Foundations of Mathematics

Foundations of Mathematics

Marvin L. Bittinger

Indiana University Purdue University Indianapolis

Judith A. Penna

Indiana University Purdue University Indianapolis

Boston San Francisco New York
London Toronto Sydney Tokyo Singapore Madrid
Mexico City Munich Paris Cape Town Hong Kong Montreal

Publisher	Greg Tobin
Editor in Chief	Maureen O'Connor
Acquisitions Editor	Jennifer Crum
Executive Project Manager	Kari Heen
Project Editor	Lauren Morse
Editorial Assistant	Katie Nopper
Managing Editor	Ron Hampton
Production Supervisor	Kathleen A. Manley
Editorial and Production Services	Martha K. Morong/Quadrata, Inc.
Art Editor and Photo Researcher	Geri Davis/The Davis Group, Inc.
Chapter Opener Art Director	Meredith Nightingale
Media Producer	Lynne Blaszak
Software Development	Alicia Anderson and Malcolm Litowitz
Marketing Manager	Dona Kenly
Marketing Coordinator	Lindsay Skay
Illustrators	Network Graphics, J. B. Woolsey Associates, Rolin Graphics, Inc., Doug Hart, and Gary Torissi
Prepress Supervisor	Caroline Fell
Compositor	Beacon Publishing Services
Cover Designer	Dennis Schaefer
Cover Photograph	Gary Conner/Index Stock Imagery
Interior Designer	Geri Davis/The Davis Group, Inc., and Susan Carsten Raymond
Print Buyer	Evelyn Beaton

Photo credits appear on page I-14.

Library of Congress Cataloging-in-Publication Data
Bittinger, Marvin L.
Foundations of mathematics /
Marvin L. Bittinger, Judith A. Penna.—1st ed.
p. cm.
ISBN 0-321-16856-9
1. Mathematics. I. Penna, Judith A. II. Title.
QA39.3.B55 2003
510—dc21 2003043361

5 6 7 8 9 10—VH—10

Contents

1 WHOLE NUMBERS

2 FRACTION NOTATION

3 DECIMAL NOTATION

4 PERCENT NOTATION

5 GEOMETRY

6 INTRODUCTION TO REAL NUMBERS AND ALGEBRAIC EXPRESSIONS

7 SOLVING EQUATIONS AND INEQUALITIES

8 GRAPHS OF LINEAR EQUATIONS

9 POLYNOMIALS: OPERATIONS

10 POLYNOMIALS: FACTORING

11 RATIONAL EXPRESSIONS AND EQUATIONS

12 GRAPHS, FUNCTIONS, AND APPLICATIONS

13 SYSTEMS OF EQUATIONS

14 MORE ON INEQUALITIES

15 RADICAL EXPRESSIONS, EQUATIONS, AND FUNCTIONS

16 QUADRATIC EQUATIONS AND FUNCTIONS

17 EXPONENTIAL AND LOGARITHMIC FUNCTIONS

APPENDIXES

Preface

Foundations of Mathematics is a new text in the Bittinger developmental mathematics series. This text covers all of the core material that is typically found in a three-semester developmental mathematics sequence—basic mathematics, introductory algebra, and intermediate algebra. Texts in the series include:

Bittinger: *Basic Mathematics,* Ninth Edition

Bittinger: *Fundamental Mathematics,* Third Edition

Bittinger/Ellenbogen: *Prealgebra,* Fourth Edition

Bittinger: *Introductory Algebra,* Ninth Edition

Bittinger: *Intermediate Algebra,* Ninth Edition

Bittinger/Beecher: *Introductory and Intermediate Algebra,* Second Edition

Bittinger/Beecher: *Developmental Mathematics,* Sixth Edition

Bittinger/Penna: *Foundations of Mathematics*

All of the topics on the Texas Higher Education Assessment test (THEA) and the majority of the topics on the state-level mathematics tests, including the CSU Entry Level Mathematics Test (ELM) and the CUNY Mathematics Skills Assessment Test, are incorporated in this text. Guidelines from many states and educational institutions were considered while planning for this text, including those for the Florida CLAST test, the Alabama College Association, and the Tennessee Board of Regents. Many of the skills required by these guidelines are covered in *Foundations of Mathematics.*

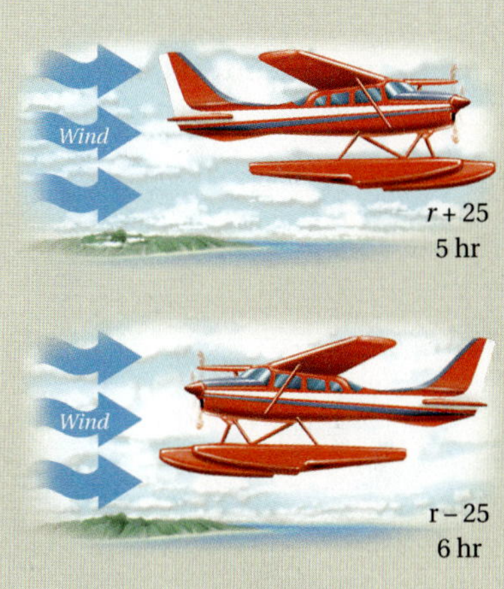

The unique approach of this text blends the following elements in order to bring students success:

- **Writing style** The authors write in a clear, easy-to-read style that helps students progress from concepts through examples and margin exercises to end-of-section exercise sets.
- **Problem-solving approach** The basis for solving problems and real-data applications is a five-step process (*Familiarize, Translate, Solve, Check,* and *State*) introduced early in the text and used consistently throughout. This problem-solving approach provides students with a consistent framework for solving applied problems. (See pages 53, 260, 554, 597, 890, and 1218.)

- **Real data** Real-data applications aid in motivating students by connecting mathematics to their everyday lives. Extensive research was conducted to find applications that relate mathematics to the real world.

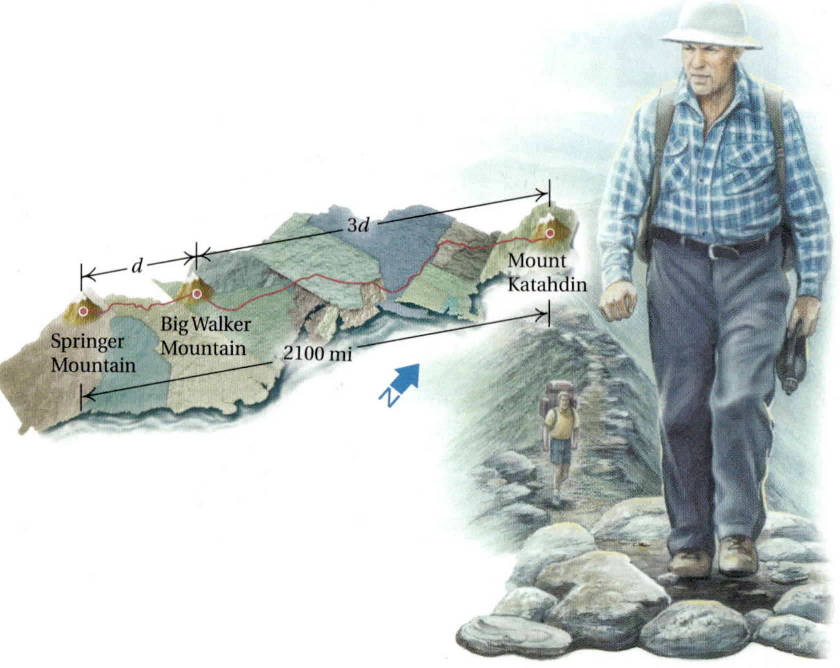

- **Art program** The art program is designed to improve the visualization of mathematical concepts and to enhance the real-data applications.
- **Reviewer feedback** The authors solicit feedback from reviewers and students to help fulfill student and instructor needs.
- **Accuracy** The manuscript is subjected to an extensive accuracy-checking process to eliminate errors.
- **Supplements package** All ancillary materials are directly tied with the text and created by members of the author team to provide a complete and consistent package for both students and instructors.

Number of Radio Stations on the Internet

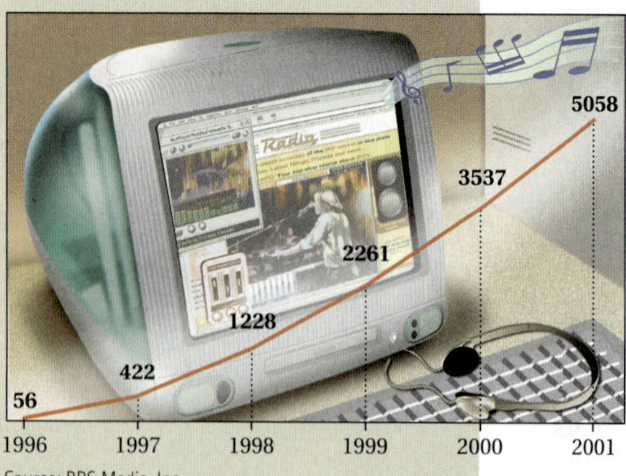

Source: BRS Media, Inc.

LET'S VISIT THE TEXT

Up-To-Date Applications Extensive research has been done to make the applications in this text up-to-date and realistic. A large number of the applications are drawn from the fields of business and economics, life and physical sciences, social sciences, and areas of general interest such as sports and daily life. To encourage students to understand the relevance of mathematics, many applications are enhanced by graphs and drawings similar to those found in today's newspapers and magazines. Many applications are also titled for quick and easy reference, and most real-data applications are authenticated with a source line. (See pages 208, 613, 889, and 928.)

Numerous Photographs An application becomes relevant when the connection to the real world is illustrated with a photograph. *Foundations of Mathematics* contains approximately 160 photos that immediately spark interest in examples and exercises. (See pages 63, 285, 820, and 1112.)

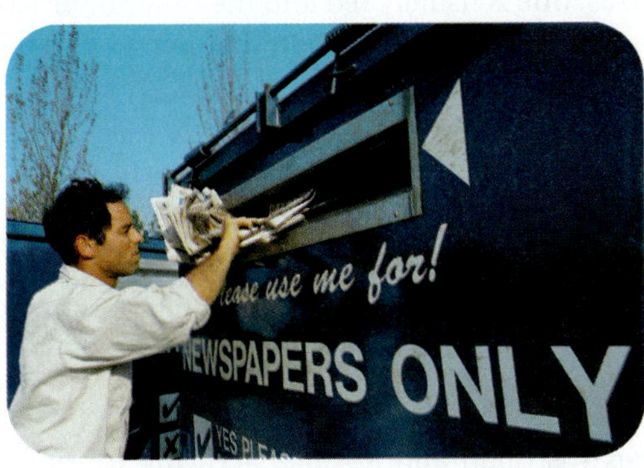

Study Tips Occurring at least twice in every chapter, these mini-lessons provide students with concrete techniques for improving their studying and test-taking skills. These features can be covered in their entirety at the beginning of the course, encouraging good study habits early on (see a complete list of study tips at the back of the book), or they can be used as they occur in the text, allowing students to encounter them gradually. (See pages 8, 137, 583, and 880.) These features can also be used in conjunction with Marvin L. Bittinger's "Math Study Skills for Students" videotape, which is free to adopters. Please contact your Addison-Wesley representative for details on how to obtain this videotape, which is also available on CD-ROM.

Calculator Corners Designed specifically for the beginning developmental-mathematics student, this optional feature includes scientific- and graphing-calculator instruction and practice exercises. (See pages 498, 809, and 991.) Answers to all Calculator Corner exercises appear at the back of the book.

Algebraic–Graphical Connections AG To give students a better visual understanding of algebra, we have included algebraic–graphical connections. (See pages 614, 877, and 1197.) This feature gives the algebra more meaning by connecting it to a graphical interpretation.

Art To enhance the emphasis on real data and applications, we have provided a significant number of pieces of technical and situational art. (See pages 212, 228, 679, and 1168.)

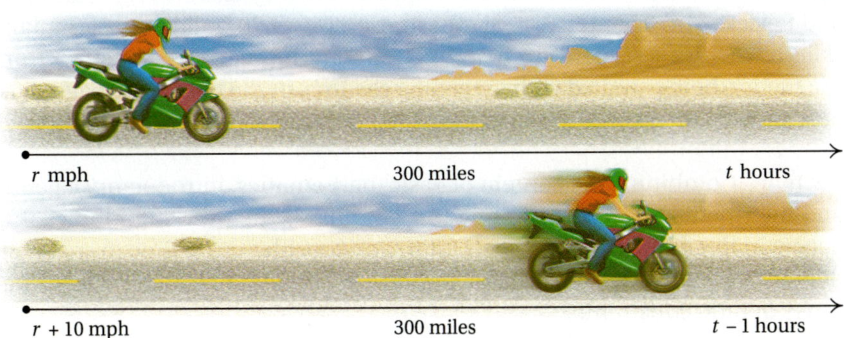

The use of color has been carried out in a methodical and precise manner so that it carries a consistent meaning, which enhances the readability of the text. For example, the use of both red and blue in mathematical art increases understanding of the concepts. When two lines are graphed using the same set of axes, one is usually red and the other blue. Note that equation labels are the same color as the corresponding line to aid in understanding.

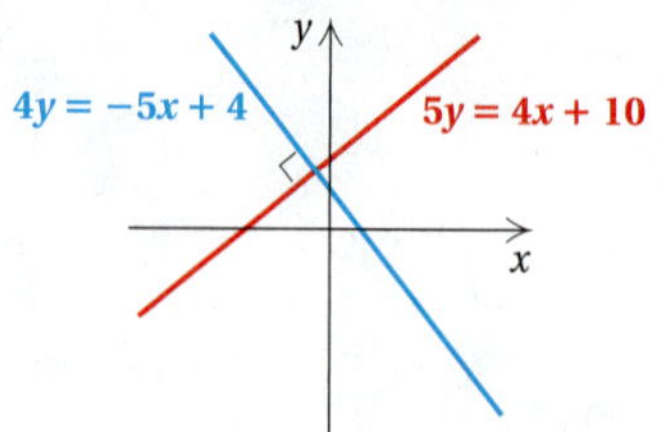

Design The design is open and flexible, allowing for an extensive art and photo package and prominent headings for the boxed definitions and rules and for the Caution boxes.

Exercises Exercises are paired, meaning that each even-numbered exercise is very much like the odd-numbered one that precedes it. This gives the instructor several options: If an instructor wants the student to have answers available, the odd-numbered exercises can be assigned; if an instructor wants the student to practice (perhaps for a test) with no answers available, then the even-numbered exercises can be assigned. In this way, each exercise set actually serves as two exercise sets. Answers to all odd-numbered exercises, with the exception of the Discussion and Writing exercises, and to *all* Skill Maintenance exercises are provided at the back of the book.

Discussion and Writing Exercises Two Discussion and Writing exercises (denoted by D_W) appear in every exercise set and Summary and Review. Designed to develop comprehension of critical concepts, these exercises encourage students both to discuss and to write about key mathematical ideas in the chapter. (See pages 193, 633, and 1083.)

Skill Maintenance Exercises The Skill Maintenance exercises review concepts from other sections of the text in order to prepare students for the Final Examination. Section and objective codes appear next to each Skill Maintenance exercise so that students can easily review each topic. Answers to *all* Skill Maintenance exercises appear at the back of the book. (See pages 157, 668, and 1228.)

Synthesis Exercises These exercises appear in every exercise set, Summary and Review, and Chapter Test. Synthesis exercises help build critical thinking skills by requiring students to synthesize or combine learning objectives from the section being studied with preceding sections in the book. (See pages 201, 605, and 1003.)

Content Combining the content of basic mathematics, introductory algebra, and intermediate algebra into one text eliminates the overlap of topics that occurs in three separate texts. This approach provides more time to spend on difficult topics as well as topics that often are eliminated from a syllabus because of time constraints.

LEARNING AIDS

Interactive Worktext Approach *Foundations of Mathematics* is designed to provide students with an interactive learning experience. The exposition, annotated examples, art, margin exercises, and exercise sets offer a clear set of learning objectives, involve students actively with the development of the material, and provide immediate and continual reinforcement and assessment.

Section objectives are keyed by letter not only to section subheadings, but also to exercises in the Pretest, exercise sets, and Summary and Review, as well as to the answers to the Chapter Test questions. This enables students to find appropriate review material easily if they are unable to do a particular exercise.

Throughout the text, students are directed to numerous *margin exercises,* which provide immediate reinforcement of the concepts covered in each section.

Objectives

a Given an equation in the form $y = mx + b$, find the slope and the y-intercept; and find an equation of a line when the slope and the y-intercept are given.

b Find an equation of a line when the slope and a point on the line are given.

c Find an equation of a line when two points on the line are given.

Review Material *Foundations of Mathematics* provides resources to help students prepare for final assessment.

A two-part *Summary and Review* appears at the end of each chapter. The first part is a checklist of Study Tips, frequently accompanied by a list of important properties and formulas. The second part provides an extensive set of review exercises. Reference codes beside each exercise or direction line allow the student to return easily to the objective being reviewed. (See pages 589, 825, 979, and 1273.)

For Extra Help Many valuable study aids accompany this text. At the beginning of each exercise set, references to appropriate videos, tutorial software, and other resources make it easy for the student to find the correct support materials.

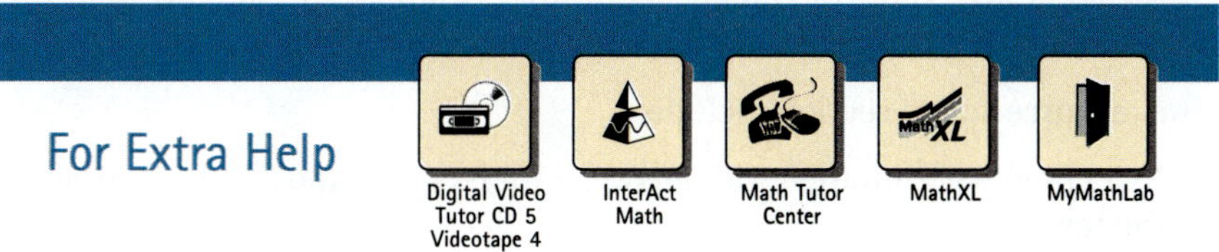

Testing The following assessment opportunities exist in the text.

Chapter Pretests can be used to place students in a specific section of the chapter, allowing them to concentrate on topics with which they have particular difficulty. (See pages 104, 512, and 1118.)

Chapter Tests allow students to review and test comprehension of chapter skills along with four objectives from earlier chapters that are retested. (See pages 100, 652, and 1114.)

In addition, a *Diagnostic Pretest,* found in the *Printed Test Bank/Instructor's Resource Guide* and in MyMathLab, can place students in the appropriate chapter for their skill level by identifying familiar material and specific trouble areas. This may be especially helpful for self-paced courses.

Answers to all Chapter Pretest and Chapter Test questions are found at the back of the book. Section and objective references for Pretest exercises are listed in blue beside each exercise or direction line preceding it. Reference codes for the Chapter Test answers are included with the answers.

SUPPLEMENTS FOR THE INSTRUCTOR

Annotated Instructor's Edition

ISBN 0-321-17248-5

The *Annotated Instructor's Edition* is a specially bound version of the student text with answers to all margin exercises and exercise sets printed in blue near the corresponding exercises.

Instructor's Solutions Manual

by Judith A. Penna
ISBN 0-321-17419-4

The *Instructor's Solutions Manual* contains brief worked-out solutions to all even-numbered exercises and answers to all Discussion and Writing exercises in the exercise sets.

Printed Test Bank/Instructor's Resource Guide

ISBN 0-321-17420-8

The test-bank section of this supplement contains the following:

- A diagnostic test that can place students in the appropriate chapter for their skill level
- Two alternate test forms for each chapter, modeled after the Chapter Tests in the text
- Two alternate test forms for each chapter designed for a 50-minute class period
- Two multiple-choice versions of each Chapter Test
- Six final examinations: two with questions organized by chapter, two with questions scrambled, and two with multiple-choice questions
- Answers for the Diagnostic Test, the Chapter Tests, and the final examinations

The resource-guide section includes:

- Extra practice exercises (with answers) for 65 of the most difficult topics in the text
- A three-column chapter summary and review for each chapter listing objectives, brief procedures, worked-out examples, multiple-choice problems similar to the examples, and answers to those problems
- Black-line masters of grids and number lines for transparency masters or test preparation
- Index to the videotapes that accompany the text

Adjunct Support Manual

ISBN 0-321-17518-2

This manual includes resources designed to help both new and adjunct faculty with course preparation and classroom management and offers helpful teaching tips.

TestGen with QuizMaster
ISBN 0-321-16929-8

Available on a dual-platform Windows/Macintosh CD-ROM, this fully networkable software enables instructors to build, edit, print, and administer tests using a computerized test bank of questions organized according to the contents of each chapter. Tests can be printed or saved for online testing via a network on the Web, and the software can generate a variety of grading reports for tests and quizzes.

InterAct MathXL® www.mathxl.com
ISBN 0-321-12986-5

MathXL is an online testing, homework, and tutorial system that uses algorithmically generated exercises correlated to the textbook.

Instructors can assign tests and homework provided by Addison-Wesley or create and customize their own tests and homework assignments. Instructors can also track students' results and tutorial work in an online gradebook. Students can take chapter tests and receive personalized study plans that will diagnose weaknesses and provide links to areas they need to study and retest. Students can also work unlimited practice problems and receive tutorial instruction for areas in which they need improvement. MathXL can be packaged with new copies of *Foundations of Mathematics*. Please contact your Addison-Wesley representative for details.

MyMathLab
MyMathLab is a complete online course for Addison-Wesley mathematics textbooks that provides interactive, multimedia instruction correlated to textbook content. MyMathLab can be easily customized to suit the needs of students and instructors and provides a comprehensive and efficient online course-management system that allows for diagnosis, assessment, and tracking of students' progress.

MyMathLab features the following:

- Fully interactive multimedia chapter and section folders from the textbook contain a wide range of instructional content, including videos, software tools, audio clips, animations, and electronic supplements.
- Hyperlinks take you directly to online testing, diagnostics, tutorials, and gradebooks in MathXL—Addison-Wesley's tutorial and testing system for mathematics and statistics.
- Instructors can create, copy, edit, assign, and track all tests for their course as well as track student tutorial and testing performance.
- With push-button ease, instructors can remove, hide, or annotate Addison-Wesley's preloaded content, add their own course documents, or change the order in which material is presented.
- Using the communication tools found in MyMathLab, instructors can hold online office hours, host a discussion board, create communication groups within their class, send e-mails, and maintain a course calendar.
- Print supplements are available online, side by side with their textbooks.

For more information, visit our Web site at www.mymathlab.com or contact your Addison-Wesley sales representative for a demonstration.

SUPPLEMENTS FOR THE STUDENT

Student's Solutions Manual
by Judith A. Penna
ISBN 0-321-17421-6

The *Student's Solutions Manual* contains fully worked-out solutions with step-by-step annotations for all the odd-numbered exercises in the exercise sets in the text, with the exception of the Discussion and Writing exercises. Students can purchase this manual from Addison-Wesley or their local college bookstore.

Videotapes
ISBN 0-321-17517-4

Digital Video Tutor CD 5 Videotape 4

This videotape series features an engaging team of mathematics instructors who present comprehensive coverage of each section of the text in a student-interactive format. The lecturers' presentations include examples and problems from the text and support an approach that emphasizes visualization and problem solving. A video symbol at the beginning of each exercise set references the appropriate videotape or CD (see *Digital Video Tutor,* below).

Digital Video Tutor
ISBN 0-321-17515-8

The videotapes for this text are also available on CD-ROM, making it easy and convenient for students to watch video segments from a computer at home or on campus. The complete digitized video set, affordable and portable for students, is ideal for distance learning or supplemental instruction.

"Math Study Skills for Students" Videotape/DVT
Videotape ISBN 0-321-11739-5/DVT ISBN 0-321-15075-9

Designed to help students make better use of their math study time, this videotape helps students improve retention of concepts and procedures taught in classes from basic mathematics through intermediate algebra. Through carefully crafted graphics and comprehensive on-camera explanation, Marvin L. Bittinger helps viewers focus on study skills that are commonly overlooked.

InterAct Math® Tutorial CD-ROM
ISBN 0-321-16927-1

InterAct Math

This interactive tutorial software provides algorithmically generated practice exercises that correlate at the objective level to the odd-numbered exercises in the text. Each practice exercise is accompanied by both an example and a guided solution designed to involve students in the solution process. Selected problems also include a video clip that helps students visualize concepts. The software recognizes common student errors and provides appropriate feedback.

InterAct MathXL® www.mathxl.com

ISBN 0-201-72611-4, stand-alone

MathXL

MathXL is an online testing, homework, and tutorial system that uses algorithmically generated exercises correlated to the textbook.

Students can take chapter tests and receive personalized study plans that will diagnose weaknesses and provide links to areas they need to study and retest. Students can also work unlimited practice problems and receive tutorial instruction for areas in which they need improvement. MathXL can be packaged with new copies of *Foundations of Mathematics.*

MyMathLab

MyMathLab

MyMathLab is a complete, online course for Addison-Wesley mathematics textbooks that provides interactive, multimedia instruction correlated to the textbook content. MyMathLab can be easily customized to suit the needs of students and instructors and provides a comprehensive and efficient online course-management system that allows for diagnosis, assessment, and tracking of students' progress.

MyMathLab features:

- Chapter and section folders in the online course mirror the textbooks' Table of Contents and contain a wide range of multimedia instruction, including video lectures, tutorial software, and electronic supplements.
- The actual pages of the textbook are loaded into MyMathLab, and as students work through a section of the online text, they can link to multimedia resources—such as video and audio clips, tutorial exercises, and interactive animations—that are correlated directly to the examples and exercises in the text.
- Hyperlinks take students directly to online testing, diagnostics, tutorials, and tracking in MathXL—Addison-Wesley's tutorial and testing system for mathematics and statistics.
- Print supplements are available online, side by side with their textbooks.

Addison-Wesley Math Tutor Center

ISBN 0-201-72170-8, stand-alone

Math Tutor Center

The Addison-Wesley Math Tutor Center is staffed by qualified mathematics instructors who provide students with tutoring on examples and odd-numbered exercises from the textbook. Tutoring is available by toll-free telephone, fax, e-mail, or the Internet. White Board technology allows tutors and students to see problems actually worked while they "talk" in real time over the Internet during tutoring sessions. An access card is required.

Acknowledgments

Many of you have helped to shape this text by reviewing and spending time with us on your campuses. Our deepest appreciation to all of you and in particular to the following:

Randall Allbritton, *Daytona Beach Community College*
Ann Arakawa, *Maui Community College*
Joaquin Armendariz, *College of Marin*
Jean Ashby, *Community College of Baltimore County—Catonsville*
Arlene Atchison, *South Seattle Community College*
Michele Bach, *Kansas City Kansas Community College*
Stan Barrick, *California State University—Sacramento*
Roseanne Benn, *Prince George's Community College*
Maria Bennett, *West Shore Community College*
Donna Bernardy, *Lane Community College*
Janis Broering, *Northern Kentucky University*
Wayne Brown, *Oklahoma State University—Oklahoma City*
Diane Christie, *University of Wisconsin—Stout*
Deirdre Collins, *Glendale College*
Patty Connelly, *Northern Kentucky University*
Joyce Couch, *Northern Kentucky University*
Karena Curtis, *Labette Community College*
Martha Daniels, *Central Oregon Community College*
Drake Dennis, *Delaware Technical & Community College*
Michael Divinia, *San Jose City College*
Jane Duncan Nesbit, *Columbia Union College*
Sharon Edgmon, *Bakersfield College*
Rafael Espericueta, *Bakersfield College*
Grace Foster, *Beaufort County Community College*
Edward A. Gallo, *Ivy Tech State College*
Bill Graesser, *Ivy Tech State College*
William Haigh, *Northern State University*
Martha Henry, *Milwaukee Area Technical College*
Celeste Hernandez, *Richland Community College*
Gerry Higdon, *Fitchburg State College*
Pat Horacek, *Pensacola Community College*
Glenn Jablonski, *Benedictine University*
Juan Jimenez, *Springfield Technical Community College*
Joe Jordan, *John Tyler Community College*
Michael Judge, *Houston Community College*
Rose Kaniper, *Burlington County Community College*

Barry King, *Okefenokee Technical School*
Lynette King, *Gadsden State Community College*
Thomas Lankston, *Ivy Tech State College—North Central*
Barbara A. Lawrence, *City University of New York, Borough of Manhattan Community College*
Jennifer Leong, *Georgia State University*
Edith Lester, *Volunteer State Community College*
Pam Lipka, *University of Wisconsin—Whitewater*
Debi Loeffler, *Community College of Baltimore County—Catonsville*
Jean-Marie Magnier, *Springfield Technical Community College*
Madeline Mahar, *Pitt Community College*
Carol A. Marinas, *Barry University*
Marianna McClymonds, *Phoenix College*
Molly Misko, *Gadsden State Community College*
Michael Montaño, *Riverside Community College—City Campus*
Valerie Morgan-Krick, *Tacoma Community College*
Linda Murphy, *Northern Essex Community College*
Rhoda Oden, *Gadsden State Community College*
Joyce Oster, *Johnson and Wales University*
Julie Pendleton, *Brookhaven Community College*
Thea Philliou, *College of Sante Fe*
Marilyn Platt, *Gaston College*
Steve Proietti, *Northern Essex Community College*
Mary Rack, *Johnson County Community College*
Greg Rosik, *Century College*
Pat Roux, *Delgado Community College*
Nelissa Rutishauser, *Mohawk Valley Community College*
Susan Santolucito, *Delgado Community College*
F. Richard Schnackenberg, *International College*
Carole Shapero, *Oakton Community College*
Cheryl Shepherd, *Cowley County Community College*
Mike Shirazi, *Germanna Community College*
Nicole Sifford, *Three Rivers Community College*
Tomesa Smith, *Wallace State Community College*
Trudy Streilein, *Northern Virginia Community College*
Sharon Testone, *Onondaga Community College*
Brad Thurmond, *Ivy Tech State College—Kokomo*
Diane Trojan, *Kutztown University*
Jimmie A. Van Alphen, *Ozarks Technical Community College*
Yvonne Wabbington, *Northern Kentucky University*
Angela Walters, *Capitol College*
Ray Weaver, *Community College of Allegheny County—Boyce Campus*
Joyce Wellington, *Southeastern Community College*
Kevin Wheeler, *Three Rivers Community College*
Annette Wiesner, *University of Wisconsin—Parkside*
Charles Wiffen, *Elon University*
Diane Williams, *Northern Kentucky University*
Jane Marie Wright, *Suffolk County Community College*

We also wish to recognize the following people who wrote scripts, presented lessons on camera, and checked the accuracy of the videotapes:

Barbara Johnson, *Indiana University Purdue University Indianapolis*
Judith A. Penna, *Indiana University Purdue University Indianapolis*
Patricia Schwarzkopf, *University of Delaware*
Clen Vance, *Houston Community College*

We wish to express our heartfelt appreciation to a number of people who have contributed in special ways to the development of this textbook. Our editor, Jennifer Crum, encouraged our vision and provided marketing insight. Kari Heen, executive project manager, deserves special recognition for overseeing every phase of the project and keeping it moving. The unwavering support of the Developmental Math group, including Lauren Morse, project editor; Katie Nopper, editorial assistant; Lynne Blaszak, media producer; Dona Kenly, senior marketing manager; Lindsay Skay, senior marketing coordinator; and Kathleen Manley, production supervisor; along with endless hours of hard work by Martha Morong and Geri Davis have led to products of which we are immensely proud. Other strong support has come from Barbara Johnson for her meticulous accuracy checking.

Study Tips

TO THE STUDENT

As your authors, we would like to welcome you to this study of *Foundations of Mathematics.*

Whatever your past experiences, we encourage you to look at this mathematics course as a fresh start and to approach it with a positive attitude. Understanding mathematics will enrich and improve your life. Mathematics is the basis for making many important decisions and for controlling your personal finances in addition to being essential for many careers.

You are the most important factor in the success of your learning experience. In earlier situations, you might have allowed yourself to sit back and let the instructor "pour in" the learning, with little or no effort on your part. One of the biggest adjustments you might have to make in college is to realize that now you must take a more assertive and proactive role. For example, as soon as possible after class, study the textbook and do the homework assigned, making use of the supplementary materials that accompany the text. Take responsibility for your own learning. This will put you in the best possible position to be successful in this course.

One of the most important suggestions we can make is that you allow yourself enough *time* to learn. You can have an outstanding instructor, an excellent textbook, and the best supplementary materials, but if you do not give yourself time to learn, how can they be of benefit? Suggestions like this are found throughout this book under the heading of *Study Tips*. You might want to read all of the Study Tips and devise a comprehensive study plan before you begin your course. An index of Study Tips is found at the back of the book.

We wish you success.

M.L.B.
J.A.P.

Feature Walkthrough

Chapter Openers

To engage students and prepare them for the upcoming chapter material, two-page gateway chapter openers are designed with exceptional artwork that is tied to a motivating real-world application, a brief overview of the material, and a table of contents for the chapter.

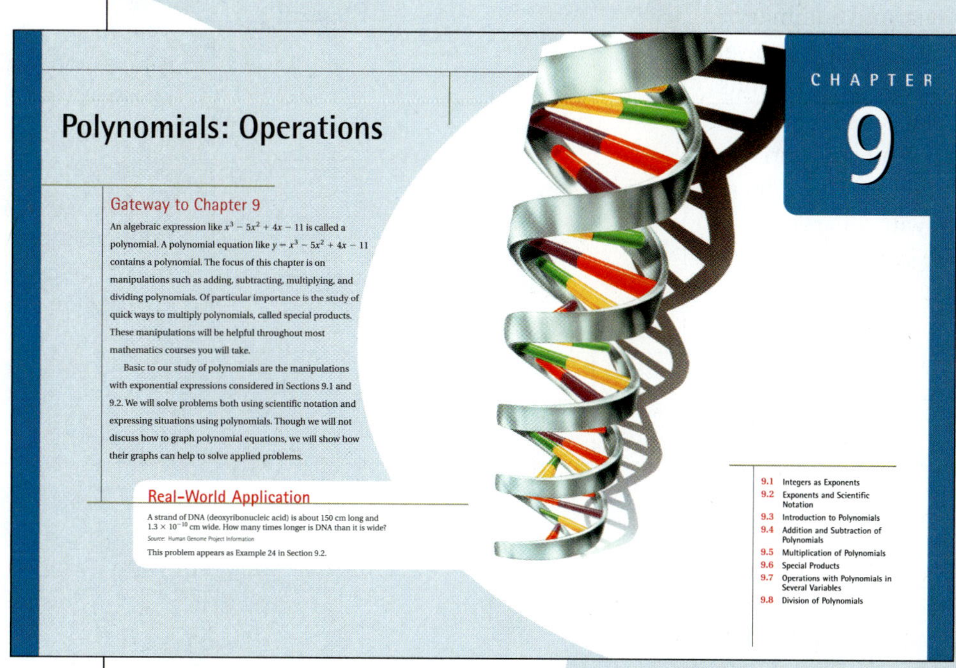

CHAPTER 9

Polynomials: Operations

Gateway to Chapter 9

An algebraic expression like $x^3 - 5x^2 + 4x - 11$ is called a polynomial. A polynomial equation like $y = x^3 - 5x^2 + 4x - 11$ contains a polynomial. The focus of this chapter is on manipulations such as adding, subtracting, multiplying, and dividing polynomials. Of particular importance is the study of quick ways to multiply polynomials, called special products. These manipulations will be helpful throughout most mathematics courses you will take.

Basic to our study of polynomials are the manipulations with exponential expressions considered in Sections 9.1 and 9.2. We will solve problems both using scientific notation and expressing situations using polynomials. Though we will not discuss how to graph polynomial equations, we will show how their graphs can help to solve applied problems.

Real-World Application

A strand of DNA (deoxyribonucleic acid) is about 150 cm long and 1.3×10^{-10} cm wide. How many times longer is DNA than it is wide?

Source: Human Genome Project Information

This problem appears as Example 24 in Section 9.2.

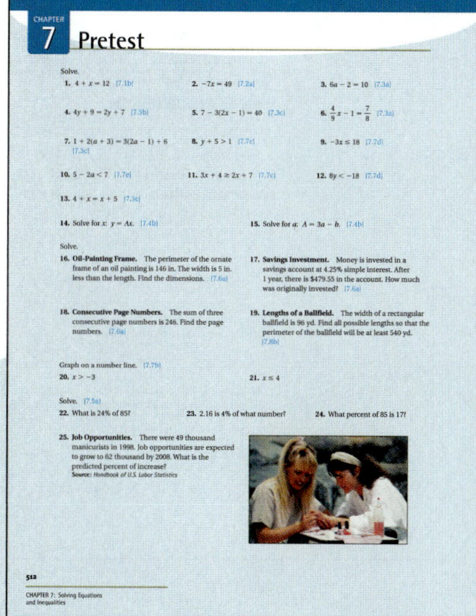

CHAPTER 7 Pretest

Solve.

1. $4 + x = 12$ [7.1b]
2. $-7x = 49$ [7.2a]
3. $6a - 2 = 10$ [7.3a]
4. $4y + 9 = 2y + 7$ [7.3b]
5. $7 - 3(2x - 1) = 40$ [7.3c]
6. $\frac{4}{9}x - 1 = \frac{7}{8}$ [7.3a]
7. $1 + 2(a + 3) = 3(2a - 1) + 6$ [7.3c]
8. $y + 5 > 1$ [7.7c]
9. $-3x \leq 18$ [7.7d]
10. $5 - 2a < 7$ [7.7e]
11. $3x + 4 \geq 2x + 7$ [7.7c]
12. $8y < -18$ [7.7d]
13. $4 + x = x + 5$ [7.3c]
14. Solve for x: $y = Ax$. [7.4b]
15. Solve for a: $A = 3a - b$. [7.4b]

Solve.

16. **Oil-Painting Frame.** The perimeter of the ornate frame of an oil painting is 146 in. The width is 5 in. less than the length. Find the dimensions. [7.6a]
17. **Savings Investment.** Money is invested in a savings account at 4.25% simple interest. After 1 year, there is \$479.55 in the account. How much was originally invested? [7.6a]
18. **Consecutive Page Numbers.** The sum of three consecutive page numbers is 246. Find the page numbers. [7.6a]
19. **Lengths of a Ballfield.** The width of a rectangular ballfield is 96 yd. Find all possible lengths so that the perimeter of the ballfield will be at least 540 yd. [7.8b]

Graph on a number line. [7.7b]

20. $x > -3$
21. $x \leq 4$

Solve. [7.5a]

22. What is 24% of 85?
23. 2.16 is 4% of what number?
24. What percent of 85 is 17?
25. **Job Opportunities.** There were 49 thousand manicurists in 1998. Job opportunities are expected to grow to 62 thousand by 2008. What is the predicted percent of increase?
Source: *Handbook of U.S. Labor Statistics*

512

CHAPTER 7: Solving Equations and Inequalities

Chapter Pretests

Allowing students to test themselves before beginning each chapter, Chapter Pretests help them to identify material that may be familiar as well as to target material that may be new or especially challenging. Instructors can use these results to assess student needs.

Objectives Boxes

At the beginning of each section, a boxed list of objectives is keyed by letter not only to section subheadings but also to the exercises in the Pretest, exercise sets, and Summary and Review, as well as answers to the Chapter Test questions. This correlation enables students to find appropriate review material easily if they need help with a particular exercise or skill.

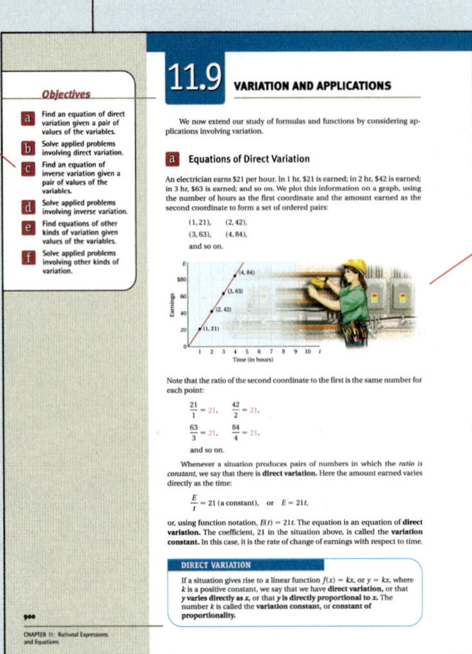

Objectives

a Find an equation of direct variation given a pair of values of the variables.

b Solve applied problems involving direct variation.

c Find an equation of inverse variation given a pair of values of the variables.

d Solve applied problems involving inverse variation.

e Find equations of other kinds of variation given values of the variables.

f Solve applied problems involving other kinds of variation.

11.9 VARIATION AND APPLICATIONS

We now extend our study of formulas and functions by considering applications involving variation.

a Equations of Direct Variation

An electrician earns \$21 per hour. In 1 hr, \$21 is earned; in 2 hr, \$42 is earned; in 3 hr, \$63 is earned; and so on. We plot this information on a graph, using the number of hours as the first coordinate and the amount earned as the second coordinate to form a set of ordered pairs:

(1, 21), (2, 42),
(3, 63), (4, 84),
and so on.

Note that the ratio of the second coordinate to the first is the same number for each point:

$$\frac{21}{1} = 21, \quad \frac{42}{2} = 21,$$
$$\frac{63}{3} = 21, \quad \frac{84}{4} = 21,$$

and so on.

Whenever a situation produces pairs of numbers in which the *ratio is constant*, we say that there is **direct variation.** Here the amount earned varies directly as the time:

$$\frac{E}{t} = 21 \text{ (a constant)}, \quad \text{or} \quad E = 21t,$$

or, using function notation, $E(t) = 21t$. The equation is an equation of **direct variation.** The coefficient, 21 in the situation above, is called the **variation constant.** In this case, it is the rate of change of earnings with respect to time.

DIRECT VARIATION

If a situation gives rise to a linear function $f(x) = kx$, or $y = kx$, where k is a positive constant, we say that we have **direct variation,** or that **y varies directly as x,** or that **y is directly proportional to x.** The number k is called the **variation constant,** or **constant of proportionality.**

900

CHAPTER 11: Rational Expressions and Equations

Art Program

Today's students are often visually oriented and their approach to a printed page is no exception. For this reason, the situational art is dynamic, and there are many photographs and art pieces overall. Where possible, mathematics is included in the art pieces to help students visualize the problem at hand.

Margin Exercises

Throughout the text, students are directed to numerous margin exercises that provide immediate practice and reinforcement of the concepts covered in each section.

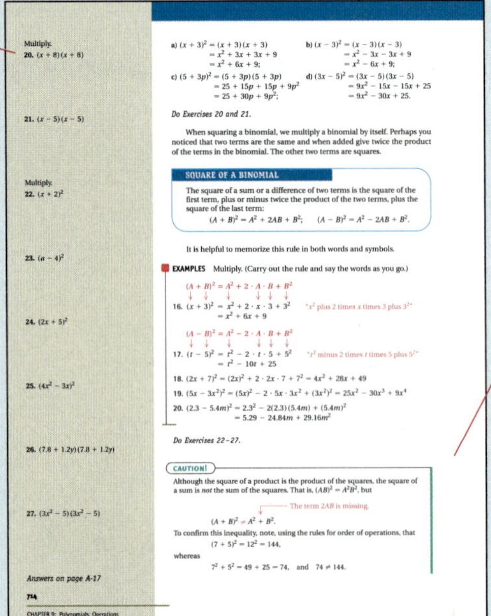

Multiply.

20. $(x + 8)(x + 8)$

21. $(x - 5)(x - 5)$

Multiply.

22. $(x + 2)^2$

23. $(a - 4)^2$

24. $(2x + 5)^2$

25. $(4x^2 - 3x)^2$

26. $(7.8 + 1.2y)(7.8 + 1.2y)$

27. $(3x^2 - 5)(3x^2 - 5)$

Answers on page A-17

a) $(x + 3)^2 = (x + 3)(x + 3) = x^2 + 3x + 3x + 9 = x^2 + 6x + 9;$

b) $(x - 3)^2 = (x - 3)(x - 3) = x^2 - 3x - 3x + 9 = x^2 - 6x + 9;$

c) $(5 + 3p)^2 = (5 + 3p)(5 + 3p) = 25 + 15p + 15p + 9p^2 = 25 + 30p + 9p^2;$

d) $(3x - 5)^2 = (3x - 5)(3x - 5) = 9x^2 - 15x - 15x + 25 = 9x^2 - 30x + 25.$

Do Exercises 20 and 21.

When squaring a binomial, we multiply a binomial by itself. Perhaps you noticed that two terms are the same and when added give twice the product of the terms in the binomial. The other two terms are squares.

SQUARE OF A BINOMIAL

The square of a sum or a difference of two terms is the square of the first term, plus or minus twice the product of the two terms, plus the square of the last term:

$$(A + B)^2 = A^2 + 2AB + B^2; \quad (A - B)^2 = A^2 - 2AB + B^2.$$

It is helpful to memorize this rule in both words and symbols.

EXAMPLES Multiply. (Carry out the rule and say the words as you go.)

$(A + B)^2 = A^2 + 2 \cdot A \cdot B + B^2$

16. $(x + 3)^2 = x^2 + 2 \cdot x \cdot 3 + 3^2 = x^2 + 6x + 9$ "x^2 plus 2 times x times 3 plus 3^2"

$(A - B)^2 = A^2 - 2 \cdot A \cdot B + B^2$

17. $(t - 5)^2 = t^2 - 2 \cdot t \cdot 5 + 5^2 = t^2 - 10t + 25$ "t^2 minus 2 times t times 5 plus 5^2"

18. $(2x + 7)^2 = (2x)^2 + 2 \cdot 2x \cdot 7 + 7^2 = 4x^2 + 28x + 49$

19. $(5x - 3x^2)^2 = (5x)^2 - 2 \cdot 5x \cdot 3x^2 + (3x^2)^2 = 25x^2 - 30x^3 + 9x^4$

20. $(2.3 - 5.4m)^2 = 2.3^2 - 2(2.3)(5.4m) + (5.4m)^2 = 5.29 - 24.84m + 29.16m^2$

Do Exercises 22–27.

CAUTION!

Although the square of a product is the product of the squares, the square of a sum is *not* the sum of the squares. That is, $(AB)^2 = A^2B^2$, but

The term $2AB$ is missing.

$$(A + B)^2 \neq A^2 + B^2.$$

To confirm this inequality, note, using the rules for order of operations, that

$$(7 + 5)^2 = 12^2 = 144,$$

whereas

$$7^2 + 5^2 = 49 + 25 = 74, \quad \text{and} \quad 74 \neq 144.$$

714

CHAPTER 9: Polynomials: Operations

Caution Boxes

Found at relevant points throughout the text, boxes with the "Caution!" heading warn students of common misconceptions or errors.

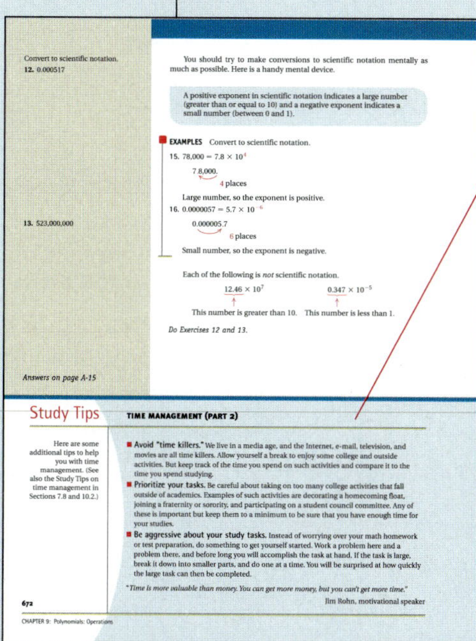

Convert to scientific notation.

12. 0.000517

13. 523,000,000

You should try to make conversions to scientific notation mentally as much as possible. Here is a handy mental device.

A positive exponent in scientific notation indicates a large number (greater than or equal to 10) and a negative exponent indicates a small number (between 0 and 1).

EXAMPLES Convert to scientific notation.

15. $78{,}000 = 7.8 \times 10^4$

7.8,000.
4 places

Large number, so the exponent is positive.

16. $0.0000057 = 5.7 \times 10^{-6}$

0.000005.7
6 places

Small number, so the exponent is negative.

Each of the following is *not* scientific notation.

12.46×10^7 — This number is greater than 10.

0.347×10^{-5} — This number is less than 1.

Do Exercises 12 and 13.

Answers on page A-15

Study Tips

TIME MANAGEMENT (PART 2)

Here are some additional tips to help you with time management. (See also the Study Tips on time management in Sections 7.8 and 10.2.)

- **Avoid "time killers."** We live in a media age, and the Internet, e-mail, television, and movies are all time killers. Allow yourself a break to enjoy some college and outside activities. But keep track of the time you spend on such activities and compare it to the time you spend studying.
- **Prioritize your tasks.** Be careful about taking on too many college activities that fall outside of academics. Examples of such activities are decorating a homecoming float, joining a fraternity or sorority, and participating on a student council committee. Any of these is important but keep them to a minimum to be sure that you have enough time for your studies.
- **Be aggressive about your study tasks.** Instead of worrying over your math homework or test preparation, do something to get yourself started. Work a problem here and a problem there, and before long you will accomplish the task at hand. If the task is large, break it down into smaller parts, and do one at a time. You will be surprised at how quickly the large task can then be completed.

"Time is more valuable than money. You can get more money, but you can't get more time."

Jim Rohn, motivational speaker

672

CHAPTER 9: Polynomials: Operations

Study Tips

A variety of Study Tips throughout the text gives students suggestions on how to develop good study habits as they progress through the course. At times very brief and at other times more lengthy, these Study Tips encourage students to get involved in the learning process.

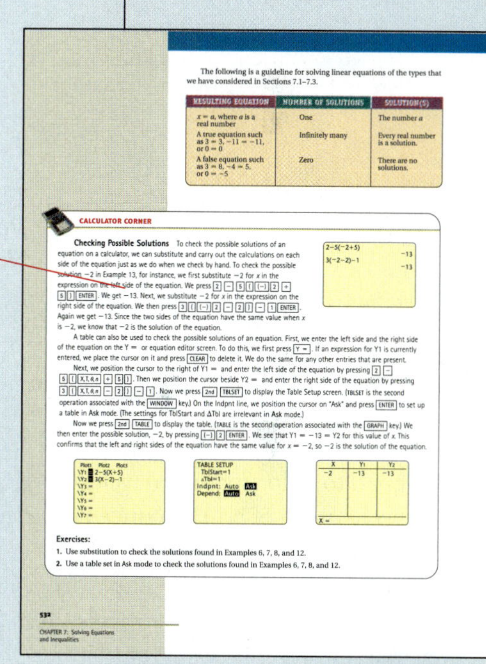

The following is a guideline for solving linear equations of the types that we have considered in Sections 7.1–7.3.

RESULTING EQUATION	NUMBER OF SOLUTIONS	SOLUTION(S)
$x = a$, where a is a real number	One	The number a
A true equation such as $3 = 3$, $-11 = -11$, or $0 = 0$	Infinitely many	Every real number is a solution.
A false equation such as $3 = 8$, $-4 = 5$, or $0 = -5$	Zero	There are no solutions.

CALCULATOR CORNER

Checking Possible Solutions To check the possible solutions of an equation on a calculator, we can substitute and carry out the calculations on each side of the equation just as we do when we check by hand. To check the possible solution −2 in Example 13, for instance, we first substitute −2 for x in the expression on the left side of the equation. We press 2 − 5 ((−) 2 + 5) ENTER. We get −13. Next, we substitute −2 for x in the expression on the right side of the equation. We then press 3 ((−) 2 − 2) − 1 ENTER. Again we get −13. Since the two sides of the equation have the same value when x is −2, we know that −2 is the solution of the equation.

A table can also be used to check the possible solutions of an equation. First, we enter the left side and the right side of the equation on the Y = or equation editor screen. To do this, we first press Y =. If an expression for Y1 is currently entered, we place the cursor on it and press CLEAR to delete it. We do the same for any other entries that are present.

Next, we position the cursor to the right of Y1 = and enter the left side of the equation by pressing 2 − 5 (X,T,θ,n + 5). Then we position the cursor beside Y2 = and enter the right side of the equation by pressing 3 (X,T,θ,n − 2) − 1. Now we press 2nd TBLSET to display the Table Setup screen. (TBLSET is the second operation associated with the WINDOW key.) On the Indpnt line, we position the cursor on "Ask" and press ENTER to set up a table in Ask mode. (The settings for TblStart and ΔTbl are irrelevant in Ask mode.)

Now we press 2nd TABLE to display the table. (TABLE is the second operation associated with the GRAPH key.) We then enter the possible solution, −2, by pressing (−) 2 ENTER. We see that Y1 = −13 = Y2 for this value of x. This confirms that the left and right sides of the equation have the same value for $x = -2$, so −2 is the solution of the equation.

Exercises:

1. Use substitution to check the solutions found in Examples 6, 7, 8, and 12.
2. Use a table set in Ask mode to check the solutions found in Examples 6, 7, 8, and 12.

532

CHAPTER 7: Solving Equations and Inequalities

Calculator Corners

Where appropriate throughout the text, students see optional Calculator Corners. This feature is designed specifically with developmental-mathematics students in mind, giving keystrokes as appropriate, and is accompanied by exercises for students to try on their own.

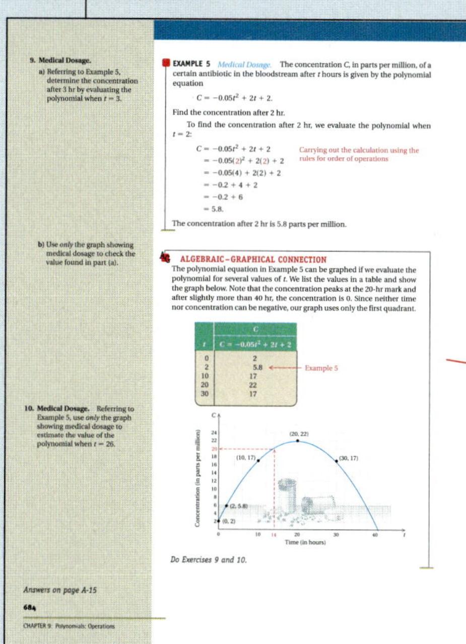

9. **Medical Dosage.**
   a) Referring to Example 5, determine the concentration after 3 hr by evaluating the polynomial when $t = 3$.
   b) Use *only* the graph showing medical dosage to check the value found in part (a).

10. **Medical Dosage.** Referring to Example 5, use *only* the graph showing medical dosage to estimate the value of the polynomial when $t = 26$.

EXAMPLE 5 *Medical Dosage.* The concentration C, in parts per million, of a certain antibiotic in the bloodstream after t hours is given by the polynomial equation

$$C = -0.05t^2 + 2t + 2.$$

Find the concentration after 2 hr.

To find the concentration after 2 hr, we evaluate the polynomial when $t = 2$:

$$\begin{aligned} C &= -0.05t^2 + 2t + 2 \\ &= -0.05(2)^2 + 2(2) + 2 \\ &= -0.05(4) + 2(2) + 2 \\ &= -0.2 + 4 + 2 \\ &= -0.2 + 6 \\ &= 5.8. \end{aligned}$$

Carrying out the calculation using the rules for order of operations

The concentration after 2 hr is 5.8 parts per million.

ALGEBRAIC–GRAPHICAL CONNECTION

The polynomial equation in Example 5 can be graphed if we evaluate the polynomial for several values of t. We list the values in a table and show the graph below. Note that the concentration peaks at the 20-hr mark and after slightly more than 40 hr, the concentration is 0. Since neither time nor concentration can be negative, our graph uses only the first quadrant.

t	C: $C = -0.05t^2 + 2t + 2$
0	2
2	5.8 ← Example 5
10	17
20	22
30	17

Do Exercises 9 and 10.

Answers on page A-15

684

CHAPTER 9: Polynomials: Operations

Algebraic–Graphical Connections

To provide a visual understanding of algebra, algebraic–graphical connections are included in each chapter beginning with Chapter 8. This feature gives the algebra more meaning by connecting the algebra to a graphical interpretation.

EXERCISE SETS

To give students the opportunity to practice what they have learned, each section is followed by an extensive exercise set designed to reinforce the section concepts. In addition, students also have the opportunity to maintain skills by reviewing material from preceding chapters and to synthesize two or more skills from the current section along with those from preceding sections.

For Extra Help

Many valuable study aids accompany this text. Located just before each exercise set, "For Extra Help" references list appropriate video, tutorial, and Web resources so students can easily find related support materials.

Exercises

Exercises are keyed by letter to the section objectives for easy review.

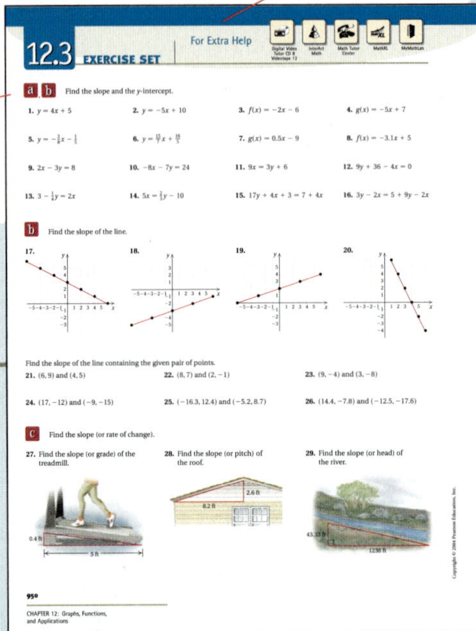

Discussion and Writing Exercises

Designed to help students develop deeper comprehension of critical concepts, Discussion and Writing exercises (indicated by the D_W symbol) are suitable for individual or group work. These exercises encourage students both to think and to write about key mathematical ideas in the chapter.

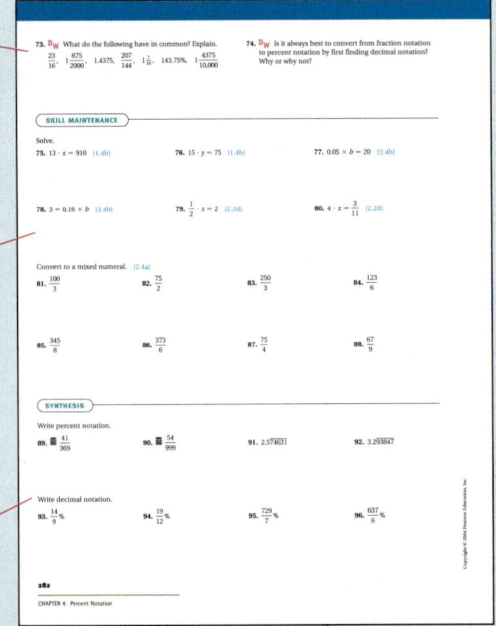

Skill Maintenance Exercises

Found in each exercise set, these exercises review concepts from other sections in the text to prepare students for the final examination. Section and objectives codes appear next to each Skill Maintenance exercise or instruction line for easy reference.

Synthesis Exercises

Synthesis exercises help build critical-thinking skills by requiring students to synthesize or combine skills taught in the current section along with those from preceding sections.

Real-Data Applications

This text encourages students to see and interpret the mathematics that appears every day in the world around them. Throughout the writing process, an energetic search for real-data applications was conducted, and the result is a variety of examples and exercises that connect the mathematical content with the real world. Most of these applications feature source lines and frequently include charts and graphs.

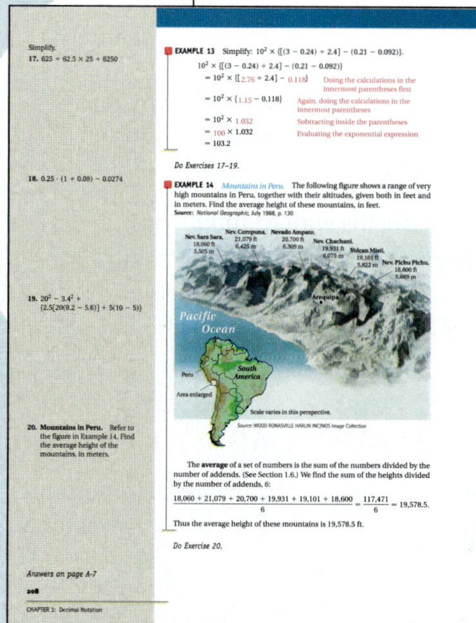

Simplify.

17. $625 \div 62.5 \times 25 \div 6250$

18. $0.25 \cdot (1 + 0.08) - 0.0274$

19. $20^2 - 3.4^2 + \{2.5[20(9.2 - 5.6)] + 5(10 - 5)\}$

20. **Mountains in Peru.** Refer to the figure in Example 14. Find the average height of the mountains, in meters.

EXAMPLE 13 Simplify: $10^2 \times \{[(3 - 0.24) \div 2.4] - (0.21 - 0.092)\}$.

$$10^2 \times \{[(3 - 0.24) \div 2.4] - (0.21 - 0.092)\}$$
$$= 10^2 \times \{[2.76 \div 2.4] - 0.118\} \quad \text{Doing the calculations in the innermost parentheses first}$$
$$= 10^2 \times \{1.15 - 0.118\} \quad \text{Again, doing the calculations in the innermost parentheses}$$
$$= 10^2 \times 1.032 \quad \text{Subtracting inside the parentheses}$$
$$= 100 \times 1.032 \quad \text{Evaluating the exponential expression}$$
$$= 103.2$$

Do Exercises 17–19.

EXAMPLE 14 *Mountains in Peru.* The following figure shows a range of very high mountains in Peru, together with their altitudes, given both in feet and in meters. Find the average height of these mountains, in feet.

Source: *National Geographic,* July 1968, p. 130

Source: WOOD RONASVILLE HARLIN INC/NGS Image Collection

The **average** of a set of numbers is the sum of the numbers divided by the number of addends. (See Section 1.6.) We find the sum of the heights divided by the number of addends, 6:

$$\frac{18{,}060 + 21{,}079 + 20{,}700 + 19{,}931 + 19{,}101 + 18{,}600}{6} = \frac{117{,}471}{6} = 19{,}578.5.$$

Thus the average height of these mountains is 19,578.5 ft.

Do Exercise 20.

Answers on page A-7

208

CHAPTER 3: Decimal Notation

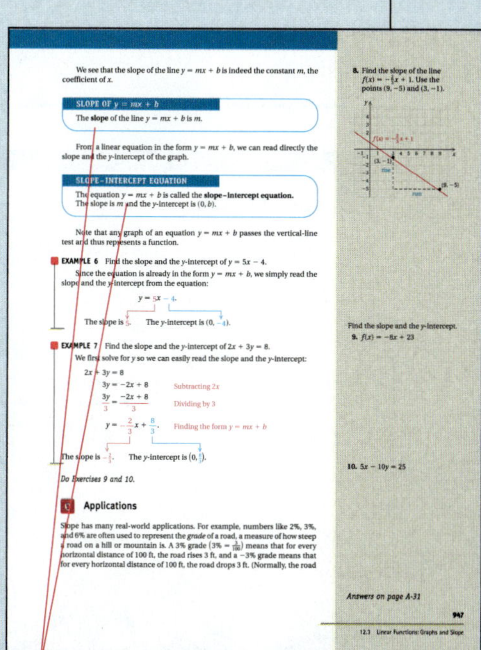

We see that the slope of the line $y = mx + b$ is indeed the constant m, the coefficient of x.

SLOPE OF $y = mx + b$

The **slope** of the line $y = mx + b$ is m.

From a linear equation in the form $y = mx + b$, we can read directly the slope and the y-intercept of the graph.

SLOPE–INTERCEPT EQUATION

The equation $y = mx + b$ is called the **slope–intercept equation.** The slope is m and the y-intercept is $(0, b)$.

Note that any graph of an equation $y = mx + b$ passes the vertical-line test and thus represents a function.

EXAMPLE 6 Find the slope and the y-intercept of $y = 5x - 4$.

Since the equation is already in the form $y = mx + b$, we simply read the slope and the y-intercept from the equation:

$$y = 5x - 4.$$

The slope is 5. The y-intercept is $(0, -4)$.

EXAMPLE 7 Find the slope and the y-intercept of $2x + 3y = 8$.

We first solve for y so we can easily read the slope and the y-intercept:

$$2x + 3y = 8$$
$$3y = -2x + 8 \quad \text{Subtracting } 2x$$
$$\frac{3y}{3} = \frac{-2x + 8}{3} \quad \text{Dividing by 3}$$
$$y = -\frac{2}{3}x + \frac{8}{3}. \quad \text{Finding the form } y = mx + b$$

The slope is $-\frac{2}{3}$. The y-intercept is $(0, \frac{8}{3})$.

Do Exercises 9 and 10.

Applications

Slope has many real-world applications. For example, numbers like 2%, 3%, and 6% are often used to represent the *grade* of a road, a measure of how steep a road on a hill or mountain is. A 3% grade $(3\% = \frac{3}{100})$ means that for every horizontal distance of 100 ft, the road rises 3 ft, and a −3% grade means that for every horizontal distance of 100 ft, the road drops 3 ft. (Normally, the road

8. Find the slope of the line $f(x) = -\frac{2}{3}x + 1$. Use the points $(9, -5)$ and $(3, -1)$.

Find the slope and the y-intercept.

9. $f(x) = -8x + 23$

10. $5x - 10y = 25$

Answers on page A-31

947

12.3 Linear Functions: Graphs and Slope

Highlighted Information

Important definitions, rules, and procedures are highlighted in titled boxes.

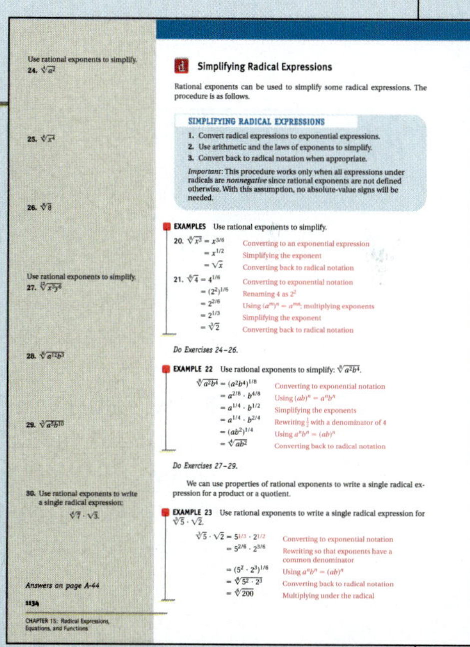

Use rational exponents to simplify.

24. $\sqrt[4]{a^2}$

25. $\sqrt[4]{x^4}$

26. $\sqrt[6]{8}$

Use rational exponents to simplify.

27. $\sqrt[12]{x^3y^6}$

28. $\sqrt[6]{a^{12}b^3}$

29. $\sqrt[5]{a^5b^{10}}$

30. Use rational exponents to write a single radical expression: $\sqrt[4]{7} \cdot \sqrt{3}$.

d Simplifying Radical Expressions

Rational exponents can be used to simplify some radical expressions. The procedure is as follows.

SIMPLIFYING RADICAL EXPRESSIONS

1. Convert radical expressions to exponential expressions.
2. Use arithmetic and the laws of exponents to simplify.
3. Convert back to radical notation when appropriate.

Important: This procedure works only when all expressions under radicals are *nonnegative* since rational exponents are not defined otherwise. With this assumption, no absolute-value signs will be needed.

EXAMPLES Use rational exponents to simplify.

20. $\sqrt[6]{x^3} = x^{3/6}$ Converting to an exponential expression
$= x^{1/2}$ Simplifying the exponent
$= \sqrt{x}$ Converting back to radical notation

21. $\sqrt[6]{4} = 4^{1/6}$ Converting to exponential notation
$= (2^2)^{1/6}$ Renaming 4 as 2^2
$= 2^{2/6}$ Using $(a^m)^n = a^{mn}$; multiplying exponents
$= 2^{1/3}$ Simplifying the exponent
$= \sqrt[3]{2}$ Converting back to radical notation

Do Exercises 24–26.

EXAMPLE 22 Use rational exponents to simplify: $\sqrt[8]{a^2b^4}$.

$\sqrt[8]{a^2b^4} = (a^2b^4)^{1/8}$ Converting to exponential notation
$= a^{2/8} \cdot b^{4/8}$ Using $(ab)^n = a^nb^n$
$= a^{1/4} \cdot b^{1/2}$ Simplifying the exponents
$= a^{1/4} \cdot b^{2/4}$ Rewriting $\frac{1}{2}$ with a denominator of 4
$= (ab^2)^{1/4}$ Using $a^nb^n = (ab)^n$
$= \sqrt[4]{ab^2}$ Converting back to radical notation

Do Exercises 27–29.

We can use properties of rational exponents to write a single radical expression for a product or a quotient.

EXAMPLE 23 Use rational exponents to write a single radical expression for $\sqrt[3]{5} \cdot \sqrt{2}$.

$\sqrt[3]{5} \cdot \sqrt{2} = 5^{1/3} \cdot 2^{1/2}$ Converting to exponential notation
$= 5^{2/6} \cdot 2^{3/6}$ Rewriting so that exponents have a common denominator
$= (5^2 \cdot 2^3)^{1/6}$ Using $a^nb^n = (ab)^n$
$= \sqrt[6]{5^2 \cdot 2^3}$ Converting back to radical notation
$= \sqrt[6]{200}$ Multiplying under the radical

Answers on page A-44

1134

CHAPTER 15: Radical Expressions, Equations, and Functions

Annotated Examples

Detailed annotations and color highlights lead the student through the structured steps of the examples.

END-OF-CHAPTER MATERIAL

At the end of each chapter, students can practice all they have learned as well as tie the current chapter material to material covered in earlier chapters.

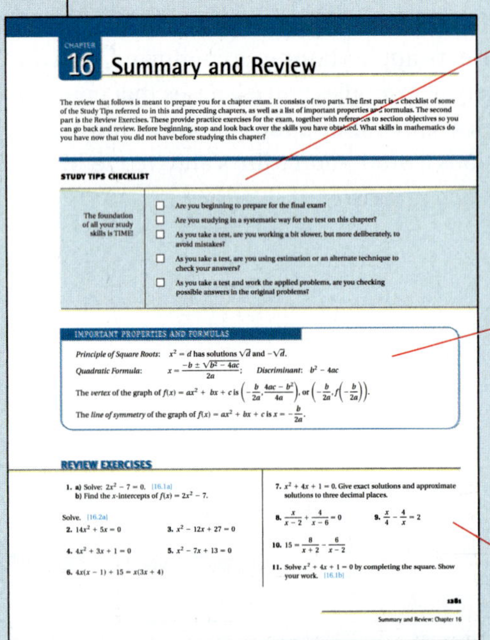

Study Tips Checklist

Each chapter review begins with a Study Tips Checklist that reviews Study Tips introduced in the current and previous chapters, making the use of these Study Tips more interactive.

Important Properties and Formulas

In many chapters, students are given a list of important properties and formulas, providing them with a valuable study aid.

Review Exercises

At the end of each chapter, students are provided with an extensive set of Review exercises. Reference codes beside each exercise or instruction line allow students to review the related objective easily.

Chapter Test

Following the Review exercises, a sample Chapter Test allows students to review and test comprehension of chapter skills prior to taking an instructor's exam.

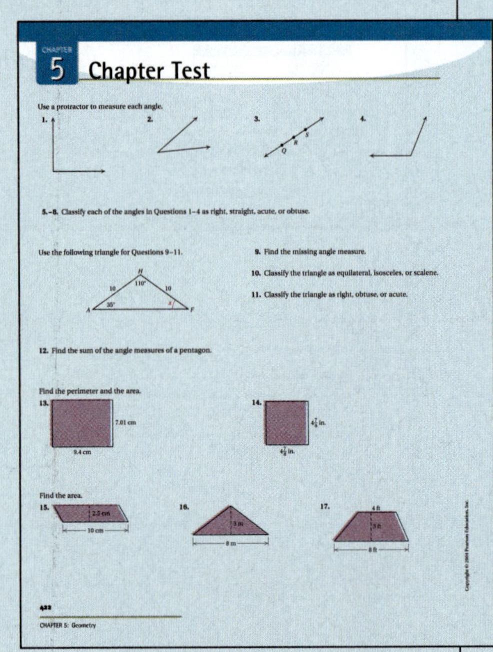

Foundations of Mathematics

Whole Numbers

Gateway to Chapter 1

In this chapter, we consider addition, subtraction, multiplication, and division of whole numbers, as well as exponential notation and order of operations. We also introduce the idea of using variables to form equations. Then we solve simple equations and use the skills of this chapter to solve applied problems.

Factorizations, divisibility, and least common multiples are also studied.

Real-World Application

Boeing Corporation builds commercial aircraft. A Boeing 767 has a seating configuration with 4 rows of 6 seats across in first class and 35 rows of 7 seats across in economy class. Find the total seating capacity of the plane.

Sources: Boeing Corporation; Delta Airlines

This problem appears as Example 8 in Section 1.5.

CHAPTER 1

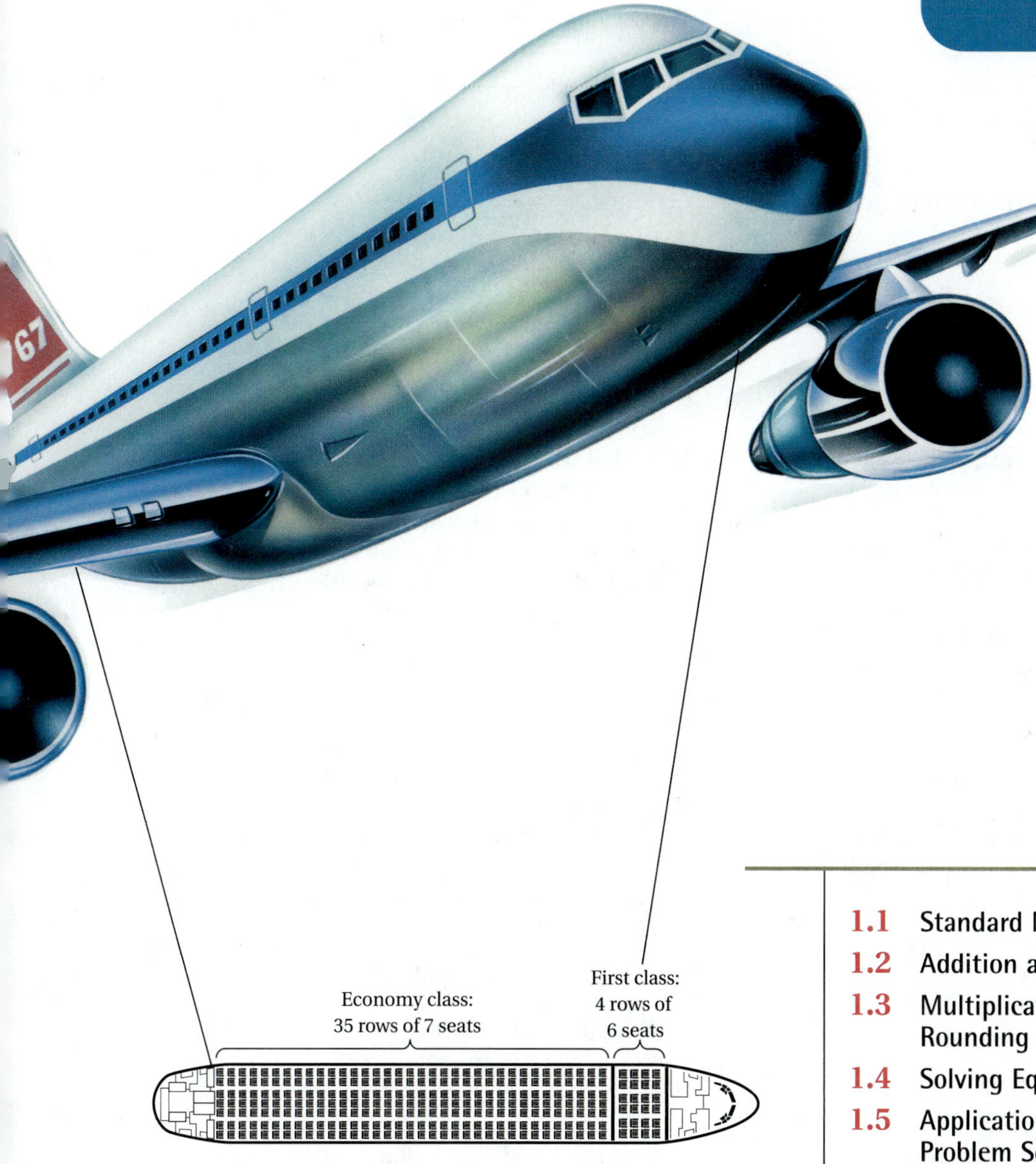

67
Economy class:
35 rows of 7 seats
First class:
4 rows of
6 seats

CHAPTER 1

Pretest

1. Write a word name: 3,078,059. [1.1c]

2. Write expanded notation: 6987. [1.1b]

3. Write standard notation: Two billion, forty-seven million, three hundred ninety-eight thousand, five hundred eighty-nine. [1.1c]

4. What does the digit 6 mean in 2,967,342? [1.1a]

5. Round 956,449 to the nearest thousand. [1.3e]

6. Estimate the product $594 \cdot 126$ by first rounding the numbers to the nearest hundred. [1.3f]

7. Add. [1.2a]

$$\begin{array}{r} 7\,3\,1\,2 \\ +\ 2\,9\,0\,4 \\ \hline \end{array}$$

8. Subtract. [1.2d]

$$\begin{array}{r} 7\,0\,1\,2 \\ -\ 2\,9\,0\,4 \\ \hline \end{array}$$

9. Multiply: $359 \cdot 64$. [1.3a]

10. Divide: $23{,}149 \div 46$. [1.3d]

Use either < or > for □ to write a true sentence. [1.1d]

11. 346 □ 364

12. 54 □ 45

Solve. [1.4b]

13. $326 \cdot 17 = m$

14. $y = 924 \div 42$

15. $19 + x = 53$

16. $34 \cdot n = 850$

Solve. [1.5a]

17. **Paper Quantity.** There are 500 sheets in a ream of paper. How many sheets are in 9 reams?

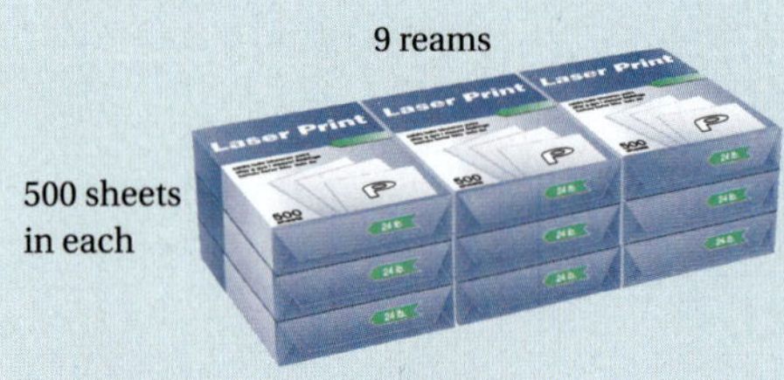

18. **Digital Cameras.** A group of 63 language students from VaMard University is planning a year abroad to study German. They decide that each of them will buy a digital camera like the one shown in the ad below. The total cost of the purchase is $18,837. What is the cost per camera?

19. **Checking Account.** You have $756 in your checking account. Using your debit card, you pay $387 for a VCR for your dorm room. How much is left in your account?

20. **College Costs.** It has been estimated that by 2012, the costs of each of the four years of college will be $7383, $7359, $7925, and $8126. Find the total cost of four years of college at that time.

21. Evaluate: 4^3. [1.6b]

22. Find the LCM of 15 and 24. [1.9a]

Simplify.

23. $8^2 \div 8 \cdot 2 - (2 + 2 \cdot 7)$ [1.6c]

24. $108 \div 9 - \{3 \cdot [18 - (5 \cdot 3)]\}$ [1.6d]

25. Determine whether 59 is prime, composite, or neither. [1.7c]

26. Find the prime factorization of 420. [1.7d]

27. Determine whether 1503 is divisible by 9. [1.8a]

28. Determine whether 768 is divisible by 6. [1.8a]

1.1 STANDARD NOTATION; ORDER

Objectives

a. Give the meaning of digits in standard notation.

b. Convert between standard notation and expanded notation.

c. Convert between standard notation and word names.

d. Use $<$ or $>$ for $\square$ to write a true sentence in a situation like $6 \square 10$.

We study mathematics in order to be able to solve problems. In this section, we study how numbers are named. We begin with the concept of place value.

a Place Value

Consider the number named in the following ad.

BURGER KING
BURGER KING® sells
1,305,716,519
WHOPPER® sandwiches each year

Source: ™ and ©2000 Burger King Brands, Inc.

A **digit** is a number 0, 1, 2, 3, 4, 5, 6, 7, 8, or 9 that names a place-value location. For large numbers, digits are separated by commas into groups of three, called **periods.** Each period has a name: *ones, thousands, millions, billions, trillions,* and so on. To understand the number in the ad, we can use a **place-value chart,** as shown below.

PLACE-VALUE CHART

Periods →	Trillions			Billions			Millions			Thousands			Ones		
						1	3	0	5	7	1	6	5	1	9
	Hundreds	Tens	Ones	Hundreds	Tens	Ones	Hundreds	Tens	Ones	Hundreds	Tens	Ones	Hundreds	Tens	Ones

1 billion, 305 million, 716 thousand, 519 ones

EXAMPLES What does the digit 8 mean in each number?

1. 278,342 — 8 thousands
2. 872,342 — 8 hundred thousands
3. 28,343,399,223 — 8 billions

Do Margin Exercises 1–4.

What does the digit 2 mean in each number?

1. 526,555

2. 265,789

3. 42,789,654

4. 24,789,654

5. **Golf Balls.** It is estimated that in one day Americans buy 486,575 golf balls. What does each digit name?
Source: U.S. Golf Association

Answers on page A-1

Write expanded notation.

6. 1895

7. $22,132, the average salary for a flight attendant in 1990

8. 3031 mi (miles), the diameter of Mercury

9. 4100 mi, the length of the Nile River, the longest in the world

10. 3860 mi, the length of the Missouri–Mississippi River, the longest in the United States

Answers on page A-1

EXAMPLE 4 *Pacific Ocean.* The area of the Pacific Ocean is 64,186,000 square miles. What does each digit name?

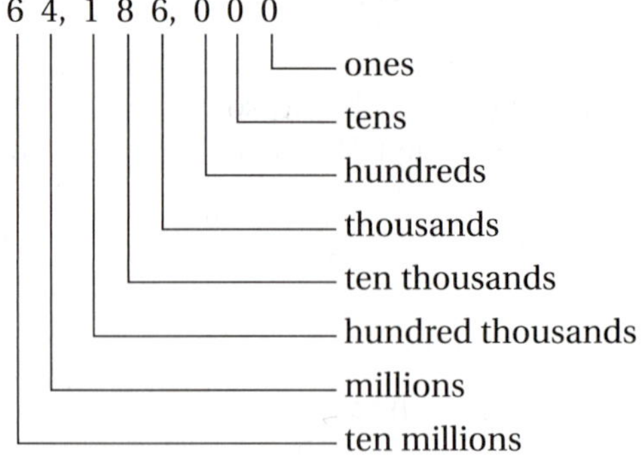

Do Exercise 5 on the preceding page.

b Converting Between Standard Notation and Expanded Notation

To answer questions such as "How many?", "How much?", and "How tall?", we use whole numbers. The set, or collection, of **whole numbers** is

0, 1, 2, 3, 4, 5, 6, 7, 8, 9, 10, 11, 12, ...

The set goes on indefinitely. There is no largest whole number, and the smallest whole number is 0. Each whole number can be named using various notations. The set 1, 2, 3, 4, 5, . . . , without 0, is called the set of **natural numbers.**

Let's look at the data from the line graph shown here.

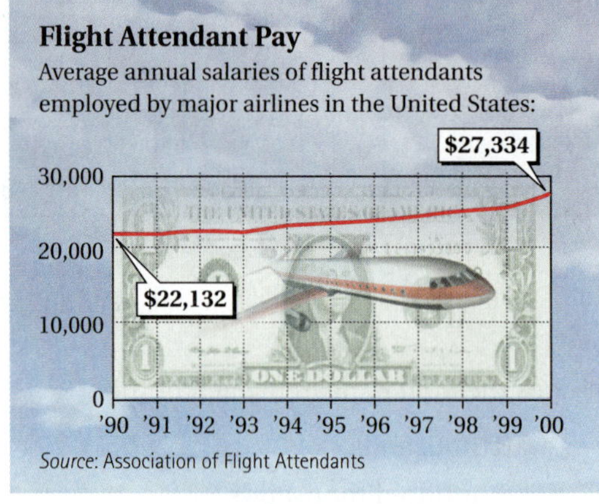

The average salary for a flight attendant in 2000 was $27,334. **Standard notation** for the salary is 27,334. We write **expanded notation** for 27,334 as follows:

27,334 = 2 ten thousands + 7 thousands
+ 3 hundreds + 3 tens + 4 ones.

EXAMPLE 5 Write expanded notation for 4218 mi, the diameter of Mars.

4218 = 4 thousands + 2 hundreds + 1 ten + 8 ones

EXAMPLE 6 Write expanded notation for 3400.

3400 = 3 thousands + 4 hundreds + 0 tens + 0 ones, or
3 thousands + 4 hundreds

Do Exercises 6–10 on the preceding page.

EXAMPLE 7 Write standard notation for 9 ten thousands + 6 thousands + 7 hundreds + 1 ten + 8 ones.

Standard notation is 96,718.

EXAMPLE 8 Write standard notation for 2 thousands + 3 tens.

Standard notation is 2030.

Do Exercises 11–13.

C Converting Between Standard Notation and Word Names

We often use **word names** for numbers. When we pronounce a number, we are speaking its word name. The People's Republic of China won 59 medals in the 2000 Summer Olympics in Sydney, Australia. A word name for 59 is "fifty-nine." Word names for some two-digit numbers like 59, 76, and 97 use hyphens. Others like 17 use only one word, "seventeen." Let's write some word names.

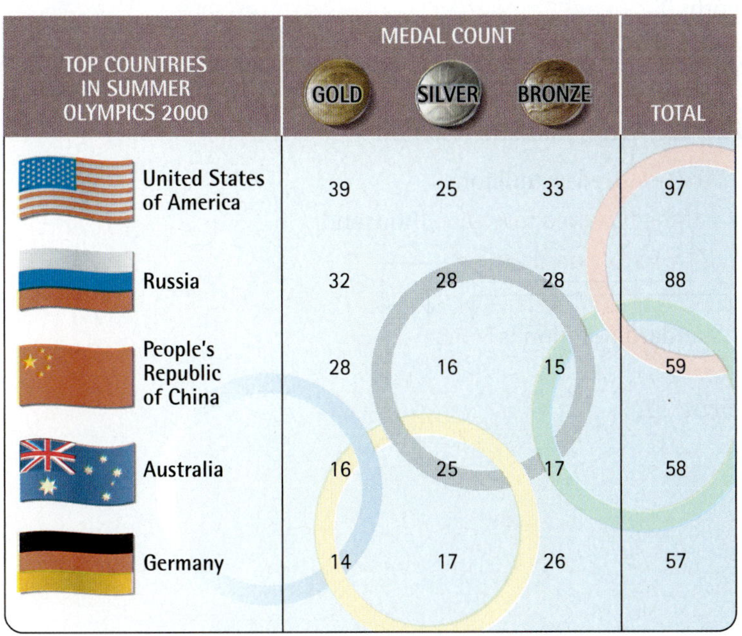

TOP COUNTRIES IN SUMMER OLYMPICS 2000	MEDAL COUNT			
	GOLD	SILVER	BRONZE	TOTAL
United States of America	39	25	33	97
Russia	32	28	28	88
People's Republic of China	28	16	15	59
Australia	16	25	17	58
Germany	14	17	26	57

Source: 2000 Olympics, Sydney, Australia

Write standard notation.

11. 5 thousands + 6 hundreds + 8 tens + 9 ones

12. 8 ten thousands + 7 thousands + 1 hundred + 2 tens + 8 ones

13. 9 thousands + 3 ones

Write a word name. (Refer to the figure at left.)

14. 88, the total number of medals won by Russia

15. 16, the number of silver medals won by the People's Republic of China

16. 32, the number of gold medals won by Russia

Answers on page A-1

Write a word name.

17. 204

18. $43,782, the average salary in 1998 for those who have a bachelor's degree
Source: U.S. Bureau of the Census

19. 1,879,204

20. 6,259,600,000, the world population in 2000
Source: U.S. Bureau of the Census

21. Write standard notation.

Two hundred thirteen million, one hundred five thousand, three hundred twenty-nine

Answers on page A-1

EXAMPLES Write a word name.

9. 97, the total number of medals won by the United States

Ninety-seven

10. 15, the number of bronze medals won by the People's Republic of China

Fifteen

Do Exercises 14–16 on the preceding page.

For word names for larger numbers, we begin at the left with the largest period. The number named in the period is followed by the name of the period; then a comma is written and the next period is named.

EXAMPLE 11 Write a word name for 46,605,314,732.

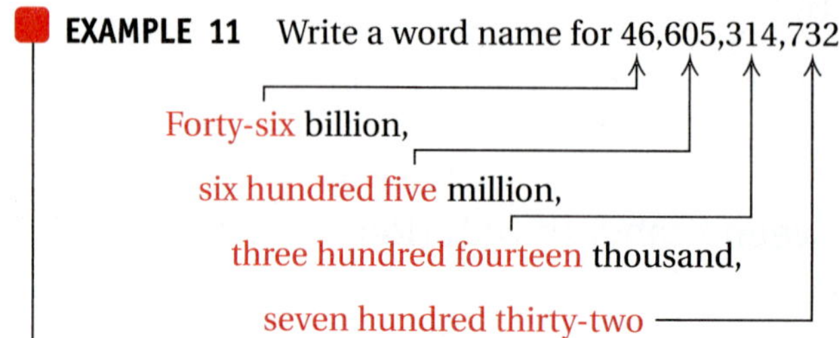

The word "and" *should not* appear in word names for whole numbers. Although we commonly hear such expressions as "two hundred *and* one," the use of "and" is not, strictly speaking, correct in word names for whole numbers. For decimal notation, it is appropriate to use "and" for the decimal point. For example, 317.4 is read as "three hundred seventeen *and* four tenths."

Do Exercises 17–20.

EXAMPLE 12 Write standard notation.

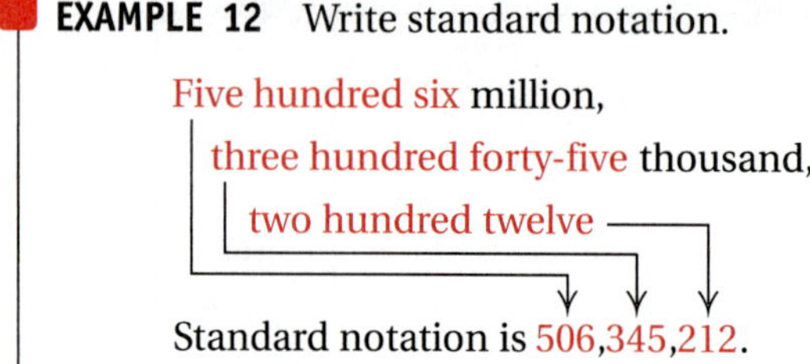

Standard notation is 506,345,212.

Do Exercise 21.

d Order

We know that 2 is not the same as 5. We express this by the sentence $2 \neq 5$. We also know that 2 is less than 5. We symbolize this by the expression $2 < 5$. We can see this order on a number line: 2 is to the left of 5.

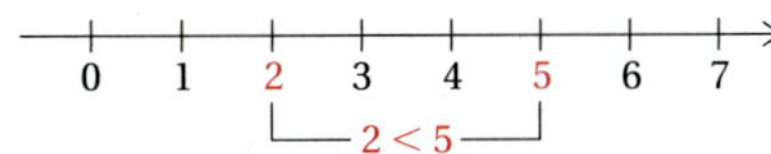

The number 0 is the smallest whole number.

ORDER OF WHOLE NUMBERS

For any whole numbers a and b:

1. $a < b$ (read "a is less than b") is true when a is to the left of b on a number line.
2. $a > b$ (read "a is greater than b") is true when a is to the right of b on a number line.

We call $<$ and $>$ **inequality symbols.**

EXAMPLE 13 Use $<$ or $>$ for $\square$ to write a true sentence: $7 \square 11$.

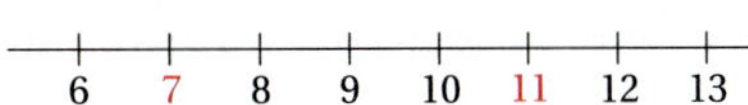

Since 7 is to the left of 11 on a number line, $7 < 11$.

EXAMPLE 14 Use $<$ or $>$ for $\square$ to write a true sentence: $92 \square 87$.

86 87 88 89 90 91 92 93

Since 92 is to the right of 87 on a number line, $92 > 87$.

A sentence like $8 + 5 = 13$ is called an **equation.** It is a *true* equation. The equation $4 + 8 = 11$ is a *false* equation. A sentence like $7 < 11$ is called an **inequality.** The sentence $7 < 11$ is a *true* inequality. The sentence $23 > 69$ is a *false* inequality.

Do Exercises 22–27.

Use $<$ or $>$ for $\square$ to write a true sentence. Draw a number line if necessary.

22. $8 \square 12$

23. $12 \square 8$

24. $76 \square 64$

25. $64 \square 76$

26. $217 \square 345$

27. $345 \square 217$

Answers on page A-1

Study Tips

Throughout this textbook, you will find a feature called *Study Tips*. These tips are intended to help improve your math study skills. On the first day of class, you should complete this form.

BASIC INFORMATION ON THE FIRST DAY OF CLASS

Instructor: Name ______________________

Office hours and location

____

Phone number ______________________

Fax number ______________________

e-mail address ______________________

Find the names of two students whom you could contact for information or study questions:

1. Name ______________________

   Phone number ______________________

   e-mail address ______________________

2. Name ______________________

   Phone number ______________________

   e-mail address ______________________

Math lab on Campus:

Location ______________________

Hours ______________________

Phone ______________________

Tutoring:

Campus location ______________________

Hours ______________________

AW Math Tutor Center ______________________

To order, call ______________________.

(See the Preface for important information concerning this tutoring.)

Important Supplements:
(See the Preface for a complete list of available supplements.)

Supplements recommended by the instructor

____

____

"I know the price of success: dedication, hard work, and an unremitting devotion to the things you want to see happen."

Frank Lloyd Wright, architect

1.1 EXERCISE SET

For Extra Help

Digital Video Tutor CD 1 Videotape 1 | InterAct Math | Math Tutor Center | MathXL | MyMathLab.com

a What does the digit 5 mean in each case?

1. 235,888

2. 253,777

3. 1,488,526

4. 500,736

Skiers. In the 1999–2000 ski season, Vail, Colorado, had 1,370,000 skiers. In the number 1,370,000, what digit names the number of:

Source: *Denver Post*

5. Ones?

6. Ten thousands?

7. Millions?

8. Hundred thousands?

b Write expanded notation.

9. 5702

10. 3097

11. 93,986

12. 38,453

Step-Climbing Races. Races in which runners climb the steps inside a building are called "run-up" races. The graph below shows the number of steps in four buildings. In Exercises 13–16, write expanded notation for the number of steps in each race.

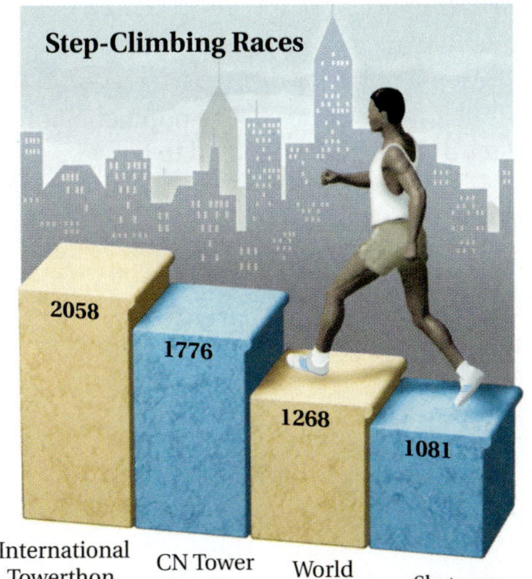

Source: New York Road Runners Club

13. 2058 steps in the International Towerthon, Kuala Lumpur, Malaysia

14. 1776 steps in the CN Tower Run-Up, Toronto, Ontario, Canada

15. 1268 steps in the World Financial Center, New York

16. 1081 steps in the Skytower Run-Up, Auckland, New Zealand

Write standard notation.

17. 2 thousands + 4 hundreds + 7 tens + 5 ones

18. 7 thousands + 9 hundreds + 8 tens + 3 ones

19. 6 ten thousands + 8 thousands + 9 hundreds + 3 tens + 9 ones

20. 1 ten thousand + 8 thousands + 4 hundreds + 6 tens + 1 one

21. 7 thousands + 3 hundreds + 0 tens + 4 ones

22. 8 thousands + 0 hundreds + 2 tens + 0 ones

23. 1 thousand + 9 ones

24. 2 thousands + 4 hundreds + 5 tens

C Write a word name.

25. 85

26. 48

27. 88,000

28. 45,987

29. 123,765

30. 111,013

31. 7,754,211,577

32. 43,550,651,808

Write standard notation.

33. Two million, two hundred thirty-three thousand, eight hundred twelve

34. Three hundred fifty-four thousand, seven hundred two

35. Eight billion

36. Seven hundred million

Write a word name for the number in each sentence.

37. *Great Pyramid.* The area of the base of the Great Pyramid in Egypt is 566,280 square feet.

38. *Population of the United States.* The population of the United States in 2000 was estimated to be 273,540,000.
Source: U.S. Bureau of the Census

39. *Monopoly.* In a recent Monopoly® game sponsored by McDonalds® restaurants, the odds of winning the grand prize were estimated to be 467,322,388 to 1.
Source: McDonald's Corporation

40. *Native American Population.* In a recent year, the population of Native Americans in Arizona was 165,385.
Source: U.S. Bureau of the Census

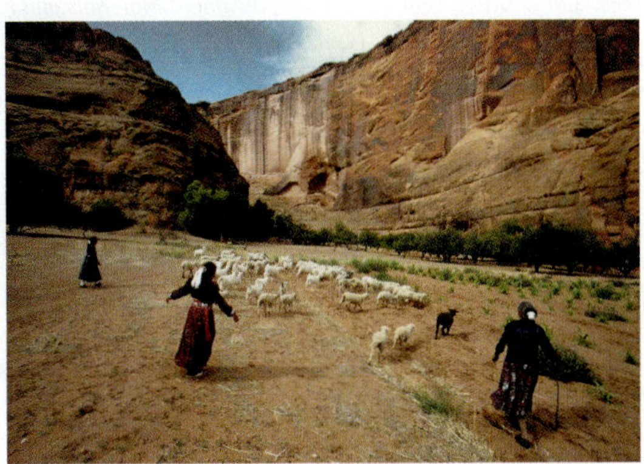

Write standard notation for the number in each sentence.

41. Light travels nine trillion, four hundred sixty billion kilometers in one year.

42. The distance from the sun to Pluto is three billion, six hundred sixty-four million miles.

43. *Pacific Ocean.* The area of the Pacific Ocean is sixty-four million, one hundred eighty-six thousand square miles.

44. *Gigabyte.* On a computer hard disk, one gigabyte is one billion, seventy-three million, seven hundred forty-one thousand, eight hundred twenty-four bytes of memory.

d Use < or > for □ to write a true sentence. Draw a number line if necessary.

45. 0 □ 17

46. 32 □ 0

47. 34 □ 12

48. 28 □ 18

49. 1000 □ 1001

50. 77 □ 117

51. 133 □ 132

52. 999 □ 997

53. 460 □ 17

54. 345 □ 456

55. 37 □ 11

56. 12 □ 32

Land-Speed Cars. Two competing jet-powered cars may soon travel faster than the speed of sound. The Thrust SCC is 54 ft long and weighs 7 tons. The Spirit of America is 47 ft long and weighs 4 tons. Use this information to answer Exercises 57 and 58.

Sources: *Car & Driver,* September 1996; *Advanced Materials and Processes,* January 1998

57. Which is longer, the Thrust SCC or the Spirit of America? Express the numbers in the situation as an inequality.

58. Which is heavier, the Thrust SCC or the Spirit of America? Express the numbers in the situation as an inequality.

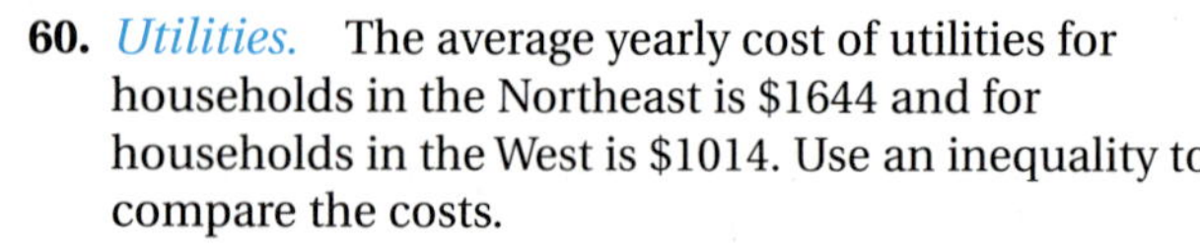

59. *Life Expectancy.* The life expectancy of a female in 2050 is predicted to be about 87 yr and of a male about 81 yr. Use an inequality to compare these life expectancies.

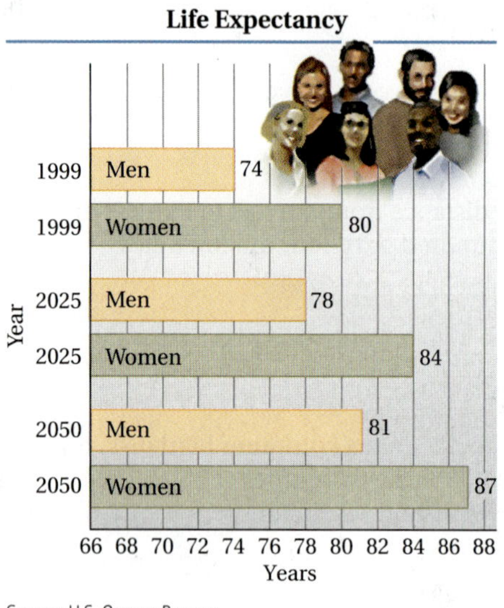

Source: U.S. Census Bureau

60. *Utilities.* The average yearly cost of utilities for households in the Northeast is \$1644 and for households in the West is \$1014. Use an inequality to compare the costs.

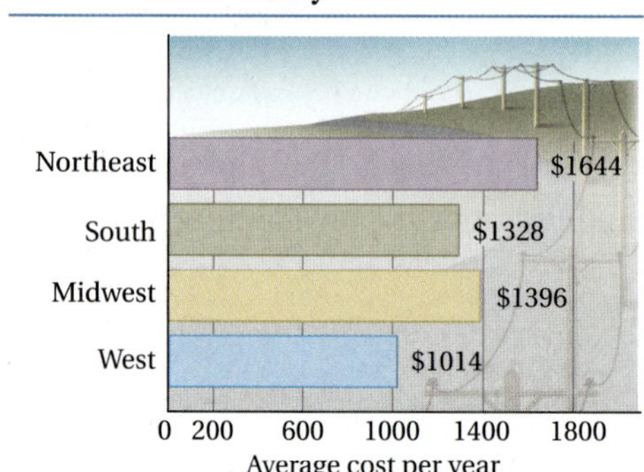

Source: Energy Information Administration

To the student and the instructor: The Discussion and Writing exercises are meant to be answered with one or more sentences. They can be discussed and answered collaboratively by the entire class or by small groups. Because of their open-ended nature, the answers to these exercises do not appear at the back of the book. They are denoted by the symbol $\mathbf{D_W}$.

61. $\mathbf{D_W}$ Explain why we use commas when writing large numbers.

62. $\mathbf{D_W}$ Write an English sentence in which the number 370,000,000 is used.

SYNTHESIS

To the student and the instructor: The Synthesis exercises found at the end of every exercise set challenge students to combine concepts or skills studied in that section or in preceding parts of the text. Exercises marked with a ▦ symbol are meant to be solved using a calculator.

63. How many whole numbers between 100 and 400 contain the digit 2 in their standard notation?

64. ▦ What is the largest number that you can name on your calculator? How many digits does that number have? How many periods?

1.2 ADDITION AND SUBTRACTION

Objectives

a Add whole numbers.

b Use addition in finding perimeter.

c Convert between addition sentences and subtraction sentences.

d Subtract whole numbers.

a Addition of Whole Numbers

Addition of whole numbers corresponds to combining or putting things together.

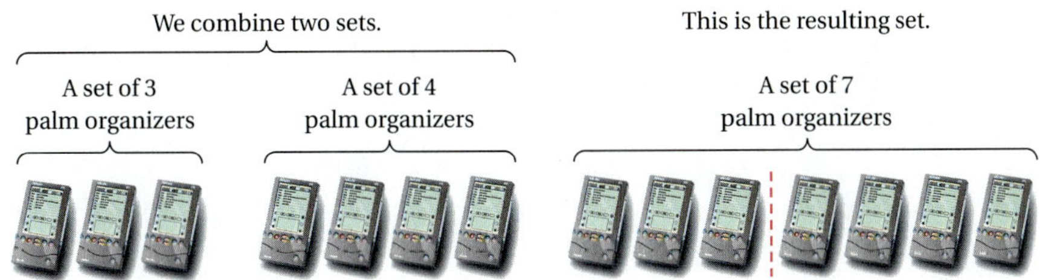

We combine two sets.
This is the resulting set.
A set of 3 palm organizers
A set of 4 palm organizers
A set of 7 palm organizers

The addition that corresponds to the figure above is $3 + 4 = 7$. The number of objects in a set can be found by counting. We count and find that the two sets have 3 palm organizers and 4 palm organizers, respectively. After combining, we count and find that there are 7 palm organizers. We say that the **sum** of 3 and 4 is 7. The numbers added are called **addends.**

Addition also corresponds to moving distances on a number line. The number line at right is marked with tick marks at equal distances of 1 *unit.* The sum $3 + 4$ is shown. We first move 3 units from 0, and then 4 more units, and end up at 7. The addition that corresponds to the situation is $3 + 4 = 7$.

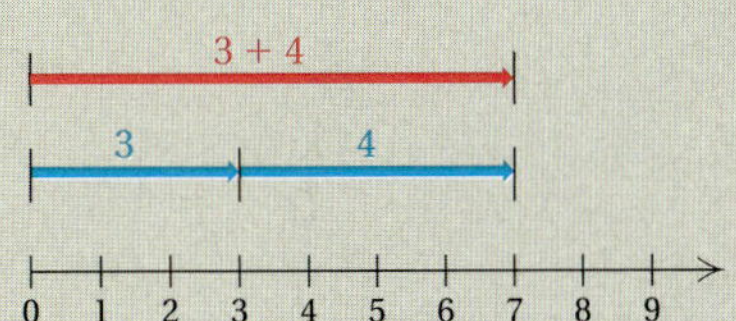

3 + 4
3
4
0 1 2 3 4 5 6 7 8 9

To add whole numbers, we add the ones digits first, then the tens, then the hundreds, then the thousands, and so on. Adding 0 to a number does not change the number: $a + 0 = 0 + a = a$. We say that 0 is the **additive identity.**

EXAMPLE 1 Add: $7312 + 2504$.

Place values are lined up in columns.

$$\begin{array}{r} 7\,3\,1\,2 \\ +\ 2\,5\,0\,4 \\ \hline 6 \end{array} \quad \text{Add ones.}$$

$$\begin{array}{r} 7\,3\,1\,2 \\ +\ 2\,5\,0\,4 \\ \hline 1\,6 \end{array} \quad \text{Add tens.}$$

$$\begin{array}{r} 7\,3\,1\,2 \\ +\ 2\,5\,0\,4 \\ \hline 8\,1\,6 \end{array} \quad \text{Add hundreds.}$$

$$\begin{array}{r} 7\,3\,1\,2 \\ +\ 2\,5\,0\,4 \\ \hline 9\,8\,1\,6 \end{array} \quad \text{Add thousands.}$$

We show you this for explanation.

You need write only this.

$$\begin{array}{r} 7\,3\,1\,2 \\ +\ 2\,5\,0\,4 \\ \hline 9\,8\,1\,6 \end{array} \quad \begin{array}{l} \leftarrow \text{Addends} \\ \\ \leftarrow \text{Sum} \end{array}$$

Do Exercises 1 and 2.

Add.

1. $74 + 23$

2. $\begin{array}{r} 6\,2\,0\,3 \\ +\ 3\,5\,4\,2 \\ \hline \end{array}$

Answers on page A-1

Add.

3.
```
  7968
+ 5497
```

4.
```
  9804
+ 6378
```

5.
```
  1932
  6723
  9878
+ 8941
```

Answers on page A-1

EXAMPLE 2 Add: 2391 + 3276 + 8789 + 1498.

```
      2
  2 3 9 1
  3 2 7 6
  8 7 8 9
+ 1 4 9 8
        4
```

Add ones. We get 24, so we have 2 tens + 4 ones. Write 4 in the ones column and 2 above the tens.

```
    3 2
  2 3 9 1
  3 2 7 6
  8 7 8 9
+ 1 4 9 8
      5 4
```

Add tens. We get 35 tens, so we have 30 tens + 5 tens. This is also 3 hundreds + 5 tens. Write 5 in the tens column and 3 above the hundreds.

```
  1 3 2
  2 3 9 1
  3 2 7 6
  8 7 8 9
+ 1 4 9 8
    9 5 4
```

Add hundreds. We get 19 hundreds, or 1 thousand + 9 hundreds. Write 9 in the hundreds column and 1 above the thousands.

```
  1 3 2
  2 3 9 1
  3 2 7 6
  8 7 8 9
+ 1 4 9 8
1 5 9 5 4
```

Add thousands. We get 15 thousands.

Do Exercises 3–5.

Study Tips

USING THIS TEXTBOOK

We began our "Study Tips" in Section 1.1. You will find many of these tips throughout the book. One of the most important ways in which to improve your math study skills is to learn the proper use of the textbook. Here we highlight a few points that we consider most helpful.

- **Be sure to note the special symbols** [a], [b], [c], **and so on, that correspond to the objectives you are to be able to perform.** The first time you see them is in the margin at the beginning of each section; the second time is in the subheadings of each section; and the third time is in the exercise set for the section. You will also find them next to the skill maintenance exercises in each exercise set and the review exercises at the end of the chapter, as well as in the answers to the chapter tests and the cumulative reviews. These objective symbols allow you to refer to the appropriate place in the text whenever you need to review a topic.
- **Read and study each step of each example.** The examples include important side comments that explain each step. These carefully chosen examples and notes prepare you for success in the exercise set.
- **Stop and do the margin exercises as you study a section.** Doing the margin exercises is one of the most effective ways to enhance your ability to learn mathematics from this text. Don't deprive yourself of this benefit!

b Finding Perimeter

Addition can be used when finding perimeter.

PERIMETER

The distance around an object is its **perimeter.**

EXAMPLE 3 A computer sales rep travels the following route to visit various electronics stores. How long is the route?

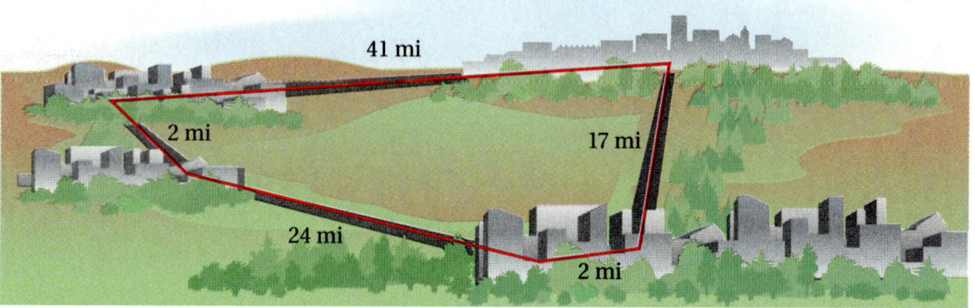

$$2 \text{ mi} + 24 \text{ mi} + 2 \text{ mi} + 17 \text{ mi} + 41 \text{ mi} = \text{Perimeter}$$

We carry out the addition as follows.

$$\begin{array}{r} \scriptstyle 1 \\ 2 \\ 24 \\ 2 \\ 17 \\ +\ 41 \\ \hline 86 \end{array}$$

The perimeter of the figure is 86 mi. The route is 86 mi long.

Do Exercises 6–8.

Solve.

6. Index Cards. Two standard sizes for index cards are 3 in. (inches) by 5 in. and 5 in. by 8 in. Find the perimeter of each card.

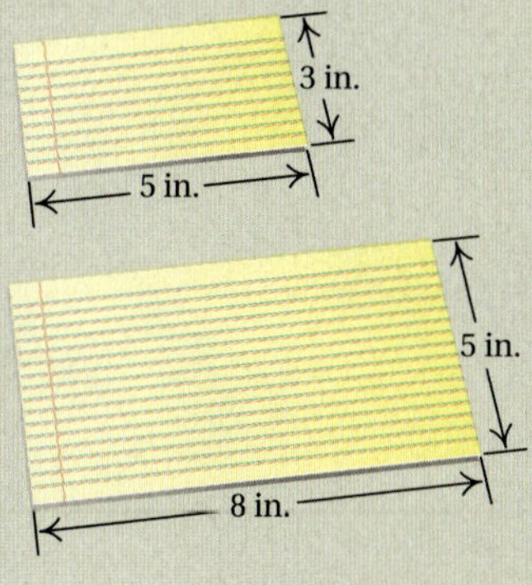

Find the perimeter of each figure.

7.

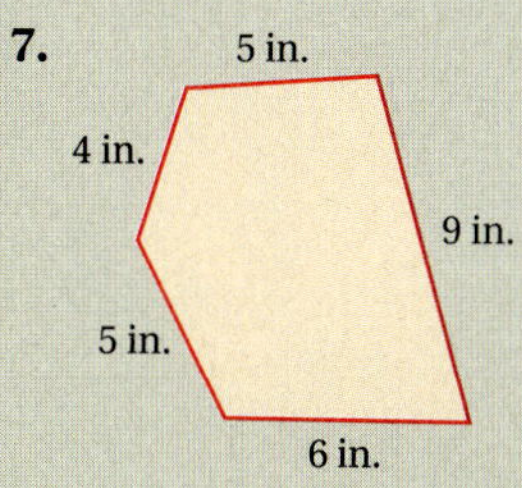

8.

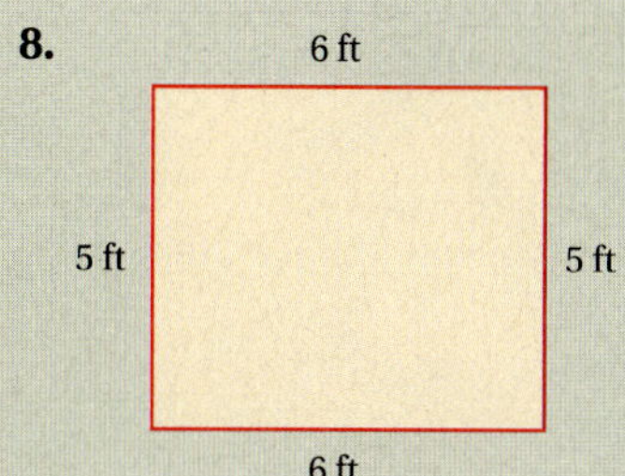

Answers on page A-1

C Subtraction and Related Sentences

TAKE AWAY

Subtraction of whole numbers applies to two kinds of situations. The first is called "take away." Consider the following example.

A bowler starts with 10 pins and knocks down 8 of them.

From 10 pins, the bowler "takes away" 8 pins. There are 2 pins left. The subtraction is $10 - 8 = 2$.

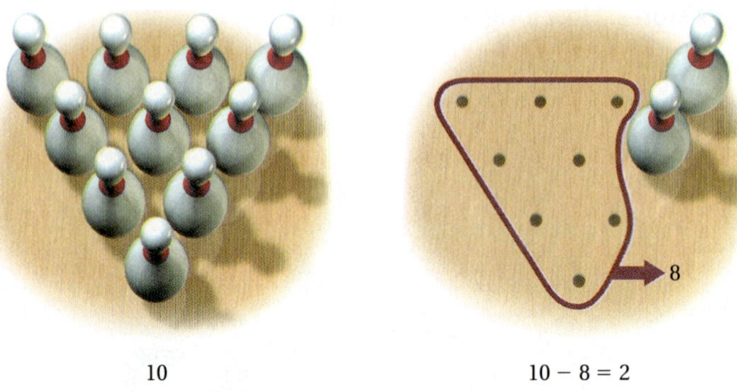

10 $\qquad$ $10 - 8 = 2$

We use the following terminology with subtraction:

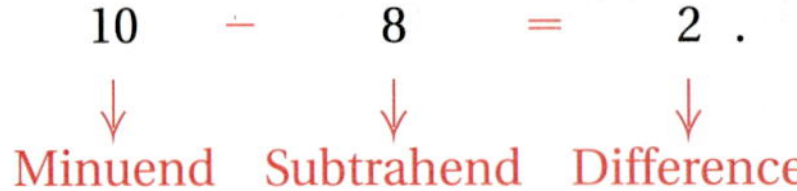

The **minuend** is the number from which another number is being subtracted. The **subtrahend** is the number being subtracted. The **difference** is the result of subtracting the subtrahend from the minuend.

Subtraction also corresponds to moving distances on a number line. The number line below is marked with tick marks at equal distances of 1 unit. The difference $10 - 8$ is shown. We first move from 0 right 10 units, and then left 8 units, and end up at 2. The subtraction that corresponds to the situation is $10 - 8 = 2$.

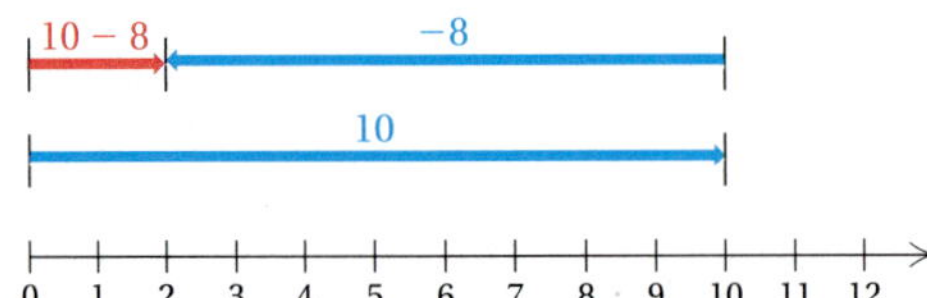

This leads us to the following definition of subtraction.

> **SUBTRACTION**
>
> The difference $a - b$ is that unique whole number c for which $a = c + b$.

RELATED SENTENCES

Subtraction is defined in terms of addition. For example, $5 - 2$ is that number which when added to 2 gives 5. Thus for the subtraction sentence

$5 - 2 = 3$, Taking away 2 from 5 gives 3.

there is a *related addition sentence*

$5 = 3 + 2$. Putting back the 2 gives 5 again.

In fact, we know that answers we find to subtractions are correct only because of the related addition, which provides a handy way to *check* a subtraction.

EXAMPLE 4 Write a related addition sentence: $8 - 5 = 3$.

$8 - 5 = 3$

This number gets added.

$8 = 3 + 5$

By the commutative law of addition, there is also another addition sentence:

$8 = 5 + 3$.

The related addition sentence is $8 = 3 + 5$.

Do Exercises 9 and 10.

EXAMPLE 5 Write two related subtraction sentences: $4 + 3 = 7$.

$4 + 3 = 7$

This addend gets subtracted from the sum.

$4 = 7 - 3$

(7 take away 3 is 4.)

$4 + 3 = 7$

This addend gets subtracted from the sum.

$3 = 7 - 4$

(7 take away 4 is 3.)

The related subtraction sentences are $4 = 7 - 3$ and $3 = 7 - 4$.

Do Exercises 11 and 12.

Write a related addition sentence.

9. $7 - 5 = 2$

10. $17 - 8 = 9$

Write two related subtraction sentences.

11. $5 + 8 = 13$

12. $11 + 3 = 14$

Answers on page A-1

Study Tips

HIGHLIGHTING

Reading and highlighting a section before your instructor lectures on it allows you to maximize your learning and understanding during the lecture.

- **Try to keep one section ahead of your syllabus.** If you study ahead of your lectures, you can concentrate on what is being explained in them, rather than trying to write everything down. You can then take notes only of special points or of questions related to what is happening in class.
- **Highlight important points.** You are probably used to highlighting key points as you study. If that works for you, continue to do so. But you will notice many design features throughout this book that already highlight important points. Thus you may not need to highlight as much as you generally do.
- **Highlight points that you do not understand.** Use a unique mark to indicate trouble spots that can lead to questions to be asked during class, in a tutoring session, or when calling or contacting the AW Math Tutor Center.

MISSING ADDEND

The second kind of situation to which subtraction can apply is called a "missing addend." You have 2 notebooks, but you need 7. You can think of this as "how many do I need to add to 2 to get 7?" Finding the answer can be thought of as finding a missing addend, and can be found by subtracting 2 from 7.

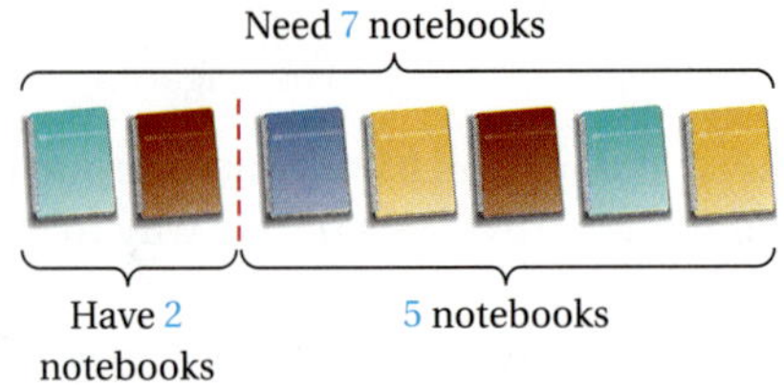

What must be added to 2 to get 7? The answer is 5.

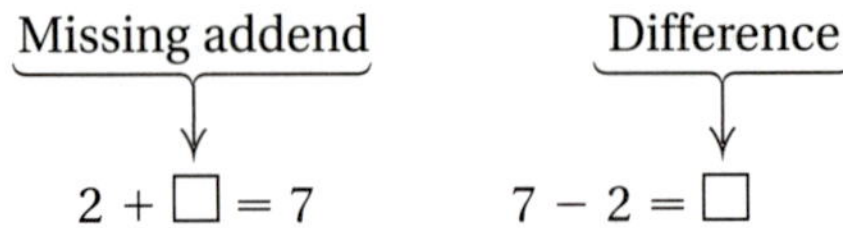

$2 + \square = 7 \qquad 7 - 2 = \square$

Let's look at the following example in which a missing addend occurs: Jason wants to buy the CD player shown in this ad. He has $30. He needs $79. How much more does he need in order to buy the CD player?

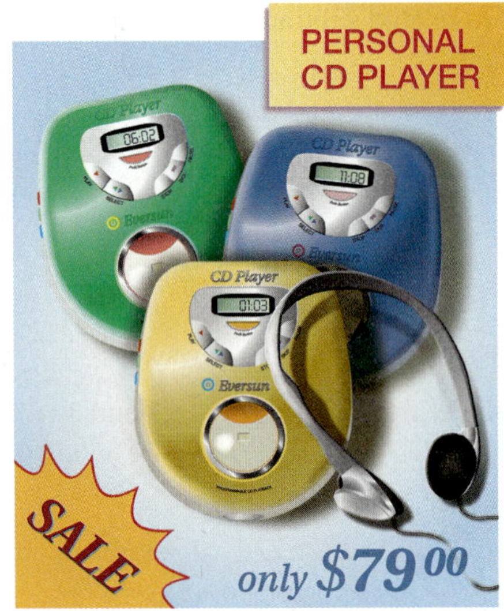

Thinking of this situation in terms of a missing addend, we have:

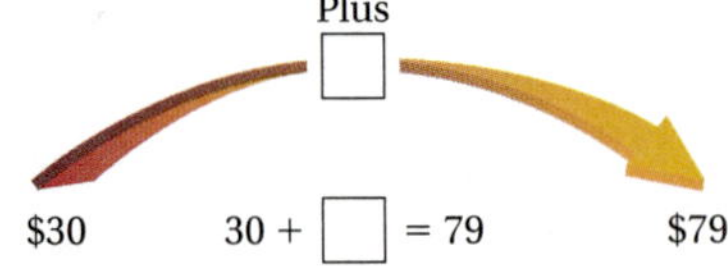

To find the answer, we think of the related subtraction sentence:

$$30 + \square = 79$$

$$\square = 79 - 30.$$

d Subtraction of Whole Numbers

To subtract numbers, we subtract the ones digits first, then the tens digits, then the hundreds, then the thousands, and so on.

EXAMPLE 6 Subtract: 9768 − 4320.

$$\begin{array}{r} 9\ 7\ 6\ 8 \\ -\ 4\ 3\ 2\ 0 \\ \hline 8 \end{array} \quad \text{Subtract ones.}$$

$$\begin{array}{r} 9\ 7\ 6\ 8 \\ -\ 4\ 3\ 2\ 0 \\ \hline 4\ 8 \end{array} \quad \text{Subtract tens.}$$

$$\begin{array}{r} 9\ 7\ 6\ 8 \\ -\ 4\ 3\ 2\ 0 \\ \hline 4\ 4\ 8 \end{array} \quad \text{Subtract hundreds.}$$

$$\begin{array}{r} 9\ 7\ 6\ 8 \\ -\ 4\ 3\ 2\ 0 \\ \hline 5\ 4\ 4\ 8 \end{array} \quad \text{Subtract thousands.}$$

This is for explanation.

$$\begin{array}{r} 9\ 7\ 6\ 8 \\ -\ 4\ 3\ 2\ 0 \\ \hline 5\ 4\ 4\ 8 \end{array}$$

You should write only this.

We have considered the subtraction 9768 − 4320 = □. That is, we have found the missing addend in the sentence 9768 = 4320 + □. If 5448 is indeed the missing addend, then if we add it to 4320, the answer should be 9768. The related addition sentence is the basis for adding as a *check*.

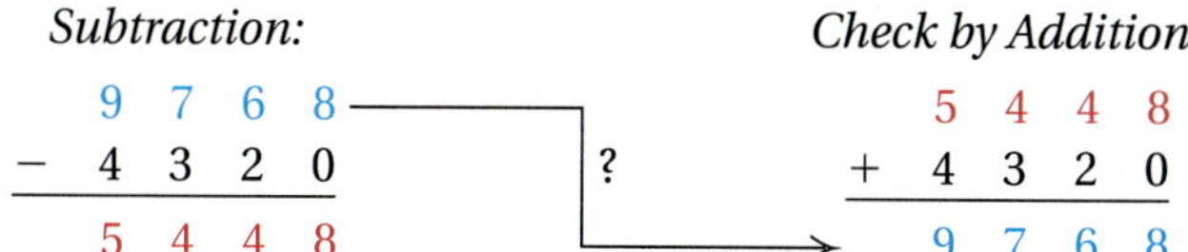

Do Exercise 13.

13. Subtract.

$$\begin{array}{r} 7\ 8\ 9\ 3 \\ -\ 4\ 0\ 9\ 2 \\ \hline \end{array}$$

Answer on page A-1

Subtract. Check by adding.

14. $\begin{array}{r} 8686 \\ -\ 2358 \\ \hline \end{array}$

15. $\begin{array}{r} 7145 \\ -\ 2398 \\ \hline \end{array}$

Subtract.

16. $\begin{array}{r} 70 \\ -\ 14 \\ \hline \end{array}$

17. $\begin{array}{r} 503 \\ -\ 298 \\ \hline \end{array}$

Subtract.

18. $\begin{array}{r} 7007 \\ -\ 6349 \\ \hline \end{array}$

19. $\begin{array}{r} 6000 \\ -\ 3149 \\ \hline \end{array}$

20. $\begin{array}{r} 9035 \\ -\ 7489 \\ \hline \end{array}$

Answers on page A-1

EXAMPLE 7 Subtract: 348 − 165.

We have

$$\begin{array}{rcl} 3 \text{ hundreds} + 4 \text{ tens} + 8 \text{ ones} & = & 2 \text{ hundreds} + 14 \text{ tens} + 8 \text{ ones} \\ -\ 1 \text{ hundred} - 6 \text{ tens} - 5 \text{ ones} & = & -\ 1 \text{ hundred} - 6 \text{ tens} - 5 \text{ ones} \\ \hline & = & 1 \text{ hundred} + 8 \text{ tens} + 3 \text{ ones} \\ & = & 183. \end{array}$$

Note that in this case, although we can subtract the ones (8 − 5 = 3), we cannot do so with the tens, because 4 − 6 is *not* a whole number. To see why, consider

$$4 - 6 = \square \quad \text{and the related addition sentence} \quad 4 = \square + 6.$$

There is no whole number that when added to 6 gives 4. To complete the subtraction, we must *borrow* 1 hundred from 3 hundreds and regroup it with 4 tens. Then we can do the subtraction 14 tens − 6 tens = 8 tens. Below we consider a shortened form.

$\begin{array}{r} 348 \\ -\ 165 \\ \hline 3 \end{array}$ Subtract ones.

$\begin{array}{r} {\scriptstyle 2\ 14} \\ \not{3}\,\not{4}\,8 \\ -\ 1\,6\,5 \\ \hline 3 \end{array}$ Borrow one hundred. That is, 1 hundred = 10 tens, and 10 tens + 4 tens = 14 tens. Write 2 above the hundreds column and 14 above the tens.

$\begin{array}{r} {\scriptstyle 2\ 14} \\ \not{3}\,\not{4}\,8 \\ -\ 1\,6\,5 \\ \hline 1\,8\,3 \end{array}$ Subtract tens; subtract hundreds.

EXAMPLE 8 Subtract: 6246 − 1879.

$\begin{array}{r} {\scriptstyle 3\ 16} \\ 6\,2\,\not{4}\,\not{6} \\ -\ 1\,8\,7\,9 \\ \hline 7 \end{array}$ We cannot subtract 9 ones from 6 ones, but we can subtract 9 ones from 16 ones. We borrow 1 ten to get 16 ones.

$\begin{array}{r} {\scriptstyle 13} \\ {\scriptstyle 1\ \not{3}\ 16} \\ 6\,\not{2}\,\not{4}\,\not{6} \\ -\ 1\,8\,7\,9 \\ \hline 6\,7 \end{array}$ We cannot subtract 7 tens from 3 tens, but we can subtract 7 tens from 13 tens. We borrow 1 hundred to get 13 tens.

$\begin{array}{r} {\scriptstyle 11\ 13} \\ {\scriptstyle 5\ \not{1}\ \not{3}\ 16} \\ \not{6}\,\not{2}\,\not{4}\,\not{6} \\ -\ 1\,8\,7\,9 \\ \hline 4\,3\,6\,7 \end{array}$ We cannot subtract 8 hundreds from 1 hundred, but we can subtract 8 hundreds from 11 hundreds. We borrow 1 thousand to get 11 hundreds.

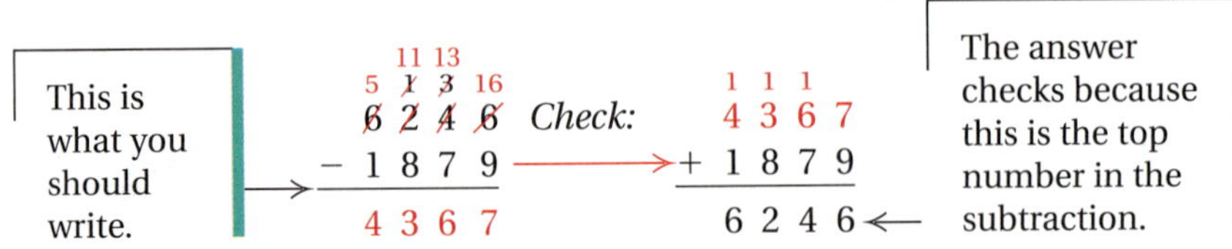

This is what you should write.
11 13
5 1 3 16
6 2 4 6
− 1 8 7 9
4 3 6 7
Check:
1 1 1
4 3 6 7
+ 1 8 7 9
6 2 4 6
The answer checks because this is the top number in the subtraction.

Do Exercises 14 and 15.

EXAMPLE 9 Subtract: 902 − 477.

```
  8 9 12
  9 0 2
− 4 7 7
  4 2 5
```

We cannot subtract 7 ones from 2 ones. We have 9 hundreds, or 90 tens. We borrow 1 ten to get 12 ones. We then have 89 tens.

Do Exercises 16 and 17 on the preceding page.

EXAMPLE 10 Subtract: 8003 − 3667.

```
  7 9 9 13
  8 0 0 3
− 3 6 6 7
  4 3 3 6
```

We have 8 thousands, or 800 tens. We borrow 1 ten to get 13 ones. We then have 799 tens.

EXAMPLES

11. Subtract: 6000 − 3762.

```
  5 9 9 10
  6 0 0 0
− 3 7 6 2
  2 2 3 8
```

12. Subtract: 6024 − 2968.

```
        11
  5 9 1 14
  6 0 2 4
− 2 9 6 8
  3 0 5 6
```

Do Exercises 18–20 on the preceding page.

CALCULATOR CORNER

Adding and Subtracting Whole Numbers *To the student and the instructor:* This is the first of a series of *optional* discussions on using a calculator. A calculator is *not* a requirement for this textbook. There are many kinds of calculators and different instructions for their usage. We have included instructions here for a minimum-cost calculator. Be sure to consult your user's manual as well. Also, check with your instructor about whether you are allowed to use a calculator in the course.

To add whole numbers on a calculator, we use the [+] and [=] keys. For example, to add 57 and 34, we press [5] [7] [+] [3] [4] [=]. The calculator displays [91], so 57 + 34 = 91. To find 314 + 259 + 478, we press [3] [1] [4] [+] [2] [5] [9] [+] [4] [7] [8] [=]. The display reads [1051], so 314 + 259 + 478 = 1051.

To subtract whole numbers on a calculator, we use the [−] and [=] keys. For example, to find 63 − 47, we press [6] [3] [−] [4] [7] [=]. The calculator displays [16], so 63 − 47 = 16. We can check this result by adding the subtrahend, 47, and the difference, 16. To do this, we press [1] [6] [+] [4] [7] [=]. The sum is the minuend, 63, so the subtraction is correct.

Exercises: Use a calculator to find each sum.

1. 925 + 677

2. 276 + 458

3.
```
  8 2 6
  4 1 5
+ 6 9 1
```

4.
```
  2 5 3
  4 9 0
+ 1 2 1
```

Use a calculator to perform each subtraction. Check by adding.

5. 145 − 78

6. 612 − 493

7.
```
  4 9 7 6
− 2 8 4 8
```

8.
```
  1 2,4 0 6
−     9 8 1 3
```

1.2 EXERCISE SET

For Extra Help

Digital Video Tutor CD 1 Videotape 1

InterAct Math

Math Tutor Center

MathXL

MyMathLab.com

a Add.

1. $\begin{array}{r} 364 \\ +\ \ 23 \\ \hline \end{array}$

2. $\begin{array}{r} 1521 \\ +\ \ 348 \\ \hline \end{array}$

3. $\begin{array}{r} 1716 \\ +\ 3482 \\ \hline \end{array}$

4. $\begin{array}{r} 7503 \\ +\ 2683 \\ \hline \end{array}$

5. 8113 + 390

6. 271 + 3338

7. 356 + 4910

8. 280 + 34,702

9. $\begin{array}{r} 99 \\ +\ \ 1 \\ \hline \end{array}$

10. $\begin{array}{r} 999 \\ +\ \ 11 \\ \hline \end{array}$

11. $\begin{array}{r} 5093 \\ +\ 3217 \\ \hline \end{array}$

12. $\begin{array}{r} 3654 \\ +\ 2700 \\ \hline \end{array}$

13. $\begin{array}{r} 4825 \\ +\ 1783 \\ \hline \end{array}$

14. $\begin{array}{r} 6775 \\ +\ 1432 \\ \hline \end{array}$

15. $\begin{array}{r} 23,443 \\ +\ 10,989 \\ \hline \end{array}$

16. $\begin{array}{r} 67,654 \\ +\ 98,786 \\ \hline \end{array}$

17. $\begin{array}{r} 12,070 \\ 2,954 \\ +\ \ 3,400 \\ \hline \end{array}$

18. $\begin{array}{r} 42,487 \\ 83,141 \\ +\ 36,712 \\ \hline \end{array}$

19. $\begin{array}{r} 327 \\ 428 \\ 569 \\ 787 \\ +\ 209 \\ \hline \end{array}$

20. $\begin{array}{r} 989 \\ 566 \\ 834 \\ 920 \\ +\ 703 \\ \hline \end{array}$

b Find the perimeter of (the distance around) each figure.

21.

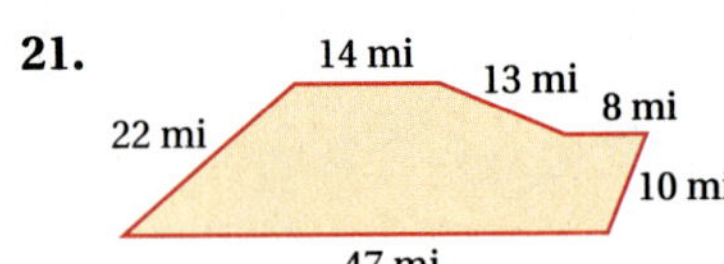

22.

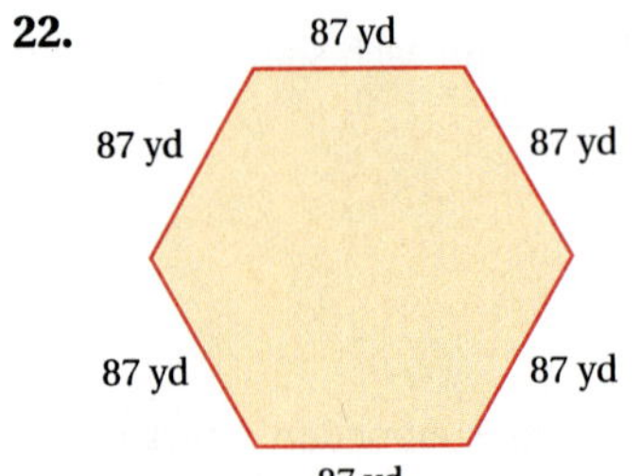

23. Find the perimeter of a standard hockey rink.

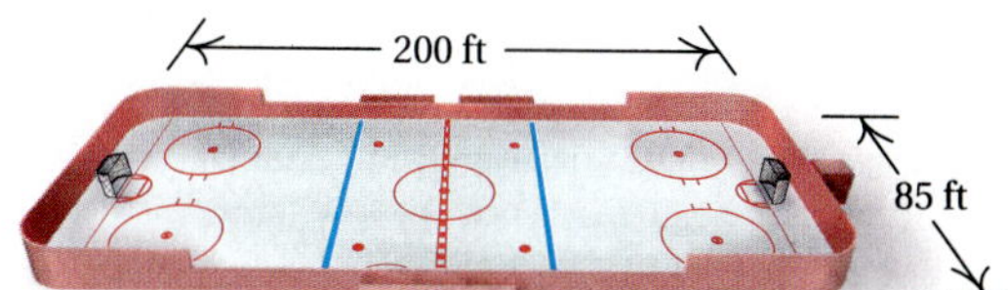

24. In major league baseball, how far does a batter travel in circling the bases when a home run has been hit?

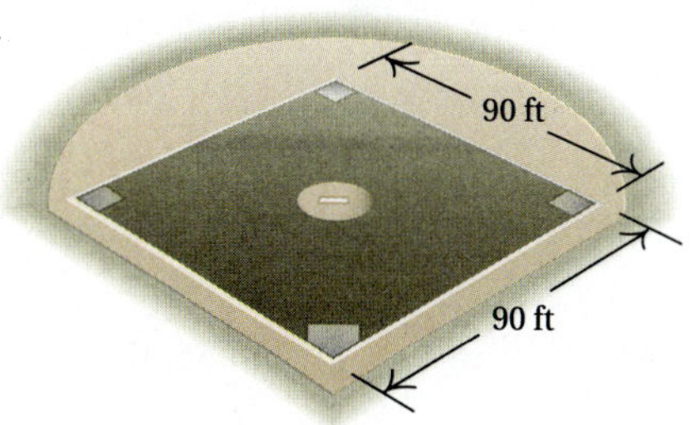

c Write a related addition sentence.

25. $7 - 4 = 3$

26. $12 - 5 = 7$

27. $43 - 16 = 27$

28. $51 - 18 = 33$

Write two related subtraction sentences.

29. $6 + 9 = 15$

30. $7 + 9 = 16$

31. $23 + 9 = 32$

32. $42 + 10 = 52$

d Subtract.

33. $\begin{array}{r} 65 \\ -\ 21 \\ \hline \end{array}$

34. $\begin{array}{r} 87 \\ -\ 34 \\ \hline \end{array}$

35. $\begin{array}{r} 4547 \\ -\ 3421 \\ \hline \end{array}$

36. $\begin{array}{r} 6875 \\ -\ 2111 \\ \hline \end{array}$

37. $\begin{array}{r} 7769 \\ -\ 2387 \\ \hline \end{array}$

38. $\begin{array}{r} 6431 \\ -\ 2896 \\ \hline \end{array}$

39. $\begin{array}{r} 7640 \\ -\ 3809 \\ \hline \end{array}$

40. $\begin{array}{r} 8003 \\ -\ 599 \\ \hline \end{array}$

41. $10,002 - 7834$

42. $23,048 - 17,592$

43. $90,237 - 47,209$

44. $84,703 - 298$

45. $\begin{array}{r} 140 \\ -\ 56 \\ \hline \end{array}$

46. $\begin{array}{r} 470 \\ -\ 188 \\ \hline \end{array}$

47. $\begin{array}{r} 690 \\ -\ 236 \\ \hline \end{array}$

48. $\begin{array}{r} 803 \\ -\ 418 \\ \hline \end{array}$

49. $\begin{array}{r} 903 \\ -\ 132 \\ \hline \end{array}$

50. $\begin{array}{r} 6408 \\ -\ 258 \\ \hline \end{array}$

51. $\begin{array}{r} 8092 \\ -\ 1073 \\ \hline \end{array}$

52. $\begin{array}{r} 6007 \\ -\ 1589 \\ \hline \end{array}$

53. 5843 − 98

54. 15,017 − 7809

55. 21,043 − 8909

56. 83,907 − 89

57. $\begin{array}{r} 7000 \\ -\ 2794 \\ \hline \end{array}$

58. $\begin{array}{r} 8001 \\ -\ 6543 \\ \hline \end{array}$

59. $\begin{array}{r} 48{,}000 \\ -\ 37{,}695 \\ \hline \end{array}$

60. $\begin{array}{r} 17{,}043 \\ -\ 11{,}598 \\ \hline \end{array}$

61. $^{\mathrm{D}}_{\mathrm{W}}$ Describe two situations that correspond to the subtraction \$20 − \$17, one "take away" and one "missing addend."

62. $^{\mathrm{D}}_{\mathrm{W}}$ Describe a situation that corresponds to this mathematical expression:

80 mi + 245 mi + 336 mi.

SKILL MAINTENANCE

The exercises that follow begin an important feature called *Skill Maintenance exercises.* These exercises provide an ongoing review of any preceding objective in the book. You will see them in virtually every exercise set. It has been found that this kind of extensive review can significantly improve your performance on a final examination.

63. What does the digit 8 mean in 486,205? [1.1a]

64. Write a word name for the number in the following sentence: [1.1c]

In a recent year, the New York Yankees topped all professional baseball teams with a total payroll of \$114,336,610.

Source: Major League Baseball

SYNTHESIS

65. A fast way to add all the numbers from 1 to 10 inclusive is to pair 1 with 9, 2 with 8, and so on. Use a similar approach to add all numbers from 1 to 100 inclusive.

66. Fill in the missing digits to make the subtraction true:

9, □48,621 − 2,097, □81 = 7,251,140.

1.3 MULTIPLICATION AND DIVISION; ROUNDING AND ESTIMATING

Objectives

a Multiply whole numbers.

b Use multiplication in finding area.

c Convert between division sentences and multiplication sentences.

d Divide whole numbers.

e Round to the nearest ten, hundred, or thousand.

f Estimate sums, differences, and products by rounding.

a Multiplication of Whole Numbers

REPEATED ADDITION

The multiplication 3×5 corresponds to this repeated addition:

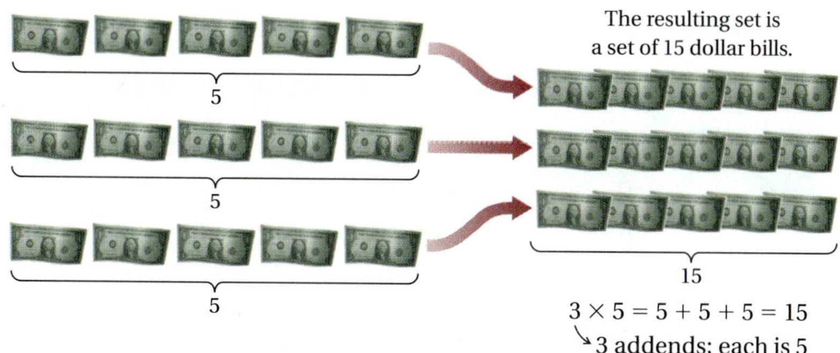

We combine 3 sets
of 5 dollar bills each.
5
5
5
The resulting set is
a set of 15 dollar bills.
15

$3 \times 5 = 5 + 5 + 5 = 15$

3 addends; each is 5

The numbers that we multiply are called **factors.** The result of the multiplication is called a **product.**

$$3 \times 5 = 15$$

Factor Factor Product

RECTANGULAR ARRAYS

Multiplications can also be thought of as rectangular arrays. Each of the following corresponds to the multiplication 3×5.

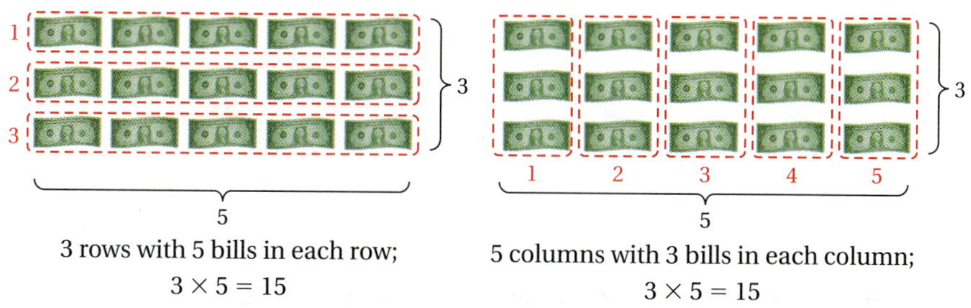

1
2
3
3
5
1
2
3
4
5
3
5

3 rows with 5 bills in each row; $3 \times 5 = 15$

5 columns with 3 bills in each column; $3 \times 5 = 15$

When you write a multiplication sentence corresponding to a real-world situation, you should think of either a rectangular array or repeated addition. In some cases, it may help to think both ways.

We have used an "$\times$" to denote multiplication. A dot "$\cdot$" is also commonly used. (Use of the dot is attributed to the German mathematician Gottfried Wilhelm von Leibniz in 1698.) Parentheses are also used to denote multiplication. For example,

$$3 \times 5 = 3 \cdot 5 = (3)(5) = 3(5) = 15.$$

Multiply.

1. $\begin{array}{r} 58 \\ \times \ \ 2 \\ \hline \end{array}$

2. $\begin{array}{r} 37 \\ \times \ \ 4 \\ \hline \end{array}$

3. $\begin{array}{r} 823 \\ \times \ \ \ 6 \\ \hline \end{array}$

4. $\begin{array}{r} 1348 \\ \times \ \ \ \ 5 \\ \hline \end{array}$

Answers on page A-2

The product of 0 and any whole number is 0: $0 \cdot a = a \cdot 0 = 0$. Multiplying a number by 1 does not change the number: $1 \cdot a = a \cdot 1 = a$. We say that 1 is the **multiplicative identity.**

EXAMPLE 1 Multiply: 5×734.

We have

$$\begin{array}{rl} 734 & \\ \times \quad 5 & \\ \hline 20 & \leftarrow \text{Multiply the 4 ones by 5: } 5 \times 4 = 20. \\ 150 & \leftarrow \text{Multiply the 3 tens by 5: } 5 \times 30 = 150. \\ 3500 & \leftarrow \text{Multiply the 7 hundreds by 5: } 5 \times 700 = 3500. \\ \hline 3670 & \leftarrow \text{Add.} \end{array}$$

Instead of writing each product on a separate line, we can use a shorter form.

$$\begin{array}{r} {}^{2} \\ 734 \\ \times \quad 5 \\ \hline 0 \end{array}$$

Multiply the ones by 5: $5 \cdot (4 \text{ ones}) = 20$ ones $=$ 2 tens $+$ 0 ones. Write 0 in the ones column and 2 above the tens.

$$\begin{array}{r} {}^{1\,2} \\ 734 \\ \times \quad 5 \\ \hline 70 \end{array}$$

Multiply the 3 tens by 5 and add 2 tens: $5 \cdot (3 \text{ tens}) = 15$ tens, 15 tens $+$ 2 tens $=$ 17 tens $=$ 1 hundred $+$ 7 tens. Write 7 in the tens column and 1 above the hundreds.

$$\begin{array}{r} {}^{1\,2} \\ 734 \\ \times \quad 5 \\ \hline 3670 \end{array}$$

Multiply the 7 hundreds by 5 and add 1 hundred: $5 \cdot (7 \text{ hundreds}) = 35$ hundreds, 35 hundreds $+$ 1 hundred $=$ 36 hundreds.

$$\begin{array}{r} {}^{1\,2} \\ 734 \\ \times \quad 5 \\ \hline 3670 \end{array}$$

You should write only this.

Do Exercises 1–4.

Let's find the product

$$\begin{array}{r} 54 \\ \times 32 \\ \hline \end{array}$$

To do this, we multiply 54 by 2, then 54 by 30, and then add.

$$\begin{array}{r} 54 \\ \times \quad 2 \\ \hline 108 \end{array} \qquad \begin{array}{r} {}^{1} \\ 54 \\ \times \ 30 \\ \hline 1620 \end{array}$$

Since we are going to add the results, let's write the work this way.

$$\begin{array}{rl} 54 & \\ \times 32 & \\ \hline 108 & \text{Multiplying by 2} \\ 1620 & \text{Multiplying by 30} \\ \hline 1728 & \text{Adding to obtain the product} \end{array}$$

EXAMPLE 2 Multiply: 43×57.

$$\begin{array}{r} {\scriptstyle 2} \\ 5\;7 \\ \times\;4\;3 \\ \hline 1\;7\;1 \end{array}$$

Multiplying by 3

$$\begin{array}{r} {\scriptstyle 2} \\ {\scriptstyle 2} \\ 5\;7 \\ \times\;4\;3 \\ \hline 1\;7\;1 \\ 2\;2\;8\;0 \end{array}$$

Multiplying by 40. (We write a 0 and then multiply 57 by 4).

You may have learned that such a 0 does not have to be written. You may omit it if you wish. If you do omit it, remember, when multiplying by tens, to put the answer in the tens place.

$$\begin{array}{r} {\scriptstyle 2} \\ {\scriptstyle 2} \\ 5\;7 \\ \times\;4\;3 \\ \hline 1\;7\;1 \\ 2\;2\;8\;0 \\ \hline 2\;4\;5\;1 \end{array}$$

Adding to obtain the product

Do Exercises 5 and 6.

EXAMPLE 3 Multiply: 457×683.

$$\begin{array}{r} {\scriptstyle 5\;\;2\;\;} \\ 6\;8\;3 \\ \times\;4\;5\;7 \\ \hline 4\;7\;8\;1 \end{array}$$

Multiplying 683 by 7

$$\begin{array}{r} {\scriptstyle 4\;\;1\;\;} \\ {\scriptstyle 5\;\;2\;\;} \\ 6\;8\;3 \\ \times\;4\;5\;7 \\ \hline 4\;7\;8\;1 \\ 3\;4\;1\;5\;0 \end{array}$$

Multiplying 683 by 50

$$\begin{array}{r} {\scriptstyle 3\;\;1\;\;} \\ {\scriptstyle 4\;\;1\;\;} \\ {\scriptstyle 5\;\;2\;\;} \\ 6\;8\;3 \\ \times\;4\;5\;7 \\ \hline 4\;7\;8\;1 \\ 3\;4\;1\;5\;0 \\ 2\;7\;3\;2\;0\;0 \\ \hline 3\;1\;2{,}1\;3\;1 \end{array}$$

Multiplying 683 by 400

Adding

Do Exercises 7 and 8.

Multiply.

5. $\begin{array}{r} 4\;5 \\ \times\;2\;3 \\ \hline \end{array}$

6. 48×63

Multiply.

7. $\begin{array}{r} 7\;4\;6 \\ \times\;\;6\;2 \\ \hline \end{array}$

8. 245×837

Multiply.

9. $\begin{array}{r} 4\;7\;2 \\ \times\;3\;0\;6 \\ \hline \end{array}$

10. 408×704

11. $\begin{array}{r} 2\;3\;4\;4 \\ \times\;6\;0\;0\;5 \\ \hline \end{array}$

Answers on page A-2

Multiply.

12. $\begin{array}{r} 472 \\ \times\ 830 \\ \hline \end{array}$

13. $\begin{array}{r} 2344 \\ \times\ 7400 \\ \hline \end{array}$

14. 100 × 562

15. 1000 × 562

CALCULATOR CORNER

Multiplying Whole Numbers To multiply whole numbers on a calculator, we use the [×] and [=] keys. For example, to find 13 × 47, we press [1] [3] [×] [4] [7] [=]. The calculator displays 611, so 13 × 47 = 611.

Exercises: Use a calculator to find each product.

1. 56 × 8
2. 845 × 26
3. 5 · 1276
4. 126(314)
5. $\begin{array}{r} 3760 \\ \times\ \ 48 \\ \hline \end{array}$
6. $\begin{array}{r} 5218 \\ \times\ 453 \\ \hline \end{array}$

Answers on page A-2

EXAMPLE 4 Multiply: 306 × 274.

Note that 306 = 3 hundreds + 6 ones.

$$\begin{array}{r} 274 \\ \times\ 306 \\ \hline 1644 \\ 82200 \\ \hline 83,844 \end{array}$$

1644 — Multiplying by 6
82200 — Multiplying by 3 hundreds. (We write 00 and then multiply 274 by 3.)
83,844 — Adding

Do Exercises 9–11 on the preceding page.

EXAMPLE 5 Multiply: 360 × 274.

Note that 360 = 3 hundreds + 6 tens.

$$\begin{array}{r} 274 \\ \times\ 360 \\ \hline 16440 \\ 82200 \\ \hline 98,640 \end{array}$$

16440 — Multiplying by 6 tens. (We write 0 and then multiply 274 by 6.)
82200 — Multiplying by 3 hundreds. (We write 00 and then multiply 274 by 3.)
98,640 — Adding

Do Exercises 12–15.

Study Tips

TIME MANAGEMENT (PART 1)

Time is the most critical factor in your success in learning mathematics. Have reasonable expectations about the time you need to study math.

- **Juggling time.** Working 40 hours per week and taking 12 credit hours is equivalent to working two full-time jobs. Can you handle such a load? Your ratio of number of work hours to number of credit hours should be about 40/3, 30/6, 20/9, 10/12, or 5/14.
- **A rule of thumb on study time.** Budget about 2–3 hours for homework and study per week for every hour of class time.
- **Scheduling your time.** Make an hour-by-hour schedule of your typical week. Include work, school, home, sleep, study, and leisure times. Try to schedule time for study when you are most alert. Choose a setting that will enable you to maximize your concentration. Plan for success and it will happen!

"You cannot increase the quality or quantity of your achievement or performance except to the degree in which you increase your ability to use time effectively."

Brian Tracy, motivational/inspirational speaker

b Finding Area

The area of a rectangular region is often considered to be the number of square units needed to fill it. Here is a rectangle 4 cm (centimeters) long and 3 cm wide. It takes 12 square centimeters (sq cm) to fill it.

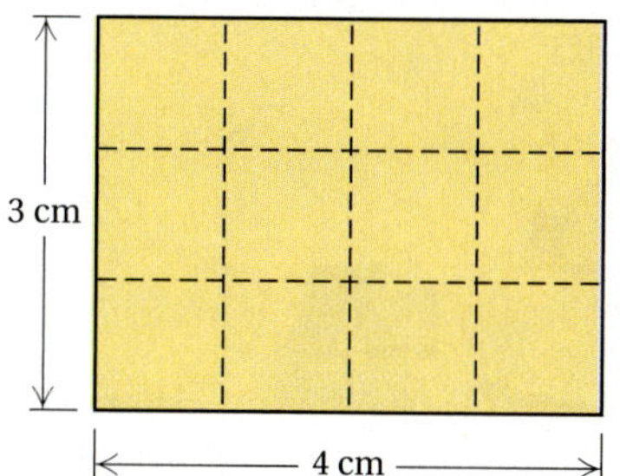

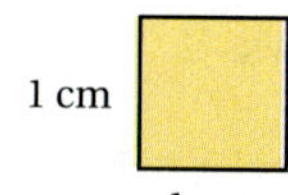

This is a square centimeter (a square unit).

In this case, we have a rectangular array of 3 rows, each of which contains 4 squares. The number of square units is given by $3 \cdot 4$, or 12.

EXAMPLE 6 *Professional Pool Table.* The playing area of a standard pool table has dimensions of 50 in. by 100 in. (There are rails 6 in. wide on the outside not included in the playing area.) Find the playing area.

If we think of filling the rectangle with square inches, we have a rectangular array. The length $l = 100$ in. and the width $w = 50$ in. Thus the area A is given by the formula

$$A = l \cdot w = 100 \cdot 50 = 5000 \text{ sq in.}$$

Do Exercise 16.

16. Table Tennis. Find the area of a standard table tennis table that has dimensions of 9 ft by 5 ft.

Professional pool player Jeanette Lee (also known as the "Black Widow")

Answer on page A-2

17. Consider $54 \div 6 = 9$. Express this division in two other ways.

Answer on page A-2

C Division and Related Sentences

REPEATED SUBTRACTION

Division of whole numbers applies to two kinds of situations. The first is repeated subtraction. Suppose we have 20 notebooks in a pile, and we want to find out how many sets of 5 there are. One way to do this is to repeatedly subtract sets of 5 as follows.

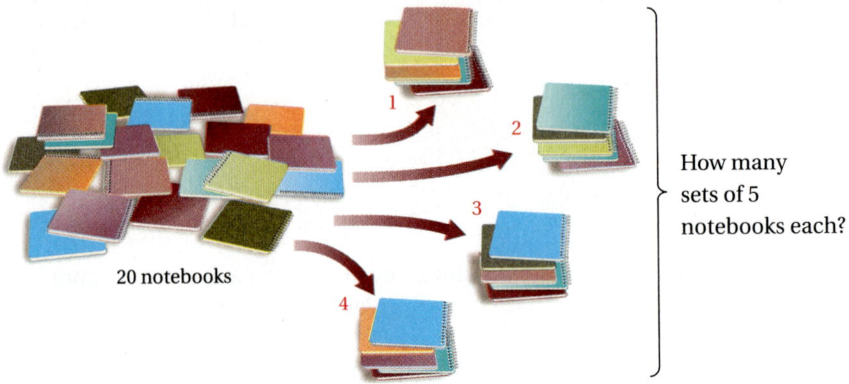
20 notebooks
1
2
3
4
How many sets of 5 notebooks each?

Since there are 4 sets of 5 notebooks each, we have

$$20 \div 5 = 4.$$

The division $20 \div 5$, read "20 divided by 5," corresponds to the figure above. We say that the **dividend** is 20, the **divisor** is 5, and the **quotient** is 4.

$$\underset{\text{Dividend}}{20} \div \underset{\text{Divisor}}{5} = \underset{\text{Quotient}}{4}$$

We divide the *dividend* by the *divisor* to get the *quotient*.

We can also express the division $20 \div 5 = 4$ as

$$\frac{20}{5} = 4 \quad \text{or} \quad 5\overline{)20}^{\,4}$$

Do Exercise 17.

Study Tips

LEARNING RESOURCES

Are you aware of all the learning resources that exist for this textbook? Many details are given in the Preface.

- The *Student's Solutions Manual* contains worked-out solutions to the odd-numbered exercises in the exercise sets.
- An extensive set of *videotapes* supplements this text. These are available on CD-ROM by calling 1-800-282-0693.
- *Tutorial software* called InterAct Math also accompanies this text. If it is not available in the campus learning center, you can order it by calling 1-800-282-0693.
- The Addison-Wesley *Math Tutor Center* is available for help with the odd-numbered exercises by experienced instructors. You can order this service by calling 1-800-824-7799.
- Extensive help is available online via MyMathLab and/or MathXL. Ask your instructor for information about these or visit MyMathLab.com and MathXL.com.

RECTANGULAR ARRAYS AND MISSING FACTORS

We can also think of division in terms of rectangular arrays. Consider again the pile of 20 notebooks and division by 5. We can arrange the notebooks in a rectangular array with 5 rows and ask, "How many are in each row?"

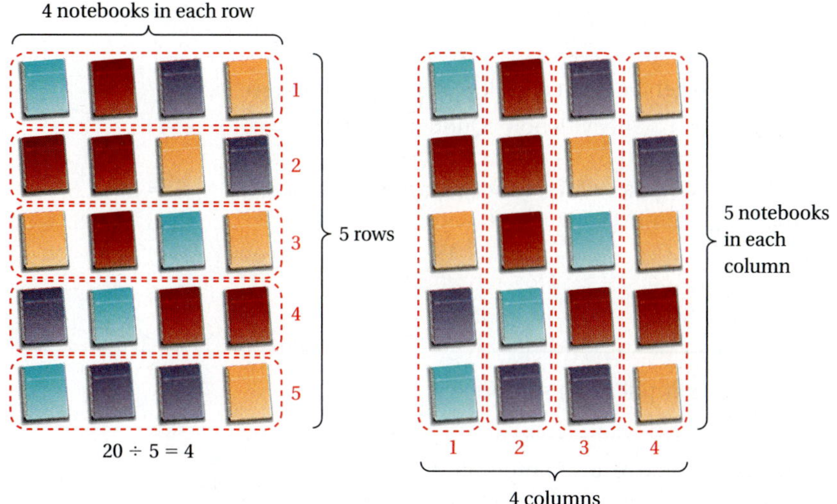

We can also consider a rectangular array with 5 notebooks in each column and ask, "How many columns are there?" The answer is still 4.

In each case, we are asking, "What do we multiply 5 by in order to get 20?"

This leads us to the following definition of division.

DIVISION

The quotient $a \div b$, where $b \neq 0$, is that unique whole number c for which $a = b \cdot c$.

RELATED SENTENCES

By looking at rectangular arrays, we can see how multiplication and division are related. The following array shows that $4 \cdot 5 = 20$.

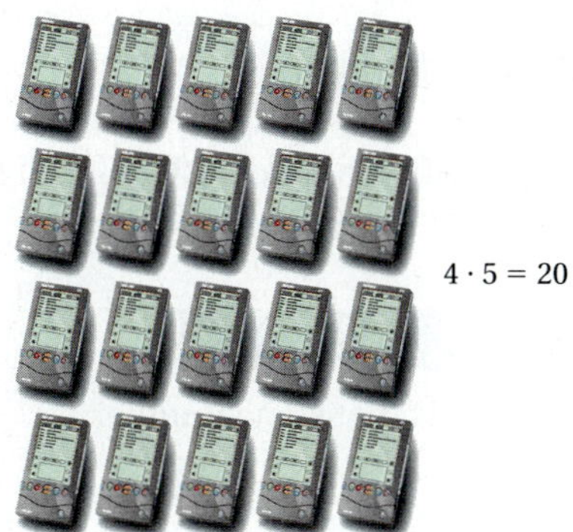

$4 \cdot 5 = 20$

The array also shows the following:

$$20 \div 5 = 4 \quad \text{and} \quad 20 \div 4 = 5.$$

Write a related multiplication sentence.

18. $15 \div 3 = 5$

19. $72 \div 8 = 9$

Write two related division sentences.

20. $6 \cdot 2 = 12$

21. $7 \cdot 6 = 42$

Answers on page A-2

The division $20 \div 5$ is defined to be the number that when multiplied by 5 gives 20. Thus, for every division sentence, there is a related multiplication sentence.

$20 \div 5 = 4$ Division sentence

$20 = 4 \cdot 5$ Related multiplication sentence

To get the related multiplication sentence, we use
Dividend = Quotient · Divisor.

EXAMPLE 7 Write a related multiplication sentence: $12 \div 6 = 2$.

We have

$12 \div 6 = 2$ Division sentence

$12 = 2 \cdot 6.$ Related multiplication sentence

The related multiplication sentence is $12 = 2 \cdot 6$.

By the commutative law of multiplication, there is also another multiplication sentence: $12 = 6 \cdot 2$.

Do Exercises 18 and 19.

For every multiplication sentence, we can write related divisions, as we can see from the preceding array.

EXAMPLE 8 Write two related division sentences: $7 \cdot 8 = 56$.

We have

$7 \cdot 8 = 56$ — This factor becomes a divisor. — $7 = 56 \div 8.$

$7 \cdot 8 = 56$ — This factor becomes a divisor. — $8 = 56 \div 7.$

The related division sentences are $7 = 56 \div 8$ and $8 = 56 \div 7$.

Do Exercises 20 and 21.

d Division of Whole Numbers

Before we consider division with remainders, let's recall four basic facts about division.

DIVIDING BY 1

Any number divided by 1 is that same number:

$$a \div 1 = \frac{a}{1} = a.$$

DIVIDING A NUMBER BY ITSELF

Any nonzero number divided by itself is 1:

$$\frac{a}{a} = 1, \quad a \neq 0.$$

DIVIDENDS OF 0

Zero divided by any nonzero number is 0:

$$\frac{0}{a} = 0, \quad a > 0.$$

EXCLUDING DIVISION BY 0

Division by 0 is not defined. (We agree not to divide by 0.)

$\frac{a}{0}$ is **not defined.**

Why can't we divide by 0? Suppose the number 4 could be divided by 0. Then if $\square$ were the answer,

$$4 \div 0 = \square$$

and since 0 times any number is 0, we would have

$$4 = \square \cdot 0 = 0. \qquad \text{False!}$$

Thus, $a \div 0$ would be some number $\square$ such that $a = \square \cdot 0 = 0$. So the only possible number that could be divided by 0 would be 0 itself.

But such a division would give us any number we wish, for

$$\left.\begin{array}{lll} 0 \div 0 = 8 & \text{because} & 0 = 8 \cdot 0; \\ 0 \div 0 = 3 & \text{because} & 0 = 3 \cdot 0; \\ 0 \div 0 = 7 & \text{because} & 0 = 7 \cdot 0. \end{array}\right\} \quad \text{All true!}$$

We avoid the preceding difficulties by agreeing to exclude division by 0.

Suppose we have 18 cans of soda and want to pack them in cartons of 6 cans each. How many cartons will we fill? We can determine this by repeated subtraction. We keep track of the number of times we subtract. We stop when the number of objects remaining, the **remainder,** is smaller than the divisor.

Divide by repeated subtraction. Then check.

22. 54 ÷ 9

23. 61 ÷ 9

24. 53 ÷ 12

25. 157 ÷ 24

Answers on page A-2

EXAMPLE 9 Divide by repeated subtraction: 18 ÷ 6.

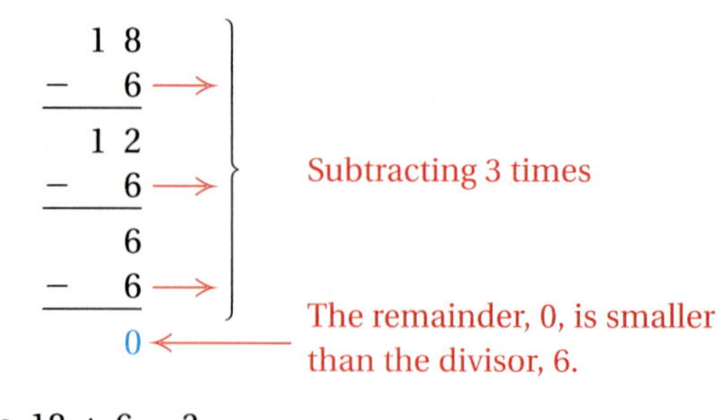

1 8
− 6
1 2
− 6
6
− 6
0
Subtracting 3 times
The remainder, 0, is smaller than the divisor, 6.

Thus, 18 ÷ 6 = 3.

Suppose we have 22 cans of soda and want to pack them in cartons of 6 cans each. We end up with 3 cartons with 4 cans left over.

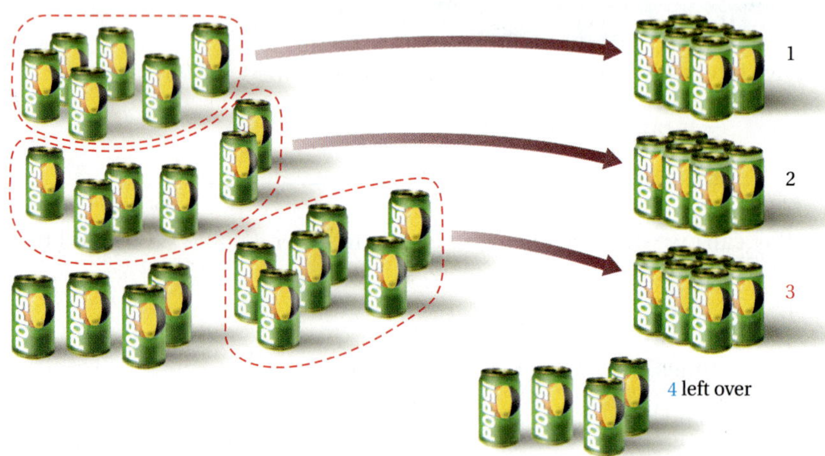

EXAMPLE 10 Divide by repeated subtraction: 22 ÷ 6.

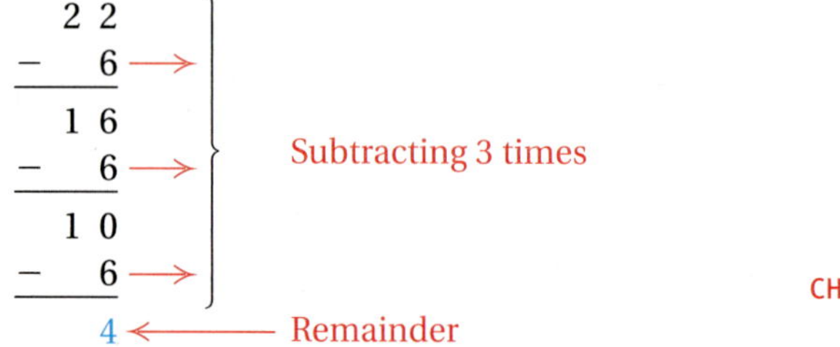

CHECK: $3 \cdot 6 = 18$,
$18 + 4 = 22$.

Note that

Quotient · Divisor + Remainder = Dividend.

We write answers to a division sentence as follows:

22 ÷ 6 = 3 R 4

Dividend Divisor Quotient Remainder

Do Exercises 22–25.

We can summarize our division procedure as follows.

To do division of whole numbers:

a) Estimate.
b) Multiply.
c) Subtract.

EXAMPLE 11 Divide and check: 3642 ÷ 5.

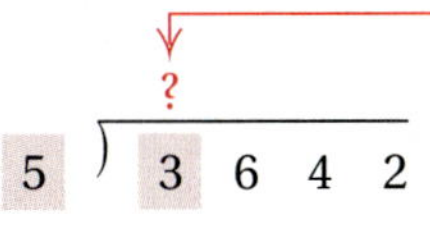

1. Find the number of thousands in the quotient. Consider 3 thousands ÷ 5 and think 3 ÷ 5. Since 3 ÷ 5 is not a whole number, move to hundreds.

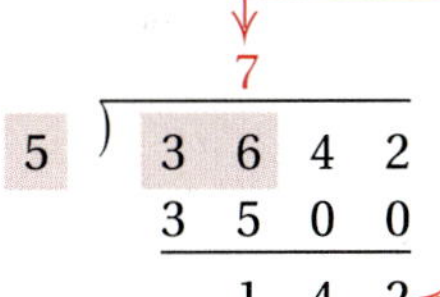

2. Find the number of hundreds in the quotient. Consider 36 hundreds ÷ 5 and think 36 ÷ 5. The estimate is about 7 hundreds. Multiply 700 by 5 and subtract.

The remainder is larger than the divisor.

3. Find the number of tens in the quotient using 142, the first remainder. Consider 14 tens ÷ 5 and think 14 ÷ 5. The estimate is about 2 tens. Multiply 20 by 5 and subtract. (If our estimate had been 3 tens, we could not have subtracted 150 from 142.)

The remainder is larger than the divisor.

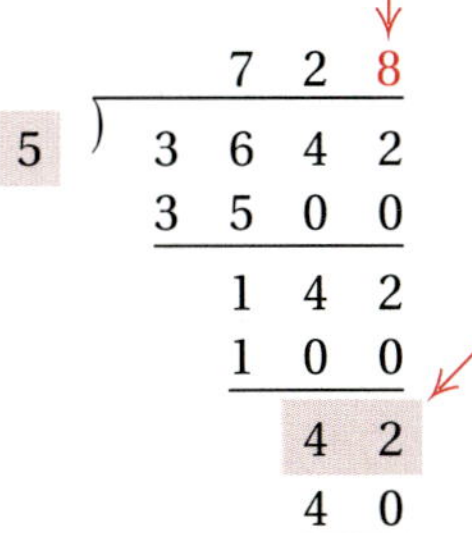

4. Find the number of ones in the quotient using 42, the second remainder. Consider 42 ones ÷ 5 and think 42 ÷ 5. The estimate is about 8 ones. Multiply 8 by 5 and subtract. The remainder, 2, is less than the divisor, 5, so we are finished.

The remainder is less than the divisor.

You may have learned to divide like this, not writing the extra zeros. You may omit them if desired.

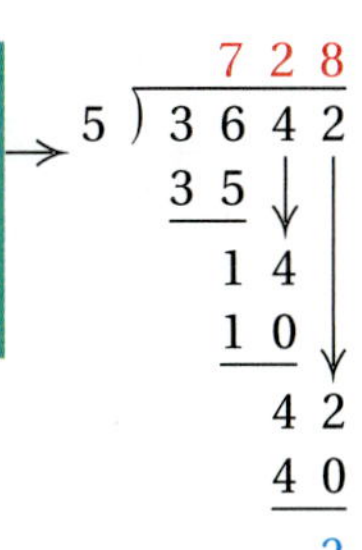

CHECK: 728 · 5 = 3640,
3640 + 2 = 3642.

The answer is 728 R 2.

Do Exercises 26–28.

Divide and check.

26. $4\overline{)2\,3\,9}$

27. $6\overline{)8\,8\,5\,5}$

28. $5\overline{)5\,0\,7\,5}$

Answers on page A-2

Divide.

29. $6\overline{)4846}$

30. $7\overline{)7616}$

Divide.

31. $27\overline{)9724}$

32. $56\overline{)44{,}847}$

Answers on page A-2

ZEROS IN QUOTIENTS

EXAMPLE 12 Divide: 6341 ÷ 7.

$$\begin{array}{r} 9 \\ 7\overline{)6341} \\ \underline{6300} \\ 41 \end{array}$$

← *Think*: 63 hundreds ÷ 7. Estimate 9 hundreds. Multiply 900 · 7 and subtract.

$$\begin{array}{r} 90 \\ 7\overline{)6341} \\ \underline{6300} \\ 41 \end{array}$$

← *Think*: 4 tens ÷ 7. There are no tens in the quotient (other than the tens in 900). We write a 0 to show this.

$$\begin{array}{r} 905 \\ 7\overline{)6341} \\ \underline{6300} \\ 41 \\ \underline{35} \\ 6 \end{array}$$

← *Think*: 41 ones ÷ 7. Estimate 5 ones. Multiply 5 · 7 and subtract.

← The remainder, 6, is less than the divisor, 7.

The answer is 905 R 6.

Do Exercises 29 and 30.

EXAMPLE 13 Divide: 8889 ÷ 37.

We round 37 to 40.

$$\begin{array}{r} 2 \\ 37\overline{)8889} \\ \underline{7400} \\ 1489 \end{array}$$

← *Think*: 37 ≈ 40; 88 hundreds ÷ 40. Estimate 2 hundreds. Multiply 200 · 37 and subtract.

$$\begin{array}{r} 24 \\ 37\overline{)8889} \\ \underline{7400} \\ 1489 \\ \underline{1480} \\ 9 \end{array}$$

← *Think*: 148 tens ÷ 40. Estimate 4 tens. Multiply 40 · 37 and subtract.

$$\begin{array}{r} 240 \\ 37\overline{)8889} \\ \underline{7400} \\ 1489 \\ \underline{1480} \\ 9 \end{array}$$

← The remainder, 9, is less than the divisor, 37.

The answer is 240 R 9.

Do Exercises 31 and 32.

CALCULATOR CORNER

Dividing Whole Numbers: Finding Remainders To divide whole numbers on a calculator, we use the [÷] and [=] keys. For example, to divide 711 by 9, we press [7] [1] [1] [÷] [9] [=]. The display reads [79], so 711 ÷ 9 = 79.

When we enter 453 ÷ 15, the display reads [30.2]. Note that the result is not a whole number. This tells us that there is a remainder. The number 30.2 is expressed in decimal notation. The symbol "." is called a decimal point. (Decimal notation will be studied in Chapter 3.) The number to the left of the decimal point, 30, is the quotient. We can use the remaining part of the result to find the remainder. To do this, first subtract 30 from 30.2. Then multiply the difference by the divisor, 15. We get 3. This is the remainder. Thus, 453 ÷ 15 = 30 R 3. The steps that we performed to find this result can be summarized as follows:

$$453 \div 15 = 30.2, \qquad 30.2 - 30 = .2, \qquad 0.2 \times 15 = 3.$$

To follow these steps on a calculator, we press [4] [5] [3] [÷] [1] [5] [=] and write the number that appears to the left of the decimal point. This is the quotient. Then we continue by pressing [−] [3] [0] [=] [×] [1] [5] [=]. The last number that appears is the remainder. In some cases, it will be necessary to round the remainder to the nearest one.

To check this result, we multiply the quotient by the divisor and then add the remainder.

$$30 \times 15 = 450, \qquad 450 + 3 = 453$$

Exercises: Use a calculator to perform each division. Check the results with a calculator also.

1. $92 \div 27$
2. $19\overline{)532}$
3. $6\overline{)746}$
4. $3817 \div 29$
5. $126\overline{)35{,}715}$
6. $308\overline{)259{,}831}$

e Rounding

We round numbers in various situations if we do not need an exact answer. For example, we might round to check if an answer to a problem is reasonable or to check a calculation done by hand or on a calculator. We might also round to see if we are being charged the correct amount in a store.

To understand how to round, we first look at some examples using number lines, even though this is not the way we generally do rounding.

EXAMPLE 14 Round 47 to the nearest ten.

Here is a part of a number line; 47 is between 40 and 50.

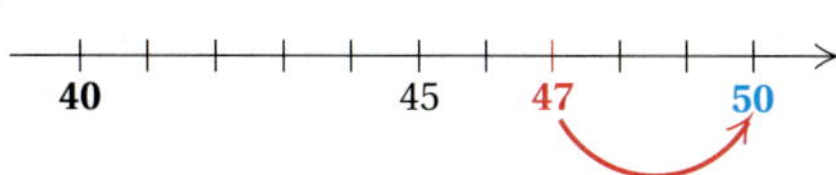

Since 47 is closer to 50, we round up to 50.

EXAMPLE 15 Round 42 to the nearest ten.

42 is between 40 and 50.

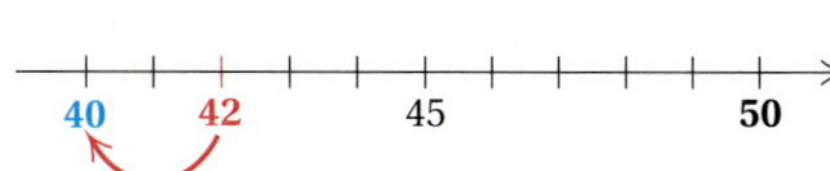

Since 42 is closer to 40, we round down to 40.

Do Exercises 33–36.

Round to the nearest ten.

33. 37

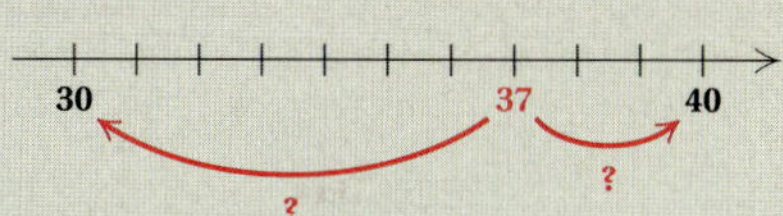

34. 52

35. 73

36. 98

Answers on page A-2

Round to the nearest ten.

37. 35

38. 75

39. 85

Round to the nearest ten.

40. 137

41. 473

42. 235

43. 285

Round to the nearest hundred.

44. 641

45. 759

46. 750

47. 9325

Round to the nearest thousand.

48. 7896

49. 8459

50. 19,343

51. 68,500

Answers on page A-2

EXAMPLE 16 Round 45 to the nearest ten.

45 is halfway between 40 and 50.

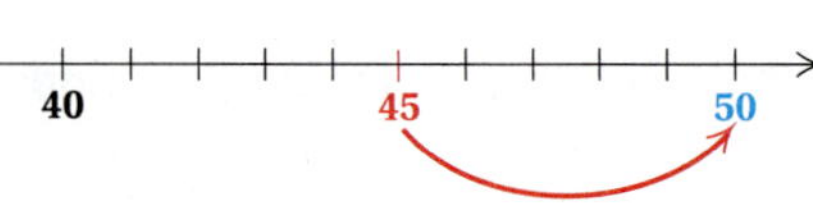

We could round 45 down to 40 or up to 50. We agree to round up to 50.

When a number is halfway between rounding numbers, round up.

Do Exercises 37–39.

Here is a rule for rounding.

ROUNDING WHOLE NUMBERS

To round to a certain place:

a) Locate the digit in that place.

b) Consider the next digit to the right.

c) If the digit to the right is 5 or higher, round up. If the digit to the right is 4 or lower, round down.

d) Change all digits to the right of the rounding location to zeros.

EXAMPLE 17 Round 6485 to the nearest (a) ten; (b) hundred; (c) thousand.

a) Locate the digit in the tens place. It is 8.

6 4 8 5
    ↑

The next digit to the right is 5, so we round up. The answer is 6490.

b) Locate the digit in the hundreds place. It is 4.

6 4 8 5
  ↑

The next digit to the right is 8, so we round up. The answer is 6500.

c) Locate the digit in the thousands place. It is 6.

6 4 8 5
↑

The next digit to the right is 4, so we round down. The answer is 6000.

Do Exercises 40–51.

CAUTION!

7000 is not a correct answer to Example 17. It is incorrect to round from the ones digit over, as follows:

6485, 6490, 6500, 7000.

There are many methods of rounding. For example, in computer applications, the rounding of 8563 to the nearest hundred might be done using a different rule called **truncating,** meaning that we simply change all digits to the right of the rounding location to zeros. Thus, 8563 would round to 8500, which is not the same answer that we would get using the rule discussed in this section.

f Estimating

In the following example, we see how estimation can be used in making a purchase.

EXAMPLE 18 *Estimating the Cost of an Automobile Purchase.* Maria and Luis Vasquez are shopping for a new car. They are considering an Oldsmobile Alero. There are three basic models of this car, and each has options beyond the basic price, as shown in the chart below. Maria and Luis have allowed themselves a budget of $20,000. They look at the list of options and want to make a quick estimate of the cost of model GL3 with all the options.

Estimate by rounding to the nearest hundred the cost of the GL3 with all the options and decide whether it will fit into their budget.

MODEL GL1 SEDAN (4 DOOR) 2.4-liter engine, 5 SPEED MANUAL TRANSMISSION	MODEL GL2 SEDAN (4 DOOR) 2.4-liter engine, 4 SPEED AUTOMATIC TRANSMISSION	MODEL GL3 SEDAN (4 DOOR) 3.4-liter engine, 4 SPEED AUTOMATIC TRANSMISSION
Base price: $17,650	*Base price:* $18,270	*Base price:* $18,875
Destination charges: $535	*Destination charges:* $535	*Destination charges:* $535

Each of these vehicles comes with several options.	
Driver's seat with 6-way power adjustment:	$305
Sunroof:	$650
Feature package: 15" aluminum wheels, remote keyless entry, foglamps, leather-wrapped steering wheel and shift knob	$585
Sport package: 16" aluminum wheels, performance radial tires, and performance suspension	$450
Rear decklid spoiler:	$225
Radio: AM/FM cassette/CD with 6-speaker dimensional sound system	$200

Source: General Motors

Refer to the chart below to answer Margin Exercises 52 and 53.

52. Suppose Maria and Luis want to buy a GL1 with all options except the sunroof and the sport package.

a) Estimate this cost by rounding to the nearest hundred.

b) Can they afford this car with a budget of $20,000?

53. By eliminating options, find a way that Luis and Maria can buy the GL3 and stay within their $20,000 budget. Answers may vary.

Answers on page A-2

First, we list the base price of the GL3 and then the cost of each of the options. We then round each number to the nearest hundred and add.

18,875	18,900
535	500
305	300
650	700
585	600
450	500
225	200
+ 200	+ 200
	21,900 Estimated answer

The estimated total cost is \$21,900. Since Maria and Luis have allowed themselves a budget of \$20,000 for their car, they will need to forego some options.

Do Exercises 52 and 53 on the preceding page.

Estimating can be done in many ways and can have many results, even though in the problems that follow we ask you to round in a specific way.

EXAMPLE 19 Estimate the difference by first rounding to the nearest thousand: 9324 − 2849.

We have

9324	9000
− 2849	− 3000
	6000

EXAMPLE 20 Estimate the following product by first rounding to the nearest ten and to the nearest hundred: 683 × 457.

Nearest ten	*Nearest hundred*	*Exact*
680	700	683
× 460	× 500	× 457
40800	350000	4781
272000		34150
312800		273200
		312131

Do Exercises 54–57.

The sentence 7 − 5 = 2 says that 7 − 5 is the same as 2. When we round, the result is rarely the same as the number we started with. Thus we use the symbol ≈ when rounding. This symbol means "**is approximately equal to.**" For example, when 687 is rounded to the nearest ten, we can write

$$687 \approx 690.$$

54. Estimate the sum by first rounding to the nearest ten. Show your work.

$$\begin{array}{r} 74 \\ 23 \\ 35 \\ +\ 66 \\ \hline \end{array}$$

55. Estimate the difference by first rounding to the nearest hundred. Show your work.

$$\begin{array}{r} 9285 \\ -\ 6739 \\ \hline \end{array}$$

56. Estimate the difference by first rounding to the nearest thousand. Show your work.

$$\begin{array}{r} 23,278 \\ -\ 11,698 \\ \hline \end{array}$$

57. Estimate the product by first rounding to the nearest ten and to the nearest hundred. Show your work.

$$\begin{array}{r} 837 \\ \times\ 245 \\ \hline \end{array}$$

Answers on page A-2

1.3 EXERCISE SET

For Extra Help

Digital Video Tutor CD 1 Videotape 1

InterAct Math

Math Tutor Center

MathXL

MyMathLab

a Multiply.

1. $\begin{array}{r} 94 \\ \times \quad 6 \\ \hline \end{array}$

2. $\begin{array}{r} 76 \\ \times \quad 9 \\ \hline \end{array}$

3. $\begin{array}{r} 2340 \\ \times 1000 \\ \hline \end{array}$

4. $\begin{array}{r} 800 \\ \times \quad 70 \\ \hline \end{array}$

5. $3 \cdot 509$

6. $7 \cdot 806$

7. 7(9229)

8. 4(7867)

9. 90(53)

10. 60(78)

11. (47)(85)

12. (34)(87)

13. $\begin{array}{r} 640 \\ \times \quad 72 \\ \hline \end{array}$

14. $\begin{array}{r} 777 \\ \times \quad 77 \\ \hline \end{array}$

15. $\begin{array}{r} 444 \\ \times \quad 33 \\ \hline \end{array}$

16. $\begin{array}{r} 509 \\ \times \quad 88 \\ \hline \end{array}$

17. $\begin{array}{r} 509 \\ \times 408 \\ \hline \end{array}$

18. $\begin{array}{r} 432 \\ \times 375 \\ \hline \end{array}$

19. $\begin{array}{r} 853 \\ \times 936 \\ \hline \end{array}$

20. $\begin{array}{r} 346 \\ \times 650 \\ \hline \end{array}$

21. $\begin{array}{r} 6428 \\ \times 3224 \\ \hline \end{array}$

22. $\begin{array}{r} 8928 \\ \times 3172 \\ \hline \end{array}$

23. $\begin{array}{r} 3482 \\ \times \quad 104 \\ \hline \end{array}$

24. $\begin{array}{r} 6408 \\ \times 6064 \\ \hline \end{array}$

25. $\begin{array}{r} 5006 \\ \times 4008 \\ \hline \end{array}$

26. $\begin{array}{r} 6789 \\ \times 2330 \\ \hline \end{array}$

27. $\begin{array}{r} 5608 \\ \times 4500 \\ \hline \end{array}$

28. $\begin{array}{r} 4560 \\ \times 7890 \\ \hline \end{array}$

b What is the area of the region?

29. 

30. 129 yd, 65 yd

31. Find the area of the region formed by the base lines on a Major League baseball diamond.

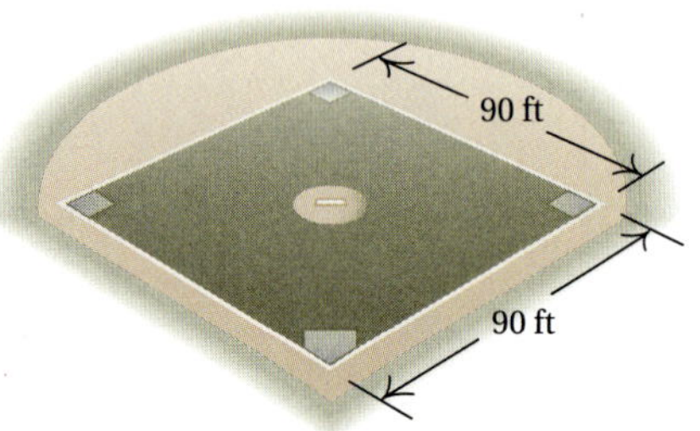

32. Find the area of a standard-sized hockey rink.

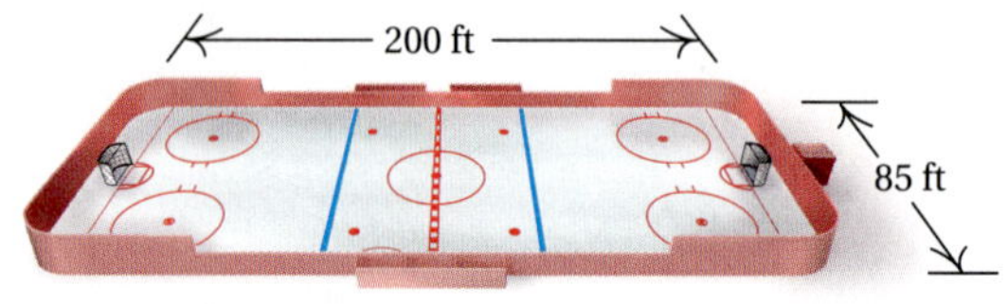

c Write a related multiplication sentence.

33. $18 \div 3 = 6$

34. $72 \div 9 = 8$

35. $22 \div 22 = 1$

36. $32 \div 1 = 32$

Write two related division sentences.

37. $9 \times 5 = 45$

38. $2 \cdot 7 = 14$

39. $37 \cdot 1 = 37$

40. $4 \cdot 12 = 48$

d Divide, if possible. If not possible, write "not defined."

41. $72 \div 6$

42. $54 \div 9$

43. $\frac{23}{23}$

44. $\frac{37}{37}$

45. $22 \div 1$

46. $\frac{56}{1}$

47. $\frac{16}{0}$

48. $74 \div 0$

Divide.

49. $277 \div 5$

50. $699 \div 3$

51. $864 \div 8$

52. $869 \div 8$

53. $4\overline{)1228}$

54. $3\overline{)2124}$

55. $738 \div 8$

56. $881 \div 6$

57. $5\overline{)8515}$

58. $3\overline{)6027}$

59. $127{,}000 \div 1000$

60. $4260 \div 10$

61. $30\overline{)875}$

62. $40\overline{)987}$

63. $852 \div 21$

64. $942 \div 23$

65. $85\overline{)7672}$

66. $54\overline{)2729}$

67. $111\overline{)3219}$

68. $102\overline{)5612}$

69. $5\overline{)5036}$

70. $7\overline{)7074}$

71. $1058 \div 46$

72. $7242 \div 24$

73. $24\overline{)8880}$

74. $36\overline{)7563}$

75. $28\overline{)17{,}067}$

76. $36\overline{)28{,}929}$

77. $285\overline{)999{,}999}$

78. $306\overline{)888{,}888}$

e Round to the nearest ten.

79. 48 **80.** 532 **81.** 467 **82.** 8945

83. 731 **84.** 17 **85.** 895 **86.** 798

Round to the nearest hundred.

87. 146 **88.** 874 **89.** 957 **90.** 650

91. 9079 **92.** 4645 **93.** 32,850 **94.** 198,402

Round to the nearest thousand.

95. 5876 **96.** 4500 **97.** 7500 **98.** 2001

99. 45,340 **100.** 735,562 **101.** 373,405 **102.** 6,713,855

 Estimate the sum or difference by first rounding to the nearest ten. Show your work.

103. $\begin{array}{r} 78 \\ +\ 97 \\ \hline \end{array}$ **104.** $\begin{array}{r} 62 \\ 97 \\ 46 \\ +\ 88 \\ \hline \end{array}$ **105.** $\begin{array}{r} 8074 \\ -\ 2347 \\ \hline \end{array}$ **106.** $\begin{array}{r} 673 \\ -\ 28 \\ \hline \end{array}$

Estimate the sum or difference by first rounding to the nearest hundred. Show your work.

107. $\begin{array}{r} 7348 \\ +\ 9247 \\ \hline \end{array}$ **108.** $\begin{array}{r} 568 \\ 472 \\ 938 \\ +\ 402 \\ \hline \end{array}$ **109.** $\begin{array}{r} 6852 \\ -\ 1748 \\ \hline \end{array}$ **110.** $\begin{array}{r} 9438 \\ -\ 2787 \\ \hline \end{array}$

Estimate the sum or difference by first rounding to the nearest thousand. Show your work.

111. $\begin{array}{r} 9643 \\ 4821 \\ 8943 \\ +\ 7004 \\ \hline \end{array}$ **112.** $\begin{array}{r} 7648 \\ 9348 \\ 7842 \\ +\ 2222 \\ \hline \end{array}$ **113.** $\begin{array}{r} 92{,}149 \\ -\ 22{,}555 \\ \hline \end{array}$ **114.** $\begin{array}{r} 84{,}890 \\ -\ 11{,}110 \\ \hline \end{array}$

Estimate the product by first rounding to the nearest ten. Show your work.

115. $\begin{array}{r} 45 \\ \times\ 67 \\ \hline \end{array}$ **116.** $\begin{array}{r} 51 \\ \times\ 78 \\ \hline \end{array}$ **117.** $\begin{array}{r} 34 \\ \times\ 29 \\ \hline \end{array}$ **118.** $\begin{array}{r} 63 \\ \times\ 54 \\ \hline \end{array}$

Estimate the product by first rounding to the nearest hundred. Show your work.

119. $\begin{array}{r} 876 \\ \times\ 345 \\ \hline \end{array}$

120. $\begin{array}{r} 355 \\ \times\ 299 \\ \hline \end{array}$

121. $\begin{array}{r} 432 \\ \times\ 199 \\ \hline \end{array}$

122. $\begin{array}{r} 789 \\ \times\ 434 \\ \hline \end{array}$

123. $^{D}_{W}$ Describe a situation that corresponds to each multiplication: $4 \cdot \$150$; $\$4 \cdot 150$.

124. $^{D}_{W}$ Suppose a student asserts that "$0 \div 0 = 0$ because nothing divided by nothing is nothing." Devise an explanation to persuade the student that the assertion is false.

SKILL MAINTENANCE

125. Write expanded notation for 7882. [1.1b]

126. Use $<$ or $>$ for $\square$ to write a true sentence: [1.1d]

$888 \square 788$.

Write a related addition sentence. [1.2c]

127. $21 - 16 = 5$

128. $56 - 14 = 42$

Write two related subtraction sentences. [1.2c]

129. $47 + 9 = 56$

130. $350 + 64 = 414$

SYNTHESIS

131. Complete the following table.

a	b	$a \cdot b$	$a + b$
	68	3672	
84			117
		32	12
		304	35

132. Find a pair of factors whose product is 36 and:

a) whose sum is 13.
b) whose difference is 0.
c) whose sum is 20.
d) whose difference is 9.

133. A group of 1231 college students is going to take buses for a field trip. Each bus can hold only 42 students. How many buses are needed?

134. 🖩 Fill in the missing digits to make the equation true:

$34{,}584{,}132 \div 76\square = 4\square{,}386$.

135. 🖩 An 18-story office building is box-shaped. Each floor measures 172 ft by 84 ft with a 20-ft by 35-ft rectangular area lost to an elevator and a stairwell. How much area is available as office space?

Objectives

a Solve simple equations by trial.

b Solve equations like $x + 28 = 54$, $28 \cdot x = 168$, and $98 \cdot 2 = y$.

Find a number that makes the sentence true.

1. $8 = 1 + \square$

2. $\square + 2 = 7$

3. Determine whether 7 is a solution of $\square + 5 = 9$.

4. Determine whether 4 is a solution of $\square + 5 = 9$.

Answers on page A-2

1.4 SOLVING EQUATIONS

a Solutions by Trial

Let's find a number that we can put in the blank to make this sentence true:

$$9 = 3 + \square.$$

We are asking "9 is 3 plus what number?" The answer is 6.

$$9 = 3 + \boxed{6}$$

Do Exercises 1 and 2.

A sentence with = is called an **equation.** A **solution** of an equation is a number that makes the sentence true. Thus, 6 is a solution of

$9 = 3 + \square$ because $9 = 3 + \boxed{6}$ is true.

However, 7 is not a solution of

$9 = 3 + \square$ because $9 = 3 + \boxed{7}$ is false.

Do Exercises 3 and 4.

We can use a letter instead of a blank. For example,

$$9 = 3 + x.$$

We call x a **variable** because it can represent any number. If a replacement for a variable makes an equation true, it is a **solution** of the equation.

SOLUTIONS OF AN EQUATION

A **solution** is a replacement for the variable that makes the equation true. When we find all the solutions, we say that we have **solved** the equation.

EXAMPLE 1 Solve $x + 12 = 27$ by trial.

We replace x with several numbers.

If we replace x with 13, we get a false equation: $13 + 12 = 27$.
If we replace x with 14, we get a false equation: $14 + 12 = 27$.
If we replace x with 15, we get a true equation: $15 + 12 = 27$.

No other replacement makes the equation true, so the solution is 15.

EXAMPLES Solve.

2. $7 + n = 22$
   (7 plus what number is 22?)
   The solution is 15.

3. $8 \cdot 23 = y$
   (8 times 23 is what?)
   The solution is 184.

Do Exercises 5–8.

b Solving Equations

We now begin to develop more efficient ways to solve certain equations. When an equation has a variable alone on one side, it is easy to see the solution or to compute it. For example, the solution of

$$x = 12$$

is 12. When a calculation is on one side and the variable is alone on the other, we can find the solution by carrying out the calculation.

EXAMPLE 4 Solve: $x = 245 \times 34$.

To solve the equation, we carry out the calculation.

$$\begin{array}{r} 2\,4\,5 \\ \times \quad 3\,4 \\ \hline 9\,8\,0 \\ 7\,3\,5\,0 \\ \hline 8\,3\,3\,0 \end{array} \qquad \begin{aligned} x &= 245 \times 34 \\ x &= 8330 \end{aligned}$$

The solution is 8330.

Do Exercises 9–12.

Look at the equation

$$x + 12 = 27.$$

We can get x alone on one side of the equation by writing a related subtraction sentence:

$x = 27 - 12$ — 12 gets subtracted to find the related subtraction sentence.

$x = 15.$ — Doing the subtraction

It is useful in our later study of algebra to think of this as "subtracting 12 *on both sides.*" Thus

$x + 12 - 12 = 27 - 12$ — Subtracting 12 on both sides

$x + 0 = 15$ — Carrying out the subtraction

$x = 15.$

SOLVING $x + a = b$

To solve $x + a = b$, subtract a on both sides.

Solve by trial.

5. $n + 3 = 8$

6. $x - 2 = 8$

7. $45 \div 9 = y$

8. $10 + t = 32$

Solve.

9. $346 \times 65 = y$

10. $x = 2347 + 6675$

11. $4560 \div 8 = t$

12. $x = 6007 - 2346$

Answers on page A-2

Solve. Be sure to check.

13. $x + 9 = 17$

14. $77 = m + 32$

15. Solve: $155 = t + 78$. Be sure to check.

Solve. Be sure to check.

16. $4566 + x = 7877$

17. $8172 = h + 2058$

Answers on page A-2

If we can get an equation in a form with the variable alone on one side, we can "see" the solution.

EXAMPLE 5 Solve: $t + 28 = 54$.

We have

$$t + 28 = 54$$
$$t + 28 - 28 = 54 - 28 \quad \text{Subtracting 28 on both sides}$$
$$t + 0 = 26$$
$$t = 26.$$

To check the answer, we substitute 26 for t in the original equation.

CHECK:
$$\begin{array}{c|l} t + 28 & 54 \\ \hline 26 + 28 \ ? & 54 \\ 54 & \end{array} \quad \text{TRUE}$$

The solution is 26.

Do Exercises 13 and 14.

EXAMPLE 6 Solve: $182 = 65 + n$.

We have

$$182 = 65 + n$$
$$182 - 65 = 65 + n - 65 \quad \text{Subtracting 65 on both sides}$$
$$117 = 0 + n \quad \text{65 plus } n \text{ minus 65 is } 0 + n.$$
$$117 = n.$$

CHECK:
$$\begin{array}{r|l} 182 & 65 + n \\ \hline 182 \ ? & 65 + 117 \\ & 182 \end{array} \quad \text{TRUE}$$

The solution is 117.

Do Exercise 15.

EXAMPLE 7 Solve: $7381 + x = 8067$.

We have

$$7381 + x = 8067$$
$$7381 + x - 7381 = 8067 - 7381 \quad \text{Subtracting 7381 on both sides}$$
$$x = 686.$$

The check is left to the student. The solution is 686.

Do Exercises 16 and 17.

We now learn to solve equations like $8 \cdot n = 96$. Look at

$$8 \cdot n = 96.$$

We can get n alone by writing a related division sentence:

$$n = 96 \div 8 = \frac{96}{8} \quad \text{96 is divided by 8.}$$

$$n = 12. \quad \text{Doing the division}$$

Note that $n = 12$ is easier to solve than $8 \cdot n = 96$. This is because we see easily that if we replace n on the left side with 12, we get a true sentence: $12 = 12$. The solution of $n = 12$ is 12, which is also the solution of $8 \cdot n = 96$.

It is useful in our later study of algebra to think of the preceding as "dividing by 8 *on both sides*." Thus,

$$\frac{8 \cdot n}{8} = \frac{96}{8} \quad \text{Dividing by 8 on both sides}$$

$$n = 12. \quad \text{8 times } n \text{ divided by 8 is } n.$$

SOLVING $a \cdot x = b$

To solve $a \cdot x = b$, divide by a on both sides.

EXAMPLE 8 Solve: $10 \cdot x = 240$.

We have

$$10 \cdot x = 240$$

$$\frac{10 \cdot x}{10} = \frac{240}{10} \quad \text{Dividing by 10 on both sides}$$

$$x = 24.$$

CHECK:

$10 \cdot x = 240$		
$10 \cdot 24$? 240		
240 |		TRUE

The solution is 24.

Do Exercises 18 and 19.

EXAMPLE 9 Solve: $5202 = 9 \cdot t$.

We have

$$5202 = 9 \cdot t$$

$$\frac{5202}{9} = \frac{9 \cdot t}{9} \quad \text{Dividing by 9 on both sides}$$

$$578 = t.$$

The check is left to the student. The solution is 578.

Do Exercise 20.

Solve. Be sure to check.

18. $8 \cdot x = 64$

19. $144 = 9 \cdot n$

20. Solve: $5152 = 8 \cdot t$.

Answers on page A-2

21. Solve: $18 \cdot y = 1728$.

22. Solve: $n \cdot 48 = 4512$.

Answers on page A-2

EXAMPLE 10 Solve: $14 \cdot y = 1092$.

We have

$$14 \cdot y = 1092$$

$$\frac{14 \cdot y}{14} = \frac{1092}{14} \quad \text{Dividing by 14 on both sides}$$

$$y = 78.$$

The check is left to the student. The solution is 78.

Do Exercise 21.

EXAMPLE 11 Solve: $n \cdot 56 = 4648$.

We have

$$n \cdot 56 = 4648$$

$$\frac{n \cdot 56}{56} = \frac{4648}{56} \quad \text{Dividing by 56 on both sides}$$

$$n = 83.$$

The check is left to the student. The solution is 83.

Do Exercise 22.

Study Tips

TIPS FROM A FORMER STUDENT

A former student of Professor Bittinger, Mike Rosenborg earned a master's degree in mathematics and now teaches mathematics. Here are some of his study tips.

- Because working problems is the best way to learn math, instructors generally assign lots of problems. Never let yourself get behind in your math homework.
- If you are struggling with a math concept, do not give up. Ask for help from your friends and your instructor. Since each concept is built on previous concepts, any gaps in your understanding will follow you through the entire course, so make sure you understand each concept as you go along.
- Read your textbook! It will often contain the help and tips you need to solve any problem with which you are struggling. It may also bring out points that you missed in class or that your instructor may not have covered.
- Learn to use scratch paper to jot down your thoughts and to draw pictures. Don't try to figure everything out "in your head." You will think more clearly and accurately this way.
- When preparing for a test, it is often helpful to work at least two problems per section as practice: one easy and one difficult. Write out all the new rules and procedures your test will cover, and then read through them twice. Doing so will enable you to both learn and retain them better.
- Most schools have classrooms set up where you can get free help from math tutors. Take advantage of this, but be sure you do the work first. Don't let your tutor do all the work for you—otherwise you'll never learn the material.
- In math, as in many other areas of life, patience and persistence are virtues—cultivate them. "Cramming" for an exam will not help you learn and retain the material.

1.4 EXERCISE SET

For Extra Help

Digital Video Tutor CD 1 Videotape 1

InterAct Math

Math Tutor Center

MathXL

MyMathLab.com

a Solve by trial.

1. $x + 0 = 14$

2. $x - 7 = 18$

3. $y \cdot 17 = 0$

4. $56 \div m = 7$

b Solve. Be sure to check.

5. $13 + x = 42$

6. $15 + t = 22$

7. $12 = 12 + m$

8. $16 = t + 16$

9. $3 \cdot x = 24$

10. $6 \cdot x = 42$

11. $112 = n \cdot 8$

12. $162 = 9 \cdot m$

13. $45 \times 23 = x$

14. $23 \times 78 = y$

15. $t = 125 \div 5$

16. $w = 256 \div 16$

17. $p = 908 - 458$

18. $9007 - 5667 = m$

19. $x = 12{,}345 + 78{,}555$

20. $5678 + 9034 = t$

21. $3 \cdot m = 96$

22. $4 \cdot y = 96$

23. $715 = 5 \cdot z$

24. $741 = 3 \cdot t$

25. $10 + x = 89$

26. $20 + x = 57$

27. $61 = 16 + y$

28. $53 = 17 + w$

29. $6 \cdot p = 1944$

30. $4 \cdot w = 3404$

31. $5 \cdot x = 3715$

32. $9 \cdot x = 1269$

33. $47 + n = 84$

34. $56 + p = 92$

35. $x + 78 = 144$

36. $z + 67 = 133$

37. $165 = 11 \cdot n$

38. $660 = 12 \cdot n$

39. $624 = t \cdot 13$

40. $784 = y \cdot 16$

41. $x + 214 = 389$

42. $x + 221 = 333$

43. $567 + x = 902$

44. $438 + x = 807$

45. $18 \cdot x = 1872$

46. $19 \cdot x = 6080$

47. $40 \cdot x = 1800$

48. $20 \cdot x = 1500$

49. $2344 + y = 6400$

50. $9281 = 8322 + t$

51. $8322 + 9281 = x$

52. $9281 - 8322 = y$

53. $234 \times 78 = y$

54. $10{,}534 \div 458 = q$

55. $58 \cdot m = 11{,}890$

56. $233 \cdot x = 22{,}135$

57. **DW** Describe a procedure that can be used to convert any equation of the form $a \cdot b = c$ to a related division equation.

58. **DW** Describe a procedure that can be used to convert any equation of the form $a + b = c$ to a related subtraction equation.

SKILL MAINTENANCE

59. Write two related subtraction sentences: $7 + 8 = 15$. [1.2c]

60. Write two related division sentences: $6 \cdot 8 = 48$. [1.3c]

Use > or < for □ to write a true sentence. [1.1d]

61. 123 □ 789

62. 342 □ 339

63. 688 □ 0

64. 0 □ 11

Divide. [1.3d]

65. $1283 \div 9$

66. $1278 \div 9$

67. $17\overline{)5678}$

68. $17\overline{)5689}$

SYNTHESIS

Solve.

69. ▦ $23{,}465 \cdot x = 8{,}142{,}355$

70. ▦ $48{,}916 \cdot x = 14{,}332{,}388$

1.5 APPLICATIONS AND PROBLEM SOLVING

Objective

a Solve applied problems involving addition, subtraction, multiplication, or division of whole numbers.

a A Problem-Solving Strategy

Applications and problem solving are the most important uses of mathematics. To solve a problem using the operations on the whole numbers, we first look at the situation. We try to translate the problem to an equation. Then we solve the equation. We check to see if the solution of the equation is a solution of the original problem. We are using the following five-step strategy.

FIVE STEPS FOR PROBLEM SOLVING

1. *Familiarize* yourself with the situation.
   a) Carefully read and reread until you understand *what* you are being asked to find.
   b) Draw a diagram or see if there is a formula that applies to the situation.
   c) Assign a letter, or *variable,* to the unknown.
2. *Translate* the problem to an equation using the letter or variable.
3. *Solve* the equation.
4. *Check* the answer in the original wording of the problem.
5. *State* the answer to the problem clearly with appropriate units.

EXAMPLE 1 *Baseball's Power Hitters.* The top three home-run hitters in the major leagues over the years from 1996 to 2000 were Sammy Sosa, Mark McGwire, and Ken Griffey, Jr. The numbers of home runs hit per year for each player are listed in the table below. Find the total number of home runs hit by Sammy Sosa over the 5-yr period.

YEAR	SAMMY SOSA	MARK MCGWIRE	KEN GRIFFEY, JR.
1996	40	52	49
1997	36	58	56
1998	66	70	56
1999	63	65	48
2000	50	32	40
Total	?	?	?

Source: Major League Baseball

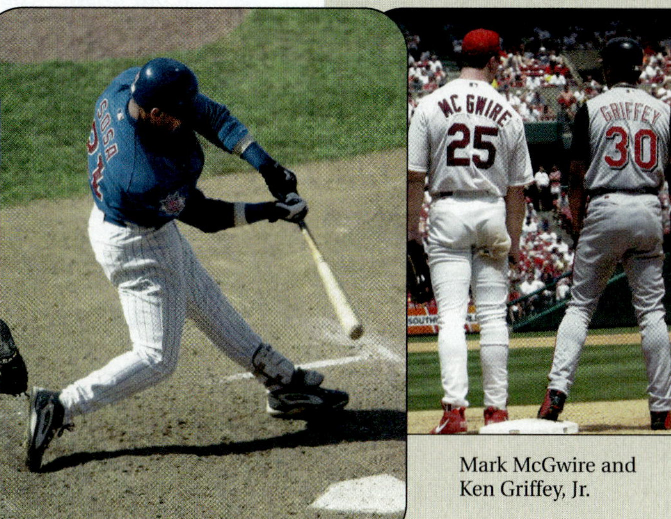

Sammy Sosa

Mark McGwire and Ken Griffey, Jr.

Refer to the table on the preceding page to answer Margin Exercises 1–3.

1. Find the total number of home runs hit by Mark McGwire from 1996 to 2000.

2. Find the total number of home runs hit by Ken Griffey, Jr., from 1996 to 2000.

3. Who hit the most home runs over the 5-yr period?

Answers on page A-2

1. **Familiarize.** We can make a drawing or at least visualize the situation.

40	+	36	+	66	+	63	+	50
in		in		in		in		in
1996		1997		1998		1999		2000

Since we are combining numbers of home runs, addition can be used. First, we define the unknown. We let $n =$ the total number of home runs hit by Sosa in the 5-yr period.

2. **Translate.** We translate to an equation:

$$40 + 36 + 66 + 63 + 50 = n.$$

3. **Solve.** We solve the equation by carrying out the addition.

$$\begin{array}{r} {}^{1} \\ 4\ 0 \\ 3\ 6 \\ 6\ 6 \\ 6\ 3 \\ +\ 5\ 0 \\ \hline 2\ 5\ 5 \end{array} \qquad \begin{aligned} 40 + 36 + 66 + 63 + 50 &= n \\ 255 &= n \end{aligned}$$

4. **Check.** We check 255 in the original problem. There are many ways in which this can be done. For example, we can repeat the calculation. (We leave this to the student.) Another way is to check whether the answer is reasonable. In this case, we would expect the total to be greater than the number of home runs in any of the individual years, which it is. We can also estimate by rounding. Here we round to the nearest ten:

$$\begin{aligned} 40 + 36 + 66 + 63 + 50 &\approx 40 + 40 + 70 + 60 + 50 \\ &= 260. \end{aligned}$$

Since $255 \approx 260$, we have a partial check. If we had an estimate like 340 or 400, we might be suspicious that our calculated answer is incorrect. Since our estimated answer is close to our calculation, we are further convinced that our answer checks.

5. **State.** The total number of home runs hit by Sammy Sosa from 1996 to 2000 was 255.

Do Exercises 1–3.

EXAMPLE 2 *Checking Account Balance.* The balance in Tyler's checking account is \$528. He uses his debit card to buy the Roto Zip Spiral Saw Combo shown in this ad. Find the new balance in his checking account.

Source: Roto Zip Tool Corporation

1. **Familiarize.** We first make a drawing or at least visualize the situation. We let $M =$ the new balance in his account. This gives us the following:

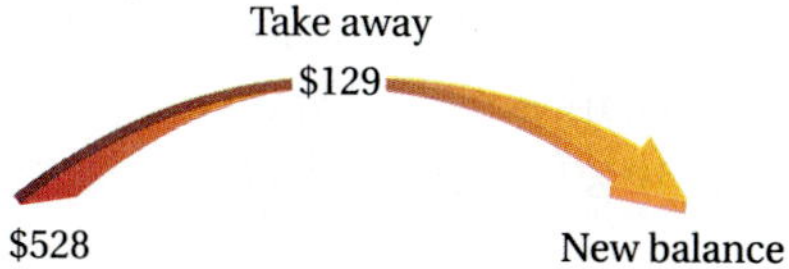

2. **Translate.** We can think of this as a "take-away" situation. We translate to an equation.

Money in the account	minus	Money spent	is	New balance
↓	↓	↓	↓	↓
528	−	129	=	M

3. **Solve.** This sentence tells us what to do. We subtract.

$$\begin{array}{r} \scriptstyle 11 \\ \scriptstyle 4\ \not{1}\ 18 \\ \not{5}\ 2\ \not{8} \\ -\ 1\ 2\ 9 \\ \hline 3\ 9\ 9 \end{array}$$

$$528 - 129 = M$$
$$399 = M$$

4. **Check.** To check our answer of \$399, we can repeat the calculation. We note that the answer should be less than the original amount, \$528, which it is. We can add the difference, 399, to the subtrahend, 129: $129 + 399 = 528$. We can also estimate:

$$528 - 129 \approx 530 - 130 = 400 \approx 399.$$

5. **State.** Tyler has a new balance of \$399 in his checking account.

Do Exercise 4.

4. **Checking Account Balance.** The balance in Heidi's checking account is \$2003. She uses her debit card to buy the same Roto Zip Spiral Saw Combo, featured in Example 2, that Tyler did. Find the new balance in her checking account.

Answer on page A-2

5. **Home Theatre Audio System.** Bernardo has \$376. He wants to purchase the Home Theatre Audio System shown in the ad below. How much more does he need?

Answer on page A-2

In the real world, problems may not be stated in written words. You must still become familiar with the situation before you can solve the problem.

EXAMPLE 3 *Travel Distance.* Vicki is driving from Indianapolis to Salt Lake City to work during the 2002 Winter Olympics. The distance from Indianapolis to Salt Lake City is 1634 mi. She travels 1154 mi to Denver. How much farther must she travel?

1. **Familiarize.** We first make a drawing or at least visualize the situation. We let $x =$ the remaining distance to Salt Lake City.

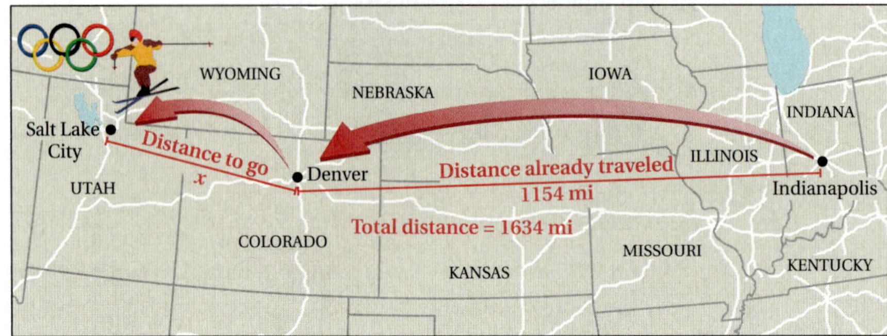

2. **Translate.** We see that this is a "missing-addend" situation. We translate to an equation.

Distance already traveled	plus	Distance to go	is	Total distance of trip
↓	↓	↓	↓	↓
1154	$+$	x	$=$	1634

3. **Solve.** To solve the equation , we subtract 1154 on both sides:

$$1154 + x = 1634$$
$$1154 + x - 1154 = 1634 - 1154$$
$$x = 480.$$

$$\begin{array}{r} \overset{5}{\cancel{6}}\ \overset{13}{\cancel{3}} \\ 1\ 6\ 3\ 4 \\ -\ 1\ 1\ 5\ 4 \\ \hline 4\ 8\ 0 \end{array}$$

4. **Check.** We check our answer of 480 mi in the original problem. This number should be less than the total distance, 1634 mi, which it is. We can add the difference, 480, to the subtrahend, 1154: $1154 + 480 = 1634$. We can also estimate:

$$1634 - 1154 \approx 1600 - 1200$$
$$= 400 \approx 480.$$

The answer, 480 mi, checks.

5. **State.** Vicki must travel 480 mi farther to Salt Lake City.

Do Exercise 5.

EXAMPLE 4 *Total Cost of DVDs.* What is the total cost of 5 DVD players if each one costs \$249?

1. **Familiarize.** We first make a drawing or at least visualize the situation. We let $T =$ the cost of 5 DVD players. Repeated addition works well in this case.

2. **Translate.** We translate to an equation.

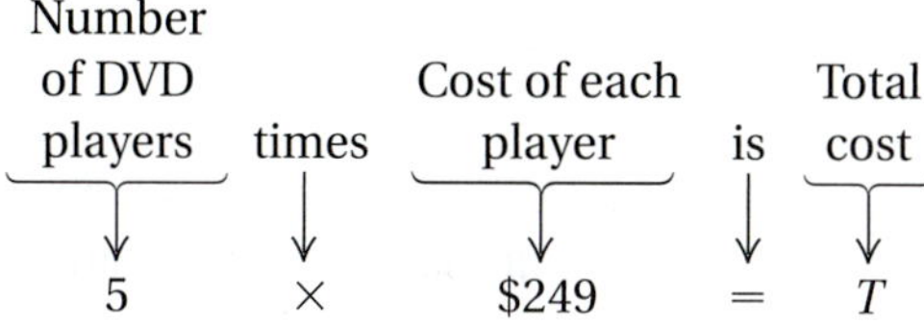

Number of DVD players	times	Cost of each player	is	Total cost
↓	↓	↓	↓	↓
5	$\times$	\$249	$=$	T

3. **Solve.** This sentence tells us what to do. We multiply.

$$\begin{array}{r} {\scriptstyle 2\,4} \\ 2\,4\,9 \\ \times\quad\;\, 5 \\ \hline 1\,2\,4\,5 \end{array} \qquad \begin{aligned} 5 \times 249 &= T \\ 1245 &= T \end{aligned}$$

4. **Check.** We have an answer, 1245, that is much greater than the cost of any individual DVD player, which is reasonable. We can repeat our calculation. We can also check by estimating:

$$5 \times 249 \approx 5 \times 250 = 1250 \approx 1245.$$

The answer checks.

5. **State.** The total cost of 5 DVD players is \$1245.

Do Exercise 6.

EXAMPLE 5 *Bed Sheets.* The dimensions of a flat sheet for a king-size bed are 108 in. by 102 in. What is the area of the sheet? (The dimension labels on sheets list width $\times$ length.)

1. **Familiarize.** We first make a drawing. We let $A =$ the area.

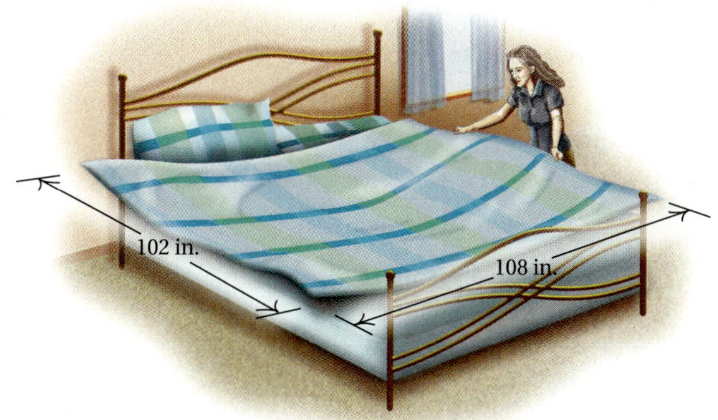

6. **Total Cost of Laptop Computers.** What is the total cost of 12 Gateway Solo 9300 laptop computers with CD ROM drives and 1 GHz processors if each one costs \$2898?
Source: Gateway Country® Stores

Answer on page A-2

7. **Bed Sheets.** The dimensions of a flat sheet for a queen-size bed are 90 in. by 102 in. What is the area of the sheet?

2. **Translate.** Using a formula for area, we have

$$A = \text{length} \cdot \text{width} = l \cdot w = 102 \cdot 108.$$

3. **Solve.** We carry out the multiplication.

$$\begin{array}{r} 1\ 0\ 8 \\ \times\quad 1\ 0\ 2 \\ \hline 2\ 1\ 6 \\ 1\ 0\ 8\ 0\ 0 \\ \hline 1\ 1\ 0\ 1\ 6 \end{array} \qquad \begin{aligned} A &= 102 \cdot 108 \\ A &= 11{,}016 \end{aligned}$$

4. **Check.** We repeat our calculation. We also note that the answer is greater than either the length or the width, which it should be. (This might not be the case if we were using fractions or decimals.) The answer checks.

5. **State.** The area of a king-size bed sheet is 11,016 sq in.

Do Exercise 7.

EXAMPLE 6 *Cartons of Soda.* A bottling company produces 3304 cans of soda. How many 12-can cartons can be filled? How many cans will be left over?

1. **Familiarize.** We first make a drawing. We let n = the number of 12-can cartons that can be filled. The problem can be considered as repeated subtraction, taking successive sets of 12 cans and putting them into n cartons.

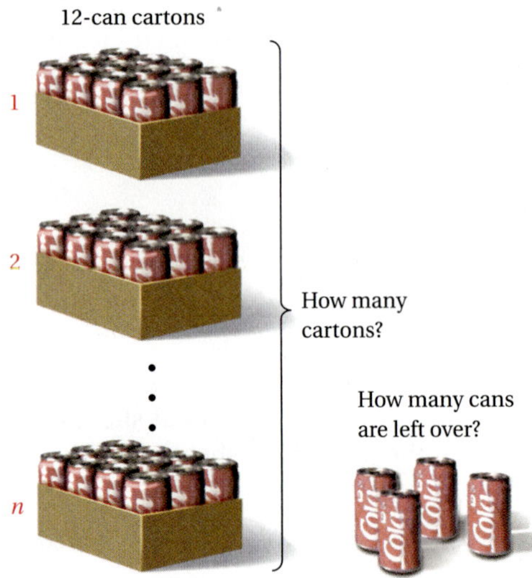
12-can cartons
1
2
n
How many cartons?
How many cans are left over?

2. **Translate.** We translate to an equation.

Number of cans	divided by	Number in each carton	is	Number of cartons
↓	↓	↓	↓	↓
3304	÷	12	=	n

Answer on page A-2

3. Solve. We solve the equation by carrying out the division.

$$\begin{array}{r} 275 \\ 12\overline{)3304} \\ 2400 \\ \hline 904 \\ 840 \\ \hline 64 \\ 60 \\ \hline 4 \end{array}$$

$$3304 \div 12 = n$$
$$275 \text{ R } 4 = n$$

4. Check. We can check by multiplying the number of cartons by 12 and adding the remainder, 4:

$12 \cdot 275 = 3300,$

$3300 + 4 = 3304.$

5. State. Thus, 275 twelve-can cartons can be filled. There will be 4 cans left over.

Do Exercise 8.

EXAMPLE 7 *Automobile Mileage.* The Chrysler PT Cruiser gets 22 miles to the gallon (mpg) in city driving. How many gallons will it use in 6028 mi of city driving?

Source: DaimlerChrysler Corporation

1. Familiarize. We first make a drawing. It is often helpful to be descriptive about how we define a variable. In this case, we let $g =$ the number of gallons ("g" comes from "gallons").

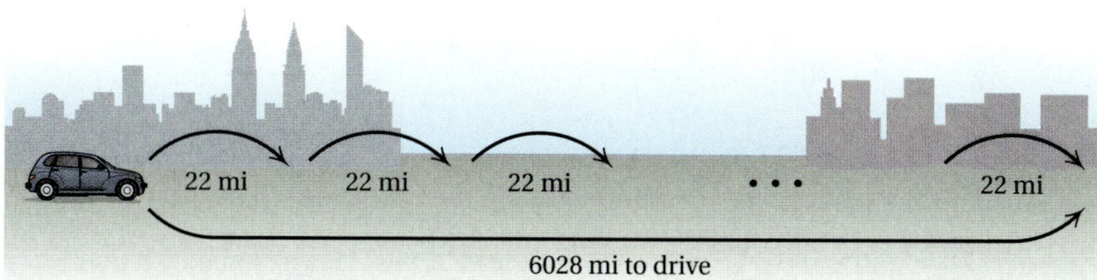

2. Translate. Repeated addition applies here. Thus the following multiplication applies to the situation.

Number of miles per gallon	times	Number of gallons needed	is	Number of miles to drive
↓	↓	↓	↓	↓
22	$\cdot$	g	$=$	6028

3. Solve. To solve the equation, we divide by 22 on both sides.

$$22 \cdot g = 6028$$
$$\frac{22 \cdot g}{22} = \frac{6028}{22}$$
$$g = 274$$

$$\begin{array}{r} 274 \\ 22\overline{)6028} \\ 4400 \\ \hline 1628 \\ 1540 \\ \hline 88 \\ 88 \\ \hline 0 \end{array}$$

8. Cartons of Soda. The bottling company in Example 6 also uses 6-can cartons. How many 6-can cartons can be filled with 2269 cans of cola? How many will be left over?

Answer on page A-2

9. Automobile Mileage. The Chrysler PT Cruiser gets 26 miles to the gallon (mpg) in highway driving. How many gallons will it take to drive 884 mi of highway driving?
Source: DaimlerChrysler Corporation

Answer on page A-2

4. Check. To check, we multiply 274 by 22: $22 \cdot 274 = 6028$.

5. State. The PT Cruiser will use 274 gal.

Do Exercise 9.

Multistep Problems

Sometimes we must use more than one operation to solve a problem, as in the following example.

EXAMPLE 8 *Aircraft Seating.* Boeing Corporation builds commercial aircraft. A Boeing 767 has a seating configuration with 4 rows of 6 seats across in first class and 35 rows of 7 seats across in economy class. Find the total seating capacity of the plane.
Sources: The Boeing Company; Delta Airlines

1. Familiarize. We first make a drawing.

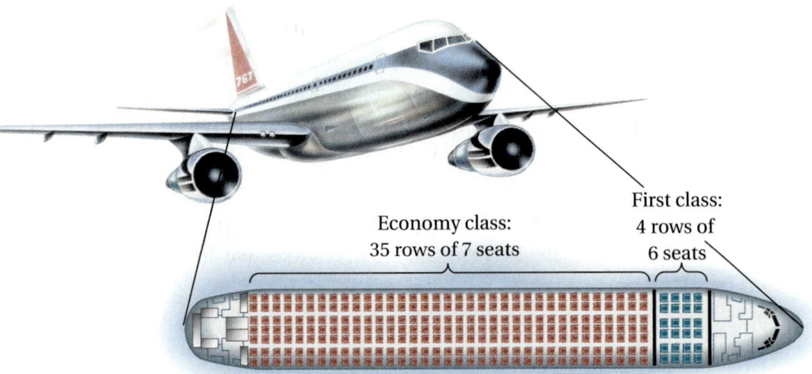

2. Translate. There are three parts to the problem. We first find the number of seats in each class. Then we add.

First-class: Repeated addition applies here. Thus the following multiplication corresponds to the situation. We let $F =$ the number of seats in first class.

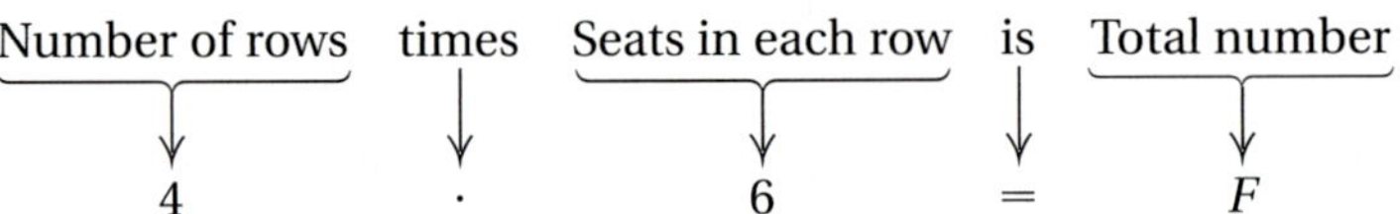

Economy class: Repeated addition applies here. Thus the following multiplication corresponds to the situation. We let $E =$ the number of seats in economy class.

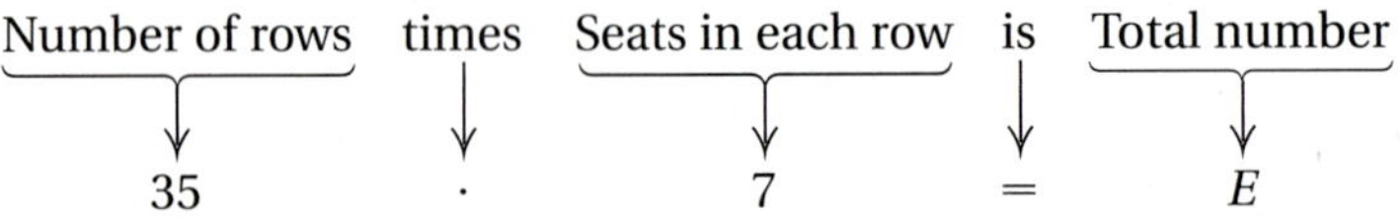

We let $T =$ the total number of seats in both classes.

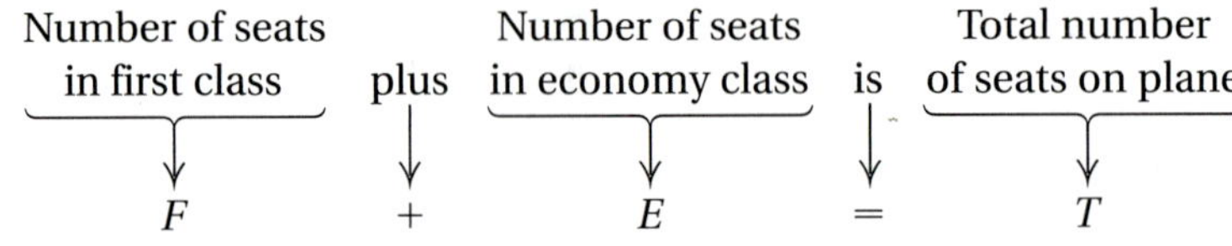

3. **Solve.** We solve each equation and add the solutions.

$$4 \cdot 6 = F \qquad 35 \cdot 7 = E \qquad F + E = T$$
$$24 = F \qquad 245 = E \qquad 24 + 245 = T$$
$$269 = T$$

4. **Check.** To check, we repeat our calculations. (We leave this to the student.) We could also check by rounding, multiplying, and adding.
5. **State.** There are 269 seats in a Boeing 767.

Do Exercise 10.

As you consider the following exercises, here are some words and phrases that may be helpful to look for when you are translating problems to equations.

KEY WORDS, PHRASES, AND CONCEPTS

Addition (+)	Subtraction (−)
add	subtract
added to	subtracted from
sum	difference
total	minus
plus	less than
more than	decreased by
increased by	take away
	how much more
	missing addend

Multiplication (·)	Division (÷)
multiply	divide
multiplied by	divided by
product	quotient
times	repeated subtraction
of	missing factor
repeated addition	finding equal quantities
rectangular arrays	

10. **Aircraft Seating.** A Boeing 767 used for foreign travel has three classes of seats. First class has 3 rows of 5 seats across; business class has 6 rows with 6 seats across and 1 row with 2 seats on each of the outside aisles. Economy class has 18 rows with 7 seats across. Find the total seating capacity of the plane.

Sources: The Boeing Company; Delta Airlines

Economy class: 18 rows of 7 seats

First class: 3 rows of 5 seats

Business class: 6 rows of 6 seats...

...with 2 seats on each outside aisle

Answer on page A-2

Study Tips

EXERCISES

- **Odd-numbered exercises.** Usually an instructor assigns some odd-numbered exercises. When you complete these, you can check your answers at the back of the book. If you miss any, check your work in the *Student's Solutions Manual* or ask your instructor for guidance.
- **Even-numbered exercises.** Whether or not your instructor assigns the even-numbered exercises, always do some on your own. As you know, there are no answers provided when you take a test, so you need to practice doing exercises without answers. Check your answers later with a friend or your instructor.

1.5 EXERCISE SET

For Extra Help

Digital Video Tutor CD 1 Videotape 1 | InterAct Math | Math Tutor Center | MathXL | MyMathLab.com

a Solve.

Top Web Properties. The bar graph below shows the four most frequently visited Web sites, in terms of the number of visits for a recent month. Use this graph for Exercises 1–4.

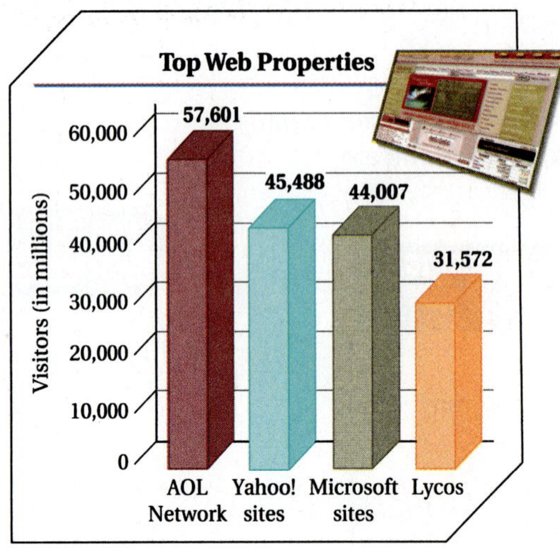

Source: Media Matrix

1. What was the total number of visits to all the sites?

2. What was the total number of visits to the three most-visited sites?

3. How many more visits were there to the AOL Network site than to the Yahoo! sites?

4. How many more visits were there to the Microsoft sites than to the Lycos site?

5. *Concorde Crash.* The Anglo-French Concorde entered service in 1976. It had its first crash 24 yr later. In what year did it have its first crash?

6. Dwight D. Eisenhower was the 34th president of the United States. He left office in 1961 and lived another 8 yr. In what year did he die?

New England. The following table lists various data about the New England states.

NEW ENGLAND STATES	TOTAL AREA (in square miles)	TOTAL INLAND WATER AREA (in square miles)	SALARY OF THE GOVERNOR	POPULATION IN 1998
Maine	33,265	2,270	$70,000	1,244,250
New Hampshire	9,279	286	86,235	1,185,048
Vermont	9,614	341	80,725	590,883
Massachusetts	8,284	460	75,000	6,147,132
Connecticut	5,018	146	78,000	3,274,069
Rhode Island	1,212	157	69,900	988,480

Source: *The New York Times Almanac*

7. Find the total area of New England.

8. Find the total area of inland water in New England.

9. Find the total amount paid in salaries to the governors of the New England states.

10. Find the total population of New England in 1998.

11. *Military Downsizing.* In 2000, there were 372,000 people in the Navy. This was down from the 583,000 who were in the Navy in 1990. How many more were in the Navy in 1990 than in 2000?

12. *Baseball Salaries.* The New York Yankees led the Major Leagues in 2000 with a total payroll of \$114,336,616. The Minnesota Twins had the lowest payroll at \$23,499,966. How much more would the Twins have to spend on payroll to equal the Yankees?
Source: Major League Baseball

13. *Longest Rivers.* The longest river in the world is the Nile in Egypt at 4100 mi. The longest river in the United States is the Missouri–Mississippi at 3860 mi. How much longer is the Nile?

14. *Speeds on Interstates.* Recently, speed limits on interstate highways in many Western states were raised from 65 mph to 75 mph. By how many miles per hour were they raised?

15. *Automobile Mileage.* The 2000 Volkswagen New Beetle GL gets 24 miles to the gallon (mpg) in city driving. How many gallons will it use in 6144 mi of city driving?
Source: Volkswagen of America, Inc.

16. *Automobile Mileage.* The 2000 Volkswagen New Beetle GL gets 31 miles to the gallon (mpg) in highway driving. How many gallons will it use in 5859 mi of highway driving?
Source: Volkswagen of America, Inc.

17. *Pixels.* A computer screen consists of small rectangular dots called *pixels.* How many pixels are there on a screen that has 600 rows with 800 pixels in each row?

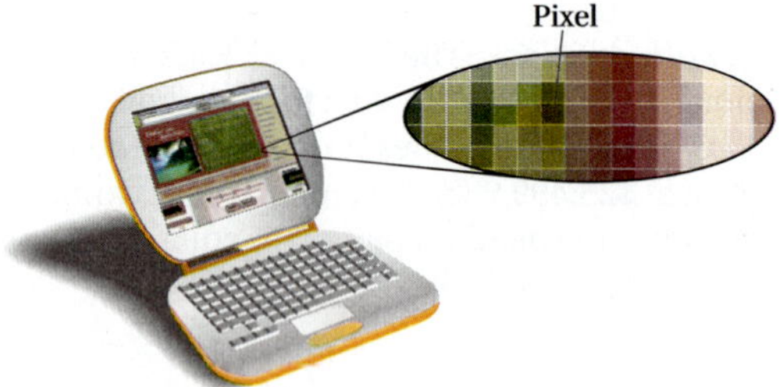

18. *Crossword.* The *USA Today* crossword puzzle is a rectangle containing 15 rows with 15 squares in each row. How many squares does the puzzle have altogether?

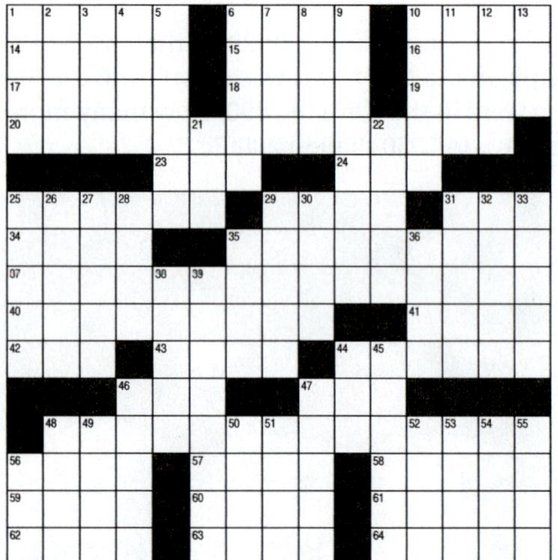

19. *Refrigerator Purchase.* Cometbucks Deli has a chain of 24 restaurants. It buys a refrigerator for each store at a cost of $499 each. Find the total cost of the purchase.

20. *Microwave Purchase.* Bridgeway College is constructing new dorms, in which each room has a small kitchen. It buys 96 microwave ovens at $88 each. Find the total cost of the purchase.

Music CD Sales. The bar graph below shows the sales of music CDs, in millions, for the years from 1995 to 1999. Use this graph for Exercises 21–24.

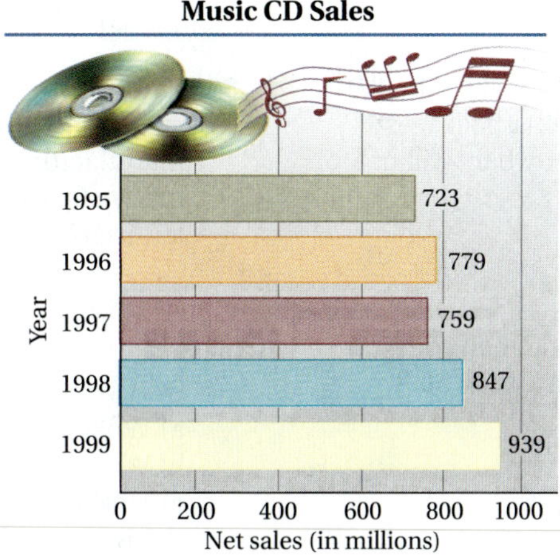

21. How many more CDs were sold in 1999 than in 1995?

22. How many more CDs were sold in 1999 than in 1998?

23. What was the total number of CDs sold from 1997 through 1999?

24. What was the total number of CDs sold from 1995 through 1999?

25. *"Seinfeld" Episodes.* "Seinfeld" is a long-running television comedy with 177 episodes created. A local station picks up the syndicated reruns. If the station runs 5 episodes per week, how many full weeks will pass before it must start over with past episodes? How many episodes will be left for the last week?

26. A lab technician separates a vial containing 70 cubic centimeters (cc) of blood into test tubes, each of which contains 3 cc of blood. How many test tubes can be filled? How much blood is left over?

27. There are 24 hours (hr) in a day and 7 days in a week. How many hours are there in a week?

28. There are 60 min in an hour and 24 hr in a day. How many minutes are there in a day?

29. Dana borrows $5928 for a used car. The loan is to be paid off in 24 equal monthly payments. How much is each payment (excluding interest)?

30. A family borrows $4824 to build a sunroom on the back of their home. The loan is to be paid off in equal monthly payments of $134 (excluding interest). How many months will it take to pay off the loan?

31. *Atlanta Population.* The population of Atlanta was 3,857,097 in 1999. This was an increase of 897,597 from its population in 1990. What was the population of Atlanta in 1990?
Source: U.S. Bureau of the Census

32. *Orlando Population.* The population of Orlando was 1,535,004 in 1999. This was an increase of 310,160 from its population in 1990. What was the population of Orlando in 1990?
Source: U.S. Bureau of the Census

33. *Crossword.* The *Los Angeles Times* crossword puzzle is a rectangle containing 441 squares arranged in 21 rows. How many columns does the puzzle have?

34. *Sheet of Stamps.* A sheet of 100 stamps typically has 10 rows of stamps. How many stamps are in each row?

35. *Hershey Bars.* Hershey Chocolate USA makes small, fun-size chocolate bars. How many 20-bar packages can be filled with 11,267 bars? How many bars will be left over?

36. *Reese's Peanut Butter Cups.* H. B. Reese Candy Co. makes small, fun-size peanut butter cups. The company manufactures 23,579 cups and fills 1025 packages. How many cups are in a package? How many cups will be left over?

37. *High School Court.* The standard basketball court used by high school players has dimensions of 50 ft by 84 ft.

a) What is its area?
b) What is its perimeter?

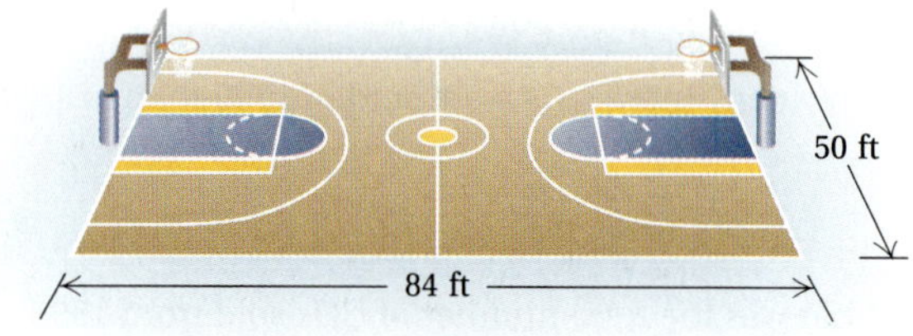

38. *NBA Court.* The standard basketball court used by college and NBA players has dimensions of 50 ft by 94 ft.

a) What is its area?
b) What is its perimeter?
c) How much greater is the area of an NBA court than a high school court? (See Exercise 37.)

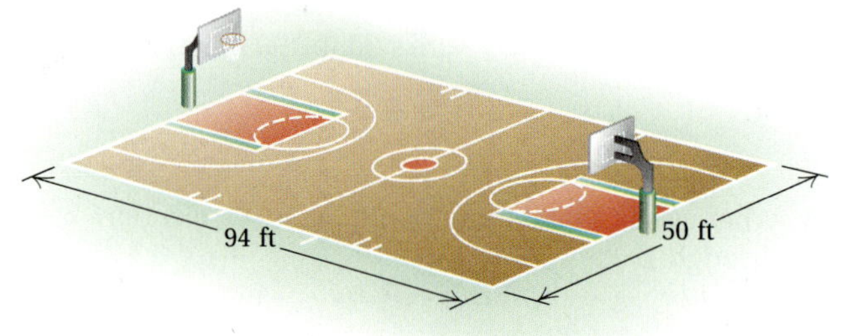

39. Copies of this book are generally shipped from the Addison-Wesley warehouse in cartons containing 24 books each. How many cartons are needed to ship 840 books?

40. According to the H. J. Heinz Company, 16-oz bottles of catsup are generally shipped in cartons containing 12 bottles each. How many cartons are needed to ship 528 bottles of catsup?

41. Copies of this book are generally shipped from the warehouse in cartons containing 24 books each. How many cartons are needed to ship 1355 books?

42. Sixteen-ounce bottles of catsup are generally shipped in cartons containing 12 bottles each. How many cartons are needed to ship 1033 bottles of catsup?

43. *Map Drawing.* A map has a scale of 64 mi to the inch. How far apart *in reality* are two cities that are 6 in. apart on the map? How far apart *on the map* are two cities that, in reality, are 1728 mi apart?

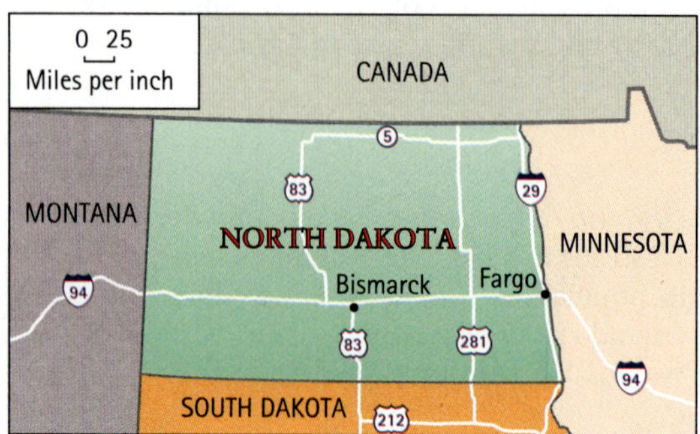

44. *Map Drawing.* A map has a scale of 150 mi to the inch. How far apart *on the map* are two cities that, in reality, are 2400 mi apart? How far apart *in reality* are two cities that are 13 in. apart on the map?

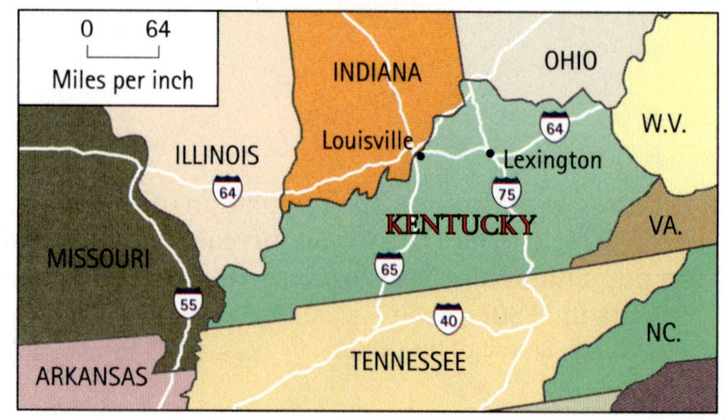

45. A carpenter drills 216 holes in a rectangular array in a pegboard. There are 12 holes in each row. How many rows are there?

46. Lou works as a CPA. He arranges 504 entries on a spreadsheet in a rectangular array that has 36 rows. How many entries are in each row?

47. Elena buys 5 video games at $64 each and pays for them with $10 bills. How many $10 bills did it take?

48. Pedro buys 5 video games at $64 each and pays for them with $20 bills. How many $20 bills did it take?

49. You have $568 in your checking account. You write checks for $46, $87, and $129. Then you deposit $94 back in the account after the return of some books. How much is left in your account?

50. The balance in your checking account is $749. You write checks for $34 and $65. Then you make a deposit of $123 from your paycheck. What is your new balance?

Weight Loss. Many Americans exercise for weight control. It is known that one must burn off about 3500 calories in order to lose one pound. The chart shown here details how much of certain types of exercise is required to burn 100 calories. Use this chart for Exercises 51–54.

To burn off 100 calories, you must:
• Run for 8 min at a brisk pace, or
• Swim for 2 min at a brisk pace, or
• Bicycle for 15 min at 9 mph, or
• Do aerobic exercises for 15 min.

51. How long must you run at a brisk pace in order to lose one pound?

52. How long must you swim in order to lose one pound?

53. How long must you do aerobic exercises in order to lose one pound?

54. How long must you bicycle at 9 mph in order to lose one pound?

55. *Bones in the Hands and Feet.* There are 27 bones in each human hand and 26 bones in each human foot. How many bones are there in all in the hands and feet?

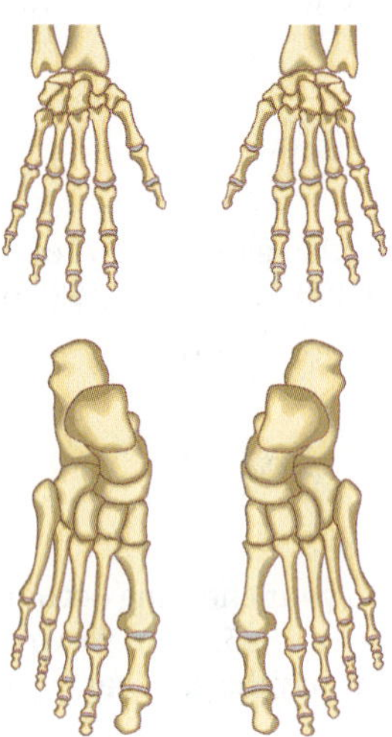

56. *Index Cards.* Index cards of dimension 3 in. by 5 in. are normally shipped in packages containing 100 cards each. How much writing area is available if one uses the front and back sides of a package of these cards?

57. Before going back to college, David buys 4 shirts at \$59 each and 6 pairs of pants at \$78 each. What is the total cost of this clothing?

58. An office for adjunct instructors at a community college has 6 bookshelves, each of which is 3 ft long. The office is moved to a new location that has dimensions of 16 ft by 21 ft. Is it possible for the bookshelves to be put side by side on the 16-ft wall?

59. D_W In the newspaper article, "When Girls Play, Knees Fail," the author discusses the fact that female athletes have six times the number of knee injuries that male athletes have. What information would be needed if you were to write a math problem based on the article? What might the problem be?
Source: *The Arizona Republic*, 2/9/00, p. C1

60. D_W Write a problem for a classmate to solve. Design the problem so that the solution is "The driver still has 329 mi to travel."

SKILL MAINTENANCE

Round 234,562 to the nearest: [1.3e]

61. Hundred.

62. Ten.

63. Thousand.

Estimate the computation by rounding to the nearest thousand. [1.3f]

64. 2783 + 4602 + 5797 + 8111

65. 28,430 − 11,977

66. 2100 + 5800

67. 5800 − 2100

Estimate the product by rounding to the nearest hundred. [1.3f]

68. 787 · 363

69. 887 · 799

70. 10,362 · 4531

SYNTHESIS

71. *Speed of Light.* Light travels about 186,000 miles per second (mi/sec) in a vacuum as in outer space. In ice it travels about 142,000 mi/sec, and in glass it travels about 109,000 mi/sec. In 18 sec, how many more miles will light travel in a vacuum than in ice? than in glass?

72. Carney Community College has 1200 students. Each professor teaches 4 classes and each student takes 5 classes. There are 30 students and 1 teacher in each classroom. How many professors are there at Carney Community College?

1.6 EXPONENTIAL NOTATION AND ORDER OF OPERATIONS

a Writing Exponential Notation

Consider the product $3 \cdot 3 \cdot 3 \cdot 3$. Such products occur often enough that mathematicians have found it convenient to create a shorter notation, called **exponential notation,** explained as follows.

$\underbrace{3 \cdot 3 \cdot 3 \cdot 3}_{4 \text{ factors}}$ is shortened to 3^4 ← exponent; 3 is the base

We read exponential notation as follows.

NOTATION	WORD DESCRIPTION
3^4	"three to the fourth power," or "the fourth power of three"
5^3	"five to the third power," or "the third power of five," or "five-cubed," or "the cube of five"
7^2	"seven to the second power," or "the second power of seven," or "seven squared," or "the square of seven"

The wording "seven squared" for 7^2 comes from the fact that a square with side s has area A given by $A = s^2$.

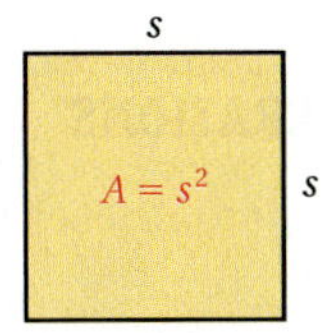

An expression like $3 \cdot 5^2$ is read "three times the square of five" or "three times five squared."

EXAMPLE 1 Write exponential notation for $10 \cdot 10 \cdot 10 \cdot 10 \cdot 10$.

Exponential notation is 10^5. 5 is the *exponent*. 10 is the *base*.

EXAMPLE 2 Write exponential notation for $2 \cdot 2 \cdot 2$.

Exponential notation is 2^3.

Do Exercises 1–4.

Objectives

a Write exponential notation for products such as $4 \cdot 4 \cdot 4$.

b Evaluate exponential notation.

c Simplify expressions using the rules for order of operations.

d Remove parentheses within parentheses.

Write exponential notation.

1. $5 \cdot 5 \cdot 5 \cdot 5$

2. $5 \cdot 5 \cdot 5 \cdot 5 \cdot 5$

3. $10 \cdot 10$

4. $10 \cdot 10 \cdot 10 \cdot 10$

Evaluate.

5. 10^4

6. 10^2

7. 8^3

8. 2^5

Answers on page A-2

b Evaluating Exponential Notation

We evaluate exponential notation by rewriting it as a product and computing the product.

EXAMPLE 3 Evaluate: 10^3.

$$10^3 = 10 \cdot 10 \cdot 10 = 1000$$

EXAMPLE 4 Evaluate: 5^4.

$$5^4 = 5 \cdot 5 \cdot 5 \cdot 5 = 625$$

CAUTION!

5^4 does not mean $5 \cdot 4$.

Do Exercises 5–8 on the preceding page.

c Simplifying Expressions

Suppose we have a calculation like the following:

$$3 + 4 \cdot 8.$$

How do we find the answer? Do we add 3 to 4 and then multiply by 8, or do we multiply 4 by 8 and then add 3? In the first case, the answer is 56. In the second, the answer is 35. We agree to compute as in the second case.

Consider the calculation

$$7 \cdot 14 - (12 + 18).$$

What do the parentheses mean? To deal with these questions, we must make some agreement regarding the order in which we perform operations. The rules are as follows.

RULES FOR ORDER OF OPERATIONS

1. Do all calculations within parentheses ( ), brackets [ ], or braces { } before operations outside.
2. Evaluate all exponential expressions.
3. Do all multiplications and divisions in order from left to right.
4. Do all additions and subtractions in order from left to right.

It is worth noting that these are the rules that computers and most scientific calculators use to do computations.

EXAMPLE 5 Simplify: $16 \div 8 \times 2$.

There are no parentheses or exponents, so we start with the third step.

$16 \div 8 \times 2 = 2 \times 2$ Doing all multiplications and divisions in order from left to right

$= 4$

CALCULATOR CORNER

Exponential Notation

Many calculators have a y^x or x^y key for raising a base to a power. To find 16^3, for example, we press 1 6 y^x 3 = or 1 6 x^y 3 =. The result is 4096.

Exercises: Use a calculator to find each of the following.

1. 3^5
2. 5^6
3. 12^4
4. 2^{11}

Simplify.

9. $93 - 14 \cdot 3$

10. $104 \div 4 + 4$

11. $25 \cdot 26 - (56 + 10)$

12. $75 \div 5 + (83 - 14)$

Answers on page A-2

EXAMPLE 6 Simplify: $7 \cdot 14 - (12 + 18)$.

$7 \cdot 14 - (12 + 18) = 7 \cdot 14 - 30$ Carrying out operations inside parentheses

$= 98 - 30$ Doing all multiplications and divisions

$= 68$ Doing all additions and subtractions

Do Exercises 9–12 on the preceding page.

EXAMPLE 7 Simplify and compare: $23 - (10 - 9)$ and $(23 - 10) - 9$.

We have

$23 - (10 - 9) = 23 - 1 = 22;$

$(23 - 10) - 9 = 13 - 9 = 4.$

We can see that $23 - (10 - 9)$ and $(23 - 10) - 9$ represent different numbers. Thus subtraction is not associative.

Do Exercises 13 and 14.

EXAMPLE 8 Simplify: $7 \cdot 2 - (12 + 0) \div 3 - (5 - 2)$.

$7 \cdot 2 - (12 + 0) \div 3 - (5 - 2) = 7 \cdot 2 - 12 \div 3 - 3$
Carrying out operations inside parentheses

$= 14 - 4 - 3$
Doing all multiplications and divisions in order from left to right

$= 7$ Doing all additions and subtractions in order from left to right

Do Exercise 15.

EXAMPLE 9 Simplify: $15 \div 3 \cdot 2 \div (10 - 8)$.

$15 \div 3 \cdot 2 \div (10 - 8) = 15 \div 3 \cdot 2 \div 2$ Carrying out operations inside parentheses

$= 5 \cdot 2 \div 2$
$= 10 \div 2$
$= 5$
Doing all multiplications and divisions in order from left to right

Do Exercises 16–18.

Simplify and compare.

13. $64 \div (32 \div 2)$ and $(64 \div 32) \div 2$

14. $(28 + 13) + 11$ and $28 + (13 + 11)$

15. Simplify:

$9 \times 4 - (20 + 4) \div 8 - (6 - 2).$

Simplify.

16. $5 \cdot 5 \cdot 5 + 26 \cdot 71 - (16 + 25 \cdot 3)$

17. $30 \div 5 \cdot 2 + 10 \cdot 20 + 8 \cdot 8 - 23$

18. $95 - 2 \cdot 2 \cdot 2 \cdot 5 \div (24 - 4)$

Answers on page A-3

Simplify.

19. $5^3 + 26 \cdot 71 - (16 + 25 \cdot 3)$

20. $(1 + 3)^3 + 10 \cdot 20 + 8^2 - 23$

21. $81 - 3^2 \cdot 2 \div (12 - 9)$

22. Simplify: $2^3 \cdot 2^8 \div 2^9$.

Answers on page A-3

EXAMPLE 10 Simplify: $4^2 \div (10 - 9 + 1)^3 \cdot 3 - 5$.

$4^2 \div (10 - 9 + 1)^3 \cdot 3 - 5$

$= 4^2 \div (1 + 1)^3 \cdot 3 - 5$ Subtracting inside parentheses

$= 4^2 \div 2^3 \cdot 3 - 5$ Adding inside parentheses

$= 16 \div 8 \cdot 3 - 5$ Evaluating exponential expressions

$= 2 \cdot 3 - 5$
$= 6 - 5$ Doing all multiplications and divisions in order from left to right

$= 1$ Subtracting

Do Exercises 19–21.

EXAMPLE 11 Simplify: $2^9 \div 2^6 \cdot 2^3$.

$2^9 \div 2^6 \cdot 2^3 = 512 \div 64 \cdot 8$ There are no parentheses. Evaluating exponential expressions

$= 8 \cdot 8$
$= 64$ Doing all multiplications and divisions in order from left to right

Do Exercise 22.

CALCULATOR CORNER

Order of Operations To determine whether a calculator is programmed to follow the rules for order of operations, we can enter a simple calculation that requires using those rules. For example, we enter [3] [+] [4] [×] [2] [=]. If the result is 11, we know that the rules for order of operations have been followed. That is, the multiplication $4 \times 2 = 8$ was performed first and then 3 was added to produce a result of 11. If the result is 14, we know that the calculator performs operations as they are entered rather than following the rules for order of operations. That means, in this case, that 3 and 4 were added first to get 7 and then that sum was multiplied by 2 to produce the result of 14. For such calculators, we would have to enter the operations in the order in which we want them performed. In this case, we would press [4] [×] [2] [+] [3] [=].

Many calculators have parenthesis keys that can be used to enter an expression containing parentheses. To enter $5(4 + 3)$, for example, we press [5] [(] [4] [+] [3] [)] [=]. The result is 35.

Exercises: Simplify.

1. $84 - 5 \cdot 7$

2. $80 + 50 \div 10$

3. $3^2 + 9^2 \div 3$

4. $4^4 \div 64 - 4$

5. $15 \cdot 7 - (23 + 9)$

6. $(4 + 3)^2$

AVERAGES

In order to find the average of a set of numbers, we use addition and then division. For example, the average of 2, 3, 6, and 9 is found as follows.

The number of addends is 4.

$$\text{Average} = \frac{2 + 3 + 6 + 9}{4} = \frac{20}{4} = 5$$

Divide by 4.

AVERAGE

The **average** of a set of numbers is the sum of the numbers divided by the number of addends.

EXAMPLE 12 *Average Height of Waterfalls.* The heights of the four highest waterfalls in the world are given in the bar graph at right. Find the average height of all four.

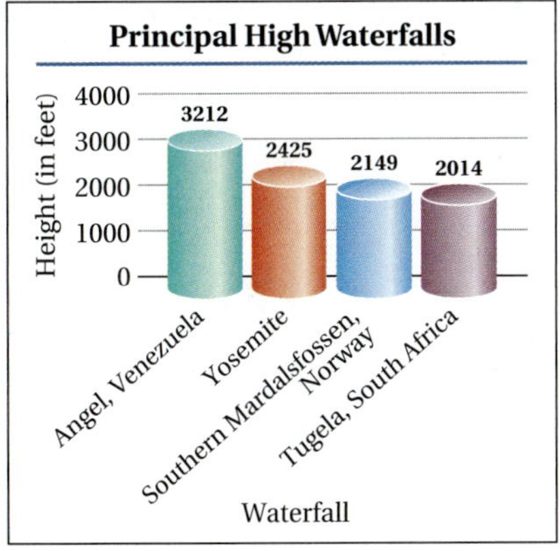

Source: World Almanac

The average is given by $\frac{3212 + 2425 + 2149 + 2014}{4} = \frac{9800}{4} = 2450.$

Thus the average height of the four highest waterfalls is 2450 ft.

Do Exercise 23.

d Removing Parentheses within Parentheses

When parentheses occur within parentheses, we can make them different shapes, such as [ ] (also called "brackets") and { } (also called "braces"). All of these have the same meaning. When parentheses occur within parentheses, computations in the innermost ones are to be done first.

EXAMPLE 13 Simplify: $[25 - (4 + 3) \times 3] \div (11 - 7)$.

$[25 - (4 + 3) \times 3] \div (11 - 7)$

$= [25 - 7 \times 3] \div (11 - 7)$ Doing the calculations in the innermost parentheses first

$= [25 - 21] \div (11 - 7)$ Doing the multiplication in the brackets

$= 4 \div 4$ Subtracting

$= 1$ Dividing

23. NBA Tall Men. The heights, in inches, of several of the tallest players in the NBA are given in the bar graph below. Find the average height of these players.

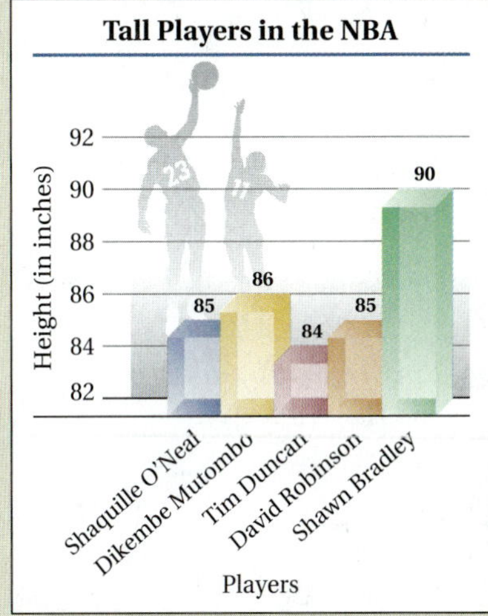

Source: NBA

Answer on page A-3

Simplify.

24. $9 \times 5 + \{6 \div [14 - (5 + 3)]\}$

25. $[18 - (2 + 7) \div 3] - (31 - 10 \times 2)$

Answers on page A-3

EXAMPLE 14 Simplify: $16 \div 2 + \{40 - [13 - (4 + 2)]\}$.

$16 \div 2 + \{40 - [13 - (4 + 2)]\}$

$= 16 \div 2 + \{40 - [13 - 6]\}$ Doing the calculations in the innermost parentheses first

$= 16 \div 2 + \{40 - 7\}$ Again, doing the calculations in the innermost parentheses

$= 16 \div 2 + 33$ Subtracting inside the braces

$= 8 + 33$ Doing all multiplications and divisions in order from left to right

$= 41$ Doing all additions and subtractions in order from left to right

Do Exercises 24 and 25.

Study Tips

TEST PREPARATION

You are probably ready to begin preparing for your first test. Here are some test-taking study tips.

- **Make up your own test questions as you study.** After you have done your homework over a particular objective, write one or two questions on your own that you think might be on a test. You will be amazed at the insight this will provide.
- **Do an overall review of the chapter, focusing on the objectives and the examples.** This should be accompanied by a study of any class notes you may have taken.
- **Do the review exercises at the end of the chapter.** Check your answers at the back of the book. If you have trouble with an exercise, use the objective symbol as a guide to go back and do further study of that objective.
- **Call the AW Math Tutor Center if you need extra help at 1-888-777-0463.**
- **Do the chapter test at the end of the chapter.** Check the answers and use the objective symbols at the back of the book as a reference for where to review.
- **Ask former students for old exams.** Working such exams can be very helpful and allows you to see what various professors think is important.
- **When taking a test, read each question carefully and try to do all the questions the first time through, but pace yourself.** Answer all the questions, and mark those to recheck if you have time at the end. Very often, your first hunch will be correct.
- **Try to write your test in a neat and orderly manner.** Very often, your instructor tries to give you partial credit when grading an exam. If your test paper is sloppy and disorderly, it is difficult to verify the partial credit. Doing your work neatly can ease such a task for the instructor.

1.6 EXERCISE SET

For Extra Help

Digital Video Tutor CD 1 Videotape 2 | InterAct Math | Math Tutor Center | MathXL | MyMathLab.com

a Write exponential notation.

1. $3 \cdot 3 \cdot 3 \cdot 3$

2. $2 \cdot 2 \cdot 2 \cdot 2 \cdot 2$

3. $5 \cdot 5$

4. $13 \cdot 13 \cdot 13$

5. $7 \cdot 7 \cdot 7 \cdot 7 \cdot 7$

6. $10 \cdot 10$

7. $10 \cdot 10 \cdot 10$

8. $1 \cdot 1 \cdot 1 \cdot 1$

b Evaluate.

9. 7^2

10. 5^3

11. 9^3

12. 10^2

13. 12^4

14. 10^5

15. 11^2

16. 6^3

c Simplify.

17. $12 + (6 + 4)$

18. $(12 + 6) + 18$

19. $52 - (40 - 8)$

20. $(52 - 40) - 8$

21. $1000 \div (100 \div 10)$

22. $(1000 \div 100) \div 10$

23. $(256 \div 64) \div 4$

24. $256 \div (64 \div 4)$

25. $(2 + 5)^2$

26. $2^2 + 5^2$

27. $(11 - 8)^2 - (18 - 16)^2$

28. $(32 - 27)^3 + (19 + 1)^3$

29. $16 \cdot 24 + 50$

30. $23 + 18 \cdot 20$

31. $83 - 7 \cdot 6$

32. $10 \cdot 7 - 4$

33. $10 \cdot 10 - 3 \cdot 4$

34. $90 - 5 \cdot 5 \cdot 2$

35. $4^3 \div 8 - 4$

36. $8^2 - 8 \cdot 2$

37. $17 \cdot 20 - (17 + 20)$

38. $1000 \div 25 - (15 + 5)$

39. $6 \cdot 10 - 4 \cdot 10$

40. $3 \cdot 8 + 5 \cdot 8$

41. $300 \div 5 + 10$

42. $144 \div 4 - 2$

43. $3 \cdot (2 + 8)^2 - 5 \cdot (4 - 3)^2$

44. $7 \cdot (10 - 3)^2 - 2 \cdot (3 + 1)^2$

45. $4^2 + 8^2 \div 2^2$

46. $6^2 - 3^4 \div 3^3$

47. $10^3 - 10 \cdot 6 - (4 + 5 \cdot 6)$

48. $7^2 + 20 \cdot 4 - (28 + 9 \cdot 2)$

49. $6 \cdot 11 - (7 + 3) \div 5 - (6 - 4)$

50. $8 \times 9 - (12 - 8) \div 4 - (10 - 7)$

51. $120 - 3^3 \cdot 4 \div (5 \cdot 6 - 6 \cdot 4)$

52. $80 - 2^4 \cdot 15 \div (7 \cdot 5 - 45 \div 3)$

53. $2^9 \cdot 2^6 \div 2^7$

54. $2^7 \div 2^5 \cdot 2^4 \div 2^2$

55. Find the average of \$64, \$97, and \$121.

56. Find the average of four test grades of 86, 92, 80, and 78.

d Simplify.

57. $8 \times 13 + \{42 \div [18 - (6 + 5)]\}$

58. $72 \div 6 - \{2 \times [9 - (4 \times 2)]\}$

59. $[14 - (3 + 5) \div 2] - [18 \div (8 - 2)]$

60. $[92 \times (6 - 4) \div 8] + [7 \times (8 - 3)]$

61. $(82 - 14) \times [(10 + 45 \div 5) - (6 \cdot 6 - 5 \cdot 5)]$

62. $(18 \div 2) \cdot \{[(9 \cdot 9 - 1) \div 2] - [5 \cdot 20 - (7 \cdot 9 - 2)]\}$

63. $4 \times \{(200 - 50 \div 5) - [(35 \div 7) \cdot (35 \div 7) - 4 \times 3]\}$

64. $15(23 - 4 \cdot 2)^3 \div (3 \cdot 25)$

65. $\{[18 - 2 \cdot 6] - [40 \div (17 - 9)]\} + \{48 - 13 \times 3 + [(50 - 7 \cdot 5) + 2]\}$

66. $(19 - 2^4)^5 - (141 \div 47)^2$

67. DW Consider the problem in Example 8 of Section 1.5. How can you translate the problem to a single equation involving what you have learned about order of operations? How does the single equation relate to how we solved the problem?

68. DW Consider the expressions $9 - (4 \cdot 2)$ and $(3 \cdot 4)^2$. Are the parentheses necessary in each case? Explain.

SKILL MAINTENANCE

Solve. [1.4b]

69. $x + 341 = 793$

70. $4197 + x = 5032$

71. $7 \cdot x = 91$

72. $1554 = 42 \cdot y$

73. $3240 = y + 898$

74. $6000 = 1102 + t$

75. $25 \cdot t = 625$

76. $10{,}000 = 100 \cdot t$

Solve. [1.5a]

77. *Colorado.* The state of Colorado is roughly the shape of a rectangle that is 270 mi by 380 mi. What is its area?

78. On a long four-day trip, a family bought the following amounts of gasoline for their motor home:

23 gallons, 24 gallons,
26 gallons, 25 gallons.

How much gasoline did they buy in all?

SYNTHESIS

Each of the answers in Exercises 79–81 is incorrect. First find the correct answer. Then place as many parentheses as needed in the expression in order to make the incorrect answer correct.

79. $1 + 5 \cdot 4 + 3 = 36$

80. $12 \div 4 + 2 \cdot 3 - 2 = 2$

81. $12 \div 4 + 2 \cdot 3 - 2 = 4$

82. Use one occurrence each of 1, 2, 3, 4, 5, 6, 7, 8, and 9 and any of the symbols +, −, ×, ÷, and ( ) to represent 100.

1.7 FACTORIZATIONS

Objectives

a Determine whether one number is a factor of another, and find the factors of a number.

b Find some multiples of a number, and determine whether a number is divisible by another.

c Given a number from 1 to 100, tell whether it is prime, composite, or neither.

d Find the prime factorization of a composite number.

Determine whether the second number is a factor of the first.

1. 72; 8

2. 2384; 28

Answers on page A-3

In Chapter 2, we will begin our work with fractions and fraction notation. Certain skills make such work easier. For example, in order to simplify

$$\frac{12}{32},$$

it is important that we be able to *factor* the 12 and the 32, as follows:

$$\frac{12}{32} = \frac{4 \cdot 3}{4 \cdot 8}.$$

Then we "remove" a factor of 1:

$$\frac{4 \cdot 3}{4 \cdot 8} = \frac{4}{4} \cdot \frac{3}{8} = 1 \cdot \frac{3}{8} = \frac{3}{8}.$$

Thus factoring is an important skill in working with fractions.

a Factors and Factorization

In Sections 1.7 and 1.8, we consider only the **natural numbers** 1, 2, 3, and so on.

Let's look at the product $3 \cdot 4 = 12$. We say that 3 and 4 are **factors** of 12.

FACTOR

- In the product $a \cdot b$, a and b are called **factors.**
- If we divide Q by d and get a remainder of 0, then the divisor d is a **factor** of the dividend Q.

EXAMPLE 1 Determine by long division whether 6 is a factor of 72.

$$\begin{array}{r} 12 \\ 6\overline{)72} \\ \underline{60} \\ 12 \\ \underline{12} \\ 0 \end{array}$$

The remainder is 0, so 6 is a factor of 72. We sometimes say that 6 divides 72 "evenly" because there is a remainder of 0.

EXAMPLE 2 Determine by long division whether 15 is a factor of 7894.

$$\begin{array}{r} 526 \\ 15\overline{)7894} \\ \underline{7500} \\ 394 \\ \underline{300} \\ 94 \\ \underline{90} \\ 4 \leftarrow \text{Not } 0 \end{array}$$

The remainder is *not* 0, so 15 is not a factor of 7894.

Do Exercises 1 and 2 on the preceding page.

Consider $12 = 3 \cdot 4$. We say that $3 \cdot 4$ is a **factorization** of 12. Similarly, $6 \cdot 2$, $12 \cdot 1$, $2 \cdot 2 \cdot 3$, and $1 \cdot 3 \cdot 4$ are also factorizations of 12. Since $a = a \cdot 1$, every number has a factorization, and every number has factors. In the case of $17 = 17 \cdot 1$, the only factors of 17 are 17 and 1.

EXAMPLE 3 Find all the factors of 70.

We find as many "two-factor" factorizations as we can. We check sequentially the numbers 1, 2, 3, and so on, to see if we can form any factorizations:

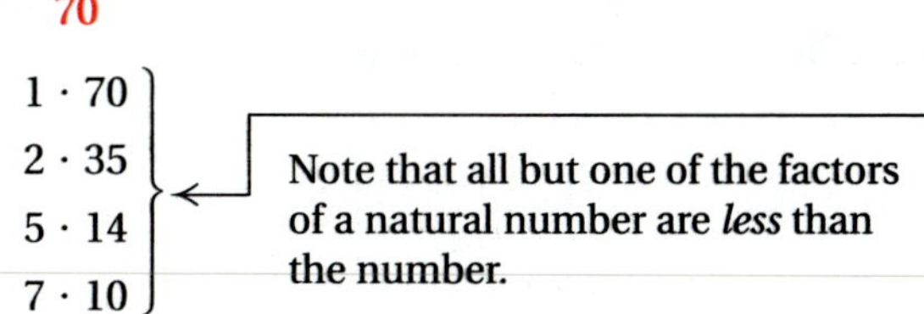

70
1 · 70
2 · 35
5 · 14
7 · 10
Note that all but one of the factors of a natural number are *less* than the number.

Note that 3, 4, and 6 are not factors. If there are additional factors, they must be between 7 and 10. Since 8 and 9 are not factors, we are finished. The factors of 70 are 1, 2, 5, 7, 10, 14, 35, and 70.

Do Exercises 3–6.

b Multiples and Divisibility

A **multiple** of a natural number is a product of it and some natural number. For example, some multiples of 2 are:

2 (because $2 = 1 \cdot 2$);
4 (because $4 = 2 \cdot 2$);
6 (because $6 = 3 \cdot 2$);
8 (because $8 = 4 \cdot 2$);
10 (because $10 = 5 \cdot 2$).

Note that all but one of the multiples of a number are *larger* than the number.

We find multiples of 2 by counting by twos: 2, 4, 6, 8, and so on. We can find multiples of 3 by counting by threes: 3, 6, 9, 12, and so on.

EXAMPLE 4 Show that each of the numbers 8, 12, 20, and 36 is a multiple of 4.

$8 = 2 \cdot 4 \qquad 12 = 3 \cdot 4 \qquad 20 = 5 \cdot 4 \qquad 36 = 9 \cdot 4$

Do Exercises 7 and 8.

EXAMPLE 5 Multiply by 1, 2, 3, and so on, to find ten multiples of 7.

$1 \cdot 7 = 7$	$6 \cdot 7 = 42$
$2 \cdot 7 = 14$	$7 \cdot 7 = 49$
$3 \cdot 7 = 21$	$8 \cdot 7 = 56$
$4 \cdot 7 = 28$	$9 \cdot 7 = 63$
$5 \cdot 7 = 35$	$10 \cdot 7 = 70$

Do Exercise 9.

Find all the factors of the number.

3. 10

4. 45

5. 62

6. 24

7. Show that each of the numbers 5, 45, and 100 is a multiple of 5.

8. Show that each of the numbers 10, 60, and 110 is a multiple of 10.

9. Multiply by 1, 2, 3, and so on, to find ten multiples of 5.

Answers on page A-3

10. Determine whether 16 is divisible by 2.

11. Determine whether 125 is divisible by 5.

12. Determine whether 125 is divisible by 6.

Answers on page A-3

DIVISIBILITY

The number a is **divisible** by another number b if there exists a number c such that $a = b \cdot c$. The statements "a is **divisible** by b," "a is a **multiple** of b," and "b is a **factor** of a" all have the same meaning.

Thus we have

27 is *divisible* by 3 because 27 is a *multiple* of 3 ($27 = 9 \cdot 3$);
27 is a *multiple* of 3 and 3 is a *factor* of 27.

EXAMPLE 6 Determine whether 45 is divisible by 9.

We divide 45 by 9:

$$\begin{array}{r} 5 \\ 9\overline{)45} \\ \underline{45} \\ 0 \end{array}$$

Because the remainder is 0, 45 is divisible by 9.

Do Exercises 10–12.

C Prime and Composite Numbers

PRIME AND COMPOSITE NUMBERS

- A natural number that has exactly two *different* factors, only itself and 1, is called a **prime number.**
- The number 1 is *not* prime.
- A natural number, other than 1, that is not prime is **composite.**

CALCULATOR CORNER

Divisibility and Factors We can use a calculator to determine whether one number is divisible by another number or whether one number is a factor of another number. For example, to determine whether 387 is divisible by 18, we first press [3] [8] [7] [÷] [1] [8] [=]. The display reads [21.5]. Note that the result is not a natural number. (The number 21.5 is in decimal notation. Decimal notation will be studied in detail in Chapter 3.) Thus we know that 387 is not a multiple of 18; that is, 387 is not divisible by 18 and 18 is not a factor of 387.

When we divide 387 by 9, the result is [43]. Since 43 is a natural number, we know that 387 is a multiple of 9; that is, $387 = 43 \cdot 9$. Thus, 387 is divisible by 9 and 9 is a factor of 387.

Exercises: For each pair of numbers, determine whether the first number is divisible by the second number.

1. 722; 19
2. 845; 7
3. 1047; 14
4. 5283; 9

For each pair of numbers, determine whether the second number is a factor of the first number.

5. 502; 8
6. 651; 21
7. 3875; 25
8. 8464; 12
9. 32,768; 256
10. 32,768; 864

EXAMPLE 7 Determine whether the numbers 1, 2, 3, 4, 5, 6, 7, 9, 10, 11, and 63 are prime, composite, or neither.

The number 1 is not prime. It does not have *two* different factors.

The number 2 is prime. It has only the factors 1 and 2.

The numbers 3, 5, 7, and 11 are prime. Each has only two factors, itself and 1.

The number 4 is not prime. It has the factors 1, 2, and 4 and is composite.

The numbers 6, 9, 10, and 63 are composite. Each has more than two factors.

Thus we have:

Prime: 2, 3, 5, 7, 11;

Composite: 4, 6, 9, 10, 63;

Neither: 1.

The number 2 is the *only* even prime number. It is also the smallest prime number. The number 0 is also neither prime nor composite, but 0 is *not* a natural number and thus is not considered here. We are considering only natural numbers.

Do Exercise 13.

13. Tell whether each number is prime, composite, or neither.

1, 2, 6, 12, 13, 19, 41, 65, 73, 99

d Prime Factorizations

To factor a composite number into a product of primes is to find a **prime factorization** of the number. To do this, we consider the primes

2, 3, 5, 7, 11, 13, 17, 19, 23, and so on,

and determine whether a given number is divisible by the primes.

EXAMPLE 8 Find the prime factorization of 39.

a) We divide by the first prime, 2.

$$\begin{array}{r} 19 \quad R = 1 \\ 2\overline{)39} \\ \underline{38} \\ 1 \end{array}$$

Because the remainder is not 0, 2 is not a factor of 39, and 39 is not divisible by 2.

b) We divide by the next prime, 3.

$$\begin{array}{r} 13 \quad R = 0 \\ 3\overline{)39} \end{array}$$

Because 13 is a prime, we are finished. The prime factorization is

$39 = 3 \cdot 13.$

The following is a table of the prime numbers from 2 to 157. There are more extensive tables, but these prime numbers will be the most helpful to you in this text.

A TABLE OF PRIMES FROM 2 TO 157

2, 3, 5, 7, 11, 13, 17, 19, 23, 29, 31, 37, 41, 43, 47, 53, 59, 61, 67, 71, 73, 79, 83, 89, 97, 101, 103, 107, 109, 113, 127, 131, 137, 139, 149, 151, 157

Answer on page A-3

Find the prime factorization of the number.

14. 6

15. 12

16. 45

17. 98

18. 126

19. 144

Answers on page A-3

EXAMPLE 9 Find the prime factorization of 76.

a) We divide by the first prime, 2.

$$2\overline{)76}\quad 38 \quad R = 0$$

b) Because 38 is composite, we start with 2 again:

$$2\overline{)38}\quad 19 \quad R = 0$$

Because 19 is a prime, we are finished. The prime factorization is

$76 = 2 \cdot 2 \cdot 19.$

We abbreviate our procedure as follows.

$$\begin{array}{r} 19 \\ 2\overline{)38} \\ 2\overline{)76} \end{array}$$

$76 = 2 \cdot 2 \cdot 19$

Multiplication is commutative so a factorization such as $2 \cdot 2 \cdot 19$ could also be expressed as $2 \cdot 19 \cdot 2$ or $19 \cdot 2 \cdot 2$ (or in exponential notation, as $2^2 \cdot 19$ or $19 \cdot 2^2$), but the prime factors are still the same. For this reason, we agree that any of these is "the" prime factorization of 76.

EXAMPLE 10 Find the prime factorization of 72.

We can do divisions "up" as follows:

$$\begin{array}{r} 3 \leftarrow \text{Prime quotient} \\ 3\overline{)\ 9} \\ 2\overline{)18} \\ 2\overline{)36} \\ 2\overline{)72} \leftarrow \text{Begin here.} \end{array}$$

Or, we can also do divisions "down":

$$\begin{array}{l} 2\underline{)72} \leftarrow \text{Begin here.} \\ 2\underline{)36} \\ 2\underline{)18} \\ 3\underline{)\ 9} \\ \quad 3 \leftarrow \text{Prime quotient} \end{array}$$

$72 = 2 \cdot 2 \cdot 2 \cdot 3 \cdot 3$

EXAMPLE 11 Find the prime factorization of 189.

We can use a string of successive divisions.

$$\begin{array}{r} 7 \\ 3\overline{)21} \\ 3\overline{)63} \\ 3\overline{)189} \end{array}$$

189 is not divisible by 2. We move to 3.

$189 = 3 \cdot 3 \cdot 3 \cdot 7$

EXAMPLE 12 Find the prime factorization of 65.

We can use a string of successive divisions.

$$\begin{array}{r} 13 \\ 5\overline{)65} \end{array}$$

65 is not divisible by 2 or 3. We move to 5.

$65 = 5 \cdot 13$

Do Exercises 14–19.

1.7 EXERCISE SET

For Extra Help

Digital Video Tutor CD 1 Videotape 2

InterAct Math

Math Tutor Center

MathXL

MyMathLab

a Determine whether the second number is a factor of the first.

1. 52; 14 **2.** 52; 13 **3.** 625; 25 **4.** 680; 16

Find all the factors of the number.

5. 18 **6.** 16 **7.** 54 **8.** 48

9. 4 **10.** 9 **11.** 7 **12.** 11

13. 1 **14.** 3 **15.** 98 **16.** 100

b Multiply by 1, 2, 3, and so on, to find ten multiples of the number.

17. 4 **18.** 11 **19.** 20 **20.** 50

21. 3 **22.** 5 **23.** 12 **24.** 13

25. 10 **26.** 6 **27.** 9 **28.** 14

29. Determine whether 26 is divisible by 6.

30. Determine whether 29 is divisible by 9.

31. Determine whether 1880 is divisible by 8.

32. Determine whether 4227 is divisible by 3.

33. Determine whether 256 is divisible by 16.

34. Determine whether 102 is divisible by 4.

35. Determine whether 4227 is divisible by 9.

36. Determine whether 200 is divisible by 25.

37. Determine whether 8650 is divisible by 16.

38. Determine whether 4143 is divisible by 7.

c Determine whether the number is prime, composite, or neither.

39. 1 **40.** 2 **41.** 9 **42.** 19

43. 11 **44.** 27 **45.** 29 **46.** 49

d Find the prime factorization of the number.

47. 8 **48.** 16 **49.** 14 **50.** 15

51. 42 **52.** 32 **53.** 25 **54.** 40

55. 50 **56.** 62 **57.** 169 **58.** 140

59. 100 **60.** 110 **61.** 35 **62.** 70

63. 72 **64.** 86 **65.** 77 **66.** 99

67. 2884 **68.** 484 **69.** 51 **70.** 91

71. DW Is every natural number a multiple of 1? Explain.

72. DW Explain a method for finding a composite number that contains exactly two factors other than itself and 1.

SKILL MAINTENANCE

Multiply. [1.3a]

73. $2 \cdot 13$ **74.** $8 \cdot 32$ **75.** $17 \cdot 25$ **76.** $25 \cdot 168$

Divide. [1.3d]

77. $0 \div 22$ **78.** $22 \div 1$ **79.** $22 \div 22$ **80.** $66 \div 22$

Solve. [1.5a]

81. Find the total cost of 7 shirts at $48 each and 4 pairs of pants at $69 each.

82. Sandy can type 62 words per minute. How long will it take her to type 12,462 words?

SYNTHESIS

83. *Factors and Sums.* To *factor* a number is to express it as a product. Since $15 = 5 \cdot 3$, we say that 15 is *factored* and that 5 and 3 are *factors* of 15. In the table below, the top number in each column has been factored in such a way that the sum of the factors is the bottom number in the column. For example, in the first column, 56 has been factored as $7 \cdot 8$, and $7 + 8 = 15$, the bottom number. Such thinking will be important in understanding the meaning of a factor and in algebra.

Product	56	63	36	72	140	96		168	110			
Factor	7									9	24	3
Factor	8						8	8		10	18	
Sum	15	16	20	38	24	20	14		21			24

Find the missing numbers in the table.

1.8 DIVISIBILITY

Objective

a Determine whether a number is divisible by 2, 3, 4, 5, 6, 8, 9, or 10.

Suppose you are asked to find the simplest fraction notation for

$$\frac{117}{225}.$$

Since the numbers are quite large, you might feel that the task is difficult. However, both the numerator and the denominator have 9 as a factor. If you knew this, you could factor and simplify quickly as follows:

$$\frac{117}{225} = \frac{9 \cdot 13}{9 \cdot 25} = \frac{9}{9} \cdot \frac{13}{25} = 1 \cdot \frac{13}{25} = \frac{13}{25}.$$

How did we know that both numbers have 9 as a factor? There are fast tests for such determinations. If the sum of the digits of a number is divisible by 9, then the number is divisible by 9; that is, it has 9 as a factor. Since $1 + 1 + 7 = 9$ and $2 + 2 + 5 = 9$, both numbers have 9 as a factor.

a Rules for Divisibility

In this section, we learn fast ways of determining whether numbers are divisible by 2, 3, 4, 5, 6, 8, 9, and 10. This will make simplifying fraction notation much easier.

DIVISIBILITY BY 2

You may already know the test for divisibility by 2.

BY 2

A number is **divisible by 2** (is *even*) if it has a ones digit of 0, 2, 4, 6, or 8 (that is, it has an even ones digit).

Let's see why. Consider 354, which is

3 hundreds + 5 tens + 4.

Hundreds and tens are both multiples of 2. If the last digit is a multiple of 2, then the entire number is a multiple of 2.

EXAMPLES Determine whether the number is divisible by 2.

1. 355 is not a multiple of 2; 5 is *not* even.
2. 4786 is a multiple of 2; 6 is even.
3. 8990 is a multiple of 2; 0 is even.
4. 4261 is not a multiple of 2; 1 is *not* even.

Do Exercises 1–4.

Determine whether the number is divisible by 2.

1. 84

2. 59

3. 998

4. 2225

Answers on page A-3

Determine whether the number is divisible by 3.

5. 111

6. 1111

7. 309

8. 17,216

Determine whether the number is divisible by 6.

9. 420

10. 106

11. 321

12. 444

Answers on page A-3

DIVISIBILITY BY 3

BY 3

A number is **divisible by 3** if the sum of its digits is divisible by 3.

EXAMPLES Determine whether the number is divisible by 3.

5. 18 $1 + 8 = 9$
6. 93 $9 + 3 = 12$
7. 201 $2 + 0 + 1 = 3$

All are divisible by 3 because the sums of their digits are divisible by 3.

8. 256 $2 + 5 + 6 = 13$

The sum, 13, is not divisible by 3, so 256 is not divisible by 3.

Do Exercises 5–8.

DIVISIBILITY BY 6

A number divisible by 6 is a multiple of 6. But $6 = 2 \cdot 3$, so the number is also a multiple of 2 and 3. Thus we have the following.

BY 6

A number is **divisible by 6** if its ones digit is 0, 2, 4, 6, or 8 (is even) and the sum of its digits is divisible by 3.

EXAMPLES Determine whether the number is divisible by 6.

9. 720

Because 720 is even, it is divisible by 2. Also, $7 + 2 + 0 = 9$, so 720 is divisible by 3. Thus, 720 is divisible by 6.

720 ↑ Even $7 + 2 + 0 = 9$ ↑ Divisible by 3

10. 73

73 is *not* divisible by 6 because it is *not* even.

73 ↑ Not even

11. 256

256 is *not* divisible by 6 because the sum of its digits is *not* divisible by 3.

$2 + 5 + 6 = 13$ ↑ Not divisible by 3

Do Exercises 9–12.

DIVISIBILITY BY 9

The test for divisibility by 9 is similar to the test for divisibility by 3.

BY 9

A number is **divisible by 9** if the sum of its digits is divisible by 9.

EXAMPLE 12 The number 6984 is divisible by 9 because

$$6 + 9 + 8 + 4 = 27$$

and 27 is divisible by 9.

EXAMPLE 13 The number 322 is *not* divisible by 9 because

$$3 + 2 + 2 = 7$$

and 7 is not divisible by 9.

Do Exercises 13–16.

DIVISIBILITY BY 10

BY 10

A number is **divisible by 10** if its ones digit is 0.

We know that this test works because the product of 10 and *any* number has a ones digit of 0.

EXAMPLES Determine whether the number is divisible by 10.

14. 3440 is divisible by 10 because the ones digit is 0.

15. 3447 is *not* divisible by 10 because the ones digit is not 0.

Do Exercises 17–20.

DIVISIBILITY BY 5

BY 5

A number is **divisible by 5** if its ones digit is 0 or 5.

EXAMPLES Determine whether the number is divisible by 5.

16. 220 is divisible by 5 because the ones digit is 0.

17. 475 is divisible by 5 because the ones digit is 5.

18. 6514 is *not* divisible by 5 because the ones digit is neither a 0 nor a 5.

Do Exercises 21–24.

Let's see why the test for 5 works. Consider 7830:

$$7830 = 10 \cdot 783 = 5 \cdot 2 \cdot 783.$$

Since 7830 is divisible by 10 and 5 is a factor of 10, 7830 is divisible by 5.

Determine whether the number is divisible by 9.

13. 16

14. 117

15. 930

16. 29,223

Determine whether the number is divisible by 10.

17. 305

18. 300

19. 847

20. 8760

Determine whether the number is divisible by 5.

21. 5780

22. 3427

23. 34,678

24. 7775

Answers on page A-3

Determine whether the number is divisible by 4.

25. 216

26. 217

27. 5865

28. 23,524

Determine whether the number is divisible by 8.

29. 7564

30. 7864

31. 17,560

32. 25,716

Answers on page A-3

Consider 6734:

6734 = 673 tens + 4.

Tens are multiples of 5, so the only number that must be checked is the ones digit. If the last digit is a multiple of 5, the entire number is. In this case, 4 is not a multiple of 5, so 6734 is *not* divisible by 5.

DIVISIBILITY BY 4

The test for divisibility by 4 is similar to the test for divisibility by 2.

BY 4

A number is **divisible by 4** if the number named by its last *two* digits is divisible by 4.

EXAMPLES Determine whether the number is divisible by 4.

19. 8212 is divisible by 4 because 12 is divisible by 4.
20. 5216 is divisible by 4 because 16 is divisible by 4.
21. 8211 is *not* divisible by 4 because 11 is *not* divisible by 4.
22. 7515 is *not* divisible by 4 because 15 is *not* divisible by 4.

Do Exercises 25–28.

To see why the test for divisibility by 4 works, consider 516:

516 = 5 hundreds + 16.

Hundreds are multiples of 4. If the number named by the last two digits is a multiple of 4, then the entire number is a multiple of 4.

DIVISIBILITY BY 8

The test for divisibility by 8 is an extension of the tests for divisibility by 2 and 4.

BY 8

A number is **divisible by 8** if the number named by its last *three* digits is divisible by 8.

EXAMPLES Determine whether the number is divisible by 8.

23. 5648 is divisible by 8 because 648 is divisible by 8.
24. 96,088 is divisible by 8 because 88 is divisible by 8.
25. 7324 is *not* divisible by 8 because 324 is *not* divisible by 8.
26. 13,420 is *not* divisible by 8 because 420 is *not* divisible by 8.

Do Exercises 29–32.

A NOTE ABOUT DIVISIBILITY BY 7

There are several tests for divisibility by 7, but all of them are more complicated than simply dividing by 7. So if you want to test for divisibility by 7, simply divide by 7, either by hand or using a calculator.

1.8 EXERCISE SET

For Extra Help

Digital Video Tutor CD 1 Videotape 2

InterAct Math

Math Tutor Center

MathXL

MyMathLab

a To answer Exercises 1–8, consider the following numbers.

46	300	85	256
224	36	711	8064
19	45,270	13,251	1867
555	4444	254,765	21,568

1. Which of the above are divisible by 2?

2. Which of the above are divisible by 3?

3. Which of the above are divisible by 4?

4. Which of the above are divisible by 5?

5. Which of the above are divisible by 6?

6. Which of the above are divisible by 8?

7. Which of the above are divisible by 9?

8. Which of the above are divisible by 10?

To answer Exercises 9–16, consider the following numbers.

56	200	75	35
324	42	812	402
784	501	2345	111,111
55,555	3009	2001	1005

9. Which of the above are divisible by 3?

10. Which of the above are divisible by 2?

11. Which of the above are divisible by 5?

12. Which of the above are divisible by 4?

13. Which of the above are divisible by 9?

14. Which of the above are divisible by 6?

15. Which of the above are divisible by 10?

16. Which of the above are divisible by 8?

17. D W How can the divisibility tests be used to find prime factorizations?

18. D W Which of the years from 2000 to 2020, if any, also happen to be prime numbers? Explain at least two ways in which you might go about solving this problem.

SKILL MAINTENANCE

Solve. [1.4b]

19. $56 + x = 194$

20. $y + 124 = 263$

21. $3008 = x + 2134$

22. $18 \cdot t = 1008$

23. $24 \cdot m = 624$

24. $338 = a \cdot 26$

Divide. [1.3d]

25. $2106 \div 9$

26. $45\overline{)180,135}$

Solve. [1.5a]

27. An automobile with a 5-speed transmission gets 33 mpg in city driving. How many gallons of gas will it use to travel 1485 mi?

28. There are 60 min in 1 hr. How many minutes are there in 72 hr?

SYNTHESIS

Find the prime factorization of the number. Use divisibility tests where applicable.

29. 7800

30. 2520

31. 2772

32. 1998

33. Fill in the missing digits of the number

95,☐☐8

so that it is divisible by 99.

34. A passenger in a taxicab asks for the driver's company number. The driver says abruptly, "Sure—you can have my number. Work it out: If you divide it by 2, 3, 4, 5, or 6, you will get a remainder of 1. If you divide it by 11, the remainder will be 0 and no driver has a company number that meets these requirements and is smaller than this one." Determine the number.

1.9 LEAST COMMON MULTIPLES

In Chapter 2, we will study addition and subtraction using fraction notation. Suppose we want to add $\frac{2}{3}$ and $\frac{1}{2}$. To do so, we rewrite the numbers using the least common multiple of the denominators: $\frac{2}{3} + \frac{1}{2} = \frac{4}{6} + \frac{3}{6}$. Then we add the numerators and keep the common denominator, 6. In order to do this, we must be able to find the **least common denominator (LCD),** or **least common multiple (LCM)** of the denominators. (A review of Section 2.1b might be helpful.)

Objective

a Find the LCM of two or more numbers.

1. By examining lists of multiples, find the LCM of 9 and 15.

a Finding Least Common Multiples

LEAST COMMON MULTIPLE, LCM

The **least common multiple,** or LCM, of two natural numbers is the smallest number that is a multiple of both.

EXAMPLE 1 Find the LCM of 20 and 30.

a) First list some multiples of 20 by multiplying 20 by 1, 2, 3, and so on:

20, 40, 60, 80, 100, 120, 140, 160, 180, 200, 220, 240, ...

b) Then list some multiples of 30 by multiplying 30 by 1, 2, 3, and so on:

30, 60, 90, 120, 150, 180, 210, 240, ...

c) Now list the numbers *common* to both lists, the common multiples:

60, 120, 180, 240, ...

d) These are the common multiples of 20 and 30. Which is the smallest? The LCM of 20 and 30 is 60.

Do Exercise 1.

Next we develop three methods that are more efficient for finding LCMs. You may choose to learn only one method (consult with your instructor), but if you are going to study algebra, you should definitely learn method 2.

METHOD 1: FINDING LCMS USING ONE LIST OF MULTIPLES

One method for finding LCMs uses *one* list of multiples. Let's consider finding the LCM of 9 and 12. The largest number, 12, is not a multiple of 9. The multiples of 12 are

12, 24, 36, 48, 60, ...

We check each multiple of 12 until we find a number that is also a multiple of 9.

$1 \cdot 12 = 12$, not a multiple of 9;

$2 \cdot 12 = 24$, not a multiple of 9;

$3 \cdot 12 = 36$, a multiple of 9: $4 \cdot 9 = 36$

The LCM of 9 and 12 is 36.

Answer on page A-3

2. By examining lists of multiples, find the LCM of 8 and 10.

Find the LCM.

3. 10, 15

4. 6, 8

5. 5, 10

6. 20, 40, 80

Answers on page A-3

Method 1. To find the LCM of a set of numbers using a list of multiples:

a) Determine whether the largest number is a multiple of the others. If it is, it is the LCM. That is, if the largest number has the others as factors, the LCM is that number.

b) If not, check multiples of the largest number until you get one that is a multiple of the others.

EXAMPLE 2 Find the LCM of 12 and 15.

a) 15 is not a multiple of 12.

b) Check multiples of 15: 15, 30, 45, and so on.

$1 \cdot 15 = 15,$ Not a multiple of 12. When we divide 15 by 12, we get a nonzero remainder.

$2 \cdot 15 = 30,$ Not a multiple of 12

$3 \cdot 15 = 45,$ Not a multiple of 12

$4 \cdot 15 = 60.$ A multiple of 12

The LCM = 60.

Do Exercise 2.

EXAMPLE 3 Find the LCM of 4 and 14.

a) 14 is not a multiple of 4.

b) Check multiples:

$1 \cdot 14 = 14,$

$2 \cdot 14 = 28.$ A multiple of 4

The LCM = 28.

EXAMPLE 4 Find the LCM of 8 and 32.

a) 32 is a multiple of 8, so the LCM = 32.

EXAMPLE 5 Find the LCM of 10, 100, and 1000.

a) 1000 is a multiple of 10 and 100, so the LCM = 1000.

Do Exercises 3–6.

METHOD 2: FINDING LCMS USING PRIME FACTORIZATIONS

A second method for finding LCMs uses prime factorizations. Consider again 20 and 30. Their prime factorizations are $20 = 2 \cdot 2 \cdot 5$ and $30 = 2 \cdot 3 \cdot 5$. Let's look at these prime factorizations in order to find the LCM. Any multiple of 20 will have to have *two* 2's as factors and *one* 5 as a factor. Any multiple of 30 will have to have *one* 2, *one* 3, and *one* 5 as factors. The smallest number satisfying these conditions is

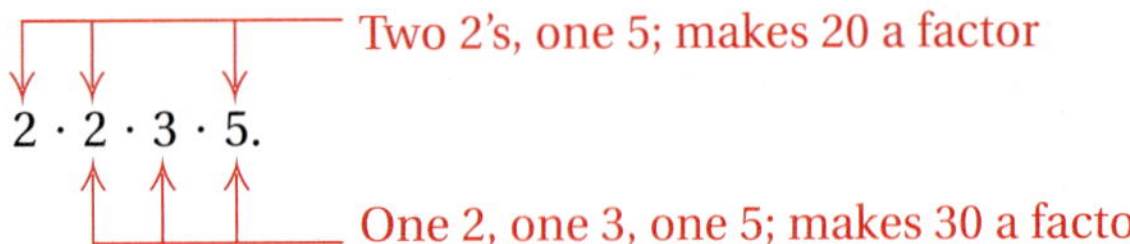
Two 2's, one 5; makes 20 a factor
2 · 2 · 3 · 5.
One 2, one 3, one 5; makes 30 a factor

The LCM must have all the factors of 20 and all the factors of 30, but the factors need not be repeated when they are common to both numbers.

The greatest number of times that a 2 occurs as a factor of either 20 or 30 is two, and the LCM has 2 as a factor twice. The greatest number of times that a 3 occurs as a factor of either 20 or 30 is one, and the LCM has 3 as a factor once. The greatest number of times that 5 occurs as a factor of either 20 or 30 is one, and the LCM has 5 as a factor once. The LCM is the product $2 \cdot 2 \cdot 3 \cdot 5$, or 60.

Method 2. To find the LCM of a set of numbers using prime factorizations:

a) Find the prime factorization of each number.

b) Create a product of factors, using each factor the greatest number of times that it occurs in any one factorization.

EXAMPLE 6 Find the LCM of 6 and 8.

a) Find the prime factorization of each number.

$6 = 2 \cdot 3, \qquad 8 = 2 \cdot 2 \cdot 2$

b) Create a product by writing factors, using each the greatest number of times that it occurs in any one factorization.

Consider the factor 2. The greatest number of times that 2 occurs in any one factorization is three. We write 2 as a factor three times.

$2 \cdot 2 \cdot 2 \cdot ?$

Consider the factor 3. The greatest number of times that 3 occurs in any one factorization is one. We write 3 as a factor one time.

$2 \cdot 2 \cdot 2 \cdot 3 \cdot ?$

Since there are no other prime factors in either factorization, the

LCM is $2 \cdot 2 \cdot 2 \cdot 3$, or 24.

EXAMPLE 7 Find the LCM of 24 and 36.

a) Find the prime factorization of each number.

$24 = 2 \cdot 2 \cdot 2 \cdot 3, \qquad 36 = 2 \cdot 2 \cdot 3 \cdot 3$

b) Create a product by writing factors, using each the greatest number of times that it occurs in any one factorization.

Consider the factor 2. The greatest number of times that 2 occurs in any one factorization is three. We write 2 as a factor three times:

$2 \cdot 2 \cdot 2 \cdot ?$

Consider the factor 3. The greatest number of times that 3 occurs in any one factorization is two. We write 3 as a factor two times:

$2 \cdot 2 \cdot 2 \cdot 3 \cdot 3 \cdot ?$

Since there are no other prime factors in either factorization, the

LCM is $2 \cdot 2 \cdot 2 \cdot 3 \cdot 3$, or 72.

Do Exercises 7–9.

Use prime factorizations to find the LCM.

7. 8, 10

8. 18, 40

9. 32, 54

Answers on page A-3

10. Find the LCM of 24, 35, and 45.

Find the LCM.

11. 3, 18

12. 12, 24

Find the LCM.

13. 4, 9

14. 5, 6, 7

Answers on page A-3

Let's compare the two methods considered so far for finding LCMs: the multiples method and the factorization method.

Method 1, the **multiples method,** can be longer than the factorization method when the LCM is large or when there are more than two numbers. But this method can be faster and easier to use mentally for two numbers.

Method 2, the **factorization method,** works well for several numbers. It is just like a method used in algebra. If you are going to study algebra, you should definitely learn the factorization method.

EXAMPLE 8 Find the LCM of 27, 90, and 84.

a) Find the prime factorization of each number.

$27 = 3 \cdot 3 \cdot 3, \quad 90 = 2 \cdot 3 \cdot 3 \cdot 5, \quad 84 = 2 \cdot 2 \cdot 3 \cdot 7$

b) Create a product by writing factors, using each the greatest number of times that it occurs in any one factorization.

Consider the factor 2. The greatest number of times that 2 occurs in any one factorization is two. We write 2 as a factor two times:

$2 \cdot 2 \cdot ?$

Consider the factor 3. The greatest number of times that 3 occurs in any one factorization is three. We write 3 as a factor three times:

$2 \cdot 2 \cdot 3 \cdot 3 \cdot 3 \cdot ?$

Consider the factor 5. The greatest number of times that 5 occurs in any one factorization is one. We write 5 as a factor one time:

$2 \cdot 2 \cdot 3 \cdot 3 \cdot 3 \cdot 5 \cdot ?$

Consider the factor 7. The greatest number of times that 7 occurs in any one factorization is one. We write 7 as a factor one time:

$2 \cdot 2 \cdot 3 \cdot 3 \cdot 3 \cdot 5 \cdot 7 \cdot ?$

Since there are no other prime factors in any of the factorizations, the

LCM is $2 \cdot 2 \cdot 3 \cdot 3 \cdot 3 \cdot 5 \cdot 7$, or 3780.

Do Exercise 10.

EXAMPLE 9 Find the LCM of 7 and 21.

We find the prime factorization of each number. Because 7 is prime, it has no prime factorization.

$7 = 7, \quad 21 = 3 \cdot 7$

Note that 7 is a factor of 21. We stated earlier that if one number is a factor of another, the LCM is the larger of the numbers. Thus the LCM is $7 \cdot 3$, or 21.

Do Exercises 11 and 12.

EXAMPLE 10 Find the LCM of 8 and 9.

We find the prime factorization of each number.

$8 = 2 \cdot 2 \cdot 2, \quad 9 = 3 \cdot 3$

Note that the two numbers, 8 and 9, have no common prime factor. When this is the case, the LCM is just the product of the two numbers. Thus the LCM is $2 \cdot 2 \cdot 2 \cdot 3 \cdot 3$, or 72.

Do Exercises 13 and 14.

1.9 EXERCISE SET

For Extra Help

Digital Video Tutor CD 1 Videotape 2

InterAct Math

Math Tutor Center

MathXL

MyMathLab

a Find the LCM of the set of numbers.

1. 2, 4

2. 3, 15

3. 10, 25

4. 10, 15

5. 20, 40

6. 8, 12

7. 18, 27

8. 9, 11

9. 30, 50

10. 24, 36

11. 30, 40

12. 21, 27

13. 18, 24

14. 12, 18

15. 60, 70

16. 35, 45

17. 16, 36

18. 18, 20

19. 32, 36

20. 36, 48

21. 2, 3, 5

22. 5, 18, 3

23. 3, 5, 7

24. 6, 12, 18

25. 24, 36, 12

26. 8, 16, 22

27. 5, 12, 15

28. 12, 18, 40

29. 9, 12, 6

30. 8, 16, 12

31. 180, 100, 450, 60

32. 18, 30, 50, 48

33. 8, 48

34. 16, 32

35. 5, 50

36. 12, 72

37. 11, 13

38. 13, 14

39. 12, 35

40. 23, 25

41. 54, 63

42. 56, 72

43. 81, 90

44. 75, 100

Applications of LCMs: Planet Orbits. The earth, Jupiter, Saturn, and Uranus all revolve around the sun. The earth takes 1 yr, Jupiter 12 yr, Saturn 30 yr, and Uranus 84 yr to make a complete revolution. On a certain night, you look at those three distant planets and wonder how many years it will take before they have the same position again. (*Hint*: To find out, you find the LCM of 12, 30, and 84. It will be that number of years.)
Source: *The Handy Science Answer Book*

45. How often will Jupiter and Saturn appear in the same direction in the night sky as seen from the earth?

46. How often will Jupiter and Uranus appear in the same direction in the night sky as seen from the earth?

47. How often will Saturn and Uranus appear in the same direction in the night sky as seen from the earth?

48. How often will Jupiter, Saturn, and Uranus appear in the same direction in the night sky as seen from the earth?

49. D_W Use both Methods 1 and 2 to find the LCM of each of the following sets of numbers.

a) 6, 8 **b)** 6, 7 **c)** 6, 21 **d)** 24, 36

Which method do you consider more efficient? Explain why.

50. D_W Is the LCM of two numbers always larger than either number? Why or why not?

SKILL MAINTENANCE

Solve.

51. Use < or > for □ to write a true sentence: [1.1d]
9001 □ 10,001.

52. *Vehicle Expense.* The most expensive cities in which to own an automobile are Los Angeles, where the yearly cost is \$9254, and Philadelphia, where the yearly cost is \$8715. How much more does it cost in Los Angeles than in Philadelphia? [1.5a]
Source: Runzheimer International

53. Add: 23,456 + 5677 + 4002. [1.2a]

54. Subtract: 10,007 − 3068. [1.2d]

55. Write expanded notation for 24,605. [1.1b]

56. Write a word name for 102,960. [1.1c]

SYNTHESIS

57. Find the LCM of 27, 90, 84, 210, 108, and 50.

58. Find the LCM of 18, 21, 24, 36, 63, 56, and 20.

59. A pencil company uses two sizes of boxes, 5 in. by 6 in. and 5 in. by 8 in. These boxes are packed in bigger cartons for shipping. Find the width and the length of the smallest carton that will accommodate boxes of either size without any room left over. (Each carton can contain only one type of box and all boxes must point in the same direction.)

60. Consider 8 and 12. Determine whether each of the following is the LCM of 8 and 12. Tell why or why not.

a) $2 \cdot 2 \cdot 3 \cdot 3$
b) $2 \cdot 2 \cdot 3$
c) $2 \cdot 3 \cdot 3$
d) $2 \cdot 2 \cdot 2 \cdot 3$

CHAPTER 1

Summary and Review

The review that follows is meant to prepare you for a chapter exam. It consists of two parts. The first part is a checklist of the Study Tips referred to in this chapter. The second part is the Review Exercises. These provide practice exercises for the exam, together with references to section objectives so you can go back and review. Before beginning, stop and look back over the skills you have obtained. What skills in mathematics do you have now that you did not have before studying this chapter?

STUDY TIPS CHECKLIST

The foundation of all your study skills is TIME!

- ☐ Have you found adequate time to study?
- ☐ Have you determined the location of the learning resource centers on your campus, such as a mathlab, tutor center, and your instructor's office?
- ☐ Are you stopping to work the margin exercises when directed to do so?
- ☐ Are you doing your homework as soon as possible after class?
- ☐ Are you making use of any of the textbook supplements, such as the Math Tutor Center, the *Student's Solutions Manual,* and the videotapes?

REVIEW EXERCISES

The review exercises that follow are for practice. Answers are given at the back of the book. If you miss an exercise, restudy the objective indicated in blue next to the exercise or direction line that precedes it.

Write expanded notation. [1.1b]

1. 2793

2. 56,078

Write standard notation. [1.1b]

3. 8 thousands + 6 hundreds + 6 tens + 9 ones

4. 9 ten thousands + 8 hundreds + 4 tens + 4 ones

Write a word name. [1.1c]

5. 67,819

6. 2,781,427

Write standard notation. [1.1c]

7. Four hundred seventy-six thousand, five hundred eighty-eight

8. *e-books.* The publishing industry predicts that sales of digital books will reach two billion, four hundred thousand by 2005.
Source: Andersen Consulting

9. What does the digit 8 mean in 4,678,952? [1.1a]

10. In 13,768,940, what digit tells the number of millions? [1.1a]

Add. [1.2a]

11. 7304 + 6968

12. 27,609 + 38,415

13. 2743 + 4125 + 6274 + 8956

14.
$$\begin{array}{r} 91,426 \\ +\ 7,495 \\ \hline \end{array}$$

15. Write a related addition sentence: [1.2c]

$10 - 6 = 4.$

16. Write two related subtraction sentences: [1.2c]

$8 + 3 = 11.$

Subtract. [1.2d]

17. $8045 - 2897$

18. $8465 - 7312$

19. $6003 - 3729$

20. $\begin{array}{r} 37,405 \\ -\ 19,648 \\ \hline \end{array}$

Round 345,759 to the nearest: [1.3e]

21. Hundred.

22. Ten.

23. Thousand.

Estimate the sum, difference, or product by first rounding to the nearest hundred. Show your work. [1.3f]

24. $41,348 + 19,749$

25. $38,652 - 24,549$

26. $396 \cdot 748$

Use $<$ or $>$ for $\square$ to write a true sentence. [1.1d]

27. $67 \square 56$

28. $1 \square 23$

Multiply. [1.3a]

29. $700 \cdot 600$

30. $7846 \cdot 800$

31. $726 \cdot 698$

32. $587 \cdot 47$

33. $\begin{array}{r} 8305 \\ \times\ 642 \\ \hline \end{array}$

34. Write a related multiplication sentence: [1.3c]

$56 \div 8 = 7.$

35. Write two related division sentences: [1.3c]

$13 \cdot 4 = 52.$

Divide. [1.3d]

36. $63 \div 5$

37. $80 \div 16$

38. $7\overline{)6394}$

39. $3073 \div 8$

40. $60\overline{)286}$

41. $4266 \div 79$

42. $38\overline{)17,176}$

43. $52,668 \div 12$

Solve. [1.4b]

44. $46 \cdot n = 368$

45. $47 + x = 92$

46. $x = 782 - 236$

47. Write exponential notation: $4 \cdot 4 \cdot 4$. [1.6a]

Evaluate. [1.6b]

48. 10^4

49. 6^2

Simplify. [1.6c, d]

50. $8 \cdot 6 + 17$

51. $10 \cdot 24 - (18 + 2) \div 4 - (9 - 7)$

52. $7 + (4 + 3)^2$

53. $7 + 4^2 + 3^2$

54. $(80 \div 16) \times [(20 - 56 \div 8) + (8 \cdot 8 - 5 \cdot 5)]$

55. Find the average of 157, 170, and 168.

Solve. [1.5a]

56. *Oak Desk.* Natasha has \$196 and wants to buy an oak computer roll-top desk for \$698. How much more does she need?

Source: Oak Express®

Desk Just...
$698
Oak Express Excalibur 48" Computer Roll-Top Desk
Accommodates most tower or desk-top computers. Slide-out mouse pad and keyboard tray. Available in light and dark finishes. Constructed of solid oak and oak veneers.

57. Tony has $406 in her checking account. She is paid $78 for a part-time job and deposits that in her checking account. How much is then in her account?

58. *Lincoln-Head Pennies.* In 1909, the first Lincoln-head pennies were minted. Seventy-three years later, these pennies were first minted with a decreased copper content. In what year was the copper content reduced?

59. A beverage company packed 222 cans of soda into 6-can cartons. How many cartons did they fill?

60. An apple farmer keeps bees in her orchard to help pollinate the apple blossoms so more apples will be produced. The bees from an average beehive can pollinate 30 surrounding trees during one growing season. A farmer has 420 trees. How many beehives does she need to pollinate them all?
Source: Jordan Orchards, Westminster, PA

61. An apartment builder bought 3 electric ranges at $299 each and 4 dishwashers at $379 each. What was the total cost?

62. A family budgets $4950 for food and clothing and $3585 for entertainment. The yearly income of the family was $28,283. How much of this income remained after these two allotments?

63. A chemist has 2753 mL of alcohol. How many 20-mL beakers can be filled? How much will be left over?

64. *Olympic Trampoline.* Shown below is an Olympic trampoline. Find the area and the perimeter of the trampoline. [1.2b], [1.3b]
Source: International Trampoline Industry Association, Inc.

Find the prime factorization of the number. [1.7d]

65. 70

66. 30

67. 45

68. 150

Determine whether: [1.8a]

69. 2432 is divisible by 6.

70. 182 is divisible by 4.

71. 4344 is divisible by 9.

72. 4344 is divisible by 8.

73. Determine whether 37 is prime, composite, or neither. [1.7c]

Find the LCM. [1.9a]

74. 12 and 18

75. 18 and 45

76. 3, 6, and 30

77. 26, 36, and 54

78. D_W Write a problem for a classmate to solve. Design the problem so that the solution is "Each of the 144 bottles will contain 8 oz of hot sauce." [1.5a]

SYNTHESIS

79. Determine the missing digit d. [1.3a]

$$\begin{array}{r} 9\,d \\ \times\ d\,2 \\ \hline 8\,0\,3\,6 \end{array}$$

80. Determine the missing digits a and b. [1.3d]

$$2\,b\,1\,\overline{)\,2\,3\,6{,}4\,2\,1}\quad\text{quotient: } 9\,a\,1$$

81. A mining company estimates that a crew must tunnel 2000 ft into a mountain to reach a deposit of copper ore. Each day the crew tunnels about 500 ft. Each night about 200 ft of loose rocks roll back into the tunnel. How many days will it take the mining company to reach the copper deposit? [1.5a]

82. A prime number that becomes a prime number when its digits are reversed is called a **palindrome prime.** For example, 17 is a palindrome prime because both 17 and 71 are primes. Which of the following numbers are palindrome primes? [1.7c]

13, 91, 16, 11, 15, 24, 29, 101, 201, 37

CHAPTER 1

Chapter Test

1. Write expanded notation: 8843.

2. Write a word name: 38,403,277.

3. In the number 546,789, which digit tells the number of hundred thousands?

Add.

4. $\begin{array}{r} 6811 \\ +\ 3178 \\ \hline \end{array}$

5. $\begin{array}{r} 45{,}889 \\ +\ 17{,}902 \\ \hline \end{array}$

6. $\begin{array}{r} 12 \\ 8 \\ 3 \\ 7 \\ +\ \ 4 \\ \hline \end{array}$

7. $\begin{array}{r} 6203 \\ +\ 4312 \\ \hline \end{array}$

Subtract.

8. $\begin{array}{r} 7983 \\ -\ 4353 \\ \hline \end{array}$

9. $\begin{array}{r} 2974 \\ -\ 1935 \\ \hline \end{array}$

10. $\begin{array}{r} 8907 \\ -\ 2059 \\ \hline \end{array}$

11. $\begin{array}{r} 23{,}067 \\ -\ 17{,}892 \\ \hline \end{array}$

Multiply.

12. $\begin{array}{r} 4568 \\ \times\ \ \ \ 9 \\ \hline \end{array}$

13. $\begin{array}{r} 8876 \\ \times\ \ 600 \\ \hline \end{array}$

14. $\begin{array}{r} 65 \\ \times\ 37 \\ \hline \end{array}$

15. $\begin{array}{r} 678 \\ \times\ 788 \\ \hline \end{array}$

Divide.

16. $15 \div 4$

17. $420 \div 6$

18. $89 \overline{)\ 8633}$

19. $44 \overline{)\ 35{,}428}$

Solve.

20. *Hostess Ding Dongs®.* Hostess packages its Ding Dong® snack products in 12-packs. It manufactures 22,231 cakes. How many 12-packs can it fill? How many will be left over?

21. *Largest States.* The following table lists the five largest states in terms of their area. Find the total area of these states.

STATE	AREA (in Square Miles)
Alaska	591,004
Texas	266,807
California	158,706
Montana	147,046
New Mexico	121,593

Source: *The New York Times Almanac*

22. *Pool Tables.* The Hartford™ pool table made by Brunswick Billiards comes in three sizes of playing area, 50 in. by 100 in., 44 in. by 88 in., and 38 in. by 76 in.

a) Find the perimeter and the area of the playing area of each table.

b) By how much area does the large table exceed the small table?

Source: Brunswick Billiards

23. *Patents Issued.* There were 169,094 patents issued in 1999. This was 70,018 more than in 1990. How many patents were issued in 1990?

Source: U.S. Patent and Trademark Office

24. A sack of oranges weighs 27 lb. A sack of apples weighs 32 lb. Find the total weight of 16 bags of oranges and 43 bags of apples.

25. A box contains 5000 staples. How many staplers can be filled from the box if each stapler holds 250 staples?

Solve.

26. $28 + x = 74$

27. $169 \div 13 = n$

28. $38 \cdot y = 532$

Round 34,578 to the nearest:

29. Thousand.

30. Ten.

31. Hundred.

Estimate the sum, difference, or product by first rounding to the nearest hundred. Show your work.

32. $\begin{array}{r} 23,649 \\ +\ 54,746 \\ \hline \end{array}$

33. $\begin{array}{r} 54,751 \\ -\ 23,649 \\ \hline \end{array}$

34. $\begin{array}{r} 824 \\ \times\ 489 \\ \hline \end{array}$

Use < or > for $\square$ to write a true sentence.

35. $34 \square 17$

36. $117 \square 157$

Evaluate.

37. 7^3

38. 2^3

39. Write exponential notation: $12 \cdot 12 \cdot 12 \cdot 12$.

40. Find the LCM of 12 and 16.

Simplify.

41. $(10 - 2)^2$

42. $10^2 - 2^2$

43. $(25 - 15) \div 5$

44. $8 \times \{(20 - 11) \cdot [(12 + 48) \div 6 - (9 - 2)]\}$

45. $2^4 + 24 \div 12$

46. Find the average of 97, 98, 87, and 86.

Find the prime factorization of the number.

47. 18

48. 60

49. Determine whether 1784 is divisible by 8.

50. Determine whether 784 is divisible by 9.

SYNTHESIS

51. An open cardboard shoe box is 8 in. wide, 12 in. long, and 6 in. high. How many square inches of cardboard are used?

52. Cara spends \$229 a month to repay her student loan. If she has already paid \$9160 on the 10-yr loan, how many payments remain?

53. Jennie scores three 90's, four 80's, and a 74 on her eight quizzes. Find her average.

54. Use trials to find the single-digit number a for which
$359 - 46 + a \div 3 \times 25 - 7^2 = 339$.

Fraction Notation

Gateway to Chapter 2

In this chapter, we consider addition, subtraction, multiplication, and division using fraction notation. Also discussed are addition, subtraction, multiplication, and division using mixed numerals. We then work with rules for order of operations, estimating, and applied problems.

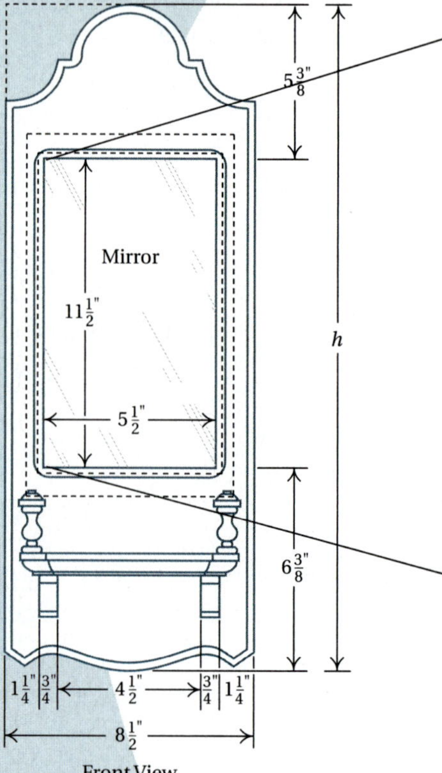

Front View

Real-World Application

The mirror-backed candle shelf, shown above with a carpenter's diagram, was designed and built by Harry Cooper. Such shelves were popular in Colonial times because the mirror provided extra lighting from the candle. A rectangular walnut board is used to make the back of the shelf. Find the area of the original board and the amount left over after the space for the mirror has been cut out.

Source: Popular Science Woodworking Projects

This problem appears as Example 8 in Section 2.5.

CHAPTER 2

CHAPTER 2

Pretest

Simplify. [2.1b, e]

1. $\frac{57}{57}$

2. $\frac{68}{1}$

3. $\frac{0}{50}$

4. $\frac{8}{32}$

5. Use < or > for $\square$ to write a true sentence: [2.3c]

$\frac{7}{9} \square \frac{4}{5}$.

6. Find the reciprocal: $\frac{7}{8}$. [2.2b]

7. Convert to fraction notation: $7\frac{5}{8}$. [2.4a]

8. Convert to a mixed numeral: $\frac{11}{2}$. [2.4a]

9. Add. Write a mixed numeral for the answer. [2.4b]

$$\begin{array}{r} 8\frac{11}{12} \\ +\ 2\frac{3}{5} \\ \hline \end{array}$$

10. Divide. Write a mixed numeral for the answer. [2.4e]

$$5\frac{5}{12} \div 3\frac{1}{4}$$

11. Multiply and simplify: $\frac{1}{3} \cdot \frac{18}{5}$. [2.2a]

12. Subtract and simplify: $\frac{2}{5} - \frac{3}{8}$. [2.3b]

Solve.

13. $\frac{7}{10} \cdot x = 21$ [2.2d]

14. $\frac{2}{3} + x = \frac{8}{9}$ [2.3d]

Solve. [2.5a]

15. At Happy Hollow Camp, the cook bought 100 lb of potatoes and used $78\frac{3}{4}$ lb. How many pounds were left over?

16. A piece of tubing $\frac{5}{8}$ m long is to be cut into 15 pieces of the same length. What is the length of each piece?

17. A courier drove $214\frac{3}{10}$ mi one day and $136\frac{9}{10}$ mi the next. How far did she travel in all?

18. A cake recipe calls for $3\frac{3}{4}$ cups of flour. How much flour would be used to make 6 cakes?

19. Simplify: $\left(\frac{3}{2}\right)^2 + 2\frac{3}{4} \div 1\frac{1}{2}$. [2.6a]

Estimate each of the following as a whole number or as a mixed numeral where the fractional part is $\frac{1}{2}$. [2.6b]

20. $10\frac{2}{17}$

21. $\frac{1}{10} + \frac{7}{8} + \frac{41}{39}$

2.1 FRACTION NOTATION AND SIMPLIFYING

Objectives

a Identify the numerator and the denominator of a fraction and write fraction notation for part of an object.

b Simplify fraction notation like n/n to 1, $0/n$ to 0, and $n/1$ to n.

c Multiply using fraction notation.

d Use multiplying by 1 to find different fraction notation for a number.

e Simplify fraction notation.

The study of arithmetic begins with the set of whole numbers

0, 1, 2, 3, 4, 5, 6, 7, 8, 9, 10, 11, and so on.

The need soon arises for fractional parts of numbers such as halves, thirds, fourths, and so on. Here are some examples:

- $\frac{1}{4}$ of the minimum daily requirement of calcium is provided by a cup of frozen yogurt.

- $\frac{1}{25}$ of the parking spaces in a commercial area in the state of Indiana are to be marked for the handicapped.

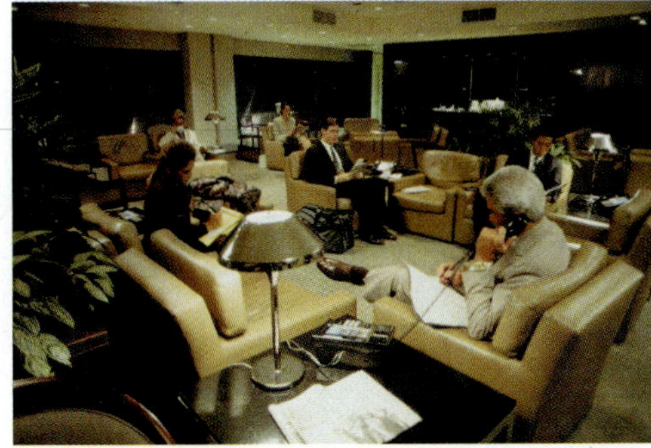

- $\frac{43}{100}$ of all corporate travel money is spent on airfares.

- About $\frac{1}{5}$ of the earth's surface is frozen.

a Fractions and the Real World

The following are some additional examples of fractions:

$$\frac{1}{2},\quad \frac{3}{4},\quad \frac{8}{5},\quad \frac{11}{23}.$$

This way of writing number names is called **fraction notation.** The top number is called the **numerator** and the bottom number is called the **denominator.**

EXAMPLE 1 Identify the numerator and the denominator.

$$\frac{7}{8} \begin{matrix}\longleftarrow \text{Numerator}\\ \longleftarrow \text{Denominator}\end{matrix}$$

Do Exercises 1–3.

Identify the numerator and the denominator.

1. $\frac{1}{6}$

2. $\frac{5}{7}$

3. $\frac{22}{3}$

Answers on page A-4

Let's look at various situations that involve fractions.

FRACTIONS AS A PARTITION OF AN OBJECT DIVIDED INTO EQUAL PARTS

Consider a candy bar divided into 5 equal sections. If you eat 2 sections, you have eaten $\frac{2}{5}$ of the candy bar.

The denominator 5 tells us the unit, $\frac{1}{5}$. The numerator 2 tells us the number of equal parts we are considering, 2.

EXAMPLE 2 What part is shaded?

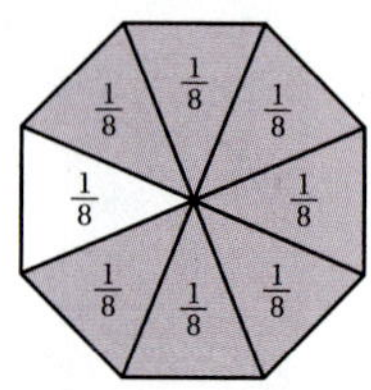

The equal parts are eighths. This tells us the unit, $\frac{1}{8}$. The *denominator* is 8. We have 7 of the units shaded. This tells us the *numerator,* 7. Thus,

$$\frac{7}{8} \begin{matrix} \leftarrow \text{7 units are shaded.} \\ \leftarrow \text{The unit is } \frac{1}{8}. \end{matrix}$$

is shaded.

The markings on a ruler use fractions.

EXAMPLE 3 What part of an inch is shaded?

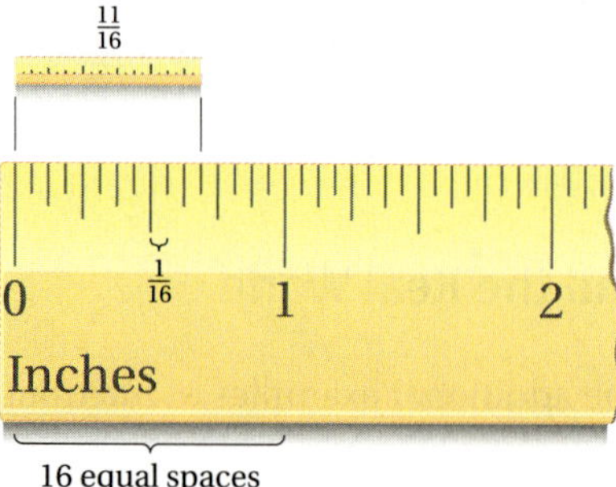

Each inch on the ruler shown above is divided into 16 equal parts. The shading extends to the 11th mark. Thus, $\frac{11}{16}$ is shaded.

Do Exercises 4–9.

What part is shaded?

4.

5.

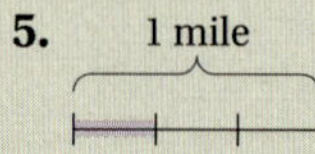

6.

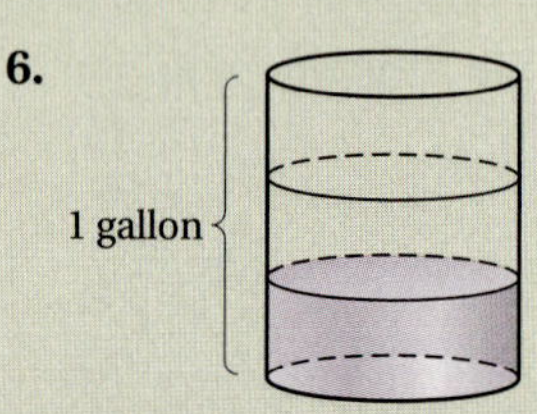

7.

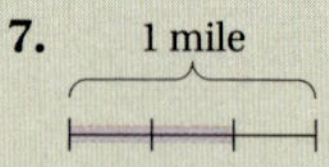

8.

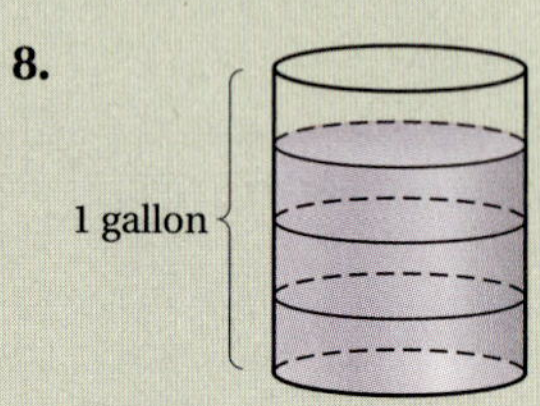

9.

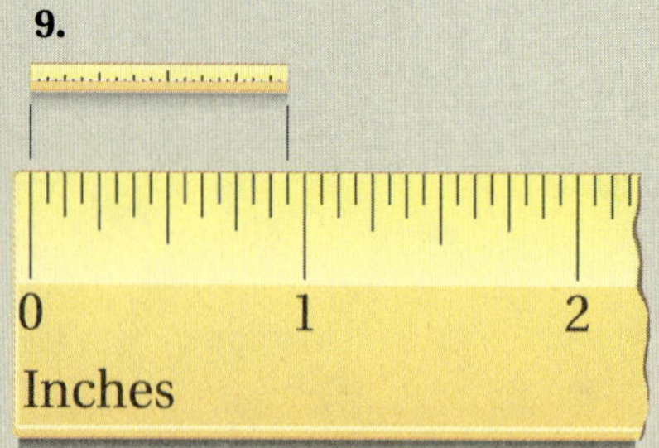

Answers on page A-4

Fractions greater than 1 correspond to situations like the following.

EXAMPLE 4 What part is shaded?

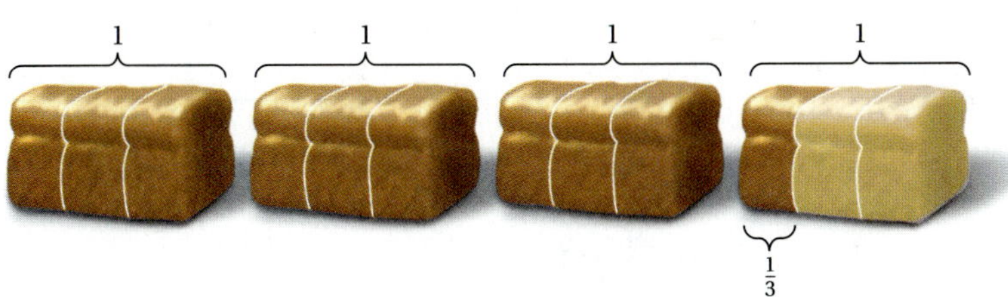

Each loaf of bread is divided into 3 equal parts. The unit is $\frac{1}{3}$. The *denominator* is 3. We have 10 of the units shaded. This tells us the *numerator* is 10. Thus, $\frac{10}{3}$ is shaded.

EXAMPLE 5 What part is shaded?

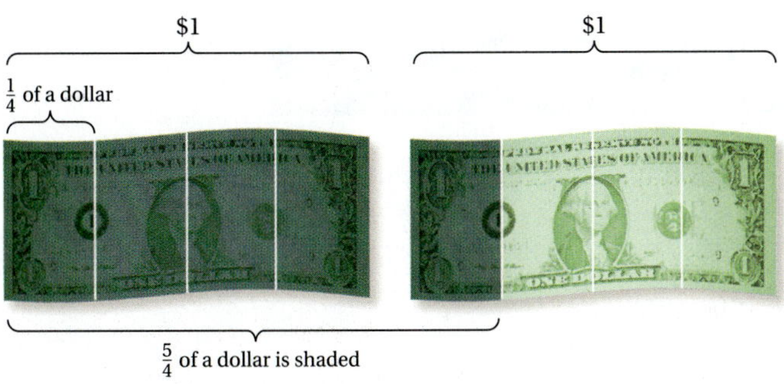

We can regard this as two objects of 4 parts each and take 5 of those parts. We have more than one whole object. Thus, $5 \cdot \frac{1}{4}$, or $\frac{5}{4}$ (also, 5 quarters) is shaded.

Do Exercises 10 and 11.

What part is shaded?

10.

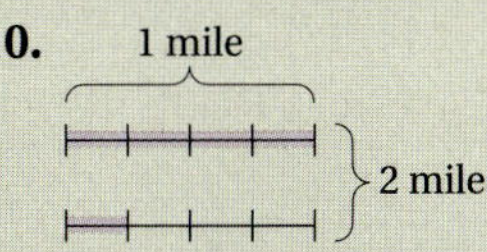

11.

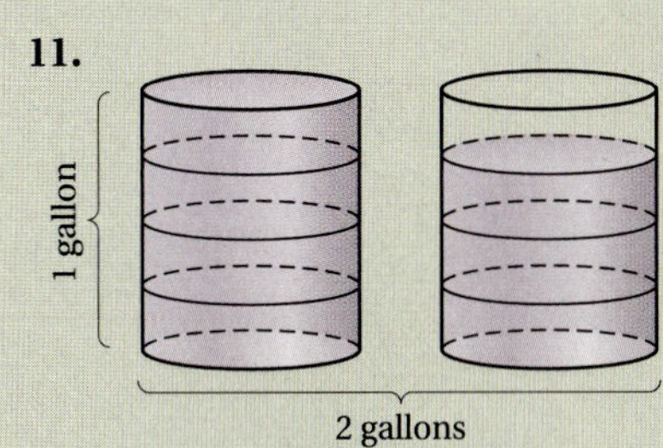

Answers on page A-4

b Some Fraction Notation for Whole Numbers

FRACTION NOTATION FOR 1

The number 1 corresponds to situations like those shown here.

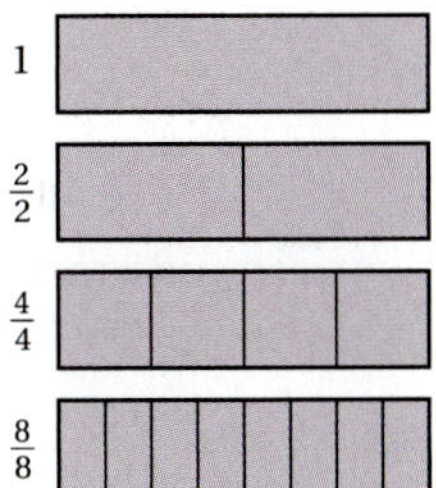

If we divide an object into n parts and take n of them, we get all of the object (1 whole object).

Simplify.

12. $\frac{1}{1}$

13. $\frac{4}{4}$

14. $\frac{34}{34}$

15. $\frac{100}{100}$

16. $\frac{2347}{2347}$

17. $\frac{103}{103}$

Simplify, if possible.

18. $\frac{0}{1}$

19. $\frac{0}{8}$

20. $\frac{0}{107}$

21. $\frac{4-4}{567}$

22. $\frac{15}{0}$

23. $\frac{0}{3-3}$

Answer on page A-4

THE NUMBER 1 IN FRACTION NOTATION

$\frac{n}{n} = 1$, for any whole number n that is not 0.

EXAMPLES Simplify.

6. $\frac{5}{5} = 1$

7. $\frac{9}{9} = 1$

8. $\frac{23}{23} = 1$

Do Exercises 12–17.

FRACTION NOTATION FOR 0

Consider the fraction $\frac{0}{4}$. This corresponds to dividing an object into 4 parts and taking none of them. We get 0.

THE NUMBER 0 IN FRACTION NOTATION

$\frac{0}{n} = 0$, for any whole number n that is not 0.

EXAMPLES Simplify.

9. $\frac{0}{1} = 0$

10. $\frac{0}{9} = 0$

11. $\frac{0}{23} = 0$

Fraction notation with a denominator of 0, such as $n/0$, is meaningless because we cannot speak of an object being divided into *zero* parts. If it is not divided at all, then we say that it is undivided and remains in one part. See also the discussion of excluding division by 0 in Section 1.3.

EXCLUDING DIVISION BY 0

$\frac{n}{0}$ is not defined for any whole number n.

Do Exercises 18–23.

OTHER WHOLE NUMBERS

Consider the fraction $\frac{4}{1}$. This corresponds to taking 4 objects and dividing each into 1 part. (We do not divide them.) We have 4 objects.

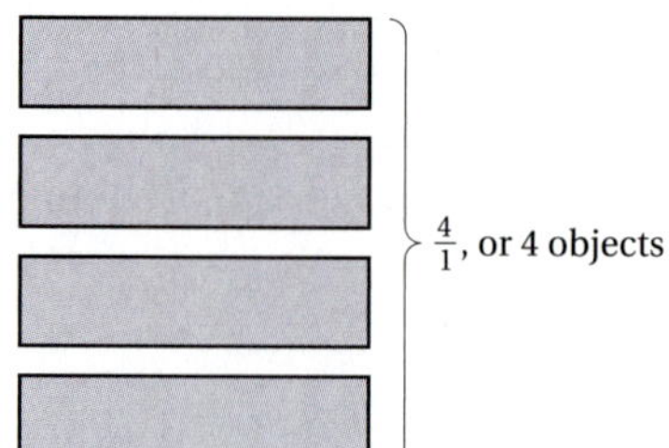

DIVISION BY

Any whole number divided by 1 is the whole number. That is,

$$\frac{n}{1} = n, \quad \text{for any whole number } n.$$

EXAMPLES Simplify.

12. $\frac{2}{1} = 2$

13. $\frac{9}{1} = 9$

14. $\frac{34}{1} = 34$

Do Exercises 24–27.

Simplify.

24. $\frac{8}{1}$

25. $\frac{10}{1}$

26. $\frac{346}{1}$

27. $\frac{24 - 1}{23 - 22}$

C Multiplication Using Fraction Notation

When neither factor is a whole number, multiplication using fraction notation does not correspond to repeated addition. Let's see how multiplication of fractions corresponds to situations in the real world. We consider the multiplication

$$\frac{3}{5} \cdot \frac{3}{4}.$$

We first consider some object and take $\frac{3}{4}$ of it. We divide it into 4 vertical parts, or columns of the same area, and take 3 of them. That is shown in the shading at right.

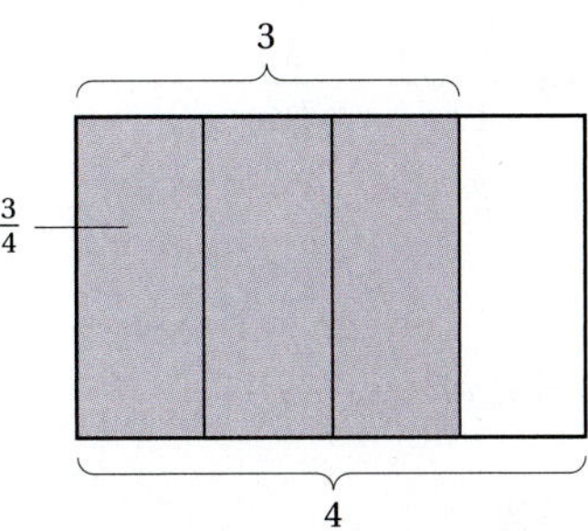

Next, we take $\frac{3}{5}$ of the result. We divide the shaded part into 5 horizontal parts, or rows of the same area, and take 3 of them. That is shown below.

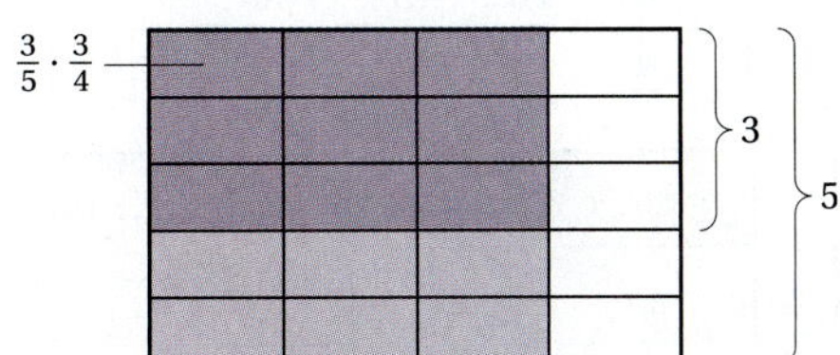

The entire object has been divided into 20 parts, and we have shaded 9 of them for a second time:

$$\frac{3}{5} \cdot \frac{3}{4} = \frac{3 \cdot 3}{5 \cdot 4} = \frac{9}{20}.$$

The figure above shows a rectangular array inside a rectangular array. The number of pieces in the entire array is $5 \cdot 4$ (the product of the denominators). The number of pieces shaded a second time is $3 \cdot 3$ (the product of the numerators). For the answer, we take 9 pieces out of a set of 20 to get $\frac{9}{20}$.

Do Exercise 28.

28. Draw a diagram like the one at left to show the multiplication $\frac{1}{3} \cdot \frac{4}{5}$.

Answers on page A-4

Multiply.

29. $\frac{3}{8} \cdot \frac{5}{7}$

30. $\frac{4}{3} \times \frac{8}{5}$

31. $\frac{3}{10} \cdot \frac{1}{10}$

32. $7 \cdot \frac{2}{3}$

Answers on page A-4

We find a product such as $\frac{9}{7} \cdot \frac{3}{4}$ as follows.

To multiply a fraction by a fraction,

a) multiply the numerators to get the new numerator, and

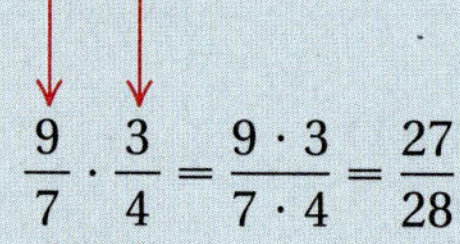

$$\frac{9}{7} \cdot \frac{3}{4} = \frac{9 \cdot 3}{7 \cdot 4} = \frac{27}{28}$$

b) multiply the denominators to get the new denominator.

EXAMPLES Multiply.

15. $\frac{5}{6} \times \frac{7}{4} = \frac{5 \times 7}{6 \times 4} = \frac{35}{24}$

Skip writing this step whenever you can.

16. $\frac{3}{5} \cdot \frac{7}{8} = \frac{3 \cdot 7}{5 \cdot 8} = \frac{21}{40}$

17. $\frac{3}{5} \cdot \frac{3}{4} = \frac{9}{20}$

18. $\frac{1}{4} \cdot \frac{1}{3} = \frac{1}{12}$

19. $6 \cdot \frac{4}{5} = \frac{6}{1} \cdot \frac{4}{5} = \frac{24}{5}$

Do Exercises 29–32.

d Multiplying by 1

Recall the following:

$$1 = \frac{1}{1} = \frac{2}{2} = \frac{3}{3} = \frac{4}{4} = \frac{10}{10} = \frac{45}{45} = \frac{100}{100} = \frac{n}{n}.$$

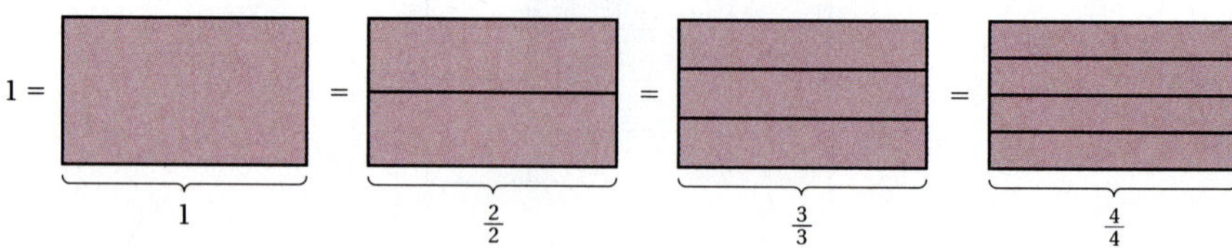

Any nonzero number divided by itself is 1. (See Section 1.3.)

Now recall the multiplicative identity from Section 1.3. For any whole number a, $1 \cdot a = a \cdot 1 = a$. This holds for numbers of arithmetic as well.

MULTIPLICATIVE IDENTITY FOR FRACTIONS

When we multiply a number by 1, we get the same number:

$$\frac{3}{5} = \frac{3}{5} \cdot 1 = \frac{3}{5} \cdot \frac{4}{4} = \frac{12}{20}.$$

Since $\frac{3}{5} = \frac{12}{20}$, we know that $\frac{3}{5}$ and $\frac{12}{20}$ are two names for the same number. We also say that $\frac{3}{5}$ and $\frac{12}{20}$ are **equivalent.**

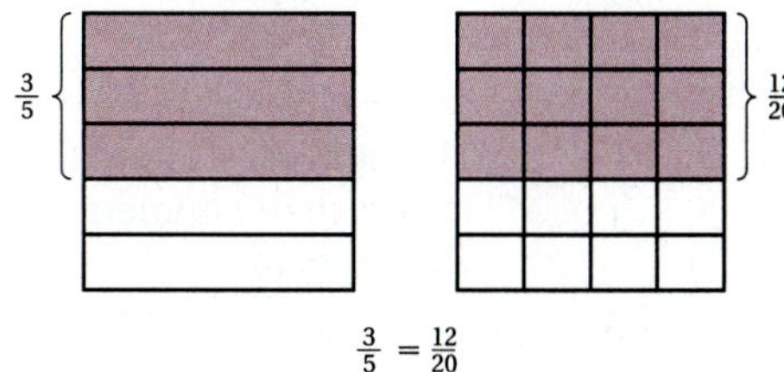

$\frac{3}{5} = \frac{12}{20}$

Do Exercises 33–36.

Suppose we want to find a name for $\frac{2}{3}$, but one that has a denominator of 9. We can multiply by 1 to find equivalent fractions:

$$\frac{2}{3} = \frac{2}{3} \cdot \frac{3}{3} = \frac{2 \cdot 3}{3 \cdot 3} = \frac{6}{9}.$$

We chose $\frac{3}{3}$ for 1 in order to get a denominator of 9.

EXAMPLE 20 Find a name for $\frac{1}{4}$ with a denominator of 24.

Since $4 \cdot 6 = 24$, we multiply by $\frac{6}{6}$:

$$\frac{1}{4} = \frac{1}{4} \cdot \frac{6}{6} = \frac{1 \cdot 6}{4 \cdot 6} = \frac{6}{24}.$$

EXAMPLE 21 Find a name for $\frac{2}{5}$ with a denominator of 35.

Since $5 \cdot 7 = 35$, we multiply by $\frac{7}{7}$:

$$\frac{2}{5} = \frac{2}{5} \cdot \frac{7}{7} = \frac{2 \cdot 7}{5 \cdot 7} = \frac{14}{35}.$$

Do Exercises 37–41.

e Simplifying Fraction Notation

All of the following are names for three-fourths:

$$\frac{3}{4}, \frac{6}{8}, \frac{9}{12}, \frac{12}{16}, \frac{15}{20}.$$

We say that $\frac{3}{4}$ is **simplest** because it has the smallest numerator and the smallest denominator. That is, the numerator and the denominator have no common factor other than 1.

To simplify, we reverse the process of multiplying by 1:

$$\frac{12}{18} = \frac{2 \cdot 6}{3 \cdot 6}$$ ← Factoring the numerator; ← Factoring the denominator

$$= \frac{2}{3} \cdot \frac{6}{6}$$ Factoring the fraction

$$= \frac{2}{3} \cdot 1$$ $\frac{6}{6} = 1$

$$= \frac{2}{3}.$$ Removing a factor of 1: $\frac{2}{3} \cdot 1 = \frac{2}{3}$

Multiply.

33. $\frac{1}{2} \cdot \frac{8}{8}$

34. $\frac{3}{5} \cdot \frac{10}{10}$

35. $\frac{13}{25} \cdot \frac{4}{4}$

36. $\frac{8}{3} \cdot \frac{25}{25}$

Find another name for the number, but with the denominator indicated. Use multiplying by 1.

37. $\frac{4}{3} = \frac{?}{9}$

38. $\frac{3}{4} = \frac{?}{24}$

39. $\frac{9}{10} = \frac{?}{100}$

40. $\frac{3}{15} = \frac{?}{45}$

41. $\frac{8}{7} = \frac{?}{49}$

Answers on page A-4

Simplify.

42. $\dfrac{2}{8}$

43. $\dfrac{10}{12}$

44. $\dfrac{40}{8}$

45. $\dfrac{24}{18}$

Simplify.

46. $\dfrac{35}{40}$

47. $\dfrac{801}{702}$

48. $\dfrac{24}{21}$

49. $\dfrac{75}{300}$

50. Simplify each fraction in this circle graph.

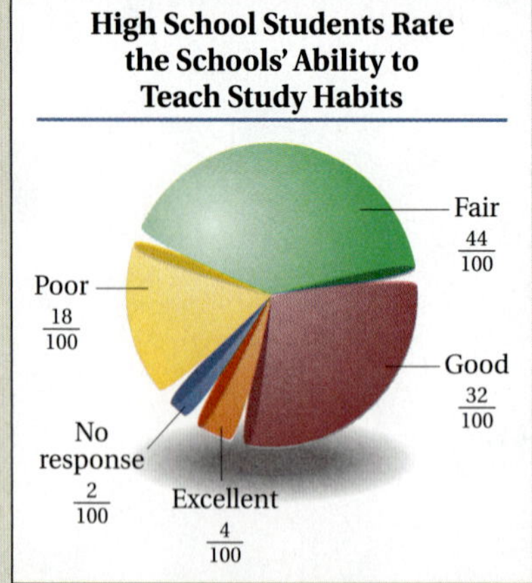

Answers on page A-4

EXAMPLES Simplify.

22. $\dfrac{8}{20} = \dfrac{2 \cdot 4}{5 \cdot 4} = \dfrac{2}{5} \cdot \dfrac{4}{4} = \dfrac{2}{5}$

23. $\dfrac{2}{6} = \dfrac{1 \cdot 2}{3 \cdot 2} = \dfrac{1}{3} \cdot \dfrac{2}{2} = \dfrac{1}{3}$

The number 1 allows for pairing of factors in the numerator and the denominator.

24. $\dfrac{30}{6} = \dfrac{5 \cdot 6}{1 \cdot 6} = \dfrac{5}{1} \cdot \dfrac{6}{6} = \dfrac{5}{1} = 5$

We could also simplify $\frac{30}{6}$ by doing the division $30 \div 6$. That is, $\frac{30}{6} = 30 \div 6 = 5$.

Do Exercises 42–45.

The use of prime factorizations can be helpful for simplifying when numerators and/or denominators are larger numbers.

EXAMPLE 25 Simplify: $\dfrac{90}{84}$.

$$\frac{90}{84} = \frac{2 \cdot 3 \cdot 3 \cdot 5}{2 \cdot 2 \cdot 3 \cdot 7}$$ Factoring the numerator and the denominator into primes

$$= \frac{2 \cdot 3 \cdot 3 \cdot 5}{2 \cdot 3 \cdot 2 \cdot 7}$$ Changing the order so that like primes are above and below each other

$$= \frac{2}{2} \cdot \frac{3}{3} \cdot \frac{3 \cdot 5}{2 \cdot 7}$$ Factoring the fraction

$$= 1 \cdot 1 \cdot \frac{3 \cdot 5}{2 \cdot 7}$$

$$= \frac{3 \cdot 5}{2 \cdot 7}$$ Removing factors of 1

$$= \frac{15}{14}$$

We could have shortened the preceding example had we recalled our tests for divisibility (Section 1.8) and noted that 6 is a factor of both the numerator and the denominator. Then

$$\frac{90}{84} = \frac{6 \cdot 15}{6 \cdot 14} = \frac{6}{6} \cdot \frac{15}{14} = \frac{15}{14}.$$

The tests for divisibility are very helpful in simplifying.

EXAMPLE 26 Simplify: $\dfrac{603}{207}$.

At first glance this looks difficult. But note, using the test for divisibility by 9 (sum of digits divisible by 9), that both the numerator and the denominator are divisible by 9. Thus we can factor 9 from both numbers:

$$\frac{603}{207} = \frac{9 \cdot 67}{9 \cdot 23} = \frac{9}{9} \cdot \frac{67}{23} = \frac{67}{23}.$$

Do Exercises 46–50.

CANCELING

Canceling is a shortcut that you may have used for removing a factor of 1 when working with fraction notation. With *great* concern, we mention it as a possibility for speeding up your work. Canceling may be done only when removing common factors in numerators and denominators. Each common factor allows us to remove a factor of 1 in a product.

Our concern is that canceling be done with care and understanding. In effect, slashes are used to indicate factors of 1 that have been removed. For instance, Example 25 might have been done faster as follows:

$$\frac{90}{84} = \frac{2 \cdot 3 \cdot 3 \cdot 5}{2 \cdot 2 \cdot 3 \cdot 7} \quad \text{Factoring the numerator and the denominator}$$

$$= \frac{\cancel{2} \cdot \cancel{3} \cdot 3 \cdot 5}{2 \cdot \cancel{2} \cdot \cancel{3} \cdot 7} \quad \text{When a factor of 1 is noted, it is "canceled" as shown: } \frac{2 \cdot 3}{2 \cdot 3} = 1.$$

$$= \frac{3 \cdot 5}{2 \cdot 7} = \frac{15}{14}.$$

CAUTION!

The difficulty with canceling is that it is often applied incorrectly in situations like the following:

$$\frac{\cancel{2} + 3}{\cancel{2}} = 3; \qquad \frac{\cancel{4} + 1}{\cancel{4} + 2} = \frac{1}{2}; \qquad \frac{1\cancel{5}}{\cancel{5}4} = \frac{1}{4}.$$

Wrong! Wrong! Wrong!

The correct answers are

$$\frac{2 + 3}{2} = \frac{5}{2}; \qquad \frac{4 + 1}{4 + 2} = \frac{5}{6}; \qquad \frac{15}{54} = \frac{5}{18}.$$

In each situation, the number canceled was not a factor of 1. Factors are parts of products. For example, in $2 \cdot 3$, 2 and 3 are factors, but in $2 + 3$, 2 and 3 are *not* factors. Canceling may not be done when sums or differences are in numerators or denominators, as shown here.

If you cannot factor, do not cancel! If in doubt, do not cancel!

CALCULATOR CORNER

Simplifying Fraction Notation Fraction calculators are equipped with a key, often labeled $a^b/_c$, that allows for simplification with fraction notation. To simplify

$$\frac{208}{256}$$

with such a fraction calculator, the following keystrokes can be used.

[2] [0] [8] [$a^b/_c$] [2] [5] [6] [=].

The display that appears

13 ⌋16.

represents simplified fraction notation $\frac{13}{16}$.

Exercises: Use a fraction calculator to simplify each of the following.

1. $\frac{84}{90}$
2. $\frac{35}{40}$
3. $\frac{690}{835}$
4. $\frac{42}{150}$

2.1 EXERCISE SET

For Extra Help

Digital Video Tutor CD 2 Videotape 3

InterAct Math

Math Tutor Center

MathXL

MyMathLab.com

a Identify the numerator and the denominator.

1. $\frac{3}{4}$

2. $\frac{9}{10}$

3. $\frac{11}{20}$

4. $\frac{18}{5}$

What part of the object is shaded? In Exercises 11–14, what part of an inch is shaded?

5. $1

6. $1

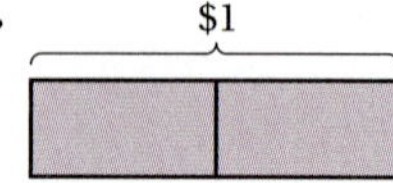

7.

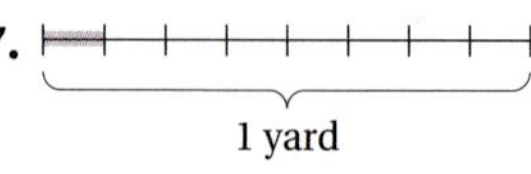

8. 1 gold bar

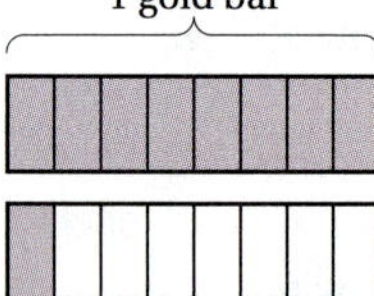

9.

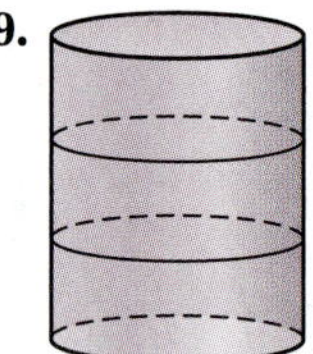

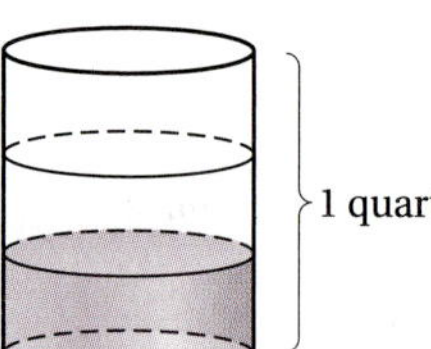

10. 1 foot

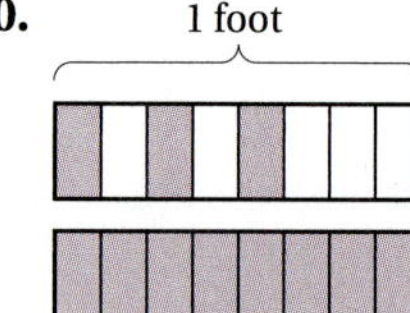

11.

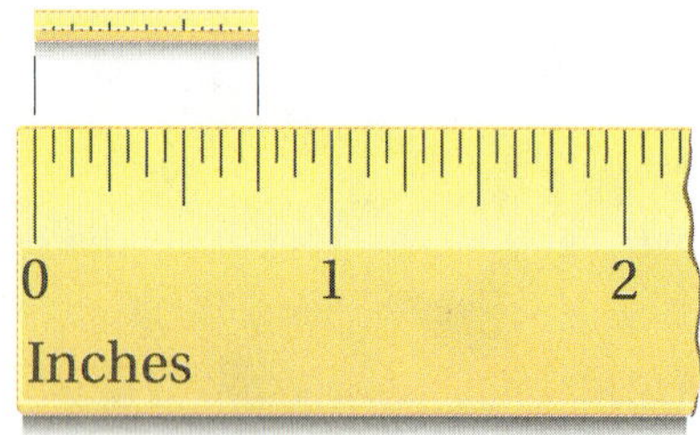

12.

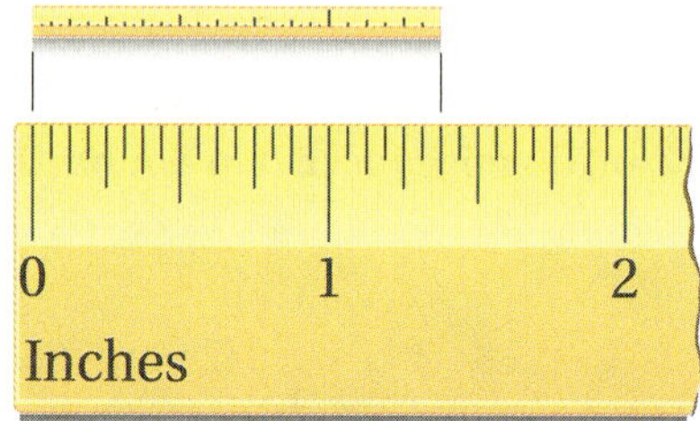

13.

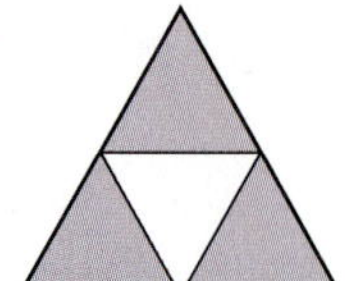

14.

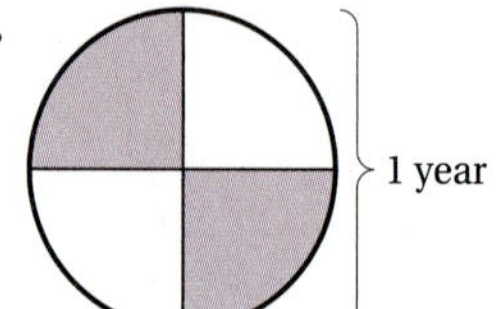

15.

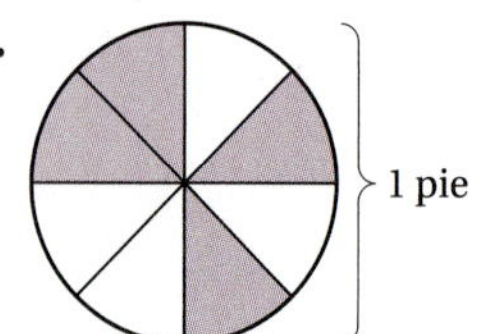

16. 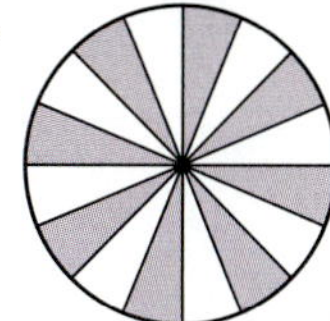

17. 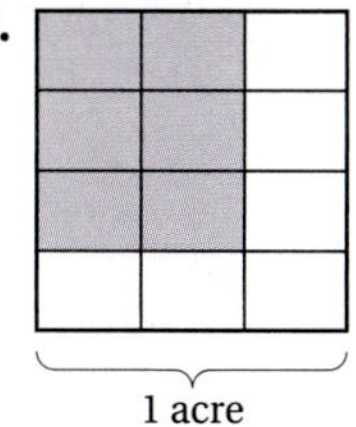

18. 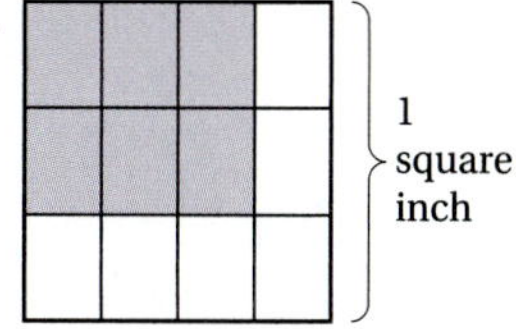

For each of Exercises 19–22, give fraction notation for the amount of gas (a) in the tank and (b) used from a full tank.

19. 

20. 

21. 

22. 

b Simplify.

23. $\frac{18}{1}$

24. $\frac{0}{16}$

25. $\frac{0}{8}$

26. $\frac{16}{1}$

27. $\frac{20}{20}$

28. $\frac{3}{3}$

29. $\frac{5}{6-6}$

30. $\frac{238}{1}$

31. $\frac{729}{0}$

32. $\frac{8-8}{1247}$

33. $\frac{87}{87}$

34. $\frac{1317}{0}$

c Multiply.

35. $\frac{1}{2} \cdot \frac{1}{3}$

36. $\frac{1}{6} \cdot \frac{1}{4}$

37. $5 \times \frac{1}{8}$

38. $4 \times \frac{1}{5}$

39. $\frac{2}{3} \times \frac{1}{5}$

40. $\frac{3}{5} \times \frac{1}{5}$

41. $\frac{2}{5} \cdot \frac{2}{3}$

42. $\frac{3}{4} \cdot \frac{3}{5}$

43. $\frac{3}{4} \cdot \frac{3}{4}$

44. $\frac{3}{7} \cdot \frac{4}{5}$

45. $\frac{2}{3} \cdot \frac{7}{13}$

46. $\frac{3}{11} \cdot \frac{4}{5}$

47. $7 \cdot \frac{3}{4}$

48. $7 \cdot \frac{2}{5}$

49. $\frac{7}{8} \cdot \frac{7}{8}$

50. $\frac{3}{10} \cdot \frac{7}{100}$

d

Find another name for the given number, but with the denominator indicated. Use multiplying by 1.

51. $\frac{1}{2} = \frac{?}{10}$

52. $\frac{1}{6} = \frac{?}{18}$

53. $\frac{5}{8} = \frac{?}{32}$

54. $\frac{2}{9} = \frac{?}{18}$

55. $\frac{5}{3} = \frac{?}{45}$

56. $\frac{11}{5} = \frac{?}{30}$

57. $\frac{7}{22} = \frac{?}{132}$

58. $\frac{10}{21} = \frac{?}{126}$

e

Simplify.

59. $\frac{6}{8}$

60. $\frac{8}{12}$

61. $\frac{3}{15}$

62. $\frac{8}{10}$

63. $\frac{24}{8}$

64. $\frac{36}{9}$

65. $\frac{18}{24}$

66. $\frac{42}{48}$

67. $\frac{14}{16}$

68. $\frac{15}{25}$

69. $\frac{150}{25}$

70. $\frac{19}{76}$

71. $\frac{17}{51}$

72. $\frac{425}{525}$

73. **D_W** Explain in your own words when it *is* possible to "cancel" and when it *is not* possible to "cancel."

74. **D_W** On p. 109, we explained, using words and pictures, why $\frac{3}{5} \cdot \frac{3}{4}$ equals $\frac{9}{20}$. Present a similar explanation of why $\frac{2}{3} \cdot \frac{4}{7}$ equals $\frac{8}{21}$.

SYNTHESIS

What part of the object is shaded?

75.

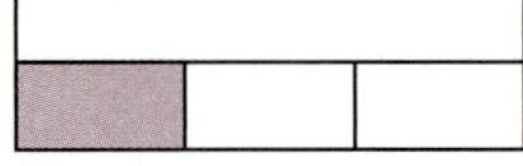

76.

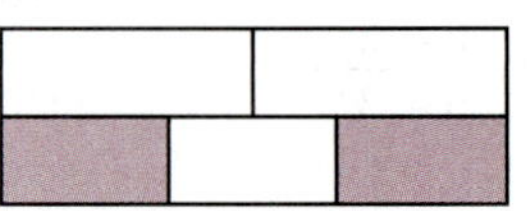

77.

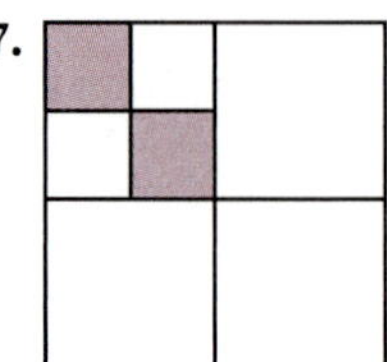

78.

2.2 MULTIPLICATION AND DIVISION

a Multiplying and Simplifying Using Fraction Notation

We usually simplify after we multiply. To make such simplifying easier, it is generally best not to carry out the products in the numerator and the denominator, but to factor and simplify before multiplying. Consider the product

$$\frac{3}{8} \cdot \frac{4}{9}.$$

We proceed as follows:

$$\frac{3}{8} \cdot \frac{4}{9} = \frac{3 \cdot 4}{8 \cdot 9}$$ We write the products in the numerator and the denominator, but we do not carry them out.

$$= \frac{3 \cdot 2 \cdot 2}{2 \cdot 2 \cdot 2 \cdot 3 \cdot 3}$$ Factoring the numerator and the denominator

$$= \frac{3 \cdot 2 \cdot 2}{3 \cdot 2 \cdot 2} \cdot \frac{1}{2 \cdot 3}$$ Factoring the fraction

$$= 1 \cdot \frac{1}{2 \cdot 3}$$

$$= \frac{1}{2 \cdot 3}$$ Removing a factor of 1

$$= \frac{1}{6}.$$

The procedure could have been shortened had we noticed that 4 is a factor of the 8 in the denominator:

$$\frac{3}{8} \cdot \frac{4}{9} = \frac{3 \cdot 4}{8 \cdot 9} = \frac{3 \cdot 4}{4 \cdot 2 \cdot 3 \cdot 3} = \frac{3 \cdot 4}{3 \cdot 4} \cdot \frac{1}{2 \cdot 3} = 1 \cdot \frac{1}{2 \cdot 3} = \frac{1}{2 \cdot 3} = \frac{1}{6}.$$

To multiply and simplify:

a) Write the products in the numerator and the denominator, but do not carry out the products.
b) Factor the numerator and the denominator.
c) Factor the fraction to remove factors of 1.
d) Carry out the remaining products.

EXAMPLES Multiply and simplify.

1. $\frac{2}{3} \cdot \frac{9}{4} = \frac{2 \cdot 9}{3 \cdot 4} = \frac{2 \cdot 3 \cdot 3}{3 \cdot 2 \cdot 2} = \frac{2 \cdot 3}{2 \cdot 3} \cdot \frac{3}{2} = 1 \cdot \frac{3}{2} = \frac{3}{2}$

2. $\frac{6}{7} \cdot \frac{5}{3} = \frac{6 \cdot 5}{7 \cdot 3} = \frac{3 \cdot 2 \cdot 5}{7 \cdot 3} = \frac{3}{3} \cdot \frac{2 \cdot 5}{7} = 1 \cdot \frac{2 \cdot 5}{7} = \frac{2 \cdot 5}{7} = \frac{10}{7}$

3. $40 \cdot \frac{7}{8} = \frac{40 \cdot 7}{8} = \frac{8 \cdot 5 \cdot 7}{8 \cdot 1} = \frac{8}{8} \cdot \frac{5 \cdot 7}{1} = 1 \cdot \frac{5 \cdot 7}{1} = \frac{5 \cdot 7}{1} = 35$

Objectives

a Multiply and simplify using fraction notation.

b Find the reciprocal of a number.

c Divide and simplify using fraction notation.

d Solve equations of the type $a \cdot x = b$ and $x \cdot a = b$, where a and b may be fractions.

Study Tips

THE SUPPLEMENTS

The new mathematical skills and concepts presented in the lectures will be of increased value to you if you begin the homework assignment as soon as possible after the lecture. Then if you still have difficulty with any of the exercises, you have time to access supplementary resources such as:

- *Student's Solutions Manual*
- Videotapes
- InterAct Math Tutorial CD-ROM
- AW Math Tutor Center
- MathXL

Multiply and simplify.

1. $\frac{2}{3} \cdot \frac{7}{8}$

2. $\frac{4}{5} \cdot \frac{5}{12}$

3. $16 \cdot \frac{3}{8}$

4. $\frac{5}{8} \cdot 4$

Find the reciprocal.

5. $\frac{2}{5}$

6. $\frac{10}{7}$

7. 9

8. $\frac{1}{5}$

Answers on page A-4

CAUTION!

Canceling can be used as follows for these examples.

1. $\frac{2}{3} \cdot \frac{9}{4} = \frac{2 \cdot 9}{3 \cdot 4} = \frac{\not{2} \cdot \not{3} \cdot 3}{\not{3} \cdot \not{2} \cdot 2} = \frac{3}{2}$ Removing a factor of 1: $\frac{2 \cdot 3}{2 \cdot 3} = 1$

2. $\frac{6}{7} \cdot \frac{5}{3} = \frac{6 \cdot 5}{7 \cdot 3} = \frac{\not{3} \cdot 2 \cdot 5}{7 \cdot \not{3}} = \frac{2 \cdot 5}{7} = \frac{10}{7}$ Removing a factor of 1: $\frac{3}{3} = 1$

3. $40 \cdot \frac{7}{8} = \frac{40 \cdot 7}{8} = \frac{\not{8} \cdot 5 \cdot 7}{\not{8} \cdot 1} = \frac{5 \cdot 7}{1} = 35$ Removing a factor of 1: $\frac{8}{8} = 1$

Remember, if you can't factor, you can't cancel!

Do Exercises 1–4.

b Reciprocals

Look at these products:

$$8 \cdot \frac{1}{8} = \frac{8 \cdot 1}{8} = \frac{8}{8} = 1; \qquad \frac{2}{3} \cdot \frac{3}{2} = \frac{2 \cdot 3}{3 \cdot 2} = \frac{6}{6} = 1.$$

RECIPROCALS

If the product of two numbers is 1, we say that they are **reciprocals** of each other. To find a reciprocal of a fraction, interchange the numerator and the denominator.

Number — $\frac{3}{4} \rightarrow \frac{4}{3}$ — Reciprocal

EXAMPLES Find the reciprocal.

4. The reciprocal of $\frac{4}{5}$ is $\frac{5}{4}$. $\quad \frac{4}{5} \cdot \frac{5}{4} = \frac{20}{20} = 1$

5. The reciprocal of $\frac{8}{7}$ is $\frac{7}{8}$. $\quad \frac{8}{7} \cdot \frac{7}{8} = \frac{56}{56} = 1$

6. The reciprocal of 8 is $\frac{1}{8}$. $\quad$ Think of 8 as $\frac{8}{1}$: $\frac{8}{1} \cdot \frac{1}{8} = \frac{8}{8} = 1.$

7. The reciprocal of $\frac{1}{3}$ is 3. $\quad \frac{1}{3} \cdot 3 = \frac{3}{3} = 1$

Do Exercises 5–8.

Does 0 have a reciprocal? If it did, it would have to be a number x such that

$$0 \cdot x = 1.$$

But 0 times any number is 0. Thus we have the following.

0 HAS NO RECIPROCAL

The number 0, or $\frac{0}{n}$, has no reciprocal. $\left(\text{Recall that } \frac{n}{0} \text{ is not defined.}\right)$

C Division

Consider the division $\frac{3}{4} \div \frac{1}{8}$. We are asking how many $\frac{1}{8}$'s are in $\frac{3}{4}$. We can answer this by looking at the figure below.

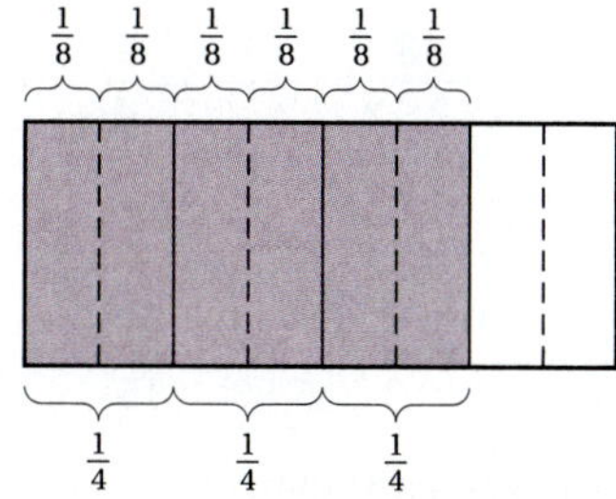

We see that there are six $\frac{1}{8}$'s in $\frac{3}{4}$. Thus,

$$\frac{3}{4} \div \frac{1}{8} = 6.$$

We can check this by multiplying:

$$6 \cdot \frac{1}{8} = \frac{6}{8} = \frac{3}{4}.$$

Here is a faster way to do this division:

$$\frac{3}{4} \div \frac{1}{8} = \frac{3}{4} \cdot \frac{8}{1} = \frac{24}{4} = 6.$$

Multiplying by the reciprocal of the divisor

To divide fractions, multiply the dividend by the reciprocal of the divisor:

$$\frac{2}{5} \div \frac{3}{4} = \frac{2}{5} \cdot \frac{4}{3} = \frac{2 \cdot 4}{5 \cdot 3} = \frac{8}{15}.$$

Multiply by the reciprocal of the divisor.

Divide and simplify.

9. $\frac{6}{7} \div \frac{3}{4}$

10. $\frac{2}{3} \div \frac{1}{4}$

11. $\frac{4}{5} \div 8$

12. $60 \div \frac{3}{5}$

13. $\frac{3}{5} \div \frac{3}{5}$

Answers on page A-4

EXAMPLES Divide and simplify.

8. $\frac{5}{6} \div \frac{2}{3} = \frac{5}{6} \cdot \frac{3}{2} = \frac{5 \cdot 3}{6 \cdot 2} = \frac{5 \cdot 3}{3 \cdot 2 \cdot 2} = \frac{3}{3} \cdot \frac{5}{2 \cdot 2} = \frac{5}{2 \cdot 2} = \frac{5}{4}$

9. $\frac{7}{8} \div \frac{1}{16} = \frac{7}{8} \cdot 16 = \frac{7 \cdot 16}{8} = \frac{7 \cdot 2 \cdot 8}{8 \cdot 1} = \frac{8}{8} \cdot \frac{7 \cdot 2}{1} = \frac{7 \cdot 2}{1} = 14$

10. $\frac{2}{5} \div 6 = \frac{2}{5} \cdot \frac{1}{6} = \frac{2 \cdot 1}{5 \cdot 6} = \frac{2 \cdot 1}{5 \cdot 2 \cdot 3} = \frac{2}{2} \cdot \frac{1}{5 \cdot 3} = \frac{1}{5 \cdot 3} = \frac{1}{15}$

11. $\frac{3}{5} \div \frac{1}{2} = \frac{3}{5} \cdot 2 = \frac{3 \cdot 2}{5} = \frac{6}{5}$

CAUTION!

Canceling can be used as follows for Examples 5–7.

8. $\frac{5}{6} \div \frac{2}{3} = \frac{5}{6} \cdot \frac{3}{2} = \frac{5 \cdot 3}{6 \cdot 2} = \frac{5 \cdot \cancel{3}}{\cancel{3} \cdot 2 \cdot 2} = \frac{5}{2 \cdot 2} = \frac{5}{4}$ Removing a factor of 1: $\frac{3}{3} = 1$

9. $\frac{7}{8} \div \frac{1}{16} = \frac{7}{8} \cdot 16 = \frac{7 \cdot 16}{8} = \frac{7 \cdot \cancel{8} \cdot 2}{\cancel{8} \cdot 1} = \frac{7 \cdot 2}{1} = 14$ Removing a factor of 1: $\frac{8}{8} = 1$

10. $\frac{2}{5} \div 6 = \frac{2}{5} \cdot \frac{1}{6} = \frac{2 \cdot 1}{5 \cdot 6} = \frac{\cancel{2} \cdot 1}{5 \cdot \cancel{2} \cdot 3} = \frac{1}{5 \cdot 3} = \frac{1}{15}$ Removing a factor of 1: $\frac{2}{2} = 1$

Remember, if you can't factor, you can't cancel!

Do Exercises 9–13.

What is the explanation for multiplying by a reciprocal when dividing? Let's consider $\frac{2}{3} \div \frac{7}{5}$. We multiply by 1. The name for 1 that we will use is (5/7)/(5/7); it comes from the reciprocal of $\frac{7}{5}$.

$$\frac{2}{3} \div \frac{7}{5} = \frac{\frac{2}{3}}{\frac{7}{5}}$$ Writing fraction notation for the division

$$= \frac{\frac{2}{3}}{\frac{7}{5}} \cdot 1$$ Multiplying by 1

$$= \frac{\frac{2}{3}}{\frac{7}{5}} \cdot \frac{\frac{5}{7}}{\frac{5}{7}}$$ Multiplying by 1; $\frac{5}{7}$ is the reciprocal of $\frac{7}{5}$ and $\frac{\frac{5}{7}}{\frac{5}{7}} = 1$

$$= \frac{\frac{2}{3} \cdot \frac{5}{7}}{\frac{7}{5} \cdot \frac{5}{7}}$$ Multiplying the numerators and the denominators

$$= \frac{\frac{2}{3} \cdot \frac{5}{7}}{1}$$ After we multiplied, we got 1 for the denominator. The numerator shows the multiplication by the reciprocal.

$$= \frac{2}{3} \cdot \frac{5}{7} = \frac{10}{21}$$

Thus,

$$\frac{2}{3} \div \frac{7}{5} = \frac{2}{3} \cdot \frac{5}{7} = \frac{10}{21}.$$

Do Exercise 14.

d Solving Equations

Now let's solve equations $a \cdot x = b$ and $x \cdot a = b$, where a and b may be fractions. We proceed as we did with equations involving whole numbers. We divide by a on both sides.

EXAMPLE 12 Solve: $\frac{4}{3} \cdot x = \frac{6}{7}$.

We have

$$\frac{4}{3} \cdot x = \frac{6}{7}$$

$$x = \frac{6}{7} \div \frac{4}{3} \quad \text{Dividing by } \tfrac{4}{3} \text{ on both sides}$$

$$= \frac{6}{7} \cdot \frac{3}{4} \quad \text{Multiplying by the reciprocal}$$

$$= \frac{2 \cdot 3 \cdot 3}{7 \cdot 2 \cdot 2} = \frac{2}{2} \cdot \frac{3 \cdot 3}{7 \cdot 2} = \frac{3 \cdot 3}{7 \cdot 2} = \frac{9}{14}.$$

The solution is $\frac{9}{14}$.

EXAMPLE 13 Solve: $t \cdot \frac{4}{5} = 80$.

Dividing by $\frac{4}{5}$ on both sides, we get

$$t = 80 \div \frac{4}{5} = 80 \cdot \frac{5}{4} = \frac{80 \cdot 5}{4} = \frac{4 \cdot 20 \cdot 5}{4 \cdot 1} = \frac{4}{4} \cdot \frac{20 \cdot 5}{1} = \frac{20 \cdot 5}{1} = 100.$$

The solution is 100.

Do Exercises 15 and 16.

14. Divide by multiplying by 1:

$$\frac{\frac{4}{5}}{\frac{6}{7}}.$$

Solve.

15. $\frac{5}{6} \cdot y = \frac{2}{3}$

16. $\frac{3}{4} \cdot n = 24$

To the student and the instructor: Recall that the Skill Maintenance exercises, which occur at the end of the exercise sets, review any skill that has been studied before in the text.

Beginning with this chapter, however, certain objectives from four particular sections, along with the material of this chapter, will be tested on the chapter test.

For this chapter, the objectives to be retested are [1.2d], [1.3d], [1.4b], *and* [1.5a].

Answers on page A-4

2.2 EXERCISE SET

For Extra Help

Digital Video Tutor CD 2 Videotape 3 | InterAct Math | Math Tutor Center | MathXL | MyMathLab

a Multiply and simplify.

Don't forget to simplify!

1. $\frac{2}{3} \cdot \frac{1}{2}$

2. $\frac{3}{8} \cdot \frac{1}{3}$

3. $\frac{1}{4} \cdot \frac{2}{3}$

4. $\frac{4}{6} \cdot \frac{1}{6}$

5. $\frac{12}{5} \cdot \frac{9}{8}$

6. $\frac{16}{15} \cdot \frac{5}{4}$

7. $\frac{10}{9} \cdot \frac{7}{5}$

8. $\frac{25}{12} \cdot \frac{4}{3}$

9. $9 \cdot \frac{1}{9}$

10. $4 \cdot \frac{1}{4}$

11. $\frac{7}{5} \cdot \frac{5}{7}$

12. $\frac{2}{11} \cdot \frac{11}{2}$

13. $24 \cdot \frac{1}{6}$

14. $16 \cdot \frac{1}{2}$

15. $12 \cdot \frac{3}{4}$

16. $18 \cdot \frac{5}{6}$

17. $\frac{7}{10} \cdot 28$

18. $\frac{5}{8} \cdot 34$

19. $240 \cdot \frac{1}{8}$

20. $150 \cdot \frac{1}{5}$

21. $\frac{4}{10} \cdot \frac{5}{10}$

22. $\frac{7}{10} \cdot \frac{34}{150}$

23. $\frac{8}{10} \cdot \frac{45}{100}$

24. $\frac{3}{10} \cdot \frac{8}{10}$

25. $\frac{11}{24} \cdot \frac{3}{5}$

26. $\frac{15}{22} \cdot \frac{4}{7}$

27. $\frac{10}{21} \cdot \frac{3}{4}$

28. $\frac{17}{18} \cdot \frac{3}{5}$

b Find the reciprocal.

29. $\frac{5}{6}$

30. $\frac{7}{8}$

31. 6

32. 4

33. $\frac{1}{6}$

34. $\frac{1}{4}$

35. $\frac{10}{3}$

36. $\frac{17}{4}$

c Divide and simplify.

Don't forget to simplify!

37. $\frac{3}{5} \div \frac{3}{4}$

38. $\frac{2}{3} \div \frac{3}{4}$

39. $\frac{3}{5} \div \frac{9}{4}$

40. $\frac{6}{7} \div \frac{3}{5}$

41. $\frac{4}{3} \div \frac{1}{3}$

42. $\frac{10}{9} \div \frac{1}{3}$

43. $\frac{1}{3} \div \frac{1}{6}$

44. $\frac{1}{4} \div \frac{1}{5}$

45. $\frac{3}{8} \div 3$

46. $\frac{5}{6} \div 5$

47. $\frac{12}{7} \div 4$

48. $\frac{18}{5} \div 2$

49. $12 \div \frac{3}{2}$

50. $24 \div \frac{3}{8}$

51. $28 \div \frac{4}{5}$

52. $40 \div \frac{2}{3}$

53. $\frac{5}{8} \div \frac{5}{8}$

54. $\frac{2}{5} \div \frac{2}{5}$

55. $\frac{8}{15} \div \frac{4}{5}$

56. $\frac{6}{13} \div \frac{3}{26}$

57. $\frac{9}{5} \div \frac{4}{5}$

58. $\frac{5}{12} \div \frac{25}{36}$

59. $120 \div \frac{5}{6}$

60. $360 \div \frac{8}{7}$

d Solve.

61. $\frac{4}{5} \cdot x = 60$

62. $\frac{3}{2} \cdot t = 90$

63. $\frac{5}{3} \cdot y = \frac{10}{3}$

64. $\frac{4}{9} \cdot m = \frac{8}{3}$

65. $x \cdot \frac{25}{36} = \frac{5}{12}$

66. $p \cdot \frac{4}{5} = \frac{8}{15}$

67. $n \cdot \frac{8}{7} = 360$

68. $y \cdot \frac{5}{6} = 120$

69. ^D_W Without performing the division, explain why $5 \div \frac{1}{7}$ is a greater number than $5 \div \frac{2}{3}$.

70. ^D_W A student incorrectly insists that $\frac{2}{5} \div \frac{3}{4}$ is $\frac{15}{8}$. What mistake is he probably making?

SKILL MAINTENANCE

Divide. [1.3d]

71. $268 \div 4$

72. $268 \div 8$

73. $6842 \div 24$

74. $8765 \div 85$

Solve. [1.4b]

75. $4 \cdot x = 268$

76. $4 + x = 268$

77. $y + 502 = 9001$

78. $56 \cdot 78 = T$

SYNTHESIS

79. If $\frac{1}{3}$ of a number is $\frac{1}{4}$, what is $\frac{1}{2}$ of the number?

80. $\left(\frac{9}{10} \div \frac{2}{5} \div \frac{3}{8}\right)^2$

2.3 ADDITION AND SUBTRACTION; ORDER

Objectives

a Add using fraction notation.

b Subtract using fraction notation.

c Use $<$ or $>$ with fraction notation to write a true sentence.

d Solve equations of the type $x + a = b$ and $a + x = b$, where a and b may be fractions.

a Addition Using Fraction Notation

LIKE DENOMINATORS

Addition using fraction notation corresponds to combining or putting like things together, just as addition with whole numbers does. For example,

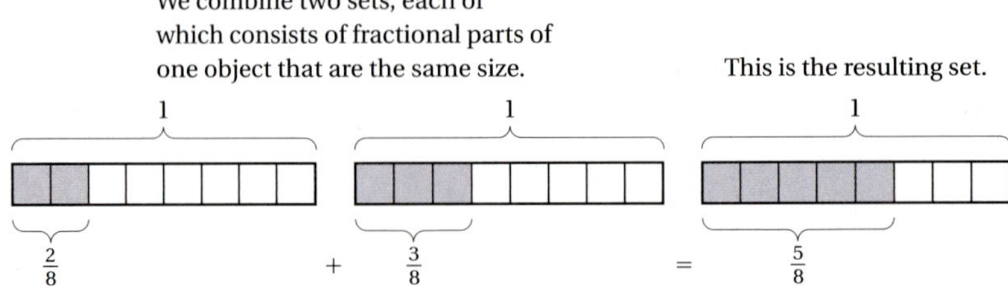

2 eighths + 3 eighths = 5 eighths,

or $\quad 2 \cdot \frac{1}{8} + 3 \cdot \frac{1}{8} = 5 \cdot \frac{1}{8}, \qquad$ or $\quad \frac{2}{8} + \frac{3}{8} = \frac{5}{8}.$

We see that to add when denominators are the same, we add the numerators, keep the denominator, and simplify, if possible.

Do Exercise 1.

1. Find $\frac{1}{5} + \frac{3}{5}$.

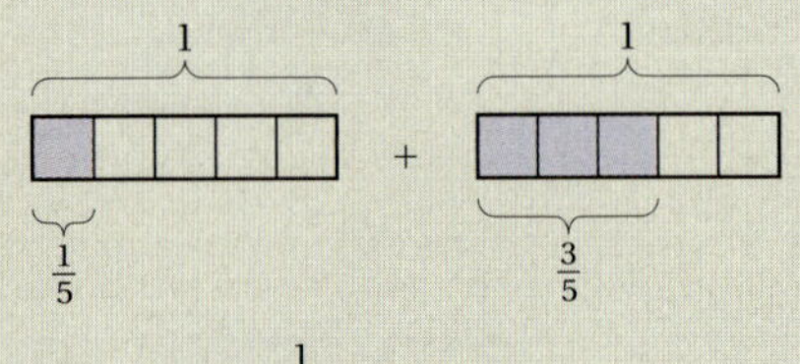

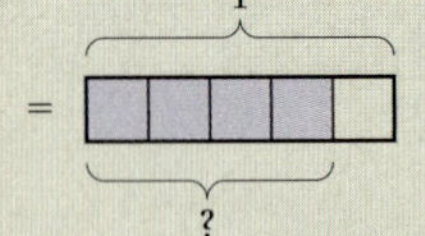

To add when denominators are the same,

a) add the numerators,

b) keep the denominator, and

c) simplify, if possible.

$$\frac{2}{6} + \frac{5}{6} = \frac{2+5}{6} = \frac{7}{6}$$

EXAMPLES Add and simplify.

1. $\frac{2}{4} + \frac{1}{4} = \frac{2+1}{4} = \frac{3}{4}$ No simplifying is possible.

2. $\frac{11}{6} + \frac{3}{6} = \frac{11+3}{6} = \frac{14}{6} = \frac{2 \cdot 7}{2 \cdot 3} = \frac{2}{2} \cdot \frac{7}{3} = 1 \cdot \frac{7}{3} = \frac{7}{3}$ Here we simplified.

3. $\frac{3}{12} + \frac{5}{12} = \frac{3+5}{12} = \frac{8}{12} = \frac{4 \cdot 2}{4 \cdot 3} = \frac{4}{4} \cdot \frac{2}{3} = 1 \cdot \frac{2}{3} = \frac{2}{3}$

Do Exercises 2–4.

Add and simplify.

2. $\frac{1}{3} + \frac{2}{3}$

3. $\frac{5}{12} + \frac{1}{12}$

4. $\frac{9}{16} + \frac{3}{16}$

DIFFERENT DENOMINATORS

What do we do when denominators are different? We can find a common denominator by multiplying by 1. Consider adding $\frac{1}{6}$ and $\frac{3}{4}$. There are many common denominators that can be obtained. Let's look at two possibilities.

Answers on page A-4

A. $\frac{1}{6} + \frac{3}{4} = \frac{1}{6} \cdot 1 + \frac{3}{4} \cdot 1$

$= \frac{1}{6} \cdot \frac{4}{4} + \frac{3}{4} \cdot \frac{6}{6}$

$= \frac{4}{24} + \frac{18}{24}$

$= \frac{22}{24}$

$= \frac{11}{12}$

B. $\frac{1}{6} + \frac{3}{4} = \frac{1}{6} \cdot 1 + \frac{3}{4} \cdot 1$

$= \frac{1}{6} \cdot \frac{2}{2} + \frac{3}{4} \cdot \frac{3}{3}$

$= \frac{2}{12} + \frac{9}{12}$

$= \frac{11}{12}$

We had to simplify in (A). We didn't have to simplify in (B). In (B), we used the least common multiple of the denominators, 12. That number is called the **least common denominator,** or **LCD.**

To add when denominators are different:

a) Find the least common multiple of the denominators. That number is the least common denominator, LCD.

b) Multiply by 1, using an appropriate notation, n/n, to express each number in terms of the LCD.

c) Add the numerators, keeping the same denominator.

d) Simplify, if possible.

EXAMPLE 4 Add: $\frac{3}{4} + \frac{1}{8}$.

The LCD is 8. 4 is a factor of 8 so the LCM of 4 and 8 is 8.

$\frac{3}{4} + \frac{1}{8} = \frac{3}{4} \cdot 1 + \frac{1}{8}$ ← This fraction already has the LCD as its denominator.

$= \frac{3}{4} \cdot \frac{2}{2} + \frac{1}{8}$ *Think*: $4 \times \square = 8$. The answer is 2, so we multiply by 1, using $\frac{2}{2}$.

$= \frac{6}{8} + \frac{1}{8} = \frac{7}{8}$

Do Exercise 5.

EXAMPLE 5 Add: $\frac{1}{9} + \frac{5}{6}$.

The LCD is 18. $9 = 3 \cdot 3$ and $6 = 2 \cdot 3$, so the LCM of 9 and 6 is $2 \cdot 3 \cdot 3$, or 18.

$\frac{1}{9} + \frac{5}{6} = \frac{1}{9} \cdot 1 + \frac{5}{6} \cdot 1 = \frac{1}{9} \cdot \frac{2}{2} + \frac{5}{6} \cdot \frac{3}{3}$

Think: $6 \times \square = 18$. The answer is 3, so we multiply by 1 using $\frac{3}{3}$.

Think: $9 \times \square = 18$. The answer is 2, so we multiply by 1 using $\frac{2}{2}$.

$= \frac{2}{18} + \frac{15}{18} = \frac{17}{18}$

Do Exercise 6.

5. Add. (Find the least common denominator.)

$\frac{2}{3} + \frac{1}{6}$

6. Add: $\frac{3}{8} + \frac{5}{6}$.

Answers on page A-4

7. Add: $\frac{1}{6} + \frac{7}{18}$.

8. Add: $\frac{4}{10} + \frac{1}{100} + \frac{3}{1000}$.

Add.

9. $\frac{7}{10} + \frac{2}{21} + \frac{1}{7}$

10. $\frac{7}{18} + \frac{5}{24} + \frac{11}{36}$

Answers on page A-4

EXAMPLE 6 Add: $\frac{5}{9} + \frac{11}{18}$.

The LCD is 18.

$$\frac{5}{9} + \frac{11}{18} = \frac{5}{9} \cdot \frac{2}{2} + \frac{11}{18} = \frac{10}{18} + \frac{11}{18}$$

$$= \frac{21}{18}$$

$$= \frac{7}{6}$$

We may still have to simplify, but it is usually easier if we have used the LCD.

Do Exercise 7.

EXAMPLE 7 Add: $\frac{1}{10} + \frac{3}{100} + \frac{7}{1000}$.

Since 10 and 100 are factors of 1000, the LCD is 1000. Then

$$\frac{1}{10} + \frac{3}{100} + \frac{7}{1000} = \frac{1}{10} \cdot \frac{100}{100} + \frac{3}{100} \cdot \frac{10}{10} + \frac{7}{1000}$$

$$= \frac{100}{1000} + \frac{30}{1000} + \frac{7}{1000} = \frac{137}{1000}.$$

Do Exercise 8.

When denominators are large, we most often use the prime factorization of each denominator. This is shown in Example 8. Using the prime factorization in this manner is similar to what is done in algebra.

EXAMPLE 8 Add: $\frac{13}{70} + \frac{11}{21} + \frac{6}{15}$.

We have

$$\frac{13}{70} + \frac{11}{21} + \frac{6}{15} = \frac{13}{2 \cdot 5 \cdot 7} + \frac{11}{3 \cdot 7} + \frac{6}{3 \cdot 5}.$$ Factoring denominators

The LCD is $2 \cdot 3 \cdot 5 \cdot 7$, or 210. Then

$$\frac{13}{70} + \frac{11}{21} + \frac{6}{15} = \frac{13}{2 \cdot 5 \cdot 7} \cdot \frac{3}{3} + \frac{11}{3 \cdot 7} \cdot \frac{2 \cdot 5}{2 \cdot 5} + \frac{6}{3 \cdot 5} \cdot \frac{7 \cdot 2}{7 \cdot 2}$$

The LCD of 70, 21, and 15 is $2 \cdot 3 \cdot 5 \cdot 7$. In each case, think of which factors are needed to get the LCD. Then multiply by 1 to obtain the LCD in each denominator.

$$= \frac{13 \cdot 3}{2 \cdot 5 \cdot 7 \cdot 3} + \frac{11 \cdot 2 \cdot 5}{3 \cdot 7 \cdot 2 \cdot 5} + \frac{6 \cdot 7 \cdot 2}{3 \cdot 5 \cdot 7 \cdot 2}$$

$$= \frac{39}{3 \cdot 5 \cdot 7 \cdot 2} + \frac{110}{3 \cdot 5 \cdot 7 \cdot 2} + \frac{84}{3 \cdot 5 \cdot 7 \cdot 2}$$

$$= \frac{233}{3 \cdot 5 \cdot 7 \cdot 2}$$

$$= \frac{233}{210}.$$ We left 210 factored until we knew we could not simplify.

Do Exercises 9 and 10.

b Subtraction Using Fraction Notation

LIKE DENOMINATORS

We can consider the difference $\frac{4}{8} - \frac{3}{8}$ as we did before, as either "take away" or "missing addend." Let's consider "take away."

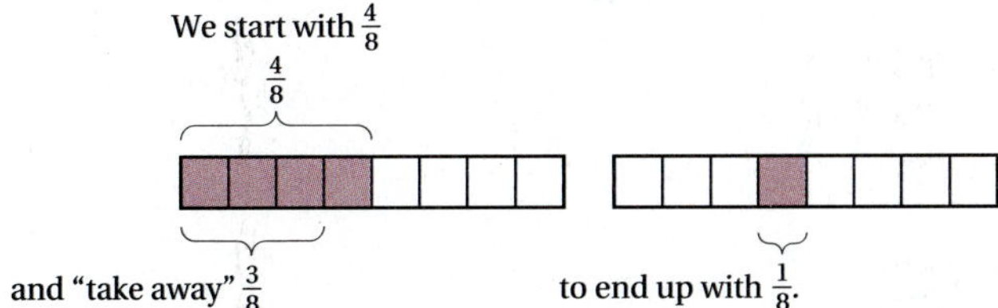

We start with $\frac{4}{8}$
$\frac{4}{8}$
and "take away" $\frac{3}{8}$
to end up with $\frac{1}{8}$.

We start with 4 eighths and take away 3 eighths:

4 eighths − 3 eighths = 1 eighth,

or $4 \cdot \frac{1}{8} - 3 \cdot \frac{1}{8} = \frac{1}{8}$, or $\frac{4}{8} - \frac{3}{8} = \frac{1}{8}$.

> To subtract when denominators are the same,
> **a)** subtract the numerators,
> **b)** keep the denominator, and
> **c)** simplify, if possible.
>
> $$\frac{7}{10} - \frac{4}{10} = \frac{7-4}{10} = \frac{3}{10}$$

EXAMPLES Subtract and simplify.

9. $\frac{7}{10} - \frac{3}{10} = \frac{7-3}{10} = \frac{4}{10} = \frac{2 \cdot 2}{5 \cdot 2} = \frac{2}{5} \cdot \frac{2}{2} = \frac{2}{5} \cdot 1 = \frac{2}{5}$

10. $\frac{8}{9} - \frac{2}{9} = \frac{8-2}{9} = \frac{6}{9} = \frac{2 \cdot 3}{3 \cdot 3} = \frac{2}{3} \cdot \frac{3}{3} = \frac{2}{3} \cdot 1 = \frac{2}{3}$

11. $\frac{32}{12} - \frac{25}{12} = \frac{32-25}{12} = \frac{7}{12}$

Do Exercises 11–13.

DIFFERENT DENOMINATORS

> To subtract when denominators are different:
> **a)** Find the least common multiple of the denominators. That number is the least common denominator, LCD.
> **b)** Multiply by 1, using an appropriate notation, n/n, to express each number in terms of the LCD.
> **c)** Subtract the numerators, keeping the same denominator.
> **d)** Simplify, if possible.

Subtract and simplify.

11. $\frac{7}{8} - \frac{3}{8}$

12. $\frac{10}{16} - \frac{4}{16}$

13. $\frac{8}{10} - \frac{3}{10}$

Answers on page A-4

Subtract.

14. $\frac{5}{6} - \frac{1}{9}$

15. $\frac{4}{5} - \frac{3}{10}$

16. Subtract: $\frac{11}{28} - \frac{5}{16}$.

17. Use < or > for □ to write a true sentence:

$\frac{3}{8} \square \frac{5}{8}$.

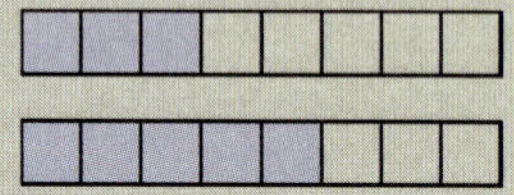

18. Use < or > for □ to write a true sentence:

$\frac{7}{10} \square \frac{6}{10}$.

Answers on page A-4

EXAMPLE 12 Subtract: $\frac{5}{6} - \frac{7}{12}$.

Since 12 is a multiple of 6, the LCM of 6 and 12 is 12. The LCD is 12.

$$\frac{5}{6} - \frac{7}{12} = \frac{5}{6} \cdot \frac{2}{2} - \frac{7}{12}$$
$$= \frac{10}{12} - \frac{7}{12} = \frac{10 - 7}{12} = \frac{3}{12}$$
$$= \frac{3 \cdot 1}{3 \cdot 4} = \frac{3}{3} \cdot \frac{1}{4} = \frac{1}{4}$$

Do Exercises 14 and 15.

EXAMPLE 13 Subtract: $\frac{17}{24} - \frac{4}{15}$.

We have

$$\frac{17}{24} - \frac{4}{15} = \frac{17}{3 \cdot 2 \cdot 2 \cdot 2} - \frac{4}{5 \cdot 3}.$$

The LCD is $3 \cdot 2 \cdot 2 \cdot 2 \cdot 5$, or 120. Then

$$\frac{17}{24} - \frac{4}{15} = \frac{17}{3 \cdot 2 \cdot 2 \cdot 2} \cdot \frac{5}{5} - \frac{4}{5 \cdot 3} \cdot \frac{2 \cdot 2 \cdot 2}{2 \cdot 2 \cdot 2}$$

The LCD of 24 and 15 is $2 \cdot 2 \cdot 2 \cdot 3 \cdot 5$. In each case, we multiply by 1 to obtain the LCD.

$$= \frac{17 \cdot 5}{3 \cdot 2 \cdot 2 \cdot 2 \cdot 5} - \frac{4 \cdot 2 \cdot 2 \cdot 2}{5 \cdot 3 \cdot 2 \cdot 2 \cdot 2}$$
$$= \frac{85}{120} - \frac{32}{120} = \frac{53}{120}.$$

Do Exercise 16.

C Order

We see from this figure that $\frac{4}{5} > \frac{3}{5}$, and $\frac{3}{5} < \frac{4}{5}$. That is, $\frac{4}{5}$ is greater than $\frac{3}{5}$, and $\frac{3}{5}$ is less than $\frac{4}{5}$.

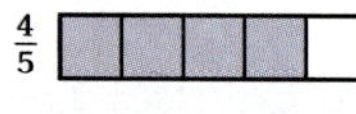

To determine which of two numbers is greater when there is a common denominator, compare the numerators:

$$\frac{4}{5}, \quad \frac{3}{5}, \qquad 4 > 3 \qquad \frac{4}{5} > \frac{3}{5}.$$

Do Exercises 17 and 18.

When denominators are different, we cannot compare numerators. We multiply by 1 to make the denominators the same.

EXAMPLE 14 Use $<$ or $>$ for $\square$ to write a true sentence:

$$\frac{2}{5} \square \frac{3}{4}.$$

We have

$$\frac{2}{5} \cdot \frac{4}{4} = \frac{8}{20};$$ We multiply by 1 using $\frac{4}{4}$ to get the LCD.

$$\frac{3}{4} \cdot \frac{5}{5} = \frac{15}{20}.$$ We multiply by 1 using $\frac{5}{5}$ to get the LCD.

Now that the denominators are the same, 20, we can compare the numerators. Since $8 < 15$, it follows that $\frac{8}{20} < \frac{15}{20}$, so

$$\frac{2}{5} < \frac{3}{4}.$$

EXAMPLE 15 Use $<$ or $>$ for $\square$ to write a true sentence: $\frac{9}{10} \square \frac{89}{100}$.

The LCD is 100.

$$\frac{9}{10} \cdot \frac{10}{10} = \frac{90}{100}$$ We multiply by $\frac{10}{10}$ to get the LCD.

Since $90 > 89$, it follows that $\frac{90}{100} > \frac{89}{100}$, so

$$\frac{9}{10} > \frac{89}{100}.$$

Do Exercises 19–21.

d Solving Equations

Now let's solve equations of the form $x + a = b$ or $a + x = b$, where a and b may be fractions. Proceeding as we have before, we subtract a on both sides of the equation.

EXAMPLE 16 Solve: $x + \frac{1}{4} = \frac{3}{5}$.

$$x + \frac{1}{4} - \frac{1}{4} = \frac{3}{5} - \frac{1}{4}$$ Subtracting $\frac{1}{4}$ on both sides

$$x + 0 = \frac{3}{5} \cdot \frac{4}{4} - \frac{1}{4} \cdot \frac{5}{5}$$ The LCD is 20. We multiply by 1 to get the LCD.

$$x = \frac{12}{20} - \frac{5}{20} = \frac{7}{20}$$

Do Exercises 22 and 23.

Use $<$ or $>$ for $\square$ to write a true sentence.

19. $\frac{2}{3} \square \frac{5}{8}$

20. $\frac{3}{4} \square \frac{8}{12}$

21. $\frac{5}{6} \square \frac{7}{8}$

Solve.

22. $x + \frac{2}{3} = \frac{5}{6}$

23. $\frac{3}{5} + t = \frac{7}{8}$

Answers on page A-4

2.3 EXERCISE SET

For Extra Help

Digital Video Tutor CD 2 Videotape 3 | InterAct Math | Math Tutor Center | MathXL | MyMathLab.com

a Add and simplify.

1. $\frac{7}{8}+\frac{1}{8}$

2. $\frac{2}{5}+\frac{3}{5}$

3. $\frac{1}{8}+\frac{5}{8}$

4. $\frac{3}{10}+\frac{3}{10}$

5. $\frac{2}{3}+\frac{5}{6}$

6. $\frac{5}{6}+\frac{1}{9}$

7. $\frac{1}{8}+\frac{1}{6}$

8. $\frac{1}{6}+\frac{3}{4}$

9. $\frac{4}{5}+\frac{7}{10}$

10. $\frac{3}{4}+\frac{1}{12}$

11. $\frac{5}{12}+\frac{3}{8}$

12. $\frac{7}{8}+\frac{1}{16}$

13. $\frac{3}{20}+\frac{3}{4}$

14. $\frac{2}{15}+\frac{2}{5}$

15. $\frac{5}{6}+\frac{7}{9}$

16. $\frac{5}{8}+\frac{5}{6}$

17. $\frac{3}{10}+\frac{1}{100}$

18. $\frac{9}{10}+\frac{3}{100}$

19. $\frac{5}{12}+\frac{4}{15}$

20. $\frac{3}{16}+\frac{1}{12}$

21. $\frac{9}{10}+\frac{99}{100}$

22. $\frac{3}{10}+\frac{27}{100}$

23. $\frac{7}{8}+\frac{0}{1}$

24. $\frac{0}{1}+\frac{5}{6}$

25. $\frac{3}{8}+\frac{1}{6}$

26. $\frac{5}{8}+\frac{1}{6}$

27. $\frac{5}{12}+\frac{7}{24}$

28. $\frac{1}{18}+\frac{7}{12}$

29. $\frac{3}{16}+\frac{5}{16}+\frac{4}{16}$

30. $\frac{3}{8}+\frac{1}{8}+\frac{2}{8}$

31. $\frac{8}{10}+\frac{7}{100}+\frac{4}{1000}$

32. $\frac{1}{10}+\frac{2}{100}+\frac{3}{1000}$

33. $\frac{3}{8}+\frac{5}{12}+\frac{8}{15}$

34. $\frac{1}{2}+\frac{3}{8}+\frac{1}{4}$

35. $\frac{15}{24}+\frac{7}{36}+\frac{91}{48}$

36. $\frac{5}{7}+\frac{25}{52}+\frac{7}{4}$

b Subtract and simplify.

37. $\frac{5}{6} - \frac{1}{6}$

38. $\frac{5}{8} - \frac{3}{8}$

39. $\frac{11}{12} - \frac{2}{12}$

40. $\frac{17}{18} - \frac{11}{18}$

41. $\frac{3}{4} - \frac{1}{8}$

42. $\frac{2}{3} - \frac{1}{9}$

43. $\frac{1}{8} - \frac{1}{12}$

44. $\frac{1}{6} - \frac{1}{8}$

45. $\frac{4}{3} - \frac{5}{6}$

46. $\frac{7}{8} - \frac{1}{16}$

47. $\frac{3}{4} - \frac{3}{28}$

48. $\frac{2}{5} - \frac{2}{15}$

49. $\frac{3}{4} - \frac{3}{20}$

50. $\frac{5}{6} - \frac{1}{2}$

51. $\frac{3}{4} - \frac{1}{20}$

52. $\frac{3}{4} - \frac{4}{16}$

53. $\frac{5}{12} - \frac{2}{15}$

54. $\frac{9}{10} - \frac{11}{16}$

55. $\frac{6}{10} - \frac{7}{100}$

56. $\frac{9}{10} - \frac{3}{100}$

57. $\frac{7}{15} - \frac{3}{25}$

58. $\frac{18}{25} - \frac{4}{35}$

59. $\frac{99}{100} - \frac{9}{10}$

60. $\frac{78}{100} - \frac{11}{20}$

61. $\frac{2}{3} - \frac{1}{8}$

62. $\frac{3}{4} - \frac{1}{2}$

63. $\frac{3}{5} - \frac{1}{2}$

64. $\frac{5}{6} - \frac{2}{3}$

65. $\frac{5}{12} - \frac{3}{8}$

66. $\frac{7}{12} - \frac{2}{9}$

67. $\frac{7}{8} - \frac{1}{16}$

68. $\frac{5}{12} - \frac{5}{16}$

69. $\frac{17}{25} - \frac{4}{15}$

70. $\frac{11}{18} - \frac{7}{24}$

71. $\frac{23}{25} - \frac{112}{150}$

72. $\frac{89}{90} - \frac{53}{120}$

c Use $<$ or $>$ for $\square$ to write a true sentence.

73. $\frac{5}{8} \square \frac{6}{8}$

74. $\frac{7}{9} \square \frac{5}{9}$

75. $\frac{1}{3} \square \frac{1}{4}$

76. $\frac{1}{8} \square \frac{1}{6}$

77. $\frac{2}{3} \square \frac{5}{7}$

78. $\frac{3}{5} \square \frac{4}{7}$

79. $\frac{4}{5} \square \frac{5}{6}$

80. $\frac{3}{2} \square \frac{7}{5}$

81. $\frac{19}{20} \square \frac{4}{5}$

82. $\frac{5}{6} \square \frac{13}{16}$

83. $\frac{19}{20} \square \frac{9}{10}$

84. $\frac{3}{4} \square \frac{11}{15}$

85. $\frac{31}{21} \square \frac{41}{13}$

86. $\frac{12}{7} \square \frac{132}{49}$

d Solve.

87. $x + \frac{1}{30} = \frac{1}{10}$

88. $y + \frac{9}{12} = \frac{11}{12}$

89. $\frac{2}{3} + t = \frac{4}{5}$

90. $\frac{2}{3} + p = \frac{7}{8}$

91. $x + \frac{1}{3} = \frac{5}{6}$

92. $m + \frac{5}{6} = \frac{9}{10}$

93. **DW** A fellow student made the following error:

$$\frac{8}{5} - \frac{8}{2} = \frac{8}{3}.$$

Find at least two ways to convince him of the mistake.

94. **DW** To add numbers with different denominators, a student consistently uses the product of the denominators as a common denominator. Is this correct? Why or why not?

SKILL MAINTENANCE

Simplify. [2.1b, e]

95. $\frac{38}{38}$

96. $\frac{38}{0}$

97. $\frac{124}{0}$

98. $\frac{124}{31}$

Holiday Expenditures. The chart at right shows average expenditures per person of consumers during the Christmas holidays of 1999 and 2000. Use these data for Exercises 99–104. [1.5a]

HOLIDAY EXPENDITURES	1999	2000
Gifts	\$1088	\$1161
Entertainment	188	197
Travel	151	154
Decorations / cards	77	88
Other holiday expenses	54	84
Total	?	?

Source: *2000 American Express Retail Index*

99. How much more was spent on gifts in 2000 than in 1999?

100. How much more was spent on decorations and cards in 2000 than in 1999?

101. How much more was spent on travel in 2000 than in 1999?

102. How much more was spent on entertainment in 2000 than in 1999?

103. What was the total expenditure in 1999?

104. What was the total expenditure in 2000?

SYNTHESIS

Simplify. Use the rules for order of operations given in Section 1.6.

105. $\frac{7}{8} - \frac{1}{10} \times \frac{5}{6}$

106. $\frac{2}{5} + \frac{1}{6} \div 3$

107. $\left(\frac{2}{3}\right)^2 + \left(\frac{3}{4}\right)^2$

108. $5 \times \frac{3}{7} - \frac{1}{7} \times \frac{4}{5}$

109. A mountain climber, beginning at sea level, climbs $\frac{3}{5}$ km, descends $\frac{1}{4}$ km, climbs $\frac{1}{3}$ km, and then descends $\frac{1}{7}$ km. At what elevation does the climber finish?

2.4 MIXED NUMERALS

a Mixed Numerals

The following figure illustrates the use of a **mixed numeral** in daily life. The bolt shown is $2\frac{3}{8}$ in. long. The length is given as a whole-number part, 2, and a fractional part less than 1, $\frac{3}{8}$. We can represent the measurement of the bolt with fraction notation as $\frac{19}{8}$, but the meaning or interpretation of such a symbol is less understandable or visual.

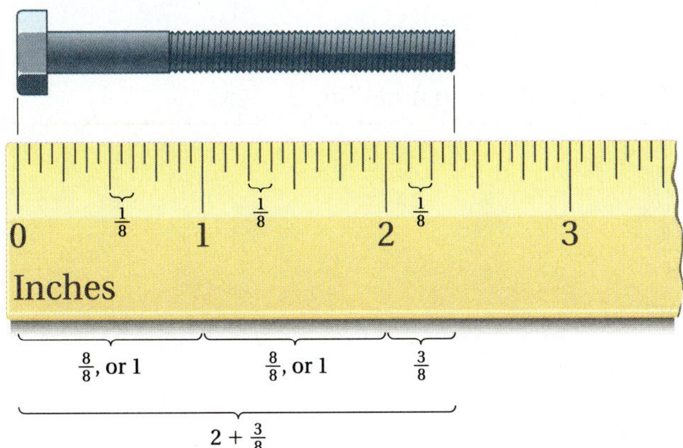

A mixed numeral $2\frac{3}{8}$ represents a sum:

$$2\frac{3}{8} \quad \text{means} \quad 2 + \frac{3}{8}$$

This is a whole number. This is a fraction less than 1.

EXAMPLES Convert to a mixed numeral.

1. $7 + \frac{2}{5} = 7\frac{2}{5}$

2. $4 + \frac{3}{10} = 4\frac{3}{10}$

Do Exercises 1–4. (Exercises 3 and 4 are on the following page.)

The notation $2\frac{3}{4}$ has a plus sign left out. To aid in understanding, we sometimes write the missing plus sign.

EXAMPLES Convert to fraction notation.

3. $2\frac{3}{4} = 2 + \frac{3}{4}$ Inserting the missing plus sign

$= \frac{2}{1} + \frac{3}{4}$ $2 = \frac{2}{1}$

$= \frac{2}{1} \cdot \frac{4}{4} + \frac{3}{4}$ Finding a common denominator

$= \frac{8}{4} + \frac{3}{4} = \frac{11}{4}$

4. $4\frac{3}{10} = 4 + \frac{3}{10} = \frac{4}{1} + \frac{3}{10} = \frac{4}{1} \cdot \frac{10}{10} + \frac{3}{10} = \frac{40}{10} + \frac{3}{10} = \frac{43}{10}$

Objectives

a Convert between mixed numerals and fraction notation.

b Add using mixed numerals.

c Subtract using mixed numerals.

d Multiply using mixed numerals.

e Divide using mixed numerals.

Convert to a mixed numeral.

1. $1 + \frac{2}{3} = \square\frac{\square}{\square}$

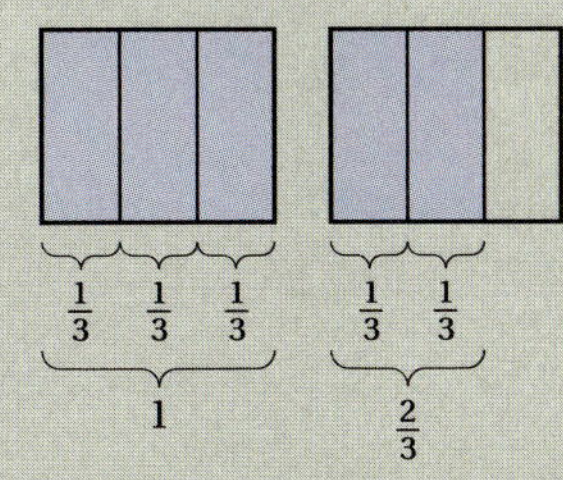

2. $2 + \frac{3}{4} = \square\frac{\square}{\square}$

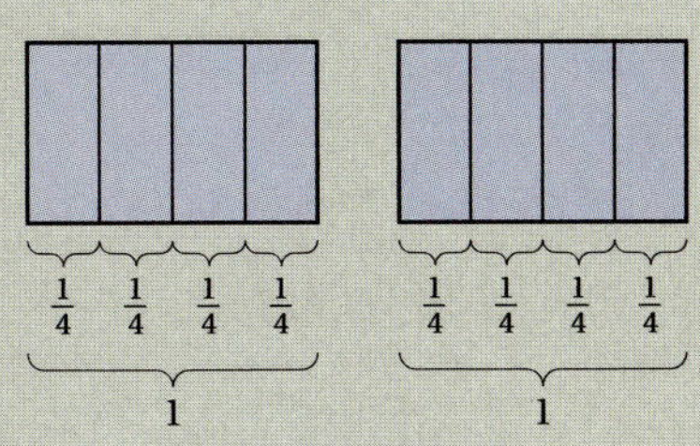

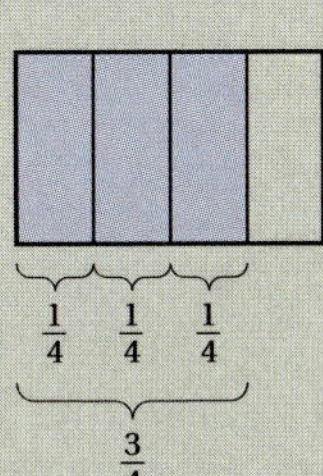

Answers on page A-5

Convert to a mixed numeral.

3. $8 + \frac{3}{4}$

4. $12 + \frac{2}{3}$

Convert to fraction notation.

5. $4\frac{2}{5}$

6. $6\frac{1}{10}$

Convert to fraction notation. Use the faster method.

7. $4\frac{5}{6}$

8. $9\frac{1}{4}$

9. $20\frac{2}{3}$

Answers on page A-5

Do Exercises 5 and 6.

Let's now consider a faster method for converting a mixed numeral to fraction notation.

To convert from a mixed numeral to fraction notation:

(a) Multiply the whole number by the denominator: $4 \cdot 10 = 40$.

(b) Add the result to the numerator: $40 + 3 = 43$.

(c) Keep the denominator.

$$4\frac{3}{10} = \frac{43}{10}$$

EXAMPLES Convert to fraction notation.

5. $6\frac{2}{3} = \frac{20}{3}$ $6 \cdot 3 = 18,\ 18 + 2 = 20$

6. $8\frac{2}{9} = \frac{74}{9}$

7. $10\frac{7}{8} = \frac{87}{8}$

Do Exercises 7–9.

WRITING MIXED NUMERALS

We can find a mixed numeral for $\frac{5}{3}$ as follows:

$$\frac{5}{3} = \frac{3}{3} + \frac{2}{3} = 1 + \frac{2}{3} = 1\frac{2}{3}.$$

In terms of objects, we can think of $\frac{5}{3}$ as $\frac{3}{3}$, or 1, plus $\frac{2}{3}$, as shown below.

$\frac{5}{3} = \frac{3}{3}$, or 1 $+$ $\frac{2}{3}$

Fraction symbols like $\frac{5}{3}$ also indicate division; $\frac{5}{3}$ means $5 \div 3$. Let's divide the numerator by the denominator.

$$\begin{array}{r} 1 \\ 3\overline{)5} \\ \underline{3} \\ 2 \end{array} \leftarrow 2 \div 3 = \frac{2}{3}$$

Thus, $\frac{5}{3} = 1\frac{2}{3}$.

To convert from fraction notation to a mixed numeral, divide.

$$\frac{13}{5} \qquad \begin{array}{r} 2 \\ 5\overline{)13} \\ \underline{10} \\ 3 \end{array} \qquad 2\frac{3}{5}$$

2 — The quotient; 5 — The divisor; 3 — The remainder

EXAMPLES Convert to a mixed numeral.

8. $\frac{69}{10}$

$$10\overline{)69} \quad \text{quotient } 6, \quad 60, \quad \text{remainder } 9$$

$$\frac{69}{10} = 6\frac{9}{10}$$

9. $\frac{122}{8}$

$$8\overline{)122} \quad \text{quotient } 15, \quad 80, \quad 42, \quad 40, \quad \text{remainder } 2$$

$$\frac{122}{8} = 15\frac{2}{8} = 15\frac{1}{4}$$

Do Exercises 10–12.

b Addition Using Mixed Numerals

To find the sum $1\frac{5}{8} + 3\frac{1}{8}$, we first add the fractions. Then we add the whole numbers.

$$\begin{array}{r} 1\frac{5}{8} \\ +\ 3\frac{1}{8} \\ \hline \frac{6}{8} \end{array} = \begin{array}{r} 1\frac{5}{8} \\ +\ 3\frac{1}{8} \\ \hline 4\frac{6}{8} \end{array} = 4\frac{3}{4}$$

Add the fractions. Add the whole numbers. Simplifying

Do Exercise 13.

EXAMPLE 10 Add: $5\frac{2}{3} + 3\frac{5}{6}$. Write a mixed numeral for the answer.

The LCD is 6.

$$\begin{array}{r} 5\frac{2}{3}\cdot\frac{2}{2} \\ +\ 3\frac{5}{6} \\ \hline \end{array} = \begin{array}{r} 5\frac{4}{6} \\ +\ 3\frac{5}{6} \\ \hline 8\frac{9}{6} \end{array}$$

$$8\frac{9}{6} = 8 + \frac{9}{6} = 8 + 1\frac{1}{2} = 9\frac{1}{2}$$

To find a mixed numeral for $\frac{9}{6}$, we divide:

$$6\overline{)9} \quad \text{quotient } 1, \quad 6, \quad \text{remainder } 3 \qquad \frac{9}{6} = 1\frac{3}{6} = 1\frac{1}{2}$$

$\frac{19}{2}$ is also a correct answer, but it is not a mixed numeral, which is what we are working with in Sections 3.4, 3.5, and 3.6.

Do Exercise 14.

Convert to a mixed numeral.

10. $\frac{7}{3}$

11. $\frac{11}{10}$

12. $\frac{110}{6}$

13. Add.

$$\begin{array}{r} 2\frac{3}{10} \\ +\ 5\frac{1}{10} \\ \hline \end{array}$$

14. Add.

$$\begin{array}{r} 8\frac{2}{5} \\ +\ 3\frac{7}{10} \\ \hline \end{array}$$

Answers on page A-5

15. Add.

$$\begin{array}{r} 9\frac{3}{4} \\ +\,3\frac{5}{6} \\ \hline \end{array}$$

Subtract.

16. $$\begin{array}{r} 10\frac{7}{8} \\ -\,9\frac{3}{8} \\ \hline \end{array}$$

17. $$\begin{array}{r} 8\frac{2}{3} \\ -\,5\frac{1}{2} \\ \hline \end{array}$$

18. Subtract.

$$\begin{array}{r} 8\frac{1}{9} \\ -\,4\frac{5}{6} \\ \hline \end{array}$$

Answers on page A-5

EXAMPLE 11 Add: $10\frac{5}{6} + 7\frac{3}{8}$.

The LCD is 24.

$$\begin{array}{rcr} 10\,\frac{5}{6}\cdot\frac{4}{4} & = & 10\frac{20}{24} \\ +\,7\,\frac{3}{8}\cdot\frac{3}{3} & = & +\,7\frac{9}{24} \\ \hline & & 17\frac{29}{24} = 18\frac{5}{24} \end{array}$$

Do Exercise 15.

C Subtraction Using Mixed Numerals

EXAMPLE 12 Subtract: $7\frac{3}{4} - 2\frac{1}{4}$.

$$\begin{array}{rcr} 7\frac{3}{4} & = & 7\frac{3}{4} \\ -\,2\frac{1}{4} & = & -\,2\frac{1}{4} \\ \hline \frac{2}{4} & & 5\frac{2}{4} = 5\frac{1}{2} \end{array}$$

Subtract the fractions. Subtract the whole numbers. Simplifying

EXAMPLE 13 Subtract: $9\frac{4}{5} - 3\frac{1}{2}$.

The LCD is 10.

$$\begin{array}{rcr} 9\,\frac{4}{5}\cdot\frac{2}{2} & = & 9\frac{8}{10} \\ -\,3\,\frac{1}{2}\cdot\frac{5}{5} & = & -\,3\frac{5}{10} \\ \hline & & 6\frac{3}{10} \end{array}$$

Do Exercises 16 and 17.

EXAMPLE 14 Subtract: $7\frac{1}{6} - 2\frac{1}{4}$.

The LCD is 12.

$$\begin{array}{rcr} 7\,\frac{1}{6}\cdot\frac{2}{2} & = & 7\frac{2}{12} \\ -\,2\,\frac{1}{4}\cdot\frac{3}{3} & = & -\,2\frac{3}{12} \\ \hline \end{array}$$

We cannot subtract $\frac{3}{12}$ from $\frac{2}{12}$. We borrow 1, or $\frac{12}{12}$, from 7: $7\frac{2}{12} = 6 + 1 + \frac{2}{12} = 6 + \frac{12}{12} + \frac{2}{12} = 6\frac{14}{12}$.

We can write this as

$$\begin{array}{rcr} 7\frac{2}{12} & = & 6\frac{14}{12} \\ -\,2\frac{3}{12} & = & -\,2\frac{3}{12} \\ \hline & & 4\frac{11}{12} \end{array}$$

Do Exercise 18.

EXAMPLE 15 Subtract: $12 - 9\frac{3}{8}$.

$$\begin{array}{rcr} 12 & = & 11\frac{8}{8} \\ -\ 9\frac{3}{8} & = & -\ 9\frac{3}{8} \\ \hline & & 2\frac{5}{8} \end{array}$$

$12 = 11 + 1 = 11 + \frac{8}{8} = 11\frac{8}{8}$

Do Exercise 19.

d Multiplication Using Mixed Numerals

Carrying out addition and subtraction with mixed numerals is usually easier if the numbers are left as mixed numerals. With multiplication and division, however, it is easier to convert the numbers first to fraction notation.

MULTIPLICATION USING MIXED NUMERALS

To multiply using mixed numerals, first convert to fraction notation. Then multiply with fraction notation and convert the answer back to a mixed numeral, if appropriate.

EXAMPLE 16 Multiply: $6 \cdot 2\frac{1}{2}$.

$$6 \cdot 2\frac{1}{2} = \frac{6}{1} \cdot \frac{5}{2} = \frac{6 \cdot 5}{1 \cdot 2} = \frac{2 \cdot 3 \cdot 5}{2 \cdot 1} = \frac{2}{2} \cdot \frac{3 \cdot 5}{1} = 15$$

Note that fraction notation is needed to carry out the multiplication.

Do Exercise 20.

EXAMPLE 17 Multiply: $3\frac{1}{2} \cdot \frac{3}{4}$.

$$3\frac{1}{2} \cdot \frac{3}{4} = \frac{7}{2} \cdot \frac{3}{4} = \frac{21}{8} = 2\frac{5}{8}$$

Here we write fraction notation.

Do Exercise 21.

19. Subtract.

$$\begin{array}{r} 5 \\ -\ 1\frac{1}{3} \\ \hline \end{array}$$

20. Multiply: $6 \cdot 3\frac{1}{3}$.

21. Multiply: $2\frac{1}{2} \cdot \frac{3}{4}$.

Answers on page A-5

Study Tips

FORMING A STUDY GROUP

Consider forming a study group with some of your fellow students. Exchange e-mail addresses, telephone numbers, and schedules so that you can coordinate study time for homework and tests.

22. Multiply: $2 \cdot 6\frac{2}{5}$.

23. Multiply: $3\frac{1}{3} \cdot 2\frac{1}{2}$.

24. Divide: $84 \div 5\frac{1}{4}$.

Answers on page A-5

EXAMPLE 18 Multiply: $8 \cdot 4\frac{2}{3}$.

$$8 \cdot 4\frac{2}{3} = \frac{8}{1} \cdot \frac{14}{3} = \frac{112}{3} = 37\frac{1}{3}$$

Do Exercise 22.

EXAMPLE 19 Multiply: $2\frac{1}{4} \cdot 3\frac{2}{5}$.

$$2\frac{1}{4} \cdot 3\frac{2}{5} = \frac{9}{4} \cdot \frac{17}{5} = \frac{153}{20} = 7\frac{13}{20}$$

CAUTION!

$2\frac{1}{4} \cdot 3\frac{2}{5} \neq 6\frac{2}{20}$. A common error is to multiply the whole numbers and then the fractions. This does not give the correct answer, $7\frac{13}{20}$, which is found by converting first to fraction notation.

Do Exercise 23.

e Division Using Mixed Numerals

The division $1\frac{1}{2} \div \frac{1}{6}$ is shown here. *Think*: "How many $\frac{1}{6}$'s are in $1\frac{1}{2}$?"

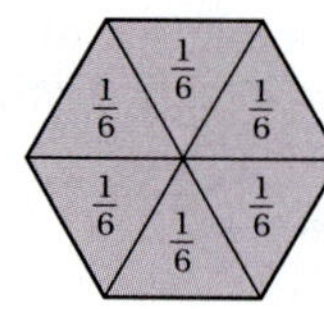

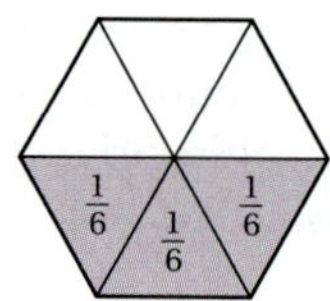

$$1\frac{1}{2} \div \frac{1}{6} = \frac{3}{2} \div \frac{1}{6} = \frac{3}{2} \cdot 6$$

$$= \frac{3 \cdot 6}{2} = \frac{3 \cdot 3 \cdot 2}{2 \cdot 1} = \frac{3 \cdot 3}{1} \cdot \frac{2}{2} = \frac{3 \cdot 3}{1} \cdot 1 = 9$$

DIVISION USING MIXED NUMERALS

To divide using mixed numerals, first write fraction notation. Then divide with fraction notation and convert the answer back to a mixed numeral, if appropriate.

EXAMPLE 20 Divide: $32 \div 3\frac{1}{5}$.

$$32 \div 3\frac{1}{5} = \frac{32}{1} \div \frac{16}{5}$$

$$= \frac{32}{1} \cdot \frac{5}{16} = \frac{32 \cdot 5}{1 \cdot 16} = \frac{2 \cdot 16 \cdot 5}{1 \cdot 16} = \frac{16}{16} \cdot \frac{2 \cdot 5}{1} = 10$$

Remember to multiply by the reciprocal.

Do Exercise 24.

EXAMPLE 21 Divide: $35 \div 4\frac{1}{3}$.

$$35 \div 4\frac{1}{3} = \frac{35}{1} \div \frac{13}{3} = \frac{35}{1} \cdot \frac{3}{13} = \frac{105}{13} = 8\frac{1}{13}$$

Do Exercise 25.

EXAMPLE 22 Divide: $2\frac{1}{3} \div 1\frac{3}{4}$.

$$2\frac{1}{3} \div 1\frac{3}{4} = \frac{7}{3} \div \frac{7}{4} = \frac{7}{3} \cdot \frac{4}{7} = \frac{7 \cdot 4}{7 \cdot 3} = \frac{7}{7} \cdot \frac{4}{3} = 1 \cdot \frac{4}{3} = \frac{4}{3} = 1\frac{1}{3}$$

CAUTION!

The reciprocal of $1\frac{3}{4}$ is *not* $1\frac{4}{3}$!

EXAMPLE 23 Divide: $1\frac{3}{5} \div 3\frac{1}{3}$.

$$1\frac{3}{5} \div 3\frac{1}{3} = \frac{8}{5} \div \frac{10}{3} = \frac{8}{5} \cdot \frac{3}{10} = \frac{2 \cdot 4 \cdot 3}{5 \cdot 2 \cdot 5} = \frac{2}{2} \cdot \frac{4 \cdot 3}{5 \cdot 5} = 1 \cdot \frac{4 \cdot 3}{5 \cdot 5} = \frac{12}{25}$$

Do Exercises 26 and 27.

25. Divide: $26 \div 3\frac{1}{2}$.

Divide.

26. $2\frac{1}{4} \div 1\frac{1}{5}$

27. $1\frac{3}{4} \div 2\frac{1}{2}$

Answers on page A-5

CALCULATOR CORNER

Operations on Fractions and Mixed Numerals Fraction calculators can add, subtract, multiply, and divide fractions and mixed numerals. The [a b/c] key is used to enter fractions and mixed numerals. To find $\frac{3}{4} + \frac{1}{2}$, for example, we press [3] [a b/c] [4] [+] [1] [a b/c] [2] [=]. Note that 3/4 and 1/2 appear on the display as [3 ⌟4] and [1 ⌟2], respectively. The result is given as the mixed numeral $1\frac{1}{4}$ and is displayed as [1 ⌟1 ⌟4]. Fraction results that are greater than 1 are always displayed as mixed numerals. To express this result as a fraction, we press [SHIFT] [d/c]. We get [5 ⌟4], or 5/4.

To find $3\frac{2}{3} \cdot 4\frac{1}{5}$, we press [3] [a b/c] [2] [a b/c] [3] [×] [4] [a b/c] [1] [a b/c] [5] [=]. The calculator displays [15 ⌟2 ⌟5], so the product is $15\frac{2}{5}$.

Some calculators are capable of displaying mixed numerals in the way in which we write them, as shown below.

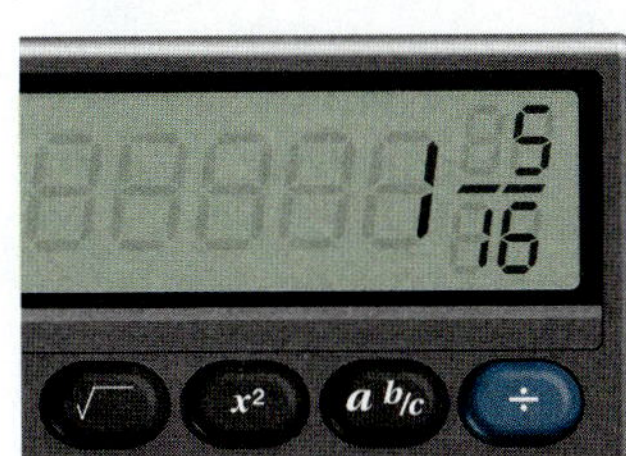

Exercises: Perform each calculation. Give the answer in fraction notation.

1. $\frac{1}{3} + \frac{1}{4}$
2. $\frac{7}{5} - \frac{3}{10}$
3. $\frac{15}{4} \cdot \frac{7}{12}$
4. $\frac{4}{5} \div \frac{8}{3}$

Perform each calculation. Give the answer as a mixed numeral.

5. $4\frac{1}{3} + 5\frac{4}{5}$
6. $9\frac{2}{7} - 8\frac{1}{4}$
7. $2\frac{1}{3} \cdot 4\frac{3}{5}$
8. $10\frac{7}{10} \div 3\frac{5}{6}$

2.4 EXERCISE SET

For Extra Help

Digital Video Tutor CD 2 Videotape 3 | InterAct Math | Math Tutor Center | MathXL | MyMathLab.com

a

1. *Garment Manufacturing.* A tailoring shop determines that for a certain size dress, it must use $3\frac{5}{8}$ yd of fabric that is 45 in. wide. To make the same dress with fabric that is 60 in. wide, it needs $2\frac{3}{4}$ yd. Convert $3\frac{5}{8}$ and $2\frac{3}{4}$ to fraction notation.

2. *Carpentry.* Dick Bonewitz, master carpenter, is making a display case according to the design below. Convert each mixed numeral to fraction notation.

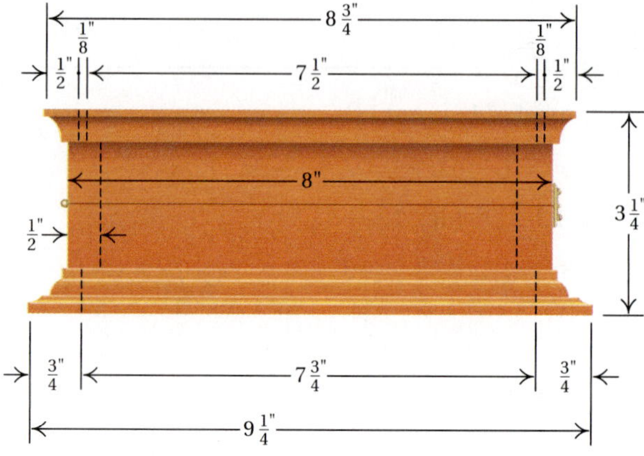

Convert to fraction notation.

3. $5\dfrac{2}{3}$

4. $20\dfrac{1}{5}$

5. $9\dfrac{5}{6}$

6. $1\dfrac{3}{5}$

7. $12\dfrac{3}{4}$

8. $33\dfrac{1}{3}$

Convert to a mixed numeral.

9. $\dfrac{18}{5}$

10. $\dfrac{17}{4}$

11. $\dfrac{57}{10}$

12. $\dfrac{50}{8}$

13. $\dfrac{345}{8}$

14. $\dfrac{467}{100}$

b Add. Write a mixed numeral for the answer.

15. $\begin{array}{r} 2\dfrac{7}{8} \\ +\ 3\dfrac{5}{8} \\ \hline \end{array}$

16. $\begin{array}{r} 4\dfrac{5}{6} \\ +\ 3\dfrac{5}{6} \\ \hline \end{array}$

17. $1\dfrac{1}{4} + 1\dfrac{2}{3}$

18. $4\dfrac{1}{3} + 5\dfrac{2}{9}$

19. $8\frac{3}{4} + 5\frac{5}{6}$

20. $4\frac{3}{8} + 6\frac{5}{12}$

21. $12\frac{4}{5} + 8\frac{7}{10}$

22. $15\frac{5}{8} + 11\frac{3}{4}$

23. $14\frac{5}{8} + 13\frac{1}{4}$

24. $16\frac{1}{4} + 15\frac{7}{8}$

25. $7\frac{1}{8} + 9\frac{2}{3} + 10\frac{3}{4}$

26. $45\frac{2}{3} + 31\frac{3}{5} + 12\frac{1}{4}$

c Subtract. Write a mixed numeral for the answer.

27. $4\frac{1}{5} - 2\frac{3}{5}$

28. $5\frac{1}{8} - 2\frac{3}{8}$

29. $6\frac{3}{5} - 2\frac{1}{2}$

30. $7\frac{2}{3} - 6\frac{1}{2}$

31. $34 - 18\frac{5}{8}$

32. $23 - 19\frac{3}{4}$

33. $21\frac{1}{6} - 13\frac{3}{4}$

34. $42\frac{1}{10} - 23\frac{7}{12}$

35. $14\frac{1}{8} - \frac{3}{4}$

36. $28\frac{1}{6} - 5$

37. $25\frac{1}{9} - 13\frac{5}{6}$

38. $23\frac{5}{16} - 14\frac{7}{12}$

d Multiply. Write a mixed numeral for the answer.

39. $8 \cdot 2\frac{5}{6}$

40. $5 \cdot 3\frac{3}{4}$

41. $3\frac{5}{8} \cdot \frac{2}{3}$

42. $6\frac{2}{3} \cdot \frac{1}{4}$

43. $3\frac{1}{2} \cdot 2\frac{1}{3}$

44. $4\frac{1}{5} \cdot 5\frac{1}{4}$

45. $3\frac{2}{5} \cdot 2\frac{7}{8}$

46. $2\frac{3}{10} \cdot 4\frac{2}{5}$

47. $4\frac{7}{10} \cdot 5\frac{3}{10}$

48. $6\frac{3}{10} \cdot 5\frac{7}{10}$

49. $20\frac{1}{2} \cdot 10\frac{1}{5} \cdot 4\frac{2}{3}$

50. $21\frac{1}{3} \cdot 11\frac{1}{3} \cdot 3\frac{5}{8}$

e Divide. Write a mixed numeral for the answer.

51. $20 \div 3\frac{1}{5}$

52. $18 \div 2\frac{1}{4}$

53. $8\frac{2}{5} \div 7$

54. $3\frac{3}{8} \div 3$

55. $4\frac{3}{4} \div 1\frac{1}{3}$

56. $5\frac{4}{5} \div 2\frac{1}{2}$

57. $1\frac{7}{8} \div 1\frac{2}{3}$

58. $4\frac{3}{8} \div 2\frac{5}{6}$

59. $5\frac{1}{10} \div 4\frac{3}{10}$

60. $4\frac{1}{10} \div 2\frac{1}{10}$

61. $20\frac{1}{4} \div 90$

62. $12\frac{1}{2} \div 50$

63. **DW** Write a problem for a classmate to solve. Design the problem so that its solution is found by performing the multiplication $4\frac{1}{2} \cdot 33\frac{1}{3}$.

64. **DW** Under what circumstances is a pair of mixed numerals more easily added than multiplied?

SKILL MAINTENANCE

65. Round to the nearest hundred: 45,765. [1.3e]

66. Round to the nearest ten: 45,765. [1.3e]

Determine whether the first number is divisible by the second. [1.8a]

67. 9993 by 3

68. 9993 by 9

69. 2345 by 9

70. 2345 by 5

71. 2335 by 10

72. 7764 by 6

73. 18,888 by 8

74. 18,888 by 4

Subtract. [1.2d]

75. $\begin{array}{r} 3004 \\ -\ 2957 \\ \hline \end{array}$

76. $\begin{array}{r} 1111 \\ -\ 222 \\ \hline \end{array}$

77. $\begin{array}{r} 10{,}013 \\ -\ 9988 \\ \hline \end{array}$

78. $\begin{array}{r} 507{,}987 \\ -\ 30{,}009 \\ \hline \end{array}$

SYNTHESIS

Multiply. Write the answer as a mixed numeral whenever possible.

79. 🖩 $15\frac{2}{11} \cdot 23\frac{31}{43}$

80. 🖩 $17\frac{23}{31} \cdot 19\frac{13}{15}$

Simplify.

81. $8 \div \frac{1}{2} + \frac{3}{4} + \left(5 - \frac{5}{8}\right)^2$

82. $\frac{7}{8} - 1\frac{1}{8} \times \frac{2}{3} + \frac{9}{10} \div \frac{3}{5}$

2.5 APPLICATIONS AND PROBLEM SOLVING

Objective

a Solve applied problems involving addition, subtraction, multiplication, and division using fraction notation and mixed numerals.

a We solve applied problems using fraction notation and mixed numerals in the same way that we do when using whole numbers. The five steps for problem solving on p. 53 should be reviewed.

Many problems that can be solved by multiplying fractions can be thought of in terms of rectangular arrays.

EXAMPLE 1 A real estate developer owns a plot of land that measures 1 square mile. He plans to use $\frac{4}{5}$ of the plot for a small strip mall and parking lot. Of this, $\frac{2}{3}$ will be needed for the parking lot. What part of the plot will be used for parking?

1. **Familiarize.** We first make a drawing to help familiarize ourselves with the problem. The land may not be rectangular. It could be in a shape like A or B below. But to think out the problem, we can think of it as a rectangle, as shown in shape C.

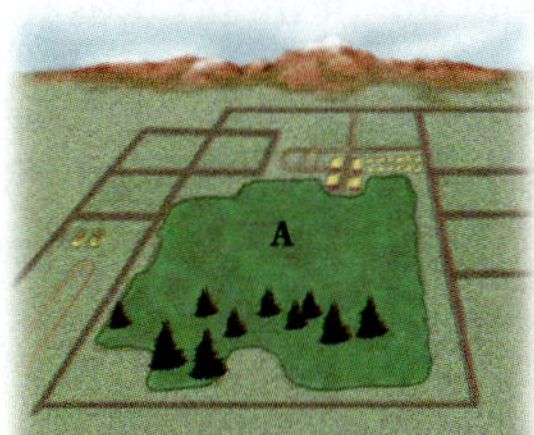

1 square mile

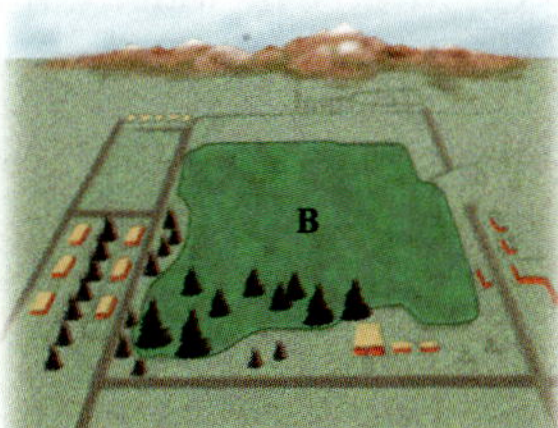

1 square mile

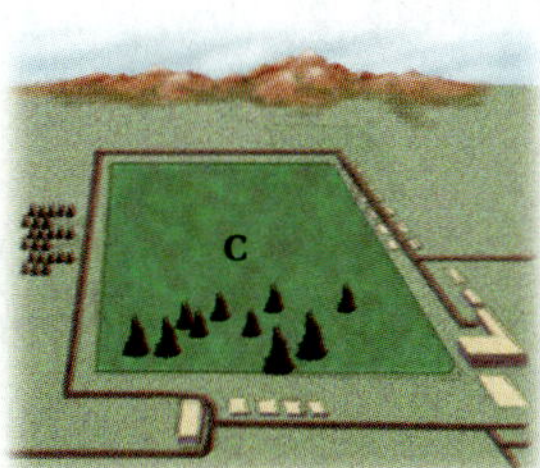

1 square mile

The strip mall including the parking lot uses $\frac{4}{5}$ of the plot. We shade $\frac{4}{5}$.

The parking lot alone takes $\frac{2}{3}$ of the preceding part. We shade that.

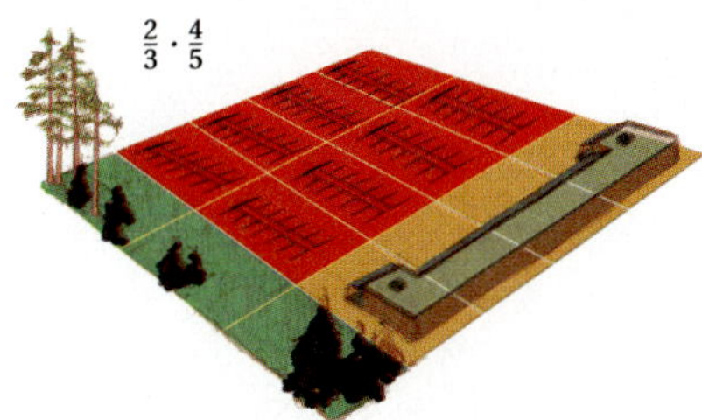

2. **Translate.** We let $n =$ the part of the plot that is used for parking. We are taking "two-thirds of four-fifths." Recall from Section 1.5 that the word "of" corresponds to multiplication. Thus the following multiplication sentence corresponds to the situation:

$$\frac{2}{3} \cdot \frac{4}{5} = n.$$

1. A resort hotel uses $\frac{3}{4}$ of its extra land for recreational purposes. Of that, $\frac{1}{2}$ is used for swimming pools. What part of the land is used for swimming pools?

2. **Area of a Fax Key.** The length of a button on a fax machine is $\frac{9}{10}$ cm. The width is $\frac{7}{10}$ cm. What is its area?

Answers on page A-5

3. **Solve.** The number sentence tells us what to do. We multiply:

$$\frac{2}{3} \cdot \frac{4}{5} = \frac{2 \cdot 4}{3 \cdot 5} = \frac{8}{15}.$$

4. **Check.** We can check partially by noting that the answer is smaller than the original area, 1, which we expect since the developer is using only part of the original plot of land. Thus, $\frac{8}{15}$ is a reasonable answer. We can also check this in the figure above, where we see that 8 of 15 parts have been shaded a second time.

5. **State.** The parking lot takes $\frac{8}{15}$ of the square mile of land.

Do Exercise 1.

Example 1 and the preceding discussion indicate that the area of a rectangular region can be found by multiplying length by width. That is true whether length and width are whole numbers or not. Remember, the area of a rectangular region is given by the formula

$$A = l \cdot w.$$

EXAMPLE 2 *Area of a Mosaic Tile.* The length of a tile on an inlaid mosaic table is $\frac{7}{10}$ in. The width is $\frac{3}{10}$ in. What is the area of one tile?

1. **Familiarize.** Recall that area is length times width. We make a drawing and let $A =$ the area of the tile.

2. **Translate.** Then we translate.

$$\underbrace{\text{Area}}_{A} \;\; \underbrace{\text{is}}_{=} \;\; \underbrace{\text{Length}}_{\frac{7}{10}} \;\; \underbrace{\text{times}}_{\times} \;\; \underbrace{\text{Width}}_{\frac{3}{10}}$$

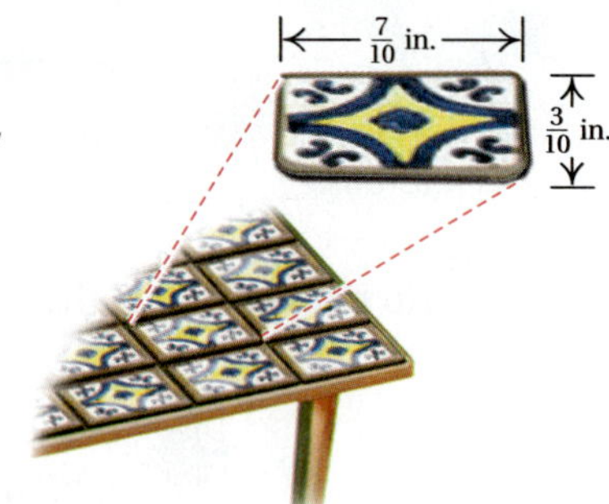

3. **Solve.** The sentence tells us what to do. We multiply:

$$\frac{7}{10} \cdot \frac{3}{10} = \frac{7 \cdot 3}{10 \cdot 10} = \frac{21}{100}.$$

4. **Check.** We check by repeating the calculation. This is left to the student.

5. **State.** The area is $\frac{21}{100}$ in^2.

Do Exercise 2.

EXAMPLE 3 *Test Tubes.* How many test tubes, each containing $\frac{3}{5}$ mL, can a nursing student fill from a container of 60 mL?

1. **Familiarize.** We are asking the question, "How many $\frac{3}{5}$'s are in 60?" Repeated addition will apply here. We make a drawing. We let $n =$ the number of test tubes in all.

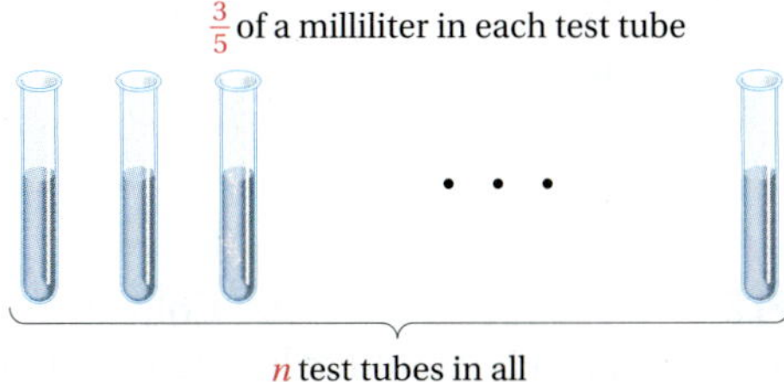

2. **Translate.** The equation that corresponds to the situation is

$$n = 60 \div \frac{3}{5}.$$

3. **Solve.** We solve the equation by carrying out the division:

$$n = 60 \div \frac{3}{5} = 60 \cdot \frac{5}{3} = \frac{60 \cdot 5}{3} = \frac{3 \cdot 20 \cdot 5}{3 \cdot 1}$$

$$= \frac{3}{3} \cdot \frac{20 \cdot 5}{1} = 100.$$

4. **Check.** We check by repeating the calculation.

5. **State.** The student can fill 100 test tubes.

Do Exercise 3.

EXAMPLE 4 Melissa Esplanah sells pharmaceutical supplies. After she had driven 210 mi, $\frac{5}{6}$ of her sales trip was completed. How long was the total trip?

1. **Familiarize.** We think: 210 mi is $\frac{5}{6}$ of the trip. We make a drawing or at least visualize the situation. We let $n =$ the length of the trip.

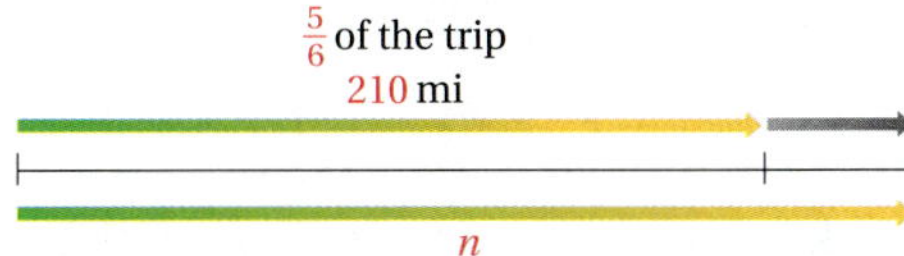

2. **Translate.** We translate to an equation.

Fraction completed	of	Total length of trip	is	Amount already traveled
↓	↓	↓	↓	↓
$\frac{5}{6}$	$\cdot$	n	$=$	210

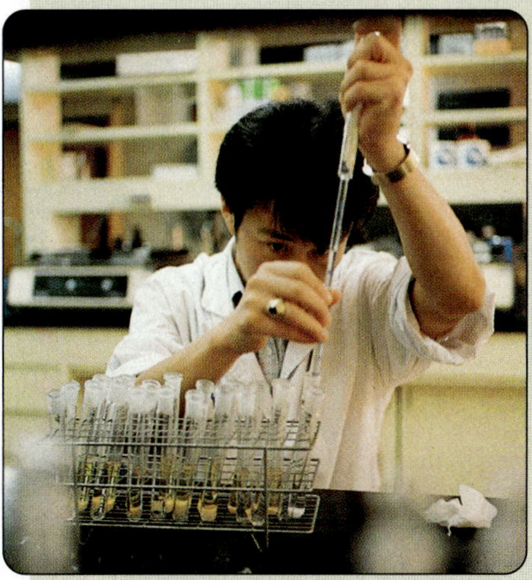

3. Each loop in a spring uses $\frac{3}{8}$ in. of wire. How many loops can be made from 120 in. of wire?

Answer on page A-5

4. A service station tank had 175 gal of oil when it was $\frac{7}{8}$ full. How much could the tank hold altogether?

3. **Solve.** The equation that corresponds to the situation is $\frac{5}{6} \cdot n = 210$. We divide by $\frac{5}{6}$ on both sides and carry out the division:

$$n = 210 \div \frac{5}{6} = 210 \cdot \frac{6}{5} = \frac{210 \cdot 6}{5} = \frac{5 \cdot 42 \cdot 6}{5 \cdot 1} = \frac{5}{5} \cdot \frac{42 \cdot 6}{1} = 252.$$

4. **Check.** We check by repeating the calculation.
5. **State.** The total trip was 252 mi.

Do Exercise 4.

Source: ©Coldwater Creek Inc. www.coldwatercreek.com

EXAMPLE 5 *Pendant Necklace.* Coldwater Creek offers the pendant necklace illustrated at left. The sterling silver capping at the top measures $\frac{11}{32}$ in. and the total length of the pendant is $\frac{7}{8}$ in. Find the length, or diameter, w of the pearl ball on the pendant.

1. **Familiarize.** We let w = the length of the pearl ball on the pendant.
2. **Translate.** We see that this is a "missing addend" situation. We can translate to an equation.

Length of silver capping	plus	Length of pearl ball	is	Total length of pendant
$\frac{11}{32}$	$+$	w	$=$	$\frac{7}{8}$

3. **Solve.** To solve the equation, we subtract $\frac{11}{32}$ on both sides:

$$\frac{11}{32} + w = \frac{7}{8}$$

$$\frac{11}{32} + w - \frac{11}{32} = \frac{7}{8} - \frac{11}{32} \qquad \text{Subtracting } \frac{11}{32} \text{ on both sides}$$

$$w + 0 = \frac{7}{8} \cdot \frac{4}{4} - \frac{11}{32} \qquad \text{The LCD is 32. We multiply by 1 to obtain the LCD.}$$

$$w = \frac{28}{32} - \frac{11}{32}$$

$$= \frac{17}{32}.$$

4. **Check.** To check, we return to the original problem and add:

$$\frac{11}{32} + \frac{17}{32} = \frac{28}{32} = \frac{7}{8} \cdot \frac{4}{4} = \frac{7}{8}.$$

5. **State.** The length of the pearl ball on the pendant is $\frac{17}{32}$ in.

Do Exercise 5.

5. Natasha has run for $\frac{2}{3}$ mi and will stop when she has run for $\frac{7}{8}$ mi. How much farther does she have to go?

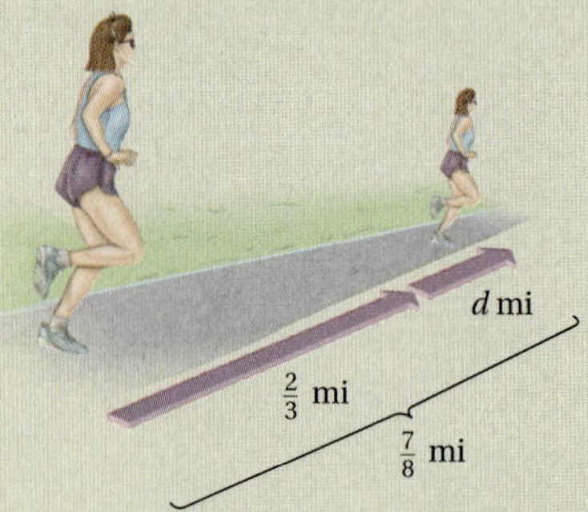

Answer on page A-5

EXAMPLE 6 *NCAA Football Goalposts.* In college football, the distance between goalposts was reduced from $23\frac{1}{3}$ ft to $18\frac{1}{2}$ ft. By how much was it reduced?

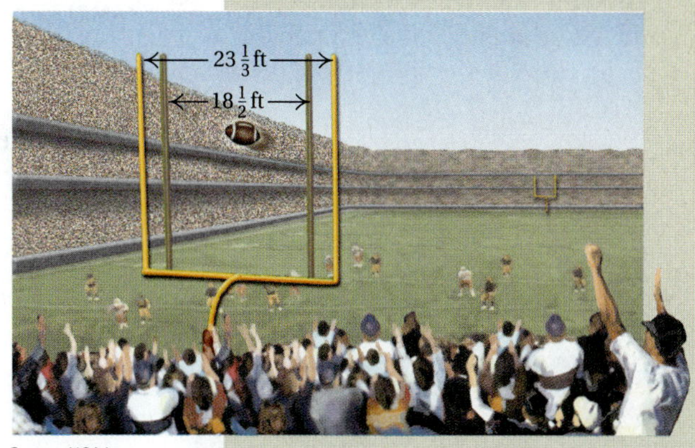

Source: NCAA

1. **Familiarize.** We let $d =$ the amount of reduction and make a drawing to illustrate the situation.
2. **Translate.** We translate as follows.

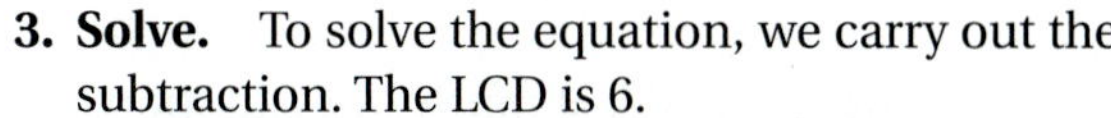

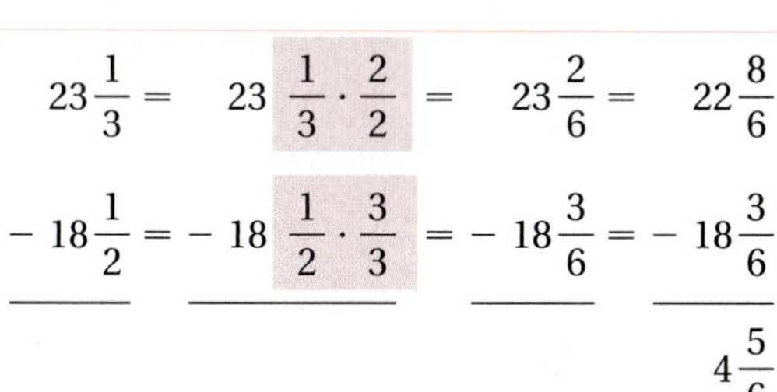

Former distance − New distance = Amount of reduction

$$23\frac{1}{3} - 18\frac{1}{2} = d$$

3. **Solve.** To solve the equation, we carry out the subtraction. The LCD is 6.

$$23\frac{1}{3} = 23\frac{1}{3}\cdot\frac{2}{2} = 23\frac{2}{6} = 22\frac{8}{6}$$
$$-\,18\frac{1}{2} = -\,18\frac{1}{2}\cdot\frac{3}{3} = -\,18\frac{3}{6} = -\,18\frac{3}{6}$$
$$4\frac{5}{6}$$

Thus, $d = 4\frac{5}{6}$ ft.

4. **Check.** To check, we add the reduction to the new distance:

$$18\frac{1}{2} + 4\frac{5}{6} = 18\frac{3}{6} + 4\frac{5}{6}$$
$$= 22\frac{8}{6}$$
$$= 23\frac{2}{6}$$
$$= 23\frac{1}{3}.$$

This checks.

5. **State.** The reduction in the goalpost distance was $4\frac{5}{6}$ ft.

Do Exercise 6.

6. **Damascus Blade.** The Damascus blade of a folding knife is $3\frac{3}{4}$ in. long. The same blade in an ATS-34 is $4\frac{1}{8}$ in. long. How many inches longer is the ATS-34 blade?
Source: *Blade Magazine* 23, no. 10, October 1996: 26–27

Answer on page A-5

7. Kyle's pickup truck travels on an interstate highway at 65 mph for $3\frac{1}{2}$ hr. How far does it travel?

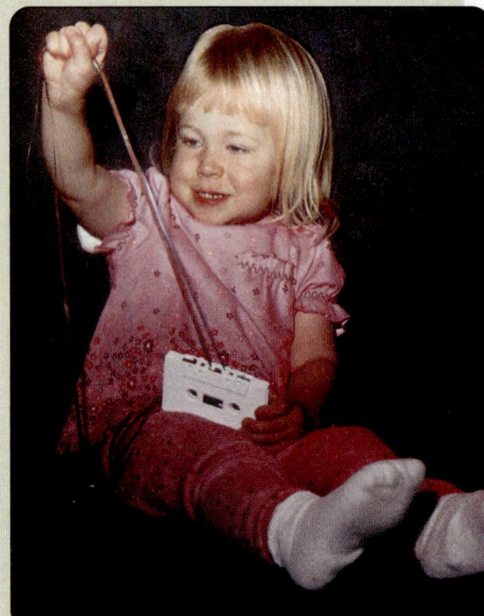

Margaret Grace Bittinger, age 2

8. Holly's minivan travels 302 mi on $15\frac{1}{10}$ gal of gas. How many miles per gallon did it get?

Answers on page A-5

EXAMPLE 7 *Cassette Tape Music.* The tape in an audio cassette is played at a rate of $1\frac{7}{8}$ in. per second. A child has destroyed 30 in. of tape. How many seconds of music have been lost?

1. **Familiarize.** We can make a drawing to help us visualize the situation.

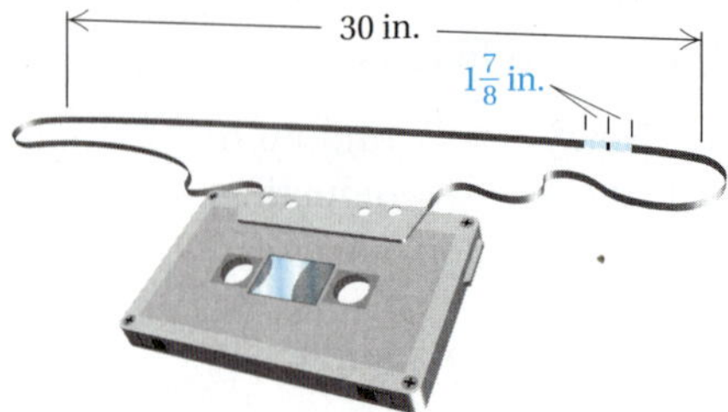

Since each $1\frac{7}{8}$ in. of tape represents 1 sec of lost music, the question can be regarded as asking how many times 30 can be divided by $1\frac{7}{8}$. We let $t =$ the number of seconds of music lost.

2. **Translate.** The situation corresponds to a division sentence:

$$t = 30 \div 1\frac{7}{8}.$$

3. **Solve.** To solve the equation, we perform the division:

$$\begin{aligned} t &= 30 \div 1\frac{7}{8} \\ &= \frac{30}{1} \div \frac{15}{8} \\ &= \frac{30}{1} \cdot \frac{8}{15} \\ &= \frac{15 \cdot 2 \cdot 8}{1 \cdot 15} \\ &= \frac{15}{15} \cdot \frac{2 \cdot 8}{1} \\ &= 16. \end{aligned}$$

4. **Check.** We check by multiplying. If 16 sec of music were lost, then

$$\begin{aligned} 16 \cdot 1\frac{7}{8} &= \frac{16}{1} \cdot \frac{15}{8} \\ &= \frac{8 \cdot 2 \cdot 15}{1 \cdot 8} \\ &= \frac{8}{8} \cdot \frac{2 \cdot 15}{1} = 30 \text{ in.} \end{aligned}$$

of tape were destroyed. A quicker, but less precise, check can be made by noting that $1\frac{7}{8} \approx 2$. Then $16 \cdot 1\frac{7}{8} \approx 16 \cdot 2 = 32 \approx 30$. Our answer checks.

5. **State.** The cassette has lost 16 sec of music.

Do Exercises 7 and 8.

EXAMPLE 8 *Mirror Area.* The mirror-backed candle shelf, shown below with a carpenter's diagram, was designed and built by Harry Cooper. Such shelves were popular in Colonial times because the mirror provided extra lighting from the candle. A rectangular walnut board is used to make the back of the shelf. Find the area of the original board and the amount left over after the mirror has been cut out.

Source: Popular Science Woodworking Projects

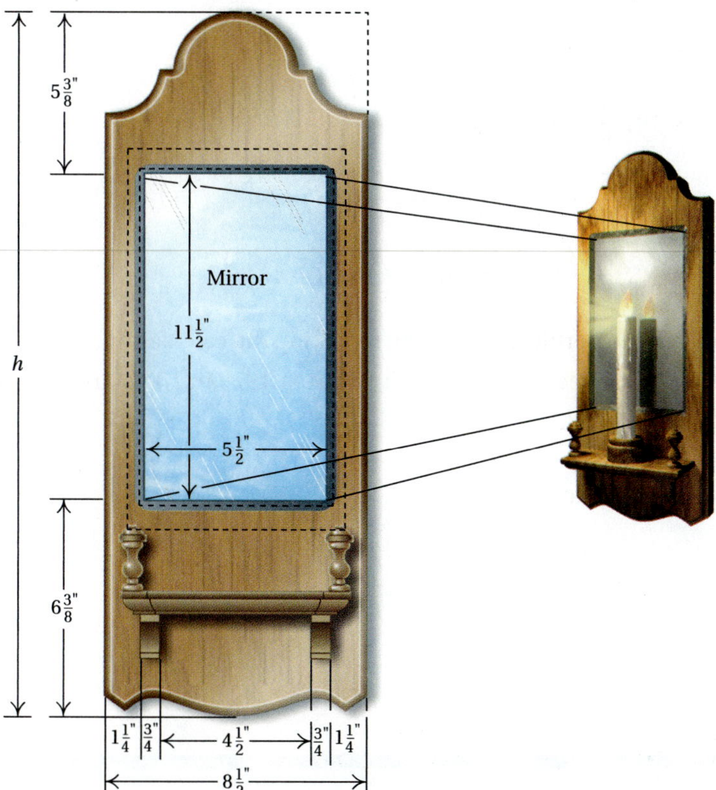

1. **Familiarize.** Refer to the figure above. We let h = the height of the back of the shelf and B = the area of the original board. We know the width of the original board, $8\frac{1}{2}$". (Remember, $8\frac{1}{2}$" means $8\frac{1}{2}$ in.) We let A = the area left over after the mirror has been cut out.
2. **Translate.** This is a multistep problem. To find B, which equals $8\frac{1}{2} \cdot h$, we first need to calculate h. We read the dimensions $5\frac{3}{8}$", $11\frac{1}{2}$", and $6\frac{3}{8}$" from the diagram and add them to find h:

$$h = 5\frac{3}{8} + 11\frac{1}{2} + 6\frac{3}{8}.$$

The dimensions of the mirror are $11\frac{1}{2}$" and $5\frac{1}{2}$". Then A is the area of the original board minus the area of the mirror. That is,

$$A = B - 11\frac{1}{2} \cdot 5\frac{1}{2}.$$

9. A room measures $22\frac{1}{2}$ ft by $15\frac{1}{2}$ ft. A 9-ft by 12-ft Oriental rug is placed in the center of the room. How much area is not covered by the rug?

Answer on page A-5

3. Solve. We carry out each calculation as follows:

$$h = 5\frac{3}{8} + 11\frac{1}{2} + 6\frac{3}{8} = 5\frac{3}{8} + 11\frac{4}{8} + 6\frac{3}{8} = 22\frac{10}{8} = 22\frac{5}{4} = 23\frac{1}{4};$$

$$B = 8\frac{1}{2} \cdot h = 8\frac{1}{2} \cdot 23\frac{1}{4} = \frac{17}{2} \cdot \frac{93}{4} = \frac{1581}{8} = 197\frac{5}{8};$$

$$\begin{aligned} A &= B - 11\frac{1}{2} \cdot 5\frac{1}{2} \\ &= 197\frac{5}{8} - 11\frac{1}{2} \cdot 5\frac{1}{2} \\ &= 197\frac{5}{8} - \frac{23}{2} \cdot \frac{11}{2} = 197\frac{5}{8} - \frac{253}{4} \\ &= 197\frac{5}{8} - 63\frac{1}{4} = 197\frac{5}{8} - 63\frac{2}{8} = 134\frac{3}{8}. \end{aligned}$$

4. Check. We perform a check by repeating the calculations.

5. State. The area of the original board is $197\frac{5}{8}$ in^2. The area left over is $134\frac{3}{8}$ in^2.

Do Exercise 9 on the preceding page.

Study Tips

BETTER TEST TAKING

How often do you make the following statement after taking a test: "I was able to do the homework, but I froze during the test"? This can be an excuse for poor study habits. Here are two tips to help you with this difficulty. Both are intended to make test taking less stressful by getting you to practice good test-taking habits on a daily basis.

- **Treat every homework exercise as if it were a test question.** If you had to work a problem at your job with no backup answer provided, what would you do? You would probably work it very deliberately, checking and rechecking every step. You might work it more than one time, or you might try to work it another way to check the result. Try to use this approach when doing your homework. Treat every exercise as though it were a test question with no answer at the back of the book.
- **Be sure that you do questions without answers as part of every homework assignment whether or not the instructor has assigned them!** One reason a test may seem such a different task is that questions on a test lack answers. That is the reason for taking a test: to see if you can do the questions without assistance. As part of your test preparation, be sure you do some exercises for which you do not have the answers. Thus when you take a test, you are doing a familiar task.

The purpose of doing your homework using these approaches is to give you more test-taking practice beforehand. Let's make a sports analogy here. At a basketball game, the players take lots of practice shots before the game. They play the first half, go to the locker room, and come out for the second half. What do they do before the second half, even though they have just played 20 minutes of basketball? They shoot baskets again! We suggest the same approach here. Create more and more situations in which you practice taking test questions by treating each homework exercise like a test question and by doing exercises for which you have no answers. Good luck!

For Extra Help

Digital Video Tutor CD 2 Videotape 3 | InterAct Math | Math Tutor Center | MathXL | MyMathLab.com

a Solve.

1. *Floor Tiling.* The floor of a room is being covered with tile. An area $\frac{3}{5}$ of the length and $\frac{3}{4}$ of the width is covered. What fraction of the floor has been tiled?

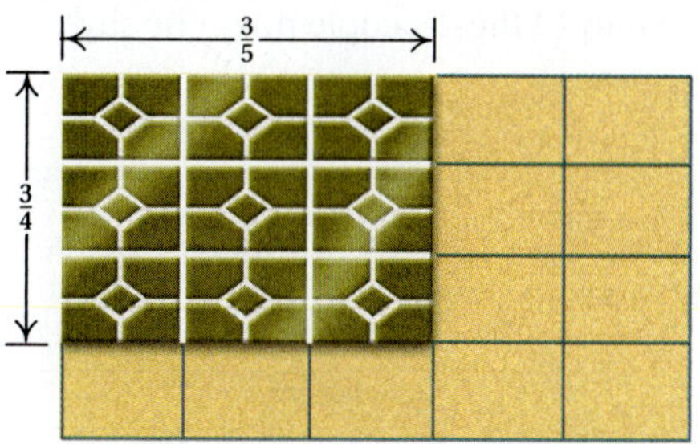

2. It takes $\frac{2}{3}$ yd of ribbon to make a bow. How much ribbon is needed to make 5 bows?

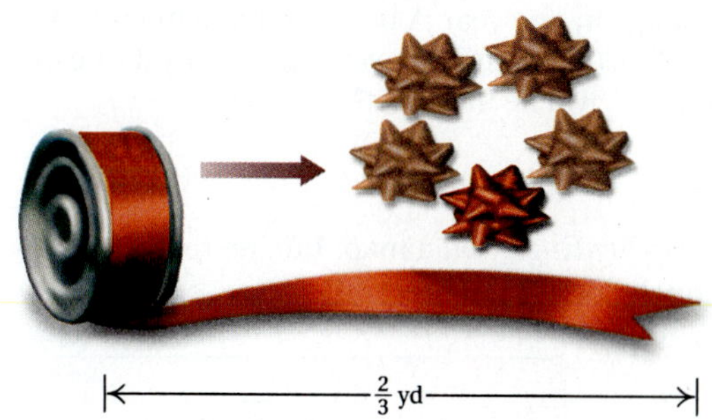

3. A rectangular table top measures $\frac{4}{5}$ m long by $\frac{3}{5}$ m wide. What is its area?

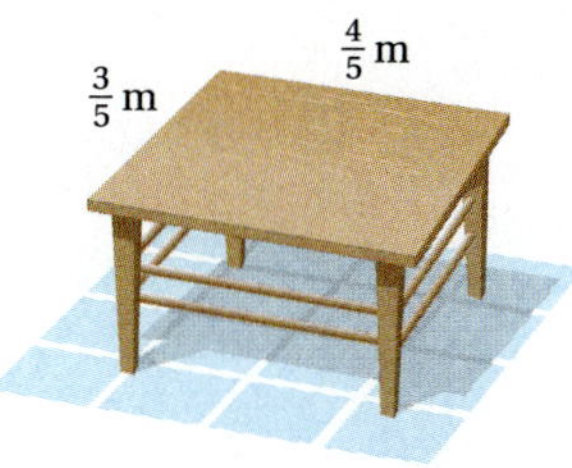

4. *Basement Carpet.* A basement floor is being covered with carpet. An area $\frac{7}{8}$ of the length and $\frac{3}{4}$ of the width is covered by lunch time. What fraction of the floor has been completed?

The *pitch* of a screw is the distance between its threads. With each complete rotation, the screw goes in or out a distance equal to its pitch. Use this information to answer Exercises 5–8.

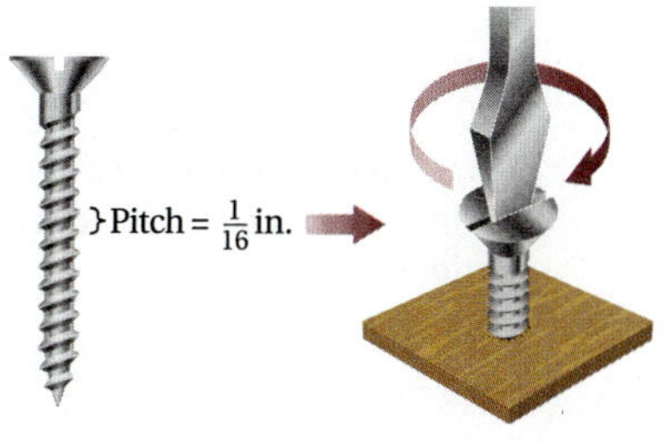

Each rotation moves the screw in or out $\frac{1}{16}$ in.

5. The pitch of a screw is $\frac{1}{16}$ in. How far will it go into a piece of oak when it is turned 10 complete rotations clockwise?

6. The pitch of a screw is $\frac{3}{32}$ in. How far will it go out of a piece of plywood when it is turned 10 complete rotations counterclockwise?

7. After a screw has been turned 8 complete rotations, it is extended $\frac{1}{2}$ in. into a piece of wallboard. What is the pitch of the screw?

8. The pitch of a screw is $\frac{3}{32}$ in. How many complete rotations are necessary to drive the screw $\frac{3}{4}$ in. into a piece of pine wood?

9. *Football: High School to Pro.* One of 39 high school football players plays college football. One of 39 college players plays professional football. What fractional part of high school players play professional football?
Source: National Football League

10. A gasoline can holds $\frac{7}{8}$ liter (L). How much will the can hold when it is $\frac{1}{2}$ full?

11. *Mailing-List Addresses.* Business people have determined that $\frac{1}{4}$ of the addresses on a mailing list will change in one year. A business has a mailing list of 2500 people. After one year, how many addresses on that list will be incorrect?

12. *Shy People.* Sociologists have determined that $\frac{2}{5}$ of the people in the world are shy. A sales manager is interviewing 650 people for an aggressive sales position. How many of these people might be shy?

13. *Map Scaling.* On a map, 1 in. represents 240 mi. How much does $\frac{2}{3}$ in. represent?

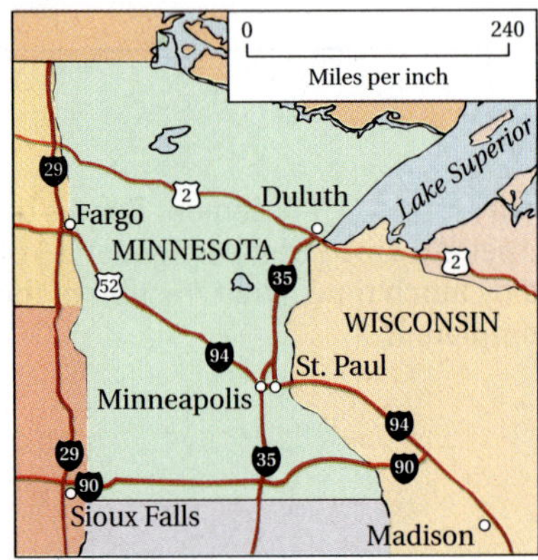

14. *Map Scaling.* On a map, 1 in. represents 120 mi. How much does $\frac{3}{4}$ in. represent?

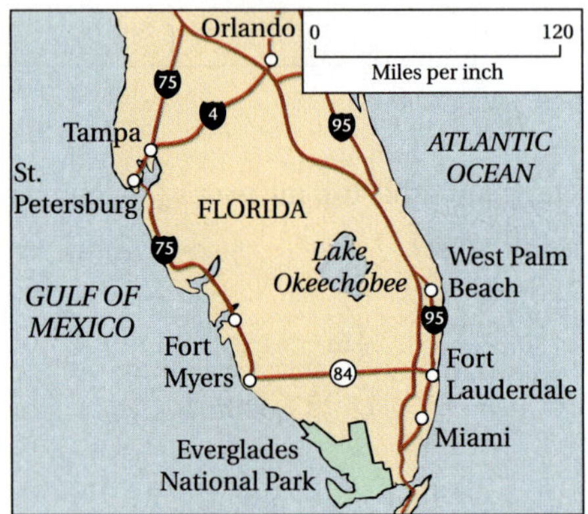

15. A recipe for piecrust calls for $\frac{2}{3}$ cup of flour. A chef is making $\frac{1}{2}$ of the recipe. How much flour should the chef use?

16. Of the students in the freshman class, $\frac{2}{5}$ have cameras; $\frac{1}{4}$ of these students also join the college photography club. What fraction of the students in the freshman class join the photography club?

17. Benny uses $\frac{2}{5}$ gram (g) of toothpaste each time he brushes his teeth. If Benny buys a 30-g tube, how many times will he be able to brush his teeth?

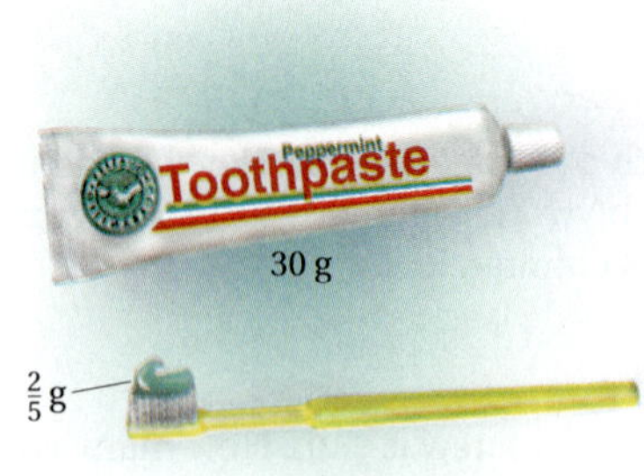

18. A piece of coaxial cable $\frac{4}{5}$ meter (m) long is to be cut into 8 pieces of the same length. What is the length of each piece?

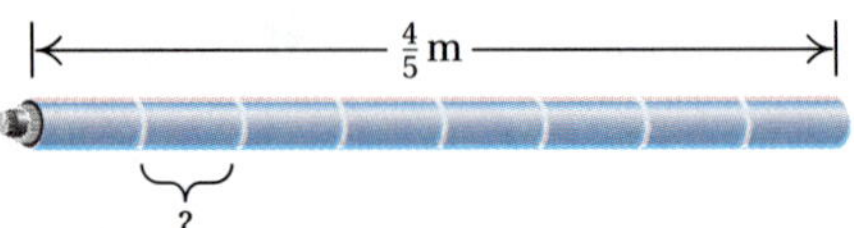

19. *Concrete Mix.* A cubic meter of concrete mix contains 420 kilograms (kg) of cement, 150 kg of stone, and 120 kg of sand. What is the total weight of the cubic meter of concrete mix? What part is cement? stone? sand? Add these amounts. What is the result?

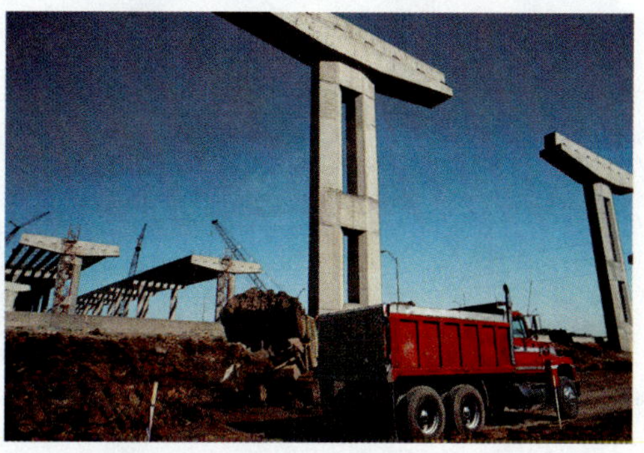

20. *Punch Recipe.* A recipe for strawberry punch calls for $\frac{1}{5}$ quart (qt) of ginger ale and $\frac{3}{5}$ qt of strawberry soda. How much liquid is needed? If the recipe is doubled, how much liquid is needed? If the recipe is halved, how much liquid is needed?

21. A tile $\frac{5}{8}$ in. thick is glued to a board $\frac{7}{8}$ in. thick. The glue is $\frac{3}{32}$ in. thick. How thick is the result?

22. A baker used $\frac{1}{2}$ lb of flour for rolls, $\frac{1}{4}$ lb for donuts, and $\frac{1}{3}$ lb for cookies. How much flour was used?

23. A pair of basketball shorts requires $\frac{3}{4}$ yd of nylon. How many pairs of shorts can be made from 24 yd of nylon?

24. A child's baseball shirt requires $\frac{5}{6}$ yd of fabric. How many shirts can be made from 25 yd of the fabric?

25. How many $\frac{2}{3}$-cup sugar bowls can be filled from 16 cups of sugar?

26. How many $\frac{2}{3}$-cup cereal bowls can be filled from 10 cups of cornflakes?

27. A bucket had 12 L of water in it when it was $\frac{3}{4}$ full. How much could it hold altogether?

28. A tank had 20 L of gasoline in it when it was $\frac{4}{5}$ full. How much could it hold altogether?

29. Rene bought $\frac{1}{3}$ lb of orange pekoe tea and $\frac{1}{2}$ lb of English cinnamon tea. How many pounds of tea did he buy?

30. Stan bought $\frac{1}{4}$ lb of gumdrops and $\frac{1}{2}$ lb of caramels. How many pounds of candy did he buy?

31. Russ walked $\frac{7}{6}$ mi to a friend's dormitory, and then $\frac{3}{4}$ mi to class. How far did he walk?

32. Elaine walked $\frac{7}{8}$ mi to the student union, and then $\frac{2}{5}$ mi to class. How far did she walk?

33. *Tire Tread.* A new long-life tire has a tread depth of $\frac{3}{8}$ in. instead of a more typical $\frac{11}{32}$ in. How much deeper is the new tread depth?
Source: *Popular Science*

34. From a $\frac{4}{5}$-lb wheel of cheese, a $\frac{1}{4}$-lb piece was served. How much cheese remained on the wheel?

35. An Arby's franchise is owned by three people. One owns $\frac{7}{12}$ of the business and the second owns $\frac{1}{6}$. What part of the business does the third person own?

36. An estate was left to four children. One received $\frac{1}{4}$ of the estate, the second $\frac{1}{16}$, and the third $\frac{3}{8}$. How much did the fourth receive?

37. A server has a bottle containing $\frac{11}{12}$ cup of olive oil. He serves $\frac{1}{4}$ cup on a plate to a customer for bread dipping. How much remains in the bottle?

38. Jovan has an $\frac{11}{10}$-lb mixture of cashews and peanuts that includes $\frac{3}{5}$ lb of cashews. How many pounds of peanuts are in the mixture?

39. *Sewing from a Pattern.* Suppose you want to make an outfit in size 8. Using 45-in. fabric, you need $1\frac{3}{8}$ yd for the dress, $\frac{5}{8}$ yd of contrasting fabric for the band at the bottom, and $3\frac{3}{8}$ yd for the jacket. How many yards in all of 45-in. fabric are needed to make the outfit?

40. *Sewing from a Pattern.* Suppose you want to make an outfit in size 12. Using 45-in. fabric, you need $2\frac{3}{4}$ yd for the dress and $3\frac{1}{2}$ yd for the jacket. How many yards in all of 45-in. fabric are needed to make the outfit?

41. For a family barbecue, Jason bought packages of hamburger weighing $1\frac{2}{3}$ lb and $5\frac{3}{4}$ lb. What was the total weight of the meat?

42. Marsha's Butcher Shop sold packages of sliced turkey breast weighing $1\frac{1}{3}$ lb and $4\frac{3}{5}$ lb. What was the total weight of the meat?

43. Kim Park is a computer technician. One day, she drove $180\frac{7}{10}$ mi away from Los Angeles for a service call. The next day, she drove $85\frac{1}{2}$ mi back toward Los Angeles for another service call. How far was she from Los Angeles?

44. Pilar is $4\frac{1}{2}$ in. taller than her daughter Teresa. Teresa is $66\frac{2}{3}$ in. tall. How tall is Pilar?

45. *Interior Design.* Sue, an interior designer, worked $10\frac{1}{2}$ hr over a three-day period. If Sue worked $2\frac{1}{2}$ hr on the first day and $4\frac{1}{5}$ hr on the second, how many hours did Sue work on the third day?

46. *Painting.* Geri had $3\frac{1}{2}$ gal of paint. It took $2\frac{3}{4}$ gal to paint the family room. It was estimated that it would take $2\frac{1}{4}$ gal to paint the living room. How much more paint was needed?

47. A car traveled 213 mi on $14\frac{2}{10}$ gal of gas. How many miles per gallon did it get?

48. *Aeronautics.* Most space shuttles orbit the earth once every $1\frac{1}{2}$ hr. How many orbits are made every 24 hr?

Find the perimeter of (distance around) the figure.

49.

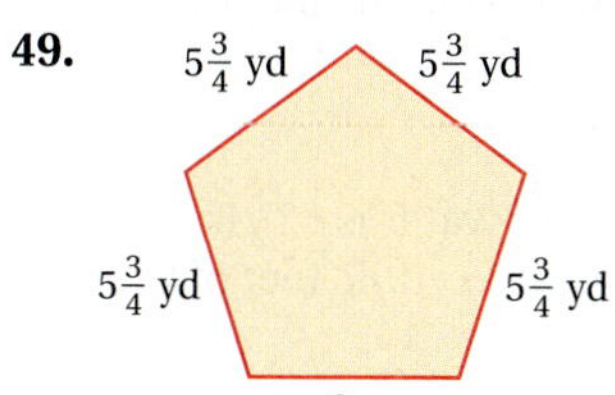

50.

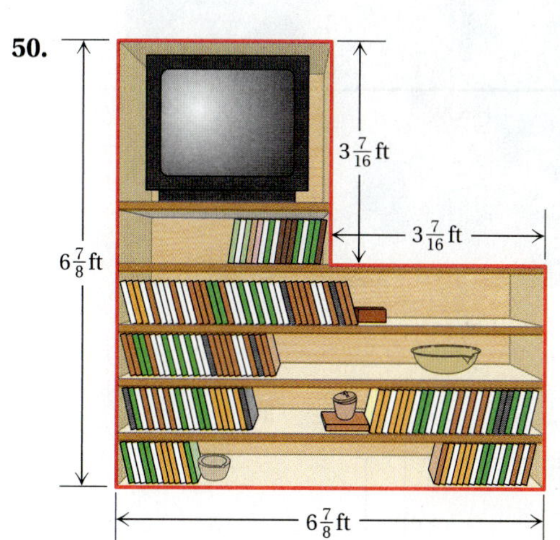

51. *Temperatures.* Fahrenheit temperature can be obtained from Celsius (centigrade) temperature by multiplying by $1\frac{4}{5}$ and adding 32°. What Fahrenheit temperature corresponds to a Celsius temperature of 20°?

52. *Carpentry.* When cutting wood with a saw, a carpenter must take into account the thickness of the saw blade. Suppose that from a piece of wood 36 in. long, a carpenter cuts a $15\frac{3}{4}$-in. length with a saw blade that is $\frac{1}{8}$ in. in thickness. How long is the piece that remains?

53. Find the length d in the figure.

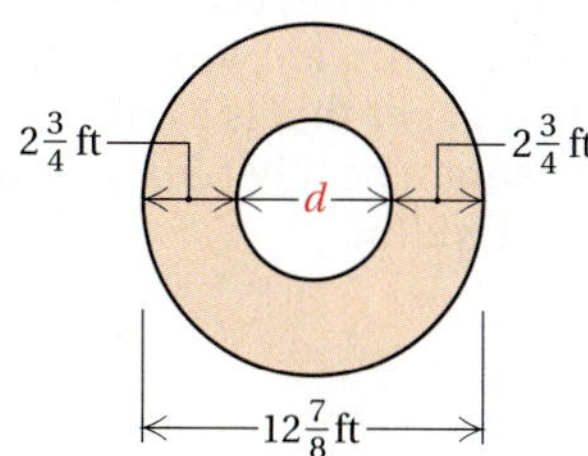

54. Find the smallest length of a bolt that will pass through a piece of tubing with an outside diameter of $\frac{1}{2}$ in., a washer $\frac{1}{16}$ in. thick, a piece of tubing with a $\frac{3}{4}$-in. outside diameter, another washer, and a nut $\frac{3}{16}$ in. thick.

55. *Sodium Consumption.* The average American woman consumes $1\frac{1}{3}$ tsp of sodium each day. How much sodium do 10 average American women consume in one day?
Source: *Nutrition Action Health Letter,* March 1994, p. 6. 1875 Connecticut Ave., N.W., Washington, DC 20009-5728

56. *Exercise.* At one point during an aerobics class at Ray's health club, Kea's bicycle wheel was completing $76\frac{2}{3}$ revolutions per minute. How many revolutions did the wheel complete in 6 min?

57. *Servings of Flounder.* A serving of filleted fish is generally considered to be about $\frac{1}{3}$ lb. How many servings can be prepared from $5\frac{1}{2}$ lb of flounder fillet?

58. *Weight of Water.* The weight of water is $62\frac{1}{2}$ lb per cubic foot. How many cubic feet would be occupied by 250 lb of water?

59. *Weight of Water.* The weight of water is $62\frac{1}{2}$ lb per cubic foot. What is the weight of $5\frac{1}{2}$ cubic feet of water?

60. *Weight of Water.* The weight of water is $62\frac{1}{2}$ lb per cubic foot. What is the weight of $2\frac{1}{4}$ cubic feet of water?

Find the area of the shaded region.

61.

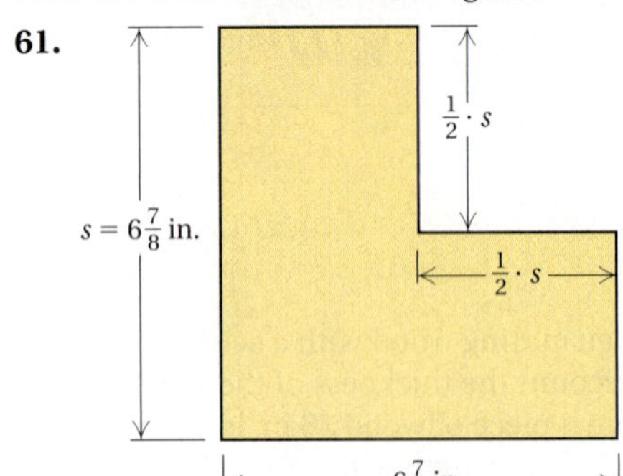

62.

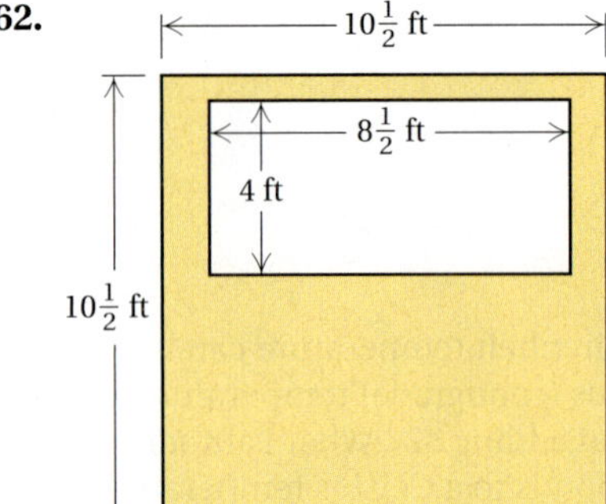

63. *Construction.* A rectangular lot has dimensions of $302\frac{1}{2}$ ft by $205\frac{1}{4}$ ft. A building with dimensions of 100 ft by $25\frac{1}{2}$ ft is built on the lot. How much area is left over?

64. *Word Processing.* Kelly wants to create a table using Microsoft® Word software for word processing. She needs to have two columns, each $1\frac{1}{2}$ in. wide, and five columns, each $\frac{3}{4}$ in. wide. Will this table fit on a piece of standard paper that is $8\frac{1}{2}$ in. wide? If so, how wide will each margin be if her margins on each side are to be of equal width?

65. **DW** Write a problem for a classmate to solve. Design the problem so that the solution is "About $\frac{1}{30}$ of the students are left-handed women."

66. **DW** Write a problem for a classmate to solve. Design the problem so the solution is "The larger package holds $4\frac{1}{2}$ oz more than the smaller package."

SKILL MAINTENANCE

Solve. [1.5a]

67. *Salaries and Education.* In 1998, the average annual salary for a person who is not a high school graduate was $16,053. The average for someone with a bachelor's degree was $43,782. How much more did the person who has a bachelor's degree earn than the person who is not a high school graduate?
Source: U.S. Bureau of the Census

68. *Gas Mileage.* The Chrysler PT Cruiser gets 22 miles to the gallon (mpg) in city driving. How many gallons will it use in 2860 mi of city driving?

69. A playing field is 78 ft long and 64 ft wide. What is its area? its perimeter?

70. A landscaper buys 13 small maple trees and 17 small oak trees for a project. A maple costs $23 and an oak costs $37. How much is spent altogether for the trees?

Subtract. [1.2d]

71. $9001 - 6798$

72. $12{,}327 - 476$

Divide. [1.3d]

73. $7140 \div 35$

74. $32{,}200 \div 46$

Solve. [1.4b]

75. $30 \cdot x = 150$

76. $10{,}947 = 123 \cdot y$

77. $5280 = 1760 + t$

78. $x + 2368 = 11{,}369$

Simplify. [1.6c]

79. $8 \cdot 12 - (63 \div 9 + 13 \cdot 3)$

80. $(10 - 3)^4 + 10^3 \cdot 4 - 10 \div 5$

SYNTHESIS

81. A guitarist's band is booked for Friday and Saturday nights at a local club. The guitarist is part of a trio on Friday and part of a quintet on Saturday. Thus the guitarist is paid one-third of one-half the weekend's pay for Friday and one-fifth of one-half the weekend's pay for Saturday. What fractional part of the band's pay did the guitarist receive for the weekend's work? If the band was paid $1200, how much did the guitarist receive?

82. *Microsoft Interview.* The following is a question taken from an employment interview with Microsoft. Try to answer it.

"Given a gold bar that can be cut exactly twice and a contractor who must be paid one-seventh of a gold bar every day for seven days, how should the bar be cut?"
Source: *Fortune Magazine*, January 22, 2001

83. *College Profile.* Of students entering a college, $\frac{7}{8}$ have completed high school and $\frac{2}{3}$ are older than 20. If $\frac{1}{7}$ of all students are left-handed, what fraction of students entering the college are left-handed high school graduates over the age of 20?

84. *College Profile.* Refer to the information in Exercise 83. If 480 students are entering the college, how many of them are left-handed high school graduates 20 yr old or younger?

2.6 ORDER OF OPERATIONS; ESTIMATION

Objectives

a Simplify expressions using the rules for order of operations.

b Estimate with fraction and mixed-numeral notation.

Simplify.

1. $\frac{2}{5} \cdot \frac{5}{8} + \frac{1}{4}$

2. $\frac{1}{3} \cdot \frac{3}{4} \div \frac{5}{8} - \frac{1}{10}$

3. Simplify: $\frac{3}{4} \cdot 16 + 8\frac{2}{3}$.

Answers on page A-5

a Order of Operations; Fraction Notation and Mixed Numerals

The rules for order of operations that we use with whole numbers (see Section 1.6) apply when we are simplifying expressions involving fraction notation and mixed numerals. For review, these rules are listed below.

RULES FOR ORDER OF OPERATIONS

1. Do all calculations within parentheses before operations outside.
2. Evaluate all exponential expressions.
3. Do all multiplications and divisions in order from left to right.
4. Do all additions and subtractions in order from left to right.

EXAMPLE 1 Simplify: $\frac{2}{3} \div \frac{1}{2} \cdot \frac{5}{8} + \frac{1}{6}$.

$$\frac{2}{3} \div \frac{1}{2} \cdot \frac{5}{8} + \frac{1}{6} = \frac{2}{3} \cdot \frac{2}{1} \cdot \frac{5}{8} + \frac{1}{6}$$ Doing the division first by multiplying by the reciprocal of $\frac{1}{2}$

$$= \frac{4 \cdot 5}{3 \cdot 8} + \frac{1}{6}$$ Doing the multiplication

$$= \frac{\not{4} \cdot 5}{3 \cdot \not{4} \cdot 2} + \frac{1}{6}$$ Factoring in order to simplify

$$= \frac{5}{3 \cdot 2} + \frac{1}{6}$$ Removing a factor of 1: $\frac{4}{4} = 1$

$$= \frac{5}{6} + \frac{1}{6}$$

$$= \frac{6}{6}, \text{ or } 1$$ Doing the addition

Do Exercises 1 and 2.

EXAMPLE 2 Simplify: $\frac{2}{3} \cdot 24 - 11\frac{1}{2}$.

$$\frac{2}{3} \cdot 24 - 11\frac{1}{2} = \frac{2 \cdot 24}{3} - 11\frac{1}{2}$$ Doing the multiplication first

$$= \frac{2 \cdot \not{3} \cdot 8}{\not{3}} - 11\frac{1}{2}$$ Factoring the numerator

$$= 2 \cdot 8 - 11\frac{1}{2}$$ Removing a factor of 1: $\frac{3}{3} = 1$

$$= 16 - 11\frac{1}{2}$$ Completing the multiplication

$$= 4\frac{1}{2}, \text{ or } \frac{9}{2}$$ Doing the subtraction

Do Exercise 3 on the preceding page.

EXAMPLE 3 Melody has had three children. Their birth weights were $7\frac{1}{2}$ lb, $7\frac{3}{4}$ lb, and $6\frac{3}{4}$ lb. What was the average weight of her babies?

Recall that to compute an **average,** we add the numbers and then divide the sum by the number of addends (see Section 1.6). We have

$$\frac{7\frac{1}{2} + 7\frac{3}{4} + 6\frac{3}{4}}{3}.$$

We first add:

$$7\frac{1}{2} + 7\frac{3}{4} + 6\frac{3}{4} = 7\frac{2}{4} + 7\frac{3}{4} + 6\frac{3}{4}$$

$$= 20\frac{8}{4} = 22. \qquad 20\tfrac{8}{4} = 20 + \tfrac{8}{4} = 20 + 2$$

Then we divide:

$$\frac{7\frac{1}{2} + 7\frac{3}{4} + 6\frac{3}{4}}{3} = \frac{22}{3} = 7\frac{1}{3}. \qquad \text{Dividing by 3}$$

The average weight of the three babies is $7\frac{1}{3}$ lb.

Do Exercises 4–6.

EXAMPLE 4 Simplify: $\left(\frac{7}{8} - \frac{1}{3}\right) \times 48 + \left(13 + \frac{4}{5}\right)^2.$

$$\left(\frac{7}{8} - \frac{1}{3}\right) \times 48 + \left(13 + \frac{4}{5}\right)^2$$

$$= \left(\frac{7}{8}\cdot\frac{3}{3} - \frac{1}{3}\cdot\frac{8}{8}\right) \times 48 + \left(13\cdot\frac{5}{5} + \frac{4}{5}\right)^2$$

Carrying out operations inside parentheses first. To do so, we first multiply by 1 to obtain the LCD.

$$= \left(\frac{21}{24} - \frac{8}{24}\right) \times 48 + \left(\frac{65}{5} + \frac{4}{5}\right)^2$$

$$= \frac{13}{24} \times 48 + \left(\frac{69}{5}\right)^2 \qquad \text{Completing the operations within parentheses}$$

$$= \frac{13}{24} \times 48 + \frac{4761}{25} \qquad \text{Evaluating exponential expressions next}$$

$$= 26 + \frac{4761}{25} \qquad \text{Doing the multiplication}$$

$$= 26 + 190\frac{11}{25} \qquad \text{Converting to a mixed numeral}$$

$$= 216\frac{11}{25}, \text{ or } \frac{5411}{25} \qquad \text{Adding}$$

Answers can be given using either fraction notation or mixed numerals as desired. Consult with your instructor.

Do Exercise 7.

4. After two weeks, Kurt's tomato seedlings measure $9\frac{1}{2}$ in., $10\frac{3}{4}$ in., $10\frac{1}{4}$ in., and 9 in. tall. Find their average height.

5. Find the average of

$$\frac{1}{2}, \frac{1}{3}, \text{ and } \frac{5}{6}.$$

6. Find the average of $\frac{3}{4}$ and $\frac{4}{5}$.

7. Simplify:

$$\left(\frac{2}{3} + \frac{3}{4}\right) \div 2\frac{1}{3} - \left(\frac{1}{2}\right)^3.$$

Answers on page A-5

Estimate each of the following as 0, $\frac{1}{2}$, or 1.

8. $\dfrac{3}{59}$

9. $\dfrac{61}{59}$

10. $\dfrac{29}{59}$

11. $\dfrac{57}{59}$

Find a number for the blank so that the fraction is close to but less than 1.

12. $\dfrac{11}{\square}$

13. $\dfrac{\square}{33}$

Answers on page A-5

Study Tips

SKILL MAINTENANCE EXERCISES

It is never too soon to begin reviewing for the final examination. The Skill Maintenance exercises found in each exercise set review and reinforce skills taught in earlier sections. Include all of these exercises in your weekly preparation. Answers to both odd-numbered and even-numbered exercises along with section references appear at the back of the book.

b Estimation with Fraction Notation and Mixed Numerals

We now estimate with fraction notation and mixed numerals.

EXAMPLES Estimate each of the following as 0, $\frac{1}{2}$, or 1.

5. $\dfrac{2}{17}$

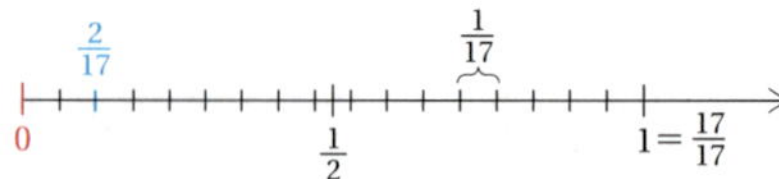

A fraction is very close to 0 when the numerator is very small in comparison to the denominator. Thus, 0 is an estimate for $\frac{2}{17}$ because 2 is very small in comparison to 17. Thus, $\frac{2}{17} \approx 0$.

6. $\dfrac{11}{23}$

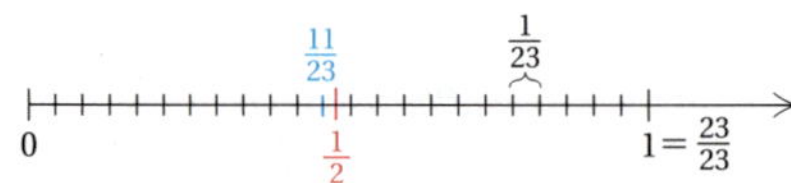

A fraction is very close to $\frac{1}{2}$ when the denominator is about twice the numerator. Thus, $\frac{1}{2}$ is an estimate for $\frac{11}{23}$ because $2 \cdot 11 = 22$ and 22 is close to 23. Thus, $\frac{11}{23} \approx \frac{1}{2}$.

7. $\dfrac{37}{38}$

A fraction is very close to 1 when the numerator is nearly equal to the denominator. Thus, 1 is an estimate for $\frac{37}{38}$ because 37 is nearly equal to 38. Thus, $\frac{37}{38} \approx 1$.

8. $\dfrac{43}{41}$

As in the preceding example, the numerator 43 is very close to the denominator 41. Thus, $\frac{43}{41} \approx 1$.

Do Exercises 8–11.

EXAMPLE 9 Find a number for the blank so that $\dfrac{9}{\square}$ is close to but less than 1. Answers may vary.

If the number in the blank were 9, we would have 1, so we increase 9 to 10. The answer is 10; $\frac{9}{10}$ is close to 1. The number 11 would also be a correct answer; $\frac{9}{11}$ is close to 1.

Do Exercises 12 and 13.

EXAMPLE 10 Find a number for the blank so that $\frac{9}{\square}$ is close to but less than $\frac{1}{2}$. Answers may vary.

If we double 9 to get 18 and use it for the blank, we have $\frac{1}{2}$. If we increase that denominator by 1, to get 19, and use it for the blank, we get a number less than $\frac{1}{2}$ but close to $\frac{1}{2}$. Thus, $\frac{9}{19} \approx \frac{1}{2}$.

Do Exercises 14 and 15.

EXAMPLE 11 Find a number for the blank so that $\frac{\square}{50}$ is close to but greater than 0.

Since 50 is rather large, any small number such as 1, 2, or 3 will make the fraction close to 0. For example, $\frac{1}{50} \approx 0$.

Do Exercises 16 and 17.

EXAMPLE 12 Estimate $16\frac{8}{9} + 11\frac{2}{13} - 4\frac{22}{43}$ as a whole number or as a mixed numeral where the fractional part is $\frac{1}{2}$.

We estimate each fraction as 0, $\frac{1}{2}$, or 1. Then we calculate:

$$16\frac{8}{9} + 11\frac{2}{13} - 4\frac{22}{43} \approx 17 + 11 - 4\frac{1}{2}$$
$$= 28 - 4\frac{1}{2}$$
$$= 23\frac{1}{2}.$$

Do Exercises 18–20.

Find a number for the blank so that the fraction is close to but less than $\frac{1}{2}$.

14. $\frac{13}{\square}$ **15.** $\frac{\square}{31}$

Find a number for the blank so that the fraction is close to but greater than 0.

16. $\frac{\square}{37}$ **17.** $\frac{13}{\square}$

Estimate each part of the following as a whole number or as a mixed numeral where the fractional part is $\frac{1}{2}$.

18. $5\frac{9}{10} + 26\frac{1}{2} - 10\frac{3}{29}$

19. $10\frac{7}{8} \cdot \left(25\frac{11}{13} - 14\frac{1}{9}\right)$

20. $\left(10\frac{4}{5} + 7\frac{5}{9}\right) \div \frac{17}{30}$

Answers on page A-5

2.6 EXERCISE SET

For Extra Help

Digital Video Tutor CD 2 Videotape 3 | InterAct Math | Math Tutor Center | MathXL | MyMathLab

a Simplify.

1. $\frac{1}{2} \cdot \frac{1}{3} \cdot \frac{1}{4}$

2. $\frac{1}{3} \cdot \frac{1}{4} \cdot \frac{1}{5}$

3. $6 \div 3 \div 5$

4. $12 \div 4 \div 8$

5. $\frac{2}{3} \div \frac{4}{3} \div \frac{7}{8}$

6. $\frac{5}{6} \div \frac{3}{4} \div \frac{2}{5}$

7. $\frac{5}{8} \div \frac{1}{4} - \frac{2}{3} \cdot \frac{4}{5}$

8. $\frac{4}{7} \cdot \frac{7}{15} + \frac{2}{3} \div 8$

9. $\frac{3}{4} - \frac{2}{3} \cdot \left(\frac{1}{2} + \frac{2}{5}\right)$

10. $\frac{3}{4} \div \frac{1}{2} \cdot \left(\frac{8}{9} - \frac{2}{3}\right)$

11. $28\frac{1}{8} - 5\frac{1}{4} + 3\frac{1}{2}$

12. $10\frac{3}{5} - 4\frac{1}{10} - 1\frac{1}{2}$

13. $\frac{7}{8} \div \frac{1}{2} \cdot \frac{1}{4}$

14. $\frac{7}{10} \cdot \frac{4}{5} \div \frac{2}{3}$

15. $\left(\frac{2}{3}\right)^2 - \frac{1}{3} \cdot 1\frac{1}{4}$

16. $\left(\frac{3}{4}\right)^2 + 3\frac{1}{2} \div 1\frac{1}{4}$

17. $\frac{1}{2} - \left(\frac{1}{2}\right)^2 + \left(\frac{1}{2}\right)^3$

18. $1 + \frac{1}{4} + \left(\frac{1}{4}\right)^2 - \left(\frac{1}{4}\right)^3$

19. Find the average of $\frac{2}{3}$ and $\frac{7}{8}$.

20. Find the average of $\frac{1}{4}$ and $\frac{1}{5}$.

21. Find the average of $\frac{1}{6}$, $\frac{1}{8}$, and $\frac{3}{4}$.

22. Find the average of $\frac{4}{5}$, $\frac{1}{2}$, and $\frac{1}{10}$.

23. Find the average of $3\frac{1}{2}$ and $9\frac{3}{8}$.

24. Find the average of $10\frac{2}{3}$ and $24\frac{5}{6}$.

25. *Birth Weights.* The Piper quadruplets of Great Britain weighed $2\frac{9}{16}$ lb, $2\frac{9}{32}$ lb, $2\frac{1}{8}$ lb, and $2\frac{5}{16}$ lb at birth. Find their average birth weight.
Source: *The Guinness Book of Records,* 1998

26. *Vertical Leaps.* Eight-year-old Zachary registered vertical leaps of $12\frac{3}{4}$ in., $13\frac{3}{4}$ in., $13\frac{1}{2}$ in., and 14 in. Find his average vertical leap.

27. *Manufacturing.* A test of five light bulbs showed that they burned for the lengths of time given on the graph below. For how many days, on average, did the bulbs burn?

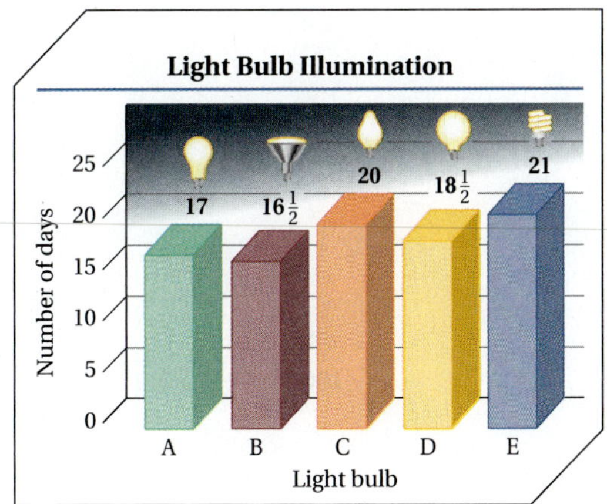

28. *Packaging.* A sample of four bags of beef jerky showed the weights given on the graph below. What was the average weight?

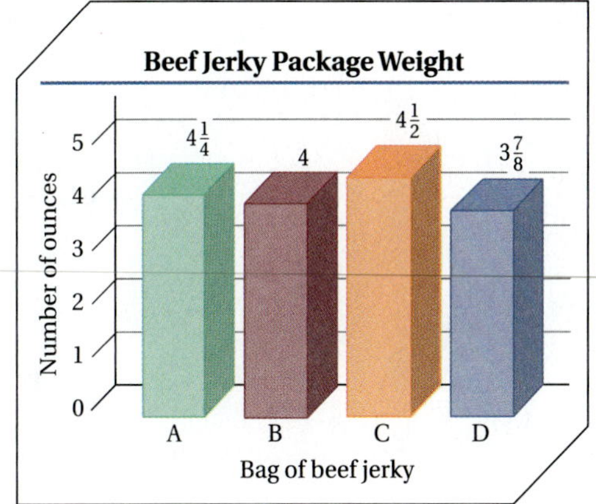

Simplify.

29. $\left(\frac{2}{3}+\frac{3}{4}\right) \div \left(\frac{5}{6}-\frac{1}{3}\right)$

30. $\left(\frac{3}{5}-\frac{1}{2}\right) \div \left(\frac{3}{4}-\frac{3}{10}\right)$

31. $\left(\frac{1}{2}+\frac{1}{3}\right)^2 \cdot 144 - \frac{5}{8} \div 10\frac{1}{2}$

32. $\left(3\frac{1}{2}-2\frac{1}{3}\right)^2 + 6 \cdot 2\frac{1}{2} \div 32$

b Estimate each of the following as 0, $\frac{1}{2}$, or 1.

33. $\frac{2}{47}$

34. $\frac{4}{5}$

35. $\frac{1}{13}$

36. $\frac{7}{8}$

37. $\frac{6}{11}$

38. $\frac{10}{13}$

39. $\frac{7}{15}$

40. $\frac{1}{16}$

41. $\frac{7}{100}$

42. $\frac{5}{9}$

43. $\frac{19}{20}$

44. $\frac{5}{12}$

Find a number for the blank so that the fraction is close to but greater than $\frac{1}{2}$. Answers may vary.

45. $\frac{\square}{11}$

46. $\frac{\square}{8}$

47. $\frac{\square}{23}$

48. $\frac{\square}{35}$

49. $\frac{10}{\square}$

50. $\frac{51}{\square}$

Find a number for the blank so that the fraction is close to but greater than 1. Answers may vary.

51. $\frac{7}{\square}$ **52.** $\frac{11}{\square}$ **53.** $\frac{13}{\square}$ **54.** $\frac{27}{\square}$ **55.** $\frac{\square}{15}$ **56.** $\frac{\square}{100}$

Estimate each part of the following as a whole number, as $\frac{1}{2}$, or as a mixed numeral where the fractional part is $\frac{1}{2}$.

57. $2\frac{7}{8}$

58. $1\frac{1}{3}$

59. $12\frac{5}{6}$

60. $26\frac{6}{13}$

61. $\frac{4}{5} + \frac{7}{8}$

62. $\frac{1}{12} \cdot \frac{7}{15}$

63. $\frac{2}{3} + \frac{7}{13} + \frac{5}{9}$

64. $\frac{8}{9} + \frac{4}{5} + \frac{11}{12}$

65. $\frac{43}{100} + \frac{1}{10} - \frac{11}{1000}$

66. $\frac{23}{24} + \frac{37}{39} + \frac{51}{50}$

67. $7\frac{29}{60} + 10\frac{12}{13} \cdot 24\frac{2}{17}$

68. $5\frac{13}{14} - 1\frac{5}{8} + 1\frac{23}{28} \cdot 6\frac{35}{74}$

69. $24 \div 7\frac{8}{9}$

70. $43\frac{16}{17} \div 11\frac{2}{13}$

71. $76\frac{3}{14} + 23\frac{19}{20}$

72. $76\frac{13}{14} \cdot 23\frac{17}{20}$

73. $16\frac{1}{5} \div 2\frac{1}{11} + 25\frac{9}{10} - 4\frac{11}{23}$

74. $96\frac{2}{13} \div 5\frac{19}{20} + 3\frac{1}{7} \cdot 5\frac{18}{21}$

75. DW A student insists that $3\frac{2}{5} \cdot 1\frac{3}{7} = 3\frac{6}{35}$. What mistake is he making and how should he have proceeded?

76. DW A student insists that $5 \cdot 3\frac{2}{7} = (5 \cdot 3) \cdot (5 \cdot \frac{2}{7})$. What mistake is she making and how should she have proceeded?

SKILL MAINTENANCE

77. Multiply: $27 \cdot 126$. [1.3a]

78. Multiply: $132 \cdot 7865$. [1.3a]

79. Divide: $7865 \div 132$. [1.3d]

Multiply and simplify. [2.2a]

80. $\frac{2}{3} \cdot 522$

81. $\frac{3}{2} \cdot 522$

Divide and simplify. [2.2c]

82. $\frac{4}{5} \div \frac{3}{10}$

83. $\frac{3}{10} \div \frac{4}{5}$

84. Classify the given numbers as prime, composite, or neither. [1.7c]

1, 5, 7, 9, 14, 23, 43

Solve.

85. *Luncheon Servings.* Ian purchased 6 lb of cold cuts for a luncheon. If Ian is to allow $\frac{3}{8}$ lb per person, how many people can he invite to the luncheon? [2.5a]

86. *Cholesterol.* A 3-oz serving of crabmeat contains 85 milligrams (mg) of cholesterol. A 3-oz serving of shrimp contains 128 mg of cholesterol. How much more cholesterol is in the shrimp? [1.5a]

SYNTHESIS

87. **a)** Find an expression for the sum of the areas of the two rectangles shown here.

b) Simplify the expression.

c) How is the computation in part (b) related to the rules for order of operations?

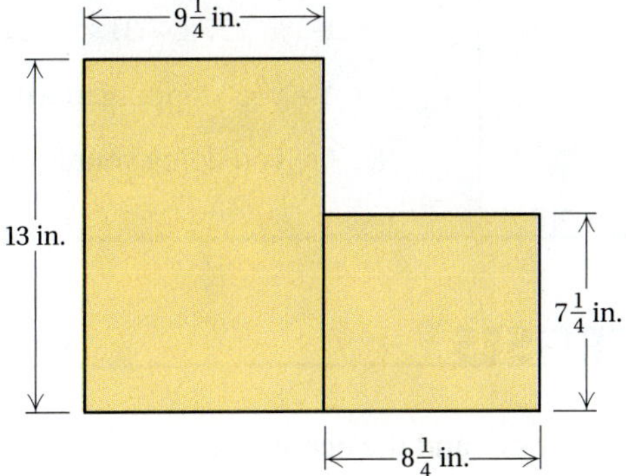

88. Find r if

$$\frac{1}{r} = \frac{1}{100} + \frac{1}{150} + \frac{1}{200}.$$

89. 🖩 In the sum below, a and b are digits. Find a and b.

$$\frac{a}{17} + \frac{1b}{23} = \frac{35a}{391}$$

90. 🖩 Consider only the numbers 3, 4, 5, and 6. Assume each can be placed in a blank in the following.

$$\square + \frac{\square}{\square} \cdot \square = ?$$

What placement of the numbers in the blanks yields the largest number?

91. 🖩 Consider only the numbers 2, 3, 4, and 5. Assume each is placed in a blank in the following.

$$\frac{\square}{\square} + \frac{\square}{\square} = ?$$

What placement of the numbers in the blanks yields the largest sum?

92. 🖩 Use a standard calculator. Arrange the following in order from smallest to largest.

$$\frac{3}{4}, \frac{17}{21}, \frac{13}{15}, \frac{7}{9}, \frac{15}{17}, \frac{13}{12}, \frac{19}{22}$$

CHAPTER 2

Summary and Review

The review that follows is meant to prepare you for a chapter exam. It consists of two parts. The first part is a checklist of the Study Tips referred to so far in this text. The second part is the Review Exercises. These provide practice exercises for the exam, together with references to section objectives so you can go back and review. Before beginning, stop and look back over the skills you have obtained. What skills in mathematics do you have now that you did not have before studying this chapter?

STUDY TIPS CHECKLIST

The foundation of all your study skills is TIME!

- ☐ Are you making use of the supplements that accompany this text?
- ☐ Are you doing some skill maintenance exercises as part of your daily assignment whether they have been assigned or not?
- ☐ Have you tried calling the Addison-Wesley Math Tutor Center at 1-888-777-0463?
- ☐ Are you stopping to work the margin exercises when directed to do so?
- ☐ Are you doing your homework as soon as possible after class?

REVIEW EXERCISES

1. Identify the numerator and the denominator of $\frac{2}{7}$. [2.1a]

2. What fractional part is shaded? [2.1a]

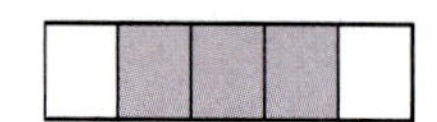

3. Simplify, if possible, the fractions on this circle graph. [2.1e]

How the Business Travel Dollar is Spent

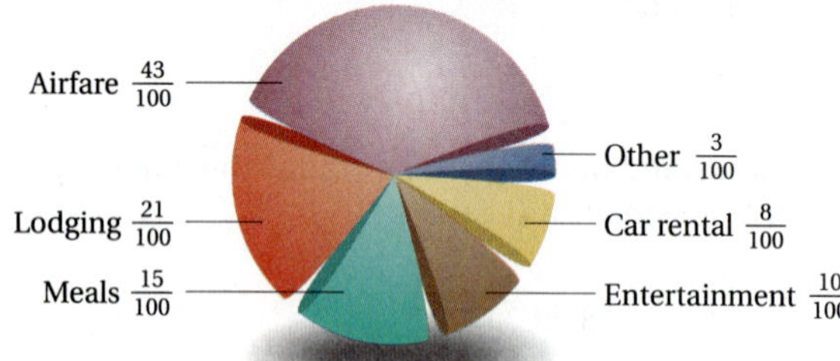

Simplify. [2.1b, e]

4. $\frac{0}{4}$

5. $\frac{23}{23}$

6. $\frac{48}{1}$

7. $\frac{48}{8}$

8. $\frac{12}{30}$

9. $\frac{18}{0}$

10. $\frac{9}{27}$

11. $\frac{7}{28}$

12. $\frac{10}{15}$

Multiply and simplify. [2.2a]

13. $4 \cdot \frac{3}{8}$

14. $\frac{6}{5} \cdot 20$

15. $\frac{5}{7} \cdot \frac{1}{10}$

16. $\frac{3}{7} \cdot \frac{14}{9}$

Find the reciprocal. [2.2b]

17. $\frac{4}{5}$

18. 3

Divide and simplify. [2.2c]

19. $6 \div \frac{4}{3}$

20. $\frac{5}{9} \div \frac{5}{18}$

21. $\frac{1}{6} \div \frac{1}{11}$

22. $\frac{3}{14} \div \frac{6}{7}$

23. $\frac{1}{4} \div \frac{1}{9}$

24. $180 \div \frac{3}{5}$

25. $\frac{23}{25} \div \frac{23}{25}$

26. $\frac{2}{3} \div \frac{3}{2}$

Add and simplify. [2.3a]

27. $\frac{6}{5} + \frac{3}{8}$

28. $\frac{5}{16} + \frac{1}{12}$

29. $\frac{6}{5} + \frac{11}{15}$

30. $\frac{5}{16} + \frac{1}{8}$

Subtract and simplify. [2.3b]

31. $\frac{5}{9} - \frac{2}{9}$

32. $\frac{7}{8} - \frac{3}{4}$

33. $\frac{11}{27} - \frac{2}{9}$

34. $\frac{5}{6} - \frac{2}{9}$

Use $<$ or $>$ for $\square$ to write a true sentence. [2.3c]

35. $\frac{4}{7} \square \frac{5}{9}$

36. $\frac{8}{9} \square \frac{11}{13}$

Convert to fraction notation. [2.4a]

37. $7\frac{1}{2}$

38. $8\frac{3}{8}$

Convert to a mixed numeral. [2.4a]

39. $\frac{7}{3}$

40. $\frac{27}{4}$

Add. Write a mixed numeral for the answer. [2.4b]

41. $5\frac{3}{5} + 4\frac{4}{5}$

42. $8\frac{1}{3} + 3\frac{2}{5}$

43. $5\frac{5}{6} + 4\frac{5}{6}$

44. $2\frac{3}{4} + 5\frac{1}{2}$

Subtract. Write a mixed numeral for the answer where appropriate. [2.4c]

45. $12 - 4\frac{2}{9}$

46. $9\frac{3}{5} - 4\frac{13}{15}$

47. $10\frac{1}{4} - 6\frac{1}{10}$

48. $24 - 10\frac{5}{8}$

Multiply. Write a mixed numeral for the answer where appropriate. [2.4d]

49. $6 \cdot 2\frac{2}{3}$

50. $5\frac{1}{4} \cdot \frac{2}{3}$

51. $2\frac{1}{5} \cdot 1\frac{1}{10}$

52. $2\frac{2}{5} \cdot 2\frac{1}{2}$

Divide. Write a mixed numeral for the answer where appropriate. [2.4e]

53. $27 \div 2\frac{1}{4}$

54. $2\frac{2}{5} \div 1\frac{7}{10}$

55. $3\frac{1}{4} \div 26$

56. $4\frac{1}{5} \div 4\frac{2}{3}$

Solve. [2.2d], [2.3d]

57. $\frac{5}{4} \cdot t = \frac{3}{8}$

58. $x \cdot \frac{2}{3} = 160$

59. $x + \frac{2}{5} = \frac{7}{8}$

60. $\frac{1}{2} + y = \frac{9}{10}$

Solve. [2.5a]

61. A road crew repaves $\frac{1}{12}$ mi of road each day. How long will it take the crew to repave a $\frac{3}{4}$-mi stretch of road?

62. After driving 60 km, the Bonewitz family has completed $\frac{3}{5}$ of their vacation. How long is the total trip?

63. Molly is making a pepper steak recipe that calls for $\frac{2}{3}$ cup of green bell peppers. How much would be needed to make $\frac{1}{2}$ recipe? 3 recipes?

64. Bernardo usually earns \$105 for working a full day. How much does he receive for working $\frac{1}{7}$ of a day?

65. *Sewing from a Pattern.* Suppose you want to make an outfit in size 12. On the back of the pattern envelope, it states that for size 12, using 60-in. fabric, you need $1\frac{5}{8}$ yd of fabric for the dress and $2\frac{5}{8}$ yd for the jacket. How many yards in all are needed to make the outfit?

66. *Turkey Servings.* Turkey contains $1\frac{1}{3}$ servings per pound. How many pounds are needed for 32 servings?

67. *Weightlifting.* In 1998, Sun Tianni of China snatched 111 kg. This amount was about $1\frac{3}{5}$ times her body weight. How much did Tianni weigh?
Source: *The Guinness Book of Records*, 2000

68. *Carpentry.* A board $\frac{9}{10}$ in. thick is glued to a board $\frac{8}{10}$ in. thick. The glue is $\frac{3}{100}$ in. thick. How thick is the result?

69. What is the sum of the areas in the figure below?

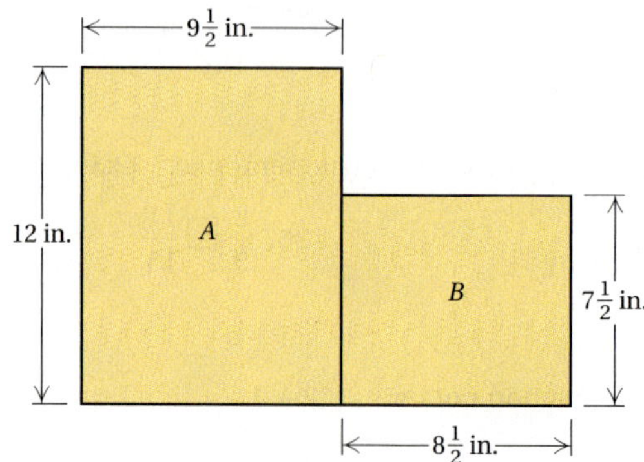

70. In the figure above, how much larger is the area of rectangle A than the area of rectangle B?

71. *Cake Recipe.* A wedding-cake recipe requires 12 cups of shortening. Being calorie-conscious, the wedding couple decides to reduce the shortening by $3\frac{5}{8}$ cups and replace it with prune purée. How many cups of shortening are used in their new recipe?

72. *Firefighters' Pie Sale.* Green River's Volunteer Fire Department recently hosted its annual ice cream social. Each of the donated 83 pies was cut into 6 pieces. At the end of the evening, the cashier said they had sold 382 pieces of pie. How many pies did they sell? How many were left over? Express your answers in mixed numerals.

73. Simplify this expression using the rules for order of operations: [2.6a]

$$\frac{1}{8} \div \frac{1}{4} + \frac{1}{2}.$$

74. Find the average of $\frac{1}{2}, \frac{1}{4}, \frac{1}{3}$, and $\frac{1}{5}$. [2.6a]

Estimate each of the following as 0, $\frac{1}{2}$, or 1. [2.6b]

75. $\frac{29}{59}$

76. $\frac{2}{59}$

77. $\frac{61}{59}$

Estimate each of the following as a whole number or as a mixed numeral where the fractional part is $\frac{1}{2}$. [2.6b]

78. $6\frac{7}{8}$

79. $10\frac{2}{17}$

80. $\frac{3}{10} + \frac{5}{6} + \frac{31}{29}$

81. $32\frac{14}{15} + 27\frac{3}{4} - 4\frac{25}{28} \cdot 6\frac{37}{76}$

82. **DW** A student claims that "taking $\frac{1}{2}$ of a number is the same as dividing by $\frac{1}{2}$." Explain the error in this reasoning. [2.2a, c]

83. **DW** Discuss the role of least common multiples in adding and subtracting with fraction notation. [2.3a, b]

SKILL MAINTENANCE

Beginning with this chapter, certain objectives from four particular sections will be retested on the chapter test. The objectives are listed with the practice problems that follow.

Solve. [1.4b]

84. $17 \cdot x = 408$

85. $765 + t = 1234$

Solve. [1.5a]

86. The balance in your checking account is $789. After purchases of $78, $97, and $102 and a deposit of $400, what is your new balance?

87. A new Beetle 1.9L TDI by Volkswagen gets 43 mpg on the highway. How far can the car be driven on a full tank of 18 gal of gasoline?
Source: 2000 Volkswagen of America

88. *Digital Tire Gauges.* A factory produces 3885 digital tire gauges per day. How long will it take to fill an order for 66,045 tire gauges?

Tire gauge

89. Divide: [1.3d]

$$36\overline{)14{,}697}$$

90. Subtract: [1.2d]

$$\begin{array}{r} 5604 \\ -\ 1997 \\ \hline \end{array}$$

SYNTHESIS

91. Place the numbers 3, 4, 5, and 6 in the boxes in order to make a true equation: [2.3a]

$$\frac{\square}{\square} + \frac{\square}{\square} = 3\frac{1}{4}.$$

92. In the division below, find a and b. [2.2c]

$$\frac{19}{24} \div \frac{a}{b} = \frac{187{,}853}{268{,}224}$$

CHAPTER 2

Chapter Test

1. What part is shaded?

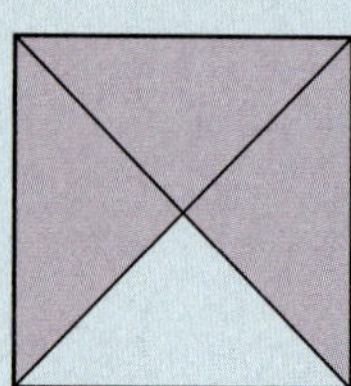

2. Use < or > for $\square$ to write a true sentence:

$$\frac{6}{7} \quad \frac{21}{25}.$$

Simplify.

3. $\frac{12}{12}$

4. $\frac{0}{16}$

5. $\frac{2}{28}$

6. $\frac{9}{0}$

Multiply and simplify.

7. $\frac{4}{3} \cdot 24$

8. $\frac{2}{3} \cdot \frac{15}{4}$

9. $\frac{3}{5} \cdot \frac{1}{6}$

Find the reciprocal.

10. $\frac{5}{8}$

11. 18

Divide and simplify.

12. $\frac{3}{8} \div \frac{5}{4}$

13. $\frac{1}{5} \div \frac{1}{8}$

14. $12 \div \frac{2}{3}$

Add and simplify.

15. $\frac{1}{2} + \frac{5}{2}$

16. $\frac{7}{8} + \frac{2}{3}$

17. $\frac{7}{10} + \frac{9}{100}$

Subtract and simplify.

18. $\frac{5}{6} - \frac{3}{6}$

19. $\frac{5}{6} - \frac{3}{4}$

20. $\frac{17}{24} - \frac{5}{8}$

Solve.

21. $\frac{7}{8} \cdot x = 56$

22. $x + \frac{2}{3} = \frac{11}{12}$

23. Convert to fraction notation: $3\frac{1}{2}$.

24. Convert to a mixed numeral: $\frac{74}{9}$.

Add, subtract, multiply, or divide. Write a mixed numeral for the answer in Exercises 25–28.

25. $\begin{array}{r} 6\frac{2}{5} \\ +\ 7\frac{4}{5} \\ \hline \end{array}$

26. $\begin{array}{r} 10\frac{1}{6} \\ -\ 5\frac{7}{8} \\ \hline \end{array}$

27. $6\frac{3}{4} \cdot \frac{2}{3}$

28. $2\frac{1}{3} \div 1\frac{1}{6}$

29. There are 7000 students at La Poloma College, and $\frac{5}{8}$ of them live in dorms. How many live in dorms?

30. A strip of taffy $\frac{9}{10}$ m long is cut into 12 equal pieces. What is the length of each piece?

31. *Book Order.* An order of books for a math course weighs 220 lb. Each book weighs $2\frac{3}{4}$ lb. How many books are in the order?

32. *Carpentry.* In carpentry, some pieces of plywood that are called "$\frac{3}{4}$-inch" plywood are actually $\frac{11}{16}$-in. thick. How much thinner is such a piece than its name indicates?

33. *Weightlifting.* In 1999, Hossein Rezazadeh of Iran did a clean and jerk of $262\frac{1}{2}$ kg. This amount was $2\frac{1}{2}$ times his body weight. How much did Rezazadeh weigh?
Source: *The Guinness Book of Records,* 2000

34. *Women's Dunks.* Only three women in the history of college basketball have been able to dunk a basketball. Their names, heights, and universities are:

Michelle Snow, $6\frac{5}{12}$ ft, Tennessee;
Charlotte Smith, $5\frac{11}{12}$ ft, North Carolina;
Georgeann Wells, $6\frac{7}{12}$ ft, West Virginia.

Find the average height of these women.
Source: *USA Today,* 11/30/00. p. 3C

35. Simplify: $\frac{2}{3} + 1\frac{1}{3} \cdot 2\frac{1}{8}$.

Estimate each of the following as 0, $\frac{1}{2}$, or 1.

36. $\frac{3}{82}$

37. $\frac{93}{91}$

Estimate each of the following as a whole number or as a mixed numeral where the fractional part is $\frac{1}{2}$.

38. $18\frac{9}{17}$

39. $256 \div 15\frac{19}{21}$

SKILL MAINTENANCE

Solve.

40. $x + 198 = 2003$

41. $47 \cdot t = 4747$

42. It is 2060 mi from San Francisco to Winnipeg, Canada. It is 1575 mi from Winnipeg to Atlanta. What is the total length of a route from San Francisco to Winnipeg to Atlanta?

43. Divide: $24\overline{)9127}$

44. Subtract: $\begin{array}{r} 8001 \\ -\ 3567 \\ \hline \end{array}$

SYNTHESIS

45. A recipe for a batch of buttermilk pancakes calls for $\frac{3}{4}$ teaspoon (tsp) of salt. Jacqueline plans to cut the amount of salt in half for each of 5 batches of pancakes. How much salt will she need?

46. Dolores runs 17 laps at her health club. Terence runs 17 laps at his health club. If the track at Dolores's health club is $\frac{1}{7}$ mi long, and the track at Terence's is $\frac{1}{8}$ mi long, who runs farther? How much farther?

Decimal Notation

Gateway to Chapter 3

In this chapter, we consider the operations of addition, subtraction, multiplication, and division with decimal notation. These skills will allow us to solve applied problems like the one about changing values of The Quaker Oats Company stock shown here. We will also study estimating sums, differences, products, and quotients. Conversion between fraction and decimal notation in which the decimal notation may be repeating will be discussed as well.

100
90
80
70
60
50
JAN.
40

Real-World Application

Over the entire 52 weeks of 2000, the price per share of The Quaker Oats Company stock ranged in value from a low of \$45.81 to a high of \$98.94. By how much did the high value differ from the low value?

Sources: The New York Stock Exchange; The Quaker Oats Company

This problem appears as Example 1 in Section 3.7.

CHAPTER 3

THE QUAKER OATS COMPANY, 2000

MARCH
MAY
JULY
SEPT.
NOV.

OATMEAL HELPS REMOVE CHOLESTEROL!
OLD FASHIONED
QUAKER OATS
100% NATURAL

QUAKER OATMEAL
Express
QUAKER
Chewy
LOW FAT GRANOLA BARS
Oatmeal Raisin
10 Moist & Chewy Bars

CHAPTER 3

Pretest

1. Write a word name: 2.347. [3.1a]

2. Write a word name, as on a check, for $3264.78. [3.1a]

Write fraction notation. [3.1b]

3. 0.21

4. 5.408

Write decimal notation. [3.1b]

5. $\frac{379}{1000}$

6. $28\frac{439}{1000}$

Which number is larger? [3.1c]

7. 3.2, 0.321

8. 0.099, 0.091

Round 21.0448 to the nearest: [3.1d]

9. Tenth.

10. Thousandth.

11. Add: [3.2a]

$$\begin{array}{r} 601.3 \\ 5.81 \\ +\ 0.109 \\ \hline \end{array}$$

12. Subtract: [3.2b]

$$\begin{array}{r} 40.0 \\ -\ 0.9099 \\ \hline \end{array}$$

Multiply. [3.3a]

13.

$$\begin{array}{r} 0.835 \\ \times\ 0.74 \\ \hline \end{array}$$

14. 0.001×324.56

Divide. [3.4a]

15. $6.6\overline{)200.64}$

16. $\frac{576.98}{1000}$

Solve.

17. $9.6 \cdot y = 808.896$ [3.4b]

18. $54.96 + q = 6400.117$ [3.2c]

Solve. [3.7a]

19. Travel Distance. On a three-day trip, a traveler drove these distances: 432.6 mi, 179.2 mi, and 469.8 mi. What is the total number of miles driven?

20. A checking account contained $434.19. After a purchase of $148.24 was made using a debit card, how much was left in the account?

21. DVD Purchase. Tanya bought 8 DVDs of the movie *The Matrix*. She paid $19.98 for each copy. What was the total cost?

22. A developer paid $47,567.89 for 14 acres of land. How much was paid for 1 acre? Round to the nearest cent.

23. Estimate the product 6.92×32.458 by rounding to the nearest one. [3.6a]

Find decimal notation. Use multiplying by 1. [3.5a]

24. $\frac{7}{5}$

25. $\frac{37}{40}$

Find decimal notation. Use division. [3.5a]

26. $\frac{11}{4}$

27. $\frac{29}{7}$

Round $4.\overline{61}$ to the nearest: [3.5b]

28. Tenth.

29. Hundredth.

30. Thousandth.

31. Convert from cents to dollars: 949¢. [3.3b]

32. Convert to standard notation: 490 trillion. [3.3b]

Calculate.

33. $(1 - 0.06)^2 + 8[5(12.1 - 7.8) + 20(17.3 - 8.7)]$ [3.4c]

34. $\frac{2}{3} \times 89.95 - \frac{5}{9} \times 3.234$ [3.5c]

3.1 DECIMAL NOTATION, ORDER, AND ROUNDING

Objectives

a Given decimal notation, write a word name, and write a word name for an amount of money.

b Convert between fraction notation and decimal notation.

c Given a pair of numbers in decimal notation, tell which is larger.

d Round decimal notation to the nearest thousandth, hundredth, tenth, one, ten, hundred, or thousand.

The set of **arithmetic numbers,** or **nonnegative rational numbers,** consists of the whole numbers 0, 1, 2, 3, 4, 5, 6, 7, 8, 9, 10, and so on, and fractions like $\frac{1}{2}, \frac{2}{3}, \frac{7}{8}, \frac{17}{10}$, and so on. We studied the use of fraction notation for arithmetic numbers in Chapter 2. In Chapter 3, we will study the use of *decimal notation.* The word *decimal* comes from the Latin word *decima,* meaning a tenth part. Although we are using different notation, we are still considering the same set of numbers. For example, instead of using fraction notation for $\frac{7}{8}$, we use decimal notation, 0.875, and instead of $48\frac{97}{100}$, we use 48.97.

a Decimal Notation and Word Names

The Razor Kick Scooter® costs \$148.97. The dot in \$148.97 is called a **decimal point.** Since 0.97, or 97¢, is $\frac{97}{100}$ of a dollar, it follows that

$$\$148.97 = 148 + \frac{97}{100} \text{ dollars.}$$

Also, since \$0.97, or 97¢, has the same value as

9 dimes + 7 cents

and 1 dime is $\frac{1}{10}$ of a dollar and 1 cent is $\frac{1}{100}$ of a dollar, we can write

$$148.97 = 1 \cdot 100 + 4 \cdot 10 + 8 \cdot 1 + 9 \cdot \frac{1}{10} + 7 \cdot \frac{1}{100}.$$

This is an extension of the expanded notation for whole numbers that we used in Chapter 1. The place values are 100, 10, 1, $\frac{1}{10}, \frac{1}{100}$, and so on. We can see this on a **place-value chart.** The value of each place is $\frac{1}{10}$ as large as the one to its left.

Let's see how to understand decimal notation using a place-value chart, using the following:

PLACE-VALUE CHART

Hundreds	Tens	Ones	Tenths	Hundredths	Thousandths	Ten-Thousandths	Hundred-Thousandths
100	10	1	$\frac{1}{10}$	$\frac{1}{100}$	$\frac{1}{1000}$	$\frac{1}{10,000}$	$\frac{1}{100,000}$
	2	9 .	5	2	9	7	

The women's record, held by Junxia Wang of China, in the 10,000-meter run is 29.5297 min.

Source: Razor USA

Study Tips

QUIZ–TEST FOLLOW-UP

You may have just completed a chapter quiz or test. Immediately after each chapter quiz or test, write out a step-by-step solution of the questions you missed. Visit your instructor or tutor for help with problems that are still giving you trouble. When the week of the final examination arrives, you will be glad to have the excellent study guide these corrected tests provide.

Write a word name for the number.

1. Each person in the United States consumed an average of 15.3 lb of seafood in a recent year.
Source: National Oceanographic and Atmospheric Administration

2. In 1999, the race horse *Charismatic* won the Kentucky Derby in a time of 2.05333 min.
Source: *The New York Times Almanac, 2000*

3. 245.89

4. 34.0064

5. 31,079.764

Answers on page A-6

The decimal notation 29.5297 means

2 tens + 9 ones + 5 tenths + 2 hundredths + 9 thousandths + 7 ten-thousandths

or

$$2 \cdot 10 + 9 \cdot 1 + 5 \cdot \frac{1}{10} + 2 \cdot \frac{1}{100} + 9 \cdot \frac{1}{1000} + 7 \cdot \frac{1}{10{,}000}$$

or

$$20 + 9 + \frac{5}{10} + \frac{2}{100} + \frac{9}{1000} + \frac{7}{10{,}000}.$$

We read both 29.5297 and $29\frac{5297}{10{,}000}$ as

"Twenty-nine and five thousand two hundred ninety-seven ten-thousandths."

When we come to the decimal point, we read it as "and." We can also read 29.5297 as

"Two nine *point* five two nine seven."

To write a word name from decimal notation,

a) write a word name for the whole number (the number named to the left of the decimal point), — 397.685 → Three hundred ninety-seven

b) write the word "and" for the decimal point, and — 397.685 → Three hundred ninety-seven and

c) write a word name for the number named to the right of the decimal point, followed by the place value of the last digit. — 397.685 → Three hundred ninety-seven and six hundred eighty-five *thousandths*

EXAMPLE 1 Write a word name for the number in this sentence: Arnold Schwarzenegger's body mass index is 27.7.

Twenty-seven and seven tenths

EXAMPLE 2 Write a word name for 410.87.

Four hundred ten and eighty-seven hundredths

EXAMPLE 3 Write a word name for the number in this sentence: The world record in the men's 800-meter run is 1.6852 min, held by Wilson Kipketer of Denmark.

One and six thousand eight hundred fifty-two ten-thousandths

EXAMPLE 4 Write a word name for 1788.405.

One thousand, seven hundred eighty-eight and four hundred five thousandths

Do Exercises 1–5 on the preceding page.

Decimal notation is also used with money. It is common on a check to write "and ninety-five cents" as "and $\frac{95}{100}$ dollars."

EXAMPLE 5 Write a word name for the amount on the check, \$5876.95.

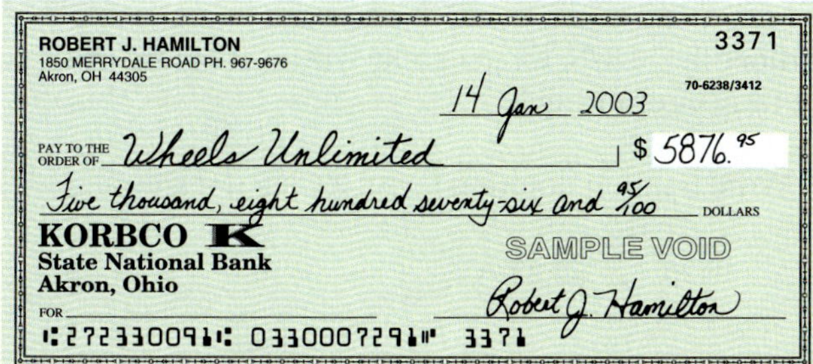

Five thousand, eight hundred seventy-six and $\frac{95}{100}$ dollars

Do Exercises 6 and 7.

Write a word name as on a check.

6. \$4217.56

7. \$13.98

b Converting Between Decimal Notation and Fraction Notation

We can find fraction notation as follows:

$$\begin{aligned} 9.875 &= 9 + \frac{8}{10} + \frac{7}{100} + \frac{5}{1000} \\ &= 9 \cdot \frac{1000}{1000} + \frac{8}{10} \cdot \frac{100}{100} + \frac{7}{100} \cdot \frac{10}{10} + \frac{5}{1000} \\ &= \frac{9000}{1000} + \frac{800}{1000} + \frac{70}{1000} + \frac{5}{1000} = \frac{9875}{1000}. \end{aligned}$$

Decimal notation → 9.875 ← 3 decimal places

Fraction notation → $\frac{9875}{1000}$ ← 3 zeros

To convert from decimal to fraction notation,

a) count the number of decimal places, — 4.98 (2 places)

b) move the decimal point that many places to the right, and — 4.98. (Move 2 places.)

c) write the answer over a denominator with a 1 followed by that number of zeros. — $\frac{498}{100}$ (2 zeros)

Answers on page A-6

Write fraction notation.

8. 0.896

9. 23.78

10. 5.6789

11. 1.9

Answers on page A-6

EXAMPLE 6 Write fraction notation for 0.876. Do not simplify.

0.876 0.876. $0.876 = \frac{876}{1000}$

3 places 3 zeros

For a number like 0.876, we generally write a 0 before the decimal point to avoid forgetting or omitting it.

EXAMPLE 7 Write fraction notation for 56.23. Do not simplify.

56.23 56.23. $56.23 = \frac{5623}{100}$

2 places 2 zeros

EXAMPLE 8 Write fraction notation for 1.5018. Do not simplify.

1.5018 1.5018. $1.5018 = \frac{15,018}{10,000}$

4 places 4 zeros

Do Exercises 8–11.

If fraction notation has a denominator that is a power of ten, such as 10, 100, 1000, and so on, we reverse the procedure we used before.

To convert from fraction notation to decimal notation when the denominator is 10, 100, 1000, and so on,

a) count the number of zeros, and $\frac{8679}{1000}$ 3 zeros

b) move the decimal point that number of places to the left. Leave off the denominator.

8.679. Move 3 places.

$\frac{8679}{1000} = 8.679$

EXAMPLE 9 Write decimal notation for $\frac{47}{10}$.

$\frac{47}{10}$ 1 zero

4.7. 1 place

$\frac{47}{10} = 4.7$

EXAMPLE 10 Write decimal notation for $\frac{123{,}067}{10{,}000}$.

$\frac{123{,}067}{10{,}000}$ ← 4 zeros 12.3067. 4 places $\frac{123{,}067}{10{,}000} = 12.3067$

EXAMPLE 11 Write decimal notation for $\frac{13}{1000}$.

$\frac{13}{1000}$ ← 3 zeros 0.013. 3 places $\frac{13}{1000} = 0.013$

EXAMPLE 12 Write decimal notation for $\frac{570}{100{,}000}$.

$\frac{570}{100{,}000}$ ← 5 zeros 0.00570. 5 places $\frac{570}{100{,}000} = 0.0057$

Do Exercises 12–17.

When denominators are numbers other than 10, 100, and so on, we will use another method for conversion. It will be considered in Section 3.5.

If a mixed numeral has a fractional part with a denominator that is a power of ten, such as 10, 100, or 1000, and so on, we first write the mixed numeral as a sum of a whole number and a fraction. Then we convert to decimal notation.

EXAMPLE 13 Write decimal notation for $23\frac{59}{100}$.

$$23\frac{59}{100} = 23 + \frac{59}{100} = 23 \text{ and } \frac{59}{100} = 23.59$$

EXAMPLE 14 Write decimal notation for $772\frac{129}{10{,}000}$.

$$772\frac{129}{10{,}000} = 772 + \frac{129}{10{,}000} = 772 \text{ and } \frac{129}{10{,}000} = 772.0129$$

Do Exercises 18–20.

Write decimal notation.

12. $\frac{743}{100}$

13. $\frac{406}{1000}$

14. $\frac{67{,}089}{10{,}000}$

15. $\frac{9}{10}$

16. $\frac{57}{1000}$

17. $\frac{830}{10{,}000}$

Write decimal notation.

18. $4\frac{3}{10}$

19. $283\frac{71}{100}$

20. $456\frac{13}{1000}$

Answers on page A-6

Which number is larger?

21. 2.04, 2.039

22. 0.06, 0.008

23. 0.5, 0.58

24. 1, 0.9999

25. 0.8989, 0.09898

26. 21.006, 21.05

Answers on page A-6

c Order

To understand how to compare numbers in decimal notation, consider 0.85 and 0.9. First note that $0.9 = 0.90$ because $\frac{9}{10} = \frac{90}{100}$. Then $0.85 = \frac{85}{100}$ and $0.90 = \frac{90}{100}$. Since $\frac{85}{100} < \frac{90}{100}$, it follows that $0.85 < 0.90$. This leads us to a quick way to compare two numbers in decimal notation.

COMPARING NUMBERS IN DECIMAL NOTATION

To compare two numbers in decimal notation, start at the left and compare corresponding digits moving from left to right. If two digits differ, the number with the larger digit is the larger of the two numbers. To ease the comparison, extra zeros can be written to the right of the last decimal place.

EXAMPLE 15 Which of 2.109 and 2.1 is larger?

Think.

2.109 ⟶ 2.109

2.1 ⟶ 2.100

Same ↑ Different; $9 > 0$

Thus, 2.109 is larger than 2.1. That is, $2.109 > 2.1$.

EXAMPLE 16 Which of 0.09 and 0.108 is larger?

Think.

0.09 ⟶ 0.090

0.108 ⟶ 0.108

Same ↑ Different; $1 > 0$

Thus, 0.108 is larger than 0.09. That is, $0.108 > 0.09$.

Do Exercises 21–26.

d Rounding

Rounding is done as for whole numbers. To understand, we first consider an example using a number line. It might help to review Section 1.3.

EXAMPLE 17 Round 0.37 to the nearest tenth.

Here is part of a number line.

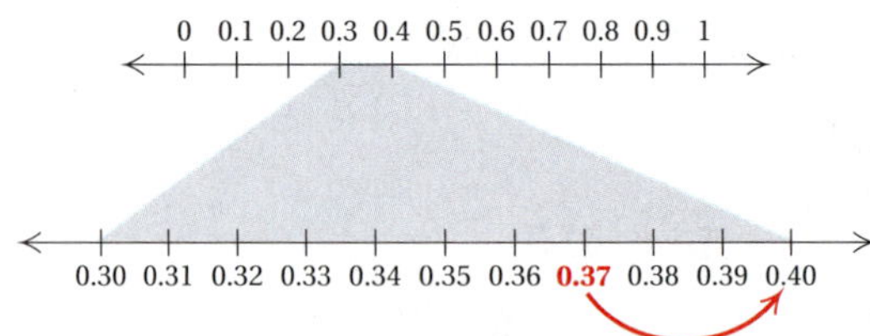

We see that 0.37 is closer to 0.40 than to 0.30. Thus, 0.37 rounded to the nearest tenth is 0.4.

ROUNDING DECIMAL NOTATION

To round to a certain place:

a) Locate the digit in that place.

b) Consider the next digit to the right.

c) If the digit to the right is 5 or higher, round up; if the digit to the right is 4 or lower, round down.

EXAMPLE 18 Round 3872.2459 to the nearest tenth.

a) Locate the digit in the tenths place, 2.

3 8 7 2.2 4 5 9
↑

b) Consider the next digit to the right, 4.

3 8 7 2.2 4 5 9
↑

c) Since that digit, 4, is less than 5, round down.

3 8 7 2.2 ← This is the answer.

CAUTION!

3872.3 is not a correct answer to Example 18. It is *incorrect* to round from the ten-thousandths digit over to the tenths digit, as follows:

3872.246→3872.25→3872.3.

EXAMPLE 19 Round 3872.2459 to the nearest thousandth, hundredth, tenth, one, ten, hundred, and thousand.

Thousandth:	3872.246	Ten:	3870
Hundredth:	3872.25	Hundred:	3900
Tenth:	3872.2	Thousand:	4000
One:	3872		

EXAMPLE 20 Round 14.8973 to the nearest hundredth.

a) Locate the digit in the hundredths place, 9. 1 4.8 9 7 3 ↑

b) Consider the next digit to the right, 7. 1 4.8 9 7 3 ↑

c) Since that digit, 7, is 5 or higher, round up. When we make the hundredths digit a 10, we carry 1 to the tenths place.

The answer is 14.90. Note that the 0 in 14.90 indicates that the answer is correct to the nearest hundredth.

EXAMPLE 21 Round 0.008 to the nearest tenth.

a) Locate the digit in the tenths place, 0. 0.0 0 8 ↑

b) Consider the next digit to the right, 0. 0.0 0 8 ↑

c) Since that digit, 0, is less than 5, round down.

The answer is 0.0.

Do Exercises 27–45.

Round to the nearest tenth.

27. 2.76 **28.** 13.85

29. 234.448 **30.** 7.009

Round to the nearest hundredth.

31. 0.636 **32.** 7.834

33. 34.675 **34.** 0.025

Round to the nearest thousandth.

35. 0.9434 **36.** 8.0038

37. 43.1119 **38.** 37.4005

Round 7459.3548 to the nearest:

39. Thousandth.

40. Hundredth.

41. Tenth.

42. One.

43. Ten. (*Caution:* "Tens" are not "tenths.")

44. Hundred.

45. Thousand.

Answers on page A-6

3.1 EXERCISE SET

For Extra Help

Digital Video Tutor CD 2 Videotape 4 | InterAct Math | Math Tutor Center | MathXL | MyMathLab

a Write a word name for the number in the sentence.

1. *MP3.* The MP3 audio format is changing the way we obtain and listen to music, allowing us to download digital music from the Internet. The cost of a Creative Labs NOMAD II Digital Audio Player is $249.94.
Source: Creative Labs

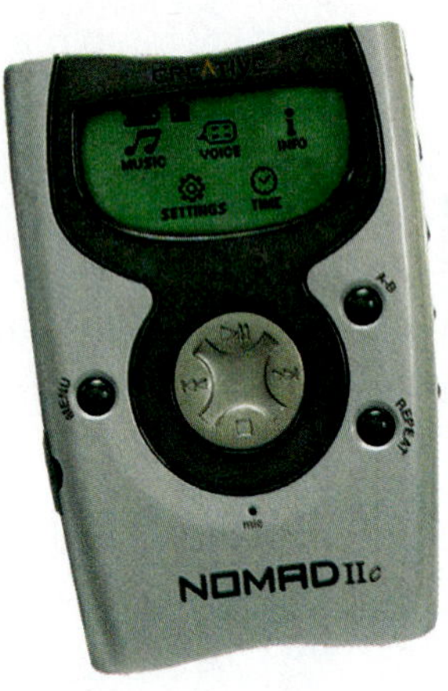

2. *Microsoft.* Recently, the stock of Microsoft sold for $43.9375 per share.

NASDAQ

NASDAQ COMPOSITE INDEX

MOST ACTIVE: SHARE VOLUME

	Vol.(000s)	Last	Change
SunMicro s	101,978	9.87	+.58
Cisco	83,072	14.94	+.52
Intel	47,702	21.96	+.41
DellCptr	42,577	22.56	+.24
Microsft	39,588	43.9375	+1.25
Oracle s	38,682	14.20	+.41
Qualcom	38,606	38.46	-3.54
JnprNtw	32,009	15.08	+1.81
WorldCom	30,837	13.33	-.60
JDS Uniph	27,145	6.92	-.10

3. *Quaker Oats.* Recently, the stock of Quaker Oats sold for $96.4375 per share.

4. *Water Weight.* One gallon of water weighs 8.35 lb.

Write a word name.

5. 34.891

6. 27.1245

Write a word name as on a check.

7. $326.48

8. $125.99

9. $36.72

10. $0.67

b Write fraction notation. Do not simplify.

11. 8.3

12. 0.17

13. 3.56

14. 203.6

15. 46.03

16. 1.509

17. 0.00013

18. 0.0109

19. 1.0008

20. 2.0114

21. 20.003

22. 4567.2

Write decimal notation.

23. $\frac{8}{10}$ **24.** $\frac{51}{10}$ **25.** $\frac{889}{100}$ **26.** $\frac{92}{100}$ **27.** $\frac{3798}{1000}$

28. $\frac{780}{1000}$ **29.** $\frac{78}{10,000}$ **30.** $\frac{56,788}{100,000}$ **31.** $\frac{19}{100,000}$ **32.** $\frac{2173}{100}$

33. $\frac{376,193}{1,000,000}$ **34.** $\frac{8,953,074}{1,000,000}$ **35.** $99\frac{44}{100}$ **36.** $4\frac{909}{1000}$ **37.** $3\frac{798}{1000}$

38. $67\frac{83}{100}$ **39.** $2\frac{1739}{10,000}$ **40.** $9243\frac{1}{10}$ **41.** $8\frac{953,073}{1,000,000}$ **42.** $2256\frac{3059}{10,000}$

c Which number is larger?

43. 0.06, 0.58 **44.** 0.008, 0.8 **45.** 0.905, 0.91 **46.** 42.06, 42.1

47. 0.0009, 0.001 **48.** 7.067, 7.054 **49.** 234.07, 235.07 **50.** 0.99999, 1

51. 0.004, $\frac{4}{100}$ **52.** $\frac{73}{10}$, 0.73 **53.** 0.432, 0.4325 **54.** 0.8437, 0.84384

d Round to the nearest tenth.

55. 0.11 **56.** 0.85 **57.** 0.49 **58.** 0.5794

59. 2.7449 **60.** 4.78 **61.** 123.65 **62.** 36.049

Round to the nearest hundredth.

63. 0.893 **64.** 0.675 **65.** 0.6666 **66.** 6.529

67. 0.995 **68.** 207.9976 **69.** 0.094 **70.** 11.4246

Round to the nearest thousandth.

71. 0.3246 **72.** 0.6666 **73.** 17.0015 **74.** 123.4562

75. 10.1011 **76.** 0.1161 **77.** 9.9989 **78.** 67.100602

Round 809.4732 to the nearest:

79. Hundred.

80. Tenth.

81. Thousandth.

82. Hundredth.

83. One.

84. Ten.

Round 34.54389 to the nearest:

85. Ten-thousandth.

86. Thousandth.

87. Hundredth.

88. Tenth.

89. One.

90. Ten.

91. DW Describe in your own words a procedure for converting from decimal notation to fraction notation.

92. DW A fellow student rounds 236.448 to the nearest one and gets 237. Explain the possible error.

SKILL MAINTENANCE

Round 6172 to the nearest: [1.3e]

93. Ten.

94. Hundred.

95. Thousand.

96. Find the LCM of 18, 27, and 54. [1.9a]

97. Subtract and simplify: $24 - 17\frac{2}{5}$. [2.4c]

Find the prime factorization. [1.7d]

98. 2000

99. 1530

100. 2002

101. 4312

SYNTHESIS

102. Arrange the following numbers in order from smallest to largest.

0.99, 0.099, 1, 0.9999, 0.89999, 1.00009, 0.909, 0.9889

103. Arrange the following numbers in order from smallest to largest.

2.1, 2.109, 2.108, 2.018, 2.0119, 2.0302, 2.000001

Truncating. There are other methods of rounding decimal notation. A computer often uses a method called **truncating.** To round using truncating, we drop off all decimal places past the rounding place, which is the same as changing all digits to the right to zeros. For example, rounding 6.78093456285102 to the ninth decimal place, using truncating, gives us 6.780934562. Use truncating to round each of the following to the fifth decimal place, that is, the hundred thousandth.

104. 6.78346123

105. 6.783461902

106. 99.999999999

107. 0.030303030303

3.2 ADDITION AND SUBTRACTION

a Addition

Adding with decimal notation is similar to adding whole numbers. First we line up the decimal points so that we can add corresponding place-value digits. Then we add digits from the right. For example, we add the thousandths, then the hundredths, and so on, carrying if necessary. If desired, we can write extra zeros to the right of the decimal point so that the number of places is the same.

EXAMPLE 1 Add: 56.314 + 17.78.

```
    5 6 . 3 1 4     Lining up the decimal points in order to add
  + 1 7 . 7 8 0     Writing an extra zero to the right
                    of the decimal point

    5 6 . 3 1 4     Adding thousandths
  + 1 7 . 7 8 0
  -------------
              4

    5 6 . 3 1 4     Adding hundredths
  + 1 7 . 7 8 0
  -------------
            9 4

      1
    5 6 . 3 1 4     Adding tenths
  + 1 7 . 7 8 0     Write a decimal point in the answer.
  -------------
        . 0 9 4     We get 10 tenths = 1 one + 0 tenths,
                    so we carry the 1 to the ones column.

    1 1
    5 6 . 3 1 4     Adding ones
  + 1 7 . 7 8 0
  -------------
      4 . 0 9 4     We get 14 ones = 1 ten + 4 ones,
                    so we carry the 1 to the tens column.

    1 1
    5 6 . 3 1 4     Adding tens
  + 1 7 . 7 8 0
  -------------
    7 4 . 0 9 4
```

Do Exercises 1 and 2.

EXAMPLE 2 Add: 3.42 + 0.237 + 14.1.

```
      3.4 2 0     Lining up the decimal points
      0.2 3 7     and writing extra zeros
  + 1 4.1 0 0
  -----------
    1 7.7 5 7     Adding
```

Do Exercises 3–5.

Objectives

a Add using decimal notation.

b Subtract using decimal notation.

c Solve equations of the type $x + a = b$ and $a + x = b$, where a and b may be in decimal notation.

d Balance a checkbook.

Add.

1.
```
     0.8 4 7
  + 1 0.0 7
```

2.
```
      2.1
      0.7 3 9
  + 3 1.3 6 8 9
```

Add.

3. 0.02 + 4.3 + 0.649

4. 0.12 + 3.006 + 0.4357

5. 0.4591 + 0.2374 + 8.70894

Answers on page A-6

Add.

6. 789 + 123.67

7. 45.78 + 2467 + 1.993

Subtract.

8. 37.428 − 26.674

9.
$$\begin{array}{r} 0.347 \\ -\ 0.008 \\ \hline \end{array}$$

Answers on page A-6

Consider the addition 3456 + 19.347. Keep in mind that any whole number has an "unwritten" decimal point at the right, with 0 fractional parts. For example, 3456 can also be written 3456.000. When adding, we can always write in that decimal point and extra zeros if desired.

EXAMPLE 3 Add: 3456 + 19.347.

$$\begin{array}{r} \scriptstyle 1 \quad\quad\quad\quad \\ 3456.000 \\ +\quad 19.347 \\ \hline 3475.347 \end{array}$$

Writing in the decimal point and extra zeros

Lining up the decimal points

Adding

Do Exercises 6 and 7.

b Subtraction

Subtracting with decimal notation is similar to subtracting whole numbers. First we line up the decimal points so that we can subtract corresponding place-value digits. Then we subtract digits from the right. For example, we subtract the thousandths, then the hundredths, the tenths, and so on, borrowing if necessary.

EXAMPLE 4 Subtract: 56.314 − 17.78.

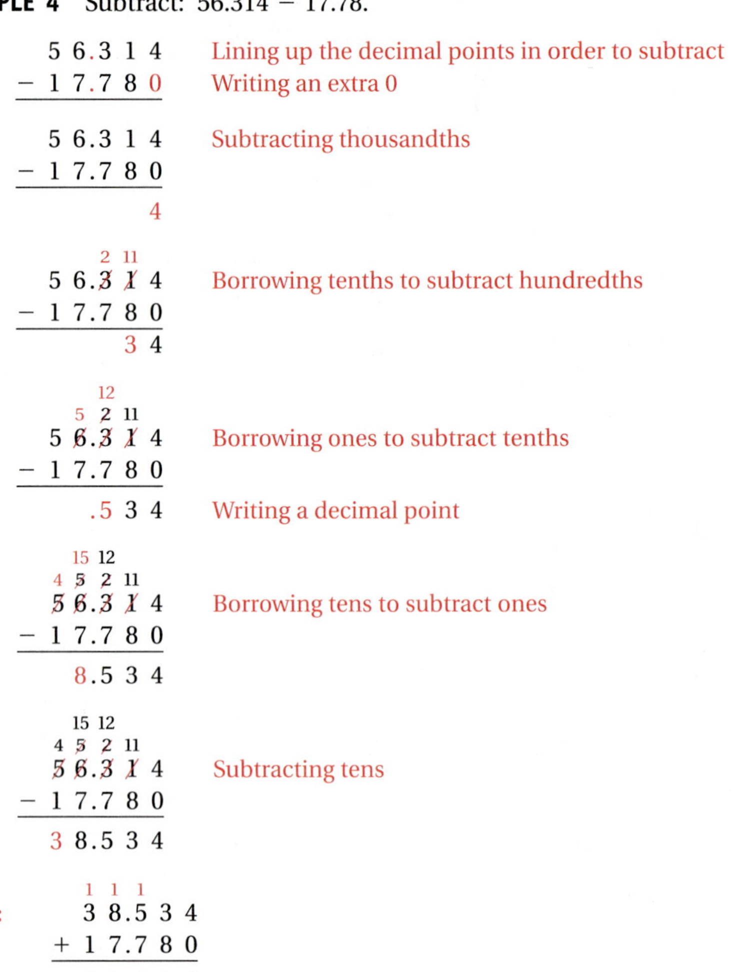

56.314
− 17.780
Lining up the decimal points in order to subtract
Writing an extra 0
Subtracting thousandths
Borrowing tenths to subtract hundredths
Borrowing ones to subtract tenths
Writing a decimal point
Borrowing tens to subtract ones
Subtracting tens
38.534
CHECK:
38.534
+ 17.780
56.314

Do Exercises 8 and 9.

EXAMPLE 5 Subtract: 13.07 − 9.205.

```
     12
      2  10 6 10
    1 3 . 0 7 0     Writing an extra zero
  −     9 . 2 0 5
  -------------
        3 . 8 6 5   Subtracting
```

EXAMPLE 6 Subtract: 23.08 − 5.0053.

```
   1 13     7  9 10
   2  3 . 0 8  0  0    Writing two extra zeros
 −    5 . 0 0  5  3
 ------------------
   1  8 . 0 7  4  7    Subtracting
```

Do Exercises 10–12.

When subtraction involves a whole number, again keep in mind that there is an "unwritten" decimal point that can be written in if desired. Extra zeros can also be written in to the right of the decimal point.

EXAMPLE 7 Subtract: 456 − 2.467.

```
        5  9 9 10
   4 5  6 . 0 0 0    Writing in the decimal point and extra zeros
 −      2 . 4 6 7
 ----------------
   4 5  3 . 5 3 3    Subtracting
```

Do Exercises 13 and 14.

CALCULATOR CORNER

Addition and Subtraction with Decimal Notation To use a calculator to add and subtract with decimal notation, we use the [·], [+], [−], and [=] keys. To find 47.046 − 28.193, for example, we press [4] [7] [·] [0] [4] [6] [−] [2] [8] [·] [1] [9] [3] [=]. The display reads [18.853], so 47.046 − 28.193 = 18.853.

Exercises:

Use a calculator to add.

1. $\begin{array}{r} 274.159 \\ +\ 43.486 \\ \hline \end{array}$

2. $\begin{array}{r} 19.805 \\ +\ 486.748 \\ \hline \end{array}$

3. 1.7 + 14.56 + 0.89

4. 3.4 + 45 + 0.68

Use a calculator to subtract.

5. $\begin{array}{r} 9.2 \\ -\ 4.8 \\ \hline \end{array}$

6. $\begin{array}{r} 52.34 \\ -\ 18.51 \\ \hline \end{array}$

7. 489 − 34.26

8. 6.09 − 5.1

Subtract.

10. 1.2345 − 0.7

11. 0.9564 − 0.4392

12. 7.37 − 0.00008

Subtract.

13. 1277 − 82.78

14. 5 − 0.0089

Answers on page A-6

Solve.

15. $x + 17.78 = 56.314$

16. $8.906 + t = 23.07$

17. Solve: $241 + y = 2374.5.$

Answers on page A-6

C Solving Equations

Now let's solve equations $x + a = b$ and $a + x = b$, where a and b may be in decimal notation. Proceeding as we have before, we subtract a on both sides.

EXAMPLE 8 Solve: $x + 28.89 = 74.567.$

We have

$$x + 28.89 - 28.89 = 74.567 - 28.89 \quad \text{Subtracting 28.89 on both sides}$$
$$x = 45.677.$$

$$\begin{array}{r} 74.567 \\ -\ 28.890 \\ \hline 45.677 \end{array}$$

The solution is 45.677.

EXAMPLE 9 Solve: $0.8879 + y = 9.0026.$

We have

$$0.8879 + y - 0.8879 = 9.0026 - 0.8879 \quad \text{Subtracting 0.8879 on both sides}$$
$$y = 8.1147.$$

$$\begin{array}{r} 9.0026 \\ -\ 0.8879 \\ \hline 8.1147 \end{array}$$

The solution is 8.1147.

Do Exercises 15 and 16.

EXAMPLE 10 Solve: $120 + x = 4380.6.$

We have

$$120 + x - 120 = 4380.6 - 120 \quad \text{Subtracting 120 on both sides}$$
$$x = 4260.6$$

$$\begin{array}{r} 4380.6 \\ -\ 120.0 \\ \hline 4260.6 \end{array}$$

The solution is 4260.6.

Do Exercise 17.

d Balancing a Checkbook

Let's use addition and subtraction with decimals to balance a checkbook.

EXAMPLE 11 Find the errors, if any, in the balances in this checkbook.

20___ RECORD ALL CHARGES OR CREDITS THAT AFFECT YOUR ACCOUNT

DATE	CHECK NUMBER	TRANSACTION DESCRIPTION	√ T	(−) PAYMENT/ DEBIT	(+ OR −) OTHER	(+) DEPOSIT/ CREDIT	BALANCE FORWARD
							8767 73
8/16	432	Burch Laundry		23 56			8744 16
8/19	433	Rogers TV		20 49			8764 65
8/20		Deposit				85 00	8848 65
8/21	434	Galaxy Records		48 60			8801 05
8/22	435	Electric Works		267 95			8533 09

There are two ways to determine whether there are errors. We assume that the amount \$8767.73 in the "Balance forward" column is correct. If we can determine that the ending balance is correct, we have some assurance that the checkbook is correct. But two errors could offset each other to give us that balance.

METHOD 1

a) We add the debits:

$$23.56 + 20.49 + 48.60 + 267.95 = 360.60.$$

b) We add the deposits/credits. In this case, there is only one deposit, 85.00.

c) We add the total of the deposits to the balance brought forward:

$$8767.73 + 85.00 = 8852.73.$$

d) We subtract the total of the debits:

$$8852.73 - 360.60 = 8492.13.$$

The result should be the ending balance, 8533.09. We see that $8492.13 \neq 8533.09$. Since the numbers are not equal, we proceed to method 2.

METHOD 2 We successively add or subtract deposit/credits and debits, and check the result in the "Balance forward" column.

$$8767.73 - 23.56 = 8744.17.$$

We have found our first error. The subtraction was incorrect. We correct it and continue, using 8744.17 as the corrected balance forward:

$$8744.17 - 20.49 = 8723.68.$$

It looks as though 20.49 was added instead of subtracted. Actually, we would have to correct this line even if it had been subtracted, because the error of 1¢ in the first step has been carried through successive calculations. We correct that balance line and continue, using 8723.68 as the balance and adding the deposit 85.00:

$$8723.68 + 85.00 = 8808.68.$$

We make the correction and continue subtracting the last two debits:

$$8808.68 - 48.60 = 8760.08.$$

Then

$$8760.08 - 267.95 = 8492.13.$$

The corrected checkbook is below.

20___ RECORD ALL CHARGES OR CREDITS THAT AFFECT YOUR ACCOUNT

DATE	CHECK NUMBER	TRANSACTION DESCRIPTION	√ T	(−) PAYMENT/ DEBIT	(+ OR −) OTHER	(+) DEPOSIT/ CREDIT	BALANCE FORWARD	
							8767 73	
8/16	432	Burch Laundry		23 56			~~8744 16~~	→ 8744.17
8/19	433	Rogers TV		20 49			~~8764 65~~	→ 8723.68
8/20		Deposit				85 00	~~8848 65~~	→ 8808.68
8/21	434	Galaxy Records		48 60			~~8801 05~~	→ 8760.08
8/22	435	Electric Works		267 95			~~8533 09~~	→ 8492.13

Do Exercise 18.

There are other ways in which errors can be made in checkbooks, such as forgetting to record a transaction or writing the amounts incorrectly, but we will not consider those here.

18. Find the errors, if any, in this checkbook.

20___ RECORD ALL CHARGES OR CREDITS THAT AFFECT YOUR ACCOUNT

DATE	CHECK NUMBER	TRANSACTION DESCRIPTION	√ T	(−) PAYMENT/ DEBIT	(+ OR −) OTHER	(+) DEPOSIT/ CREDIT	BALANCE FORWARD
							3078 92
12/1	888	H.H. Gregg Appliances		340 69			2738 23
12/3	889	Marie Callendar's Pies		78 56			2659 66
12/5		Deposit <Paycheck>				230 80	2890 46
12/6	890	Chili's Restaurant		13 14			2877 32
12/8	891	Stonycreek Golf Course		48 00			2829 32
12/8		Deposit <Molly>				39 58	2868 90
12/10	892	Galyan's Trading Post		102 87			2766 83
12/14	893	Goody's Music		68 59			2697 45
12/15	894	Salvation Army		100 00			2497 45

Answer on page A-6

3.2 EXERCISE SET

For Extra Help

Digital Video Tutor CD 2 Videotape 4 | InterAct Math | Math Tutor Center | MathXL | MyMathLab

a Add.

1. $\begin{array}{r} 316.25 \\ +\ 18.12 \\ \hline \end{array}$

2. $\begin{array}{r} 641.803 \\ +\ 14.935 \\ \hline \end{array}$

3. $\begin{array}{r} 659.403 \\ +\ 916.812 \\ \hline \end{array}$

4. $\begin{array}{r} 4203.28 \\ +\ 3.39 \\ \hline \end{array}$

5. $\begin{array}{r} 9.104 \\ +\ 123.456 \\ \hline \end{array}$

6. $\begin{array}{r} 6.1528 \\ +\ 5.2777 \\ \hline \end{array}$

7. $\begin{array}{r} 81.008 \\ +\ 3.409 \\ \hline \end{array}$

8. $0.8096 + 0.7856$

9. $20.0124 + 30.0124$

10. $0.687 + 0.9$

11. $39 + 1.007$

12. $0.845 + 10.02$

13. $0.34 + 3.5 + 0.127 + 768$

14. $2.3 + 0.729 + 23$

15. $17 + 3.24 + 0.256 + 0.3689$

16. $\begin{array}{r} 47.8 \\ 219.852 \\ 43.59 \\ +\ 666.713 \\ \hline \end{array}$

17. $\begin{array}{r} 2.703 \\ 78.33 \\ 28.0009 \\ +\ 118.4341 \\ \hline \end{array}$

18. $\begin{array}{r} 13.72 \\ 9.112 \\ 6542.7908 \\ +\ 23.901 \\ \hline \end{array}$

19. $99.6001 + 7285.18 + 500.042 + 870$

20. $65.987 + 9.4703 + 6744.02 + 1.0003 + 200.895$

b Subtract.

21. $\begin{array}{r} 5.2 \\ -\ 3.9 \\ \hline \end{array}$

22. $\begin{array}{r} 44.345 \\ -\ 3.105 \\ \hline \end{array}$

23. $\begin{array}{r} 51.31 \\ -\ 2.29 \\ \hline \end{array}$

24. $\begin{array}{r} 87.46 \\ -\ 6.32 \\ \hline \end{array}$

25. $\begin{array}{r} 48.76 \\ -\ 3.15 \\ \hline \end{array}$

26. $\begin{array}{r} 97.01 \\ -\ 3.15 \\ \hline \end{array}$

27. $\begin{array}{r} 92.341 \\ -\ 6.42 \\ \hline \end{array}$

28. $\begin{array}{r} 0.8468 \\ -\ 0.034 \\ \hline \end{array}$

29. $\begin{array}{r} 2.5 \\ -\ 0.0025 \\ \hline \end{array}$

30. $\begin{array}{r} 39.0 \\ -\ 0.28 \\ \hline \end{array}$

31. $\begin{array}{r} 3.4 \\ -\ 0.003 \\ \hline \end{array}$

32. $\begin{array}{r} 2.8 \\ -\ 2.08 \\ \hline \end{array}$

33. 28.2 − 19.35

34. 100.16 − 0.118

35. 34.07 − 30.7

36. 36.2 − 16.28

37. 8.45 − 7.405

38. 3.801 − 2.81

39. 6.003 − 2.3

40. 9.087 − 8.807

41. 1 − 0.0098

42. 2 − 1.0908

43. 100 − 0.34

44. 624 − 18.79

45. 7.48 − 2.6

46. 18.4 − 5.92

47. 3 − 2.006

48. 263.7 − 102.08

49. 19 − 1.198

50. 2548.98 − 2.007

51. 65 − 13.87

52. 45 − 0.999

53. 3.907 − 1.416

54. 70.0009 − 23.0567

55. $\begin{array}{r} 32.7978 \\ -\ \ 0.0592 \\ \hline \end{array}$

56. $\begin{array}{r} 0.49634 \\ -\ 0.12678 \\ \hline \end{array}$

57. $\begin{array}{r} 3.0074 \\ -\ 1.3408 \\ \hline \end{array}$

58. $\begin{array}{r} 6.07 \\ -\ 2.0078 \\ \hline \end{array}$

59. $\begin{array}{r} 2345.90786 \\ -\ \ \ \ 0.999 \\ \hline \end{array}$

60. $\begin{array}{r} 1.0 \\ -\ 0.9999 \\ \hline \end{array}$

c Solve.

61. $x + 17.5 = 29.15$

62. $t + 50.7 = 54.07$

63. $3.205 + m = 22.456$

64. $4.26 + q = 58.32$

65. $17.95 + p = 402.63$

66. $w + 1.3004 = 47.8$

67. $13{,}083.3 = x + 12{,}500.33$

68. $100.23 = 67.8 + z$

69. $x + 2349 = 17{,}684.3$

70. $1830.4 + t = 23{,}067$

d Find the errors, if any, in each checkbook.

71.

20___ RECORD ALL CHARGES OR CREDITS THAT AFFECT YOUR ACCOUNT

DATE	CHECK NUMBER	TRANSACTION DESCRIPTION	√ T	(−) PAYMENT/ DEBIT	(+ OR −) OTHER	(+) DEPOSIT/ CREDIT	BALANCE FORWARD
							9704 56
8/8	342	Bill Rydman		27 44			9677 12
8/9		Deposit <Beauty Contest>				1000 00	10,677 12
8/12	343	Jason Jordan		123 95			10,553 17
8/14	344	Jennifer Crum		124 02			10,677 19
8/22	345	Neon Johnny's Pizza		12 43			10,664 76
8/24		Deposit <Bowling Tournament>				2500 00	13,164 76
8/29	346	Border's Bookstore		137 78			13,302 54
9/2		Deposit <Bodybuilder Contest>				18 88	13,283 66
9/3	347	Fireman's Fund		2800 00			10,483 66

72.

20___ RECORD ALL CHARGES OR CREDITS THAT AFFECT YOUR ACCOUNT

DATE	CHECK NUMBER	TRANSACTION DESCRIPTION	√ T	(−) PAYMENT/ DEBIT	(+ OR −) OTHER	(+) DEPOSIT/ CREDIT	BALANCE FORWARD
							1876 43
4/1	500	Ed Moura		500 12			1376 31
4/3	501	Jim Lawler		28 56			1347 75
4/3		Deposit <State Lottery>				10,000 00	11,347 75
4/3	502	Victoria Montoya		464 00			10,883 75
4/3		Deposit <Jewelry Sale>.				2500 00	8383 75
4/4	503	Baskin & Robbins		1600 00			6783 75
4/8	504	Golf Galaxy		1349 98			5433 77
4/12	505	Don Mitchell Pro Shops		658 97			4774 80
4/13		Deposit <Publisher's Clearing House>				100000 00	104,774 80
4/15	506	American Airlines		6885 58			98,889 22

73. DW Explain the error in the following:

Add.

$$\begin{array}{r} 13.07 \\ +\ \ 9.205 \\ \hline 10.512 \end{array}$$

74. DW Explain the error in the following:

Subtract.

$$\begin{array}{r} 73.089 \\ -\ \ 5.0061 \\ \hline 2.3028 \end{array}$$

SKILL MAINTENANCE

75. Find the LCM of 32 and 85. [1.9a]

76. Find the prime factorization of 228. [1.7d]

Subtract.

77. $\frac{13}{24} - \frac{3}{8}$ [2.3b]

78. $\frac{8}{9} - \frac{2}{15}$ [2.3b]

79. 8805 − 2639 [1.2d]

80. 8005 − 2639 [1.2d]

Solve. [2.5a]

81. A serving of filleted fish is generally considered to be about $\frac{1}{3}$ lb. How many servings can be prepared from $5\frac{1}{2}$ lb of flounder fillet?

82. A photocopier technician drove $125\frac{7}{10}$ mi away from Scottsdale for a repair call. The next day he drove $65\frac{1}{2}$ mi back toward Scottsdale for another service call. How far was the technician from Scottsdale?

SYNTHESIS

83. A student presses the wrong button when using a calculator and adds 235.7 instead of subtracting it. The incorrect answer is 817.2. What is the correct answer?

Objectives

a Multiply using decimal notation.

b Convert from notation like 45.7 million to standard notation, and from dollars to cents and cents to dollars.

3.3 MULTIPLICATION

a Multiplication

Let's find the product

$2.3 \times 1.12.$

To understand how we find such a product, we first convert each factor to fraction notation. Next, we multiply the whole numbers 23 and 112, and then divide by 1000.

$$2.3 \times 1.12 = \frac{23}{10} \times \frac{112}{100} = \frac{23 \times 112}{10 \times 100} = \frac{2576}{1000} = 2.576$$

Note the number of decimal places.

$$\begin{array}{r} 1.12 \\ \times\ \ 2.3 \\ \hline 2.576 \end{array} \quad \begin{array}{l} \text{(2 decimal places)} \\ \text{(1 decimal place)} \\ \text{(3 decimal places)} \end{array}$$

Now consider

$$0.011 \times 15.0002 = \frac{11}{1000} \times \frac{150{,}002}{10{,}000} = \frac{1{,}650{,}022}{10{,}000{,}000} = 0.1650022.$$

Note the number of decimal places.

$$\begin{array}{r} 15.0002 \\ \times\ \ 0.011 \\ \hline 0.1650022 \end{array} \quad \begin{array}{l} \text{(4 decimal places)} \\ \text{(3 decimal places)} \\ \text{(7 decimal places)} \end{array}$$

To multiply using decimals:

a) Ignore the decimal points and multiply as though both factors were whole numbers.

b) Then place the decimal point in the result. The number of decimal places in the product is the sum of the numbers of places in the factors (count places from the right).

0.8×0.43

$$\begin{array}{r} \overset{2}{0.4}3 \\ \times\ \ 0.8 \\ \hline 344 \end{array} \quad \text{Ignore the decimal points for now.}$$

$$\begin{array}{r} 0.43 \\ \times\ \ 0.8 \\ \hline 0.344 \end{array} \quad \begin{array}{l} \text{(2 decimal places)} \\ \text{(1 decimal place)} \\ \text{(3 decimal places)} \end{array}$$

EXAMPLE 1 Multiply: 8.3×74.6.

a) Ignore the decimal points and multiply as though factors were whole numbers:

$$\begin{array}{r} \scriptstyle 3\ 4 \\ \scriptstyle 1\ 1 \\ 74.6 \\ \times\ \ 8.3 \\ \hline 2238 \\ 59680 \\ \hline 61918 \end{array}$$

b) Place the decimal point in the result. The number of decimal places in the product is the sum, 1 + 1, of the number of places in the factors.

```
      7 4.6    (1 decimal place)
×       8.3    (1 decimal place)
    2 2 3 8
  5 9 6 8 0
  6 1 9.1 8    (2 decimal places)
```

Do Exercise 1.

EXAMPLE 2 Multiply: 0.0032×2148.

As we catch on to the skill, we can combine the two steps.

```
      2 1 4 8    (0 decimal places)
× 0.0 0 3 2      (4 decimal places)
      4 2 9 6
    6 4 4 4 0
    6.8 7 3 6    (4 decimal places)
```

EXAMPLE 3 Multiply: 0.14×0.867.

```
    0.8 6 7    (3 decimal places)
×     0.1 4    (2 decimal places)
    3 4 6 8
    8 6 7 0
0.1 2 1 3 8    (5 decimal places)
```

Do Exercises 2 and 3.

MULTIPLYING BY 0.1, 0.01, 0.001, AND SO ON

Now let's consider some special kinds of products. The first involves multiplying by a tenth, hundredth, thousandth, or ten-thousandth. Let's look at those products.

$$0.1 \times 38 = \frac{1}{10} \times 38 = \frac{38}{10} = 3.8$$

$$0.01 \times 38 = \frac{1}{100} \times 38 = \frac{38}{100} = 0.38$$

$$0.001 \times 38 = \frac{1}{1000} \times 38 = \frac{38}{1000} = 0.038$$

$$0.0001 \times 38 = \frac{1}{10{,}000} \times 38 = \frac{38}{10{,}000} = 0.0038$$

Note in each case that the product is *smaller* than 38.

1. Multiply.

```
  8 5.4
×   6.2
```

Multiply.

2.
```
      1 2 3 4
× 0.0 0 4 1
```

3.
```
    4 2.6 5
× 0.8 0 4
```

Answers on page A-7

Multiply.

4. 0.1×3.48

5. 0.01×3.48

6. 0.001×3.48

7. 0.0001×3.48

Multiply.

8. 10×3.48

9. 100×3.48

10. 1000×3.48

11. $10{,}000 \times 3.48$

Answers on page A-7

To multiply any number by 0.1, 0.01, 0.001, and so on,

a) count the number of decimal places in the tenth, hundredth, or thousandth, and so on, and
b) move the decimal point that many places to the left.

0.001×34.45678 → 3 places

$0.001 \times 34.45678 = 0.034.45678$

Move 3 places to the left.

$0.001 \times 34.45678 = 0.03445678$

EXAMPLES Multiply.

4. $0.1 \times 14.605 = 1.4605$ 1.4.605
5. $0.01 \times 14.605 = 0.14605$
6. $0.001 \times 14.605 = 0.014605$ — We write an extra zero.
7. $0.0001 \times 14.605 = 0.0014605$ — We write two extra zeros.

Do Exercises 4–7.

MULTIPLYING BY 10, 100, 1000, AND SO ON

Next, let's consider multiplying by 10, 100, 1000, and so on. Let's look at those products.

$$10 \times 97.34 = 973.4$$
$$100 \times 97.34 = 9734$$
$$1000 \times 97.34 = 97{,}340$$
$$10{,}000 \times 97.34 = 973{,}400$$

Note in each case that the product is *larger* than 97.34.

To multiply any number by 10, 100, 1000, and so on,

a) count the number of zeros, and
b) move the decimal point that many places to the right.

1000×34.45678 → 3 zeros

$1000 \times 34.45678 = 34.456.78$

Move 3 places to the right.

$1000 \times 34.45678 = 34{,}456.78$

EXAMPLES Multiply.

8. $10 \times 14.605 = 146.05$ 14.6.05
9. $100 \times 14.605 = 1460.5$
10. $1000 \times 14.605 = 14{,}605$
11. $10{,}000 \times 14.605 = 146{,}050$ 14.6050.

Do Exercises 8–11.

b Applications Using Multiplication with Decimal Notation

NAMING LARGE NUMBERS

We often see notation like the following in newspapers and magazines and on television.

> The largest building in the world is the Pentagon, which has 3.7 million square feet of floor space.
>
> By 2004, it is expected that \$7.3 trillion dollars worth of business will be transacted over the Internet.
>
> In 1999, the U. S. Mint produced 11.6 billion pennies.

Source: U.S. Mint

To understand such notation, consider the information in the following table.

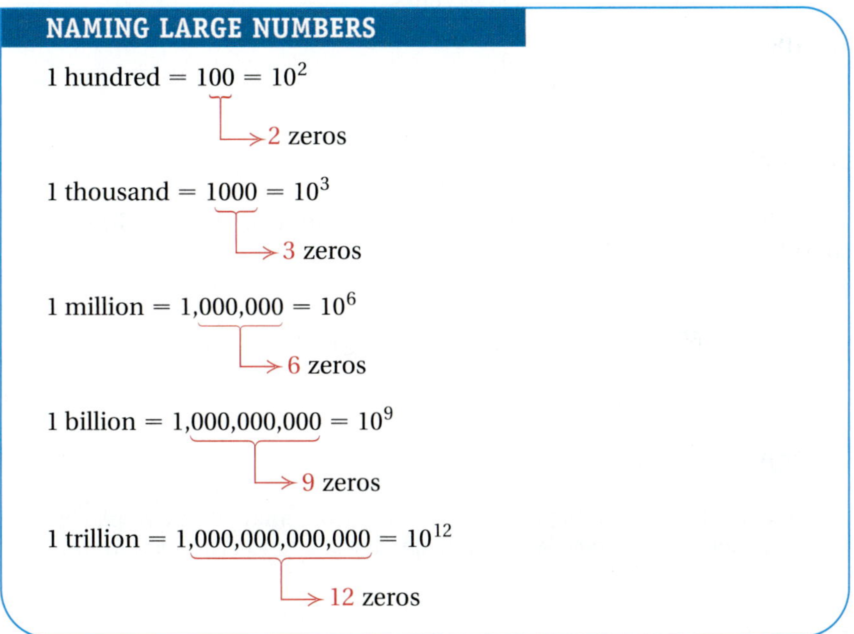

NAMING LARGE NUMBERS

1 hundred = 100 = 10^2 → 2 zeros

1 thousand = 1000 = 10^3 → 3 zeros

1 million = 1,000,000 = 10^6 → 6 zeros

1 billion = 1,000,000,000 = 10^9 → 9 zeros

1 trillion = 1,000,000,000,000 = 10^{12} → 12 zeros

CALCULATOR CORNER

Multiplication with Decimal Notation To use a calculator to multiply with decimal notation, we use the [·], [×], and [=] keys. To find 4.78×0.34, for example, we press [4] [·] [7] [8] [×] [·] [3] [4] [=]. The display reads [1.6252], so $4.78 \times 0.34 = 1.6252$.

Exercises: Use a calculator to multiply.

1. 5.4×9
2. 415×16.7
3. 17.63×8.1
4. 0.04×12.69
5. 586.4×13.5
6. 4.003×5.1

Convert the number in the sentence to standard notation.

12. In 1999, the U.S. Mint produced 4.4 billion quarters.
Source: U.S. Mint

13. The largest building in the world is the Pentagon, which has 3.7 million square feet of floor space.

Convert from dollars to cents.

14. \$15.69

15. \$0.17

Convert from cents to dollars.

16. 35¢

17. 577¢

Answers on page A-7

To convert a large number to standard notation, we proceed as follows.

EXAMPLE 12 Convert the number in this sentence to standard notation: In 1999, the U.S. Mint produced 11.6 billion pennies.
Source: U.S. Mint

$$\begin{aligned} 11.6 \text{ billion} &= 11.6 \times 1 \text{ billion} \\ &= 11.6 \times \underbrace{1{,}000{,}000{,}000}_{\text{9 zeros}} \\ &= 11{,}600{,}000{,}000 \end{aligned}$$

Do Exercises 12 and 13.

MONEY CONVERSION

Converting from dollars to cents is like multiplying by 100. To see why, consider \$19.43.

$\$19.43 = 19.43 \times \$1$	We think of \$19.43 as 19.43 × 1 dollar, or 19.43 × \$1.
$= 19.43 \times 100¢$	Substituting 100¢ for \$1: \$1 = 100¢
$= 1943¢$	Multiplying

DOLLARS TO CENTS

To convert from dollars to cents, move the decimal point two places to the right and change from the \$ sign in front to the ¢ sign at the end.

EXAMPLES Convert from dollars to cents.

13. \$189.64 = 18,964¢
14. \$0.75 = 75¢

Do Exercises 14 and 15.

Converting from cents to dollars is like multiplying by 0.01. To see why, consider 65¢.

$65¢ = 65 \times 1¢$	We think of 65¢ as 65 × 1 cent, or 65 × 1¢.
$= 65 \times \$0.01$	Substituting \$0.01 for 1¢: 1¢ = \$0.01
$= \$0.65$	Multiplying

CENTS TO DOLLARS

To convert from cents to dollars, move the decimal point two places to the left and change from the ¢ sign at the end to the \$ sign in front.

EXAMPLES Convert from cents to dollars.

15. 395¢ = \$3.95
16. 8503¢ = \$85.03

Do Exercises 16 and 17.

3.3 EXERCISE SET

For Extra Help

Digital Video Tutor CD 2 Videotape 4

InterAct Math

Math Tutor Center

MathXL

MyMathLab

a Multiply.

1. $\begin{array}{r} 8.6 \\ \times \quad 7 \\ \hline \end{array}$

2. $\begin{array}{r} 5.7 \\ \times 0.8 \\ \hline \end{array}$

3. $\begin{array}{r} 0.84 \\ \times \quad 8 \\ \hline \end{array}$

4. $\begin{array}{r} 9.4 \\ \times 0.6 \\ \hline \end{array}$

5. $\begin{array}{r} 6.3 \\ \times 0.04 \\ \hline \end{array}$

6. $\begin{array}{r} 9.8 \\ \times 0.08 \\ \hline \end{array}$

7. $\begin{array}{r} 87 \\ \times 0.006 \\ \hline \end{array}$

8. $\begin{array}{r} 18.4 \\ \times 0.07 \\ \hline \end{array}$

9. 10×23.76

10. 100×3.8798

11. 1000×583.686852

12. 0.34×1000

13. 7.8×100

14. 0.00238×10

15. 0.1×89.23

16. 0.01×789.235

17. 0.001×97.68

18. 8976.23×0.001

19. 78.2×0.01

20. 0.0235×0.1

21. $\begin{array}{r} 32.6 \\ \times \quad 16 \\ \hline \end{array}$

22. $\begin{array}{r} 9.28 \\ \times \quad 8.6 \\ \hline \end{array}$

23. $\begin{array}{r} 0.984 \\ \times \quad 3.3 \\ \hline \end{array}$

24. $\begin{array}{r} 8.489 \\ \times \quad 7.4 \\ \hline \end{array}$

25. $\begin{array}{r} 374 \\ \times \quad 2.4 \\ \hline \end{array}$

26. $\begin{array}{r} 865 \\ \times 1.08 \\ \hline \end{array}$

27. $\begin{array}{r} 749 \\ \times 0.43 \\ \hline \end{array}$

28. $\begin{array}{r} 978 \\ \times 20.5 \\ \hline \end{array}$

29. $\begin{array}{r} 0.87 \\ \times \quad 64 \\ \hline \end{array}$

30. $\begin{array}{r} 7.25 \\ \times \quad 60 \\ \hline \end{array}$

31. $\begin{array}{r} 46.50 \\ \times \quad 75 \\ \hline \end{array}$

32. $\begin{array}{r} 8.24 \\ \times 703 \\ \hline \end{array}$

33. $\begin{array}{r} 81.7 \\ \times\ 0.612 \\ \hline \end{array}$

34. $\begin{array}{r} 31.82 \\ \times\ 7.15 \\ \hline \end{array}$

35. $\begin{array}{r} 10.105 \\ \times\ 11.324 \\ \hline \end{array}$

36. $\begin{array}{r} 151.2 \\ \times\ 4.555 \\ \hline \end{array}$

37. $\begin{array}{r} 12.3 \\ \times\ 1.08 \\ \hline \end{array}$

38. $\begin{array}{r} 7.82 \\ \times\ 0.024 \\ \hline \end{array}$

39. $\begin{array}{r} 32.4 \\ \times\ 2.8 \\ \hline \end{array}$

40. $\begin{array}{r} 8.09 \\ \times\ 0.0075 \\ \hline \end{array}$

41. $\begin{array}{r} 0.00342 \\ \times\ 0.84 \\ \hline \end{array}$

42. $\begin{array}{r} 2.0056 \\ \times\ 3.8 \\ \hline \end{array}$

43. $\begin{array}{r} 0.347 \\ \times\ 2.09 \\ \hline \end{array}$

44. $\begin{array}{r} 2.532 \\ \times\ 1.067 \\ \hline \end{array}$

45. $\begin{array}{r} 3.005 \\ \times\ 0.623 \\ \hline \end{array}$

46. $\begin{array}{r} 16.34 \\ \times\ 0.000512 \\ \hline \end{array}$

47. 1000×45.678

48. 0.001×45.678

b Convert from dollars to cents.

49. $28.88

50. $67.43

51. $0.66

52. $1.78

Convert from cents to dollars.

53. 34¢

54. 95¢

55. 3445¢

56. 933¢

Convert the number in the sentence to standard notation.

57. The average distance from the earth to the sun is 93 million miles. (This was a $1 million question on the TV quiz show "Who Wants to Be a Millionaire?")

58. In 2001, 3.5 million sport utility vehicles were sold.
Source: Autodata

59. In 1999, total box office sales at the movies was $7.2 billion.
Source: Motion Picture Association of America

60. By 2003, it is expected that $3.5 trillion dollars worth of business will be transacted over the Internet.

61. **DW** If two rectangles have the same perimeter, will they also have the same area? Experiment with different dimensions. Be sure to use decimals. Explain your answer.

62. **DW** A student insists that $346.708 \times 0.1 = 3467.08$. How could you convince him that a mistake had been made without checking on a calculator?

SKILL MAINTENANCE

Calculate.

63. $2\frac{1}{3} \cdot 4\frac{4}{5}$ [2.4d]

64. $2\frac{1}{3} \div 4\frac{4}{5}$ [2.4e]

65. $4\frac{4}{5} - 2\frac{1}{3}$ [2.4c]

66. $4\frac{4}{5} + 2\frac{1}{3}$ [2.4b]

Divide. [1.3d]

67. $24\overline{)8208}$

68. $4\overline{)348}$

69. $7\overline{)31{,}962}$

70. $18\overline{)22{,}626}$

71. $40\overline{)3480}$

72. $17\overline{)20{,}006}$

SYNTHESIS

Consider the following names for large numbers in addition to those already discussed in this section:

$$1 \text{ quadrillion} = 1{,}000{,}000{,}000{,}000{,}000 = 10^{15};$$
$$1 \text{ quintillion} = 1{,}000{,}000{,}000{,}000{,}000{,}000 = 10^{18};$$
$$1 \text{ sextillion} = 1{,}000{,}000{,}000{,}000{,}000{,}000{,}000 = 10^{21};$$
$$1 \text{ septillion} = 1{,}000{,}000{,}000{,}000{,}000{,}000{,}000{,}000 = 10^{24}.$$

Find each of the following. Express the answer with a name that is a power of 10.

73. (1 trillion) · (1 billion)

74. (1 million) · (1 billion)

75. (1 trillion) · (1 trillion)

76. Is a billion millions the same as a million billions? Explain.

Objectives

a Divide using decimal notation.

b Solve equations of the type $a \cdot x = b$, where a and b may be in decimal notation.

c Simplify expressions using the rules for order of operations.

Divide.

1. $9\,\overline{)\,5.4}$

2. $15\,\overline{)\,22.5}$

3. $82\,\overline{)\,38.54}$

Answers on page A-7

3.4 DIVISION

a Division

WHOLE-NUMBER DIVISORS

Compare these divisions by a whole number.

$$\frac{588}{7} = 84$$

$$\frac{58.8}{7} = 8.4$$

$$\frac{5.88}{7} = 0.84$$

$$\frac{0.588}{7} = 0.084$$

When we are dividing by a whole number, the number of decimal places in the *quotient* is the same as the number of decimal places in the *dividend*.

These examples lead us to this method for dividing by a whole number.

To divide by a whole number,

a) place the decimal point directly above the decimal point in the dividend, and

b) divide as though dividing whole numbers.

```
               0.8 4 ← Quotient
Divisor → 7 ) 5.8 8 ← Dividend
              5 6 0
                2 8
                2 8
                  0 ← Remainder
```

EXAMPLE 1 Divide: $379.2 \div 8$.

```
      4 7.4       ← Place the decimal point.
  8 ) 3 7 9.2
      3 2 0 0
        5 9 2
        5 6 0     Divide as though dividing whole numbers.
          3 2
          3 2
            0
```

EXAMPLE 2 Divide: $82.08 \div 24$.

```
          3.4 2     ← Place the decimal point.
  2 4 ) 8 2.0 8
        7 2 0 0
        1 0 0 8
          9 6 0     Divide as though dividing whole numbers.
            4 8
            4 8
              0
```

Do Exercises 1–3 on the preceding page.

Sometimes it helps to write some extra zeros to the right of the decimal point. They don't change the number.

EXAMPLE 3 Divide: $30 \div 8$.

$$\begin{array}{r} 3. \\ 8\,\overline{)\,30.} \\ \underline{24} \\ 6 \end{array}$$

Place the decimal point and divide to find how many ones

$$\begin{array}{r} 3. \\ 8\,\overline{)\,30.0} \\ \underline{24} \\ 60 \end{array}$$

Write an extra zero.

$$\begin{array}{r} 3.7 \\ 8\,\overline{)\,30.0} \\ \underline{24} \\ 60 \\ \underline{56} \\ 4 \end{array}$$

Divide to find how many tenths.

$$\begin{array}{r} 3.7 \\ 8\,\overline{)\,30.00} \\ \underline{24} \\ 60 \\ \underline{56} \\ 40 \end{array}$$

Write an extra zero.

$$\begin{array}{r} 3.75 \\ 8\,\overline{)\,30.00} \\ \underline{24} \\ 60 \\ \underline{56} \\ 40 \\ \underline{40} \\ 0 \end{array}$$

Divide to find how many hundredths.

EXAMPLE 4 Divide: $4 \div 25$.

$$\begin{array}{r} 0.16 \\ 25\,\overline{)\,4.00} \\ \underline{25} \\ 150 \\ \underline{150} \\ 0 \end{array}$$

Do Exercises 4–6.

Divide.

4. $25\,\overline{)\,8}$

5. $4\,\overline{)\,15}$

6. $86\,\overline{)\,21.5}$

Answers on page A-7

7. a) Complete.

$$\frac{3.75}{0.25} = \frac{3.75}{0.25} \times \frac{100}{100} = \frac{(\ )}{25}$$

b) Divide.

$0.25 \overline{)3.75}$

Divide.

8. $0.83 \overline{)4.067}$

9. $3.5 \overline{)44.8}$

Answers on page A-7

DIVISORS THAT ARE NOT WHOLE NUMBERS

Consider the division

$0.24 \overline{)8.208}$

We write the division as $\frac{8.208}{0.24}$. Then we multiply by 1 to change to a whole-number divisor:

$$\frac{8.208}{0.24} = \frac{8.208}{0.24} \times \frac{100}{100} = \frac{820.8}{24}.$$

The division $0.24\overline{)8.208}$ is the same as $24\overline{)820.8}$.

The divisor is now a whole number.

To divide when the divisor is not a whole number,

a) move the decimal point (multiply by 10, 100, and so on) to make the divisor a whole number;

$0.24 \overline{)8.208}$

Move 2 places to the right.

b) move the decimal point (multiply the same way) in the dividend the same number of places; and

$0.24 \overline{)8.208}$

Move 2 places to the right.

c) place the decimal point directly above the new decimal point in the dividend and divide as though dividing whole numbers.

```
          3 4.2
0.2 4 ) 8.2 0^8
        7 2 0 0
        1 0 0 8
          9 6 0
            4 8
            4 8
              0
```

(The new decimal point in the dividend is indicated by a caret.)

EXAMPLE 5 Divide: 5.848 ÷ 8.6.

$8.6 \overline{)5.848}$

Multiply the divisor by 10 (move the decimal point 1 place). Multiply the same way in the dividend (move 1 place).

```
        0.6 8
8.6 ) 5.8^4 8
      5 1 6 0
        6 8 8
        6 8 8
            0
```

Place a decimal point above the new decimal point and then divide.

Note: $\frac{5.848}{8.6} = \frac{5.848}{8.6} \cdot \frac{10}{10} = \frac{58.48}{86}$.

Do Exercises 7–9.

Suppose the dividend is a whole number. We can think of it as having a decimal point at the end with as many 0's as we wish after the decimal point. For example,

$$12 = 12. = 12.0 = 12.00 = 12.000, \text{ and so on.}$$

EXAMPLE 6 Divide: $12 \div 0.64$.

$0.64\overline{)12.}$ Place a decimal point at the end of the whole number.

$0.64\overline{)12.00}$ Multiply the divisor by 100 (move the decimal point 2 places). Multiply the same way in the dividend (move 2 places).

Place a decimal point above and then divide.

$$
\begin{array}{r}
18.75 \\
0.64\overline{)12.00\,00} \\
\underline{6\,4\,0} \\
5\,6\,0 \\
\underline{5\,1\,2} \\
4\,8\,0 \\
\underline{4\,4\,8} \\
3\,2\,0 \\
\underline{3\,2\,0} \\
0
\end{array}
$$

Do Exercise 10.

DIVIDING BY 10, 100, 1000, AND SO ON

It is often helpful to be able to divide quickly by a ten, hundred, or thousand, or by a tenth, hundredth, or thousandth. Each procedure we use is based on multiplying by 1. Consider the following example:

$$\frac{23.789}{1000} = \frac{23.789}{1000} \cdot \frac{1000}{1000} = \frac{23{,}789}{1{,}000{,}000} = 0.023789.$$

We are dividing by a number greater than 1: The result is *smaller* than 23.789.

To divide by 10, 100, 1000, and so on,

a) count the number of zeros in the divisor, and

$\frac{713.49}{100}$ — 2 zeros

b) move the decimal point that number of places to the left.

$\frac{713.49}{100}$, 7.13.49, $\frac{713.49}{100} = 7.1349$

2 places to the left

EXAMPLE 7 Divide: $\frac{0.0104}{10}$.

$\frac{0.0104}{10}$, 0.0.0104, $\frac{0.0104}{10} = 0.00104$

1 zero — 1 place to the left

10. Divide.

$1.6\overline{)25}$

CALCULATOR CORNER

Division with Decimal Notation To use a calculator to divide with decimal notation, we use the [·], [÷], and [=] keys. To find $237.12 \div 5.2$, for example, we press [2] [3] [7] [·] [1] [2] [÷] [5] [·] [2] [=]. The display reads [45.6], so $237.12 \div 5.2 = 45.6$.

Exercises: Use a calculator to divide.

1. $12.4\overline{)177.32}$
2. $49\overline{)125.44}$
3. $3.2\overline{)640}$
4. $16\overline{)12}$
5. $14 \div 0.7$
6. $1.6 \div 25$
7. $474.14 \div 30.2$
8. $518.472 \div 6.84$

Answer on page A-7

Study Tips

HOMEWORK TIPS

Prepare for your homework assignment by reading the explanations of concepts and following the step-by-step solutions of examples in the text. The time you spend preparing will save valuable time when you do your assignment.

Divide.

11. $\dfrac{0.1278}{0.01}$

12. $\dfrac{0.1278}{100}$

13. $\dfrac{98.47}{1000}$

14. $\dfrac{6.7832}{0.1}$

Answers on page A-7

DIVIDING BY 0.1, 0.01, 0.001, AND SO ON

Now consider the following example:

$$\frac{23.789}{0.01} = \frac{23.789}{0.01} \cdot \frac{100}{100} = \frac{2378.9}{1} = 2378.9.$$

We are dividing by a number less than 1: The result is *larger* than 23.789.

We use the following procedure.

To divide by 0.1, 0.01, 0.001, and so on,

a) count the number of decimal places in the divisor, and

$\dfrac{713.49}{0.001}$ → 3 places

b) move the decimal point that number of places to the right.

$\dfrac{713.49}{0.001}$, 713.490. $\dfrac{713.49}{0.001} = 713{,}490$

3 places to the right

EXAMPLE 8 Divide: $\dfrac{23.738}{0.001}$.

$\dfrac{23.738}{0.001}$, 23.738. $\dfrac{23.738}{0.001} = 23{,}738$

3 places 3 places to the right to change 0.001 to 1

Do Exercises 11–14.

b Solving Equations

Now let's solve equations of the type $a \cdot x = b$, where a and b may be in decimal notation. Proceeding as before, we divide by a on both sides.

EXAMPLE 9 Solve: $8 \cdot x = 27.2$.

We have

$$\frac{8 \cdot x}{8} = \frac{27.2}{8} \qquad \text{Dividing by 8 on both sides}$$

$$x = 3.4. \qquad \begin{array}{r} 3.4 \\ 8\overline{)\,27.2} \\ \underline{24\,0} \\ 3\,2 \\ \underline{3\,2} \\ 0 \end{array}$$

The solution is 3.4.

EXAMPLE 10 Solve: $2.9 \cdot t = 0.14616$.

We have

$$\frac{2.9 \cdot t}{2.9} = \frac{0.14616}{2.9} \quad \text{Dividing by 2.9 on both sides}$$

$$t = 0.0504.$$

```
           0.0 5 0 4
   2.9 ) 0.1^4 6 1 6
           1 4 5 0 0
               1 1 6
               1 1 6
                   0
```

The solution is 0.0504.

Do Exercises 15 and 16.

Solve.

15. $100 \cdot x = 78.314$

16. $0.25 \cdot y = 276.4$

C Order of Operations: Decimal Notation

The same rules for order of operations used with whole numbers and fraction notation apply when simplifying expressions with decimal notation.

RULES FOR ORDER OF OPERATIONS

1. Do all calculations within grouping symbols before operations outside.
2. Evaluate all exponential expressions.
3. Do all multiplications and divisions in order from left to right.
4. Do all additions and subtractions in order from left to right.

EXAMPLE 11 Simplify: $2.56 \times 25.6 \div 25{,}600 \times 256$.

There are no exponents or parentheses, so we multiply and divide from left to right:

$$\begin{aligned} 2.56 \times 25.6 \div 25{,}600 \times 256 &= 65.536 \div 25{,}600 \times 256 && \text{Doing all multiplications and divisions in order from left to right} \\ &= 0.00256 \times 256 \\ &= 0.65536. \end{aligned}$$

EXAMPLE 12 Simplify: $(5 - 0.06) \div 2 + 3.42 \times 0.1$.

$$\begin{aligned} (5 - 0.06) \div 2 + 3.42 \times 0.1 &= 4.94 \div 2 + 3.42 \times 0.1 && \text{Carrying out operations inside parentheses} \\ &= 2.47 + 0.342 && \text{Doing all multiplications and divisions in order from left to right} \\ &= 2.812 \end{aligned}$$

Answers on page A-7

Simplify.

17. $625 \div 62.5 \times 25 \div 6250$

18. $0.25 \cdot (1 + 0.08) - 0.0274$

19. $20^2 - 3.4^2 + \{2.5[20(9.2 - 5.6)] + 5(10 - 5)\}$

20. **Mountains in Peru.** Refer to the figure in Example 14. Find the average height of the mountains, in meters.

Answers on page A-7

EXAMPLE 13 Simplify: $10^2 \times \{[(3 - 0.24) \div 2.4] - (0.21 - 0.092)\}$.

$10^2 \times \{[(3 - 0.24) \div 2.4] - (0.21 - 0.092)\}$

$= 10^2 \times \{[2.76 \div 2.4] - 0.118\}$ Doing the calculations in the innermost parentheses first

$= 10^2 \times \{1.15 - 0.118\}$ Again, doing the calculations in the innermost parentheses

$= 10^2 \times 1.032$ Subtracting inside the parentheses

$= 100 \times 1.032$ Evaluating the exponential expression

$= 103.2$

Do Exercises 17–19.

EXAMPLE 14 *Mountains in Peru.* The following figure shows a range of very high mountains in Peru, together with their altitudes, given both in feet and in meters. Find the average height of these mountains, in feet.

Source: *National Geographic*, July 1968, p. 130

Source: WOOD RONASVILLE HARLIN INC/NGS Image Collection

The **average** of a set of numbers is the sum of the numbers divided by the number of addends. (See Section 1.6.) We find the sum of the heights divided by the number of addends, 6:

$$\frac{18{,}060 + 21{,}079 + 20{,}700 + 19{,}931 + 19{,}101 + 18{,}600}{6} = \frac{117{,}471}{6} = 19{,}578.5.$$

Thus the average height of these mountains is 19,578.5 ft.

Do Exercise 20.

3.4 EXERCISE SET

For Extra Help

Digital Video Tutor CD 2 Videotape 4

InterAct Math

Math Tutor Center

MathXL

MyMathLab

a Divide.

1. $2\overline{)5.98}$

2. $5\overline{)18}$

3. $4\overline{)95.12}$

4. $8\overline{)25.92}$

5. $12\overline{)89.76}$

6. $23\overline{)25.07}$

7. $33\overline{)237.6}$

8. $12.4 \div 4$

9. $9.144 \div 8$

10. $4.5 \div 9$

11. $12.123 \div 3$

12. $7\overline{)5.6}$

13. $5\overline{)0.35}$

14. $0.04\overline{)1.68}$

15. $0.12\overline{)8.4}$

16. $0.36\overline{)2.88}$

17. $3.4\overline{)68}$

18. $0.25\overline{)5}$

19. $15\overline{)6}$

20. $12\overline{)1.8}$

21. $36\overline{)14.76}$

22. $52\overline{)119.6}$

23. $3.2\overline{)27.2}$

24. $8.5\overline{)27.2}$

25. $4.2\overline{)39.06}$

26. $4.8\overline{)0.1104}$

27. $8\overline{)5}$

28. $8\overline{)3}$

29. $0.47\overline{)0.1222}$

30. $1.08\overline{)0.54}$

31. $4.8\overline{)75}$

32. $0.28\overline{)63}$

33. $0.032\overline{)0.07488}$

34. $0.017\overline{)1.581}$

35. $82\overline{)38.54}$

36. $34\overline{)0.1462}$

37. $\frac{213.4567}{1000}$

38. $\frac{213.4567}{100}$

39. $\frac{213.4567}{10}$

40. $\frac{100.7604}{0.1}$

41. $\frac{1.0237}{0.001}$

42. $\frac{1.0237}{0.01}$

b Solve.

43. $4.2 \cdot x = 39.06$

44. $36 \cdot y = 14.76$

45. $1000 \cdot y = 9.0678$

46. $789.23 = 0.25 \cdot q$

47. $1048.8 = 23 \cdot t$

48. $28.2 \cdot x = 423$

c Simplify.

49. $14 \times (82.6 + 67.9)$

50. $(26.2 - 14.8) \times 12$

51. $0.003 + 3.03 \div 0.01$

52. $9.94 + 4.26 \div (6.02 - 4.6) - 0.9$

53. $42 \times (10.6 + 0.024)$

54. $(18.6 - 4.9) \times 13$

55. $4.2 \times 5.7 + 0.7 \div 3.5$

56. $123.3 - 4.24 \times 1.01$

57. $9.0072 + 0.04 \div 0.1^2$

58. $12 \div 0.03 - 12 \times 0.03^2$

59. $(8 - 0.04)^2 \div 4 + 8.7 \times 0.4$

60. $(5 - 2.5)^2 \div 100 + 0.1 \times 6.5$

61. $86.7 + 4.22 \times (9.6 - 0.03)^2$

62. $2.48 \div (1 - 0.504) + 24.3 - 11 \times 2$

63. $4 \div 0.4 + 0.1 \times 5 - 0.1^2$

64. $6 \times 0.9 + 0.1 \div 4 - 0.2^3$

65. $5.5^2 \times [(6 - 4.2) \div 0.06 + 0.12]$

66. $12^2 \div (12 + 2.4) - [(2 - 1.6) \div 0.8]$

67. $200 \times \{[(4 - 0.25) \div 2.5] - (4.5 - 4.025)\}$

68. $0.03 \times \{1 \times 50.2 - [(8 - 7.5) \div 0.05]\}$

69. Find the average of \$1276.59, \$1350.49, \$1123.78, and \$1402.58.

70. Find the average weight of two wrestlers who weigh 308 lb and 296.4 lb.

71. *Porsche Sales.* Because of the exchange rate of the dollar to the Euro (European Monetary Unit), sales of Porsches have soared in the United States, as shown in the bar graph below. Find the average number of sales per year over the 5-yr period.

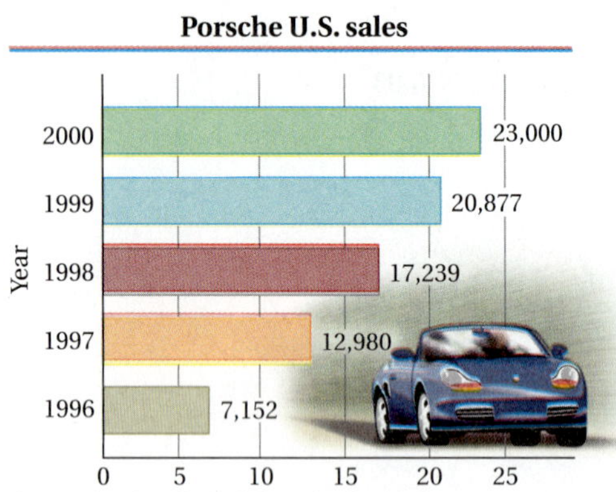

Sources: Autodata; Bridge Information Services

72. *Apples.* Americans are growing and eating more apples each year. The following graph shows the number of apples, in millions of bushels, in storage on May 1 of four recent years. Find the average number of apples in storage per year.

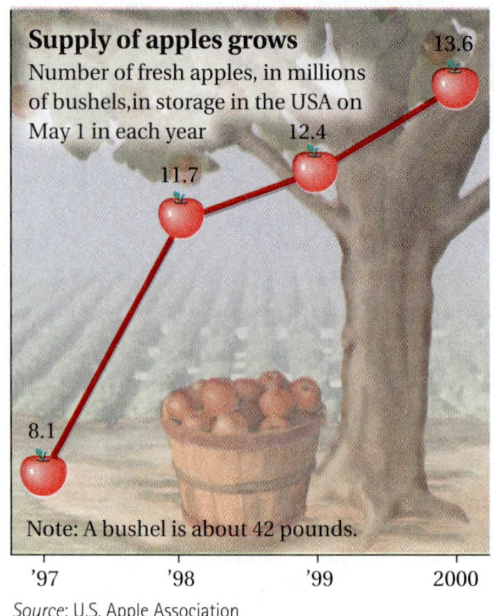

Source: U.S. Apple Association

73. D_W How is division with decimal notation similar to division of whole numbers? How is it different?

74. D_W A student made these two computational mistakes:

$0.247 \div 0.1 = 0.0247; \qquad 0.247 \div 10 = 2.47.$

In each case, how could you convince her that a mistake has been made?

SKILL MAINTENANCE

Simplify. [2.1e]

75. $\frac{36}{42}$

76. $\frac{56}{64}$

77. $\frac{38}{146}$

78. $\frac{114}{438}$

Find the prime factorization. [1.7d]

79. 684

80. 162

81. 2007

82. 2005

83. Add: $10\frac{1}{2} + 4\frac{5}{8}$. [2.4b]

84. Subtract: $10\frac{1}{2} - 4\frac{5}{8}$. [2.4c]

SYNTHESIS

Simplify.

85. ▦ $9.0534 - 2.041^2 \times 0.731 \div 1.043^2$

86. ▦ $23.042(7 - 4.037 \times 1.46 - 0.932^2)$

In Exercises 87–90, find the missing value.

87. $439.57 \times 0.01 \div 1000 \times \square = 4.3957$

88. $5.2738 \div 0.01 \times 1000 \div \square = 52.738$

89. $0.0329 \div 0.001 \times 10^4 \div \square = 3290$

90. $0.0047 \times 0.01 \div 10^4 \times \square = 4.7$

3.5 CONVERTING FROM FRACTION NOTATION TO DECIMAL NOTATION

Objectives

a. Convert from fraction notation to decimal notation.

b. Round numbers named by repeating decimals in problem solving.

c. Calculate using fraction and decimal notation together.

a Fraction Notation to Decimal Notation

When a denominator has no prime factors other than 2's and 5's, we can find decimal notation by multiplying by 1. We multiply to get a denominator that is a power of ten, like 10, 100, or 1000.

EXAMPLE 1 Find decimal notation for $\frac{3}{5}$.

$$\frac{3}{5} = \frac{3}{5} \cdot \frac{2}{2} = \frac{6}{10} = 0.6$$

We use $\frac{2}{2}$ for 1 to get a denominator of 10.

EXAMPLE 2 Find decimal notation for $\frac{7}{20}$.

$$\frac{7}{20} = \frac{7}{20} \cdot \frac{5}{5} = \frac{35}{100} = 0.35$$

We use $\frac{5}{5}$ for 1 to get a denominator of 100.

EXAMPLE 3 Find decimal notation for $\frac{87}{25}$.

$$\frac{87}{25} = \frac{87}{25} \cdot \frac{4}{4} = \frac{348}{100} = 3.48$$

We use $\frac{4}{4}$ for 1 to get a denominator of 100.

EXAMPLE 4 Find decimal notation for $\frac{9}{40}$.

$$\frac{9}{40} = \frac{9}{40} \cdot \frac{25}{25} = \frac{225}{1000} = 0.225$$

We use $\frac{25}{25}$ for 1 to get a denominator of 1000.

Do Exercises 1–4.

We can also divide to find decimal notation.

EXAMPLE 5 Find decimal notation for $\frac{3}{5}$.

$$\frac{3}{5} = 3 \div 5 \qquad \begin{array}{r} 0.6 \\ 5\,\overline{)\,3.0} \\ \underline{3\,0} \\ 0 \end{array} \qquad \frac{3}{5} = 0.6$$

EXAMPLE 6 Find decimal notation for $\frac{7}{8}$.

$$\frac{7}{8} = 7 \div 8 \qquad \begin{array}{r} 0.875 \\ 8\,\overline{)\,7.000} \\ \underline{6\,4} \\ 6\,0 \\ \underline{5\,6} \\ 4\,0 \\ \underline{4\,0} \\ 0 \end{array} \qquad \frac{7}{8} = 0.875$$

Do Exercises 5 and 6.

Find decimal notation. Use multiplying by 1.

1. $\frac{4}{5}$

2. $\frac{9}{20}$

3. $\frac{11}{40}$

4. $\frac{33}{25}$

Find decimal notation.

5. $\frac{2}{5}$

6. $\frac{3}{8}$

Answers on page A-7

Find decimal notation.

7. $\frac{1}{6}$

8. $\frac{2}{3}$

Find decimal notation.

9. $\frac{5}{11}$

10. $\frac{12}{11}$

Answers on page A-7

In Examples 5 and 6, the division *terminated,* meaning that eventually we got a remainder of 0. A **terminating decimal** occurs when the denominator has only 2's or 5's, or both, as factors, as in $\frac{17}{25}$, $\frac{5}{8}$, or $\frac{83}{100}$. This assumes that the fraction notation has been simplified.

Consider a different situation:

$$\frac{5}{6}, \quad \text{or} \quad \frac{5}{2 \cdot 3}.$$

Since 6 has a 3 as a factor, the division will not terminate. Although we can still use division to get decimal notation, the answer will be a **repeating decimal,** as follows.

EXAMPLE 7 Find decimal notation for $\frac{5}{6}$.

$$\frac{5}{6} = 5 \div 6 \qquad \begin{array}{r|l} & 0.8\,3\,3 \\ 6 & 5.0\,0\,0 \\ & 4\,8 \\ \hline & 2\,0 \\ & 1\,8 \\ \hline & 2\,0 \\ & 1\,8 \\ \hline & 2 \end{array}$$

Since 2 keeps reappearing as a remainder, the digits repeat and will continue to do so; therefore,

$$\frac{5}{6} = 0.83333\ldots.$$

The red dots indicate an endless sequence of digits in the quotient. When there is a repeating pattern, the dots are often replaced by a bar to indicate the repeating part—in this case, only the 3:

$$\frac{5}{6} = 0.8\overline{3}.$$

Do Exercises 7 and 8.

EXAMPLE 8 Find decimal notation for $\frac{4}{11}$.

$$\frac{4}{11} = 4 \div 11 \qquad \begin{array}{r|l} & 0.3\,6\,3\,6 \\ 11 & 4.0\,0\,0\,0 \\ & 3\,3 \\ \hline & 7\,0 \\ & 6\,6 \\ \hline & 4\,0 \\ & 3\,3 \\ \hline & 7\,0 \\ & 6\,6 \\ \hline & 4 \end{array}$$

Since 7 and 4 keep repeating as remainders, the sequence of digits "36" repeats in the quotient, and

$$\frac{4}{11} = 0.363636\ldots, \quad \text{or} \quad 0.\overline{36}.$$

Do Exercises 9 and 10.

EXAMPLE 9 Find decimal notation for $\frac{5}{7}$.

$$\begin{array}{r} 0.714285 \\ 7\,)\overline{5.000000} \\ \underline{4\,9} \\ 1\,0 \\ \underline{7} \\ 3\,0 \\ \underline{2\,8} \\ 2\,0 \\ \underline{1\,4} \\ 6\,0 \\ \underline{5\,6} \\ 4\,0 \\ \underline{3\,5} \\ 5 \end{array}$$

Since 5 appears as a remainder, the sequence of digits "714285" repeats in the quotient, and

$$\frac{5}{7} = 0.714285714285\ldots, \quad \text{or} \quad 0.\overline{714285}.$$

The length of a repeating part can be very long—too long to find on a calculator. An example is $\frac{5}{97}$, which has a repeating part of 96 digits.

Do Exercise 11.

11. Find decimal notation for $\frac{3}{7}$.

b Rounding in Problem Solving

In applied problems, repeating decimals are rounded to get approximate answers. To round a repeating decimal, we can extend the decimal notation at least one place past the rounding digit, and then round as before.

EXAMPLES Round each of the following to the nearest tenth, hundredth, and thousandth.

	Nearest tenth	*Nearest hundredth*	*Nearest thousandth*
10. $0.8\overline{3} = 0.83333\ldots$	0.8	0.83	0.833
11. $0.\overline{09} = 0.090909\ldots$	0.1	0.09	0.091
12. $0.\overline{714285} = 0.714285714285\ldots$	0.7	0.71	0.714

Do Exercises 12–14.

Round each to the nearest tenth, hundredth, and thousandth.

12. $0.\overline{6}$

13. $0.\overline{80}$

14. $6.\overline{245}$

CONVERTING RATIOS TO DECIMAL NOTATION

When solving applied problems, we often convert ratios to decimal notation.

EXAMPLE 13 *Forest Fires.* The National Forest Service reports that in a recent year, 6.4 million acres were burned by 73,000 fires. Find the ratio of number of acres burned to number of fires and convert it to decimal notation. Round to the nearest thousandth.

Source: National Forest Service

Answers on page A-7

15. **Coin Tossing.** A coin is tossed 51 times. It lands heads 26 times. Find the ratio of heads to tosses and convert it to decimal notation. Round to the nearest thousandth. (This is also the experimental probability of getting heads.)

16. **Gas Mileage.** A car goes 380 mi on 15.7 gal of gasoline. Find the gasoline mileage and convert the ratio to decimal notation rounded to the nearest tenth.

17. **SUV Models.** Refer to the data in bold on the bar graph in Example 15. Find the average number of SUV models available per year for the 5-yr period. Round to the nearest tenth.

Answers on page A-7

We have

$$\frac{\text{Acres burned}}{\text{Number of fires}} = \frac{6{,}400{,}000 \text{ acres}}{73{,}000 \text{ fires}} \approx 87.671.$$

There were about 87.671 acres burned per fire.

EXAMPLE 14 *Gas Mileage.* A car goes 457 mi on 16.4 gal of gasoline. The ratio of number of miles driven to amount of gasoline used is *gas mileage.* Find the gas mileage and convert the ratio to decimal notation rounded to the nearest tenth.

$$\frac{\text{Miles driven}}{\text{Gasoline used}} = \frac{457}{16.4} \approx 27.9$$

The gas mileage is 27.9 miles to the gallon.

Do Exercises 15 and 16.

AVERAGES

When finding an average, we may at times need to round an answer.

EXAMPLE 15 *Sport Utility Vehicles.* Sport utility vehicles have experienced a great explosion in sales. The following bar graph shows total sales, in millions, and the number of models available (in bold). Find the average number of vehicles sold per year for the period from 1995 to 2001. Round the answer to the nearest hundredth.

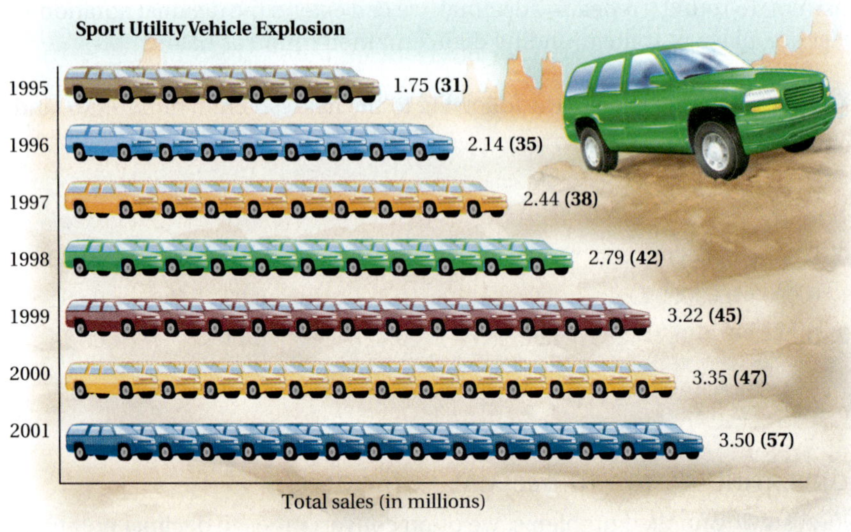

Source: Autodata

We add the sales totals shown on the bar graph and divide by the number of addends, 7. Since all the units are in millions, we need not convert them to standard notation. The average is

$$\frac{1.75 + 2.14 + 2.44 + 2.79 + 3.22 + 3.35 + 3.50}{7} = \frac{19.19}{7} = 2.7414\ldots \approx 2.74.$$

The average number of SUVs sold per year for the 7-yr period is about 2.74 million.

Do Exercise 17.

C Calculations with Fraction and Decimal Notation Together

In certain kinds of calculations, fraction and decimal notation might occur together. In such cases, there are at least three ways in which we might proceed.

EXAMPLE 16 Calculate: $\frac{2}{3} \times 0.576$.

METHOD 1 One way to do this calculation is to convert the fraction notation to decimal notation so that both numbers are in decimal notation. Since $\frac{2}{3}$ converts to repeating decimal notation, it is first rounded to some chosen decimal place. We choose three decimal places. Then, using decimal notation, we multiply.

$$\frac{2}{3} \times 0.576 = 0.\overline{6} \times 0.576 \approx 0.667 \times 0.576 = 0.384192$$

METHOD 2 A second way to do this calculation is to convert the decimal notation to fraction notation so that both numbers are in fraction notation. The answer can be left in fraction notation and simplified, or we can convert back to decimal notation and round, if appropriate.

$$\begin{aligned}\frac{2}{3} \times 0.576 &= \frac{2}{3} \cdot \frac{576}{1000} = \frac{2 \cdot 576}{3 \cdot 1000} \\ &= \frac{2 \cdot 2 \cdot 2 \cdot 2 \cdot 2 \cdot 2 \cdot 2 \cdot 3 \cdot 3}{2 \cdot 2 \cdot 2 \cdot 3 \cdot 5 \cdot 5 \cdot 5} \\ &= \frac{2 \cdot 2 \cdot 2 \cdot 3}{2 \cdot 2 \cdot 2 \cdot 3} \cdot \frac{2 \cdot 2 \cdot 2 \cdot 2 \cdot 3}{5 \cdot 5 \cdot 5} \\ &= 1 \cdot \frac{2 \cdot 2 \cdot 2 \cdot 2 \cdot 3}{5 \cdot 5 \cdot 5} \\ &= \frac{2 \cdot 2 \cdot 2 \cdot 2 \cdot 3}{5 \cdot 5 \cdot 5} = \frac{48}{125}, \text{ or } 0.384\end{aligned}$$

METHOD 3 A third way to do this calculation is to treat 0.576 as $\frac{0.576}{1}$. Then we multiply 0.576 by 2, and divide the result by 3.

$$\frac{2}{3} \times 0.576 = \frac{2}{3} \times \frac{0.576}{1} = \frac{2 \times 0.576}{3} = \frac{1.152}{3} = 0.384$$

Do Exercise 18.

EXAMPLE 17 Calculate: $\frac{2}{3} \times 0.576 + 3.287 \div \frac{4}{5}$.

We use the rules for order of operations, doing first the multiplication and then the division. Then we add.

$$\begin{aligned}\frac{2}{3} \times 0.576 + 3.287 \div \frac{4}{5} &= 0.384 + 3.287 \cdot \frac{5}{4} \\ &= 0.384 + 4.10875 \\ &= 4.49275\end{aligned}$$

Method 3:
$\frac{2}{3} \times \frac{0.576}{1} = 0.384$;
$\frac{3.287}{1} \times \frac{5}{4} = 4.10875$

Do Exercises 19 and 20.

18. Calculate: $\frac{5}{6} \times 0.864$.

Calculate.

19. $\frac{1}{3} \times 0.384 + \frac{5}{8} \times 0.6784$

20. $\frac{5}{6} \times 0.864 + 14.3 \div \frac{8}{5}$

Answers on page A-7

3.5 EXERCISE SET

For Extra Help

Digital Video Tutor CD 3 Videotape 4 | InterAct Math | Math Tutor Center | MathXL | MyMathLab

a Find decimal notation.

1. $\frac{23}{100}$ **2.** $\frac{9}{100}$ **3.** $\frac{3}{5}$ **4.** $\frac{19}{20}$ **5.** $\frac{13}{40}$ **6.** $\frac{3}{16}$

7. $\frac{1}{5}$ **8.** $\frac{4}{5}$ **9.** $\frac{17}{20}$ **10.** $\frac{11}{20}$ **11.** $\frac{3}{8}$ **12.** $\frac{7}{8}$

13. $\frac{39}{40}$ **14.** $\frac{31}{40}$ **15.** $\frac{13}{25}$ **16.** $\frac{61}{125}$ **17.** $\frac{2502}{125}$ **18.** $\frac{181}{200}$

19. $\frac{1}{4}$ **20.** $\frac{1}{2}$ **21.** $\frac{29}{25}$ **22.** $\frac{37}{25}$ **23.** $\frac{19}{16}$ **24.** $\frac{5}{8}$

25. $\frac{4}{15}$ **26.** $\frac{7}{9}$ **27.** $\frac{1}{3}$ **28.** $\frac{1}{9}$ **29.** $\frac{4}{3}$ **30.** $\frac{8}{9}$

31. $\frac{7}{6}$ **32.** $\frac{7}{11}$ **33.** $\frac{4}{7}$ **34.** $\frac{14}{11}$ **35.** $\frac{11}{12}$ **36.** $\frac{5}{12}$

b

37.–47. Odds. Round each answer of the odd-numbered Exercises 25–35 to the nearest tenth, hundredth, and thousandth.

38.–48. Evens. Round each answer of the even-numbered Exercises 26–36 to the nearest tenth, hundredth, and thousandth.

Round each to the nearest tenth, hundredth, and thousandth.

49. $0.\overline{18}$ **50.** $0.\overline{83}$ **51.** $0.2\overline{7}$ **52.** $3.5\overline{4}$

53. For this set of people, what is the ratio, in decimal notation rounded to the nearest thousandth, where appropriate, of:

a) women to the total number of people?
b) women to men?
c) men to the total number of people?
d) men to women?

54. For this set of nuts and bolts, what is the ratio, in decimal notation rounded to the nearest thousandth, where appropriate, of:

a) nuts to bolts?
b) bolts to nuts?
c) nuts to the total?
d) total number to nuts?

Gas Mileage. In each of Exercises 55–58, find the gas mileage rounded to the nearest tenth.

55. 285 mi; 18 gal

56. 396 mi; 17 gal

57. 324.8 mi; 18.2 gal

58. 264.8 mi; 12.7 gal

59. *Windy Cities.* Although nicknamed the Windy City, Chicago is not the windiest city in the United States. Listed in the table below are the six windiest cities and their average wind speeds. Find the average of these wind speeds and round your answer to the nearest tenth.
Source: *The Handy Geography Answer Book*

CITY	AVERAGE WIND SPEED (in miles per hour)
Mt. Washington, NH	35.3
Boston, MA	12.5
Honolulu, HI	11.3
Dallas, TX	10.7
Kansas City, MO	10.7
Chicago, IL	10.4

60. *Areas of the New England States.* The table below lists the areas of the New England states. Find the average area and round your answer to the nearest tenth.
Source: *The New York Times Almanac*

STATE	TOTAL AREA (in square miles)
Maine	33,265
New Hampshire	9,279
Vermont	9,614
Massachusetts	8,284
Connecticut	5,018
Rhode Island	1,211

Stock Prices. At one time stock prices were given using mixed numerals involving halves, fourths, eighths, and, more recently, sixteenths. The Securities and Exchange Commission has mandated the use of decimal notation. Thus a price of $\$23\frac{13}{16}$ is now converted to decimal notation rounded to the nearest hundredth, that is, \$23.81. Complete the following table.

Sources: *The Indianapolis Star*, 1/30/01; www.yahoo.com

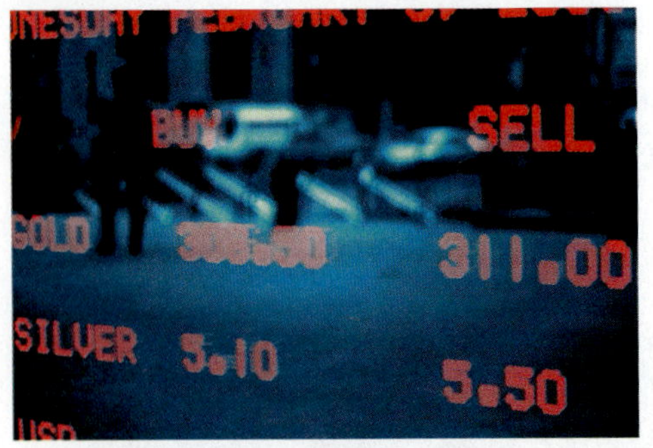

	STOCK	PRICE PER SHARE	DECIMAL NOTATION	ROUNDED TO NEAREST HUNDREDTH
61.	General Mills	$\$41\frac{11}{16}$		
62.	Quaker Oats	$\$98\frac{15}{16}$		
63.	Kellogg	$\$25\frac{7}{8}$		
64.	Dillard's	$\$20\frac{5}{8}$		
65.	Hudson's Bay	$\$19\frac{3}{64}$		
66.	Abercrombie & Fitch	$\$31\frac{47}{64}$		

C Calculate.

67. $\frac{7}{8} \times 12.64$

68. $\frac{4}{5} \times 384.8$

69. $2\frac{3}{4} + 5.65$

70. $4\frac{4}{5} + 3.25$

71. $\frac{47}{9} \times 79.95$

72. $\frac{7}{11} \times 2.7873$

73. $\frac{1}{2} - 0.5$

74. $3\frac{1}{8} - 2.75$

75. $4.875 - 2\frac{1}{16}$

76. $55\frac{3}{5} - 12.22$

77. $\frac{5}{6} \times 0.0765 + \frac{5}{4} \times 0.1124$

78. $\frac{3}{5} \times 6384.1 - \frac{3}{8} \times 156.56$

79. $\frac{4}{5} \times 384.8 + 24.8 \div \frac{8}{3}$

80. $102.4 \div \frac{2}{5} - 12 \times \frac{5}{6}$

81. $\frac{7}{8} \times 0.86 - 0.76 \times \frac{3}{4}$

82. $17.95 \div \frac{5}{8} + \frac{3}{4} \times 16.2$

83. $3.375 \times 5\frac{1}{3}$

84. $2.5 \times 3\frac{5}{8}$

85. $6.84 \div 2\frac{1}{2}$

86. $8\frac{1}{2} \div 2.125$

87. D_W When is long division *not* the fastest way to convert from fraction notation to decimal notation?

88. D_W Examine Example 16 of this section. How could the problem be changed so that method 1 would give a result that is completely accurate?

SKILL MAINTENANCE

Multiply. [2.4d]

89. $9 \cdot 2\frac{1}{3}$

90. $10\frac{1}{2} \cdot 22\frac{3}{4}$

Divide. [2.4e]

91. $84 \div 8\frac{2}{5}$

92. $8\frac{3}{5} \div 10\frac{2}{5}$

Add. [2.4b]

93. $17\frac{5}{6} + 32\frac{3}{8}$

94. $14\frac{3}{5} + 16\frac{1}{10}$

Subtract. [2.4c]

95. $16\frac{1}{10} - 14\frac{3}{5}$

96. $32\frac{3}{8} - 17\frac{5}{6}$

Solve. [2.5a]

97. A recipe for bread calls for $\frac{2}{3}$ cup of water, $\frac{1}{4}$ cup of milk, and $\frac{1}{8}$ cup of oil. How many cups of liquid ingredients does the recipe call for?

98. A board $\frac{7}{10}$ in. thick is glued to a board $\frac{3}{5}$ in. thick. The glue is $\frac{3}{100}$ in. thick. How thick is the result?

Find the LCM. [1.9a]

99. 15, 27, and 30

100. 8, 11, and 36

SYNTHESIS

Find decimal notation.

101. $\frac{1}{7}$

102. $\frac{2}{7}$

103. $\frac{3}{7}$

104. $\frac{4}{7}$

105. $\frac{5}{7}$

106. From the pattern of Exercises 101–105, guess the decimal notation for $\frac{6}{7}$. Check on your calculator.

Find decimal notation.

107. $\frac{1}{9}$

108. $\frac{1}{99}$

109. $\frac{1}{999}$

110. From the pattern of Exercises 107–109, guess the decimal notation for $\frac{1}{9999}$. Check on your calculator.

3.6 ESTIMATING

Objective

a Estimate sums, differences, products, and quotients.

1. Estimate by rounding to the nearest ten the total cost of one TV and one vacuum cleaner. Which of the following is an appropriate estimate?

a) $5700   b) $570
c) $500   d) $57

2. About how much more does the TV cost than the vacuum cleaner? Estimate by rounding to the nearest ten. Which of the following is an appropriate estimate?

a) $130   b) $1300
c) $580   d) $13

Answers on page A-7

Study Tips

ASKING QUESTIONS

Don't be afraid to ask questions in class. Most instructors welcome this and encourage students to ask them. Other students probably have the same questions you do.

"Better to ask twice than lose your way once."

Danish Proverb

a Estimating Sums, Differences, Products, and Quotients

Estimating has many uses. It can be done before a problem is even attempted in order to get an idea of the answer. It can be done afterward as a check, even when we are using a calculator. In many situations, an estimate is all we need. We usually estimate by rounding the numbers so that there are one or two nonzero digits, depending on how accurate we want our estimate. Consider the following advertisements for Examples 1–4.

EXAMPLE 1 Estimate by rounding to the nearest ten the total cost of one fax machine and one TV.

We are estimating the sum

$$\$149.95 + \$346.95 = \text{Total cost}.$$

The estimate found by rounding the addends to the nearest ten is

$$\$150 + \$350 = \$500. \quad \text{(Estimated total cost)}$$

Do Exercise 1.

EXAMPLE 2 About how much more does the TV cost than the fax machine? Estimate by rounding to the nearest ten.

We are estimating the difference

$$\$346.95 - \$149.95 = \text{Price difference}.$$

The estimate to the nearest ten is

$$\$350 - \$150 = \$200. \quad \text{(Estimated price difference)}$$

Do Exercise 2.

EXAMPLE 3 Estimate the total cost of 4 vacuum cleaners.

We are estimating the product

$4 \times \$219.95 =$ Total cost.

The estimate is found by rounding $219.95 to the nearest ten:

$4 \times \$220 = \$880.$

Do Exercise 3.

EXAMPLE 4 About how many fax machines can be purchased for $1480?

We estimate the quotient

$\$1480 \div \$149.95.$

Since we want a whole-number estimate, we choose our rounding appropriately. Rounding $149.95 to the nearest one, we get $150. Since $1480 is close to $1500, which is a multiple of 150, we estimate

$\$1500 \div \$150,$

so the answer is 10.

Do Exercise 4.

EXAMPLE 5 Estimate: 4.8×52. Do not find the actual product. Which of the following is an appropriate estimate?

a) 25 b) 250 c) 2500 d) 360

We have

$5 \times 50 = 250.$ (Estimated product)

We rounded 4.8 to the nearest one and 52 to the nearest ten. Thus an appropriate estimate is (b).

Compare these estimates for the product 4.94×38:

$5 \times 40 = 200,$ $5 \times 38 = 190,$ $4.9 \times 40 = 196.$

The first estimate was the easiest. You could probably do it mentally. The others had more nonzero digits.

Do Exercises 5–10.

3. Estimate the total cost of 6 TVs. Which of the following is an appropriate estimate?

a) $4400 b) $350
c) $21,000 d) $2100

4. About how many vacuum cleaners can be purchased for $1100? Which of the following is an appropriate estimate?

a) 8 b) 5
c) 11 d) 124

Estimate the product. Do not find the actual product. Which of the following is an appropriate estimate?

5. 2.4×8

a) 16 b) 34
c) 125 d) 5

6. 24×0.6

a) 200 b) 5
c) 110 d) 20

7. 0.86×0.432

a) 0.04 b) 0.4
c) 1.1 d) 4

8. 0.82×0.1

a) 800 b) 8
c) 0.08 d) 80

9. 0.12×18.248

a) 180 b) 1.8
c) 0.018 d) 18

10. 24.234×5.2

a) 200 b) 125
c) 12.5 d) 234

Answers on page A-7

Estimate the quotient. Which of the following is an appropriate estimate?

11. 59.78 ÷ 29.1

a) 200 **b)** 20
c) 2 **d)** 0.2

12. 82.08 ÷ 2.4

a) 40 **b)** 4.0
c) 400 **d)** 0.4

13. 0.1768 ÷ 0.08

a) 8 **b)** 10
c) 2 **d)** 20

14. Estimate: 0.0069 ÷ 0.15. Which of the following is an appropriate estimate?

a) 0.5 **b)** 50
c) 0.05 **d)** 0.004

Answers on page A-7

EXAMPLE 6 Estimate: 82.08 ÷ 24. Which of the following is an appropriate estimate?

a) 400 b) 16 c) 40 d) 4

This is about 80 ÷ 20, so the answer is about 4. Thus an appropriate estimate is (d).

EXAMPLE 7 Estimate: 94.18 ÷ 3.2. Which of the following is an appropriate estimate?

a) 30 b) 300 c) 3 d) 60

This is about 90 ÷ 3, so the answer is about 30. Thus an appropriate estimate is (a).

EXAMPLE 8 Estimate: 0.0156 ÷ 1.3. Which of the following is an appropriate estimate?

a) 0.2 b) 0.002 c) 0.02 d) 20

This is about 0.02 ÷ 1, so the answer is about 0.02. Thus an appropriate estimate is (c).

Do Exercises 11–13.

In some cases, it is easier to estimate a quotient directly rather than by rounding the divisor and the dividend.

EXAMPLE 9 Estimate: 0.0074 ÷ 0.23. Which of the following is an appropriate estimate?

a) 0.3 b) 0.03 c) 300 d) 3

We estimate 3 for a quotient. We check by multiplying.

$0.23 \times 3 = 0.69$

We make the estimate smaller. We estimate 0.3 and check by multiplying.

$0.23 \times 0.3 = 0.069$

We make the estimate smaller. We estimate 0.03 and check by multiplying.

$0.23 \times 0.03 = 0.0069$

This is about 0.0074, so the quotient is about 0.03. Thus an appropriate estimate is (b).

Do Exercise 14.

3.6 EXERCISE SET

For Extra Help

Digital Video Tutor CD 3 Videotape 4 | InterAct Math | Math Tutor Center | MathXL | MyMathLab

a Consider the following advertisements for Exercises 1–8. Estimate the sums, differences, products, or quotients involved in these problems. Indicate which of the choices is an appropriate estimate.

1. Estimate the total cost of one entertainment center and one sound system.
a) \$36 **b)** \$72 **c)** \$3.60 **d)** \$360

2. Estimate the total cost of one entertainment center and one TV.
a) \$410 **b)** \$820 **c)** \$41 **d)** \$4.10

3. About how much more does the TV cost than the sound system?
a) \$500 **b)** \$80 **c)** \$50 **d)** \$5

4. About how much more does the TV cost than the entertainment center?
a) \$100 **b)** \$190 **c)** \$250 **d)** \$150

5. Estimate the total cost of 9 TVs.
a) \$2700 **b)** \$27 **c)** \$270 **d)** \$540

6. Estimate the total cost of 16 sound systems.
a) \$5010 **b)** \$4000 **c)** \$40 **d)** \$410

7. About how many TVs can be purchased for \$1700?
a) 600 **b)** 72 **c)** 6 **d)** 60

8. About how many sound systems can be purchased for \$1300?
a) 10 **b)** 5 **c)** 50 **d)** 500

Estimate by rounding as directed.

9. $0.02 + 1.31 + 0.34$; nearest tenth

10. $0.88 + 2.07 + 1.54$; nearest one

11. $6.03 + 0.007 + 0.214$; nearest one

12. $1.11 + 8.888 + 99.94$; nearest one

13. $52.367 + 1.307 + 7.324$; nearest one

14. $12.9882 + 1.0115$; nearest tenth

15. 2.678 − 0.445; nearest tenth

16. 12.9882 − 1.0115; nearest one

17. 198.67432 − 24.5007; nearest ten

Estimate. Choose a rounding digit that gives one or two nonzero digits. Indicate which of the choices is an appropriate estimate.

18. 234.12321 − 200.3223

a) 600 **b)** 60
c) 300 **d)** 30

19. 49 × 7.89

a) 400 **b)** 40
c) 4 **d)** 0.4

20. 7.4 × 8.9

a) 95 **b)** 63
c) 124 **d)** 6

21. 98.4 × 0.083

a) 80 **b)** 12
c) 8 **d)** 0.8

22. 78 × 5.3

a) 400 **b)** 800
c) 40 **d)** 8

23. 3.6 ÷ 4

a) 10 **b)** 1
c) 0.1 **d)** 0.01

24. 0.0713 ÷ 1.94

a) 4 **b)** 0.4
c) 0.04 **d)** 40

25. 74.68 ÷ 24.7

a) 9 **b)** 3
c) 12 **d)** 120

26. 914 ÷ 0.921

a) 10 **b)** 100
c) 1000 **d)** 1

27. *Palm VIIxe and the Sears Tower.* The Palm VIIxe PDA (Personal Digital Assistant) is 4.7 in. (about 0.39167 ft) high. Estimate how many PDAs it would take, if placed end to end, to reach from the ground to the top of the Sears Tower, which is 1454 ft tall. Round to the nearest one.
Source: www. yahoo.com

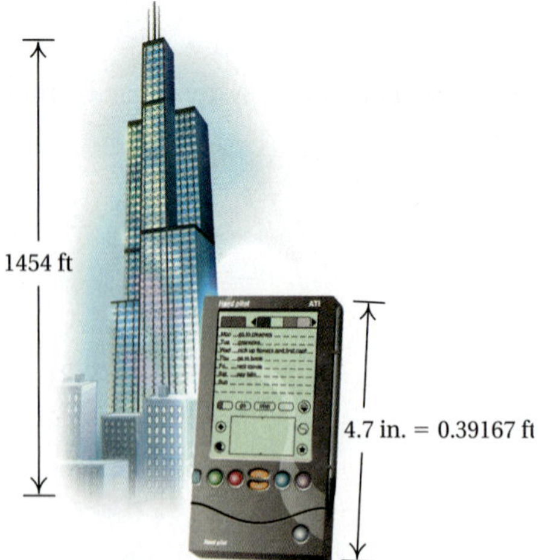

28. *Ticketmaster.* Recently, Ticketmaster stock sold for $8.63 per share. Estimate how many shares can be purchased for $27,000.

29. $^{\mathrm{D}}{}_{\mathrm{W}}$ Describe a situation in which an estimation is made by rounding to the nearest 10,000 and then multiplying.

30. $^{\mathrm{D}}{}_{\mathrm{W}}$ A roll of fiberglass insulation costs \$21.95. Describe two situations involving estimating and the cost of fiberglass insulation. Devise one situation so that \$21.95 is rounded to \$22. Devise the other situation so that \$21.95 is rounded to \$20.

SKILL MAINTENANCE

Find the prime factorization. [1.7d]

31. 108

32. 400

33. 325

34. 666

35. 1728

Simplify. [2.1e]

36. $\frac{125}{400}$

37. $\frac{3225}{6275}$

38. $\frac{72}{81}$

39. $\frac{325}{625}$

40. $\frac{625}{475}$

SYNTHESIS

The following were done on a calculator. Estimate to determine whether the decimal point was placed correctly.

41. $178.9462 \times 61.78 = 11{,}055.29624$

42. $14{,}973.35 \div 298.75 = 501.2$

43. $19.7236 - 1.4738 \times 4.1097 = 1.366672414$

44. $28.46901 \div 4.9187 - 2.5081 = 3.279813473$

45. 🖩 Use one of $+$, $-$, $\times$, and $\div$ in each blank to make a true sentence.

a) $(0.37 \square 18.78) \square 2^{13} = 156{,}876.8$
b) $2.56 \square 6.4 \square 51.2 \square 17.4 = 312.84$

46. 🖩 In the subtraction below, a and b are digits. Find a and b.

$$\begin{array}{r} b876.a4321 \\ -1234.a678b \\ \hline 8641.b7a32 \end{array}$$

3.7 APPLICATIONS AND PROBLEM SOLVING

Objective

a Solve applied problems involving decimals.

1. **Body Temperature.** Normal body temperature is 98.6°F. When fevered, most people will die if their bodies reach 107°F. This is a rise of how many degrees?

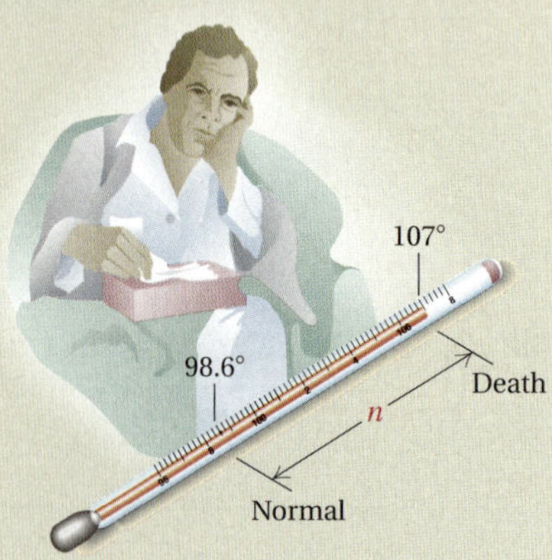

Answer on page A-7

a Solving Applied Problems

Solving applied problems with decimals is like solving applied problems with whole numbers. We translate first to an equation that corresponds to the situation. Then we solve the equation.

EXAMPLE 1 *Quaker Oats Stock Prices.* The Quaker Oats Company is a manufacturer of hot cereals, pancake syrups, grain-based snacks, pancake mixes, and pasta products. Over the entire 52 weeks in 2000, the price per share of its stock ranged in value from a low of \$45.81 to a high of \$98.94. By how much did the high value differ from the low value?

Sources: The New York Stock Exchange; The Quaker Oats Company

1. **Familiarize.** The stock prices are charted in the graph above. We let c = the amount that the price per share rose over the 52-week period.
2. **Translate.** This is a "missing-addend" situation. We translate as follows, using the given information.

Price at the start	plus	Amount of increase	is	Price at the end
\$45.81	+	c	=	\$98.94

3. **Solve.** We solve the equation by subtracting \$45.81 on both sides:

$$45.81 + c = 98.94$$
$$45.81 + c - 45.81 = 98.94 - 45.81$$
$$c = 53.13.$$

$$\begin{array}{r} 98.94 \\ -\ 45.81 \\ \hline 53.13 \end{array}$$

4. **Check.** We can check by adding 53.13 to 45.81 to get 98.94.
5. **State.** The stock rose by \$53.13 per share over the 52-week period.

Do Exercise 1.

EXAMPLE 2 *Injections of Medication.* A patient was given injections of 2.8 mL, 1.35 mL, 2.0 mL, and 1.88 mL over a 24-hr period. What was the total amount of the injections?

1. **Familiarize.** We make a drawing or at least visualize the situation. We let $t =$ the amount of the injections.

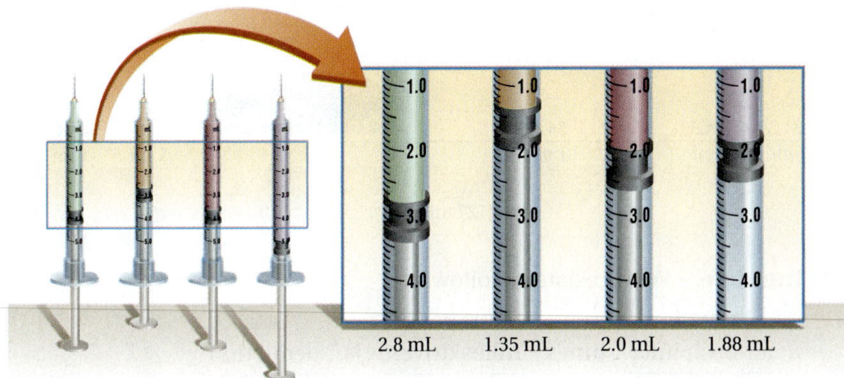

2. **Translate.** Amounts are being combined. We translate to an equation:

$$\underbrace{\text{First}}_{2.8} \quad \underbrace{\text{plus}}_{+} \quad \underbrace{\text{second}}_{1.35} \quad \underbrace{\text{plus}}_{+} \quad \underbrace{\text{third}}_{2.0} \quad \underbrace{\text{plus}}_{+} \quad \underbrace{\text{fourth}}_{1.88} \quad \underbrace{\text{is}}_{=} \quad \underbrace{\text{total.}}_{t}$$

3. **Solve.** To solve, we carry out the addition.

$$\begin{array}{r} \scriptstyle 2\ 1 \\ 2.8\ 0 \\ 1.3\ 5 \\ 2.0\ 0 \\ +\ 1.8\ 8 \\ \hline 8.0\ 3 \end{array}$$

Thus, $t = 8.03$.

4. **Check.** We can check by repeating our addition. We can also see whether our answer is reasonable by first noting that it is indeed larger than any of the numbers being added. We can also partially check by rounding:

$$2.8 + 1.35 + 2.0 + 1.88 \approx 3 + 1 + 2 + 2$$
$$= 8 \approx 8.03.$$

If we had gotten an answer like 80.3 or 0.803, then our estimate, 8, would have told us that we did something wrong, like not lining up the decimal points.

5. **State.** The total amount of the injections was 8.03 mL.

Do Exercise 2.

2. **Liquid Consumption.** Each year, the average American drinks about 49.0 gal of soft drinks, 41.2 gal of water, 25.3 gal of milk, 24.8 gal of coffee, and 7.8 gal of fruit juice. What is the total amount that the average American drinks?

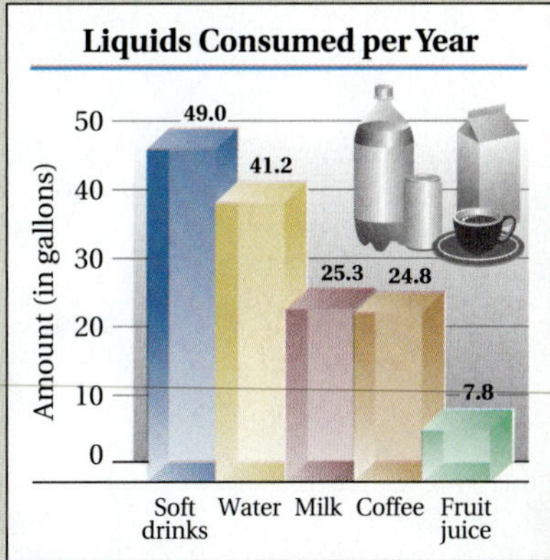

Source: U.S. Department of Agriculture

Answer on page A-7

3. Printing Costs. At a printing company, the cost of copying is 11 cents per page. How much, in dollars, would it cost to make 466 copies?

EXAMPLE 3 *IRS Driving Allowance.* In 2001, the Internal Revenue Service allowed a tax deduction of 34.5¢ per mile for mileage driven for business purposes. What deduction, in dollars, would be allowed for driving 127 mi?
Source: Internal Revenue Service

1. **Familiarize.** We first make a drawing or at least visualize the situation. Repeated addition fits this situation. We let $d =$ the deduction, in dollars, allowed for driving 127 mi.

2. **Translate.** We translate as follows.

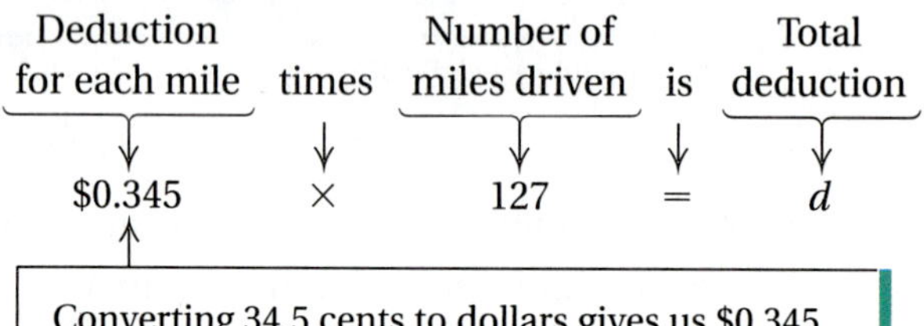

Deduction for each mile
times
Number of miles driven
is
Total deduction
$0.345
×
127
=
d

Converting 34.5 cents to dollars gives us $0.345.

3. **Solve.** To solve the equation, we carry out the multiplication.

$$\begin{array}{r} 1\,2\,7 \\ \times \quad 0.3\,4\,5 \\ \hline 6\,3\,5 \\ 5\,0\,8\,0 \\ 3\,8\,1\,0\,0 \\ \hline 4\,3.8\,1\,5 \end{array}$$

Thus, $d = 43.815 \approx \$43.82$.

4. **Check.** We can obtain a partial check by rounding and estimating:

$$127 \times 0.345 \approx 130 \times 0.3$$
$$= 39 \approx 43.82.$$

5. **State.** The total allowable deduction would be $43.82.

Do Exercise 3.

EXAMPLE 4 *Loan Payments.* A car loan of $7382.52 is to be paid off in 36 monthly payments. How much is each payment?

1. **Familiarize.** We first make a drawing. We let $n =$ the amount of each payment.

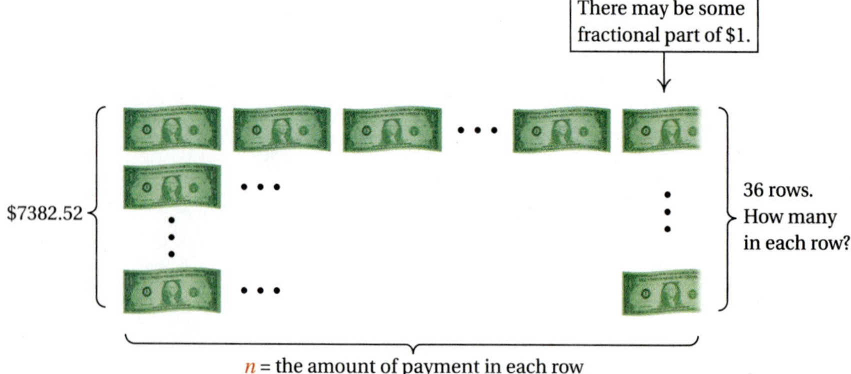

There may be some fractional part of $1.

36 rows. How many in each row?

Answer on page A-7

2. **Translate.** The problem can be translated to the following equation, thinking that

$$(\text{Total loan}) \div (\text{Number of payments}) = \text{Amount of each payment}$$
$$\$7382.52 \div 36 = n.$$

3. **Solve.** To solve the equation, we carry out the division.

$$\begin{array}{r} 205.07 \\ 36\overline{)\,7382.52} \\ \underline{7200000} \\ 18252 \\ \underline{18000} \\ 252 \\ \underline{252} \\ 0 \end{array}$$

Thus, $n = 205.07$.

4. **Check.** A partial check can be obtained by estimating the quotient: $\$7382.56 \div 36 \approx 8000 \div 40 = 200 \approx 205.07$. The estimate checks.

5. **State.** Each payment is \$205.07.

Do Exercise 4.

EXAMPLE 5 *Jackie Robinson Poster.* A special limited-edition poster was painted by sports artist Leroy Neiman. Commissioned by Barton L. Kaufman, it commemorates the entrance of the first African-American, Jackie Robinson, into major league baseball in 1947. The dimensions of the poster are 19.3 in. by 27.4 in. Find the area.

Source: Barton L. Kaufman, private collection

1. **Familiarize.** We first make a drawing. We let A = the area.

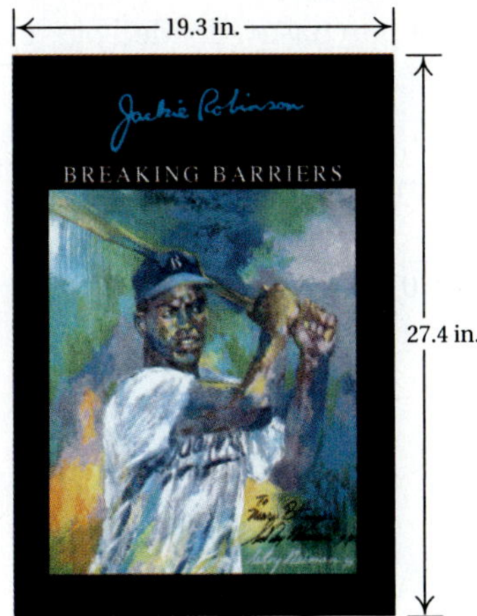

4. **Loan Payments.** A loan of \$4425 is to be paid off in 12 monthly payments. How much is each payment?

Answer on page A-7

Study Tips

FIVE STEPS FOR PROBLEM SOLVING

Are you remembering to use the following five steps for problem solving that were developed in Section 1.5?

1. **Familiarize** yourself with the situation.
   a) Carefully read and reread until you understand *what* you are being asked to find.
   b) Draw a diagram or see if there is a formula that applies.
   c) Assign a letter, or *variable*, to the unknown.
2. **Translate** the problem to an equation using the letter or variable.
3. **Solve** the equation.
4. **Check** the answer in the original wording of the problem.
5. **State** the answer to the problem clearly with appropriate units.

5. **Index Cards.** A standard-size index card measures 12.7 cm by 7.6 cm. Find its area.

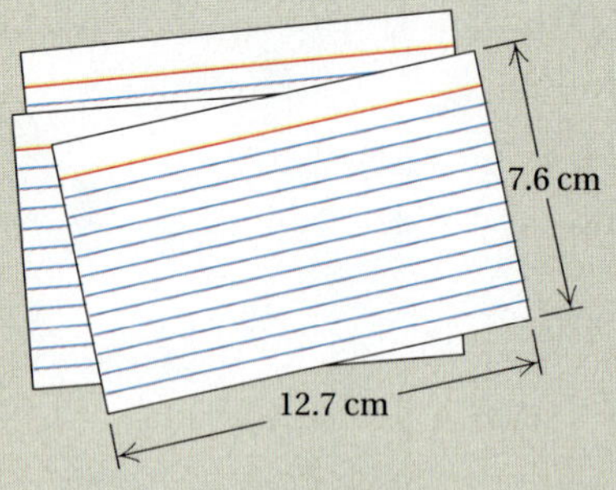

2. **Translate.** We use the formula $A = l \cdot w$ and substitute.

$$A = l \cdot w$$
$$A = 27.4 \times 19.3.$$

3. **Solve.** We solve by carrying out the multiplication.

```
    2 7.4
  × 1 9.3
  -------
    8 2 2
  2 4 6 6 0
  2 7 4 0 0
  ---------
  5 2 8.8 2
```

4. **Check.** We obtain a partial check by estimating the product:

$$A = 27.4 \times 19.3 \approx 30 \times 20 = 600.$$

Since this estimate is not too close to 528.82, we might repeat our calculation or change our estimate by rounding to the nearest one. This is left to the student. We see that 528.82 checks.

5. **State.** The area of the Jackie Robinson poster is 528.82 in^2.

Do Exercise 5.

6. One pound of lean boneless ham contains 4.5 servings. It costs $5.99 per pound. What is the cost per serving? Round to the nearest cent.

EXAMPLE 6 *Digital Camera Purchase.* Kelly Real Estate spends $11,998.80 on a set of 24 Olympus D-490 Digital Zoom cameras, so that its realtors can make instant photos and place them on the firm's website. How much did each camera cost?

Source: d-store™: Olympus Camera and Accessories Store

1. **Familiarize.** We let c = the cost of each camera.

2. **Translate.** We translate as follows.

Cost of each camera	is	Total cost of purchase	divided by	Number of cameras purchased
↓	↓	↓	↓	↓
c	=	$11,998.80	÷	24

3. **Solve.** To solve, we carry out the division.

```
          4 9 9.9 5
2 4 ) 1 1,9 9 8.8 0
        9 6
        -----
        2 3 9
        2 1 6
        -----
          2 3 8
          2 1 6
          -----
            2 2 8
            2 1 6
            -----
              1 2 0
              1 2 0
              -----
                  0
```

4. **Check.** We check by estimating $11,998.80 \div 24 \approx 12,000 \div 25 = 480$. Since 480 is close to 499.95, the answer is probably correct.

5. **State.** The cost of each camera was $499.95.

Do Exercise 6.

Answers on page A-7

Multistep Problems

EXAMPLE 7 *Gas Mileage.* A driver filled the gasoline tank and noted that the odometer read 67,507.8. After the next filling, the odometer read 68,006.1. It took 16.5 gal to fill the tank. How many miles per gallon did the driver get?

1. Familiarize. We first make a drawing.

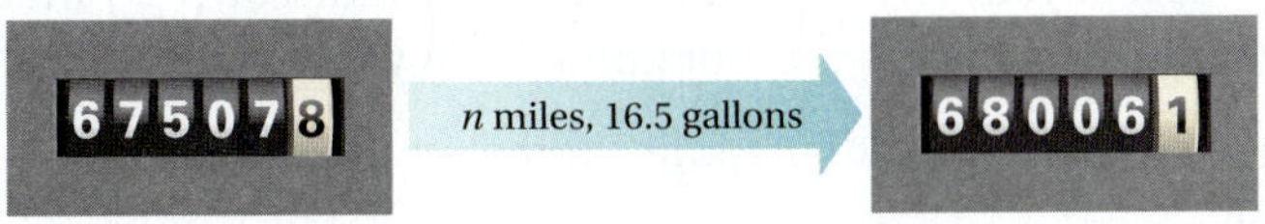

This is a two-step problem. First, we find the number of miles that have been driven between fillups. We let $n =$ the number of miles driven.

2., 3. Translate and **Solve.** This is a "missing-addend" situation. We translate and solve as follows.

First odometer reading	plus	Number of miles driven	is	Second odometer reading
↓	↓	↓	↓	↓
67,507.8	+	n	=	68,006.1

To solve the equation, we subtract 67,507.8 on both sides:

$$n = 68{,}006.1 - 67{,}507.8 = 498.3.$$

$$\begin{array}{r} 68{,}006.1 \\ -\ 67{,}507.8 \\ \hline 498.3 \end{array}$$

Second, we divide the total number of miles driven by the number of gallons. This gives us $m =$ the number of miles per gallon—that is, the mileage. The division that corresponds to the situation is

$$498.3 \div 16.5 = m.$$

To find the number m, we divide.

$$\begin{array}{r} 30.2 \\ 16.5\overline{)\,498.3_{\wedge}0} \\ \underline{4950} \\ 330 \\ \underline{330} \\ 0 \end{array}$$

Thus, $m = 30.2$.

4. Check. To check, we first multiply the number of miles per gallon times the number of gallons:

$$16.5 \times 30.2 = 498.3.$$

Then we add 498.3 to 67,507.8:

$$67{,}507.8 + 498.3 = 68{,}006.1.$$

The mileage 30.2 checks.

5. State. The driver got 30.2 miles per gallon.

Do Exercise 7.

7. Gas Mileage. A driver filled the gasoline tank and noted that the odometer read 38,320.8. After the next filling, the odometer read 38,735.5. It took 14.5 gal to fill the tank. How many miles per gallon did the driver get?

Answer on page A-7

EXAMPLE 8 *Home-Cost Comparison.* Suppose you own a home like the one shown here and it is valued at $250,000 in Indianapolis, Indiana. What would it cost to buy a similar (replacement) home in Palo Alto, California? To find out, we can use an index table prepared by Coldwell Banker Real Estate Corporation. (For a complete index table, contact your local representative.) We use the following formula:

$$\begin{pmatrix}\text{Cost of your}\\ \text{home in new city}\end{pmatrix} = \begin{pmatrix}\text{Value of}\\ \text{your home}\end{pmatrix} \div \begin{pmatrix}\text{Index of}\\ \text{your city}\end{pmatrix} \times \begin{pmatrix}\text{Index of}\\ \text{new city}\end{pmatrix}.$$

Find the cost of your Indianapolis home in Palo Alto. Round to the nearest one.

Source: Coldwell Banker Real Estate Corporation

STATE	CITY	INDEX
California	San Francisco	310
	Palo Alto	398
	Hollywood Hills	271
Indiana	Indianapolis	63
	Fort Wayne	52
Arizona	Phoenix	79
	Tucson	77
Illinois	Barrington	184
	Naperville	99
Texas	Austin	88
	Dallas	73
	Houston	80
Florida	Miami	112
	Orlando	77
	Tampa	74
Minnesota	Minneapolis	112
	St. Paul	94
Georgia	Atlanta	97
New York	Queens/North Shore	159
	Albany	80

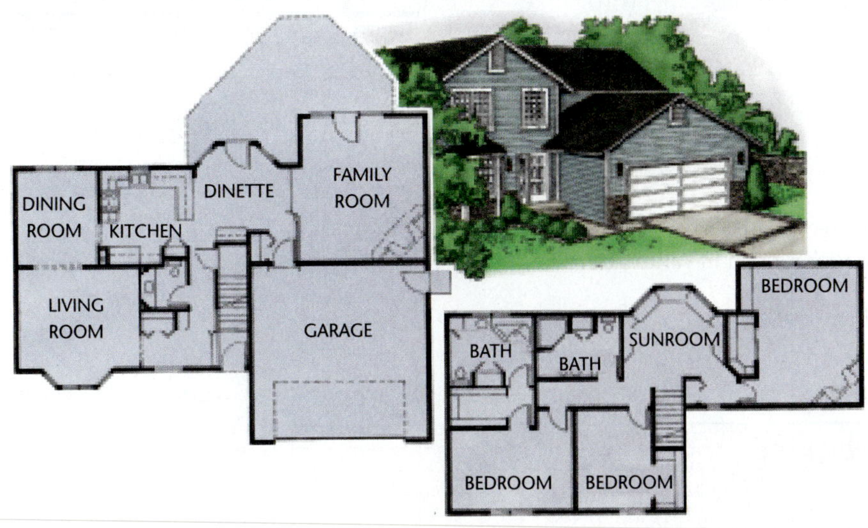

1. **Familiarize.** We let $C =$ the cost of the home in Palo Alto. We use the table and look up the indexes of the city in which you now live and the city to which you are moving.
2. **Translate.** Using the formula, we translate to the following equation:

   $C = \$250{,}000 \div 63 \times 398.$
3. **Solve.** To solve, we carry out the computations using the rules for order of operations (see Section 3.4):

   $C = \$250{,}000 \div 63 \times 398$

   $\approx \$3968.254 \times 398$ — Carrying out the division first

   $\approx \$1{,}579{,}365.$ — Carrying out the multiplication and rounding to the nearest one

   On a calculator, the computation could be done in one step.
4. **Check.** We can repeat our computations.
5. **State.** A home that sells for \$250,000 in Indianapolis would cost about \$1,579,365 in Palo Alto.

Do Exercises 8 and 9.

Refer to the table in Example 8 to answer Margin Exercises 8 and 9.

8. **Home-Cost Comparison.** Find the cost of a \$250,000 home in Indianapolis if you were to try to replace it when moving to Dallas. Round to the nearest one.

9. Find the cost of a \$250,000 home in Phoenix if you were to try to replace it when moving to Barrington. Round to the nearest one.

Answers on page A-7

For Extra Help

Digital Video Tutor CD 3 Videotape 4 | InterAct Math | Math Tutor Center |  MathXL |  MyMathLab

a Solve.

1. *Sherwin-Williams® Stock.* Sherwin-Williams Company specializes in many kinds of home-improvement items. Over the 52 weeks in 2000, the price per share of its stock ranged in value from a low of $17.13 to a high of $27.63. By how much did the high value differ from the low value?
Sources: The New York Stock Exchange; Sherwin-Williams Company

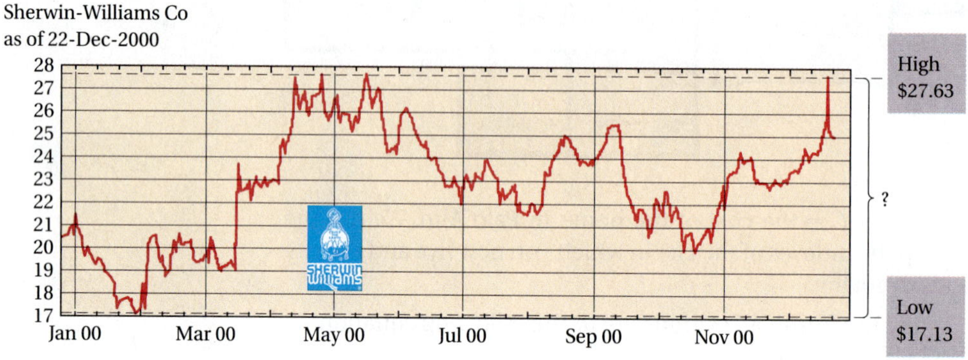

2. *Intel Corporation® Stock.* Intel is a corporation that specializes in microprocessors and other semiconductor products, such as computer chips. Over the 52 weeks in 2000, the price per share of its stock ranged in value from a low of $31.25 to a high of $75.81. By how much did the high value differ from the low value?
Sources: NasdaqNM; Intel Corporation

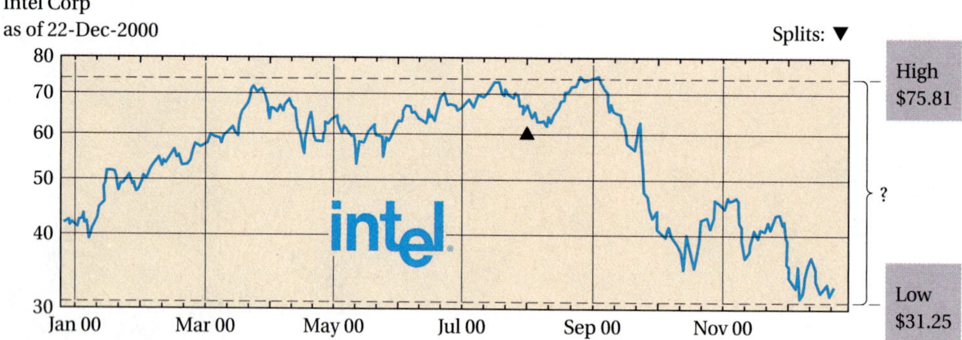

3. Roberto bought the CD "No Strings Attached" by *NSYNC for $14.99 plus $1.14 sales tax. He paid for it with a $20 bill. How much change did he receive?

4. Hannah bought a DVD of the movie *Gladiator* for $25.87 plus $1.55 sales tax. She paid for it with a $50 bill. How much change did she receive?

Russell Crowe holds his Oscar for Best Actor for his role in *Gladiator* at the 73rd Annual Academy Awards.

5. *Body Temperature.* Normal body temperature is 98.6°F. During an illness, a patient's temperature rose 4.2°. What was the new temperature?

6. *Gasoline Cost.* What is the cost, in dollars, of 20.4 gal of gasoline at 159.9 cents per gallon? Round the answer to the nearest cent.

7. *Lottery Winnings.* In Texas, one of the state lotteries is called "Cash 5." In a recent weekly game, the lottery prize of $127,315 was shared equally by 6 winners. How much was each winner's share? Round to the nearest cent.
Source: Texas Lottery

8. *Lunch Costs.* A group of 4 students pays $47.84 for lunch. What is each person's share?

9. *Stamp.* Find the area and the perimeter of the stamp shown here.

10. *Pole Vault Pit.* Find the area and the perimeter of the landing area and the pole vault pit shown here.

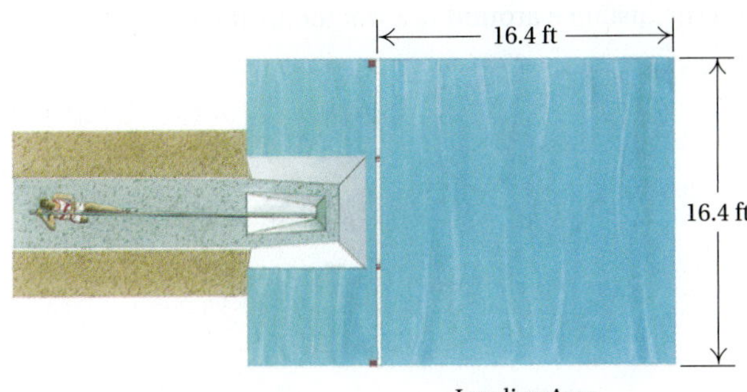

11. *Odometer Reading.* A family checked the odometer before starting a trip. It read 22,456.8 and they know that they will be driving 234.7 mi. What will the odometer read at the end of the trip?

12. *Miles Driven.* Petra bought gasoline when the odometer read 14,296.3. At the next gasoline purchase, the odometer read 14,515.8. How many miles had been driven?

13. *Gas Mileage.* Peggy filled her van's gas tank and noted that the odometer read 26,342.8. After the next filling, the odometer read 26,736.7. It took 19.5 gal to fill the tank. How many miles per gallon did the van get?

14. *Gas Mileage.* Peter filled his Honda's gas tank and noted that the odometer read 18,943.2. After the next filling, the odometer read 19,306.2. It took 13.2 gal to fill the tank. How many miles per gallon did the car get?

15. *Cost of Video Game.* A certain video game costs 25 cents and runs for 1.5 min. Assuming a player does not win any free games and plays continuously, how much money, in dollars, does it cost to play the video game for 1 hr?

16. *Property Taxes.* The Colavitos own a house with an assessed value of $184,500. For every $1000 of assessed value, they pay $7.68 in taxes. How much do they pay in taxes?

17. *Chemistry.* The water in a filled tank weighs 748.45 lb. One cubic foot of water weighs 62.5 lb. How many cubic feet of water does the tank hold?

18. *Highway Routes.* You can drive from home to work using either of two routes:

Route A: Via interstate highway, 7.6 mi, with a speed limit of 65 mph.
Route B: Via a country road, 5.6 mi, with a speed limit of 50 mph.

Assuming you drive at the posted speed limit, which route takes less time? (Use the formula *Distance* = *Speed* $\times$ *Time*.)

Find the distance around (perimeter of) the figure.

19.

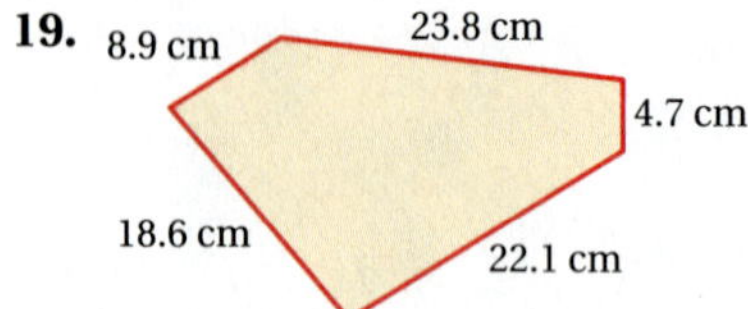

20.

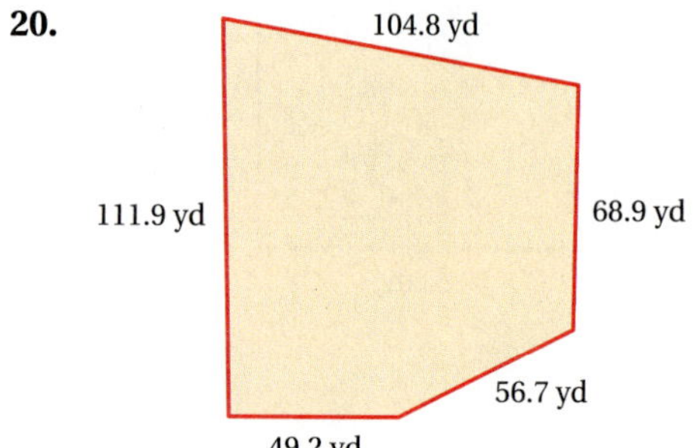

21.

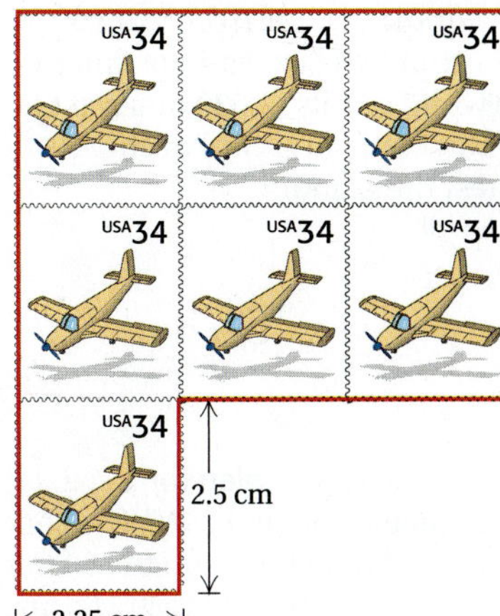

22.

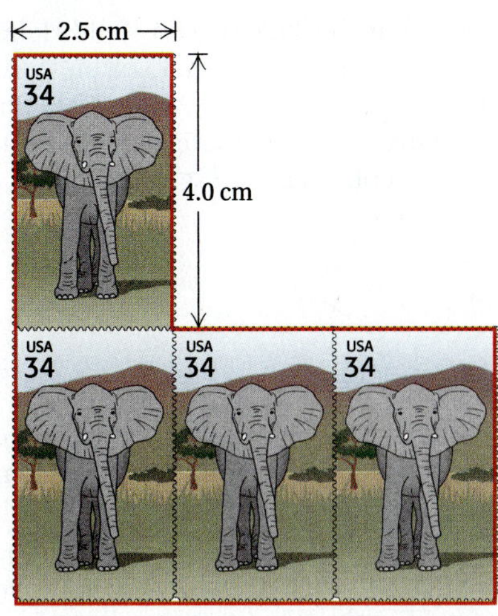

Find the length d in the figure.

23.

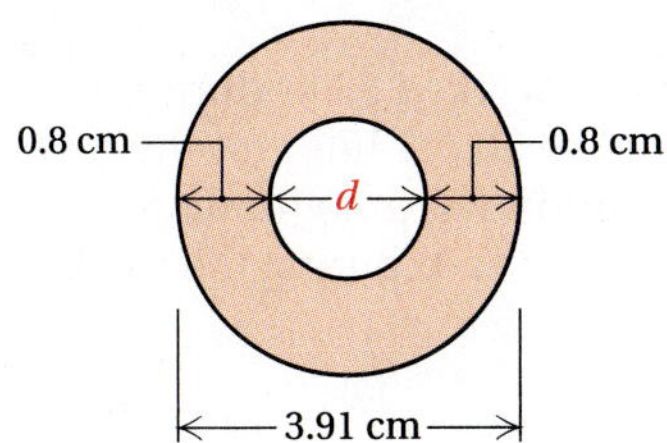

24.

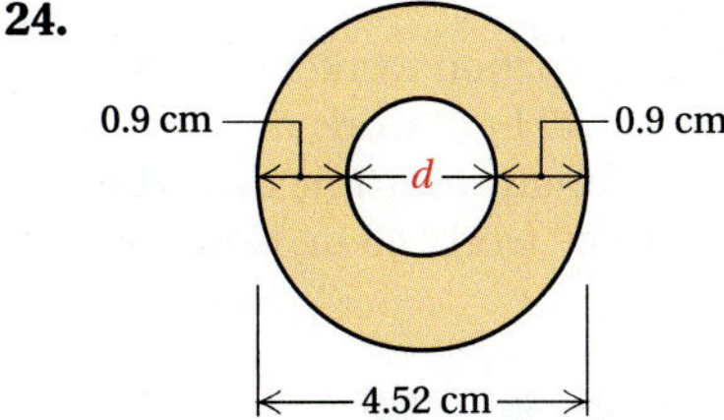

25. *Calories Burned Mowing.* A person weighing 150 lb burns 7.3 calories per minute while mowing a lawn with a power lawnmower. How many calories would be burned in 2 hr of mowing?
Source: *The Handy Science Answer Book*

26. Lot A measures 250.1 ft by 302.7 ft. Lot B measures 389.4 ft by 566.2 ft. What is the total area of the two lots?

27. Holly had \$1123.56 in her checking account. She used her debit card to pay bills of \$23.82, \$507.88, and \$98.32. She then deposited a bonus check of \$678.20. How much is in her account after these changes?

28. Natalie had \$185.00 to spend for fall clothes: \$44.95 was spent on shoes, \$71.95 for a jacket, and \$55.35 for pants. How much was left?

29. A rectangular yard is 20 ft by 15 ft. The yard is covered with grass except for an 8.5-ft square flower garden. How much grass is in the yard?

30. Rita earns a gross paycheck (before deductions) of \$495.72. Her deductions are \$59.60 for federal income tax, \$29.00 for FICA, and \$29.00 for medical insurance. What is her take-home paycheck?

31. *Batting Averages.* For the 2000 season, Todd Helton of the Colorado Rockies won the National League batting title with 216 hits in 580 times at bat. What part of his at-bats were hits? Give decimal notation to the nearest thousandth. (This is a player's *batting average.*)
Source: Major League Baseball

32. *Batting Averages.* For the 2000 season, Nomar Garciaparra of the Boston Red Sox won the American League batting title with 197 hits in 529 times at bat. What part of his at-bats were hits? Give decimal notation to the nearest thousandth.
Source: Major League Baseball

33. *CellularOne® Rates.* One recent plan for a cellular phone in Indiana was called "Indiana 400." The charge was $39.99 per month, and it included up to 400 min of statewide calling time. Minutes over 400 were charged at a rate of $0.25 per minute. One month Maggie used her cell phone for 517 min. What was the charge?
Source: CellularOne® from Bell South

34. *CellularOne® Rates.* One recent plan for a cellular phone in Indiana was called "Indiana 700." The charge was $64.99 per month, and it included up to 700 min of calling time. Minutes over 700 were charged at a rate of $0.25 per minute. One month Dave used his cell phone for 946 min. What was the charge?
Source: CellularOne® from Bell South

35. *Construction Pay.* A construction worker is paid $18.50 per hour for the first 40 hr of work, and time and a half, or $27.75 per hour, for any overtime exceeding 40 hr per week. One week she works 46 hr. How much is her pay?

36. *Summer Work.* Zachary worked 53 hr during a week one summer. He earned $6.50 per hour for the first 40 hr and $9.75 per hour for overtime (hours exceeding 40). How much did Zachary earn during the week?

37. *Egg Costs.* A restaurant owner bought 20 dozen eggs for $13.80. Find the cost of each egg to the nearest tenth of a cent (thousandth of a dollar).

38. *Weight Loss.* A person weighing 170 lb burns 8.6 calories per minute while mowing a lawn. One must burn about 3500 calories in order to lose 1 lb. How many pounds would be lost by mowing for 2 hr? Round to the nearest tenth.

39. *Field Dimensions.* The dimensions of a World Cup soccer field are 114.9 yd by 74.4 yd. The dimensions of a standard football field are 120 yd by 53.3 yd. How much greater is the area of a World Cup soccer field?

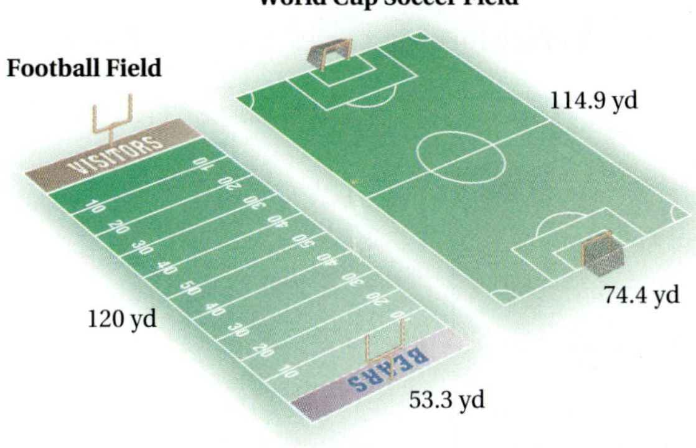

40. *Loan Payment.* In order to make money on loans, financial institutions are paid back more money than they loan. You borrow \$120,000 to buy a house and agree to make monthly payments of \$880.52 for 30 yr. How much do you pay back altogether? How much more do you pay back than the amount of the loan?

41. *World Population.* Using the information in the following bar graph, determine the average population of the world for the years 1950 through 2000.

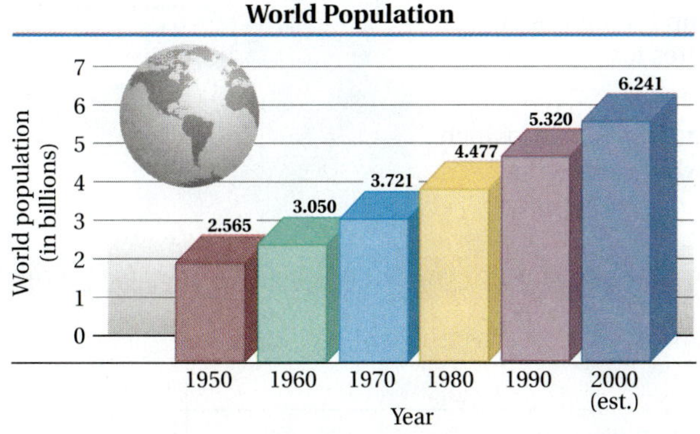

Source: Francis Urban and Philip Rose. *World Population by Country and Region, 1950–86*, and *Projections to 2050*, U.S. Dept. of Agriculture.

42. *Sleep Aid Prescriptions.* The following bar graph shows the number of sleep aid prescriptions written for recent years. Find the average number of prescriptions written each year for that period.
Source: IMS Health

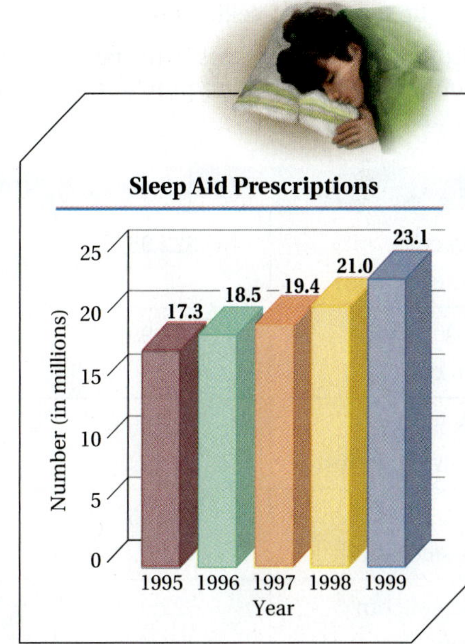

43. *Body Temperature.* Normal body temperature is 98.6°F. A baby's bath water should be 100°F. How many degrees above normal body temperature is this?

44. *Body Temperature.* Normal body temperature is 98.6°F. The lowest temperature at which a patient has survived is 69°F. How many degrees below normal is this?

Home-Cost Comparison. Use the table and formula from Example 8. In each of the following cases, find the value of the house in the new location.
Source: Coldwell Banker Real Estate Corporation

	VALUE	PRESENT LOCATION	NEW LOCATION	NEW VALUE
45.	$125,000	Hollywood Hills	San Francisco	
46.	$180,000	Barrington	Palo Alto	
47.	$96,000	Indianapolis	Tampa	
48.	$300,000	Miami	Queens/North Shore	
49.	$240,000	San Francisco	Atlanta	
50.	$160,000	St. Paul	Phoenix	

Comparison Shopping. The Internet now provides many sites to shop for a product rather than shopping in a local store. The following lists various web sites for purchasing the recent best-selling novel *The Rescue* by Nicholas Sparks.

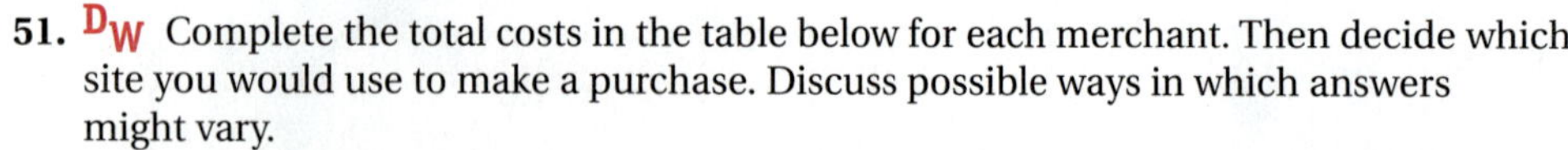

51. DW Complete the total costs in the table below for each merchant. Then decide which site you would use to make a purchase. Discuss possible ways in which answers might vary.

MERCHANT	RETAIL PRICE	OUR PRICE	SHIPPING AND HANDLING	SALES TAX	TOTAL COST
Amazon.com www.amazon.com	$22.95	$13.77	$4.48	$0	?
Barnes & Noble www.bn.com	$22.95	$13.77	$3.49 per order, plus $0.99 per item	$0	?
Powell's Books www.powells.com	$22.95	$15.00	$4.00 per order, plus $1.00 per item	$0	?
Costco Wholesale www.costco.com	$22.95	$12.99	$2.45	$0	?
1bookstreet.com www.1bookstreet.com	$22.95	$16.07	Free if the order is over $15	$0	?
Borders Bookstore, local store	$22.95	$22.95	$0	$1.38	?

Sources: www.yahoo.com; Borders Bookstore, Indianapolis

52. $\mathbf{D_W}$ *Internet Project.* Consider using the Internet to buy a copy of the book *The Bear and the Dragon* by Tom Clancy. Then compare your costs with buying it at a local bookstore. Decide which way you would make a purchase. Discuss possible ways in which answers might vary.

SKILL MAINTENANCE

Add.

53. 4569 + 1766 [1.2a]

54. $\frac{2}{3} + \frac{5}{8}$ [2.3a]

55. $4\frac{1}{3} + 2\frac{1}{2}$ [2.4b]

Subtract.

56. 4569 − 1766 [1.2d]

57. $\frac{2}{3} - \frac{5}{8}$ [2.3b]

58. $4\frac{1}{3} - 2\frac{1}{2}$ [2.4c]

Multiply. [2.4d]

59. $2\frac{2}{7} \cdot 3\frac{1}{2}$

60. $10 \cdot 1\frac{1}{10}$

61. $6\frac{4}{5} \cdot \frac{1}{2}$

Divide. [2.4e]

62. $20 \div 1\frac{1}{20}$

63. $8\frac{2}{3} \div 3$

64. $4\frac{4}{9} \div 1\frac{1}{9}$

Solve. [2.5a]

65. If a water wheel made 469 revolutions at a rate of $16\frac{3}{4}$ revolutions per minute, how long did it rotate?

66. If a bicycle wheel made 480 revolutions at a rate of $66\frac{2}{3}$ revolutions per minute, how long did it rotate?

SYNTHESIS

67. You buy a half-dozen packs of basketball cards with a dozen cards in each pack. The cost is twelve dozen cents for each half-dozen cards. How much do you pay for the cards?

CHAPTER 3

Summary and Review

The review that follows is meant to prepare you for a chapter exam. It consists of two parts. The first part is a checklist of some of the Study Tips referred to in this text. The second part is the Review Exercises. These provide practice exercises for the exam, together with references to section objectives so you can go back and review. Before beginning, stop and look back over the skills you have obtained. What powers in mathematics do you have now that you did not have before studying this chapter?

STUDY TIPS CHECKLIST

The foundation of all your study skills is TIME!

- ☐ Have you been trying to take the primary responsibility for your learning?
- ☐ Have you been asking questions in class?
- ☐ Have you been using the five steps for problem solving?
- ☐ Are you doing your homework as soon as possible after class?
- ☐ Have you formed a study group with some of your fellow students?

REVIEW EXERCISES

Convert the number in the sentence to standard notation.

1. Russia has the largest total area of any country in the world, at 6.59 million square miles. [3.3b]

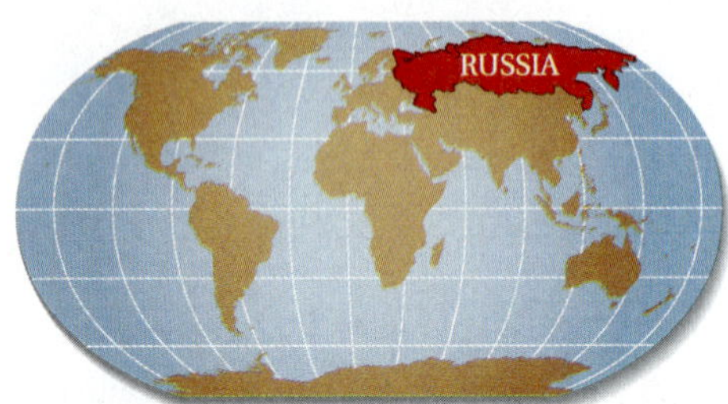

2. The total weight of the turkeys consumed by Americans during the Thanksgiving holidays is about 6.9 million pounds. [3.3b]

Write a word name. [3.1a]

3. 3.47
4. 0.031

Write a word name as on a check. [3.1a]

5. $597.25
6. $0.96

Write fraction notation. [3.1b]

7. 0.09
8. 4.561
9. 0.089
10. 3.0227

Write decimal notation. [3.1b]

11. $\frac{34}{1000}$
12. $\frac{42,603}{10,000}$
13. $27\frac{91}{100}$
14. $867\frac{6}{1000}$

Which number is larger? [3.1c]

15. 0.034, 0.0185
16. 0.91, 0.19
17. 0.741, 0.6943
18. 1.038, 1.041

Round 17.4287 to the nearest: [3.1d]

19. Tenth.
20. Hundredth.
21. Thousandth.
22. One.

Add. [3.2a]

23.
$$\begin{array}{r} 2.048 \\ 65.371 \\ +\ 507.1 \\ \hline \end{array}$$

24.
$$\begin{array}{r} 0.6 \\ 0.004 \\ 0.07 \\ +0.0098 \\ \hline \end{array}$$

25. $219.3 + 2.8 + 7$

26. $0.41 + 4.1 + 41 + 0.041$

Subtract. [3.2b]

27.
$$\begin{array}{r} 30.0 \\ -\ 0.7908 \\ \hline \end{array}$$

28.
$$\begin{array}{r} 845.08 \\ -\ 54.79 \\ \hline \end{array}$$

29. $37.645 - 8.497$

30. $70.8 - 0.0109$

Multiply. [3.3a]

31.
$$\begin{array}{r} 48 \\ \times\ 0.27 \\ \hline \end{array}$$

32.
$$\begin{array}{r} 0.174 \\ \times\ 0.83 \\ \hline \end{array}$$

33. 100×0.043

34. 0.001×24.68

Divide. [3.4a]

35. $8\overline{)60}$

36. $52\overline{)23.4}$

37. $2.6\overline{)117.52}$

38. $2.14\overline{)2.18708}$

39. $\dfrac{276.3}{1000}$

40. $\dfrac{13.892}{0.01}$

Solve. [3.2c], [3.4b]

41. $x + 51.748 = 548.0275$

42. $3 \cdot x = 20.85$

43. $10 \cdot y = 425.4$

44. $0.0089 + y = 5$

Solve. [3.7a]

45. *Tea Consumption.* The average person drinks about 3.48 cups of tea per day. How many cups of tea does the average person drink in a week? in a 30-day month?
Source: Tom Parker, *In One Day.* Boston: Houghton Mifflin, 1984.

46. Stacia, a coronary intensive care nurse, earned $620.74 during a recent 40-hr week. What was her hourly wage? Round to the nearest cent.

47. Derek had $6274.35 in his checking account. He used $485.79 to buy a Palm Digital Assistant with his debit card. How much was left in his account?

48. *CellularOne® Rates.* One recent plan for a cellular phone in Indiana was called "Indiana 1600." The charge was $124.99 per month, and it included up to 1600 min of calling time. Minutes over 1600 were charged at a rate of $0.25 per minute. One month Maria used her cell phone for 2000 min. What was the charge?
Source: CellularOne® from Bell South

49. *Gas Mileage.* A driver wants to estimate gas mileage per gallon. At 36,057.1 mi, the tank is filled with 10.7 gal. At 36,217.6 mi, the tank is filled with 11.1 gal. Find the mileage per gallon. Round to the nearest tenth.

50. *Seafood Consumption.* The following graph shows the annual consumption, in pounds, of seafood per person in the United States in recent years.

a) Find the total per capita consumption for the four years.
b) Find the average per capita consumption.

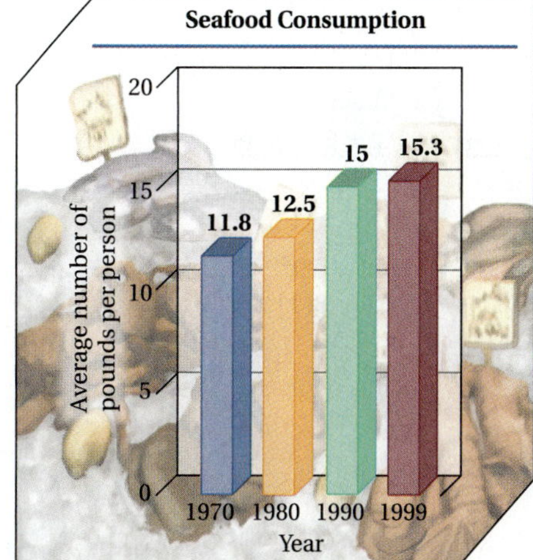

Source: National Oceanographic and Atmospheric Administration

Estimate each of the following. [3.6a]

51. The product 7.82×34.487 by rounding to the nearest one

52. The difference $219.875 - 4.478$ by rounding to the nearest one

53. The quotient $82.304 \div 17.287$ by rounding to the nearest ten

54. The sum $45.78 + $78.99 by rounding to the nearest one

Find decimal notation. Use multiplying by 1. [3.5a]

55. $\frac{13}{5}$ **56.** $\frac{32}{25}$ **57.** $\frac{11}{4}$

Find decimal notation. Use division. [3.5a]

58. $\frac{13}{4}$ **59.** $\frac{7}{6}$ **60.** $\frac{17}{11}$

Round the answer to Exercise 60 to the nearest: [3.5b]

61. Tenth. **62.** Hundredth. **63.** Thousandth.

Convert from cents to dollars. [3.3b]

64. 8273¢ **65.** 487¢

Convert from dollars to cents. [3.3b]

66. $24.93 **67.** $9.86

Calculate. [3.4c], [3.5c]

68. $(8 - 1.23) \div 4 + 5.6 \times 0.02$

69. $(1 + 0.07)^2 + 10^3 \div 10^2 + [4(10.1 - 5.6) + 8(11.3 - 7.8)]$

70. $\frac{3}{4} \times 20.85$

71. $\frac{1}{3} \times 123.7 + \frac{4}{9} \times 0.684$

72. DW Consider finding decimal notation for $\frac{44}{125}$. Discuss as many ways as you can for finding such notation and give the answer. [3.5a]

73. DW Explain how we can use fraction notation to understand why we count decimal places when multiplying with decimal notation. [3.3a]

SKILL MAINTENANCE

Certain objectives from four particular sections will be retested on the chapter test. The objectives are listed with the practice problems that follow.

74. Divide: $20 \div 5\frac{1}{3}$. [2.4e]

75. Subtract: $24 - 17\frac{2}{5}$. [2.4c]

76. Simplify: $\frac{28}{56}$. [2.1e]

77. Find the prime factorization of 192. [1.7d]

78. Find the LCM of 20, 33, and 75. [1.9a]

SYNTHESIS

79. In each of the following, use one of $+$, $-$, $\times$, and $\div$ in each blank to make a true sentence. [3.4c]

a) $2.56 \square 6.4 \square 51.2 \square 17.4 \square 89.7 = 72.62$
b) $(11.12 \square 0.29) \square 3^4 = 877.23$

80. Find repeating decimal notation for 1 and explain. Use the following hints. [3.5a]

$$\frac{1}{3} = 0.33333333\ldots,$$
$$\frac{2}{3} = 0.66666666\ldots$$

81. Find repeating decimal notation for 2. [3.5a]

CHAPTER 3

Chapter Test

Convert the number in the sentence to standard notation.

1. The annual sales of antibiotics in the United States is \$8.9 billion.
Source: IMS Health

2. There are 3.756 million people enrolled in bowling organizations in the United States.
Source: *Bowler's Journal International*, December 2000

3. Write a word name: 2.34.

4. Write a word name, as on a check, for \$1234.78.

Write fraction notation.

5. 0.91

6. 2.769

Write decimal notation.

7. $\frac{74}{1000}$

8. $\frac{37{,}047}{10{,}000}$

9. $756\frac{9}{100}$

10. $91\frac{703}{1000}$

Which number is larger?

11. 0.07, 0.162

12. 0.078, 0.06

13. 0.09, 0.9

Round 5.6783 to the nearest:

14. One.

15. Hundredth.

16. Thousandth.

17. Tenth.

Calculate.

18.
$$\begin{array}{r} 0.7 \\ 0.08 \\ 0.009 \\ +\ 0.0012 \\ \hline \end{array}$$

19. $102.4 + 6.1 + 78$

20. $0.93 + 9.3 + 93 + 930$

21.
$$\begin{array}{r} 52.678 \\ -\ 4.321 \\ \hline \end{array}$$

22.
$$\begin{array}{r} 20.0 \\ -\ 0.9099 \\ \hline \end{array}$$

23. $234.6788 - 81.7854$

24.
$$\begin{array}{r} 0.125 \\ \times\ 0.24 \\ \hline \end{array}$$

25. 0.001×213.45

26. 1000×73.962

27. $4\overline{)19}$

28. $3.3\overline{)100.32}$

29. $82\overline{)15.58}$

30. $\dfrac{346.89}{1000}$

31. $\dfrac{346.89}{0.01}$

Solve.

32. $4.8 \cdot y = 404.448$

33. $x + 0.018 = 9$

34. *CellularOne® Rates.* One recent plan for a cellular phone in Indiana was called "Indiana 1000." The charge was $84.99 per month, and it included up to 1000 min of calling time. Minutes over 1000 were charged at a rate of $0.25 per minute. One month Ramon used his cell phone for 1142 min. What was the charge?
Source: CellularOne® from Bell South

35. *Gas Mileage.* Tina wants to estimate the gas mileage per gallon in her economy car. At 76,843 mi, the tank is filled with 14.3 gal of gasoline. At 77,310 mi, the tank is filled with 16.5 gal of gasoline. Find the mileage per gallon. Round to the nearest tenth.

36. *Checking Account Balance.* Nicholas has a balance of $10,200 in his checking account before making purchases of $123.89, $56.68, and $3446.98 with his debit card. What was the balance after making the purchases?

37. *MP3 Players.* Matt buys 6 Compaq iPAQ 64 Mb Personal Audio Players at $199.99 each. What is the total cost?
Source: Compaq

38. *Airport Passengers.* The following graph shows the number of passengers in a recent year who traveled through the country's busiest airports. Find the average number of passengers through these airports.

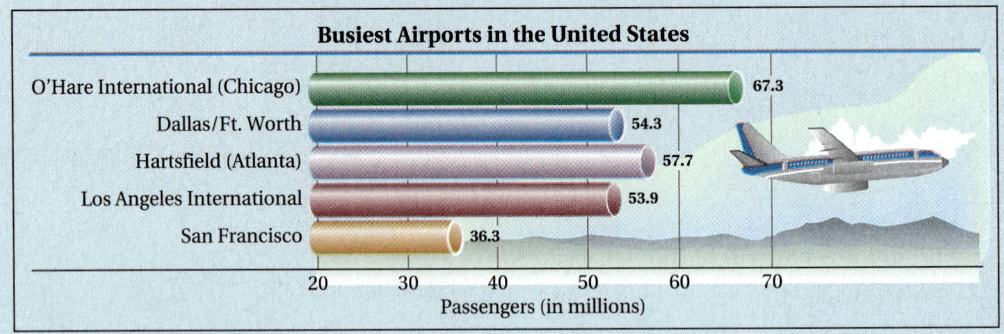

Source: Air Transport Association of America

Estimate each of the following.

39. The product 8.91×22.457 by rounding to the nearest one

40. The quotient $78.2209 \div 16.09$ by rounding to the nearest ten

Find decimal notation. Use multiplying by 1.

41. $\frac{8}{5}$

42. $\frac{22}{25}$

43. $\frac{21}{4}$

Find decimal notation. Use division.

44. $\frac{3}{4}$

45. $\frac{11}{9}$

46. $\frac{15}{7}$

Round the answer to Question 46 to the nearest:

47. Tenth.

48. Hundredth.

49. Thousandth.

50. Convert from cents to dollars: 949 cents.

Calculate.

51. $256 \div 3.2 \div 2 - 1.56 + 78.325 \times 0.02$

52. $(1 - 0.08)^2 + 6[5(12.1 - 8.7) + 10(14.3 - 9.6)]$

53. $\frac{7}{8} \times 345.6$

54. $\frac{2}{3} \times 79.95 - \frac{7}{9} \times 1.235$

SKILL MAINTENANCE

55. Subtract: $28\frac{2}{3} - 2\frac{1}{6}$.

56. Divide: $3\frac{3}{8} \div 3$.

57. Simplify: $\frac{33}{54}$.

58. Find the prime factorization of 360.

59. Find the LCM of 15, 36, and 40.

SYNTHESIS

60. The Silver's Health Club generally charges a \$79 membership fee and \$42.50 a month. Allise has a coupon that will allow her to join the club for \$299 for six months. How much will Allise save if she uses the coupon?

61. Arrange from smallest to largest.

$\frac{2}{3}, \frac{15}{19}, \frac{11}{13}, \frac{5}{7}, \frac{13}{15}, \frac{17}{20}$

Percent Notation

Gateway to Chapter 4

This chapter introduces percent notation. We will see that $\frac{3}{8}$ (fraction notation), 0.375 (decimal notation), and 37.5% (percent notation) are all names for the same number. Percent notation has extensive applications in everyday life, in such diverse areas as business, sports, science, and medicine. We consider as well applications involving sales tax, commission, discount, interest, and interest rates on credit cards and loans.

Also introduced are the topics of ratio and proportion.

Real-World Application

George W. Bush was inaugurated as the 43rd president of the United States in 2001. Since Grover Cleveland was both the 22nd and the 24th presidents, there have been only 42 different presidents. Of these 42 presidents, 8 have died in office: William Henry Harrison, Zachary Taylor, Abraham Lincoln, James A. Garfield, William McKinley, Warren G. Harding, Franklin D. Roosevelt, and John F. Kennedy. What percent have died in office?

This problem appears as Example 1 in Section 4.6.

CHAPTER 4

CHAPTER 4

Pretest

Write fraction notation for the ratio. [4.1a]

1. 35 to 43

2. 0.079 to 1.043

Solve.

3. $\frac{5}{6} = \frac{x}{27}$ [4.1d]

4. What is the rate in miles per gallon? [4.1b]
   408 miles, 16 gallons

5. Juan's digital car clock loses 5 min in 10 hr. At this rate, how much will it lose in 24 hr? [4.1e]

6. If 4 packs of gum cost $5.16, how many packs of gum can you buy for $28.38? [4.1e]

7. Insurance costs account for 13.3% of the annual cost of owning and operating an automobile. Find decimal notation for 13.3%. [4.2b]
   Source: Runzheimer International

8. Depreciation and interest charges on a car loan account for 0.504 of the annual cost of owning and operating an automobile. Find percent notation for 0.504. [4.2b]
   Source: Runzheimer International

9. Tire costs account for $\frac{1}{25}$ of the annual cost of owning and operating an automobile. Find percent notation for $\frac{1}{25}$. [4.3a]
   Source: Runzheimer International

10. Fuel costs account for 19% of the annual cost of owning and operating an automobile. Find fraction notation for 19%. [4.3b]
    Source: Runzheimer International

11. Translate to a percent equation. Then solve.
    What is 60% of 75? [4.4a, b]

12. Translate to a proportion. Then solve.
    What percent of 50 is 35? [4.5a, b]

Solve.

13. **Weight of Muscles.** The weight of muscles in a human body is 40% of total body weight. A person weighs 225 lb. What do the muscles weigh? [4.6a]

14. **Ticket Price Increase.** In 2001, the Indianapolis Colts raised the price of a ticket from $125 to $149 for a seat between the 30-yd lines on the lower level. What was the percent of increase? [4.6b]
    Source: The Indianapolis Colts

15. **Massachusetts Sales Tax.** The sales tax rate in Massachusetts is 5%. How much tax is charged on a purchase of $286? What is the total price? [4.7a]

16. A salesperson's commission rate is 28%. What is the commission from the sale of $18,400 worth of merchandise? [4.7b]

17. The marked price of a home theater system is $4450. The system is on sale at Lowland Appliances for 25% off. What are the discount and the sale price? [4.7c]

18. What is the simple interest on $1200 principal at the interest rate of 8.3% for 1 year? [4.7d]

19. What is the simple interest on $500 at 8% for $\frac{1}{2}$ year? [4.7d]

20. Interest is compounded annually. Find the amount in an account if $6000 is invested at 9% for 2 years. [4.7e]

4.1 RATIO AND PROPORTION

Objectives

a Find fraction notation for ratios.

b Give the ratio of two different measures as a rate.

c Determine whether two pairs of numbers are proportional.

d Solve proportions.

e Solve applied problems involving proportions.

a Ratios

RATIO

A **ratio** is the quotient of two quantities.

In January 2001, the Atlanta Hawks basketball team averaged 86.5 points per game and allowed their opponents an average of 90.8 points per game. The *ratio* of points earned to points allowed is given by the fraction notation

Points earned → 86.5
Points allowed → 90.8

$\frac{86.5}{90.8}$ or by the colon notation 86.5 : 90.8.

We read both forms of notation as "the ratio of 86.5 to 90.8," listing the numerator first and the denominator second.

RATIO NOTATION

The **ratio** of a to b is given by the fraction notation $\frac{a}{b}$, where a is the numerator and b is the denominator, or by the colon notation $a:b$.

EXAMPLE 1 Find the ratio of 31.4 to 100.

The ratio is $\frac{31.4}{100}$, or 31.4 : 100.

Do Exercises 1–3.

In most of our work, we will use fraction notation for ratios.

EXAMPLE 2 *Batting.* In the 2000 season, Gary Sheffield of the Los Angeles Dodgers got 163 hits in 501 at-bats. What was the ratio of hits to at-bats? of at-bats to hits?
Source: Major League Baseball

The ratio of hits to at-bats is

$$\frac{163}{501}.$$

The ratio of at-bats to hits is

$$\frac{501}{163}.$$

Do Exercises 4–6. (Exercises 5 and 6 are on the following page.)

1. Find the ratio of 5 to 11.

2. Find the ratio of 57.3 to 86.1.

3. Find the ratio of $6\frac{3}{4}$ to $7\frac{2}{5}$.

4. **Rainfall.** The greatest amount of rainfall ever recorded for a 12-month period was 739 in. in Kukui, Maui, Hawaii, from December 1981 to December 1982. Find the ratio of rainfall to time in months.
Source: *The Handy Science Answer Book*

Answers on page A-8

5. **Fat Grams.** In one serving ($\frac{1}{2}$-cup) of fried scallops, there is 12 g of fat. In one serving ($\frac{1}{2}$-cup) of fried oysters, there is 14 g of fat. What is the ratio of grams of fat in one serving of scallops to grams of fat in one serving of oysters?
Source: *Better Homes and Gardens: A New Cook Book*

6. **Earned Runs.** In the 2000 season, Randy Johnson of the Arizona Diamondbacks gave up 73 earned runs in $248\frac{2}{3}$ innings pitched. What was the ratio of earned runs to innings pitched? of innings pitched to earned runs?
Source: Major League Baseball

7. In the triangle below, what is the ratio of the length of the shortest side to the length of the longest side?

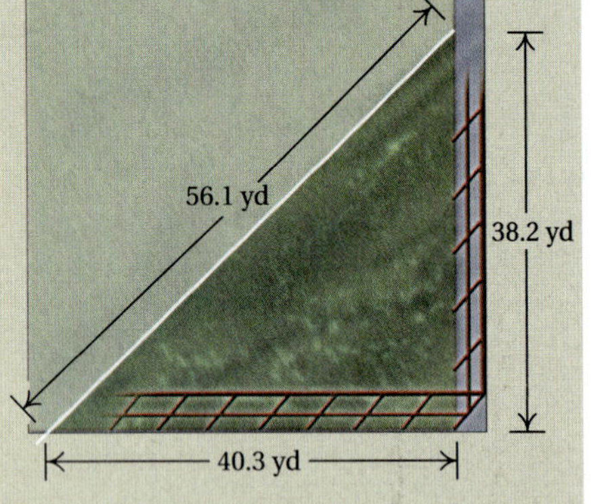

Answers on page A-8

EXAMPLE 3 Refer to the triangle below.

a) What is the ratio of the length of the longest side to the length of the shortest side?

$$\frac{5}{3}$$

b) What is the ratio of the length of the shortest side to the length of the longest side?

$$\frac{3}{5}$$

Do Exercise 7.

EXAMPLE 4 Find the ratio of 2.4 to 10. Then simplify and find two other numbers in the same ratio.

We first write the ratio in fraction notation. Next, we multiply by 1 to clear the decimal from the numerator. Then we simplify.

$$\frac{2.4}{10} = \frac{2.4}{10} \cdot \frac{10}{10} = \frac{24}{100} = \frac{4 \cdot 6}{4 \cdot 25} = \frac{4}{4} \cdot \frac{6}{25} = \frac{6}{25}$$

Thus, 2.4 is to 10 as 6 is to 25.

Do Exercises 8–10 on the following page.

EXAMPLE 5 A standard television screen with a width of 32 in. has a height of 24 in. Find the ratio of width to height and simplify.

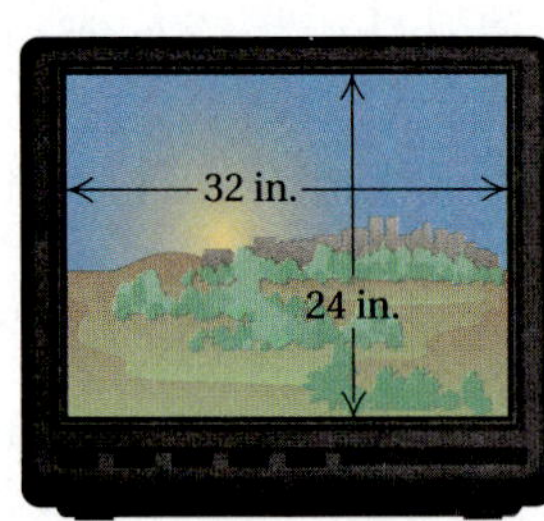

The ratio is $\frac{32}{24} = \frac{8 \cdot 4}{8 \cdot 3} = \frac{8}{8} \cdot \frac{4}{3} = \frac{4}{3}$.

Thus we can say that the ratio of width to height is 4 to 3.

Do Exercise 11 on the following page.

b Rates

A 2001 Honda Civic can go 464 miles on 16 gallons of gasoline. Let's consider the ratio of miles to gallons:
Source: *Consumer Reports*

$$\frac{464 \text{ mi}}{16 \text{ gal}} = \frac{464}{16} \frac{\text{miles}}{\text{gallon}} = \frac{29}{1} \frac{\text{miles}}{\text{gallon}}$$
$$= 29 \text{ miles per gallon} = 29 \text{ mpg}.$$

"per" means "division," or "for each."

The ratio $\frac{464 \text{ mi}}{16 \text{ gal}}$, or $\frac{464}{16} \frac{\text{mi}}{\text{gal}}$, or 29 mpg is called a **rate.**

RATE

When a ratio is used to compare two different kinds of measure, we call it a **rate.**

Suppose Alyssa says her car goes 462.4 mi on 15.8 gal of gasoline. Is the mpg (mileage) of her car better than that of the Civic above? To determine this, it helps to convert to decimal notation and perhaps round. Then we have

$$\frac{462.4 \text{ miles}}{15.8 \text{ gallons}} = \frac{462.4}{15.8} \text{ mpg} \approx 29.266 \text{ mpg}.$$

Since $29.266 > 29$, Alyssa's car gets better mileage than the Civic does.

EXAMPLE 6 It takes 60 oz of grass seed to seed 3000 sq ft of lawn. What is the rate in ounces per square foot?

$$\frac{60 \text{ oz}}{3000 \text{ sq ft}} = \frac{1}{50} \frac{\text{oz}}{\text{sq ft}}, \quad \text{or} \quad 0.02 \frac{\text{oz}}{\text{sq ft}}$$

EXAMPLE 7 A cook buys 10 lb of potatoes for \$3.69. What is the rate in cents per pound?

$$\frac{\$3.69}{10 \text{ lb}} = \frac{369 \text{ cents}}{10 \text{ lb}}, \quad \text{or} \quad 36.9 \frac{\text{cents}}{\text{lb}}$$

EXAMPLE 8 A pharmacy student working as a pharmacist's assistant earned \$3690 for working 3 months one summer. What was the rate of pay per month?

The rate of pay is the ratio of money earned per length of time worked, or

$$\frac{\$3690}{3 \text{ mo}} = 1230 \frac{\text{dollars}}{\text{month}}, \quad \text{or}$$
\$1230 per month.

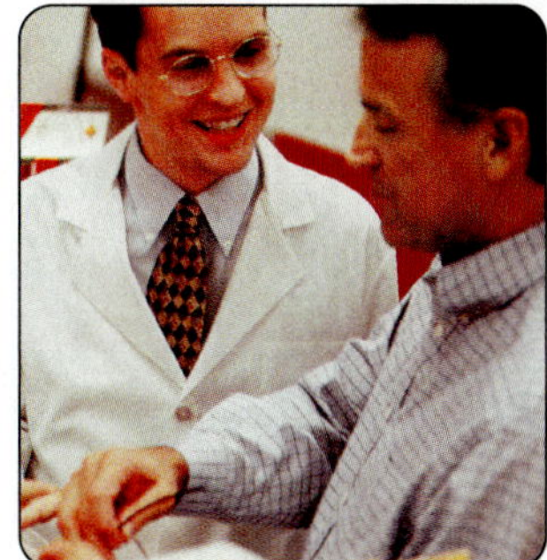

Do Exercises 12–19 on the following page.

8. Find the ratio of 18 to 27. Then simplify and find two other numbers in the same ratio.

9. Find the ratio of 3.6 to 12. Then simplify and find two other numbers in the same ratio.

10. Find the ratio of 1.2 to 1.5. Then simplify and find two other numbers in the same ratio.

11. In Example 5, find the ratio of the width of the shortest side of the television screen to the width of the longest side and simplify.

Answers on page A-8

A ratio of distance traveled to time is called *speed.* What is the rate, or speed, in miles per hour?

12. 45 mi, 9 hr

13. 120 mi, 10 hr

14. 89 km, 13 hr (Round to the nearest hundredth.)

What is the rate, or speed, in feet per second?

15. 2200 ft, 2 sec

16. 52 ft, 13 sec

17. 242 ft, 16 sec

18. A well-hit golf ball can travel 500 ft in 2 sec. What is the rate, or speed, of the golf ball in feet per second?

19. A leaky faucet can lose 14 gal of water in a week. What is the rate in gallons per day?

Answers on page A-8

C Proportions

Suppose we want to compare $\frac{2}{4}$ and $\frac{3}{6}$. We find a common denominator and compare numerators. To do this, we multiply by 1 using symbols for 1 formed by looking at contrasting denominators.

The "unit" is $\frac{1}{6}$.

$$\frac{3}{6}=\frac{3}{6}\cdot\frac{4}{4}=\frac{3\cdot 4}{6\cdot 4}=\frac{12}{24}$$

$$\frac{2}{4}=\frac{2}{4}\cdot\frac{6}{6}=\frac{2\cdot 6}{4\cdot 6}=\frac{12}{24}$$

The "unit" is $\frac{1}{4}$.

Both "units" are $\frac{1}{24}$.

We see that $\frac{3}{6}=\frac{2}{4}$.

Note in the preceding that if

$$\frac{3}{6}=\frac{2}{4},\quad \text{then}\quad 3\cdot 4=6\cdot 2.$$

We need to check only the products $3\cdot 4$ and $6\cdot 2$ to compare the fractions.

A TEST FOR EQUALITY

We multiply these two numbers: $3\cdot 4$.

We multiply these two numbers: $6\cdot 2$.

$$\frac{3}{6}\quad\frac{2}{4}$$

We call $3\cdot 4$ and $6\cdot 2$ **cross products.** Since the cross products are the same, that is, $3\cdot 4=6\cdot 2$, we know that

$$\frac{3}{6}=\frac{2}{4}.$$

When two pairs of numbers (such as 3, 6 and 2, 4) have the same ratio, we say that they are **proportional.** The equation

$$\frac{3}{6}=\frac{2}{4}$$

states that the pairs 3, 6 and 2, 4 are proportional. Such an equation is called a **proportion.** We sometimes read $\frac{3}{6}=\frac{2}{4}$ as "3 is to 6 as 2 is to 4."

EXAMPLE 9 Determine whether 1, 2, and 3, 6 are proportional.

We can use cross products:

$$1\cdot 6=6\qquad \frac{1}{2}\stackrel{?}{=}\frac{3}{6}\qquad 2\cdot 3=6.$$

Since the cross products are the same, $6=6$, we know that $\frac{1}{2}=\frac{3}{6}$, so the numbers are proportional.

EXAMPLE 10 Determine whether 2, 5 and 4, 7 are proportional.

We can use cross products:

$$2 \cdot 7 = 14 \qquad \frac{2}{5} \stackrel{?}{=} \frac{4}{7} \qquad 5 \cdot 4 = 20.$$

Since the cross products are not the same, $14 \neq 20$, we know that $\frac{2}{5} \neq \frac{4}{7}$, so the numbers are not proportional.

Do Exercises 20–22.

Determine whether the two pairs of numbers are proportional.

20. 3, 4 and 6, 8

21. 1, 4 and 10, 39

22. 1, 2 and 20, 39

d Solving Proportions

Let's now look at solving proportions. Consider the proportion

$$\frac{x}{3} = \frac{4}{6}.$$

One way to solve a proportion is to use cross products. Then we can divide on both sides to get the variable alone:

$$x \cdot 6 = 3 \cdot 4 \qquad \text{Equating cross products (finding cross products and setting them equal)}$$

$$\frac{x \cdot 6}{6} = \frac{3 \cdot 4}{6} \qquad \text{Dividing by 6 on both sides}$$

$$x = \frac{3 \cdot 4}{6} = \frac{12}{6} = 2.$$

We can check that 2 is the solution by replacing x with 2 and using cross products:

$$2 \cdot 6 = 12 \qquad \frac{2}{3} \stackrel{?}{=} \frac{4}{6} \qquad 3 \cdot 4 = 12$$

Since the cross products are the same, it follows that $\frac{2}{3} = \frac{4}{6}$; so the numbers 2, 3 and 4, 6 are proportional, and 2 is the solution of the equation.

SOLVING PROPORTIONS

To solve $\frac{x}{a} = \frac{c}{d}$, equate *cross products* and divide on both sides to get x alone.

Do Exercise 23.

23. Solve: $\frac{x}{63} = \frac{2}{9}$.

Answers on page A-8

24. Solve: $\frac{x}{9} = \frac{5}{4}$.

25. Solve: $\frac{21}{5} = \frac{n}{2.5}$.

Answers on page A-8

Study Tips

WRITING ALL THE STEPS

Take the time to include all the steps when working your homework problems. Doing so will help you organize your thinking and avoid computational errors. If you find a wrong answer, having all the steps allows easier checking of your work. It will also give you complete, step-by-step solutions of the exercises that can be used to study for an exam.

Writing down all the steps and keeping your work organized may also give you a better chance of getting partial credit.

"Success comes before work only in the dictionary."

Anonymous

EXAMPLE 11 Solve: $\frac{x}{7} = \frac{5}{3}$. Write a mixed numeral for the answer.

We have

$$\frac{x}{7} = \frac{5}{3}$$

$$x \cdot 3 = 7 \cdot 5 \qquad \text{Equating cross products}$$

$$\frac{x \cdot 3}{3} = \frac{7 \cdot 5}{3} \qquad \text{Dividing by 3}$$

$$x = \frac{7 \cdot 5}{3}$$

$$= \frac{35}{3}, \text{ or } 11\frac{2}{3}.$$

The solution is $11\frac{2}{3}$.

Do Exercise 24.

EXAMPLE 12 Solve: $\frac{7.7}{15.4} = \frac{y}{2.2}$.

We have

$$\frac{7.7}{15.4} = \frac{y}{2.2}$$

$$7.7 \times 2.2 = 15.4 \times y \qquad \text{Equating cross products}$$

$$\frac{7.7 \times 2.2}{15.4} = \frac{15.4 \times y}{15.4}. \qquad \text{Dividing by 15.4}$$

$$\frac{7.7 \times 2.2}{15.4} = y$$

$$\frac{16.94}{15.4} = y \qquad \text{Multiplying}$$

$$1.1 = y. \qquad \text{Dividing:}$$

```
         1.1
15.4 ) 16.9^4
       15 4 0
         1 5 4
         1 5 4
             0
```

The solution is 1.1.

Do Exercise 25.

EXAMPLE 13 Solve: $\frac{8}{x} = \frac{5}{3}$. Write decimal notation for the answer.

We have

$$\frac{8}{x} = \frac{5}{3}$$

$$8 \cdot 3 = x \cdot 5 \qquad \text{Equating cross products}$$

$$\frac{8 \cdot 3}{5} = \frac{x \cdot 5}{5}. \qquad \text{Dividing by 5}$$

Then

$$\frac{8 \cdot 3}{5} = x$$

$$\frac{24}{5} = x \qquad \text{Multiplying}$$

$$4.8 = x. \qquad \text{Simplifying}$$

The solution is 4.8.

Do Exercise 26.

EXAMPLE 14 Solve: $\dfrac{3.4}{4.93} = \dfrac{10}{n}$.

We have

$$\frac{3.4}{4.93} = \frac{10}{n}$$

$$3.4 \times n = 4.93 \times 10 \qquad \text{Equating cross products}$$

$$\frac{3.4 \times n}{3.4} = \frac{4.93 \times 10}{3.4} \qquad \text{Dividing by 3.4}$$

$$n = \frac{4.93 \times 10}{3.4}$$

$$= \frac{49.3}{3.4} \qquad \text{Multiplying}$$

$$= 14.5. \qquad \text{Dividing}$$

The solution is 14.5.

Do Exercise 27.

26. Solve: $\dfrac{6}{x} = \dfrac{25}{11}$.

27. Solve: $\dfrac{0.4}{0.9} = \dfrac{4.8}{t}$.

Answers on page A-8

CALCULATOR CORNER

Solving Proportions Note in Examples 11–14 that when we solve a proportion, we equate cross products and then we divide on both sides to isolate the variable on one side of the equation. We can use a calculator to do the calculations in this situation. In Example 14, for instance, after equating cross products and dividing by 3.4 on both sides, we have

$$n = \frac{4.93 \times 10}{3.4}.$$

To find n on a calculator, we can press [4] [.] [9] [3] [×] [1] [0] [÷] [3] [.] [4] [=]. The result is 14.5, so $n = 14.5$.

Exercises:

1. Use a calculator to solve each of the proportions in Examples 11–13.
2. Use a calculator to solve each of the proportions in Margin Exercises 23–27.

Solve each proportion.

3. $\dfrac{15.75}{20} = \dfrac{a}{35}$
4. $\dfrac{32}{x} = \dfrac{25}{20}$
5. $\dfrac{t}{57} = \dfrac{17}{64}$
6. $\dfrac{71.2}{a} = \dfrac{42.5}{23.9}$
7. $\dfrac{29.6}{3.15} = \dfrac{x}{4.23}$
8. $\dfrac{a}{3.01} = \dfrac{1.7}{0.043}$

28. Determining Paint Needs. Lowell and Chris run a summer painting company to support their college expenses. They can paint 1600 ft^2 of clapboard with 4 gal of paint. How much paint would be needed for a building with 6000 ft^2 of clapboard?

Answer on page A-8

e Applications and Problem Solving

Proportions have applications in such diverse fields as business, chemistry, health sciences, and home economics, as well as to many areas of daily life. Proportions are useful in making predictions.

EXAMPLE 15 *Recommended Dosage.* To control a fever, a doctor suggests that a child who weighs 28 kg be given 420 mg of Tylenol. If the dosage is proportional to the child's weight, how much Tylenol is recommended for a child who weighs 35 kg?

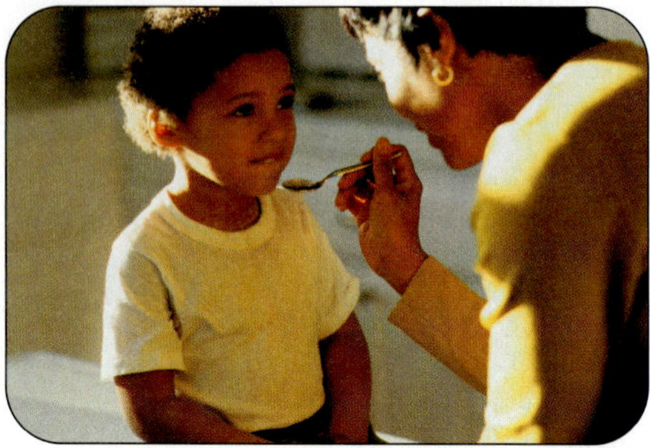

1. **Familiarize.** We let $t =$ the number of milligrams of Tylenol.
2. **Translate.** We translate to a proportion, keeping the amount of Tylenol in the numerators.

$$\begin{array}{r} \text{Tylenol suggested} \rightarrow \\ \text{Child's weight} \rightarrow \end{array} \frac{420}{28} = \frac{t}{35} \begin{array}{l} \leftarrow \text{Tylenol suggested} \\ \leftarrow \text{Child's weight} \end{array}$$

3. **Solve.** Next, we solve the proportion:

$$420 \cdot 35 = 28 \cdot t \qquad \text{Equating cross products}$$

$$\frac{420 \cdot 35}{28} = \frac{28 \cdot t}{28} \qquad \text{Dividing by 28 on both sides}$$

$$\frac{420 \cdot 35}{28} = t$$

$$525 = t. \qquad \text{Multiplying and dividing}$$

4. **Check.** We substitute into the proportion and check cross products:

$$\frac{420}{28} = \frac{525}{35};$$

$$420 \cdot 35 = 14{,}700; \qquad 28 \cdot 525 = 14{,}700.$$

The cross products are the same.

5. **State.** The dosage for a child who weighs 35 kg is 525 mg.

Do Exercise 28.

EXAMPLE 16 *Construction Plans.* Architects make blueprints of projects being constructed. These are scale drawings in which lengths are in proportion to actual sizes. The Hennesseys are constructing a rectangular deck just outside their house. The architectural blueprints are rendered such that $\frac{3}{4}$ in. on the drawing is actually 2.25 ft on the deck. The width of the deck on the drawing is 4.3 in. How wide is the deck in reality?

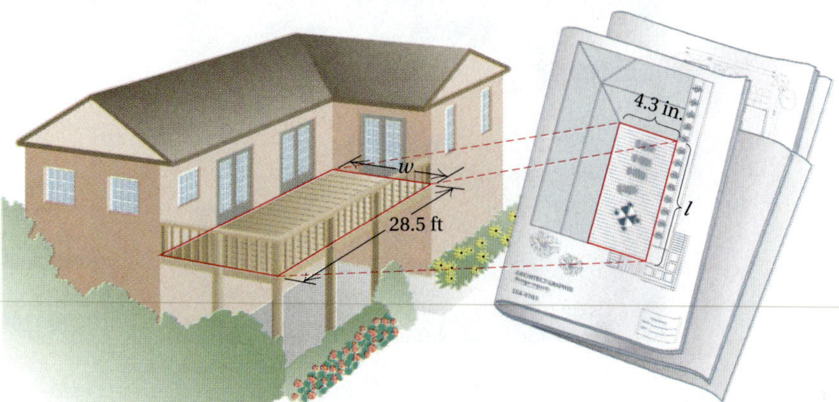

1. **Familiarize.** We let w = the width of the deck.
2. **Translate.** Then we translate to a proportion, using 0.75 for $\frac{3}{4}$ in.

$$\text{Measure on drawing} \rightarrow \frac{0.75}{2.25} = \frac{4.3}{w} \leftarrow \text{Width of drawing}$$
$$\text{Measure on deck} \rightarrow \qquad \leftarrow \text{Width of deck}$$

3. **Solve.** Next, we solve the proportion:

$$0.75 \times w = 2.25 \times 4.3 \qquad \text{Equating cross products}$$
$$\frac{0.75 \times w}{0.75} = \frac{2.25 \times 4.3}{0.75} \qquad \text{Dividing by 0.75 on both sides}$$
$$w = \frac{2.25 \times 4.3}{0.75}$$
$$w = 12.9.$$

4. **Check.** We substitute into the proportion and check cross products:

$$\frac{0.75}{2.25} = \frac{4.3}{12.9};$$
$$0.75 \times 12.9 = 9.675; \qquad 2.25 \times 4.3 = 9.675.$$

The cross products are the same.

5. **State.** The width of the deck is 12.9 ft.

Do Exercise 29.

29. **Construction Plans.** In Example 16, the length of the actual deck is 28.5 ft. What is the length of the deck on the blueprints?

Answer on page A-8

30. Estimating a Deer Population. To determine the number of deer in a forest, a conservationist catches 612 deer, tags them, and releases them. Later, 244 deer are caught, and it is found that 72 of them are tagged. Estimate how many deer are in the forest.

EXAMPLE 17 *Estimating a Wildlife Population.* To determine the number of fish in a lake, a conservationist catches 225 fish, tags them, and throws them back into the lake. Later, 108 fish are caught, and it is found that 15 of them are tagged. Estimate how many fish are in the lake.

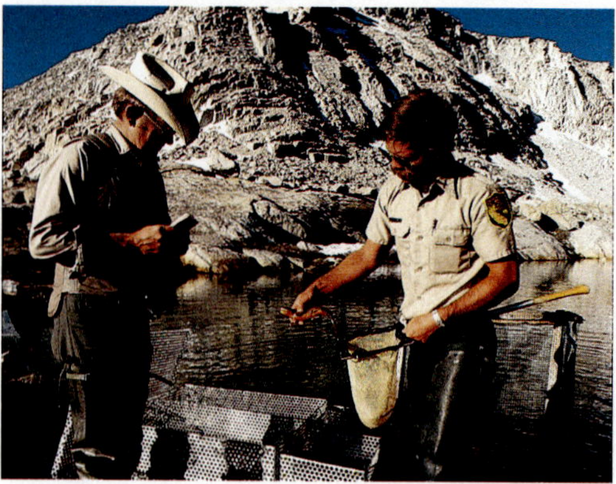

1. **Familiarize.** We let $F =$ the number of fish in the lake.
2. **Translate.** We translate to a proportion as follows:

$$\text{Fish tagged originally} \rightarrow \frac{225}{F} = \frac{15}{108} \leftarrow \text{Tagged fish caught later}$$
$$\text{Fish in lake} \rightarrow \qquad\qquad \leftarrow \text{Fish caught later}$$

3. **Solve.** Next, we solve the proportion:

$$225 \cdot 108 = F \cdot 15 \qquad \text{Equating cross products}$$
$$\frac{225 \cdot 108}{15} = \frac{F \cdot 15}{15} \qquad \text{Dividing by 15 on both sides}$$
$$\frac{225 \cdot 108}{15} = F$$
$$1620 = F. \qquad \text{Multiplying and dividing}$$

4. **Check.** We substitute into the proportion and check cross products:

$$\frac{225}{1620} = \frac{15}{108};$$
$$225 \cdot 108 = 24{,}300; \qquad 1620 \cdot 15 = 24{,}300.$$

The cross products are the same.

5. **State.** We estimate that there are 1620 fish in the lake.

Do Exercise 30.

Answer on page A-8

4.1 EXERCISE SET

For Extra Help

Digital Video Tutor CD 3 Videotape 5

InterAct Math

Math Tutor Center

MathXL

MyMathLab

a Find fraction notation for the ratio. You need not simplify.

1. 4 to 5

2. 329 to 967

3. 56.78 to 98.35

4. $10\frac{1}{2}$ to $43\frac{1}{4}$

5. *Corvette Accidents.* Of every 5 fatal accidents involving a Corvette, 4 do not involve another vehicle. Find the ratio of fatal accidents involving just a Corvette to those involving a Corvette and at least one other vehicle.
Source: *Harper's Magazine*

6. *Cancer Deaths.* In the state of Texas, of every 1000 people, 122.8 will die of cancer. Find the ratio of those who die of cancer to every 1000 people.
Source: "Reforming the Health Care System; State Profiles 1999," AARP

Find the ratio of the first number to the second and simplify.

7. 18 to 24

8. 5.6 to 10

9. 2.8 to 3.6

10. 0.32 to 0.96

11. In this rectangle, find the ratios of length to width and of width to length.

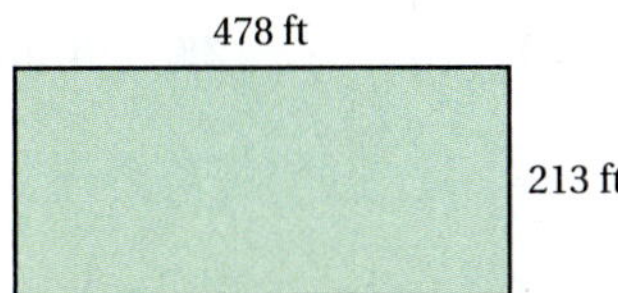

12. In this right triangle, find the ratios of shortest length to longest length and of longest length to shortest length.

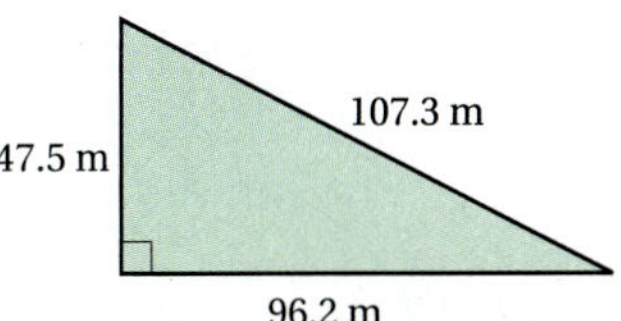

b In Exercises 13–16, find the rate, or speed, as a ratio of distance to time. Round to the nearest hundredth where appropriate.

13. 120 km, 3 hr

14. 18 mi, 9 hr

15. 217 mi, 29 sec

16. 443 m, 48 sec

17. *BMW 330Ci Convertible.* A 2001 BMW 330Ci Convertible will go 434 mi on 15.5 gal of gasoline in highway driving. What is the rate in miles per gallon?
Source: BMW

18. *Population Density of Monaco.* Monaco is a tiny country on the Mediterranean coast of France. It has an area of 1.21 square miles and a population of 32,149 people. What is the rate of number of people per square mile? The rate per square mile is called the *population density.* Monaco has the highest population density in the world.
Sources: *The New York Times Almanac; The Handy Geography Answer Book*

19. *Lawn Watering.* To water a lawn adequately requires 623 gal of water for every 1000 ft^2. What is the rate in gallons per square foot?

20. A car is driven 200 km on 40 L of gasoline. What is the rate in kilometers per liter?

21. Impulses in nerve fibers travel 310 km in 2.5 hr. What is the rate, or speed, in kilometers per hour?

22. A black racer snake can travel 4.6 km in 2 hr. What is its rate, or speed, in kilometers per hour?

23. *Elephant Heartbeat.* The heart of an elephant, at rest, will beat an average of 1500 beats in 60 min. What is the rate in beats per minute?
Source: *The Handy Science Answer Book*

24. *Human Heartbeat.* The heart of a human, at rest, will beat an average of 4200 beats in 60 min. What is the rate in beats per minute?
Source: *The Handy Science Answer Book*

c Determine whether the two pairs of numbers are proportional.

25. 5, 6 and 7, 9

26. 7, 5 and 6, 4

27. 1, 2 and 10, 20

28. 7, 3 and 21, 9

29. 2.4, 3.6 and 1.8, 2.7

30. 4.5, 3.8 and 6.7, 5.2

31. $5\frac{1}{3}, 8\frac{1}{4}$ and $2\frac{1}{5}, 9\frac{1}{2}$

32. $2\frac{1}{3}, 3\frac{1}{2}$ and 14, 21

d Solve.

33. $\frac{18}{4} = \frac{x}{10}$

34. $\frac{x}{45} = \frac{20}{25}$

35. $\frac{t}{12} = \frac{5}{6}$

36. $\frac{12}{4} = \frac{x}{3}$

37. $\frac{2}{5} = \frac{8}{n}$

38. $\frac{10}{6} = \frac{5}{x}$

39. $\frac{16}{12} = \frac{24}{x}$

40. $\frac{7}{11} = \frac{2}{x}$

41. $\frac{t}{0.16} = \frac{0.15}{0.40}$

42. $\frac{x}{11} = \frac{7.1}{2}$

43. $\frac{100}{25} = \frac{20}{n}$

44. $\frac{35}{125} = \frac{7}{m}$

45. $\frac{\frac{1}{4}}{\frac{1}{2}} = \frac{\frac{1}{2}}{x}$

46. $\frac{5\frac{1}{5}}{6\frac{1}{6}} = \frac{y}{3\frac{1}{2}}$

47. $\frac{1.28}{3.76} = \frac{4.28}{y}$

48. $\frac{10.4}{12.4} = \frac{6.76}{t}$

e Solve.

49. *Overweight Americans.* A study recently confirmed that of every 100 Americans, 60 are considered overweight. There were 281 million Americans in 2001. How many would be considered overweight?
Source: U.S. Centers for Disease Control

50. *Cancer Death Rate in Illinois.* It is predicted that for every 1000 people in the state of Illinois, 130.9 will die of cancer. The population of Chicago is about 2,721,547. How many of these people will die of cancer?
Source: *2001 New York Times Almanac*

51. *Gasoline Mileage.* Nancy's van traveled 84 mi on 6.5 gal of gasoline. At this rate, how many gallons would be needed to travel 126 mi?

52. *Bicycling.* Roy bicycled 234 mi in 14 days. At this rate, how far would Roy travel in 42 days?

53. *Quality Control.* A quality-control inspector examined 100 lightbulbs and found 7 of them to be defective. At this rate, how many defective bulbs will there be in a lot of 2500?

54. *Grading.* A professor must grade 32 essays in a literature class. She can grade 5 essays in 40 min. At this rate, how long will it take her to grade all 32 essays?

55. *Painting.* Fred uses 3 gal of paint to cover 1275 ft^2 of siding. How much siding can Fred paint with 7 gal of paint?

56. *Waterproofing.* Bonnie can waterproof 450 ft^2 of decking with 2 gal of sealant. How many gallons should Bonnie buy for a 1200-ft^2 deck?

57. *Exchanging Money.* On 22 December 2000, 1 U.S. dollar was worth about 1.80 Australian dollars.

a) How much would 250 U.S. dollars be worth in Australian dollars?

b) Derek was traveling in Australia and bought a sweatshirt that cost 50 Australian dollars. How much would it cost in U.S. dollars?

58. *Coffee Production.* Coffee beans from 14 trees are required to produce the 17 lb of coffee that the average person in the United States drinks each year. How many trees are required to produce 375 lb of coffee?

59. *Gas Mileage.* A 2001 BMW 330Ci Convertible will go 434 mi on 15.5 gal of gasoline in highway driving.

a) How many gallons of gasoline will it take to drive 2690 mi from Boston to Phoenix?

b) How far can the car be driven on 140 gal of gasoline?

Source: BMW

60. *Gas Mileage.* A 2001 Mercedes-Benz Cabriolet will go 396 mi on 16.5 gal of gasoline in highway driving.

a) How many gallons of gasoline will it take to drive 1650 mi from Pittsburgh to Albuquerque?

b) How far can the car be driven on 130 gal of gasoline?

Source: Mercedes-Benz

61. *Painting.* Helen can paint 950 ft^2 with 2 gal of paint. How many 1-gal cans does she need in order to paint a 30,000-ft^2 wall?

62. *Snow to Water.* Under typical conditions, $1\frac{1}{2}$ ft of snow will melt to 2 in. of water. To how many inches of water will $5\frac{1}{2}$ ft of snow melt?

63. *Estimating a Deer Population.* To determine the number of deer in a game preserve, a forest ranger catches 318 deer, tags them, and releases them. Later, 168 deer are caught, and it is found that 56 of them are tagged. Estimate how many deer are in the game preserve.

64. *Estimating a Trout Population.* To determine the number of trout in a lake, a conservationist catches 112 trout, tags them, and throws them back into the lake. Later, 82 trout are caught, and it is found that 32 of them are tagged. Estimate how many trout there are in the lake.

65. *Map Scaling.* On a road atlas map, 1 in. represents 16.6 mi. If two cities are 3.5 in. apart on the map, how far apart are they in reality?

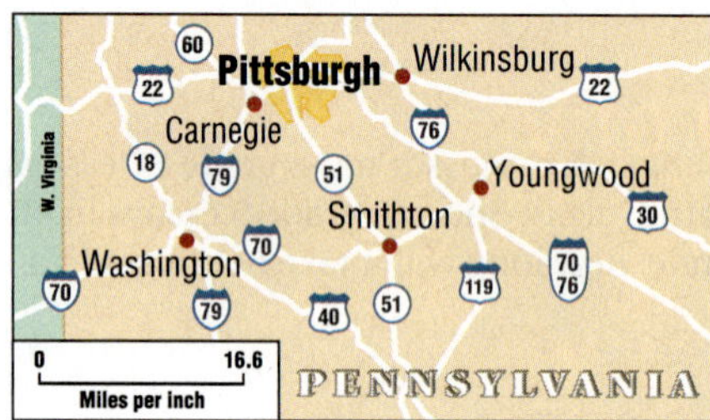

66. *Map Scaling.* On a map, $\frac{1}{4}$ in. represents 50 mi. If two cities are $3\frac{1}{4}$ in. apart on the map, how far apart are they in reality?

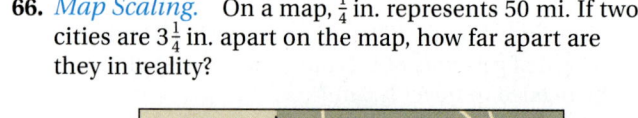

67. *Metallurgy.* In a metal alloy, the ratio of zinc to copper is 3 to 13. If there are 520 lb of copper, how many pounds of zinc are there?

68. *Class Size.* A college advertises that its student-to-faculty ratio is 14 to 1. If 56 students register for Introductory Spanish, how many sections of the course would you expect to see offered?

69. ^{D}W An instructor predicts that a student's test grade will be proportional to the amount of time the student spends studying. What is meant by this? Write an example of a proportion that involves the grades of two students and their study times.

70. ^{D}W *Earned Run Average.* In baseball, the average number of runs given up by a pitcher in nine innings is his *earned run average,* or *ERA.* Set up a formula for determining a player's ERA. Then use your formula to determine the ERA of Randy Johnson of the Arizona Diamondbacks, who gave up 73 earned runs in $248\frac{2}{3}$ innings. Is a low ERA considered good or bad?

Source: Major League Baseball

SKILL MAINTENANCE

Divide. Write decimal notation for the answer. [3.4a]

71. $260 \div 4$

72. $395 \div 10$

73. $4648 \div 16$

74. $3427 \div 2.25$

SYNTHESIS

75. ▦ Carney College is expanding from 850 to 1050 students. To avoid any rise in the student-to-faculty ratio, the faculty of 69 professors must also increase. How many new faculty positions should be created?

76. ▦ In recognition of her outstanding work, Sheri's salary has been increased from $26,000 to $29,380. Tim is earning $23,000 and is requesting a proportional raise. How much more should he ask for?

4.2 PERCENT NOTATION

Objectives

a Write three kinds of notation for a percent.

b Convert between percent notation and decimal notation.

a Understanding Percent Notation

Of all the surface area of the earth, 70% of it is covered by water. What does this mean? It means that of every 100 square miles of the earth's surface area, 70 square miles are covered by water. Thus, 70% is a ratio of 70 to 100, or $\frac{70}{100}$.
Source: *The Handy Geography Answer Book*

70 of 100 squares are shaded.

70% or $\frac{70}{100}$ or 0.70 of the large square is shaded.

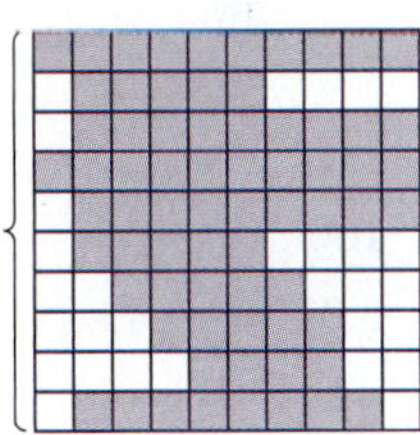

Percent notation is used extensively in our everyday lives. Here are some examples:

63% of all aluminum used in the United States is recycled.

46% of the people at a major-league baseball game are women.

33% of all Americans say the day they dread the most is the day they go to the dentist.

20% of the time that people declare as sick leave is actually used for personal needs.

60% of the vehicles involved in a rollover fatality are sport utility vehicles.

0.08% blood alcohol level is a standard used by some states as the legal limit for drunk driving.

Percent notation is often represented in pie charts to show how the parts of a quantity are related. For example, the chart below relates the amounts of different kinds of juices that are sold.

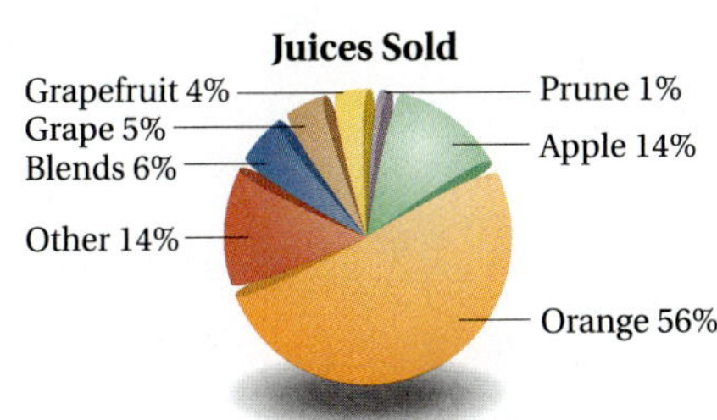

Source: Beverage Marketing Corporation

PERCENT NOTATION

The notation ***n*%** means "*n* per hundred."

Write three kinds of notation as in Examples 1 and 2.

1. 70%

2. 23.4%

3. 100%

It is thought that the Roman emperor Augustus began percent notation by taxing goods sold at a rate of $\frac{1}{100}$. In time, the symbol "%" evolved by interchanging the parts of the symbol "100" to "0/0" and then to "%."

Answers on page A-9

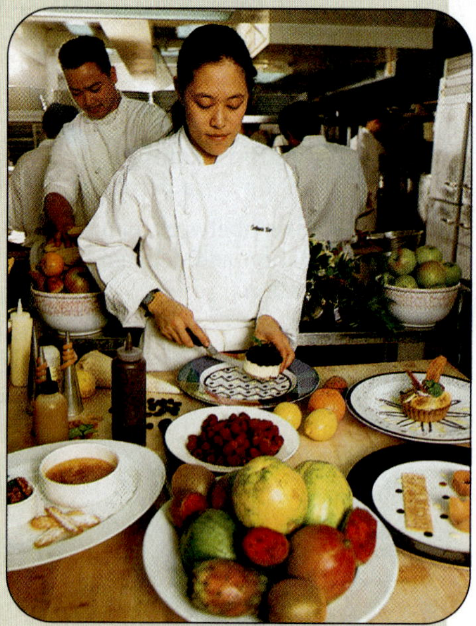

From 1998 to 2008, the number of jobs for professional chefs will increase by 13.4%.
Source: *Handbook of U.S. Labor Statistics*

CALCULATOR CORNER

Converting from Percent Notation to Decimal Notation Many calculators have a [%] key that can be used to convert from percent notation to decimal notation. This is often the second operation associated with a particular key and is accessed by first pressing a [2nd] or [SHIFT] key. To convert 57.6% to decimal notation, for example, you might press [5] [7] [.] [6] [2nd] [%] or [5] [7] [.] [6] [SHIFT] [%]. The display would read [0.576], so 57.6% = 0.576. Read the user's manual to determine whether your calculator can do this conversion.

Exercises: Use a calculator to find decimal notation.

1. 14% **2.** 0.069% **3.** 43.8% **4.** 125%

This definition leads us to the following equivalent ways of defining percent notation.

NOTATION FOR $n\%$

Percent notation, $n\%$, can be expressed using:

ratio → $n\% = \text{the ratio of } n \text{ to } 100 = \dfrac{n}{100}$,

fraction notation → $n\% = n \times \dfrac{1}{100}$, or

decimal notation → $n\% = n \times 0.01$.

EXAMPLE 1 Write three kinds of notation for 35%.

Using ratio: $35\% = \dfrac{35}{100}$ A ratio of 35 to 100

Using fraction notation: $35\% = 35 \times \dfrac{1}{100}$ Replacing % with $\times \dfrac{1}{100}$

Using decimal notation: $35\% = 35 \times 0.01$ Replacing % with $\times$ 0.01

EXAMPLE 2 Write three kinds of notation for 67.8%.

Using ratio: $67.8\% = \dfrac{67.8}{100}$ A ratio of 67.8 to 100

Using fraction notation: $67.8\% = 67.8 \times \dfrac{1}{100}$ Replacing % with $\times \dfrac{1}{100}$

Using decimal notation: $67.8\% = 67.8 \times 0.01$ Replacing % with $\times$ 0.01

Do Exercises 1–3 on the preceding page.

b Converting Between Percent Notation and Decimal Notation

Consider 78%. To convert to decimal notation, we can think of percent notation as a ratio and write

$$78\% = \frac{78}{100} \quad \text{Using the definition of percent as a ratio}$$
$$= 0.78. \quad \text{Dividing}$$

Similarly,

$$4.9\% = \frac{4.9}{100} \quad \text{Using the definition of percent as a ratio}$$
$$= 0.049. \quad \text{Dividing}$$

We could also convert 78% to decimal notation by replacing "%" with "× 0.01" and write

$$78\% = 78 \times 0.01 \quad \text{Replacing \% with} \times 0.01$$
$$= 0.78. \quad \text{Multiplying}$$

Similarly,

$$4.9\% = 4.9 \times 0.01 \quad \text{Replacing \% with} \times 0.01$$
$$= 0.049. \quad \text{Multiplying}$$

Dividing by 100 amounts to moving the decimal point two places to the left, which is the same as multiplying by 0.01. This leads us to a quick way to convert from percent notation to decimal notation: We drop the percent symbol and move the decimal point two places to the left.

To convert from percent notation to decimal notation,	36.5%	
a) replace the percent symbol % with $\times$ 0.01, and	36.5×0.01	
b) multiply by 0.01, which means move the decimal point two places to the left.	0.36.5	Move 2 places to the left.
	$36.5\% = 0.365$	

EXAMPLE 3 Find decimal notation for 99.44%.

a) Replace the percent symbol with $\times$ 0.01. $\quad 99.44 \times 0.01$

b) Move the decimal point two places to the left. $\quad$ 0.99.44

Thus, $99.44\% = 0.9944$.

EXAMPLE 4 The interest rate on a $2\frac{1}{2}$-year certificate of deposit is $6\frac{3}{8}\%$. Find decimal notation for $6\frac{3}{8}\%$.

a) Convert $6\frac{3}{8}$ to decimal notation and replace the percent symbol with $\times$ 0.01. $\quad 6\frac{3}{8}\%$; 6.375×0.01

b) Move the decimal point two places to the left. $\quad$ 0.06.375

Thus, $6\frac{3}{8}\% = 0.06375$.

Do Exercises 4–8.

To convert 0.38 to percent notation, we can first write fraction notation, as follows:

$$0.38 = \frac{38}{100} \qquad \text{Converting to fraction notation}$$
$$= 38\%. \qquad \text{Using the definition of percent as a ratio}$$

Note that $100\% = 100 \times 0.01 = 1$. Thus to convert 0.38 to percent notation, we can multiply by 1, using 100% as a symbol for 1. Then

$$\begin{aligned} 0.38 &= 0.38 \times 1 \\ &= 0.38 \times 100\% \\ &= 0.38 \times 100 \times 0.01 && \text{Replacing 100\% with } 100 \times 0.01 \\ &= (0.38 \times 100) \times 0.01 && \text{Using the associative law of multiplication} \\ &= 38 \times 0.01 \\ &= 38\%. && \text{Replacing "}\times 0.01\text{" with the \% symbol} \end{aligned}$$

Even more quickly, since $0.38 = 0.38 \times 100\%$, we can simply multiply 0.38 by 100 and write the % symbol.

Find decimal notation.

4. 34%

5. 78.9%

6. $6\frac{5}{8}\%$

Find decimal notation for the percent notation in the sentence.

7. Of all aluminum used in the United States, 63% is recycled.

8. A blood alcohol level of 0.08% is a standard used by some states as the legal limit for drunk driving.

Answers on page A-9

Find percent notation.

9. 0.24

10. 3.47

11. 1

Find percent notation for the decimal notation in the sentence.

12. Of all vehicles involved in a rollover fatality, 0.6 are sport utility vehicles.
Source: National Highway Traffic Safety Administration

13. Of those who play golf, 0.253 play 25–49 rounds per year.
Source: U.S. Golf Association

Answers on page A-9

To convert from decimal notation to percent notation, we multiply by 100%—that is, we move the decimal point two places to the right and write a percent symbol.

To convert from decimal notation to percent notation, multiply by 100%. That is,	$0.675 = 0.675 \times 100\%$	
a) move the decimal point two places to the right, and	0.67.5	Move 2 places to the right.
b) write a % symbol.	67.5% 0.675 = 67.5%	

EXAMPLE 5 Find percent notation for 1.27.

a) Move the decimal point two places to the right. 1.27.

b) Write a % symbol. 127%

Thus, 1.27 = 127%.

EXAMPLE 6 Of the time that people declare as sick leave, 0.21 is actually used for family issues. Find percent notation for 0.21.
Source: CCH Inc.

a) Move the decimal point two places to the right. 0.21.

b) Write a % symbol. 21%

Thus, 0.21 = 21%.

EXAMPLE 7 Find percent notation for 5.6.

a) Move the decimal point two places to the right, adding an extra zero. 5.60.

b) Write a % symbol. 560%

Thus, 5.6 = 560%.

EXAMPLE 8 Of those who play golf, 0.149 play 8–24 rounds per year. Find percent notation for 0.149.
Source: U.S. Golf Association

a) Move the decimal point two places to the right. 0.14.9

b) Write a % symbol. 14.9%

Thus, 0.149 = 14.9%.

Do Exercises 9–13.

4.2 EXERCISE SET

For Extra Help

Digital Video Tutor CD 3 Videotape 5 | InterAct Math | Math Tutor Center | MathXL | MyMathLab

a Write three kinds of notation as in Examples 1 and 2 on p. 268.

1. 90%

2. 58.7%

3. 12.5%

4. 130%

b Find decimal notation.

5. 67%

6. 17%

7. 45.6%

8. 76.3%

9. 59.01%

10. 30.02%

11. 10%

12. 80%

13. 1%

14. 100%

15. 200%

16. 300%

17. 0.1%

18. 0.4%

19. 0.09%

20. 0.12%

21. 0.18%

22. 5.5%

23. 23.19%

24. 87.99%

25. $14\frac{7}{8}\%$

26. $93\frac{1}{8}\%$

27. $56\frac{1}{2}\%$

28. $61\frac{3}{4}\%$

Find decimal notation for the percent notation in the sentence.

29. Of the people who declare time off as sick leave, 40% actually have a personal illness.
Source: CCH, Inc.

30. Of those who play golf, 39% play 50–99 rounds per year.
Source: U.S. Golf Association

31. Of those who play golf, 18.6% play 100 or more rounds per year.
Source: U.S. Golf Association

32. Recently, the average interest rate on a 30-yr mortgage loan was 6.89%.
Source: Freddie Mac

33. According to a recent survey, 29% of those asked to name their favorite ice cream chose vanilla.
Source: International Ice Cream Association

34. According to a recent survey, 95.1% of those asked to name what sports they participate in chose swimming.
Source: Sporting Goods Manufacturers

Find percent notation.

35. 0.47

36. 0.87

37. 0.03

38. 0.01

39. 8.7

40. 4

41. 0.334

42. 0.889

43. 0.75

44. 0.99

45. 0.4

46. 0.5

47. 0.006

48. 0.008

49. 0.017

50. 0.024

51. 0.2718

52. 0.8911

53. 0.0239

54. 0.00073

Find percent notation for the decimal notation in the sentence.

55. According to a recent survey, 0.526 of those asked to name what sports they participate in chose bowling.
Source: Sporting Goods Manufacturers

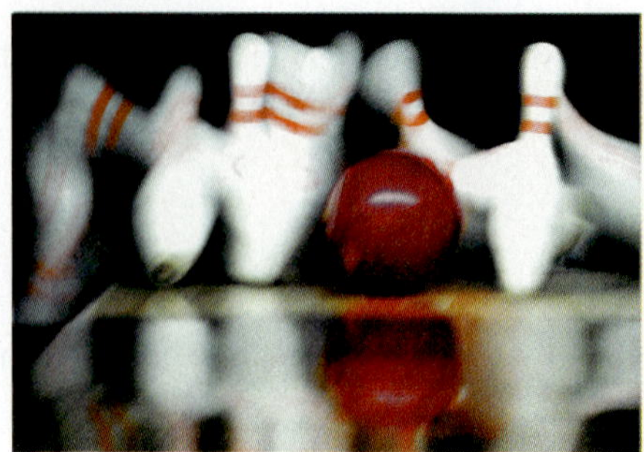

56. On average, churchgoers donate 0.03 of their income to their churches.
Source: Lutheran Brotherhood

57. In 2000, the cost of college to a middle-income family was 0.17 of their income.
Source: College Board

58. About 0.69 of all newspapers are recycled.
Sources: American Forest and Paper Association; Newspaper Association of America

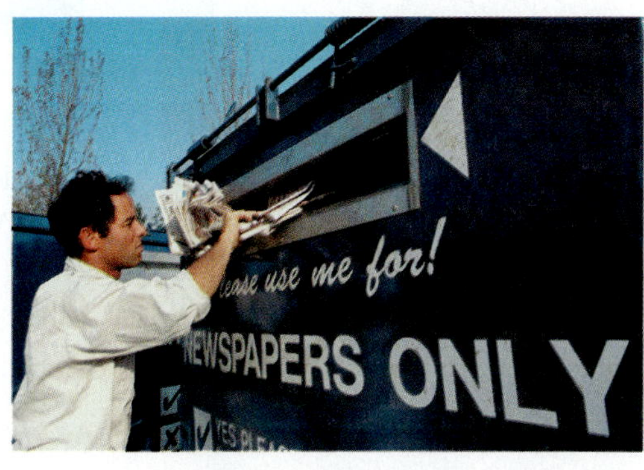

59. At one point in the 2000–2001 NBA season, Allen Iverson of the Philadelphia 76ers had made 0.411 of his field goals. His shooting percentage was 0.411.
Source: National Basketball Association

60. Of those people living in North Carolina, 0.1134 will die of heart disease.
Source: American Association of Retired Persons

61. ^{D}W *Winning Percentage.* During the 2000 regular baseball season, the New York Yankees won 87 of 162 games and went on to win the World Series. Find the ratio of number of wins to total number of games played in the regular season and convert it to decimal notation. Such a rate is often called a "winning percentage." Explain why.

62. ^{D}W Athletes sometimes speak of "giving 110%" effort. Does this make sense? Explain.

SKILL MAINTENANCE

Convert to a mixed numeral. [2.4a]

63. $\frac{100}{3}$

64. $\frac{75}{2}$

65. $\frac{75}{8}$

66. $\frac{297}{16}$

67. $\frac{567}{98}$

68. $\frac{2345}{21}$

Convert to decimal notation. [3.5a]

69. $\frac{2}{3}$

70. $\frac{1}{3}$

71. $\frac{5}{6}$

72. $\frac{17}{12}$

73. $\frac{8}{3}$

74. $\frac{15}{16}$

Objectives

a Convert from fraction notation to percent notation.

b Convert from percent notation to fraction notation.

4.3 PERCENT AND FRACTION NOTATION

a Converting from Fraction Notation to Percent Notation

Consider the fraction notation $\frac{7}{8}$. To convert to percent notation, we use two skills we already have. We first find decimal notation by dividing:

$$\frac{7}{8} = 0.875$$

$$\begin{array}{r} 0.875 \\ 8\overline{)7.000} \\ \underline{64} \\ 60 \\ \underline{56} \\ 40 \\ \underline{40} \\ 0 \end{array}$$

Then we convert the decimal notation to percent notation. We move the decimal point two places to the right

0.87.5

and write a % symbol:

$$\frac{7}{8} = 87.5\%, \text{ or } 87\frac{1}{2}\%.$$

To convert from fraction notation to percent notation,

a) find decimal notation by division, and

b) convert the decimal notation to percent notation.

$\frac{3}{5}$ Fraction notation

$$\begin{array}{r} 0.6 \\ 5\overline{)3.0} \\ \underline{30} \\ 0 \end{array}$$

$0.6 = 0.60 = 60\%$ Percent notation

$\frac{3}{5} = 60\%$

EXAMPLE 1 Find percent notation for $\frac{9}{16}$.

a) We first find decimal notation by division.

$$\begin{array}{r} 0.5625 \\ 16\overline{)9.0000} \\ \underline{80} \\ 100 \\ \underline{96} \\ 40 \\ \underline{32} \\ 80 \\ \underline{80} \\ 0 \end{array}$$

$$\frac{9}{16} = 0.5625$$

Study Tips

BEING A TUTOR

Try being a tutor for a fellow student. Understanding and retention of concepts can be maximized for yourself if you explain the material to someone else.

b) Next, we convert the decimal notation to percent notation. We move the decimal point two places to the right and write a % symbol.

0.56.25

$\frac{9}{16} = 56.25\%$, or $56\frac{1}{4}\%$

Don't forget the % symbol.

Do Exercises 1 and 2.

Find percent notation.

1. $\frac{1}{4}$

2. $\frac{5}{8}$

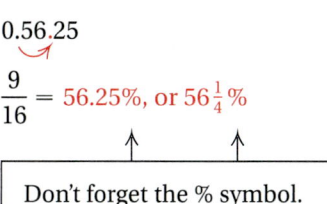

CALCULATOR CORNER

Converting from Fraction Notation to Percent Notation A calculator can be used to convert from fraction notation to percent notation. We simply perform the division on the calculator and then use the percent key. To convert $\frac{17}{40}$ to percent notation, for example, we press [1] [7] [÷] [4] [0] [2nd] [%], or [1] [7] [÷] [4] [0] [SHIFT] [%]. The display reads [42.5], so $\frac{17}{40} = 42.5\%$. Read the user's manual to determine whether your calculator can do this conversion.

Exercises: Use a calculator to find percent notation. Round to the nearest hundredth of a percent.

1. $\frac{13}{25}$

2. $\frac{5}{13}$

3. $\frac{43}{39}$

4. $\frac{12}{7}$

5. $\frac{217}{364}$

6. $\frac{2378}{8401}$

EXAMPLE 2 *Death from Heart Attack.* Of all those who suffer a heart attack, $\frac{1}{3}$ will die. Find percent notation for $\frac{1}{3}$.
Source: American Heart Association

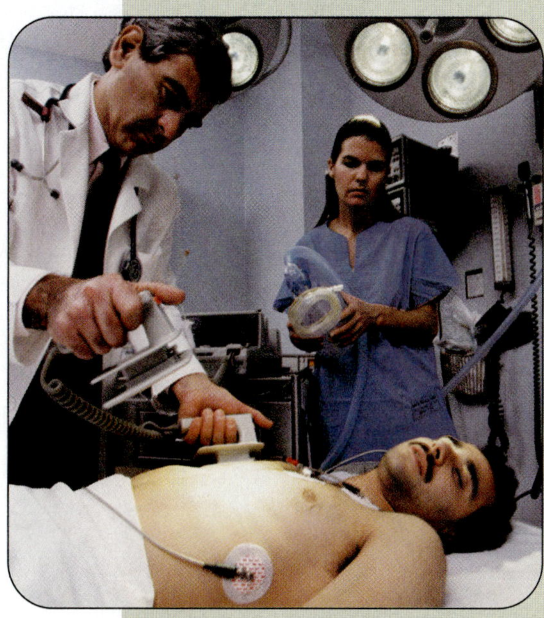

a) Find decimal notation by division.

$$\begin{array}{r} 0.3\,3\,3 \\ 3\,\overline{)\,1.0\,0\,0} \\ \underline{9} \\ 1\,0 \\ \underline{9} \\ 1\,0 \\ \underline{9} \\ 1 \end{array}$$

We get a repeating decimal: $0.33\overline{3}$.

b) Convert the answer to percent notation.

$0.33.\overline{3}$

$\frac{1}{3} = 33.\overline{3}\%$, or $33\frac{1}{3}\%$

Answers on page A-9

3. Water is the single most abundant chemical in the body. The human body is about $\frac{2}{3}$ water. Find percent notation for $\frac{2}{3}$.

4. Find percent notation: $\frac{5}{6}$.

Find percent notation.

5. $\frac{57}{100}$

6. $\frac{19}{25}$

Answers on page A-9

Do Exercises 3 and 4.

In some cases, division is not the fastest way to convert. The following are some optional ways in which conversion might be done.

EXAMPLE 3 Find percent notation for $\frac{69}{100}$.

We use the definition of percent as a ratio.

$$\frac{69}{100} = 69\%$$

EXAMPLE 4 Find percent notation for $\frac{17}{20}$.

We multiply by 1 to get 100 in the denominator. We think of what we have to multiply 20 by in order to get 100. That number is 5, so we multiply by 1 using $\frac{5}{5}$.

$$\frac{17}{20} \cdot \frac{5}{5} = \frac{85}{100} = 85\%$$

Note that this shortcut works only when the denominator is a factor of 100.

Do Exercises 5 and 6.

b Converting from Percent Notation to Fraction Notation

To convert from percent notation to fraction notation,

a) use the definition of percent as a ratio, and

b) simplify, if possible.

30% Percent notation

$\frac{30}{100}$

$\frac{3}{10}$ Fraction notation

EXAMPLE 5 Find fraction notation for 75%.

$$75\% = \frac{75}{100} \quad \text{Using the definition of percent}$$

$$= \frac{3 \cdot 25}{4 \cdot 25} = \frac{3}{4} \cdot \frac{25}{25}$$

$$= \frac{3}{4} \quad \text{Simplifying}$$

EXAMPLE 6 Find fraction notation for 62.5%.

$$62.5\% = \frac{62.5}{100} \quad \text{Using the definition of percent}$$

$$= \frac{62.5}{100} \times \frac{10}{10} \quad \text{Multiplying by 1 to eliminate the decimal point in the numerator}$$

$$= \frac{625}{1000}$$

$$= \frac{5 \cdot 125}{8 \cdot 125} = \frac{5}{8} \cdot \frac{125}{125}$$

$$= \frac{5}{8} \quad \text{Simplifying}$$

EXAMPLE 7 Find fraction notation for $16\frac{2}{3}\%$.

$$16\frac{2}{3}\% = \frac{50}{3}\% \quad \text{Converting from the mixed numeral to fraction notation}$$

$$= \frac{50}{3} \times \frac{1}{100} \quad \text{Using the definition of percent}$$

$$= \frac{50 \cdot 1}{3 \cdot 50 \cdot 2} = \frac{1}{6} \cdot \frac{50}{50}$$

$$= \frac{1}{6} \quad \text{Simplifying}$$

The table on the inside front cover lists decimal, fraction, and percent equivalents used so often that it would speed up your work if you memorized them. For example, $\frac{1}{3} = 0.\overline{3}$, so we say that the **decimal equivalent** of $\frac{1}{3}$ is $0.\overline{3}$, or that $0.\overline{3}$ has the **fraction equivalent** $\frac{1}{3}$.

EXAMPLE 8 Find fraction notation for $16.\overline{6}\%$.

We can use the table on the inside front cover or recall that $16.\overline{6}\% = 16\frac{2}{3}\% = \frac{1}{6}$. We can also recall from our work with repeating decimals in Chapter 3 that $0.\overline{6} = \frac{2}{3}$. Then we have $16.\overline{6}\% = 16\frac{2}{3}\%$ and can proceed as in Example 7.

Do Exercises 7–10.

Find fraction notation.

7. 60%

8. 3.25%

9. $66\frac{2}{3}\%$

10. Complete this table.

Fraction Notation	$\frac{1}{5}$		
Decimal Notation		$0.83\overline{3}$	
Percent Notation			$37\frac{1}{2}\%$

Answers on page A-9

Study Tips

MEMORIZING

Memorizing is a very helpful tool in the study of mathematics. Don't underestimate its power as you memorize the table of decimal, fraction, and percent notation on the inside front cover. We will discuss memorizing more later.

CALCULATOR CORNER

Applications of Ratio and Percent: The Price–Earnings Ratio and Stock Yields

The Price–Earnings Ratio If the total earnings of a company one year were $5,000,000 and 100,000 shares of stock were issued, the earnings per share was $50. At one time, the price per share of Coca-Cola was $58.125 and the earnings per share was $0.76. The **price-earnings ratio,** *P/E,* is the price of the stock divided by the earnings per share. For the Coca-Cola stock, the price–earnings ratio, *P/E,* is given by

$$\frac{P}{E} = \frac{58.125}{0.76} \approx 76.48. \quad \text{Dividing, using a calculator, and rounding to the nearest hundredth}$$

Stock Yields At one time, the price per share of Coca-Cola stock was $58.125 and the company was paying a yearly dividend of $0.68 per share. It is helpful to those interested in stocks to know what percent the dividend is of the price of the stock. The percent is called the **yield.** For the Coca-Cola stock, the yield is given by

$$\text{Yield} = \frac{\text{Dividend}}{\text{Price per share}} = \frac{0.68}{58.125} \approx 0.0117 \quad \text{Dividing and rounding to the nearest ten-thousandth}$$

$$= 1.17\% \quad \text{Converting to percent notation}$$

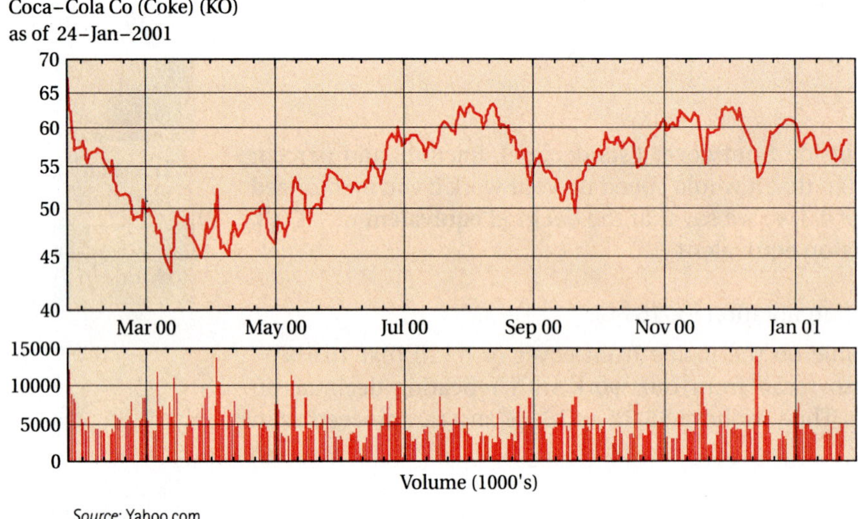

Source: Yahoo.com

Exercises: Compute the price–earnings ratio and the yield for each stock listed below.

	STOCK	PRICE PER SHARE	EARNINGS	DIVIDEND	P/E	YIELD
1.	Pepsi (PEP)	$42.75	$1.40	$0.56		
2.	Pearson (PSO)	$25.00	$0.78	$0.30		
3.	Quaker Oats (OAT)	$92.375	$2.68	$1.10		
4.	Texas Insts (TEX)	$42.875	$1.62	$0.43		
5.	Ford Motor Co (F)	$27.5625	$2.30	$1.19		
6.	Wendy's Intl (WEN)	$25.75	$1.47	$0.23		

4.3 EXERCISE SET

For Extra Help

Digital Video Tutor CD 3 Videotape 5 | InterAct Math | Math Tutor Center | MathXL | MyMathLab

a Find percent notation.

1. $\frac{41}{100}$ **2.** $\frac{36}{100}$ **3.** $\frac{5}{100}$ **4.** $\frac{1}{100}$ **5.** $\frac{2}{10}$ **6.** $\frac{7}{10}$

7. $\frac{3}{10}$ **8.** $\frac{9}{10}$ **9.** $\frac{1}{2}$ **10.** $\frac{3}{4}$ **11.** $\frac{7}{8}$ **12.** $\frac{1}{8}$

13. $\frac{4}{5}$ **14.** $\frac{2}{5}$ **15.** $\frac{2}{3}$ **16.** $\frac{1}{3}$ **17.** $\frac{1}{6}$ **18.** $\frac{5}{6}$

19. $\frac{3}{16}$ **20.** $\frac{11}{16}$ **21.** $\frac{13}{16}$ **22.** $\frac{7}{16}$ **23.** $\frac{4}{25}$ **24.** $\frac{17}{25}$

25. $\frac{1}{20}$ **26.** $\frac{31}{50}$ **27.** $\frac{17}{50}$ **28.** $\frac{3}{20}$

Find percent notation for the fraction notation in the sentence.

29. Of all people, $\frac{2}{25}$ dread their birthday.
Source: Yankelovich Partners for Lutheran Brotherhood

30. Of all the water taken into the body, $\frac{3}{5}$ of it comes from beverages.

In Exercises 31–34, write percent notation for the fractions in this pie chart.

Engagement Times of Married Couples

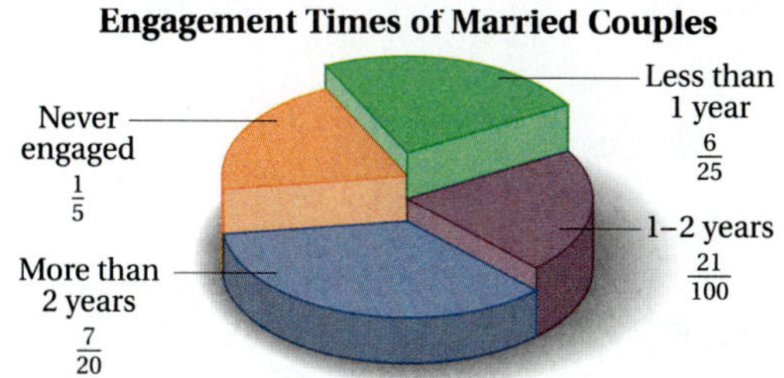

31. $\frac{21}{100}$ **32.** $\frac{1}{5}$

33. $\frac{6}{25}$ **34.** $\frac{7}{20}$

b Find fraction notation. Simplify.

35. 85%

36. 55%

37. 62.5%

38. 12.5%

39. $33\frac{1}{3}\%$

40. $83\frac{1}{3}\%$

41. $16.\overline{6}\%$

42. $66.\overline{6}\%$

43. 7.25%

44. 4.85%

45. 0.8%

46. 0.2%

47. $25\frac{3}{8}\%$

48. $48\frac{7}{8}\%$

49. $78\frac{2}{9}\%$

50. $16\frac{5}{9}\%$

51. $64\frac{7}{11}\%$

52. $73\frac{3}{11}\%$

53. 150%

54. 110%

55. 0.0325%

56. 0.419%

57. $33.\overline{3}\%$

58. $83.\overline{3}\%$

Find fraction notation for the percent notation in the following bar graph.

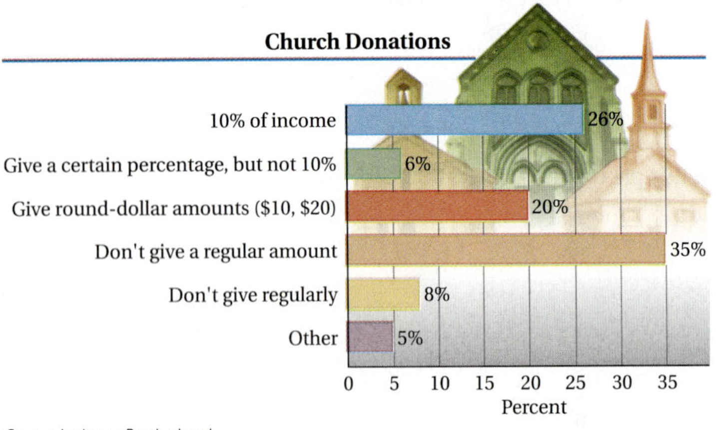

Source: Lutheran Brotherhood

59. 26%

60. 20%

61. 5%

62. 8%

63. 6%

64. 35%

Find fraction notation for the percent notation in the sentence.

65. A 1.1-oz serving of Complete® cereal with $\frac{1}{2}$ cup of skim milk satisfies 45% of the minimum daily requirements for iron.
Source: Kellogg's USA, Inc.

66. A 1-cup serving of Wheaties® cereal with $\frac{1}{2}$ cup of skim milk satisfies 15% of the minimum daily requirements for calcium.
Source: General Mills Sales, Inc.

67. Of all those who are 85 or older, 47% have Alzheimer's disease.
Source: Alzheimer's Association

68. In 1998, 24.1% of Americans 18 and older smoked cigarettes.
Source: U.S. Centers for Disease Control

Complete the table.

69.

FRACTION NOTATION	DECIMAL NOTATION	PERCENT NOTATION
$\frac{1}{8}$		12.5%, or $12\frac{1}{2}\%$
$\frac{1}{6}$		
		20%
	0.25	
		$33.\overline{3}\%$, or $33\frac{1}{3}\%$
		37.5%, or $37\frac{1}{2}\%$
		40%
$\frac{1}{2}$		

70.

FRACTION NOTATION	DECIMAL NOTATION	PERCENT NOTATION
$\frac{3}{5}$		
	0.625	
$\frac{2}{3}$		
	0.75	75%
$\frac{4}{5}$		
$\frac{5}{6}$		$83.\overline{3}\%$, or $83\frac{1}{3}\%$
$\frac{7}{8}$		87.5%, or $87\frac{1}{2}\%$
		100%

71.

FRACTION NOTATION	DECIMAL NOTATION	PERCENT NOTATION
	0.5	
$\frac{1}{3}$		
		25%
		$16.\overline{6}\%$, or $16\frac{2}{3}\%$
	0.125	
$\frac{3}{4}$		
	$0.8\overline{3}$	
$\frac{3}{8}$		

72.

FRACTION NOTATION	DECIMAL NOTATION	PERCENT NOTATION
		40%
		62.5%, or $62\frac{1}{2}\%$
	0.875	
$\frac{1}{1}$		
	0.6	
	$0.\overline{6}$	
$\frac{1}{5}$		

73. **DW** What do the following have in common? Explain.

$\frac{23}{16}$, $1\frac{875}{2000}$, 1.4375, $\frac{207}{144}$, $1\frac{7}{16}$, 143.75%, $1\frac{4375}{10{,}000}$

74. **DW** Is it always best to convert from fraction notation to percent notation by first finding decimal notation? Why or why not?

SKILL MAINTENANCE

Solve.

75. $13 \cdot x = 910$ [1.4b]

76. $15 \cdot y = 75$ [1.4b]

77. $0.05 \times b = 20$ [3.4b]

78. $3 = 0.16 \times b$ [3.4b]

79. $\frac{1}{2} \cdot x = 2$ [2.2d]

80. $4 \cdot x = \frac{3}{11}$ [2.2d]

Convert to a mixed numeral. [2.4a]

81. $\frac{100}{3}$

82. $\frac{75}{2}$

83. $\frac{250}{3}$

84. $\frac{123}{6}$

85. $\frac{345}{8}$

86. $\frac{373}{6}$

87. $\frac{75}{4}$

88. $\frac{67}{9}$

SYNTHESIS

Write percent notation.

89. 🖩 $\frac{41}{369}$

90. 🖩 $\frac{54}{999}$

91. $2.5\overline{74631}$

92. $3.2\overline{93847}$

Write decimal notation.

93. $\frac{14}{9}\%$

94. $\frac{19}{12}\%$

95. $\frac{729}{7}\%$

96. $\frac{637}{6}\%$

4.4 SOLVING PERCENT PROBLEMS USING PERCENT EQUATIONS

Objectives

a Translate percent problems to percent equations.

b Solve basic percent problems.

a Translating to Equations

To solve a problem involving percents, it is helpful to translate first to an equation. To distinguish the method in Section 4.4 from that of Section 4.5, we will call these *percent equations.*

EXAMPLE 1 Translate:

23%	of	5	is	what?
↓	↓	↓	↓	↓
23%	$\cdot$	5	$=$	a

This is a *percent equation.*

Translate to an equation. Do not solve.

1. 12% of 50 is what?

KEY WORDS IN PERCENT TRANSLATIONS

"**Of**" translates to "$\cdot$", or "$\times$".

"**What**" translates to any letter.

"**Is**" translates to "$=$".

"**%**" translates to "$\times \frac{1}{100}$" or "$\times$ 0.01".

EXAMPLE 2 Translate:

What	is	11%	of	49?
↓	↓	↓	↓	↓
a	$=$	11%	$\cdot$	49

Any letter can be used.

2. What is 40% of 60?

Do Exercises 1 and 2.

EXAMPLE 3 Translate:

3	is	10%	of	what?
↓	↓	↓	↓	↓
3	$=$	10%	$\cdot$	b

Translate to an equation. Do not solve.

3. 45 is 20% of what?

EXAMPLE 4 Translate:

45%	of	what	is	23?
↓	↓	↓	↓	↓
45%	$\times$	b	$=$	23

Do Exercises 3 and 4.

4. 120% of what is 60?

EXAMPLE 5 Translate:

10	is	what percent	of	20?
↓	↓	↓	↓	↓
10	$=$	p	$\times$	20

Answers on page A-10

Translate to an equation. Do not solve.

5. 16 is what percent of 40?

6. What percent of 84 is 10.5?

EXAMPLE 6 Translate:

What percent	of	50	is	7?
↓	↓	↓	↓	↓
p	$\cdot$	50	$=$	7

Do Exercises 5 and 6.

b Solving Percent Problems

In solving percent problems, we use the *Translate* and *Solve* steps in the problem-solving strategy used throughout this text.

Percent problems are actually of three different types. Although the method we present does *not* require that you be able to identify which type you are solving, it is helpful to know them.

We know that

15 is 25% of 60, or

$15 = 25\% \times 60.$

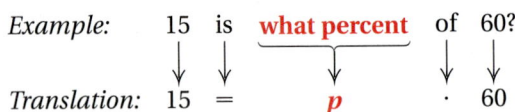

We can think of this as:

Amount = Percent number × Base.

Each of the three types of percent problems depend on which of the three pieces of information is missing.

1. **Finding the *amount* (the result of taking the percent)**

Example:	**What**	is	25%	of	60?
	↓	↓	↓	↓	↓
Translation:	a	$=$	25%	$\cdot$	60

2. **Finding the *base* (the number you are taking the percent of)**

Example:	15	is	25%	of	**what number?**
	↓	↓	↓	↓	↓
Translation:	15	$=$	25%	$\cdot$	b

3. **Finding the *percent number* (the percent itself)**

Example:	15	is	**what percent**	of	60?
	↓	↓	↓	↓	↓
Translation:	15	$=$	p	$\cdot$	60

7. Solve:

What is 12% of 50?

John's Used Cars currently has 60 used cars for sale; 15% of those cars are at least 8 yr old. How many cars are at least 8 yr old?

FINDING THE AMOUNT

EXAMPLE 7 What is 15% of 60?

Translate: $a = 15\% \times 60.$

Solve: The letter is by itself. To solve the equation, we just convert 15% to decimal notation and multiply:

$$a = 15\% \times 60 = 0.15 \times 60 = 9.$$

Thus, 9 is 15% of 60. The answer is 9.

Do Exercise 7.

Answers on page A-10

EXAMPLE 8 120% of \$42 is what?

Translate: $120\% \times 42 = a.$

Solve: The letter is by itself. To solve the equation, we carry out the calculation:

$$\begin{aligned} a &= 120\% \times 42 \\ &= 1.2 \times 42 \\ &= 50.4. \end{aligned}$$

Thus, 120% of \$42 is \$50.40. The answer is \$50.40.

Do Exercise 8.

8. Solve:

64% of \$55 is what?

FINDING THE BASE

EXAMPLE 9 5% of what is 20?

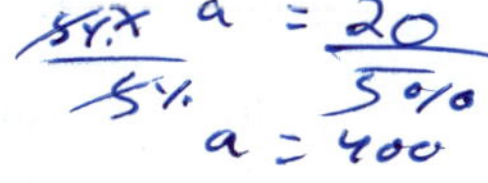

Translate: $5\% \times b = 20.$

Solve: This time the letter is *not* by itself. To solve the equation, we divide by 5% on both sides:

$$\frac{5\% \times b}{5\%} = \frac{20}{5\%} \quad \text{Dividing by 5\% on both sides}$$

$$b = \frac{20}{0.05} \quad 5\% = 0.05$$

$$b = 400.$$

Thus, 5% of 400 is 20. The answer is 400.

In a survey of a group of people, it was found that 5%, or 20 people, chose strawberry as their favorite ice cream. How many people were surveyed?
Source: International Ice Cream Association

EXAMPLE 10 \$3 is 16% of what?

Translate: \$3 is 16% of what?

$$\begin{array}{ccccc} \$3 & \text{is} & 16\% & \text{of} & \text{what?} \\ \downarrow & \downarrow & \downarrow & \downarrow & \downarrow \\ 3 & = & 16\% & \times & b. \end{array}$$

Solve: Again, the letter is *not* by itself. To solve the equation, we divide by 16% on both sides:

$$\frac{3}{16\%} = \frac{16\% \times b}{16\%} \quad \text{Dividing by 16\% on both sides}$$

$$\frac{3}{0.16} = b \quad 16\% = 0.16$$

$$18.75 = b.$$

Thus, \$3 is 16% of \$18.75. The answer is \$18.75.

Do Exercises 9 and 10.

Solve.

9. 20% of what is 45?

10. \$60 is 120% of what?

Answers on page A-10

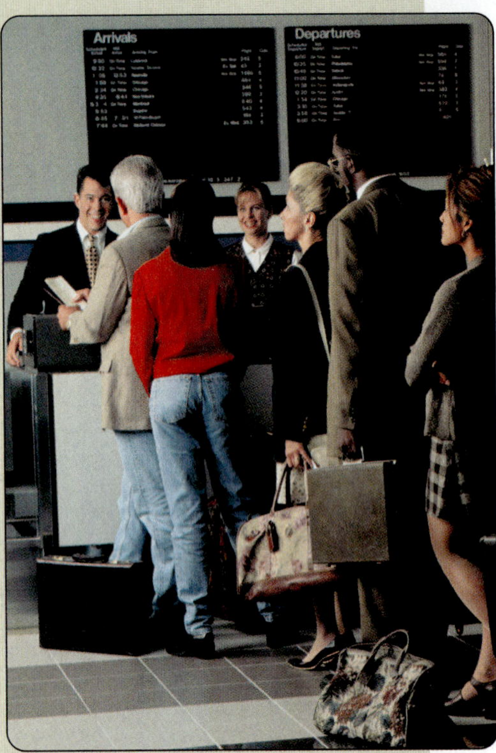

Of every 20 people who travel, 10 will feel stress while waiting in line. What percent will feel stress?
Source: TNS Intersearch

11. Solve:

16 is what percent of 40?

12. Solve:

What percent of \$84 is \$10.50?

Answers on page A-10

FINDING THE PERCENT NUMBER

In solving these problems, you *must* remember to convert to percent notation after you have solved the equation.

EXAMPLE 11 10 is what percent of 20?

Translate: 10 is what percent of 20?

$$\underbrace{10}_{10} \quad \underbrace{\text{is}}_{=} \quad \underbrace{\text{what percent}}_{p} \quad \underbrace{\text{of}}_{\times} \quad \underbrace{20?}_{20.}$$

Solve: To solve the equation, we divide by 20 on both sides and convert the result to percent notation:

$$p \cdot 20 = 10$$

$$\frac{p \cdot 20}{20} = \frac{10}{20} \qquad \text{Dividing by 20 on both sides}$$

$$p = 0.50 = 50\%. \qquad \text{Converting to percent notation}$$

Thus, 10 is 50% of 20. The answer is 50%.

Do Exercise 11.

EXAMPLE 12 What percent of \$50 is \$16?

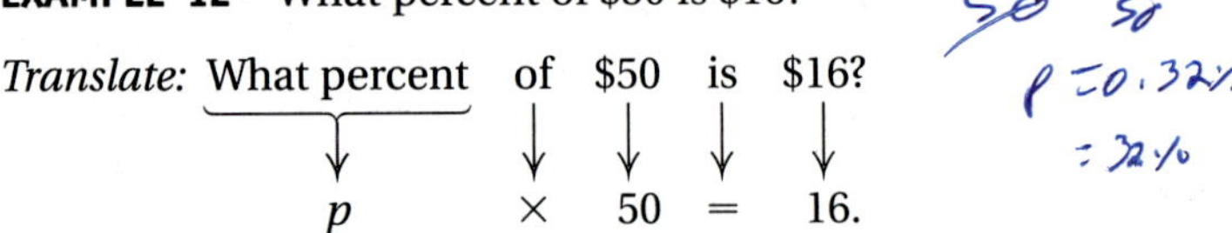

Translate: What percent of \$50 is \$16?

$$\underbrace{\text{What percent}}_{p} \quad \underbrace{\text{of}}_{\times} \quad \underbrace{\$50}_{50} \quad \underbrace{\text{is}}_{=} \quad \underbrace{\$16?}_{16.}$$

Solve: To solve the equation, we divide by 50 on both sides and convert the answer to percent notation:

$$\frac{p \times 50}{50} = \frac{16}{50} \qquad \text{Dividing by 50 on both sides}$$

$$p = \frac{16}{50}$$

$$p = 0.32$$

$$p = 32\%. \qquad \text{Converting to percent notation}$$

Thus, 32% of \$50 is \$16. The answer is 32%.

Do Exercise 12.

CAUTION!

When a question asks "what percent?", be sure to give the answer in percent notation.

CALCULATOR CORNER

Using Percents in Computations Many calculators have a [%] key that can be used in computations. (See the Calculator Corner on page 268.) For example, to find 11% of 49, we press [1] [1] [2nd] [%] [×] [4] [9] [=] or [4] [9] [×] [1] [1] [SHIFT] [%]. The display reads [5.39], so 11% of 49 is 5.39.

In Example 9, we perform the computation 20/5%. To use the [%] key in this computation, we press [2] [0] [÷] [5] [2nd] [%] [=], or [2] [0] [÷] [5] [SHIFT] [%]. The result is 400.

We can also use the [%] key to find the percent number in a problem. In Example 11, for instance, we answer the question "10 is what percent of 20?" On a calculator, we press [1] [0] [÷] [2] [0] [2nd] [%] [=], or [1] [0] [÷] [2] [0] [SHIFT] [%]. The result is 50, so 10 is 50% of 20.

Exercises: Use a calculator to find each of the following.

1. What is 5% of 24?
2. What is 12.6% of $40?
3. What is 19% of 256?
4. 140% of $16 is what?
5. 0.04% of 28 is what?
6. 33% of $90 is what?
7. Use the percent key on a calculator to perform the computations in Example 10 and Margin Exercises 9 and 10.
8. Use the percent key on a calculator to perform the computations in Example 12 and Margin Exercises 11 and 12.

Study Tips

TIME MANAGEMENT (PART 2)

Here are some additional tips to help you with time management. (See also the Study Tips on time management in Sections 1.3 and 10.2.)

- **Avoid "time killers."** We live in a media age, and the Internet, e-mail, television, and movies all are time killers. Allow yourself a break to enjoy some college and outside activities. But keep track of the time you spend on such activities and compare it to the time you spend studying.
- **Prioritize your tasks.** Be careful about taking on too many college activities that fall outside of academics. Examples of such activities are decorating a homecoming float, joining a fraternity or sorority, and participating on a student council committee. Any of these is important but keep them to a minimum to be sure that you have enough time for your studies.
- **Be aggressive about your study tasks.** Instead of worrying over your math homework or test preparation, do something to get yourself started. Work a problem here and a problem there, and before long you will accomplish the task at hand. If the task is large, break it down into smaller parts, and do one at a time. You will be surprised at how quickly the large task can then be completed.

4.4 EXERCISE SET

For Extra Help

Digital Video Tutor CD 3 Videotape 5

InterAct Math

Math Tutor Center

MathXL

MyMathLab

a Translate to an equation. Do not solve.

1. What is 32% of 78?

2. 98% of 57 is what?

3. 89 is what percent of 99?

4. What percent of 25 is 8?

5. 13 is 25% of what?

6. 21.4% of what is 20?

b Solve.

7. What is 85% of 276?

8. What is 74% of 53?

9. 150% of 30 is what?

10. 100% of 13 is what?

11. What is 6% of $300?

12. What is 4% of $45?

13. 3.8% of 50 is what?

14. $33\frac{1}{3}\%$ of 480 is what?
(*Hint:* $33\frac{1}{3}\% = \frac{1}{3}$.)

15. $39 is what percent of $50?

16. $16 is what percent of $90?

17. 20 is what percent of 10?

18. 60 is what percent of 20?

19. What percent of $300 is $150?

20. What percent of $50 is $40?

21. What percent of 80 is 100?

22. What percent of 60 is 15?

23. 20 is 50% of what?

24. 57 is 20% of what?

25. 40% of what is \$16?

26. 100% of what is \$74?

27. 56.32 is 64% of what?

28. 71.04 is 96% of what?

29. 70% of what is 14?

30. 70% of what is 35?

31. What is $62\frac{1}{2}\%$ of 10?

32. What is $35\frac{1}{4}\%$ of 1200?

33. What is 8.3% of \$10,200?

34. What is 9.2% of \$5600?

35. D_W Write a question that could be translated to the equation

$$25 = 4\% \times b.$$

36. D_W Suppose we know that 40% of 92 is 36.8. What is a quick way to find 4% of 92? 400% of 92? Explain.

SKILL MAINTENANCE

Write fraction notation. [3.1b]

37. 0.09

38. 1.79

39. 0.875

40. 0.125

41. 0.9375

42. 0.6875

Write decimal notation. [3.1b]

43. $\frac{89}{100}$

44. $\frac{7}{100}$

45. $\frac{3}{10}$

46. $\frac{17}{1000}$

SYNTHESIS

Solve.

47. What is 7.75% of \$10,880?
Estimate ____________
Calculate ____________

48. 50,951.775 is what percent of 78,995?
Estimate ____________
Calculate ____________

49. \$2496 is 24% of what amount?
Estimate ____________
Calculate ____________

50. What is 38.2% of \$52,345.79?
Estimate ____________
Calculate ____________

51. 40% of $18\frac{3}{4}\%$ of \$25,000 is what?

4.5 SOLVING PERCENT PROBLEMS USING PROPORTIONS*

Objectives

a Translate percent problems to proportions.

b Solve basic percent problems.

A survey has found that 75% of all people watch TV in bed before they go to sleep. The city of San Francisco has 745,780 people. This means that 559,335 of them watch TV in bed before they go to sleep.
Sources: Bruskin-Goldring Research for Serta; *The New York Times Almanac*

**Note: This section presents an alternative method for solving basic percent problems. You can use either equations or proportions to solve percent problems, but you might prefer one method over the other, or your instructor may direct you to use one method over the other.*

a Translating to Proportions

A percent is a ratio of some number to 100. For example, 75% is the ratio $\frac{75}{100}$. The numbers 559,335 and 745,780 have the same ratio as 75 and 100. The numbers 3 and 4 also have the same ratio.

$$\frac{75}{100} = \frac{559{,}335}{745{,}780} = \frac{3}{4}$$

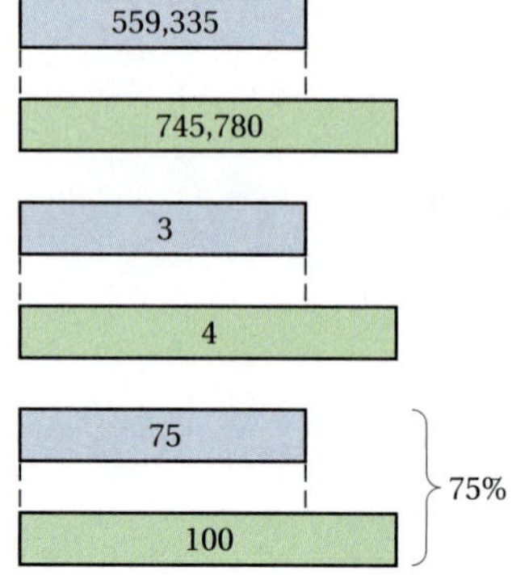

To solve a percent problem using a proportion, we translate as follows:

$$\frac{P}{100} = \frac{a}{b}$$

Number → P; 100 → 100; a ← Amount; b ← Base

You might find it helpful to read this as "part is to whole as part is to whole."

For example, 60% of 25 is 15 translates to

$$\frac{60}{100} = \frac{15}{25}.$$

15 ← Amount; 25 ← Base

A clue in translating is that the base, b, corresponds to 100 and usually follows the wording "percent of." Also, $P\%$ always translates to $P/100$. Another aid in translating is to make a comparison drawing. To do this, we start with the percent side and list 0% at the top and 100% near the bottom. Then we estimate where the specified percent—in this case, 60%—is located. The corresponding quantities are then filled in. The base—in this case, 25—always corresponds to 100% and the amount—in this case, 15—corresponds to the specified percent.

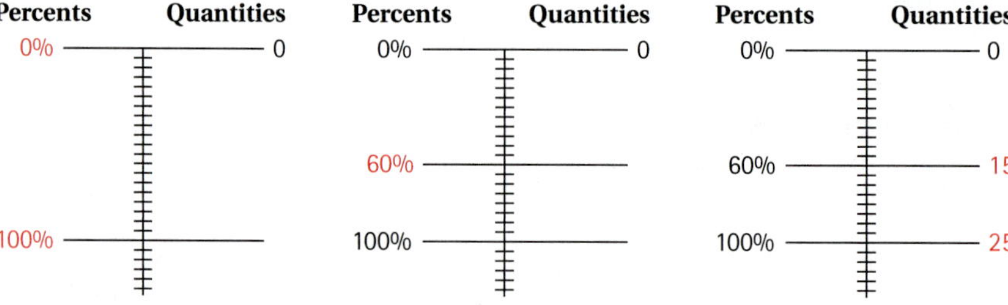

The proportion can then be read easily from the drawing: $\frac{60}{100} = \frac{15}{25}$.

EXAMPLE 1 Translate to a proportion.

23% of 5 is what?

$$\frac{23}{100} = \frac{a}{5}$$

Percents | Quantities
0% — 0
23% — a
100% — 5

EXAMPLE 2 Translate to a proportion.

What is 124% of 49?

$$\frac{124}{100} = \frac{a}{49}$$

Percents | Quantities
0% — 0
100% — 49
124% — a

Do Exercises 1–3.

EXAMPLE 3 Translate to a proportion.

3 is 10% of what?

$$\frac{10}{100} = \frac{3}{b}$$

Percents | Quantities
0% — 0
10% — 3
100% — b

EXAMPLE 4 Translate to a proportion.

45% of what is 23?

$$\frac{45}{100} = \frac{23}{b}$$

Percents | Quantities
0% — 0
45% — 23
100% — b

Do Exercises 4 and 5.

EXAMPLE 5 Translate to a proportion.

10 is what percent of 20?

$$\frac{P}{100} = \frac{10}{20}$$

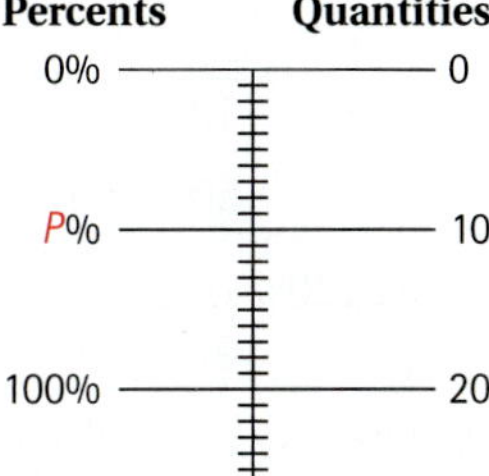

Translate to a proportion. Do not solve.

1. 12% of 50 is what?

2. What is 40% of 60?

3. 130% of 72 is what?

Translate to a proportion. Do not solve.

4. 45 is 20% of what?

5. 120% of what is 60?

Answers on page A-10

Translate to a proportion. Do not solve.

6. 16 is what percent of 40?

7. What percent of 84 is 10.5?

EXAMPLE 6 Translate to a proportion.

What percent of 50 is 7?

$$\frac{P}{100} = \frac{7}{50}$$

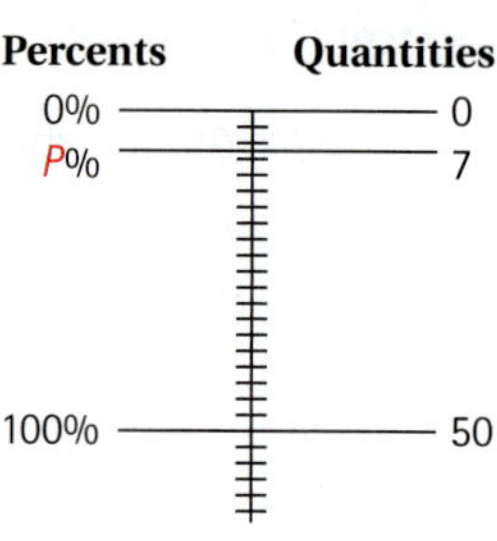

Do Exercises 6 and 7.

b Solving Percent Problems

After a percent problem has been translated to a proportion, we solve as in Section 4.1.

8. Solve:

20% of what is $45?

EXAMPLE 7 5% of what is $20?

Translate: $\frac{5}{100} = \frac{20}{b}$

Solve: $5 \cdot b = 100 \cdot 20$ Equating cross products

$\frac{5 \cdot b}{5} = \frac{100 \cdot 20}{5}$ Dividing by 5

$b = \frac{2000}{5}$

$b = 400$ Simplifying

Thus, 5% of $400 is $20. The answer is $400.

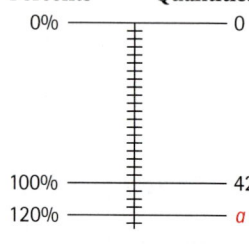

Do Exercise 8.

Solve.

9. 64% of 55 is what?

10. What is 12% of 50?

EXAMPLE 8 120% of 42 is what?

Translate: $\frac{120}{100} = \frac{a}{42}$

Solve: $120 \cdot 42 = 100 \cdot a$ Equating cross products

$\frac{120 \cdot 42}{100} = \frac{100 \cdot a}{100}$ Dividing by 100

$\frac{5040}{100} = a$

$50.4 = a$ Simplifying

Thus, 120% of 42 is 50.4. The answer is 50.4.

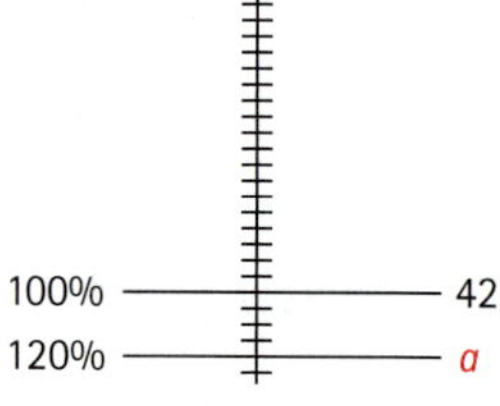

Do Exercises 9 and 10.

Answers on page A-10

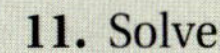

EXAMPLE 9 3 is 16% of what?

Percents	Quantities
0%	0
16%	3
100%	b

Translate: $\frac{3}{b} = \frac{16}{100}$

Solve:

$3 \cdot 100 = b \cdot 16$ — Equating cross products

$\frac{3 \cdot 100}{16} = \frac{b \cdot 16}{16}$ — Dividing by 16

$\frac{300}{16} = b$ — Multiplying and simplifying

$18.75 = b$ — Dividing

Thus, 3 is 16% of 18.75. The answer is 18.75.

Do Exercise 11.

EXAMPLE 10 $10 is what percent of $20?

Percents	Quantities
0%	0
P%	$10
100%	$20

Translate: $\frac{10}{20} = \frac{P}{100}$

Solve:

$10 \cdot 100 = 20 \cdot P$ — Equating cross products

$\frac{10 \cdot 100}{20} = \frac{20 \cdot P}{20}$ — Dividing by 20

$\frac{1000}{20} = P$ — Multiplying and simplifying

$50 = P$ — Dividing

Thus, $10 is 50% of $20. The answer is 50%.

Do Exercise 12.

EXAMPLE 11 What percent of 50 is 16?

Percents	Quantities
0%	0
P%	16
100%	50

Translate: $\frac{P}{100} = \frac{16}{50}$

Solve:

$50 \cdot P = 100 \cdot 16$ — Equating cross products

$\frac{50 \cdot P}{50} = \frac{100 \cdot 16}{50}$ — Dividing by 50

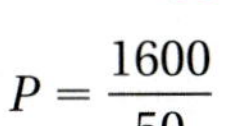

$P = \frac{1600}{50}$ — Multiplying and simplifying

$P = 32$ — Dividing

Thus, 32% of 50 is 16. The answer is 32%.

Do Exercise 13.

11. Solve:

60 is 120% of what?

12. Solve:

$12 is what percent of $40?

13. Solve:

What percent of 84 is 10.5?

Answers on page A-10

4.5 EXERCISE SET

For Extra Help

Digital Video Tutor CD 3 Videotape 5

InterAct Math

Math Tutor Center

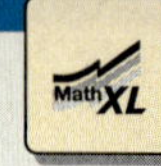

MathXL

MyMathLab

a Translate to a proportion. Do not solve.

1. What is 37% of 74?

2. 66% of 74 is what?

3. 4.3 is what percent of 5.9?

4. What percent of 6.8 is 5.3?

5. 14 is 25% of what?

6. 133% of what is 40?

b Solve.

7. What is 76% of 90?

8. What is 32% of 70?

9. 70% of 660 is what?

10. 80% of 920 is what?

11. What is 4% of 1000?

12. What is 6% of 2000?

13. 4.8% of 60 is what?

14. 63.1% of 80 is what?

15. \$24 is what percent of \$96?

16. \$14 is what percent of \$70?

17. 102 is what percent of 100?

18. 103 is what percent of 100?

19. What percent of \$480 is \$120?

20. What percent of \$80 is \$60?

21. What percent of 160 is 150?

22. What percent of 33 is 11?

23. \$18 is 25% of what?

24. \$75 is 20% of what?

25. 60% of what is 54?

26. 80% of what is 96?

27. 65.12 is 74% of what?

28. 63.7 is 65% of what?

29. 80% of what is 16?

30. 80% of what is 10?

31. What is $62\frac{1}{2}\%$ of 40?

32. What is $43\frac{1}{4}\%$ of 2600?

33. What is 9.4% of $8300?

34. What is 8.7% of $76,000?

35. DW In your own words, list steps that a classmate could use to solve any percent problem in this section.

36. DW In solving Example 10, a student simplifies $\frac{10}{20}$ before solving. Is this a good idea? Why or why not?

SKILL MAINTENANCE

Solve. [4.1d]

37. $\frac{x}{188} = \frac{2}{47}$

38. $\frac{15}{x} = \frac{3}{800}$

39. $\frac{4}{7} = \frac{x}{14}$

40. $\frac{612}{t} = \frac{72}{244}$

41. $\frac{5000}{t} = \frac{3000}{60}$

42. $\frac{75}{100} = \frac{n}{20}$

43. $\frac{x}{1.2} = \frac{36.2}{5.4}$

44. $\frac{y}{1\frac{1}{2}} = \frac{2\frac{3}{4}}{22}$

Solve. [2.5a]

45. A recipe for muffins calls for $\frac{1}{2}$ qt of buttermilk, $\frac{1}{3}$ qt of skim milk, and $\frac{1}{16}$ qt of oil. How many quarts of liquid ingredients does the recipe call for?

46. The Ferristown School District purchased $\frac{3}{4}$ ton (T) of clay. If the clay is to be shared equally among the district's 6 art departments, how much will each art department receive?

SYNTHESIS

Solve.

47. ▦ What is 8.85% of $12,640?
Estimate ______________
Calculate ______________

48. ▦ 78.8% of what is 9809.024?
Estimate ______________
Calculate ______________

Objectives

a Solve applied problems involving percent.

b Solve applied problems involving percent of increase or decrease.

4.6 APPLICATIONS OF PERCENT

a Applied Problems Involving Percent

Applied problems involving percent are not always stated in a manner easily translated to an equation. In such cases, it is helpful to rephrase the problem before translating. Sometimes it also helps to make a drawing.

EXAMPLE 1 *Presidential Deaths in Office.* George W. Bush was inaugurated as the 43rd President of the United States in 2001. Since Grover Cleveland was both the 22nd and the 24th presidents, there have been only 42 different presidents. Of the 42 presidents, 8 have died in office: William Henry Harrison, Zachary Taylor, Abraham Lincoln, James A. Garfield, William McKinley, Warren G. Harding, Franklin D. Roosevelt, and John F. Kennedy. What percent have died in office?

1. **Familiarize.** The question asks for a percent of the presidents who have died in office. We note that 42 is approximately 40 and 8 is $\frac{1}{5}$, or 20%, of 40, so our answer is close to 20%. We let $p =$ the percent who have died in office.

2. **Translate.** There are two ways in which we can translate this problem.

   Percent equation (see Section 4.4):

   8 is what percent of 42?

   $$8 = p \cdot 42$$

   Proportion (see Section 4.5):

   $$\frac{P}{100} = \frac{8}{42}$$

   For proportions, $P\% = p$.

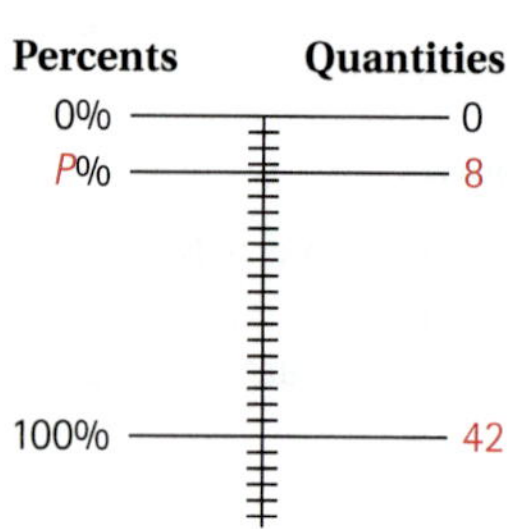

Study Tips

MAKING APPLICATIONS REAL

Newspapers and magazines are full of mathematical applications. Some of the easiest ones to find in the area of Basic Mathematics are about percent. Find such an application and share it with your class. As you obtain more skills in mathematics, you will find yourself observing the world from a different perspective, seeing mathematics everywhere. Math courses become more interesting when we connect the concepts to the real world.

3. **Solve.** We now have two ways in which to solve this problem.

Percent equation (see Section 4.4):

$$8 = p \cdot 42$$

$$\frac{8}{42} = \frac{p \cdot 42}{42} \quad \text{Dividing by 42 on both sides}$$

$$\frac{8}{42} = p$$

$$0.190 \approx p \quad \text{Finding decimal notation and rounding to the nearest thousandth}$$

$$19.0\% \approx p \quad \text{Remember to find percent notation.}$$

Note here that the solution, p, includes the % symbol.

Proportion (see Section 4.5):

$$\frac{P}{100} = \frac{8}{42}$$

$$P \cdot 42 = 100 \cdot 8 \quad \text{Equating cross products}$$

$$\frac{P \cdot 42}{42} = \frac{800}{42} \quad \text{Dividing by 42 on both sides}$$

$$P = \frac{800}{42}$$

$$P \approx 19.0 \quad \text{Dividing and rounding to the nearest tenth}$$

We use the solution of the proportion to express the answer to the problem as 19.0%. Note that in the proportion method, $P\% = p$.

4. **Check.** To check, we note that the answer 19.0% is close to 20%, as estimated in the *Familiarize* step.

5. **State.** About 19.0% of the U.S. presidents have died in office.

Do Exercise 1.

EXAMPLE 2 *Water Ingested as Beverages.* The average adult ingests 2500 milliliters (mL) of water each day, that is, about 2.5 qt. About 60% of this comes from beverages. How many milliliters of water does the average adult ingest as beverages in one day?

Source: Elaine N. Meriab, *Essentials of Anatomy & Physiology*, 6th ed. San Francisco: Benjamin/Cummings Science Publishing, 2000

1. **Presidential Assassinations in Office.** Of the 42 U.S. presidents, 4 have been assassinated in office. These were Garfield, McKinley, Lincoln, and Kennedy. What percent have been assassinated in office?

Answer on page A-10

2. Water Ingested as Food. The average adult ingests 2500 mL of water each day. About 30% of this comes from foods. How many milliliters of water does the average adult ingest as food in one day?

Source: Elaine N. Meriab, *Essentials of Anatomy & Physiology*, 6th ed. San Francisco: Benjamin/Cummings Science Publishing, 2000

what number is 30% of 2500?

A = 30% × 2500

Answer on page A-10

1. **Familiarize.** We can make a drawing of a pie chart to help familiarize ourselves with the problem. We let b = the total number of milliliters of water ingested through beverages.

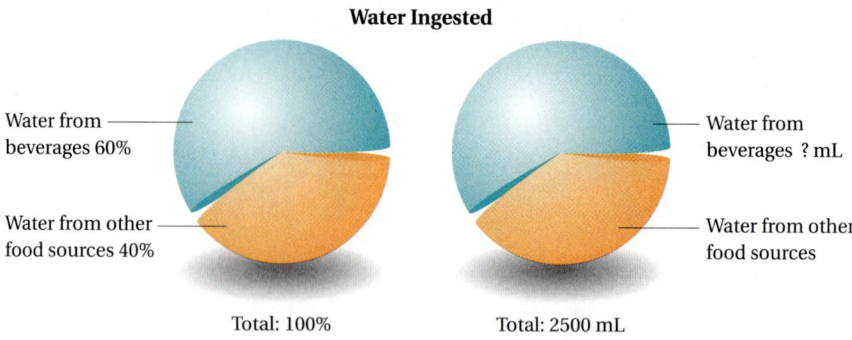

2. **Translate.** There are two ways in which we can translate this problem.

Percent equation:

What number is 60% of 2500?

$$b = 60\% \cdot 2500$$

Proportion:

$$\frac{60}{100} = \frac{b}{2500}$$

Percents	Quantities
0%	0
60%	b
100%	2500

3. **Solve.** We now have two ways in which to solve this problem.

Percent equation:

$$b = 60\% \cdot 2500$$

We convert 60% to decimal notation and multiply:

$$b = 60\% \cdot 2500 = 0.60 \times 2500 = 1500.$$

Proportion:

$$\frac{60}{100} = \frac{b}{2500}$$

$$60 \cdot 2500 = 100 \cdot b \quad \text{Equating cross products}$$

$$\frac{60 \cdot 2500}{100} = \frac{100 \cdot b}{100} \quad \text{Dividing by 100}$$

$$\frac{150{,}000}{100} = b$$

$$1500 = b \quad \text{Simplifying}$$

4. **Check.** To check, we can repeat the calculations. We can also think about our answer. Since we are taking 60% of 2500, we would expect 1500 to be smaller than 2500 and exactly three-fifths of 2500, which it is.

5. **State.** The amount of water ingested through beverages by the average adult in one day is 1500 mL.

Do Exercise 2.

b Percent of Increase or Decrease

Percent is often used to state increase or decrease. Let's consider an example of each, using the price of a car as the original number.

PERCENT OF INCREASE

One year a car sold for \$20,455. The manufacturer decides to raise the price of the following year's model by 6%. The increase is $0.06 \times \$20,455$, or \$1227.30. The new price is $\$20,455 + \$1227.30$, or \$21,682.30. The *percent of increase* is 6%.

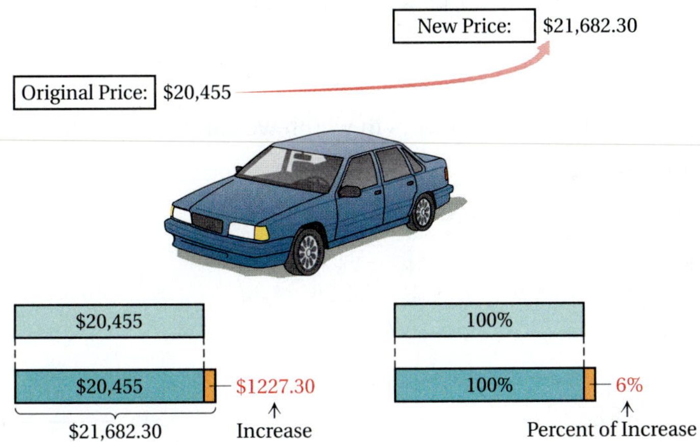

PERCENT OF DECREASE

Lisa buys the car listed above for \$20,455. After one year, the car depreciates in value by 25%. This is $0.25 \times \$20,455$, or \$5113.75. This lowers the value of the car to

$$\$20,455 - \$5113.75, \quad \text{or} \quad \$15,341.25.$$

Note that the new price is thus 75% of the original price. If Lisa decides to sell the car after a year, \$15,341.25 might be the most she could expect to get for it. The *percent of decrease* is 25%, and the decrease is \$5113.75.

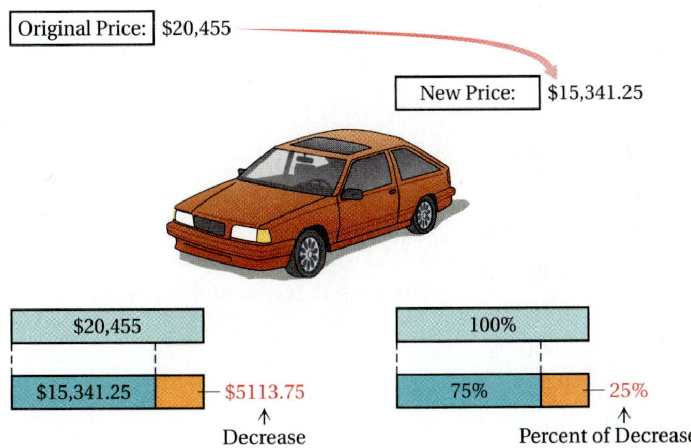

Do Exercises 3 and 4.

3. **Percent of Increase.** The value of a car is \$36,875. The price is increased by 4%.
   a) How much is the increase?
   b) What is the new price?

4. **Percent of Decrease.** The value of a car is \$36,875. The car depreciates in value by 25% after one year.
   a) How much is the decrease?
   b) What is the depreciated value of the car?

Answers on page A-10

5. Book Price. Recently Barnes & Noble sold the book *The 9 Steps to Financial Freedom* by Suze Orman on its Web site. The retail price of $13.95 was decreased to a sale price of $9.76. What was the percent of decrease?
Source: Barnes & Noble

Answer on page A-10

When a quantity is decreased by a certain percent, we say we have **percent of decrease.**

EXAMPLE 3 *DVD Price.* Barnes & Noble recently sold the DVD *When Harry Met Sally* on its Web site. The retail price of $24.98 was decreased to a sale price of $19.98. What was the percent of decrease?
Source: Barnes & Noble

1. Familiarize. We find the amount of decrease and then make a drawing.

$$\begin{array}{r} 24.98 \\ -\ 19.98 \\ \hline 5.00 \end{array} \quad \begin{array}{l} \text{Retail price} \\ \text{Sale price} \\ \text{Decrease} \end{array}$$

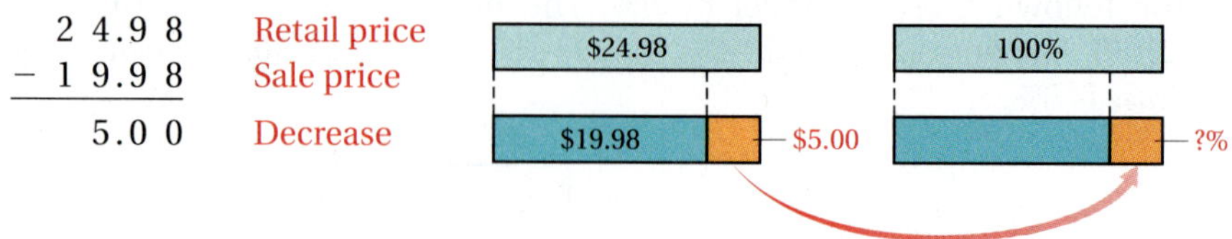

2. Translate. There are two ways in which we can translate this problem.

Percent equation:

5.00 is what percent of 24.98?

$$5.00 = p \times 24.98$$

Proportion:

$$\frac{P}{100} = \frac{5.00}{24.98}$$

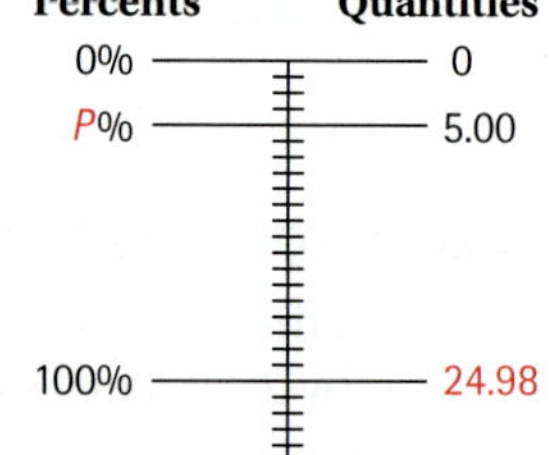

For proportions, $P\% = p$.

3. Solve. We have two ways in which to solve this problem.

Percent equation:

$$5.00 = p \times 24.98$$

$$\frac{5.00}{24.98} = \frac{p \times 24.98}{24.98} \quad \text{Dividing by 24.98 on both sides}$$

$$\frac{5.00}{24.98} = p$$

$$0.20 \approx p$$

$$20\% \approx p \quad \text{Converting to percent notation}$$

Proportion:

$$\frac{P}{100} = \frac{5.00}{24.98}$$

$$24.98 \times P = 100 \times 5 \quad \text{Equating cross products}$$

$$\frac{24.98 \times P}{24.98} = \frac{100 \times 5}{24.98} \quad \text{Dividing by 24.98 on both sides}$$

$$P = \frac{500}{24.98}$$

$$P \approx 20$$

We use the solution of the proportion to express the answer to the problem as 20%.

4. **Check.** To check, we note that, with a 20% decrease, the reduced (or sale) price should be 80% of the retail (or original) price. Since

$$80\% \times 24.98 = 0.80 \times 24.98 = 19.984 \approx 19.98,$$

our answer checks.

5. **State.** The percent of decrease in the price of the DVD was 20%.

Do Exercise 5 on the preceding page.

When a quantity is increased by a certain percent, we say we have **percent of increase.**

EXAMPLE 4 *Doctor Visits.* The average length of a doctor's visit paid for by the patient or insurance increased from 16.4 min in 1989 to 18.5 min in 1998. What was the percent of increase in the time of an average visit?

Source: *The New England Journal of Medicine*

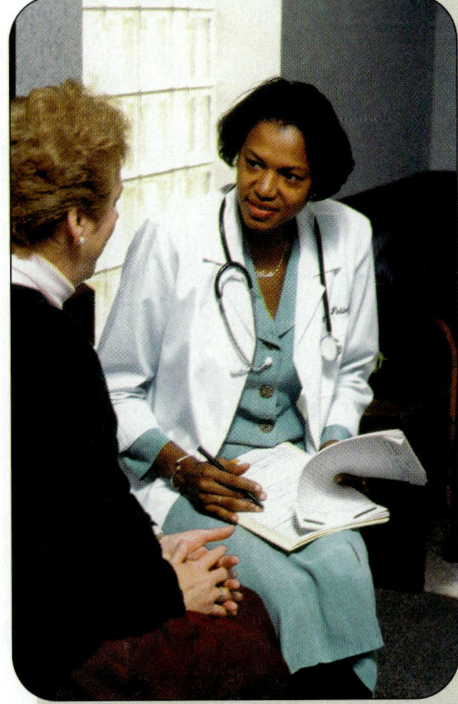

1. **Familiarize.** We note that the increase in time was 18.5 − 16.4, or 2.1 min. A drawing can help us visualize the situation. We let $p =$ the percent of increase.

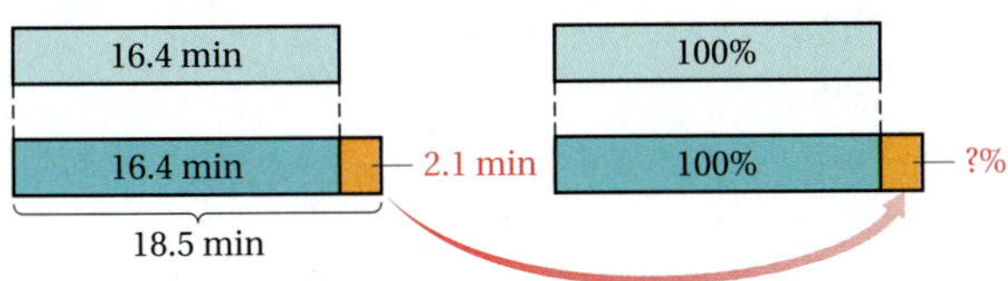

2. **Translate.** There are two ways in which we can translate this problem.

Percent equation:

2.1 min	is	what percent	of	16.4?
↓	↓	↓	↓	↓
2.1	=	p	·	16.4

Proportion:

$$\frac{P}{100} = \frac{2.1}{16.4}$$

For proportions, $P\% = p$.

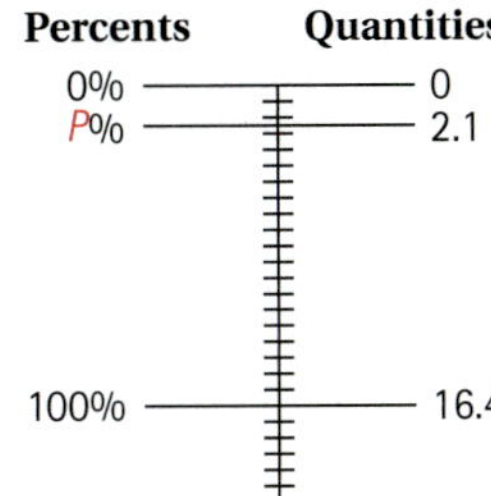

6. Doctor Visits. The average length of a visit paid for by an HMO increased from 15.4 min in 1989 to 17.9 min in 1998. What was the percent of increase in the time of an average visit?
Source: *The New England Journal of Medicine*

Answer on page A-10

Study Tips

TAPING YOUR LECTURES

Consider recording your notes and playing them back when convenient, say, while commuting to campus. It can even be advantageous to record math lectures. (Be sure to get permission from your instructor before doing so, however.) Important points can be emphasized verbally. We consider this idea so worthwhile that we provide a series of audiotapes that accompany the book. (See the Preface for more information.)

3. Solve. We have two ways in which to solve this problem.

Percent equation:

$$2.1 = p \cdot 16.4$$

$$\frac{2.1}{16.4} = \frac{p \times 16.4}{16.4} \qquad \text{Dividing by 16.4 on both sides}$$

$$\frac{2.1}{16.4} = p$$

$$0.128 \approx p$$

$$12.8\% \approx p \qquad \text{Converting to percent notation}$$

Proportion:

$$\frac{P}{100} = \frac{2.1}{16.4}$$

$$16.4 \times P = 100 \times 2.1 \qquad \text{Equating cross products}$$

$$\frac{16.4 \times P}{16.4} = \frac{100 \times 2.1}{16.4} \qquad \text{Dividing by 16.4 on both sides}$$

$$P = \frac{210}{16.4}$$

$$P \approx 12.8$$

We use the solution of the proportion to express the answer to the problem as 12.8%.

4. Check. To check, we take 12.8% of 16.4:

$$12.8\% \times 16.4 = 0.128 \times 16.4 = 2.0992.$$

Since we rounded the percent, this approximation is close enough to 2.1 to be a good check.

5. State. The percent of increase in the average length of a visit to a doctor was 12.8%.

Do Exercise 6.

4.6 EXERCISE SET

For Extra Help

Digital Video Tutor CD 3 Videotape 5 | InterAct Math | Math Tutor Center | MathXL | MyMathLab

a Solve.

1. *Panda Survival.* Breeding the much-loved panda bear in captivity has been quite difficult for zookeepers.
   a) From 1964 to 1997, of 133 panda cubs born in captivity, only 90 lived to be one month old. What percent lived to be one month old?
   b) In 1999, Mark Edwards of the San Diego Zoo developed a nutritional formula on which 18 of 20 newborns lived to be one month old. What percent lived to be one month old?

2. *Batting Averages.* Nomar Garciaparra of the Boston Red Sox won the 2000 American League baseball batting title with 196 hits in 586 at-bats. What percent of his at-bats were hits?
   Source: Major League Baseball

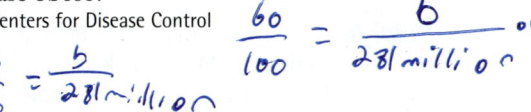

3. *Pass Completions.* Trent Dilfer, quarterback of the Baltimore Ravens, completed 59.3% of his passes in the 2000 NFL season. He attempted 226 passes. How many did he complete?
   Source: National Football League

4. *Pass Completions.* Kerry Collins, quarterback of the New York Giants, completed 58.8% of his passes in the 2000 NFL season. He attempted 529 passes. How many did he complete?
   Source: National Football League

5. *Overweight and Obese.* Of the 281 million people in the United States, 60% are considered overweight and 25% are considered obese. How many are overweight? How many are obese?
   Source: U.S. Centers for Disease Control

6. *Smoking and Diabetes.* Of the 281 million people in the United States, 25% are smokers and 6.5% have diabetes. How many are smokers? How many have diabetes?
   Source: U.S. Centers for Disease Control

7. A lab technician has 680 mL of a solution of water and acid; 3% is acid. How many milliliters are acid? water?

8. A lab technician has 540 mL of a solution of alcohol and water; 8% is alcohol. How many milliliters are alcohol? water?

9. *Field Goals.* At one point in the 2000–2001 NBA season, Vince Carter of the Toronto Raptors had successfully completed 45.2% of his field goals. He made 288 field goals. How many did he attempt?
Source: National Basketball Association

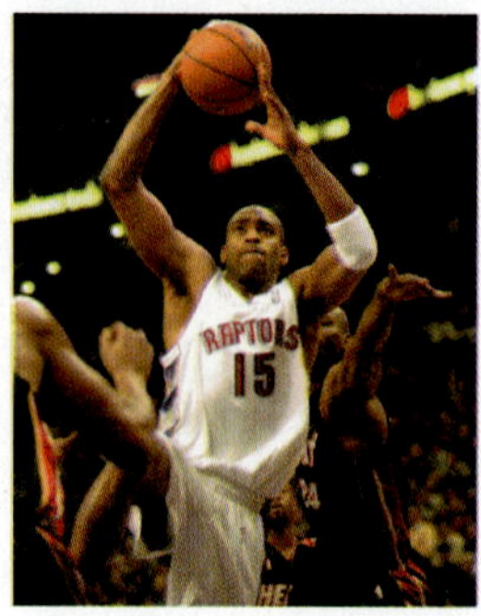

10. *Field Goals.* At one point in the 2000–2001 NBA season, Glenn Robinson of the Milwaukee Bucks had completed 45.4% of his field goals. He made 269 field goals. How many did he attempt?
Source: National Basketball Association

11. *Test Results.* On a test of 80 items, Antonio got 76 correct. What percent were correct? incorrect?

12. *Test Results.* On a test of 40 items, Cole got 33 correct. What percent were correct? incorrect?

13. *Test Results.* On a test of 40 items, Christina got 91% correct. (There was partial credit on some items.) How many items did she get correct? incorrect?

14. *Test Results.* On a test of 80 items, Pedro got 93% correct. (There was partial credit on some items.) How many items did he get correct? incorrect?

15. *Test Results.* On a test, Maj Ling got 86%, or 81.7, of the items correct. (There was partial credit on some items.) How many items were on the test?

16. *Test Results.* On a test, Juan got 85%, or 119, of the items correct. How many items were on the test?

17. *TV Usage.* Of the 8760 hr in a year, most television sets are on for 2190 hr. What percent is this?

18. *Colds from Kissing.* In a medical study, it was determined that if 800 people kiss someone who has a cold, only 56 will actually catch a cold. What percent is this?
Source: U.S. Centers for Disease Control

19. *Maximum Heart Rate.* Treadmill tests are often administered to diagnose heart ailments. A guideline in such a test is to try to get you to reach your *maximum heart rate,* in beats per minute. The maximum heart rate is found by subtracting your age from 220 and then multiplying by 85%. What is the maximum heart rate of someone whose age is 25? 36? 48? 55? 76? Round to the nearest one.

20. It costs an oil company $40,000 a day to operate two refineries. Refinery A accounts for 37.5% of the cost, and refinery B for the rest of the cost.

a) What percent of the cost does it take to run refinery B?

b) What is the cost of operating refinery A? refinery B?

b Solve.

21. *Savings Increase.* The amount in a savings account increased from $200 to $216. What was the percent of increase?

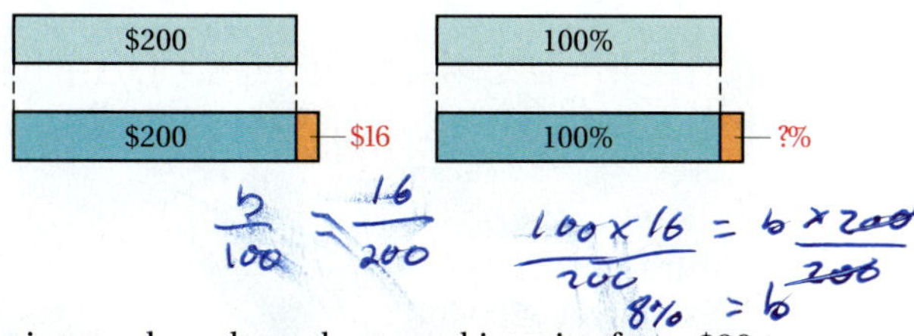

22. *Population Increase.* The population of a small mountain town increased from 840 to 882. What was the percent of increase?

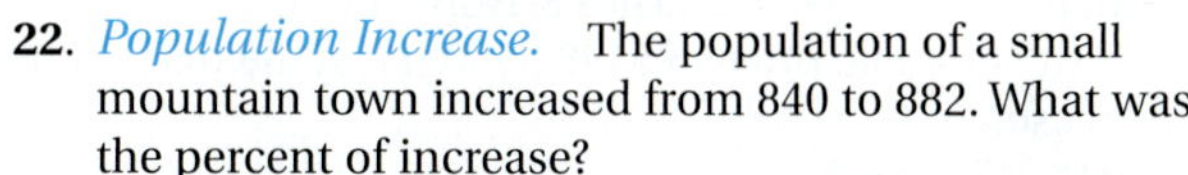

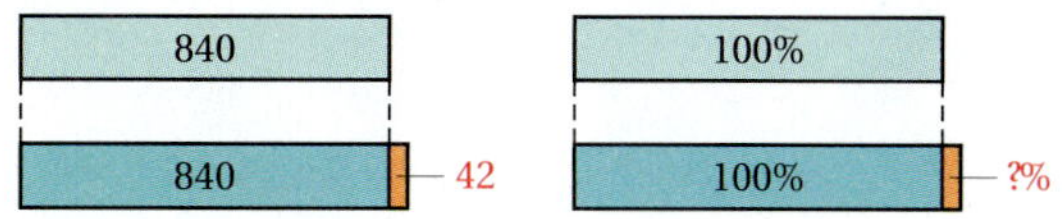

23. During a sale, a dress decreased in price from $90 to $72. What was the percent of decrease?

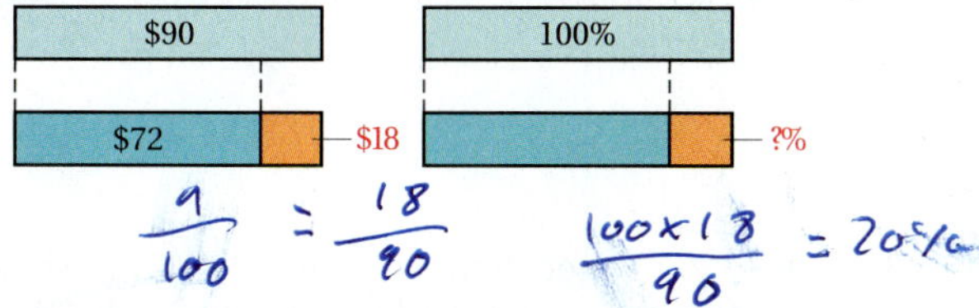

24. A person on a diet goes from a weight of 125 lb to a weight of 110 lb. What is the percent of decrease?

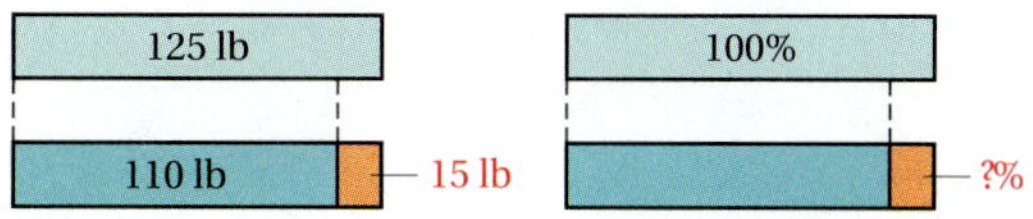

25. *Population Increase.* The population of the state of Colorado increased from 3,294,394 in 1990 to 4,301,261 in 2000. What is the percent of increase?
Source: U.S. Bureau of the Census

26. *Population Increase.* The population of the state of Utah increased from 1,722,850 in 1990 to 2,233,169 in 2000. What is the percent of increase?
Source: U.S. Bureau of the Census

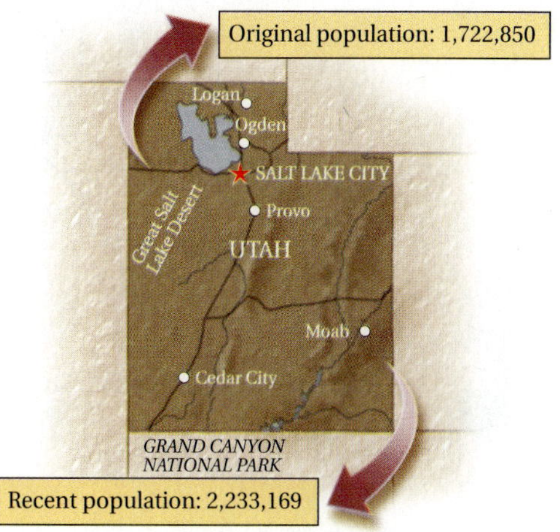

27. A person earns $28,600 one year and receives a 5% raise in salary. What is the new salary?

28. A person earns $20,400 one year and receives an 8% raise in salary. What is the new salary?

29. *Car Depreciation.* Irwin buys a car for $21,566. It depreciates 25% each year that he owns it. What is the depreciated value of the car after 1 yr? after 2 yr?

30. *Car Depreciation.* Janice buys a car for $22,688. It depreciates 25% each year that she owns it. What is the depreciated value of the car after 1 yr? after 2 yr?

31. *DVD Price.* The set of DVDs *Ken Burns: Jazz* has a retail price of $199.98. Barnes & Noble offers it on its Web site at a sale price of $149.99. What is the percent of decrease?
Source: Barnes & Noble

32. *Portable DVD Player.* A Sharp Portable DVD video player has a retail price of $1,499.95. Amazon.com offers it on its Web site at a sale price of $899.88. What is the percent of decrease?
Sources: Sharp Electronics Corporation; Amazon.com

33. *Two-by-Four.* A cross-section of a standard or nominal "two-by-four" board actually measures $1\frac{1}{2}$ in. by $3\frac{1}{2}$ in. The rough board is 2 in. by 4 in. but is planed and dried to the finished size. What percent of the wood is removed in planing and drying?

34. *Tipping.* Diners frequently add a 15% tip when charging a meal to a credit card. What is the total amount charged if the cost of the meal, without tip, is $18? $34? $49?

35. *Population Decrease.* Between 1990 and 2000, the population of Washington, D. C., decreased from 606,900 to 572,059.

a) What is the percent of decrease?

b) If this percent of decrease repeated itself in the following decade, what would the population be in 2010?

Source: U.S. Bureau of the Census

36. *World Population.* World population is increasing by 1.6% each year. In 2000, it was 6.26 billion. How much will it be in 2003? 2005? 2008?

Life Insurance Rates for Smokers and Nonsmokers. The following table provides data showing how yearly rates (premiums) for a $500,000 term life insurance policy are increased for smokers. Complete the missing numbers in the table.

TYPICAL INSURANCE PREMIUMS (DOLLARS)

	AGE	RATE FOR NONSMOKER	RATE FOR SMOKER	PERCENT INCREASE FOR SMOKER
	35	$ 345	$ 630	83%
37.	40	$ 430	$ 735	
38.	45	$ 565		84%
39.	50	$ 780		100%
40.	55	$ 985		117%
41.	60	$1645	$2955	
42.	65	$2943	$5445	

Source: Pacific Life PL Protector Term Life Portfolio, OYT Rates

Population Increase. The following table provides data showing how the populations of various states increased from 1990 to 2000. Complete the missing numbers in the table.

	STATE	POPULATION IN 1990	POPULATION IN 2000	CHANGE	PERCENT CHANGE
43.	Vermont	562,758	608,827		
44.	Virginia	6,187,358	7,078,515		
45.	Washington	4,866,692	5,894,121		
46.	West Virginia	1,793,477		14,867	
47.	Wisconsin		5,363,675	471,906	
48.	Wyoming	453,588		40,194	

Source: U.S. Bureau of the Census

49. *Car Depreciation.* A car generally depreciates 25% of its original value in the first year. A car is worth $27,300 after the first year. What was its original cost?

50. *Car Depreciation.* Given normal use, an American-made car will depreciate 25% of its original cost the first year and 14% of its remaining value in the second year. What is the value of a car at the end of the second year if its original cost was $36,400? $28,400? $26,800?

51. *Strike Zone.* In baseball, the *strike zone* is normally a 17-in. by 30-in. rectangle. Some batters give the pitcher an advantage by swinging at pitches thrown out of the strike zone. By what percent is the area of the strike zone increased if a 2-in. border is added to the outside?
Source: Major League Baseball

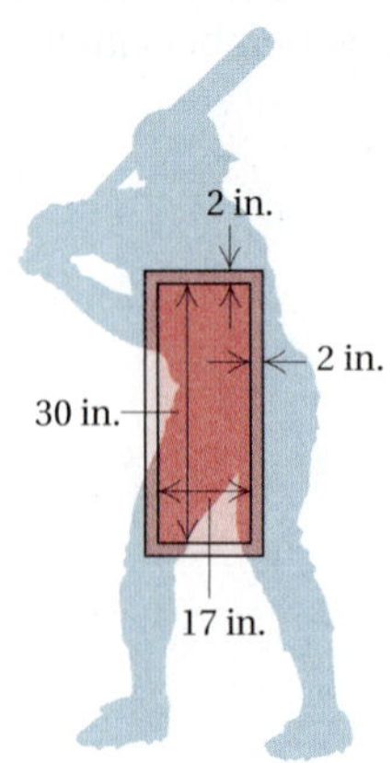

52. Tony is planting grass on a 24-ft by 36-ft area in his back yard. He installs a 6-ft by 8-ft garden. By what percent has he reduced the area he has to mow?

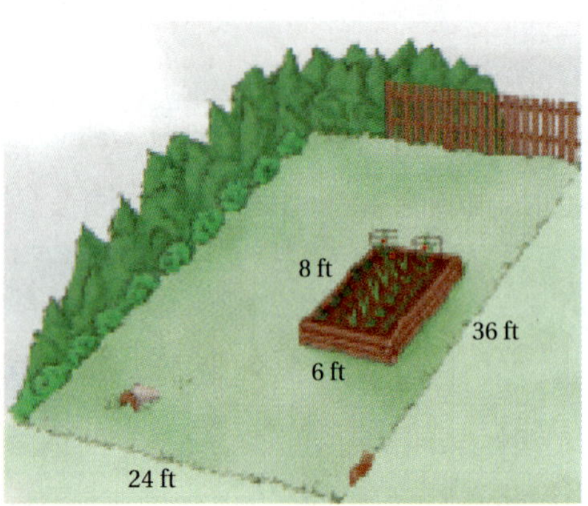

53. **D_W** Which is better for a wage earner, and why: a 10% raise followed by a 5% raise a year later, or a 5% raise followed by a 10% raise a year later?

54. **D_W** A worker receives raises of 3%, 6%, and then 9%. By what percent has the original salary increased? Explain.

SKILL MAINTENANCE

Convert to decimal notation. [3.1b], [3.5a]

55. $\frac{25}{11}$

56. $\frac{11}{25}$

57. $\frac{27}{8}$

58. $\frac{43}{9}$

59. $\frac{23}{25}$

60. $\frac{20}{24}$

61. $\frac{14}{32}$

62. $\frac{2317}{1000}$

63. $\frac{34,809}{10,000}$

64. $\frac{27}{40}$

SYNTHESIS

65. *Adult Height.* It has been determined that at the age of 10, a girl has reached 84.4% of her final adult growth. Cynthia is 4 ft, 8 in. at the age of 10. What will be her final adult height?
Source: *Dunlop Illustrated Encyclopedia of Facts.* New York: Sterling Publishing, 1970.

66. *Adult Height.* It has been determined that at the age of 15, a boy has reached 96.1% of his final adult height. Claude is 6 ft, 4 in. at the age of 15. What will be his final adult height?
Source: *Dunlop Illustrated Encyclopedia of Facts.* New York: Sterling Publishing, 1970.

67. If p is 120% of q, then q is what percent of p?

68. A coupon allows a couple to have dinner and then have $10 subtracted from the bill. Before subtracting $10, however, the restaurant adds a tip of 15%. If the couple is presented with a bill for $44.05, how much would the dinner (without tip) have cost without the coupon?

4.7 SALES TAX, COMMISSION, DISCOUNT, AND INTEREST

Objectives

a Solve applied problems involving sales tax and percent.

b Solve applied problems involving commission and percent.

c Solve applied problems involving discount and percent.

d Solve applied problems involving simple interest.

e Solve applied problems involving compound interest.

a Sales Tax

Sales tax computations represent a special type of percent of increase problem. The sales tax rate in Maryland is 5%. This means that the tax is 5% of the purchase price. Suppose the purchase price on a coat is $124.95. The sales tax is then 5% of $124.95, or 0.05×124.95, or 6.2475, or about $6.25.

The total that you pay is the price plus the sales tax:

$124.95 + $6.25, or $131.20.

BILL:		
Purchase price	=	$124.95
Sales tax (5% of $124.95)	=	+ 6.25
Total price		$131.20

SALES TAX

Sales tax = Sales tax rate × Purchase price
Total price = Purchase price + Sales tax

EXAMPLE 1 *Connecticut Sales Tax.* The sales tax rate in Connecticut is 6%. How much tax is charged on the purchase of 3 copies of the DVD *The Matrix* at $19.98 each? What is the total price?

a) We first find the cost of the DVDs. It is

$3 \times \$19.98 = \$59.94$.

b) The sales tax on items costing $59.94 is

Sales tax rate × Purchase price
↓ ↓
6% × $59.94

or 0.06×59.94, or 3.5964. Thus the tax is $3.60 (rounded to the nearest cent).

c) The total price is given by the purchase price plus the sales tax:

$59.94 + $3.60, or $63.54.

To check, note that the total price is the purchase price plus 6% of the purchase price. Thus the total price is 106% of the purchase price. Since $1.06 \times 59.94 \approx 63.54$, we have a check. The sales tax is $3.60 and the total price is $63.54.

Do Exercises 1 and 2.

1. **California Sales Tax.** The sales tax rate in California is 8%. How much tax is charged on the purchase of a refrigerator that sells for $668.95? What is the total price?

2. **Illinois Sales Tax.** Maggie buys 5 hardcover copies of Dean Koontz's novel *From the Corner of His Eye* for $26.95 each. The sales tax rate in Illinois is 7%. How much sales tax will be charged? What is the total price?

Answers on page A-10

3. The sales tax is $50.94 on the purchase of a night table that costs $849. What is the sales tax?

4. The sales tax on a television is $25.20 and the sales tax rate is 6%. Find the purchase price (the price before taxes are added).

Answers on page A-10

EXAMPLE 2 The sales tax is $83.96 on the purchase of this lingerie chest, which costs $2099. What is the sales tax rate?

Rephrase: Sales tax is what percent of purchase price?

Translate: $83.96 = r × $2099

To solve the equation, we divide by 2099 on both sides:

$$\frac{83.96}{2099} = \frac{r \times 2099}{2099}$$

$$\frac{83.96}{2099} = r$$

$$0.04 = r$$

$$4\% = r.$$

The sales tax rate is 4%.

Do Exercise 3.

EXAMPLE 3 The sales tax on an inkjet printer is $17.19 and the sales tax rate is 5%. Find the purchase price (the price before taxes are added).

Rephrase: Sales tax is 5% of what?

Translate: 17.19 = 5% × b, or $17.19 = 0.05 \times b$.

To solve, we divide by 0.05 on both sides:

$$\frac{17.19}{0.05} = \frac{0.05 \times b}{0.05}$$

$$\frac{17.19}{0.05} = b$$

$$343.80 = b.$$

The purchase price is $343.80.

Do Exercise 4.

b Commission

When you work for a **salary,** you receive the same amount of money each week or month. When you work for a **commission,** you are paid a percentage of the total sales for which you are responsible.

COMMISSION

Commission = Commission rate × Sales

EXAMPLE 4 *Stereo Equipment Sales.* A salesperson's commission rate is 20%. What is the commission from the sale of $25,560 worth of stereophonic equipment?

Commission	=	*Commission rate*	×	*Sales*
C	=	20%	×	25,560
C	=	0.20	×	25,560
C	=	5112		

The commission is $5112.

Do Exercise 5.

EXAMPLE 5 *Farm Machinery Sales.* Dawn earns a commission of $30,000 selling $600,000 worth of farm machinery. What is the commission rate?

Commission	=	*Commission rate*	×	*Sales*
30,000	=	r	×	600,000

5. Raul's commission rate is 30%. What is the commission from the sale of $18,760 worth of air conditioners?

Answer on page A-10

6. Liz earns a commission of $3000 selling $24,000 worth of *NSYNC concert tickets. What is the commission rate?

To solve this equation, we divide by 600,000 on both sides:

$$\frac{30{,}000}{600{,}000} = \frac{r \times 600{,}000}{600{,}000}$$

$$\frac{1}{20} = r$$

$$0.05 = r$$

$$5\% = r.$$

The commission rate is 5%.

Do Exercise 6.

EXAMPLE 6 *Motorcycle Sales.* Joyce's commission rate is 25%. She receives a commission of $425 on the sale of a motorcycle. How much did the motorcycle cost?

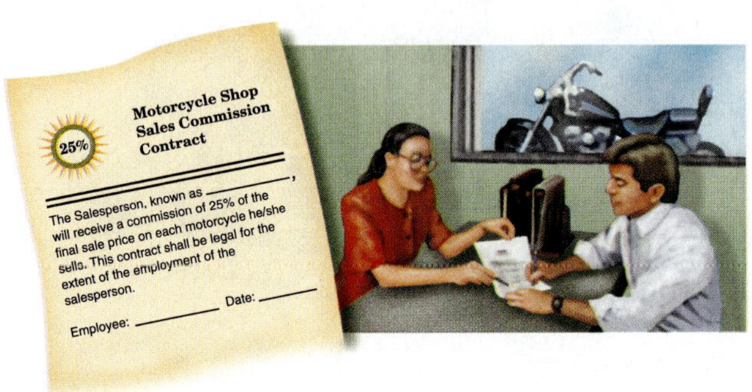

7. Ben's commission rate is 16%. He receives a commission of $268 from sales of clothing. How many dollars worth of clothing were sold?

$$\textit{Commission} = \textit{Commission rate} \times \textit{Sales}$$

$$425 = 25\% \times S, \quad \text{or } 425 = 0.25 \times S$$

To solve this equation, we divide by 0.25 on both sides:

$$\frac{425}{0.25} = \frac{0.25 \times S}{0.25}$$

$$\frac{4.25}{0.25} = S$$

$$1700 = S.$$

The motorcycle cost $1700.

Do Exercise 7.

C Discount

Suppose that the regular price of a rug is $60, and the rug is on sale at 25% off. Since 25% of $60 is $15, the sale price is $60 − $15, or $45. We call $60 the **original,** or **marked price,** 25% the **rate of discount,** $15 the **discount,** and $45 the **sale price.** Note that discount problems are a type of percent of decrease problem.

DISCOUNT AND SALE PRICE

Discount = Rate of discount × Original price
Sale price = Original price − Discount

Answers on page A-10

EXAMPLE 7 A rug marked \$240 is on sale at $33\frac{1}{3}\%$ off. What is the discount? the sale price?

a) *Discount* = *Rate of discount* × *Original price*

$$D = 33\frac{1}{3}\% \times 240$$

$$D = \frac{1}{3} \times 240$$

$$D = \frac{240}{3} = 80$$

b) *Sale price* = *Original price* − *Discount*

$$S = 240 - 80$$

$$S = 160$$

The discount is \$80 and the sale price is \$160.

Do Exercise 8.

EXAMPLE 8 *Antique Pricing.* An antique table is marked down from \$620 to \$527. What is the rate of discount?

We first find the discount by subtracting the sale price from the original price:

$$620 - 527 = 93.$$

The discount is \$93.

Next, we use the equation for discount:

Discount = *Rate of discount* × *Original price*

$$93 = r \times 620.$$

To solve, we divide by 620 on both sides:

$$\frac{93}{620} = \frac{r \times 620}{620}$$

$$\frac{93}{620} = r$$

$$0.15 = r$$

$$15\% = r.$$

The discount rate is 15%.

To check, note that a 15% discount rate means that 85% of the original price is paid:
$0.85 \times 620 = 527.$

Do Exercise 9.

8. A suit marked \$540 is on sale at $33\frac{1}{3}\%$ off. What is the discount? the sale price?

9. A pair of hiking boots is reduced from \$75 to \$60. Find the rate of discount.

Answers on page A-10

10. What is the simple interest on \$4300 invested at an interest rate of 14% for 1 year?

11. What is the simple interest on a principal of \$4300 invested at an interest rate of 14% for $\frac{3}{4}$ year?

Answers on page A-10

d Simple Interest

Suppose you put \$1000 into an investment for 1 year. The \$1000 is called the **principal.** If the **interest rate** is 8%, in addition to the principal, you get back 8% of the principal, which is

8% of \$1000, or 0.08×1000, or \$80.00.

The \$80.00 is called the **simple interest.** It is, in effect, the price that a financial institution pays for the use of the money over time.

SIMPLE INTEREST FORMULA

The **simple interest *I*** on principal P, invested for t years at interest rate r, is given by

$$I = P \cdot r \cdot t.$$

EXAMPLE 9 What is the simple interest on \$2500 invested at an interest rate of 6% for 1 year?

We use the formula $I = P \cdot r \cdot t$:

$$\begin{aligned} I = P \cdot r \cdot t &= \$2500 \times 6\% \times 1 \\ &= \$2500 \times 0.06 \\ &= \$150. \end{aligned}$$

The simple interest for 1 year is \$150.

Do Exercise 10.

EXAMPLE 10 What is the simple interest on a principal of \$2500 invested at an interest rate of 6% for $\frac{1}{4}$ year?

We use the formula $I = P \cdot r \cdot t$:

$$\begin{aligned} I = P \cdot r \cdot t &= \$2500 \times 6\% \times \frac{1}{4} \\ &= \frac{\$2500 \times 0.06}{4} \\ &= \$37.50. \end{aligned}$$

We could instead have found $\frac{1}{4}$ of 6% and then multiplied by 2500.

The simple interest for $\frac{1}{4}$ year is \$37.50.

Do Exercise 11.

When time is given in days, we generally divide it by 365 to express the time as a fractional part of a year.

EXAMPLE 11 To pay for a shipment of tee shirts, New Wave Designs borrows \$8000 at $9\frac{3}{4}\%$ for 60 days. Find (a) the amount of simple interest that is due and (b) the total amount that must be paid after 60 days.

a) We express 60 days as a fractional part of a year:

$$I = P \cdot r \cdot t = \$8000 \times 9\frac{3}{4}\% \times \frac{60}{365}$$

$$= \$8000 \times 0.0975 \times \frac{60}{365}$$

$$\approx \$128.22.$$

The interest due for 60 days is \$128.22.

b) The total amount to be paid after 60 days is the principal plus the interest:

$$\$8000 + \$128.22 = \$8128.22$$

The total amount due is \$8128.22.

Do Exercise 12.

12. The Glass Nook borrows \$4800 at $9\frac{1}{2}\%$ for 30 days. Find (a) the amount of simple interest due and (b) the total amount that must be paid after 30 days.

e Compound Interest

When interest is paid *on interest,* we call it **compound interest.** This is the type of interest usually paid on investments. Suppose you have \$5000 in a savings account at 6%. In 1 year, the account will contain the original \$5000 plus 6% of \$5000. Thus the total in the account after 1 year will be

106% of \$5000, or 1.06 × \$5000, or \$5300.

Now suppose that the total of \$5300 remains in the account for another year. At the end of this second year, the account will contain the \$5300 plus 6% of \$5300. The total in the account would thus be

106% of \$5300, or 1.06 × \$5300, or \$5618.

Note that in the second year, interest is earned on the first year's interest. When this happens, we say that interest is **compounded annually.**

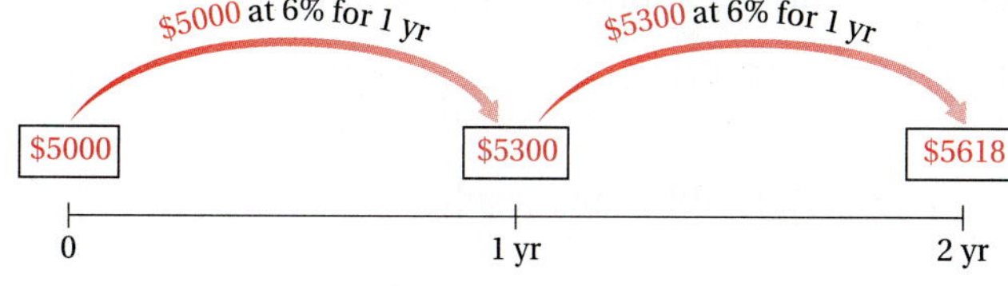

Answer on page A-10

13. Find the amount in an account if \$2000 is invested at 11%, compounded annually, for 2 years.

Answer on page A-10

EXAMPLE 12 Find the amount in an account if \$2000 is invested at 8%, compounded annually, for 2 years.

a) After 1 year, the account will contain 108% of \$2000:

$$1.08 \times \$2000 = \$2160.$$

b) At the end of the second year, the account will contain 108% of \$2160:

$$1.08 \times \$2160 = \$2332.80.$$

The amount in the account after 2 years is \$2332.80.

Do Exercise 13.

Suppose that the interest in Example 12 were **compounded semiannually**—that is, every half year. Interest would then be calculated twice a year at a rate of 8% ÷ 2, or 4% each time. The approach used in Example 4 can then be adapted, as follows.

After the first $\frac{1}{2}$ year, the account will contain 104% of \$2000:

$$1.04 \times \$2000 = \$2080.$$

After a second $\frac{1}{2}$ year (1 full year), the account will contain 104% of \$2080:

$$1.04 \times \$2080 = \$2163.20.$$

After a third $\frac{1}{2}$ year $\left(1\frac{1}{2}\text{ full years}\right)$, the account will contain 104% of \$2163.20:

$$1.04 \times \$2163.20 = \$2249.728$$
$$\approx \$2249.73. \quad \text{Rounding to the nearest cent}$$

Finally, after a fourth $\frac{1}{2}$ year (2 full years), the account will contain 104% of \$2249.73:

$$1.04 \times \$2249.73 = \$2339.7192$$
$$\approx \$2339.72. \quad \text{Rounding to the nearest cent}$$

Let's summarize our results and look at them another way:

End of 1st $\frac{1}{2}$ year $\rightarrow 1.04 \times 2000 = 2000 \times (1.04)^1$;
End of 2nd $\frac{1}{2}$ year $\rightarrow 1.04 \times (1.04 \times 2000) = 2000 \times (1.04)^2$;
End of 3rd $\frac{1}{2}$ year $\rightarrow 1.04 \times (1.04 \times 1.04 \times 2000) = 2000 \times (1.04)^3$;
End of 4th $\frac{1}{2}$ year $\rightarrow 1.04 \times (1.04 \times 1.04 \times 1.04 \times 2000) = 2000 \times (1.04)^4$.

Note that each multiplication was by 1.04 and that

$$\$2000 \times 1.04^4 \approx \$2339.72. \quad \text{Using a calculator and rounding to the nearest cent}$$

We have illustrated the following result.

COMPOUND INTEREST FORMULA

If a principal P has been invested at interest rate r, compounded n times a year, in t years it will grow to an amount A given by

$$A = P \cdot \left(1 + \frac{r}{n}\right)^{n \cdot t}.$$

Let's apply this formula to confirm our preceding discussion, where the amount invested is $P = \$2000$, the number of years is $t = 2$, and the number of compounding periods each year is $n = 2$. Substituting into the compound interest formula, we have

$$A = P \cdot \left(1 + \frac{r}{n}\right)^{n \cdot t} = 2000 \cdot \left(1 + \frac{8\%}{2}\right)^{2 \cdot 2}$$
$$= 2000 \cdot \left(1 + \frac{0.08}{2}\right)^4 = 2000(1.04)^4$$
$$= 2000 \times 1.16985856 \approx \$2339.72.$$

If you are using a calculator, you could perform this computation in one step.

EXAMPLE 13 The Ibsens invest \$4000 in an account paying $8\frac{5}{8}\%$, compounded quarterly. Find the amount in the account after $2\frac{1}{2}$ years.

The compounding is quarterly, so n is 4. We substitute \$4000 for P, $8\frac{5}{8}\%$, or 0.08625, for r, 4 for n, and $2\frac{1}{2}$, or $\frac{5}{2}$, for t and compute A:

$$A = P \cdot \left(1 + \frac{r}{n}\right)^{n \cdot t} = \$4000 \cdot \left(1 + \frac{8\frac{5}{8}\%}{4}\right)^{4 \cdot 5/2}$$
$$= \$4000 \cdot \left(1 + \frac{0.08625}{4}\right)^{10}$$
$$= \$4000(1.0215625)^{10}$$
$$\approx \$4951.19.$$

The amount in the account after $2\frac{1}{2}$ years is \$4951.19.

Do Exercise 14.

14. A couple invests \$7000 in an account paying $10\frac{3}{8}\%$, compounded semiannually. Find the amount in the account after $1\frac{1}{2}$ years.

Answer on page A-10

CALCULATOR CORNER

Compound Interest A calculator is useful in computing compound interest. Not only does it do computations quickly but it also eliminates the need to round until the computation is completed. This minimizes "round-off errors" that occur when rounding is done at each stage of the computation. We must keep order of operations in mind when computing compound interest.

To find the amount due on a \$20,000 loan made for 25 days at 11% interest, compounded daily, we would compute $20{,}000\left(1 + \frac{0.11}{365}\right)^{25}$. To do this on a calculator, we press [2] [0] [0] [0] [0] [×] [(] [1] [+] [.] [1] [1] [÷] [3] [6] [5] [)] [y^x] (or [x^y]) [2] [5] [=]. Without parentheses, we would first find $1 + \frac{0.11}{365}$, raise this result to the 25th power, and then multiply by 20,000. To do this, we press [1] [+] [.] [1] [1] [÷] [3] [6] [5] [=] [y^x] (or [x^y]) [2] [5] [=] [×] [2] [0] [0] [0] [0]. In either case, the result is 20,151.23, rounded to the nearest cent.

Some calculators have business keys that allow such computations to be done more quickly.

Exercises:

1. Find the amount due on a \$16,000 loan made for 62 days at 13% interest, compounded daily.
2. An investment of \$12,500 is made for 90 days at 8.5% interest, compounded daily. How much is the investment worth after 90 days?

4.7 EXERCISE SET

For Extra Help

Digital Video Tutor CD 4 Videotape 5

InterAct Math

Math Tutor Center

MathXL

MyMathLab

a Solve.

1. *Fort Worth, Texas, Sales Tax.* The sales tax rate in Fort Worth, Texas, is 8.25%. How much sales tax would be charged on a copy of J. K. Rowling's novel, *Harry Potter and the Goblet of Fire,* which sells for $25.95?

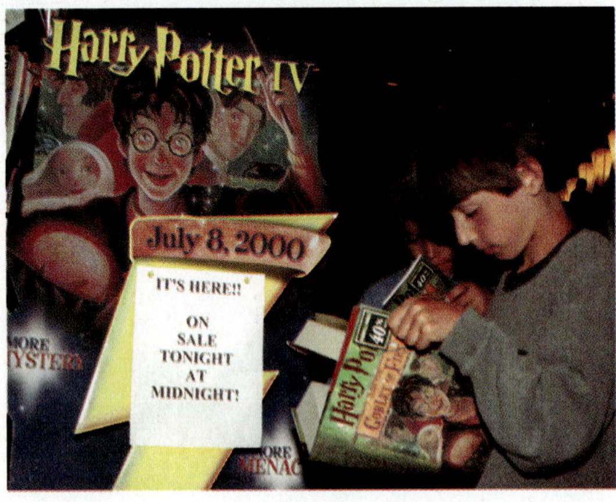

2. *Fort Worth, Texas, Sales Tax.* The sales tax rate in Fort Worth, Texas, is 8.25%. How much sales tax would be charged on a copy of John Grisham's novel *A Painted House,* which sells for $27.95?
Source: Borders Bookstore; Andrea Sutcliffe, *Numbers*

3. The sales tax is $48 on the purchase of a dining room set that sells for $960. What is the sales tax rate?

4. The sales tax is $15 on the purchase of a diamond ring that sells for $500. What is the sales tax rate?

5. *Illinois Sales Tax.* The sales tax rate in Illinois is 7%. How much tax is charged on a purchase of 5 telephones at $53 apiece? What is the total price?

6. *Kentucky Sales Tax.* The sales tax rate in Kentucky is 6%. How much tax is charged on a purchase of 5 teapots at $37.99 apiece? What is the total price?

7. The sales tax is $35.80 on the purchase of a refrigerator–freezer that sells for $895. What is the sales tax rate?

8. The sales tax is $9.12 on the purchase of a patio set that sells for $456. What is the sales tax rate?

9. The sales tax on a used car is $100 and the sales tax rate is 5%. Find the purchase price (the price before taxes are added).

10. The sales tax on the purchase of a new boat is $112 and the sales tax rate is 2%. Find the purchase price.

11. The sales tax on a dining room set is $28 and the sales tax rate is 3.5%. Find the purchase price.

12. The sales tax on a portable CD player is $66 and the sales tax rate is 5.5%. Find the purchase price.

13. The sales tax rate in Austin is 2% for the city and county and 6.25% for the state. Find the total amount paid for 2 shower units at $332.50 apiece.

14. The sales tax rate in Omaha is 1.5% for the city and 5% for the state. Find the total amount paid for 3 air conditioners at $260 apiece.

15. The sales tax is $1030.40 on an automobile purchase of $18,400. What is the sales tax rate?

16. The sales tax is $979.60 on an automobile purchase of $15,800. What is the sales tax rate?

b Solve.

17. Katrina's commission rate is 6%. What is the commission from the sale of $45,000 worth of furnaces?

18. Jose's commission rate is 32%. What is the commission from the sale of $12,500 worth of sailboards?

19. Vince earns $120 selling $2400 worth of television sets in a consignment shop. What is the commission rate?

20. Donna earns $408 selling $3400 worth of shoes. What is the commission rate?

21. An art gallery's commission rate is 40%. They receive a commission of $392. How many dollars worth of artwork were sold?

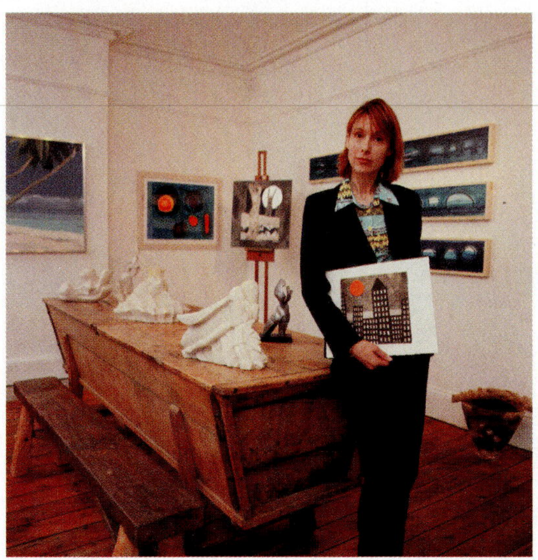

22. A real estate agent's commission rate is 7%. She receives a commission of $5600 on the sale of a home. How much did the home sell for?

23. A real estate commission is 6%. What is the commission on the sale of a $98,000 home?

24. A real estate commission is 8%. What is the commission on the sale of a piece of land for $68,000?

25. Bonnie earns $280.80 selling $2340 worth of tee shirts. What is the commission rate?

26. Chuck earns $1147.50 selling $7650 worth of ski passes. What is the commission rate?

27. Miguel's commission is increased according to how much he sells. He receives a commission of 5% for the first $2000 and 8% on the amount over $2000. What is the total commission on sales of $6000?

28. Lucinda earns a salary of $500 a month, plus a 2% commission on sales. One month, she sold $990 worth of encyclopedias. What were her wages that month?

c Solve.

29. Find the discount and the rate of discount for the ring in this ad.

30. Find the marked price and the rate of discount for the cedar chest in this ad.

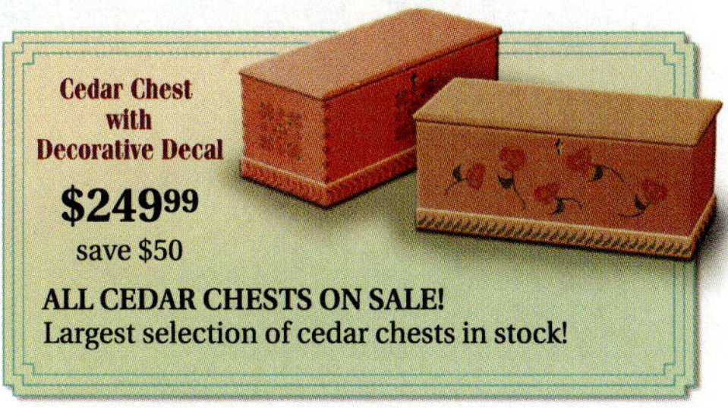

Find what is missing.

	MARKED PRICE	RATE OF DISCOUNT	DISCOUNT	SALE PRICE
31.	$300	10%		
32.	$2000	40%		
33.	$17	15%		
34.	$20	25%		
35.		10%	$12.50	
36.		15%	$65.70	
37.	$600		$240	
38.	$12,800		$1920	

d Find the simple interest.

	PRINCIPAL	RATE OF INTEREST	TIME	SIMPLE INTEREST
39.	$200	8%	1 year	
40.	$450	6%	1 year	
41.	$2,000	8.4%	$\frac{1}{2}$ year	
42.	$200	7.7%	$\frac{1}{2}$ year	
43.	$4,300	10.56%	$\frac{1}{4}$ year	
44.	$8,000	9.42%	$\frac{1}{6}$ year	
45.	$20,000	$7\frac{5}{8}$%	1 year	
46.	$100,000	$8\frac{7}{8}$%	1 year	
47.	$50,000	$5\frac{3}{8}$%	$\frac{1}{4}$ year	
48.	$80,000	$6\frac{3}{4}$%	$\frac{1}{12}$ year	

49. Animal Instinct, a pet supply shop, borrows $6500 at 8% for 90 days. Find (a) the amount of interest due and (b) the total amount that must be paid after 90 days.

50. Andante's Cafe borrows $4500 at 9% for 60 days. Find (a) the amount of interest due and (b) the total amount that must be paid after 60 days.

e Interest is compounded annually. Find the amount in the account after the given length of time. Round to the nearest cent.

	PRINCIPAL	RATE OF INTEREST	TIME	AMOUNT IN THE ACCOUNT
51.	\$400	10%	2 years	
52.	\$450	8%	2 years	
53.	\$2,000	8.8%	4 years	
54.	\$4,000	7.7%	4 years	
55.	\$4,300	10.56%	6 years	
56.	\$8,000	9.42%	6 years	
57.	\$20,000	$7\frac{5}{8}\%$	25 years	
58.	\$100,000	$8\frac{7}{8}\%$	30 years	

Interest is compounded semiannually. Find the amount in the account after the given length of time. Round to the nearest cent.

	PRINCIPAL	RATE OF INTEREST	TIME	AMOUNT IN THE ACCOUNT
59.	\$4,000	7%	1 year	
60.	\$1,000	5%	1 year	
61.	\$20,000	8.8%	4 years	
62.	\$40,000	7.7%	4 years	
63.	\$5,000	10.56%	6 years	
64.	\$8,000	9.42%	8 years	
65.	\$20,000	$7\frac{5}{8}\%$	25 years	
66.	\$100,000	$8\frac{7}{8}\%$	30 years	

Solve.

67. A family invests \$4000 in an account paying 6%, compounded monthly. How much is in the account after 5 months?

68. A couple invests \$2500 in an account paying 9%, compounded monthly. How much is in the account after 6 months?

69. ^D_W Which is a better investment and why: \$1000 invested at $14\frac{3}{4}\%$ simple interest for 1 year, or \$1000 invested at 14% compounded monthly for 1 year?

70. ^D_W A firm must choose between borrowing \$5000 at 10% for 30 days and borrowing \$10,000 at 8% for 60 days. Give arguments in favor of and against each option.

SKILL MAINTENANCE

71. Write fraction notation: 0.93. [3.1b]

72. Solve: $2.3 \times y = 85.1$. [3.4b]

73. Convert to decimal notation: $\frac{13}{11}$. [3.5a]

74. Convert to a mixed numeral: $\frac{29}{11}$. [2.4a]

4.8 INTEREST RATES ON CREDIT CARDS AND LOANS

Objective

a Solve applied problems involving interest rates on credit cards and loans.

a Credit Cards and Loans

Look at the following graphs. They offer good reason for a study of the real-world applications of percent, interest, loans, and credit cards.

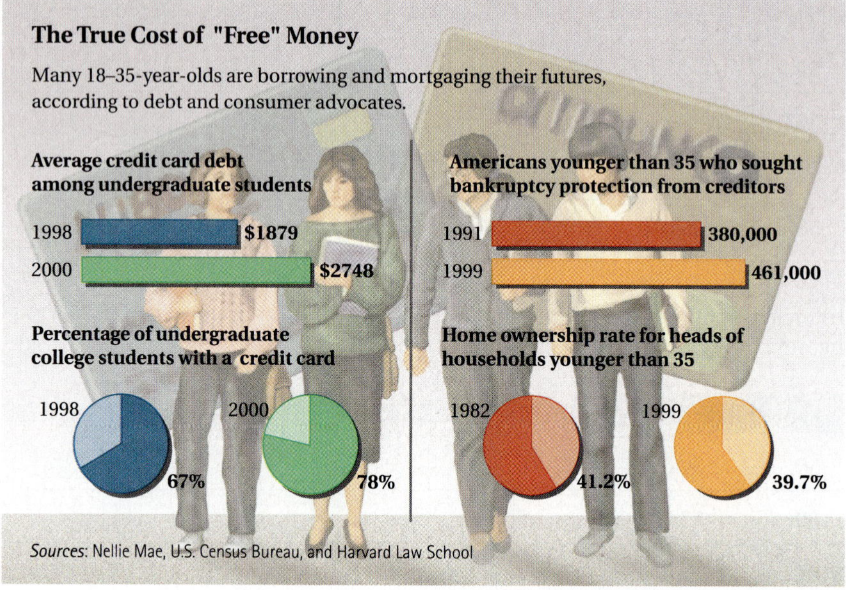

The True Cost of "Free" Money
Many 18–35-year-olds are borrowing and mortgaging their futures, according to debt and consumer advocates.
Average credit card debt among undergraduate students
1998 $1879
2000 $2748
Americans younger than 35 who sought bankruptcy protection from creditors
1991 380,000
1999 461,000
Percentage of undergraduate college students with a credit card
1998 67%
2000 78%
Home ownership rate for heads of households younger than 35
1982 41.2%
1999 39.7%
Sources: Nellie Mae, U.S. Census Bureau, and Harvard Law School

Comparing interest rates is essential if one is to become financially responsible. A small change in an interest rate can make a *large* difference in the cost of a loan. When you make a payment on a loan, do you know how much of that payment is interest and how much is applied to reducing the principal?

We begin with an example involving credit cards. A balance carried on a credit card is a type of loan. Last year in the United States, 100,000 young adults declared bankruptcy because of excessive credit card debt. The money you obtain through the use of a credit card is not "free" money. There is a price (interest) to be paid for the privilege.

EXAMPLE 1 *Credit Cards.* After the holidays, Sarah has a balance of $3216.28 on a credit card with an annual percentage rate (APR) of 19.7%. She decides to not make any additional purchases with this card until she has paid off the balance.

a) Many credit cards require a minimum monthly payment of 2% of the balance. What is Sarah's minimum payment on a balance of $3216.28? Round the answer to the nearest dollar.

b) Find the amount of interest and the amount applied to reduce the principal in the minimum payment found in part (a).

c) If Sarah had transferred her balance to a card with an APR of 12.5%, how much of her first payment would be interest and how much would be applied to reduce the principal?

d) Compare the amounts for 12.5% from part (c) with the amounts for 19.7% from part (b).

We solve as follows.

a) We multiply the balance of \$3216.28 by 2%:

$0.02 \times \$3216.28 = \$64.3256.$ Sarah's minimum payment, rounded to the nearest dollar, is \$64.

b) The amount of interest on \$3216.28 at 19.7% for one month* is given by

$$I = P \cdot r \cdot t = \$3216.28 \times 0.197 \times \frac{1}{12} \approx \$52.80.$$

We subtract to find the amount applied to reduce the principal in the first payment:

$$\begin{aligned}\text{Amount applied to reduce the principal} &= \text{Minimum payment} - \text{Interest for the month}\\ &= \$64 - \$52.80\\ &= \$11.20.\end{aligned}$$

Thus the principal of \$3216.28 is decreased by only \$11.20 with the first payment. (Sarah still owes \$3205.08.)

c) The amount of interest on \$3216.28 at 12.5% for one month is

$$I = P \cdot r \cdot t = \$3216.28 \times 0.125 \times \frac{1}{12} \approx \$33.50.$$

We subtract to find the amount applied to reduce the principal in the first payment:

$$\begin{aligned}\text{Amount applied to reduce the principal} &= \text{Minimum payment} - \text{Interest for the month}\\ &= \$64 - \$33.50\\ &= \$30.50.\end{aligned}$$

Thus the principal of \$3216.28 is decreased by \$30.50 with the first payment. (Sarah still owes \$3185.78.)

d) Let's organize the information for both rates in the following table.

BALANCE BEFORE FIRST PAYMENT	FIRST MONTH'S PAYMENT	% APR	AMOUNT OF INTEREST	AMOUNT APPLIED TO PRINCIPAL	BALANCE AFTER FIRST PAYMENT
\$3216.28	\$64	19.7%	\$52.80	\$11.20	\$3205.08
3216.28	64	12.5	33.50	30.50	3185.78

Difference in balance after first payment → \$19.30

At 19.7%, the interest is \$52.80 and the principal is decreased by \$11.20. At 12.5%, the interest is \$33.50 and the principal is decreased by \$30.50. Thus the principal is decreased by \$30.50 − \$11.20, or \$19.30 more with the 12.5% rate than with the 19.7% rate. Thus the interest at 19.7% is \$19.30 greater than the interest at 12.5%.

*Actually, the interest on a credit card is computed daily with a rate called a daily percentage rate (DPR). The DPR for Example 1 would be $19.7\%/365 \approx 0.054\%$. When no payments or additional purchases are made during the month, the difference in total interest for the month is minimal and we will not deal with it here.

1. Credit Cards. After the holidays, Jamal has a balance of \$4867.59 on a credit card with an annual percentage rate (APR) of 21.3%. He decides to not make any additional purchases with this card until he has paid off the balance.

a) Many credit cards require a minimum monthly payment of 2% of the balance. What is Jamal's minimum payment on a balance of \$4867.59? Round the answer to the nearest dollar.

b) Find the amount of interest and the amount applied to reduce the principal in the minimum payment found in part (a).

c) If Jamal had transferred his balance to a card with an APR of 13.6%, how much of his first payment would be interest and how much would be applied to reduce the principal?

d) Compare the amounts for 13.6% from part (c) with the amounts for 21.3% from part (b).

Answers on page A-11

Do Exercise 1 on the preceding page.

Even though the mathematics of the information in the chart below is beyond the scope of this text, it is interesting to compare how long it takes to pay off the balance of Example 1 if Sarah continues to pay $64 for each payment with how long it takes if she pays double that amount, $128, for each payment. Financial consultants frequently tell clients that if they want to take control of their debt, they should pay double the minimum payment.

RATE	PAYMENT PER MONTH	NUMBER OF PAYMENTS TO PAY OFF DEBT	TOTAL PAID BACK	ADDITIONAL COST OF PURCHASES
19.7%	$64	107, or 8 yr 11 mo	$6848	$3631.72
19.7	128	33, or 2 yr 9 mo	4224	1007.72
12.5	64	72, or 6 yr	4608	1391.72
12.5	128	29, or 2 yr 5 mo	3712	495.72

As with most loans, if you pay an extra amount toward the principal with each payment, the length of the loan can be greatly reduced. Note that at the rate of 19.7%, it will take Sarah almost 9 yr to pay off her debt if she pays only $64 per month and does not make additional purchases. If she transfers her balance to a card with a 12.5% rate and pays $128 per month, she could eliminate her debt in approximately $2\frac{1}{2}$ yr. You can see how debt can get out of control if you continue to make purchases and pay only the minimum payment. The debt will never be eliminated.

The Federal Stafford Loan program provides educational loans to students at interest rates (usually from 6.5% to 8.5%) that are much lower than those on credit cards. Payments on a loan do not begin until 6 months after graduation. At that time, the student has 10 years, or 120 monthly payments, to pay off the loan.

EXAMPLE 2 *Federal Stafford Loans.* After graduation, the balance on Taylor's Stafford loan is $28,650. If the rate on his loan is 7%, he will make 120 payments of approximately $333 each to pay off the loan.

a) Find the amount of interest and the amount of principal in the first payment.

b) If the interest rate were 8.25%, he would make 120 monthly payments of approximately $351 each. How much more of the first payment is interest if the loan is 8.25% rather than 7%?

c) Compare the total amount of interest on the loan at 7% with the amount on the loan at 8.25%. How much more would Taylor pay in interest on the 8.25% loan than on the 7% loan?

We solve as follows.

a) We use the formula $I = P \cdot r \cdot t$, substituting $28,650 for P, 0.07 for r, and 1/12 for t:

$$I = \$28{,}650 \times 0.07 \times \frac{1}{12}$$

$$\approx \$167.13.$$

The amount of interest in the first payment is $167.13. The payment is $333. We subtract to determine the amount applied to the principal:

$$\$333 - \$167.13 = \$165.87.$$

With the first payment, the principal will be reduced by $165.87.

b) The interest at 8.25% would be

$$I = \$28{,}650 \times 0.0825 \times \frac{1}{12}$$

$$\approx \$196.97.$$

At the rate of 8.25%, the additional interest in the first payment is

$$\$196.97 - \$167.13 = \$29.84.$$

The higher interest rate results in an additional $29.84 in interest in the first payment.

c) For the 7% loan, there will be 120 payments of $333 each:

$$120 \times \$333 = \$39{,}960.$$

The total amount of interest at this rate is

$$\$39{,}960 - \$28{,}650 = \$11{,}310.$$

For the 8.25% loan, there will be 120 payments of $351 each:

$$120 \times \$351 = \$42{,}120.$$

The total amount of interest at this rate is

$$\$42{,}120 - \$28{,}650 = \$13{,}470.$$

At the rate of 8.25%, Taylor would pay

$$\$13{,}470 - \$11{,}310 = \$2160$$

more in interest than at the rate of 7%.

Do Exercise 2.

EXAMPLE 3 *Home Loans.* The Sawyers recently purchased their first home. They borrowed $123,000 at $8\frac{7}{8}$% for 30 years (360 payments). Their monthly payment (excluding insurance and taxes) is $978.64.

a) How much of the first payment is interest and how much is applied to reduce the principal?

b) If the Sawyers pay the entire 360 payments, how much interest will be paid on the loan?

We solve as follows.

a) To find the amount of interest paid in the first payment, we use the formula $I = P \cdot r \cdot t$:

$$I = P \cdot r \cdot t = \$123{,}000 \times 0.08875 \times \frac{1}{12} \approx \$909.69.$$

The amount applied to the principal is

$$\$978.64 - \$909.69, \text{ or } \$68.95.$$

b) Over the 30-year period, the total paid will be

$$360 \times \$978.64, \text{ or } \$352{,}310.40.$$

The total amount of interest paid over the lifetime of the loan is

$$\$352{,}310.40 - \$123{,}000, \text{ or } \$229{,}310.40.$$

Do Exercises 3 and 4 on the following page.

2. **Federal Stafford Loans.** After graduation, the balance on Maggie's Stafford loan is $32,680. To pay off the loan at 7.25%, she will make 120 payments of approximately $384 each.

   a) Find the amount of interest and the amount of principal in the first payment.

   b) If the interest rate were 8.5%, she would make 120 payments of approximately $405 each. How much more of the first payment is interest if the loan is 8.5% rather than 7.25%?

   c) Compare the total amount of interest on the loan at 7.25% with the amount of interest on the loan at 8.5%. How much more would Maggie pay in interest on the 8.5% loan than on the 7.25% loan?

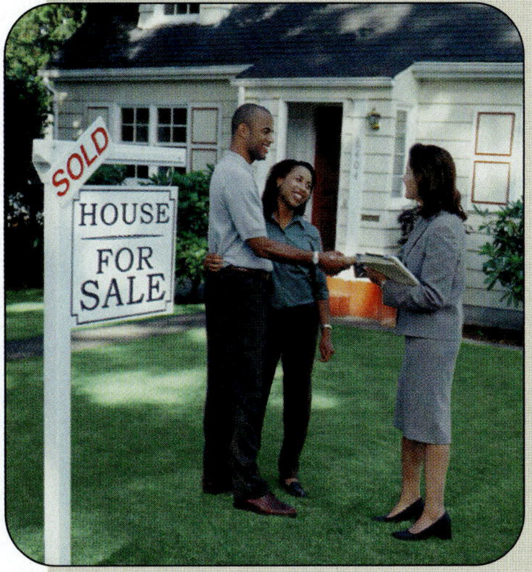

Answers on page A-11

Refer to Example 3 for Margin Exercises 3 and 4.

3. **Home Loans.** Since the principal has been reduced by the first payment, at the time of the second payment of the Sawyers' 30-year loan, the new principal is the decreased principal

$123,000 − $68.95,

or

$122,931.05.

Use $122,931.05 as the principal, and determine how much of the second payment is interest and how much is applied to reduce the principal. (In effect, repeat Example 3(a) using the new principal.)

4. **Home Loans.** The Sawyers decide to change the period of their home loan from 30 years to 15 years. Their monthly payment increases to $1238.42.

a) How much of the first payment is interest and how much is applied to reduce the principal?

b) If the Sawyers pay the entire 180 payments, how much interest will be paid on this loan?

c) Compare the amount of interest to pay off the 15-yr loan with the amount of interest to pay off the 30-yr loan.

Answers on page A-11

AMORTIZATION TABLES

If we make 360 calculations as in Example 3(a) and continue with a decreased principal as in Margin Exercise 4, we can create an *amortization table,* part of which is shown below. Such tables are also found in reference books. The beginning, middle, and last part of the loan described are shown. Look over the table and note how small a portion of the payment reduces the principal at the beginning of the loan and how that portion increases throughout the lifetime of the loan. Do you see again why a loan is not "free"?

MORTGAGE AMORTIZATION PROGRAM

MORTGAGE AMOUNT: $123,000
INTEREST RATE: 8.875%
NUMBER OF YEARS: 30
MONTHLY PAYMENTS ARE: $978.64

PAYMENT	PRINCIPAL	INTEREST	BALANCE	
1	$ 68.95	$909.69	$122,931.05	← Example 3
2	69.46	909.18	122,861.59	← Margin Exercise 4
3	69.98	908.66	122,791.61	
4	70.49	908.15	122,721.12	
5	71.02	907.62	122,650.10	
6	71.54	907.10	122,578.56	
7	72.07	906.57	122,506.49	
8	72.60	906.04	122,433.89	
9	73.14	905.50	122,360.75	
10	73.68	904.96	122,287.07	
11	74.23	904.41	122,212.84	
12	74.77	903.87	122,138.07	
⋮	⋮	⋮ ↳	Interest for 12 periods =	$10,881.75
175	248.53	730.11	98,470.83	
176	250.37	728.27	98,220.46	
177	252.22	726.42	97,968.24	
178	254.08	724.56	97,714.16	
179	255.96	722.68	97,458.20	
180	257.86	720.78	97,200.34	
181	259.76	718.88	96,940.58	
182	261.68	716.96	96,678.90	
183	263.62	715.02	96,415.28	
184	265.57	713.07	96,149.71	
185	267.53	711.11	95,882.18	
186	269.51	709.13	95,612.67	
⋮	⋮	⋮ ↳	Interest for 12 periods =	$8636.99
349	895.78	82.86	10,307.76	
350	902.41	76.23	9,405.35	
351	909.08	69.56	8,496.27	
352	915.80	62.84	7,580.47	
353	922.58	56.06	6,657.89	
354	929.40	49.24	5,728.49	
355	936.27	42.37	4,792.22	
356	943.20	35.44	3,849.02	
357	950.17	28.47	2,898.85	
358	957.20	21.44	1,941.65	
359	964.28	14.36	977.37	
360	971.41	7.23	5.96	
		↳	Interest for 12 periods =	$546.10

Total Interest for 360 Periods = $229,316.36

EXAMPLE 4 *Refinancing a Home Loan.* Refer to Example 3. Ten months after the Sawyers buy their home financed at a rate of $8\frac{7}{8}\%$, the rates drop to $6\frac{1}{2}\%$. After much consideration, they decide to refinance even though the new loan will cost them \$1200 in refinance charges. They have reduced the principal a small amount in the 10 payments they have made, but they decide to again borrow \$123,000 for 30 years at the new rate. Their new monthly payment is \$777.44.

a) How much of the first payment is interest and how much is applied to the principal?

b) Compare the amounts at $6\frac{1}{2}\%$ found in part (a) with the amounts at $8\frac{7}{8}\%$ found in Example 3(a).

c) With the lower house payment, how long will it take the Sawyers to recoup the refinance charge of \$1200?

d) If the Sawyers pay the entire 360 payments, how much interest will be paid on this loan? How much less is the total interest at $6\frac{1}{2}\%$ than at $8\frac{7}{8}\%$?

We solve as follows.

a) To find the interest paid in the first payment, we use the formula $I = P \cdot r \cdot t$:

$$I = P \cdot r \cdot t = \$123{,}000 \times 0.065 \times \frac{1}{12} = \$662.25.$$

The amount applied to the principal is

\$777.44 − \$666.25, or \$111.19.

b) We compare the amount found in part (a) with the amount found in Example 3(a):

Rate	*Monthly payment*	*Interest in first payment*	*Amount applied to principal*
$8\frac{7}{8}\%$	\$978.64	\$909.69	\$68.95
$6\frac{1}{2}\%$	\$777.44	\$666.25	\$111.19

At $6\frac{1}{2}\%$, the amount of interest in the payment is \$909.69 − \$666.25, or \$243.44, less than at $8\frac{7}{8}\%$. The amount applied to the principal is \$111.19 − \$68.95, or \$42.24, more.

c) The monthly payment at $6\frac{1}{2}\%$ is \$978.64 − \$777.44, or \$201.20 less than the payment at $8\frac{7}{8}\%$. The total savings each month is approximately \$200. We can divide the cost of the refinancing by this monthly savings to determine the number of months it will take to recoup the \$1200 refinancing charge: \$1200 ÷ \$200 = 6. It will take the Sawyers approximately 6 months to break even.

d) Over the 30-year period, the total paid will be

360 × \$777.44, or \$279,878.40.

The total amount of interest paid over the lifetime of the loan is

\$279,878.40 − \$123,000, or \$156,878.40.

The total interest paid at $6\frac{1}{2}\%$ is

\$229,310.40 (see Example 3) − \$156,878.40, or \$72,432

less than the total interest paid at $8\frac{7}{8}\%$. Thus the $6\frac{1}{2}\%$ loan saves the Sawyers approximately \$70,000 in interest charges over the 30 years.

Do Exercise 5.

5. Refinancing a Home Loan. Consider Example 4 for a 15-yr loan. The new monthly payment is \$1071.46.

a) How much of the first payment is interest and how much is applied to reduce the principal?

b) If the Sawyers pay the entire 180 payments, how much interest will be paid on this loan?

c) Compare the amount of interest to pay off the 15-yr loan at $6\frac{1}{2}\%$ with the amount of interest to pay off the 15-yr loan at $8\frac{7}{8}\%$ in Margin Exercise 4.

Answer on page A-11

For Extra Help

Digital Video Tutor CD 4 Videotape 5

InterAct Math

Math Tutor Center

MathXL

MyMathLab.com

a Solve.

1. *Credit Cards.* At the end of his freshman year of college, Antonio has a balance of $4876.54 on a credit card with an annual percentage rate (APR) of 21.3%. He decides to not make any additional purchases with his card until he has paid off the balance.
   a) Many credit cards require a minimum monthly payment of 2% of the balance. What is Antonio's minimum payment on a balance of $4876.54? Round the answer to the nearest dollar.
   b) Find the amount of interest and the amount applied to reduce the principal in the minimum payment found in part (a).
   c) If Antonio had transferred his balance to a card with an APR of 12.6%, how much of his first payment would be interest and how much would be applied to reduce the principal?
   d) Compare the amounts for 12.6% from part (c) with the amounts for 21.3% from part (b).

2. *Credit Cards.* At the end of her junior year of college, Becky had a balance of $5328.88 on a credit card with an annual percentage rate (APR) of 18.7%. She decides to not make any additional purchases with this card until she has paid off the balance.
   a) Many credit cards require a minimum monthly payment of 2% of the balance. What is Becky's minimum payment on a balance of $5328.88? Round the answer to the nearest dollar.
   b) Find the amount of interest and the amount applied to reduce the principal in the minimum payment found in part (a).
   c) If Becky had transferred her balance to a card with an APR of 13.2%, how much of her first payment would be interest and how much would be applied to reduce the principal?
   d) Compare the amounts for 13.2% from part (c) with the amounts for 18.7% from part (b).

3. *Federal Stafford Loans.* After graduation, the balance on Grace's Stafford loan is $44,560. To pay off the loan at 6.5%, she will make 120 payments of approximately $505.97 each.
   a) Find the amount of interest and the amount applied to reduce the principal in the first payment.
   b) If the interest rate were 8.5%, she would make 120 monthly payments of approximately $552.48 each. How much more of the first payment is interest if the loan is 8.5% rather than 6.5%?
   c) Compare the total amount of interest on a loan at 6.5% with the amount on the loan at 8.5%. How much more would Grace pay on the 8.5% loan than on the 6.5% loan?

4. *Federal Stafford Loans.* After graduation, the balance on Ricky's Stafford loan is $38,970. To pay off the loan at 8.2%, he will make 120 payments of approximately $476.94 each.
   a) Find the amount of interest and the amount applied to reduce the principal in the first payment.
   b) If the interest rate were 7.4%, he would make 120 monthly payments of approximately $460.55 each. How much less of the first payment is interest if the loan is 7.4% rather than 8.2%?
   c) Compare the total amount of interest on the loan at 8.2% with the amount on the loan at 7.4%. How much more would Ricky pay on the 8.2% loan than on the 7.4% loan?

5. *Home Loan.* The Martinez family recently purchased a home. They borrowed $150,000 at 6.98% for 30 years (360 payments). Their monthly payment (excluding insurance and taxes) is $995.94.
   a) How much of the first payment is interest and how much is applied to reduce the principal?
   b) If this family pays the entire 360 payments, how much interest will be paid on the loan?
   c) Determine the new principal after the first payment. Use that new principal to determine how much of the second payment is interest and how much is applied to reduce the principal.

6. *Home Loan.* The Kaufmans recently purchased a home. They borrowed $180,000 at 8.36% for 30 years (360 payments). Their monthly payment (excluding insurance and taxes) is $1366.22.
   a) How much of the first payment is interest and how much is applied to reduce the principal?
   b) If the Kaufmans pay the entire 360 payments, how much interest will be paid on the loan?
   c) Determine the new principal after the first payment. Use that new principal to determine how much of the second payment is interest and how much is applied to reduce the principal.

7. *Refinancing a Home Loan.* Refer to Exercise 5. The Martinez decide to change the period of their home loan to 15 years. Their monthly payment increased to $1346.57.

a) How much of the first payment is interest and how much is applied to reduce the principal?
b) If the Martinez pay the entire 180 payments, how much interest will be paid on the loan?
c) Compare the amount of interest to pay off the 15-yr loan with the amount of interest to pay off the 30-yr loan.

8. *Refinancing a Home Loan.* Refer to Exercise 6. The Kaufmans decide to change the period of their home loan to 15 years. Their monthly payment increased to $1757.79.

a) How much of the first payment is interest and how much is applied to reduce the principal?
b) If the Kaufmans pay the entire 180 payments, how much interest will be paid on the loan?
c) Compare the amount of interest to pay off the 15-yr loan with the amount of interest to pay off the 30-yr loan.

Complete the following table, assuming monthly payments as given.

	INTEREST RATE	HOME MORTGAGE	TIME OF LOAN	MONTHLY PAYMENT	PRINCIPAL AFTER FIRST PAYMENT	PRINCIPAL AFTER SECOND PAYMENT
9.	6.98%	$100,000	360 mos	$663.96		
10.	6.98%	$100,000	180 mos	$897.71		
11.	8.04%	$100,000	180 mos	$957.96		
12.	8.04%	$100,000	360 mos	$736.55		
13.	7.24%	$150,000	360 mos	$1022.25		
14.	7.24%	$75,000	180 mos	$684.22		
15.	7.24%	$200,000	180 mos	$1824.60		
16.	7.24%	$180,000	360 mos	$1226.70		

17. *New-Car Loan.* After working at her first job for 2 years, Janice buys a new Saturn for $16,385. She makes a down payment of $1385 and finances $15,000 for 4 years at a new-car loan rate of 8.99%. Her monthly payment is $373.20.

a) How much of her first payment is interest and how much is applied to reduce the principal?
b) Find the principal balance at the beginning of the second month and determine how much less interest she will pay in the second payment than in the first.
c) What is the total interest cost of the loan if she pays all of the 48 payments?

18. *Manufacturer's Car Loan Offer.* For a trip to Colorado, Michael and Rebecca buy a new Jeep Cherokee Classic whose selling price is $22,085. For financing, they accept the promotion from the manufacturer that offers a 36-month loan at 1.9% with 20% down. Their monthly payment is $505.29.

a) What is the down payment? the amount borrowed?
b) How much of the first payment is interest and how much is applied to reduce the principal?
c) What is the total interest cost of the loan if they pay all of the 36 payments?

19. *Used-Car Loan.* Twin brothers, Jerry and Terry, each take a job at the college cafeteria in order to have the money to make payments on the purchase of a used 1997 Ford Taurus for $7900. They make a down payment of 10% and finance the remainder at 12.49% for 3 years. (Used-car loan rates are generally higher than new-car loan rates.) Their monthly payment is $237.82.

a) What is the down payment? the amount borrowed?
b) How much of the first payment is interest and how much is applied to reduce the principal?
c) If they pay all 36 payments, how much interest will they pay for the loan?

20. *Used-Car Loan.* For his construction job, Clint buys a 1994 Chevrolet S-10 truck for $5350. He makes a down payment of $550 and finances the remainder for 2 years at 11.3%. The monthly payment is $224.39.

a) How much is financed?
b) How much of the first payment is interest and how much is applied to reduce the principal?
c) If he pays all 24 payments, how much interest will he pay for thc loan?

21. **DW** Based on the skills of mathematics you have obtained in this section, discuss the significant new ideas you now have about interest rates and credit cards that you didn't have before.

22. **DW** Examine the information in the graphs at the beginning of the section. Discuss how a knowledge of this section might have been of help to some of these students.

23. **DW** Compare the following two purchases and describe a situation in which each purchase is the best choice.

Purchase A: A new car for $15,145. The loan is for $14,500 at 6.9% for 4 years. The monthly payment is $346.55.

Purchase B: A used car for $10,600. The loan is for $9300 at 12.5% for 3 years. The monthly payment is $311.12.

24. **DW** Look over the examples and exercises in this section. What seems to happen to the monthly payment on a loan if the time of payment changes from 30 years to 15 years, assuming the interest rate stays the same? Discuss the pros and cons of both time periods.

SKILL MAINTENANCE

Solve. [4.1d]

25. $\frac{x}{12} = \frac{24}{16}$

26. $\frac{7}{2} = \frac{11}{x}$

Solve. [3.4b]

27. $0.64 \cdot x = 170$

28. $28.5 = 25.6 \times y$

Find decimal notation. [3.5a]

29. $\frac{5}{9}$

30. $\frac{23}{11}$

31. $\frac{11}{12}$

32. $\frac{13}{7}$

33. $\frac{15}{7}$

34. $\frac{19}{12}$

Convert to standard notation. [3.3b]

35. $4.03 trillion

36. 5.8 million

37. 42.7 million

38. 6.09 trillion

CHAPTER 4

Summary and Review

The review that follows is meant to prepare you for a chapter exam. It consists of two parts. The first part is a checklist of some of the Study Tips referred to in this and preceding chapters, as well as a list of important properties and formulas. The second part is the Review Exercises. These provide practice exercises for the exam, together with references to section objectives so you can go back and review. Before beginning, stop and look back over the skills you have obtained. What skills in mathematics do you have now that you did not have before studying this chapter?

STUDY TIPS CHECKLIST

The foundation of all your study skills is TIME!

- ☐ Have you tried *memorizing* the formulas and the three types of percent problems?
- ☐ Have you tried being a tutor to a fellow student?
- ☐ Have you found any applications of percent in a newspaper or magazine?
- ☐ Are you stopping to work the margin exercises when directed to do so?
- ☐ Are you doing your homework as soon as possible after class?

IMPORTANT PROPERTIES AND FORMULAS

Commission = Commission rate × Sales
Discount = Rate of discount × Original price
Sale price = Original price − Discount

Simple Interest: $I = P \cdot r \cdot t$
Compound Interest: $A = P \cdot \left(1 + \dfrac{r}{n}\right)^{n \cdot t}$

REVIEW EXERCISES

Write fraction notation for the ratio. Do not simplify. [4.1a]

1. 47 to 84

2. 46 to 1.27

3. 83 to 100

4. 0.72 to 197

Find the ratio of the first number to the second number and simplify. [4.1a]

5. 9 to 12

6. 3.6 to 6.4

7. *Gas Mileage.* The Chrysler PT Cruiser will go 377 mi on 14.5 gal of gasoline in highway driving. What is the rate in miles per gallon? [4.1b]
Source: DaimlerChrysler Corporation

8. *CD-ROM Spin Rate.* A 12x CD-ROM on a computer will spin 472,500 revolutions if left running for 75 min. What is the rate of its spin in revolutions per minute (rpm)? [4.1b]
Source: *Electronic Engineering Times*, June 1997

9. A lawn requires 319 gal of water for every 500 ft^2. What is the rate in gallons per square foot? [4.1b]

10. *Turkey Servings.* A 25-lb turkey serves 18 people. Find the rate in servings per pound. [4.1b]

Determine whether the two pairs of numbers are proportional. [4.1c]

11. 9, 15 and 36, 59

12. 24, 37 and 40, 46.25

Solve. [4.1d]

13. $\frac{8}{9} = \frac{x}{36}$

14. $\frac{6}{x} = \frac{48}{56}$

15. $\frac{120}{\frac{3}{7}} = \frac{7}{x}$

16. $\frac{4.5}{120} = \frac{0.9}{x}$

Solve. [4.1e]

17. If 3 dozen eggs cost \$2.67, how much will 5 dozen eggs cost?

18. *Quality Control.* A factory manufacturing computer circuits found 39 defective circuits in a lot of 65 circuits. At this rate, how many defective circuits can be expected in a lot of 585 circuits?

19. *Exchanging Money.* On 22 December 2000, 1 U.S. dollar was worth about 1.08 European Monetary Units (Euros).

a) How much would 250 U.S. dollars be worth in Euros?

b) Jamal was traveling in France and saw a sweatshirt that cost 50 Euros. How much would it cost in U.S. dollars?

20. A train travels 448 mi in 7 hr. At this rate, how far will it travel in 13 hr?

21. Fifteen acres are required to produce 54 bushels of tomatoes. At this rate, how many acres are required to produce 97.2 bushels of tomatoes?

22. *Garbage Production.* It is known that 5 people produce 13 kg of garbage in one day. San Diego, California, has 1,220,666 people. How many kilograms of garbage are produced in San Diego in one day?

23. *Snow to Water.* Under typical conditions, $1\frac{1}{2}$ ft of snow will melt to 2 in. of water. To how many inches of water will $4\frac{1}{2}$ ft of snow melt?

24. *Lawyers in Michigan.* In Michigan, there are 2.3 lawyers for every 1000 people. The population of Detroit is 4,307,000. How many lawyers would you expect there to be in Detroit?
Source: U.S. Bureau of the Census

Find percent notation for the decimal notation in the sentence in Exercises 25 and 26. [4.2b]

25. Of all the vehicles in Mexico City, 0.017 of them are taxis.
Source: *The Handy Geography Answer Book*

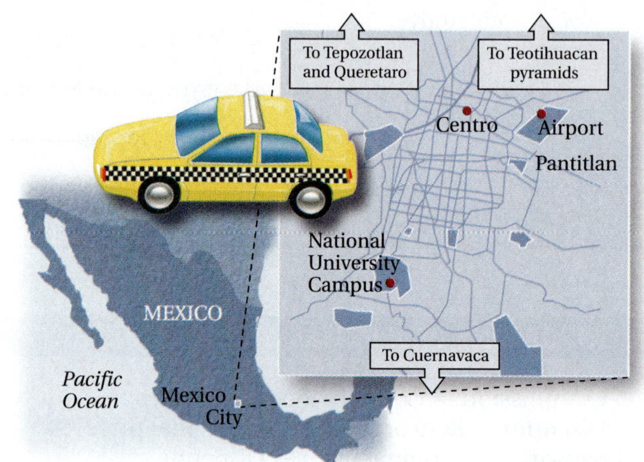

26. Of all the snacks eaten on Super Bowl Sunday, 0.56 of them are chips and salsa.
Source: Korbel Research and Pace Foods

Find percent notation. [4.3a]

27. $\frac{3}{8}$

28. $\frac{1}{3}$

Find decimal notation. [4.2b]

29. 73.5%

30. $6\frac{1}{2}\%$

Find fraction notation. [4.3b]

31. 24%

32. 6.3%

Translate to a percent equation. Then solve. [4.4a, b]

33. 30.6 is what percent of 90?

34. 63 is 84 percent of what?

35. What is $38\frac{1}{2}\%$ of 168?

Translate to a proportion. Then solve. [4.5a, b]

36. 24 percent of what is 16.8?

37. 42 is what percent of 30?

38. What is 10.5% of 84?

Solve. [4.6a, b]

39. *Favorite Ice Creams.* According to a recent survey, 8.9% of those interviewed chose chocolate as their favorite ice cream flavor and 4.2% chose butter pecan. Of the 2500 students in a freshman class, how many would choose chocolate as their favorite ice cream? butter pecan?
Source: International Ice Cream Association

40. *Prescriptions.* Of the 281 million people in the United States, 123.64 million take at least one kind of prescription drug per day. What percent take at least one kind of prescription drug per day?
Source: American Society of Health-System Pharmacies

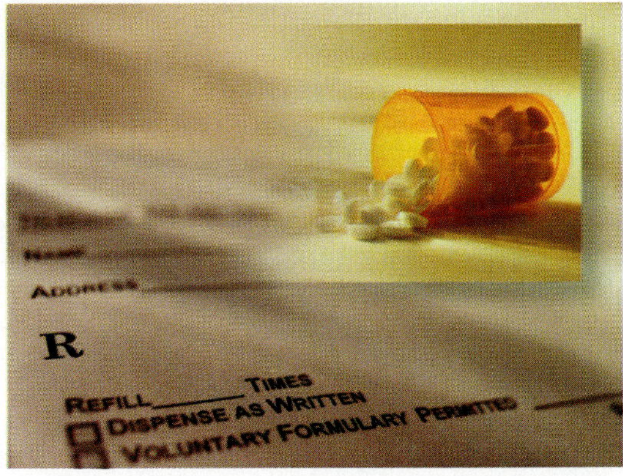

41. *Water Output.* The average person expels 200 mL of water per day by sweating. This is 8% of the total output of water from the body. How much is the total output of water?
Source: Elaine N. Marieb, *Essentials of Human Anatomy and Physiology*, 6th ed. Boston: Addison Wesley Longman, Inc., 2000

42. *Test Scores.* Jason got a 75 on a math test. He was allowed to go to the math lab and take a retest. He increased his score to 84. What was the percent of increase?

43. *Test Scores.* Jenny got an 81 on a math test. By taking a retest in the math lab, she increased her score by 15%. What was her new score?

Solve. [4.7a, b, c]

44. A state charges a meals tax of $4\frac{1}{2}\%$. What is the meals tax charged on a dinner party costing \$320?

45. In a certain state, a sales tax of \$378 is collected on the purchase of a used car for \$7560. What is the sales tax rate?

46. Kim earns \$753.50 selling \$6850 worth of televisions. What is the commission rate?

47. An air conditioner has a marked price of \$350. It is placed on sale at 12% off. What are the discount and the sale price?

48. A fax machine priced at \$305 is discounted at the rate of 14%. What are the discount and the sale price?

49. An insurance salesperson receives a 7% commission. If \$42,000 worth of life insurance is sold, what is the commission?

50. Find the rate of discount.

Solve. [4.7d, e]

51. What is the simple interest on \$1800 at 6% for $\frac{1}{3}$ year?

52. The Dress Shack borrows \$24,000 at 10% simple interest for 60 days. Find (a) the amount of interest due and (b) the total amount that must be paid after 60 days.

53. What is the simple interest on \$2200 principal at the interest rate of 5.5% for 1 year?

54. The Kleins invest \$7500 in an investment account paying an annual interest rate of 12%, compounded monthly. How much is in the account after 3 months?

55. Find the amount in an investment account if \$8000 is invested at 9%, compounded annually, for 2 years.

Solve. [6.8a]

56. *Credit Cards.* At the end of her junior year of college, Judy has a balance of \$6428.74 on a credit card with an annual percentage rate (APR) of 18.7%. She decides to not make any additional purchases with this card until she has paid off the balance.

a) Many credit cards require a minimum payment of 2% of the balance. What is Judy's minimum payment on a balance of \$6428.74? Round the answer to the nearest dollar.

b) Find the amount of interest and the amount applied to reduce the principal in the minimum payment found in part (a).

c) If Judy had transferred her balance to a card with an APR of 13.2%, how much of her first payment would be interest and how much would be applied to reduce the principal?

d) Compare the amounts for 13.2% from part (c) with the amounts for 18.7% from part (b).

57. D_W Ollie buys a microwave oven during a 10%-off sale. The sale price that Ollie paid was \$162. To find the original price, Ollie calculates 10% of \$162 and adds that to \$162. Is this correct? Why or why not? [4.7c]

58. D_W Which is the better deal for a consumer and why: a discount of 40% or a discount of 20% followed by another of 22%? [4.7c]

SKILL MAINTENANCE

Certain objectives from four particular sections will be retested on the chapter test. The objectives are listed with the practice problems that follow.

Write fraction notation. [3.1b]

59. 3.107

60. 0.29

Solve. [3.4b]

61. $10.4 \times y = 665.6$

62. $100 \cdot x = 761.23$

Convert to decimal notation. [3.5a]

63. $\frac{11}{3}$

64. $\frac{11}{7}$

Convert to a mixed numeral. [2.4a]

65. $\frac{11}{3}$

66. $\frac{121}{7}$

SYNTHESIS

67. It takes Yancy Martinez 10 min to type two-thirds of a page of his term paper. At this rate, how long will it take him to type a 7-page term paper? [4.1e]

68. Rhonda's Dress Shop reduces the price of a dress by 40% during a sale. By what percent must the store increase the sale price, after the sale, to get back to the original price? [4.7c]

69. A \$200 coat is marked up 20%. After 30 days, it is marked down 30% and sold. What was the final selling price of the coat? [4.7c]

CHAPTER 4

Chapter Test

Write fraction notation for the ratio. Do not simplify.

1. 85 to 97

2. 0.34 to 124

3. *Gas Mileage.* The 2000 Volkswagen New Beetle GL will go 341 mi on 14.5 gal of gasoline in city driving. What is the rate in miles per gallon?
Source: Volkswagen of America, Inc.

4. *Ham Servings.* A 12-lb shankless ham contains 16 servings. What is the rate in servings per pound?

Determine whether the two pairs of numbers are proportional.

5. 7, 8 and 63, 72

6. 1.3, 3.4 and 5.6, 15.2

Solve.

7. $\frac{9}{4} = \frac{27}{x}$

8. $\frac{150}{2.5} = \frac{x}{6}$

9. *Map Scaling.* On a map, 3 in. represents 225 mi. If two cities are 7 in. apart on the map, how far are they apart in reality?

10. *Time Loss.* A watch loses 2 min in 10 hr. At this rate, how much will it lose in 24 hr?

11. *Exchanging Money.* On 22 December 2000, it was known that 1 U.S. dollar was worth about 1.52 Canadian dollars.

a) How much would 450 U.S. dollars be worth in Canadian dollars?

b) Mitchell was traveling in Toronto and saw a DVD player that cost 560 Canadian dollars. How much would it cost in U.S. dollars?

12. *Police Department Arrests.* In a recent year, the Indianapolis Police Department employed 1088 officers and made 37,493 arrests. At this rate, how many arrests could be made if the number of officers were increased to 2500?
Source: *Indianapolis Star*, 12-31-00

13. During a recent month, 0.905 of all the flights of Aloha Airlines arrived on time. This was the highest percentage in the airline industry. Find percent notation for 0.905.
Source: U.S. Department of Transportation

14. Stephen King's novel *Dream Catcher* was sold on the Barnes & Noble Web site at a 20% discount. Find decimal notation for 20%.
Source: Barnes & Noble

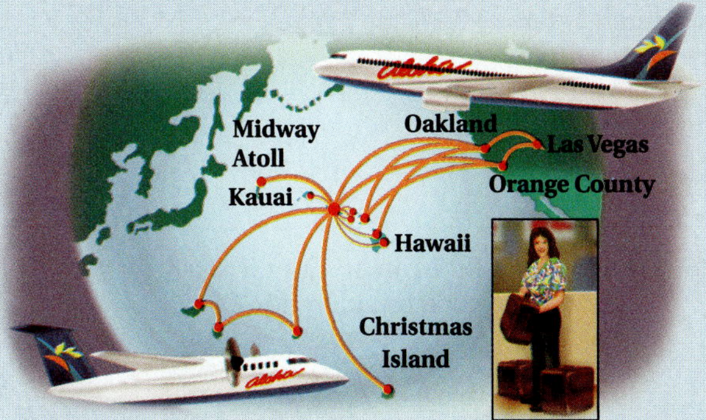

15. Find percent notation for $\frac{11}{8}$.

16. Find fraction notation for 65%.

17. Translate to a percent equation. Then solve.

What is 40% of 55?

18. Translate to a proportion. Then solve.

What percent of 80 is 65?

Solve.

19. *Cruise Ship Passengers.* Of the passengers on a typical cruise ship, 16% are in the 25–34 age group and 23% are in the 35–44 age group. A cruise ship has 2500 passengers. How many are in the 25–34 age group? the 35–44 age group?
Source: Polk

20. *Batting Averages.* Luis Castillo, second baseman for the Florida Marlins, got 180 hits during the 2000 baseball season. This was about 33.4% of his at-bats. How many at-bats did he have?
Source: Major League Baseball

21. *Airline Profits.* Profits of the entire U. S. Airline industry decreased from $5.5 billion in 1999 to $2.7 billion in 2000. Find the percent of decrease.
Source: Air Transport Association

22. There are 6.6 billion people living in the world today. It is estimated that the total number who have ever lived is about 120 billion. What percent of people who have ever lived are alive today?
Source: *The Handy Geography Answer Book*

23. *Maine Sales Tax.* The sales tax rate in Maine is 5%. How much tax is charged on a purchase of $324? What is the total price?

24. Gwen's commission rate is 15%. What is the commission from the sale of $4200 worth of merchandise?

25. The marked price of a CD player is $200 and the item is on sale at 20% off. What are the discount and the sale price?

26. What is the simple interest on a principal of $120 at the interest rate of 7.1% for 1 year?

27. The Burnham Parents–Teachers Association invests $5200 at 6% simple interest. How much is in the account after $\frac{1}{2}$ year?

28. Find the amount in an account if $1000 is invested at $5\frac{3}{8}\%$, compounded annually, for 2 years.

29. The Suarez family invests $10,000 at an annual interest rate of 9%, compounded monthly. How much is in the account after 3 years?

30. *Job Opportunities.* The table below lists job opportunities, in thousands, in 1998 and projected increases to 2008. Find the missing numbers.

OCCUPATION	NUMBER OF JOBS IN 1998 (in thousands)	NUMBER OF JOBS IN 2008 (in thousands)	CHANGE	PERCENT OF INCREASE
Restaurant waitstaff	2019	2322	303	15.0%
Dental assistant	229		97	
Nurse psychiatric aide	1461	1794		
Child-care worker		1141	236	
Hairdresser/hairstylist/cosmetologist		670		10.2%

Source: Handbook of U.S. Labor Statistics

31. Find the discount and the discount rate of the bed in this ad.

WHITE IRON DAYBED
WITH BRASS ACCENTS
100 TO SELL
FANTASTIC VALUE!
MARKET VALUE
$249.95
CHOICE OF FINISH!
$118
Springs Included!

32. *Home Loan.* Complete the following table, assuming the monthly payment as given.

Interest Rate	7.4%
Mortgage	$120,000
Time of Loan	360 mos
Monthly Payment	$830.86
Principal after First Payment	
Principal after Second Payment	

SKILL MAINTENANCE

33. Solve: $8.4 \times y = 1864.8$.

34. Write fraction notation for 44.7.

35. Convert to decimal notation: $\frac{17}{12}$.

36. Convert to a mixed numeral: $\frac{153}{44}$.

SYNTHESIS

37. By selling a home without using a realtor, Juan and Marie can avoid paying a 7.5% commission. They receive an offer of $180,000 from a potential buyer. In order to give a comparable offer, for what price would a realtor need to sell the house? Round to the nearest hundred.

38. Nancy Morano-Smith wants to win a season football ticket from the local bookstore. Her goal is to guess the number of marbles in an 8-gal jar. She knows that there are 128 oz in a gallon. She goes home and fills an 8-oz jar with 46 marbles. How many marbles should she guess are in the larger jar?

Geometry

Gateway to Chapter 5

In this chapter, we introduce basic geometric figures, such as segments, rays, lines, and angles. Measures considered in this chapter are perimeter, area, and volume. Relationships between angle measures, congruent and similar triangles, and properties of parallelograms are also studied.

Real-World Application

Major league baseball underwent a rule change between the 2000 and 2001 seasons. Over the years before 2001, the strike zone evolved to something other than what was defined in the rule book. The zone in the rule book is described by rectangle *ABCD* in the illustration. The zone used before 2001 is described by the region *AQRST*. By what percent has the area of the strike zone been increased by the change?

Sources: *The Cincinnati Enquirer;* Major League Baseball; Gannett News Service; *The Sporting News Official Baseball Rules Book*

This problem appears as Example 10 in Section 5.3.

CHAPTER 5

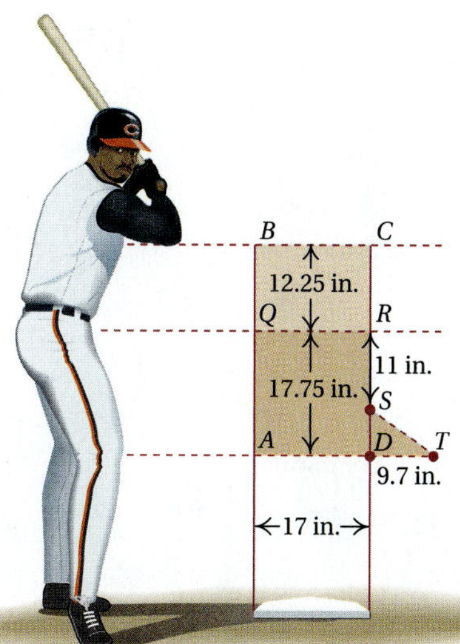

CHAPTER 5

Pretest

1. Find the missing angle measure. [5.1f]

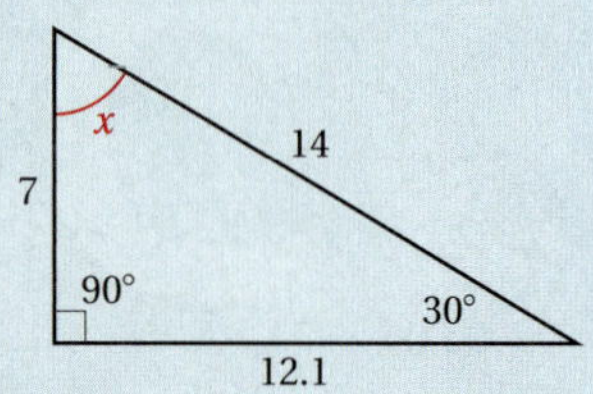

2. Find the perimeter. [5.2a]

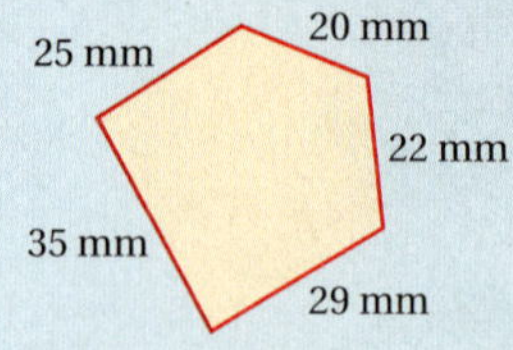

3. Find the area of the shaded region. [5.3c]

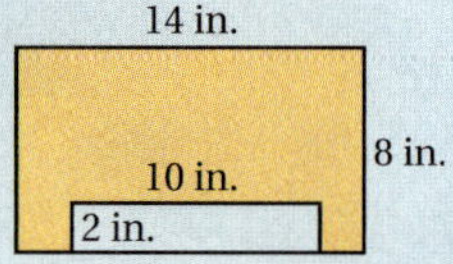

Find the area. [5.3b]

4.

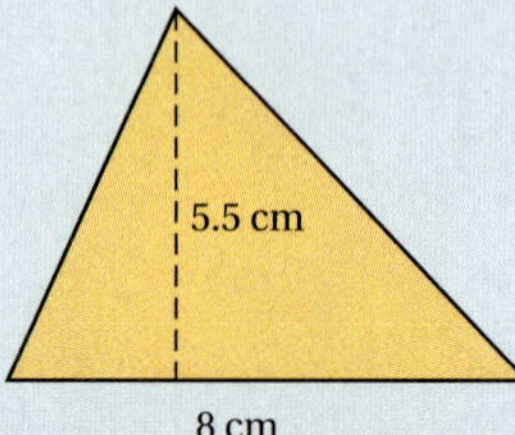

5.

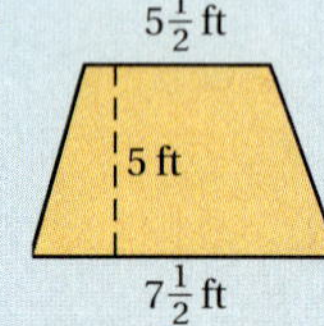

6.

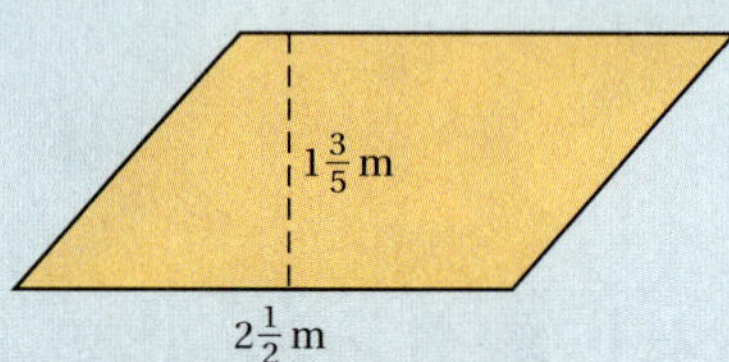

7. Find the length of a diameter of a circle with a radius of 4.8 m. [5.4a]

8. Find the circumference and the area of the circle in Question 7. Use 3.14 for π. [5.4b, c]

Find the volume. Use 3.14 for π.

9. [5.5a]

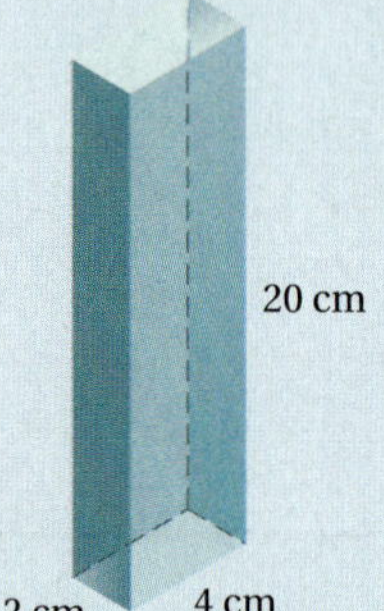

10. [5.5b]

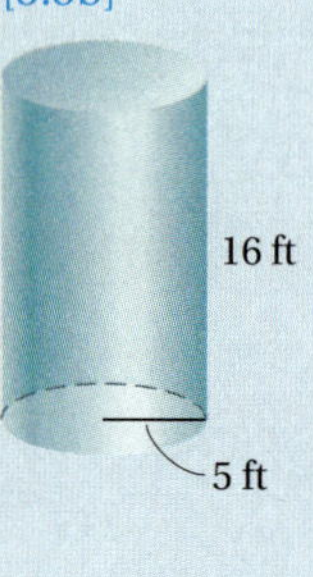

11. [5.5c]

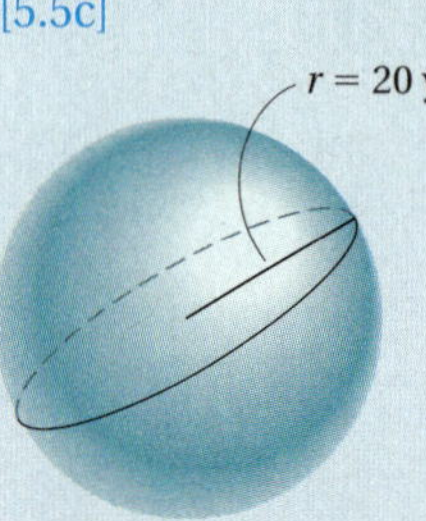

12. [5.5d]

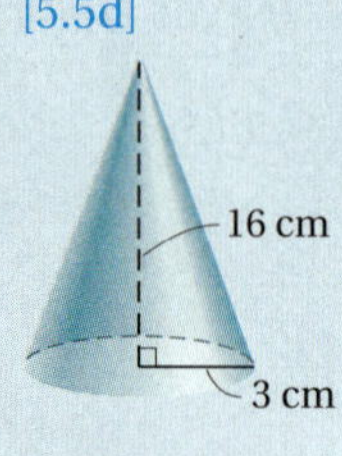

13. If $m \parallel n$ and $m\angle 8 = 29°$, what are the measures of the other angles? [5.6d]

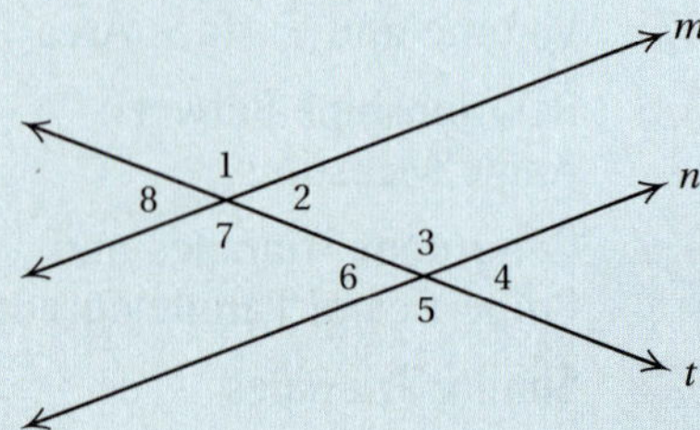

14. Given that $\triangle PQR \cong \triangle STV$, list the congruent corresponding parts. [5.7a]

15. Given that $\triangle MAC \sim \triangle GET$, find MA and GT. [5.8b]

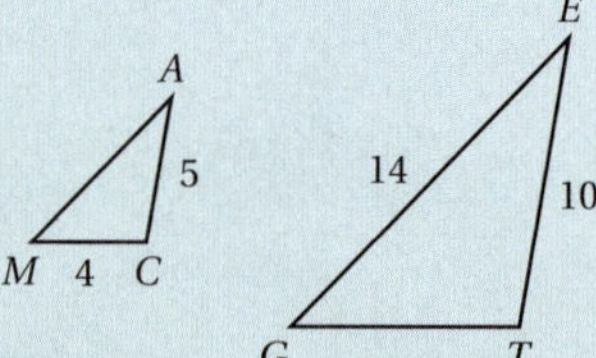

5.1 BASIC GEOMETRIC FIGURES

In geometry we study sets of points. A **geometric figure** (or *figure*) is simply a set of points. Thus a figure can be a set with one point, a set with two points, or sets that look like those below.

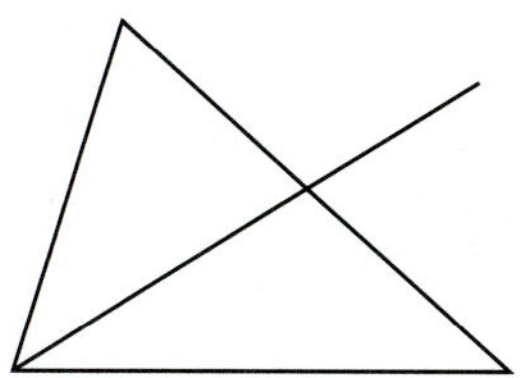

Objectives

a. Draw and name segments, rays, and lines. Also, identify endpoints, if they exist.

b. Name an angle in five different ways, and given an angle, measure it with a protractor.

c. Classify an angle as right, straight, acute, or obtuse.

d. Identify perpendicular lines.

e. Classify a triangle as equilateral, isosceles, or scalene and as right, obtuse, or acute. Given a polygon of twelve, ten, or fewer sides, classify it as a dodecagon, decagon, and so on.

f. Given a polygon of n sides, find the sum of its angle measures using the formula $(n - 2) \cdot 180°$.

a Segments, Rays, and Lines

A **segment** is a geometric figure consisting of two points, called *endpoints*, and all points between them. The segment whose endpoints are A and B is shown below. It can be named $\overline{AB}$ or $\overline{BA}$.

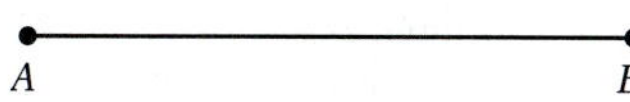

Do Exercise 1.

We get an idea of a geometric figure called a ray by thinking of a ray of light. A **ray** consists of a segment, say $\overline{AB}$, and all points X such that B is between A and X: that is, $\overline{AB}$ and all points "beyond" B.

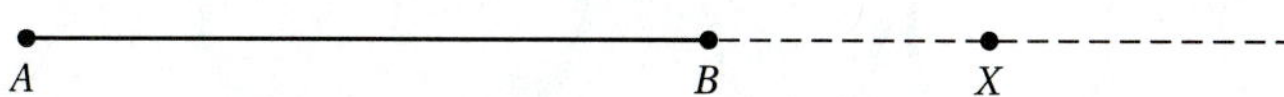

A ray is usually drawn as shown below. It has just one endpoint. The arrow indicates that it extends forever in one direction.

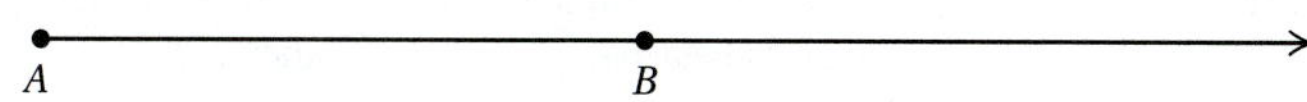

A ray is named $\overrightarrow{AB}$, where B is some point on the ray other than A. The endpoint is always listed first. Thus rays $\overrightarrow{AB}$ and $\overrightarrow{BA}$ are different.

Do Exercises 2–5.

Two rays such as $\overrightarrow{PQ}$ and $\overrightarrow{QP}$ make up what is known as a **line.** A line can be named with a smaller letter m, as shown below, or it can be named by two points P and Q on the line as $\overleftrightarrow{PQ}$.

Do Exercises 6–11 on the following page.

1. a) Draw a segment.

b) Label its endpoints E and F.

c) Name this segment in two ways.

2. Draw two points P and Q.

3. Draw $\overline{PQ}$.

4. Draw $\overrightarrow{PQ}$. What is its endpoint?

5. Use a colored pencil to draw $\overrightarrow{QP}$. What is its endpoint?

Answers on page A-12

6. Draw two points R and S.

7. Draw $\overline{RS}$. What are its endpoints?

8. Draw $\overrightarrow{RS}$. What is its endpoint?

9. Draw $\overrightarrow{SR}$. What is its endpoint?

10. Draw $\overleftrightarrow{RS}$. What are its endpoints?

11. Name this line in seven different ways.

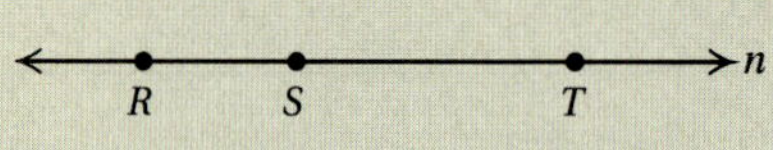

Name the angle in five different ways.

12.

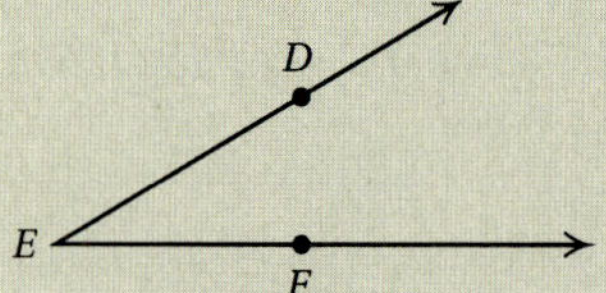

13.

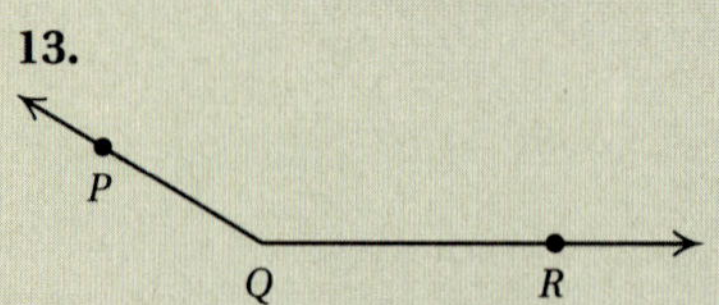

Answers on page A-12

Lines in the same plane are called **coplanar.** Coplanar lines that do not intersect are called **parallel.** For example, lines l and m below are *parallel* ($l \parallel m$).

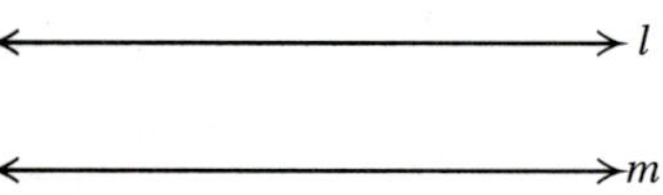

The figure below shows two lines that cross. Their *intersection* is D. They are also called **intersecting lines.**

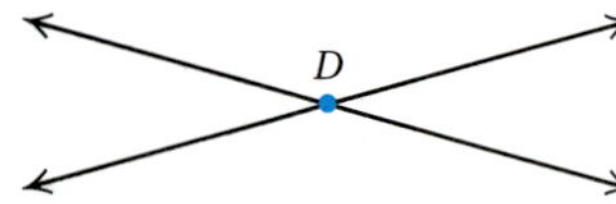

b Angles

We see a real-world application of *angles* of various types in the spokes of these bicycles and the different back postures of the riders.

Style of Biking Determines Cycling Posture

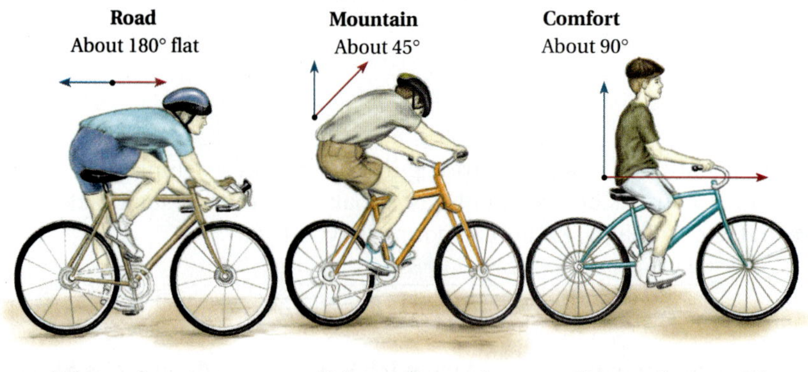

Riders prefer a more aerodynamic flat-back position.

Riders prefer a semi-upright position to help lift the front wheel over obstacles.

Riders prefer an upright position that lessens stress on the lower back and neck.

Source: USA TODAY research

An **angle** is a set of points consisting of two **rays,** or half-lines, with a common endpoint. The endpoint is called the **vertex.**

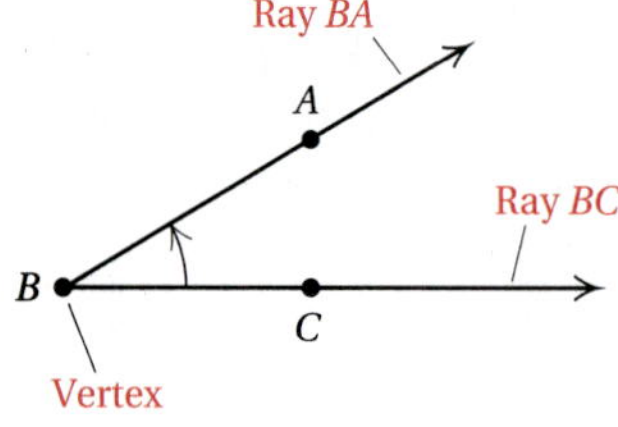

The rays are called the *sides*. The angle above can be named

angle ABC, angle CBA, $\angle ABC$, $\angle CBA$, or $\angle B$.

Note that the name of the vertex is either in the middle or, if no confusion results, listed by itself.

Do Exercises 12 and 13.

Measuring angles is similar to measuring segments. To measure angles, we start with some arbitrary angle and assign to it a measure of 1. We call it a *unit angle.* Suppose that $\angle U$, shown below, is a unit angle. Let's measure $\angle DEF$. If we made 3 copies of $\angle U$, they would "fill up" $\angle DEF$. Thus the measure of $\angle DEF$ would be 3.

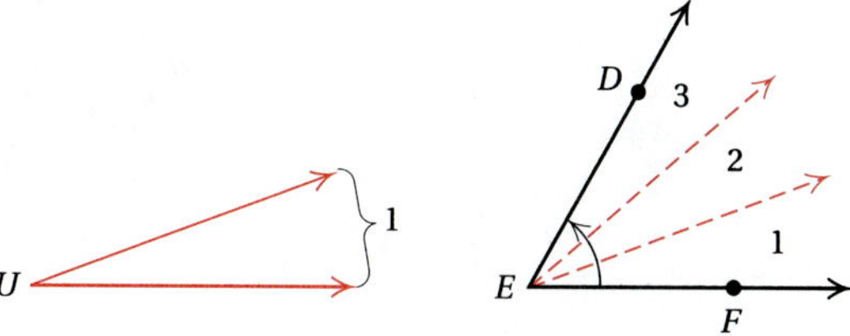

The unit most commonly used for angle measure is the **degree.** Below is such a unit. Its measure is 1 degree, or 1°.

A 1° angle:

Here are some other angles with their degree measures.

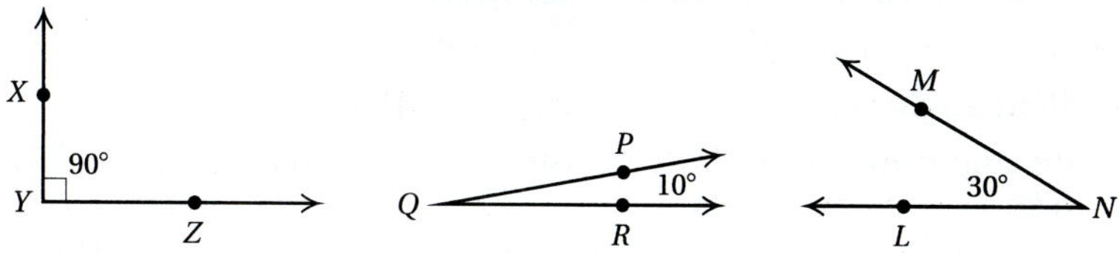

To indicate the *measure* of $\angle XYZ$, we write $m\angle XYZ = 90°$. The symbol ⌉ is sometimes drawn on a figure to indicate a 90° angle.

A device called a **protractor** is used to measure angles. Protractors have two scales. To measure an angle like $\angle Q$ below, we place the protractor's ▲ at the vertex and line up one of the angle's sides at 0°. Then we check where the angle's other side crosses the scale. In the figure below, 0° is on the inside scale, so we check where the angle's other side crosses the inside scale. We see that $m\angle Q = 145°$. The notation $m\angle Q$ is read "the measure of angle Q."

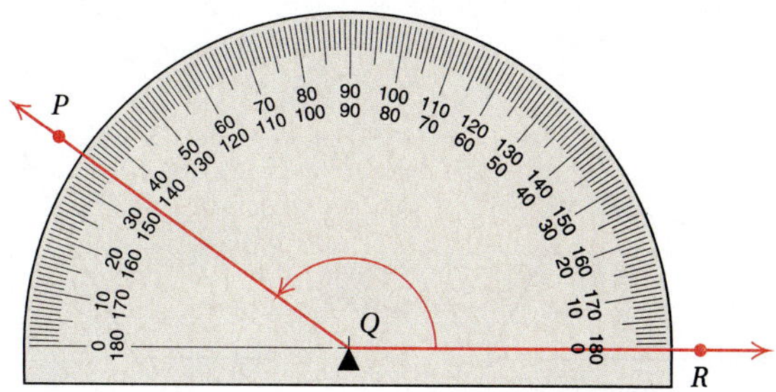

Do Exercise 14.

14. Use a protractor to measure this angle.

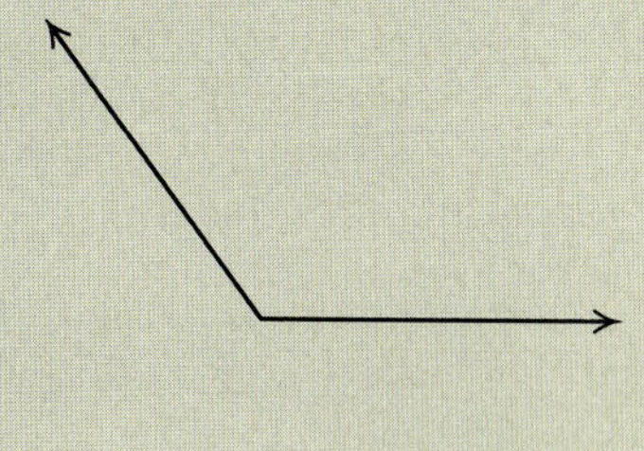

Answer on page A-12

Study Tips

WORKING WITH A CLASSMATE

If you are finding it difficult to master a particular topic or concept, try talking about it with a classmate. Verbalizing your questions about the material might help clarify it. If your classmate is also finding the material difficult, it is possible that the majority of the people in your class are confused and you can ask your instructor to explain the concept again.

15. Use a protractor to measure this angle.

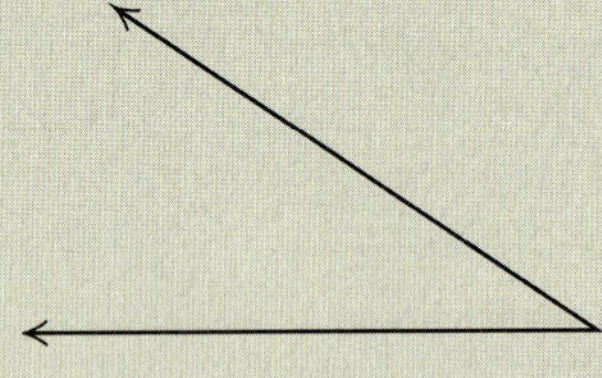

Classify the angle as right, straight, acute, or obtuse. Use a protractor if necessary.

16.

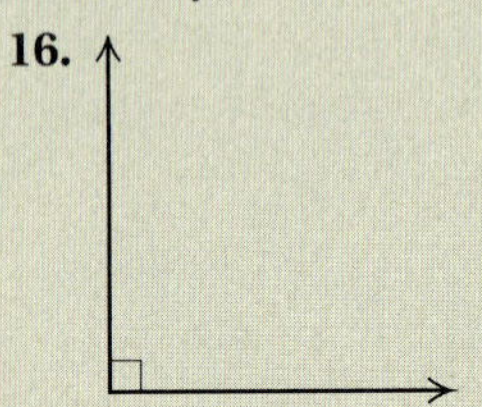

17.

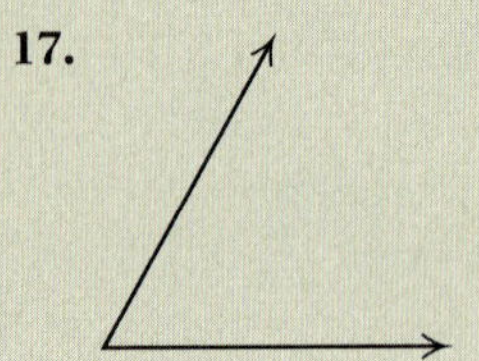

18.

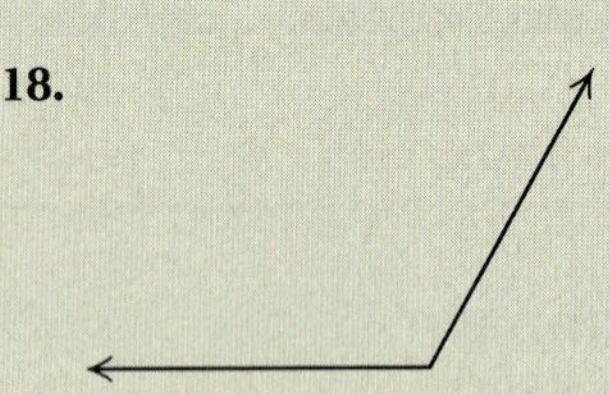

19.

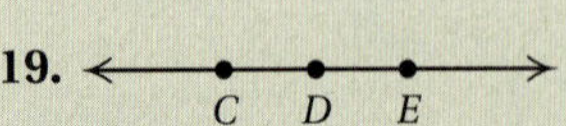

Answers on page A-12

Let's find the measure of $\angle ABC$. This time we will use the 0° on the outside scale. We see that $m\angle ABC = 42°$.

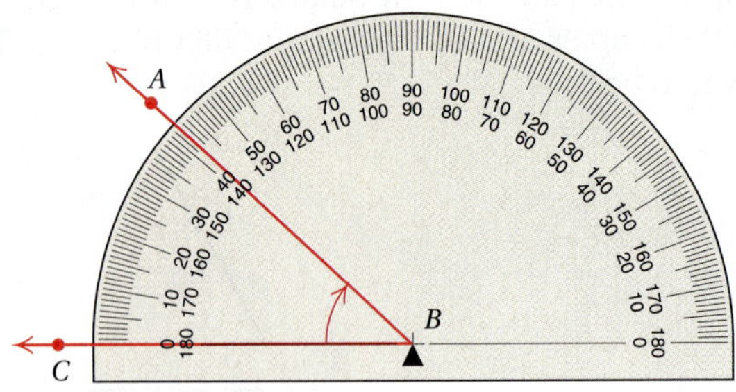

Do Exercise 15.

C Classifying Angles

The following are ways in which we classify angles.

TYPES OF ANGLES

Right angle: An angle whose measure is 90°.

Straight angle: An angle whose measure is 180°.

Acute angle: An angle whose measure is greater than 0° and less than 90°.

Obtuse angle: An angle whose measure is greater than 90° and less than 180°.

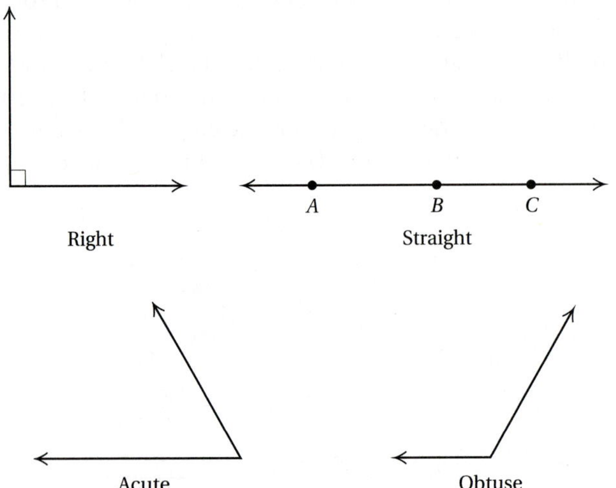

Do Exercises 16–19.

d Perpendicular Lines

Two lines are **perpendicular** if they intersect to form a right angle.

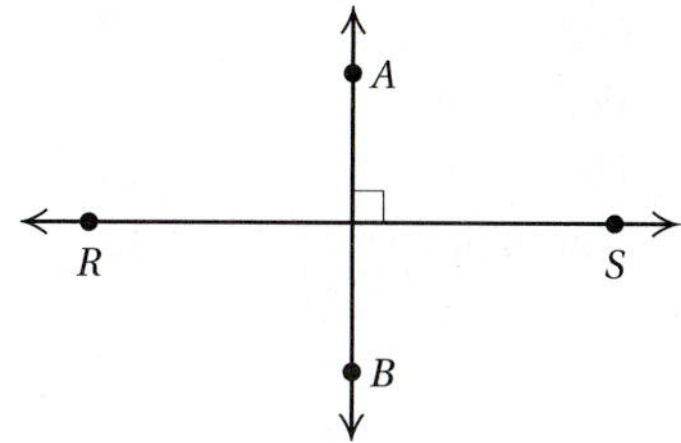

To say that $\overleftrightarrow{AB}$ is perpendicular to $\overleftrightarrow{RS}$, we write $\overleftrightarrow{AB} \perp \overleftrightarrow{RS}$. If two lines intersect to form one right angle, they form four right angles.

Do Exercises 20 and 21.

e Polygons

The figures below are examples of **polygons.**

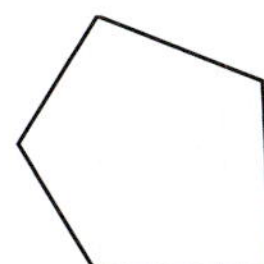
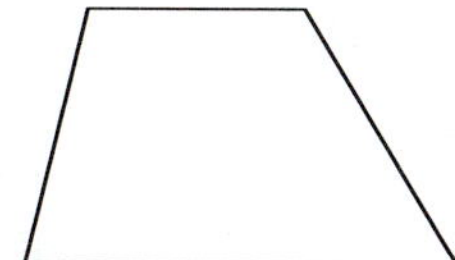
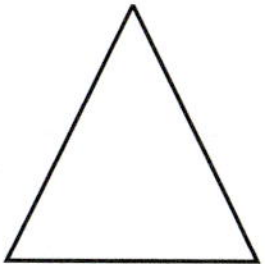

A **triangle** is a polygon made up of three segments, or sides. Consider these triangles. The triangle with vertices A, B, and C can be named $\triangle ABC$.

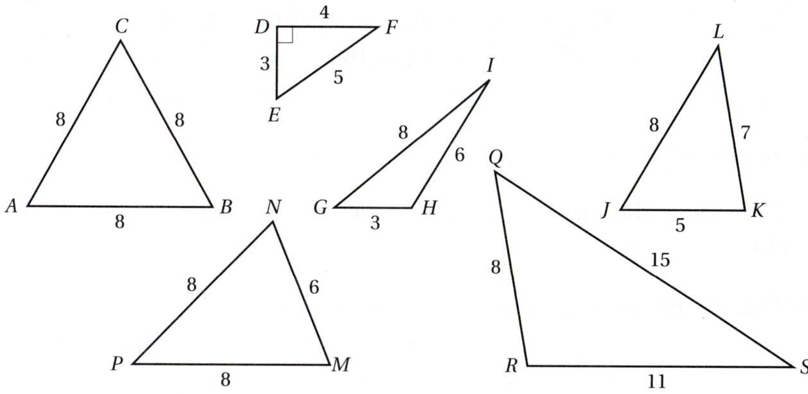

We can classify triangles according to sides and according to angles.

TYPES OF TRIANGLES

Equilateral triangle: All sides are the same length.
Isosceles triangle: Two or more sides are the same length.
Scalene triangle: All sides are of different lengths.
Right triangle: One angle is a right angle.
Obtuse triangle: One angle is an obtuse angle.
Acute triangle: All three angles are acute.

Do Exercises 22–25.

Determine whether the pair of lines is perpendicular. Use a protractor.

20.

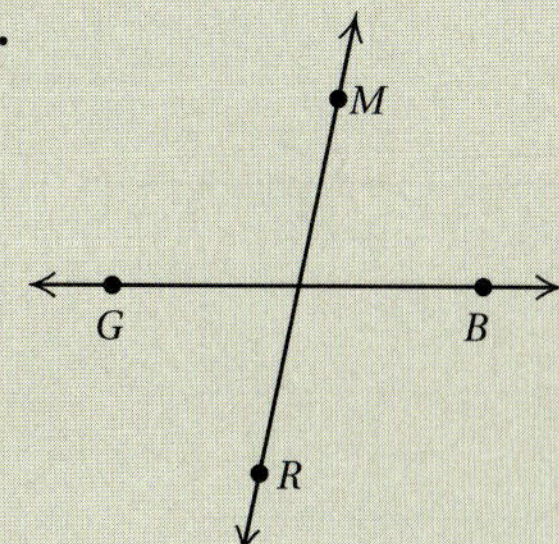

21.

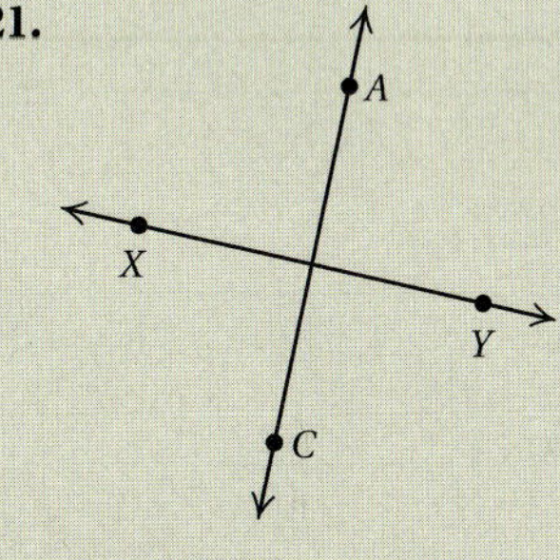

22. Which triangles at left are:
a) equilateral?
b) isosceles?
c) scalene?

23. Are all equilateral triangles isosceles?

24. Are all isosceles triangles equilateral?

25. Which triangles at left are:
a) right triangles?
b) obtuse triangles?
c) acute triangles?

Answers on page A-12

Classify the polygon by name.

26.

27.

28.

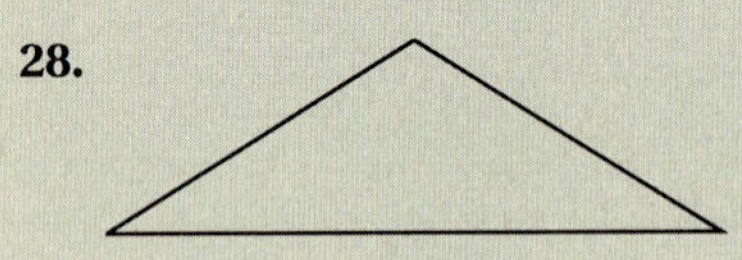

29.

30.

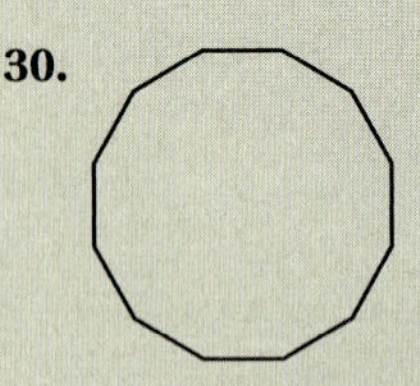

31.

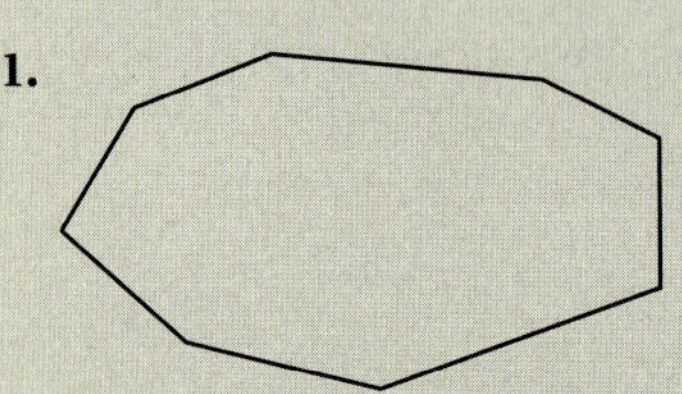

32. Find $m(\angle P) + m(\angle Q) + m(\angle R)$.

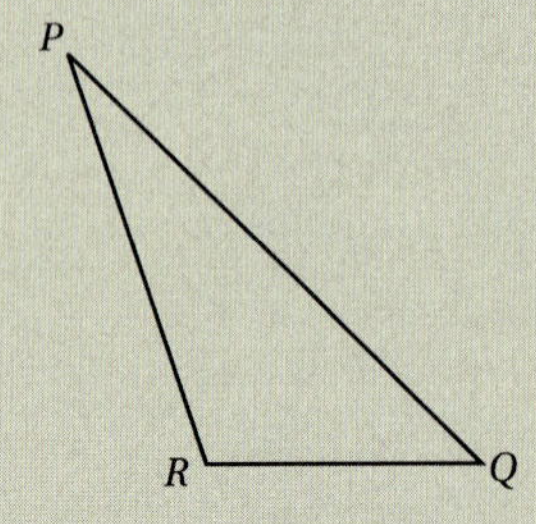

Answers on page A-12

We can further classify polygons as follows.

NUMBER OF SIDES	POLYGON	NUMBER OF SIDES	POLYGON
4	Quadrilateral	8	Octagon
5	Pentagon	9	Nonagon
6	Hexagon	10	Decagon
7	Heptagon	12	Dodecagon

Do Exercises 26–31.

f Sum of the Angle Measures of a Polygon

The sum of the angle measures of a triangle is 180°. To see this, note that we can think of cutting apart a triangle as shown on the left below. If we re-assemble the pieces, we see that a straight angle is formed.

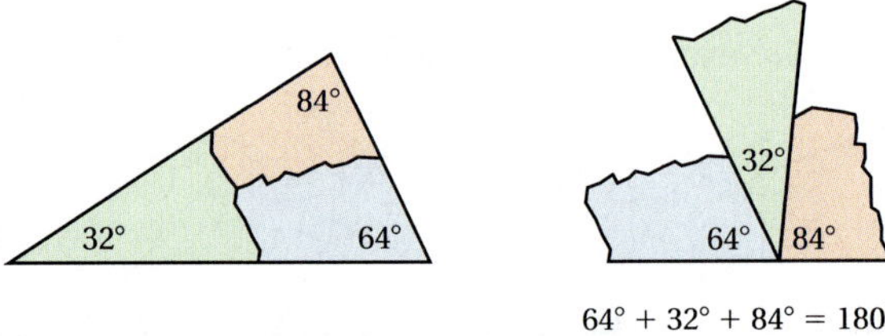

$64° + 32° + 84° = 180°$

SUM OF THE ANGLE MEASURES OF A TRIANGLE

In any triangle *ABC*, the sum of the measures of the angles is 180°:

$$m(\angle A) + m(\angle B) + m(\angle C) = 180°.$$

Do Exercise 32.

If we know the measures of two angles of a triangle, we can calculate the third.

EXAMPLE 1 Find the missing angle measure.

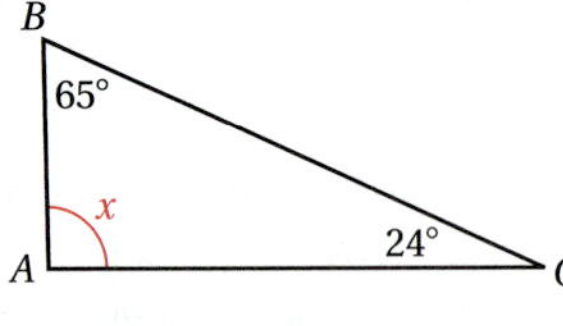

$$m(\angle A) + m(\angle B) + m(\angle C) = 180°$$
$$x + 65° + 24° = 180°$$
$$x + 89° = 180°$$
$$x = 180° - 89°$$
$$x = 91°$$

Do Exercise 33 on the following page.

Now let's use this idea to find the sum of the measures of the angles of a polygon of n sides. First let's consider a four-sided figure:

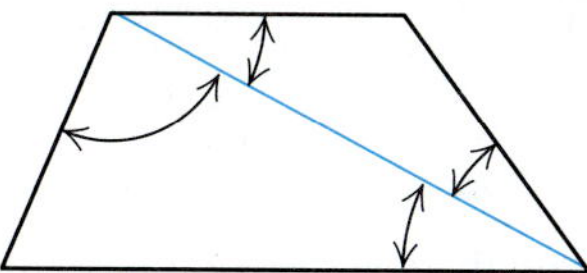

We can divide the figure into two triangles. The sum of the angle measures of each triangle is 180°. We have two triangles, so the sum of the angle measures of the figure is $2 \cdot 180°$, or 360°.

Do Exercise 34.

If a polygon has n sides, it can be divided into $n - 2$ triangles, each having 180° as the sum of its angle measures. Thus the sum of the angle measures of the polygon is $(n - 2) \cdot 180°$.

SUM OF ANGLE MEASURES

If a polygon has n sides, then the sum of its angle measures is $(n - 2) \cdot 180°$.

EXAMPLE 2 What is the sum of the angle measures of a hexagon?

A hexagon has 6 sides. We use the formula $(n - 2) \cdot 180°$:

$$\begin{aligned}(n - 2) \cdot 180° &= (6 - 2) \cdot 180° \\ &= 4 \cdot 180° \\ &= 720°.\end{aligned}$$

Do Exercises 35 and 36.

33. Find the missing angle measure.

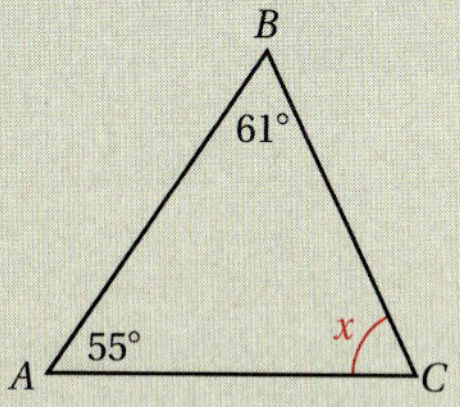

34. Consider a five-sided figure:

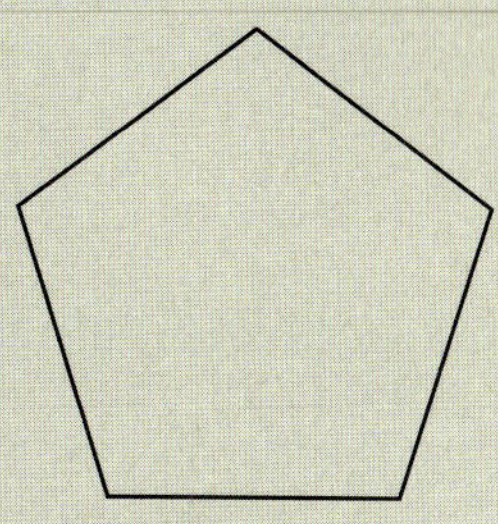

Complete.

a) The figure can be divided into ______ triangles.

b) The sum of the angle measures of each triangle is ______.

c) The sum of the angle measures of the polygon is ______ · 180°, or ______.

35. What is the sum of the angle measures of an octagon?

36. What is the sum of the angle measures of a 25-sided figure?

Answers on page A-12

5.1 EXERCISE SET

For Extra Help

Digital Video Tutor CD 4 Videotape 6 | InterAct Math | Math Tutor Center | MathXL | MyMathLab

a

1. Draw the segment whose endpoints are G and H. Name the segment in two ways.

G H

2. Draw the segment whose endpoints are C and D. Name the segment in two ways.

C D

3. Draw the ray with endpoint Q. Name the ray.

Q D

4. Draw the ray with endpoint D. Name the ray.

Q D

Name the line in seven different ways.

5.
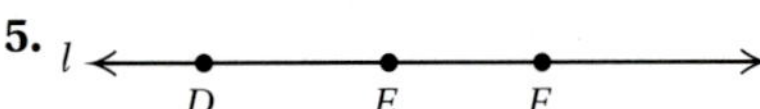

6.
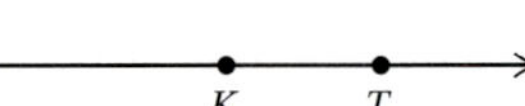

b

Name the angle in five different ways.

7.
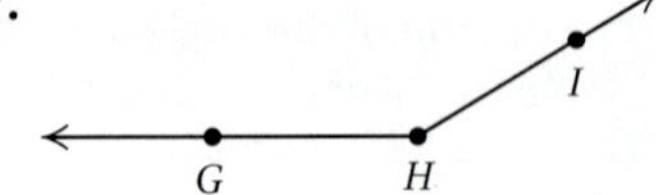

8.
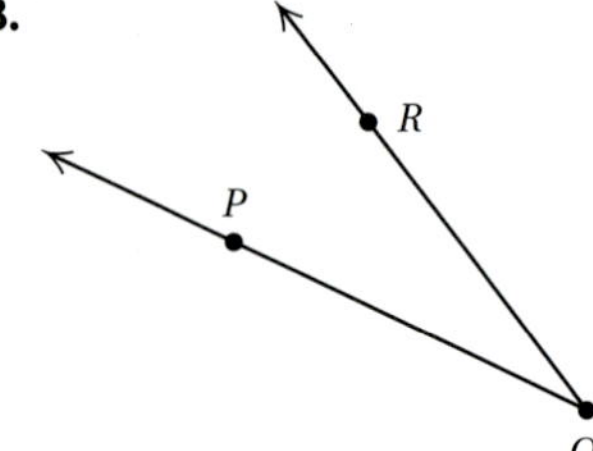

Use a protractor to measure the angle.

9.

10.
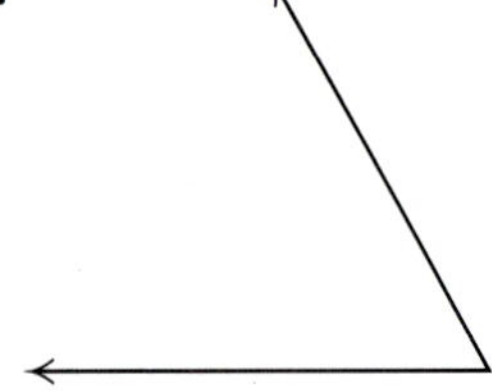

11.
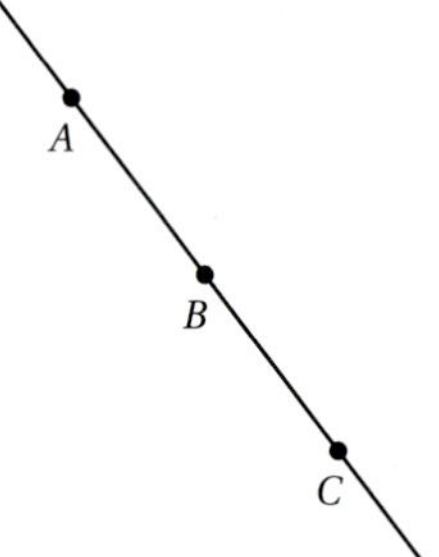

12.

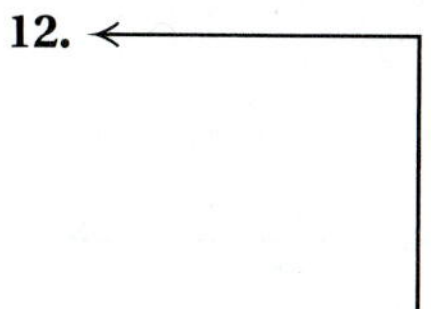

13.

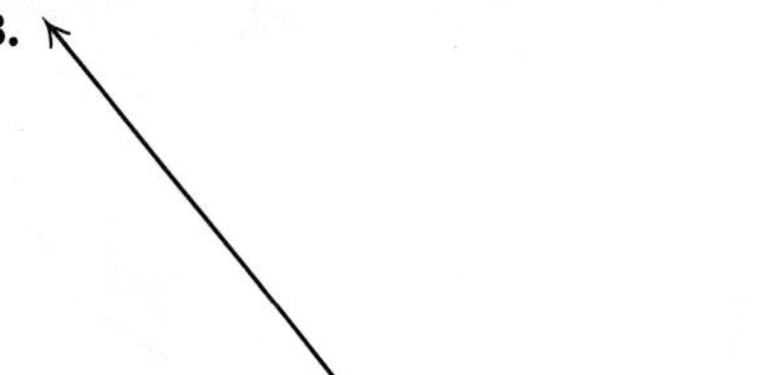

14.

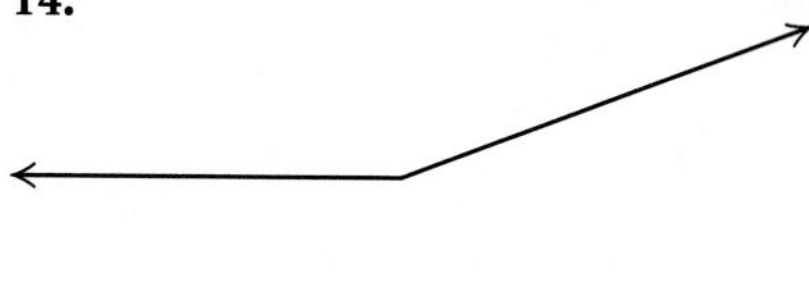

c

15.–22. Classify each of the angles in Exercises 7–14 as right, straight, acute, or obtuse.

23.–26. Classify each of the angles in Margin Exercises 12–15 as right, straight, acute, or obtuse.

d

Determine whether the pair of lines is perpendicular. Use a protractor.

27.

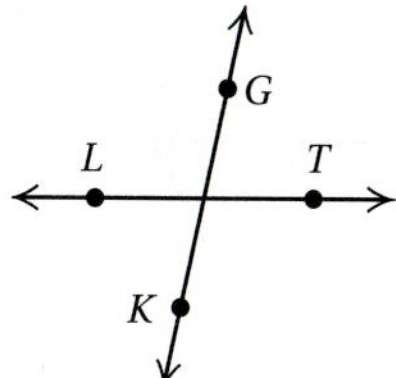

28.

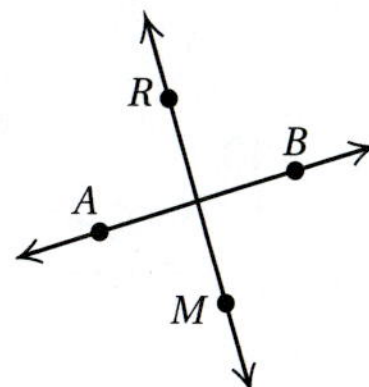

29.

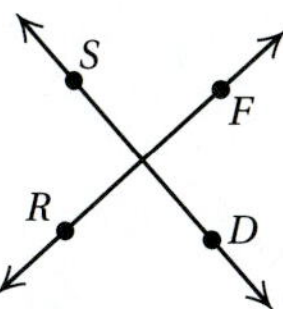

30.

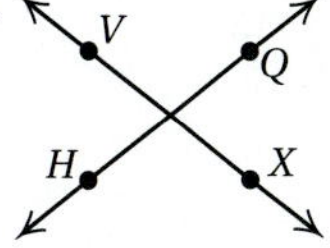

e

Classify the triangle as equilateral, isosceles, or scalene. Then classify it as right, obtuse, or acute.

31.

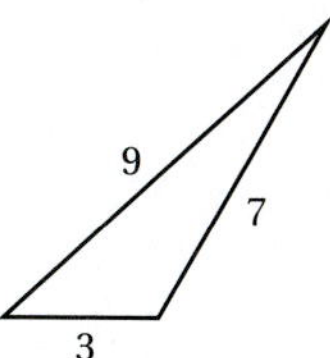

32.

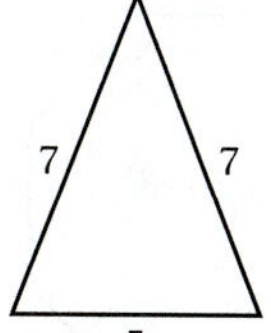

33.

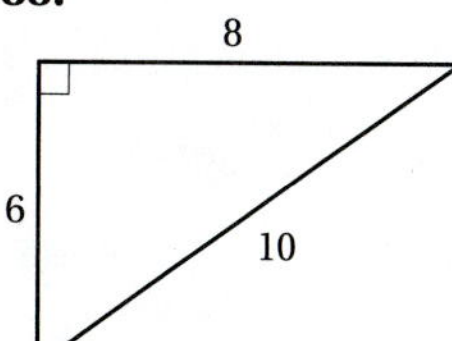

34.

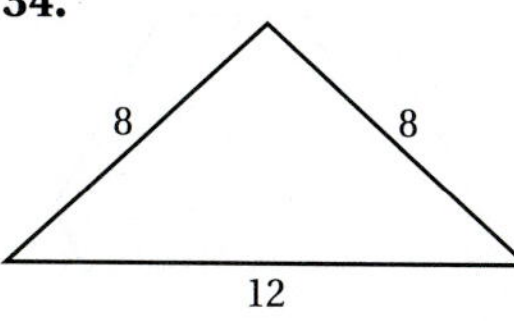

35.

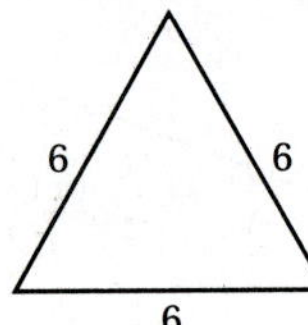

36.

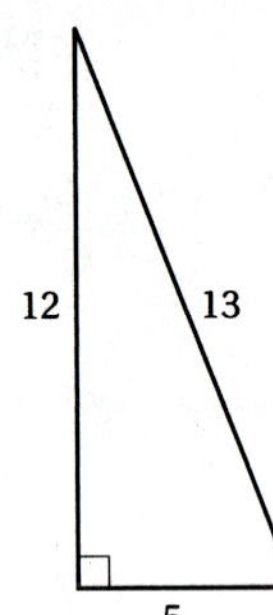

37.

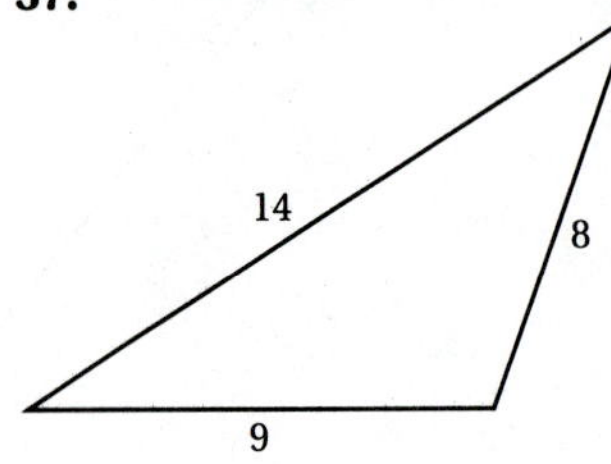

38.

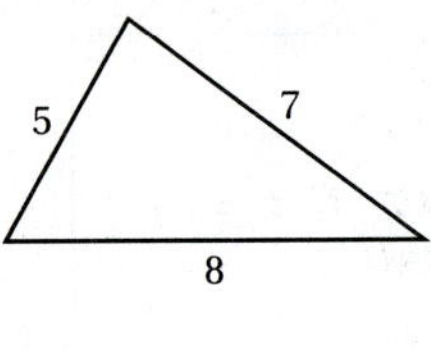

Classify the polygon by name.

39.

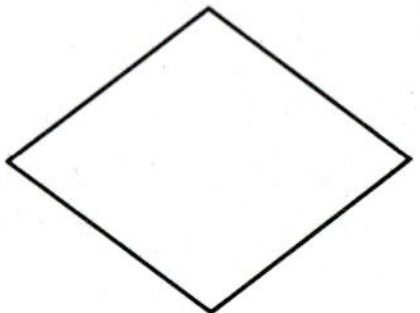

40.

41.

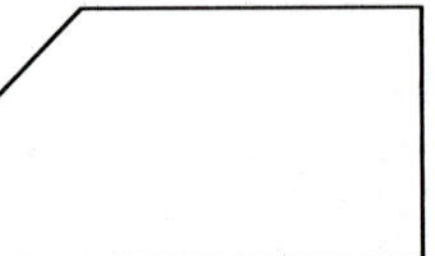

42.

43.

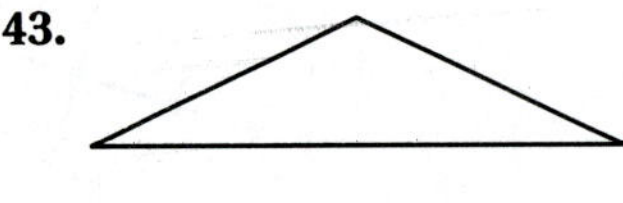

44.

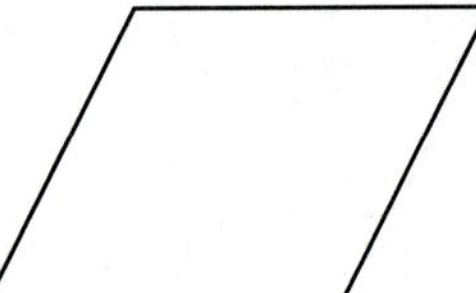

45.

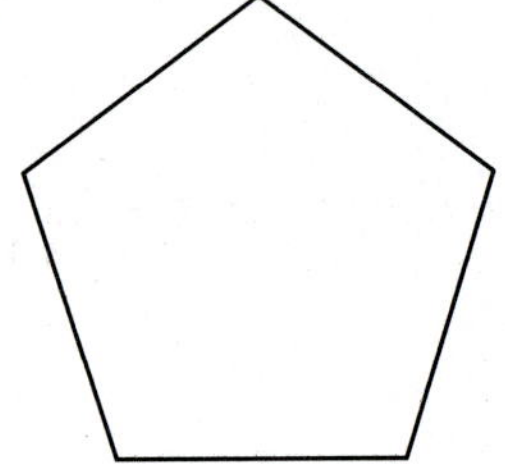

46.

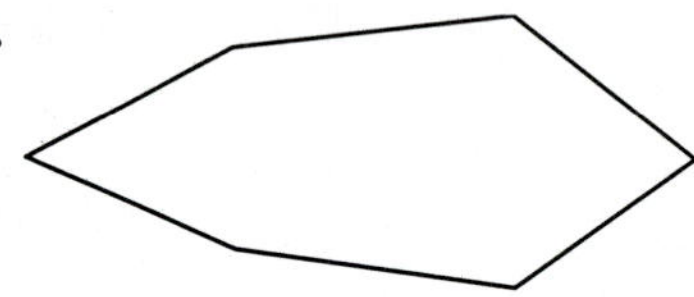

47.

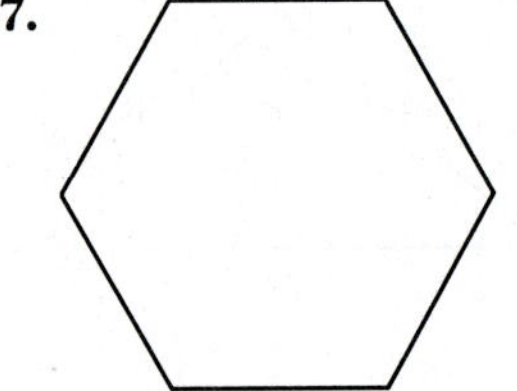

48.

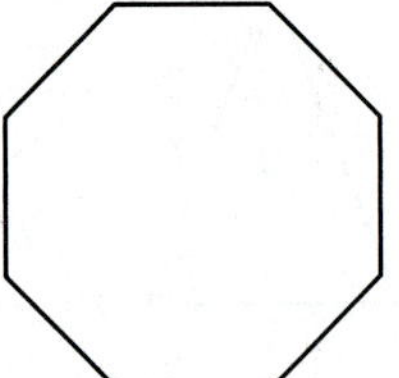

Find the sum of the angle measures of each of the following.

49. A decagon

50. A quadrilateral

51. A heptagon

52. A nonagon

53. A 14-sided polygon

54. A 17-sided polygon

55. A 20-sided polygon

56. A 32-sided polygon

Find the missing angle measure.

57.

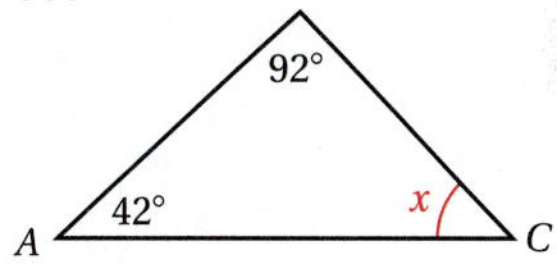

58.

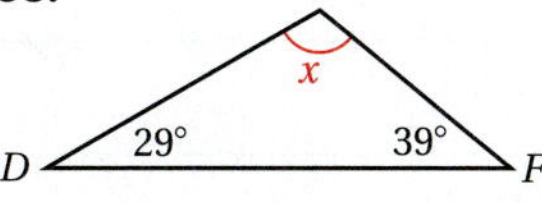

59.

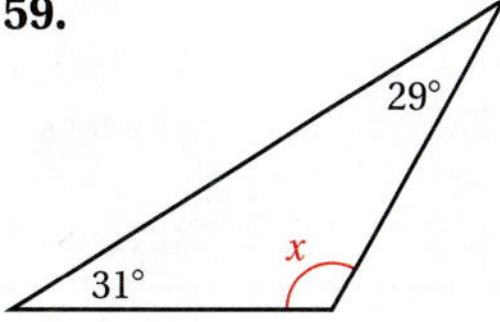

60.

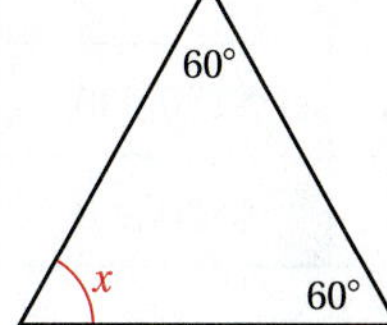

61. In $\triangle RST$, $m(\angle S) = 58°$ and $m(\angle T) = 79°$. Find $m(\angle R)$.

62. In $\triangle KNP$, $m(\angle K) = 137°$ and $m(\angle P) = 12°$. Find $m(\angle N)$.

63. DW Explain how you might use triangles to find the sum of the angle measures of this figure.

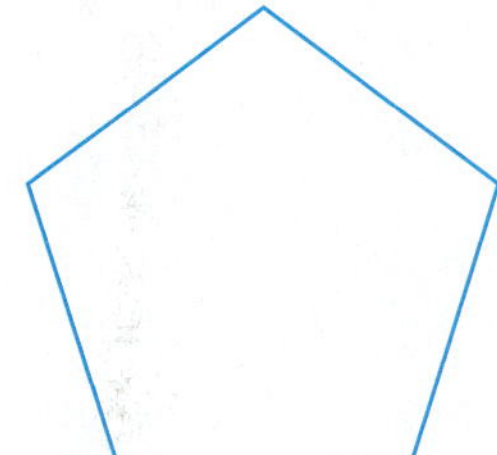

64. DW Determine whether the following statement is true or false and explain your answer.

All equilateral triangles are isosceles, but not all isosceles triangles are equilateral.

SKILL MAINTENANCE

Find the simple interest. [4.7d]

	PRINCIPAL	RATE OF INTEREST	TIME	SIMPLE INTEREST
65.	\$2000	8%	1 year	
66.	\$750	6%	$\frac{1}{2}$ year	
67.	\$4000	7.4%	$\frac{1}{2}$ year	
68.	\$200,000	6.7%	$\frac{1}{12}$ year	

Interest is compounded semiannually. Find the amount in the account after the given length of time. Round to the nearest cent. [4.7e]

	PRINCIPAL	RATE OF INTEREST	TIME	AMOUNT IN THE ACCOUNT
69.	\$25,000	6%	5 years	
70.	\$150,000	$6\frac{7}{8}\%$	15 years	
71.	\$150,000	7.4%	20 years	
72.	\$160,000	7.4%	20 years	

SYNTHESIS

73. In the figure, $m\angle 1 = 79.8°$ and $m\angle 6 = 33.07°$. Find $m\angle 2$, $m\angle 3$, $m\angle 4$, and $m\angle 5$.

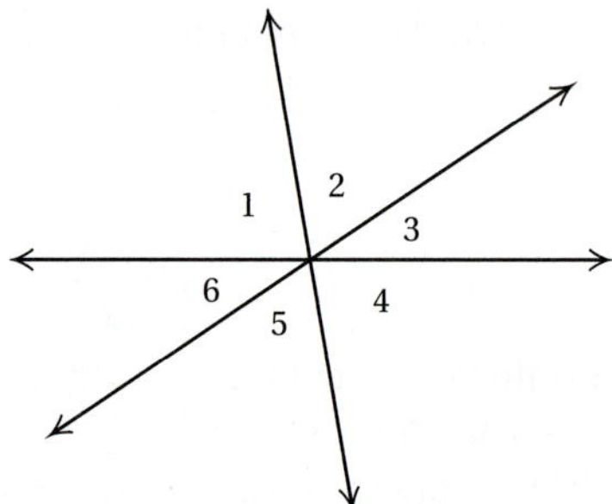

74. In the figure, $m\angle 2 = 42.17°$ and $m\angle 3 = 81.9°$. Find $m\angle 1$, $m\angle 4$, $m\angle 5$, and $m\angle 6$.

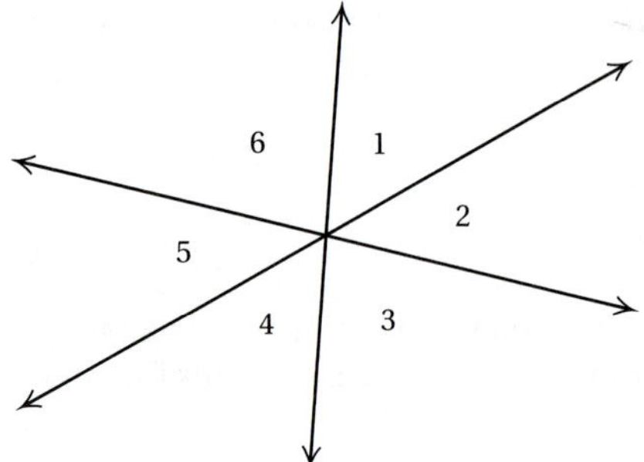

75. Find $m\angle ACB$, $m\angle CAB$, $m\angle EBC$, $m\angle EBA$, $m\angle AEB$, and $m\angle ADB$ in the rectangle shown below.

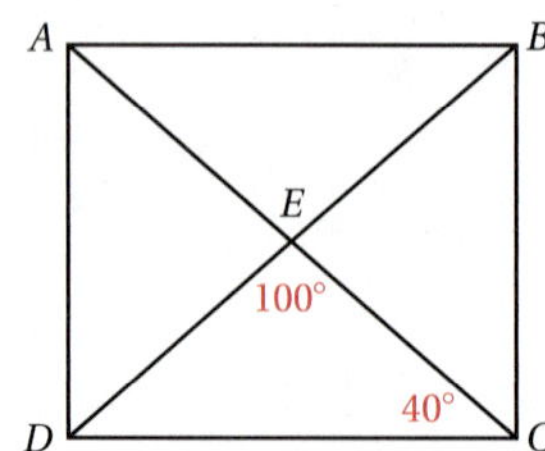

5.2 PERIMETER

a Finding Perimeters

PERIMETER OF A POLYGON

A **polygon** is a geometric figure with three or more sides. The **perimeter** of a **polygon** is the distance around it, or the sum of the lengths of its sides.

EXAMPLE 1 Find the perimeter of this polygon.

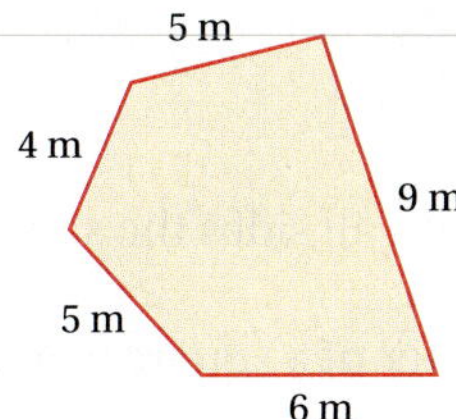

We add the lengths of the sides. Since all units are the same, we add the numbers, keeping meters (m) as the unit.

$$\begin{aligned}\text{Perimeter} &= 6\text{ m} + 5\text{ m} + 4\text{ m} + 5\text{ m} + 9\text{ m}\\ &= (6 + 5 + 4 + 5 + 9)\text{ m}\\ &= 29\text{ m}\end{aligned}$$

Do Exercises 1 and 2.

A **rectangle** is a figure with four sides and four 90°-angles, like the one shown in Example 2.

EXAMPLE 2 Find the perimeter of a rectangle that is 3 cm by 4 cm.

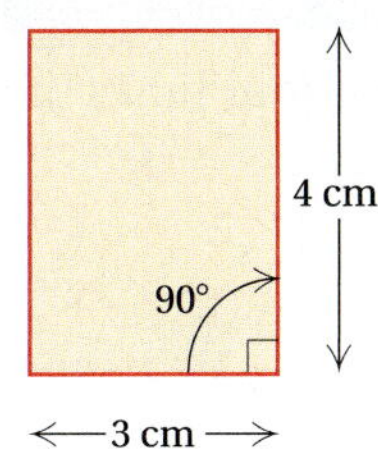

$$\begin{aligned}\text{Perimeter} &= 3\text{ cm} + 3\text{ cm} + 4\text{ cm} + 4\text{ cm}\\ &= (3 + 3 + 4 + 4)\text{ cm}\\ &= 14\text{ cm}\end{aligned}$$

Do Exercise 3.

Objectives

a Find the perimeter of a polygon.

b Solve applied problems involving perimeter.

Find the perimeter of the polygon.

1.

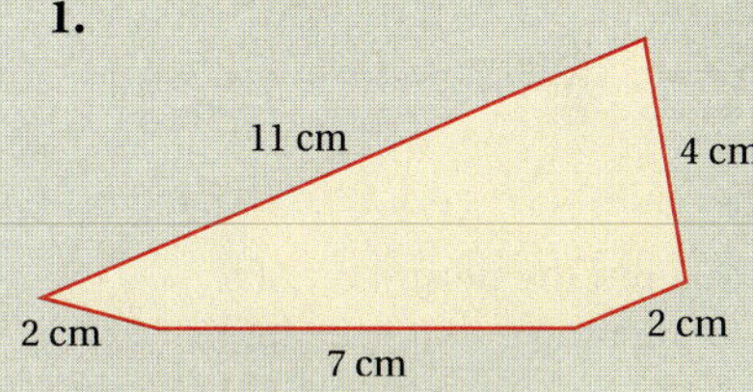

2.

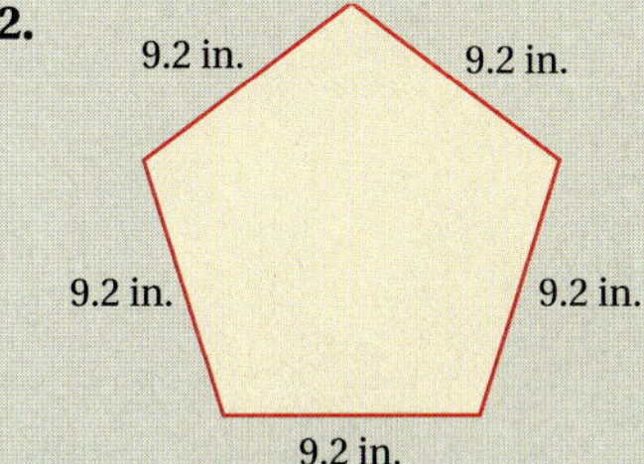

3. Find the perimeter of a rectangle that is 2 cm by 4 cm.

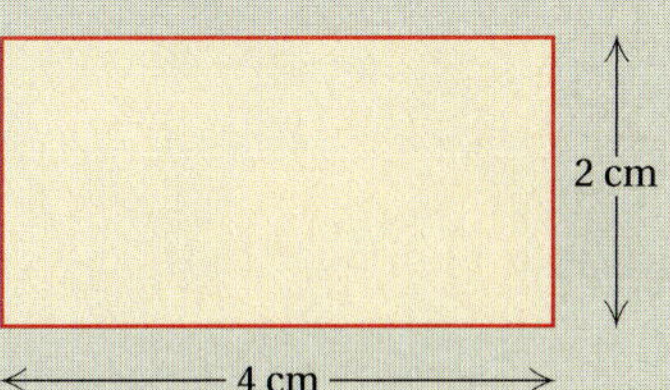

Answers on page A-12

4. Find the perimeter of a rectangle that is 5.25 yd by 3.5 yd.

5. Find the perimeter of a rectangle that is $8\frac{1}{4}$ in. by $5\frac{2}{3}$ in.

6. Find the perimeter of a square with sides of length 10 km.

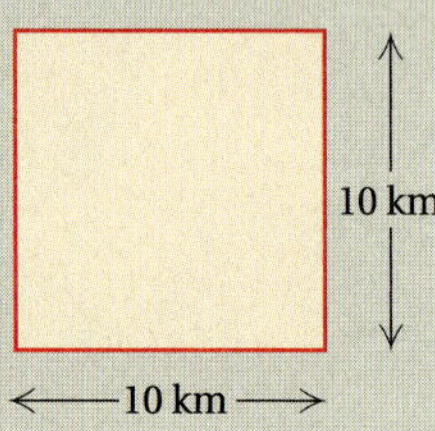

Answers on page A-12

PERIMETER OF A RECTANGLE

The **perimeter of a rectangle** is twice the sum of the length and the width, or 2 times the length plus 2 times the width:

$$\mathbf{P = 2 \cdot (l + w),} \quad \text{or} \quad \mathbf{P = 2 \cdot l + 2 \cdot w.}$$

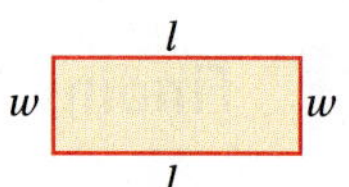

EXAMPLE 3 Find the perimeter of a rectangle that is 4.3 ft by 7.8 ft.

$$\begin{aligned} P &= 2 \cdot (l + w) \\ &= 2 \cdot (4.3 \text{ ft} + 7.8 \text{ ft}) \\ &= 2 \cdot (12.1 \text{ ft}) \\ &= 24.2 \text{ ft} \end{aligned}$$

Do Exercises 4 and 5.

A **square** is a rectangle with all sides the same length.

EXAMPLE 4 Find the perimeter of a square whose sides are 9 mm long.

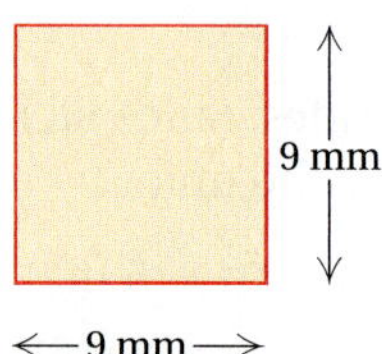

$$\begin{aligned} P &= 9 \text{ mm} + 9 \text{ mm} + 9 \text{ mm} + 9 \text{ mm} \\ &= (9 + 9 + 9 + 9) \text{ mm} \\ &= 36 \text{ mm} \end{aligned}$$

Do Exercise 6.

PERIMETER OF A SQUARE

The **perimeter of a square** is four times the length of a side:

$$\mathbf{P = 4 \cdot s.}$$

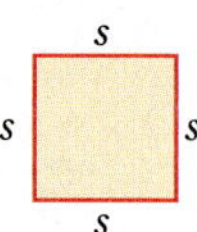

EXAMPLE 5 Find the perimeter of a square whose sides are $20\frac{1}{8}$ in. long.

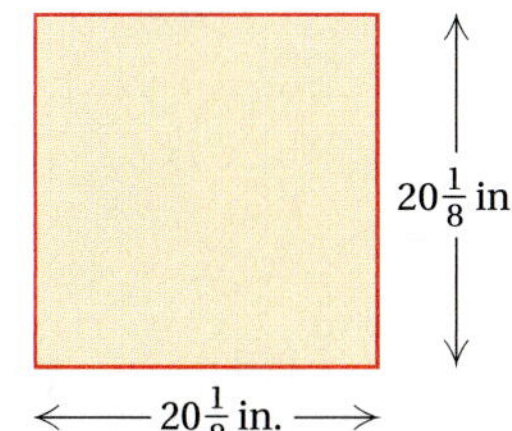

$$P = 4 \cdot s = 4 \cdot 20\frac{1}{8} \text{ in.}$$

$$= 4 \cdot \frac{161}{8} \text{ in.} = \frac{4 \cdot 161}{4 \cdot 2} \text{ in.}$$

$$= \frac{161}{2} \cdot \frac{4}{4} \text{ in.} = 80\frac{1}{2} \text{ in.}$$

Do Exercises 7 and 8.

7. Find the perimeter of a square with sides of length $5\frac{1}{4}$ yd.

8. Find the perimeter of a square with sides of length 7.8 km.

b Solving Applied Problems

EXAMPLE 6 A vegetable garden is 20 ft by 15 ft. A fence is to be built around the garden. How many feet of fence will be needed? If fencing sells for $2.95 per foot, what will the fencing cost?

1. **Familiarize.** We make a drawing and let $P =$ the perimeter.

2. **Translate.** The perimeter of the garden is given by

$$P = 2 \cdot (l + w) = 2 \cdot (20 \text{ ft} + 15 \text{ ft}).$$

3. **Solve.** We calculate the perimeter as follows:

$$P = 2 \cdot (20 \text{ ft} + 15 \text{ ft}) = 2 \cdot (35 \text{ ft}) = 70 \text{ ft}$$

Then we multiply by $2.95 to find the cost of the fencing:

$$\text{Cost} = \$2.95 \times \text{Perimeter} = \$2.95 \times 70 \text{ ft} = \$206.50.$$

4. **Check.** The check is left to the student.
5. **State.** The 70 ft of fencing that is needed will cost $206.50.

Do Exercise 9.

9. A play area is 25 ft by 10 ft. A fence is to be built around the play area. How many feet of fencing will be needed? If fencing costs $4.95 per foot, what will the fencing cost?

Answers on page A-12

5.2 EXERCISE SET

For Extra Help

Digital Video Tutor CD 4 Videotape 6 | InterAct Math | Math Tutor Center | MathXL | MyMathLab

a Find the perimeter of the polygon.

1.

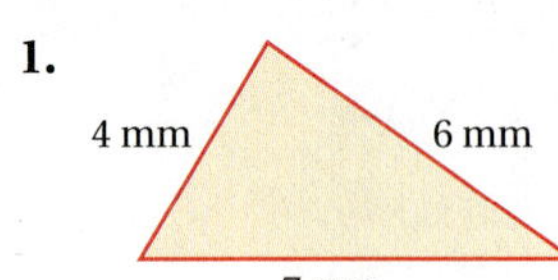

2.

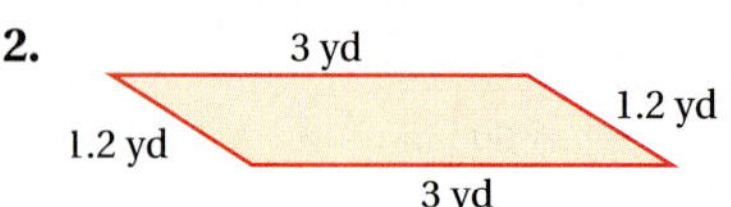

3.

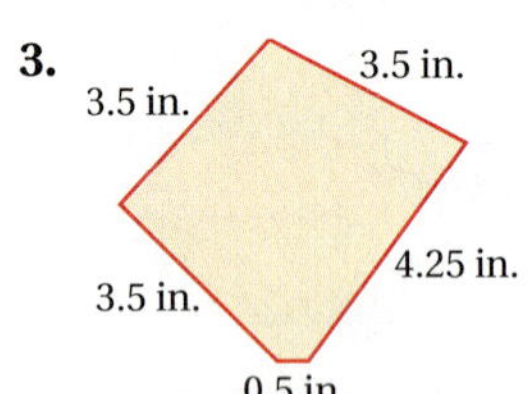

4.

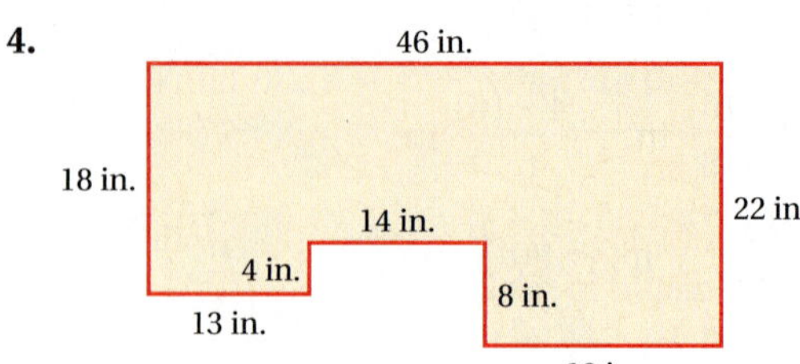

5.

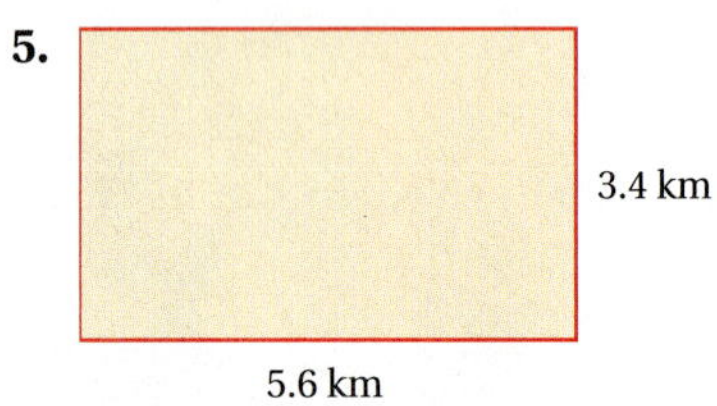

6.

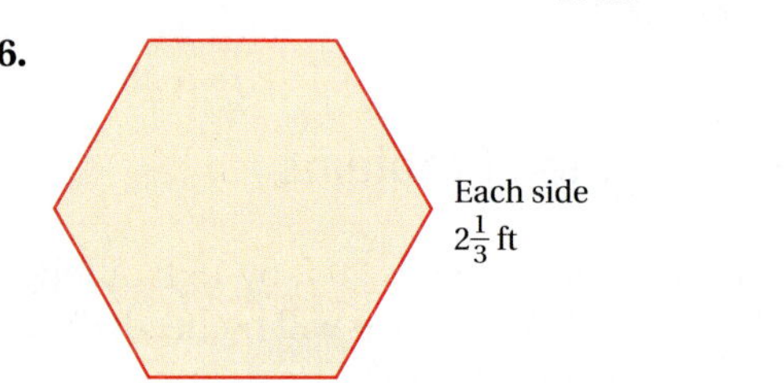

Find the perimeter of the rectangle.

7. 5 ft by 10 ft

8. 2.5 m by 100 m

9. 34.67 cm by 4.9 cm

10. $3\frac{1}{2}$ yd by $4\frac{1}{2}$ yd

Find the perimeter of the square.

11. 22 ft on a side

12. 56.9 km on a side

13. 45.5 mm on a side

14. $3\frac{1}{8}$ yd on a side

Solve.

15. A security fence is to be built around a 173-m by 240-m rectangular field. What is the perimeter of the field? If fence wire costs $1.45 per meter, what will the fencing cost?

16. *Softball Diamond.* A standard-sized slow-pitch softball diamond is a square with sides of length 65 ft. What is the perimeter of this softball diamond? (This is the distance you would have to run if you hit a home run.)
Source: American Softball Association

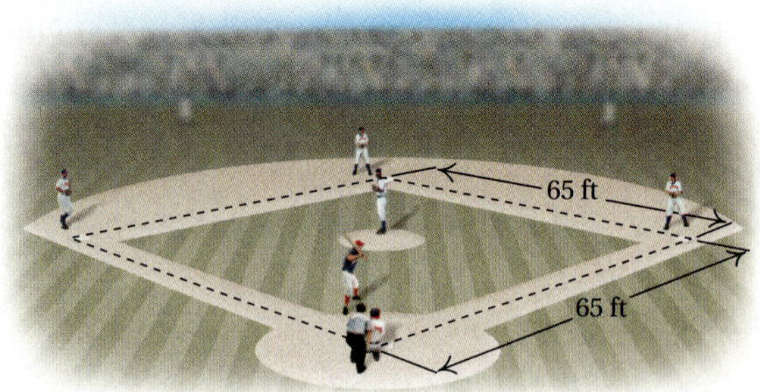

17. A piece of flooring tile is a square with sides of length 30.5 cm. What is the perimeter of a piece of tile?

18. A rectangular posterboard is 61.8 cm by 87.9 cm. What is the perimeter of the board?

19. A rain gutter is to be installed around the house shown in the figure.

a) Find the perimeter of the house.
b) If the gutter costs $4.59 per foot, what is the total cost of the gutter?

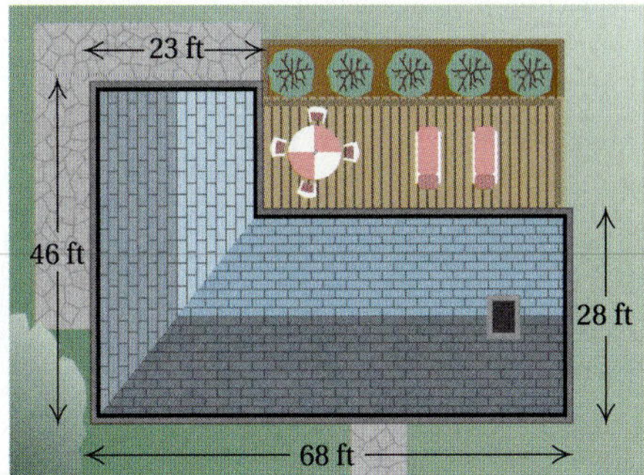

20. A carpenter is to build a fence around a 9-m by 12-m garden.

a) The posts are 3 m apart. How many posts will be needed?
b) The posts cost $2.40 each. How much will the posts cost?
c) The fence will surround all but 3 m of the garden, which will be a gate. How long will the fence be?
d) The fence costs $2.85 per meter. What will the cost of the fence be?
e) The gate costs $9.95. What is the total cost of the materials?

21. DW Create for a fellow student a development of the formula

$$P = 2 \cdot (l + w) = 2 \cdot l + 2 \cdot w$$

for the perimeter of a rectangle.

22. DW Create for a fellow student a development of the formula

$$P = 4 \cdot s$$

for the perimeter of a square.

SKILL MAINTENANCE

23. Convert to decimal notation: 56.1%. [4.2b]

24. Convert to percent notation: 0.6734. [4.2b]

25. Convert to percent notation: $\frac{9}{8}$. [4.3a]

Evaluate. [1.6b]

26. 5^2

27. 10^2

28. 31^2

Convert the number in the sentence to standard notation. [3.3b]

29. It is estimated that 4.7 million fax machines were sold in a recent year.

30. In a recent year, 4.3 billion CDs were sold.

SYNTHESIS

Find the perimeter, in feet, of the figure.

31.

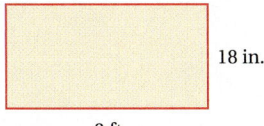

32.

5.3 AREA

Objectives

a Find the area of a rectangle and a square.

b Find the area of a parallelogram, a triangle, and a trapezoid.

c Solve applied problems involving areas of rectangles, squares, parallelograms, triangles, and trapezoids.

1. What is the area of this region? Count the number of square centimeters.

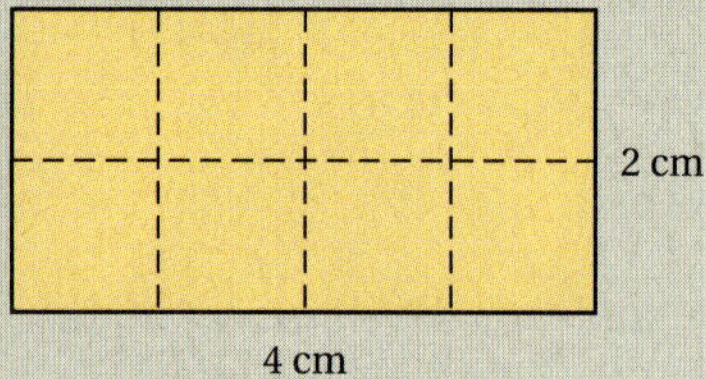

2. Find the area of a rectangle that is 7 km by 8 km.

3. Find the area of a rectangle that is $5\frac{1}{4}$ yd by $3\frac{1}{2}$ yd.

Answers on page A-12

a Rectangles and Squares

A polygon and its interior form a plane region. We can find the area of a *rectangular region,* or *rectangle,* by filling it in with square units. Two such units, a *square inch* and a *square centimeter,* are shown below.

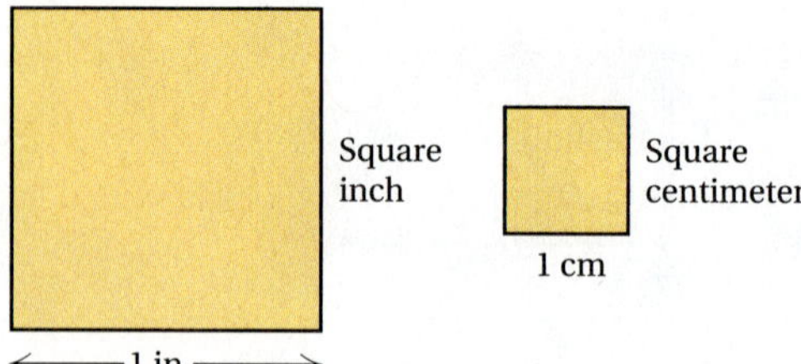

EXAMPLE 1 What is the area of this region?

We have a rectangular array. Since the region is filled with 12 square centimeters, its area is 12 square centimeters (sq cm), or 12 cm^2. The number of units is 3×4, or 12.

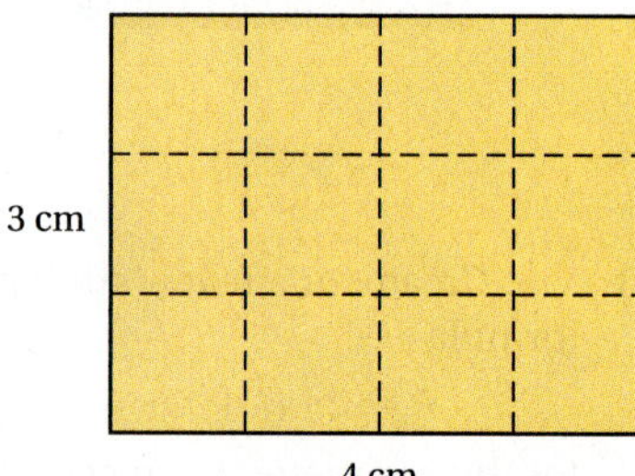

Do Exercise 1.

AREA OF A RECTANGLE

The **area of a rectangle** is the product of the length l and the width w:

$$A = l \cdot w.$$

w

l

EXAMPLE 2 Find the area of a rectangle that is 7 yd by 4 yd.

$$A = l \cdot w = 7\text{ yd} \cdot 4\text{ yd}$$
$$= 7 \cdot 4 \cdot \text{yd} \cdot \text{yd} = 28\text{ yd}^2$$

We think of yd · yd as $(\text{yd})^2$ and denote it yd^2. Thus we read "28 yd^2" as "28 square yards."

Do Exercises 2 and 3.

EXAMPLE 3 Find the area of a square with sides of length 9 mm.

$$A = (9\text{ mm}) \cdot (9\text{ mm})$$
$$= 9 \cdot 9 \cdot \text{mm} \cdot \text{mm}$$
$$= 81\text{ mm}^2$$

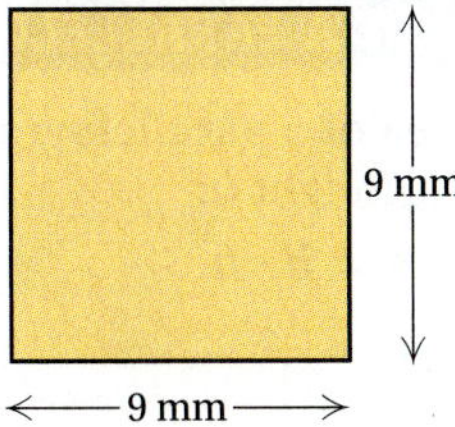

Do Exercise 4.

4. Find the area of a square with sides of length 12 km.

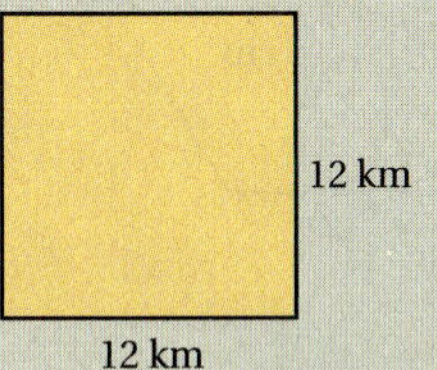

AREA OF A SQUARE

The **area of a square** is the square of the length of a side:

$\boldsymbol{A = s \cdot s}$, or $\boldsymbol{A = s^2}$.

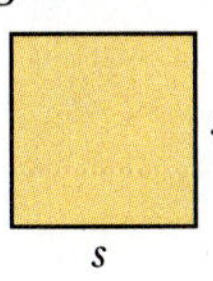

EXAMPLE 4 Find the area of a square with sides of length 20.3 m.

$A = s \cdot s = 20.3\text{ m} \times 20.3\text{ m} = 20.3 \times 20.3 \times \text{m} \times \text{m} = 412.09\text{ m}^2$

Do Exercises 5 and 6.

5. Find the area of a square with sides of length 10.9 m.

6. Find the area of a square with sides of length $3\frac{1}{2}$ yd.

Answers on page A-12

b Finding Other Areas

PARALLELOGRAMS

A **parallelogram** is a four-sided figure with two pairs of parallel sides, as shown below.

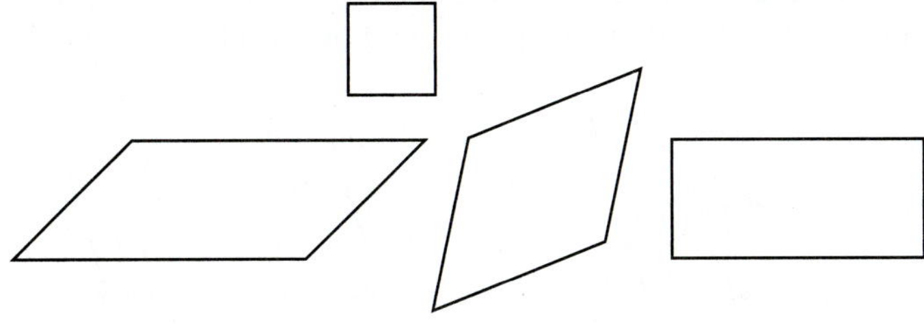

To find the area of a parallelogram, consider the one below.

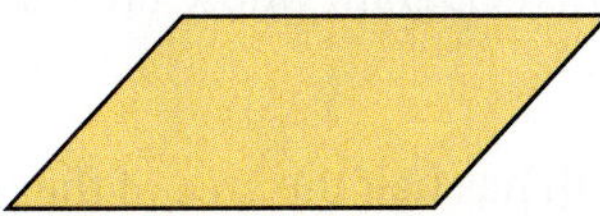

If we cut off a piece and move it to the other end, we get a rectangle.

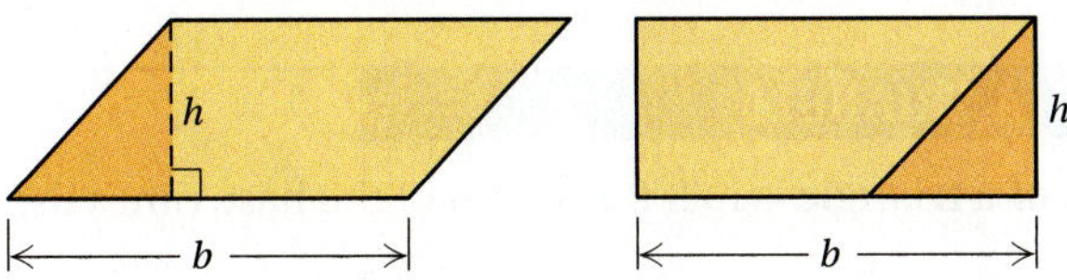

We can find the area by multiplying the length b, called a **base,** by h, called the **height.**

Study Tips

STUDYING THE ART PIECES

When you study a section of a mathematics text, read it slowly, observing all the details of the corresponding art pieces that are discussed in the paragraphs. Also note the precise color markings in the art that enhances the learning process. These tips apply especially to this chapter because geometry, by its nature, is quite visual.

Find the area.

7.

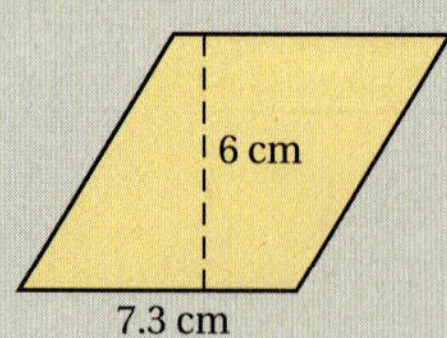

8.

5.5 km

2.25 km

Answers on page A-12

AREA OF A PARALLELOGRAM

The **area of a parallelogram** is the product of the length of a base b and the height h:

$$A = b \cdot h.$$

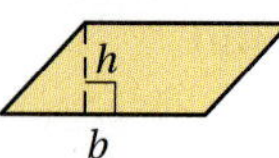

EXAMPLE 5 Find the area of this parallelogram.

$$A = b \cdot h$$
$$= 7 \text{ km} \cdot 5 \text{ km}$$
$$= 35 \text{ km}^2$$

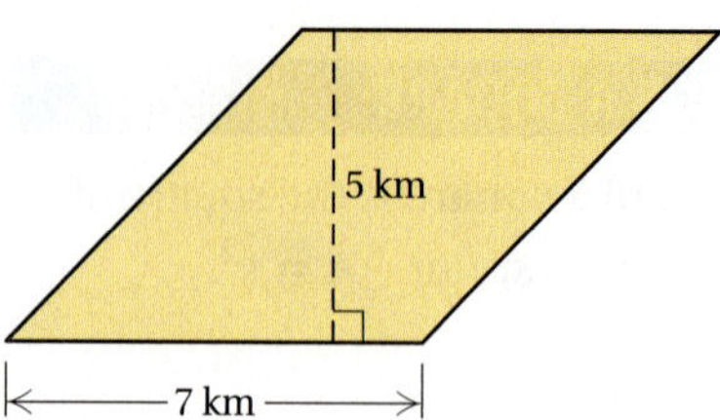

EXAMPLE 6 Find the area of this parallelogram.

$$A = b \cdot h$$
$$= 1.2 \text{ m} \times 6 \text{ m}$$
$$= 7.2 \text{ m}^2$$

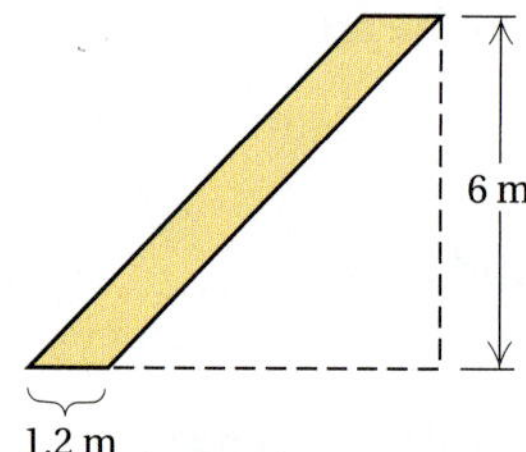

Do Exercises 7 and 8.

TRIANGLES

To find the area of a triangle like the one shown on the left below, think of cutting out another just like it and placing it as shown on the right below.

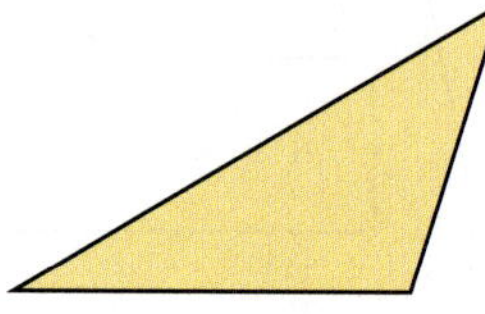

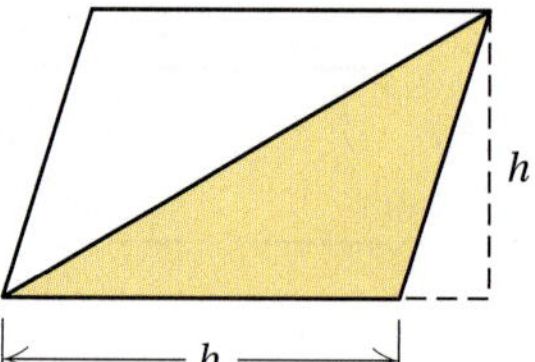

The resulting figure is a parallelogram whose area is

$$b \cdot h.$$

The triangle we started with has half the area of the parallelogram, or

$$\frac{1}{2} \cdot b \cdot h.$$

AREA OF A TRIANGLE

The **area of a triangle** is half the length of the base times the height:

$$A = \frac{1}{2} \cdot b \cdot h.$$

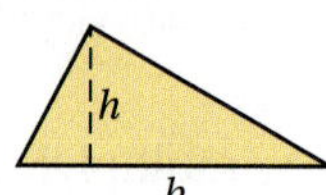

EXAMPLE 7 Find the area of this triangle.

$$A = \frac{1}{2} \cdot b \cdot h$$
$$= \frac{1}{2} \cdot 9\text{ m} \cdot 6\text{ m}$$
$$= \frac{9 \cdot 6}{2}\text{ m}^2$$
$$= 27\text{ m}^2$$

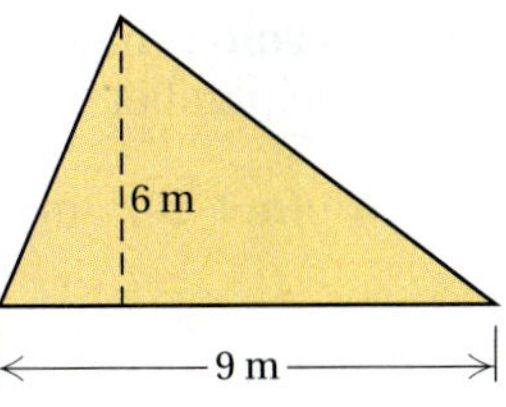

EXAMPLE 8 Find the area of this triangle.

$$A = \frac{1}{2} \cdot b \cdot h$$
$$= \frac{1}{2} \times 6.25\text{ cm} \times 5.5\text{ cm}$$
$$= 0.5 \times 6.25 \times 5.5\text{ cm}^2$$
$$= 17.1875\text{ cm}^2$$

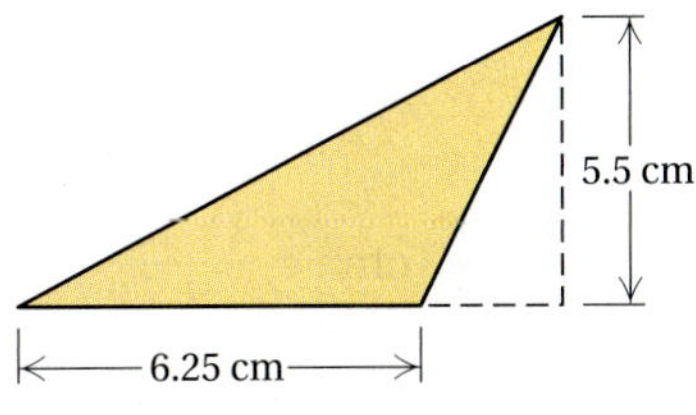

Do Exercises 9 and 10.

Find the area.

9.

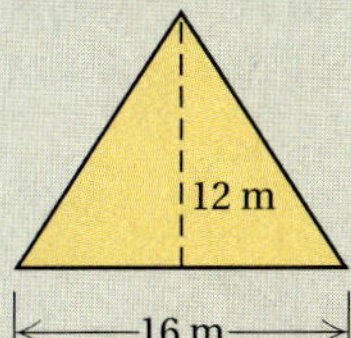

10.

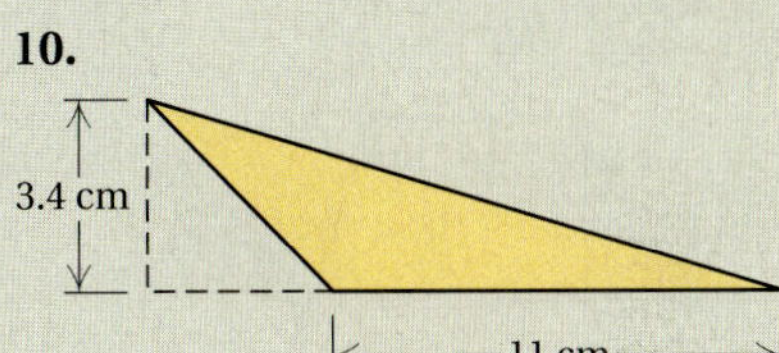

Answers on page A-12

TRAPEZOIDS

A **trapezoid** is a polygon with four sides, two of which, the **bases,** are parallel to each other.

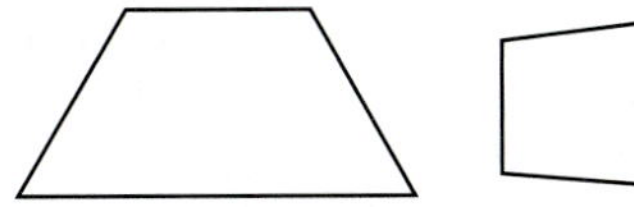

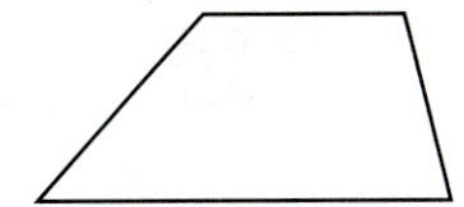

To find the area of a trapezoid, think of cutting out another just like it.

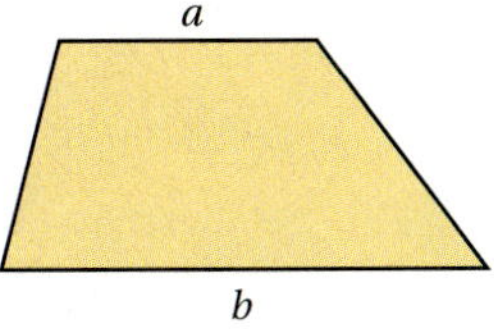

Then place the second one like this.

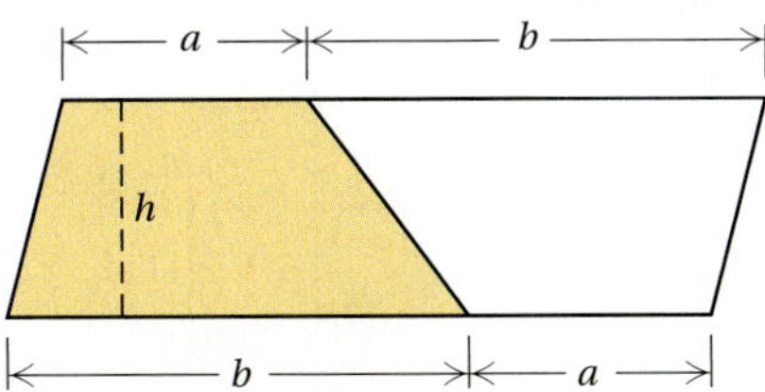

The resulting figure is a parallelogram whose area is

$h \cdot (a + b)$. The base is $a + b$.

The trapezoid we started with has half the area of the parallelogram, or

$$\frac{1}{2} \cdot h \cdot (a + b).$$

AREA OF A TRAPEZOID

The **area of a trapezoid** is half the product of the height and the sum of the lengths of the parallel sides (bases):

$$A = \frac{1}{2} \cdot h \cdot (a + b), \quad \text{or} \quad A = \frac{a + b}{2} \cdot h.$$

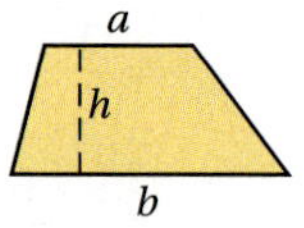

EXAMPLE 9 Find the area of this trapezoid.

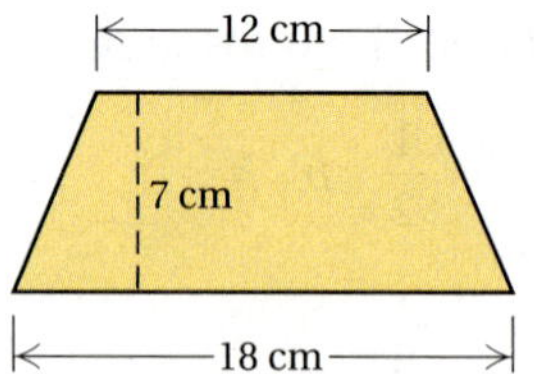

$$A = \frac{1}{2} \cdot h \cdot (a + b)$$

$$= \frac{1}{2} \cdot 7\text{ cm} \cdot (12 + 18)\text{ cm}$$

$$= \frac{7 \cdot 30}{2} \cdot \text{cm}^2 = \frac{7 \cdot 15 \cdot 2}{1 \cdot 2}\text{ cm}^2$$

$$= \frac{7 \cdot 15}{1} \cdot \frac{2}{2}\text{ cm}^2 = 105\text{ cm}^2$$

Do Exercises 11 and 12.

C Solving Applied Problems

EXAMPLE 10 *Baseball's Strike Zones.* Major league baseball underwent a rule change between the 2000 and 2001 seasons. Over the years before 2001, the strike zone evolved to something other than what was defined in the rule book. The zone in the rule book is described by rectangle *ABCD* below. The zone used before 2001 is described by the region *AQRST*. The figure shown here represents the zones for a normal-sized player, but they vary depending on the height of the player.

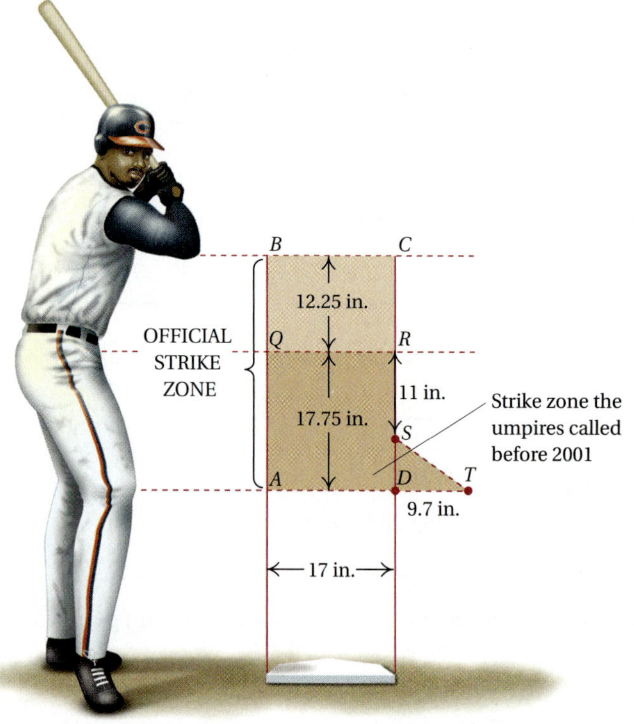

Find the area.

11.

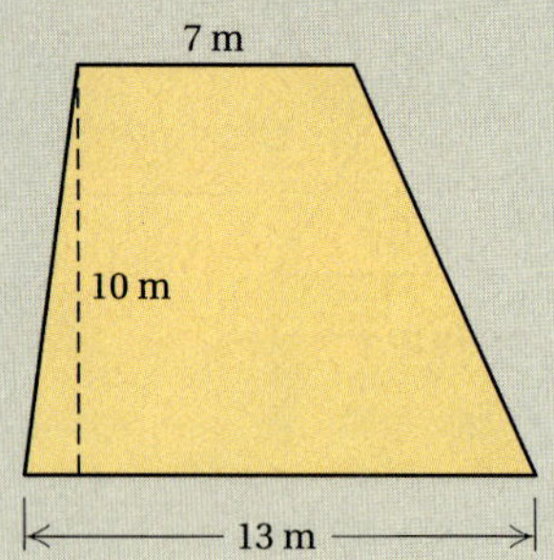

12.

6 cm

11 cm

10 cm

Answers on page A-12

a) Find the area of the new zone.

b) Find the area of the former zone.

c) How much larger is the rule-book zone than the former zone?

d) By what percent has the area of the strike zone been increased by the change?

Sources: *The Cincinnati Inquirer*; Major League Baseball; Gannett News Service; *The Sporting News Official Baseball Rules Book*

This is a multistep problem that makes use of many of the skills we have learned in this book.

a) The area of rectangle *ABCD* (the rule-book zone), denoted A_1, is the length times width:

$$A_1 = l \cdot w = (17.75 \text{ in.} + 12.25 \text{ in.}) \cdot (17 \text{ in.}) = (30 \text{ in.}) \cdot (17 \text{ in.}) = 510 \text{ in}^2.$$

b) The area of the region *AQRST* (the former zone), denoted A_2, is shown shaded in the figure. To find that area, we add the area of rectangle *AQRD* to the area of triangle *SDT*. We first determine

$$\text{Area of } AQRD = l \cdot w = (17.75 \text{ in.}) \cdot (17 \text{ in.}) = 301.75 \text{ in}^2.$$

To find the area of triangle *SDT*, we first note that the length of the base is given as 9.7 in. To find the height that is the length of segment *SD*, we subtract 11 in. from 17.75 in.: 17.75 in. − 11 in. = 6.75 in. Then

$$\text{Area of triangle } SDT = \frac{1}{2} \cdot b \cdot h = \frac{1}{2}(9.7 \text{ in.}) \cdot (6.75 \text{ in.}) = 32.7375 \text{ in}^2.$$

The area of the former zone A_2 is the sum of the areas of the triangle *SDT* and the rectangle *AQRD*:

$$A_2 = 301.75 \text{ in}^2 + 32.7375 \text{ in}^2 = 334.4875 \text{ in}^2 \approx 334.5 \text{ in}^2.$$

c) To find the increase in the area, we subtract A_2 from A_1:

$$510 \text{ in}^2 - 334.5 \text{ in}^2 = 175.5 \text{ in}^2.$$

d) To determine the percent of increase, note that we are asking "What percent of the former area is the increase?" We translate this to an equation as follows:

What percent of 334.5 is 175.5?

$$p \cdot 334.5 = 175.5$$

We solve the equation:

$$p \cdot 334.5 = 175.5$$

$$\frac{p \cdot 334.5}{334.5} = \frac{175.5}{334.5}$$

$$p = \frac{175.5}{334.5} \approx 0.5247 \approx 52\%.$$

There was an increase of approximately 52% in the strike zone.

Do Exercise 13.

13. Find the area of this kite.

Answer on page A-12

5.3 EXERCISE SET

For Extra Help

Digital Video Tutor CD 4 Videotape 6

InterAct Math

Math Tutor Center

MathXL

MyMathLab

a Find the area.

1.

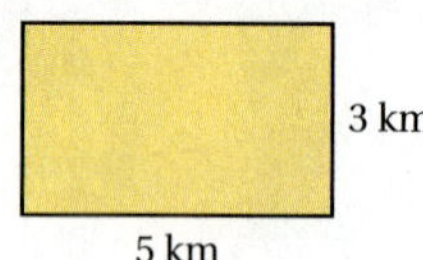

2.

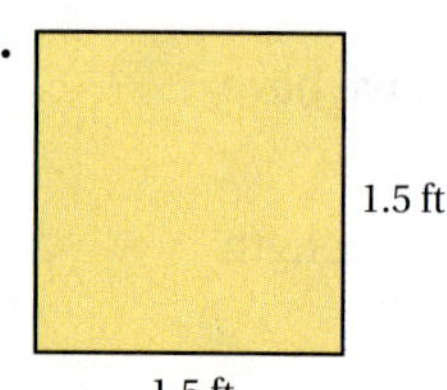

3.

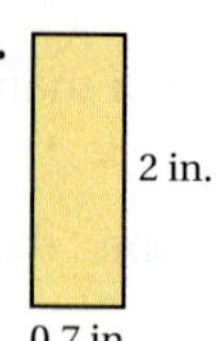

4.

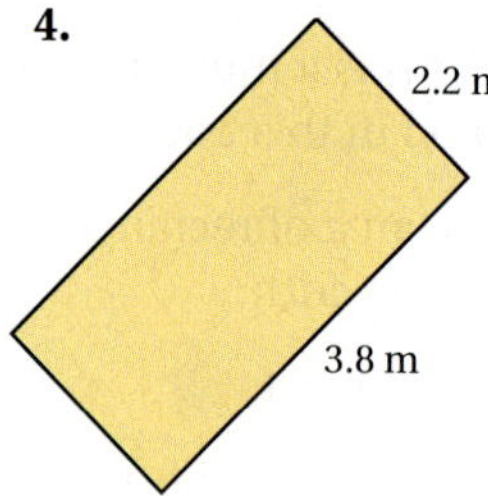

5.

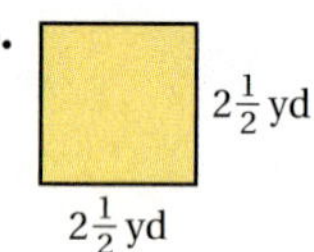

6.

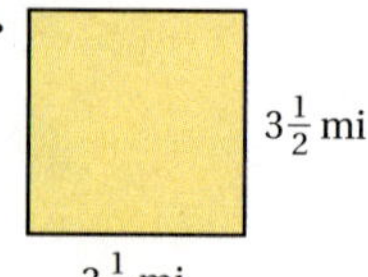

7.

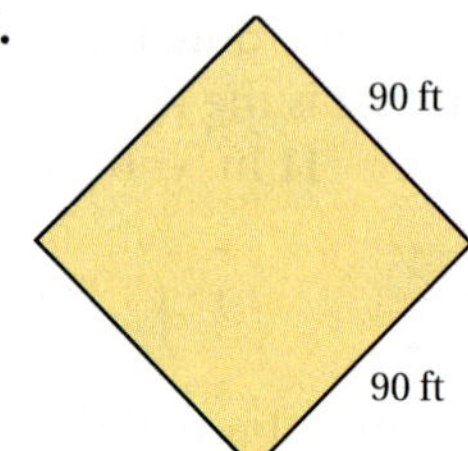

8.

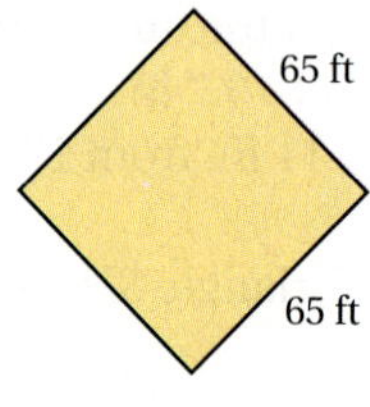

Find the area of the rectangle.

9. 5 ft by 10 ft

10. 14 yd by 8 yd

11. 34.67 cm by 4.9 cm

12. 2.45 km by 100 km

13. $4\frac{2}{3}$ in. by $8\frac{5}{6}$ in.

14. $10\frac{1}{3}$ mi by $20\frac{2}{3}$ mi

Find the area of the square.

15. 22 ft on a side

16. 18 yd on a side

17. 56.9 km on a side

18. 45.5 m on a side

19. $5\frac{3}{8}$ yd on a side

20. $7\frac{2}{3}$ ft on a side

b Find the area.

21.

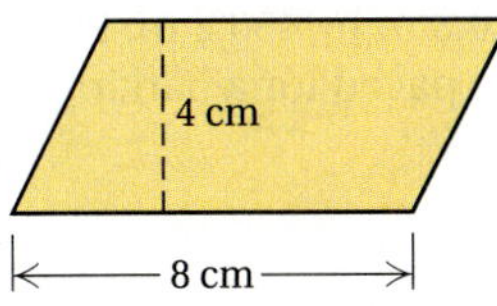

22.

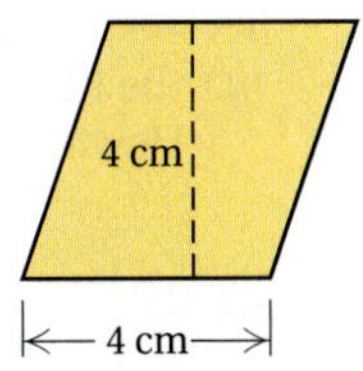

23.

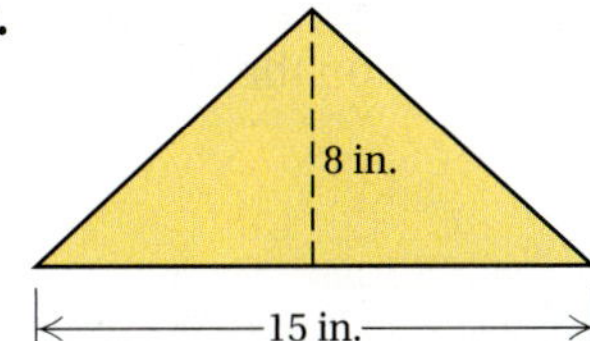

24.

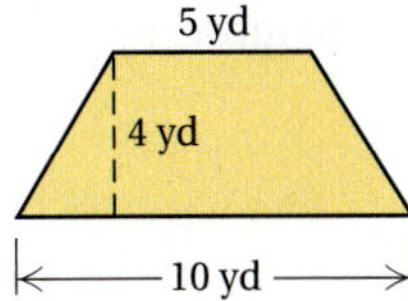

25.

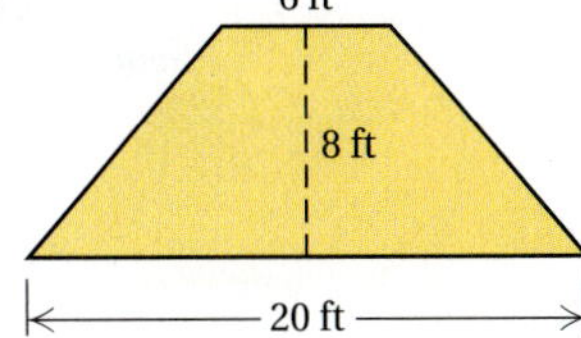

26.

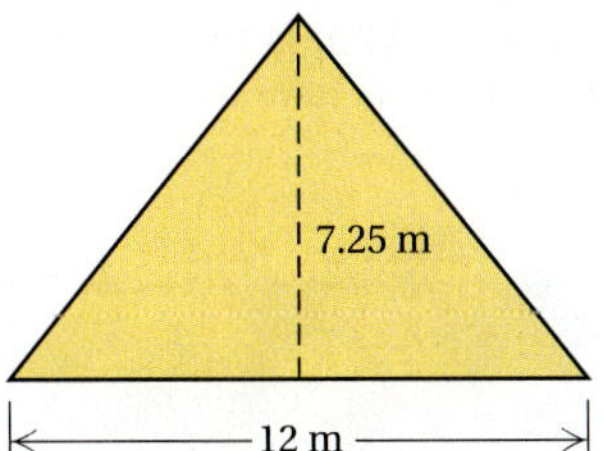

27.

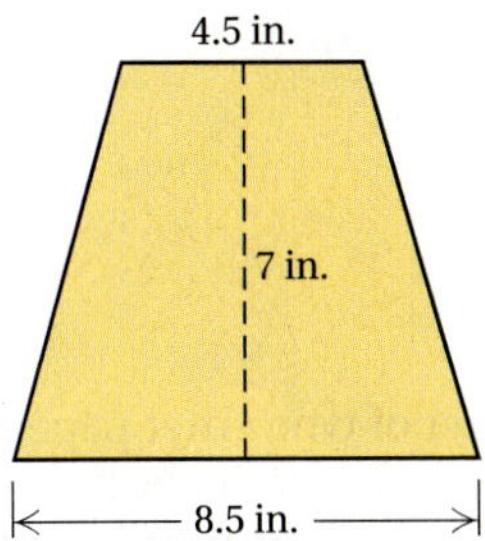

28.

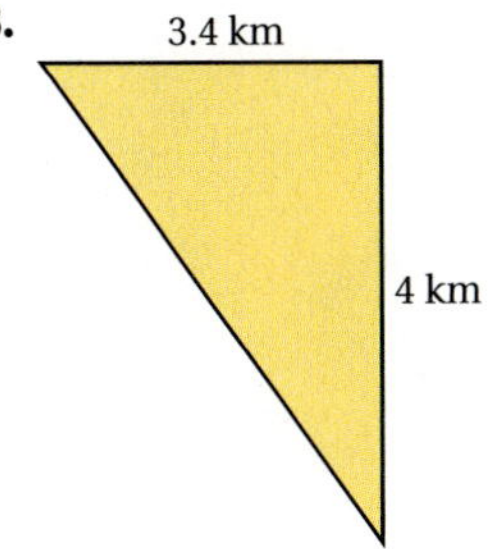

29.

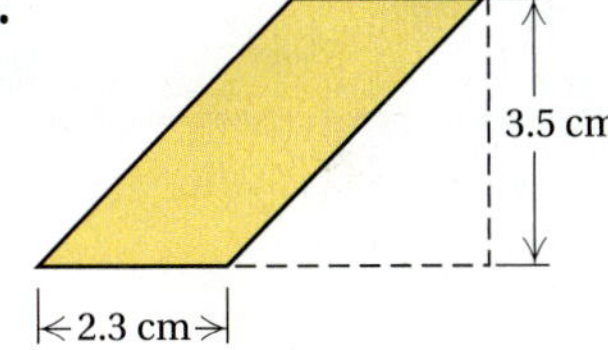

30.

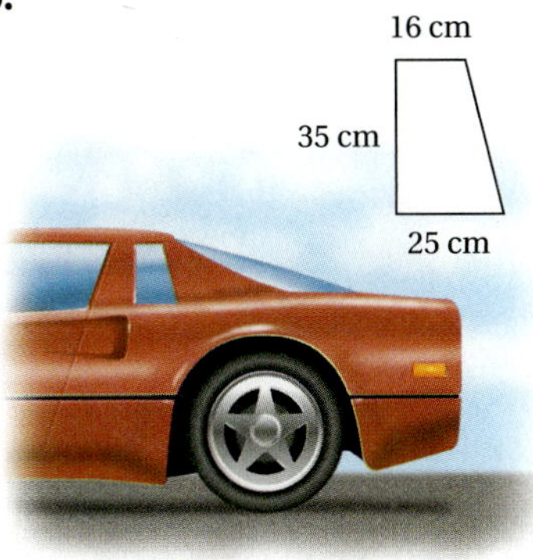

31.

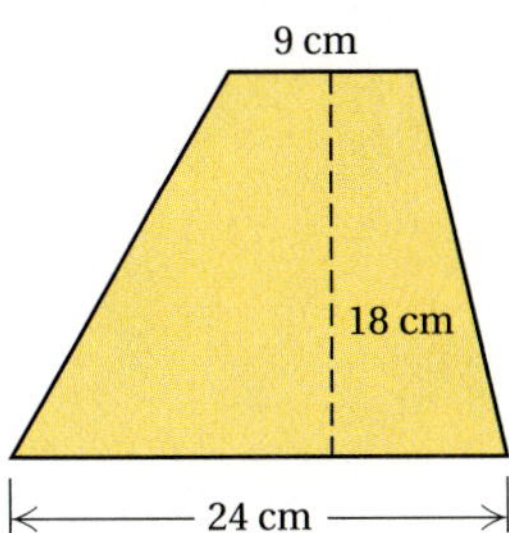

32.

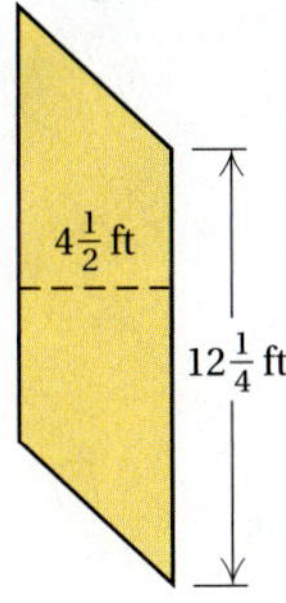

33.

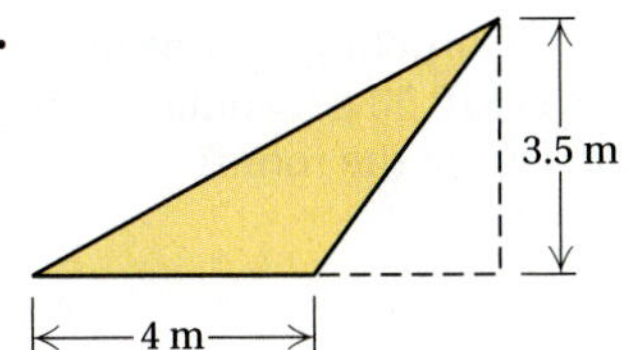

34.

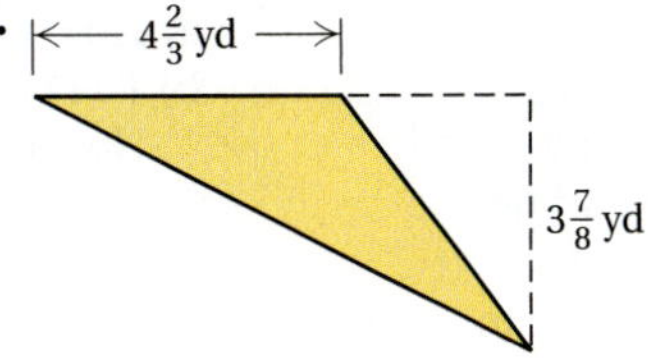

C Solve.

35. *Area of a Lawn.* A lot is 40 m by 36 m. A house 27 m by 9 m is built on the lot. How much area is left over for a lawn?

36. *Area of a Field.* A field is 240.8 m by 450.2 m. Part of the field, 160.4 m by 90.6 m, is paved for a parking lot. How much area is unpaved?

37. *Mowing Expense.* A square sandbox 4.5 ft on a side is placed on a 60-ft by $93\frac{2}{3}$-ft lawn.

a) Find the area of the lawn.

b) It costs $0.008 per square foot to have the lawn mowed. What is the total cost of the mowing?

Source: Jackson's Lawn Care, Carmel IN

38. *Mowing Expense.* A square flower bed 10.5 ft on a side is dug on a 90-ft by $67\frac{1}{4}$-ft lawn.

a) Find the area of the lawn.

b) It costs $0.03 per square foot to have the lawn mowed. What is the total cost of the mowing?

Source: Jackson's Lawn Care, Carmel IN

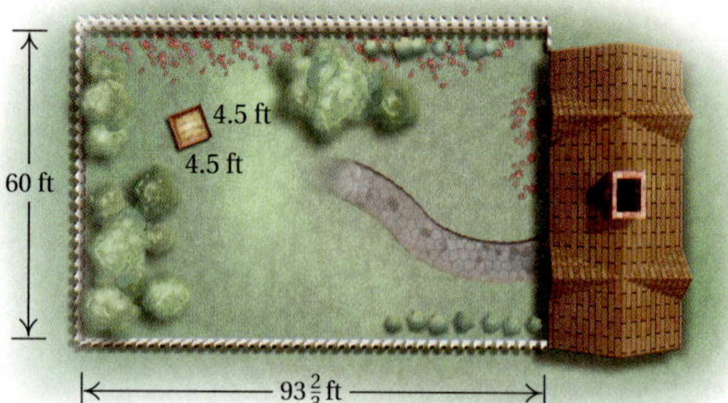

39. *Area of a Sidewalk.* Franklin Construction Company builds a sidewalk around two sides of the Municipal Trust Bank building, as shown in the figure. What is the area of the sidewalk?

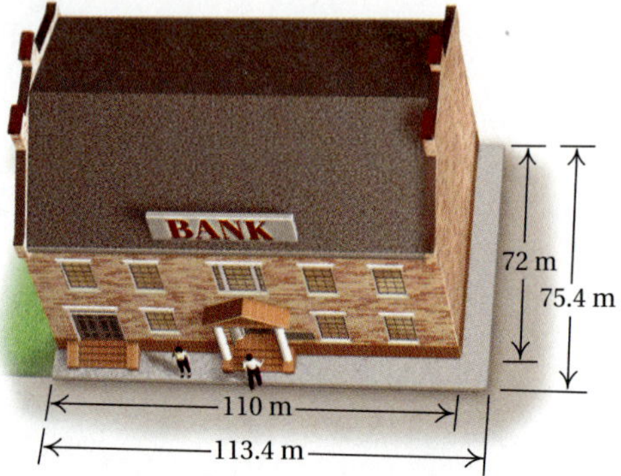

40. *Margin Area.* A standard sheet of typewriter paper is $8\frac{1}{2}$ in. by 11 in. We generally type on a $7\frac{1}{2}$-in. by 9-in. area of the paper. What is the area of the margin?

41. *Painting Costs.* A room is 15 ft by 20 ft. The ceiling is 8 ft above the floor. There are two windows in the room, each 3 ft by 4 ft. The door is $2\frac{1}{2}$ ft by $6\frac{1}{2}$ ft.

a) What is the total area of the walls and the ceiling?

b) A gallon of paint will cover 86.625 ft^2. How many gallons of paint are needed for the room, including the ceiling?

c) Paint costs $17.95 a gallon. How much will it cost to paint the room?

42. *Carpeting Costs.* A restaurant owner wants to carpet a 15-yd by 20-yd room.

a) How many square yards of carpeting are needed?

b) The carpeting she wants is $18.50 per square yard. How much will it cost to carpet the room?

Find the area of the shaded region.

43.

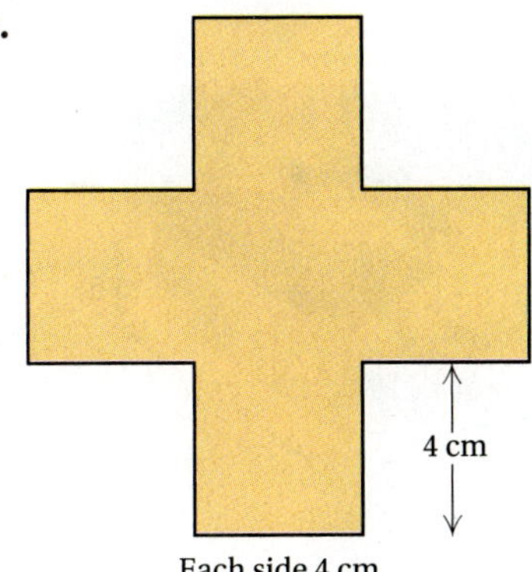

44.

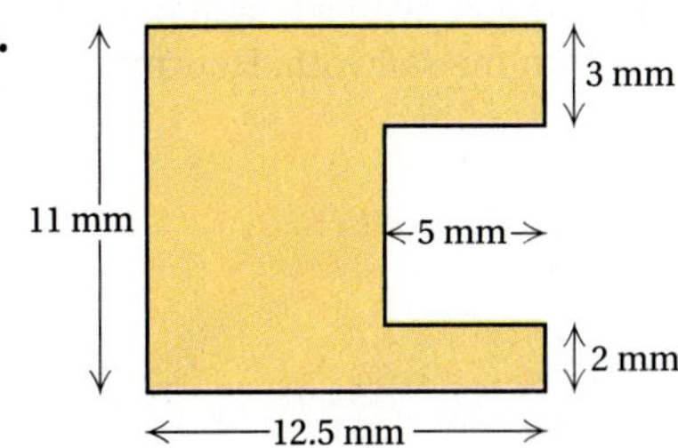

45.

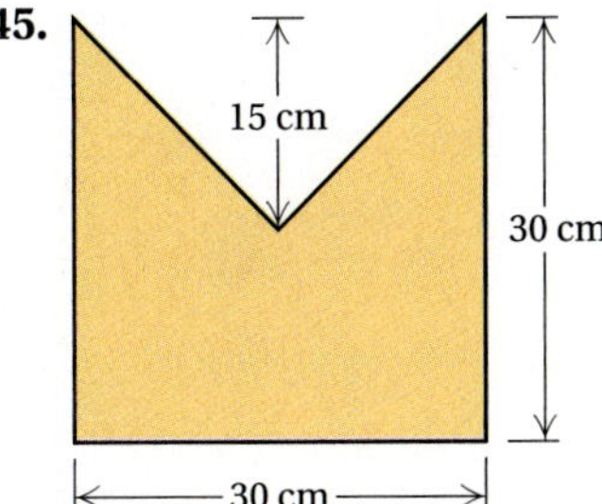

46.

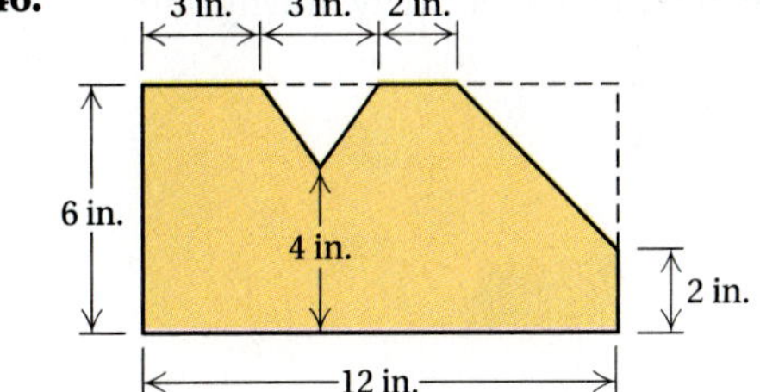

47.

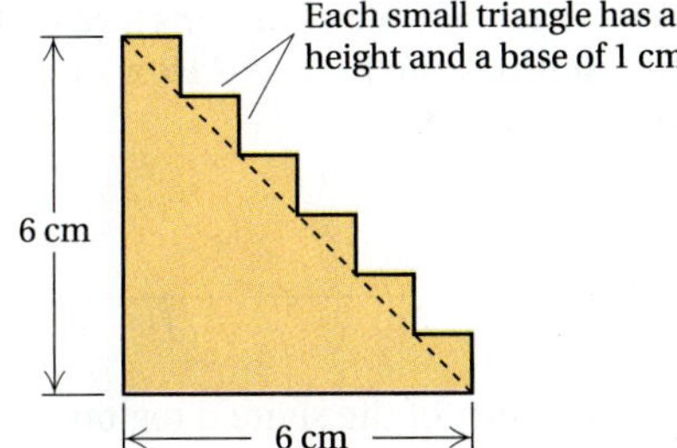

48.

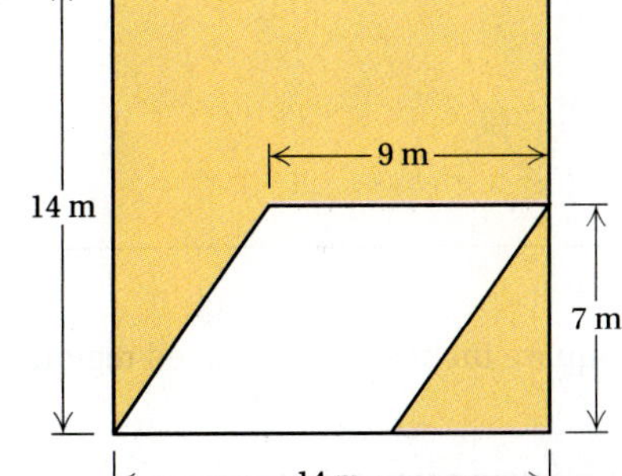

49. *Triangular Sail.* A rectangular piece of sailcloth is 36 ft by 24 ft. A triangular area with a height of 4.6 ft and a base of 5.2 ft is cut from the sailcloth. How much area is left over?

50. *Building Area.* Find the total area of the sides and ends of the building.

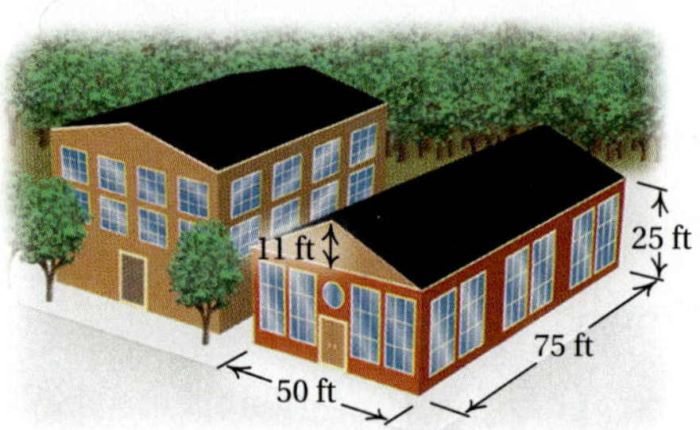

51. DW The length and the width of one rectangle are each three times the length and the width of another rectangle. Is the area of the first rectangle three times the area of the other rectangle? Why or why not?

52. DW Explain how the area of a triangle can be found by considering the area of a parallelogram.

SKILL MAINTENANCE

Convert to fraction notation. [4.3b]

53. 35%

54. 85.5%

55. $37\frac{1}{2}\%$

56. $66.\overline{6}\%$

57. $83.\overline{3}\%$

58. $16\frac{2}{3}\%$

Solve. [1.5a]

59. A ream of paper contains 500 sheets. How many sheets are there in 15 reams?

60. A lab technician separates a vial containing 140 cc of blood into test tubes, each of which contains 3 cc of blood. How many test tubes can be filled? How much blood is left over?

SYNTHESIS

61. Find the area, in square inches, of the shaded region.

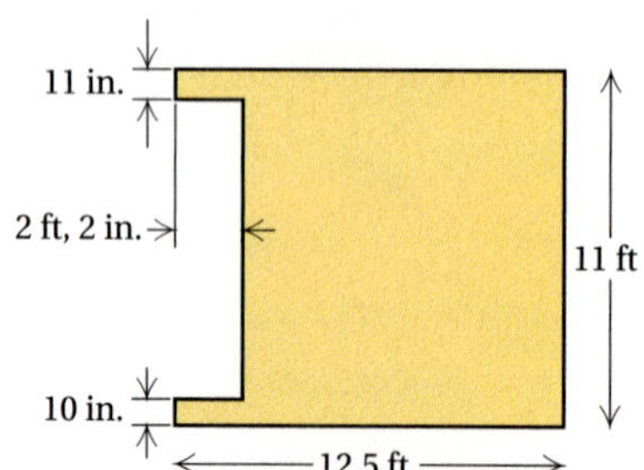

62. Find the area, in square feet, of the shaded region.

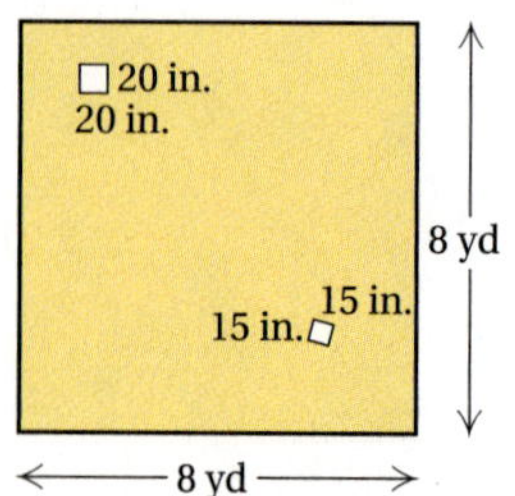

5.4 CIRCLES

a Radius and Diameter

Shown below is a circle with center O. Segment $\overline{AC}$ is a *diameter*. A **diameter** is a segment that passes through the center of the circle and has endpoints on the circle. Segment $\overline{OB}$ is called a *radius*. A **radius** is a segment with one endpoint on the center and the other endpoint on the circle.

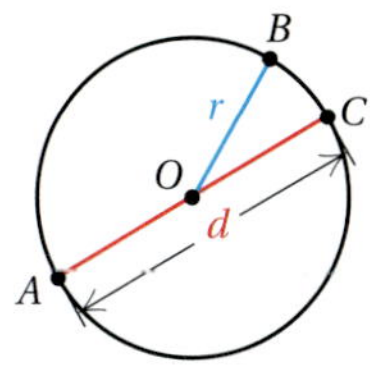

DIAMETER AND RADIUS

Suppose that d is the diameter of a circle and r is the radius. Then

$$d = 2 \cdot r \quad \text{and} \quad r = \frac{d}{2}.$$

EXAMPLE 1 Find the length of a radius of this circle.

$$r = \frac{d}{2} = \frac{12\text{ m}}{2} = 6\text{ m}$$

12 m

The radius is 6 m.

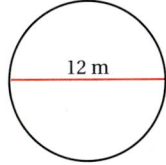

EXAMPLE 2 Find the length of a diameter of this circle.

$$d = 2 \cdot r = 2 \cdot \frac{1}{4}\text{ ft} = \frac{1}{2}\text{ ft}$$

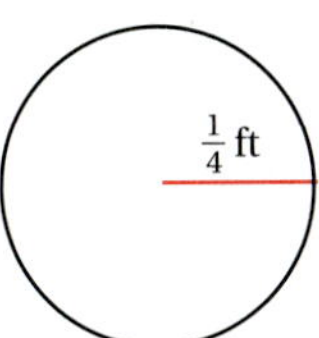

The diameter is $\frac{1}{2}$ ft.

Do Exercises 1 and 2.

Objectives

a Find the length of a radius of a circle given the length of a diameter, and find the length of a diameter given the length of a radius.

b Find the circumference of a circle given the length of a diameter or a radius.

c Find the area of a circle given the length of a radius.

d Solve applied problems involving circles.

1. Find the length of a radius.

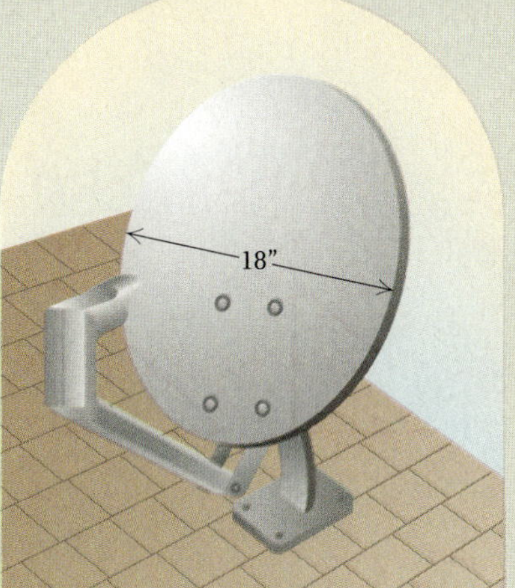

2. Find the length of a diameter.

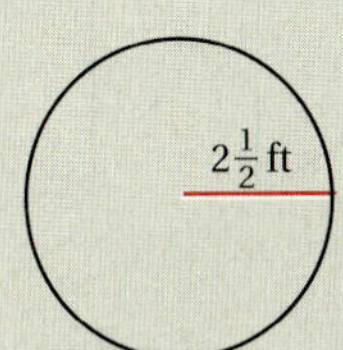

Answers on page A-12

3. Find the circumference of this circle. Use 3.14 for π.

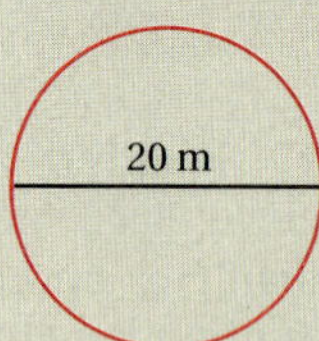

Answer on page A-12

b Circumference

The **circumference** of a circle is the distance around it. Calculating circumference is similar to finding the perimeter of a polygon.

To find a formula for the circumference of any circle given its diameter, we first need to consider the ratio C/d. Take a 12-oz soda can and measure the circumference C with a tape measure. Also measure the diameter d. The results are shown in the figure. Then

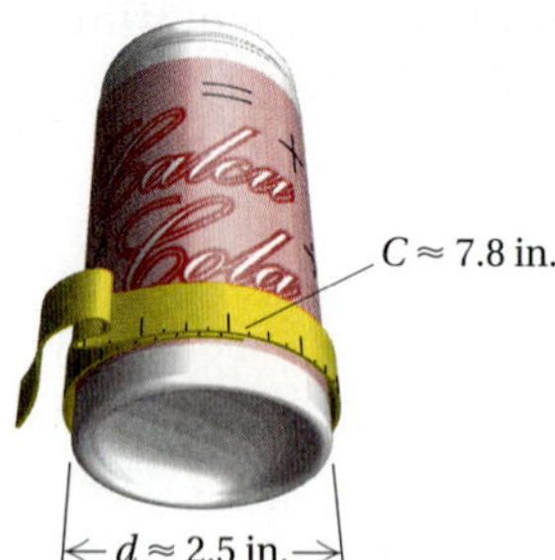

$$\frac{C}{d} = \frac{7.8 \text{ in.}}{2.5 \text{ in.}} \approx 3.1.$$

Suppose we did this with cans and circles of several sizes. We would get a number close to 3.1. For any circle, if we divide the circumference C by the diameter d, we get the same number. We call this number π (pi).

CIRCUMFERENCE AND DIAMETER

The circumference C of a circle of diameter d is given by

$$C = \pi \cdot d.$$

The number π is about 3.14, or about $\frac{22}{7}$.

EXAMPLE 3 Find the circumference of this circle. Use 3.14 for π.

$$\begin{aligned} C &= \pi \cdot d \\ &\approx 3.14 \times 6 \text{ cm} \\ &= 18.84 \text{ cm} \end{aligned}$$

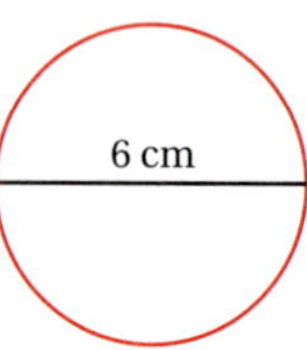

The circumference is about 18.84 cm.

Do Exercise 3.

Since $d = 2 \cdot r$, where r is the length of a radius, it follows that

$$C = \pi \cdot d = \pi \cdot (2 \cdot r).$$

CIRCUMFERENCE AND RADIUS

The circumference C of a circle of radius r is given by

$$\mathbf{C = 2 \cdot \pi \cdot r}.$$

EXAMPLE 4 Find the circumference of this circle. Use $\frac{22}{7}$ for π.

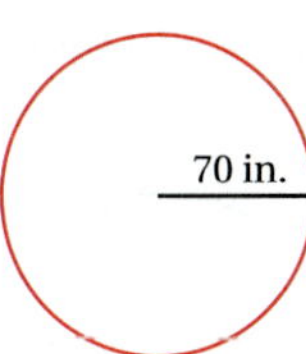

$$C = 2 \cdot \pi \cdot r$$
$$\approx 2 \cdot \frac{22}{7} \cdot 70 \text{ in.}$$
$$= 2 \cdot 22 \cdot \frac{70}{7} \text{ in.}$$
$$= 44 \cdot 10 \text{ in.}$$
$$= 440 \text{ in.}$$

The circumference is about 440 in.

EXAMPLE 5 Find the perimeter of this figure. Use 3.14 for π.

We let $P =$ the perimeter. We see that we have half a circle attached to a square. Thus we add half the circumference of the circle to the lengths of the three line segments.

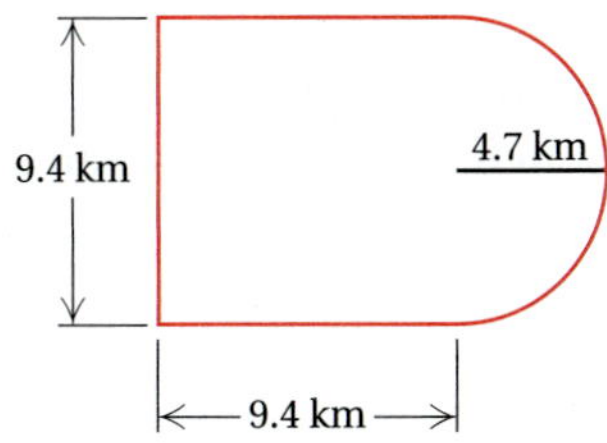

$P =$ Length of three sides of the square + Half of the circumference of the circle

$$= 3 \times 9.4 \text{ km} + \frac{1}{2} \times 2 \times \pi \times 4.7 \text{ km}$$
$$= 28.2 \text{ km} + 3.14 \times 4.7 \text{ km}$$
$$= 28.2 \text{ km} + 14.758 \text{ km}$$
$$= 42.958 \text{ km}$$

The perimeter is about 42.958 km.

Do Exercises 4 and 5.

4. Find the circumference of this circle. Use $\frac{22}{7}$ for π.

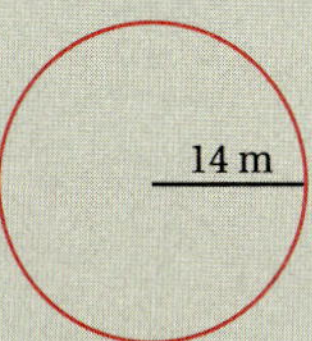

5. Find the perimeter of this figure. Use 3.14 for π.

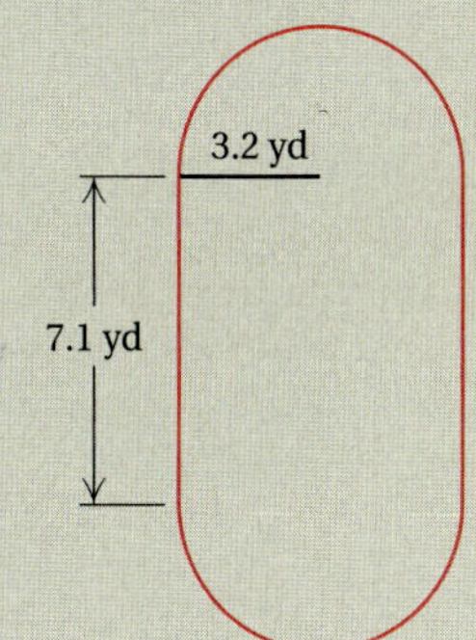

Answers on page A-12

6. Find the area of this circle. Use $\frac{22}{7}$ for π.

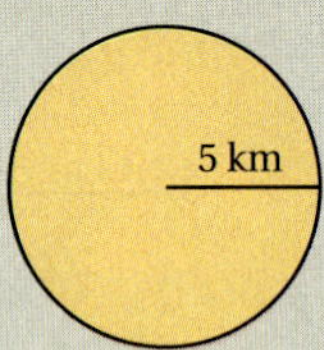

C Area

Below is a circle of radius r.

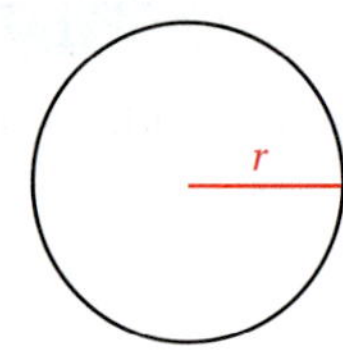

Think of cutting half the circular region into small pieces and arranging them as shown below.

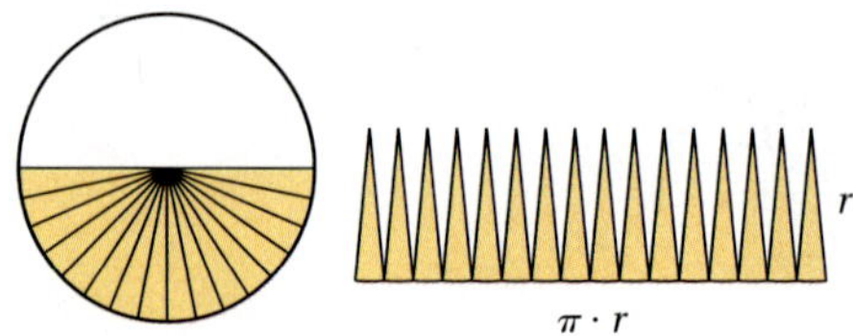

Then imagine cutting the other half of the circular region and arranging the pieces in with the others as shown below.

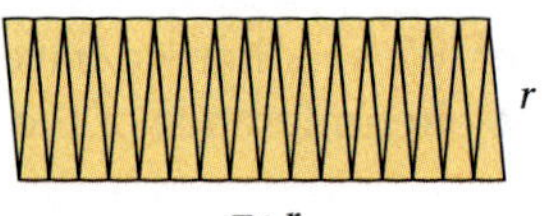

This is almost a parallelogram. The base has length $\frac{1}{2} \cdot 2 \cdot \pi \cdot r$, or $\pi \cdot r$ (half the circumference) and the height is r. Thus the area is

$$(\pi \cdot r) \cdot r.$$

This is the area of a circle.

AREA OF A CIRCLE

The **area of a circle** with radius of length r is given by

$$A = \pi \cdot r \cdot r, \quad \text{or} \quad A = \pi \cdot r^2.$$

EXAMPLE 6 Find the area of this circle. Use $\frac{22}{7}$ for π.

$$\begin{aligned} A &= \pi \cdot r \cdot r \\ &\approx \frac{22}{7} \cdot 14\text{ cm} \cdot 14\text{ cm} \\ &= \frac{22}{7} \cdot 196\text{ cm}^2 \\ &= 616\text{ cm}^2 \end{aligned}$$

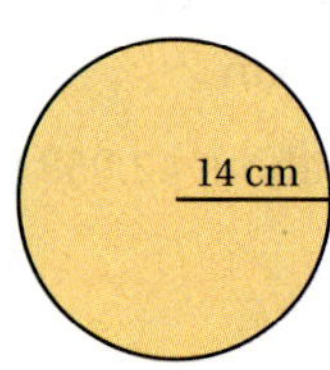

The area is about 616 cm^2.

Do Exercise 6.

Answer on page A-12

EXAMPLE 7 Find the area of this circle. Use 3.14 for π. Round to the nearest hundredth.

$$A = \pi \cdot r \cdot r$$
$$\approx 3.14 \times 2.1 \text{ m} \times 2.1 \text{ m}$$
$$= 3.14 \times 4.41 \text{ m}^2$$
$$= 13.8474 \text{ m}^2$$
$$\approx 13.85 \text{ m}^2$$

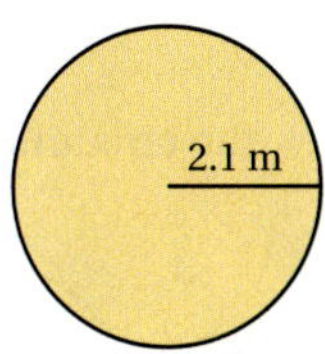

The area is about 13.85 m^2.

CAUTION!

Remember that circumference is always measured in linear units like ft, m, cm, yd, and so on. But area is measured in square units like ft^2, m^2, cm^2, yd^2, and so on.

Do Exercise 7.

7. Find the area of this circle. Use 3.14 for π. Round to the nearest hundredth.

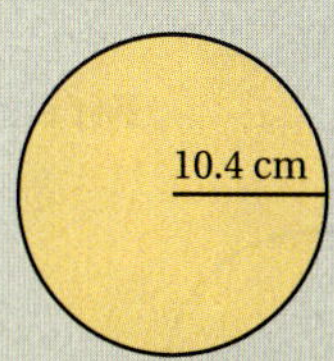

d Solving Applied Problems

EXAMPLE 8 *Area of Pizza Pans.* Which is larger and by how much: a pizza made in a 16-in. square pizza pan or a pizza made in a 16-in. diameter circular pan?

First, we make a drawing of each.

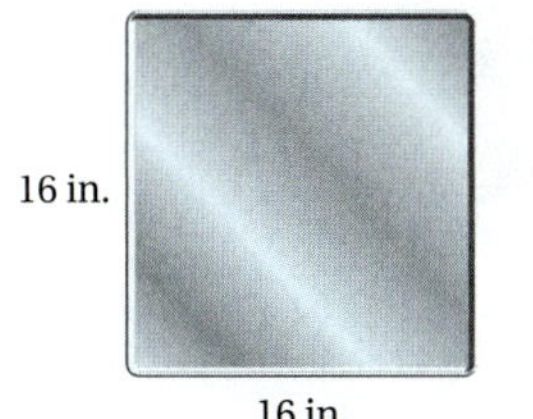

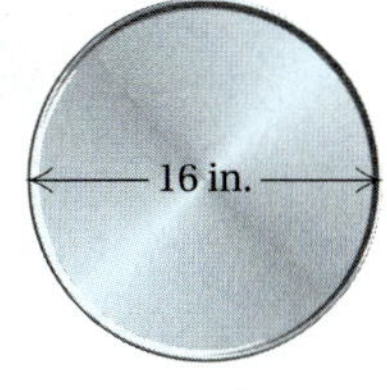

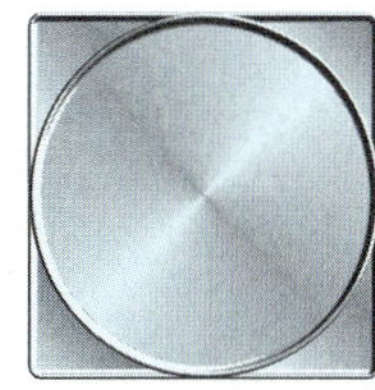

Then we compute areas.

The area of the square is

$$A = s \cdot s$$
$$= 16 \text{ in.} \times 16 \text{ in.} = 256 \text{ in}^2.$$

The diameter of the circle is 16 in., so the radius is 16 in./2, or 8 in. The area of the circle is

$$A = \pi \cdot r \cdot r$$
$$\approx 3.14 \times 8 \text{ in.} \times 8 \text{ in.} = 200.96 \text{ in}^2.$$

We see that the square pizza is larger by about

$$256 \text{ in}^2 - 200.96 \text{ in}^2, \quad \text{or} \quad 55.04 \text{ in}^2.$$

Thus the pizza made in the square pan is larger, by about 55.04 in^2.

Do Exercise 8.

8. Which is larger and by how much: a 10-ft square flower bed or a 12-ft diameter flower bed?

Answers on page A-12

5.4 EXERCISE SET

For Extra Help

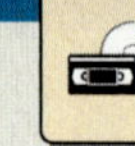

Digital Video Tutor CD 4 Videotape 6

InterAct Math

Math Tutor Center

MathXL

MyMathLab.com

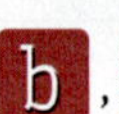

For each circle, find the length of a diameter, the circumference, and the area. Use $\frac{22}{7}$ for π.

1.
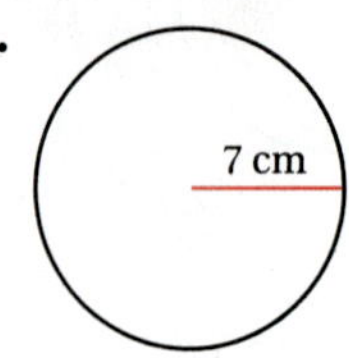

2.
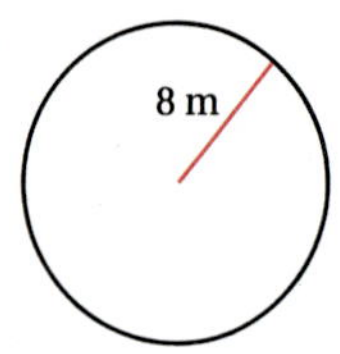

3.
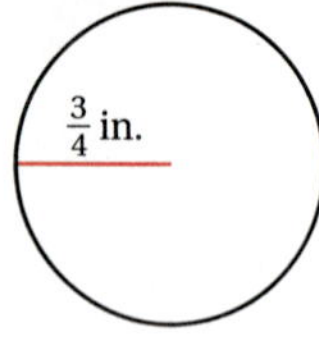

4.
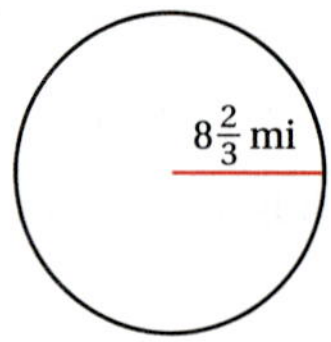

For each circle, find the length of a radius, the circumference, and the area. Use 3.14 for π.

5.
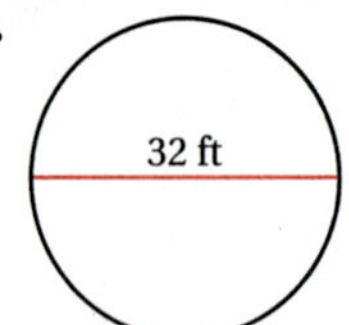

6.
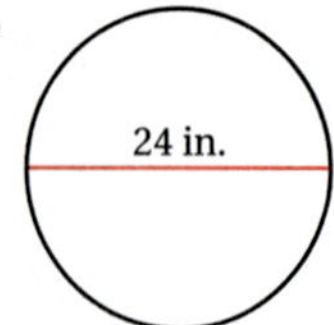

7.
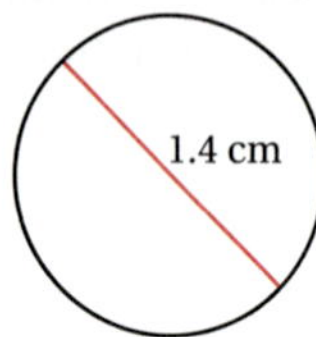

8.
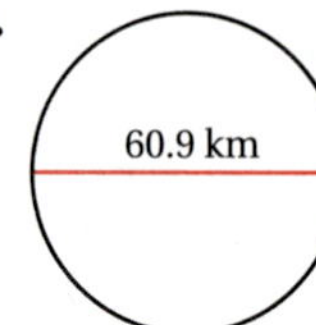

d

Solve. Use 3.14 for π.

9. *Soda-Can Top.* The top of a soda can has a 6-cm diameter. What is its radius? its circumference? its area?

10. *Penny.* A penny has a 1-cm radius. What is its diameter? its circumference? its area?

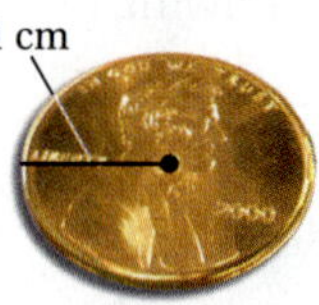

11. *Trampoline.* The standard backyard trampoline has a diameter of 14 ft. What is its area?

Source: International Trampoline Industry Association, Inc.

14 ft

Frame height: 36 in.

12. *Area of Pizza Pans.* Which is larger and by how much: a pizza made in a 12-in square pizza pan or a pizza made in a 12-in diameter circular pan?

13. *Dimensions of a Quarter.* The circumference of a quarter is 7.85 cm. What is the diameter? the radius? the area?

14. *Dimensions of a Dime.* The circumference of a dime is 2.23 in. What is the diameter? the radius? the area?

15. *Gypsy-Moth Tape.* To protect an elm tree in your backyard, you need to attach gypsy moth caterpillar tape around the trunk. The tree has a 1.1-ft diameter. What length of tape is needed?

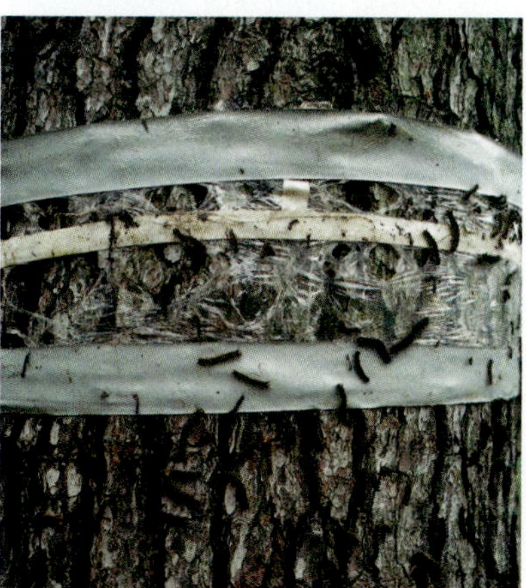

16. *Earth.* The diameter of the earth at the equator is 7926.41 mi. What is the circumference of the earth at the equator?
Source: *The Handy Geography Answer Book*

17. *Swimming-Pool Walk.* You want to install a 1-yd–wide walk around a circular swimming pool. The diameter of the pool is 20 yd. What is the area of the walk?

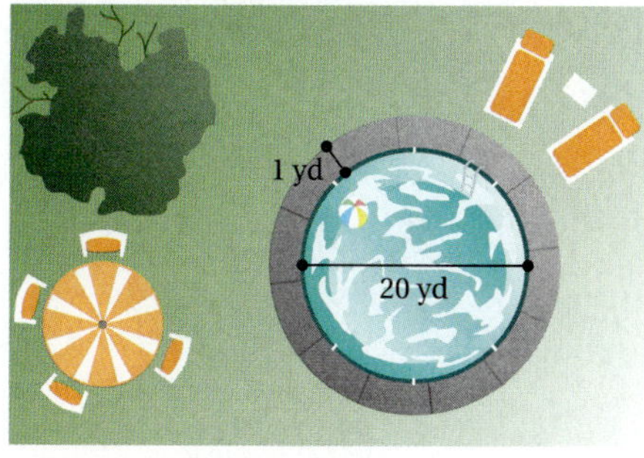

18. *Roller-Rink Floor.* A roller rink floor is shown below. What is its area? If hardwood flooring costs $10.50 per square meter, how much will the flooring cost?

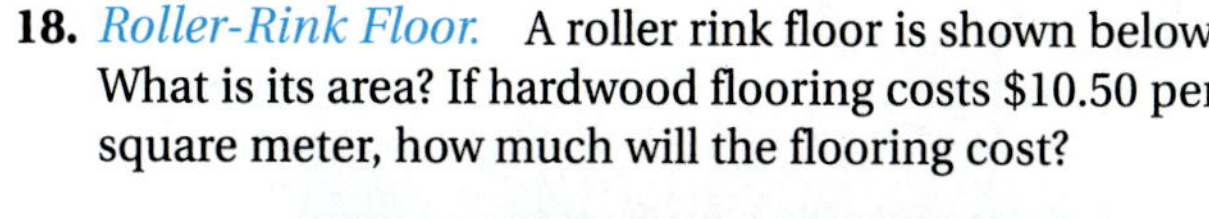

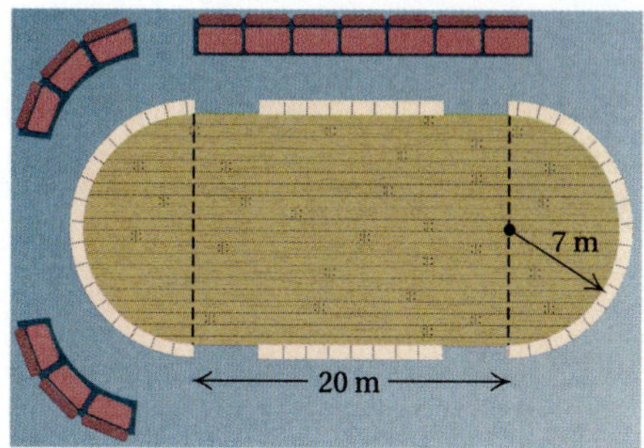

Find the perimeter. Use 3.14 for π.

19.

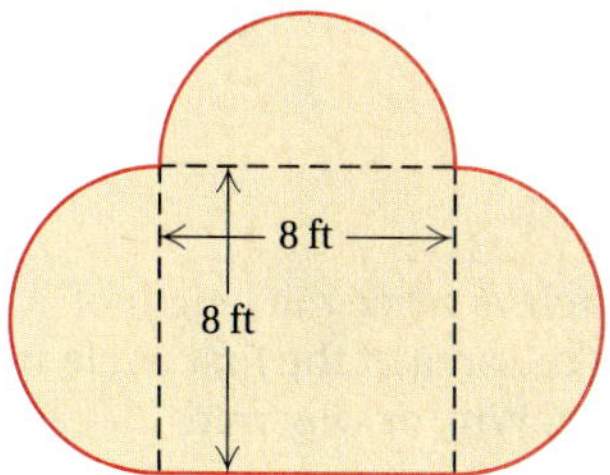

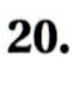

20.

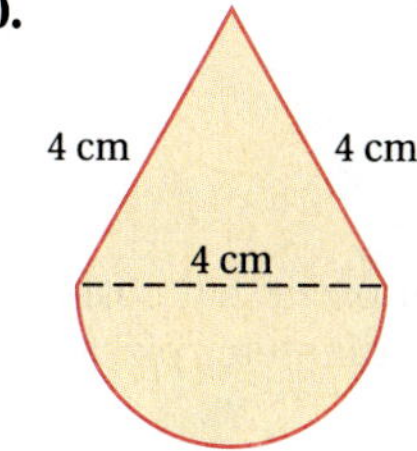

21. 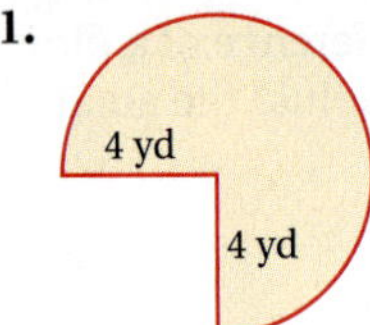

22. 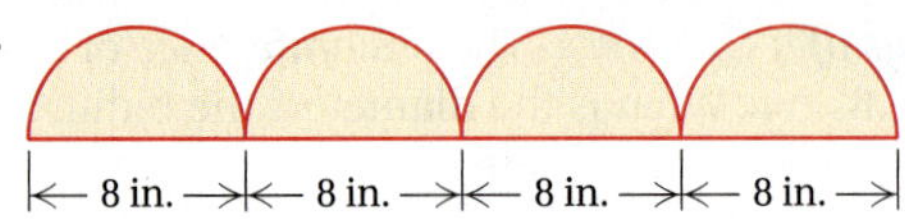

23. 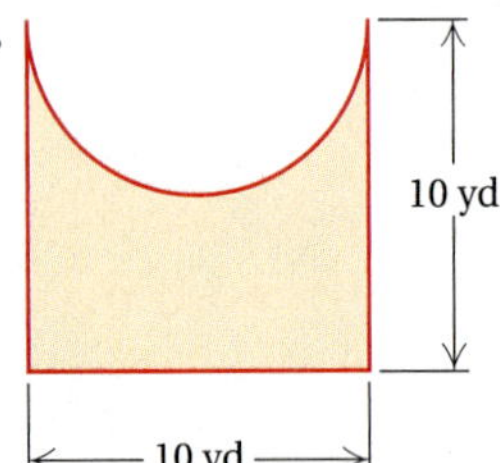

24. 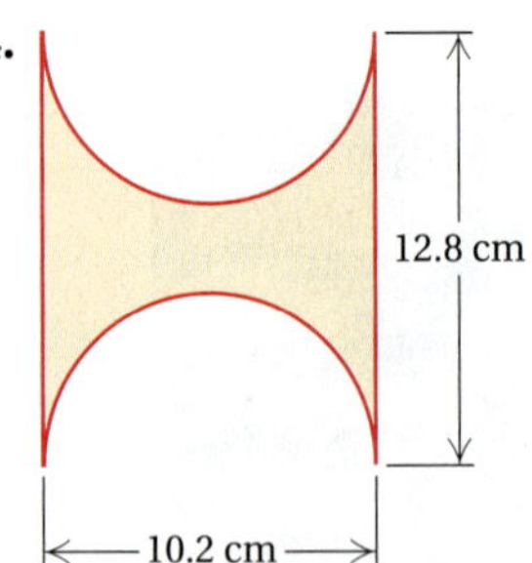

Find the area of the shaded region. Use 3.14 for π.

25. 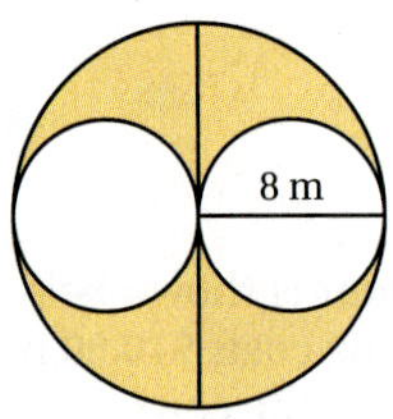

26. 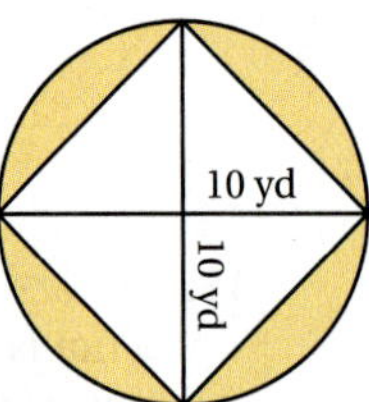

27. 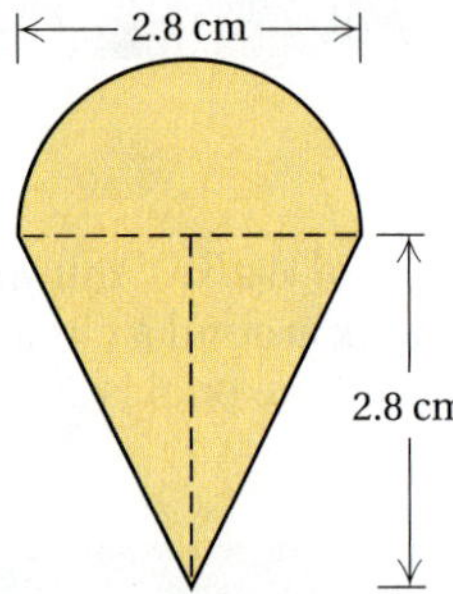

28. 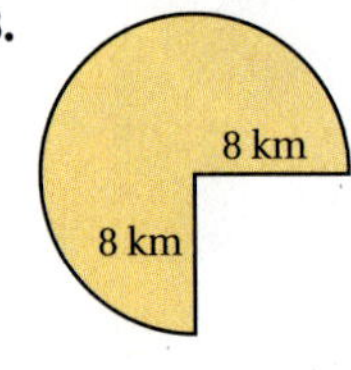

29. 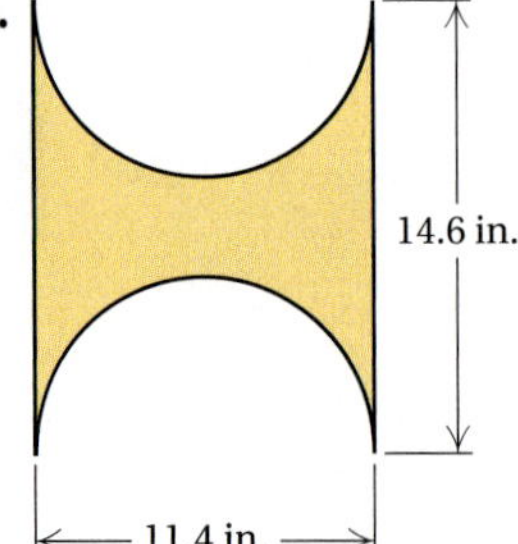

30. 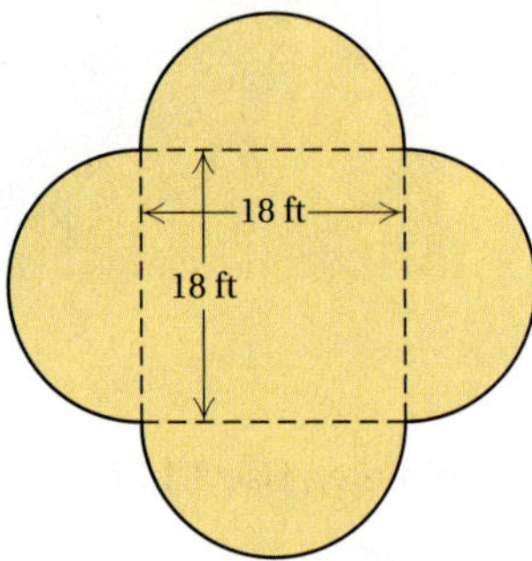

31. **D**W Explain why a 16-in.–diameter pizza that costs $16.25 is a better buy than a 10-in.–diameter pizza that costs $7.85.

32. **D**W The radius of one circle is twice the length of another circle's radius. Is the area of the first circle twice the area of the other circle? Why or why not?

SKILL MAINTENANCE

Convert to percent notation. [4.2b]

33. 0.875

34. 0.58

35. $0.\overline{6}$

36. 0.4361

Convert to percent notation. [4.3a]

37. $\frac{3}{8}$

38. $\frac{5}{8}$

39. $\frac{2}{3}$

40. $\frac{1}{5}$

Estimate each of the following as a whole number, as $\frac{1}{2}$, or as a mixed numeral where the fractional part is $\frac{1}{2}$. [2.6b]

41. $3\frac{7}{8}$

42. $8\frac{1}{3}$

43. $13\frac{1}{6}$

44. $39\frac{7}{13}$

45. $\frac{4}{5} + 3\frac{7}{8}$

46. $\frac{1}{11} \cdot \frac{7}{15}$

47. $\frac{2}{3} + \frac{7}{15} + \frac{8}{9}$

48. $\frac{8}{9} + \frac{4}{5} + \frac{13}{14}$

49. $\frac{57}{100} - \frac{1}{10} + \frac{9}{1000}$

50. $\frac{23}{24} + \frac{38}{39} + \frac{61}{60}$

51. $11\frac{29}{80} + 10\frac{14}{15} \cdot 24\frac{2}{17}$

52. $\frac{13}{14} + 9\frac{5}{8} - 1\frac{23}{28} \cdot 1\frac{36}{73}$

SYNTHESIS

53. 🖩 $\pi \approx \frac{3927}{1250}$ is another approximation for π. Find decimal notation using a calculator.

54. 🖩 The distance from Kansas City to Indianapolis is 500 mi. A car was driven this distance using tires with a radius of 14 in. How many revolutions of each tire occurred on the trip? Use $\frac{22}{7}$ for π.

55. *Tennis Balls.* Tennis balls are generally packed vertically three in a can, one on top of another. Suppose the diameter of a tennis ball is d. Find the height of the stack of balls. Find the circumference of one ball. Which is greater? Explain.

5.5 VOLUME AND SURFACE AREA

Objectives

a Find the volume and the surface area of a rectangular solid.

b Given the radius and the height, find the volume of a circular cylinder.

c Given the radius, find the volume of a sphere.

d Given the radius and the height, find the volume of a circular cone.

e Solve applied problems involving volume of rectangular solids, circular cylinders, spheres, and cones.

1. Find the volume.

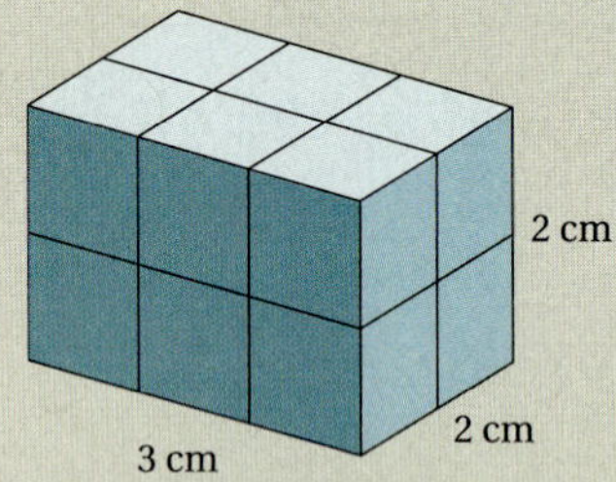

a Rectangular Solids

The **volume** of a **rectangular solid** is the number of unit cubes needed to fill it.

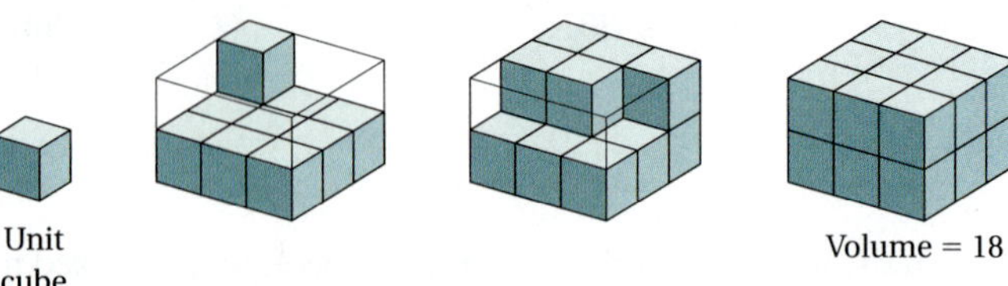

Two other units are shown below.

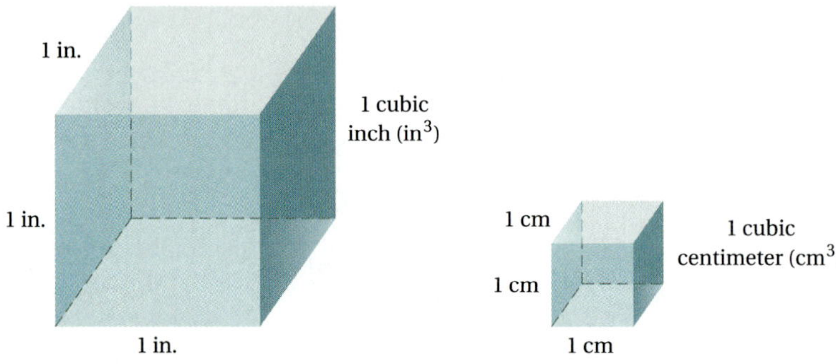

EXAMPLE 1 Find the volume.

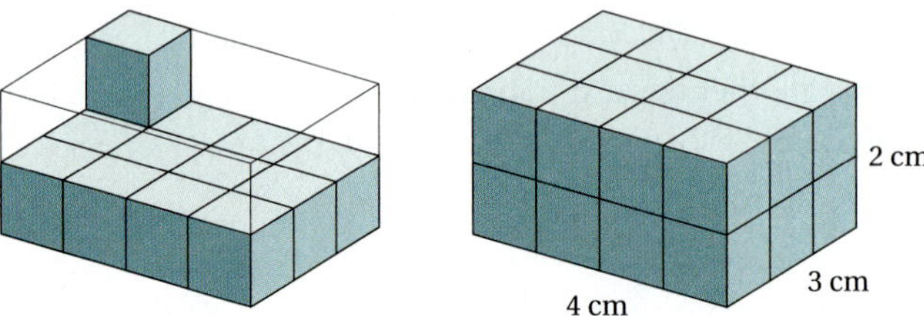

The figure is made up of 2 layers of 12 cubes each, so its volume is 24 cubic centimeters (cm^3).

Do Exercise 1.

VOLUME OF A RECTANGULAR SOLID

The **volume of a rectangular solid** is found by multiplying length by width by height:

$$V = l \cdot w \cdot h.$$

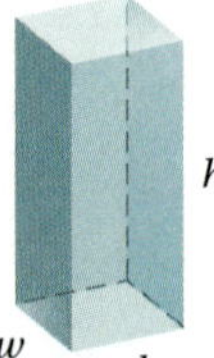

Answer on page A-12

EXAMPLE 2 *Carry-on Luggage.* The largest piece of luggage that you can carry on an airplane measures 23 in. by 10 in. by 13 in. Find the volume of this solid.

$$V = l \cdot w \cdot h$$
$$= 23 \text{ in.} \cdot 10 \text{ in.} \cdot 13 \text{ in.}$$
$$= 230 \cdot 13 \text{ in}^3$$
$$= 2990 \text{ in}^3$$

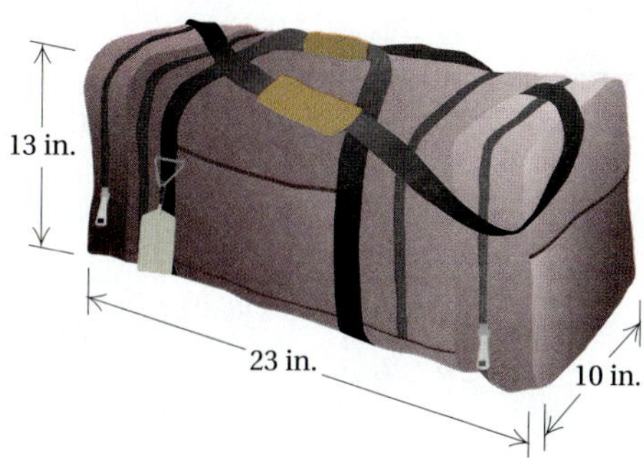

Do Exercises 2 and 3.

The **surface area** of a rectangular solid is the total area of the six rectangles that form the surface of the solid. For the rectangular solid below, we can show the six rectangles with a diagram.

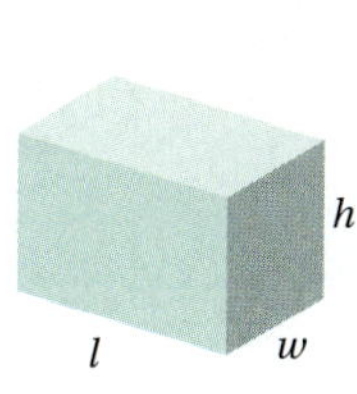

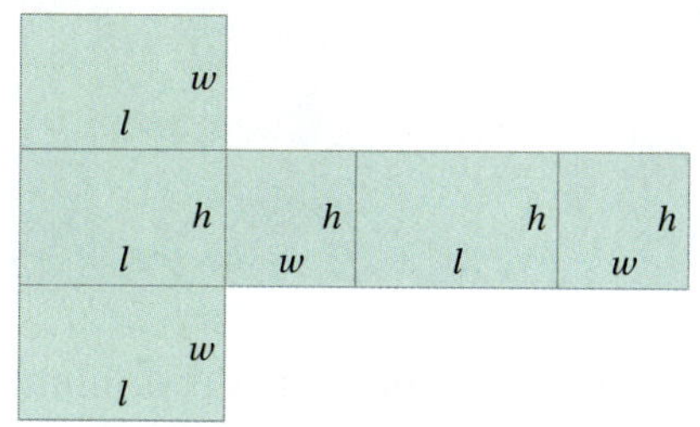

SA = 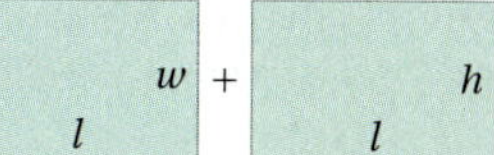 +  + lh + wh + lh + wh

$$= lw + lw + lh + wh + lh + wh$$
$$= 2lw + 2lh + 2wh, \quad \text{or} \quad 2(lw + lh + wh)$$

SURFACE AREA OF A RECTANGULAR SOLID

The surface area of a rectangular solid with length l, width w, and height h is given by the formula

$$\text{SA} = 2lw + 2lh + 2wh, \quad \text{or} \quad 2(lw + lh + wh).$$

EXAMPLE 3 Find the surface area of this rectangular solid.

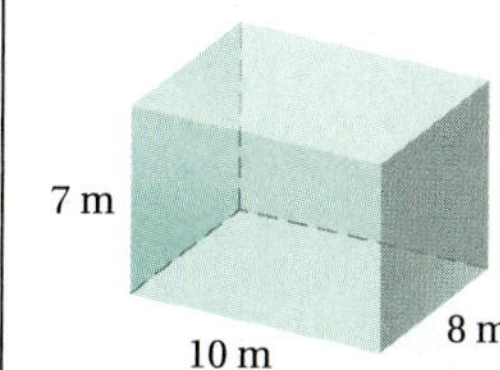

$$\text{SA} = 2lw + 2lh + 2wh$$
$$= 2 \cdot 10 \text{ m} \cdot 8 \text{ m} + 2 \cdot 10 \text{ m} \cdot 7 \text{ m} + 2 \cdot 8 \text{ m} \cdot 7 \text{ m}$$
$$= 160 \text{ m}^2 + 140 \text{ m}^2 + 112 \text{ m}^2$$
$$= 412 \text{ m}^2$$

The units used for area are square units.
The units used for volume are cubic units.

2. **Popcorn.** In a recent year, people in the United States bought enough unpopped popcorn to provide every person in the country with a bag of popped corn measuring 2 ft by 2 ft by 5 ft. Find the volume of such a bag.

3. **Cord of Wood.** A cord of wood measures 4 ft by 4 ft by 8 ft. What is the volume of a cord of wood?

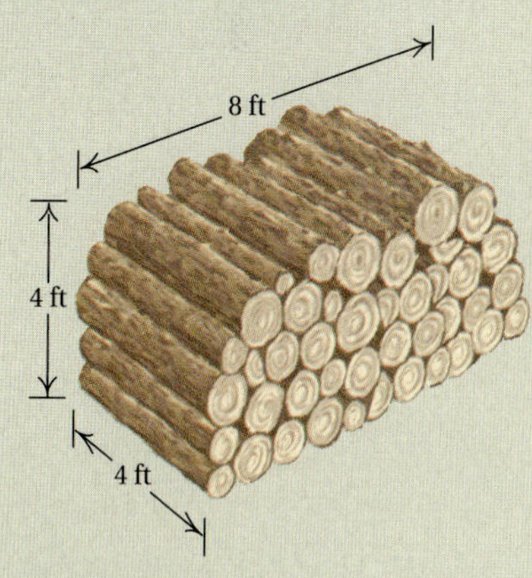

Answers on page A-12

Find the volume and the surface area of the rectangular solid.

4.

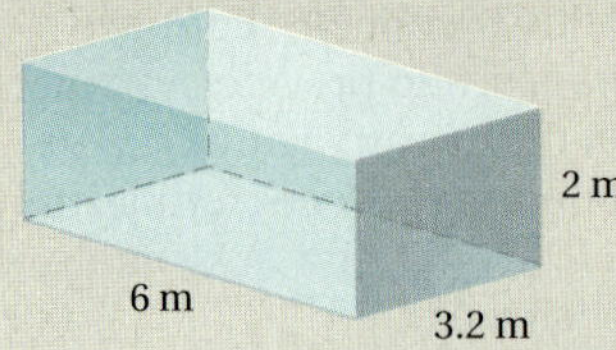

5.

$2\frac{1}{2}$ ft

1 ft

$\frac{3}{4}$ ft

Answers on page A-12

Do Exercises 4 and 5.

b Cylinders

A rectangular solid is shown below. Note that we can think of the volume as the product of the area of the base times the height:

$$\begin{aligned} V &= l \cdot w \cdot h \\ &= (l \cdot w) \cdot h \\ &= (\text{Area of the base}) \cdot h \\ &= B \cdot h, \end{aligned}$$

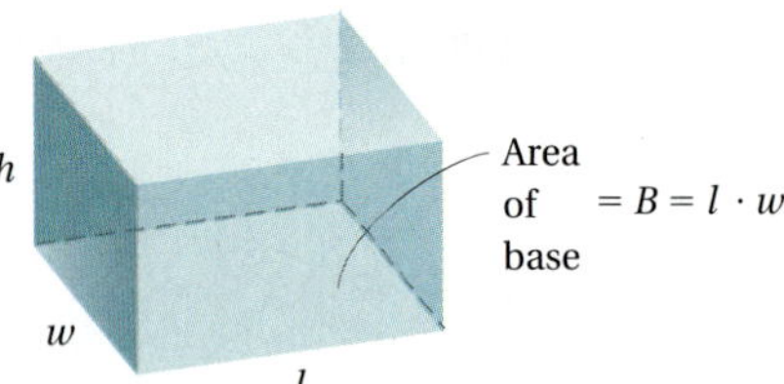

where B represents the area of the base.

Like rectangular solids, **circular cylinders** have bases of equal area that lie in parallel planes. The bases of circular cylinders are circular regions.

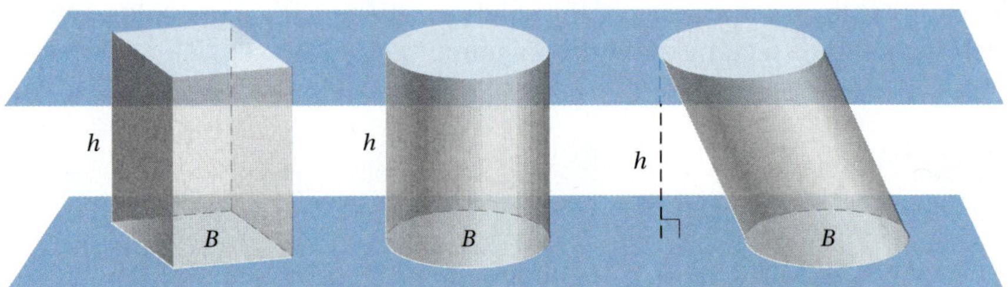

The volume of a circular cylinder is found in a manner similar to finding the volume of a rectangular solid. The volume is the product of the area of the base times the height. The height is always measured perpendicular to the base.

VOLUME OF A CIRCULAR CYLINDER

The **volume of a circular cylinder** is the product of the area of the base B and the height h:

$$\boldsymbol{V = B \cdot h, \quad \text{or} \quad V = \pi \cdot r^2 \cdot h.}$$

EXAMPLE 4 Find the volume of this circular cylinder. Use 3.14 for π.

$$V = Bh = \pi \cdot r^2 \cdot h$$
$$\approx 3.14 \times 4\text{ cm} \times 4\text{ cm} \times 12\text{ cm}$$
$$= 602.88\text{ cm}^3$$

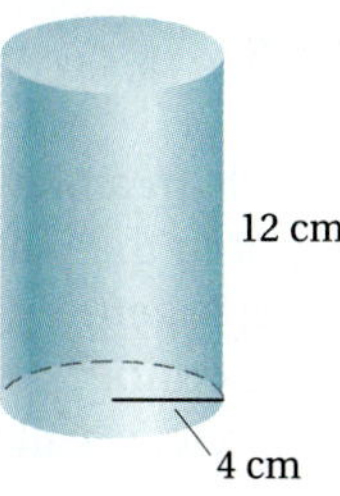

Do Exercises 6 and 7.

C Spheres

A **sphere** is the three-dimensional counterpart of a circle. It is the set of all points in space that are a given distance (the radius) from a given point (the center).

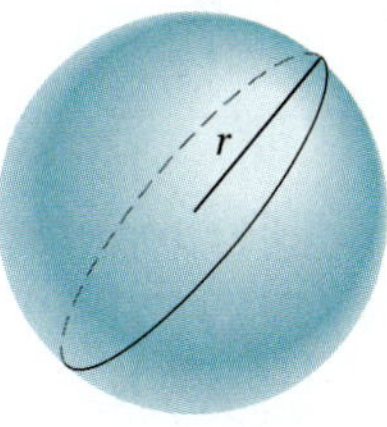

We find the volume of a sphere as follows.

VOLUME OF A SPHERE

The **volume of a sphere** of radius r is given by

$$V = \frac{4}{3} \cdot \pi \cdot r^3.$$

EXAMPLE 5 *Bowling Ball.* The radius of a standard-sized bowling ball is 4.2915 in. Find the volume of a standard-sized bowling ball. Round to the nearest hundredth of a cubic inch. Use 3.14 for π.

$$V = \frac{4}{3} \cdot \pi \cdot r^3 \approx \frac{4}{3} \times 3.14 \times (4.2915\text{ in.})^3$$
$$\approx \frac{4 \times 3.14 \times 79.0364\text{ in}^3}{3} \approx 330.90\text{ in}^3 \quad \text{Using a calculator}$$

Do Exercises 8 and 9.

6. Find the volume of the cylinder. Use 3.14 for π.

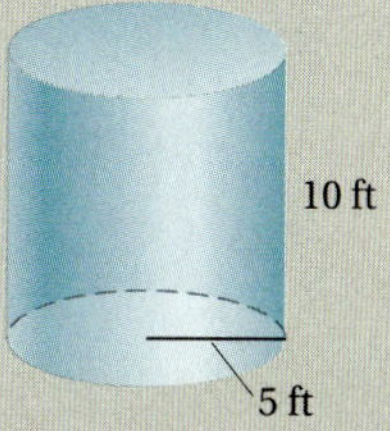

7. Find the volume of the cylinder. Use $\frac{22}{7}$ for π.

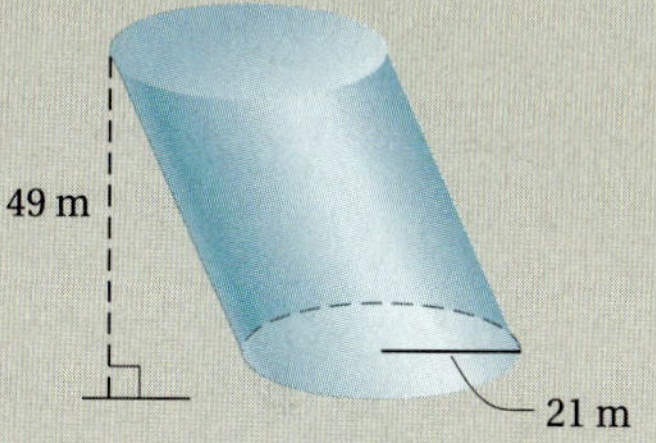

8. Find the volume of the sphere. Use $\frac{22}{7}$ for π.

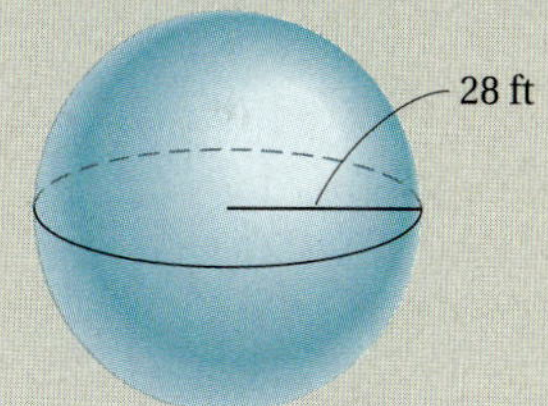

9. The radius of a standard-sized golf ball is 2.1 cm. Find its volume. Use 3.14 for π.

Answers on page A-12

10. Find the volume of this cone. Use 3.14 for π.

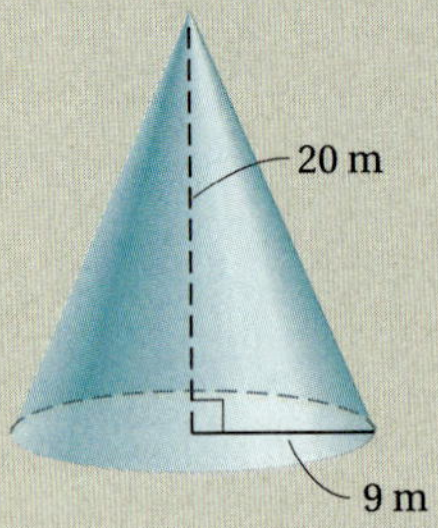

11. Find the volume of this cone. Use $\frac{22}{7}$ for π.

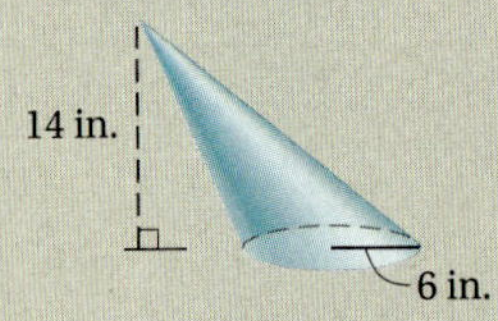

Answers on page A-12

d Cones

Consider a circle in a plane and choose any point P not in the plane. The circular region, together with the set of all segments connecting P to a point on the circle, is called a **circular cone.**

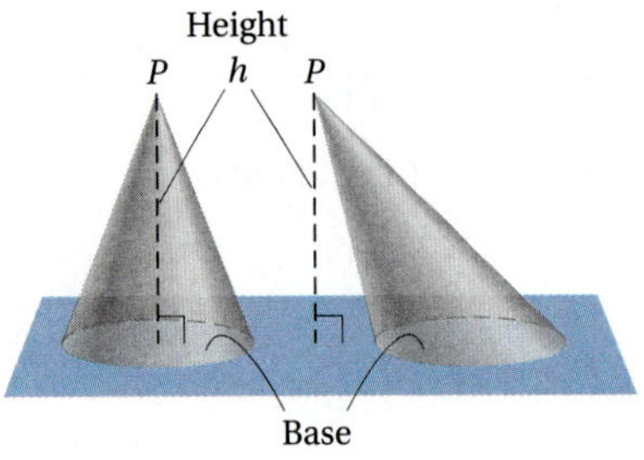

We find the volume of a cone as follows.

VOLUME OF A CIRCULAR CONE

The **volume of a circular cone** with base radius r is one-third the product of the base area and the height:

$$V = \frac{1}{3} \cdot B \cdot h = \frac{1}{3} \pi \cdot r^2 \cdot h.$$

EXAMPLE 6 Find the volume of this circular cone. Use 3.14 for π.

$$V = \tfrac{1}{3} \pi \cdot r^2 \cdot h$$
$$\approx \tfrac{1}{3} \times 3.14 \times 3 \text{ cm} \times 3 \text{ cm} \times 7 \text{ cm}$$
$$= 65.94 \text{ cm}^3$$

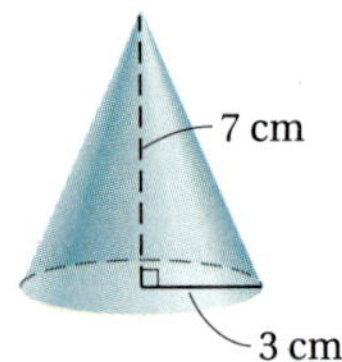

Do Exercises 10 and 11.

CALCULATOR CORNER

The [π] Key Many calculators have a [π] key that can be used to enter the value of π in a computation. It might be necessary to press a [2nd] or [SHIFT] key before pressing the [π] key on some calculators. Since 3.14 is a rounded value for π, results obtained using the [π] key might be slightly different from those obtained when 3.14 is used for the value of π in a computation.

To find the volume of the circular cylinder in Example 4, we press [2nd] [π] [×] [4] [×] [4] [×] [1] [2] [=] or [SHIFT] [π] [×] [4] [×] [4] [×] [1] [2] [=]. The result is approximately 603.19. Note that this is slightly different from the result found using 3.14 for the value of π.

Exercises

1. Use a calculator with a [π] key to perform the computations in Examples 4 and 5.
2. Use a calculator with a [π] key to perform the computations in Margin Exercises 5–10.

e Solving Applied Problems

EXAMPLE 7 *Propane Gas Tank.* A propane gas tank is shaped like a circular cylinder with half of a sphere at each end. Find the volume of the tank if the cylindrical section is 5 ft long with a 4-ft diameter. Use 3.14 for π.

1. **Familiarize.** We first make a drawing.

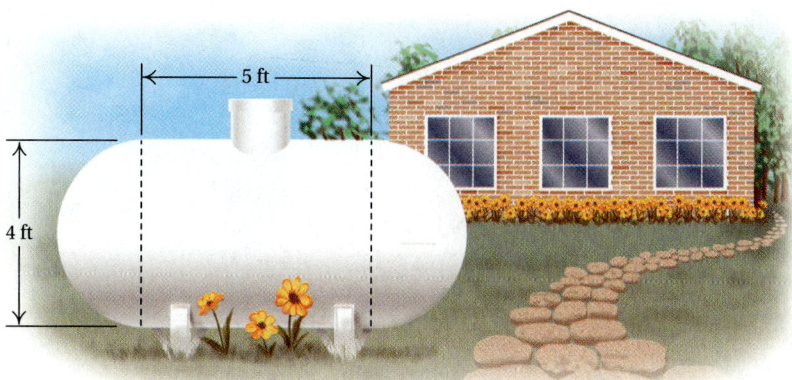

2. **Translate.** This is a two-step problem. We first find the volume of the cylindrical portion. Then we find the volume of the two ends and add. Note that the radius is 2 ft and that together the two ends make a sphere. We let

$$V = \pi \cdot r^2 \cdot h + \frac{4}{3} \cdot \pi \cdot r^3,$$

where V is the total volume. Then

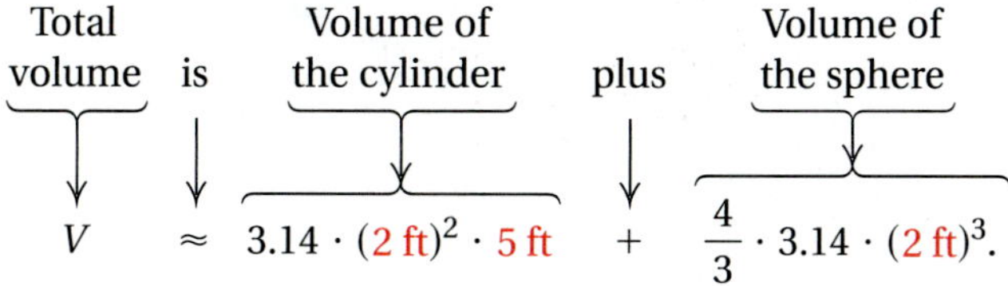

$$V \approx 3.14 \cdot (2\text{ ft})^2 \cdot 5\text{ ft} + \frac{4}{3} \cdot 3.14 \cdot (2\text{ ft})^3.$$

3. **Solve.** The volume of the cylinder is approximately

$$\begin{aligned} 3.14 \cdot (2\text{ ft})^2 \cdot 5\text{ ft} &= 3.14 \cdot 2\text{ ft} \cdot 2\text{ ft} \cdot 5\text{ ft} \\ &= 62.8\text{ ft}^3. \end{aligned}$$

The volume of the two ends is approximately

$$\begin{aligned} \frac{4}{3} \cdot 3.14 \cdot (2\text{ ft})^3 &= \frac{4}{3} \cdot 3.14 \cdot 2\text{ ft} \cdot 2\text{ ft} \cdot 2\text{ ft} \\ &\approx 33.5\text{ ft}^3. \end{aligned}$$

The total volume is about

$$62.8\text{ ft}^3 + 33.5\text{ ft}^3 = 96.3\text{ ft}^3.$$

4. **Check.** The check is left to the student.
5. **State.** The volume of the tank is about 96.3 ft³.

Do Exercise 12.

12. **Medicine Capsule.** A cold capsule is 8 mm long and 4 mm in diameter. Find the volume of the capsule. Use 3.14 for π. (*Hint*: First find the length of the cylindrical section.)

Answer on page A-12

5.5 EXERCISE SET

For Extra Help

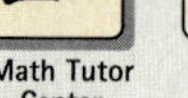

Digital Video Tutor CD 4 Videotape 6 | InterAct Math | Math Tutor Center | MathXL | MyMathLab

a Find the volume and the surface area of the rectangular solid.

1.

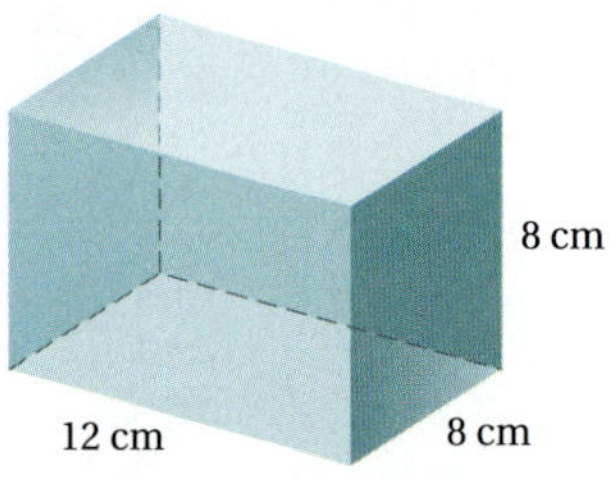

2.

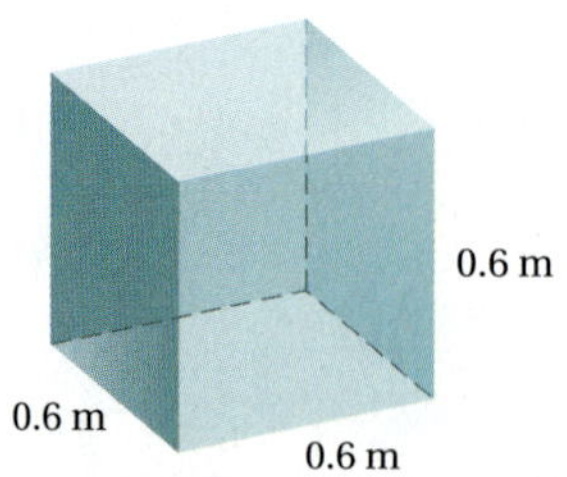

3.

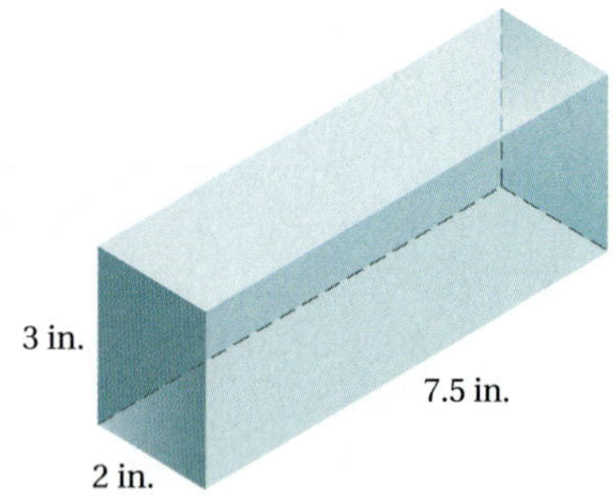

4.

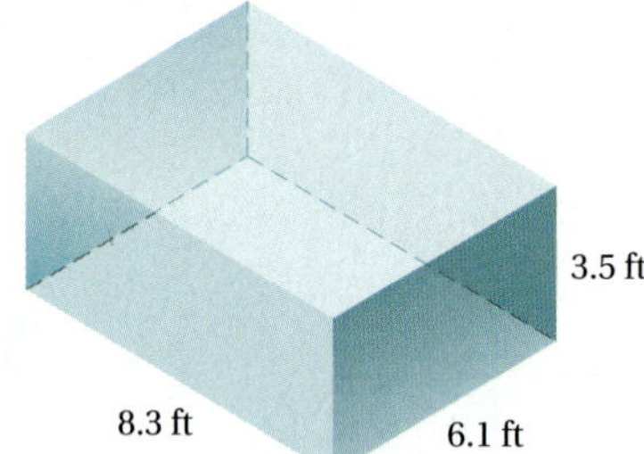

5.

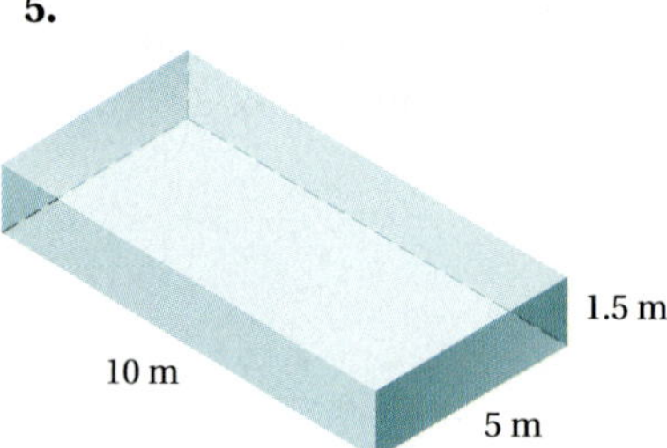

6.

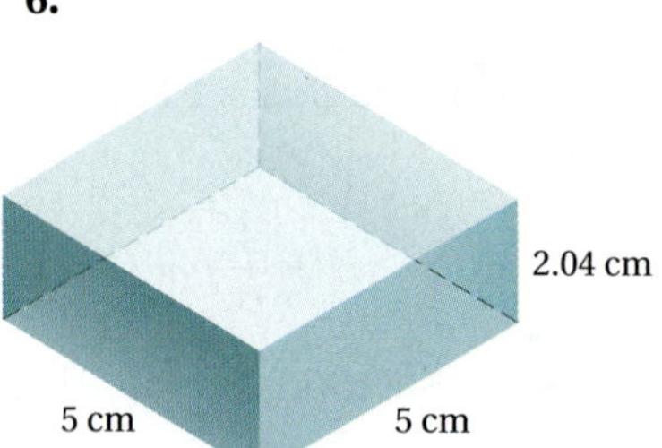

7.

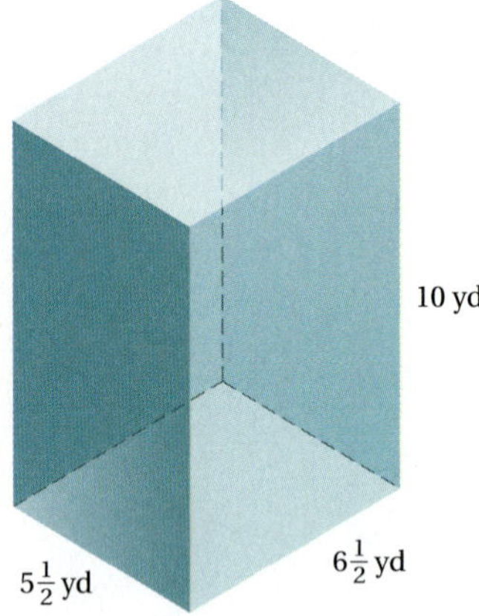

8.

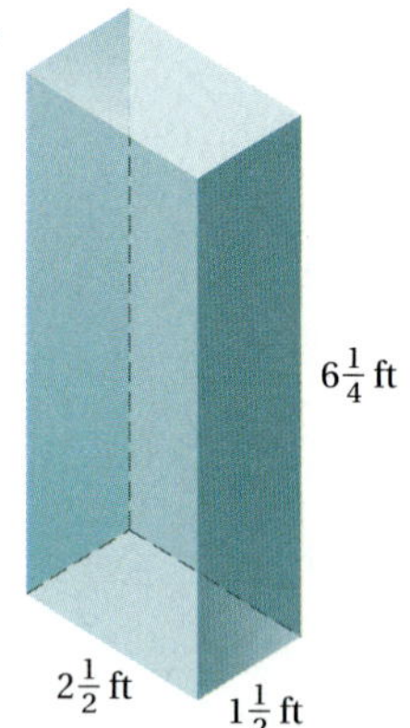

b Find the volume of the circular cylinder. Use 3.14 for π in Exercises 9–12. Use $\frac{22}{7}$ for π in Exercises 13 and 14.

9.

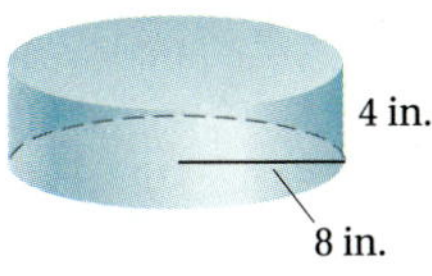

10.

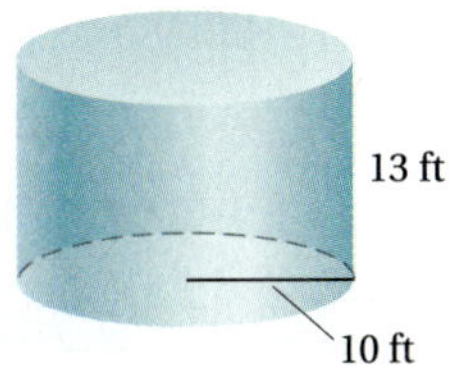

11.

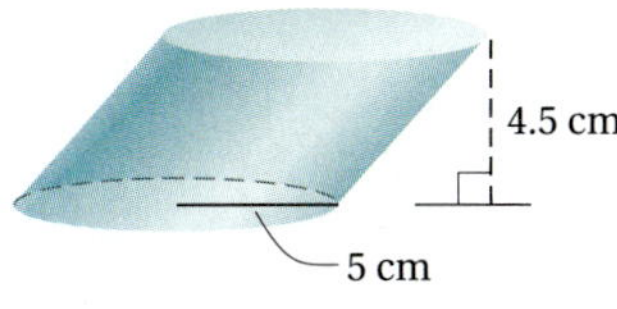

12.

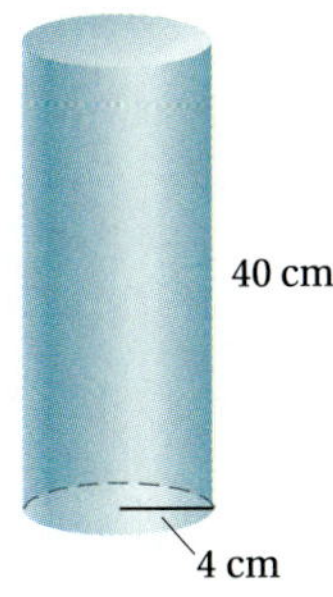

13.

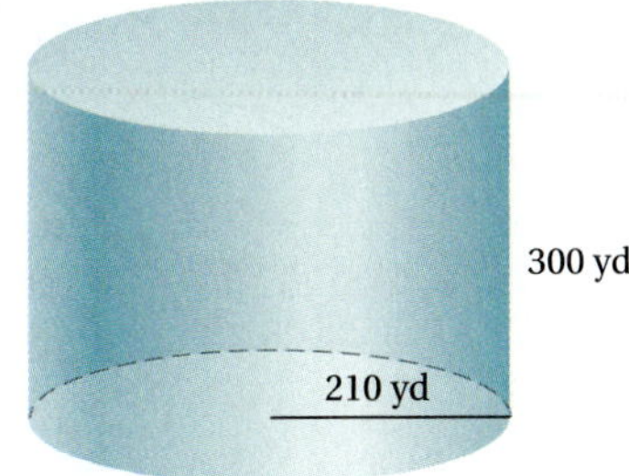

14.

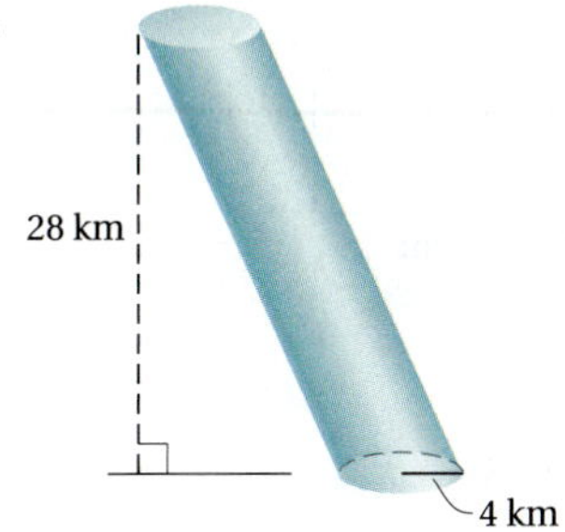

c Find the volume of the sphere. Use 3.14 for π in Exercises 15–18. Use $\frac{22}{7}$ for π in Exercises 19 and 20.

15.

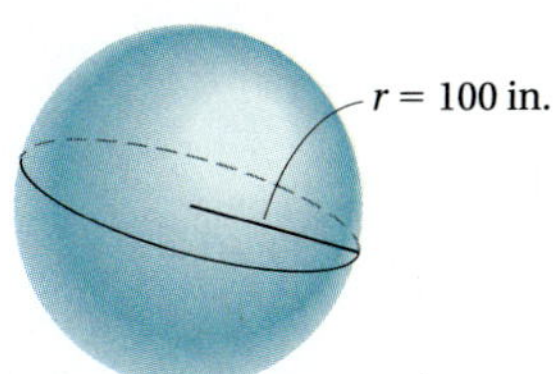

16.

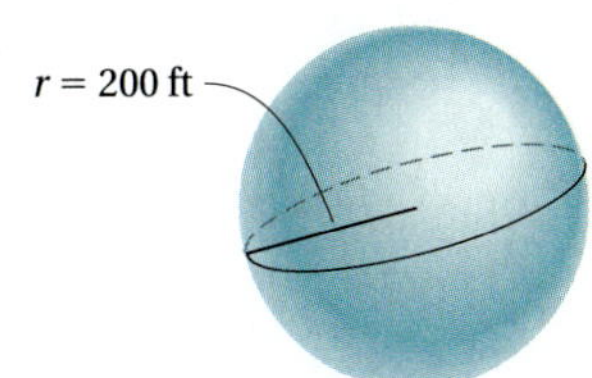

17.

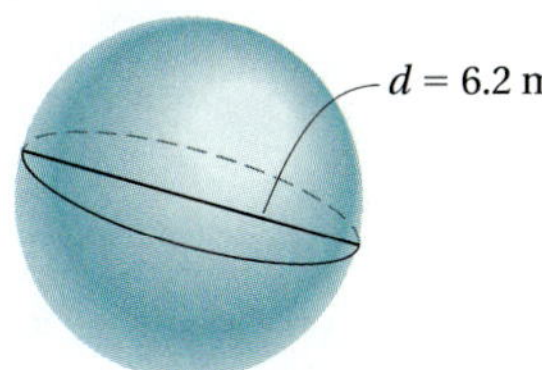

18.

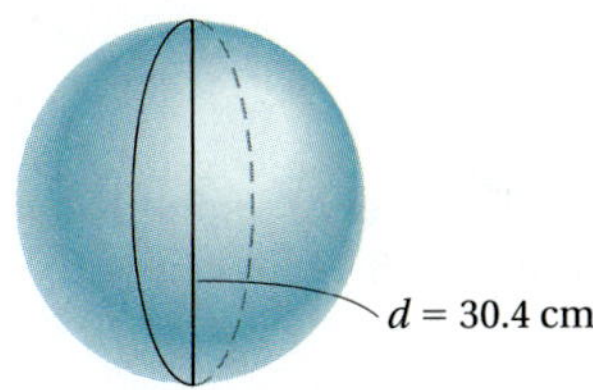

19.

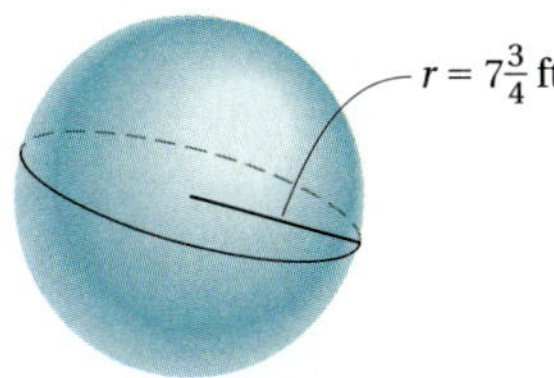

20.

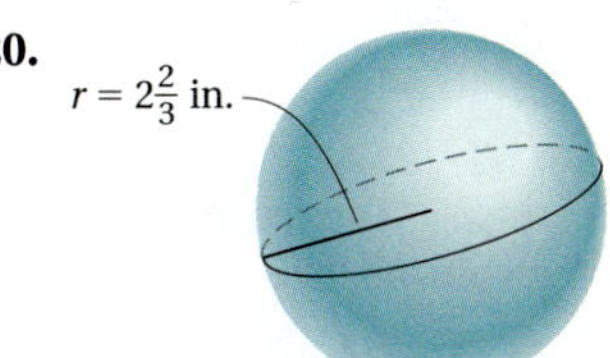

d Find the volume of the circular cone. Use 3.14 for π in Exercises 21 and 22. Use $\frac{22}{7}$ for π in Exercises 23 and 24.

21.

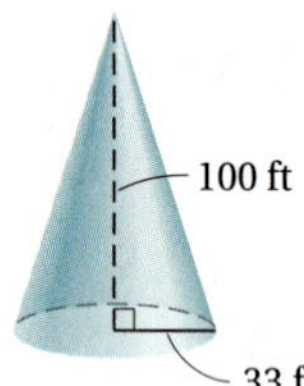

22.

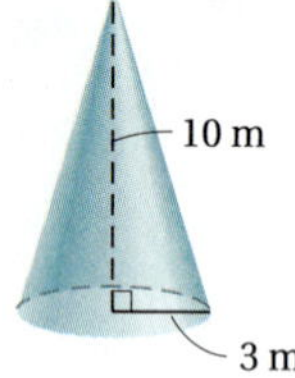

23.

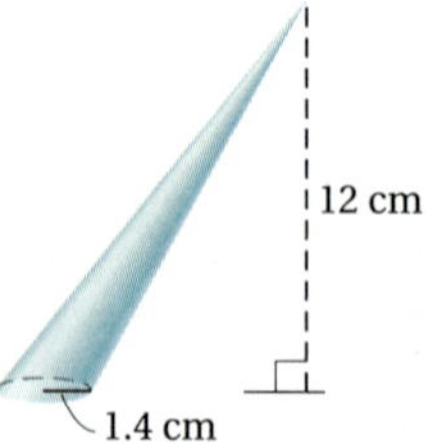

24.

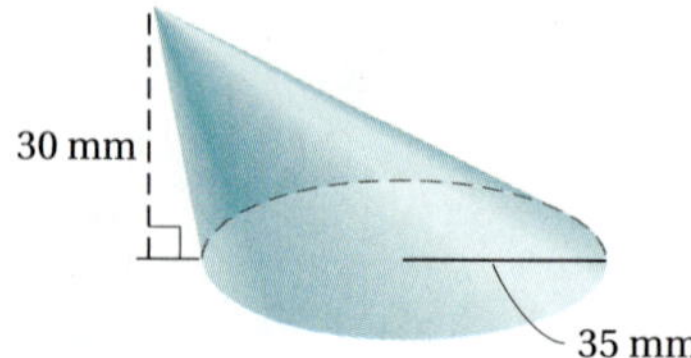

e Solve.

25. *Volume of a Trash Can.* The diameter of the base of a cylindrical trash can is 0.7 yd. The height is 1.1 yd. Find the volume. Use 3.14 for π.

26. *Ladder Rung.* A rung of a ladder is 2 in. in diameter and 16 in. long. Find the volume. Use 3.14 for π.

27. *Barn Silo.* A barn silo, excluding the top, is a circular cylinder. The silo is 6 m in diameter and the height is 13 m. Find the volume of the silo. Use 3.14 for π.

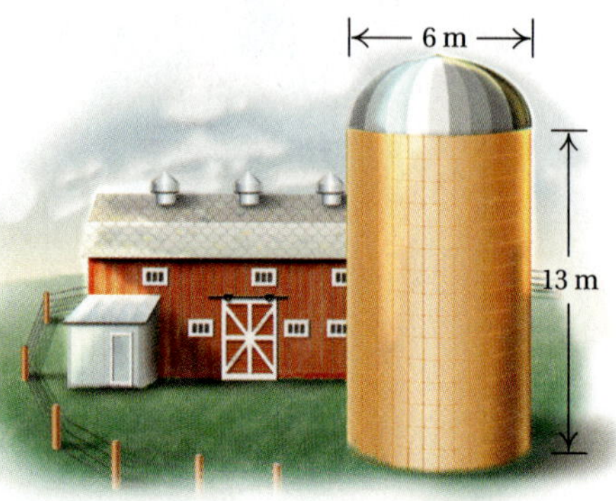

28. *Oak Log.* An oak log has a diameter of 12 cm and a length (height) of 42 cm. Find the volume. Use 3.14 for π.

29. *Tennis Ball.* The diameter of a tennis ball is 6.5 cm. Find the volume. Use 3.14 for π.

30. *Spherical Gas Tank.* The diameter of a spherical gas tank is 6 m. Find the volume. Use 3.14 for π.

31. *Volume of Earth.* The diameter of the earth is about 3980 mi. Find the volume of the earth. Use 3.14 for π. Round to the nearest ten thousand cubic miles.

32. *Astronomy.* The radius of Pluto's moon is about 500 km. Find the volume of this satellite. Use $\frac{22}{7}$ for π.

33. *Tennis-Ball Packaging.* Tennis balls are generally packaged in circular cylinders that hold 3 balls each. The diameter of a tennis ball is 6.5 cm. Find the volume of a can of tennis balls. Use 3.14 for π.

34. *Golf-Ball Packaging.* The box shown is just big enough to hold 3 golf balls. If the radius of a golf ball is 2.1 cm, how much air surrounds the three balls? Use 3.14 for π.

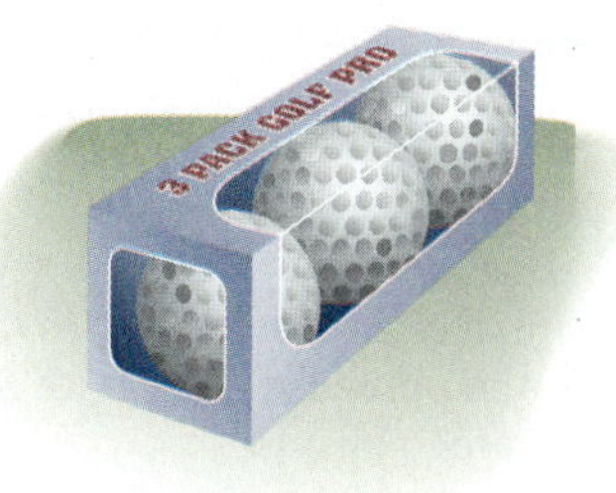

35. *Water Storage.* A water storage tank is a right circular cylinder with a radius of 14 cm and a height of 100 cm. What is the tank's volume? Use $\frac{22}{7}$ for π.

36. *Oceanography.* A research submarine is capsule-shaped. Find the volume of the submarine if it has a length of 10 m and a diameter of 8 m. Use 3.14 for π. (*Hint*: First find the length of the cylindrical section.)

37. *Metallurgy.* If all the gold in the world could be gathered together, it would form a cube 18 yd on a side. Find the volume of the world's gold.

38. The volume of a ball is $36\pi\ \text{cm}^3$. Find the dimensions of a rectangular box that is just large enough to hold the ball.

39. **DW** How could you use the volume formulas given in this section to help estimate the volume of an egg?

40. **DW** The design of a modern home includes a cylindrical tower that will be capped with either a 10-ft–high dome or a 10-ft–high cone. Which type of cap will be more energy-efficient and why?

SKILL MAINTENANCE

41. Find the simple interest on \$600 at 6.4% for $\frac{1}{2}$ yr. [4.7d]

42. Find the simple interest on \$600 at 8% for 2 yr. [4.7d]

Evaluate. [1.6b]

43. 10^3

44. 15^2

45. 7^2

46. 4^3

Solve.

47. *Sales Tax.* In a certain state, a sales tax of \$878 is collected on the purchase of a car for \$17,560. What is the sales tax rate? [4.7a]

48. *Commission Rate.* Rich earns \$1854.60 selling \$16,860 worth of cellular phones. What is the commission rate? [4.7b]

SYNTHESIS

49. The width of a dollar bill is 2.3125 in., the length is 6.0625 in., and the thickness is 0.0041 in. Find the volume occupied by one million one-dollar bills.

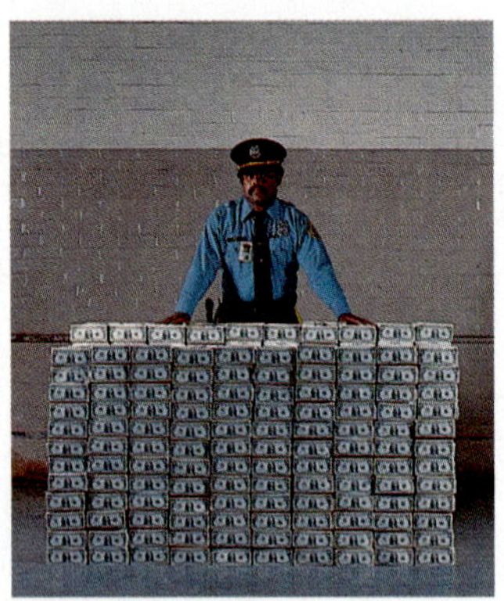

© 1998 AL SATTERWHITE

50. Audio-cassette cases are typically 7 cm by 10.75 cm by 1.5 cm and contain 90 min of music. Compact-disc cases are typically 12.4 cm by 14.1 cm by 1 cm and contain 50 min of music. Which container holds the most music per cubic centimeter?

51. A 2-cm–wide stream of water passes through a 30-m–long garden hose. At the instant that the water is turned off, how many liters of water are in the hose? Use 3.141593 for π.

52. The volume of a basketball is 2304π cm^3. Find the volume of a cube-shaped box that is just large enough to hold a ball.

53. *Circumference of Earth.* The circumference of the earth at the equator is about 24,901.55 mi. Due to the irregular shape of the earth, the circumference of a circle of longitude wrapped around the earth between the north and south poles is about 24,859.82 mi. Describe and carry out a procedure for estimating the volume of the earth.
Source: *The Handy Geography Answer Book*

54. A sphere with diameter 1 m is circumscribed by a cube. How much more volume is in the cube?

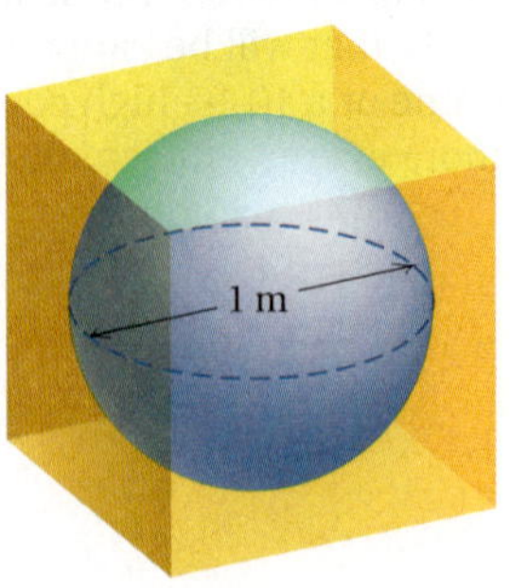

55. A cube is circumscribed by a sphere with a 1-m diameter. How much more volume is in the sphere?

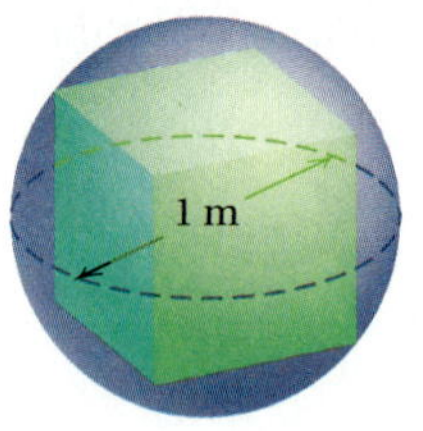

5.6 RELATIONSHIPS BETWEEN ANGLE MEASURES

a Complementary and Supplementary Angles

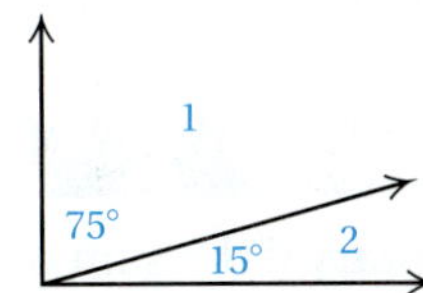

∠1 and ∠2 above are **complementary** angles.

$$m\angle 1 + m\angle 2 = 90^\circ$$
$$75^\circ + 15^\circ = 90^\circ$$

COMPLEMENTARY ANGLES

Two angles are **complementary** if the sum of their measures is 90°. Each angle is called a **complement** of the other.

If two angles are complementary, each is an acute angle. When complementary angles are adjacent to each other, they form a right angle.

EXAMPLE 1 Identify each pair of complementary angles.

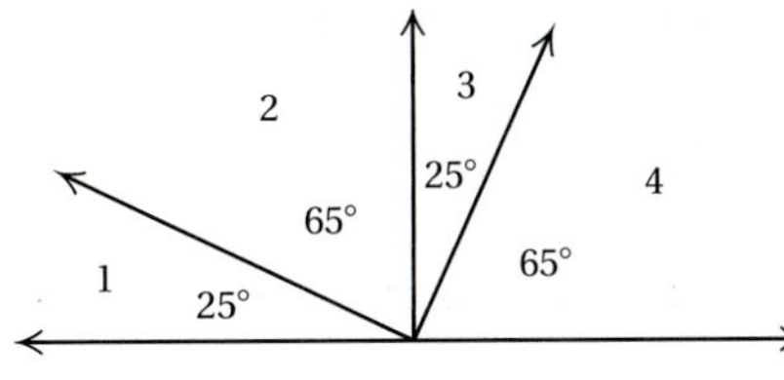

∠1 and ∠2 $25^\circ + 65^\circ = 90^\circ$ ∠2 and ∠3
∠1 and ∠4 ∠3 and ∠4

Do Exercise 1.

EXAMPLE 2 Find the measure of a complement of an angle of 39°.

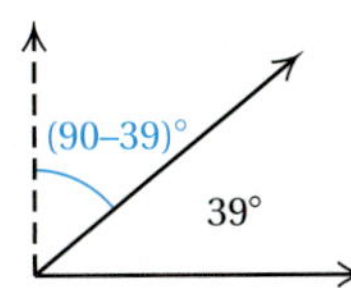

$$90^\circ - 39^\circ = 51^\circ$$

The measure of a complement is 51°.

Do Exercises 2–4.

Objectives

a Identify complementary and supplementary angles and find the measure of a complement or a supplement of a given angle.

b Determine whether segments are congruent and whether angles are congruent.

c Use the Vertical Angle Property to find measures of angles.

d Identify pairs of corresponding angles, interior angles, and alternate interior angles and apply properties of transversals and parallel lines to find measures of angles.

1. Identify each pair of complementary angles.

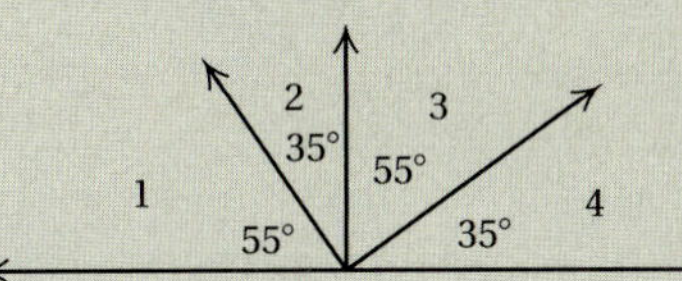

Find the measure of a complement of the angle.

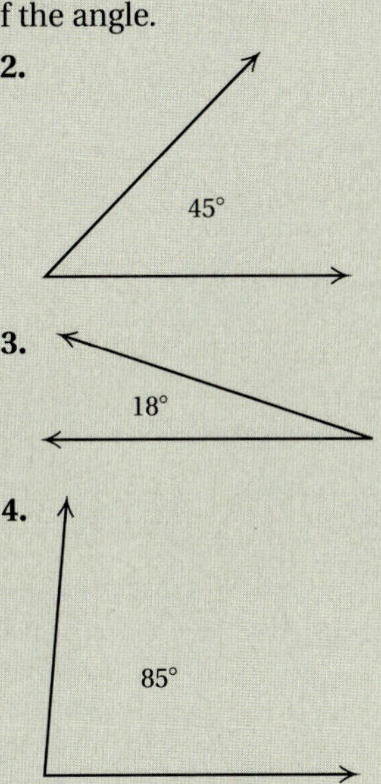

Answers on page A-13

Next, consider ∠1 and ∠2 as shown below. Because the sum of their measures is 180°, ∠1 and ∠2 are said to be **supplementary.** Note that when supplementary angles are adjacent, they form a straight angle.

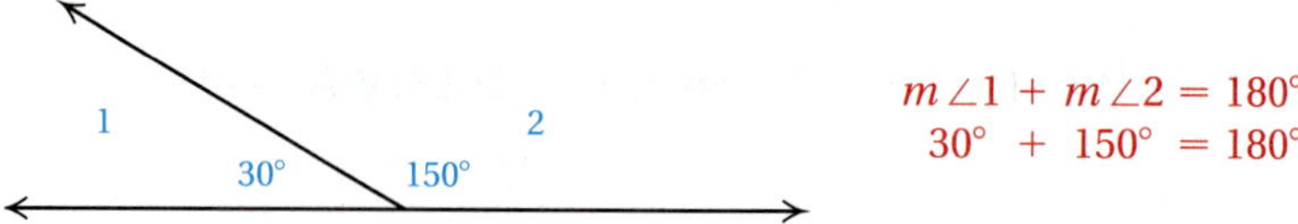

SUPPLEMENTARY ANGLES

Two angles are **supplementary** if the sum of their measures is 180°. Each angle is called a **supplement** of the other.

EXAMPLE 3 Identify each pair of supplementary angles.

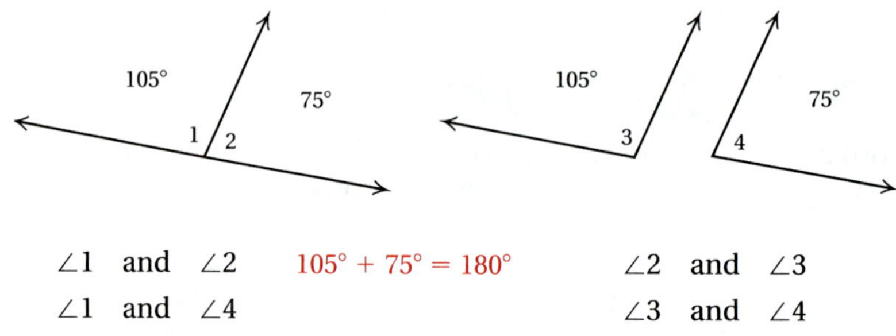

Do Exercise 5.

EXAMPLE 4 Find the measure of a supplement of an angle of 112°.

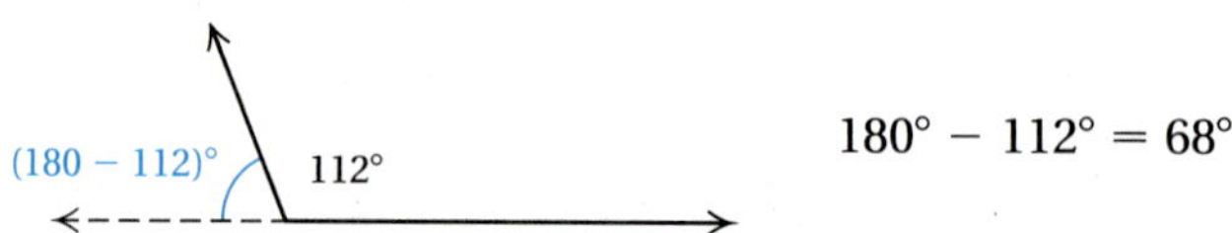

The measure of a supplement is 68°.

Do Exercises 6–8.

5. Identify each pair of supplementary angles.

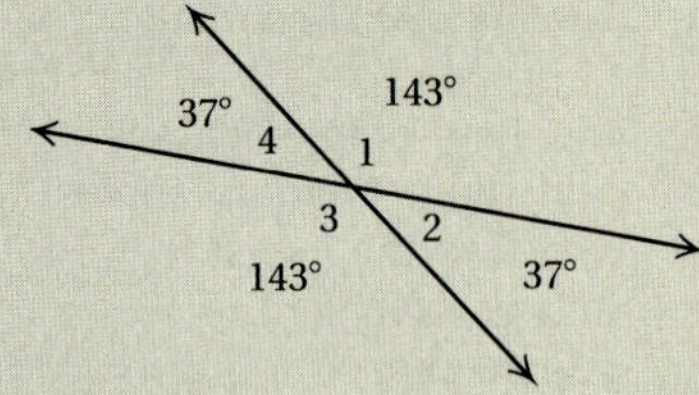

Find the measure of a supplement of an angle with the given measure.

6. 38°

7. 157°

8. 90°

Answers on page A-13

b Congruent Segments and Angles

Congruent figures have the same size and shape. They fit together exactly.

CONGRUENT SEGMENTS

Two segments are **congruent** if and only if they have the same length.

EXAMPLE 5 Use a ruler to show that $\overline{PQ}$ and $\overline{RS}$ are congruent.

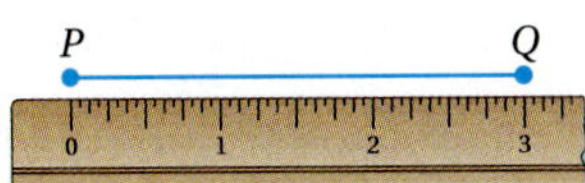

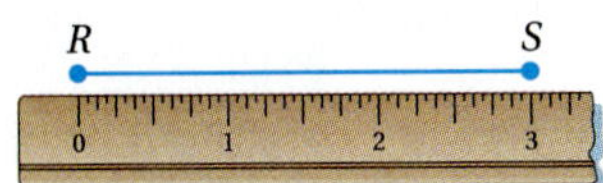

Since both segments have the same length, $\overline{PQ}$ and $\overline{RS}$ are congruent. To say that $\overline{PQ}$ and $\overline{RS}$ are congruent, we write

$$\overline{PQ} \cong \overline{RS}.$$

EXAMPLE 6 Which pairs of segments are congruent? Use a ruler.

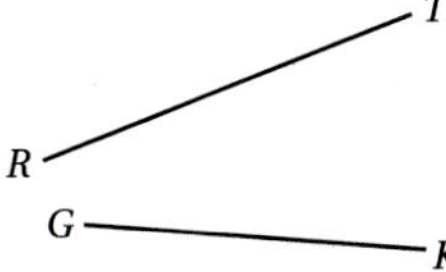

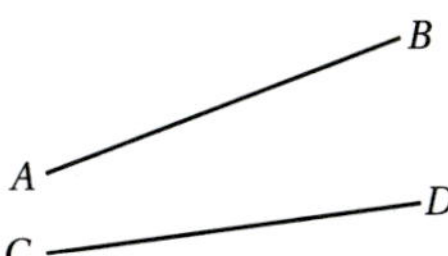

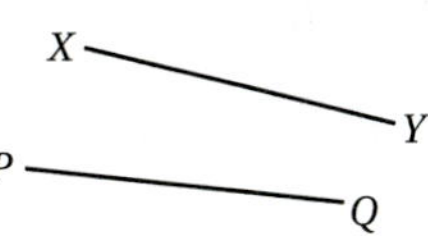

$\overline{AB} \cong \overline{CD}$ and $\overline{PQ} \cong \overline{XY}$.

Do Exercises 9 and 10.

CONGRUENT ANGLES

Two angles are **congruent** if and only if they have the same measure.

EXAMPLE 7 Use a protractor to show that $\angle P$ and $\angle Q$ are congruent.

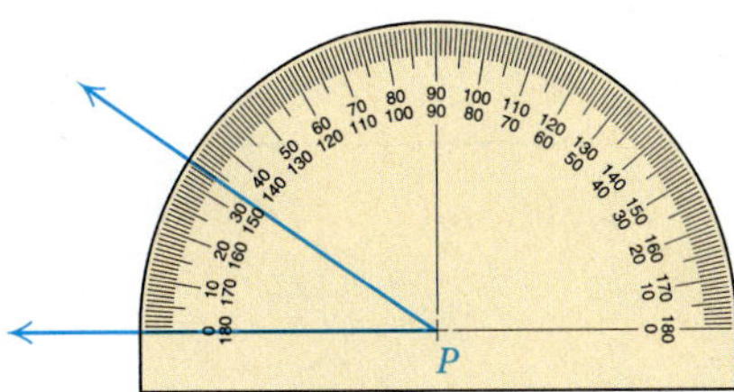

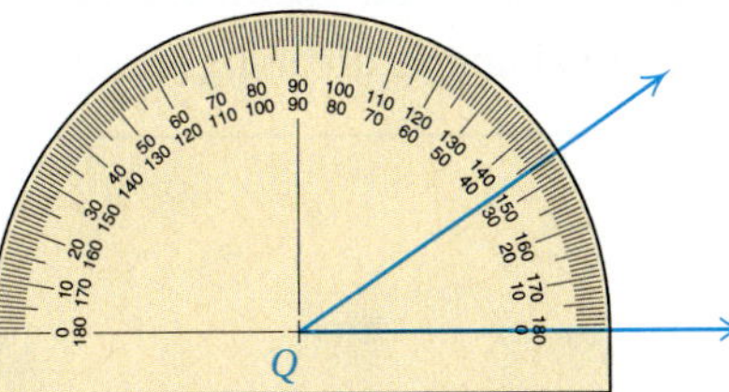

Since $m\angle P = m\angle Q = 34°$, $\angle P$ and $\angle Q$ are congruent. To say that $\angle P$ and $\angle Q$ are congruent, we write

$$\angle P \cong \angle Q.$$

Which pairs of segments are congruent? Use a ruler.

9.

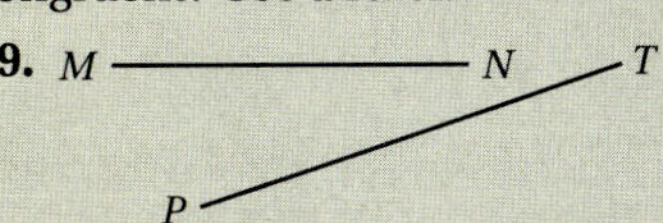

10.

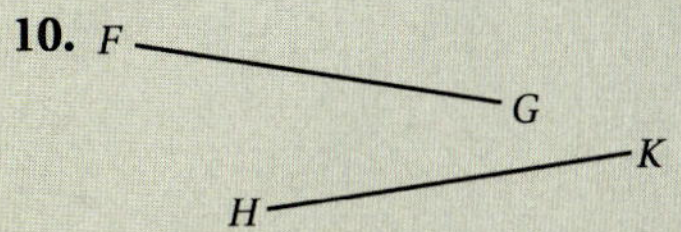

Answers on page A-13

Which pairs of angles are congruent? Use a protractor.

11.

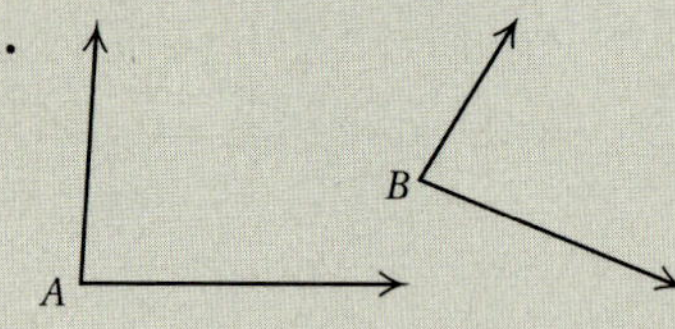

12.

EXAMPLE 8 Which pairs of angles are congruent? Use a protractor.

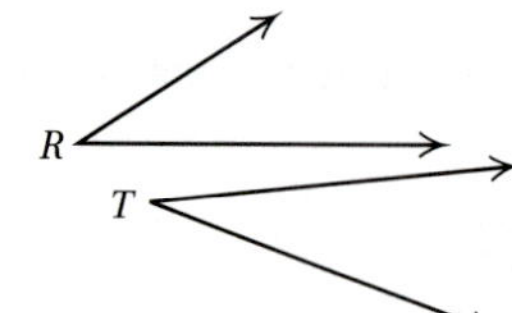

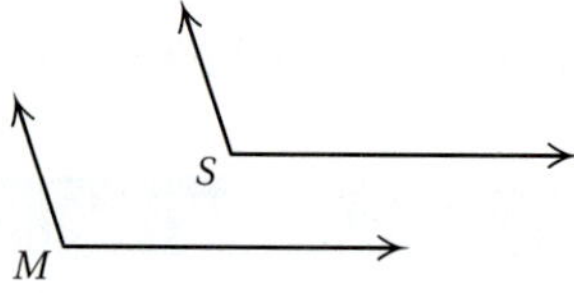

$$\angle M \cong \angle S \quad \text{since} \quad m\angle M = m\angle S = 108°.$$

Do Exercises 11 and 12.

If two angles are congruent, then their supplements are congruent and their complements are congruent.

C Vertical Angles

When $\overleftrightarrow{RT}$ intersects $\overleftrightarrow{SQ}$ at P, four angles are formed:

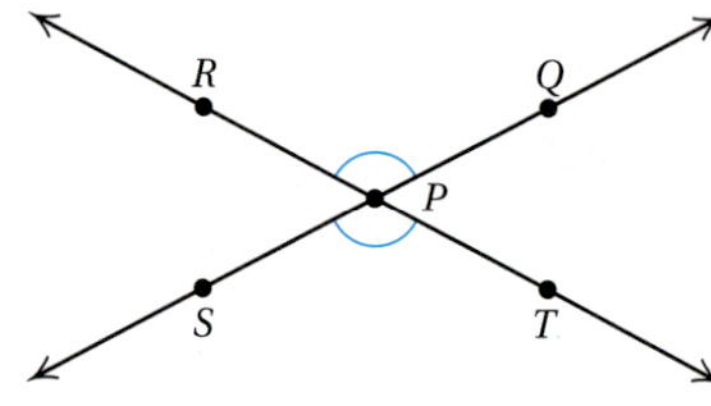

$\angle SPT$

$\angle RPQ$

$\angle SPR$

$\angle QPT$

Pairs of angles such as $\angle RPQ$ and $\angle SPT$ are called **vertical angles.**

VERTICAL ANGLES

Two nonstraight angles are **vertical angles** if and only if their sides form two pairs of opposite rays.

Vertical angles are supplements of the same angle. Thus they are congruent.

THE VERTICAL ANGLE PROPERTY

Vertical angles are congruent.

Answers on page A-13

EXAMPLE 9 In the figure below, $m\angle 1 = 23°$ and $m\angle 3 = 34°$. Find $m\angle 2$, $m\angle 4$, $m\angle 5$, and $m\angle 6$.

Since $\angle 1$ and $\angle 4$ are vertical angles, $m\angle 4 = 23°$. Likewise, $\angle 3$ and $\angle 6$ are vertical angles, so $m\angle 6 = 34°$.

$$m\angle 1 + m\angle 2 + m\angle 3 = 180$$
$$23 + m\angle 2 + 34 = 180 \quad \text{Substituting}$$
$$m\angle 2 = 180 - 57$$
$$m\angle 2 = 123°$$

Since $\angle 2$ and $\angle 5$ are vertical angles, $m\angle 5 = 123°$.

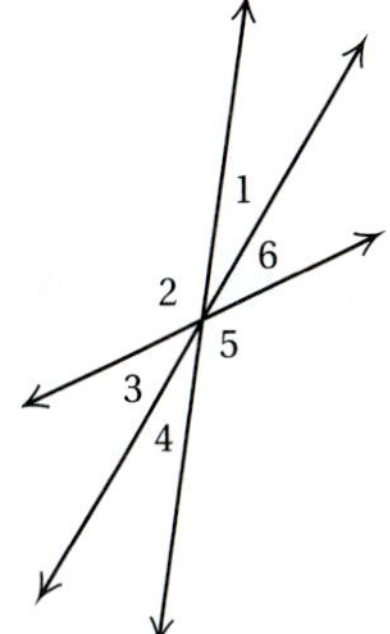

Do Exercise 13.

13. In the figure below, $m\angle 2 = 41°$ and $m\angle 4 = 10°$. Find $m\angle 1$, $m\angle 3$, $m\angle 5$, and $m\angle 6$.

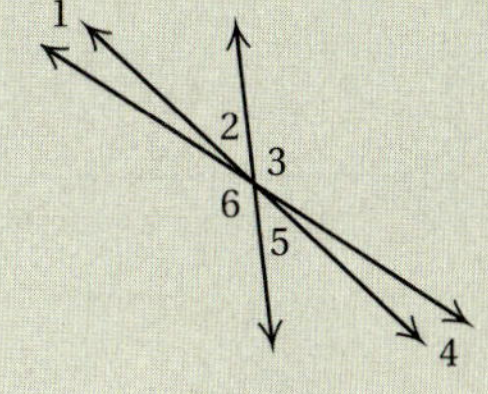

d Transversals and Angles

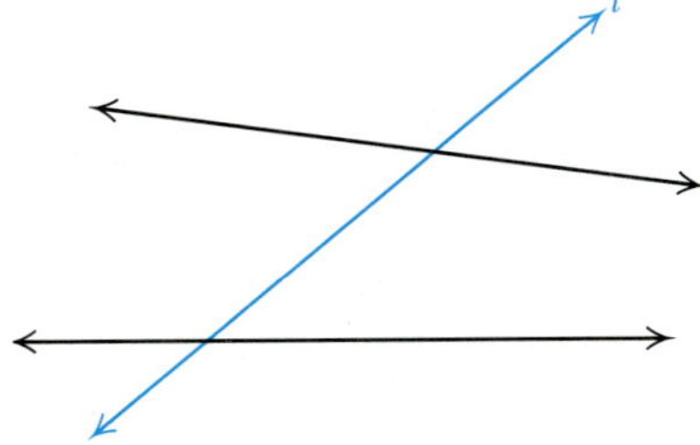

TRANSVERSAL

A **transversal** is a line that intersects two or more coplanar lines in different points.

When a transversal intersects a pair of lines, eight angles are formed. Certain pairs of these angles have special names.

CORRESPONDING ANGLES

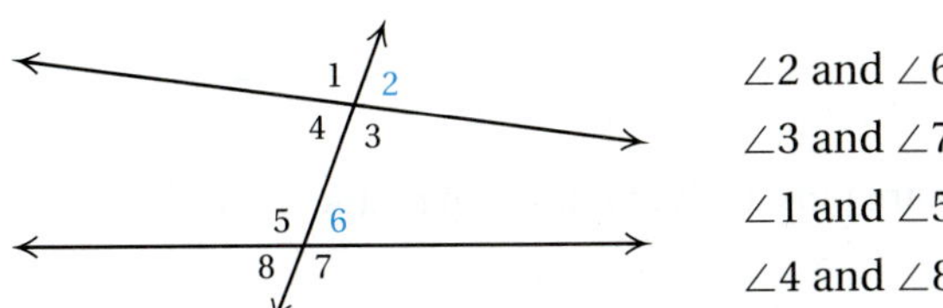

$\angle 2$ and $\angle 6$
$\angle 3$ and $\angle 7$
$\angle 1$ and $\angle 5$
$\angle 4$ and $\angle 8$

Answers on page A-13

Use the following figure to answer Margin Exercises 14–16.

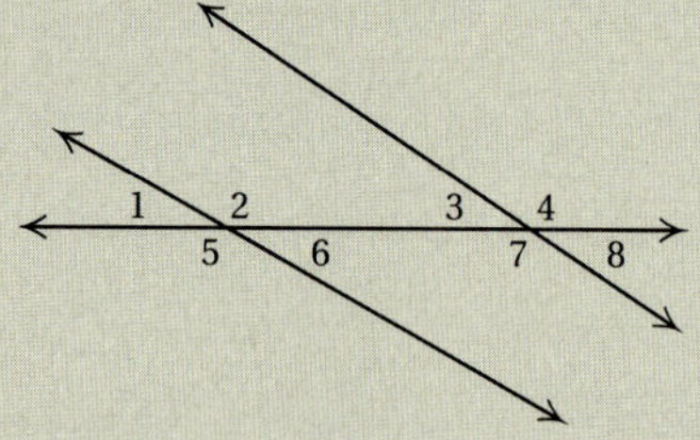

14. Identify all pairs of corresponding angles.

15. Identify all interior angles.

16. Identify all pairs of alternate interior angles.

Answers on page A-13

INTERIOR ANGLES

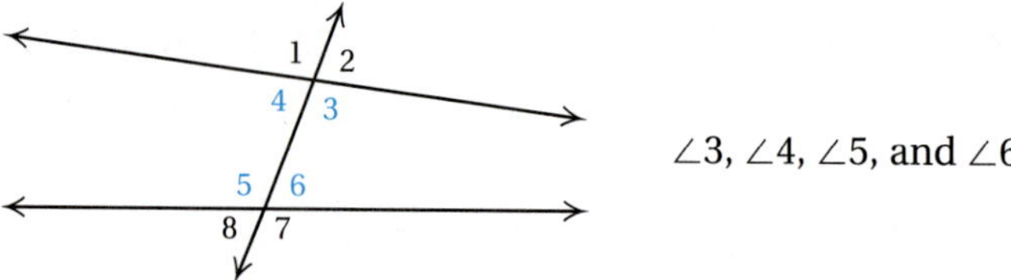

$\angle 3$, $\angle 4$, $\angle 5$, and $\angle 6$

ALTERNATE INTERIOR ANGLES

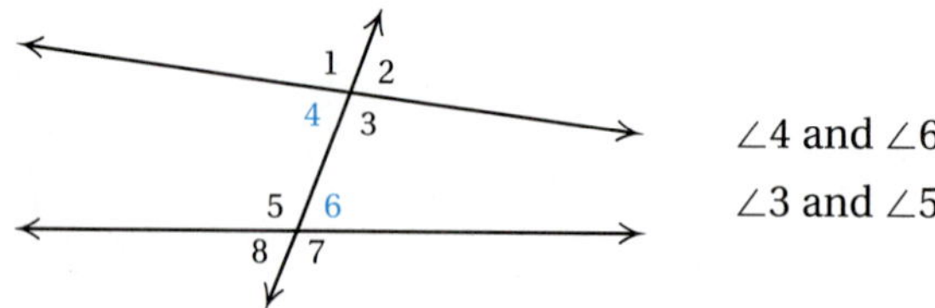

$\angle 4$ and $\angle 6$

$\angle 3$ and $\angle 5$

EXAMPLE 10 Identify all pairs of corresponding angles, all interior angles, and all pairs of alternate interior angles.

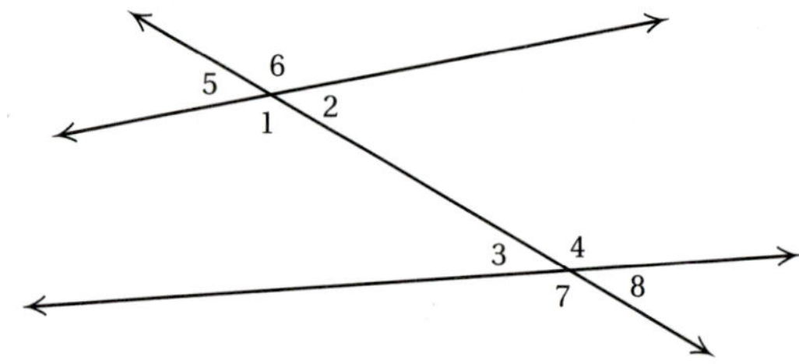

Corresponding angles: $\angle 6$ and $\angle 4$, $\angle 2$ and $\angle 8$, $\angle 5$ and $\angle 3$, $\angle 1$ and $\angle 7$

Interior angles: $\angle 1$, $\angle 2$, $\angle 3$, $\angle 4$

Alternate interior angles: $\angle 1$ and $\angle 4$, $\angle 2$ and $\angle 3$

Do Exercises 14–16.

Given a line l and a point P not on l, there is at most one line that contains P and is parallel to l.

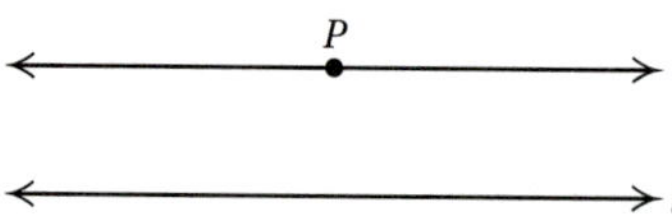

If two lines are parallel, the following relations hold.

PROPERTIES OF PARALLEL LINES

1. If a transversal intersects two parallel lines, then the corresponding angles are congruent.

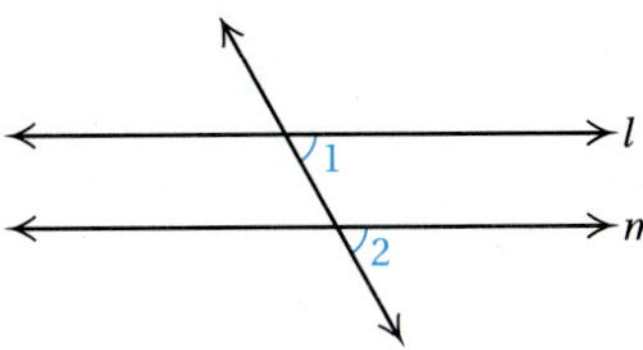

If $l \parallel m$, then $\angle 1 \cong \angle 2$.

2. If a transversal intersects two parallel lines, then the alternate interior angles are congruent.

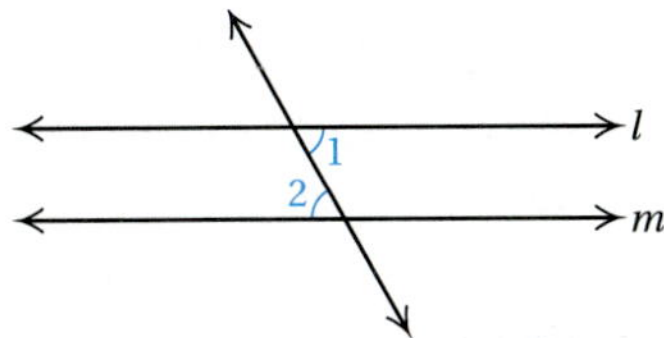

If $l \parallel m$, then $\angle 1 \cong \angle 2$.

3. In a plane, if two lines are parallel to a third line, then the two lines are parallel to each other.

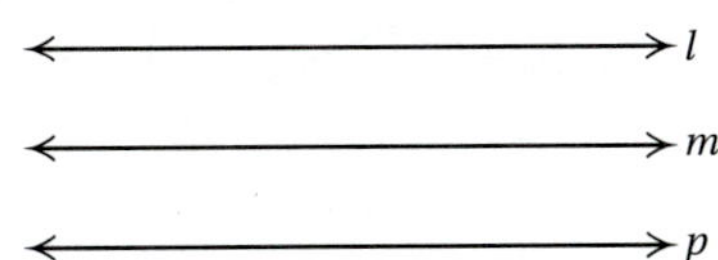

If $l \parallel p$ and $m \parallel p$, then $l \parallel m$.

4. If a transversal intersects two parallel lines, then the interior angles on the same side of the transversal are supplementary.

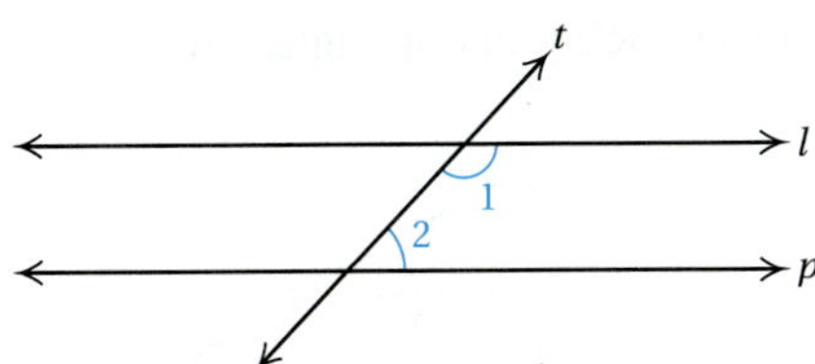

If $l \parallel p$, then $m\angle 1 + m\angle 2 = 180°$.

5. If a transversal is perpendicular to one of two parallel lines, then it is perpendicular to the other.

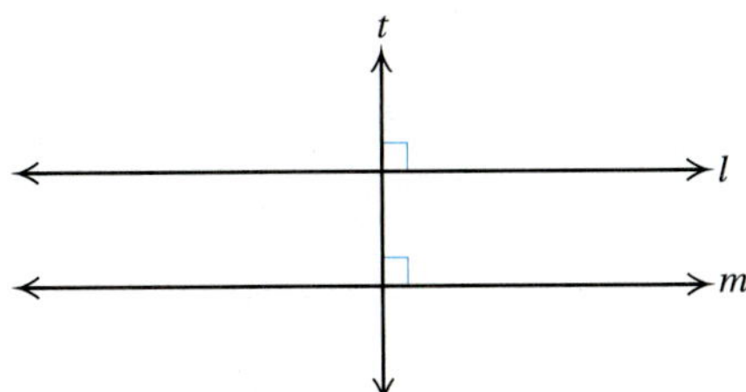

17. If $l \parallel m$ and $m\angle 3 = 51°$, what are the measures of the other angles?

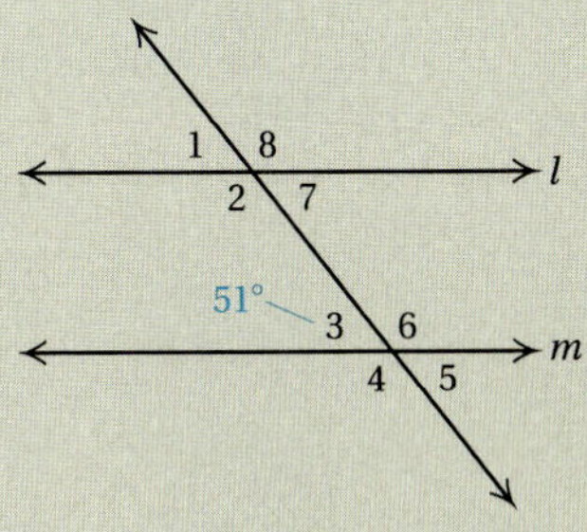

Answers on page A-13

18. If $\overline{AB} \parallel \overline{CD}$, which pairs of angles are congruent?

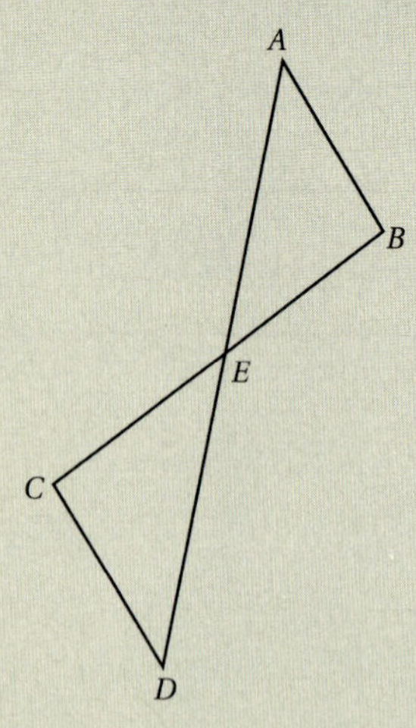

19. If $\overline{PQ} \parallel \overline{RS}$, which pairs of angles are congruent?

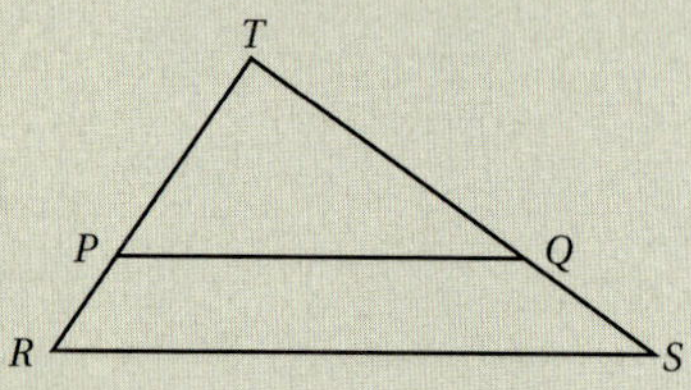

Answers on page A-13

EXAMPLE 11 If $l \parallel m$ and $m\angle 1 = 40°$, what are the measures of the other angles?

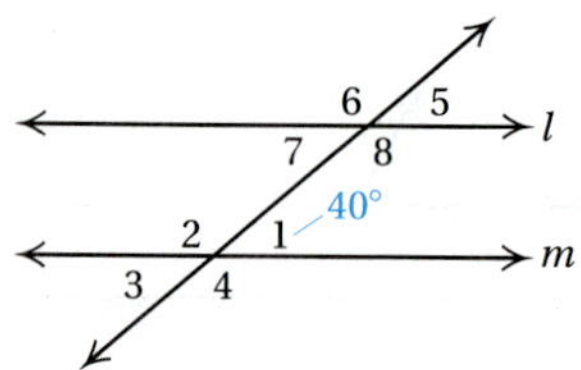

$m\angle 7 = 40°$	Using Property 2
$m\angle 5 = 40°$	Using Property 1
$m\angle 8 = 140°$	Using Property 4
$m\angle 3 = 40°$	$\angle 1$ and $\angle 3$ are vertical angles
$m\angle 4 = 140°$	Using Property 1 and $m\angle 8 = 140°$
$m\angle 2 = 140°$	$\angle 2$ and $\angle 4$ are vertical angles and $m\angle 4 = 140°$
$m\angle 6 = 140°$	$\angle 6$ and $\angle 8$ are vertical angles and $m\angle 8 = 140°$

Do Exercise 17 on the preceding page.

EXAMPLE 12 If $\overline{PT} \parallel \overline{SR}$, which pairs of angles are congruent?

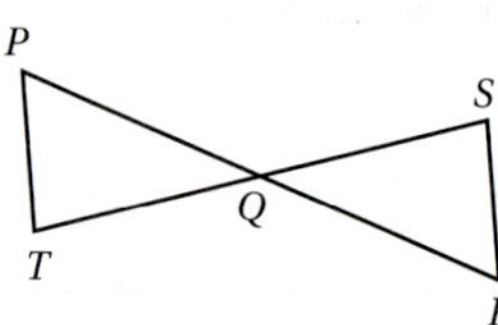

$\angle TPQ \cong \angle SRQ$ and $\angle PTQ \cong \angle RSQ$ Using Property 2

$\angle PQT \cong \angle RQS$ and $\angle PQS \cong \angle RQT$ Vertical angles

Do Exercise 18.

EXAMPLE 13 If $\overline{DE} \parallel \overline{BC}$, which pairs of angles are congruent?

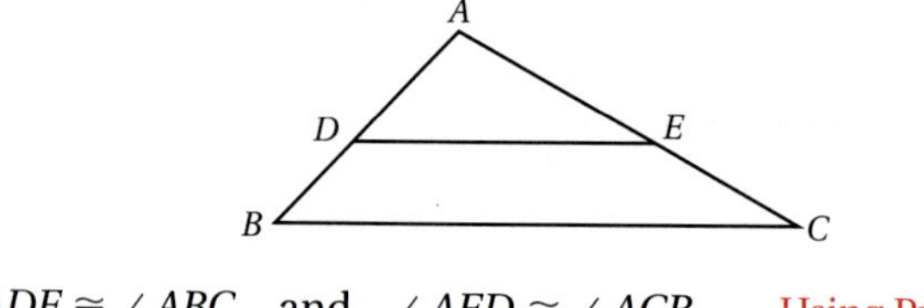

$\angle ADE \cong \angle ABC$ and $\angle AED \cong \angle ACB$ Using Property 1

Do Exercise 19.

5.6 EXERCISE SET

For Extra Help

Digital Video Tutor CD 4 Videotape 6 | InterAct Math | Math Tutor Center | MathXL | MyMathLab

a Find the measure of a complement of an angle with the given measure.

1. 11°

2. 83°

3. 67°

4. 5°

5. 58°

6. 32°

7. 29°

8. 54°

Find the measure of a supplement of an angle with the given measure.

9. 3°

10. 54°

11. 139°

12. 13°

13. 85°

14. 129°

15. 102°

16. 45°

b Determine whether the pair of segments is congruent. Use a ruler.

17.

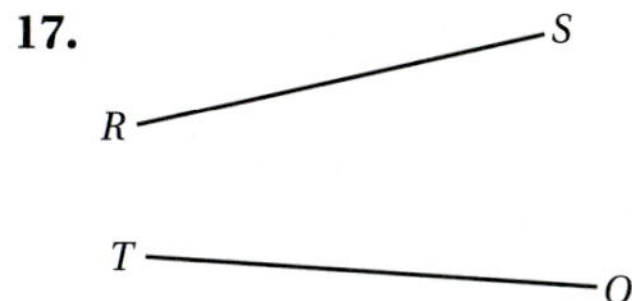

18. H J K V

Determine whether the pair of angles is congruent. Use a protractor.

19.

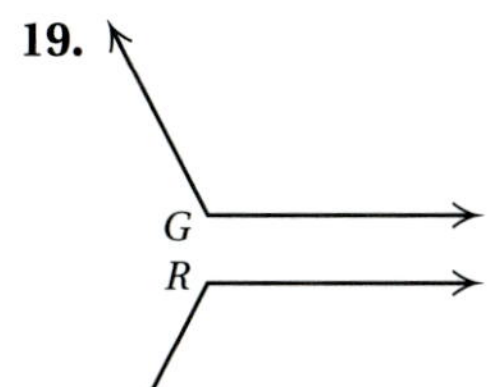

20.

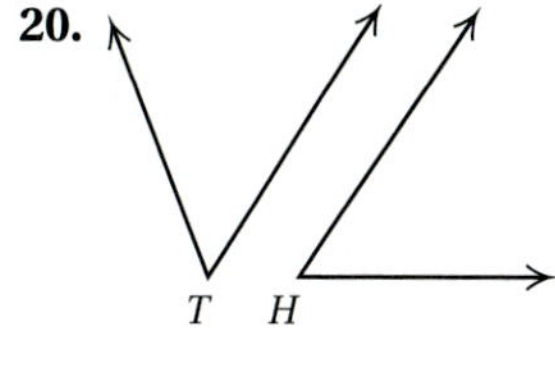

c

21. In the figure, $m\angle 1 = 80°$ and $m\angle 5 = 67°$. Find $m\angle 2$, $m\angle 3$, $m\angle 4$, and $m\angle 6$.

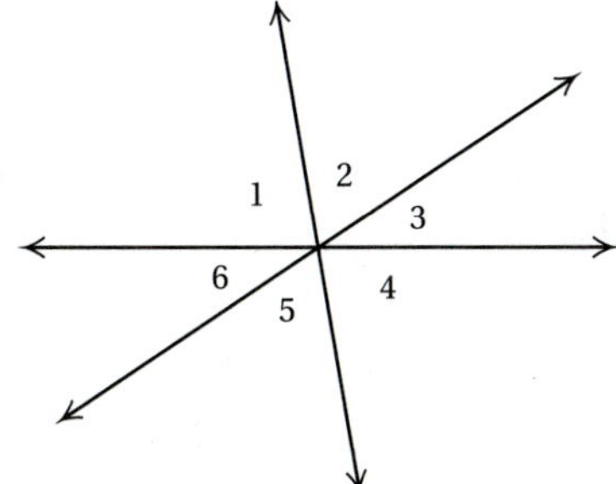

22. In the figure, $m\angle 2 = 42°$ and $m\angle 4 = 56°$. Find $m\angle 1$, $m\angle 3$, $m\angle 5$, and $m\angle 6$.

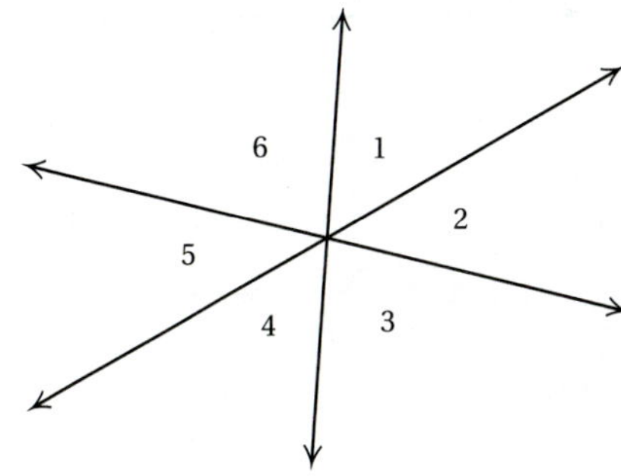

d In Exercises 23 and 24, **(a)** identify all pairs of corresponding angles, **(b)** identify all interior angles, and **(c)** identify all pairs of alternate interior angles.

23.

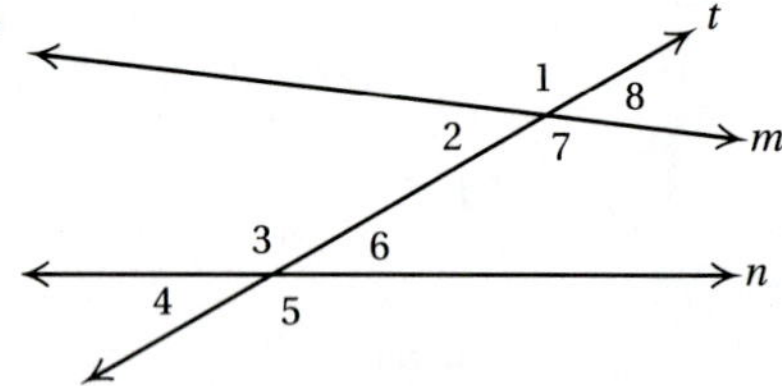

Lines m and n
Transversal t

24.

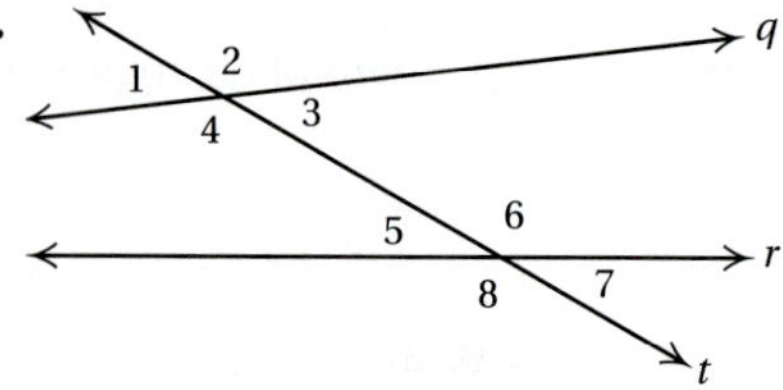

Lines q and r
Transversal t

25. If $m \parallel n$ and $m\angle 4 = 125°$, what are the measures of the other angles?

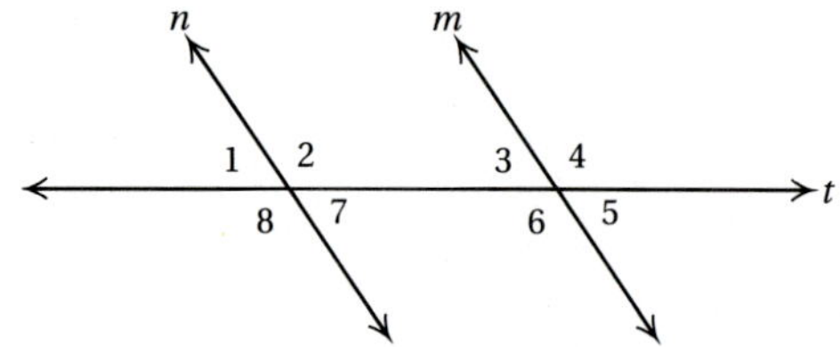

26. If $m \parallel n$ and $m\angle 8 = 34°$, what are the measures of the other angles?

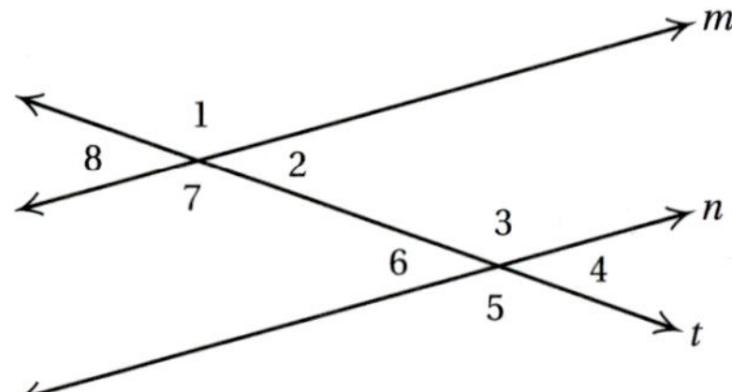

In each figure, $\overline{AB} \parallel \overline{CD}$. Identify pairs of congruent angles. When possible, give the measures of the angles.

27.

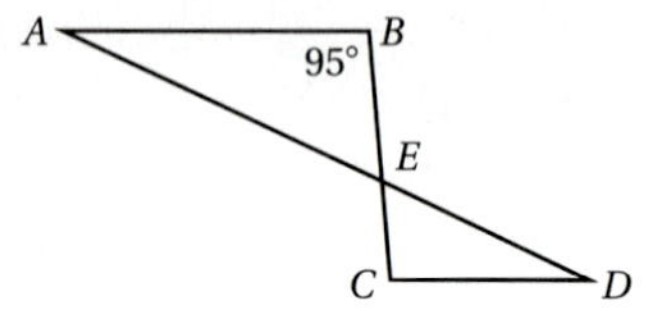

28.

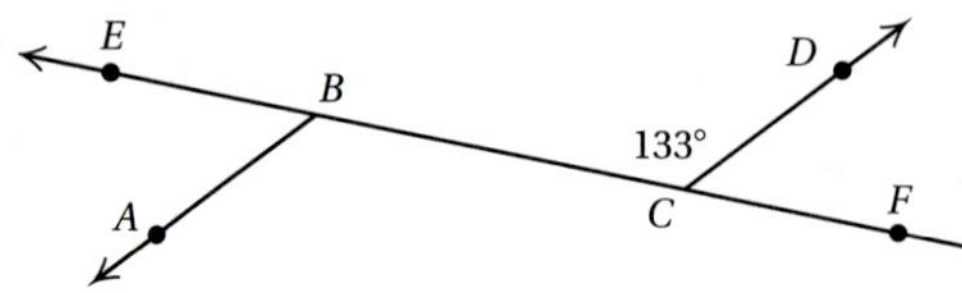

29.

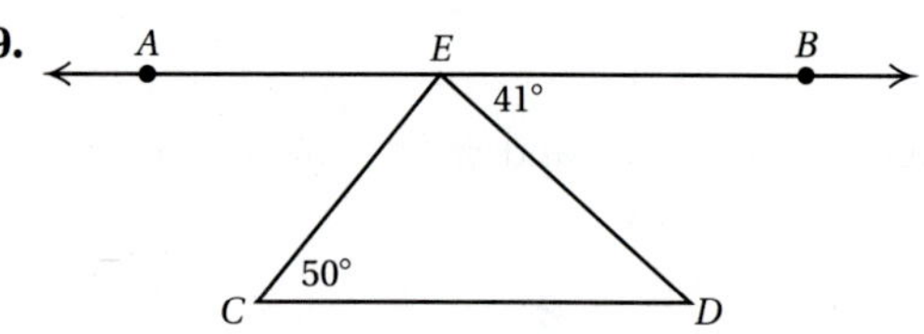

30.

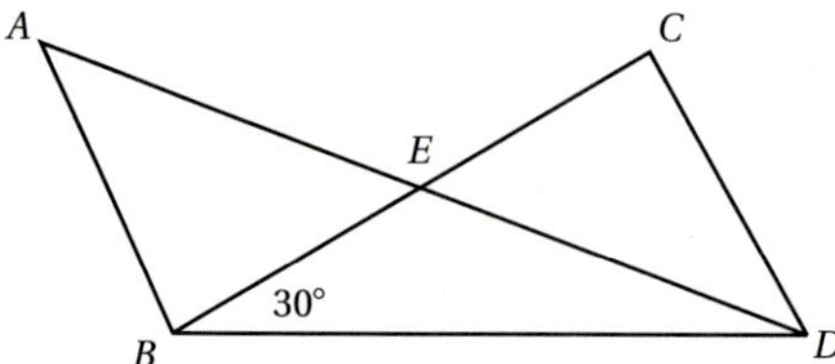

SKILL MAINTENANCE

Multiply. [2.4d]

31. $6 \times 1\frac{7}{8}$

32. $10\frac{3}{4} \times 1\frac{1}{2}$

33. $8\frac{3}{7} \times 14$

34. $2\frac{2}{3} \times 5\frac{1}{2}$

5.7 CONGRUENT TRIANGLES AND PROPERTIES OF PARALLELOGRAMS

Objectives

a Identify the corresponding parts of congruent triangles and show why triangles are congruent using SAS, SSS, and ASA.

b Use properties of parallelograms to find lengths of sides and measures of angles of parallelograms.

a Congruent Triangles

Triangles can be classified by their angles.

Acute:	All angles acute
Right:	One right angle
Obtuse:	One obtuse angle
Equiangular:	All angles congruent

Triangles can also be classified by their sides.

Equilateral:	All sides congruent
Isosceles:	At least two sides congruent
Scalene:	No sides congruent

We know that congruent figures fit together exactly.

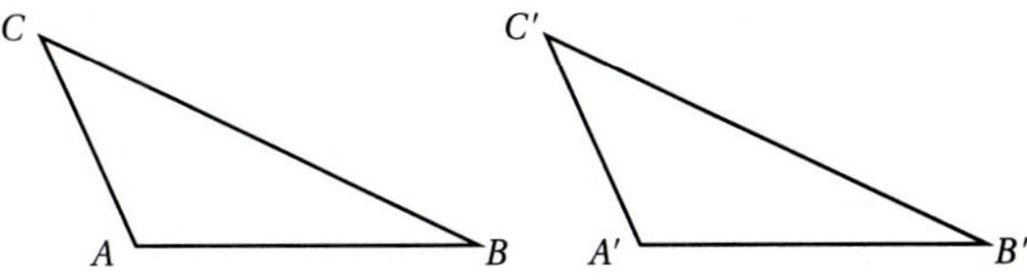

B' is read "B prime."

These triangles will fit together exactly if we match A with A', B with B', and C with C'. On the other hand, if we match A with B', B with C', and C with A', the triangles will not fit together exactly. The matching of vertices determines corresponding sides and angles.

EXAMPLES Consider $\triangle ABC$ and $\triangle A'B'C'$ above.

1. If we match A with A', B with B', and C with C', what are the corresponding sides?

$\overline{AB} \leftrightarrow \overline{A'B'}$

$\overline{BC} \leftrightarrow \overline{B'C'}$ $\quad$ $\leftrightarrow$ means "corresponds to."

$\overline{AC} \leftrightarrow \overline{A'C'}$

2. If we match A with B', B with C', and C with A', what are the corresponding angles?

$\angle A \leftrightarrow \angle B'$ $\qquad$ $\angle B \leftrightarrow \angle C'$ $\qquad$ $\angle C \leftrightarrow \angle A'$

If $A \leftrightarrow A'$, $B \leftrightarrow B'$, and $C \leftrightarrow C'$, then we write $ABC \leftrightarrow A'B'C'$.

CONGRUENT TRIANGLES

Two triangles are **congruent** if and only if their vertices can be matched so that the corresponding angles and sides are congruent.

The corresponding sides and angles of two congruent triangles are called *corresponding parts* of congruent triangles. Corresponding parts of congruent triangles are always congruent.

1. Suppose that $\triangle ABC \cong \triangle DEF$. What are the congruent corresponding parts?

We write $\triangle ABC \cong \triangle A'B'C'$ to say that $\triangle ABC$ and $\triangle A'B'C'$ are congruent. We agree that this symbol also tells us the way in which the vertices are matched.

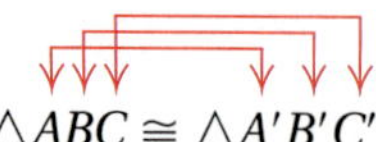

$\triangle ABC \cong \triangle A'B'C'$

$\triangle ABC \cong \triangle A'B'C'$ means that

$\angle A \cong \angle A'$ and $\overline{AB} \cong \overline{A'B'}$

$\angle B \cong \angle B'$ $\quad \overline{AC} \cong \overline{A'C'}$

$\angle C \cong \angle C'$ $\quad \overline{BC} \cong \overline{B'C'}$.

EXAMPLE 3 Suppose that $\triangle PQR \cong \triangle STV$. What are the congruent corresponding parts?

Angles	*Sides*
$\angle P \cong \angle S$	$\overline{PQ} \cong \overline{ST}$
$\angle Q \cong \angle T$	$\overline{PR} \cong \overline{SV}$
$\angle R \cong \angle V$	$\overline{QR} \cong \overline{TV}$

Do Exercise 1.

2. Name the corresponding parts of these congruent triangles.

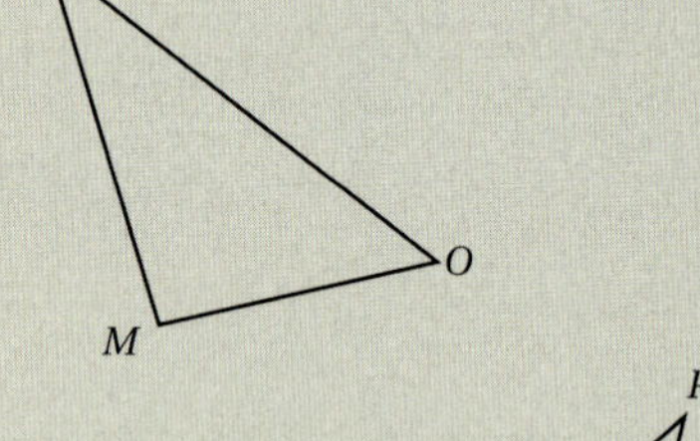

P

Q

R

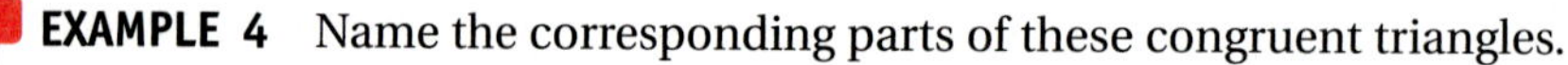

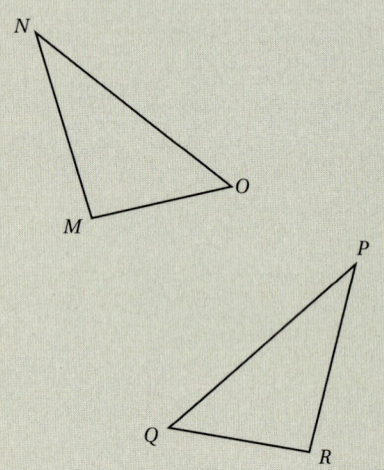

EXAMPLE 4 Name the corresponding parts of these congruent triangles.

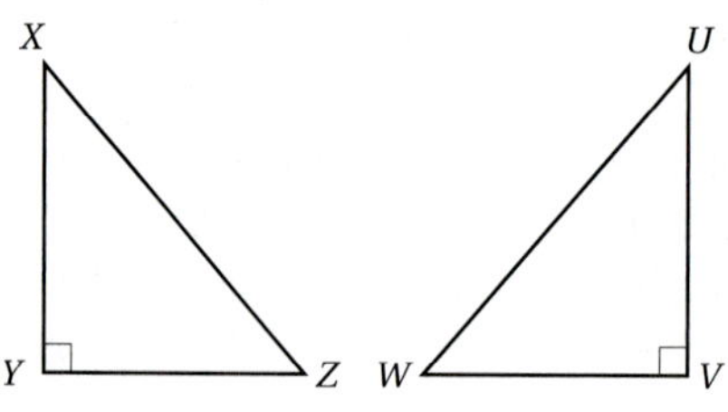

Angles	*Sides*
$\angle X \cong \angle U$	$\overline{XY} \cong \overline{UV}$
$\angle Y \cong \angle V$	$\overline{YZ} \cong \overline{VW}$
$\angle Z \cong \angle W$	$\overline{ZX} \cong \overline{WU}$

Do Exercise 2.

Sometimes we can show that triangles are congruent without already knowing that all six corresponding parts are congruent.

On a full sheet of paper, draw $\triangle ABC$. On another sheet of paper, make a copy of $\angle A$. Label the copy $\angle D$. On the sides of $\angle D$, copy $\overline{AB}$ and $\overline{AC}$. Label the copy $\overline{DE}$ and $\overline{DF}$. Draw $\overline{EF}$. Cut out $\triangle DEF$ and $\triangle ABC$ and place them together. What do you conclude?

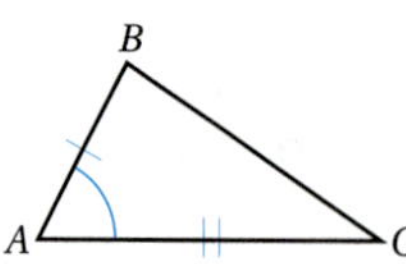

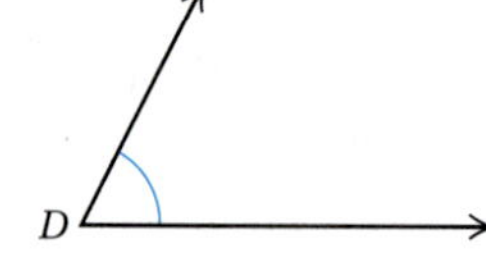

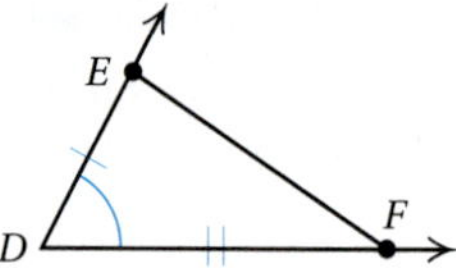

Answers on page A-13

THE SIDE–ANGLE–SIDE (SAS) PROPERTY

Two triangles are congruent if two sides and the included angle of one triangle are congruent to two sides and the included angle of the other triangle.

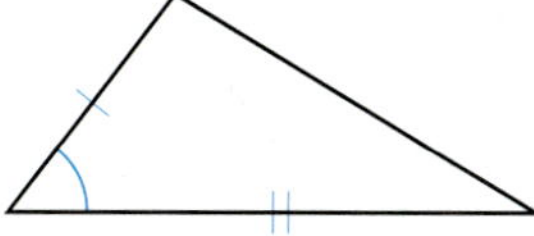

EXAMPLE 5 Which pairs of triangles are congruent by the SAS property?

a)

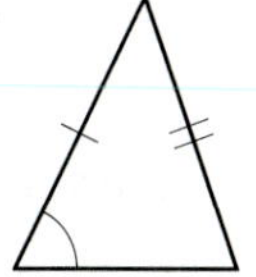

b)

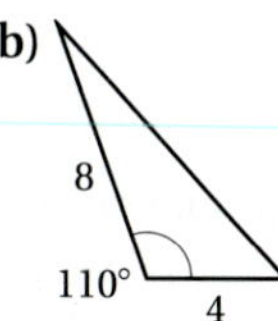

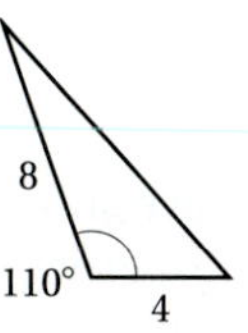

c)

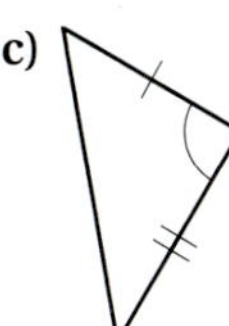

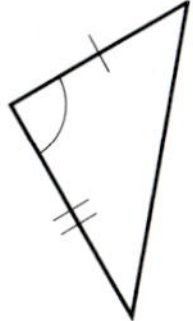

d)

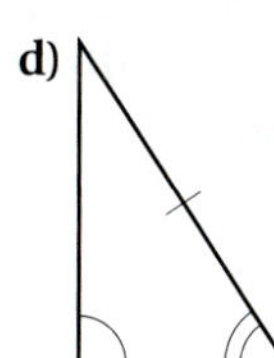

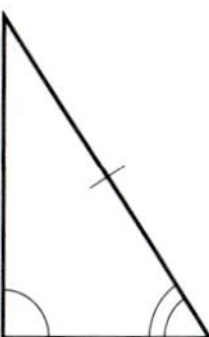

Pairs (b) and (c) are congruent by the SAS property.

Do Exercise 3.

On a sheet of paper, draw a triangle. Then copy this triangle by copying each of its sides. Cut both triangles out and place them together. This suggests the following property.

THE SIDE–SIDE–SIDE (SSS) PROPERTY

If three sides of one triangle are congruent to three sides of another triangle, then the triangles are congruent.

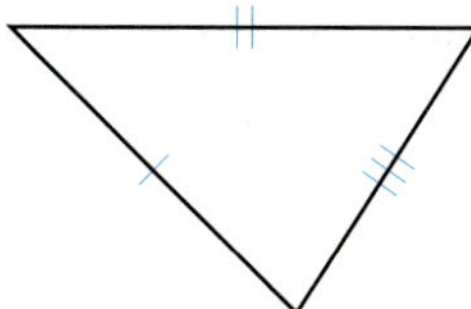

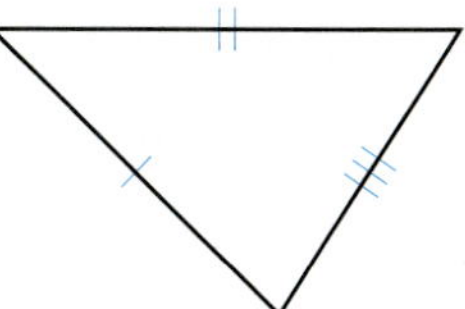

3. Which pairs of triangles are congruent by the SAS property?

a)

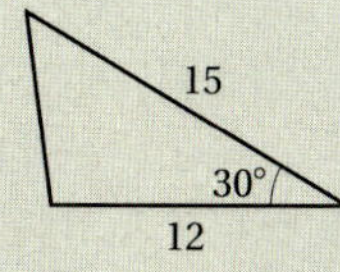

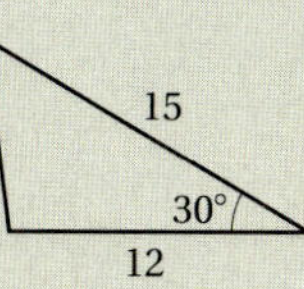

b)

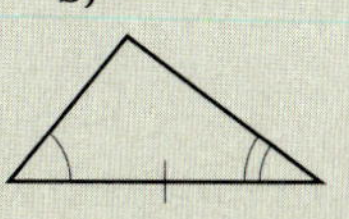

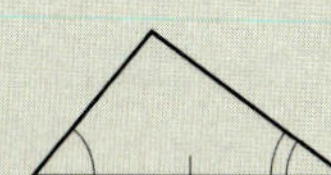

c)

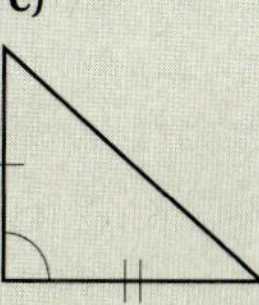

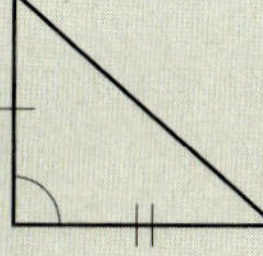

d)

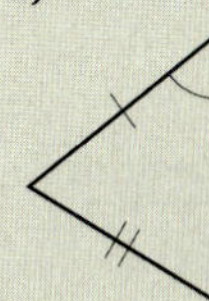

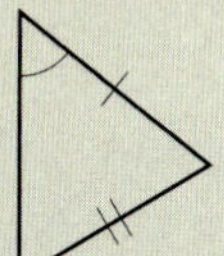

Answers on page A-13

4. Which pairs of triangles are congruent by the SSS property?

a)

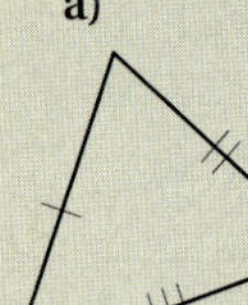

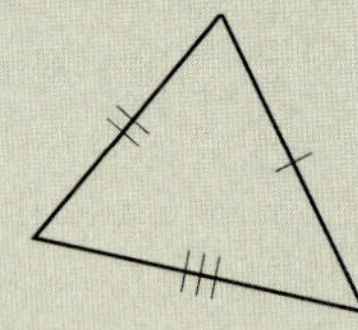

b)

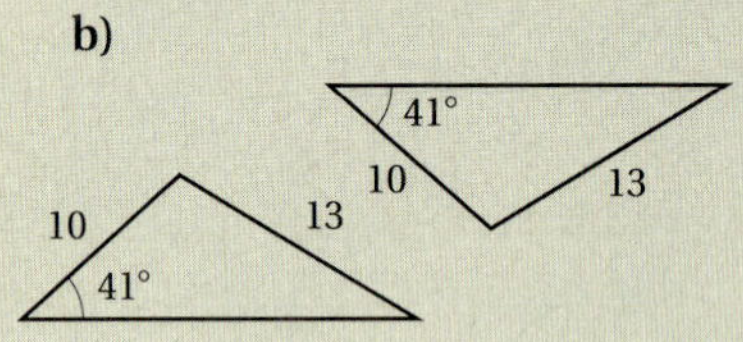

5. Which pairs of triangles are congruent by the ASA property?

a)

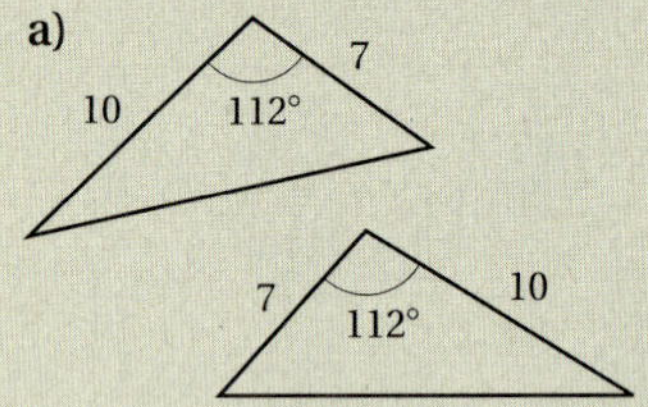

b)

c)

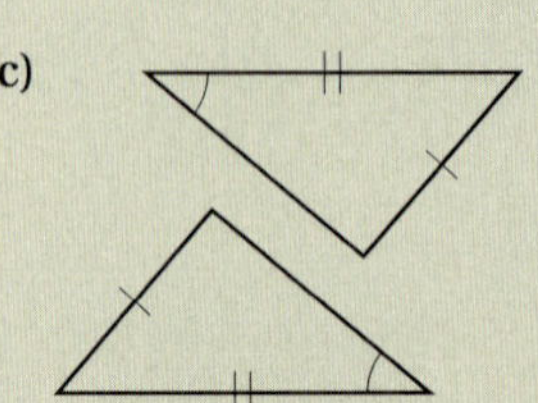

Answers on page A-13

EXAMPLE 6 Which pairs of triangles are congruent by the SSS property?

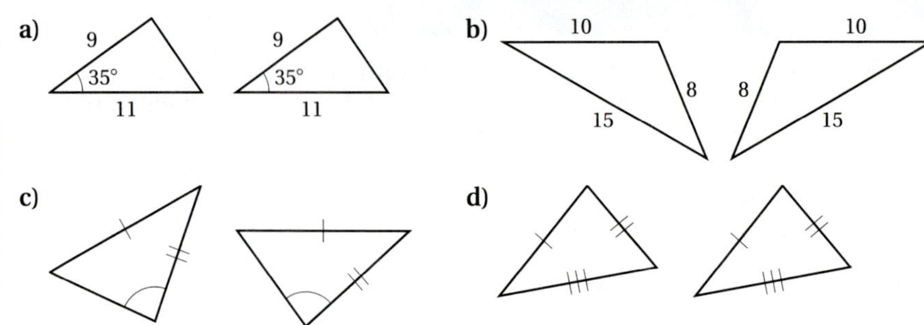

Pairs (b) and (d) are congruent by the SSS property.

Do Exercise 4.

We have shown triangles to be congruent using SAS and SSS. A third way to show congruence is shown below.

On a full sheet of paper, draw a triangle, $\triangle ABC$. On another sheet of paper, draw a segment $\overline{DE}$ so that $DE = AB$*. At D, make a copy of $\angle A$. At E, make a copy of $\angle B$. Label the third vertex of the copy F. Cut out $\triangle ABC$ and $\triangle DEF$ and place them together. What do you conclude?

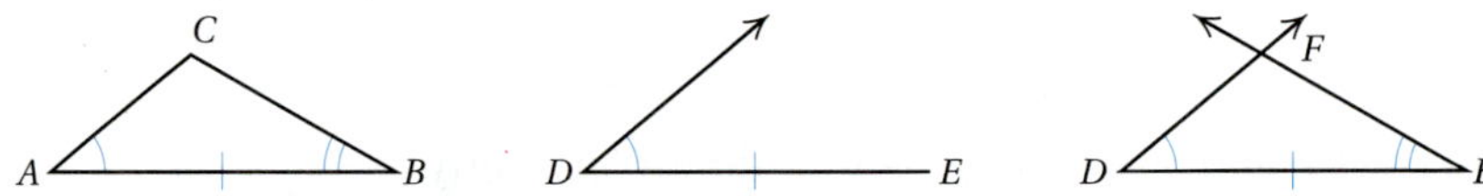

THE ANGLE–SIDE–ANGLE (ASA) PROPERTY

If two angles and the included side of a triangle are congruent to two angles and the included side of another triangle, then the triangles are congruent.

EXAMPLE 7 Which pairs of triangles are congruent by the ASA property?

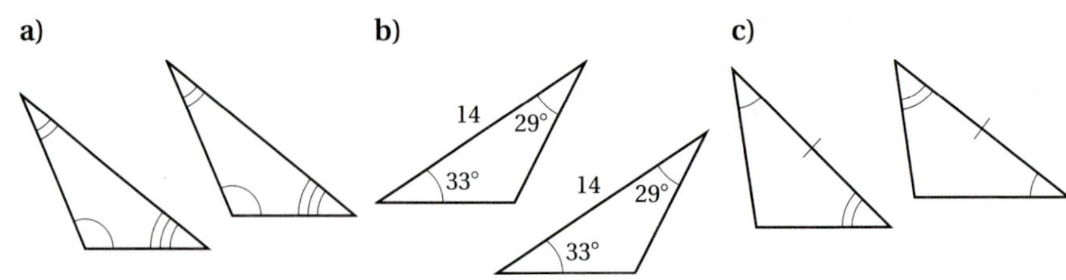

Pairs (b) and (c) are congruent by the ASA property.

Do Exercise 5.

*$\overline{DE}$ denotes the segment with endpoints D and E. DE denotes the length of $\overline{DE}$.

EXAMPLES Which property (if any) should be used to show that these pairs of triangles are congruent?

8.

Use SAS.

9.

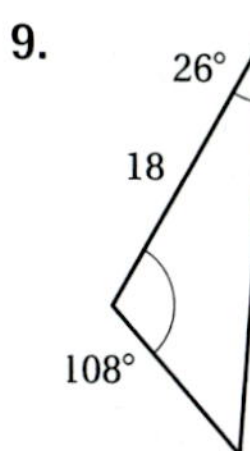

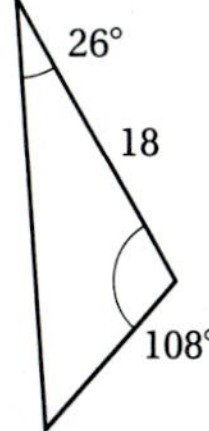

Use ASA.

10.

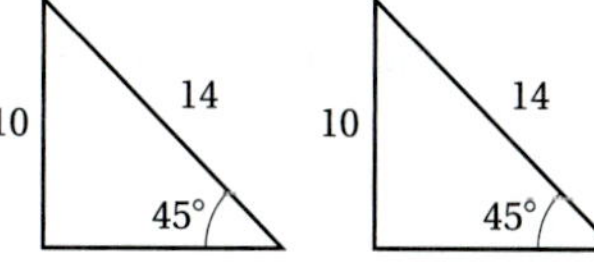

None.

11.

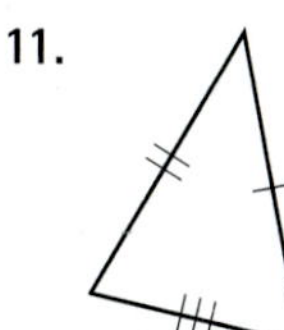

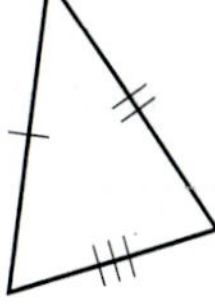

Use SSS.

Do Exercises 6–9.

It is important to be able to explain why triangles are congruent.

EXAMPLE 12 In $\triangle ABC$ and $\triangle DEF$, $\overline{AB} \cong \overline{DE}$, $\overline{AC} \cong \overline{DF}$, and $\angle A \cong \angle D$. Explain why the triangles are congruent.

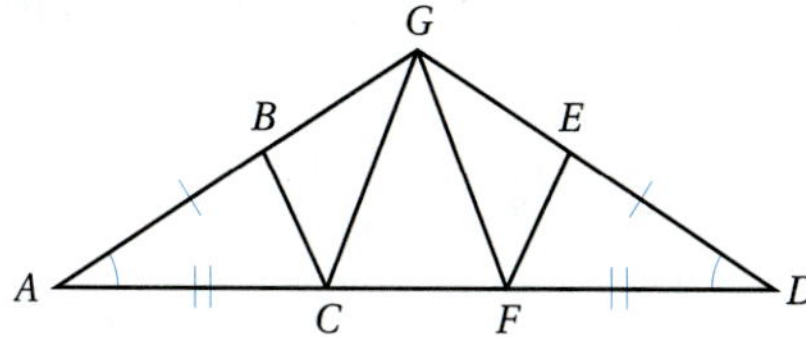

We have two sides and an included angle of $\triangle ABC$ congruent to the corresponding parts of $\triangle DEF$. Thus, $\triangle ABC \cong \triangle DEF$ by SAS.

EXAMPLE 13 In $\triangle CPD$ and $\triangle EQD$, $\overline{CP} \perp \overline{QP}$ and $\overline{EQ} \perp \overline{QP}$. Also, $\angle QDE \cong \angle PDC$ and D is the midpoint of $\overline{QP}$. Explain why $\triangle CPD \cong \triangle EQD$.

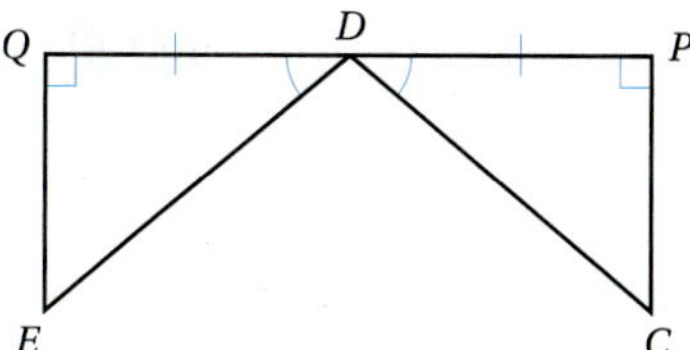

The perpendicular sides form right angles, which are congruent. Since D is the midpoint of $\overline{QP}$, we know that $\overline{QD} \cong \overline{PD}$. With $\angle QDE \cong \angle PDC$, we have $\triangle CPD \cong \triangle EQD$ by ASA.

Do Exercise 10.

Which property (if any) should be used to show that these pairs of triangles are congruent?

6.

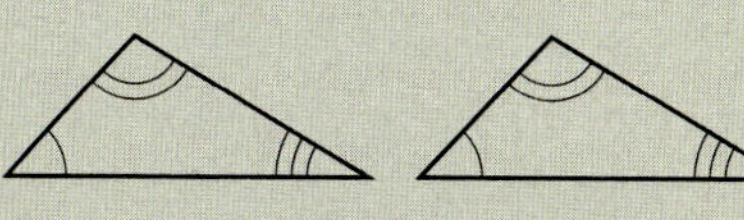

7.

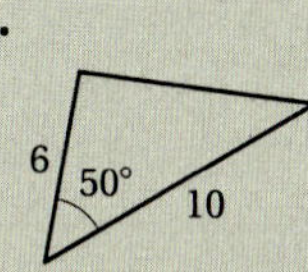

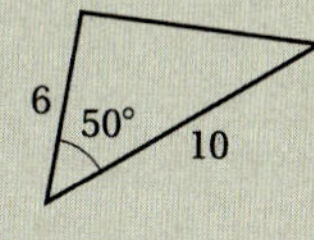

8.

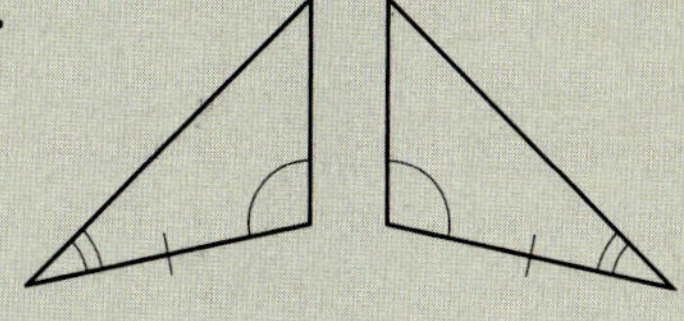

9.

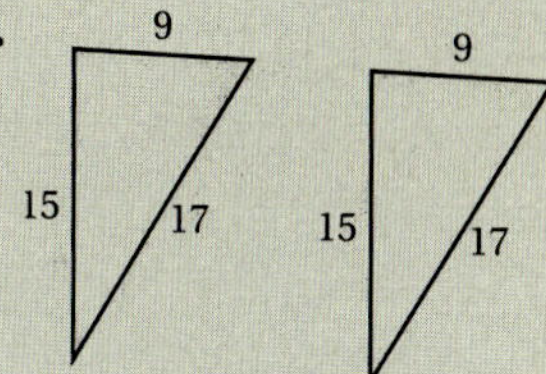

10. In this figure, $\overline{AB} \perp \overline{ED}$ and B is the midpoint of $\overline{ED}$. Explain why $\triangle ABD \cong \triangle ABE$.

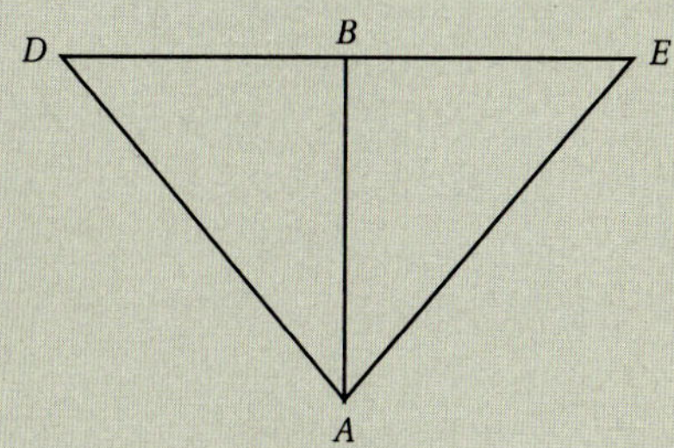

Answers on page A-13

11. $\angle R \cong \angle T$, $\angle W \cong \angle V$, and $\overline{RW} \cong \overline{TV}$. What can you conclude about this figure?

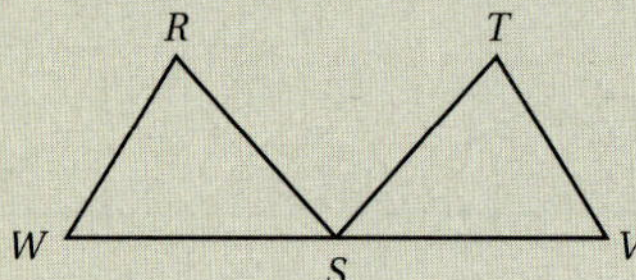

12. On a pair of pinking shears, the indicated angles and sides are congruent. How do you know that P is the midpoint of $\overline{GR}$?

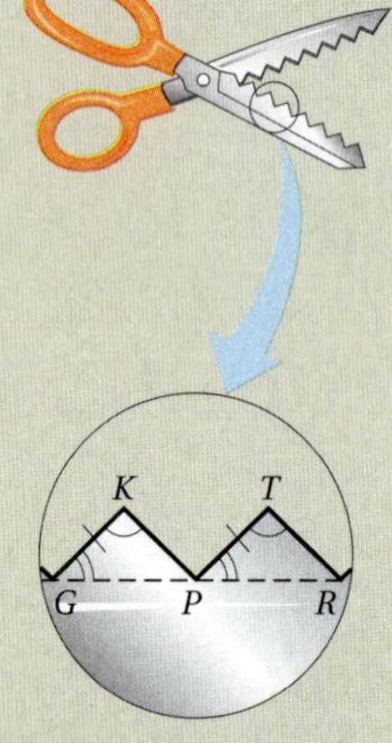

Answers on page A-13

Sometimes we can conclude that angles and segments are congruent by first showing that triangles are congruent.

EXAMPLE 14 $\overline{AB} \cong \overline{BC}$ and $\overline{EB} \cong \overline{DB}$. What can you conclude?

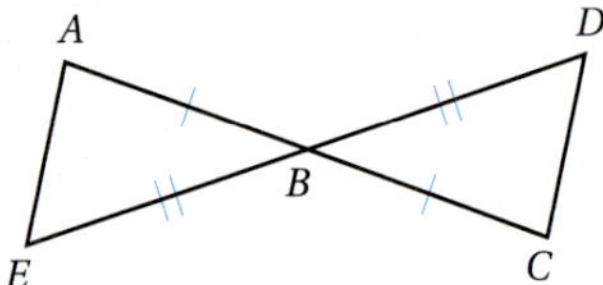

Since $\angle ABE$ and $\angle CBD$ are vertical angles, $\angle ABE \cong \angle CBD$. Thus, $\triangle ABE \cong \triangle CBD$ by SAS. As corresponding parts, $\overline{AE} \cong \overline{CD}$, $\angle A \cong \angle C$, and $\angle E \cong \angle D$.

EXAMPLE 15 Explain how you can use congruent triangles to find the distance across a marsh.

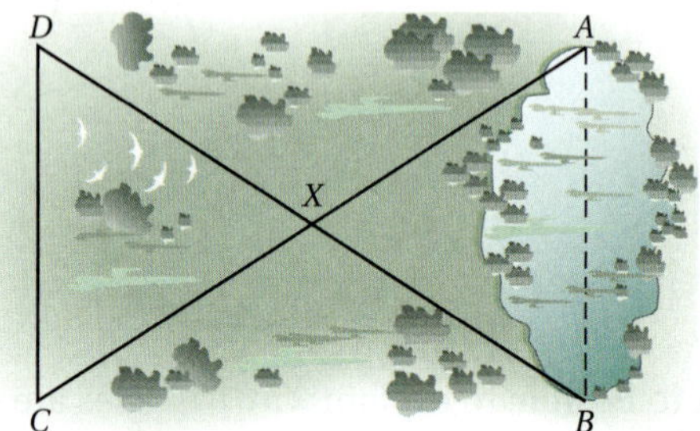

Mark off distances AX and BX. Extend $\overline{AX}$ and $\overline{BX}$ so that point X becomes the midpoint of $\overline{AC}$ and $\overline{BD}$. Then $\triangle ABX \cong \triangle CDX$ by SAS. Thus, $\overline{DC} \cong \overline{AB}$ as corresponding parts. Then we can measure $\overline{DC}$ knowing that $DC = AB$.

Do Exercises 11 and 12.

b Properties of Parallelograms

A quadrilateral is a polygon with four sides. A **diagonal** of a quadrilateral is a segment that joins two opposite vertices.

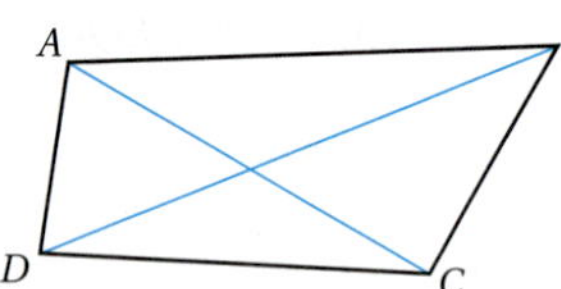

$\overline{AC}$ and $\overline{BD}$ are diagonals.

The sum of the measures of the angles of a quadrilateral is 360°.

A parallelogram is a quadrilateral with two pairs of parallel sides.

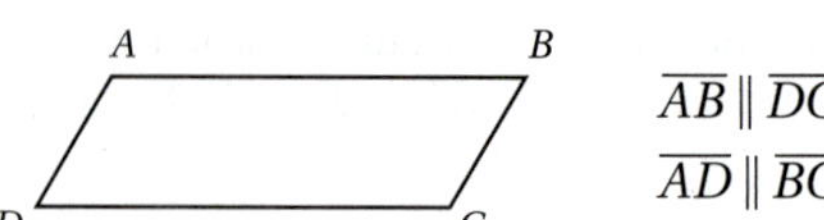

$\overline{AB} \parallel \overline{DC}$

$\overline{AD} \parallel \overline{BC}$

Draw two pairs of parallel lines to form parallelogram $ABCD$. Compare the lengths of opposite sides. Compare the measures of opposite angles. Compare the measures of consecutive angles. Draw diagonal $\overline{AC}$. How are $\triangle ADC$ and $\triangle CBA$ related? Draw diagonal $\overline{BD}$, intersecting $\overline{AC}$ at point E. What is special about point E?

Using the comparisons and the fact that corresponding parts of congruent triangles are congruent, we can list the following properties of parallelograms.

PROPERTIES OF PARALLELOGRAMS

1. A diagonal of a parallelogram determines two congruent triangles.
2. The opposite angles of a parallelogram are congruent.
3. The opposite sides of a parallelogram are congruent.
4. Consecutive angles of a parallelogram are supplementary.
5. The diagonals of a parallelogram bisect each other.

EXAMPLE 16 If $m\angle A = 120°$, find the measures of the other angles of parallelogram $ABCD$.

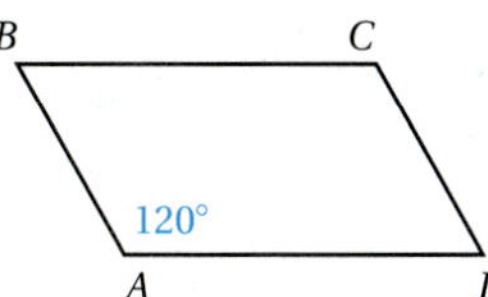

$m\angle C = 120°$ Using Property 2

$m\angle B = 60°$ Using Property 4

$m\angle D = 60°$ Using Property 2

EXAMPLE 17 Find AB and BC.

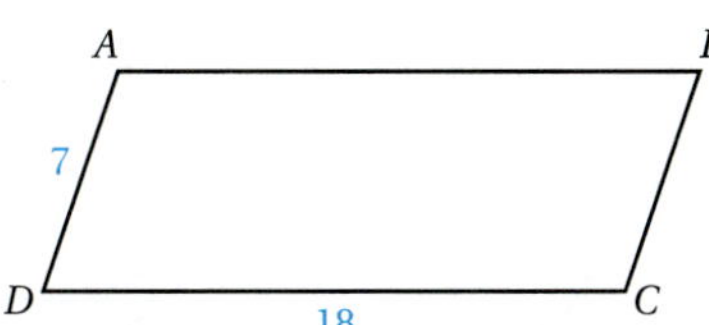

$AB = 18$ and $BC = 7$ Using Property 3

Do Exercises 13–16.

Find the measure of each angle.

13.

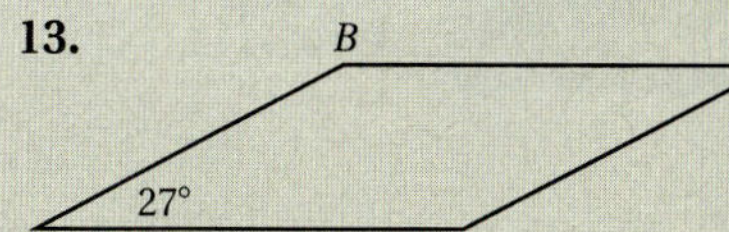

14.

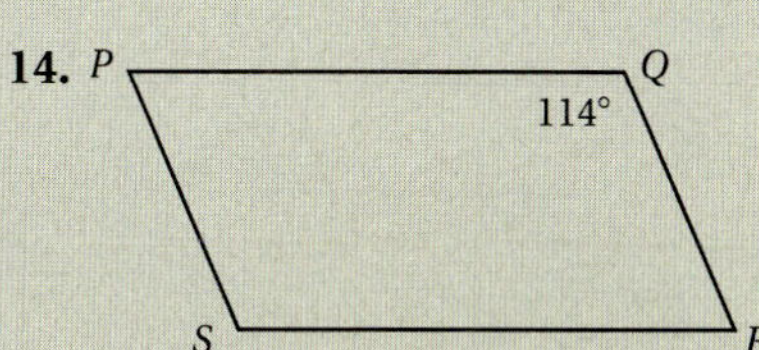

Find the length of each side.

15.

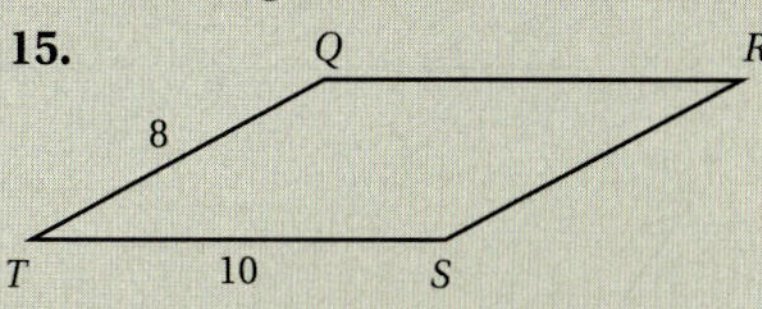

16. The perimeter of ▱$DEFG$ is 68.

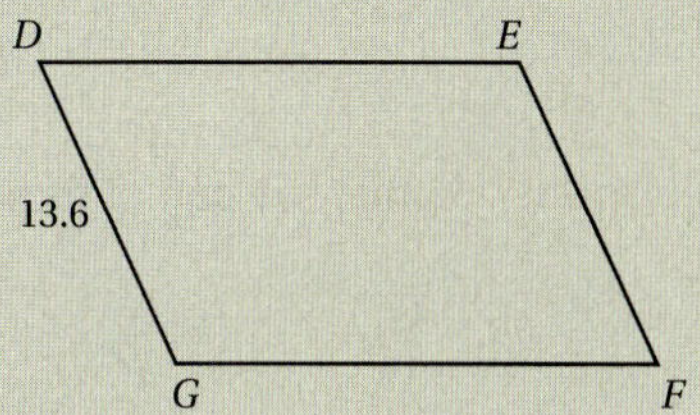

Answers on page A-13

5.7 EXERCISE SET

For Extra Help

Digital Video Tutor CD 4 Videotape 6 | InterAct Math | Math Tutor Center | MathXL | MyMathLab

a Name the corresponding parts of the congruent triangles.

1. $\triangle ABC \cong \triangle RST$

2. $\triangle MNQ \cong \triangle HJK$

3. $\triangle DEF \cong \triangle GHK$

4. $\triangle ABC \cong \triangle ABC$

5. $\triangle XYZ \cong \triangle UVW$

6. $\triangle ABC \cong \triangle ACB$

Name the corresponding parts of the congruent triangles.

7.

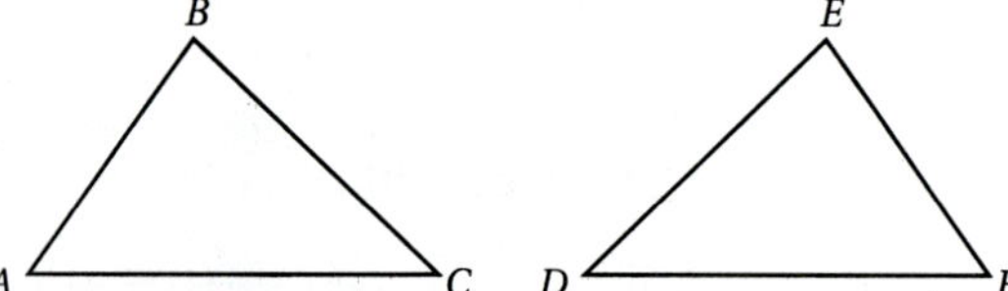

8.

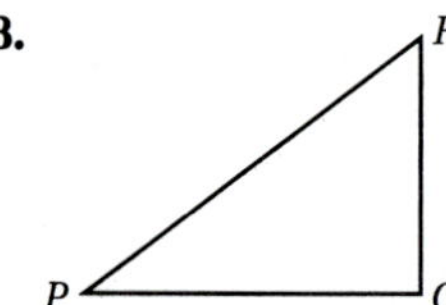

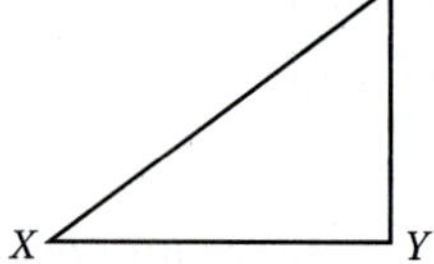

9.

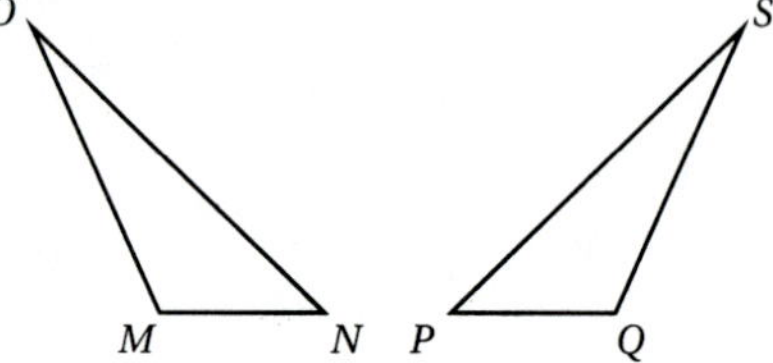

10.

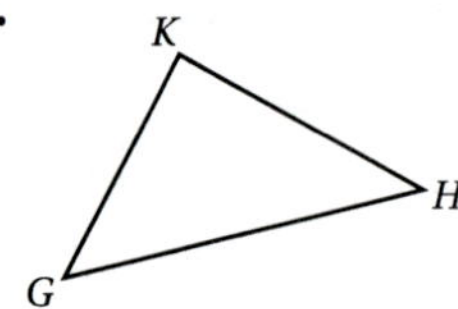

Determine whether the pair of triangles is congruent by the SAS property.

11.

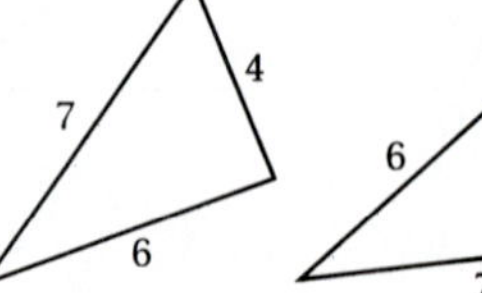

12.

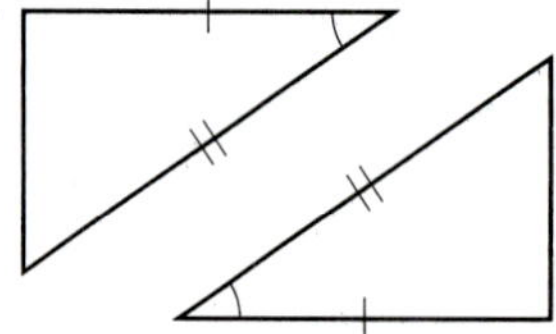

13.

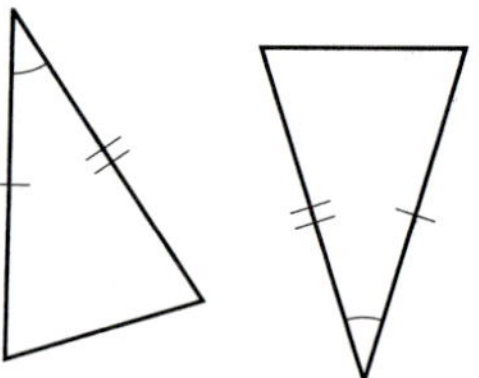

14.  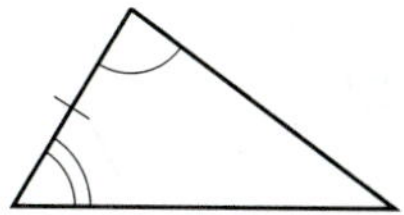

15. 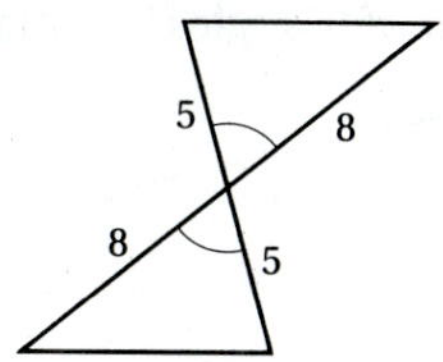

16. 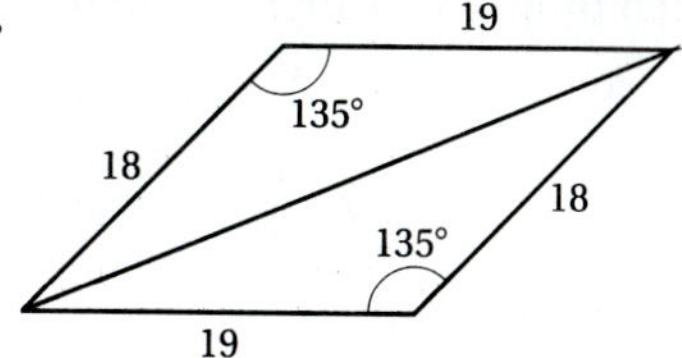

Determine whether the pair of triangles is congruent by the SSS property.

17. 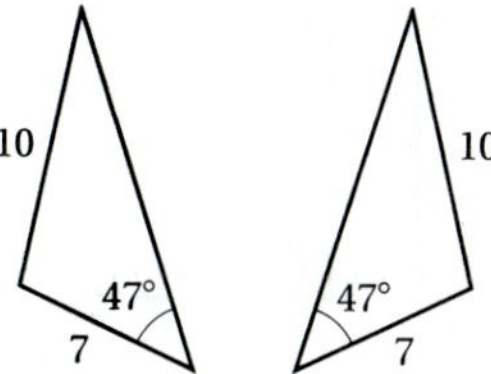

18. 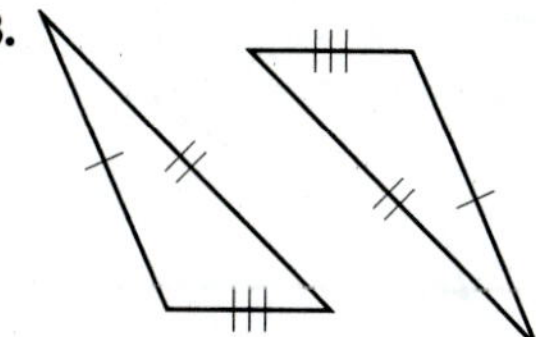

19. 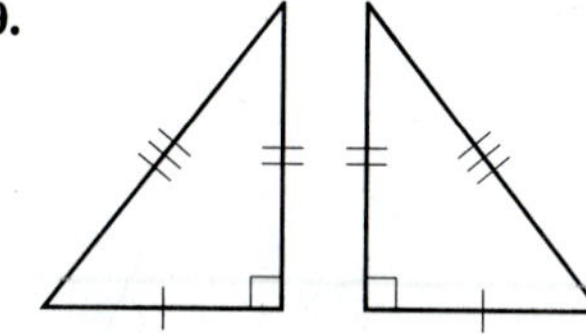

20. 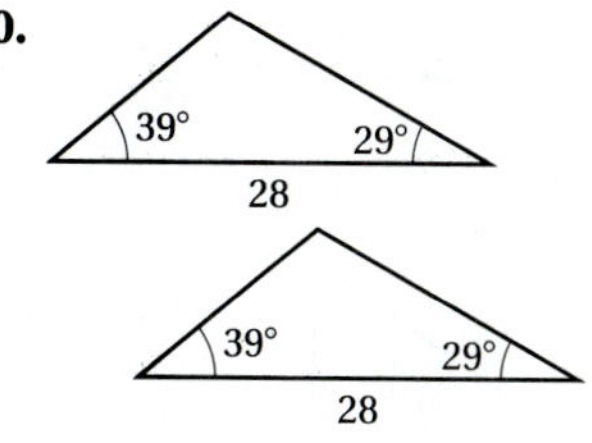

21. 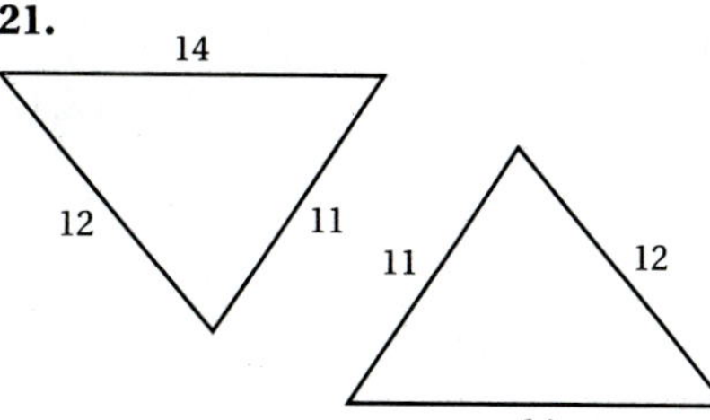

22. 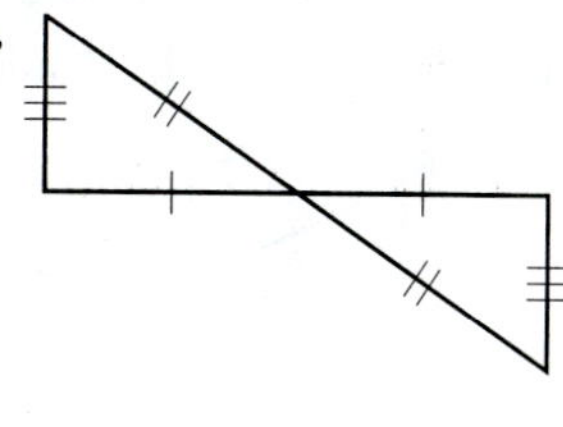

Determine whether the pair of triangles is congruent by the ASA property.

23. 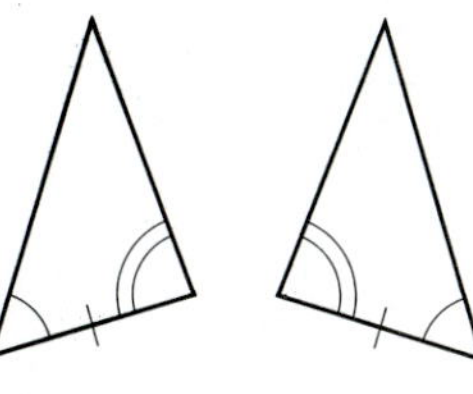

24. 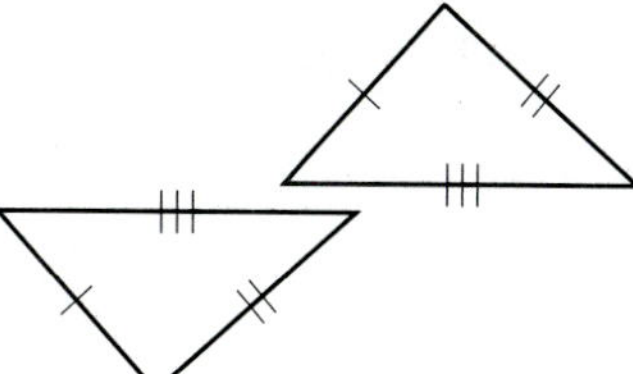

25. 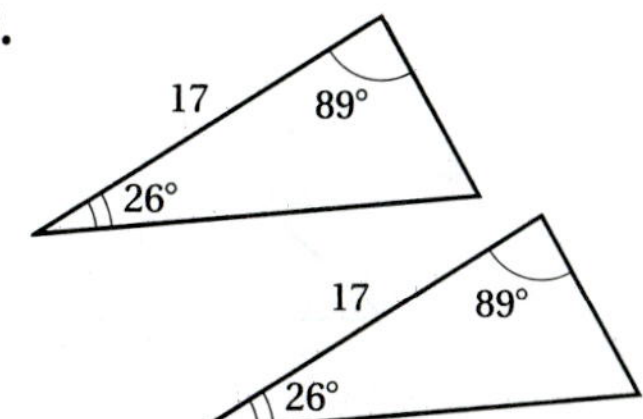

26. 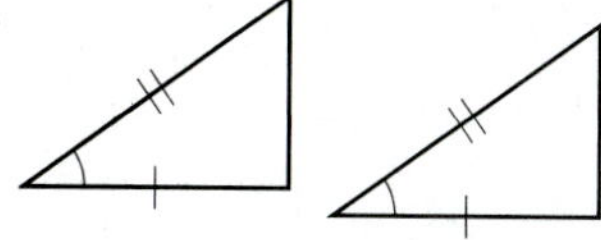

27. 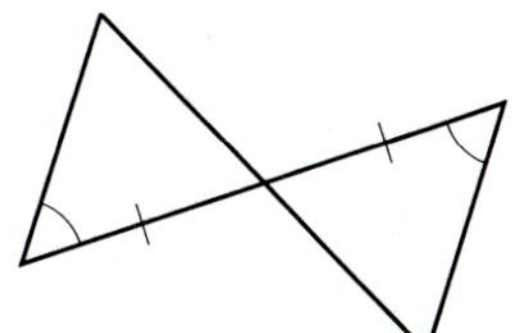

28. 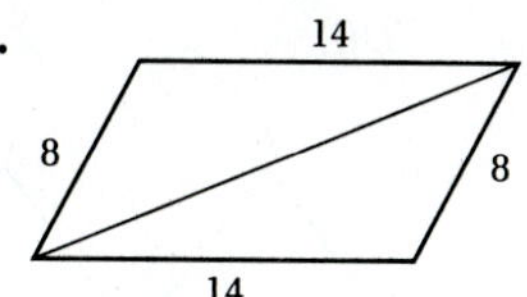

Which property (if any) should be used to show that the pair of triangles is congruent?

29.

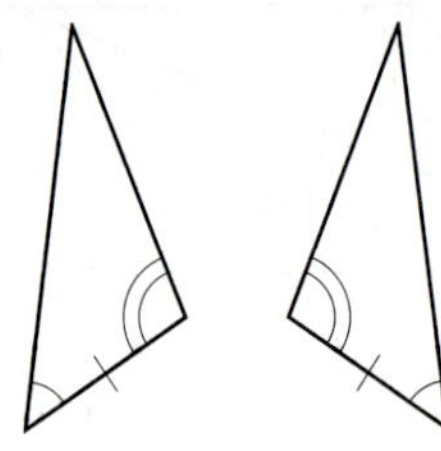

30.

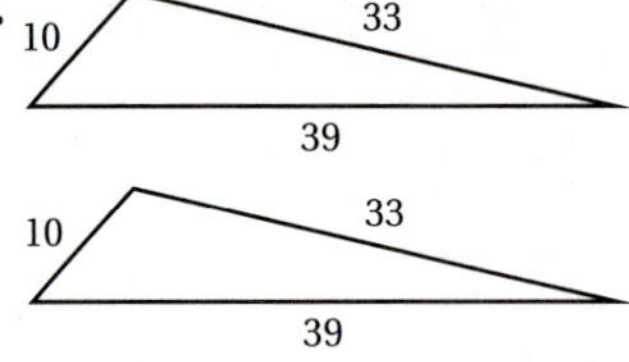

31.

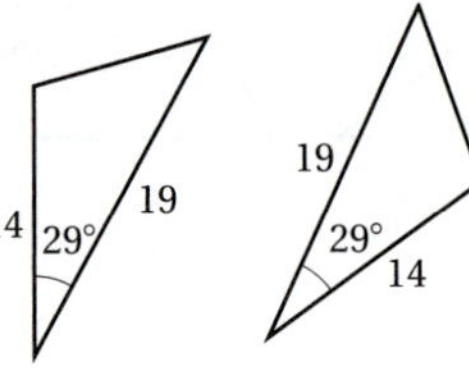

32.

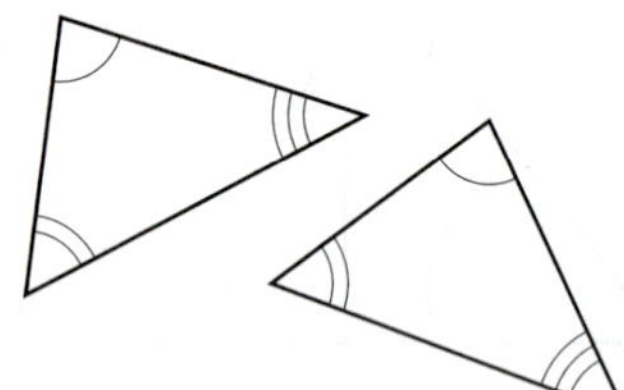

33.

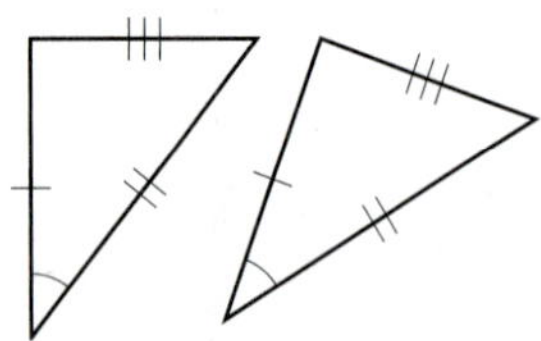

34.

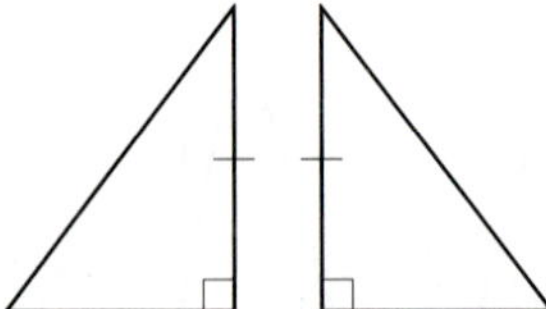

Explain why the triangles indicated in parentheses are congruent.

35. R is the midpoint of both $\overline{PT}$ and $\overline{QS}$. ($\triangle PRQ \cong \triangle TRS$)

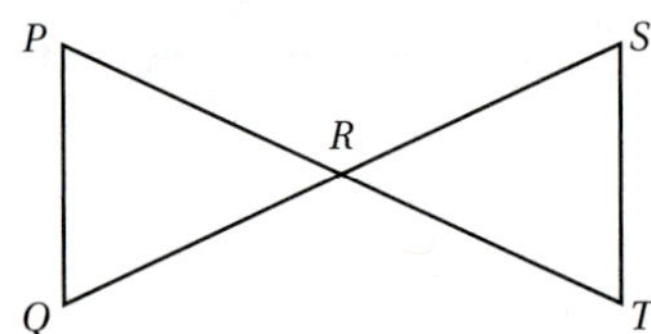

36. $\angle 1$ and $\angle 2$ are right angles, X is the midpoint of $\overline{AY}$, and $\overline{XB} \cong \overline{YZ}$. ($\triangle ABX \cong \triangle XZY$)

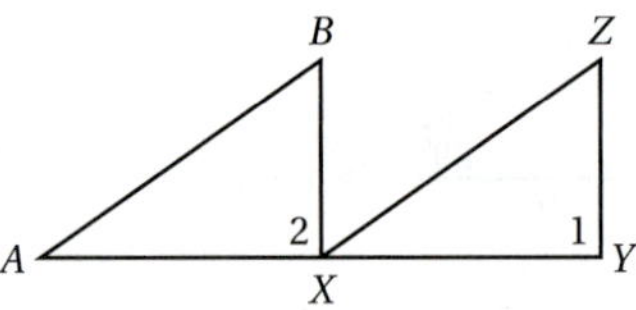

37. L is the midpoint of $\overline{KM}$ and $\overline{GL} \perp \overline{KM}$. ($\triangle KLG \cong \triangle MLG$)

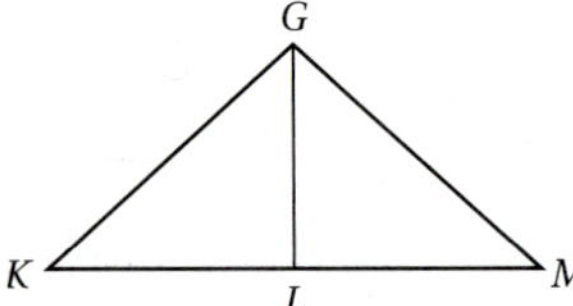

38. X is the midpoint of $\overline{QS}$ and $\overline{RP}$ with $RQ = SP$. ($\triangle RQX \cong \triangle PSX$)

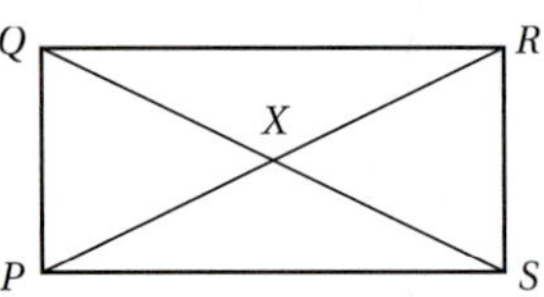

39. $\triangle AEB$ and $\triangle CDB$ are isosceles with $\overline{AE} \cong \overline{AB} \cong \overline{CB} \cong \overline{CD}$. Also, B is the midpoint of $\overline{ED}$. ($\triangle AEB \cong \triangle CDB$)

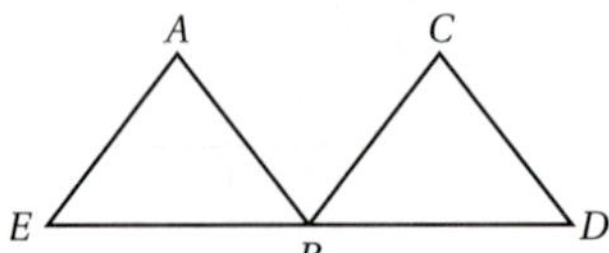

40. $\overline{AB} \perp \overline{BE}$ and $\overline{DE} \perp \overline{BE}$. $\overline{AB} \cong \overline{DE}$ and $\angle BAC \cong \angle EDC$. ($\triangle ABC \cong \triangle DEC$)

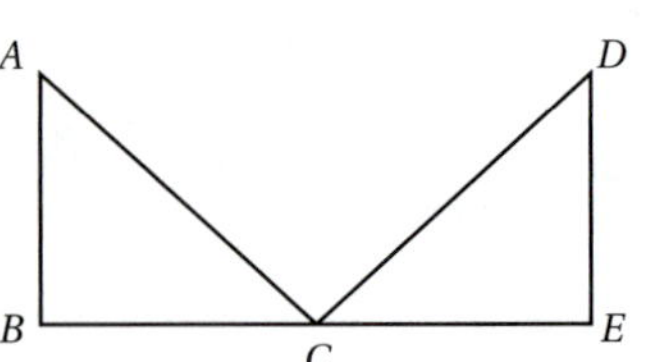

What can you conclude about each figure using the given information?

41. $\overline{GK} \perp \overline{LJ}$, $\overline{HK} \cong \overline{KJ}$, and $\overline{GK} \cong \overline{LK}$

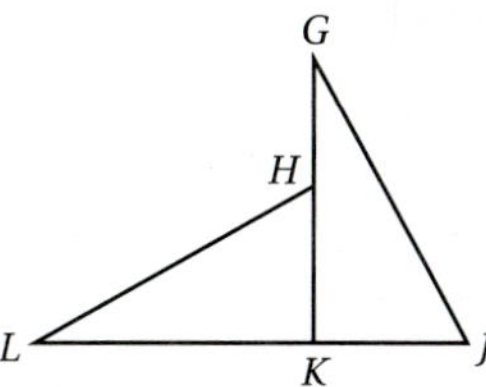

42. $\overline{AB} \cong \overline{DC}$ and $\angle BAC \cong \angle DCA$

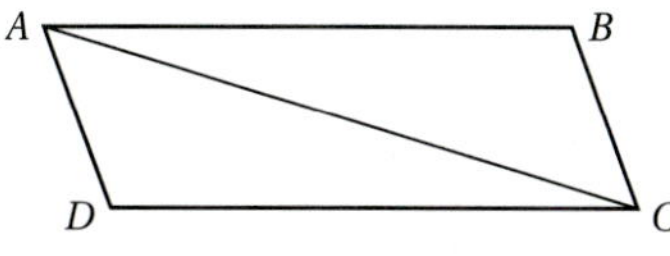

Use corresponding parts to solve Exercises 43 and 44.

43. On this national flag, the indicated segments and angles are congruent. Explain why P is the midpoint of $\overline{EF}$.

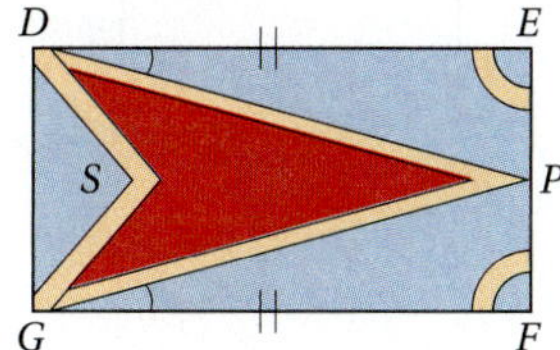

44. The indicated sides of a kite are congruent. Explain how you know that $\angle 1 \cong \angle 2$.

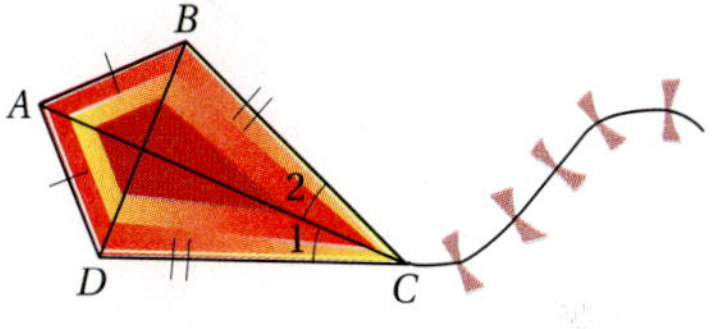

b Find the measures of the angles of the parallelogram.

45.

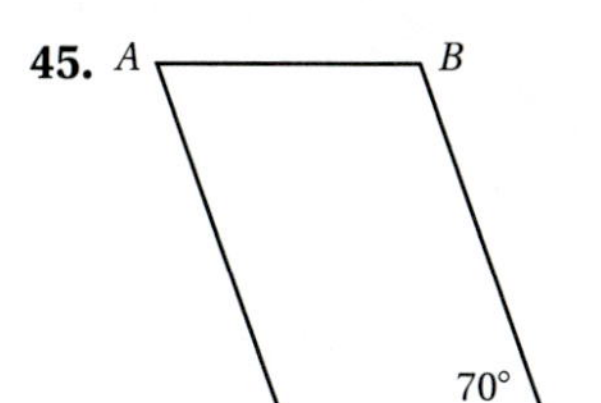

46.

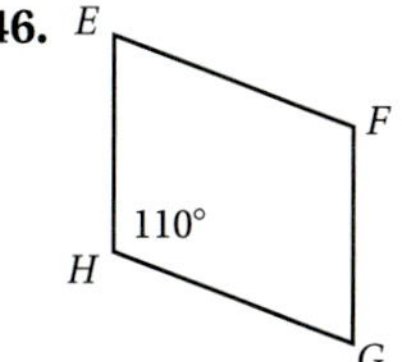

47.

J K

71°

M L

48.

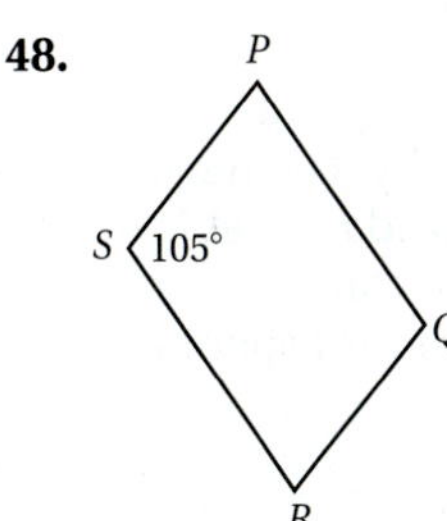

Find the lengths of the sides of the parallelogram.

49.

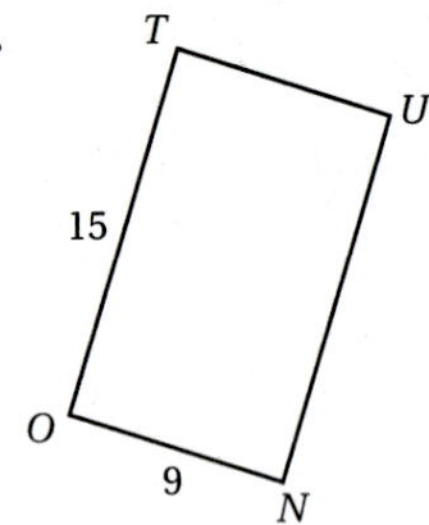

50.

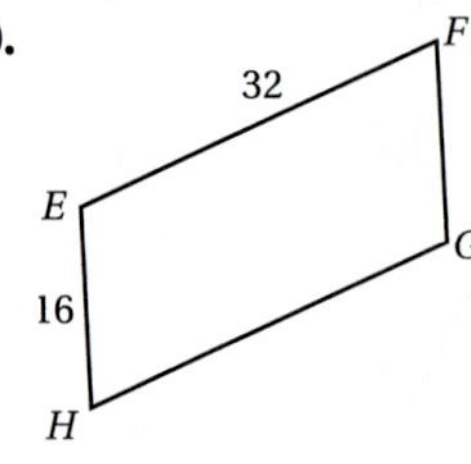

51. The perimeter of ▱*JKLM* is 22.

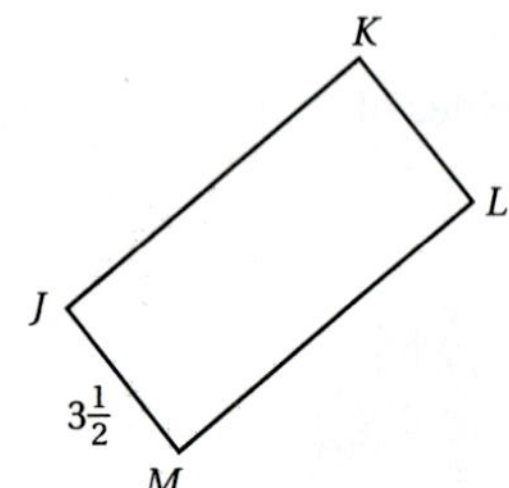

52. The perimeter of ▱*WXYZ* is 248.

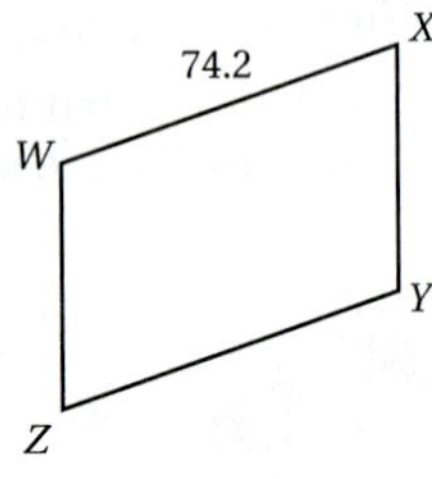

53. $AB = 14$ and $BD = 19$. Find the length of each diagonal.

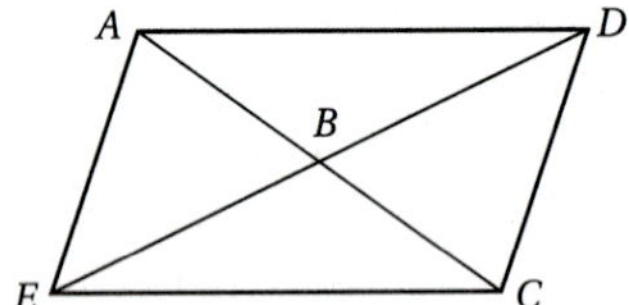

54. $EJ = 23$ and $GJ = 13$. Find the length of each diagonal.

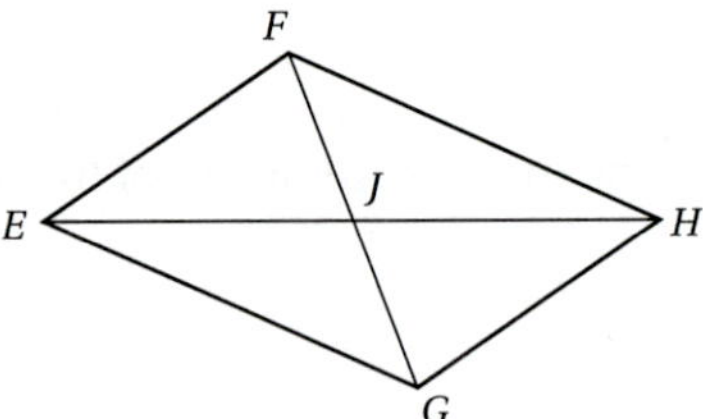

SKILL MAINTENANCE

Convert to percent notation. [4.2b], [4.3a]

55. 0.452

56. $\frac{1}{3}$

57. $\frac{11}{20}$

58. $\frac{22}{25}$

59. *Tourist Spending.* Foreign tourists spend $13.1 billion in this country annually. The most money, $2.7 billion, is spent in Florida. What is the ratio of amount spent in Florida to total amount spent? What is the ratio of total amount spent to amount spent in Florida? [4.1a]

60. One person in four plays a musical instrument. In a given group of people, what is the ratio of those who play an instrument to total number of people? What is the ratio of those who do not play an instrument to total number of people? [4.1a]

Divide. Find decimal notation for the answer. [3.4a]

61. $21 \div 12$

62. $23.4 \div 10$

63. $23.4 \div 100$

64. $23.4 \div 1000$

65. Multiply 3.14×4.41. Round to the nearest hundredth. [3.3a], [3.1d]

5.8 SIMILAR TRIANGLES

a Proportions and Similar Triangles

We know that congruent figures have the same shape and size. *Similar figures* have the same shape, but are not necessarily the same size.

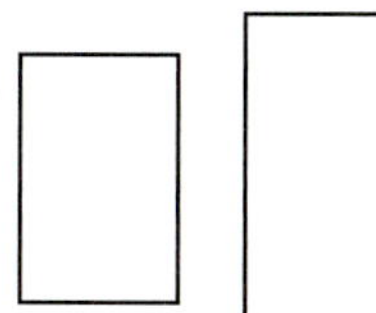

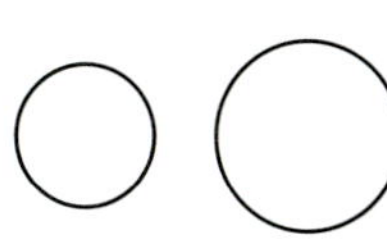
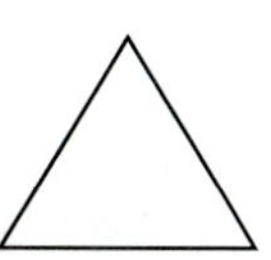
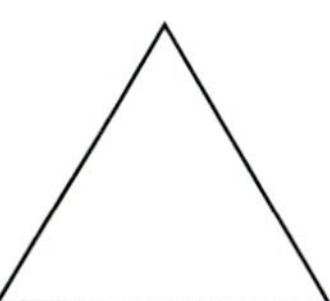

Similar figures

EXAMPLE 1 Which pairs of triangles appear to be similar?

a)
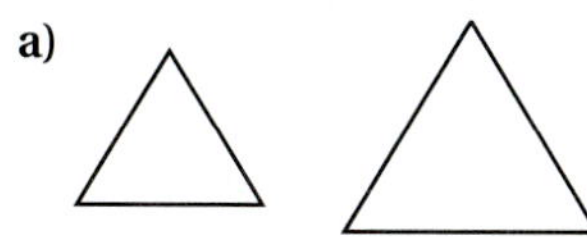

b)

c)
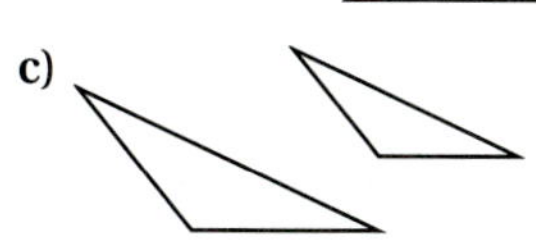

d)
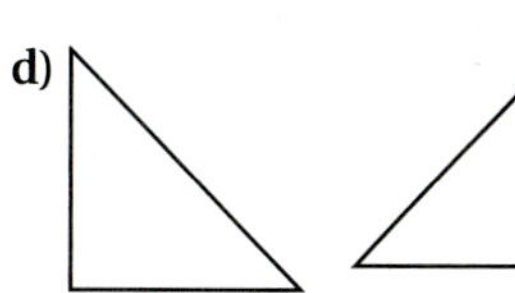

Pairs (a), (c), and (d) appear to be similar.

Do Exercise 1.

Similar triangles have corresponding sides and angles.

EXAMPLE 2 $\triangle ABC$ and $\triangle DEF$ are similar. Name their corresponding sides and angles.

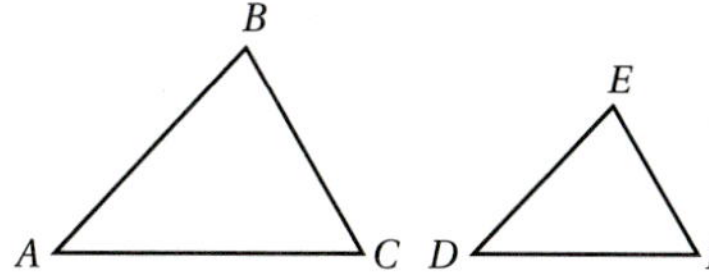

$\overline{AB} \leftrightarrow \overline{DE}$ $\quad \angle A \leftrightarrow \angle D$

$\overline{AC} \leftrightarrow \overline{DF}$ $\quad \angle B \leftrightarrow \angle E$

$\overline{BC} \leftrightarrow \overline{EF}$ $\quad \angle C \leftrightarrow \angle F$

Do Exercise 2.

Objectives

a Identify the corresponding parts of similar triangles and determine which sides of a given pair of triangles have lengths that are proportional.

b Find lengths of sides of similar triangles using proportions.

1. Which pairs of triangles appear to be similar?

a)
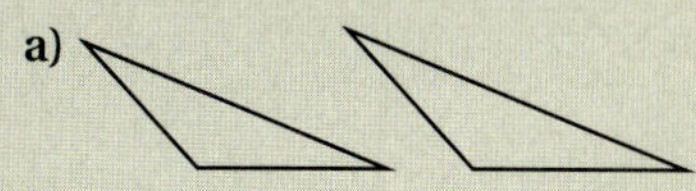

b)

c)
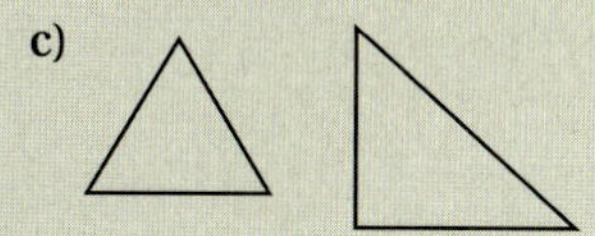

d)
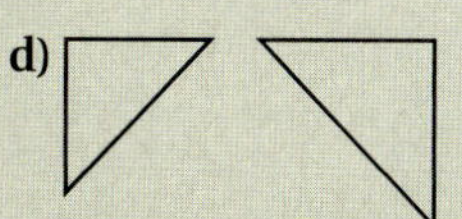

2. $\triangle PQR$ and $\triangle GHK$ are similar. Name their corresponding sides and angles.

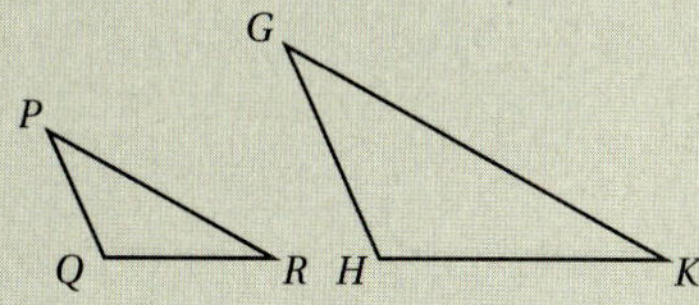

Answers on page A-13

3. Suppose that $\triangle JKL \sim \triangle ABC$. Which angles are congruent? Which sides are proportional?

4. These triangles are similar. Which sides are proportional?

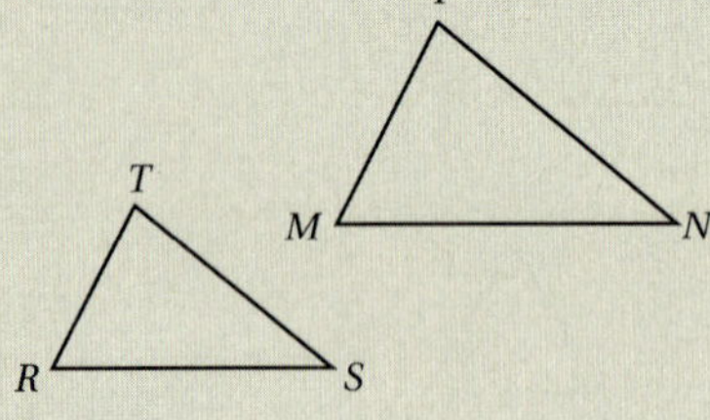

Answers on page A-13

SIMILAR TRIANGLES

Two triangles are **similar** if and only if their vertices can be matched so that the corresponding angles are congruent and the lengths of corresponding sides are proportional.

To say that $\triangle ABC$ and $\triangle DEF$ are similar, we write "$\triangle ABC \sim \triangle DEF$." We will agree that this symbol also tells us the way in which the vertices are matched.

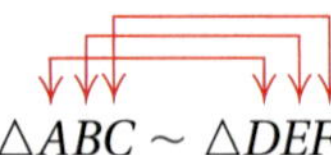

$\triangle ABC \sim \triangle DEF$

Thus, $\triangle ABC \sim \triangle DEF$ means that

$$\angle A \cong \angle D, \quad \angle B \cong \angle E, \quad \angle C \cong \angle F \quad \text{and} \quad \frac{AB}{DE} = \frac{AC}{DF} = \frac{BC}{EF}.$$

EXAMPLE 3 Suppose that $\triangle PQR \sim \triangle STV$. Which angles are congruent? Which sides are proportional?

$$\angle P \cong \angle S, \quad \angle Q \cong \angle T, \quad \angle R \cong \angle V \quad \text{and} \quad \frac{PQ}{ST} = \frac{PR}{SV} = \frac{QR}{TV}.$$

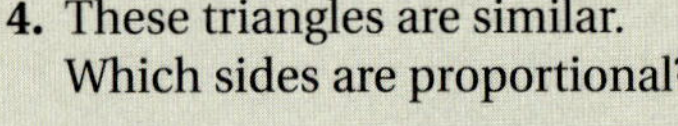

Do Exercise 3.

EXAMPLE 4 These triangles are similar. Which sides are proportional?

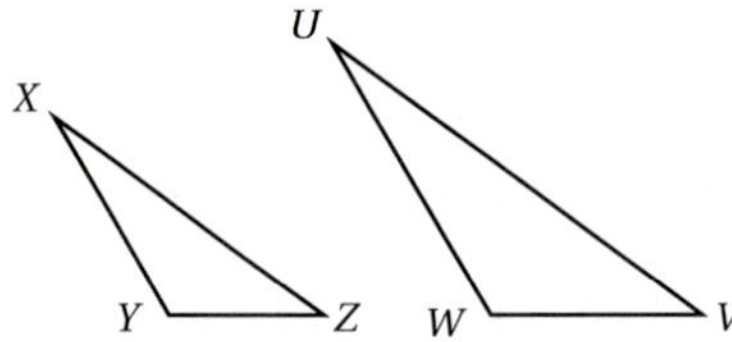

It appears that if we match X with U, Y with W, and Z with V, the corresponding angles will be congruent. Thus,

$$\frac{XY}{UW} = \frac{XZ}{UV} = \frac{YZ}{WV}.$$

Do Exercise 4.

b Proportions and Similar Triangles

We can find lengths of sides in similar triangles.

EXAMPLE 5 If $\triangle RAE \sim \triangle GQL$, find QL and GL.

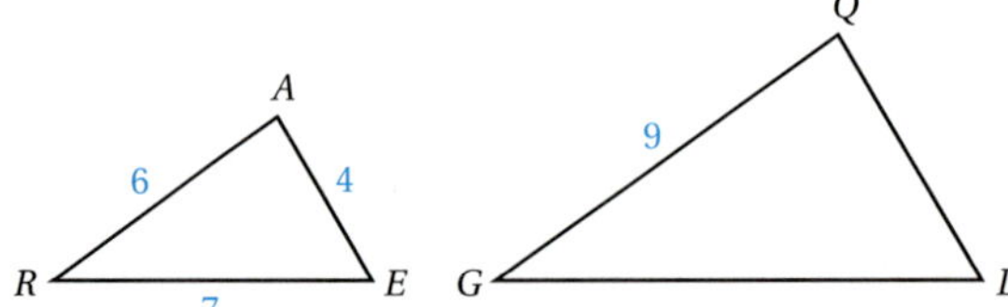

Since $\triangle RAE \sim \triangle GQL$, the corresponding sides are proportional. Thus,

$$\frac{6}{9} = \frac{4}{QL}$$

$6(QL) = 9 \cdot 4$ Equating cross products

$6(QL) = 36$

$QL = 6$ Dividing both sides by 6

and

$$\frac{6}{9} = \frac{7}{GL}$$

$6(GL) = 9 \cdot 7$

$6(GL) = 63$

$GL = 10\frac{1}{2}.$

Do Exercise 5.

5. If $\triangle WNE \sim \triangle CBT$, find BT and CT.

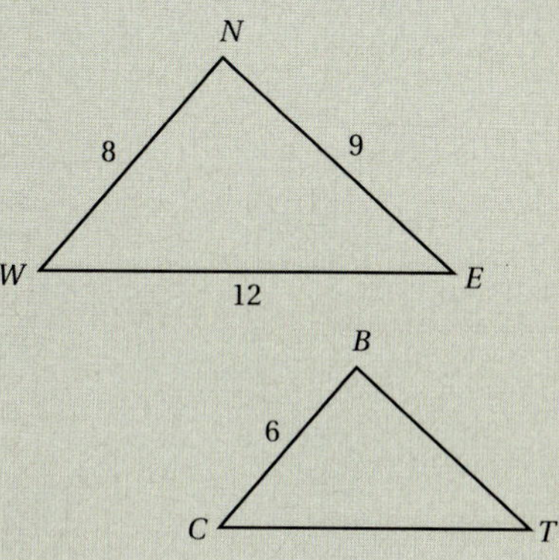

EXAMPLE 6 If $\overline{AB} \parallel \overline{CD}$, find CD.

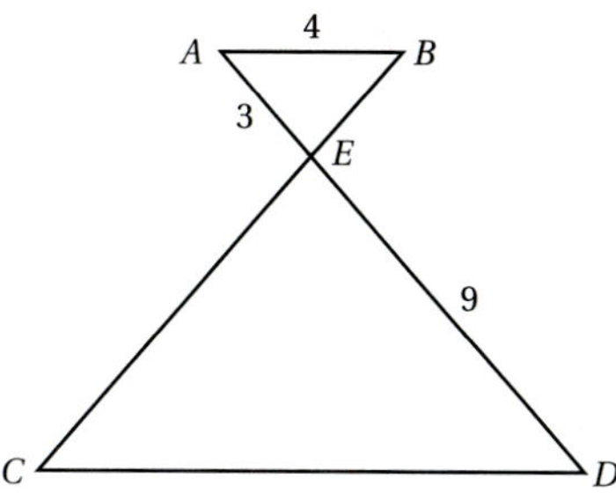

Recall that if a transversal intersects two parallel lines, then the alternate interior angles are congruent (Section 6.6). Thus,

$$\angle A \cong \angle D \quad \text{and} \quad \angle C \cong \angle B,$$

because they are pairs of alternate interior angles. Since $\angle AEB$ and $\angle DEC$ are vertical angles, they are congruent. Thus by definition

$$\triangle AEB \sim \triangle DEC$$

and the lengths of the corresponding sides are proportional. Thus,

$$\frac{AE}{DE} = \frac{AB}{CD}.$$

Solve: $\frac{3}{9} = \frac{4}{CD}$ Substituting

$3(CD) = 9 \cdot 4$ Equating cross products

$3(CD) = 36$

$CD = 12.$ Dividing both sides by 3

Do Exercise 6.

6. If $\overline{QR} \parallel \overline{ST}$, find QR.

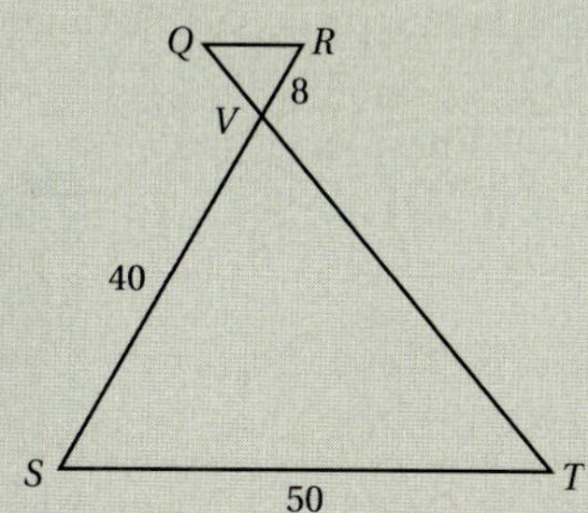

Similar triangles and proportions can often be used to find lengths that would ordinarily be difficult to measure. For example, we could find the height of a flagpole without climbing it or the distance across a river without crossing it.

Answers on page A-13

7. How high is a flagpole that casts a 45-ft shadow at the same time that a 5.5-ft woman casts a 10-ft shadow?

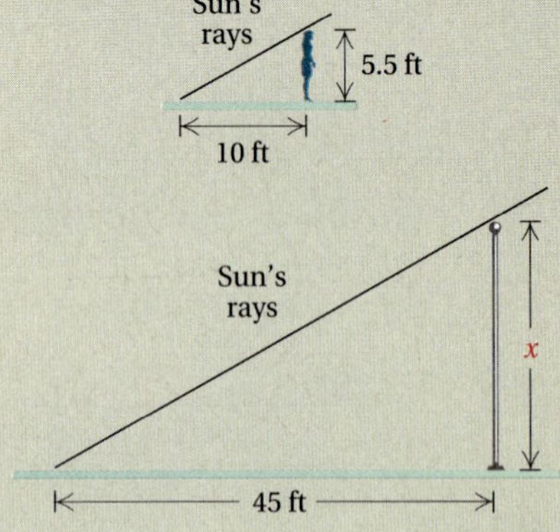

EXAMPLE 7 How high is a flagpole that casts a 56-ft shadow at the same time that a 6-ft man casts a 5-ft shadow?

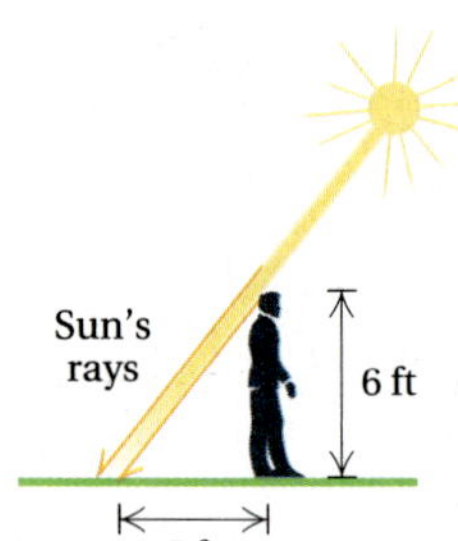

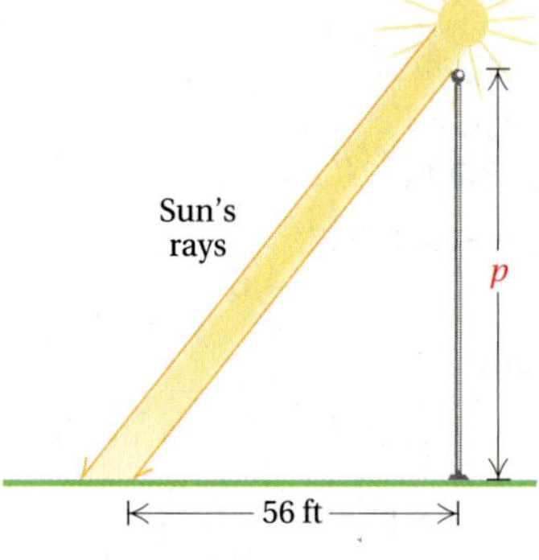

If we use the sun's rays to represent the third side of the triangle in our drawing of the situation, we see that we have similar triangles. Let $p =$ the height of the flagpole. The ratio of 6 to p is the same as the ratio of 5 to 56. Thus we have the proportion

Height of man → 6, Height of pole → p; Length of shadow of man ← 5, Length of shadow of pole ← 56

$$\frac{6}{p} = \frac{5}{56}.$$

Solve:

$6 \cdot 56 = 5 \cdot p$	Equating cross products
$\frac{6 \cdot 56}{5} = p$	Dividing both sides by 5
$67.2 = p$	Simplifying

The height of the flagpole is 67.2 ft.

Do Exercise 7.

8. **F-106 Blueprint.** Referring to Example 8, find the length x of the wing.

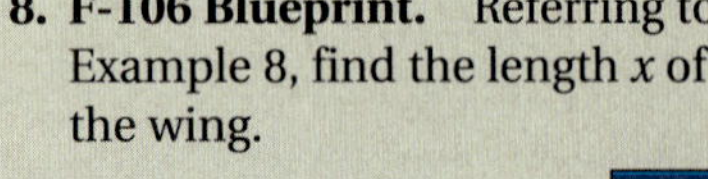

EXAMPLE 8 *F-106 Blueprint.* A blueprint for an F-106 Delta Dart fighter plane is a scale drawing. Each wing of the plane has a triangular shape. The blueprint shows similar triangles. Find the length of side a of the wing.

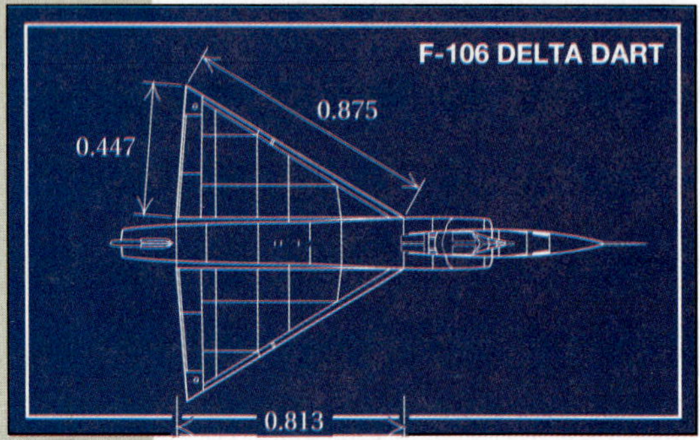

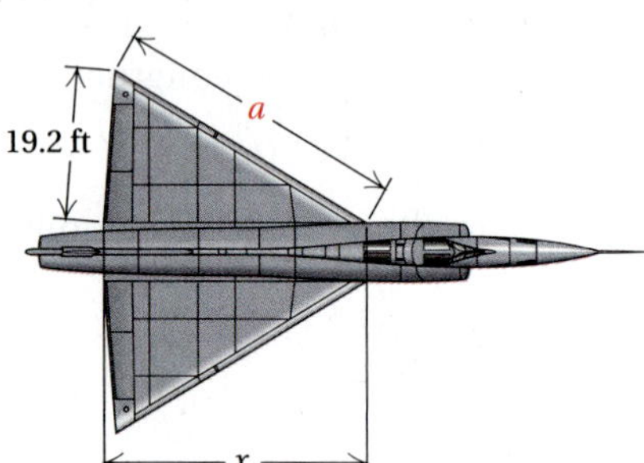

We let $a =$ the length of the wing. Thus we have the proportion

Length on the blueprint → 0.447, Length of the wing → 19.2; Length on the blueprint ← 0.875, Length of the wing ← a

$$\frac{0.447}{19.2} = \frac{0.875}{a}.$$

Solve:

$0.447 \cdot a = 19.2 \cdot 0.875$	Equating cross products
$a = \frac{19.2 \cdot 0.875}{0.447}$	Dividing both sides by 0.447
≈ 37.6 ft	

The length of side a of the wing is about 37.6 ft.

Do Exercise 8.

Answers on page A-13

5.8 EXERCISE SET

For Extra Help

Digital Video Tutor CD 5 Videotape 6

InterAct Math

Math Tutor Center

MathXL

MyMathLab

a For each pair of similar triangles, name the corresponding sides and angles.

1.

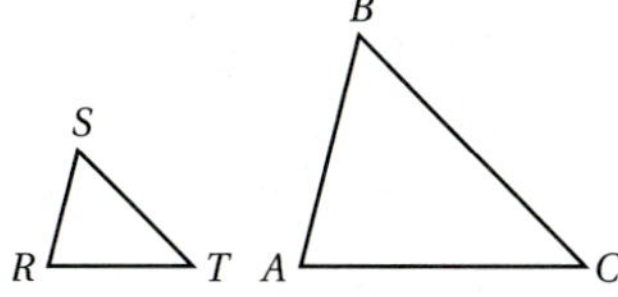

2.

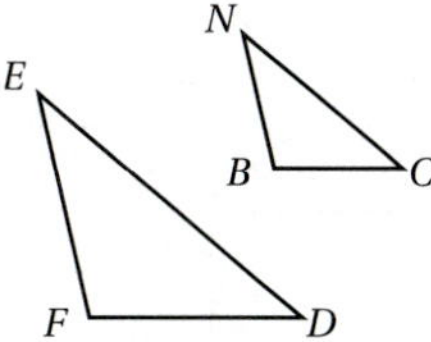

3.

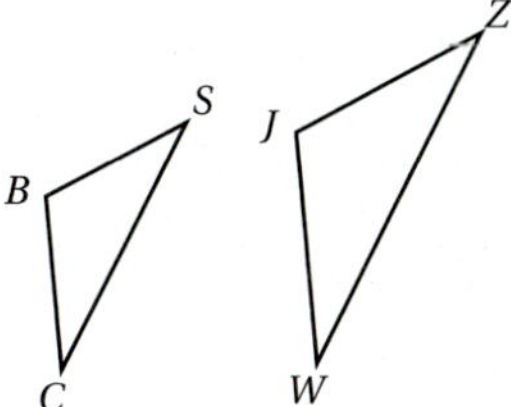

4.

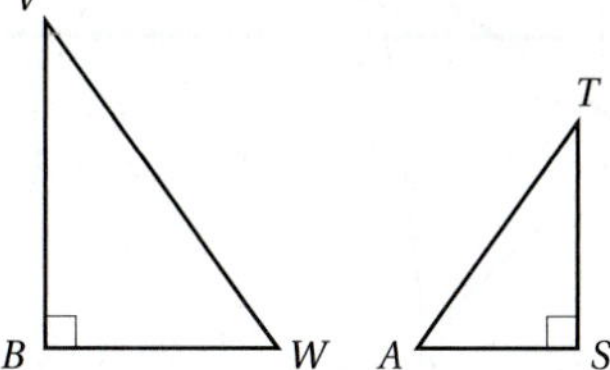

For each pair of similar triangles, name the congruent angles and proportional sides.

5. $\triangle ABC \sim \triangle RST$

6. $\triangle PQR \sim \triangle STV$

7. $\triangle MES \sim \triangle CLF$

8. $\triangle SMH \sim \triangle WLK$

Name the proportional sides in these similar triangles.

9.

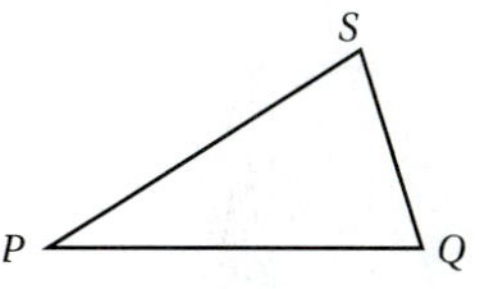

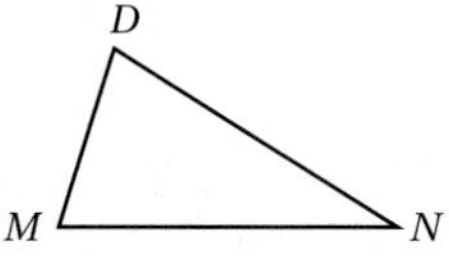

10.

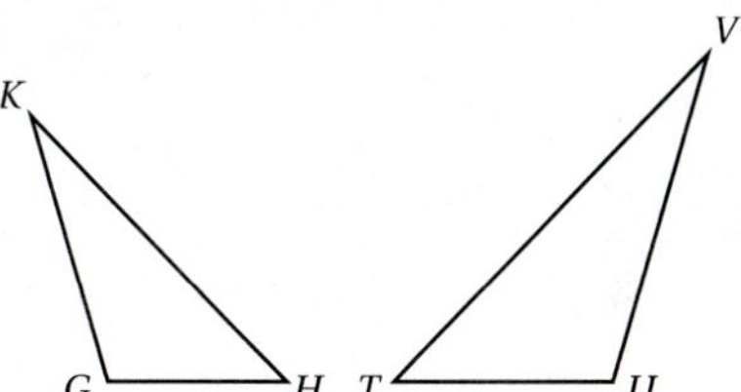

11.

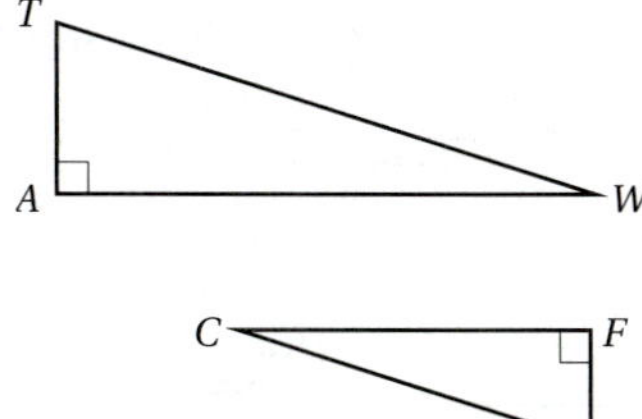

12.

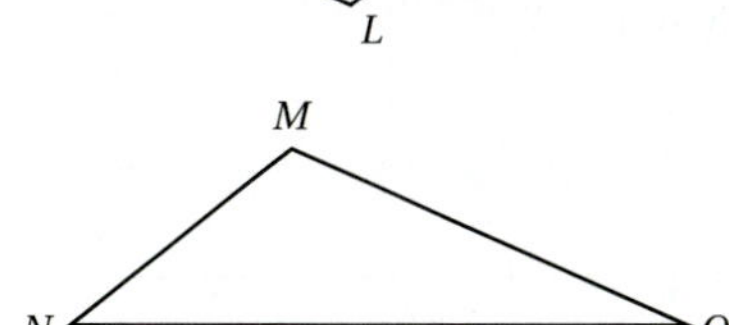

b Find the missing lengths.

13. If $\triangle ABC \sim \triangle PQR$, find QR and PR.

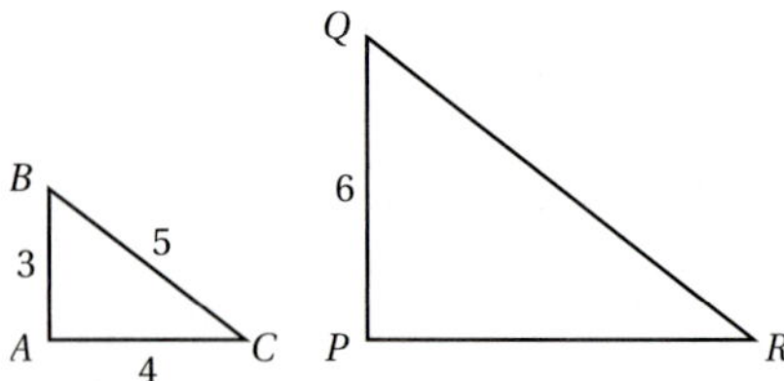

14. If $\triangle MAC \sim \triangle GET$, find AM and GT.

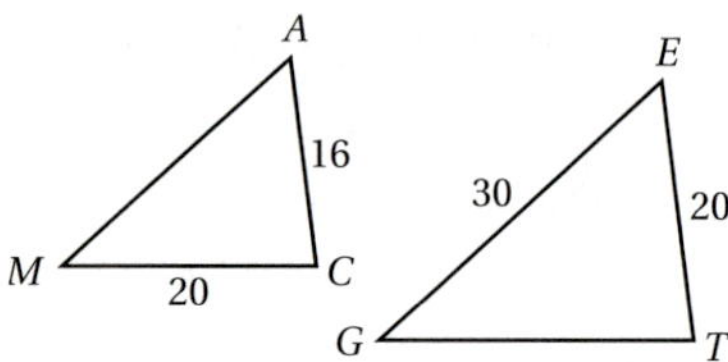

15. If $\overline{AD} \parallel \overline{CB}$, find EC.

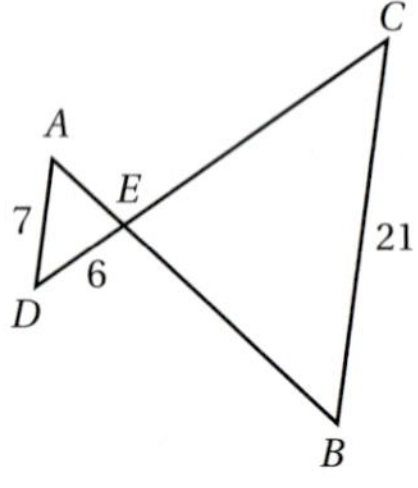

16. If $\overline{LN} \parallel \overline{PM}$, find QM.

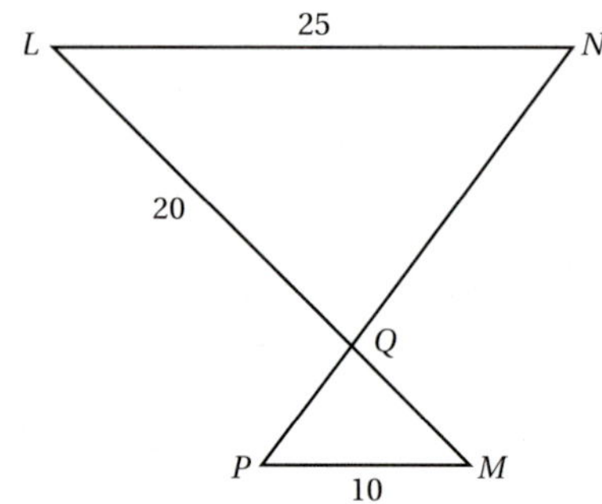

17. How high is a tree that casts a 27-ft shadow at the same time that a 4-ft fence post casts a 3-ft shadow?

18. How high is a flagpole that casts a 42-ft shadow at the same time that a $5\frac{1}{2}$-ft woman casts a 7-ft shadow?

19. Find the distance across the river. Assume that the ratio of d to 25 ft is the same as the ratio of 40 ft to 10 ft.

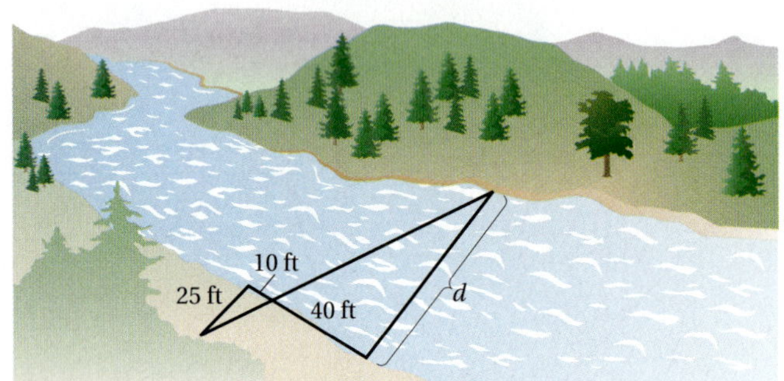

20. To measure the height of a hill, a string is drawn tight from level ground to the top of the hill. A 3-ft yardstick is placed under the string, touching it at point P, a distance of 5 ft from point G, where the string touches the ground. The string is then detached and found to be 120 ft long. How high is the hill?

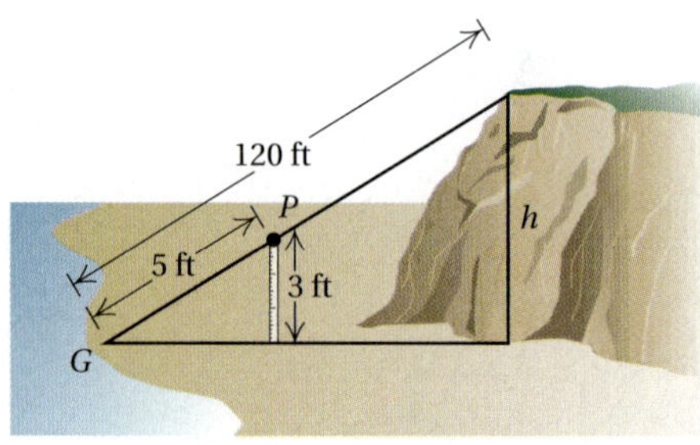

21. **DW** Is it possible for two triangles to have two pairs of sides that are proportional without the triangles being similar? Why or why not?

22. **DW** Design for a classmate a problem involving similar triangles for which

$$\frac{18}{128.95} = \frac{x}{789.89}.$$

SKILL MAINTENANCE

Multiply. [2.4d], [3.3a]

23. $2\frac{4}{5} \times 10\frac{1}{2}$

24. 3.05×0.08

25. $8 \times 9\frac{3}{4}$

26. 10.01×6.11

CHAPTER 5 Summary and Review

The review that follows is meant to prepare you for a chapter exam. It consists of two parts. The first part is a checklist of some of the Study Tips referred to in this and preceding chapters, as well as a list of important properties and formulas. The second part is the Review Exercises. These provide practice exercises for the exam, together with references to section objectives so you can go back and review. Before beginning, stop and look back over the skills you have obtained. What skills in mathematics do you have now that you did not have before studying this chapter?

STUDY TIPS CHECKLIST

The foundation of all your study skills is TIME!

- ☐ Are you still working the margin exercises when directed to do so?
- ☐ Are you working to improve your time-management skills?
- ☐ Did you carefully analyze the art pieces as you studied this chapter?
- ☐ Are you keeping one section ahead in your syllabus?

IMPORTANT PROPERTIES AND FORMULAS

Perimeter of a Rectangle:	$P = 2 \cdot (l + w)$, or $P = 2 \cdot l + 2 \cdot w$
Perimeter of a Square:	$P = 4 \cdot s$
Area of a Rectangle:	$A = l \cdot w$
Area of a Square:	$A = s \cdot s$, or $A = s^2$
Area of a Parallelogram:	$A = b \cdot h$
Area of a Triangle:	$A = \frac{1}{2} \cdot b \cdot h$
Area of a Trapezoid:	$A = \frac{1}{2} \cdot h \cdot (a + b)$
Radius and Diameter of a Circle:	$d = 2 \cdot r$, or $r = \frac{d}{2}$
Circumference of a Circle:	$C = \pi \cdot d$, or $C = 2 \cdot \pi \cdot r$
Area of a Circle:	$A = \pi \cdot r \cdot r$, or $A = \pi \cdot r^2$
Volume of a Rectangular Solid:	$V = l \cdot w \cdot h$
Surface Area of a Rectangular Solid:	$SA = 2lw + 2lh + 2wh$, or $SA = 2(lw + lh + wh)$
Volume of a Circular Cylinder:	$V = \pi \cdot r^2 \cdot h$
Volume of a Sphere:	$V = \frac{4}{3} \cdot \pi \cdot r^3$
Volume of a Cone:	$V = \frac{1}{3} \cdot \pi \cdot r^2 \cdot h$
Sum of Angle Measures of a Triangle:	$m(\angle A) + m(\angle B) + m(\angle C) = 180°$
Pythagorean Equation:	$a^2 + b^2 = c^2$

REVIEW EXERCISES

Use a protractor to measure each angle. [5.1b]

1.

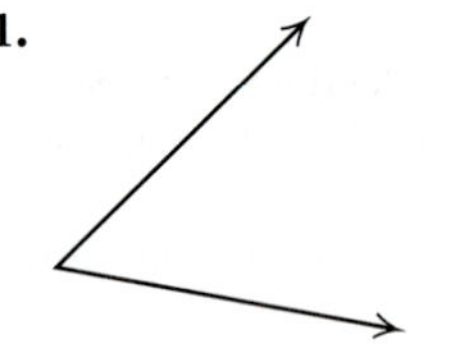

2.

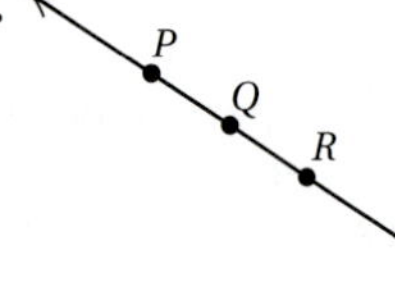

3.

4.

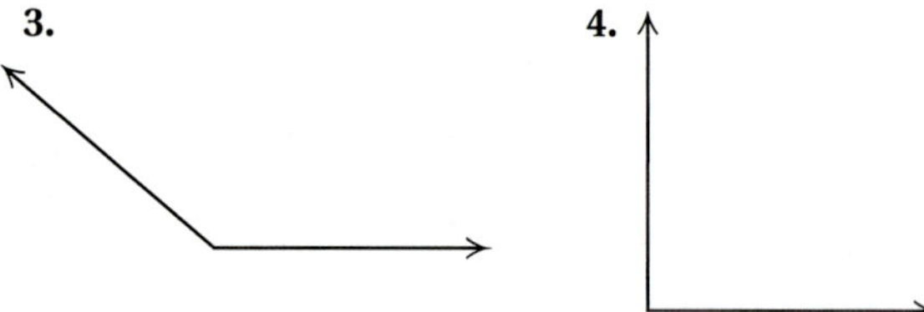

5.–8. Classify each of the angles in Exercises 1–4 as right, straight, acute, or obtuse. [5.1c]

Use the following triangle for Exercises 9–11.

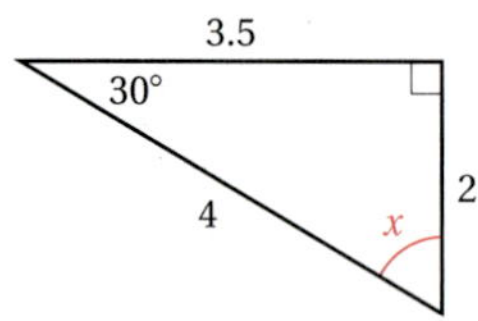

9. Find the missing angle measure. [5.1f]

10. Classify the triangle as equilateral, isosceles, or scalene. [5.1e]

11. Classify the triangle as right, obtuse, or acute. [5.1e]

12. Find the sum of the angle measures of a hexagon. [5.1f]

Find the perimeter. [5.2a]

13.

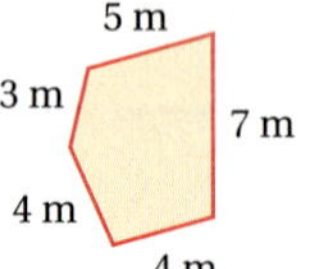

14.

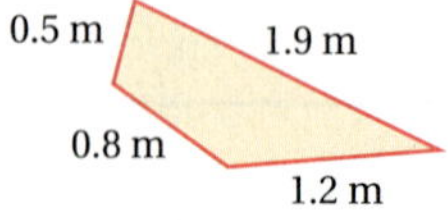

15. *Tennis Court.* The dimensions of a standard-sized tennis court are 78 ft by 36 ft. Find the perimeter and the area of the tennis court. [5.2b], [5.3c]

Find the perimeter and the area. [5.2a], [5.3a]

16.

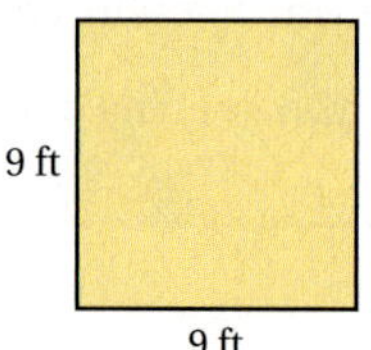

17.

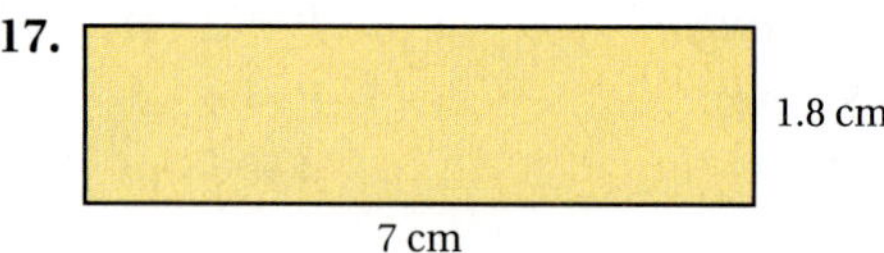

Find the area. [5.3b]

18.

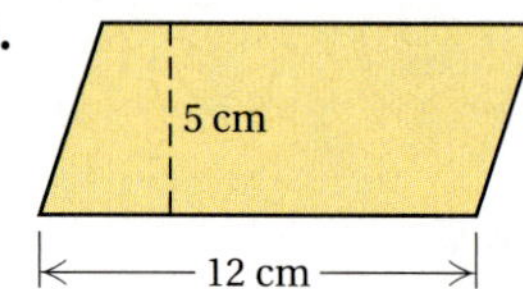

19.

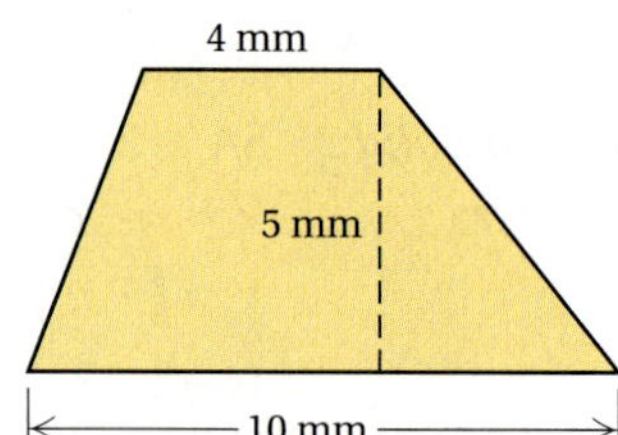

20.

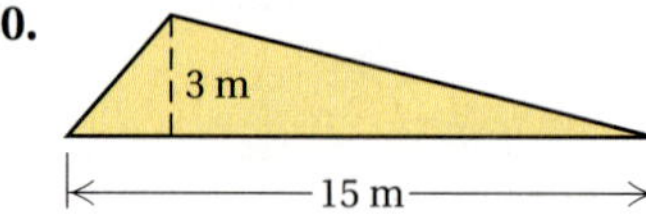

21.

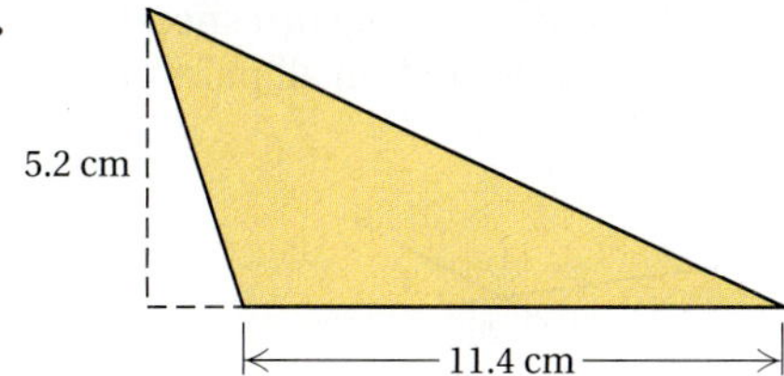

22.

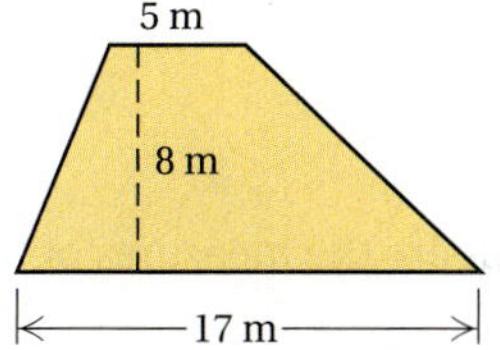

23.

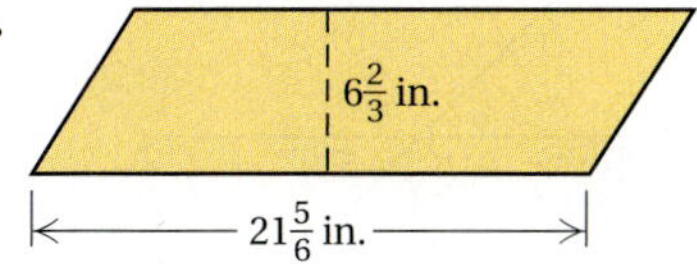

24. *Seeded Area.* A grassy area is to be seeded around three sides of a building and has equal width on the three sides, as shown below. What is the seeded area? [5.3c]

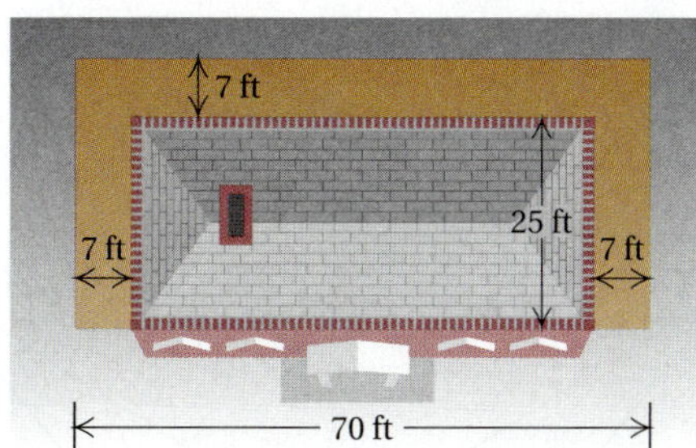

Find the length of a radius of the circle. [5.4a]

25.

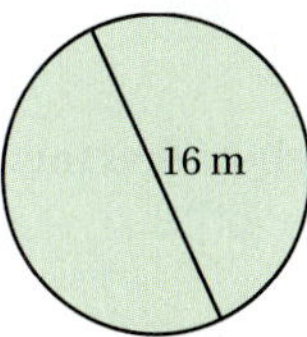

26.

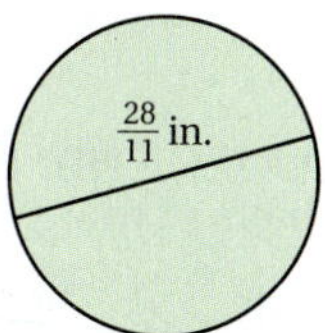

Find the length of a diameter of the circle. [5.4a]

27.

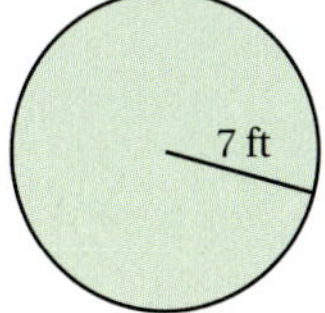

28.

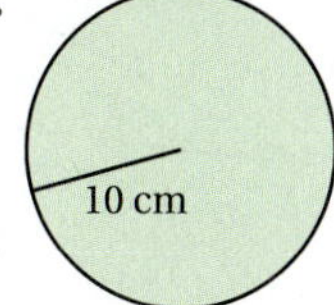

29. Find the circumference of the circle in Exercise 25. Use 3.14 for π. [5.4b]

30. Find the circumference of the circle in Exercise 26. Use $\frac{22}{7}$ for π. [5.4b]

31. Find the area of the circle in Exercise 25. Use 3.14 for π. [5.4c]

32. Find the area of the circle in Exercise 26. Use $\frac{22}{7}$ for π. [5.4c]

33. Find the area of the shaded region. Use 3.14 for π. [5.4d]

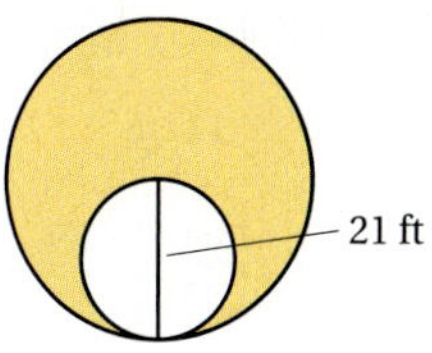

Find the volume and the surface area. [5.5a]

34.

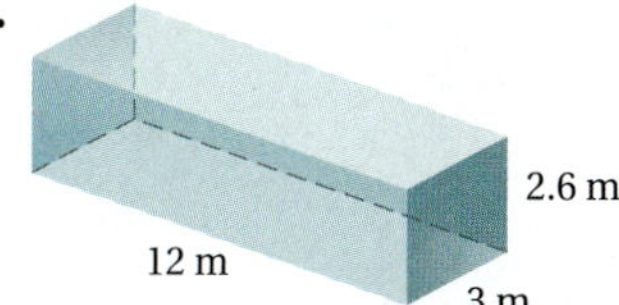

35.

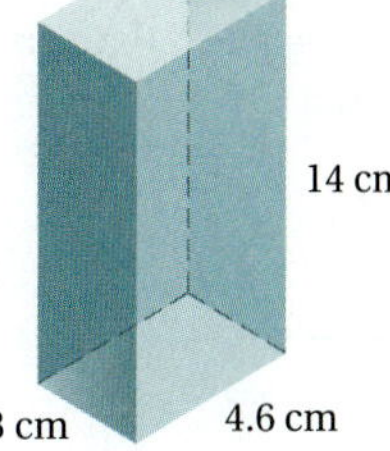

Find the volume. Use 3.14 for π.

36. [5.5b]

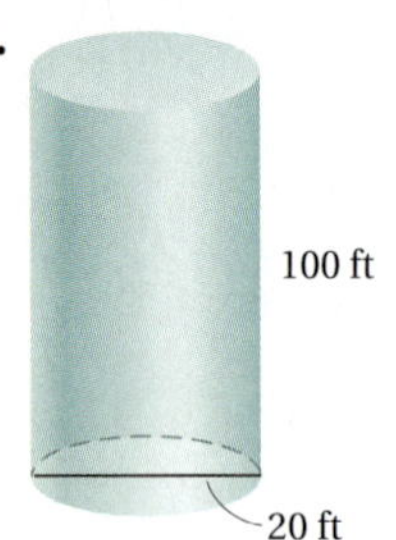

37. [5.5d]

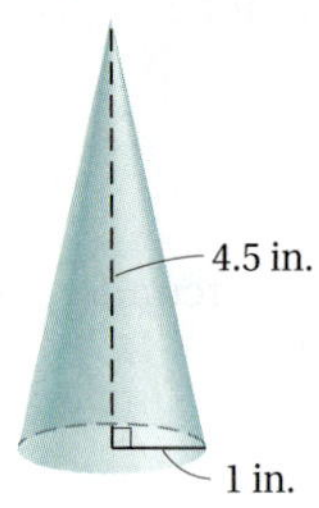

38. [5.5c]

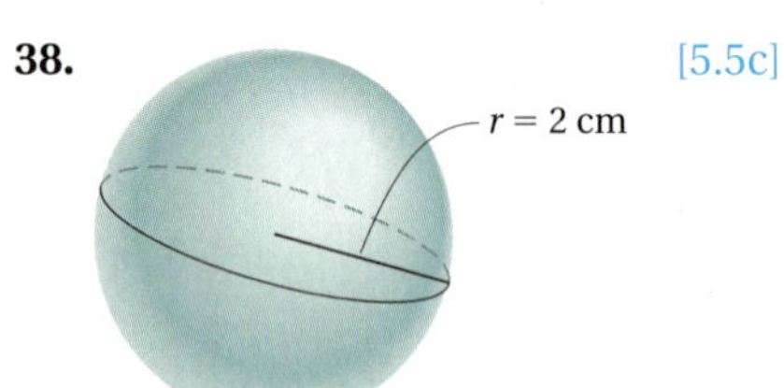

39. [5.5b]

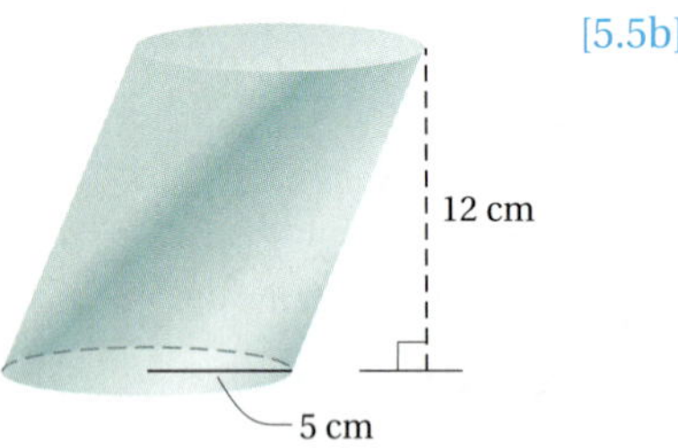

Find the measure of a complement of an angle with the given measure. [5.6a]

40. 82°

41. 5°

Find the measure of a supplement of an angle with the given measure. [5.6a]

42. 33°

43. 133°

44. In this figure, $m\angle 1 = 38°$ and $m\angle 5 = 105°$. Find $m\angle 2$, $m\angle 3$, $m\angle 4$, and $m\angle 6$. [5.1f], [5.6c]

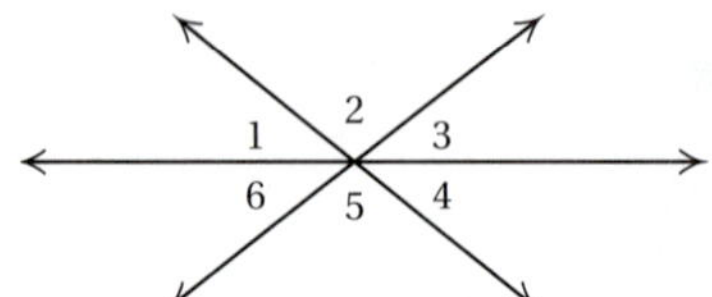

45. In this figure, identify **(a)** all pairs of corresponding angles, **(b)** all interior angles, and **(c)** all pairs of alternate interior angles. [5.6d]

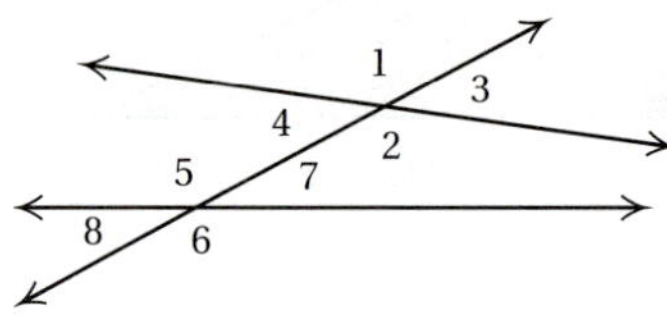

46. If $m \parallel n$ and $m\angle 4 = 135°$, what are the measures of the other angles? [5.6d]

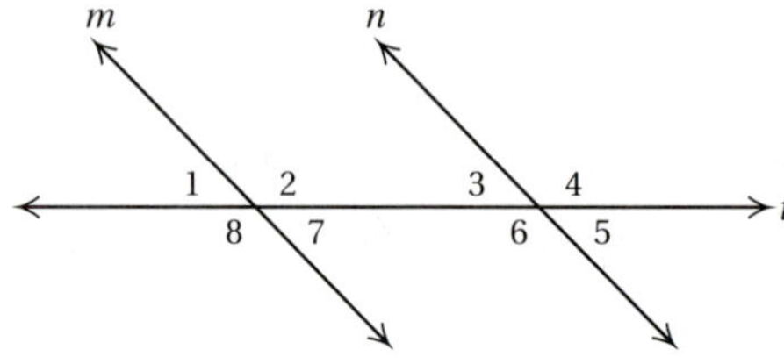

Name the corresponding parts of these congruent triangles. [5.7a]

47. $\triangle DHJ \cong \triangle RZK$

48.

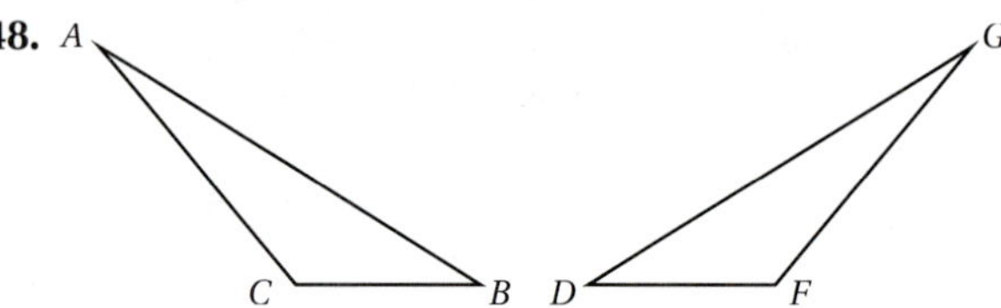

Which property (if any) should be used to show that the following pairs of triangles are congruent? [5.7a]

49.

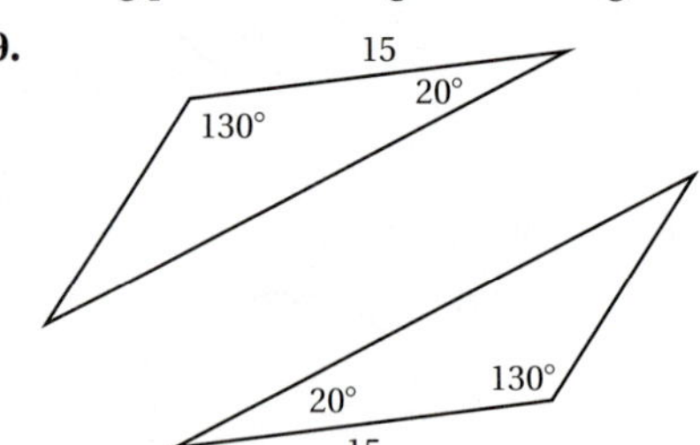

50.

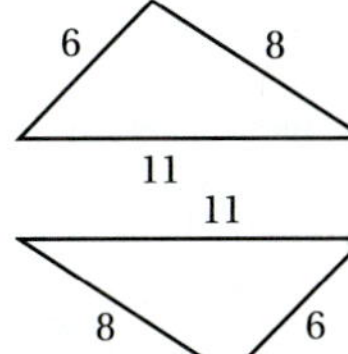

51.

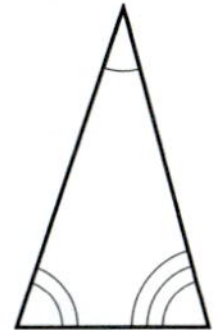

52. J is the midpoint of $\overline{IK}$ and $\overline{HI} \parallel \overline{KL}$. Explain why $\triangle JIH \cong \triangle JKL$. [5.7a]

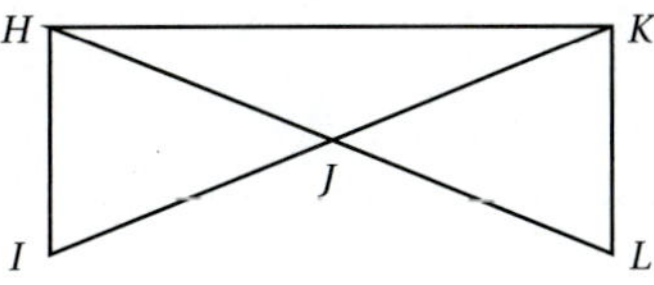

53. Find the measures of the angles and the lengths of the sides of this parallelogram. [5.7b]

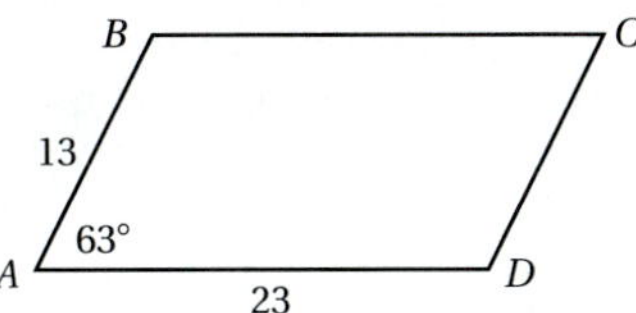

54. If $\triangle CQW \sim \triangle FAS$, name the congruent angles and the proportional sides. [5.8a]

55. If $\triangle NMO \sim \triangle STR$, find MO. [5.8b]

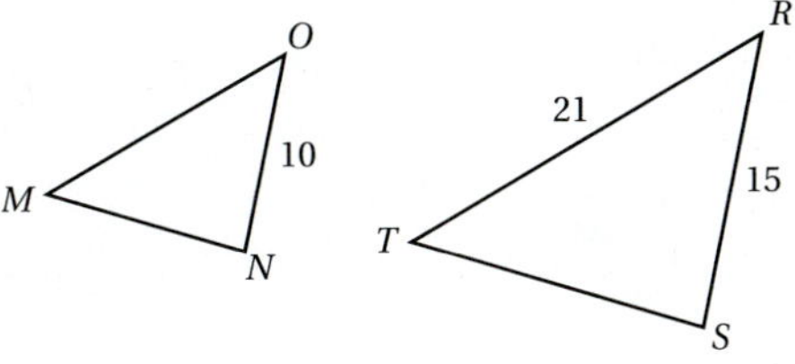

56. D$_W$ Which occupies more volume: two spheres, each with radius r, or one sphere with radius $2r$? Explain why. [5.5c]

57. D$_W$ Describe the difference among linear, area, and volume units of measure. [5.2a], [5.3a], [5.5a]

SKILL MAINTENANCE

58. Multiply: $5\frac{3}{4} \times 9\frac{1}{2}$. [2.4d]

Evaluate. [1.6b]

59. 4.7^3

60. $\left(\frac{1}{2}\right)^4$

61. Convert to fraction notation: 73%. [4.3b]

62. Convert to percent notation: 0.47. [4.2b]

63. Convert to percent notation: $\frac{23}{25}$. [4.3a]

SYNTHESIS

64. A square is cut in half so that the perimeter of each of the resulting rectangles is 30 ft. Find the area of the original square. [5.2a], [5.3a]

65. Find the area, in square meters, of the shaded region. [5.3c]

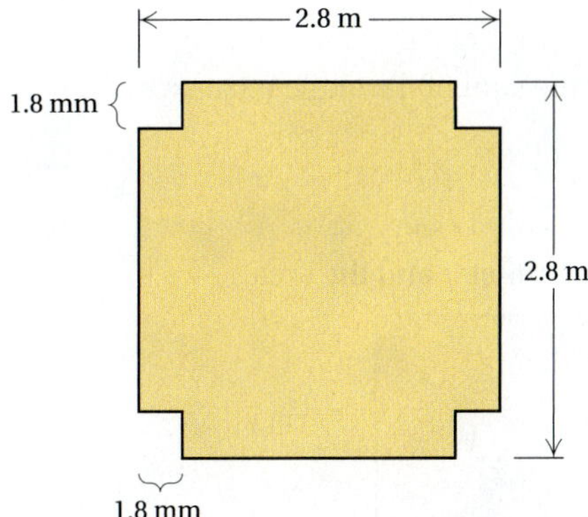

66. Find the area, in square centimeters, of the shaded region. [5.3c]

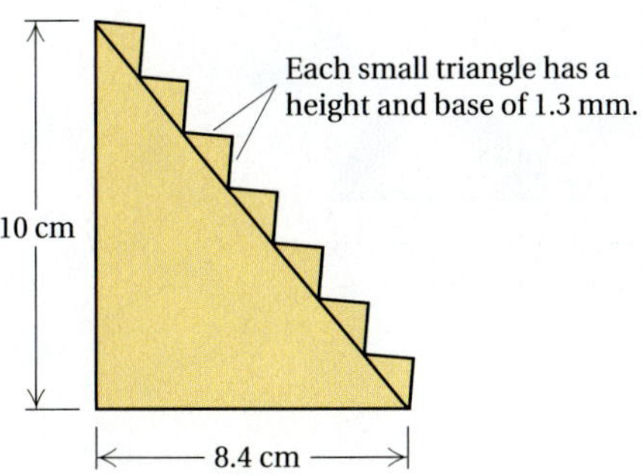

CHAPTER 5

Chapter Test

Use a protractor to measure each angle.

1.

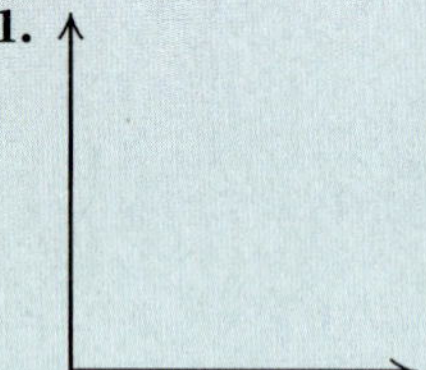

2.

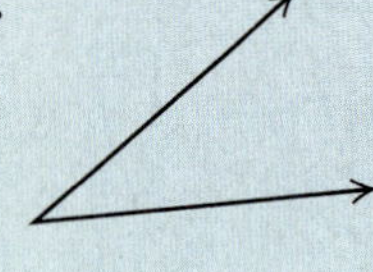

3.

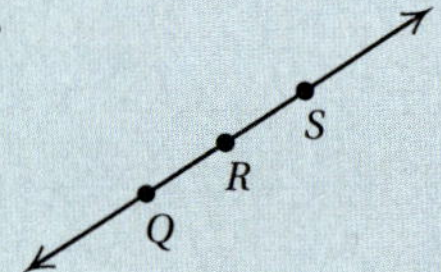

4.

5.–8. Classify each of the angles in Questions 1–4 as right, straight, acute, or obtuse.

Use the following triangle for Questions 9–11.

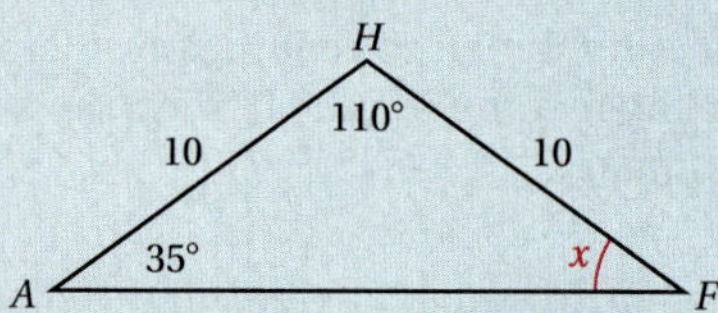

9. Find the missing angle measure.

10. Classify the triangle as equilateral, isosceles, or scalene.

11. Classify the triangle as right, obtuse, or acute.

12. Find the sum of the angle measures of a pentagon.

Find the perimeter and the area.

13.

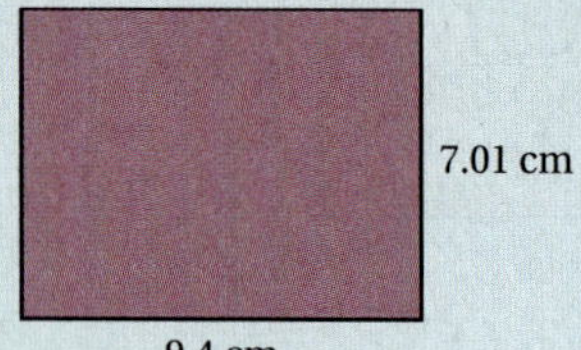

14.

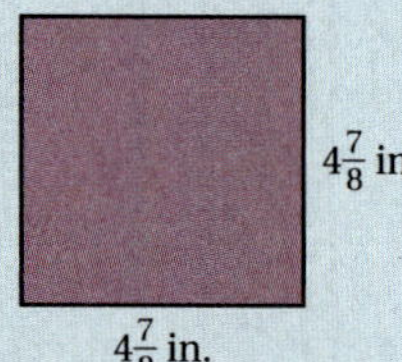

Find the area.

15.

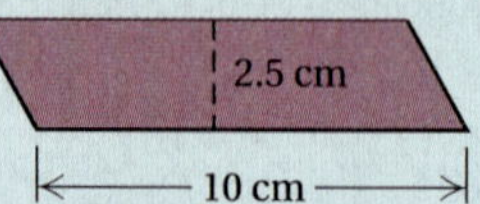

16.

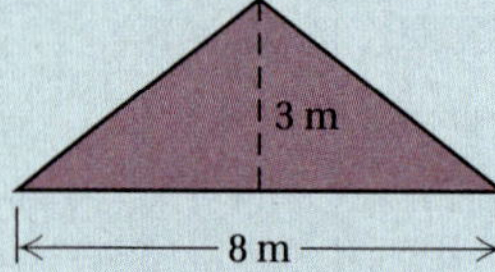

17.

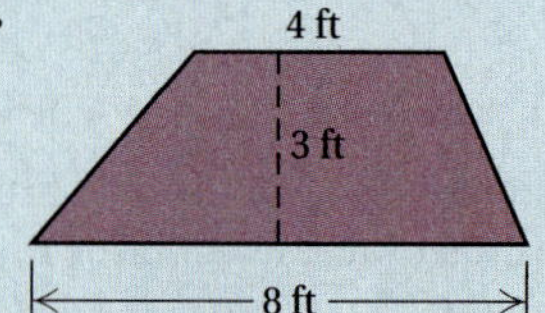

18. Find the length of a diameter of this circle.

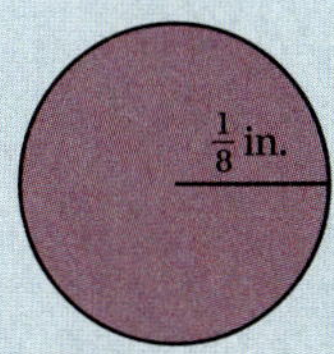

19. Find the length of a radius of this circle.

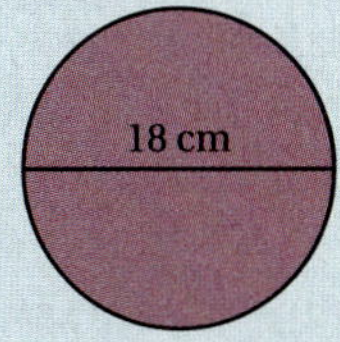

20. Find the circumference of the circle in Question 18. Use $\frac{22}{7}$ for π.

21. Find the area of the circle in Question 19. Use 3.14 for π.

22. Find the perimeter and the area of the shaded region. Use 3.14 for π.

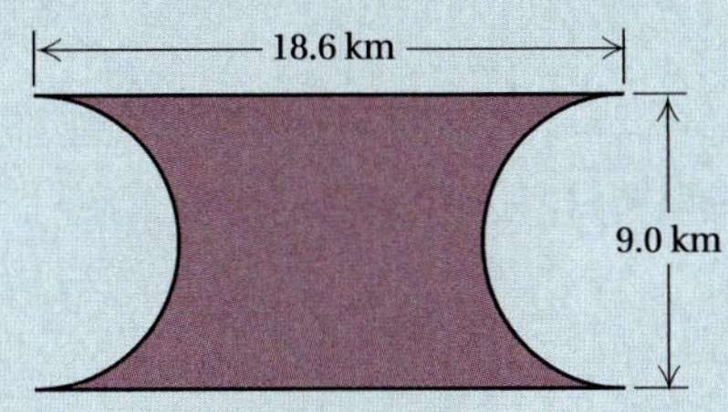

23. Find the volume and the surface area.

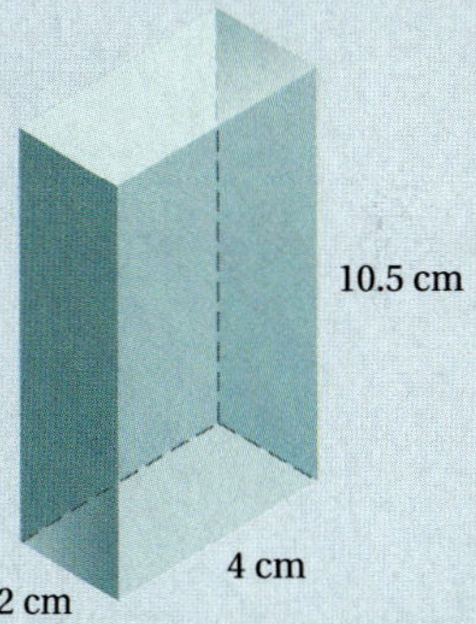

24. A twelve-box carton of 12-oz juice boxes comes in a rectangular box $10\frac{1}{2}$ in. by 8 in. by 5 in. What is the volume of the carton?

Find the volume. Use 3.14 for π.

25.

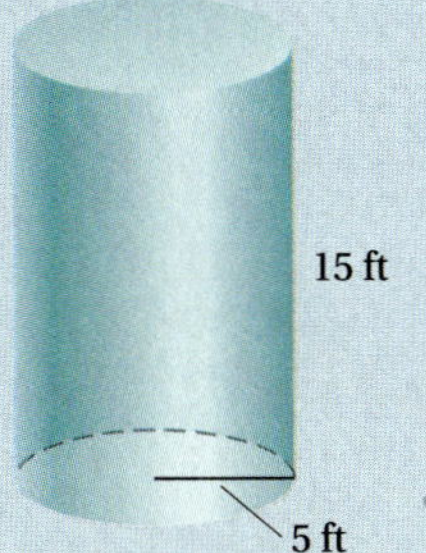

26.

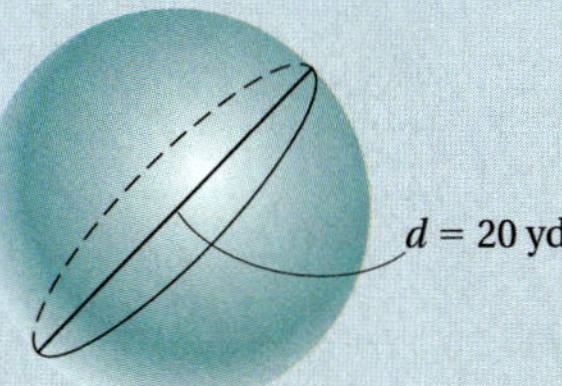

27.

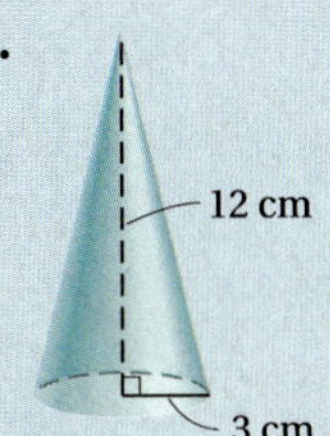

28. Find the measure of a supplement of an angle of 31°.

29. Find the measure of a complement of an angle of 79°.

30. In the figure, $m\angle 1 = 62°$ and $m\angle 5 = 110°$. Find $m\angle 2$, $m\angle 3$, $m\angle 4$, and $m\angle 6$.

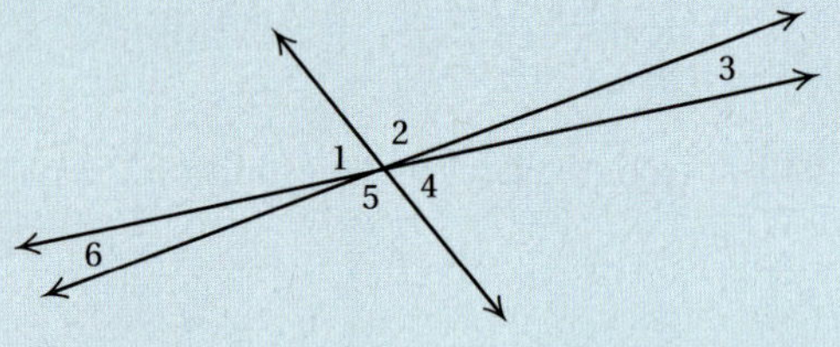

31. If $m \parallel n$ and $m\angle 4 = 120°$, what are the measures of the other angles?

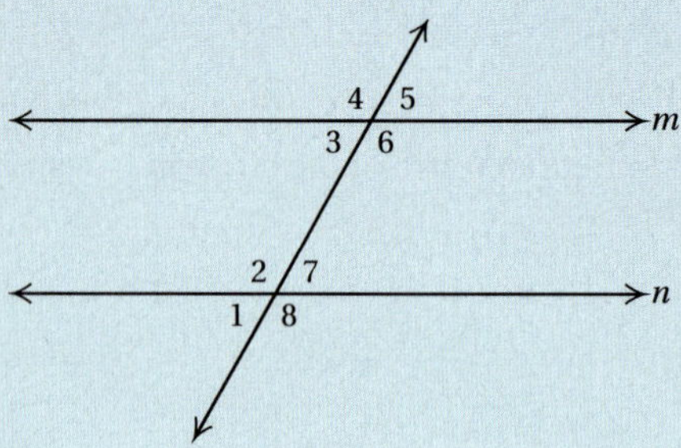

32. Name the corresponding parts of these congruent triangles: $\triangle CWS \cong \triangle ATZ$.

Which property (if any) would you use to show that $\triangle RST \cong \triangle DEF$ with the given information?

33. $\overline{RS} \cong \overline{DE}$, $\overline{RT} \cong \overline{DF}$, and $\angle R \cong \angle D$

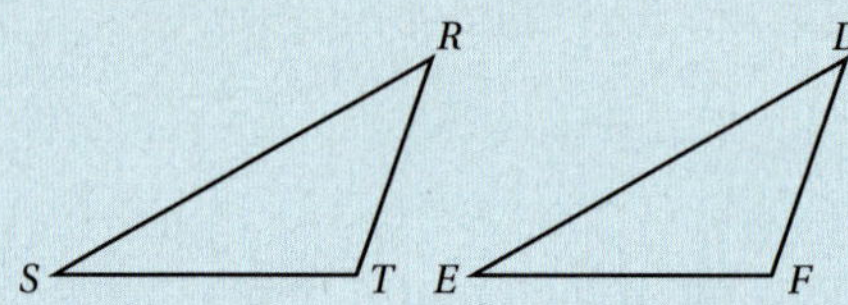

34. $\angle R \cong \angle D$, $\angle S \cong \angle E$, and $\angle T \cong \angle F$

35. $\overline{RS} \cong \overline{DE}$, $\angle R \cong \angle D$, and $\angle S \cong \angle E$

36. $\angle R \cong \angle D$, $\overline{RT} \cong \overline{DF}$, and $\overline{ST} \cong \overline{EF}$

37. The perimeter of $\square DEFG$ is 62. Find the measures of the angles and the lengths of the sides.

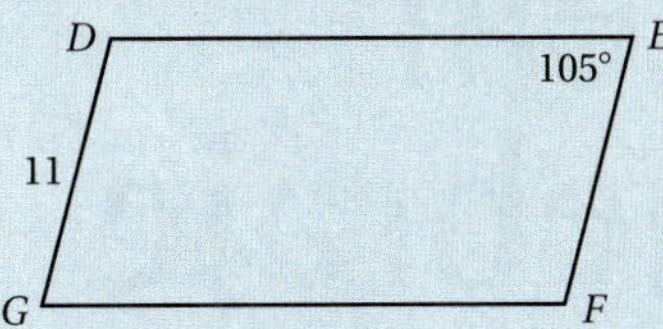

38. In $\square JKLM$, $JN = 3.2$ and $KN = 3$. Find the lengths of the diagonals, $\overline{LJ}$ and $\overline{KM}$.

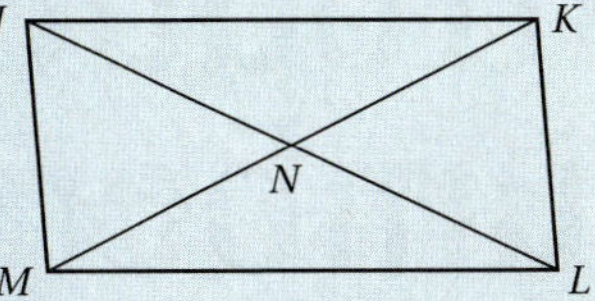

39. If $\triangle ERS \sim \triangle TGF$, name the congruent angles and the proportional sides.

40. If $\triangle GTR \sim \triangle ZEK$, find EK and ZK.

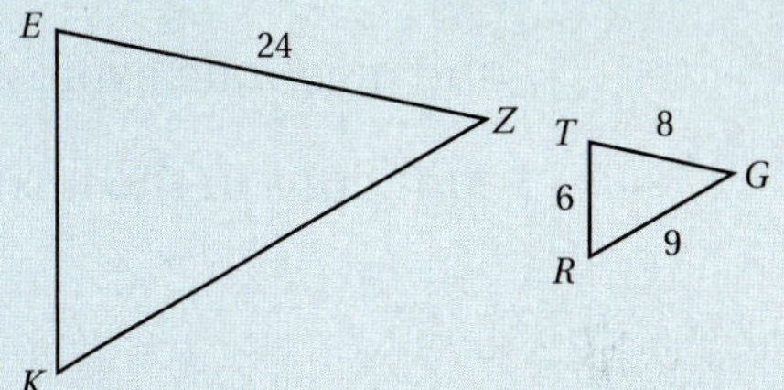

SKILL MAINTENANCE

Evaluate.

41. 10^3

42. $\left(\frac{1}{4}\right)^2$

43. Convert to percent notation: $\frac{13}{16}$.

44. Convert to decimal notation: 93.2%.

45. Convert to fraction notation: $33\frac{1}{3}\%$.

46. Multiply: $8\frac{1}{4} \times 2\frac{2}{3}$.

SYNTHESIS

47. Find the area of the shaded region. Give the answer in square feet.

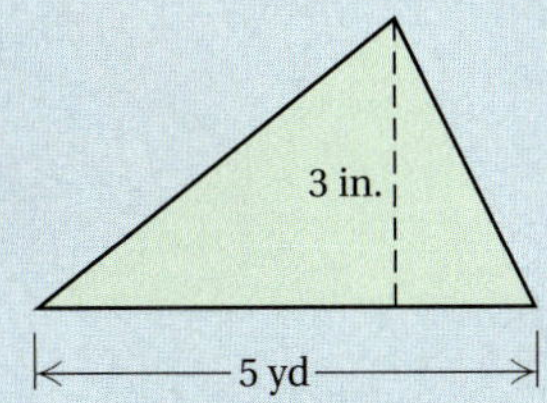

48. Find the volume of the solid. Give the answer in cubic feet. (Note that the solid is not drawn in perfect proportion.)

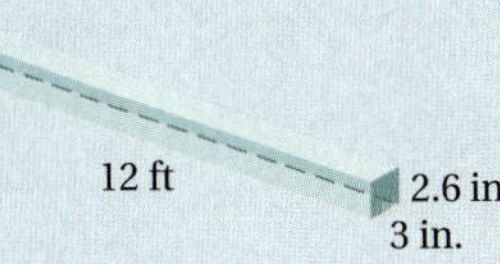

Introduction to Real Numbers and Algebraic Expressions

Gateway to Chapter 6

In this chapter, we emphasize skills involving real numbers and algebraic expressions. We will use these skills to solve equations in the next chapter.

Real-World Application

The Viking 2 Lander spacecraft has determined that temperatures on Mars range from $-125°C$ (Celsius) to $25°C$. Find the difference between the highest value and the lowest value in this temperature range.

Source: The Lunar and Planetary Institute

This problem appears as Example 12 in Section 6.4.

CHAPTER 6

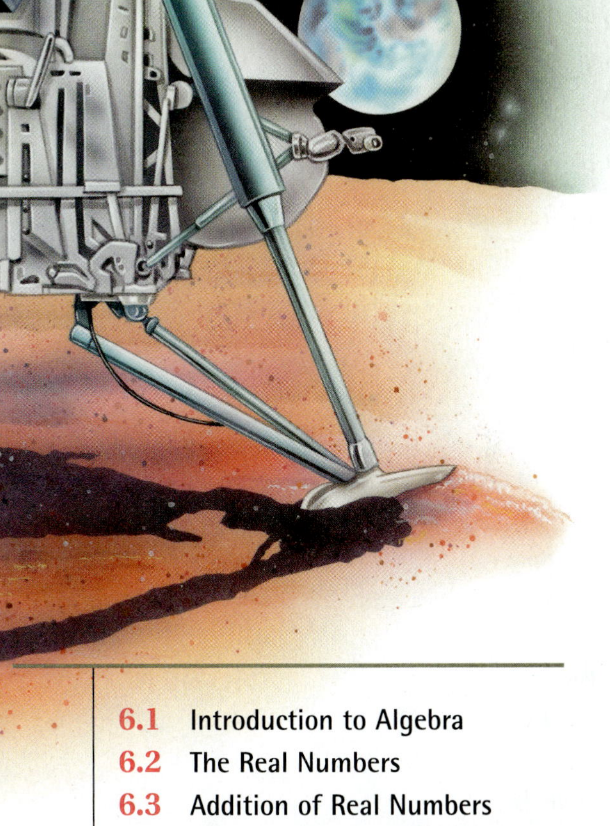

CHAPTER 6

Pretest

1. Evaluate $x/2y$ when $x = 5$ and $y = 8$. [6.1a]

2. Write an algebraic expression: Seventy-eight percent of some number. [6.1b]

3. Find the area of a rectangle when the length is 22.5 ft and the width is 16 ft. [6.1a]

4. Find $-x$ when $x = -12$. [6.3b]

Use either < or > for □ to write a true sentence. [6.2d]

5. $0 \square -5$

6. $10 \square -5$

7. $-35 \square -45$

8. $-\frac{2}{3} \square \frac{4}{5}$

Find the absolute value. [6.2e]

9. $|-12|$

10. $|2.3|$

11. $|0|$

Find the opposite, or additive inverse. [6.3b]

12. 5.4

13. $-\frac{2}{3}$

Find the reciprocal. [6.6b]

14. 10

15. $-\frac{2}{3}$

Compute and simplify.

16. $-9 + (-8)$ [6.3a]

17. $20.2 - (-18.4)$ [6.4a]

18. $-\frac{5}{6} - \frac{3}{10}$ [6.4a]

19. $-11.5 + 6.5$ [6.3a]

20. $-9(-7)$ [6.5a]

21. $\frac{5}{8}\left(-\frac{2}{3}\right)$ [6.5a]

22. $-19.6 \div 0.2$ [6.6c]

23. $-56 \div (-7)$ [6.6a]

24. $12 - (-6) + 14 - 8$ [6.4a]

25. $20 - 10 \div 5 + 2^3$ [6.8d]

Multiply. [6.7c]

26. $9(z - 2)$

27. $-2(2a + b - 5c)$

Factor. [6.7d]

28. $4x - 12$

29. $6y - 9z - 18$

Simplify.

30. $3y - 7 - 2(2y + 3)$ [6.8b]

31. $2[3(y + 1) - 4] - [5(y - 3) - 5]$ [6.8c]

32. Write an inequality with the same meaning as $x > 12$. [6.2d]

33. **Temperature Extremes.** In Churchill, Manitoba, Canada, the average daily low temperature in January is −31°C. The average daily low temperature in Key West, Florida, is 19°C. How much higher is the average daily low temperature in Key West, Florida? [6.4b]

6.1 INTRODUCTION TO ALGEBRA

The study of algebra involves the use of equations to solve problems. Equations are constructed from algebraic expressions. The purpose of this section is to introduce you to the types of expressions encountered in algebra.

Objectives

a Evaluate algebraic expressions by substitution.

b Translate phrases to algebraic expressions.

a Evaluating Algebraic Expressions

In arithmetic, you have worked with expressions such as

$$49 + 75, \quad 8 \times 6.07, \quad 29 - 14, \quad \text{and} \quad \frac{5}{6}.$$

In algebra, we use certain letters for numbers and work with *algebraic expressions* such as

$$x + 75, \quad 8 \times y, \quad 29 - t, \quad \text{and} \quad \frac{a}{b}.$$

Sometimes a letter can represent various numbers. In that case, we call the letter a **variable.** Let a = your age. Then a is a variable since a changes from year to year. Sometimes a letter can stand for just one number. In that case, we call the letter a **constant.** Let b = your date of birth. Then b is a constant.

Where do algebraic expressions occur? Most often we encounter them when we are solving applied problems. For example, consider the bar graph shown at right, one that we might find in a book or magazine. Suppose we want to know how many more moons Saturn has than Jupiter. Using arithmetic, we might simply subtract. But let's see how we might find this out using algebra. We translate the problem into a statement of equality, an equation. It might be done as follows:

Number of moons of Jupiter	plus	How many more	is	Number of moons of Saturn
↓	↓	↓	↓	↓
17	+	x	=	28

Note that we have an algebraic expression, $17 + x$, on the left of the equals sign. To find the number x, we can subtract 17 on both sides of the equation:

$$17 + x = 28$$
$$17 + x - 17 = 28 - 17$$
$$x = 11.$$

The value of x gives the answer, 11 moons.

We call $17 + x$ an *algebraic expression* and $17 + x = 28$ an *algebraic equation*. Note that there is no equals sign, =, in an algebraic expression.

In arithmetic, you probably would do this subtraction without ever considering an equation. *In algebra, more complex problems are difficult to solve without first solving an equation.*

Do Exercise 1.

1. Translate this problem to an equation. Use the graph below.

How many more moons does Uranus have than Neptune?

Moons of planets

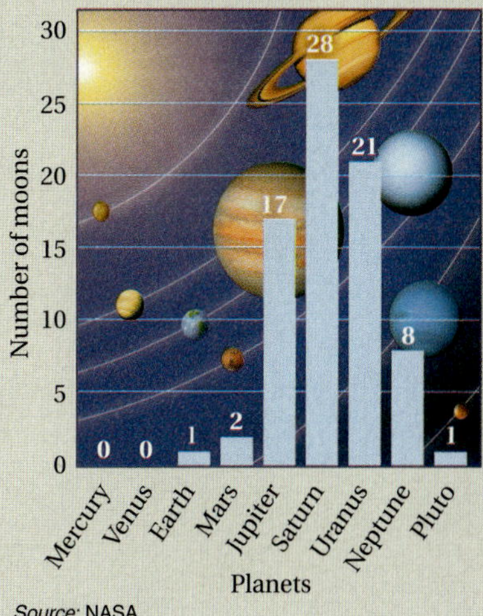

Source: NASA

Answer on page A-14

2. Evaluate $a + b$ when $a = 38$ and $b = 26$.

3. Evaluate $x - y$ when $x = 57$ and $y = 29$.

4. Evaluate $4t$ when $t = 15$.

5. Find the area of a rectangle when l is 24 ft and w is 8 ft.

6. Evaluate a/b when $a = 200$ and $b = 8$.

7. Evaluate $10p/q$ when $p = 40$ and $q = 25$.

Answers on page A-14

An **algebraic expression** consists of variables, constants, numerals, and operation signs. When we replace a variable with a number, we say that we are **substituting** for the variable. This process is called **evaluating the expression.**

EXAMPLE 1 Evaluate $x + y$ when $x = 37$ and $y = 29$.

We substitute 37 for x and 29 for y and carry out the addition:

$$x + y = 37 + 29 = 66.$$

The number 66 is called the **value** of the expression.

Algebraic expressions involving multiplication can be written in several ways. For example, "8 times a" can be written as $8 \times a$, $8 \cdot a$, $8(a)$, or simply $8a$.

Two letters written together without an operation symbol, such as ab, also indicate a multiplication.

EXAMPLE 2 Evaluate $3y$ when $y = 14$.

$$3y = 3(14) = 42$$

Do Exercises 2–4.

EXAMPLE 3 *Area of a Rectangle.* The area A of a rectangle of length l and width w is given by the formula $A = lw$. Find the area when l is 24.5 in. and w is 16 in.

We substitute 24.5 in. for l and 16 in. for w and carry out the multiplication:

$$\begin{aligned} A = lw &= (24.5 \text{ in.})(16 \text{ in.}) \\ &= (24.5)(16)(\text{in.})(\text{in.}) \\ &= 392 \text{ in}^2, \text{ or } 392 \text{ square inches.} \end{aligned}$$

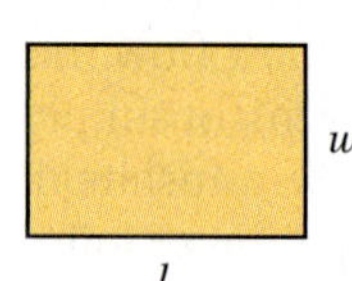

Do Exercise 5.

Algebraic expressions involving division can also be written in several ways. For example, "8 divided by t" can be written as $8 \div t$, $\frac{8}{t}$, $8/t$, or $8 \cdot \frac{1}{t}$, where the fraction bar is a division symbol.

EXAMPLE 4 Evaluate $\frac{a}{b}$ when $a = 63$ and $b = 9$.

We substitute 63 for a and 9 for b and carry out the division:

$$\frac{a}{b} = \frac{63}{9} = 7.$$

EXAMPLE 5 Evaluate $\frac{12m}{n}$ when $m = 8$ and $n = 16$.

$$\frac{12m}{n} = \frac{12 \cdot 8}{16} = \frac{96}{16} = 6$$

Do Exercises 6 and 7.

EXAMPLE 6 *Motorcycle Travel.* Ed takes a trip on his motorcycle. He wants to travel 660 mi on a particular day. The time t, in hours, that it takes to travel 660 mi is given by

$$t = \frac{660}{r},$$

where r is the speed of Ed's motorcycle. Find the time of travel if the speed r is 60 mph.

We substitute 60 for r and carry out the division:

$$t = \frac{660}{r} = \frac{660}{60} = 11 \text{ hr.}$$

Do Exercise 8.

8. Motorcycle Travel. Find the time it takes to travel 660 mi if the speed is 55 mph.

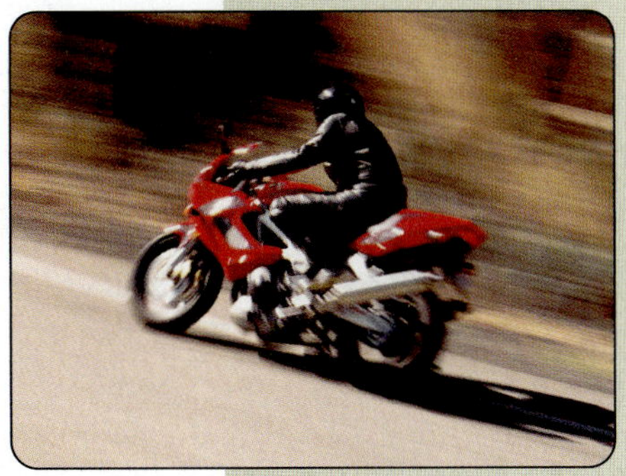

CALCULATOR CORNER

Evaluating Algebraic Expressions *To the student and the instructor*: This book contains a series of *optional* discussions on using a calculator. A calculator is *not* a requirement for this textbook. There are many kinds of calculators and different instructions for their usage. For the remainder of the text, we include instructions for the scientific keys on a graphing calculator such as a TI-83 or a TI-83 Plus. Be sure to consult your user's manual as well. Also, check with your instructor about whether you are allowed to use a calculator in the course.

Note that there are options above the keys as well as on them. To access the option written on a key, simply press the key. The options written in yellow above the keys are accessed by first pressing the yellow [2nd] key and then pressing the key corresponding to the desired option. The green options are accessed by first pressing the green [ALPHA] key.

To turn the calculator on, press the [ON] key at the bottom left-hand corner of the keypad. You should see a blinking rectangle, or cursor, on the screen. If you do not see the cursor, try adjusting the display contrast. To do this, first press [2nd] and then press and hold [△] to increase the contrast of [▽] to decrease the contrast.

To turn the calculator off, press [2nd] [OFF]. (OFF is the second operation associated with the [ON] key.) The calculator will turn itself off automatically after about five minutes of no activity.

We can evaluate algebraic expressions on a calculator by making the appropriate substitutions, keeping in mind the rules for order of operations, and then carrying out the resulting calculations. To evaluate $12m/n$ when $m = 8$ and $n = 16$, as in Example 5, we enter $12 \cdot 8/16$ by pressing [1] [2] [×] [8] [÷] [1] [6] [ENTER]. The result is 6.

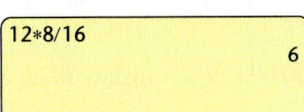

Exercises: Evaluate.

1. $\frac{12m}{n}$, when $m = 42$ and $n = 9$
2. $a + b$, when $a = 8.2$ and $b = 3.7$
3. $b - a$, when $a = 7.6$ and $b = 9.4$
4. $27xy$, when $x = 12.7$ and $y = 100.4$
5. $3a + 2b$, when $a = 2.9$ and $b = 5.7$
6. $2a + 3b$, when $a = 7.3$ and $b = 5.1$

Answer on page A-14

Translate to an algebraic expression.

9. Eight less than some number

10. Eight more than some number

11. Four less than some number

12. Half of a number

13. Six more than eight times some number

14. The difference of two numbers

15. Fifty-nine percent of some number

16. Two hundred less than the product of two numbers

17. The sum of two numbers

Answers on page A-14

b Translating to Algebraic Expressions

In algebra, we translate problems to equations. The different parts of an equation are translations of word phrases to algebraic expressions. It is easier to translate if we know that certain words often translate to certain operation symbols.

KEY WORDS, PHRASES, AND CONCEPTS

ADDITION (+)	SUBTRACTION (−)	MULTIPLICATION (·)	DIVISION (÷)
add	subtract	multiply	divide
added to	subtracted from	multiplied by	divided by
sum	difference	product	quotient
total	minus	times	
plus	less than	of	
more than	decreased by		
increased by	take away		

EXAMPLE 7 Translate to an algebraic expression:

Twice (or two times) some number.

Think of some number, say, 8. What number is twice 8? It is 16. How did you get 16? You multiplied by 2. Do the same thing using a variable. We can use any variable we wish, such as x, y, m, or n. Let's use y to stand for some number. If we multiply by 2, we get an expression

$$y \times 2, \quad 2 \times y, \quad 2 \cdot y, \quad \text{or} \quad 2y.$$

In algebra, $2y$ is the expression generally used.

EXAMPLE 8 Translate to an algebraic expression:

Thirty-eight percent of some number.

The word "of" translates to a multiplication symbol, so we get the following expressions as a translation:

$$38\% \cdot n, \quad 0.38 \times n, \quad \text{or} \quad 0.38n.$$

EXAMPLE 9 Translate to an algebraic expression:

Seven less than some number.

We let

x represent the number.

Now if the number were 23, then the translation would be "7 subtracted from 23," or $23 - 7$. If we knew the number to be 345, then the translation would be $345 - 7$. If the number is x, then the translation is

$$x - 7.$$

CAUTION!

Note that $7 - x$ is *not* a correct translation of the expression in Example 9. The expression $7 - x$ is a translation of "seven minus some number" or "some number less than seven."

EXAMPLE 10 Translate to an algebraic expression:

Eighteen more than a number.

We let

$t =$ the number.

Now if the number were 26, then the translation would be $26 + 18$, or $18 + 26$. If we knew the number to be 174, then the translation would be $174 + 18$, or $18 + 174$. If the number is t, then the translation is

$$t + 18, \quad \text{or} \quad 18 + t.$$

EXAMPLE 11 Translate to an algebraic expression:

A number divided by 5.

We let

$m =$ the number.

Now if the number were 76, then the translation would be $76 \div 5$, or $76/5$, or $\frac{76}{5}$. If the number were 213, then the translation would be $213 \div 5$, or $213/5$, or $\frac{213}{5}$. If the number is m, then the translation is

$$m \div 5, \quad m/5, \quad \text{or} \quad \frac{m}{5}.$$

EXAMPLE 12 Translate each phrase to an algebraic expression.

PHRASE	ALGEBRAIC EXPRESSION
Five more than some number	$n + 5$, or $5 + n$
Half of a number	$\frac{1}{2}t$, $\frac{t}{2}$, or $t/2$
Five more than three times some number	$3p + 5$, or $5 + 3p$
The difference of two numbers	$x - y$
Six less than the product of two numbers	$mn - 6$
Seventy-six percent of some number	$76\%z$, or $0.76z$
Four less than twice some number	$2x - 4$

Do Exercises 9–17 on the preceding page.

Study Tips

USING THIS TEXTBOOK

From time to time we will repeat Study Tips that appeared earlier in the textbook because we think it will be helpful to you to read them again. This is one such Study Tip.

Knowing how to derive the full benefit of the organization and features in your textbook will contribute significantly to your understanding of the material in this course. Here we highlight a few points that should contribute to your learning.

- **Be sure to note the special symbols [a], [b], [c], and so on, that correspond to the objectives you are to be able to perform.** The first time you see them is in the margin at the beginning of each section; the second time is in the subheadings of each section; and the third time is in the exercise set for the section. You will also find them next to the skill maintenance exercises in each exercise set and in the review exercises at the end of the chapter, as well as in the answers to the chapter tests and the cumulative reviews. These objective symbols allow you to refer to the appropriate place in the text whenever you need to review a topic.
- **Read and study each step of each example.** The examples include important side comments that explain each step. These carefully chosen examples and notes prepare you for success in the exercise set.
- **Stop and do the margin exercises as you study a section.** Doing the margin exercises is one of the most effective ways to enhance your ability to learn mathematics from this text. Don't deprive yourself of its benefits!
- **Note the icons listed at the top of each exercise set.** These refer to the many distinctive multimedia study aids that accompany the book.
- **Odd-numbered exercises.** Usually an instructor assigns some odd-numbered exercises. When you complete these, you can check your answers at the back of the book. If you miss any, check your work in the *Student's Solutions Manual* or ask your instructor for guidance.
- **Even-numbered exercises.** Whether or not your instructor assigns the even-numbered exercises, always do some on your own. Remember, there are no answers given for the chapter tests, so you need to practice doing exercises without answers. Check your answers later with a friend or your instructor.

6.1 EXERCISE SET

For Extra Help

Digital Video Tutor CD 5 Videotape 7

InterAct Math

Math Tutor Center

MathXL

MyMathLab

a Substitute to find values of the expressions in each of the following applied problems.

1. *Enrollment Costs.* At Emmett Community College, it costs \$600 to enroll in the 8 A.M. section of Elementary Algebra. Suppose that the variable n stands for the number of students who enroll. Then $600n$ stands for the total amount of money collected for this course. How much is collected if 34 students enroll? 78 students? 250 students?

2. *Commuting Time.* It takes Erin 24 min less time to commute to work than it does George. Suppose that the variable x stands for the time it takes George to get to work. Then $x - 24$ stands for the time it takes Erin to get to work. How long does it take Erin to get to work if it takes George 56 min? 93 min? 105 min?

3. *Area of a Triangle.* The area A of a triangle with base b and height h is given by $A = \frac{1}{2}bh$. Find the area when $b = 45$ m (meters) and $h = 86$ m.

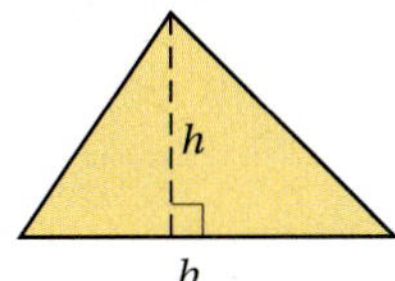

4. *Area of a Parallelogram.* The area A of a parallelogram with base b and height h is given by $A = bh$. Find the area of the parallelogram when the height is 15.4 cm (centimeters) and the base is 6.5 cm.

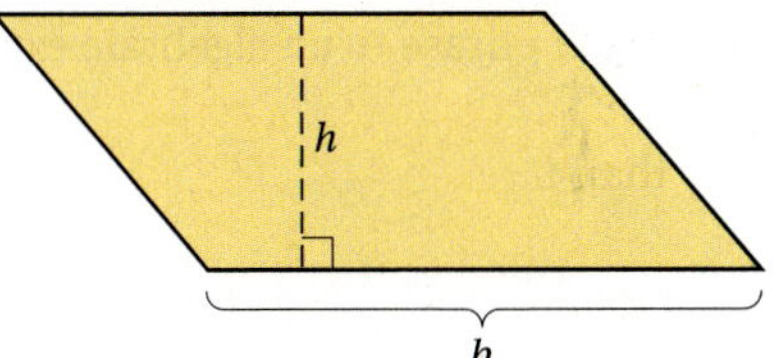

5. *Distance Traveled.* A driver who drives at a speed of r mph for t hr will travel a distance d mi given by $d = rt$ mi. How far will a driver travel at a speed of 65 mph for 4 hr?

6. *Simple Interest.* The simple interest I on a principal of P dollars at interest rate r for time t, in years, is given by $I = Prt$. Find the simple interest on a principal of \$4800 at 9% for 2 yr. (*Hint*: 9% = 0.09.)

7. *Hockey Goal.* The front of a regulation hockey goal is 6 ft wide and 4 ft high. Find its area.
Source: National Hockey League

8. *Zoology.* A great white shark has triangular teeth. Each tooth measures about 5 cm across the base and has a height of 6 cm. Find the surface area of one side of one tooth. (See Exercise 3.)

Evaluate.

9. $8x$, when $x = 7$

10. $6y$, when $y = 7$

11. $\frac{a}{b}$, when $a = 24$ and $b = 3$

12. $\frac{p}{q}$, when $p = 16$ and $q = 2$

13. $\frac{3p}{q}$, when $p = 2$ and $q = 6$

14. $\frac{5y}{z}$, when $y = 15$ and $z = 25$

15. $\frac{x + y}{5}$, when $x = 10$ and $y = 20$

16. $\frac{p + q}{2}$, when $p = 2$ and $q = 16$

17. $\frac{x - y}{8}$, when $x = 20$ and $y = 4$

18. $\frac{m - n}{5}$, when $m = 16$ and $n = 6$

b Translate each phrase to an algebraic expression.

19. Seven more than b

20. Nine more than t

21. Twelve less than c

22. Fourteen less than d

23. Four increased by q

24. Thirteen increased by z

25. b more than a

26. c more than d

27. x divided by y

28. c divided by h

29. x plus w

30. s added to t

31. m subtracted from n

32. p subtracted from q

33. The sum of x and y

34. The sum of a and b

35. Twice z

36. Three times q

37. Three multiplied by m

38. The product of 8 and t

39. The product of 89% and your salary

40. 67% of the women attending

41. Danielle drove at a speed of 65 mph for t hours. How far did Danielle travel?

42. Juan has d dollars before spending $19.95 on a DVD of the movie *Castaway*. How much did Juan have after the purchase?

43. Lisa had $50 before spending x dollars on pizza. How much money remains?

44. Dino drove his pickup truck at 55 mph for t hours. How far did he travel?

45. $\mathbf{D}_\mathbf{W}$ If the length of a rectangle is doubled, does the area double? Why or why not?

46. $\mathbf{D}_\mathbf{W}$ If the height and the base of a triangle are doubled, what happens to the area? Explain.

SKILL MAINTENANCE

Find the area. Use 3.14 for π.

47. [5.3a]

48. [5.3b]

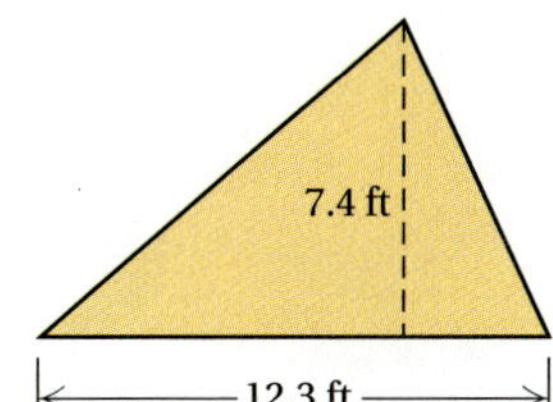

49. [5.3a]

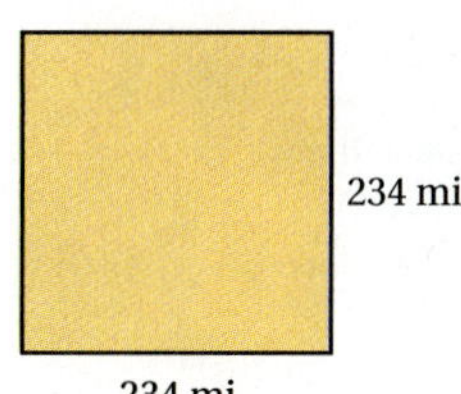

50. [5.3b]

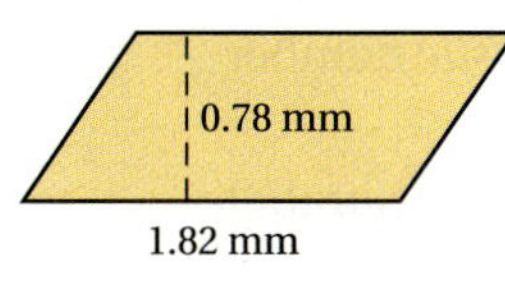

51. [5.3b]

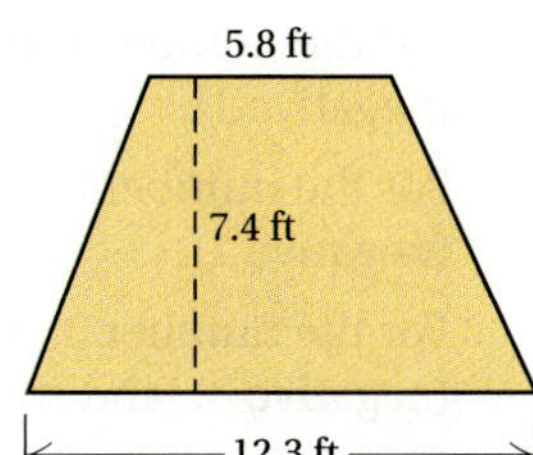

52. [5.4c]

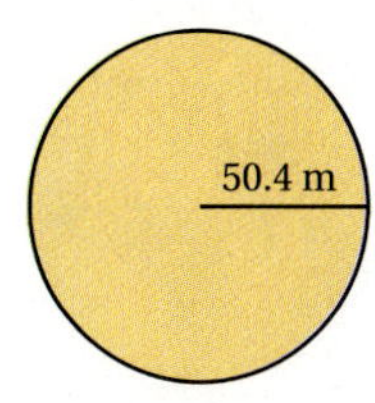

SYNTHESIS

Translate to an algebraic expression.

53. Some number x plus three times y

54. Some number a plus 2 plus b

55. A number that is 3 less than twice x

56. Your age in 5 years, if you are a years old now

Objectives

a State the integer that corresponds to a real-world situation.

b Graph rational numbers on a number line.

c Convert from fraction notation to decimal notation for a rational number.

d Determine which of two real numbers is greater and indicate which, using < or >; given an inequality like $a > b$, write another inequality with the same meaning. Determine whether an inequality like $-3 \leq 5$ is true or false.

e Find the absolute value of a real number.

6.2 THE REAL NUMBERS

A **set** is a collection of objects. For our purposes, we will most often be considering sets of numbers. One way to name a set uses what is called **roster notation.** For example, roster notation for the set containing the numbers 0, 2, and 5 is $\{0, 2, 5\}$.

Sets that are part of other sets are called **subsets.** In this section, we become acquainted with the set of *real numbers* and its various subsets.

Two important subsets of the real numbers are listed below using roster notation.

NATURAL NUMBERS

The set of **natural numbers** $= \{1, 2, 3, \ldots\}$. These are the numbers used for counting.

WHOLE NUMBERS

The set of **whole numbers** $= \{0, 1, 2, 3, \ldots\}$. This is the set of natural numbers with 0 included.

We can represent these sets on a number line. The natural numbers are those to the right of zero. The whole numbers are the natural numbers and zero.

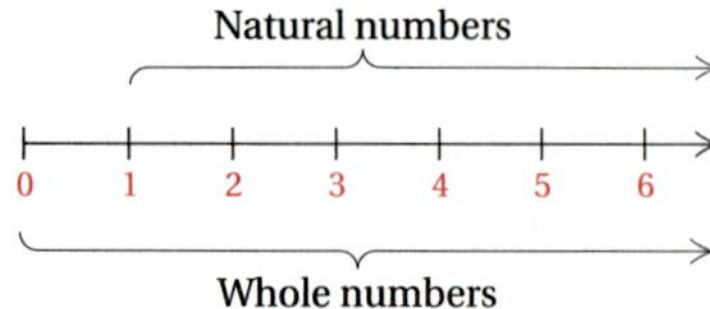

We create a new set, called the *integers,* by starting with the whole numbers, 0, 1, 2, 3, and so on. For each natural number 1, 2, 3, and so on, we obtain a new number to the left of zero on the number line:

For the number 1, there will be an *opposite* number -1 (negative 1).

For the number 2, there will be an *opposite* number -2 (negative 2).

For the number 3, there will be an *opposite* number -3 (negative 3), and so on.

The **integers** consist of the whole numbers and these new numbers.

INTEGERS

The set of **integers** $= \{\ldots, -5, -4, -3, -2, -1, 0, 1, 2, 3, 4, 5, \ldots\}$.

We picture the integers on a number line as follows.

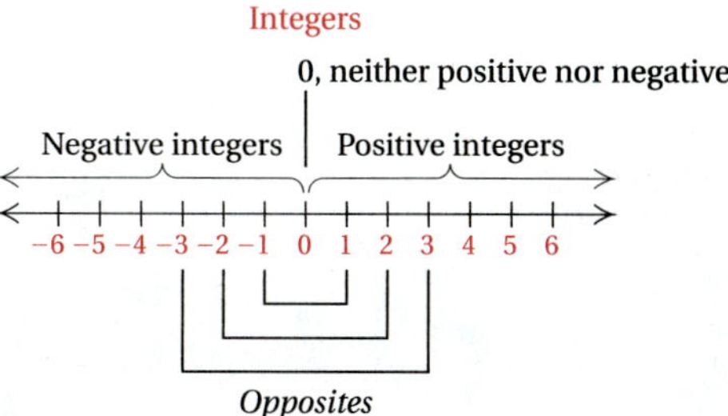

We call these new numbers to the left of 0 **negative integers.** The natural numbers are also called **positive integers.** Zero is neither positive nor negative. We call -1 and 1 **opposites** of each other. Similarly, -2 and 2 are opposites, -3 and 3 are opposites, -100 and 100 are opposites, and 0 is its own opposite. Pairs of opposite numbers like -3 and 3 are the same distance from 0. The integers extend infinitely on the number line to the left and right of zero.

a Integers and the Real World

Integers correspond to many real-world problems and situations. The following examples will help you get ready to translate problem situations that involve integers to mathematical language.

EXAMPLE 1 Tell which integer corresponds to this situation: The temperature is 3 degrees below zero.

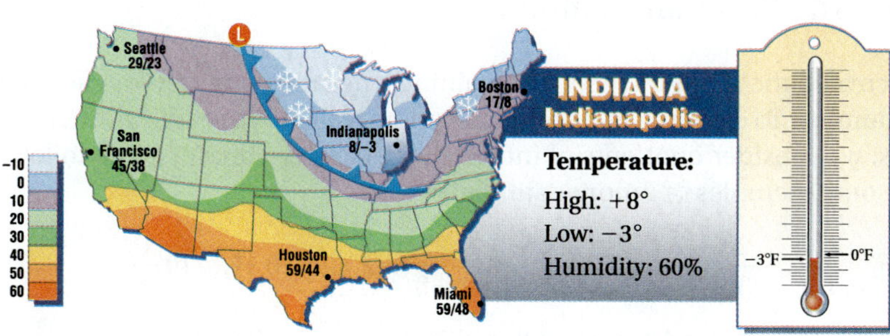

The integer -3 corresponds to the situation. The temperature is $-3°$.

EXAMPLE 2 *Jeopardy.* Tell which integer corresponds to this situation: A contestant missed a \$600 question on the television game show "Jeopardy."

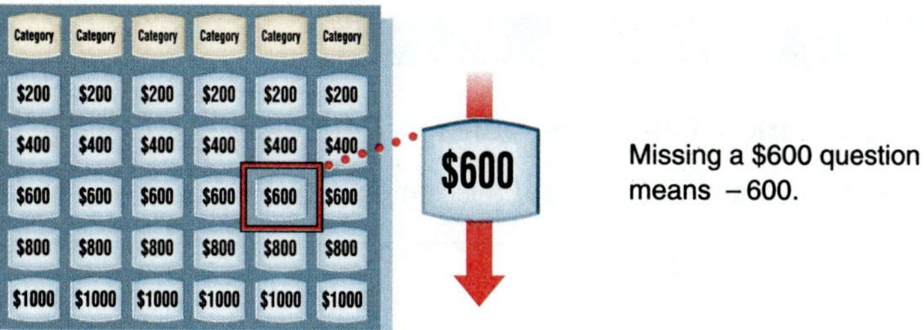

Missing a \$600 question causes a \$600 loss on the score—that is, the contestant earns -600 dollars.

State the integers that correspond to the given situation.

1. The halfback gained 8 yd on the first down. The quarterback was sacked for a 5-yd loss on the second down.

2. **Temperature High.** The highest man-made temperature on record is 950,000,000°F. It was created on May 27, 1994, at the Tokamak Fusion Test Reactor at the Princeton Plasma Physics Laboratory in New Jersey.
Source: *The Guinness Book of Records*

3. **Stock Decrease.** The price of Sherwin Williams Co. stock decreased from \$25 per share to \$19 per share over a recent time period.
Source: The New York Stock Exchange

4. At 10 sec before liftoff, ignition occurs. At 156 sec after liftoff, the first stage is detached from the rocket.

5. A submarine dove 120 ft, rose 50 ft, and then dove 80 ft.

Answers on page A-14

EXAMPLE 3 *Elevation.* Tell which integer corresponds to this situation: The lowest point in New Orleans is 8 ft below sea level.

The integer −8 corresponds to the situation. The elevation is −8 ft.

EXAMPLE 4 *Stock Price Change.* Tell which integers correspond to this situation: The price of Pearson Education stock decreased from \$24 per share to \$17 per share over a recent time period. The price of Sherwin Williams Co. stock increased from \$21 per share to \$25 per share over a recent time period.
Source: The New York Stock Exchange

The integer −7 corresponds to the decrease in the stock value. The integer 4 represents the increase in stock value.

Do Exercises 1–5.

b The Rational Numbers

We created the set of integers by obtaining a negative number for each natural number. To create a larger number system, called the set of **rational numbers,** we consider quotients of integers with nonzero divisors. The following are some examples of rational numbers:

$$\frac{2}{3},\quad -\frac{2}{3},\quad \frac{7}{1},\quad 4,\quad -3,\quad 0,\quad \frac{23}{-8},\quad 2.4,\quad -0.17,\quad 10\frac{1}{2}.$$

The number $-\frac{2}{3}$ (read "negative two-thirds") can also be named $\frac{2}{-3}$ or $\frac{-2}{3}$. The number 2.4 can be named $\frac{24}{10}$ or $\frac{12}{5}$, and −0.17 can be named $-\frac{17}{100}$.

Note that this new set of numbers, the rational numbers, contains the whole numbers, the integers, and the arithmetic numbers (also called the nonnegative rational numbers). We can describe the set of rational numbers as follows.

RATIONAL NUMBERS

The set of **rational numbers** = the set of numbers $\frac{a}{b}$, where a and b are integers and b is not equal to 0 ($b \neq 0$).

We picture the rational numbers on a number line as follows.

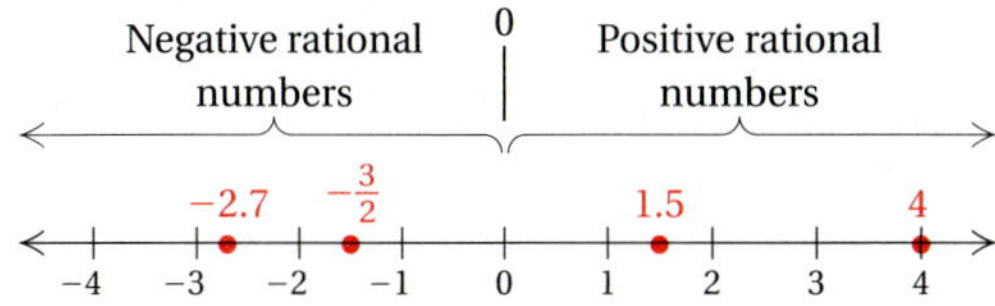

To **graph** a number means to find and mark its point on the number line. Some rational numbers are graphed in the preceding figure.

EXAMPLE 5 Graph: $\frac{5}{2}$.

The number $\frac{5}{2}$ can be named $2\frac{1}{2}$, or 2.5. Its graph is halfway between 2 and 3.

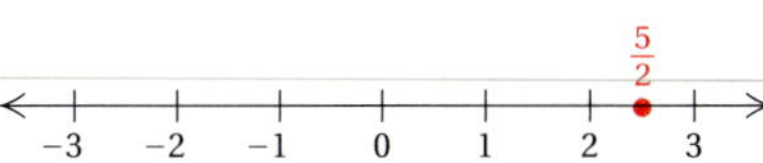

EXAMPLE 6 Graph: −3.2.

The graph of −3.2 is $\frac{2}{10}$ of the way from −3 to −4.

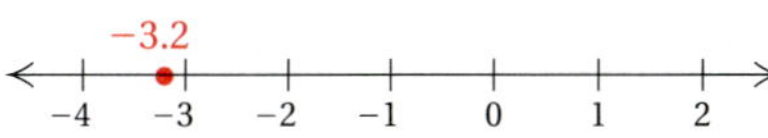

EXAMPLE 7 Graph: $\frac{13}{8}$.

The number $\frac{13}{8}$ can be named $1\frac{5}{8}$, or 1.625. The graph is about $\frac{6}{10}$ of the way from 1 to 2.

13/8

−3 −2 −1 0 1 2 3

Do Exercises 6–8.

C Notation for Rational Numbers

Each rational number can be named using fraction or decimal notation.

EXAMPLE 8 Convert to decimal notation: $-\frac{5}{8}$.

We first find decimal notation for $\frac{5}{8}$. Since $\frac{5}{8}$ means $5 \div 8$, we divide.

```
    0.6 2 5
8 ) 5.0 0 0
    4 8
      2 0
      1 6
        4 0
        4 0
          0
```

Thus, $\frac{5}{8} = 0.625$, so $-\frac{5}{8} = -0.625$.

Graph on a number line.

6. $-\frac{7}{2}$

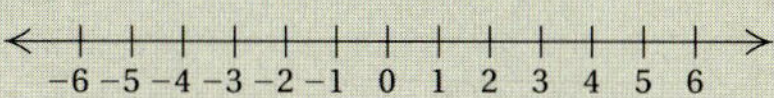

7. −1.4

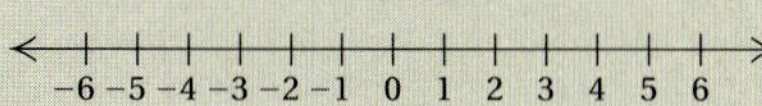

8. $\frac{11}{4}$

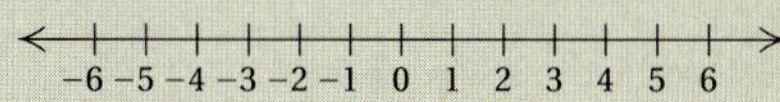

Answers on page A-14

CALCULATOR CORNER

Negative Numbers on a Calculator; Converting to Decimal Notation We use the opposite key [(−)] to enter negative numbers on a graphing calculator. Note that this is different from the [−] key, which is used for the operation of subtraction. To convert $-\frac{5}{8}$ to decimal notation, as in Example 8, we press [(−)] [5] [÷] [8] [ENTER]. The result is −0.625.

Exercises: Convert each of the following negative numbers to decimal notation.

1. $-\frac{3}{4}$

2. $-\frac{9}{20}$

3. $-\frac{1}{8}$

4. $-\frac{9}{5}$

5. $-\frac{27}{40}$

6. $-\frac{11}{16}$

7. $-\frac{7}{2}$

8. $-\frac{19}{25}$

Convert to decimal notation.

9. $-\frac{3}{8}$

10. $-\frac{6}{11}$

11. $\frac{4}{3}$

Answers on page A-14

CALCULATOR CORNER

Approximating Square Roots and π Square roots are found by pressing [2nd] [$\sqrt{\ }$]. ($\sqrt{\ }$ is the second operation associated with the [x^2] key.)

To find an approximation for $\sqrt{48}$, we press [2nd] [$\sqrt{\ }$] [4] [8] [ENTER]. The approximation 6.92820323 is displayed.

To find $8 \cdot \sqrt{13}$, we press [8] [2nd] [$\sqrt{\ }$] [1] [3] [ENTER]. The approximation 28.8444102 is displayed. The number π is used widely enough to have its own key. (π is the second operation associated with the [^] key.)

To approximate π, we press [2nd] [π] [ENTER]. The approximation 3.141592654 is displayed.

Exercises: Approximate.

1. $\sqrt{76}$
2. $\sqrt{317}$
3. $15 \cdot \sqrt{20}$
4. $29 + \sqrt{42}$
5. π
6. $29 \cdot \pi$
7. $\pi \cdot 13^2$
8. $5 \cdot \pi + 8 \cdot \sqrt{237}$

Decimal notation for $-\frac{5}{8}$ is -0.625. We consider -0.625 to be a **terminating decimal.** Decimal notation for some numbers repeats.

EXAMPLE 9 Convert to decimal notation: $\frac{7}{11}$.

We divide.

$$
\begin{array}{r}
0.6363\ldots \\
11\,\overline{)\,7.0000} \\
\underline{6\,6} \\
4\,0 \\
\underline{3\,3} \\
7\,0 \\
\underline{6\,6} \\
4\,0 \\
\underline{3\,3} \\
7
\end{array}
$$

We can abbreviate repeating decimal notation by writing a bar over the repeating part—in this case, $0.\overline{63}$. Thus, $\frac{7}{11} = 0.\overline{63}$.

The following are other examples to show how each rational number can be named using fraction or decimal notation:

$$0 = \frac{0}{8}, \quad \frac{27}{100} = 0.27, \quad -8\frac{3}{4} = -8.75, \quad -\frac{13}{6} = -2.1\overline{6}.$$

Do Exercises 9–11.

d The Real Numbers and Order

Every rational number has a point on the number line. However, there are some points on the line for which there is no rational number. These points correspond to what are called **irrational numbers.**

What kinds of numbers are irrational? One example is the number π, which is used in finding the area and the circumference of a circle: $A = \pi r^2$ and $C = 2\pi r$.

Another example of an irrational number is the square root of 2, named $\sqrt{2}$.

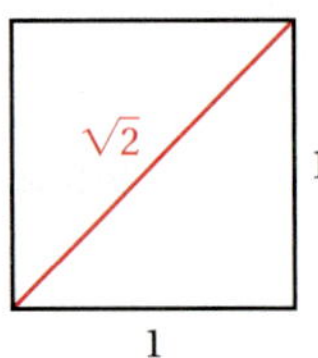

It is the length of the diagonal of a square with sides of length 1. It is also the number that when multiplied by itself gives 2, that is, $\sqrt{2} \cdot \sqrt{2} = 2$. There is no rational number that can be multiplied by itself to get 2. But the following are rational *approximations*:

1.4 is an approximation of $\sqrt{2}$ because $(1.4)^2 = 1.96$;

1.41 is a better approximation because $(1.41)^2 = 1.9881$;

1.4142 is an even better approximation because $(1.4142)^2 = 1.99996164$.

We can find rational approximations for square roots using a calculator.

Decimal notation for rational numbers *either* terminates *or* repeats. Decimal notation for irrational numbers *neither* terminates *nor* repeats. Some other examples of irrational numbers are $\sqrt{3}$, $-\sqrt{8}$, $\sqrt{11}$, and 0.121221222122221. . . . Whenever we take the square root of a number that is not a perfect square, we will get an irrational number.

The rational numbers and the irrational numbers together correspond to all the points on a number line and make up what is called the **real-number system.**

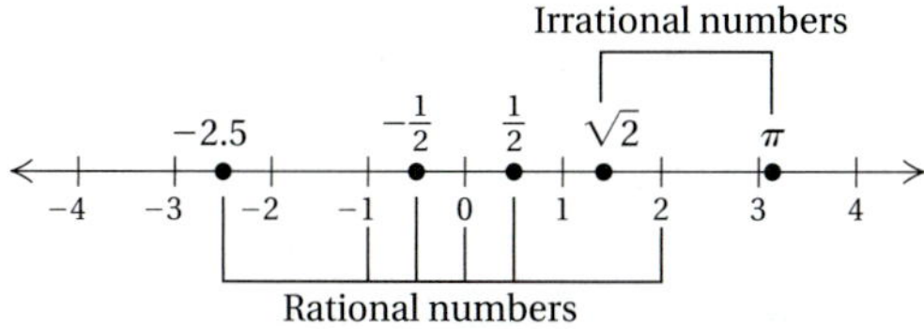

REAL NUMBERS

The set of **real numbers** = The set of all numbers corresponding to points on the number line.

The real numbers consist of the rational numbers and the irrational numbers. The following figure shows the relationships among various kinds of numbers.

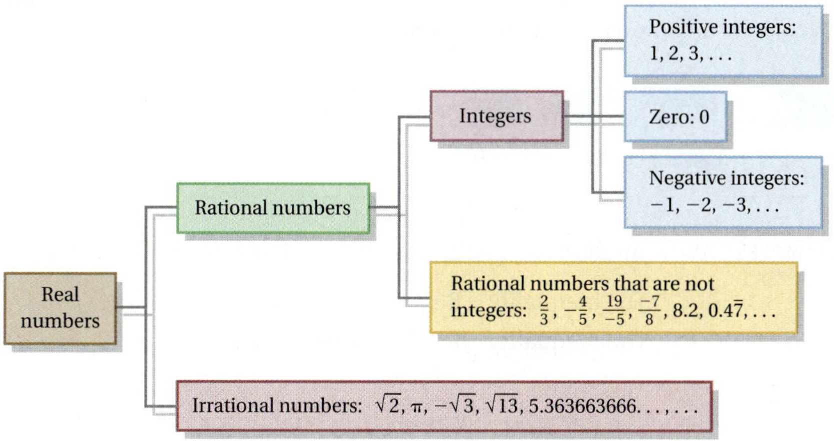

ORDER

Real numbers are named in order on the number line, with larger numbers named farther to the right. For any two numbers on the line, the one to the left is less than the one to the right.

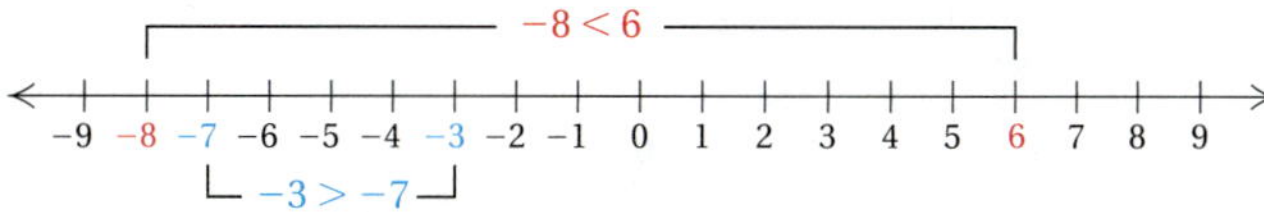

We use the symbol $<$ to mean "**is less than.**" The sentence $-8 < 6$ means "-8 is less than 6." The symbol $>$ means "**is greater than.**" The sentence $-3 > -7$ means "-3 is greater than -7." The sentences $-8 < 6$ and $-3 > -7$ are **inequalities.**

Use either $<$ or $>$ for $\square$ to write a true sentence.

12. $-3 \square 7$

13. $-8 \square -5$

14. $7 \square -10$

15. $3.1 \square -9.5$

16. $-\frac{2}{3} \square -1$

17. $-\frac{11}{8} \square \frac{23}{15}$

18. $-\frac{2}{3} \square -\frac{5}{9}$

19. $-4.78 \square -5.01$

Answers on page A-14

Write another inequality with the same meaning.

20. $-5 < 7$

21. $x > 4$

Write true or false.

22. $-4 \le -6$

23. $7.8 \ge 7.8$

24. $-2 \le \dfrac{3}{8}$

Answers on page A-14

EXAMPLES Use either < or > for □ to write a true sentence.

10. $2 \square 9$	Since 2 is to the left of 9, 2 is less than 9, so $2 < 9$.
11. $-7 \square 3$	Since -7 is to the left of 3, we have $-7 < 3$.
12. $6 \square -12$	Since 6 is to the right of -12, then $6 > -12$.
13. $-18 \square -5$	Since -18 is to the left of -5, we have $-18 < -5$.
14. $-2.7 \square -\frac{3}{2}$	The answer is $-2.7 < -\frac{3}{2}$.

−2.7 −3/2 1.5
−3 −2 −1 0 1 2 3 4

15. $1.5 \square -2.7$	The answer is $1.5 > -2.7$.
16. $1.38 \square 1.83$	The answer is $1.38 < 1.83$.
17. $-3.45 \square 1.32$	The answer is $-3.45 < 1.32$.
18. $-4 \square 0$	The answer is $-4 < 0$.
19. $5.8 \square 0$	The answer is $5.8 > 0$.
20. $\frac{5}{8} \square \frac{7}{11}$	We convert to decimal notation: $\frac{5}{8} = 0.625$ and $\frac{7}{11} = 0.6363\ldots$. Thus, $\frac{5}{8} < \frac{7}{11}$.

Do Exercises 12–19 on the preceding page.

Note that both $-8 < 6$ and $6 > -8$ are true. Every true inequality yields another true inequality when we interchange the numbers or variables and reverse the direction of the inequality sign.

ORDER; >, <

$a < b$ also has the meaning $b > a$.

EXAMPLES Write another inequality with the same meaning.

21. $-3 > -8$ The inequality $-8 < -3$ has the same meaning.

22. $a < -5$ The inequality $-5 > a$ has the same meaning.

A helpful mental device is to think of an inequality sign as an "arrow" with the arrow pointing to the smaller number.

Do Exercises 20 and 21.

Note that all positive real numbers are greater than zero and all negative real numbers are less than zero.

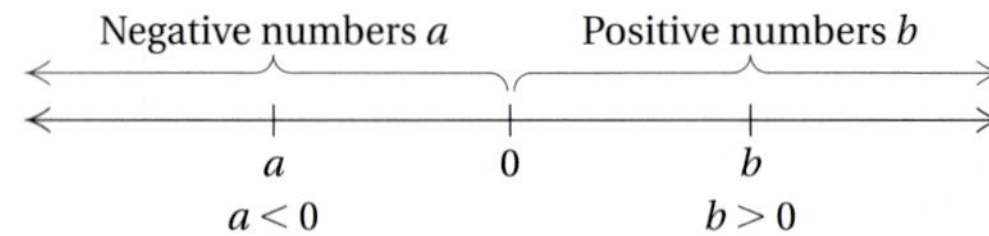

If b is a positive real number, then $b > 0$.
If a is a negative real number, then $a < 0$.

Expressions like $a \le b$ and $b \ge a$ are also inequalities. We read $\boldsymbol{a \le b}$ as "**a is less than or equal to b.**" We read $\boldsymbol{a \ge b}$ as "**a is greater than or equal to b.**"

EXAMPLES Write true or false for each statement.

23. $-3 \le 5.4$ True since $-3 < 5.4$ is true
24. $-3 \le -3$ True since $-3 = -3$ is true
25. $-5 \ge 1\frac{2}{3}$ False since neither $-5 > 1\frac{2}{3}$ nor $-5 = 1\frac{2}{3}$ is true

Do Exercises 22–24 on the preceding page.

e Absolute Value

From the number line, we see that numbers like 4 and -4 are the same distance from zero. Distance is always a nonnegative number. We call the distance of a number from zero on a number line the **absolute value** of the number.

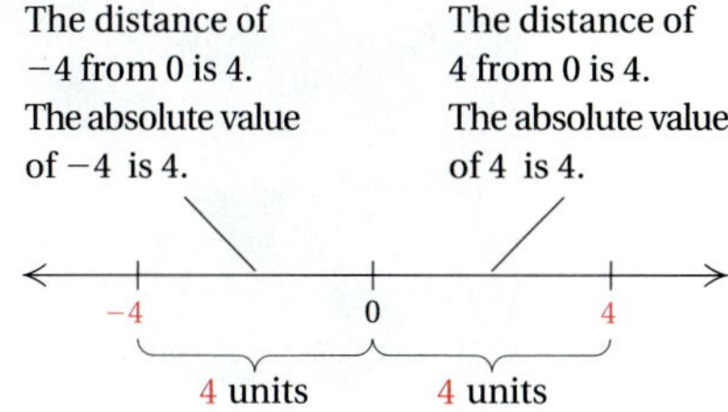

The distance of −4 from 0 is 4. The absolute value of −4 is 4.
The distance of 4 from 0 is 4. The absolute value of 4 is 4.
−4
0
4
4 units
4 units

ABSOLUTE VALUE

The **absolute value** of a number is its distance from zero on a number line. We use the symbol $|x|$ to represent the absolute value of a number x.

FINDING ABSOLUTE VALUE

a) If a number is negative, its absolute value is positive.
b) If a number is positive or zero, its absolute value is the same as the number.

EXAMPLES Find the absolute value.

26. $|-7|$ The distance of -7 from 0 is 7, so $|-7| = 7$.
27. $|12|$ The distance of 12 from 0 is 12, so $|12| = 12$.
28. $|0|$ The distance of 0 from 0 is 0, so $|0| = 0$.
29. $\left|\frac{3}{2}\right| = \frac{3}{2}$
30. $|-2.73| = 2.73$

Do Exercises 25–28.

Find the absolute value.

25. $|8|$
26. $|-9|$
27. $\left|-\frac{2}{3}\right|$
28. $|5.6|$

Answers on page A-14

CALCULATOR CORNER

Absolute Value The absolute-value operation is the first item in the Catalog on the TI-83 Plus graphing calculator. To find $|-7|$, as in Example 26, we first press [2nd] [CATALOG] [ENTER] to copy "abs(" to the home screen. (CATALOG is the second operation associated with the [0] numeric key.) Then we press [(−)] [7] [)] [ENTER]. The result is 7. To find $\left|-\frac{1}{2}\right|$ and express the result as a fraction, we press [2nd] [CATALOG] [ENTER] [(−)] [1] [÷] [2] [)] [MATH] [1] [ENTER]. The result is $\frac{1}{2}$.

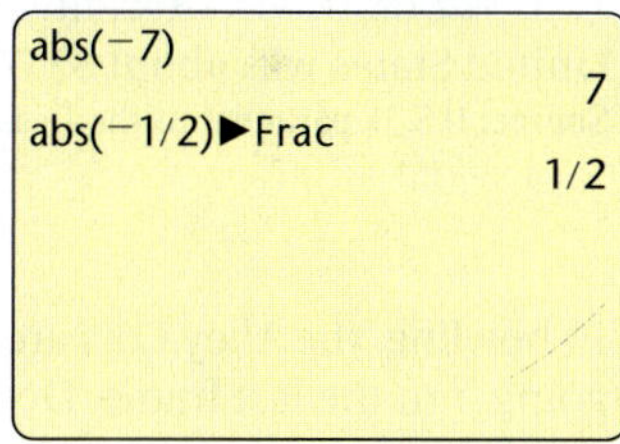

abs(−7)
7
abs(−1/2)►Frac
1/2

Exercises: Find the absolute value.

1. $|-5|$
2. $|17|$
3. $|0|$
4. $|6.48|$
5. $|-12.7|$
6. $|-0.9|$
7. $\left|-\frac{5}{7}\right|$
8. $\left|\frac{4}{3}\right|$

6.2 EXERCISE SET

For Extra Help

Digital Video Tutor CD 5 Videotape 7 | InterAct Math | Math Tutor Center | MathXL | MyMathLab

a State the integers that correspond to the situation.

1. *Elevations.* The Dead Sea, between Jordan and Israel, is 1286 ft below sea level. Mount Rainier in Washington State is 14,410 ft above sea level.
Sources: *The Handy Geography Answer Book; The New York Times Almanac*

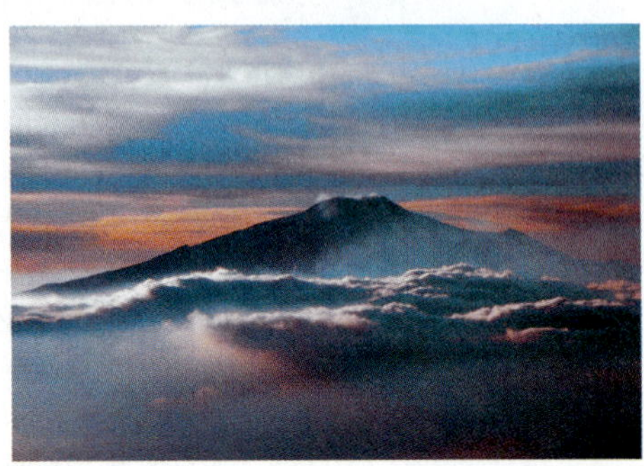

2. *Golf Score.* Tiger Woods' score in winning the 2000 PGA Championship was 18 under par.
Source: U.S. Golf Association

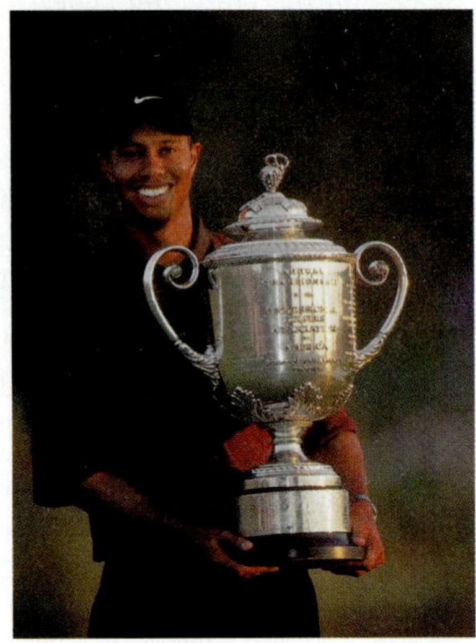

3. On Wednesday, the temperature was 24° above zero. On Thursday, it was 2° below zero.

4. A student deposited her tax refund of $750 in a savings account. Two weeks later, she withdrew $125 to pay sorority fees.

5. *U.S. Public Debt.* Recently, the total public debt of the United States was about $5,600,000,000,000.
Source: U.S. Department of the Treasury

6. *Birth and Death Rates.* Recently, the world birth rate was 270 per ten thousand. The death rate was 97 per ten thousand.
Source: United Nations Population Fund

7. In bowling, the Alley Cats are 34 pins behind the Strikers going into the last frame. Describe the situation of each team.

8. During a video game, Maggie intercepted a missile worth 20 points, lost a starship worth 150 points, and captured a landing base worth 300 points.

b Graph the number on the number line.

9. $\frac{10}{3}$

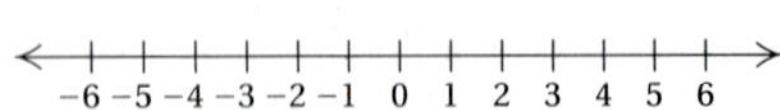

10. $-\frac{17}{4}$

−6 −5 −4 −3 −2 −1 0 1 2 3 4 5 6

11. -5.2

−6 −5 −4 −3 −2 −1 0 1 2 3 4 5 6

12. 4.78

−6 −5 −4 −3 −2 −1 0 1 2 3 4 5 6

c Convert to decimal notation.

13. $-\frac{7}{8}$

14. $-\frac{1}{8}$

15. $\frac{5}{6}$

16. $\frac{5}{3}$

17. $-\frac{7}{6}$

18. $-\frac{5}{12}$

19. $\frac{2}{3}$

20. $\frac{1}{4}$

21. $-\frac{1}{2}$

22. $\frac{5}{8}$

23. $\frac{1}{10}$

24. $-\frac{7}{20}$

d Use either $<$ or $>$ for $\square$ to write a true sentence.

25. $8 \square 0$

26. $3 \square 0$

27. $-8 \square 3$

28. $6 \square -6$

29. $-8 \square 8$

30. $0 \square -9$

31. $-8 \square -5$

32. $-4 \square -3$

33. $-5 \square -11$

34. $-3 \square -4$

35. $-6 \square -5$

36. $-10 \square -14$

37. $2.14 \square 1.24$

38. $-3.3 \square -2.2$

39. $-14.5 \square 0.011$

40. $17.2 \square -1.67$

41. $-12.88 \square -6.45$

42. $-14.34 \square -17.88$

43. $\frac{5}{12} \square \frac{11}{25}$

44. $-\frac{13}{16} \square -\frac{5}{9}$

Write true or false.

45. $-3 \geq -11$

46. $5 \leq -5$

47. $0 \geq 8$

48. $-5 \leq 7$

Write an inequality with the same meaning.

49. $-6 > x$

50. $x < 8$

51. $-10 \leq y$

52. $12 \geq t$

e Find the absolute value.

53. $|-3|$

54. $|-7|$

55. $|10|$

56. $|11|$

57. $|0|$

58. $|-4|$

59. $|-24|$

60. $|325|$

61. $\left|-\frac{2}{3}\right|$

62. $\left|-\frac{10}{7}\right|$

63. $\left|\frac{0}{4}\right|$

64. $|14.8|$

65. $\left|-3\frac{5}{8}\right|$

66. $\left|-7\frac{4}{5}\right|$

67. D_W When Jennifer's calculator gives a decimal approximation for $\sqrt{2}$ and that approximation is promptly squared, the result is 2. Yet, when that same approximation is entered by hand and then squared, the result is not exactly 2. Why do you suppose this happens?

68. D_W How many rational numbers are there between 0 and 1? Why?

SKILL MAINTENANCE

Find the LCM of the set of numbers. [1.9a]

69. 10, 35

70. 24, 42

71. 4, 10, 18

Add and simplify, if possible. [2.3a]

72. $\frac{1}{2} + \frac{2}{3}$

73. $\frac{5}{8} + \frac{7}{12}$

74. $\frac{7}{4} + \frac{3}{10}$

SYNTHESIS

List in order from the least to the greatest.

75. $-\frac{2}{3}, \frac{1}{2}, -\frac{3}{4}, -\frac{5}{6}, \frac{3}{8}, \frac{1}{6}$

76. $-8\frac{7}{8}, 7^1, -5, |-6|, 4, |3|, -8\frac{5}{8}, -100, 0, 1^7, \frac{14}{4}, -\frac{67}{8}$

Given that $0.\overline{3} = \frac{1}{3}$ and $0.\overline{6} = \frac{2}{3}$, express each of the following as a quotient or ratio of two integers.

77. $0.\overline{1}$

78. $0.\overline{9}$

79. $5.\overline{5}$

6.3 ADDITION OF REAL NUMBERS

In this section, we consider addition of real numbers. First, to gain an understanding, we add using a number line. Then we consider rules for addition.

ADDITION ON A NUMBER LINE

To do the addition $a + b$ on a number line, we start at 0. Then we move to a and then move according to b.

a) If b is positive, we move to the right.
b) If b is negative, we move to the left.
c) If b is 0, we stay at a.

EXAMPLE 1 Add: $3 + (-5)$.

We start at 0 and move 3 units right since 3 is positive. Then we move 5 units left since -5 is negative.

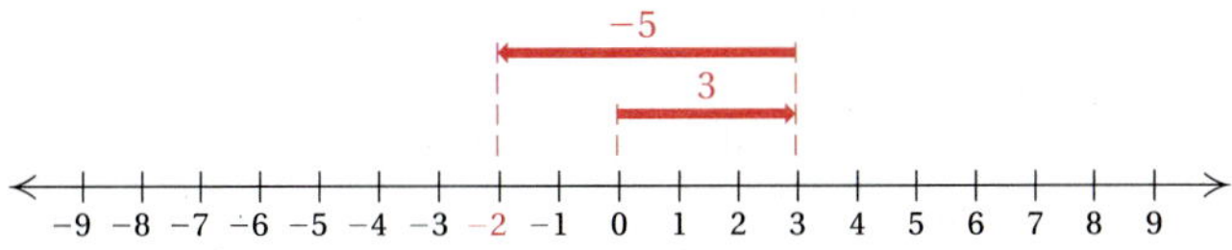

$3 + (-5) = -2$

EXAMPLE 2 Add: $-4 + (-3)$.

We start at 0 and move 4 units left since -4 is negative. Then we move 3 units further left since -3 is negative.

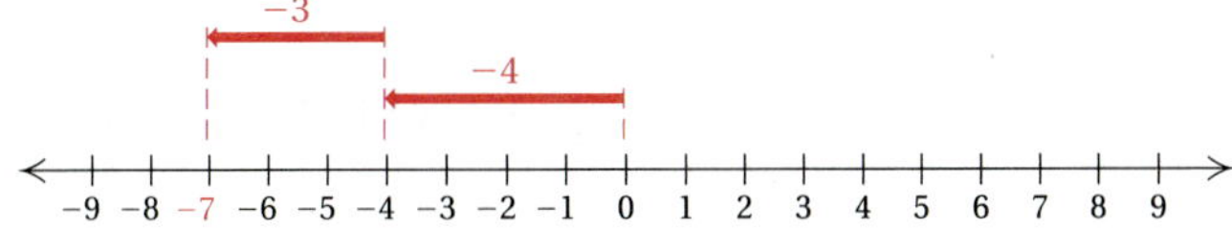

$-4 + (-3) = -7$

EXAMPLE 3 Add: $-4 + 9$.

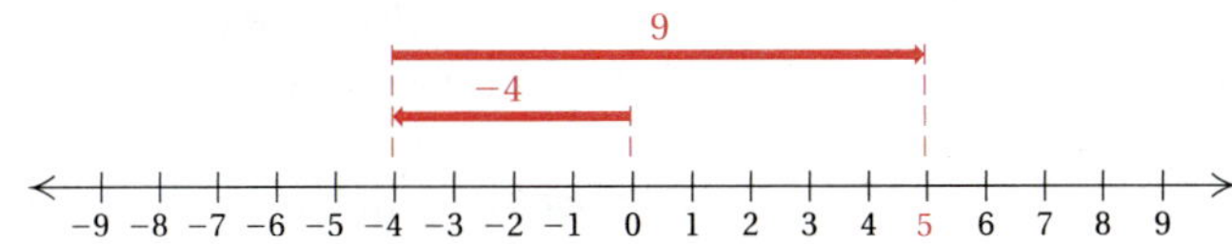

$-4 + 9 = 5$

Objectives

a Add real numbers without using a number line.

b Find the opposite, or additive inverse, of a real number.

c Solve applied problems involving addition of real numbers.

Add using a number line.

1. $0 + (-3)$

2. $1 + (-4)$

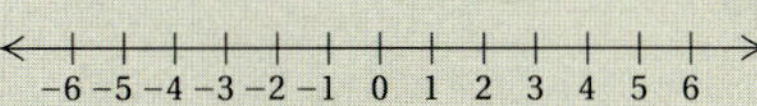

3. $-3 + (-2)$

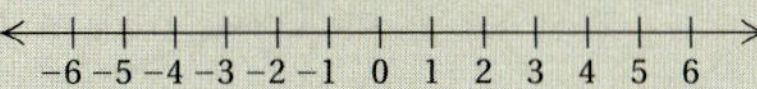

4. $-3 + 7$

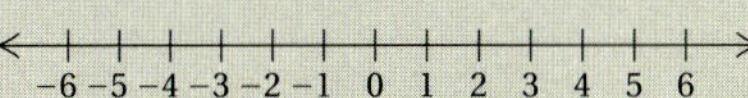

5. $-2.4 + 2.4$

6. $-\frac{5}{2} + \frac{1}{2}$

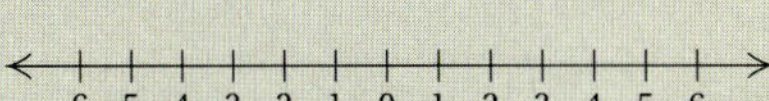

Answers on page A-15

Add without using a number line.

7. $-5 + (-6)$

8. $-9 + (-3)$

9. $-4 + 6$

10. $-7 + 3$

11. $5 + (-7)$

12. $-20 + 20$

13. $-11 + (-11)$

14. $10 + (-7)$

15. $-0.17 + 0.7$

16. $-6.4 + 8.7$

17. $-4.5 + (-3.2)$

18. $-8.6 + 2.4$

19. $\frac{5}{9} + \left(-\frac{7}{9}\right)$

20. $-\frac{1}{5} + \left(-\frac{3}{4}\right)$

Answers on page A-15

EXAMPLE 4 Add: $-5.2 + 0$.

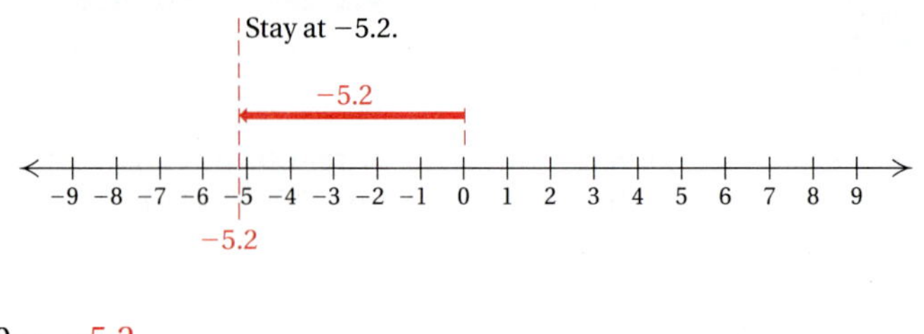

$-5.2 + 0 = -5.2$

Do Exercises 1–6 on the preceding page.

a Adding Without a Number Line

You may have noticed some patterns in the preceding examples. These lead us to rules for adding without using a number line that are more efficient for adding larger numbers.

RULES FOR ADDITION OF REAL NUMBERS

1. *Positive numbers*: Add the same as arithmetic numbers. The answer is positive.
2. *Negative numbers*: Add absolute values. The answer is negative.
3. *A positive and a negative number*: Subtract the smaller absolute value from the larger. Then:
   a) If the positive number has the greater absolute value, the answer is positive.
   b) If the negative number has the greater absolute value, the answer is negative.
   c) If the numbers have the same absolute value, the answer is 0.
4. *One number is zero*: The sum is the other number.

Rule 4 is known as the **identity property of 0.** It says that for any real number a, $a + 0 = a$.

EXAMPLES Add without using a number line.

5. $-12 + (-7) = -19$ Two negatives. Add the absolute values: $|-12| + |-7| = 12 + 7 = 19$. Make the answer *negative*: -19.

6. $-1.4 + 8.5 = 7.1$ One negative, one positive. Find the absolute values: $|-1.4| = 1.4$; $|8.5| = 8.5$. Subtract the smaller absolute value from the larger: $8.5 - 1.4 = 7.1$. The *positive* number, 8.5, has the larger absolute value, so the answer is *positive*, 7.1.

7. $-36 + 21 = -15$ One negative, one positive. Find the absolute values: $|-36| = 36$, $|21| = 21$. Subtract the smaller absolute value from the larger: $36 - 21 = 15$. The negative number, -36, has the larger absolute value, so the answer is *negative*, -15.

8. $1.5 + (-1.5) = 0$ The numbers have the same absolute value. The sum is 0.

9. $-\frac{7}{8} + 0 = -\frac{7}{8}$ One number is zero. The sum is $-\frac{7}{8}$.

10. $-9.2 + 3.1 = -6.1$

11. $-\frac{3}{2} + \frac{9}{2} = \frac{6}{2} = 3$

12. $-\frac{2}{3} + \frac{5}{8} = -\frac{16}{24} + \frac{15}{24} = -\frac{1}{24}$

Do Exercises 7–20 on the preceding page.

Suppose we want to add several numbers, some positive and some negative, as follows. How can we proceed?

$$15 + (-2) + 7 + 14 + (-5) + (-12)$$

We can change grouping and order as we please when adding. For instance, we can group the positive numbers together and the negative numbers together and add them separately. Then we add the two results.

EXAMPLE 13 Add: $15 + (-2) + 7 + 14 + (-5) + (-12)$.

a) $15 + 7 + 14 = 36$ Adding the positive numbers

b) $-2 + (-5) + (-12) = -19$ Adding the negative numbers

$36 + (-19) = 17$ Adding (a) and (b)

We can also add the numbers in any other order we wish, say, from left to right as follows:

$$\begin{aligned} 15 + (-2) + 7 + 14 + (-5) + (-12) &= 13 + 7 + 14 + (-5) + (-12) \\ &= 20 + 14 + (-5) + (-12) \\ &= 34 + (-5) + (-12) \\ &= 29 + (-12) \\ &= 17 \end{aligned}$$

Do Exercises 21–24.

b Opposites, or Additive Inverses

Suppose we add two numbers that are **opposites,** such as 6 and −6. The result is 0. When opposites are added, the result is always 0. Such numbers are also called **additive inverses.** Every real number has an opposite, or additive inverse.

OPPOSITES, OR ADDITIVE INVERSES

Two numbers whose sum is 0 are called **opposites,** or **additive inverses,** of each other.

Add.

21. $(-15) + (-37) + 25 + 42 + (-59) + (-14)$

22. $42 + (-81) + (-28) + 24 + 18 + (-31)$

23. $-2.5 + (-10) + 6 + (-7.5)$

24. $-35 + 17 + 14 + (-27) + 31 + (-12)$

Find the opposite, or additive inverse, of each of the following.

25. -4

26. 8.7

27. -7.74

28. $-\frac{8}{9}$

29. 0

30. 12

Answers on page A-15

Evaluate $-x$ and $-(-x)$ when:

31. $x = 14$.

32. $x = 1$.

33. $x = -19$.

34. $x = -1.6$.

35. $x = \frac{2}{3}$.

36. $x = -\frac{9}{8}$.

Answers on page A-15

EXAMPLES Find the opposite, or additive inverse, of each number.

14. 34 The opposite of 34 is -34 because $34 + (-34) = 0$.

15. -8 The opposite of -8 is 8 because $-8 + 8 = 0$.

16. 0 The opposite of 0 is 0 because $0 + 0 = 0$.

17. $-\frac{7}{8}$ The opposite of $-\frac{7}{8}$ is $\frac{7}{8}$ because $-\frac{7}{8} + \frac{7}{8} = 0$.

Do Exercises 25–30 on the preceding page.

To name the opposite, we use the symbol $-$, as follows.

SYMBOLIZING OPPOSITES

The opposite, or additive inverse, of a number a can be named $-a$ (read "the opposite of a," or "the additive inverse of a").

Note that if we take a number, say, 8, and find its opposite, -8, and then find the opposite of the result, we will have the original number, 8, again.

THE OPPOSITE OF AN OPPOSITE

The **opposite of the opposite** of a number is the number itself. (The additive inverse of the additive inverse of a number is the number itself.) That is, for any number a,

$$-(-a) = a.$$

EXAMPLE 18 Evaluate $-x$ and $-(-x)$ when $x = 16$.

If $x = 16$, then $-x = -16$. The opposite of 16 is -16.

If $x = 16$, then $-(-x) = -(-16) = 16$. The opposite of the opposite of 16 is 16.

EXAMPLE 19 Evaluate $-x$ and $-(-x)$ when $x = -3$.

If $x = -3$, then $-x = -(-3) = 3$.

If $x = -3$, then $-(-x) = -(-(-3)) = -3$.

Note that in Example 19 we used a second set of parentheses to show that we are substituting the negative number -3 for x. Symbolism like $--x$ is not considered meaningful.

Do Exercises 31–36.

A symbol such as -8 is usually read "negative 8." It could be read "the additive inverse of 8," because the additive inverse of 8 is negative 8. It could also be read "the opposite of 8," because the opposite of 8 is -8. Thus a symbol like -8 can be read in more than one way. It is never correct to read -8 as "minus 8."

CAUTION!

A symbol like $-x$, which has a variable, should be read "the opposite of x" or "the additive inverse of x" and *not* "negative x," because we do not know whether x represents a positive number, a negative number, or 0. You can check this in Examples 18 and 19.

We can use the symbolism $-a$ to restate the definition of opposite, or additive inverse.

THE SUM OF OPPOSITES

For any real number a, the **opposite,** or **additive inverse,** of a, expressed as $-a$, is such that

$$a + (-a) = (-a) + a = 0.$$

SIGNS OF NUMBERS

A negative number is sometimes said to have a "negative sign." A positive number is said to have a "positive sign." When we replace a number with its opposite, we can say that we have "changed its sign."

EXAMPLES Find the opposite. (Change the sign.)

20. -3 $\quad -(-3) = 3$

21. -10 $\quad -(-10) = 10$

22. 0 $\quad -(0) = 0$

23. 14 $\quad -(14) = -14$

Do Exercises 37–40.

C Applications and Problem Solving

Addition of real numbers occurs in many real-world situations.

EXAMPLE 24 *Lake Level.* In the course of one four-month period, the water level of Lake Champlain went down 2 ft, up 1 ft, down 5 ft, and up 3 ft. How much had the lake level changed at the end of the four months?

We let $T =$ the total change in the level of the lake. Then the problem translates to a sum:

Total change	is	1st change	plus	2nd change	plus	3rd change	plus	4th change.
↓	↓	↓	↓	↓	↓	↓	↓	↓
T	$=$	-2	$+$	1	$+$	(-5)	$+$	3

Adding from left to right, we have

$$\begin{aligned} T = -2 + 1 + (-5) + 3 &= -1 + (-5) + 3 \\ &= -6 + 3 \\ &= -3. \end{aligned}$$

The lake level has dropped 3 ft at the end of the four-month period.

Do Exercise 41.

Find the opposite. (Change the sign.)

37. -4

38. -13.4

39. 0

40. $\frac{1}{4}$

41. **Class Size.** There were 27 students in Eliza's algebra class when the semester began. During the first two weeks, 5 students withdrew, 8 students enrolled in the class, and 6 students were dropped as "no shows." How many were in the class after two weeks?

Answers on page A-15

6.3 EXERCISE SET

For Extra Help

Digital Video Tutor CD 5 Videotape 7 | InterAct Math | Math Tutor Center | MathXL | MyMathLab

a Add. Do not use a number line except as a check.

1. $2 + (-9)$

2. $-5 + 2$

3. $-11 + 5$

4. $4 + (-3)$

5. $-6 + 6$

6. $8 + (-8)$

7. $-3 + (-5)$

8. $-4 + (-6)$

9. $-7 + 0$

10. $-13 + 0$

11. $0 + (-27)$

12. $0 + (-35)$

13. $17 + (-17)$

14. $-15 + 15$

15. $-17 + (-25)$

16. $-24 + (-17)$

17. $18 + (-18)$

18. $-13 + 13$

19. $-28 + 28$

20. $11 + (-11)$

21. $8 + (-5)$

22. $-7 + 8$

23. $-4 + (-5)$

24. $10 + (-12)$

25. $13 + (-6)$

26. $-3 + 14$

27. $-25 + 25$

28. $50 + (-50)$

29. $53 + (-18)$

30. $75 + (-45)$

31. $-8.5 + 4.7$

32. $-4.6 + 1.9$

33. $-2.8 + (-5.3)$

34. $-7.9 + (-6.5)$

35. $-\frac{3}{5} + \frac{2}{5}$

36. $-\frac{4}{3} + \frac{2}{3}$

37. $-\frac{2}{9} + \left(-\frac{5}{9}\right)$

38. $-\frac{4}{7} + \left(-\frac{6}{7}\right)$

39. $-\frac{5}{8} + \frac{1}{4}$

40. $-\frac{5}{6} + \frac{2}{3}$

41. $-\frac{5}{8} + \left(-\frac{1}{6}\right)$

42. $-\frac{5}{6} + \left(-\frac{2}{9}\right)$

43. $-\frac{3}{8} + \frac{5}{12}$

44. $-\frac{7}{16} + \frac{7}{8}$

45. $76 + (-15) + (-18) + (-6)$

46. $29 + (-45) + 18 + 32 + (-96)$

47. $-44 + \left(-\frac{3}{8}\right) + 95 + \left(-\frac{5}{8}\right)$

48. $24 + 3.1 + (-44) + (-8.2) + 63$

49. $98 + (-54) + 113 + (-998) + 44 + (-612)$

50. $-458 + (-124) + 1025 + (-917) + 218$

b Find the opposite, or additive inverse.

51. 24

52. -64

53. -26.9

54. 48.2

Evaluate $-x$ when:

55. $x = 8$.

56. $x = -27$.

57. $x = -\frac{13}{8}$.

58. $x = \frac{1}{236}$.

Evaluate $-(-x)$ when:

59. $x = -43$.

60. $x = 39$.

61. $x = \frac{4}{3}$.

62. $x = -7.1$.

Find the opposite. (Change the sign.)

63. -24

64. -12.3

65. $-\frac{3}{8}$

66. 10

Solve.

67. *Tallest Mountain.* The tallest mountain in the world, when measured from base to peak, is Mauna Kea (White Mountain) in Hawaii. From its base 19,684 ft below sea level in the Hawaiian Trough, it rises 33,480 ft. What is the elevation of the peak above sea level?
Source: *The Guinness Book of Records*

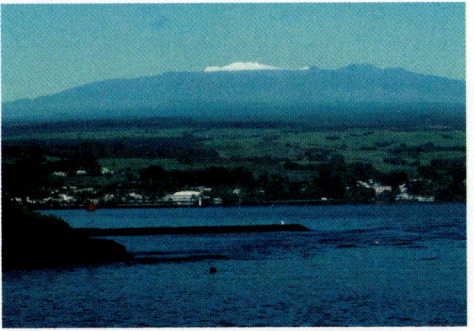

68. *Telephone Bills.* Erika's cell-phone bill for July was \$82. She sent a check for \$50 and then made \$37 worth of calls in August. How much did she then owe on her cell-phone bill?

69. *Temperature Changes.* One day the temperature in Lawrence, Kansas, is 32°F at 6:00 A.M. It rises 15° by noon, but falls 50° by midnight when a cold front moves in. What is the final temperature?

70. *Stock Changes.* On a recent day, the price of Quaker Oats stock opened at a value of \$61.38. During the day, it rose \$4.75, dropped \$7.38, and rose \$5.13. Find the value of the stock at the end of the day.
Source: The New York Stock Exchange

71. *Profits and Losses.* A business expresses a profit as a positive number and refers to it as operating "in the black." A loss is expressed as a negative number and is referred to as operating "in the red." The profits and losses of Xponent Corporation over various years are shown in the bar graph below. Find the sum of the profits and losses.

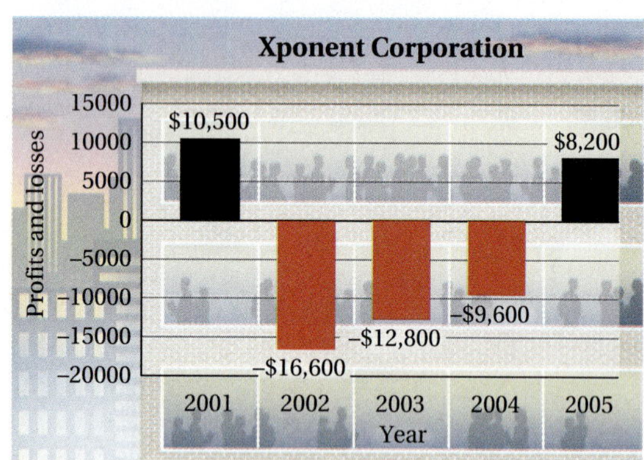

72. *Football Yardage.* In a college football game, the quarterback attempted passes with the following results. Find the total gain or loss.

TRY	GAIN OR LOSS
1st	13-yd gain
2nd	12-yd loss
3rd	21-yd gain

73. *Account Balance.* Leah has $460 in a checking account. She writes a check for $530, makes a deposit of $75, and then writes a check for $90. What is the balance in her account?

74. *Credit Card Bills.* On August 1, Lyle's credit card bill shows that he owes $470. During the month of August, Lyle sends a check for $45 to the credit card company, charges another $160 in merchandise, and then pays off another $500 of his bill. What is the new balance of Lyle's account at the end of August?

75. ^{D}W Without actually performing the addition, explain why the sum of all integers from -50 to 50 is 0.

76. ^{D}W Explain in your own words why the sum of two negative numbers is always negative.

SKILL MAINTENANCE

Round 87,452 to the nearest: [1.3e]

77. Ten.

78. Hundred.

79. Thousand.

Subtract and simplify, if possible. [2.3b]

80. $\frac{12}{5} - \frac{7}{10}$

81. $\frac{9}{4} - \frac{5}{6}$

82. $\frac{7}{8} - \frac{1}{3}$

SYNTHESIS

83. For what numbers x is $-x$ negative?

84. For what numbers x is $-x$ positive?

For each of Exercises 85 and 86, choose the correct answer from the selections given.

85. If a is positive and b is negative, then $-a + b$ is:

a) Positive.
b) Negative.
c) 0.
d) Cannot be determined without more information

86. If $a = b$ and a and b are negative, then $-a + (-b)$ is:

a) Positive.
b) Negative.
c) 0.
d) Cannot be determined without more information

6.4 SUBTRACTION OF REAL NUMBERS

Objectives

a Subtract real numbers and simplify combinations of additions and subtractions.

b Solve applied problems involving subtraction of real numbers.

a Subtraction

We now consider subtraction of real numbers.

SUBTRACTION

The difference $a - b$ is the number c for which $a = b + c$.

Consider, for example, $45 - 17$. *Think*: What number can we add to 17 to get 45? Since $45 = 17 + 28$, we know that $45 - 17 = 28$. Let's consider an example whose answer is a negative number.

EXAMPLE 1 Subtract: $3 - 7$.

Think: What number can we add to 7 to get 3? The number must be negative. Since $7 + (-4) = 3$, we know the number is -4: $3 - 7 = -4$. That is, $3 - 7 = -4$ because $7 + (-4) = 3$.

Do Exercises 1–3.

The definition above does not provide the most efficient way to do subtraction. We can develop a faster way to subtract. As a rationale for the faster way, let's compare $3 + 7$ and $3 - 7$ on a number line.

To find $3 + 7$ on a number line, we move 3 units to the right from 0 since 3 is positive. Then we move 7 units farther to the right since 7 is positive.

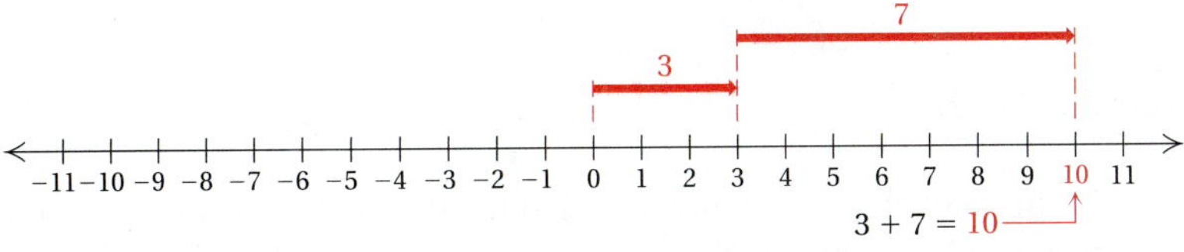

To find $3 - 7$, we do the "opposite" of adding 7: We move 7 units to the *left* to do the subtracting. This is the same as *adding* the opposite of 7, -7, to 3.

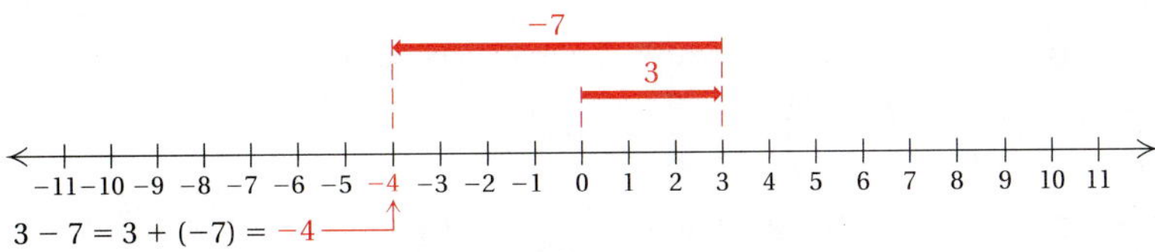

Do Exercises 4–6.

Look for a pattern in the examples shown at right.

SUBTRACTIONS	ADDING AN OPPOSITE
$5 - 8 = -3$	$5 + (-8) = -3$
$-6 - 4 = -10$	$-6 + (-4) = -10$
$-7 - (-2) = -5$	$-7 + 2 = -5$

Subtract.

1. $-6 - 4$

Think: What number can be added to 4 to get -6:

$$\square + 4 = -6?$$

2. $-7 - (-10)$

Think: What number can be added to -10 to get -7:

$$\square + (-10) = -7?$$

3. $-7 - (-2)$

Think: What number can be added to -2 to get -7:

$$\square + (-2) = -7?$$

Subtract. Use a number line, doing the "opposite" of addition.

4. $-4 - (-3)$

−6 −5 −4 −3 −2 −1 0 1 2 3 4 5 6

5. $-4 - (-6)$

6. $5 - 9$

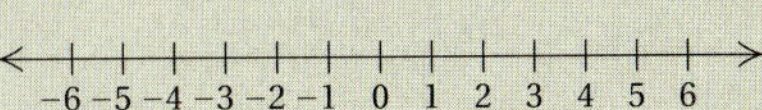

Answers on page A-15

Complete the addition and compare with the subtraction.

7. $4 - 6 = -2$;
$4 + (-6) =$ ______

8. $-3 - 8 = -11$;
$-3 + (-8) =$ ______

9. $-5 - (-9) = 4$;
$-5 + 9 =$ ______

10. $-5 - (-3) = -2$;
$-5 + 3 =$ ______

Subtract.

11. $2 - 8$

12. $-6 - 10$

13. $12.4 - 5.3$

14. $-8 - (-11)$

15. $-8 - (-8)$

16. $\frac{2}{3} - \left(-\frac{5}{6}\right)$

Answers on page A-15

Do Exercises 7–10.

Perhaps you have noticed that we can subtract by adding the opposite of the number being subtracted. This can always be done.

SUBTRACTING BY ADDING THE OPPOSITE

For any real numbers a and b,

$$a - b = a + (-b).$$

(To subtract, add the opposite, or additive inverse, of the number being subtracted.)

This is the method generally used for quick subtraction of real numbers.

EXAMPLES Subtract.

2. $2 - 6 = 2 + (-6) = -4$ — The opposite of 6 is -6. We change the subtraction to addition and add the opposite. *Check*: $-4 + 6 = 2$.

3. $4 - (-9) = 4 + 9 = 13$ — The opposite of -9 is 9. We change the subtraction to addition and add the opposite. *Check*: $13 + (-9) = 4$.

4. $-4.2 - (-3.6) = -4.2 + 3.6 = -0.6$ — Adding the opposite. *Check*: $-0.6 + (-3.6) = -4.2$.

5. $-\frac{1}{2} - \left(-\frac{3}{4}\right) = -\frac{1}{2} + \frac{3}{4} = \frac{1}{4}$ — Adding the opposite. *Check*: $\frac{1}{4} + \left(-\frac{3}{4}\right) = -\frac{1}{2}$.

Do Exercises 11–16.

EXAMPLES Read each of the following. Then subtract by adding the opposite of the number being subtracted.

6. $3 - 5$ — Read "three minus five is three plus the opposite of five"

$3 - 5 = 3 + (-5) = -2$

7. $\frac{1}{8} - \frac{7}{8}$ — Read "one-eighth minus seven-eighths is one-eighth plus the opposite of seven-eighths"

$\frac{1}{8} - \frac{7}{8} = \frac{1}{8} + \left(-\frac{7}{8}\right) = -\frac{6}{8}$, or $-\frac{3}{4}$

8. $-4.6 - (-9.8)$ — Read "negative four point six minus negative nine point eight is negative four point six plus the opposite of negative nine point eight"

$-4.6 - (-9.8) = -4.6 + 9.8 = 5.2$

9. $-\frac{3}{4} - \frac{7}{5}$ — Read "negative three-fourths minus seven-fifths is negative three-fourths plus the opposite of seven-fifths"

$-\frac{3}{4} - \frac{7}{5} = -\frac{3}{4} + \left(-\frac{7}{5}\right) = -\frac{15}{20} + \left(-\frac{28}{20}\right) = -\frac{43}{20}$

Do Exercises 17–21 on the following page.

When several additions and subtractions occur together, we can make them all additions.

EXAMPLES Simplify.

10. $8 - (-4) - 2 - (-4) + 2 = 8 + 4 + (-2) + 4 + 2$ Adding the opposite
$= 16$

11. $8.2 - (-6.1) + 2.3 - (-4) = 8.2 + 6.1 + 2.3 + 4 = 20.6$

Do Exercises 22–24.

b Applications and Problem Solving

Let's now see how we can use subtraction of real numbers to solve applied problems.

EXAMPLE 12 *Temperatures on Mars.* The Viking 2 Lander spacecraft has determined that temperatures on Mars range from −125°C (Celsius) to 25°C. Find the difference between the highest value and the lowest value in this temperature range.
Source: The Lunar and Planetary Institute

We let D = the difference in the temperatures. Then the problem translates to the following subtraction:

Difference in temperature	is	Highest temperature	minus	Lowest temperature.
↓	↓	↓	↓	↓
D	$=$	25	$-$	(-125)
D	$= 25 + 125 = 150$			

The difference in the temperatures is 150°C.

Do Exercise 25.

Read each of the following. Then subtract by adding the opposite of the number being subtracted.

17. $3 - 11$

18. $12 - 5$

19. $-12 - (-9)$

20. $-12.4 - 10.9$

21. $-\frac{4}{5} - \left(-\frac{4}{5}\right)$

Simplify.

22. $-6 - (-2) - (-4) - 12 + 3$

23. $9 - (-6) + 7 - 11 - 14 - (-20)$

24. $-9.6 + 7.4 - (-3.9) - (-11)$

25. Temperature Extremes. The highest temperature ever recorded in the United States is 134°F in Greenland Ranch, California, on July 10, 1913. The lowest temperature ever recorded is −80°F in Prospect Creek, Alaska, on January 23, 1971. How much higher was the temperature in Greenland Ranch than that in Prospect Creek?
Source: National Oceanographic and Atmospheric Administration

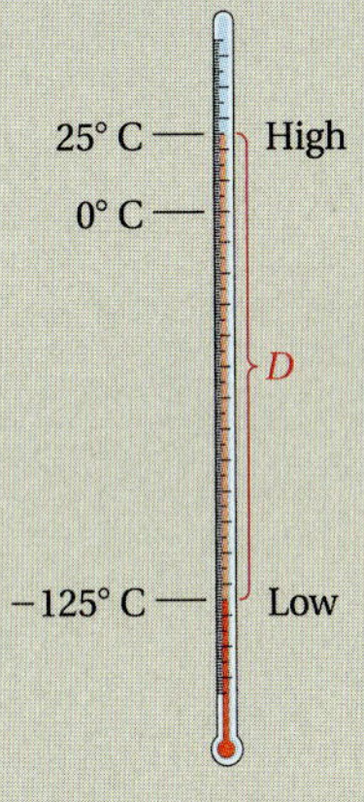

Answers on page A-15

6.4 EXERCISE SET

For Extra Help

Digital Video Tutor CD 5 Videotape 7

InterAct Math

Math Tutor Center

MathXL

MyMathLab

a Subtract.

1. $2 - 9$

2. $3 - 8$

3. $0 - 4$

4. $0 - 9$

5. $-8 - (-2)$

6. $-6 - (-8)$

7. $-11 - (-11)$

8. $-6 - (-6)$

9. $12 - 16$

10. $14 - 19$

11. $20 - 27$

12. $30 - 4$

13. $-9 - (-3)$

14. $-7 - (-9)$

15. $-40 - (-40)$

16. $-9 - (-9)$

17. $7 - 7$

18. $9 - 9$

19. $7 - (-7)$

20. $4 - (-4)$

21. $8 - (-3)$

22. $-7 - 4$

23. $-6 - 8$

24. $6 - (-10)$

25. $-4 - (-9)$

26. $-14 - 2$

27. $1 - 8$

28. $2 - 8$

29. $-6 - (-5)$

30. $-4 - (-3)$

31. $8 - (-10)$

32. $5 - (-6)$

33. $0 - 10$

34. $0 - 18$

35. $-5 - (-2)$

36. $-3 - (-1)$

37. $-7 - 14$

38. $-9 - 16$

39. $0 - (-5)$

40. $0 - (-1)$

41. $-8 - 0$

42. $-9 - 0$

43. $7 - (-5)$

44. $7 - (-4)$

45. $2 - 25$

46. $18 - 63$

47. $-42 - 26$

48. $-18 - 63$

49. $-71 - 2$

50. $-49 - 3$

51. $24 - (-92)$

52. $48 - (-73)$

53. $-50 - (-50)$

54. $-70 - (-70)$

55. $-\frac{3}{8} - \frac{5}{8}$

56. $\frac{3}{9} - \frac{9}{9}$

57. $\frac{3}{4} - \frac{2}{3}$

58. $\frac{5}{8} - \frac{3}{4}$

59. $-\frac{3}{4} - \frac{2}{3}$

60. $-\frac{5}{8} - \frac{3}{4}$

61. $-\frac{5}{8} - \left(-\frac{3}{4}\right)$

62. $-\frac{3}{4} - \left(-\frac{2}{3}\right)$

63. $6.1 - (-13.8)$

64. $1.5 - (-3.5)$

65. $-2.7 - 5.9$

66. $-3.2 - 5.8$

67. $0.99 - 1$

68. $0.87 - 1$

69. $-79 - 114$

70. $-197 - 216$

71. $0 - (-500)$

72. $500 - (-1000)$

73. $-2.8 - 0$

74. $6.04 - 1.1$

75. $7 - 10.53$

76. $8 - (-9.3)$

77. $\frac{1}{6} - \frac{2}{3}$

78. $-\frac{3}{8} - \left(-\frac{1}{2}\right)$

79. $-\frac{4}{7} - \left(-\frac{10}{7}\right)$

80. $\frac{12}{5} - \frac{12}{5}$

81. $-\frac{7}{10} - \frac{10}{15}$

82. $-\frac{4}{18} - \left(-\frac{2}{9}\right)$

83. $\frac{1}{5} - \frac{1}{3}$

84. $-\frac{1}{7} - \left(-\frac{1}{6}\right)$

Simplify.

85. $18 - (-15) - 3 - (-5) + 2$

86. $22 - (-18) + 7 + (-42) - 27$

87. $-31 + (-28) - (-14) - 17$

88. $-43 - (-19) - (-21) + 25$

89. $-34 - 28 + (-33) - 44$

90. $39 + (-88) - 29 - (-83)$

91. $-93 - (-84) - 41 - (-56)$

92. $84 + (-99) + 44 - (-18) - 43$

93. $-5 - (-30) + 30 + 40 - (-12)$

94. $14 - (-50) + 20 - (-32)$

95. $132 - (-21) + 45 - (-21)$

96. $81 - (-20) - 14 - (-50) + 53$

Solve.

97. *Ocean Depth.* The deepest point in the Pacific Ocean is the Marianas Trench, with a depth of 11,033 m. The deepest point in the Atlantic Ocean is the Puerto Rico Trench, with a depth of 8648 m. What is the difference in the elevation of the two trenches?
Source: *The Handy Geography Answer Book*

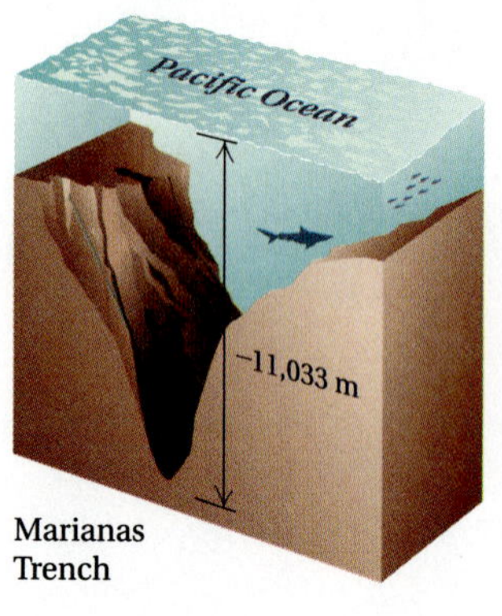

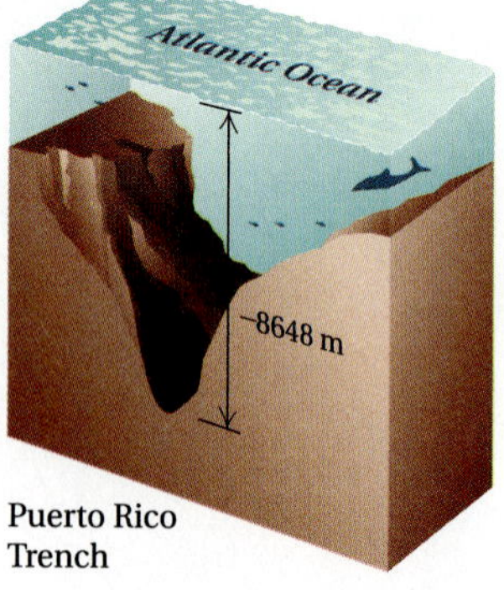

98. *Depth of Offshore Oil Wells.* In 1993, the elevation of the world's deepest offshore oil well was −2860 ft. By 1998, the deepest well was expected to be 360 ft deeper. What was the elevation of the deepest well in 1998?

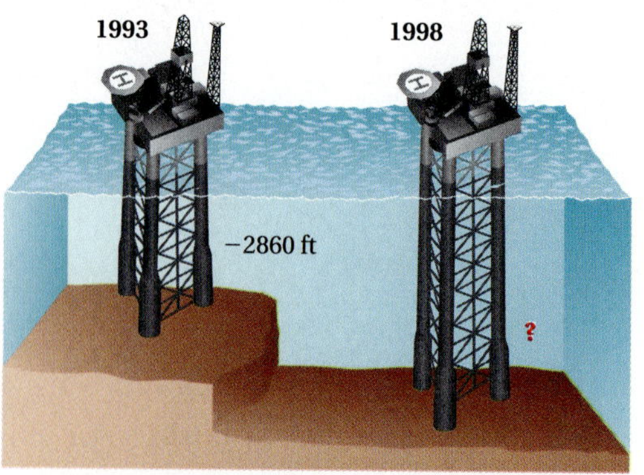

99. Laura has a charge of \$476.89 on her credit card, but she then returns a sweater that cost \$128.95. How much does she now owe on her credit card?

100. Chris has \$720 in a checking account. He writes a check for \$970 to pay for a sound system. What is the balance in his checking account?

101. *Home-Run Differential.* In baseball, the difference between the number of home runs hit by a team's players and the number allowed by its pitchers is called the *home-run differential,* that is,

$$\text{Home run differential} = \begin{matrix}\text{Number of}\\ \text{home runs hit}\end{matrix} - \begin{matrix}\text{Number of home}\\ \text{runs allowed}\end{matrix}.$$

Teams strive for a positive home-run differential.

a) In a recent year, Atlanta hit 197 home runs and allowed 120. Find its home-run differential.

b) In a recent year, San Francisco hit 153 home runs and allowed 194. Find its home-run differential.

Source: Major League Baseball

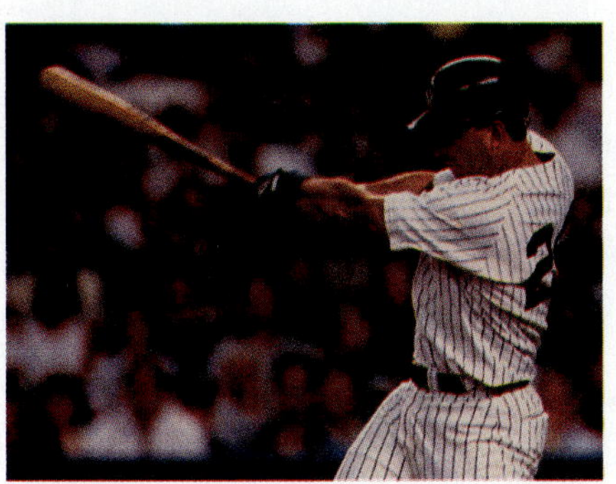

102. *Temperature Records.* The greatest recorded temperature change in one day occurred in Browning, Montana, where the temperature fell from 44°F to −56°F. By how much did the temperature drop?

Source: *The Guinness Book of Records*

103. *Low Points on Continents.* The lowest point in Africa is Lake Assal, which is 515 ft below sea level. The lowest point in South America is the Valdes Peninsula, which is 132 ft below sea level. How much lower is Lake Assal than the Valdes Peninsula?

Source: National Geographic Society

104. *Elevation Changes.* The lowest elevation in Asia, the Dead Sea, is 1286 ft below sea level. The highest elevation in Asia, Mount Everest, is 29,028 ft. Find the difference in elevation between the highest point and the lowest.

Source: *The Handy Geography Answer Book*

105. D_W If a negative number is subtracted from a positive number, will the result always be positive? Why or why not?

106. D_W Write a problem for a classmate to solve. Design the problem so that the solution is "The temperature dropped to −9°."

SKILL MAINTENANCE

Multiply and simplify, if possible. [2.2a]

107. $\frac{3}{4} \cdot \frac{8}{9}$

108. $5 \cdot \frac{1}{30}$

109. $\frac{15}{16} \cdot \frac{6}{25}$

Find percent notation. [4.3a]

110. $\frac{3}{20}$

111. $\frac{18}{25}$

112. $\frac{2}{3}$

SYNTHESIS

Subtract.

113. 🖩 123,907 − 433,789

114. 🖩 23,011 − (−60,432)

Tell whether the statement is true or false for all integers a and b. If false, show why.

115. $a - 0 = 0 - a$

116. $0 - a = a$

117. If $a \neq b$, then $a - b \neq 0$.

118. If $a = -b$, then $a + b = 0$.

119. If $a + b = 0$, then a and b are opposites.

120. If $a - b = 0$, then $a = -b$.

121. *Blackjack Counting System.* The casino game of blackjack makes use of many card-counting systems to give players a winning edge if the count becomes negative. One such system is called *High–Low,* first developed by Harvey Dubner in 1963. Each card counts as −1, 0, or 1 as follows:

2, 3, 4, 5, 6	count as +1;
7, 8, 9	count as 0;
10, J, Q, K, A	count as −1.

Source: Jerry L. Patterson, *Casino Gambling.* New York: Perigee, 1982

a) Find the final count on the sequence of cards

K, A, 2, 4, 5, 10, J, 8, Q, K, 5.

b) Does the player have a winning edge?

6.5 MULTIPLICATION OF REAL NUMBERS

a Multiplication

Multiplication of real numbers is very much like multiplication of arithmetic numbers. The only difference is that we must determine whether the answer is positive or negative.

MULTIPLICATION OF A POSITIVE NUMBER AND A NEGATIVE NUMBER

To see how to multiply a positive number and a negative number, consider the pattern of the following.

This number decreases by 1 each time. This number decreases by 5 each time.

$$\begin{aligned} 4 \cdot 5 &= 20 \\ 3 \cdot 5 &= 15 \\ 2 \cdot 5 &= 10 \\ 1 \cdot 5 &= 5 \\ 0 \cdot 5 &= 0 \\ -1 \cdot 5 &= -5 \\ -2 \cdot 5 &= -10 \\ -3 \cdot 5 &= -15 \end{aligned}$$

Do Exercise 1.

According to this pattern, it looks as though the product of a negative number and a positive number is negative. That is the case, and we have the first part of the rule for multiplying numbers.

THE PRODUCT OF A POSITIVE AND A NEGATIVE NUMBER

To multiply a positive number and a negative number, multiply their absolute values. The answer is negative.

EXAMPLES Multiply.

1. $8(-5) = -40$
2. $-\frac{1}{3} \cdot \frac{5}{7} = -\frac{5}{21}$
3. $(-7.2)5 = -36$

Do Exercises 2–7.

MULTIPLICATION OF TWO NEGATIVE NUMBERS

How do we multiply two negative numbers? Again, we look for a pattern.

This number decreases by 1 each time. This number increases by 5 each time.

$$\begin{aligned} 4 \cdot (-5) &= -20 \\ 3 \cdot (-5) &= -15 \\ 2 \cdot (-5) &= -10 \\ 1 \cdot (-5) &= -5 \\ 0 \cdot (-5) &= 0 \\ -1 \cdot (-5) &= 5 \\ -2 \cdot (-5) &= 10 \\ -3 \cdot (-5) &= 15 \end{aligned}$$

Objectives

a Multiply real numbers.

b Solve applied problems involving multiplication of real numbers.

1. Complete, as in the example.

$$\begin{aligned} 4 \cdot 10 &= 40 \\ 3 \cdot 10 &= 30 \\ 2 \cdot 10 &= \\ 1 \cdot 10 &= \\ 0 \cdot 10 &= \\ -1 \cdot 10 &= \\ -2 \cdot 10 &= \\ -3 \cdot 10 &= \end{aligned}$$

Multiply.

2. $-3 \cdot 6$

3. $20 \cdot (-5)$

4. $4 \cdot (-20)$

5. $-\frac{2}{3} \cdot \frac{5}{6}$

6. $-4.23(7.1)$

7. $\frac{7}{8}\left(-\frac{4}{5}\right)$

8. Complete, as in the example.

$$\begin{aligned} 3 \cdot (-10) &= -30 \\ 2 \cdot (-10) &= -20 \\ 1 \cdot (-10) &= \\ 0 \cdot (-10) &= \\ -1 \cdot (-10) &= \\ -2 \cdot (-10) &= \\ -3 \cdot (-10) &= \end{aligned}$$

Answers on page A-15

Multiply.

9. $-9 \cdot (-3)$

10. $-16 \cdot (-2)$

11. $-7 \cdot (-5)$

12. $-\frac{4}{7}\left(-\frac{5}{9}\right)$

13. $-\frac{3}{2}\left(-\frac{4}{9}\right)$

14. $-3.25(-4.14)$

Multiply.

15. $5(-6)$

16. $(-5)(-6)$

17. $(-3.2) \cdot 0$

18. $\left(-\frac{4}{5}\right)\left(\frac{10}{3}\right)$

19. $0 \cdot (-34.2)$

20. $23 \cdot 0 \cdot \left(-4\frac{2}{3}\right)$

Answers on page A-15

Do Exercise 8 on the preceding page.

According to the pattern, it appears that the product of two negative numbers is positive. That is actually so, and we have the second part of the rule for multiplying real numbers.

THE PRODUCT OF TWO NEGATIVE NUMBERS

To multiply two negative numbers, multiply their absolute values. The answer is positive.

Do Exercises 9–14.

The following is another way to consider the rules we have for multiplication.

To multiply two nonzero real numbers:

a) Multiply the absolute values.

b) If the signs are the same, the answer is positive.

c) If the signs are different, the answer is negative.

MULTIPLICATION BY ZERO

The only case that we have not considered is multiplying by zero. As with other numbers, the product of any real number and 0 is 0.

THE MULTIPLICATION PROPERTY OF ZERO

For any real number a,

$$a \cdot 0 = 0 \cdot a = 0.$$

(The product of 0 and any real number is 0.)

EXAMPLES Multiply.

4. $(-3)(-4) = 12$
5. $-1.6(2) = -3.2$
6. $-19 \cdot 0 = 0$
7. $\left(-\frac{5}{6}\right)\left(-\frac{1}{9}\right) = \frac{5}{54}$
8. $0 \cdot (-452) = 0$
9. $23 \cdot 0 \cdot \left(-8\frac{2}{3}\right) = 0$

Do Exercises 15–20.

MULTIPLYING MORE THAN TWO NUMBERS

When multiplying more than two real numbers, we can choose order and grouping as we please.

EXAMPLES Multiply.

10. $-8 \cdot 2(-3) = -16(-3)$ Multiplying the first two numbers
$= 48$

11. $-8 \cdot 2(-3) = 24 \cdot 2$ Multiplying the negatives. Every pair of negative numbers gives a positive product.
$= 48$

12. $-3(-2)(-5)(4) = 6(-5)(4)$ Multiplying the first two numbers
$= (-30)4$
$= -120$

13. $\left(-\frac{1}{2}\right)(8)\left(-\frac{2}{3}\right)(-6) = (-4)4$ Multiplying the first two numbers and the last two numbers
$= -16$

14. $-5 \cdot (-2) \cdot (-3) \cdot (-6) = 10 \cdot 18$
$= 180$

15. $(-3)(-5)(-2)(-3)(-6) = (-30)(18)$
$= -540$

Considering that the product of a pair of negative numbers is positive, we see the following pattern.

> The product of an even number of negative numbers is positive.
> The product of an odd number of negative numbers is negative.

Do Exercises 21–26.

Let's compare the expressions $(-x)^2$ and $-x^2$.

EXAMPLE 16 Evaluate $(-x)^2$ and $-x^2$ when $x = 5$.

$(-x)^2 = (-5)^2 = (-5)(-5) = 25;$ Substitute 5 for x. Then evaluate the power.

$-x^2 = -(5)^2 = -25$ Substitute 5 for x. Evaluate the power. Then find the opposite.

The expressions $(-x)^2$ and $-x^2$ are *not* equivalent. That is, they do not have the same value for every allowable replacement of the variable by a real number. To find $(-x)^2$, we take the opposite and then square. To find $-x^2$, we find the square and then take the opposite.

Do Exercises 27 and 28.

EXAMPLE 17 Evaluate $2x^2$ when $x = 3$ and $x = -3$.

$2x^2 = 2(3)^2 = 2(9) = 18;$

$2x^2 = 2(-3)^2 = 2(9) = 18$

Do Exercise 29.

Multiply.

21. $5 \cdot (-3) \cdot 2$

22. $-3 \times (-4.1) \times (-2.5)$

23. $-\frac{1}{2} \cdot \left(-\frac{4}{3}\right) \cdot \left(-\frac{5}{2}\right)$

24. $-2 \cdot (-5) \cdot (-4) \cdot (-3)$

25. $(-4)(-5)(-2)(-3)(-1)$

26. $(-1)(-1)(-2)(-3)(-1)(-1)$

27. Evaluate $(-x)^2$ and $-x^2$ when $x = 2$.

28. Evaluate $(-x)^2$ and $-x^2$ when $x = 3$.

29. Evaluate $3x^2$ when $x = 4$ and when $x = -4$.

Answers on page A-15

30. Chemical Reaction. During a chemical reaction, the temperature in the beaker increased by 3°C every minute until 1:34 P.M. If the temperature was −17°C at 1:10 P.M., when the reaction began, what was the temperature at 1:34 P.M.?

Answer on page A-15

b Applications and Problem Solving

We now consider multiplication of real numbers in real-world applications.

EXAMPLE 18 *Chemical Reaction.* During a chemical reaction, the temperature in the beaker decreased by 2°C every minute until 10:23 A.M. If the temperature was 17°C at 10:00 A.M., when the reaction began, what was the temperature at 10:23 A.M.?

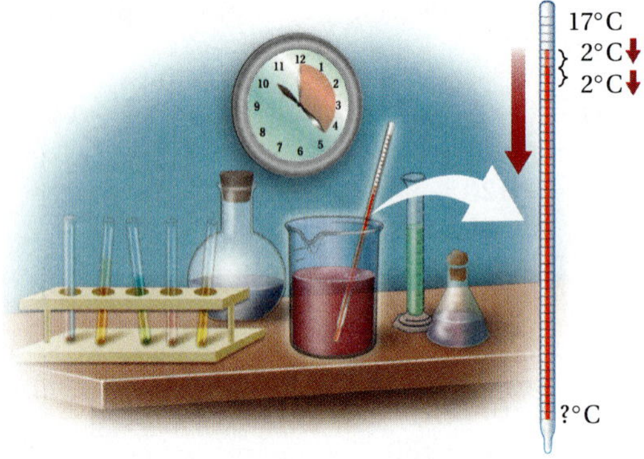

This is a multistep problem. We first find the total number of degrees that the temperature dropped, using −2° for each minute. Since it dropped 2° for each of the 23 minutes, we know that the total drop d is given by

$$d = 23 \cdot (-2) = -46.$$

To determine the temperature after this time period, we find the sum of 17 and −46, or

$$T = 17 + (-46) = -29.$$

Thus the temperature at 10:23 A.M. was −29°C.

Do Exercise 30.

Study Tips

HIGHLIGHTING

Reading and highlighting a section before your instructor lectures on it allows you to maximize your learning and understanding during the lecture.

- **Try to keep one section ahead of your syllabus.** If you study ahead of your lectures, you can concentrate on what is being explained in them, rather than trying to write everything down. You can then take notes only of special points or of questions related to what is happening in class.
- **Highlight important points.** You are probably used to highlighting key points as you study. If that works for you, continue to do so. But you will notice many design features throughout this book that already highlight important points. Thus you may not need to highlight as much as you generally do.
- **Highlight points that you do not understand.** Use a unique mark to indicate trouble spots that can lead to questions to be asked during class, in a tutoring session, or when calling or contacting the AW Math Tutor Center.

6.5 EXERCISE SET

For Extra Help

Digital Video Tutor CD 5 Videotape 7 | InterAct Math | Math Tutor Center | MathXL | MyMathLab

a Multiply.

1. $-4 \cdot 2$

2. $-3 \cdot 5$

3. $-8 \cdot 6$

4. $-5 \cdot 2$

5. $8 \cdot (-3)$

6. $9 \cdot (-5)$

7. $-9 \cdot 8$

8. $-10 \cdot 3$

9. $-8 \cdot (-2)$

10. $-2 \cdot (-5)$

11. $-7 \cdot (-6)$

12. $-9 \cdot (-2)$

13. $15 \cdot (-8)$

14. $-12 \cdot (-10)$

15. $-14 \cdot 17$

16. $-13 \cdot (-15)$

17. $-25 \cdot (-48)$

18. $39 \cdot (-43)$

19. $-3.5 \cdot (-28)$

20. $97 \cdot (-2.1)$

21. $9 \cdot (-8)$

22. $7 \cdot (-9)$

23. $4 \cdot (-3.1)$

24. $3 \cdot (-2.2)$

25. $-5 \cdot (-6)$

26. $-6 \cdot (-4)$

27. $-7 \cdot (-3.1)$

28. $-4 \cdot (-3.2)$

29. $\frac{2}{3} \cdot \left(-\frac{3}{5}\right)$

30. $\frac{5}{7} \cdot \left(-\frac{2}{3}\right)$

31. $-\frac{3}{8} \cdot \left(-\frac{2}{9}\right)$

32. $-\frac{5}{8} \cdot \left(-\frac{2}{5}\right)$

33. -6.3×2.7

34. -4.1×9.5

35. $-\frac{5}{9} \cdot \frac{3}{4}$

36. $-\frac{8}{3} \cdot \frac{9}{4}$

37. $7 \cdot (-4) \cdot (-3) \cdot 5$

38. $9 \cdot (-2) \cdot (-6) \cdot 7$

39. $-\frac{2}{3} \cdot \frac{1}{2} \cdot \left(-\frac{6}{7}\right)$

40. $-\frac{1}{8} \cdot \left(-\frac{1}{4}\right) \cdot \left(-\frac{3}{5}\right)$

41. $-3 \cdot (-4) \cdot (-5)$

42. $-2 \cdot (-5) \cdot (-7)$

43. $-2 \cdot (-5) \cdot (-3) \cdot (-5)$

44. $-3 \cdot (-5) \cdot (-2) \cdot (-1)$

45. $\frac{1}{5}\left(-\frac{2}{9}\right)$

46. $-\frac{3}{5}\left(-\frac{2}{7}\right)$

47. $-7 \cdot (-21) \cdot 13$

48. $-14 \cdot (34) \cdot 12$

49. $-4 \cdot (-1.8) \cdot 7$

50. $-8 \cdot (-1.3) \cdot (-5)$

51. $-\frac{1}{9}\left(-\frac{2}{3}\right)\left(\frac{5}{7}\right)$

52. $-\frac{7}{2}\left(-\frac{5}{7}\right)\left(-\frac{2}{5}\right)$

53. $4 \cdot (-4) \cdot (-5) \cdot (-12)$

54. $-2 \cdot (-3) \cdot (-4) \cdot (-5)$

55. $0.07 \cdot (-7) \cdot 6 \cdot (-6)$

56. $80 \cdot (-0.8) \cdot (-90) \cdot (-0.09)$

57. $\left(-\frac{5}{6}\right)\left(\frac{1}{8}\right)\left(-\frac{3}{7}\right)\left(-\frac{1}{7}\right)$

58. $\left(\frac{4}{5}\right)\left(-\frac{2}{3}\right)\left(-\frac{15}{7}\right)\left(\frac{1}{2}\right)$

59. $(-14) \cdot (-27) \cdot 0$

60. $7 \cdot (-6) \cdot 5 \cdot (-4) \cdot 3 \cdot (-2) \cdot 1 \cdot 0$

61. $(-8)(-9)(-10)$

62. $(-7)(-8)(-9)(-10)$

63. $(-6)(-7)(-8)(-9)(-10)$

64. $(-5)(-6)(-7)(-8)(-9)(-10)$

65. Evaluate $(-3x)^2$ and $-3x^2$ when $x = 7$.

66. Evaluate $(-2x)^2$ and $-2x^2$ when $x = 3$.

67. Evaluate $5x^2$ when $x = 2$ and when $x = -2$.

68. Evaluate $2x^2$ when $x = 5$ and when $x = -5$.

b Solve.

69. *Lost Weight.* Dave lost 2 lb each week for a period of 10 weeks. Express his total weight change as an integer.

70. *Stock Loss.* Michelle lost \$3 each day for a period of 5 days in the value of a stock she owned. Express her total loss as an integer.

71. *Chemical Reaction.* The temperature of a chemical compound was 0°C at 11:00 A.M. During a reaction, it dropped 3°C per minute until 11:18 A.M. What was the temperature at 11:18 A.M.?

72. *Chemical Reaction.* The temperature in a chemical compound was −5°C at 3:20 P.M. During a reaction, it increased 2°C per minute until 3:52 P.M. What was the temperature at 3:52 P.M.?

73. *Stock Price.* The price of ePDQ.com began the day at \$23.75 per share and dropped \$1.38 per hour for 8 hr. What was the price of the stock after 8 hr?

74. *Population Decrease.* The population of a rural town was 12,500. It decreased 380 each year for 4 yr. What was the population of the town after 4 yr?

75. *Diver's Position.* After diving 95 m below the sea level, a diver rises at a rate of 7 meters per minute for 9 min. Where is the diver in relation to the surface?

76. *Checking Account Balance.* Karen had \$68 in her checking account. After writing checks to make seven purchases at \$13 each, what was the balance in her checking account?

77. ^D_W Multiplication can be thought of as repeated addition. Using this concept and a number line, explain why $3 \cdot (-5) = -15$.

78. ^D_W What rule have we developed that would tell you the sign of $(-7)^8$ and $(-7)^{11}$ without doing the computations? Explain.

SKILL MAINTENANCE

Divide and simplify, if possible. [2.2c]

79. $12 \div \frac{3}{4}$

80. $\frac{4}{5} \div \frac{8}{9}$

81. $\frac{5}{6} \div 30$

Find percent notation. [4.2c]

82. 0.23

83. 1.4

84. 0.046

SYNTHESIS

For each of Exercises 85 and 86, choose the correct answer from the selections given.

85. If a is positive and b is negative, then $-ab$ is:
a) Positive.
b) Negative.
c) 0.
d) Cannot be determined without more information

86. If a is positive and b is negative, then $(-a)(-b)$ is:
a) Positive.
b) Negative.
c) 0.
d) Cannot be determined without more information

87. Below is a number line showing 0 and two positive numbers x and y. Use a compass or ruler and locate as best you can the following:

$2x, \quad 3x, \quad 2y, \quad -x, \quad -y, \quad x + y, \quad x - y, \quad x - 2y.$

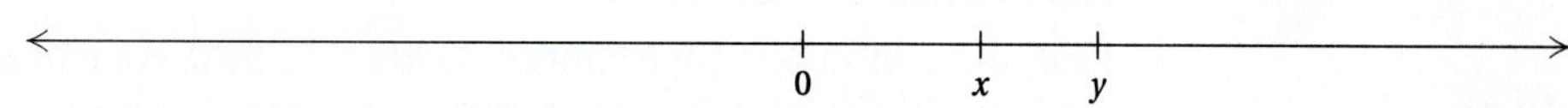

88. Below is a number line showing 0 and two negative numbers x and y. Use a compass or ruler and locate as best you can the following:

$2x, \quad 3x, \quad -x, \quad -y, \quad -3y, \quad x + y, \quad x - y, \quad 2x - y.$

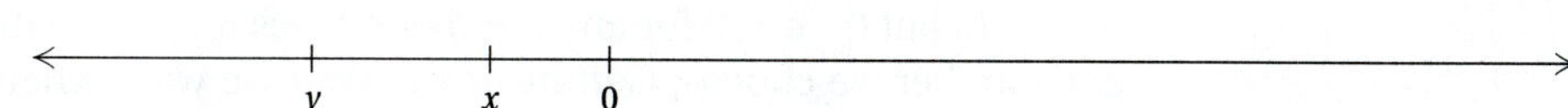

6.6 DIVISION OF REAL NUMBERS

Objectives

a Divide integers.

b Find the reciprocal of a real number.

c Divide real numbers.

d Solve applied problems involving division of real numbers.

We now consider division of real numbers. The definition of division results in rules for division that are the same as those for multiplication.

a Division of Integers

DIVISION

The quotient $a \div b$, or $\frac{a}{b}$, where $b \neq 0$, is that unique real number c for which $a = b \cdot c$.

Let's use the definition to divide integers.

EXAMPLES Divide, if possible. Check your answer.

1. $14 \div (-7) = -2$ — *Think*: What number multiplied by -7 gives 14? That number is -2. *Check*: $(-2)(-7) = 14$.

2. $\frac{-32}{-4} = 8$ — *Think*: What number multiplied by -4 gives -32? That number is 8. *Check*: $8(-4) = -32$.

3. $\frac{-10}{7} = -\frac{10}{7}$ — *Think*: What number multiplied by 7 gives -10? That number is $-\frac{10}{7}$. *Check*: $-\frac{10}{7} \cdot 7 = -10$.

4. $\frac{-17}{0}$ **is not defined.** — *Think*: What number multiplied by 0 gives -17? There is no such number because the product of 0 and *any* number is 0.

The rules for division are the same as those for multiplication.

To multiply or divide two real numbers (where the divisor is nonzero):

a) Multiply or divide the absolute values.

b) If the signs are the same, the answer is positive.

c) If the signs are different, the answer is negative.

Do Exercises 1–6.

EXCLUDING DIVISION BY 0

Example 4 shows why we cannot divide -17 by 0. We can use the same argument to show why we cannot divide any nonzero number b by 0. Consider $b \div 0$. We look for a number that when multiplied by 0 gives b. There is no such number because the product of 0 and any number is 0. Thus we cannot divide a nonzero number b by 0.

On the other hand, if we divide 0 by 0, we look for a number c such that $0 \cdot c = 0$. But $0 \cdot c = 0$ for any number c. Thus it appears that $0 \div 0$ could be any number we choose. Getting any answer we want when we divide 0 by 0 would be very confusing. Thus we agree that division by zero is not defined.

Divide.

1. $6 \div (-3)$

   Think: What number multiplied by -3 gives 6?

2. $\frac{-15}{-3}$

   Think: What number multiplied by -3 gives -15?

3. $-24 \div 8$

   Think: What number multiplied by 8 gives -24?

4. $\frac{-48}{-6}$

5. $\frac{30}{-5}$

6. $\frac{30}{-7}$

Answers on page A-16

EXCLUDING DIVISION BY 0

Division by 0 is not defined.

$a \div 0$, or $\frac{a}{0}$, is not defined for all real numbers a.

DIVIDING 0 BY OTHER NUMBERS

Note that

$0 \div 8 = 0$ because $0 = 0 \cdot 8$; $\quad \frac{0}{-5} = 0$ because $0 = 0 \cdot (-5)$.

DIVIDENDS OF 0

Zero divided by any nonzero real number is 0:

$\frac{0}{a} = 0; \qquad a \neq 0.$

EXAMPLES Divide.

5. $0 \div (-6) = 0$

6. $\frac{0}{12} = 0$

7. $\frac{-3}{0}$ is not defined.

Do Exercises 7 and 8.

b Reciprocals

When two numbers like $\frac{1}{2}$ and 2 are multiplied, the result is 1. Such numbers are called **reciprocals** of each other. Every nonzero real number has a reciprocal, also called a **multiplicative inverse.**

RECIPROCALS

Two numbers whose product is 1 are called **reciprocals,** or **multiplicative inverses,** of each other.

EXAMPLES Find the reciprocal.

8. $\frac{7}{8}$ The reciprocal of $\frac{7}{8}$ is $\frac{8}{7}$ because $\frac{7}{8} \cdot \frac{8}{7} = 1$.

9. -5 The reciprocal of -5 is $-\frac{1}{5}$ because $-5\left(-\frac{1}{5}\right) = 1$.

10. 3.9 The reciprocal of 3.9 is $\frac{1}{3.9}$ because $3.9\left(\frac{1}{3.9}\right) = 1$.

11. $-\frac{1}{2}$ The reciprocal of $-\frac{1}{2}$ is -2 because $\left(-\frac{1}{2}\right)(-2) = 1$.

12. $-\frac{2}{3}$ The reciprocal of $-\frac{2}{3}$ is $-\frac{3}{2}$ because $\left(-\frac{2}{3}\right)\left(-\frac{3}{2}\right) = 1$.

13. $\frac{1}{3/4}$ The reciprocal of $\frac{1}{3/4}$ is $\frac{3}{4}$ because $\left(\frac{1}{3/4}\right)\left(\frac{3}{4}\right) = 1$.

Divide, if possible.

7. $\frac{-5}{0}$

8. $\frac{0}{-3}$

Find the reciprocal.

9. $\frac{2}{3}$

10. $-\frac{5}{4}$

11. -3

12. $-\frac{1}{5}$

13. 1.6

14. $\frac{1}{2/3}$

Answers on page A-16

15. Complete the following table.

NUMBER	OPPOSITE	RECIPROCAL
$\frac{2}{3}$		
$-\frac{5}{4}$		
0		
1		
-8		
-4.5		

Answers on page A-16

Study Tips

TAKE THE TIME!

The foundation of all your study skills is *time*! If you invest your time, we will help you achieve success.

"Nine-tenths of wisdom is being wise in time."

Theodore Roosevelt

RECIPROCAL PROPERTIES

For $a \neq 0$, the reciprocal of a can be named $\frac{1}{a}$ and the reciprocal of $\frac{1}{a}$ is a.

The reciprocal of a nonzero number $\frac{a}{b}$ can be named $\frac{b}{a}$.

The number 0 has no reciprocal.

Do Exercises 9–14 on the preceding page.

The reciprocal of a positive number is also a positive number, because their product must be the positive number 1. The reciprocal of a negative number is also a negative number, because their product must be the positive number 1.

THE SIGN OF A RECIPROCAL

The reciprocal of a number has the same sign as the number itself.

CAUTION!

It is important *not* to confuse *opposite* with *reciprocal.* Keep in mind that the opposite, or additive inverse, of a number is what we add to the number to get 0. The reciprocal, or multiplicative inverse, is what we multiply the number by to get 1.

Compare the following.

NUMBER	OPPOSITE (Change the sign.)	RECIPROCAL (Invert but do not change the sign.)
$-\frac{3}{8}$	$\frac{3}{8}$	$-\frac{8}{3}$
19	-19	$\frac{1}{19}$
$\frac{18}{7}$	$-\frac{18}{7}$	$\frac{7}{18}$
-7.9	7.9	$-\frac{1}{7.9}$, or $-\frac{10}{79}$
0	0	Not defined

$\left(-\frac{3}{8}\right)\left(-\frac{8}{3}\right) = 1$

$-\frac{3}{8} + \frac{3}{8} = 0$

Do Exercise 15.

C Division of Real Numbers

We know that we can subtract by adding an opposite. Similarly, we can divide by multiplying by a reciprocal.

RECIPROCALS AND DIVISION

For any real numbers a and b, $b \neq 0$,

$$a \div b = \frac{a}{b} = a \cdot \frac{1}{b}.$$

(To divide, multiply by the reciprocal of the divisor.)

EXAMPLES Rewrite the division as a multiplication.

14. $-4 \div 3$ $\quad$ $-4 \div 3$ is the same as $-4 \cdot \frac{1}{3}$

15. $\frac{6}{-7}$ $\quad$ $\frac{6}{-7} = 6\left(-\frac{1}{7}\right)$

16. $\frac{x+2}{5}$ $\quad$ $\frac{x+2}{5} = (x+2)\frac{1}{5}$ $\quad$ Parentheses are necessary here.

17. $\frac{-17}{1/b}$ $\quad$ $\frac{-17}{1/b} = -17 \cdot b$

18. $\frac{3}{5} \div \left(-\frac{9}{7}\right)$ $\quad$ $\frac{3}{5} \div \left(-\frac{9}{7}\right) = \frac{3}{5}\left(-\frac{7}{9}\right)$

Do Exercises 16–20.

When actually doing division calculations, we sometimes multiply by a reciprocal and we sometimes divide directly. With fraction notation, it is usually better to multiply by a reciprocal. With decimal notation, it is usually better to divide directly.

EXAMPLES Divide by multiplying by the reciprocal of the divisor.

19. $\frac{2}{3} \div \left(-\frac{5}{4}\right) = \frac{2}{3} \cdot \left(-\frac{4}{5}\right) = -\frac{8}{15}$

20. $-\frac{5}{6} \div \left(-\frac{3}{4}\right) = -\frac{5}{6} \cdot \left(-\frac{4}{3}\right) = \frac{20}{18} = \frac{10 \cdot 2}{9 \cdot 2} = \frac{10}{9} \cdot \frac{2}{2} = \frac{10}{9}$

CAUTION!

Be careful not to change the sign when taking a reciprocal!

21. $-\frac{3}{4} \div \frac{3}{10} = -\frac{3}{4} \cdot \left(\frac{10}{3}\right) = -\frac{30}{12} = -\frac{5}{2} \cdot \frac{6}{6} = -\frac{5}{2}$

Rewrite the division as a multiplication.

16. $\frac{4}{7} \div \left(-\frac{3}{5}\right)$

17. $\frac{5}{-8}$

18. $\frac{a-b}{7}$

19. $\frac{-23}{1/a}$

20. $-5 \div 7$

Divide by multiplying by the reciprocal of the divisor.

21. $\frac{4}{7} \div \left(-\frac{3}{5}\right)$

22. $-\frac{8}{5} \div \frac{2}{3}$

23. $-\frac{12}{7} \div \left(-\frac{3}{4}\right)$

24. Divide: $21.7 \div (-3.1)$.

Answers on page A-16

Find two equal expressions for the number with negative signs in different places.

25. $\frac{-5}{6}$

26. $-\frac{8}{7}$

27. $\frac{10}{-3}$

Answers on page A-16

With decimal notation, it is easier to carry out long division than to multiply by the reciprocal.

EXAMPLES Divide.

22. $-27.9 \div (-3) = \frac{-27.9}{-3} = 9.3$ Do the long division $3\overline{)27.9}$ (quotient 9.3). The answer is positive.

23. $-6.3 \div 2.1 = -3$ Do the long division $2.1.\overline{)6.3_\wedge}$ (quotient 3.). The answer is negative.

Do Exercises 21–24 on the preceding page.

Consider the following:

1. $\frac{2}{3} = \frac{2}{3} \cdot 1 = \frac{2}{3} \cdot \frac{-1}{-1} = \frac{2(-1)}{3(-1)} = \frac{-2}{-3}$. Thus, $\frac{2}{3} = \frac{-2}{-3}$.

   (A negative number divided by a negative number is positive.)

2. $-\frac{2}{3} = -1 \cdot \frac{2}{3} = \frac{-1}{1} \cdot \frac{2}{3} = \frac{-1 \cdot 2}{1 \cdot 3} = \frac{-2}{3}$. Thus, $-\frac{2}{3} = \frac{-2}{3}$.

   (A negative number divided by a positive number is negative.)

3. $\frac{-2}{3} = \frac{-2}{3} \cdot 1 = \frac{-2}{3} \cdot \frac{-1}{-1} = \frac{-2(-1)}{3(-1)} = \frac{2}{-3}$. Thus, $-\frac{2}{3} = \frac{2}{-3}$.

   (A positive number divided by a negative number is negative.)

We can use the following properties to make sign changes in fraction notation.

SIGN CHANGES IN FRACTION NOTATION

For any numbers a and b, $b \neq 0$:

1. $\frac{-a}{-b} = \frac{a}{b}$

(The opposite of a number a divided by the opposite of another number b is the same as the quotient of the two numbers a and b.)

2. $\frac{-a}{b} = \frac{a}{-b} = -\frac{a}{b}$

(The opposite of a number a divided by another number b is the same as the number a divided by the opposite of the number b, and both are the same as the opposite of *a divided by b*.)

Do Exercises 25–27.

d Applications and Problem Solving

EXAMPLE 24 *Chemical Reaction.* During a chemical reaction, the temperature in the beaker decreased every minute by the same number of degrees. The temperature was 56°F at 10:10 A.M. By 10:42 A.M., the temperature had dropped to −12°F. By how many degrees did it change each minute?

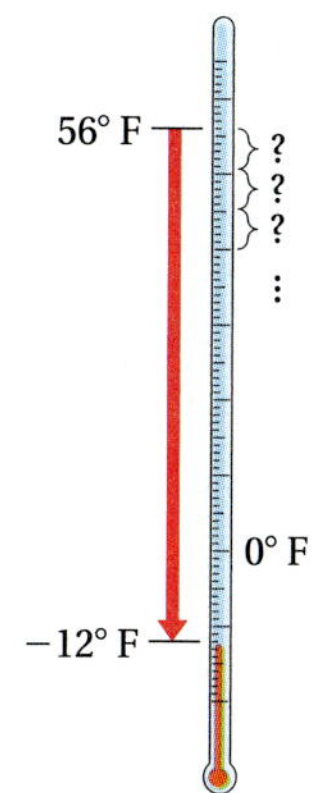

We first determine by how many degrees d the temperature changed altogether. We subtract −12 from 56:

$$d = 56 - (-12) = 56 + 12 = 68.$$

The temperature changed a total of 68°. We can express this as −68° since the temperature dropped.

The amount of time t that passed was 42 − 10, or 32 min. Thus the number of degrees T that the temperature dropped each minute is given by

$$T = \frac{d}{t} = \frac{-68}{32} = -2.125.$$

The change was −2.125°F per minute.

Do Exercise 28.

28. **Chemical Reaction.** During a chemical reaction, the temperature in the beaker decreased every minute by the same number of degrees. The temperature was 71°F at 2:12 P.M. By 2:37 P.M., the temperature had changed to −14°F. By how many degrees did it change each minute?

Answer on page A-16

CALCULATOR CORNER

Operations on the Real Numbers We can perform operations on the real numbers on a graphing calculator. Recall that negative numbers are entered using the opposite key, [(−)], rather than the subtraction operation key, [−]. Consider the sum −5 + (−3.8). We use parentheses when we write this sum in order to separate the addition symbol and the "opposite of" symbol and thus make the expression more easily read. When we enter this calculation on a graphing calculator, however, the parentheses are not necessary. We can press [(−)] [5] [+] [(−)] [3] [.] [8] [ENTER]. The result is −8.8. Note that it is not incorrect to enter the parentheses. The result will be the same if this is done.

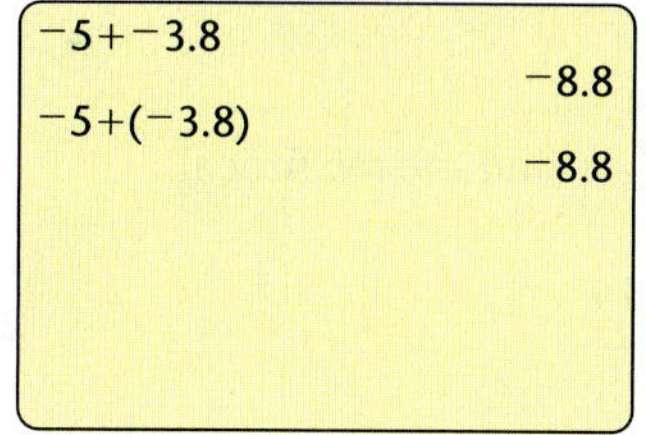

To find the difference 10 − (−17), we press [1] [0] [−] [(−)] [1] [7] [ENTER]. The result is 27. We can also multiply and divide real numbers. To find −5 · (−7), we press [(−)] [5] [×] [(−)] [7] [ENTER], and to find 45 ÷ (−9), we press [4] [5] [÷] [(−)] [9] [ENTER]. Note that it is not necessary to use parentheses in any of these calculations.

```
10−−17
                 27
−5*−7
                 35
45/−9
                 −5
```

Exercises: Use a calculator to perform the operation.

1. −8 + 4
2. 1.2 + (−1.5)
3. −7 + (−5)
4. −7.6 + (−1.9)
5. −8 − 4
6. 1.2 − (−1.5)
7. −7 − (−5)
8. −7.6 − (−1.9)
9. −8 · 4
10. 1.2 · (−1.5)
11. −7 · (−5)
12. −7.6 · (−1.9)
13. −8 ÷ 4
14. 1.2 ÷ (−1.5)
15. −7 ÷ (−5)
16. −7.6 ÷ (−1.9)

6.6 EXERCISE SET

For Extra Help

Digital Video Tutor CD 5 Videotape 7 | InterAct Math | Math Tutor Center | MathXL | MyMathLab

a Divide, if possible. Check each answer.

1. $48 \div (-6)$

2. $\dfrac{42}{-7}$

3. $\dfrac{28}{-2}$

4. $24 \div (-12)$

5. $\dfrac{-24}{8}$

6. $-18 \div (-2)$

7. $\dfrac{-36}{-12}$

8. $-72 \div (-9)$

9. $\dfrac{-72}{9}$

10. $\dfrac{-50}{25}$

11. $-100 \div (-50)$

12. $\dfrac{-200}{8}$

13. $-108 \div 9$

14. $\dfrac{-63}{-7}$

15. $\dfrac{200}{-25}$

16. $-300 \div (-16)$

17. $\dfrac{75}{0}$

18. $\dfrac{0}{-5}$

19. $\dfrac{-23}{-2}$

20. $\dfrac{-23}{0}$

b Find the reciprocal.

21. $\dfrac{15}{7}$

22. $\dfrac{3}{8}$

23. $-\dfrac{47}{13}$

24. $-\dfrac{31}{12}$

25. 13

26. -10

27. 4.3

28. -8.5

29. $\dfrac{1}{-7.1}$

30. $\dfrac{1}{-4.9}$

31. $\dfrac{p}{q}$

32. $\dfrac{s}{t}$

33. $\dfrac{1}{4y}$

34. $\dfrac{-1}{8a}$

35. $\dfrac{2a}{3b}$

36. $\dfrac{-4y}{3x}$

C Rewrite the division as a multiplication.

37. $4 \div 17$

38. $5 \div (-8)$

39. $\frac{8}{-13}$

40. $-\frac{13}{47}$

41. $\frac{13.9}{-1.5}$

42. $-\frac{47.3}{21.4}$

43. $\frac{x}{\frac{1}{y}}$

44. $\frac{13}{x}$

45. $\frac{3x + 4}{5}$

46. $\frac{4y - 8}{-7}$

47. $\frac{5a - b}{5a + b}$

48. $\frac{2x + x^2}{x - 5}$

Divide.

49. $\frac{3}{4} \div \left(-\frac{2}{3}\right)$

50. $\frac{7}{8} \div \left(-\frac{1}{2}\right)$

51. $-\frac{5}{4} \div \left(-\frac{3}{4}\right)$

52. $-\frac{5}{9} \div \left(-\frac{5}{6}\right)$

53. $-\frac{2}{7} \div \left(-\frac{4}{9}\right)$

54. $-\frac{3}{5} \div \left(-\frac{5}{8}\right)$

55. $-\frac{3}{8} \div \left(-\frac{8}{3}\right)$

56. $-\frac{5}{8} \div \left(-\frac{6}{5}\right)$

57. $-6.6 \div 3.3$

58. $-44.1 \div (-6.3)$

59. $\frac{-11}{-13}$

60. $\frac{-1.9}{20}$

61. $\frac{48.6}{-3}$

62. $\frac{-17.8}{3.2}$

63. $\frac{-9}{17 - 17}$

64. $\frac{-8}{-5 + 5}$

d *Percent of Increase or Decrease in Employment.* A percent of increase is generally positive and a percent of decrease is generally negative. The following table lists estimates of the number of job opportunities for various occupations in 1998 and 2008. In Exercises 65–68, find the missing numbers.

	OCCUPATION	NUMBER OF JOBS IN 1998 (in thousands)	NUMBER OF JOBS IN 2008 (in thousands)	CHANGE	PERCENT OF INCREASE OR DECREASE
	Court clerk	100	112	12	12%
	Bank teller	560	529	−31	−5.5%
65.	Barber	54	50	−4	
66.	Child-care worker in private household	306	209	−97	
67.	Dental assistant	229	326	97	
68.	Cook (short-order and fast-food)	677	801	124	

Source: Handbook of U.S. Labor Statistics

69. $^{\text{D}}\text{W}$ Explain how multiplication can be used to justify why a negative number divided by a positive number is negative.

70. $^{\text{D}}\text{W}$ Explain how multiplication can be used to justify why a negative number divided by a negative number is positive.

SKILL MAINTENANCE

Convert to decimal notation. [3.5a]

71. $\frac{1}{11}$

72. $\frac{11}{12}$

73. Find the amount in an account if $3000 is invested at 3%, compounded annually, for 3 years. [4.7e]

74. Find the perimeter of a rectangle that is 25.9 cm by 16.4 cm. [5.2a]

SYNTHESIS

75. Find the reciprocal of −10.5. What happens if you take the reciprocal of the result?

76. Determine those real numbers a for which the opposite of a is the same as the reciprocal of a.

Tell whether the expression represents a positive number or a negative number when a and b are negative.

77. $\frac{-a}{b}$

78. $\frac{-a}{-b}$

79. $-\left(\frac{a}{-b}\right)$

80. $-\left(\frac{-a}{b}\right)$

81. $-\left(\frac{-a}{-b}\right)$

6.7 PROPERTIES OF REAL NUMBERS

a Equivalent Expressions

In solving equations and doing other kinds of work in algebra, we manipulate expressions in various ways. For example, instead of

$x + x,$

we might write

$2x,$

knowing that the two expressions represent the same number for any allowable replacement of x. In that sense, the expressions $x + x$ and $2x$ are **equivalent**, as are $\frac{3}{x}$ and $\frac{3x}{x^2}$, even though 0 is not an allowable replacement because division by 0 is not defined.

EQUIVALENT EXPRESSIONS

Two expressions that have the same value for all allowable replacements are called **equivalent.**

The expressions $x + 3x$ and $5x$ are *not* equivalent.

Do Exercises 1 and 2.

In this section, we will consider several laws of real numbers that will allow us to find equivalent expressions. The first two laws are the *identity properties of 0 and 1.*

THE IDENTITY PROPERTY OF 0

For any real number a,

$a + 0 = 0 + a = a.$

(The number 0 is the *additive identity.*)

THE IDENTITY PROPERTY OF 1

For any real number a,

$a \cdot 1 = 1 \cdot a = a.$

(The number 1 is the *multiplicative identity.*)

We often refer to the use of the identity property of 1 as "multiplying by 1." We can use this method to find equivalent fractional expressions. Recall from arithmetic that to multiply with fraction notation, we multiply numerators and denominators.

Objectives

a Find equivalent fractional expressions and simplify fractional expressions.

b Use the commutative and associative laws to find equivalent expressions.

c Use the distributive laws to multiply expressions like 8 and $x - y$.

d Use the distributive laws to factor expressions like $4x - 12 + 24y$.

e Collect like terms.

Complete the table by evaluating each expression for the given values.

1.

VALUE	$x + x$	$2x$
$x = 3$		
$x = -6$		
$x = 4.8$		

2.

VALUE	$x + 3x$	$5x$
$x = 2$		
$x = -6$		
$x = 4.8$		

Answers on page A-16

3. Write a fractional expression equivalent to $\frac{3}{4}$ with a denominator of 8.

4. Write a fractional expression equivalent to $\frac{3}{4}$ with a denominator of $4t$.

Simplify.

5. $\dfrac{3y}{4y}$

6. $-\dfrac{16m}{12m}$

7. Evaluate $x + y$ and $y + x$ when $x = -2$ and $y = 3$.

8. Evaluate xy and yx when $x = -2$ and $y = 5$.

Answers on page A-16

EXAMPLE 1 Write a fractional expression equivalent to $\frac{2}{3}$ with a denominator of $3x$.

Note that $3x = 3 \cdot x$. We want fraction notation for $\frac{2}{3}$ that has a denominator of $3x$, but the denominator 3 is missing a factor of x. Thus we multiply by 1, using x/x as an equivalent expression for 1:

$$\frac{2}{3} = \frac{2}{3} \cdot 1 = \frac{2}{3} \cdot \frac{x}{x} = \frac{2x}{3x}.$$

The expressions $2/3$ and $2x/3x$ are equivalent. They have the same value for any allowable replacement. Note that $2x/3x$ is not defined for a replacement of 0, but for all nonzero real numbers, the expressions $2/3$ and $2x/3x$ have the same value.

Do Exercises 3 and 4.

In algebra, we consider an expression like $2/3$ to be "simplified" from $2x/3x$. To find such simplified expressions, we use the identity property of 1 to remove a factor of 1.

EXAMPLE 2 Simplify: $-\dfrac{20x}{12x}$.

$$-\frac{20x}{12x} = -\frac{5 \cdot 4x}{3 \cdot 4x} \quad \text{We look for the largest factor common to both the numerator and the denominator and factor each.}$$

$$= -\frac{5}{3} \cdot \frac{4x}{4x} \quad \text{Factoring the fractional expression}$$

$$= -\frac{5}{3} \cdot 1 \quad \frac{4x}{4x} = 1$$

$$= -\frac{5}{3} \quad \text{Removing a factor of 1 using the identity property of 1}$$

Do Exercises 5 and 6.

b The Commutative and Associative Laws

Let's examine the expressions $x + y$ and $y + x$, as well as xy and yx.

EXAMPLE 3 Evaluate $x + y$ and $y + x$ when $x = 4$ and $y = 3$.

We substitute 4 for x and 3 for y in both expressions:

$$x + y = 4 + 3 = 7; \qquad y + x = 3 + 4 = 7.$$

EXAMPLE 4 Evaluate xy and yx when $x = 23$ and $y = 12$.

We substitute 23 for x and 12 for y in both expressions:

$$xy = 23 \cdot 12 = 276; \qquad yx = 12 \cdot 23 = 276.$$

Do Exercises 7 and 8.

Note that the expressions

$$x + y \quad \text{and} \quad y + x$$

have the same values no matter what the variables stand for. Thus they are equivalent. Therefore, when we add two numbers, the order in which we add does not matter. Similarly, the expressions xy and yx are equivalent. They also have the same values, no matter what the variables stand for. Therefore, when we multiply two numbers, the order in which we multiply does not matter.

The following are examples of general patterns or laws.

THE COMMUTATIVE LAWS

Addition. For any numbers a and b,

$$a + b = b + a.$$

(We can change the order when adding without affecting the answer.)

Multiplication. For any numbers a and b,

$$ab = ba.$$

(We can change the order when multiplying without affecting the answer.)

Using a commutative law, we know that $x + 2$ and $2 + x$ are equivalent. Similarly, $3x$ and $x(3)$ are equivalent. Thus, in an algebraic expression, we can replace one with the other and the result will be equivalent to the original expression.

EXAMPLE 5 Use the commutative laws to write an expression equivalent to $y + 5$, ab, and $7 + xy$.

An expression equivalent to $y + 5$ is $5 + y$ by the commutative law of addition.

An expression equivalent to ab is ba by the commutative law of multiplication.

An expression equivalent to $7 + xy$ is $xy + 7$ by the commutative law of addition. Another expression equivalent to $7 + xy$ is $7 + yx$ by the commutative law of multiplication.

Do Exercises 9–11.

THE ASSOCIATIVE LAWS

Now let's examine the expressions $a + (b + c)$ and $(a + b) + c$. Note that these expressions involve the use of parentheses as *grouping* symbols, and they also involve three numbers. Calculations within parentheses are to be done first.

EXAMPLE 6 Calculate and compare: $3 + (8 + 5)$ and $(3 + 8) + 5$.

$3 + (8 + 5) = 3 + 13$ Calculating within parentheses first; adding the 8 and 5

$= 16;$

$(3 + 8) + 5 = 11 + 5$ Calculating within parentheses first; adding the 3 and 8

$= 16$

Use a commutative law to write an equivalent expression.

9. $x + 9$

10. pq

11. $xy + t$

Answers on page A-16

Study Tips

THE AW MATH TUTOR CENTER

The AW Math Tutor Center is staffed by highly qualified mathematics instructors who provide students with tutoring on text examples and odd-numbered exercises. Tutoring is provided free to students who have bought a new textbook with a special access card bound with the book. Tutoring is available by telephone (toll-free), fax, and e-mail. White-board technology allows tutors and students to actually see problems worked while they talk live during the tutoring sessions. If you purchased a book without this card, you can purchase an access code through your bookstore using ISBN# 0-201-72170-8. (This is also discussed in the Preface.)

12. Calculate and compare:

$8 + (9 + 2)$ and $(8 + 9) + 2$.

13. Calculate and compare:

$10 \cdot (5 \cdot 3)$ and $(10 \cdot 5) \cdot 3$.

Use an associative law to write an equivalent expression.

14. $r + (s + 7)$

15. $9(ab)$

Answers on page A-16

The two expressions in Example 6 name the same number. Moving the parentheses to group the additions differently does not affect the value of the expression.

EXAMPLE 7 Calculate and compare: $3 \cdot (4 \cdot 2)$ and $(3 \cdot 4) \cdot 2$.

$$3 \cdot (4 \cdot 2) = 3 \cdot 8 = 24; \qquad (3 \cdot 4) \cdot 2 = 12 \cdot 2 = 24$$

Do Exercises 12 and 13.

You may have noted that when only addition is involved, parentheses can be placed any way we please without affecting the answer. When only multiplication is involved, parentheses also can be placed any way we please without affecting the answer.

THE ASSOCIATIVE LAWS

Addition. For any numbers a, b, and c,

$$a + (b + c) = (a + b) + c.$$

(Numbers can be grouped in any manner for addition.)

Multiplication. For any numbers a, b, and c,

$$a \cdot (b \cdot c) = (a \cdot b) \cdot c.$$

(Numbers can be grouped in any manner for multiplication.)

EXAMPLE 8 Use an associative law to write an expression equivalent to $(y + z) + 3$ and $8(xy)$.

An expression equivalent to $(y + z) + 3$ is $y + (z + 3)$ by the associative law of addition.

An expression equivalent to $8(xy)$ is $(8x)y$ by the associative law of multiplication.

Do Exercises 14 and 15.

The associative laws say parentheses can be placed any way we please when only additions or only multiplications are involved. Thus we often omit them. For example,

$$x + (y + 2) \text{ means } x + y + 2, \quad \text{and} \quad (lw)h \text{ means } lwh.$$

USING THE COMMUTATIVE AND ASSOCIATIVE LAWS TOGETHER

EXAMPLE 9 Use the commutative and associative laws to write at least three expressions equivalent to $(x + 5) + y$.

a) $(x + 5) + y = x + (5 + y)$ Using the associative law first and then using the commutative law

$= x + (y + 5)$

b) $(x + 5) + y = y + (x + 5)$ Using the commutative law first and then the commutative law again

$= y + (5 + x)$

c) $(x + 5) + y = (5 + x) + y$ Using the commutative law first and then the associative law

$= 5 + (x + y)$

EXAMPLE 10 Use the commutative and associative laws to write at least three expressions equivalent to $(3x)y$.

a) $(3x)y = 3(xy)$ Using the associative law first and then using the commutative law

$= 3(yx)$

b) $(3x)y = y(3x)$ Using the commutative law twice

$= y(x3)$

c) $(3x)y = (x3)y$ Using the commutative law, and then the associative law, and then the commutative law again

$= x(3y)$

$= x(y3)$

Do Exercises 16 and 17.

Use the commutative and associative laws to write at least three equivalent expressions.

16. $4(tu)$

17. $r + (2 + s)$

C The Distributive Laws

The *distributive laws* are the basis of many procedures in both arithmetic and algebra. They are probably the most important laws that we use to manipulate algebraic expressions. The distributive law of multiplication over addition involves two operations: addition and multiplication.

Let's begin by considering a multiplication problem from arithmetic:

```
    4 5
  ×   7
  -----
    3 5 ← This is 7 · 5.
  2 8 0 ← This is 7 · 40.
  -----
  3 1 5 ← This is the sum 7 · 40 + 7 · 5.
```

To carry out the multiplication, we actually added two products. That is,

$$7 \cdot 45 = 7(40 + 5) = 7 \cdot 40 + 7 \cdot 5.$$

Let's examine this further. If we wish to multiply a sum of several numbers by a factor, we can either add and then multiply, or multiply and then add.

EXAMPLE 11 Compute in two ways: $5 \cdot (4 + 8)$.

a) $5 \cdot (4 + 8)$ Adding within parentheses first, and then multiplying

$= 5 \cdot 12$

$= 60$

b) $(5 \cdot 4) + (5 \cdot 8)$ Distributing the multiplication to terms within parentheses first and then adding

$= 20 + 40$

$= 60$

Do Exercises 18–20.

Compute.

18. a) $7 \cdot (3 + 6)$

b) $(7 \cdot 3) + (7 \cdot 6)$

19. a) $2 \cdot (10 + 30)$

b) $(2 \cdot 10) + (2 \cdot 30)$

20. a) $(2 + 5) \cdot 4$

b) $(2 \cdot 4) + (5 \cdot 4)$

Answers on page A-16

THE DISTRIBUTIVE LAW OF MULTIPLICATION OVER ADDITION

For any numbers a, b, and c,

$$a(b + c) = ab + ac.$$

Calculate.

21. a) $4(5-3)$

b) $4 \cdot 5 - 4 \cdot 3$

22. a) $-2 \cdot (5-3)$

b) $-2 \cdot 5 - (-2) \cdot 3$

23. a) $5 \cdot (2-7)$

b) $5 \cdot 2 - 5 \cdot 7$

What are the terms of the expression?

24. $5x - 8y + 3$

25. $-4y - 2x + 3z$

Multiply.

26. $3(x-5)$

27. $5(x+1)$

28. $\frac{3}{5}(p+q-t)$

Answers on page A-16

In the statement of the distributive law, we know that in an expression such as $ab + ac$, the multiplications are to be done first according to the rules for order of operations. So, instead of writing $(4 \cdot 5) + (4 \cdot 7)$, we can write $4 \cdot 5 + 4 \cdot 7$. However, in $a(b + c)$, we cannot omit the parentheses. If we did, we would have $ab + c$, which means $(ab) + c$. For example, $3(4 + 2) = 18$, but $3 \cdot 4 + 2 = 14$.

There is another distributive law that relates multiplication and subtraction. This law says that to multiply by a difference, we can either subtract and then multiply, or multiply and then subtract.

THE DISTRIBUTIVE LAW OF MULTIPLICATION OVER SUBTRACTION

For any numbers a, b, and c,

$$a(b-c) = ab - ac.$$

We often refer to "*the* distributive law" when we mean *either* or *both* of these laws.

Do Exercises 21–23.

What do we mean by the *terms* of an expression? **Terms** are separated by addition signs. If there are subtraction signs, we can find an equivalent expression that uses addition signs.

EXAMPLE 12 What are the terms of $3x - 4y + 2z$?

We have

$$3x - 4y + 2z = 3x + (-4y) + 2z. \qquad \text{Separating parts with + signs}$$

The terms are $3x$, $-4y$, and $2z$.

Do Exercises 24 and 25.

The distributive laws are a basis for a procedure in algebra called **multiplying.** In an expression like $8(a + 2b - 7)$, we multiply each term inside the parentheses by 8:

$$8(a + 2b - 7) = 8 \cdot a + 8 \cdot 2b - 8 \cdot 7 = 8a + 16b - 56.$$

EXAMPLES Multiply.

13. $9(x-5) = 9x - 9(5)$ Using the distributive law of multiplication over subtraction

$= 9x - 45$

14. $\frac{2}{3}(w+1) = \frac{2}{3} \cdot w + \frac{2}{3} \cdot 1$ Using the distributive law of multiplication over addition

$= \frac{2}{3}w + \frac{2}{3}$

15. $\frac{4}{3}(s - t + w) = \frac{4}{3}s - \frac{4}{3}t + \frac{4}{3}w$ Using both distributive laws

Do Exercises 26–28.

EXAMPLE 16 Multiply: $-4(x - 2y + 3z)$.

$$\begin{aligned} -4(x - 2y + 3z) &= -4 \cdot x - (-4)(2y) + (-4)(3z) && \text{Using both distributive laws} \\ &= -4x - (-8y) + (-12z) && \text{Multiplying} \\ &= -4x + 8y - 12z \end{aligned}$$

We can also do this problem by first finding an equivalent expression with all plus signs and then multiplying:

$$\begin{aligned} -4(x - 2y + 3z) &= -4[x + (-2y) + 3z] \\ &= -4 \cdot x + (-4)(-2y) + (-4)(3z) \\ &= -4x + 8y - 12z. \end{aligned}$$

Do Exercises 29–31.

Multiply.

29. $-2(x - 3)$

30. $5(x - 2y + 4z)$

31. $-5(x - 2y + 4z)$

d Factoring

Factoring is the reverse of multiplying. To factor, we can use the distributive laws in reverse:

$$ab + ac = a(b + c) \quad \text{and} \quad ab - ac = a(b - c).$$

FACTORING

To **factor** an expression is to find an equivalent expression that is a product.

Look at Example 13. To *factor* $9x - 45$, we find an equivalent expression that is a product, $9(x - 5)$. When all the terms of an expression have a factor in common, we can "factor it out" using the distributive laws. Note the following.

$9x$ has the factors $9, -9, 3, -3, 1, -1, x, -x, 3x, -3x, 9x, -9x$;

-45 has the factors $1, -1, 3, -3, 5, -5, 9, -9, 15, -15, 45, -45$

We generally remove the largest common factor. In this case, that factor is 9. Thus,

$$\begin{aligned} 9x - 45 &= 9 \cdot x - 9 \cdot 5 \\ &= 9(x - 5). \end{aligned}$$

Remember that an expression has been factored when we have found an equivalent expression that is a product.

EXAMPLES Factor.

17. $5x - 10 = 5 \cdot x - 5 \cdot 2$ Try to do this step mentally.

$= 5(x - 2)$ You can check by multiplying.

18. $ax - ay + az = a(x - y + z)$

19. $9x + 27y - 9 = 9 \cdot x + 9 \cdot 3y - 9 \cdot 1 = 9(x + 3y - 1)$

Answers on page A-16

Factor.

32. $6x - 12$

33. $3x - 6y + 9$

34. $bx + by - bz$

35. $16a - 36b + 42$

36. $\frac{3}{8}x - \frac{5}{8}y + \frac{7}{8}$

37. $-12x + 32y - 16z$

Answers on page A-16

Note in Example 19 that you might, at first, just factor out a 3, as follows:

$$\begin{aligned} 9x + 27y - 9 &= 3 \cdot 3x + 3 \cdot 9y - 3 \cdot 3 \\ &= 3(3x + 9y - 3). \end{aligned}$$

At this point, the mathematics is correct, but the answer is not because there is another factor of 3 that can be factored out, as follows:

$$\begin{aligned} 3 \cdot 3x + 3 \cdot 9y - 3 \cdot 3 &= 3(3x + 9y - 3) \\ &= 3(3 \cdot x + 3 \cdot 3y - 3 \cdot 1) \\ &= 3 \cdot 3(x + 3y - 1) \\ &= 9(x + 3y - 1). \end{aligned}$$

We now have a correct answer, but it took more work than we did in Example 19. Thus it is better to look for the greatest common factor at the outset.

EXAMPLES Factor. Try to write just the answer, if you can.

20. $5x - 5y = 5(x - y)$

21. $-3x + 6y - 9z = -3(x - 2y + 3z)$

We usually factor out a negative factor when the first term is negative. The way we factor can depend on the situation in which we are working. We might also factor the expression in Example 21 as follows:

$$-3x + 6y - 9z = 3(-x + 2y - 3z).$$

22. $18z - 12x - 24 = 6(3z - 2x - 4)$

23. $\frac{1}{2}x + \frac{3}{2}y - \frac{1}{2} = \frac{1}{2}(x + 3y - 1)$

Remember that you can always check factoring by multiplying. Keep in mind that an expression is factored when it is written as a product.

Do Exercises 32–37.

e Collecting Like Terms

Terms such as $5x$ and $-4x$, whose variable factors are exactly the same, are called **like terms.** Similarly, numbers, such as -7 and 13, are like terms. Also, $3y^2$ and $9y^2$ are like terms because the variables are raised to the same power. Terms such as $4y$ and $5y^2$ are not like terms, and $7x$ and $2y$ are not like terms.

The process of **collecting like terms** is also based on the distributive laws. We can apply the distributive law when a factor is on the right because of the commutative law of multiplication.

Later in this text, terminology like "collecting like terms" and "combining like terms" will also be referred to as "simplifying."

EXAMPLES Collect like terms. Try to write just the answer, if you can.

24. $4x + 2x = (4 + 2)x = 6x$ Factoring out the x using a distributive law

25. $2x + 3y - 5x - 2y = 2x - 5x + 3y - 2y$
$= (2 - 5)x + (3 - 2)y = -3x + y$

26. $3x - x = 3x - 1x = (3 - 1)x = 2x$

27. $x - 0.24x = 1 \cdot x - 0.24x = (1 - 0.24)x = 0.76x$

28. $x - 6x = 1 \cdot x - 6 \cdot x = (1 - 6)x = -5x$

29. $4x - 7y + 9x - 5 + 3y - 8 = 13x - 4y - 13$

30. $\frac{2}{3}a - b + \frac{4}{5}a + \frac{1}{4}b - 10 = \frac{2}{3}a - 1 \cdot b + \frac{4}{5}a + \frac{1}{4}b - 10$
$= \left(\frac{2}{3} + \frac{4}{5}\right)a + \left(-1 + \frac{1}{4}\right)b - 10$
$= \left(\frac{10}{15} + \frac{12}{15}\right)a + \left(-\frac{4}{4} + \frac{1}{4}\right)b - 10$
$= \frac{22}{15}a - \frac{3}{4}b - 10$

Do Exercises 38–44.

Collect like terms.

38. $6x - 3x$

39. $7x - x$

40. $x - 9x$

41. $x - 0.41x$

42. $5x + 4y - 2x - y$

43. $3x - 7x - 11 + 8y + 4 - 13y$

44. $-\frac{2}{3} - \frac{3}{5}x + y + \frac{7}{10}x - \frac{2}{9}y$

Answers on page A-16

Study Tips

LEARNING RESOURCES

Are you aware of all the learning resources that exist for this textbook? Many details are given in the Preface.

- The *Student's Solutions Manual* contains worked-out solutions to the odd-numbered exercises in the exercise sets. You can order this through the bookstore or by calling 1-800-282-0693.
- An extensive set of *videotapes* supplements this text. These are available on CD-ROM by calling 1-800-282-0693.
- *Tutorial software* called InterAct Math also accompanies this text. If it is not available in the campus learning center, you can order it by calling 1-800-282-0693.
- The Addison-Wesley *Math Tutor Center* is available for help with the odd-numbered exercises. You can order this service by calling 1-800-824-7799.
- Extensive help is available online via MyMathLab and/or MathXL. Ask your instructor for information about these or visit MyMathLab.com and MathXL.com.

6.7 EXERCISE SET

For Extra Help

Digital Video Tutor CD 5 Videotape 7 | InterAct Math | Math Tutor Center | MathXL | MyMathLab

a Find an equivalent expression with the given denominator.

1. $\frac{3}{5}$; $5y$

2. $\frac{5}{8}$; $8t$

3. $\frac{2}{3}$; $15x$

4. $\frac{6}{7}$; $14y$

Simplify.

5. $-\frac{24a}{16a}$

6. $-\frac{42t}{18t}$

7. $-\frac{42ab}{36ab}$

8. $-\frac{64pq}{48pq}$

b Write an equivalent expression. Use a commutative law.

9. $y + 8$

10. $x + 3$

11. mn

12. ab

13. $9 + xy$

14. $11 + ab$

15. $ab + c$

16. $rs + t$

Write an equivalent expression. Use an associative law.

17. $a + (b + 2)$

18. $3(vw)$

19. $(8x)y$

20. $(y + z) + 7$

21. $(a + b) + 3$

22. $(5 + x) + y$

23. $3(ab)$

24. $(6x)y$

Use the commutative and associative laws to write three equivalent expressions.

25. $(a + b) + 2$

26. $(3 + x) + y$

27. $5 + (v + w)$

28. $6 + (x + y)$

29. $(xy)3$

30. $(ab)5$

31. $7(ab)$

32. $5(xy)$

c Multiply.

33. $2(b + 5)$

34. $4(x + 3)$

35. $7(1 + t)$

36. $4(1 + y)$

37. $6(5x + 2)$

38. $9(6m + 7)$

39. $7(x + 4 + 6y)$

40. $4(5x + 8 + 3p)$

41. $7(x-3)$

42. $15(y-6)$

43. $-3(x-7)$

44. $1.2(x-2.1)$

45. $\frac{2}{3}(b-6)$

46. $\frac{5}{8}(y+16)$

47. $7.3(x-2)$

48. $5.6(x-8)$

49. $-\frac{3}{5}(x-y+10)$

50. $-\frac{2}{3}(a+b-12)$

51. $-9(-5x-6y+8)$

52. $-7(-2x-5y+9)$

53. $-4(x-3y-2z)$

54. $8(2x-5y-8z)$

55. $3.1(-1.2+3.2y-1.1)$

56. $-2.1(-4.2x-4.3y-2.2)$

List the terms of the expression.

57. $4x+3z$

58. $8x-1.4y$

59. $7x+8y-9z$

60. $8a+10b-18c$

d Factor. Check by multiplying.

61. $2x+4$

62. $5y+20$

63. $30+5y$

64. $7x+28$

65. $14x+21y$

66. $18a+24b$

67. $5x+10+15y$

68. $9a+27b+81$

69. $8x-24$

70. $10x-50$

71. $32-4y$

72. $24-6m$

73. $8x + 10y - 22$

74. $9a + 6b - 15$

75. $ax - a$

76. $by - 9b$

77. $ax - ay - az$

78. $cx + cy - cz$

79. $18x - 12y + 6$

80. $-14x + 21y + 7$

81. $\frac{2}{3}x - \frac{5}{3}y + \frac{1}{3}$

82. $\frac{3}{5}a + \frac{4}{5}b - \frac{1}{5}$

e Collect like terms.

83. $9a + 10a$

84. $12x + 2x$

85. $10a - a$

86. $-16x + x$

87. $2x + 9z + 6x$

88. $3a - 5b + 7a$

89. $7x + 6y^2 + 9y^2$

90. $12m^2 + 6q + 9m^2$

91. $41a + 90 - 60a - 2$

92. $42x - 6 - 4x + 2$

93. $23 + 5t + 7y - t - y - 27$

94. $45 - 90d - 87 - 9d + 3 + 7d$

95. $\frac{1}{2}b + \frac{1}{2}b$

96. $\frac{2}{3}x + \frac{1}{3}x$

97. $2y + \frac{1}{4}y + y$

98. $\frac{1}{2}a + a + 5a$

99. $11x - 3x$

100. $9t - 17t$

101. $6n - n$

102. $100t - t$

103. $y - 17y$

104. $3m - 9m + 4$

105. $-8 + 11a - 5b + 6a - 7b + 7$

106. $8x - 5x + 6 + 3y - 2y - 4$

107. $9x + 2y - 5x$

108. $8y - 3z + 4y$

109. $11x + 2y - 4x - y$

110. $13a + 9b - 2a - 4b$

111. $2.7x + 2.3y - 1.9x - 1.8y$

112. $6.7a + 4.3b - 4.1a - 2.9b$

113. $\frac{13}{2}a + \frac{9}{5}b - \frac{2}{3}a - \frac{3}{10}b - 42$

114. $\frac{11}{4}x + \frac{2}{3}y - \frac{4}{5}x - \frac{1}{6}y + 12$

115. ^{D}W The distributive law was introduced before the discussion on collecting like terms. Why do you think this was done?

116. ^{D}W Find two different expressions for the total area of the two rectangles shown below. Explain the equivalence of the expressions in terms of the distributive law.

SKILL MAINTENANCE

Solve. [1.4b]

117. $x + 27 = 36$

118. $34 \cdot y = 408$

Evaluate. [1.6b]

119. 4^3

120. 2^5

121. 15^2

122. 3^4

SYNTHESIS

Tell whether the expressions are equivalent. Explain.

123. $3t + 5$ and $3 \cdot 5 + t$

124. $4x$ and $x + 4$

125. $5m + 6$ and $6 + 5m$

126. $(x + y) + z$ and $z + (x + y)$

127. Factor: $q + qr + qrs + qrst$.

128. Collect like terms:
$21x + 44xy + 15y - 16x - 8y - 38xy + 2y + xy$.

6.8 SIMPLIFYING EXPRESSIONS; ORDER OF OPERATIONS

Objectives

a Find an equivalent expression for an opposite without parentheses, where an expression has several terms.

b Simplify expressions by removing parentheses and collecting like terms.

c Simplify expressions with parentheses inside parentheses.

d Simplify expressions using rules for order of operations.

We now expand our ability to manipulate expressions by first considering opposites of sums and differences. Then we simplify expressions involving parentheses.

a Opposites of Sums

What happens when we multiply a real number by -1? Consider the following products:

$$-1(7) = -7, \qquad -1(-5) = 5, \qquad -1(0) = 0.$$

From these examples, it appears that when we multiply a number by -1, we get the opposite, or additive inverse, of that number.

THE PROPERTY OF −1

For any real number a,

$$-1 \cdot a = -a.$$

(Negative one times a is the opposite, or additive inverse, of a.)

The property of -1 enables us to find certain expressions equivalent to opposites of sums.

EXAMPLES Find an equivalent expression without parentheses.

1. $-(3 + x) = -1(3 + x)$ Using the property of -1

$= -1 \cdot 3 + (-1)x$ Using a distributive law, multiplying each term by -1

$= -3 + (-x)$ Using the property of -1

$= -3 - x$

2. $-(3x + 2y + 4) = -1(3x + 2y + 4)$ Using the property of -1

$= -1(3x) + (-1)(2y) + (-1)4$ Using a distributive law

$= -3x - 2y - 4$ Using the property of -1

Do Exercises 1 and 2.

Suppose we want to remove parentheses in an expression like

$$-(x - 2y + 5).$$

We can first rewrite any subtractions inside the parentheses as additions. Then we take the opposite of each term:

$$-(x - 2y + 5) = -[x + (-2y) + 5]$$
$$= -x + 2y - 5.$$

The most efficient method for removing parentheses is to replace each term in the parentheses with its opposite ("change the sign of every term"). Doing so for $-(x - 2y + 5)$, we obtain $-x + 2y - 5$ as an equivalent expression.

Find an equivalent expression without parentheses.

1. $-(x + 2)$

2. $-(5x + 2y + 8)$

Answers on page A-17

EXAMPLES Find an equivalent expression without parentheses.

3. $-(5 - y) = -5 + y = y + (-5) = y - 5$ Changing the sign of each term

4. $-(2a - 7b - 6) = -2a + 7b + 6$

5. $-(-3x + 4y + z - 7w - 23) = 3x - 4y - z + 7w + 23$

Do Exercises 3–6.

b Removing Parentheses and Simplifying

When a sum is added, as in $5x + (2x + 3)$, we can simply remove, or drop, the parentheses and collect like terms because of the associative law of addition:

$$5x + (2x + 3) = 5x + 2x + 3 = 7x + 3.$$

On the other hand, when a sum is subtracted, as in $3x - (4x + 2)$, no "associative" law applies. However, we can subtract by adding an opposite. We then remove parentheses by changing the sign of each term inside the parentheses and collecting like terms.

EXAMPLE 6 Remove parentheses and simplify.

$3x - (4x + 2) = 3x + [-(4x + 2)]$ Adding the opposite of $(4x + 2)$

$= 3x + (-4x - 2)$ Changing the sign of each term inside the parentheses

$= 3x - 4x - 2$

$= -x - 2$ Collecting like terms

Do Exercises 7 and 8.

In practice, the first three steps of Example 6 are usually combined by changing the sign of each term in parentheses and then collecting like terms.

EXAMPLES Remove parentheses and simplify.

7. $5y - (3y + 4) = 5y - 3y - 4$ Removing parentheses by changing the sign of every term inside the parentheses

$= 2y - 4$ Collecting like terms

8. $3x - 2 - (5x - 8) = 3x - 2 - 5x + 8$

$= -2x + 6$, or $6 - 2x$

9. $(3a + 4b - 5) - (2a - 7b + 4c - 8)$

$= 3a + 4b - 5 - 2a + 7b - 4c + 8$

$= a + 11b - 4c + 3$

Do Exercises 9–11.

Find an equivalent expression without parentheses. Try to do this in one step.

3. $-(6 - t)$

4. $-(x - y)$

5. $-(-4a + 3t - 10)$

6. $-(18 - m - 2n + 4z)$

Remove parentheses and simplify.

7. $5x - (3x + 9)$

8. $5y - 2 - (2y - 4)$

Remove parentheses and simplify.

9. $6x - (4x + 7)$

10. $8y - 3 - (5y - 6)$

11. $(2a + 3b - c) - (4a - 5b + 2c)$

Answers on page A-17

Remove parentheses and simplify.

12. $y - 9(x + y)$

13. $5a - 3(7a - 6)$

14. $4a - b - 6(5a - 7b + 8c)$

15. $5x - \frac{1}{4}(8x + 28)$

Simplify.

16. $12 - (8 + 2)$

17. $\{9 - [10 - (13 + 6)]\}$

18. $[24 \div (-2)] \div (-2)$

19. $5(3 + 4) - \{8 - [5 - (9 + 6)]\}$

Answers on page A-17

Next, consider subtracting an expression consisting of several terms multiplied by a number other than 1 or -1.

EXAMPLE 10 Remove parentheses and simplify.

$$\begin{aligned} x - 3(x + y) &= x + [-3(x + y)] && \text{Adding the opposite of } 3(x + y) \\ &= x + [-3x - 3y] && \text{Multiplying } x + y \text{ by } -3 \\ &= x - 3x - 3y \\ &= -2x - 3y && \text{Collecting like terms} \end{aligned}$$

EXAMPLES Remove parentheses and simplify.

11. $3y - 2(4y - 5) = 3y - 8y + 10$ — Multiplying each term in parentheses by -2

$= -5y + 10$

12. $(2a + 3b - 7) - 4(-5a - 6b + 12)$

$= 2a + 3b - 7 + 20a + 24b - 48$

$= 22a + 27b - 55$

13. $2y - \frac{1}{3}(9y - 12) = 2y - 3y + 4 = -y + 4$

Do Exercises 12–15.

C Parentheses Within Parentheses

In addition to parentheses, some expressions contain other grouping symbols such as brackets [ ] and braces { }.

> When more than one kind of grouping symbol occurs, do the computations in the innermost ones first. Then work from the inside out.

EXAMPLES Simplify.

14. $[3 - (7 + 3)] = [3 - 10]$ — Computing $7 + 3$

$= -7$

15. $\{8 - [9 - (12 + 5)]\} = \{8 - [9 - 17]\}$ — Computing $12 + 5$

$= \{8 - [-8]\}$ — Computing $9 - 17$

$= 8 + 8 = 16$

16. $\left[(-4) \div \left(-\frac{1}{4}\right)\right] \div \frac{1}{4} = [(-4) \cdot (-4)] \div \frac{1}{4}$ — Working within the brackets; computing $(-4) \div \left(-\frac{1}{4}\right)$

$= 16 \div \frac{1}{4}$

$= 16 \cdot 4 = 64$

17. $4(2 + 3) - \{7 - [4 - (8 + 5)]\}$

$= 4 \cdot 5 - \{7 - [4 - 13]\}$ — Working with the innermost parentheses first

$= 20 - \{7 - [-9]\}$ — Computing $4 \cdot 5$ and $4 - 13$

$= 20 - 16$ — Computing $7 - [-9]$

$= 4$

Do Exercises 16–19.

EXAMPLE 18 Simplify.

$$[5(x+2)-3x]-[3(y+2)-7(y-3)]$$
$= [5x + 10 - 3x] - [3y + 6 - 7y + 21]$ Working with the innermost parentheses first

$= [2x + 10] - [-4y + 27]$ Collecting like terms within brackets

$= 2x + 10 + 4y - 27$ Removing brackets

$= 2x + 4y - 17$ Collecting like terms

Do Exercise 20.

d Order of Operations

When several operations are to be done in a calculation or a problem, we apply the following rules.

RULES FOR ORDER OF OPERATIONS

1. Do all calculations within grouping symbols before operations outside.
2. Evaluate all exponential expressions.
3. Do all multiplications and divisions in order from left to right.
4. Do all additions and subtractions in order from left to right.

These rules are consistent with the way in which most computers and scientific calculators perform calculations.

EXAMPLE 19 Simplify: $-34 \cdot 56 - 17$.

There are no parentheses or powers, so we start with the third step.

$-34 \cdot 56 - 17 = -1904 - 17$ Doing all multiplications and divisions in order from left to right

$= -1921$ Doing all additions and subtractions in order from left to right

EXAMPLE 20 Simplify: $25 \div (-5) + 50 \div (-2)$.

There are no calculations inside parentheses or powers. The parentheses with (-5) and (-2) are used only to represent the negative numbers. We begin by doing all multiplications and divisions.

$25 \div (-5) + 50 \div (-2)$

$= -5 + (-25)$ Doing all multiplications and divisions in order from left to right

$= -30$ Doing all additions and subtractions in order from left to right.

Do Exercises 21–23.

20. Simplify:

$[3(x+2)+2x]-$
$[4(y+2)-3(y-2)]$.

Simplify.

21. $23 - 42 \cdot 30$

22. $32 \div 8 \cdot 2$

23. $-24 \div 3 - 48 \div (-4)$

Answers on page A-17

EXAMPLE 21 Simplify: $2^4 + 51 \cdot 4 - (37 + 23 \cdot 2)$.

$$2^4 + 51 \cdot 4 - (37 + 23 \cdot 2)$$

$= 2^4 + 51 \cdot 4 - (37 + 46)$	Following the rules for order of operations within the parentheses first
$= 2^4 + 51 \cdot 4 - 83$	Completing the addition inside parentheses
$= 16 + 51 \cdot 4 - 83$	Evaluating exponential expressions
$= 16 + 204 - 83$	Doing all multiplications
$= 220 - 83$	Doing all additions and subtractions in order from left to right
$= 137$	

CALCULATOR CORNER

Order of Operations and Grouping Symbols Parentheses are necessary in some calculations in order to ensure that operations are performed in the desired order. To simplify $-5(3 - 6) - 12$, we press [(−)] [5] [(] [3] [−] [6] [)] [−] [1] [2] [ENTER]. The result is 3. Without parentheses, the computation is $-5 \cdot 3 - 6 - 12$, and the result is -33.

```
-5(3-6)-12
                3
-5*3-6-12
              -33
```

When a negative number is raised to an even power, parentheses must also be used. To find $(-3)^4$, we press [(] [(−)] [3] [)] [^] [4] [ENTER]. The result is 81. Without parentheses, the computation is $-3^4 = -1 \cdot 3^4 = -1 \cdot 81 = -81$.

```
(-3)^4
               81
-3^4
              -81
```

To simplify an expression like $\dfrac{49 - 104}{7 + 4}$, we must enter it as $(49 - 104) \div (7 + 4)$. We press [(] [4] [9] [−] [1] [0] [4] [)] [÷] [(] [7] [+] [4] [)] [ENTER]. The result is -5.

```
(49-104)/(7+4)
               -5
```

Exercises: Calculate.

1. $-8 + 4(7 - 9) + 5$
2. $-3[2 + (-5)]$
3. $7[4 - (-3)] + 5[3^2 - (-4)]$
4. $(-7)^6$
5. $(-17)^5$
6. $(-104)^3$
7. -7^6
8. -17^5
9. -104^3
10. $\dfrac{38 - 178}{5 + 30}$
11. $\dfrac{311 - 17^2}{2 - 13}$
12. $785 - \dfrac{285 - 5^4}{17 + 3 \cdot 51}$

A fraction bar can play the role of a grouping symbol, although such a symbol is not as evident as the others.

EXAMPLE 22 Simplify: $\dfrac{-64 \div (-16) \div (-2)}{2^3 - 3^2}$.

An equivalent expression with brackets as grouping symbols is

$$[-64 \div (-16) \div (-2)] \div [2^3 - 3^2].$$

This shows, in effect, that we do the calculations in the numerator and then in the denominator, and divide the results:

$$\frac{-64 \div (-16) \div (-2)}{2^3 - 3^2} = \frac{4 \div (-2)}{8 - 9} = \frac{-2}{-1} = 2.$$

Do Exercises 24 and 25.

Simplify.

24. $52 \cdot 5 + 5^3 - (4^2 - 48 \div 4)$

25. $\dfrac{5 - 10 - 5 \cdot 23}{2^3 + 3^2 - 7}$

Answers on page A-17

Study Tips

TEST PREPARATION

You are probably ready to begin preparing for the test on this chapter. Here are some test-taking study tips.

- **Make up your own test questions as you study.** After you have done your homework over a particular objective, write one or two questions on your own that you think might be on a test. You will be amazed at the insight this will provide.
- **Do an overall review of the chapter, focusing on the objectives and the examples.** This should be accompanied by a study of any class notes you may have taken.
- **Do the review exercises at the end of the chapter.** Check your answers at the back of the book. If you have trouble with an exercise, use the objective symbol as a guide to go back and do further study of that objective.
- **If you need extra help, call the AW Math Tutor Center at 1-888-777-0463.**
- **Do the chapter test at the end of the chapter.** Check the answers and use the objective symbols at the back of the book as a reference for where to review.
- **Ask former students for old exams.** Working such exams can be very helpful and allows you to see what various professors think is important.
- **When taking a test, read each question carefully and try to do all the questions the first time through, but pace yourself.** Answer all the questions, and mark those to recheck if you have time at the end. Very often, your first hunch will be correct.
- **Try to write your test in a neat and orderly manner.** Very often, your instructor tries to give you partial credit when grading an exam. If your test paper is sloppy and disorderly, it is difficult to verify the partial credit. Doing your work neatly can ease such a task for the instructor.

6.8 EXERCISE SET

For Extra Help

Digital Video Tutor CD 6 Videotape 7 | InterAct Math | Math Tutor Center | MathXL | MyMathLab

a Find an equivalent expression without parentheses.

1. $-(2x + 7)$

2. $-(8x + 4)$

3. $-(8 - x)$

4. $-(a - b)$

5. $-(4a - 3b + 7c)$

6. $-(x - 4y - 3z)$

7. $-(6x - 8y + 5)$

8. $-(4x + 9y + 7)$

9. $-(3x - 5y - 6)$

10. $-(6a - 4b - 7)$

11. $-(-8x - 6y - 43)$

12. $-(-2a + 9b - 5c)$

b Remove parentheses and simplify.

13. $9x - (4x + 3)$

14. $4y - (2y + 5)$

15. $2a - (5a - 9)$

16. $12m - (4m - 6)$

17. $2x + 7x - (4x + 6)$

18. $3a + 2a - (4a + 7)$

19. $2x - 4y - 3(7x - 2y)$

20. $3a - 9b - 1(4a - 8b)$

21. $15x - y - 5(3x - 2y + 5z)$

22. $4a - b - 4(5a - 7b + 8c)$

23. $(3x + 2y) - 2(5x - 4y)$

24. $(-6a - b) - 5(2b + a)$

25. $(12a - 3b + 5c) - 5(-5a + 4b - 6c)$

26. $(-8x + 5y - 12) - 6(2x - 4y - 10)$

c Simplify.

27. $[9 - 2(5 - 4)]$

28. $[6 - 5(8 - 4)]$

29. $8[7 - 6(4 - 2)]$

30. $10[7 - 4(7 - 5)]$

31. $[4(9 - 6) + 11] - [14 - (6 + 4)]$

32. $[7(8 - 4) + 16] - [15 - (7 + 8)]$

33. $[10(x + 3) - 4] + [2(x - 1) + 6]$

34. $[9(x + 5) - 7] + [4(x - 12) + 9]$

35. $[7(x + 5) - 19] - [4(x - 6) + 10]$

36. $[6(x + 4) - 12] - [5(x - 8) + 14]$

37. $3\{[7(x - 2) + 4] - [2(2x - 5) + 6]\}$

38. $4\{[8(x - 3) + 9] - [4(3x - 2) + 6]\}$

39. $4\{[5(x - 3) + 2] - 3[2(x + 5) - 9]\}$

40. $3\{[6(x - 4) + 5] - 2[5(x + 8) - 3]\}$

d Simplify.

41. $8 - 2 \cdot 3 - 9$

42. $8 - (2 \cdot 3 - 9)$

43. $(8 - 2 \cdot 3) - 9$

44. $(8 - 2)(3 - 9)$

45. $[(-24) \div (-3)] \div \left(-\frac{1}{2}\right)$

46. $[32 \div (-2)] \div (-2)$

47. $16 \cdot (-24) + 50$

48. $10 \cdot 20 - 15 \cdot 24$

49. $2^4 + 2^3 - 10$

50. $40 - 3^2 - 2^3$

51. $5^3 + 26 \cdot 71 - (16 + 25 \cdot 3)$

52. $4^3 + 10 \cdot 20 + 8^2 - 23$

53. $4 \cdot 5 - 2 \cdot 6 + 4$

54. $4 \cdot (6 + 8)/(4 + 3)$

55. $4^3/8$

56. $5^3 - 7^2$

57. $8(-7) + 6(-5)$

58. $10(-5) + 1(-1)$

59. $19 - 5(-3) + 3$

60. $14 - 2(-6) + 7$

61. $9 \div (-3) + 16 \div 8$

62. $-32 - 8 \div 4 - (-2)$

63. $-4^2 + 6$

64. $-5^2 + 7$

65. $-8^2 - 3$

66. $-9^2 - 11$

67. $12 - 20^3$

68. $20 + 4^3 \div (-8)$

69. $2 \cdot 10^3 - 5000$

70. $-7(3^4) + 18$

71. $6[9 - (3 - 4)]$

72. $8[(6 - 13) - 11]$

73. $-1000 \div (-100) \div 10$

74. $256 \div (-32) \div (-4)$

75. $8 - (7 - 9)$

76. $(8 - 7) - 9$

77. $\dfrac{10 - 6^2}{9^2 + 3^2}$

78. $\dfrac{5^2 - 4^3 - 3}{9^2 - 2^2 - 1^5}$

79. $\dfrac{3(6-7)-5\cdot 4}{6\cdot 7-8(4-1)}$

80. $\dfrac{20(8-3)-4(10-3)}{10(2-6)-2(5+2)}$

81. $\dfrac{2^3-3^2+12\cdot 5}{-32\div(-16)\div(-4)}$

82. $\dfrac{|3-5|^2-|7-13|}{|12-9|+|11-14|}$

83. DW Jake keys in 18/2 · 3 on his calculator and expects the result to be 3. What mistake is he making?

84. DW Determine whether $|-x|$ and $|x|$ are equivalent. Explain.

SKILL MAINTENANCE

Find fraction notation. Simplify. [4.3b]

85. 65%

86. 150%

87. 37.5%

88. $33\frac{1}{3}\%$

For the circle with the given radius, find the diameter, the circumference, and the area. Use 3.14 for π. [5.4a, b, c]

89. $r = 15$ yd

90. $r = 8.2$ m

91. $r = 9\frac{1}{2}$ mi

92. $r = 2400$ cm

SYNTHESIS

Find an equivalent expression by enclosing the last three terms in parentheses preceded by a minus sign.

93. $6y + 2x - 3a + c$

94. $x - y - a - b$

95. $6m + 3n - 5m + 4b$

Simplify.

96. $z - \{2z - [3z - (4z - 5z) - 6z] - 7z\} - 8z$

97. $\{x - [f - (f - x)] + [x - f]\} - 3x$

98. $x - \{x - 1 - [x - 2 - (x - 3 - \{x - 4 - [x - 5 - (x - 6)]\})]\}$

99. Use your calculator to do the following.

a) Evaluate $x^2 + 3$ when $x = 7$, when $x = -7$, and when $x = -5.013$.

b) Evaluate $1 - x^2$ when $x = 5$, when $x = -5$, and when $x = -10.455$.

100. Express $3^3 + 3^3 + 3^3$ as a power of 3.

Find the average.

101. $-15,\ 20,\ 50,\ -82,\ -7,\ -2$

102. $-1,\ 1,\ 2,\ -2,\ 3,\ -8,\ -10$

CHAPTER

6 Summary and Review

The review that follows is meant to prepare you for a chapter exam. It consists of two parts. The first part is a checklist of some of the Study Tips referred to in this chapter, as well as a list of important properties and formulas. The second part is the Review Exercises. These provide practice exercises for the exam, together with references to section objectives so you can go back and review. Before beginning, stop and look back over the skills you have obtained. What skills in mathematics do you have now that you did not have before studying this chapter?

STUDY TIPS CHECKLIST

The foundation of all your study skills is TIME!

- ☐ Are you making use of the textbook supplements, such as the Math Tutor Center, the *Student's Solutions Manual*, and the videotapes?
- ☐ Have you determined the location of the learning resource centers on your campus, such as a math lab, tutor center, and your instructor's office?
- ☐ Are you stopping to work the margin exercises when directed to do so?
- ☐ Are you keeping one section ahead in your syllabus?

IMPORTANT PROPERTIES AND FORMULAS

PROPERTIES OF THE REAL-NUMBER SYSTEM

The Commutative Laws: $a + b = b + a, \quad ab = ba$

The Associative Laws: $a + (b + c) = (a + b) + c, \quad a(bc) = (ab)c$

The Identity Properties: $a + 0 = 0 + a = a, \quad a \cdot 1 = 1 \cdot a = a$

The Inverse Properties: For any real number a, there is an opposite $-a$ such that $a + (-a) = (-a) + a = 0$.
For any nonzero real number a, there is a reciprocal $\frac{1}{a}$ such that $a \cdot \frac{1}{a} = \frac{1}{a} \cdot a = 1$.

The Distributive Laws: $a(b + c) = ab + ac, \quad a(b - c) = ab - ac$

REVIEW EXERCISES

1. Evaluate $\frac{x - y}{3}$ when $x = 17$ and $y = 5$. [6.1a]

2. Translate to an algebraic expression: [6.1b]
Nineteen percent of some number.

3. Tell which integers correspond to this situation: [6.2a]
David has a debt of \$45 and Joe has \$72 in his savings account.

4. Find: $|-38|$. [6.2e]

Graph the number on a number line. [6.2b]

5. -2.5

6. $\frac{8}{9}$

Use either $<$ or $>$ for $\square$ to write a true sentence. [6.2d]

7. $-3 \square 10$

8. $-1 \square -6$

9. $0.126 \square -12.6$

10. $-\frac{2}{3} \square -\frac{1}{10}$

Find the opposite. [6.3b]

11. 3.8

12. $-\frac{3}{4}$

Find the reciprocal. [6.6b]

13. $\frac{3}{8}$

14. -7

15. Evaluate $-x$ when $x = -34$. [6.3b]

16. Evaluate $-(-x)$ when $x = 5$. [6.3b]

Compute and simplify.

17. $4 + (-7)$ [6.3a]

18. $6 + (-9) + (-8) + 7$ [6.3a]

19. $-3.8 + 5.1 + (-12) + (-4.3) + 10$ [6.3a]

20. $-3 - (-7)$ [6.4a]

21. $-\frac{9}{10} - \frac{1}{2}$ [6.4a]

22. $-3.8 - 4.1$ [6.4a]

23. $-9 \cdot (-6)$ [6.5a]

24. $-2.7(3.4)$ [6.5a]

25. $\frac{2}{3} \cdot \left(-\frac{3}{7}\right)$ [6.5a]

26. $3 \cdot (-7) \cdot (-2) \cdot (-5)$ [6.5a]

27. $35 \div (-5)$ [6.6a]

28. $-5.1 \div 1.7$ [6.6c]

29. $-\frac{3}{11} \div \left(-\frac{4}{11}\right)$ [6.6c]

30. $(-3.4 - 12.2) - 8(-7)$ [6.8d]

31. $\frac{-12(-3) - 2^3 - (-9)(-10)}{3 \cdot 10 + 1}$ [6.8d]

32. $-16 \div 4 - 30 \div (-5)$ [6.8d]

33. $9[(7 - 14) - 13]$ [6.8d]

Solve.

34. On the first, second, and third downs, a football team had these gains and losses: 5-yd gain, 12-yd loss, and 15-yd gain, respectively. Find the total gain (or loss). [6.3c]

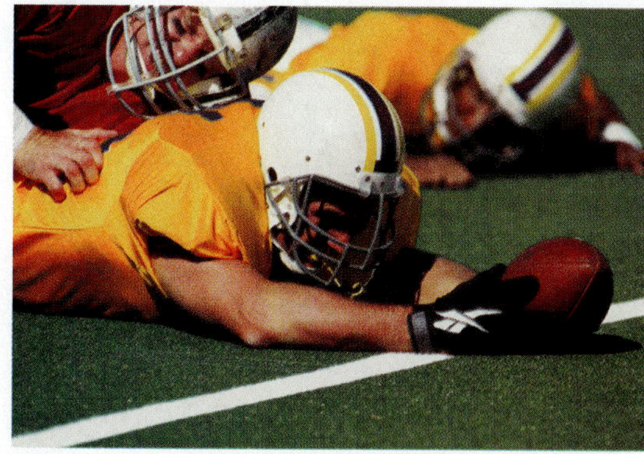

35. Kaleb's total assets are \$170. He borrows \$300. What are his total assets now? [6.4b]

36. *Stock Price.* The value of EFX Corp. began the day at \$17.68 per share and dropped \$1.63 per hour for 8 hr. What was the price of the stock after 8 hr? [6.5b]

37. *Checking Account Balance.* Yuri had \$68 in his checking account. After writing checks to make seven purchases of DVDs at the same price for each, the balance in his account was −\$64.65. What was the price of each DVD? [6.4b], [6.6d]

Multiply. [6.7c]

38. $5(3x - 7)$

39. $-2(4x - 5)$

40. $10(0.4x + 1.5)$

41. $-8(3 - 6x)$

Factor. [6.7d]

42. $2x - 14$

43. $6x - 6$

44. $5x + 10$

45. $12 - 3x$

Collect like terms. [6.7e]

46. $11a + 2b - 4a - 5b$

47. $7x - 3y - 9x + 8y$

48. $6x + 3y - x - 4y$

49. $-3a + 9b + 2a - b$

Remove parentheses and simplify.

50. $2a - (5a - 9)$ [6.8b]

51. $3(b + 7) - 5b$ [6.8b]

52. $3[11 - 3(4 - 1)]$ [6.8c]

53. $2[6(y - 4) + 7]$ [6.8c]

54. $[8(x + 4) - 10] - [3(x - 2) + 4]$ [6.8c]

55. $5\{[6(x - 1) + 7] - [3(3x - 4) + 8]\}$ [6.8c]

Write true or false. [6.2d]

56. $-9 \leq 11$

57. $-11 \geq -3$

58. Write another inequality with the same meaning as $-3 < x$. [6.2d]

59. D$_W$ Explain the notion of the opposite of a number in as many ways as possible. [6.3b]

60. D$_W$ Is the absolute value of a number always positive? Why or why not? [6.2e]

SYNTHESIS

Simplify. [6.2e], [6.4a], [6.6a], [6.8d]

61. $-\left|\frac{7}{8} - \left(-\frac{1}{2}\right) - \frac{3}{4}\right|$

62. $(|2.7 - 3| + 3^2 - |-3|) \div (-3)$

63. $2000 - 1990 + 1980 - 1970 + \cdots - 20 + 10$

64. Find a formula for the perimeter of the following figure. [6.7e]

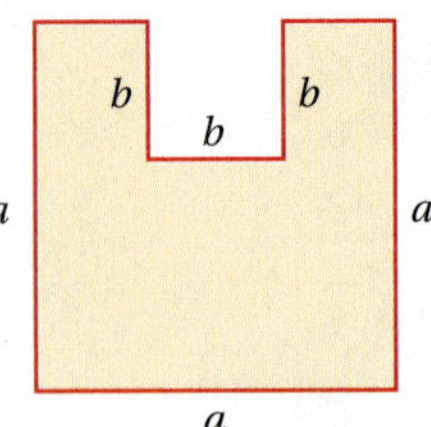

CHAPTER 6

Chapter Test

1. Evaluate $\frac{3x}{y}$ when $x = 10$ and $y = 5$.

2. Write an algebraic expression: Nine less than some number.

3. Find the area of a triangle when the height h is 30 ft and the base b is 16 ft.

Use either < or > for □ to write a true sentence.

4. $-4 \square 0$

5. $-3 \square -8$

6. $-0.78 \square -0.87$

7. $-\frac{1}{8} \square \frac{1}{2}$

Find the absolute value.

8. $|-7|$

9. $\left|\frac{9}{4}\right|$

10. $|-2.7|$

Find the opposite.

11. $\frac{2}{3}$

12. -1.4

13. Evaluate $-x$ when $x = -8$.

Find the reciprocal.

14. -2

15. $\frac{4}{7}$

Compute and simplify.

16. $3.1 - (-4.7)$

17. $-8 + 4 + (-7) + 3$

18. $-\frac{1}{5} + \frac{3}{8}$

19. $2 - (-8)$

20. $3.2 - 5.7$

21. $\frac{1}{8} - \left(-\frac{3}{4}\right)$

22. $4 \cdot (-12)$

23. $-\frac{1}{2} \cdot \left(-\frac{3}{8}\right)$

24. $-45 \div 5$

25. $-\frac{3}{5} \div \left(-\frac{4}{5}\right)$

26. $4.864 \div (-0.5)$

27. $-2(16) - |2(-8) - 5^3|$

28. $-20 \div (-5) + 36 \div (-4)$

29. *Antarctica Highs and Lows.* The continent of Antarctica, which lies in the southern hemisphere, experiences winter in July. The average high temperature is $-67°F$ and the average low temperature is $-81°F$. How much higher is the average high than the average low?
Source: National Climatic Data Center

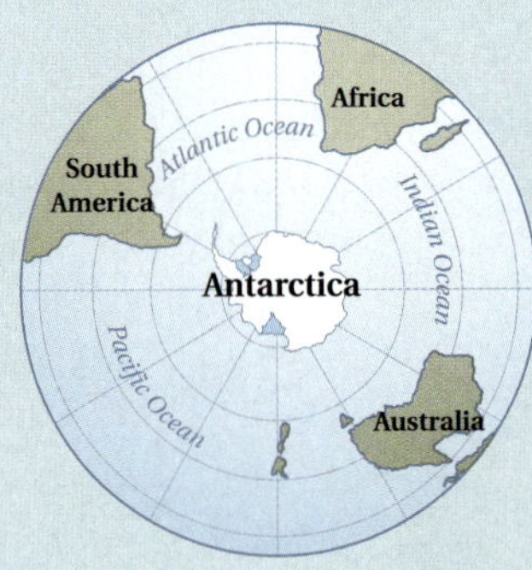

30. Maureen is a stockbroker. She kept track of the changes in the stock market over a period of 5 weeks. By how many points had the market risen or fallen over this time?

WEEK 1	WEEK 2	WEEK 3	WEEK 4	WEEK 5
Down 13 pts	Down 16 pts	Up 36 pts	Down 11 pts	Up 19 pts

31. *Population Decrease.* The population of a city was 18,600. It dropped 420 each year for 6 yr. What was the population of the city after 6 yr?

32. *Chemical Experiment.* During a chemical reaction, the temperature in the beaker decreased every minute by the same number of degrees. The temperature was 16°C at 11:08 A.M. By 11:43 A.M., the temperature had dropped to $-17°C$. By how many degrees did it drop each minute?

Multiply.

33. $3(6 - x)$

34. $-5(y - 1)$

Factor.

35. $12 - 22x$

36. $7x + 21 + 14y$

Simplify.

37. $6 + 7 - 4 - (-3)$

38. $5x - (3x - 7)$

39. $4(2a - 3b) + a - 7$

40. $4\{3[5(y - 3) + 9] + 2(y + 8)\}$

41. $256 \div (-16) \div 4$

42. $2^3 - 10[4 - (-2 + 18)3]$

43. Write an inequality with the same meaning as $x \leq -2$.

SKILL MAINTENANCE

44. Add: $\frac{1}{10} + \frac{7}{15}$.

45. Jenna's portion of the rent on the apartment she shares with her roommates is \$225 per month. The landlord decides to raise the rent 4%. What will Jenna's monthly portion of the rent be after the increase?

46. Find the perimeter of a square whose sides are $5\frac{3}{8}$ in. long.

47. Find the area of the figure.

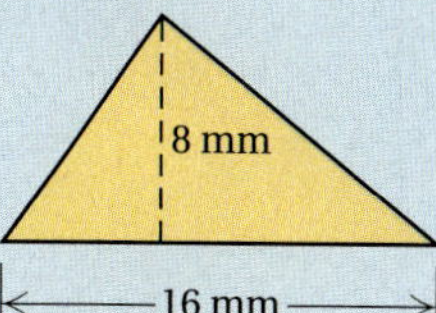

SYNTHESIS

Simplify.

48. $|-27 - 3(4)| - |-36| + |-12|$

49. $a - \{3a - [4a - (2a - 4a)]\}$

50. Find a formula for the perimeter of the figure shown here.

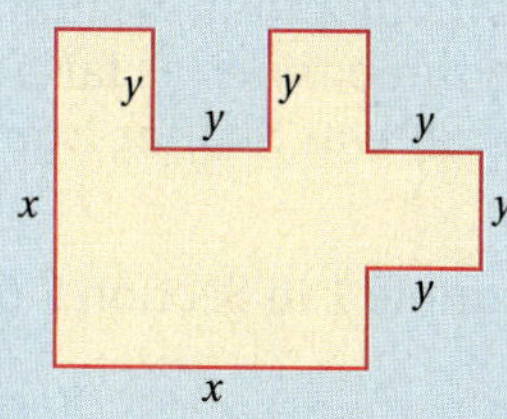

Solving Equations and Inequalities

Gateway to Chapter 7

In this chapter, we use the manipulations considered in Chapter 6 to solve equations and inequalities. We then use equations and inequalities to solve applied problems.

Real-World Application

In 1998, at age 79, Earl Shaffer became the oldest person to hike all 2100 miles of the Appalachian Trail—from Springer Mountain, Georgia, to Mount Katahdin, Maine. At one point, Shaffer stood atop Big Walker Mountain, Virginia, which is three times as far from the northern end as from the southern end. How far was Shaffer from each end of the trail?

Source: Appalachian Trail Conference

This problem appears as Example 1 in Section 7.6.

CHAPTER 7

CHAPTER 7

Pretest

Solve.

1. $4 + x = 12$ [7.1b]

2. $-7x = 49$ [7.2a]

3. $6a - 2 = 10$ [7.3a]

4. $4y + 9 = 2y + 7$ [7.3b]

5. $7 - 3(2x - 1) = 40$ [7.3c]

6. $\frac{4}{9}x - 1 = \frac{7}{8}$ [7.3a]

7. $1 + 2(a + 3) = 3(2a - 1) + 6$ [7.3c]

8. $y + 5 > 1$ [7.7c]

9. $-3x \leq 18$ [7.7d]

10. $5 - 2a < 7$ [7.7e]

11. $3x + 4 \geq 2x + 7$ [7.7c]

12. $8y < -18$ [7.7d]

13. $4 + x = x + 5$ [7.3c]

14. Solve for x: $y = Ax$. [7.4b]

15. Solve for a: $A = 3a - b$. [7.4b]

Solve.

16. **Oil-Painting Frame.** The perimeter of the ornate frame of an oil painting is 146 in. The width is 5 in. less than the length. Find the dimensions. [7.6a]

17. **Savings Investment.** Money is invested in a savings account at 4.25% simple interest. After 1 year, there is \$479.55 in the account. How much was originally invested? [7.6a]

18. **Consecutive Page Numbers.** The sum of three consecutive page numbers is 246. Find the page numbers. [7.6a]

19. **Lengths of a Ballfield.** The width of a rectangular ballfield is 96 yd. Find all possible lengths so that the perimeter of the ballfield will be at least 540 yd. [7.8b]

Graph on a number line. [7.7b]

20. $x > -3$

21. $x \leq 4$

Solve. [7.5a]

22. What is 24% of 85?

23. 2.16 is 4% of what number?

24. What percent of 85 is 17?

25. **Job Opportunities.** There were 49 thousand manicurists in 1998. Job opportunities are expected to grow to 62 thousand by 2008. What is the predicted percent of increase?
Source: *Handbook of U.S. Labor Statistics*

7.1 SOLVING EQUATIONS: THE ADDITION PRINCIPLE

Objectives

a Determine whether a given number is a solution of a given equation.

b Solve equations using the addition principle.

a Equations and Solutions

In order to solve problems, we must learn to solve equations.

EQUATION

An **equation** is a number sentence that says that the expressions on either side of the equals sign, =, represent the same number.

Here are some examples:

$$3 + 2 = 5, \quad 14 - 10 = 1 + 3, \quad x + 6 = 13, \quad 3x - 2 = 7 - x.$$

Equations have expressions on each side of the equals sign. The sentence "$14 - 10 = 1 + 3$" asserts that the expressions $14 - 10$ and $1 + 3$ name the same number.

Some equations are true. Some are false. Some are neither true nor false.

EXAMPLES Determine whether the equation is true, false, or neither.

1. $3 + 2 = 5$ The equation is *true.*
2. $7 - 2 = 4$ The equation is *false.*
3. $x + 6 = 13$ The equation is *neither* true nor false, because we do not know what number x represents.

Do Exercises 1–3.

SOLUTION OF AN EQUATION

Any replacement for the variable that makes an equation true is called a **solution** of the equation. To solve an equation means to find *all* of its solutions.

One way to determine whether a number is a solution of an equation is to evaluate the expression on each side of the equals sign by substitution. If the values are the same, then the number is a solution.

EXAMPLE 4 Determine whether 7 is a solution of $x + 6 = 13$.

We have

$x + 6 = 13$	Writing the equation
$7 + 6 \ ? \ 13$	Substituting 7 for x
$13 \mid$	TRUE

Since the left-hand and the right-hand sides are the same, we have a solution. No other number makes the equation true, so the only solution is the number 7.

Determine whether the equation is true, false, or neither.

1. $5 - 8 = -4$

false

2. $12 + 6 = 18$

true

3. $x + 6 = 7 - x$

Answers on page A-17

Determine whether the given number is a solution of the given equation.

4. 8; $x + 4 = 12$

true

5. 0; $x + 4 = 12$

False

6. -3; $7 + x = -4$

False

7. Solve using the addition principle:

$x + 2 = 11.$

$x+2-2 = 11-2$

$x = 9$

Answers on page A-17

EXAMPLE 5 Determine whether 19 is a solution of $7x = 141$.

We have

$7x = 141$	Writing the equation
$7(19) \stackrel{?}{=} 141$	Substituting 19 for x
133	FALSE

Since the left-hand and the right-hand sides are not the same, we do not have a solution.

Do Exercises 4–6.

b Using the Addition Principle

Consider the equation

$$x = 7.$$

We can easily see that the solution of this equation is 7. If we replace x with 7, we get

$$7 = 7, \quad \text{which is true.}$$

Now consider the equation of Example 4:

$$x + 6 = 13.$$

In Example 4, we discovered that the solution of this equation is also 7, but the fact that 7 is the solution is not as obvious. We now begin to consider principles that allow us to start with an equation like $x + 6 = 13$ and end up with an *equivalent equation,* like $x = 7$, in which the variable is alone on one side and for which the solution is easier to find.

EQUIVALENT EQUATIONS

Equations with the same solutions are called **equivalent equations.**

One of the principles that we use in solving equations involves adding. An equation $a = b$ says that a and b stand for the same number. Suppose this is true, and we add a number c to the number a. We get the same answer if we add c to b, because a and b are the same number.

THE ADDITION PRINCIPLE

For any real numbers a, b, and c,

$a = b$ is equivalent to $a + c = b + c$.

Let's again solve the equation $x + 6 = 13$ using the addition principle. We want to get x alone on one side. To do so, we use the addition principle, choosing to add -6 because $6 + (-6) = 0$:

$x + 6 = 13$	
$x + 6 + (-6) = 13 + (-6)$	Using the addition principle: adding -6 on both sides
$x + 0 = 7$	Simplifying
$x = 7.$	Identity property of 0: $x + 0 = x$

The solution of $x + 6 = 13$ is 7.

Do Exercise 7 on the preceding page.

When we use the addition principle, we sometimes say that we "add the same number on both sides of the equation." This is also true for subtraction, since we can express every subtraction as an addition. That is, since

$$a - c = b - c \quad \text{is equivalent to} \quad a + (-c) = b + (-c),$$

the addition principle tells us that we can "subtract the same number on both sides of the equation."

EXAMPLE 6 Solve: $x + 5 = -7$.

We have

$$\begin{aligned} x + 5 &= -7 \\ x + 5 - 5 &= -7 - 5 && \text{Using the addition principle: adding } -5 \text{ on both sides or subtracting 5 on both sides} \\ x + 0 &= -12 && \text{Simplifying} \\ x &= -12. && \text{Identity property of 0} \end{aligned}$$

We can see that the solution of $x = -12$ is the number -12. To check the answer, we substitute -12 in the original equation.

CHECK:

$$\begin{array}{c|c} x + 5 & -7 \\ \hline -12 + 5 \;?\; & -7 \\ -7 & \end{array} \quad \text{TRUE}$$

The solution of the original equation is -12.

In Example 6, to get x alone, we used the addition principle and subtracted 5 on both sides. This eliminated the 5 on the left. We started with $x + 5 = -7$, and, using the addition principle, we found a simpler equation $x = -12$ for which it was easy to *"see"* the solution. The equations $x + 5 = -7$ and $x = -12$ are *equivalent.*

Do Exercise 8.

Now we use the addition principle to solve an equation that involves a subtraction.

EXAMPLE 7 Solve: $a - 4 = 10$.

We have

$$\begin{aligned} a - 4 &= 10 \\ a - 4 + 4 &= 10 + 4 && \text{Using the addition principle: adding 4 on both sides} \\ a + 0 &= 14 && \text{Simplifying} \\ a &= 14. && \text{Identity property of 0} \end{aligned}$$

CHECK:

$$\begin{array}{c|c} a - 4 & 10 \\ \hline 14 - 4 \;?\; & 10 \\ 10 & \end{array} \quad \text{TRUE}$$

The solution is 14.

Do Exercise 9.

8. Solve using the addition principle, subtracting 5 on both sides:

$$x + 5 = -8.$$

9. Solve: $t - 3 = 19$.

Answers on page A-17

Study Tips

VIDEOTAPES

(ISBN 0-321-17517-4)

Developed and produced especially for this text, the videotapes feature an engaging team of instructors, who present material and concepts by using examples and exercises from every section of the text.

DIGITAL VIDEO TUTOR

(ISBN 0-321-17515-8)

The videotapes for this text are also available on CD-ROM, making it easy and convenient for you to watch video segments from a computer at home or on campus. The complete digitized video set, both affordable and portable, is ideal for distance learning or supplemental instruction.

Solve.

10. $8.7 = n - 4.5$

11. $y + 17.4 = 10.9$

Solve.

12. $x + \frac{1}{2} = -\frac{3}{2}$

13. $t - \frac{13}{4} = \frac{5}{8}$

Answers on page A-17

EXAMPLE 8 Solve: $-6.5 = y - 8.4$.

We have

$$-6.5 = y - 8.4$$

$$-6.5 + 8.4 = y - 8.4 + 8.4 \qquad \text{Using the addition principle: adding 8.4 on both sides to eliminate } -8.4 \text{ on the right}$$

$$1.9 = y.$$

CHECK:

$$\begin{array}{c|l} -6.5 & y - 8.4 \\ \hline -6.5 \;?\; & 1.9 - 8.4 \\ & -6.5 \end{array} \qquad \text{TRUE}$$

The solution is 1.9.

Note that equations are reversible. That is, if $a = b$ is true, then $b = a$ is true. Thus when we solve $-6.5 = y - 8.4$, we can reverse it and solve $y - 8.4 = -6.5$ if we wish.

Do Exercises 10 and 11.

EXAMPLE 9 Solve: $-\frac{2}{3} + x = \frac{5}{2}$.

We have

$$-\frac{2}{3} + x = \frac{5}{2}$$

$$\frac{2}{3} - \frac{2}{3} + x = \frac{2}{3} + \frac{5}{2} \qquad \text{Adding } \tfrac{2}{3} \text{ on both sides}$$

$$x = \frac{2}{3} \cdot \frac{2}{2} + \frac{5}{2} \cdot \frac{3}{3} \qquad \text{Multiplying by 1 to obtain equivalent fractional expressions with the least common denominator 6}$$

$$x = \frac{4}{6} + \frac{15}{6}$$

$$x = \frac{19}{6}.$$

CHECK:

$$\begin{array}{l|l} -\frac{2}{3} + x & \frac{5}{2} \\ \hline -\frac{2}{3} + \frac{19}{6} \;?\; & \frac{5}{2} \\ -\frac{4}{6} + \frac{19}{6} & \\ \frac{15}{6} & \\ \frac{5}{2} & \end{array} \qquad \text{TRUE}$$

The solution is $\frac{19}{6}$.

Do Exercises 12 and 13.

7.1 EXERCISE SET

For Extra Help

Digital Video Tutor CD 6 Videotape 8 | InterAct Math | Math Tutor Center | MathXL | MyMathLab

a Determine whether the given number is a solution of the given equation.

1. 15; $x + 17 = 32$

2. 35; $t + 17 = 53$

3. 21; $x - 7 = 12$

4. 36; $a - 19 = 17$

5. -7; $6x = 54$

6. -9; $8y = -72$

7. 30; $\frac{x}{6} = 5$

8. 49; $\frac{y}{8} = 6$

9. 19; $5x + 7 = 107$

10. 9; $9x + 5 = 86$

11. -11; $7(y - 1) = 63$

12. -18; $x + 3 = 3 + x$

b Solve using the addition principle. Don't forget to check!

13. $x + 2 = 6$

CHECK: $x + 2 = 6$?

14. $y + 4 = 11$

CHECK: $y + 4 = 11$?

15. $x + 15 = -5$

CHECK: $x + 15 = -5$?

16. $t + 10 = 44$

CHECK: $t + 10 = 44$?

17. $x + 6 = -8$

CHECK: $x + 6 = -8$?

18. $z + 9 = -14$

19. $x + 16 = -2$

20. $m + 18 = -13$

21. $x - 9 = 6$

22. $x - 11 = 12$

23. $x - 7 = -21$

24. $x - 3 = -14$

25. $5 + t = 7$

26. $8 + y = 12$

27. $-7 + y = 13$

28. $-8 + y = 17$

29. $-3 + t = -9$

30. $-8 + t = -24$

31. $x + \frac{1}{2} = 7$

32. $24 = -\frac{7}{10} + r$

33. $12 = a - 7.9$

34. $2.8 + y = 11$

35. $r + \frac{1}{3} = \frac{8}{3}$

36. $t + \frac{3}{8} = \frac{5}{8}$

37. $m + \frac{5}{6} = -\frac{11}{12}$

38. $x + \frac{2}{3} = -\frac{5}{6}$

39. $x - \frac{5}{6} = \frac{7}{8}$

40. $y - \frac{3}{4} = \frac{5}{6}$

41. $-\frac{1}{5} + z = -\frac{1}{4}$

42. $-\frac{1}{8} + y = -\frac{3}{4}$

43. $7.4 = x + 2.3$

44. $8.4 = 5.7 + y$

45. $7.6 = x - 4.8$

46. $8.6 = x - 7.4$

47. $-9.7 = -4.7 + y$

48. $-7.8 = 2.8 + x$

49. $5\frac{1}{6} + x = 7$

50. $5\frac{1}{4} = 4\frac{2}{3} + x$

51. $q + \frac{1}{3} = -\frac{1}{7}$

52. $52\frac{3}{8} = -84 + x$

53. **DW** Explain the difference between equivalent expressions and equivalent equations.

54. **DW** When solving an equation using the addition principle, how do you determine which number to add or subtract on both sides of the equation?

SKILL MAINTENANCE

55. Add: $-3 + (-8)$. [6.3a]

56. Subtract: $-3 - (-8)$. [6.4a]

57. Multiply: $-\frac{2}{3} \cdot \frac{5}{8}$. [6.5a]

58. Divide: $-\frac{3}{7} \div \left(-\frac{9}{7}\right)$. [6.6c]

59. Divide: $\frac{2}{3} \div \left(-\frac{4}{9}\right)$. [6.6c]

60. Add: $-8.6 + 3.4$. [6.3a]

61. Subtract: $-\frac{2}{3} - \left(-\frac{5}{8}\right)$. [6.4a]

62. Multiply: $(-25.4)(-6.8)$. [6.5a]

Translate to an algebraic expression. [6.1b]

63. Jane had $83 before paying x dollars for a pair of tennis shoes. How much does she have left?

64. Justin drove his S-10 pickup truck 65 mph for t hours. How far did he drive?

SYNTHESIS

Solve.

65. 🖩 $-356.788 = -699.034 + t$

66. $-\frac{4}{5} + \frac{7}{10} = x - \frac{3}{4}$

67. $x + \frac{4}{5} = -\frac{2}{3} - \frac{4}{15}$

68. $8 - 25 = 8 + x - 21$

69. $16 + x - 22 = -16$

70. $x + x = x$

71. $x + 3 = 3 + x$

72. $x + 4 = 5 + x$

73. $-\frac{3}{2} + x = -\frac{5}{17} - \frac{3}{2}$

74. $|x| = 5$

75. $|x| + 6 = 19$

7.2 SOLVING EQUATIONS: THE MULTIPLICATION PRINCIPLE

a Using the Multiplication Principle

Suppose that $a = b$ is true, and we multiply a by some number c. We get the same number if we multiply b by c, because a and b are the same number.

THE MULTIPLICATION PRINCIPLE

For any real numbers a, b, and c, $c \neq 0$,

$a = b$ is equivalent to $a \cdot c = b \cdot c$.

When using the multiplication principle, we sometimes say that we "multiply on both sides of the equation by the same number."

EXAMPLE 1 Solve: $5x = 70$.

To get x alone, we multiply by the *multiplicative inverse,* or *reciprocal,* of 5. Then we get the *multiplicative identity* 1 times x, or $1 \cdot x$, which simplifies to x. This allows us to eliminate 5 on the left.

$5x = 70$	The reciprocal of 5 is $\frac{1}{5}$.
$\frac{1}{5} \cdot 5x = \frac{1}{5} \cdot 70$	Multiplying by $\frac{1}{5}$ to get $1 \cdot x$ and eliminate 5 on the left
$1 \cdot x = 14$	Simplifying
$x = 14$	Identity property of 1: $1 \cdot x = x$

CHECK:

$$\begin{array}{c|l} 5x & = 70 \\ \hline 5 \cdot 14 & ?\ 70 \\ 70 & \end{array} \quad \text{TRUE}$$

The solution is 14.

The multiplication principle also tells us that we can "divide on both sides of the equation by a nonzero number." This is because division is the same as multiplying by a reciprocal. That is,

$$\frac{a}{c} = \frac{b}{c} \text{ is equivalent to } a \cdot \frac{1}{c} = b \cdot \frac{1}{c}, \quad \text{when } c \neq 0.$$

In an expression like $5x$ in Example 1, the number 5 is called the **coefficient.** Example 1 could be done as follows, dividing by 5, the coefficient of x, on both sides.

EXAMPLE 2 Solve: $5x = 70$.

We have

$5x = 70$	
$\frac{5x}{5} = \frac{70}{5}$	Dividing by 5 on both sides
$1 \cdot x = 14$	Simplifying
$x = 14.$	Identity property of 1

Objective

a Solve equations using the multiplication principle.

1. Solve. Multiply on both sides.

$6x = 90$

2. Solve. Divide on both sides.

$4x = -7$

Answers on page A-18

3. Solve: $-6x = 108$.

Do Exercises 1 and 2 on the preceding page.

EXAMPLE 3 Solve: $-4x = 92$.

We have

$$\begin{aligned} -4x &= 92 \\ \frac{-4x}{-4} &= \frac{92}{-4} && \text{Using the multiplication principle. Dividing by } -4 \text{ on both sides is the same as multiplying by } -\tfrac{1}{4}. \\ 1 \cdot x &= -23 && \text{Simplifying} \\ x &= -23. && \text{Identity property of 1} \end{aligned}$$

CHECK:

$$\begin{array}{c|l} -4x & = 92 \\ \hline -4(-23) & ?\ 92 \\ 92 & \end{array} \quad \text{TRUE}$$

The solution is -23.

4. Solve: $-x = -10$.

Do Exercise 3.

EXAMPLE 4 Solve: $-x = 9$.

We have

$$\begin{aligned} -x &= 9 \\ -1 \cdot x &= 9 && \text{Using the property of } -1\text{: } -x = -1 \cdot x \\ \frac{-1 \cdot x}{-1} &= \frac{9}{-1} && \text{Dividing by } -1 \text{ on both sides} \\ 1 \cdot x &= -9 \\ x &= -9. \end{aligned}$$

CHECK:

$$\begin{array}{c|l} -x & = 9 \\ \hline -(-9) & ?\ 9 \\ 9 & \end{array} \quad \text{TRUE}$$

The solution is -9.

5. Solve: $-x = -10$.

Do Exercise 4.

We can also solve the equation $-x = 9$ by multiplying as follows.

EXAMPLE 5 Solve: $-x = 9$.

We have

$$\begin{aligned} -x &= 9 \\ -1(-x) &= -1 \cdot 9 && \text{Multiplying by } -1 \text{ on both sides} \\ -1 \cdot (-1) \cdot x &= -9 \\ 1 \cdot x &= -9 \\ x &= -9. \end{aligned}$$

The solution is -9.

Do Exercise 5.

Answers on page A-18

In practice, it is generally more convenient to divide on both sides of the equation if the coefficient of the variable is in decimal notation or is an integer. If the coefficient is in fraction notation, it is more convenient to multiply by a reciprocal.

EXAMPLE 6 Solve: $\frac{3}{8} = -\frac{5}{4}x$.

$$\frac{3}{8} = -\frac{5}{4}x$$

The reciprocal of $-\frac{5}{4}$ is $-\frac{4}{5}$. There is no sign change.

$$-\frac{4}{5} \cdot \frac{3}{8} = -\frac{4}{5} \cdot \left(-\frac{5}{4}x\right)$$ Multiplying by $-\frac{4}{5}$ to get $1 \cdot x$ and eliminate $-\frac{5}{4}$ on the right

$$-\frac{12}{40} = 1 \cdot x$$

$$-\frac{3}{10} = 1 \cdot x$$ Simplifying

$$-\frac{3}{10} = x$$ Identity property of 1

CHECK:

$$\frac{3}{8} = -\frac{5}{4}x$$

$$\frac{3}{8} \; ? \; -\frac{5}{4}\left(-\frac{3}{10}\right)$$

$$\frac{3}{8}$$ TRUE

The solution is $-\frac{3}{10}$.

Note that equations are reversible. That is, if $a = b$ is true, then $b = a$ is true. Thus when we solve $\frac{3}{8} = -\frac{5}{4}x$, we can reverse it and solve $-\frac{5}{4}x = \frac{3}{8}$ if we wish.

Do Exercise 6.

EXAMPLE 7 Solve: $1.16y = 9744$.

$$1.16y = 9744$$

$$\frac{1.16y}{1.16} = \frac{9744}{1.16}$$ Dividing by 1.16 on both sides

$$y = \frac{9744}{1.16}$$

$$y = 8400$$ Using a calculator to divide

CHECK:

$$1.16y = 9744$$

$$1.16(8400) \; ? \; 9744$$

$$9744$$ TRUE

The solution is 8400.

Do Exercises 7 and 8.

6. Solve: $\frac{2}{3} = -\frac{5}{6}y$.

Solve.

7. $1.12x = 8736$

8. $6.3 = -2.1y$

Answers on page A-18

9. Solve: $-14 = \frac{-y}{2}$.

Answer on page A-18

Now we use the multiplication principle to solve an equation that involves division.

EXAMPLE 8 Solve: $\frac{-y}{9} = 14$.

$$\frac{-y}{9} = 14$$

$$9 \cdot \frac{-y}{9} = 9 \cdot 14 \quad \text{Multiplying by 9 on both sides}$$

$$-y = 126$$

$$-1 \cdot (-y) = -1 \cdot 126 \quad \text{Multiplying by } -1 \text{ on both sides}$$

$$y = -126$$

CHECK:

$$\frac{-y}{9} = 14$$

$\frac{-(-126)}{9}$	? 14
$\frac{126}{9}$	
14	TRUE

The solution is -126.

Do Exercise 9.

Study Tips

ATTITUDE AND THE POWER OF YOUR CHOICES

Making the right choices can give you the power to succeed in learning mathematics.

You can choose to improve your attitude and raise the academic goals that you have set for yourself. Projecting a positive attitude toward your study of mathematics and expecting a positive outcome can make it easier for you to learn and to perform well in this course.

Here are some positive choices you can make:

- Choose to allocate the proper amount of time to learn.
- Choose to place the primary responsibility for learning on yourself.
- Choose to establish a learning relationship with your instructor.
- Choose to make a strong commitment to learning.

Well-known American psychologist William James once said, "The one thing that will guarantee the successful conclusion of a doubtful undertaking is faith in the beginning that you can do it." Having a positive attitude and making the right choices will add to your confidence in this course and multiply your successes.

7.2 EXERCISE SET

For Extra Help

Digital Video Tutor CD 6 Videotape 8

InterAct Math

Math Tutor Center

MathXL

MyMathLab

a Solve using the multiplication principle. Don't forget to check!

1. $6x = 36$

CHECK: $6x = 36$?

2. $3x = 51$

CHECK: $3x = 51$?

3. $5x = 45$

CHECK: $5x = 45$?

4. $8x = 72$

CHECK: $8x = 72$?

5. $84 = 7x$

6. $63 = 9x$

7. $-x = 40$

8. $53 = -x$

9. $-x = -1$

10. $-47 = -t$

11. $7x = -49$

12. $8x = -56$

13. $-12x = 72$

14. $-15x = 105$

15. $-21x = -126$

16. $-13x = -104$

17. $\frac{t}{7} = -9$

18. $\frac{y}{-8} = 11$

19. $\frac{3}{4}x = 27$

20. $\frac{4}{5}x = 16$

21. $\frac{-t}{3} = 7$

22. $\frac{-x}{6} = 9$

23. $-\frac{m}{3} = \frac{1}{5}$

24. $\frac{1}{8} = -\frac{y}{5}$

25. $-\frac{3}{5}r = \frac{9}{10}$

26. $\frac{2}{5}y = -\frac{4}{15}$

27. $-\frac{3}{2}r = -\frac{27}{4}$

28. $-\frac{3}{8}x = -\frac{15}{16}$

29. $6.3x = 44.1$

30. $2.7y = 54$

31. $-3.1y = 21.7$

32. $-3.3y = 6.6$

33. $38.7m = 309.6$

34. $29.4m = 235.2$

35. $-\frac{2}{3}y = -10.6$

36. $-\frac{9}{7}y = 12.06$

37. $\frac{-x}{5} = 10$

38. $\frac{-x}{8} = -16$

39. $-\frac{t}{2} = 7$

40. $\frac{m}{-3} = 10$

41. **DW** When solving an equation using the multiplication principle, how do you determine by what number to multiply or divide on both sides of the equation?

42. **DW** Are the equations $x = 5$ and $x^2 = 25$ equivalent? Why or why not?

SKILL MAINTENANCE

Collect like terms. [6.7e]

43. $3x + 4x$

44. $6x + 5 - 7x$

45. $-4x + 11 - 6x + 18x$

46. $8y - 16y - 24y$

Remove parentheses and simplify. [6.8b]

47. $3x - (4 + 2x)$

48. $2 - 5(x + 5)$

49. $8y - 6(3y + 7)$

50. $-2a - 4(5a - 1)$

Translate to an algebraic expression. [6.1b]

51. Patty drives her van for 8 hr at a speed of r mph. How far does she drive?

52. A triangle has a height of 10 meters and a base of b meters. What is the area of the triangle?

SYNTHESIS

Solve.

53. $-0.2344m = 2028.732$

54. $0 \cdot x = 0$

55. $0 \cdot x = 9$

56. $4|x| = 48$

57. $2|x| = -12$

Solve for x.

58. $ax = 5a$

59. $3x = \frac{b}{a}$

60. $cx = a^2 + 1$

61. $\frac{a}{b}x = 4$

62. A student makes a calculation and gets an answer of 22.5. On the last step, she multiplies by 0.3 when a division by 0.3 should have been done. What is the correct answer?

EXAMPLE 6 Solve: $2x - 2 = -3x + 3$.

$$\begin{aligned} 2x - 2 &= -3x + 3 \\ 2x - 2 + 2 &= -3x + 3 + 2 && \text{Adding 2} \\ 2x &= -3x + 5 && \text{Collecting like terms} \\ 2x + 3x &= -3x + 3x + 5 && \text{Adding } 3x \\ 5x &= 5 && \text{Simplifying} \\ \frac{5x}{5} &= \frac{5}{5} && \text{Dividing by 5} \\ x &= 1 && \text{Simplifying} \end{aligned}$$

CHECK:

$$\begin{array}{c|l} 2x - 2 & = -3x + 3 \\ \hline 2 \cdot 1 - 2 & ?\ -3 \cdot 1 + 3 \\ 2 - 2 & -3 + 3 \\ 0 & 0 \end{array} \qquad \text{TRUE}$$

The solution is 1.

Do Exercises 9 and 10.

In Example 6, we used the addition principle to get all terms with a variable on one side and all numbers on the other side. Then we collected like terms and proceeded as before. If there are like terms on one side at the outset, they should be collected before proceeding.

EXAMPLE 7 Solve: $6x + 5 - 7x = 10 - 4x + 3$.

$$\begin{aligned} 6x + 5 - 7x &= 10 - 4x + 3 \\ -x + 5 &= 13 - 4x && \text{Collecting like terms} \\ 4x - x + 5 &= 13 - 4x + 4x && \text{Adding } 4x \text{ to get all terms with a variable on one side} \\ 3x + 5 &= 13 && \text{Simplifying; that is, collecting like terms} \\ 3x + 5 - 5 &= 13 - 5 && \text{Subtracting 5} \\ 3x &= 8 && \text{Simplifying} \\ \frac{3x}{3} &= \frac{8}{3} && \text{Dividing by 3} \\ x &= \frac{8}{3} && \text{Simplifying} \end{aligned}$$

The number $\frac{8}{3}$ checks, so it is the solution.

Do Exercises 11 and 12.

Clearing Fractions and Decimals

In general, equations are easier to solve if they do not contain fractions or decimals. Consider, for example,

$$\frac{1}{2}x + 5 = \frac{3}{4} \quad \text{and} \quad 2.3x + 7 = 5.4.$$

Solve.

9. $7y + 5 = 2y + 10$

10. $5 - 2y = 3y - 5$

Solve.

11. $7x - 17 + 2x = 2 - 8x + 15$

12. $3x - 15 = 5x + 2 - 4x$

Answers on page A-18

If we multiply by 4 on both sides of the first equation and by 10 on both sides of the second equation, we have

$$4\left(\frac{1}{2}x + 5\right) = 4 \cdot \frac{3}{4} \quad \text{and} \quad 10(2.3x + 7) = 10 \cdot 5.4$$

or

$$4 \cdot \frac{1}{2}x + 4 \cdot 5 = 4 \cdot \frac{3}{4} \quad \text{and} \quad 10 \cdot 2.3x + 10 \cdot 7 = 10 \cdot 5.4$$

or

$$2x + 20 = 3 \quad \text{and} \quad 23x + 70 = 54.$$

The first equation has been "cleared of fractions" and the second equation has been "cleared of decimals." Both resulting equations are equivalent to the original equations and are easier to solve. *It is your choice* whether to clear fractions or decimals, but doing so often eases computations.

The easiest way to clear an equation of fractions is to multiply *every term on both sides* by the **least common multiple of all the denominators.**

EXAMPLE 8 Solve: $\frac{2}{3}x - \frac{1}{6} + \frac{1}{2}x = \frac{7}{6} + 2x$.

The number 6 is the least common multiple of all the denominators. We multiply by 6 on both sides.

$$6\left(\frac{2}{3}x - \frac{1}{6} + \frac{1}{2}x\right) = 6\left(\frac{7}{6} + 2x\right) \qquad \text{Multiplying by 6 on both sides}$$

$$6 \cdot \frac{2}{3}x - 6 \cdot \frac{1}{6} + 6 \cdot \frac{1}{2}x = 6 \cdot \frac{7}{6} + 6 \cdot 2x \qquad \text{Using the distributive law (\textit{Caution}! Be sure to multiply \textit{all} the terms by 6.)}$$

$$4x - 1 + 3x = 7 + 12x \qquad \text{Simplifying. Note that the fractions are cleared.}$$

$$7x - 1 = 7 + 12x \qquad \text{Collecting like terms}$$

$$7x - 1 - 12x = 7 + 12x - 12x \qquad \text{Subtracting } 12x$$

$$-5x - 1 = 7 \qquad \text{Collecting like terms}$$

$$-5x - 1 + 1 = 7 + 1 \qquad \text{Adding 1}$$

$$-5x = 8 \qquad \text{Collecting like terms}$$

$$\frac{-5x}{-5} = \frac{8}{-5} \qquad \text{Dividing by } -5$$

$$x = -\frac{8}{5}$$

CHECK:

$$\begin{array}{r|l}
\multicolumn{2}{c}{\frac{2}{3}x - \frac{1}{6} + \frac{1}{2}x = \frac{7}{6} + 2x} \\
\hline
\frac{2}{3}\left(-\frac{8}{5}\right) - \frac{1}{6} + \frac{1}{2}\left(-\frac{8}{5}\right) \ ? & \frac{7}{6} + 2\left(-\frac{8}{5}\right) \\
-\frac{16}{15} - \frac{1}{6} - \frac{8}{10} & \frac{7}{6} - \frac{16}{5} \\
-\frac{32}{30} - \frac{5}{30} - \frac{24}{30} & \frac{35}{30} - \frac{96}{30} \\
\frac{-32 - 5 - 24}{30} & -\frac{61}{30} \\
-\frac{61}{30} &
\end{array}$$

TRUE

Study Tips

SMALL STEPS LEAD TO GREAT SUCCESS (PART 1)

What is your long-term goal for getting an education? How does math help you to attain that goal? As you begin this course, approach each short-term task, such as going to class, asking questions, using your time wisely, and doing your homework, as part of the framework of your long-term goal.

"What man actually needs is not a tensionless state but rather the struggling and striving for a worthwhile goal, a freely chosen task."

Victor Frankl

The solution is $-\frac{8}{5}$.

Do Exercise 13.

To illustrate clearing decimals, we repeat Example 4, but this time we clear the equation of decimals first. Compare both methods.

To clear an equation of decimals, we count the greatest number of decimal places in any one number. If the greatest number of decimal places is 1, we multiply by 10; if it is 2, we multiply by 100; and so on.

EXAMPLE 9 Solve: $16.3 - 7.2y = -8.18$.

The greatest number of decimal places in any one number is *two.* Multiplying by 100, which has *two* 0's, will clear all decimals.

$100(16.3 - 7.2y) = 100(-8.18)$	Multiplying by 100 on both sides
$100(16.3) - 100(7.2y) = 100(-8.18)$	Using the distributive law
$1630 - 720y = -818$	Simplifying
$1630 - 720y - 1630 = -818 - 1630$	Subtracting 1630
$-720y = -2448$	Collecting like terms
$\frac{-720y}{-720} = \frac{-2448}{-720}$	Dividing by -720
$y = \frac{17}{5}$, or 3.4	

The number $\frac{17}{5}$, or 3.4, checks, so it is the solution.

Do Exercise 14.

C Equations Containing Parentheses

To solve certain kinds of equations that contain parentheses, we first use the distributive laws to remove the parentheses. Then we proceed as before.

EXAMPLE 10 Solve: $4x = 2(12 - 2x)$.

$4x = 2(12 - 2x)$	
$4x = 24 - 4x$	Using the distributive laws to multiply and remove parentheses
$4x + 4x = 24 - 4x + 4x$	Adding $4x$ to get all the x-terms on one side
$8x = 24$	Collecting like terms
$\frac{8x}{8} = \frac{24}{8}$	Dividing by 8
$x = 3$	

The number 3 checks, so the solution is 3.

Do Exercises 15 and 16.

13. Solve: $\frac{7}{8}x - \frac{1}{4} + \frac{1}{2}x = \frac{3}{4} + x$.

14. Solve: $41.68 = 4.7 - 8.6y$.

Solve.

15. $2(2y + 3) = 14$

16. $5(3x - 2) = 35$

Answers on page A-18

Solve.

17. $3(7 + 2x) = 30 + 7(x - 1)$

18. $4(3 + 5x) - 4 = 3 + 2(x - 2)$

Determine whether the given number is a solution of the given equation.

19. 10; $3 + x = x + 3$

20. -7; $3 + x = x + 3$

21. $\frac{1}{2}$; $3 + x = x + 3$

22. 0; $3 + x = x + 3$

Answers on page A-18

Here is a procedure for solving the types of equation discussed in this section.

AN EQUATION-SOLVING PROCEDURE

1. Multiply on both sides to clear the equation of fractions or decimals. (This is optional, but it can ease computations.)
2. If parentheses occur, multiply to remove them using the *distributive laws.*
3. Collect like terms on each side, if necessary.
4. Get all terms with variables on one side and all numbers (constant terms) on the other side, using the *addition principle.*
5. Collect like terms again, if necessary.
6. Multiply or divide to solve for the variable, using the *multiplication principle.*
7. Check all possible solutions in the original equation.

EXAMPLE 11 Solve: $2 - 5(x + 5) = 3(x - 2) - 1$.

$$2 - 5(x + 5) = 3(x - 2) - 1$$

$$2 - 5x - 25 = 3x - 6 - 1 \quad \text{Using the distributive laws to multiply and remove parentheses}$$

$$-5x - 23 = 3x - 7 \quad \text{Collecting like terms}$$

$$-5x - 23 + 5x = 3x - 7 + 5x \quad \text{Adding } 5x$$

$$-23 = 8x - 7 \quad \text{Collecting like terms}$$

$$-23 + 7 = 8x - 7 + 7 \quad \text{Adding 7}$$

$$-16 = 8x \quad \text{Collecting like terms}$$

$$\frac{-16}{8} = \frac{8x}{8} \quad \text{Dividing by 8}$$

$$-2 = x$$

CHECK:

$2 - 5(x + 5)$	$= 3(x - 2) - 1$
$2 - 5(-2 + 5)$	? $3(-2 - 2) - 1$
$2 - 5(3)$	$3(-4) - 1$
$2 - 15$	$-12 - 1$
-13	-13 TRUE

The solution is -2.

Do Exercises 17 and 18.

EQUATIONS WITH INFINITELY MANY SOLUTIONS

The types of equations we have considered thus far in Sections 7.1–7.3 have all had exactly one solution. We now look at two other possibilities.

Consider

$$3 + x = x + 3.$$

Let's explore the solutions in Margin Exercises 19–22.

Do Exercises 19–22.

We know by the commutative law that this equation holds for any replacement of x with a real number. (See Section 6.7.) We have confirmed some of these solutions in Margin Exercises 19–22. Suppose we try to solve this equation using the addition principle:

$$3 + x = x + 3$$

$$-x + 3 + x = -x + x + 3 \qquad \text{Adding } -x$$

$$3 = 3. \qquad \text{TRUE}$$

We end with a true equation. The original equation holds for all real-number replacements. Thus the number of solutions is **infinite.**

EXAMPLE 12 Solve: $7x - 17 = 4 + 7(x - 3)$.

$$7x - 17 = 4 + 7(x - 3)$$

$$7x - 17 = 4 + 7x - 21 \qquad \text{Using the distributive law to multiply and remove parentheses}$$

$$7x - 17 = 7x - 17 \qquad \text{Collecting like terms}$$

$$-7x + 7x - 17 = -7x + 7x - 17 \qquad \text{Adding } -7x$$

$$-17 = -17 \qquad \text{TRUE}$$

Every real number is a solution. There are infinitely many solutions.

EQUATIONS WITH NO SOLUTION

Now consider

$$3 + x = x + 8.$$

Let's explore the solutions in Margin Exercises 23–26.

Do Exercises 23–26.

None of the replacements in Margin Exercises 23–26 are solutions of the given equation. In fact, there are no solutions. Let's try to solve this equation using the addition principle:

$$3 + x = x + 8$$

$$-x + 3 + x = -x + x + 8 \qquad \text{Adding } -x$$

$$3 = 8. \qquad \text{FALSE}$$

We end with a false equation. The original equation is false for all real-number replacements. Thus it has **no** solutions.

EXAMPLE 13 Solve: $3x + 4(x + 2) = 11 + 7x$.

$$3x + 4(x + 2) = 11 + 7x$$

$$3x + 4x + 8 = 11 + 7x \qquad \text{Using the distributive law to multiply and remove parentheses}$$

$$7x + 8 = 11 + 7x \qquad \text{Collecting like terms}$$

$$7x + 8 - 7x = 11 + 7x - 7x \qquad \text{Subtracting } 7x$$

$$8 = 11 \qquad \text{FALSE}$$

There are no solutions.

Do Exercises 27 and 28.

Determine whether the given number is a solution of the given equation.

23. 10; $3 + x = x + 8$

24. -7; $3 + x = x + 8$

25. $\frac{1}{2}$; $3 + x = x + 8$

26. 0; $3 + x = x + 8$

Solve.

27. $30 + 5(x + 3) = -3 + 5x + 48$

28. $2x + 7(x - 4) = 13 + 9x$

Answers on page A-18

The following is a guideline for solving linear equations of the types that we have considered in Sections 7.1–7.3.

RESULTING EQUATION	NUMBER OF SOLUTIONS	SOLUTION(S)
$x = a$, where a is a real number	One	The number a
A true equation such as $3 = 3$, $-11 = -11$, or $0 = 0$	Infinitely many	Every real number is a solution.
A false equation such as $3 = 8$, $-4 = 5$, or $0 = -5$	Zero	There are no solutions.

CALCULATOR CORNER

Checking Possible Solutions To check the possible solutions of an equation on a calculator, we can substitute and carry out the calculations on each side of the equation just as we do when we check by hand. To check the possible solution -2 in Example 13, for instance, we first substitute -2 for x in the expression on the left side of the equation. We press [2] [−] [5] [(] [(−)] [2] [+] [5] [)] [ENTER]. We get -13. Next, we substitute -2 for x in the expression on the right side of the equation. We then press [3] [(] [(−)] [2] [−] [2] [)] [−] [1] [ENTER]. Again we get -13. Since the two sides of the equation have the same value when x is -2, we know that -2 is the solution of the equation.

```
2-5(-2+5)
                 -13
3(-2-2)-1
                 -13
```

A table can also be used to check the possible solutions of an equation. First, we enter the left side and the right side of the equation on the Y = or equation editor screen. To do this, we first press [Y =]. If an expression for Y1 is currently entered, we place the cursor on it and press [CLEAR] to delete it. We do the same for any other entries that are present.

Next, we position the cursor to the right of Y1 = and enter the left side of the equation by pressing [2] [−] [5] [(] [X,T,θ,n] [+] [5] [)]. Then we position the cursor beside Y2 = and enter the right side of the equation by pressing [3] [(] [X,T,θ,n] [−] [2] [)] [−] [1]. Now we press [2nd] [TBLSET] to display the Table Setup screen. (TBLSET is the second operation associated with the [WINDOW] key.) On the Indpnt line, we position the cursor on "Ask" and press [ENTER] to set up a table in Ask mode. (The settings for TblStart and ΔTbl are irrelevant in Ask mode.)

Now we press [2nd] [TABLE] to display the table. (TABLE is the second operation associated with the [GRAPH] key.) We then enter the possible solution, -2, by pressing [(−)] [2] [ENTER]. We see that Y1 $= -13 =$ Y2 for this value of x. This confirms that the left and right sides of the equation have the same value for $x = -2$, so -2 is the solution of the equation.

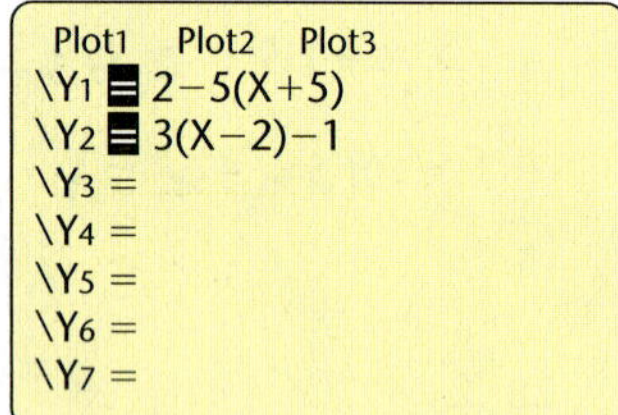

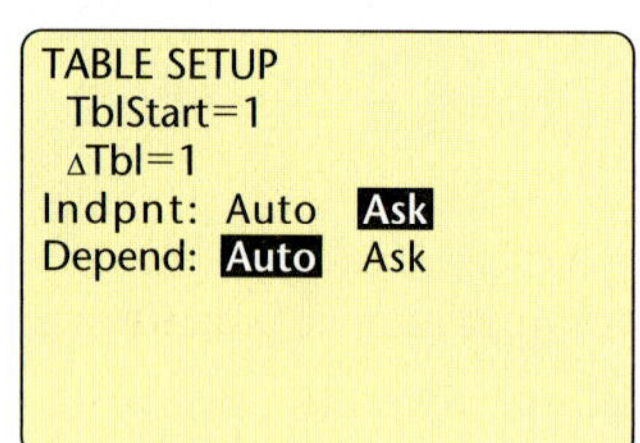

X	Y1	Y2
−2	−13	−13

X =

Exercises:

1. Use substitution to check the solutions found in Examples 6, 7, 8, and 12.
2. Use a table set in Ask mode to check the solutions found in Examples 6, 7, 8, and 12.

7.3 EXERCISE SET

For Extra Help

Digital Video Tutor CD 6 Videotape 8

InterAct Math

Math Tutor Center

MathXL

MyMathLab

a Solve. Don't forget to check!

1. $5x + 6 = 31$

CHECK: $5x + 6 = 31$?

2. $7x + 6 = 13$

CHECK: $7x + 6 = 13$?

3. $8x + 4 = 68$

CHECK: $8x + 4 = 68$?

4. $4y + 10 = 46$

CHECK: $4y + 10 = 46$?

5. $4x - 6 = 34$

6. $5y - 2 = 53$

7. $3x - 9 = 33$

8. $4x - 19 = 5$

9. $7x + 2 = -54$

10. $5x + 4 = -41$

11. $-45 = 3 + 6y$

12. $-91 = 9t + 8$

13. $-4x + 7 = 35$

14. $-5x - 7 = 108$

15. $-7x - 24 = -129$

16. $-6z - 18 = -132$

b Solve.

17. $5x + 7x = 72$

CHECK: $5x + 7x = 72$?

18. $8x + 3x = 55$

CHECK: $8x + 3x = 55$?

19. $8x + 7x = 60$

CHECK: $8x + 7x = 60$?

20. $8x + 5x = 104$

CHECK: $8x + 5x = 104$?

21. $4x + 3x = 42$

22. $7x + 18x = 125$

23. $-6y - 3y = 27$

24. $-5y - 7y = 144$

25. $-7y - 8y = -15$

26. $-10y - 3y = -39$

27. $x + \frac{1}{3}x = 8$

28. $x + \frac{1}{4}x = 10$

29. $10.2y - 7.3y = -58$

30. $6.8y - 2.4y = -88$

31. $8y - 35 = 3y$

32. $4x - 6 = 6x$

33. $8x - 1 = 23 - 4x$

34. $5y - 2 = 28 - y$

35. $2x - 1 = 4 + x$

36. $4 - 3x = 6 - 7x$

37. $6x + 3 = 2x + 11$

38. $14 - 6a = -2a + 3$

39. $5 - 2x = 3x - 7x + 25$

40. $-7z + 2z - 3z - 7 = 17$

41. $4 + 3x - 6 = 3x + 2 - x$

42. $5 + 4x - 7 = 4x - 2 - x$

43. $4y - 4 + y + 24 = 6y + 20 - 4y$

44. $5y - 7 + y = 7y + 21 - 5y$

Solve. Clear fractions or decimals first.

45. $\frac{7}{2}x + \frac{1}{2}x = 3x + \frac{3}{2} + \frac{5}{2}x$

46. $\frac{7}{8}x - \frac{1}{4} + \frac{3}{4}x = \frac{1}{16} + x$

47. $\frac{2}{3} + \frac{1}{4}t = \frac{1}{3}$

48. $-\frac{3}{2} + x = -\frac{5}{6} - \frac{4}{3}$

49. $\frac{2}{3} + 3y = 5y - \frac{2}{15}$

50. $\frac{1}{2} + 4m = 3m - \frac{5}{2}$

51. $\frac{5}{3} + \frac{2}{3}x = \frac{25}{12} + \frac{5}{4}x + \frac{3}{4}$

52. $1 - \frac{2}{3}y = \frac{9}{5} - \frac{y}{5} + \frac{3}{5}$

53. $2.1x + 45.2 = 3.2 - 8.4x$

54. $0.96y - 0.79 = 0.21y + 0.46$

55. $1.03 - 0.62x = 0.71 - 0.22x$

56. $1.7t + 8 - 1.62t = 0.4t - 0.32 + 8$

57. $\frac{2}{7}x - \frac{1}{2}x = \frac{3}{4}x + 1$

58. $\frac{5}{16}y + \frac{3}{8}y = 2 + \frac{1}{4}y$

C Solve.

59. $3(2y - 3) = 27$

60. $8(3x + 2) = 30$

61. $40 = 5(3x + 2)$

62. $9 = 3(5x - 2)$

63. $-23 + y = y + 25$

64. $17 - t = -t + 68$

65. $-23 + x = x - 23$

66. $y - \frac{2}{3} = -\frac{2}{3} + y$

67. $2(3 + 4m) - 9 = 45$

68. $5x + 5(4x - 1) = 20$

69. $5r - (2r + 8) = 16$

70. $6b - (3b + 8) = 16$

71. $6 - 2(3x - 1) = 2$

72. $10 - 3(2x - 1) = 1$

73. $5x + 5 - 7x = 15 - 12x + 10x - 10$

74. $3 - 7x + 10x - 14 = 9 - 6x + 9x - 20$

75. $22x - 5 - 15x + 3 = 10x - 4 - 3x + 11$

76. $11x - 6 - 4x + 1 = 9x - 8 - 2x + 12$

77. $5(d + 4) = 7(d - 2)$

78. $3(t - 2) = 9(t + 2)$

79. $8(2t + 1) = 4(7t + 7)$

80. $7(5x - 2) = 6(6x - 1)$

81. $3(r - 6) + 2 = 4(r + 2) - 21$

82. $5(t + 3) + 9 = 3(t - 2) + 6$

83. $19 - (2x + 3) = 2(x + 3) + x$

84. $13 - (2c + 2) = 2(c + 2) + 3c$

85. $2[4 - 2(3 - x)] - 1 = 4[2(4x - 3) + 7] - 25$

86. $5[3(7 - t) - 4(8 + 2t)] - 20 = -6[2(6 + 3t) - 4]$

87. $11 - 4(x + 1) - 3 = 11 + 2(4 - 2x) - 16$

88. $6(2x - 1) - 12 = 7 + 12(x - 1)$

89. $22x - 1 - 12x = 5(2x - 1) + 4$

90. $2 + 14x - 9 = 7(2x + 1) - 14$

91. $0.7(3x + 6) = 1.1 - (x + 2)$

92. $0.9(2x + 8) = 20 - (x + 5)$

93. $^D W$ What procedure would you follow to solve an equation like $0.23x + \frac{17}{3} = -0.8 + \frac{3}{4}x$? Could your procedure be streamlined? If so, how?

94. $^D W$ You are trying to explain to a classmate how equations can arise with infinitely many solutions and with no solutions. Give such an explanation. Does having no solution mean that 0 is a solution? Explain.

SKILL MAINTENANCE

95. Divide: $-22.1 \div 3.4$. [6.6c]

96. Multiply: $-22.1(3.4)$. [6.5a]

97. Factor: $7x - 21 - 14y$. [6.7d]

98. Factor: $8y - 88x + 8$. [6.7d]

SYNTHESIS

Solve.

99. $0.008 + 9.62x - 42.8 = 0.944x + 0.0083 - x$

100. $\frac{1}{4}(8y + 4) - 17 = -\frac{1}{2}(4y - 8)$

101. $\frac{2}{3}\left(\frac{7}{8} - 4x\right) - \frac{5}{8} = \frac{3}{8}$

102. $\frac{4 - 3x}{7} = \frac{2 + 5x}{49} - \frac{x}{14}$

7.4 FORMULAS

a Evaluating Formulas

A **formula** is a "recipe" for doing a certain type of calculation. Formulas are often given as equations. When we replace the variables in an equation with numbers and calculate the result, we are **evaluating** the formula. We did some evaluating in Section 6.1.

Let's consider another example. A formula that has to do with weather is $M = \frac{1}{5}t$. You see a flash of lightning. After a few seconds you hear the thunder associated with that flash. How far away was the lightning?

Your distance from the storm is M miles. You can find that distance by counting the number of seconds t that it takes the sound of the thunder to reach you and then multiplying by $\frac{1}{5}$.

EXAMPLE 1 *Storm Distance.* Consider the formula $M = \frac{1}{5}t$. It takes 10 sec for the sound of thunder to reach you after you have seen a flash of lightning. How far away is the storm?

We substitute 10 for t and calculate M: $M = \frac{1}{5}t = \frac{1}{5}(10) = 2$. The storm is 2 mi away.

EXAMPLE 2 *Distance, Rate, and Time.* The distance d that a car will travel at a rate, or speed, r in time t is given by

$$d = rt.$$

A car travels at 75 miles per hour (mph) for 4.5 hr. How far will it travel?

We substitute 75 for r, 4.5 for t, and calculate d:

$$d = rt = (75)(4.5) = 337.5 \text{ mi.}$$

The car will travel 337.5 mi.

Do Exercises 1 and 2.

Objectives

a Evaluate a formula.

b Solve a formula for a specified letter.

1. Storm Distance. Suppose that it takes the sound of thunder 14 sec to reach you. How far away is the storm?

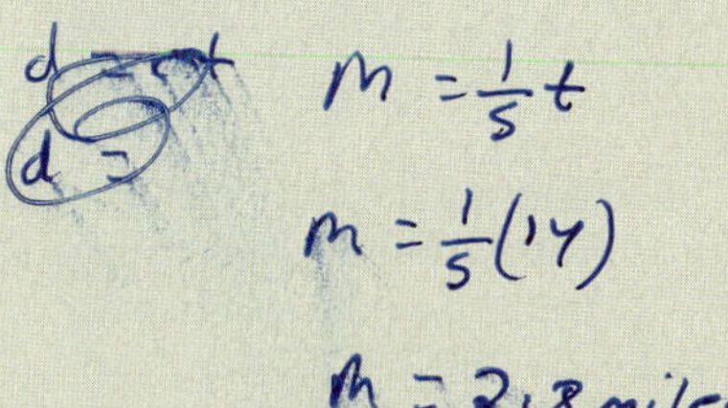

2. Distance, Rate, and Time. A car travels at 55 mph for 6.2 hr. How far will it travel?

Answers on page A-18

3. Solve for q: $B = \frac{1}{3}q$.

4. **Distance, Rate, and Time.** Solve for r: $d = rt$.

5. **Electricity.** Solve for I: $E = IR$. (This formula relates voltage E, current I, and resistance R.)

Solve for x.

6. $y = x + 5$

7. $y = x - 7$

8. $y = x - b$

Answers on page A-18

b Solving Formulas

Refer to Example 1. Suppose that we think we know how far we are from the storm and want to check by calculating the number of seconds it should take the sound of the thunder to reach us. We could substitute a number for M—say, 2—and solve for t:

$$2 = \tfrac{1}{5}t$$
$$10 = t. \quad \text{Multiplying by 5}$$

However, if we wanted to do this repeatedly, it might be easier to solve for t by getting it alone on one side. We "solve" the formula for t.

EXAMPLE 3 Solve for t: $M = \frac{1}{5}t$.

$M = \frac{1}{5}t$ — We want this letter alone.

$5 \cdot M = 5 \cdot \frac{1}{5}t$ — Multiplying by 5 on both sides

$5M = t$

In the above situation for $M = 2$, $t = 5M = 5(2)$, or 10.

EXAMPLE 4 *Distance, Rate, and Time.* Solve for t: $d = rt$.

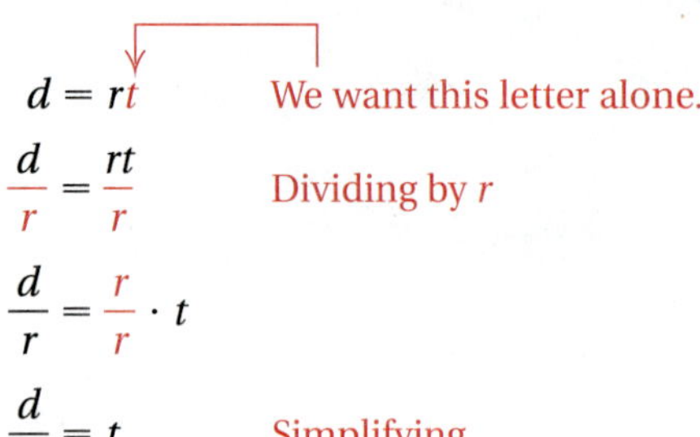

$d = rt$ — We want this letter alone.

$\frac{d}{r} = \frac{rt}{r}$ — Dividing by r

$\frac{d}{r} = \frac{r}{r} \cdot t$

$\frac{d}{r} = t$ — Simplifying

Do Exercises 3–5.

EXAMPLE 5 Solve for x: $y = x + 3$.

$y = x + 3$ — We want this letter alone.

$y - 3 = x + 3 - 3$ — Subtracting 3

$y - 3 = x$ — Simplifying

EXAMPLE 6 Solve for x: $y = x - a$.

$y = x - a$ — We want this letter alone.

$y + a = x - a + a$ — Adding a

$y + a = x$ — Simplifying

Do Exercises 6–8.

EXAMPLE 7 Solve for y: $6y = 3x$.

$$6y = 3x \qquad \text{We want this letter alone.}$$

$$\frac{6y}{6} = \frac{3x}{6} \qquad \text{Dividing by 6}$$

$$y = \frac{1}{2}x \qquad \text{Simplifying}$$

EXAMPLE 8 Solve for y: $by = ax$.

$$by = ax \qquad \text{We want this letter alone.}$$

$$\frac{by}{b} = \frac{ax}{b} \qquad \text{Dividing by } b$$

$$y = \frac{ax}{b} \qquad \text{Simplifying}$$

Do Exercises 9 and 10.

To see how the addition and multiplication principles apply to formulas, compare the following.

A. *Solve.* We carry this out as we did in Sections 7.1–7.3.

$$5x + 2 = 12 \qquad \text{We want this letter alone.}$$

$$5x + 2 - 2 = 12 - 2 \qquad \text{Subtracting 2}$$

$$5x = 10 \qquad \text{Simplifying}$$

$$\frac{5x}{5} = \frac{10}{5} \qquad \text{Dividing by 5}$$

$$x = 2 \qquad \text{Simplifying}$$

B. *Solve.* We carry this out as we did in Sections 7.1–7.3, but we do not do as much simplifying or collecting like terms.

$$5x + 2 = 12$$

$$5x + 2 - 2 = 12 - 2$$

$$5x = 12 - 2$$

$$\frac{5x}{5} = \frac{12 - 2}{5}$$

$$x = \frac{12 - 2}{5}$$

C. *Solve for x:* $ax + b = c$. In this case, we cannot carry out any calculations because we have unknown letters.

$$ax + b = c \qquad \text{We want this letter alone.}$$

$$ax + b - b = c - b \qquad \text{Subtracting } b$$

$$ax = c - b \qquad \text{Simplifying}$$

$$\frac{ax}{a} = \frac{c - b}{a} \qquad \text{Dividing by } a$$

$$x = \frac{c - b}{a} \qquad \text{Simplifying}$$

9. Solve for y: $9y = 5x$.

10. Solve for p: $ap = bq$.

11. Solve for x: $y = mx + b$.

12. Solve for Q: $tQ - p = a$.

Answers on page A-18

13. Circumference. Solve for D:

$$C = \pi D.$$

(This is a formula for the circumference C of a circle of diameter D.)

14. Averages. Solve for c:

$$A = \frac{a + b + c + d}{4}.$$

Answers on page A-18

Do Exercises 11 and 12 on the preceding page.

Solving Formulas

To solve a formula for a given letter, identify the letter and:

1. Multiply on both sides to clear fractions or decimals, if that is needed.
2. Collect like terms on each side, if necessary.
3. Get all terms with the letter to be solved for on one side of the equation and all other terms on the other side.
4. Collect like terms again, if necessary.
5. Solve for the letter in question.

EXAMPLE 9 *Circumference.* Solve for r: $C = 2\pi r$. This is a formula for the circumference C of a circle of radius r.

$$C = 2\pi r \quad \text{We want this letter alone.}$$

$$\frac{C}{2\pi} = \frac{2\pi r}{2\pi} \quad \text{Dividing by } 2\pi$$

$$\frac{C}{2\pi} = r$$

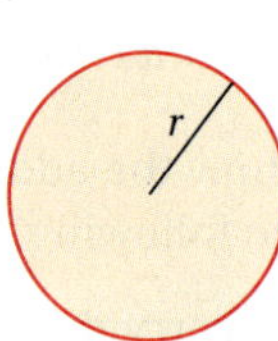

EXAMPLE 10 *Averages.* Solve for a: $A = \dfrac{a + b + c}{3}$. This is a formula for the average A of three numbers a, b, and c.

$$A = \frac{a + b + c}{3} \quad \text{We want the letter } a \text{ alone.}$$

$$3A = a + b + c \quad \text{Multiplying by 3 to clear the fraction}$$

$$3A - b - c = a \quad \text{Subtracting } b \text{ and } c$$

Do Exercises 13 and 14.

7.4 EXERCISE SET

For Extra Help

Digital Video Tutor CD 6 Videotape 8 | InterAct Math | Math Tutor Center  | MathXL  | MyMathLab 

a, **b** Solve.

1. *Furnace Output.* The formula

$$B = 30a$$

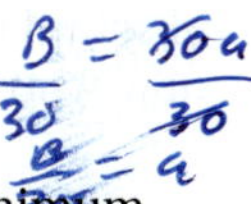

is used in New England to estimate the minimum furnace output B, in Btu's, for a modern house with a square feet of flooring.

a) Determine the minimum furnace output for a 1900-ft^2 modern house.

b) Solve for a.

Source: U.S. Department of Energy

2. *Furnace Output.* The formula

$$B = 50a$$

is used in New England to estimate the minimum furnace output B, in Btu's, for an old, poorly insulated house with a square feet of flooring.

a) Determine the minimum furnace output for a 3200-ft^2 old, poorly insulated house.

b) Solve for a.

Source: U.S. Department of Energy

3. *Distance from a Storm.* The formula

$$M = \tfrac{1}{5}t$$

can be used to determine how far M, in miles, you are from lightning when its thunder takes t seconds to reach your ears.

a) It takes 8 sec for the sound of thunder to reach you after you have seen the lightning. How far away is the storm?

b) Solve for t.

4. *Electrical Power.* The power rating P, in watts, of an electrical appliance is determined by

$$P = I \cdot V,$$

where I is the current, in amperes, and V is measured in volts.

a) A kitchen requires 30 amps of current and the voltage in the house is 115 volts. What is the wattage of the kitchen?

b) Solve for I; for V.

5. *College Enrollment.* At many colleges, the number of "full-time-equivalent" students f is given by

$$f = \frac{n}{15},$$

where n is the total number of credits for which students have enrolled in a given semester.

a) Determine the number of full-time-equivalent students on a campus in which students registered for a total of 21,345 credits.

b) Solve for n.

6. *Surface Area of a Cube.* The surface area A of a cube with side s is given by

$$A = 6s^2.$$

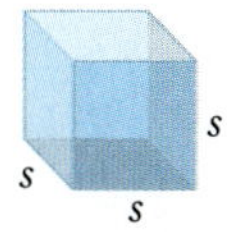

a) Find the surface area of a cube with sides of 3 in.

b) Solve for s^2.

7. *Calorie Density.* The calorie density D, in calories per ounce, of a food that contains c calories and weighs w ounces is given by

$$D = \frac{c}{w}.$$

Eight ounces of fat-free milk contains 84 calories. Find the calorie density of fat-free milk.

Source: *Nutrition Action Healthletter,* March 2000, p. 9. Center for Science in the Public Interest, Suite 300; 1875 Connecticut Ave NW, Washington, D.C. 20008.

8. *Wavelength of a Musical Note.* The wavelength w, in meters per cycle, of a musical note is given by

$$w = \frac{r}{f},$$

where r is the speed of the sound, in meters per second, and f is the frequency, in cycles per second. The speed of sound in air is 344 m/sec. What is the wavelength of a note whose frequency in air is 24 cycles per second?

9. *Size of a League Schedule.* When all n teams in a league play every other team twice, a total of N games are played, where

$$N = n^2 - n.$$

A soccer league has 7 teams and all teams play each other twice. How many games are played?

10. *Size of a League Schedule.* When all n teams in a league play every other team twice, a total of N games are played, where

$$N = n^2 - n.$$

A basketball league has 11 teams and all teams play each other twice. How many games are played?

b Solve for the indicated letter.

11. $y = 5x$, for x

12. $d = 55t$, for t

13. $a = bc$, for c

14. $y = mx$, for x

15. $y = 13 + x$, for x

16. $y = x - \frac{2}{3}$, for x

17. $y = x + b$, for x

18. $y = x - A$, for x

19. $y = 5 - x$, for x

20. $y = 10 - x$, for x

21. $y = a - x$, for x

22. $y = q - x$, for x

23. $8y = 5x$, for y

24. $10y = -5x$, for y

25. $By = Ax$, for x

26. $By = Ax$, for y

27. $W = mt + b$, for t

28. $W = mt - b$, for t

29. $y = bx + c$, for x

30. $y = bx - c$, for x

31. $A = \dfrac{a + b + c}{3}$, for b

32. $A = \dfrac{a + b + c}{3}$, for c

33. $A = at + b$, for t

34. $S = rx + s$, for x

35. *Area of a Parallelogram*:

$A = bh$, for h

(Area A, base b, height h)

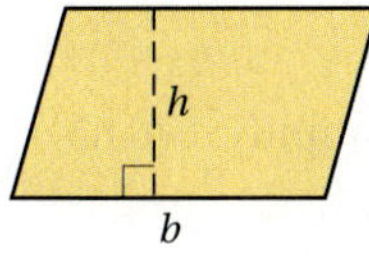

36. 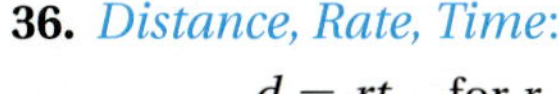 *Distance, Rate, Time*:

$d = rt$, for r

(Distance d, speed r, time t)

37.  *Perimeter of a Rectangle*:

$P = 2l + 2w$, for w

(Perimeter P, length l, width w)

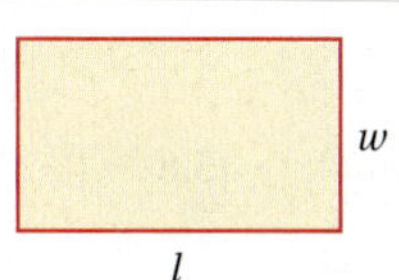

38. *Area of a Circle*:

$A = \pi r^2$, for r^2

(Area A, radius r)

39.  *Average of Two Numbers*:

$$A = \frac{a + b}{2}, \quad \text{for } a$$

40.  *Area of a Triangle*:

$$A = \frac{1}{2}bh, \quad \text{for } b$$

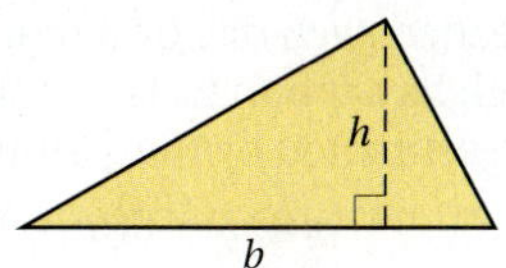

41. *Force*:

$F = ma$, for a

(Force F, mass m, acceleration a)

42. *Simple Interest*:

$I = Prt$, for P

(Interest I, principal P, interest rate r, time t)

43. *Relativity*:

$E = mc^2$, for c^2

(Energy E, mass m, speed of light c)

44. $Q = \dfrac{p - q}{2}$, for p

45. $Ax + By = c$, for x

46. $Ax + By = c$, for y

47. $v = \dfrac{3k}{t}$, for t

48. $P = \dfrac{ab}{c}$, for c

49. ^{D}W Devise an application in which it would be useful to solve the equation $d = rt$ for r. (See Exercise 36.)

50. ^{D}W The equations

$$P = 2l + 2w \quad \text{and} \quad w = \frac{P}{2} - l$$

are equivalent formulas involving the perimeter P, the length l, and the width w of a rectangle. (See Exercise 37.) Devise a problem for which the second of the two formulas would be more useful.

SKILL MAINTENANCE

51. Evaluate $2a - b$ for $a = 2$ and $b = 3$. [6.1a]

52. Add: $-23 + (-67)$. [6.3a]

53. Add: $0.082 + (-9.407)$. [6.3a]

54. Subtract: $-23 - (-67)$. [6.4a]

55. Subtract: $-45.8 - (-32.6)$. [6.4a]

56. Remove parentheses and simplify: [6.8b]

$$4a - 8b - 5(5a - 4b).$$

57. Add: $-\frac{2}{3} + \frac{5}{6}$. [6.3a]

58. Subtract: $-\frac{2}{3} - \frac{5}{6}$. [6.4a]

SYNTHESIS

59. *Female Caloric Needs.* The number of calories K needed each day by a moderately active woman who weighs w pounds, is h inches tall, and is a years old can be estimated by the formula

$$K = 917 + 6(w + h - a).$$

a) Elaine is moderately active, weighs 120 lb, is 67 in. tall, and is 23 yr old. What are her caloric needs?
b) Solve the formula for a; for h; for w.

Source: Parker, M., *She Does Math.* Mathematical Association of America, p. 96

60. *Male Caloric Needs.* The number of calories K needed each day by a moderately active man who weighs w kilograms, is h centimeters tall, and is a years old can be estimated by the formula

$$K = 19.18w + 7h - 9.52a + 92.4.$$

a) Marv is moderately active, weighs 97 kg, is 185 cm tall, and is 55 yr old. What are his caloric needs?
b) Solve the formula for a; for h; for w.

Source: Parker, M., *She Does Math.* Mathematical Association of America, p. 96

Solve.

61. $A = \frac{1}{2}ah + \frac{1}{2}bh$, for b; for h

62. $P = 4m + 7mn$, for m

63. In $A = lw$, l and w both double. What is the effect on A?

64. In $P = 2a + 2b$, P doubles. Do a and b necessarily both double?

65. In $A = \frac{1}{2}bh$, b increases by 4 units and h does not change. What happens to A?

66. Solve for F: $D = \frac{1}{E + F}$.

7.5 APPLICATIONS OF PERCENT

a Translating and Solving

Many applied problems involve percent. Here we begin to see how equation solving can enhance our problem-solving skills.

In solving percent problems, we first *translate* the problem to an equation. Then we *solve* the equation using the techniques discussed in Sections 7.1–7.3. The key words in the translation are as follows.

KEY WORDS IN PERCENT TRANSLATIONS

"**Of**" translates to "·" or "×". "**Is**" translates to "=".

"**What**" translates to any letter. **%** translates to "$\times \frac{1}{100}$" or "× 0.01".

EXAMPLE 1 Translate:

28%	of	5	is	what?
↓	↓	↓	↓	↓
28%	·	5	=	a

This is a percent equation.

EXAMPLE 2 Translate:

45%	of	what	is	28?
↓	↓	↓	↓	↓
45%	×	b	=	28

EXAMPLE 3 Translate:

What percent	of	90	is	7?
↓	↓	↓	↓	↓
n	·	90	=	7

Do Exercises 1–6.

Percent problems are actually of three different types. Although the method we present does *not* require that you be able to identify which type we are studying, it is helpful to know them.

We know that

15 is 25% of 60, or

15 = 25% × 60.

We can think of this as:

Amount = Percent number × Base.

Objective

a Solve applied problems involving percent.

Translate to an equation. Do not solve.

1. 13% of 80 is what?

2. What is 60% of 70?

3. 43 is 20% of what?

4. 110% of what is 30?

5. 16 is what percent of 80?

6. What percent of 94 is 10.5?

Answers on page A-18

7. What is 2.4% of 80?

8. 25.3 is 22% of what number?

Answers on page A-18

Study Tips

USING THE SUPPLEMENTS

The new mathematical skills and concepts presented in the lectures will be of increased value to you if you begin the homework assignment as soon as possible after the lecture. Then if you still have difficulty with any of the exercises, you have time to access supplementary resources such as:

- *Student's Solutions Manual*
- Videotapes
- InterAct Math Tutorial CD-ROM
- AW Math Tutor Center
- MyMathLab
- MathXL

Each of the three types of percent problems depends on which of the three pieces of information is missing.

1. Finding the ***amount*** (the result of taking the percent)

   Example: What is 25% of 60?

   Translation: $y = 25\% \cdot 60$

2. Finding the ***base*** (the number you are taking the percent of)

   Example: 15 is 25% of what number?

   Translation: $15 = 25\% \cdot y$

3. Finding the ***percent number*** (the percent itself)

   Example: 15 is what percent of 60?

   Translation: $15 = y \cdot 60$

FINDING THE AMOUNT

EXAMPLE 4 What is 11% of 49?

What is 11% of 49?

Translate: $a = 11\% \times 49$

Solve: The letter is by itself. To solve the equation, we need only convert 11% to decimal notation and multiply:

$$a = 11\% \times 49 = 0.11 \times 49 = 5.39.$$

Thus, 5.39 is 11% of 49. The answer is 5.39.

Do Exercise 7.

FINDING THE BASE

EXAMPLE 5 3 is 16% of what?

3 is 16% of what number?

Translate: $3 = 16\% \times b$

$3 = 0.16 \times b$ Converting 16% to decimal notation

Solve: In this case, the letter is not by itself. To solve the equation, we divide by 0.16 on both sides:

$$3 = 0.16 \times b$$

$$\frac{3}{0.16} = \frac{0.16 \times b}{0.16} \quad \text{Dividing by 0.16}$$

$$18.75 = b. \quad \text{Simplifying}$$

The answer is 18.75.

Do Exercise 8.

FINDING THE PERCENT NUMBER

In solving these problems, you *must* remember to convert to percent notation after you have solved the equation.

EXAMPLE 6 $32 is what percent of $50?

$32 is what percent of $50?

Translate: $32 = p \times 50$

Solve: To solve the equation, we divide by 50 on both sides and convert the answer to percent notation:

$$32 = p \times 50$$

$$\frac{32}{50} = \frac{p \times 50}{50} \quad \text{Dividing by 50}$$

$$0.64 = p$$

$$64\% = p. \quad \text{Converting to percent notation}$$

Thus, 64% of $50 is $32. The answer is 64%.

Do Exercise 9.

EXAMPLE 7 *Coronary Heart Disease.* In 2001, there were 281 million people in the United States. About 2.5% of them had heart disease. How many had heart disease?

Source: American Heart Association

To solve the problem, we first reword and then translate. We let a = the number of people in the United States with heart disease.

Rewording: What is 2.5% of 281?

Translate: $a = 2.5\% \times 281$

Solve: The letter is by itself. To solve the equation, we need only convert 2.5% to decimal notation and multiply:

$$a = 2.5\% \times 281 = 0.025 \times 281 = 7.025.$$

Thus, 7.025 million is 2.5% of 281 million, so in 2001 about 7.025 million people in the United States had heart disease.

Do Exercise 10.

EXAMPLE 8 *DVD Players.* At one time, Amazon.com had a Sharp DVD video player on sale for $899.98. This was 60% of the list price. What was the list price?

To solve the problem, we first reword and then translate. We let L = the list price.

Rewording: $899.98 is 60% of what number?

Translate: $899.98 = 60\% \times L$

9. What percent of $50 is $18?

10. Areas of Alaska and Arizona. The area of Arizona is 19% of the area of Alaska. The area of Alaska is 586,400 mi^2. What is the area of Arizona?

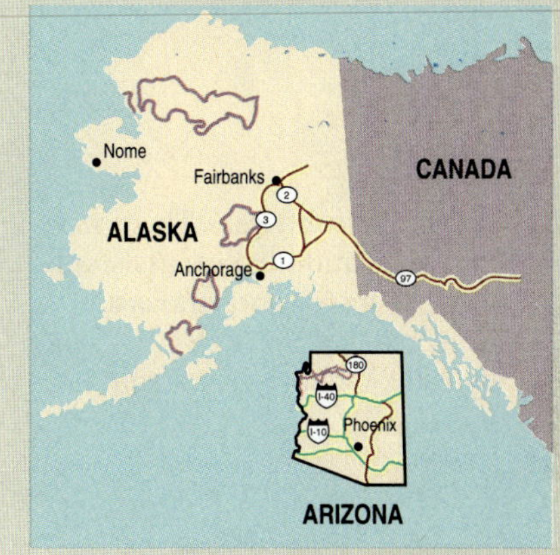

Answers on page A-18

11. **Population of Arizona.** The population of Arizona was 5.1 million in 2000. This was 130.6% of its population in 1990. What was the population in 1990?
Source: U.S. Bureau of the Census

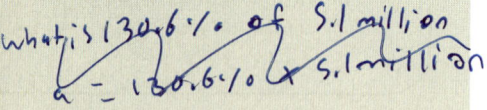

12. **Job Opportunities.** There were 252 thousand medical assistants in 1998. Job opportunities are expected to grow to 398 thousand by 2008. What is the percent of increase?
Source: *Handbook of U.S. Labor Statistics*

Answers on page A-18

Solve: To solve the equation, we convert 60% to decimal notation and divide by 0.60 on both sides:

$$899.98 = 60\% \times L$$

$$899.98 = 0.60 \times L \qquad \text{Converting to decimal notation}$$

$$\frac{899.98}{0.60} = \frac{0.60 \times L}{0.60} \qquad \text{Dividing by 0.60}$$

$$1499.97 \approx L. \qquad \text{Simplifying using a calculator and rounding to the nearest cent}$$

The list price was about $1499.97.

Do Exercise 11.

EXAMPLE 9 *Compaq PAP.* A Compaq iPAQ PA-1 64 MB Personal Audio Player (PAP) was on sale on the Internet for $199.99, decreased from a normal list price of $249.99. What was the percent of decrease?

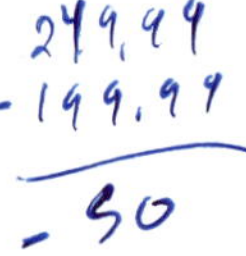

To solve the problem, we must first determine the amount of decrease from the original price:

Original price	minus	Sale price	=	Decrease
↓	↓	↓	↓	↓
$249.99	−	$199.99	=	$50.00.

Using the $50 decrease, we reword and translate. We let $p =$ the percent of decrease. We want to know, "What percent of the *original* price is $50?"

Rewording:	$50	is	what percent	of	$249.99?
	↓	↓	↓	↓	↓
Translate:	50	=	p	×	249.99

Solve: To solve the equation, we divide by 249.99 on both sides and convert the answer to percent notation:

$$50 = p \times 249.99$$

$$\frac{50}{249.99} = \frac{p \times 249.99}{249.99} \qquad \text{Dividing by 249.99}$$

$$0.20 \approx p$$

$$20\% = p. \qquad \text{Simplifying and converting to percent notation}$$

Thus the percent of decrease was about 20%.

Do Exercise 12.

7.5 EXERCISE SET

For Extra Help

Digital Video Tutor CD 6 Videotape 8

InterAct Math

Math Tutor Center

MathXL

MyMathLab

a Solve.

1. What percent of 180 is 36?

2. What percent of 76 is 19?

3. 45 is 30% of what number?

4. 20.4 is 24% of what number?

5. What number is 65% of 840?

6. What is 50% of 50? (This was a $500.00 question on the "Who Wants To Be a Millionaire?" television quiz show.)

7. 30 is what percent of 125?

8. 57 is what percent of 300?

9. 12% of what number is 0.3?

10. 7 is 175% of what number?

11. 2 is what percent of 40?

12. 40 is 2% of what number?

13. What percent of 68 is 17?

14. What percent of 150 is 39?

15. What number is 35% of 240?

16. What number is 1% of one million?

17. What percent of 125 is 30?

18. What percent of 60 is 75?

19. What percent of 300 is 48?

20. What percent of 70 is 70?

21. 14 is 30% of what number?

22. 54 is 24% of what number?

23. What is 2% of 40?

24. What is 40% of 2?

25. 0.8 is 16% of what number?

26. 25 is what percent of 50?

27. 54 is 135% of what number?

28. 8 is 2% of what number?

Costs of Owning a Dog. The American Pet Products Manufacturers Association estimates that the total cost of owning a dog for its lifetime is $6600. The following circle graph shows the relative costs of raising a dog from birth to death.

Costs of Owning a Dog

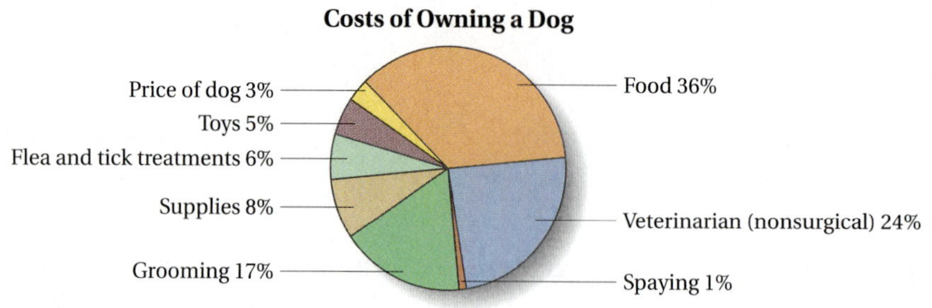

Source: The American Pet Products Manufacturers Association

Complete the following table of costs of owning a dog for its lifetime.

	EXPENSE ITEM	COST		EXPENSE ITEM	COST
29.	Price of dog		**30.**	Food	
31.	Veterinarian		**32.**	Grooming	
33.	Supplies		**34.**	Flea and tick treatments	

35. *Auto Sales.* In 2000, 17.4 million cars were sold in the United States. Of these, 11.9 million were manufactured in the United States, 4.5 million in Asia, and 1 million in Europe. What percent were manufactured in each region?
Source: *Autodata*

36. *Auto Sales.* In 1997, 15.2 million cars were sold in the United States. Of these, 10.8 million were manufactured in the United States, 3.7 million in Asia, and 0.7 million in Europe. What percent were manufactured in each region?
Source: *Autodata*

37. *Batting Average.* At one point in a recent season, Sammy Sosa of the Chicago Cubs had 193 hits. His batting average was 0.320, or 32%. That is, of the total number of at-bats, 32% were hits. How many at-bats did he have?
Source: Major League Baseball

38. *Pass Completions.* At one point in a recent season, Peyton Manning of the Indianapolis Colts had completed 357 passes. This was 62.5% of his attempts. How many attempts did he make?
Source: National Football League

39. *Student Loans.* To finance her community college education, Sarah takes out a Stafford loan for $3500. After a year, Sarah decides to pay off the interest, which is 8% of $3500. How much will she pay?

40. *Student Loans.* Paul takes out a subsidized federal Stafford loan for $2400. After a year, Paul decides to pay off the interest, which is 7% of $2400. How much will he pay?

41. *Tipping.* Leon left a $4 tip for a meal that cost $25.

a) What percent of the cost of the meal was the tip?
b) What was the total cost of the meal including the tip?

42. *Tipping.* Selena left a $12.76 tip for a meal that cost $58.

a) What percent of the cost of the meal was the tip?
b) What was the total cost of the meal including the tip?

43. *Tipping.* Leon left a 15% tip for a meal that cost $25.

a) How much was the tip?

b) What was the total cost of the meal including the tip?

44. *Tipping.* Sam, Selena, Rachel, and Clement left a 15% tip for a meal that cost $58.

a) How much was the tip?

b) What was the total cost of the meal including the tip?

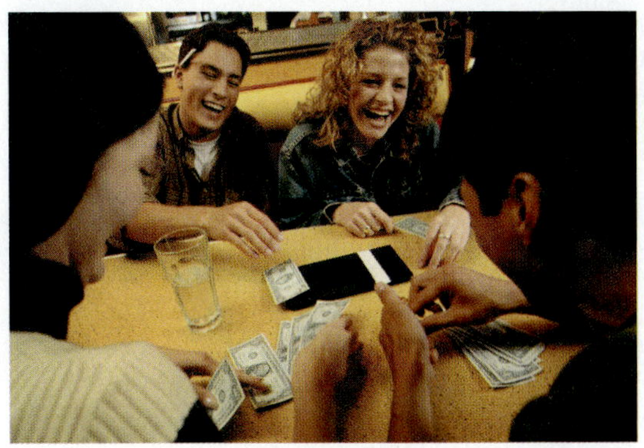

45. *Tipping.* Leon left a 15% tip of $4.32 for a meal.

a) What was the cost of the meal before the tip?

b) What was the total cost of the meal including the tip?

46. *Tipping.* Selena left a 15% tip of $8.40 for a meal.

a) What was the cost of the meal before the tip?

b) What was the total cost of the meal including the tip?

47. In a medical study of a group of pregnant women with "poor" diets, 16 of the women, or 8%, had babies who were in good or excellent health. How many women were in the original study?

48. In a medical study of a group of pregnant women with "good-to-excellent" diets, 285 of the women, or 95%, had babies who were in good or excellent health. How many women were in the original study?

49. *Body Fat.* The author of this text exercises regularly at a local YMCA that recently offered a body-fat percentage test to its members. The device used measures the passage of a very low voltage of electricity through the body. The author's body-fat percentage was found to be 16.5% and he weighs 191 lb. What part, in pounds, of his body weight is fat?

50. *Junk Mail.* The U.S. Postal Service reports that we open and read 78% of the junk mail that we receive. A sports instructional videotape company sends out 10,500 advertising brochures.

a) How many of the brochures can it expect to be opened and read?

b) The company sells videos to 189 of the people who receive the brochure. What percent of the 10,500 people who receive the brochure buy the video?

Source: U.S. Postal Service

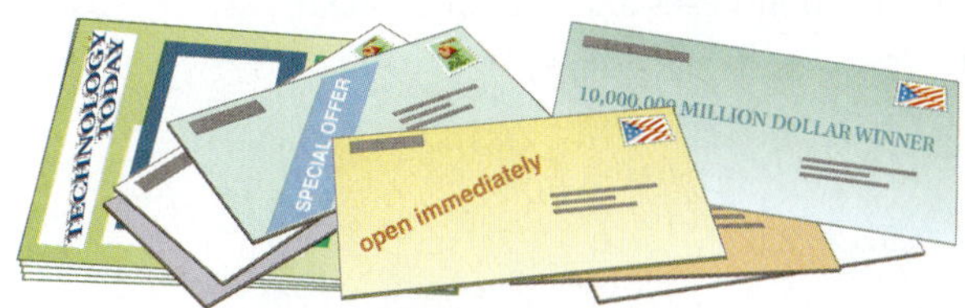

Life Insurance Rates for Smokers and Nonsmokers. The data in the following table illustrate how yearly rates (premiums) for a $500,000 term life insurance policy are increased for smokers. Complete the table by finding the missing numbers. Round to the nearest percent and dollar.

TYPICAL INSURANCE PREMIUMS (DOLLARS)

	AGE	RATE FOR NONSMOKER	RATE FOR SMOKER	RATE INCREASE	PERCENT OF INCREASE FOR SMOKER
	35	$345	$630	$285	83%
51.	40	$430	$735		
52.	45	$565			84%
53.	50	$780			100%
54.	55	$985	$2137		
55.	60	$1645	$2955		
56.	65	$2943			85%

Source: Pacific Life PL Protector Term Life Portfolio, OYT Rates

57. ^{D}W The 80/20 rule is commonly quoted in the field of business. It asserts that 80% of your results will come from 20% of your activities. Discuss how this might affect you as a student and as an employee.

58. ^{D}W Comment on the following quote by Yogi Berra, a famous Major League Hall of Fame baseball player: "Ninety percent of hitting is mental. The other half is physical."

SKILL MAINTENANCE

Simplify.

59. $-3 - 8$ [6.4a]

60. $5 \cdot (-20)$ [6.5a]

61. $-\frac{3}{5} + \frac{1}{5}$ [6.3a]

62. $-12 \div (-6)$ [6.6a]

Remove parentheses and simplify. [6.8b]

63. $-5a + 3c - 2(c - 3a)$

64. $4(x - 2y) - (y - 3x)$

Add. [6.3a]

65. $-6.5 + 2.6$

66. $-\frac{3}{8} + (-5) + \frac{1}{4} + (-1)$

SYNTHESIS

67. It has been determined that at the age of 15, a boy has reached 96.1% of his final adult height. Jaraan is 6 ft 4 in. at the age of 15. What will his final adult height be?

68. It has been determined that at the age of 10, a girl has reached 84.4% of her final adult height. Dana is 4 ft 8 in. at the age of 10. What will her final adult height be?

7.6 APPLICATIONS AND PROBLEM SOLVING

Objective

a Solve applied problems by translating to equations.

a Five Steps for Solving Problems

We have discussed many new equation-solving tools in this chapter and used them for applications and problem solving. Here we review the five-step strategy for solving problems that was introduced in Section 1.5.

FIVE STEPS FOR PROBLEM SOLVING IN ALGEBRA

1. *Familiarize* yourself with the problem situation.
2. *Translate* the problem to an equation.
3. *Solve* the equation.
4. *Check* the answer in the original problem.
5. *State* the answer to the problem clearly.

Of the five steps, the most important is probably the first one: becoming familiar with the problem situation. The table below lists some hints for familiarization.

TO FAMILIARIZE YOURSELF WITH A PROBLEM:

- If a problem is given in words, read it carefully. Reread the problem, perhaps aloud. Try to verbalize the problem as if you were explaining it to someone else.
- Choose a variable (or variables) to represent the unknown and clearly state what the variable represents. Be descriptive! For example, let $L =$ the length, $d =$ the distance, and so on.
- Make a drawing and label it with known information, using specific units if given. Also, indicate unknown information.
- Find further information. Look up formulas or definitions with which you are not familiar. (Geometric formulas appear at the front of this text.) Consult a reference librarian or the Internet.
- Create a table that lists all the information you have available. Look for patterns that may help in the translation to an equation.
- Think of a possible answer and check the guess. Note the manner in which the guess is checked.

EXAMPLE 1 *Hiking.* In 1998, at age 79, Earl Shaffer became the oldest person to hike all 2100 miles of the Appalachian Trail—from Springer Mountain, Georgia, to Mount Katahdin, Maine. At one point, Shaffer stood atop Big Walker Mountain, Virginia, which is three times as far from the northern end as from the southern end. How far was Shaffer from each end of the trail?

Source: Appalachian Trail Conference

2100 total 3x

1. Running. In 1997, Yiannis Kouros of Australia set the record for the greatest distance run in 24 hr by running 188 mi. After 8 hr, he was approximately twice as far from the finish line as he was from the start. How far had he run?

Source: *Guinness World Records 2000 Millennium Edition*

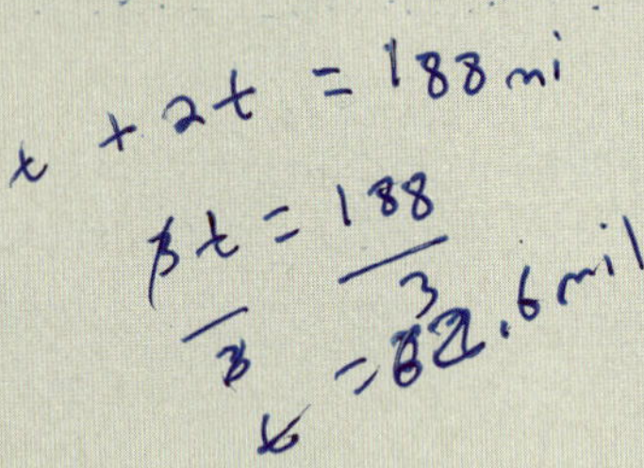

Answer on page A-19

1. Familiarize. Let's consider a drawing.

To become familiar with the problem, let's guess a possible distance that Shaffer stood from Springer Mountain—say, 600 mi. Three times 600 mi is 1800 mi. Since 600 mi + 1800 mi = 2400 mi and 2400 mi is greater than 2100 mi, we see that our guess is too large. Rather than guess again, let's use the skills we have obtained in the ability to solve equations. We let

d = distance, in miles, to the southern end,

and

$3d$ = the distance, in miles, to the northern end.

(We could also let x = the distance to the northern end and $\frac{1}{3}x$ = the distance to the southern end.)

2. Translate. From the drawing, we see that the lengths of the two parts of the trail must add up to 2100 mi. This leads to our translation.

Distance to southern end	plus	Distance to northern end	is	2100 mi
↓	↓	↓	↓	↓
d	$+$	$3d$	$=$	2100

3. Solve. We solve the equation:

$$d + 3d = 2100$$

$$4d = 2100 \quad \text{Collecting like terms}$$

$$\frac{4d}{4} = \frac{2100}{4} \quad \text{Dividing by 4}$$

$$d = 525.$$

4. Check. As expected, d is less than 600 mi. If $d = 525$ mi, then $3d = 1575$ mi. Since 525 mi + 1575 mi = 2100 mi, we have a check.

5. State. Atop Big Walker Mountain, Shaffer stood 525 mi from Springer Mountain and 1575 mi from Mount Katahdin.

Do Exercise 1 on the preceding page.

EXAMPLE 2 *Gourmet Sandwiches.* A gourmet sandwich shop located near a college campus specializes in sandwiches prepared in buns of length 18 in. Suppose Jenny, Demi, and Sarah buy one of these sandwiches and take it back to their apartment. Since they have different appetites, Jenny cuts the sandwich in such a way that Demi gets half of what Jenny gets and Sarah gets three-fourths of what Jenny gets. Find the length of each person's sandwich.

1. Familiarize. We first make a drawing.

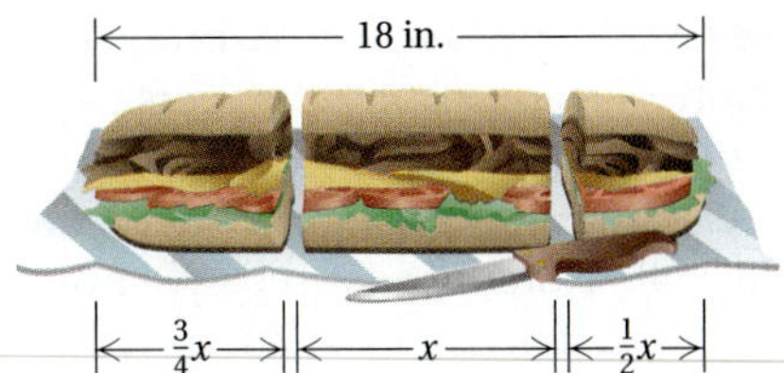

Because the sandwich lengths are expressed in terms of Jenny's sandwich, we let

$x =$ the length of Jenny's sandwich.

Then $\frac{1}{2}x =$ the length of Demi's sandwich

and $\frac{3}{4}x =$ the length of Sarah's sandwich.

2. Translate. From the statement of the problem and the drawing, we see that the lengths add up to 18 in. That gives us our translation:

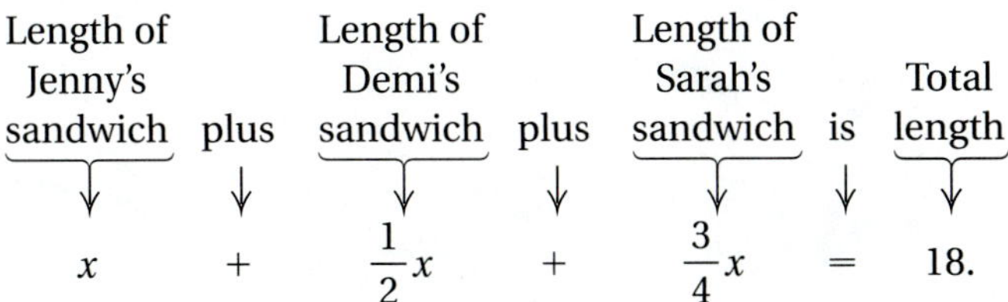

3. Solve. We begin by clearing fractions as follows:

$$x + \frac{1}{2}x + \frac{3}{4}x = 18 \qquad \text{The LCM of all the denominators is 4.}$$

$$4\left(x + \frac{1}{2}x + \frac{3}{4}x\right) = 4 \cdot 18 \qquad \text{Multiplying by the LCM, 4}$$

$$4 \cdot x + 4 \cdot \frac{1}{2}x + 4 \cdot \frac{3}{4}x = 4 \cdot 18 \qquad \text{Using the distributive law}$$

$$4x + 2x + 3x = 72 \qquad \text{Simplifying}$$

$$9x = 72 \qquad \text{Collecting like terms}$$

$$\frac{9x}{9} = \frac{72}{9} \qquad \text{Dividing by 9}$$

$$x = 8.$$

4. Check. Do we have an answer to the *problem*? If the length of Jenny's sandwich is 8 in., then the length of Demi's sandwich is $\frac{1}{2} \cdot 8$ in., or 4 in., and the length of Sarah's sandwich is $\frac{3}{4} \cdot 8$ in., or 6 in. These lengths add up to 18 in. Our answer checks.

2. Rocket Sections. A rocket is divided into three sections: the payload and navigation section in the top, the fuel section in the middle, and the rocket engine section in the bottom. The top section is one-sixth the length of the bottom section. The middle section is one-half the length of the bottom section. The total length is 240 ft. Find the length of each section.

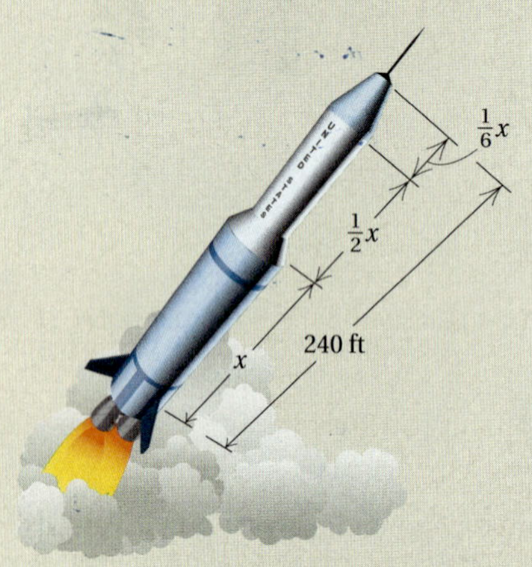

5. **State.** The length of Jenny's sandwich is 8 in., the length of Demi's sandwich is 4 in., and the length of Sarah's sandwich is 6 in.

Do Exercise 2.

Recall that the

$$\text{Set of integers} = \{\ldots, -5, -4, -3, -2, -1, 0, 1, 2, 3, 4, 5, \ldots\}.$$

Before we solve the next problem, we need to learn some additional terminology regarding integers.

The following are examples of **consecutive integers:** 16, 17, 18, 19, 20; and −31, −30, −29, −28. Note that consecutive integers can be represented in the form x, $x + 1$, $x + 2$, and so on.

The following are examples of **consecutive even integers:** 16, 18, 20, 22, 24; and −52, −50, −48, −46. Note that consecutive even integers can be represented in the form x, $x + 2$, $x + 4$, and so on.

The following are examples of **consecutive odd integers:** 21, 23, 25, 27, 29; and −71, −69, −67, −65. Note that consecutive odd integers can be represented in the form x, $x + 2$, $x + 4$, and so on.

EXAMPLE 3 *Interstate Mile Markers.* If you are traveling on a U.S. interstate highway, you will notice numbered markers every mile to tell your location in case of an accident or other emergency. In many states, the numbers on the markers increase from west to east. The sum of two consecutive mile markers on I-70 in Kansas is 559. Find the numbers on the markers.
Source: Federal Highway Administration, Ed Rotalewski

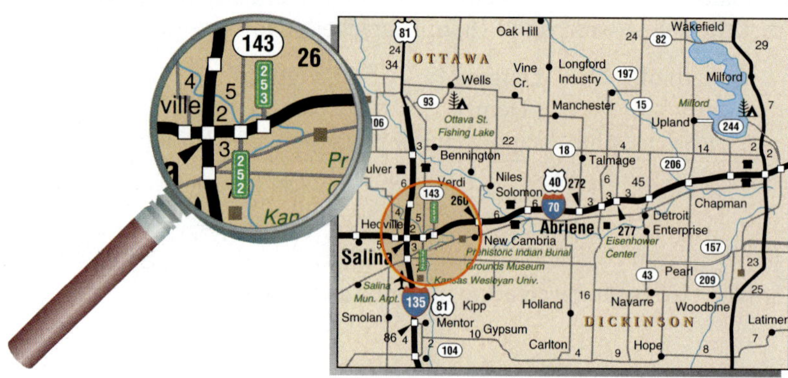

1. **Familiarize.** The numbers on the mile markers are consecutive positive integers. Thus if we let x = the smaller number, then $x + 1$ = the larger number.

   To become familiar with the problem, we can make a table. First, we guess a value for x; then we find $x + 1$. Finally, we add the two numbers and check the sum.

x	$x + 1$	SUM OF x AND $x + 1$
114	115	229
252	253	505
302	303	605

Answer on page A-19

From the table, we see that the first marker should be between 252 and 302. You might actually solve the problem this way, but let's work on developing our algebra skills.

2. **Translate.** We reword the problem and translate as follows.

First integer	plus	Second integer	is	559	Rewording
x	$+$	$(x + 1)$	$=$	559	Translating

3. **Solve.** We solve the equation:

$$x + (x + 1) = 559$$

$$2x + 1 = 559 \quad \text{Collecting like terms}$$

$$2x + 1 - 1 = 559 - 1 \quad \text{Subtracting 1}$$

$$2x = 558$$

$$\frac{2x}{2} = \frac{558}{2} \quad \text{Dividing by 2}$$

$$x = 279.$$

If x is 279, then $x + 1$ is 280.

4. **Check.** Our possible answers are 279 and 280. These are consecutive positive integers and $279 + 280 = 559$, so the answers check.

5. **State.** The mile markers are 279 and 280.

Do Exercise 3.

EXAMPLE 4 *IKON Copiers.* IKON Office Solutions rents a Canon IR330 copier for \$225 per month plus 1.2¢ per copy. A law firm needs to lease a copy machine for use during a special case that they anticipate will take 3 months. If they allot a budget of \$1100, how many copies can they make?

Source: IKON Office Solutions, Nathan DuMond, Sales Manager

1. **Familiarize.** Suppose that the law firm makes 20,000 copies. Then the cost is monthly charges plus copy charges, or

3(\$225)	plus	Cost per copy	times	Number of copies
\$675	$+$	\$0.012	$\cdot$	20,000,

3. **Interstate Mile Markers.** The sum of two consecutive mile markers on I-90 in upstate New York is 627. (On I-90 in New York, the marker numbers increase from east to west.) Find the numbers on the markers.
Source: New York State Department of Transportation

Answer on page A-19

4. IKON Copiers. The law firm in Example 4 decides to raise its budget to $1400 for the 3-month period. How many copies can they make for $1400?

which is $915. This process familiarizes us with the way in which a calculation is made. Note that we convert 1.2¢ to $0.012 so that all information is in the same unit, dollars. Otherwise, we will not get the correct answer.

We let $c =$ the number of copies that can be made for $1100.

2. Translate. We reword the problem and translate as follows.

Monthly cost	plus	Cost per copy	times	Number of copies	is	Cost
↓	↓	↓	↓	↓	↓	↓
3($225)	+	$0.012	·	c	=	$1100

3. Solve. We solve the equation:

$$3(225) + 0.012c = 1100$$
$$675 + 0.012c = 1100$$
$$0.012c = 425 \quad \text{Subtracting 675}$$
$$\frac{0.012c}{0.012} = \frac{425}{0.012} \quad \text{Dividing by 0.012}$$
$$c \approx 35{,}417. \quad \text{Rounding to the nearest one}$$

4. Check. We check in the original problem. The cost for 35,417 pages is 35,417($0.012) = $425.004. The rental for 3 months is 3($225) = $675. The total cost is then $425.004 + $675 ≈ $1100, which is the $1100 that was allotted.

5. State. The law firm can make 35,417 copies on the copy rental allotment of $1100.

Do Exercise 4.

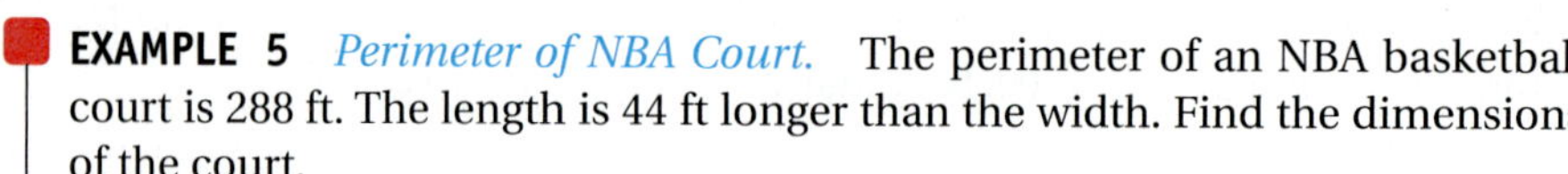

EXAMPLE 5 *Perimeter of NBA Court.* The perimeter of an NBA basketball court is 288 ft. The length is 44 ft longer than the width. Find the dimensions of the court.

Source: National Basketball Association

1. Familiarize. We first make a drawing.

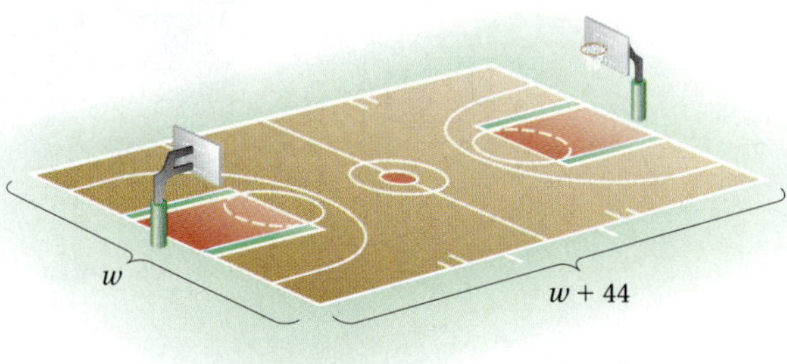

We let $w =$ the width of the rectangle. Then $w + 44 =$ the length. The perimeter P of a rectangle is the distance around the rectangle and is given by the formula $2l + 2w = P$, where

$l =$ the length and $w =$ the width.

Answer on page A-19

2. **Translate.** To translate the problem, we substitute $w + 44$ for l and 288 for P:

$$2l + 2w = P$$
$$2(w + 44) + 2w = 288.$$

CAUTION! Parentheses are important here.

3. **Solve.** We solve the equation:

$$2(w + 44) + 2w = 288$$
$$2 \cdot w + 2 \cdot 44 + 2w = 288 \quad \text{Using the distributive law}$$
$$4w + 88 = 288 \quad \text{Collecting like terms}$$
$$4w + 88 - 88 = 288 - 88 \quad \text{Subtracting 88}$$
$$4w = 200$$
$$\frac{4w}{4} = \frac{200}{4} \quad \text{Dividing by 4}$$
$$w = 50.$$

Thus possible dimensions are

$w = 50$ ft and $l = w + 44 = 50 + 44$, or 94 ft.

4. **Check.** If the width is 50 ft and the length is 94 ft, then the perimeter is $2(50\text{ ft}) + 2(94\text{ ft})$, or 288 ft. This checks.

5. **State.** The width is 50 ft and the length is 94 ft.

Do Exercise 5.

5. **Perimeter of High School Basketball Court.** The perimeter of a standard high school basketball court is 268 ft. The length is 34 ft longer than the width. Find the dimensions of the court.

Source: Indiana High School Athletic Association

Answer on page A-19

EXAMPLE 6 *Cross Section of a Roof.* In a triangular cross section of a roof, the second angle is twice as large as the first angle. The measure of the third angle is 20° greater than that of the first angle. How large are the angles?

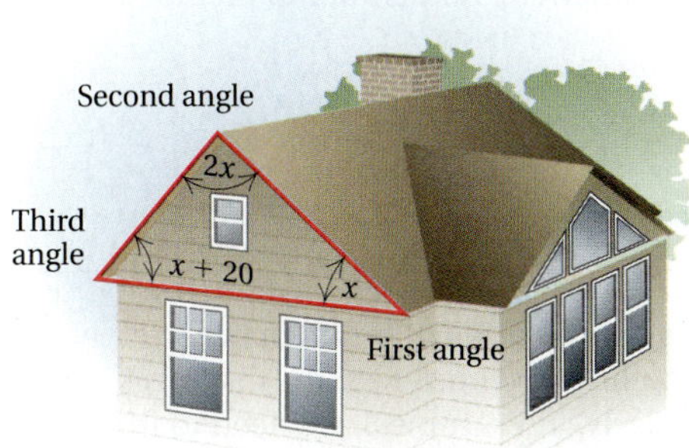

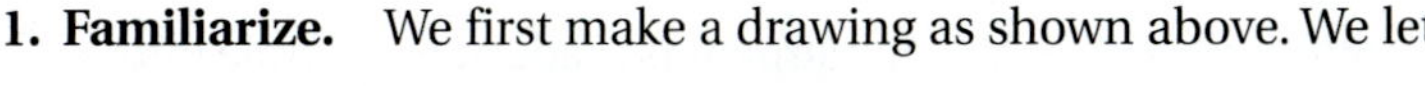

1. **Familiarize.** We first make a drawing as shown above. We let

measure of first angle $= x$.

Then measure of second angle $= 2x$

and measure of third angle $= x + 20$.

6. The second angle of a triangle is three times as large as the first. The third angle measures 30° more than the first angle. Find the measures of the angles.

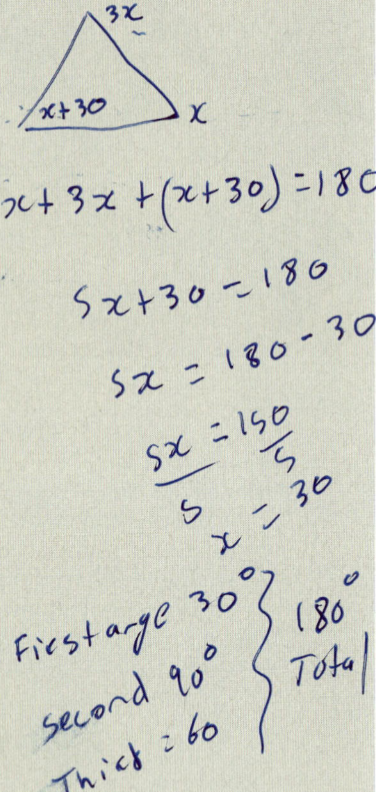

Answer on page A-19

2. **Translate.** To translate, we need to recall a geometric fact. (You might, as part of step 1, look it up in a geometry book or in the list of formulas at the front of this text.) Remember, the measures of the angles of a triangle total 180°.

Measure of first angle	plus	Measure of second angle	plus	Measure of third angle	is	180°
↓	↓	↓	↓	↓	↓	↓
x	$+$	$2x$	$+$	$(x + 20)$	$=$	$180°$

3. **Solve.** We solve the equation:

$$x + 2x + (x + 20) = 180$$
$$4x + 20 = 180$$
$$4x + 20 - 20 = 180 - 20$$
$$4x = 160$$
$$\frac{4x}{4} = \frac{160}{4}$$
$$x = 40.$$

Possible measures for the angles are as follows:

First angle: $x = 40°$;

Second angle: $2x = 2(40) = 80°$;

Third angle: $x + 20 = 40 + 20 = 60°$.

4. **Check.** Consider our answers: 40°, 80°, and 60°. The second is twice the first and the third is 20° greater than the first. The sum is 180°. The angles check.

5. **State.** The measures of the angles are 40°, 80°, and 60°.

CAUTION!

Units are important in answers. Remember to include them, where appropriate.

Do Exercise 6.

CAUTION!

Always be sure to answer the original problem completely. For instance, in Example 1, we need to find *two* numbers: the distances from *each* end of the trail to the hiker. Similarly, in Example 3, we need to find two mile markers, and in Example 5, we need to find two dimensions, not just the width.

EXAMPLE 7 *Simple Interest.* An investment is made at 6% simple interest for 1 year. It grows to $768.50. How much was originally invested (the principal)?

1. **Familiarize.** Suppose that $100 was invested. Recalling the formula for simple interest, $I = Prt$, we know that the interest for 1 year on $100 at 6% simple interest is given by $I = \$100 \cdot 0.06 \cdot 1 = \$6$. Then, at the end of the year, the amount in the account is found by adding the principal and the interest:

Principal	+	Interest	=	Amount
↓		↓		↓
$100	+	$6	=	$106.

In this problem, we are working backward. We are trying to find the principal, which is the original investment. We let $x =$ the principal.

2. **Translate.** We reword the problem and then translate.

$$\begin{array}{ccccc} \text{Principal} & + & \text{Interest} & = & \text{Amount} \\ \downarrow & & \downarrow & & \downarrow \\ x & + & 6\%x & = & 768.50 \end{array}$$

Interest is 6% of the principal.

3. **Solve.** We solve the equation:

$x + 6\%x = 768.50$

$x + 0.06x = 768.50$ Converting to decimal notation

$1x + 0.06x = 768.50$ Identity property of 1

$1.06x = 768.50$ Collecting like terms

$\frac{1.06x}{1.06} = \frac{768.50}{1.06}$ Dividing by 1.06

$x = 725.$

4. **Check.** We check by taking 6% of \$725 and adding it to \$725:

$6\% \times \$725 = 0.06 \times 725 = \$43.50.$

Then $\$725 + \$43.50 = \$768.50$, so \$725 checks.

5. **State.** The original investment was \$725.

Do Exercise 7.

EXAMPLE 8 *Selling a home.* The Landers are planning to sell their home. If they want to be left with \$117,500 after paying 6% of the selling price to a realtor as a commission, for how much must they sell the house?

1. **Familiarize.** Suppose the Landers sell the house for \$120,000. A 6% commission can be determined by finding 6% of \$120,000:

6% of $\$120{,}000 = 0.06(\$120{,}000) = \$7200.$

Subtracting this commission from \$120,000 would leave the Landers with

$\$120{,}000 - \$7200 = \$112{,}800.$

This shows that in order for the Landers to clear \$117,500, the house must sell for more than \$120,000. To determine what the sale price must be, we could check more guesses. Instead, we let $x =$ the selling price, in dollars. With a 6% commission, the realtor would receive $0.06x$.

2. **Translate.** We reword the problem and translate as follows.

$$\begin{array}{ccccc} \underbrace{\text{Selling price}} & \text{less} & \underbrace{\text{Commission}} & \text{is} & \underbrace{\text{Amount remaining.}} \\ \downarrow & \downarrow & \downarrow & \downarrow & \downarrow \\ x & - & 0.06x & = & 117{,}500 \end{array}$$

7. **Simple Interest.** An investment is made at 7% simple interest for 1 year. It grows to \$8988. How much was originally invested (the principal)?

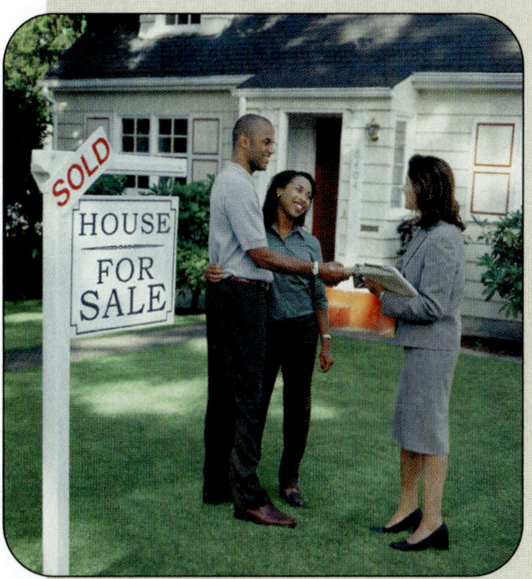

Answer on page A-19

8. Price Before Sale. The price of a suit was decreased to a sale price of \$526.40. This was a 20% reduction. What was the former price?

Answer on page A-19

3. Solve. We solve the equation:

$$x - 0.06x = 117{,}500$$

$$1x - 0.06x = 117{,}500$$

$$0.94x = 117{,}500 \qquad \text{Collecting like terms. Had we noted that after the commission has been paid, 94\% remains, we could have begun with this equation.}$$

$$\frac{094x}{0.94} = \frac{117{,}500}{0.94} \qquad \text{Dividing by 0.94}$$

$$x = 125{,}000.$$

4. Check. To check, we first find 6% of \$125,000:

$$6\% \text{ of } \$125{,}000 = 0.06(\$125{,}000) = \$7500. \qquad \text{This is the commission.}$$

Next, we subtract the commission to find the remaining amount:

$$\$125{,}000 - \$7500 = \$117{,}500.$$

Since, after the commission, the Landers are left with \$117,500, our answer checks. Note that the \$125,000 selling price is greater than \$120,000, as predicted in the *Familiarize* step.

5. State. To be left with \$117,500, the Landers must sell the house for \$125,000.

Do Exercise 8.

CAUTION!

The problem in Example 8 is easy to solve with algebra. Without algebra, it is not. A common error in such a problem is to take 6% of the price after commission and then subtract or add. Note that 6% of the selling price ($6\% \cdot \$125{,}000 = \$7500$) is not equal to 6% of the amount that the Landers wanted to be left with ($6\% \cdot \$117{,}500 = \$7050$).

Study Tips

PROBLEM-SOLVING TIPS

The more problems you solve, the more your skills will improve.

1. Look for patterns when solving problems. Each time you study an example in a text, you may observe a pattern for problems that you will encounter later in the exercise sets or in other practical situations.
2. When translating in mathematics, consider the dimensions of the variables and constants in the equation. The variables that represent length should all be in the same unit, those that represent money should all be in dollars or all in cents, and so on.
3. Make sure that units appear in the answer whenever appropriate and that you have completely answered the original problem.

7.6 EXERCISE SET

For Extra Help

Digital Video Tutor CD 6 Videotape 8

InterAct Math

Math Tutor Center

MathXL

MyMathLab

a Solve. *Even though you might find the answer quickly in some other way, practice using the five-step problem-solving strategy.*

1. *Pipe Cutting.* A 240-in. pipe is cut into two pieces. One piece is three times the length of the other. Find the lengths of the pieces.

2. *Board Cutting.* A 72-in. board is cut into two pieces. One piece is 2 in. longer than the other. Find the lengths of the pieces.

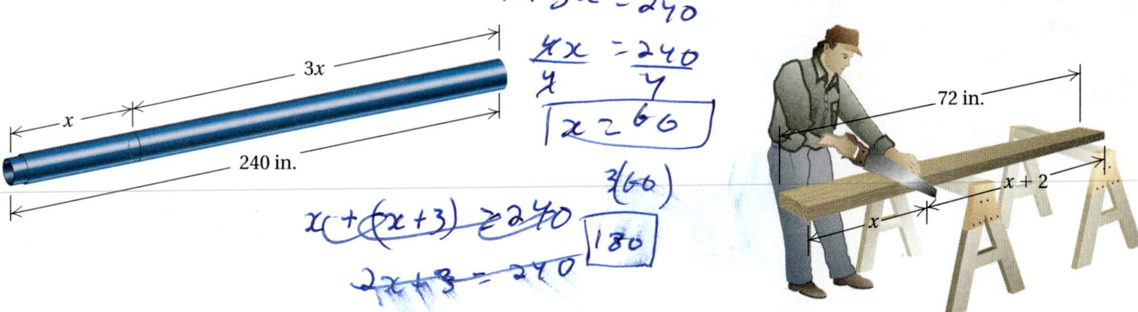

3. *Wheaties.* Recently, the cost of four 18-oz boxes of Wheaties cereal was \$14.68. What was the cost of one box?

4. *Area of Lake Ontario.* The area of Lake Superior is about four times the area of Lake Ontario. The area of Lake Superior is 30,172 mi^2. What is the area of Lake Ontario?

5. *Women's Dresses.* In a recent year, the total amount spent on women's blouses was \$6.5 billion. This was \$0.2 billion more than what was spent on women's dresses. How much was spent on women's dresses?

6. *Statue of Liberty.* The height of the Eiffel Tower is 974 ft, which is about 669 ft higher than the Statue of Liberty. What is the height of the Statue of Liberty?

7. *Iditarod Race.* The Iditarod sled dog race extends for 1049 mi from Anchorage to Nome. If a musher is twice as far from Anchorage as from Nome, how much of the race has the musher completed?
Source: Iditarod Trail Commission

8. *Home Remodeling.* In a recent year, Americans spent a total of \$35 billion to remodel bathrooms and kitchens. Twice as much was spent on kitchens as bathrooms. How much was spent on each?

9. *Consecutive Post Office Box Numbers.* The sum of the numbers on two consecutive post office boxes is 547. What are the numbers?

10. *Consecutive Page Numbers.* The sum of the page numbers on the facing pages of a book is 573. What are the page numbers?

11. *Consecutive Ticket Numbers.* The numbers on Sam's three raffle tickets are consecutive integers. The sum of the numbers is 126. What are the numbers?

12. *Consecutive Ages.* The ages of Whitney, Wesley, and Wanda are consecutive integers. The sum of their ages is 108. What are their ages?

13. *Consecutive Odd Integers.* The sum of three consecutive odd integers is 189. What are the integers?

14. *Consecutive Integers.* Three consecutive integers are such that the first plus one-half the second plus seven less than twice the third is 2101. What are the integers?

15. *Standard Billboard Sign.* A standard rectangular highway billboard sign has a perimeter of 124 ft. The length is 6 ft more than three times the width. Find the dimensions of the sign.

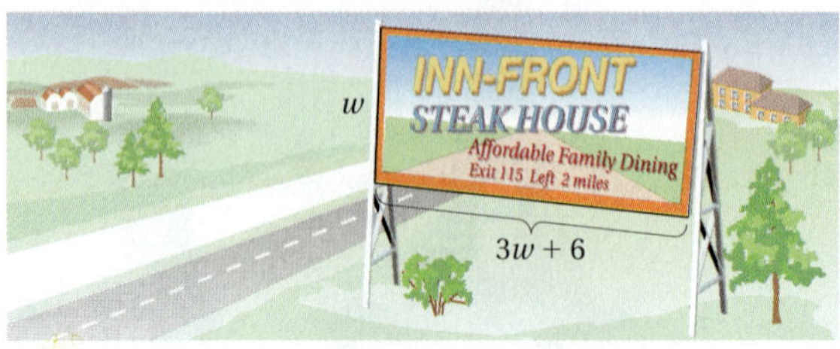

16. *Two-by-Four.* The perimeter of a cross section of a "two-by-four" piece of lumber is $10\frac{1}{2}$ in. The length is twice the width. Find the actual dimensions of the cross section of a two-by-four.

17. *Price of Sneakers.* Amy paid \$63.75 for a pair of New Balance 903 running shoes during a 15%-off sale. What was the regular price?

18. *Price of a CD Player.* Doug paid \$72 for a shockproof portable CD player during a 20%-off sale. What was the regular price?

19. *Price of a Textbook.* Evelyn paid \$89.25, including 5% tax, for her biology textbook. How much did the book itself cost?

20. *Price of a Printer.* Jake paid \$100.70, including 6% tax, for a color printer. How much did the printer itself cost?

21. *Parking Costs.* A hospital parking lot charges \$1.50 for the first hour or part thereof, and \$1.00 for each additional hour or part thereof. A weekly pass costs \$27.00 and allows unlimited parking for 7 days. Suppose that each visit Ed makes to the hospital lasts $1\frac{1}{2}$ hr. What is the minimum number of times that Ed would have to visit per week to make it worthwhile for him to buy the pass?

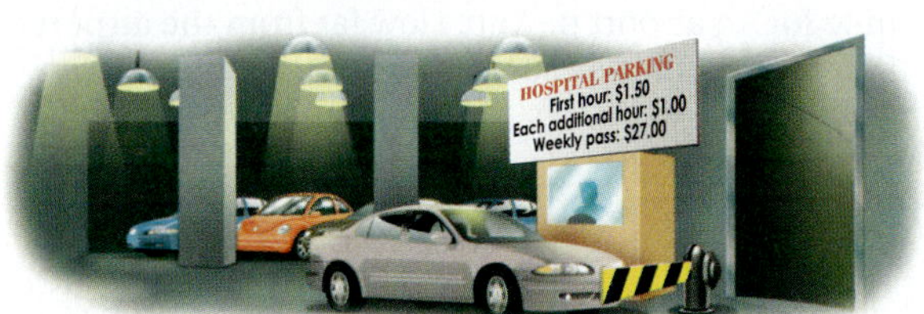

22. *Van Rental.* Value Rent-A-Car rents vans at a daily rate of \$84.95 plus 60 cents per mile. Molly rents a van to deliver electrical parts to her customers. She is allotted a daily budget of \$320. How many miles can she drive for \$320?

23. *Triangular Field.* The second angle of a triangular field is three times as large as the first angle. The third angle is 40° greater than the first angle. How large are the angles?

24. *Triangular Parking Lot.* The second angle of a triangular parking lot is four times as large as the first angle. The third angle is 45° less than the sum of the other two angles. How large are the angles?

25. *Triangular Backyard.* A home has a triangular backyard. The second angle of the triangle is 5° more than the first angle. The third angle is 10° more than three times the first angle. Find the angles of the triangular yard.

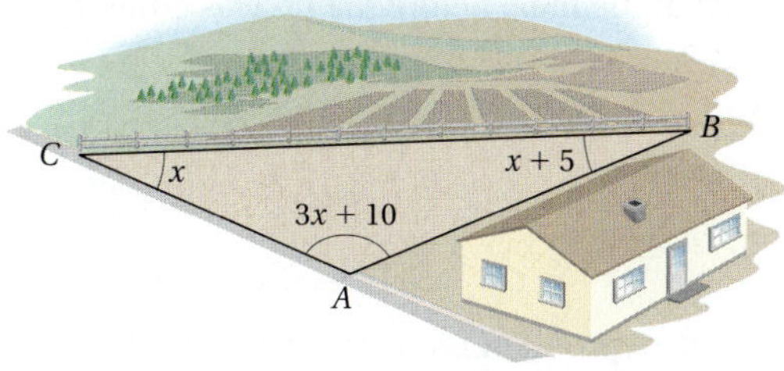

26. *Boarding Stable.* A rancher needs to form a triangular horse pen using ropes next to a stable. The second angle is three times the first angle. The third angle is 15° less than the first angle. Find the angles of the triangular pen.

27. *Stock Prices.* Sarah's investment in AOL/Time Warner stock grew 28% to \$448. How much did she invest?

28. *Savings Interest.* Sharon invested money in a savings account at a rate of 6% simple interest. After 1 yr, she has \$6996 in the account. How much did Sharon originally invest?

29. *Credit Cards.* The balance in Will's Mastercard® account grew 2%, to \$870, in one month. What was his balance at the beginning of the month?

30. *Loan Interest.* Alvin borrowed money from a cousin at a rate of 10% simple interest. After 1 yr, \$7194 paid off the loan. How much did Alvin borrow?

31. *Taxi Fares.* In Beniford, taxis charge \$3 plus 75¢ per mile for an airport pickup. How far from the airport can Courtney travel for \$12?

32. *Taxi Fares.* In Cranston, taxis charge \$4 plus 90¢ per mile for an airport pickup. How far from the airport can Ralph travel for \$17.50?

33. *Tipping.* Leon left a 15% tip for a meal. The total cost of the meal, including the tip, was \$41.40. What was the cost of the meal before the tip was added?

34. *Tipping.* Selena left an 18% tip for a meal. The total cost of the meal, including the tip, was \$40.71. What was the cost of the meal before the tip was added?

35. D$_W$ Erin returns a tent that she bought during a storewide 35% off sale that has ended. She is offered store credit for 125% of what she paid (not to be used on sale items). Is this fair to Erin? Why or why not?

36. D$_W$ Write a problem for a classmate to solve so that it can be translated to the equation

$$\frac{2}{3}x + (x + 5) + x = 375.$$

SKILL MAINTENANCE

Calculate.

37. $-\frac{4}{5} - \frac{3}{8}$ [6.4a]

38. $-\frac{4}{5} + \frac{3}{8}$ [6.3a]

39. $-\frac{4}{5} \cdot \frac{3}{8}$ [6.5a]

40. $-\frac{4}{5} \div \frac{3}{8}$ [6.6c]

41. $\frac{1}{10} \div \left(-\frac{1}{100}\right)$ [6.6c]

42. $-25.6 \div (-16)$ [6.6c]

43. $-25.6(-16)$ [6.5a]

44. $-25.6 - (-16)$ [6.4a]

45. $-25.6 + (-16)$ [6.3a]

46. $(-0.02) \div (-0.2)$ [6.6c]

SYNTHESIS

47. Apples are collected in a basket for six people. One-third, one-fourth, one-eighth, and one-fifth are given to four people, respectively. The fifth person gets ten apples with one apple remaining for the sixth person. Find the original number of apples in the basket.

48. A student scored 78 on a test that had 4 seven-point fill-ins and 24 three-point multiple-choice questions. The student had one fill-in wrong. How many multiple-choice questions did the student answer correctly?

49. The area of this triangle is 2.9047 in^2. Find x.

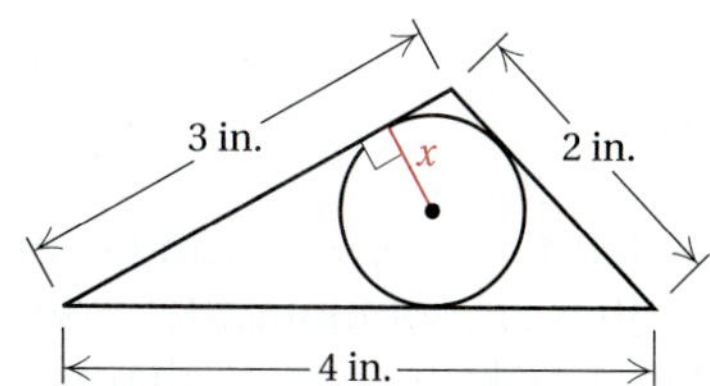

50. A storekeeper goes to the bank to get $10 worth of change. She requests twice as many quarters as half dollars, twice as many dimes as quarters, three times as many nickels as dimes, and no pennies or dollars. How many of each coin did the storekeeper get?

51. In one city, a sales tax of 9% was added to the price of gasoline as registered on the pump. Suppose a driver asked for $10 worth of gas. The attendant filled the tank until the pump read $9.10 and charged the driver $10. Something was wrong. Use algebra to correct the error.

7.7 SOLVING INEQUALITIES

Objectives

a Determine whether a given number is a solution of an inequality.

b Graph an inequality on a number line.

c Solve inequalities using the addition principle.

d Solve inequalities using the multiplication principle.

e Solve inequalities using the addition and multiplication principles together.

We now extend our equation-solving principles to the solving of inequalities.

a Solutions of Inequalities

In Section 6.2, we defined the symbols $>$ (is greater than), $<$ (is less than), $\geq$ (is greater than or equal to), and $\leq$ (is less than or equal to). For example, $3 \leq 4$ and $3 \leq 3$ are both true, but $-3 \leq -4$ and $0 \geq 2$ are both false.

An **inequality** is a number sentence with $>$, $<$, $\geq$, or $\leq$ as its verb—for example,

$$-4 > t, \quad x < 3, \quad 2x + 5 \geq 0, \quad \text{and} \quad -3y + 7 \leq -8.$$

Some replacements for a variable in an inequality make it true and some make it false.

SOLUTION

A replacement that makes an inequality true is called a **solution.** The set of all solutions is called the **solution set.** When we have found the set of all solutions of an inequality, we say that we have **solved** the inequality.

EXAMPLES Determine whether the number is a solution of $x < 2$.

1. -2.7 Since $-2.7 < 2$ is true, -2.7 is a solution.
2. 2 Since $2 < 2$ is false, 2 is not a solution.

EXAMPLES Determine whether the number is a solution of $y \geq 6$.

3. 6 Since $6 \geq 6$ is true, 6 is a solution.
4. $-\frac{4}{3}$ Since $-\frac{4}{3} \geq 6$ is false, $-\frac{4}{3}$ is not a solution.

Do Exercises 1 and 2.

b Graphs of Inequalities

Some solutions of $x < 2$ are $-3, 0, 1, 0.45, -8.9, -\pi, \frac{5}{8}$, and so on. In fact, there are infinitely many real numbers that are solutions. Because we cannot list them all individually, it is helpful to make a drawing that represents all the solutions.

A **graph** of an inequality is a drawing that represents its solutions. An inequality in one variable can be graphed on a number line. An inequality in two variables can be graphed on a coordinate plane; we will study such graphs in Chapter 14.

Determine whether each number is a solution of the inequality.

1. $x > 3$

a) 2 **b)** 0

c) -5 **d)** 15.4

e) 3 **f)** $-\frac{2}{5}$

2. $x \leq 6$

a) 6 **b)** 0

c) -4.3 **d)** 25

e) -6 **f)** $\frac{5}{8}$

Answers on page A-19

EXAMPLE 5 Graph: $x < 2$.

The solutions of $x < 2$ are all those numbers less than 2. They are shown on the graph by shading all points to the left of 2. The open circle at 2 indicates that 2 is *not* part of the graph.

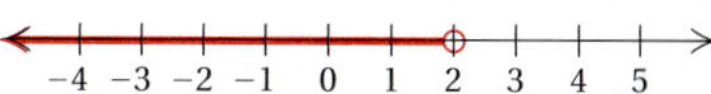

EXAMPLE 6 Graph: $x \geq -3$.

The solutions of $x \geq -3$ are shown on the number line by shading the point for -3 and all points to the right of -3. The closed circle at -3 indicates that -3 *is* part of the graph.

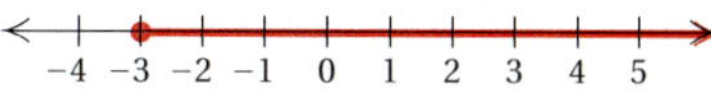

EXAMPLE 7 Graph: $-3 \leq x < 2$.

The inequality $-3 \leq x < 2$ is read "-3 is less than or equal to x *and* x is less than 2," or "x is greater than or equal to -3 *and* x is less than 2." In order to be a solution of this inequality, a number must be a solution of both $-3 \leq x$ and $x < 2$. The number 1 is a solution, as are -1.7, 0, 1.5, and $\frac{3}{8}$. We can see from the graphs below that the solution set consists of the numbers that overlap in the two solution sets in Examples 5 and 6:

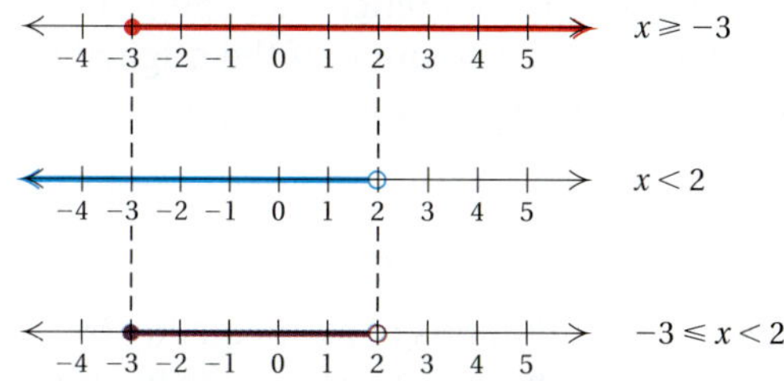

The open circle at 2 means that 2 is *not* part of the graph. The closed circle at -3 means that -3 *is* part of the graph. The other solutions are shaded.

Do Exercises 3–5.

Graph.

3. $x \leq 4$

4. $x > -2$

5. $-2 < x \leq 4$

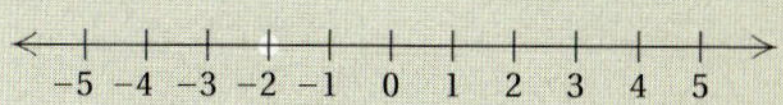

Answers on page A-19

C Solving Inequalities Using the Addition Principle

Consider the true inequality $3 < 7$. If we add 2 on both sides, we get another true inequality:

$$3 + 2 < 7 + 2, \quad \text{or} \quad 5 < 9.$$

Similarly, if we add -4 on both sides of $x + 4 < 10$, we get an *equivalent* inequality:

$$x + 4 + (-4) < 10 + (-4),$$

or

$$x < 6.$$

To say that $x + 4 < 10$ and $x < 6$ are **equivalent** is to say that they have the same solution set. For example, the number 3 is a solution of $x + 4 < 10$. It is also a solution of $x < 6$. The number -2 is a solution of $x < 6$. It is also a solution of $x + 4 < 10$. Any solution of one is a solution of the other—they are equivalent.

Solve. Then graph.

6. $x + 3 > 5$

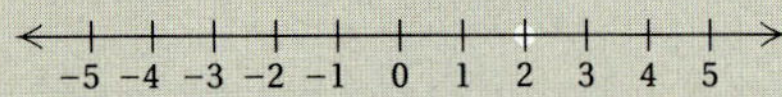

7. $x - 1 \leq 2$

8. $5x + 1 < 4x - 2$

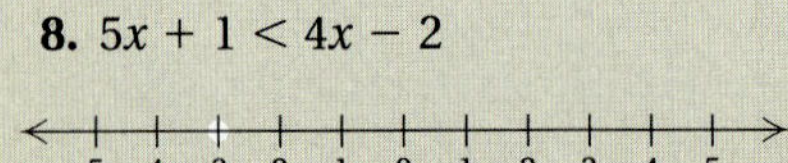

Answers on page A-19

THE ADDITION PRINCIPLE FOR INEQUALITIES

For any real numbers a, b, and c:

$a < b$ is equivalent to $a + c < b + c$;

$a > b$ is equivalent to $a + c > b + c$;

$a \leq b$ is equivalent to $a + c \leq b + c$;

$a \geq b$ is equivalent to $a + c \geq b + c$.

In other words, when we add or subtract the same number on both sides of an inequality, the direction of the inequality symbol is not changed.

As with equation solving, when solving inequalities, our goal is to isolate the variable on one side. Then it is easier to determine the solution set.

EXAMPLE 8 Solve: $x + 2 > 8$. Then graph.

We use the addition principle, subtracting 2 on both sides:

$$x + 2 - 2 > 8 - 2$$
$$x > 6.$$

From the inequality $x > 6$, we can determine the solutions directly. Any number greater than 6 makes the last sentence true and is a solution of that sentence. Any such number is also a solution of the original sentence. Thus the inequality is solved. The graph is as follows:

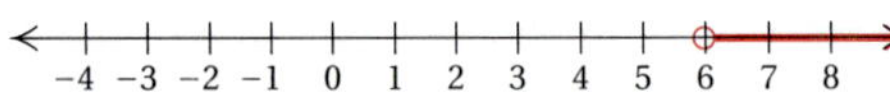

We cannot check all the solutions of an inequality by substitution, as we can check solutions of equations, because there are too many of them. A partial check can be done by substituting a number greater than 6—say, 7—into the original inequality:

$x + 2 > 8$	
$7 + 2$	8
9	

TRUE

Since $9 > 8$ is true, 7 is a solution. Any number greater than 6 is a solution.

EXAMPLE 9 Solve: $3x + 1 \leq 2x - 3$. Then graph.

We have

$$3x + 1 \leq 2x - 3$$
$$3x + 1 - 1 \leq 2x - 3 - 1 \quad \text{Subtracting 1}$$
$$3x \leq 2x - 4 \quad \text{Simplifying}$$
$$3x - 2x \leq 2x - 4 - 2x \quad \text{Subtracting } 2x$$
$$x \leq -4. \quad \text{Simplifying}$$

The graph is as follows:

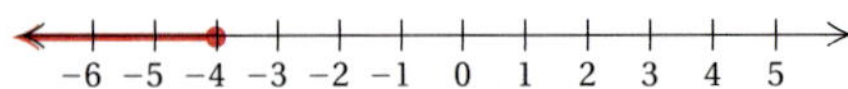

Remember that the graph is a drawing that represents the solutions of the original inequality.

In Example 9, any number less than or equal to -4 is a solution. The following are some solutions:

$$-4, \quad -5, \quad -6, \quad -\frac{13}{3}, \quad -204.5, \quad \text{and} \quad -18\pi.$$

Besides drawing a graph, we can also describe all the solutions of an inequality using **set notation.** We could just begin to list them in a set using roster notation (see p. 438), as follows:

$$\left\{-4, -5, -6, -\frac{13}{3}, -204.5, -18\pi, \dots\right\}.$$

We can never list them all this way, however. Seeing this set without knowing the inequality makes it difficult for us to know what real numbers we are considering. There is, however, another kind of notation that we can use. It is

$$\{x \mid x \le -4\},$$

which is read

"The set of all x such that x is less than or equal to -4."

This shorter notation for sets is called **set-builder notation.**

From now on, we will use this notation when solving inequalities.

Do Exercises 6–8 on the preceding page.

Solve.

9. $x + \frac{2}{3} \ge \frac{4}{5}$

10. $5y + 2 \le -1 + 4y$

EXAMPLE 10 Solve: $x + \frac{1}{3} > \frac{5}{4}$.

We have

$$x + \tfrac{1}{3} > \tfrac{5}{4}$$

$$x + \tfrac{1}{3} - \tfrac{1}{3} > \tfrac{5}{4} - \tfrac{1}{3} \qquad \text{Subtracting } \tfrac{1}{3}$$

$$x > \tfrac{5}{4} \cdot \tfrac{3}{3} - \tfrac{1}{3} \cdot \tfrac{4}{4} \qquad \text{Multiplying by 1 to obtain a common denominator}$$

$$x > \tfrac{15}{12} - \tfrac{4}{12}$$

$$x > \tfrac{11}{12}.$$

Any number greater than $\frac{11}{12}$ is a solution. The solution set is

$$\{x \mid x > \tfrac{11}{12}\},$$

which is read

"The set of all x such that x is greater than $\frac{11}{12}$."

When solving inequalities, you may obtain an answer like $\frac{11}{12} < x$. Recall from Chapter 6 that this has the same meaning as $x > \frac{11}{12}$. Thus the solution set in Example 10 can be described as $\{x \mid \frac{11}{12} < x\}$ or as $\{x \mid x > \frac{11}{12}\}$. The latter is used most often.

Do Exercises 9 and 10.

d Solving Inequalities Using the Multiplication Principle

There is a multiplication principle for inequalities that is similar to that for equations, but it must be modified. When we are multiplying on both sides by a negative number, the direction of the inequality symbol must be changed.

Answers on page A-19

Solve. Then graph.

11. $8x < 64$

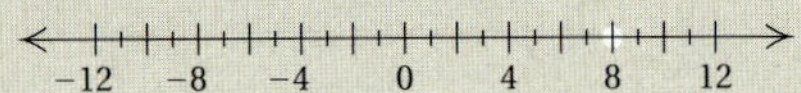

12. $5y \geq 160$

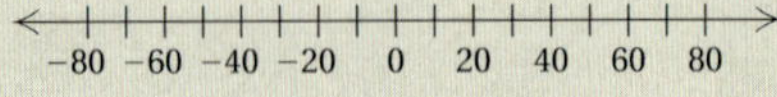

Answers on page A-19

Let's see what happens. Consider the true inequality $3 < 7$. If we multiply on both sides by a *positive* number, like 2, we get another true inequality:

$$3 \cdot 2 < 7 \cdot 2, \quad \text{or} \quad 6 < 14. \qquad \text{True}$$

If we multiply on both sides by a *negative* number, like -2, and we do not change the direction of the inequality symbol, we get a *false* inequality:

$$3 \cdot (-2) < 7 \cdot (-2), \quad \text{or} \quad -6 < -14. \qquad \text{False}$$

The fact that $6 < 14$ is true but $-6 < -14$ is false stems from the fact that the negative numbers, in a sense, mirror the positive numbers. That is, whereas 14 is to the *right* of 6 on a number line, the number -14 is to the *left* of -6. Thus, if we reverse (change the direction of) the inequality symbol, we get a *true* inequality: $-6 > -14$.

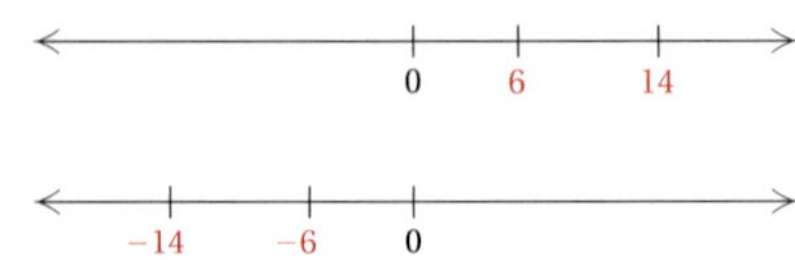

THE MULTIPLICATION PRINCIPLE FOR INEQUALITIES

For any real numbers a and b, and any *positive* number c:

$a < b$ is equivalent to $ac < bc$;

$a > b$ is equivalent to $ac > bc$.

For any real numbers a and b, and any *negative* number c:

$a < b$ is equivalent to $ac > bc$;

$a > b$ is equivalent to $ac < bc$.

Similar statements hold for $\leq$ and $\geq$.

In other words, when we multiply or divide by a positive number on both sides of an inequality, the direction of the inequality symbol stays the same. When we multiply or divide by a negative number on both sides of an inequality, the direction of the inequality symbol is reversed.

EXAMPLE 11 Solve: $4x < 28$. Then graph.

We have

$$4x < 28$$

$$\frac{4x}{4} < \frac{28}{4} \qquad \text{Dividing by 4}$$

The symbol stays the same.

$$x < 7. \qquad \text{Simplifying}$$

The solution set is $\{x \mid x < 7\}$. The graph is as follows:

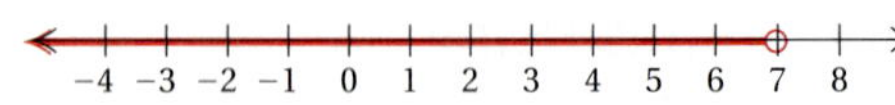

Do Exercises 11 and 12.

EXAMPLE 12 Solve: $-2y < 18$. Then graph.

We have

$$-2y < 18$$

$$\frac{-2y}{-2} > \frac{18}{-2}$$ Dividing by -2

The symbol must be reversed!

$$y > -9.$$ Simplifying

The solution set is $\{y \mid y > -9\}$. The graph is as follows:

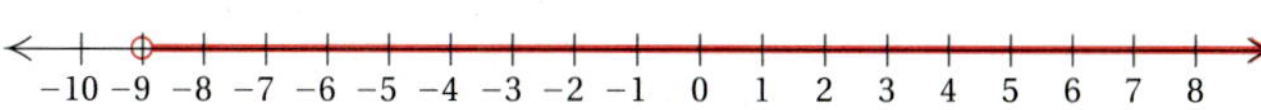

Do Exercises 13 and 14.

Solve.

13. $-4x \le 24$

14. $-5y > 13$

e Using the Principles Together

All of the equation-solving techniques used in Sections 7.1–7.3 can be used with inequalities provided we remember to reverse the inequality symbol when multiplying or dividing on both sides by a negative number.

EXAMPLE 13 Solve: $6 - 5y > 7$.

We have

$$6 - 5y > 7$$

$$-6 + 6 - 5y > -6 + 7$$ Adding -6. The symbol stays the same.

$$-5y > 1$$ Simplifying

$$\frac{-5y}{-5} < \frac{1}{-5}$$ Dividing by -5

The symbol must be reversed because we are dividing by a *negative* number, -5.

$$y < -\frac{1}{5}.$$ Simplifying

The solution set is $\{y \mid y < -\frac{1}{5}\}$.

Do Exercise 15.

15. Solve: $7 - 4x < 8$.

EXAMPLE 14 Solve: $8y - 5 > 17 - 5y$.

$$-17 + 8y - 5 > -17 + 17 - 5y$$ Adding -17. The symbol stays the same.

$$8y - 22 > -5y$$ Simplifying

$$-8y + 8y - 22 > -8y - 5y$$ Adding $-8y$

$$-22 > -13y$$ Simplifying

$$\frac{-22}{-13} < \frac{-13y}{-13}$$ Dividing by -13

The symbol must be reversed because we are dividing by a *negative* number, -13.

$$\frac{22}{13} < y$$

The solution set is $\{y \mid \frac{22}{13} < y\}$, or $\{y \mid y > \frac{22}{13}\}$.

Answers on page A-19

16. Solve: $24 - 7y \leq 11y - 14$.

17. Solve. Use a method like the one used in Example 15.

$24 - 7y \leq 11y - 14$

18. Solve:

$3(7 + 2x) \leq 30 + 7(x - 1)$.

Answers on page A-19

We can often solve inequalities in such a way as to avoid having to reverse the inequality symbol. We add so that after like terms have been collected, the coefficient of the variable term is positive. We show this by solving the inequality in Example 14 a different way.

EXAMPLE 15 Solve: $8y - 5 > 17 - 5y$.

Note that if we add $5y$ on both sides, the coefficient of the y-term will be positive after like terms have been collected.

$$8y - 5 + 5y > 17 - 5y + 5y \quad \text{Adding } 5y$$
$$13y - 5 > 17 \quad \text{Simplifying}$$
$$13y - 5 + 5 > 17 + 5 \quad \text{Adding 5}$$
$$13y > 22 \quad \text{Simplifying}$$
$$\frac{13y}{13} > \frac{22}{13} \quad \text{Dividing by 13. We leave the inequality symbol the same because we are dividing by a positive number.}$$
$$y > \frac{22}{13}$$

The solution set is $\{y \mid y > \frac{22}{13}\}$.

Do Exercises 16 and 17.

EXAMPLE 16 Solve: $3(x - 2) - 1 < 2 - 5(x + 6)$.

$$3(x - 2) - 1 < 2 - 5(x + 6)$$
$$3x - 6 - 1 < 2 - 5x - 30 \quad \text{Using the distributive law to multiply and remove parentheses}$$
$$3x - 7 < -5x - 28 \quad \text{Simplifying}$$
$$3x + 5x < -28 + 7 \quad \text{Adding } 5x \text{ and 7 to get all } x\text{-terms on one side and all other terms on the other side}$$
$$8x < -21 \quad \text{Simplifying}$$
$$x < \frac{-21}{8}, \text{ or } -\frac{21}{8} \quad \text{Dividing by 8}$$

The solution set is $\{x \mid x < -\frac{21}{8}\}$.

Do Exercise 18.

EXAMPLE 17 Solve: $16.3 - 7.2p \le -8.18$.

The greatest number of decimal places in any one number is *two*. Multiplying by 100, which has two 0's, will clear decimals. Then we proceed as before.

$$16.3 - 7.2p \le -8.18$$

$$100(16.3 - 7.2p) \le 100(-8.18) \quad \text{Multiplying by 100}$$

$$100(16.3) - 100(7.2p) \le 100(-8.18) \quad \text{Using the distributive law}$$

$$1630 - 720p \le -818 \quad \text{Simplifying}$$

$$1630 - 720p - 1630 \le -818 - 1630 \quad \text{Subtracting 1630}$$

$$-720p \le -2448 \quad \text{Simplifying}$$

$$\frac{-720p}{-720} \ge \frac{-2448}{-720} \quad \text{Dividing by } -720$$

The symbol must be reversed.

$$p \ge 3.4$$

The solution set is $\{p \mid p \ge 3.4\}$.

Do Exercise 19.

EXAMPLE 18 Solve: $\frac{2}{3}x - \frac{1}{6} + \frac{1}{2}x > \frac{7}{6} + 2x$.

The number 6 is the least common multiple of all the denominators. Thus we multiply by 6 on both sides.

$$\frac{2}{3}x - \frac{1}{6} + \frac{1}{2}x > \frac{7}{6} + 2x$$

$$6\left(\frac{2}{3}x - \frac{1}{6} + \frac{1}{2}x\right) > 6\left(\frac{7}{6} + 2x\right) \quad \text{Multiplying by 6 on both sides}$$

$$6 \cdot \frac{2}{3}x - 6 \cdot \frac{1}{6} + 6 \cdot \frac{1}{2}x > 6 \cdot \frac{7}{6} + 6 \cdot 2x \quad \text{Using the distributive law}$$

$$4x - 1 + 3x > 7 + 12x \quad \text{Simplifying}$$

$$7x - 1 > 7 + 12x \quad \text{Collecting like terms}$$

$$7x - 1 - 12x > 7 + 12x - 12x \quad \text{Subtracting } 12x$$

$$-5x - 1 > 7 \quad \text{Collecting like terms}$$

$$-5x - 1 + 1 > 7 + 1 \quad \text{Adding 1}$$

$$-5x > 8 \quad \text{Simplifying}$$

$$\frac{-5x}{-5} < \frac{8}{-5} \quad \text{Dividing by } -5$$

The symbol must be reversed.

$$x < -\frac{8}{5}$$

The solution set is $\{x \mid x < -\frac{8}{5}\}$.

Do Exercise 20.

19. Solve:

$2.1x + 43.2 \ge 1.2 - 8.4x$.

20. Solve.

$\frac{3}{4} + x < \frac{7}{8}x - \frac{1}{4} + \frac{1}{2}x.$

Answers on page A-19

7.7 EXERCISE SET

For Extra Help

Digital Video Tutor CD 6 Videotape 8 | InterAct Math | Math Tutor Center | MathXL | MyMathLab

a Determine whether each number is a solution of the given inequality.

1. $x > -4$
- **a)** 4
- **b)** 0
- **c)** -4
- **d)** 6
- **e)** 5.6

2. $x \leq 5$
- **a)** 0
- **b)** 5
- **c)** -1
- **d)** -5
- **e)** $7\frac{1}{4}$

3. $x \geq 6.8$
- **a)** -6
- **b)** 0
- **c)** 6
- **d)** 8
- **e)** $-3\frac{1}{2}$

4. $x < 8$
- **a)** 8
- **b)** -10
- **c)** 0
- **d)** 11
- **e)** -4.7

b Graph on a number line.

5. $x > 4$

−5 −4 −3 −2 −1 0 1 2 3 4 5

6. $x < 0$

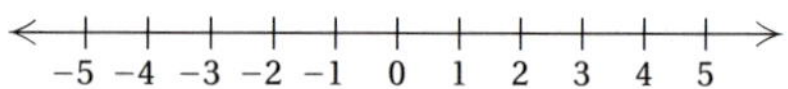

7. $t < -3$

−5 −4 −3 −2 −1 0 1 2 3 4 5

8. $y > 5$

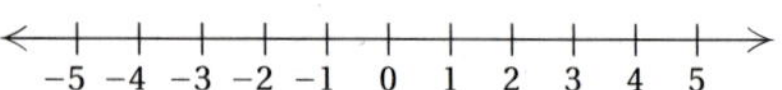

9. $m \geq -1$

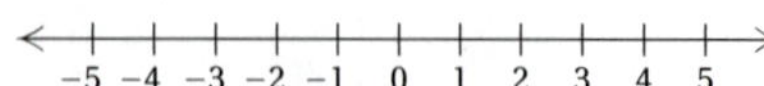

10. $x \leq -2$

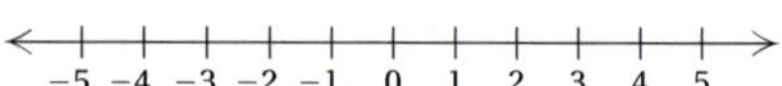

11. $-3 < x \leq 4$

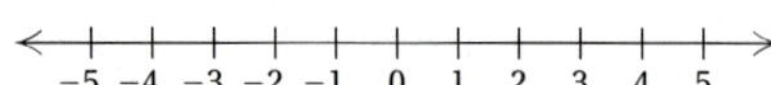

12. $-5 \leq x < 2$

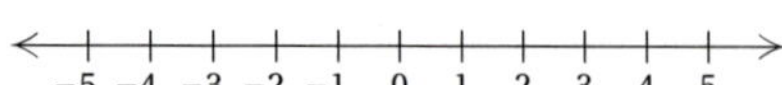

13. $0 < x < 3$

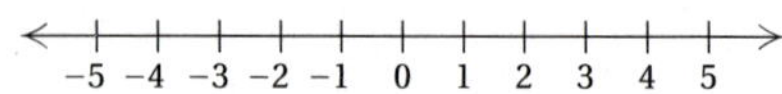

14. $-5 \leq x \leq 0$

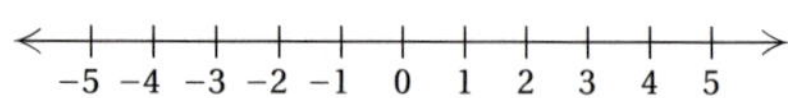

c Solve using the addition principle. Then graph.

15. $x + 7 > 2$

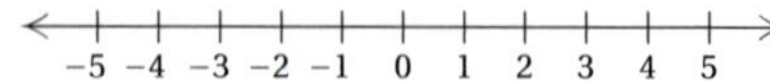

16. $x + 5 > 2$

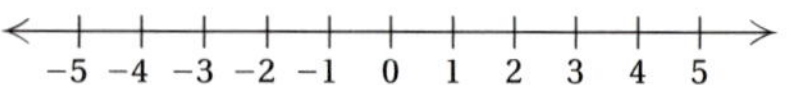

17. $x + 8 \leq -10$

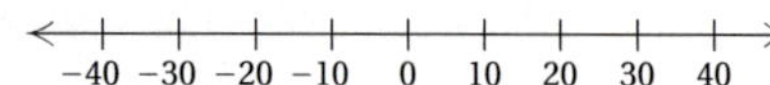

18. $x + 8 \leq -11$

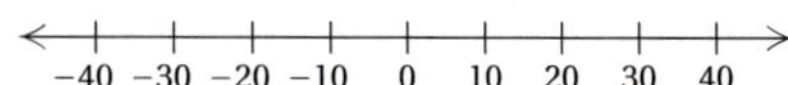

Solve using the addition principle.

19. $y - 7 > -12$

20. $y - 9 > -15$

21. $2x + 3 > x + 5$

22. $2x + 4 > x + 7$

23. $3x + 9 \leq 2x + 6$

24. $3x + 18 \leq 2x + 16$

25. $5x - 6 < 4x - 2$

26. $9x - 8 < 8x - 9$

27. $-9 + t > 5$

28. $-8 + p > 10$

29. $y + \frac{1}{4} \leq \frac{1}{2}$

30. $x - \frac{1}{3} \leq \frac{5}{6}$

31. $x - \frac{1}{3} > \frac{1}{4}$

32. $x + \frac{1}{8} > \frac{1}{2}$

d Solve using the multiplication principle. Then graph.

33. $5x < 35$

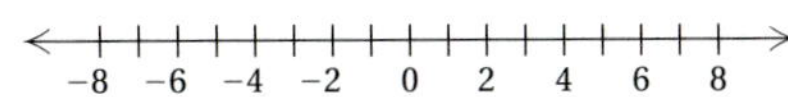

34. $8x \geq 32$

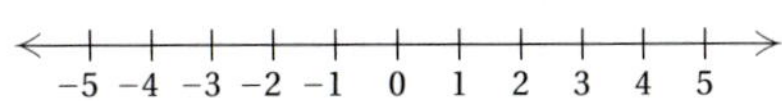

35. $-12x > -36$

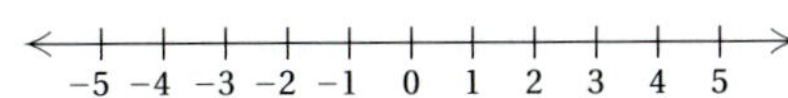

36. $-16x > -64$

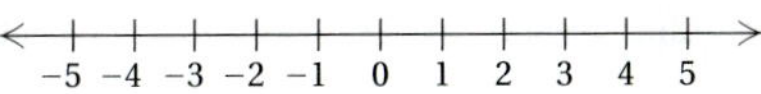

Solve using the multiplication principle.

37. $5y \geq -2$

38. $3x < -4$

39. $-2x \leq 12$

40. $-3x \leq 15$

41. $-4y \geq -16$

42. $-7x < -21$

43. $-3x < -17$

44. $-5y > -23$

45. $-2y > \frac{1}{7}$

46. $-4x \leq \frac{1}{9}$

47. $-\frac{6}{5} \leq -4x$

48. $-\frac{7}{9} > 63x$

e Solve using the addition and multiplication principles.

49. $4 + 3x < 28$

50. $3 + 4y < 35$

51. $3x - 5 \leq 13$

52. $5y - 9 \leq 21$

53. $13x - 7 < -46$

54. $8y - 6 < -54$

55. $30 > 3 - 9x$

56. $48 > 13 - 7y$

57. $4x + 2 - 3x \leq 9$

58. $15x + 5 - 14x \leq 9$

59. $-3 < 8x + 7 - 7x$

60. $-8 < 9x + 8 - 8x - 3$

61. $6 - 4y > 4 - 3y$

62. $9 - 8y > 5 - 7y + 2$

63. $5 - 9y \leq 2 - 8y$

64. $6 - 18x \leq 4 - 12x - 5x$

65. $19 - 7y - 3y < 39$

66. $18 - 6y - 4y < 63 + 5y$

67. $2.1x + 45.2 > 3.2 - 8.4x$

68. $0.96y - 0.79 \leq 0.21y + 0.46$

69. $\frac{x}{3} - 2 \leq 1$

70. $\frac{2}{3} + \frac{x}{5} < \frac{4}{15}$

71. $\frac{y}{5} + 1 \leq \frac{2}{5}$

72. $\frac{3x}{4} - \frac{7}{8} \geq -15$

73. $3(2y - 3) < 27$

74. $4(2y - 3) > 28$

75. $2(3 + 4m) - 9 \geq 45$

76. $3(5 + 3m) - 8 \leq 88$

77. $8(2t + 1) > 4(7t + 7)$

78. $7(5y - 2) > 6(6y - 1)$

79. $3(r - 6) + 2 < 4(r + 2) - 21$

80. $5(x + 3) + 9 \leq 3(x - 2) + 6$

81. $0.8(3x + 6) \geq 1.1 - (x + 2)$

82. $0.4(2x + 8) \geq 20 - (x + 5)$

83. $\frac{5}{3} + \frac{2}{3}x < \frac{25}{12} + \frac{5}{4}x + \frac{3}{4}$

84. $1 - \frac{2}{3}y \geq \frac{9}{5} - \frac{y}{5} + \frac{3}{5}$

85. DW Are the inequalities $3x - 4 < 10 - 4x$ and $2(x - 5) > 3(2x - 6)$ equivalent? Why or why not?

86. DW Explain in your own words why it is necessary to reverse the inequality symbol when multiplying on both sides of an inequality by a negative number.

SKILL MAINTENANCE

Add or subtract. [6.3a], [6.4a]

87. $-56 + (-18)$

88. $-2.3 + 7.1$

89. $-\frac{3}{4} + \frac{1}{8}$

90. $8.12 - 9.23$

91. $-56 - (-18)$

92. $-\frac{3}{4} - \frac{1}{8}$

93. $-2.3 - 7.1$

94. $-8.12 + 9.23$

Simplify.

95. $5 - 3^2 + (8 - 2)^2 \cdot 4$ [6.8d]

96. $10 \div 2 \cdot 5 - 3^2 + (-5)^2$ [6.8d]

97. $5(2x - 4) - 3(4x + 1)$ [6.8b]

98. $9(3 + 5x) - 4(7 + 2x)$ [6.8b]

SYNTHESIS

99. Determine whether each number is a solution of the inequality $|x| < 3$.

a) 0
b) −2
c) −3
d) 4
e) 3
f) 1.7
g) −2.8

100. Graph $|x| < 3$ on a number line.

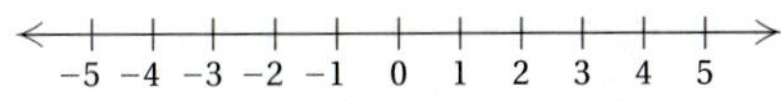

Solve.

101. $x + 3 \leq 3 + x$

102. $x + 4 < 3 + x$

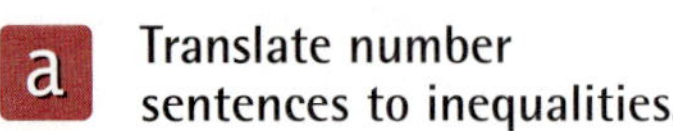

7.8 APPLICATIONS AND PROBLEM SOLVING WITH INEQUALITIES

Objectives

a Translate number sentences to inequalities.

b Solve applied problems using inequalities.

The five steps for problem solving can be used for problems involving inequalities.

a Translating to Inequalities

Before solving problems that involve inequalities, we list some important phrases to look for. Sample translations are listed as well.

IMPORTANT WORDS	SAMPLE SENTENCE	TRANSLATION
is at least	Bill is at least 21 years old.	$b \geq 21$
is at most	At most 5 students dropped the course.	$n \leq 5$
cannot exceed	To qualify, earnings cannot exceed \$12,000.	$r \leq 12{,}000$
must exceed	The speed must exceed 15 mph.	$s > 15$
is less than	Tucker's weight is less than 50 lb.	$w < 50$
is more than	Boston is more than 200 miles away.	$d > 200$
is between	The film was between 90 and 100 minutes long.	$90 < t < 100$
no more than	Bing weighs no more than 90 lb.	$w \leq 90$
no less than	Valerie scored no less than 8.3.	$s \geq 8.3$

The following phrases deserve special attention.

TRANSLATING "AT LEAST" AND "AT MOST"

A quantity x is at least some amount q: $x \geq q$.
(If x is at least q, it cannot be less than q.)

A quantity x is at most some amount q: $x \leq q$.
(If x is at most q, it cannot be more than q.)

Do Exercises 1–10.

Translate.

1. Maggie scored no less than 92 on her English exam.

2. The average credit card holder is at least \$4000 in debt.

3. The price of that PT Cruiser is at most \$21,900.

4. The time of the test was between 45 and 55 min.

5. Normandale Community College is more than 15 mi away.

6. Tania's weight is less than 110 lb.

7. That number is greater than −2.

8. The costs of production of that CD-ROM cannot exceed \$12,500.

9. At most, 11.4% of all deaths in Arizona are from cancer.

10. Yesterday, at least 23 people got tickets for speeding.

Answers on page A-19

b Solving Problems

EXAMPLE 1 *Catering costs.* To cater a party, Curtis' Barbeque charges a \$50 setup fee plus \$15 per person. The cost of Hotel Pharmacy's end-of-season softball party cannot exceed \$450. How many people can attend the party?
Source: Curtis' All American Barbeque, Putney, Vermont

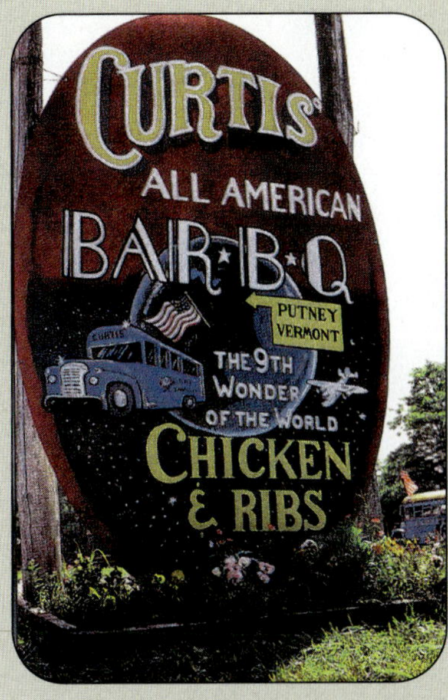

1. **Familiarize.** Suppose that 20 people were to attend the party. The cost would then be $\$50 + \$15 \cdot 20$, or \$350. This shows that more than 20 people could attend without exceeding \$450. Instead of making another guess, we let $n =$ the number of people in attendance.
2. **Translate.** The cost of the party will be \$50 for the setup fee plus \$15 times the number of people attending. We can reword and translate to an inequality as follows:

Rewording:	The setup fee	plus	the cost of the meals	cannot exceed	\$450.
Translating:	50	+	$15 \cdot n$	$\le$	450

3. **Solve.** We solve the inequality for n:

$$50 + 15n \le 450$$
$$50 + 15n - 50 \le 450 - 50 \quad \text{Subtracting 50}$$
$$15n \le 400 \quad \text{Simplifying}$$
$$\frac{15n}{15} \le \frac{400}{15} \quad \text{Dividing by 15}$$
$$n \le \frac{400}{15}$$
$$n \le 26\tfrac{2}{3}. \quad \text{Simplifying}$$

4. **Check.** Although the solution set of the inequality is all numbers less than or equal to $26\frac{2}{3}$, since $n =$ the number of people in attendance, we round *down* to 26. If 26 people attend, the cost will be $\$50 + \$15 \cdot 26$, or \$440, and if 27 attend, the cost will exceed \$450.
5. **State.** At most 26 people can attend the party.

Do Exercise 11.

CAUTION!

Solutions of problems should always be checked using the original wording of the problem. In some cases, answers might need to be whole numbers or integers or rounded off in a particular direction.

11. **Butter Temperatures.** Butter stays solid at Fahrenheit temperatures below 88°. The formula

$$F = \tfrac{9}{5}C + 32$$

can be used to convert Celsius temperatures C to Fahrenheit temperatures F. Determine (in terms of an inequality) those Celsius temperatures for which butter stays solid.

Answer on page A-19

EXAMPLE 2 *Nutrition.* The U.S. Department of Health and Human Services and the Department of Agriculture recommend that for a typical 2000-calorie daily diet, no more than 65 g of fat be consumed. In the first three days of a four-day vacation, Phil consumed 70 g, 62 g, and 80 g of fat. Determine (in terms of an inequality) how many grams of fat Phil can consume on the fourth day if he is to average no more than 65 g of fat per day.

Sources: U.S. Department of Health and Human Services and Department of Agriculture

1. **Familiarize.** Suppose Phil consumed 64 g of fat on the fourth day. His daily average for the vacation would then be

$$\frac{70\text{ g} + 62\text{ g} + 80\text{ g} + 64\text{ g}}{4} = 69\text{ g}.$$

This shows that Phil cannot consume 64 g of fat on the fourth day, if he is to average no more than 65 g of fat per day. We let x = the number of grams of fat that Phil consumes on the fourth day.

2. **Translate.** We reword the problem and translate to an inequality as follows:

Rewording:	The average consumption of fat	should be no more than	65 g.
Translating:	$\frac{70 + 62 + 80 + x}{4}$	$\leq$	65

3. **Solve.** Because of the fraction expression, it is convenient to use the multiplication principle first to solve the inequality:

$$\frac{70 + 62 + 80 + x}{4} \leq 65$$

$$4\left(\frac{70 + 62 + 80 + x}{4}\right) \leq 4 \cdot 65 \quad \text{Multiplying by 4}$$

$$70 + 62 + 80 + x \leq 260$$

$$212 + x \leq 260 \quad \text{Simplifying}$$

$$x \leq 48. \quad \text{Subtracting 212}$$

4. **Check.** As a partial check, we show that Phil can consume 48 g of fat on the fourth day and not exceed a 65-g average for the four days:

$$\frac{70 + 62 + 80 + 48}{4} = \frac{260}{4} = 65.$$

5. **State.** Phil's average fat intake for the vacation will not exceed 65 g per day if he consumes no more than 48 g of fat on the fourth day.

Do Exercise 12.

Translate to an inequality and solve.

12. **Test Scores.** A pre-med student is taking a chemistry course in which four tests are to be given. To get an A, she must average at least 90 on the four tests. The student got scores of 91, 86, and 89 on the first three tests. Determine (in terms of an inequality) what scores on the last test will allow her to get an A.

Answer on page A-19

Study Tips

TIME MANAGEMENT (PART 1)

Time is the most critical factor in your success in learning mathematics. Have reasonable expectations about the time you need to study math. (See also the Study Tips on time management in Sections 9.2 and 10.2.)

- **Juggling time.** Working 40 hours per week and taking 12 credit hours is equivalent to working two full-time jobs. Can you handle such a load? Your ratio of number of work hours to number of credit hours should be about 40/3, 30/6, 20/9, 10/12, or 5/14.
- **A rule of thumb on study time.** Budget about 2–3 hours for homework and study per week for every hour of class time.
- **Scheduling your time.** Make an hour-by-hour schedule of your typical week. Include work, school, home, sleep, study, and leisure times. Try to schedule time for study when you are most alert. Choose a setting that will enable you to maximize your concentration. Plan for success and it will happen!

"You cannot increase the quality or quantity of your achievement or performance except to the degree in which you increase your ability to use time effectively."

Brian Tracy, motivational speaker

7.8 EXERCISE SET

For Extra Help

Digital Video Tutor CD 6 Videotape 8 | InterAct Math | Math Tutor Center | MathXL | MyMathLab.com

a Translate to an inequality.

1. A number is at least 7.

2. A number is greater than or equal to 5.

3. The baby weighs more than 2 kilograms (kg).

4. Between 75 and 100 people attended the concert.

5. The speed of the train was between 90 and 110 mph.

6. At least 400,000 people attended the Million Man March.

7. At most 1,200,000 people attended the Million Man March.

8. The amount of acid is not to exceed 40 liters (L).

9. The cost of gasoline is no less than $1.50 per gallon.

10. The temperature is at most $-2°$.

11. A number is greater than 8.

12. A number is less than 5.

13. A number is less than or equal to -4.

14. A number is greater than or equal to 18.

15. The number of people is at least 1300.

16. The cost is at most $4857.95.

17. The amount of acid is not to exceed 500 liters.

18. The cost of gasoline is no less than 94 cents per gallon.

19. Two more than three times a number is less than 13.

20. Five less than one-half a number is greater than 17.

b Solve.

21. *Test Scores.* A student is taking a literature course in which four tests are to be given. To get a B, he must average at least 80 on the four tests. The student got scores of 82, 76, and 78 on the first three tests. Determine (in terms of an inequality) what scores on the last test will allow him to get at least a B.

22. *Test Scores.* Your quiz grades are 73, 75, 89, and 91. Determine (in terms of an inequality) what scores on the last quiz will allow you to get an average quiz grade of at least 85.

23. *Gold Temperatures.* Gold stays solid at Fahrenheit temperatures below 1945.4°. Determine (in terms of an inequality) those Celsius temperatures for which gold stays solid. Use the formula given in Margin Exercise 11.

24. *Body Temperatures.* The human body is considered to be fevered when its temperature is higher than 98.6°F. Using the formula given in Margin Exercise 11, determine (in terms of an inequality) those Celsius temperatures for which the body is fevered.

25. *World Records in the 1500-m Run.* The formula

$$R = -0.075t + 3.85$$

can be used to predict the world record in the 1500-m run t years after 1930. Determine (in terms of an inequality) those years for which the world record will be less than 3.5 min.

26. *World Records in the 200-m Dash.* The formula

$$R = -0.028t + 20.8$$

can be used to predict the world record in the 200-m dash t years after 1920. Determine (in terms of an inequality) those years for which the world record will be less than 19.0 sec.

27. *Sizes of Envelopes.* Rhetoric Advertising is a direct-mail company. It determines that for a particular campaign, it can use any envelope with a fixed width of $3\frac{1}{2}$ in. and an area of at least $17\frac{1}{2}$ in^2. Determine (in terms of an inequality) those lengths that will satisfy the company constraints.

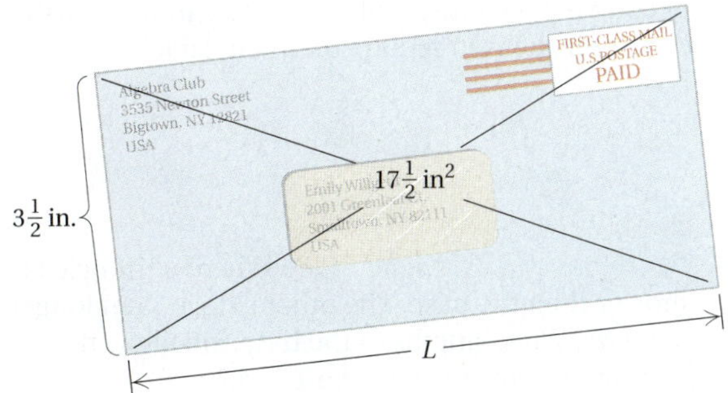

28. *Sizes of Packages.* An overnight delivery service accepts packages of up to 165 in. in length and girth combined. (Girth is the distance around the package.) A package has a fixed girth of 53 in. Determine (in terms of an inequality) those lengths for which a package is acceptable.

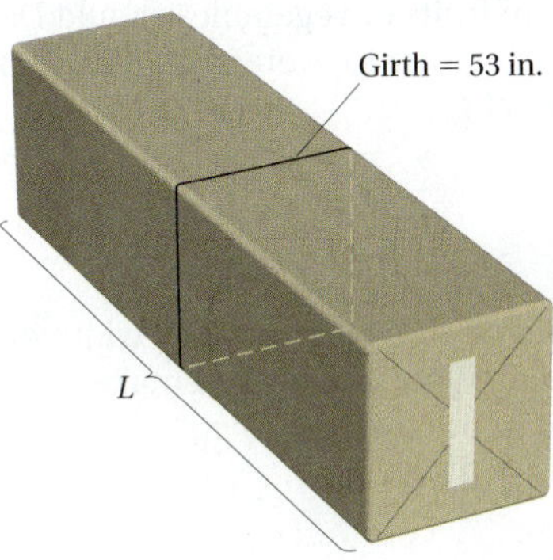

29. *Blueprints.* To make copies of blueprints, Vantage Reprographics charges a $5 setup fee plus $4 per copy. Myra can spend no more than $65 for the copying. What numbers of copies will allow her to stay within budget?

30. *Banquet Costs.* The women's volleyball team can spend at most $450 for its awards banquet at a local restaurant. If the restaurant charges a $40 setup fee plus $16 per person, at most how many can attend?

31. *Phone Costs.* Simon claims that it costs him at least $3.00 every time he calls an overseas customer. If his typical call costs 75¢ plus 45¢ for each minute, how long do his calls typically last?

32. *Parking Costs.* Laura is certain that every time she parks in the municipal garage it costs her at least $2.20. If the garage charges 45¢ plus 25¢ for each half hour, for how long is Laura's car generally parked?

33. *College Tuition.* Angelica's financial aid stipulates that her tuition not exceed $1000. If her local community college charges a $35 registration fee plus $375 per course, what is the greatest number of courses for which Angelica can register?

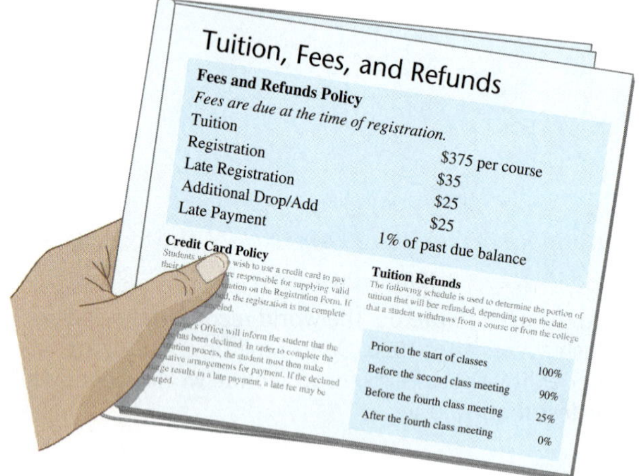

34. *Furnace Repairs.* RJ's Plumbing and Heating charges $25 plus $30 per hour for emergency service. Gary remembers being billed over $100 for an emergency call. How long was RJ's there?

35. *Nutrition.* Following the guidelines of the Food and Drug Administration, Dale tries to eat at least 5 servings of fruits or vegetables each day. For the first six days of one week, he had 4, 6, 7, 4, 6, and 4 servings. How many servings of fruits or vegetables should Dale eat on Saturday, in order to average at least 5 servings per day for the week?

36. *College Course Load.* To remain on financial aid, Millie needs to complete an average of at least 7 credits per quarter each year. In the first three quarters of 2001, Millie completed 5, 7, and 8 credits. How many credits of course work must Millie complete in the fourth quarter if she is to remain on financial aid?

37. *Perimeter of a Rectangle.* The width of a rectangle is fixed at 8 ft. What lengths will make the perimeter at least 200 ft? at most 200 ft?

38. *Perimeter of a Triangle.* One side of a triangle is 2 cm shorter than the base. The other side is 3 cm longer than the base. What lengths of the base will allow the perimeter to be greater than 19 cm?

39. *Area of a Rectangle.* The width of a rectangle is fixed at 4 cm. For what lengths will the area be less than 86 cm^2?

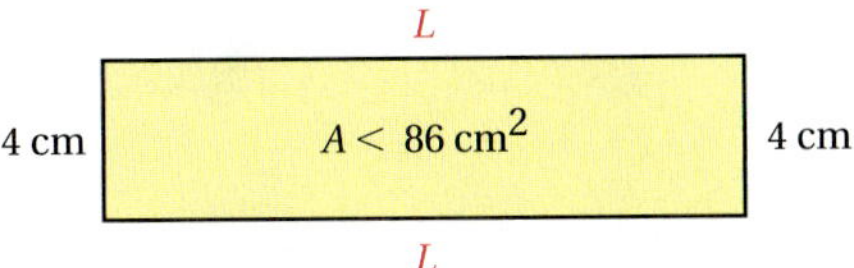

40. *Area of a Rectangle.* The width of a rectangle is fixed at 16 yd. For what lengths will the area be at least 264 yd^2?

41. *Insurance-covered Repairs.* Most insurance companies will replace a vehicle if an estimated repair exceeds 80% of the "blue-book" value of the vehicle. Michelle's insurance company paid \$8500 for repairs to her Subaru after an accident. What can be concluded about the blue-book value of the car?

42. *Insurance-covered Repairs.* Following an accident, Jeff's Ford pickup was replaced by his insurance company because the damage was so extensive. Before the damage, the blue-book value of the truck was \$21,000. How much would it have cost to repair the truck? (See Exercise 41.)

43. *Fat Content in Foods.* Reduced Fat Skippy® peanut butter contains 12 g of fat per serving. In order for a food to be labeled "reduced fat," it must have at least 25% less fat than the regular item. What can you conclude about the number of grams of fat in a serving of the regular Skippy peanut butter?
Source: Best Foods

44. *Fat Content in Foods.* Reduced Fat Chips Ahoy!® cookies contain 5 g of fat per serving. What can you conclude about the number of grams of fat in regular Chips Ahoy! cookies (see Exercise 43)?
Source: Nabisco Brands, Inc.

45. *Pond Depth.* On July 1, Garrett's Pond was 25 ft deep. Since that date, the water level has dropped $\frac{2}{3}$ ft per week. For what dates will the water level not exceed 21 ft?

46. *Weight Gain.* A 3-lb puppy is gaining weight at a rate of $\frac{3}{4}$ lb per week. When will the puppy's weight exceed $22\frac{1}{2}$ lb?

47. *Area of a Triangular Flag.* As part of an outdoor education course, Wanda needs to make a bright-colored triangular flag with an area of at least 3 ft^2. What heights can the triangle be if the base is $1\frac{1}{2}$ ft?

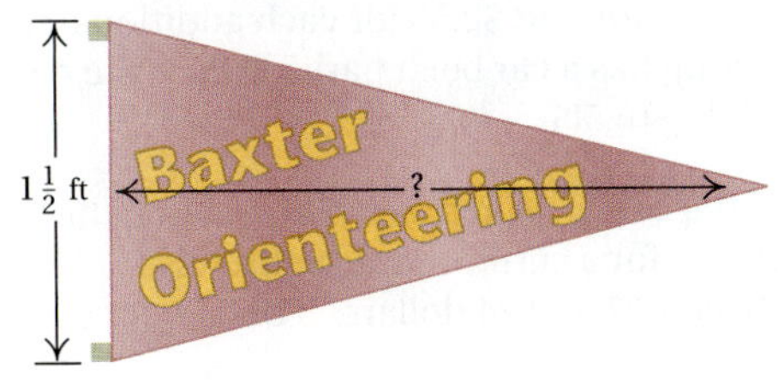

48. *Area of a Triangular Sign.* Zoning laws in Harrington prohibit displaying signs with areas exceeding 12 ft^2. If Flo's Marina is ordering a triangular sign with an 8-ft base, how tall can the sign be?

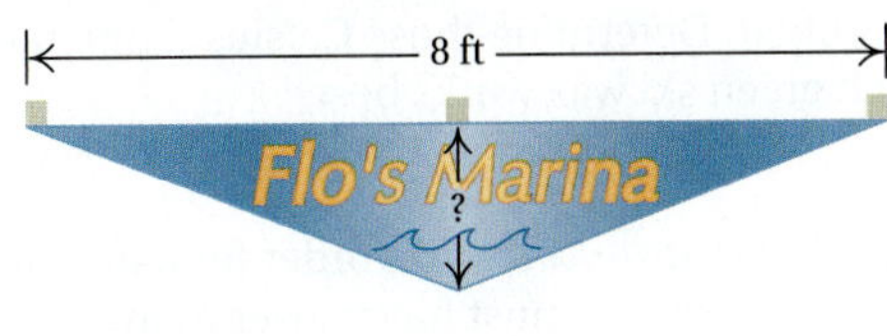

49. *Electrician Visits.* Dot's Electric made 17 customer calls last week and 22 calls this week. How many calls must be made next week in order to maintain an average of at least 20 for the three-week period?

50. *Volunteer Work.* George and Joan do volunteer work at a hospital. Joan worked 3 more hr than George, and together they worked more than 27 hr. What possible numbers of hours did each work?

51. **DW** If f represents Fran's age and t represents Todd's age, write a sentence that would translate to $t + 3 < f$.

52. **DW** Explain how the meanings of "Five more than a number" and "Five is more than a number" differ.

SKILL MAINTENANCE

Simplify.

53. $-3 + 2(-5)^2(-3) - 7$ [6.8d]

54. $3x + 2[4 - 5(2x - 1)]$ [6.8c]

55. $23(2x - 4) - 15(10 - 3x)$ [6.8b]

56. $256 \div 64 \div 4^2$ [6.8d]

SYNTHESIS

57. *Ski Wax.* Green ski wax works best between 5° and 15° Fahrenheit. Determine those Celsius temperatures for which green ski wax works best.

58. *Parking Fees.* Mack's Parking Garage charges \$4.00 for the first hour and \$2.50 for each additional hour. For how long has a car been parked when the charge exceeds \$16.50?

59. *Nutritional Standards.* In order for a food to be labeled "lowfat," it must have fewer than 3 g of fat per serving. Reduced fat Tortilla Pops® contain 60% less fat than regular nacho cheese tortilla chips, but still cannot be labeled lowfat. What can you conclude about the fat content of a serving of nacho cheese tortilla chips?

60. *Parking Fees.* When asked how much the parking charge is for a certain car (see Exercise 58), Mack replies "between 14 and 24 dollars." For how long has the car been parked?

CHAPTER 7

Summary and Review

The review that follows is meant to prepare you for a chapter exam. It consists of two parts. The first part is a checklist of some of the Study Tips referred to in this and preceding chapters, as well as a list of important properties and formulas. The second part is the Review Exercises. These provide practice exercises for the exam, together with references to section objectives so you can go back and review. Before beginning, stop and look back over the skills you have obtained. What skills in mathematics do you have now that you did not have before studying this chapter?

STUDY TIPS CHECKLIST

The foundation of all your study skills is TIME!

- ☐ Are you making progress in learning to manage your time?
- ☐ Are you practicing the five-step problem-solving strategy?
- ☐ Are you doing the homework as soon as possible after class?
- ☐ Are you stopping to work the margin exercises when directed to do so?

IMPORTANT PROPERTIES AND FORMULAS

The Addition Principle for Equations: For any real numbers a, b, and c: $a = b$ is equivalent to $a + c = b + c$.

The Multiplication Principle for Equations: For any real numbers a, b, and c, $c \neq 0$: $a = b$ is equivalent to $a \cdot c = b \cdot c$.

The Addition Principle for Inequalities: For any real numbers a, b, and c:
$a < b$ is equivalent to $a + c < b + c$;
$a > b$ is equivalent to $a + c > b + c$;
$a \leq b$ is equivalent to $a + c \leq b + c$;
$a \geq b$ is equivalent to $a + c \geq b + c$.

The Multiplication Principle for Inequalities: For any real numbers a and b, and any *positive* number c:
$a < b$ is equivalent to $ac < bc$; $a > b$ is equivalent to $ac > bc$.
For any real numbers a and b, and any *negative* number c:
$a < b$ is equivalent to $ac > bc$; $a > b$ is equivalent to $ac < bc$.

REVIEW EXERCISES

Solve. [7.1b]

1. $x + 5 = -17$

2. $n - 7 = -6$

3. $x - 11 = 14$

4. $y - 0.9 = 9.09$

Solve. [7.2a]

5. $-\frac{2}{3}x = -\frac{1}{6}$

6. $-8x = -56$

7. $-\frac{x}{4} = 48$

8. $15x = -35$

9. $\frac{4}{5}y = -\frac{3}{16}$

Solve. [7.3a]

10. $5 - x = 13$

11. $\frac{1}{4}x - \frac{5}{8} = \frac{3}{8}$

Solve. [7.3b, c]

12. $5t + 9 = 3t - 1$

13. $7x - 6 = 25x$

14. $14y = 23y - 17 - 10$

15. $0.22y - 0.6 = 0.12y + 3 - 0.8y$

16. $\frac{1}{4}x - \frac{1}{8}x = 3 - \frac{1}{16}x$

17. $14y + 17 + 7y = 9 + 21y + 8$

Solve. [7.3c]

18. $4(x + 3) = 36$

19. $3(5x - 7) = -66$

20. $8(x - 2) - 5(x + 4) = 20 + x$

21. $-5x + 3(x + 8) = 16$

22. $6(x - 2) - 16 = 3(2x - 5) + 11$

Determine whether the given number is a solution of the inequality $x \leq 4$. [7.7a]

23. -3

24. 7

25. 4

Solve. Write set notation for the answers. [7.7c, d, e]

26. $y + \frac{2}{3} \geq \frac{1}{6}$

27. $9x \geq 63$

28. $2 + 6y > 14$

29. $7 - 3y \geq 27 + 2y$

30. $3x + 5 < 2x - 6$

31. $-4y < 28$

32. $4 - 8x < 13 + 3x$

33. $-4x \leq \frac{1}{3}$

Graph on a number line. [7.7b, e]

34. $4x - 6 < x + 3$

−5 −4 −3 −2 −1 0 1 2 3 4 5

35. $-2 < x \leq 5$

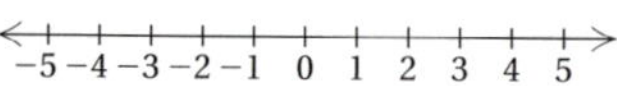

36. $y > 0$

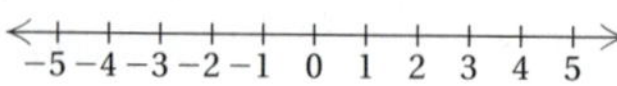

Solve. [7.4b]

37. $C = \pi d$, for d

38. $V = \frac{1}{3}Bh$, for B

39. $A = \frac{a + b}{2}$, for a

40. $y = mx + b$, for x

Solve. [7.6a]

41. *Dimensions of Wyoming.* The state of Wyoming is roughly in the shape of a rectangle whose perimeter is 1280 mi. The length is 90 mi more than the width. Find the dimensions.

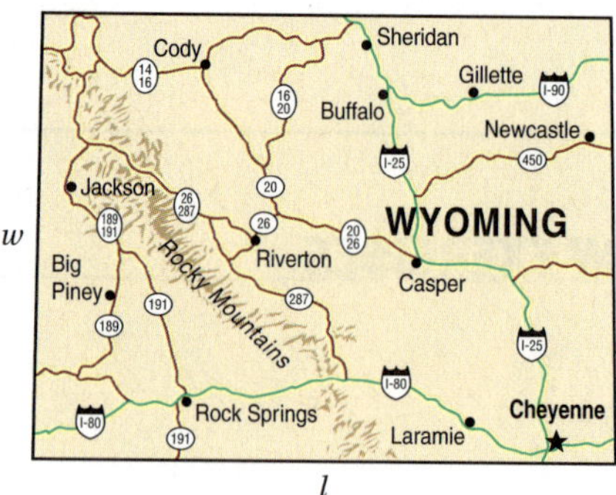

42. *Interstate Mile Markers.* The sum of two consecutive mile markers on I-5 in California is 691. Find the numbers on the markers.

43. An entertainment center sold for $2449 in June. This was $332 more than the cost in February. Find the cost in February.

44. Ty is paid a commission of $4 for each appliance he sells. One week, he received $108 in commissions. How many appliances did he sell?

45. The measure of the second angle of a triangle is 50° more than that of the first angle. The measure of the third angle is 10° less than twice the first angle. Find the measures of the angles.

Solve. [7.5a]

46. What is 20% of 75?

47. Fifteen is what percent of 80?

48. 18 is 3% of what number?

49. *Job Opportunities.* There were 905 thousand child-care workers in 1998. Job opportunities are expected to grow to 1141 thousand by 2008. What is the percent of increase?
Source: *Handbook of U.S. Labor Statistics*

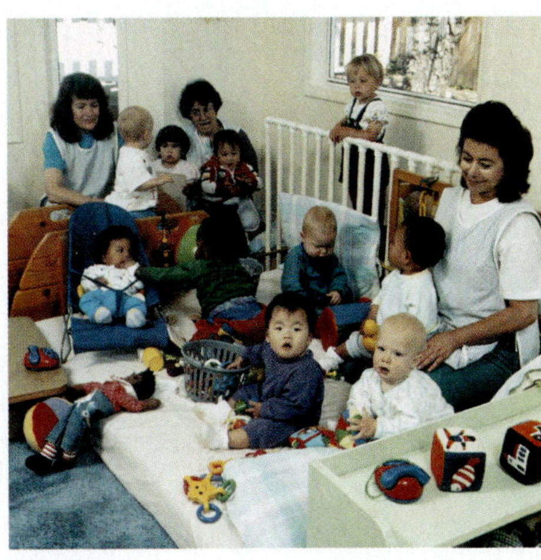

Solve. [7.6a]

50. After a 30% reduction, a bread maker is on sale for $154. What was the marked price (the price before the reduction)?

51. A hotel manager's salary is $61,410, which is a 15% increase over the previous year's salary. What was the previous salary?

52. A tax-exempt charity received a bill of $145.90 for a sump pump. The bill incorrectly included sales tax of 5%. How much does the charity actually owe?

Solve. [7.8b]

53. *Test Scores.* Your test grades are 71, 75, 82, and 86. What is the lowest grade that you can get on the next test and still have an average test score of at least 80?

54. The length of a rectangle is 43 cm. What widths will make the perimeter greater than 120 cm?

55. $\mathbf{D_W}$ Would it be better to receive a 5% raise and then an 8% raise or the other way around? Why? [7.5a]

56. $\mathbf{D_W}$ Are the inequalities $x > -5$ and $-x < 5$ equivalent? Why or why not? [7.7d]

SKILL MAINTENANCE

Certain objectives from four particular sections will be retested on the chapter test. The objectives are listed with the practice problems that follow.

57. Evaluate $\frac{a + b}{4}$ when $a = 16$ and $b = 25$. [6.1a]

58. Translate to an algebraic expression: [6.1b]

Tricia drives her car at 58 mph for t hours. How far has she driven?

59. Add: $-12 + 10 + (-19) + (-24)$. [6.3a]

60. Remove parentheses and simplify: $5x - 8(6x - y)$. [6.8b]

SYNTHESIS

Solve.

61. $2|x| + 4 = 50$ [6.2e], [7.3a]

62. $|3x| = 60$ [6.2e], [7.2a]

63. $y = 2a - ab + 3$, for a [7.4b]

CHAPTER 7 Chapter Test

Solve.

1. $x + 7 = 15$
2. $t - 9 = 17$
3. $3x = -18$
4. $-\frac{4}{7}x = -28$
5. $3t + 7 = 2t - 5$
6. $\frac{1}{2}x - \frac{3}{5} = \frac{2}{5}$
7. $8 - y = 16$
8. $-\frac{2}{5} + x = -\frac{3}{4}$
9. $3(x + 2) = 27$
10. $-3x - 6(x - 4) = 9$
11. $0.4p + 0.2 = 4.2p - 7.8 - 0.6p$
12. $4(3x - 1) + 11 = 2(6x + 5) - 8$
13. $-2 + 7x + 6 = 5x + 4 + 2x$

Solve. Write set notation for the answers.

14. $x + 6 \leq 2$
15. $14x + 9 > 13x - 4$
16. $12x \leq 60$
17. $-2y \geq 26$
18. $-4y \leq -32$
19. $-5x \geq \frac{1}{4}$
20. $4 - 6x > 40$
21. $5 - 9x \geq 19 + 5x$

Graph on a number line.

22. $y \leq 9$

−12 −8 −4 0 4 8 12

23. $6x - 3 < x + 2$

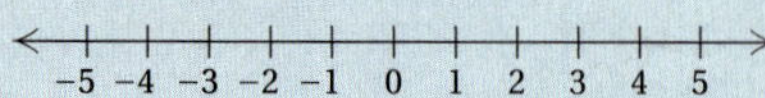

24. $-2 \leq x \leq 2$

−5 −4 −3 −2 −1 0 1 2 3 4 5

Solve.

25. What is 24% of 75?
26. 15.84 is what percent of 96?
27. 800 is 2% of what number?
28. *Job Opportunities.* Job opportunities for flight attendants are expected to increase from 99 thousand in 1998 to 129 thousand by 2008. What is the percent of increase?
Source: *Handbook of U.S. Labor Statistics*

29. *Perimeter of a Photograph.* The perimeter of a rectangular photograph is 36 cm. The length is 4 cm greater than the width. Find the width and the length.

30. *Charitable Contributions.* About 53% of all charitable contributions are made to religious organizations. In 1997, \$75 billion was given to religious organizations. How much was given to charities in general?
Source: *Statistical Abstract of the United States, 1999*

31. *Raffle Tickets.* The numbers on three raffle tickets are consecutive integers whose sum is 7530. Find the integers.

32. *Savings Account.* Money is invested in a savings account at 5% simple interest. After 1 year, there is \$924 in the account. How much was originally invested?

33. *Board Cutting.* An 8-m board is cut into two pieces. One piece is 2 m longer than the other. How long are the pieces?

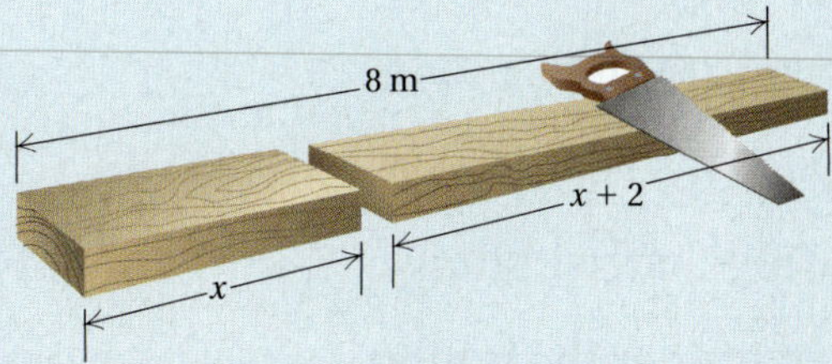

34. *Lengths of a Rectangle.* The width of a rectangle is 96 yd. Find all possible lengths such that the perimeter of the rectangle will be at least 540 yd.

35. *Budgeting.* Jason has budgeted an average of \$95 a month for entertainment. For the first five months of the year, he has spent \$98, \$89, \$110, \$85, and \$83. How much can Jason spend in the sixth month without exceeding his average budget?

36. *IKON Copiers.* IKON Office Solutions rents a Canon IR330 copier for \$225 per month plus 1.2¢ per copy. A catalog publisher needs to lease a copy machine for use during a special project that they anticipate will take 3 months. They decide to rent the copier, but must stay within a budget of \$2400 for copies. Determine (in terms of an inequality) the number of copies they can make and still remain within budget.
Source: Ikon Office Solutions, Nathan DuMond, Sales Manager

37. Solve $A = 2\pi rh$ for r.

38. Solve $y = 8x + b$ for x.

SKILL MAINTENANCE

39. Add: $\frac{2}{3} + \left(-\frac{8}{9}\right)$.

40. Evaluate $\frac{4x}{y}$ when $x = 2$ and $y = 3$.

41. Translate to an algebraic expression: Seventy-three percent of p.

42. Simplify: $2x - 3y - 5(4x - 8y)$.

SYNTHESIS

43. Solve $c = \frac{1}{a - d}$ for d.

44. Solve: $3|w| - 8 = 37$.

45. A movie theater had a certain number of tickets to give away. Five people got the tickets. The first got one-third of the tickets, the second got one-fourth of the tickets, and the third got one-fifth of the tickets. The fourth person got eight tickets, and there were five tickets left for the fifth person. Find the total number of tickets given away.

Graphs of Linear Equations

Gateway to Chapter 8

We now begin a study of graphs. First, we examine graphs as they commonly appear in newspapers or magazines and develop some terminology. Following that, we study graphs of linear equations. Finally, we consider the notion of the slope of a line and connect it to the concept of rate of change.

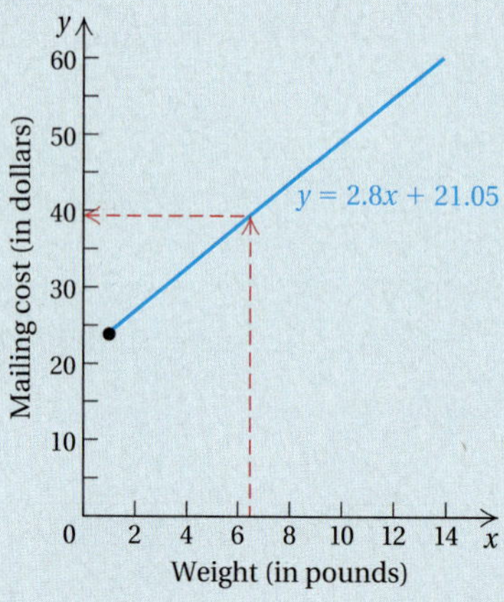

Real-World Application

The cost y, in dollars, of shipping a FedEx Priority Overnight package weighing 1 lb or more a distance of 1001 to 1400 mi is given by $y = 2.8x + 21.05$, where x is the number of pounds. Graph the equation and then use the graph to find the cost of shipping a $6\frac{1}{2}$-lb package.

Source: Federal Express Corporation

This problem appears as Example 8 in Section 8.2.

CHAPTER 8

CHAPTER 8 Pretest

Graph on a plane.

1. $y = -x$ [8.2b]

2. $x = -4$ [8.3b]

3. $4x - 5y = 20$ [8.2b]

4. $y = \frac{2}{3}x - 1$ [8.2b]

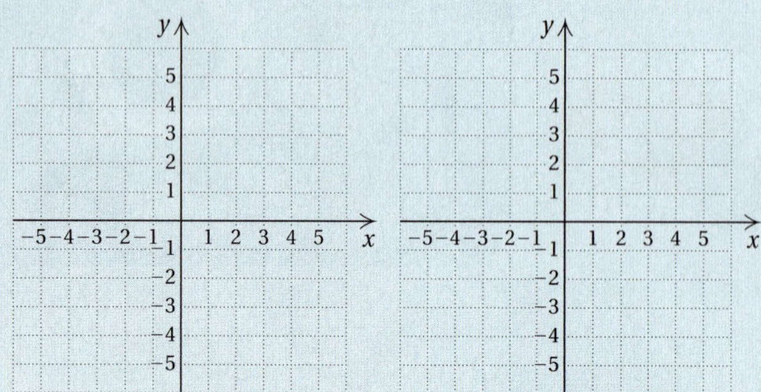

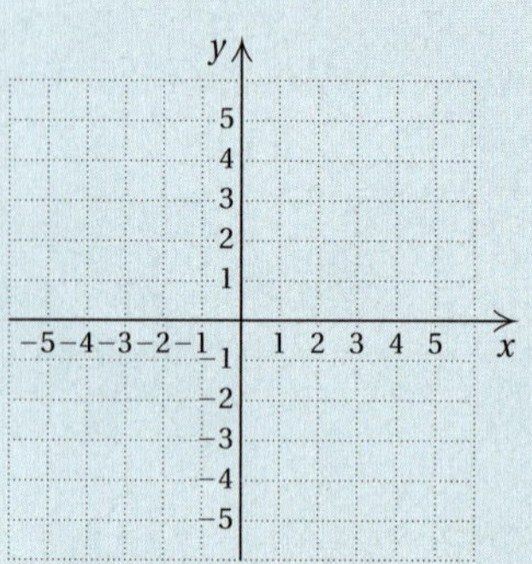

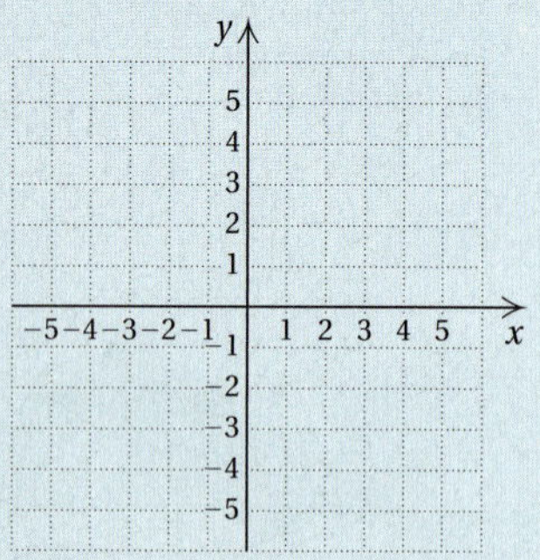

5. In which quadrant is the point $(-4, -1)$ located? [8.1c]

6. Determine whether the ordered pair $(-4, -1)$ is a solution of $4x - 5y = 20$. [8.2a]

7. Find the intercepts of the graph of $4x - 5y = 20$. [8.3a]

8. Find the y-intercept of $y = 3x - 8$. [8.2b]

9. Price of Printing. The price P, in cents, of a photocopied and bound lab manual is given by

$$P = \frac{7}{2}n + 20,$$

where n is the number of pages in the manual. Graph the equation and then use the graph to estimate the price of an 85-page manual. [8.2c]

10. Find the slope of the line. [8.4a]

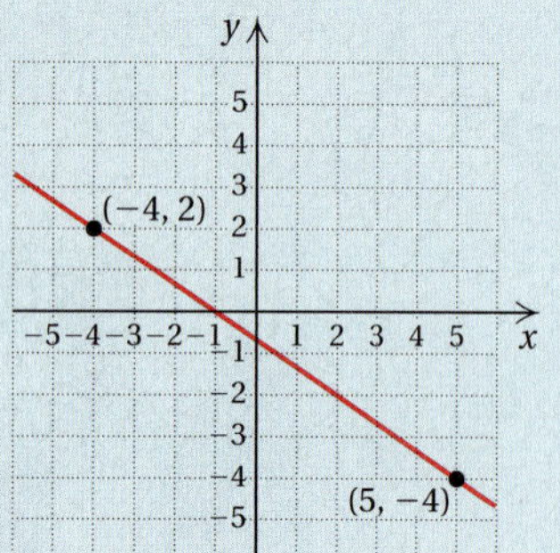

11. Find the rate of change. [8.4b]

TIME	NUMBER OF CARS PRODUCED
9:00 A.M.	11
11:00 A.M.	16

12. Find the rate at which a runner burns calories. [8.4b]

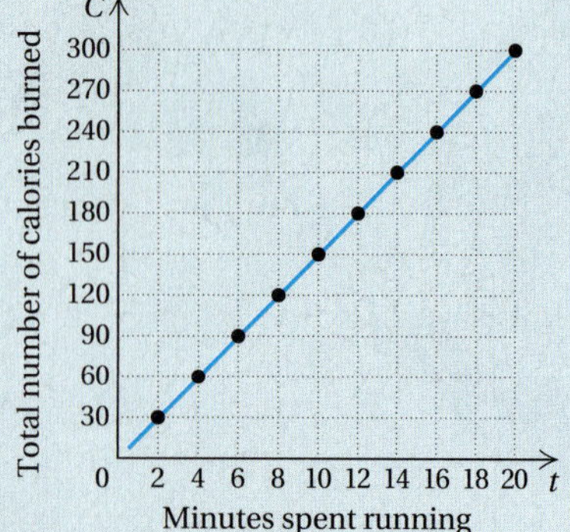

13. Find the slope, if it exists. [8.4c]

$$3x - 5y = 15$$

8.1 GRAPHS AND APPLICATIONS

Objectives

a Solve applied problems involving circle, bar, and line graphs.

b Plot points associated with ordered pairs of numbers.

c Determine the quadrant in which a point lies.

d Find the coordinates of a point on a graph.

Often data are available regarding an application in mathematics. We can use graphs to show the data and extract information about the data that can lead to making analyses and predictions.

Today's print and electronic media make extensive use of graphs. This is due in part to the ease with which some graphs can be prepared by computer and in part to the large quantity of information that a graph can display. We first consider applications with circle, bar, and line graphs.

a Applications with Graphs

CIRCLE GRAPHS

Circle graphs and *pie graphs,* or *charts,* are often used to show what percent of a whole each particular item in a group represents.

EXAMPLE 1 *Careers of Women Who Travel.* Consider all the businesswomen who travel in their careers. The following circle graph shows the percentage of these women in specific careers.

Jobs of Women Who Travel

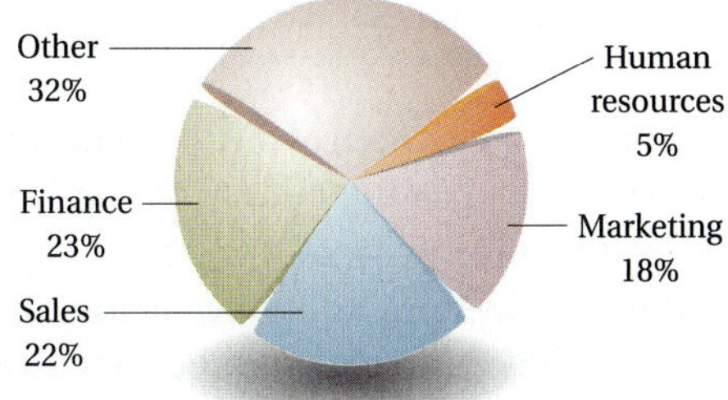

Source: Runzheimer Reports of Travel Management

In a sample of 1500 women travelers, how many of them can be expected to be in sales?

1. **Familiarize.** The circle graph shows that 22% of these women are in sales. We let $y =$ the number of women who are in sales.
2. **Translate.** We reword and translate the problem as follows.

   What is 22% of 1500?   Rewording

   $y = 22\% \cdot 1500$ Translating

3. **Solve.** We solve the equation by carrying out the computation on the right:

   $y = 22\% \cdot 1500 = 0.22 \cdot 1500 = 330.$

4. **Check.** We note that 330 is less than 1500, which we would expect. We can also repeat the calculation.
5. **State.** We expect that 330 of the women travelers are in sales.

Do Exercise 1.

1. **Careers of Women Who Travel.** Referring to the graph in Example 1, determine how many of the 1500 travelers are in marketing.

Answer on page A-20

2. Tornado Touchdowns. Referring to the graph in Example 2, determine the following.

a) During which interval did the least number of touchdowns occur?

b) During which intervals was the number of touchdowns less than 60?

Answers on page A-20

BAR GRAPHS

Bar graphs are convenient for showing comparisons. In every bar graph, certain categories are paired with certain numbers. Example 2 pairs intervals of time with the total number of reported cases of tornado touchdowns.

EXAMPLE 2 *Tornado Touchdowns.* The following bar graph shows the total number of tornado touchdowns by time of day in Indiana from 1950–1994.

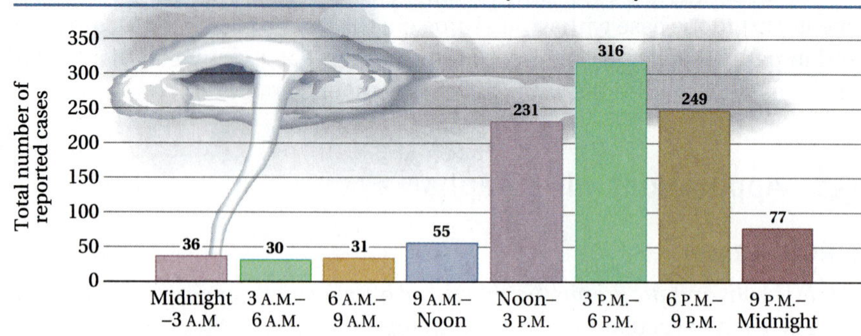

Source: National Weather Service

a) During which interval of time did the greatest number of tornado touchdowns occur?

b) During which intervals was the number of tornado touchdowns greater than 200?

We solve as follows.

a) In this bar graph, the values are written at the top of the bars. We see that 316 is the greatest number. We look at the bottom of that bar on the horizontal scale and see that the time interval of greatest occurrence is 3 P.M.–6 P.M.

b) We locate 200 on the vertical scale and move across the graph or draw a horizontal line. We note that the value on three bars exceeds 200. Then we look down at the horizontal scale and see that the corresponding time intervals are noon–3 P.M., 3 P.M.–6 P.M., and 6 P.M.–9 P.M.

Do Exercise 2.

LINE GRAPHS

Line graphs are often used to show change over time. Certain points are plotted to represent given information. When segments are drawn to connect the points, a line graph is formed.

Sometimes it is important to begin the labeling of horizontal or vertical values on the x- and y-axes with zero. When the values are large, as in Example 3, the symbol ⦚ can be used to indicate a break in the list of values.

EXAMPLE 3 *Exercise and Pulse Rate.* The following line graph shows the relationship between a person's resting pulse rate and months of regular exercise.

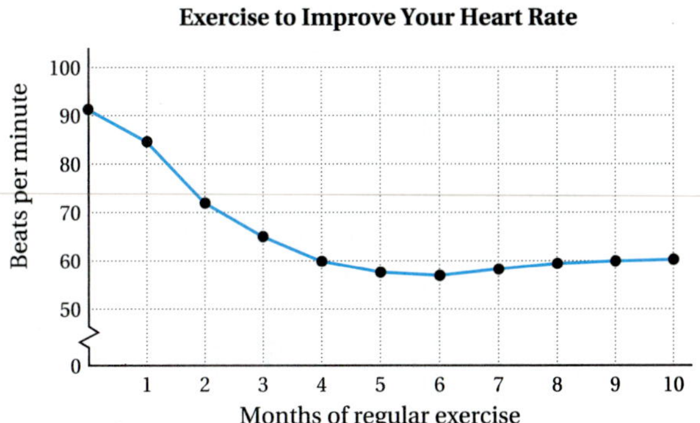

Source: Hughes, Martin, *Body Clock*. New York: Facts on File, Inc., p. 60

a) How many months of regular exercise are required to lower the pulse rate to its lowest point?

b) How many months of regular exercise are needed to achieve a pulse rate of 65 beats per minute?

We solve as follows.

a) The lowest point on the graph occurs above the number 6. Thus after 6 months of regular exercise, the pulse rate has been lowered as much as possible.

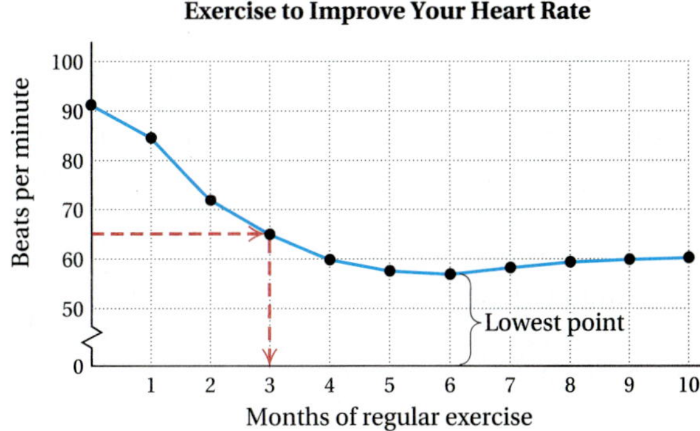

b) We locate 65 on the vertical scale and then move right until we reach the line. At that point, we move down to the horizontal scale and read the information we are seeking. The pulse rate is 65 beats per minute after 3 months of regular exercise.

Do Exercise 3.

3. Exercise and Pulse Rate. Referring to the graph in Example 3, determine the following.

a) About how many months of regular exercise are needed to achieve a pulse rate of about 72 beats per minute?

b) What pulse rate has been achieved after 10 months of exercise?

Answers on page A-20

Plot these points on the graph below.

4. $(4, 5)$ **5.** $(5, 4)$

6. $(-2, 5)$ **7.** $(-3, -4)$

8. $(5, -3)$ **9.** $(-2, -1)$

10. $(0, -3)$ **11.** $(2, 0)$

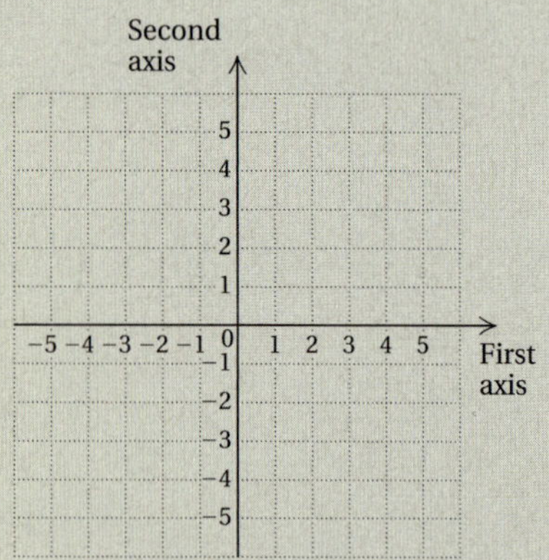

Answers on page A-20

b Plotting Ordered Pairs

The line graph in Example 3 is formed from a collection of points. Each point pairs a number of months of exercise with a pulse rate.

In Chapter 7, we graphed numbers and inequalities in one variable on a line. To enable us to graph an equation that contains two variables, we now learn to graph number pairs on a plane.

On a number line, each point is the graph of a number. On a plane, each point is the graph of a number pair. We use two perpendicular number lines called **axes.** They cross at a point called the **origin.** The arrows show the positive directions.

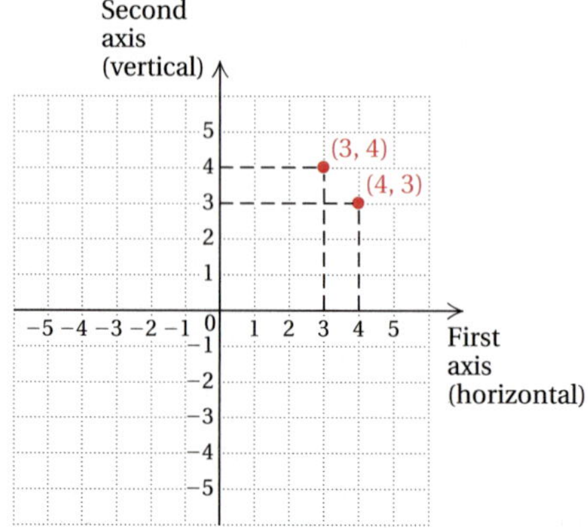

Consider the ordered pair $(3, 4)$. The numbers in an ordered pair are called **coordinates.** In $(3, 4)$, the **first coordinate (abscissa)** is 3 and the **second coordinate (ordinate)** is 4. To plot $(3, 4)$, we start at the origin and move horizontally to the 3. Then we move up vertically 4 units and make a "dot."

The point $(4, 3)$ is also plotted. Note that $(3, 4)$ and $(4, 3)$ give different points. The order of the numbers in the pair is indeed important. They are called **ordered pairs** because it makes a difference which number comes first. The coordinates of the origin are $(0, 0)$.

EXAMPLE 4 Plot the point $(-5, 2)$.

The first number, -5, is negative. Starting at the origin, we move -5 units in the horizontal direction (5 units to the left). The second number, 2, is positive. We move 2 units in the vertical direction (up).

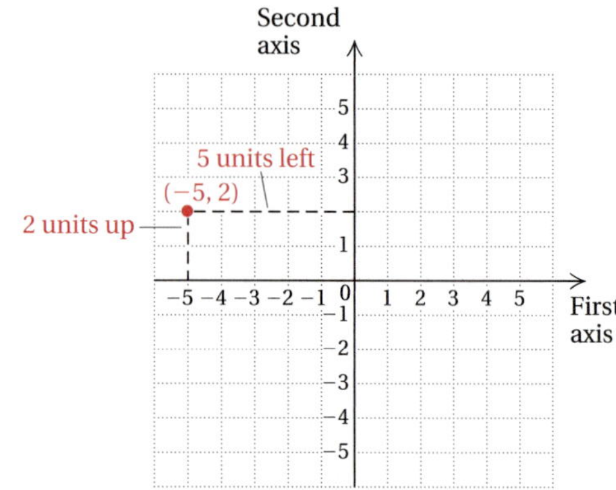

CAUTION!

The *first* coordinate of an ordered pair is always graphed in a *horizontal* direction and the *second* coordinate is always graphed in a *vertical* direction.

Do Exercises 4–11.

c Quadrants

The figure below shows some points and their coordinates. In region I (the *first quadrant*), both coordinates of any point are positive. In region II (the *second quadrant*), the first coordinate is negative and the second positive. In region III (the *third quadrant*), both coordinates are negative. In region IV (the *fourth quadrant*), the first coordinate is positive and the second is negative.

EXAMPLE 5 In which quadrant, if any, are the points $(-4, 5)$, $(5, -5)$, $(2, 4)$, $(-2, -5)$, and $(-5, 0)$ located?

The point $(-4, 5)$ is in the second quadrant. The point $(5, -5)$ is in the fourth quadrant. The point $(2, 4)$ is in the first quadrant. The point $(-2, -5)$ is in the third quadrant. The point $(-5, 0)$ is on an axis and is *not* in any quadrant.

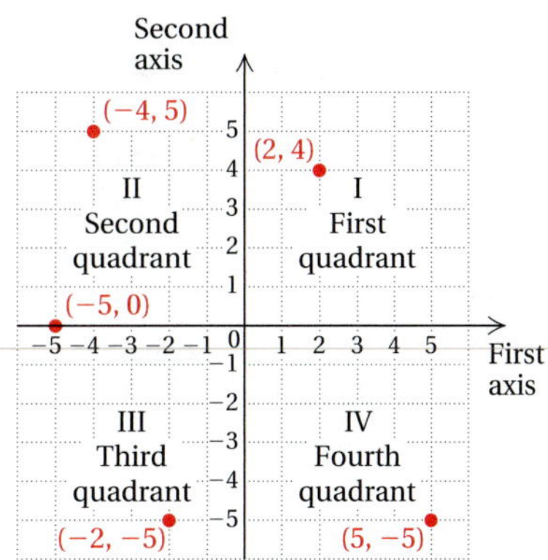

Do Exercises 12–18.

d Finding Coordinates

To find the coordinates of a point, we see how far to the right or left of zero it is located and how far up or down.

EXAMPLE 6 Find the coordinates of points *A*, *B*, *C*, *D*, *E*, *F*, and *G*.

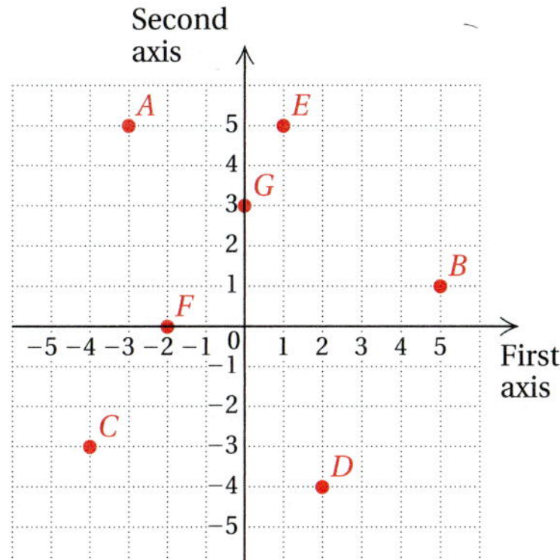

Point *A* is 3 units to the left (horizontal direction) and 5 units up (vertical direction). Its coordinates are $(-3, 5)$. Point *D* is 2 units to the right and 4 units down. Its coordinates are $(2, -4)$. The coordinates of the other points are as follows:

B: $(5, 1)$; *C*: $(-4, -3)$;
E: $(1, 5)$; *F*: $(-2, 0)$; *G*: $(0, 3)$.

Do Exercise 19.

12. What can you say about the coordinates of a point in the third quadrant?

13. What can you say about the coordinates of a point in the fourth quadrant?

In which quadrant, if any, is the point located?

14. $(5, 3)$

15. $(-6, -4)$

16. $(10, -14)$

17. $(-13, 9)$

18. $(0, -3)$

19. Find the coordinates of points *A*, *B*, *C*, *D*, *E*, *F*, and *G* on the graph below.

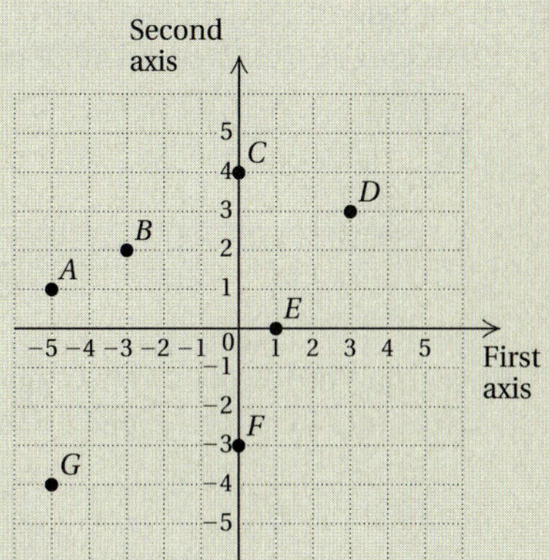

Answers on page A-20

8.1 EXERCISE SET

For Extra Help

Digital Video Tutor CD 6 Videotape 8

InterAct Math

Math Tutor Center

MathXL

MyMathLab

a Solve.

Causes of Death. The following circle graphs show the leading causes of death for women and men ages 65 and over. Use the circle graphs for Exercises 1–8.

Leading Causes of Death Among Persons 65 Years of Age and Over

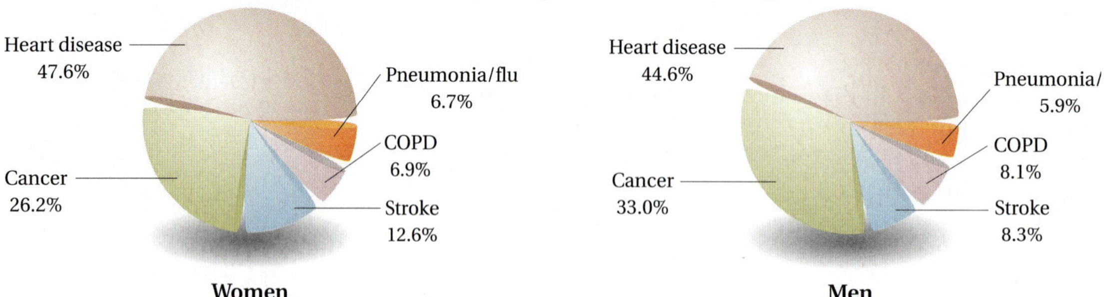

Sources: Centers for Disease Control and Prevention, National Center for Health Statistics, National Vital Statistics System, 1998.
Note: COPD is chronic obstructive pulmonary diseases such as emphysema, chronic bronchitis, and so on, often the result of cigarette smoking.

1. What percent of women age 65 and over die of heart disease?

2. What percent of men age 65 and over die of a stroke?

3. What percent of men age 65 and over die of heart disease?

4. What percent of women age 65 and over die of cancer?

5. In a group of 150,000 men age 65 and over, how many of them can be expected to die of cancer?

6. In a group of 150,000 women age 65 and over, how many of them can be expected to die of heart disease?

7. In a group of 150,000 women age 65 and over, how many of them can be expected to die of a stroke?

8. In a group of 150,000 men age 65 and over, how many of them can be expected to die of chronic obstructive pulmonary disease?

Driving While Intoxicated (DWI). State laws have determined that a blood alcohol level of at least 0.10% or higher indicates that an individual has consumed too much alcohol to drive safely. The following bar graph shows the number of drinks that a person of a certain weight would need to consume in order to reach a blood alcohol level of 0.10%. A 12-oz beer, a 5-oz glass of wine, or a cocktail containing $1\frac{1}{2}$ oz of distilled liquor all count as one drink. Use the bar graph for Exercises 9–14.

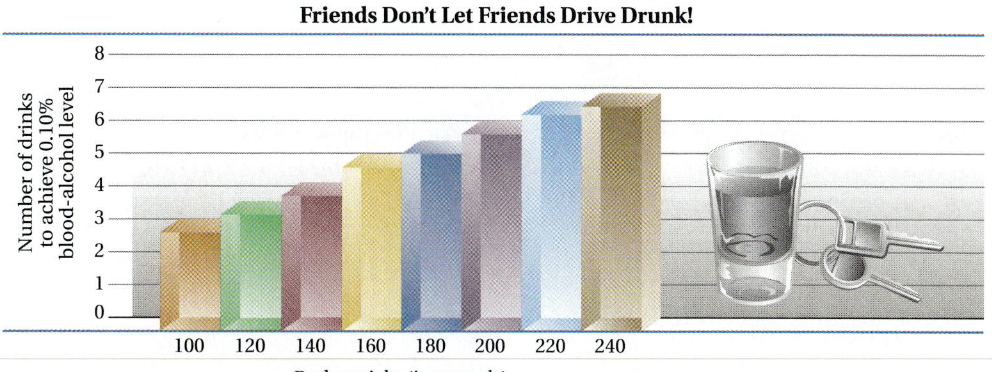

Source: Neighborhood Digest, 7, no. 12

9. Approximately how many drinks would a 200-lb person have consumed if he or she had a blood alcohol level of 0.10%?

10. What can be concluded about the weight of someone who can consume 4 drinks without reaching a blood alcohol level of 0.10%?

11. What can be concluded about the weight of someone who can consume 6 drinks without reaching a blood alcohol level of 0.10%?

12. Approximately how many drinks would a 160-lb person have consumed if he or she had a blood alcohol level of 0.10%?

13. What can be concluded about the weight of someone who has consumed $3\frac{1}{2}$ drinks without reaching a blood alcohol level of 0.10%?

14. What can be concluded about the weight of someone who has consumed $4\frac{1}{2}$ drinks without reaching a blood alcohol level of 0.10%?

Alcohol-Related Deaths. The data in the following line graph show the number of deaths from 1990 to 1998. Use the line graph for Exercises 15–20.

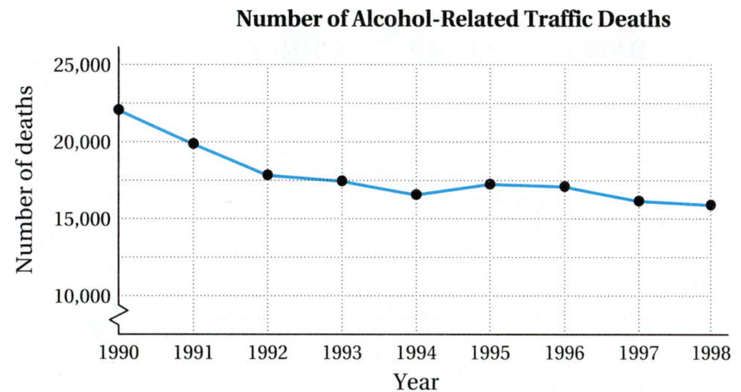

Source: National Highway Traffic Safety Administration

15. About how many alcohol-related deaths occurred in 1995?

16. About how many alcohol-related deaths occurred in 1998?

17. In what year did the lowest number of deaths occur?

18. In what years did fewer than 18,000 deaths occur?

19. By how much did the number of alcohol-related deaths decrease from 1995 to 1998?

20. By how much did the number of alcohol-related deaths increase from 1994 to 1995?

21. Plot these points.

$(2, 5)$ $(-1, 3)$ $(3, -2)$ $(-2, -4)$
$(0, 4)$ $(0, -5)$ $(5, 0)$ $(-5, 0)$

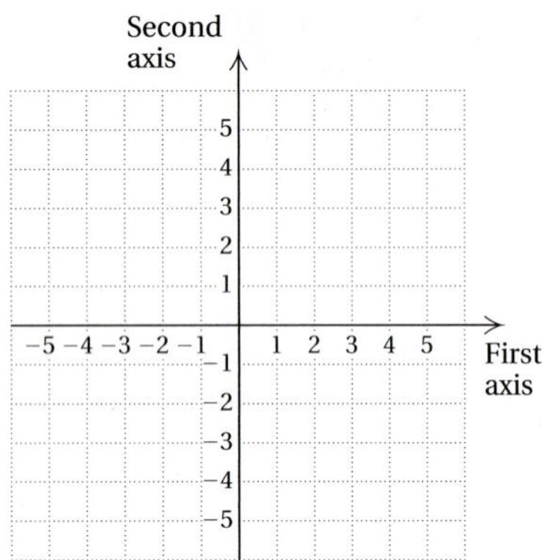

22. Plot these points.

$(4, 4)$ $(-2, 4)$ $(5, -3)$ $(-5, -5)$
$(0, 4)$ $(0, -4)$ $(3, 0)$ $(-4, 0)$

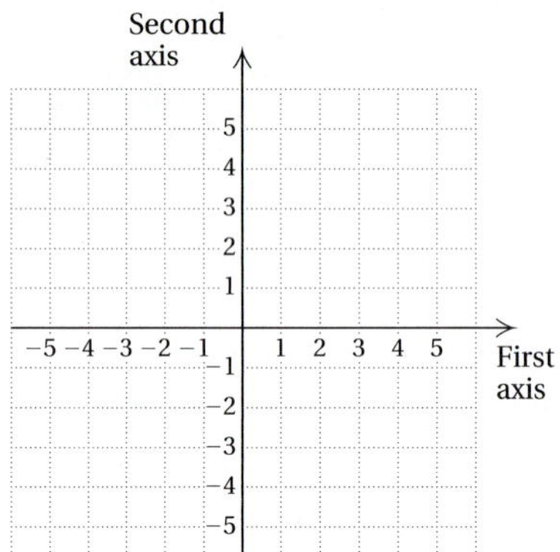

c In which quadrant is the point located?

23. $(-5, 3)$

24. $(1, -12)$

25. $(100, -1)$

26. $(-2.5, 35.6)$

27. $(-6, -29)$

28. $(3.6, 105.9)$

29. $(3.8, 9.2)$

30. $(-895, -492)$

31. $\left(-\frac{1}{3}, \frac{15}{7}\right)$

32. $\left(-\frac{2}{3}, -\frac{9}{8}\right)$

33. $\left(12\frac{7}{8}, -1\frac{1}{2}\right)$

34. $\left(23\frac{5}{8}, 81.74\right)$

For each of Exercises 35–38, complete the table regarding the signs of coordinates in certain quadrants.

	QUADRANT	FIRST COORDINATES	SECOND COORDINATES
35.		Positive	Positive
36.	III		Negative
37.	II	Negative	
38.		Positive	Negative

In which quadrant(s) can the point described be located?

39. The first coordinate is positive.

40. The second coordinate is negative.

41. The first and second coordinates are equal.

42. The first coordinate is the additive inverse of the second coordinate.

d

43. Find the coordinates of points A, B, C, D, and E.

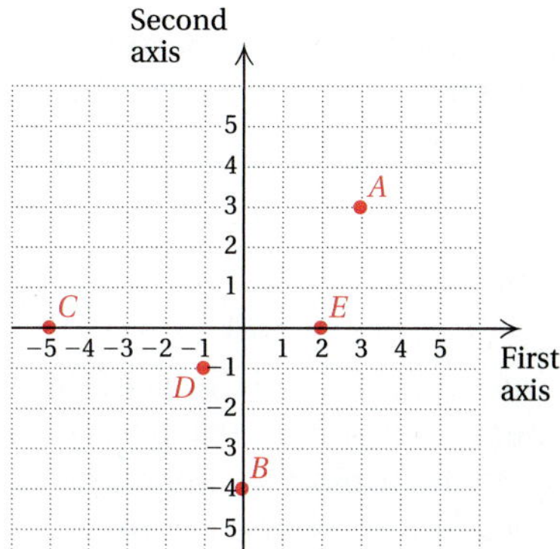

44. Find the coordinates of points A, B, C, D, and E.

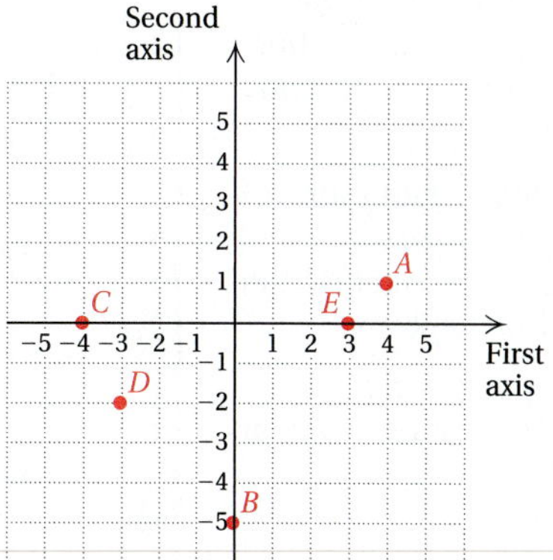

45. DW The sales of snow skis are highest in the winter months and lowest in the summer months. Sketch a line graph that might show the sales of a ski store and explain how an owner might use such a graph in decision making.

46. DW The graph in Example 3 tends to flatten out. Explain why the graph does not continue to decrease downward.

SKILL MAINTENANCE

Find the absolute value. [6.2e]

47. $|-12|$

48. $|4.89|$

49. $|0|$

50. $\left|-\frac{4}{5}\right|$

51. $|-3.4|$

52. $|\sqrt{2}|$

53. $\left|\frac{2}{3}\right|$

54. $\left|-\frac{7}{8}\right|$

Solve. [7.5a]

55. *Baseball Ticket Prices.* In 2001, the average price of a ticket to a Boston Red Sox baseball game was \$36.08, the highest in the major leagues. This price was an increase of 27.4% over the price in 2000. What was the price in 2000?
Source: Major League Baseball

56. *Tipping.* Erin left a 15% tip for a meal. The total cost of the meal, including the tip, was \$21.16. What was the cost of the meal before the tip was added?

SYNTHESIS

57. The points $(-1, 1)$, $(4, 1)$, and $(4, -5)$ are three vertices of a rectangle. Find the coordinates of the fourth vertex.

58. Three parallelograms share the vertices $(-2, -3)$, $(-1, 2)$, and $(4, -3)$. Find the fourth vertex of each parallelogram.

59. Graph eight points such that the sum of the coordinates in each pair is 6.

60. Graph eight points such that the first coordinate minus the second coordinate is 1.

61. Find the perimeter of a rectangle whose vertices have coordinates $(5, 3)$, $(5, -2)$, $(-3, -2)$, and $(-3, 3)$.

62. Find the area of a triangle whose vertices have coordinates $(0, 9)$, $(0, -4)$, and $(5, -4)$.

8.2 GRAPHING LINEAR EQUATIONS

Objectives

a. Determine whether an ordered pair is a solution of an equation with two variables.

b. Graph linear equations of the type $y = mx + b$ and $Ax + By = C$, identifying the y-intercept.

c. Solve applied problems involving graphs of linear equations.

We have seen how circle, bar, and line graphs can be used to represent the data in an application. Now we begin to learn how graphs can be used to represent solutions of equations.

a Solutions of Equations

When an equation contains two variables, the solutions of the equation are *ordered pairs* in which each number in the pair corresponds to a letter in the equation. Unless stated otherwise, to determine whether a pair is a solution, we use the first number in each pair to replace the variable that occurs first *alphabetically.*

EXAMPLE 1 Determine whether each of the following pairs is a solution of $4q - 3p = 22$: $(2, 7)$ and $(-1, 6)$.

For $(2, 7)$, we substitute 2 for p and 7 for q (using alphabetical order of variables):

$$\begin{array}{r|l} 4q - 3p & = 22 \\ \hline 4 \cdot 7 - 3 \cdot 2 & ?\ 22 \\ 28 - 6 & \\ 22 & \quad \text{TRUE} \end{array}$$

Thus, $(2, 7)$ is a solution of the equation.

For $(-1, 6)$, we substitute -1 for p and 6 for q:

$$\begin{array}{r|l} 4q - 3p & = 22 \\ \hline 4 \cdot 6 - 3 \cdot (-1) & ?\ 22 \\ 24 + 3 & \\ 27 & \quad \text{FALSE} \end{array}$$

Thus, $(-1, 6)$ is *not* a solution of the equation.

Do Exercises 1 and 2.

EXAMPLE 2 Show that the pairs $(3, 7)$, $(0, 1)$, and $(-3, -5)$ are solutions of $y = 2x + 1$. Then graph the three points and use the graph to determine another pair that is a solution.

To show that a pair is a solution, we substitute, replacing x with the first coordinate and y with the second coordinate of each pair:

$$\begin{array}{l|l} y & = 2x + 1 \\ \hline 7 & ?\ 2 \cdot 3 + 1 \\ & 6 + 1 \\ & 7 \quad \text{TRUE} \end{array} \qquad \begin{array}{l|l} y & = 2x + 1 \\ \hline 1 & ?\ 2 \cdot 0 + 1 \\ & 0 + 1 \\ & 1 \quad \text{TRUE} \end{array}$$

$$\begin{array}{l|l} y & = 2x + 1 \\ \hline -5 & ?\ 2(-3) + 1 \\ & -6 + 1 \\ & -5 \quad \text{TRUE} \end{array}$$

In each of the three cases, the substitution results in a true equation. Thus the pairs are all solutions.

1. Determine whether $(2, -4)$ is a solution of $4q - 3p = 22$.

2. Determine whether $(2, -4)$ is a solution of $7a + 5b = -6$.

Answers on page A-21

We plot the points as shown at right. The order of the points follows the alphabetical order of the variables. That is, x comes before y, so x-values are first coordinates and y-values are second coordinates. Similarly, we also label the horizontal axis as the x-axis and the vertical axis as the y-axis.

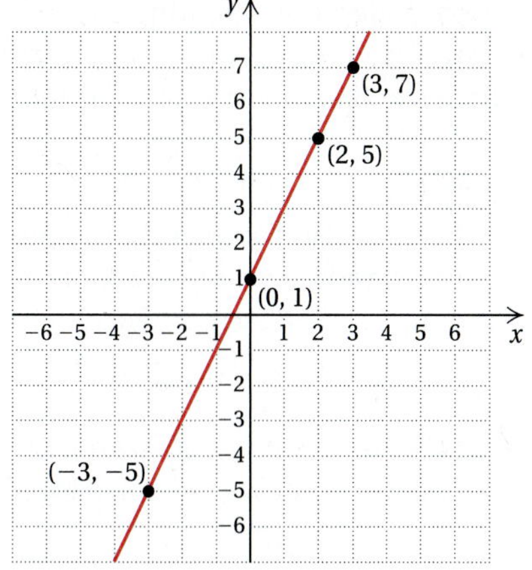

Note that the three points appear to "line up." That is, they appear to be on a straight line. Will other points that line up with these points also represent solutions of $y = 2x + 1$? To find out, we use a straightedge and lightly sketch a line passing through $(3, 7)$, $(0, 1)$, and $(-3, -5)$.

The line appears to also pass through $(2, 5)$. Let's see if this pair is a solution of $y = 2x + 1$:

$$\begin{array}{c|l} y & = 2x + 1 \\ \hline 5 & ?\ 2 \cdot 2 + 1 \\ & 4 + 1 \\ & 5 \qquad \text{TRUE} \end{array}$$

Thus, $(2, 5)$ is a solution.

Do Exercise 3.

Example 2 leads us to suspect that any point on the line that passes through $(3, 7)$, $(0, 1)$, and $(-3, -5)$ represents a solution of $y = 2x + 1$. In fact, every solution of $y = 2x + 1$ is represented by a point on that line and every point on that line represents a solution. The line is the *graph* of the equation.

GRAPH OF AN EQUATION

The **graph** of an equation is a drawing that represents all its solutions.

b Graphs of Linear Equations

Equations like $y = 2x + 1$ and $4q - 3p = 22$ are said to be **linear** because the graph of each equation is a straight line. In general, any equation equivalent to one of the form $y = mx + b$ or $Ax + By = C$, where m, b, A, B, and C are constants (not variables) and A and B are not both 0, is linear.

To graph a linear equation:

1. Select a value for one variable and calculate the corresponding value of the other variable. Form an ordered pair using alphabetical order as indicated by the variables.
2. Repeat step (1) to obtain at least two other ordered pairs. Two points are essential to determine a straight line. A third point serves as a check.
3. Plot the ordered pairs and draw a straight line passing through the points.

3. Use the graph in Example 2 to find at least two more points that are solutions of $y = 2x + 1$.

Answer on page A-21

Complete the table and graph.

4. $y = -2x$

x	y	(x, y)
-3		
-1		
0		
1		
3		

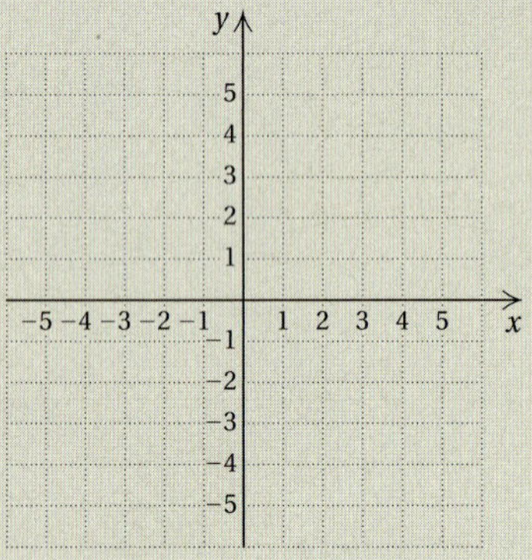

5. $y = \frac{1}{2}x$

x	y	(x, y)
4		
2		
0		
-2		
-4		
-1		

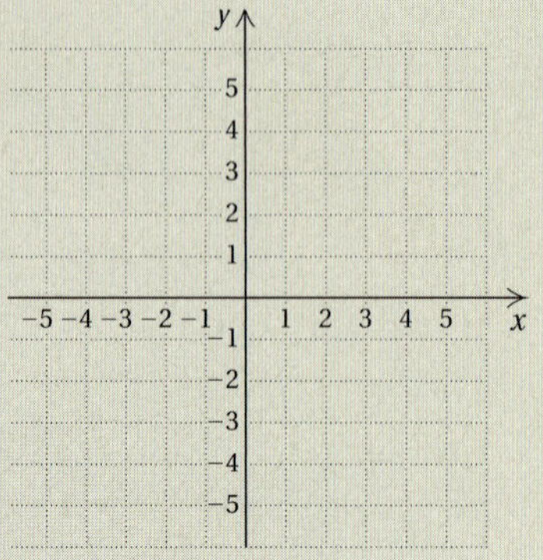

Answers on page A-21

In general, calculating three (or more) ordered pairs is not difficult for equations of the form $y = mx + b$. We simply substitute values for x and calculate the corresponding values for y.

EXAMPLE 3 Graph: $y = 2x$.

First, we find some ordered pairs that are solutions. We choose *any* number for x and then determine y by substitution. Since $y = 2x$, we find y by doubling x. Suppose that we choose 3 for x. Then

$$y = 2x = 2 \cdot 3 = 6.$$

We get a solution: the ordered pair $(3, 6)$.

Suppose that we choose 0 for x. Then

$$y = 2x = 2 \cdot 0 = 0.$$

We get another solution: the ordered pair $(0, 0)$.

For a third point, we make a negative choice for x. We now have enough points to plot the line, but if we wish, we can compute more. If a number takes us off the graph paper, we either do not use it or we use larger paper or rescale the axes. Continuing in this manner, we create a table like the one shown below.

Now we plot these points. We draw the line, or graph, with a straightedge and label it $y = 2x$.

x	y $y = 2x$	(x, y)
3	6	$(3, 6)$
1	2	$(1, 2)$
0	0	$(0, 0)$
-2	-4	$(-2, -4)$
-3	-6	$(-3, -6)$

(1) Choose x.
(2) Compute y.
(3) Form the pair (x, y).
(4) Plot the points.

CAUTION!

Keep in mind that you can choose *any* number for x and then compute y. Our choice of certain numbers in the examples does not dictate the ones you can choose.

Do Exercises 4 and 5.

EXAMPLE 4 Graph: $y = -3x + 1$.

We select a value for x, compute y, and form an ordered pair. Then we repeat the process for other choices of x.

If $x = 2$, then $y = -3 \cdot 2 + 1 = -5$, and $(2, -5)$ is a solution.
If $x = 0$, then $y = -3 \cdot 0 + 1 = 1$, and $(0, 1)$ is a solution.
If $x = -1$, then $y = -3 \cdot (-1) + 1 = 4$, and $(-1, 4)$ is a solution.

Results are often listed in a table, as shown below. The points corresponding to each pair are then plotted.

x	y $y = -3x + 1$	(x, y)
2	−5	(2, −5)
0	1	(0, 1)
−1	4	(−1, 4)

(1) Choose x.
(2) Compute y.
(3) Form the pair (x, y).
(4) Plot the points.

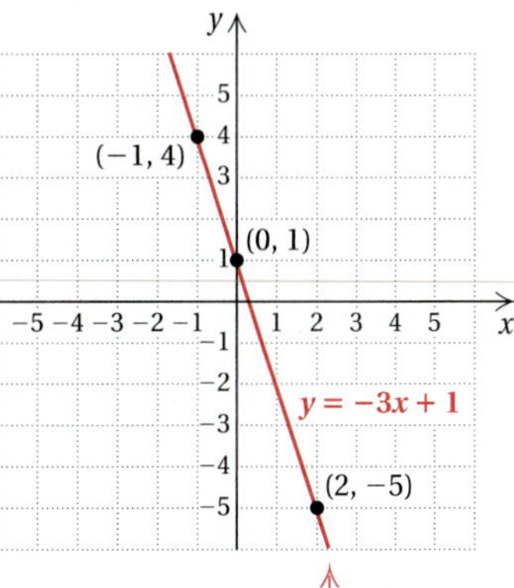

Note that all three points line up. If they did not, we would know that we had made a mistake. When only two points are plotted, a mistake is harder to detect. We use a ruler or other straightedge to draw a line through the points. Every point on the line represents a solution of $y = -3x + 1$.

Do Exercises 6 and 7.

In Example 3, we saw that $(0, 0)$ is a solution of $y = 2x$. It is also the point at which the graph crosses the y-axis. Similarly, in Example 4, we saw that $(0, 1)$ is a solution of $y = -3x + 1$. It is also the point at which the graph crosses the y-axis. A generalization can be made: If x is replaced with 0 in the equation $y = mx + b$, then the corresponding y-value is $m \cdot 0 + b$, or b. Thus any equation of the form $y = mx + b$ has a graph that passes through the point $(0, b)$. Since $(0, b)$ is the point at which the graph crosses the y-axis, it is called the ***y*-intercept.** Sometimes, for convenience, we simply refer to b as the y-intercept.

y-INTERCEPT

The graph of the equation $y = mx + b$ passes through the ***y*-intercept** $(0, b)$.

Graph.

6. $y = 2x + 3$

x	y	(x, y)

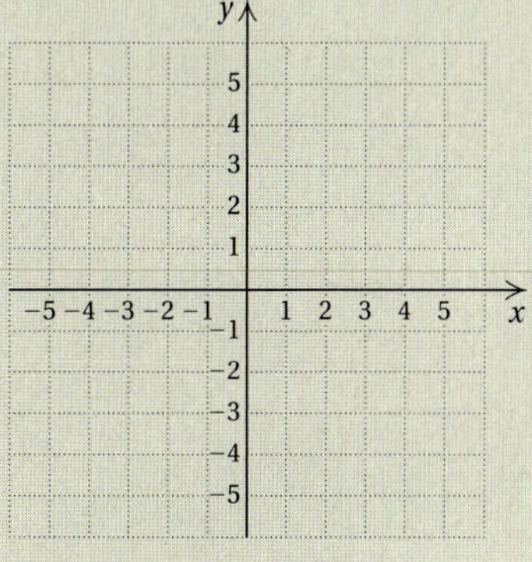

7. $y = -\frac{1}{2}x - 3$

x	y	(x, y)

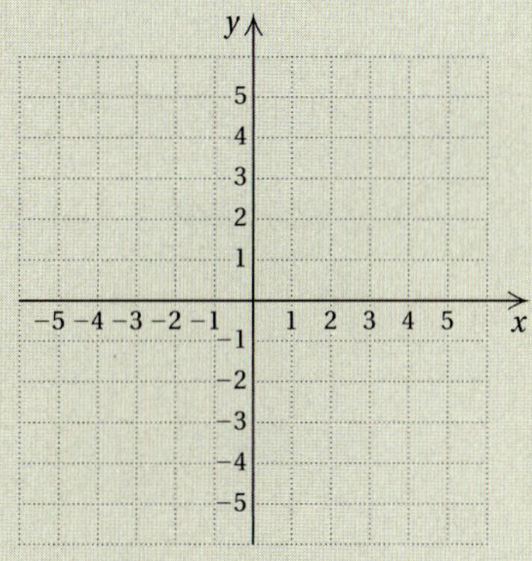

Answers on page A-21

CALCULATOR CORNER

Finding Solutions of Equations A table of values representing ordered pairs that are solutions of an equation can be displayed on a graphing calculator. To do this for the equation in Example 4, $y = -3x + 1$, we first press [Y =] to access the equation-editor screen. Then we clear any equations that are present. (See the Calculator Corner in Section 7.3 for the procedure for doing this.) Next, we enter the equation by positioning the cursor beside "Y1 =" and pressing [(–)] [3] [X, T, θ, n] [+] [1]. Now we press [2nd] [TblSet] to display the table set-up screen. (TblSet is the second function associated with the [WINDOW] key.) You can choose to supply the x-values yourself or you can set the calculator to supply them. To supply them yourself, follow the procedure for selecting ASK mode on p. 532. To have the calculator supply the x-values, set "Indpnt" to "Auto" by positioning the cursor over "Auto" and pressing [ENTER]. "Depend" should also be set to "Auto."

When "Indpnt" is set to "Auto," the graphing calculator will supply values of x, beginning with the value specified as TblStart and continuing by adding the value of △Tbl to the preceding value for x. Below, we show a table of values that starts with $x = -2$ and adds 1 to the preceding x-value. We press [(–)] [2] [▽] [1] or [(–)] [2] [ENTER] [1] to select a minimum x-value of -2 and an increment of 1. To display the table, we press [2nd] [TABLE]. (TABLE is the second operation associated with the [GRAPH] key.)

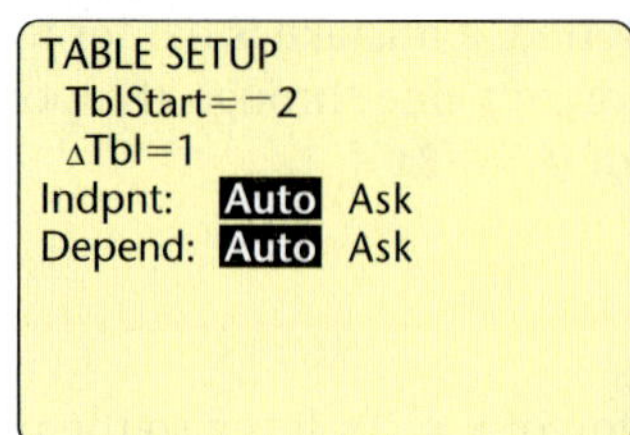

X	Y1
−2	7
−1	4
0	1
1	−2
2	−5
3	−8
4	−11

X = −2

We can use the [△] and [▽] keys to scroll up and down through the table to see other solutions of the equation.

Exercise:

1. Create a table of ordered pairs that are solutions of the equations in Examples 3 and 5.

EXAMPLE 5 Graph $y = \frac{2}{5}x + 4$ and identify the y-intercept.

We select a value for x, compute y, and form an ordered pair. Then we repeat the process for other choices of x. In this case, using multiples of 5 avoids fractions. We try to avoid graphing ordered pairs with fractions because they are difficult to graph accurately.

If $x = 0$, then $y = \frac{2}{5} \cdot 0 + 4 = 4$, and $(0, 4)$ is a solution.

If $x = 5$, then $y = \frac{2}{5} \cdot 5 + 4 = 6$, and $(5, 6)$ is a solution.

If $x = -5$, then $y = \frac{2}{5} \cdot (-5) + 4 = 2$, and $(-5, 2)$ is a solution.

The following table lists these solutions. Next, we plot the points and see that they form a line. Finally, we draw and label the line.

x	y $y = \frac{2}{5}x + 4$	(x, y)
0	4	$(0, 4)$
5	6	$(5, 6)$
−5	2	$(-5, 2)$

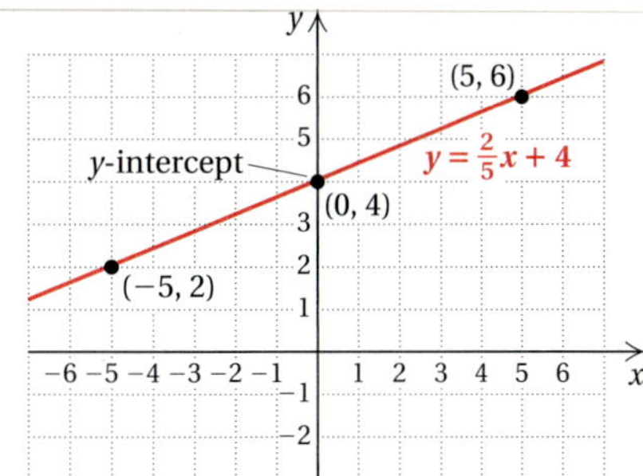

We see that $(0, 4)$ is a solution of $y = \frac{2}{5}x + 4$. It is the y-intercept. Because the equation is in the form $y = mx + b$, we can read the y-intercept directly from the equation as follows:

$$y = \frac{2}{5}x + 4 \qquad (0, 4) \text{ is the } y\text{-intercept.}$$

Do Exercises 8 and 9.

Calculating ordered pairs is generally easiest when y is isolated on one side of the equation, as in $y = mx + b$. To graph an equation in which y is not isolated, we can use the addition and multiplication principles to solve for y (see Section 7.3).

EXAMPLE 6 Graph $3y + 5x = 0$ and identify the y-intercept.

To find an equivalent equation in the form $y = mx + b$, we solve for y:

$$3y + 5x = 0$$

$$3y + 5x - 5x = 0 - 5x \qquad \text{Subtracting } 5x$$

$$3y = -5x \qquad \text{Collecting like terms}$$

$$\frac{3y}{3} = \frac{-5x}{3} \qquad \text{Dividing by 3}$$

$$y = -\frac{5}{3}x.$$

Graph the equation and identify the y-intercept.

8. $y = \frac{3}{5}x + 2$

x	y	(x, y)

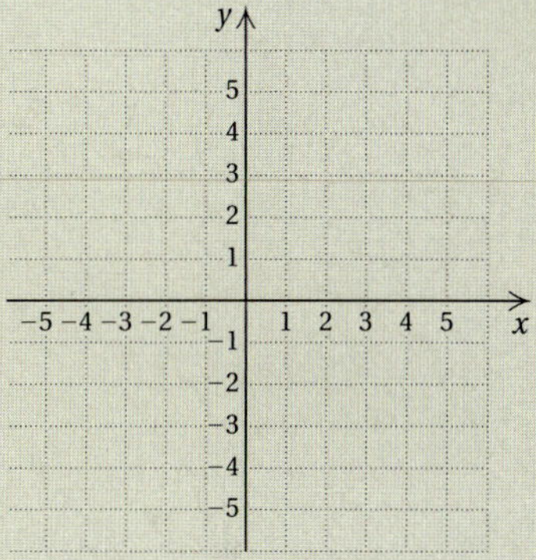

9. $y = -\frac{3}{5}x - 1$

x	y	(x, y)

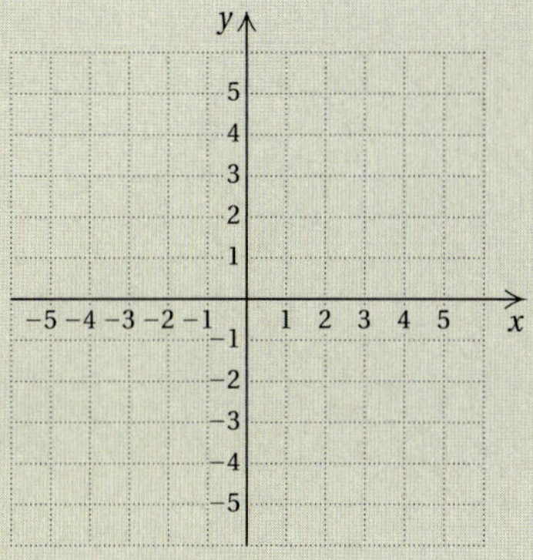

Answers on page A-21

Graph the equation and identify the y-intercept.

10. $5y + 4x = 0$

x	y	
0		← y-intercept

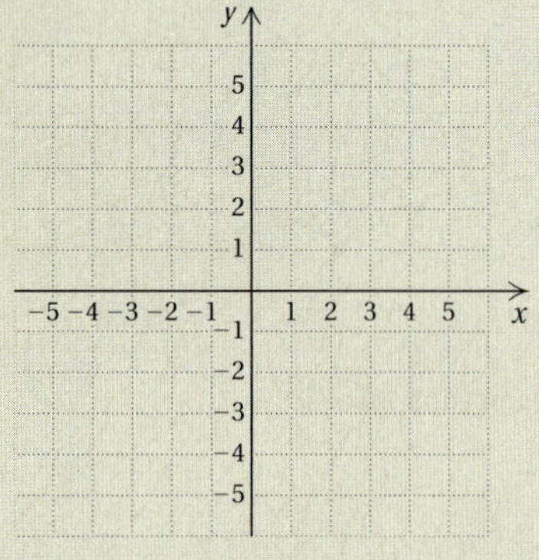

11. $4y = 3x$

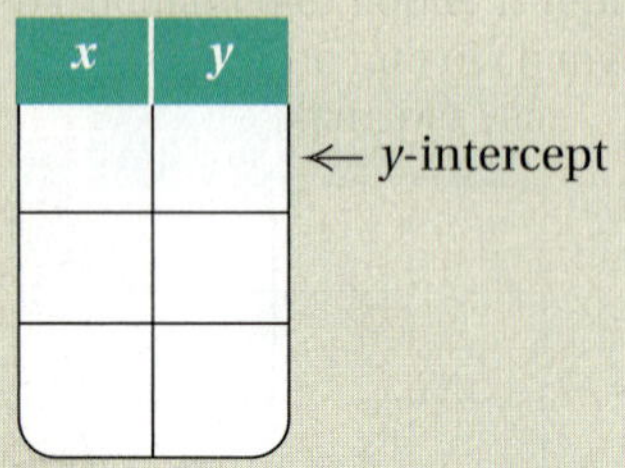

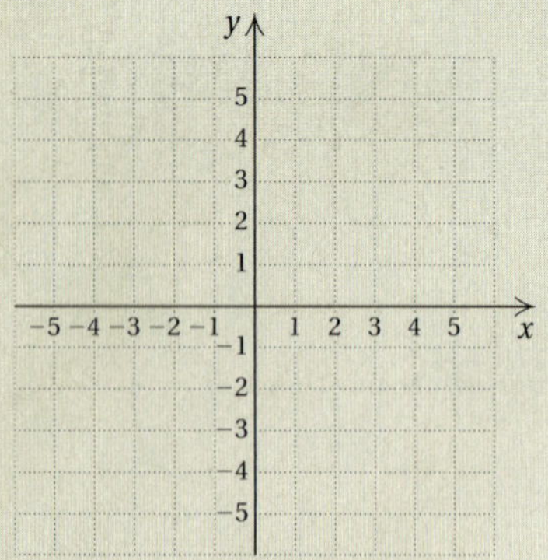

Answers on page A-21

Because all the equations above are equivalent, we can use $y = -\frac{5}{3}x$ to draw the graph of $3y + 5x = 0$. To graph $y = -\frac{5}{3}x$, we select x-values and compute y-values. In this case, if we select multiples of 3, we can avoid fractions.

$$\text{If } x = 0, \quad \text{then } y = -\frac{5}{3} \cdot 0 = 0.$$

$$\text{If } x = 3, \quad \text{then } y = -\frac{5}{3} \cdot 3 = -5.$$

$$\text{If } x = -3, \quad \text{then } y = -\frac{5}{3} \cdot (-3) = 5.$$

We list these solutions in a table. Next, we plot the points and see that they form a line. Finally, we draw and label the line. The y-intercept is $(0, 0)$.

x	y
0	0
3	−5
−3	5

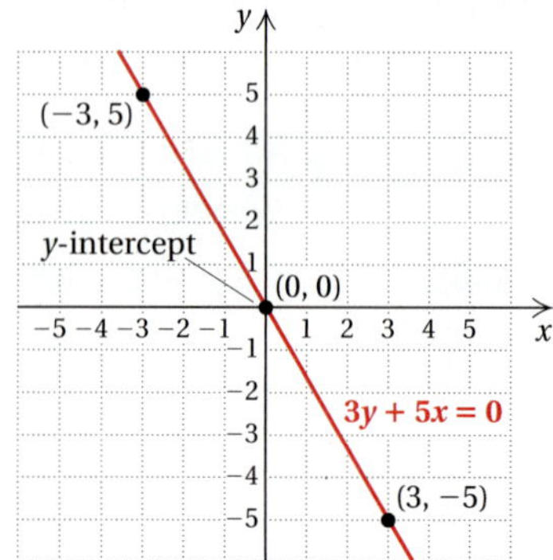

Do Exercises 10 and 11.

EXAMPLE 7 Graph $4y + 3x = -8$ and identify the y-intercept.

To find an equivalent equation in the form $y = mx + b$, we solve for y:

$$4y + 3x = -8$$

$$4y + 3x - 3x = -8 - 3x \qquad \text{Subtracting } 3x$$

$$4y = -3x - 8 \qquad \text{Simplifying}$$

$$\frac{1}{4} \cdot 4y = \frac{1}{4} \cdot (-3x - 8) \qquad \text{Multiplying by } \tfrac{1}{4} \text{ or dividing by 4}$$

$$y = \frac{1}{4} \cdot (-3x) - \frac{1}{4} \cdot 8 \qquad \text{Using the distributive law}$$

$$y = -\frac{3}{4}x - 2. \qquad \text{Simplifying}$$

Thus, $4y + 3x = -8$ is equivalent to $y = -\frac{3}{4}x - 2$. The y-intercept is $(0, -2)$. We find two other pairs using multiples of 4 for x to avoid fractions. We then complete and label the graph as shown.

x	y
0	−2
4	−5
−4	1

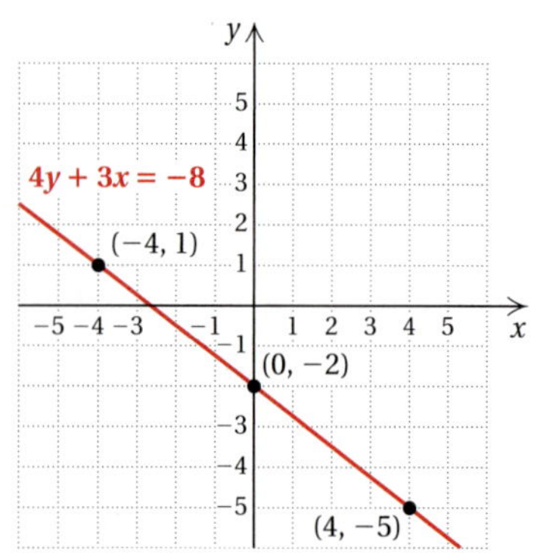

Do Exercises 12 and 13.

C Applications of Linear Equations

Mathematical concepts become more understandable through visualization. Throughout this text, you will occasionally see the heading AG Algebraic–Graphical Connection, as in Example 8, which follows. In this feature, the algebraic approach is enhanced and expanded with a graphical connection. Relating a solution of an equation to a graph can often give added meaning to the algebraic solution.

EXAMPLE 8 *FedEx Mailing Costs.* The cost y, in dollars, of shipping a FedEx Priority Overnight package weighing 1 lb or more a distance of 1001 to 1400 mi is given by

$$y = 2.8x + 21.05,$$

where x is the number of pounds.
Source: Federal Express Corporation

a) Find the cost of shipping packages weighing 2 lb, 5 lb, and 7 lb.
b) Graph the equation and then use the graph to estimate the cost of shipping a $6\frac{1}{2}$-lb package.
c) If a package costs \$239.45 to ship, how much does it weigh?

We solve as follows.

a) We substitute 2, 5, and 7 for x and then calculate y:

$$y = 2.8(2) + 21.05 = \$26.65;$$
$$y = 2.8(5) + 21.05 = \$35.05;$$
$$y = 2.8(7) + 21.05 = \$40.65.$$

It costs \$26.65, \$35.05, and \$40.65 to ship packages that weigh 2 lb, 5 lb, and 7 lb, respectively.

Graph the equation and identify the y-intercept.

12. $5y - 3x = -10$

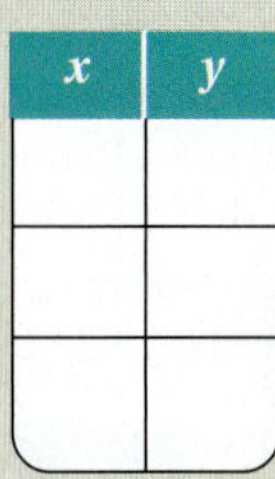

x	y

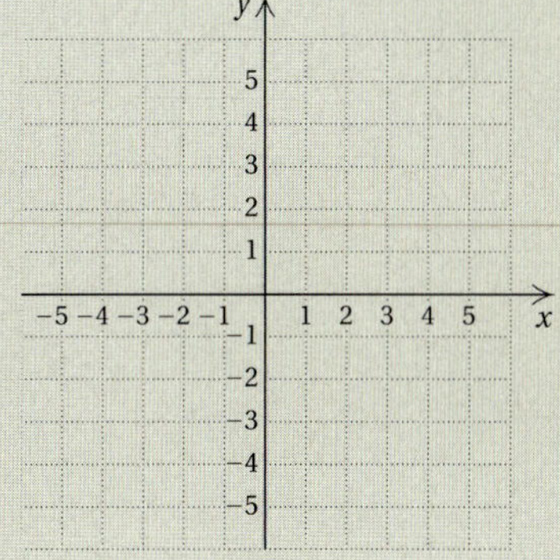

13. $5y + 3x = 20$

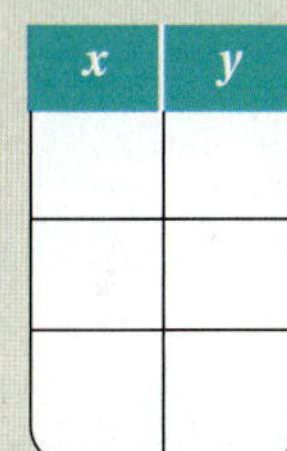

x	y

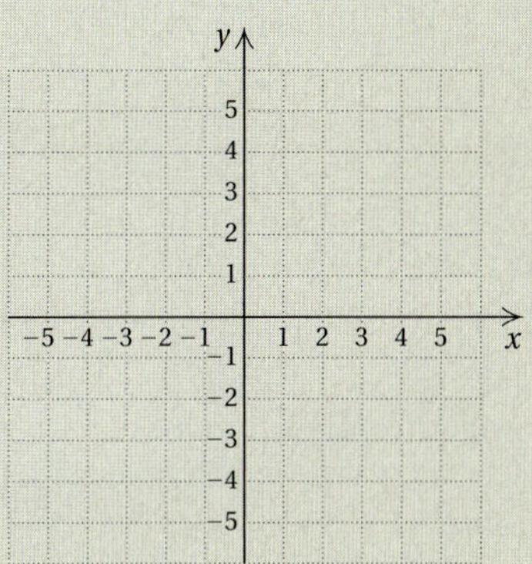

Answers on page A-21

14. Value of a Color Copier. The value of Dupliographic's color copier is given by

$$v = -0.68t + 3.4,$$

where v is the value, in thousands of dollars, t years from the date of purchase.

a) Find the value after 1 yr, 2 yr, 4 yr, and 5 yr.

t	v
1	
2	
4	
5	

b) Graph the equation and use the graph to estimate the value of the copier after $2\frac{1}{2}$ yr.

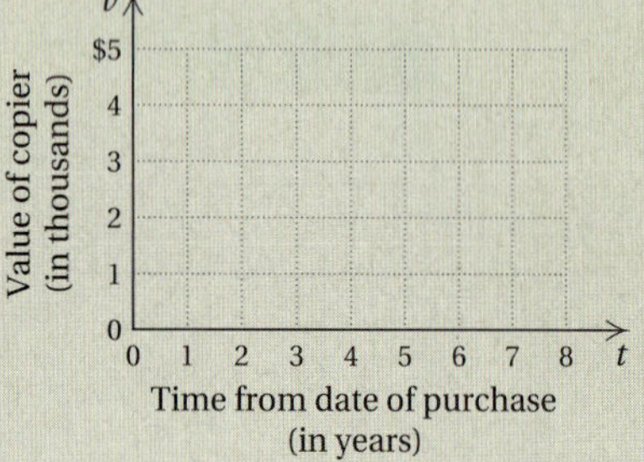

c) After what amount of time is the value of the copier \$1500?

Answers on page A-21

ALGEBRAIC–GRAPHICAL CONNECTION

b) We have three ordered pairs from part (a). We plot these points and see that they line up. Thus our calculations are probably correct. Since this formula, $y = 2.8x + 21.05$, gives the cost of mailing packages that weigh 1 lb or more, we begin at $(1, 23.85)$ when drawing the graph.

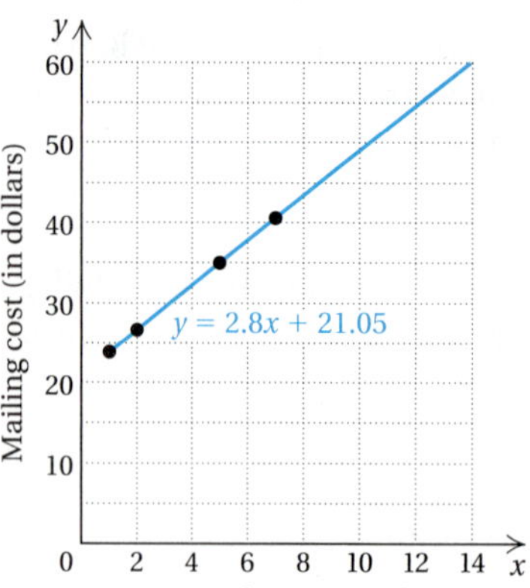

To estimate the cost of shipping a $6\frac{1}{2}$-lb package, we need to determine what y-value is paired with $x = 6\frac{1}{2}$. We locate the point on the line that is above $6\frac{1}{2}$ and then find the value on the y-axis that corresponds to that point. It appears that the cost of shipping a $6\frac{1}{2}$-lb package is about \$39.

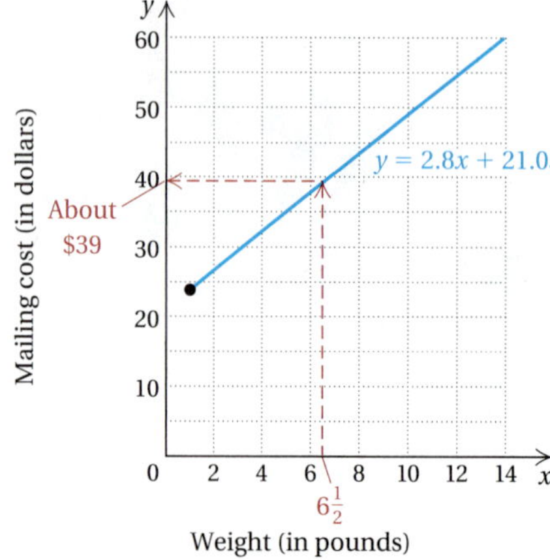

To obtain a more accurate cost, we can simply substitute into the equation:

$$y = 2.8(6.5) + 21.05 = \$39.25.$$

c) We substitute \$239.45 for y and then solve for x:

$$y = 2.8x + 21.05$$
$$239.45 = 2.8x + 21.05 \quad \text{Substituting}$$
$$218.40 = 2.8x \quad \text{Subtracting 21.05}$$
$$78 = x. \quad \text{Dividing by 2.8}$$

A package that costs \$239.45 to ship weighs 78 lb.

Do Exercise 14.

Many equations in two variables have graphs that are not straight lines. Three such graphs are shown below. As before, each graph represents the solutions of the given equation. We are not going to develop methods of doing such graphing at this time, although such *nonlinear graphs* can be created very easily using a graphing calculator. We will cover such graphs in the optional Calculator Corners throughout the text and in Chapter 16.

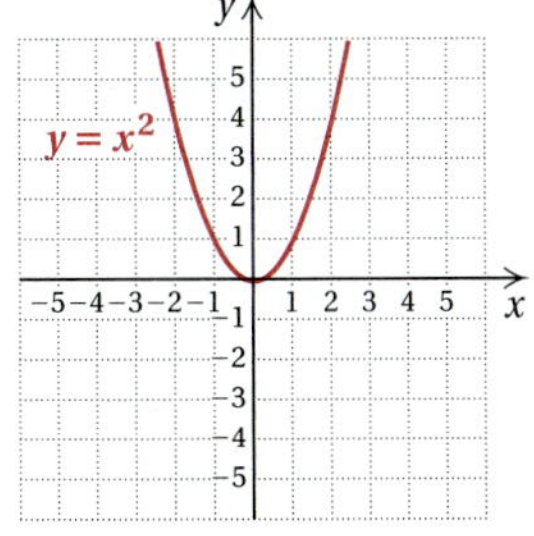

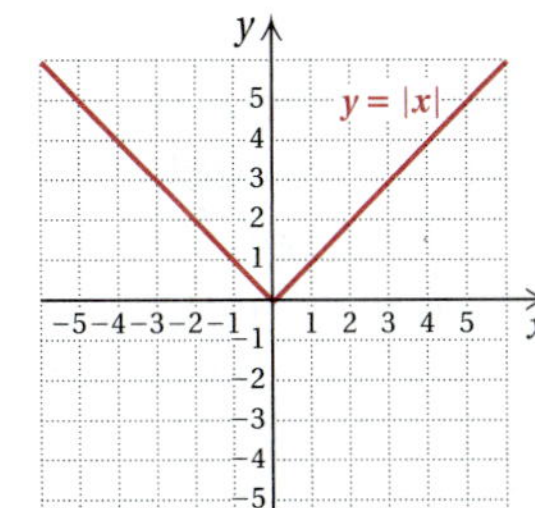

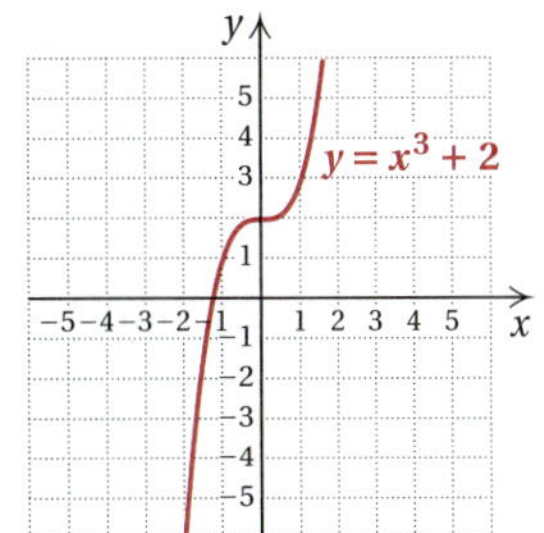

Study Tips

BETTER TEST TAKING

How often do you make the following statement after taking a test: "I was able to do the homework, but I froze during the test"? This can be an excuse for poor study habits. Here are two tips to help you with this difficulty. Both are intended to make test taking less stressful by getting you to practice good test-taking habits on a daily basis.

- **Treat every homework exercise as if it were a test question.** If you had to work a problem at your job with no backup answer provided, what would you do? You would probably work it very deliberately, checking and rechecking every step. You might work it more than one time, or you might try to work it another way to check the result. Try to use this approach when doing your homework. Treat every exercise as though it were a test question with no answer at the back of the book.
- **Be sure that you do questions without answers as part of every homework assignment whether or not the instructor has assigned them!** One reason a test may seem such a different task is that questions on a test lack answers. That is the reason for taking a test: to see if you can do the questions without assistance. As part of your test preparation, be sure you do some exercises for which you do not have the answers. Thus when you take a test, you are doing a more familiar task.

The purpose of doing your homework using these approaches is to give you more test-taking practice beforehand. Let's use a sports analogy: At a basketball game, the players take lots of practice shots before the game. They play the first half, go to the locker room, and come out for the second half. What do they do before the second half, even though they have just played 20 minutes of basketball? They shoot baskets again! We suggest the same approach here. Create more and more situations in which you practice taking test questions by treating each homework exercise like a test question and by doing exercises for which you have no answers. Good luck!

"He who does not venture has no luck."

Mexican proverb

CALCULATOR CORNER

Graphing Equations Graphs of equations are displayed in the **viewing window** of a graphing calculator. The viewing window is the portion of the coordinate plane that appears on the calculator's screen. It is defined by the minimum and maximum values of x and y: Xmin, Xmax, Ymin, and Ymax. The notation [Xmin, Xmax, Ymin, Ymax] is used to represent these window settings or dimensions. For example, $[-12, 12, -8, 8]$ denotes a window that displays the portion of the x-axis from -12 to 12 and the portion of the y-axis from -8 to 8. In addition, the distance between tick marks on the axes is defined by the settings Xscl and Yscl. The Xres setting indicates the pixel resolution. We usually select Xres = 1. The window corresponding to the settings $[-20, 30, -12, 20]$, Xscl = 5, Yscl = 2, Xres = 1, is shown on the left below. Press [WINDOW] on the top row of the keypad of your calculator to display the current window settings. The settings for the **standard viewing window** are shown on the right below.

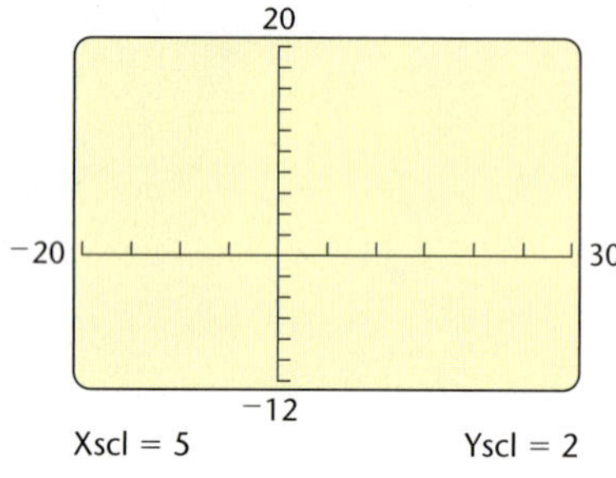

Xscl = 5 Yscl = 2

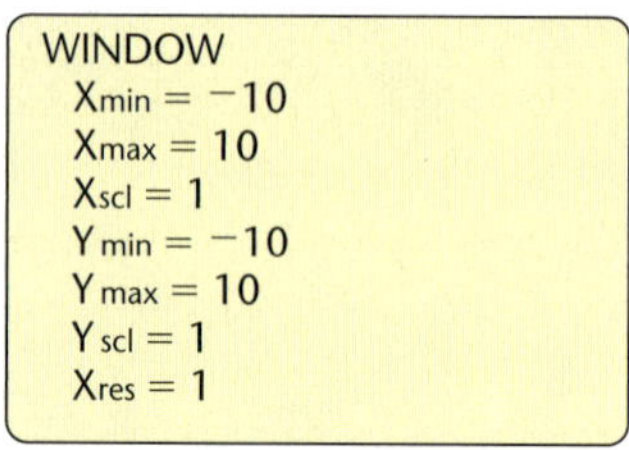
WINDOW
Xmin = −10
Xmax = 10
Xscl = 1
Ymin = −10
Ymax = 10
Yscl = 1
Xres = 1

To change a setting, we position the cursor beside the setting we wish to change and enter the new value. For example, to change from the standard settings to $[-20, 30, -12, 20]$, Xscl = 5, Yscl = 2, on the WINDOW screen, we press [(−)] [2] [0] [ENTER] [3] [0] [ENTER] [5] [ENTER] [(−)] [1] [2] [ENTER] [2] [0] [ENTER] [2] [ENTER]. The [▽] key can be used instead of [ENTER] after typing each window setting. To see the window, we press [GRAPH] on the top row of the keypad. To return quickly to the standard window setting $[-10, 10, -10, 10]$, Xscl = 1, Yscl = 1, we press [ZOOM] [6].

Equations must be solved for y before they can be graphed on the TI-83 Plus. Consider the equation $3x + 2y = 6$. Solving for y, we get $y = \dfrac{6 - 3x}{2}$. We enter this equation as $y_1 = (6 - 3x)/2$ on the equation-editor screen as described in the Calculator Corner in Section 7.3 (see p. 532). Then we press [ZOOM] [6] to select the standard viewing window and display the graph.

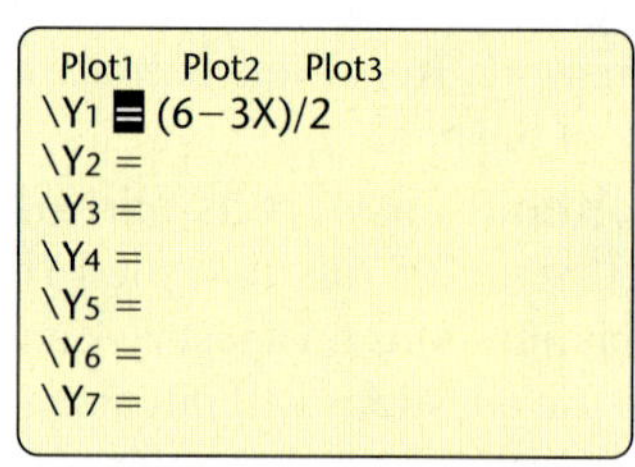
Plot1 Plot2 Plot3
\Y1 = (6−3X)/2
\Y2 =
\Y3 =
\Y4 =
\Y5 =
\Y6 =
\Y7 =

$y = (6 - 3x)/2$

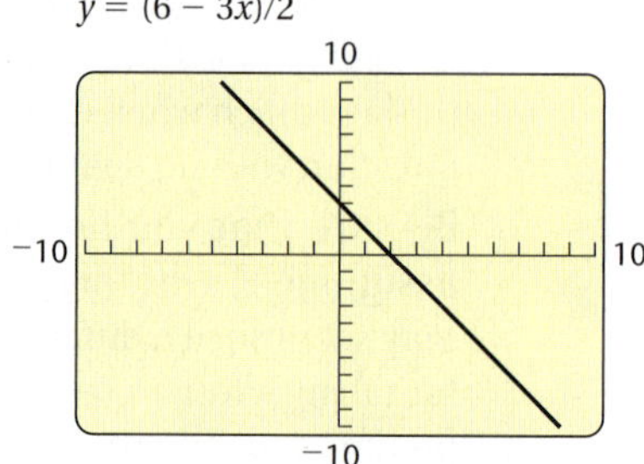

Exercises: Graph each equation in the standard viewing window $[-10, 10, -10, 10]$, Xscl = 1, Yscl = 1.

1. $y = 2x + 1$
2. $y = -3x + 1$
3. $y = -5x + 3$
4. $y = 4x - 5$
5. $y = \dfrac{4}{5}x + 2$
6. $y = -\dfrac{3}{5}x - 1$
7. $y = 2.085x + 5.08$
8. $y = -3.45x - 1.68$

8.2 EXERCISE SET

For Extra Help

Digital Video Tutor CD 6 Videotape 8 | InterAct Math | Math Tutor Center | MathXL | MyMathLab

a Determine whether the given ordered pair is a solution of the equation.

1. $(2, 9)$; $y = 3x - 1$

2. $(1, 7)$; $y = 2x + 5$

3. $(4, 2)$; $2x + 3y = 12$

4. $(0, 5)$; $5x - 3y = 15$

5. $(3, -1)$; $3a - 4b = 13$

6. $(-5, 1)$; $2p - 3q = -13$

In Exercises 7–12, an equation and two ordered pairs are given. Show that each pair is a solution. Then use the graph of the two points to determine another solution. Answers may vary.

7. $y = x - 5$; $(4, -1)$ and $(1, -4)$

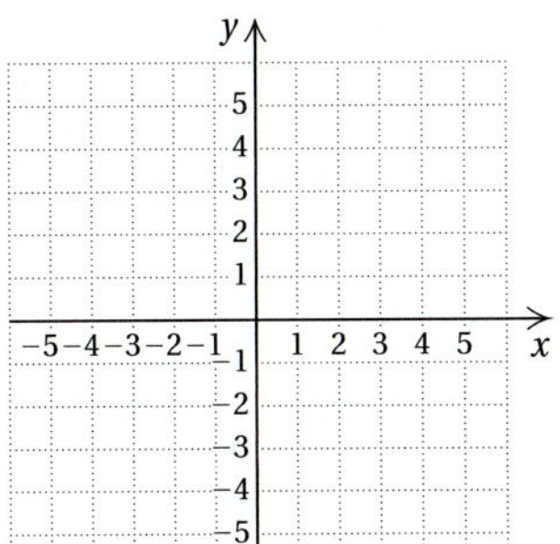

8. $y = x + 3$; $(-1, 2)$ and $(3, 6)$

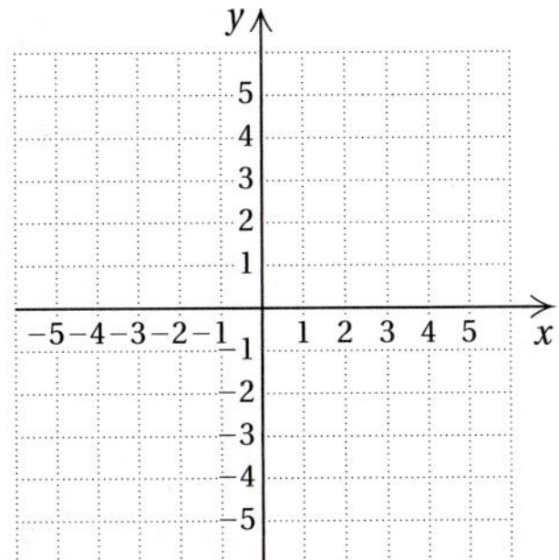

9. $y = \frac{1}{2}x + 3$; $(4, 5)$ and $(-2, 2)$

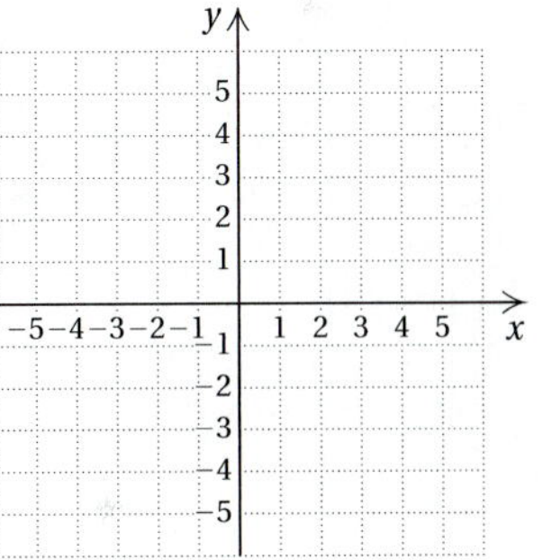

10. $3x + y = 7$; $(2, 1)$ and $(4, -5)$

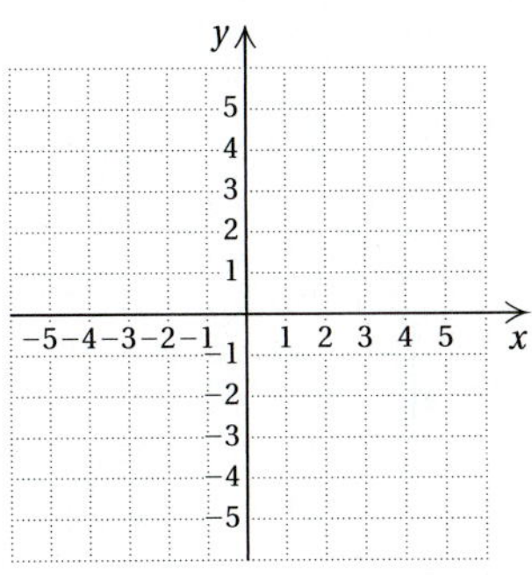

11. $4x - 2y = 10$; $(0, -5)$ and $(4, 3)$

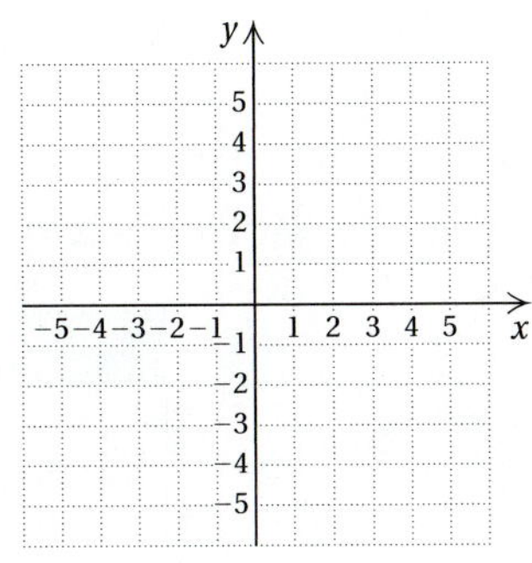

12. $6x - 3y = 3$; $(1, 1)$ and $(-1, -3)$

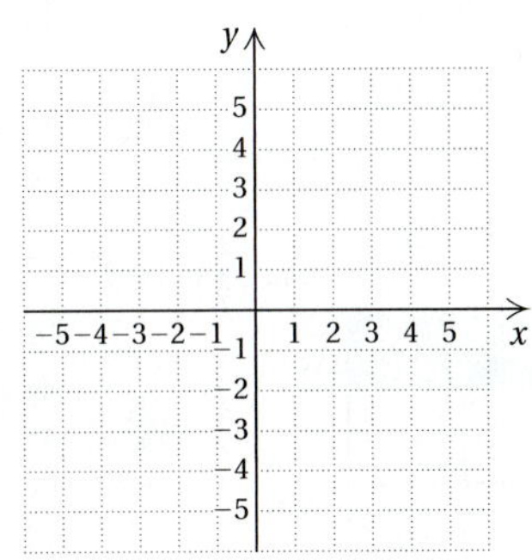

b Graph the equation and identify the y-intercept.

13. $y = x + 1$

x	y
-2	
-1	
0	
1	
2	
3	

14. $y = x - 1$

x	y
-2	
-1	
0	
1	
2	
3	

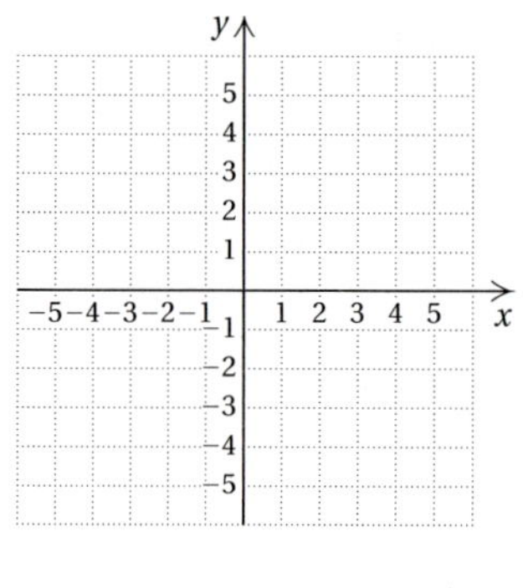

15. $y = x$

x	y
-2	
-1	
0	
1	
2	
3	

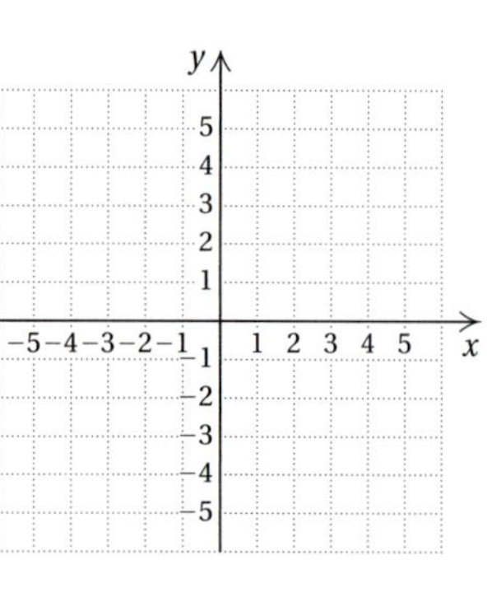

16. $y = -x$

x	y
-2	
-1	
0	
1	
2	
3	

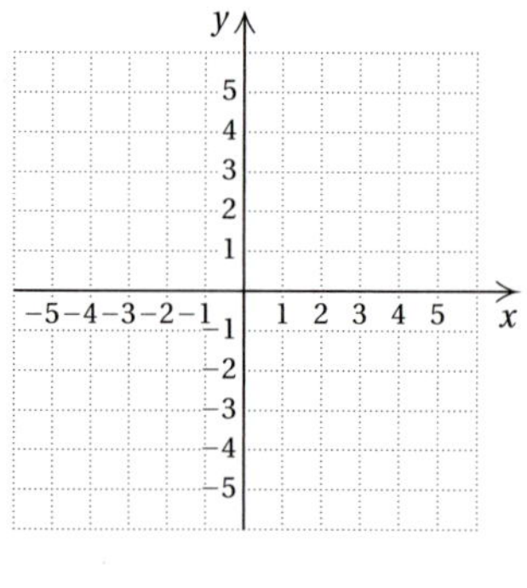

17. $y = \frac{1}{2}x$

x	y
-2	
0	
4	

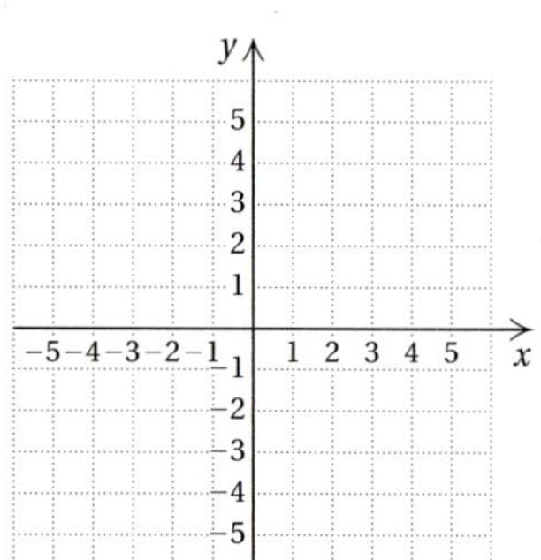

18. $y = \frac{1}{3}x$

x	y
-6	
0	
3	

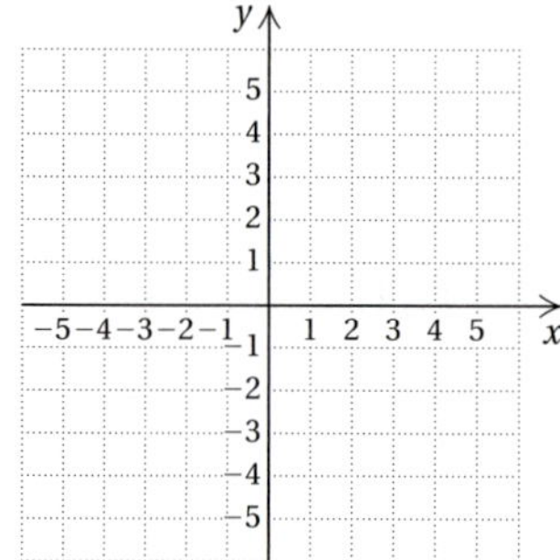

19. $y = x - 3$

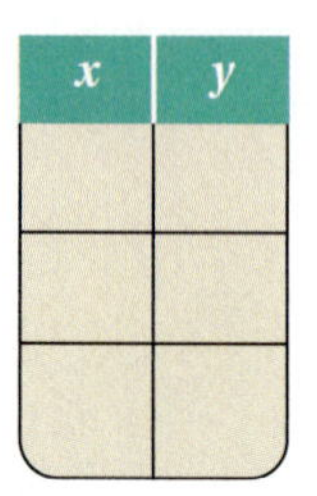

x	y

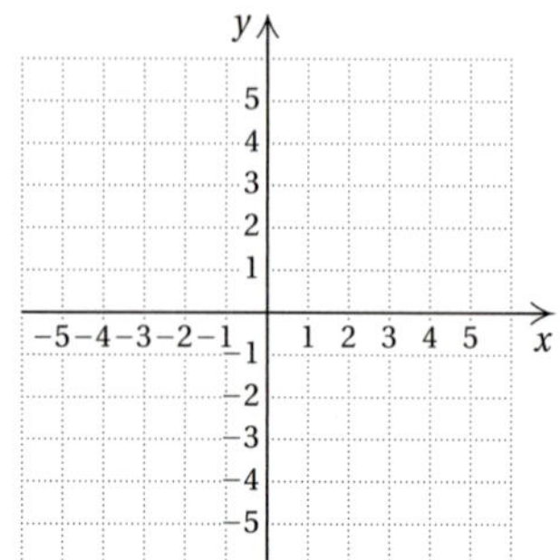

20. $y = x + 3$

x	y

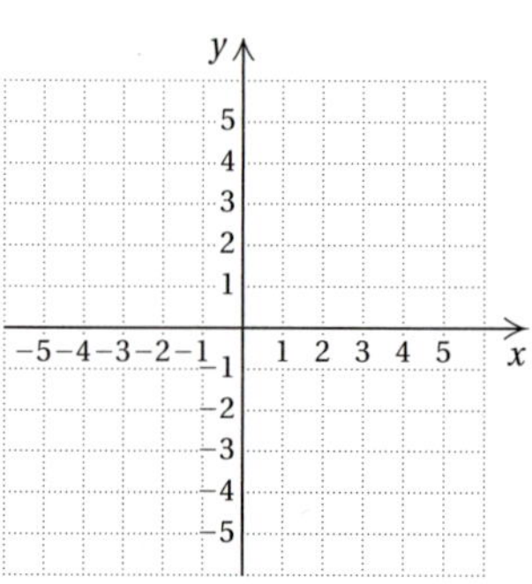

21. $y = 3x - 2$

x	y

22. $y = 2x + 2$

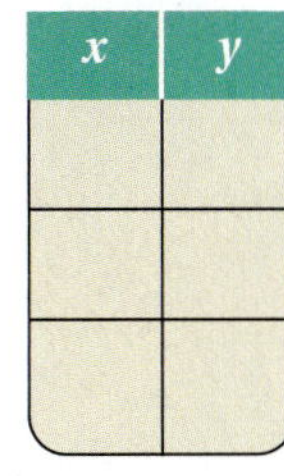

x	y

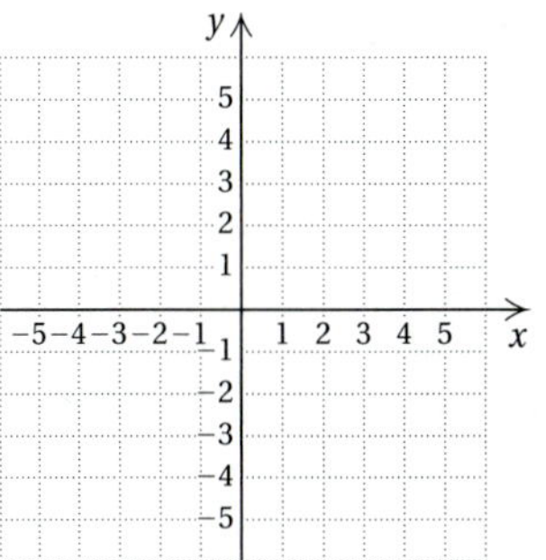

23. $y = \frac{1}{2}x + 1$

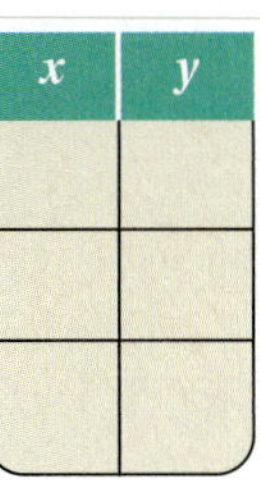

x	y

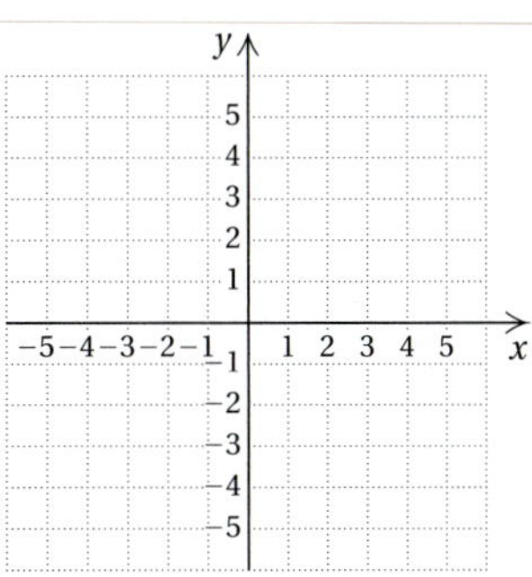

24. $y = \frac{1}{3}x - 4$

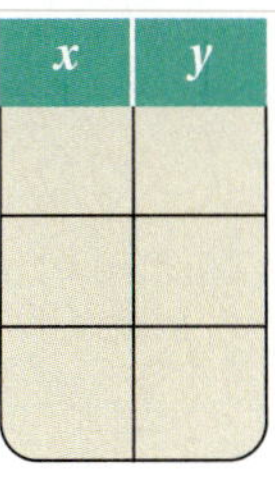

x	y

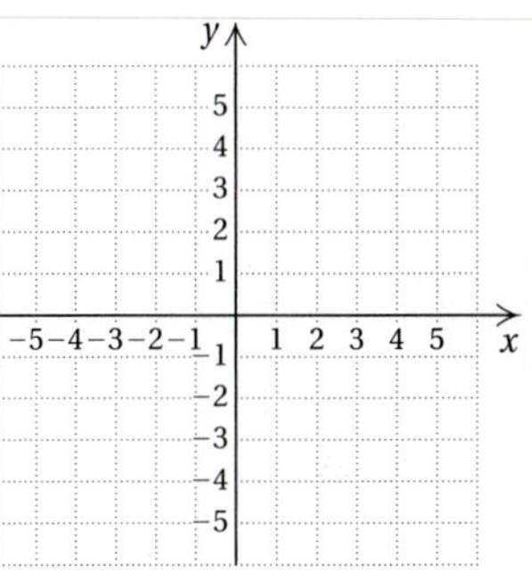

25. $x + y = -5$

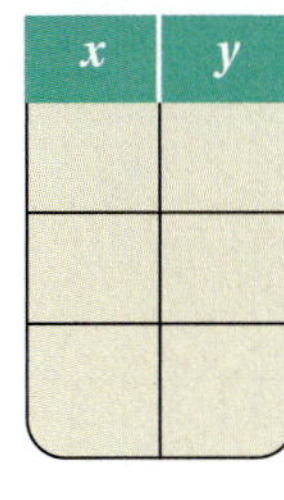

x	y

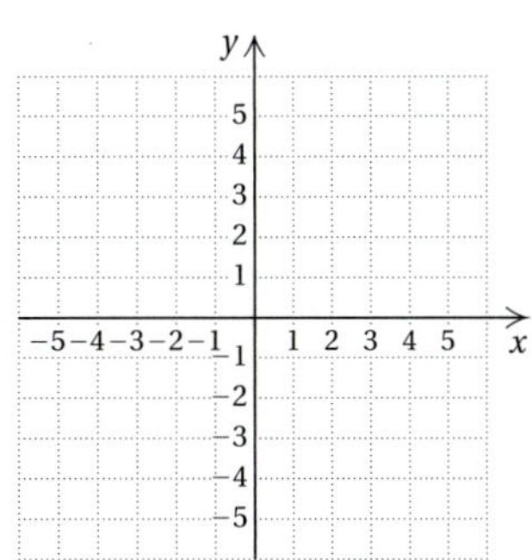

26. $x + y = 4$

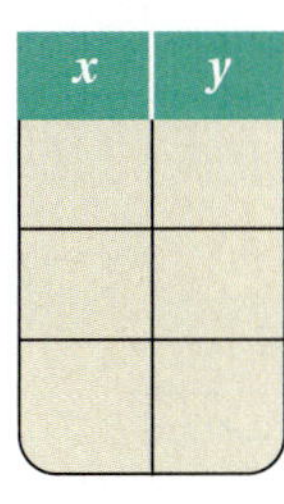

x	y

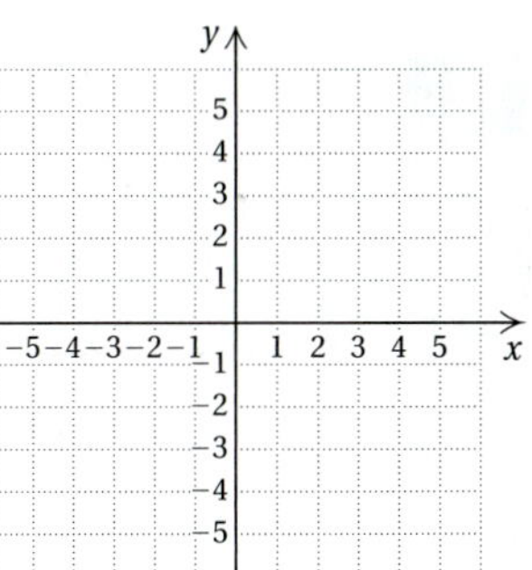

27. $y = \frac{5}{3}x - 2$

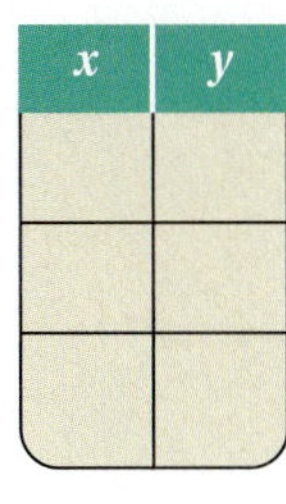

x	y

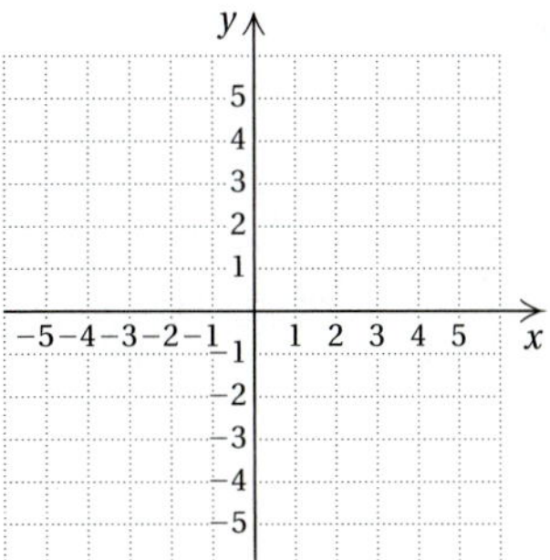

28. $y = \frac{5}{2}x + 3$

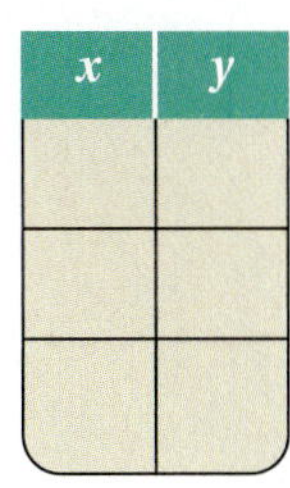

x	y

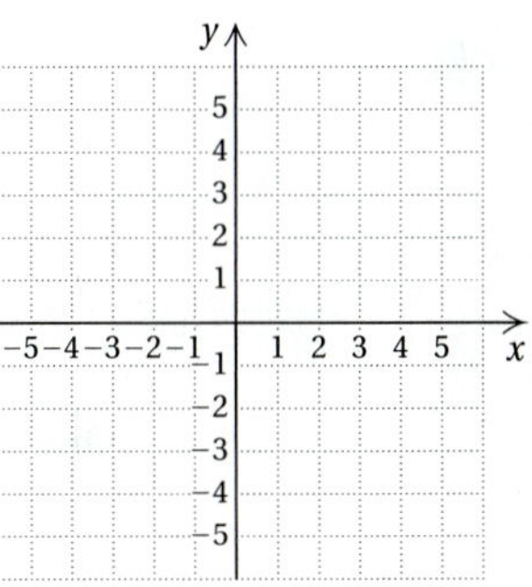

29. $x + 2y = 8$

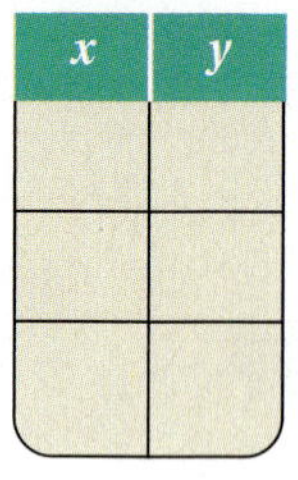

30. $x + 2y = -6$

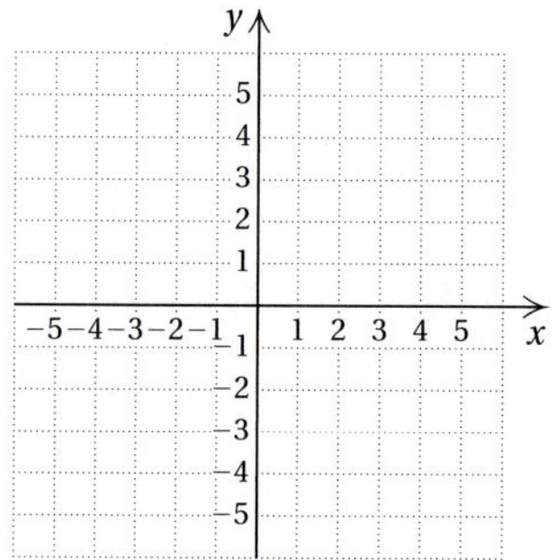

31. $y = \frac{3}{2}x + 1$

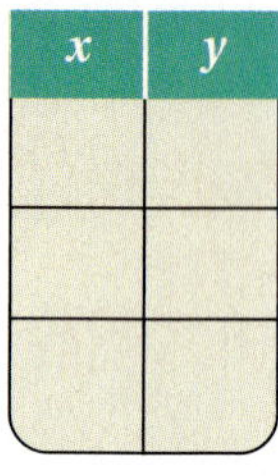

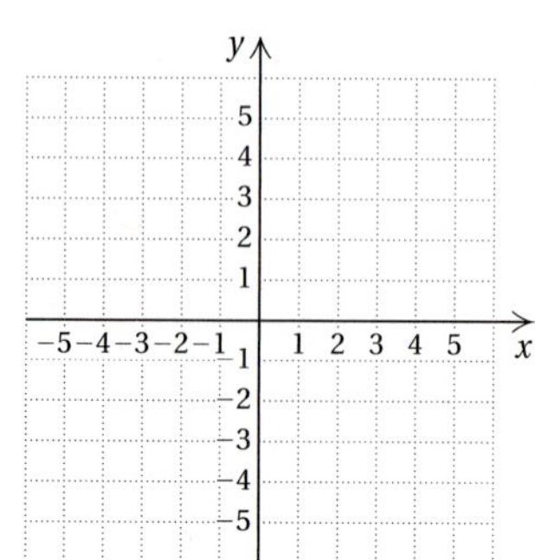

32. $y = -\frac{1}{2}x - 3$

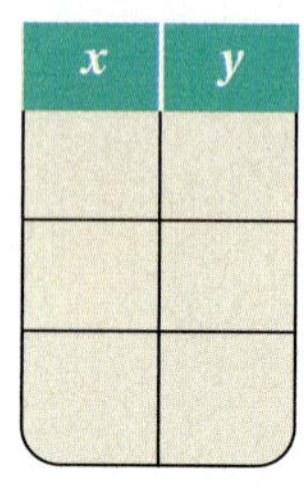

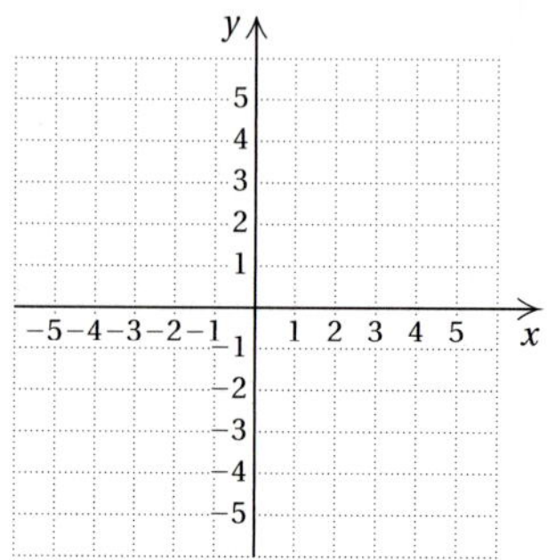

33. $8x - 2y = -10$

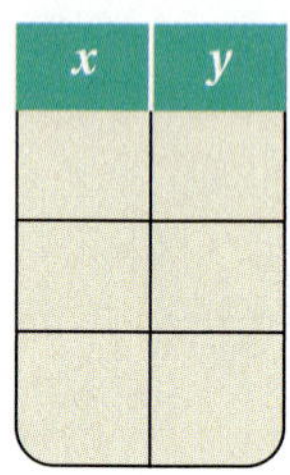

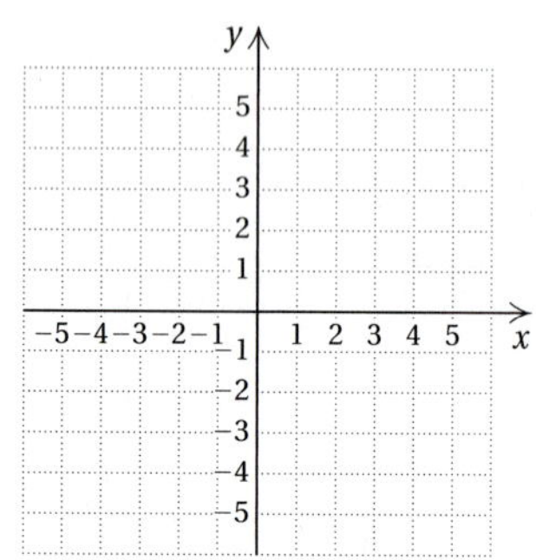

34. $6x - 3y = 9$

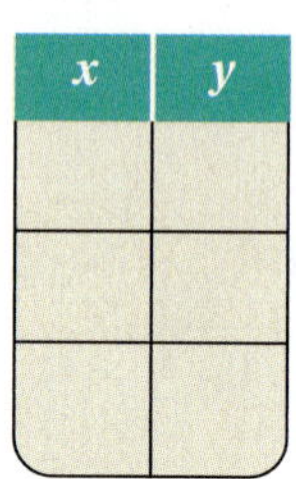

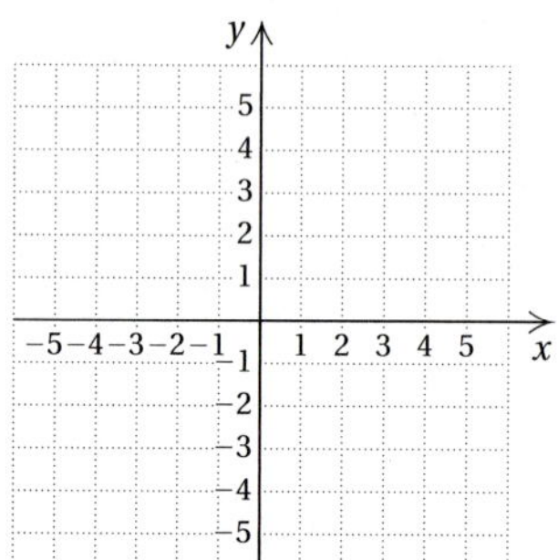

35. $8y + 2x = -4$

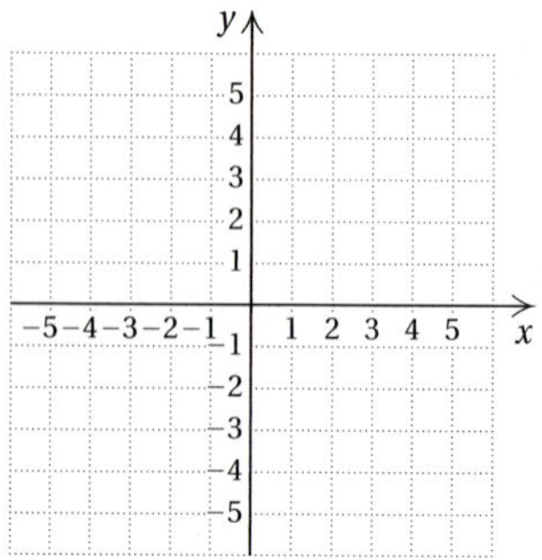

36. $6y + 2x = 8$

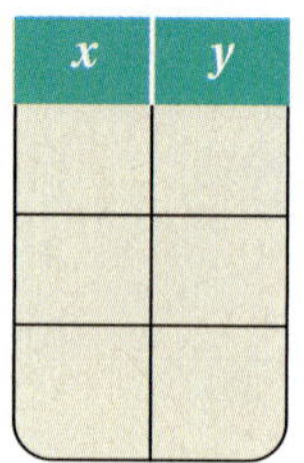

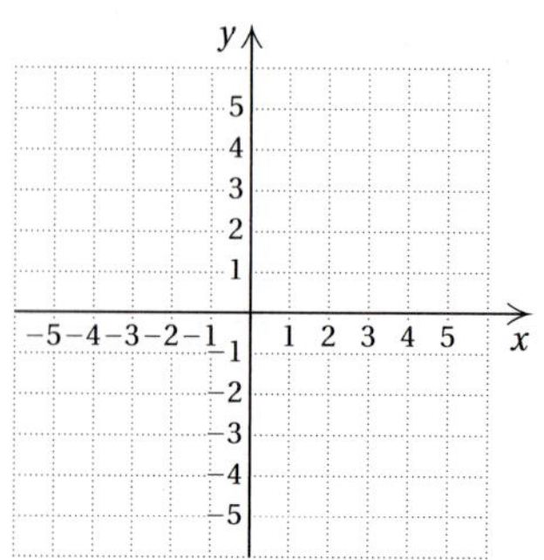

C Solve.

37. *Value of Computer Software.* The value V, in dollars, of a shopkeeper's inventory software program is given by $V = -50t + 300$, where t is the number of years since the shopkeeper first bought the program.

a) Find the value of the software after 0 yr, 4 yr, and 6 yr.

b) Graph the equation and then use the graph to estimate the value of the software after 5 yr.

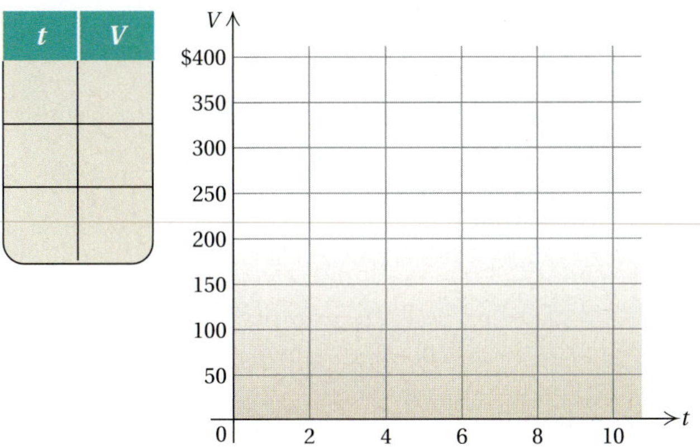

c) After how many years is the value of the software $150?

38. *College Costs.* The cost T, in dollars, of tuition and fees at many community colleges can be approximated by $T = 120c + 100$, where c is the number of credits for which a student registers.
Source: Community College of Vermont

a) Find the cost of tuition for a student who takes 8 hr, 12 hr, and 15 hr.

b) Graph the equation and then use the graph to estimate the cost of tuition and fees for 9 hr.

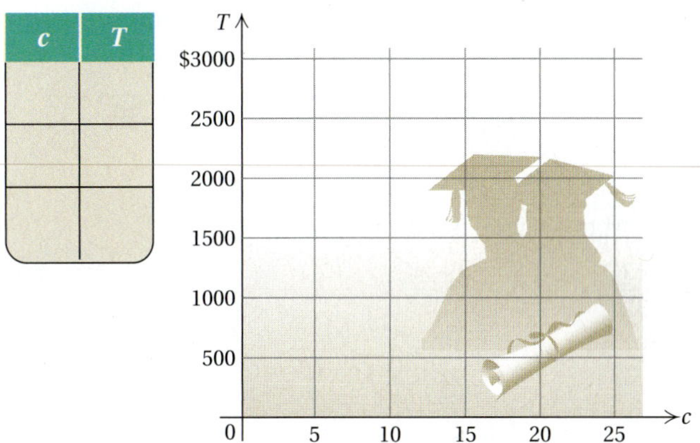

c) How many hours can a student take for $1420?

39. *Tea Consumption.* The number of gallons N of tea consumed each year by the average U.S. consumer can be approximated by $N = 0.1d + 7$, where d is the number of years since 1991.
Source: *Statistical Abstract of the United States*

a) Find the number of gallons of tea consumed in 1992 ($d = 1$), 1996, 2001, and 2011.

b) Graph the equation and use the graph to estimate what the tea consumption was in 1997.

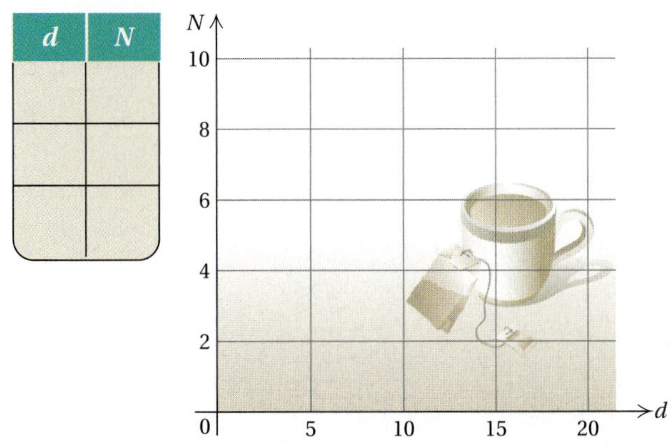

c) In what year will tea consumption be about 8.5 gal?

40. *Record Temperature Drop.* On 22 January 1943, the temperature T, in degrees Fahrenheit, in Spearfish, South Dakota, could be approximated by $T = -2.15m + 54$, where m is the number of minutes since 9:00 that morning.
Source: *Information Please Almanac*

a) Find the temperature at 9:01 A.M., 9:08 A.M., and 9:20 A.M.

b) Graph the equation and use the graph to estimate the temperature at 9:15 A.M.

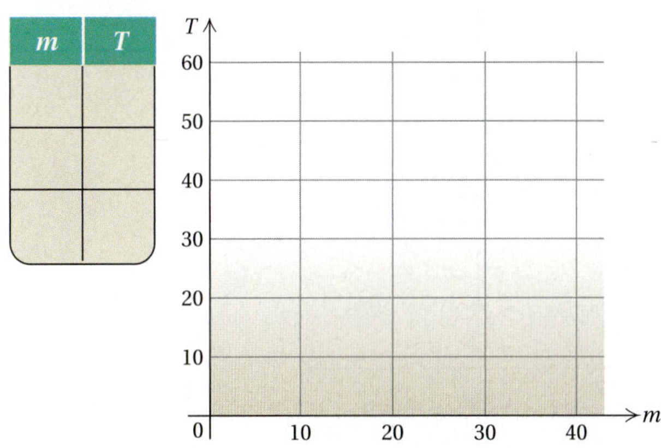

c) The temperature stopped dropping when it reached −4°F. At what time did this occur?

41. **DW** The equations $3x + 4y = 8$ and $y = -\frac{3}{4}x + 2$ are equivalent. Which equation is easier to graph and why?

42. **DW** Referring to Exercise 40, discuss why the linear equation no longer described the temperature after the temperature reached $-4°$.

SKILL MAINTENANCE

Solve. [7.2a]

43. $63 = 9x$

44. $\frac{2}{3}y = -\frac{1}{3}$

45. $13x = -52$

46. $\frac{a}{8} = -2$

47. $\frac{1}{10}x = \frac{2}{5}$

Convert to decimal notation. [6.2c]

48. $\frac{23}{32}$

49. $-\frac{7}{8}$

50. $-\frac{27}{12}$

51. $\frac{117}{64}$

52. $-\frac{17}{16}$

SYNTHESIS

In Exercises 53–56, find an equation for the graph shown.

53.

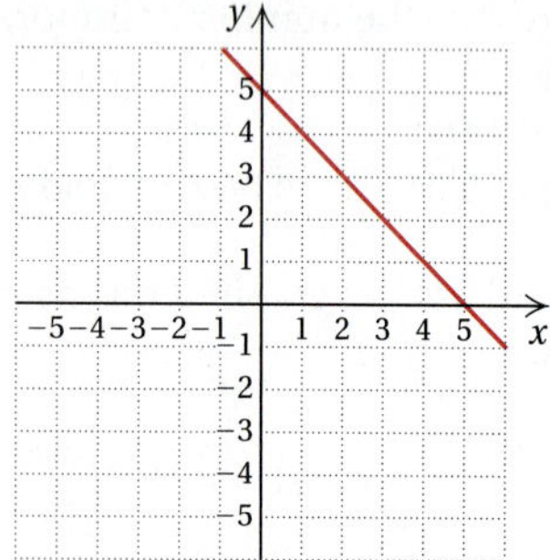

54.

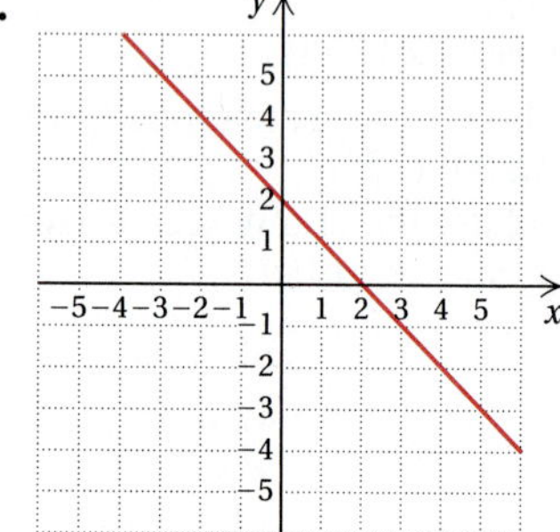

55.

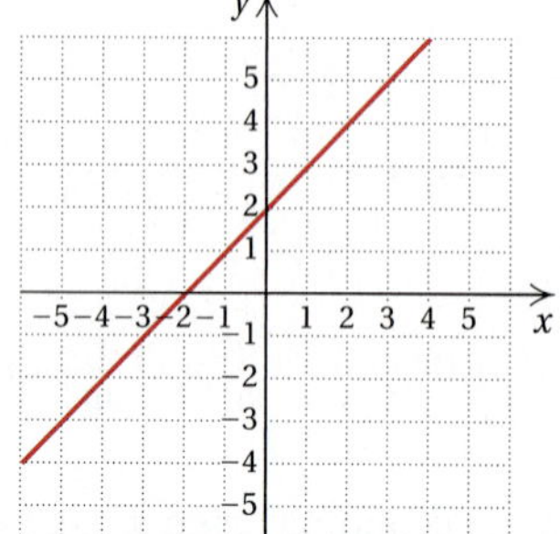

56.

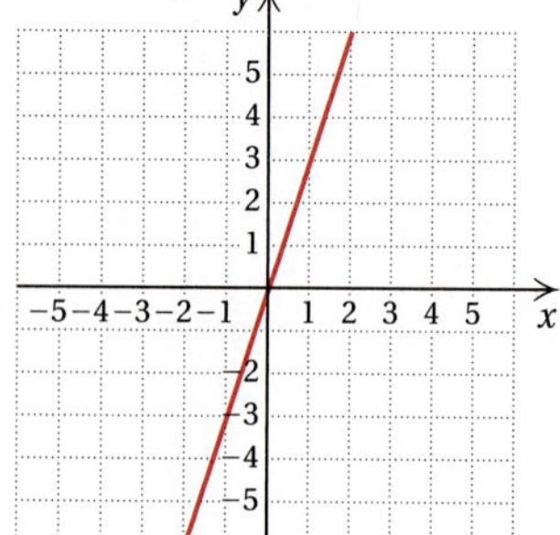

8.3 MORE WITH GRAPHING AND INTERCEPTS

Objectives

a Find the intercepts of a linear equation, and graph using intercepts.

b Graph equations equivalent to those of the type $x = a$ and $y = b$.

a Graphing Using Intercepts

In Section 8.2, we graphed linear equations of the form $Ax + By = C$ by first solving for y to find an equivalent equation in the form $y = mx + b$. We did so because it is then easier to calculate the y-value that corresponds to a given x-value. Another convenient way to graph $Ax + By = C$ is to use **intercepts.** Look at the graph of $-2x + y = 4$ shown below.

The y-intercept is $(0, 4)$. It occurs where the line crosses the y-axis and thus will always have 0 as the first coordinate. The x-intercept is $(-2, 0)$. It occurs where the line crosses the x-axis and thus will always have 0 as the second coordinate.

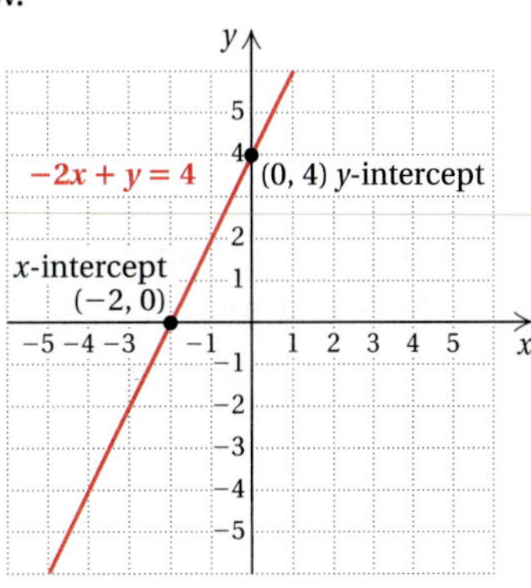

Do Exercise 1.

We find intercepts as follows.

INTERCEPTS

The **y-intercept** is $(0, b)$. To find b, let $x = 0$ and solve the original equation for y.

The **x-intercept** is $(a, 0)$. To find a, let $y = 0$ and solve the original equation for x.

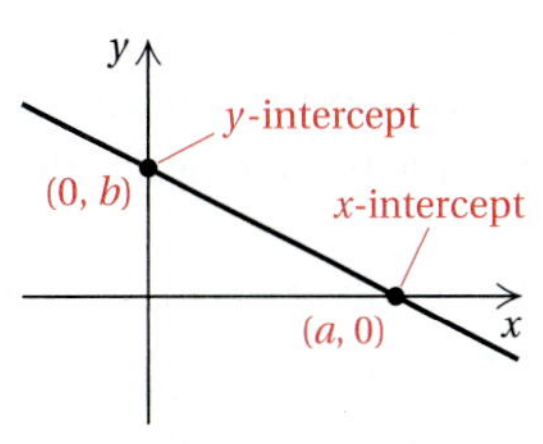

Now let's draw a graph using intercepts.

EXAMPLE 1 Consider $4x + 3y = 12$. Find the intercepts. Then graph the equation using the intercepts.

To find the y-intercept, we let $x = 0$. Then we solve for y:

$$4 \cdot 0 + 3y = 12$$
$$3y = 12$$
$$y = 4.$$

Thus, $(0, 4)$ is the y-intercept. Note that finding this intercept amounts to covering up the x-term and solving the rest of the equation.

To find the x-intercept, we let $y = 0$. Then we solve for x:

$$4x + 3 \cdot 0 = 12$$
$$4x = 12$$
$$x = 3.$$

1. Look at the graph shown below.

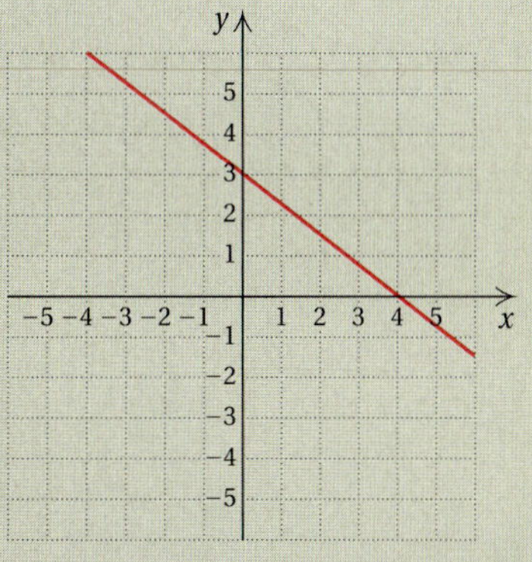

a) Find the coordinates of the y-intercept.

b) Find the coordinates of the x-intercept.

Answers on page A-23

For each equation, find the intercepts. Then graph the equation using the intercepts.

2. $2x + 3y = 6$

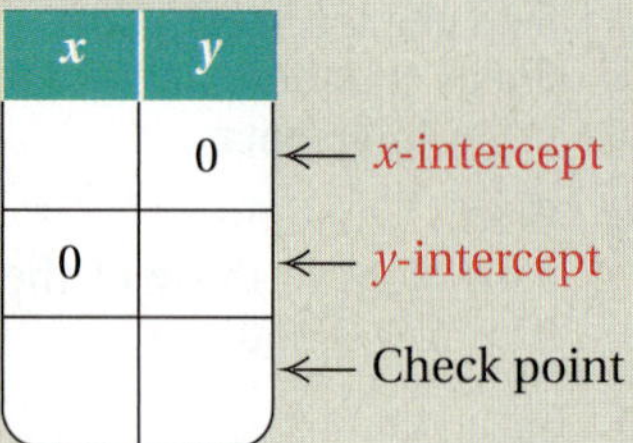

x	y	
	0	← x-intercept
0		← y-intercept
		← Check point

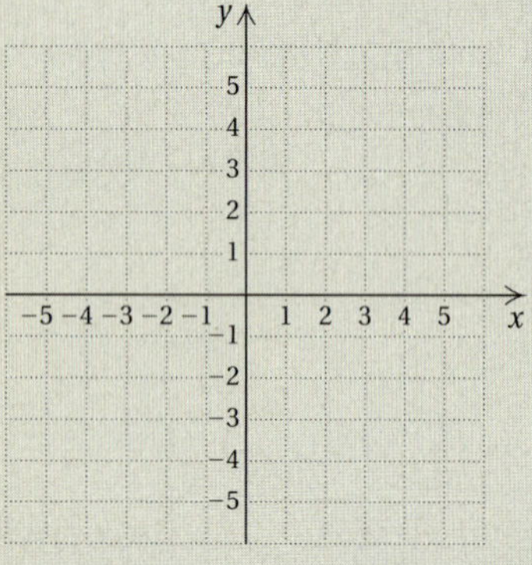

3. $3y - 4x = 12$

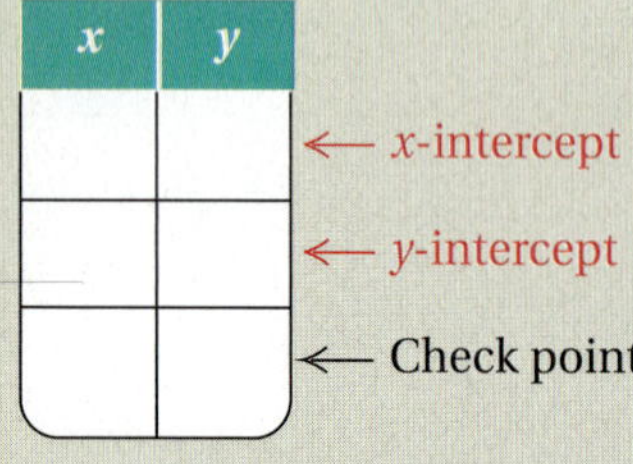

x	y	
		← x-intercept
		← y-intercept
		← Check point

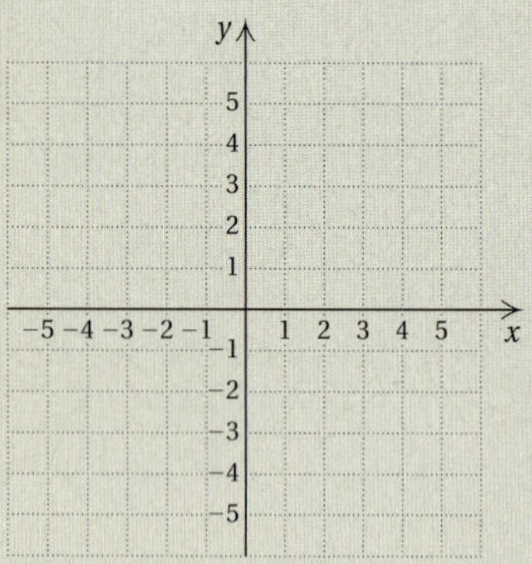

Answers on page A-23

Thus, $(3, 0)$ is the x-intercept. Note that finding this intercept amounts to covering up the y-term and solving the rest of the equation.

We plot these points and draw the line, or graph.

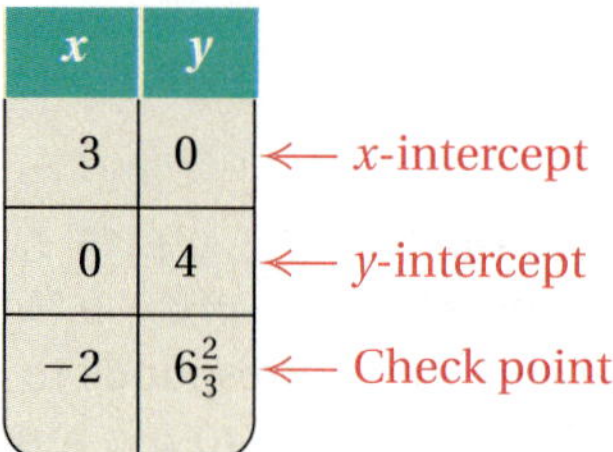

x	y	
3	0	← x-intercept
0	4	← y-intercept
−2	$6\frac{2}{3}$	← Check point

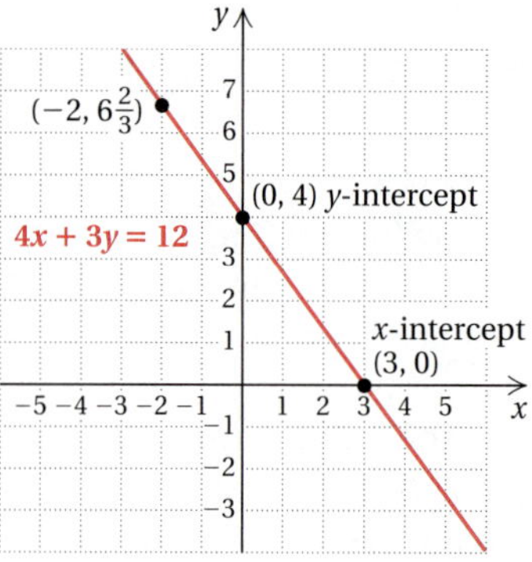

A third point should be used as a check. We substitute any convenient value for x and solve for y. In this case, we choose $x = -2$. Then

$$4(-2) + 3y = 12 \qquad \text{Substituting } -2 \text{ for } x$$
$$-8 + 3y = 12$$
$$3y = 12 + 8 = 20$$
$$y = \frac{20}{3}, \text{ or } 6\frac{2}{3}. \qquad \text{Solving for } y$$

It appears that the point $\left(-2, 6\frac{2}{3}\right)$ is on the graph, though graphing fraction values can be inexact. The graph is probably correct.

Do Exercises 2 and 3.

Graphs of equations of the type $y = mx$ pass through the origin. Thus the x-intercept and the y-intercept are the same, $(0, 0)$. In such cases, we must calculate another point in order to complete the graph. Another point would also have to be calculated if a check is desired.

EXAMPLE 2 Graph: $y = 3x$.

We know that $(0, 0)$ is both the x-intercept and the y-intercept. We calculate values at two other points and complete the graph, knowing that it passes through the origin $(0, 0)$.

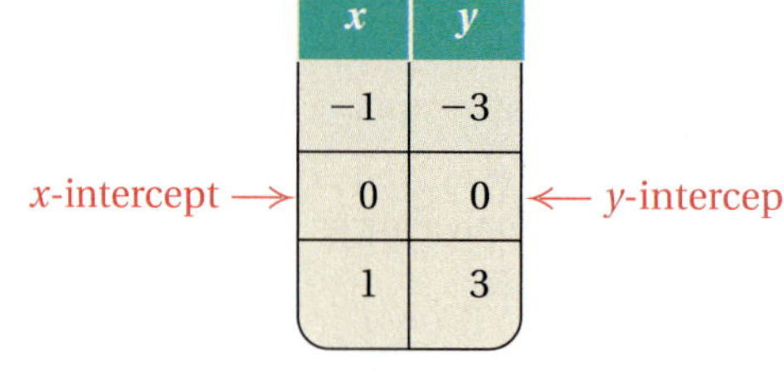

	x	y	
	−1	−3	
x-intercept →	0	0	← y-intercept
	1	3	

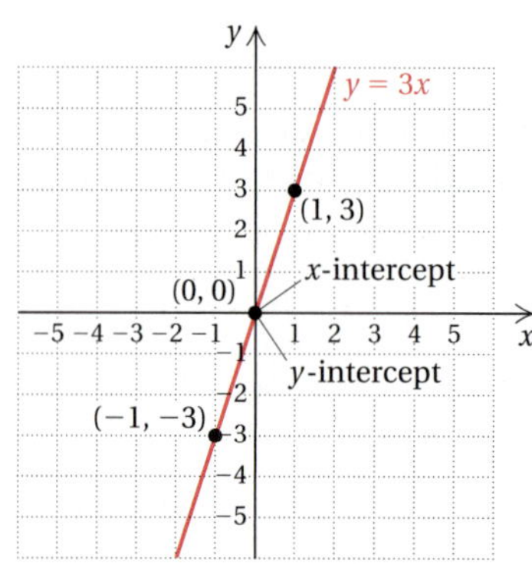

Do Exercises 4 and 5 on the following page.

CALCULATOR CORNER

Viewing the Intercepts Knowing the intercepts of a linear equation helps us to determine a good viewing window for the graph of the equation. For example, when we graph the equation $y = -x + 15$ in the standard window, we see only a small portion of the graph in the upper righthand corner of the screen, as shown on the left below.

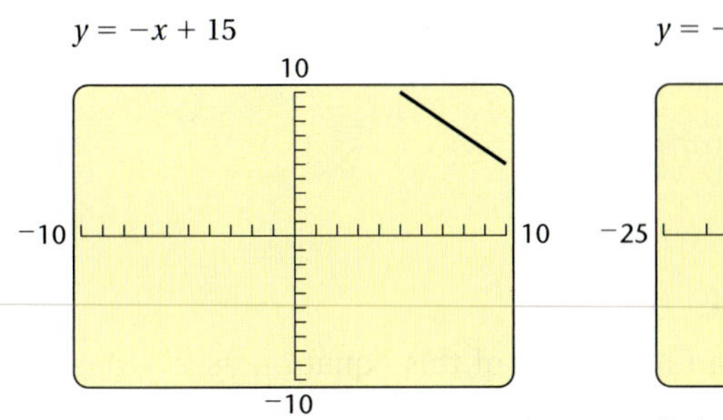

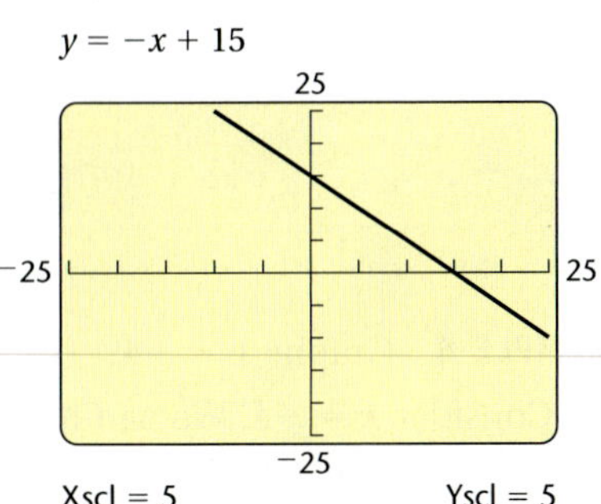

Using algebra, as we did in Example 1, we find that the intercepts of the graph of this equation are (0, 15) and (15, 0). This tells us that, if we are to see more of the graph than is shown on the left above, both Xmax and Ymax should be greater than 15. We can try different window settings until we find one that suits us. One good choice is [−25, 25, −25, 25], Xscl = 5, Yscl = 5, shown on the right above.

Exercises: Find the intercepts of each equation algebraically. Then graph the equation on a graphing calculator, choosing window settings that allow the intercepts to be seen clearly. (Settings may vary.)

1. $y = -7.5x - 15$

2. $y - 2.15x = 43$

3. $6x - 5y = 150$

4. $y = 0.2x - 4$

5. $y = 1.5x - 15$

6. $5x - 4y = 2$

b Equations Whose Graphs Are Horizontal or Vertical Lines

EXAMPLE 3 Graph: $y = 3$.

Consider $y = 3$. We can also think of this equation as $0 \cdot x + y = 3$. No matter what number we choose for x, we find that y is 3. We make up a table with all 3's in the y-column.

x	y
	3
	3
	3

y must be 3.

Choose any number for x. →

x	y
−2	3
0	3
4	3

← y-intercept (row 0, 3)

Graph.

4. $y = 2x$

x	y
−1	
0	
1	

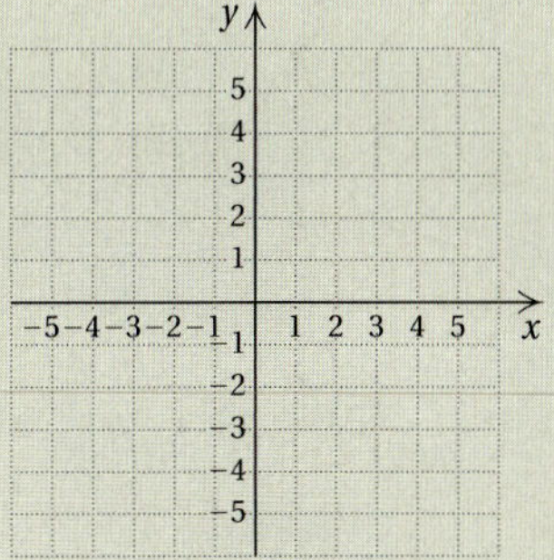

5. $y = -\frac{2}{3}x$

x	y

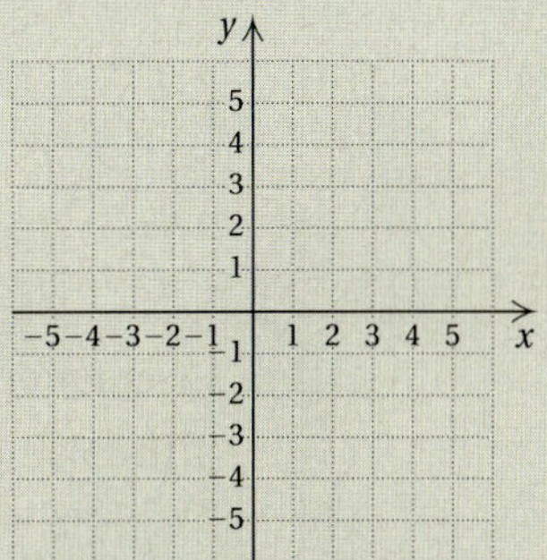

Answers on page A-23

Graph.

6. $x = 5$

x	y
5	
5	
5	

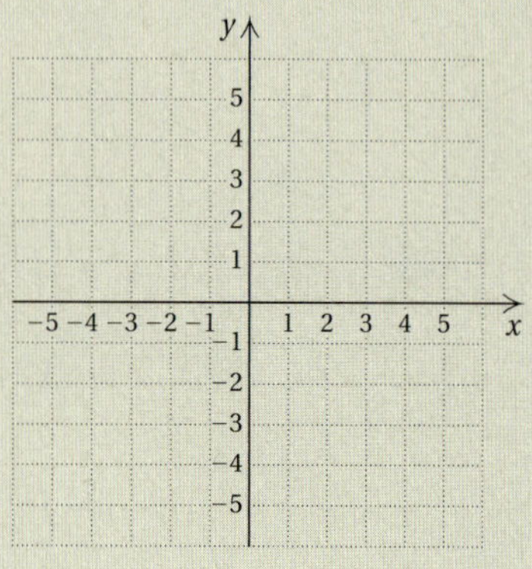

7. $y = -2$

x	y
	-2
	-2
	-2

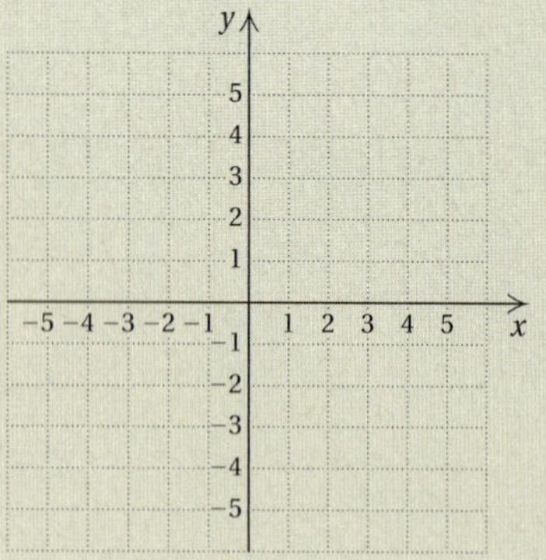

Answers on page A-23

When we plot the ordered pairs $(-2, 3)$, $(0, 3)$, and $(4, 3)$ and connect the points, we will obtain a horizontal line. Any ordered pair $(x, 3)$ is a solution. So the line is parallel to the x-axis with y-intercept $(0, 3)$.

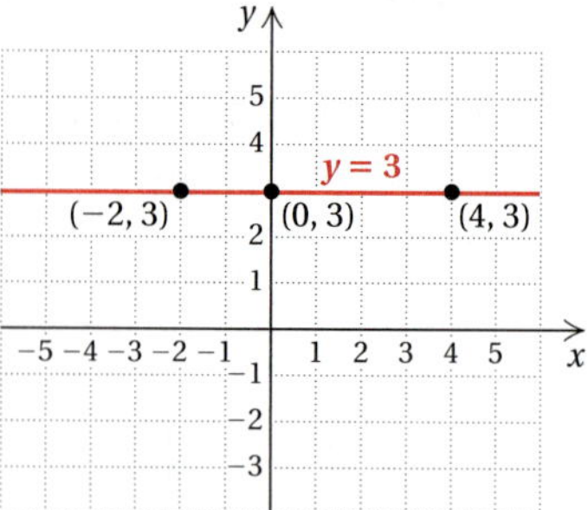

EXAMPLE 4 Graph: $x = -4$.

Consider $x = -4$. We can also think of this equation as $x + 0 \cdot y = -4$. We make up a table with all -4's in the x-column.

x	y
-4	
-4	
-4	
-4	

x must be -4.

Choose any number for y. →

x	y
-4	-5
-4	1
-4	3
-4	0

x-intercept → (last row)

When we plot the ordered pairs $(-4, -5)$, $(-4, 1)$, $(-4, 3)$, and $(-4, 0)$ and connect the points, we will obtain a vertical line. Any ordered pair $(-4, y)$ is a solution. So the line is parallel to the y-axis with x-intercept $(-4, 0)$.

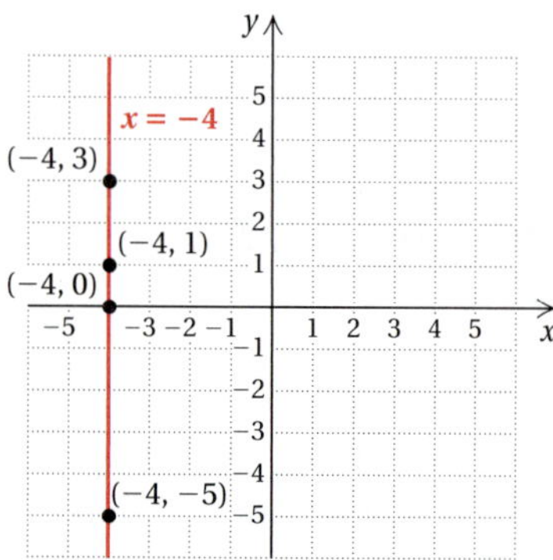

HORIZONTAL AND VERTICAL LINES

The graph of $y = b$ is a **horizontal line.** The y-intercept is $(0, b)$.
The graph of $x = a$ is a **vertical line.** The x-intercept is $(a, 0)$.

Do Exercises 6–9. (Exercises 8 and 9 are on the following page.)

The following is a general procedure for graphing linear equations.

GRAPHING LINEAR EQUATIONS

1. If the equation is of the type $x = a$ or $y = b$, the graph will be a line parallel to an axis; $x = a$ is vertical and $y = b$ is horizontal.

Examples.

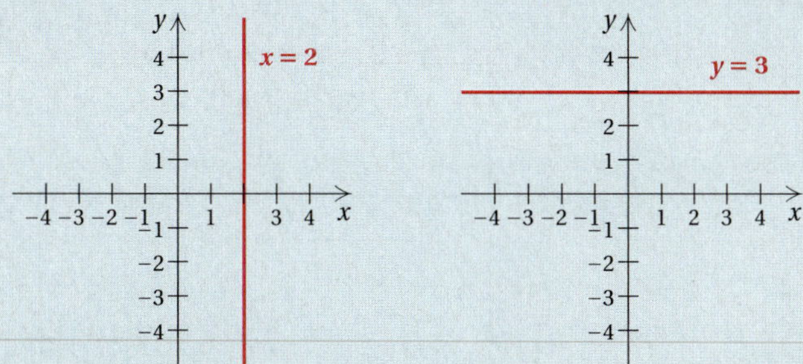

2. If the equation is of the type $y = mx$, both intercepts are the origin, $(0, 0)$. Plot $(0, 0)$ and two other points.

Example.

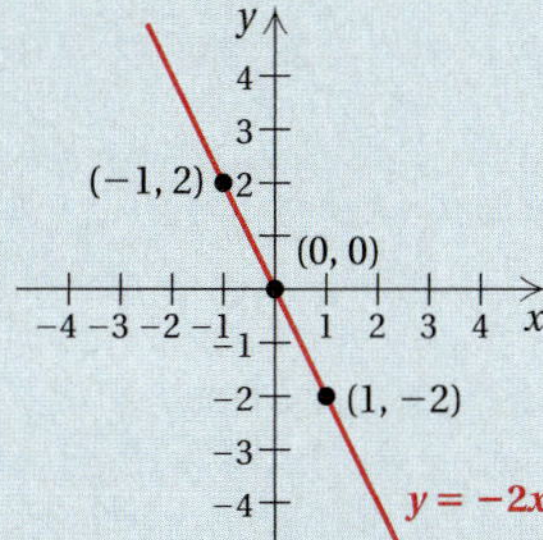

3. If the equation is of the type $y = mx + b$, plot the y-intercept $(0, b)$ and two other points.

Example.

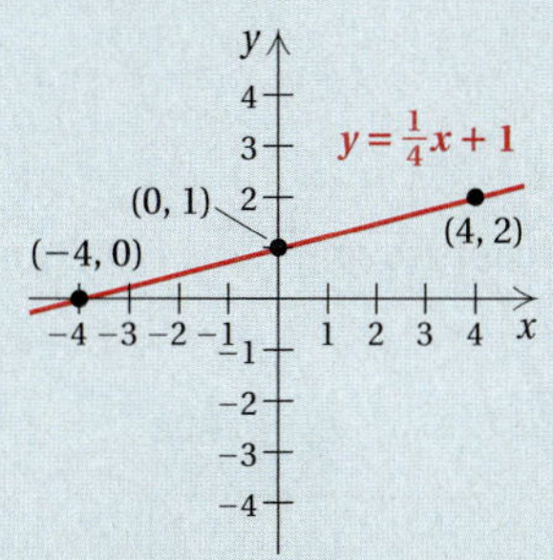

4. If the equation is of the type $Ax + By = C$, but not of the type $x = a$, $y = b$, $y = mx$, or $y = mx + b$, then either solve for y and proceed as with the equation $y = mx + b$, or graph using intercepts. If the intercepts are too close together, choose another point or points farther from the origin.

Examples.

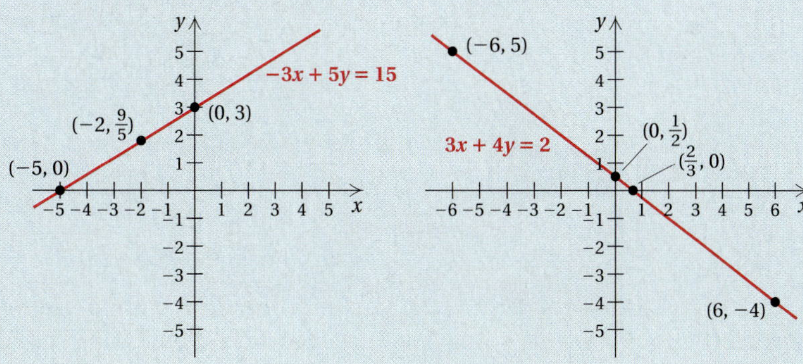

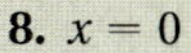

8. $x = 0$

x	y

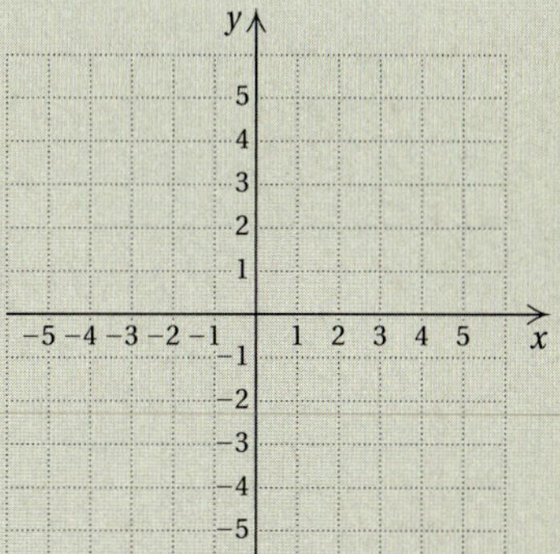

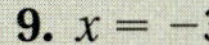

9. $x = -3$

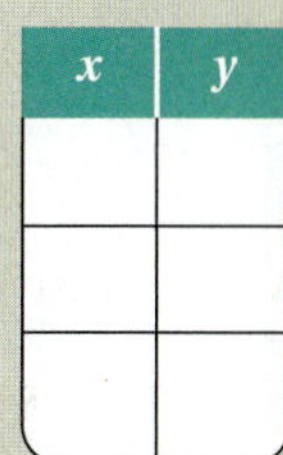

x	y

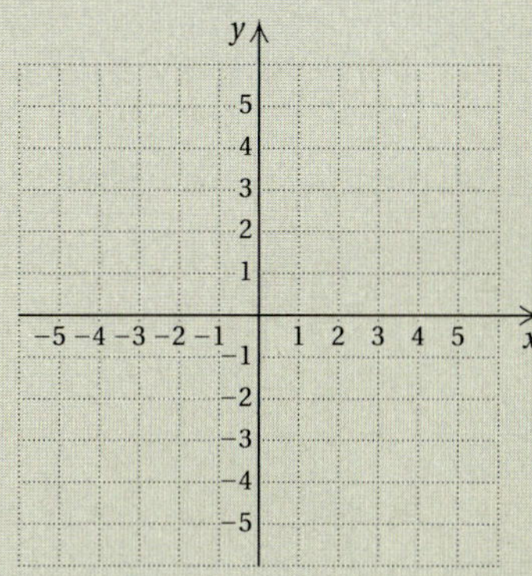

Answers on page A-23

8.3 EXERCISE SET

For Extra Help

Digital Video Tutor CD 7 Videotape 8 | InterAct Math | Math Tutor Center | MathXL | MyMathLab

a For Exercises 1–4, find (a) the coordinates of the y-intercept and (b) the coordinates of the x-intercept.

1.

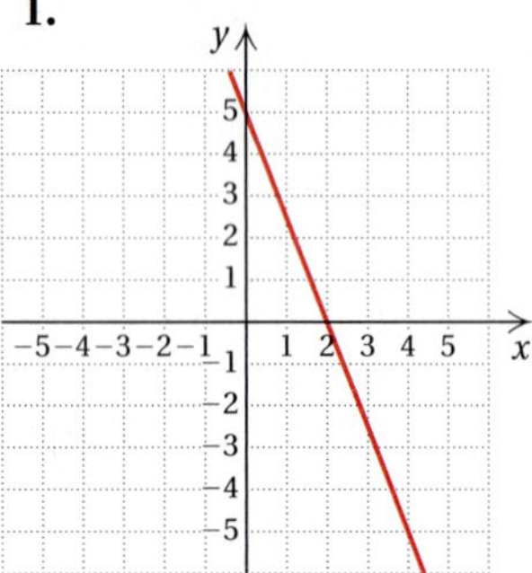

2.

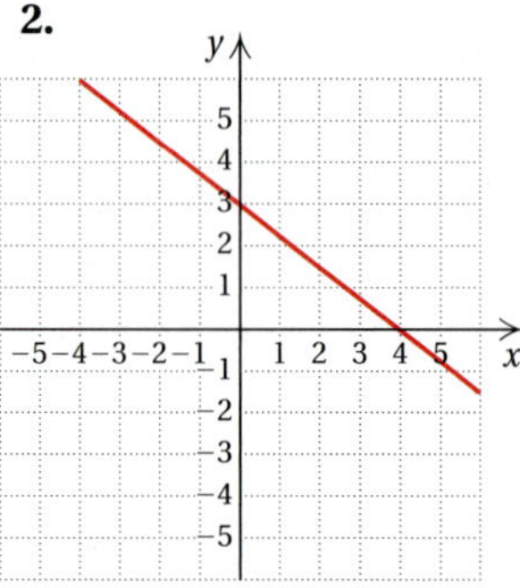

3.

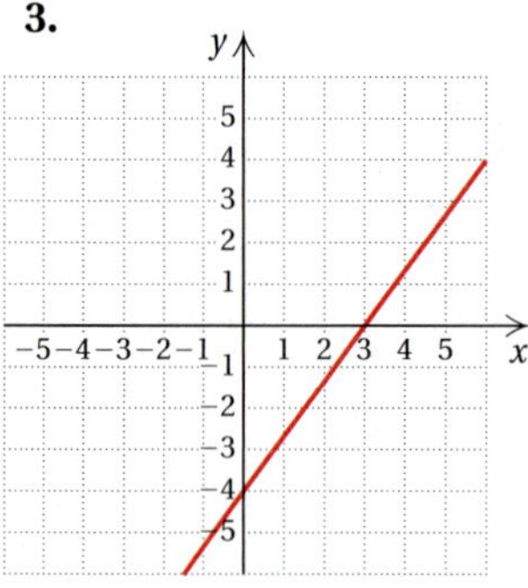

4.

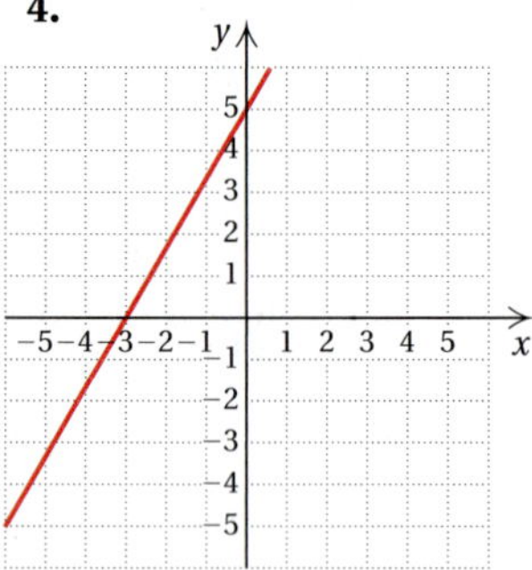

For Exercises 5–12, find (a) the coordinates of the y-intercept and (b) the coordinates of the x-intercept. Do not graph.

5. $3x + 5y = 15$

6. $5x + 2y = 20$

7. $7x - 2y = 28$

8. $3x - 4y = 24$

9. $-4x + 3y = 10$

10. $-2x + 3y = 7$

11. $6x - 3 = 9y$

12. $4y - 2 = 6x$

For each equation, find the intercepts. Then use the intercepts to graph the equation.

13. $x + 3y = 6$

x	y	
0		← y-intercept
	0	← x-intercept

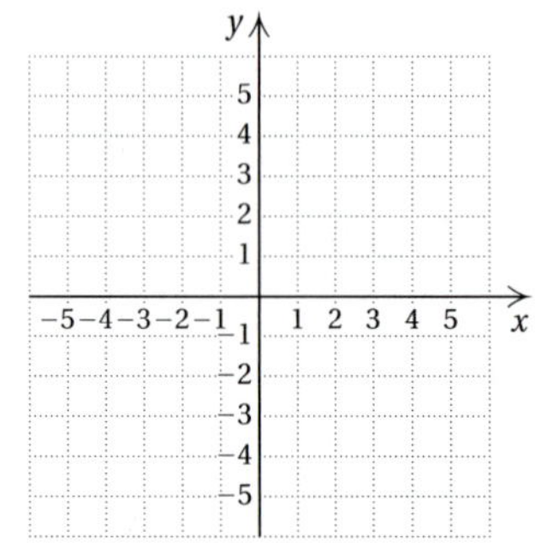

14. $x + 2y = 2$

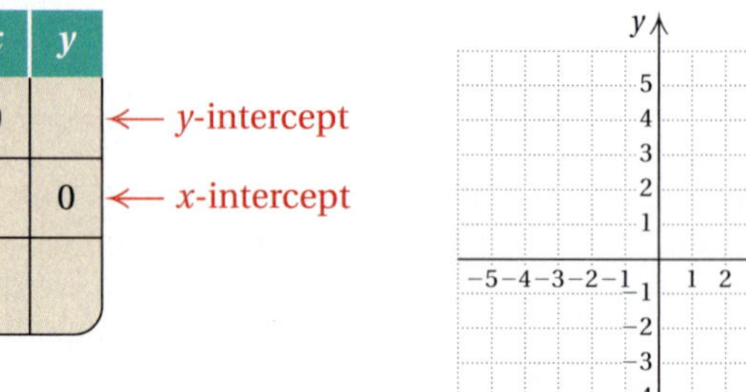

x	y	
0		← y-intercept
	0	← x-intercept

15. $-x + 2y = 4$

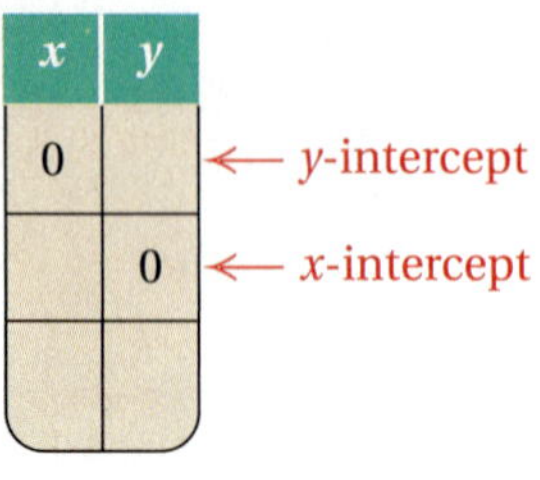

x	y	
0		← y-intercept
	0	← x-intercept

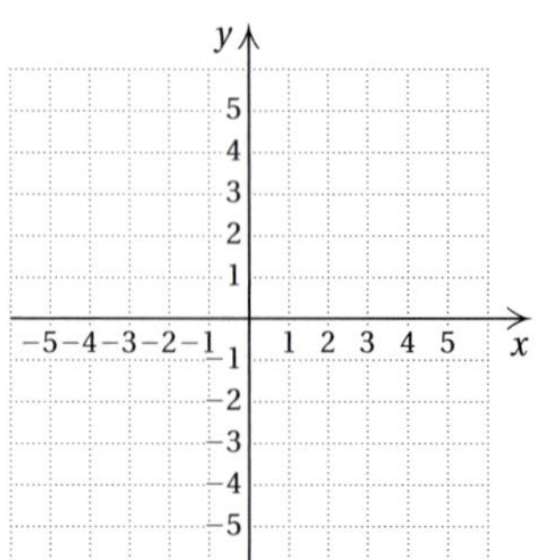

16. $-x + y = 5$

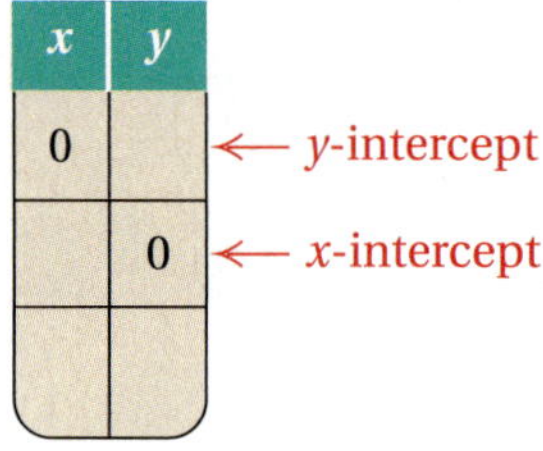

x	y	
0		← y-intercept
	0	← x-intercept

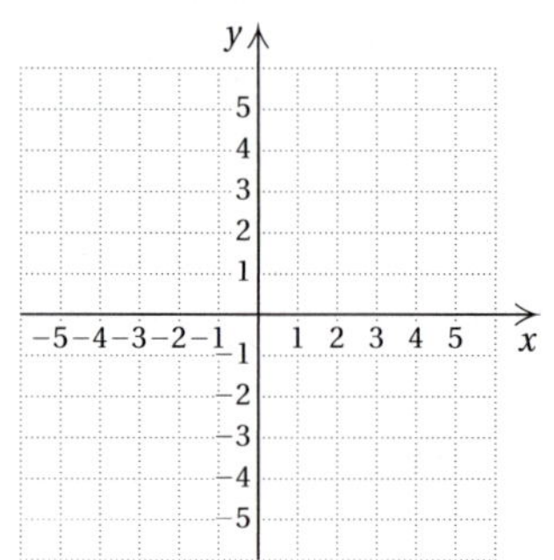

17. $3x + y = 6$

x	y	
	0	← x-intercept
0		← y-intercept

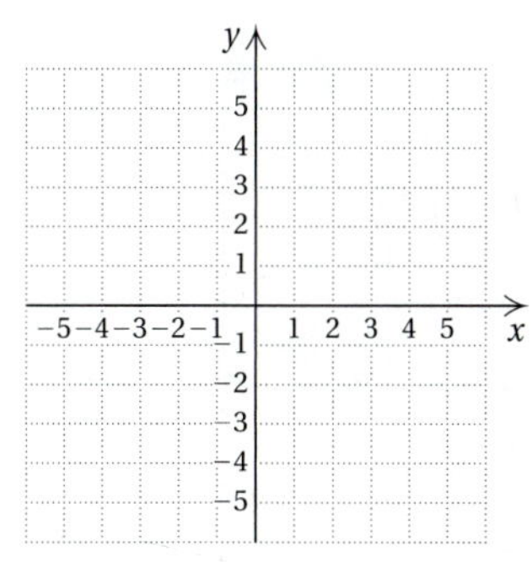

18. $2x + y = 6$

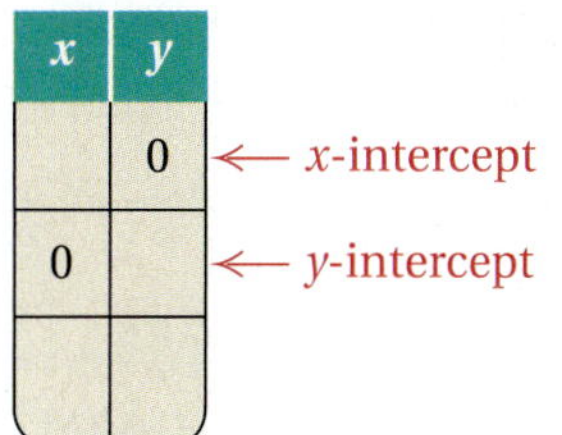

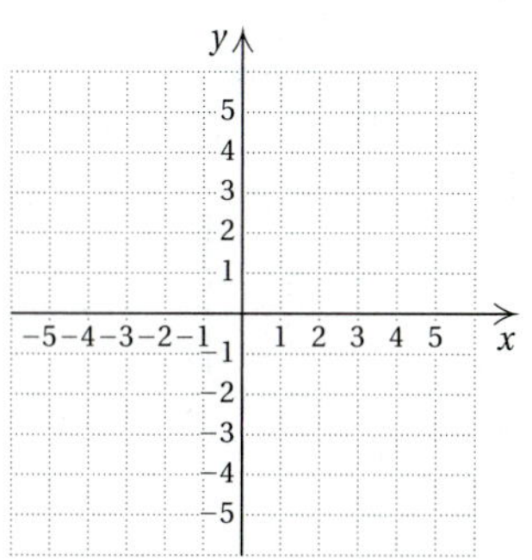

19. $2y - 2 = 6x$

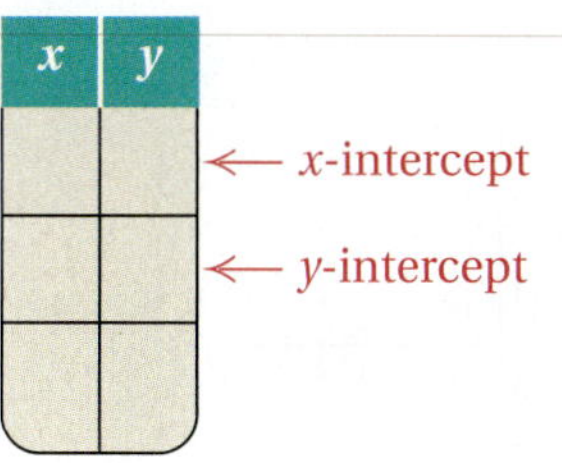

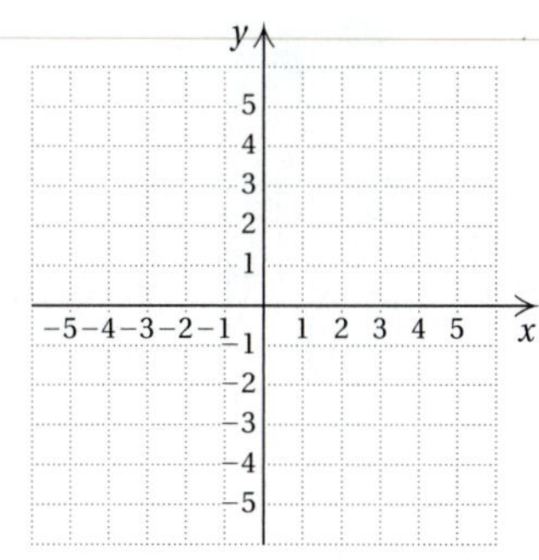

20. $3y - 6 = 9x$

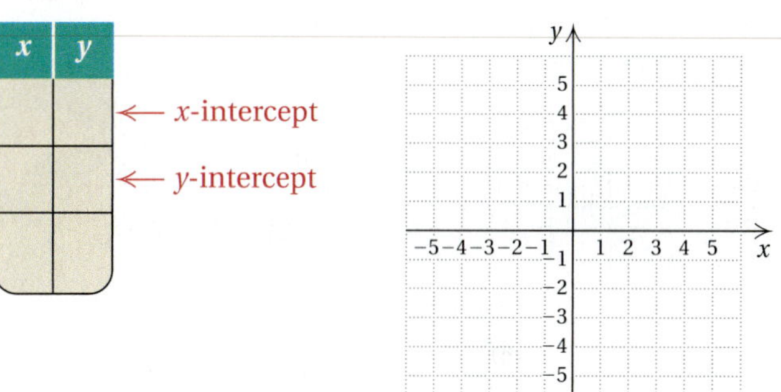

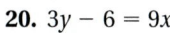

21. $3x - 9 = 3y$

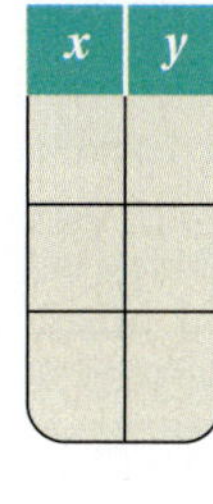

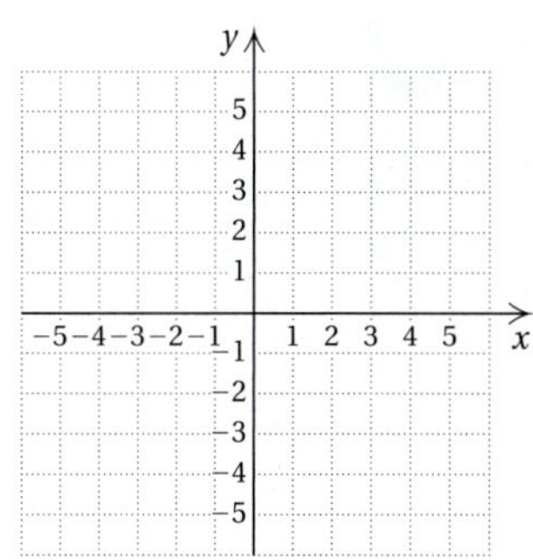

22. $5x - 10 = 5y$

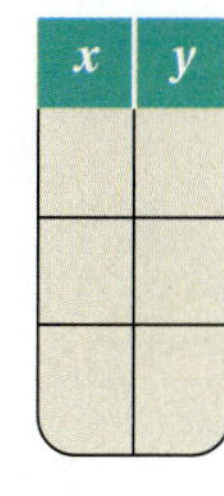

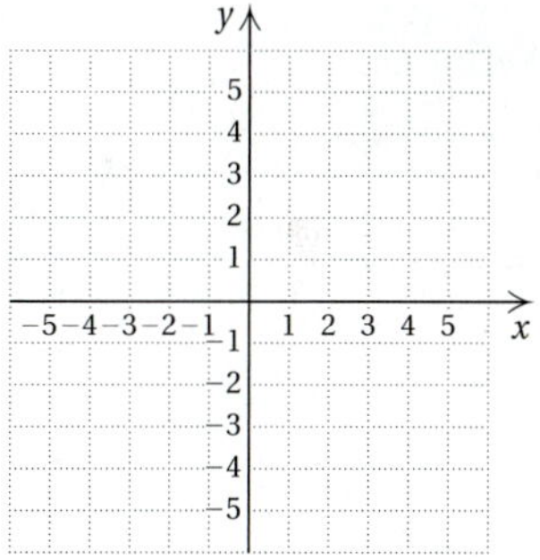

23. $2x - 3y = 6$

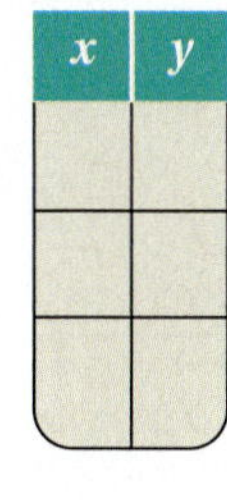

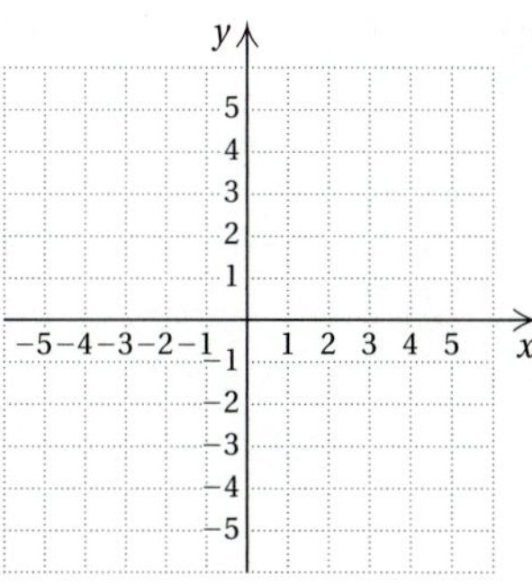

24. $2x - 5y = 10$

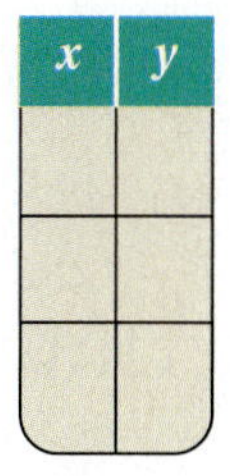

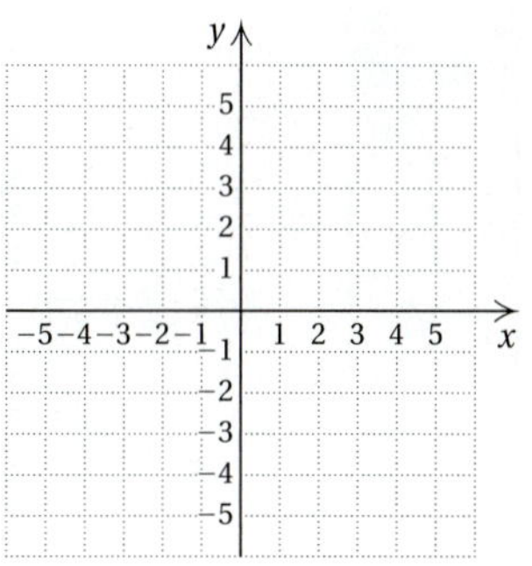

25. $4x + 5y = 20$

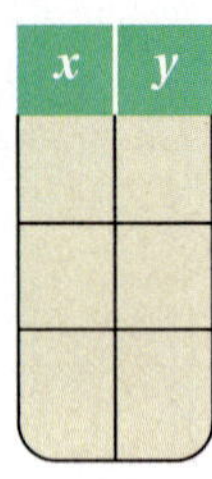

26. $2x + 6y = 12$

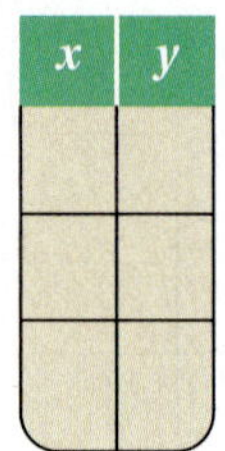

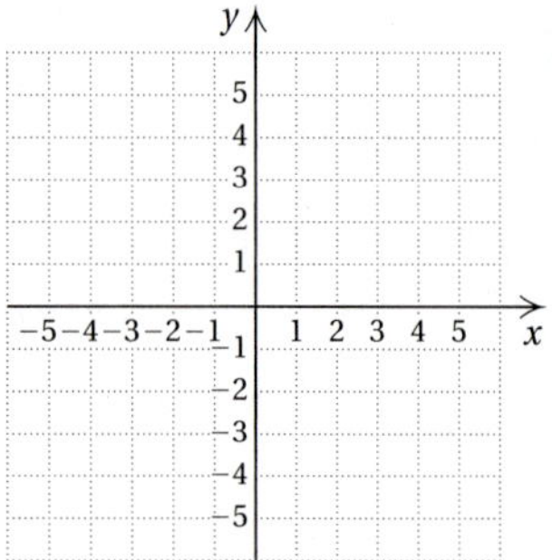

27. $2x + 3y = 8$

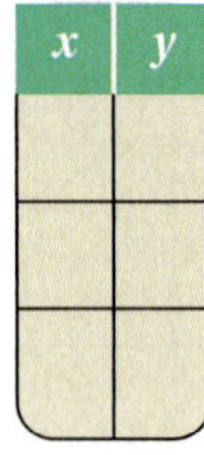

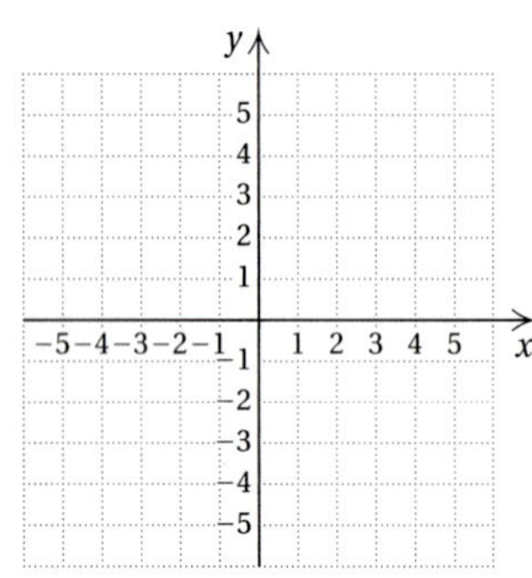

28. $x - 1 = y$

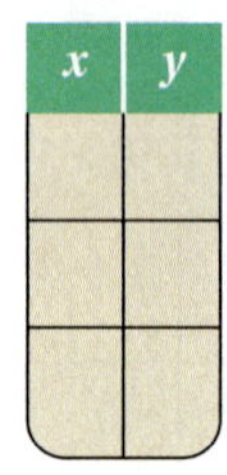

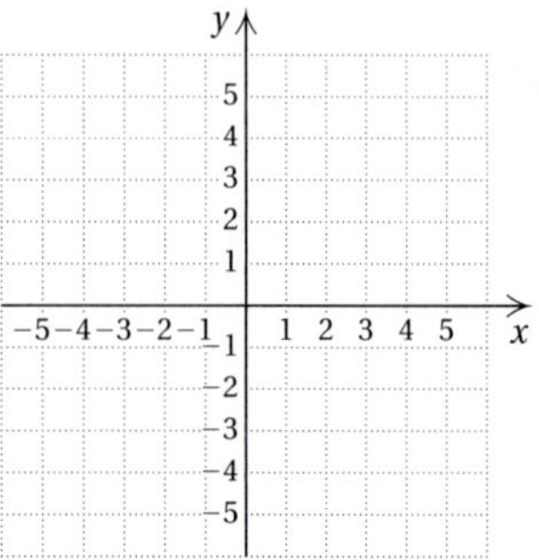

29. $x - 3 = y$

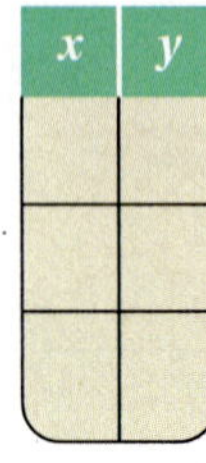

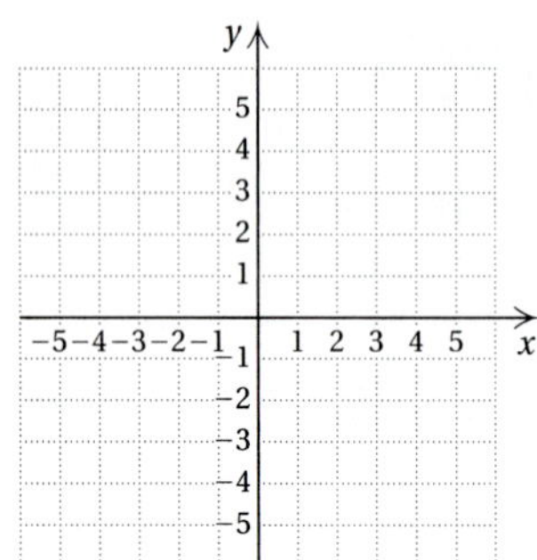

30. $2x - 1 = y$

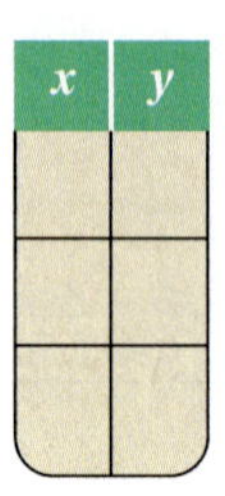

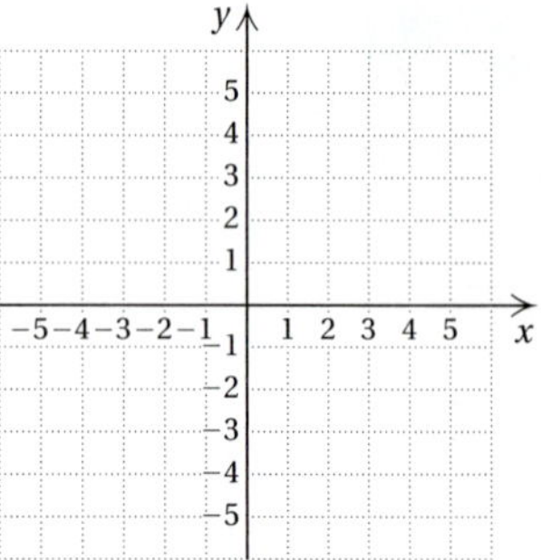

31. $3x - 2 = y$

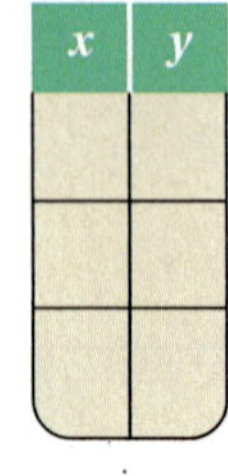

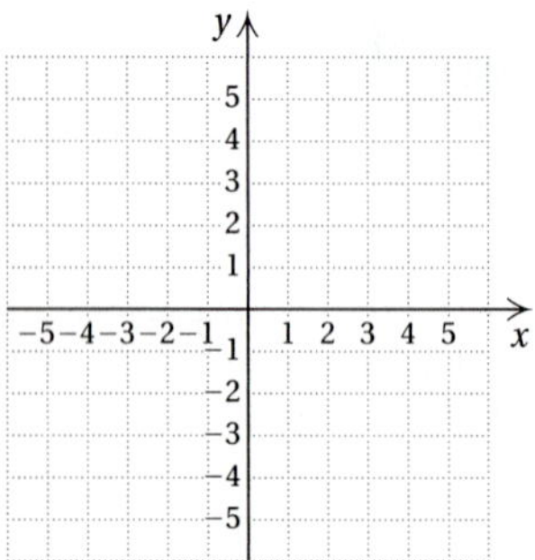

32. $4x - 3y = 12$

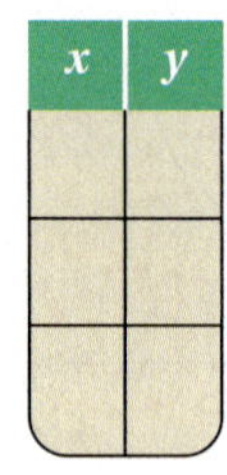

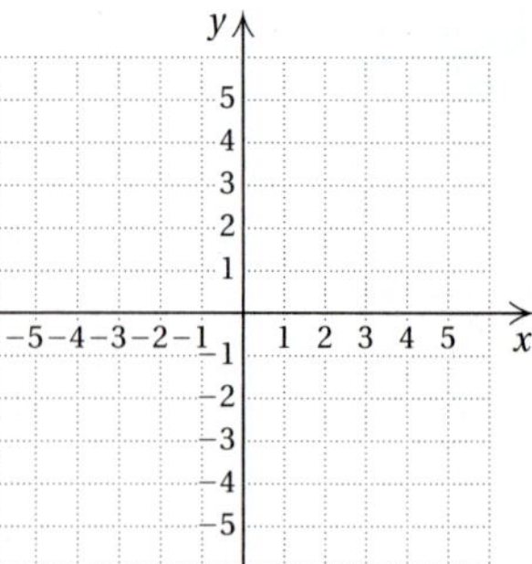

33. $6x - 2y = 12$

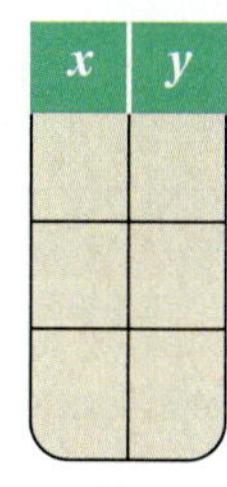

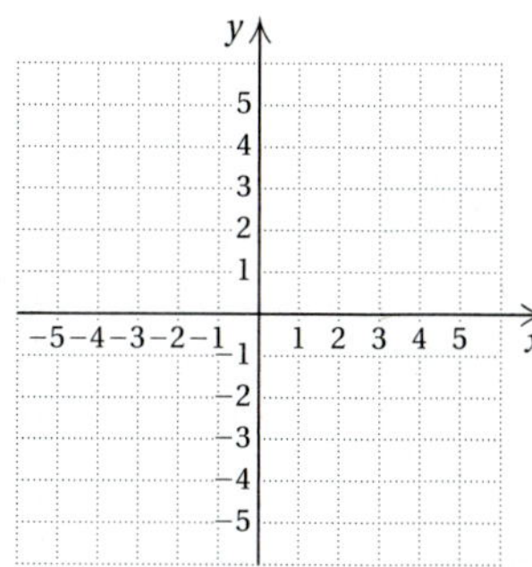

34. $7x + 2y = 6$

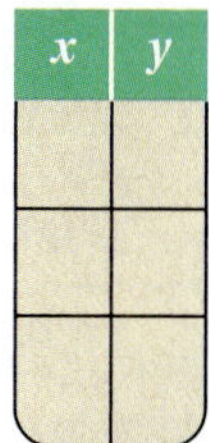

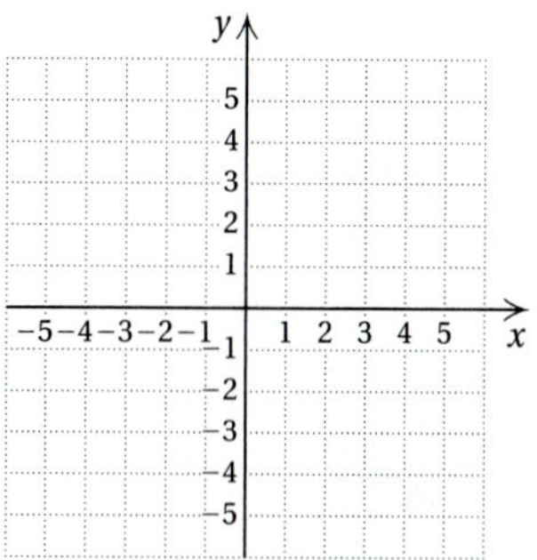

35. $3x + 4y = 5$

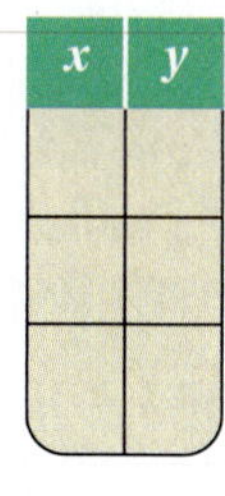

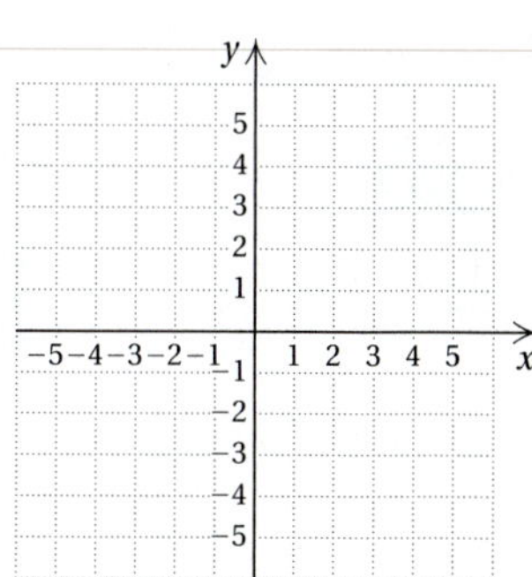

36. $y = -4 - 4x$

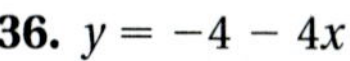

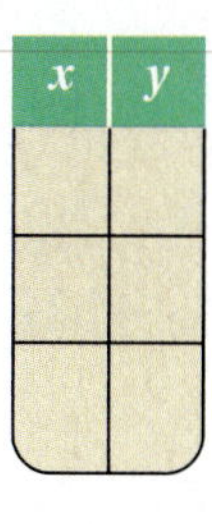

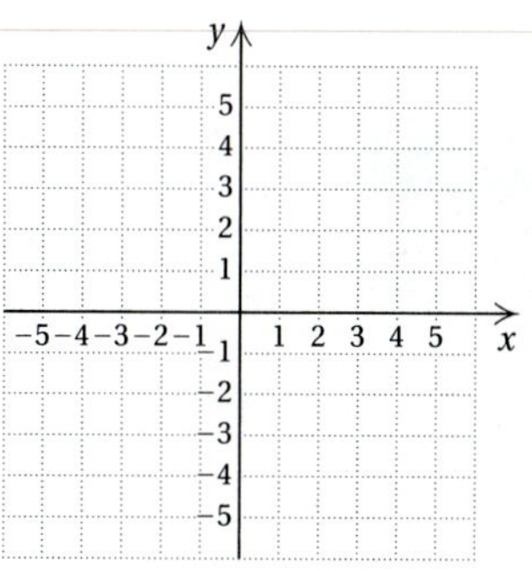

37. $y = -3 - 3x$

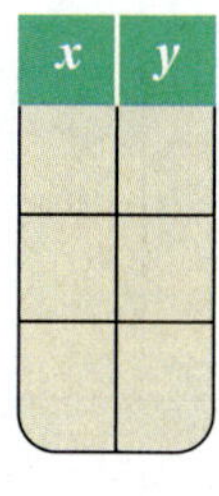

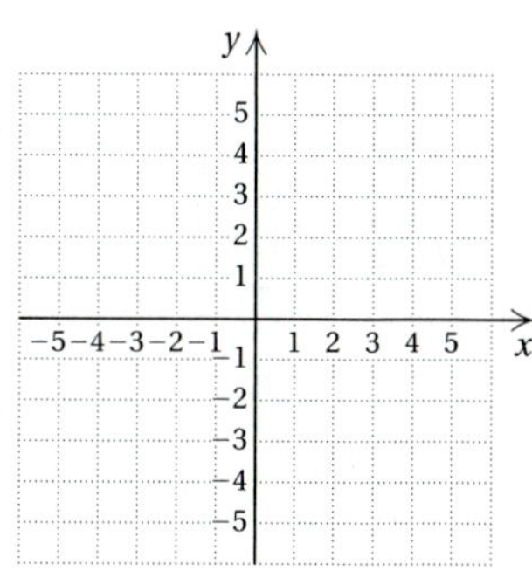

38. $-3x = 6y - 2$

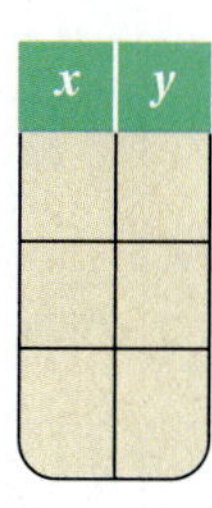

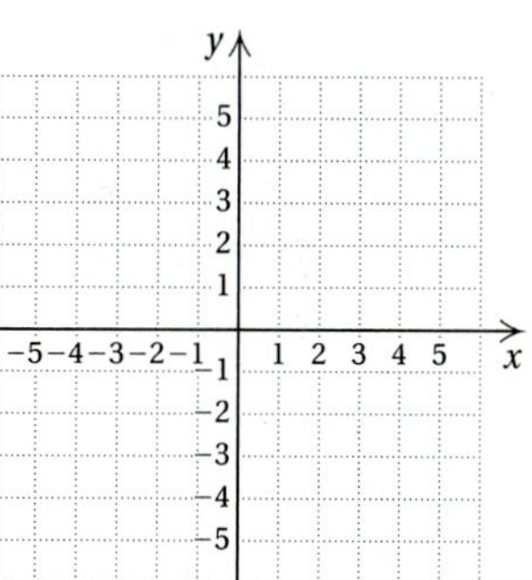

39. $y - 3x = 0$

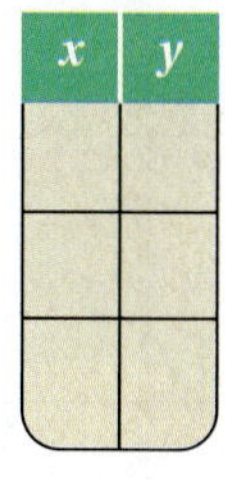

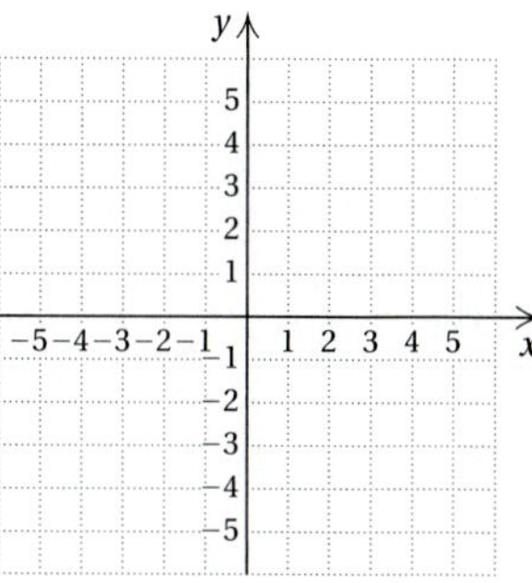

40. $x + 2y = 0$

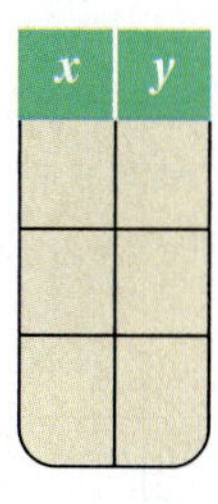

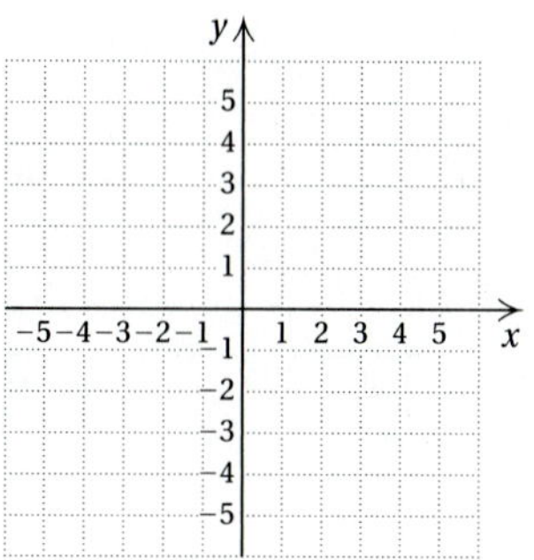

b Graph.

41. $x = -2$

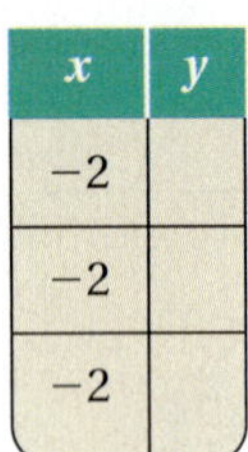

x	y
−2	
−2	
−2	

42. $x = 1$

x	y
1	
1	
1	

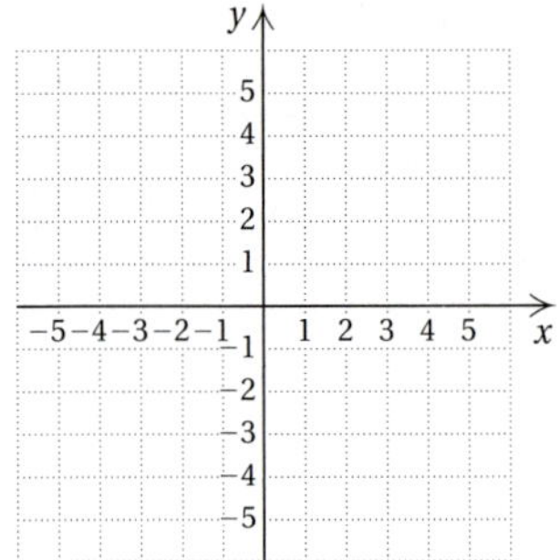

43. $y = 2$

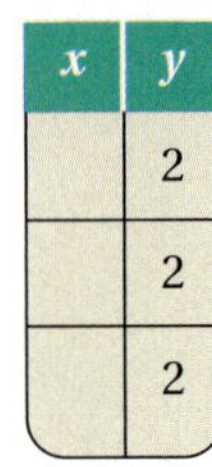

x	y
	2
	2
	2

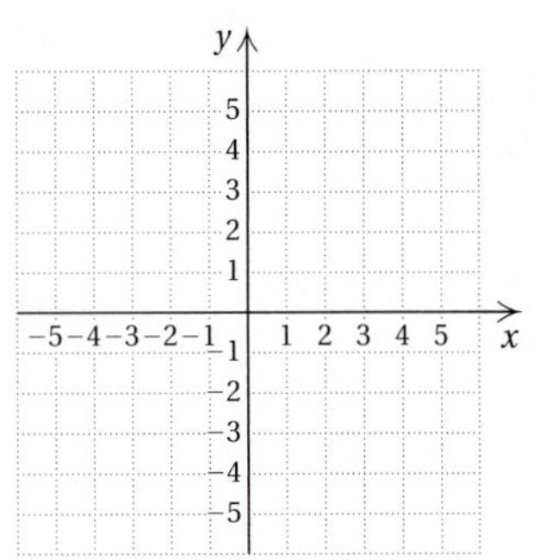

44. $y = -4$

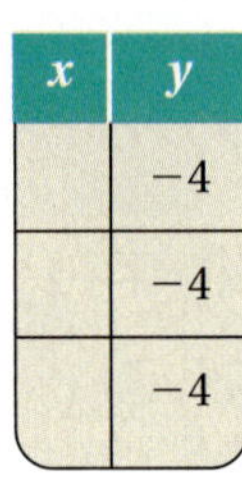

x	y
	−4
	−4
	−4

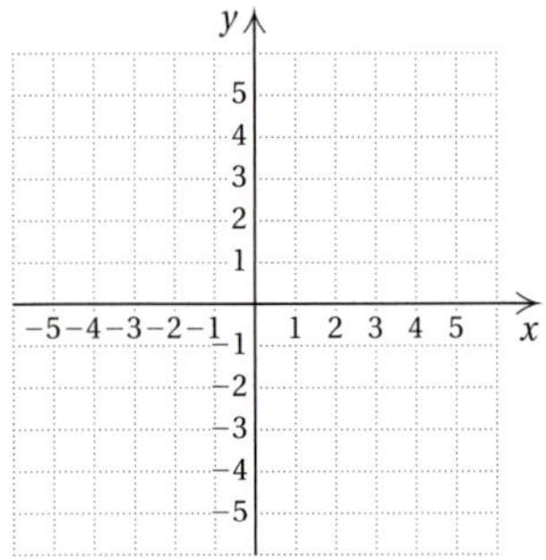

45. $x = 2$

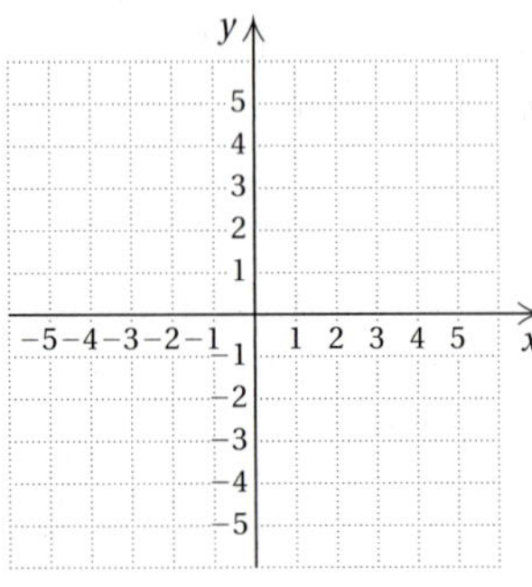

46. $x = 3$

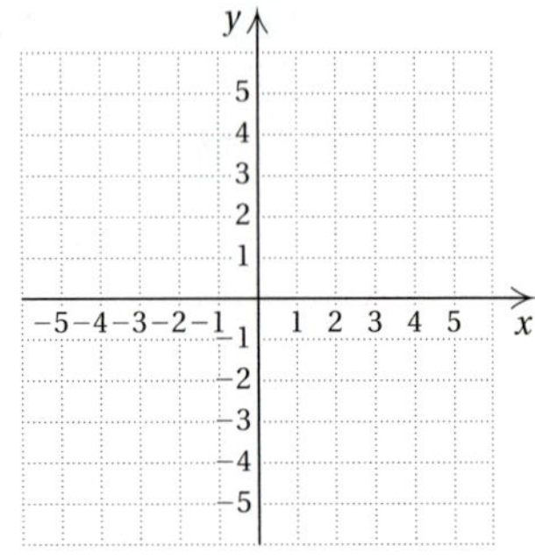

47. $y = 0$

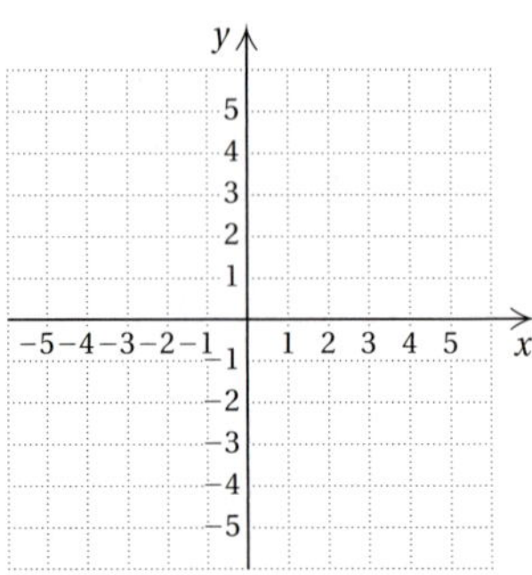

48. $y = -1$

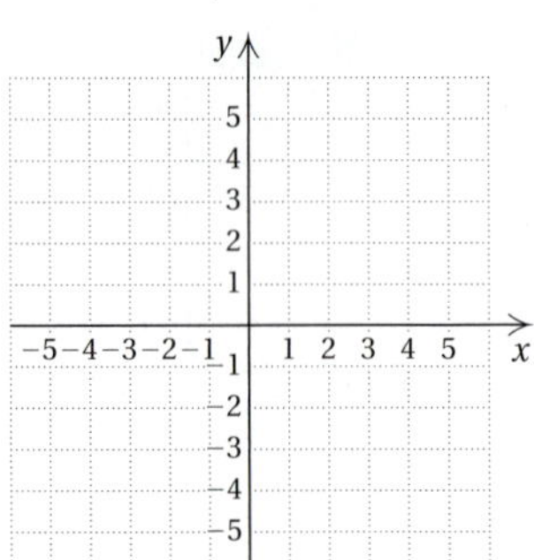

49. $x = \frac{3}{2}$

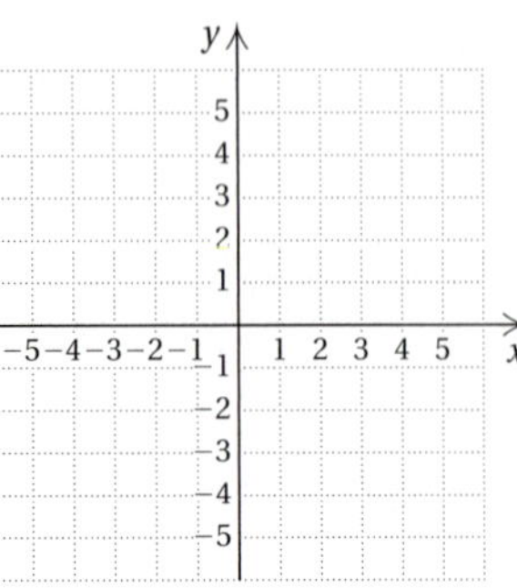

50. $x = -\frac{5}{2}$

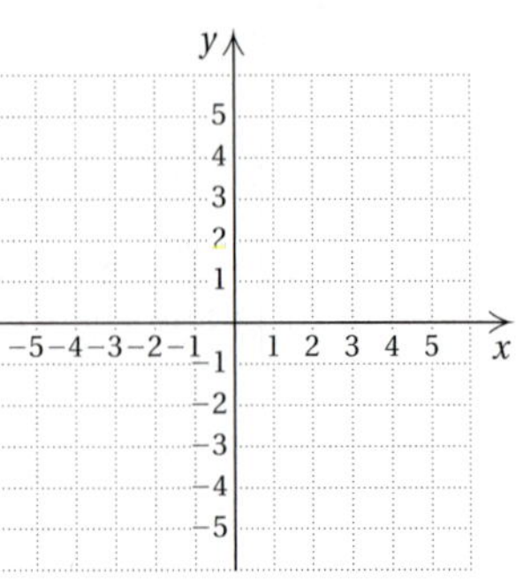

51. $3y = -5$

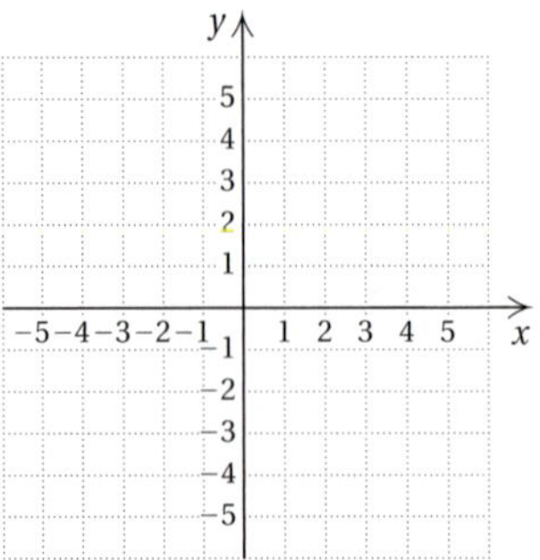

52. $12y = 45$

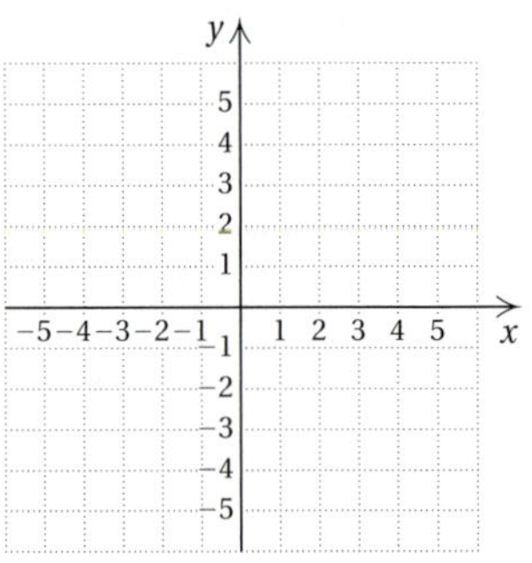

53. $4x + 3 = 0$

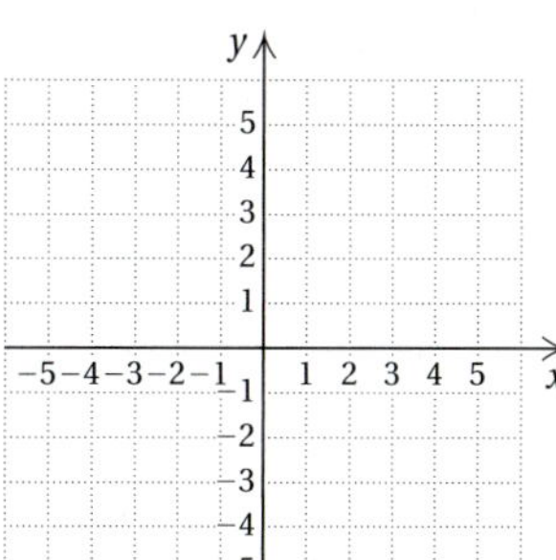

54. $-3x + 12 = 0$

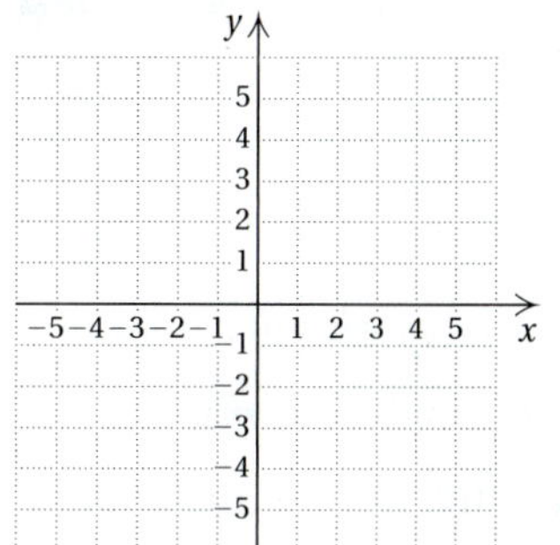

55. $48 - 3y = 0$

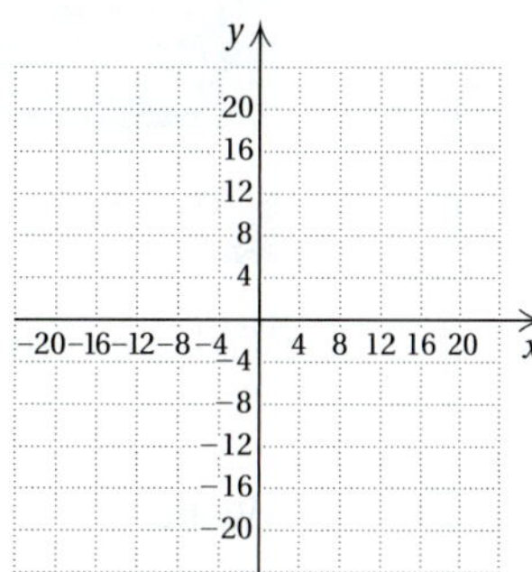

56. $63 + 7y = 0$

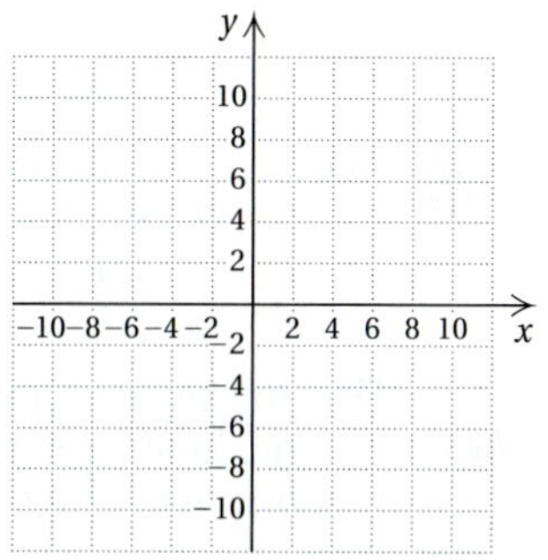

Write an equation for the graph shown.

57.

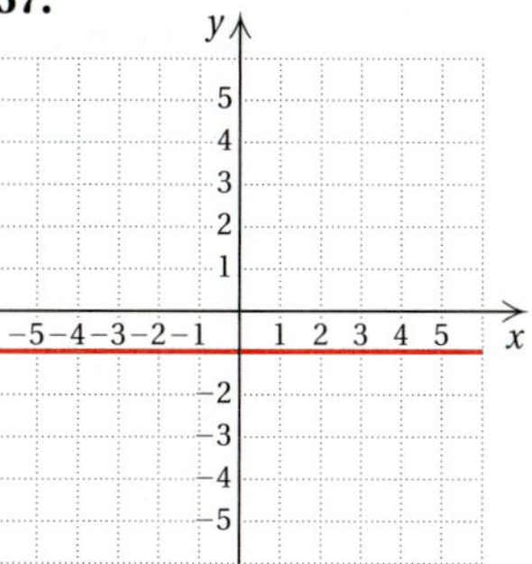

58.

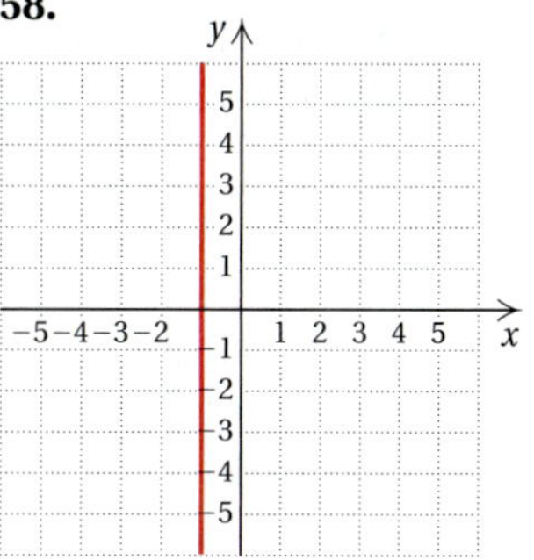

59.

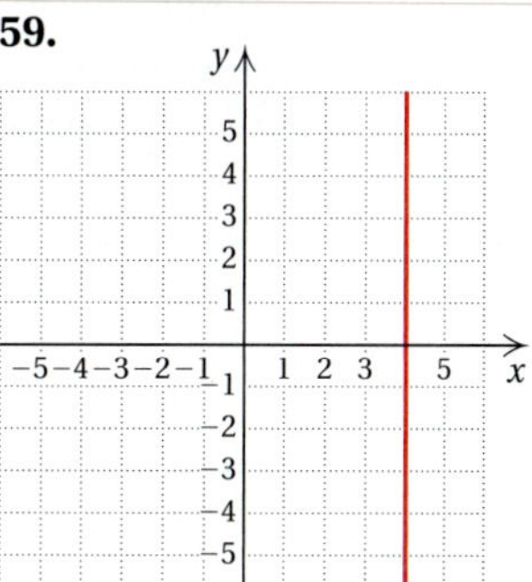

60.

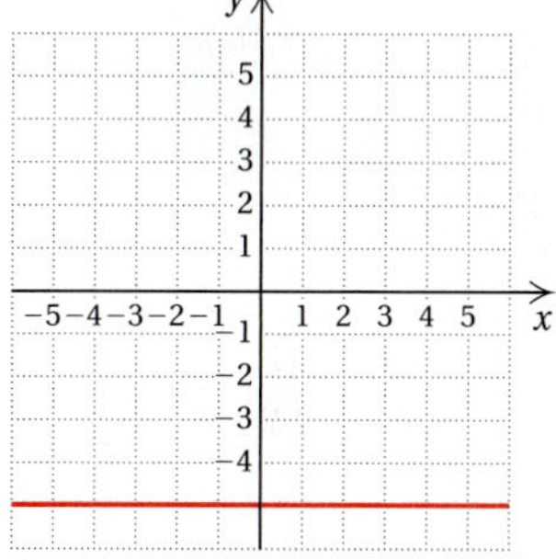

61. D$_W$ If the graph of the equation $Ax + By = C$ is a horizontal line, what can you conclude about A? Why?

62. D$_W$ Explain in your own words why the graph of $x = 7$ is a vertical line.

SKILL MAINTENANCE

Solve. [7.5a]

63. *Desserts.* If a restaurant sells 250 desserts in an evening, it is typical that 40 of them will be pie. What percent of the desserts sold will be pie?

64. *Tipping.* Harry left a 20% tip of \$6.50 for a meal. What was the cost of the meal before the tip?

Solve. [7.7d, e]

65. $-1.6x < 64$

66. $-12x - 71 \geq 13$

67. $x + (x - 1) < (x + 2) - (x + 1)$

68. $6 - 18x \leq 4 - 12x - 5x$

SYNTHESIS

69. Write an equation of a line parallel to the x-axis and passing through $(-3, -4)$.

70. Find the value of m such that the graph of $y = mx + 6$ has an x-intercept of $(2, 0)$.

71. Find the value of k such that the graph of $3x + k = 5y$ has an x-intercept of $(-4, 0)$.

72. Find the value of k such that the graph of $4x = k - 3y$ has a y-intercept of $(0, -8)$.

Objectives

a Given the coordinates of two points on a line, find the slope of the line, if it exists.

b Find the slope, or rate of change, in an applied problem involving slope.

c Find the slope of a line from an equation.

8.4 SLOPE AND APPLICATIONS

a Slope

We have considered two forms of a linear equation,

$$Ax + By = C \quad \text{and} \quad y = mx + b.$$

We found that from the form of the equation $y = mx + b$, we know certain information—namely, that the y-intercept of the line is $(0, b)$.

$$y = mx + b$$

? ← m The y-intercept is $(0, b)$.

What about the constant m? Does it give us certain information about the line? Look at the following graphs and see if you can make any connection between the constant m and the "slant" of the line.

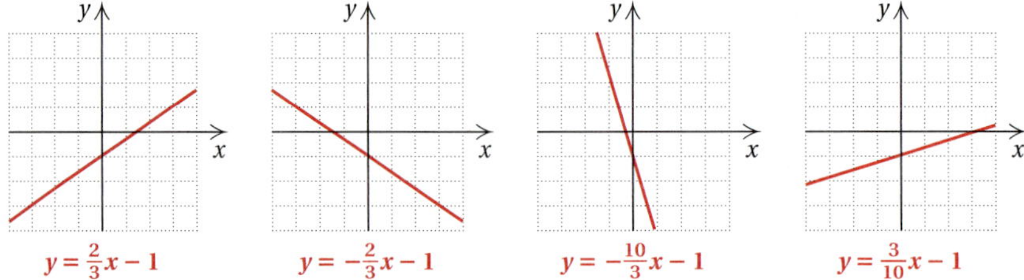

The graphs of some linear equations slant upward from left to right. Others slant downward. Some are vertical and some are horizontal. Some slant more steeply than others. We now look for a way to describe such possibilities with numbers.

Consider a line with two points marked P and Q. As we move from P to Q, the y-coordinate changes from 1 to 3 and the x-coordinate changes from 2 to 6. The change in y is $3 - 1$, or 2. The change in x is $6 - 2$, or 4.

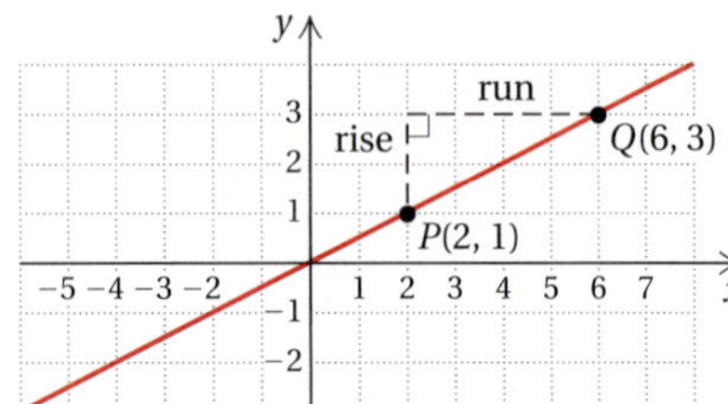

We call the change in y the **rise** and the change in x the **run.** The ratio rise/run is the same for any two points on a line. We call this ratio the **slope.** Slope describes the slant of a line. The slope of the line in the graph above is given by

$$\frac{\text{rise}}{\text{run}} = \frac{\text{the change in } y}{\text{the change in } x}, \text{ or } \frac{2}{4}, \text{ or } \frac{1}{2}.$$

SLOPE

The **slope** of a line containing points (x_1, y_1) and (x_2, y_2) is given by

$$m = \frac{\text{rise}}{\text{run}} = \frac{\text{the change in } y}{\text{the change in } x} = \frac{y_2 - y_1}{x_2 - x_1}.$$

In the preceding definition, (x_1, y_1) and (x_2, y_2)—read "x sub-one, y sub-one and x sub-two, y sub-two"— represent two different points on a line. It does not matter which point is considered (x_1, y_1) and which is considered (x_2, y_2) so long as coordinates are subtracted in the same order in both the numerator and the denominator — for example,

$$\frac{y_2 - y_1}{x_2 - x_1} = \frac{y_1 - y_2}{x_1 - x_2}.$$

EXAMPLE 1 Graph the line containing the points $(-4, 3)$ and $(2, -6)$ and find the slope.

The graph is shown below. We consider (x_1, y_1) to be $(-4, 3)$ and (x_2, y_2) to be $(2, -6)$. From $(-4, 3)$ and $(2, -6)$, we see that the change in y, or the rise, is $-6 - 3$, or -9. The change in x, or the run, is $2 - (-4)$, or 6.

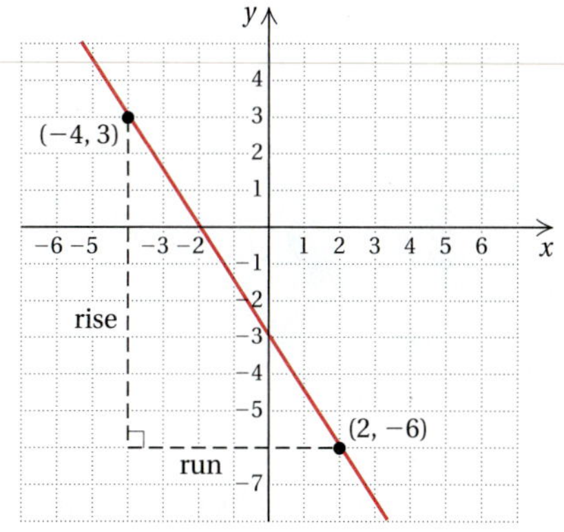

$$\text{Slope} = \frac{\text{rise}}{\text{run}} = \frac{\text{change in } y}{\text{change in } x}$$

$$= \frac{y_2 - y_1}{x_2 - x_1}$$

$$= \frac{-6 - 3}{2 - (-4)}$$

$$= \frac{-9}{6} = -\frac{9}{6}, \text{ or } -\frac{3}{2}.$$

When we use the formula

$$m = \frac{y_2 - y_1}{x_2 - x_1},$$

we can subtract in two ways. We must remember, however, to subtract the y-coordinates in the same order that we subtract the x-coordinates. Let's redo Example 1, where we consider (x_1, y_1) to be $(2, -6)$ and (x_2, y_2) to be $(-4, 3)$:

$$\text{Slope} = \frac{\text{change in } y}{\text{change in } x} = \frac{3 - (-6)}{-4 - 2} = \frac{9}{-6} = -\frac{3}{2}.$$

The slope of a line tells how it slants. A line with positive slope slants up from left to right. The larger the slope, the steeper the slant. A line with negative slope slants downward from left to right.

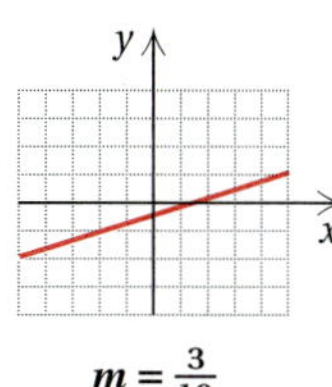

$m = \frac{3}{10}$

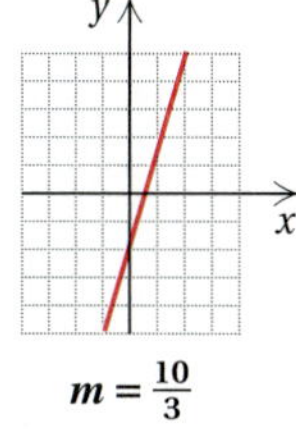

$m = \frac{10}{3}$

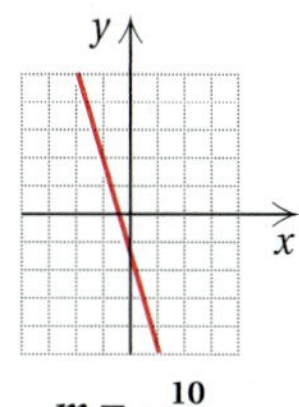

$m = -\frac{10}{3}$

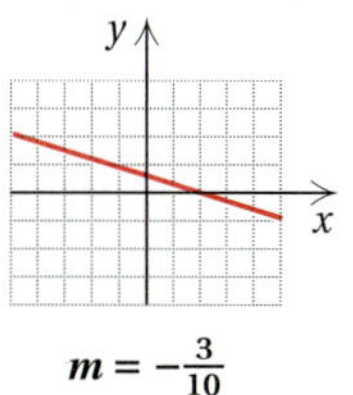

$m = -\frac{3}{10}$

Do Exercises 1 and 2.

Graph the line containing the points and find the slope in two different ways.

1. $(-2, 3)$ and $(3, 5)$

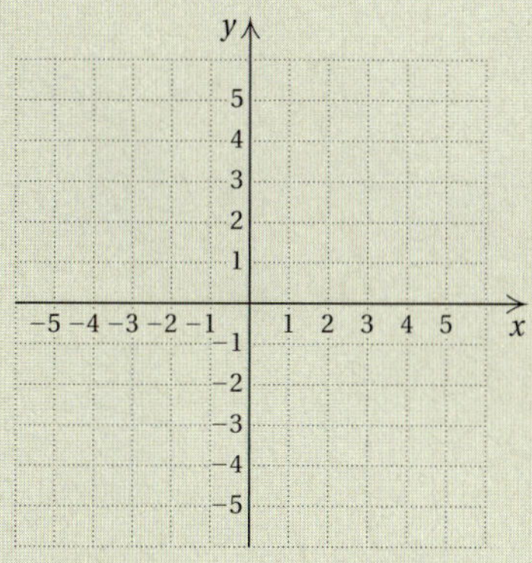

2. $(0, -3)$ and $(-3, 2)$

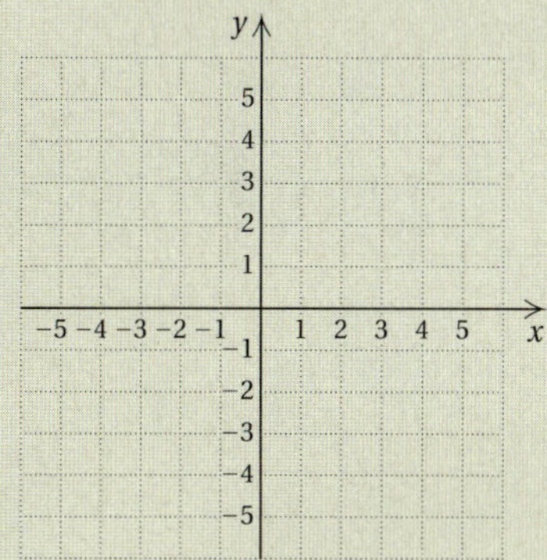

Answers on page A-25

3. **Construction.** Public buildings regularly include steps with 7-in. risers and 11-in. treads. Find the grade of such a stairway.

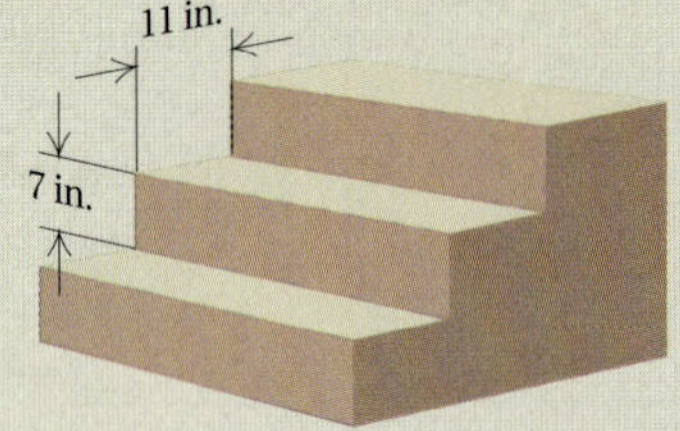

b Applications of Slope; Rates of Change

Slope has many real-world applications. For example, numbers like 2%, 3%, and 6% are often used to represent the *grade* of a road, a measure of how steep a road on a hill or mountain is. For example, a 3% grade $\left(3\% = \frac{3}{100}\right)$ means that for every horizontal distance of 100 ft, the road rises 3 ft, and a -3% grade means that for every horizontal distance of 100 ft, the road drops 3 ft. (Road signs do not include negative signs. It's usually obvious whether you are climbing or descending.) The concept of grade also occurs in skiing or snowboarding, where a 4% grade is considered very tame, but a 40% grade is considered extremely steep. And in cardiology, a physician may change the grade of a treadmill to measure its effect on heartbeat.

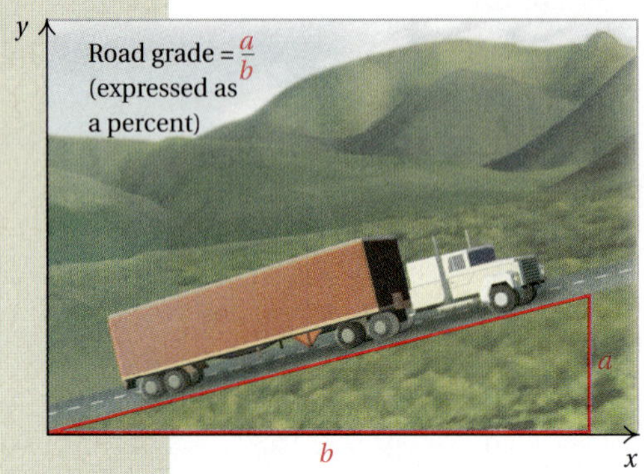

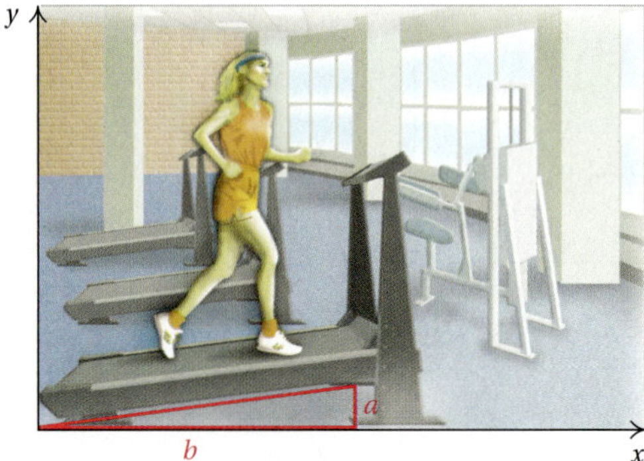

Architects and carpenters use slope when designing and building stairs, ramps, or roof pitches. Another application occurs in hydrology. When a river flows, the strength or force of the river depends on how far the river falls vertically compared to how far it flows horizontally.

EXAMPLE 2 *Skiing.* Among the steepest skiable terrain in North America, the Headwall on Mount Washington, in New Hampshire, drops 720 ft over a horizontal distance of 900 ft. Find the grade of the Headwall.

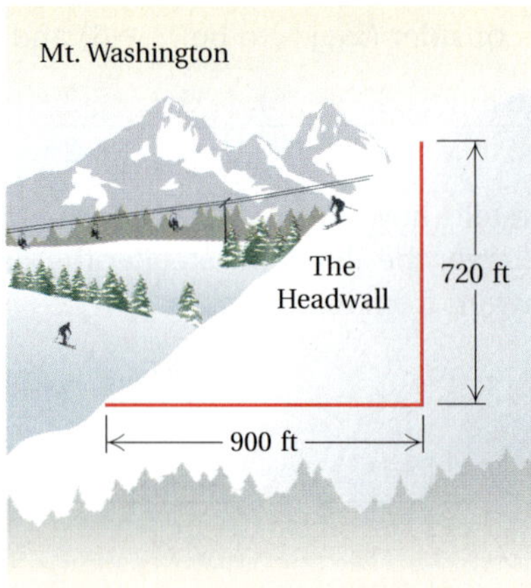

The grade of the Headwall is its slope, expressed as a percent:

$$m = \frac{720}{900} \begin{array}{l}\leftarrow \text{Vertical change}\\ \leftarrow \text{Horizontal change}\end{array}$$

$$= \frac{8}{10} = 80\%.$$

Do Exercise 3.

Answer on page A-25

Slope can also be considered as a **rate of change.**

EXAMPLE 3 *Haircutting.* Gary's Barber Shop has a graph displaying data from a recent day's work. Use the graph to determine the slope, or the rate of change, of the number of haircuts with respect to time.

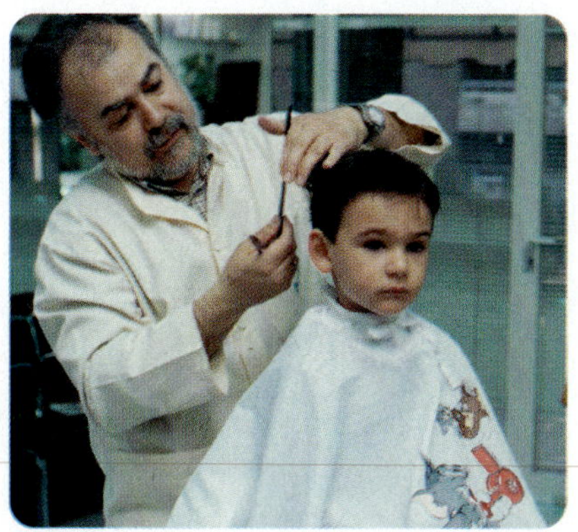

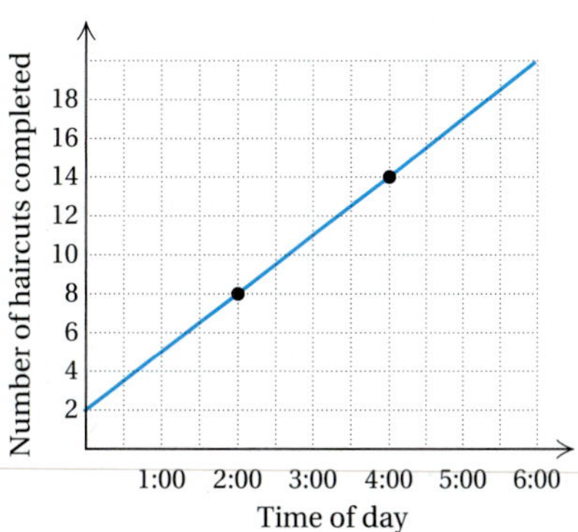

The vertical axis of the graph shows the number of haircuts and the horizontal axis the time, in units of one hour. We can describe the rate of change in the number of haircuts with respect to time as

$$\frac{\text{Haircuts}}{\text{Hour}}, \quad \text{or} \quad \textit{number of haircuts per hour.}$$

This value is the slope of the line. We determine two ordered pairs on the graph—in this case,

(2:00, 8 haircuts) and (4:00, 14 haircuts).

This tells us that in the 2 hr between 2:00 and 4:00, 6 haircuts were completed. Thus,

$$\text{Rate of change} = \frac{14 \text{ haircuts} - 8 \text{ haircuts}}{4{:}00 - 2{:}00} = \frac{6 \text{ haircuts}}{2 \text{ hours}} = 3 \text{ haircuts per hour.}$$

Do Exercise 4.

EXAMPLE 4 *Defense Spending.* Each year the United States spends a smaller percent of its annual budget on defense. Use the following graph to determine the slope, or rate of change, of the percent of the budget spent on defense with respect to time.
Source: U.S. Office of Management and Budget

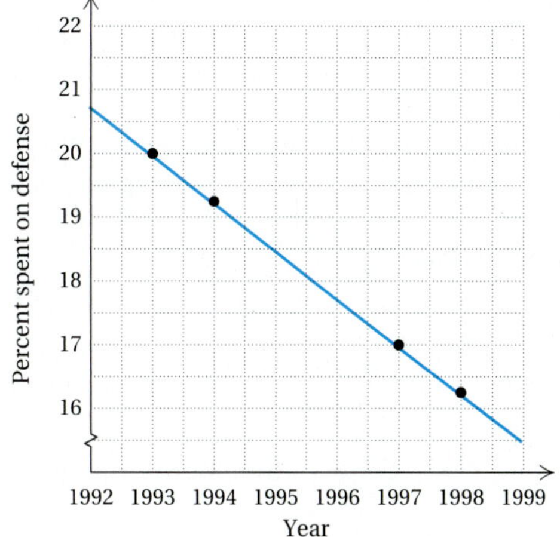

4. **Cost of a Telephone Call.** The following graph shows data concerning recent MCI phone calls from San Francisco to Pittsburgh. At what rate was the customer being billed?
Source: MCI

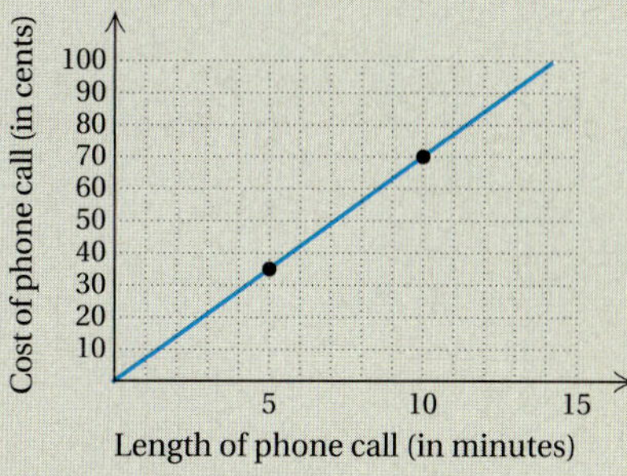

Answer on page A-25

5. **Unemployment.** Find the rate of change in the percent of U.S. workers that are unemployed.
Source: U.S. Bureau of Labor Statistics

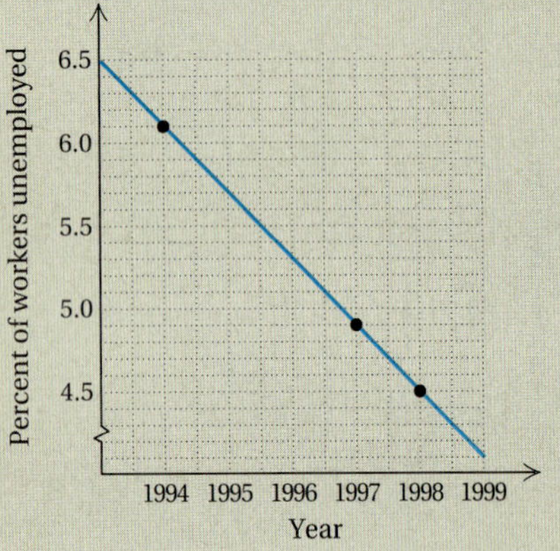

Answer on page A-25

The vertical axis of the graph shows the percent of the budget spent on defense and the horizontal axis the years. We can describe the rate of change in the percent spent with respect to time as

$$\frac{\text{Percent spent}}{\text{Years}}, \quad \text{or} \quad \textit{percent spent per year.}$$

This value is the slope of the line. We determine two ordered pairs on the graph—in this case,

$$(1993, 20) \quad \text{and} \quad (1997, 17).$$

This tells us that in the 4 yr from 1993 to 1997, the percent spent dropped from 20% to 17%. Thus,

$$\text{Rate of change} = \frac{17\% - 20\%}{1997 - 1993} = \frac{-3\%}{4 \text{ yr}} = -\tfrac{3}{4}\% \text{ per year.}$$

Do Exercise 5.

Study Tips

SMALL STEPS LEAD TO GREAT SUCCESS (PART 2)

Chris Widener is a popular motivational speaker and writer. In his article "A Little Equation That Creates Big Results," he proposes the following equation: "Your Short-Term Actions Multiplied By Time = Your Long-Term Accomplishments."

Think of the major or career toward which you are working as a long-term accomplishment. We (your authors and instructors) are at a point in life where we realize the long-term benefits of learning mathematics. For you as students, it may be more difficult to see those long-term results. But make an effort to do so.

Widener goes on to say, "We need to take action on our dreams and beliefs every day." Think of the long-term goal as you do the short-term tasks of homework in math, studying for tests, and completing this course so you can move on to what it prepares you for.

Who writes best-selling novels? The person who only dreams of becoming a best-selling author or the one who also spends 4 hours a day doing research and working at a computer?

Who loses weight? The person who thinks about being thin or the one who also plans a healthy diet and runs 3 miles a day?

Who is successful at math? The person who only knows all the benefits of math or the one who also spends 2 hours studying outside of class for every hour spent inside class?

"The purpose of man is in action, not thought."

Thomas Carlyle, British historian/essayist

"Prepare for your success in little ways and you will eventually see results in big ways. It's almost magical."

Tom Morris, public philosopher/speaker

C Finding the Slope from an Equation

It is possible to find the slope of a line from its equation. Let's consider the equation $y = 2x + 3$, which is in the form $y = mx + b$. We can find two points by choosing convenient values for x—say, 0 and 1—and substituting to find the corresponding y-values. We find the two points on the line to be $(0, 3)$ and $(1, 5)$. The slope of the line is found using the definition of slope:

$$m = \frac{\text{change in } y}{\text{change in } x} = \frac{5-3}{1-0} = \frac{2}{1} = 2.$$

The slope is 2. Note that this is also the coefficient of the x-term in the equation $y = 2x + 3$.

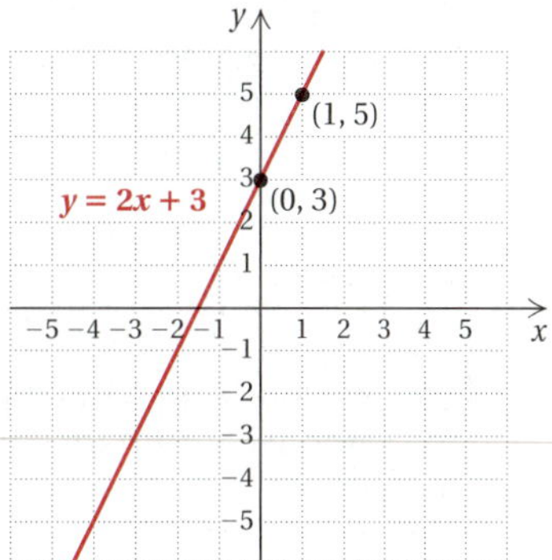

CALCULATOR CORNER

Visualizing Slope

Exercises: Graph each of the following sets of equations using the window settings $[-6, 6, -4, 4]$, Xscl = 1, Yscl = 1.

1. $y = x$, $y = 2x$, $y = 5x$, $y = 10x$

   What do you think the graph of $y = 123x$ will look like?

2. $y = x$, $y = \frac{3}{4}x$, $y = 0.38x$, $y = \frac{5}{32}x$

   What do you think the graph of $y = 0.000043x$ will look like?

DETERMINING SLOPE FROM THE EQUATION $y = mx + b$

The slope of the line $y = mx + b$ is m. To find the slope of a nonvertical line, solve the linear equation in x and y for y and get the resulting equation in the form $y = mx + b$. The coefficient of the x-term, m, is the slope of the line.

EXAMPLES Find the slope of the line.

5. $y = -3x + \frac{2}{9}$

   $\to m = -3 = \text{Slope}$

6. $y = \frac{4}{5}x$

   $\to m = \frac{4}{5} = \text{Slope}$

7. $y = x + 6$

   $\to m = 1 = \text{Slope}$

8. $y = -0.6x - 3.5$

   $\to m = -0.6 = \text{Slope}$

Do Exercises 6–9.

To find slope from an equation, we may have to first find an equivalent form of the equation.

EXAMPLE 9 Find the slope of the line $2x + 3y = 7$.

We solve for y to get the equation in the form $y = mx + b$:

$$2x + 3y = 7$$
$$3y = -2x + 7$$
$$y = \frac{-2x + 7}{3}$$
$$y = -\frac{2}{3}x + \frac{7}{3}. \qquad \text{This is } y = mx + b.$$

The slope is $-\frac{2}{3}$.

Find the slope of the line.

6. $y = 4x + 11$

7. $y = -17x + 8$

8. $y = -x + \frac{1}{2}$

9. $y = \frac{2}{3}x - 1$

Find the slope of the line.

10. $4x + 4y = 7$

11. $5x - 4y = 8$

Answers on page A-25

CALCULATOR CORNER

Visualizing Slope

Exercises: Graph each of the following sets of equations using the window settings $[-6, 6, -4, 4]$, Xscl $= 1$, Yscl $= 1$.

1. $y = -x, y = -2x,$
$y = -5x, y = -10x$
What do you think the graph of $y = -123x$ will look like?

2. $y = -x, y = -\frac{3}{4}x,$
$y = -0.38x, y = -\frac{5}{32}x$
What do you think the graph of $y = -0.000043x$ will look like?

Find the slope, if it exists, of the line.

12. $x = 7$

13. $y = -5$

Answers on page A-25

Do Exercises 10 and 11 on the preceding page.

What about the slope of a horizontal or a vertical line?

EXAMPLE 10 Find the slope of the line $y = 5$.

We can think of $y = 5$ as $y = 0x + 5$. Then from this equation, we see that $m = 0$. Consider the points $(-3, 5)$ and $(4, 5)$, which are on the line. The change in $y = 5 - 5$, or 0. The change in $x = -3 - 4$, or -7. We have

$$m = \frac{5 - 5}{-3 - 4} = \frac{0}{-7} = 0.$$

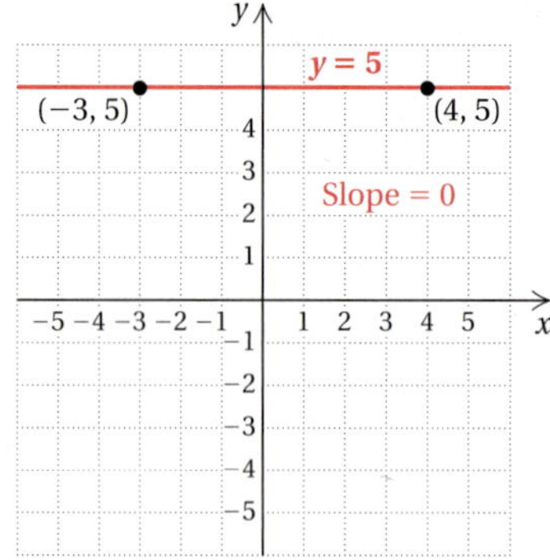

Any two points on a horizontal line have the same y-coordinate. Thus the change in y is 0.

EXAMPLE 11 Find the slope of the line $x = -4$.

Consider the points $(-4, 3)$ and $(-4, -2)$, which are on the line. The change in $y = 3 - (-2)$, or 5. The change in $x = -4 - (-4)$, or 0. We have

$$m = \frac{3 - (-2)}{-4 - (-4)} = \frac{5}{0}. \quad \text{Not defined}$$

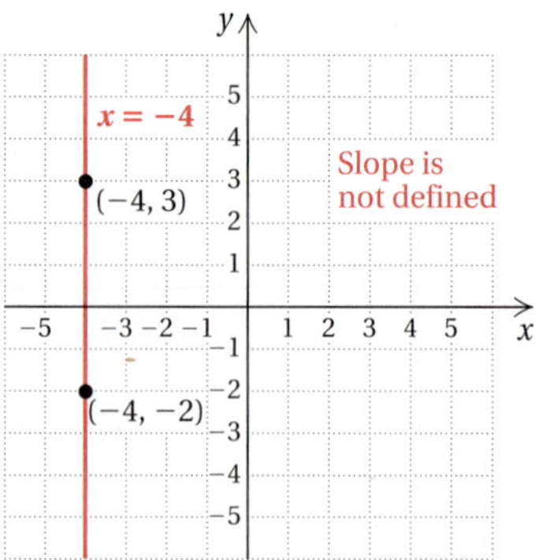

Since division by 0 is not defined, the slope of this line is not defined. The answer in this example is "The slope of this line is not defined."

SLOPE 0; SLOPE NOT DEFINED

The slope of a horizontal line is 0. The slope of a vertical line is not defined.

Do Exercises 12 and 13.

We will consider slope again and use it in graphing in Chapter 12.

For Extra Help

InterAct Math

a Find the slope, if it exists, of the line.

1.

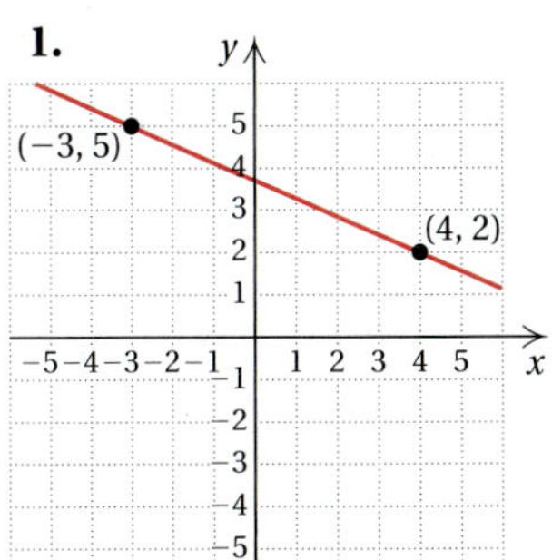

2.

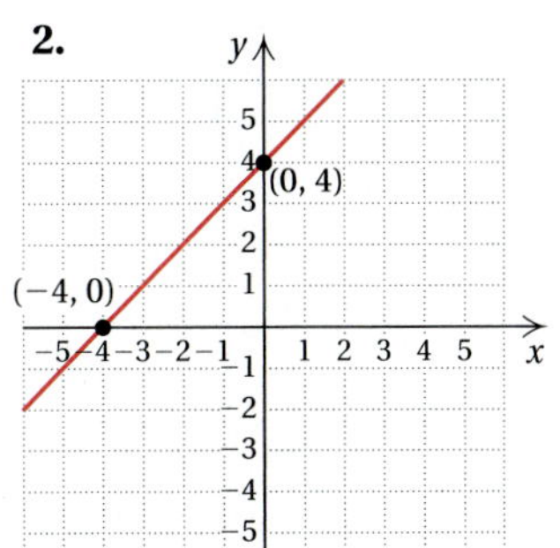

3.

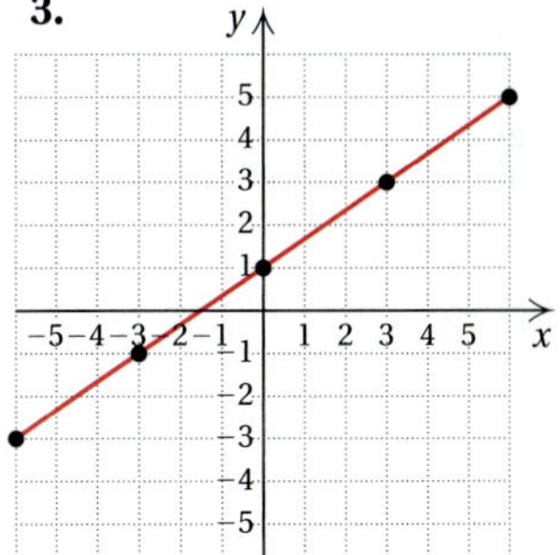

4.

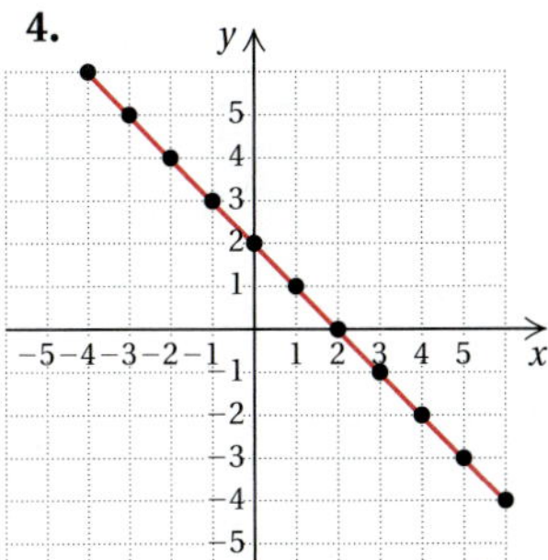

5.

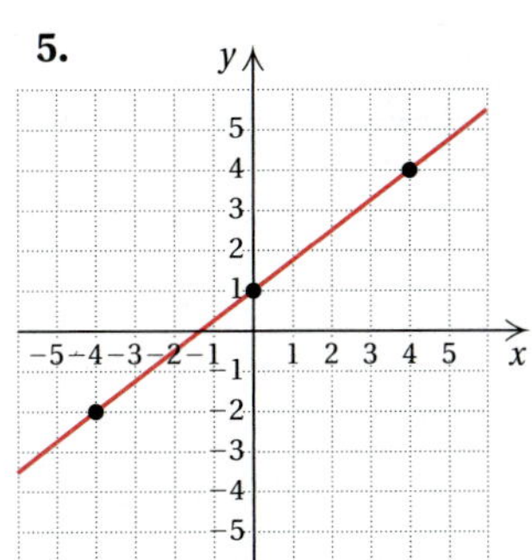

6.

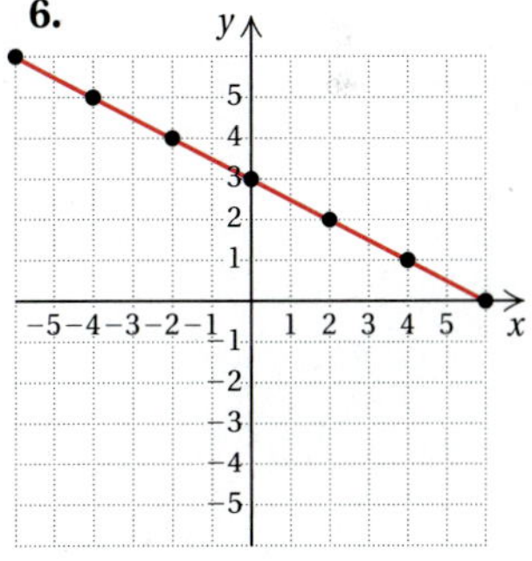

7.

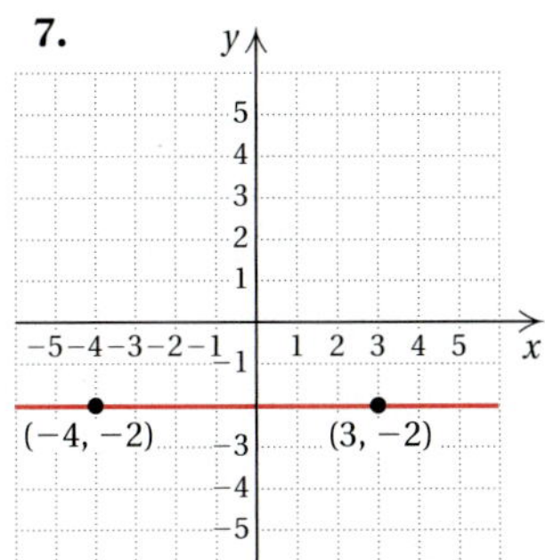

8.

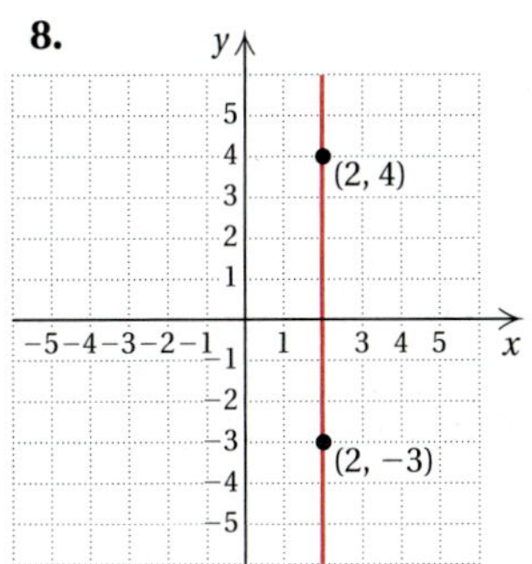

Graph the line containing the given pair of points and find the slope.

9. $(-2, 4), (3, 0)$

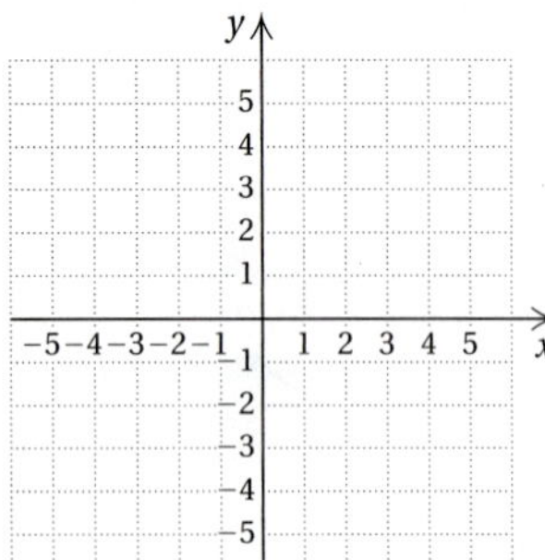

10. $(2, -4), (-3, 2)$

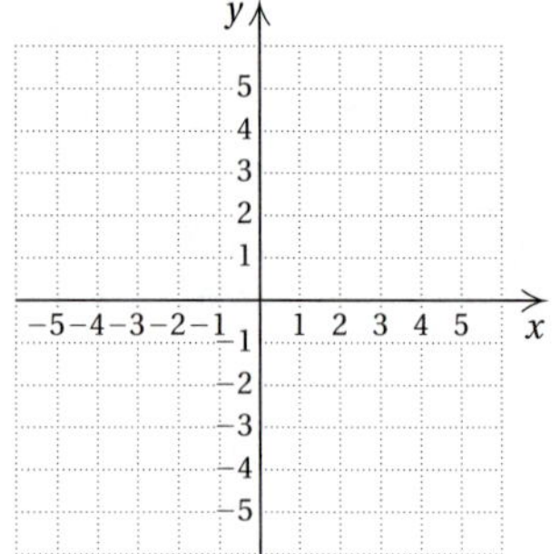

11. $(-4, 0), (-5, -3)$

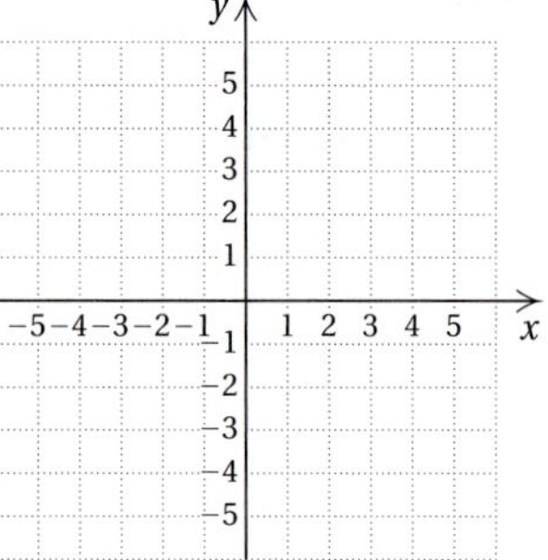

12. $(-3, 0), (-5, -2)$

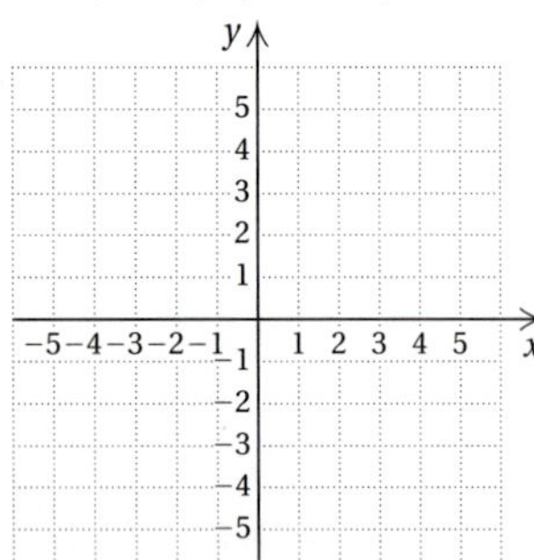

13. $(-4, 2), (2, -3)$

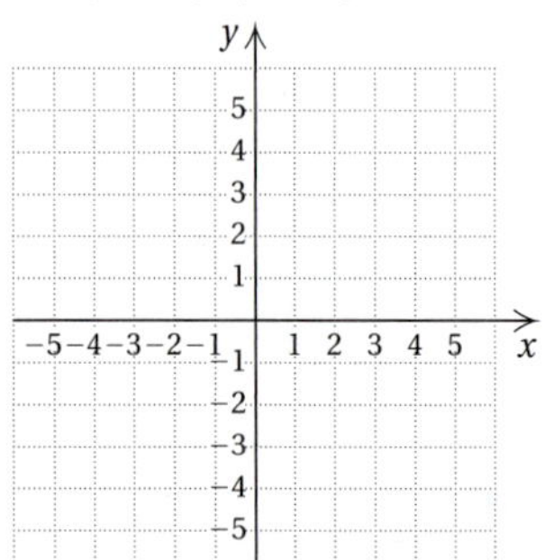

14. $(-3, 5), (4, -3)$

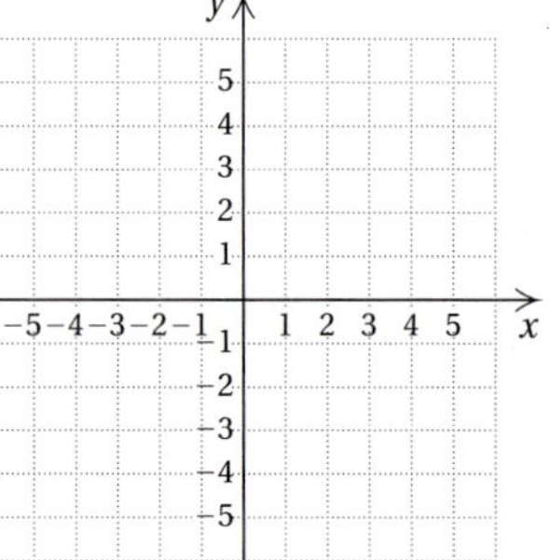

15. $(5, 3), (-3, -4)$

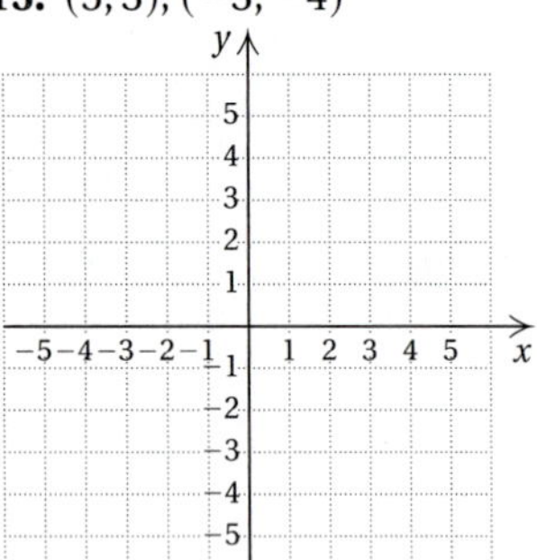

16. $(-4, -3), (2, 5)$

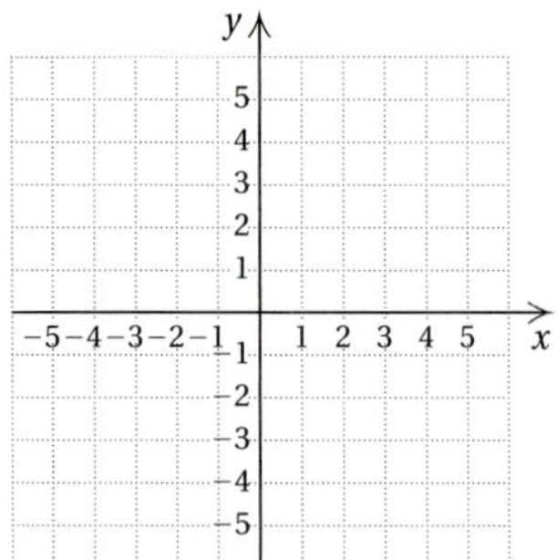

Find the slope, if it exists, of the line containing the given pair of points.

17. $\left(2, -\frac{1}{2}\right), \left(5, \frac{3}{2}\right)$

18. $\left(\frac{2}{3}, -1\right), \left(\frac{5}{3}, 2\right)$

19. $(4, -2), (4, 3)$

20. $(4, -3), (-2, -3)$

b In Exercises 21–24, find the slope (or rate of change).

21. Find the slope (or pitch) of the roof.

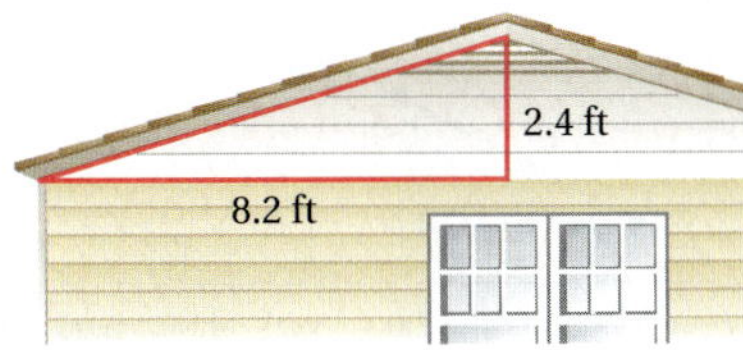

22. Find the slope (or grade) of the road.

23. Find the slope of the river.

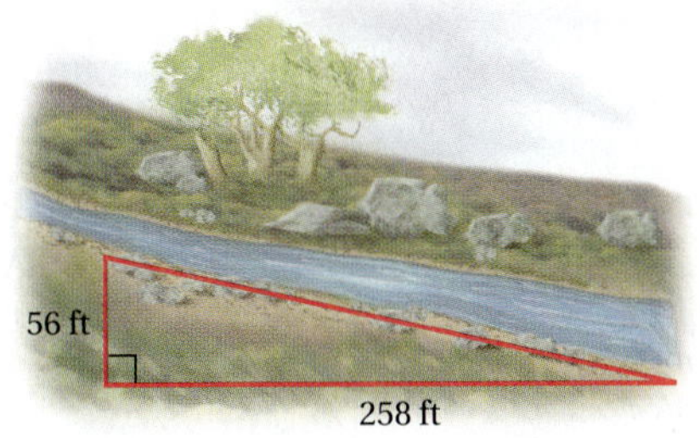

24. Find the slope of the treadmill.

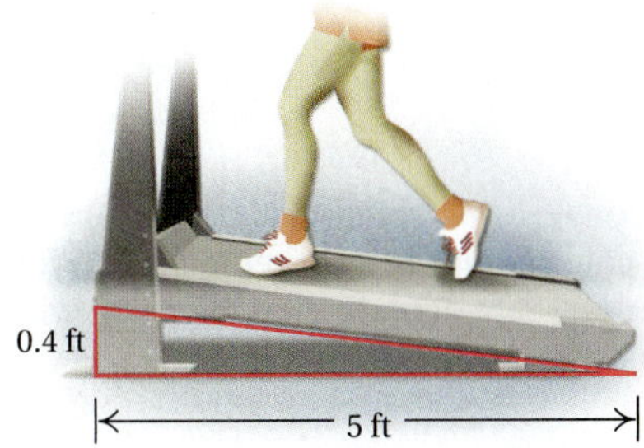

25. *Slope of Long's Peak.* From a base elevation of 9600 ft, Long's Peak in Colorado rises to a summit elevation of 14,255 ft over a horizontal distance of 15,840 ft. Find the grade of Long's Peak.

26. *Ramps for the Disabled.* In order to meet federal standards, a wheelchair ramp must not rise more than 1 ft over a horizontal distance of 12 ft. Express this slope as a grade.

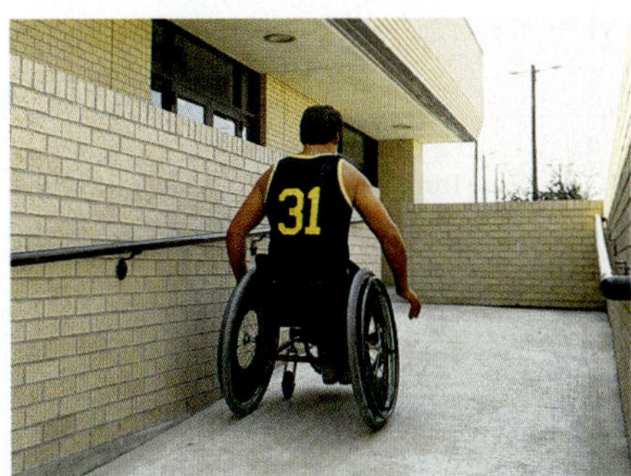

In Exercises 27–32, use the graph to calculate a rate of change in which the units of the horizontal axis are used in the denominator.

27. *Gas Mileage.* The following graph shows data for a Honda Odyssey driven on interstate highways. Find the rate of change in miles per gallon, that is, the gas mileage.
Source: Honda Motor Company

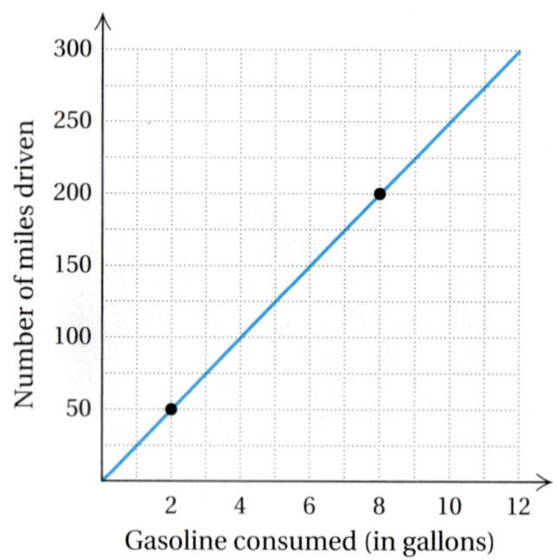

28. *Hairdresser.* Eve's Custom Cuts has a graph displaying data from a recent day of work. Find the rate of change of the number of haircuts with respect to time.

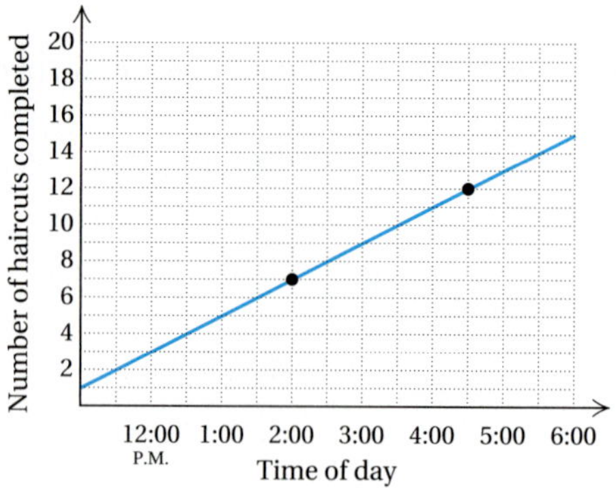

29. *Depreciation of an Office Machine.* The value of a particular color copier is represented in the following graph. Find the rate of change of the value with respect to time, in dollars per year.

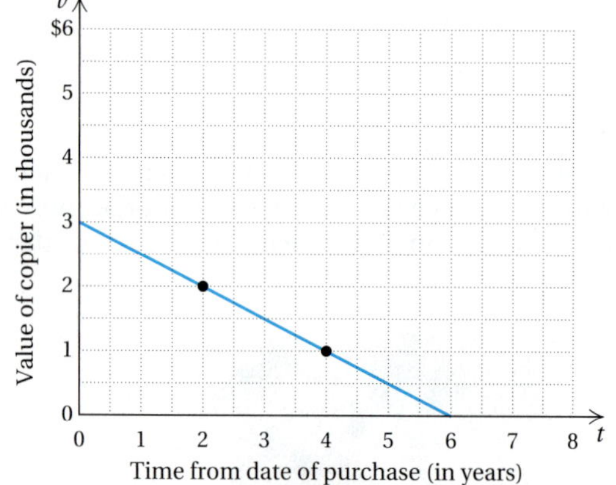

30. *NASA Spending.* The amount spent by the National Aeronautics and Space Administration (NASA) is represented in the following graph. Find the rate of change of spending with respect to time, in dollars per year.
Source: NASA

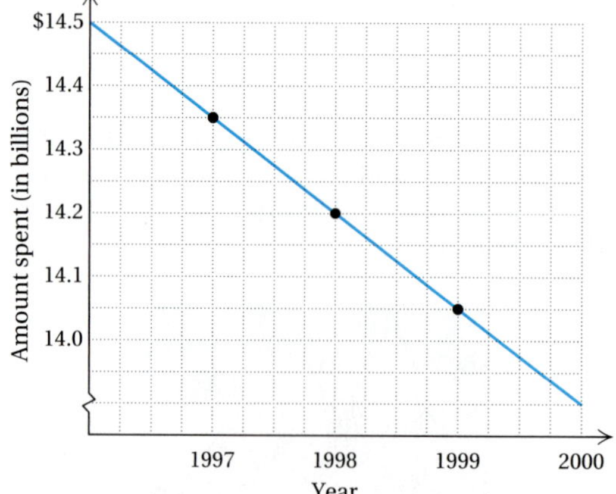

31. *Population Growth of Alaska.* The population of Alaska is shown in the following graph. Find the rate of change of the population with respect to time, in number of people per year.

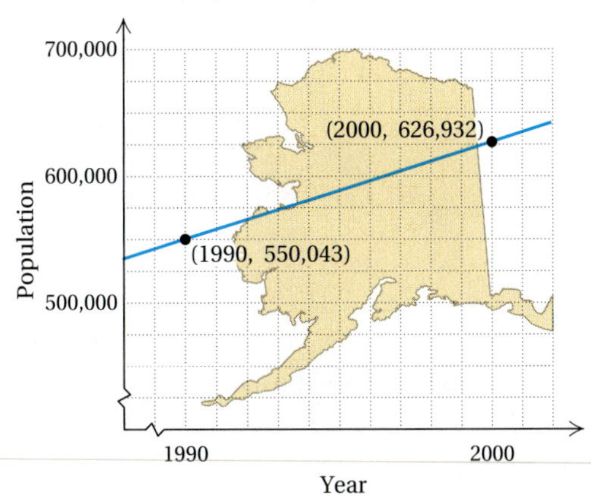

32. *Population Growth of Utah.* The population of Utah is shown in the following graph. Find the rate of change of the population, in number of people per year.

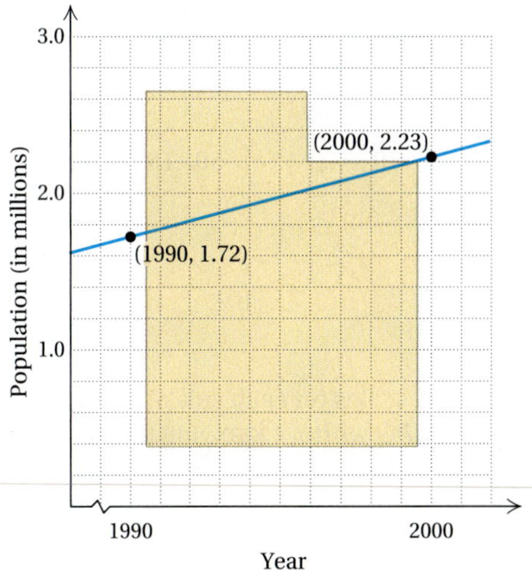

C Find the slope, if it exists.

33. $y = -10x + 7$

34. $y = \frac{10}{3}x - \frac{5}{7}$

35. $y = 3.78x - 4$

36. $y = -\frac{3}{5}x + 28$

37. $3x - y = 4$

38. $-2x + y = 8$

39. $x + 5y = 10$

40. $x - 4y = 8$

41. $3x + 2y = 6$

42. $2x - 4y = 8$

43. $5x - 7y = 14$

44. $3x - 6y = 10$

45. $y = -2.74x$

46. $y = \frac{219}{298}x - 6.7$

47. $9x = 3y + 5$

48. $4y = 9x - 7$

49. $5x - 4y + 12 = 0$

50. $16 + 2x - 8y = 0$

51. $y = 4$

52. $x = -3$

Assuming that the scales on each axis of each graph are the same, explain how you can estimate the slope of the line that contains segment PQ without knowing the coordinates of the points P and Q.

53. DW

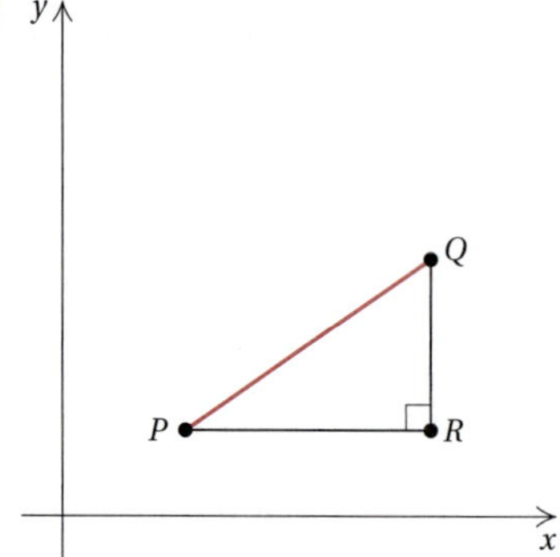

54. DW

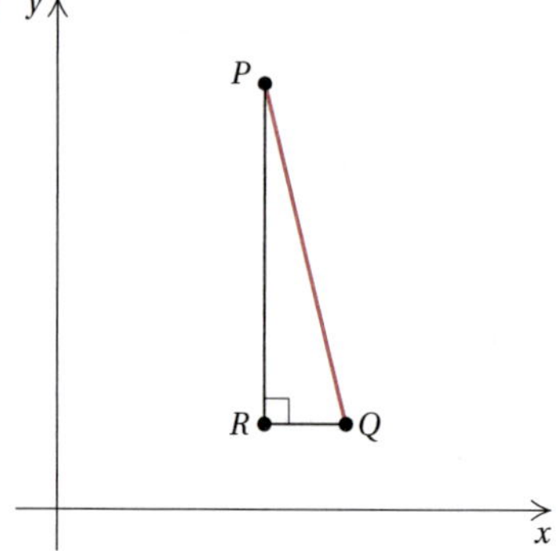

SKILL MAINTENANCE

Solve. [7.2a]

55. $15x = -60$

56. $-\frac{1}{2}x = 10$

57. $-x = 37$

58. $\frac{1}{4} = -\frac{x}{3}$

Solve. [7.5a]

59. What is 15% of $23.80?

60. $7.29 is 15% of what number?

61. Jennifer left an $8.50 tip for a meal that cost $42.50. What percent of the cost of the meal was the tip?

62. Kristen left an 18% tip of $3.24 for a meal. What was the cost of the meal before the tip?

63. Juan left a 15% tip for a meal. The total cost of the meal, including the tip, was $51.92. What was the cost of the meal before the tip was added?

64. After a 25% reduction, a sweater is on sale for $41.25. What was the original price?

SYNTHESIS

Graph the equation using the standard viewing window. Then construct a table of y-values for x-values starting at $x = -10$ with $\Delta\text{Tbl} = 0.1$.

65. $y = 0.35x - 7$

66. $y = 5.6 - x^2$

67. $y = x^3 - 5$

68. $y = 4 + 3x - x^2$

CHAPTER 8 Summary and Review

The review that follows is meant to prepare you for a chapter exam. It consists of two parts. The first part is a checklist of some of the Study Tips referred to in this and preceding chapters. The second part is the Review Exercises. These provide practice exercises for the exam, together with references to section objectives so you can go back and review. Before beginning, stop and look back over the skills you have obtained. What skills in mathematics do you have now that you did not have before studying this chapter?

STUDY TIPS CHECKLIST

The foundation of all your study skills is TIME!	☐ Are you doing exercises without answers at the back of the book as part of every homework assignment to better prepare you to take tests? ☐ Have you been taking the primary responsibility for your learning? ☐ Have you established a learning relationship with your instructor? ☐ Have you used the videotapes to supplement your learning? ☐ Are you preparing for each homework assignment by reading the explanations and following the step-by-step examples in the text?

REVIEW EXERCISES

1. *Federal Spending.* The following pie chart shows how our federal income tax dollars are used. As a freelance graphic artist, Jennifer pays $3525 in taxes. How much of Jennifer's tax payment goes toward defense? toward social programs? [8.1a]

Where Your Tax Dollars Are Spent

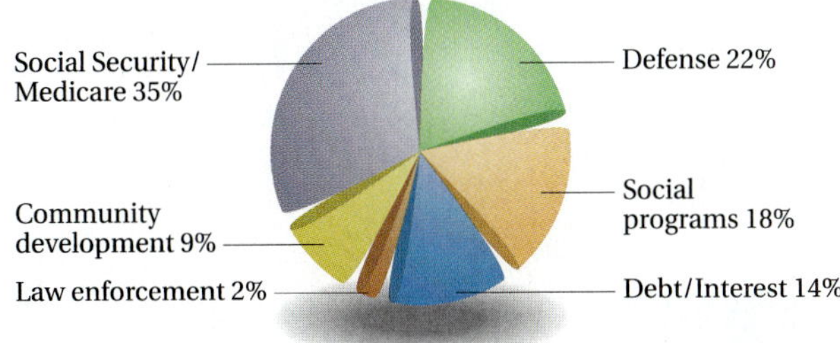

Source: U.S. Department of the Treasury

Chicken Consumption. The following line graph shows average chicken consumption from 1980 to 2000. Use the line graph for Exercises 2–6. [8.1a]

Chicken Consumption

Number of pounds per person: 10, 20, 30, 40, 50, 60, 70, 80

Year: 1980, 1985, 1990, 1995, 2000

2. About how many pounds of chicken were consumed per person in 1980?

3. About how many pounds of chicken were consumed per person in 2000?

4. By what amount did chicken consumption increase from 1980 to 2000?

5. In what year did the consumption of chicken exceed 70 lb per person?

6. In what 5-yr period was the difference in consumption the greatest?

Water Usage. The following bar graph shows water usage, in gallons, for various tasks. Use the bar graph for Exercises 7–10. [8.1a]

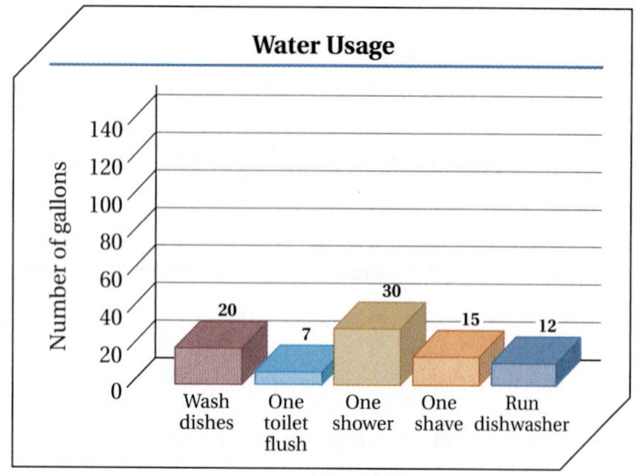

Source: American Water Works Association

7. Which task requires the most water?

8. Which task requires the least water?

9. Which tasks require 15 or more gallons?

10. Which task requires 7 gallons?

Find the coordinates of the point. [8.1d]

11. *A* **12.** *B* **13.** *C*

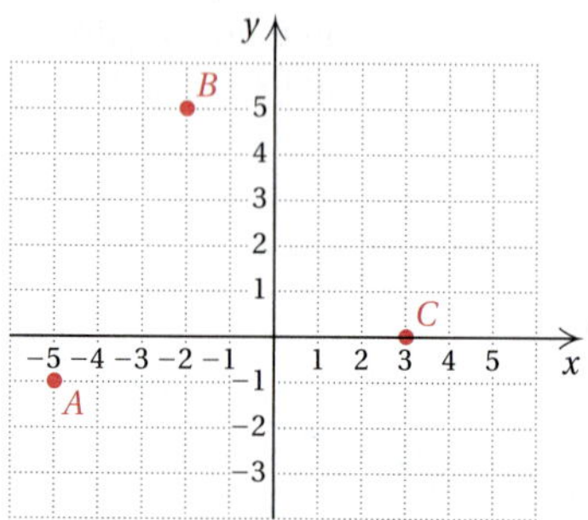

Plot the point. [8.1b]

14. $(2, 5)$ **15.** $(0, -3)$ **16.** $(-4, -2)$

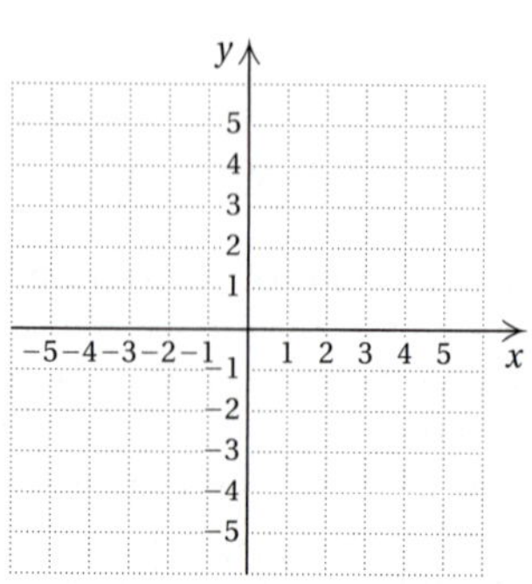

In which quadrant is the point located? [8.1c]

17. $(3, -8)$ **18.** $(-20, -14)$ **19.** $(4.9, 1.3)$

Determine whether the ordered pair is a solution of $2y - x = 10$. [8.2a]

20. $(2, -6)$ **21.** $(0, 5)$

22. Show that the ordered pairs $(0, -3)$ and $(2, 1)$ are solutions of the equation $2x - y = 3$. Then use the graph of the two points to determine another solution. Answers may vary. [8.2a]

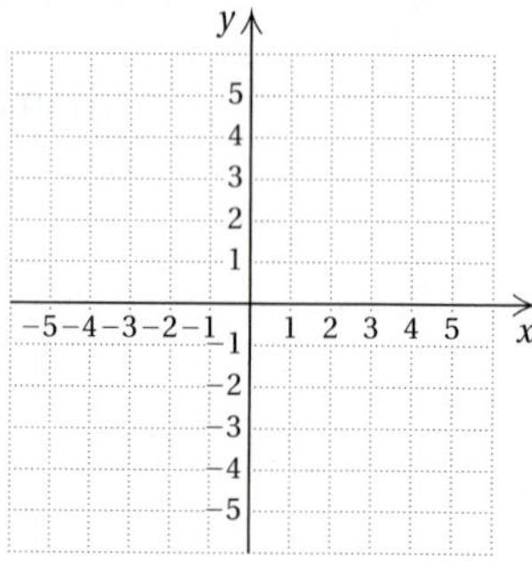

Graph the equation, identifying the y-intercept. [8.2b]

23. $y = 2x - 5$

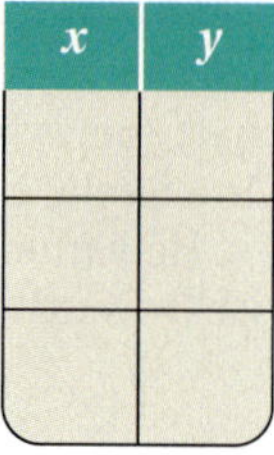

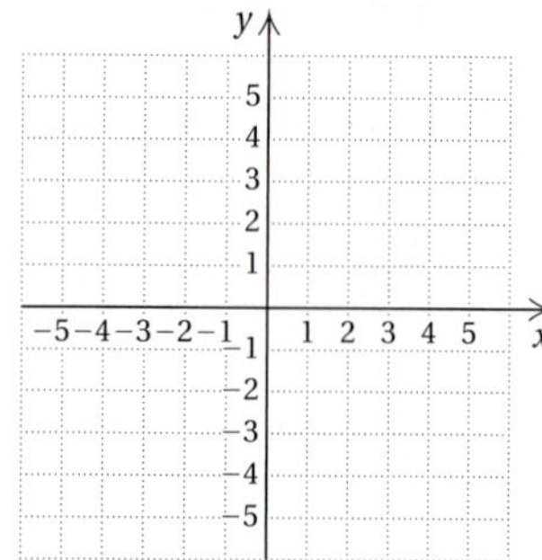

24. $y = -\frac{3}{4}x$

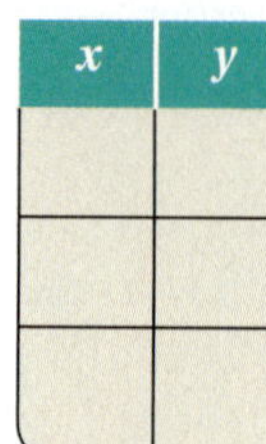

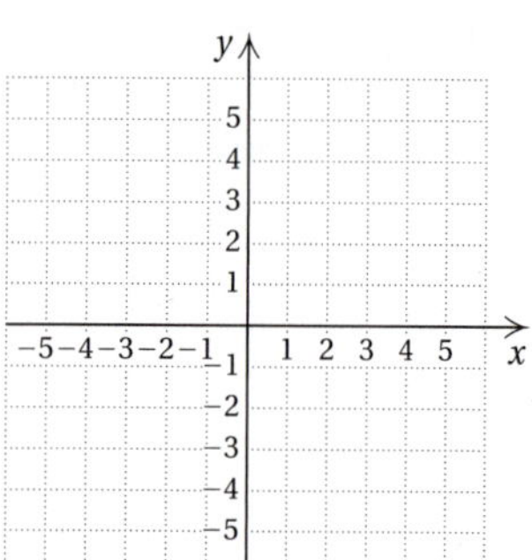

25. $y = -x + 4$

x	y

26. $y = 3 - 4x$

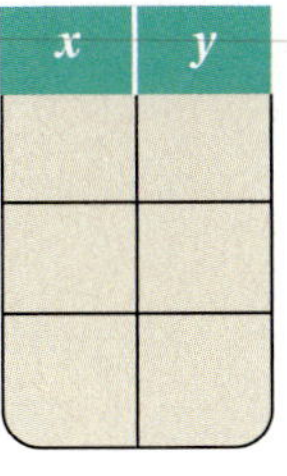

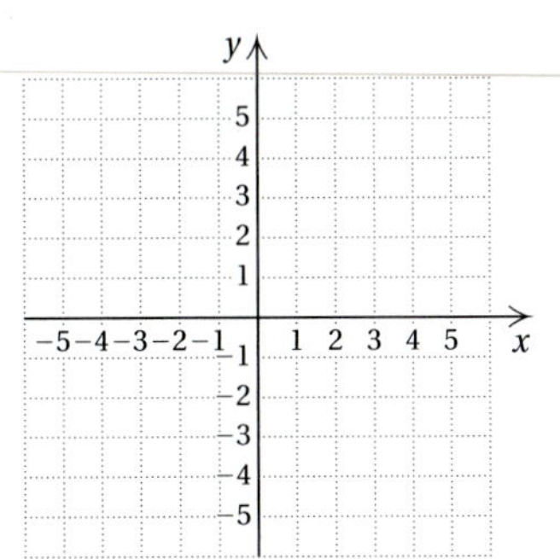

Graph the equation. [8.3b]

27. $y = 3$

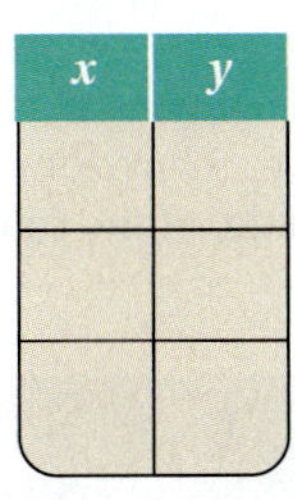

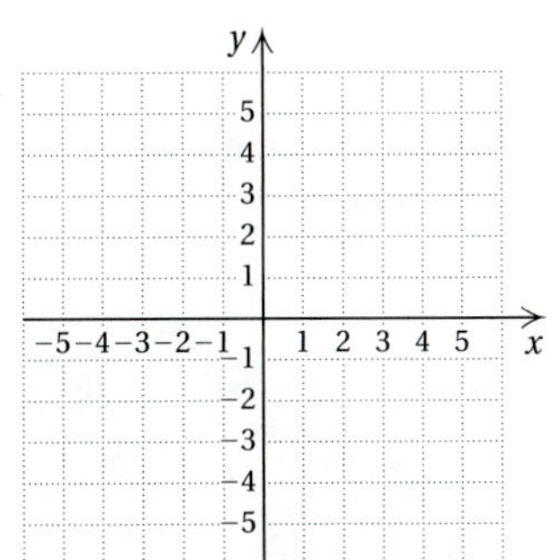

28. $5x - 4 = 0$

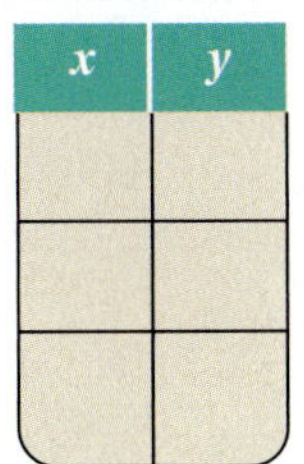

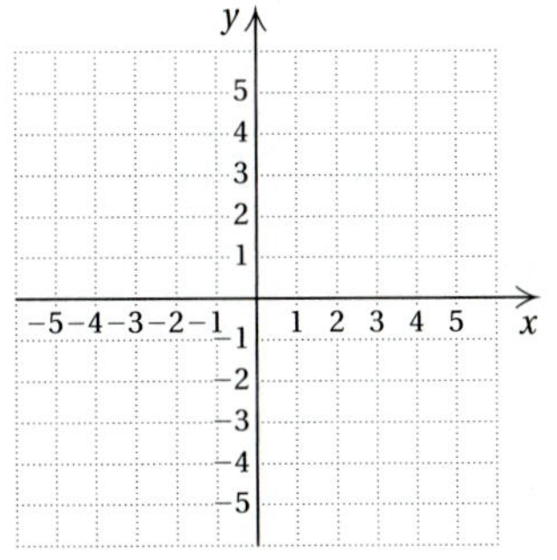

Find the intercepts of the equation. Then graph the equation. [8.3a]

29. $x - 2y = 6$

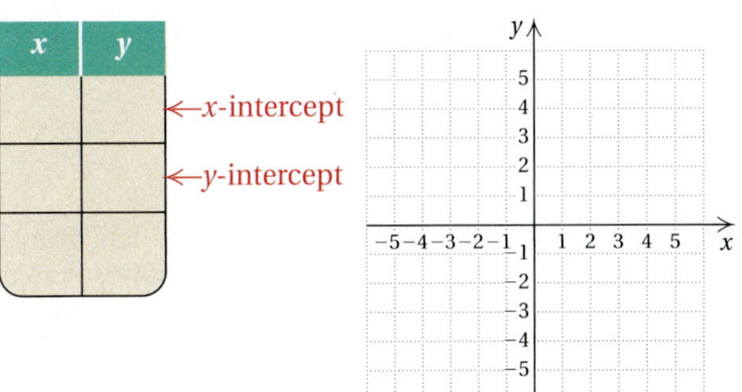

30. $5x - 2y = 10$

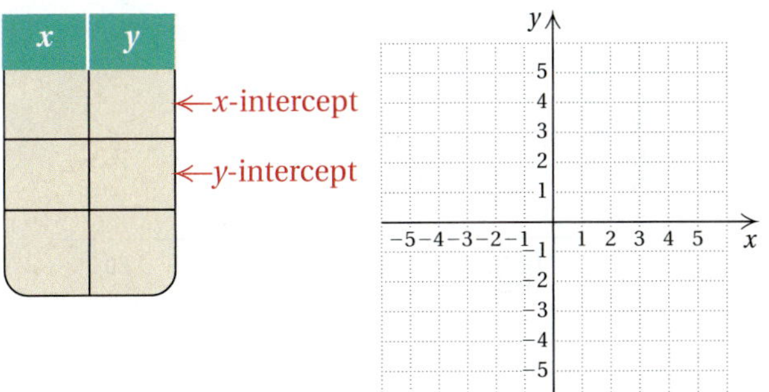

Solve.

31. *Kitchen Design.* Kitchen designers recommend that a refrigerator be selected on the basis of the number of people n in the household. The appropriate size S, in cubic feet, is given by

$$S = \frac{3}{2}n + 13.$$

a) Determine the recommended size of a refrigerator if the number of people is 1, 2, 5, and 10.

b) Graph the equation and use the graph to estimate the recommended size of a refrigerator for 3 people sharing an apartment.

c) A refrigerator is 22 ft^3. For how many residents is it the recommended size? [8.2c]

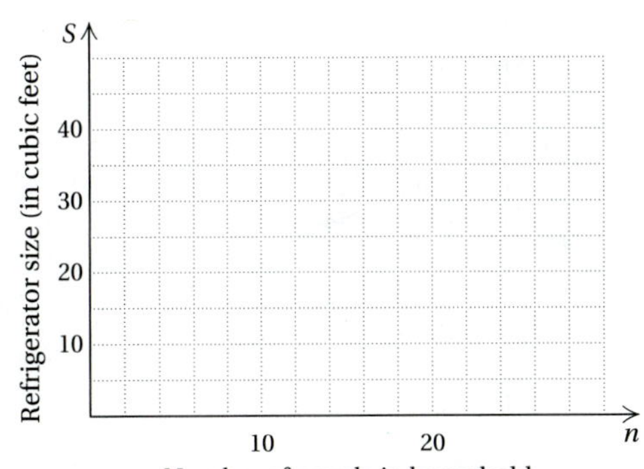

32. *Snow Removal.* By 3:00 P.M., Erin had plowed 7 driveways and by 5:30 P.M. had completed 13.

a) Find Erin's plowing rate in number of driveways per hour.

b) Find Erin's plowing rate in minutes per driveway. [8.4b]

33. *Manicures.* The following graph shows data from a recent day's work at the O'Hara School of Cosmetology. What is the rate of change, in number of manicures per hour? [8.4b]

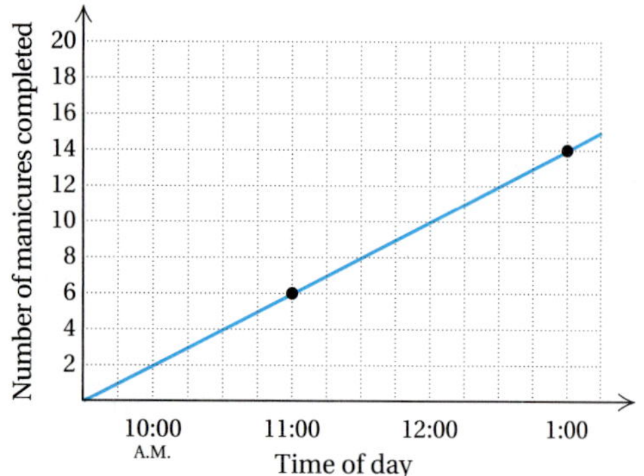

Find the slope. [8.4a]

34.

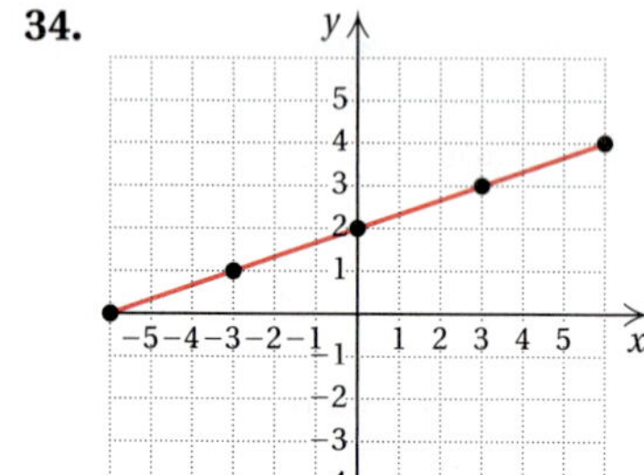

35.

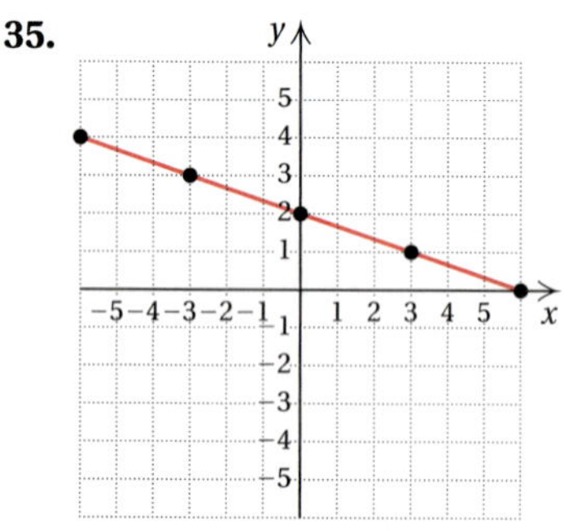

Graph the line containing the given pair of points and find the slope. [8.4a]

36. $(-5, -2), (5, 4)$

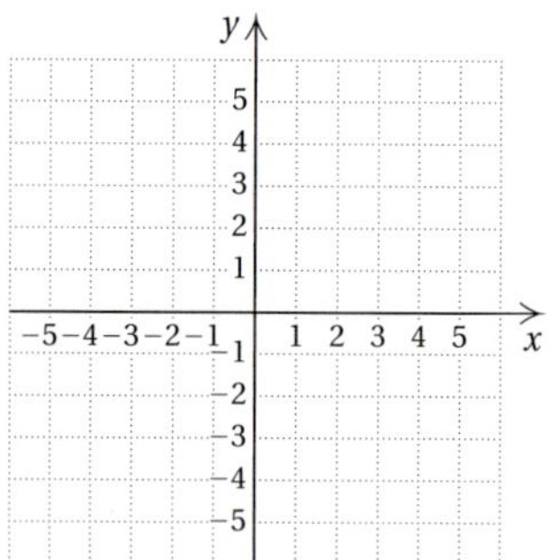

37. $(-5, 5), (4, -4)$

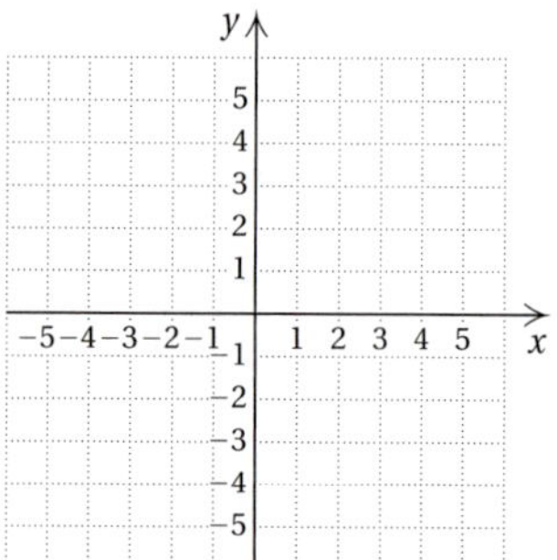

38. *Road Grade.* At one point, Beartooth Highway in Yellowstone National Park rises 315 ft over a horizontal distance of 4500 ft. Find the slope, or grade, of the road. [8.4b]

Find the slope, if it exists. [8.4c]

39. $y = -\frac{5}{8}x - 3$

40. $2x - 4y = 8$

41. $x = -2$

42. $y = 9$

43. DW Describe two ways in which a small business might make use of graphs. [8.1a], [8.2c]

44. DW Explain why the first coordinate of the y-intercept is always 0. [8.2b]

SKILL MAINTENANCE

Certain objectives from four particular sections will be retested on the chapter test. The objectives are listed with the practice problems that follow.

Convert to decimal notation. [6.2c]

45. $-\frac{11}{32}$

46. $\frac{8}{9}$

Find the absolute value. [6.2e]

47. $|-3.2|$

48. $\left|\frac{17}{19}\right|$

Solve. [7.2a]

49. $-5x = -55$

50. $\frac{4}{7}a = \frac{3}{7}$

Solve. [7.5a]

51. An investment was made at 6% simple interest for 1 year. It grows to \$10,340.40. How much was originally invested?

52. After a 20% reduction, a pair of slacks is on sale for \$63.96. What was the original price (that is, the price before reduction)?

SYNTHESIS

53. Find the value of m in $y = mx + 3$ such that $(-2, 5)$ is on the graph. [8.2a]

54. Find the area and the perimeter of a rectangle for which $(-2, 2)$, $(7, 2)$, and $(7, -3)$ are three of the vertices. [8.1b]

55. *Mountaineering.* As part of an ill-fated expedition to climb Mount Everest in 1996, author Jon Krakauer departed "The Balcony," elevation 27,600 ft, at 7:00 A.M. and reached the summit, elevation 29,028 ft, at 1:25 P.M. [8.4b]

a) Find Krakauer's rate of ascent in feet per minute.
b) Find Krakauer's rate of ascent in minutes per foot.

Source: Jon Krakauer, *Into Thin Air.* New York: Villard, 1998.

CHAPTER 8

Chapter Test

Toothpaste Sales. The following pie chart shows the percentages of sales of various toothpaste brands in the United States. In a recent year, total sales of toothpaste were $1,500,000,000. Use the pie chart for Questions 1–4.

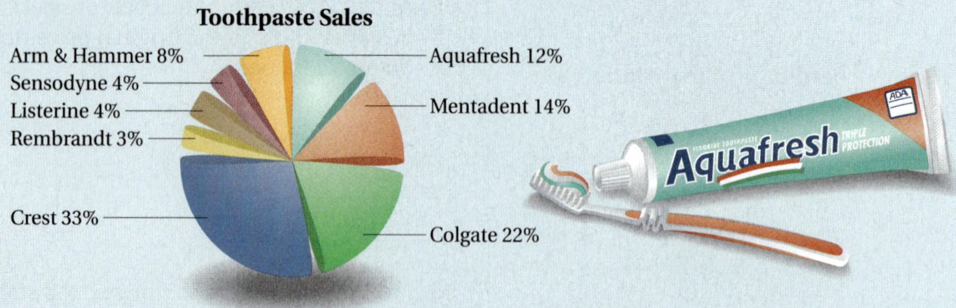

1. What were the total sales of Crest?

2. Which two brands together accounted for over half the sales?

3. Which brand had the greatest sales?

4. Which brand had sales of $120,000,000?

Tornado Touchdowns. The following bar graph shows the total number of tornado touchdowns by month in Indiana from 1950–1994. Use the bar graph for Questions 5–8.

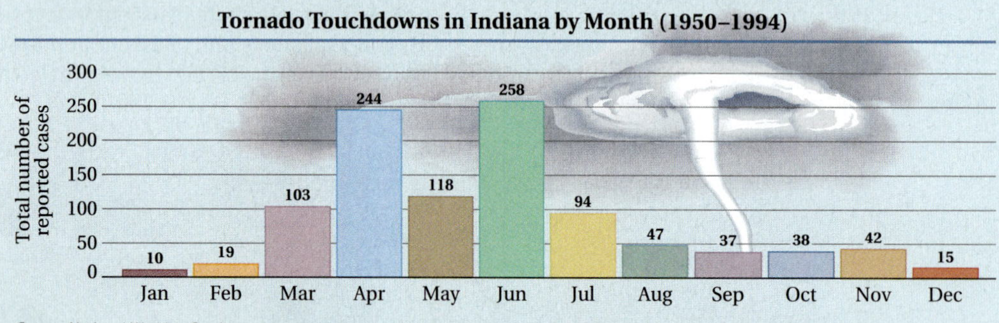

Source: National Weather Service

5. In which month did the greatest number of touchdowns occur?

6. In which month did the least number of touchdowns occur?

7. In which months was the number of touchdowns greater than 90?

8. In which month were there 47 touchdowns?

Radio Stations on the Internet. The line graph at right shows the number of radio stations in various years that were transmitting over the Internet. Use the line graph for Questions 9–14.

Number of Radio Stations on the Internet

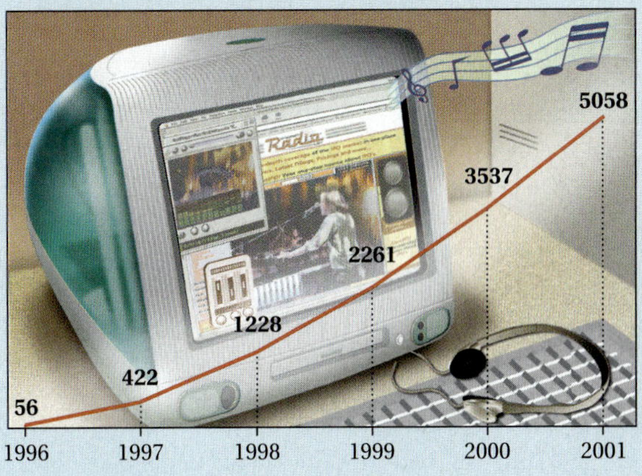

Source: BRS Media, Inc.

9. In which year were the greatest number of stations transmitting over the Internet?

10. In which year were the least number of stations transmitting over the Internet?

11. What is the difference between the greatest and least number of stations?

12. Between which two years was the increase in the number of stations the greatest?

13. By how much did the number of stations increase from 1996 to 2000?

14. By how much did the number of stations increase from 1998 to 2000?

In which quadrant is the point located?

15. $\left(-\frac{1}{2}, 7\right)$

16. $(-5, -6)$

Find the coordinates of the point.

17. A

18. B

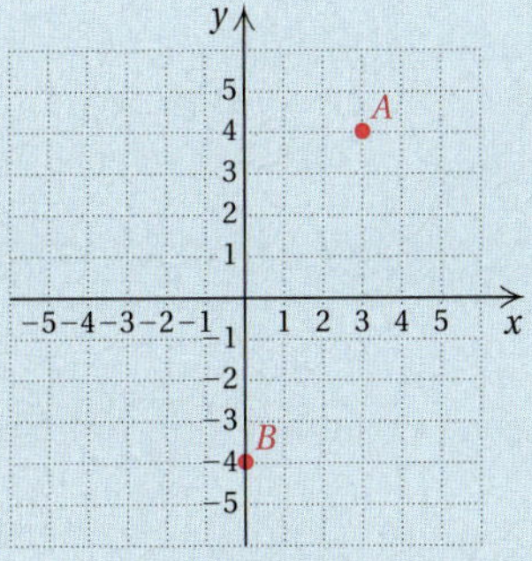

19. Show that the ordered pairs $(-4, -3)$ and $(-1, 3)$ are solutions of the equation $y - 2x = 5$. Then use the graph of the straight line containing the two points to determine another solution. Answers may vary.

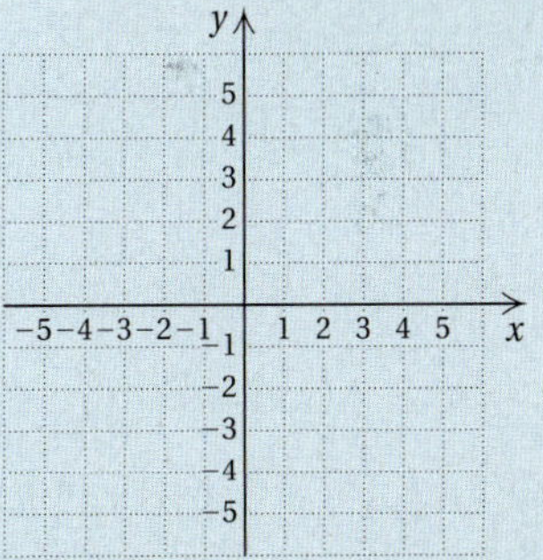

Graph the equation. Identify the y-intercept.

20. $y = 2x - 1$

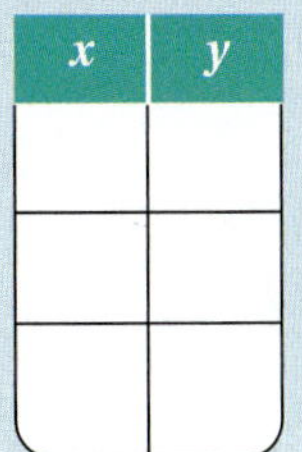

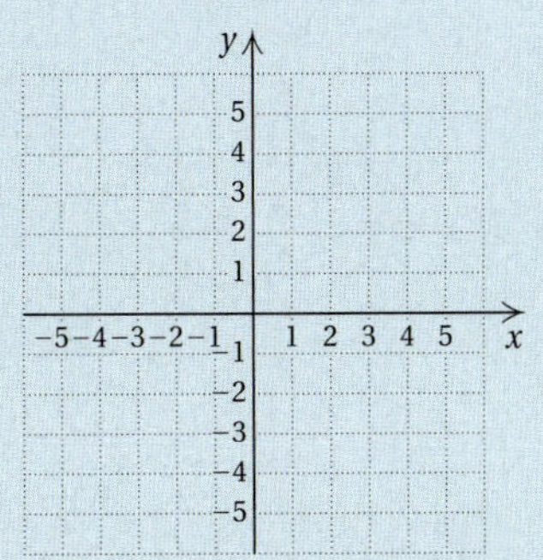

21. $y = -\frac{3}{2}x$

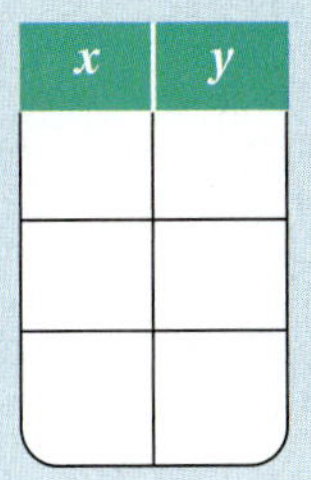

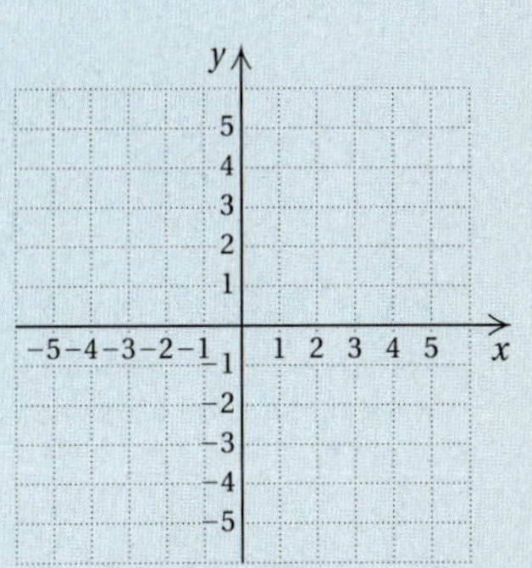

Graph the equation.

22. $2x + 8 = 0$

x	y

23. $y = 5$

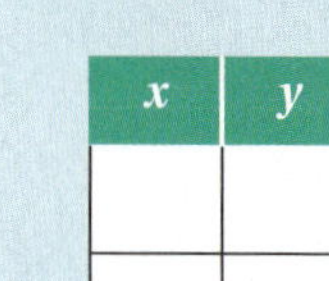

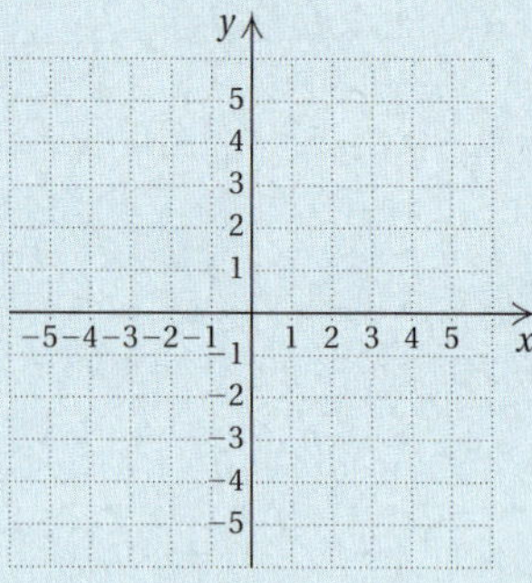

Find the intercepts of the equation. Then graph the equation.

24. $2x - 4y = -8$

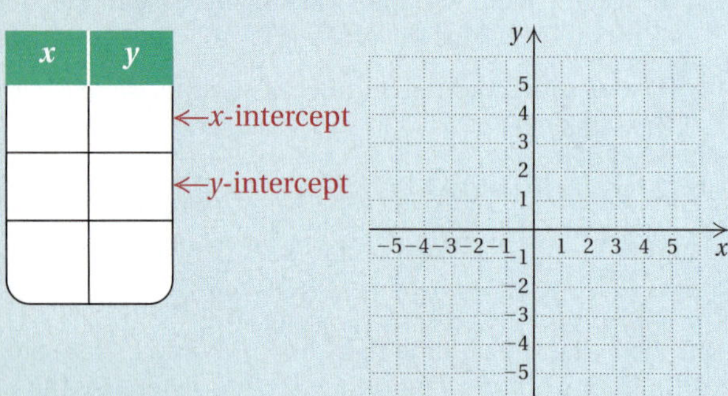

25. $2x - y = 3$

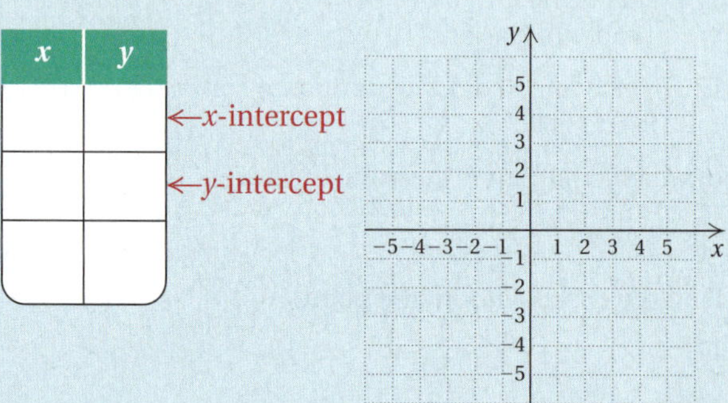

26. *Private-College Costs.* The cost T, in thousands of dollars, of tuition and fees at a private college (all expenses) can be approximated by

$$T = \frac{4}{5}n + 17,$$

where n is the number of years since 1992. That is, $n = 0$ corresponds to 1992, $n = 7$ corresponds to 1999, and so on.

a) Find the cost of tuition in 1992, 1995, 1999, and 2001.

b) Graph the equation and then use the graph to estimate the cost of tuition in 2005.

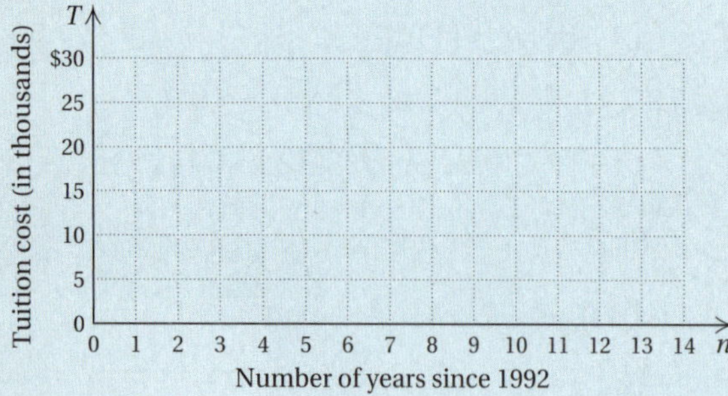

c) Estimate the year in which the cost of tuition will be \$25,000.

Source: *Statistical Abstract of the United States*

27. *Elevators.* At 2:38, Serge entered an elevator on the 34th floor of the Regency Hotel. At 2:40, he stepped off at the 5th floor.

a) Find the elevator's average rate of travel in number of floors per minute.

b) Find the elevator's average rate of travel in seconds per floor.

28. *Train Travel.* The following graph shows data concerning a recent train ride from Denver to Kansas City. At what rate did the train travel?

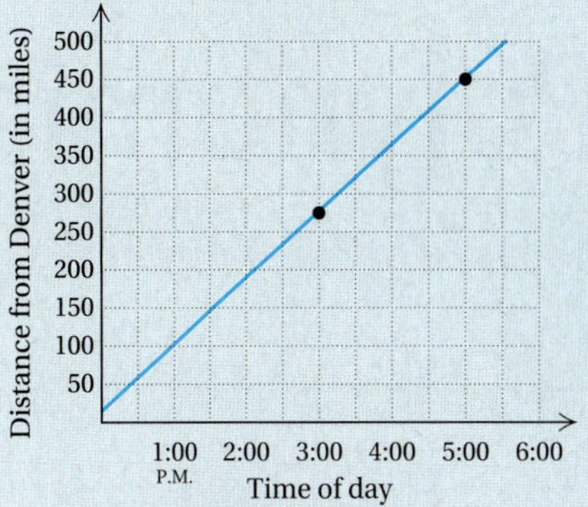

29. Find the slope.

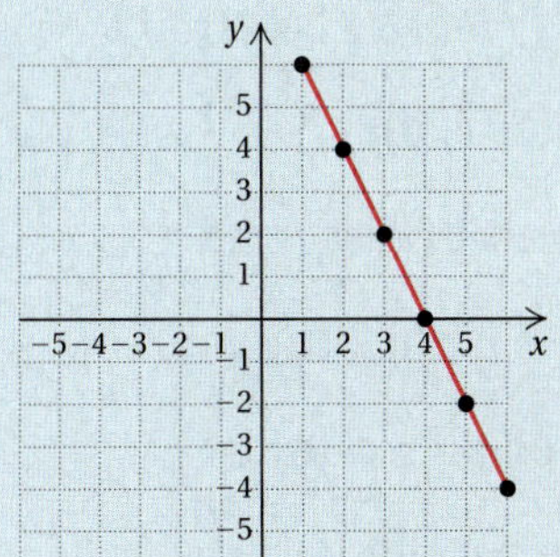

30. Graph the line containing $(-3, 1)$ and $(5, 4)$ and find the slope.

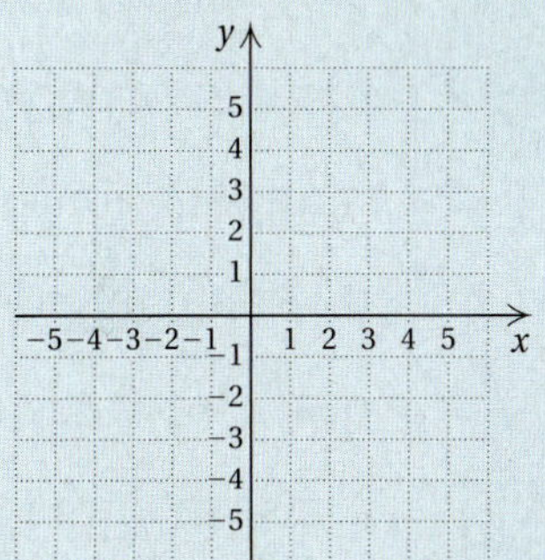

31. *Navigation.* Capital Rapids drops 54 ft vertically over a horizontal distance of 1080 ft. What is the slope of the rapids?

32. Find the slope, if it exists.

a) $2x - 5y = 10$

b) $x = -2$

SKILL MAINTENANCE

Convert to decimal notation.

33. $\frac{39}{40}$

34. $-\frac{13}{12}$

Find the absolute value.

35. $|71.2|$

36. $\left|-\frac{13}{47}\right|$

Solve.

37. $-\frac{5}{8}x = \frac{3}{8}$

38. $39 = 3t$

Solve.

39. After a 24% reduction, a software game is on sale for \$64.22. What was the original price (that is, the price before reduction)?

40. An investment was made at 7% simple interest for 1 year. It grows to \$38,948. How much was originally invested?

SYNTHESIS

41. A diagonal of a square connects the points $(-3, -1)$ and $(2, 4)$. Find the area and the perimeter of the square.

42. Write an equation of a line parallel to the x-axis and 3 units above it.

Polynomials: Operations

Gateway to Chapter 9

An algebraic expression like $x^3 - 5x^2 + 4x - 11$ is called a polynomial. A polynomial equation like $y = x^3 - 5x^2 + 4x - 11$ contains a polynomial. The focus of this chapter is on manipulations such as adding, subtracting, multiplying, and dividing polynomials. Of particular importance is the study of quick ways to multiply polynomials, called special products. These manipulations will be helpful throughout most mathematics courses you will take.

Basic to our study of polynomials are the manipulations with exponential expressions considered in Sections 9.1 and 9.2. We will solve problems both using scientific notation and expressing situations using polynomials. Though we will not discuss how to graph polynomial equations, we will show how their graphs can help to solve applied problems.

Real-World Application

A strand of DNA (deoxyribonucleic acid) is about 150 cm long and 1.3×10^{-10} cm wide. How many times longer is DNA than it is wide?

Source: Human Genome Project Information

This problem appears as Example 24 in Section 9.2.

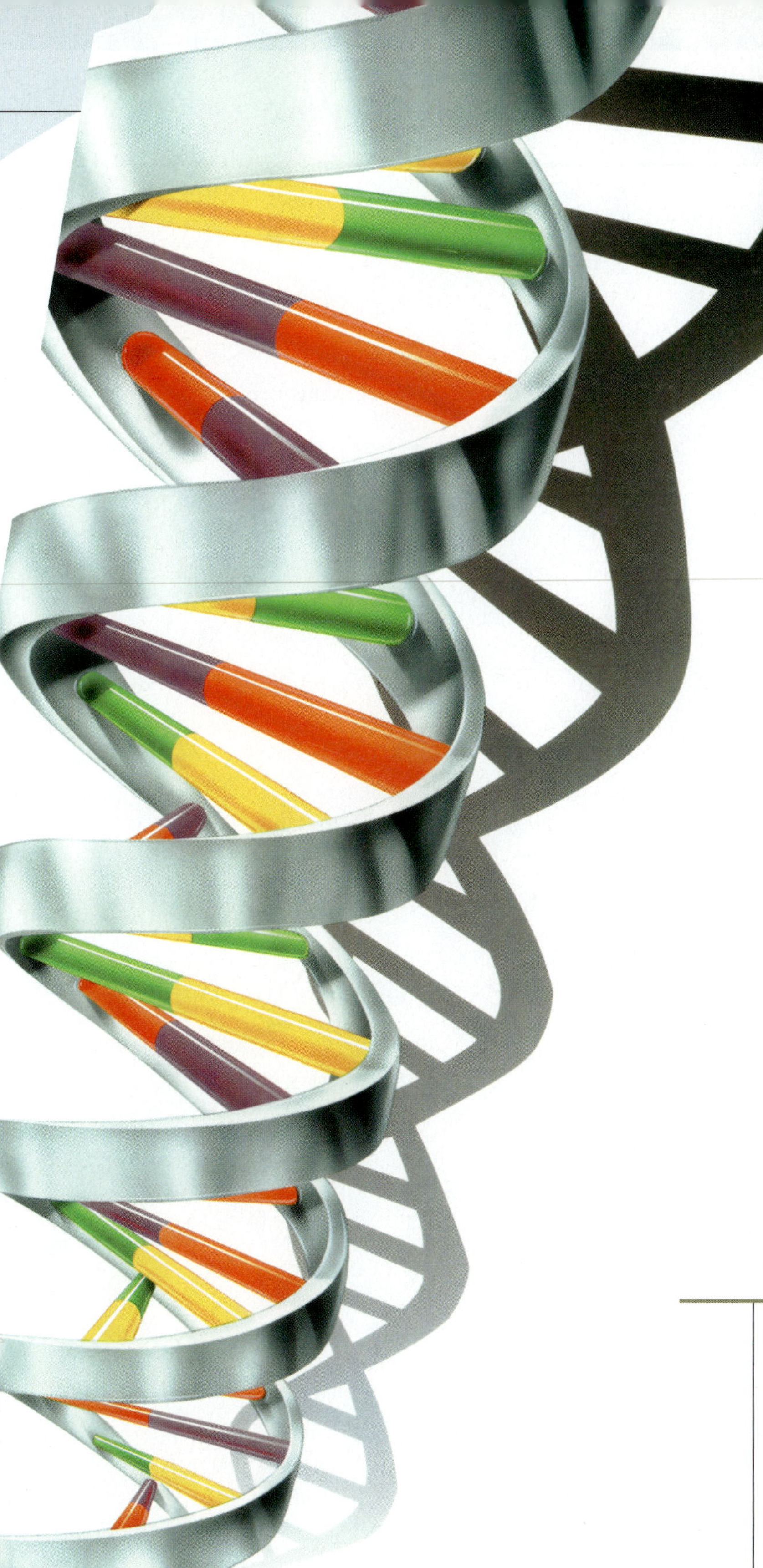

CHAPTER 9

CHAPTER 9 Pretest

1. Multiply: $x^{-3} \cdot x^5$. [9.1d]

2. Divide: $\frac{x^{-2}}{x^5}$. [9.1e]

3. Simplify: $(-4x^2y^{-3})^2$. [9.2b]

4. Express using a positive exponent: p^{-3}. [9.1f]

5. Convert to scientific notation: 0.000347. [9.2c]

6. Convert to decimal notation: 3.4×10^6. [9.2c]

Multiply and divide and express your results using scientific notation. [9.2d]

7. $(3.1 \times 10^5)(4.5 \times 10^{-3})$

8. $\frac{6.4 \times 10^{-7}}{8.0 \times 10^{-6}}$

9. Identify the degree of each term and the degree of the polynomial: [9.3g]

$2x^3 - 4x^2 + 3x - 5$.

10. Collect like terms: [9.3e]

$2a^3b - a^2b^2 + ab^3 + 9 - 5a^3b - a^2b^2 + 12b^3$.

11. Add: [9.4a]

$(5x^2 - 7x + 8) + (6x^2 + 11x - 19)$.

12. Subtract: [9.4c]

$(5x^2 - 7x + 8) - (6x^2 + 11x - 19)$.

Multiply.

13. $5x^2(3x^2 - 4x + 1)$ [9.5b]

14. $(x + 5)^2$ [9.6c]

15. $(x - 5)(x + 5)$ [9.6b]

16. $(x^3 + 6)(4x^3 - 5)$ [9.6a]

17. $(2x - 3y)(2x - 3y)$ [9.6c]

18. Divide: $(x^3 - x^2 + x + 2) \div (x - 2)$. [9.8b]

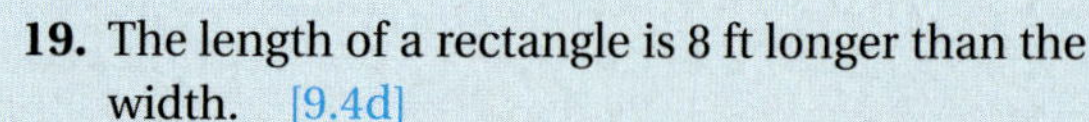

19. The length of a rectangle is 8 ft longer than the width. [9.4d]

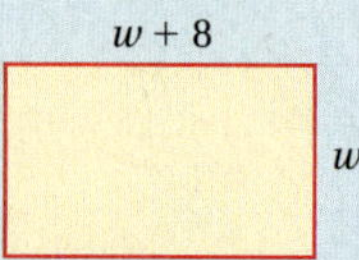

a) Find a polynomial for the perimeter.
b) Find a polynomial for the area.

9.1 INTEGERS AS EXPONENTS

Objectives

a Tell the meaning of exponential notation.

b Evaluate exponential expressions with exponents of 0 and 1.

c Evaluate algebraic expressions containing exponents.

d Use the product rule to multiply exponential expressions with like bases.

e Use the quotient rule to divide exponential expressions with like bases.

f Express an exponential expression involving negative exponents with positive exponents.

a Exponential Notation

An exponent of 2 or greater tells how many times the base is used as a factor. For example,

$$a \cdot a \cdot a \cdot a = a^4.$$

In this case, the **exponent** is 4 and the **base** is a. An expression for a power is called **exponential notation.**

a^n ← This is the exponent.
↑
This is the base.

EXAMPLE 1 What is the meaning of 3^5? of n^4? of $(2n)^3$? of $50x^2$? of $(-n)^3$?

3^5 means $3 \cdot 3 \cdot 3 \cdot 3 \cdot 3$; n^4 means $n \cdot n \cdot n \cdot n$;
$(2n)^3$ means $2n \cdot 2n \cdot 2n$; $50x^2$ means $50 \cdot x \cdot x$;
$(-n)^3$ means $(-n) \cdot (-n) \cdot (-n)$

Do Exercises 1–5.

We read exponential notation as follows: a^n is read the ***n*th power of *a*,** or simply ***a* to the *n*th,** or ***a* to the *n*.** We often read x^2 as "***x*-squared.**" The reason for this is that the area of a square of side x is $x \cdot x$, or x^2. We often read x^3 as "***x*-cubed.**" The reason for this is that the volume of a cube with length, width, and height x is $x \cdot x \cdot x$, or x^3.

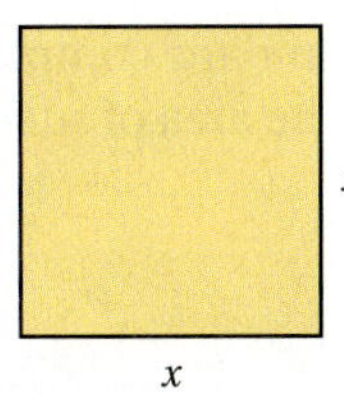

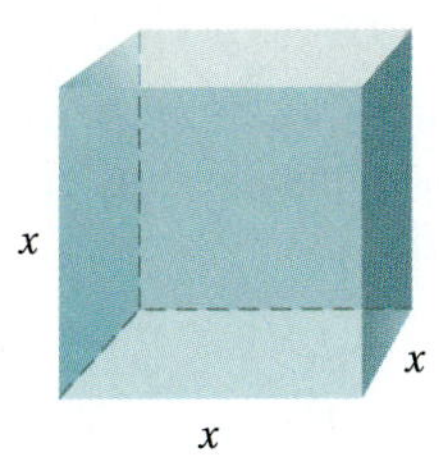

What is the meaning of each of the following?

1. 5^4

2. x^5

3. $(3t)^2$

4. $3t^2$

5. $(-x)^4$

b One and Zero as Exponents

Look for a pattern in the following:

On each side, we divide by 8 at each step.

$$8 \cdot 8 \cdot 8 \cdot 8 = 8^4$$
$$8 \cdot 8 \cdot 8 = 8^3$$
$$8 \cdot 8 = 8^2$$
$$8 = 8^?$$
$$1 = 8^?.$$

On this side, the exponents decrease by 1.

To continue the pattern, we would say that

$$8 = 8^1$$

and

$$1 = 8^0.$$

Answers on page A-27

Evaluate.

6. 6^1

7. 7^0

8. $(8.4)^1$

9. 8654^0

Answers on page A-27

Study Tips

AUDIOTAPES

Your instructor can request a complete set of audiotapes designed to help lead you through each section of this textbook. If you have difficulty reading or if you want extra review, these tapes explain solution steps for examples, caution you about errors, give instructions to work margin exercises, and then review the solutions to the margin exercises. Ideal for use outside of class, these tapes can be played anywhere you have a tape player. To obtain these audiotapes, consult with your instructor and refer to the Preface of this text.

We make the following definition.

EXPONENTS OF 0 AND 1

$a^1 = a$, for any number a;
$a^0 = 1$, for any nonzero number a

We consider 0^0 to be not defined. We will explain why later in this section.

EXAMPLE 2 Evaluate 5^1, 8^1, 3^0, $(-7.3)^0$, and $(186{,}892{,}046)^0$.

$5^1 = 5;\quad 8^1 = 8;\quad 3^0 = 1;$
$(-7.3)^0 = 1;\quad (186{,}892{,}046)^0 = 1$

Do Exercises 6–9.

C Evaluating Algebraic Expressions

Algebraic expressions can involve exponential notation. For example, the following are algebraic expressions:

$$x^4,\qquad (3x)^3 - 2,\qquad a^2 + 2ab + b^2.$$

We evaluate algebraic expressions by replacing variables with numbers and following the rules for order of operations.

EXAMPLE 3 Evaluate x^4 when $x = 2$.

$x^4 = 2^4$ Substituting
$= 2 \cdot 2 \cdot 2 \cdot 2 = 16$

EXAMPLE 4 *Area of a Compact Disc.* The standard compact disc used for software and music has a radius of 6 cm. Find the area of such a CD (ignoring the hole in the middle).

$A = \pi r^2$
$= \pi \cdot (6\text{ cm})^2$
$\approx 3.14 \times 36\text{ cm}^2$
$= 113.04\text{ cm}^2$

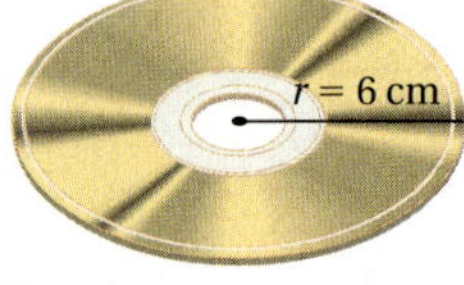

In Example 4, "cm^2" means "square centimeters" and "$\approx$" means "is approximately equal to."

EXAMPLE 5 Evaluate $(5x)^3$ when $x = -2$.

When we evaluate with a negative number, we often use extra parentheses to show the substitution.

$(5x)^3 = [5 \cdot (-2)]^3$ Substituting
$= [-10]^3$ Multiplying within brackets first
$= [-10] \cdot [-10] \cdot [-10]$
$= -1000$ Evaluating the power

EXAMPLE 6 Evaluate $5x^3$ when $x = -2$.

$$5x^3 = 5 \cdot (-2)^3 \quad \text{Substituting}$$
$$= 5(-8) \quad \text{Evaluating the power first}$$
$$= -40$$

Recall that two expressions are equivalent if they have the same value for all meaningful replacements. Note that Examples 5 and 6 show that $(5x)^3$ and $5x^3$ are *not* equivalent—that is, $(5x)^3 \neq 5x^3$.

Do Exercises 10–14.

d Multiplying Powers with Like Bases

There are several rules for manipulating exponential notation to obtain equivalent expressions. We first consider multiplying powers with like bases:

$$a^3 \cdot a^2 = \underbrace{(a \cdot a \cdot a)}_{3 \text{ factors}}\underbrace{(a \cdot a)}_{2 \text{ factors}} = \underbrace{a \cdot a \cdot a \cdot a \cdot a}_{5 \text{ factors}} = a^5.$$

Since an integer exponent greater than 1 tells how many times we use a base as a factor, then $(a \cdot a \cdot a)(a \cdot a) = a \cdot a \cdot a \cdot a \cdot a = a^5$ by the associative law. Note that the exponent in a^5 is the sum of those in $a^3 \cdot a^2$. That is, $3 + 2 = 5$. Likewise,

$$b^4 \cdot b^3 = (b \cdot b \cdot b \cdot b)(b \cdot b \cdot b) = b^7, \quad \text{where} \quad 4 + 3 = 7.$$

Adding the exponents gives the correct result.

THE PRODUCT RULE

For any number a and any positive integers m and n,

$$a^m \cdot a^n = a^{m+n}.$$

(When multiplying with exponential notation, if the bases are the same, keep the base and add the exponents.)

EXAMPLES Multiply and simplify. By simplify, we mean write the expression as one number to a nonnegative power.

7. $8^4 \cdot 8^3 = 8^{4+3}$ Adding exponents: $a^m \cdot a^n = a^{m+n}$
$= 8^7$

8. $x^2 \cdot x^9 = x^{2+9} = x^{11}$

9. $m^5 m^{10} m^3 = m^{5+10+3} = m^{18}$

10. $x \cdot x^8 = x^1 \cdot x^8$ Writing x as x^1
$= x^{1+8}$
$= x^9$

11. $(a^3b^2)(a^3b^5) = (a^3a^3)(b^2b^5)$
$= a^6b^7$

Do Exercises 15–19.

10. Evaluate t^3 when $t = 5$.

11. Find the area of a circle when $r = 32$ cm. Use 3.14 for π.

12. Evaluate $200 - a^4$ when $a = 3$.

13. Evaluate $t^1 - 4$ and $t^0 - 4$ when $t = 7$.

14. a) Evaluate $(4t)^2$ when $t = -3$.

b) Evaluate $4t^2$ when $t = -3$.

c) Determine whether $(4t)^2$ and $4t^2$ are equivalent.

Multiply and simplify.

15. $3^5 \cdot 3^5$

16. $x^4 \cdot x^6$

17. $p^4p^{12}p^8$

18. $x \cdot x^4$

19. $(a^2b^3)(a^7b^5)$

Answers on page A-27

Divide and simplify.

20. $\dfrac{4^5}{4^2}$

21. $\dfrac{y^6}{y^2}$

22. $\dfrac{p^{10}}{p}$

23. $\dfrac{a^7b^6}{a^3b^4}$

Answers on page A-27

e Dividing Powers with Like Bases

The following suggests a rule for dividing powers with like bases, such as a^5/a^2:

$$\frac{a^5}{a^2} = \frac{a \cdot a \cdot a \cdot a \cdot a}{a \cdot a} = \frac{a \cdot a \cdot a \cdot a \cdot a}{1 \cdot a \cdot a} = \frac{a \cdot a \cdot a}{1} \cdot \frac{a \cdot a}{a \cdot a} = \frac{a \cdot a \cdot a}{1} \cdot 1$$

$$= a \cdot a \cdot a = a^3.$$

Note that the exponent in a^3 is the difference of those in $a^5 \div a^2$, that is, $5 - 2 = 3$. In a similar way, we have

$$\frac{t^9}{t^4} = \frac{t \cdot t \cdot t \cdot t \cdot t \cdot t \cdot t \cdot t \cdot t}{t \cdot t \cdot t \cdot t} = t^5, \quad \text{where } 9 - 4 = 5.$$

Subtracting exponents gives the correct answer.

THE QUOTIENT RULE

For any nonzero number a and any positive integers m and n,

$$\frac{a^m}{a^n} = a^{m-n}.$$

(When dividing with exponential notation, if the bases are the same, keep the base and subtract the exponent of the denominator from the exponent of the numerator.)

EXAMPLES Divide and simplify. By simplify, we mean write the expression as one number to a nonnegative power.

12. $\dfrac{6^5}{6^3} = 6^{5-3}$ Subtracting exponents

$= 6^2$

13. $\dfrac{x^8}{x^2} = x^{8-2}$

$= x^6$

14. $\dfrac{t^{12}}{t} = \dfrac{t^{12}}{t^1} = t^{12-1}$

$= t^{11}$

15. $\dfrac{p^5q^7}{p^2q^5} = \dfrac{p^5}{p^2} \cdot \dfrac{q^7}{q^5} = p^{5-2}q^{7-5}$

$= p^3q^2$

The quotient rule can also be used to explain the definition of 0 as an exponent. Consider the expression a^4/a^4, where a is nonzero:

$$\frac{a^4}{a^4} = \frac{a \cdot a \cdot a \cdot a}{a \cdot a \cdot a \cdot a} = 1.$$

This is true because the numerator and the denominator are the same. Now suppose we apply the rule for dividing powers with the same base:

$$\frac{a^4}{a^4} = a^{4-4} = a^0 = 1.$$

Since both expressions a^4/a^4 and a^{4-4} are equivalent to 1, it follows that $a^0 = 1$, when $a \neq 0$.

We can explain why we do not define 0^0 using the quotient rule. We know that 0^0 is 0^{1-1}. But 0^{1-1} is also equal to $0/0$. We have already seen that division by 0 is not defined, so 0^0 is also not defined.

Do Exercises 20–23.

Negative Integers as Exponents

We can use the rule for dividing powers with like bases to lead us to a definition of exponential notation when the exponent is a negative integer. Consider $5^3/5^7$ and first simplify it using procedures we have learned for working with fractions:

$$\frac{5^3}{5^7} = \frac{5 \cdot 5 \cdot 5}{5 \cdot 5 \cdot 5 \cdot 5 \cdot 5 \cdot 5 \cdot 5} = \frac{5 \cdot 5 \cdot 5 \cdot 1}{5 \cdot 5 \cdot 5 \cdot 5 \cdot 5 \cdot 5 \cdot 5}$$

$$= \frac{5 \cdot 5 \cdot 5}{5 \cdot 5 \cdot 5} \cdot \frac{1}{5 \cdot 5 \cdot 5 \cdot 5} = \frac{1}{5^4}.$$

Now we apply the rule for dividing exponential expressions with the same bases. Then

$$\frac{5^3}{5^7} = 5^{3-7} = 5^{-4}.$$

From these two expressions for $5^3/5^7$, it follows that

$$5^{-4} = \frac{1}{5^4}.$$

This leads to our definition of negative exponents.

NEGATIVE EXPONENT

For any real number a that is nonzero and any integer n,

$$a^{-n} = \frac{1}{a^n}.$$

In fact, the numbers a^n and a^{-n} are reciprocals of each other because

$$a^n \cdot a^{-n} = a^n \cdot \frac{1}{a^n} = \frac{a^n}{a^n} = 1.$$

EXAMPLES Express using positive exponents. Then simplify.

16. $4^{-2} = \frac{1}{4^2} = \frac{1}{16}$

17. $(-3)^{-2} = \frac{1}{(-3)^2} = \frac{1}{(-3)(-3)} = \frac{1}{9}$

18. $m^{-3} = \frac{1}{m^3}$

19. $ab^{-1} = a\left(\frac{1}{b^1}\right) = a\left(\frac{1}{b}\right) = \frac{a}{b}$

20. $\frac{1}{x^{-3}} = x^{-(-3)} = x^3$

21. $3c^{-5} = 3\left(\frac{1}{c^5}\right) = \frac{3}{c^5}$

Example 20 might also be done as follows:

$$\frac{1}{x^{-3}} = \frac{1}{\frac{1}{x^3}} = 1 \cdot \frac{x^3}{1} = x^3.$$

CAUTION!

As shown in Examples 16 and 17, a negative exponent does not necessarily mean that an expression is negative.

Do Exercises 24–29.

Express with positive exponents. Then simplify.

24. 4^{-3}

25. 5^{-2}

26. 2^{-4}

27. $(-2)^{-3}$

28. $4p^{-3}$

29. $\frac{1}{x^{-2}}$

Answers on page A-27

Simplify.

30. $5^{-2} \cdot 5^4$

31. $x^{-3} \cdot x^{-4}$

32. $\dfrac{7^{-2}}{7^3}$

33. $\dfrac{b^{-2}}{b^{-3}}$

34. $\dfrac{t}{t^{-5}}$

Answers on page A-27

The rules for multiplying and dividing powers with like bases still hold when exponents are 0 or negative. We will state them in a summary at the end of this section.

EXAMPLES Simplify. By simplify, we generally mean write the expression as one number or variable to a nonnegative power.

22. $7^{-3} \cdot 7^6 = 7^{-3+6} = 7^3$ Adding exponents

23. $x^4 \cdot x^{-3} = x^{4+(-3)} = x^1 = x$

24. $\dfrac{5^4}{5^{-2}} = 5^{4-(-2)} = 5^{4+2} = 5^6$ Subtracting exponents

25. $\dfrac{x}{x^7} = x^{1-7} = x^{-6} = \dfrac{1}{x^6}$

26. $\dfrac{b^{-4}}{b^{-5}} = b^{-4-(-5)} = b^{-4+5} = b^1 = b$

27. $y^{-4} \cdot y^{-8} = y^{-4+(-8)} = y^{-12} = \dfrac{1}{y^{12}}$

Do Exercises 30–34.

The following is another way to arrive at the definition of negative exponents.

On each side, we divide by 5 at each step.

$$5 \cdot 5 \cdot 5 \cdot 5 = 5^4$$
$$5 \cdot 5 \cdot 5 = 5^3$$
$$5 \cdot 5 = 5^2$$
$$5 = 5^1$$
$$1 = 5^0$$
$$\frac{1}{5} = 5^?$$
$$\frac{1}{25} = 5^?$$

On this side, the exponents decrease by 1.

To continue the pattern, it should follow that

$$\frac{1}{5} = \frac{1}{5^1} = 5^{-1} \quad \text{and} \quad \frac{1}{25} = \frac{1}{5^2} = 5^{-2}.$$

The following is a summary of the definitions and rules for exponents that we have considered in this section.

DEFINITIONS AND RULES FOR EXPONENTS

1 as an exponent:	$a^1 = a$
0 as an exponent:	$a^0 = 1, a \neq 0$
Negative integers as exponents:	$a^{-n} = \dfrac{1}{a^n}, \dfrac{1}{a^{-n}} = a^n; a \neq 0$
Product Rule:	$a^m \cdot a^n = a^{m+n}$
Quotient Rule:	$\dfrac{a^m}{a^n} = a^{m-n}, a \neq 0$

9.1 EXERCISE SET

For Extra Help

Digital Video Tutor CD 7 Videotape 9

InterAct Math

Math Tutor Center

MathXL

MyMathLab

a What is the meaning of each of the following?

1. 3^4

2. 4^3

3. $(-1.1)^5$

4. $(87.2)^6$

5. $\left(\frac{2}{3}\right)^4$

6. $\left(-\frac{5}{8}\right)^3$

7. $(7p)^2$

8. $(11c)^3$

9. $8k^3$

10. $17x^2$

b Evaluate.

11. $a^0, a \neq 0$

12. $t^0, t \neq 0$

13. b^1

14. c^1

15. $\left(\frac{2}{3}\right)^0$

16. $\left(-\frac{5}{8}\right)^0$

17. 8.38^0

18. 8.38^1

19. $(ab)^1$

20. $(ab)^0, a, b \neq 0$

21. ab^1

22. ab^0

c Evaluate.

23. m^3, when $m = 3$

24. x^6, when $x = 2$

25. p^1, when $p = 19$

26. x^{19}, when $x = 0$

27. x^4, when $x = 4$

28. y^{15}, when $y = 1$

29. $y^2 - 7$, when $y = -10$

30. $z^5 + 5$, when $z = -2$

31. $x^1 + 3$ and $x^0 + 3$, when $x = 7$

32. $y^0 - 8$ and $y^1 - 8$, when $y = -3$

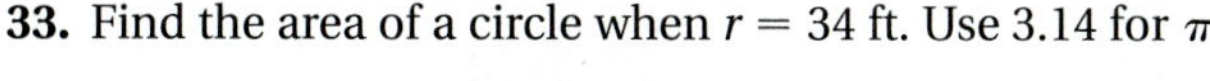

33. Find the area of a circle when $r = 34$ ft. Use 3.14 for π.

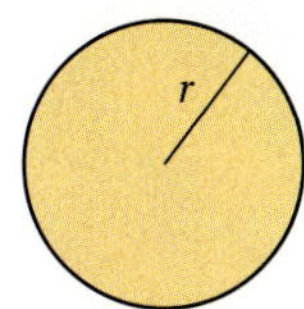

34. The area A of a square with sides of length s is given by $A = s^2$. Find the area of a square with sides of length 24 m.

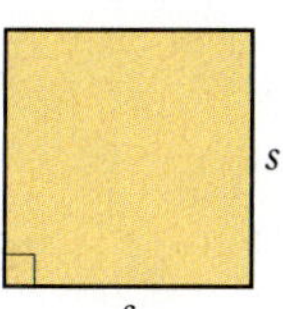

f Express using positive exponents. Then simplify.

35. 3^{-2}

36. 2^{-3}

37. 10^{-3}

38. 5^{-4}

39. 7^{-3}

40. 5^{-2}

41. a^{-3}

42. x^{-2}

43. $\frac{1}{8^{-2}}$

44. $\frac{1}{2^{-5}}$

45. $\frac{1}{y^{-4}}$

46. $\frac{1}{t^{-7}}$

47. $\frac{1}{z^{-n}}$

48. $\frac{1}{h^{-n}}$

Express using negative exponents.

49. $\frac{1}{4^3}$

50. $\frac{1}{5^2}$

51. $\frac{1}{x^3}$

52. $\frac{1}{y^2}$

53. $\frac{1}{a^5}$

54. $\frac{1}{b^7}$

d, **f** Multiply and simplify.

55. $2^4 \cdot 2^3$

56. $3^5 \cdot 3^2$

57. $8^5 \cdot 8^9$

58. $n^3 \cdot n^{20}$

59. $x^4 \cdot x^3$

60. $y^7 \cdot y^9$

61. $9^{17} \cdot 9^{21}$

62. $t^0 \cdot t^{16}$

63. $(3y)^4(3y)^8$

64. $(2t)^8(2t)^{17}$

65. $(7y)^1(7y)^{16}$

66. $(8x)^0(8x)^1$

67. $3^{-5} \cdot 3^8$

68. $5^{-8} \cdot 5^9$

69. $x^{-2} \cdot x$

70. $x \cdot x^{-1}$

71. $x^{14} \cdot x^3$

72. $x^9 \cdot x^4$

73. $x^{-7} \cdot x^{-6}$

74. $y^{-5} \cdot y^{-8}$

75. $a^{11} \cdot a^{-3} \cdot a^{-18}$

76. $a^{-11} \cdot a^{-3} \cdot a^{-7}$

77. $t^8 \cdot t^{-8}$

78. $m^{10} \cdot m^{-10}$

e, **f** Divide and simplify.

79. $\frac{7^5}{7^2}$

80. $\frac{5^8}{5^6}$

81. $\frac{8^{12}}{8^6}$

82. $\frac{8^{13}}{8^2}$

83. $\frac{y^9}{y^5}$

84. $\frac{x^{11}}{x^9}$

85. $\frac{16^2}{16^8}$

86. $\frac{7^2}{7^9}$

87. $\frac{m^6}{m^{12}}$

88. $\frac{a^3}{a^4}$

89. $\frac{(8x)^6}{(8x)^{10}}$

90. $\frac{(8t)^4}{(8t)^{11}}$

91. $\frac{(2y)^9}{(2y)^9}$

92. $\frac{(6y)^7}{(6y)^7}$

93. $\frac{x}{x^{-1}}$

94. $\frac{y^8}{y}$

95. $\frac{x^7}{x^{-2}}$

96. $\frac{t^8}{t^{-3}}$

97. $\frac{z^{-6}}{z^{-2}}$

98. $\frac{x^{-9}}{x^{-3}}$

99. $\frac{x^{-5}}{x^{-8}}$

100. $\frac{y^{-2}}{y^{-9}}$

101. $\frac{m^{-9}}{m^{-9}}$

102. $\frac{x^{-7}}{x^{-7}}$

Matching. In Exercises 103 and 104, match each item in the first column with the appropriate item in the second column by drawing connecting lines.

103.

5^2	$-\frac{1}{10}$
5^{-2}	$\frac{1}{10}$
$\left(\frac{1}{5}\right)^2$	$-\frac{1}{25}$
$\left(\frac{1}{5}\right)^{-2}$	10
-5^2	25
$(-5)^2$	-25
$-\left(-\frac{1}{5}\right)^2$	$\frac{1}{25}$
$\left(-\frac{1}{5}\right)^{-2}$	-10

104.

$-\left(\frac{1}{8}\right)^2$	16
$\left(\frac{1}{8}\right)^{-2}$	-16
8^{-2}	64
8^2	-64
-8^2	$\frac{1}{64}$
$(-8)^2$	$-\frac{1}{64}$
$\left(-\frac{1}{8}\right)^{-2}$	$-\frac{1}{16}$
$\left(-\frac{1}{8}\right)^2$	$\frac{1}{16}$

105. DW Suppose that the width of a square is three times the width of a second square. How do the areas of the squares compare? Why?

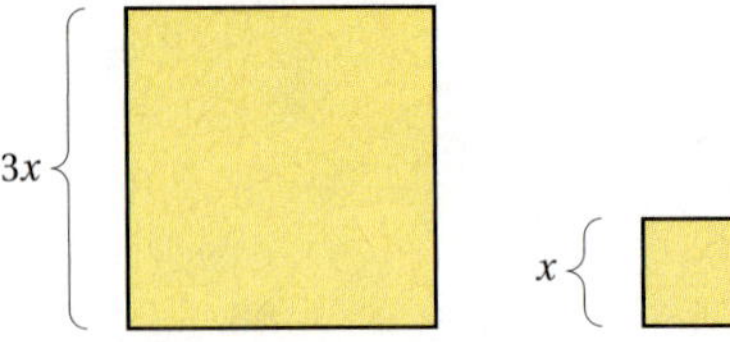

106. DW Suppose that the width of a cube is twice the width of a second cube. How do the volumes of the cubes compare? Why?

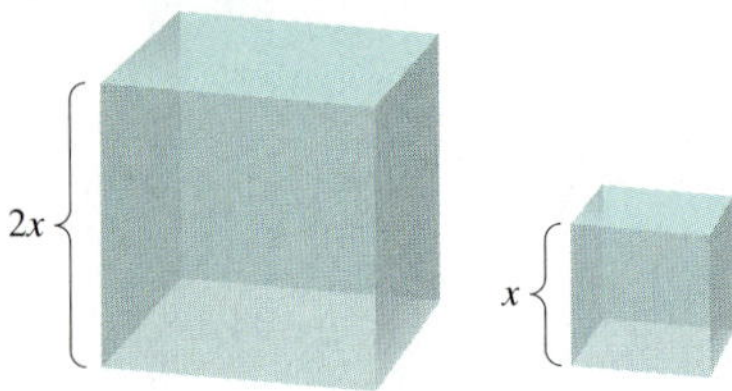

SKILL MAINTENANCE

107. Translate to an algebraic expression: Sixty-four percent of t. [6.1b]

108. Evaluate $\frac{3x}{y}$ when $x = 4$ and $y = 12$. [6.1a]

109. Divide: $1555.2 \div 24.3$. [6.6c]

110. Add: $1555.2 + 24.3$. [6.3a]

111. Solve: $3x - 4 + 5x - 10x = x - 8$. [7.3b]

112. Factor: $8x - 56$. [6.7d]

Solve. [7.6a]

113. *Cutting a Submarine Sandwich.* A 12-in. submarine sandwich is cut into two pieces. One piece is twice as long as the other. How long are the pieces?

114. *Book Pages.* A book is opened. The sum of the page numbers on the facing pages is 457. Find the page numbers.

SYNTHESIS

Determine whether each of the following is correct.

115. $(x + 1)^2 = x^2 + 1$

116. $(x - 1)^2 = x^2 - 2x + 1$

117. $(5x)^0 = 5x^0$

118. $\frac{x^3}{x^5} = x^2$

Simplify.

119. $(y^{2x})(y^{3x})$

120. $a^{5k} \div a^{3k}$

121. $\frac{a^{6t}(a^{7t})}{a^{9t}}$

122. $\frac{\left(\frac{1}{2}\right)^4}{\left(\frac{1}{2}\right)^5}$

123. $\frac{(0.8)^5}{(0.8)^3(0.8)^2}$

124. Determine whether $(a + b)^2$ and $a^2 + b^2$ are equivalent. (*Hint*: Choose values for a and b and evaluate.)

Use >, <, or = for □ to write a true sentence.

125. $3^5 \square 3^4$

126. $4^2 \square 4^3$

127. $4^3 \square 5^3$

128. $4^3 \square 3^4$

Find a value of the variable that shows that the two expressions are *not* equivalent.

129. $3x^2$; $(3x)^2$

130. $\frac{x + 2}{2}$; x

9.2 EXPONENTS AND SCIENTIFIC NOTATION

We now enhance our ability to manipulate exponential expressions by considering three more rules. The rules are also applied to a new way to name numbers called *scientific notation.*

a Raising Powers to Powers

Consider an expression like $(3^2)^4$. We are raising 3^2 to the fourth power:

$$(3^2)^4 = (3^2)(3^2)(3^2)(3^2)$$
$$= (3 \cdot 3)(3 \cdot 3)(3 \cdot 3)(3 \cdot 3)$$
$$= 3 \cdot 3 \cdot 3 \cdot 3 \cdot 3 \cdot 3 \cdot 3 \cdot 3$$
$$= 3^8.$$

Note that in this case we could have multiplied the exponents:

$$(3^2)^4 = 3^{2 \cdot 4} = 3^8.$$

Likewise, $(y^8)^3 = (y^8)(y^8)(y^8) = y^{24}$. Once again, we get the same result if we multiply the exponents:

$$(y^8)^3 = y^{8 \cdot 3} = y^{24}.$$

THE POWER RULE

For any real number a and any integers m and n,

$$(a^m)^n = a^{mn}.$$

(To raise a power to a power, multiply the exponents.)

EXAMPLES Simplify. Express the answers using positive exponents.

1. $(3^5)^4 = 3^{5 \cdot 4}$ Multiplying exponents
$= 3^{20}$

2. $(2^2)^5 = 2^{2 \cdot 5} = 2^{10}$

3. $(y^{-5})^7 = y^{-5 \cdot 7} = y^{-35} = \dfrac{1}{y^{35}}$

4. $(x^4)^{-2} = x^{4(-2)} = x^{-8} = \dfrac{1}{x^8}$

5. $(a^{-4})^{-6} = a^{(-4)(-6)} = a^{24}$

Do Exercises 1–4.

b Raising a Product or a Quotient to a Power

When an expression inside parentheses is raised to a power, the inside expression is the base. Let's compare $2a^3$ and $(2a)^3$:

$2a^3 = 2 \cdot a \cdot a \cdot a;$ The base is a.

$(2a)^3 = (2a)(2a)(2a)$ The base is $2a$.

$= (2 \cdot 2 \cdot 2)(a \cdot a \cdot a)$ Using the associative and commutative laws of multiplication to regroup the factors

$= 2^3a^3$

$= 8a^3.$

Objectives

a Use the power rule to raise powers to powers.

b Raise a product to a power and a quotient to a power.

c Convert between scientific notation and decimal notation.

d Multiply and divide using scientific notation.

e Solve applied problems using scientific notation.

Simplify. Express the answers using positive exponents.

1. $(3^4)^5$

2. $(x^{-3})^4$

3. $(y^{-5})^{-3}$

4. $(x^4)^{-8}$

Answers on page A-28

Simplify.

5. $(2x^5y^{-3})^4$

6. $(5x^5y^{-6}z^{-3})^2$

7. $[(-x)^{37}]^2$

8. $(3y^{-2}x^{-5}z^8)^3$

Simplify.

9. $\left(\frac{x^6}{5}\right)^2$

10. $\left(\frac{2t^5}{w^4}\right)^3$

11. $\left(\frac{x^4}{3}\right)^{-2}$

Do this two ways.

Answers on page A-28

We see that $2a^3$ and $(2a)^3$ are *not* equivalent. We also see that we can evaluate the power $(2a)^3$ by raising each factor to the power 3. This leads us to the following rule for raising a product to a power.

RAISING A PRODUCT TO A POWER

For any real numbers a and b and any integer n,

$$(ab)^n = a^n b^n.$$

(To raise a product to the nth power, raise each factor to the nth power.)

EXAMPLES Simplify.

6. $(4x^2)^3 = (4^1x^2)^3$ Since $4 = 4^1$

$= (4^1)^3 \cdot (x^2)^3$ Raising each factor to the third power

$= 4^3 \cdot x^6 = 64x^6$

7. $(5x^3y^5z^2)^4 = 5^4(x^3)^4(y^5)^4(z^2)^4$ Raising each factor to the fourth power

$= 625x^{12}y^{20}z^8$

8. $(-5x^4y^3)^3 = (-5)^3(x^4)^3(y^3)^3$

$= -125x^{12}y^9$

9. $[(-x)^{25}]^2 = (-x)^{50}$ Using the power rule

$= (-1 \cdot x)^{50}$ Using the property of -1 (Section 6.8)

$= (-1)^{50}x^{50}$

$= 1 \cdot x^{50}$ The product of an even number of negative factors is positive.

$= x^{50}$

10. $(5x^2y^{-2})^3 = 5^3(x^2)^3(y^{-2})^3 = 125x^6y^{-6}$ Be sure to raise *each* factor to the third power.

$= \frac{125x^6}{y^6}$

11. $(3x^3y^{-5}z^2)^4 = 3^4(x^3)^4(y^{-5})^4(z^2)^4 = 81x^{12}y^{-20}z^8 = \frac{81x^{12}z^8}{y^{20}}$

Do Exercises 5–8.

There is a similar rule for raising a quotient to a power.

RAISING A QUOTIENT TO A POWER

For any real numbers a and b, $b \neq 0$, and any integer n,

$$\left(\frac{a}{b}\right)^n = \frac{a^n}{b^n}.$$

(To raise a quotient to the nth power, raise both the numerator and the denominator to the nth power.) Also,

$$\left(\frac{a}{b}\right)^{-n} = \left(\frac{b}{a}\right)^n = \frac{b^n}{a^n},\ a \neq 0.$$

EXAMPLES Simplify.

12. $\left(\frac{x^2}{4}\right)^3 = \frac{(x^2)^3}{4^3} = \frac{x^6}{64}$

13. $\left(\frac{3a^4}{b^3}\right)^2 = \frac{(3a^4)^2}{(b^3)^2} = \frac{3^2(a^4)^2}{b^{3\cdot 2}} = \frac{9a^8}{b^6}$

14. $\left(\frac{y^3}{5}\right)^{-2} = \frac{(y^3)^{-2}}{5^{-2}} = \frac{y^{-6}}{5^{-2}} = \frac{\frac{1}{y^6}}{\frac{1}{5^2}} = \frac{1}{y^6} \div \frac{1}{5^2} = \frac{1}{y^6} \cdot \frac{5^2}{1} = \frac{25}{y^6}$

Example 14 might also be done as follows:

$$\left(\frac{y^3}{5}\right)^{-2} = \left(\frac{5}{y^3}\right)^2 = \frac{5^2}{(y^3)^2} = \frac{25}{y^6}.$$

Do Exercises 9–11 on the preceding page.

C Scientific Notation

There are many kinds of symbols, or notation, for numbers. You are already familiar with fraction notation, decimal notation, and percent notation. Now we study another, **scientific notation,** which makes use of exponential notation. Scientific notation is especially useful when calculations involve very large or very small numbers. The following are examples of scientific notation:

①

②

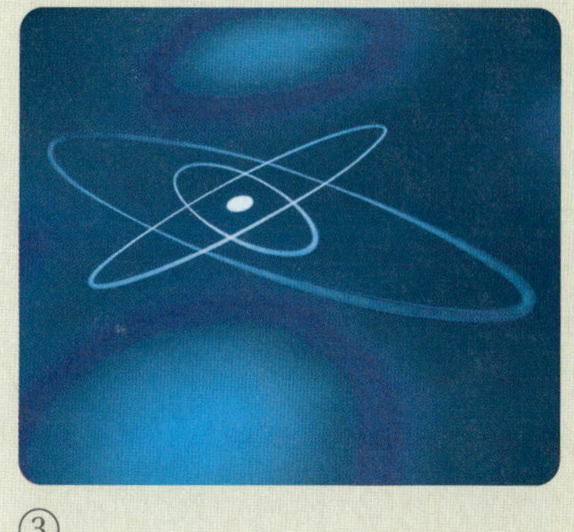

③

① *Niagara Falls*: On the Canadian side, during the summer the amount of water that spills over the falls in 1 day is about

4.9793×10^{10} gal $= 49{,}793{,}000{,}000$ gal.

② *The mass of the earth*:

$6.615 \times 10^{21} = 6{,}615{,}000{,}000{,}000{,}000{,}000{,}000$ tons.

③ *The mass of a hydrogen atom*:

1.7×10^{-24} g $= 0.0000000000000000000000017$ g.

SCIENTIFIC NOTATION

Scientific notation for a number is an expression of the type

$M \times 10^n$,

where n is an integer, M is greater than or equal to 1 and less than 10 ($1 \leq M < 10$), and M is expressed in decimal notation. 10^n is also considered to be scientific notation when $M = 1$.

Convert to scientific notation.

12. 0.000517

13. 523,000,000

Answers on page A-28

You should try to make conversions to scientific notation mentally as much as possible. Here is a handy mental device.

> A positive exponent in scientific notation indicates a large number (greater than or equal to 10) and a negative exponent indicates a small number (between 0 and 1).

EXAMPLES Convert to scientific notation.

15. $78{,}000 = 7.8 \times 10^4$

7.8,000.

4 places

Large number, so the exponent is positive.

16. $0.0000057 = 5.7 \times 10^{-6}$

0.000005.7

6 places

Small number, so the exponent is negative.

Each of the following is *not* scientific notation.

12.46×10^7 — This number is greater than 10.

0.347×10^{-5} — This number is less than 1.

Do Exercises 12 and 13.

Study Tips

TIME MANAGEMENT (PART 2)

Here are some additional tips to help you with time management. (See also the Study Tips on time management in Sections 7.8 and 10.2.)

- **Avoid "time killers."** We live in a media age, and the Internet, e-mail, television, and movies are all time killers. Allow yourself a break to enjoy some college and outside activities. But keep track of the time you spend on such activities and compare it to the time you spend studying.
- **Prioritize your tasks.** Be careful about taking on too many college activities that fall outside of academics. Examples of such activities are decorating a homecoming float, joining a fraternity or sorority, and participating on a student council committee. Any of these is important but keep them to a minimum to be sure that you have enough time for your studies.
- **Be aggressive about your study tasks.** Instead of worrying over your math homework or test preparation, do something to get yourself started. Work a problem here and a problem there, and before long you will accomplish the task at hand. If the task is large, break it down into smaller parts, and do one at a time. You will be surprised at how quickly the large task can then be completed.

"Time is more valuable than money. You can get more money, but you can't get more time."

Jim Rohn, motivational speaker

EXAMPLES Convert mentally to decimal notation.

17. $7.893 \times 10^5 = 789{,}300$

7.89300.
5 places

Positive exponent, so the answer is a large number.

18. $4.7 \times 10^{-8} = 0.000000047$

.00000004.7
8 places

Negative exponent, so the answer is a small number.

Do Exercises 14 and 15.

Convert to decimal notation.

14. 6.893×10^{11}

15. 5.67×10^{-5}

d Multiplying and Dividing Using Scientific Notation

MULTIPLYING

Consider the product

$$400 \cdot 2000 = 800{,}000.$$

In scientific notation, this is

$$(4 \times 10^2) \cdot (2 \times 10^3) = (4 \cdot 2)(10^2 \cdot 10^3) = 8 \times 10^5.$$

By applying the commutative and associative laws, we can find this product by multiplying $4 \cdot 2$, to get 8, and $10^2 \cdot 10^3$, to get 10^5 (we do this by adding the exponents).

EXAMPLE 19 Multiply: $(1.8 \times 10^6) \cdot (2.3 \times 10^{-4})$.

We apply the commutative and associative laws to get

$$\begin{aligned}(1.8 \times 10^6) \cdot (2.3 \times 10^{-4}) &= (1.8 \cdot 2.3) \times (10^6 \cdot 10^{-4}) \\ &= 4.14 \times 10^{6+(-4)} \\ &= 4.14 \times 10^2.\end{aligned}$$

We get 4.14 by multiplying 1.8 and 2.3. We get 10^2 by adding the exponents 6 and -4.

EXAMPLE 20 Multiply: $(3.1 \times 10^5) \cdot (4.5 \times 10^{-3})$.

We have

$$\begin{aligned}(3.1 \times 10^5) \cdot (4.5 \times 10^{-3}) &= (3.1 \times 4.5)(10^5 \cdot 10^{-3}) \\ &= 13.95 \times 10^2 && \text{Not scientific notation. 13.95 is greater than 10.} \\ &= (1.395 \times 10^1) \times 10^2 && \text{Substituting } 1.395 \times 10^1 \text{ for 13.95} \\ &= 1.395 \times (10^1 \times 10^2) && \text{Associative law} \\ &= 1.395 \times 10^3. && \text{Adding exponents. The answer is now in scientific notation.}\end{aligned}$$

Do Exercises 16 and 17.

Multiply and write scientific notation for the result.

16. $(1.12 \times 10^{-8})(5 \times 10^{-7})$

17. $(9.1 \times 10^{-17})(8.2 \times 10^3)$

Answers on page A-28

Divide and write scientific notation for the result.

18. $\dfrac{4.2 \times 10^5}{2.1 \times 10^2}$

19. $\dfrac{1.1 \times 10^{-4}}{2.0 \times 10^{-7}}$

Answers on page A-28

DIVIDING

Consider the quotient

$$800{,}000 \div 400 = 2000.$$

In scientific notation, this is

$$(8 \times 10^5) \div (4 \times 10^2) = \frac{8 \times 10^5}{4 \times 10^2} = \frac{8}{4} \times \frac{10^5}{10^2} = 2 \times 10^3.$$

We can find this product by dividing 8 by 4, to get 2, and 10^5 by 10^2, to get 10^3 (we do this by subtracting the exponents.)

EXAMPLE 21 Divide: $(3.41 \times 10^5) \div (1.1 \times 10^{-3})$.

We have

$$\begin{aligned}(3.41 \times 10^5) \div (1.1 \times 10^{-3}) &= \frac{3.41 \times 10^5}{1.1 \times 10^{-3}} = \frac{3.41}{1.1} \times \frac{10^5}{10^{-3}} \\ &= 3.1 \times 10^{5-(-3)} \\ &= 3.1 \times 10^8.\end{aligned}$$

CALCULATOR CORNER

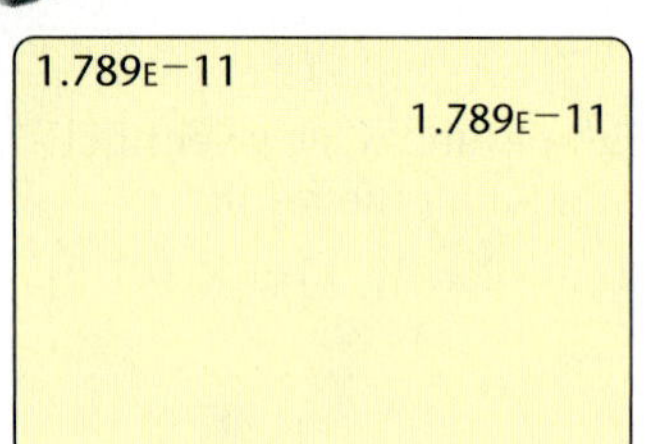

Normal Sci Eng
Float 0123456789
Radian Degree
Func Par Pol Seq
Connected Dot
Sequential Simul
Real a+bi re^θi
Full Horiz G−T

1.8E6∗2.3E−4
4.14E2

Scientific Notation To enter a number in scientific notation on a graphing calculator, we first type the decimal portion of the number and then press [2nd] [EE]. (EE is the second operation associated with the [·] key.) Finally, we type the exponent, which can be at most two digits. For example, to enter 1.789×10^{-11} in scientific notation, we press [1] [.] [7] [8] [9] [2nd] [EE] [(−)] [1] [1] [ENTER]. The decimal portion of the number appears before a small E and the exponent follows the E.

The graphing calculator can be used to perform computations using scientific notation. To find the product in Example 19 and express the result in scientific notation, we first set the calculator in Scientific mode by pressing [MODE], positioning the cursor over Sci on the first line, and pressing [ENTER]. Then we press [2nd] [QUIT] to go to the home screen and enter the computation by pressing [1] [.] [8] [2nd] [EE] [6] [×] [2] [.] [3] [2nd] [EE] [(−)] [4] [ENTER].

Exercises: Multiply or divide and express the answer in scientific notation.

1. $(3.15 \times 10^7)(4.3 \times 10^{-12})$

2. $(4.76 \times 10^{-5})(1.9 \times 10^{10})$

3. $(8 \times 10^9)(4 \times 10^{-5})$

4. $(4 \times 10^4)(9 \times 10^7)$

5. $\dfrac{4.5 \times 10^6}{1.5 \times 10^{12}}$

6. $\dfrac{6.4 \times 10^{-5}}{1.6 \times 10^{-10}}$

7. $\dfrac{4 \times 10^{-9}}{5 \times 10^{16}}$

8. $\dfrac{9 \times 10^{11}}{3 \times 10^{-2}}$

EXAMPLE 22 Divide: $(6.4 \times 10^{-7}) \div (8.0 \times 10^6)$.

We have

$$\begin{aligned}(6.4 \times 10^{-7}) \div (8.0 \times 10^6) &= \frac{6.4 \times 10^{-7}}{8.0 \times 10^6} \\ &= \frac{6.4}{8.0} \times \frac{10^{-7}}{10^6} \\ &= 0.8 \times 10^{-7-6} \\ &= 0.8 \times 10^{-13} && \text{Not scientific notation. 0.8 is less than 1.} \\ &= (8.0 \times 10^{-1}) \times 10^{-13} && \text{Substituting } 8.0 \times 10^{-1} \text{ for } 0.8 \\ &= 8.0 \times (10^{-1} \times 10^{-13}) && \text{Associative law} \\ &= 8.0 \times 10^{-14}. && \text{Adding exponents}\end{aligned}$$

Do Exercises 18 and 19 on the preceding page.

e Applications with Scientific Notation

EXAMPLE 23 *Distance from the Sun to Earth.* Light from the sun traveling at a rate of 300,000 kilometers per second (km/s) reaches Earth in 499 sec. Find the distance, expressed in scientific notation, from the sun to Earth.

The time t that it takes for light to reach Earth from the sun is 4.99×10^2 sec (s). The speed is 3.0×10^5 km/s. Recall that distance can be expressed in terms of speed and time as

$$\text{Distance} = \text{Speed} \cdot \text{Time}$$
$$d = rt.$$

We substitute 3.0×10^5 for r and 4.99×10^2 for t:

$$\begin{aligned}d &= rt \\ &= (3.0 \times 10^5)(4.99 \times 10^2) && \text{Substituting} \\ &= 14.97 \times 10^7 \\ &= (1.497 \times 10^1) \times 10^7 \\ &= 1.497 \times (10^1 \times 10^7) \\ &= 1.497 \times 10^8 \text{ km}. && \text{Converting to scientific notation}\end{aligned}$$

Thus the distance from the sun to Earth is 1.497×10^8 km.

Do Exercise 20.

20. Niagara Falls Water Flow. On the Canadian side, during the summer the amount of water that spills over the falls in 1 min is about

$$1.3088 \times 10^8 \text{ L}.$$

How much water spills over the falls in one day? Express the answer in scientific notation.
Source: *Collier's Encyclopedia*, 1997, Vol. 17

Answer on page A-28

Study Tips

WRITING ALL THE STEPS

Take the time to include all the steps when working your homework problems. Doing so will help you organize your thinking and avoid computational errors. If you find a wrong answer, having all the steps allows easier checking of your work. It will also give you complete, step-by-step solutions of the exercises that can be used to study for an exam.

Writing down all the steps and keeping your work organized may also give you a better chance of getting partial credit.

"Success comes before work only in the dictionary."

Anonymous

21. Earth vs. Saturn. The mass of Earth is about 6×10^{21} metric tons. The mass of Saturn is about 5.7×10^{23} metric tons. About how many times the mass of Earth is the mass of Saturn? Express the answer in scientific notation.

Answer on page A-28

EXAMPLE 24 *DNA.* A strand of DNA (deoxyribonucleic acid) is about 150 cm long and 1.3×10^{-10} cm wide. How many times longer is DNA than it is wide?

Source: Human Genome Project Information

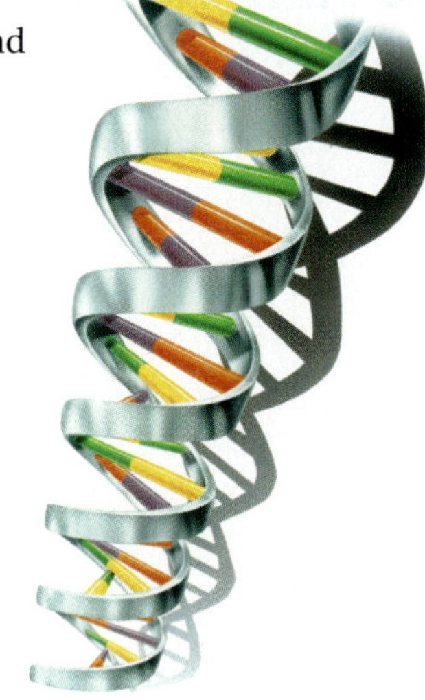

To determine how many times longer (N) DNA is than it is wide, we divide the length by the width:

$$\begin{aligned} N = \frac{150}{1.3 \times 10^{-10}} &= \frac{150}{1.3} \times \frac{1}{10^{-10}} \\ &\approx 115.385 \times 10^{10} \\ &= (1.15385 \times 10^2) \times 10^{10} \\ &= 1.15385 \times 10^{12}. \end{aligned}$$

Thus the length of DNA is about 1.15385×10^{12} times its width.

Do Exercise 21.

The following is a summary of the definitions and rules for exponents that we have considered in this section and the preceding one.

DEFINITIONS AND RULES FOR EXPONENTS

Exponent of 1:	$a^1 = a$
Exponent of 0:	$a^0 = 1, a \neq 0$
Negative exponents:	$a^{-n} = \dfrac{1}{a^n}, \dfrac{1}{a^{-n}} = a^n, a \neq 0$
Product Rule:	$a^m \cdot a^n = a^{m+n}$
Quotient Rule:	$\dfrac{a^m}{a^n} = a^{m-n}, a \neq 0$
Power Rule:	$(a^m)^n = a^{mn}$
Raising a product to a power:	$(ab)^n = a^n b^n$
Raising a quotient to a power:	$\left(\dfrac{a}{b}\right)^n = \dfrac{a^n}{b^n}, b \neq 0;$
	$\left(\dfrac{a}{b}\right)^{-n} = \dfrac{b^n}{a^n}, b \neq 0, a \neq 0$
Scientific notation:	$M \times 10^n$, or 10^n, where $1 \leq M < 10$

9.2 EXERCISE SET

For Extra Help

Digital Video Tutor CD 7 Videotape 9

InterAct Math

Math Tutor Center

MathXL

MyMathLab

a, **b** Simplify.

1. $(2^3)^2$

2. $(5^2)^4$

3. $(5^2)^{-3}$

4. $(7^{-3})^5$

5. $(x^{-3})^{-4}$

6. $(a^{-5})^{-6}$

7. $(a^{-2})^9$

8. $(x^{-5})^6$

9. $(t^{-3})^{-6}$

10. $(a^{-4})^{-7}$

11. $(t^4)^{-3}$

12. $(t^5)^{-2}$

13. $(x^{-2})^{-4}$

14. $(t^{-6})^{-5}$

15. $(ab)^3$

16. $(xy)^2$

17. $(ab)^{-3}$

18. $(xy)^{-6}$

19. $(mn^2)^{-3}$

20. $(x^3y)^{-2}$

21. $(4x^3)^2$

22. $4(x^3)^2$

23. $(3x^{-4})^2$

24. $(2a^{-5})^3$

25. $(x^4y^5)^{-3}$

26. $(t^5x^3)^{-4}$

27. $(x^{-6}y^{-2})^{-4}$

28. $(x^{-2}y^{-7})^{-5}$

29. $(a^{-2}b^7)^{-5}$

30. $(q^5r^{-1})^{-3}$

31. $(5r^{-4}t^3)^2$

32. $(4x^5y^{-6})^3$

33. $(a^{-5}b^7c^{-2})^3$

34. $(x^{-4}y^{-2}z^9)^2$

35. $(3x^3y^{-8}z^{-3})^2$

36. $(2a^2y^{-4}z^{-5})^3$

37. $\left(\frac{y^3}{2}\right)^2$

38. $\left(\frac{a^5}{3}\right)^3$

39. $\left(\frac{a^2}{b^3}\right)^4$

40. $\left(\frac{x^3}{y^4}\right)^5$

41. $\left(\frac{y^2}{2}\right)^{-3}$

42. $\left(\frac{a^4}{3}\right)^{-2}$

43. $\left(\frac{7}{x^{-3}}\right)^2$

44. $\left(\frac{3}{a^{-2}}\right)^3$

45. $\left(\frac{x^2y}{z}\right)^3$

46. $\left(\frac{m}{n^4p}\right)^3$

47. $\left(\frac{a^2b}{cd^3}\right)^{-2}$

48. $\left(\frac{2a^2}{3b^4}\right)^{-3}$

c Convert to scientific notation.

49. 28,000,000,000

50. 4,900,000,000,000

51. 907,000,000,000,000,000

52. 168,000,000,000,000

53. 0.00000304

54. 0.000000000865

55. 0.000000018

56. 0.00000000002

57. 100,000,000,000

58. 0.0000001

Convert the number in the sentence to scientific notation.

59. *Population of the United States.* After the 2000 census, the population of the United States was 281 million (1 million $= 10^6$).
Source: U.S. Bureau of the Census

60. *NASCAR.* Total revenue of NASCAR (National Association of Stock Car Automobile Racing) is expected to be $3423 million by 2006.
Source: NASCAR

61. *State Lottery.* Typically, the probability of winning a state lottery is about 1/10,000,000.

62. *Cancer Death Rate.* In Michigan, the death rate due to cancer is about 127.1/1000.
Source: AARP

Convert to decimal notation.

63. 8.74×10^7

64. 1.85×10^8

65. 5.704×10^{-8}

66. 8.043×10^{-4}

67. 10^7

68. 10^6

69. 10^{-5}

70. 10^{-8}

d Multiply or divide and write scientific notation for the result.

71. $(3 \times 10^4)(2 \times 10^5)$

72. $(3.9 \times 10^8)(8.4 \times 10^{-3})$

73. $(5.2 \times 10^5)(6.5 \times 10^{-2})$

74. $(7.1 \times 10^{-7})(8.6 \times 10^{-5})$

75. $(9.9 \times 10^{-6})(8.23 \times 10^{-8})$

76. $(1.123 \times 10^4) \times 10^{-9}$

77. $\dfrac{8.5 \times 10^{8}}{3.4 \times 10^{-5}}$

78. $\dfrac{5.6 \times 10^{-2}}{2.5 \times 10^{5}}$

79. $(3.0 \times 10^{6}) \div (6.0 \times 10^{9})$

80. $(1.5 \times 10^{-3}) \div (1.6 \times 10^{-6})$

81. $\dfrac{7.5 \times 10^{-9}}{2.5 \times 10^{12}}$

82. $\dfrac{4.0 \times 10^{-3}}{8.0 \times 10^{20}}$

Solve.

83. *River Discharge.* The average discharge at the mouths of the Amazon River is 4,200,000 cubic feet per second. How much water is discharged from the Amazon River in 1 yr? Express the answer in scientific notation.

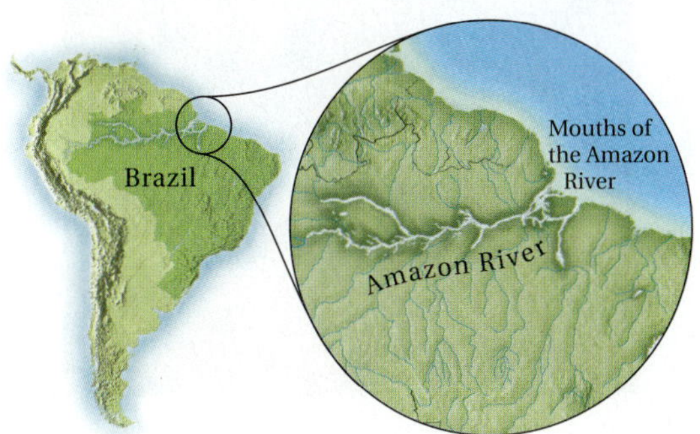

84. *Computers.* A gigabyte is a measure of a computer's storage capacity. One gigabyte holds about one billion bytes of information. If a firm's computer network contains 2500 gigabytes of memory, how many bytes are in the network? Express the answer in scientific notation.

85. *Earth vs. Jupiter.* The mass of Earth is about 6×10^{21} metric tons. The mass of Jupiter is about 1.908×10^{24} metric tons. About how many times the mass of Earth is the mass of Jupiter? Express the answer in scientific notation.

86. *Water Contamination.* In the United States, 200 million gal of used motor oil is improperly disposed of each year. One gallon of used oil can contaminate one million gallons of drinking water. How many gallons of drinking water can 200 million gallons of oil contaminate? Express the answer in scientific notation.

Source: *The Macmillan Visual Almanac*

87. *Stars.* It is estimated that there are 10 billion trillion stars in the known universe. Express the number of stars in scientific notation (1 billion $= 10^9$; 1 trillion $= 10^{12}$).

88. *Closest Star.* Excluding the sun, the closest star to Earth is Proxima Centauri, which is 4.3 light-years away (one light-year $= 5.88 \times 10^{12}$ mi). How far, in miles, is Proxima Centauri from Earth? Express the answer in scientific notation.

89. *Earth vs. Sun.* The mass of Earth is about 6×10^{21} metric tons. The mass of the sun is about 1.998×10^{27} metric tons. About how many times the mass of Earth is the mass of the sun? Express the answer in scientific notation.

90. *Red Light.* The wavelength of light is given by the velocity divided by the frequency. The velocity of red light is 300,000,000 m/sec, and its frequency is 400,000,000,000,000 cycles per second. What is the wavelength of red light? Express the answer in scientific notation.

Space Travel. Use the following information for Exercises 91 and 92.

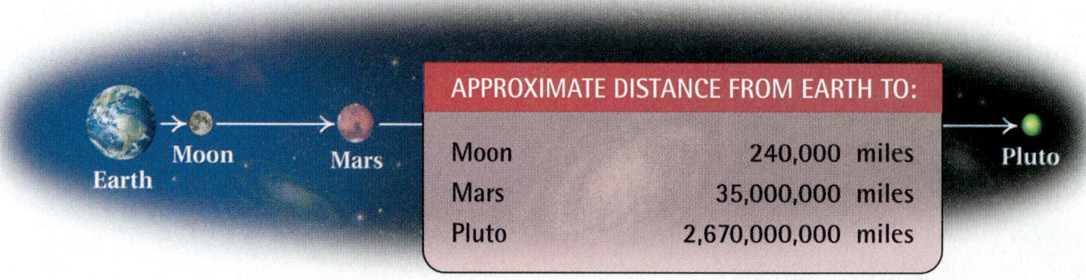

Earth
Moon
Mars
Pluto
APPROXIMATE DISTANCE FROM EARTH TO:
Moon 240,000 miles
Mars 35,000,000 miles
Pluto 2,670,000,000 miles

91. *Time to Reach Mars.* Suppose that it takes about 3 days for a space vehicle to travel from Earth to the moon. About how long would it take the same space vehicle traveling at the same speed to reach Mars? Express the answer in scientific notation.

92. *Time to Reach Pluto.* Suppose that it takes about 3 days for a space vehicle to travel from Earth to the moon. About how long would it take the same space vehicle traveling at the same speed to reach Pluto? Express the answer in scientific notation.

93. DW Explain in your own words when exponents should be added and when they should be multiplied.

94. DW Without performing actual computations, explain why 3^{-29} is smaller than 2^{-29}.

SKILL MAINTENANCE

Factor. [6.7d]

95. $9x - 36$

96. $4x - 2y + 16$

97. $3s + 3t + 24$

98. $-7x - 14$

Solve. [7.3b]

99. $2x - 4 - 5x + 8 = x - 3$

100. $8x + 7 - 9x = 12 - 6x + 5$

Solve. [7.3c]

101. $8(2x + 3) - 2(x - 5) = 10$

102. $4(x - 3) + 5 = 6(x + 2) - 8$

Graph. [8.2b], [8.3a]

103. $y = x - 5$

104. $2x + y = 8$

SYNTHESIS

105. Carry out the indicated operations. Express the result in scientific notation.

$$\frac{(5.2 \times 10^6)(6.1 \times 10^{-11})}{1.28 \times 10^{-3}}$$

106. Find the reciprocal and express it in scientific notation.

$$6.25 \times 10^{-3}$$

Simplify.

107. $\frac{(5^{12})^2}{5^{25}}$

108. $\frac{a^{22}}{(a^2)^{11}}$

109. $\frac{(3^5)^4}{3^5 \cdot 3^4}$

110. $\left(\frac{5x^{-2}}{3y^{-2}z}\right)^0$

111. $\frac{49^{18}}{7^{35}}$

112. $\left(\frac{1}{a}\right)^{-n}$

113. $\frac{(0.4)^5}{[(0.4^3]^2}$

114. $\left(\frac{4a^3b^{-2}}{5c^{-3}}\right)^1$

Determine whether each of the following is true for any pairs of integers m and n and any positive numbers x and y.

115. $x^m \cdot y^n = (xy)^{mn}$

116. $x^m \cdot y^m = (xy)^{2m}$

117. $(x - y)^m = x^m - y^m$

9.3 INTRODUCTION TO POLYNOMIALS

Objectives

a Evaluate a polynomial for a given value of the variable.

b Identify the terms of a polynomial.

c Identify the like terms of a polynomial.

d Identify the coefficients of a polynomial.

e Collect the like terms of a polynomial.

f Arrange a polynomial in descending order, or collect the like terms and then arrange in descending order.

g Identify the degree of each term of a polynomial and the degree of the polynomial.

h Identify the missing terms of a polynomial.

i Classify a polynomial as a monomial, binomial, trinomial, or none of these.

We have already learned to evaluate and to manipulate certain kinds of algebraic expressions. We will now consider algebraic expressions called *polynomials.*

The following are examples of *monomials in one variable*:

$$3x^2, \quad 2x, \quad -5, \quad 37p^4, \quad 0.$$

Each expression is a constant or a constant times some variable to a nonnegative integer power.

MONOMIAL

A **monomial** is an expression of the type ax^n, where a is a real-number constant and n is a nonnegative integer.

Algebraic expressions like the following are **polynomials:**

$$\tfrac{3}{4}y^5, \quad -2, \quad 5y + 3, \quad 3x^2 + 2x - 5, \quad -7a^3 + \tfrac{1}{2}a, \quad 6x, \quad 37p^4, \quad x, \quad 0.$$

POLYNOMIAL

A **polynomial** is a monomial or a combination of sums and/or differences of monomials.

The following algebraic expressions are *not* polynomials:

$$\textbf{(1)}\ \frac{x+3}{x-4}, \qquad \textbf{(2)}\ 5x^3 - 2x^2 + \frac{1}{x}, \qquad \textbf{(3)}\ \frac{1}{x^3-2}.$$

Expressions (1) and (3) are not polynomials because they represent quotients, not sums. Expression (2) is not a polynomial because

$$\frac{1}{x} = x^{-1},$$

and this is not a monomial because the exponent is negative.

Do Exercise 1.

1. Write three polynomials.

a Evaluating Polynomials and Applications

When we replace the variable in a polynomial with a number, the polynomial then represents a number called a **value** of the polynomial. Finding that number, or value, is called **evaluating the polynomial.** We evaluate a polynomial using the rules for order of operations (Section 6.8).

EXAMPLE 1 Evaluate the polynomial when $x = 2$.

a) $3x + 5 = 3 \cdot 2 + 5$
$\quad = 6 + 5$
$\quad = 11$

b) $2x^2 - 7x + 3 = 2 \cdot 2^2 - 7 \cdot 2 + 3$
$\quad = 2 \cdot 4 - 7 \cdot 2 + 3$
$\quad = 8 - 14 + 3$
$\quad = -3$

Answer on page A-28

EXAMPLE 2 Evaluate the polynomial when $x = -4$.

a) $2 - x^3 = 2 - (-4)^3 = 2 - (-64)$
$= 2 + 64$
$= 66$

b) $-x^2 - 3x + 1 = -(-4)^2 - 3(-4) + 1$
$= -16 + 12 + 1$
$= -3$

Do Exercises 2–5.

ALGEBRAIC–GRAPHICAL CONNECTION

An equation like $y = 2x - 2$, which has a polynomial on one side and y on the other, is called a **polynomial equation.** Here and in many places throughout the book, we will connect graphs to related concepts.

Recall from Chapter 8 that in order to plot points before graphing an equation, we choose values for x and compute the corresponding y-values. If the equation has y on one side and a polynomial involving x on the other, then determining y is the same as evaluating the polynomial. Once the graph of such an equation has been drawn, we can evaluate the polynomial for a given x-value by finding the y-value that is paired with it on the graph.

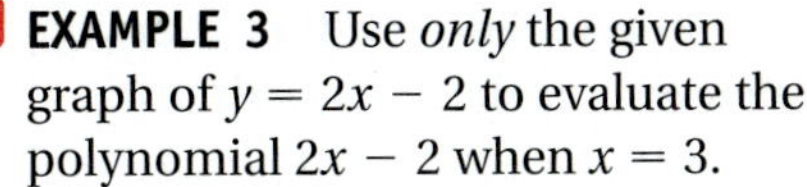

EXAMPLE 3 Use *only* the given graph of $y = 2x - 2$ to evaluate the polynomial $2x - 2$ when $x = 3$.

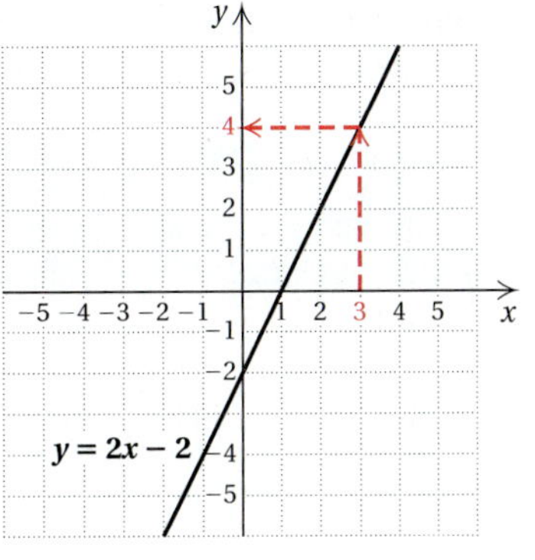

First, we locate 3 on the x-axis. From there we move vertically to the graph of the equation and then horizontally to the y-axis. There we locate the y-value that is paired with 3. Although our drawing may not be precise, it appears that the y-value 4 is paired with 3. Thus the value of $2x - 2$ is 4 when $x = 3$.

Do Exercise 6.

Polynomial equations can be used to model many real-world situations.

EXAMPLE 4 *Games in a Sports League.* In a sports league of x teams in which each team plays every other team twice, the total number of games N to be played is given by the polynomial equation

$$N = x^2 - x.$$

A women's slow-pitch softball league has 10 teams. What is the total number of games to be played?

We evaluate the polynomial when $x = 10$:

$$N = x^2 - x = 10^2 - 10 = 100 - 10 = 90.$$

The league plays 90 games.

Do Exercises 7 and 8.

Evaluate the polynomial when $x = 3$.

2. $-4x - 7$

3. $-5x^3 + 7x + 10$

Evaluate the polynomial when $x = -5$.

4. $5x + 7$

5. $2x^2 + 5x - 4$

6. Use *only* the graph shown in Example 3 to evaluate the polynomial $2x - 2$ when $x = 4$ and when $x = -1$.

7. Referring to Example 4, determine the total number of games to be played in a league of 12 teams.

8. Perimeter of a Baseball Diamond. The perimeter P of a square of side x is given by the polynomial equation $P = 4x$.

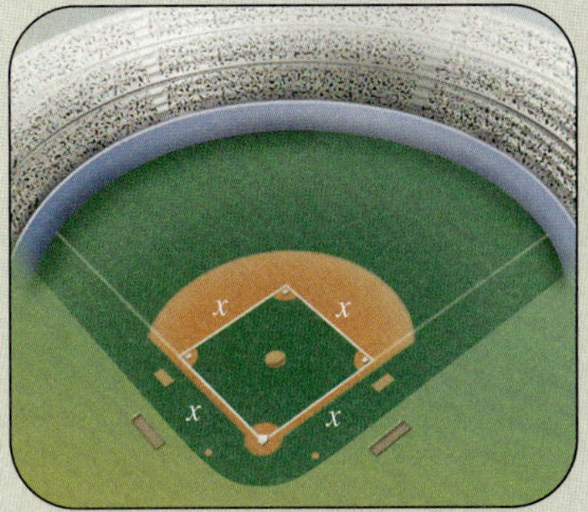

A baseball diamond is a square 90 ft on a side. Find the perimeter of a baseball diamond.

Answers on page A-28

9. Medical Dosage.

a) Referring to Example 5, determine the concentration after 3 hr by evaluating the polynomial when $t = 3$.

b) Use *only* the graph showing medical dosage to check the value found in part (a).

10. Medical Dosage. Referring to Example 5, use *only* the graph showing medical dosage to estimate the value of the polynomial when $t = 26$.

Answers on page A-28

EXAMPLE 5 *Medical Dosage.* The concentration C, in parts per million, of a certain antibiotic in the bloodstream after t hours is given by the polynomial equation

$$C = -0.05t^2 + 2t + 2.$$

Find the concentration after 2 hr.

To find the concentration after 2 hr, we evaluate the polynomial when $t = 2$:

$$\begin{aligned} C &= -0.05t^2 + 2t + 2 \\ &= -0.05(2)^2 + 2(2) + 2 \\ &= -0.05(4) + 2(2) + 2 \\ &= -0.2 + 4 + 2 \\ &= -0.2 + 6 \\ &= 5.8. \end{aligned}$$

Carrying out the calculation using the rules for order of operations

The concentration after 2 hr is 5.8 parts per million.

ALGEBRAIC–GRAPHICAL CONNECTION

The polynomial equation in Example 5 can be graphed if we evaluate the polynomial for several values of t. We list the values in a table and show the graph below. Note that the concentration peaks at the 20-hr mark and after slightly more than 40 hr, the concentration is 0. Since neither time nor concentration can be negative, our graph uses only the first quadrant.

t	C $C = -0.05t^2 + 2t + 2$
0	2
2	5.8 ← Example 5
10	17
20	22
30	17

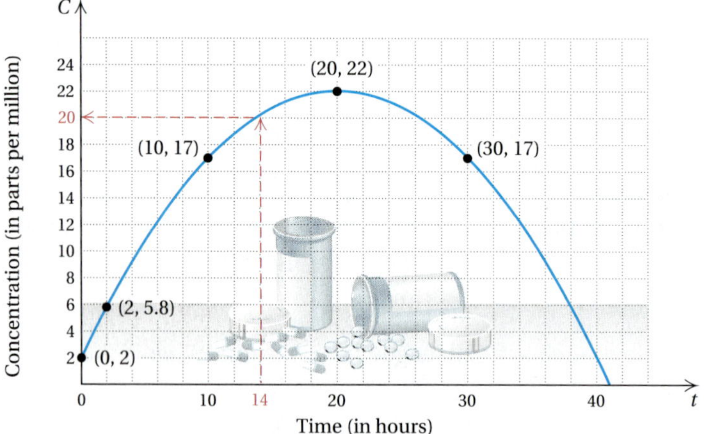

Do Exercises 9 and 10.

CALCULATOR CORNER

Evaluating Polynomials (*Note*: If you set your graphing calculator in Sci (scientific) mode to do the exercises in Section 9.2, return it to Normal mode now.)

There are several ways to evaluate polynomials on a graphing calculator. One method uses a table. To evaluate the polynomial in Example 2(b), $-x^2 - 3x + 1$, when $x = -4$, we first enter $y_1 = -x^2 - 3x + 1$ on the equation-editor screen. Then we set up a table in ASK mode (see p. 532) and enter the value -4 for x. We see that when $x = -4$, the value of Y1 is -3. This is the value of the polynomial when $x = -4$.

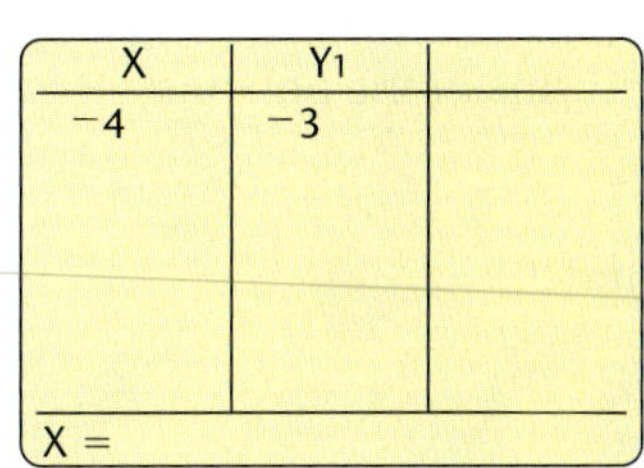

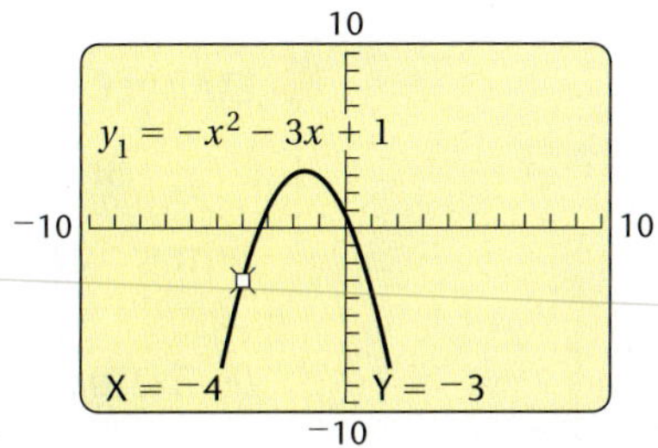

We can also use the Value feature from the CALC menu to evaluate this polynomial. First, we graph $y_1 = -x^2 - 3x + 1$ in a window that includes the x-value -4. We will use the standard window (see p. 616). Then we press [2nd] [CALC] [1] or [2nd] [CALC] [ENTER] to access the CALC menu and select item 1, Value. Now we supply the desired x-value by pressing [(−)] [4]. We then press [ENTER] to see X = −4, Y = −3 at the bottom of the screen. Thus, when $x = -4$, the value of $-x^2 - 3x + 1$ is -3.

Exercises: Use the Value feature to evaluate the polynomial for the given values of x.

1. $-x^2 - 3x + 1$, when $x = -2$, $x = -0.5$, and $x = 4$
2. $3x^2 - 5x + 2$, when $x = -3$, $x = 1$, and $x = 2.6$
3. $2x^2 - x - 8$, when $x = -3$, $x = 1.8$, and $x = 3$
4. $-5x^2 + 3x + 7$, when $x = -1$, $x = 2$, and $x = 3.4$

b Identifying Terms

As we saw in Section 6.4, subtractions can be rewritten as additions. For any polynomial that has some subtractions, we can find an equivalent polynomial using only additions.

EXAMPLES Find an equivalent polynomial using only additions.

6. $-5x^2 - x = -5x^2 + (-x)$
7. $4x^5 - 2x^6 - 4x + 7 = 4x^5 + (-2x^6) + (-4x) + 7$

Do Exercises 11 and 12.

Find an equivalent polynomial using only additions.

11. $-9x^3 - 4x^5$

12. $-2y^3 + 3y^7 - 7y$

Answers on page A-28

Identify the terms of the polynomial.

13. $3x^2 + 6x + \frac{1}{2}$

14. $-4y^5 + 7y^2 - 3y - 2$

Identify the like terms in the polynomial.

15. $4x^3 - x^3 + 2$

16. $4t^4 - 9t^3 - 7t^4 + 10t^3$

17. $5x^2 + 3x - 10 + 7x^2 - 8x + 11$

18. Identify the coefficient of each term in the polynomial

$$2x^4 - 7x^3 - 8.5x^2 + 10x - 4.$$

Answers on page A-28

When a polynomial has only additions, the monomials being added are called **terms.** In Example 6, the terms are $-5x^2$ and $-x$. In Example 7, the terms are $4x^5$, $-2x^6$, $-4x$, and 7.

EXAMPLE 8 Identify the terms of the polynomial

$$4x^7 + 3x + 12 + 8x^3 + 5x.$$

Terms: $4x^7$, $3x$, 12, $8x^3$, and $5x$.

If there are subtractions, you can *think* of them as additions without rewriting.

EXAMPLE 9 Identify the terms of the polynomial

$$3t^4 - 5t^6 - 4t + 2.$$

Terms: $3t^4$, $-5t^6$, $-4t$, and 2.

Do Exercises 13 and 14.

c Like Terms

When terms have the same variable and the variable is raised to the same power, we say that they are **like terms.**

EXAMPLES Identify the like terms in the polynomials.

10. $4x^3 + 5x - 4x^2 + 2x^3 + x^2$

Like terms: $4x^3$ and $2x^3$ — Same variable and exponent

Like terms: $-4x^2$ and x^2 — Same variable and exponent

11. $6 - 3a^2 + 8 - a - 5a$

Like terms: 6 and 8 — Constant terms are like terms because $6 = 6x^0$ and $8 = 8x^0$.

Like terms: $-a$ and $-5a$

Do Exercises 15–17.

d Coefficients

The coefficient of the term $5x^3$ is 5. In the following polynomial, the color numbers are the **coefficients,** 3, −2, 5, and 4:

$$3x^5 - 2x^3 + 5x + 4.$$

EXAMPLE 12 Identify the coefficient of each term in the polynomial

$$3x^4 - 4x^3 + 7x^2 + x - 8.$$

The coefficient of the first term is 3.

The coefficient of the second term is −4.

The coefficient of the third term is 7.

The coefficient of the fourth term is 1.

The coefficient of the fifth term is −8.

Do Exercise 18.

e Collecting Like Terms

We can often simplify polynomials by **collecting like terms,** or **combining like terms.** To do this, we use the distributive laws. We factor out the variable expression and add or subtract the coefficients. We try to do this mentally as much as possible.

EXAMPLES Collect like terms.

13. $2x^3 - 6x^3 = (2 - 6)x^3 = -4x^3$ Using a distributive law

14. $5x^2 + 7 + 4x^4 + 2x^2 - 11 - 2x^4 = (5 + 2)x^2 + (4 - 2)x^4 + (7 - 11)$
$= 7x^2 + 2x^4 - 4$

Note that using the distributive laws in this manner allows us to collect like terms by adding or subtracting the coefficients. Often the middle step is omitted and we add or subtract mentally, writing just the answer. In collecting like terms, we may get 0.

EXAMPLES Collect like terms.

15. $5x^3 - 5x^3 = (5 - 5)x^3 = 0x^3 = 0$

16. $3x^4 + 2x^2 - 3x^4 + 8 = (3 - 3)x^4 + 2x^2 + 8$
$= 0x^4 + 2x^2 + 8 = 2x^2 + 8$

Do Exercises 19–24.

Expressing a term like x^2 by showing 1 as a factor may make it easier to understand how to factor or collect like terms.

EXAMPLES Collect like terms.

17. $5x^2 + x^2 = 5x^2 + 1x^2$ Replacing x^2 with $1x^2$
$= (5 + 1)x^2$ Using a distributive law
$= 6x^2$

18. $5x^4 - 6x^3 - x^4 = 5x^4 - 6x^3 - 1x^4$ $x^4 = 1x^4$
$= (5 - 1)x^4 - 6x^3$
$= 4x^4 - 6x^3$

19. $\frac{2}{3}x^4 - x^3 - \frac{1}{6}x^4 + \frac{2}{5}x^3 - \frac{3}{10}x^3 = \left(\frac{2}{3} - \frac{1}{6}\right)x^4 + \left(-1 + \frac{2}{5} - \frac{3}{10}\right)x^3$
$= \left(\frac{4}{6} - \frac{1}{6}\right)x^4 + \left(-\frac{10}{10} + \frac{4}{10} - \frac{3}{10}\right)x^3$
$= \frac{3}{6}x^4 - \frac{9}{10}x^3 = \frac{1}{2}x^4 - \frac{9}{10}x^3$

Do Exercises 25–28.

f Descending and Ascending Order

Note in the following polynomial that the exponents decrease from left to right. We say that the polynomial is arranged in **descending order:**

$$2x^4 - 8x^3 + 5x^2 - x + 3.$$

The term with the largest exponent is first. The term with the next largest exponent is second, and so on. The associative and commutative laws allow us to arrange the terms of a polynomial in descending order.

Collect like terms.

19. $3x^2 + 5x^2$

20. $4x^3 - 2x^3 + 2 + 5$

21. $\frac{1}{2}x^5 - \frac{3}{4}x^5 + 4x^2 - 2x^2$

22. $24 - 4x^3 - 24$

23. $5x^3 - 8x^5 + 8x^5$

24. $-2x^4 + 16 + 2x^4 + 9 - 3x^5$

Collect like terms.

25. $7x - x$

26. $5x^3 - x^3 + 4$

27. $\frac{3}{4}x^3 + 4x^2 - x^3 + 7$

28. $8x^2 - x^2 + x^3 - 1 - 4x^2 + 10$

Answers on page A-28

Arrange the polynomial in descending order.

29. $x + 3x^5 + 4x^3 + 5x^2 + 6x^7 - 2x^4$

30. $4x^2 - 3 + 7x^5 + 2x^3 - 5x^4$

31. $-14 + 7t^2 - 10t^5 + 14t^7$

Collect like terms and then arrange in descending order.

32. $3x^2 - 2x + 3 - 5x^2 - 1 - x$

33. $-x + \frac{1}{2} + 14x^4 - 7x - 1 - 4x^4$

34. Identify the degree of each term and the degree of the polynomial

$$-6x^4 + 8x^2 - 2x + 9.$$

Answers on page A-28

EXAMPLES Arrange the polynomial in descending order.

20. $6x^5 + 4x^7 + x^2 + 2x^3 = 4x^7 + 6x^5 + 2x^3 + x^2$

21. $\frac{2}{3} + 4x^5 - 8x^2 + 5x - 3x^3 = 4x^5 - 3x^3 - 8x^2 + 5x + \frac{2}{3}$

Do Exercises 29–31.

EXAMPLE 22 Collect like terms and then arrange in descending order:

$$2x^2 - 4x^3 + 3 - x^2 - 2x^3.$$

$2x^2 - 4x^3 + 3 - x^2 - 2x^3 = x^2 - 6x^3 + 3$ Collecting like terms

$= -6x^3 + x^2 + 3$ Arranging in descending order

Do Exercises 32 and 33.

We usually arrange polynomials in descending order, but not always. The opposite order is called **ascending order.** Generally, if an exercise is written in a certain order, we give the answer in that same order.

g Degrees

The **degree** of a term is the exponent of the variable. The degree of the term $5x^3$ is 3.

EXAMPLE 23 Identify the degree of each term of $8x^4 + 3x + 7$.

The degree of $8x^4$ is 4.

The degree of $3x$ is 1. Recall that $x = x^1$.

The degree of 7 is 0. Think of 7 as $7x^0$. Recall that $x^0 = 1$.

The **degree of a polynomial** is the largest of the degrees of the terms, unless it is the polynomial 0. The polynomial 0 is a special case. We agree that it has *no* degree either as a term or as a polynomial. This is because we can express 0 as $0 = 0x^5 = 0x^7$, and so on, using any exponent we wish.

EXAMPLE 24 Identify the degree of the polynomial $5x^3 - 6x^4 + 7$.

$5x^3 - 6x^4 + 7.$ The largest exponent is 4.

The degree of the polynomial is 4.

Do Exercise 34.

Let's summarize the terminology that we have learned, using the polynomial $3x^4 - 8x^3 + 5x^2 + 7x - 6$.

TERM	COEFFICIENT	DEGREE OF THE TERM	DEGREE OF THE POLYNOMIAL
$3x^4$	3	4	
$-8x^3$	−8	3	
$5x^2$	5	2	4
$7x$	7	1	
−6	−6	0	

h Missing Terms

If a coefficient is 0, we generally do not write the term. We say that we have a **missing term.**

EXAMPLE 25 Identify the missing terms in the polynomial

$$8x^5 - 2x^3 + 5x^2 + 7x + 8.$$

There is no term with x^4. We say that the x^4-term is missing.

Do Exercises 35–38.

For certain skills or manipulations, we can write missing terms with zero coefficients or leave space.

EXAMPLE 26 Write the polynomial $x^4 - 6x^3 + 2x - 1$ in two ways: with its missing terms and by leaving space for them.

a) $x^4 - 6x^3 + 2x - 1 = x^4 - 6x^3 + 0x^2 + 2x - 1$

b) $x^4 - 6x^3 + 2x - 1 = x^4 - 6x^3 \qquad + 2x - 1$

EXAMPLE 27 Write the polynomial $y^5 - 1$ in two ways: with its missing terms and by leaving space for them.

a) $y^5 - 1 = y^5 + 0y^4 + 0y^3 + 0y^2 + 0y - 1$

b) $y^5 - 1 = y^5 \qquad\qquad\qquad\qquad - 1$

Do Exercises 39 and 40.

i Classifying Polynomials

Polynomials with just one term are called **monomials.** Polynomials with just two terms are called **binomials.** Those with just three terms are called **trinomials.** Those with more than three terms are generally not specified with a name.

EXAMPLE 28

MONOMIALS	BINOMIALS	TRINOMIALS	NONE OF THESE
$4x^2$	$2x + 4$	$3x^3 + 4x + 7$	$4x^3 - 5x^2 + x - 8$
9	$3x^5 + 6x$	$6x^7 - 7x^2 + 4$	$z^5 + 2z^4 - z^3 + 7z + 3$
$-23x^{19}$	$-9x^7 - 6$	$4x^2 - 6x - \frac{1}{2}$	$4x^6 - 3x^5 + x^4 - x^3 + 2x - 1$

Do Exercises 41–44.

Identify the missing terms in the polynomial.

35. $2x^3 + 4x^2 - 2$

36. $-3x^4$

37. $x^3 + 1$

38. $x^4 - x^2 + 3x + 0.25$

Write the polynomial in two ways: with its missing terms and by leaving space for them.

39. $2x^3 + 4x^2 - 2$

40. $a^4 + 10$

Classify the polynomial as a monomial, binomial, trinomial, or none of these.

41. $5x^4$

42. $4x^3 - 3x^2 + 4x + 2$

43. $3x^2 + x$

44. $3x^2 + 2x - 4$

Answers on page A-28

9.3 EXERCISE SET

For Extra Help

Digital Video Tutor CD 7 Videotape 9 | InterAct Math | Math Tutor Center | MathXL | MyMathLab

a Evaluate the polynomial when $x = 4$ and when $x = -1$.

1. $-5x + 2$

2. $-8x + 1$

3. $2x^2 - 5x + 7$

4. $3x^2 + x - 7$

5. $x^3 - 5x^2 + x$

6. $7 - x + 3x^2$

Evaluate the polynomial when $x = -2$ and when $x = 0$.

7. $3x + 5$

8. $8 - 4x$

9. $x^2 - 2x + 1$

10. $5x + 6 - x^2$

11. $-3x^3 + 7x^2 - 3x - 2$

12. $-2x^3 + 5x^2 - 4x + 3$

13. *Skydiving.* During the first 13 sec of a jump, the number of feet S that a skydiver falls in t seconds can be approximated by the polynomial equation

$$S = 11.12t^2.$$

Approximately how far has a skydiver fallen 10 sec after having jumped from a plane?

14. *Skydiving.* For jumps that exceed 13 sec, the polynomial equation

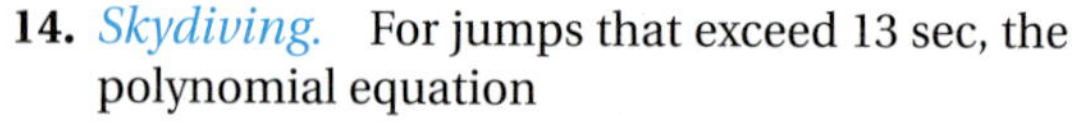

$$S = 173t - 369$$

can be used to approximate the distance S, in feet, that a skydiver has fallen in t seconds. Approximately how far has a skydiver fallen 20 sec after having jumped from a plane?

15. *Electricity Consumption.* The consumption of electricity in the United States can be approximated by the polynomial equation

$$E = 0.19t + 3.93,$$

where E is the electricity consumption, in millions of gigawatt hours, and t is the number of years since 2000—that is, $t = 0$ corresponds to 2000, $t = 5$ corresponds to 2005, and so on.

Source: Cambridge Energy Research Associates

a) Approximate the electricity consumption in 2000, 2001, 2003, 2005, 2008, and 2010.

b) Check the results of part (a) using the graph below.

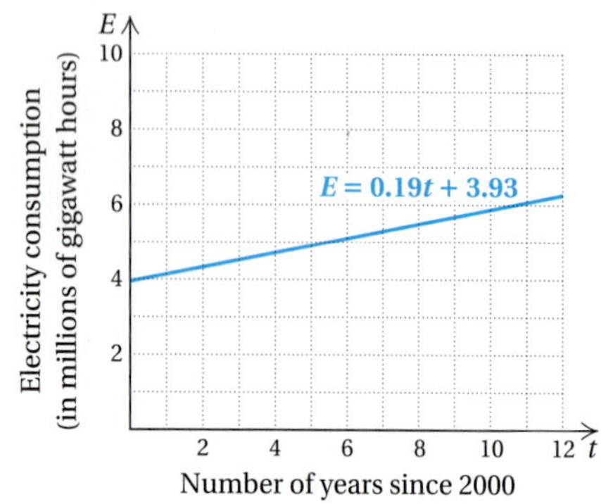

16. *Consumer Debt.* The amount of debt D, in trillions of dollars, held by consumers in the United States is given by the polynomial equation

$$D = 0.15t + 1.42,$$

where t is the number of years since 2000—that is, $t = 0$ corresponds to 2000, $t = 5$ corresponds to 2005, and so on.

Source: Federal Reserve

a) Find consumer debt in 2000, 2001, 2003, 2005, 2008, and 2010.

b) Check the results of part (a) using the graph below.

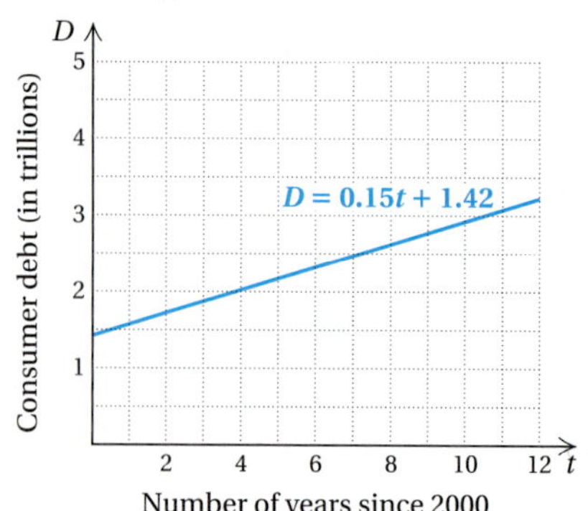

17. *Total Revenue.* Hadley Electronics is marketing a new kind of high-density TV. The firm determines that when it sells x TVs, its total revenue R (the total amount of money taken in) will be

$$R = 280x - 0.4x^2 \text{ dollars.}$$

What is the total revenue from the sale of 75 TVs? 100 TVs?

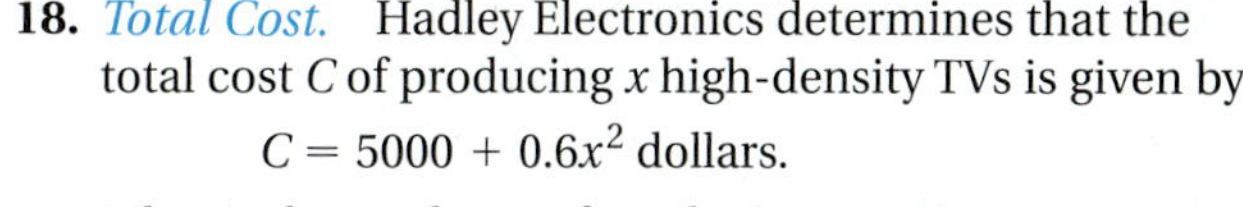

18. *Total Cost.* Hadley Electronics determines that the total cost C of producing x high-density TVs is given by

$$C = 5000 + 0.6x^2 \text{ dollars.}$$

What is the total cost of producing 500 TVs? 650 TVs?

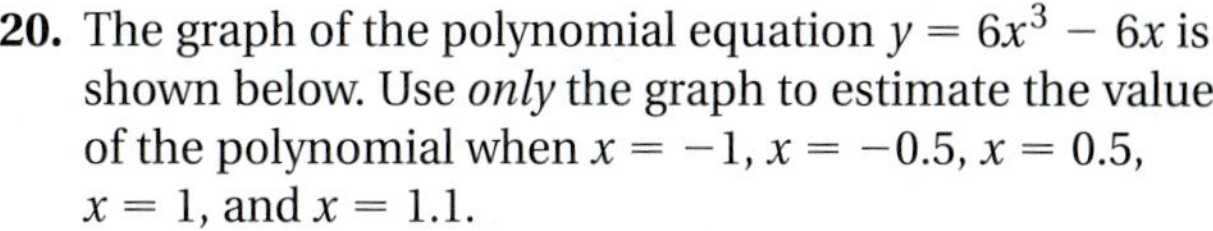

19. The graph of the polynomial equation $y = 5 - x^2$ is shown below. Use *only* the graph to estimate the value of the polynomial when $x = -3$, $x = -1$, $x = 0$, $x = 1.5$, and $x = 2$.

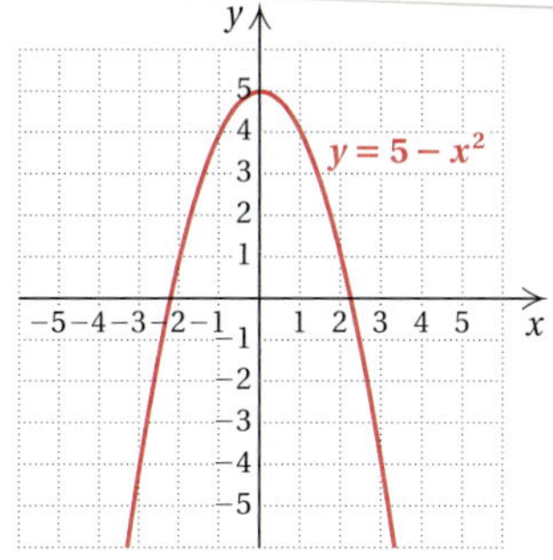

20. The graph of the polynomial equation $y = 6x^3 - 6x$ is shown below. Use *only* the graph to estimate the value of the polynomial when $x = -1$, $x = -0.5$, $x = 0.5$, $x = 1$, and $x = 1.1$.

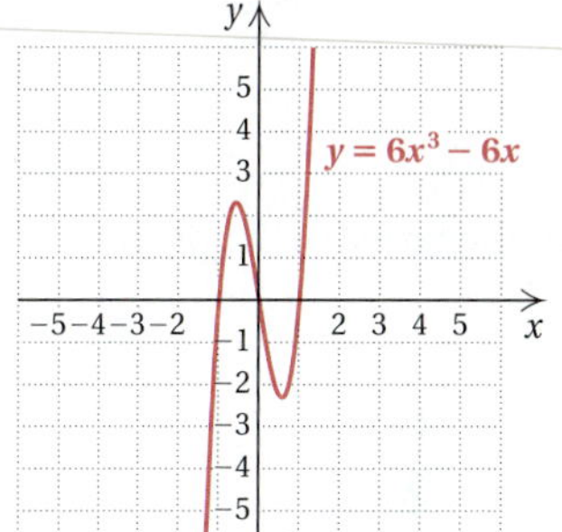

Hearing-Impaired Americans. The number N, in millions, of hearing-impaired Americans of age x can be approximated by the polynomial equation

$$N = -0.00006x^3 + 0.006x^2 - 0.1x + 1.9.$$

The graph of this equation is shown at right. Use either the graph or the polynomial equation for Exercises 21 and 22.

Source: American Speech-Language Hearing Association

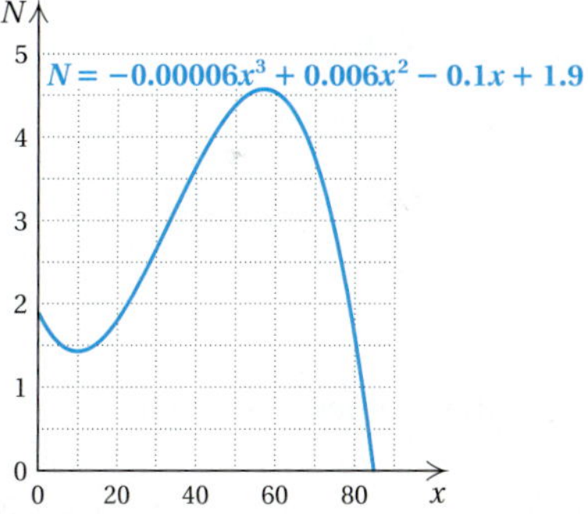

21. Approximate the number of hearing-impaired Americans of ages 20 and 40.

22. Approximate the number of hearing-impaired Americans of ages 50 and 60.

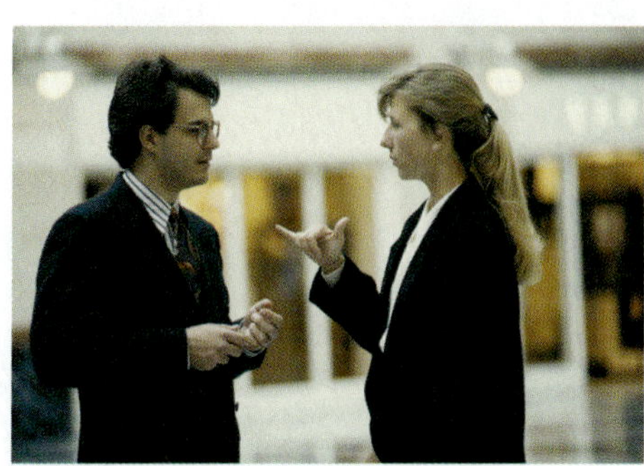

Memorizing words. Participants in a psychology experiment were able to memorize an average of M words in t minutes, where $M = -0.001t^3 + 0.1t^2$. Use the graph below for Exercises 23–28.

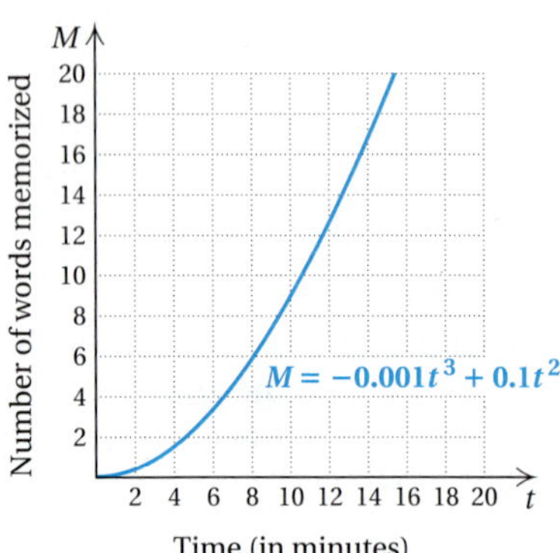

23. Estimate the number of words memorized after 10 min.

24. Estimate the number of words memorized after 14 min.

25. Find the approximate value of M for $t = 8$.

26. Find the approximate value of M for $t = 12$.

27. Estimate the value of M when t is 13.

28. Estimate the value of M when t is 7.

b Identify the terms of the polynomial.

29. $2 - 3x + x^2$

30. $2x^2 + 3x - 4$

c Identify the like terms in the polynomial.

31. $5x^3 + 6x^2 - 3x^2$

32. $3x^2 + 4x^3 - 2x^2$

33. $2x^4 + 5x - 7x - 3x^4$

34. $-3t + t^3 - 2t - 5t^3$

35. $3x^5 - 7x + 8 + 14x^5 - 2x - 9$

36. $8x^3 + 7x^2 - 11 - 4x^3 - 8x^2 - 29$

d Identify the coefficient of each term of the polynomial.

37. $-3x + 6$

38. $2x - 4$

39. $5x^2 + 3x + 3$

40. $3x^2 - 5x + 2$

41. $-5x^4 + 6x^3 - 3x^2 + 8x - 2$

42. $7x^3 - 4x^2 - 4x + 5$

e Collect like terms.

43. $2x - 5x$

44. $2x^2 + 8x^2$

45. $x - 9x$

46. $x - 5x$

47. $5x^3 + 6x^3 + 4$

48. $6x^4 - 2x^4 + 5$

49. $5x^3 + 6x - 4x^3 - 7x$

50. $3a^4 - 2a + 2a + a^4$

51. $6b^5 + 3b^2 - 2b^5 - 3b^2$

52. $2x^2 - 6x + 3x + 4x^2$

53. $\frac{1}{4}x^5 - 5 + \frac{1}{2}x^5 - 2x - 37$

54. $\frac{1}{3}x^3 + 2x - \frac{1}{6}x^3 + 4 - 16$

55. $6x^2 + 2x^4 - 2x^2 - x^4 - 4x^2$

56. $8x^2 + 2x^3 - 3x^3 - 4x^2 - 4x^2$

57. $\frac{1}{4}x^3 - x^2 - \frac{1}{6}x^2 + \frac{3}{8}x^3 + \frac{5}{16}x^3$

58. $\frac{1}{5}x^4 + \frac{1}{5} - 2x^2 + \frac{1}{10} - \frac{3}{15}x^4 + 2x^2 - \frac{3}{10}$

f Arrange the polynomial in descending order.

59. $x^5 + x + 6x^3 + 1 + 2x^2$

60. $3 + 2x^2 - 5x^6 - 2x^3 + 3x$

61. $5y^3 + 15y^9 + y - y^2 + 7y^8$

62. $9p - 5 + 6p^3 - 5p^4 + p^5$

Collect like terms and then arrange in descending order.

63. $3x^4 - 5x^6 - 2x^4 + 6x^6$

64. $-1 + 5x^3 - 3 - 7x^3 + x^4 + 5$

65. $-2x + 4x^3 - 7x + 9x^3 + 8$

66. $-6x^2 + x - 5x + 7x^2 + 1$

67. $3x + 3x + 3x - x^2 - 4x^2$

68. $-2x - 2x - 2x + x^3 - 5x^3$

69. $-x + \frac{3}{4} + 15x^4 - x - \frac{1}{2} - 3x^4$

70. $2x - \frac{5}{6} + 4x^3 + x + \frac{1}{3} - 2x$

g Identify the degree of each term of the polynomial and the degree of the polynomial.

71. $2x - 4$

72. $6 - 3x$

73. $3x^2 - 5x + 2$

74. $5x^3 - 2x^2 + 3$

75. $-7x^3 + 6x^2 + 3x + 7$

76. $5x^4 + x^2 - x + 2$

77. $x^2 - 3x + x^6 - 9x^4$

78. $8x - 3x^2 + 9 - 8x^3$

79. Complete the following table for the polynomial $-7x^4 + 6x^3 - 3x^2 + 8x - 2$.

TERM	COEFFICIENT	DEGREE OF THE TERM	DEGREE OF THE POLYNOMIAL
$-7x^4$			
$6x^3$	6		
		2	
$8x$		1	
	-2		

80. Complete the following table for the polynomial $3x^2 + 8x^5 - 46x^3 + 6x - 2.4 - \frac{1}{2}x^4$.

TERM	COEFFICIENT	DEGREE OF THE TERM	DEGREE OF THE POLYNOMIAL
		5	
$-\frac{1}{2}x^4$		4	
	-46		
$3x^2$		2	
	6		
-2.4			

h Identify the missing terms in the polynomial.

81. $x^3 - 27$

82. $x^5 + x$

83. $x^4 - x$

84. $5x^4 - 7x + 2$

85. $2x^3 - 5x^2 + x - 3$

86. $-6x^3$

Write the polynomial in two ways: with its missing terms and by leaving space for them.

87. $x^3 - 27$

88. $x^5 + x$

89. $x^4 - x$

90. $5x^4 - 7x + 2$

91. $2x^3 - 5x^2 + x - 3$

92. $-6x^3$

i Classify the polynomial as a monomial, binomial, trinomial, or none of these.

93. $x^2 - 10x + 25$

94. $-6x^4$

95. $x^3 - 7x^2 + 2x - 4$

96. $x^2 - 9$

97. $4x^2 - 25$

98. $2x^4 - 7x^3 + x^2 + x - 6$

99. $40x$

100. $4x^2 + 12x + 9$

101. DW Is it better to evaluate a polynomial before or after like terms have been collected? Why?

102. DW Explain why an understanding of the rules for order of operations is essential when evaluating polynomials.

SKILL MAINTENANCE

103. Three tired campers stopped for the night. All they had to eat was a bag of apples. During the night, one awoke and ate one-third of the apples. Later, a second camper awoke and ate one-third of the apples that remained. Much later, the third camper awoke and ate one-third of those apples yet remaining after the other two had eaten. When they got up the next morning, 8 apples were left. How many apples did they begin with? [7.6a]

Subtract. [6.4a]

104. $1 - 20$

105. $\frac{1}{8} - \frac{5}{6}$

106. $\frac{3}{8} - \left(-\frac{1}{4}\right)$

107. $5.6 - 8.2$

108. Solve: $3(x + 2) = 5x - 9$. [7.3c]

109. Solve $C = ab - r$ for b. [7.4b]

110. A nut dealer has 1800 lb of peanuts, 1500 lb of cashews, and 700 lb of almonds. What percent of the total is peanuts? cashews? almonds? [7.5a]

111. Factor: $3x - 15y + 63$. [6.7d]

SYNTHESIS

Collect like terms.

112. $\frac{9}{2}x^8 + \frac{1}{9}x^2 + \frac{1}{2}x^9 + \frac{9}{2}x + \frac{9}{2}x^9 + \frac{8}{9}x^2 + \frac{1}{2}x - \frac{1}{2}x^8$

113. $(3x^2)^3 + 4x^2 \cdot 4x^4 - x^4(2x)^2 + ((2x)^2)^3 - 100x^2(x^2)^2$

114. Construct a polynomial in x (meaning that x is the variable) of degree 5 with four terms and coefficients that are integers.

115. What is the degree of $(5m^5)^2$?

116. A polynomial in x has degree 3. The coefficient of x^2 is 3 less than the coefficient of x^3. The coefficient of x is three times the coefficient of x^2. The remaining coefficient is 2 more than the coefficient of x^3. The sum of the coefficients is -4. Find the polynomial.

Use the CALC feature and choose VALUE on your graphing calculator to find the values in each of the following.

117. Exercise 19

118. Exercise 20

119. Exercise 27

120. Exercise 28

9.4 ADDITION AND SUBTRACTION OF POLYNOMIALS

Objectives

a Add polynomials.

b Simplify the opposite of a polynomial.

c Subtract polynomials.

d Use polynomials to represent perimeter and area.

a Addition of Polynomials

To add two polynomials, we can write a plus sign between them and then collect like terms. Depending on the situation, you may see polynomials written in descending order, ascending order, or neither. Generally, if an exercise is written in a particular order, we write the answer in that same order.

EXAMPLE 1 Add: $(-3x^3 + 2x - 4) + (4x^3 + 3x^2 + 2)$.

$$(-3x^3 + 2x - 4) + (4x^3 + 3x^2 + 2)$$
$$= (-3 + 4)x^3 + 3x^2 + 2x + (-4 + 2) \quad \text{Collecting like terms (\textit{No} signs are changed.)}$$
$$= x^3 + 3x^2 + 2x - 2$$

EXAMPLE 2 Add:

$$\left(\tfrac{2}{3}x^4 + 3x^2 - 2x + \tfrac{1}{2}\right) + \left(-\tfrac{1}{3}x^4 + 5x^3 - 3x^2 + 3x - \tfrac{1}{2}\right).$$

We have

$$\left(\tfrac{2}{3}x^4 + 3x^2 - 2x + \tfrac{1}{2}\right) + \left(-\tfrac{1}{3}x^4 + 5x^3 - 3x^2 + 3x - \tfrac{1}{2}\right)$$
$$= \left(\tfrac{2}{3} - \tfrac{1}{3}\right)x^4 + 5x^3 + (3 - 3)x^2 + (-2 + 3)x + \left(\tfrac{1}{2} - \tfrac{1}{2}\right) \quad \text{Collecting like terms}$$
$$= \tfrac{1}{3}x^4 + 5x^3 + x.$$

We can add polynomials as we do because they represent numbers. After some practice, you will be able to add mentally.

Do Exercises 1–4.

EXAMPLE 3 Add: $(3x^2 - 2x + 2) + (5x^3 - 2x^2 + 3x - 4)$.

$$(3x^2 - 2x + 2) + (5x^3 - 2x^2 + 3x - 4)$$
$$= 5x^3 + (3 - 2)x^2 + (-2 + 3)x + (2 - 4) \quad \text{You might do this step mentally.}$$
$$= 5x^3 + x^2 + x - 2 \quad \text{Then you would write only this.}$$

Do Exercises 5 and 6.

We can also add polynomials by writing like terms in columns.

EXAMPLE 4 Add: $9x^5 - 2x^3 + 6x^2 + 3$ and $5x^4 - 7x^2 + 6$ and $3x^6 - 5x^5 + x^2 + 5$.

We arrange the polynomials with the like terms in columns.

$$\begin{array}{rrrrrr} & 9x^5 & & -\,2x^3 & +\,6x^2 & +\;3 \\ & & 5x^4 & & -\,7x^2 & +\;6 \\ 3x^6 & -\,5x^5 & & & +\;\;x^2 & +\;5 \\ \hline 3x^6 & +\,4x^5 & +\,5x^4 & -\,2x^3 & & +\,14 \end{array}$$

We leave spaces for missing terms.

Adding

We write the answer as $3x^6 + 4x^5 + 5x^4 - 2x^3 + 14$ without the space.

Add.

1. $(3x^2 + 2x - 2) + (-2x^2 + 5x + 5)$

2. $(-4x^5 + x^3 + 4) + (7x^4 + 2x^2)$

3. $(31x^4 + x^2 + 2x - 1) + (-7x^4 + 5x^3 - 2x + 2)$

4. $(17x^3 - x^2 + 3x + 4) + \left(-15x^3 + x^2 - 3x - \dfrac{2}{3}\right)$

Add mentally. Try to write just the answer.

5. $(4x^2 - 5x + 3) + (-2x^2 + 2x - 4)$

6. $(3x^3 - 4x^2 - 5x + 3) + \left(5x^3 + 2x^2 - 3x - \dfrac{1}{2}\right)$

Answers on page A-29

Do Exercises 7 and 8.

b Opposites of Polynomials

In Section 6.8, we used the property of -1 to show that we can find the opposite of an expression like

$$-(x - 2y + 5)$$

by changing the sign of every term:

$$-(x - 2y + 5) = -x + 2y - 5.$$

This applies to polynomials as well.

OPPOSITES OF POLYNOMIALS

To find an equivalent polynomial for the **opposite,** or **additive inverse,** of a polynomial, change the sign of every term. This is the same as multiplying by -1.

EXAMPLE 5 Simplify: $-(x^2 - 3x + 4)$.

$$-(x^2 - 3x + 4) = -x^2 + 3x - 4$$

EXAMPLE 6 Simplify: $-(-t^3 - 6t^2 - t + 4)$.

$$-(-t^3 - 6t^2 - t + 4) = t^3 + 6t^2 + t - 4$$

EXAMPLE 7 Simplify: $-\left(-7x^4 - \frac{5}{9}x^3 + 8x^2 - x + 67\right)$.

$$-\left(-7x^4 - \tfrac{5}{9}x^3 + 8x^2 - x + 67\right) = 7x^4 + \tfrac{5}{9}x^3 - 8x^2 + x - 67$$

Do Exercises 9–11.

c Subtraction of Polynomials

Recall that we can subtract a real number by adding its opposite, or additive inverse: $a - b = a + (-b)$. This allows us to subtract polynomials.

EXAMPLE 8 Subtract:

$$(9x^5 + x^3 - 2x^2 + 4) - (2x^5 + x^4 - 4x^3 - 3x^2).$$

We have

$$(9x^5 + x^3 - 2x^2 + 4) - (2x^5 + x^4 - 4x^3 - 3x^2)$$

$= 9x^5 + x^3 - 2x^2 + 4 + [-(2x^5 + x^4 - 4x^3 - 3x^2)]$ Adding the opposite

$= 9x^5 + x^3 - 2x^2 + 4 - 2x^5 - x^4 + 4x^3 + 3x^2$ Finding the opposite by changing the sign of *each* term

$= 7x^5 - x^4 + 5x^3 + x^2 + 4.$ Adding (collecting like terms)

Do Exercises 12 and 13.

Add.

7.
$$\begin{array}{r} -2x^3 + 5x^2 - 2x + 4 \\ x^4 \qquad\quad + 6x^2 + 7x - 10 \\ -9x^4 + 6x^3 + x^2 \qquad\quad - 2 \\ \hline \end{array}$$

8. $-3x^3 + 5x + 2$ and
$x^3 + x^2 + 5$ and
$x^3 - 2x - 4$

Simplify.

9. $-(4x^3 - 6x + 3)$

10. $-(5x^4 + 3x^2 + 7x - 5)$

11. $-\left(14x^{10} - \frac{1}{2}x^5 + 5x^3 - x^2 + 3x\right)$

Subtract.

12. $(7x^3 + 2x + 4) - (5x^3 - 4)$

13. $(-3x^2 + 5x - 4) - (-4x^2 + 11x - 2)$

Answers on page A-29

Subtract.

14. $(-6x^4 + 3x^2 + 6) - (2x^4 + 5x^3 - 5x^2 + 7)$

15. $\left(\frac{3}{2}x^3 - \frac{1}{2}x^2 + 0.3\right) - \left(\frac{1}{2}x^3 + \frac{1}{2}x^2 + \frac{4}{3}x + 1.2\right)$

Write in columns and subtract.

16. $(4x^3 + 2x^2 - 2x - 3) - (2x^3 - 3x^2 + 2)$

17. $(2x^3 + x^2 - 6x + 2) - (x^5 + 4x^3 - 2x^2 - 4x)$

Answers on page A-29

As with similar work in Section 6.8, we combine steps by changing the sign of each term of the polynomial being subtracted and collecting like terms. Try to do this mentally as much as possible.

EXAMPLE 9 Subtract: $(9x^5 + x^3 - 2x) - (-2x^5 + 5x^3 + 6)$.

$(9x^5 + x^3 - 2x) - (-2x^5 + 5x^3 + 6)$

$= 9x^5 + x^3 - 2x + 2x^5 - 5x^3 - 6$ Finding the opposite by changing the sign of each term

$= 11x^5 - 4x^3 - 2x - 6$ Adding (collecting like terms)

Do Exercises 14 and 15.

We can use columns to subtract. We replace coefficients with their opposites, as shown in Example 8.

EXAMPLE 10 Write in columns and subtract:

$(5x^2 - 3x + 6) - (9x^2 - 5x - 3)$.

a) $\begin{array}{r} 5x^2 - 3x + 6 \\ -(9x^2 - 5x - 3) \\ \hline \end{array}$ Writing like terms in columns

b) $\begin{array}{r} 5x^2 - 3x + 6 \\ -9x^2 + 5x + 3 \\ \hline \end{array}$ Changing signs

c) $\begin{array}{r} 5x^2 - 3x + 6 \\ -9x^2 + 5x + 3 \\ \hline -4x^2 + 2x + 9 \end{array}$ Adding

If you can do so without error, you can arrange the polynomials in columns and write just the answer, remembering to change the signs and add.

EXAMPLE 11 Write in columns and subtract:

$(x^3 + x^2 + 2x - 12) - (-2x^3 + x^2 - 3x)$.

$\begin{array}{r} x^3 + x^2 + 2x - 12 \\ -(-2x^3 + x^2 - 3x \qquad) \\ \hline 3x^3 \qquad + 5x - 12. \end{array}$ Leaving space for the missing term; Adding

Do Exercises 16 and 17.

d Polynomials and Geometry

EXAMPLE 12 Find a polynomial for the sum of the areas of these rectangles.

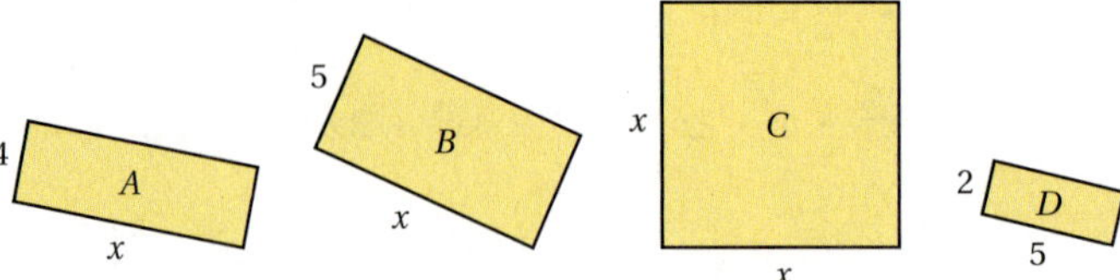

Recall that the area of a rectangle is the product of the length and the width. The sum of the areas is a sum of products. We find these products and then collect like terms.

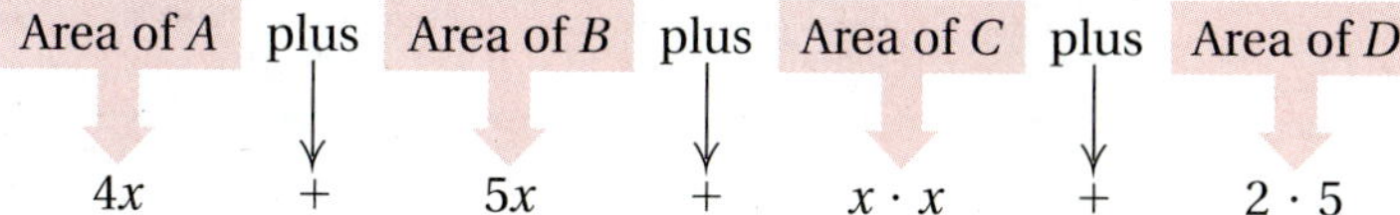

Area of A plus Area of B plus Area of C plus Area of D
$4x$ + $5x$ + $x \cdot x$ + $2 \cdot 5$

We collect like terms:

$$4x + 5x + x^2 + 10 = x^2 + 9x + 10.$$

Do Exercise 18.

18. Find a polynomial for the sum of the perimeters and the areas of the rectangles.

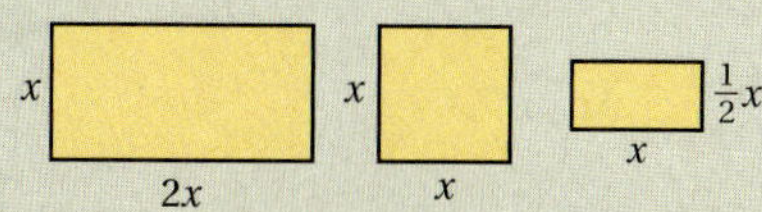

EXAMPLE 13 *Lawn Area.* A water fountain with a 4-ft by 4-ft square base is placed on a square grassy park area that is x ft on a side. To determine the amount of grass seed needed for the lawn, find a polynomial for the grassy area.

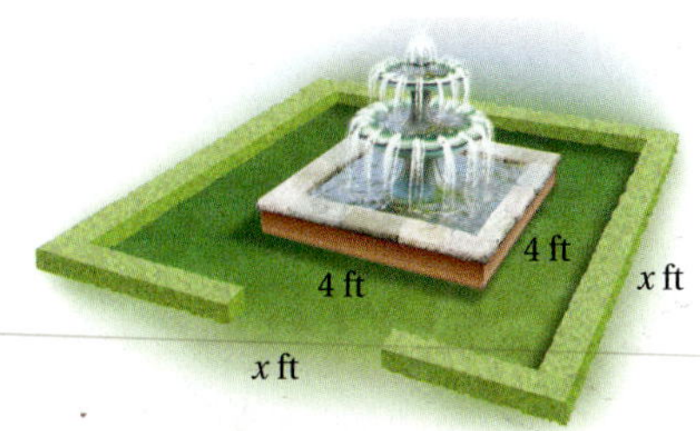

We make a drawing of the situation as shown here. We then reword the problem and write the polynomial as follows.

Area of park	$-$	Area of base of fountain	$=$	Area left over
$x \cdot x$	$-$	$4 \cdot 4$	$=$	Area left over

Then $x^2 - 16$ ft^2 = Area left over.

Do Exercise 19.

19. Lawn Area. An 8-ft by 8-ft shed is placed on a lawn x ft on a side. Find a polynomial for the remaining area.

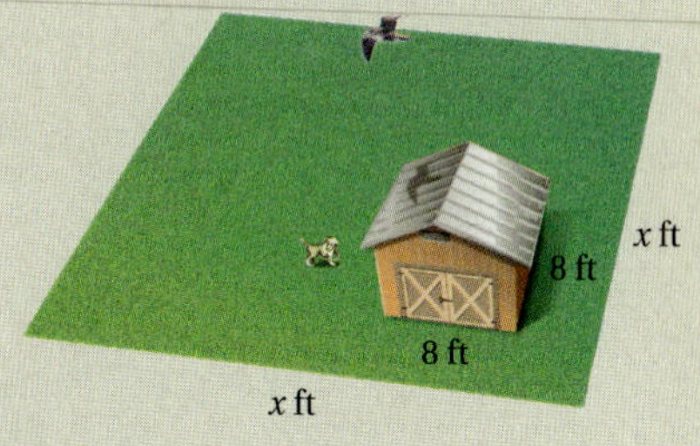

Answers on page A-29

CALCULATOR CORNER

Checking Addition and Subtraction of Polynomials A table set in AUTO mode can be used to perform a partial check that polynomials have been added or subtracted correctly. To check Example 3, we enter $y_1 = (3x^2 - 2x + 2) + (5x^3 - 2x^2 + 3x - 4)$ and $y_2 = 5x^3 + x^2 + x - 2$. If the addition has been done correctly, the values of y_1 and y_2 will be the same regardless of the table settings used.

X	Y1	Y2
−2	−40	−40
−1	−7	−7
0	−2	−2
1	5	5
2	44	44
3	145	145
4	338	338
X = −2		

A graph can also be used to check addition and subtraction. See the Calculator Corner on p. 707 for the procedure.

Exercises: Use a table to determine whether the sum or difference is correct.

1. $(-3x^3 + 2x - 4) + (4x^3 + 3x^2 + 2) = x^3 + 3x^2 + 2x - 2$
2. $(x^3 - 2x^2 + 3x - 7) + (3x^2 - 4x + 5) = x^3 + x^2 - x - 2$
3. $(5x^2 - 7x + 4) + (2x^2 + 3x - 6) = 7x^2 + 4x - 2$
4. $(9x^5 + x^3 - 2x) - (-2x^5 + 5x^3 + 6) = 11x^5 - 4x^3 - 2x - 6$
5. $(3x^4 - 2x^2 - 1) - (2x^4 - 3x^2 - 4) = x^4 + x^2 - 5$
6. $(-2x^3 + 3x^2 - 4x + 5) - (3x^2 + 2x + 8) = -2x^3 - 6x - 3$

9.4 EXERCISE SET

For Extra Help

Digital Video Tutor CD 7 Videotape 9 | InterAct Math | Math Tutor Center | MathXL | MyMathLab

a Add.

1. $(3x + 2) + (-4x + 3)$

2. $(6x + 1) + (-7x + 2)$

3. $(-6x + 2) + (x^2 + x - 3)$

4. $(x^2 - 5x + 4) + (8x - 9)$

5. $(x^2 - 9) + (x^2 + 9)$

6. $(x^3 + x^2) + (2x^3 - 5x^2)$

7. $(3x^2 - 5x + 10) + (2x^2 + 8x - 40)$

8. $(6x^4 + 3x^3 - 1) + (4x^2 - 3x + 3)$

9. $(1.2x^3 + 4.5x^2 - 3.8x) + (-3.4x^3 - 4.7x^2 + 23)$

10. $(0.5x^4 - 0.6x^2 + 0.7) + (2.3x^4 + 1.8x - 3.9)$

11. $(1 + 4x + 6x^2 + 7x^3) + (5 - 4x + 6x^2 - 7x^3)$

12. $(3x^4 - 6x - 5x^2 + 5) + (6x^2 - 4x^3 - 1 + 7x)$

13. $\left(\frac{1}{4}x^4 + \frac{2}{3}x^3 + \frac{5}{8}x^2 + 7\right) + \left(-\frac{3}{4}x^4 + \frac{3}{8}x^2 - 7\right)$

14. $\left(\frac{1}{3}x^9 + \frac{1}{5}x^5 - \frac{1}{2}x^2 + 7\right) +$
$\left(-\frac{1}{5}x^9 + \frac{1}{4}x^4 - \frac{3}{5}x^5 + \frac{3}{4}x^2 + \frac{1}{2}\right)$

15. $(0.02x^5 - 0.2x^3 + x + 0.08) +$
$(-0.01x^5 + x^4 - 0.8x - 0.02)$

16. $(0.03x^6 + 0.05x^3 + 0.22x + 0.05) +$
$\left(\frac{7}{100}x^6 - \frac{3}{100}x^3 + 0.5\right)$

17. $(9x^8 - 7x^4 + 2x^2 + 5) + (8x^7 + 4x^4 - 2x) +$
$(-3x^4 + 6x^2 + 2x - 1)$

18. $(4x^5 - 6x^3 - 9x + 1) + (6x^3 + 9x^2 + 9x) +$
$(-4x^3 + 8x^2 + 3x - 2)$

19.
$$\begin{array}{rrrrr}
0.15x^4 & + 0.10x^3 & - \ 0.9x^2 & & \\
 & - 0.01x^3 & + 0.01x^2 & + x & \\
1.25x^4 & & + 0.11x^2 & & + 0.01 \\
 & 0.27x^3 & & & + 0.99 \\
\hline
-0.35x^4 & & + \ 15x^2 & & - 0.03
\end{array}$$

20.
$$\begin{array}{rrrrr}
0.05x^4 & + 0.12x^3 & - \ 0.5x^2 & & \\
 & - 0.02x^3 & + 0.02x^2 & + 2x & \\
1.5x^4 & & + 0.01x^2 & & + 0.15 \\
 & 0.25x^3 & & & + 0.85 \\
\hline
-0.25x^4 & & + \ 10x^2 & & - 0.04
\end{array}$$

b Simplify.

21. $-(-5x)$

22. $-(x^2 - 3x)$

23. $-(-x^2 + 10x - 2)$

24. $-(-4x^3 - x^2 - x)$

25. $-(12x^4 - 3x^3 + 3)$

26. $-(4x^3 - 6x^2 - 8x + 1)$

27. $-(3x - 7)$

28. $-(-2x + 4)$

29. $-(4x^2 - 3x + 2)$

30. $-(-6a^3 + 2a^2 - 9a + 1)$

31. $-\left(-4x^4 + 6x^2 + \frac{3}{4}x - 8\right)$

32. $-(-5x^4 + 4x^3 - x^2 + 0.9)$

c Subtract.

33. $(3x + 2) - (-4x + 3)$

34. $(6x + 1) - (-7x + 2)$

35. $(-6x + 2) - (x^2 + x - 3)$

36. $(x^2 - 5x + 4) - (8x - 9)$

37. $(x^2 - 9) - (x^2 + 9)$

38. $(x^3 + x^2) - (2x^3 - 5x^2)$

39. $(6x^4 + 3x^3 - 1) - (4x^2 - 3x + 3)$

40. $(-4x^2 + 2x) - (3x^3 - 5x^2 + 3)$

41. $(1.2x^3 + 4.5x^2 - 3.8x) - (-3.4x^3 - 4.7x^2 + 23)$

42. $(0.5x^4 - 0.6x^2 + 0.7) - (2.3x^4 + 1.8x - 3.9)$

43. $\left(\frac{5}{8}x^3 - \frac{1}{4}x - \frac{1}{3}\right) - \left(-\frac{1}{8}x^3 + \frac{1}{4}x - \frac{1}{3}\right)$

44. $\left(\frac{1}{5}x^3 + 2x^2 - 0.1\right) - \left(-\frac{2}{5}x^3 + 2x^2 + 0.01\right)$

45. $(0.08x^3 - 0.02x^2 + 0.01x) - (0.02x^3 + 0.03x^2 - 1)$

46. $(0.8x^4 + 0.2x - 1) - \left(\frac{7}{10}x^4 + \frac{1}{5}x - 0.1\right)$

Subtract.

47. $\begin{array}{r} x^2 + 5x + 6 \\ -(x^2 + 2x) \\ \hline \end{array}$

48. $\begin{array}{r} x^3 \qquad + 1 \\ -(x^3 + x^2 \qquad) \\ \hline \end{array}$

49. $\begin{array}{r} 5x^4 + 6x^3 - 9x^2 \qquad\qquad \\ -(-6x^4 - 6x^3 \qquad + 8x + 9) \\ \hline \end{array}$

50. $\begin{array}{r} 5x^4 \qquad + 6x^2 - 3x + 6 \\ -(\quad 6x^3 + 7x^2 - 8x - 9) \\ \hline \end{array}$

51. $\begin{array}{r} x^5 \qquad\qquad\qquad\qquad - 1 \\ -(x^5 - x^4 + x^3 - x^2 + x - 1) \\ \hline \end{array}$

52. $\begin{array}{r} x^5 + x^4 - x^3 + x^2 - x + 2 \\ -(x^5 - x^4 + x^3 - x^2 - x + 2) \\ \hline \end{array}$

Solve.

53. Find a polynomial for the sum of the areas of these rectangles.

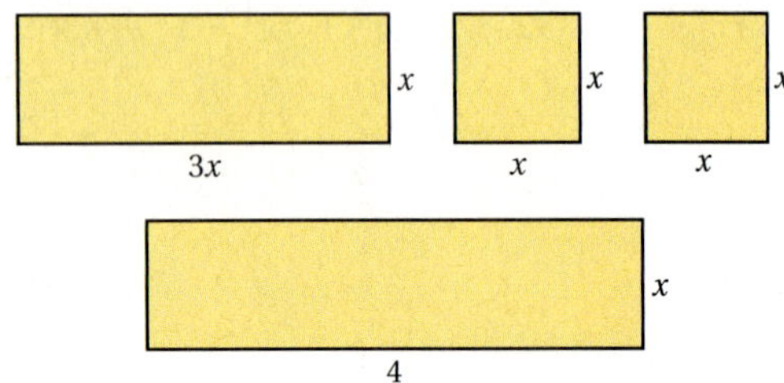

54. Find a polynomial for the sum of the areas of these circles.

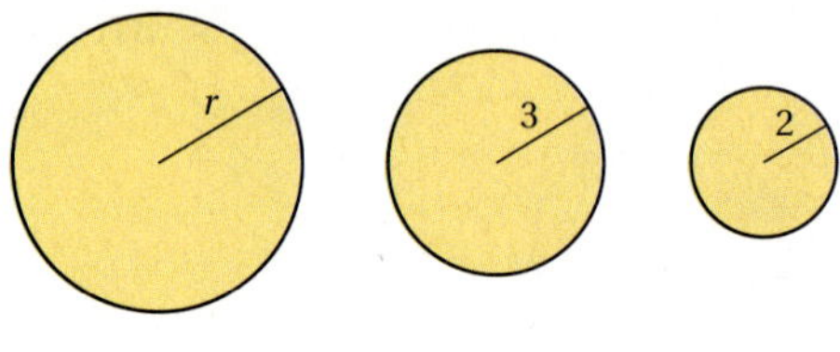

Find a polynomial for the perimeter of the figure.

55.

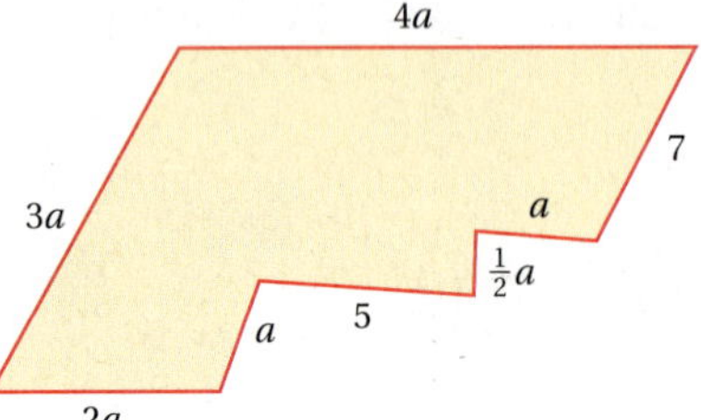

56.

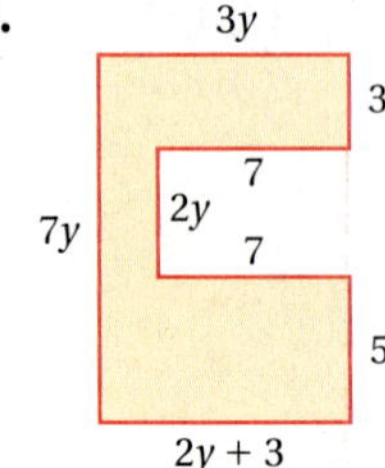

Find two algebraic expressions for the area of each figure. First, regard the figure as one large rectangle, and then regard the figure as a sum of four smaller rectangles.

57.

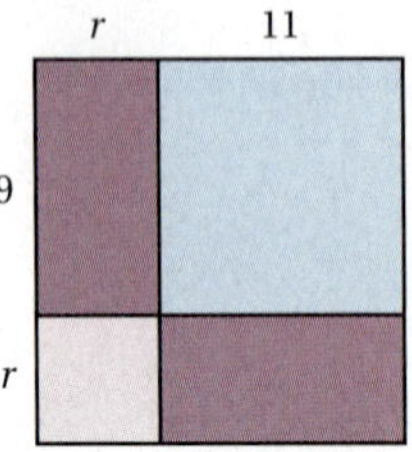

58.

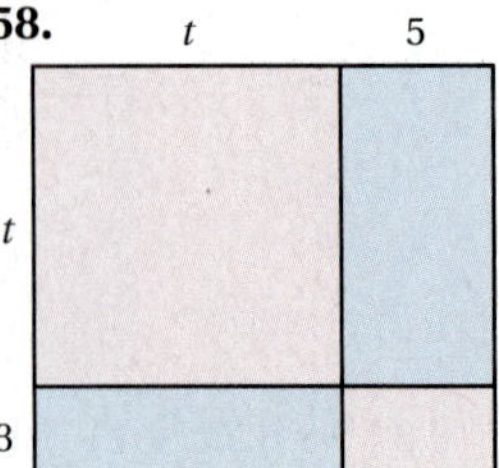

59.

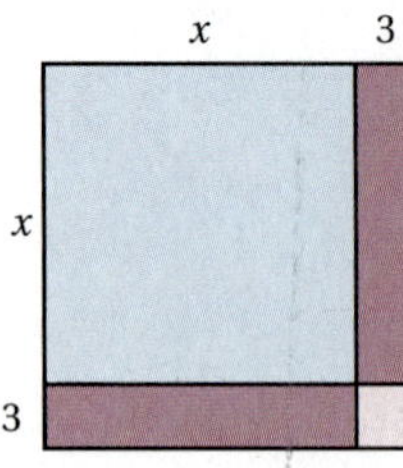

60.

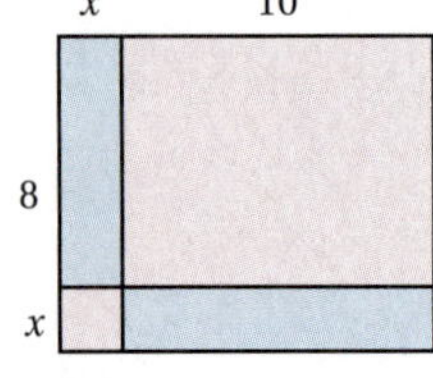

Find a polynomial for the shaded area of each figure.

61.

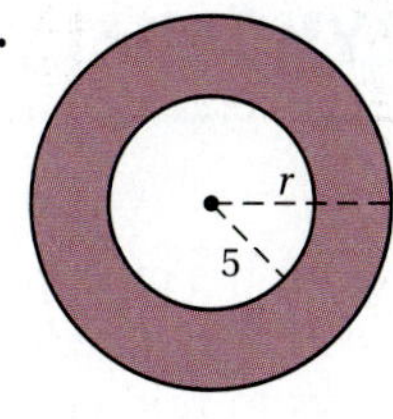

62.

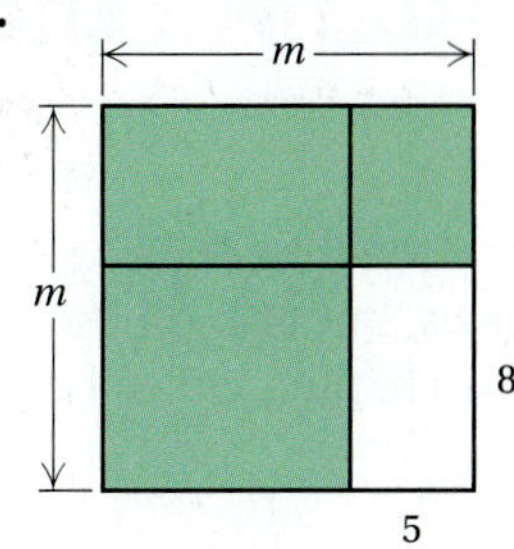

63.

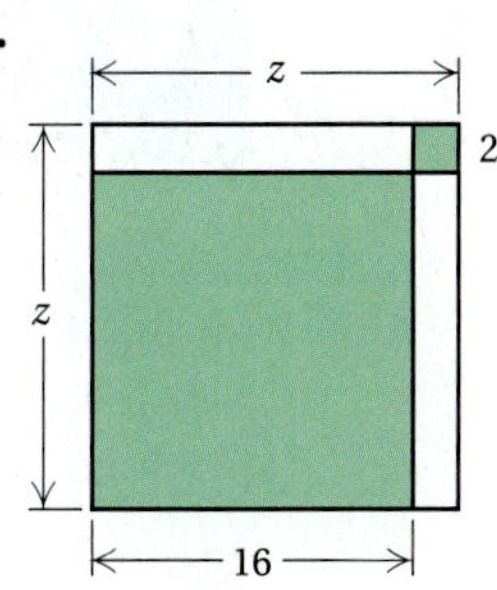

64.

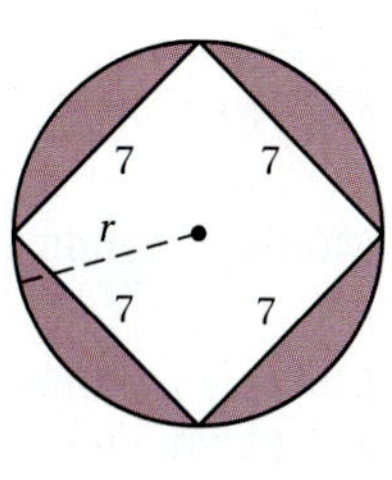

65. DW Is the sum of two binomials ever a trinomial? Why or why not?

66. DW Which, if any, of the commutative, associative, and distributive laws are needed for adding polynomials? Why?

SKILL MAINTENANCE

Solve. [7.3b]

67. $8x + 3x = 66$

68. $5x - 7x = 38$

69. $\frac{3}{8}x + \frac{1}{4} - \frac{3}{4}x = \frac{11}{16} + x$

70. $5x - 4 = 26 - x$

71. $1.5x - 2.7x = 22 - 5.6x$

72. $3x - 3 = -4x + 4$

Solve. [7.3c]

73. $6(y - 3) - 8 = 4(y + 2) + 5$

74. $8(5x + 2) = 7(6x - 3)$

Solve. [7.7e]

75. $3x - 7 \leq 5x + 13$

76. $2(x - 4) > 5(x - 3) + 7$

SYNTHESIS

Find a polynomial for the surface area of the right rectangular solid.

77.

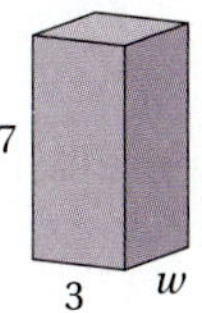

78.

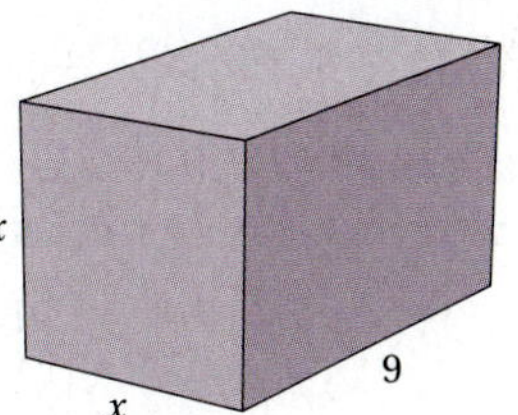

79.

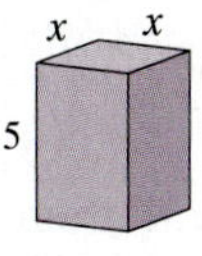

80.

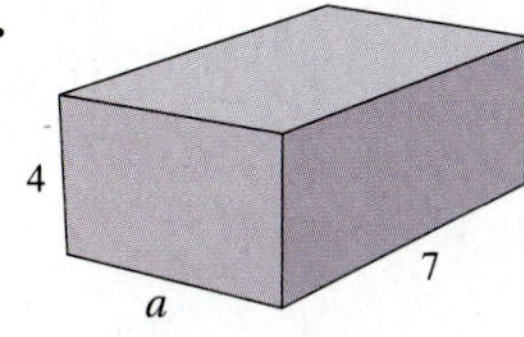

81. Find $(y - 2)^2$ using the four parts of this square.

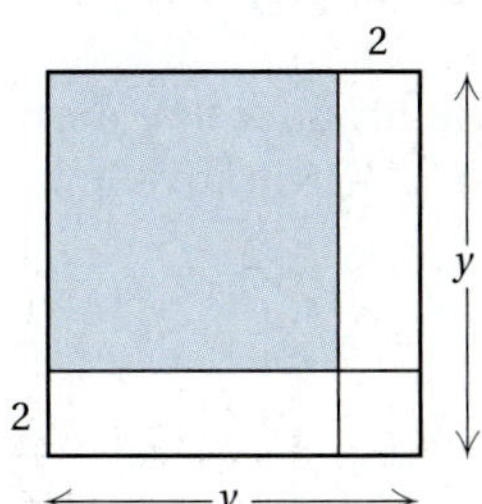

Simplify.

82. $(3x^2 - 4x + 6) - (-2x^2 + 4) + (-5x - 3)$

83. $(7y^2 - 5y + 6) - (3y^2 + 8y - 12) + (8y^2 - 10y + 3)$

84. $(-4 + x^2 + 2x^3) - (-6 - x + 3x^3) - (-x^2 - 5x^3)$

85. $(-y^4 - 7y^3 + y^2) + (-2y^4 + 5y - 2) - (-6y^3 + y^2)$

9.5 MULTIPLICATION OF POLYNOMIALS

Objectives

a. Multiply monomials.

b. Multiply a monomial and any polynomial.

c. Multiply two binomials.

d. Multiply any two polynomials.

Multiply.

1. $(3x)(-5)$

2. $(-x) \cdot x$

3. $(-x)(-x)$

4. $(-x^2)(x^3)$

5. $3x^5 \cdot 4x^2$

6. $(4y^5)(-2y^6)$

7. $(-7y^4)(-y)$

8. $7x^5 \cdot 0$

Answers on page A-29

We now multiply polynomials using techniques based, for the most part, on the distributive laws, but also on the associative and commutative laws. As we proceed in this chapter, we will develop special ways to find certain products.

a Multiplying Monomials

Consider $(3x)(4x)$. We multiply as follows:

$$\begin{aligned}(3x)(4x) &= 3 \cdot x \cdot 4 \cdot x && \text{By the associative law of multiplication}\\ &= 3 \cdot 4 \cdot x \cdot x && \text{By the commutative law of multiplication}\\ &= (3 \cdot 4)(x \cdot x) && \text{By the associative law}\\ &= 12x^2. && \text{Using the product rule for exponents}\end{aligned}$$

MULTIPLYING MONOMIALS

To find an equivalent expression for the product of two monomials, multiply the coefficients and then multiply the variables using the product rule for exponents.

EXAMPLES Multiply.

1. $5x \cdot 6x = (5 \cdot 6)(x \cdot x)$ — By the associative and commutative laws
   $= 30x^2$ — Multiplying the coefficients and multiplying the variables

2. $(3x)(-x) = (3x)(-1x)$
   $= (3)(-1)(x \cdot x) = -3x^2$

3. $(-7x^5)(4x^3) = (-7 \cdot 4)(x^5 \cdot x^3)$
   $= -28x^{5+3}$ — Adding the exponents
   $= -28x^8$ — Simplifying

After some practice, you can do this mentally. Multiply the coefficients and then the variables by keeping the base and adding the exponents. Write only the answer.

Do Exercises 1–8.

b Multiplying a Monomial and Any Polynomial

To find an equivalent expression for the product of a monomial, such as $2x$, and a binomial, such as $5x + 3$, we use a distributive law and multiply each term of $5x + 3$ by $2x$.

EXAMPLE 4 Multiply: $2x(5x + 3)$.

$$\begin{aligned}2x(5x + 3) &= (2x)(5x) + (2x)(3) && \text{Using a distributive law}\\ &= 10x^2 + 6x && \text{Multiplying the monomials}\end{aligned}$$

EXAMPLE 5 Multiply: $5x(2x^2 - 3x + 4)$.

$$5x(2x^2 - 3x + 4) = (5x)(2x^2) - (5x)(3x) + (5x)(4)$$
$$= 10x^3 - 15x^2 + 20x$$

MULTIPLYING A MONOMIAL AND A POLYNOMIAL

To multiply a monomial and a polynomial, multiply each term of the polynomial by the monomial.

EXAMPLE 6 Multiply: $-2x^2(x^3 - 7x^2 + 10x - 4)$.

$$-2x^2(x^3 - 7x^2 + 10x - 4) = -2x^5 + 14x^4 - 20x^3 + 8x^2$$

Do Exercises 9–11.

C Multiplying Two Binomials

To find an equivalent expression for the product of two binomials, we use the distributive laws more than once. In Example 7, we use a distributive law three times.

EXAMPLE 7 Multiply: $(x + 5)(x + 4)$.

$$(x + 5)(x + 4) = x(x + 4) + 5(x + 4) \quad \text{Using a distributive law}$$
$$= x \cdot x + x \cdot 4 + 5 \cdot x + 5 \cdot 4 \quad \text{Using a distributive law on each part}$$
$$= x^2 + 4x + 5x + 20 \quad \text{Multiplying the monomials}$$
$$= x^2 + 9x + 20 \quad \text{Collecting like terms}$$

To visualize the product in Example 7, consider a rectangle of length $x + 5$ and width $x + 4$.

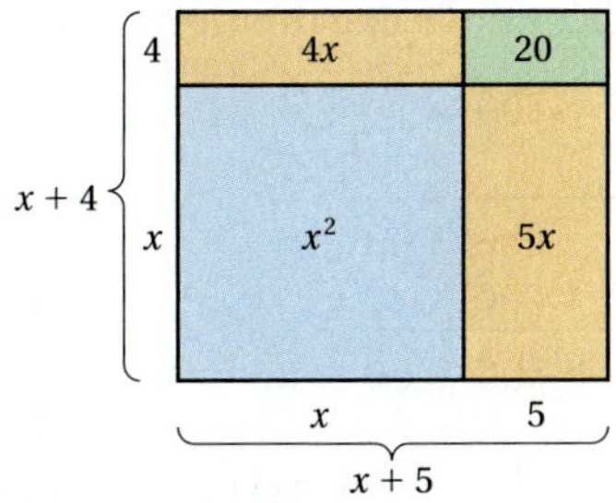

The total area can be expressed as $(x + 5)(x + 4)$ or, by adding the four smaller areas, $x^2 + 5x + 4x + 20$.

Do Exercises 12–14.

Multiply.

9. $4x(2x + 4)$

10. $3t^2(-5t + 2)$

11. $-5x^3(x^3 + 5x^2 - 6x + 8)$

12. Multiply: $(y + 2)(y + 7)$.

a) Fill in the blanks in the steps of the solution below.

$(y + 2)(y + 7)$

$= y \cdot$ ______ $+ 2 \cdot$ ______

$= y \cdot$ ______ $+ y \cdot$ ______

$+ 2 \cdot$ ______ $+ 2 \cdot$ ______

$=$ ______ $+$ ______

$+$ ______ $+$ ______

$= y^2 +$ ______ $+ 14$

b) Write an algebraic expression that represents the area of the four smaller rectangles in the figure shown here.

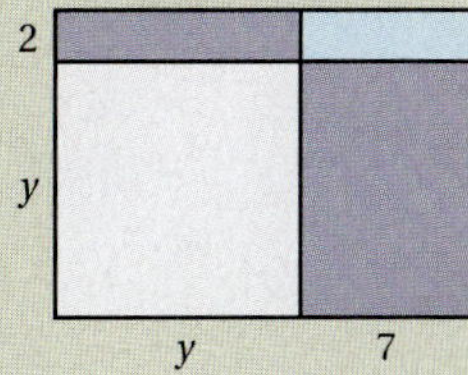

Multiply.

13. $(x + 8)(x + 5)$

14. $(x + 5)(x - 4)$

Answers on page A-29

Multiply.

15. $(5x + 3)(x - 4)$

16. $(2x - 3)(3x - 5)$

Multiply.

17. $(x^2 + 3x - 4)(x^2 + 5)$

18. $(3y^2 - 7)(2y^3 - 2y + 5)$

Answers on page A-29

EXAMPLE 8 Multiply: $(4x + 3)(x - 2)$.

$$
\begin{aligned}
(4x + 3)(x - 2) &= 4x(x - 2) + 3(x - 2) && \text{Using a distributive law} \\
&= 4x \cdot x - 4x \cdot 2 + 3 \cdot x - 3 \cdot 2 && \text{Using a distributive law on each part} \\
&= 4x^2 - 8x + 3x - 6 && \text{Multiplying the monomials} \\
&= 4x^2 - 5x - 6 && \text{Collecting like terms}
\end{aligned}
$$

Do Exercises 15 and 16.

d Multiplying Any Two Polynomials

Let's consider the product of a binomial and a trinomial. We use a distributive law four times. You may see ways to skip some steps and do the work mentally.

EXAMPLE 9 Multiply: $(x^2 + 2x - 3)(x^2 + 4)$.

$$
\begin{aligned}
(x^2 + 2x - 3)(x^2 + 4) &= x^2(x^2 + 4) + 2x(x^2 + 4) - 3(x^2 + 4) \\
&= x^2 \cdot x^2 + x^2 \cdot 4 + 2x \cdot x^2 + 2x \cdot 4 - 3 \cdot x^2 - 3 \cdot 4 \\
&= x^4 + 4x^2 + 2x^3 + 8x - 3x^2 - 12 \\
&= x^4 + 2x^3 + x^2 + 8x - 12
\end{aligned}
$$

Do Exercises 17 and 18.

PRODUCT OF TWO POLYNOMIALS

To multiply two polynomials P and Q, select one of the polynomials—say, P. Then multiply each term of P by every term of Q and collect like terms.

To use columns for long multiplication, multiply each term in the top row by every term in the bottom row. We write like terms in columns, and then add the results. Such multiplication is like multiplying with whole numbers:

$$
\begin{array}{r}
321 \\
\times \quad 12 \\
\hline
642 \\
321 \\
\hline
3852
\end{array}
\qquad
\begin{array}{rl}
300 + 20 + 1 & \\
\times \qquad 10 + 2 & \\
\hline
600 + 40 + 2 & \text{Multiplying the top row by 2} \\
3000 + 200 + 10 & \text{Multiplying the top row by 10} \\
\hline
3000 + 800 + 50 + 2 & \text{Adding}
\end{array}
$$

EXAMPLE 10 Multiply: $(4x^3 - 2x^2 + 3x)(x^2 + 2x)$.

$$
\begin{array}{rrrrl}
 & 4x^3 & -\ 2x^2 & +\ 3x & \\
 & & x^2 & +\ 2x & \\
\hline
 & 8x^4 & -\ 4x^3 & +\ 6x^2 & \text{Multiplying the top row by } 2x \\
4x^5 & -\ 2x^4 & +\ 3x^3 & & \text{Multiplying the top row by } x^2 \\
\hline
4x^5 & +\ 6x^4 & -\ x^3 & +\ 6x^2 & \text{Collecting like terms}
\end{array}
$$

Line up like terms in columns.

EXAMPLE 11 Multiply: $(5x^3 - 3x + 4)(-2x^2 - 3)$.

When missing terms occur, it helps to leave spaces for them and align like terms as we multiply.

$$\begin{array}{rrrrrl}
 & 5x^3 & & -3x & +4 & \\
 & & -2x^2 & & -3 & \\
\hline
 & -15x^3 & & +9x & -12 & \quad \text{Multiplying by } -3 \\
-10x^5 & +6x^3 & -8x^2 & & & \quad \text{Multiplying by } -2x^2 \\
\hline
-10x^5 & -9x^3 & -8x^2 & +9x & -12 & \quad \text{Collecting like terms}
\end{array}$$

Do Exercises 19 and 20.

EXAMPLE 12 Multiply: $(2x^2 + 3x - 4)(2x^2 - x + 3)$.

$$\begin{array}{rrrrrl}
 & & 2x^2 & +3x & -4 & \\
 & & 2x^2 & -x & +3 & \\
\hline
 & & 6x^2 & +9x & -12 & \quad \text{Multiplying by } 3 \\
 & -2x^3 & -3x^2 & +4x & & \quad \text{Multiplying by } -x \\
4x^4 & +6x^3 & -8x^2 & & & \quad \text{Multiplying by } 2x^2 \\
\hline
4x^4 & +4x^3 & -5x^2 & +13x & -12 & \quad \text{Collecting like terms}
\end{array}$$

Do Exercise 21.

Multiply.

19. $$\begin{array}{r} 3x^2 - 2x + 4 \\ x + 5 \\ \hline \end{array}$$

20. $$\begin{array}{r} -5x^2 + 4x + 2 \\ -4x^2 - 8 \\ \hline \end{array}$$

21. Multiply.

$$\begin{array}{r} 3x^2 - 2x - 5 \\ 2x^2 + x - 2 \\ \hline \end{array}$$

Answers on page A-29

CALCULATOR CORNER

Checking Multiplication of Polynomials A partial check of multiplication of polynomials can be performed graphically on the TI-83 Plus graphing calculator. Consider the product $(x + 3)(x - 2) = x^2 + x - 6$. We will use two graph styles to determine whether this product is correct. First, we press MODE to determine whether Sequential mode is selected. If it is not, we position the blinking cursor over Sequential and then press ENTER. Next, on the Y = screen, we enter $y_1 = (x + 3)(x - 2)$ and $y_2 = x^2 + x - 6$. We will select the line-graph style for y_1 and the path style for y_2. To select these graph styles, we use ◁ to position the cursor over the icon to the left of the equation and press ENTER repeatedly until the desired style of icon appears, as shown below.

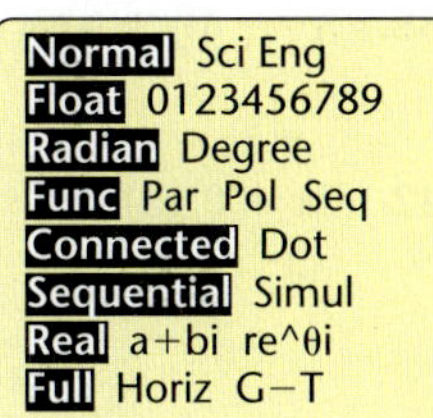

Plot1 Plot2 Plot3
\Y1 = (X+3)(X−2)
-0Y2 = X2+X−6
\Y3 =
\Y4 =
\Y5 =
\Y6 =
\Y7 =

$y_1 = (x + 3)(x - 2)$, $y_2 = x^2 + x - 6$

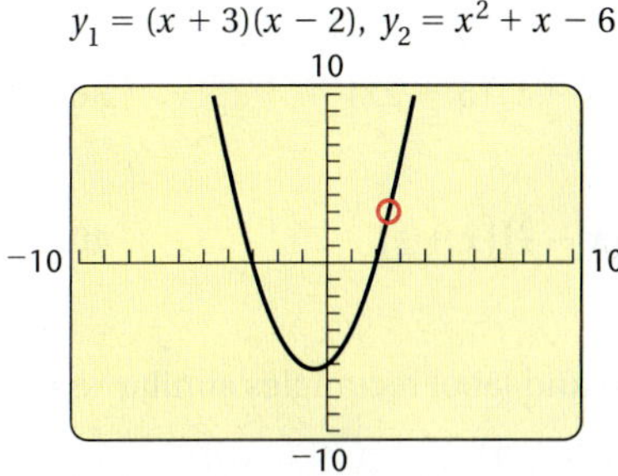

The graphing calculator will graph y_1 first as a solid line. Then it will graph y_2 as the circular cursor traces the leading edge of the graph, allowing us to determine visually whether the graphs coincide. In this case, the graphs appear to coincide, so the factorization is probably correct.

A table can also be used to perform a partial check of a product. See the Calculator Corner on p. 699 for the procedure.

Exercises: Determine graphically whether the following products are correct.

1. $(x + 5)(x + 4) = x^2 + 9x + 20$

2. $(4x + 3)(x - 2) = 4x^2 - 5x - 6$

3. $(5x + 3)(x - 4) = 5x^2 + 17x - 12$

4. $(2x - 3)(3x - 5) = 6x^2 - 19x - 15$

9.5 EXERCISE SET

For Extra Help

Digital Video Tutor CD 7 Videotape 10 | InterAct Math | Math Tutor Center | MathXL | MyMathLab

a Multiply.

1. $(8x^2)(5)$

2. $(4x^2)(-2)$

3. $(-x^2)(-x)$

4. $(-x^3)(x^2)$

5. $(8x^5)(4x^3)$

6. $(10a^2)(2a^2)$

7. $(0.1x^6)(0.3x^5)$

8. $(0.3x^4)(-0.8x^6)$

9. $\left(-\frac{1}{5}x^3\right)\left(-\frac{1}{3}x\right)$

10. $\left(-\frac{1}{4}x^4\right)\left(\frac{1}{5}x^8\right)$

11. $(-4x^2)(0)$

12. $(-4m^5)(-1)$

13. $(3x^2)(-4x^3)(2x^6)$

14. $(-2y^5)(10y^4)(-3y^3)$

b Multiply.

15. $2x(-x + 5)$

16. $3x(4x - 6)$

17. $-5x(x - 1)$

18. $-3x(-x - 1)$

19. $x^2(x^3 + 1)$

20. $-2x^3(x^2 - 1)$

21. $3x(2x^2 - 6x + 1)$

22. $-4x(2x^3 - 6x^2 - 5x + 1)$

23. $(-6x^2)(x^2 + x)$

24. $(-4x^2)(x^2 - x)$

25. $(3y^2)(6y^4 + 8y^3)$

26. $(4y^4)(y^3 - 6y^2)$

c Multiply.

27. $(x + 6)(x + 3)$

28. $(x + 5)(x + 2)$

29. $(x + 5)(x - 2)$

30. $(x + 6)(x - 2)$

31. $(x - 4)(x - 3)$

32. $(x - 7)(x - 3)$

33. $(x + 3)(x - 3)$

34. $(x + 6)(x - 6)$

35. $(5 - x)(5 - 2x)$

36. $(3 + x)(6 + 2x)$

37. $(2x + 5)(2x + 5)$

38. $(3x - 4)(3x - 4)$

39. $\left(x - \frac{5}{2}\right)\left(x + \frac{2}{5}\right)$

40. $\left(x + \frac{4}{3}\right)\left(x + \frac{3}{2}\right)$

41. $(x - 2.3)(x + 4.7)$

42. $(2x + 0.13)(2x - 0.13)$

Draw and label rectangles similar to the one following Example 7 to illustrate each product.

43. $x(x + 5)$

44. $x(x + 2)$

45. $(x + 1)(x + 2)$

46. $(x + 3)(x + 1)$

47. $(x + 5)(x + 3)$

48. $(x + 4)(x + 6)$

49. $(3x + 2)(3x + 2)$

50. $(5x + 3)(5x + 3)$

d Multiply.

51. $(x^2 + x + 1)(x - 1)$

52. $(x^2 + x - 2)(x + 2)$

53. $(2x + 1)(2x^2 + 6x + 1)$

54. $(3x - 1)(4x^2 - 2x - 1)$

55. $(y^2 - 3)(3y^2 - 6y + 2)$

56. $(3y^2 - 3)(y^2 + 6y + 1)$

57. $(x^3 + x^2)(x^3 + x^2 - x)$

58. $(x^3 - x^2)(x^3 - x^2 + x)$

59. $(-5x^3 - 7x^2 + 1)(2x^2 - x)$

60. $(-4x^3 + 5x^2 - 2)(5x^2 + 1)$

61. $(1 + x + x^2)(-1 - x + x^2)$

62. $(1 - x + x^2)(1 - x + x^2)$

63. $(2t^2 - t - 4)(3t^2 + 2t - 1)$

64. $(3a^2 - 5a + 2)(2a^2 - 3a + 4)$

65. $(x - x^3 + x^5)(x^2 - 1 + x^4)$

66. $(x - x^3 + x^5)(3x^2 + 3x^6 + 3x^4)$

67. $(x^3 + x^2 + x + 1)(x - 1)$

68. $(x + 2)(x^3 - x^2 + x - 2)$

69. $(x + 1)(x^3 + 7x^2 + 5x + 4)$

70. $(x + 2)(x^3 + 5x^2 + 9x + 3)$

71. $\left(x - \frac{1}{2}\right)\left(2x^3 - 4x^2 + 3x - \frac{2}{5}\right)$

72. $\left(x + \frac{1}{3}\right)\left(6x^3 - 12x^2 - 5x + \frac{1}{2}\right)$

73. ^D_W Under what conditions will the product of two binomials be a trinomial?

74. ^D_W How can the following figure be used to show that $(x + 3)^2 \neq x^2 + 9$?

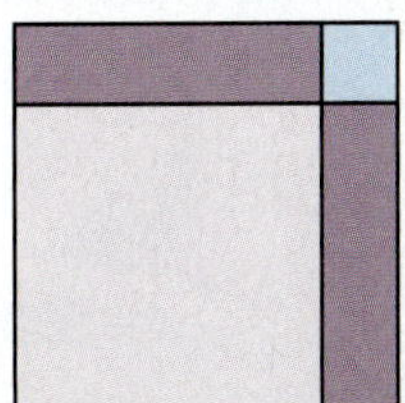

SKILL MAINTENANCE

Simplify.

75. $-\frac{1}{4} - \frac{1}{2}$ [6.4a]

76. $-3.8 - (-10.2)$ [6.4a]

77. $(10 - 2)(10 + 2)$ [6.8d]

78. $10 - 2 + (-6)^2 \div 3 \cdot 2$ [6.8d]

Factor. [6.7d]

79. $15x - 18y + 12$

80. $16x - 24y + 36$

81. $-9x - 45y + 15$

82. $100x - 100y + 1000a$

83. Graph: $y = \frac{1}{2}x - 3$. [8.2b]

84. Solve: $4(x - 3) = 5(2 - 3x) + 1$. [7.3c]

SYNTHESIS

Find a polynomial for the shaded area of each figure.

85.

14y − 5
3y
6y
3y + 5

86.

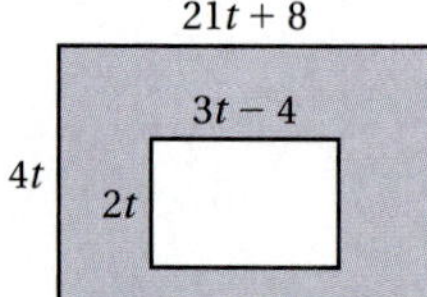

87. A box with a square bottom is to be made from a 12-in.-square piece of cardboard. Squares with side x are cut out of the corners and the sides are folded up. Find the polynomials for the volume and the outside surface area of the box.

x x
x x
12
x x
x x
12

For each figure, determine what the missing number must be in order for the figure to have the given area.

88. Area $= x^2 + 7x + 10$

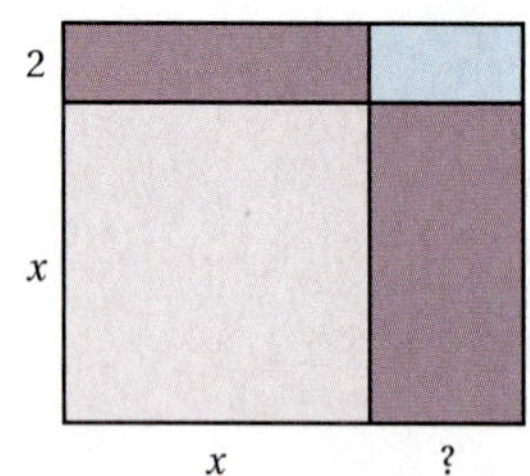

89. Area $= x^2 + 8x + 15$

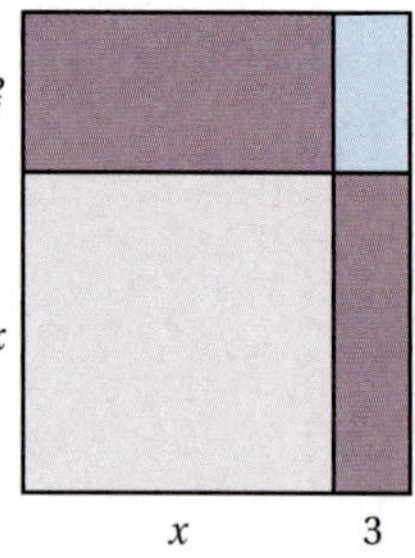

90. An open wooden box is a cube with side x cm. The box, including its bottom, is made of wood that is 1 cm thick. Find a polynomial for the interior volume of the cube.

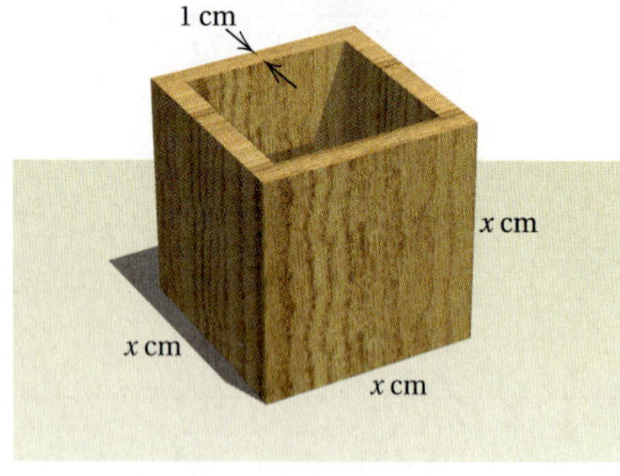

91. Find a polynomial for the volume of the solid shown below.

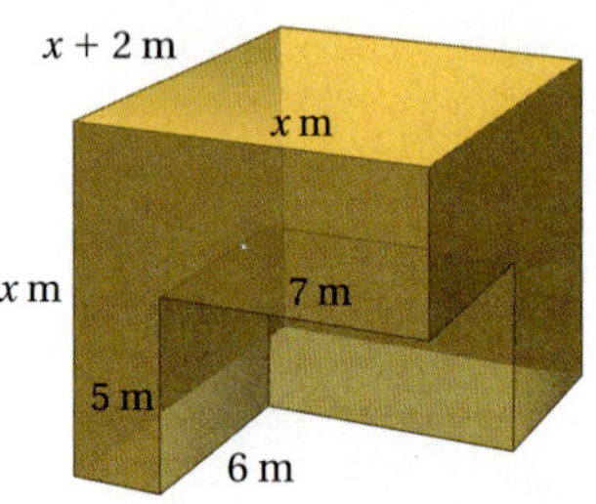

Compute and simplify.

92. $(x + 3)(x + 6) + (x + 3)(x + 6)$

93. $(x - 2)(x - 7) - (x - 7)(x - 2)$

94. $(x + 5)^2 - (x - 3)^2$

95. Extend the pattern and simplify

$$(x - a)(x - b)(x - c)(x - d) \cdots (x - z).$$

96. Use a graphing calculator to check your answers to Exercises 15, 29, and 51. Use graphs, tables, or both, as directed by your instructor.

9.6 SPECIAL PRODUCTS

Objectives

- **a** Multiply two binomials mentally using the FOIL method.
- **b** Multiply the sum and the difference of two terms mentally.
- **c** Square a binomial mentally.
- **d** Find special products when polynomial products are mixed together.

We encounter certain products so often that it is helpful to have faster methods of computing. We now consider special ways of multiplying any two binomials. Such techniques are called *special products.*

a Products of Two Binomials Using FOIL

To multiply two binomials, we can select one binomial and multiply each term of that binomial by every term of the other. Then we collect like terms. Consider the product $(x + 5)(x + 4)$:

$$\begin{aligned}(x + 5)(x + 4) &= x \cdot x + 5 \cdot x + x \cdot 4 + 5 \cdot 4\\ &= x^2 + 5x + 4x + 20\\ &= x^2 + 9x + 20.\end{aligned}$$

We can rewrite the first line of this product to show a special technique for finding the product of two binomials:

First terms, Outside terms, Inside terms, Last terms

$$(x + 5)(x + 4) = x \cdot x + 4 \cdot x + 5 \cdot x + 5 \cdot 4.$$

To remember this method of multiplying, we use the initials **FOIL.**

THE FOIL METHOD

To multiply two binomials, $A + B$ and $C + D$, multiply the First terms AC, the Outside terms AD, the Inside terms BC, and then the Last terms BD. Then collect like terms, if possible.

$$(A + B)(C + D) = AC + AD + BC + BD$$

1. Multiply First terms: AC.
2. Multiply Outside terms: AD.
3. Multiply Inside terms: BC.
4. Multiply Last terms: BD.

FOIL

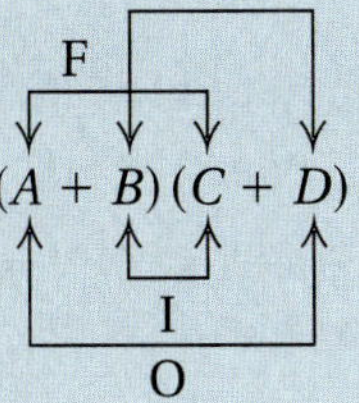

EXAMPLE 1 Multiply: $(x + 8)(x^2 - 5)$.

We have

$$\begin{aligned}(x + 8)(x^2 - 5) &= \overset{F}{x \cdot x^2} + \overset{O}{x \cdot (-5)} + \overset{I}{8 \cdot x^2} + \overset{L}{8(-5)}\\ &= x^3 - 5x + 8x^2 - 40\\ &= x^3 + 8x^2 - 5x - 40.\end{aligned}$$

Since each of the original binomials is in descending order, we write the product in descending order, as is customary, but this is not a "must."

Multiply mentally, if possible. If you need extra steps, be sure to use them.

1. $(x + 3)(x + 4)$

2. $(x + 3)(x - 5)$

3. $(2x - 1)(x - 4)$

4. $(2x^2 - 3)(x - 2)$

5. $(6x^2 + 5)(2x^3 + 1)$

6. $(y^3 + 7)(y^3 - 7)$

7. $(t + 5)(t + 3)$

8. $(2x^4 + x^2)(-x^3 + x)$

Multiply.

9. $\left(x + \frac{4}{5}\right)\left(x - \frac{4}{5}\right)$

10. $(x^3 - 0.5)(x^2 + 0.5)$

11. $(2 + 3x^2)(4 - 5x^2)$

12. $(6x^3 - 3x^2)(5x^2 - 2x)$

Answers on page A-30

Often we can collect like terms after we have multiplied.

EXAMPLES Multiply.

2. $(x + 6)(x - 6) = x^2 - 6x + 6x - 36$ Using FOIL
$= x^2 - 36$ Collecting like terms

3. $(x + 7)(x + 4) = x^2 + 4x + 7x + 28$
$= x^2 + 11x + 28$

4. $(y - 3)(y - 2) = y^2 - 2y - 3y + 6$
$= y^2 - 5y + 6$

5. $(x^3 - 5)(x^3 + 5) = x^6 + 5x^3 - 5x^3 - 25$
$= x^6 - 25$

Do Exercises 1–8.

EXAMPLES Multiply.

6. $(4t^3 + 5)(3t^2 - 2) = 12t^5 - 8t^3 + 15t^2 - 10$

7. $\left(x - \frac{2}{3}\right)\left(x + \frac{2}{3}\right) = x^2 + \frac{2}{3}x - \frac{2}{3}x - \frac{4}{9}$
$= x^2 - \frac{4}{9}$

8. $(x^2 - 0.3)(x^2 - 0.3) = x^4 - 0.3x^2 - 0.3x^2 + 0.09$
$= x^4 - 0.6x^2 + 0.09$

9. $(3 - 4x)(7 - 5x^3) = 21 - 15x^3 - 28x + 20x^4$
$= 21 - 28x - 15x^3 + 20x^4$

(*Note*: If the original polynomials are in ascending order, it is natural to write the product in ascending order, but this is not a "must.")

10. $(5x^4 + 2x^3)(3x^2 - 7x) = 15x^6 - 35x^5 + 6x^5 - 14x^4$
$= 15x^6 - 29x^5 - 14x^4$

Do Exercises 9–12.

We can show the FOIL method geometrically as follows.

The area of the large rectangle is $(A + B)(C + D)$.

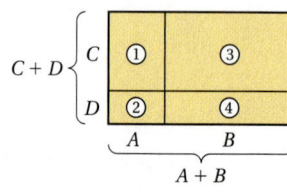

The area of rectangle ① is AC.
The area of rectangle ② is AD.
The area of rectangle ③ is BC.
The area of rectangle ④ is BD.

The area of the large rectangle is the sum of the areas of the smaller rectangles. Thus,

$$(A + B)(C + D) = AC + AD + BC + BD.$$

b Multiplying Sums and Differences of Two Terms

Consider the product of the sum and the difference of the same two terms, such as

$$(x + 2)(x - 2).$$

Since this is the product of two binomials, we can use FOIL. This type of product occurs so often, however, that it would be valuable if we could use an even faster method. To find a faster way to compute such a product, look for a pattern in the following:

a) $(x + 2)(x - 2) = x^2 - 2x + 2x - 4$
$= x^2 - 4;$

b) $(3x - 5)(3x + 5) = 9x^2 + 15x - 15x - 25$
$= 9x^2 - 25.$

Do Exercises 13 and 14.

Multiply.

13. $(x + 5)(x - 5)$

14. $(2x - 3)(2x + 3)$

Perhaps you discovered in each case that when you multiply the two binomials, two terms are opposites, or additive inverses, which add to 0 and "drop out."

PRODUCT OF THE SUM AND THE DIFFERENCE

The product of the sum and the difference of the same two terms is the square of the first term minus the square of the second term:

$$(A + B)(A - B) = A^2 - B^2.$$

It is helpful to memorize this rule in both words and symbols. (If you do forget it, you can, of course, use FOIL.)

EXAMPLES Multiply. (Carry out the rule and say the words as you go.)

$(A + B)(A - B) = A^2 - B^2$

11. $(x + 4)(x - 4) = x^2 - 4^2$ "The square of the first term, x^2, minus the square of the second, 4^2"
$= x^2 - 16$ Simplifying

12. $(5 + 2w)(5 - 2w) = 5^2 - (2w)^2$
$= 25 - 4w^2$

13. $(3x^2 - 7)(3x^2 + 7) = (3x^2)^2 - 7^2$
$= 9x^4 - 49$

14. $(-4x - 10)(-4x + 10) = (-4x)^2 - 10^2$
$= 16x^2 - 100$

15. $\left(x + \frac{3}{8}\right)\left(x - \frac{3}{8}\right) = x^2 - \left(\frac{3}{8}\right)^2 = x^2 - \frac{9}{64}$

Do Exercises 15–19.

Multiply.

15. $(x + 2)(x - 2)$

16. $(x - 7)(x + 7)$

17. $(6 - 4y)(6 + 4y)$

18. $(2x^3 - 1)(2x^3 + 1)$

19. $\left(x - \frac{2}{5}\right)\left(x + \frac{2}{5}\right)$

C Squaring Binomials

Consider the square of a binomial, such as $(x + 3)^2$. This can be expressed as $(x + 3)(x + 3)$. Since this is the product of two binomials, we can again use FOIL. But again, this type of product occurs so often that we would like to use an even faster method. Look for a pattern in the following:

Answers on page A-30

Multiply.

20. $(x + 8)(x + 8)$

21. $(x - 5)(x - 5)$

Multiply.

22. $(x + 2)^2$

23. $(a - 4)^2$

24. $(2x + 5)^2$

25. $(4x^2 - 3x)^2$

26. $(7.8 + 1.2y)(7.8 + 1.2y)$

27. $(3x^2 - 5)(3x^2 - 5)$

Answers on page A-30

a) $(x + 3)^2 = (x + 3)(x + 3) = x^2 + 3x + 3x + 9 = x^2 + 6x + 9;$

b) $(x - 3)^2 = (x - 3)(x - 3) = x^2 - 3x - 3x + 9 = x^2 - 6x + 9;$

c) $(5 + 3p)^2 = (5 + 3p)(5 + 3p) = 25 + 15p + 15p + 9p^2 = 25 + 30p + 9p^2;$

d) $(3x - 5)^2 = (3x - 5)(3x - 5) = 9x^2 - 15x - 15x + 25 = 9x^2 - 30x + 25.$

Do Exercises 20 and 21.

When squaring a binomial, we multiply a binomial by itself. Perhaps you noticed that two terms are the same and when added give twice the product of the terms in the binomial. The other two terms are squares.

SQUARE OF A BINOMIAL

The square of a sum or a difference of two terms is the square of the first term, plus or minus twice the product of the two terms, plus the square of the last term:

$$(A + B)^2 = A^2 + 2AB + B^2; \qquad (A - B)^2 = A^2 - 2AB + B^2.$$

It is helpful to memorize this rule in both words and symbols.

EXAMPLES Multiply. (Carry out the rule and say the words as you go.)

$$(A + B)^2 = A^2 + 2 \cdot A \cdot B + B^2$$

16. $(x + 3)^2 = x^2 + 2 \cdot x \cdot 3 + 3^2 = x^2 + 6x + 9$ "x^2 plus 2 times x times 3 plus 3^2"

$$(A - B)^2 = A^2 - 2 \cdot A \cdot B + B^2$$

17. $(t - 5)^2 = t^2 - 2 \cdot t \cdot 5 + 5^2 = t^2 - 10t + 25$ "t^2 minus 2 times t times 5 plus 5^2"

18. $(2x + 7)^2 = (2x)^2 + 2 \cdot 2x \cdot 7 + 7^2 = 4x^2 + 28x + 49$

19. $(5x - 3x^2)^2 = (5x)^2 - 2 \cdot 5x \cdot 3x^2 + (3x^2)^2 = 25x^2 - 30x^3 + 9x^4$

20. $(2.3 - 5.4m)^2 = 2.3^2 - 2(2.3)(5.4m) + (5.4m)^2 = 5.29 - 24.84m + 29.16m^2$

Do Exercises 22–27.

CAUTION!

Although the square of a product is the product of the squares, the square of a sum is *not* the sum of the squares. That is, $(AB)^2 = A^2B^2$, but

The term $2AB$ is missing.

$$(A + B)^2 \neq A^2 + B^2.$$

To confirm this inequality, note, using the rules for order of operations, that

$$(7 + 5)^2 = 12^2 = 144,$$

whereas

$$7^2 + 5^2 = 49 + 25 = 74, \quad \text{and} \quad 74 \neq 144.$$

We can look at the rule for finding $(A + B)^2$ geometrically as follows. The area of the large square is

$$(A + B)(A + B) = (A + B)^2.$$

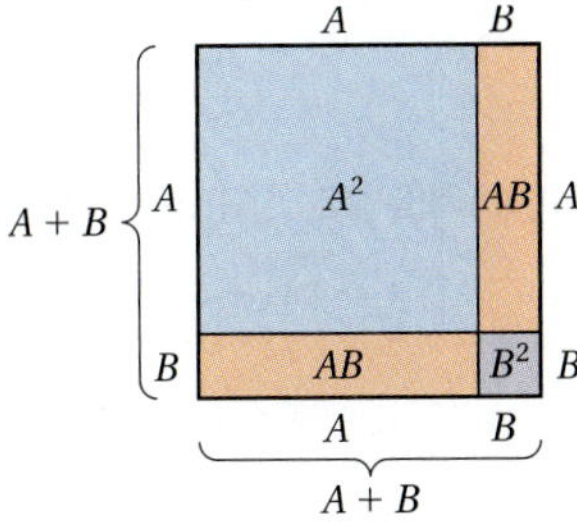

This is equal to the sum of the areas of the smaller rectangles:

$$A^2 + AB + AB + B^2 = A^2 + 2AB + B^2.$$

Thus, $(A + B)^2 = A^2 + 2AB + B^2$.

d Multiplication of Various Types

We have considered how to quickly multiply certain kinds of polynomials. Let's now try several types of multiplications mixed together so that we can learn to sort them out. When you multiply, first see what kind of multiplication you have. Then use the best method. The formulas you should know and the questions you should ask yourself are as follows:

MULTIPLYING TWO POLYNOMIALS

1. Is it the product of a monomial and a polynomial? If so, multiply each term of the polynomial by the monomial.
   Example: $5x(x + 7) = 5x \cdot x + 5x \cdot 7 = 5x^2 + 35x$
2. Is it the product of the sum and the difference of the *same* two terms? If so, use the following:
   $$(A + B)(A - B) = A^2 - B^2.$$
   The product of the sum and the difference of the same two terms is the difference of the squares. [The answer has 2 terms.]
   Example: $(x + 7)(x - 7) = x^2 - 7^2 = x^2 - 49$
3. Is the product the square of a binomial? If so, use the following:
   $$(A + B)(A + B) = (A + B)^2 = A^2 + 2AB + B^2,$$
   or $(A - B)(A - B) = (A - B)^2 = A^2 - 2AB + B^2$.
   The square of a binomial is the square of the first term, plus or minus *twice* the product of the two terms, plus the square of the last term. [The answer has 3 terms.]
   Example: $(x + 7)(x + 7) = (x + 7)^2$
   $= x^2 + 2 \cdot x \cdot 7 + 7^2 = x^2 + 14x + 49$
4. Is it the product of two binomials other than those above? If so, use FOIL. [The answer will have 3 or 4 terms.]
   Example: $(x + 7)(x - 4) = x^2 - 4x + 7x - 28 = x^2 + 3x - 28$
5. Is it the product of two polynomials other than those above? If so, multiply each term of one by every term of the other. Use columns if you wish. [The answer will have 2 or more terms, usually more than 2 terms.]
   Example:
   $$\begin{aligned}(x^2 - 3x + 2)(x + 7) &= x^2(x + 7) - 3x(x + 7) + 2(x + 7)\\ &= x^2 \cdot x + x^2 \cdot 7 - 3x \cdot x - 3x \cdot 7\\ &\quad + 2 \cdot x + 2 \cdot 7\\ &= x^3 + 7x^2 - 3x^2 - 21x + 2x + 14\\ &= x^3 + 4x^2 - 19x + 14\end{aligned}$$

Remember that FOIL will *always* work for two binomials. You can use it instead of either of rules 2 and 3, but those rules will make your work go faster.

Study Tips

MEMORIZING FORMULAS

Memorizing can be a very helpful tool in the study of mathematics. Don't underestimate its power as you consider the special products. Consider putting the rules, in words and in math symbols, on index cards and go over them many times.

Multiply.

28. $(x + 5)(x + 6)$

29. $(t - 4)(t + 4)$

30. $4x^2(-2x^3 + 5x^2 + 10)$

31. $(9x^2 + 1)^2$

32. $(2a - 5)(2a + 8)$

33. $\left(5x + \frac{1}{2}\right)^2$

34. $\left(2x - \frac{1}{2}\right)^2$

35. $(x^2 - x + 4)(x - 2)$

Answers on page A-30

EXAMPLE 21 Multiply: $(x + 3)(x - 3)$.

$(x + 3)(x - 3) = x^2 - 9$ Using method 2 (the product of the sum and the difference of two terms)

EXAMPLE 22 Multiply: $(t + 7)(t - 5)$.

$(t + 7)(t - 5) = t^2 + 2t - 35$ Using method 4 (the product of two binomials, but neither the square of a binomial nor the product of the sum and the difference of two terms)

EXAMPLE 23 Multiply: $(x + 6)(x + 6)$.

$$(x + 6)(x + 6) = x^2 + 2(6)x + 36$$ Using method 3 (the square of a binomial sum)

$$= x^2 + 12x + 36$$

EXAMPLE 24 Multiply: $2x^3(9x^2 + x - 7)$.

$2x^3(9x^2 + x - 7) = 18x^5 + 2x^4 - 14x^3$ Using method 1 (the product of a monomial and a trinomial; multiplying each term of the trinomial by the monomial)

EXAMPLE 25 Multiply: $(5x^3 - 7x)^2$.

$$(5x^3 - 7x)^2 = 25x^6 - 2(5x^3)(7x) + 49x^2$$ Using method 3 (the square of a binomial difference)

$$= 25x^6 - 70x^4 + 49x^2$$

EXAMPLE 26 Multiply: $\left(3x + \frac{1}{4}\right)^2$.

$$\left(3x + \tfrac{1}{4}\right)^2 = 9x^2 + 2(3x)\left(\tfrac{1}{4}\right) + \tfrac{1}{16}$$ Using method 3 (the square of a binomial sum. To get the middle term, we multiply $3x$ by $\frac{1}{4}$ and double.)

$$= 9x^2 + \tfrac{3}{2}x + \tfrac{1}{16}$$

EXAMPLE 27 Multiply: $\left(4x - \frac{3}{4}\right)^2$.

$$\left(4x - \tfrac{3}{4}\right)^2 = 16x^2 - 2(4x)\left(\tfrac{3}{4}\right) + \tfrac{9}{16}$$ Using method 3 (the square of a binomial difference)

$$= 16x^2 - 6x + \tfrac{9}{16}$$

EXAMPLE 28 Multiply: $(p + 3)(p^2 + 2p - 1)$.

$$\begin{array}{r} p^2 + 2p - 1 \\ p + 3 \\ \hline 3p^2 + 6p - 3 \\ p^3 + 2p^2 - p \\ \hline p^3 + 5p^2 + 5p - 3 \end{array}$$

Using method 5 (the product of two polynomials)

Multiplying by 3

Multiplying by p

Do Exercises 28–35.

9.6 EXERCISE SET

For Extra Help

Digital Video Tutor CD 7 Videotape 10 | InterAct Math | Math Tutor Center | MathXL | MyMathLab

a Multiply. Try to write only the answer. If you need more steps, be sure to use them.

1. $(x + 1)(x^2 + 3)$

2. $(x^2 - 3)(x - 1)$

3. $(x^3 + 2)(x + 1)$

4. $(x^4 + 2)(x + 10)$

5. $(y + 2)(y - 3)$

6. $(a + 2)(a + 3)$

7. $(3x + 2)(3x + 2)$

8. $(4x + 1)(4x + 1)$

9. $(5x - 6)(x + 2)$

10. $(x - 8)(x + 8)$

11. $(3t - 1)(3t + 1)$

12. $(2m + 3)(2m + 3)$

13. $(4x - 2)(x - 1)$

14. $(2x - 1)(3x + 1)$

15. $\left(p - \frac{1}{4}\right)\left(p + \frac{1}{4}\right)$

16. $\left(q + \frac{3}{4}\right)\left(q + \frac{3}{4}\right)$

17. $(x - 0.1)(x + 0.1)$

18. $(x + 0.3)(x - 0.4)$

19. $(2x^2 + 6)(x + 1)$

20. $(2x^2 + 3)(2x - 1)$

21. $(-2x + 1)(x + 6)$

22. $(3x + 4)(2x - 4)$

23. $(a + 7)(a + 7)$

24. $(2y + 5)(2y + 5)$

25. $(1 + 2x)(1 - 3x)$

26. $(-3x - 2)(x + 1)$

27. $(x^2 + 3)(x^3 - 1)$

28. $(x^4 - 3)(2x + 1)$

29. $(3x^2 - 2)(x^4 - 2)$

30. $(x^{10} + 3)(x^{10} - 3)$

31. $(2.8x - 1.5)(4.7x + 9.3)$

32. $\left(x - \frac{3}{8}\right)\left(x + \frac{4}{7}\right)$

33. $(3x^5 + 2)(2x^2 + 6)$

34. $(1 - 2x)(1 + 3x^2)$

35. $(8x^3 + 1)(x^3 + 8)$

36. $(4 - 2x)(5 - 2x^2)$

37. $(4x^2 + 3)(x - 3)$

38. $(7x - 2)(2x - 7)$

39. $(4y^4 + y^2)(y^2 + y)$

40. $(5y^6 + 3y^3)(2y^6 + 2y^3)$

b Multiply mentally, if possible. If you need extra steps, be sure to use them.

41. $(x + 4)(x - 4)$

42. $(x + 1)(x - 1)$

43. $(2x + 1)(2x - 1)$

44. $(x^2 + 1)(x^2 - 1)$

45. $(5m - 2)(5m + 2)$

46. $(3x^4 + 2)(3x^4 - 2)$

47. $(2x^2 + 3)(2x^2 - 3)$

48. $(6x^5 - 5)(6x^5 + 5)$

49. $(3x^4 - 4)(3x^4 + 4)$

50. $(t^2 - 0.2)(t^2 + 0.2)$

51. $(x^6 - x^2)(x^6 + x^2)$

52. $(2x^3 - 0.3)(2x^3 + 0.3)$

53. $(x^4 + 3x)(x^4 - 3x)$

54. $\left(\frac{3}{4} + 2x^3\right)\left(\frac{3}{4} - 2x^3\right)$

55. $(x^{12} - 3)(x^{12} + 3)$

56. $(12 - 3x^2)(12 + 3x^2)$

57. $(2y^8 + 3)(2y^8 - 3)$

58. $\left(m - \frac{2}{3}\right)\left(m + \frac{2}{3}\right)$

59. $\left(\frac{5}{8}x - 4.3\right)\left(\frac{5}{8}x + 4.3\right)$

60. $(10.7 - x^3)(10.7 + x^3)$

c Multiply mentally, if possible. If you need extra steps, be sure to use them.

61. $(x + 2)^2$

62. $(2x - 1)^2$

63. $(3x^2 + 1)^2$

64. $\left(3x + \frac{3}{4}\right)^2$

65. $\left(a - \frac{1}{2}\right)^2$

66. $\left(2a - \frac{1}{5}\right)^2$

67. $(3 + x)^2$

68. $(x^3 - 1)^2$

69. $(x^2 + 1)^2$

70. $(8x - x^2)^2$

71. $(2 - 3x^4)^2$

72. $(6x^3 - 2)^2$

73. $(5 + 6t^2)^2$

74. $(3p^2 - p)^2$

75. $\left(x - \frac{5}{8}\right)^2$

76. $(0.3y + 2.4)^2$

d Multiply mentally, if possible.

77. $(3 - 2x^3)^2$

78. $(x - 4x^3)^2$

79. $4x(x^2 + 6x - 3)$

80. $8x(-x^5 + 6x^2 + 9)$

81. $\left(2x^2 - \frac{1}{2}\right)\left(2x^2 - \frac{1}{2}\right)$

82. $(-x^2 + 1)^2$

83. $(-1 + 3p)(1 + 3p)$

84. $(-3q + 2)(3q + 2)$

85. $3t^2(5t^3 - t^2 + t)$

86. $-6x^2(x^3 + 8x - 9)$

87. $(6x^4 + 4)^2$

88. $(8a + 5)^2$

89. $(3x + 2)(4x^2 + 5)$

90. $(2x^2 - 7)(3x^2 + 9)$

91. $(8 - 6x^4)^2$

92. $\left(\frac{1}{5}x^2 + 9\right)\left(\frac{3}{5}x^2 - 7\right)$

93. $(t - 1)(t^2 + t + 1)$

94. $(y + 5)(y^2 - 5y + 25)$

Compute each of the following and compare.

95. $3^2 + 4^2; (3 + 4)^2$

96. $6^2 + 7^2; (6 + 7)^2$

97. $9^2 - 5^2; (9 - 5)^2$

98. $11^2 - 4^2; (11 - 4)^2$

Find the total area of all the shaded rectangles.

99.

1

a

a

1

100.

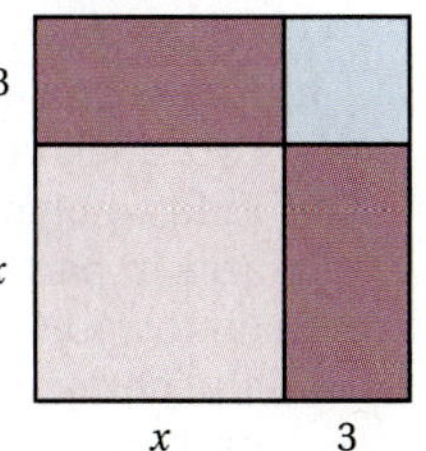

101.

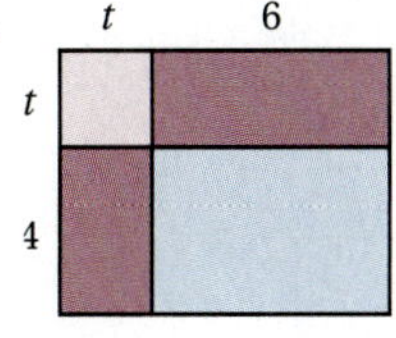

102.

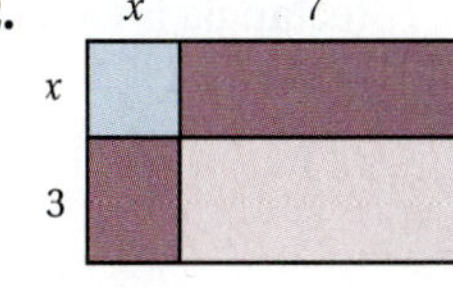

103. DW Under what conditions is the product of two binomials a binomial?

104. DW Brittney feels that since the FOIL method can be used to find the product of any two binomials, she needn't study the other special products. What advice would you give her?

SKILL MAINTENANCE

105. *Electricity Usage.* In apartment 3B, lamps, an air conditioner, and a television set are all operating at the same time. The lamps use 10 times as many watts of electricity as the television set, and the air conditioner uses 40 times as many watts as the television set. The total wattage used in the apartment is 2550. How many watts are used by each appliance? [7.6a]

Solve. [7.3c]

106. $3x - 8x = 4(7 - 8x)$

107. $3(x - 2) = 5(2x + 7)$

108. $5(2x - 3) - 2(3x - 4) = 20$

Solve. [7.4b]

109. $3x - 2y = 12$, for y

110. $3a - 5d = 4$, for a

SYNTHESIS

Multiply.

111. $5x(3x - 1)(2x + 3)$

112. $[(2x - 3)(2x + 3)](4x^2 + 9)$

113. $[(a - 5)(a + 5)]^2$

114. $(a - 3)^2(a + 3)^2$
(*Hint*: Examine Exercise 113.)

115. $(3t^4 - 2)^2(3t^4 + 2)^2$
(*Hint*: Examine Exercise 113.)

116. $[3a - (2a - 3)][3a + (2a - 3)]$

Solve.

117. $(x + 2)(x - 5) = (x + 1)(x - 3)$

118. $(2x + 5)(x - 4) = (x + 5)(2x - 4)$

119. *Factors and Sums.* To *factor* a number is to express it as a product. Since $12 = 4 \cdot 3$, we say that 12 is *factored* and that 4 and 3 are *factors* of 12. In the following table, the top number has been factored in such a way that the sum of the factors is the bottom number. For example, in the first column, 40 has been factored as $5 \cdot 8$, and $5 + 8 = 13$, the bottom number. Such thinking is important in algebra when we factor trinomials of the type $x^2 + bx + c$. Find the missing numbers in the table.

Product	40	63	36	72	−140	−96	48	168	110			
Factor	5									−9	−24	−3
Factor	8									−10	18	
Sum	13	16	−20	−38	−4	4	−14	−29	−21			18

120. A factored polynomial for the shaded area in this rectangle is $(A + B)(A - B)$.

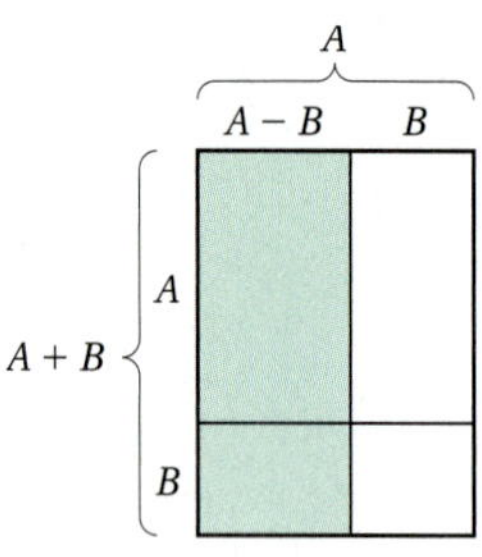

a) Find a polynomial for the area of the entire rectangle.

b) Find a polynomial for the sum of the areas of the two small unshaded rectangles.

c) Find a polynomial for the area in part (a) minus the area in part (b).

d) Find a polynomial for the area of the shaded region and compare this with the polynomial found in part (c).

Use the TABLE or GRAPH feature to check whether each of the following is correct.

121. $(x - 1)^2 = x^2 - 2x + 1$

122. $(x - 2)^2 = x^2 - 4x - 4$

123. $(x - 3)(x + 3) = x^2 - 6$

124. $(x - 3)(x + 2) = x^2 - x - 6$

9.7 OPERATIONS WITH POLYNOMIALS IN SEVERAL VARIABLES

Objectives

a. Evaluate a polynomial in several variables for given values of the variables.

b. Identify the coefficients and the degrees of the terms of a polynomial and the degree of a polynomial.

c. Collect like terms of a polynomial.

d. Add polynomials.

e. Subtract polynomials.

f. Multiply polynomials.

The polynomials that we have been studying have only one variable. A **polynomial in several variables** is an expression like those you have already seen, but with more than one variable. Here are two examples:

$$3x + xy^2 + 5y + 4, \qquad 8xy^2z - 2x^3z - 13x^4y^2 + 15.$$

a Evaluating Polynomials

EXAMPLE 1 Evaluate the polynomial $4 + 3x + xy^2 + 8x^3y^3$ when $x = -2$ and $y = 5$.

We replace x with -2 and y with 5:

$$\begin{aligned} 4 + 3x + xy^2 + 8x^3y^3 &= 4 + 3(-2) + (-2) \cdot 5^2 + 8(-2)^3 \cdot 5^3 \\ &= 4 - 6 - 50 - 8000 \\ &= -8052. \end{aligned}$$

EXAMPLE 2 *Male Caloric Needs.* The number of calories needed each day by a moderately active man who weighs w kilograms, is h centimeters tall, and is a years old can be estimated by the polynomial

$$19.18w + 7h - 9.52a + 92.4.$$

Marv Bittinger, an author of this text, is moderately active, weighs 87 kg, is 185 cm tall, and is 59 yr old. What are his daily caloric needs?

Source: Parker, M., *She Does Math.* Mathematical Association of America

We evaluate the polynomial for $w = 87$, $h = 185$, and $a = 59$:

$$\begin{aligned} &19.18w + 7h - 9.52a + 92.4 \\ &\quad = 19.18(87) + 7(185) - 9.52(59) + 92.4 \qquad \text{Substituting} \\ &\quad = 2494.38. \end{aligned}$$

His daily caloric need is about 2494 calories.

Do Exercises 1–3.

1. Evaluate the polynomial

$$4 + 3x + xy^2 + 8x^3y^3$$

when $x = 2$ and $y = -5$.

2. Evaluate the polynomial

$$8xy^2 - 2x^3z - 13x^4y^2 + 5$$

when $x = -1$, $y = 3$, and $z = 4$.

3. Female Caloric Needs. The number of calories needed each day by a moderately active woman who weighs w pounds, is h inches tall, and is a years old can be estimated by the polynomial

$$917 + 6w + 6h - 6a.$$

Christine is moderately active, weighs 125 lb, is 64 in. tall, and is 27 yr old. What are her daily caloric needs?

Source: Parker, M., *She Does Math.* Mathematical Association of America

Answers on page A-30

4. Identify the coefficient of each term:

$-3xy^2 + 3x^2y - 2y^3 + xy + 2.$

5. Identify the degree of each term and the degree of the polynomial

$4xy^2 + 7x^2y^3z^2 - 5x + 2y + 4.$

Collect like terms.

6. $4x^2y + 3xy - 2x^2y$

7. $-3pq - 5pqr^3 - 12 + 8pq + 5pqr^3 + 4$

Answers on page A-30

b Coefficients and Degrees

The **degree** of a term is the sum of the exponents of the variables. The **degree of a polynomial** is the degree of the term of highest degree.

EXAMPLE 3 Identify the coefficient and the degree of each term and the degree of the polynomial

$$9x^2y^3 - 14xy^2z^3 + xy + 4y + 5x^2 + 7.$$

TERM	COEFFICIENT	DEGREE	DEGREE OF THE POLYNOMIAL
$9x^2y^3$	9	5	
$-14xy^2z^3$	-14	6	6
xy	1	2	
$4y$	4	1	Think: $4y = 4y^1$.
$5x^2$	5	2	
7	7	0	Think: $7 = 7x^0$, or $7x^0y^0z^0$.

Do Exercises 4 and 5.

c Collecting Like Terms

Like terms have exactly the same variables with exactly the same exponents. For example,

$3x^2y^3$ and $-7x^2y^3$ are like terms;
$9x^4z^7$ and $12x^4z^7$ are like terms.

But

$13xy^5$ and $-2x^2y^5$ are *not* like terms, because the x-factors have different exponents;

and

$3xyz^2$ and $4xy$ are *not* like terms, because there is no factor of z^2 in the second expression.

Collecting like terms is based on the distributive laws.

EXAMPLES Collect like terms.

4. $5x^2y + 3xy^2 - 5x^2y - xy^2 = (5 - 5)x^2y + (3 - 1)xy^2 = 2xy^2$
5. $8a^2 - 2ab + 7b^2 + 4a^2 - 9ab - 17b^2 = 12a^2 - 11ab - 10b^2$
6. $7xy - 5xy^2 + 3xy^2 - 7 + 6x^3 + 9xy - 11x^3 + y - 1$
$\quad = -2xy^2 + 16xy - 5x^3 + y - 8$

Do Exercises 6 and 7.

d Addition

We can find the sum of two polynomials in several variables by writing a plus sign between them and then collecting like terms.

EXAMPLE 7 Add: $(-5x^3 + 3y - 5y^2) + (8x^3 + 4x^2 + 7y^2)$.

$$\begin{aligned}&(-5x^3 + 3y - 5y^2) + (8x^3 + 4x^2 + 7y^2)\\&\quad = (-5 + 8)x^3 + 4x^2 + 3y + (-5 + 7)y^2\\&\quad = 3x^3 + 4x^2 + 3y + 2y^2\end{aligned}$$

EXAMPLE 8 Add:

$$(5xy^2 - 4x^2y + 5x^3 + 2) + (3xy^2 - 2x^2y + 3x^3y - 5).$$

We first look for like terms. They are $5xy^2$ and $3xy^2$, $-4x^2y$ and $-2x^2y$, and 2 and -5. We collect these. Since there are no more like terms, the answer is

$$8xy^2 - 6x^2y + 5x^3 + 3x^3y - 3.$$

Do Exercises 8–10.

e Subtraction

We subtract a polynomial by adding its opposite, or additive inverse. The opposite of the polynomial $4x^2y - 6x^3y^2 + x^2y^2 - 5y$ is

$$-(4x^2y - 6x^3y^2 + x^2y^2 - 5y) = -4x^2y + 6x^3y^2 - x^2y^2 + 5y.$$

EXAMPLE 9 Subtract:

$$(4x^2y + x^3y^2 + 3x^2y^3 + 6y + 10) - (4x^2y - 6x^3y^2 + x^2y^2 - 5y - 8).$$

We have

$$\begin{aligned}&(4x^2y + x^3y^2 + 3x^2y^3 + 6y + 10) - (4x^2y - 6x^3y^2 + x^2y^2 - 5y - 8)\\&\quad = 4x^2y + x^3y^2 + 3x^2y^3 + 6y + 10 - 4x^2y + 6x^3y^2 - x^2y^2 + 5y + 8\end{aligned}$$

Finding the opposite by changing the sign of each term

$$= 7x^3y^2 + 3x^2y^3 - x^2y^2 + 11y + 18.$$

Collecting like terms. (Try to write just the answer!)

Do Exercises 11 and 12.

Add.

8. $(4x^3 + 4x^2 - 8y - 3) + (-8x^3 - 2x^2 + 4y + 5)$

9. $(13x^3y + 3x^2y - 5y) + (x^3y + 4x^2y - 3xy + 3y)$

10. $(-5p^2q^4 + 2p^2q^2 + 3q) + (6pq^2 + 3p^2q + 5)$

Subtract.

11. $(-4s^4t + s^3t^2 + 2s^2t^3) - (4s^4t - 5s^3t^2 + s^2t^2)$

12. $(-5p^4q + 5p^3q^2 - 3p^2q^3 - 7q^4 - 2) - (4p^4q - 4p^3q^2 + p^2q^3 + 2q^4 - 7)$

Answers on page A-30

Multiply.

13. $(x^2y^3 + 2x)(x^3y^2 + 3x)$

14. $(p^4q - 2p^3q^2 + 3q^3)(p + 2q)$

Multiply.

15. $(3xy + 2x)(x^2 + 2xy^2)$

16. $(x - 3y)(2x - 5y)$

17. $(4x + 5y)^2$

18. $(3x^2 - 2xy^2)^2$

19. $(2xy^2 + 3x)(2xy^2 - 3x)$

20. $(3xy^2 + 4y)(-3xy^2 + 4y)$

21. $(3y + 4 - 3x)(3y + 4 + 3x)$

22. $(2a + 5b + c)(2a - 5b - c)$

Answers on page A-30

f Multiplication

To multiply polynomials in several variables, we can multiply each term of one by every term of the other. We can use columns for long multiplications as with polynomials in one variable. We multiply each term at the top by every term at the bottom. We write like terms in columns, and then we add the results.

EXAMPLE 10 Multiply: $(3x^2y - 2xy + 3y)(xy + 2y)$.

$$\begin{array}{rrrrl} & 3x^2y & - 2xy & + 3y & \\ & & xy & + 2y & \\ \hline & 6x^2y^2 & - 4xy^2 & + 6y^2 & \text{Multiplying by } 2y \\ 3x^3y^2 & - 2x^2y^2 & + 3xy^2 & & \text{Multiplying by } xy \\ \hline 3x^3y^2 & + 4x^2y^2 & - xy^2 & + 6y^2 & \text{Adding} \end{array}$$

Do Exercises 13 and 14.

Where appropriate, we use the special products that we have learned.

EXAMPLES Multiply.

11. $(x^2y + 2x)(xy^2 + y^2) = x^3y^3 + x^2y^3 + 2x^2y^2 + 2xy^2$ (F O I L)

12. $(p + 5q)(2p - 3q) = 2p^2 - 3pq + 10pq - 15q^2$
$= 2p^2 + 7pq - 15q^2$

$(A + B)^2 = A^2 + 2 \cdot A \cdot B + B^2$

13. $(3x + 2y)^2 = (3x)^2 + 2(3x)(2y) + (2y)^2$
$= 9x^2 + 12xy + 4y^2$

$(A - B)^2 = A^2 - 2 \cdot A \cdot B + B^2$

14. $(2y^2 - 5x^2y)^2 = (2y^2)^2 - 2(2y^2)(5x^2y) + (5x^2y)^2$
$= 4y^4 - 20x^2y^3 + 25x^4y^2$

$(A + B)(A - B) = A^2 - B^2$

15. $(3x^2y + 2y)(3x^2y - 2y) = (3x^2y)^2 - (2y)^2$
$= 9x^4y^2 - 4y^2$

16. $(-2x^3y^2 + 5t)(2x^3y^2 + 5t) = (5t - 2x^3y^2)(5t + 2x^3y^2)$
$= (5t)^2 - (2x^3y^2)^2$
$= 25t^2 - 4x^6y^4$

$(A - B)(A + B) = A^2 - B^2$

17. $(2x + 3 - 2y)(2x + 3 + 2y) = (2x + 3)^2 - (2y)^2$
$= 4x^2 + 12x + 9 - 4y^2$

Do Exercises 15–22.

9.7 EXERCISE SET

For Extra Help

Digital Video Tutor CD 8 Videotape 10 | InterAct Math | Math Tutor Center | MathXL | MyMathLab

a Evaluate the polynomial when $x = 3$, $y = -2$, and $z = -5$.

1. $x^2 - y^2 + xy$
2. $x^2 + y^2 - xy$
3. $x^2 - 3y^2 + 2xy$
4. $x^2 - 4xy + 5y^2$
5. $8xyz$
6. $-3xyz^2$
7. $xyz^2 - z$
8. $xy - xz + yz$

Lung Capacity. The polynomial equation

$$C = 0.041h - 0.018A - 2.69$$

can be used to estimate the lung capacity C, in liters, of a female of height h, in centimeters, and age A, in years.

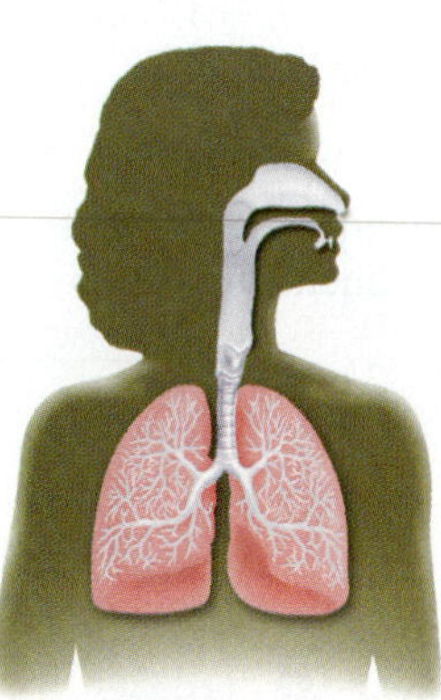

9. Find the lung capacity of a 20-yr-old woman who is 165 cm tall.

10. Find the lung capacity of a 50-yr-old woman who is 160 cm tall.

Altitude of a Launched Object. The altitude h, in meters, of a launched object is given by the polynomial equation

$$h = h_0 + vt - 4.9t^2,$$

where h_0 is the height, in meters, from which the launch occurs, v is the initial upward speed (or velocity), in meters per second (m/s), and t is the number of seconds for which the object is airborne.

11. A model rocket is launched from the top of the Leaning Tower of Pisa, 50 m above the ground. The upward speed is 40 m/s. How high will the rocket be 2 sec after the blastoff?

12. A golf ball is thrown upward with an initial speed of 30 m/s by a golfer atop the Washington Monument, which is 160 m above the ground. How high above the ground will the ball be after 3 sec?

Surface Area of a Right Circular Cylinder. The surface area S of a right circular cylinder is given by the polynomial equation

$$S = 2\pi rh + 2\pi r^2,$$

where h is the height and r is the radius of the base.

13. A 12-oz beverage can has a height of 4.7 in. and a radius of 1.2 in. Evaluate the polynomial when $h = 4.7$ and $r = 1.2$ to find the area of the can. Use 3.14 for π.

14. A 26-oz coffee can has a height of 6.5 in. and a radius of 2.5 in. Evaluate the polynomial when $h = 6.5$ and $r = 2.5$ to find the area of the can. Use 3.14 for π.

Surface Area of a Silo. A silo is a structure that is shaped like a right circular cylinder with a half sphere on top. The surface area S of a silo of height h and radius r (including the area of the base) is given by the polynomial equation $S = 2\pi rh + \pi r^2$.

15. A container of tennis balls is silo-shaped, with a height of $7\frac{1}{2}$ in. and a radius of $1\frac{1}{4}$ in. Find the surface area of the container. Use 3.14 for π.

16. A $1\frac{1}{2}$-oz bottle of roll-on deodorant has a height of 4 in. and a radius of $\frac{3}{4}$ in. Find the surface area of the bottle if the bottle is shaped like a silo. Use 3.14 for π.

b Identify the coefficient and the degree of each term of the polynomial. Then find the degree of the polynomial.

17. $x^3y - 2xy + 3x^2 - 5$

18. $5y^3 - y^2 + 15y + 1$

19. $17x^2y^3 - 3x^3yz - 7$

20. $6 - xy + 8x^2y^2 - y^5$

c Collect like terms.

21. $a + b - 2a - 3b$

22. $y^2 - 1 + y - 6 - y^2$

23. $3x^2y - 2xy^2 + x^2$

24. $m^3 + 2m^2n - 3m^2 + 3mn^2$

25. $6au + 3av + 14au + 7av$

26. $3x^2y - 2z^2y + 3xy^2 + 5z^2y$

27. $2u^2v - 3uv^2 + 6u^2v - 2uv^2$

28. $3x^2 + 6xy + 3y^2 - 5x^2 - 10xy - 5y^2$

d Add.

29. $(2x^2 - xy + y^2) + (-x^2 - 3xy + 2y^2)$

30. $(2z - z^2 + 5) + (z^2 - 3z + 1)$

31. $(r - 2s + 3) + (2r + s) + (s + 4)$

32. $(ab - 2a + 3b) + (5a - 4b) + (3a + 7ab - 8b)$

33. $(b^3a^2 - 2b^2a^3 + 3ba + 4) + (b^2a^3 - 4b^3a^2 + 2ba - 1)$

34. $(2x^2 - 3xy + y^2) + (-4x^2 - 6xy - y^2) + (x^2 + xy - y^2)$

e Subtract.

35. $(a^3 + b^3) - (a^2b - ab^2 + b^3 + a^3)$

36. $(x^3 - y^3) - (-2x^3 + x^2y - xy^2 + 2y^3)$

37. $(xy - ab - 8) - (xy - 3ab - 6)$

38. $(3y^4x^2 + 2y^3x - 3y - 7) - (2y^4x^2 + 2y^3x - 4y - 2x + 5)$

39. $(-2a + 7b - c) - (-3b + 4c - 8d)$

40. Find the sum of $2a + b$ and $3a - b$. Then subtract $5a + 2b$.

f Multiply.

41. $(3z - u)(2z + 3u)$

42. $(a - b)(a^2 + b^2 + 2ab)$

43. $(a^2b - 2)(a^2b - 5)$

44. $(xy + 7)(xy - 4)$

45. $(a^3 + bc)(a^3 - bc)$

46. $(m^2 + n^2 - mn)(m^2 + mn + n^2)$

47. $(y^4x + y^2 + 1)(y^2 + 1)$

48. $(a - b)(a^2 + ab + b^2)$

49. $(3xy - 1)(4xy + 2)$

50. $(m^3n + 8)(m^3n - 6)$

51. $(3 - c^2d^2)(4 + c^2d^2)$

52. $(6x - 2y)(5x - 3y)$

53. $(m^2 - n^2)(m + n)$

54. $(pq + 0.2)(0.4pq - 0.1)$

55. $(xy + x^5y^5)(x^4y^4 - xy)$

56. $(x - y^3)(2y^3 + x)$

57. $(x + h)^2$

58. $(3a + 2b)^2$

59. $(r^3t^2 - 4)^2$

60. $(3a^2b - b^2)^2$

61. $(p^4 + m^2n^2)^2$

62. $(2ab - cd)^2$

63. $\left(2a^3 - \frac{1}{2}b^3\right)^2$

64. $-3x(x + 8y)^2$

65. $3a(a - 2b)^2$

66. $(a^2 + b + 2)^2$

67. $(2a - b)(2a + b)$

68. $(x - y)(x + y)$

69. $(c^2 - d)(c^2 + d)$

70. $(p^3 - 5q)(p^3 + 5q)$

71. $(ab + cd^2)(ab - cd^2)$

72. $(xy + pq)(xy - pq)$

73. $(x + y - 3)(x + y + 3)$

74. $(p + q + 4)(p + q - 4)$

75. $[x + y + z][x - (y + z)]$

76. $[a + b + c][a - (b + c)]$

77. $(a + b + c)(a - b - c)$

78. $(3x + 2 - 5y)(3x + 2 + 5y)$

79. D_W Is it possible for a polynomial in four variables to have a degree less than 4? Why or why not?

80. D_W Can the sum of two trinomials in several variables be a trinomial in one variable? Why or why not?

SKILL MAINTENANCE

In which quadrant is the point located? [8.1c]

81. $(2, -5)$

82. $(-8, -9)$

83. $(16, 23)$

84. $(-3, 2)$

Graph. [8.3b]

85. $2x = -10$

86. $y = -4$

87. $8y - 16 = 0$

88. $x = 4$

SYNTHESIS

Find a polynomial for the shaded area. (Leave results in terms of π where appropriate.)

89.

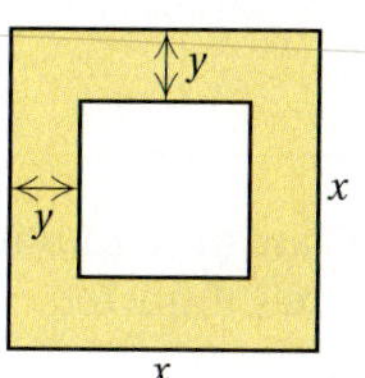

90.

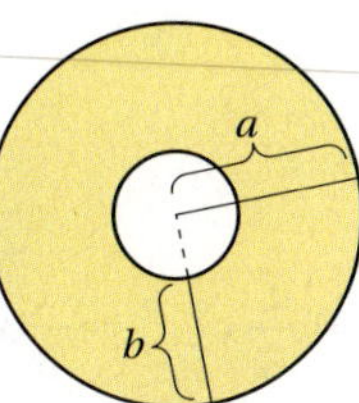

91.

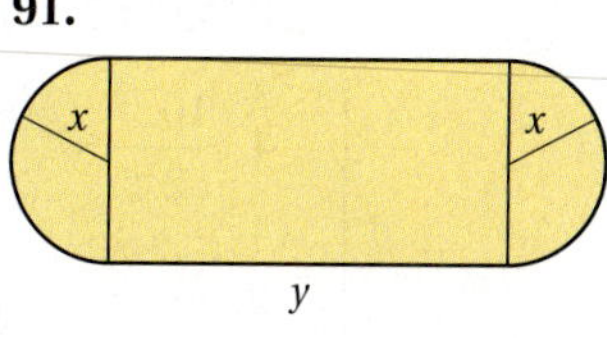

92.

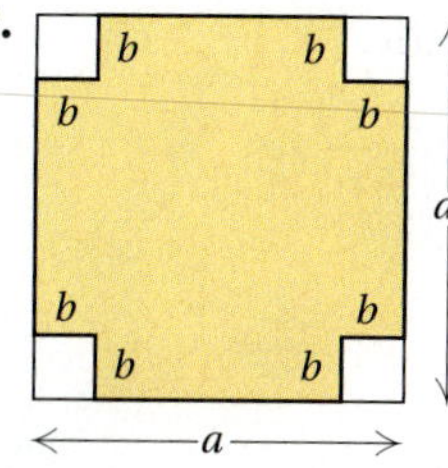

Find a formula for the surface area of the solid object. Leave results in terms of π.

93.

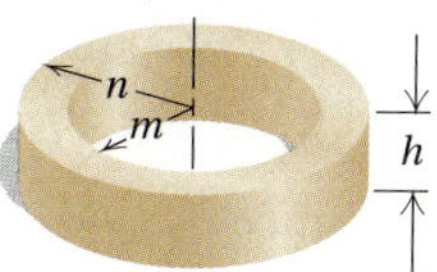

94.

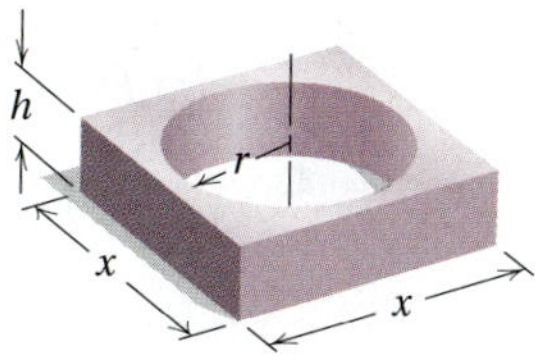

95. *Observatory Paint Costs.* The observatory at Danville University is shaped like a silo that is 40 ft high and 30 ft wide (see Exercise 15). The Heavenly Bodies Astronomy Club is to paint the exterior of the observatory using paint that covers 250 ft^2 per gallon. How many gallons should they purchase?

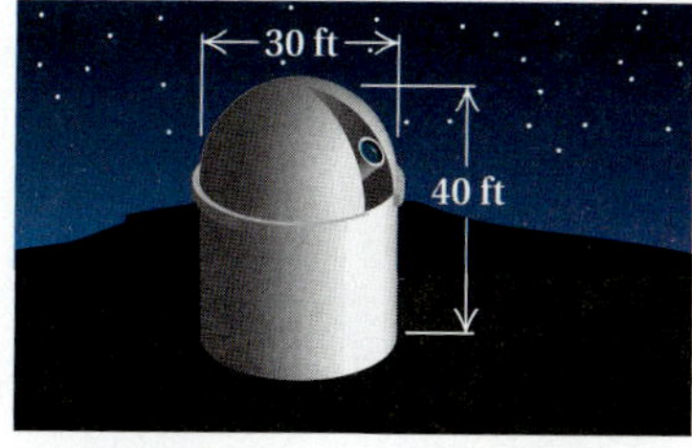

96. *Interest Compounded Annually.* An amount of money P that is invested at the yearly interest rate r grows to the amount

$$P(1 + r)^t$$

after t years. Find a polynomial that can be used to determine the amount to which P will grow after 2 yr.

97. Suppose that \$10,400 is invested at 8.5% compounded annually. How much is in the account at the end of 5 yr? (See Exercise 96.)

98. Multiply: $(x + a)(x - b)(x - a)(x + b)$.

9.8 DIVISION OF POLYNOMIALS

Objectives

a Divide a polynomial by a monomial.

b Divide a polynomial by a divisor that is a binomial.

c Use synthetic division to divide a polynomial by a binomial of the type $x - a$.

Divide.

1. $\dfrac{20x^3}{5x}$

2. $\dfrac{-28x^{14}}{4x^3}$

3. $\dfrac{-56p^5q^7}{2p^2q^6}$

4. $\dfrac{x^5}{4x}$

Answers on page A-31

In this section, we consider division of polynomials. You will see that such division is similar to what is done in arithmetic.

a Divisor a Monomial

We first consider division by a monomial. When dividing a monomial by a monomial, we use the quotient rule of Section 9.1 to subtract exponents when the bases are the same. We also divide the coefficients.

EXAMPLES Divide.

1. $\dfrac{10x^2}{2x} = \dfrac{10}{2} \cdot \dfrac{x^2}{x} = 5x^{2-1} = 5x$

2. $\dfrac{x^9}{3x^2} = \dfrac{1x^9}{3x^2} = \dfrac{1}{3} \cdot \dfrac{x^9}{x^2} = \dfrac{1}{3}x^{9-2} = \dfrac{1}{3}x^7$

3. $\dfrac{-18x^{10}}{3x^3} = \dfrac{-18}{3} \cdot \dfrac{x^{10}}{x^3} = -6x^{10-3} = -6x^7$

4. $\dfrac{42a^2b^5}{-3ab^2} = \dfrac{42}{-3} \cdot \dfrac{a^2}{a} \cdot \dfrac{b^5}{b^2} = -14a^{2-1}b^{5-2} = -14ab^3$

CAUTION!

The coefficients are divided but the exponents are subtracted.

Do Exercises 1–4.

To divide a polynomial by a monomial, we note that since

$$\frac{A}{C} + \frac{B}{C} = \frac{A+B}{C},$$

it follows that

$$\frac{A+B}{C} = \frac{A}{C} + \frac{B}{C}.$$ Switching the left and right sides of the equation

This is actually the procedure we use when performing divisions like $86 \div 2$. Although we might write

$$\frac{86}{2} = 43,$$

we could also calculate as follows:

$$\frac{86}{2} = \frac{80+6}{2} = \frac{80}{2} + \frac{6}{2} = 40 + 3 = 43.$$

Similarly, to divide a polynomial by a monomial, we divide each term by the monomial.

EXAMPLE 5 Divide: $(9x^8 + 12x^6) \div 3x^2$.

We have

$$(9x^8 + 12x^6) \div 3x^2 = \frac{9x^8 + 12x^6}{3x^2}$$

$$= \frac{9x^8}{3x^2} + \frac{12x^6}{3x^2}.$$ To see this, add and get the original expression.

We now perform the separate divisions:

$$\frac{9x^8}{3x^2} + \frac{12x^6}{3x^2} = \frac{9}{3} \cdot \frac{x^8}{x^2} + \frac{12}{3} \cdot \frac{x^6}{x^2}$$
$$= 3x^{8-2} + 4x^{6-2}$$
$$= 3x^6 + 4x^4.$$

CAUTION!

The coefficients are *divided*, but the exponents are *subtracted*.

To check, we multiply the quotient $3x^6 + 4x^4$ by the divisor $3x^2$:

$$3x^2(3x^6 + 4x^4) = (3x^2)(3x^6) + (3x^2)(4x^4) = 9x^8 + 12x^6.$$

This is the polynomial that was being divided, so our answer is $3x^6 + 4x^4$.

Do Exercises 5–7.

EXAMPLE 6 Divide and check: $(10a^5b^4 - 2a^3b^2 + 6a^2b) \div (2a^2b)$.

$$\frac{10a^5b^4 - 2a^3b^2 + 6a^2b}{2a^2b} = \frac{10a^5b^4}{2a^2b} - \frac{2a^3b^2}{2a^2b} + \frac{6a^2b}{2a^2b}$$
$$= \frac{10}{2}a^{5-2}b^{4-1} - \frac{2}{2}a^{3-2}b^{2-1} + \frac{6}{2}$$
$$= 5a^3b^3 - ab + 3$$

CHECK:

$$2a^2b(5a^3b^3 - ab + 3) = 2a^2b \cdot 5a^3b^3 - 2a^2b \cdot ab + 2a^2b \cdot 3$$
$$= 10a^5b^4 - 2a^3b^2 + 6a^2b$$

Our answer, $5a^3b^3 - ab + 3$, checks.

To divide a polynomial by a monomial, divide each term by the monomial.

Do Exercises 8 and 9.

b Divisor a Binomial

Let's first consider long division as it is performed in arithmetic. When we divide, we repeat the following procedure.

To carry out long division:

1. Divide,
2. Multiply,
3. Subtract, and
4. Bring down the next term.

We review this by considering the division $3711 \div 8$.

```
     4      ← ① Divide: 37 ÷ 8 ≈ 4.
8 ) 3 7 1 1
    3 2     ← ② Multiply: 4 × 8 = 32.
      5 1     ③ Subtract: 37 − 32 = 5.
              ④ Bring down the 1.
```

```
      4 6 3
8 ) 3 7 1 1
    3 2
      5 1
      4 8
        3 1
        2 4
          7
```

5. Divide: $(28x^7 + 32x^5) \div 4x^3$. Check the result.

6. Divide: $(2x^3 + 6x^2 + 4x) \div 2x$. Check the result.

7. Divide: $(6x^2 + 3x - 2) \div 3$. Check the result.

Divide and check.

8. $(8x^2 - 3x + 1) \div 2$

9. $\dfrac{2x^4y^6 - 3x^3y^4 + 5x^2y^3}{x^2y^2}$

Answers on page A-31

10. Divide and check:

$(x^2 + x - 6) \div (x + 3)$.

Next, we repeat the process two more times. We obtain the complete division as shown on the right above. The quotient is 463. The remainder is 7, expressed as R = 7. We write the answer as

$$463 \text{ R } 7 \qquad \text{or} \qquad 463 + \frac{7}{8} = 463\frac{7}{8}.$$

We check by multiplying the quotient, 463, by the divisor, 8, and adding the remainder, 7:

$$8 \cdot 463 + 7 = 3704 + 7 = 3711.$$

Now let's look at long division with polynomials. We use this procedure when the divisor is not a monomial. We write polynomials in descending order and then write in missing terms.

EXAMPLE 7 Divide $x^2 + 5x + 6$ by $x + 2$.

$$\begin{array}{r l}
x & \leftarrow \text{Divide the first term by the first term: } x^2/x = x. \\
x + 2\overline{)x^2 + 5x + 6} & \quad \text{Ignore the term 2.} \\
\underline{x^2 + 2x} & \leftarrow \text{Multiply } x \text{ above by the divisor, } x + 2. \\
3x & \leftarrow \text{Subtract: } (x^2 + 5x) - (x^2 + 2x) = x^2 + 5x - x^2 - 2x \\
 & \qquad = 3x.
\end{array}$$

We now "bring down" the next term of the dividend—in this case, 6.

$$\begin{array}{r l}
x + 3 & \leftarrow \text{Divide the first term by the first term: } 3x/x = 3. \\
x + 2\overline{)x^2 + 5x + 6} & \\
\underline{x^2 + 2x} & \\
3x + 6 & \leftarrow \text{The 6 has been "brought down."} \\
\underline{3x + 6} & \leftarrow \text{Multiply 3 by the divisor, } x + 2. \\
0 & \leftarrow \text{Subtract: } (3x + 6) - (3x + 6) = 3x + 6 - 3x - 6 = 0.
\end{array}$$

The quotient is $x + 3$. The remainder is 0, expressed as R = 0. A remainder of 0 is generally not listed in an answer.

To check, we multiply the quotient by the divisor and add the remainder, if any, to see if we get the dividend:

$$\underbrace{(x + 2)}_{\text{Divisor}} \cdot \underbrace{(x + 3)}_{\text{Quotient}} + \underbrace{0}_{\text{Remainder}} = \underbrace{x^2 + 5x + 6}_{\text{Dividend}}. \qquad \text{The division checks.}$$

Do Exercise 10.

EXAMPLE 8 Divide and check: $(x^2 + 2x - 12) \div (x - 3)$.

$$\begin{array}{r l}
x & \leftarrow \text{Divide the first term by the first term: } x^2/x = x. \\
x - 3\overline{)x^2 + 2x - 12} & \\
\underline{x^2 - 3x} & \leftarrow \text{Multiply } x \text{ above by the divisor, } x - 3. \\
5x & \leftarrow \text{Subtract: } (x^2 + 2x) - (x^2 - 3x) = x^2 + 2x - x^2 + 3x \\
 & \qquad = 5x.
\end{array}$$

We now "bring down" the next term of the dividend—in this case, -12.

$$\begin{array}{r l}
x + 5 & \leftarrow \text{Divide the first term by the first term: } 5x/x = 5. \\
x - 3\overline{)x^2 + 2x - 12} & \\
\underline{x^2 - 3x} & \\
5x - 12 & \leftarrow \text{Bring down the } -12. \\
\underline{5x - 15} & \leftarrow \text{Multiply 5 above by the divisor, } x - 3. \\
3 & \leftarrow \text{Subtract:} \\
 & \quad (5x - 12) - (5x - 15) = 5x - 12 - 5x + 15 \\
 & \qquad = 3.
\end{array}$$

Answer on page A-31

Study Tips

FORMING A STUDY GROUP

Consider forming a study group with some of your fellow students. Exchange e-mail addresses, telephone numbers, and schedules so that you can coordinate study time for homework and tests.

The answer is $x + 5$ with R = 3, or

$$\underbrace{x + 5}_{\text{Quotient}} + \frac{3 \leftarrow \text{Remainder}}{\underbrace{x - 3}_{\text{Divisor}}}$$

(This is the way answers will be given at the back of the book.)

CHECK: We can check by multiplying the divisor by the quotient and adding the remainder, as follows:

$$\begin{aligned}(x - 3)(x + 5) + 3 &= x^2 + 2x - 15 + 3 \\ &= x^2 + 2x - 12.\end{aligned}$$

When dividing, an answer may "come out even" (that is, have a remainder of 0, as in Example 7), or it may not (as in Example 8). If a remainder is not 0, we continue dividing until the degree of the remainder is less than the degree of the divisor. Check this in each of Examples 7 and 8.

Do Exercises 11 and 12.

Divide and check.

11. $x - 2\overline{)x^2 + 2x - 8}$

12. $x + 3\overline{)x^2 + 7x + 10}$

EXAMPLE 9 Divide and check: $(x^3 + 1) \div (x + 1)$.

$$\begin{array}{r} x^2 - x + 1 \\ x + 1\overline{)x^3 + 0x^2 + 0x + 1} \\ \underline{x^3 + x^2} \\ -x^2 + 0x \\ \underline{-x^2 - x} \\ x + 1 \\ \underline{x + 1} \\ 0 \end{array}$$

← Fill in the missing terms (see Section 9.3).

← Subtract: $x^3 - (x^3 + x^2) = -x^2$.

← Subtract: $-x^2 - (-x^2 - x) = x$.

← Subtract: $(x + 1) - (x + 1) = 0$.

The answer is $x^2 - x + 1$. The check is left to the student.

EXAMPLE 10 Divide and check: $(x^4 - 3x^2 + 1) \div (x - 4)$.

$$\begin{array}{r} x^3 + 4x^2 + 13x + 52 \\ x - 4\overline{)x^4 + 0x^3 - 3x^2 + 0x + 1} \\ \underline{x^4 - 4x^3} \\ 4x^3 - 3x^2 \\ \underline{4x^3 - 16x^2} \\ 13x^2 + 0x \\ \underline{13x^2 - 52x} \\ 52x + 1 \\ \underline{52x - 208} \\ 209 \end{array}$$

← Fill in the missing terms.

← Subtract: $x^4 - (x^4 - 4x^3) = 4x^3$.

← Subtract: $(4x^3 - 3x^2) - (4x^3 - 16x^2) = 13x^2$.

← Subtract: $13x^2 - (13x^2 - 52x) = 52x$.

← Subtract: $(52x + 1) - (52x - 208) = 209$.

The answer is $x^3 + 4x^2 + 13x + 52$, with R = 209, or

$$x^3 + 4x^2 + 13x + 52 + \frac{209}{x - 4}.$$

CHECK: $(x - 4)(x^3 + 4x^2 + 13x + 52) + 209$

$$= -4x^3 - 16x^2 - 52x - 208 + x^4 + 4x^3 + 13x^2 + 52x + 209$$
$$= x^4 - 3x^2 + 1$$

Do Exercise 13.

13. Divide and check:

$$(x^3 - 1) \div (x - 1).$$

Answers on page A-31

C Synthetic Division

To divide a polynomial by a binomial of the type $x - a$, we can streamline the general procedure by a process called **synthetic division.**

Compare the following. In **A,** we perform a division. In **B,** we also divide but we do not write the variables.

A.

$$\begin{array}{r} 4x^2 + 5x + 11 \\ x - 2\overline{)4x^3 - 3x^2 + x + 7} \\ \underline{4x^3 - 8x^2} \\ 5x^2 + x \\ \underline{5x^2 - 10x} \\ 11x + 7 \\ \underline{11x - 22} \\ 29 \end{array}$$

B.

$$\begin{array}{r} 4 + 5 + 11 \\ 1 - 2\overline{)4 - 3 + 1 + 7} \\ \underline{4 - 8} \\ 5 + 1 \\ \underline{5 - 10} \\ 11 + 7 \\ \underline{11 - 22} \\ 29 \end{array}$$

In **B,** there is still some duplication of writing. Also, since we can subtract by adding the opposite, we can use 2 instead of -2 and then add instead of subtracting.

C. *Synthetic Division*

a) $\underline{2}|\, 4 \; -3 \; +1 \; +7$

$\quad 4$

Write the 2, the opposite of -2 in the divisor $x - 2$, and the coefficients of the dividend.

Bring down the first coefficient.

b) $\underline{2}|\, 4 \; -3 \; +1 \; +7$

$\quad\quad 8$

$\quad 4 \quad 5$

Multiply 4 by 2 to get 8. Add 8 and -3.

c) $\underline{2}|\, 4 \; -3 \; +1 \; +7$

$\quad\quad 8 \quad 10$

$\quad 4 \quad 5 \quad 11$

Multiply 5 by 2 to get 10. Add 10 and 1.

d) $\underline{2}|\, 4 \; -3 \; +1 \; +7$

$\quad\quad 8 \quad 10 \quad 22$

$\quad 4 \quad 5 \quad 11 \;|\; 29$

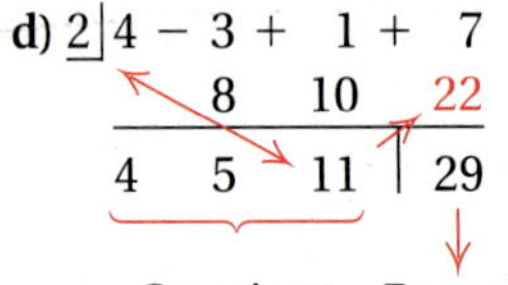

Multiply 11 by 2 to get 22. Add 22 and 7.

The last number, 29, is the remainder. The other numbers are the coefficients of the quotient, with that of the term of highest degree, which is 1 less than the degree of the dividend, first, as follows.

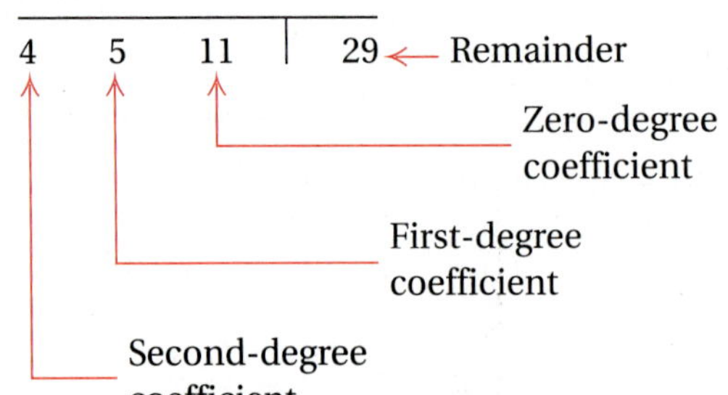

The answer is $4x^2 + 5x + 11$, R 29, or $4x^2 + 5x + 11 + \dfrac{29}{x - 2}$.

It is important to remember that in order for synthetic division to work, the divisor must be of the form $x - a$, that is, a variable minus a constant. The coefficient of the variable must be 1.

EXAMPLE 11 Use synthetic division to divide:

$$(x^3 + 6x^2 - x - 30) \div (x - 2).$$

We have

$$\begin{array}{r|rrrr} 2 & 1 & 6 & -1 & -30 \\ & & 2 & 16 & 30 \\ \hline & 1 & 8 & 15 & 0 \end{array}$$

The answer is $x^2 + 8x + 15$, R 0, or just $x^2 + 8x + 15$.

Do Exercise 14.

When there are missing terms, be sure to write 0's for their coefficients.

EXAMPLES Use synthetic division to divide.

12. $(2x^3 + 7x^2 - 5) \div (x + 3)$

There is no x-term, so we must write a 0 for its coefficient. Note that $x + 3 = x - (-3)$, so we write -3 at the left.

$$\begin{array}{r|rrrr} -3 & 2 & 7 & 0 & -5 \\ & & -6 & -3 & 9 \\ \hline & 2 & 1 & -3 & 4 \end{array}$$

The answer is $2x^2 + x - 3$, R 4, or $2x^2 + x - 3 + \frac{4}{x + 3}$.

13. $(x^4 - 1) \div (x - 1)$

The divisor is $x - 1$, so we write 1 at the left.

$$\begin{array}{r|rrrrr} 1 & 1 & 0 & 0 & 0 & -1 \\ & & 1 & 1 & 1 & 1 \\ \hline & 1 & 1 & 1 & 1 & 0 \end{array}$$

The answer is $x^3 + x^2 + x + 1$.

14. $(8x^5 - 6x^3 + x - 8) \div (x + 2)$

Note that $x + 2 = x - (-2)$, so we write -2 at the left.

$$\begin{array}{r|rrrrrr} -2 & 8 & 0 & -6 & 0 & 1 & -8 \\ & & -16 & 32 & -52 & 104 & -210 \\ \hline & 8 & -16 & 26 & -52 & 105 & -218 \end{array}$$

The answer is $8x^4 - 16x^3 + 26x^2 - 52x + 105$, R -218, or

$$8x^4 - 16x^3 + 26x^2 - 52x + 105 + \frac{-218}{x + 2}.$$

Do Exercises 15 and 16.

14. Use synthetic division to divide:

$(2x^3 - 4x^2 + 8x - 8) \div (x - 3)$.

Use synthetic division to divide.

15. $(x^3 - 2x^2 + 5x - 4) \div (x + 2)$

16. $(y^3 + 1) \div (y + 1)$

Answer on page A-31

9.8 EXERCISE SET

For Extra Help

Digital Video Tutor CD 8 Videotape 10 | InterAct Math | Math Tutor Center | MathXL | MyMathLab

a Divide and check.

1. $\dfrac{24x^4}{8}$

2. $\dfrac{-2u^2}{u}$

3. $\dfrac{25x^3}{5x^2}$

4. $\dfrac{16x^7}{-2x^2}$

5. $\dfrac{-54x^{11}}{-3x^8}$

6. $\dfrac{-75a^{10}}{3a^2}$

7. $\dfrac{64a^5b^4}{16a^2b^3}$

8. $\dfrac{-34p^{10}q^{11}}{-17pq^9}$

9. $\dfrac{24x^4 - 4x^3 + x^2 - 16}{8}$

10. $\dfrac{12a^4 - 3a^2 + a - 6}{6}$

11. $\dfrac{u - 2u^2 - u^5}{u}$

12. $\dfrac{50x^5 - 7x^4 + x^2}{x}$

13. $(15t^3 + 24t^2 - 6t) \div (3t)$

14. $(25t^3 + 15t^2 - 30t) \div (5t)$

15. $(20x^6 - 20x^4 - 5x^2) \div (-5x^2)$

16. $(24x^6 + 32x^5 - 8x^2) \div (-8x^2)$

17. $(24x^5 - 40x^4 + 6x^3) \div (4x^3)$

18. $(18x^6 - 27x^5 - 3x^3) \div (9x^3)$

19. $\dfrac{18x^2 - 5x + 2}{2}$

20. $\dfrac{15x^2 - 30x + 6}{3}$

21. $\dfrac{12x^3 + 26x^2 + 8x}{2x}$

22. $\dfrac{2x^4 - 3x^3 + 5x^2}{x^2}$

23. $\dfrac{9r^2s^2 + 3r^2s - 6rs^2}{3rs}$

24. $\dfrac{4x^4y - 8x^6y^2 + 12x^8y^6}{4x^4y}$

b Divide.

25. $(x^2 + 4x + 4) \div (x + 2)$

26. $(x^2 - 6x + 9) \div (x - 3)$

27. $(x^2 - 10x - 25) \div (x - 5)$

28. $(x^2 + 8x - 16) \div (x + 4)$

29. $(x^2 + 4x - 14) \div (x + 6)$

30. $(x^2 + 5x - 9) \div (x - 2)$

31. $\dfrac{x^2 - 9}{x + 3}$

32. $\dfrac{x^2 - 25}{x - 5}$

33. $\dfrac{x^5 + 1}{x + 1}$

34. $\dfrac{x^5 - 1}{x - 1}$

35. $\dfrac{8x^3 - 22x^2 - 5x + 12}{4x + 3}$

36. $\dfrac{2x^3 - 9x^2 + 11x - 3}{2x - 3}$

37. $(x^6 - 13x^3 + 42) \div (x^3 - 7)$

38. $(x^6 + 5x^3 - 24) \div (x^3 - 3)$

39. $(x^4 - 16) \div (x - 2)$

40. $(x^4 - 81) \div (x - 3)$

41. $(t^3 - t^2 + t - 1) \div (t - 1)$

42. $(t^3 - t^2 + t - 1) \div (t + 1)$

C Use synthetic division to divide.

43. $(x^3 - 2x^2 + 2x - 5) \div (x - 1)$

44. $(x^3 - 2x^2 + 2x - 5) \div (x + 1)$

45. $(a^2 + 11a - 19) \div (a + 4)$

46. $(a^2 + 11a - 19) \div (a - 4)$

47. $(x^3 - 7x^2 - 13x + 3) \div (x - 2)$

48. $(x^3 - 7x^2 - 13x + 3) \div (x + 2)$

49. $(3x^3 + 7x^2 - 4x + 3) \div (x + 3)$

50. $(3x^3 + 7x^2 - 4x + 3) \div (x - 3)$

51. $(y^3 - 3y + 10) \div (y - 2)$

52. $(x^3 - 2x^2 + 8) \div (x + 2)$

53. $(3x^4 - 25x^2 - 18) \div (x - 3)$

54. $(6y^4 + 15y^3 + 28y + 6) \div (y + 3)$

55. $(x^3 - 8) \div (x - 2)$

56. $(y^3 + 125) \div (y + 5)$

57. $(y^4 - 16) \div (y - 2)$

58. $(x^5 - 32) \div (x - 2)$

59. **DW** How is the distributive law used when dividing a polynomial by a binomial?

60. **DW** On an assignment, Emma *incorrectly* writes

$$\frac{12x^3 - 6x}{3x} = 4x^2 - 6x.$$

What mistake do you think she is making and how might you convince her that a mistake has been made?

SKILL MAINTENANCE

Subtract. [6.4a]

61. $17 - 45$

62. $-14 - 45$

63. $-2.3 - (-9.1)$

64. $-\frac{5}{8} - \frac{3}{4}$

Solve. [7.6a]

65. The perimeter of a rectangle is 640 ft. The length is 15 ft more than the width. Find the area of the rectangle.

66. The first angle of a triangle is 24° more than the second. The third angle is twice the first. Find the measures of the angles of the triangle.

Solve. [7.3c]

67. $-6(2 - x) + 10(5x - 7) = 10$

68. $-10(x - 4) = 5(2x + 5) - 7$

Factor. [6.7d]

69. $4x - 12 + 24y$

70. $256 - 2a - 4b$

SYNTHESIS

Divide.

71. $(x^4 + 9x^2 + 20) \div (x^2 + 4)$

72. $(y^4 + a^2) \div (y + a)$

73. $(5a^3 + 8a^2 - 23a - 1) \div (5a^2 - 7a - 2)$

74. $(15y^3 - 30y + 7 - 19y^2) \div (3y^2 - 2 - 5y)$

75. $(6x^5 - 13x^3 + 5x + 3 - 4x^2 + 3x^4) \div (3x^3 - 2x - 1)$

76. $(5x^7 - 3x^4 + 2x^2 - 10x + 2) \div (x^2 - x + 1)$

77. $(a^6 - b^6) \div (a - b)$

78. $(x^5 + y^5) \div (x + y)$

If the remainder is 0 when one polynomial is divided by another, the divisor is a *factor* of the dividend. Find the value(s) of c for which $x - 1$ is a factor of the polynomial.

79. $x^2 + 4x + c$

80. $2x^2 + 3cx - 8$

81. $c^2x^2 - 2cx + 1$

CHAPTER 9

Summary and Review

The review that follows is meant to prepare you for a chapter exam. It consists of two parts. The first part is a checklist of some of the Study Tips referred to in this and preceding chapters, as well as a list of important properties and formulas. The second part is the Review Exercises. These provide practice exercises for the exam, together with references to section objectives so you can go back and review. Before beginning, stop and look back over the skills you have obtained. What skills in mathematics do you have now that you did not have before studying this chapter?

STUDY TIPS CHECKLIST

The foundation of all your study skills is TIME!

- [ ] Have you tried using the audiotapes?
- [ ] Are you taking the time to include all the steps when working your homework and the tests?
- [ ] Are you using the time-management suggestions we have given so you have the proper amount of time to study mathematics?
- [ ] Have you been using the supplements for the text such as the *Student's Solutions Manual* and the Math Tutor Center?
- [ ] Have you memorized the rules for special products of polynomials and for manipulating expressions with exponents?

IMPORTANT PROPERTIES AND FORMULAS

FOIL: $(A + B)(C + D) = AC + AD + BC + BD$

Square of a Sum: $(A + B)(A + B) = (A + B)^2 = A^2 + 2AB + B^2$

Square of a Difference: $(A - B)(A - B) = (A - B)^2 = A^2 - 2AB + B^2$

Product of a Sum and a Difference: $(A + B)(A - B) = A^2 - B^2$

Definitions and Rules for Exponents: See p. 676.

REVIEW EXERCISES

Multiply and simplify. [9.1d, f]

1. $7^2 \cdot 7^{-4}$

2. $y^7 \cdot y^3 \cdot y$

3. $(3x)^5 \cdot (3x)^9$

4. $t^8 \cdot t^0$

Divide and simplify. [9.1e, f]

5. $\dfrac{4^5}{4^2}$

6. $\dfrac{a^5}{a^8}$

7. $\dfrac{(7x)^4}{(7x)^4}$

Simplify.

8. $(3t^4)^2$ [9.2a, b]

9. $(2x^3)^2(-3x)^2$ [9.1d], [9.2a, b]

10. $\left(\dfrac{2x}{y}\right)^{-3}$ [9.2b]

11. Express using a negative exponent: $\dfrac{1}{t^5}$. [9.1f]

12. Express using a positive exponent: y^{-4}. [9.1f]

13. Convert to scientific notation: 0.0000328. [9.2c]

14. Convert to decimal notation: 8.3×10^6. [9.2c]

Multiply or divide and write scientific notation for the result. [9.2d]

15. $(3.8 \times 10^4)(5.5 \times 10^{-1})$

16. $\dfrac{1.28 \times 10^{-8}}{2.5 \times 10^{-4}}$

17. *Diet-Drink Consumption.* It has been estimated that there will be 292 million people in the United States by 2005 and that on average, each of them will drink 15.3 gal of diet drinks that year. How many gallons of diet drinks will be consumed by the entire population in 2005? Express the answer in scientific notation. [9.2e]
Source: U.S. Department of Agriculture

18. Evaluate the polynomial $x^2 - 3x + 6$ when $x = -1$. [9.3a]

19. Identify the terms of the polynomial $-4y^5 + 7y^2 - 3y - 2$. [9.3b]

20. Identify the missing terms in $x^3 + x$. [9.3h]

21. Identify the degree of each term and the degree of the polynomial $4x^3 + 6x^2 - 5x + \frac{5}{3}$. [9.3g]

Classify the polynomial as a monomial, binomial, trinomial, or none of these. [9.3i]

22. $4x^3 - 1$

23. $4 - 9t^3 - 7t^4 + 10t^2$

24. $7y^2$

Collect like terms and then arrange in descending order. [9.3f]

25. $3x^2 - 2x + 3 - 5x^2 - 1 - x$

26. $-x + \frac{1}{2} + 14x^4 - 7x^2 - 1 - 4x^4$

Add. [9.4a]

27. $(3x^4 - x^3 + x - 4) + (x^5 + 7x^3 - 3x^2 - 5) + (-5x^4 + 6x^2 - x)$

28. $(3x^5 - 4x^4 + x^3 - 3) + (3x^4 - 5x^3 + 3x^2) + (-5x^5 - 5x^2) + (-5x^4 + 2x^3 + 5)$

Subtract. [9.4c]

29. $(5x^2 - 4x + 1) - (3x^2 + 1)$

30. $(3x^5 - 4x^4 + 3x^2 + 3) - (2x^5 - 4x^4 + 3x^3 + 4x^2 - 5)$

31. Find a polynomial for the perimeter and for the area. [9.4d], [9.5b]

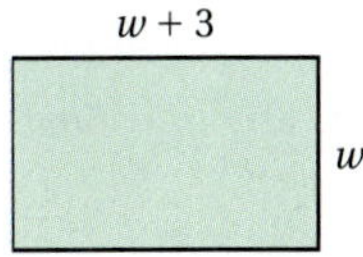

32. Find two algebraic expressions for the area of this figure. First, regard the figure as one large rectangle, and then regard the figure as a sum of four smaller rectangles. [9.4d]

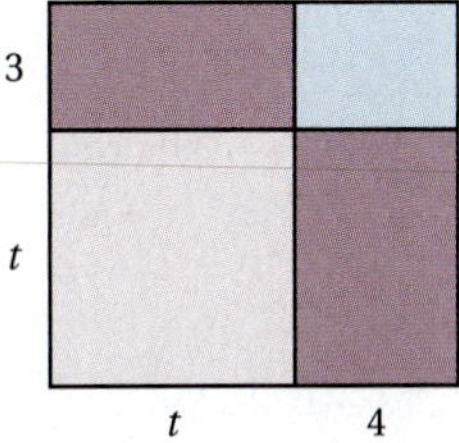

Multiply.

33. $\left(x + \frac{2}{3}\right)\left(x + \frac{1}{2}\right)$ [9.6a]

34. $(7x + 1)^2$ [9.6c]

35. $(4x^2 - 5x + 1)(3x - 2)$ [9.5d]

36. $(3x^2 + 4)(3x^2 - 4)$ [9.6b]

37. $5x^4(3x^3 - 8x^2 + 10x + 2)$ [9.5b]

38. $(x + 4)(x - 7)$ [9.6a]

39. $(3y^2 - 2y)^2$ [9.6c]

40. $(2t^2 + 3)(t^2 - 7)$ [9.6a]

41. Evaluate the polynomial

$$2 - 5xy + y^2 - 4xy^3 + x^6$$

when $x = -1$ and $y = 2$. [9.7a]

42. Identify the coefficient and the degree of each term of the polynomial

$$x^5y - 7xy + 9x^2 - 8.$$

Then find the degree of the polynomial. [9.7b]

Collect like terms. [9.7c]

43. $y + w - 2y + 8w - 5$

44. $m^6 - 2m^2n + m^2n^2 + n^2m - 6m^3 + m^2n^2 + 7n^2m$

45. Add: [9.7d]
$(5x^2 - 7xy + y^2) + (-6x^2 - 3xy - y^2) + (x^2 + xy - 2y^2)$.

46. Subtract: [9.7e]
$(6x^3y^2 - 4x^2y - 6x) - (-5x^3y^2 + 4x^2y + 6x^2 - 6)$.

Multiply. [9.7f]

47. $(p - q)(p^2 + pq + q^2)$

48. $\left(3a^4 - \frac{1}{3}b^3\right)^2$

Divide.

49. $(10x^3 - x^2 + 6x) \div (2x)$ [9.8a]

50. $(6x^3 - 5x^2 - 13x + 13) \div (2x + 3)$ [9.8b]

Divide using synthetic division. Show your work. [9.8c]

51. $(x^3 + 5x^2 + 4x - 7) \div (x - 4)$

52. $(3x^4 - 5x^3 + 2x - 7) \div (x + 1)$

53. The graph of the polynomial equation $y = 10x^3 - 10x$ is shown below. Use *only* the graph to estimate the value of the polynomial when $x = -1$, $x = -0.5$, $x = 0.5$, $x = 1$, and $x = 1.1$. [9.3a]

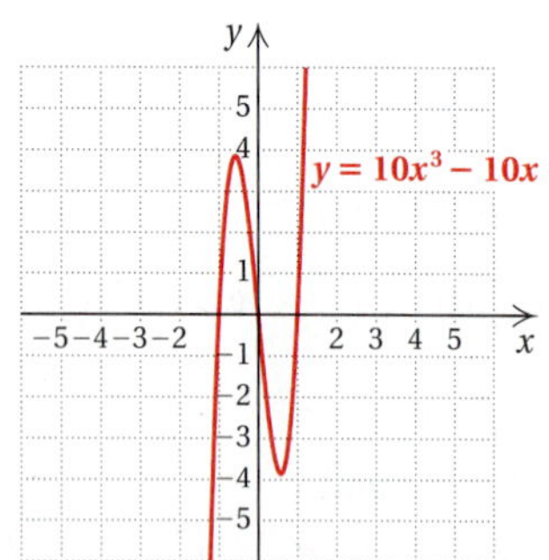

54. **DW** Explain why the expression 578.6×10^{-7} is not in scientific notation. [9.2c]

55. **DW** Write a short explanation of the difference between a monomial, a binomial, a trinomial, and a general polynomial. [9.3i]

SKILL MAINTENANCE

Certain objectives from four particular sections will be retested on the chapter test. The objectives are listed with the practice problems that follow.

56. Factor: $25t - 50 + 100m$. [6.7d]

57. Solve: $7x + 6 - 8x = 11 - 5x + 4$. [7.3b]

58. Solve: $3(x - 2) + 6 = 5(x + 3) + 9$. [7.3c]

59. Subtract: $-3.4 - 7.8$. [6.4a]

60. The perimeter of a rectangle is 540 m. The width is 19 m less than the length. Find the width and the length. [7.6a]

SYNTHESIS

Find a polynomial for the shaded area. [9.4d], [9.6b]

61.

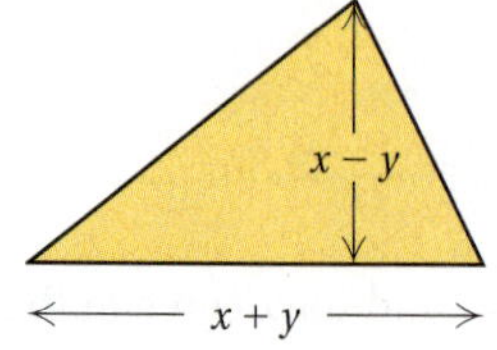

62.

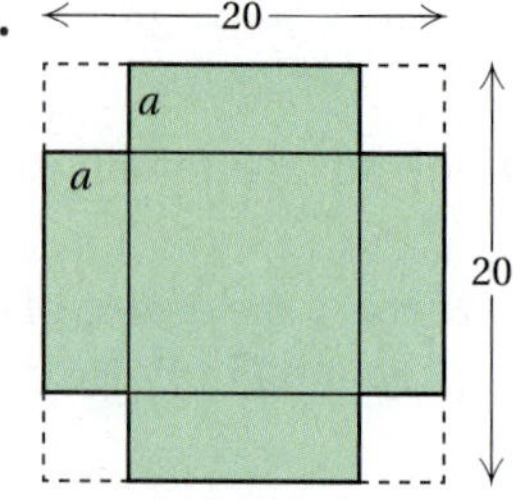

63. Collect like terms: [9.1d], [9.2a], [9.3e]
$-3x^5 \cdot 3x^3 - x^6(2x)^2 + (3x^4)^2 + (2x^2)^4 - 40x^2(x^3)^2$.

64. Solve: [7.3b], [9.6a]
$(x - 7)(x + 10) = (x - 4)(x - 6)$.

65. The product of two polynomials is $x^5 - 1$. One of the polynomials is $x - 1$. Find the other. [9.8b]

66. A rectangular garden is twice as long as it is wide and is surrounded by a sidewalk that is 4 ft wide (see the figure below). The area of the sidewalk is 256 ft^2. Find the dimensions of the garden. [7.3b], [9.4d], [9.5a], [9.6a]

CHAPTER 9

Chapter Test

Multiply and simplify.

1. $6^{-2} \cdot 6^{-3}$

2. $x^6 \cdot x^2 \cdot x$

3. $(4a)^3 \cdot (4a)^8$

Divide and simplify.

4. $\frac{3^5}{3^2}$

5. $\frac{x^3}{x^8}$

6. $\frac{(2x)^5}{(2x)^5}$

Simplify.

7. $(x^3)^2$

8. $(-3y^2)^3$

9. $(2a^3b)^4$

10. $\left(\frac{ab}{c}\right)^3$

11. $(3x^2)^3(-2x^5)^3$

12. $3(x^2)^3(-2x^5)^3$

13. $2x^2(-3x^2)^4$

14. $(2x)^2(-3x^2)^4$

15. Express using a positive exponent:

5^{-3}.

16. Express using a negative exponent:

$\frac{1}{y^8}$.

17. Convert to scientific notation:

3,900,000,000.

18. Convert to decimal notation:

5×10^{-8}.

Multiply or divide and write scientific notation for the answer.

19. $\frac{5.6 \times 10^6}{3.2 \times 10^{-11}}$

20. $(2.4 \times 10^5)(5.4 \times 10^{16})$

21. *CD-ROM Memory.* A CD-ROM can contain about 600 million pieces of information (bytes). How many sound files, each containing 40,000 bytes, can a CD-ROM hold? Express the answer in scientific notation.

22. Evaluate the polynomial $x^5 + 5x - 1$ when $x = -2$.

23. Identify the coefficient of each term of the polynomial $\frac{1}{3}x^5 - x + 7$.

24. Identify the degree of each term and the degree of the polynomial $2x^3 - 4 + 5x + 3x^6$.

25. Classify the polynomial $7 - x$ as a monomial, a binomial, a trinomial, or none of these.

Collect like terms.

26. $4a^2 - 6 + a^2$

27. $y^2 - 3y - y + \frac{3}{4}y^2$

28. Collect like terms and then arrange in descending order:

$3 - x^2 + 2x^3 + 5x^2 - 6x - 2x + x^5.$

Add.

29. $(3x^5 + 5x^3 - 5x^2 - 3) +$
$(x^5 + x^4 - 3x^3 - 3x^2 + 2x - 4)$

30. $\left(x^4 + \frac{2}{3}x + 5\right) + \left(4x^4 + 5x^2 + \frac{1}{3}x\right)$

Subtract.

31. $(2x^4 + x^3 - 8x^2 - 6x - 3) - (6x^4 - 8x^2 + 2x)$

32. $(x^3 - 0.4x^2 - 12) - (x^5 + 0.3x^3 + 0.4x^2 + 9)$

Multiply.

33. $-3x^2(4x^2 - 3x - 5)$

34. $\left(x - \frac{1}{3}\right)^2$

35. $(3x + 10)(3x - 10)$

36. $(3b + 5)(b - 3)$

37. $(x^6 - 4)(x^8 + 4)$

38. $(8 - y)(6 + 5y)$

39. $(2x + 1)(3x^2 - 5x - 3)$

40. $(5t + 2)^2$

41. Collect like terms: $x^3y - y^3 + xy^3 + 8 - 6x^3y - x^2y^2 + 11.$

42. Subtract: $(8a^2b^2 - ab + b^3) - (-6ab^2 - 7ab - ab^3 + 5b^3).$

43. Multiply: $(3x^5 - 4y^5)(3x^5 + 4y^5).$

Divide.

44. $(12x^4 + 9x^3 - 15x^2) \div (3x^2)$

45. $(6x^3 - 8x^2 - 14x + 13) \div (3x + 2)$

Divide using synthetic division. Show your work.

46. $(x^3 + 3x^2 + 2x - 6) \div (x - 3)$

47. $(4x^3 - 6x^2 - 9) \div (x + 5)$

48. The graph of the polynomial equation $y = x^3 - 5x - 1$ is shown at right. Use *only* the graph to estimate the value of the polynomial when $x = -1$, $x = -0.5$, $x = 0.5$, $x = 1$, and $x = 1.1$.

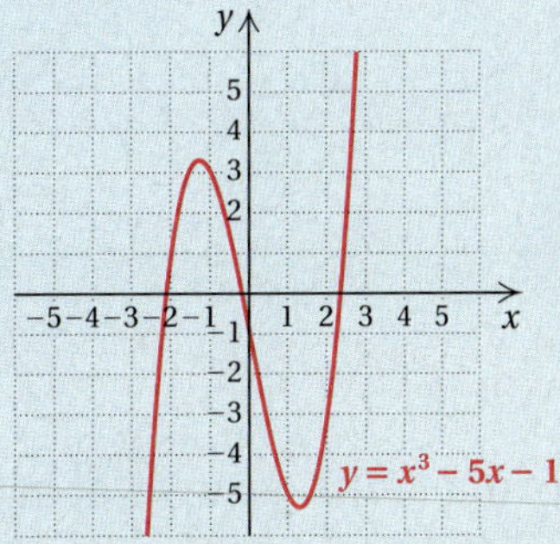

49. Find a polynomial for the surface area of this right rectangular solid.

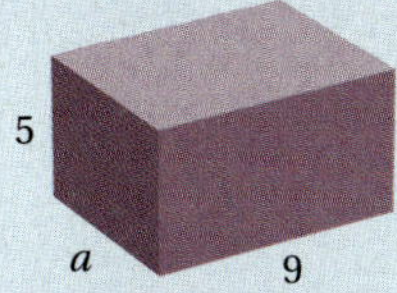

50. Find two algebraic expressions for the area of this figure. First, regard the figure as one large rectangle, and then regard the figure as a sum of four smaller rectangles.

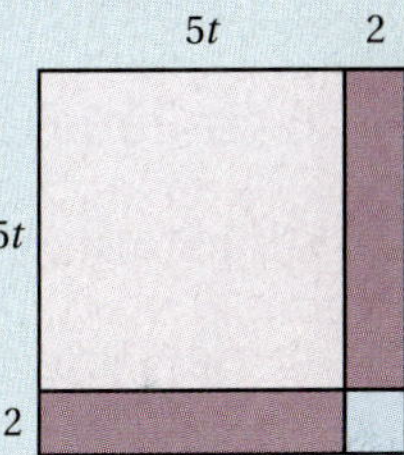

SKILL MAINTENANCE

51. Solve: $7x - 4x - 2 = 37$.

52. Solve: $4(x + 2) - 21 = 3(x - 6) + 2$.

53. Factor: $64t - 32m + 16$.

54. Subtract: $\frac{2}{5} - \left(-\frac{3}{4}\right)$.

55. The first angle of a triangle is four times as large as the second. The measure of the third angle is 30° greater than that of the second. How large are the angles?

SYNTHESIS

56. The height of a box is 1 less than its length, and the length is 2 more than its width. Find the volume in terms of the length.

57. Solve: $(x - 5)(x + 5) = (x + 6)^2$.

Polynomials: Factoring

Gateway to Chapter 10

Factoring is the reverse of multiplying. To factor a polynomial or other algebraic expression is to find an equivalent expression that is a product. In this chapter, we study the very important skill of factoring polynomials. To learn to factor quickly, we use the quick methods for multiplication that we learned in Chapter 9.

In this chapter, we introduce a new equation-solving technique to be used to solve equations involving quadratic, or second-degree, polynomials. This leads us to new ways of problem solving in Section 10.8. Problems we consider there could not have been solved easily without these new skills.

Real-World Application

The height of a triangular sail on a racing sailboat is 9 ft more than the base. The area of the triangle is 110 ft^2. Find the height and the base of the sail.

Source: Whitney Gladstone, North Graphics, San Diego, California

This problem appears as Example 2 in Section 10.8.

CHAPTER 10

CHAPTER 10

Pretest

1. Find three factorizations of $-20x^6$. [10.1a]

Factor.

2. $2x^2 + 4x + 2$ [10.5b]

3. $x^2 + 6x + 8$ [10.2a]

4. $8a^5 + 4a^3 - 20a$ [10.1b]

5. $-6 + 5x^2 - 13x$ [10.3a], [10.4a]

6. $81 - z^4$ [10.5d]

7. $y^6 - 4y^3 + 4$ [10.5b]

8. $3x^3 + 2x^2 + 12x + 8$ [10.1c]

9. $p^2 - p - 30$ [10.2a]

10. $x^4y^2 - 64$ [10.5d]

11. $2p^2 + 7pq - 4q^2$ [10.3a], [10.4a]

12. $y^3 - 64$ [10.6a]

Solve.

13. $x^2 - 5x = 0$ [10.8b]

14. $(x - 4)(5x - 3) = 0$ [10.8a]

15. $3x^2 + 10x - 8 = 0$ [10.8b]

16. $(x + 2)(x - 2) = 5$ [10.8b]

Solve.

17. Dimensions of a Triangle. The height of a triangle is 3 cm longer than the base. The area of the triangle is 44 cm^2. Find the base and the height. [10.9a]

18. Framing. A rectangular picture frame is twice as long as it is wide. The area of the frame is 162 in^2. Find the dimensions of the frame. [10.9a]

19. Right-Triangle Geometry. Find the length of the missing side of this right triangle. [10.9b]

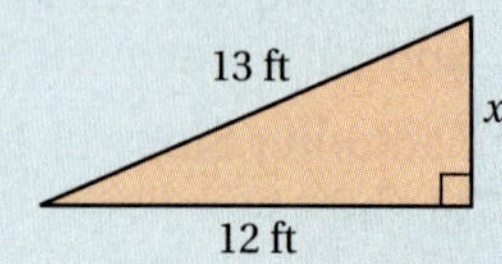

10.1 INTRODUCTION TO FACTORING

Objectives

a Factor monomials.

b Factor polynomials when the terms have a common factor, factoring out the largest common factor.

c Factor certain expressions with four terms using factoring by grouping.

To solve certain types of algebraic equations involving polynomials of second degree, we must learn to factor polynomials.

Consider the product $15 = 3 \cdot 5$. We say that 3 and 5 are **factors** of 15 and that $3 \cdot 5$ is a **factorization** of 15. Since $15 = 15 \cdot 1$, we also know that 15 and 1 are factors of 15 and that $15 \cdot 1$ is a factorization of 15. We use the word "factor" as both a verb and a noun.

FACTOR; FACTORIZATION

To **factor** a polynomial is to express it as a product.

A **factor** of a polynomial P is a polynomial that can be used to express P as a product.

A **factorization** of a polynomial is an expression that names that polynomial as a product.

1. a) Multiply: $(3x)(4x)$.

b) Factor: $12x^2$.

2. a) Multiply: $(2x)(-8x^2)$.

b) Factor: $-16x^3$.

a Factoring Monomials

To factor a monomial, we find two monomials whose product is equivalent to the original monomial. Compare.

	Multiplying	*Factoring*
a)	$(4x)(5x) = 20x^2$	$20x^2 = (4x)(5x)$
b)	$(2x)(10x) = 20x^2$	$20x^2 = (2x)(10x)$
c)	$(-4x)(-5x) = 20x^2$	$20x^2 = (-4x)(-5x)$
d)	$(x)(20x) = 20x^2$	$20x^2 = (x)(20x)$

In each case, we have expressed $20x^2$ as a *product.*

You can see that the monomial $20x^2$ has many factorizations. There are still other ways to factor $20x^2$.

Do Exercises 1 and 2.

EXAMPLE 1 Find three factorizations of $15x^3$.

a) $15x^3 = (3 \cdot 5)(x \cdot x^2)$
$= (3x)(5x^2)$ ← This is a *product.*

b) $15x^3 = (3 \cdot 5)(x^2 \cdot x)$
$= (3x^2)(5x)$

c) $15x^3 = (-15)(-1)x^3$
$= (-15)(-x^3)$

Do Exercises 3–5.

Find three factorizations of the monomial.

3. $8x^4$

4. $-21x^2$

5. $6x^5$

b Factoring When Terms Have a Common Factor

To multiply a monomial and a polynomial with more than one term, we multiply each term of the polynomial by the monomial using the distributive laws,

$$a(b + c) = ab + ac \quad \text{and} \quad a(b - c) = ab - ac.$$

Answers on page A-32

6. a) Multiply: $3(x + 2)$.

b) Factor: $3x + 6$.

7. a) Multiply: $2x(x^2 + 5x + 4)$.

b) Factor: $2x^3 + 10x^2 + 8x$.

Answers on page A-32

To factor, we do the reverse. We express a polynomial as a product using the distributive laws in reverse:

$$ab + ac = a(b + c) \quad \text{and} \quad ab - ac = a(b - c).$$

Compare.

Multiply	*Factor*
$3x(x^2 + 2x - 4)$	$3x^3 + 6x^2 - 12x$
$= 3x \cdot x^2 + 3x \cdot 2x - 3x \cdot 4$	$= 3x \cdot x^2 + 3x \cdot 2x - 3x \cdot 4$
$= 3x^3 + 6x^2 - 12x$	$= 3x(x^2 + 2x - 4)$

Do Exercises 6 and 7.

CAUTION!

Consider the following:

$$3x^3 + 6x^2 - 12x = 3 \cdot x \cdot x \cdot x + 2 \cdot 3 \cdot x \cdot x - 2 \cdot 2 \cdot 3 \cdot x.$$

The terms of the polynomial, $3x^3$, $6x^2$, and $-12x$, have been factored but the polynomial itself has not been factored. This is not what we mean by a factorization of the polynomial. The *factorization* is

$$3x(x^2 + 2x - 4).$$

The expressions $3x$ and $x^2 + 2x - 4$ are *factors* of $3x^3 + 6x^2 - 12x$.

To factor, we first try to find a factor common to all terms. There may not be one other than 1. When there is, we generally use the factor with the largest possible coefficient and the largest possible exponent.

EXAMPLE 2 Factor: $7x^2 + 14$.

We have

$$7x^2 + 14 = 7 \cdot x^2 + 7 \cdot 2 \qquad \text{Factoring each term}$$
$$= 7(x^2 + 2). \qquad \text{Factoring out the common factor 7}$$

CHECK: We multiply to check:

$$7(x^2 + 2) = 7 \cdot x^2 + 7 \cdot 2 = 7x^2 + 14.$$

EXAMPLE 3 Factor: $16x^3 + 20x^2$.

$$16x^3 + 20x^2 = (4x^2)(4x) + (4x^2)(5) \qquad \text{Factoring each term}$$
$$= 4x^2(4x + 5) \qquad \text{Factoring out the common factor } 4x^2$$

Suppose in Example 3 that you had not recognized the largest common factor and removed only part of it, as follows:

$$16x^3 + 20x^2 = (2x^2)(8x) + (2x^2)(10)$$
$$= 2x^2(8x + 10).$$

Note that $8x + 10$ still has a common factor of 2. You need not begin again. Just continue factoring out common factors, as follows, until finished:

$$= 2x^2[2(4x + 5)]$$
$$= 4x^2(4x + 5).$$

EXAMPLE 4 Factor: $15x^5 - 12x^4 + 27x^3 - 3x^2$.

$$15x^5 - 12x^4 + 27x^3 - 3x^2 = (3x^2)(5x^3) - (3x^2)(4x^2) + (3x^2)(9x) - (3x^2)(1)$$
$$= 3x^2(5x^3 - 4x^2 + 9x - 1) \quad \text{Factoring out } 3x^2$$

CAUTION! Don't forget the term -1.

CHECK: We multiply to check:

$$3x^2(5x^3 - 4x^2 + 9x - 1)$$
$$= (3x^2)(5x^3) - (3x^2)(4x^2) + (3x^2)(9x) - (3x^2)(1)$$
$$= 15x^5 - 12x^4 + 27x^3 - 3x^2.$$

As you become more familiar with factoring, you will be able to spot the largest common factor without factoring each term. Then you can write just the answer.

EXAMPLES Factor.

5. $8m^3 - 16m = 8m(m^2 - 2)$
6. $14p^2y^3 - 8py^2 + 2py = 2py(7py^2 - 4y + 1)$
7. $\frac{4}{5}x^2 + \frac{1}{5}x + \frac{2}{5} = \frac{1}{5}(4x^2 + x + 2)$
8. $2.4x^2 + 1.2x - 3.6 = 1.2(2x^2 + x - 3)$

Do Exercises 8–13.

There are two important points to keep in mind as we study this chapter.

TIPS FOR FACTORING

- Before doing any other kind of factoring, first try to factor out the largest common factor.
- Always check the result of factoring by multiplying.

C Factoring by Grouping: Four Terms

Certain polynomials with four terms can be factored using a method called *factoring by grouping.*

EXAMPLE 9 Factor: $x^2(x + 1) + 2(x + 1)$.

The binomial $x + 1$ is common to both terms:

$$x^2(x + 1) + 2(x + 1) = (x^2 + 2)(x + 1).$$

The factorization is $(x^2 + 2)(x + 1)$.

Do Exercises 14 and 15.

Factor. Check by multiplying.

8. $x^2 + 3x$

9. $3y^6 - 5y^3 + 2y^2$

10. $9x^4 - 15x^3 + 3x^2$

11. $\frac{3}{4}t^3 + \frac{5}{4}t^2 + \frac{7}{4}t + \frac{1}{4}$

12. $35x^7 - 49x^6 + 14x^5 - 63x^3$

13. $8.4x^2 - 5.6x + 2.8$

Factor.

14. $x^2(x + 7) + 3(x + 7)$

15. $x^2(a + b) + 2(a + b)$

Answers on page A-32

Factor by grouping.

16. $x^3 + 7x^2 + 3x + 21$

17. $8t^3 + 2t^2 + 12t + 3$

18. $3m^5 - 15m^3 + 2m^2 - 10$

19. $3x^3 - 6x^2 - x + 2$

20. $4x^3 - 6x^2 - 6x + 9$

21. $y^4 - 2y^3 - 2y - 10$

Answers on page A-32

Consider the four-term polynomial

$$x^3 + x^2 + 2x + 2.$$

There is no factor other than 1 that is common to all the terms. We can, however, factor $x^3 + x^2$ and $2x + 2$ separately:

$x^3 + x^2 = x^2(x + 1)$; Factoring $x^3 + x^2$

$2x + 2 = 2(x + 1)$. Factoring $2x + 2$

We have grouped certain terms and factored each polynomial separately:

$$\begin{aligned} x^3 + x^2 + 2x + 2 &= (x^3 + x^2) + (2x + 2) \\ &= x^2(x + 1) + 2(x + 1) \\ &= (x^2 + 2)(x + 1), \end{aligned}$$

as in Example 9. This method is called **factoring by grouping.** We began with a polynomial with four terms. After grouping and removing common factors, we obtained a polynomial with two parts, each having a common factor $x + 1$. Not all polynomials with four terms can be factored by this procedure, but it does give us a method to try.

EXAMPLES Factor by grouping.

10. $6x^3 - 9x^2 + 4x - 6$

$= (6x^3 - 9x^2) + (4x - 6)$

$= 3x^2(2x - 3) + 2(2x - 3)$ Factoring each binomial

$= (3x^2 + 2)(2x - 3)$ Factoring out the common factor $2x - 3$

We think through this process as follows:

$$6x^3 - 9x^2 + 4x - 6 = 3x^2(2x - 3) \; \square \; (2x - 3)$$

(1) Factor the first two terms.

(2) This factor, $2x - 3$, gives us a hint to the factorization of the last two terms.

(3) Now we ask ourselves, "What needs to be here to enable us to get $4x - 6$ when we multiply?"

11. $x^3 + x^2 + x + 1 = (x^3 + x^2) + (x + 1)$

$= x^2(x + 1) + 1(x + 1)$ Factoring each binomial

$= (x^2 + 1)(x + 1)$ Factoring out the common factor $x + 1$

CAUTION! Don't forget the 1.

12. $2x^3 - 6x^2 - x + 3$

$= (2x^3 - 6x^2) + (-x + 3)$

$= 2x^2(x - 3) - 1(x - 3)$ *Check:* $-1(x - 3) = -x + 3$.

$= (2x^2 - 1)(x - 3)$ Factoring out the common factor $x - 3$

13. $12x^5 + 20x^2 - 21x^3 - 35 = 4x^2(3x^3 + 5) - 7(3x^3 + 5)$

$= (4x^2 - 7)(3x^3 + 5)$

14. $x^3 + x^2 + 2x - 2 = x^2(x + 1) + 2(x - 1)$

This polynomial is not factorable using factoring by grouping. It may be factorable, but not by methods that we will consider in this text.

Do Exercises 16–21.

10.1 EXERCISE SET

For Extra Help

Digital Video Tutor CD 8 Videotape 11 | InterAct Math | Math Tutor Center | MathXL | MyMathLab

a Find three factorizations for the monomial.

1. $8x^3$

2. $6x^4$

3. $-10a^6$

4. $-8y^5$

5. $24x^4$

6. $15x^5$

b Factor. Check by multiplying.

7. $x^2 - 6x$

8. $x^2 + 5x$

9. $2x^2 + 6x$

10. $8y^2 - 8y$

11. $x^3 + 6x^2$

12. $3x^4 - x^2$

13. $8x^4 - 24x^2$

14. $5x^5 + 10x^3$

15. $2x^2 + 2x - 8$

16. $8x^2 - 4x - 20$

17. $17x^5y^3 + 34x^3y^2 + 51xy$

18. $16p^6q^4 + 32p^5q^3 - 48pq^2$

19. $6x^4 - 10x^3 + 3x^2$

20. $5x^5 + 10x^2 - 8x$

21. $x^5y^5 + x^4y^3 + x^3y^3 - x^2y^2$

22. $x^9y^6 - x^7y^5 + x^4y^4 + x^3y^3$

23. $2x^7 - 2x^6 - 64x^5 + 4x^3$

24. $8y^3 - 20y^2 + 12y - 16$

25. $1.6x^4 - 2.4x^3 + 3.2x^2 + 6.4x$

26. $2.5x^6 - 0.5x^4 + 5x^3 + 10x^2$

27. $\frac{5}{3}x^6 + \frac{4}{3}x^5 + \frac{1}{3}x^4 + \frac{1}{3}x^3$

28. $\frac{5}{9}x^7 + \frac{2}{9}x^5 - \frac{4}{9}x^3 - \frac{1}{9}x$

c Factor.

29. $x^2(x + 3) + 2(x + 3)$

30. $3z^2(2z + 1) + (2z + 1)$

31. $5a^3(2a - 7) - (2a - 7)$

32. $m^4(8 - 3m) - 7(8 - 3m)$

Factor by grouping.

33. $x^3 + 3x^2 + 2x + 6$

34. $6z^3 + 3z^2 + 2z + 1$

35. $2x^3 + 6x^2 + x + 3$

36. $3x^3 + 2x^2 + 3x + 2$

37. $8x^3 - 12x^2 + 6x - 9$

38. $10x^3 - 25x^2 + 4x - 10$

39. $12x^3 - 16x^2 + 3x - 4$

40. $18x^3 - 21x^2 + 30x - 35$

41. $5x^3 - 5x^2 - x + 1$

42. $7x^3 - 14x^2 - x + 2$

43. $x^3 + 8x^2 - 3x - 24$

44. $2x^3 + 12x^2 - 5x - 30$

45. $2x^3 - 8x^2 - 9x + 36$

46. $20g^3 - 4g^2 - 25g + 5$

47. D$_W$ Josh says that there is no need to print answers for Exercises 1–46 at the back of the book. Is he correct in saying this? Why or why not?

48. D$_W$ Explain how one could construct a polynomial with four terms that can be factored by grouping.

SKILL MAINTENANCE

Solve.

49. $-2x < 48$ [7.7d]

50. $4x - 8x + 16 \geq 6(x - 2)$ [7.7e]

51. Divide: $\frac{-108}{-4}$. [6.6a]

52. Solve $A = \frac{p + q}{2}$ for p. [7.4b]

Multiply. [9.6d]

53. $(y + 5)(y + 7)$

54. $(y + 7)^2$

55. $(y + 7)(y - 7)$

56. $(y - 7)^2$

Find the intercepts of the equation. Then graph the equation. [8.3a]

57. $x + y = 4$

58. $x - y = 3$

59. $5x - 3y = 15$

60. $y - 3x = 6$

SYNTHESIS

Factor.

61. $4x^5 + 6x^3 + 6x^2 + 9$

62. $x^6 + x^4 + x^2 + 1$

63. $x^{12} + x^7 + x^5 + 1$

64. $x^3 - x^2 - 2x + 5$

65. $p^3 + p^2 - 3p + 10$

10.2 FACTORING TRINOMIALS OF THE TYPE $x^2 + bx + c$

a Factoring $x^2 + bx + c$

We now begin a study of the factoring of trinomials. We first factor trinomials like

$$x^2 + 5x + 6 \quad \text{and} \quad x^2 + 3x - 10$$

by a refined *trial-and-error process*. In this section, we restrict our attention to trinomials of the type $ax^2 + bx + c$, where $a = 1$. The coefficient a is often called the **leading coefficient.**

To understand the factoring that follows, compare the following multiplications:

$$\begin{aligned}
& \quad\;\;\; \text{F} \quad\;\; \text{O} \quad\;\; \text{I} \quad\;\;\; \text{L} \\
(x + 2)(x + 5) &= x^2 + 5x + 2x + 2 \cdot 5 \\
&= x^2 + \quad 7x \quad + \; 10; \\
(x - 2)(x - 5) &= x^2 - 5x - 2x + 2 \cdot 5 \\
&= x^2 - \quad 7x \quad + \; 10; \\
(x + 3)(x - 7) &= x^2 - 7x + 3x + 3(-7) \\
&= x^2 - \quad 4x \quad - \; 21; \\
(x - 3)(x + 7) &= x^2 + 7x - 3x + (-3)7 \\
&= x^2 + \quad 4x \quad - \; 21.
\end{aligned}$$

Note that for all four products:

- The product of the two binomials is a trinomial.
- The coefficient of x in the trinomial is the sum of the constant terms in the binomials.
- The constant term in the trinomial is the product of the constant terms in the binomials.

These observations lead to a method for factoring certain trinomials. The first type we consider has a positive constant term, just as in the first two multiplications above.

CONSTANT TERM POSITIVE

To factor $x^2 + 7x + 10$, we think of FOIL in reverse. We multiplied x times x to get the first term of the trinomial, so we know that the first term of each binomial factor is x. Next, we look for numbers p and q such that

$$x^2 + 7x + 10 = (x + p)(x + q).$$

To get the middle term and the last term of the trinomial, we look for two numbers p and q whose product is 10 and whose sum is 7. Those numbers are 2 and 5. Thus the factorization is

$$(x + 2)(x + 5).$$

CHECK:
$$\begin{aligned}(x + 2)(x + 5) &= x^2 + 5x + 2x + 10 \\ &= x^2 + 7x + 10.\end{aligned}$$

Objective

a Factor trinomials of the type $x^2 + bx + c$ by examining the constant term c.

1. Consider the trinomial $x^2 + 7x + 12$.

a) Complete the following table.

PAIRS OF FACTORS	SUMS OF FACTORS
1, 12	13
−1, −12	
2, 6	
−2, −6	
3, 4	
−3, −4	

b) Explain why you need to consider only positive factors, as in the following table.

PAIRS OF FACTORS	SUMS OF FACTORS
1, 12	
2, 6	
3, 4	

c) Factor: $x^2 + 7x + 12$.

2. Factor: $x^2 + 13x + 36$.

Answers on page A-33

3. Explain why you would *not* consider the pairs of factors listed below in factoring $y^2 - 8y + 12$.

PAIRS OF FACTORS	SUMS OF FACTORS
1, 12	
2, 6	
3, 4	

Factor.

4. $x^2 - 8x + 15$

5. $t^2 - 9t + 20$

Answers on page A-33

EXAMPLE 1 Factor: $x^2 + 5x + 6$.

Think of FOIL in reverse. The first term of each factor is x: $(x + \square)(x + \square)$. Next, we look for two numbers whose product is 6 and whose sum is 5. All the pairs of factors of 6 are shown in the table on the left below. Since both the product, 6, and the sum, 5, of the pair of numbers must be positive, we need consider only the positive factors, listed in the table on the right.

PAIRS OF FACTORS	SUMS OF FACTORS
1, 6	7
−1, −6	−7
2, 3	5
−2, −3	−5

PAIRS OF FACTORS	SUMS OF FACTORS
1, 6	7
2, 3	5

The numbers we need are 2 and 3.

The factorization is $(x + 2)(x + 3)$. We can check by multiplying to see whether we get the original trinomial.

CHECK: $(x + 2)(x + 3) = x^2 + 3x + 2x + 6 = x^2 + 5x + 6$.

Do Exercises 1 and 2 on the preceding page.

Compare these multiplications:

$$(x - 2)(x - 5) = x^2 - 5x - 2x + 10 = x^2 - 7x + 10;$$
$$(x + 2)(x + 5) = x^2 + 5x + 2x + 10 = x^2 + 7x + 10.$$

TO FACTOR $x^2 + bx + c$ WHEN c IS POSITIVE

When the constant term of a trinomial is positive, look for two numbers with the same sign. The sign is that of the middle term:

$$x^2 - 7x + 10 = (x - 2)(x - 5);$$

$$x^2 + 7x + 10 = (x + 2)(x + 5).$$

EXAMPLE 2 Factor: $y^2 - 8y + 12$.

Since the constant term, 12, is positive and the coefficient of the middle term, −8, is negative, we look for a factorization of 12 in which both factors are negative. Their sum must be −8.

PAIRS OF FACTORS	SUMS OF FACTORS
−1, −12	−13
−2, −6	−8
−3, −4	−7

The numbers we need are −2 and −6.

The factorization is $(y - 2)(y - 6)$.

Do Exercises 3–5.

CONSTANT TERM NEGATIVE

As we saw in two of the multiplications earlier in this section, the product of two binomials can have a negative constant term:

$$(x + 3)(x - 7) = x^2 - 4x - 21$$

and

$$(x - 3)(x + 7) = x^2 + 4x - 21.$$

Note that when the signs of the constants in the binomials are reversed, only the sign of the middle term in the product changes.

EXAMPLE 3 Factor: $x^2 - 8x - 20$.

The constant term, -20, must be expressed as the product of a negative number and a positive number. Since the sum of these two numbers must be negative (specifically, -8), the negative number must have the greater absolute value.

PAIRS OF FACTORS	SUMS OF FACTORS
1, −20	−19
2, −10	−8
4, −5	−1
5, −4	1
10, −2	8
20, −1	19

The numbers we need are 2 and −10.

Because these sums are all positive, for this problem all of the corresponding pairs can be disregarded. Note that in all three pairs, the positive number has the greater absolute value.

The numbers that we are looking for are 2 and −10. The factorization is $(x + 2)(x - 10)$.

CHECK: $(x + 2)(x - 10) = x^2 - 10x + 2x - 20$
$= x^2 - 8x - 20.$

TO FACTOR $x^2 + bx + c$ WHEN c IS NEGATIVE

When the constant term of a trinomial is negative, look for two numbers whose product is negative. One must be positive and the other negative:

$$x^2 - 4x - 21 = (x + 3)(x - 7);$$

$$x^2 + 4x - 21 = (x - 3)(x + 7).$$

Consider pairs of numbers for which the number with the larger absolute value has the same sign as b, the coefficient of the middle term.

Do Exercises 6 and 7. (Exercise 7 is on the following page.)

6. Consider $x^2 - 5x - 24$.

a) Explain why you would *not* consider the pairs of factors listed below in factoring $x^2 - 5x - 24$.

PAIRS OF FACTORS	SUMS OF FACTORS
−1, 24	
−2, 12	
−3, 8	
−4, 6	

b) Explain why you *would* consider the pairs of factors listed below in factoring $x^2 - 5x - 24$.

PAIRS OF FACTORS	SUMS OF FACTORS
1, −24	
2, −12	
3, −8	
4, −6	

c) Factor: $x^2 - 5x - 24$.

Answers on page A-33

7. Consider $x^2 + 10x - 24$.

a) Explain why you would *not* consider the pairs of factors listed below in factoring $x^2 + 10x - 24$.

PAIRS OF FACTORS	SUMS OF FACTORS
1, −24	
2, −12	
3, −8	
4, −6	

b) Explain why you *would* consider the pairs of factors listed below in factoring $x^2 + 10x - 24$.

PAIRS OF FACTORS	SUMS OF FACTORS
−1, 24	
−2, 12	
−3, 8	
−4, 6	

c) Factor: $x^2 + 10x - 24$.

Factor.

8. $a^2 - 24 + 10a$

9. $-24 - 10t + t^2$

Answers on page A-33

EXAMPLE 4 Factor: $t^2 - 24 + 5t$.

It helps to first write the trinomial in descending order: $t^2 + 5t - 24$. Since the constant term, -24, is negative, we look for a factorization of -24 in which one factor is positive and one factor is negative. Their sum must be 5, so the positive factor must have the larger absolute value. Thus we consider only pairs of factors in which the positive term has the larger absolute value.

PAIRS OF FACTORS	SUMS OF FACTORS
−1, 24	23
−2, 12	10
−3, 8	5
−4, 6	2

The numbers we need are −3 and 8.

The factorization is $(t - 3)(t + 8)$. The check is left to the student.

Do Exercises 8 and 9.

EXAMPLE 5 Factor: $x^4 - x^2 - 110$.

Consider this trinomial as $(x^2)^2 - x^2 - 110$. We look for numbers p and q such that

$$x^4 - x^2 - 110 = (x^2 + p)(x^2 + q).$$

Since the constant term, -110, is negative, we look for a factorization of -110 in which one factor is positive and one factor is negative. Their sum must be -1. The middle-term coefficient, -1, is small compared to -110. This tells us that the desired factors are close to each other in absolute value. The numbers we want are 10 and -11. The factorization is

$$(x^2 + 10)(x^2 - 11).$$

EXAMPLE 6 Factor: $a^2 + 4ab - 21b^2$.

We consider the trinomial in the equivalent form

$$a^2 + 4ba - 21b^2.$$

This way we think of $-21b^2$ as the "constant" term and $4b$ as the "coefficient" of the middle term. Then we try to express $-21b^2$ as a product of two factors whose sum is $4b$. Those factors are $-3b$ and $7b$. The factorization is $(a - 3b)(a + 7b)$.

CHECK: $(a - 3b)(a + 7b) = a^2 + 7ab - 3ba - 21b^2$
$= a^2 + 4ab - 21b^2$.

There are polynomials that are not factorable.

EXAMPLE 7 Factor: $x^2 - x + 5$.

Since 5 has very few factors, we can easily check all possibilities.

PAIRS OF FACTORS	SUMS OF FACTORS
5, 1	6
−5, −1	−6

There are no factors whose sum is -1. Thus the polynomial is *not* factorable into factors that are polynomials.

In this text, a polynomial like $x^2 - x + 5$ that cannot be factored further is said to be **prime.** In more advanced courses, polynomials like $x^2 - x + 5$ can be factored and are not considered prime.

Do Exercises 10–12.

Often factoring requires two or more steps. In general, when told to factor, we should *factor completely.* This means that the final factorization should not contain any factors that can be factored further.

EXAMPLE 8 Factor: $2x^3 - 20x^2 + 50x$.

Always look first for a common factor. This time there is one, $2x$, which we factor out first:

$$2x^3 - 20x^2 + 50x = 2x(x^2 - 10x + 25).$$

Now consider $x^2 - 10x + 25$. Since the constant term is positive and the coefficient of the middle term is negative, we look for a factorization of 25 in which both factors are negative. Their sum must be -10.

PAIRS OF FACTORS	SUMS OF FACTORS
$-25, -1$	-26
$-5, -5$	-10 ← The numbers we need are -5 and -5.

The factorization of $x^2 - 10x + 25$ is $(x - 5)(x - 5)$, or $(x - 5)^2$. The final factorization is $2x(x - 5)^2$.

Do Exercises 13–15.

Once any common factors have been factored out, the following summary can be used to factor $x^2 + bx + c$.

TO FACTOR $x^2 + bx + c$

1. First arrange in descending order.
2. Use a trial-and-error process that looks for factors of c whose sum is b.
3. If c is positive, the signs of the factors are the same as the sign of b.
4. If c is negative, one factor is positive and the other is negative. If the sum of two factors is the opposite of b, changing the sign of each factor will give the desired factors whose sum is b.
5. Check by multiplying.

Factor.

10. $y^2 - 12 - 4y$

11. $t^4 + 5t^2 - 14$

12. $x^2 + 2x + 7$

Factor.

13. $x^3 + 4x^2 - 12x$

14. $p^2 - pq - 3pq^2$

15. $3x^3 + 24x^2 + 48x$

Answers on page A-33

Factor.

16. $14 + 5x - x^2$

17. $-x^2 + 3x + 18$

Answers on page A-33

LEADING COEFFICIENT −1

EXAMPLE 9 Factor: $10 - 3x - x^2$.

Note that the polynomial is written in ascending order. When we write it in descending order, we get

$$-x^2 - 3x + 10,$$

which has a leading coefficient of -1. Before factoring in such a case, we can factor out a -1, as follows:

$$-x^2 - 3x + 10 = -1(x^2 + 3x - 10).$$

Then we proceed to factor $x^2 + 3x - 10$. We get

$$-x^2 - 3x + 10 = -1(x^2 + 3x - 10) = -1(x + 5)(x - 2).$$

We can also express this answer in two other ways by multiplying either binomial by -1. Thus each of the following is a correct answer:

$$-x^2 - 3x + 10 = -1(x + 5)(x - 2)$$
$$= (-x - 5)(x - 2) \quad \text{Multiplying } x + 5 \text{ by } -1$$
$$= (x + 5)(-x + 2). \quad \text{Multiplying } x - 2 \text{ by } -1$$

Do Exercises 16 and 17.

Study Tips

Here are some additional tips to help you with time management. (See also the Study Tips on time management in Sections 7.8 and 9.2.)

TIME MANAGEMENT (PART 3)

- **Are you a morning or an evening person?** If you are an evening person, it might be best to avoid scheduling early-morning classes. If you are a morning person, do the opposite, but go to bed earlier to compensate. Nothing can drain your study time and effectiveness like fatigue.
- **Keep on schedule.** Your course syllabus provides a plan for the semester's schedule. Use a write-on calendar, daily planner, laptop computer, or personal digital assistant to outline your time for the semester. Be sure to note deadlines involving term papers and exams so you can begin a task early, breaking it down into smaller segments that can be accomplished more easily.
- **Balance your class schedule.** You may be someone who prefers large blocks of time for study on the off days. In that case, it might be advantageous for you to take courses that meet only three days a week. Keep in mind, however, that this might be a problem when tests in more than one course are scheduled for the same day.

"Time is our most important asset, yet we tend to waste it, kill it, and spend it rather than invest it."

Jim Rohn, motivational speaker

10.2 EXERCISE SET

For Extra Help

Digital Video Tutor CD 8 Videotape 11 | InterAct Math | Math Tutor Center | MathXL | MyMathLab

a Factor. Remember that you can check by multiplying.

1. $x^2 + 8x + 15$

PAIRS OF FACTORS	SUMS OF FACTORS

2. $x^2 + 5x + 6$

PAIRS OF FACTORS	SUMS OF FACTORS

3. $x^2 + 7x + 12$

PAIRS OF FACTORS	SUMS OF FACTORS

4. $x^2 + 9x + 8$

PAIRS OF FACTORS	SUMS OF FACTORS

5. $x^2 - 6x + 9$

PAIRS OF FACTORS	SUMS OF FACTORS

6. $y^2 - 11y + 28$

PAIRS OF FACTORS	SUMS OF FACTORS

7. $x^2 - 5x - 14$

PAIRS OF FACTORS	SUMS OF FACTORS

8. $a^2 + 7a - 30$

PAIRS OF FACTORS	SUMS OF FACTORS

9. $b^2 + 5b + 4$

PAIRS OF FACTORS	SUMS OF FACTORS

10. $z^2 - 8z + 7$

PAIRS OF FACTORS	SUMS OF FACTORS

11. $x^2 + \frac{2}{3}x + \frac{1}{9}$

PAIRS OF FACTORS	SUMS OF FACTORS

12. $x^2 - \frac{2}{5}x + \frac{1}{25}$

PAIRS OF FACTORS	SUMS OF FACTORS

13. $d^2 - 7d + 10$

14. $t^2 - 12t + 35$

15. $y^2 - 11y + 10$

16. $x^2 - 4x - 21$

17. $x^2 + x + 1$

18. $x^2 + 5x + 3$

19. $x^2 - 7x - 18$

20. $y^2 - 3y - 28$

21. $x^3 - 6x^2 - 16x$

22. $x^3 - x^2 - 42x$

23. $y^3 - 4y^2 - 45y$

24. $x^3 - 7x^2 - 60x$

25. $-2x - 99 + x^2$

26. $x^2 - 72 + 6x$

27. $c^4 + c^2 - 56$

28. $b^4 + 5b^2 - 24$

29. $a^4 + 2a^2 - 35$

30. $x^4 - x^2 - 6$

31. $x^2 + x - 42$

32. $x^2 + 2x - 15$

33. $7 - 2p + p^2$

34. $11 - 3w + w^2$

35. $x^2 + 20x + 100$

36. $a^2 + 19a + 88$

37. $30 + 7x - x^2$

38. $45 + 4x - x^2$

39. $24 - a^2 - 10a$

40. $-z^2 + 36 - 9z$

41. $x^4 - 21x^3 - 100x^2$

42. $x^4 - 20x^3 + 96x^2$

43. $x^2 - 21x - 72$

44. $4x^2 + 40x + 100$

45. $x^2 - 25x + 144$

46. $y^2 - 21y + 108$

47. $a^2 + a - 132$

48. $a^2 + 9a - 90$

49. $120 - 23x + x^2$

50. $96 + 22d + d^2$

51. $108 - 3x - x^2$

52. $112 + 9y - y^2$

53. $y^2 - 0.2y - 0.08$

54. $t^2 - 0.3t - 0.10$

55. $p^2 + 3pq - 10q^2$

56. $a^2 + 2ab - 3b^2$

57. $84 - 8t - t^2$

58. $72 - 6m - m^2$

59. $m^2 + 5mn + 4n^2$

60. $x^2 + 11xy + 24y^2$

61. $s^2 - 2st - 15t^2$

62. $p^2 + 5pq - 24q^2$

63. $6a^{10} - 30a^9 - 84a^8$

64. $7x^9 - 28x^8 - 35x^7$

65. ^D_W Gwyneth factors $x^3 - 8x^2 + 15x$ as $(x^2 - 5x)(x - 3)$. Is she wrong? Why or why not? What advice would you offer?

66. ^D_W When searching for a factorization, why do we list pairs of numbers with the correct *product* instead of pairs of numbers with the correct *sum*?

67. ^D_W Without multiplying $(x - 17)(x - 18)$, explain why it cannot possibly be a factorization of $x^2 + 35x + 306$.

68. ^D_W What is the advantage of writing out the prime factorization of c when factoring $x^2 + bx + c$ with a large value of c?

SKILL MAINTENANCE

Multiply. [9.6d]

69. $8x(2x^2 - 6x + 1)$

70. $(7w + 6)(4w - 11)$

71. $(7w + 6)^2$

72. $(4w - 11)^2$

73. $(4w - 11)(4w + 11)$

74. Simplify: $(3x^4)^3$. [9.2a, b]

Solve. [7.3a]

75. $3x - 8 = 0$

76. $2x + 7 = 0$

Solve.

77. *Arrests for Counterfeiting.* In a recent year, 29,200 people were arrested for counterfeiting. This number was down 1.2% from the preceding year. How many people were arrested the preceding year? [7.5a]

78. The first angle of a triangle is four times as large as the second. The measure of the third angle is 30° greater than that of the second. Find the angle measures. [7.6a]

SYNTHESIS

79. Find all integers m for which $y^2 + my + 50$ can be factored.

80. Find all integers b for which $a^2 + ba - 50$ can be factored.

Factor completely.

81. $x^2 - \frac{1}{2}x - \frac{3}{16}$

82. $x^2 - \frac{1}{4}x - \frac{1}{8}$

83. $x^2 + \frac{30}{7}x - \frac{25}{7}$

84. $\frac{1}{3}x^3 + \frac{1}{3}x^2 - 2x$

85. $b^{2n} + 7b^n + 10$

86. $a^{2m} - 11a^m + 28$

Find a polynomial in factored form for the shaded area. (Leave answers in terms of π.)

87.

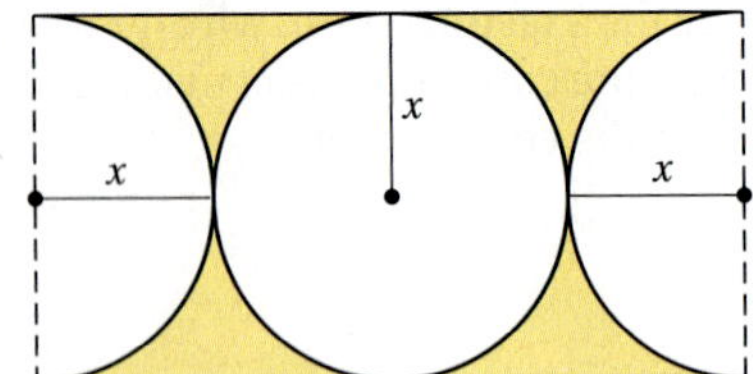

88.

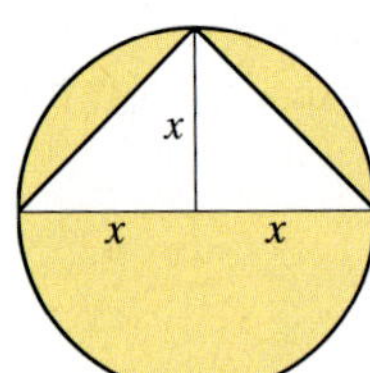

89. A census taker asks a woman, "How many children do you have?" "Three," she answers. "What are their ages?" She responds, "The product of their ages is 36. The sum of their ages is the house number next door." The math-savvy census taker walks next door, reads the house number, appears puzzled, and returns to the woman, asking, "Is there something you forgot to tell me?" "Oh yes," says the woman. "I'm sorry. The oldest child is at the park." The census taker records the three ages, thanks the woman for her time, and leaves. How old is each child? Explain how you reached this conclusion. (*Hint*: Consider factorizations.)

Source: Adapted from Anita Harnadek, *Classroom Quickies.* Pacific Grove, CA: Critical Thinking Press and Software

10.3 FACTORING $ax^2 + bx + c, a \neq 1$: THE FOIL METHOD

Objective

a Factor trinomials of the type $ax^2 + bx + c, a \neq 1$, using the FOIL method.

In Section 10.2, we learned a trial-and-error method to factor trinomials of the type $x^2 + bx + c$. In this section, we factor trinomials in which the coefficient of the leading term x^2 is not 1. The procedure we learn is a refined trial-and-error method.

a The FOIL Method

We want to factor trinomials of the type $ax^2 + bx + c$. Consider the following multiplication:

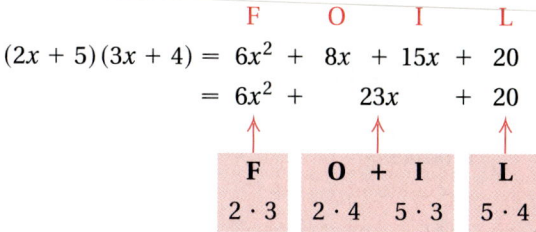

$$\begin{aligned}(2x+5)(3x+4) &= \overset{F}{6x^2} + \overset{O}{8x} + \overset{I}{15x} + \overset{L}{20} \\ &= 6x^2 + 23x + 20\end{aligned}$$

F	O + I	L
$2 \cdot 3$	$2 \cdot 4 \quad 5 \cdot 3$	$5 \cdot 4$

To factor $6x^2 + 23x + 20$, we reverse the above multiplication, using what we might call an "unFOIL" process. We look for two binomials $rx + p$ and $sx + q$ whose product is $(rx + p)(sx + q) = 6x^2 + 23x + 20$. The product of the First terms must be $6x^2$. The product of the Outside terms plus the product of the Inside terms must be $23x$. The product of the Last terms must be 20. We know from the preceding discussion that the answer is $(2x + 5)(3x + 4)$. Generally, however, finding such an answer is a refined trial-and-error process. It turns out that $(-2x - 5)(-3x - 4)$ is also a correct answer, but we generally choose an answer in which the first coefficients are positive.

We will use the following trial-and-error method.

THE FOIL METHOD

To factor $ax^2 + bx + c, a \neq 1$, using the FOIL method:

1. Factor out the largest common factor, if one exists.
2. Find two First terms whose product is ax^2.

   $(\square x + \quad)(\square x + \quad) = ax^2 + bx + c.$

   FOIL

3. Find two Last terms whose product is c:

   $(\quad x + \square)(\quad x + \square) = ax^2 + bx + c.$

   FOIL

4. Look for Outer and Inner products resulting from steps (2) and (3) for which the sum is bx:

   $(\square x + \square)(\square x + \square) = ax^2 + bx + c.$

   I, O — FOIL

5. Always check by multiplying.

To the student: In Section 10.4, we will consider an alternative method for the same kind of factoring. It involves factoring by grouping and is called the ac-method.

To the instructor: We present two ways to factor general trinomials in Sections 10.3 and 10.4: the FOIL method in Section 10.3 and the ac-method in Section 10.4. You can teach both methods and let the student use the one that he or she prefers or you can select just one. In the latter case, the exercise set that is not studied can be used for extra practice.

Factor.

1. $2x^2 - x - 15$

2. $12x^2 - 17x - 5$

Answers on page A-34

Study Tips

READING EXAMPLES

A careful study of the examples in these sections on factoring is critical. *Read them carefully* to ensure success!

EXAMPLE 1 Factor: $3x^2 - 10x - 8$.

1) First, we check for a common factor. Here there is none (other than 1 or -1).

2) Find two **F**irst terms whose product is $3x^2$.

The only possibilities for the **F**irst terms are $3x$ and x, so any factorization must be of the form

$$(3x + \square)(x + \square).$$

3) Find two **L**ast terms whose product is -8.

Possible factorizations of -8 are

$$(-8) \cdot 1, \quad 8 \cdot (-1), \quad (-2) \cdot 4, \quad \text{and} \quad 2 \cdot (-4).$$

Since the First terms are not identical, we must also consider

$$1 \cdot (-8), \quad (-1) \cdot 8, \quad 4 \cdot (-2), \quad \text{and} \quad (-4) \cdot 2.$$

4) Inspect the **O**uter and **I**nner products resulting from steps (2) and (3). Look for a combination in which the sum of the products is the middle term, $-10x$:

Trial	*Product*	
$(3x - 8)(x + 1)$	$3x^2 + 3x - 8x - 8$ $= 3x^2 - 5x - 8$	← Wrong middle term
$(3x + 8)(x - 1)$	$3x^2 - 3x + 8x - 8$ $= 3x^2 + 5x - 8$	← Wrong middle term
$(3x - 2)(x + 4)$	$3x^2 + 12x - 2x - 8$ $= 3x^2 + 10x - 8$	← Wrong middle term
$(3x + 2)(x - 4)$	$3x^2 - 12x + 2x - 8$ $= 3x^2 - 10x - 8$	← **Correct middle term!**
$(3x + 1)(x - 8)$	$3x^2 - 24x + x - 8$ $= 3x^2 - 23x - 8$	← Wrong middle term
$(3x - 1)(x + 8)$	$3x^2 + 24x - x - 8$ $= 3x^2 + 23x - 8$	← Wrong middle term
$(3x + 4)(x - 2)$	$3x^2 - 6x + 4x - 8$ $= 3x^2 - 2x - 8$	← Wrong middle term
$(3x - 4)(x + 2)$	$3x^2 + 6x - 4x - 8$ $= 3x^2 + 2x - 8$	← Wrong middle term

The correct factorization is $(3x + 2)(x - 4)$.

5) CHECK: $(3x + 2)(x - 4) = 3x^2 - 10x - 8$.

Two observations can be made from Example 1. First, we listed all possible trials even though we could have stopped after having found the correct factorization. We did this to show that each trial differs only in the middle term of the product. **Second, note that as in Section 10.2, only the sign of the middle term changes when the signs in the binomials are reversed:**

Plus Minus

$$(3x + 4)(x - 2) = 3x^2 - 2x - 8$$

Minus Plus ← Middle term changes sign

$$(3x - 4)(x + 2) = 3x^2 + 2x - 8.$$

Do Exercises 1 and 2.

EXAMPLE 2 Factor: $24x^2 - 76x + 40$.

1) First, we factor out the largest common factor, 4:

$$4(6x^2 - 19x + 10).$$

Now we factor the trinomial $6x^2 - 19x + 10$.

2) Because $6x^2$ can be factored as $3x \cdot 2x$ or $6x \cdot x$, we have these possibilities for factorizations:

$$(3x + \square)(2x + \square) \quad \text{or} \quad (6x + \square)(x + \square).$$

3) There are four pairs of factors of 10 and they each can be listed in two ways:

$$10, 1 \qquad -10, -1 \qquad 5, 2 \qquad -5, -2$$

and

$$1, 10 \qquad -1, -10 \qquad 2, 5 \qquad -2, -5.$$

4) The two possibilities from step (2) and the eight possibilities from step (3) give $2 \cdot 8$, or 16 possibilities for factorizations. We look for **O**uter and **I**nner products resulting from steps (2) and (3) for which the sum is the middle term, $-19x$. Since the sign of the middle term is negative, but the sign of the last term, 10, is positive, the two factors of 10 must both be negative. This means only four pairings from step (3) need be considered. We first try these factors with $(3x + \square)(2x + \square)$. If none gives the correct factorization, we will consider $(6x + \square)(x + \square)$.

Trial	*Product*	
$(3x - 10)(2x - 1)$	$6x^2 - 3x - 20x + 10$	
	$= 6x^2 - 23x + 10$	← Wrong middle term
$(3x - 1)(2x - 10)$	$6x^2 - 30x - 2x + 10$	
	$= 6x^2 - 32x + 10$	← Wrong middle term
$(3x - 5)(2x - 2)$	$6x^2 - 6x - 10x + 10$	
	$= 6x^2 - 16x + 10$	← Wrong middle term
$(3x - 2)(2x - 5)$	$6x^2 - 15x - 4x + 10$	
	$= 6x^2 - 19x + 10$	← **Correct middle term!**

Since we have a correct factorization, we need not consider

$$(6x + \square)(x + \square).$$

The factorization of $6x^2 - 19x + 10$ is $(3x - 2)(2x - 5)$, but *do not forget the common factor*! We must include it in order to factor the original trinomial:

$$24x^2 - 76x + 40 = 4(6x^2 - 19x + 10)$$
$$= 4(3x - 2)(2x - 5).$$

5) CHECK: $4(3x - 2)(2x - 5) = 4(6x^2 - 19x + 10) = 24x^2 - 76x + 40$.

CAUTION!

When factoring any polynomial, always look for a common factor. Failure to do so is such a common error that this caution bears repeating.

In Example 2, look again at the possibility $(3x - 5)(2x - 2)$. Without multiplying, we can reject such a possibility. To see why, consider the following:

$$(3x - 5)(2x - 2) = 2(3x - 5)(x - 1).$$

Factor.

3. $3x^2 - 19x + 20$

4. $20x^2 - 46x + 24$

5. Factor: $6x^2 + 7x + 2$.

Answers on page A-34

The expression $2x - 2$ has a common factor, 2. But we removed the *largest* common factor in the first step. If $2x - 2$ were one of the factors, then 2 would have to be a common factor in addition to the original 4. Thus, $(2x - 2)$ cannot be part of the factorization of the original trinomial.

Given that the largest common factor is factored out at the outset, we need not consider factorizations that have a common factor.

Do Exercises 3 and 4.

EXAMPLE 3 Factor: $10x^2 + 37x + 7$.

1) There is no common factor (other than 1 or -1).
2) Because $10x^2$ factors as $10x \cdot x$ or $5x \cdot 2x$, we have these possibilities for factorizations:

$$(10x + \square)(x + \square) \quad \text{or} \quad (5x + \square)(2x + \square).$$

3) There are two pairs of factors of 7 and they each can be listed in two ways:

$$1, 7 \qquad -1, -7 \qquad \text{and} \qquad 7, 1 \qquad -7, -1.$$

4) From steps (2) and (3), we see that there are 8 possibilities for factorizations. Look for **O**uter and **I**nner products for which the sum is the middle term. Because all coefficients in $10x^2 + 37x + 7$ are positive, we need consider only positive factors of 7. The possibilities are

$$(10x + 1)(x + 7) = 10x^2 + 71x + 7,$$
$$(10x + 7)(x + 1) = 10x^2 + 17x + 7,$$
$$(5x + 7)(2x + 1) = 10x^2 + 19x + 7,$$
$$(5x + 1)(2x + 7) = 10x^2 + 37x + 7. \quad \leftarrow \text{Correct middle term}$$

The factorization is $(5x + 1)(2x + 7)$.

5) CHECK: $(5x + 1)(2x + 7) = 10x^2 + 37x + 7$.

Keep in mind that this method of factoring trinomials of the type $ax^2 + bx + c$ involves *trial and error.* As you practice, you will find that you can make better and better guesses.

Do Exercise 5.

TIPS FOR FACTORING $ax^2 + bx + c$, $a \neq 1$

- Always factor out the largest common factor, if one exists.
- Once the common factor has been factored out of the original trinomial, no binomial factor can contain a common factor (other than 1 or -1).
- If c is positive, then the signs in both binomial factors must match the sign of b. (This assumes that $a > 0$.)
- Reversing the signs in the binomials reverses the sign of the middle term of their product.
- Organize your work so that you can keep track of which possibilities have or have not been checked.
- Always check by multiplying.

EXAMPLE 4 Factor: $10x + 8 - 3x^2$.

An important problem-solving strategy is to find a way to make new problems look like problems we already know how to solve. (See Example 9 in Section 10.2.) The factoring tips above apply only to trinomials of the form $ax^2 + bx + c$, with $a > 0$. This leads us to rewrite $10x + 8 - 3x^2$ in descending order:

$$10x + 8 - 3x^2 = -3x^2 + 10x + 8. \quad \text{Writing in descending order}$$

Although $-3x^2 + 10x + 8$ looks similar to the trinomials we have factored, the tips above require a positive leading coefficient. This can be attained by factoring out -1:

$$-3x^2 + 10x + 8 = -1(3x^2 - 10x - 8) \quad \text{Factoring out } -1 \text{ changes the signs of the coefficients.}$$

$$= -1(3x + 2)(x - 4). \quad \text{Using the result from Example 1}$$

The factorization of $10x + 8 - 3x^2$ is $-1(3x + 2)(x - 4)$. Other correct answers are

$$10x + 8 - 3x^2 = (3x + 2)(-x + 4) \quad \text{Multiplying } x - 4 \text{ by } -1$$

$$= (-3x - 2)(x - 4). \quad \text{Multiplying } 3x + 2 \text{ by } -1$$

Do Exercises 6 and 7.

EXAMPLE 5 Factor: $6p^2 - 13pq - 28q^2$.

1) Factor out a common factor, if any.

   There is none (other than 1 or -1).

2) Factor the first term, $6p^2$.

   Possibilities are $2p$, $3p$ and $6p$, p. We have these as possibilities for factorizations:

   $$(2p + \quad)(3p + \quad) \quad \text{or} \quad (6p + \quad)(p + \quad).$$

3) Factor the last term, $-28q^2$, which has a negative coefficient.

   The possibilities are $-14q$, $2q$ and $14q$, $-2q$; $-28q$, q and $28q$, $-q$; and $-7q$, $4q$ and $7q$, $-4q$.

4) The coefficient of the middle term is negative, so we look for combinations of factors from steps (2) and (3) such that the sum of their products has a negative coefficient. We try some possibilities:

   $$(2p + q)(3p - 28q) = 6p^2 - 53pq - 28q^2,$$
   $$(2p - 7q)(3p + 4q) = 6p^2 - 13pq - 28q^2. \quad \leftarrow \textbf{Correct middle term}$$

   The factorization of $6p^2 - 13pq - 28q^2$ is $(2p - 7q)(3p + 4q)$.

5) The check is left to the student.

Do Exercises 8 and 9.

Factor.

6. $2 - x - 6x^2$

7. $2x + 8 - 6x^2$

Factor.

8. $6a^2 - 5ab + b^2$

9. $6x^2 + 15xy + 9y^2$

Answers on page A-34

CALCULATOR CORNER

Checking Factorizations A partial check of a factorization can be performed using a table or a graph. To check the factorization $6x^3 - 9x^2 + 4x - 6 = (3x^2 + 2)(2x - 3)$, for example, we enter $y_1 = 6x^3 - 9x^2 + 4x - 6$ and $y_2 = (3x^2 + 2)(2x - 3)$ on the equation-editor screen (see page 532). Then we set up a table in AUTO mode (see page 610). If the factorization is correct, the values of y_1 and y_2 will be the same regardless of the table settings used.

X	Y1	Y2
−3	−261	−261
−2	−98	−98
−1	−25	−25
0	−6	−6
1	−5	−5
2	14	14
3	87	87
X = −3		

We can also graph $y_1 = 6x^3 - 9x^2 + 4x - 6$ and $y_2 = (3x^2 + 2)(2x - 3)$. If the graphs appear to coincide, the factorization is probably correct.

$y_1 = 6x^3 - 9x^2 + 4x - 6,$
$y_2 = (3x^2 + 2)(2x - 3)$

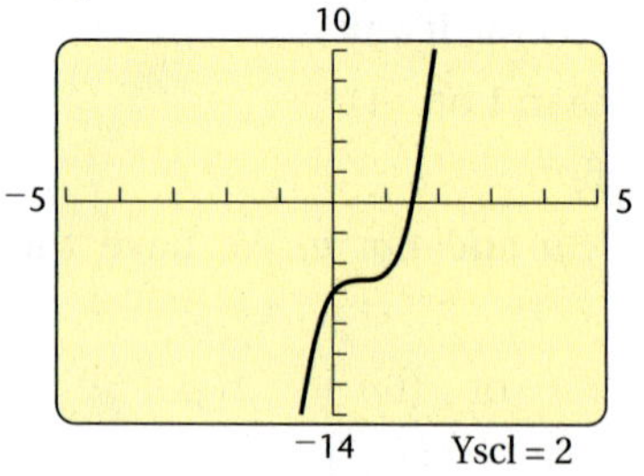

10

−5 5

−14 Yscl = 2

Keep in mind that these procedures provide only a partial check since we cannot view all possible values of x in a table nor can we see the entire graph.

Exercises: Use a table or a graph to determine whether the factorization is correct.

1. $24x^2 - 76x + 40 = 4(3x - 2)(2x - 5)$
2. $4x^2 - 5x - 6 = (4x + 3)(x - 2)$
3. $5x^2 + 17x - 12 = (5x + 3)(x - 4)$
4. $10x^2 + 37x + 7 = (5x - 1)(2x + 7)$
5. $12x^2 - 17x - 5 = (6x + 1)(2x - 5)$
6. $12x^2 - 17x - 5 = (4x + 1)(3x - 5)$
7. $x^2 - 4 = (x - 2)(x - 2)$
8. $x^2 - 4 = (x + 2)(x - 2)$

Study Tips

SKILL MAINTENANCE EXERCISES

It is never too soon to begin reviewing for the final examination. The Skill Maintenance exercises found in each exercise set review and reinforce skills taught in earlier sections. Include all of these exercises in your weekly preparation. Answers to both odd-numbered and even-numbered exercises along with section references appear at the back of the book.

10.3 EXERCISE SET

For Extra Help

Digital Video Tutor CD 8 Videotape 11 | InterAct Math | Math Tutor Center | MathXL | MyMathLab

a Factor.

1. $2x^2 - 7x - 4$

2. $3x^2 - x - 4$

3. $5x^2 - x - 18$

4. $4x^2 - 17x + 15$

5. $6x^2 + 23x + 7$

6. $6x^2 - 23x + 7$

7. $3x^2 + 4x + 1$

8. $7x^2 + 15x + 2$

9. $4x^2 + 4x - 15$

10. $9x^2 + 6x - 8$

11. $2x^2 - x - 1$

12. $15x^2 - 19x - 10$

13. $9x^2 + 18x - 16$

14. $2x^2 + 5x + 2$

15. $3x^2 - 5x - 2$

16. $18x^2 - 3x - 10$

17. $12x^2 + 31x + 20$

18. $15x^2 + 19x - 10$

19. $14x^2 + 19x - 3$

20. $35x^2 + 34x + 8$

21. $9x^2 + 18x + 8$

22. $6 - 13x + 6x^2$

23. $49 - 42x + 9x^2$

24. $16 + 36x^2 + 48x$

25. $24x^2 + 47x - 2$

26. $16p^2 - 78p + 27$

27. $35x^2 - 57x - 44$

28. $9a^2 + 12a - 5$

29. $20 + 6x - 2x^2$

30. $15 + x - 2x^2$

31. $12x^2 + 28x - 24$

32. $6x^2 + 33x + 15$

33. $30x^2 - 24x - 54$

34. $18t^2 - 24t + 6$

35. $4y + 6y^2 - 10$

36. $-9 + 18x^2 - 21x$

37. $3x^2 - 4x + 1$

38. $6t^2 + 13t + 6$

39. $12x^2 - 28x - 24$

40. $6x^2 - 33x + 15$

41. $-1 + 2x^2 - x$

42. $-19x + 15x^2 + 6$

43. $9x^2 - 18x - 16$

44. $14y^2 + 35y + 14$

45. $15x^2 - 25x - 10$

46. $18x^2 + 3x - 10$

47. $12p^3 + 31p^2 + 20p$

48. $15x^3 + 19x^2 - 10x$

49. $16 + 18x - 9x^2$

50. $33t - 15 - 6t^2$

51. $-15x^2 + 19x - 6$

52. $1 + p - 2p^2$

53. $14x^4 + 19x^3 - 3x^2$

54. $70x^4 + 68x^3 + 16x^2$

55. $168x^3 - 45x^2 + 3x$

56. $144x^5 + 168x^4 + 48x^3$

57. $15x^4 - 19x^2 + 6$

58. $9x^4 + 18x^2 + 8$

59. $25t^2 + 80t + 64$

60. $9x^2 - 42x + 49$

61. $6x^3 + 4x^2 - 10x$

62. $18x^3 - 21x^2 - 9x$

63. $25x^2 + 79x + 64$

64. $9y^2 + 42y + 47$

65. $6x^2 - 19x - 5$

66. $2x^2 + 11x - 9$

67. $12m^2 - mn - 20n^2$

68. $12a^2 - 17ab + 6b^2$

69. $6a^2 - ab - 15b^2$

70. $3p^2 - 16pq - 12q^2$

71. $9a^2 + 18ab + 8b^2$

72. $10s^2 + 4st - 6t^2$

73. $35p^2 + 34pq + 8q^2$

74. $30a^2 + 87ab + 30b^2$

75. $18x^2 - 6xy - 24y^2$

76. $15a^2 - 5ab - 20b^2$

77. D$_W$ Explain how the factoring in Exercise 21 can be used to aid the factoring in Exercise 71.

78. D$_W$ A student presents the following work:

$$4x^2 + 28x + 48 = (2x + 6)(2x + 8)$$
$$= 2(x + 3)(x + 4).$$

Is it correct? Explain.

SKILL MAINTENANCE

Solve. [7.4b]

79. $A = pq - 7$, for q

80. $y = mx + b$, for x

81. $3x + 2y = 6$, for y

82. $p - q + r = 2$, for q

Solve. [7.7e]

83. $5 - 4x < -11$

84. $2x - 4(x + 3x) \geq 6x - 8 - 9x$

85. Graph: $y = \frac{2}{5}x - 1$. [8.2b]

86. Divide: $\frac{y^{12}}{y^4}$. [9.1e]

Multiply. [9.6d]

87. $(3x - 5)(3x + 5)$

88. $(4a - 3)^2$

SYNTHESIS

Factor.

89. $20x^{2n} + 16x^n + 3$

90. $-15x^{2m} + 26x^m - 8$

91. $3x^{6a} - 2x^{3a} - 1$

92. $x^{2n+1} - 2x^{n+1} + x$

93.–102. Use the TABLE feature to check the factoring in Exercises 15–24.

Objective

a Factor trinomials of the type $ax^2 + bx + c$, $a \neq 1$, using the ac-method.

10.4 FACTORING $ax^2 + bx + c$, $a \neq 1$: THE *ac*-METHOD

a The *ac*-Method

Another method for factoring trinomials of the type $ax^2 + bx + c$, $a \neq 1$, involves the product, ac, of the leading coefficient a and the last term c. It is called the ***ac*-method**. Because it uses factoring by grouping, it is also referred to as the **grouping method**.

We know how to factor the trinomial $x^2 + 5x + 6$. We look for factors of the constant term, 6, whose sum is the coefficient of the middle term, 5:

$$x^2 + 5x + 6.$$

(1) Factor: $6 = 2 \cdot 3$

(2) Sum: $2 + 3 = 5$

What happens when the leading coefficient is not 1? To factor a trinomial like $3x^2 - 10x - 8$, we can use a method similar to what we used for $x^2 + 5x + 6$, but we need two more steps. That method is outlined as follows.

THE *ac*-METHOD

To factor $ax^2 + bx + c$, $a \neq 1$, using the ac-method:

1. Factor out a common factor, if any.
2. Multiply the leading coefficient a and the constant c.
3. Try to factor the product ac so that the sum of the factors is b. That is, find integers p and q such that $pq = ac$ and $p + q = b$.
4. Split the middle term. That is, write it as a sum using the factors found in step (3).
5. Factor by grouping.
6. Check by multiplying.

EXAMPLE 1 Factor: $3x^2 - 10x - 8$.

1) First, we factor out a common factor, if any. There is none (other than 1 or -1).

2) We multiply the leading coefficient, 3, and the constant, -8:

$$3(-8) = -24.$$

3) Then we look for a factorization of -24 in which the sum of the factors is the coefficient of the middle term, -10.

PAIRS OF FACTORS	SUMS OF FACTORS
$-1,\ 24$	23
$1,\ -24$	-23
$-2,\ 12$	10
$2,\ -12$	-10 ← $2 + (-12) = -10$
$-3,\ 8$	5
$3,\ -8$	-5
$-4,\ 6$	2
$4,\ -6$	-2

4) Next, we split the middle term as a sum or a difference using the factors found in step (3): $-10x = 2x - 12x$.

5) Finally, we factor by grouping, as follows:

$$3x^2 - 10x - 8 = 3x^2 + 2x - 12x - 8 \quad \text{Substituting } 2x - 12x \text{ for } -10x$$
$$= x(3x + 2) - 4(3x + 2) \quad \text{Factoring by grouping}$$
$$= (x - 4)(3x + 2).$$

We can also split the middle term as $-12x + 2x$. We still get the same factorization, although the factors may be in a different order. Note the following:

$$3x^2 - 10x - 8 = 3x^2 - 12x + 2x - 8 \quad \text{Substituting } -12x + 2x \text{ for } -10x$$
$$= 3x(x - 4) + 2(x - 4) \quad \text{Factoring by grouping}$$
$$= (3x + 2)(x - 4).$$

6) CHECK: $(3x + 2)(x - 4) = 3x^2 - 10x - 8$.

Do Exercises 1 and 2.

Factor.

1. $6x^2 + 7x + 2$

2. $12x^2 - 17x - 5$

EXAMPLE 2 Factor: $8x^2 + 8x - 6$.

1) First, we factor out a common factor, if any. The number 2 is common to all three terms, so we factor it out: $2(4x^2 + 4x - 3)$.

2) Next, we factor the trinomial $4x^2 + 4x - 3$. We multiply the leading coefficient and the constant, 4 and -3: $4(-3) = -12$.

3) We try to factor -12 so that the sum of the factors is 4.

PAIRS OF FACTORS	SUMS OF FACTORS
$-1,\ 12$	11
$1,\ -12$	-11
$-2,\ 6$	4 ← $-2 + 6 = 4$
$2,\ -6$	-4
$-3,\ 4$	1
$3,\ -4$	-1

4) Then we split the middle term, $4x$, as follows: $4x = -2x + 6x$.

5) Finally, we factor by grouping:

$$4x^2 + 4x - 3 = 4x^2 - 2x + 6x - 3 \quad \text{Substituting } -2x + 6x \text{ for } 4x$$
$$= 2x(2x - 1) + 3(2x - 1) \quad \text{Factoring by grouping}$$
$$= (2x + 3)(2x - 1).$$

The factorization of $4x^2 + 4x - 3$ is $(2x + 3)(2x - 1)$. But don't forget the common factor! We must include it to get a factorization of the original trinomial: $8x^2 + 8x - 6 = 2(2x + 3)(2x - 1)$.

6) CHECK: $2(2x + 3)(2x - 1) = 2(4x^2 + 4x - 3) = 8x^2 + 8x - 6$.

Do Exercises 3 and 4.

Factor.

3. $6x^2 + 15x + 9$

4. $20x^2 - 46x + 24$

Answers on page A-34

10.4 EXERCISE SET

For Extra Help

Digital Video Tutor CD 8 Videotape 11

InterAct Math

Math Tutor Center

MathXL

MyMathLab

a Factor. Note that the middle term has already been split.

1. $x^2 + 2x + 7x + 14$

2. $x^2 + 3x + x + 3$

3. $x^2 - 4x - x + 4$

4. $a^2 + 5a - 2a - 10$

5. $6x^2 + 4x + 9x + 6$

6. $3x^2 - 2x + 3x - 2$

7. $3x^2 - 4x - 12x + 16$

8. $24 - 18y - 20y + 15y^2$

9. $35x^2 - 40x + 21x - 24$

10. $8x^2 - 6x - 28x + 21$

11. $4x^2 + 6x - 6x - 9$

12. $2x^4 - 6x^2 - 5x^2 + 15$

13. $2x^4 + 6x^2 + 5x^2 + 15$

14. $9x^4 - 6x^2 - 6x^2 + 4$

Factor by grouping.

15. $2x^2 + 7x - 4$

16. $5x^2 + x - 18$

17. $3x^2 - 4x - 15$

18. $3x^2 + x - 4$

19. $6x^2 + 23x + 7$

20. $6x^2 + 13x + 6$

21. $3x^2 - 4x + 1$

22. $7x^2 - 15x + 2$

23. $4x^2 - 4x - 15$

24. $9x^2 - 6x - 8$

25. $2x^2 + x - 1$

26. $15x^2 + 19x - 10$

27. $9x^2 - 18x - 16$

28. $2x^2 - 5x + 2$

29. $3x^2 + 5x - 2$

30. $18x^2 + 3x - 10$

31. $12x^2 - 31x + 20$

32. $15x^2 - 19x - 10$

33. $14x^2 - 19x - 3$

34. $35x^2 - 34x + 8$

35. $9x^2 + 18x + 8$

36. $6 - 13x + 6x^2$

37. $49 - 42x + 9x^2$

38. $25x^2 + 40x + 16$

39. $24x^2 - 47x - 2$

40. $16a^2 + 78a + 27$

41. $5 - 9a^2 - 12a$

42. $17x - 4x^2 + 15$

43. $20 + 6x - 2x^2$

44. $15 + x - 2x^2$

45. $12x^2 + 28x - 24$

46. $6x^2 + 33x + 15$

47. $30x^2 - 24x - 54$

48. $18t^2 - 24t + 6$

49. $4y + 6y^2 - 10$

50. $-9 + 18x^2 - 21x$

51. $3x^2 - 4x + 1$

52. $6t^2 + t - 15$

53. $12x^2 - 28x - 24$

54. $6x^2 - 33x + 15$

55. $-1 + 2x^2 - x$

56. $-19x + 15x^2 + 6$

57. $9x^2 + 18x - 16$

58. $14y^2 + 35y + 14$

59. $15x^2 - 25x - 10$

60. $18x^2 + 3x - 10$

61. $12p^3 + 31p^2 + 20p$

62. $15x^3 + 19x^2 - 10x$

63. $4 - x - 5x^2$

64. $1 - p - 2p^2$

65. $33t - 15 - 6t^2$

66. $-15x^2 - 19x - 6$

67. $14x^4 + 19x^3 - 3x^2$

68. $70x^4 + 68x^3 + 16x^2$

69. $168x^3 - 45x^2 + 3x$

70. $144x^5 + 168x^4 + 48x^3$

71. $15x^4 - 19x^2 + 6$

72. $9x^4 + 18x^2 + 8$

73. $25t^2 + 80t + 64$

74. $9x^2 - 42x + 49$

75. $6x^3 + 4x^2 - 10x$

76. $18x^3 - 21x^2 - 9x$

77. $25x^2 + 79x + 64$

78. $9y^2 + 42y + 47$

79. $6x^2 - 19x - 5$

80. $2x^2 + 11x - 9$

81. $12m^2 - mn - 20n^2$

82. $12a^2 - 17ab + 6b^2$

83. $6a^2 - ab - 15b^2$

84. $3p^2 - 16pq - 12q^2$

85. $9a^2 - 18ab + 8b^2$

86. $10s^2 + 4st - 6t^2$

87. $35p^2 + 34pq + 8q^2$

88. $30a^2 + 87ab + 30b^2$

89. $18x^2 - 6xy - 24y^2$

90. $15a^2 - 5ab - 20b^2$

91. $60x + 18x^2 - 6x^3$

92. $60x + 4x^2 - 8x^3$

93. $35x^5 - 57x^4 - 44x^3$

94. $15x^3 + 33x^4 + 6x^5$

95. D$_W$ If you have studied both the FOIL and the *ac*-methods of factoring $ax^2 + bx + c, a \neq 1$, decide which method you think is better and explain why.

96. D$_W$ Explain factoring $ax^2 + bx + c, a \neq 1$, using the *ac*-method as though you were teaching a fellow student.

SKILL MAINTENANCE

Solve. [7.7d, e]

97. $-10x > 1000$

98. $-3.8x \leq -824.6$

99. $6 - 3x \geq -18$

100. $3 - 2x - 4x > -9$

101. $\frac{1}{2}x - 6x + 10 \leq x - 5x$

102. $-2(x + 7) > -4(x - 5)$

103. $3x - 6x + 2(x - 4) > 2(9 - 4x)$

104. $-6(x - 4) + 8(4 - x) \leq 3(x - 7)$

Solve. [7.6a]

105. The earth is a sphere (or ball) that is about 40,000 km in circumference. Find the radius of the earth, in kilometers and in miles. Use 3.14 for π. (*Hint*: 1 km $\approx$ 0.62 mi.)

106. The second angle of a triangle is 10° less than twice the first. The third angle is 15° more than four times the first. Find the measure of the second angle.

SYNTHESIS

Factor.

107. $9x^{10} - 12x^5 + 4$

108. $24x^{2n} + 22x^n + 3$

109. $16x^{10} + 8x^5 + 1$

110. $(a + 4)^2 - 2(a + 4) + 1$

111.–120. Use the TABLE feature to check the factoring in Exercises 15–24.

Objectives

a Recognize trinomial squares.

b Factor trinomial squares.

c Recognize differences of squares.

d Factor differences of squares, being careful to factor completely.

10.5 FACTORING TRINOMIAL SQUARES AND DIFFERENCES OF SQUARES

In this section, we first learn to factor trinomials that are squares of binomials. Then we factor binomials that are differences of squares.

a Recognizing Trinomial Squares

Some trinomials are squares of binomials. For example, the trinomial $x^2 + 10x + 25$ is the square of the binomial $x + 5$. To see this, we can calculate $(x + 5)^2$. It is $x^2 + 2 \cdot x \cdot 5 + 5^2$, or $x^2 + 10x + 25$. A trinomial that is the square of a binomial is called a **trinomial square,** or a **perfect-square trinomial.**

In Chapter 9, we considered squaring binomials as special-product rules:

$$(A + B)^2 = A^2 + 2AB + B^2;$$
$$(A - B)^2 = A^2 - 2AB + B^2.$$

We can use these equations in reverse to factor trinomial squares.

TRINOMIAL SQUARES

$$A^2 + 2AB + B^2 = (A + B)^2;$$
$$A^2 - 2AB + B^2 = (A - B)^2$$

How can we recognize when an expression to be factored is a trinomial square? Look at $A^2 + 2AB + B^2$ and $A^2 - 2AB + B^2$. In order for an expression to be a trinomial square:

a) The two expressions A^2 and B^2 must be squares, such as

$$4, \quad x^2, \quad 25x^4, \quad 16t^2.$$

When the coefficient is a perfect square and the power(s) of the variable(s) is (are) even, then the expression is a perfect square.

b) There must be no minus sign before A^2 or B^2.

c) If we multiply A and B and double the result, we get either the remaining term $2 \cdot A \cdot B$, or its opposite, $-2 \cdot A \cdot B$.

EXAMPLE 1 Determine whether $x^2 + 6x + 9$ is a trinomial square.

a) We know that x^2 and 9 are squares.

b) There is no minus sign before x^2 or 9.

c) If we multiply the square roots, x and 3, and double the product, we get the remaining term: $2 \cdot x \cdot 3 = 6x$.

Thus, $x^2 + 6x + 9$ is the square of a binomial. In fact, $x^2 + 6x + 9 = (x + 3)^2$.

EXAMPLE 2 Determine whether $x^2 + 6x + 11$ is a trinomial square.

The answer is no, because only one term is a square.

EXAMPLE 3 Determine whether $16x^2 + 49 - 56x$ is a trinomial square.

It helps to first write the trinomial in descending order:

$$16x^2 - 56x + 49.$$

a) We know that $16x^2$ and 49 are squares.

b) There is no minus sign before $16x^2$ or 49.

c) If we multiply the square roots, $4x$ and 7, and double the product, we get the opposite of the remaining term: $2 \cdot 4x \cdot 7 = 56x$; $56x$ is the opposite of $-56x$.

Thus, $16x^2 + 49 - 56x$ is a trinomial square. In fact, $16x^2 - 56x + 49 = (4x - 7)^2$.

Do Exercises 1–8.

Determine whether each is a trinomial square. Write "yes" or "no."

1. $x^2 + 8x + 16$
2. $25 - x^2 + 10x$
3. $t^2 - 12t + 4$
4. $25 + 20y + 4y^2$
5. $5x^2 + 16 - 14x$
6. $16x^2 + 40x + 25$
7. $p^2 + 6p - 9$
8. $25a^2 + 9 - 30a$

b Factoring Trinomial Squares

We can use the trial-and-error or grouping methods from Sections 10.2–10.4 to factor trinomial squares, but there is a faster method using the following equations.

FACTORING TRINOMIAL SQUARES

$A^2 + 2AB + B^2 = (A + B)^2$;
$A^2 - 2AB + B^2 = (A - B)^2$

We consider 3 to be a square root of 9 because $3^2 = 9$. Similarly, A is a square root of A^2. We use square roots of the squared terms and the sign of the remaining term to factor a trinomial square.

EXAMPLE 4 Factor: $x^2 + 6x + 9$.

$$x^2 + 6x + 9 = x^2 + 2 \cdot x \cdot 3 + 3^2 = (x + 3)^2$$

$$A^2 + 2\ A\ B + B^2 = (A + B)^2$$

The sign of the middle term is positive.

EXAMPLE 5 Factor: $x^2 + 49 - 14x$.

$x^2 + 49 - 14x = x^2 - 14x + 49$ Changing to descending order

$= x^2 - 2 \cdot x \cdot 7 + 7^2$ The sign of the middle term is negative.

$= (x - 7)^2$

EXAMPLE 6 Factor: $16x^2 - 40x + 25$.

$$16x^2 - 40x + 25 = (4x)^2 - 2 \cdot 4x \cdot 5 + 5^2 = (4x - 5)^2$$

$$A^2\ - 2\ A\ B + B^2 = (A - B)^2$$

Do Exercises 9–13.

Factor.

9. $x^2 + 2x + 1$
10. $1 - 2x + x^2$
11. $4 + t^2 + 4t$
12. $25x^2 - 70x + 49$
13. $49 - 56y + 16y^2$

Answers on page A-35

Factor.

14. $48m^2 + 75 + 120m$

15. $p^4 + 18p^2 + 81$

16. $4z^5 - 20z^4 + 25z^3$

17. $9a^2 + 30ab + 25b^2$

Answers on page A-35

EXAMPLE 7 Factor: $t^4 + 20t^2 + 100$.

$$t^4 + 20t^2 + 100 = (t^2)^2 + 2(t^2)(10) + 10^2$$
$$= (t^2 + 10)^2$$

EXAMPLE 8 Factor: $75m^3 + 210m^2 + 147m$.

Always look first for a common factor. This time there is one, $3m$:

$$75m^3 + 210m^2 + 147m = 3m[25m^2 + 70m + 49]$$
$$= 3m[(5m)^2 + 2(5m)(7) + 7^2]$$
$$= 3m(5m + 7)^2.$$

EXAMPLE 9 Factor: $4p^2 - 12pq + 9q^2$.

$$4p^2 - 12pq + 9q^2 = (2p)^2 - 2(2p)(3q) + (3q)^2$$
$$= (2p - 3q)^2$$

Do Exercises 14–17.

C Recognizing Differences of Squares

The following polynomials are *differences of squares*:

$$x^2 - 9, \quad 4t^2 - 49, \quad a^2 - 25b^2.$$

To factor a difference of squares such as $x^2 - 9$, think about the formula we used in Chapter 9:

$$\mathbf{(A + B)(A - B) = A^2 - B^2.}$$

Equations are reversible, so we also know the following.

DIFFERENCE OF SQUARES

$$A^2 - B^2 = (A + B)(A - B)$$

Thus,

$$x^2 - 9 = (x + 3)(x - 3).$$

To use this formula, we must be able to recognize when it applies. A **difference of squares** is an expression like the following:

$$A^2 - B^2.$$

How can we recognize such expressions? Look at $A^2 - B^2$. In order for a binomial to be a difference of squares:

a) There must be two expressions, both squares, such as

$$4x^2, \quad 9, \quad 25t^4, \quad 1, \quad x^6, \quad 49y^8.$$

b) The terms must have different signs.

32. $1 - 2a^5 + a^{10}$

33. $4p^2 + 12pq + 9q^2$

34. $25m^2 + 20mn + 4n^2$

35. $a^2 - 6ab + 9b^2$

36. $x^2 - 14xy + 49y^2$

37. $81a^2 - 18ab + b^2$

38. $64p^2 + 16pq + q^2$

39. $36a^2 + 96ab + 64b^2$

40. $16m^2 - 40mn + 25n^2$

c Determine whether each of the following is a difference of squares.

41. $x^2 - 4$

42. $x^2 - 36$

43. $x^2 + 25$

44. $x^2 + 9$

45. $x^2 - 45$

46. $x^2 - 80y^2$

47. $16x^2 - 25y^2$

48. $-1 + 36x^2$

d Factor completely. Remember to look first for a common factor.

49. $y^2 - 4$

50. $q^2 - 1$

51. $p^2 - 9$

52. $x^2 - 36$

53. $-49 + t^2$

54. $-64 + m^2$

55. $a^2 - b^2$

56. $p^2 - q^2$

57. $25t^2 - m^2$

58. $w^2 - 49z^2$

59. $100 - k^2$

60. $81 - w^2$

61. $16a^2 - 9$

62. $25x^2 - 4$

63. $4x^2 - 25y^2$

64. $9a^2 - 16b^2$

65. $8x^2 - 98$

66. $24x^2 - 54$

67. $36x - 49x^3$

68. $16x - 81x^3$

69. $49a^4 - 81$

70. $25a^4 - 9$

71. $a^4 - 16$

72. $y^4 - 1$

73. $5x^4 - 405$

74. $4x^4 - 64$

75. $1 - y^8$

76. $x^8 - 1$

77. $x^{12} - 16$

78. $x^8 - 81$

79. $y^2 - \frac{1}{16}$

80. $x^2 - \frac{1}{25}$

81. $25 - \frac{1}{49}x^2$

82. $\frac{1}{4} - 9q^2$

83. $16m^4 - t^4$

84. $p^4q^4 - 1$

85. DW Explain in your own words how to determine whether a polynomial is a trinomial square.

86. DW Spiro concludes that since $x^2 - 9 = (x - 3)(x + 3)$, it must follow that $x^2 + 9 = (x + 3)(x + 3)$. What mistake is the student making? How would you go about correcting the misunderstanding?

SKILL MAINTENANCE

Divide. [6.6a, c]

87. $(-110) \div 10$

88. $-1000 \div (-2.5)$

89. $\left(-\frac{2}{3}\right) \div \frac{4}{5}$

90. $8.1 \div (-9)$

91. $-64 \div (-32)$

92. $-256 \div 1.6$

Find a polynomial for the shaded area. (Leave results in terms of π where appropriate.) [9.4d]

93.

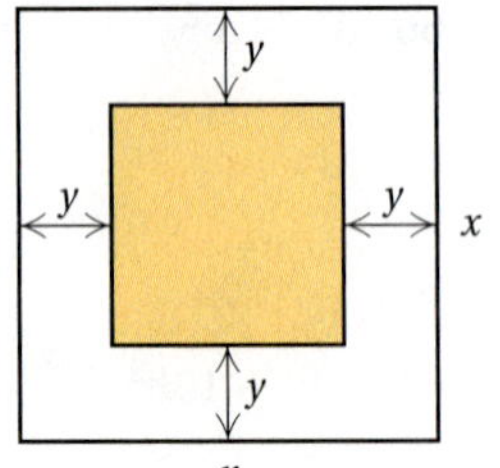

94.

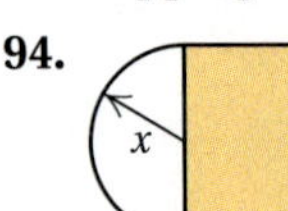

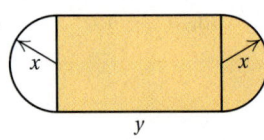

Simplify.

95. $y^5 \cdot y^7$ [9.1d]

96. $(5a^2b^3)^2$ [9.2a, b]

Find the intercepts. Then graph the equation. [8.3a]

97. $y - 6x = 6$

98. $3x - 5y = 15$

SYNTHESIS

Factor completely, if possible.

99. $49x^2 - 216$

100. $27x^3 - 13x$

101. $x^2 + 22x + 121$

102. $x^2 - 5x + 25$

103. $18x^3 + 12x^2 + 2x$

104. $162x^2 - 82$

105. $x^8 - 2^8$

106. $4x^4 - 4x^2$

107. $3x^5 - 12x^3$

108. $3x^2 - \frac{1}{3}$

109. $18x^3 - \frac{8}{25}x$

110. $x^2 - 2.25$

111. $0.49p - p^3$

112. $3.24x^2 - 0.81$

113. $0.64x^2 - 1.21$

114. $1.28x^2 - 2$

115. $(x + 3)^2 - 9$

116. $(y - 5)^2 - 36q^2$

117. $x^2 - \left(\frac{1}{x}\right)^2$

118. $a^{2n} - 49b^{2n}$

119. $81 - b^{4k}$

120. $9x^{18} + 48x^9 + 64$

121. $9b^{2n} + 12b^n + 4$

122. $(x + 7)^2 - 4x - 24$

123. $(y + 3)^2 + 2(y + 3) + 1$

124. $49(x + 1)^2 - 42(x + 1) + 9$

Find c such that the polynomial is the square of a binomial.

125. $cy^2 + 6y + 1$

126. $cy^2 - 24y + 9$

Use the TABLE feature to determine whether the factorization is correct.

127. $x^2 + 9 = (x + 3)(x + 3)$

128. $x^2 - 49 = (x - 7)(x + 7)$

129. $x^2 + 9 = (x + 3)^2$

130. $x^2 - 49 = (x - 7)^2$

Objective

a Factor sums or differences of two cubes.

10.6 FACTORING SUMS OR DIFFERENCES OF CUBES

a Sums or Differences of Cubes

We can factor the sum or the difference of two expressions that are cubes. Consider the following products:

$$(A + B)(A^2 - AB + B^2) = A(A^2 - AB + B^2) + B(A^2 - AB + B^2)$$
$$= A^3 - A^2B + AB^2 + A^2B - AB^2 + B^3$$
$$= A^3 + B^3$$

and

$$(A - B)(A^2 + AB + B^2) = A(A^2 + AB + B^2) - B(A^2 + AB + B^2)$$
$$= A^3 + A^2B + AB^2 - A^2B - AB^2 - B^3$$
$$= A^3 - B^3.$$

The above equations (reversed) show how we can factor a sum or a difference of two cubes.

SUM OR DIFFERENCE OF CUBES

$A^3 + B^3 = (A + B)(A^2 - AB + B^2)$;
$A^3 - B^3 = (A - B)(A^2 + AB + B^2)$

Note that what we are considering here is a sum or a difference of cubes. We are not cubing a binomial. For example, $(A + B)^3$ is *not* the same as $A^3 + B^3$. The table of cubes in the margin is helpful.

N	N^3
0.2	0.008
0.1	0.001
0	0
1	1
2	8
3	27
4	64
5	125
6	216
7	343
8	512
9	729
10	1000

EXAMPLE 1 Factor: $x^3 - 27$.

We have

$$\begin{array}{c} A^3 - B^3 \\ \downarrow \quad \downarrow \\ x^3 - 27 = x^3 - 3^3. \end{array}$$

In one set of parentheses, we write the cube root of the first term, x. Then we write the cube root of the second term, -3. This gives us the expression $x - 3$:

$$(x - 3)(\qquad\qquad).$$

To get the next factor, we think of $x - 3$ and do the following:

Square the first term: $x \cdot x = x^2$.

Multiply the terms, $x(-3) = -3x$, and then change the sign: $3x$.

Square the second term: $(-3)^2 = 9$.

$$(x - 3)(x^2 + 3x + 9).$$
$$(A - B)(A^2 + AB + B^2)$$

Note that we cannot factor $x^2 + 3x + 9$. It is not a trinomial square nor can it be factored by trial and error. Check this on your own.

Do Exercises 1 and 2.

Factor.

1. $x^3 - 8$

2. $64 - y^3$

EXAMPLE 2 Factor: $125x^3 + y^3$.

We have

$$125x^3 + y^3 = (5x)^3 + y^3.$$

In one set of parentheses, we write the cube root of the first term, $5x$. Then we write a plus sign, and then the cube root of the second term, y:

$$(5x + y)(\qquad\qquad\qquad).$$

To get the next factor, we think of $5x + y$ and do the following:

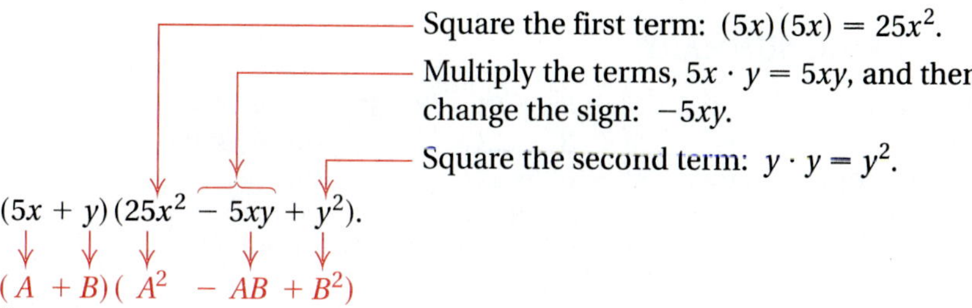

Square the first term: (5x)(5x) = 25x².
Multiply the terms, 5x · y = 5xy, and then change the sign: −5xy.
Square the second term: y · y = y².
(5x + y)(25x² − 5xy + y²).
(A + B)(A² − AB + B²)

Do Exercises 3 and 4.

Factor.

3. $27x^3 + y^3$

4. $8y^3 + z^3$

EXAMPLE 3 Factor: $128y^7 - 250x^6y$.

We first look for the largest common factor:

$$\begin{aligned} 128y^7 - 250x^6y &= 2y(64y^6 - 125x^6) \\ &= 2y[(4y^2)^3 - (5x^2)^3] \\ &= 2y(4y^2 - 5x^2)(16y^4 + 20x^2y^2 + 25x^4). \end{aligned}$$

EXAMPLE 4 Factor: $a^6 - b^6$.

We can express this polynomial as a difference of squares:

$$a^6 - b^6 = (a^3)^2 - (b^3)^2.$$

We factor as follows:

$$a^6 - b^6 = (a^3 + b^3)(a^3 - b^3).$$

One factor is a sum of two cubes, and the other factor is a difference of two cubes. We factor them:

$$a^6 - b^6 = (a + b)(a^2 - ab + b^2)(a - b)(a^2 + ab + b^2).$$

We have now factored completely.

In Example 4, had we thought of factoring first as a difference of two cubes, we would have had

$$\begin{aligned} (a^2)^3 - (b^2)^3 &= (a^2 - b^2)(a^4 + a^2b^2 + b^4) \\ &= (a + b)(a - b)(a^4 + a^2b^2 + b^4). \end{aligned}$$

In this case, we might have missed some factors; $a^4 + a^2b^2 + b^4$ can be factored as $(a^2 - ab + b^2)(a^2 + ab + b^2)$, but we probably would not have known to do such factoring.

Answers on page A-35

Factor.

5. $m^6 - n^6$

6. $16x^7y + 54xy^7$

7. $729x^6 - 64y^6$

8. $x^3 - 0.027$

Answers on page A-35

EXAMPLE 5 Factor: $64a^6 - 729b^6$.

We have

$$\begin{aligned} 64a^6 - 729b^6 &= (8a^3)^2 - (27b^3)^2 \\ &= (8a^3 - 27b^3)(8a^3 + 27b^3) \quad \text{Factoring a difference of squares} \\ &= [(2a)^3 - (3b)^3][(2a)^3 + (3b)^3]. \end{aligned}$$

Each factor is a sum or a difference of cubes. We factor each:

$$= (2a - 3b)(4a^2 + 6ab + 9b^2)(2a + 3b)(4a^2 - 6ab + 9b^2).$$

FACTORING SUMMARY

Sum of cubes: $A^3 + B^3 = (A + B)(A^2 - AB + B^2)$;
Difference of cubes: $A^3 - B^3 = (A - B)(A^2 + AB + B^2)$;
Difference of squares: $A^2 - B^2 = (A + B)(A - B)$;
Sum of squares: $A^2 + B^2$ cannot be factored using real numbers if the largest common factor has been removed.

Do Exercises 5–8.

Study Tips

TROUBLE SPOTS

By now you have probably encountered certain topics that gave you more difficulty than others. It is important to know that this happens to every person who studies mathematics. Unfortunately, frustration is often part of the learning process and it is important not to give up when difficulty arises.

Below are some study tips that might be useful if and when troubles arise.

- **Realize that everyone—even your instructor—has been stumped at times when studying math.** You are not the first person, nor will you be the last, to encounter a "roadblock."
- **Whether working alone or with a classmate, try to allow enough study time so that you won't need to constantly glance at a clock.** Difficult material is best mastered when your mind is completely focused on the subject matter. Thus, if you are tired, it is usually best to study early the next morning or to take a ten-minute "power-nap" in order to make the most productive use of your time.
- **Talk about your trouble spot with a classmate.** It is possible that she or he is also having difficulty with the same material. If that is the case, perhaps the majority of your class is confused and your instructor's coverage of the topic is not yet finished. If your classmate *does* understand the topic that is troubling you, patiently allow him or her to explain it to you. By verbalizing the math in question, your classmate may help clarify the material for both of you. Perhaps you will be able to return the favor for your classmate when he or she is struggling with a topic that you understand.
- **When working on difficult material, it is often helpful to "back up" and review the most recent material that did make sense.** This can build your confidence and create a momentum that can often carry you through the roadblock. Sometimes a small piece of information that appeared in a previous section is all that is needed for your problem spot to disappear. When the difficult material is finally mastered, try to make use of what is fresh in your mind by taking a "sneak preview" of what your next topic for study will be.

10.6 EXERCISE SET

For Extra Help

Digital Video Tutor CD 8 Videotape 12 | InterAct Math | Math Tutor Center | MathXL | MyMathLab

a Factor.

1. $z^3 + 27$
2. $a^3 + 8$
3. $x^3 - 1$
4. $c^3 - 64$
5. $y^3 + 125$
6. $x^3 + 1$
7. $8a^3 + 1$
8. $27x^3 + 1$
9. $y^3 - 8$
10. $p^3 - 27$
11. $8 - 27b^3$
12. $64 - 125x^3$
13. $64y^3 + 1$
14. $125x^3 + 1$
15. $8x^3 + 27$
16. $27y^3 + 64$
17. $a^3 - b^3$
18. $x^3 - y^3$
19. $a^3 + \frac{1}{8}$
20. $b^3 + \frac{1}{27}$
21. $2y^3 - 128$
22. $3z^3 - 3$
23. $24a^3 + 3$
24. $54x^3 + 2$
25. $rs^3 + 64r$
26. $ab^3 + 125a$
27. $5x^3 - 40z^3$

28. $2y^3 - 54z^3$

29. $x^3 + 0.001$

30. $y^3 + 0.125$

31. $64x^6 - 8t^6$

32. $125c^6 - 8d^6$

33. $2y^4 - 128y$

34. $3z^5 - 3z^2$

35. $z^6 - 1$

36. $t^6 + 1$

37. $t^6 + 64y^6$

38. $p^6 - q^6$

SKILL MAINTENANCE

Simplify. [9.2b]

39. $(7y^{-5})^3$

40. $(a^{-4}b^{-9})^{-2}$

41. $\left(\frac{x^3}{4}\right)^{-2}$

Multiply.

42. $(2y^5 + 3)(2y^5 - 3)$ [9.6b]

43. $\left(w - \frac{1}{3}\right)^2$ [9.6c]

44. $(x - 0.1)(x + 0.5)$ [9.6a]

SYNTHESIS

Consider these polynomials:

$$(a + b)^3;\quad a^3 + b^3;\quad (a + b)(a^2 - ab + b^2);$$
$$(a + b)(a^2 + ab + b^2);\quad (a + b)(a + b)(a + b).$$

45. Evaluate each polynomial when $a = -2$ and $b = 3$.

46. Evaluate each polynomial when $a = 4$ and $b = -1$.

Factor. Assume that variables in exponents represent natural numbers.

47. $x^{6a} + y^{3b}$

48. $a^3x^3 - b^3y^3$

49. $3x^{3a} + 24y^{3b}$

50. $\frac{8}{27}x^3 + \frac{1}{64}y^3$

51. $\frac{1}{24}x^3y^3 + \frac{1}{3}z^3$

52. $7x^3 - \frac{7}{8}$

53. $(x + y)^3 - x^3$

54. $(1 - x)^3 + (x - 1)^6$

55. $(a + 2)^3 - (a - 2)^3$

56. $y^4 - 8y^3 - y + 8$

10.7 FACTORING: A GENERAL STRATEGY

Objective

a Factor polynomials completely using any of the methods considered in this chapter.

a We now combine all of our factoring techniques and consider a general strategy for factoring polynomials. Here we will encounter polynomials of all the types we have considered, in random order, so you will have the opportunity to determine which method to use.

FACTORING STRATEGY

To factor a polynomial:

a) Always look first for a common factor. If there is one, factor out the largest common factor.

b) Then look at the number of terms.

Two terms: Try factoring as a difference of squares first. Next, try factoring as a sum or a difference of cubes. Do not try to factor a sum of squares: $A^2 + B^2$.

Three terms: Determine whether the trinomial is a square. If it is, you know how to factor. If not, try trial and error, using FOIL or the *ac*-method.

Four terms: Try factoring by grouping.

c) *Always factor completely.* If a factor with more than one term can still be factored, you should factor it. When no factor can be factored further, you have finished.

d) Check by multiplying.

EXAMPLE 1 Factor: $5t^4 - 80$.

a) We look for a common factor:

$$5t^4 - 80 = 5(t^4 - 16).$$

b) The factor $t^4 - 16$ has only two terms. It is a difference of squares: $(t^2)^2 - 4^2$. We factor $t^4 - 16$ and then include the common factor:

$$5(t^2 + 4)(t^2 - 4).$$

c) We see that one of the factors is again a difference of squares. We factor it:

$$5(t^2 + 4)(t + 2)(t - 2).$$

↑ This is a sum of squares. It cannot be factored!

We have factored completely because no factor with more than one term can be factored further.

d) CHECK: $5(t^2 + 4)(t + 2)(t - 2) = 5(t^2 + 4)(t^2 - 4)$
$= 5(t^4 - 16)$
$= 5t^4 - 80.$

Factor.

1. $3m^4 - 3$

2. $x^6 + 8x^3 + 16$

3. $2x^4 + 8x^3 + 6x^2$

4. $3x^3 + 12x^2 - 2x - 8$

5. $8x^3 - 200x$

Answers on page A-36

EXAMPLE 2 Factor: $2x^3 + 10x^2 + x + 5$.

a) We look for a common factor. There isn't one.

b) There are four terms. We try factoring by grouping:

$$\begin{aligned} &2x^3 + 10x^2 + x + 5 \\ &\quad = (2x^3 + 10x^2) + (x + 5) && \text{Separating into two binomials} \\ &\quad = 2x^2(x + 5) + 1(x + 5) && \text{Factoring each binomial} \\ &\quad = (2x^2 + 1)(x + 5). && \text{Factoring out the common factor } x + 5 \end{aligned}$$

c) None of these factors can be factored further, so we have factored completely.

d) CHECK: $(2x^2 + 1)(x + 5) = 2x^2 \cdot x + 2x^2 \cdot 5 + 1 \cdot x + 1 \cdot 5$
$= 2x^3 + 10x^2 + x + 5.$

EXAMPLE 3 Factor: $x^5 - 2x^4 - 35x^3$.

a) We look first for a common factor. This time there is one, x^3:

$$x^5 - 2x^4 - 35x^3 = x^3(x^2 - 2x - 35).$$

b) The factor $x^2 - 2x - 35$ has three terms, but it is not a trinomial square. We factor it using trial and error (FOIL):

$$x^5 - 2x^4 - 35x^3 = x^3(x^2 - 2x - 35) = x^3(x - 7)(x + 5).$$

Don't forget to include the common factor in the final answer!

c) No factor with more than one term can be factored further, so we have factored completely.

d) CHECK: $x^3(x - 7)(x + 5) = x^3(x^2 - 2x - 35) = x^5 - 2x^4 - 35x^3.$

EXAMPLE 4 Factor: $x^4 - 10x^2 + 25$.

a) We look first for a common factor. There isn't one.

b) There are three terms. We see that this polynomial is a trinomial square. We factor it:

$$x^4 - 10x^2 + 25 = (x^2)^2 - 2 \cdot x^2 \cdot 5 + 5^2 = (x^2 - 5)^2.$$

We could use FOIL if we have not recognized that we have a trinomial square.

c) Since $x^2 - 5$ cannot be factored further, we have factored completely.

d) CHECK: $(x^2 - 5)^2 = (x^2)^2 - 2(x^2)(5) + 5^2 = x^4 - 10x^2 + 25.$

Do Exercises 1–5.

EXAMPLE 5 Factor: $6x^2y^4 - 21x^3y^5 + 3x^2y^6$.

a) We look first for a common factor:

$$6x^2y^4 - 21x^3y^5 + 3x^2y^6 = 3x^2y^4(2 - 7xy + y^2).$$

b) There are three terms in $2 - 7xy + y^2$. We determine whether the trinomial is a square. Since only y^2 is a square, we do not have a trinomial square. Can the trinomial be factored by trial and error? A key to the answer is that x is only in the term $-7xy$. The polynomial might be in a form like $(1 - y)(2 + y)$, but there would be no x in the middle term. Thus, $2 - 7xy + y^2$ cannot be factored.

c) Have we factored completely? Yes because no factor with more than one term can be factored further.

d) The check is left to the student.

EXAMPLE 6 Factor: $(p + q)(x + 2) + (p + q)(x + y)$.

a) We look for a common factor:

$$\begin{aligned}(p + q)(x + 2) + (p + q)(x + y) &= (p + q)[(x + 2) + (x + y)]\\ &= (p + q)(2x + y + 2).\end{aligned}$$

b) There are three terms in $2x + y + 2$, but this trinomial cannot be factored further.

c) Neither factor can be factored further, so we have factored completely.

d) The check is left to the student.

EXAMPLE 7 Factor: $px + py + qx + qy$.

a) We look first for a common factor. There isn't one.

b) There are four terms. We try factoring by grouping:

$$\begin{aligned}px + py + qx + qy &= p(x + y) + q(x + y)\\ &= (p + q)(x + y).\end{aligned}$$

c) Have we factored completely? Since neither factor can be factored further, we have factored completely.

d) CHECK: $(p + q)(x + y) = px + py + qx + qy$.

EXAMPLE 8 Factor: $25x^2 + 20xy + 4y^2$.

a) We look first for a common factor. There isn't one.

b) There are three terms. We determine whether the trinomial is a square. The first term and the last term are squares:

$$25x^2 = (5x)^2 \quad \text{and} \quad 4y^2 = (2y)^2.$$

Since twice the product of $5x$ and $2y$ is the other term,

$$2 \cdot 5x \cdot 2y = 20xy,$$

the trinomial is a perfect square.

We factor by writing the square roots of the square terms and the sign of the middle term:

$$25x^2 + 20xy + 4y^2 = (5x + 2y)^2.$$

c) Since $5x + 2y$ cannot be factored further, we have factored completely.

d) CHECK: $(5x + 2y)^2 = (5x)^2 + 2(5x)(2y) + (2y)^2$
$= 25x^2 + 20xy + 4y^2.$

EXAMPLE 9 Factor: $p^2q^2 + 7pq + 12$.

a) We look first for a common factor. There isn't one.

b) There are three terms. We determine whether the trinomial is a square. The first term is a square, but neither of the other terms is a square, so we do not have a trinomial square. We factor, thinking of the product pq as a single variable. We consider this possibility for factorization:

$$(pq + \square)(pq + \square).$$

We factor the last term, 12. All the signs are positive, so we consider only positive factors. Possibilities are 1, 12 and 2, 6 and 3, 4. The pair 3, 4 gives a sum of 7 for the coefficient of the middle term. Thus,

$$p^2q^2 + 7pq + 12 = (pq + 3)(pq + 4).$$

c) No factor with more than one term can be factored further, so we have factored completely.

d) CHECK: $(pq + 3)(pq + 4) = (pq)(pq) + 4 \cdot pq + 3 \cdot pq + 3 \cdot 4$
$= p^2q^2 + 7pq + 12.$

EXAMPLE 10 Factor: $8x^4 - 20x^2y - 12y^2$.

a) We look first for a common factor:

$$8x^4 - 20x^2y - 12y^2 = 4(2x^4 - 5x^2y - 3y^2).$$

b) There are three terms in $2x^4 - 5x^2y - 3y^2$. We determine whether the trinomial is a square. Since none of the terms is a square, we do not have a trinomial square. We factor $2x^4$. Possibilities are $2x^2$, x^2 and $2x$, x^3 and others. We also factor the last term, $-3y^2$. Possibilities are $3y$, $-y$ and $-3y$, y and others. We look for factors such that the sum of their products is the middle term. The x^2 in the middle term, $-5x^2y$, should lead us to try $(2x^2)(x^2)$. We try some possibilities:

$$(2x^2 - y)(x^2 + 3y) = 2x^4 + 5x^2y - 3y^2,$$
$$(2x^2 + y)(x^2 - 3y) = 2x^4 - 5x^2y - 3y^2.$$

c) No factor with more than one term can be factored further, so we have factored completely. The factorization, including the common factor, is

$$4(2x^2 + y)(x^2 - 3y).$$

d) CHECK: $4(2x^2 + y)(x^2 - 3y) = 4[(2x^2)(x^2) + 2x^2(-3y) + yx^2 + y(-3y)]$
$= 4[2x^4 - 6x^2y + x^2y - 3y^2]$
$= 4(2x^4 - 5x^2y - 3y^2)$
$= 8x^4 - 20x^2y - 12y^2.$

EXAMPLE 11 Factor: $a^4 - 16b^4$.

a) We look first for a common factor. There isn't one.

b) There are two terms. Since $a^4 = (a^2)^2$ and $16b^4 = (4b^2)^2$, we see that we do have a difference of squares. Thus,

$$a^4 - 16b^4 = (a^2 + 4b^2)(a^2 - 4b^2).$$

c) The last factor can be factored further. It is also a difference of squares. Thus,

$$a^4 - 16b^4 = (a^2 + 4b^2)(a + 2b)(a - 2b).$$

d) CHECK: $(a^2 + 4b^2)(a + 2b)(a - 2b) = (a^2 + 4b^2)(a^2 - 4b^2)$
$= a^4 - 16b^4.$

EXAMPLE 12 Factor: $40t^3 - 5s^3$.

a) We look first for a common factor:

$$40t^3 - 5s^3 = 5(8t^3 - s^3).$$

b) The factor $8t^3 - s^3$ has only two terms. It is a difference of two cubes. We factor as follows:

$$(2t - s)(4t^2 + 2ts + s^2).$$

c) No factor with more than one term can be factored further, so we have factored completely. The factorization, including the common factor, is

$$5(2t - s)(4t^2 + 2ts + s^2).$$

Do Exercises 6–13.

Factor.

6. $x^4y^2 + 2x^3y + 3x^2y$

7. $10p^6q^2 + 4p^5q^3 + 2p^4q^4$

8. $(a - b)(x + 5) + (a - b)(x + y^2)$

9. $ax^2 + ay + bx^2 + by$

10. $x^4 + 2x^2y^2 + y^4$

11. $x^2y^2 + 5xy + 4$

12. $p^4 - 81q^4$

13. $15a^3 - 120b^3$

Answers on page A-36

Study Tips

SPECIAL VIDEOTAPES

In addition to the videotaped lectures for the books, there are two special tapes, *Math Study Skills for Students* and *Math Problem Solving in the Real World*. Check with your instructor to see whether these tapes are available on your campus.

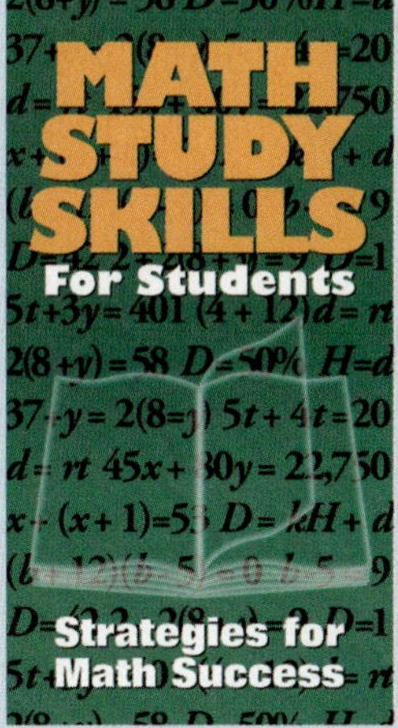

10.7 EXERCISE SET

For Extra Help

Digital Video Tutor CD 9 Videotape 12 | InterAct Math | Math Tutor Center | MathXL | MyMathLab

a Factor completely.

1. $3x^2 - 192$

2. $2t^2 - 18$

3. $a^2 + 25 - 10a$

4. $y^2 + 49 + 14y$

5. $2x^2 - 11x + 12$

6. $8y^2 - 18y - 5$

7. $x^3 + 24x^2 + 144x$

8. $x^3 - 18x^2 + 81x$

9. $x^3 + 3x^2 - 4x - 12$

10. $x^3 - 5x^2 - 25x + 125$

11. $48x^2 - 3$

12. $50x^2 - 32$

13. $9x^3 + 12x^2 - 45x$

14. $20x^3 - 4x^2 - 72x$

15. $x^2 + 4$

16. $t^2 + 25$

17. $x^4 + 7x^2 - 3x^3 - 21x$

18. $m^4 + 8m^3 + 8m^2 + 64m$

19. $x^5 - 14x^4 + 49x^3$

20. $2x^6 + 8x^5 + 8x^4$

21. $20 - 6x - 2x^2$

22. $45 - 3x - 6x^2$

23. $x^2 - 6x + 1$

24. $x^2 + 8x + 5$

25. $4x^4 - 64$

26. $5x^5 - 80x$

27. $1 - y^8$

28. $t^8 - 1$

29. $x^5 - 4x^4 + 3x^3$

30. $x^6 - 2x^5 + 7x^4$

31. $\frac{1}{81}x^6 - \frac{8}{27}x^3 + \frac{16}{9}$

32. $36a^2 - 15a + \frac{25}{16}$

33. $mx^2 + my^2$

34. $125a^3 - 8b^3$

35. $9x^2y^2 - 36xy$

36. $x^2y - xy^2$

37. $2\pi rh + 2\pi r^2$

38. $10p^4q^4 + 35p^3q^3 + 10p^2q^2$

39. $(a + b)(x - 3) + (a + b)(x + 4)$

40. $5c(a^3 + b) - (a^3 + b)$

41. $(x - 1)(x + 1) - y(x + 1)$

42. $3(p - q) - q^2(p - q)$

43. $n^2 + 2n + np + 2p$

44. $a^2 - 3a + ay - 3y$

45. $6q^2 - 3q + 2pq - p$

46. $2x^2 - 4x + xy - 2y$

47. $4b^2 + a^2 - 4ab$

48. $x^2 + y^2 - 2xy$

49. $16x^2 + 24xy + 9y^2$

50. $9c^2 + 6cd + d^2$

51. $49m^4 - 112m^2n + 64n^2$

52. $4x^2y^2 + 12xyz + 9z^2$

53. $y^4 + 10y^2z^2 + 25z^4$

54. $0.01x^4 - 0.1x^2y^2 + 0.25y^4$

55. $\frac{1}{4}a^2 + \frac{1}{3}ab + \frac{1}{9}b^2$

56. $4p^2q + pq^2 + 4p^3$

57. $a^2 - ab - 2b^2$

58. $3b^2 - 17ab - 6a^2$

59. $2mn - 360n^2 + m^2$

60. $15 + x^2y^2 + 8xy$

61. $m^2n^2 - 4mn - 32$

62. $p^2q^2 + 7pq + 6$

63. $r^5s^2 - 10r^4s + 16r^3$

64. $p^5q^2 + 3p^4q - 10p^3$

65. $a^5 + 4a^4b - 5a^3b^2$

66. $2s^6t^2 + 10s^3t^3 + 12t^4$

67. $a^2 - \frac{1}{25}b^2$

68. $p^2 - \frac{1}{49}b^2$

69. $7x^6 - 7y^6$

70. $16p^3 + 54q^3$

71. $16 - p^4q^4$

72. $15a^4 - 15b^4$

73. $1 - 16x^{12}y^{12}$

74. $81a^4 - b^4$

75. $q^3 + 8q^2 - q - 8$

76. $m^3 - 7m^2 - 4m + 28$

77. $112xy + 49x^2 + 64y^2$

78. $4ab^5 - 32b^4 + a^2b^6$

79. DW Kelly factored $16 - 8x + x^2$ as $(x - 4)^2$, while Tony factored it as $(4 - x)^2$. Evaluate each expression for several values of x. Then explain why both answers are correct.

80. DW Describe in your own words a strategy that can be used to factor polynomials.

SKILL MAINTENANCE

CD Sales. The line graph below charts data concerning the number of music CDs sold in recent years. Use it for Exercises 81–86. [8.1a]

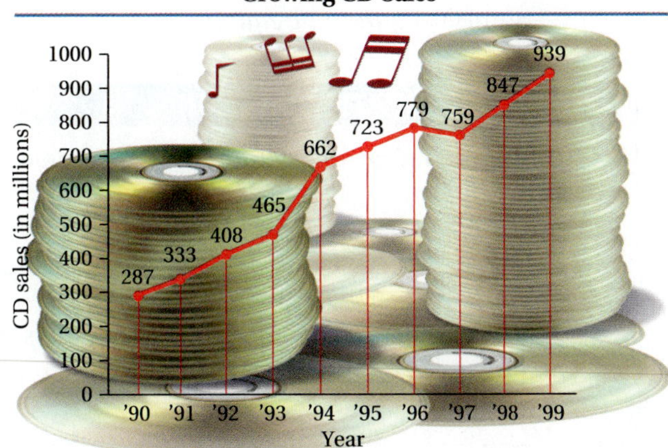

Source: Recording Industry Association of America

81. In which year were CD sales highest?

82. In which year were CD sales lowest?

83. In which year were CD sales 779 million?

84. What were CD sales in 1998?

85. What was the percent of increase in sales from 1998 to 1999? [7.5a]

86. What was the percent of decrease in sales from 1996 to 1997? [7.5a]

87. Divide: $\frac{7}{5} \div \left(-\frac{11}{10}\right)$. [6.6c]

88. Multiply: $(5x - t)^2$. [9.6c]

89. Solve $A = aX + bX - 7$ for X. [7.4b]

90. Solve: $4(x - 9) - 2(x + 7) < 14$. [7.7e]

SYNTHESIS

Factor completely.

91. $a^4 - 2a^2 + 1$

92. $x^4 + 9$

93. $12.25x^2 - 7x + 1$

94. $\frac{1}{5}x^2 - x + \frac{4}{5}$

95. $5x^2 + 13x + 7.2$

96. $x^3 - (x - 3x^2) - 3$

97. $18 + y^3 - 9y - 2y^2$

98. $-(x^4 - 7x^2 - 18)$

99. $a^3 + 4a^2 + a + 4$

100. $x^3 + x^2 - (4x + 4)$

101. $x^3 - x^2 - 4x + 4$

102. $3x^4 - 15x^2 + 12$

103. $y^2(y - 1) - 2y(y - 1) + (y - 1)$

104. $y^2(y + 1) - 4y(y + 1) - 21(y + 1)$

105. $(y + 4)^2 + 2x(y + 4) + x^2$

106. $6(x - 1)^2 + 7y(x - 1) - 3y^2$

Objectives

a Solve equations (already factored) using the principle of zero products.

b Solve quadratic equations by factoring and then using the principle of zero products.

10.8 SOLVING QUADRATIC EQUATIONS BY FACTORING

Second-degree equations like $x^2 + x - 156 = 0$ and $9 - x^2 = 0$ are examples of *quadratic equations*.

QUADRATIC EQUATION

A **quadratic equation** is an equation equivalent to an equation of the type

$$ax^2 + bx + c = 0, \quad a \neq 0.$$

In order to solve quadratic equations, we need a new equation-solving principle.

a The Principle of Zero Products

The product of two numbers is 0 if one or both of the numbers is 0. Furthermore, *if any product is* 0, *then a factor must be* 0. For example:

If $7x = 0$, then we know that $x = 0$.

If $x(2x - 9) = 0$, then we know that $x = 0$ or $2x - 9 = 0$.

If $(x + 3)(x - 2) = 0$, then we know that $x + 3 = 0$ or $x - 2 = 0$.

In a product such as $ab = 24$, we cannot conclude with certainty that a is 24 or that b is 24, but if $ab = 0$, we can conclude that $a = 0$ or $b = 0$.

EXAMPLE 1 Solve: $(x + 3)(x - 2) = 0$.

We have a product of 0. This equation will be true when either factor is 0. Thus it is true when

$$x + 3 = 0 \quad \text{or} \quad x - 2 = 0.$$

Here we have two simple equations that we know how to solve:

$$x = -3 \quad \text{or} \quad x = 2.$$

Each of the numbers -3 and 2 is a solution of the original equation, as we can see in the following checks.

CHECK: For -3:

$$\begin{array}{r|l} (x + 3)(x - 2) & = 0 \\ \hline (-3 + 3)(-3 - 2) & ?\ 0 \\ 0(-5) & \\ 0 & \end{array} \quad \text{TRUE}$$

For 2:

$$\begin{array}{r|l} (x + 3)(x - 2) & = 0 \\ \hline (2 + 3)(2 - 2) & ?\ 0 \\ 5(0) & \\ 0 & \end{array} \quad \text{TRUE}$$

We now have a principle to help in solving quadratic equations.

THE PRINCIPLE OF ZERO PRODUCTS

An equation $ab = 0$ is true if and only if $a = 0$ is true or $b = 0$ is true, or both are true. (A product is 0 if and only if one or both of the factors is 0.)

Study Tips

WORKING WITH A CLASSMATE

If you are finding it difficult to master a particular topic or concept, try talking about it with a classmate. Verbalizing your questions about the material might help clarify it. If your classmate is also finding the material difficult, it is possible that the majority of the people in your class are confused and you can ask your instructor to explain the concept again.

EXAMPLE 2 Solve: $(5x+1)(x-7)=0$.

$$(5x+1)(x-7)=0$$

$5x+1=0$ *or* $x-7=0$ Using the principle of zero products

$5x=-1$ *or* $x=7$ Solving the two equations separately

$x=-\frac{1}{5}$ *or* $x=7$

CHECK: For $-\frac{1}{5}$:

$$\begin{array}{c|l} (5x+1)(x-7) & 0 \\ \hline \left(5\left(-\frac{1}{5}\right)+1\right)\left(-\frac{1}{5}-7\right) & ?\ 0 \\ (-1+1)\left(-7\frac{1}{5}\right) & \\ 0\left(-7\frac{1}{5}\right) & \\ 0 & \end{array} \quad \text{TRUE}$$

For 7:

$$\begin{array}{c|l} (5x+1)(x-7) & 0 \\ \hline (5(7)+1)(7-7) & ?\ 0 \\ (35+1)\cdot 0 & \\ 36\cdot 0 & \\ 0 & \end{array} \quad \text{TRUE}$$

The solutions are $-\frac{1}{5}$ and 7.

When you solve an equation using the principle of zero products, a check by substitution, as in Examples 1 and 2, will detect errors in solving.

Do Exercises 1–3.

When some factors have only one term, you can still use the principle of zero products.

EXAMPLE 3 Solve: $x(2x-9)=0$.

We have

$$x(2x-9)=0$$

$x=0$ *or* $2x-9=0$ Using the principle of zero products

$x=0$ *or* $2x=9$

$x=0$ *or* $x=\frac{9}{2}$.

The solutions are 0 and $\frac{9}{2}$. The check is left to the student.

Do Exercise 4.

Solve using the principle of zero products.

1. $(x-3)(x+4)=0$

2. $(x-7)(x-3)=0$

3. $(4t+1)(3t-2)=0$

4. Solve: $y(3y-17)=0$.

Answers on page A-36

5. Solve: $x^2 - x - 6 = 0$.

Solve.

6. $x^2 - 3x = 28$

7. $x^2 = 6x - 9$

Solve.

8. $x^2 - 4x = 0$

9. $9x^2 = 16$

Answers on page A-36

b Using Factoring to Solve Equations

Using factoring and the principle of zero products, we can solve some new kinds of equations. Thus we have extended our equation-solving abilities.

EXAMPLE 4 Solve: $x^2 + 5x + 6 = 0$.

Compare this equation to those that we know how to solve from Chapter 7. There are no like terms to collect, and we have a squared term. We first factor the polynomial. Then we use the principle of zero products.

$$x^2 + 5x + 6 = 0$$

$$(x + 2)(x + 3) = 0 \quad \text{Factoring}$$

$$x + 2 = 0 \quad or \quad x + 3 = 0 \quad \text{Using the principle of zero products}$$

$$x = -2 \quad or \quad x = -3$$

CHECK: For -2:

$$\begin{array}{c|c} x^2 + 5x + 6 & 0 \\ \hline (-2)^2 + 5(-2) + 6 \ ? & 0 \\ 4 - 10 + 6 & \\ -6 + 6 & \\ 0 & \end{array} \quad \text{TRUE}$$

For -3:

$$\begin{array}{c|c} x^2 + 5x + 6 & 0 \\ \hline (-3)^2 + 5(-3) + 6 \ ? & 0 \\ 9 - 15 + 6 & \\ -6 + 6 & \\ 0 & \end{array} \quad \text{TRUE}$$

The solutions are -2 and -3.

CAUTION!

Keep in mind that you *must* have 0 on one side of the equation before you can use the principle of zero products. Get all nonzero terms on one side and 0 on the other.

Do Exercise 5.

EXAMPLE 5 Solve: $x^2 - 8x = -16$.

We first add 16 to get a 0 on one side:

$$x^2 - 8x = -16$$

$$x^2 - 8x + 16 = 0 \quad \text{Adding 16}$$

$$(x - 4)(x - 4) = 0 \quad \text{Factoring}$$

$$x - 4 = 0 \quad or \quad x - 4 = 0 \quad \text{Using the principle of zero products}$$

$$x = 4 \quad or \quad x = 4. \quad \text{Solving each equation}$$

There is only one solution, 4. The check is left to the student.

Do Exercises 6 and 7.

EXAMPLE 6 Solve: $x^2 + 5x = 0$.

$$x^2 + 5x = 0$$

$$x(x + 5) = 0 \quad \text{Factoring out a common factor}$$

$$x = 0 \quad or \quad x + 5 = 0 \quad \text{Using the principle of zero products}$$

$$x = 0 \quad or \quad x = -5$$

The solutions are 0 and -5. The check is left to the student.

EXAMPLE 7 Solve: $4x^2 = 25$.

$$4x^2 = 25$$
$$4x^2 - 25 = 0 \quad \text{Subtracting 25 on both sides to get 0 on one side}$$
$$(2x - 5)(2x + 5) = 0 \quad \text{Factoring a difference of squares}$$
$$2x - 5 = 0 \quad or \quad 2x + 5 = 0$$
$$2x = 5 \quad or \quad 2x = -5 \quad \text{Solving each equation}$$
$$x = \frac{5}{2} \quad or \quad x = -\frac{5}{2}$$

The solutions are $\frac{5}{2}$ and $-\frac{5}{2}$. The check is left to the student.

Do Exercises 8 and 9 on the preceding page.

EXAMPLE 8 Solve: $-5x^2 + 2x + 3 = 0$.

In this case, the leading coefficient of the trinomial is negative. Thus we first multiply by -1 and then proceed as we have in Examples 1–7.

$$-5x^2 + 2x + 3 = 0$$
$$-1(-5x^2 + 2x + 3) = -1 \cdot 0 \quad \text{Multiplying by } -1$$
$$5x^2 - 2x - 3 = 0 \quad \text{Simplifying}$$
$$(5x + 3)(x - 1) = 0 \quad \text{Factoring}$$
$$5x + 3 = 0 \quad or \quad x - 1 = 0 \quad \text{Using the principle of zero products}$$
$$5x = -3 \quad or \quad x = 1$$
$$x = -\frac{3}{5} \quad or \quad x = 1$$

The solutions are $-\frac{3}{5}$ and 1. The check is left to the student.

Do Exercises 10 and 11.

EXAMPLE 9 Solve: $(x + 2)(x - 2) = 5$.

Be careful with an equation like this one! It might be tempting to set each factor equal to 5. Remember: We must have a 0 on one side. We first carry out the product on the left. Then we subtract 5 on both sides to get 0 on one side. Then we proceed with the principle of zero products.

$$(x + 2)(x - 2) = 5$$
$$x^2 - 4 = 5 \quad \text{Multiplying on the left}$$
$$x^2 - 4 - 5 = 5 - 5 \quad \text{Subtracting 5}$$
$$x^2 - 9 = 0 \quad \text{Simplifying}$$
$$(x + 3)(x - 3) = 0 \quad \text{Factoring}$$
$$x + 3 = 0 \quad or \quad x - 3 = 0 \quad \text{Using the principle of zero products}$$
$$x = -3 \quad or \quad x = 3$$

The solutions are -3 and 3. The check is left to the student.

Do Exercise 12.

Solve.

10. $-2x^2 + 13x - 21 = 0$

11. $10 - 3x - x^2 = 0$

12. Solve: $(x + 1)(x - 1) = 8$.

Answers on page A-36

13. Find the x-intercepts of the graph shown below.

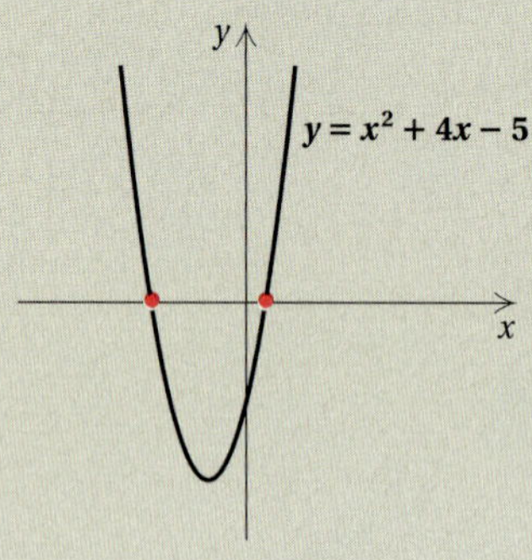

14. Use *only* the graph shown below to solve $3x - x^2 = 0$.

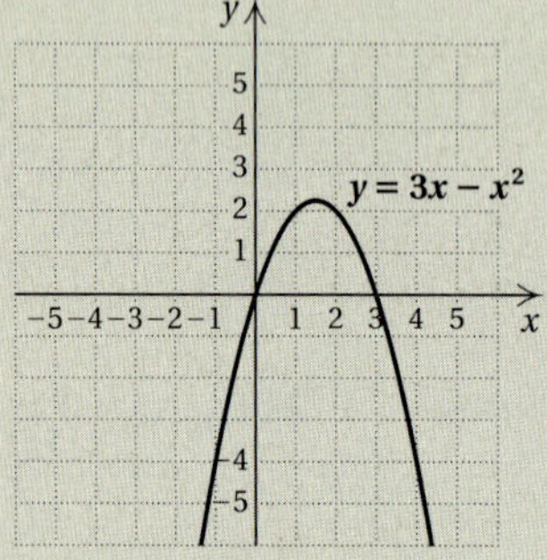

Answers on page A-36

AG ALGEBRAIC–GRAPHICAL CONNECTION

In Chapter 8, we graphed linear equations of the type $y = mx + b$ and $Ax + By = C$. Recall that to find the x-intercept, we replaced y with 0 and solved for x. This procedure can also be used to find the x-intercepts when an equation of the form $y = ax^2 + bx + c$, $a \neq 0$, is to be graphed. Although the details of creating such graphs will be left to Chapter 16, we consider them briefly here from the standpoint of finding the x-intercepts. The graphs are shaped like the following curves. Note that each x-intercept represents a solution of $ax^2 + bx + c = 0$.

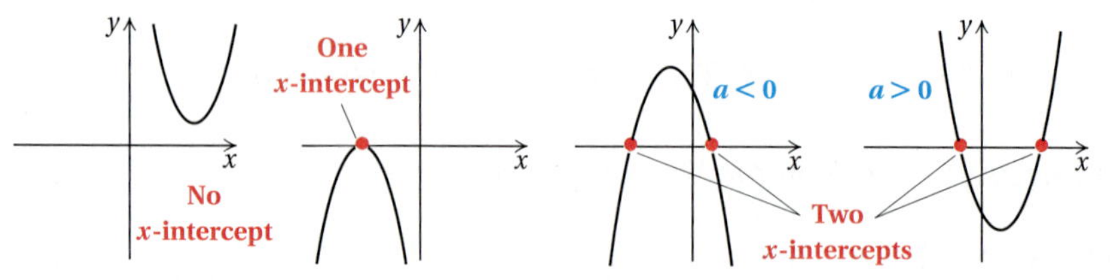

EXAMPLE 10 Find the x-intercepts of the graph of $y = x^2 - 4x - 5$ shown at right. (The grid is intentionally not included.)

To find the x-intercepts, we let $y = 0$ and solve for x:

$0 = x^2 - 4x - 5$ Substituting 0 for y

$0 = (x - 5)(x + 1)$ Factoring

$x - 5 = 0$ *or* $x + 1 = 0$ Using the principle of zero products

$x = 5$ *or* $x = -1$.

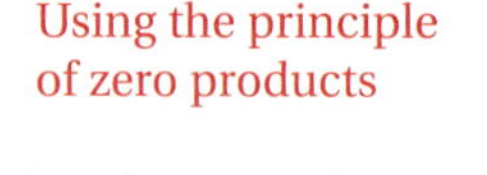

The x-intercepts are $(5, 0)$ and $(-1, 0)$. We can now label them on the graph.

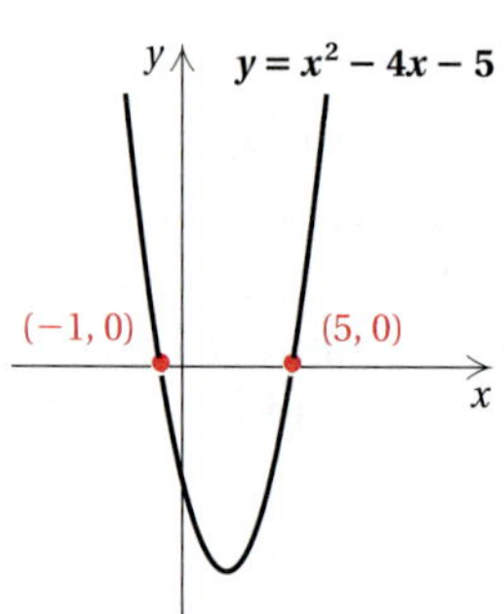

Do Exercises 13 and 14.

CALCULATOR CORNER

Solving Quadratic Equations We can solve quadratic equations graphically. Consider the equation $x^2 + 2x = 8$. First, we must write the equation with 0 on one side. To do this, we subtract 8 on both sides of the equation; we get $x^2 + 2x - 8 = 0$. Next, we graph $y = x^2 + 2x - 8$ in a window that shows the x-intercepts. The standard window works well in this case.

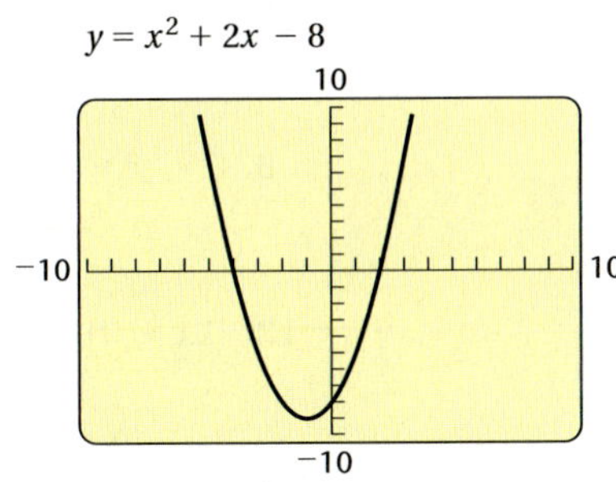

The solutions of the equation are the values of x for which $x^2 + 2x - 8 = 0$. These are also the first coordinates of the x-intercepts of the graph. We use the ZERO feature from the CALC menu to find these numbers. To find the solution corresponding to the leftmost x-intercept, we first press [2nd] [CALC] [2] to select the ZERO feature. The prompt "Left Bound?" appears. Next, we use the [◁] or the [▷] key to move the cursor to the left of the intercept and press [ENTER]. Now the prompt "Right Bound?" appears. Then we move the cursor to the right of the intercept and press [ENTER]. The prompt "Guess?" appears. We move the cursor close to the intercept and press [ENTER] again. We now see the cursor positioned at the leftmost x-intercept and the coordinates of that point, $x = -4$, $y = 0$, are displayed. Thus, $x^2 + 2x - 8 = 0$ when $x = -4$. This is one solution of the equation.

We can repeat this procedure to find the first coordinate of the other x-intercept. We see that $x = 2$ at that point. Thus the solutions of the equation are -4 and 2.

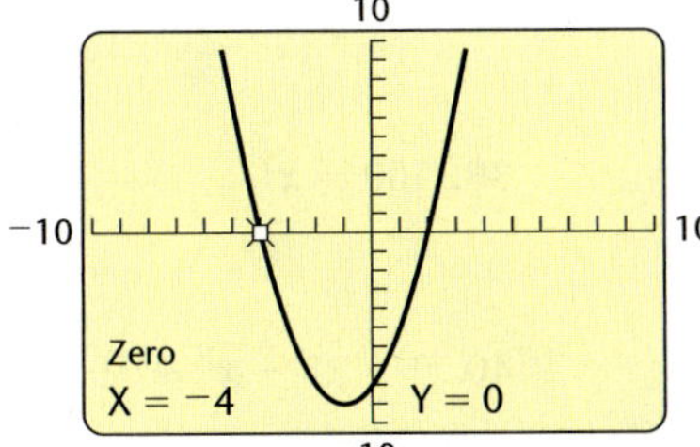

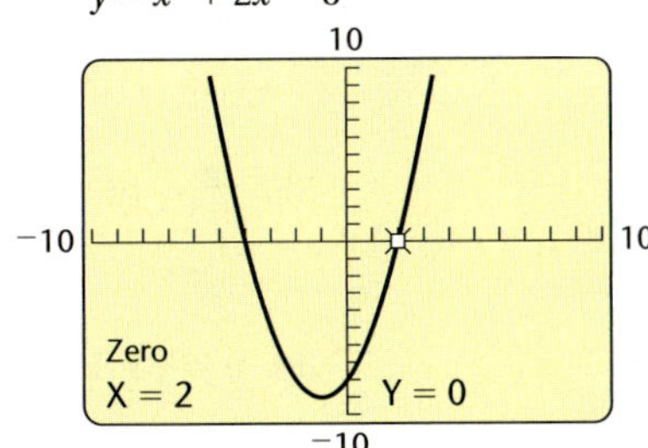

Exercises:

1. Solve each of the equations in Examples 4–8 graphically.

10.8 EXERCISE SET

For Extra Help

Digital Video Tutor CD 9 Videotape 12 | InterAct Math | Math Tutor Center | MathXL | MyMathLab

a Solve using the principle of zero products.

1. $(x + 4)(x + 9) = 0$

2. $(x + 2)(x - 7) = 0$

3. $(x + 3)(x - 8) = 0$

4. $(x + 6)(x - 8) = 0$

5. $(x + 12)(x - 11) = 0$

6. $(x - 13)(x + 53) = 0$

7. $x(x + 3) = 0$

8. $y(y + 5) = 0$

9. $0 = y(y + 18)$

10. $0 = x(x - 19)$

11. $(2x + 5)(x + 4) = 0$

12. $(2x + 9)(x + 8) = 0$

13. $(5x + 1)(4x - 12) = 0$

14. $(4x + 9)(14x - 7) = 0$

15. $(7x - 28)(28x - 7) = 0$

16. $(13x + 14)(6x - 5) = 0$

17. $2x(3x - 2) = 0$

18. $55x(8x - 9) = 0$

19. $\left(\frac{1}{5} + 2x\right)\left(\frac{1}{9} - 3x\right) = 0$

20. $\left(\frac{7}{4}x - \frac{1}{16}\right)\left(\frac{2}{3}x - \frac{16}{15}\right) = 0$

21. $(0.3x - 0.1)(0.05x + 1) = 0$

22. $(0.1x + 0.3)(0.4x - 20) = 0$

23. $9x(3x - 2)(2x - 1) = 0$

24. $(x + 5)(x - 75)(5x - 1) = 0$

b Solve by factoring and using the principle of zero products. Remember to check.

25. $x^2 + 6x + 5 = 0$

26. $x^2 + 7x + 6 = 0$

27. $x^2 + 7x - 18 = 0$

28. $x^2 + 4x - 21 = 0$

29. $x^2 - 8x + 15 = 0$

30. $x^2 - 9x + 14 = 0$

31. $x^2 - 8x = 0$

32. $x^2 - 3x = 0$

33. $x^2 + 18x = 0$

34. $x^2 + 16x = 0$

35. $x^2 = 16$

36. $100 = x^2$

37. $9x^2 - 4 = 0$

38. $4x^2 - 9 = 0$

39. $0 = 6x + x^2 + 9$

40. $0 = 25 + x^2 + 10x$

41. $x^2 + 16 = 8x$

42. $1 + x^2 = 2x$

43. $5x^2 = 6x$

44. $7x^2 = 8x$

45. $6x^2 - 4x = 10$

46. $3x^2 - 7x = 20$

47. $12y^2 - 5y = 2$

48. $2y^2 + 12y = -10$

49. $t(3t + 1) = 2$

50. $x(x - 5) = 14$

51. $100y^2 = 49$

52. $64a^2 = 81$

53. $x^2 - 5x = 18 + 2x$

54. $3x^2 + 8x = 9 + 2x$

55. $10x^2 - 23x + 12 = 0$

56. $12x^2 + 17x - 5 = 0$

Find the x-intercepts for the graph of the equation. (The grids are intentionally not included.)

57.

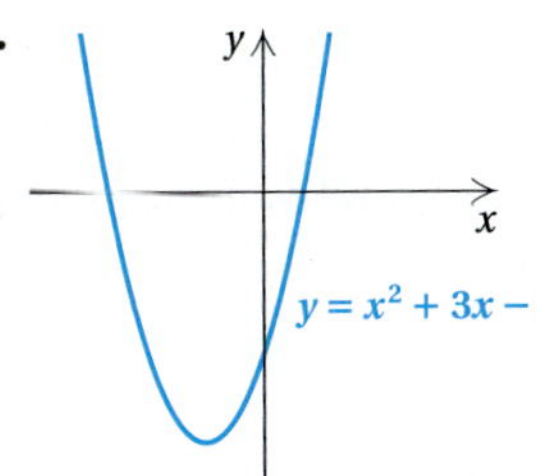

58.

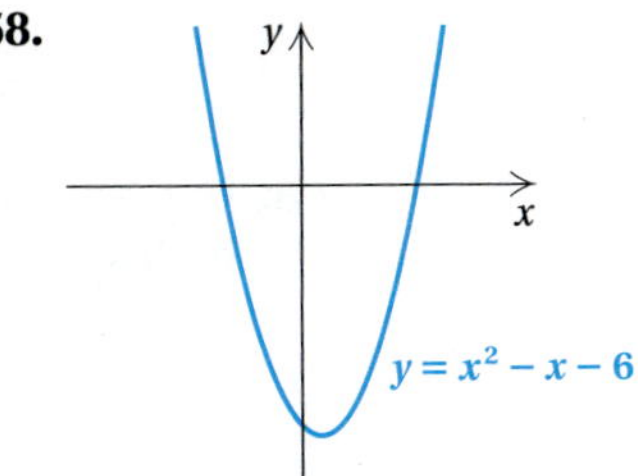

59.

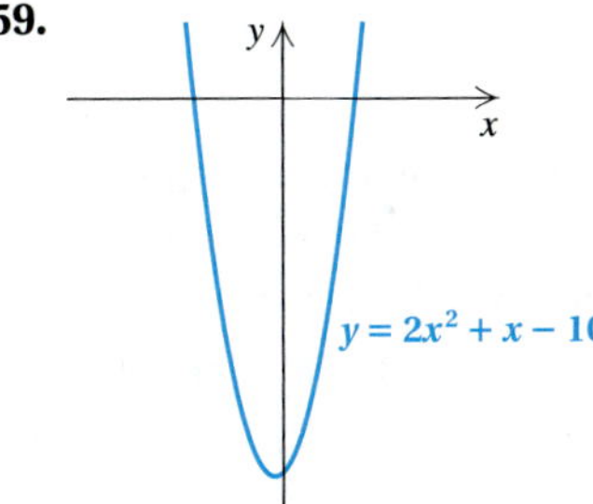

60.

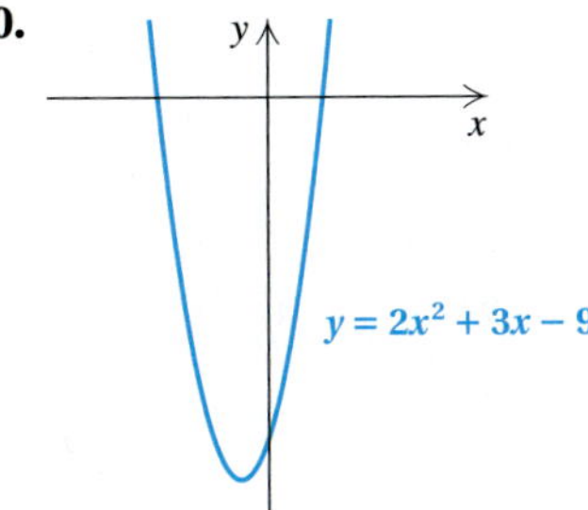

61.

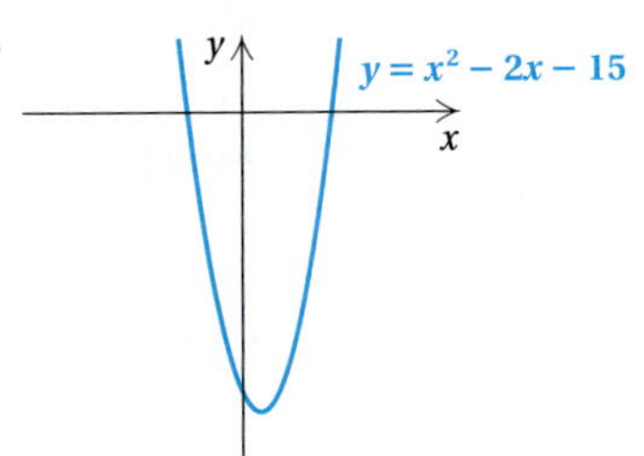

62.

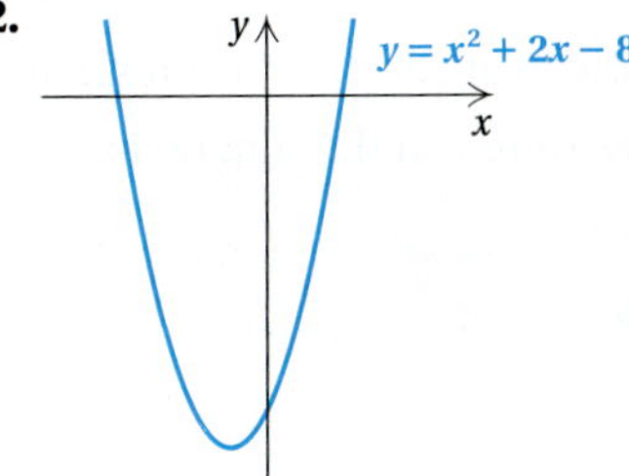

63. Use the following graph to solve $x^2 - 3x - 4 = 0$.

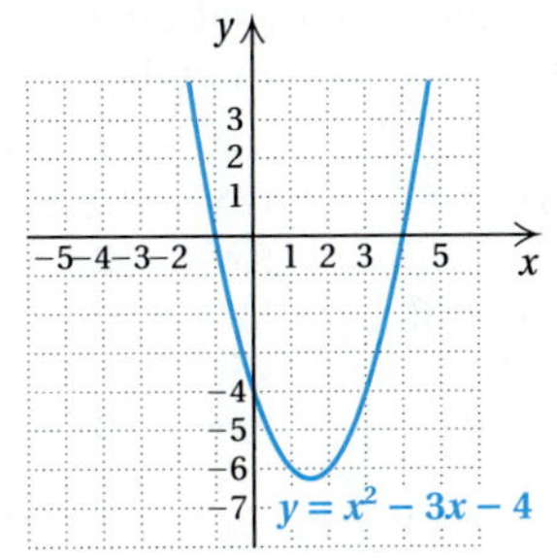

64. Use the following graph to solve $x^2 + x - 6 = 0$.

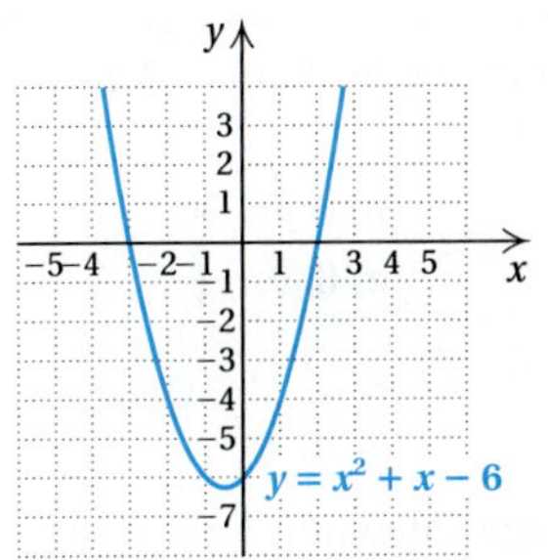

65. Use the following graph to solve $-x^2 + 2x + 3 = 0$.

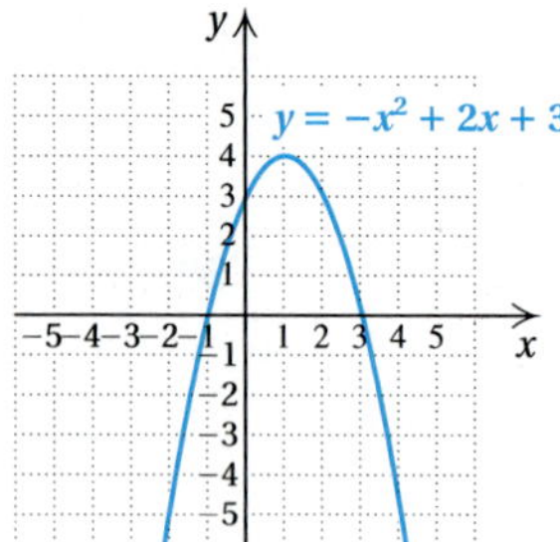

66. Use the following graph to solve $-x^2 - x + 6 = 0$.

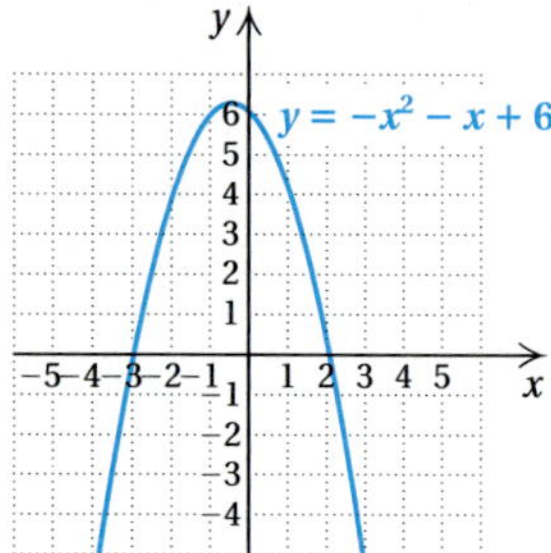

67. **DW** What is wrong with the following? Explain the correct method of solution.

$$(x - 3)(x + 4) = 8$$
$$x - 3 = 8 \quad or \quad x + 4 = 8$$
$$x = 11 \quad or \quad x = 4$$

68. **DW** What is incorrect about solving $x^2 = 3x$ by dividing both sides by x?

SKILL MAINTENANCE

Translate to an algebraic expression. [6.1b]

69. The square of the sum of a and b

70. The sum of the squares of a and b

Divide. [6.6a, c]

71. $144 \div (-9)$

72. $-24.3 \div 5.4$

73. $-\frac{5}{8} \div \frac{3}{16}$

74. $-\frac{3}{16} \div \left(-\frac{5}{8}\right)$

SYNTHESIS

Solve.

75. $b(b + 9) = 4(5 + 2b)$

76. $y(y + 8) = 16(y - 1)$

77. $(t - 3)^2 = 36$

78. $(t - 5)^2 = 2(5 - t)$

79. $x^2 - \frac{1}{64} = 0$

80. $x^2 - \frac{25}{36} = 0$

81. $\frac{5}{16}x^2 = 5$

82. $\frac{27}{25}x^2 = \frac{1}{3}$

83. Find an equation that has the given numbers as solutions. For example, 3 and -2 are solutions to $x^2 - x - 6 = 0$.

a) $-3, 4$ **b)** $-3, -4$ **c)** $\frac{1}{2}, \frac{1}{2}$

d) $5, -5$ **e)** $0, 0.1, \frac{1}{4}$

84. *Matching.* Match each equation in the first column with the equivalent equation in the second column.

$x^2 + 10x - 2 = 0$	$4x^2 + 8x + 36 = 0$
$(x - 6)(x + 3) = 0$	$(2x + 8)(2x - 5) = 0$
$5x^2 - 5 = 0$	$9x^2 - 12x + 24 = 0$
$(2x - 5)(x + 4) = 0$	$(x + 1)(5x - 5) = 0$
$x^2 + 2x + 9 = 0$	$x^2 - 3x - 18 = 0$
$3x^2 - 4x + 8 = 0$	$2x^2 + 20x - 4 = 0$

Use a graphing calculator to find the solutions of each equation. Round solutions to the nearest hundredth.

85. $x^2 - 9.10x + 15.77 = 0$

86. $x^2 + 1.80x - 5.69 = 0$

87. $x^2 + 13.74x + 42.00 = 0$

88. $-x^2 + 0.63x + 0.22 = 0$

89. $0.84x^2 - 2.30x = 0$

90. $6.4x^2 - 8.45x - 94.06 = 0$

10.9 APPLICATIONS OF QUADRATIC EQUATIONS

Objective

a Solve applied problems involving quadratic equations that can be solved by factoring.

b Solve applied problems involving the Pythagorean theorem and quadratic equations that can be solved by factoring.

a Applied Problems, Quadratic Equations, and Factoring

We can now use our new method for solving quadratic equations and the five steps for solving problems.

EXAMPLE 1 *Manufacturing.* Wooden Work, Ltd., builds cutting boards that are twice as long as they are wide. The most popular board that Wooden Work makes has an area of 800 cm^2. What are the dimensions of the board?

1. **Familiarize.** We first make a drawing. Recall that the area of any rectangle is Length · Width. We let $x =$ the width of the board, in centimeters. The length is then $2x$.

2. **Translate.** We reword and translate as follows:

Rewording: The area of the rectangle is 800 cm^2.

Translating: $2x \cdot x = 800$

3. **Solve.** We solve the equation as follows:

$$2x \cdot x = 800$$
$$2x^2 = 800$$
$$2x^2 - 800 = 0 \quad \text{Subtracting 800 to get 0 on one side}$$
$$2(x^2 - 400) = 0 \quad \text{Removing a common factor of 2}$$
$$2(x - 20)(x + 20) = 0 \quad \text{Factoring a difference of squares}$$
$$(x - 20)(x + 20) = 0 \quad \text{Dividing by 2}$$
$$x - 20 = 0 \quad or \quad x + 20 = 0 \quad \text{Using the principle of zero products}$$
$$x = 20 \quad or \quad x = -20. \quad \text{Solving each equation}$$

4. **Check.** The solutions of the equation are 20 and -20. Since the width must be positive, -20 cannot be a solution. To check 20 cm, we note that if the width is 20 cm, then the length is $2 \cdot 20$ cm $= 40$ cm and the area is 20 cm · 40 cm = 800 cm^2. Thus the solution 20 checks.

5. **State.** The cutting board is 20 cm wide and 40 cm long.

Do Exercise 1.

1. **Framing.** A rectangular picture frame is twice as long as it is wide. If the area of the frame is 288 in^2, find its dimensions.

Answer on page A-36

2. Dimensions of a Sail. The mainsail of Stacey's lightning-styled sailboat has an area of 125 ft^2. The sail is 15 ft taller than it is wide. Find the height and the width of the sail.

Answer on page A-36

EXAMPLE 2 *Racing Sailboat.* The height of a triangular sail on a racing sailboat is 9 ft more than the base. The area of the triangle is 110 ft^2. Find the height and the base of the sail.
Source: Whitney Gladstone, North Graphics, San Diego, CA

1. **Familiarize.** We first make a drawing. If you don't remember the formula for the area of a triangle, look it up on the inside front cover of this book or in a geometry book. The area is $\frac{1}{2}$(base)(height).
   We let b = the base of the triangle. Then $b + 9$ = the height.

2. **Translate.** It helps to reword this problem before translating:

$\frac{1}{2}$	times	Base	times	Height	is	110.	Rewording
↓	↓	↓	↓	↓	↓	↓	
$\frac{1}{2}$	$\cdot$	b	$\cdot$	$(b+9)$	$=$	110	Translating

3. **Solve.** We solve the equation as follows:

$$\frac{1}{2}\cdot b\cdot(b+9) = 110$$
$$\frac{1}{2}(b^2+9b) = 110 \quad \text{Multiplying}$$
$$2\cdot\frac{1}{2}(b^2+9b) = 2\cdot 110 \quad \text{Multiplying by 2}$$
$$b^2+9b = 220 \quad \text{Simplifying}$$
$$b^2+9b-220 = 220-220 \quad \text{Subtracting 220 to get 0 on one side}$$
$$b^2+9b-220 = 0$$
$$(b-11)(b+20) = 0 \quad \text{Factoring}$$
$$b-11 = 0 \quad or \quad b+20 = 0 \quad \text{Using the principle of zero products}$$
$$b = 11 \quad or \quad b = -20.$$

4. **Check.** The base of a triangle cannot have a negative length, so -20 cannot be a solution. Suppose the base is 11 ft. The height is 9 ft more than the base, so the height is 20 ft and the area is $\frac{1}{2}(11)(20)$, or 110 ft^2. These numbers check in the original problem.
5. **State.** The height is 20 ft and the base is 11 ft.

Do Exercise 2.

EXAMPLE 3 *Games in a Sports League.* In a sports league of x teams in which each team plays every other team twice, the total number N of games to be played is given by

$$x^2 - x = N.$$

Maggie's basketball league plays a total of 240 games. How many teams are in the league?

1., 2. Familiarize and **Translate.** We are given that x is the number of teams in a league and N is the number of games. To familiarize yourself with this problem, reread Example 4 in Section 9.3 where we first considered it. To find the number of teams x in a league in which 240 games are played, we substitute 240 for N in the equation:

$$x^2 - x = 240. \quad \text{Substituting 240 for } N$$

3. Solve. We solve the equation as follows:

$$x^2 - x = 240$$

$$x^2 - x - 240 = 240 - 240 \quad \text{Subtracting 240 to get 0 on one side}$$

$$x^2 - x - 240 = 0$$

$$(x - 16)(x + 15) = 0 \quad \text{Factoring}$$

$$x - 16 = 0 \quad or \quad x + 15 = 0 \quad \text{Using the principle of zero products}$$

$$x = 16 \quad or \quad x = -15.$$

4. Check. The solutions of the equation are 16 and -15. Since the number of teams cannot be negative, -15 cannot be a solution. But 16 checks, since $16^2 - 16 = 256 - 16 = 240$.

5. State. There are 16 teams in the league.

Do Exercise 3.

3. Use $N = x^2 - x$ for the following.

a) Volleyball League. Amy's volleyball league has 19 teams. What is the total number of games to be played?

b) Softball League. Barry's slow-pitch softball league plays a total of 72 games. How many teams are in the league?

Answers on page A-36

Study Tips

FIVE STEPS FOR PROBLEM SOLVING

Are you remembering to use the five steps for problem solving that were developed in Section 7.6?

1. **Familiarize** yourself with the situation.
   a) Carefully read and reread until you understand *what* you are being asked to find.
   b) Draw a diagram or see if there is a formula that applies.
   c) Assign a letter, or *variable,* to the unknown.
2. **Translate** the problem to an equation using the letter or variable.
3. **Solve** the equation.
4. **Check** the answer in the original wording of the problem.
5. **State** the answer to the problem clearly with appropriate units.

"Most worthwhile achievements are the result of many little things done in a simple direction."

Nido Quebin, speaker/entrepreneur

4. **Page Numbers.** The product of the page numbers on two facing pages of a book is 506. Find the page numbers.

Answer on page A-36

Study Tips

ASKING QUESTIONS

Don't be afraid to ask questions in class. Most instructors welcome this and encourage students to ask them. Other students probably have the same questions you do.

"Better to ask twice than lose your way once."

Danish proverb

EXAMPLE 4 *Athletic Numbers.* The product of the numbers of two consecutive entrants in a marathon race is 156. Find the numbers.

1. **Familiarize.** The numbers are consecutive integers. Recall that consecutive integers are next to each other, such as 49 and 50, or -6 and -5. Let $x =$ the smaller integer; then $x + 1 =$ the larger integer.
2. **Translate.** It helps to reword the problem before translating:

First integer	times	Second integer	is	156.	Rewording
↓	↓	↓	↓	↓	
x	$\cdot$	$(x + 1)$	$=$	156	Translating

3. **Solve.** We solve the equation as follows:

$$x(x + 1) = 156$$
$$x^2 + x = 156 \quad \text{Multiplying}$$
$$x^2 + x - 156 = 156 - 156 \quad \text{Subtracting 156 to get 0 on one side}$$
$$x^2 + x - 156 = 0 \quad \text{Simplifying}$$
$$(x - 12)(x + 13) = 0 \quad \text{Factoring}$$
$$x - 12 = 0 \quad or \quad x + 13 = 0 \quad \text{Using the principle of zero products}$$
$$x = 12 \quad or \quad x = -13.$$

4. **Check.** The solutions of the equation are 12 and -13. When x is 12, then $x + 1$ is 13, and $12 \cdot 13 = 156$. The numbers 12 and 13 are consecutive integers that are solutions to the problem. When x is -13, then $x + 1$ is -12, and $(-13)(-12) = 156$. The numbers -13 and -12 are consecutive integers, but they are not solutions of the problem because negative numbers are not used as entry numbers.
5. **State.** The entry numbers are 12 and 13.

Do Exercise 4.

b The Pythagorean Theorem

The following problems involve the Pythagorean theorem, which relates the lengths of the sides of a *right* triangle. A triangle is a **right triangle** if it has a 90°, or *right*, angle. The side opposite the 90° angle is called the **hypotenuse**. The other sides are called **legs**.

THE PYTHAGOREAN THEOREM

In any right triangle, if a and b are the lengths of the legs and c is the length of the hypotenuse, then

$$a^2 + b^2 = c^2.$$

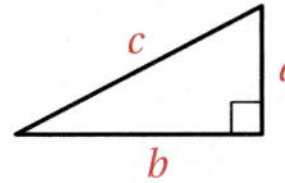

The symbol ⌐ denotes a 90° angle.

EXAMPLE 5 *Lookout Tower.* The diagonal braces in a lookout tower are 15 ft long and span a distance of 12 ft. How high does each brace reach vertically?

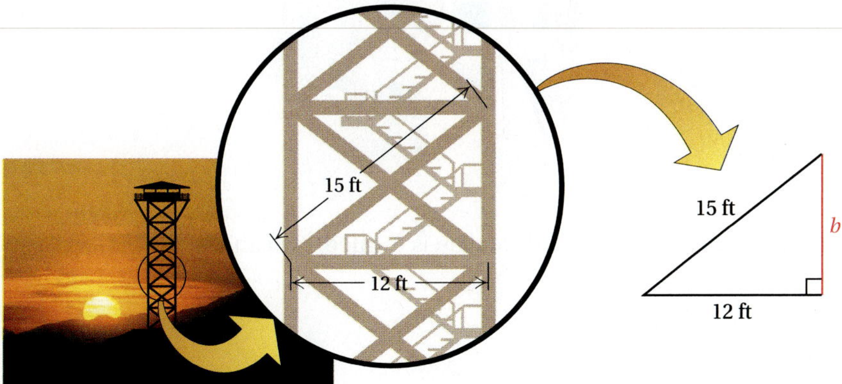

1. **Familiarize.** We make a drawing as shown above. We let b = the height of the vertical part of the brace.
2. **Translate.** Since a right triangle is formed, we can use the Pythagorean theorem:

$$a^2 + b^2 = c^2$$

$$12^2 + b^2 = 15^2. \quad \text{Substituting}$$

3. **Solve.** We solve the equation as follows:

$$12^2 + b^2 = 15^2$$

$$144 + b^2 = 225 \quad \text{Squaring 12 and 15}$$

$$b^2 - 81 = 0 \quad \text{Subtracting 225}$$

$$(b - 9)(b + 9) = 0 \quad \text{Factoring}$$

$$b - 9 = 0 \quad or \quad b + 9 = 0 \quad \text{Using the principle of zero products}$$

$$b = 9 \quad or \quad b = -9.$$

4. **Check.** Since the height cannot be negative, -9 cannot be a solution. If the height is 9 ft, we have $12^2 + 9^2 = 144 + 81 = 225$, which is 15^2. Thus, 9 checks and is a solution.
5. **Solve.** The vertical height of the brace is 9 ft.

Do Exercise 5.

5. **Reach of a Ladder.** Twila has a 26-ft ladder leaning against her house. If the bottom of the ladder is 10 ft from the base of the house, how high does the ladder reach?

Answer on page A-36

6. **Right-Triangle Geometry.** The length of one leg of a right triangle is 1 m longer than the other. The length of the hypotenuse is 5 m. Find the lengths of the legs.

Answer on page A-36

EXAMPLE 6 *Ladder Settings.* A ladder of length 13 ft is placed against a building in such a way that the distance from the top of the ladder to the ground is 7 ft more than the distance from the bottom of the ladder to the building. Find both distances.

1. **Familiarize.** We first make a drawing. The ladder and the missing distances form the hypotenuse and legs of a right triangle. We let $x =$ the length of the side (leg) across the bottom. Then $x + 7 =$ the length of the other side (leg). The hypotenuse has length 13 ft.

2. **Translate.** Since a right triangle is formed, we can use the Pythagorean theorem:

$$a^2 + b^2 = c^2$$

$$x^2 + (x + 7)^2 = 13^2. \quad \text{Substituting}$$

3. **Solve.** We solve the equation as follows:

$x^2 + (x^2 + 14x + 49) = 169$	Squaring the binomial and 13
$2x^2 + 14x + 49 = 169$	Collecting like terms
$2x^2 + 14x + 49 - 169 = 169 - 169$	Subtracting 169 to get 0 on one side
$2x^2 + 14x - 120 = 0$	Simplifying
$2(x^2 + 7x - 60) = 0$	Factoring out a common factor
$x^2 + 7x - 60 = 0$	Dividing by 2
$(x + 12)(x - 5) = 0$	Factoring
$x + 12 = 0 \quad or \quad x - 5 = 0$	Using the principle of zero products
$x = -12 \quad or \quad x = 5.$	

4. **Check.** The negative integer -12 cannot be the length of a side. When $x = 5$, $x + 7 = 12$, and $5^2 + 12^2 = 13^2$. So 5 and 12 check.

5. **State.** The distance from the top of the ladder to the ground is 12 ft. The distance from the bottom of the ladder to the building is 5 ft.

Do Exercise 6.

a Solve.

1. *Furnishings.* A rectangular table in Arlo's House of Tunes is six times as long as it is wide. The area of the table is 24 ft^2. Find the length and the width of the table.

2. *Framing.* A rectangular picture frame is three times as long as it is wide. The area of the frame is 588 in^2. Find the dimensions of the frame.

3. *Design.* The keypad and viewing window of the TI83 graphing calculator is rectangular. The length of the rectangle is 2 cm more than twice the width, and the area of the rectangle is 144 cm^2. Find the length and the width.

4. *Area of a Garden.* The length of a rectangular garden is 4 m greater than the width. The area of the garden is 96 m^2. Find the length and the width.

5. *Dimensions of a Triangle.* A triangle is 10 cm wider than it is tall. The area is 28 cm^2. Find the height and the base.

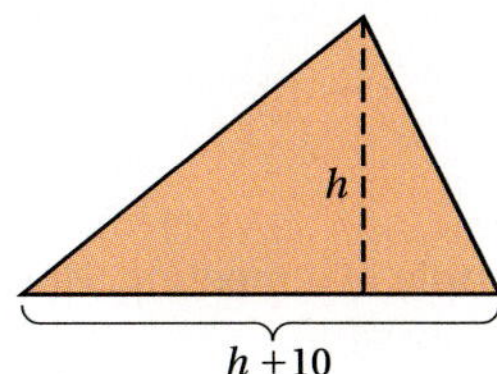

6. *Dimensions of a Triangle.* The height of a triangle is 3 cm less than the length of the base. The area of the triangle is 35 cm^2. Find the height and the length of the base.

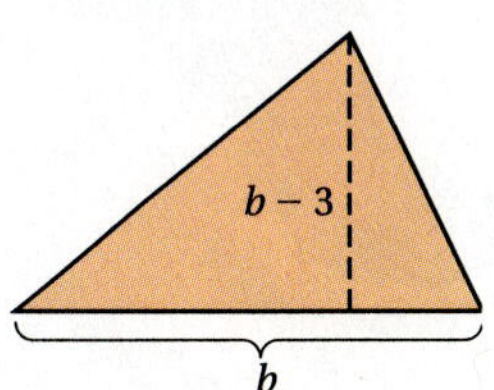

7. *Road Design.* A triangular traffic island has a base half as long as its height. The island has an area of 64 m^2. Find the base and the height.

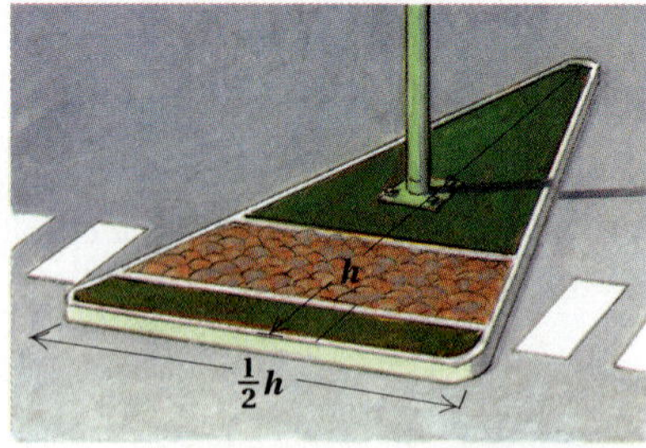

8. *Dimensions of a Sail.* The height of the jib sail on a Lightning sailboat is 5 ft greater than the length of its "foot." The area of the sail is 42 ft^2. Find the length of the foot and the height of the sail.

Games in a League. Use $x^2 - x = N$ for Exercises 9–12.

9. A chess league has 14 teams. What is the total number of games to be played?

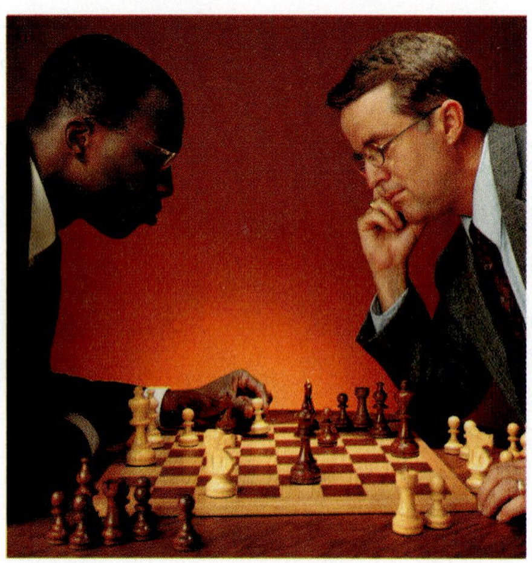

10. A women's volleyball league has 23 teams. What is the total number of games to be played?

11. A slow-pitch softball league plays a total of 132 games. How many teams are in the league?

12. A basketball league plays a total of 90 games. How many teams are in the league?

Handshakes. A researcher wants to investigate the potential spread of germs by contact. She knows that the number of possible handshakes within a group of x people is given by

$$N = \frac{1}{2}(x^2 - x).$$

13. There are 100 people at a party. How many handshakes are possible?

14. There are 40 people at a meeting. How many handshakes are possible?

15. Everyone at a meeting shook hands. There were 300 handshakes in all. How many people were at the meeting?

16. Everyone at a party shook hands. There were 153 handshakes in all. How many people were at the party?

17. *Toasting.* During a toast at a party, there were 190 "clicks" of glasses. How many people took part in the toast?

18. *High-fives.* After winning the championship, all Los Angeles Laker teammates exchanged "high-fives." Altogether there were 66 high-fives. How many players were there?

19. *Consecutive Page Numbers.* The product of the page numbers on two facing pages of a book is 210. Find the page numbers.

20. *Consecutive Page Numbers.* The product of the page numbers on two facing pages of a book is 420. Find the page numbers.

21. The product of two consecutive even integers is 168. Find the integers. (See Section 7.6.)

22. The product of two consecutive even integers is 224. Find the integers. (See Section 7.6.)

23. The product of two consecutive odd integers is 255. Find the integers.

24. The product of two consecutive odd integers is 143. Find the integers.

25. *Right-Triangle Geometry.* The length of one leg of a right triangle is 8 ft. The length of the hypotenuse is 2 ft longer than the other leg. Find the length of the hypotenuse and the other leg.

26. *Right-Triangle Geometry.* The length of one leg of a right triangle is 24 ft. The length of the other leg is 16 ft shorter than the hypotenuse. Find the length of the hypotenuse and the other leg.

27. *Roadway Design.* Elliott Street is 24 ft wide when it ends at Main Street in Brattleboro, Vermont. A 40-ft long diagonal crosswalk allows pedestrians to cross Main Street to or from either corner of Elliott Street (see the figure). Determine the width of Main Street.

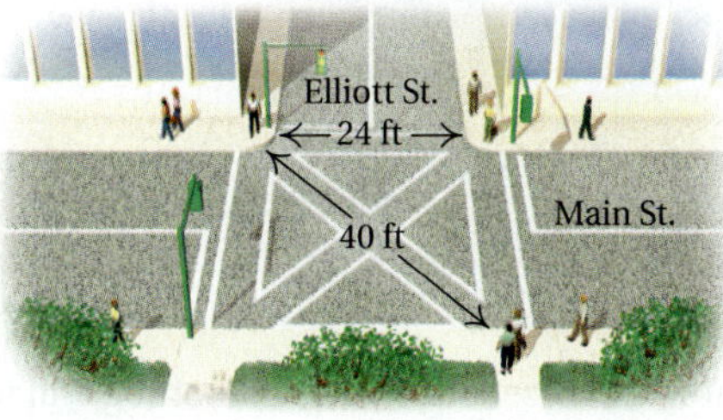

28. *Sailing.* The mainsail of a Lightning sailboat is a right triangle in which the hypotenuse is called the leech. If a 24-ft tall mainsail has a leech length of 26 ft and if Dacron® sailcloth costs $10 per square foot, find the cost of a new mainsail.

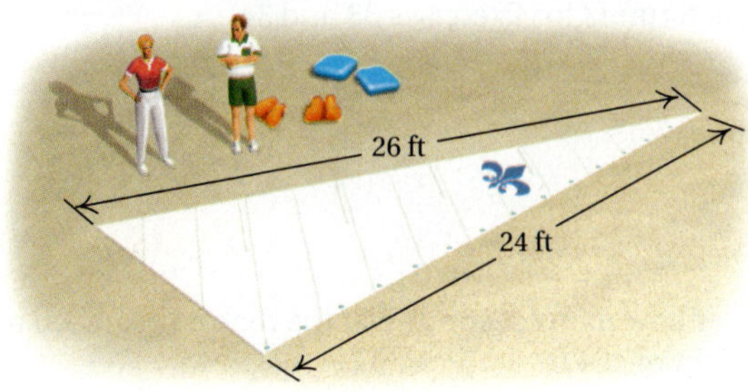

29. *Physical Education.* An outdoor-education ropes course includes a cable that slopes downward from a height of 37 ft to a height of 30 ft. The trees that the cable connects are 24 ft apart. How long is the cable?

30. *Aviation.* Engine failure forced Geraldine to pilot her Cessna 150 to an emergency landing. To land, Geraldine's plane glided 17,000 ft over a 15,000-ft stretch of deserted highway. From what altitude did the descent begin?

31. *Architecture.* An architect has allocated a rectangular space of 264 ft^2 for a square dining room and a 10-ft wide kitchen, as shown in the figure. Find the dimensions of each room.

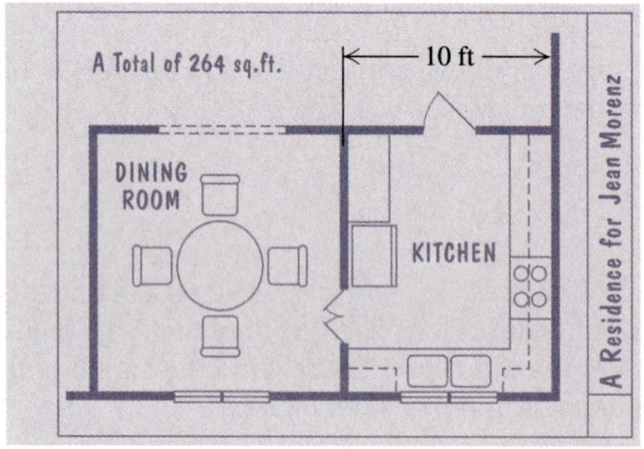

32. *Guy Wire.* The guy wire on a TV antenna is 1 m longer than the height of the antenna. If the guy wire is anchored 3 m from the foot of the antenna, how tall is the antenna?

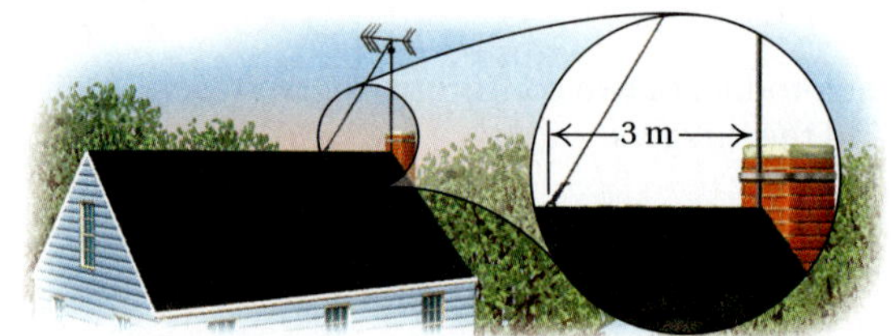

Rocket Launch. A model water rocket is launched with an initial velocity of 180 ft/sec. Its height h, in feet, after t seconds is given by the formula

$$h = 180t - 16t^2.$$

Use this formula for Exercises 33 and 34.

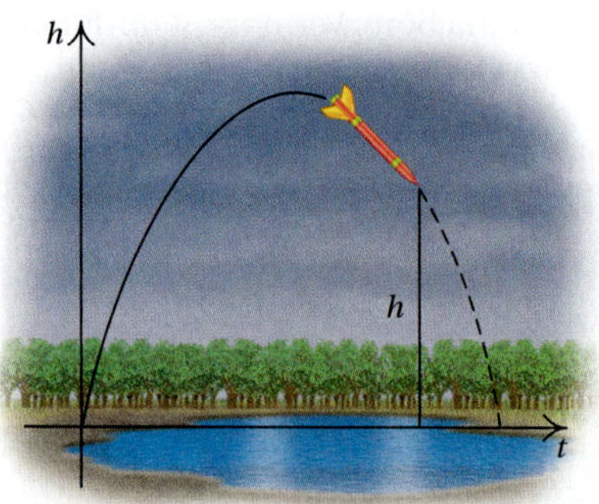

33. After how many seconds will the rocket first reach a height of 464 ft?

34. After how many seconds from launching will the rocket again be at that same height of 464 ft? (See Exercise 33.)

35. The sum of the squares of two consecutive odd positive integers is 74. Find the integers.

36. The sum of the squares of two consecutive odd positive integers is 130. Find the integers.

37. D$_W$ An archaeologist has measuring sticks of 3 ft, 4 ft, and 5 ft. Explain how she could draw a 7-ft by 9-ft rectangle on a piece of land being excavated.

38. D$_W$ Write a problem for a classmate to solve such that only one of the two solutions of a quadratic equation can be used as an answer.

SKILL MAINTENANCE

Multiply. [9.6d], [9.7f]

39. $(3x - 5y)(3x + 5y)$

40. $(3x - 5y)^2$

41. $(3x + 5y)^2$

42. $(3x - 5y)(2x + 7y)$

Find the intercepts of the equation. [8.3a]

43. $4x - 16y = 64$

44. $4x + 16y = 64$

45. $x - 1.3y = 6.5$

46. $\frac{2}{3}x + \frac{5}{8}y = \frac{5}{12}$

47. $y = 4 - 5x$

48. $y = 2x - 5$

SYNTHESIS

49. *Telephone Service.* Use the information in the figure below to determine the height of the telephone pole.

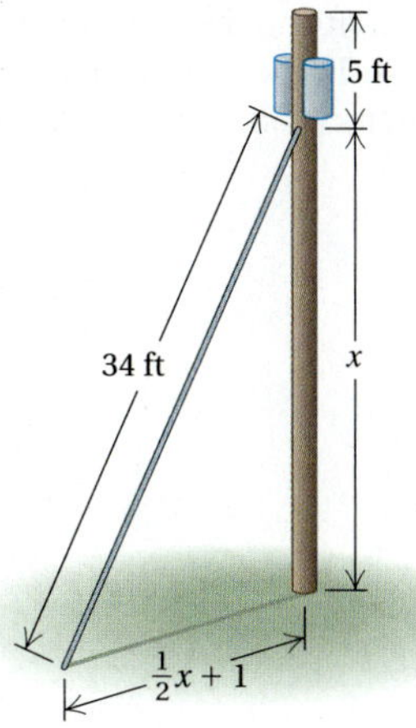

50. *Roofing.* A *square* of shingles covers 100 ft^2 of surface area. How many squares will be needed to reshingle the house shown?

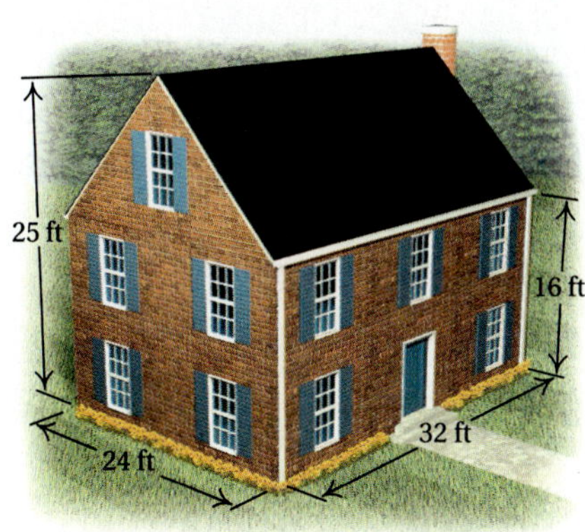

51. *Pool Sidewalk.* A cement walk of constant width is built around a 20-ft by 40-ft rectangular pool. The total area of the pool and the walk is 1500 ft^2. Find the width of the walk.

52. *Rain-gutter Design.* An open rectangular gutter is made by turning up the sides of a piece of metal 20 in. wide. The area of the cross-section of the gutter is 50 in^2. Find the depth of the gutter.

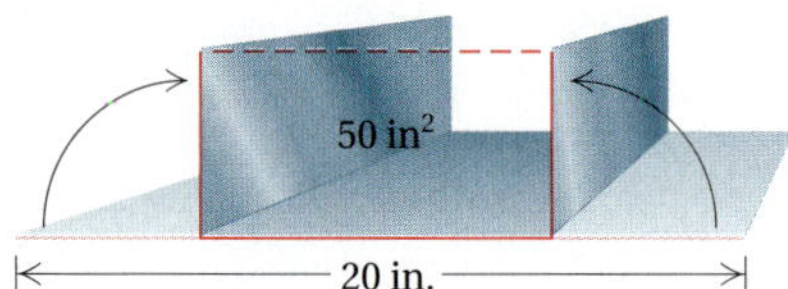

53. *Dimensions of an Open Box.* A rectangular piece of cardboard is twice as long as it is wide. A 4-cm square is cut out of each corner, and the sides are turned up to make a box with an open top. The volume of the box is 616 cm^3. Find the original dimensions of the cardboard.

54. *Dimensions of a Closed Box.* The total surface area of a closed box is 350 m^2. The box is 9 m high and has a square base and lid. Find the length of a side of the base.

55. Solve for x.

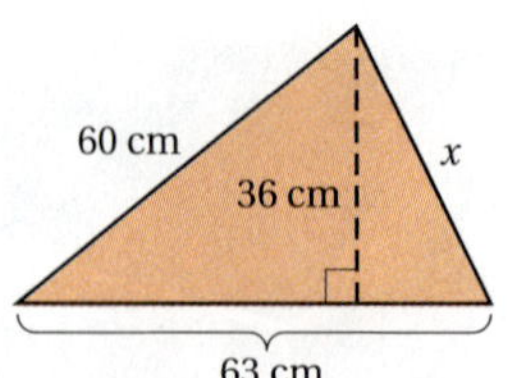

56. The ones digit of a number less than 100 is 4 greater than the tens digit. The sum of the number and the product of the digits is 58. Find the number.

CHAPTER

10 Summary and Review

The review that follows is meant to prepare you for a chapter exam. It consists of two parts. The first part is a checklist of some of the Study Tips referred to in this and preceding chapters, as well as a list of important properties and formulas. The second part is the Review Exercises. These provide practice exercises for the exam, together with references to section objectives so you can go back and review. Before beginning, stop and look back over the skills you have obtained. What skills in mathematics do you have now that you did not have before studying this chapter?

STUDY TIPS CHECKLIST

The foundation of all your study skills is TIME!

- ☐ Are you staying on schedule and on time for class and adapting your study time and class schedule to your personality?
- ☐ Did you study the examples in this chapter carefully?
- ☐ Did you use the five steps for problem solving as you did the applications in Section 10.9?
- ☐ Are you asking questions at appropriate times in class and with your tutors?
- ☐ Are you doing exercises without answers as part of every homework assignment to prepare you for tests?

IMPORTANT PROPERTIES AND FORMULAS

Factoring Formulas:
$A^2 - B^2 = (A + B)(A - B)$, $A^3 + B^3 = (A + B)(A^2 - AB + B^2)$,
$A^2 + 2AB + B^2 = (A + B)^2$, $A^3 - B^3 = (A - B)(A^2 + AB + B^2)$
$A^2 - 2AB + B^2 = (A - B)^2$,

The Principle of Zero Products: An equation $ab = 0$ is true if and only if $a = 0$ is true or $b = 0$ is true, or both are true.

Pythagorean Theorem: $a^2 + b^2 = c^2$

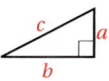

c a b

REVIEW EXERCISES

Find three factorizations of the monomial. [10.1a]

1. $-10x^2$

2. $36x^5$

Factor completely. [10.7a]

3. $5 - 20x^6$

4. $x^2 - 3x$

5. $9x^2 - 4$

6. $x^2 + 4x - 12$

7. $x^2 + 14x + 49$

8. $6x^3 + 12x^2 + 3x$

9. $x^3 + x^2 + 3x + 3$

10. $6x^2 - 5x + 1$

11. $x^4 - 81$

12. $9x^3 + 12x^2 - 45x$

13. $2x^2 - 50$

14. $x^4 + 4x^3 - 2x - 8$

15. $16x^4 - 1$

16. $8x^6 - 32x^5 + 4x^4$

17. $75 + 12x^2 + 60x$

18. $x^2 + 9$

19. $x^3 - x^2 - 30x$

20. $8x^3 - 125$

21. $9x^2 + 25 - 30x$

22. $6x^2 - 28x - 48$

23. $x^2 - 6x + 9$

24. $2x^2 - 7x - 4$

25. $18x^2 - 12x + 2$

26. $3x^2 - 27$

27. $15 - 8x + x^2$

28. $25x^2 - 20x + 4$

29. $49b^{10} + 4a^8 - 28a^4b^5$

30. $x^2y^2 + xy - 12$

31. $12a^2 + 84ab + 147b^2$

32. $m^2 + 5m + mt + 5t$

33. $32x^4 - 128y^4z^4$

Solve. [10.8a], [10.8b]

34. $(x - 1)(x + 3) = 0$

35. $x^2 + 2x - 35 = 0$

36. $x^2 + x - 12 = 0$

37. $3x^2 + 2 = 5x$

38. $2x^2 + 5x = 12$

39. $16 = x(x - 6)$

Solve. [10.9a, b]

40. *Sharks' Teeth.* Sharks' teeth are shaped like triangles. The height of a tooth of a great white shark is 1 cm longer than the base. The area is 15 cm^2. Find the height and the base.

41. The product of two consecutive even integers is 288. Find the integers.

42. The product of two consecutive odd integers is 323. Find the integers.

43. *Antenna Guy Wire.* The guy wires for a television antenna are 2 m longer than the height of the antenna. The guy wires are anchored 4 m from the foot of the antenna. How tall is the antenna?

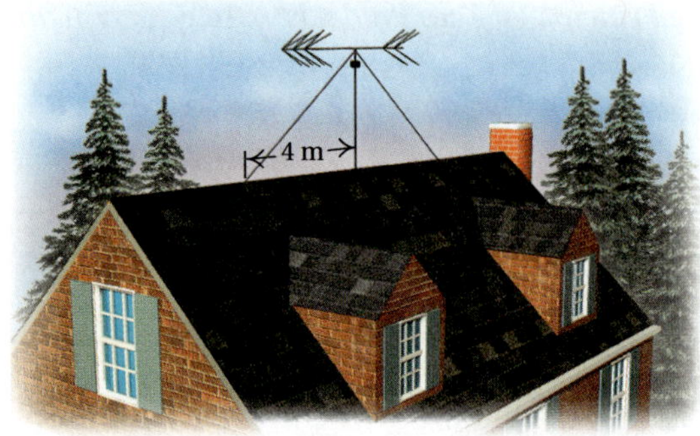

44. If the sides of a square are lengthened by 3 km, the area becomes 81 km^2. Find the length of a side of the original square.

Find the x-intercepts for the graph of the equation. [10.8b]

45. $y = x^2 + 9x + 20$

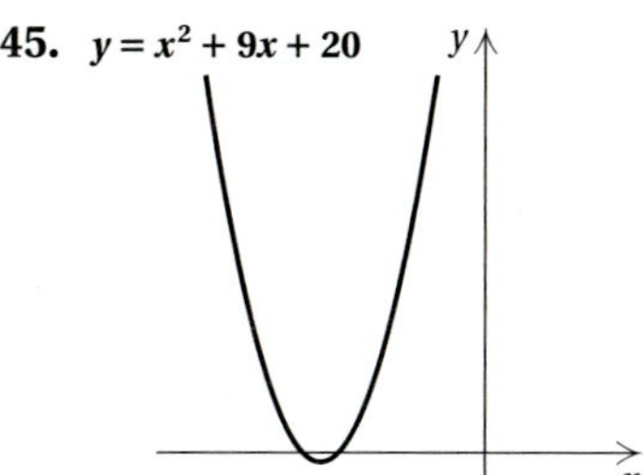

46. $y = 2x^2 - 7x - 15$

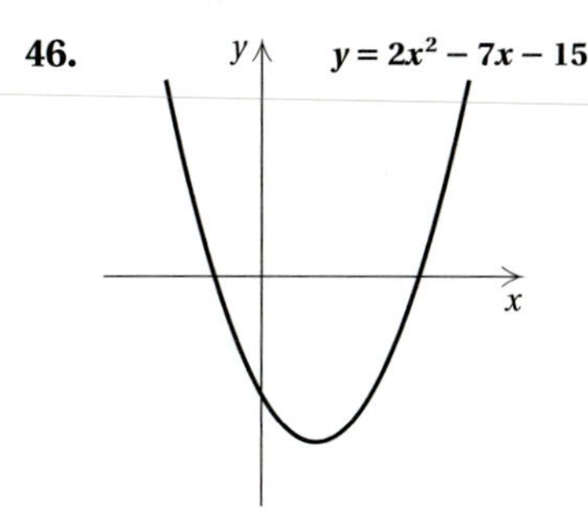

47. DW On a quiz, Sheri writes the factorization of $4x^2 - 100$ as $(2x - 10)(2x + 10)$. If this were a 10-point question, how many points would you give Sheri? Why? [10.5d]

48. DW How do the equations solved in this chapter differ from those solved in previous chapters? [10.8b]

SKILL MAINTENANCE

Certain objectives from four particular sections will be retested on the chapter test. The objectives are listed with the practice problems that follow.

49. Divide: $-\frac{12}{25} \div \left(-\frac{21}{10}\right)$. [6.6c]

50. Solve: $20 - (3x + 2) \geq 2(x + 5) + x$. [7.7e]

51. Multiply: $(2a - 3)(2a + 3)$. [9.6d]

52. Find the intercepts. Then graph the equation. [8.3a]

$$3y - 4x = -12$$

SYNTHESIS

Solve. [10.9a]

53. The pages of a book measure 15 cm by 20 cm. Margins of equal width surround the printing on each page and constitute one-half of the area of the page. Find the width of the margins.

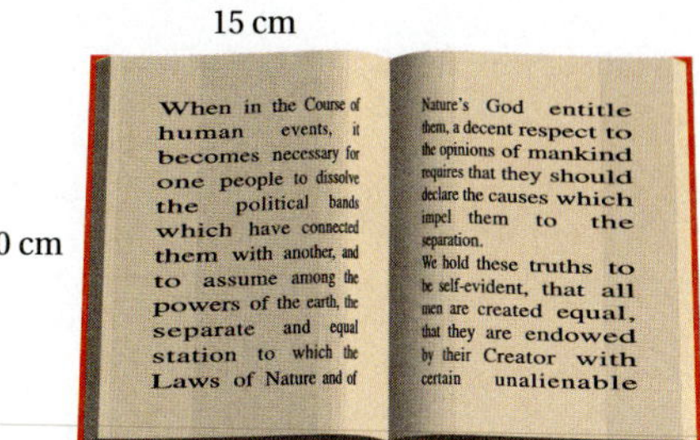

54. The cube of a number is the same as twice the square of the number. Find all such numbers.

55. The length of a rectangle is two times its width. When the length is increased by 20 and the width decreased by 1, the area is 160. Find the original length and width.

Solve. [10.8a, b]

56. $x^2 + 25 = 0$

57. $(x - 2)(x + 3)(2x - 5) = 0$

58. $(x - 3)4x^2 + 3x(x - 3) - (x - 3)10 = 0$

59. Find a polynomial for the shaded area in the figure below. [9.4d], [10.9b]

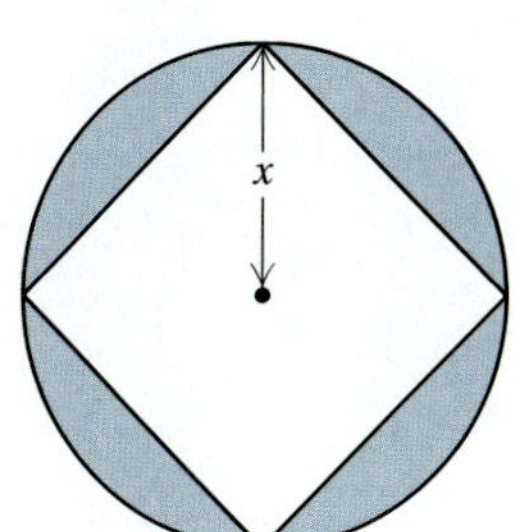

CHAPTER 10

Chapter Test

1. Find three factorizations of $4x^3$.

Factor completely.

2. $x^2 - 7x + 10$
3. $x^2 + 25 - 10x$
4. $6y^2 - 8y^3 + 4y^4$
5. $x^3 + x^2 + 2x + 2$
6. $x^2 - 5x$
7. $x^3 + 2x^2 - 3x$
8. $28x - 48 + 10x^2$
9. $4x^2 - 9$
10. $x^2 - x - 12$
11. $6m^3 + 9m^2 + 3m$
12. $3w^2 - 75$
13. $60x + 45x^2 + 20$
14. $3x^4 - 48$
15. $49x^2 - 84x + 36$
16. $5x^2 - 26x + 5$
17. $x^4 + 2x^3 - 3x - 6$
18. $80 - 5x^4$
19. $4x^2 - 4x - 15$
20. $6t^3 + 9t^2 - 15t$
21. $3m^2 - 9mn - 30n^2$
22. $1000a^3 - 27b^3$

Solve.

23. $x^2 - x - 20 = 0$
24. $2x^2 + 7x = 15$
25. $x(x - 3) = 28$

Solve.

26. The length of a rectangle is 2 m more than the width. The area of the rectangle is 48 m^2. Find the length and the width.

27. The base of a triangle is 6 cm greater than twice the height. The area is 28 cm^2. Find the height and the base.

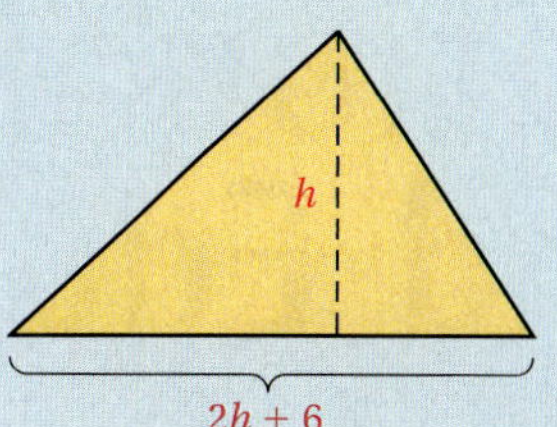

28. *Masonry Corner.* A mason wants to be sure she has a right corner in a building's foundation. She marks a point 3 ft from the corner along one wall and another point 4 ft from the corner along the other wall. If the corner is a right angle, what should the distance be between the two marked points?

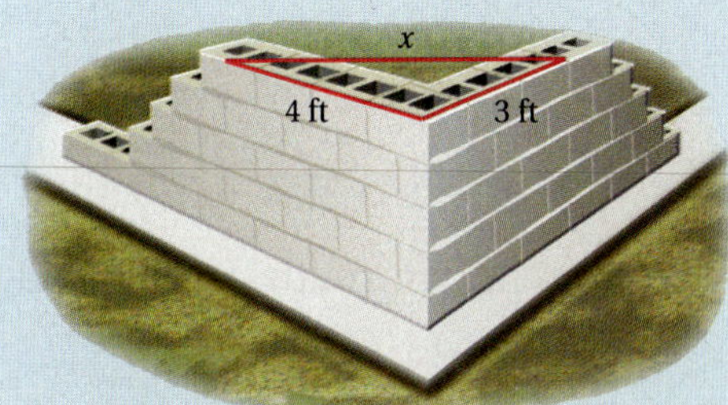

Find the x-intercepts for the graph of the equation.

29.

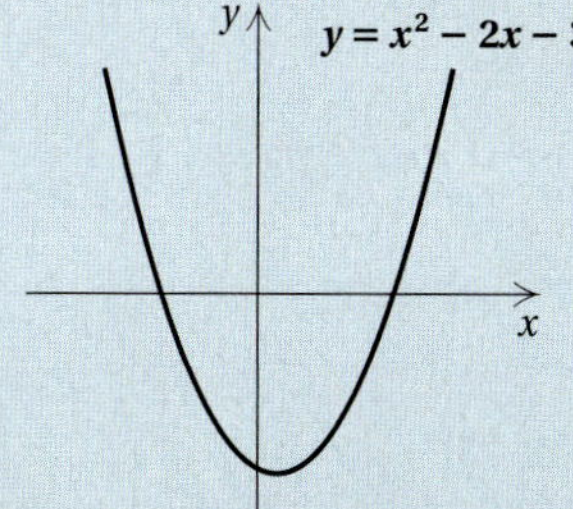

30.

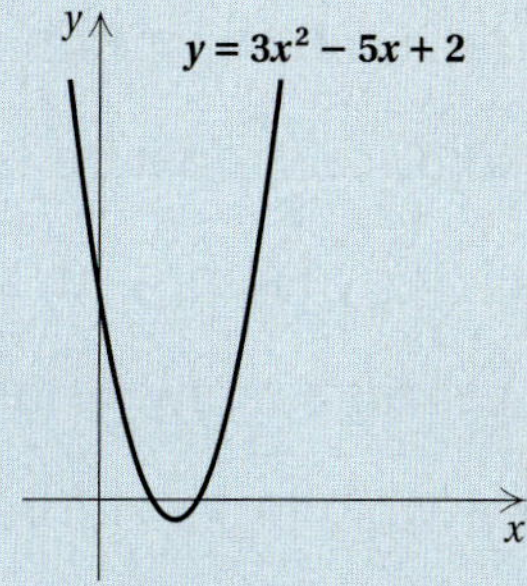

SKILL MAINTENANCE

31. Divide: $\frac{5}{8} \div \left(-\frac{11}{16}\right)$.

32. Solve: $10(x - 3) < 4(x + 2)$.

33. Find the intercepts. Then graph the equation.

$$2y - 5x = 10$$

34. Multiply: $(5x^2 - 7)^2$.

SYNTHESIS

35. The length of a rectangle is five times its width. When the length is decreased by 3 and the width is increased by 2, the area of the new rectangle is 60. Find the original length and width.

36. Factor: $(a + 3)^2 - 2(a + 3) - 35$.

37. Solve: $20x(x + 2)(x - 1) = 5x^3 - 24x - 14x^2$.

38. If $x + y = 4$ and $x - y = 6$, then $x^2 - y^2 = ?$

a) 2 **b)** 10
c) 34 **d)** 24

Rational Expressions and Equations

Gateway to Chapter 11

A rational expression is the ratio or quotient of two polynomials. In this chapter, we learn to add, subtract, multiply, and divide rational expressions. Then we apply these manipulative skills to the solving of rational equations. These equations require careful thinking because certain numbers that seem to be solutions must be checked carefully.

Finally, we apply our new equation-solving skills to new kinds of applications, such as work and motion problems, and to applications involving ratio and proportion. You may have studied ratio and proportion problems in earlier mathematics courses.

Real-World Application

A cheetah can run 20 mph faster than a lion. A cheetah can run 7 mi in the same time that a lion can run 5 mi. Find the speed of each animal.

Source: Barbara Ann Kipfer, *The Order of Things*. New York: Random House, 1998.

This problem appears as Example 2 in Section 11.8.

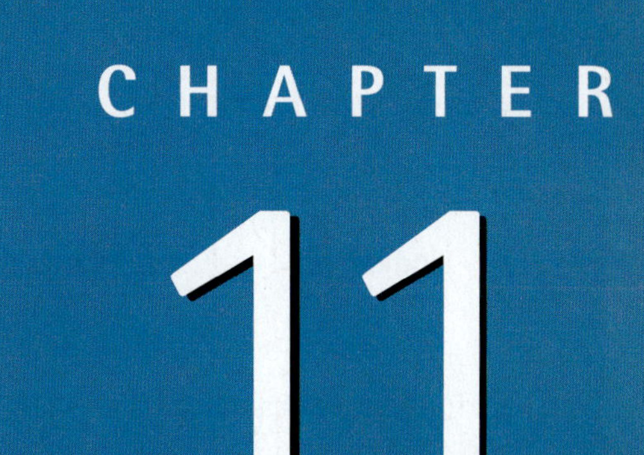

CHAPTER 11

CHAPTER 11

Pretest

1. Find the LCM of $x^2 + 5x + 6$ and $x^2 + 6x + 9$. [11.3c]

Perform the indicated operations and simplify.

2. $\dfrac{b-1}{2-b} + \dfrac{b^2-3}{b^2-4}$ [11.4a]

3. $\dfrac{4y-4}{y^2-y-2} - \dfrac{3y-5}{y^2-y-2}$ [11.5a]

4. $\dfrac{4}{a+2} + \dfrac{3}{a}$ [11.4a]

5. $\dfrac{x}{x+1} - \dfrac{x}{x-1} + \dfrac{2x^2}{x^2-1}$ [11.5b]

6. $\dfrac{4x+8}{x+1} \cdot \dfrac{x^2-2x-3}{2x^2-8}$ [11.1d]

7. $\dfrac{x+3}{x^2-9} \div \dfrac{x+3}{x^2-6x+9}$ [11.2b]

8. Simplify: $\dfrac{\dfrac{1}{x}+\dfrac{1}{y}}{\dfrac{1}{x}-\dfrac{1}{y}}$. [11.6a]

Solve. [11.7a]

9. $\dfrac{1}{x+4} = \dfrac{5}{x}$

10. $\dfrac{3}{x-2} + \dfrac{x}{2} = \dfrac{6}{2x-4}$

11. **Mercedes-Benz Cabriolet Gas Mileage.** A Mercedes-Benz Cabriolet travels 297 miles on 16.5 gal of gas. How much gas would it take to drive 1000 mi? [11.8b]
Source: Mercedes-Benz

12. **Paper Delivery.** It takes 6 hr for a paper carrier to deliver 200 papers. At this rate, how long would it take to deliver 350 papers? [11.8b]

13. **Data Entry.** One data-entry clerk can key in a report in 6 hr. Another can key in the same report in 5 hr. How long would it take them, working together, to key in the same report? [11.8a]

14. **Car Speeds.** One car travels 20 mph faster than another. While one car travels 300 mi, the other travels 400 mi. Find the speed of each car. [11.8a]

11.1 MULTIPLYING AND SIMPLIFYING RATIONAL EXPRESSIONS

a Rational Expressions and Replacements

Rational numbers are quotients of integers. Some examples are

$$\frac{2}{3}, \quad \frac{4}{-5}, \quad \frac{-8}{17}, \quad \frac{563}{1}.$$

The following are called **rational expressions** or **fractional expressions.** They are quotients, or ratios, of polynomials:

$$\frac{3}{4}, \quad \frac{z}{6}, \quad \frac{5}{x+2}, \quad \frac{t^2+3t-10}{7t^2-4}.$$

A rational expression is also a division. For example,

$$\frac{3}{4} \text{ means } 3 \div 4 \text{ and } \frac{x-8}{x+2} \text{ means } (x-8) \div (x+2).$$

Because rational expressions indicate division, we must be careful to avoid denominators of zero. When a variable is replaced with a number that produces a denominator equal to zero, the rational expression is not defined. For example, in the expression

$$\frac{x-8}{x+2},$$

when x is replaced with -2, the denominator is 0, and the expression is *not* defined:

$$\frac{x-8}{x+2} = \frac{-2-8}{-2+2} = \frac{-10}{0} \leftarrow \text{Division by 0 is not defined.}$$

When x is replaced with a number other than -2, such as 3, the expression *is* defined because the denominator is nonzero:

$$\frac{x-8}{x+2} = \frac{3-8}{3+2} = \frac{-5}{5} = -1.$$

EXAMPLE 1 Find all numbers for which the rational expression

$$\frac{x+4}{x^2-3x-10}$$

is not defined.

The value of the numerator has no bearing on whether or not a rational expression is defined. To determine which numbers make the rational expression not defined, we set the *denominator* equal to 0 and solve:

$$x^2 - 3x - 10 = 0$$
$$(x-5)(x+2) = 0 \qquad \text{Factoring}$$
$$x-5=0 \quad or \quad x+2=0 \qquad \text{Using the principle of zero products (see Section 5.8)}$$
$$x=5 \quad or \quad x=-2.$$

The expression is not defined for the replacement numbers 5 and -2.

Do Exercises 1–3.

Objectives

a Find all numbers for which a rational expression is not defined.

b Multiply a rational expression by 1, using an expression such as A/A.

c Simplify rational expressions by factoring the numerator and the denominator and removing factors of 1.

d Multiply rational expressions and simplify.

Find all numbers for which the rational expression is not defined.

1. $\frac{16}{x-3}$

2. $\frac{2x-7}{x^2+5x-24}$

3. $\frac{x+5}{8}$

Answers on page A-37

Multiply.

4. $\dfrac{2x+1}{3x-2} \cdot \dfrac{x}{x}$

5. $\dfrac{x+1}{x-2} \cdot \dfrac{x+2}{x+2}$

6. $\dfrac{x-8}{x-y} \cdot \dfrac{-1}{-1}$

Answers on page A-37

b Multiplying by 1

We multiply rational expressions in the same way that we multiply fraction notation in arithmetic. For a review, see Section 2.2. We know that

$$\frac{3}{7} \cdot \frac{2}{5} = \frac{3 \cdot 2}{7 \cdot 5} = \frac{6}{35}.$$

MULTIPLYING RATIONAL EXPRESSIONS

To multiply rational expressions, multiply numerators and multiply denominators:

$$\frac{A}{B} \cdot \frac{C}{D} = \frac{AC}{BD}.$$

For example,

$$\frac{x-2}{3} \cdot \frac{x+2}{x+7} = \frac{(x-2)(x+2)}{3(x+7)}.$$ Multiplying the numerators and the denominators

Note that we leave the numerator, $(x-2)(x+2)$, and the denominator, $3(x+7)$, in factored form because it is easier to simplify if we do not multiply. In order to learn to simplify, we first need to consider multiplying the rational expression by 1.

Any rational expression with the same numerator and denominator is a symbol for 1:

$$\frac{19}{19} = 1, \quad \frac{x+8}{x+8} = 1, \quad \frac{3x^2-4}{3x^2-4} = 1, \quad \frac{-1}{-1} = 1.$$

EQUIVALENT EXPRESSIONS

Expressions that have the same value for all allowable (or meaningful) replacements are called **equivalent expressions.**

We can multiply by 1 to obtain an *equivalent expression*. At this point, we select expressions for 1 arbitrarily. Later, we will have a system for our choices when we add and subtract.

EXAMPLES Multiply.

2. $\dfrac{3x+2}{x+1} \cdot 1 = \dfrac{3x+2}{x+1} \cdot \dfrac{2x}{2x} = \dfrac{(3x+2)2x}{(x+1)2x}$ Using the identity property of 1. We arbitrarily choose $2x/2x$ as a symbol for 1.

3. $\dfrac{x+2}{x-7} \cdot \dfrac{x+3}{x+3} = \dfrac{(x+2)(x+3)}{(x-7)(x+3)}$ We arbitrarily choose $(x+3)/(x+3)$ as a symbol for 1.

4. $\dfrac{2+x}{2-x} \cdot \dfrac{-1}{-1} = \dfrac{(2+x)(-1)}{(2-x)(-1)}$

Do Exercises 4–6.

C Simplifying Rational Expressions

Simplifying rational expressions is similar to simplifying fractional expressions in arithmetic. For a review, see Section 2.1. We know, for example, that an expression like $\frac{15}{40}$ can be simplified as follows:

$$\frac{15}{40} = \frac{3 \cdot 5}{8 \cdot 5}$$ Factoring the numerator and the denominator. Note the common factor of 5.

$$= \frac{3}{8} \cdot \frac{5}{5}$$ Factoring the fractional expression

$$= \frac{3}{8} \cdot 1$$ $\frac{5}{5} = 1$

$$= \frac{3}{8}.$$ Using the identity property of 1, or "removing a factor of 1"

Similar steps are followed when simplifying rational expressions: We factor and remove a factor of 1, using the fact that

$$\frac{ab}{cb} = \frac{a}{c} \cdot \frac{b}{b} = \frac{a}{c} \cdot 1 = \frac{a}{c}.$$

In algebra, instead of simplifying

$$\frac{15}{40},$$

we may need to simplify an expression like

$$\frac{x^2 - 16}{x + 4}.$$

Just as factoring is important in simplifying in arithmetic, so too is it important in simplifying rational expressions. The factoring we use most is the factoring of polynomials, which we studied in Chapter 10.

To simplify, we can do the reverse of multiplying. We factor the numerator and the denominator and "remove" a factor of 1.

EXAMPLE 5 Simplify: $\frac{8x^2}{24x}$.

$$\frac{8x^2}{24x} = \frac{8 \cdot x \cdot x}{3 \cdot 8 \cdot x}$$ Factoring the numerator and the denominator

$$= \frac{8x}{8x} \cdot \frac{x}{3}$$ Factoring the rational expression

$$= 1 \cdot \frac{x}{3}$$ $\frac{8x}{8x} = 1$

$$= \frac{x}{3}$$ We removed a factor of 1.

Do Exercises 7 and 8.

Simplify.

7. $\frac{5y}{y}$

8. $\frac{9x^2}{36x}$

Answers on page A-37

Simplify.

9. $\dfrac{2x^2 + x}{3x^2 + 2x}$

10. $\dfrac{x^2 - 1}{2x^2 - x - 1}$

11. $\dfrac{7x + 14}{7}$

12. $\dfrac{12y + 24}{48}$

Answers on page A-37

EXAMPLES Simplify.

6. $\dfrac{5a + 15}{10} = \dfrac{5(a + 3)}{5 \cdot 2}$ Factoring the numerator and the denominator

$= \dfrac{5}{5} \cdot \dfrac{a + 3}{2}$ Factoring the rational expression

$= 1 \cdot \dfrac{a + 3}{2}$ $\dfrac{5}{5} = 1$

$= \dfrac{a + 3}{2}$ Removing a factor of 1

7. $\dfrac{6a + 12}{7a + 14} = \dfrac{6(a + 2)}{7(a + 2)}$ Factoring the numerator and the denominator

$= \dfrac{6}{7} \cdot \dfrac{a + 2}{a + 2}$ Factoring the rational expression

$= \dfrac{6}{7} \cdot 1$ $\dfrac{a + 2}{a + 2} = 1$

$= \dfrac{6}{7}$ Removing a factor of 1

8. $\dfrac{6x^2 + 4x}{2x^2 + 2x} = \dfrac{2x(3x + 2)}{2x(x + 1)}$ Factoring the numerator and the denominator

$= \dfrac{2x}{2x} \cdot \dfrac{3x + 2}{x + 1}$ Factoring the rational expression

$= 1 \cdot \dfrac{3x + 2}{x + 1}$ $\dfrac{2x}{2x} = 1$

$= \dfrac{3x + 2}{x + 1}$ Removing a factor of 1

9. $\dfrac{x^2 + 3x + 2}{x^2 - 1} = \dfrac{(x + 2)(x + 1)}{(x + 1)(x - 1)}$

$= \dfrac{x + 1}{x + 1} \cdot \dfrac{x + 2}{x - 1}$

$= 1 \cdot \dfrac{x + 2}{x - 1}$

$= \dfrac{x + 2}{x - 1}$

CAUTION!

Note in this step that you *cannot* remove the x's because x is not a factor of the entire numerator and the entire denominator.

CANCELING

You may have encountered canceling when working with rational expressions. With great concern, we mention it as a possible way to speed up your work. Our concern is that canceling be done with care and understanding. Example 9 might have been done faster as follows:

$$\frac{x^2+3x+2}{x^2-1} = \frac{(x+2)(x+1)}{(x+1)(x-1)}$$ Factoring the numerator and the denominator

$$= \frac{(x+2)\cancel{(x+1)}}{\cancel{(x+1)}(x-1)}$$ When a factor of 1 is noted, it is canceled, as shown: $\frac{x+1}{x+1} = 1$.

$$= \frac{x+2}{x-1}.$$ Simplifying

CAUTION!

The difficulty with canceling is that it is often applied incorrectly, as in the following situations:

$$\frac{\cancel{x}+3}{\cancel{x}} = 3; \qquad \frac{\cancel{4}+1}{\cancel{4}+2} = \frac{1}{2}; \qquad \frac{1\cancel{5}}{\cancel{5}4} = \frac{1}{4}.$$

Wrong! Wrong! Wrong!

In each of these situations, the expressions canceled were *not* factors of 1. Factors are parts of products. For example, in $2 \cdot 3$, 2 and 3 are factors, but in $2 + 3$, 2 and 3 are *not* factors. If you can't factor, you can't cancel. If in doubt, don't cancel!

Do Exercises 9–12 on the preceding page.

OPPOSITES IN RATIONAL EXPRESSIONS

Expressions of the form $a - b$ and $b - a$ are opposites of each other. When either of these binomials is multiplied by -1, the result is the other binomial:

$$-1(a-b) = -a + b = b + (-a) = b - a;$$
$$-1(b-a) = -b + a = a + (-b) = a - b.$$

Multiplication by -1 reverses the order in which subtraction occurs.

Consider, for example,

$$\frac{x-4}{4-x}.$$

At first glance, it appears as though the numerator and the denominator do not have any common factors other than 1. But $x - 4$ and $4 - x$ are opposites, or additive inverses, of each other. Thus we can rewrite one as the opposite of the other by factoring out a -1.

EXAMPLE 10 Simplify: $\frac{x-4}{4-x}$.

$$\frac{x-4}{4-x} = \frac{x-4}{-(x-4)} = \frac{x-4}{-1(x-4)}$$ $4 - x = -(x - 4)$; $4 - x$ and $x - 4$ are opposites.

$$= -1 \cdot \frac{x-4}{x-4}$$

$$= -1 \cdot 1$$

$$= -1$$

Do Exercises 13–15.

Simplify.

13. $\frac{x-8}{8-x}$

14. $\frac{c-d}{d-c}$

15. $\frac{-x-7}{x+7}$

Answers on page A-37

16. Multiply and simplify:

$$\frac{a^2 - 4a + 4}{a^2 - 9} \cdot \frac{a + 3}{a - 2}.$$

17. Multiply and simplify:

$$\frac{x^2 - 25}{6} \cdot \frac{3}{x + 5}.$$

Answers on page A-37

d Multiplying and Simplifying

We try to simplify after we multiply. That is why we leave the numerator and the denominator in factored form.

EXAMPLE 11 Multiply and simplify: $\frac{5a^3}{4} \cdot \frac{2}{5a}$.

$$\frac{5a^3}{4} \cdot \frac{2}{5a} = \frac{5a^3(2)}{4(5a)}$$ Multiplying the numerators and the denominators

$$= \frac{2 \cdot 5 \cdot a \cdot a \cdot a}{2 \cdot 2 \cdot 5 \cdot a}$$ Factoring the numerator and the denominator

$$= \frac{\not{2} \cdot \not{5} \cdot \not{a} \cdot a \cdot a}{\not{2} \cdot 2 \cdot \not{5} \cdot \not{a}}$$ Removing a factor of 1: $\frac{2 \cdot 5 \cdot a}{2 \cdot 5 \cdot a} = 1$

$$= \frac{a^2}{2}$$ Simplifying

EXAMPLE 12 Multiply and simplify: $\frac{x^2 + 6x + 9}{x^2 - 4} \cdot \frac{x - 2}{x + 3}$.

$$\frac{x^2 + 6x + 9}{x^2 - 4} \cdot \frac{x - 2}{x + 3} = \frac{(x^2 + 6x + 9)(x - 2)}{(x^2 - 4)(x + 3)}$$ Multiplying the numerators and the denominators

$$= \frac{(x + 3)(x + 3)(x - 2)}{(x + 2)(x - 2)(x + 3)}$$ Factoring the numerator and the denominator

$$= \frac{\cancel{(x + 3)}(x + 3)\cancel{(x - 2)}}{(x + 2)\cancel{(x - 2)}\cancel{(x + 3)}}$$ Removing a factor of 1: $\frac{(x + 3)(x - 2)}{(x + 3)(x - 2)} = 1$

$$= \frac{x + 3}{x + 2}$$ Simplifying

Do Exercise 16.

EXAMPLE 13 Multiply and simplify: $\frac{x^2 + x - 2}{15} \cdot \frac{5}{2x^2 - 3x + 1}$.

$$\frac{x^2 + x - 2}{15} \cdot \frac{5}{2x^2 - 3x + 1} = \frac{(x^2 + x - 2)5}{15(2x^2 - 3x + 1)}$$ Multiplying the numerators and the denominators

$$= \frac{(x + 2)(x - 1)5}{5(3)(x - 1)(2x - 1)}$$ Factoring the numerator and the denominator

$$= \frac{(x + 2)\cancel{(x - 1)}\cancel{5}}{\cancel{5}(3)\cancel{(x - 1)}(2x - 1)}$$ Removing a factor of 1: $\frac{(x - 1)5}{(x - 1)5} = 1$

$$= \frac{x + 2}{3(2x - 1)}$$ Simplifying

You need not carry out this multiplication.

Do Exercise 17.

CALCULATOR CORNER

Checking Multiplication and Simplification We can use the TABLE feature as a partial check that rational expressions have been multiplied and/or simplified correctly. To check the simplification in Example 9,

$$\frac{x^2 + 3x + 2}{x^2 - 1} = \frac{x + 2}{x - 1},$$

we first enter $y_1 = (x^2 + 3x + 2)/(x^2 - 1)$ and $y_2 = (x + 2)/(x - 1)$. Then, using AUTO mode, we look at a table of values of y_1 and y_2. If the simplification is correct, the values should be the same for all allowable replacements.

X	Y1	Y2
−4	.4	.4
−3	.25	.25
−2	0	0
−1	ERROR	−.5
0	−2	−2
1	ERROR	ERROR
2	4	4
X = −4		

The ERROR messages indicate that −1 and 1 are not allowable replacements in the first expression, and 1 is not an allowable replacement in the second. For all other numbers, we see that y_1 and y_2 are the same, so the simplification appears to be correct. Remember, this is only a partial check since we cannot check all possible values.

Exercises: Use the TABLE feature to determine whether each of the following appears to be correct.

1. $\frac{8x^2}{24x} = \frac{x}{3}$
2. $\frac{5x + 15}{10} = \frac{x + 3}{2}$
3. $\frac{x + 3}{x} = 3$
4. $\frac{x^2 + 3x - 4}{x^2 - 16} = \frac{x - 1}{x + 4}$
5. $\frac{x^2 + 2x - 3}{x^2 - 4} \cdot \frac{4x - 8}{x + 3} = \frac{4x - 1}{x + 2}$
6. $\frac{x^2 - 25}{6} \cdot \frac{3}{x + 5} = \frac{x - 5}{3}$
7. $\frac{x^2 + 6x + 9}{x^2 - 4} \cdot \frac{x - 2}{x + 3} = \frac{x + 3}{x + 2}$
8. $\frac{x^2}{x^2 - 3x} \cdot \frac{x^2 - 9}{3} = \frac{x(x + 3)}{3}$

11.1 EXERCISE SET

For Extra Help

Digital Video Tutor CD 9 Videotape 13 | InterAct Math | Math Tutor Center | MathXL | MyMathLab

a Find all numbers for which the rational expression is not defined.

1. $\frac{-3}{2x}$

2. $\frac{24}{-8y}$

3. $\frac{5}{x-8}$

4. $\frac{y-4}{y+6}$

5. $\frac{3}{2y+5}$

6. $\frac{x^2-9}{4x-12}$

7. $\frac{x^2+11}{x^2-3x-28}$

8. $\frac{p^2-9}{p^2-7p+10}$

9. $\frac{m^3-2m}{m^2-25}$

10. $\frac{7-3x+x^2}{49-x^2}$

11. $\frac{x-4}{3}$

12. $\frac{x^2-25}{14}$

b Multiply. Do not simplify. Note that in each case you are multiplying by 1.

13. $\frac{4x}{4x} \cdot \frac{3x^2}{5y}$

14. $\frac{5x^2}{5x^2} \cdot \frac{6y^3}{3z^4}$

15. $\frac{2x}{2x} \cdot \frac{x-1}{x+4}$

16. $\frac{2a-3}{5a+2} \cdot \frac{a}{a}$

17. $\frac{3-x}{4-x} \cdot \frac{-1}{-1}$

18. $\frac{x-5}{5-x} \cdot \frac{-1}{-1}$

19. $\frac{y+6}{y+6} \cdot \frac{y-7}{y+2}$

20. $\frac{x^2+1}{x^3-2} \cdot \frac{x-4}{x-4}$

c Simplify.

21. $\frac{8x^3}{32x}$

22. $\frac{4x^2}{20x}$

23. $\frac{48p^7q^5}{18p^5q^4}$

24. $\frac{-76x^8y^3}{-24x^4y^3}$

25. $\frac{4x-12}{4x}$

26. $\frac{5a-40}{5}$

27. $\dfrac{3m^2 + 3m}{6m^2 + 9m}$

28. $\dfrac{4y^2 - 2y}{5y^2 - 5y}$

29. $\dfrac{a^2 - 9}{a^2 + 5a + 6}$

30. $\dfrac{t^2 - 25}{t^2 + t - 20}$

31. $\dfrac{a^2 - 10a + 21}{a^2 - 11a + 28}$

32. $\dfrac{x^2 - 2x - 8}{x^2 - x - 6}$

33. $\dfrac{x^2 - 25}{x^2 - 10x + 25}$

34. $\dfrac{x^2 + 8x + 16}{x^2 - 16}$

35. $\dfrac{a^2 - 1}{a - 1}$

36. $\dfrac{t^2 - 1}{t + 1}$

37. $\dfrac{x^2 + 1}{x + 1}$

38. $\dfrac{m^2 + 9}{m + 3}$

39. $\dfrac{6x^2 - 54}{4x^2 - 36}$

40. $\dfrac{8x^2 - 32}{4x^2 - 16}$

41. $\dfrac{6t + 12}{t^2 - t - 6}$

42. $\dfrac{4x + 32}{x^2 + 9x + 8}$

43. $\dfrac{2t^2 + 6t + 4}{4t^2 - 12t - 16}$

44. $\dfrac{3a^2 - 9a - 12}{6a^2 + 30a + 24}$

45. $\dfrac{w^3 - z^3}{w^2 - z^2}$

46. $\dfrac{a^2 - b^2}{a^3 + b^3}$

47. $\dfrac{6-x}{x-6}$

48. $\dfrac{t-3}{3-t}$

49. $\dfrac{a-b}{b-a}$

50. $\dfrac{y-x}{-x+y}$

51. $\dfrac{6t-12}{2-t}$

52. $\dfrac{5a-15}{3-a}$

53. $\dfrac{x^2-1}{1-x}$

54. $\dfrac{a^2-b^2}{b^2-a^2}$

d Multiply and simplify.

55. $\dfrac{4x^3}{3x} \cdot \dfrac{14}{x}$

56. $\dfrac{18}{x^3} \cdot \dfrac{5x^2}{6}$

57. $\dfrac{3c}{d^2} \cdot \dfrac{4d}{6c^3}$

58. $\dfrac{3x^2y}{2} \cdot \dfrac{4}{xy^3}$

59. $\dfrac{x^2-3x-10}{x^2-4x+4} \cdot \dfrac{x-2}{x-5}$

60. $\dfrac{t^2}{t^2-4} \cdot \dfrac{t^2-5t+6}{t^2-3t}$

61. $\dfrac{a^2-9}{a^2} \cdot \dfrac{a^2-3a}{a^2+a-12}$

62. $\dfrac{x^2+10x-11}{x^2-1} \cdot \dfrac{x+1}{x+11}$

63. $\dfrac{4a^2}{3a^2-12a+12} \cdot \dfrac{3a-6}{2a}$

64. $\dfrac{5v+5}{v-2} \cdot \dfrac{v^2-4v+4}{v^2-1}$

65. $\dfrac{t^4-16}{t^4-1} \cdot \dfrac{t^2+1}{t^2+4}$

66. $\dfrac{x^4-1}{x^4-81} \cdot \dfrac{x^2+9}{x^2+1}$

67. $\dfrac{(x+4)^3}{(x+2)^3} \cdot \dfrac{x^2+4x+4}{x^2+8x+16}$

68. $\dfrac{(t-2)^3}{(t-1)^3} \cdot \dfrac{t^2-2t+1}{t^2-4t+4}$

69. $\dfrac{5a^2-180}{10a^2-10} \cdot \dfrac{20a+20}{2a-12}$

70. $\dfrac{2t^2-98}{4t^2-4} \cdot \dfrac{8t+8}{16t-112}$

71. $\dfrac{c^3+8}{c^2-4} \cdot \dfrac{c^2-4c+4}{c^2-2c+4}$

72. $\dfrac{x^2-y^2}{x^3-y^3} \cdot \dfrac{x^2+xy+y^2}{x^2+2xy+y^2}$

73. **DW** How is the process of canceling related to the identity property of 1?

74. **DW** Explain how a rational expression can be formed for which -3 and 4 are not allowable replacements.

SKILL MAINTENANCE

Solve.

75. *Consecutive Even Integers.* The product of two consecutive even integers is 360. Find the integers. [10.9a]

76. *Chemistry.* About 5 L of oxygen can be dissolved in 100 L of water at 0°C. This is 1.6 times the amount that can be dissolved in the same volume of water at 20°C. How much oxygen can be dissolved in 100 L at 20°C? [7.6a]

Factor. [10.7a]

77. $x^2 - x - 56$

78. $a^2 - 16a + 64$

79. $x^5 - 2x^4 - 35x^3$

80. $2y^3 - 10y^2 + y - 5$

81. $16 - t^4$

82. $10x^2 + 80x + 70$

83. $x^2 - 9x + 14$

84. $x^2 + x + 7$

85. $16x^2 - 40xy + 25y^2$

86. $a^2 - 9ab + 14b^2$

SYNTHESIS

Simplify.

87. $\dfrac{x^4-16y^4}{(x^2+4y^2)(x-2y)}$

88. $\dfrac{(a-b)^2}{b^2-a^2}$

89. $\dfrac{t^4-1}{t^4-81} \cdot \dfrac{t^2-9}{t^2+1} \cdot \dfrac{(t-9)^2}{(t+1)^2}$

90. $\dfrac{(t+2)^3}{(t+1)^3} \cdot \dfrac{t^2+2t+1}{t^2+4t+4} \cdot \dfrac{t+1}{t+2}$

91. $\dfrac{x^2-y^2}{(x-y)^2} \cdot \dfrac{x^2-2xy+y^2}{x^2-4xy-5y^2}$

92. $\dfrac{x-1}{x^2+1} \cdot \dfrac{x^4-1}{(x-1)^2} \cdot \dfrac{x^2-1}{x^4-2x^2+1}$

93. Select any number x, multiply by 2, add 5, multiply by 5, subtract 25, and divide by 10. What do you get? Explain how this procedure can be used for a number trick.

11.2 DIVISION AND RECIPROCALS

Objectives

a Find the reciprocal of a rational expression.

b Divide rational expressions and simplify.

Find the reciprocal.

1. $\frac{7}{2}$

2. $\frac{x^2 + 5}{2x^3 - 1}$

3. $x - 5$

4. $\frac{1}{x^2 - 3}$

5. Divide: $\frac{3}{5} \div \frac{7}{10}$.

Answers on page A-38

There is a similarity between what we do with rational expressions and what we do with rational numbers. In fact, after variables have been replaced with rational numbers, a rational expression represents a rational number.

a Finding Reciprocals

Two expressions are reciprocals of each other if their product is 1. The reciprocal of a rational expression is found by interchanging the numerator and the denominator.

EXAMPLES

1. The reciprocal of $\frac{2}{5}$ is $\frac{5}{2}$. $\left(\text{This is because } \frac{2}{5} \cdot \frac{5}{2} = \frac{10}{10} = 1.\right)$
2. The reciprocal of $\frac{2x^2 - 3}{x + 4}$ is $\frac{x + 4}{2x^2 - 3}$.
3. The reciprocal of $x + 2$ is $\frac{1}{x + 2}$. $\left(\text{Think of } x + 2 \text{ as } \frac{x + 2}{1}.\right)$

Do Exercises 1–4.

b Division

We divide rational expressions in the same way that we divide fraction notation in arithmetic. For a review, see Section 2.2.

DIVIDING RATIONAL EXPRESSIONS

To divide by a rational expression, multiply by its reciprocal:

$$\frac{A}{B} \div \frac{C}{D} = \frac{A}{B} \cdot \frac{D}{C} = \frac{AD}{BC}.$$

Then factor and, if possible, simplify.

EXAMPLE 4 Divide: $\frac{3}{4} \div \frac{9}{5}$.

$$\frac{3}{4} \div \frac{9}{5} = \frac{3}{4} \cdot \frac{5}{9}$$ Multiplying by the reciprocal of the divisor

$$= \frac{3 \cdot 5}{4 \cdot 9} = \frac{3 \cdot 5}{2 \cdot 2 \cdot 3 \cdot 3}$$ Factoring

$$= \frac{\cancel{3} \cdot 5}{2 \cdot 2 \cdot \cancel{3} \cdot 3}$$ Removing a factor of 1: $\frac{3}{3} = 1$

$$= \frac{5}{12}$$ Simplifying

Do Exercise 5.

EXAMPLE 5 Divide: $\frac{2}{x} \div \frac{3}{x}$.

$$\frac{2}{x} \div \frac{3}{x} = \frac{2}{x} \cdot \frac{x}{3}$$ Multiplying by the reciprocal of the divisor

$$= \frac{2 \cdot x}{x \cdot 3}$$

$$= \frac{2 \cdot \not{x}}{\not{x} \cdot 3}$$ Removing a factor of 1: $\frac{x}{x} = 1$

$$= \frac{2}{3}$$

Do Exercise 6.

6. Divide: $\frac{x}{8} \div \frac{x}{5}$.

EXAMPLE 6 Divide: $\frac{x+1}{x+2} \div \frac{x-1}{x+3}$.

$$\frac{x+1}{x+2} \div \frac{x-1}{x+3} = \frac{x+1}{x+2} \cdot \frac{x+3}{x-1}$$ Multiplying by the reciprocal of the divisor

$$= \frac{(x+1)(x+3)}{(x+2)(x-1)}$$

We usually do not carry out the multiplication in the numerator or the denominator. It is not wrong to do so, but the factored form is often more useful.

Do Exercise 7.

7. Divide:

$$\frac{x-3}{x+5} \div \frac{x+5}{x-2}.$$

Answers on page A-38

EXAMPLE 7 Divide and simplify: $\frac{x+1}{x^2-1} \div \frac{x+1}{x^2-2x+1}$.

$$\frac{x+1}{x^2-1} \div \frac{x+1}{x^2-2x+1}$$

$$= \frac{x+1}{x^2-1} \cdot \frac{x^2-2x+1}{x+1}$$ Multiplying by the reciprocal

$$= \frac{(x+1)(x^2-2x+1)}{(x^2-1)(x+1)}$$

$$= \frac{(x+1)(x-1)(x-1)}{(x-1)(x+1)(x+1)}$$ Factoring the numerator and the denominator

$$= \frac{\cancel{(x+1)}\cancel{(x-1)}(x-1)}{\cancel{(x-1)}\cancel{(x+1)}(x+1)}$$ Removing a factor of 1: $\frac{(x+1)(x-1)}{(x+1)(x-1)} = 1$

$$= \frac{x-1}{x+1}$$

CALCULATOR CORNER

Checking Division Use the TABLE feature, as described on p. 839, to check the divisions in Examples 5–8. Then check your answers to Margin Exercises 7–9.

EXAMPLE 8 Divide and simplify: $\frac{x^2-2x-3}{x^2-4} \div \frac{x+1}{x+5}$.

$$\frac{x^2-2x-3}{x^2-4} \div \frac{x+1}{x+5}$$

$$= \frac{x^2-2x-3}{x^2-4} \cdot \frac{x+5}{x+1}$$ Multiplying by the reciprocal

Divide and simplify.

8. $\dfrac{x-3}{x+5} \div \dfrac{x+2}{x+5}$

9. $\dfrac{x^2-5x+6}{x+5} \div \dfrac{x+2}{x+5}$

10. $\dfrac{y^2-1}{y+1} \div \dfrac{y^2-2y+1}{y+1}$

Answers on page A-38

Then

$$= \frac{(x^2-2x-3)(x+5)}{(x^2-4)(x+1)}$$

$$= \frac{(x-3)(x+1)(x+5)}{(x-2)(x+2)(x+1)} \quad \text{Factoring the numerator and the denominator}$$

$$= \frac{(x-3)\cancel{(x+1)}(x+5)}{(x-2)(x+2)\cancel{(x+1)}} \quad \text{Removing a factor of 1: } \frac{x+1}{x+1} = 1$$

$$= \frac{(x-3)(x+5)}{(x-2)(x+2)}.$$

You need not carry out the multiplications in the numerator and the denominator.

Do Exercises 8–10.

Study Tips

TURNING NEGATIVES INTO POSITIVES (PART 1)

Dusty Baker, presently manager of the Chicago Cubs, was cut from Little League by his father because he had a poor attitude. Dusty turned that negative into many positives, eventually earning his way back into Little League. He went on to an outstanding Major League career primarily with the Atlanta Braves and Los Angeles Dodgers. Dusty was the manager of the San Francisco Giants from 1993 to 2002, winning the National League Manager of the Year Award in 1993, 1997, and 2000.

Dusty has amazing people skills. He speaks fluent Spanish and easily relates to players from a variety of backgrounds. Rod Beck, former relief pitcher with the Giants, once said of Dusty, "If you can't play for Dusty Baker, you can't play for anybody."

Dusty is an avid reader, finding inspiration in books on positive thinking, because he wants all the creative aids he can find to counsel his ballplayers. He refuses to let his players consider the word "slump," referring to such experiences as "unfortunate time periods." Major-league players face many failures on a daily basis. Think about it: A player hitting .300 gets 3 hits in 10 at-bats, but also fails 7 out of 10 times at bat! A .300 hitter is considered highly successful.

What might Dusty say to you as a math student? You may have had some negative experiences, or "unfortunate time periods," in your past with mathematics, or maybe even as you take this course. What can you do? One answer is to keep picking yourself up with an intense resolve to keep on with the task, not letting negative thinking drag you down. Think about how Dusty and others overcame obstacles to achieve positive results.

In 1993, one of your authors, Marv Bittinger, had the pleasure of coauthoring with Dusty a book, a three-part video, and a CD-ROM titled *You Can Teach Hitting*. For more on Dusty, visit www.dustybaker.com.

11.2 EXERCISE SET

For Extra Help

Digital Video Tutor CD 9 Videotape 13 | InterAct Math | Math Tutor Center | MathXL | MyMathLab

a Find the reciprocal.

1. $\frac{4}{x}$

2. $\frac{a+3}{a-1}$

3. $x^2 - y^2$

4. $x^2 - 5x + 7$

5. $\frac{1}{a+b}$

6. $\frac{x^2}{x^2-3}$

7. $\frac{x^2+2x-5}{x^2-4x+7}$

8. $\frac{(a-b)(a+b)}{(a+4)(a-5)}$

b Divide and simplify.

9. $\frac{2}{5} \div \frac{4}{3}$

10. $\frac{3}{10} \div \frac{3}{2}$

11. $\frac{2}{x} \div \frac{8}{x}$

12. $\frac{t}{3} \div \frac{t}{15}$

13. $\frac{a}{b^2} \div \frac{a^2}{b^3}$

14. $\frac{x^2}{y} \div \frac{x^3}{y^3}$

15. $\frac{a+2}{a-3} \div \frac{a-1}{a+3}$

16. $\frac{x-8}{x+9} \div \frac{x+2}{x-1}$

17. $\frac{x^2-1}{x} \div \frac{x+1}{x-1}$

18. $\frac{4y-8}{y+2} \div \frac{y-2}{y^2-4}$

19. $\frac{x+1}{6} \div \frac{x+1}{3}$

20. $\frac{a}{a-b} \div \frac{b}{a-b}$

21. $\frac{5x-5}{16} \div \frac{x-1}{6}$

22. $\frac{4y-12}{12} \div \frac{y-3}{3}$

23. $\frac{-6+3x}{5} \div \frac{4x-8}{25}$

24. $\frac{-12+4x}{4} \div \frac{-6+2x}{6}$

25. $\frac{a+2}{a-1} \div \frac{3a+6}{a-5}$

26. $\frac{t-3}{t+2} \div \frac{4t-12}{t+1}$

27. $\frac{x^2-4}{x} \div \frac{x-2}{x+2}$

28. $\frac{x+y}{x-y} \div \frac{x^2+y}{x^2-y^2}$

29. $\frac{x^2-9}{4x+12} \div \frac{x-3}{6}$

30. $\frac{a-b}{2a} \div \frac{a^2-b^2}{8a^3}$

31. $\frac{c^2+3c}{c^2+2c-3} \div \frac{c}{c+1}$

32. $\frac{y+5}{2y} \div \frac{y^2-25}{4y^2}$

33. $\dfrac{2y^2 - 7y + 3}{2y^2 + 3y - 2} \div \dfrac{6y^2 - 5y + 1}{3y^2 + 5y - 2}$

34. $\dfrac{x^2 + x - 20}{x^2 - 7x + 12} \div \dfrac{x^2 + 10x + 25}{x^2 - 6x + 9}$

35. $\dfrac{x^3 - 64}{x^3 + 64} \div \dfrac{x^2 - 16}{x^2 - 4x + 16}$

36. $\dfrac{8y^3 + 27}{64y^3 - 1} \div \dfrac{4y^2 - 9}{16y^2 + 4y + 1}$

37. **D**W Is the reciprocal of a product the product of the reciprocals? Why or why not?

38. **D**W Explain why 5, −1, and 7 are *not* allowable replacements in the division

$$\frac{x + 3}{x - 5} \div \frac{x - 7}{x + 1}.$$

SKILL MAINTENANCE

Solve.

39. Bonnie is taking an astronomy course. In order to receive an A, she must average at least 90 after four exams. Bonnie scored 96, 98, and 89 on the first three tests. Determine (in terms of an inequality) what scores on the last test will earn her an A. [7.8b]

40. *Triangle Dimensions.* The base of a triangle is 6 cm greater than twice the height. The area is 28 cm^2. Find the height and the base. [10.9a]

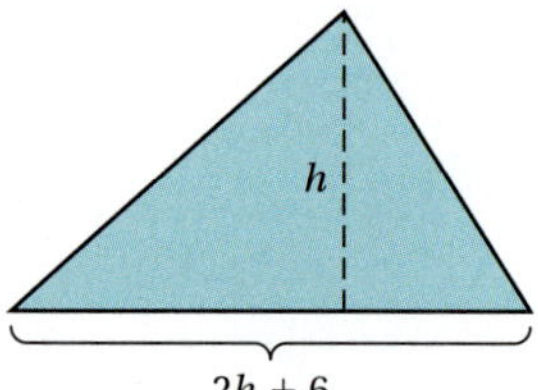

Subtract. [9.4c]

41. $(8x^3 - 3x^2 + 7) - (8x^2 + 3x - 5)$

42. $(3p^2 - 6pq + 7q^2) - (5p^2 - 10pq + 11q^2)$

Simplify. [9.2a, b]

43. $(2x^{-3}y^4)^2$

44. $(5x^6y^{-4})^3$

45. $\left(\dfrac{2x^3}{y^5}\right)^2$

46. $\left(\dfrac{a^{-3}}{b^4}\right)^5$

SYNTHESIS

Simplify.

47. $\dfrac{3a^2 - 5ab - 12b^2}{3ab + 4b^2} \div (3b^2 - ab)$

48. $\dfrac{3x + 3y + 3}{9x} \div \dfrac{x^2 + 2xy + y^2 - 1}{x^4 + x^2}$

49. The volume of this rectangular solid is $x - 3$. What is its height?

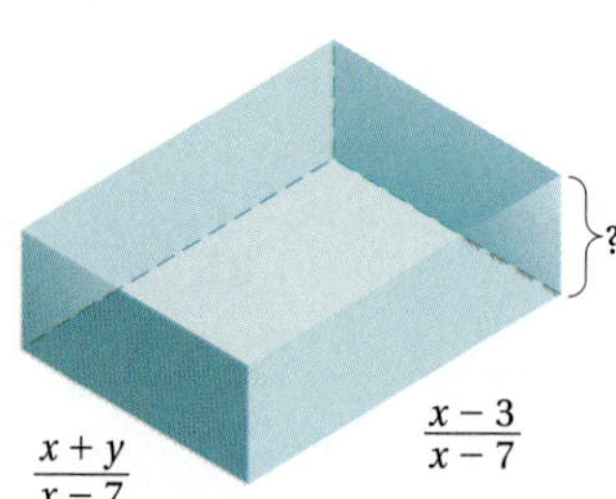

11.3 LEAST COMMON MULTIPLES AND DENOMINATORS

Objectives

a Find the LCM of several numbers by factoring.

b Add fractions, first finding the LCD.

c Find the LCM of algebraic expressions by factoring.

a Least Common Multiples

To add when denominators are different, we first find a common denominator. For a review, see Section 2.3. We know, for example, that to add $\frac{5}{12}$ and $\frac{7}{30}$, we first look for the **least common multiple, LCM,** of both 12 and 30. That number becomes the **least common denominator, LCD.** To find the LCM of 12 and 30, we factor:

$$12 = 2 \cdot 2 \cdot 3;$$
$$30 = 2 \cdot 3 \cdot 5.$$

The LCM is the number that has 2 as a factor twice, 3 as a factor once, and 5 as a factor once:

$$\text{LCM} = 2 \cdot 2 \cdot 3 \cdot 5 = 60.$$

12 is a factor of the LCM.

30 is a factor of the LCM.

FINDING LCMS

To find the LCM, use each factor the greatest number of times that it appears in any one factorization.

EXAMPLE 1 Find the LCM of 24 and 36.

$$\left.\begin{aligned} 24 &= 2 \cdot 2 \cdot 2 \cdot 3 \\ 36 &= 2 \cdot 2 \cdot 3 \cdot 3 \end{aligned}\right\} \quad \text{LCM} = 2 \cdot 2 \cdot 2 \cdot 3 \cdot 3, \text{ or } 72$$

Do Exercises 1–4.

Find the LCM by factoring.

1. 16, 18

2. 6, 12

3. 2, 5

4. 24, 30, 20

b Adding Using the LCD

Let's finish adding $\frac{5}{12}$ and $\frac{7}{30}$:

$$\frac{5}{12} + \frac{7}{30} = \frac{5}{2 \cdot 2 \cdot 3} + \frac{7}{2 \cdot 3 \cdot 5}.$$

The least common denominator, LCD, is $2 \cdot 2 \cdot 3 \cdot 5$. To get the LCD in the first denominator, we need a 5. To get the LCD in the second denominator, we need another 2. We get these numbers by multiplying by 1:

$$\frac{5}{12} + \frac{7}{30} = \frac{5}{2 \cdot 2 \cdot 3} \cdot \frac{5}{5} + \frac{7}{2 \cdot 3 \cdot 5} \cdot \frac{2}{2} \qquad \text{Multiplying by 1}$$

$$= \frac{25}{2 \cdot 2 \cdot 3 \cdot 5} + \frac{14}{2 \cdot 3 \cdot 5 \cdot 2} \qquad \text{The denominators are now the LCD.}$$

$$= \frac{39}{2 \cdot 2 \cdot 3 \cdot 5} \qquad \text{Adding the numerators and keeping the LCD}$$

$$= \frac{\cancel{3} \cdot 13}{2 \cdot 2 \cdot \cancel{3} \cdot 5} \qquad \text{Factoring the numerator and removing a factor of 1: } \frac{3}{3} = 1$$

$$= \frac{13}{20}. \qquad \text{Simplifying}$$

Answers on page A-38

Add, first finding the LCD. Simplify if possible.

5. $\frac{3}{16} + \frac{1}{18}$

6. $\frac{1}{6} + \frac{1}{12}$

7. $\frac{1}{2} + \frac{3}{5}$

8. $\frac{1}{24} + \frac{1}{30} + \frac{3}{20}$

Find the LCM.

9. $12xy^2,\ 15x^3y$

10. $y^2 + 5y + 4,\ y^2 + 2y + 1$

11. $t^2 + 16,\ t - 2,\ 7$

12. $x^2 + 2x + 1,\ 3x^2 - 3x,\ x^2 - 1$

Answers on page A-38

EXAMPLE 2 Add: $\frac{5}{12} + \frac{11}{18}$.

$$\left.\begin{aligned}12 &= 2 \cdot 2 \cdot 3\\ 18 &= 2 \cdot 3 \cdot 3\end{aligned}\right\} \quad \text{LCD} = 2 \cdot 2 \cdot 3 \cdot 3\text{, or } 36$$

$$\frac{5}{12} + \frac{11}{18} = \frac{5}{2 \cdot 2 \cdot 3} \cdot \frac{3}{3} + \frac{11}{2 \cdot 3 \cdot 3} \cdot \frac{2}{2} = \frac{15 + 22}{2 \cdot 2 \cdot 3 \cdot 3} = \frac{37}{36}$$

Do Exercises 5–8.

C LCMs of Algebraic Expressions

To find the LCM of two or more algebraic expressions, we factor them. Then we use each factor the greatest number of times that it occurs in any one expression. In Section 11.4, each LCM will become an LCD used to add rational expressions.

EXAMPLE 3 Find the LCM of $12x$, $16y$, and $8xyz$.

$$\left.\begin{aligned}12x &= 2 \cdot 2 \cdot 3 \cdot x\\ 16y &= 2 \cdot 2 \cdot 2 \cdot 2 \cdot y\\ 8xyz &= 2 \cdot 2 \cdot 2 \cdot x \cdot y \cdot z\end{aligned}\right\} \quad \begin{aligned}\text{LCM} &= 2 \cdot 2 \cdot 2 \cdot 2 \cdot 3 \cdot x \cdot y \cdot z\\ &= 48xyz\end{aligned}$$

EXAMPLE 4 Find the LCM of $x^2 + 5x - 6$ and $x^2 - 1$.

$$\left.\begin{aligned}x^2 + 5x - 6 &= (x + 6)(x - 1)\\ x^2 - 1 &= (x + 1)(x - 1)\end{aligned}\right\} \quad \text{LCM} = (x + 6)(x - 1)(x + 1)$$

EXAMPLE 5 Find the LCM of $x^2 + 4$, $x + 1$, and 5.

These expressions do not share a common factor other than 1, so the LCM is their product:

$$5(x^2 + 4)(x + 1).$$

EXAMPLE 6 Find the LCM of $x^2 - 25$ and $2x - 10$.

$$\left.\begin{aligned}x^2 - 25 &= (x + 5)(x - 5)\\ 2x - 10 &= 2(x - 5)\end{aligned}\right\} \quad \text{LCM} = 2(x + 5)(x - 5)$$

EXAMPLE 7 Find the LCM of $x^2 - 4y^2$, $x^2 - 4xy + 4y^2$, and $x - 2y$.

$$\left.\begin{aligned}x^2 - 4y^2 &= (x - 2y)(x + 2y)\\ x^2 - 4xy + 4y^2 &= (x - 2y)(x - 2y)\\ x - 2y &= x - 2y\end{aligned}\right\} \quad \begin{aligned}\text{LCM} &= (x + 2y)(x - 2y)(x - 2y)\\ &= (x + 2y)(x - 2y)^2\end{aligned}$$

Do Exercises 9–12.

11.3 EXERCISE SET

For Extra Help

Digital Video Tutor CD 9 Videotape 13 | InterAct Math | Math Tutor Center | MathXL | MyMathLab

a Find the LCM.

1. 12, 27

2. 10, 15

3. 8, 9

4. 12, 18

5. 6, 9, 21

6. 8, 36, 40

7. 24, 36, 40

8. 4, 5, 20

9. 10, 100, 500

10. 28, 42, 60

b Add, first finding the LCD. Simplify if possible.

11. $\frac{7}{24} + \frac{11}{18}$

12. $\frac{7}{60} + \frac{2}{25}$

13. $\frac{1}{6} + \frac{3}{40}$

14. $\frac{5}{24} + \frac{3}{20}$

15. $\frac{1}{20} + \frac{1}{30} + \frac{2}{45}$

16. $\frac{2}{15} + \frac{5}{9} + \frac{3}{20}$

c Find the LCM.

17. $6x^2,\ 12x^3$

18. $2a^2b,\ 8ab^3$

19. $2x^2,\ 6xy,\ 18y^2$

20. $p^3q,\ p^2q,\ pq^2$

21. $2(y - 3),\ 6(y - 3)$

22. $5(m + 2),\ 15(m + 2)$

23. $t,\ t + 2,\ t - 2$

24. $y,\ y - 5,\ y + 5$

25. $x^2 - 4,\ x^2 + 5x + 6$

26. $x^2 - 4,\ x^2 - x - 2$

27. $t^3 + 4t^2 + 4t,\ t^2 - 4t$

28. $m^4 - m^2,\ m^3 - m^2$

29. $a + 1,\ (a - 1)^2,\ a^2 - 1$

30. $a^2 - 2ab + b^2,\ a^2 - b^2,\ 3a + 3b$

31. $m^2 - 5m + 6,\ m^2 - 4m + 4$

32. $2x^2 + 5x + 2,\ 2x^2 - x - 1$

33. $2 + 3x,\ 4 - 9x^2,\ 2 - 3x$

34. $9 - 4x^2,\ 3 + 2x,\ 3 - 2x$

35. $10v^2 + 30v,\ 5v^2 + 35v + 60$

36. $12a^2 + 24a,\ 4a^2 + 20a + 24$

37. $9x^3 - 9x^2 - 18x,\ 6x^5 - 24x^4 + 24x^3$

38. $x^5 - 4x^3,\ x^3 + 4x^2 + 4x$

39. $x^5 + 4x^4 + 4x^3,\ 3x^2 - 12,\ 2x + 4$

40. $x^5 + 2x^4 + x^3,\ 2x^3 - 2x,\ 5x - 5$

41. **DW** If the LCM of a binomial and a trinomial is the trinomial, what relationship exists between the two expressions?

42. **DW** Explain why the product of two numbers is not always their least common multiple.

SKILL MAINTENANCE

Factor. [10.7a]

43. $x^2 - 6x + 9$

44. $6x^2 + 4x$

45. $x^2 - 9$

46. $x^2 + 4x - 21$

47. $x^2 + 6x + 9$

48. $x^2 - 4x - 21$

Divorce Rate. The graph at right is that of the equation

$$D = 0.00509x^2 - 19.17x + 18{,}065.305$$

for values of x ranging from 1900 to 2010. It shows the percentage of couples who are married in a given year, x, whose marriages, it is predicted, will end in divorce. Use *only* the graph to answer the questions in Exercises 49–54. [8.1a], [9.3a]

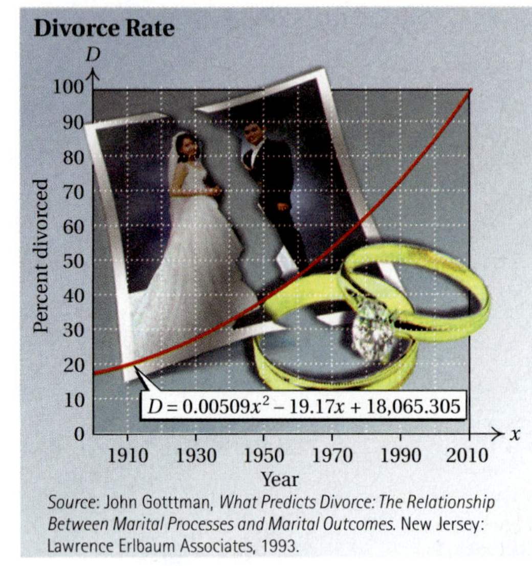

Source: John Gotttman, *What Predicts Divorce: The Relationship Between Marital Processes and Marital Outcomes*. New Jersey: Lawrence Erlbaum Associates, 1993.

49. Estimate the divorce percentage of those married in 1970.

50. Estimate the divorce percentage of those married in 1980.

51. Estimate the divorce percentage of those married in 1990.

52. Estimate the divorce percentage of those married in 2010. Does this seem reasonable?

53. In what year was the divorce percentage about 50%?

54. In what year was the divorce percentage about 84%?

SYNTHESIS

55. *Running.* Pedro and Maria leave the starting point of a fitness loop at the same time. Pedro jogs a lap in 6 min and Maria jogs one in 8 min. Assuming they continue to run at the same pace, when will they next meet at the starting place?

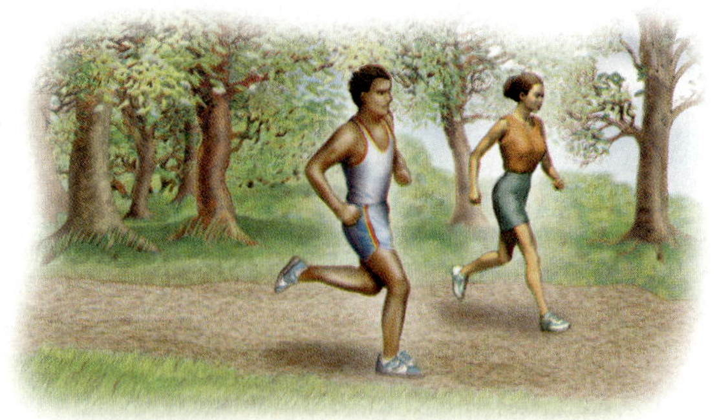

11.4 ADDING RATIONAL EXPRESSIONS

a Adding Rational Expressions

We add rational expressions as we do rational numbers.

ADDING RATIONAL EXPRESSIONS WITH LIKE DENOMINATORS

To add when the denominators are the same, add the numerators and keep the same denominator. Then simplify if possible.

EXAMPLES Add.

1. $\dfrac{x}{x+1} + \dfrac{2}{x+1} = \dfrac{x+2}{x+1}$

2. $\dfrac{2x^2+3x-7}{2x+1} + \dfrac{x^2+x-8}{2x+1} = \dfrac{(2x^2+3x-7)+(x^2+x-8)}{2x+1}$

$$= \frac{3x^2+4x-15}{2x+1}$$

3. $\dfrac{x-5}{x^2-9} + \dfrac{2}{x^2-9} = \dfrac{(x-5)+2}{x^2-9} = \dfrac{x-3}{x^2-9}$

$$= \frac{x-3}{(x-3)(x+3)} \quad \text{Factoring}$$

$$= \frac{\cancel{x-3}}{\cancel{(x-3)}(x+3)} \quad \text{Removing a factor of 1: } \frac{x-3}{x-3} = 1$$

$$= \frac{1}{x+3} \quad \text{Simplifying}$$

As in Example 3, simplifying should be done if possible after adding.

Do Exercises 1–3.

When denominators are different, we find the least common denominator, LCD. The procedure we use is as follows.

ADDING RATIONAL EXPRESSIONS WITH DIFFERENT DENOMINATORS

To add rational expressions with different denominators:

1. Find the LCM of the denominators. This is the least common denominator (LCD).
2. For each rational expression, find an equivalent expression with the LCD. To do so, multiply by 1 using an expression for 1 made up of factors of the LCD that are missing from the original denominator.
3. Add the numerators. Write the sum over the LCD.
4. Simplify if possible.

Objective

a Add rational expressions.

Add.

1. $\dfrac{5}{9} + \dfrac{2}{9}$

2. $\dfrac{3}{x-2} + \dfrac{x}{x-2}$

3. $\dfrac{4x+5}{x-1} + \dfrac{2x-1}{x-1}$

Answers on page A-38

Add.

4. $\dfrac{3x}{16} + \dfrac{5x^2}{24}$

5. $\dfrac{3}{16x} + \dfrac{5}{24x^2}$

6. Add:

$$\frac{3}{x^3 - x} + \frac{4}{x^2 + 2x + 1}.$$

Answers on page A-38

EXAMPLE 4 Add: $\dfrac{5x^2}{8} + \dfrac{7x}{12}$.

First, we find the LCD:

$$\left.\begin{array}{l} 8 = 2 \cdot 2 \cdot 2 \\ 12 = 2 \cdot 2 \cdot 3 \end{array}\right\} \quad \text{LCD} = 2 \cdot 2 \cdot 2 \cdot 3, \text{ or } 24.$$

Compare the factorization $8 = 2 \cdot 2 \cdot 2$ with the factorization of the LCD, $24 = 2 \cdot 2 \cdot 2 \cdot 3$. The factor of 24 that is missing from 8 is 3. Compare $12 = 2 \cdot 2 \cdot 3$ and $24 = 2 \cdot 2 \cdot 2 \cdot 3$. The factor of 24 that is missing from 12 is 2.

We multiply by a symbol for 1 to get the LCD in each expression, and then add and, if possible, simplify:

$$\frac{5x^2}{8} + \frac{7x}{12} = \frac{5x^2}{2 \cdot 2 \cdot 2} + \frac{7x}{2 \cdot 2 \cdot 3}$$

$$= \frac{5x^2}{2 \cdot 2 \cdot 2} \cdot \frac{3}{3} + \frac{7x}{2 \cdot 2 \cdot 3} \cdot \frac{2}{2} \quad \text{Multiplying by 1 to get the same denominators}$$

$$= \frac{15x^2}{24} + \frac{14x}{24} = \frac{15x^2 + 14x}{24}.$$

EXAMPLE 5 Add: $\dfrac{3}{8x} + \dfrac{5}{12x^2}$.

First, we find the LCD:

$$\left.\begin{array}{l} 8x = 2 \cdot 2 \cdot 2 \cdot x \\ 12x^2 = 2 \cdot 2 \cdot 3 \cdot x \cdot x \end{array}\right\} \quad \text{LCD} = 2 \cdot 2 \cdot 2 \cdot 3 \cdot x \cdot x, \text{ or } 24x^2.$$

The factors of the LCD missing from $8x$ are 3 and x. The factor of the LCD missing from $12x^2$ is 2. We multiply by 1 to get the LCD in each expression, and then add and, if possible, simplify:

$$\frac{3}{8x} + \frac{5}{12x^2} = \frac{3}{8x} \cdot \frac{3 \cdot x}{3 \cdot x} + \frac{5}{12x^2} \cdot \frac{2}{2}$$

$$= \frac{9x}{24x^2} + \frac{10}{24x^2} = \frac{9x + 10}{24x^2}.$$

Do Exercises 4 and 5.

EXAMPLE 6 Add: $\dfrac{2a}{a^2 - 1} + \dfrac{1}{a^2 + a}$.

First, we find the LCD:

$$\left.\begin{array}{l} a^2 - 1 = (a - 1)(a + 1) \\ a^2 + a = a(a + 1) \end{array}\right\} \quad \text{LCD} = a(a - 1)(a + 1).$$

We multiply by 1 to get the LCD in each expression, and then add and simplify:

$$\frac{2a}{(a - 1)(a + 1)} \cdot \frac{a}{a} + \frac{1}{a(a + 1)} \cdot \frac{a - 1}{a - 1}$$

$$= \frac{2a^2}{a(a - 1)(a + 1)} + \frac{a - 1}{a(a - 1)(a + 1)}$$

$$= \frac{2a^2 + a - 1}{a(a - 1)(a + 1)}$$

$$= \frac{(a + 1)(2a - 1)}{a(a - 1)(a + 1)}. \quad \text{Factoring the numerator in order to simplify}$$

Then

$$= \frac{(a+1)(2a-1)}{a(a-1)(a+1)} \quad \text{Removing a factor of 1: } \frac{a+1}{a+1} = 1$$

$$= \frac{2a-1}{a(a-1)}.$$

Do Exercise 6 on the preceding page.

EXAMPLE 7 Add: $\frac{x+4}{x-2} + \frac{x-7}{x+5}$.

First, we find the LCD. It is just the product of the denominators:

LCD $= (x-2)(x+5)$.

We multiply by 1 to get the LCD in each expression, and then add and simplify:

$$\frac{x+4}{x-2} \cdot \frac{x+5}{x+5} + \frac{x-7}{x+5} \cdot \frac{x-2}{x-2}$$

$$= \frac{(x+4)(x+5)}{(x-2)(x+5)} + \frac{(x-7)(x-2)}{(x-2)(x+5)}$$

$$= \frac{x^2+9x+20}{(x-2)(x+5)} + \frac{x^2-9x+14}{(x-2)(x+5)}$$

$$= \frac{x^2+9x+20+x^2-9x+14}{(x-2)(x+5)} = \frac{2x^2+34}{(x-2)(x+5)}.$$

Do Exercise 7.

EXAMPLE 8 Add: $\frac{x}{x^2+11x+30} + \frac{-5}{x^2+9x+20}$.

$$\frac{x}{x^2+11x+30} + \frac{-5}{x^2+9x+20}$$

$$= \frac{x}{(x+5)(x+6)} + \frac{-5}{(x+5)(x+4)} \quad \text{Factoring the denominators in order to find the LCD. The LCD is } (x+4)(x+5)(x+6).$$

$$= \frac{x}{(x+5)(x+6)} \cdot \frac{x+4}{x+4} + \frac{-5}{(x+5)(x+4)} \cdot \frac{x+6}{x+6} \quad \text{Multiplying by 1}$$

$$= \frac{x(x+4)+(-5)(x+6)}{(x+4)(x+5)(x+6)} = \frac{x^2+4x-5x-30}{(x+4)(x+5)(x+6)}$$

$$= \frac{x^2-x-30}{(x+4)(x+5)(x+6)}$$

$$= \frac{(x-6)(x+5)}{(x+4)(x+5)(x+6)}$$

$$= \frac{(x-6)}{(x+4)(x+6)}$$

Always simplify at the end if possible: $\frac{x+5}{x+5} = 1$.

Do Exercise 8.

DENOMINATORS THAT ARE OPPOSITES

When one denominator is the opposite of the other, we can first multiply either expression by 1 using $-1/-1$.

7. Add:

$$\frac{x-2}{x+3} + \frac{x+7}{x+8}.$$

8. Add:

$$\frac{5}{x^2+17x+16} + \frac{3}{x^2+9x+8}.$$

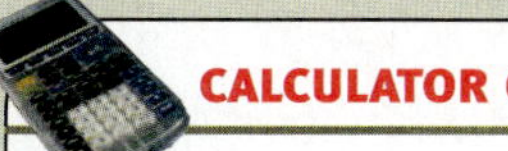

CALCULATOR CORNER

Checking Addition Use the TABLE feature, as described on p. 839, to check the additions in Examples 7 and 9. Then check your answers to Margin Exercises 7–9.

Answers on page A-38

Add.

9. $\dfrac{x}{4} + \dfrac{5}{-4}$

10. $\dfrac{2x+1}{x-3} + \dfrac{x+2}{3-x}$

11. Add:

$$\frac{x+3}{x^2-16} + \frac{5}{12-3x}.$$

Answers on page A-38

Study Tips

WORKING WITH RATIONAL EXPRESSIONS

The procedures covered in this chapter are by their nature rather long. It may help to write out lots of steps as you do the problems. If you have difficulty, consider taking a clean sheet of paper and starting over. Don't squeeze your work into a small amount of space. When using lined paper, consider using two spaces at a time, with the paper's line representing the fraction bar.

EXAMPLES

9. $\dfrac{x}{2} + \dfrac{3}{-2} = \dfrac{x}{2} + \dfrac{3}{-2} \cdot \dfrac{-1}{-1}$ Multiplying by 1 using $\dfrac{-1}{-1}$

$= \dfrac{x}{2} + \dfrac{-3}{2}$ The denominators are now the same.

$= \dfrac{x+(-3)}{2} = \dfrac{x-3}{2}$

10. $\dfrac{3x+4}{x-2} + \dfrac{x-7}{2-x} = \dfrac{3x+4}{x-2} + \dfrac{x-7}{2-x} \cdot \dfrac{-1}{-1}$

We could have chosen to multiply this expression by $-1/-1$. We multiply only one expression, *not* both.

$= \dfrac{3x+4}{x-2} + \dfrac{-x+7}{x-2}$ *Note:* $(2-x)(-1) = -2 + x = x - 2.$

$= \dfrac{(3x+4)+(-x+7)}{x-2} = \dfrac{2x+11}{x-2}$

Do Exercises 9 and 10.

FACTORS THAT ARE OPPOSITES

Suppose that when we factor to find the LCD, we find factors that are opposites. The easiest way to handle this is to first go back and multiply by $-1/-1$ appropriately to change factors so that they are not opposites.

EXAMPLE 11 Add: $\dfrac{x}{x^2-25} + \dfrac{3}{10-2x}$.

First, we factor to find the LCD:

$x^2 - 25 = (x-5)(x+5);$

$10 - 2x = 2(5 - x).$

We note that there is an $x - 5$ as one factor of $x^2 - 25$ and a $5 - x$ as one factor of $10 - 2x$. If the denominator of the second expression were $2x - 10$, this situation would not occur. To rewrite the second expression with a denominator of $2x - 10$, we multiply by 1 using $-1/-1$, and then continue as before:

$$\frac{x}{x^2-25} + \frac{3}{10-2x} = \frac{x}{(x-5)(x+5)} + \frac{3}{10-2x} \cdot \frac{-1}{-1}$$

$$= \frac{x}{(x-5)(x+5)} + \frac{-3}{2x-10}$$

$$= \frac{x}{(x-5)(x+5)} + \frac{-3}{2(x-5)} \qquad \text{LCD} = 2(x-5)(x+5)$$

$$= \frac{x}{(x-5)(x+5)} \cdot \frac{2}{2} + \frac{-3}{2(x-5)} \cdot \frac{x+5}{x+5}$$

$$= \frac{2x}{2(x-5)(x+5)} + \frac{-3(x+5)}{2(x-5)(x+5)}$$

$$= \frac{2x-3(x+5)}{2(x-5)(x+5)} = \frac{2x-3x-15}{2(x-5)(x+5)}$$

$$= \frac{-x-15}{2(x-5)(x+5)}. \qquad \text{Collecting like terms}$$

Do Exercise 11.

11.4 EXERCISE SET

For Extra Help

Digital Video Tutor CD 9 Videotape 13 | InterAct Math | Math Tutor Center | MathXL | MyMathLab

a Add. Simplify if possible.

1. $\frac{5}{8} + \frac{3}{8}$

2. $\frac{3}{16} + \frac{5}{16}$

3. $\frac{1}{3 + x} + \frac{5}{3 + x}$

4. $\frac{4x + 6}{2x - 1} + \frac{5 - 8x}{-1 + 2x}$

5. $\frac{x^2 + 7x}{x^2 - 5x} + \frac{x^2 - 4x}{x^2 - 5x}$

6. $\frac{4}{x + y} + \frac{9}{y + x}$

7. $\frac{2}{x} + \frac{5}{x^2}$

8. $\frac{3}{y^2} + \frac{6}{y}$

9. $\frac{5}{6r} + \frac{7}{8r}$

10. $\frac{13}{18x} + \frac{7}{24x}$

11. $\frac{4}{xy^2} + \frac{6}{x^2y}$

12. $\frac{8}{ab^3} + \frac{3}{a^2b}$

13. $\frac{2}{9t^3} + \frac{1}{6t^2}$

14. $\frac{5}{c^2d^3} + \frac{-4}{7cd^2}$

15. $\frac{x + y}{xy^2} + \frac{3x + y}{x^2y}$

16. $\frac{2c - d}{c^2d} + \frac{c + d}{cd^2}$

17. $\frac{3}{x - 2} + \frac{3}{x + 2}$

18. $\frac{2}{y + 1} + \frac{2}{y - 1}$

19. $\frac{3}{x + 1} + \frac{2}{3x}$

20. $\frac{4}{5y} + \frac{7}{y - 2}$

21. $\frac{2x}{x^2 - 16} + \frac{x}{x - 4}$

22. $\dfrac{4x}{x^2 - 25} + \dfrac{x}{x + 5}$

23. $\dfrac{5}{z + 4} + \dfrac{3}{3z + 12}$

24. $\dfrac{t}{t - 3} + \dfrac{5}{4t - 12}$

25. $\dfrac{3}{x - 1} + \dfrac{2}{(x - 1)^2}$

26. $\dfrac{8}{(y + 3)^2} + \dfrac{5}{y + 3}$

27. $\dfrac{4a}{5a - 10} + \dfrac{3a}{10a - 20}$

28. $\dfrac{9x}{6x - 30} + \dfrac{3x}{4x - 20}$

29. $\dfrac{x + 4}{x} + \dfrac{x}{x + 4}$

30. $\dfrac{a}{a - 3} + \dfrac{a - 3}{a}$

31. $\dfrac{4}{a^2 - a - 2} + \dfrac{3}{a^2 + 4a + 3}$

32. $\dfrac{a}{a^2 - 2a + 1} + \dfrac{1}{a^2 - 5a + 4}$

33. $\dfrac{x + 3}{x - 5} + \dfrac{x - 5}{x + 3}$

34. $\dfrac{3x}{2y - 3} + \dfrac{2x}{3y - 2}$

35. $\dfrac{a}{a^2 - 1} + \dfrac{2a}{a^2 - a}$

36. $\dfrac{3x + 2}{3x + 6} + \dfrac{x - 2}{x^2 - 4}$

37. $\dfrac{7}{8} + \dfrac{5}{-8}$

38. $\dfrac{5}{-3} + \dfrac{11}{3}$

39. $\dfrac{3}{t} + \dfrac{4}{-t}$

40. $\dfrac{5}{-a} + \dfrac{8}{a}$

41. $\dfrac{2x + 7}{x - 6} + \dfrac{3x}{6 - x}$

42. $\dfrac{2x - 7}{5x - 8} + \dfrac{6 + 10x}{8 - 5x}$

43. $\dfrac{y^2}{y-3}+\dfrac{9}{3-y}$

44. $\dfrac{t^2}{t-2}+\dfrac{4}{2-t}$

45. $\dfrac{b-7}{b^2-16}+\dfrac{7-b}{16-b^2}$

46. $\dfrac{a-3}{a^2-25}+\dfrac{a-3}{25-a^2}$

47. $\dfrac{a^2}{a-b}+\dfrac{b^2}{b-a}$

48. $\dfrac{x^2}{x-7}+\dfrac{49}{7-x}$

49. $\dfrac{x+3}{x-5}+\dfrac{2x-1}{5-x}+\dfrac{2(3x-1)}{x-5}$

50. $\dfrac{3(x-2)}{2x-3}+\dfrac{5(2x+1)}{2x-3}+\dfrac{3(x+1)}{3-2x}$

51. $\dfrac{2(4x+1)}{5x-7}+\dfrac{3(x-2)}{7-5x}+\dfrac{-10x-1}{5x-7}$

52. $\dfrac{5(x-2)}{3x-4}+\dfrac{2(x-3)}{4-3x}+\dfrac{3(5x+1)}{4-3x}$

53. $\dfrac{x+1}{(x+3)(x-3)}+\dfrac{4(x-3)}{(x-3)(x+3)}+\dfrac{(x-1)(x-3)}{(3-x)(x+3)}$

54. $\dfrac{2(x+5)}{(2x-3)(x-1)}+\dfrac{3x+4}{(2x-3)(1-x)}+\dfrac{x-5}{(3-2x)(x-1)}$

55. $\dfrac{6}{x-y}+\dfrac{4x}{y^2-x^2}$

56. $\dfrac{a-2}{3-a}+\dfrac{4-a^2}{a^2-9}$

57. $\dfrac{4-a}{25-a^2}+\dfrac{a+1}{a-5}$

58. $\dfrac{x+2}{x-7}+\dfrac{3-x}{49-x^2}$

59. $\dfrac{2}{t^2+t-6}+\dfrac{3}{t^2-9}$

60. $\dfrac{10}{a^2-a-6}+\dfrac{3a}{a^2+4a+4}$

61. DW Explain why the expressions

$$\frac{1}{3-x} \quad \text{and} \quad \frac{1}{x-3}$$

are opposites.

62. DW A student insists on finding a common denominator by always multiplying the denominators of the expressions being added. How could this approach be improved?

SKILL MAINTENANCE

Subtract. [9.4c]

63. $(x^2 + x) - (x + 1)$

64. $(4y^3 - 5y^2 + 7y - 24) - (-9y^3 + 9y^2 - 5y + 49)$

Simplify. [9.2a, b]

65. $(2x^4y^3)^{-3}$

66. $\left(\dfrac{x^3}{5y}\right)^2$

67. $\left(\dfrac{x^{-4}}{y^7}\right)^3$

68. $(5x^{-2}y^{-3})^2$

Graph.

69. $y = \dfrac{1}{2}x - 5$ [8.2b], [8.3a]

70. $2y + x + 10 = 0$ [8.2b], [8.3a]

71. $y = 3$ [8.3b]

72. $x = -5$ [8.3b]

Solve.

73. $3x - 7 = 5x + 9$ [7.3b]

74. $2a + 8 = 13 - 4a$ [7.3b]

75. $x^2 - 8x + 15 = 0$ [10.8b]

76. $x^2 - 7x = 18$ [10.8b]

SYNTHESIS

Find the perimeter and the area of the figure.

77.

$\dfrac{y+4}{3}$

$\dfrac{y-2}{5}$

78.

$\dfrac{3}{x+4}$

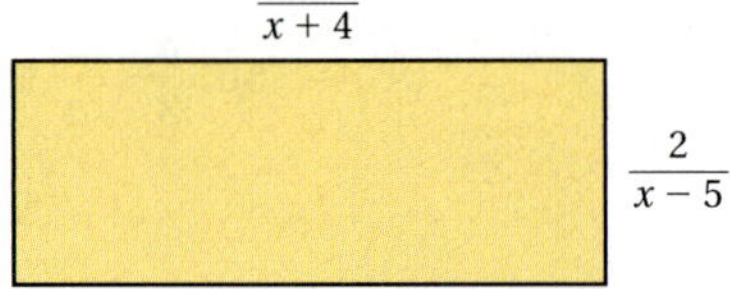

$\dfrac{2}{x-5}$

Add. Simplify if possible.

79. $\dfrac{5}{z+2} + \dfrac{4z}{z^2-4} + 2$

80. $\dfrac{-2}{y^2-9} + \dfrac{4y}{(y-3)^2} + \dfrac{6}{3-y}$

81. $\dfrac{3z^2}{z^4-4} + \dfrac{5z^2-3}{2z^4+z^2-6}$

82. Find an expression equivalent to

$$\frac{a-3b}{a-b}$$

that is a sum of two rational expressions. Answers may vary.

83.–86. Use the TABLE feature to check the additions in Exercises 29–32.

11.5 SUBTRACTING RATIONAL EXPRESSIONS

Objectives

a Subtract rational expressions.

b Simplify combined additions and subtractions of rational expressions.

a Subtracting Rational Expressions

We subtract rational expressions as we do rational numbers.

SUBTRACTING RATIONAL EXPRESSIONS WITH LIKE DENOMINATORS

To subtract when the denominators are the same, subtract the numerators and keep the same denominator. Then simplify if possible.

EXAMPLE 1 Subtract: $\frac{8}{x} - \frac{3}{x}$.

$$\frac{8}{x} - \frac{3}{x} = \frac{8-3}{x} = \frac{5}{x}$$

EXAMPLE 2 Subtract: $\frac{3x}{x+2} - \frac{x-2}{x+2}$.

$$\frac{3x}{x+2} - \frac{x-2}{x+2} = \frac{3x-(x-2)}{x+2}$$

CAUTION! The parentheses are important to make sure that you subtract the entire numerator.

$$= \frac{3x-x+2}{x+2} \quad \text{Removing parentheses}$$

$$= \frac{2x+2}{x+2}$$

Do Exercises 1–3.

To subtract rational expressions with different denominators, we use a procedure similar to what we used for addition, except that we subtract numerators and write the difference over the LCD.

SUBTRACTING RATIONAL EXPRESSIONS WITH DIFFERENT DENOMINATORS

To subtract rational expressions with different denominators:

1. Find the LCM of the denominators. This is the least common denominator (LCD).
2. For each rational expression, find an equivalent expression with the LCD. To do so, multiply by 1 using a symbol for 1 made up of factors of the LCD that are missing from the original denominator.
3. Subtract the numerators. Write the difference over the LCD.
4. Simplify if possible.

Subtract.

1. $\frac{7}{11} - \frac{3}{11}$

2. $\frac{7}{y} - \frac{2}{y}$

3. $\frac{2x^2+3x-7}{2x+1} - \frac{x^2+x-8}{2x+1}$

Answers on page A-39

4. Subtract:

$$\frac{x-2}{3x}-\frac{2x-1}{5x}.$$

5. Subtract:

$$\frac{x}{x^2+15x+56}-\frac{6}{x^2+13x+42}.$$

Answers on page A-39

EXAMPLE 3 Subtract: $\frac{x+2}{x-4}-\frac{x+1}{x+4}$.

The LCD $=(x-4)(x+4)$.

$$\frac{x+2}{x-4}\cdot\frac{x+4}{x+4}-\frac{x+1}{x+4}\cdot\frac{x-4}{x-4}$$ Multiplying by 1

$$=\frac{(x+2)(x+4)}{(x-4)(x+4)}-\frac{(x+1)(x-4)}{(x-4)(x+4)}$$

$$=\frac{x^2+6x+8}{(x-4)(x+4)}-\frac{x^2-3x-4}{(x-4)(x+4)}$$

$$=\frac{x^2+6x+8-(x^2-3x-4)}{(x-4)(x+4)}$$ Subtracting this numerator. Don't forget the parentheses.

$$=\frac{x^2+6x+8-x^2+3x+4}{(x-4)(x+4)}$$ Removing parentheses

$$=\frac{9x+12}{(x-4)(x+4)}$$

Do Exercise 4.

EXAMPLE 4 Subtract: $\frac{x}{x^2+5x+6}-\frac{2}{x^2+3x+2}$.

$$\frac{x}{x^2+5x+6}-\frac{2}{x^2+3x+2}$$

$$=\frac{x}{(x+2)(x+3)}-\frac{2}{(x+2)(x+1)}$$ LCD $=(x+1)(x+2)(x+3)$

$$=\frac{x}{(x+2)(x+3)}\cdot\frac{x+1}{x+1}-\frac{2}{(x+2)(x+1)}\cdot\frac{x+3}{x+3}$$

$$=\frac{x^2+x}{(x+1)(x+2)(x+3)}-\frac{2x+6}{(x+1)(x+2)(x+3)}$$

$$=\frac{x^2+x-(2x+6)}{(x+1)(x+2)(x+3)}$$ Subtracting this numerator. Don't forget the parentheses.

$$=\frac{x^2+x-2x-6}{(x+1)(x+2)(x+3)}$$

$$=\frac{x^2-x-6}{(x+1)(x+2)(x+3)}$$

$$=\frac{(x+2)(x-3)}{(x+1)(x+2)(x+3)}$$

$$=\frac{\cancel{(x+2)}(x-3)}{(x+1)\cancel{(x+2)}(x+3)}$$ Simplifying by removing a factor of 1: $\frac{x+2}{x+2}=1$

$$=\frac{x-3}{(x+1)(x+3)}.$$

Do Exercise 5.

DENOMINATORS THAT ARE OPPOSITES

When one denominator is the opposite of the other, we can first multiply one expression by $-1/-1$ to obtain a common denominator.

EXAMPLE 5 Subtract: $\frac{x}{5} - \frac{3x - 4}{-5}$.

$$\frac{x}{5} - \frac{3x - 4}{-5} = \frac{x}{5} - \frac{3x - 4}{-5} \cdot \frac{-1}{-1} \quad \text{Multiplying by 1 using } \frac{-1}{-1}$$

This is equal to 1 (not -1).

$$= \frac{x}{5} - \frac{(3x - 4)(-1)}{(-5)(-1)}$$

$$= \frac{x}{5} - \frac{4 - 3x}{5}$$

$$= \frac{x - (4 - 3x)}{5} \quad \text{Remember the parentheses!}$$

$$= \frac{x - 4 + 3x}{5} = \frac{4x - 4}{5}$$

EXAMPLE 6 Subtract: $\frac{5y}{y - 5} - \frac{2y - 3}{5 - y}$.

$$\frac{5y}{y - 5} - \frac{2y - 3}{5 - y} = \frac{5y}{y - 5} - \frac{2y - 3}{5 - y} \cdot \frac{-1}{-1}$$

$$= \frac{5y}{y - 5} - \frac{(2y - 3)(-1)}{(5 - y)(-1)}$$

$$= \frac{5y}{y - 5} - \frac{3 - 2y}{y - 5}$$

$$= \frac{5y - (3 - 2y)}{y - 5} \quad \text{Remember the parentheses!}$$

$$= \frac{5y - 3 + 2y}{y - 5} = \frac{7y - 3}{y - 5}$$

Do Exercises 6 and 7.

FACTORS THAT ARE OPPOSITES

Suppose that when we factor to find the LCD, we find factors that are opposites. Then we multiply by $-1/-1$ appropriately to change factors so that they are not opposites.

EXAMPLE 7 Subtract: $\frac{p}{64 - p^2} - \frac{5}{p - 8}$.

Factoring $64 - p^2$, we get $(8 - p)(8 + p)$. Note that the factors $8 - p$ in the first denominator and $p - 8$ in the second denominator are opposites. We multiply the first expression by $-1/-1$ to avoid this situation. Then we proceed as before.

$$\frac{p}{64 - p^2} - \frac{5}{p - 8} = \frac{p}{64 - p^2} \cdot \frac{-1}{-1} - \frac{5}{p - 8}$$

$$= \frac{-p}{p^2 - 64} - \frac{5}{p - 8}$$

$$= \frac{-p}{(p - 8)(p + 8)} - \frac{5}{p - 8} \quad \text{LCD} = (p - 8)(p + 8)$$

$$= \frac{-p}{(p - 8)(p + 8)} - \frac{5}{p - 8} \cdot \frac{p + 8}{p + 8}$$

Subtract.

6. $\frac{x}{3} - \frac{2x - 1}{-3}$

7. $\frac{3x}{x - 2} - \frac{x - 3}{2 - x}$

Answers on page A-39

CALCULATOR CORNER

Checking Subtraction

Use the TABLE feature, as described on p. 839, to check the subtractions in Examples 5 and 6. Then check your answers to Margin Exercises 6 and 7.

8. Subtract:

$$\frac{y}{16 - y^2} - \frac{7}{y - 4}.$$

9. Perform the indicated operations and simplify:

$$\frac{x + 2}{x^2 - 9} - \frac{x - 7}{9 - x^2} + \frac{-8 - x}{x^2 - 9}.$$

10. Perform the indicated operations and simplify:

$$\frac{1}{x} - \frac{5}{3x} + \frac{2x}{x + 1}.$$

Answers on page A-39

Then

$$= \frac{-p}{(p-8)(p+8)} - \frac{5p + 40}{(p-8)(p+8)}$$

Subtracting this numerator. Don't forget the parentheses.

$$= \frac{-p - (5p + 40)}{(p-8)(p+8)}$$

$$= \frac{-p - 5p - 40}{(p-8)(p+8)} = \frac{-6p - 40}{(p-8)(p+8)}.$$

Do Exercise 8.

b Combined Additions and Subtractions

Now let's look at some combined additions and subtractions.

EXAMPLE 8 Perform the indicated operations and simplify:

$$\frac{x + 9}{x^2 - 4} + \frac{5 - x}{4 - x^2} - \frac{2 + x}{x^2 - 4}.$$

$$\frac{x + 9}{x^2 - 4} + \frac{5 - x}{4 - x^2} - \frac{2 + x}{x^2 - 4}$$

$$= \frac{x + 9}{x^2 - 4} + \frac{5 - x}{4 - x^2} \cdot \frac{-1}{-1} - \frac{2 + x}{x^2 - 4}$$

$$= \frac{x + 9}{x^2 - 4} + \frac{x - 5}{x^2 - 4} - \frac{2 + x}{x^2 - 4} = \frac{(x + 9) + (x - 5) - (2 + x)}{x^2 - 4}$$

$$= \frac{x + 9 + x - 5 - 2 - x}{x^2 - 4} = \frac{x + 2}{x^2 - 4}$$

$$= \frac{\cancel{(x + 2)} \cdot 1}{\cancel{(x + 2)}(x - 2)} = \frac{1}{x - 2}.$$

Do Exercise 9.

EXAMPLE 9 Perform the indicated operations and simplify:

$$\frac{1}{x} - \frac{1}{x^2} + \frac{2}{x + 1}.$$

The LCD $= x \cdot x(x + 1)$, or $x^2(x + 1)$.

$$\frac{1}{x} \cdot \frac{x(x + 1)}{x(x + 1)} - \frac{1}{x^2} \cdot \frac{(x + 1)}{(x + 1)} + \frac{2}{x + 1} \cdot \frac{x^2}{x^2}$$

$$= \frac{x(x + 1)}{x^2(x + 1)} - \frac{x + 1}{x^2(x + 1)} + \frac{2x^2}{x^2(x + 1)}$$

Subtracting this numerator. Don't forget the parentheses.

$$= \frac{x(x + 1) - (x + 1) + 2x^2}{x^2(x + 1)}$$

$$= \frac{x^2 + x - x - 1 + 2x^2}{x^2(x + 1)} \qquad \text{Removing parentheses}$$

$$= \frac{3x^2 - 1}{x^2(x + 1)}$$

Do Exercise 10.

11.5 EXERCISE SET

For Extra Help

Digital Video Tutor CD 9 Videotape 13 | InterAct Math | Math Tutor Center | MathXL | MyMathLab

a Subtract. Simplify if possible.

1. $\frac{7}{x} - \frac{3}{x}$

2. $\frac{5}{a} - \frac{8}{a}$

3. $\frac{y}{y-4} - \frac{4}{y-4}$

4. $\frac{t^2}{t+5} - \frac{25}{t+5}$

5. $\frac{2x-3}{x^2+3x-4} - \frac{x-7}{x^2+3x-4}$

6. $\frac{x+1}{x^2-2x+1} - \frac{5-3x}{x^2-2x+1}$

7. $\frac{a-2}{10} - \frac{a+1}{5}$

8. $\frac{y+3}{2} - \frac{y-4}{4}$

9. $\frac{4z-9}{3z} - \frac{3z-8}{4z}$

10. $\frac{a-1}{4a} - \frac{2a+3}{a}$

11. $\frac{4x+2t}{3xt^2} - \frac{5x-3t}{x^2t}$

12. $\frac{5x+3y}{2x^2y} - \frac{3x+4y}{xy^2}$

13. $\frac{5}{x+5} - \frac{3}{x-5}$

14. $\frac{3t}{t-1} - \frac{8t}{t+1}$

15. $\frac{3}{2t^2-2t} - \frac{5}{2t-2}$

16. $\frac{11}{x^2-4} - \frac{8}{x+2}$

17. $\frac{2s}{t^2-s^2} - \frac{s}{t-s}$

18. $\frac{3}{12+x-x^2} - \frac{2}{x^2-9}$

19. $\frac{y-5}{y}-\frac{3y-1}{4y}$

20. $\frac{3x-2}{4x}-\frac{3x+1}{6x}$

21. $\frac{a}{x+a}-\frac{a}{x-a}$

22. $\frac{a}{a-b}-\frac{a}{a+b}$

23. $\frac{11}{6}-\frac{5}{-6}$

24. $\frac{5}{9}-\frac{7}{-9}$

25. $\frac{5}{a}-\frac{8}{-a}$

26. $\frac{8}{x}-\frac{3}{-x}$

27. $\frac{4}{y-1}-\frac{4}{1-y}$

28. $\frac{5}{a-2}-\frac{3}{2-a}$

29. $\frac{3-x}{x-7}-\frac{2x-5}{7-x}$

30. $\frac{t^2}{t-2}-\frac{4}{2-t}$

31. $\frac{a-2}{a^2-25}-\frac{6-a}{25-a^2}$

32. $\frac{x-8}{x^2-16}-\frac{x-8}{16-x^2}$

33. $\frac{4-x}{x-9}-\frac{3x-8}{9-x}$

34. $\frac{4x-6}{x-5}-\frac{7-2x}{5-x}$

35. $\frac{5x}{x^2-9}-\frac{4}{3-x}$

36. $\frac{8x}{16-x^2}-\frac{5}{x-4}$

37. $\frac{t^2}{2t^2 - 2t} - \frac{1}{2t - 2}$

38. $\frac{4}{5a^2 - 5a} - \frac{2}{5a - 5}$

39. $\frac{x}{x^2 + 5x + 6} - \frac{2}{x^2 + 3x + 2}$

40. $\frac{a}{a^2 + 11a + 30} - \frac{5}{a^2 + 9a + 20}$

b Perform the indicated operations and simplify.

41. $\frac{3(2x + 5)}{x - 1} - \frac{3(2x - 3)}{1 - x} + \frac{6x - 1}{x - 1}$

42. $\frac{a - 2b}{b - a} - \frac{3a - 3b}{a - b} + \frac{2a - b}{a - b}$

43. $\frac{x - y}{x^2 - y^2} + \frac{x + y}{x^2 - y^2} - \frac{2x}{x^2 - y^2}$

44. $\frac{x - 3y}{2(y - x)} + \frac{x + y}{2(x - y)} - \frac{2x - 2y}{2(x - y)}$

45. $\frac{2(x - 1)}{2x - 3} - \frac{3(x + 2)}{2x - 3} - \frac{x - 1}{3 - 2x}$

46. $\frac{5(2y + 1)}{2y - 3} - \frac{3(y - 1)}{3 - 2y} - \frac{3(y - 2)}{2y - 3}$

47. $\frac{10}{2y - 1} - \frac{6}{1 - 2y} + \frac{y}{2y - 1} + \frac{y - 4}{1 - 2y}$

48. $\frac{(x + 1)(2x - 1)}{(2x - 3)(x - 3)} - \frac{(x - 3)(x + 1)}{(3 - x)(3 - 2x)} + \frac{(2x + 1)(x + 3)}{(3 - 2x)(x - 3)}$

49. $\frac{a + 6}{4 - a^2} - \frac{a + 3}{a + 2} + \frac{a - 3}{2 - a}$

50. $\frac{4t}{t^2 - 1} - \frac{2}{t} - \frac{2}{t + 1}$

51. $\frac{2z}{1 - 2z} + \frac{3z}{2z + 1} - \frac{3}{4z^2 - 1}$

52. $\frac{1}{x - y} - \frac{2x}{x^2 - y^2} + \frac{1}{x + y}$

53. $\frac{1}{x + y} - \frac{1}{x - y} + \frac{2x}{x^2 - y^2}$

54. $\frac{2b}{a^2 - b^2} - \frac{1}{a + b} + \frac{1}{a - b}$

55. **DW** Are parentheses as important when adding rational expressions as they are when subtracting? Why or why not?

56. **DW** Is it possible to add or subtract rational expressions without knowing how to factor? Why or why not?

SKILL MAINTENANCE

Simplify.

57. $\frac{x^8}{x^3}$ [9.1e]

58. $3x^4 \cdot 10x^8$ [9.1d]

59. $(a^2b^{-5})^{-4}$ [9.2a, b]

60. $\frac{54x^{10}}{3x^7}$ [9.1e]

61. $\frac{66x^2}{11x^5}$ [9.1e, f]

62. $5x^{-7} \cdot 2x^4$ [9.1d]

Find a polynomial for the shaded area of the figure. [9.4d]

63.

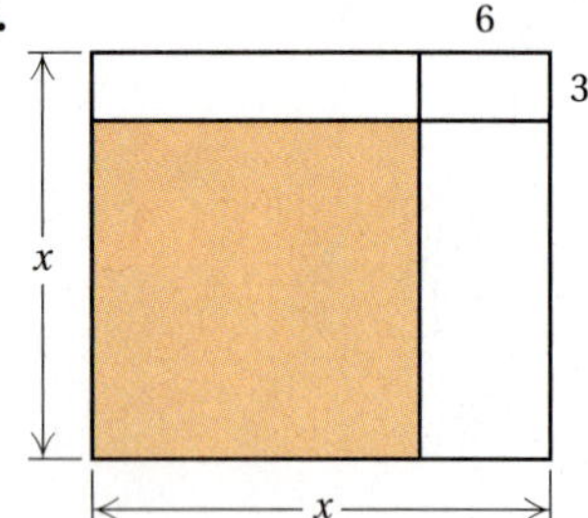

64.

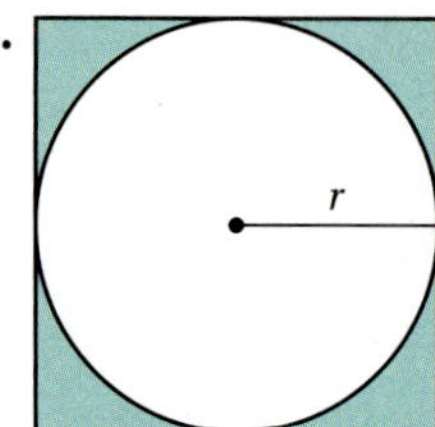

SYNTHESIS

Perform the indicated operations and simplify.

65. $\frac{2x+11}{x-3} \cdot \frac{3}{x+4} + \frac{2x+1}{4+x} \cdot \frac{3}{3-x}$

66. $\frac{x^2}{3x^2-5x-2} - \frac{2x}{3x+1} \cdot \frac{1}{x-2}$

67. $\frac{x}{x^4-y^4} - \left(\frac{1}{x+y}\right)^2$

68. $\left(\frac{a}{a-b} + \frac{b}{a+b}\right)\left(\frac{1}{3a+b} + \frac{2a+6b}{9a^2-b^2}\right)$

69. The perimeter of the following right triangle is $2a + 5$. Find the length of the missing side and the area.

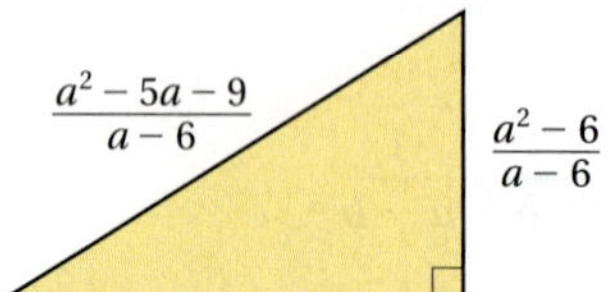

70.–73. Use the TABLE feature to check the subtractions in Exercises 15, 16, 19, and 20.

11.6 COMPLEX RATIONAL EXPRESSIONS

Objective

a Simplify complex rational expressions.

a Simplifying Complex Rational Expressions

A **complex rational expression,** or **complex fractional expression,** is a rational expression that has one or more rational expressions within its numerator or denominator. Here are some examples:

$$\frac{1+\frac{2}{x}}{3}, \quad \frac{\frac{x+y}{2}}{\frac{2x}{x+1}}, \quad \frac{\frac{1}{3}+\frac{1}{5}}{\frac{2}{x}-\frac{x}{y}}.$$

These are rational expressions within the complex rational expression.

There are two methods to simplify complex rational expressions. We will consider them both.

METHOD 1

MULTIPLYING BY THE LCM OF ALL THE DENOMINATORS

To simplify a complex rational expression:

1. First, find the LCM of all the denominators of all the rational expressions occurring *within* both the numerator and the denominator of the complex rational expression.
2. Then multiply by 1 using LCM/LCM.
3. If possible, simplify by removing a factor of 1.

To the instructor and the student: Students can be instructed to either try both methods and then choose the one that works best for them, or use the method chosen by the instructor.

EXAMPLE 1 Simplify: $\dfrac{\frac{1}{2}+\frac{3}{4}}{\frac{5}{6}-\frac{3}{8}}$.

We have

$$\frac{\frac{1}{2}+\frac{3}{4}}{\frac{5}{6}-\frac{3}{8}}$$

The denominators *within* the complex rational expression are 2, 4, 6, and 8. The LCM of these denominators is 24. We multiply by 1 using $\frac{24}{24}$. This amounts to multiplying both the numerator *and* the denominator by 24.

$$= \frac{\frac{1}{2}+\frac{3}{4}}{\frac{5}{6}-\frac{3}{8}} \cdot \frac{24}{24}$$ Multiplying by 1

$$= \frac{\left(\frac{1}{2}+\frac{3}{4}\right)24}{\left(\frac{5}{6}-\frac{3}{8}\right)24}$$

← Multiplying the numerator by 24

← Multiplying the denominator by 24

1. Simplify. Use method 1.

$$\frac{\frac{1}{3}+\frac{4}{5}}{\frac{7}{8}-\frac{5}{6}}$$

2. Simplify. Use method 1.

$$\frac{\frac{x}{2}+\frac{2x}{3}}{\frac{1}{x}-\frac{x}{2}}$$

3. Simplify. Use method 1.

$$\frac{1+\frac{1}{x}}{1-\frac{1}{x^2}}$$

Answers on page A-39

Using the distributive laws, we carry out the multiplications:

$$= \frac{\frac{1}{2}(24)+\frac{3}{4}(24)}{\frac{5}{6}(24)-\frac{3}{8}(24)}$$

$$= \frac{12+18}{20-9} \quad \text{Simplifying}$$

$$= \frac{30}{11}.$$

Multiplying in this manner has the effect of clearing fractions in both the top and the bottom of the complex rational expression.

Do Exercise 1.

EXAMPLE 2 Simplify: $\dfrac{\frac{3}{x}+\frac{1}{2x}}{\frac{1}{3x}-\frac{3}{4x}}$.

The denominators within the complex expression are x, $2x$, $3x$, and $4x$. The LCM of these denominators is $12x$. We multiply by 1 using $12x/12x$.

$$\frac{\frac{3}{x}+\frac{1}{2x}}{\frac{1}{3x}-\frac{3}{4x}}\cdot\frac{12x}{12x} = \frac{\left(\frac{3}{x}+\frac{1}{2x}\right)12x}{\left(\frac{1}{3x}-\frac{3}{4x}\right)12x} = \frac{\frac{3}{x}(12x)+\frac{1}{2x}(12x)}{\frac{1}{3x}(12x)-\frac{3}{4x}(12x)}$$

$$= \frac{36+6}{4-9} = -\frac{42}{5}$$

Do Exercise 2.

EXAMPLE 3 Simplify: $\dfrac{1-\frac{1}{x}}{1-\frac{1}{x^2}}$.

The denominators within the complex expression are x and x^2. The LCM of these denominators is x^2. We multiply by 1 using x^2/x^2. Then, after obtaining a single rational expression, we simplify:

$$\frac{1-\frac{1}{x}}{1-\frac{1}{x^2}}\cdot\frac{x^2}{x^2} = \frac{\left(1-\frac{1}{x}\right)x^2}{\left(1-\frac{1}{x^2}\right)x^2} = \frac{1(x^2)-\frac{1}{x}(x^2)}{1(x^2)-\frac{1}{x^2}(x^2)} = \frac{x^2-x}{x^2-1}$$

$$= \frac{x\cancel{(x-1)}}{(x+1)\cancel{(x-1)}} = \frac{x}{x+1}.$$

Do Exercise 3.

METHOD 2

ADDING IN THE NUMERATOR AND THE DENOMINATOR

To simplify a complex rational expression:

1. Add or subtract, as necessary, to get a single rational expression in the numerator.
2. Add or subtract, as necessary, to get a single rational expression in the denominator.
3. Divide the numerator by the denominator.
4. If possible, simplify by removing a factor of 1.

We will redo Examples 1–3 using this method.

EXAMPLE 4 Simplify: $\dfrac{\frac{1}{2}+\frac{3}{4}}{\frac{5}{6}-\frac{3}{8}}$.

The LCM of 2 and 4 in the numerator is 4. The LCM of 6 and 8 in the denominator is 24. We have

$$\frac{\frac{1}{2}+\frac{3}{4}}{\frac{5}{6}-\frac{3}{8}} = \frac{\frac{1}{2}\cdot\frac{2}{2}+\frac{3}{4}}{\frac{5}{6}\cdot\frac{4}{4}-\frac{3}{8}\cdot\frac{3}{3}}$$

← Multiplying the $\frac{1}{2}$ by 1 to get the common denominator, 4

← Multiplying the $\frac{5}{6}$ and the $\frac{3}{8}$ by 1 to get the common denominator, 24

$$= \frac{\frac{2}{4}+\frac{3}{4}}{\frac{20}{24}-\frac{9}{24}}$$

$$= \frac{\frac{5}{4}}{\frac{11}{24}}$$ Adding in the numerator; subtracting in the denominator

$$= \frac{5}{4}\cdot\frac{24}{11}$$ Multiplying by the reciprocal of the divisor

$$= \frac{5\cdot 3\cdot 2\cdot 2\cdot 2}{2\cdot 2\cdot 11}$$ Factoring

$$= \frac{5\cdot 3\cdot 2\cdot \cancel{2}\cdot \cancel{2}}{\cancel{2}\cdot \cancel{2}\cdot 11}$$ Removing a factor of 1: $\frac{2\cdot 2}{2\cdot 2} = 1$

$$= \frac{30}{11}.$$

Do Exercise 4.

4. Simplify. Use method 2.

$$\frac{\frac{1}{3}+\frac{4}{5}}{\frac{7}{8}-\frac{5}{6}}$$

Answer on page A-39

5. Simplify. Use method 2.

$$\frac{\frac{x}{2}+\frac{2x}{3}}{\frac{1}{x}-\frac{x}{2}}$$

6. Simplify. Use method 2.

$$\frac{1+\frac{1}{x}}{1-\frac{1}{x^2}}$$

Answers on page A-39

EXAMPLE 5 Simplify: $\dfrac{\frac{3}{x}+\frac{1}{2x}}{\frac{1}{3x}-\frac{3}{4x}}$.

We have

$$\frac{\frac{3}{x}+\frac{1}{2x}}{\frac{1}{3x}-\frac{3}{4x}} = \frac{\frac{3}{x}\cdot\frac{2}{2}+\frac{1}{2x}}{\frac{1}{3x}\cdot\frac{4}{4}-\frac{3}{4x}\cdot\frac{3}{3}}$$

Finding the LCD, $2x$, and multiplying by 1 in the numerator

Finding the LCD, $12x$, and multiplying by 1 in the denominator

$$= \frac{\frac{6}{2x}+\frac{1}{2x}}{\frac{4}{12x}-\frac{9}{12x}} = \frac{\frac{7}{2x}}{\frac{-5}{12x}}$$

Adding in the numerator and subtracting in the denominator

$$= \frac{7}{2x}\cdot\frac{12x}{-5}$$

Multiplying by the reciprocal of the divisor

$$= \frac{7}{2x}\cdot\frac{6(2x)}{-5}$$

Factoring

$$= \frac{7}{\cancel{2x}}\cdot\frac{6(\cancel{2x})}{-5}$$

Removing a factor of 1: $\frac{2x}{2x}=1$

$$= \frac{42}{-5} = -\frac{42}{5}.$$

Do Exercise 5.

EXAMPLE 6 Simplify: $\dfrac{1-\frac{1}{x}}{1-\frac{1}{x^2}}$.

We have

$$\frac{1-\frac{1}{x}}{1-\frac{1}{x^2}} = \frac{1\cdot\frac{x}{x}-\frac{1}{x}}{1\cdot\frac{x^2}{x^2}-\frac{1}{x^2}}$$

Finding the LCD, x, and multiplying by 1 in the numerator

Finding the LCD, x^2, and multiplying by 1 in the denominator

$$= \frac{\frac{x-1}{x}}{\frac{x^2-1}{x^2}}$$

Subtracting in the numerator and subtracting in the denominator

$$= \frac{x-1}{x}\cdot\frac{x^2}{x^2-1}$$

Multiplying by the reciprocal of the divisor

$$= \frac{(x-1)x\cdot x}{x(x-1)(x+1)}$$

Factoring

$$= \frac{\cancel{(x-1)}\cancel{x}\cdot x}{\cancel{x}\cancel{(x-1)}(x+1)}$$

Removing a factor of 1: $\frac{x(x-1)}{x(x-1)}=1$

$$= \frac{x}{x+1}.$$

Do Exercise 6.

11.6 EXERCISE SET

For Extra Help

Digital Video Tutor CD 10 Videotape 13 | InterAct Math | Math Tutor Center | MathXL | MyMathLab

a Simplify.

1. $\dfrac{1+\frac{9}{16}}{1-\frac{3}{4}}$

2. $\dfrac{6-\frac{3}{8}}{4+\frac{5}{6}}$

3. $\dfrac{1-\frac{3}{5}}{1+\frac{1}{5}}$

4. $\dfrac{2+\frac{2}{3}}{2-\frac{2}{3}}$

5. $\dfrac{\frac{1}{2}+\frac{3}{4}}{\frac{5}{8}-\frac{5}{6}}$

6. $\dfrac{\frac{3}{4}+\frac{7}{8}}{\frac{2}{3}-\frac{5}{6}}$

7. $\dfrac{\frac{1}{x}+3}{\frac{1}{x}-5}$

8. $\dfrac{2-\frac{1}{a}}{4+\frac{1}{a}}$

9. $\dfrac{4-\frac{1}{x^2}}{2-\frac{1}{x}}$

10. $\dfrac{\frac{2}{y}+\frac{1}{2y}}{y+\frac{y}{2}}$

11. $\dfrac{8+\frac{8}{d}}{1+\frac{1}{d}}$

12. $\dfrac{3+\frac{2}{t}}{3-\frac{2}{t}}$

13. $\dfrac{\frac{x}{8}-\frac{8}{x}}{\frac{1}{8}+\frac{1}{x}}$

14. $\dfrac{\frac{2}{m}+\frac{m}{2}}{\frac{m}{3}-\frac{3}{m}}$

15. $\dfrac{1+\frac{1}{y}}{1-\frac{1}{y^2}}$

16. $\dfrac{\frac{1}{q^2}-1}{\frac{1}{q}+1}$

17. $\dfrac{\frac{1}{5}-\frac{1}{a}}{\frac{5-a}{5}}$

18. $\dfrac{\frac{4}{t}}{4+\frac{1}{t}}$

19. $\dfrac{\frac{1}{a}+\frac{1}{b}}{\frac{1}{a^2}-\frac{1}{b^2}}$

20. $\dfrac{\frac{1}{x^2}-\frac{1}{y^2}}{\frac{2}{x}-\frac{2}{y}}$

21. $\dfrac{\frac{p}{q}+\frac{q}{p}}{\frac{1}{p}+\frac{1}{q}}$

22. $\dfrac{x-3+\frac{2}{x}}{x-4+\frac{3}{x}}$

23. $\dfrac{\frac{2}{a}+\frac{4}{a^2}}{\frac{5}{a^3}-\frac{3}{a}}$

24. $\dfrac{\frac{5}{x^3}-\frac{1}{x^2}}{\frac{2}{x}+\frac{3}{x^2}}$

25. $\dfrac{\dfrac{2}{7a^4} - \dfrac{1}{14a}}{\dfrac{3}{5a^2} + \dfrac{2}{15a}}$

26. $\dfrac{\dfrac{5}{4x^3} - \dfrac{3}{8x}}{\dfrac{3}{2x} + \dfrac{3}{4x^3}}$

27. $\dfrac{\dfrac{a}{b} + \dfrac{c}{d}}{\dfrac{b}{a} + \dfrac{d}{c}}$

28. $\dfrac{\dfrac{a}{b} - \dfrac{c}{d}}{\dfrac{b}{a} - \dfrac{d}{c}}$

29. $\dfrac{\dfrac{x}{5y^3} + \dfrac{3}{10y}}{\dfrac{3}{10y} + \dfrac{x}{5y^3}}$

30. $\dfrac{\dfrac{a}{6b^3} + \dfrac{4}{9b^2}}{\dfrac{5}{6b} - \dfrac{1}{9b^3}}$

31. $\dfrac{\dfrac{1}{z^2} - \dfrac{1}{w^2}}{\dfrac{1}{z^3} + \dfrac{1}{w^3}}$

32. $\dfrac{\dfrac{1}{b^2} - \dfrac{1}{c^2}}{\dfrac{1}{b^3} - \dfrac{1}{c^3}}$

33. **DW** Why is factoring an important skill when simplifying complex rational expressions?

34. **DW** Why is the distributive law especially important when using method 1 of this section?

SKILL MAINTENANCE

Add. [9.4a]

35. $(2x^3 - 4x^2 + x - 7) + (4x^4 + x^3 + 4x^2 + x)$

36. $(2x^3 - 4x^2 + x - 7) + (-2x^3 + 4x^2 - x + 7)$

Factor. [10.7a]

37. $p^2 - 10p + 25$

38. $p^2 + 10p + 25$

39. $50p^2 - 100$

40. $5p^2 - 40p - 100$

Solve.

41. *Perimeter of a Rectangle.* The length of a rectangle is 3 yd greater than the width. The area of the rectangle is 10 yd^2. Find the perimeter. [10.9a]

42. *Ladder Distances.* A ladder of length 13 ft is placed against a building in such a way that the distance from the top of the ladder to the ground is 7 ft more than the distance from the bottom of the ladder to the building. Find these distances. [10.9b]

SYNTHESIS

43. Find the reciprocal of $\dfrac{2}{x-1} - \dfrac{1}{3x-2}$.

Simplify.

44. $\left[\dfrac{\dfrac{x+1}{x-1} + 1}{\dfrac{x+1}{x-1} - 1}\right]^5$

45. $1 + \dfrac{1}{1 + \dfrac{1}{1 + \dfrac{1}{1 + \dfrac{1}{x}}}}$

46. $\dfrac{\dfrac{z}{1 - \dfrac{z}{2 + 2z}} - 2z}{\dfrac{2z}{5z - 2} - 3}$

11.7 SOLVING RATIONAL EQUATIONS

a Rational Equations

In Sections 11.1–11.6, we studied operations with *rational expressions.* These expressions have no equals signs. We can add, subtract, multiply, or divide and simplify expressions, but we cannot solve if there are no equals signs—as, for example, in

$$\frac{x^2+6x+9}{x^2-4}\cdot\frac{x-2}{x+3},\qquad \frac{x+y}{x-y}\div\frac{x^2+y}{x^2-y^2},\quad\text{and}\quad \frac{a+3}{a^2-16}+\frac{5}{12-3a}.$$

Operation signs occur. There are no equals signs!

Most often, the result of our calculation is another rational expression that has not been cleared of fractions.

Equations *do have* equals signs, and we can clear them of fractions as we did in Section 7.3. A **rational,** or **fractional, equation** is an equation containing one or more rational expressions. Here are some examples:

$$\frac{2}{3}+\frac{5}{6}=\frac{x}{9},\qquad x+\frac{6}{x}=-5,\quad\text{and}\quad \frac{x^2}{x-1}=\frac{1}{x-1}.$$

There are equals signs as well as operation signs.

SOLVING RATIONAL EQUATIONS

To solve a rational equation, the first step is to clear the equation of fractions. To do this, multiply all terms on both sides of the equation by the LCM of all the denominators. Then carry out the equation-solving process as we learned it in Chapter 7.

When clearing an equation of fractions, we use the terminology LCM instead of LCD because we are *not* adding or subtracting rational expressions.

EXAMPLE 1 Solve: $\frac{2}{3}+\frac{5}{6}=\frac{x}{9}$.

The LCM of all denominators is $2\cdot3\cdot3$, or 18. We multiply all terms on both sides by 18:

$$18\left(\frac{2}{3}+\frac{5}{6}\right)=18\cdot\frac{x}{9}$$ Multiplying both sides by the LCM

$$18\cdot\frac{2}{3}+18\cdot\frac{5}{6}=18\cdot\frac{x}{9}$$ Multiplying each term by the LCM to remove parentheses

$$12+15=2x$$ Simplifying. Note that we have now cleared fractions.

$$27=2x$$

$$\frac{27}{2}=x.$$

The solution is $\frac{27}{2}$.

Do Exercise 1.

Objective

a Solve rational equations.

1. Solve: $\frac{3}{4}+\frac{5}{8}=\frac{x}{12}$.

Answer on page A-39

2. Solve: $\frac{x}{4} - \frac{x}{6} = \frac{1}{8}$.

3. Solve: $\frac{1}{x} = \frac{1}{6 - x}$.

Answers on page A-39

EXAMPLE 2 Solve: $\frac{x}{6} - \frac{x}{8} = \frac{1}{12}$.

The LCM is 24. We multiply all terms on both sides by 24:

$$\frac{x}{6} - \frac{x}{8} = \frac{1}{12}$$

$$24\left(\frac{x}{6} - \frac{x}{8}\right) = 24 \cdot \frac{1}{12} \quad \text{Multiplying both sides by the LCM}$$

$$24 \cdot \frac{x}{6} - 24 \cdot \frac{x}{8} = 24 \cdot \frac{1}{12} \quad \text{Multiplying to remove parentheses}$$

Be sure to multiply *each* term by the LCM.

$$4x - 3x = 2 \quad \text{Simplifying}$$

$$x = 2.$$

CHECK:

$$\frac{x}{6} - \frac{x}{8} = \frac{1}{12}$$

$$\frac{2}{6} - \frac{2}{8} \;\Big|\; \frac{1}{12}$$

$$\frac{1}{3} - \frac{1}{4}$$

$$\frac{4}{12} - \frac{3}{12}$$

$$\frac{1}{12} \qquad \text{TRUE}$$

This checks, so the solution is 2.

Do Exercise 2.

EXAMPLE 3 Solve: $\frac{1}{x} = \frac{1}{4 - x}$.

The LCM is $x(4 - x)$. We multiply all terms on both sides by $x(4 - x)$:

$$\frac{1}{x} = \frac{1}{4 - x}$$

$$x(4 - x) \cdot \frac{1}{x} = x(4 - x) \cdot \frac{1}{4 - x} \quad \text{Multiplying both sides by the LCM}$$

$$4 - x = x \quad \text{Simplifying}$$

$$4 = 2x$$

$$x = 2.$$

CHECK:

$$\frac{1}{x} = \frac{1}{4 - x}$$

$$\frac{1}{2} \;\Big|\; \frac{1}{4 - 2}$$

$$\frac{1}{2} \qquad \text{TRUE}$$

This checks, so the solution is 2.

Do Exercise 3.

ALGEBRAIC–GRAPHICAL CONNECTION

We can obtain a visual check of the solutions of a rational equation by graphing. For example, consider the equation

$$\frac{x}{4} + \frac{x}{2} = 6.$$

We can examine the solution by graphing the equations

$$y = \frac{x}{4} + \frac{x}{2} \quad \text{and} \quad y = 6$$

using the same set of axes.

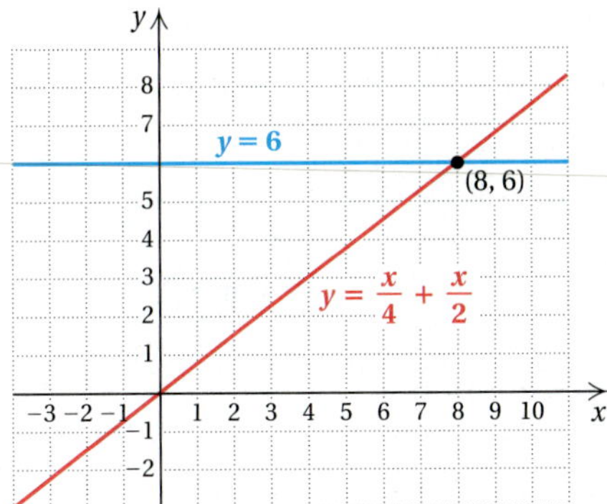

The y-values for each equation will be the same where the graphs intersect. The x-value of that point will yield that value, so it will be the solution of the equation. It appears from the graph that when $x = 8$, the value of $x/4 + x/2$ is 6. We can check by substitution:

$$\frac{x}{4} + \frac{x}{2} = \frac{8}{4} + \frac{8}{2} = 2 + 4 = 6.$$

Thus the solution is 8.

EXAMPLE 4 Solve: $\frac{2}{3x} + \frac{1}{x} = 10.$

The LCM is $3x$. We multiply all terms on both sides by $3x$:

$$\frac{2}{3x} + \frac{1}{x} = 10$$

$$3x\left(\frac{2}{3x} + \frac{1}{x}\right) = 3x \cdot 10 \qquad \text{Multiplying both sides by the LCM}$$

$$3x \cdot \frac{2}{3x} + 3x \cdot \frac{1}{x} = 3x \cdot 10 \qquad \text{Multiplying to remove parentheses}$$

$$2 + 3 = 30x \qquad \text{Simplifying}$$

$$5 = 30x$$

$$\frac{5}{30} = x$$

$$\frac{1}{6} = x.$$

The check is left to the student. The solution is $\frac{1}{6}$.

Do Exercise 4.

4. Solve: $\frac{1}{2x} + \frac{1}{x} = -12.$

Answer on page A-39

5. Solve: $x + \frac{1}{x} = 2.$

Answer on page A-39

EXAMPLE 5 Solve: $x + \frac{6}{x} = -5.$

The LCM is x. We multiply all terms on both sides by x:

$$x + \frac{6}{x} = -5$$

$$x\left(x + \frac{6}{x}\right) = -5x \qquad \text{Multiplying both sides by } x$$

$$x \cdot x + x \cdot \frac{6}{x} = -5x \qquad \text{Note that each rational expression on the left is now multiplied by } x.$$

$$x^2 + 6 = -5x \qquad \text{Simplifying}$$

$$x^2 + 5x + 6 = 0 \qquad \text{Adding } 5x \text{ to get a 0 on one side}$$

$$(x + 3)(x + 2) = 0 \qquad \text{Factoring}$$

$$x + 3 = 0 \quad or \quad x + 2 = 0 \qquad \text{Using the principle of zero products}$$

$$x = -3 \quad or \quad x = -2.$$

CHECK:

For -3:

$$x + \frac{6}{x} = -5$$

$-3 + \frac{6}{-3}$	-5
$-3 - 2$	
-5	TRUE

For -2:

$$x + \frac{6}{x} = -5$$

$-2 + \frac{6}{-2}$	-5
$-2 - 3$	
-5	TRUE

Both of these check, so there are two solutions, -3 and -2.

Do Exercise 5.

CALCULATOR CORNER

Checking Solutions of Rational Equations We can use a table to check possible solutions of rational equations. Consider the equation in Example 6,

$$\frac{x^2}{x - 1} = \frac{1}{x - 1},$$

and the possible solutions that were found, 1 and -1. To check these solutions, we enter $y_1 = x^2/(x - 1)$ and $y_2 = 1/(x - 1)$ on the equation-editor screen. Then, with a table set in ASK mode, we enter $x = 1$. The ERROR messages indicate that 1 is not a solution because it is not an allowable replacement for x in the equation. Next, we enter $x = -1$. Since y_1 and y_2 have the same value, we know that the equation is true when $x = -1$, and thus -1 is a solution.

X	Y1	Y2
1	ERROR	ERROR
−1	−.5	−.5
X =		

Exercises: Use a graphing calculator to check the possible solutions in each of the following.

1. Examples 1, 3, 5, and 7
2. Margin Exercises 1, 3, 6, and 7

CHECKING POSSIBLE SOLUTIONS

When we multiply both sides of an equation by the LCM, the resulting equation might have solutions that are *not* solutions of the original equation. Thus we must *always* check possible solutions in the original equation.

1. If you have carried out all algebraic procedures correctly, you need only check if a number makes a denominator 0 in the original equation. If it does make a denominator 0, it is *not* a solution.
2. To be sure that no computational errors have been made and that you indeed have a solution, a complete check is necessary, as we did in Chapter 7.

Example 6 illustrates the importance of checking all possible solutions.

EXAMPLE 6 Solve: $\dfrac{x^2}{x-1} = \dfrac{1}{x-1}$.

The LCM is $x - 1$. We multiply all terms on both sides by $x - 1$:

$$\frac{x^2}{x-1} = \frac{1}{x-1}$$

$$(x-1)\cdot\frac{x^2}{x-1} = (x-1)\cdot\frac{1}{x-1} \quad \text{Multiplying both sides by } x-1$$

$$x^2 = 1 \quad \text{Simplifying}$$

$$x^2 - 1 = 0 \quad \text{Subtracting 1 to get a 0 on one side}$$

$$(x-1)(x+1) = 0 \quad \text{Factoring}$$

$$x - 1 = 0 \quad or \quad x + 1 = 0 \quad \text{Using the principle of zero products}$$

$$x = 1 \quad or \quad x = -1.$$

The numbers 1 and -1 are possible solutions. We look at the original equation and see that 1 makes a denominator 0 and is therefore not a solution. The number -1 checks and is a solution.

Do Exercise 6.

EXAMPLE 7 Solve: $\dfrac{3}{x-5} + \dfrac{1}{x+5} = \dfrac{2}{x^2-25}$.

The LCM is $(x-5)(x+5)$. We multiply all terms on both sides by $(x-5)(x+5)$:

$$(x-5)(x+5)\left(\frac{3}{x-5} + \frac{1}{x+5}\right) = (x-5)(x+5)\left(\frac{2}{x^2-25}\right)$$

Multiplying both sides by the LCM

$$(x-5)(x+5)\cdot\frac{3}{x-5} + (x-5)(x+5)\cdot\frac{1}{x+5} = (x-5)(x+5)\cdot\frac{2}{x^2-25}$$

$$3(x+5) + (x-5) = 2 \quad \text{Simplifying}$$

$$3x + 15 + x - 5 = 2 \quad \text{Removing parentheses}$$

$$4x + 10 = 2$$

$$4x = -8$$

$$x = -2.$$

The check is left to the student. The number -2 checks and is the solution.

Do Exercise 7.

6. Solve: $\dfrac{x^2}{x+2} = \dfrac{4}{x+2}$.

7. Solve: $\dfrac{4}{x-2} + \dfrac{1}{x+2} = \dfrac{26}{x^2-4}$.

CAUTION!

We have introduced a new use of the LCM in this section. We previously used the LCM in adding or subtracting rational expressions. *Now* we have equations with equals signs. We clear fractions by multiplying both sides of the equation by the LCM. This eliminates the denominators. Do *not* make the mistake of trying to clear fractions when you do not have an equation.

Answers on page A-39

Study Tips

ARE YOU CALCULATING OR SOLVING?

One of the common difficulties with this chapter is knowing for sure the task at hand. Are you combining expressions using operations to get another *rational expression,* or are you solving equations for which the results are numbers that are *solutions* of an equation? To learn to make these decisions, complete the following list by writing in the blank the type of answer you should get: "Rational expression" or "Solutions." You need not complete the mathematical operations.

Task	*Answer (Just write "Rational expression" or "Solutions.")*
1. Add: $\frac{4}{x-2} + \frac{1}{x+2}$.	
2. Solve: $\frac{4}{x-2} = \frac{1}{x+2}$.	
3. Subtract: $\frac{4}{x-2} - \frac{1}{x+2}$.	
4. Multiply: $\frac{4}{x-2} \cdot \frac{1}{x+2}$.	
5. Divide: $\frac{4}{x-2} \div \frac{1}{x+2}$.	
6. Solve: $\frac{4}{x-2} + \frac{1}{x+2} = \frac{26}{x^2-4}$.	
7. Perform the indicated operations and simplify: $\frac{4}{x-2} + \frac{1}{x+2} - \frac{26}{x^2-4}$.	
8. Solve: $\frac{x^2}{x-1} = \frac{1}{x-1}$.	
9. Solve: $\frac{2}{y^2-25} = \frac{3}{y-5} + \frac{1}{y-5}$.	
10. Solve: $\frac{x}{x+4} - \frac{4}{x-4} = \frac{x^2+16}{x^2-16}$.	
11. Perform the indicated operations and simplify: $\frac{x}{x+4} - \frac{4}{x-4} - \frac{x^2+16}{x^2-16}$.	
12. Solve: $\frac{5}{y-3} - \frac{30}{y^2-9} = 1$.	
13. Add: $\frac{5}{y-3} + \frac{30}{y^2-9} + 1$.	

11.7 EXERCISE SET

For Extra Help

Digital Video Tutor CD 10 Videotape 13 | InterAct Math | Math Tutor Center | MathXL | MyMathLab.com

a Solve. Don't forget to check!

1. $\frac{4}{5} - \frac{2}{3} = \frac{x}{9}$

2. $\frac{x}{20} = \frac{3}{8} - \frac{4}{5}$

3. $\frac{3}{5} + \frac{1}{8} = \frac{1}{x}$

4. $\frac{2}{3} + \frac{5}{6} = \frac{1}{x}$

5. $\frac{3}{8} + \frac{4}{5} = \frac{x}{20}$

6. $\frac{3}{5} + \frac{2}{3} = \frac{x}{9}$

7. $\frac{1}{x} = \frac{2}{3} - \frac{5}{6}$

8. $\frac{1}{x} = \frac{1}{8} - \frac{3}{5}$

9. $\frac{1}{6} + \frac{1}{8} = \frac{1}{t}$

10. $\frac{1}{8} + \frac{1}{12} = \frac{1}{t}$

11. $x + \frac{4}{x} = -5$

12. $\frac{10}{x} - x = 3$

13. $\frac{x}{4} - \frac{4}{x} = 0$

14. $\frac{x}{5} - \frac{5}{x} = 0$

15. $\frac{5}{x} = \frac{6}{x} - \frac{1}{3}$

16. $\frac{4}{x} = \frac{5}{x} - \frac{1}{2}$

17. $\frac{5}{3x} + \frac{3}{x} = 1$

18. $\frac{5}{2y} + \frac{8}{y} = 1$

19. $\frac{t-2}{t+3} = \frac{3}{8}$

20. $\frac{x-7}{x+2} = \frac{1}{4}$

21. $\frac{2}{x+1} = \frac{1}{x-2}$

22. $\frac{8}{y-3} = \frac{6}{y+4}$

23. $\frac{x}{6} - \frac{x}{10} = \frac{1}{6}$

24. $\frac{x}{8} - \frac{x}{12} = \frac{1}{8}$

25. $\frac{t+2}{5} - \frac{t-2}{4} = 1$

26. $\frac{x+1}{3} - \frac{x-1}{2} = 1$

27. $\frac{5}{x-1} = \frac{3}{x+2}$

28. $\frac{x-7}{x-9} = \frac{2}{x-9}$

29. $\frac{a-3}{3a+2} = \frac{1}{5}$

30. $\frac{x+7}{8x-5} = \frac{2}{3}$

31. $\frac{x-1}{x-5} = \frac{4}{x-5}$

32. $\frac{y+11}{y+8} = \frac{3}{y+8}$

33. $\frac{2}{x+3} = \frac{5}{x}$

34. $\frac{6}{y} = \frac{5}{y-8}$

35. $\frac{x-2}{x-3} = \frac{x-1}{x+1}$

36. $\frac{t+5}{t-2} = \frac{t-2}{t+4}$

37. $\frac{1}{x+3} + \frac{1}{x-3} = \frac{1}{x^2-9}$

38. $\frac{4}{x-3} + \frac{2x}{x^2-9} = \frac{1}{x+3}$

39. $\frac{x}{x+4} - \frac{4}{x-4} = \frac{x^2+16}{x^2-16}$

40. $\frac{5}{y-3} - \frac{30}{y^2-9} = 1$

41. $\frac{4-a}{8-a} = \frac{4}{a-8}$

42. $\frac{3}{x-7} = \frac{x+10}{x-7}$

43. $2 - \frac{a-2}{a+3} = \frac{a^2-4}{a+3}$

44. $\frac{5}{x-1} + x + 1 = \frac{5x+4}{x-1}$

45. **DW** Why is it especially important to check the possible solutions to a rational equation?

46. **DW** How can a graph be used to determine how many solutions an equation has?

SKILL MAINTENANCE

Simplify.

47. $(a^2b^5)^{-3}$ [9.2a, b]

48. $(x^{-2}y^{-3})^{-4}$ [9.2a, b]

49. $\left(\frac{2x}{t^2}\right)^4$ [9.2a, b]

50. $\left(\frac{y^3}{w^2}\right)^{-2}$ [9.2a, b]

51. $4x^{-5} \cdot 8x^{11}$ [9.1d]

52. $(8x^5y^{-4})^2$ [9.2a, b]

Find the intercepts. Then graph the equation. [8.3a]

53. $5x + 10y = 20$

54. $2x - 4y = 8$

55. $10y - 4x = -20$

56. $y - 5x = 5$

SYNTHESIS

Solve.

57. $\frac{4}{y-2} - \frac{2y-3}{y^2-4} = \frac{5}{y+2}$

58. $\frac{x}{x^2+3x-4} + \frac{x+1}{x^2+6x+8} = \frac{2x}{x^2+x-2}$

59. $\frac{x+1}{x+2} = \frac{x+3}{x+4}$

60. $\frac{x^2}{x^2-4} = \frac{x}{x+2} - \frac{2x}{2-x}$

61. $4a - 3 = \frac{a+13}{a+1}$

62. $\frac{3x-9}{x-3} = \frac{5x-4}{2}$

63. $\frac{y^2-4}{y+3} = 2 - \frac{y-2}{y+3}$

64. $\frac{3a-5}{a^2+4a+3} + \frac{2a+2}{a+3} = \frac{a-3}{a+1}$

65. Use a graphing calculator to check the solutions to Exercises 1–4.

66. Use a graphing calculator to check the solutions to Exercises 13, 15, and 25.

11.8 APPLICATIONS USING RATIONAL EQUATIONS AND PROPORTIONS

Objectives

a Solve applied problems using rational equations.

b Solve proportion problems.

In many areas of study, applications involving rates, proportions, or reciprocals translate to rational equations. By using the five steps for problem solving and the skills of Sections 11.1–11.7, we can now solve such problems.

a Solving Applied Problems

PROBLEMS INVOLVING WORK

EXAMPLE 1 *Recyclable Work.* Erin and Nick work as volunteers at a community recycling depot. Erin can sort a morning's accumulation of recyclables in 4 hr, while Nick requires 6 hr to do the same job. How long would it take them, working together, to sort the recyclables?

1. **Familiarize.** We familiarize ourselves with the problem by considering two *incorrect* ways of translating the problem to mathematical language.

   a) A common *incorrect* way to translate the problem is to add the two times: $4\text{ hr} + 6\text{ hr} = 10\text{ hr}$. Let's think about this. Erin can do the job alone in 4 hr. If Erin and Nick work together, whatever time it takes them should be *less* than 4 hr. Thus we reject 10 hr as a solution, but we do have a partial check on any answer we get. The answer should be less than 4 hr.

   b) Another *incorrect* way to translate the problem is as follows. Suppose the two people split up the sorting job in such a way that Erin does half the sorting and Nick does the other half. Then

   Erin sorts $\frac{1}{2}$ the recyclables in $\frac{1}{2}(4\text{ hr})$, or 2 hr,

   and Nick sorts $\frac{1}{2}$ the recyclables in $\frac{1}{2}(6\text{ hr})$, or 3 hr.

   But time is wasted since Erin would finish 1 hr earlier than Nick. In effect, they have not worked together to get the job done as fast as possible. If Erin helps Nick after completing her half, the entire job could be done in a time somewhere between 2 hr and 3 hr.

   We proceed to a translation by considering how much of the job is finished in 1 hr, 2 hr, 3 hr, and so on. It takes Erin 4 hr to do the sorting job alone. Then, in 1 hr, she can do $\frac{1}{4}$ of the job. It takes Nick 6 hr to do the job alone. Then, in 1 hr, he can do $\frac{1}{6}$ of the job. Working together, they can do

   $$\frac{1}{4} + \frac{1}{6}, \text{ or } \frac{5}{12} \text{ of the job in 1 hr.}$$

   In 2 hr, Erin can do $2\left(\frac{1}{4}\right)$ of the job and Nick can do $2\left(\frac{1}{6}\right)$ of the job. Working together, they can do

   $$2\left(\frac{1}{4}\right) + 2\left(\frac{1}{6}\right), \text{ or } \frac{5}{6} \text{ of the job in 2 hr.}$$

Study Tips

BEING A TUTOR

Try being a tutor for a fellow student. You can maximize your understanding and retention of concepts if you explain the material to someone else.

Continuing this reasoning, we can create a table like the following one.

TIME	FRACTION OF THE JOB COMPLETED		
	Erin	Nick	Together
1 hr	$\frac{1}{4}$	$\frac{1}{6}$	$\frac{1}{4}+\frac{1}{6}$, or $\frac{5}{12}$
2 hr	$2\left(\frac{1}{4}\right)$	$2\left(\frac{1}{6}\right)$	$2\left(\frac{1}{4}\right)+2\left(\frac{1}{6}\right)$, or $\frac{5}{6}$
3 hr	$3\left(\frac{1}{4}\right)$	$3\left(\frac{1}{6}\right)$	$3\left(\frac{1}{4}\right)+3\left(\frac{1}{6}\right)$, or $1\frac{1}{4}$
t hr	$t\left(\frac{1}{4}\right)$	$t\left(\frac{1}{6}\right)$	$t\left(\frac{1}{4}\right)+t\left(\frac{1}{6}\right)$

From the table, we see that if they work 3 hr, the fraction of the job completed is $1\frac{1}{4}$, which is more of the job than needs to be done. We see again that the answer is somewhere between 2 hr and 3 hr. What we want is a number t such that the fraction of the job that gets completed is 1; that is, the job is just completed.

2. **Translate.** From the table, we see that the time we want is some number t for which

$$t\left(\frac{1}{4}\right)+t\left(\frac{1}{6}\right)=1, \quad \text{or} \quad \frac{t}{4}+\frac{t}{6}=1,$$

where 1 represents the idea that the entire job is completed in time t.

3. **Solve.** We solve the equation:

$$12\left(\frac{t}{4}+\frac{t}{6}\right)=12\cdot 1 \quad \text{Multiplying by the LCM, which is } 2\cdot 2\cdot 3\text{, or } 12$$

$$12\cdot\frac{t}{4}+12\cdot\frac{t}{6}=12$$

$$3t+2t=12$$

$$5t=12$$

$$t=\frac{12}{5}, \text{ or } 2\frac{2}{5} \text{ hr.}$$

4. **Check.** The check can be done by recalculating:

$$\frac{12}{5}\left(\frac{1}{4}\right)+\frac{12}{5}\left(\frac{1}{6}\right)=\frac{3}{5}+\frac{2}{5}=\frac{5}{5}=1.$$

We also have another check in what we learned from the *Familiarize* step. The answer, $2\frac{2}{5}$ hr, is between 2 hr and 3 hr (see the table), and it is less than 4 hr, the time it takes Erin working alone.

5. **State.** It takes $2\frac{2}{5}$ hr for them to do the sorting, working together.

THE WORK PRINCIPLE

Suppose a = the time it takes A to do a job, b = the time it takes B to do the same job, and t = the time it takes them to do the same job working together. Then

$$\frac{t}{a} + \frac{t}{b} = 1, \quad \text{or} \quad \frac{1}{a} + \frac{1}{b} = \frac{1}{t}.$$

Do Exercise 1.

1. **Wall Construction.** By checking work records, a contractor finds that it takes Eduardo 6 hr to construct a wall of a certain size. It takes Yolanda 8 hr to construct the same wall. How long would it take if they worked together?

Answer on page A-40

PROBLEMS INVOLVING MOTION

Problems that deal with distance, speed (or rate), and time are called **motion problems.** Translation of these problems involves the distance formula, $d = r \cdot t$, and/or the equivalent formulas $r = d/t$ and $t = d/r$.

MOTION FORMULAS

The following are the formulas for motion problems:

$d = rt$; Distance = Rate · Time (basic formula)

$r = \dfrac{d}{t}$; Rate = Distance/Time

$t = \dfrac{d}{r}$. Time = Distance/Rate

EXAMPLE 2 *Animal Speeds.* A cheetah can run 20 mph faster than a lion. A cheetah can run 7 mi in the same time that a lion can run 5 mi. Find the speed of each animal.

Source: Barbara Ann Kipfer, *The Order of Things.* New York: Random House, 1998.

1. **Familiarize.** We first make a drawing. Let r = the speed of the lion. Then $r + 20$ = the speed of the cheetah.

Recall that sometimes we need to find a formula in order to solve an application. A formula that relates the notions of distance, speed, and time is $d = rt$, or

Distance = Speed · Time.

(Indeed, you may need to look up such a formula.)

Study Tips

TEST TAKING: MORE ON DOING EVEN-NUMBERED EXERCISES

In an earlier study tip (p. 615), as a way to improve your test-taking skills, we encouraged you to build some even-numbered exercises into your homework. Here we explore this issue further.

Working a test is different from working your homework, when the answers are provided. When taking the test, you are "on your own," so to speak. Keep the following tips in mind when taking your next test or quiz.

1. Work a bit slower and deliberately, taking a fresh piece of paper to redo the problem. Check your work against your previous work to see if there is a difference and why. This is especially helpful if you finish the test early and have extra time.
2. Use estimation techniques to solve the problem as a check.
3. Do the checks to applied problems that we so often discuss in the book.

2. Driving speed. Nancy drives 20 mph faster than her father, Greg. In the same time that Nancy travels 180 mi, her father travels 120 mi. Find their speeds.

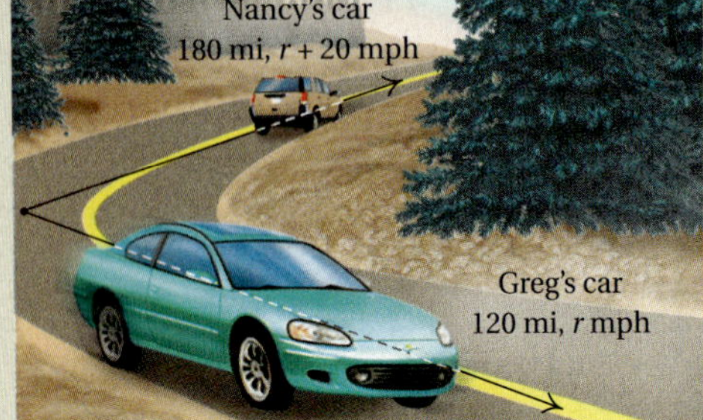

Answer on page A-40

Since each animal travels the same length of time, we can use just t for time. We organize the information in a chart, as follows.

$$d = r \cdot t$$

	DISTANCE	SPEED	TIME	
Lion	5	r	t	$\rightarrow 5 = rt$
Cheetah	7	$r + 20$	t	$\rightarrow 7 = (r + 20)t$

2. Translate. We can apply the formula $d = rt$ along the rows of the table to obtain two equations:

$$5 = rt, \qquad (1)$$
$$7 = (r + 20)t. \qquad (2)$$

We know that the animals travel for the same length of time. Thus if we solve each equation for t and set the results equal to each other, we get an equation in terms of r.

Solving $5 = rt$ for t: $\quad t = \frac{5}{r}$

Solving $7 = (r + 20)t$ for t: $\quad t = \frac{7}{r + 20}$

Since the times are the same, we have the following equation:

$$\frac{5}{r} = \frac{7}{r + 20}.$$

3. Solve. To solve the equation, we first multiply both sides by the LCM, which is $r(r + 20)$:

$$r(r + 20) \cdot \frac{5}{r} = r(r + 20) \cdot \frac{7}{r + 20} \qquad \text{Multiplying both sides by the LCM, which is } r(r + 20)$$
$$5(r + 20) = 7r \qquad \text{Simplifying}$$
$$5r + 100 = 7r \qquad \text{Removing parentheses}$$
$$100 = 2r$$
$$50 = r.$$

We now have a possible solution. The speed of the lion is 50 mph, and the speed of the cheetah is $r + 20 = 50 + 20$, or 70 mph.

4. Check. We first reread the problem to see what we were to find. We check the speeds of 50 for the lion and 70 for the cheetah. The cheetah does travel 20 mph faster than the lion and will travel farther than the lion, which runs at a slower speed. If the cheetah runs 7 mi at 70 mph, the time it has traveled is $\frac{7}{70}$, or $\frac{1}{10}$ hr. If the lion runs 5 mi at 50 mph, the time it has traveled is $\frac{5}{50}$, or $\frac{1}{10}$ hr. Since the times are the same, the speeds check.

5. State. The speed of the lion is 50 mph and the speed of the cheetah is 70 mph.

Do Exercise 2.

b Applications Involving Proportions

We now consider applications with proportions. A **proportion** involves ratios. A **ratio** of two quantities is their quotient. For example, 73% is the ratio of 73 to 100, $\frac{73}{100}$. The ratio of two different kinds of measure is called a **rate.** Suppose an animal travels 720 ft in 2.5 hr. Its **rate,** or **speed,** is then

$$\frac{720 \text{ ft}}{2.5 \text{ hr}} = 288 \frac{\text{ft}}{\text{hr}}.$$

Do Exercises 3–6.

PROPORTION

An equality of ratios, $A/B = C/D$, is called a **proportion.** The numbers within a proportion are said to be **proportional** to each other.

Proportions can be used to solve a variety of applied problems.

EXAMPLE 3 *Mileage.* A Honda Insight is a gasoline–electric car that travels 280 mi on 4 gal of gas. Find the amount of gas required for a 700-mi trip.
Source: American Honda Motor Company

1. **Familiarize.** We know that the Honda can travel 280 mi on 4 gal of gas. Thus we can set up a ratio, letting x = the amount of gas required to drive 700 mi.
2. **Translate.** We assume that the car uses gas at the same rate throughout the 700-mi trip. Thus the ratios are the same and we can write a proportion. Note that the units of *mileage* are in the numerators and the units of *gasoline* are in the denominators.

$$\begin{array}{c} \text{Miles} \rightarrow \\ \text{Gas} \rightarrow \end{array} \frac{280}{4} = \frac{700}{x} \begin{array}{c} \leftarrow \text{Miles} \\ \leftarrow \text{Gas} \end{array}$$

3. **Solve.** To solve for x, we multiply both sides by the LCM, which is $4x$:

$$4x \cdot \frac{280}{4} = 4x \cdot \frac{700}{x} \quad \text{Multiplying by } 4x$$
$$280x = 2800 \quad \text{Simplifying}$$
$$\frac{280x}{280} = \frac{2800}{280} \quad \text{Dividing by 280}$$
$$x = 10. \quad \text{Simplifying}$$

We can also use cross products to solve the proportion:

$$\frac{280}{4} = \frac{700}{x} \quad 280x \text{ and } 4 \cdot 700 \text{ are cross products.}$$

$$280x = 4 \cdot 700 \quad \text{Equating cross products}$$
$$\frac{280x}{280} = \frac{4 \cdot 700}{280} \quad \text{Dividing by 280}$$
$$x = 10.$$

4. **Check.** The check is left for the student.
5. **State.** The Honda Insight will require 10 gal of gas for 700 mi of driving.

---

3. Find the ratio of 145 km to 2.5 liters (L).

4. **Batting Average.** Recently, a baseball player got 7 hits in 25 times at bat. What was the rate, or batting average, in hits per times at bat?

5. Impulses in nerve fibers travel 310 km in 2.5 hr. What is the rate, or speed, in kilometers per hour?

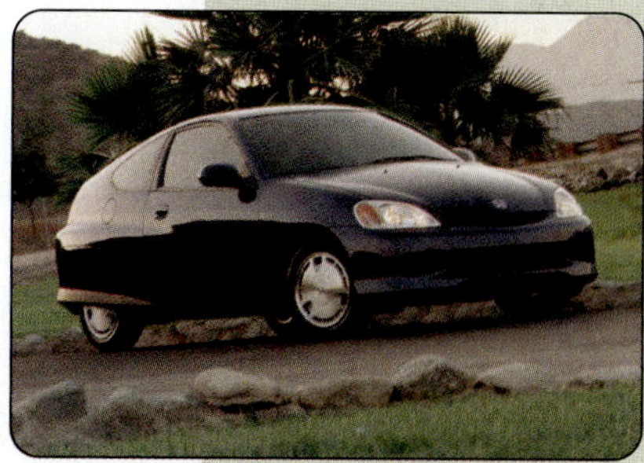

6. A lake of area 550 yd^2 contains 1320 fish. What is the population density of the lake in number of fish per square yard?

7. **Automotive Mileage.** In highway driving, a Chrysler PT Cruiser will travel 377 mi on 14.5 gal of gasoline. How much gas will be required for a 900-mi trip?
Source: DaimlerChrysler Corporation

Answers on page A-40

8. Environmental Science. To determine the number of humpback whales in a pod, a marine biologist, using tail markings, identifies 27 members of the pod. Several weeks later, 40 whales from the pod are randomly sighted. Of the 40 sighted, 12 are from the 27 originally identified. Estimate the number of whales in the pod.

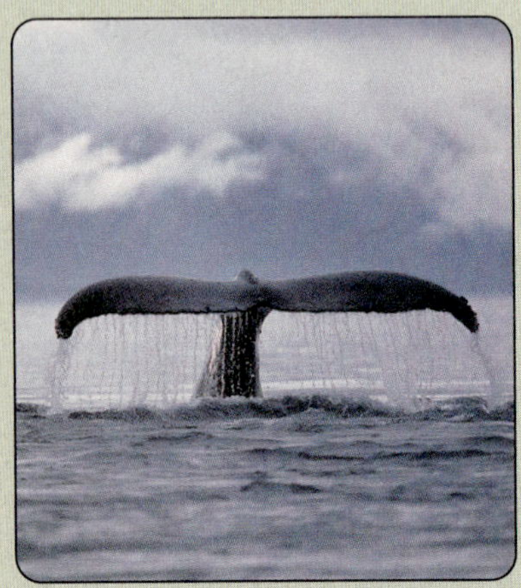

Answer on page A-40

Do Exercise 7 on the preceding page.

EXAMPLE 4 *Environmental Science.* To determine the number of fish in a lake, a park ranger catches 225 fish, tags them, and throws them back into the lake. Later, 108 fish are caught, and 15 of them are found to be tagged. Estimate how many fish are in the lake.

1. **Familiarize.** The ratio of the number of fish tagged to the total number of fish in the lake, F, is $\frac{225}{F}$. Of the 108 fish caught later, 15 fish were tagged. The ratio of fish tagged to fish caught is $\frac{15}{108}$.
2. **Translate.** Assuming that the two ratios are the same, we can translate to a proportion.

$$\frac{225}{F} = \frac{15}{108}$$

Fish tagged originally → 225; Fish in lake → F; Tagged fish caught later ← 15; Fish caught later ← 108

3. **Solve.** We solve the proportion. We multiply by the LCM, which is $108F$:

$$108F \cdot \frac{225}{F} = 108F \cdot \frac{15}{108} \quad \text{Multiplying by } 108F$$
$$108 \cdot 225 = F \cdot 15$$
$$\frac{108 \cdot 225}{15} = F \quad \text{Dividing by 15}$$
$$1620 = F.$$

4. **Check.** The check is left to the student.
5. **State.** We estimate that there are about 1620 fish in the lake.

Do Exercise 8.

In the following example, we predict whether an important home-run record can be broken.

EXAMPLE 5 *Pursuit of Baseball's Home-Run Record.* Mark McGwire hit 70 home runs in 1998 to claim the major-league season home-run record. In 2001, Barry Bonds of the San Francisco Giants hit 31 home runs in the first 58 games of the season, which consists of 162 games. At this rate, could it be predicted that Bonds would break McGwire's record?
Source: Major League Baseball

1. **Familiarize.** Let's assume that Bond's rate of hitting 31 home runs in 58 games will continue for the entire 162-game season. We let $H =$ the number of home runs that Bonds can hit in 162 games.

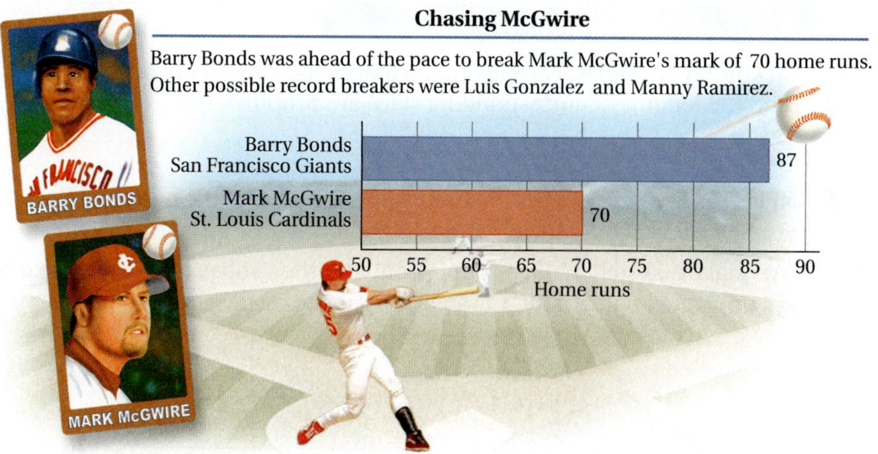

2. **Translate.** Assuming the rate continues, the ratios are the same, and we have the proportion

$$\text{Number of home runs} \longrightarrow \frac{H}{162} = \frac{31}{58}. \longleftarrow \text{Number of home runs}$$
$$\text{Number of games} \longrightarrow \qquad\qquad \longleftarrow \text{Number of games}$$

3. **Solve.** We solve the equation:

$$\frac{H}{162} = \frac{31}{58}$$

$$58H = 162 \cdot 31 \quad \text{Equating cross products}$$

$$\frac{58H}{58} = \frac{162 \cdot 31}{58} \quad \text{Dividing by 58}$$

$$H \approx 87.$$

4. **Check.** The check is left to the student.

5. **State.** We can indeed predict that Bonds will hit 87 home runs and break McGwire's record. (Bonds actually completed the season with 73 home runs and broke McGwire's record.)

Do Exercise 9.

9. Baseball's Home-Run Record. In 2001, Luis Gonzalez of the Arizona Diamondbacks hit 22 home runs in the first 59 games. The season consists of 162 games. At this rate, could it be predicted that Gonzalez would break McGwire's record?
Source: Major League Baseball

SIMILAR TRIANGLES

Proportions arise in geometry when we are studying *similar triangles.* If two triangles are **similar,** then their corresponding angles have the same measure and their corresponding sides are proportional. To illustrate, if triangle ABC is similar to triangle RST, then angles A and R have the same measure, angles B and S have the same measure, angles C and T have the same measure, and

$$\frac{a}{r} = \frac{b}{s} = \frac{c}{t}.$$

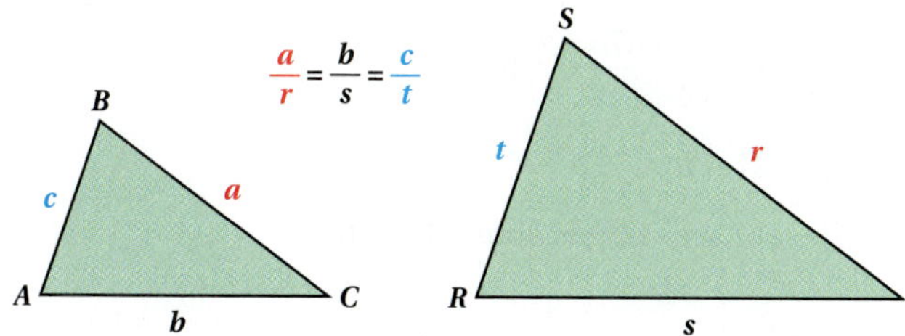

Answer on page A-40

10. Height of a Flagpole. How high is a flagpole that casts a 45-ft shadow at the same time that a 5.5-ft woman casts a 10-ft shadow?

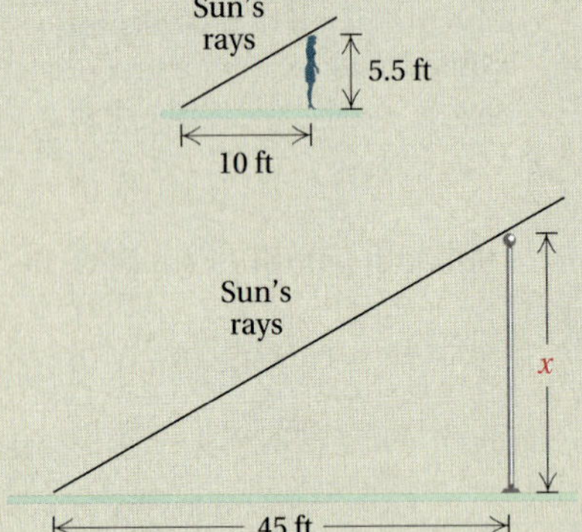

SIMILAR TRIANGLES

In **similar triangles,** corresponding angles have the same measure and the lengths of corresponding sides are proportional.

EXAMPLE 6 *Similar Triangles.* Triangles ABC and XYZ below are similar triangles. Solve for z if $x = 10$, $a = 8$, and $c = 5$.

We make a sketch, write a proportion, and then solve. Note that side a is always opposite angle A, side x is always opposite angle X, and so on.

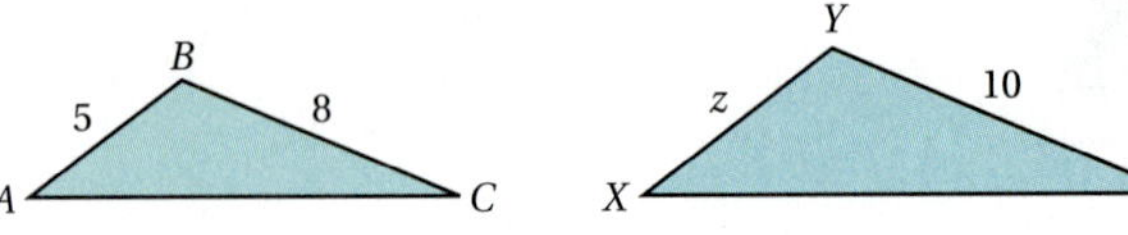

We have

$$\frac{z}{5} = \frac{10}{8}$$ The proportion $\frac{5}{z} = \frac{8}{10}$ could also be used.

$$40 \cdot \frac{z}{5} = 40 \cdot \frac{10}{8}$$ Multiplying by 40

$$8z = 50$$

$$z = \frac{50}{8}, \text{ or } 6.25.$$ Dividing by 8

EXAMPLE 7 *F-106 Blueprint.* A blueprint for an F-106 Delta Dart fighter plane is a scale drawing, as shown below. Each wing has a triangular shape. The blueprint shows similar triangles. Find the length of side a of the wing.

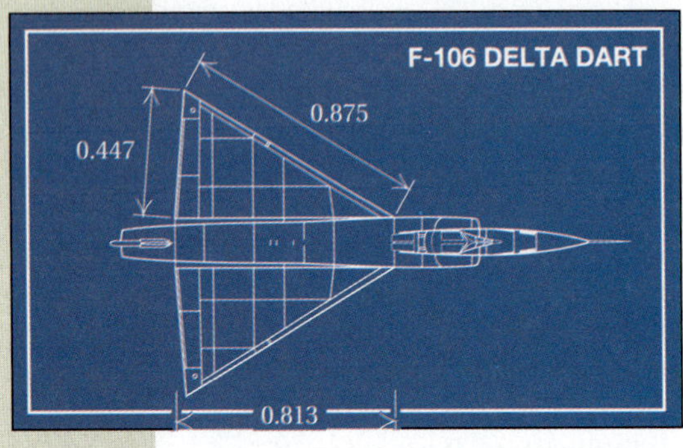

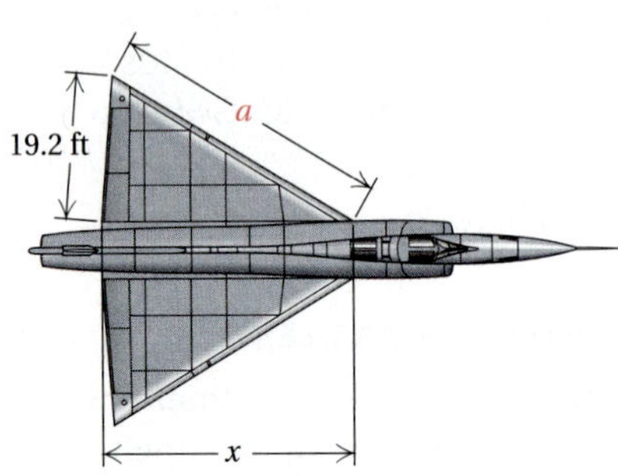

11. F-106 Blueprint. Referring to Example 7, find the length x on the plane.

We let $a =$ the length of the wing. Thus we have the proportion

Length on the blueprint → $\frac{0.447}{19.2} = \frac{0.875}{a}$. ← Length on the blueprint
Length of the wing → ← Length of the wing

Solve: $0.447 \cdot a = 19.2 \cdot 0.875$ Equating cross products

$$a = \frac{19.2 \cdot 0.875}{0.447}$$ Dividing by 0.447

$$a \approx 37.6 \text{ ft}$$

The length of side a of the wing is about 37.6 ft.

Do Exercises 10 and 11.

Answers on page A-40

11.8 EXERCISE SET

For Extra Help

Digital Video Tutor CD 10 Videotape 13

InterAct Math

Math Tutor Center

MathXL

MyMathLab

a Solve.

1. *Construction.* It takes Mandy 4 hr to put up paneling in a room. Omar takes 5 hr to do the same job. How long would it take them, working together, to panel the room?

2. *Carpentry.* By checking work records, a carpenter finds that Juanita can build a small shed in 12 hr. Anton can do the same job in 16 hr. How long would it take if they worked together?

3. *Shoveling.* Vern can shovel the snow from his driveway in 45 min. Nina can do the same job in 60 min. How long would it take Nina and Vern to shovel the driveway if they worked together?

4. *Raking.* Zoë can rake her yard in 4 hr. Steffi does the same job in 3 hr. How long would it take the two of them, working together, to rake the yard?

5. *Wiring.* By checking work records, a contractor finds that Kenny Dewitt can wire a room addition in 9 hr. It takes Betty Wohnt 7 hr to wire the same room. How long would it take if they worked together?

6. *Plumbing.* By checking work records, a plumber finds that Raul can plumb a house in 48 hr. Mira can do the same job in 36 hr. How long would it take if they worked together?

7. *Gardening.* Nicole can weed her vegetable garden in 50 min. Glen can weed the same garden in 40 min. How long would it take if they worked together?

8. *Harvesting.* Bobbi can pick a quart of raspberries in 20 min. Blanche can pick a quart in 25 min. How long would it take if Bobbi and Blanche worked together?

9. *Computer Printers.* The HP OfficeJetG85 printer can copy Charlotte's dissertation in 12 min. The HP LaserJet 3200se can copy the same document in 20 min. If the two machines work together, how long would they take to copy the dissertation?

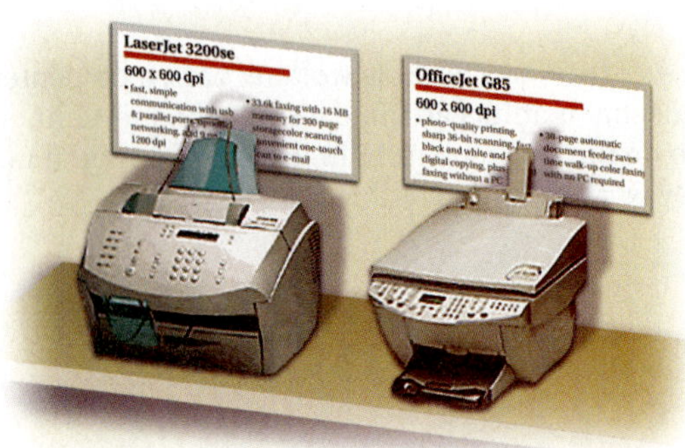

10. *Fax Machines.* The Brother MFC4500® can fax a year-end report in 10 min while the Xerox 850® can fax the same report in 8 min. How long would it take the two machines, working together, to fax the report? (Assume that the recipient has at least two machines for incoming faxes.)

11. *Car Speed.* Rick drives his four-wheel-drive truck 40 km/h faster than Sarah drives her Saturn. While Sarah travels 150 km, Rick travels 350 km. Find their speeds.

Complete this table and the equations as part of the *Familiarize* step.

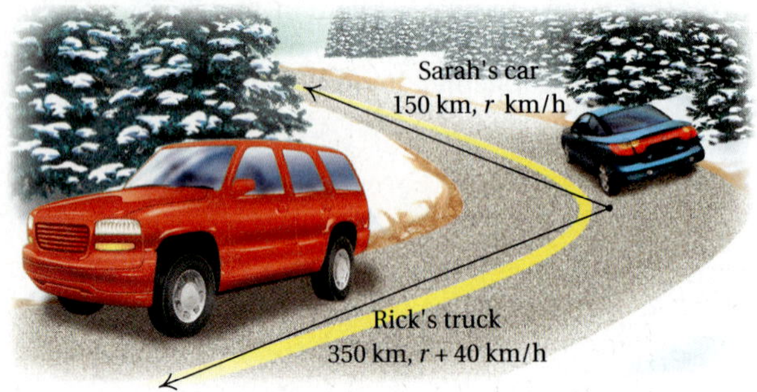

	d = DISTANCE	r · SPEED	t TIME	
Car	150	r		→ $150 = r(\ )$
Truck	350		t	→ $350 = (\ )t$

12. *Car Speed.* A passenger car travels 30 km/h faster than a delivery truck. While the car goes 400 km, the truck goes 250 km. Find their speeds.

13. *Train Speed.* The speed of a B & M freight train is 14 mph slower than the speed of an Amtrak passenger train. The freight train travels 330 mi in the same time that it takes the passenger train to travel 400 mi. Find the speed of each train.

Complete this table and the equations as part of the *Familiarize* step.

	d = DISTANCE	r · SPEED	t TIME	
B & M	330		t	→ $330 = (\ )t$
Amtrak	400	r		→ $400 = r(\ )$

14. *Train Speed.* The speed of a freight train is 15 mph slower than the speed of a passenger train. The freight train travels 390 mi in the same time that it takes the passenger train to travel 480 mi. Find the speed of each train.

15. *Trucking Speed.* A long-distance trucker traveled 120 mi in one direction during a snowstorm. The return trip in rainy weather was accomplished at double the speed and took 3 hr less time. Find the speed going.

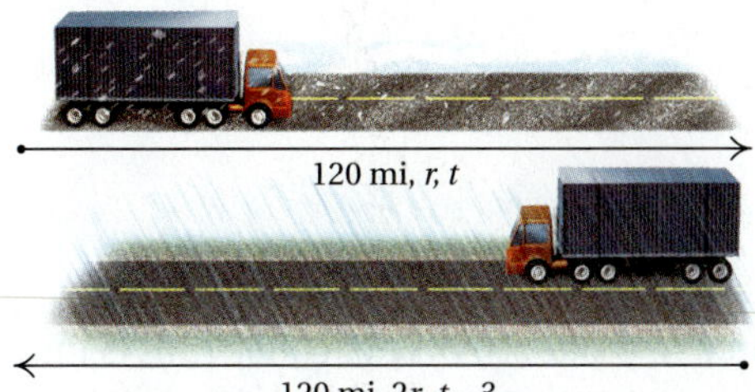

16. *Car Speed.* After making a trip of 126 mi, a person found that the trip would have taken 1 hr less time by increasing the speed by 8 mph. What was the actual speed?

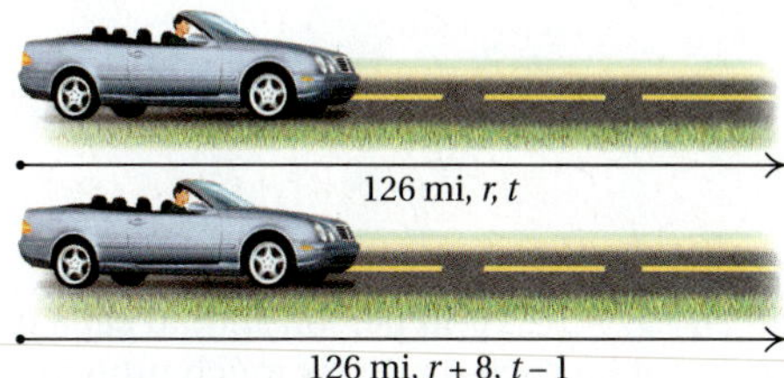

17. *Bicycle Speed.* Hank bicycles 5 km/h slower than Kelly. In the time that it takes Hank to bicycle 42 km, Kelly can bicycle 57 km. How fast does each bicyclist travel?

18. *Driving Speed.* Hillary's Lexus travels 30 mph faster than Bill's Harley. In the same time that Bill travels 75 mi, Hillary travels 120 mi. Find their speeds.

19. *Walking Speed.* Bonnie power walks 3 km/h faster than Ralph. In the time that it takes Ralph to walk 7.5 km, Bonnie walks 12 km. Find their speeds.

20. *Cross-Country Skiing.* Gerard cross-country skis 4 km/h faster than Sally. In the time that it takes Sally to ski 18 km, Gerard skis 24 km. Find their speeds.

21. *Tractor Speed.* Manley's tractor is just as fast as Caledonia's. It takes Manley 1 hr more than it takes Caledonia to drive to town. If Manley is 20 mi from town and Caledonia is 15 mi from town, how long does it take Caledonia to drive to town?

22. *Boat Speed.* Tory and Emilio's motorboats both travel at the same speed. Tory pilots her boat 40 km before docking. Emilio continues for another 2 hr, traveling a total of 100 km before docking. How long did it take Tory to navigate the 40 km?

b Find the ratio of the following. Simplify if possible.

23. 10 divorces, 18 marriages

24. 800 mi, 50 gal

25. *Speed of Black Racer.* A black racer snake travels 4.6 km in 2 hr. What is the speed in kilometers per hour?

26. *Speed of Light.* Light travels 558,000 mi in 3 sec. What is the speed in miles per second?

Solve.

27. *Protein Needs.* A 120-lb person should eat a minimum of 44 g of protein each day. How much protein should a 180-lb person eat each day?

28. *Coffee Beans.* The coffee beans from 14 trees are required to produce 7.7 kg of coffee (this is the average amount that each person in the United States drinks each year). How many trees are required to produce 320 kg of coffee?

29. *Hemoglobin.* A normal 10-cc specimen of human blood contains 1.2 g of hemoglobin. How much hemoglobin would 16 cc of the same blood contain?

30. *Walking Speed.* Wanda walked 234 km in 14 days. At this rate, how far would she walk in 42 days?

31. *Honey Bees.* Making 1 lb of honey requires 20,000 trips by bees to flowers to gather nectar. How many pounds of honey would 35,000 trips produce?
Source: Tom Turpin, Professor of Entomology, Purdue University

32. *Cockroaches and Horses.* A cockroach can run about 2 mi/hr (mph). The average body length of a cockroach is 1 in. The average body length of a horse is 8 ft (96 in.). If we assume that a horse's speed-to-length ratio is the same as that of a cockroach, how fast can a horse run?
Source: Tom Turpin, Professor of Entomology, Purdue University

Professor Turpin founded the annual cockroach race at Purdue University.

33. *Money.* The ratio of the weight of copper to the weight of zinc in a U.S. penny is $\frac{1}{39}$. If 50 kg of zinc is being turned into pennies, how much copper is needed?

34. *Baking.* In a potato bread recipe, the ratio of milk to flour is $\frac{3}{13}$. If 5 cups of milk are used, how many cups of flour are used?

35. *Ichiro Suzuki.* In the 2001 major-league baseball season, Ichiro Suzuki, a rookie from Japan playing for the Seattle Mariners, led the American League in hitting by collecting 96 hits in 266 at-bats in the first 58 games.

  a) The ratio of number of hits to number of at-bats, rounded to the nearest thousandth, is a player's *batting average*. What was Suzuki's batting average?
  b) Based on the ratio of number of hits to number of games, how many hits would he get in the 162-game season?
  c) Based on the ratio of number of hits to number of at-bats and assuming he bats 560 times in 2001, how many hits would he get?

36. *Rich Aurilia.* In the 2001 major-league baseball season, Rich Aurilia, playing for the San Francisco Giants, led the National League in hitting by collecting 79 hits in 213 at-bats in the first 55 games.

  a) The ratio of number of hits to number of at-bats, rounded to the nearest thousandth, is a player's *batting average*. What was Aurilia's batting average?
  b) Based on the ratio of number of hits to number of games, how many hits would he get in the 162-game season?
  c) Based on the ratio of number of hits to number of at-bats and assuming he bats 550 times in 2001, how many hits would he get?

Hat Sizes. Hat sizes are determined by measuring the circumference of one's head in either inches or centimeters. Use ratio and proportion to complete the missing parts of the following table.

	HAT SIZE	HEAD CIRCUMFERENCE (in inches)	HEAD CIRCUMFERENCE (in centimeters)
	$6\frac{3}{4}$	$21\frac{1}{5}$ in.	53.8 cm
37.	7		
38.			56.8 cm
39.		$22\frac{4}{5}$ in.	
40.	$7\frac{3}{8}$		
41.			59.8 cm
42.		24 in.	

43. *Estimating Whale Population.* To determine the number of blue whales in the world's oceans, marine biologists tag 500 blue whales in various parts of the world. Later, 400 blue whales are checked, and it is found that 20 of them are tagged. Estimate the blue whale population.

44. *Estimating Trout Population.* To determine the number of trout in a lake, a conservationist catches 112 trout, tags them, and throws them back into the lake. Later, 82 trout are caught; 32 of them are tagged. Estimate the number of trout in the lake.

45. *Weight on Mars.* The ratio of the weight of an object on Mars to the weight of an object on Earth is 0.4 to 1.

a) How much would a 12-ton rocket weigh on Mars?
b) How much would a 120-lb astronaut weigh on Mars?

46. *Weight on Moon.* The ratio of the weight of an object on the moon to the weight of an object on Earth is 0.16 to 1.

a) How much would a 12-ton rocket weigh on the moon?
b) How much would a 180-lb astronaut weigh on the moon?

47. *Quality Control.* A sample of 144 firecrackers contained 9 "duds." How many duds would you expect in a sample of 3200 firecrackers?

48. *Grass Seed.* It takes 60 oz of grass seed to seed 3000 ft^2 of lawn. At this rate, how much would be needed to seed 5000 ft^2 of lawn?

Geometry. For each pair of similar triangles, find the length of the indicated side.

49. b:

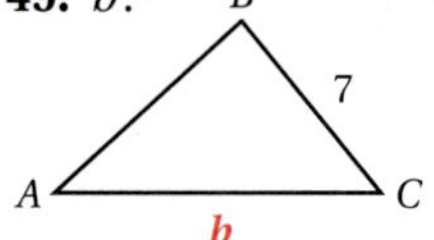

50. a:

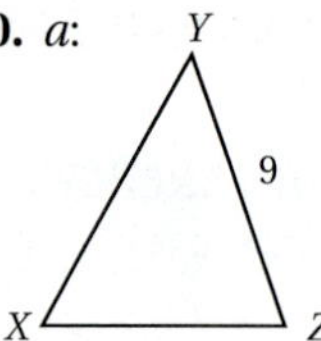

51. f:

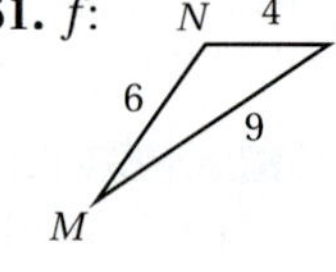

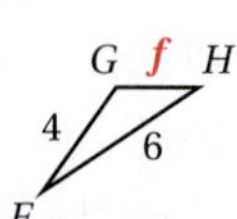

52. r:

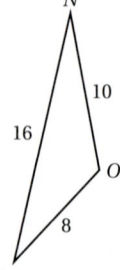

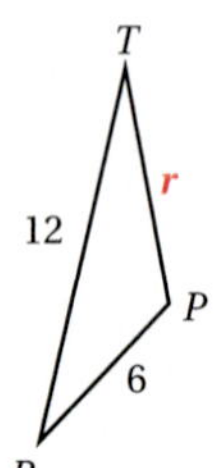

53. h:

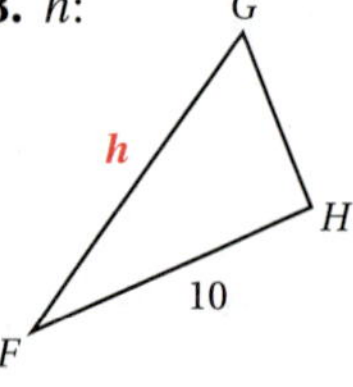

54. n:

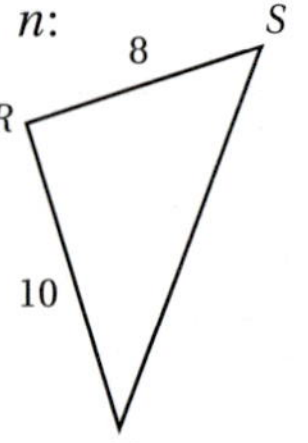

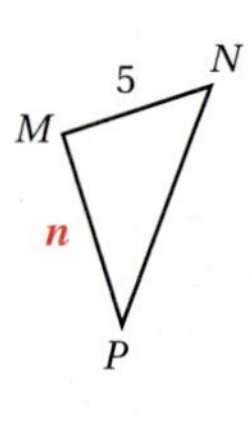

55. l:

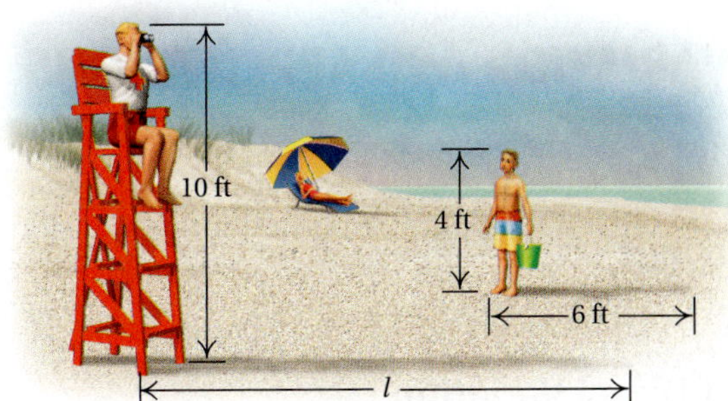

56. h:

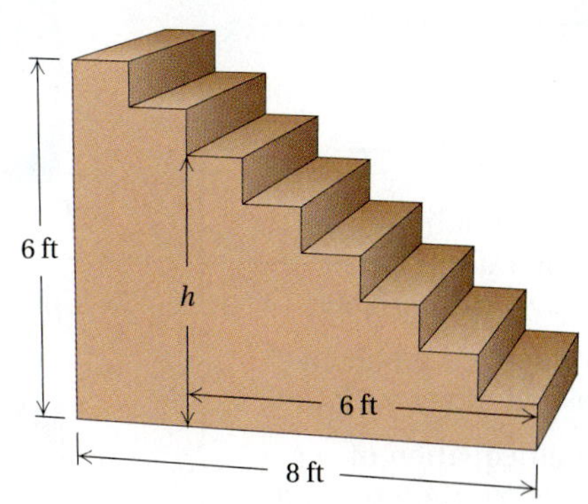

57. DW Explain why it is incorrect to assume that two workers can complete a task twice as quickly as one person working alone.

58. DW Write a problem similar to Example 1 or Margin Exercise 1 for a classmate to solve.

SKILL MAINTENANCE

Simplify. [9.1d]

59. $x^5 \cdot x^6$

60. $x^{-5} \cdot x^6$

61. $x^{-5} \cdot x^{-6}$

62. $x^5 \cdot x^{-6}$

Graph. [8.2b], [8.3a]

63. $y = 2x - 6$

64. $y = -2x + 6$

65. $3x + 2y = 12$

66. $x - 3y = 6$

67. $y = -\frac{3}{4}x + 2$

68. $y = \frac{2}{5}x - 4$

SYNTHESIS

69. Ann and Betty work together and complete a sales report in 4 hr. It would take Betty 6 hr longer, working alone, to do the job than it would Ann. How long would it take each of them to do the job working alone?

70. Express 100 as the sum of two numbers for which the ratio of one number, increased by 5, to the other number, decreased by 5, is 4.

71. How soon after 5 o'clock will the hands on a clock first be together?

72. Rachel allows herself 1 hr to reach a sales appointment 50 mi away. After she has driven 30 mi, she realizes that she must increase her speed by 15 mph in order to get there on time. What was her speed for the first 30 mi?

73. Solve $\frac{t}{a} + \frac{t}{b} = 1$ for t.

Objectives

- **a** Find an equation of direct variation given a pair of values of the variables.
- **b** Solve applied problems involving direct variation.
- **c** Find an equation of inverse variation given a pair of values of the variables.
- **d** Solve applied problems involving inverse variation.
- **e** Find equations of other kinds of variation given values of the variables.
- **f** Solve applied problems involving other kinds of variation.

11.9 VARIATION AND APPLICATIONS

We now extend our study of formulas and functions by considering applications involving variation.

a Equations of Direct Variation

An electrician earns $21 per hour. In 1 hr, $21 is earned; in 2 hr, $42 is earned; in 3 hr, $63 is earned; and so on. We plot this information on a graph, using the number of hours as the first coordinate and the amount earned as the second coordinate to form a set of ordered pairs:

$(1, 21)$, $(2, 42)$,
$(3, 63)$, $(4, 84)$,
and so on.

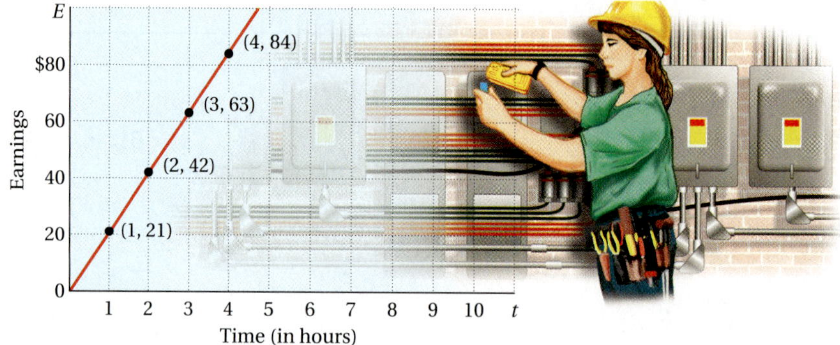

Note that the ratio of the second coordinate to the first is the same number for each point:

$$\frac{21}{1} = 21, \qquad \frac{42}{2} = 21,$$

$$\frac{63}{3} = 21, \qquad \frac{84}{4} = 21,$$

and so on.

Whenever a situation produces pairs of numbers in which the *ratio is constant,* we say that there is **direct variation.** Here the amount earned varies directly as the time:

$$\frac{E}{t} = 21 \text{ (a constant)}, \quad \text{or} \quad E = 21t,$$

or, using function notation, $E(t) = 21t$. The equation is an equation of **direct variation.** The coefficient, 21 in the situation above, is called the **variation constant.** In this case, it is the rate of change of earnings with respect to time.

DIRECT VARIATION

If a situation gives rise to a linear function $f(x) = kx$, or $y = kx$, where k is a positive constant, we say that we have **direct variation,** or that **y varies directly as x,** or that **y is directly proportional to x.** The number k is called the **variation constant,** or **constant of proportionality.**

EXAMPLE 1 Find the variation constant and an equation of variation in which y varies directly as x, and $y = 32$ when $x = 2$.

We know that $(2, 32)$ is a solution of $y = kx$. Thus,

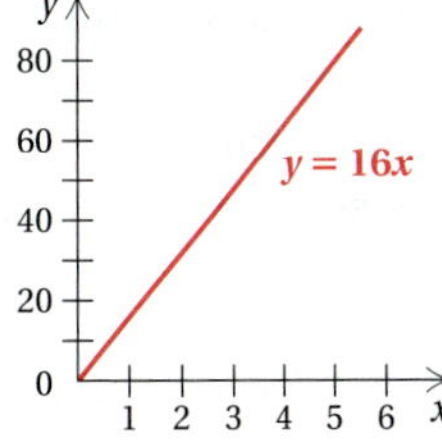

$y = kx$

$32 = k \cdot 2$ Substituting

$\frac{32}{2} = k$, or $k = 16$. Solving for k

The variation constant, 16, is the rate of change of y with respect to x. The equation of variation is $y = 16x$.

The graph of $y = kx$, $k > 0$, always goes through the origin and rises from left to right. Note that as x increases, y increases. The constant k is also the slope of the line.

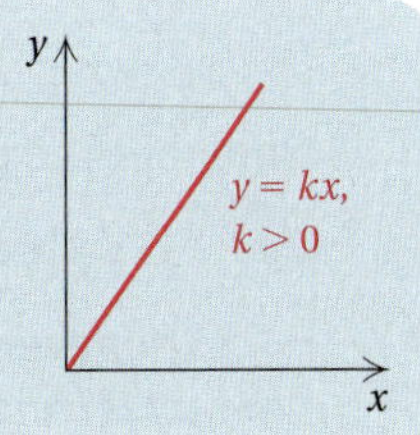

Do Exercises 1 and 2.

b Applications of Direct Variation

EXAMPLE 2 *Water from Melting Snow.* The number of centimeters W of water produced from melting snow varies directly as S, the number of centimeters of snow. Meteorologists have found that 150 cm of snow will melt to 16.8 cm of water. To how many centimeters of water will 200 cm of snow melt?

We first find the variation constant using the data and then find an equation of variation:

$W = kS$ W varies directly as S.

$16.8 = k \cdot 150$ Substituting

$\frac{16.8}{150} = k$ Solving for k

$0.112 = k$. This is the variation constant.

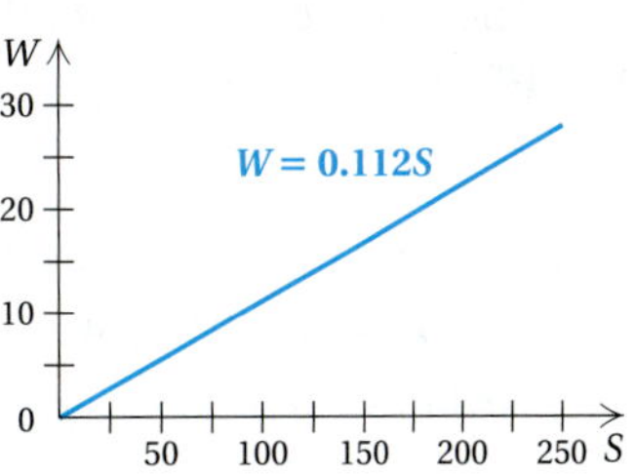

The equation of variation is $W = 0.112S$.

Next, we use the equation to find how many centimeters of water will result from melting 200 cm of snow:

$W = 0.112S$

$= 0.112(200)$ Substituting

$= 22.4$.

Thus, 200 cm of snow will melt to 22.4 cm of water.

Do Exercises 3 and 4. (Exercise 4 is on the following page.)

1. Find the variation constant and an equation of variation in which y varies directly as x, and $y = 8$ when $x = 20$.

2. Find the variation constant and an equation of variation in which y varies directly as x, and $y = 5.6$ when $x = 8$.

3. **Ohm's Law.** Ohm's Law states that the voltage V in an electric circuit varies directly as the number of amperes I of electric current in the circuit. If the voltage is 10 volts when the current is 3 amperes, what is the voltage when the current is 15 amperes?

Answers on page A-40

4. **An Ecology Problem.** The amount of garbage G produced in the United States varies directly as the number of people N who produce the garbage. It is known that 50 tons of garbage is produced by 200 people in 1 year. The population of the San Francisco–Oakland–San Jose area is 6,300,000. How much garbage is produced by this area in 1 year?

C Equations of Inverse Variation

A bus is traveling a distance of 20 mi. At a speed of 5 mph, the trip will take 4 hr; at 20 mph, it will take 1 hr; at 40 mph, it will take $\frac{1}{2}$ hr; and so on. We plot this information on a graph, using speed as the first coordinate and time as the second coordinate to determine a set of ordered pairs:

$(5, 4)$, $(20, 1)$, $\left(40, \frac{1}{2}\right)$, and so on.

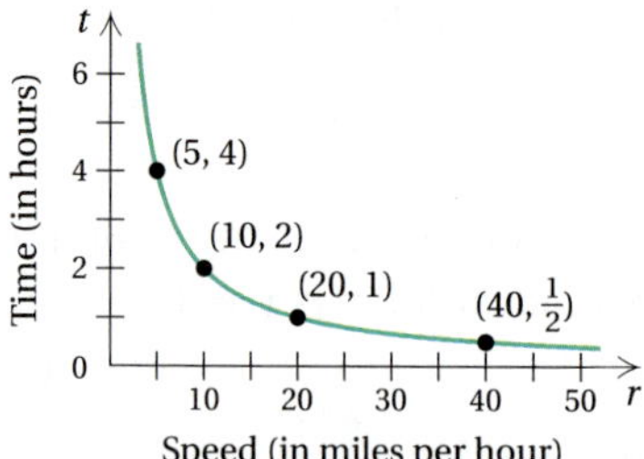

Note that the products of the coordinates are all the same number:

$$5 \cdot 4 = 20, \qquad 20 \cdot 1 = 20, \qquad 40 \cdot \tfrac{1}{2} = 20, \quad \text{and so on.}$$

Whenever a situation produces pairs of numbers in which the *product is constant,* we say that there is **inverse variation.** Here the time varies inversely as the speed:

$$rt = 20 \text{ (a constant)}, \quad \text{or} \quad t = \frac{20}{r}.$$

The equation is an equation of **inverse variation.** The coefficient, 20 in the situation above, is called the **variation constant.** Note that as the first number (speed) increases, the second number (time) decreases.

INVERSE VARIATION

If a situation gives rise to a function $f(x) = k/x$, or $y = k/x$, where k is a positive constant, we say that we have **inverse variation,** or that **y varies inversely as x,** or that **y is inversely proportional to x.** The number k is called the **variation constant,** or **constant of proportionality.**

EXAMPLE 3 Find the variation constant and an equation of variation in which y varies inversely as x, and $y = 32$ when $x = 0.2$.

We know that $(0.2, 32)$ is a solution of $y = k/x$. We substitute:

$$y = \frac{k}{x}$$

$$32 = \frac{k}{0.2} \qquad \text{Substituting}$$

$$(0.2)32 = k \qquad \text{Solving for } k$$

$$6.4 = k.$$

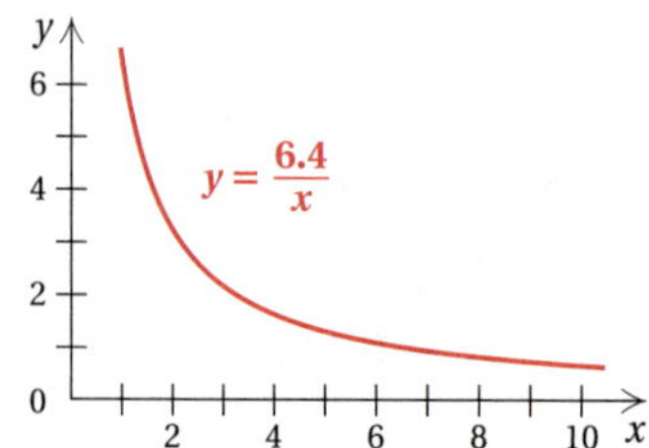

The variation constant is 6.4. The equation of variation is $y = \dfrac{6.4}{x}$.

Answer on page A-40

It is helpful to look at the graph of $y = k/x$, $k > 0$. The graph is like the one shown at right for positive values of x. Note that as x increases, y decreases.

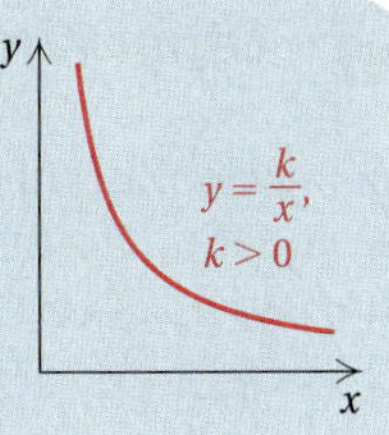

Do Exercise 5.

5. Find the variation constant and an equation of variation in which y varies inversely as x, and $y = 0.012$ when $x = 50$.

d Applications of Inverse Variation

EXAMPLE 4 *Ultraviolet Index.* The ultraviolet index, UV, is a measure issued daily by the National Weather Service. It indicates the strength of the sun rays in a particular locale. For those people whose skin is quite sensitive, a UV rating of 7 will cause a sunburn after 10 min. Given that the number of minutes it takes to burn t varies inversely as the UV rating U, how long will it take a person with highly sensitive skin to burn on a day with a UV rating of 2?
Source: *Los Angeles Times*, 3/24/98

We first find the variation constant using the data given and then find an equation of variation:

$t = \frac{k}{U}$ t varies inversely as U.

$10 = \frac{k}{7}$ Substituting

$70 = k$.   Solving for k, the variation constant

The equation of variation is $t = \frac{70}{U}$.

Next, we use the equation to find the time it would take a person with highly sensitive skin to burn on a day with a UV rating of 2:

$t = \frac{70}{U}$ Equation of variation

$t = \frac{70}{2}$ Substituting

$t = 35$.

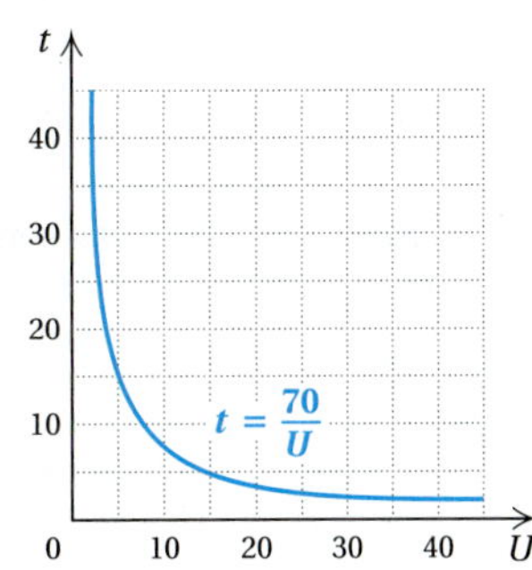

It would take 35 min for a person with highly sensitive skin to burn on a day when the UV rating is 2.

Do Exercise 6.

6. **Building a Shed.** The time t required to do a job varies inversely as the number of people P who work on the job (assuming that all work at the same rate). It takes 4 hr for 12 people to build a woodshed. How long would it take 3 people to complete the same job?

Answers on page A-40

7. Find an equation of variation in which y varies directly as the square of x, and $y = 175$ when $x = 5$.

e Other Kinds of Variation

We now look at other kinds of variation. Consider the equation for the area of a circle, in which A and r are variables and π is a constant:

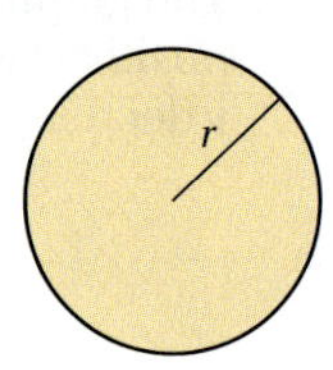

$$A = \pi r^2, \quad \text{or, as a function,} \quad A(r) = \pi r^2.$$

We say that the area *varies directly* as the square of the radius.

> y varies directly as the nth power of x if there is some positive constant k such that $y = kx^n$.

EXAMPLE 5 Find an equation of variation in which y varies directly as the square of x, and $y = 12$ when $x = 2$.

We write an equation of variation and find k:

$$y = kx^2$$
$$12 = k \cdot 2^2$$
$$12 = k \cdot 4$$
$$3 = k.$$

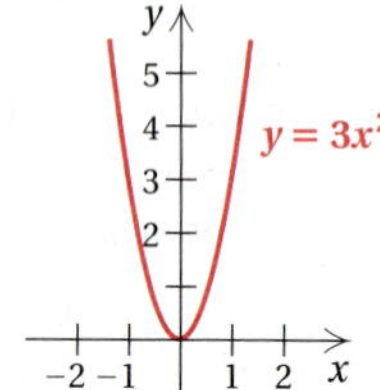

Thus, $y = 3x^2$.

Do Exercise 7.

From the law of gravity, we know that the weight W of an object *varies inversely* as the square of its distance d from the center of the earth:

$$W = \frac{k}{d^2}.$$

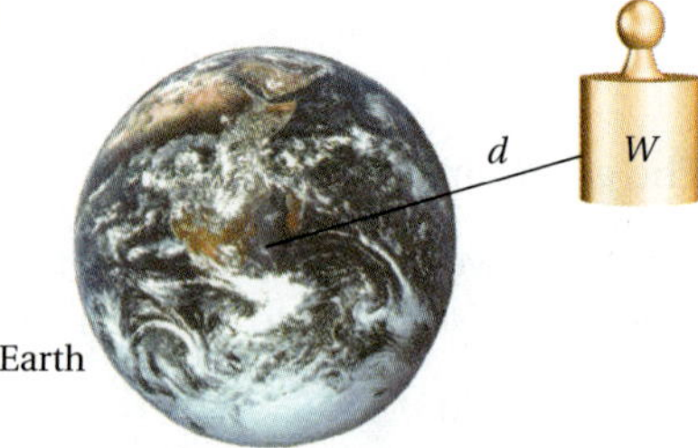

> y varies inversely as the nth power of x if there is some positive constant k such that
> $$y = \frac{k}{x^n}.$$

Answer on page A-40

EXAMPLE 6 Find an equation of variation in which W varies inversely as the square of d, and $W = 3$ when $d = 5$.

$$W = \frac{k}{d^2}$$

$$3 = \frac{k}{5^2} \qquad \text{Substituting}$$

$$3 = \frac{k}{25}$$

$$75 = k$$

Thus, $W = \dfrac{75}{d^2}$.

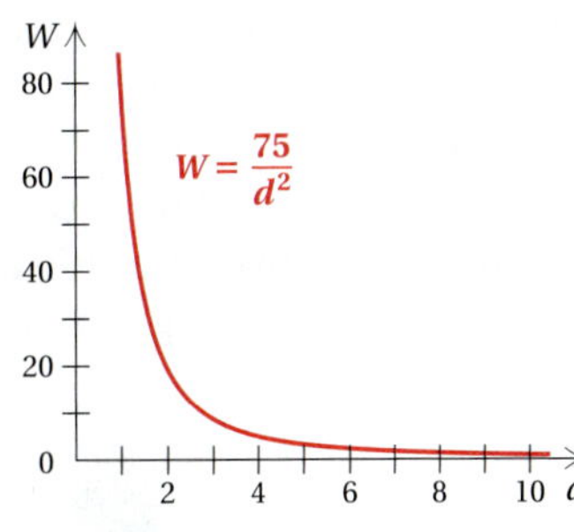

Do Exercise 8.

8. Find an equation of variation in which y varies inversely as the square of x, and $y = \frac{1}{4}$ when $x = 6$.

Consider the equation for the area A of a triangle with height h and base b: $A = \frac{1}{2}bh$. We say that the area *varies jointly* as the height and the base.

> y varies jointly as x and z if there is some positive constant k such that
> $y = kxz$.

EXAMPLE 7 Find an equation of variation in which y varies jointly as x and z, and $y = 42$ when $x = 2$ and $z = 3$.

$$y = kxz$$

$$42 = k \cdot 2 \cdot 3 \qquad \text{Substituting}$$

$$42 = k \cdot 6$$

$$7 = k$$

Thus, $y = 7xz$.

Do Exercise 9.

9. Find an equation of variation in which y varies jointly as x and z, and $y = 65$ when $x = 10$ and $z = 13$.

The equation

$$y = k \cdot \frac{xz^2}{w}$$

asserts that y varies jointly as x and the square of z, and inversely as w.

EXAMPLE 8 Find an equation of variation in which y varies jointly as x and z and inversely as the square of w, and $y = 105$ when $x = 3$, $z = 20$, and $w = 2$.

$$y = k \cdot \frac{xz}{w^2}$$

$$105 = k \cdot \frac{3 \cdot 20}{2^2} \qquad \text{Substituting}$$

$$105 = k \cdot 15$$

$$7 = k$$

Thus, $y = 7 \cdot \dfrac{xz}{w^2}$.

Do Exercise 10.

10. Find an equation of variation in which y varies jointly as x and the square of z and inversely as w, and $y = 80$ when $x = 4$, $z = 10$, and $w = 25$.

Answers on page A-40

11. **Distance of a Dropped Object.** The distance s that an object falls when dropped from some point above the ground varies directly as the square of the time t that it falls. If the object falls 19.6 m in 2 sec, how far will the object fall in 10 sec?

12. **Electrical Resistance.** At a fixed temperature, the resistance R of a wire varies directly as the length l and inversely as the square of its diameter d. If the resistance is 0.1 ohm when the diameter is 1 mm and the length is 50 cm, what is the resistance when the length is 2000 cm and the diameter is 2 mm?

Answers on page A-40

f Other Applications of Variation

Many problem situations can be described with equations of variation.

EXAMPLE 9 *Volume of a Tree.* The volume of wood V in a tree varies jointly as the height h and the square of the girth g (girth is distance around). If the volume of a redwood tree is 216 m^3 when the height is 30 m and the girth is 1.5 m, what is the height of a tree whose volume is 960 m^3 and girth is 2 m?

We first find k using the first set of data. Then we solve for h using the second set of data.

$$V = khg^2$$
$$216 = k \cdot 30 \cdot 1.5^2$$
$$3.2 = k$$

Then the equation of variation is $V = 3.2hg^2$. We substitute the second set of data into the equation:

$$960 = 3.2 \cdot h \cdot 2^2$$
$$75 = h.$$

Therefore, the height of the tree is 75 m.

EXAMPLE 10 *TV Signal.* The intensity I of a TV signal varies inversely as the square of the distance d from the transmitter. If the intensity is 23 watts per square meter (W/m^2) at a distance of 2 km, what is the intensity at a distance of 6 km?

We first find k using the first set of data. Then we solve for I using the second set of data.

$$I = \frac{k}{d^2}$$
$$23 = \frac{k}{2^2}$$
$$92 = k$$

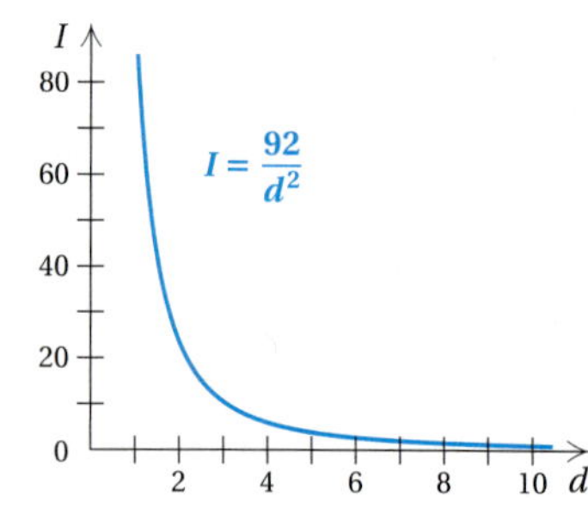

Then the equation of variation is $I = 92/d^2$. We substitute the second distance into the equation:

$$I = \frac{92}{d^2} = \frac{92}{6^2} \approx 2.56. \quad \text{Rounded to the nearest hundredth}$$

Therefore, at 6 km, the intensity is about 2.56 W/m^2.

Do Exercises 11 and 12.

11.9 EXERCISE SET

For Extra Help

Digital Video Tutor CD 10 Videotape 13 | InterAct Math | Math Tutor Center | MathXL | MyMathLab

a Find the variation constant and an equation of variation in which y varies directly as x and the following are true.

1. $y = 40$ when $x = 8$

2. $y = 54$ when $x = 12$

3. $y = 4$ when $x = 30$

4. $y = 3$ when $x = 33$

5. $y = 0.9$ when $x = 0.4$

6. $y = 0.8$ when $x = 0.2$

b Solve.

7. *Aluminum Usage.* The number N of aluminum cans used each year varies directly as the number of people using the cans. If 250 people use 60,000 cans in one year, how many cans are used each year in Dallas, which has a population of 1,189,000?

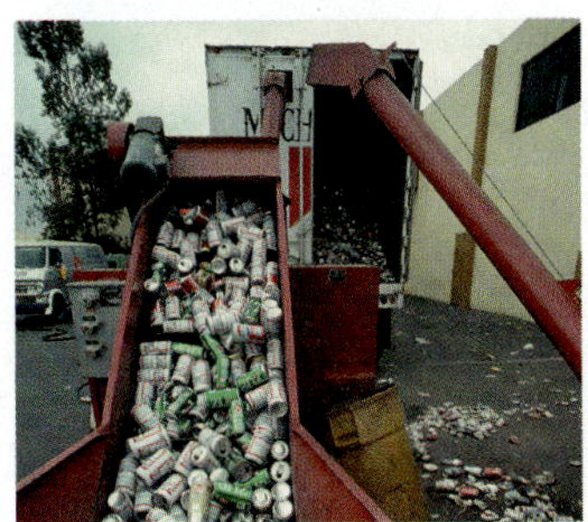

8. *Weekly Allowance.* The average weekly allowance A of children varies directly as their grade level G. It is known that the average allowance of a 9th-grade student is \$9.66 per week. What then is the average weekly allowance of a 4th-grade student?

Source: Fidelity Investments *Investment Vision Magazine*

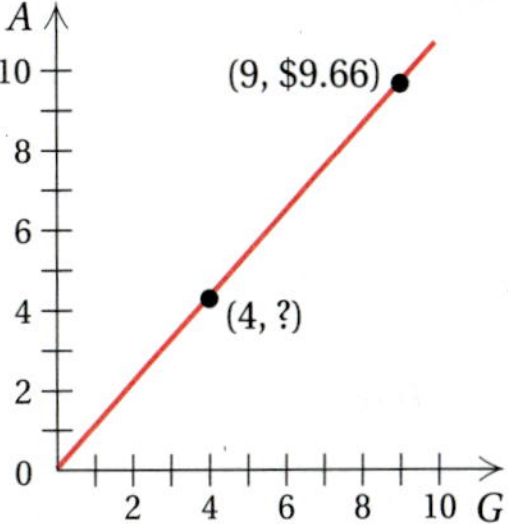

9. *Hooke's Law.* Hooke's law states that the distance d that a spring is stretched by a hanging object varies directly as the weight w of the object. If a spring is stretched 40 cm by a 3-kg barbell, what is the distance stretched by a 5-kg barbell?

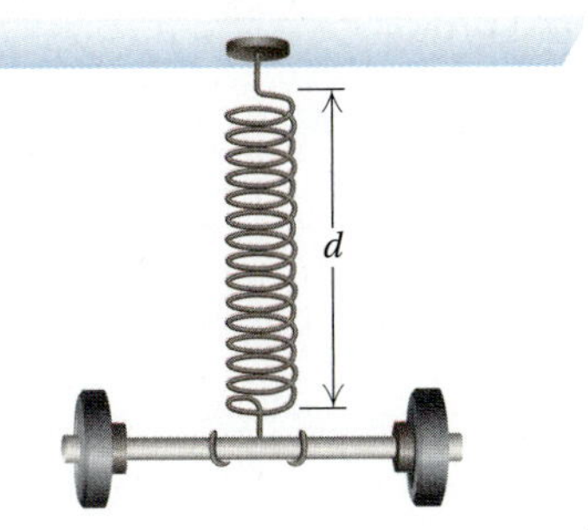

10. *Lead Pollution.* The average U.S. community of population 12,500 released about 385 tons of lead into the environment in a recent year. How many tons were released nationally? Use 281,000,000 as the U.S. population.

Source: *Conservation Matters,* Autumn 1995 issue. Boston: Conservation Law Foundation, p. 30

11. *Fat Intake.* The maximum number of grams of fat that should be in a diet varies directly as a person's weight. A person weighing 120 lb should have no more than 60 g of fat per day. What is the maximum daily fat intake for a person weighing 180 lb?

12. *Relative Aperture.* The relative aperture, or f-stop, of a 23.5-mm diameter lens is directly proportional to the focal length F of the lens. If a 150-mm focal length has an f-stop of 6.3, find the f-stop of a 23.5-mm diameter lens with a focal length of 80 mm.

13. *Mass of Water in Body.* The number of kilograms W of water in a human body varies directly as the mass of the body. A 96-kg person contains 64 kg of water. How many kilograms of water are in a 60-kg person?

14. *Weight on Mars.* The weight M of an object on Mars varies directly as its weight E on Earth. A person who weighs 95 lb on Earth weighs 38 lb on Mars. How much would a 100-lb person weigh on Mars?

c Find the variation constant and an equation of variation in which y varies inversely as x and the following are true.

15. $y = 14$ when $x = 7$

16. $y = 1$ when $x = 8$

17. $y = 3$ when $x = 12$

18. $y = 12$ when $x = 5$

19. $y = 0.1$ when $x = 0.5$

20. $y = 1.8$ when $x = 0.3$

d Solve.

21. *Work Rate.* The time T required to do a job varies inversely as the number of people P working. It takes 5 hr for 7 bricklayers to build a park wall. How long will it take 10 bricklayers to complete the job?

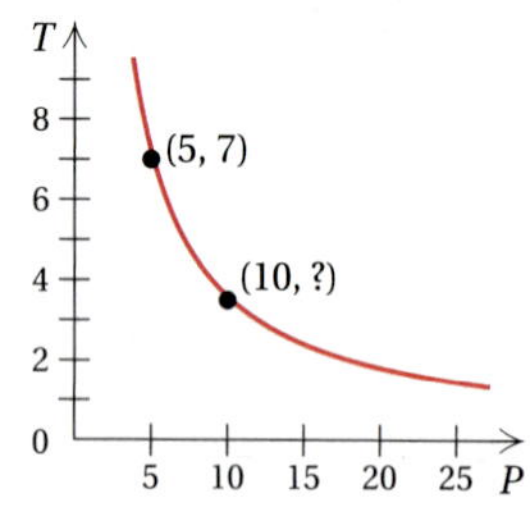

22. *Pumping Rate.* The time t required to empty a tank varies inversely as the rate r of pumping. If a pump can empty a tank in 45 min at the rate of 600 kL/min, how long will it take the pump to empty the same tank at the rate of 1000 kL/min?

23. *Current and Resistance.* The current I in an electrical conductor varies inversely as the resistance R of the conductor. If the current is $\frac{1}{2}$ ampere when the resistance is 240 ohms, what is the current when the resistance is 540 ohms?

24. *Wavelength and Frequency.* The wavelength W of a radio wave varies inversely as its frequency F. A wave with a frequency of 1200 kilohertz has a length of 300 meters. What is the length of a wave with a frequency of 800 kilohertz?

25. *Musical Pitch.* The pitch P of a musical tone varies inversely as its wavelength W. One tone has a pitch of 330 vibrations per second and a wavelength of 3.2 ft. Find the wavelength of another tone that has a pitch of 550 vibrations per second.

26. *Beam Weight.* The weight W that a horizontal beam can support varies inversely as the length L of the beam. Suppose that an 8-m beam can support 1200 kg. How many kilograms can a 14-m beam support?

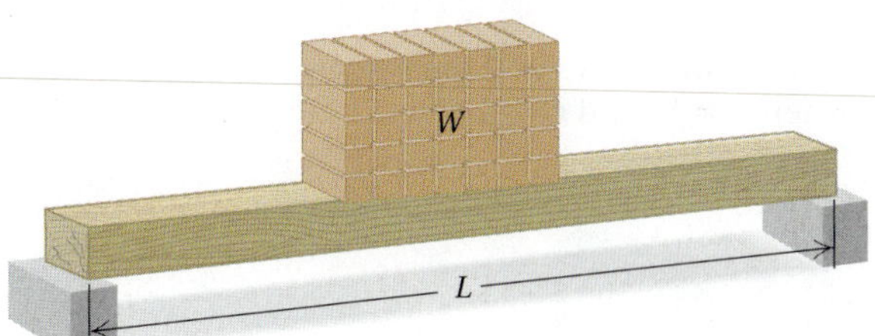

27. *Volume and Pressure.* The volume V of a gas varies inversely as the pressure P upon it. The volume of a gas is 200 cm^3 under a pressure of 32 kg/cm^2. What will be its volume under a pressure of 40 kg/cm^2?

28. *Rate of Travel.* The time t required to drive a fixed distance varies inversely as the speed r. It takes 5 hr at a speed of 80 km/h to drive a fixed distance. How long will it take to drive the same distance at a speed of 70 km/h?

e Find an equation of variation in which the following are true.

29. y varies directly as the square of x, and $y = 0.15$ when $x = 0.1$

30. y varies directly as the square of x, and $y = 6$ when $x = 3$

31. y varies inversely as the square of x, and $y = 0.15$ when $x = 0.1$

32. y varies inversely as the square of x, and $y = 6$ when $x = 3$

33. y varies jointly as x and z, and $y = 56$ when $x = 7$ and $z = 8$

34. y varies directly as x and inversely as z, and $y = 4$ when $x = 12$ and $z = 15$

35. y varies jointly as x and the square of z, and $y = 105$ when $x = 14$ and $z = 5$

36. y varies jointly as x and z and inversely as w, and $y = \frac{3}{2}$ when $x = 2$, $z = 3$, and $w = 4$

37. y varies jointly as x and z and inversely as the product of w and p, and $y = \frac{3}{28}$ when $x = 3$, $z = 10$, $w = 7$, and $p = 8$

38. y varies jointly as x and z and inversely as the square of w, and $y = \frac{12}{5}$ when $x = 16$, $z = 3$, and $w = 5$

f Solve.

39. *Stopping Distance of a Car.* The stopping distance d of a car after the brakes have been applied varies directly as the square of the speed r. If a car traveling 60 mph can stop in 200 ft, how fast can a car travel and still stop in 72 ft?

40. *Boyle's Law.* The volume V of a given mass of a gas varies directly as the temperature T and inversely as the pressure P. If $V = 231$ cm^3 when $T = 42°$ and $P = 20$ kg/cm^2, what is the volume when $T = 30°$ and $P = 15$ kg/cm^2?

41. *Intensity of Light.* The intensity I of light from a light bulb varies inversely as the square of the distance d from the bulb. Suppose that I is 90 W/m^2 (watts per square meter) when the distance is 5 m. How much *further* would it be to a point where the intensity is 40 W/m^2?

42. *Weight of an Astronaut.* The weight W of an object varies inversely as the square of the distance d from the center of the earth. At sea level (3978 mi from the center of the earth), an astronaut weighs 220 lb. Find his weight when he is 200 mi above the surface of the earth and the spacecraft is not in motion.

43. *Earned-Run Average.* A pitcher's earned-run average E varies directly as the number R of earned runs allowed and inversely as the number I of innings pitched. In a recent year, Shawn Estes of the San Francisco Giants had an earned-run average of 3.18. He gave up 71 earned runs in 201 innings. How many earned runs would he have given up had he pitched 300 innings with the same average? Round to the nearest whole number.

44. *Atmospheric Drag.* Wind resistance, or atmospheric drag, tends to slow down moving objects. Atmospheric drag varies jointly as an object's surface area A and velocity v. If a car traveling at a speed of 40 mph with a surface area of 37.8 ft^2 experiences a drag of 222 N (Newtons), how fast must a car with 51 ft^2 of surface area travel in order to experience a drag force of 430 N?

45. *Water Flow.* The amount Q of water emptied by a pipe varies directly as the square of the diameter d. A pipe 5 in. in diameter will empty 225 gal of water over a fixed time period. If we assume the same kind of flow, how many gallons of water are emptied in the same amount of time by a pipe that is 9 in. in diameter?

46. *Weight of a Sphere.* The weight W of a sphere of a given material varies directly as its volume V, and its volume V varies directly as the cube of its diameter.

a) Find an equation of variation relating the weight W to the diameter d.

b) An iron ball that is 5 in. in diameter is known to weigh 25 lb. Find the weight of an iron ball that is 8 in. in diameter.

47. **DW** Write a variation problem for a classmate to solve. Design the problem so that the answer is "When Simone studies for 8 hr a week, her quiz score is 92."

48. **DW** If y varies directly as x and x varies inversely as z, how does y vary with regard to z? Why?

SKILL MAINTENANCE

Factor completely. [10.7a]

49. $x^2 - x - 56$

50. $a^2 - 16a + 64$

51. $x^5 - 2x^4 - 35x^3$

52. $2y^3 - 10y^2 + y - 5$

53. $16 - t^4$

54. $10x^2 + 80x + 70$

55. $x^2 - 9x + 14$

56. $x^2 + x + 7$

57. $16x^2 - 40xy + 25y^2$

58. $a^2 - 9ab + 14b^2$

59. $3x^3 - 3y^3$

60. $w^6 - t^6$

SYNTHESIS

61. *Area of a Circle.* The area of a circle varies directly as the square of the length of a diameter. What is the variation constant?

62. In each of the following equations, state whether y varies directly as x, inversely as x, or neither directly nor inversely as x.

a) $7xy = 14$

b) $x - 2y = 12$

c) $-2x + 3y = 0$

d) $x = \frac{3}{4}y$

e) $\frac{x}{y} = 2$

Describe, in words, the variation given by the equation.

63. $Q = \frac{kp^2}{q^3}$

64. $W = \frac{km_1M_1}{d^2}$

65. *Volume and Cost.* A peanut butter jar in the shape of a right circular cylinder is 4 in. high and 3 in. in diameter and sells for \$1.20. If we assume that cost is proportional to volume, how much should a jar 6 in. high and 6 in. in diameter cost?

CHAPTER

11 Summary and Review

The review that follows is meant to prepare you for a chapter exam. It consists of two parts. The first part is a checklist of some of the Study Tips referred to in this and preceding chapters. The second part is the Review Exercises. These provide practice exercises for the exam, together with references to section objectives so you can go back and review. Before beginning, stop and look back over the skills you have obtained. What skills in mathematics do you have now that you did not have before studying this chapter?

STUDY TIPS CHECKLIST

The foundation of all your study skills is TIME!

- [ ] Are you finding quiet, nondistracting places to study?
- [ ] Have you tried tutoring a fellow student?
- [ ] Did you use the five-step problem-solving strategy when doing the applications in Section 11.8?
- [ ] Are you doing more even-numbered exercises and using strategies to check your work, such as doing the problem a second time, checking, or estimating?

REVIEW EXERCISES

Find all numbers for which the rational expression is not defined. [11.1a]

1. $\dfrac{3}{x}$

2. $\dfrac{4}{x-6}$

3. $\dfrac{x+5}{x^2-36}$

4. $\dfrac{x^2-3x+2}{x^2+x-30}$

5. $\dfrac{-4}{(x+2)^2}$

6. $\dfrac{x-5}{x^3-8x^2+15x}$

Simplify. [11.1c]

7. $\dfrac{4x^2-8x}{4x^2+4x}$

8. $\dfrac{14x^2-x-3}{2x^2-7x+3}$

9. $\dfrac{a^2+2a+4}{a^3-8}$

Multiply and simplify. [11.1d]

10. $\dfrac{a^2-36}{10a}\cdot\dfrac{2a}{a+6}$

11. $\dfrac{6t-6}{2t^2+t-1}\cdot\dfrac{t^2-1}{t^2-2t+1}$

12. $\dfrac{x^3-64}{x^2-16}\cdot\dfrac{x^2+7x+12}{x^2+5x+6}$

Divide and simplify. [11.2b]

13. $\dfrac{10-5t}{3}\div\dfrac{t-2}{12t}$

14. $\dfrac{4x^4}{x^2-1}\div\dfrac{2x^3}{x^2-2x+1}$

Find the LCM. [11.3c]

15. $3x^2,\ 10xy,\ 15y^2$

16. $a - 2,\ 4a - 8$

17. $y^2 - y - 2,\ y^2 - 4$

Add and simplify. [11.4a]

18. $\dfrac{x+8}{x+7} + \dfrac{10-4x}{x+7}$

19. $\dfrac{3}{3x-9} + \dfrac{x-2}{3-x}$

20. $\dfrac{2a}{a+1} + \dfrac{4a}{a^2-1}$

21. $\dfrac{d^2}{d-c} + \dfrac{c^2}{c-d}$

Subtract and simplify. [11.5a]

22. $\dfrac{6x-3}{x^2-x-12} - \dfrac{2x-15}{x^2-x-12}$

23. $\dfrac{3x-1}{2x} - \dfrac{x-3}{x}$

24. $\dfrac{x+3}{x-2} - \dfrac{x}{2-x}$

25. $\dfrac{1}{x^2-25} - \dfrac{x-5}{x^2-4x-5}$

26. Perform the indicated operations and simplify: [11.5b]

$$\frac{3x}{x+2} - \frac{x}{x-2} + \frac{8}{x^2-4}.$$

Simplify. [11.6a]

27. $\dfrac{\dfrac{1}{z}+1}{\dfrac{1}{z^2}-1}$

28. $\dfrac{\dfrac{c}{d}-\dfrac{d}{c}}{\dfrac{1}{c}+\dfrac{1}{d}}$

Solve. [11.7a]

29. $\dfrac{3}{y} - \dfrac{1}{4} = \dfrac{1}{y}$

30. $\dfrac{15}{x} - \dfrac{15}{x+2} = 2$

31. Find an equation of variation in which y varies directly as x, and $y = 100$ when $x = 25$. [11.9a]

32. Find an equation of variation in which y varies inversely as x, and $y = 100$ when $x = 25$. [11.9c]

33. *Pumping Time.* The time t required to empty a tank varies inversely as the rate r of pumping. If a pump can empty a tank in 35 min at the rate of 800 kL per minute, how long will it take the pump to empty the same tank at the rate of 1400 kL per minute? [11.9d]

Solve. [11.8a]

34. *Highway Work.* In checking records, a contractor finds that crew A can pave a certain length of highway in 9 hr, while crew B can do the same job in 12 hr. How long would it take if they worked together?

35. *Train Speed.* A manufacturer is testing two high-speed trains. One train travels 40 km/h faster than the other. While one train travels 70 km, the other travels 60 km. Find the speed of each train.

36. *Airplane Speed.* One plane travels 80 mph faster than another. While one travels 1750 mi, the other travels 950 mi. Find the speed of each plane.

Solve. [11.8b]

37. *Quality Control.* A sample of 250 calculators contained 8 defective calculators. How many defective calculators would you expect to find in a sample of 5000?

38. *Pizza Proportions.* A certain kind of pizza at Finnelli's Pizzeria uses the following ratio: 5 parts sausage to 7 parts cheese, 6 parts onion to 13 parts green pepper, and 9 parts pepperoni to 14 parts cheese.

a) Finnelli's makes several pizzas with green pepper and onion. They use 2 cups of green pepper. How much onion would they use?
b) Finnelli's makes several pizzas with sausage and cheese. They use 3 cups of sausage. How much cheese would they use?
c) Finnelli's makes several pizzas with pepperoni and cheese. They use 6 cups of pepperoni. How much cheese would they use?

39. *Frog Population.* To estimate how many frogs there are in a rain forest, a research team tags 600 frogs and then releases them. Later the team catches 300 frogs and notes that 25 of them have been tagged. Estimate the total frog population in the rain forest.

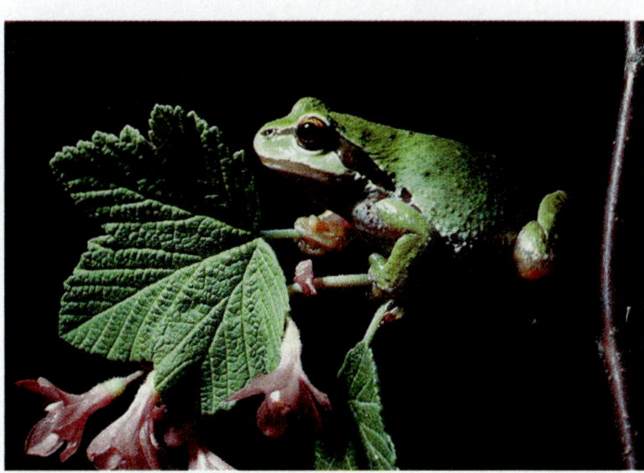

40. Triangles ABC and XYZ below are similar. Find the value of x.

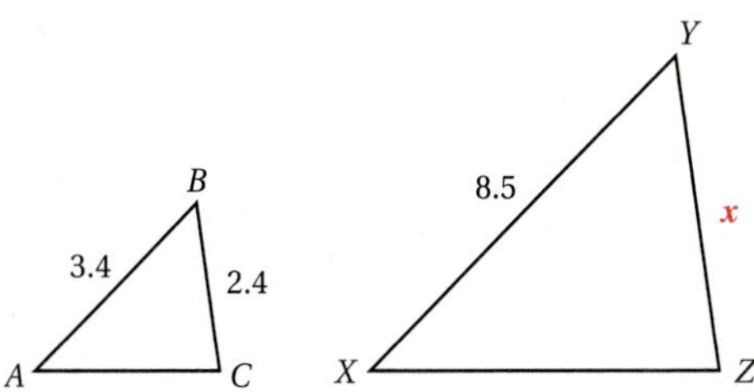

D_W Carry out the direction for each of the following. Explain the use of the LCM in each case.

41. Add: $\frac{4}{x-2} + \frac{1}{x+2}$. [11.4a]

42. Subtract: $\frac{4}{x-2} - \frac{1}{x+2}$. [11.5a]

43. Solve: $\frac{4}{x-2} + \frac{1}{x+2} = \frac{26}{x^2-4}$. [11.7a]

44. Simplify: $\dfrac{1 - \frac{2}{x}}{1 + \frac{x}{4}}$. [11.6a]

SKILL MAINTENANCE

Certain objectives from four particular sections will be retested on the chapter test. The objectives are listed with the practice problems that follow.

45. Factor: $5x^3 + 20x^2 - 3x - 12$. [10.7a]

46. Simplify: $(5x^3y^2)^{-3}$. [9.2a, b]

47. Subtract: [9.4c]
$$(5x^3 - 4x^2 + 3x - 4) - (7x^3 - 7x^2 - 9x + 14).$$

48. *Rectangle Dimensions.* The width of a rectangle is 2 cm less than the length. The area is 15 cm^2. Find the dimensions and the perimeter of the rectangle. [10.9a]

SYNTHESIS

Simplify.

49. $\frac{2a^2 + 5a - 3}{a^2} \cdot \frac{5a^3 + 30a^2}{2a^2 + 7a - 4} \div \frac{a^2 + 6a}{a^2 + 7a + 12}$
[11.1d], [11.2b]

50. $\frac{12a}{(a-b)(b-c)} - \frac{2a}{(b-a)(c-b)}$ [11.5a]

51. Compare
$$\frac{A+B}{B} = \frac{C+D}{D}$$
with the proportion
$$\frac{A}{B} = \frac{C}{D}.$$
[11.8b]

CHAPTER 11

Chapter Test

Find all numbers for which the rational expression is not defined.

1. $\dfrac{8}{2x}$

2. $\dfrac{5}{x + 8}$

3. $\dfrac{x - 7}{x^2 - 49}$

4. $\dfrac{x^2 + x - 30}{x^2 - 3x + 2}$

5. $\dfrac{11}{(x - 1)^2}$

6. $\dfrac{x + 2}{x^3 + 8x^2 + 15x}$

7. Simplify:
$$\frac{6x^2 + 17x + 7}{2x^2 + 7x + 3}.$$

8. Multiply and simplify:
$$\frac{a^2 - 25}{6a} \cdot \frac{3a}{a - 5}.$$

9. Divide and simplify:
$$\frac{25x^2 - 1}{9x^2 - 6x} \div \frac{5x^2 + 9x - 2}{3x^2 + x - 2}.$$

10. Find the LCM:
$$y^2 - 9,\ y^2 + 10y + 21,\ y^2 + 4y - 21.$$

Add or subtract. Simplify if possible.

11. $\dfrac{16 + x}{x^3} + \dfrac{7 - 4x}{x^3}$

12. $\dfrac{5 - t}{t^2 + 1} - \dfrac{t - 3}{t^2 + 1}$

13. $\dfrac{x - 4}{x - 3} + \dfrac{x - 1}{3 - x}$

14. $\dfrac{x - 4}{x - 3} - \dfrac{x - 1}{3 - x}$

15. $\dfrac{5}{t - 1} + \dfrac{3}{t}$

16. $\dfrac{1}{x^2 - 16} - \dfrac{x + 4}{x^2 - 3x - 4}$

17. $\dfrac{1}{x - 1} + \dfrac{4}{x^2 - 1} - \dfrac{2}{x^2 - 2x + 1}$

18. Simplify: $\dfrac{9 - \dfrac{1}{y^2}}{3 - \dfrac{1}{y}}.$

Solve.

19. $\dfrac{7}{y} - \dfrac{1}{3} = \dfrac{1}{4}$

20. $\dfrac{15}{x} - \dfrac{15}{x - 2} = -2$

Solve.

21. *Quality Control.* A sample of 125 spark plugs contained 4 defective spark plugs. How many defective spark plugs would you expect to find in a sample of 500?

22. *Zebra Population.* A game warden catches, tags, and then releases 15 zebras. A month later, a sample of 20 zebras is collected and 6 of them have tags. Use this information to estimate the size of the zebra population in that area.

23. *Copying Time.* Kopy Kwik has 2 copiers. One can copy a year-end report in 20 min. The other can copy the same document in 30 min. How long would it take both machines, working together, to copy the report?

24. *Driving Speed.* Craig drives 20 km/h faster than Marilyn. In the same time that Marilyn drives 225 km, Craig drives 325 km. Find the speed of each car.

25. Find an equation of variation in which Q varies jointly as x and y, and $Q = 25$ when $x = 2$ and $y = 5$.

26. Find an equation of variation in which y varies inversely as x, and $y = 10$ when $x = 25$.

27. *Income vs. Time.* Dean's income I varies directly as the time t worked. He gets a job that pays $275 for 40 hr of work. What is he paid for working 72 hr, assuming that there is no change in payscale for overtime?

28. This pair of triangles is similar. Find the missing length x.

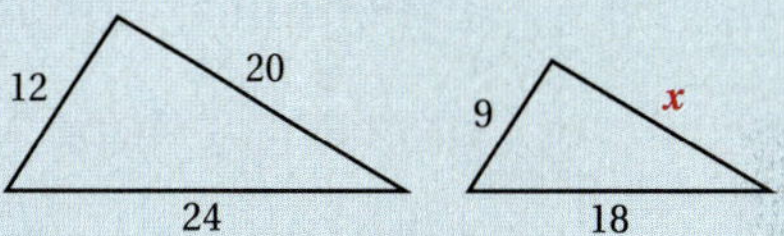

SKILL MAINTENANCE

29. Factor: $16a^2 - 49$.

30. Simplify: $\left(\frac{3x^2}{y^3}\right)^{-4}$.

31. Subtract:

$$(5x^2 - 19x + 34) - (-8x^2 + 10x - 42).$$

32. The product of two consecutive integers is 462. Find the integers.

SYNTHESIS

33. Reggie and Rema work together to mulch the flower beds around an office complex in $2\frac{6}{7}$ hr. Working alone, it would take Reggie 6 hr more than it would take Rema. How long would it take each of them to complete the landscaping working alone?

34. Simplify: $1 + \cfrac{1}{1 + \cfrac{1}{1 + \cfrac{1}{a}}}$.

Graphs, Functions, and Applications

Gateway to Chapter 12

Graphs help us to see relationships among quantities. In this chapter, we graph many types of equations, emphasizing those whose graphs are straight lines. These are called linear equations.

We also consider the concept of a function. Functions are the backbone of any future study in mathematics. Here we introduce both graphs and applications of functions.

For linear equations and functions, we develop the concepts of slope and intercepts. Finally, we bring the ideas of this chapter together with applications.

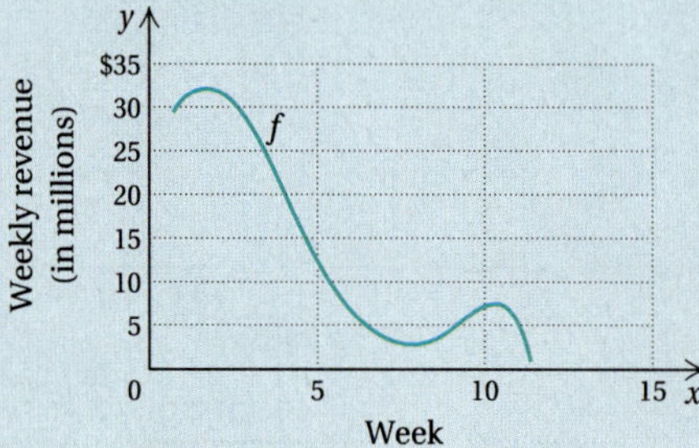

Real-World Application

The graph shown here approximates the weekly revenue from the movie *The Matrix*. The weekly revenue is a function f of the number of weeks x since the movie was released. No equation is given for the function. What was the movie revenue for week 2? That is, find $f(2)$.

Source: Motion Picture Association of America

This problem appears as Example 9 in Section 12.1.

CHAPTER 12

CHAPTER 12
Pretest

Graph on a plane.

1. $2x - 5y = 20$ [12.4a]
2. $x = 4$ [12.4c]
3. $y = x - 2$ [12.4b]
4. $f(x) = -2$ [12.4c]
5. $f(x) = 3 - x^2$ [12.1c]

6. Determine whether each of the following is the graph of a function. [12.1d]

a)

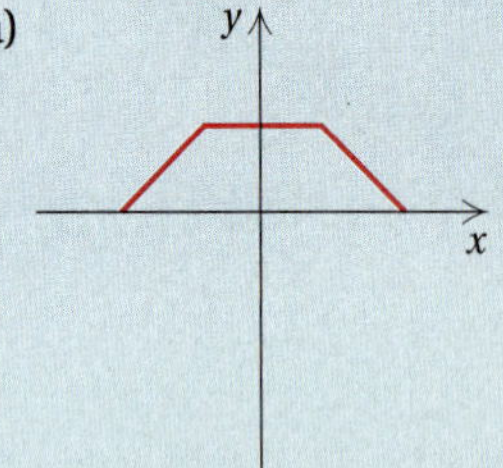

b)

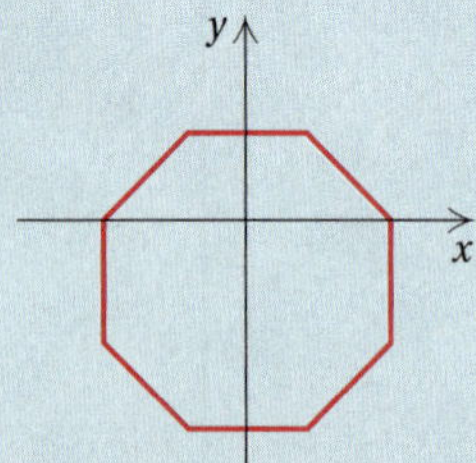

7. Find the slope and the y-intercept: $y = 5x - 3$. [12.3b]

8. Find the slope, if it exists, of the line containing the points $(7, 4)$ and $(-5, 4)$. [12.3b]

9. Find an equation of the line containing the points $(-3, 7)$ and $(-8, 2)$. [12.5c]

10. Find an equation of the line having the given slope and containing the given point. [12.5b]

$m = 2$; $(-2, 3)$

11. Find an equation of the line containing the given point and parallel to the given line. [12.5d]

$(2, 5)$; $2x - 7y = 10$

12. Find an equation of the line containing the given point and perpendicular to the given line. [12.5d]

$(2, 5)$; $2x - 7y = 10$

Determine whether the graphs of the pair of lines are parallel or perpendicular. [12.4d]

13. $3y - 2x = 21,$
$3x + 2y = 8$

14. $y = 3x + 7,$
$y = 3x - 4$

15. Find the intercepts of $2x - 3y = 12$. [12.4a]

16. For the function f given by $f(x) = |x| - 3$, find $f(0)$, $f(-2)$, and $f(4)$. [12.1b]

17. Find the domain: [12.2a]

$$f(x) = \frac{3}{2x - 5}.$$

18. **Spending on Recorded Music.** Consider the graph at right, which shows the average amount of spending per person per year on recorded music. [12.5e]

a) Use the two points $(0, \$37.73)$ and $(10, \$71.56)$ to find a linear function that fits the data.

b) Use the function to predict the average amount of spending on recorded music in 2011.

Average Amount Spent per Person

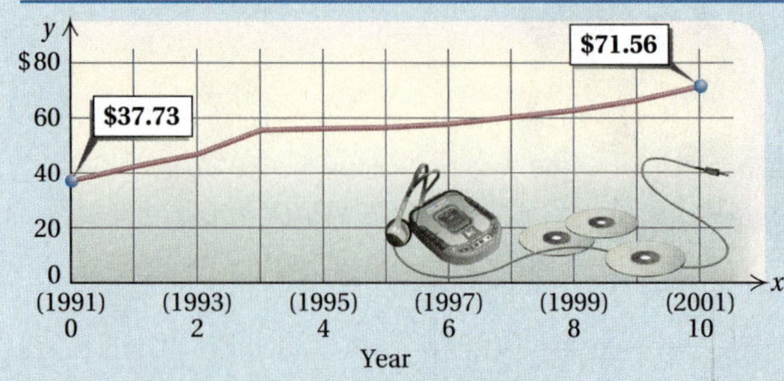

Source: Veronis, Suhler & Associates, Industry Sources

12.1 FUNCTIONS AND GRAPHS

a Identifying Functions

Consider the equation $y = 2x - 3$. If we substitute a value for x—say 5—we get a value for y, 7:

$$y = 2x - 3 = 2(5) - 3 = 10 - 3 = 7.$$

The equation $y = 2x - 3$ is an example of a *function*. We now develop the concept of a *function*, one of the most important concepts in mathematics.

In much the same way that ordered pairs form correspondences between first and second coordinates, a *function* is a correspondence from one set to another. For example:

To each student in a college, there corresponds his or her student ID.

To each item in a store, there corresponds its price.

To each real number, there corresponds the cube of that number.

In each case, the first set is called the **domain** and the second set is called the **range.** This kind of correspondence is called a **function.** Given a member of the domain, there is *just one* member of the range to which it corresponds.

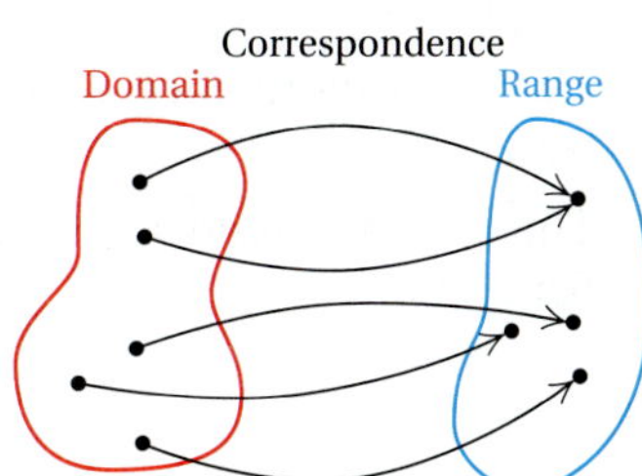

EXAMPLE 1 Determine whether the correspondence is a function.

f:

Domain	*Range*
1 →	$107.4
2 →	$ 34.1
3 →	$ 29.6
4 →	$ 19.6

g:

Domain	*Range*
3 →	5
4 →	9
5 →	−7
6 →	−7

h:

Domain	*Range*
Chicago →	Cubs
Chicago →	White Sox
Baltimore →	Orioles
San Diego →	Padres

p:

Domain	*Range*
Cubs →	Chicago
White Sox →	Chicago
Orioles →	Baltimore
Padres →	San Diego

The correspondence f *is* a function because each member of the domain is matched to *only one* member of the range.

The correspondence g *is* also a function because each member of the domain is matched to *only one* member of the range.

The correspondence h *is not* a function because one member of the domain, Chicago, is matched to *more than one* member of the range.

The correspondence p *is* a function because each member of the domain is matched to *only one* member of the range.

Objectives

a Determine whether a correspondence is a function.

b Given a function described by an equation, find function values (outputs) for specified values (inputs).

c Draw the graph of a function.

d Determine whether a graph is that of a function using the vertical-line test.

e Solve applied problems involving functions and their graphs.

Determine whether the correspondence is a function.

1. *Domain* *Range*

   Cheetah → 70 mph
   Human → 28 mph
   Lion → 50 mph
   Chicken → 9 mph

2. *Domain* *Range*

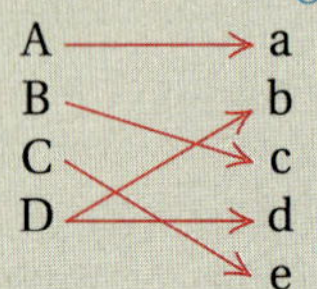

3. *Domain* *Range*

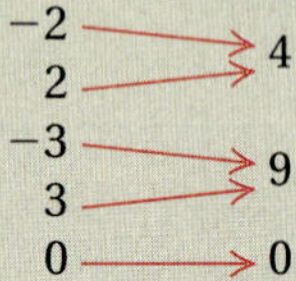

4. *Domain* *Range*

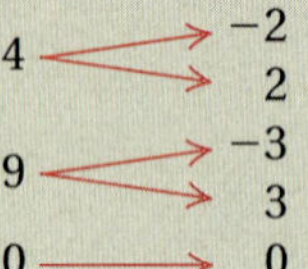

Determine whether each of the following is a function.

5. *Domain*
   A set of numbers
   Correspondence
   Square each number and subtract 10.
   Range
   A set of numbers

6. *Domain*
   A set of polygons
   Correspondence
   Find the area of each polygon.
   Range
   A set of numbers

Answers on page A-42

FUNCTION; DOMAIN; RANGE

A **function** is a correspondence between a first set, called the **domain,** and a second set, called the **range,** such that each member of the domain corresponds to **exactly one** member of the range.

Do Exercises 1–4.

EXAMPLE 2 Determine whether the correspondence is a function.

	Domain	*Correspondence*	*Range*
a)	A family	Each person's weight	A set of positive numbers
b)	The integers	Each number's square	A set of nonnegative integers
c)	The set of all states	Each state's members of the U.S. Senate	The set of U.S. Senators

a) The correspondence *is* a function because each person has *only one* weight.

b) The correspondence *is* a function because each integer has *only one* square.

c) The correspondence *is not* a function because each state has two U.S. Senators.

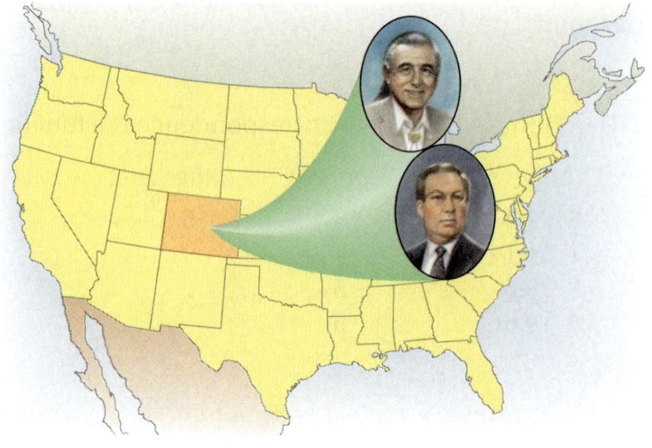

Do Exercises 5 and 6.

When a correspondence between two sets is not a function, it is still an example of a **relation.**

RELATION

A **relation** is a correspondence between a first set, called the **domain,** and a second set, called the **range,** such that each member of the domain corresponds to **at least one** member of the range.

Thus, although the correspondences of Examples 1 and 2 are not all functions, they *are* all relations. A function is a special type of relation—one in which each member of the domain is paired with *exactly one* member of the range.

b Finding Function Values

Most functions considered in mathematics are described by equations like $y = 2x + 3$ or $y = 4 - x^2$. We graph the function $y = 2x + 3$ by first performing calculations like the following:

for $x = 4$, $y = 2x + 3 = 2 \cdot 4 + 3 = 11$;
for $x = -5$, $y = 2x + 3 = 2 \cdot (-5) + 3 = -7$;
for $x = 0$, $y = 2x + 3 = 2 \cdot 0 + 3 = 3$; and so on.

For $y = 2x + 3$, the **inputs** (members of the domain) are values of x substituted into the equation. The **outputs** (members of the range) are the resulting values of y. If we call the function f, we can use x to represent an arbitrary *input* and $f(x)$—read "f of x," or "f at x," or "the value of f at x"—to represent the corresponding *output*. In this notation, the function given by $y = 2x + 3$ is written as $f(x) = 2x + 3$ and the calculations above can be written more concisely as follows:

$y = f(4) = 2 \cdot 4 + 3 = 11$;
$y = f(-5) = 2 \cdot (-5) + 3 = -7$;
$y = f(0) = 2 \cdot 0 + 3 = 3$; and so on.

Thus instead of writing "when $x = 4$, the value of y is 11," we can simply write "$f(4) = 11$," which can also be read as "f of 4 is 11" or "for the input 4,the output of f is 11."

It helps to think of a function as a machine. Think of $f(4) = 11$ as putting a member of the domain (an input), 4, into the machine. The machine knows the correspondence $f(x) = 2x + 3$, multiplies 4 by 2 and adds 3, and gives out a member of the range (the output), 11.

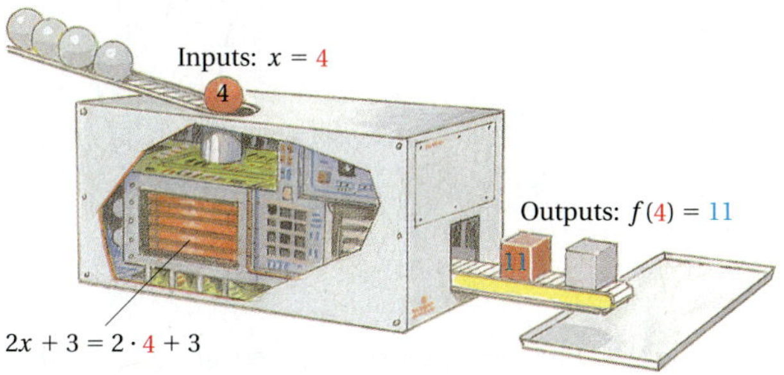

CAUTION!

The notation $f(x)$ *does not mean* "f times x" and should not be read that way.

EXAMPLE 3 A function f is given by $f(x) = 3x^2 - 2x + 8$. Find each of the indicated function values.

a) $f(0)$ **b)** $f(1)$ **c)** $f(-5)$ **d)** $f(7a)$

One way to find function values when a formula is given is to think of the formula with blanks, or placeholders, as follows:

$$f(\square) = 3\square^2 - 2\square + 8.$$

To find an output for a given input, we think: "Whatever goes in the blank on the left goes in the blank(s) on the right." With this in mind, let's complete the example.

a) $f(0) = 3 \cdot 0^2 - 2 \cdot 0 + 8 = 8$

7. Find the indicated function values for the following function:

$$f(x) = 2x^2 + 3x - 4.$$

a) $f(0)$

b) $f(8)$

c) $f(-5)$

d) $f(7a)$

8. Find the indicated function value.

a) $f(-6)$, for $f(x) = 5x - 3$

b) $g(-1)$, for $g(r) = 5r^2 + 3r$

c) $h(55)$, for $h(x) = 7$

d) $F(a + 2)$, for $F(x) = 5x - 8$

Answers on page A-42

b) $f(1) = 3 \cdot 1^2 - 2 \cdot 1 + 8 = 3 \cdot 1 - 2 + 8 = 3 - 2 + 8 = 9$

c) $f(-5) = 3(-5)^2 - 2 \cdot (-5) + 8 = 3 \cdot 25 + 10 + 8 = 75 + 10 + 8 = 93$

d) $f(7a) = 3(7a)^2 - 2(7a) + 8 = 3 \cdot 49a^2 - 14a + 8 = 147a^2 - 14a + 8$

Do Exercise 7.

EXAMPLE 4 Find the indicated function value.

a) $f(5)$, for $f(x) = 3x + 2$

b) $g(-2)$, for $g(r) = 5r^2 + 3r$

c) $h(4)$, for $h(x) = 7$

d) $F(a + 1)$, for $F(x) = 5x - 8$

a) $f(5) = 3 \cdot 5 + 2 = 15 + 2 = 17$

b) $g(-2) = 5(-2)^2 + 3(-2) = 5(4) - 6 = 20 - 6 = 14$

c) For the function given by $h(x) = 7$, all inputs share the same output, 7. Thus, $h(4) = 7$. The function h is an example of a **constant function.**

d) $F(a + 1) = 5(a + 1) - 8 = 5a + 5 - 8 = 5a - 3$

Do Exercise 8.

CALCULATOR CORNER

Finding Function Values We can find function values on a graphing calculator. One method is to substitute inputs directly into the formula. Consider the function $f(x) = x^2 + 3x - 4$. To find $f(-5)$, we press [(] [(–)] [5] [)] [x^2] [+] [3] [(] [(–)] [5] [)] [–] [4] [ENTER]. We find that $f(-5) = 6$.

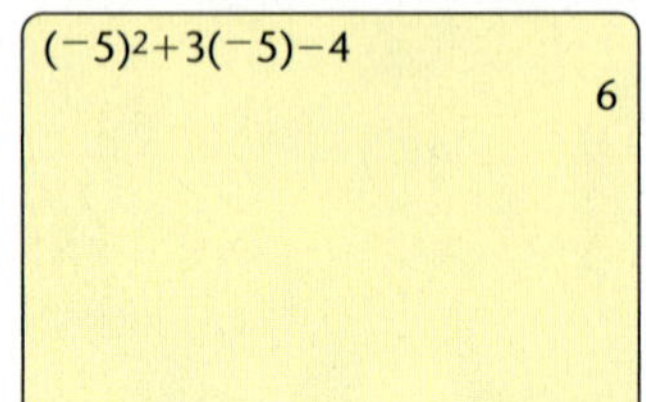

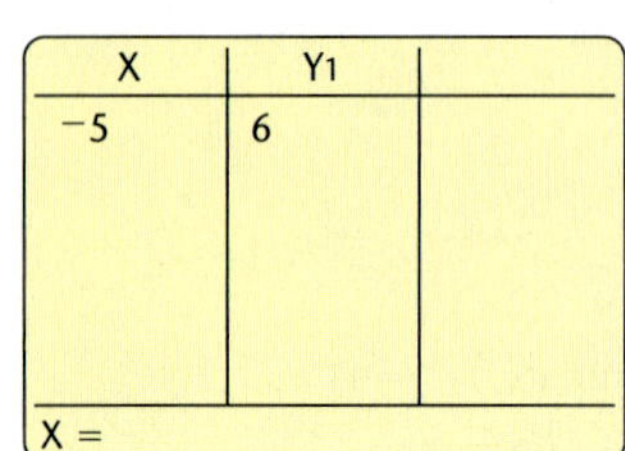

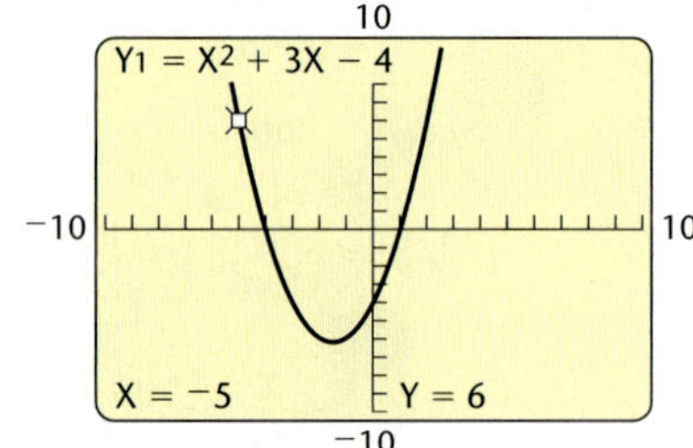

After we have entered the function as $y_1 = x^2 + 3x - 4$ on the equation-editor screen, there are several other methods that we can use to find function values. We can use a table set in ASK mode and enter $x = -5$. (See p. 532.) We see that the function value, y_1, is 6. We can also use the VALUE feature to evaluate the function. To do this, we first graph the function. Then we press [2nd] [CALC] [1] to access the VALUE feature. Next, we supply the desired x-value by pressing [(–)] [5]. Finally, we press [ENTER] to see X = −5, Y = 6 at the bottom of the screen. Again we see that the function value is 6. Note that, when the VALUE feature is used to find a function value, the x-value must be in the viewing window.

A fourth method for finding function values uses the TRACE feature. With the function graphed in a window that includes the x-value −5, we press [TRACE]. The coordinates of the point where the blinking cursor is positioned on the graph are displayed at the bottom of the screen. To move the cursor to the point with x-coordinate −5, we press [(–)] [5] [ENTER]. Now we see X = −5, Y = 6 displayed at the bottom of the screen. This tells us that $f(-5) = 6$. The final calculator display for this method is the same as the one shown above for the VALUE feature. There are other ways to find function values, but we will not discuss them here.

Exercises: Find the function values.

1. $f(-5.1)$, for $f(x) = 3x + 2$
2. $f(4)$, for $f(x) = -3.6x$
3. $f(-3)$, for $f(x) = x^2 + 5$
4. $f(3)$, for $f(x) = 4x^2 + x - 5$

C Graphs of Functions

To graph a function, we find ordered pairs (x, y) or $(x, f(x))$, plot them, and connect the points. Note that y and $f(x)$ are used interchangeably—that is, $y = f(x)$—when we are working with functions and their graphs.

EXAMPLE 5 Graph: $f(x) = x + 2$.

A list of some function values is shown in this table. We plot the points and connect them. The graph is a straight line. The "y" on the vertical axis could also be labeled "$f(x)$".

x	$f(x)$
−4	−2
−3	−1
−2	0
−1	1
0	2
1	3
2	4
3	5
4	6

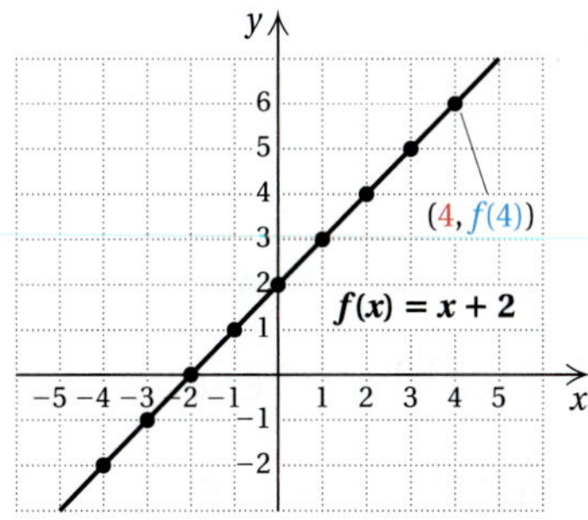

Do Exercise 9.

EXAMPLE 6 Graph: $g(x) = 4 - x^2$.

We calculate some function values and draw the curve.

$$g(0) = 4 - 0^2 = 4 - 0 = 4,$$
$$g(-1) = 4 - (-1)^2 = 4 - 1 = 3,$$
$$g(2) = 4 - 2^2 = 4 - 4 = 0,$$
$$g(-3) = 4 - (-3)^2 = 4 - 9 = -5$$

x	$g(x)$
−3	−5
−2	0
−1	3
0	4
1	3
2	0
3	−5

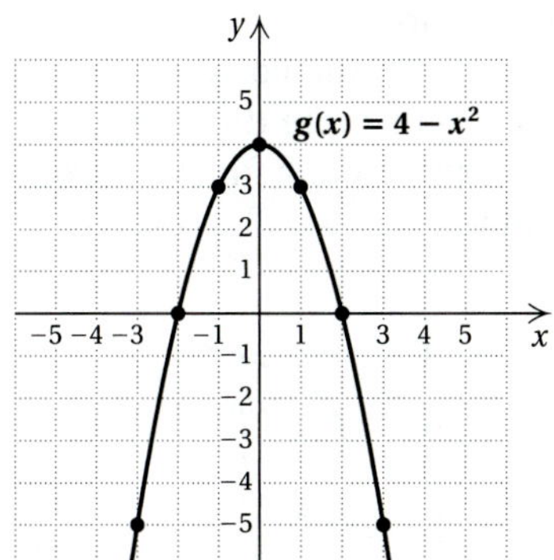

Do Exercise 10.

Graph.

9. $f(x) = x - 4$

x	$f(x)$

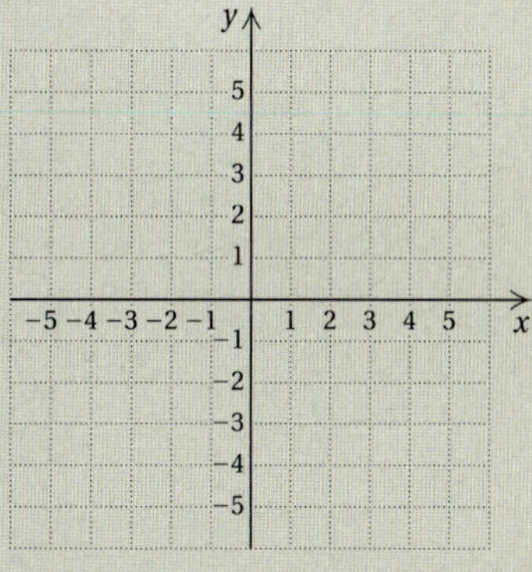

10. $g(x) = 5 - x^2$

x	$g(x)$

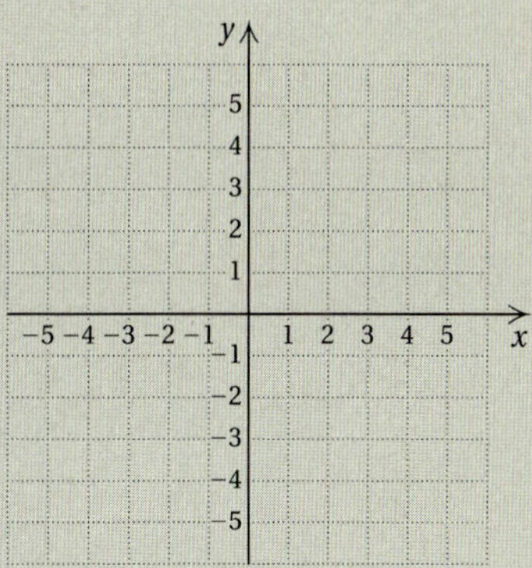

Answers on page A-42

11. $t(x) = 3 - |x|$

x	$t(x)$

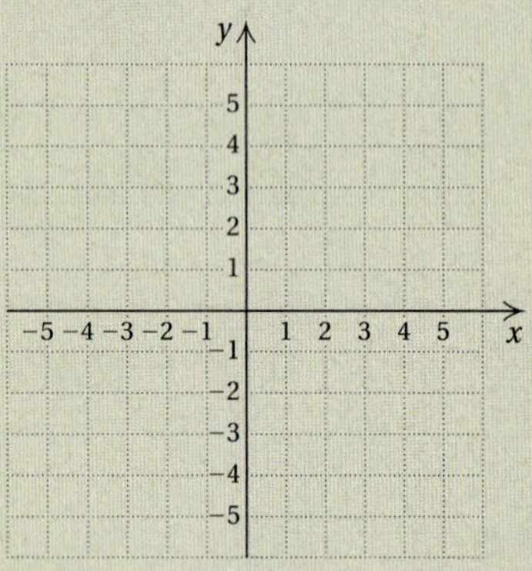

Answer on page A-42

EXAMPLE 7 Graph: $h(x) = |x|$.

A list of some function values is shown in the following table. We plot the points and connect them. The graph is a V-shaped "curve" that rises on either side of the vertical axis.

x	$h(x)$
−3	3
−2	2
−1	1
0	0
1	1
2	2
3	3

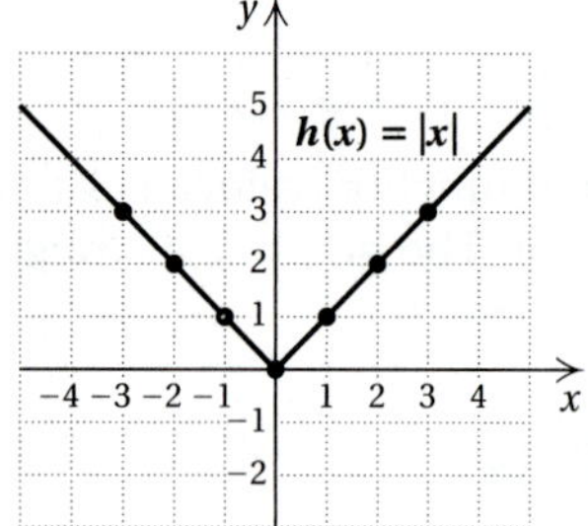

Do Exercise 11.

d The Vertical-Line Test

Consider the graph of the function f described by $f(x) = x^2 - 5$. Its graph is shown at right. It is also the graph of the equation $y = x^2 - 5$.

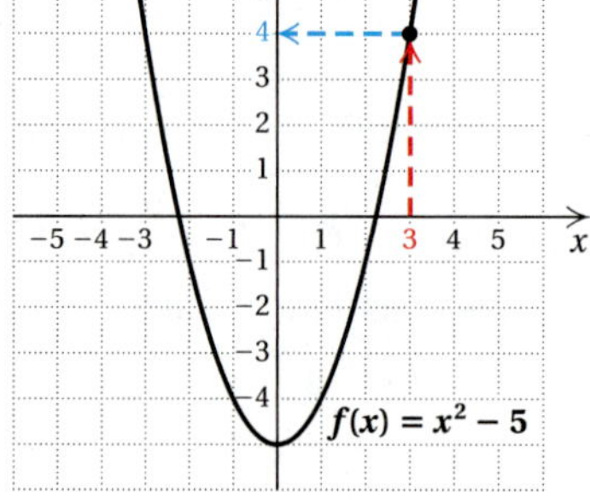

To find a function value, like $f(3)$, from a graph, we locate the input on the horizontal axis, move directly up or down to the graph of the function, and then move left or right to find the output on the vertical axis. Thus, $f(3) = 4$. Keep in mind that members of the domain are found on the horizontal axis, members of the range are found on the vertical axis, and the y on the vertical axis could also be labeled $f(x)$.

When one member of the domain is paired with two or more different members of the range, the correspondence is not a function. Thus, when a graph contains two or more different points with the same first coordinate, the graph cannot represent a function. Points sharing a common first coordinate are vertically above or below each other (see the following graph).

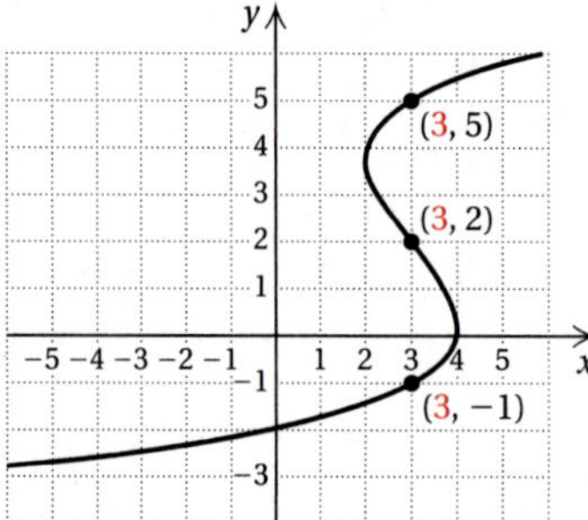

Since 3 is paired with more than one member of the range, the graph does not represent a function.

This observation leads to the *vertical-line test.*

THE VERTICAL-LINE TEST

A graph represents a function if it is impossible to draw a vertical line that intersects the graph more than once.

EXAMPLE 8 Determine whether each of the following is the graph of a function.

a) 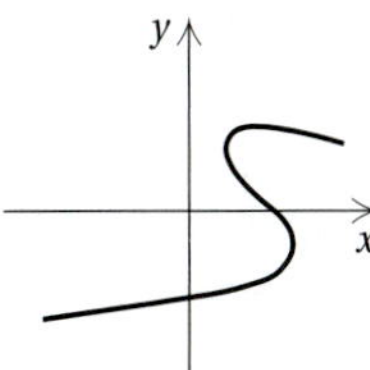

b) 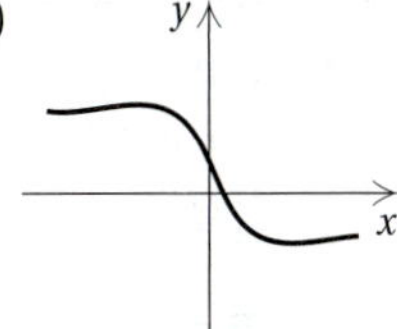

c) 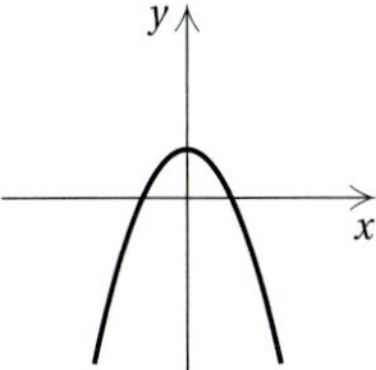

d) 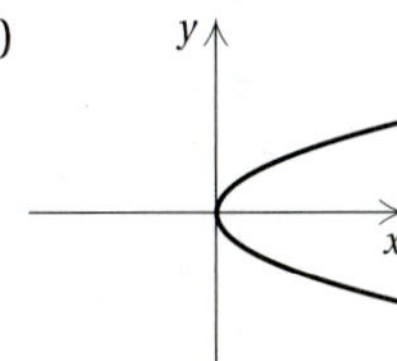

a) The graph *is not* that of a function because a vertical line can cross the graph at more than one point.

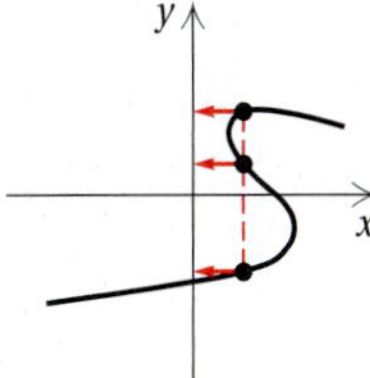

b) The graph *is* that of a function because no vertical line can cross the graph at more than one point. This can be confirmed with a ruler or straightedge.

c) The graph *is* that of a function.

d) The graph *is not* that of a function. A vertical line can cross the graph more than once.

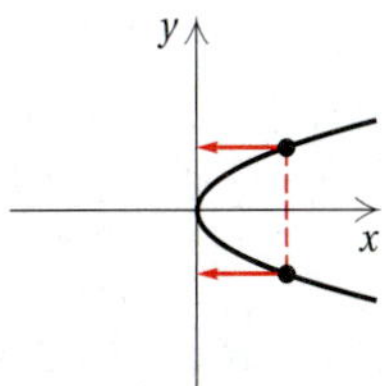

Do Exercises 12–15 on the following page.

e Applications of Functions and Their Graphs

Functions are often described by graphs, whether or not an equation is given. To use a graph in an application, we note that each point on the graph represents a pair of values.

CALCULATOR CORNER

Graphing Functions To graph a function using a graphing calculator, we replace the function notation with y and proceed as described in the Calculator Corner on p. 616. To graph $f(x) = 2x^2 + x$ in the standard window, for example, we replace $f(x)$ with y and enter $y_1 = 2x^2 + x$ on the Y = screen and then press ZOOM 6.

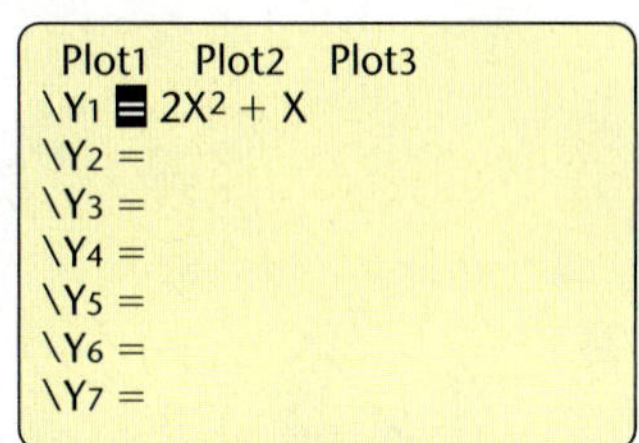

$y = 2x^2 + x$

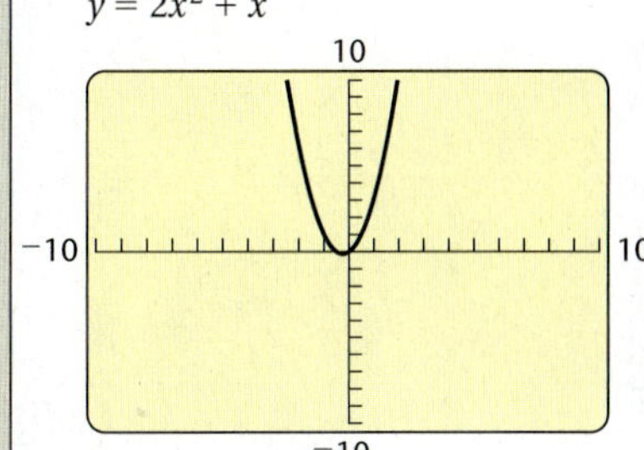

Exercises: Graph the function.

1. $f(x) = x - 4$
2. $f(x) = -2x - 3$
3. $g(x) = -3x + 4$
4. $f(x) = x^2 + 2$
5. $h(x) = 1 - x^2$
6. $f(x) = 3x^2 - 4x + 1$
7. $f(x) = x^3$
8. $f(x) = |x + 3|$
9. $g(x) = x^4 + 2$

Determine whether each of the following is the graph of a function.

12.

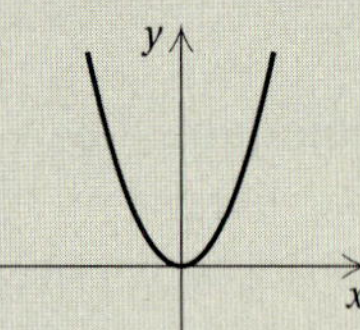

13.

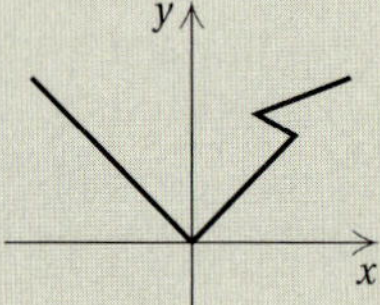

14.

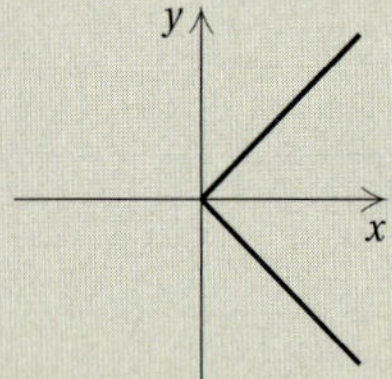

15.

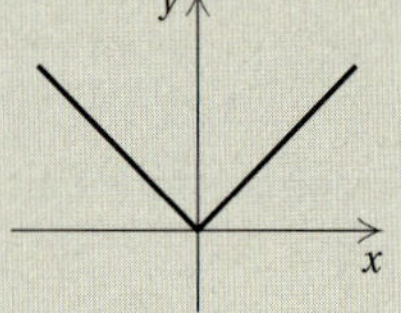

Referring to the graph in Example 9:

16. What was the movie revenue for week 3?

17. What was the movie revenue for week 6?

Answers on page A-42

EXAMPLE 9 *Movie Revenue.* The following graph approximates the weekly revenue from the movie *The Matrix*. The weekly revenue is a function f of the number of weeks x since the movie was released. No equation is given for the function.

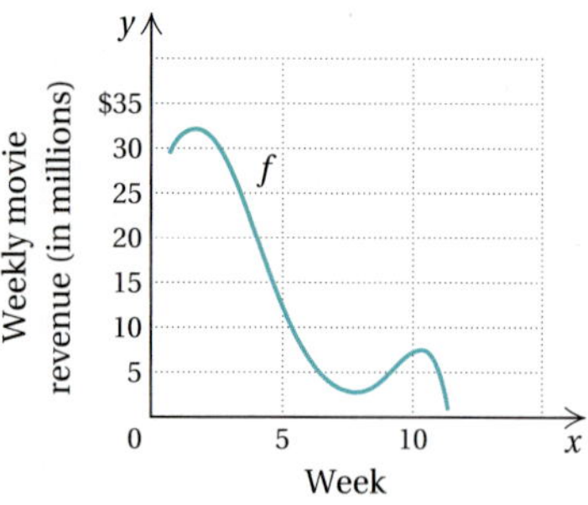

Source: Motion Picture Association of America

a) What was the movie revenue for week 2? That is, find $f(2)$.

b) What was the movie revenue for week 5? That is, find $f(5)$.

a) To estimate the weekly revenue for week 2, we locate 2 on the horizontal axis and move directly up until we reach the graph. Then we move across to the vertical axis. We estimate that value to be about \$32 million—that is, $f(2) = 32$.

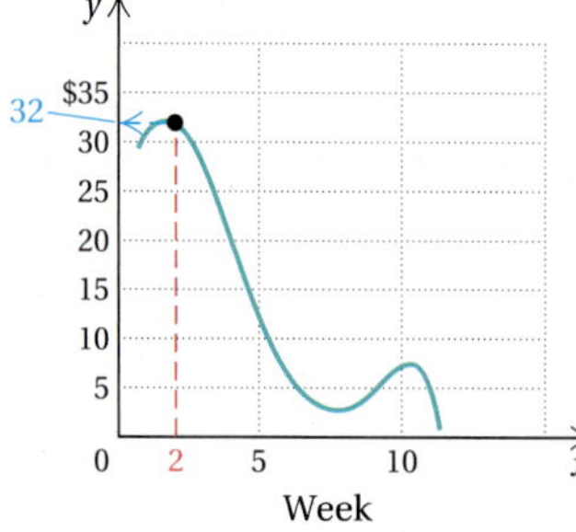

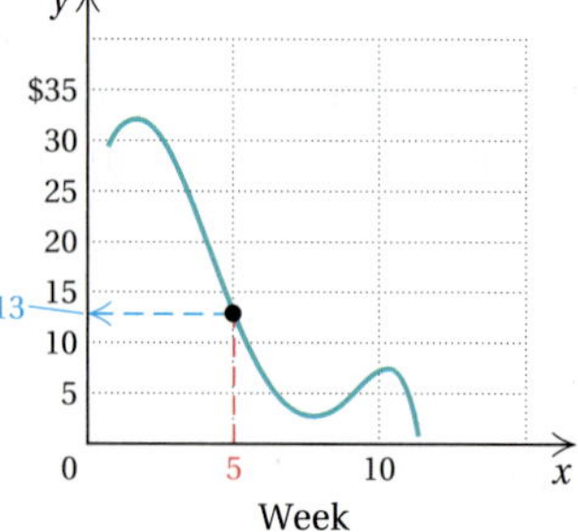

b) To estimate the weekly revenue for week 5, we locate 5 on the horizontal axis and move directly up until we reach the graph. Then we move across to the vertical axis. We estimate that value to be about \$13 million—that is, $f(5) = 13$.

Do Exercises 16 and 17.

12.1 EXERCISE SET

For Extra Help

Digital Video Tutor CD 10 Videotape 14

InterAct Math

Math Tutor Center

MathXL

MyMathLab

a Determine whether the correspondence is a function.

1. *Domain* *Range*

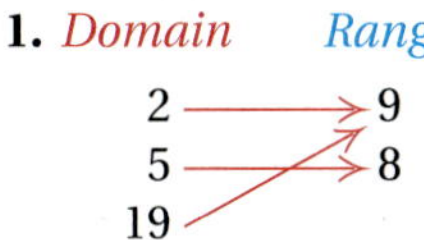

2. *Domain* *Range*

5 → 3
−3 → 7
7 → 7
−7 → 7

3. *Domain* *Range*

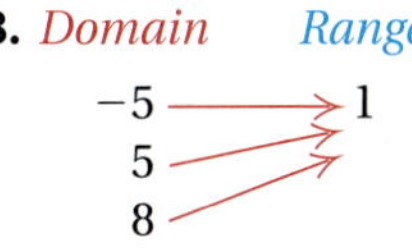

4. *Domain* *Range*

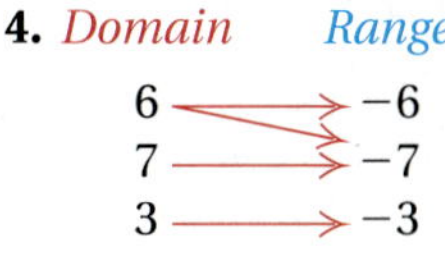

5. *Domain* *Range*

STATE	BLOOD ALCOHOL CONTENT THAT CONSTITUTES THE THRESHOLD OF LEGAL INTOXICATION
Alabama	0.08
Alaska	0.10
Arizona	0.10
Arkansas	0.10
California	0.08

Source: National Safety Council

6. *Domain* *Range*

YEAR	NUMBER OF HISPANIC COMPANIES IN THE UNITED STATES
1982	284,011
1987	489,973
1992	862,605
1997	1,121,387

Source: U.S. Small Business Administration

7. *Domain* *Range*

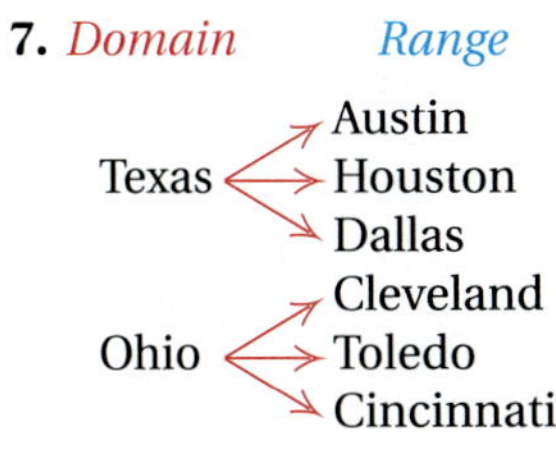

8. *Domain* *Range*

Austin → Texas
Houston → Texas
Dallas → Texas
Cleveland → Ohio
Toledo → Ohio
Cincinnati → Ohio

	Domain	*Correspondence*	*Range*
9.	A family	Each person's height, in inches	A set of positive numbers
10.	A textbook	An even-numbered page in the book	A set of pages
11.	A set of avenues	An intersecting road	A set of cross streets
12.	A math class	Each person's seat number	A set of numbers
13.	A set of numbers	Square each number and then add 4.	A set of positive numbers
14.	A set of shapes	The perimeter of each shape	A set of positive numbers

b Find the function values.

15. $f(x) = x + 5$

a) $f(4)$ **b)** $f(7)$
c) $f(-3)$ **d)** $f(0)$
e) $f(2.4)$ **f)** $f(\frac{2}{3})$

16. $g(t) = t - 6$

a) $g(0)$ **b)** $g(6)$
c) $g(13)$ **d)** $g(-1)$
e) $g(-1.08)$ **f)** $g(\frac{7}{8})$

17. $h(p) = 3p$

a) $h(-7)$ **b)** $h(5)$
c) $h(14)$ **d)** $h(0)$
e) $h(\frac{2}{3})$ **f)** $h(a + 1)$

18. $f(x) = -4x$

a) $f(6)$ **b)** $f(-\frac{1}{2})$
c) $f(a - 1)$ **d)** $f(11.8)$
e) $f(0)$ **f)** $f(-1)$

19. $g(s) = 3s + 4$

a) $g(1)$ **b)** $g(-7)$
c) $g(6.7)$ **d)** $g(0)$
e) $g(-10)$ **f)** $g(\frac{2}{3})$

20. $h(x) = 19$, a constant function

a) $h(4)$ **b)** $h(-6)$
c) $h(12.5)$ **d)** $h(0)$
e) $h(\frac{2}{3})$ **f)** $h(1234)$

21. $f(x) = 2x^2 - 3x$

a) $f(0)$ **b)** $f(-1)$
c) $f(2)$ **d)** $f(10)$
e) $f(-5)$ **f)** $f(4a)$

22. $f(x) = 3x^2 - 2x + 1$

a) $f(0)$ **b)** $f(1)$
c) $f(-1)$ **d)** $f(10)$
e) $f(2a)$ **f)** $f(-3)$

23. $f(x) = |x| + 1$

a) $f(0)$ **b)** $f(-2)$
c) $f(2)$ **d)** $f(-3)$
e) $f(-10)$ **f)** $f(a - 1)$

24. $g(t) = |t - 1|$

a) $g(4)$ **b)** $g(-2)$
c) $g(-1)$ **d)** $g(100)$
e) $g(-50)$ **f)** $g(a + 1)$

25. $f(x) = x^3$

a) $f(0)$ **b)** $f(-1)$
c) $f(2)$ **d)** $f(10)$
e) $f(-5)$ **f)** $f(-10)$

26. $f(x) = x^4 - 3$

a) $f(1)$ **b)** $f(-1)$
c) $f(0)$ **d)** $f(2)$
e) $f(-2)$ **f)** $f(10)$

27. *Archaeology.* The function H described by

$$H(x) = 2.75x + 71.48$$

can be used to determine the height, in centimeters, of a woman whose *humerus* (the bone from the elbow to the shoulder) is x cm long. If a humerus is known to be from a female, how tall was she if the bone is **(a)** 32 cm long? **(b)** 35 cm long?

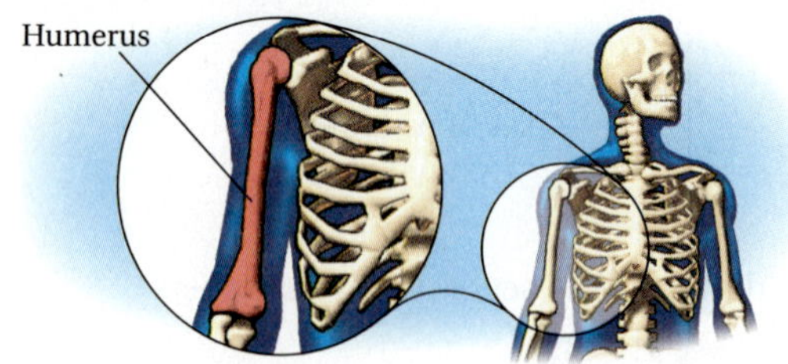

28. Refer to Exercise 27. When a humerus is from a male, the function

$$M(x) = 2.89x + 70.64$$

can be used to find the male's height, in centimeters. If a humerus is known to be from a male, how tall was the male if the bone is **(a)** 30 cm long? **(b)** 35 cm long?

29. *Pressure at Sea Depth.* The function $P(d) = 1 + (d/33)$ gives the pressure, in *atmospheres* (atm), at a depth of d feet in the sea. Note that $P(0) = 1$ atm, $P(33) = 2$ atm, and so on. Find the pressure at 20 ft, 30 ft, and 100 ft.

30. *Temperature as a Function of Depth.* The function $T(d) = 10d + 20$ gives the temperature, in degrees Celsius, inside the earth as a function of the depth d, in kilometers. Find the temperature at 5 km, 20 km, and 1000 km.

31. *Melting Snow.* The function $W(d) = 0.112d$ approximates the amount, in centimeters, of water that results from d centimeters of snow melting. Find the amount of water that results from snow melting from depths of 16 cm, 25 cm, and 100 cm.

32. *Temperature Conversions.* The function $C(F) = \frac{5}{9}(F - 32)$ determines the Celsius temperature that corresponds to F degrees Fahrenheit. Find the Celsius temperature that corresponds to 62°F, 77°F, and 23°F.

C Graph the function.

33. $f(x) = 3x - 1$

x	$f(x)$
−1	
0	
1	
2	

34. $g(x) = 2x + 5$

x	$g(x)$
−3	
−2	
−1	
0	

35. $g(x) = -2x + 3$

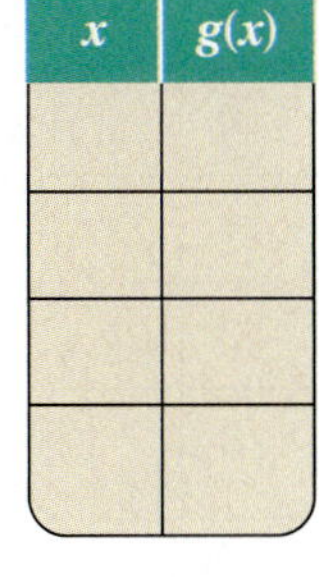

x	$g(x)$

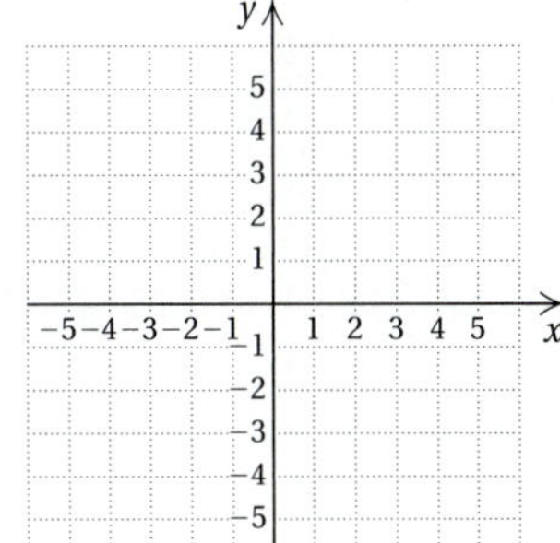

36. $f(x) = -\frac{1}{2}x + 2$

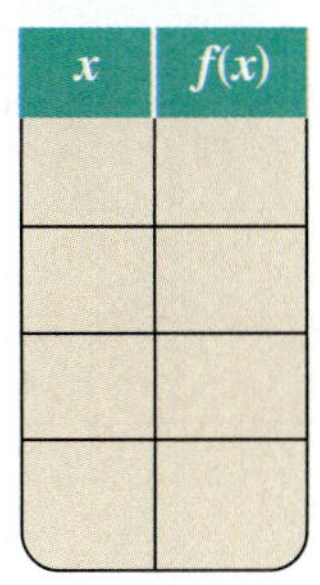

x	$f(x)$

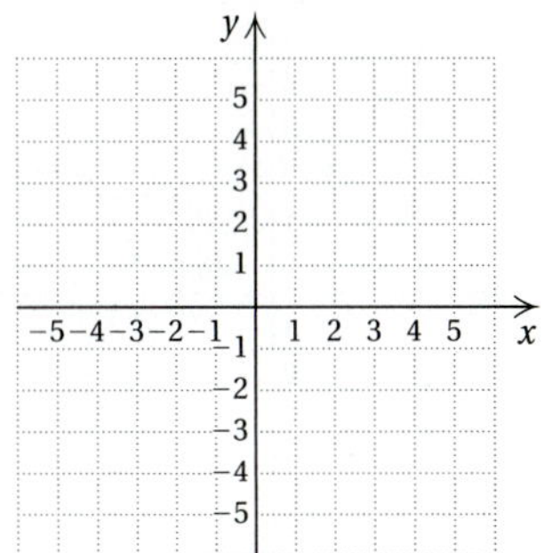

37. $f(x) = \frac{1}{2}x + 1$

x	$f(x)$

38. $f(x) = -\frac{3}{4}x - 2$

x	$f(x)$

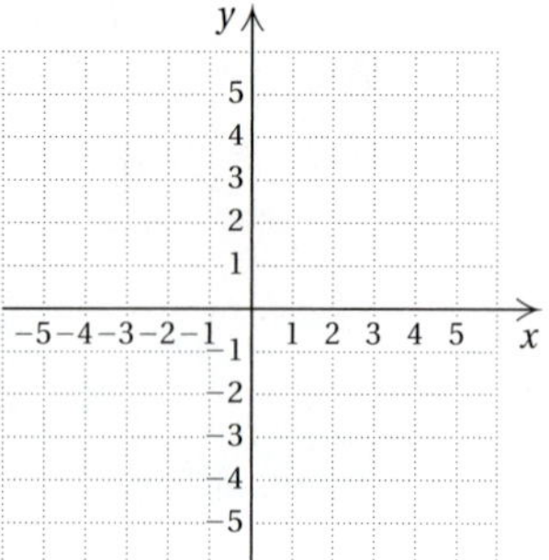

39. $f(x) = 2 - |x|$

x	$f(x)$
−3	
−2	
−1	
0	
1	
2	
3	

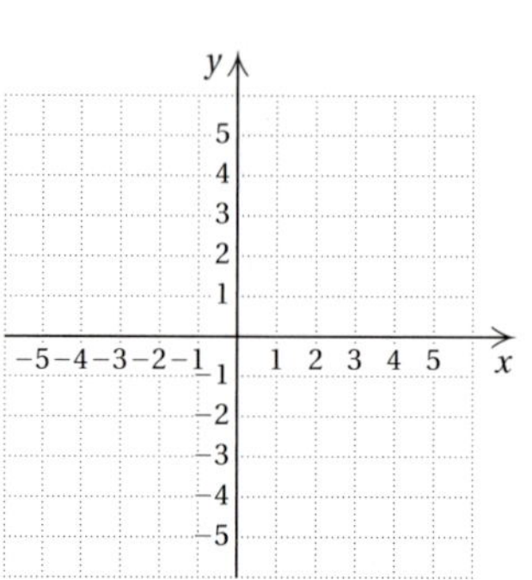

40. $f(x) = |x| - 4$

x	$f(x)$
−3	
−2	
−1	
0	
1	
2	
3	

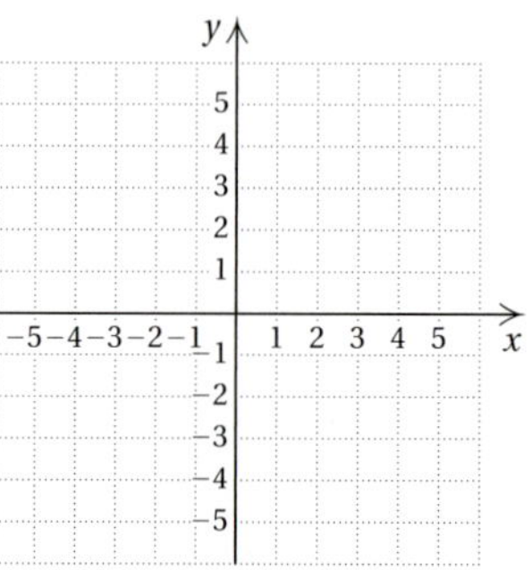

41. $f(x) = x^2$

x	$f(x)$
−3	
−2	
−1	
0	
1	
2	
3	

42. $f(x) = x^2 - 1$

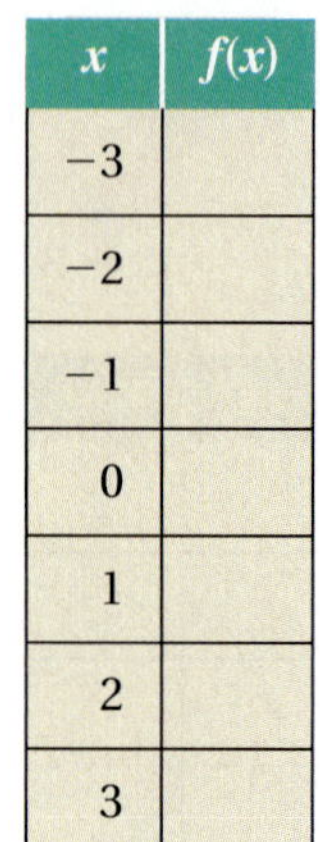

x	$f(x)$
−3	
−2	
−1	
0	
1	
2	
3	

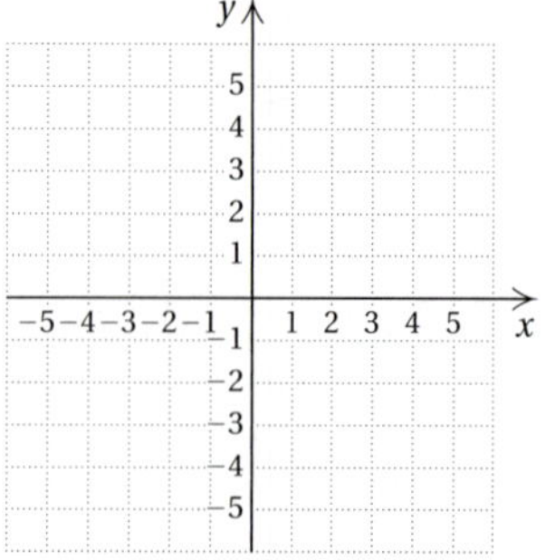

43. $f(x) = x^2 - x - 2$

x	$f(x)$
−3	
−2	
−1	
0	
1	
2	
3	

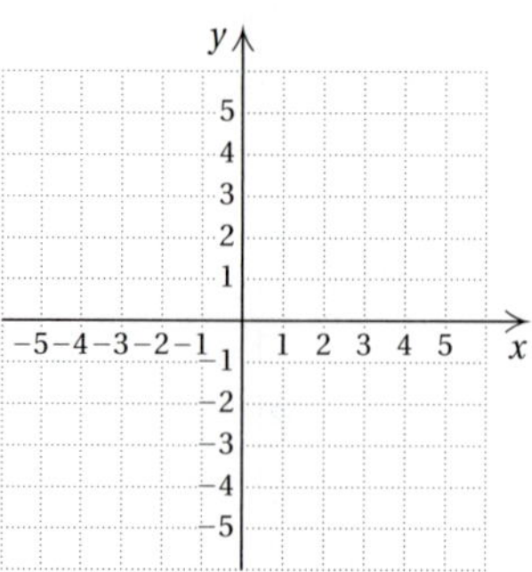

44. $f(x) = x^2 + 6x + 5$

x	$f(x)$
−5	
−4	
−3	
−2	
−1	
0	
1	

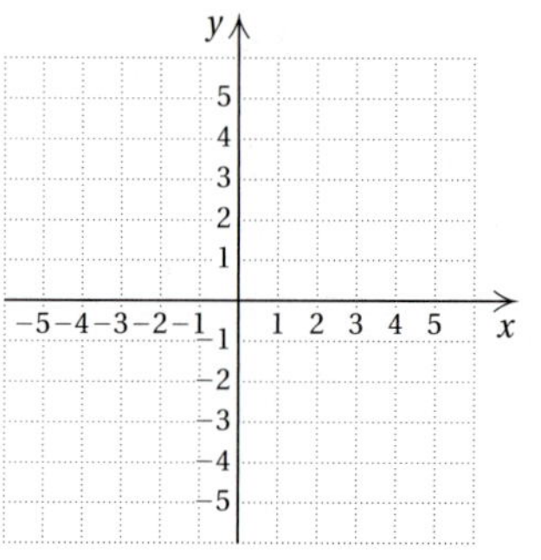

d Determine whether each of the following is the graph of a function.

45.

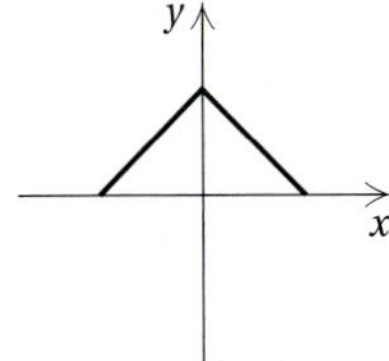

46.

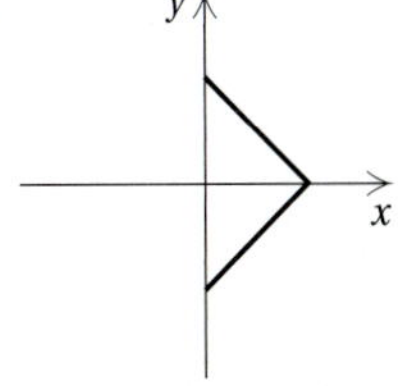

47.

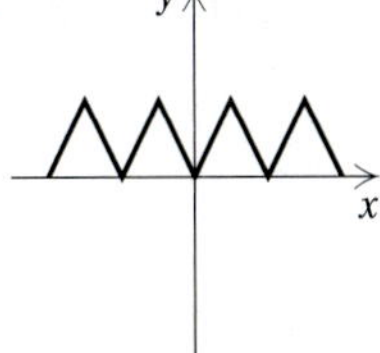

48.

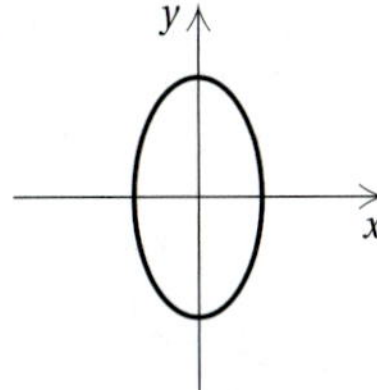

49.

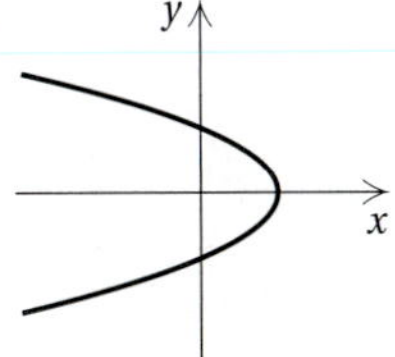

50.

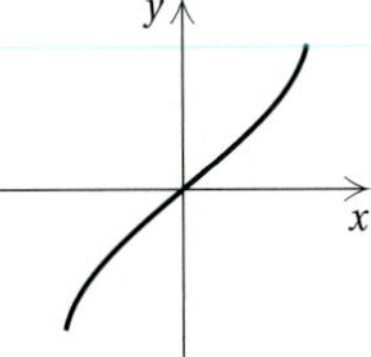

51.

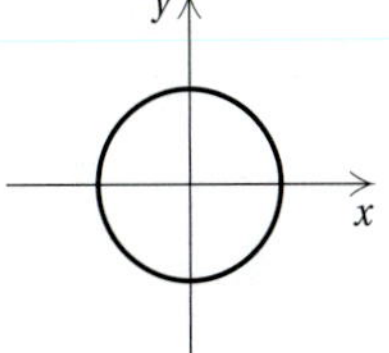

52.

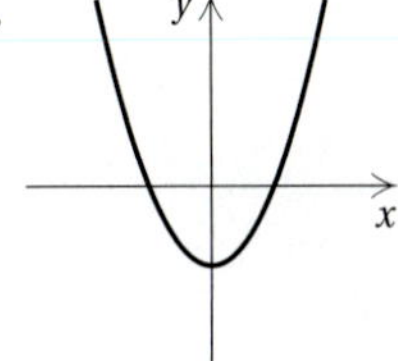

e

Cholesterol Level and Risk of a Heart Attack. The following graph shows the annual heart attack rate per 10,000 men as a function of blood cholesterol level.

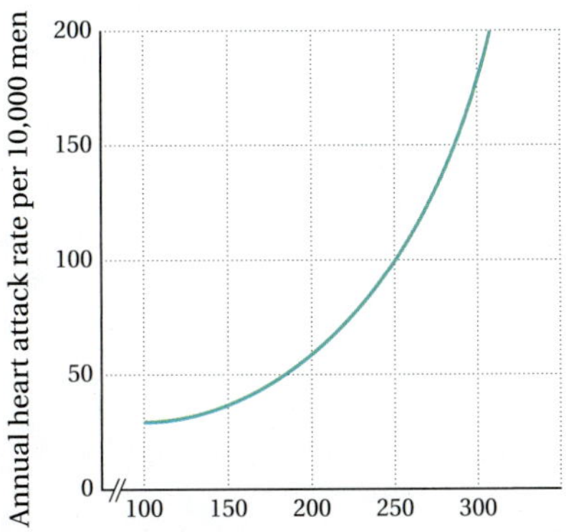

Source: Copyright 1989, CSPI. Adapted from *Nutrition Action Healthletter* (1875 Connecticut Avenue, N.W., Suite 300, Washington, DC 20009-5728)

53. Approximate the annual heart attack rate per 10,000 men for those whose blood cholesterol level is 225 mg/dl.

54. Approximate the annual heart attack rate per 10,000 men for those whose blood cholesterol level is 275 mg/dl.

Movie Revenue. The following graph approximates the weekly revenue, in millions of dollars, from the movie *Pearl Harbor*. The weekly revenue is a function f of the number of weeks x since the movie was released. No equation is given for the function.

Source: Motion Picture Association of America

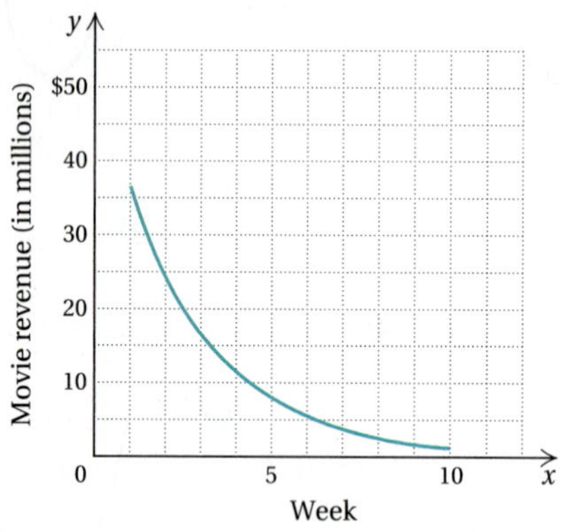

55. What was the movie revenue for week 2? That is, find $f(2)$.

56. What was the movie revenue for week 5? That is, find $f(5)$.

Love Theories. Researchers at Yale University have suggested that the following graphs may represent three different aspects of love.

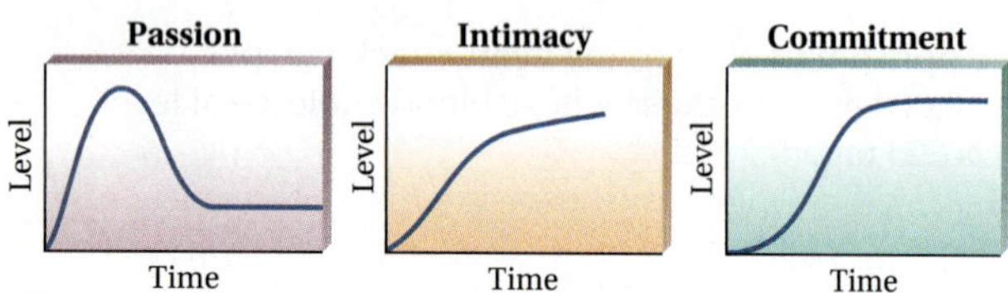

Source: A Triangular Theory of Love, by R. J. Sternberg, 1986, *Psychological Review*, 93(2), 119–135. Copyright 1986 by the American Psychological Association, Inc. Reprinted by permission.

57. D$_W$ In what unit would you measure time if the horizontal length of each graph were 10 units? Why?

58. D$_W$ Do you agree with the researchers that these graphs should be shaped as they are? Why or why not?

59. D$_W$ Is it possible for a function to have more numbers as outputs than as inputs? Why or why not?

60. D$_W$ Look up the word "function" in a dictionary. Explain how that definition might be related to the mathematical one given in this section.

SKILL MAINTENANCE

Solve.

61. *Consecutive Even Integers.* Find three consecutive even integers such that the sum of the first, twice the second, and three times the third is 124. [7.6a]

62. *Wire Cutting.* A piece of wire 32.8 ft long is to be cut into two pieces and each of those pieces is to be bent into a square. The length of a side of one square is to be 2.2 ft longer than the length of a side of the other square. How should the wire be cut? [7.6a]

63. The surface area of a rectangular solid of length l, width w, and height h is given by

$$S = 2lh + 2lw + 2wh.$$

Solve for l. [7.4b]

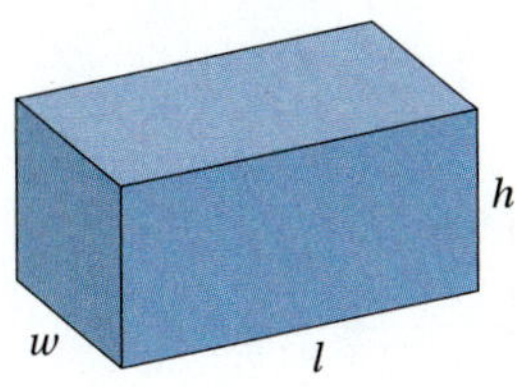

64. Solve the formula in Exercise 63 for w. [7.4b]

Convert to decimal notation. [9.2c]

65. 9.3×10^{-9}

66. 3.04×10^{6}

Convert to scientific notation. [9.2c]

67. 1,075,000,000

68. 0.00703

Factor. [10.6a]

69. $w^3 + \frac{1}{27}$

70. $a^3 - b^3$

71. $xy^3 + 64x$

72. $c^3 - 0.008$

SYNTHESIS

For Exercises 73 and 74, let $f(x) = 3x^2 - 1$ and $g(x) = 2x + 5$.

73. Find $f(g(-4))$ and $g(f(-4))$.

74. Find $f(g(-1))$ and $g(f(-1))$.

75. Suppose that a function g is such that $g(-1) = -7$ and $g(3) = 8$. Find a formula for g if $g(x)$ is of the form $g(x) = mx + b$, where m and b are constants.

Objective

a Find the domain and the range of a function.

1. Find the domain and the range of the function f whose graph is shown below.

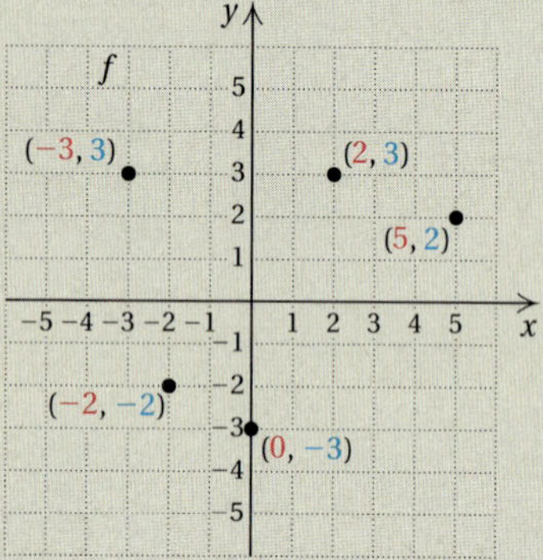

Answer on page A-43

Study Tips

HOMEWORK TIPS

Prepare for your homework assignment by reading the explanations of concepts and following the step-by-step solutions of examples in the text. The time you spend preparing will save valuable time when you do your assignment.

"I will prepare and some day my chance will come."

Abraham Lincoln

12.2 FINDING DOMAIN AND RANGE

a Finding Domain and Range

The solutions of an equation in two variables consist of a set of ordered pairs. An arbitrary set of ordered pairs is called a **relation.** When a set of ordered pairs is such that no two different pairs share a common first coordinate, we have a **function.** The **domain** is the set of all first coordinates and the **range** is the set of all second coordinates.

EXAMPLE 1 Find the domain and the range of the function f whose graph is shown below.

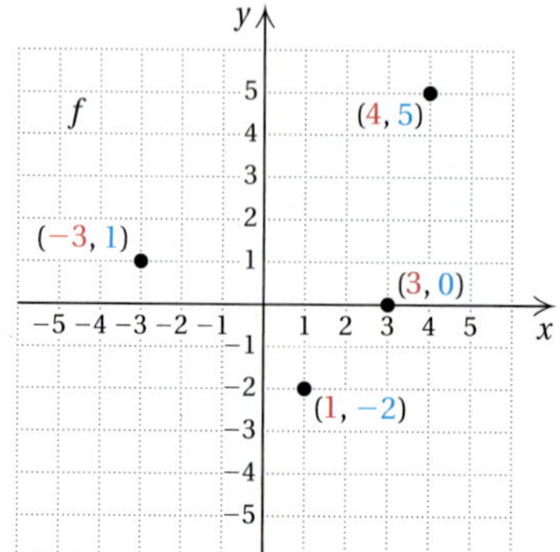

This function contains just four ordered pairs and it can be written as

$$\{(-3, 1), (1, -2), (3, 0), (4, 5)\}.$$

We can determine the domain and the range by reading the x- and y-values directly from the graph.

The domain is the set of all first coordinates, $\{-3, 1, 3, 4\}$. The range is the set of all second coordinates, $\{1, -2, 0, 5\}$.

Do Exercise 1.

EXAMPLE 2 For the function f whose graph is shown below, determine each of the following.

a) The number in the range that is paired with 1 (from the domain). That is, find $f(1)$.

b) The domain of f

c) The numbers in the domain that are paired with 1 (from the range). That is, find all x such that $f(x) = 1$.

d) The range of f

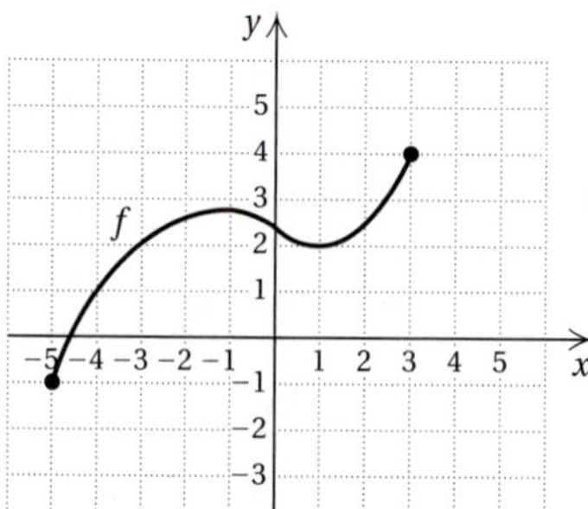

a) To determine which number in the range is paired with 1 in the domain, we locate 1 on the horizontal axis. Next, we find the point on the graph of f for which 1 is the first coordinate. From that point, we can look to the vertical axis to find the corresponding y-coordinate, 2. The input 1 has the output 2—that is, $f(1) = 2$.

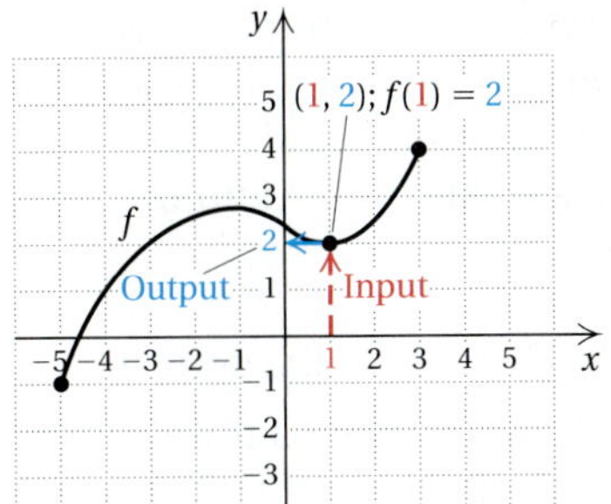

b) The domain of the function is the set of all x-values, or inputs, of the points on the graph. These extend from -5 to 3 and can be viewed as the curve's shadow, or projection, onto the x-axis. Thus the domain is the set $\{x | -5 \le x \le 3\}$.

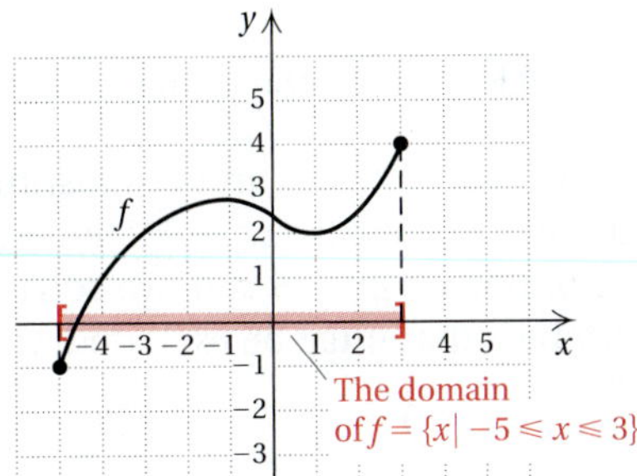

c) To determine which numbers in the domain are paired with 1 in the range, we locate 1 on the vertical axis. From there, we look left and right to the graph of f to find any points for which 1 is the second coordinate (output). One such point exists, $(-4, 1)$. For this function, we note that $x = -4$ is the only member of the domain paired with 1. For other functions, there might be more than one member of the domain paired with a member of the range.

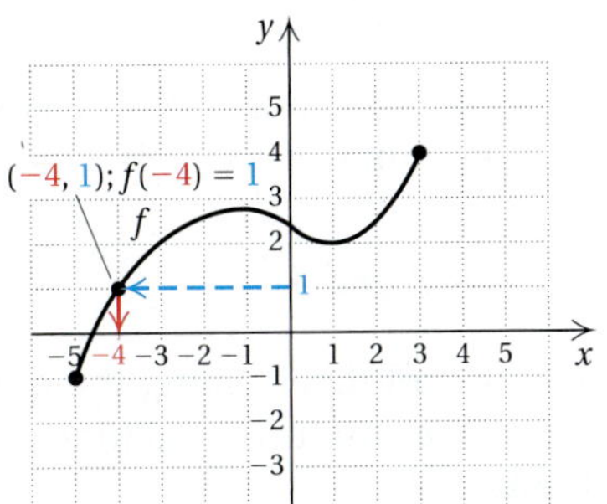

d) The range of the function is the set of all y-values, or outputs, of the points on the graph. These extend from -1 to 4 and can be viewed as the curve's shadow, or projection, onto the y-axis. Thus the range is the set $\{y | -1 \le y \le 4\}$.

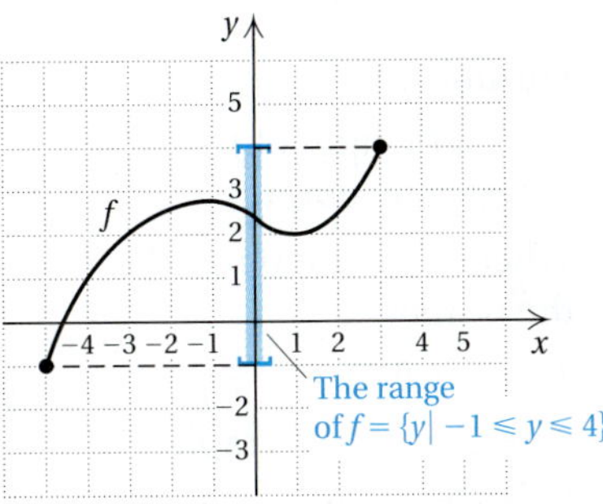

Do Exercise 2.

2. For the function f whose graph is shown below, determine each of the following.

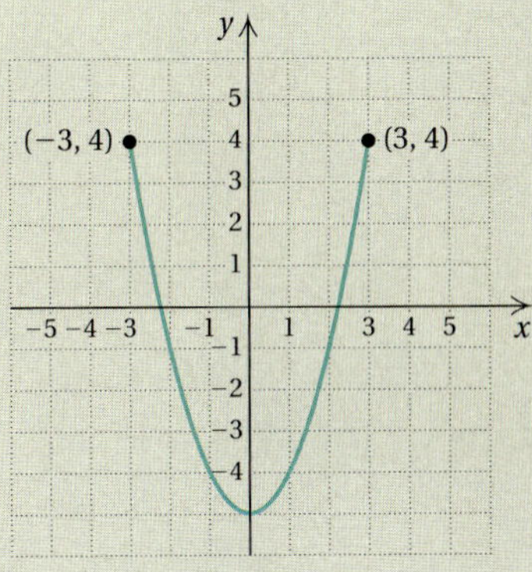

a) The number in the range that is paired with the input 1. That is, find $f(1)$.

b) The domain of f

c) The numbers in the domain that are paired with 4

d) The range of f

Answers on page A-43

Find the domain.

3. $f(x) = x^3 - |x|$

4. $f(x) = \dfrac{4}{3x + 2}$

Answers on page A-43

When a function is given by an equation or formula, the domain is understood to be the largest set of real numbers (inputs) for which function values (outputs) can be calculated. That is, the domain is the set of all possible allowable inputs into the formula. To find the domain, think, "What can we substitute?"

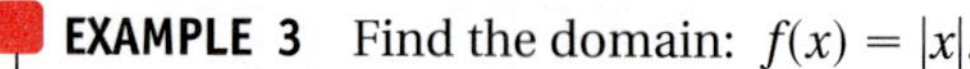

EXAMPLE 3 Find the domain: $f(x) = |x|$.

We ask, "What can we substitute?" Is there any number x for which we cannot calculate $|x|$? The answer is no. Thus the domain of f is the set of all real numbers.

EXAMPLE 4 Find the domain: $f(x) = \dfrac{3}{2x - 5}$.

We ask, "What can we substitute?" Is there any number x for which we cannot calculate $3/(2x - 5)$? Since $3/(2x - 5)$ cannot be calculated when the denominator $2x - 5$ is 0, we solve the following equation to find those real numbers that must be excluded from the domain of f:

$2x - 5 = 0$	Setting the denominator equal to 0
$2x = 5$	Adding 5
$x = \frac{5}{2}$.	Dividing by 2

Thus $\frac{5}{2}$ is not in the domain, whereas all other real numbers are.

The domain of f is $\{x \mid x \text{ is a real number } and\ x \neq \frac{5}{2}\}$.

Do Exercises 3 and 4.

The task of determining the domain and the range of a function is one that we will return to several times as we consider other types of functions in this book.

FUNCTIONS: A REVIEW

The following is a review of the function concepts considered in Sections 12.1 and 12.2.

Function Concepts

- Formula for f: $f(x) = x^2 - 7$
- For every input of f, there is exactly one output.
- When 1 is the input, -6 is the output.
- $f(1) = -6$
- $(1, -6)$ is on the graph.
- Domain = The set of all inputs
  = The set of all real numbers
- Range = The set of all outputs
  $= \{y \mid y \geq -7\}$

Graph

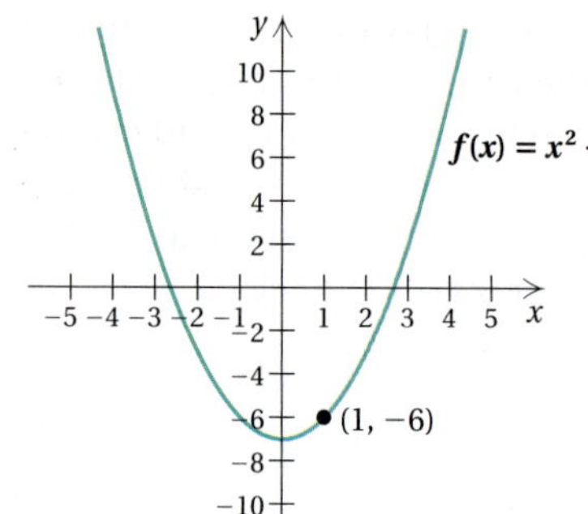

For Extra Help

Digital Video Tutor CD 10 Videotape 14 | InterAct Math | Math Tutor Center | MathXL | MyMathLab

a In Exercises 1–12, the graph is that of a function. Determine for each one **(a)** $f(1)$; **(b)** the domain; **(c)** all x-values such that $f(x) = 2$; and **(d)** the range. An open dot indicates that the point does not belong to the graph.

1.
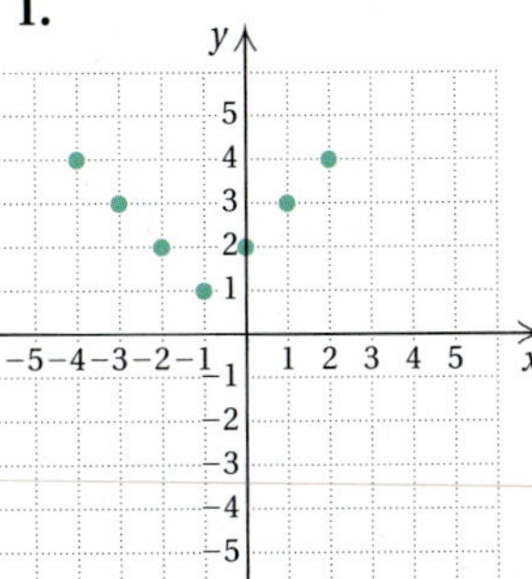

2.
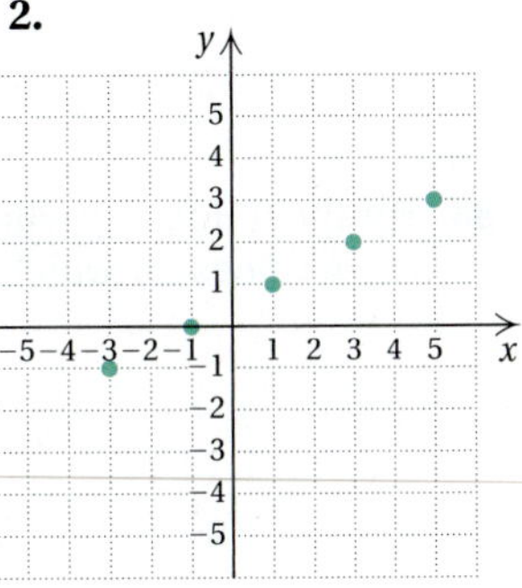

3.
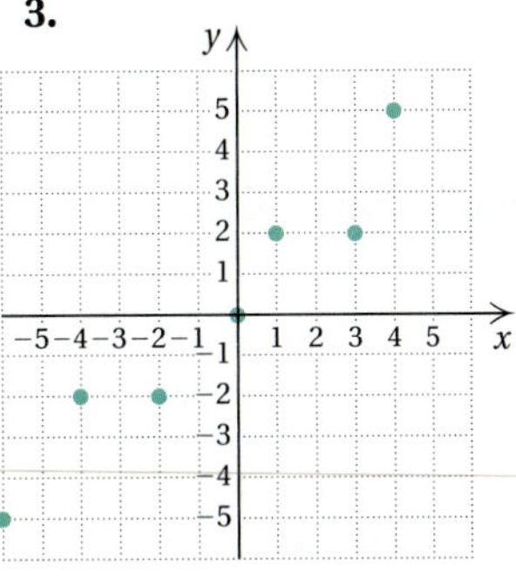

4.
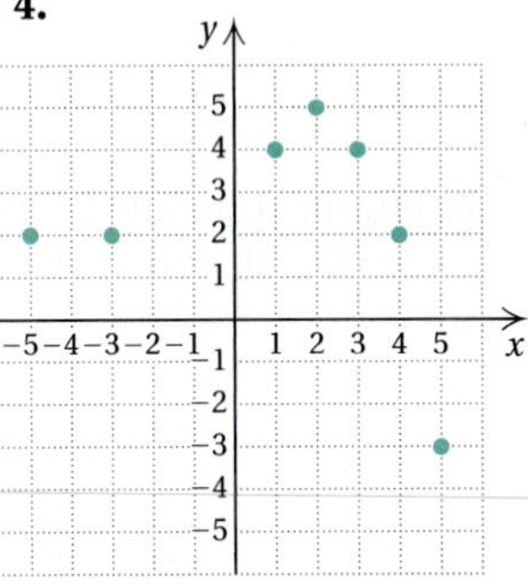

5.
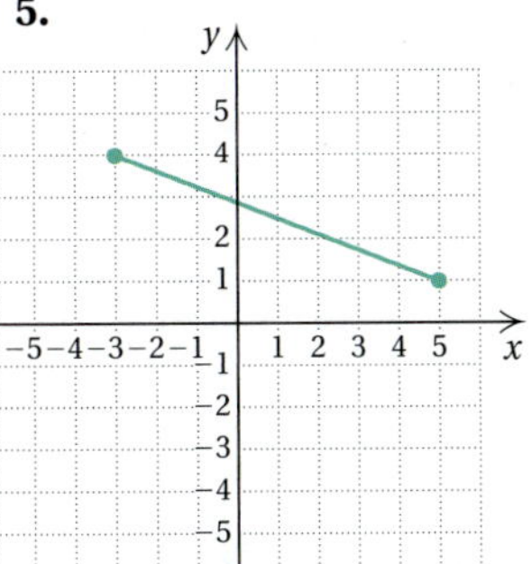

6.
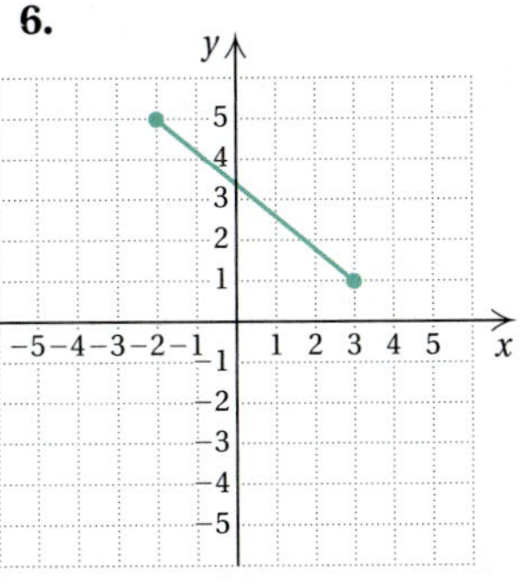

7.
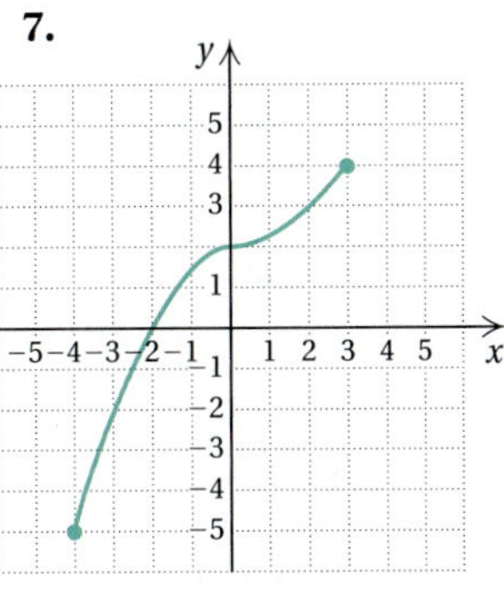

8.
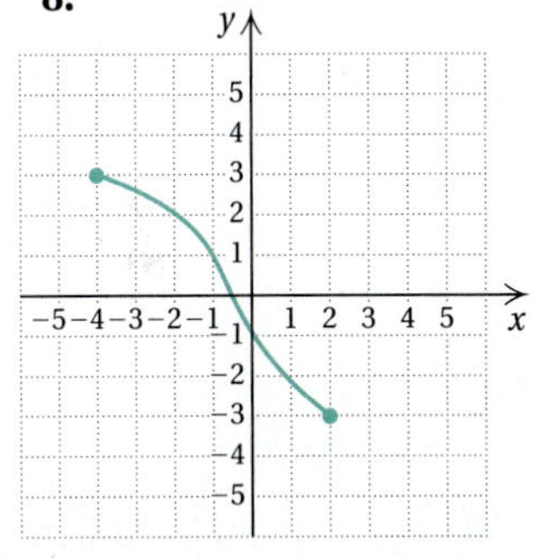

9.
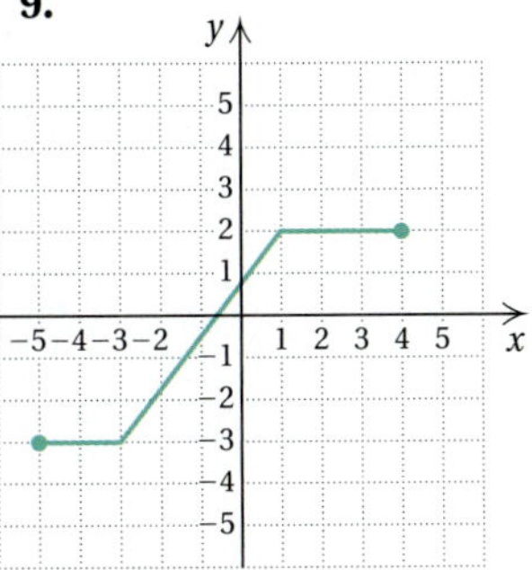

10.
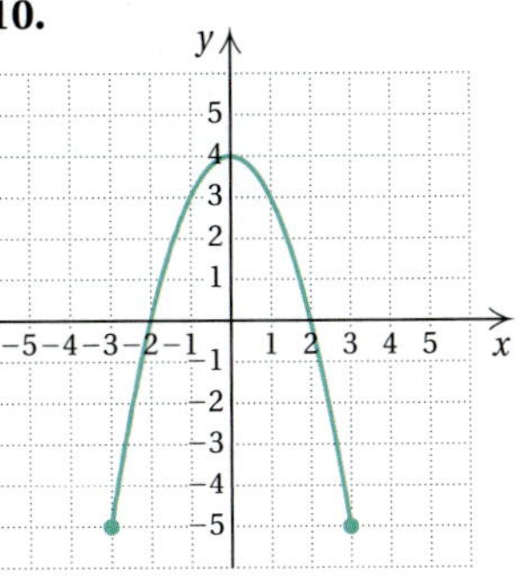

11.
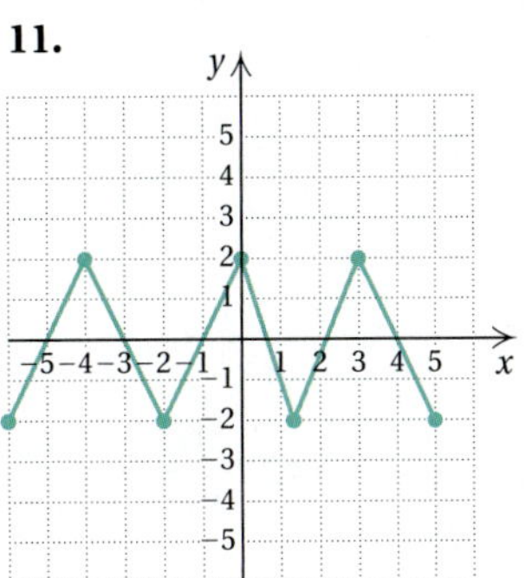

12.
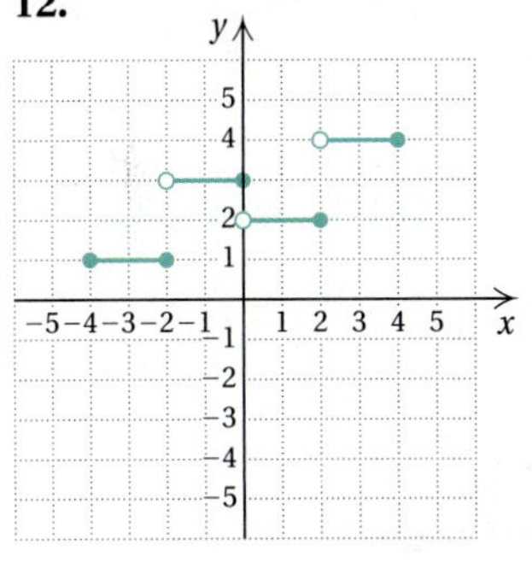

Find the domain.

13. $f(x) = \dfrac{2}{x+3}$

14. $f(x) = \dfrac{7}{5-x}$

15. $f(x) = 2x + 1$

16. $f(x) = 4 - 5x$

17. $f(x) = x^2 + 3$

18. $f(x) = x^2 - 2x + 3$

19. $f(x) = \dfrac{8}{5x - 14}$

20. $f(x) = \dfrac{x-2}{3x+4}$

21. $f(x) = |x| - 4$

22. $f(x) = |x - 4|$

23. $f(x) = \dfrac{4}{|2x - 3|}$

24. $f(x) = \dfrac{x^2 - 3x}{|4x - 7|}$

25. $g(x) = \dfrac{1}{x-1}$

26. $g(x) = \dfrac{-11}{4+x}$

27. $g(x) = x^2 - 2x + 1$

28. $g(x) = 8 - x^2$

29. $g(x) = x^3 - 1$

30. $g(x) = 4x^3 + 5x^2 - 2x$

31. $g(x) = \dfrac{7}{20 - 8x}$

32. $g(x) = \dfrac{2x - 3}{6x - 12}$

33. $g(x) = |x + 7|$

34. $g(x) = |x| + 1$

35. $g(x) = \dfrac{-2}{|4x + 5|}$

36. $g(x) = \dfrac{x^2 + 2x}{|10x - 20|}$

37. For the function f whose graph is shown below, find $f(-1)$, $f(0)$, and $f(1)$.

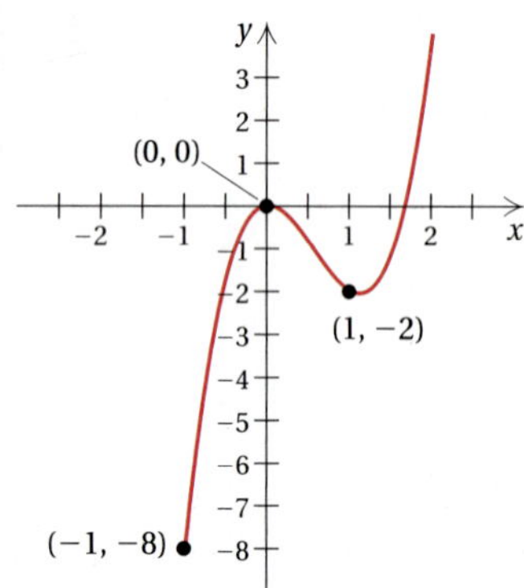

38. For the function g whose graph is shown below, find all the x-values for which $g(x) = 1$.

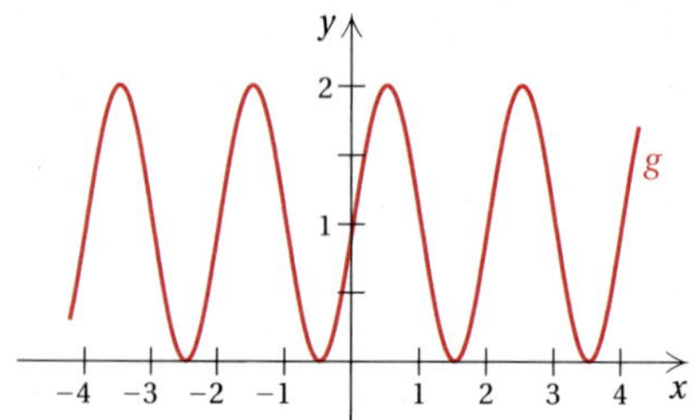

39. DW Explain the difference between the domain and the range of a function.

40. DW For a given function f, it is known that $f(2) = -3$. Give as many interpretations of this fact as you can.

SKILL MAINTENANCE

Simplify. [11.1c]

41. $\dfrac{a^2 - 1}{a + 1}$

42. $\dfrac{10y^2 + 10y - 20}{35y^2 + 210y - 245}$

43. $\dfrac{5x - 15}{x^2 - x - 6}$

Divide. [9.8b, c]

44. $(t^2 + 3t - 28) \div (t + 7)$

45. $(w^2 + 4w + 5) \div (w + 3)$

46. $(x^6 + x^5 - 3x^4 + x + 5) \div (x^2 - 1)$

Multiply. [9.6d]

47. $(7x - 3)(2x + 9)$

48. $(a - 1)(a + 1)$

49. $(9y + 10)^2$

50. $\left(2w - \frac{1}{2}\right)\left(4w + \frac{1}{2}\right)$

SYNTHESIS

51. Determine the range of each of the functions in Exercises 13, 18, 21, and 22.

52. Determine the range of each of the functions in Exercises 26, 27, 28, and 34.

12.3 LINEAR FUNCTIONS: GRAPHS AND SLOPE

Objectives

a Find the y-intercept of a line from the equation $y = mx + b$ or $f(x) = mx + b$.

b Given two points on a line, find the slope; given a linear equation, derive the equivalent slope–intercept equation and determine the slope and the y-intercept.

c Solve applied problems involving slope.

We now turn our attention to functions whose graphs are straight lines. Such functions are called **linear** and can be written in the form $f(x) = mx + b$.

LINEAR FUNCTION

A **linear function** f is any function that can be described by $f(x) = mx + b$.

Compare the two equations $7y + 2x = 11$ and $y = 3x + 5$. Both are linear equations because their graphs are straight lines. Each can be expressed in an equivalent form that is a linear function.

The equation $y = 3x + 5$ can be expressed as $f(x) = mx + b$, where $m = 3$ and $b = 5$.

The equation $7y + 2x = 11$ also has an equivalent form $f(x) = mx + b$. To see this, we solve for y:

$$7y + 2x = 11$$

$$7y + 2x - 2x = -2x + 11 \qquad \text{Adding } -2x$$

$$7y = -2x + 11$$

$$\frac{7y}{7} = \frac{-2x + 11}{7} \qquad \text{Dividing by 7}$$

$$y = -\frac{2}{7}x + \frac{11}{7}. \qquad \text{Simplifying}$$

(It might be helpful to review the discussion on solving formulas in Section 7.4.) We now have an equivalent equation in the form

$$f(x) = -\frac{2}{7}x + \frac{11}{7}, \quad \text{where} \quad m = -\frac{2}{7} \quad \text{and} \quad b = \frac{11}{7}.$$

In this section, we consider the effects of the constants m and b on the graphs of linear functions.

a The Constant b: The y-Intercept

Let's first explore the effect of the constant b.

EXAMPLE 1 Graph $y = 2x$ and $y = 2x + 3$ using the same set of axes. Compare the graphs.

We first make a table of solutions of both equations.

	y	y
x	$y = 2x$	$y = 2x + 3$
0	0	3
1	2	5
−1	−2	1
2	4	7
−2	−4	−1

CALCULATOR CORNER

Exploring b We can use a graphing calculator to explore the effect of the constant b on the graph of a function of the form $f(x) = mx + b$. We graph $y_1 = x$ in the standard $[-10, 10, -10, 10]$ viewing window. Then we graph $y_2 = x + 4$, followed by $y_3 = x - 3$, in the same viewing window.

1. Compare the graph of y_2 with the graph of y_1.
2. Compare the graph of y_3 with the graph of y_1.
3. Visualize the graph of $y = x + 8$. Compare it with the graph of y_1.
4. Visualize the graph of $y = x - 5$. Compare it with the graph of y_1.

1. Graph $y = 3x$ and $y = 3x - 6$ using the same set of axes. Compare the graphs.

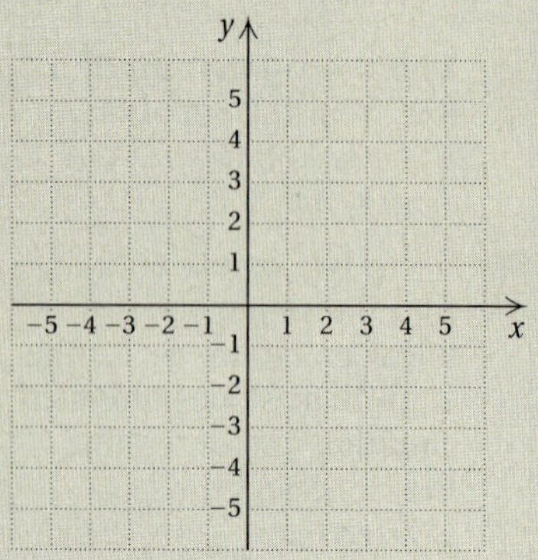

2. Graph $y = -2x$ and $y = -2x + 3$ using the same set of axes. Compare the graphs.

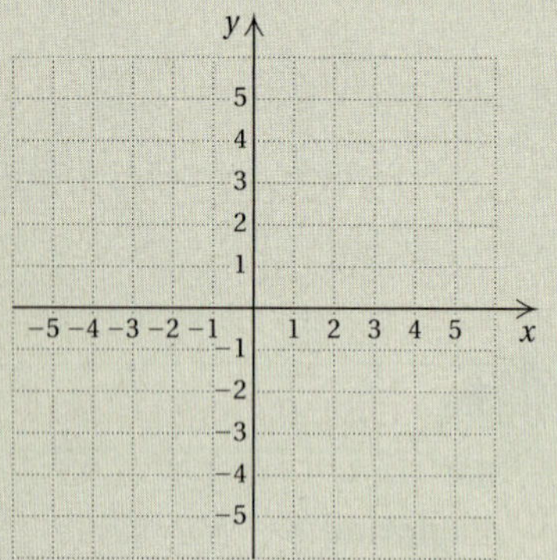

3. Graph $f(x) = \frac{1}{3}x$ and $g(x) = \frac{1}{3}x + 2$ using the same set of axes. Compare the graphs.

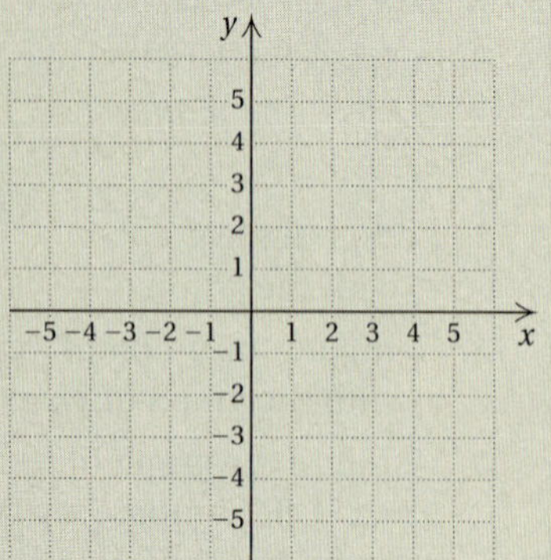

Answers on page A-43

Next, we plot these points. Drawing a red line for $y = 2x$ and a blue line for $y = 2x + 3$, we note that the graph of $y = 2x + 3$ is simply the graph of $y = 2x$ shifted, or *translated*, up 3 units. The lines are parallel.

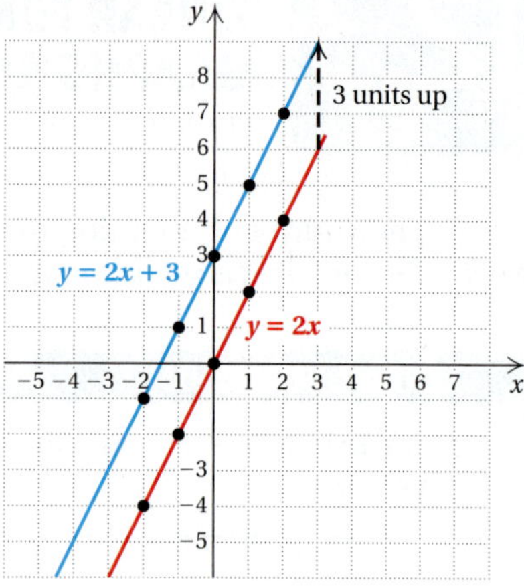

Do Exercises 1 and 2.

EXAMPLE 2 Graph $f(x) = \frac{1}{3}x$ and $g(x) = \frac{1}{3}x - 2$ using the same set of axes. Compare the graphs.

We first make a table of solutions of both equations. By choosing multiples of 3, we can avoid fractions.

	$f(x)$	$g(x)$
x	$f(x) = \frac{1}{3}x$	$g(x) = \frac{1}{3}x - 2$
0	0	−2
3	1	−1
−3	−1	−3
6	2	0

We then plot these points. Drawing a red line for $f(x) = \frac{1}{3}x$ and a blue line for $g(x) = \frac{1}{3}x - 2$, we see that the graph of $g(x) = \frac{1}{3}x - 2$ is simply the graph of $f(x) = \frac{1}{3}x$ shifted, or translated, down 2 units.

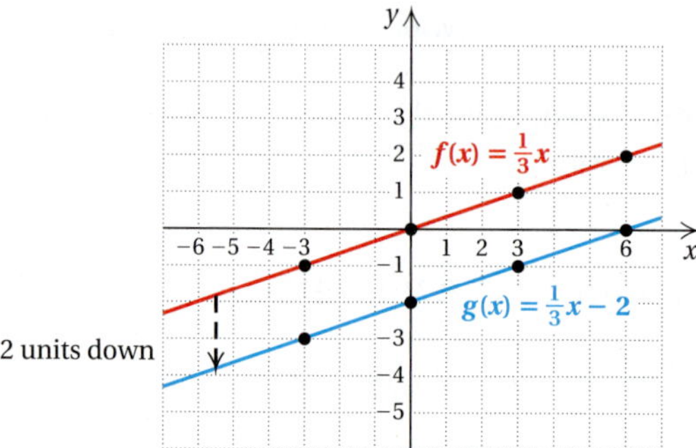

Note that in Example 1, the graph of $y = 2x + 3$ passed through the point $(0, 3)$ and in Example 2, the graph of $g(x) = \frac{1}{3}x - 2$ passed through the point $(0, -2)$. In general, the graph of $y = mx + b$ is a line parallel to $y = mx$, passing through the point $(0, b)$. The point $(0, b)$ is called the ***y*-intercept** because it is the point at which the graph crosses the *y*-axis. Often it is convenient to refer to the number b as the *y*-intercept. The constant b has the effect of moving the graph of $y = mx$ up or down $|b|$ units to obtain the graph of $y = mx + b$.

Do Exercise 3 on the preceding page.

y-INTERCEPT OF $f(x) = mx + b$

The ***y*-intercept** of the graph of $f(x) = mx + b$ is the point $(0, b)$ or, simply, b.

y
f(x) = mx + b
(0, b)
x

EXAMPLE 3 Find the *y*-intercept: $y = -5x + 4$.

$y = -5x + 4$ $\quad$ $(0, 4)$ is the *y*-intercept.

EXAMPLE 4 Find the *y*-intercept: $f(x) = 6.3x - 7.8$.

$f(x) = 6.3x - 7.8$ $\quad$ $(0, -7.8)$ is the *y*-intercept.

Do Exercises 4 and 5.

Find the *y*-intercept.

4. $y = 7x + 8$

5. $f(x) = -6x - \frac{2}{3}$

Answers on page A-43

b The Constant m: Slope

Look again at the graphs in Examples 1 and 2. Note that the slant of each red line seems to match the slant of each blue line. This leads us to believe that the number m in the equation $y = mx + b$ is related to the slant of the line. Let's consider some examples.

Graphs with $m < 0$:

$m = -1$

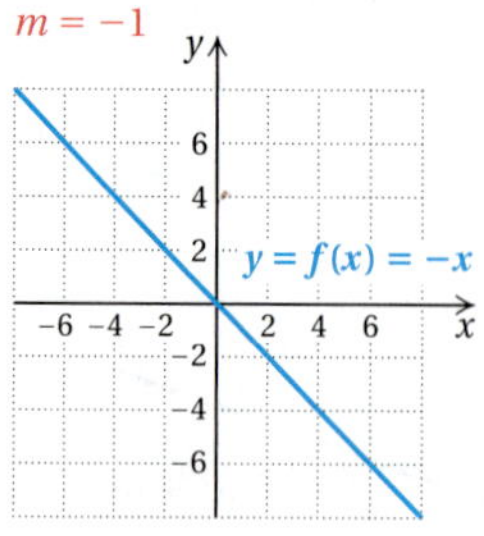

$m = -5$

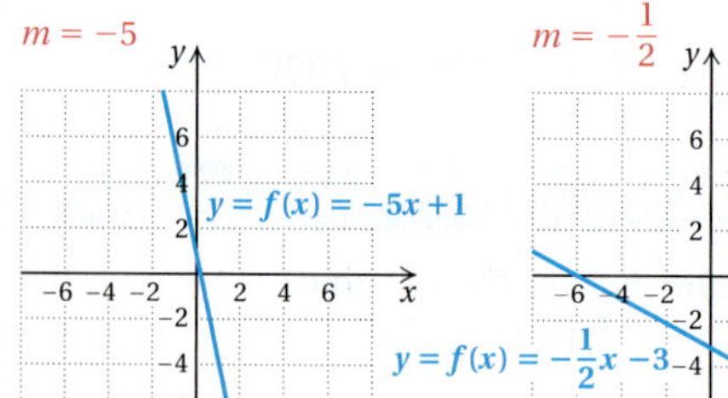

$m = -\frac{1}{2}$

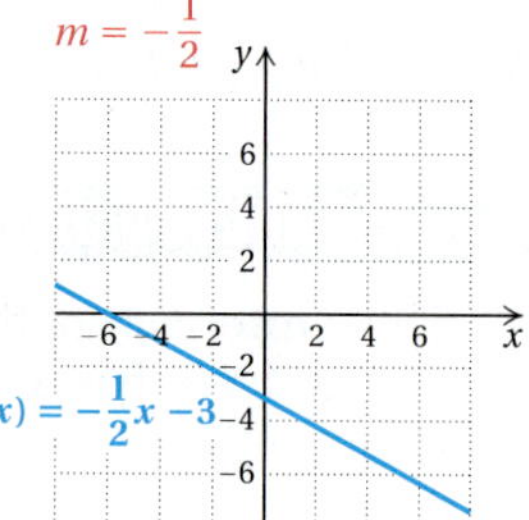

Graphs with $m = 0$:

$m = 0$

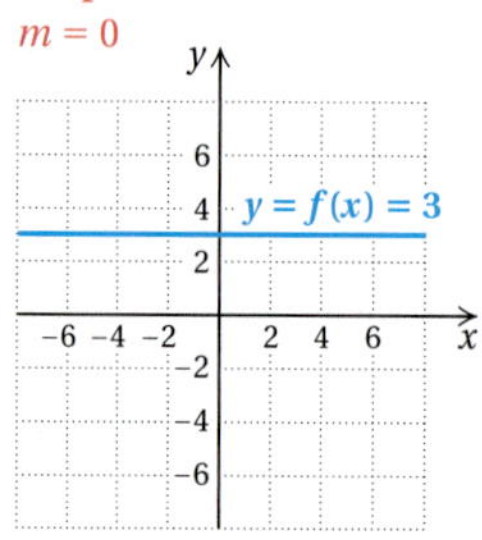

$m = 0$

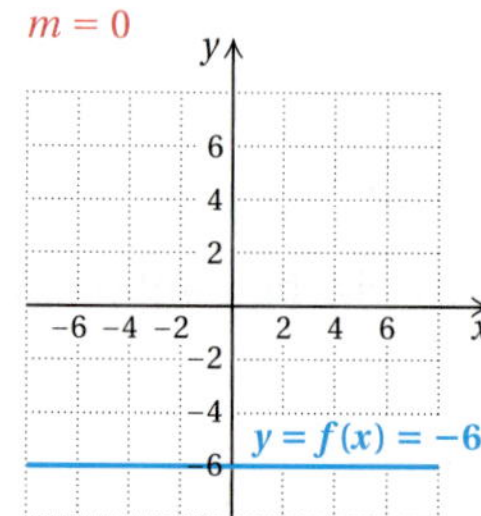

$m = 0$

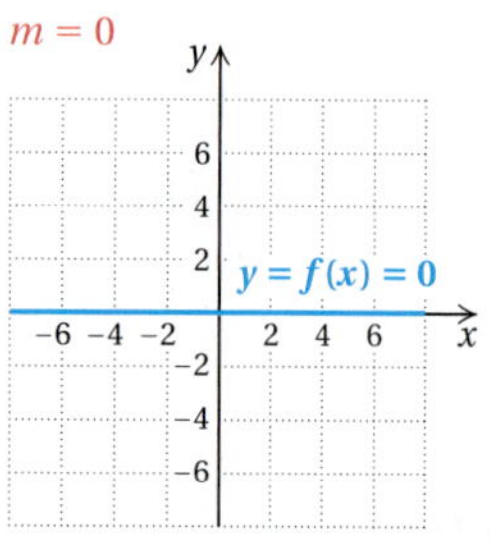

Graphs with $m > 0$:

$m = 1$

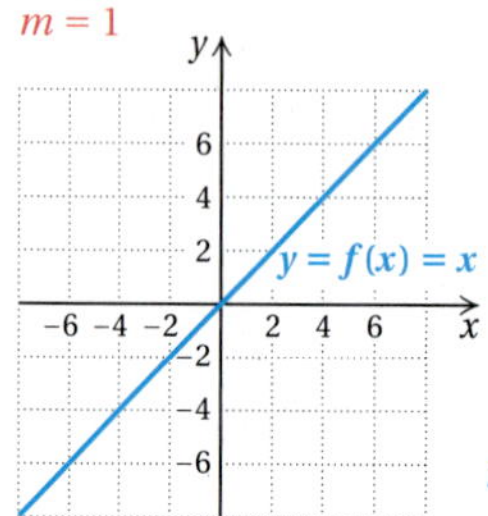

$m = 6$

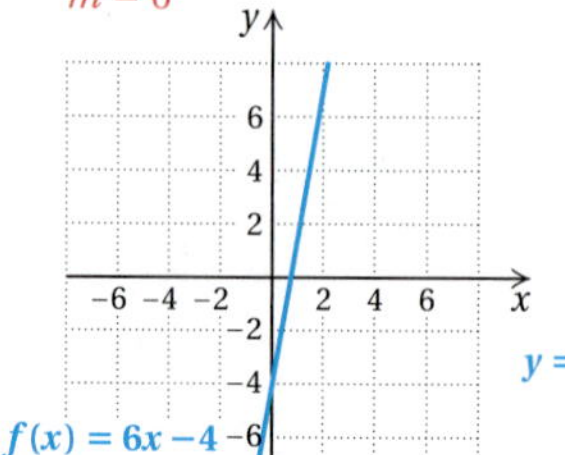

$m = \frac{1}{3}$

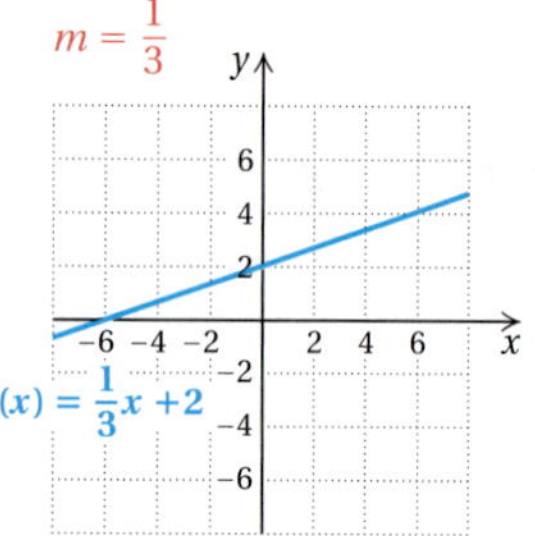

Note that

$m < 0 \longrightarrow$ The graph slants down from left to right;

$m = 0 \longrightarrow$ the graph is horizontal; and

$m > 0 \longrightarrow$ the graph slants up from left to right.

The following definition enables us to visualize the slant and attach a number, a geometric ratio, or *slope*, to the line.

SLOPE

The **slope** of a line containing points (x_1, y_1) and (x_2, y_2) is given by

$$m = \frac{\text{rise}}{\text{run}}$$

$$= \frac{\text{change in } y}{\text{change in } x} = \frac{y_2 - y_1}{x_2 - x_1} = \frac{y_1 - y_2}{x_1 - x_2}.$$

Consider a line with two points marked P_1 and P_2, as follows. As we move from P_1 to P_2, the y-coordinate changes from 1 to 3 and the x-coordinate changes from 2 to 7. The change in y is $3 - 1$, or 2. The change in x is $7 - 2$, or 5.

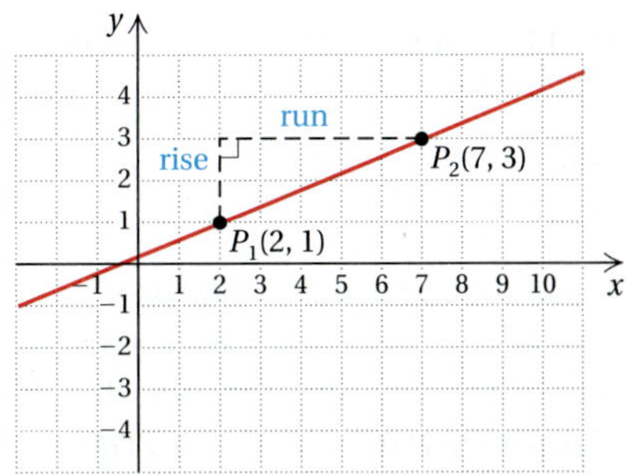

We call the change in y the **rise** and the change in x the **run.** The ratio rise/run is the same for any two points on a line. We call this ratio the **slope.** Slope describes the slant of a line. The slope of the line in the graph above is given by

$$\frac{\text{rise}}{\text{run}}, \quad \text{or} \quad \frac{\text{change in } y}{\text{change in } x}, \quad \text{or} \quad \frac{2}{5}.$$

Whenever x increases by 5 units, y increases by 2 units. Equivalently, whenever x increases by 1 unit, y increases by $\frac{2}{5}$ unit.

EXAMPLE 5 Graph the line containing the points $(-4, 3)$ and $(2, -5)$ and find the slope.

The graph is shown below. Going from $(-4, 3)$ to $(2, -5)$, we see that the change in y, or the rise, is $-5 - 3$, or -8. The change in x, or the run, is $2 - (-4)$, or 6.

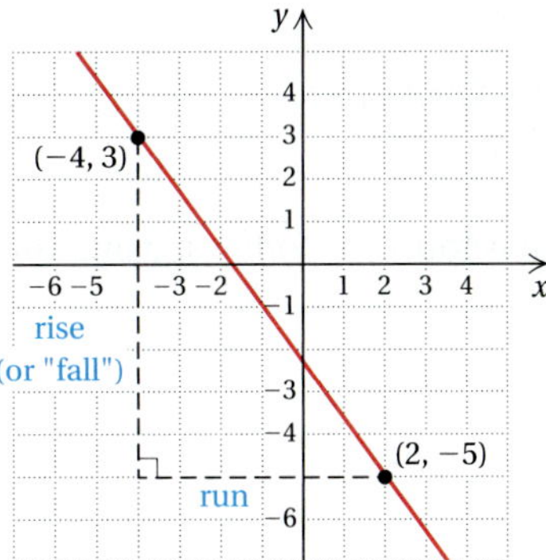

$$\text{Slope} = \frac{\text{rise}}{\text{run}} = \frac{\text{change in } y}{\text{change in } x}$$
$$= \frac{-5 - 3}{2 - (-4)}$$
$$= \frac{-8}{6} = -\frac{8}{6}, \text{ or } -\frac{4}{3}$$

CALCULATOR CORNER

Visualizing Slope For Exercises 1–4, use the window settings $[-6, 6, -4, 4]$, with Xscl = 1 and Yscl = 1.

1. Graph $y = x$, $y = 2x$, and $y = 5x$ in the same window. What do you think the graph of $y = 10x$ will look like?
2. Graph $y = x$, $y = \frac{1}{2}x$, and $y = 0.1x$ in the same window. What do you think the graph of $y = 0.005x$ will look like?
3. Graph $y = -x$, $y = -2x$, and $y = -5x$ in the same window. What do you think the graph of $y = -10x$ will look like?
4. Graph $y = -x$, $y = -\frac{1}{2}x$, and $y = -0.1x$ in the same window. What do you think the graph of $y = -0.005x$ will look like?

Graph the line through the given points and find its slope.

6. $(-1, -1)$ and $(2, -4)$

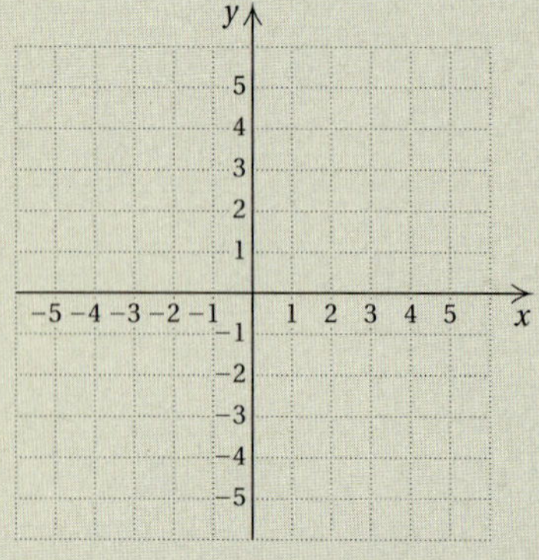

7. $(0, 2)$ and $(3, 1)$

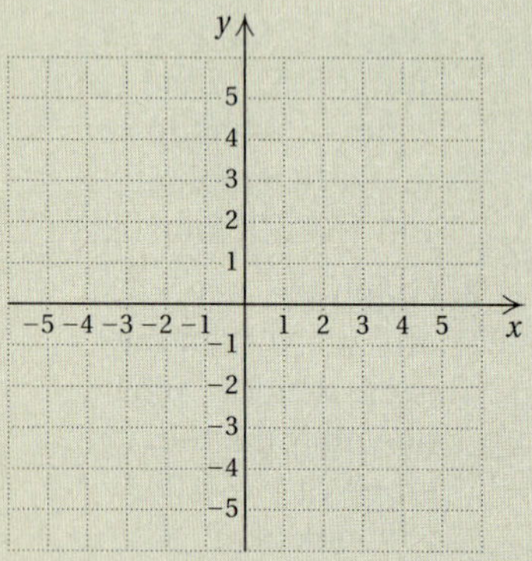

Answers on page A-43

The formula

$$m = \frac{y_2 - y_1}{x_2 - x_1} = \frac{y_1 - y_2}{x_1 - x_2}$$

tells us that we can subtract in two ways. We must remember, however, to subtract the x-coordinates in the same order that we subtract the y-coordinates.

Let's do Example 5 again:

$$\text{Slope} = \frac{\text{change in } y}{\text{change in } x} = \frac{3 - (-5)}{-4 - 2} = \frac{8}{-6} = -\frac{8}{6} = -\frac{4}{3}.$$

We see that both ways give the same value for the slope.

The slope of a line tells how it slants. A line with positive slope slants up from left to right. The larger the positive number, the steeper the slant. A line with negative slope slants downward from left to right. The smaller the negative number, the steeper the line.

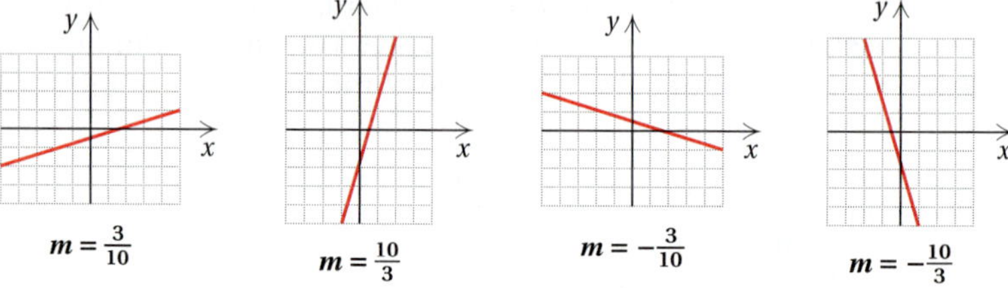

Do Exercises 6 and 7.

How can we find the slope from a given equation? Let's consider the equation $y = 2x + 3$, which is in the form $y = mx + b$. We can find two points by choosing convenient values for x, say 0 and 1, and substituting to find the corresponding y-values.

If $x = 0$, $y = 2 \cdot 0 + 3 = 3$.

If $x = 1$, $y = 2 \cdot 1 + 3 = 5$.

We find two points on the line to be

$(0, 3)$ and $(1, 5)$.

The slope of the line is found as follows, using the definition of slope:

$$m = \frac{\text{change in } y}{\text{change in } x} = \frac{5 - 3}{1 - 0} = \frac{2}{1} = 2.$$

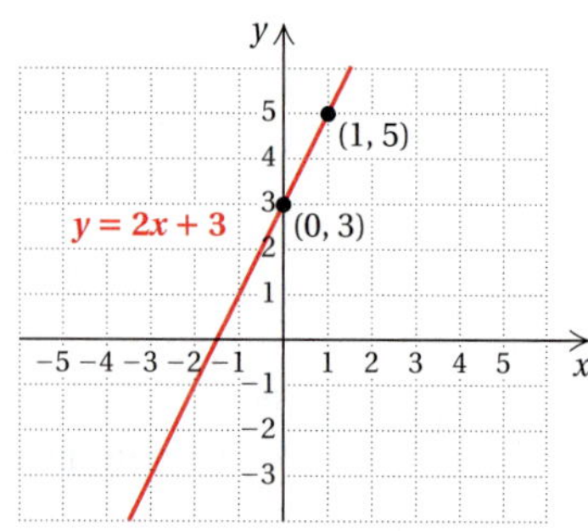

The slope is 2. Note that this is the coefficient of the x-term in the equation $y = 2x + 3$.

If we had chosen different points on the line—say $(-2, -1)$ and $(4, 11)$—the slope would still be 2, as we see in the following calculation:

$$m = \frac{11 - (-1)}{4 - (-2)} = \frac{11 + 1}{4 + 2} = \frac{12}{6} = 2.$$

Do Exercise 8 on the following page.

We see that the slope of the line $y = mx + b$ is indeed the constant m, the coefficient of x.

SLOPE OF $y = mx + b$

The **slope** of the line $y = mx + b$ is m.

From a linear equation in the form $y = mx + b$, we can read directly the slope and the y-intercept of the graph.

SLOPE–INTERCEPT EQUATION

The equation $y = mx + b$ is called the **slope–intercept equation.** The slope is m and the y-intercept is $(0, b)$.

Note that any graph of an equation $y = mx + b$ passes the vertical-line test and thus represents a function.

EXAMPLE 6 Find the slope and the y-intercept of $y = 5x - 4$.

Since the equation is already in the form $y = mx + b$, we simply read the slope and the y-intercept from the equation:

$$y = 5x - 4.$$

The slope is 5. The y-intercept is $(0, -4)$.

EXAMPLE 7 Find the slope and the y-intercept of $2x + 3y = 8$.

We first solve for y so we can easily read the slope and the y-intercept:

$$2x + 3y = 8$$

$$3y = -2x + 8 \quad \text{Subtracting } 2x$$

$$\frac{3y}{3} = \frac{-2x + 8}{3} \quad \text{Dividing by 3}$$

$$y = -\frac{2}{3}x + \frac{8}{3}. \quad \text{Finding the form } y = mx + b$$

The slope is $-\frac{2}{3}$. The y-intercept is $\left(0, \frac{8}{3}\right)$.

Do Exercises 9 and 10.

C Applications

Slope has many real-world applications. For example, numbers like 2%, 3%, and 6% are often used to represent the *grade* of a road, a measure of how steep a road on a hill or mountain is. A 3% grade $\left(3\% = \frac{3}{100}\right)$ means that for every horizontal distance of 100 ft, the road rises 3 ft, and a -3% grade means that for every horizontal distance of 100 ft, the road drops 3 ft. (Normally, the road

8. Find the slope of the line $f(x) = -\frac{2}{3}x + 1$. Use the points $(9, -5)$ and $(3, -1)$.

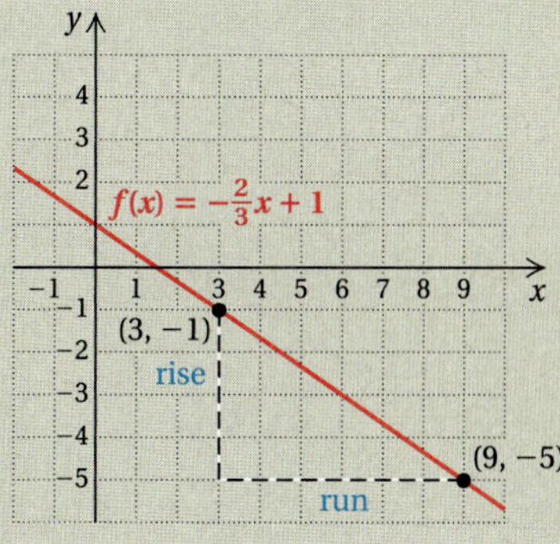

Find the slope and the y-intercept.

9. $f(x) = -8x + 23$

10. $5x - 10y = 25$

Answers on page A-44

signs do not include negative signs, since it is obvious whether you are climbing or descending.) An athlete might change the grade of a treadmill during a workout. An escape ramp on an airliner might have a slope of about -0.6.

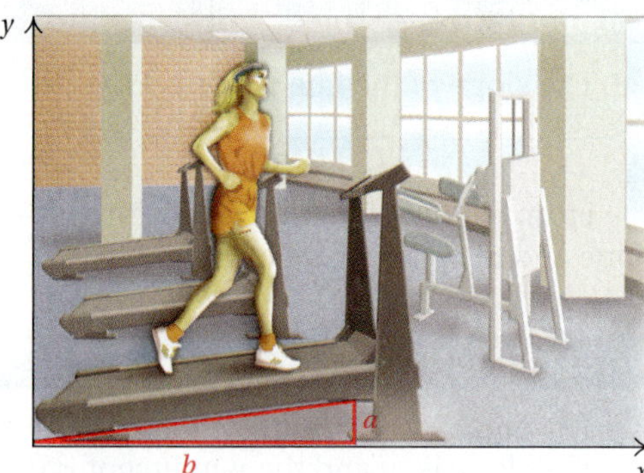

11. Haircutting. Gary's Barber Shop has a graph displaying data from a recent day's work. Use the graph to determine the rate of change in the number of haircuts with respect to time.

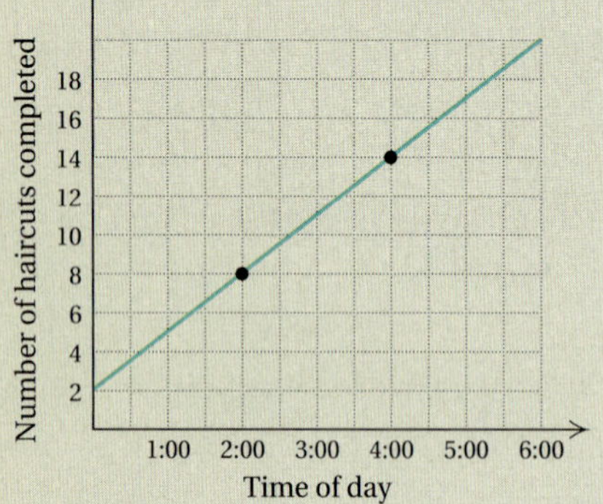

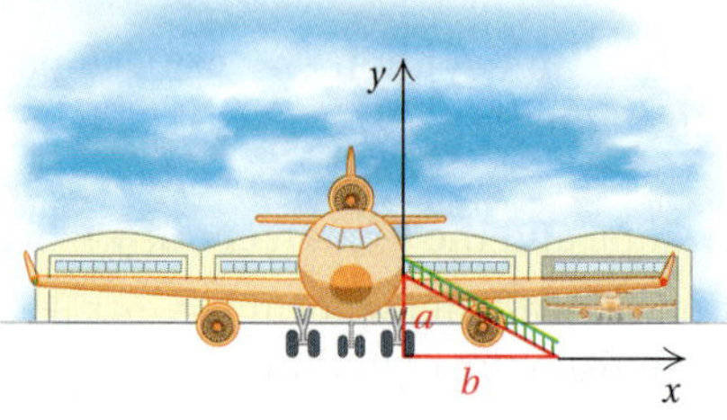

Architects and carpenters use slope when designing and building stairs, ramps, or roof pitches. Another application occurs in hydrology. The strength or force of a river depends on how far the river falls vertically compared to how far it flows horizontally. Slope can also be considered as a **rate of change.**

EXAMPLE 8 *Online Travel Plans.* More and more Americans are making their travel arrangements using the Internet. Use the graph below to find the rate at which this number is growing. Note that the jagged "break" on the vertical axis is used to avoid including a large portion of unused grid.

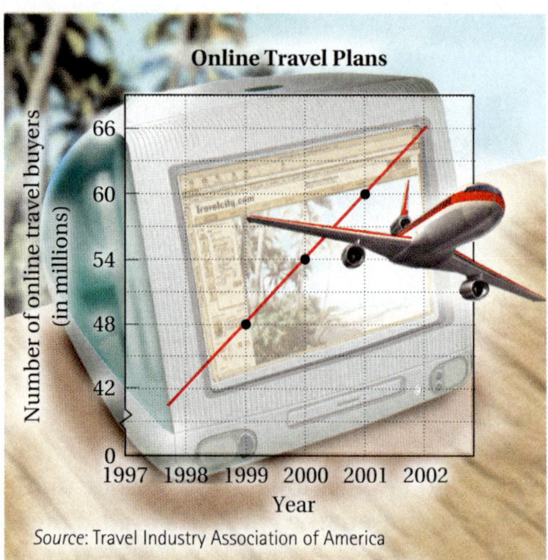

Answer on page A-44

Since the graph is linear, we can use any pair of points to determine the rate of change. We choose the points (1999, 48 million) and (2001, 60 million), which gives us

$$\text{Rate of change} = \frac{60 \text{ million} - 48 \text{ million}}{2001 - 1999}$$
$$= \frac{12 \text{ million}}{2 \text{ years}} = 6 \text{ million per year.}$$

The number of Americans making online travel arrangements is growing at the rate of approximately 6 million users per year.

Do Exercise 11 on the preceding page.

EXAMPLE 9 *First-Class Mail.* The amount of first-class mail is expected to decrease steadily from 2000 to 2005. (This is probably due to the increasing use of e-mail.) In 2000, there were 700 million pieces of first-class mail delivered in the United States. In 2005, it is expected that the number will decrease to 583 million. Find the rate of change of the number of first-class pieces of mail delivered with respect to time, in years.

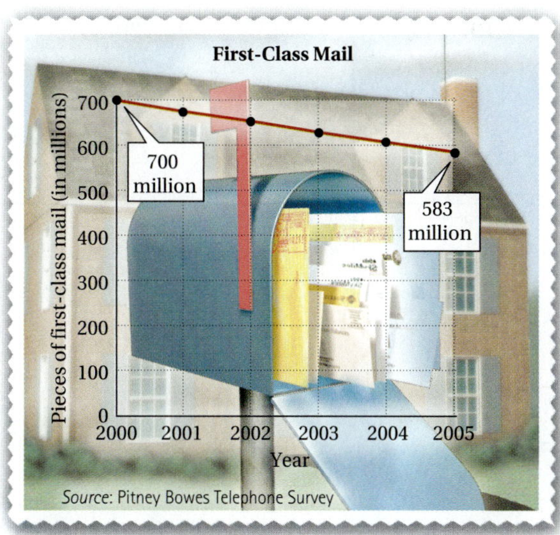

The rate of change of the number of first-class pieces of mail delivered with respect to time, in years, is given by

$$\text{Rate of change} = \frac{583 \text{ million pieces} - 700 \text{ million pieces}}{2005 - 2000}$$
$$= \frac{-117 \text{ million pieces}}{5 \text{ years}}$$
$$= -23.4 \text{ million pieces per year.}$$

Do Exercise 12.

12. Defense Spending. Each year, the United States spends a smaller percent of its annual budget on defense. In 1993, defense spending was 20% of the annual budget. In 1997, defense spending was 17% of the annual budget. Find the rate of change of the percent of the budget spent on defense with respect to time, in years.
Source: U.S. Office of Management and Budget

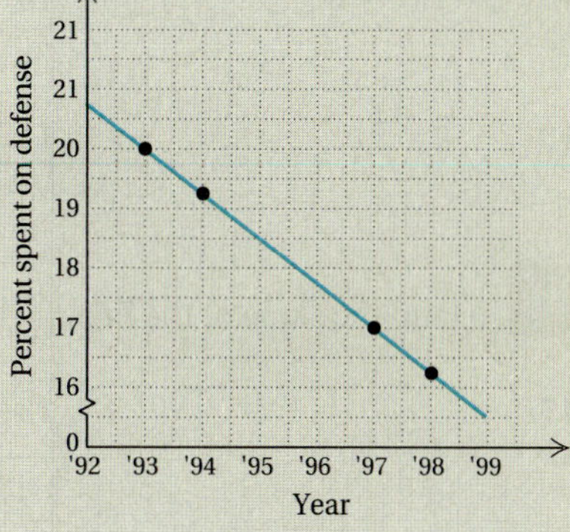

Answer on page A-44

12.3 EXERCISE SET

For Extra Help

Digital Video Tutor CD 10 Videotape 14 | InterAct Math | Math Tutor Center | MathXL | MyMathLab

a, b Find the slope and the y-intercept.

1. $y = 4x + 5$

2. $y = -5x + 10$

3. $f(x) = -2x - 6$

4. $g(x) = -5x + 7$

5. $y = -\frac{3}{8}x - \frac{1}{5}$

6. $y = \frac{15}{7}x + \frac{16}{5}$

7. $g(x) = 0.5x - 9$

8. $f(x) = -3.1x + 5$

9. $2x - 3y = 8$

10. $-8x - 7y = 24$

11. $9x = 3y + 6$

12. $9y + 36 - 4x = 0$

13. $3 - \frac{1}{4}y = 2x$

14. $5x = \frac{2}{3}y - 10$

15. $17y + 4x + 3 = 7 + 4x$

16. $3y - 2x = 5 + 9y - 2x$

b Find the slope of the line.

17.

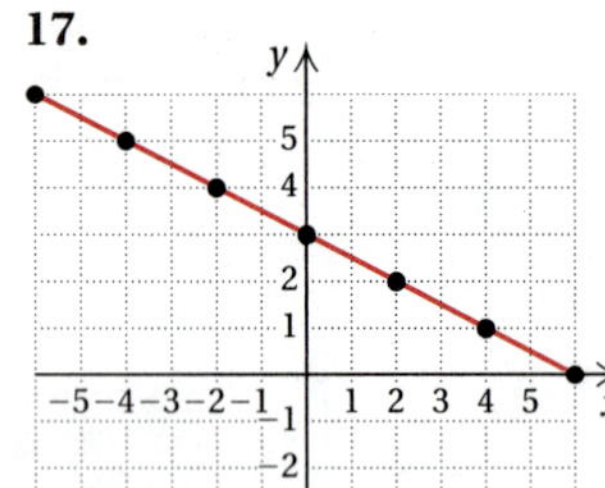

18.

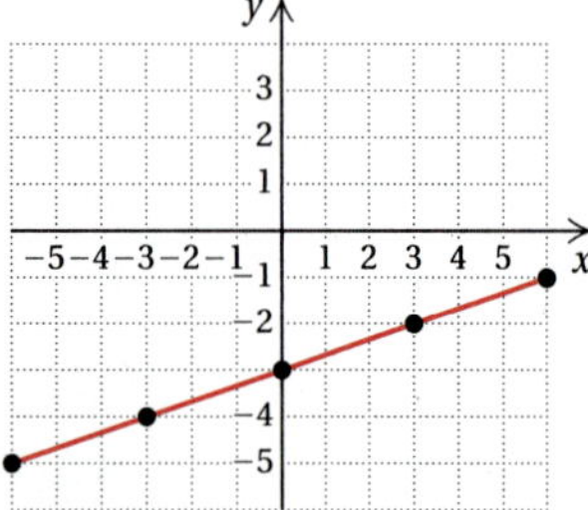

19.

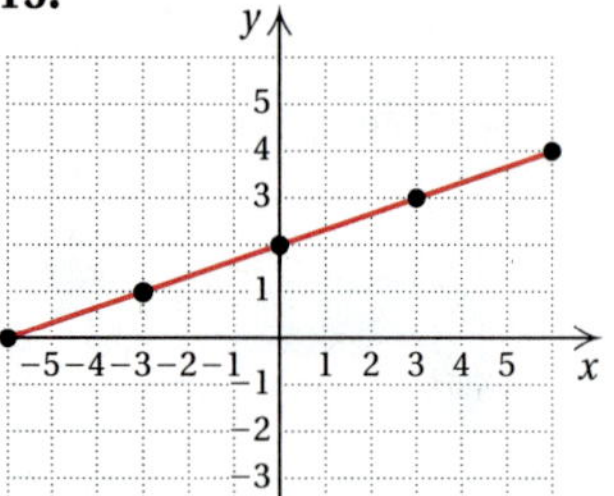

20.

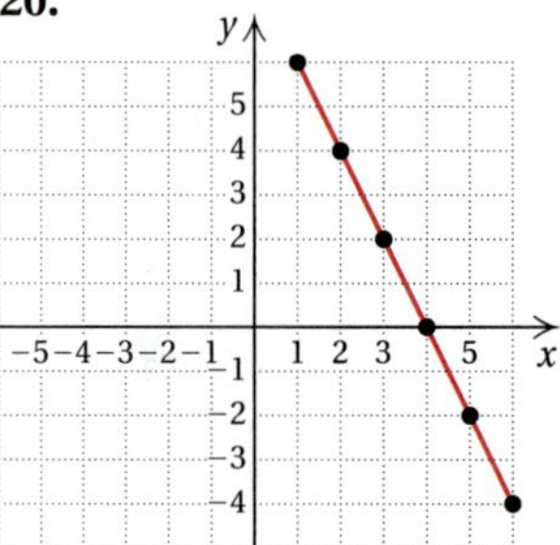

Find the slope of the line containing the given pair of points.

21. $(6, 9)$ and $(4, 5)$

22. $(8, 7)$ and $(2, -1)$

23. $(9, -4)$ and $(3, -8)$

24. $(17, -12)$ and $(-9, -15)$

25. $(-16.3, 12.4)$ and $(-5.2, 8.7)$

26. $(14.4, -7.8)$ and $(-12.5, -17.6)$

c Find the slope (or rate of change).

27. Find the slope (or grade) of the treadmill.

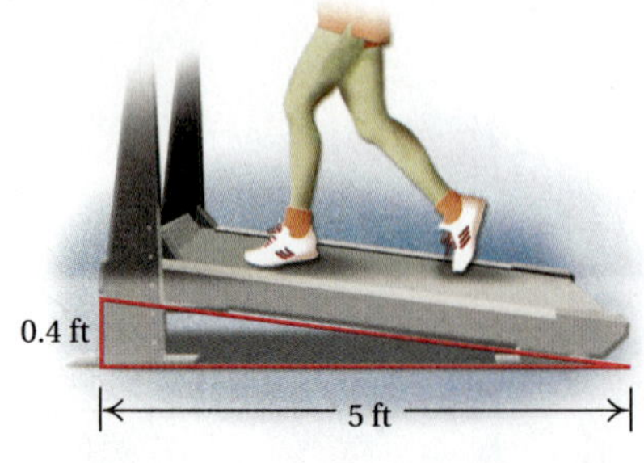

28. Find the slope (or pitch) of the roof.

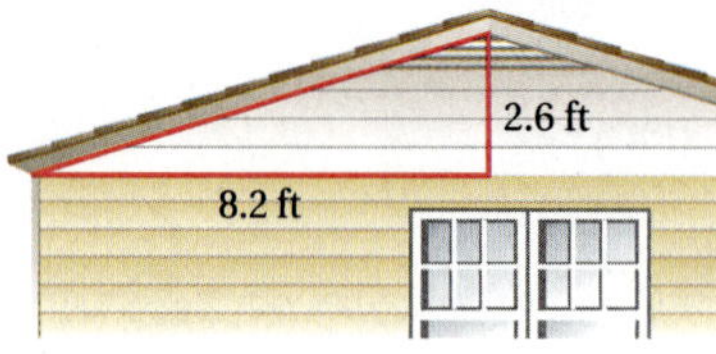

29. Find the slope (or head) of the river.

30. Public buildings regularly include steps with 7-in. risers and 11-in. treads. Find the grade of such a stairway.

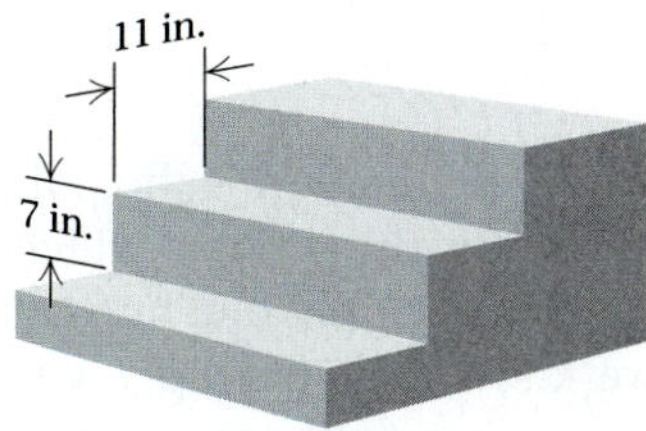

31. *E-mail.* Find the rate of change of the average daily volume of e-mail messages, in billions, in North America, with respect to time, in years.

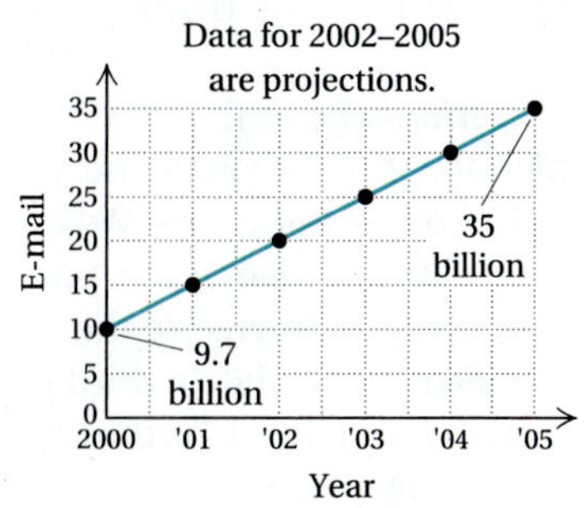

Source: Pitney Bowes Telephone Survey

32. Find the rate of change of the cost of a formal wedding.

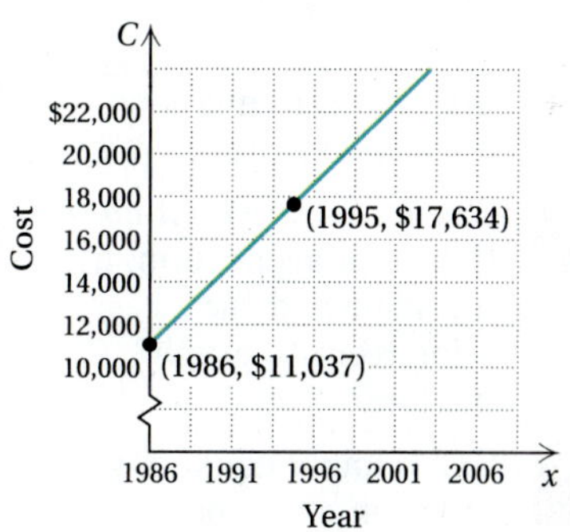

Source: *Modern Bride Magazine*

Find the rate of change.

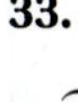

33.

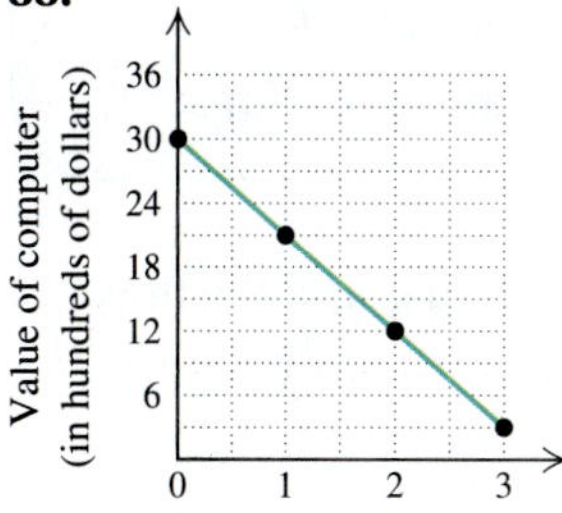

34.

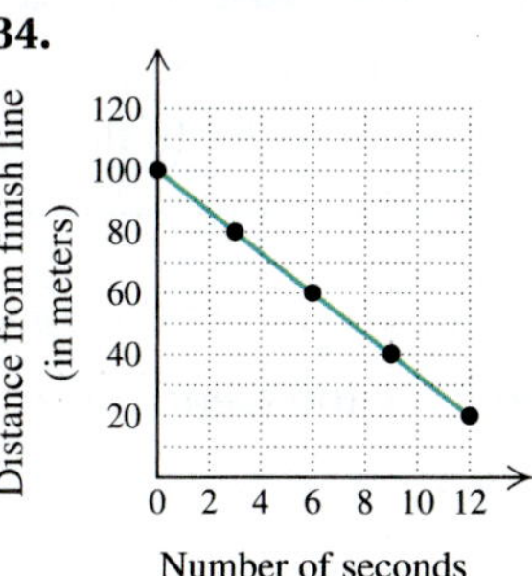

35.

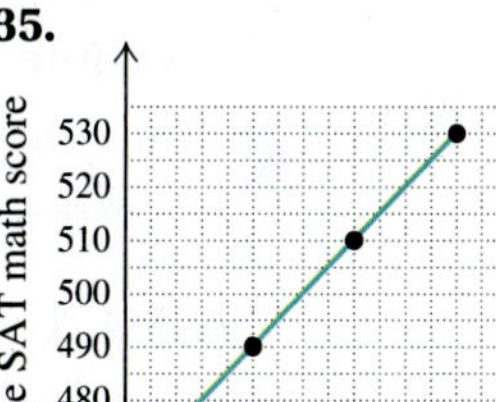

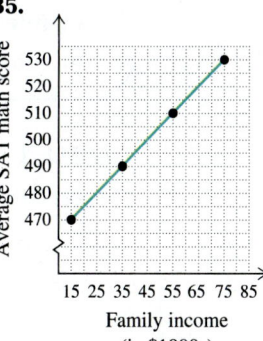

Source: College Board Online

36.

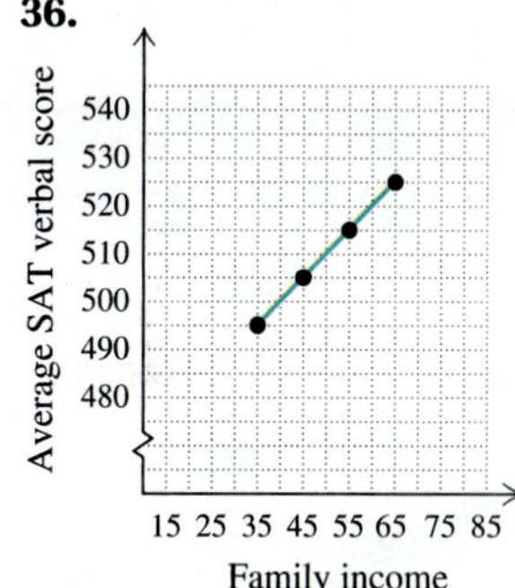

Source: College Board Online

37. **DW** A student makes a mistake when using a graphing calculator to draw $4x + 5y = 12$ and the following screen appears. Use algebra to show that a mistake has been made. What do you think the mistake was?

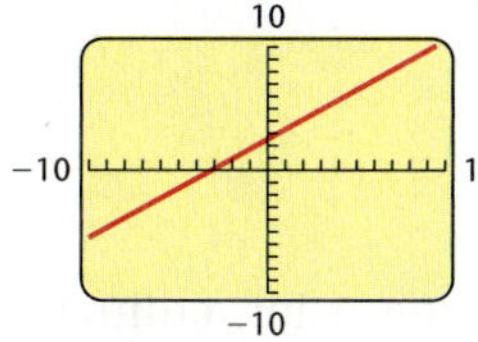

38. **DW** A student makes a mistake when using a graphing calculator to draw $5x - 2y = 3$ and the following screen appears. Use algebra to show that a mistake has been made. What do you think the mistake was?

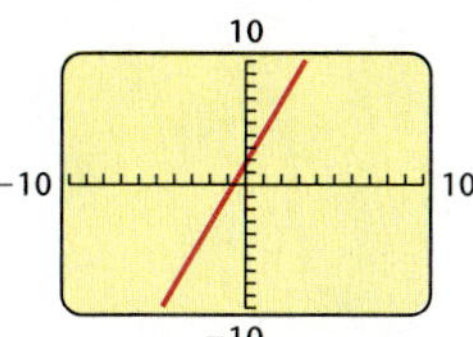

SKILL MAINTENANCE

Simplify. [6.8d], [9.1b]

39. $3^2 - 24 \cdot 56 + 144 \div 12$

40. $9\{2x - 3[5x + 2(-3x + y^0 - 2)]\}$

41. $10\{2x + 3[5x - 2(-3x + y^1 - 2)]\}$

42. $5^4 \div 625 \div 5^2 \cdot 5^7 \div 5^3$

Solve. [7.6a]

43. One side of a square is 5 yd less than a side of an equilateral triangle. If the perimeter of the square is the same as the perimeter of the triangle, what is the length of a side of the square? of the triangle?

Factor. [10.6a]

44. $8 - 125x^3$

45. $c^6 - d^6$

46. $56x^3 - 7$

47. Divide: $(a^2 - 11a + 6) \div (a - 1)$. [9.8b, c]

Objectives

a Graph linear equations using intercepts.

b Given a linear equation in slope–intercept form, use the slope and the y-intercept to graph the line.

c Graph linear equations of the form $x = a$ or $y = b$.

d Given the equations of two lines, determine whether their graphs are parallel or whether they are perpendicular.

12.4 MORE ON GRAPHING LINEAR EQUATIONS

a Graphing Using Intercepts

The ***x*-intercept** of the graph of a linear equation or function is the point at which the graph crosses the x-axis. The ***y*-intercept** is the point at which the graph crosses the y-axis. We know from geometry that only one line can be drawn through two given points. Thus, if we know the intercepts, we can graph the line. To ensure that a computation error has not been made, it is a good idea to calculate a third point as a check.

Many equations of the type $Ax + By = C$ can be graphed conveniently using intercepts.

x- AND y-INTERCEPTS

A ***y*-intercept** is a point $(0, b)$. To find b, let $x = 0$ and solve for y.

An ***x*-intercept** is a point $(a, 0)$. To find a, let $y = 0$ and solve for x.

EXAMPLE 1 Find the intercepts of $3x + 2y = 12$ and then graph the line.

***y*-intercept:** To find the y-intercept, we let $x = 0$ and solve for y:

$$\begin{aligned} 3x + 2y &= 12 \\ 3 \cdot 0 + 2y &= 12 \quad \text{Substituting 0 for } x \\ 2y &= 12 \\ y &= 6. \end{aligned}$$

The y-intercept is $(0, 6)$.

***x*-intercept:** To find the x-intercept, we let $y = 0$ and solve for x:

$$\begin{aligned} 3x + 2y &= 12 \\ 3x + 2 \cdot 0 &= 12 \quad \text{Substituting 0 for } y \\ 3x &= 12 \\ x &= 4. \end{aligned}$$

The x-intercept is $(4, 0)$.

We plot these points and draw the line, using a third point as a check. We choose $x = 6$ and solve for y:

$$\begin{aligned} 3(6) + 2y &= 12 \\ 18 + 2y &= 12 \\ 2y &= -6 \\ y &= -3. \end{aligned}$$

We plot $(6, -3)$ and note that it is on the line.

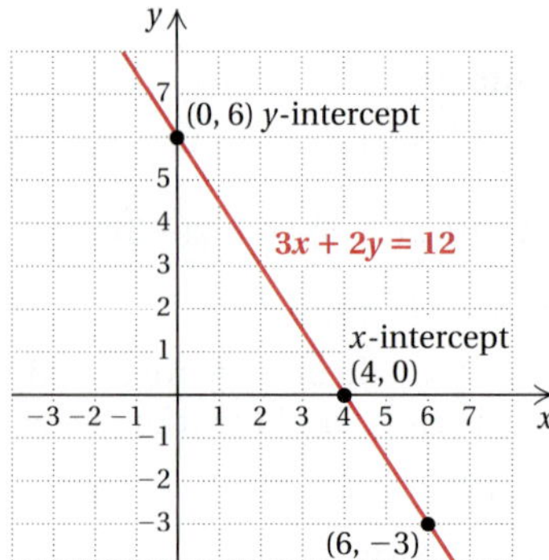

When both the x- and y-intercepts are $(0, 0)$, as is the case with an equation such as $y = 2x$, whose graph passes through the origin, another point would have to be calculated and a third point used as a check.

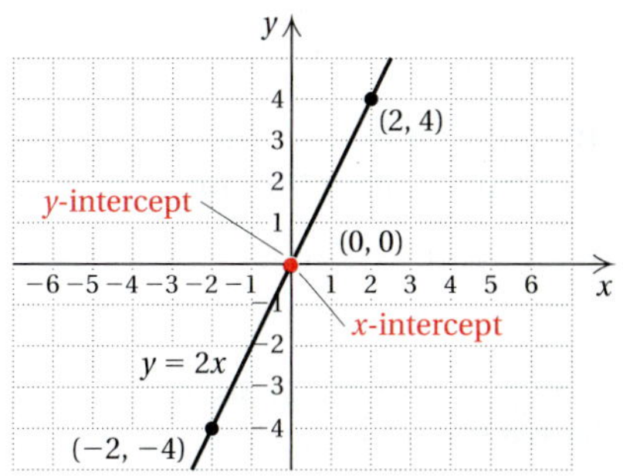

Do Exercise 1.

1. Find the intercepts of $4y - 12 = -6x$ and then graph the line.

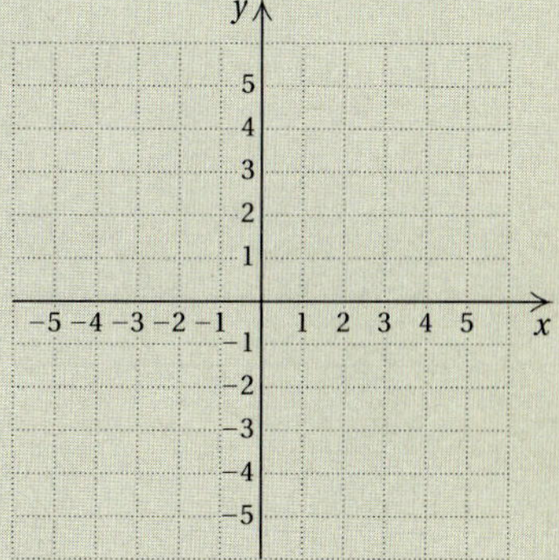

CALCULATOR CORNER

Viewing the Intercepts Knowing the intercepts of a linear equation helps us determine a good viewing window for the graph of the equation. For example, when we graph the equation $y = -x + 15$ in the standard window, we see only a small portion of the graph in the upper right-hand corner of the screen, as shown on the left below.

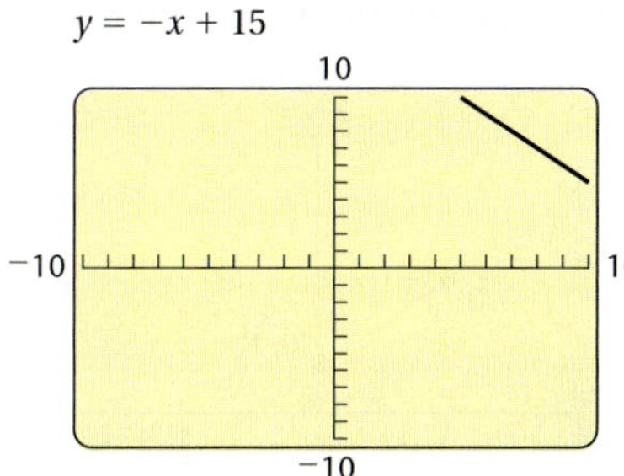

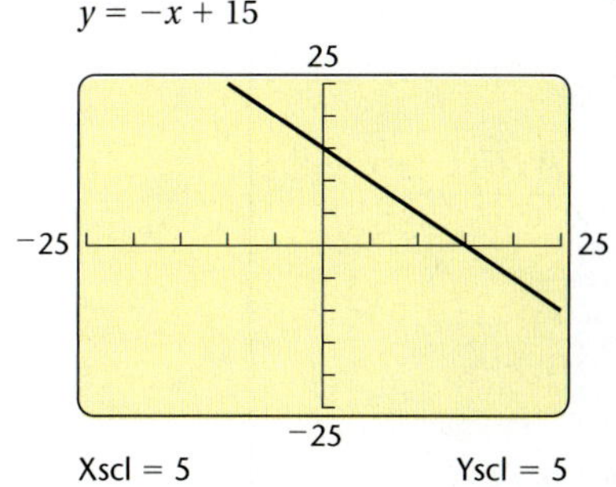

Using algebra, as we did in Example 1, we can find that the intercepts of the graph of this equation are $(0, 15)$ and $(15, 0)$. This tells us that, if we are to see more of the graph than is shown above, both Xmax and Ymax should be greater than 15. We can try different window settings until we find one that suits us. One good choice, shown on the right above, is $[-25, 25, -25, 25]$, with Xscl $= 5$ and Yscl $= 5$.

Exercises: Find the intercepts of the equation algebraically. Then graph the equation on a graphing calculator, choosing window settings that allow the intercepts to be seen clearly. (Settings may vary.)

1. $y = -3.2x - 16$
2. $y - 4.25x = 85$
3. $6x + 5y = 90$
4. $5x - 6y = 30$
5. $8x + 3y = 9$
6. $y = 0.4x - 5$
7. $y = 1.2x - 12$
8. $4x - 5y = 2$

Answer on page A-44

Graph using the slope and the y-intercept.

2. $y = \frac{3}{2}x + 1$

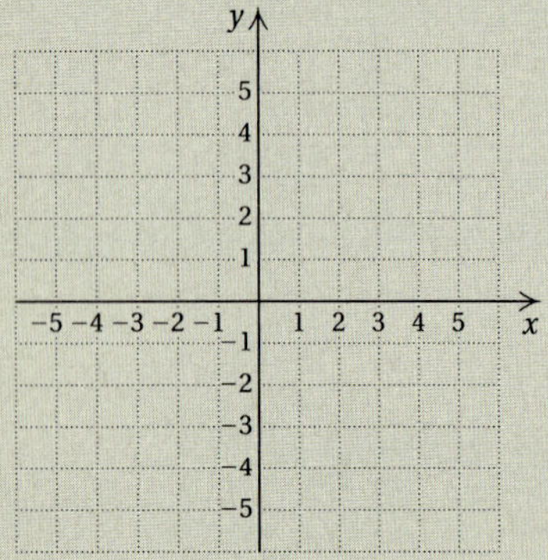

3. $f(x) = \frac{3}{4}x - 2$

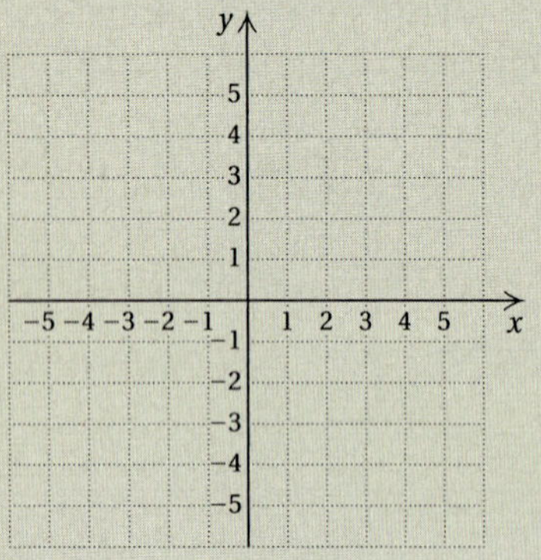

4. $g(x) = -\frac{3}{5}x + 5$

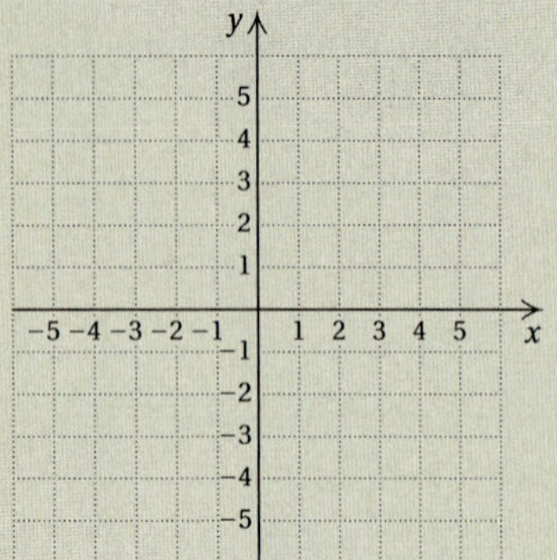

Answers on page A-44

b Graphing Using the Slope and the y-intercept

We can also graph a line using its slope and y-intercept.

EXAMPLE 2 Graph: $y = -\frac{2}{3}x + 1$.

This equation is in slope–intercept form, $y = mx + b$. The y-intercept is $(0, 1)$. We plot $(0, 1)$. We can think of the slope $\left(m = -\frac{2}{3}\right)$ as $\frac{-2}{3}$.

$$m = \frac{\text{Rise}}{\text{Run}} = \frac{-2}{3} \quad \begin{matrix}\text{Move 2 units down.}\\ \text{Move 3 units right.}\end{matrix}$$

Starting at the y-intercept and using the slope, we find another point by moving 2 units down (since the numerator is *negative* and corresponds to the change in y) and 3 units to the right (since the denominator is *positive* and corresponds to the change in x). We get to a new point, $(3, -1)$. In a similar manner, we can move from the point $(3, -1)$ to find another point, $(6, -3)$.

We could also think of the slope $\left(m = -\frac{2}{3}\right)$ as $\frac{2}{-3}$.

$$m = \frac{\text{Rise}}{\text{Run}} = \frac{2}{-3} \quad \begin{matrix}\text{Move 2 units up.}\\ \text{Move 3 units left.}\end{matrix}$$

Then we can start again at $(0, 1)$, but this time we move 2 units up (since the numerator is *positive* and corresponds to the change in y) and 3 units to the left (since the denominator is *negative* and corresponds to the change in x).We get another point on the graph, $(-3, 3)$, and from it we can obtain $(-6, 5)$ and others in a similar manner. We plot the points and draw the line.

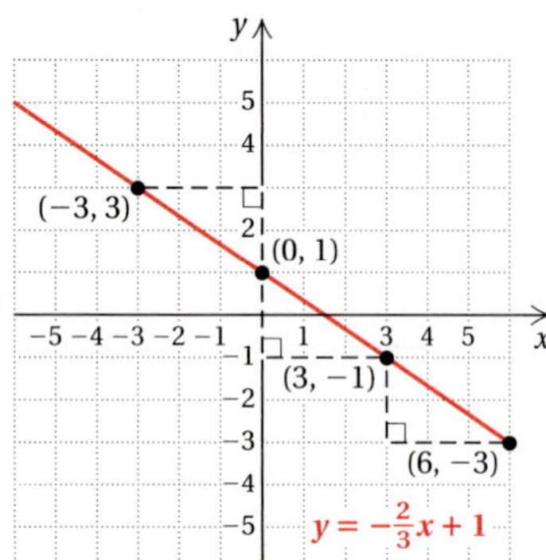

EXAMPLE 3 Graph: $f(x) = \frac{2}{5}x + 4$.

First, we plot the y-intercept, $(0, 4)$. We then consider the slope $\frac{2}{5}$. Starting at the y-intercept and using the slope, we find another point by moving 2 units up (since the numerator is *positive* and corresponds to the change in y) and 5 units to the right (since the denominator is *positive* and corresponds to the change in x). We get to a new point, $(5, 6)$.

We can also think of the slope $\frac{2}{5}$ as $\frac{-2}{-5}$. We again start at the y-intercept, $(0, 4)$. We move 2 units down (since the numerator is *negative* and corresponds to the change in y) and 5 units to the left (since the denominator is *negative* and corresponds to the change in x). We get to another new point, $(-5, 2)$. We plot the points and draw the line.

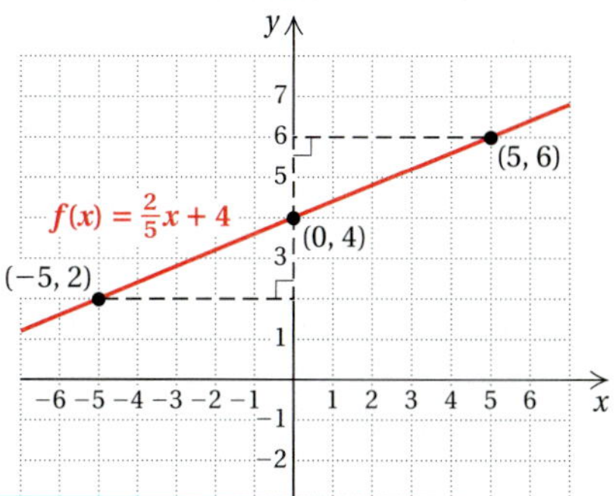

Do Exercises 2–5. (Exercises 2–4 are on the preceding page.)

5. $y = -\dfrac{5}{3}x - 4$

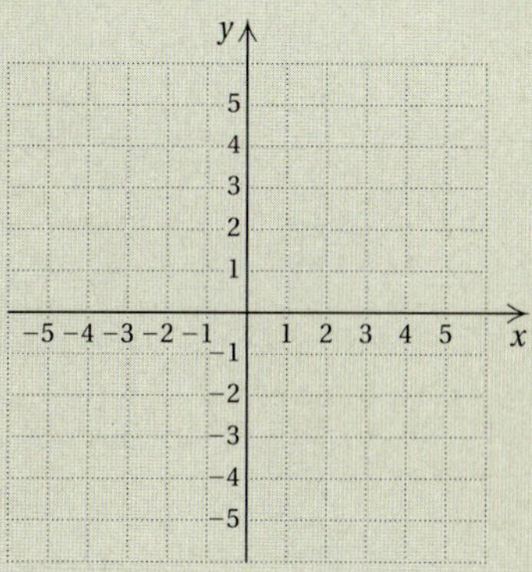

C Horizontal and Vertical Lines

Some equations have graphs that are parallel to one of the axes. This happens when either A or B is 0 in $Ax + By = C$. These equations have a missing variable, that is, there is only one variable in the equation. In the following example, x is missing.

EXAMPLE 4 Graph: $y = 3$.

Since x is missing, any number for x will do. Thus all ordered pairs $(x, 3)$ are solutions. The graph is a **horizontal line** parallel to the x-axis.

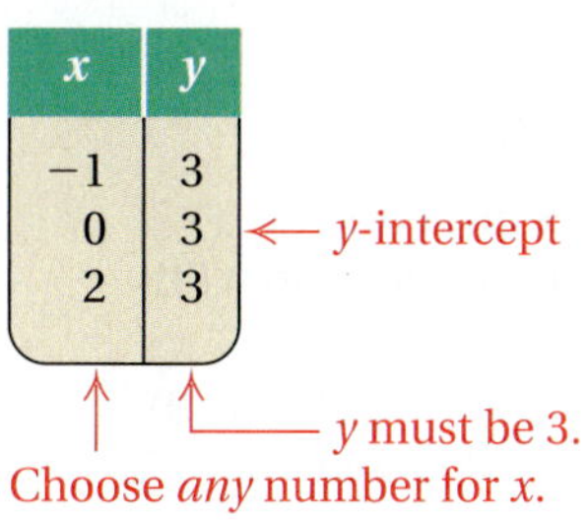

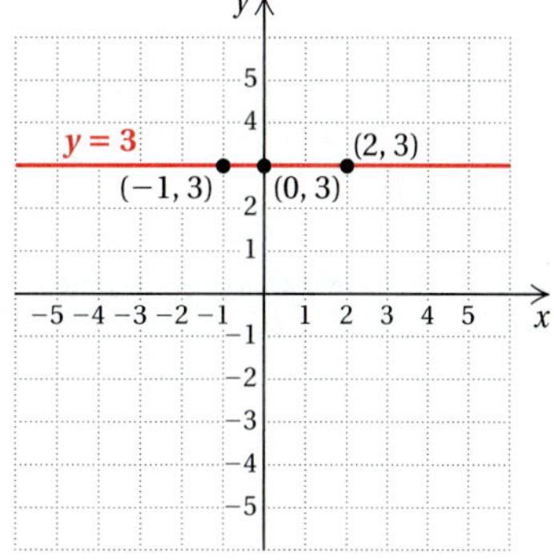

What about the slope of a horizontal line? In Example 4, consider the points $(-1, 3)$ and $(2, 3)$, which are on the line $y = 3$. The change in y is $3 - 3$, or 0. The change in x is $-1 - 2$, or -3. Thus,

$$m = \frac{3 - 3}{-1 - 2} = \frac{0}{-3} = 0.$$

Any two points on a horizontal line have the same y-coordinate. Thus the change in y is always 0, so the slope is 0.

We can also determine the slope by noting that $y = 3$ can be written in slope–intercept form as $y = 0x + 3$, or $f(x) = 0x + 3$. From this equation, we read that the slope is 0. A function of this type is called a **constant function.** We can express it in the form $y = b$, or $f(x) = b$. Its graph is a horizontal line that crosses the y-axis at $(0, b)$.

Do Exercises 6 and 7.

Graph and determine the slope.

6. $f(x) = -4$

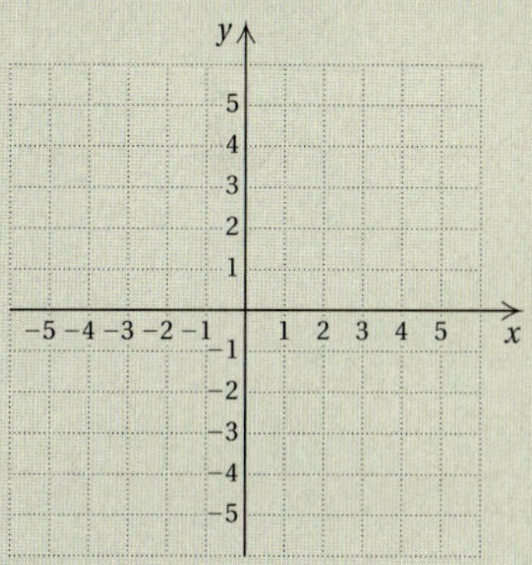

7. $y = 3.6$

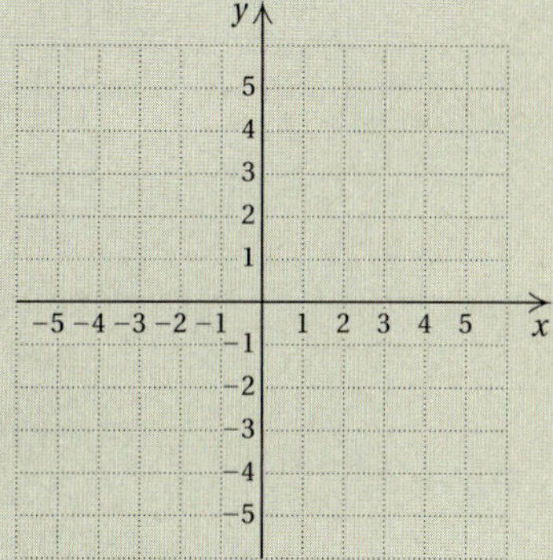

Answers on page A-44

Graph.

8. $x = -5$

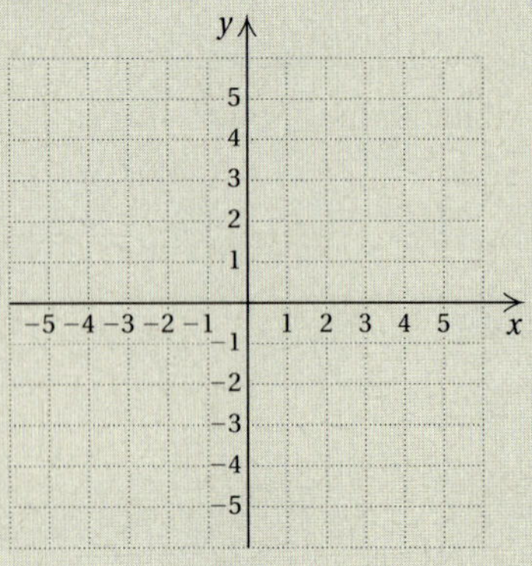

9. $8x - 5 = 19$ (*Hint*: Solve for x.)

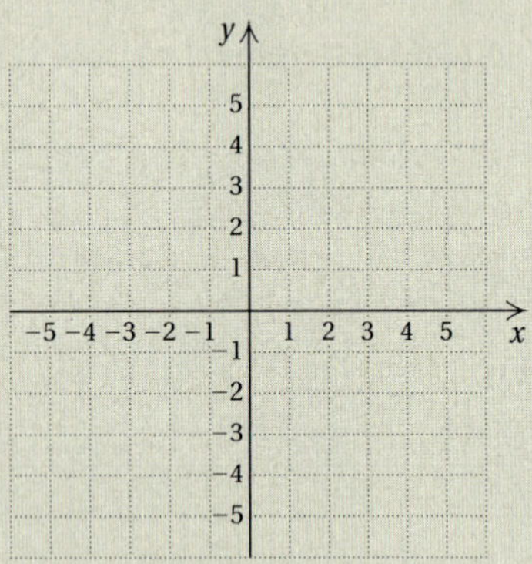

10. Determine, if possible, the slope of each line.

a) $x = -12$

b) $y = 6$

c) $2y + 7 = 11$

d) $x = 0$

e) $y = -\frac{3}{4}$

f) $10 - 5x = 15$

Answers on page A-44

In the following example, y is missing and the graph is parallel to the y-axis.

EXAMPLE 5 Graph: $x = -2$.

Since y is missing, any number for y will do. Thus all ordered pairs $(-2, y)$ are solutions. The graph is a **vertical line** parallel to the y-axis.

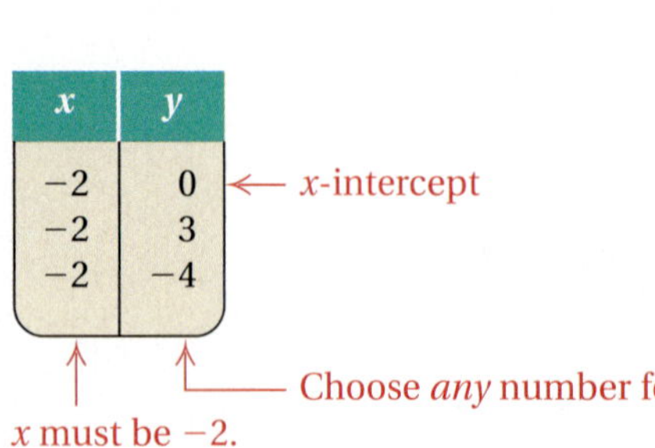

x	y
−2	0
−2	3
−2	−4

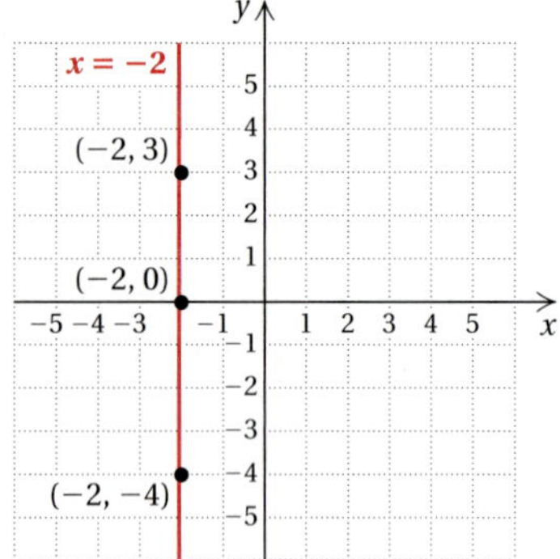

This graph is not the graph of a function because it fails the vertical-line test. The vertical line itself crosses the graph more than once.

What about the slope of a vertical line? In Example 5, consider the points $(-2, 3)$ and $(-2, -4)$, which are on the line $x = -2$. The change in y is $3 - (-4)$, or 7. The change in x is $-2 - (-2)$, or 0. Thus,

$$m = \frac{3 - (-4)}{-2 - (-2)} = \frac{7}{0}. \quad \text{Not defined}$$

Since division by 0 is not defined, the slope of this line is not defined. Any two points on a vertical line have the same x-coordinate. Thus the change in x is always 0, so the slope of any vertical line is not defined.

The following summarizes horizontal and vertical lines and their equations.

HORIZONTAL LINE; VERTICAL LINE

The graph of $y = b$, or $f(x) = b$, is a **horizontal line** with y-intercept $(0, b)$. It is the graph of a constant function with slope 0.

The graph of $x = a$ is a **vertical line** through the point $(a, 0)$. The slope is not defined. It is not the graph of a function.

Do Exercises 8–10.

We have graphed linear equations in several ways in this chapter. Although, in general, you can use any method that works best for you, we list some guidelines in the margin at right.

To graph a linear equation:

1. Is the equation of the type $x = a$ or $y = b$? If so, the graph will be a line parallel to an axis; $x = a$ is vertical and $y = b$ is horizontal.
2. If the line is of the type $y = mx$, both intercepts are the origin $(0, 0)$. Plot $(0, 0)$ and one other point.
3. If the line is of the type $y = mx + b$, plot the y-intercept and one other point.
4. If the equation is of the form $Ax + By = C$, graph using intercepts. If the intercepts are too close together, choose another point farther from the origin.
5. In all cases, use a third point as a check.

d Parallel and Perpendicular Lines

PARALLEL LINES

If two lines are vertical, they are parallel. How can we tell whether nonvertical lines are parallel? We examine their slopes and y-intercepts.

PARALLEL LINES

Two nonvertical lines are **parallel** if they have the *same* slope and *different* y-intercepts.

EXAMPLE 6 Determine whether the graphs of

$$y - 3x = 1 \quad \text{and} \quad 3x + 2y = -2$$

are parallel.

To determine whether lines are parallel, we first find their slopes. Thus we first find the slope–intercept form of each equation by solving for y:

$$\begin{aligned} y - 3x &= 1 \\ y &= 3x + 1; \end{aligned} \qquad \begin{aligned} 3x + 2y &= -2 \\ 2y &= -3x - 2 \\ y &= \tfrac{1}{2}(-3x - 2) \\ &= \tfrac{1}{2}(-3x) - \tfrac{1}{2}(2) \\ &= -\tfrac{3}{2}x - 1. \end{aligned}$$

The slopes, 3 and $-\frac{3}{2}$, are different. Thus the lines are not parallel, as the graphs at right confirm.

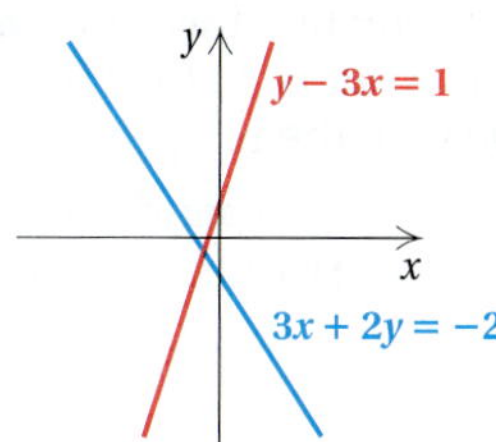

EXAMPLE 7 Determine whether the graphs of

$$3x - y = -5 \quad \text{and} \quad y - 3x = -2$$

are parallel.

We first find the slope–intercept form of each equation by solving for y:

$$\begin{aligned} 3x - y &= -5 \\ -y &= -3x - 5 \\ y &= 3x + 5; \end{aligned} \qquad \begin{aligned} y - 3x &= -2 \\ y &= 3x - 2. \end{aligned}$$

The slopes, 3, are the same. The y-intercepts, $(0, 5)$ and $(0, -2)$, are different. Thus the lines are parallel, as the graphs appear to confirm.

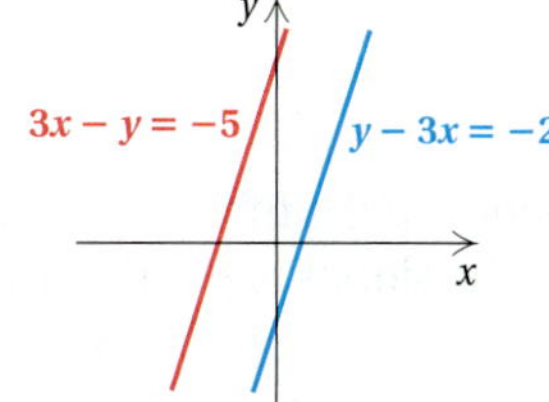

Determine whether the graphs of the given pair of lines are parallel.

11. $x + 4 = y,$
$y - x = -3$

12. $y + 4 = 3x,$
$4x - y = -7$

13. $y = 4x + 5,$
$2y = 8x + 10$

Determine whether the graphs of the given pair of lines are perpendicular.

14. $2y - x = 2,$
$y + 2x = 4$

15. $3y = 2x + 15,$
$2y = 3x + 10$

Answers on page A-44

Do Exercises 11–13.

PERPENDICULAR LINES

If one line is vertical and another is horizontal, they are perpendicular. Otherwise, how can we tell whether two lines are perpendicular?

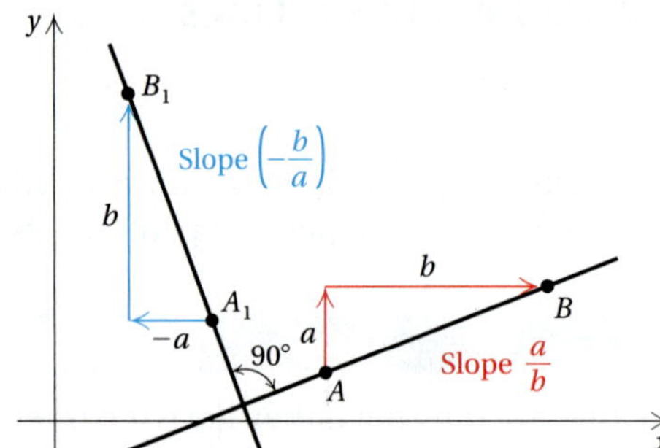

Consider a line $\overleftrightarrow{AB}$, as shown in the figure above, with slope a/b. Then think of rotating the line 90° to get a line $\overleftrightarrow{A_1B_1}$ perpendicular to $\overleftrightarrow{AB}$. For the new line, the rise and the run are interchanged, but the run is now negative. Thus the slope of the new line is $-b/a$, which is the opposite of the reciprocal of the slope of the first line. Also note that when we multiply the slopes, we get

$$\frac{a}{b}\left(-\frac{b}{a}\right) = -1.$$

This is the condition under which lines will be perpendicular.

> **PERPENDICULAR LINES**
>
> Two nonvertical lines are **perpendicular** if the product of their slopes is -1. (If one line has slope m, the slope of a line perpendicular to it is $-1/m$. That is, to find the slope of a line perpendicular to a given line, we take the reciprocal of the given slope and change the sign.)
>
> Lines are also perpendicular if one of them is vertical ($x = a$) and one of them is horizontal ($y = b$).

EXAMPLE 8 Determine whether the graphs of $5y = 4x + 10$ and $4y = -5x + 4$ are perpendicular.

To determine whether the lines are perpendicular, we determine whether the product of their slopes is -1. We first find the slope–intercept form of each equation by solving for y.

We have

$$\begin{aligned} 5y &= 4x + 10 \\ y &= \tfrac{1}{5}(4x + 10) \\ &= \tfrac{1}{5}(4x) + \tfrac{1}{5}(10) \\ &= \tfrac{4}{5}x + 2; \end{aligned}$$

$$\begin{aligned} 4y &= -5x + 4 \\ y &= \tfrac{1}{4}(-5x + 4) \\ &= \tfrac{1}{4}(-5x) + \tfrac{1}{4}(4) \\ &= -\tfrac{5}{4}x + 1. \end{aligned}$$

$4y = -5x + 4$
$5y = 4x + 10$

The slope of the first line is $\frac{4}{5}$, and the slope of the second line is $-\frac{5}{4}$. The product of the slopes is -1; that is, $\frac{4}{5} \cdot \left(-\frac{5}{4}\right) = -1$. Thus the lines are perpendicular.

Do Exercises 14 and 15.

CALCULATOR CORNER

Squaring the Viewing Window; Visualizing Perpendicular Lines Consider the standard $[-10, 10, -10, 10]$ viewing window on the left below. Note that the distance between tick marks is not the same on both axes. In fact, the distance between tick marks on the y-axis is about two-thirds the distance between tick marks on the x-axis. If we change the dimensions of the window to $[-6, 6, -4, 4]$, we get a window in which the distance between tick marks is about the same on both axes, as shown in the window on the right below. Such a window is called a **square viewing window.** In a square window, the length of the portion of the y-axis shown is two-thirds the length of the portion of the x-axis shown. One way to obtain a square window, then, is to choose window dimensions in which the length of the y-axis is two-thirds the length of the x-axis. Other examples of square windows are $[-3, 3, -2, 2]$ and $[-9, 9, -6, 6]$.

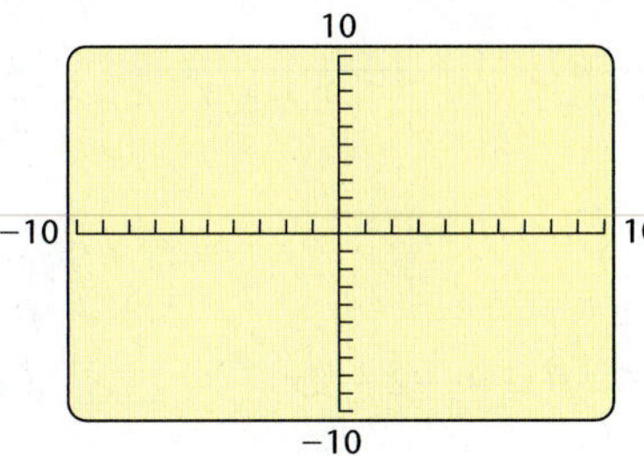

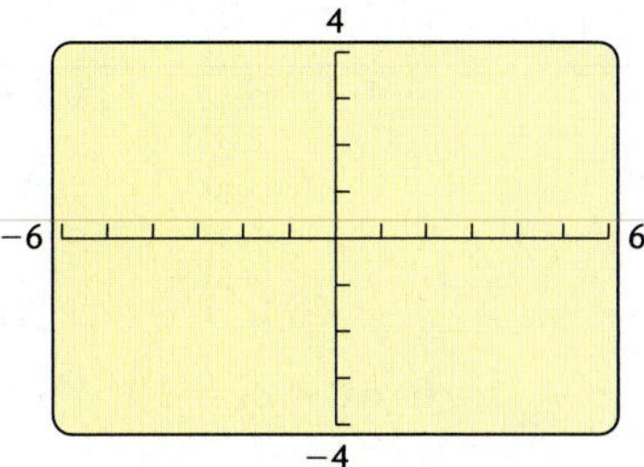

When the viewing window is square, we get an accurate representation of the slope of a line. This is important when we want to visualize perpendicular lines. The graphs of the perpendicular lines $y = x + 1$ and $y = -x + 2$ are shown in the standard window on the left below and in the square window $[-6, 6, -4, 4]$ on the right. Note that the lines do not appear to be perpendicular in the standard window, whereas they do in the square window.

$y_1 = x + 1,\ y_2 = -x + 2$

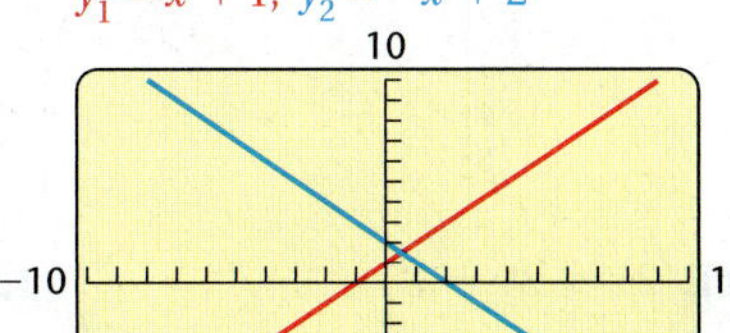

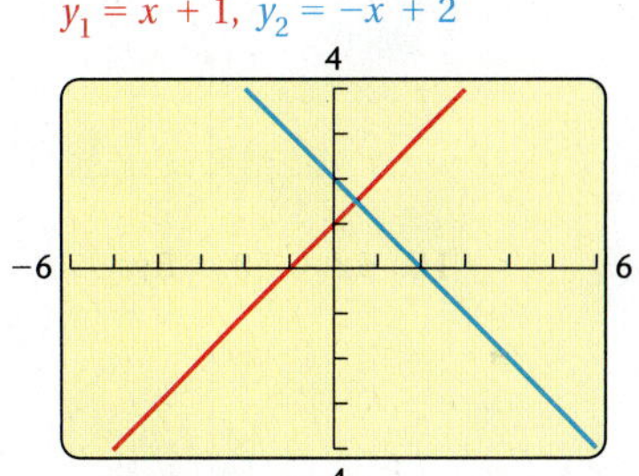

Exercise:

1. Graph each pair of equations in Margin Exercises 14 and 15 using a square viewing window. Note that each equation must be solved for y before it is entered on the equation-editor screen. Check visually whether the lines appear to be perpendicular.

12.4 EXERCISE SET

For Extra Help

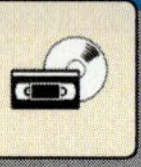
Digital Video Tutor CD 10 Videotape 14

InterAct Math

Math Tutor Center

MathXL

MyMathLab

a Find the intercepts and then graph the line.

1. $x - 2 = y$

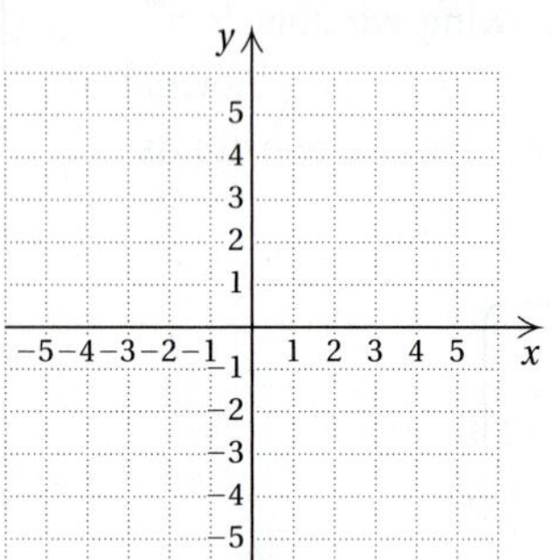

2. $x + 3 = y$

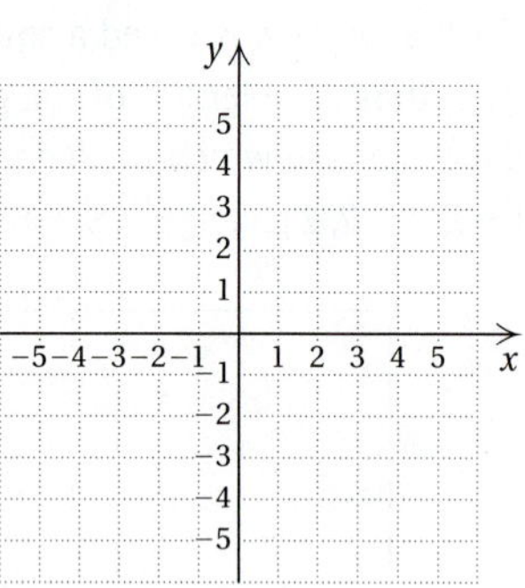

3. $x + 3y = 6$

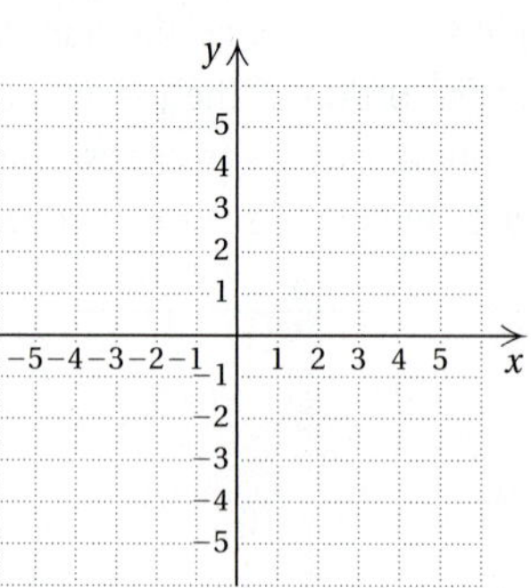

4. $x - 2y = 4$

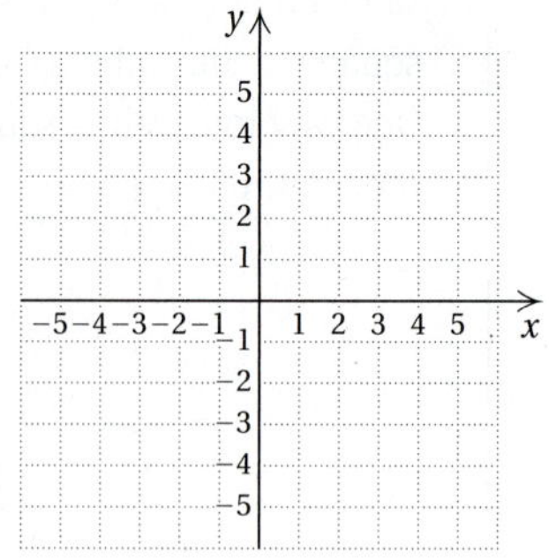

5. $2x + 3y = 6$

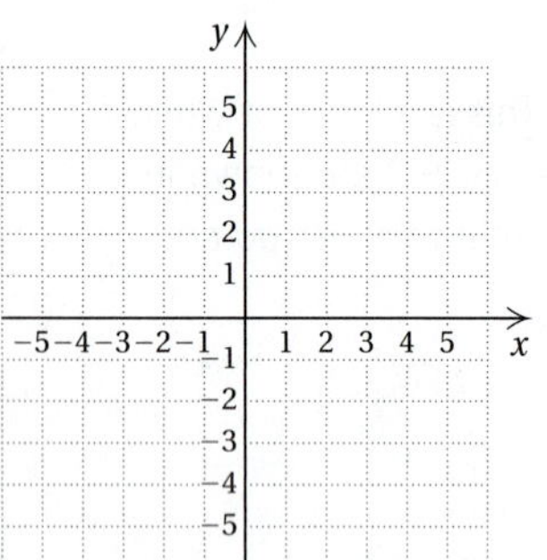

6. $5x - 2y = 10$

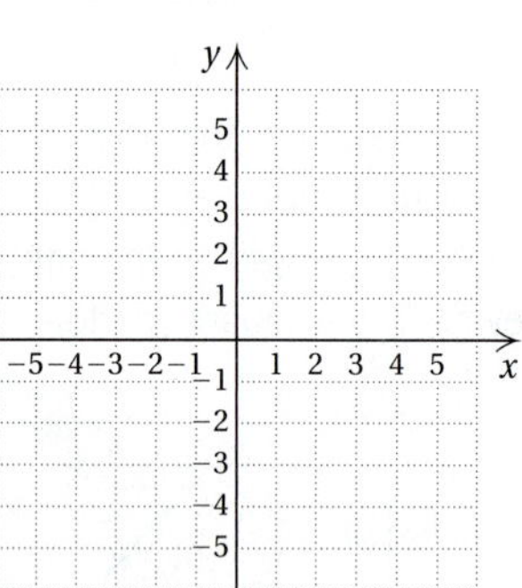

7. $f(x) = -2 - 2x$

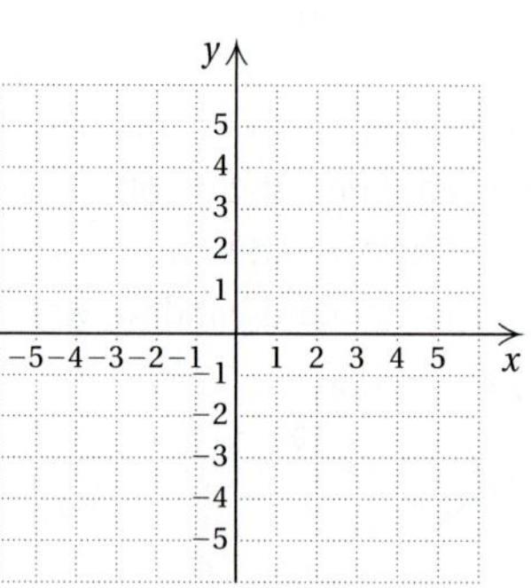

8. $g(x) = 5x - 5$

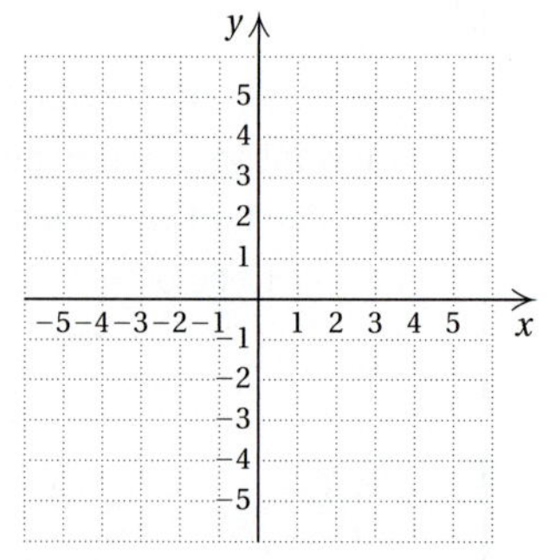

9. $5y = -15 + 3x$

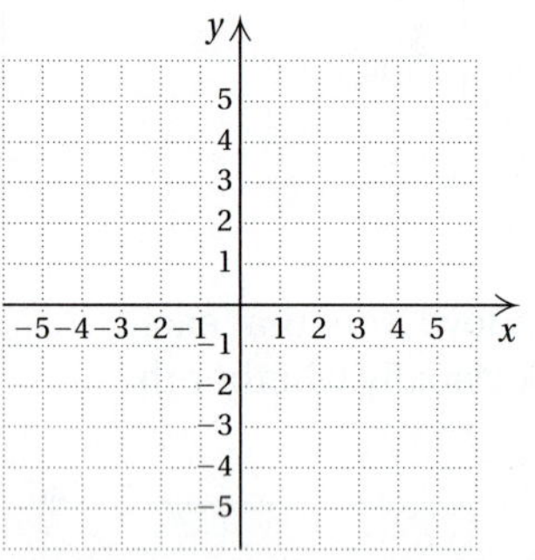

10. $5x - 10 = 5y$

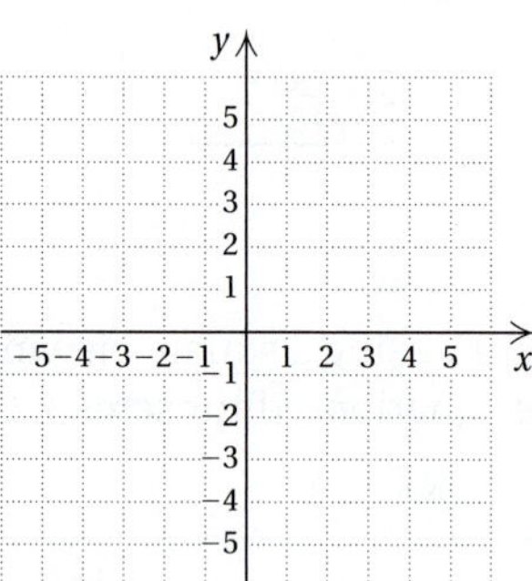

11. $2x - 3y = 6$

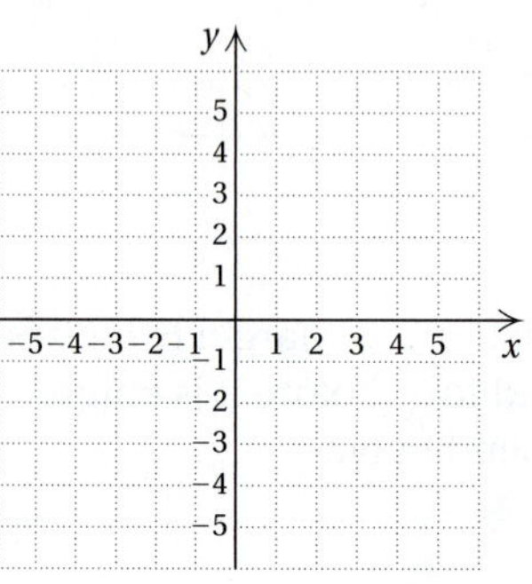

12. $4x + 5y = 20$

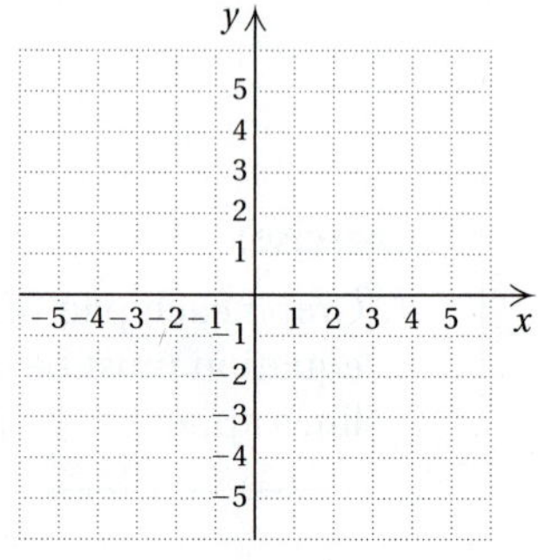

13. $2.8y - 3.5x = -9.8$

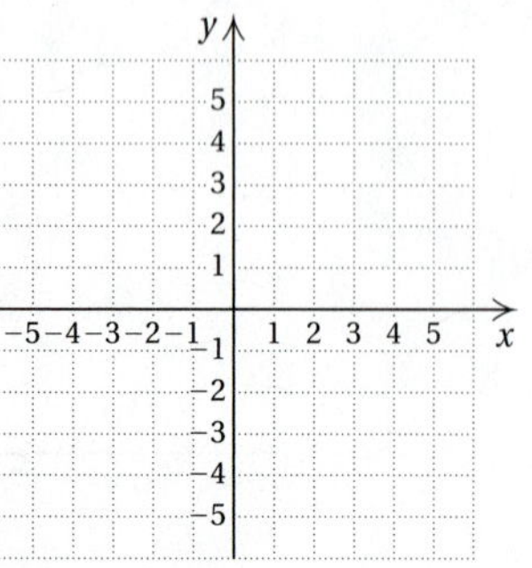

14. $10.8x - 22.68 = 4.2y$

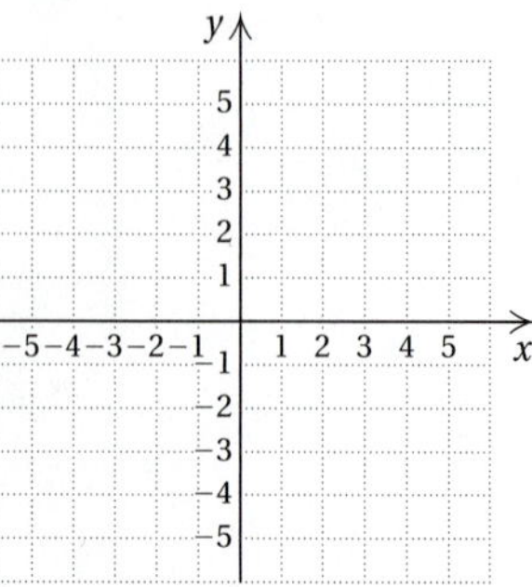

15. $5x + 2y = 7$

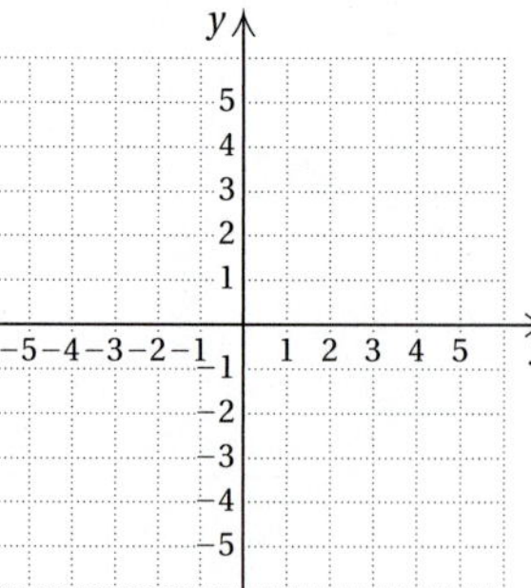

16. $3x - 4y = 10$

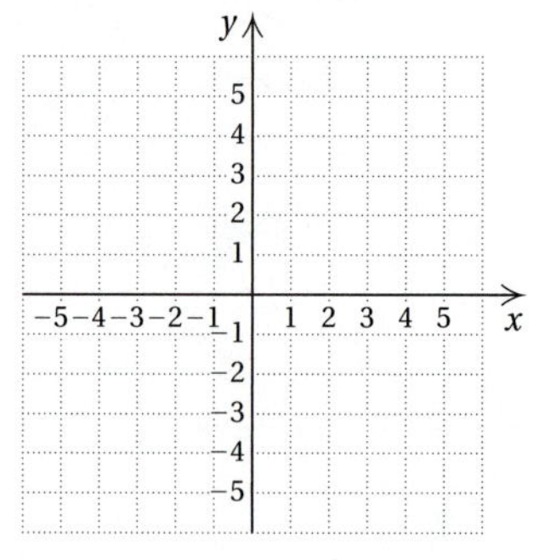

b Graph using the slope and the y-intercept.

17. $y = \frac{5}{2}x + 1$

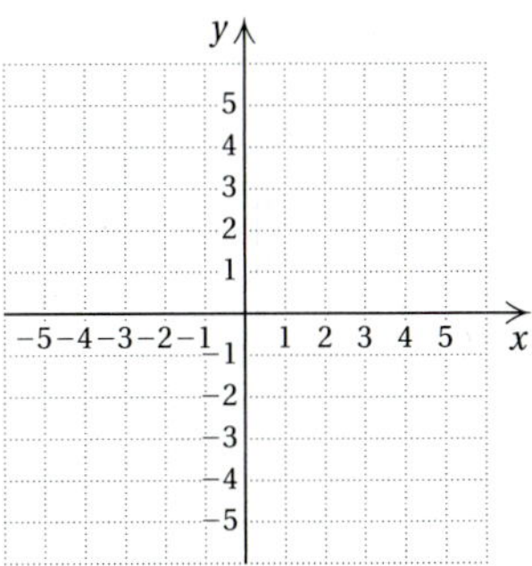

18. $y = \frac{2}{5}x - 4$

19. $f(x) = -\frac{5}{2}x - 4$

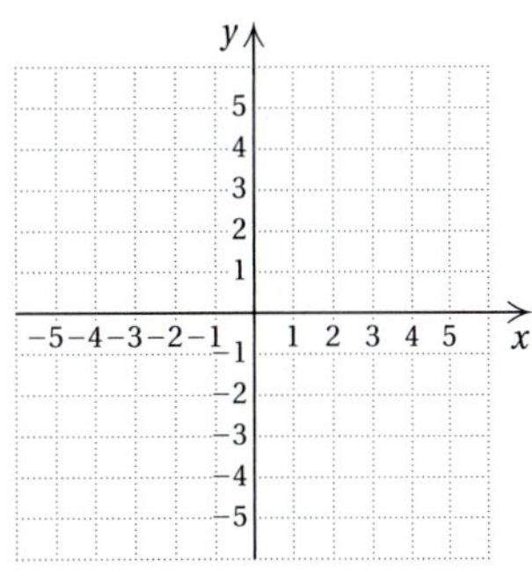

20. $f(x) = \frac{2}{5}x + 3$

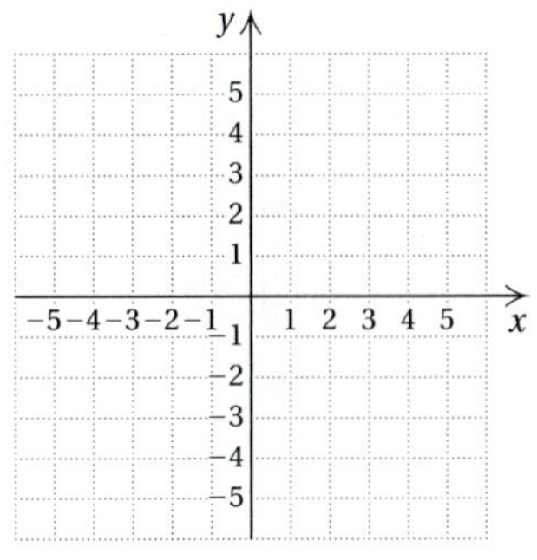

21. $x + 2y = 4$

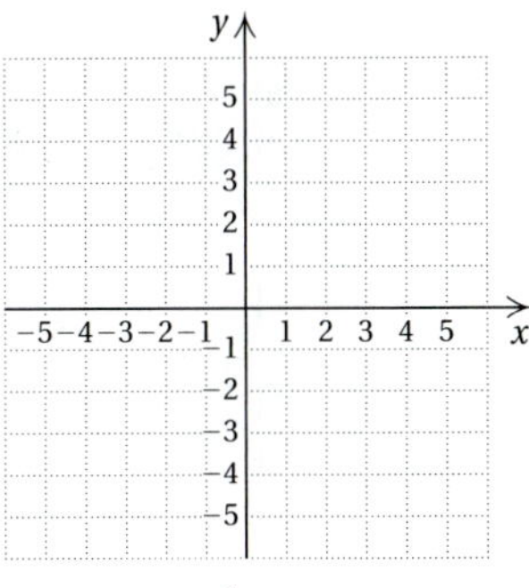

22. $x - 3y = 6$

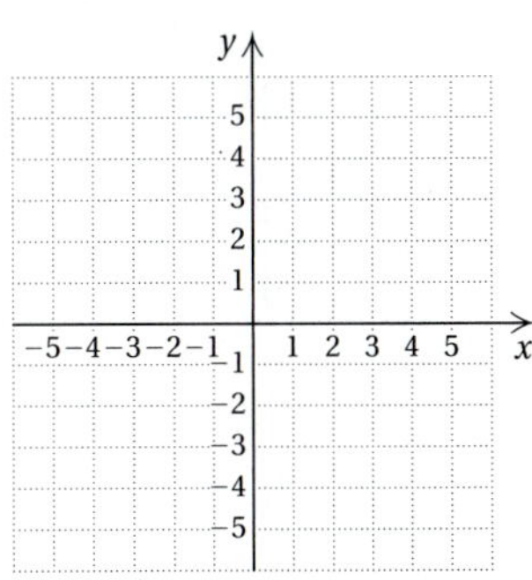

23. $4x - 3y = 12$

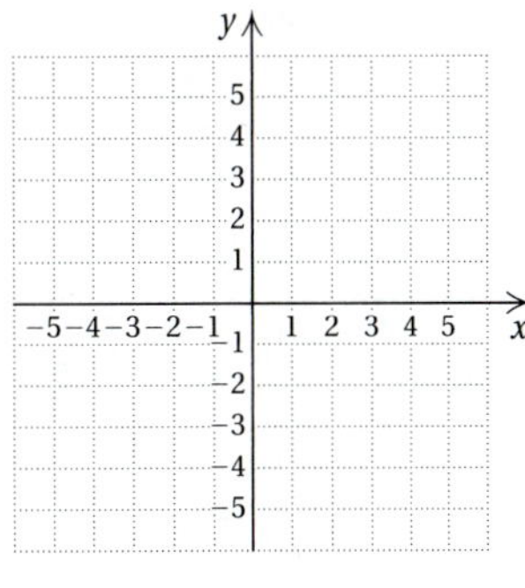

24. $2x + 6y = 12$

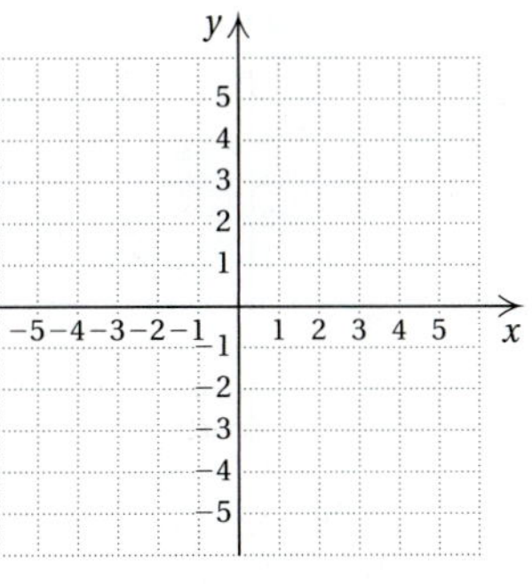

25. $f(x) = \frac{1}{3}x - 4$

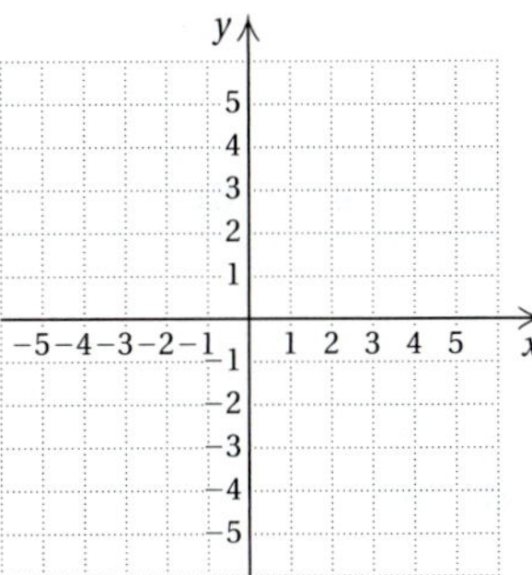

26. $g(x) = -0.25x + 2$

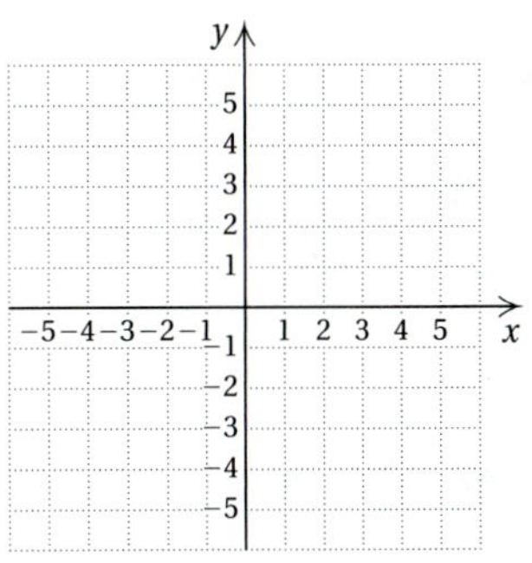

27. $5x + 4 \cdot f(x) = 4$
(*Hint*: Solve for $f(x)$.)

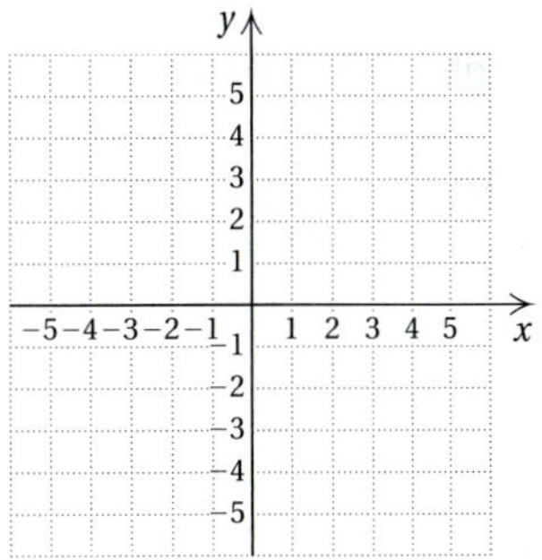

28. $3 \cdot f(x) = 4x + 6$

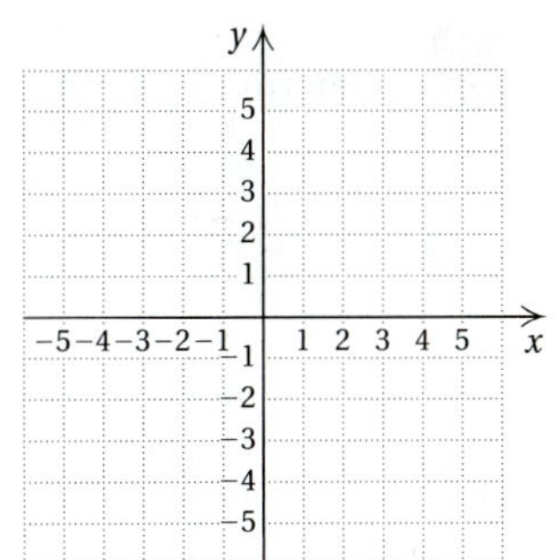

c Graph and, if possible, determine the slope.

29. $x = 1$

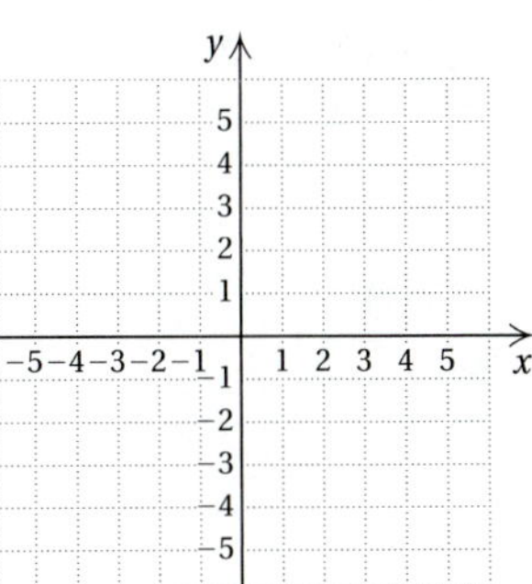

30. $x = -4$

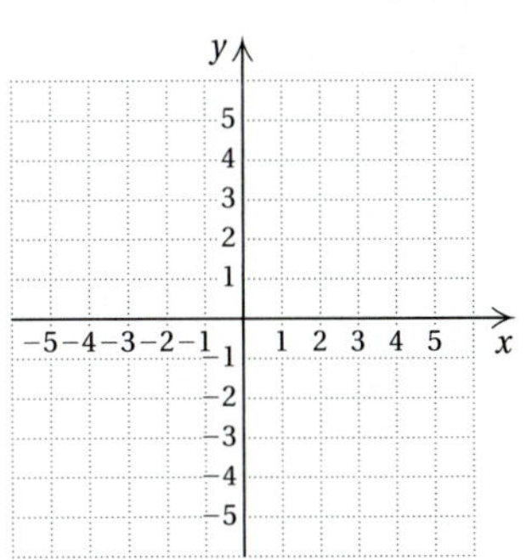

31. $y = -1$

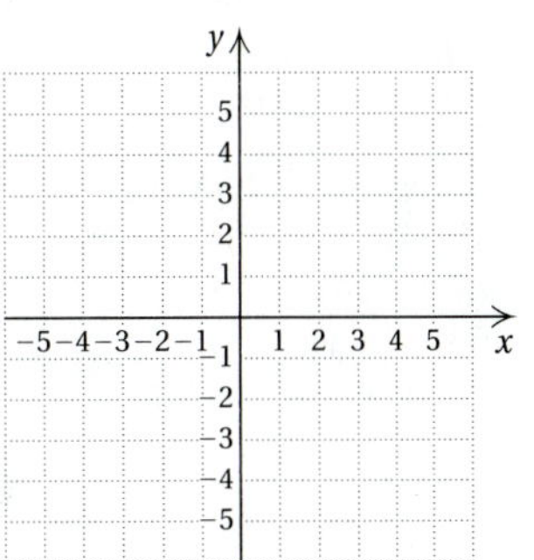

32. $y = \frac{3}{2}$

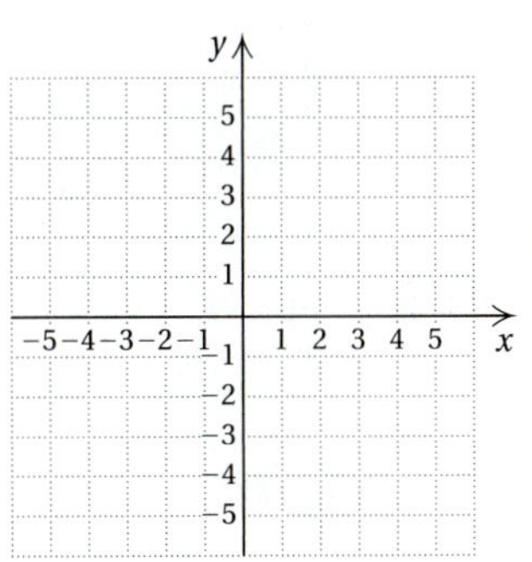

33. $f(x) = -6$

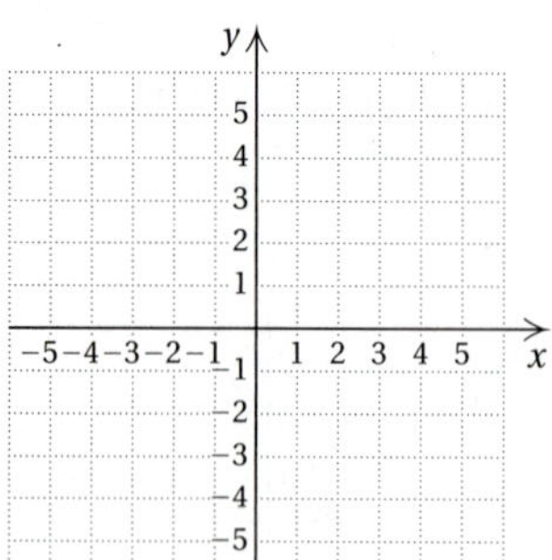

34. $f(x) = 2$

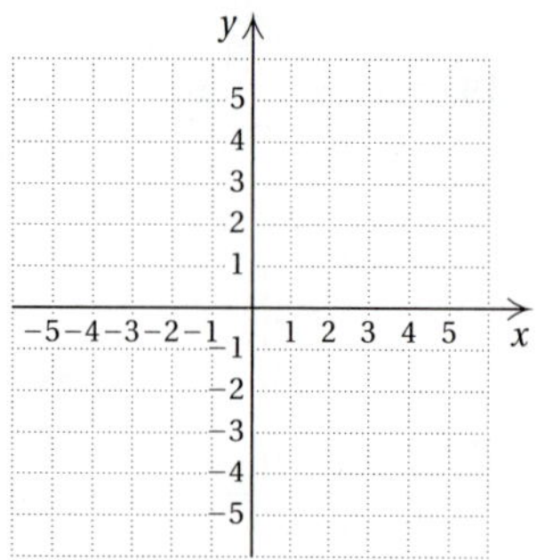

35. $y = 0$

36. $x = 0$

37. $2 \cdot f(x) + 5 = 0$

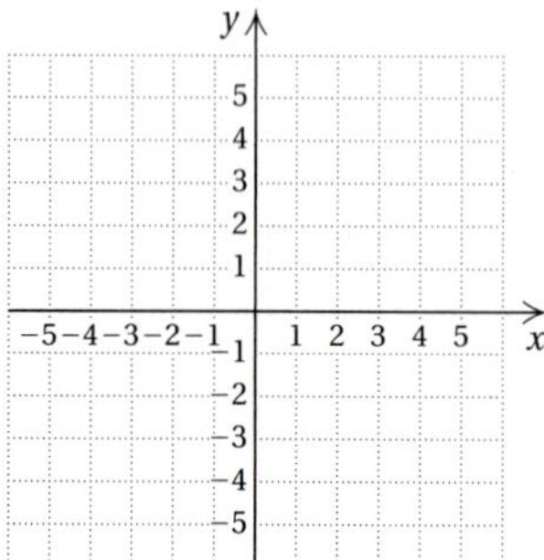

38. $4 \cdot g(x) + 3x = 12 + 3x$

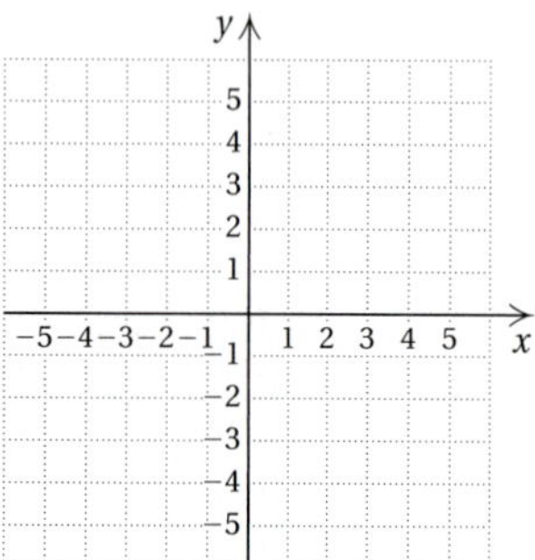

39. $7 - 3x = 4 + 2x$

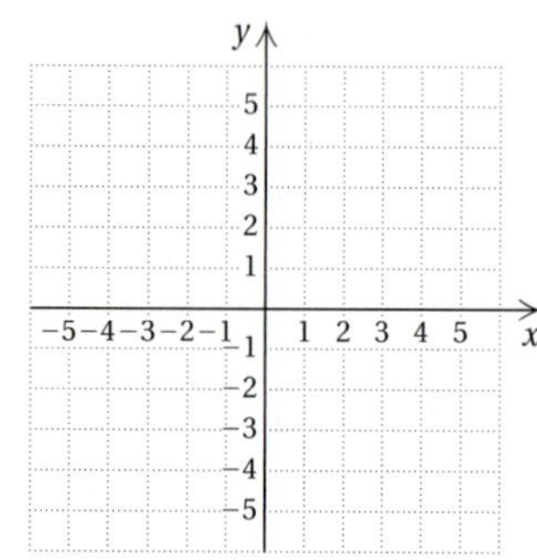

40. $3 - f(x) = 2$

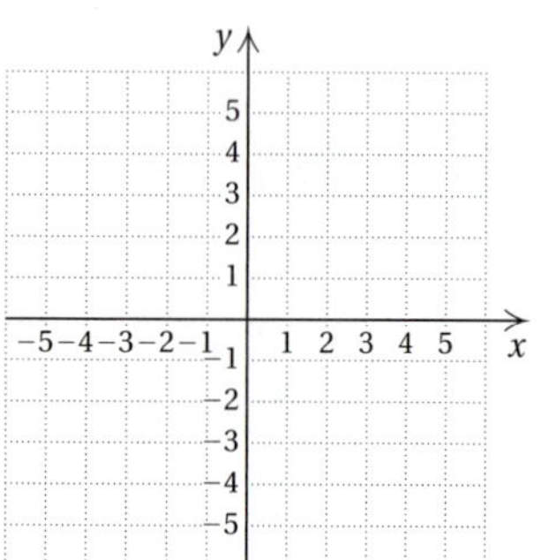

d Determine whether the graphs of the given pair of lines are parallel.

41. $x + 6 = y,$
$y - x = -2$

42. $2x - 7 = y,$
$y - 2x = 8$

43. $y + 3 = 5x,$
$3x - y = -2$

44. $y + 8 = -6x,$
$-2x + y = 5$

45. $y = 3x + 9,$
$2y = 6x - 2$

46. $y + 7x = -9,$
$-3y = 21x + 7$

47. $12x = 3,$
$-7x = 10$

48. $5y = -2,$
$\frac{3}{4}x = 16$

Determine whether the graphs of the given pair of lines are perpendicular.

49. $y = 4x - 5,$
$4y = 8 - x$

50. $2x - 5y = -3,$
$2x + 5y = 4$

51. $x + 2y = 5,$
$2x + 4y = 8$

52. $y = -x + 7,$
$y = x + 3$

53. $2x - 3y = 7,$
$2y - 3x = 10$

54. $x = y,$
$y = -x$

55. $2x = 3,$
$-3y = 6$

56. $-5y = 10,$
$y = -\frac{4}{9}$

57. D_W Under what conditions will the x- and the y-intercepts of a line be the same? What would the equation for such a line look like?

58. D_W Explain why the slope of a vertical line is not defined but the slope of a horizontal line is 0.

SKILL MAINTENANCE

Write in scientific notation. [9.2c]

59. 53,000,000,000

60. 0.000047

61. 0.018

62. 99,902,000

Write in decimal notation. [9.2c]

63. 2.13×10^{-5}

64. 9.01×10^{8}

65. 2×10^{4}

66. 8.5677×10^{-2}

Factor. [10.1b]

67. $9x - 15y$

68. $12a + 21ab$

69. $21p - 7pq + 14p$

70. $64x - 128y + 256$

SYNTHESIS

71. Find an equation of a horizontal line that passes through the point $(-2, 3)$.

72. Find an equation of a vertical line that passes through the point $(-2, 3)$.

73. Find the value of a such that the graphs of $5y = ax + 5$ and $\frac{1}{4}y = \frac{1}{10}x - 1$ are parallel.

74. Find the value of k such that the graphs of $x + 7y = 70$ and $y + 3 = kx$ are perpendicular.

75. Write an equation of the line that has x-intercept $(-3, 0)$ and y-intercept $\left(0, \frac{2}{5}\right)$.

76. Find the coordinates of the point of intersection of the graphs of the equations $x = -4$ and $y = 5$.

77. Write an equation for the x-axis. Is this equation a function?

78. Write an equation for the y-axis. Is this equation a function?

79. Find the value of m in $y = mx + 3$ so that the x-intercept of its graph will be $(4, 0)$.

80. Find the value of b in $2y = -7x + 3b$ so that the y-intercept of its graph will be $(0, -13)$.

81. Match each sentence with the most appropriate graph.

a) The rate at which fluids were given intravenously was doubled after 3 hr.

b) The rate at which fluids were given intravenously was gradually reduced to 0.

c) The rate at which fluids were given intravenously remained constant for 5 hr.

d) The rate at which fluids were given intravenously was gradually increased.

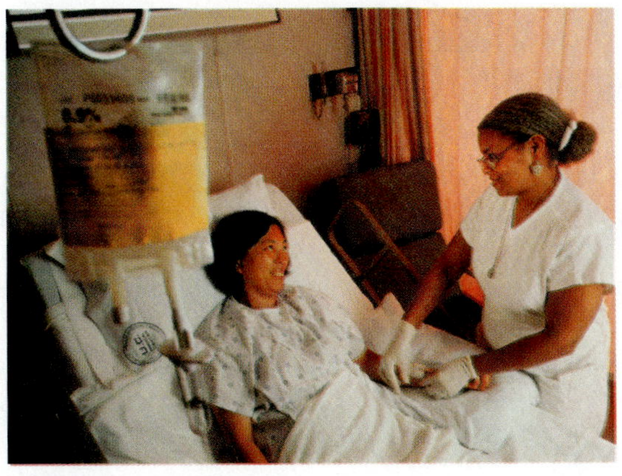

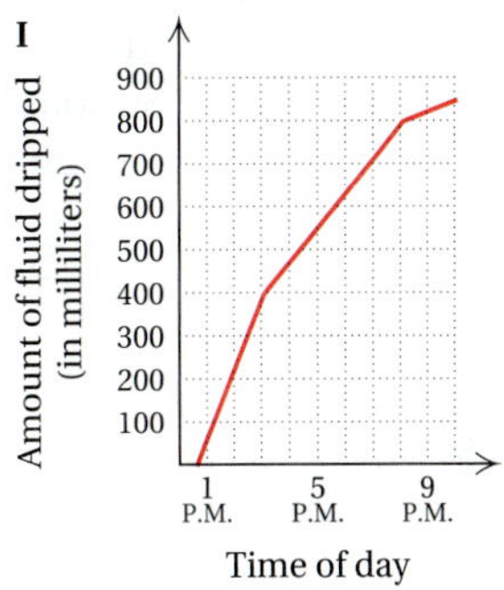

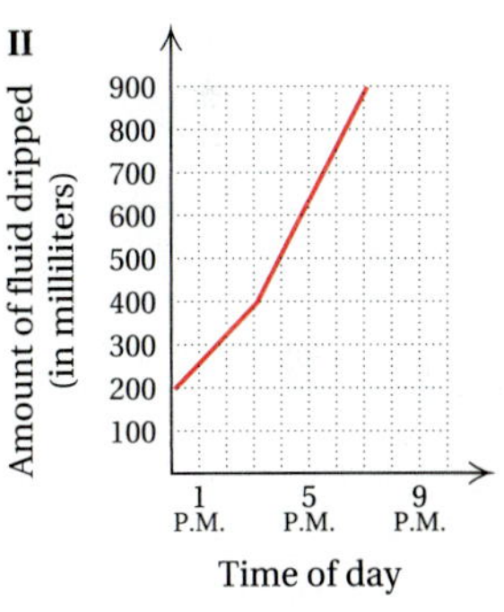

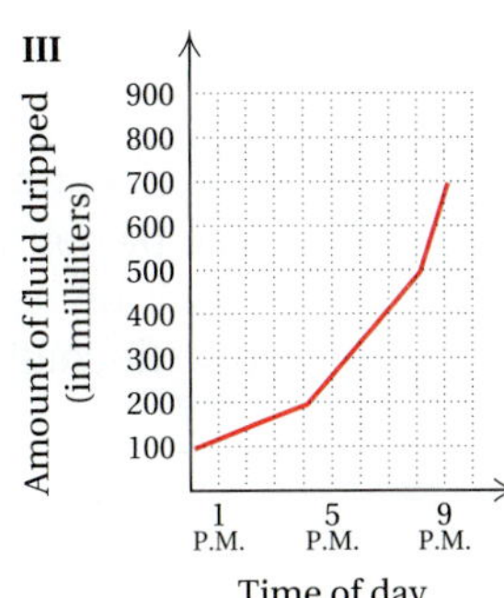

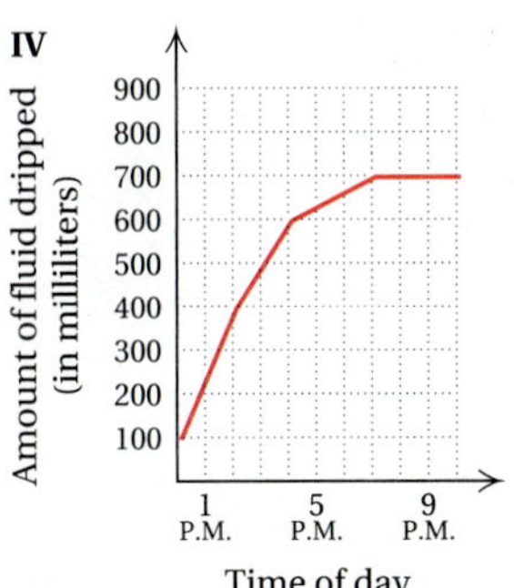

12.5 FINDING EQUATIONS OF LINES; APPLICATIONS

Objectives

a Find an equation of a line when the slope and the y-intercept are given.

b Find an equation of a line when the slope and a point are given.

c Find an equation of a line when two points are given.

d Given a line and a point not on the given line, find an equation of the line parallel to the line and containing the point, and find an equation of the line perpendicular to the line and containing the point.

e Solve applied problems involving linear functions.

In this section, we will learn to find a linear equation for a line for which we have been given two pieces of information.

a Finding an Equation of a Line When the Slope and the y-Intercept Are Given

If we know the slope and the y-intercept of a line, we can find an equation of the line using the slope–intercept equation $y = mx + b$.

EXAMPLE 1 A line has slope -0.7 and y-intercept $(0, 13)$. Find an equation of the line.

We use the slope–intercept equation and substitute -0.7 for m and 13 for b:

$$y = mx + b$$

$$y = -0.7x + 13.$$

Do Exercise 1 on the following page.

b Finding an Equation of a Line When the Slope and a Point Are Given

Suppose we know the slope of a line and the coordinates of one point on the line. We can use the slope–intercept equation to find an equation of the line. Or, we can use what is called a **point–slope equation.** We first develop a formula for such a line.

Suppose that a line of slope m passes through the point (x_1, y_1). For any other point (x, y) to lie on this line, we must have

$$\frac{y - y_1}{x - x_1} = m.$$

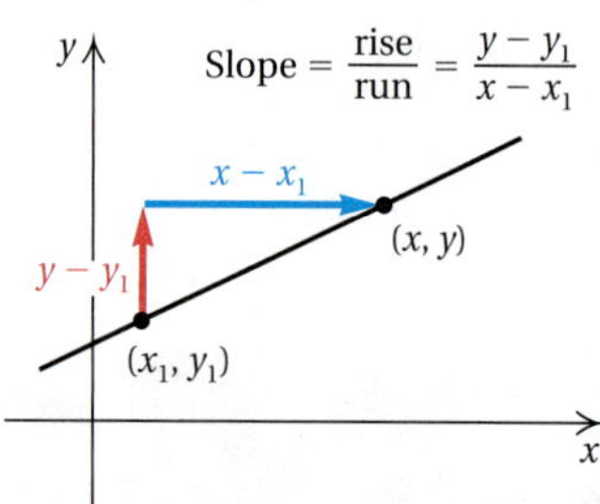

It is tempting to use this last equation as an equation of the line of slope m that passes through (x_1, y_1). The only problem with this form is that when x and y are replaced with x_1 and y_1, we have $\frac{0}{0} = m$, a false equation. To avoid this difficulty, we multiply both sides by $x - x_1$ and simplify:

$$(x - x_1)\frac{y - y_1}{x - x_1} = m(x - x_1) \qquad \text{Multiplying both sides by } x - x_1$$

$$y - y_1 = m(x - x_1). \qquad \text{Removing a factor of 1: } \frac{x - x_1}{x - x_1} = 1$$

This is the *point–slope* form of a linear equation.

POINT–SLOPE EQUATION

The **point–slope equation** of a line with slope m, passing through (x_1, y_1), is

$$y - y_1 = m(x - x_1).$$

If we know the slope of a line and a certain point on the line, we can find an equation of the line using either the point–slope equation,

$$y - y_1 = m(x - x_1),$$

or the slope–intercept equation,

$$y = mx + b.$$

EXAMPLE 2 Find an equation of the line with slope 5 and containing the point $\left(\frac{1}{2}, -1\right)$.

Using the Point–Slope Equation: We consider $\left(\frac{1}{2}, -1\right)$ to be (x_1, y_1) and 5 to be the slope m, and substitute:

$$\begin{aligned} y - y_1 &= m(x - x_1) \\ y - (-1) &= 5\left(x - \tfrac{1}{2}\right) && \text{Substituting} \\ y + 1 &= 5x - \tfrac{5}{2} && \text{Simplifying} \\ y &= 5x - \tfrac{5}{2} - 1 \\ y &= 5x - \tfrac{5}{2} - \tfrac{2}{2} \\ y &= 5x - \tfrac{7}{2}. \end{aligned}$$

Using the Slope–Intercept Equation: The point $\left(\frac{1}{2}, -1\right)$ is on the line, so it is a solution. Thus we can substitute $\frac{1}{2}$ for x and -1 for y in $y = mx + b$. We also substitute 5 for m, the slope. Then we solve for b:

$$\begin{aligned} y &= mx + b \\ -1 &= 5 \cdot \left(\tfrac{1}{2}\right) + b && \text{Substituting} \\ -1 &= \tfrac{5}{2} + b \\ -1 - \tfrac{5}{2} &= b \\ -\tfrac{2}{2} - \tfrac{5}{2} &= b \\ -\tfrac{7}{2} &= b. && \text{Solving for } b \end{aligned}$$

We then use the equation $y = mx + b$ and substitute 5 for m and $-\frac{7}{2}$ for b:

$$y = 5x - \tfrac{7}{2}.$$

Do Exercises 2–5.

1. A line has slope 3.4 and y-intercept $(0, -8)$. Find an equation of the line.

Find an equation of the line with the given slope and containing the given point.

2. $m = -5,\ (-4, 2)$

3. $m = 3,\ (1, -2)$

4. $m = 8,\ (3, 5)$

5. $m = -\dfrac{2}{3},\ (1, 4)$

Answers on page A-46

6. Find an equation of the line containing the points $(4, -3)$ and $(1, 2)$.

7. Find an equation of the line containing the points $(-3, -5)$ and $(-4, 12)$.

C Finding an Equation of a Line When Two Points Are Given

We can also use the slope–intercept equation or the point–slope equation to find an equation of a line when two points are given.

EXAMPLE 3 Find an equation of the line containing the points $(2, 3)$ and $(-6, 1)$.

First, we find the slope:

$$m = \frac{3 - 1}{2 - (-6)} = \frac{2}{8}, \text{ or } \frac{1}{4}.$$

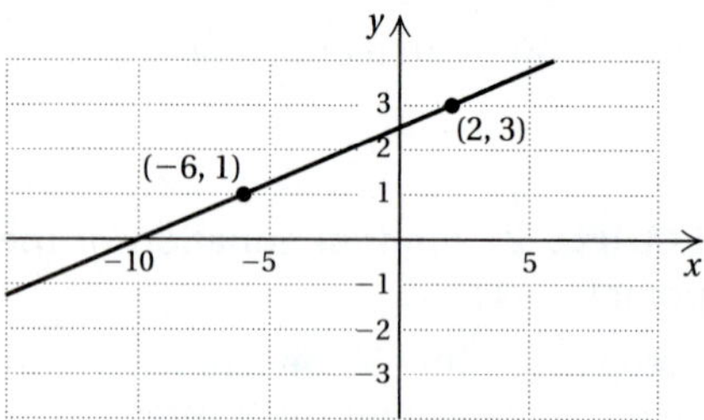

Now we have the slope and two points. We then proceed as we did in Example 2, using either point, and either the point–slope equation or the slope–intercept equation.

Using the Point–Slope Equation: We choose $(2, 3)$ and substitute 2 for x_1, 3 for y_1, and $\frac{1}{4}$ for m:

$$\begin{aligned} y - y_1 &= m(x - x_1) \\ y - 3 &= \tfrac{1}{4}(x - 2) \qquad \text{Substituting} \\ y - 3 &= \tfrac{1}{4}x - \tfrac{1}{2} \\ y &= \tfrac{1}{4}x - \tfrac{1}{2} + 3 \\ y &= \tfrac{1}{4}x - \tfrac{1}{2} + \tfrac{6}{2} \\ y &= \tfrac{1}{4}x + \tfrac{5}{2}. \end{aligned}$$

Using the Slope–Intercept Equation: We choose $(2, 3)$ and substitute 2 for x, 3 for y, and $\frac{1}{4}$ for m:

$$\begin{aligned} y &= mx + b \\ 3 &= \tfrac{1}{4} \cdot 2 + b \qquad \text{Substituting} \\ 3 &= \tfrac{1}{2} + b \\ 3 - \tfrac{1}{2} &= \tfrac{1}{2} + b - \tfrac{1}{2} \\ \tfrac{6}{2} - \tfrac{1}{2} &= b \\ \tfrac{5}{2} &= b. \qquad \text{Solving for } b \end{aligned}$$

Finally, we use the equation $y = mx + b$ and substitute $\frac{1}{4}$ for m and $\frac{5}{2}$ for b:

$$y = \tfrac{1}{4}x + \tfrac{5}{2}.$$

Do Exercises 6 and 7.

Answers on page A-46

d Finding an Equation of a Line Parallel or Perpendicular to a Given Line Through a Point Off the Line

We can also use the methods of Example 2 to find equations of lines through a point off the line parallel or perpendicular to a given line.

EXAMPLE 4 Find an equation of the line containing the point $(-1, 3)$ and parallel to the line $2x + y = 10$.

An equation parallel to the given line $2x + y = 10$ must have the same slope. To find that slope, we first find the slope–intercept equation by solving for y:

$$2x + y = 10$$
$$y = -2x + 10.$$

Thus the new line through $(-1, 3)$ must also have slope -2.

Using the Point–Slope Equation: We use the point $(-1, 3)$ and the slope -2, substituting -1 for x_1, 3 for y_1, and -2 for m:

$$y - y_1 = m(x - x_1)$$
$$y - 3 = -2(x - (-1)) \qquad \text{Substituting}$$
$$y - 3 = -2(x + 1) \qquad \text{Simplifying}$$
$$y - 3 = -2x + (-2) \cdot 1$$
$$y - 3 = -2x - 2$$
$$y = -2x - 2 + 3$$
$$y = -2x + 1.$$

Using the Slope–Intercept Equation: We substitute -1 for x and 3 for y in $y = mx + b$, and -2 for m, the slope. Then we solve for b:

$$y = mx + b$$
$$3 = -2(-1) + b \qquad \text{Substituting}$$
$$3 = 2 + b$$
$$1 = b. \qquad \text{Solving for } b$$

We then use the equation $y = mx + b$ and substitute -2 for m and 1 for b:

$$y = -2x + 1.$$

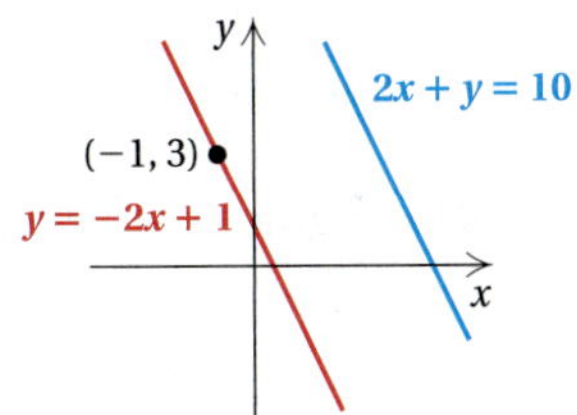

The given line $2x + y = 10$, or $y = -2x + 10$, and the new line $y = -2x + 1$ have the same slope but different y-intercepts. Thus their graphs are parallel.

Do Exercise 8.

8. Find an equation of the line containing the point $(2, -1)$ and parallel to the line $8x = 7y - 24$.

Answer on page A-46

9. Find an equation of the line containing the point $(5, 4)$ and perpendicular to the line $2x - 4y = 12$.

EXAMPLE 5 Find an equation of the line containing the point $(2, -3)$ and perpendicular to the line $4y - x = 20$.

To find the slope of the given line, we first find its slope–intercept form by solving for y:

$$4y - x = 20$$
$$4y = x + 20$$
$$\frac{4y}{4} = \frac{x + 20}{4}$$
$$y = \tfrac{1}{4}x + 5.$$

We know that the slope of the perpendicular line must be the opposite of the reciprocal of $\frac{1}{4}$. Thus the new line through $(2, -3)$ must have slope -4.

Using the Point–Slope Equation: We use the point $(2, -3)$ and the slope -4, substituting 2 for x_1, -3 for y_1, and -4 for m:

$$y - y_1 = m(x - x_1)$$
$$y - (-3) = -4(x - 2) \quad \text{Substituting}$$
$$y + 3 = -4(x - 2) \quad \text{Simplifying}$$
$$y + 3 = -4x + (-4)(-2)$$
$$y + 3 = -4x + 8$$
$$y = -4x + 8 - 3$$
$$y = -4x + 5.$$

Using the Slope–Intercept Equation: We now substitute 2 for x and -3 for y in $y = mx + b$. We also substitute -4 for m, the slope. Then we solve for b:

$$y = mx + b$$
$$-3 = -4(2) + b \quad \text{Substituting}$$
$$-3 = -8 + b$$
$$5 = b. \quad \text{Solving for } b$$

Finally, we use the equation $y = mx + b$ and substitute -4 for m and 5 for b:

$$y = -4x + 5.$$

The product of the slopes of the lines $4y - x = 20$ and $y = -4x + 5$ is -1. That is, $\frac{1}{4} \cdot (-4) = -1$. Thus their graphs are perpendicular.

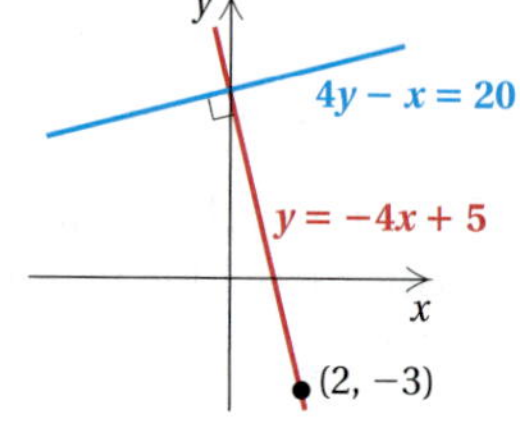

Do Exercise 9.

Answer on page A-46

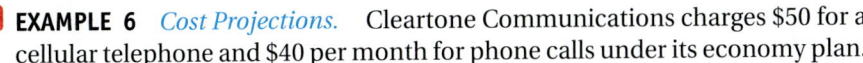

e Applications of Linear Functions

When the essential parts of a problem are described in mathematical language, we say that we have a **mathematical model.** We have already studied many kinds of mathematical models in this text—for example, the formulas in Section 7.4 and the functions in Section 12.1. Here we study linear functions as models.

EXAMPLE 6 *Cost Projections.* Cleartone Communications charges \$50 for a cellular telephone and \$40 per month for phone calls under its economy plan.

a) Formulate a linear function for total cost $C(t)$, where t is the number of months of telephone usage.

b) Graph the model.

c) Use the model to determine the cost of $3\frac{1}{2}$ months of service.

a) The problem describes a situation in which a monthly fee is charged after an initial purchase has been made. After 1 month of service, the total cost is

$$\$50 + \$40 \cdot 1 = \$90.$$

After 2 months of service, the total cost is

$$\$50 + \$40 \cdot 2 = \$130.$$

We can generalize that after t months of service, the total cost $C(t)$ is $C(t) = 50 + 40t$, where $t \geq 0$ (since there cannot be a negative number of months). Note that $C(t)$ is a way of saying that the cost C of the phone is a function of time t.

b) Before graphing, we rewrite the model in slope–intercept form:

$$C(t) = 40t + 50.$$

We note that the y-intercept is $(0, 50)$ and the slope, or rate of change, is \$40 per month. We plot $(0, 50)$, and from there we count \$40 *up* and 1 month *to the right.* This takes us to $(1, 90)$. We then draw a line through the points, calculating a third value as a check:

$$C(4) = 40 \cdot 4 + 50 = 210.$$

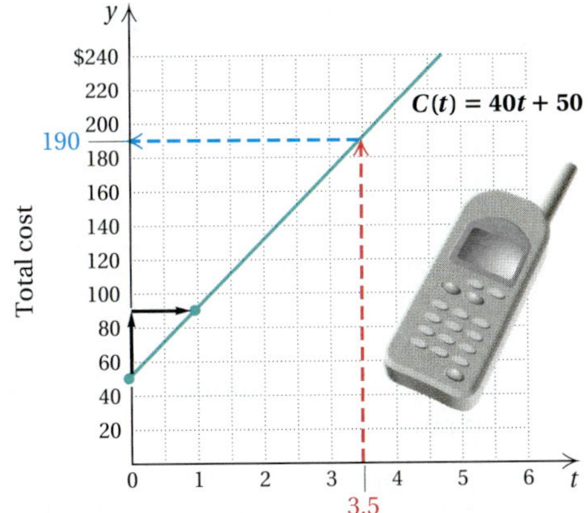

10. Cable TV Service. Clear County Cable TV Service charges a $25 installation fee and $20 per month for basic service.

a) Formulate a linear function for cost $C(t)$ for t months of cable TV service.

b) Graph the model.

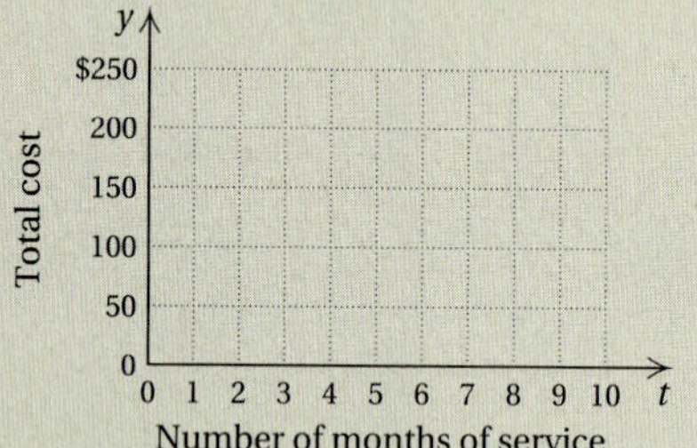

c) Use the model to determine the cost of $8\frac{1}{2}$ months of service.

Answers on page A-46

c) To find the cost for $3\frac{1}{2}$ months, we determine $C(3.5)$:

$$C(3.5) = 40(3.5) + 50 = 190.$$

Thus it would cost $190 for $3\frac{1}{2}$ months of service.

Do Exercise 10.

In the following example, we use two points and find an equation for the linear function through these points. Then we use the equation to make a prediction.

EXAMPLE 7 *Hat Size as a Function of Head Circumference.* The table below lists data relating hat size to head circumference.

HEAD CIRCUMFERENCE C (in inches)	HAT SIZE H
21.2	$6\frac{3}{4}$
22	7

Source: Shushan's New Orleans

a) Assuming a constant rate of change, use the two data points to find a linear function that fits the data.

b) If a person has a head circumference of 24.8 in., what is his or her hat size?

c) If a person has a hat size of 8, what is the circumference of his or her head?

a) A linear function can be used to model the data. We let

H = a person's hat size and

C = the circumference of his or her head.

This gives us two ordered pairs (21.2, 6.75) and (22, 7). After choosing suitable scales on the axes, we can draw a graph as shown below.

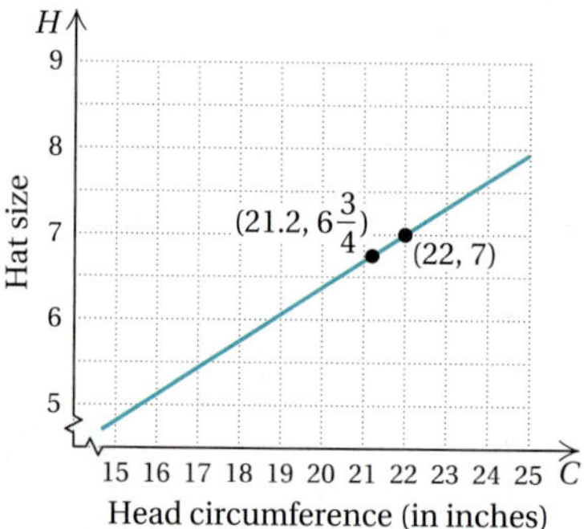

Next, we find the slope. This corresponds to the rate of change of hat size with respect to head circumference:

$$m = \frac{7 \text{ size} - 6.75 \text{ size}}{22 \text{ in.} - 21.2 \text{ in.}} = \frac{0.25 \text{ size}}{0.8 \text{ in.}} = 0.3125 \text{ sizes per in.}$$

Next, we find the linear equation $H = mC + b$, which fits the data. We use either the point–slope equation or the slope–intercept equation.

Using the Point–Intercept Equation: We choose one of the points—say, (22, 7)—and use the slope in the point–slope equation:

$$H - H_1 = m(C - C_1)$$

$$H - 7 = 0.3125(C - 22) \quad \text{Substituting}$$

$$H - 7 = 0.3125C - 6.875 \quad \text{Using a distributive law to remove parentheses}$$

$$H = 0.3125C + 0.125. \quad \text{Adding 7}$$

Using function notation in either case, we have

$$H(C) = 0.3125C + 0.125, \quad \text{or} \quad H(C) = \frac{5}{16}C + \frac{1}{8}.$$

Using the Slope–Intercept Equation: We choose one of the points—say, (22, 7). We substitute 22 for C, 7 for H, and 0.3125 for m in $H = mC + b$, and solve for b:

$$H = mC + b$$

$$7 = 0.3125(22) + b \quad \text{Substituting}$$

$$7 = 6.875 + b$$

$$0.125 = b. \quad \text{Solving for } b$$

Finally, we use the equation $H = mC + b$ and substitute 0.3125 for m and 0.125 for b:

$$H = 0.3125C + 0.125, \quad \text{or} \quad H = \frac{5}{16}C + \frac{1}{8}.$$

b) We have a linear function

$$H(C) = 0.3125C + 0.125.$$

To predict the hat size, we substitute the head circumference, 24.8, for C:

$$H(C) = 0.3125C + 0.125$$

$$= 0.3125(24.8) + 0.125 \quad \text{Substituting}$$

$$= 7.75 + 0.125$$

$$= 7.875, \text{ or } 7\tfrac{5}{8}.$$

The hat size is $7\frac{5}{8}$.

c) To find the circumference of his or her head, we substitute 8 for H and solve for C:

$$H(C) = 0.3125C + 0.125$$

$$8 = 0.3125C + 0.125 \quad \text{Substituting}$$

$$7.875 = 0.3125C$$

$$\frac{7.875}{0.3125} = C$$

$$25.2 = C.$$

The head circumference is 25.2 in.

Do Exercise 11.

11. Ring Size. The table below lists data regarding the correspondence between the diameter of a person's finger, in inches, and his or her ring size (U.S.) Let x = the diameter of a finger, in inches, and R = the U.S. ring size.

FINGER DIAMETER, x (in inches)	RING SIZE, R (in U.S. units)
0.586	4
0.783	10

Source: Oromi, Inc.

a) Use the two data points to find a linear function that fits the data.

b) Use the function to determine the ring size of a person whose finger has a diameter of 0.619 in.

c) Cliff has a ring size of $11\frac{1}{2}$. What is the diameter of his finger, in inches?

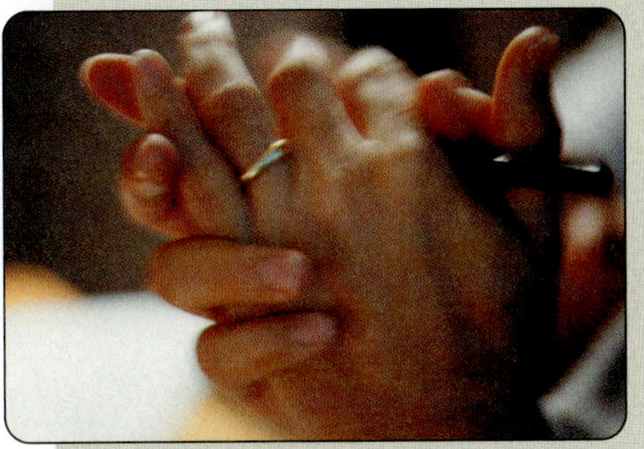

Answers on page A-46

CALCULATOR CORNER

Linear Regression: Fitting a Linear Equation to a Set of Data We can use the LINEAR REGRESSION feature on a graphing calculator to fit a linear equation to a set of data. The following table shows the correspondence between the diameter of a person's finger and his or her ring size, in U.S. units.

FINGER DIAMETER (in inches)	RING SIZE (in U.S. Units)
0.487	1
0.528	$2\frac{1}{4}$
0.577	$3\frac{3}{4}$
0.619	5
0.676	$6\frac{3}{4}$
0.693	$7\frac{1}{4}$
0.734	$8\frac{1}{2}$
0.783	10
0.825	$11\frac{1}{4}$

Source: Oromi, Inc.

a) Fit a regression line to the data using the LINEAR REGRESSION feature on a graphing calculator.

b) Use the linear model to determine the ring size of a person with a finger diameter of 0.685 in.

a) We will enter the data on the STAT list editor screen as ordered pairs in which the first coordinate is the finger diameter and the second is the corresponding ring size. First, we clear any existing lists by pressing [STAT] 4 [2nd] [L_1] [,] [2nd] [L_2] [,] [2nd] [L_3] [,] [2nd] [L_4] [,] [2nd] [L_5] [,] [2nd] [L_6] [ENTER]. (L_1 through L_6 are the second operations associated with the numeric keys 1 through 6.)

Once the lists have been cleared, we can enter the coordinates of the points. We press [STAT] [1] and enter the first coordinates in L_1 and the second coordinates in L_2. We position the cursor at the top of column L_1, below the L_1 heading. To enter 0.487, we press [.] [4] [8] [7] [ENTER], or [.] [4] [8] [7] [▽]. We then continue typing the first coordinates in order, each followed by [ENTER] or [▽]. Next, we press [▷] to move to the top of column L_2, and type the second coordinates in succession, each followed by [ENTER] or [▽]. (We enter the mixed numerals in decimal notation, as shown in the window below.) Note that the coordinates of each point must be in the same position in both lists.

L1	L2	L3 2
.487	1	------
.528	2.25	
.577	3.75	
.619	5	
.676	6.75	
.693	7.25	
.734	8.5	

L2 (1) = 1

(continued)

Now we will plot the data points to determine if a linear equation would be a good model for these data. First, we turn on the STAT PLOT feature. To access the STAT PLOT screen, we press [2nd] [STAT PLOT]. (STAT PLOT is the second operation associated with the [Y=] key in the upper left-hand corner of the keypad.)

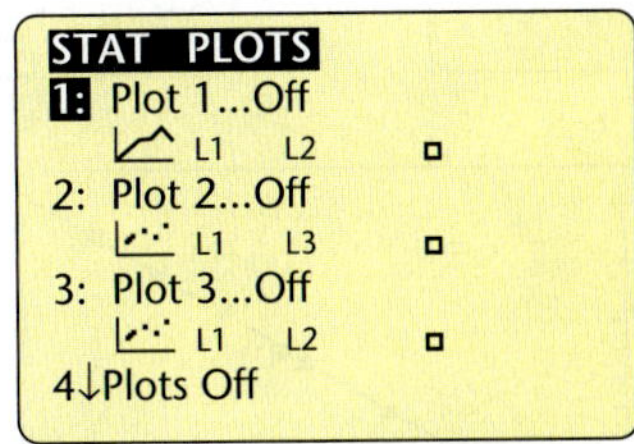

We will use Plot 1. We access it by highlighting 1 and pressing [ENTER], or simply by pressing 1. Next, we position thc cursor over On and press [ENTER] to turn on Plot 1. The entries Type, Xlist, and Ylist should be as shown below. The last item, Mark, allows us to choose a box, a cross, or a dot for each point. Here we have selected a box. To select Type and Mark, we position the cursor over the appropriate selection and press [ENTER]. We use the L_1 and L_2 keys (associated with the 1 and 2 numeric keys) to select Xlist and Ylist.

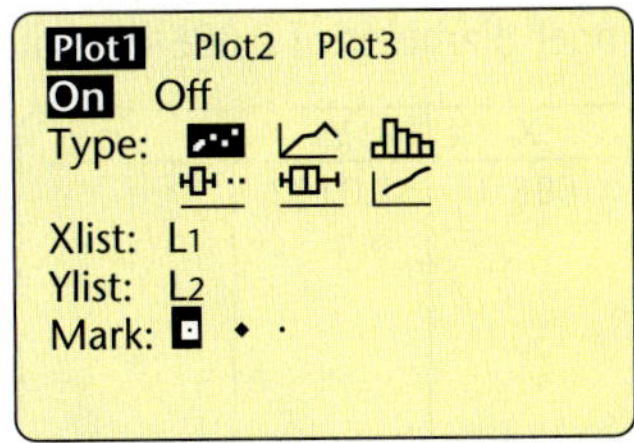

Note that there should be no equations entered on the "Y = " screen. If there are entries present, they should be cleared before we proceed. (See p. 532.) To see the plotted points, we press [ZOOM] [9]. These keystrokes activate the ZOOMSTAT operation, which automatically defines a viewing window that displays all the points.

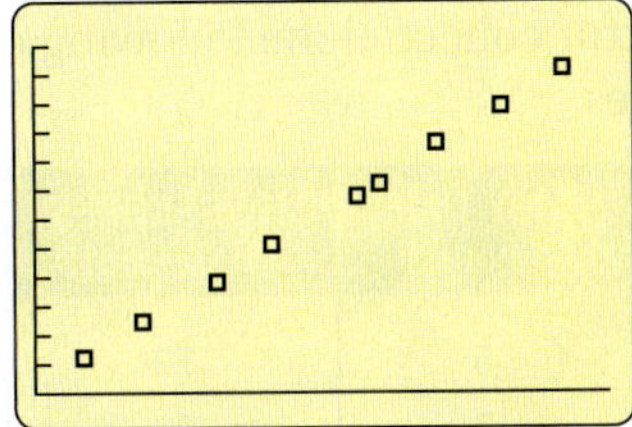

The data points appear to fall on or near a straight line, so we will fit a linear equation to these data. We press [STAT] [▷] [4] [VARS] [▷] [1] [1] [ENTER]. The first three keystrokes select LinReg($ax + b$) from the STAT CALC menu and display the coefficients a and b of the regression equation $y = ax + b$. The keystrokes [VARS] [▷] [1] [1] copy the regression equation to the equation-editor screen as y_1. We see that the regression equation is $y = 30.34160781x - 13.7703335$. If we press [ZOOM] 9 again, we will see the regression equation graphed with the data points.

(continued)

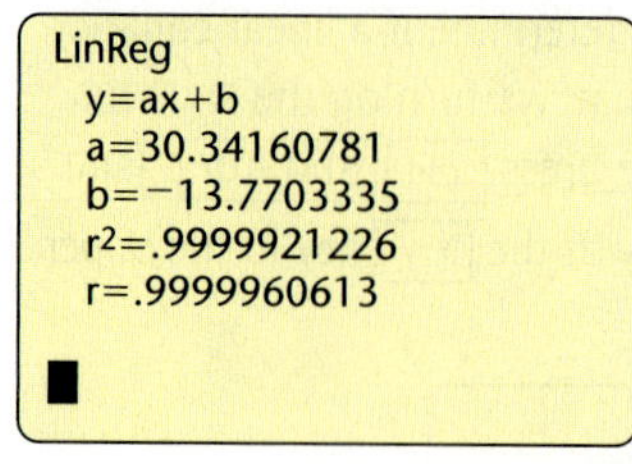

Plot1 Plot2 Plot3
\Y1 = 30.341607812
53X+−13.77033349
6201
\Y2 =
\Y3 =
\Y4 =
\Y5 =

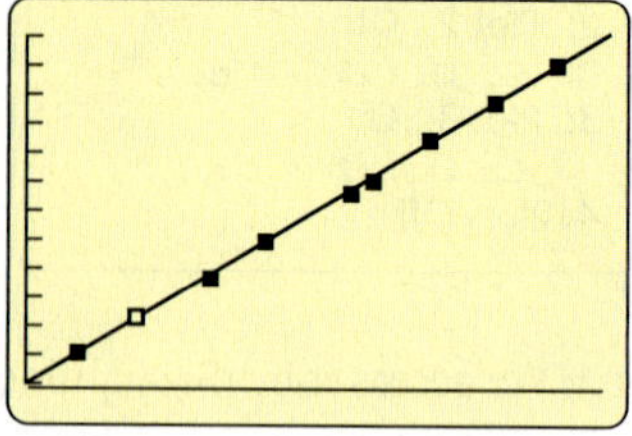

Before graphing other equations, we turn off the STAT PLOT by pressing [Y=], positioning the cursor over Plot 1 at the top of the screen, and then pressing [ENTER]. Plot 1 is now no longer highlighted, indicating that it is not turned on.

b) We can use a table set in ASK mode to determine the ring size of a person with a finger diameter of 0.685 in. (See p. 924.) When $x = 0.685$, $y \approx 7$, so a person with a finger diameter of 0.685 in. would wear a size 7 ring.

X	Y1	
.685	7.0137	

X =

Exercises:

1. Find the ring size of a person with a finger diameter of 0.619 in. Compare your answer with the answer you found in Margin Exercise 11(b).
2. An algebra instructor collected the following data relating study time and test scores.

STUDY TIME (in hours)	TEST SCORES (in percent)
5	73
6	80
8	88
11	92
12	94

a) Fit a regression line to the data using the LINEAR REGRESSION feature on a graphing calculator.

b) Use the linear model to estimate the test score of a student who studies for 10 hr.

12.5 EXERCISE SET

For Extra Help

Digital Video Tutor CD 11 Videotape 14 | InterAct Math | Math Tutor Center | MathXL | MyMathLab

a Find an equation of the line having the given slope and y-intercept.

1. Slope: -8; y-intercept: $(0, 4)$

2. Slope: 5; y-intercept: $(0, -3)$

3. Slope: 2.3; y-intercept: $(0, -1)$

4. Slope: -9.1; y-intercept: $(0, 2)$

Find a linear function $f(x) = mx + b$ whose graph has the given slope and y-intercept.

5. Slope: $-\frac{7}{3}$; y-intercept: $(0, -5)$

6. Slope: $\frac{4}{5}$; y-intercept: $(0, 28)$

7. Slope: $\frac{2}{3}$; y-intercept: $\left(0, \frac{5}{8}\right)$

8. Slope: $-\frac{7}{8}$; y-intercept: $\left(0, -\frac{7}{11}\right)$

b Find an equation of the line having the given slope and containing the given point.

9. $m = 5$, $(4, 3)$

10. $m = 4$, $(5, 2)$

11. $m = -3$, $(9, 6)$

12. $m = -2$, $(2, 8)$

13. $m = 1$, $(-1, -7)$

14. $m = 3$, $(-2, -2)$

15. $m = -2$, $(8, 0)$

16. $m = -3$, $(-2, 0)$

17. $m = 0$, $(0, -7)$

18. $m = 0$, $(0, 4)$

19. $m = \frac{2}{3}$, $(1, -2)$

20. $m = -\frac{4}{5}$, $(2, 3)$

c Find an equation of the line containing the given pair of points.

21. $(1, 4)$ and $(5, 6)$

22. $(2, 5)$ and $(4, 7)$

23. $(-3, -3)$ and $(2, 2)$

24. $(-1, -1)$ and $(9, 9)$

25. $(-4, 0)$ and $(0, 7)$

26. $(0, -5)$ and $(3, 0)$

27. $(-2, -3)$ and $(-4, -6)$

28. $(-4, -7)$ and $(-2, -1)$

29. $(0, 0)$ and $(6, 1)$

30. $(0, 0)$ and $(-4, 7)$

31. $\left(\frac{1}{4}, -\frac{1}{2}\right)$ and $\left(\frac{3}{4}, 6\right)$

32. $\left(\frac{2}{3}, \frac{3}{2}\right)$ and $\left(-3, \frac{5}{6}\right)$

d Write an equation of the line containing the given point and parallel to the given line.

33. $(3, 7)$; $x + 2y = 6$

34. $(0, 3)$; $2x - y = 7$

35. $(2, -1)$; $5x - 7y = 8$

36. $(-4, -5)$; $2x + y = -3$

37. $(-6, 2)$; $3x = 9y + 2$

38. $(-7, 0)$; $2y + 5x = 6$

Write an equation of the line containing the given point and perpendicular to the given line.

39. $(2, 5)$; $2x + y = 3$

40. $(4, 1)$; $x - 3y = 9$

41. $(3, -2)$; $3x + 4y = 5$

42. $(-3, -5)$; $5x - 2y = 4$

43. $(0, 9)$; $2x + 5y = 7$

44. $(-3, -4)$; $6y - 3x = 2$

Solve.

45. *Moving Costs.* Musclebound Movers charges \$85 plus \$40 an hour to move households across town.

a) Formulate a linear function for total cost $C(t)$ for t hours of moving.
b) Graph the model.
c) Use the model to determine the cost of $6\frac{1}{2}$ hr of moving service.

46. *Deluxe Cable TV Service.* Twin Cities Cable TV Service charges a \$35 installation fee and \$20 per month for basic service.

a) Formulate a linear function for total cost $C(t)$ for t months of cable TV service.
b) Graph the model.
c) Use the model to determine the cost of 9 months of service.

47. *Value of a Fax Machine.* FaxMax bought a multifunction fax machine for \$750. The value $V(t)$ of the machine depreciates (declines) at a rate of \$25 per month.

a) Formulate a linear function for the value $V(t)$ of the machine after t months.
b) Graph the model.
c) Use the model to determine the value of the machine after 13 months.

48. *Value of a Computer.* SendUp Graphics bought a computer for \$3800. The value $V(t)$ of the computer depreciates at a rate of \$50 per month.

a) Formulate a linear function for the value $V(t)$ of the computer after t months.
b) Graph the model.
c) Use the model to determine the value of the computer after $10\frac{1}{2}$ months.

In Exercises 49–54, assume that a constant rate of change exists for each model formed.

49. *NHL Salaries.* The table below lists data regarding the average salaries of players in the National Hockey League in 1991 and 2001.

YEAR	AVERAGE SALARY
1991	\$ 271,000
2001	\$1,430,000

Source: National Hockey League

a) Use the two data points to find a linear function that fits the data. Let $x =$ the number of years since 1990 and $S =$ the average salary x years from 1990.
b) Use the function to estimate the average salary of players in the National Hockey League in 2005 and in 2010.

50. *Median Age of Women at First Marriage.* The table below lists data regarding the median age of women at first marriage in 1970 and 1998.

YEAR	MEDIAN AGE OF WOMEN AT FIRST MARRIAGE
1970	20.8
1998	25.0

Source: U.S. Bureau of the Census

a) Use the two data points to find a linear function that fits the data. Let $x =$ the number of years since 1970 and $A =$ the median age at first marriage x years from 1970.
b) Use the function to predict the median age of women at first marriage in 2008 and in 2015.

51. *Records in the 400-Meter Run.* In 1930, the record for the 400-m run was 46.8 sec. In 1970, it was 43.8 sec. Let $R(t) =$ the record in the 400-m run and $t =$ the number of years since 1930.

a) Find a linear function that fits the data.
b) Use the function of part (a) to estimate the record in 2003; in 2006.
c) When will the record be 40 sec?

52. *Recycling.* In 1993, Americans recycled 43.8 million tons of solid waste. In 1997, the figure grew to 60.8 million tons. Let $N(t) =$ the number of tons recycled, in millions, and $t =$ the number of years since 1993.

Source: *Statistical Abstract of the United States*

a) Find a linear function that fits the data.
b) Use the function of part (a) to estimate the amount recycled in 2005.

53. *Life Expectancy of Males in the United States.* In 1990, the life expectancy of males was 71.8 yr. In 1997, it was 73.6 yr. Let $E(t) =$ life expectancy and $t =$ the number of years since 1990.
Source: *Statistical Abstract of the United States*

a) Find a linear function that fits the data.
b) Use the function of part (a) to predict the life expectancy of males in 2007.

54. *PAC Contributions.* In 1992, Political Action Committees (PACs) contributed $178.6 million to congressional candidates. In 2000, the figure rose to $243.1 million. Let $A(t) =$ the amount of PAC contributions, in millions, and $t =$ the number of years since 1992.
Source: Congressional Research Service and Federal Election Commission

a) Find a linear function that fits the data.
b) Use the function of part (a) to estimate the amount of PAC contributions in 2006.

Determine m and b in each application and explain their meaning.

55. **DW** *Hair Growth.* After Tina gets a "buzz cut," the length $L(t)$ of her hair, in inches, is given by $L(t) = \frac{1}{2}t + 1$, where t is the number of months after she gets the haircut.

56. **DW** *Cost of a Taxi Ride.* The cost $C(d)$, in dollars, of a taxi ride in Pelham is given by

$$C(d) = 0.75d + 2,$$

where d is the number of miles traveled.

SKILL MAINTENANCE

Simplify. [11.1c]

57. $\dfrac{w - t}{t - w}$

58. $\dfrac{b^2 - 1}{b - 1}$

59. $\dfrac{3x^2 + 15x - 72}{6x^2 + 18x - 240}$

60. $\dfrac{4y + 32}{y^2 - y - 72}$

SYNTHESIS

For Exercises 61 and 62, assume that a linear equation models each situation.

61. *Depreciation of a Computer.* After 6 mos of use, the value of Pearl's computer had dropped to $900. After 8 mos, the value had gone down to $750. How much did the computer cost originally?

62. *Cellular Phone Charges.* The total cost of Mel's cellular phone was $230 after 5 mos of service and $390 after 9 mos. What cost had Mel already incurred when his service just began?

CHAPTER 12

Summary and Review

The review that follows is meant to prepare you for a chapter exam. It consists of two parts. The first part is a checklist of some of the Study Tips referred to in this text, as well as a list of important properties and formulas. The second part is the Review Exercises. These provide practice exercises for the exam, together with references to section objectives so you can go back and review. Before beginning, stop and look back over the skills you have obtained. What skills in mathematics do you have now that you did not have before studying this chapter?

STUDY TIPS CHECKLIST

The foundation of all your study skills is TIME!

- ☐ Are you taking the time to include all the steps when working your homework and your tests?
- ☐ Are you asking questions at appropriate times in class and with your tutors?
- ☐ Are you doing each short-term task such as going to class, asking questions, using your time wisely, and doing your homework as if it exists within the framework of your long-term goal?
- ☐ Have you been using the supplements for the text such as the *Student's Solutions Manual*, MyMathLab, and the Math Tutor Center?

IMPORTANT PROPERTIES AND FORMULAS

Slope $= m = \dfrac{y_2 - y_1}{x_2 - x_1}$, or $\dfrac{y_1 - y_2}{x_1 - x_2}$

EQUATIONS OF LINES AND LINEAR FUNCTIONS

Horizontal Line: $f(x) = b$, or $y = b$; slope $= 0$

Vertical Line: $x = a$, slope is not defined.

Slope–Intercept Equation: $f(x) = mx + b$, or $y = mx + b$

Point–Slope Equation: $y - y_1 = m(x - x_1)$

Parallel Lines: $m_1 = m_2, b_1 \neq b_2$

Perpendicular Lines: $m_1 = -\dfrac{1}{m_2}, m_1, m_2 \neq 0$

REVIEW EXERCISES

Determine whether the correspondence is a function. [12.1a]

1. 3 → a; 5 → b; 7 → c; 7 → d; 9 → c; 3 → e

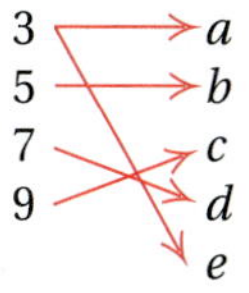

2. 1 → a; 2 → a; 3 → b; 4 → d; 5 → d

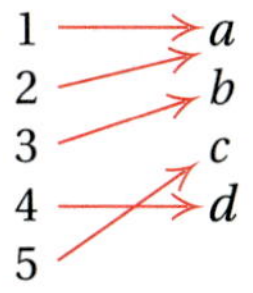

Find the function values. [12.1b]

3. $g(x) = -2x + 5$; $g(0)$ and $g(-1)$

4. $f(x) = 3x^2 - 2x + 7$; $f(0)$ and $f(-1)$

5. *Tuition Cost.* The function described by $C(t) = 645t + 7865$ can be used to estimate the average cost of tuition at a state university t years after 2000. Estimate the average cost of tuition at a state university in 2010. [12.1b]

Determine whether each of the following is the graph of a function. [12.1d]

6.

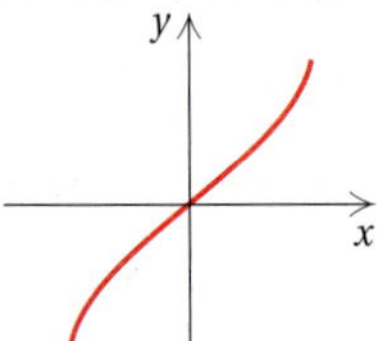

7.

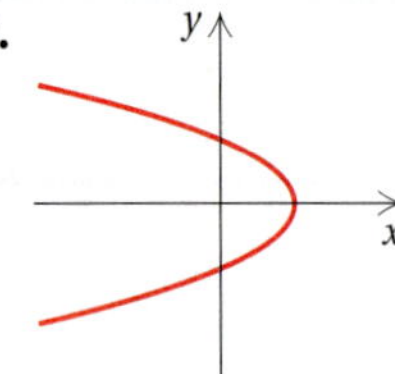

8. For the following graph of a function f, determine **(a)** $f(2)$; **(b)** the domain; **(c)** all x-values such that $f(x) = 2$; and **(d)** the range. [12.2a]

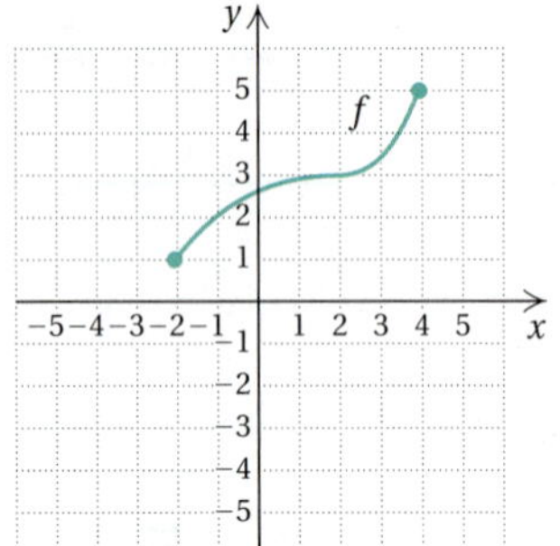

Find the domain. [12.2a]

9. $f(x) = \dfrac{5}{x - 4}$

10. $g(x) = x - x^2$

Find the slope and the y-intercept. [12.3a, b]

11. $y = -3x + 2$

12. $4y + 2x = 8$

13. Find the slope, if it exists, of the line containing the points $(13, 7)$ and $(10, -4)$. [12.3b]

Find the intercepts. Then graph the equation. [12.4a]

14. $2y + x = 4$

15. $2y = 6 - 3x$

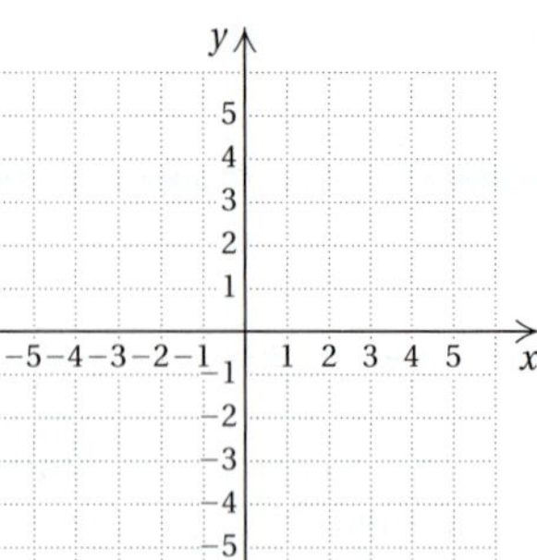

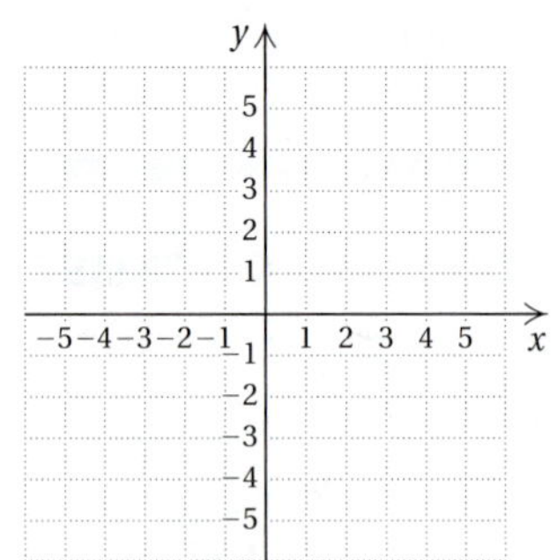

Graph using the slope and the y-intercept. [12.4b]

16. $g(x) = -\frac{2}{3}x - 4$

17. $f(x) = \frac{5}{2}x + 3$

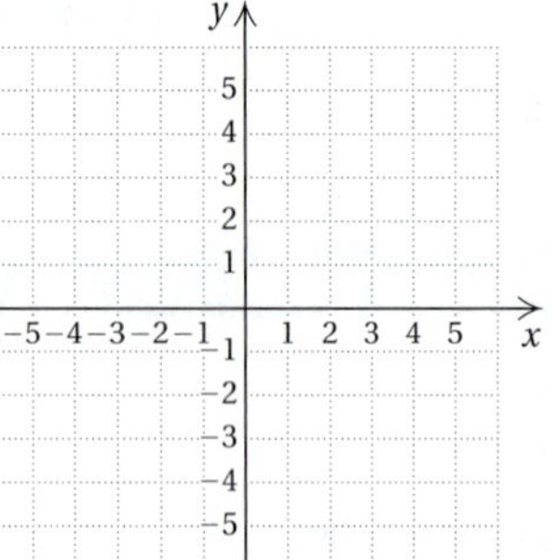

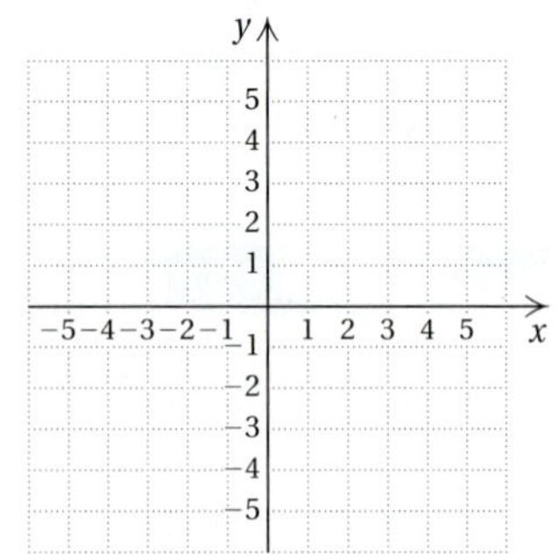

Graph. [12.4c]

18. $x = -3$

19. 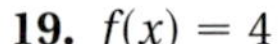 $f(x) = 4$

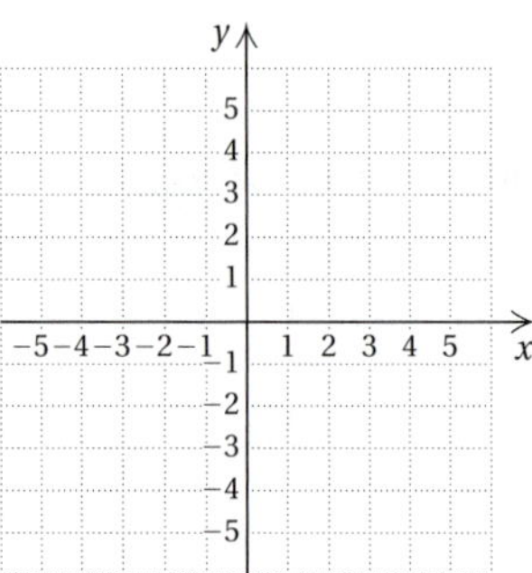

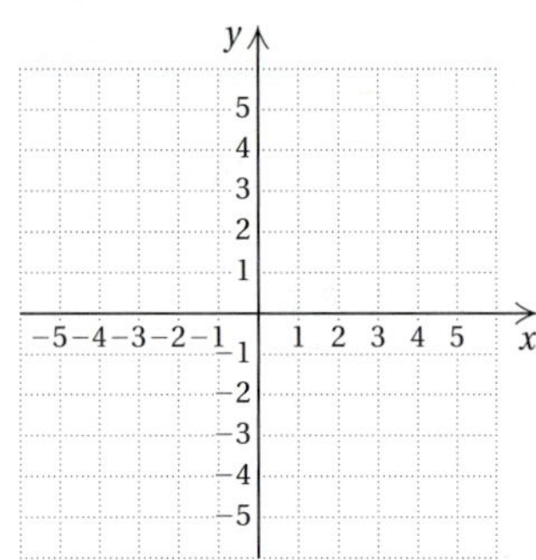

Determine whether the graphs of the given pair of lines are parallel or perpendicular. [12.4d]

20. $y + 5 = -x,$
$x - y = 2$

21. $3x - 5 = 7y,$
$7y - 3x = 7$

22. $4y + x = 3,$
$2x + 8y = 5$

23. $x = 4,$
$y = -3$

24. Find a linear function $f(x) = mx + b$ whose graph has the given slope and y-intercept: [12.5a]

slope: 4.7; y-intercept: $(0, -23)$.

25. Find an equation of the line having the given slope and containing the given point: [12.5b]

$m = -3$; $(3, -5)$.

26. Find an equation of the line containing the given pair of points: [12.5c]

$(-2, 3)$ and $(-4, 6)$.

27. Find an equation of the line containing the given point and parallel to the given line: [12.5d]

$(14, -1)$; $5x + 7y = 8$.

28. Find an equation of the line containing the given point and perpendicular to the given line: [12.5d]

$(5, 2)$; $3x + y = 5$.

29. *Records in the 400-Meter Run.* The table below lists data regarding the world records in the 400-meter run in 1970 and in 2000. [12.5e]

YEAR	RECORDS IN THE 400-M RUN (in seconds)
1970	46.8
2001	44.63

a) Use the two data points to find a linear function that fits the data. Let x = the number of years since 1970 and R = the world record x years from 1970.
b) Use the function to predict the world record in the 400-m run in 2005 and in 2010.

30. DW Explain the usefulness of the slope concept when describing a line. [12.3b, c], [12.4b], [12.5a, b, c, d]

31. DW Explain the meaning of the notation $f(x)$ when considering a function in as many ways as you can. [12.1b]

SKILL MAINTENANCE

Certain objectives from four particular sections will be retested on the chapter test. The objectives are listed with the practice problems that follow.

32. Convert 3.3×10^{-8} to decimal notation. [9.2c]

33. Convert 13,700,000,000 to scientific notation. [9.2c]

Divide. [9.8b]

34. $(6x^2 + 13x - 5) \div (2x + 5)$

35. $(r^2 - 3r + 6) \div (r - 4)$

Factor. [10.6a]

36. $81y^3 - 3000$

37. $9 + 72z^3$

38. Simplify: $\dfrac{2a^2 - 5a - 12}{6a^2 + 7a - 3}$. [11.1c]

SYNTHESIS

39. Homespun Jellies charges \$2.49 for each jar of preserves. Shipping charges are \$3.75 for handling, plus \$0.60 per jar. Find a linear function for determining the cost of shipping x jars of preserves. [12.5e]

CHAPTER 12

Chapter Test

Graph.

1. $f(x) = -\frac{3}{5}x$

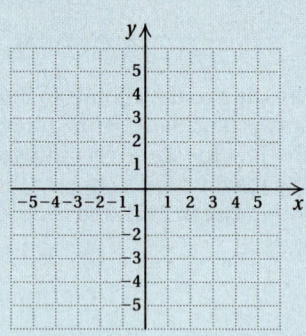

2. $g(x) = 2 - |x|$

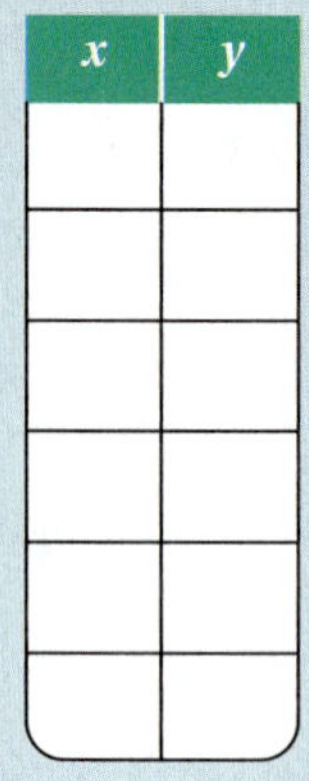

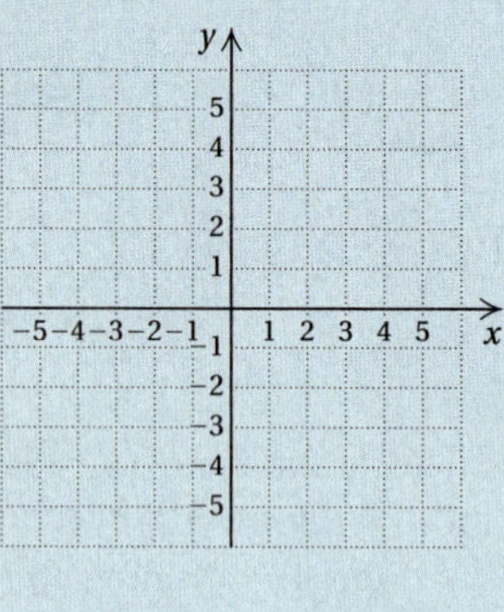

3. *Median Age of Cars.* The function

$$A(t) = 0.233t + 5.87$$

can be used to estimate the median age of cars in the United States t years after 1990. (In this context, we mean that if the median age of cars is 3 yr, then half the cars are older than 3 yr and half are younger.)

Source: The Polk Co.

a) Find the median age of cars in 2002.

b) In what year was the median age of cars 7.734 yr?

Determine whether the correspondence is a function.

4. cat → fish; fish → worm; dog → dog; tiger → cat; teacher → student

Domain	Range
cat	dog
fish	worm
dog	cat
tiger	fish
teacher	student

5. Lake Placid → 1980, 1932; Oslo → 1952; Squaw Valley → 1960; Innsbruck → 1976

Domain	Range
Lake Placid	1980
Oslo	1976
Squaw Valley	1960
Innsbruck	1952
	1932

Find the function values.

6. $f(x) = -3x - 4$; $f(0)$ and $f(-2)$

7. $g(x) = x^2 + 7$; $g(0)$ and $g(-1)$

Determine whether each of the following is the graph of a function.

8.

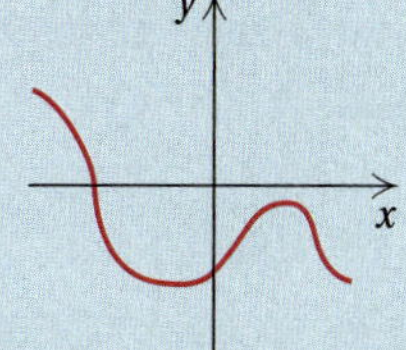

9.

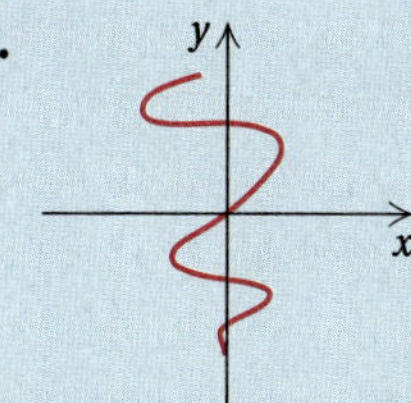

10. *Movie Revenue.* The following graph approximates the weekly revenue, in millions of dollars, from the movie *Jurassic Park—The Lost World*. The revenue is given as a function of the week.

a) What was the movie revenue for week 1?

b) What was the movie revenue for week 5?

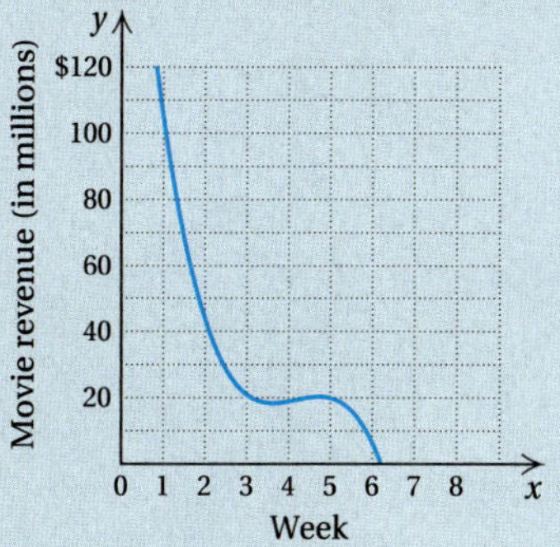

Source: Exhibitor Relations Co., Inc.

11. For the following graph of function f, determine **(a)** $f(2)$; **(b)** the domain; **(c)** all x-values such that $f(x) = 2$; and **(d)** the range.

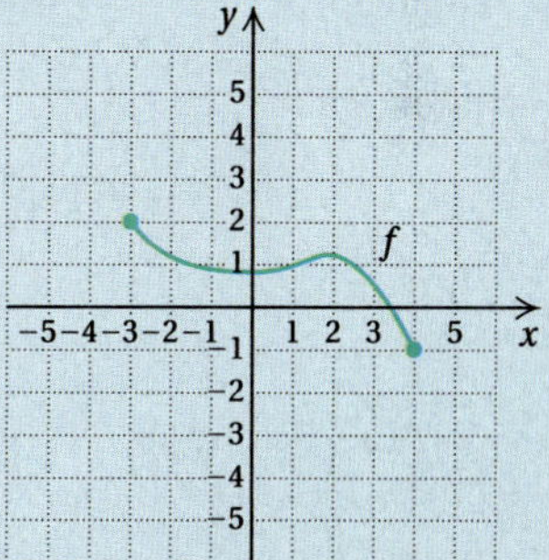

Find the domain.

12. $g(x) = 5 - x^2$

13. $f(x) = \dfrac{8}{2x + 3}$

Find the slope and the y-intercept.

14. $f(x) = -\frac{3}{5}x + 12$

15. $-5y - 2x = 7$

Find the slope, if it exists, of the line containing the following points.

16. $(-2, -2)$ and $(6, 3)$

17. $(-3.1, 5.2)$ and $(-4.4, 5.2)$

18. Find the slope, or rate of change, of the graph at right.

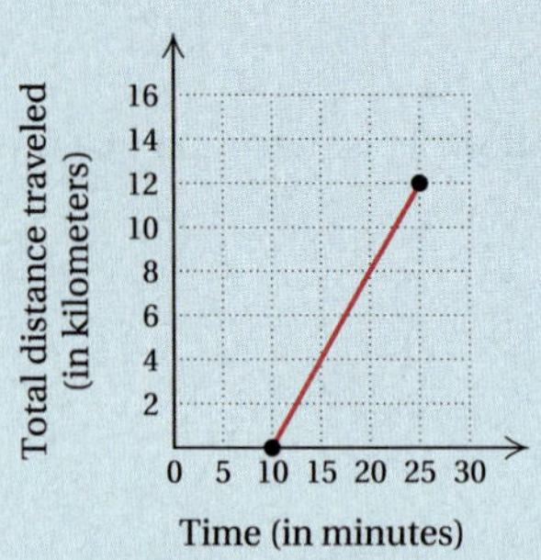

19. Find the intercepts. Then graph the equation.

$2x + 3y = 6$

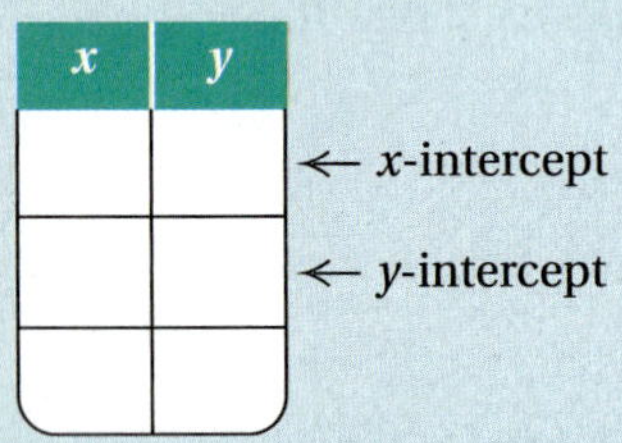

x	y	
		← x-intercept
		← y-intercept

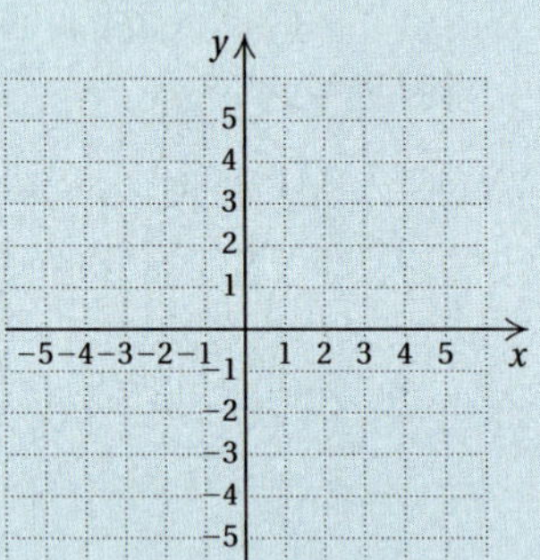

20. Graph using the slope and the y-intercept:

$$f(x) = -\frac{2}{3}x - 1.$$

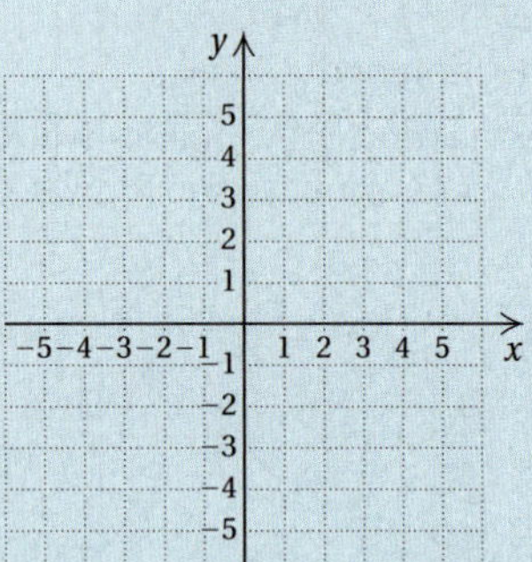

Graph.

21. $y = f(x) = -3$

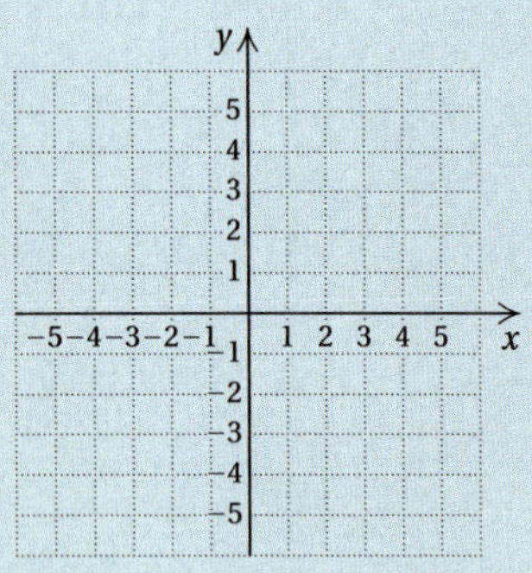

22. $2x = -4$

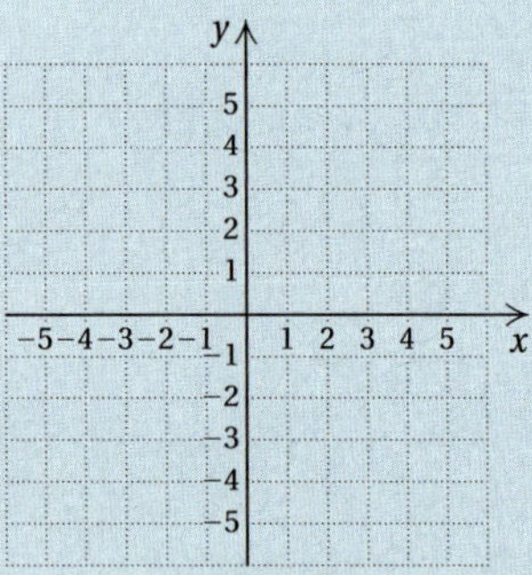

Determine whether the graphs of the given pair of lines are parallel or perpendicular.

23. $4y + 2 = 3x,$
$-3x + 4y = -12$

24. $y = -2x + 5,$
$2y - x = 6$

25. Find an equation of the line that has the given characteristics:

slope: -3; y-intercept: $(0, 4.8)$.

26. Find a linear function $f(x) = mx + b$ whose graph has the given slope and y-intercept:

slope: 5.2; y-intercept: $\left(0, -\frac{5}{8}\right)$.

27. Find an equation of the line having the given slope and containing the given point:

$m = -4$; $(1, -2)$.

28. Find an equation of the line containing the given pair of points:

$(4, -6)$ and $(-10, 15)$.

29. Find an equation of the line containing the given point and parallel to the given line:

$(4, -1)$; $x - 2y = 5$.

30. Find an equation of the line containing the given point and perpendicular to the given line:

$(2, 5)$; $x + 3y = 2$.

31. *Median Age of Men at First Marriage.* The table below lists data regarding the median age of men at first marriage in 1970 and in 1998. [2.6e]

YEAR	MEDIAN AGE OF MEN AT FIRST MARRIAGE
1970	23.2
1998	26.7

Source: U.S. Bureau of the Census

a) Use the two data points to find a linear function that fits the data. Let $x =$ the number of years since 1970 and $A =$ the median age at first marriage x years from 1970.
b) Use the function to predict the median age of men at first marriage in 2008 and in 2015.

SKILL MAINTENANCE

32. Convert 0.0000008204 to scientific notation.

33. Divide: $(y^2 + 6y - 3) \div (y + 4)$.

34. Factor: $125a^3 + 27$.

35. Simplify: $\dfrac{3m^2 + m - 10}{7m^2 + 18m + 8}$.

SYNTHESIS

36. Find k such that the line $3x + ky = 17$ is perpendicular to the line $8x - 5y = 26$.

37. Find a formula for a function f for which $f(-2) = 3$.

Systems of Equations

Gateway to Chapter 13

As you may have noted, the most difficult and time-consuming step in solving an applied problem is generally the translation of the situation to mathematical language. Often the translation step can be eased by using more than one variable and more than one equation, thus obtaining a system of equations. In this chapter, you will learn to solve systems of equations and then apply that skill to the solving of applied problems.

Real-World Application

Tara's website, Garden Edibles, specializes in the sale of herbs and flowers for colorful meals and garnishes. Tara sells packets of nasturtium seeds for \$0.95 each and packets of Johnny-jump-up seeds for \$1.43 each. She decides to offer a 16-packet spring-garden combination, combining packets of both types of seeds at \$1.10 per packet. How many packets of each type of seed should be put in her garden mix?

This problem appears as Example 2 in Section 13.4.

CHAPTER 13

CHAPTER

13 Pretest

1. Solve this system by graphing: [13.1a]

$y = x + 1,$

$y + x = 3.$

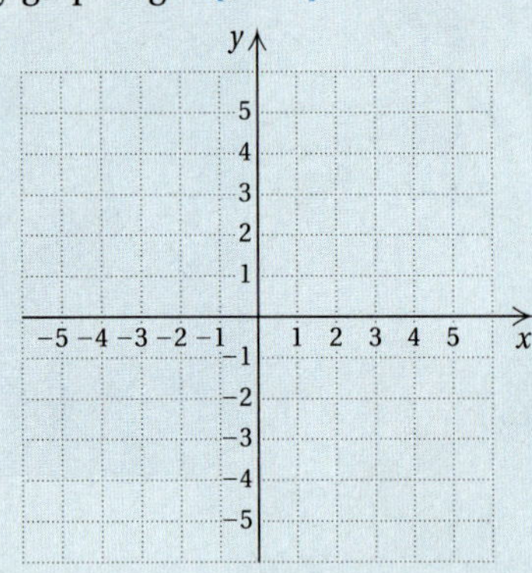

2. Solve the system of equations in Question 1 by the substitution method. [13.2a]

3. Solve this system by the elimination method: [13.3a]

$3x + 5y = 1,$

$4x + 3y = -6.$

4. Classify the system of equations in Question 1 as consistent or inconsistent. [13.1a]

5. Classify the system of equations in Question 1 as dependent or independent. [13.1a]

6. Solve: [13.5a]

$3x + 5y - 2z = 7,$

$2x + y - 3z = -5,$

$4x - 2y + z = 3.$

Solve.

7. Mixed Nuts. The Nutty Professor sells cashews for \$6.75 per pound and Brazil nuts for \$5.00 per pound. How much of each type should be used to make a 50-lb mixture that sells for \$5.70 per pound? [13.4a]

8. Investments. Two investments are made totaling \$7500. For a certain year, the investments yielded \$516 in simple interest. Part of the \$7500 is invested at 8% and part at 6%. Find the amount invested at each rate. [13.4a]

9. Salt-Water Mixtures. Mixture A is 32% salt and the rest water. Mixture B is 58% salt and the rest water. How many pounds of each mixture should be combined in order to obtain 120 lb of a mixture that is 44% salt? [13.4a]

10. Marine Travel. A motorboat took 4 hr to make a trip downstream with a 6-mph current. The return trip against the same current took 5 hr. Find the speed of the boat in still water. [13.4b]

11. For the following total-cost and total-revenue functions, find **(a)** the total-profit function and **(b)** the break-even point. [13.7a]

$C(x) = 90{,}000 + 15x,$

$R(x) = 26x$

13.1 SYSTEMS OF EQUATIONS IN TWO VARIABLES

Objective

a Solve a system of two linear equations or two functions by graphing and determine whether a system is consistent or inconsistent and whether the equations in a system are dependent or independent.

We can solve many applied problems more easily by translating to two or more equations in two or more variables than by translating to a single equation. Let's look at such a problem.

Scholastic Aptitude Test. Many high-school students take the Scholastic Aptitude Test. Each student receives two scores, a *verbal* score and a *math* score. In 1999–2000, the average total score of students was 1019, with the average math score exceeding the verbal score by 9 points. What was the average verbal score and what was the average math score?
Source: National Center for Education Statistics

To solve, we first let

$x =$ the average verbal score and

$y =$ the average math score.

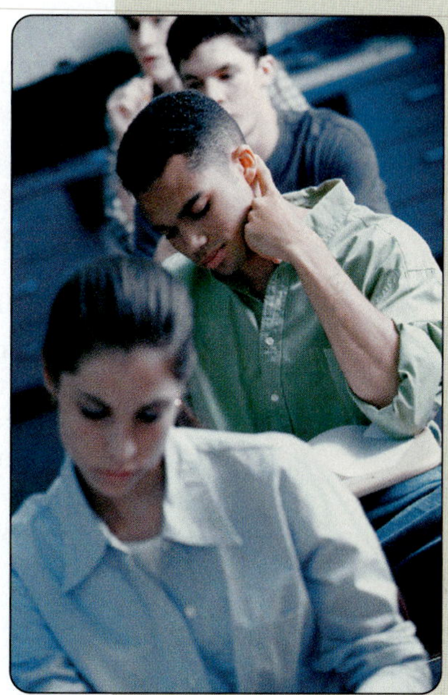

The problem contains two statements to translate. First, we look at the total of the two scores:

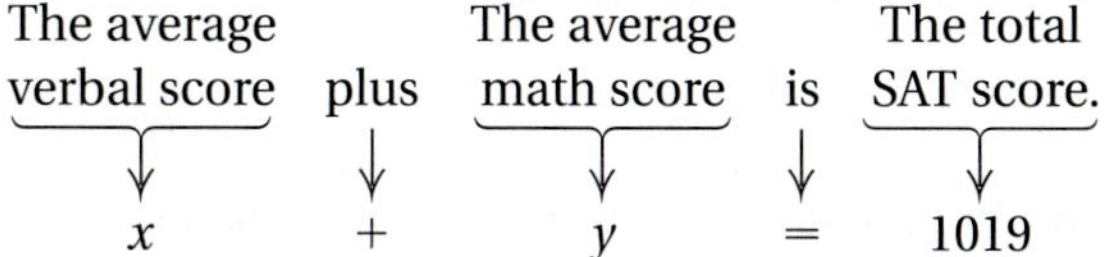

The second statement compares the two scores.

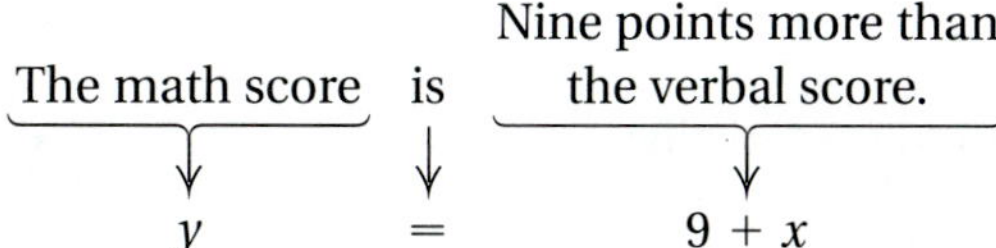

The math score
is
Nine points more than the verbal score.
y
=
9 + x

We have now translated the problem to a **pair,** or **system, of equations:**

$$x + y = 1019,$$
$$y = 9 + x.$$

We can also write this system in function notation by solving each equation for y:

$$x + y = 1019 \rightarrow y = 1019 - x \rightarrow f(x) = 1019 - x,$$
$$y = 9 + x \rightarrow g(x) = 9 + x.$$

Solve the system graphically.

1. $-2x + y = 1,$
$3x + y = 1$

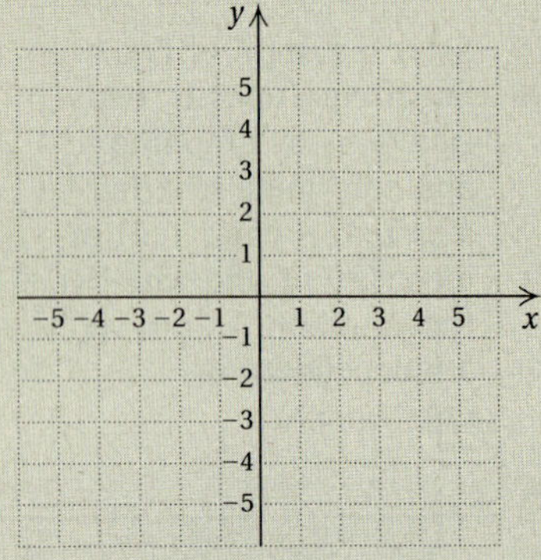

2. $f(x) = \frac{1}{2}x,$
$g(x) = -\frac{1}{4}x + \frac{3}{2}$

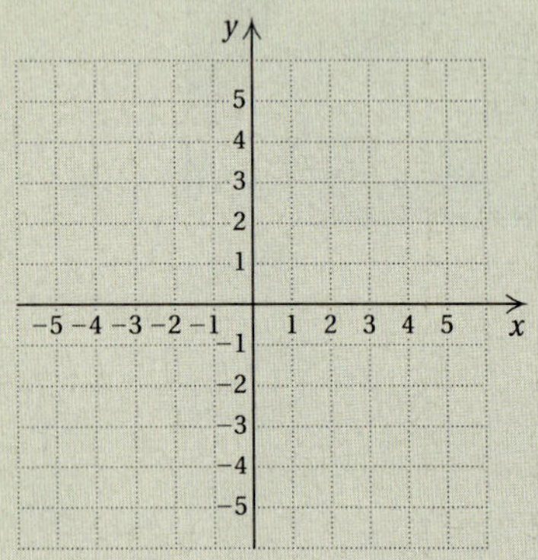

Answers on page A-48

To solve this system, whether it is written in function notation or not, we graph each equation and look for the point of intersection of the graphs. (We may need to use detailed graphing paper to determine the point of intersection exactly.)

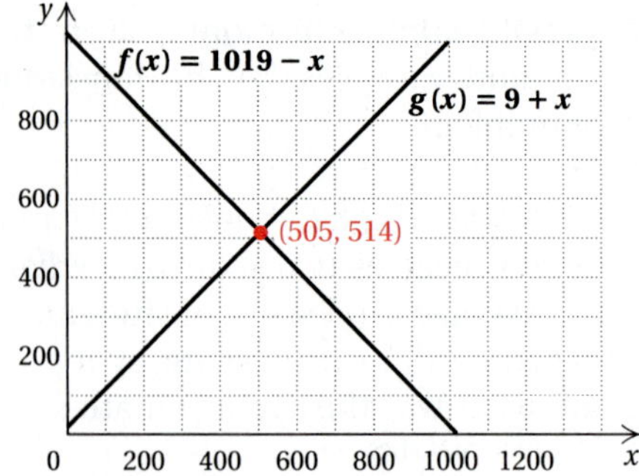

As we see in the graph above, the ordered pair (505, 514) is the intersection and thus the solution—that is, $x = 505$ and $y = 514$. This tells us that the average verbal score was 505 and the average math score was 514.

a Solving Systems of Equations Graphically

ONE SOLUTION

A **solution** of a system of two equations in two variables is an ordered pair that makes *both* equations true. If we graph a system of equations or functions, the point at which the graphs intersect will be a solution of *both* equations or functions.

EXAMPLE 1 Solve this system graphically:

$$y - x = 1,$$
$$y + x = 3.$$

We draw the graph of each equation using any method studied in Chapter 7.

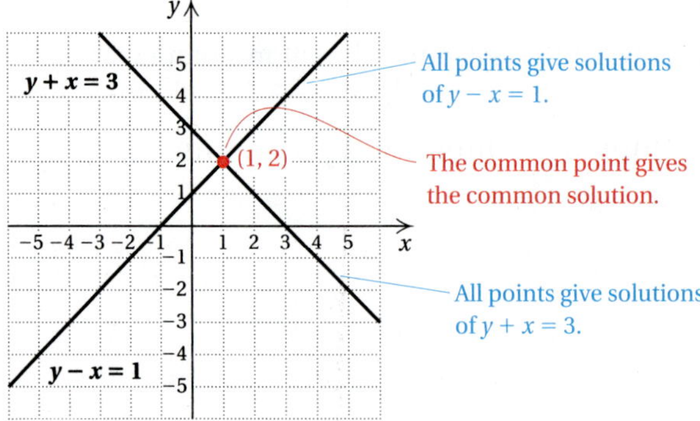

The point of intersection has coordinates that make *both* equations true. The solution seems to be the point (1, 2). However, since graphing alone is not perfectly accurate, solving by graphing may give only approximate answers. Thus we check the pair (1, 2) as follows.

CHECK:

$$\begin{array}{c|c} y - x & 1 \\ \hline 2 - 1 & ?\ 1 \\ 1 & \end{array} \quad \text{TRUE} \qquad \begin{array}{c|c} y + x & 3 \\ \hline 2 + 1 & ?\ 3 \\ 3 & \end{array} \quad \text{TRUE}$$

The solution is $(1, 2)$.

Do Exercises 1 and 2 on the preceding page.

CALCULATOR CORNER

Solving Systems of Equations We can solve a system of two equations in two variables on a graphing calculator. Consider the system of equations in Example 1:

$$y - x = 1,$$
$$y + x = 3.$$

First, we solve the equations for y, obtaining $y = x + 1$ and $y = -x + 3$. Next, we enter $y_1 = x + 1$ and $y_2 = -x + 3$ on the equation-editor screen and graph the equations. We can use the standard viewing window, $[-10, 10, -10, 10]$. (If you turned on Plot 1 to do the Calculator Corner in Section 12.5 and haven't turned it off yet, do so now. See p. 974 for the procedure.)

We will use the INTERSECT feature to find the coordinates of the point of intersection of the lines. To access this feature, we press [2nd] [CALC] [5]. (CALC is the second operation associated with the [TRACE] key.) The query "First curve?" appears on the graph screen. The blinking cursor is positioned on the graph of y_1. We press [ENTER] to indicate that this is the first curve involved in the intersection. Next, the query "Second curve?" appears and the blinking cursor is positioned on the graph of y_2. We press [ENTER] to indicate that this is the second curve. Now the query "Guess?" appears. We use the [▷] and [◁] keys to move the cursor close to the point of intersection and press [ENTER]. The coordinates of the point of intersection of the graphs, $x = 1$, $y = 2$, appear at the bottom of the screen. Thus the solution of the system of equations is $(1, 2)$.

$y_1 = x + 1,\ y_2 = -x + 3$

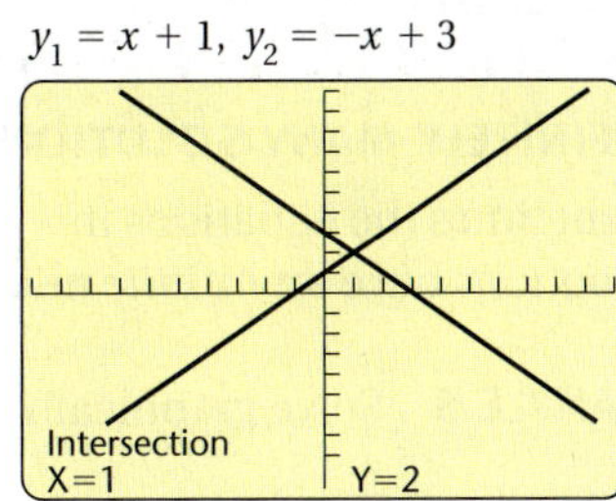

Exercises: Use a graphing calculator to solve the system of equations.

1. $x + y = 5,$
$y = x + 1$

2. $y = x + 3,$
$2x - y = -7$

3. $x - y = -6,$
$y = 2x + 7$

4. $x + 4y = -1,$
$x - y = 4$

3. Solve graphically:

$$y + 2x = 3,$$
$$y + 2x = -4.$$

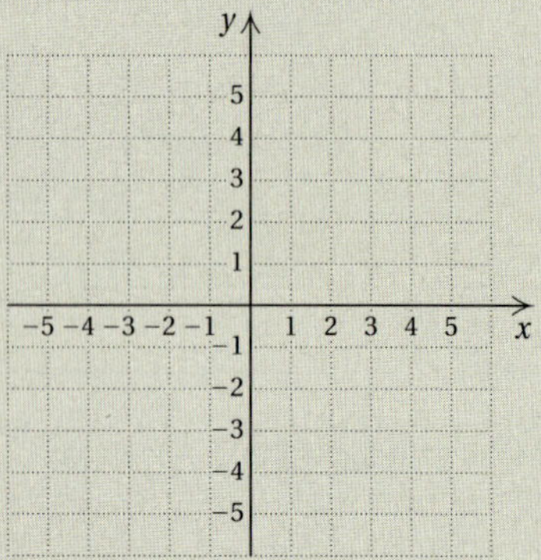

4. Classify each of the systems in Margin Exercises 1–3 as consistent or inconsistent.

Answers on page A-48

NO SOLUTION

Sometimes the equations in a system have graphs that are parallel lines.

EXAMPLE 2 Solve graphically:

$$f(x) = -3x + 5,$$
$$g(x) = -3x - 2.$$

We graph the functions. The graphs have the same slope, -3, and different y-intercepts, so they are parallel. There is no point at which they cross, so the system has no solution. No matter what point we try, it will *not* check in *both* equations. The solution set is thus the empty set, denoted $\varnothing$ or $\{ \}$.

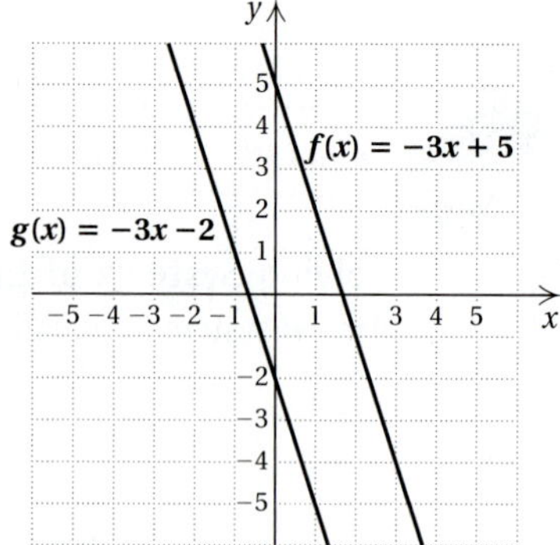

CONSISTENT AND INCONSISTENT SYSTEMS

A system of equations has at least one solution. → It is **consistent.**

A system of equations has no solution. → It is **inconsistent.**

The system in Example 1 is consistent. The system in Example 2 is inconsistent.

Do Exercises 3 and 4.

INFINITELY MANY SOLUTIONS

Sometimes the equations in a system have the same graph. In such a case, the equations have an *infinite* number of solutions in common.

EXAMPLE 3 Solve graphically:

$$3y - 2x = 6,$$
$$-12y + 8x = -24.$$

We graph the equations and see that the graphs are the same. Thus any solution of one of the equations is a solution of the other. Each equation has an infinite number of solutions, two of which are shown on the graph.

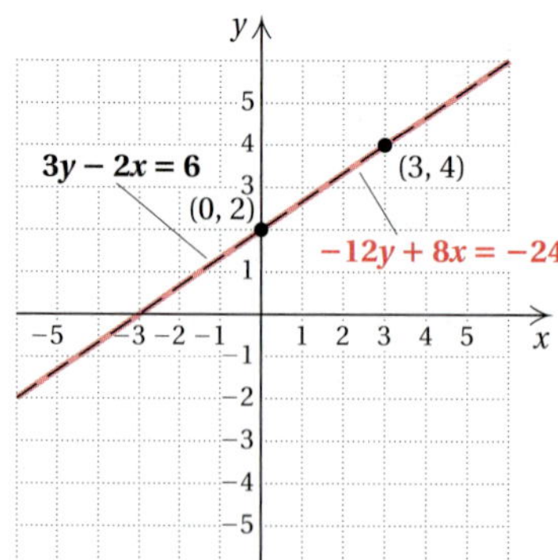

We check one such solution, $(0, 2)$, which is the y-intercept of each equation.

CHECK:

$$\begin{array}{c|c} 3y - 2x = 6 & \\ \hline 3(2) - 2(0) \; ? \; 6 & \\ 6 - 0 & \\ 6 & \text{TRUE} \end{array} \qquad \begin{array}{c|c} -12y + 8x = -24 & \\ \hline -12(2) + 8(0) \; ? \; -24 & \\ -24 + 0 & \\ -24 & \text{TRUE} \end{array}$$

On your own, check that $(3, 4)$ is a solution of both equations. If $(0, 2)$ and $(3, 4)$ are solutions, then all points on the line containing them will be solutions. The system has an infinite number of solutions.

DEPENDENT AND INDEPENDENT EQUATIONS

A system of two equations in two variables:

has infinitely many solutions. → The equations are **dependent.**

has one solution or no solutions. → The equations are **independent.**

The equations in Example 3 are dependent. The equations in Example 1 and in Example 2 are independent.

When we graph a system of two equations, one of the following three things can happen.

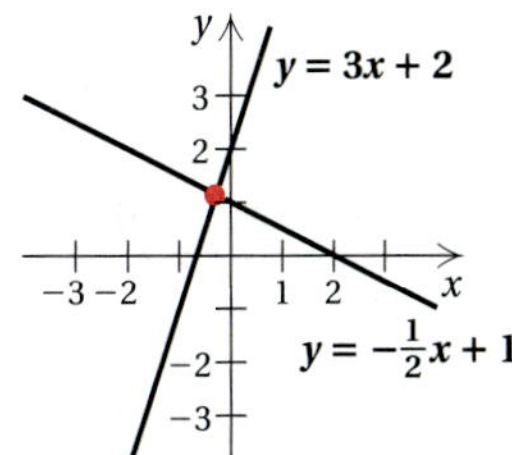

One solution.
Graphs intersect.
The system is consistent and *the equations are independent.*

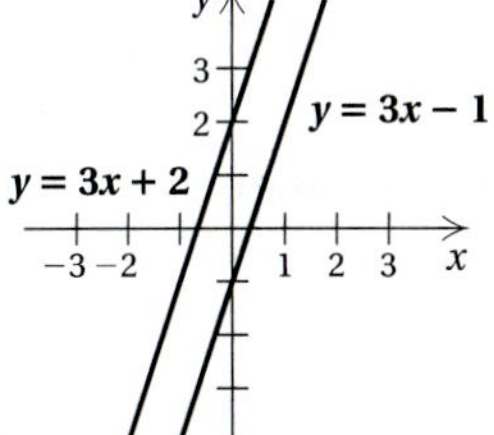

No solution.
Graphs are parallel.
The system is inconsistent and *the equations are independent.*

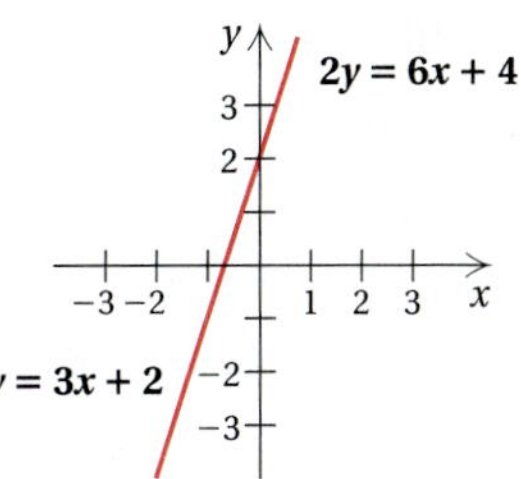

Infinitely many solutions.
Equations have the same graph. *The system is consistent* and *the equations are dependent.*

Do Exercises 5 and 6.

ALGEBRAIC–GRAPHICAL CONNECTION

Consider the equation $-2x + 13 = 4x - 17$. Let's solve it algebraically as we did in Chapter 7:

$$\begin{aligned} -2x + 13 &= 4x - 17 & & \\ 13 &= 6x - 17 & & \text{Adding } 2x \\ 30 &= 6x & & \text{Adding } 17 \\ 5 &= x. & & \text{Dividing by } 6 \end{aligned}$$

Could we also solve the equation graphically? The answer is yes, as we see in the following two methods.

5. Solve graphically:

$2x - 5y = 10,$
$-6x + 15y = -30.$

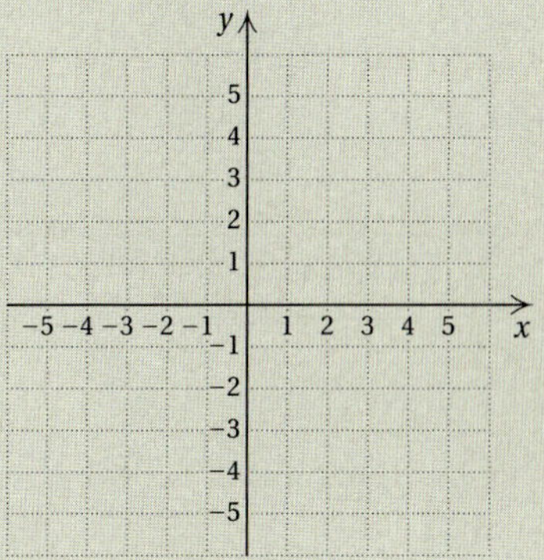

6. Classify the equations in Margin Exercises 1, 2, 3, and 5 as dependent or independent.

7. a) Solve $x + 1 = \frac{2}{3}x$ algebraically.

b) Solve $x + 1 = \frac{2}{3}x$ graphically using Method 1.

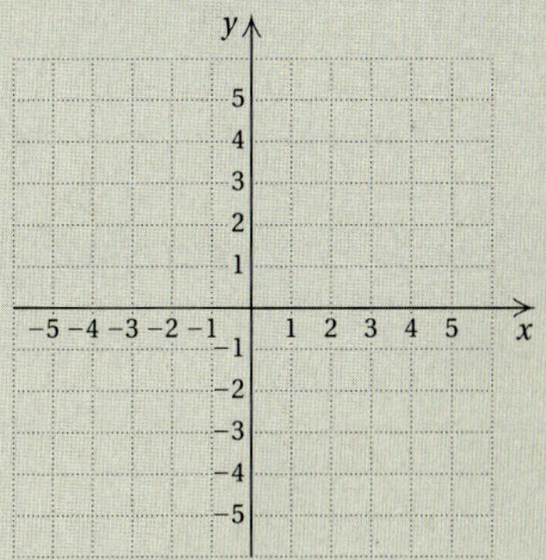

c) Compare your answers to parts (a) and (b).

Answers on page A-48

8. Solve $\frac{1}{2}x + 3 = 2$ graphically using Method 1.

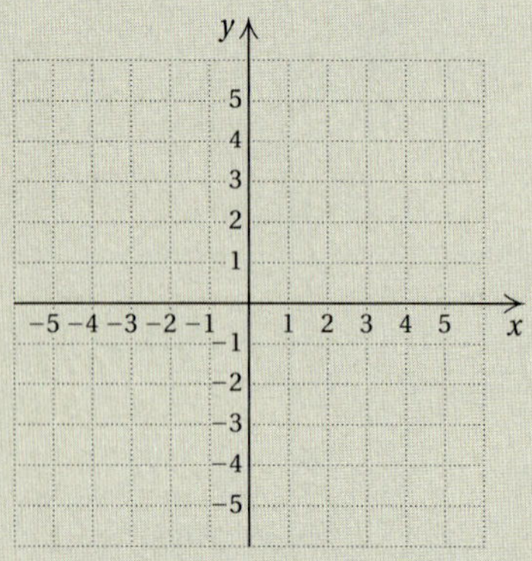

9. a) Solve $x + 1 = \frac{2}{3}x$ graphically using Method 2.

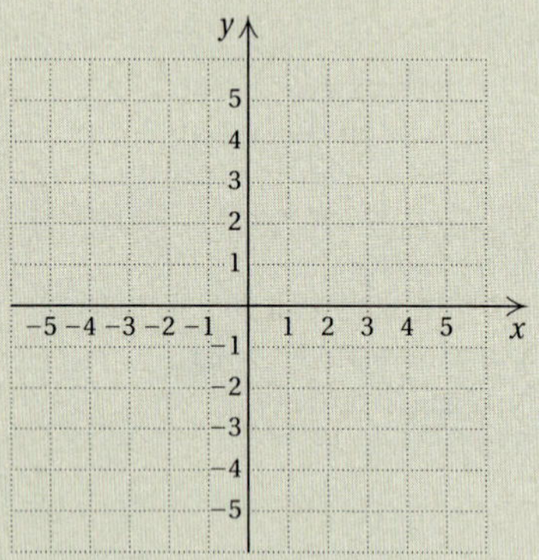

b) Compare your answers to Margin Exercises 7(a), 7(b), and 9(a).

10. Solve $\frac{1}{2}x + 3 = 2$ graphically using Method 2.

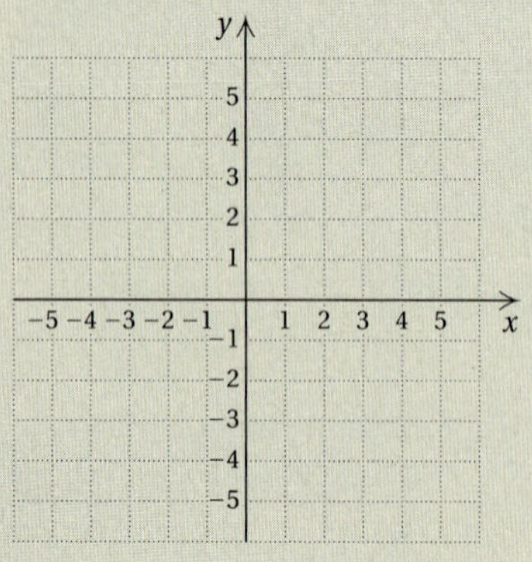

Answers on page A-48

METHOD 1 Solve $-2x + 13 = 4x - 17$ graphically.

We let

$$f(x) = -2x + 13 \quad \text{and} \quad g(x) = 4x - 17.$$

Graphing the system of equations, we get the graph shown at right.

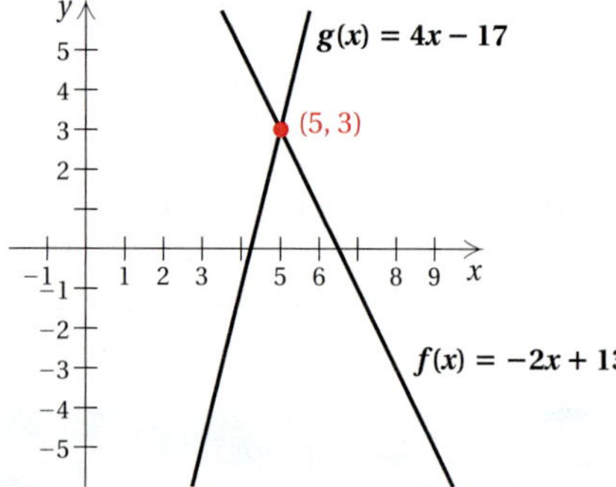

The point of intersection of the two graphs is $(5, 3)$. Note that the x-coordinate of the intersection is 5. This value for x is the solution of the equation $-2x + 13 = 4x - 17$.

Do Exercises 7 and 8. (Exercise 7 is on the preceding page.)

METHOD 2 Solve $-2x + 13 = 4x - 17$ graphically.

Adding $-4x$ and 17 on both sides, we obtain the form

$$-6x + 30 = 0.$$

This time we let

$$f(x) = -6x + 30 \quad \text{and} \quad g(x) = 0.$$

Since $g(x) = 0$, or $y = 0$, is the x-axis, we need only graph $f(x) = -6x + 30$ and see where it crosses the x-axis.

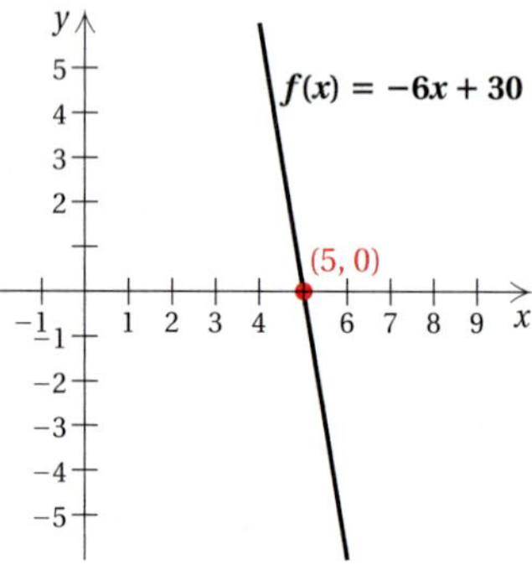

Note that the x-intercept of $f(x) = -6x + 30$ is $(5, 0)$, or just 5. This x-value is the solution of the equation $-2x + 13 = 4x - 17$.

Do Exercise 9.

Let's compare the two methods. Using Method 1, we graph two functions. The solution of the original equation is the x-coordinate of the point of intersection. Using Method 2, we find that the solution of the original equation is the x-intercept of the graph.

Do Exercise 10.

13.1 EXERCISE SET

For Extra Help

Digital Video Tutor CD 11 Videotape 15 | InterAct Math | Math Tutor Center | MathXL | MyMathLab

a Solve the system of equations graphically. Then classify the system as consistent or inconsistent and the equations as dependent or independent. Complete the check for Exercises 1–4.

1. $x + y = 4,$
$x - y = 2$

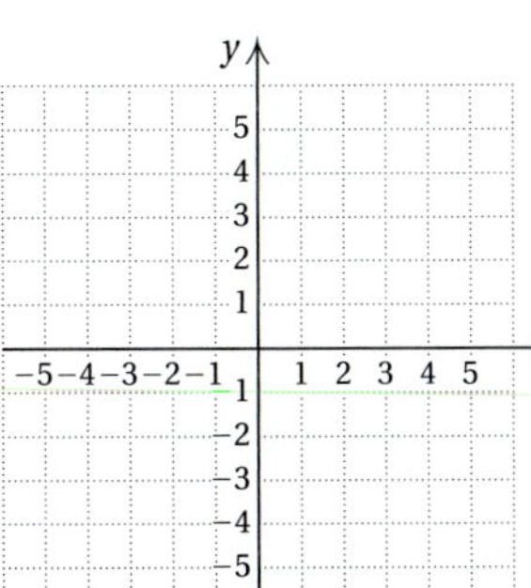

CHECK:

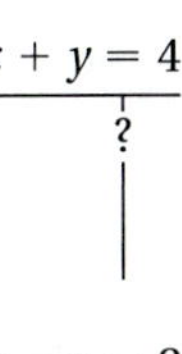

2. $x - y = 3,$
$x + y = 5$

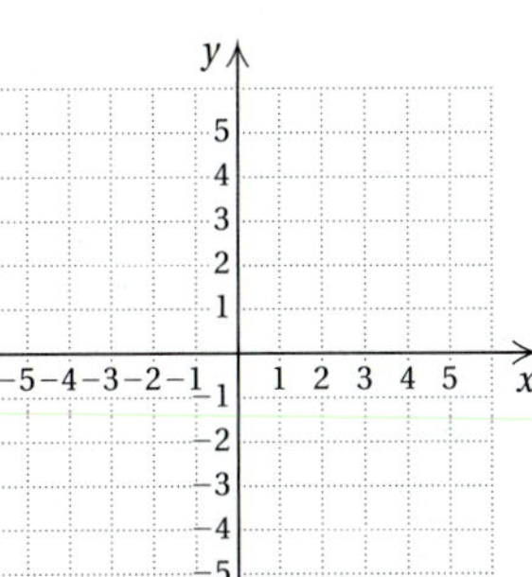

CHECK:

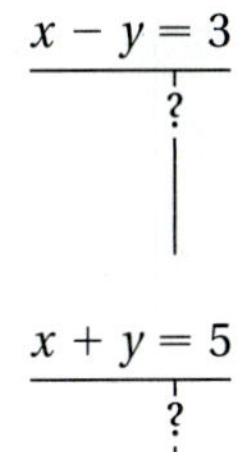

3. $2x - y = 4,$
$2x + 3y = -4$

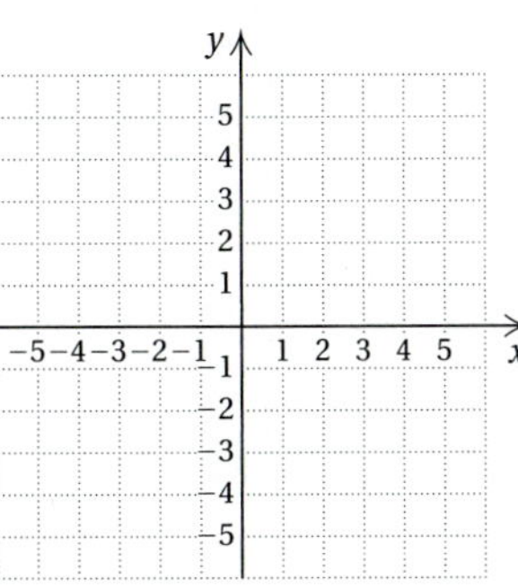

CHECK:

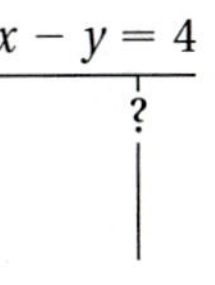

4. $3x + y = 5,$
$x - 2y = 4$

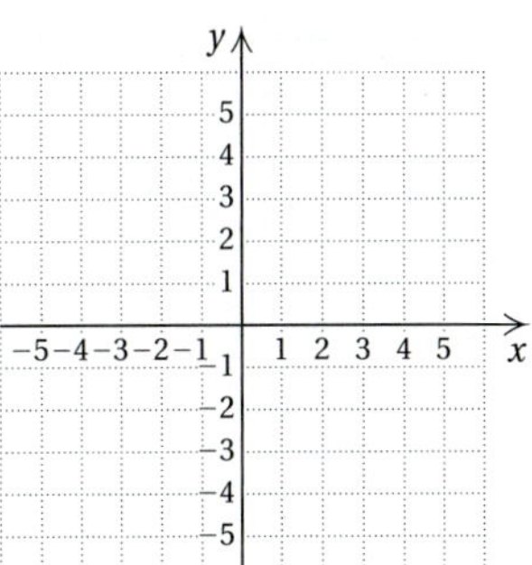

CHECK:

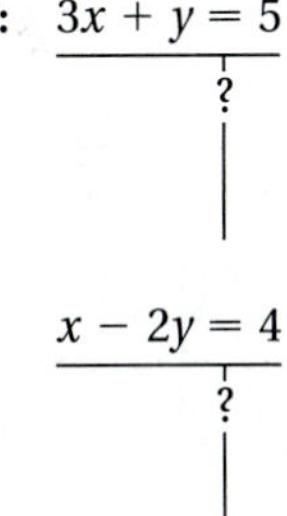

5. $2x + y = 6,$
$3x + 4y = 4$

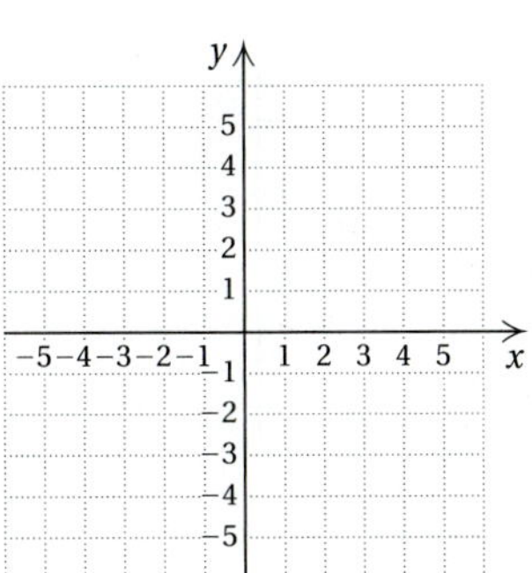

6. $2y = 6 - x,$
$3x - 2y = 6$

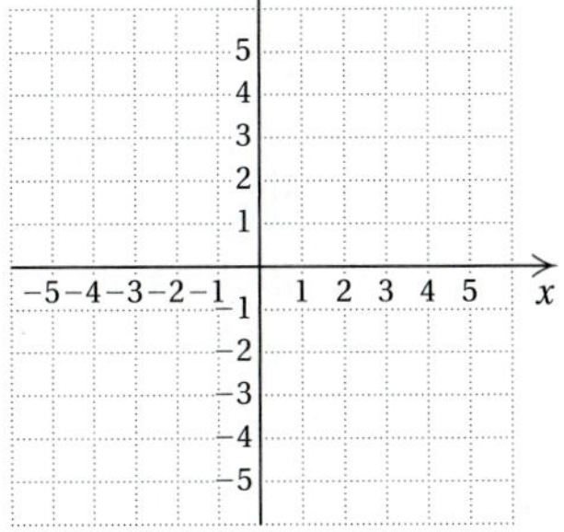

7. $f(x) = x - 1,$
$g(x) = -2x + 5$

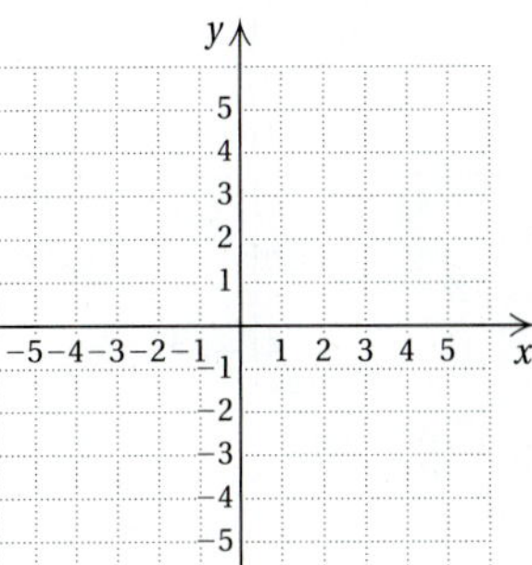

8. $f(x) = x + 1,$
$g(x) = \frac{2}{3}x$

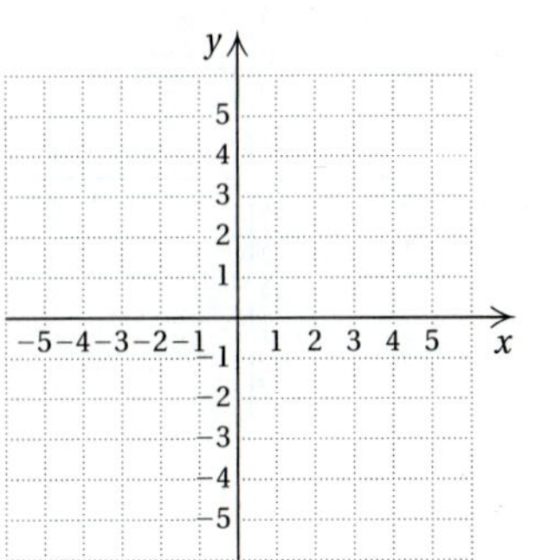

9. $2u + v = 3,$
$2u = v + 7$

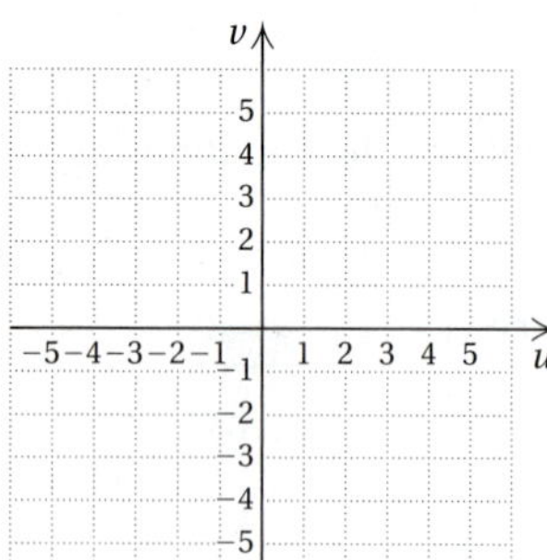

10. $2b + a = 11,$
$a - b = 5$

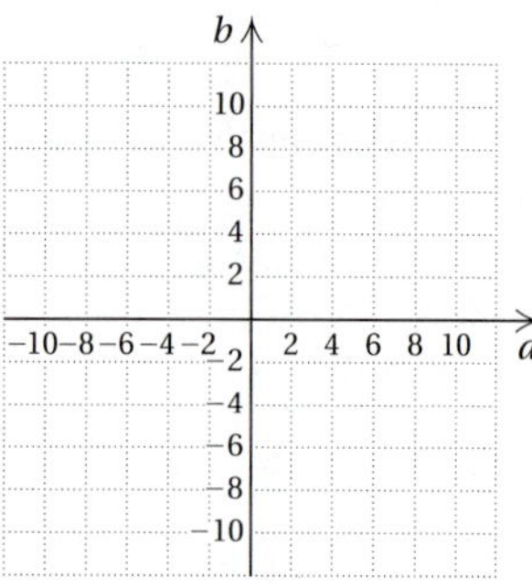

11. $f(x) = -\frac{1}{3}x - 1,$
$g(x) = \frac{4}{3}x - 6$

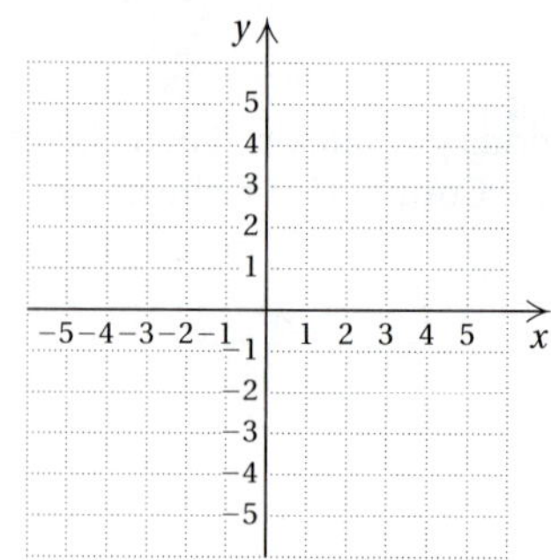

12. $f(x) = -\frac{1}{4}x + 1,$
$g(x) = \frac{1}{2}x - 2$

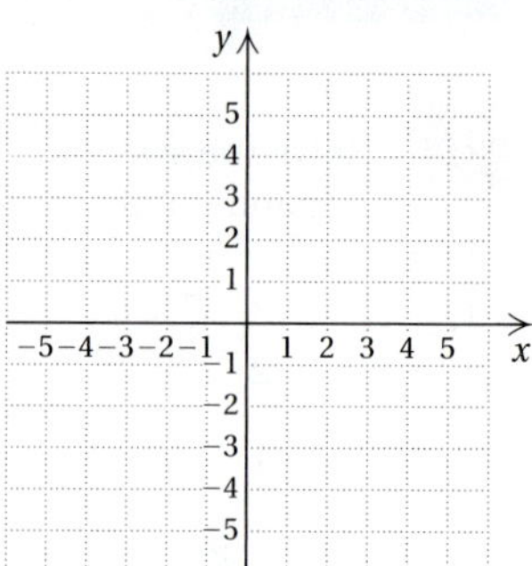

13. $6x - 2y = 2,$
$9x - 3y = 1$

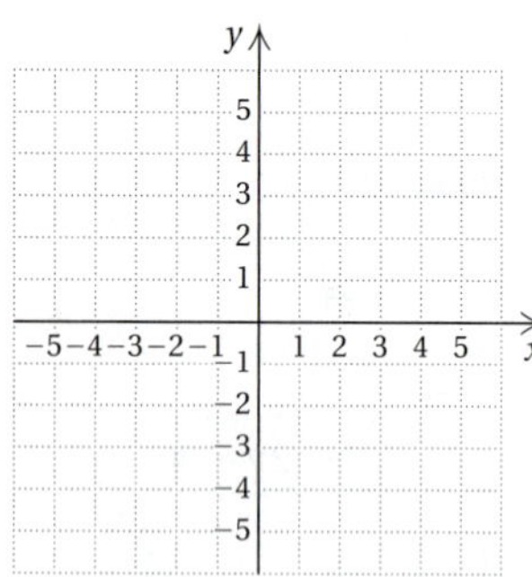

14. $y - x = 5,$
$2x - 2y = 10$

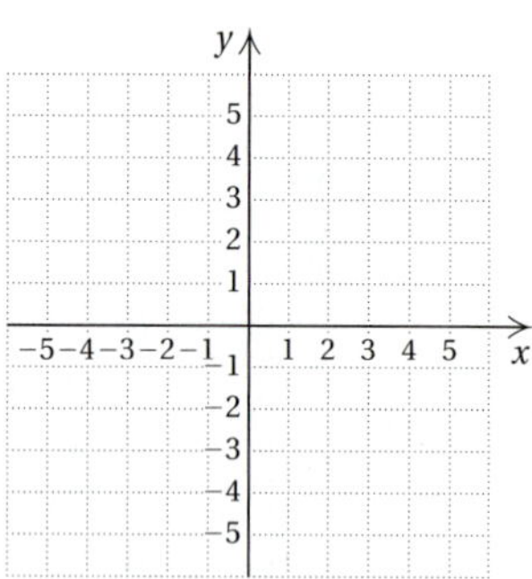

15. $2x - 3y = 6,$
$3y - 2x = -6$

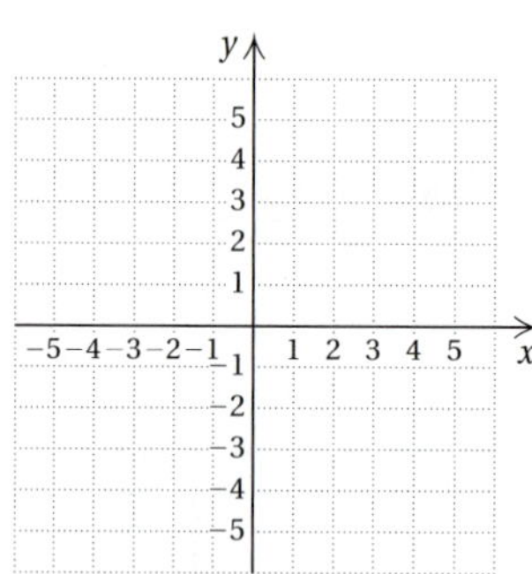

16. $y = 3 - x,$
$2x + 2y = 6$

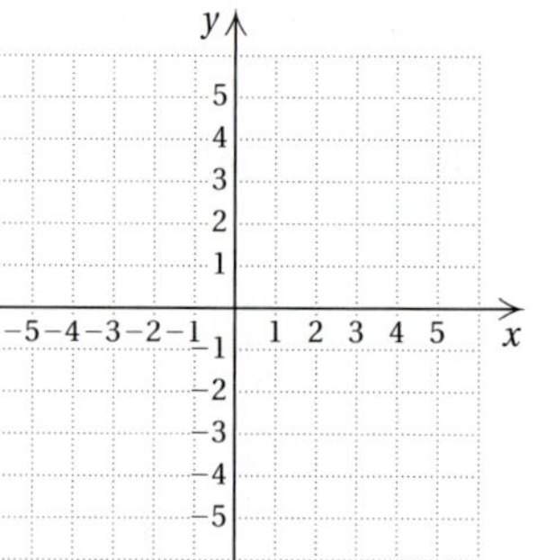

17. $x = 4,$
$y = -5$

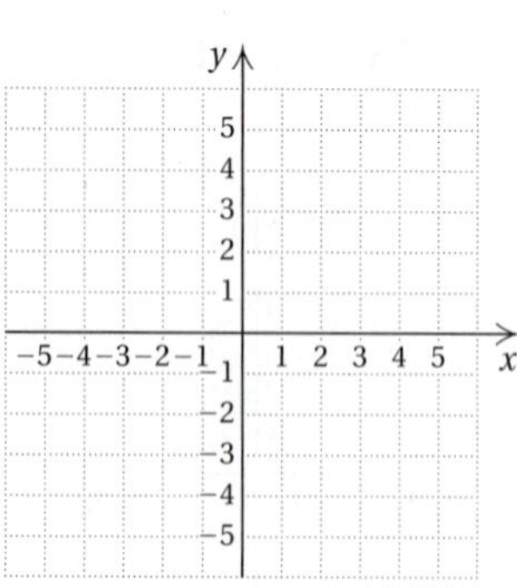

18. $x = -3,$
$y = 2$

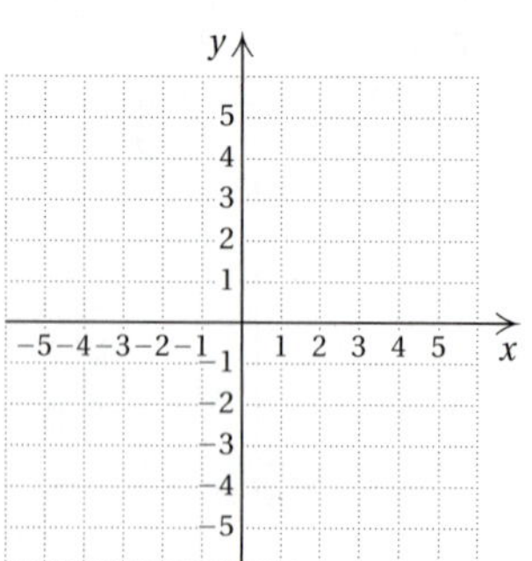

19. $y = -x - 1,$
$4x - 3y = 17$

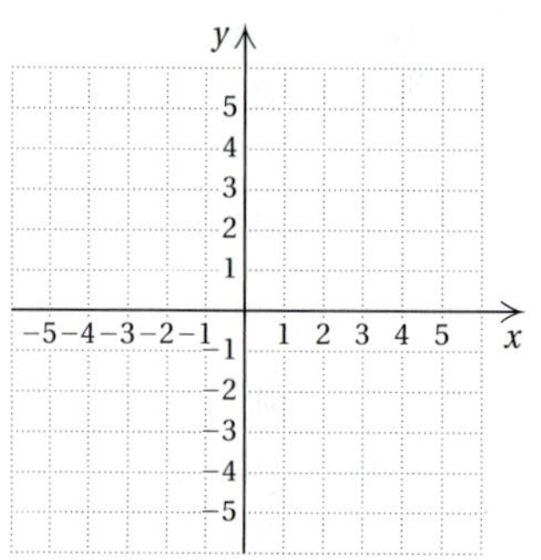

20. $a + 2b = -3,$
$b - a = 6$

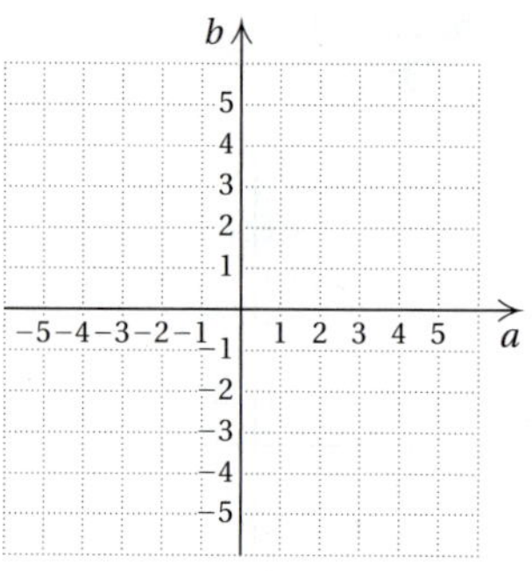

Matching. Each of Exercises 21–26 shows the graph of a system and its solution. First, classify the system as consistent or inconsistent and the equations as dependent or independent. Then match it with one of the appropriate systems of equations (A)–(F), which follow.

21. Solution: $(3, 3)$

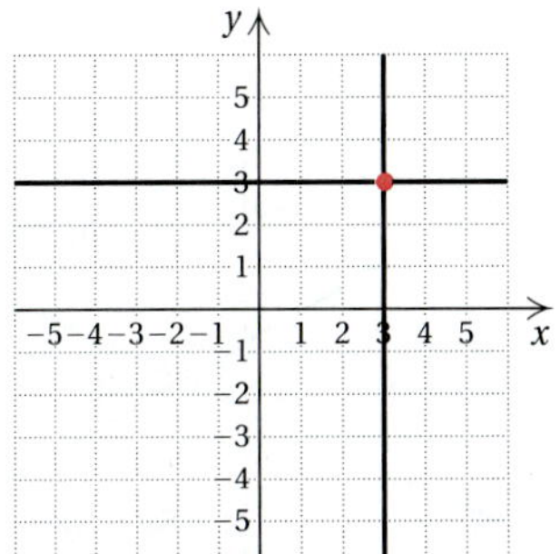

22. Solution: $(1, 1)$

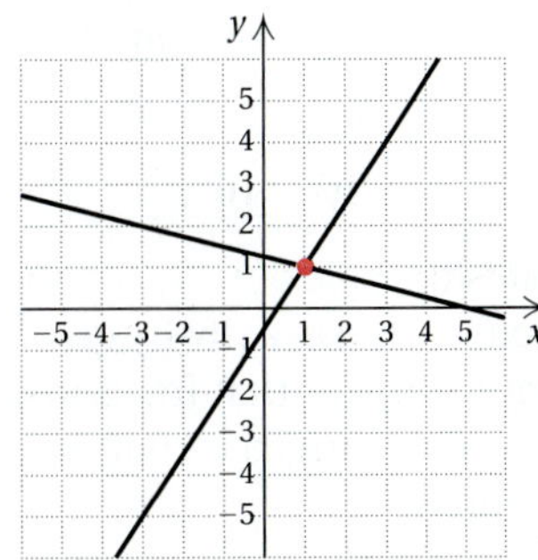

23. Solutions: Infinitely many

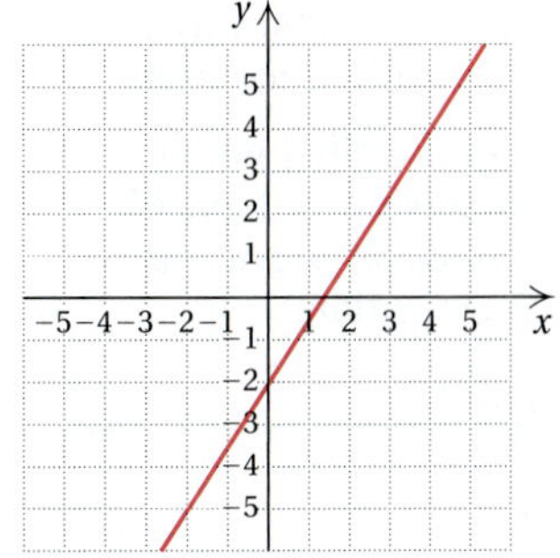

24. Solution: $(4, -3)$

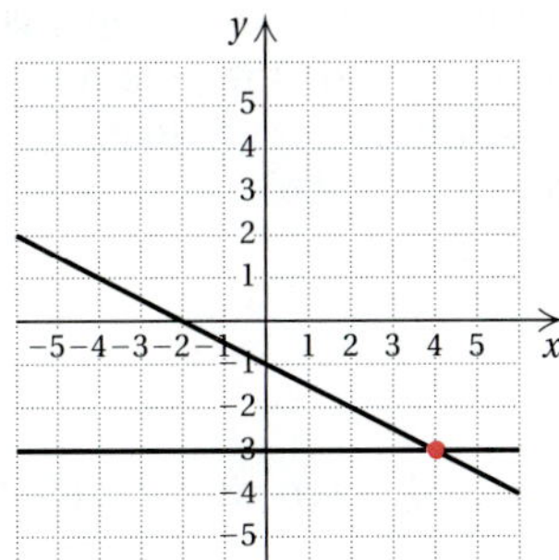

25. Solution: No solution

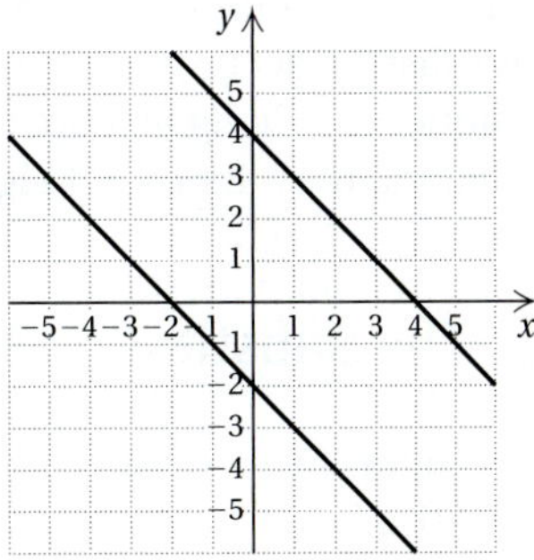

26. Solution: $(-1, 3)$

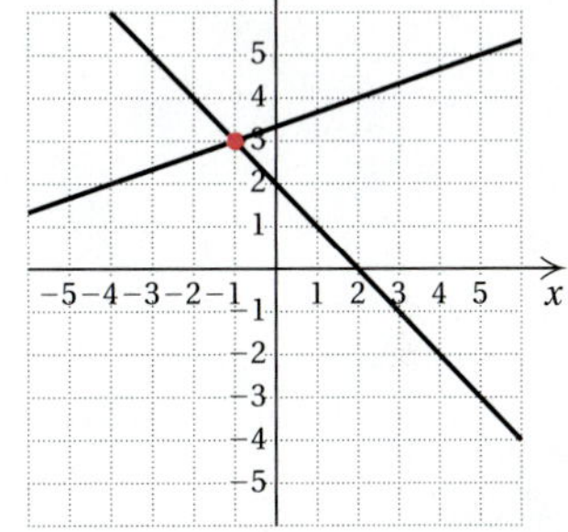

A. $3y - x = 10,$
$x = -y + 2$

B. $9x - 6y = 12,$
$y = \frac{3}{2}x - 2$

C. $2y - 3x = -1,$
$x + 4y = 5$

D. $x + y = 4,$
$y = -x - 2$

E. $\frac{1}{2}x + y = -1,$
$y = -3$

F. $x = 3,$
$y = 3$

27. D$_W$ Explain how to find the solution of $\frac{3}{4}x + 2 = \frac{2}{5}x - 5$ in two ways graphically and in two ways algebraically.

28. D$_W$ Write a system of equations with the given solution. Answers may vary.

a) $(4, -3)$ **b)** No solution
c) Infinitely many solutions

SKILL MAINTENANCE

Write an equation of the line containing the given point and parallel to the given line. [12.5d]

29. $(-4, 2)$; $3x = 5y - 4$

30. $(-6, 0)$; $8y - 3x = 2$

SYNTHESIS

Use a graphing calculator to find the point of intersection of the pair of equations. Round all answers to the nearest hundredth. You may need to solve for y first.

31. $2.18x + 7.81y = 13.78,$
$5.79x - 3.45y = 8.94$

32. $f(x) = 123.52x + 89.32,$
$g(x) = -89.22x + 33.76$

Solve graphically.

33. $y = |x|,$
$x + 4y = 15$

34. $x - y = 0,$
$y = x^2$

13.2 SOLVING BY SUBSTITUTION

Objectives

a Solve systems of equations in two variables by the substitution method.

b Solve applied problems by solving systems of two equations using substitution.

Consider this system of equations:

$$5x + 9y = 2,$$
$$4x - 9y = 10.$$

What is the solution? It is rather difficult to tell exactly by graphing. It would appear that fractions are involved. It turns out that the solution is

$$\left(\frac{4}{3}, -\frac{14}{27}\right).$$

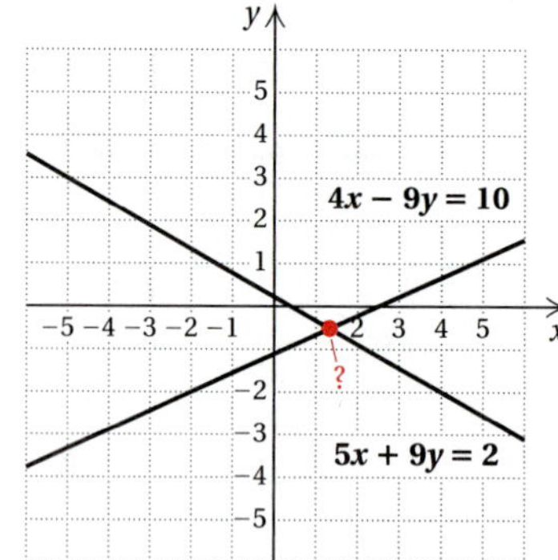

Solving by graphing, though useful in many applied situations, is not always fast or accurate in cases where solutions are not integers. We need techniques involving algebra to determine the solution exactly. Because they use algebra, they are called **algebraic methods.**

a The Substitution Method

One nongraphical method for solving systems is known as the **substitution method.**

EXAMPLE 1 Solve this system:

$$x + y = 4, \quad (1)$$
$$x = y + 1. \quad (2)$$

Equation (2) says that x and $y + 1$ name the same number. Thus we can substitute $y + 1$ for x in equation (1):

$x + y = 4$ Equation (1)

$(y + 1) + y = 4.$ Substituting $y + 1$ for x

Since this equation has only one variable, we can solve for y using methods learned earlier:

$(y + 1) + y = 4$

$2y + 1 = 4$ Removing parentheses and collecting like terms

$2y = 3$ Subtracting 1

$y = \frac{3}{2}.$ Dividing by 2

We return to the original pair of equations and substitute $\frac{3}{2}$ for y in *either* equation so that we can solve for x. Calculation will be easier if we choose equation (2) since it is already solved for x:

$x = y + 1$ Equation (2)

$= \frac{3}{2} + 1$ Substituting $\frac{3}{2}$ for y

$= \frac{3}{2} + \frac{2}{2} = \frac{5}{2}.$

We obtain the ordered pair $\left(\frac{5}{2}, \frac{3}{2}\right)$. Even though we solved for y *first*, it is still the *second* coordinate since x is before y alphabetically. We check to be sure that the ordered pair is a solution.

Study Tips

TUNE OUT DISTRACTIONS

Do you generally study in noisy places? If there is constant noise in your home, dorm, or other study area, consider finding a quiet place in the library—maybe an uncrowded place so you are not distracted with people-watching!

"The ability to concentrate and use your time well is everything."

Lee Iacocca, former CEO of Chrysler Corporation

CHECK:

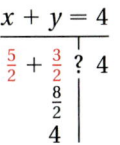

$$\begin{array}{c|c} x + y = 4 & \\ \hline \frac{5}{2} + \frac{3}{2} \ ? \ 4 & \\ \frac{8}{2} & \\ 4 & \text{TRUE} \end{array} \qquad \begin{array}{c|c} x = y + 1 & \\ \hline \frac{5}{2} \ ? & \frac{3}{2} + 1 \\ & \frac{3}{2} + \frac{2}{2} \\ & \frac{5}{2} \quad \text{TRUE} \end{array}$$

Since $\left(\frac{5}{2}, \frac{3}{2}\right)$ checks, it is the solution. Even though exact fraction solutions are difficult to determine graphically, a graph can help us to visualize whether the solution is reasonable.

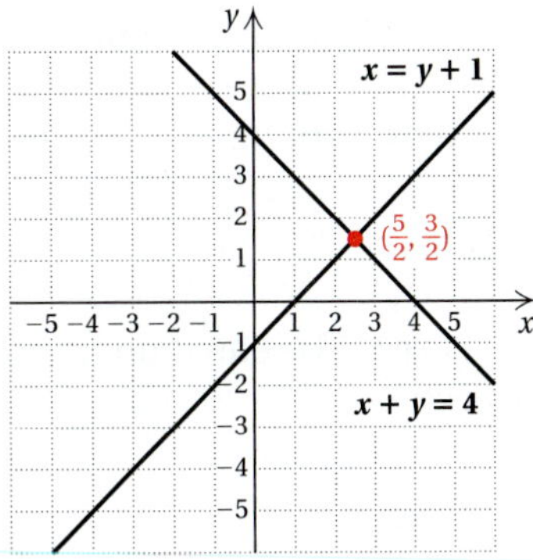

Do Exercises 1 and 2.

Suppose neither equation of a pair has a variable alone on one side. We then solve one equation for one of the variables.

EXAMPLE 2 Solve this system:

$$2x + y = 6, \quad (1)$$
$$3x + 4y = 4. \quad (2)$$

First, we solve one equation for one variable. Since the coefficient of y is 1 in equation (1), it is the easier one to solve for y:

$$y = 6 - 2x. \quad (3)$$

Next, we substitute $6 - 2x$ for y in equation (2) and solve for x:

$$3x + 4(6 - 2x) = 4 \quad \text{Substituting } 6 - 2x \text{ for } y$$

Caution! Remember to use parentheses when you substitute. Then remove them carefully.

$$3x + 24 - 8x = 4 \quad \text{Multiplying to remove parentheses}$$
$$24 - 5x = 4 \quad \text{Collecting like terms}$$
$$-5x = -20 \quad \text{Subtracting 24}$$
$$x = 4. \quad \text{Dividing by } -5$$

In order to find y, we return to either of the original equations, (1) or (2), or equation (3), which we solved for y. It is generally easier to use an equation like (3), where we have solved for the specific variable. We substitute 4 for x in equation (3) and solve for y:

$$y = 6 - 2x = 6 - 2(4) = 6 - 8 = -2.$$

We obtain the ordered pair $(4, -2)$.

CHECK:

$$\begin{array}{c|l} 2x + y = 6 & \\ \hline 2(4) + (-2) \ ? & 6 \\ 8 - 2 & \\ 6 & \quad \text{TRUE} \end{array} \qquad \begin{array}{c|l} 3x + 4y = 4 & \\ \hline 3(4) + 4(-2) \ ? & 4 \\ 12 - 8 & \\ 4 & \quad \text{TRUE} \end{array}$$

Since $(4, -2)$ checks, it is the solution.

Do Exercises 3 and 4.

Solve by the substitution method.

1. $x + y = 6,$
$y = x + 2$

2. $y = 7 - x,$
$2x - y = 8$

(*Caution*: Use parentheses when you substitute, being careful about removing them. Remember to solve for both variables.)

Solve by the substitution method.

3. $2y + x = 1,$
$y - 2x = 8$

4. $8x + 5y = 184,$
$x - y = -3$

Answers on page A-48

CALCULATOR CORNER

Solving Systems of Equations Use the INTERSECT feature to solve the systems of equations in Margin Exercises 1, 2, and 3. (See the Calculator Corner on p. 991 for the procedure.)

5. a) Solve this system of equations algebraically using substitution:

$$y + 2x = 3,$$
$$y + 2x = -4.$$

b) Check your answer in part (a) with the one you found graphically in Margin Exercise 3 of Section 13.1.

EXAMPLE 3 Solve this system of equations:

$$y = -3x + 5, \quad \textbf{(1)}$$
$$y = -3x - 2. \quad \textbf{(2)}$$

We solved this system graphically in Example 2 of Section 13.1. We found that the graphs are parallel and the system has no solution. Let's try to solve this system algebraically using substitution.

We substitute $-3x - 2$ for y in equation (1):

$$-3x - 2 = -3x + 5 \quad \text{Substituting } -3x - 2 \text{ for } y$$
$$-2 = 5. \quad \text{Adding } 3x$$

We have a false equation. The equation has **no solution.** (See also Example 13 of Section 7.3.)

Do Exercise 5.

b Solving Applied Problems Involving Two Equations

Many applied problems are easier to solve if we first translate to a system of two equations rather than to a single equation. Using substitution, we will solve a few fairly easy problems here. Then in Section 13.4, we will consider more complicated problems.

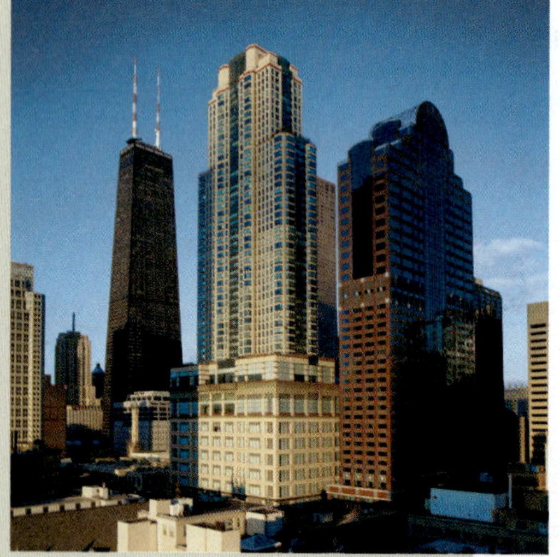

EXAMPLE 4 *Architecture.* The architects who designed the John Hancock Building in Chicago created a visually appealing building that slants on the sides. Thus the ground floor is a rectangle that is larger than the rectangle formed by the top floor. The ground floor has a perimeter of 860 ft. The length is 100 ft more than the width. Find the length and the width.

1. **Familiarize.** We first make a drawing and label it. We recall, or look up, the formula for perimeter: $P = 2l + 2w$. This formula can be found at the back of the book.

2. **Translate.** We translate as follows:

The perimeter is 860 ft.

$$2l + 2w = 860$$

We then translate the second statement:

The length is 100 ft more than the width.

$$l = w + 100$$

We now have a system of equations:

$$2l + 2w = 860, \quad \textbf{(1)}$$
$$l = w + 100. \quad \textbf{(2)}$$

Answers on page A-48

3. Solve. We substitute $w + 100$ for l in equation (1):

$$2(w + 100) + 2w = 860 \quad \text{Substituting in equation (1)}$$
$$2w + 200 + 2w = 860 \quad \text{Multiplying to remove parentheses on the left}$$
$$4w + 200 = 860 \quad \text{Collecting like terms}$$
$$\left.\begin{aligned} 4w &= 660 \\ w &= 165. \end{aligned}\right\} \quad \text{Solving for } w$$

Next, we substitute 165 for w in equation (2) and solve for l:

$$l = 165 + 100 = 265.$$

4. Check. Consider the dimensions 265 ft and 165 ft. The length is 100 ft more than the width. The perimeter is 2(265 ft) + 2(165 ft), or 860 ft. The dimensions 265 ft and 165 ft check in the original problem.

5. State. The length is 265 ft and the width is 165 ft.

Do Exercise 6.

6. Architecture. The top floor of the John Hancock Building is also a rectangle, but its perimeter is 520 ft. The width is 60 ft less than the length. Find the length and the width.

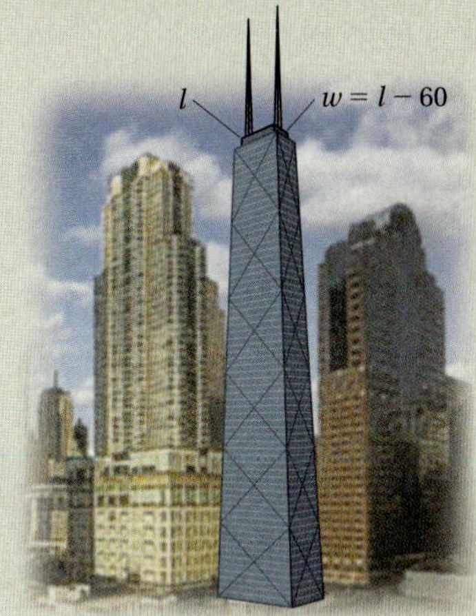

Answer on page A-48

Study Tips

TURNING NEGATIVES INTO POSITIVES (PART 2)

B. C. Forbes said, "History has demonstrated that notable *winners* usually encountered heartbreaking obstacles before they triumphed. They won because they refused to become discouraged by their defeats."

Here are some anecdotes about well-known people who turned what could have been a negative experience into a positive outcome.

- *Richard Bach* sold more than 7 million copies of his story about a "soaring" seagull, Jonathan Livingston Seagull. His work was turned down by 18 publishers before Macmillan finally published it in 1970.
- *Walt Disney* was once fired by a newspaper for what they said was his "lack of ideas." He went bankrupt several times before he built an entertainment empire that now includes Disneyland and Disney World.
- *Hank Aaron* holds the all-time Major League home run record with a total of 755, topping the former record holder, Babe Ruth, who had 714. But Aaron also held the all-time record for many years for striking out 1383 times, also topping Babe Ruth, who struck out 1330 times!

In an article entitled "*Mistakes—Important Teacher,*" Josh Hinds writes, "Another approach (to negative experiences) is to remind ourselves that failures are not always failures, rather they are lessons. I would challenge you to find one occurrence in your own life where you have learned from a past mistake. While it can be true that we don't gain direct rewards from them, we still gain something of great importance. Therefore we need to explore our failures and take the time to use them as our teachers ..."

13.2 EXERCISE SET

For Extra Help

Digital Video Tutor CD 11 Videotape 15 | InterAct Math | Math Tutor Center | MathXL | MyMathLab

a Solve the system by the substitution method.

1. $y = 5 - 4x,$
$2x - 3y = 13$

2. $x = 8 - 4y,$
$3x + 5y = 3$

3. $2y + x = 9,$
$x = 3y - 3$

4. $9x - 2y = 3,$
$3x - 6 = y$

5. $3s - 4t = 14,$
$5s + t = 8$

6. $m - 2n = 16,$
$4m + n = 1$

7. $9x - 2y = -6,$
$7x + 8 = y$

8. $t = 4 - 2s,$
$t + 2s = 6$

9. $-5s + t = 11,$
$4s + 12t = 4$

10. $5x + 6y = 14,$
$-3y + x = 7$

11. $2x + 2y = 2,$
$3x - y = 1$

12. $4p - 2q = 16,$
$5p + 7q = 1$

13. $3a - b = 7,$
$2a + 2b = 5$

14. $5x + 3y = 4,$
$x - 4y = 3$

15. $2x - 3 = y,$
$y - 2x = 1$

16. $4x + 13y = 5,$
$-6x + y = 13$

b Solve.

17. *Racquetball Court.* A regulation racquetball court has a perimeter of 120 ft, with a length that is twice the width. Find the length and the width of such a court.

Court width

Court length

18. *Soccer Field.* The perimeter of a soccer field is 340 m. The length exceeds the width by 50 m. Find the length and the width.

19. *Supplementary Angles.* **Supplementary angles** are angles whose sum is 180°. Two supplementary angles are such that one angle is 12° less than three times the other. Find the measures of the angles.

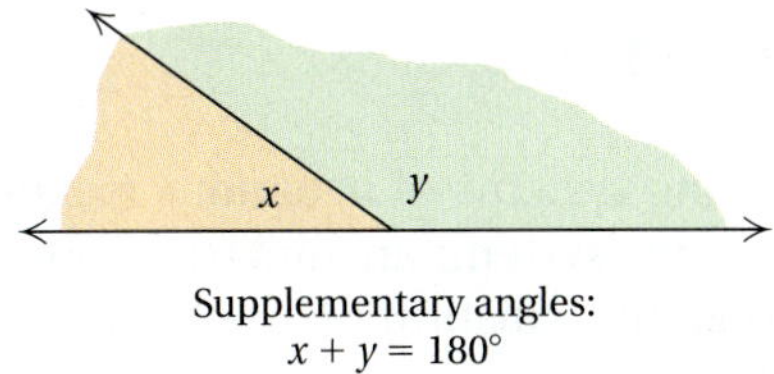

Supplementary angles:
$x + y = 180°$

20. *Complementary Angles.* **Complementary angles** are angles whose sum is 90°. Two complementary angles are such that one angle is 6° more than five times the other. Find the measures of the angles.

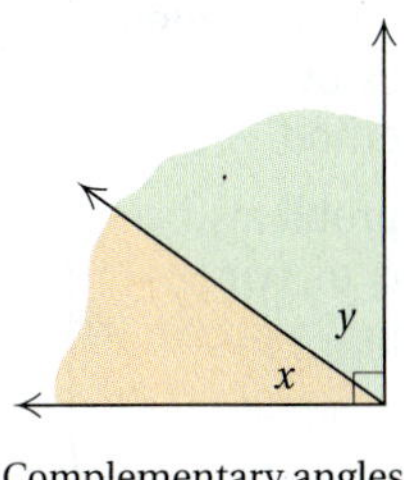

Complementary angles:
$x + y = 90°$

21. *Hockey Points.* Hockey teams receive two points when they win a game and one point when they tie. One season, a team won a championship with 60 points. They won 9 more games than they tied. How many wins and how many ties did the team have?

22. *Airplane Seating.* An airplane has a total of 152 seats. The number of coach-class seats is 5 more than six times the number of first-class seats. How many of each type of seat are there on the plane?

23. **DW** Describe a method that could be used to create inconsistent systems of equations.

24. **DW** Write a problem for a classmate to solve that requires writing a system of two equations. Devise the problem so that the solution is "The Lakers made 6 three-point baskets and 31 two-point baskets."

SKILL MAINTENANCE

25. Find the slope of the line $y = 1.3x - 7$. [12.3b]

26. Simplify: $-9(y + 7) - 6(y - 4)$. [6.8b]

27. Solve $A = \frac{pq}{7}$ for p. [7.4b]

28. Find the slope of the line containing the points $(-2, 3)$ and $(-5, -4)$. [12.3b]

SYNTHESIS

29. For $y = mx + b$, two solutions are $(1, 2)$ and $(-3, 4)$. Find m and b.

30. Solve for x and y in terms of a and b:

$$5x + 2y = a,$$
$$x - y = b.$$

31. *Design.* A piece of posterboard has a perimeter of 156 in. If you cut 6 in. off the width, the length becomes four times the width. What are the dimensions of the original piece of posterboard?

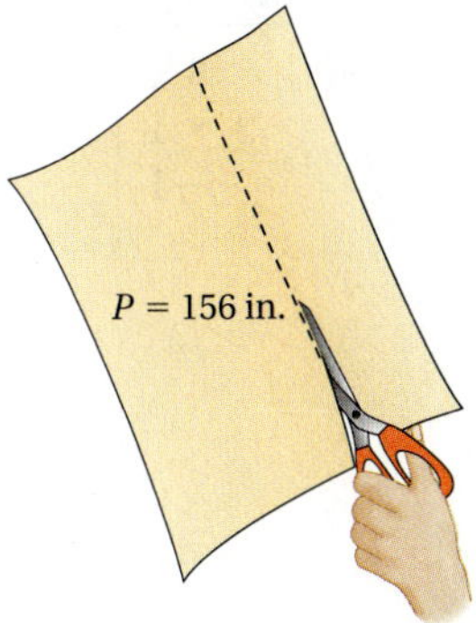

32. *Nontoxic Scouring Powder.* A nontoxic scouring powder is made up of 4 parts baking soda and 1 part vinegar. How much of each ingredient is needed for a 16-oz mixture?

Objectives

a Solve systems of equations in two variables by the elimination method.

b Solve applied problems by solving systems of two equations using elimination.

13.3 SOLVING BY ELIMINATION

a The Elimination Method

The **elimination method** for solving systems of equations makes use of the *addition principle* for equations. Some systems are much easier to solve using the elimination method rather than the substitution method.

EXAMPLE 1 Solve this system:

$$2x - 3y = 0, \qquad (1)$$
$$-4x + 3y = -1. \qquad (2)$$

The key to the advantage of the elimination method in this case is the $-3y$ in one equation and the $3y$ in the other. These terms are opposites. If we add them, these terms will add to 0, and in effect, the variable y will have been "eliminated."

We will use the addition principle for equations, adding the same number on both sides of the equation. According to equation (2), $-4x + 3y$ and -1 are the same number. Thus we can use a vertical form and add $-4x + 3y$ to the left side of equation (1) and -1 to the right side:

$$\begin{array}{rl} 2x - 3y = 0 & (1) \\ \underline{-4x + 3y = -1} & (2) \\ -2x + 0y = -1 & \text{Adding} \\ -2x + \quad 0 = -1 & \\ -2x = -1. & \end{array}$$

We have eliminated the variable y, which is why we call this the *elimination method.** We now have an equation with just one variable, which we solve for x:

$$-2x = -1$$
$$x = \tfrac{1}{2}.$$

Next, we substitute $\frac{1}{2}$ for x in either equation and solve for y:

$$\begin{array}{rl} 2 \cdot \frac{1}{2} - 3y = 0 & \text{Substituting in equation (1)} \\ 1 - 3y = 0 & \\ -3y = -1 & \text{Subtracting 1} \\ y = \frac{1}{3}. & \text{Dividing by } -3 \end{array}$$

We obtain the ordered pair $\left(\frac{1}{2}, \frac{1}{3}\right)$.

CHECK:

$2x - 3y = 0$	
$2\left(\frac{1}{2}\right) - 3\left(\frac{1}{3}\right) \; ? \; 0$	
$1 - 1$	
0	TRUE

$-4x + 3y = -1$	
$-4\left(\frac{1}{2}\right) + 3\left(\frac{1}{3}\right) \; ? \; -1$	
$-2 + 1$	
-1	TRUE

*Also called the *addition method.*

Since $\left(\frac{1}{2}, \frac{1}{3}\right)$ checks, it is the solution. We can also see this in the graph shown at right.

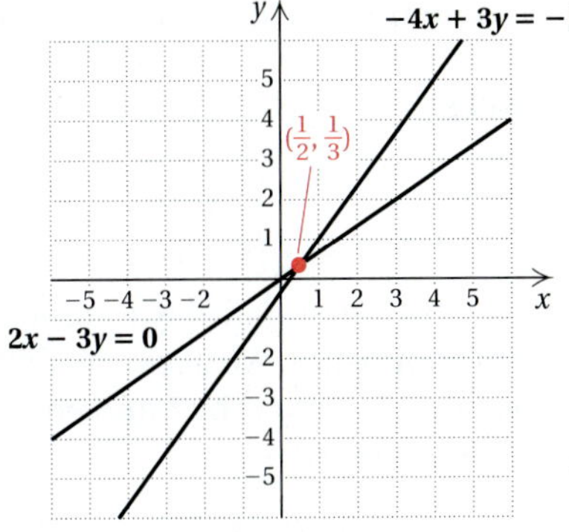

Do Exercises 1 and 2.

In order to eliminate a variable, we sometimes use the multiplication principle to multiply one or both of the equations by a particular number before adding.

EXAMPLE 2 Solve this system:

$$3x + 3y = 15, \quad (1)$$
$$2x + 6y = 22. \quad (2)$$

If we add directly, we get $5x + 9y = 37$, but we have not eliminated a variable. However, note that if the $3y$ in equation (1) were $-6y$, we could eliminate y. Thus we multiply both sides of equation (1) by -2 and add:

$$-6x - 6y = -30 \qquad \text{Multiplying both sides of equation (1) by } -2$$
$$2x + 6y = 22 \qquad \text{Equation (2)}$$
$$-4x + 0 = -8 \qquad \text{Adding}$$
$$-4x = -8$$
$$x = 2. \qquad \text{Solving for } x$$

Then

$$2 \cdot 2 + 6y = 22 \qquad \text{Substituting 2 for } x \text{ in equation (2)}$$
$$\left.\begin{aligned} 4 + 6y &= 22 \\ 6y &= 18 \\ y &= 3. \end{aligned}\right\} \text{Solving for } y$$

We obtain $(2, 3)$, or $x = 2$, $y = 3$. This checks, so it is the solution.

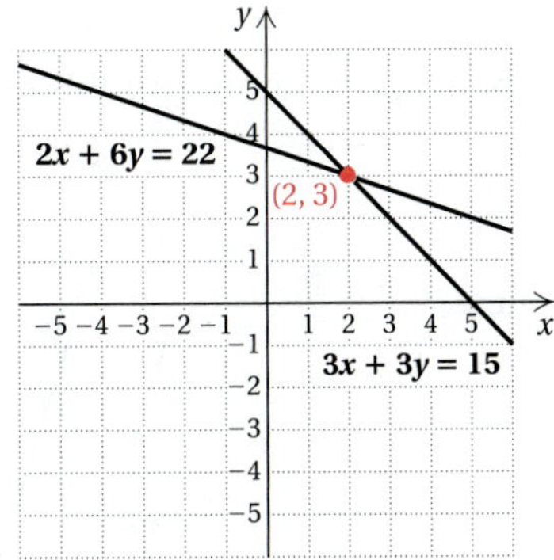

Do Exercise 3.

Solve by the elimination method.

1. $5x + 3y = 17,$
$-5x + 2y = 3$

2. $-3a + 2b = 0,$
$3a - 4b = -1$

3. Solve by the elimination method:

$$2y + 3x = 12,$$
$$-4y + 5x = -2.$$

Answers on page A-48

Solve by the elimination method.

4. $4x + 5y = -8,$
$7x + 9y = 11$

5. $4x - 5y = 38,$
$7x - 8y = -22$

Answers on page A-48

Sometimes we must multiply twice in order to make two terms opposites.

EXAMPLE 3 Solve this system:

$$2x + 3y = 17, \quad (1)$$
$$5x + 7y = 29. \quad (2)$$

We must first multiply in order to make one pair of terms with the same variable opposites. We decide to do this with the x-terms in each equation. We multiply equation (1) by 5 and equation (2) by -2. Then we get $10x$ and $-10x$, which are opposites.

From equation (1):	$10x + 15y = 85$	Multiplying by 5
From equation (2):	$-10x - 14y = -58$	Multiplying by -2
	$0 + y = 27$	Adding
	$y = 27.$	Solving for y

Then

$$2x + 3 \cdot 27 = 17 \qquad \text{Substituting 27 for } y \text{ in equation (1)}$$
$$\left.\begin{aligned} 2x + 81 &= 17 \\ 2x &= -64 \\ x &= -32. \end{aligned}\right\} \text{ Solving for } x$$

We check the ordered pair $(-32, 27)$.

CHECK:

$2x + 3y = 17$		$5x + 7y = 29$	
$2(-32) + 3(27) \ ? \ 17$		$5(-32) + 7(27) \ ? \ 29$	
$-64 + 81$		$-160 + 189$	
17	TRUE	29	TRUE

We obtain $(-32, 27)$, or $x = -32$, $y = 27$, as the solution.

Do Exercises 4 and 5.

Some systems have no solution, as we saw graphically in Section 13.1 and algebraically in Example 3 of Section 13.2. How do we recognize such systems if we are solving using elimination?

EXAMPLE 4 Solve this system:

$$y + 3x = 5, \quad (1)$$
$$y + 3x = -2. \quad (2)$$

If we find the slope–intercept equations for this system, we get

$$y = -3x + 5,$$
$$y = -3x - 2.$$

The graphs are parallel lines. The system has no solution.

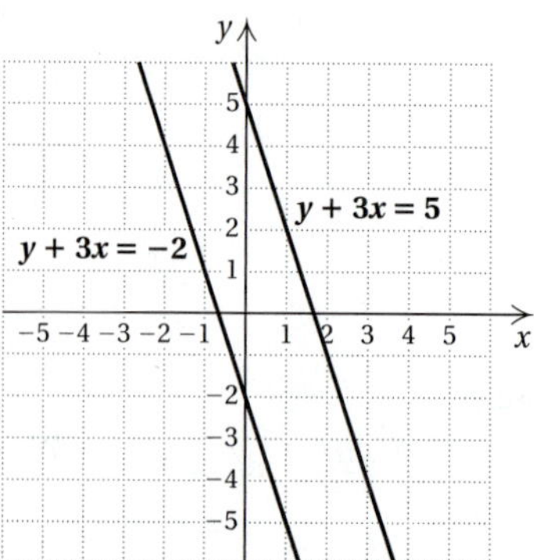

Let's see what happens if we attempt to solve the system by the elimination method. We multiply both sides of equation (2) by -1 and add:

$$\begin{aligned} y + 3x &= 5 \quad &&\text{Equation (1)} \\ -y - 3x &= 2 \quad &&\text{Multiplying equation (2) by } -1 \\ \hline 0 &= 7. \quad &&\text{Adding, we obtain a false equation.} \end{aligned}$$

The x-terms and the y-terms are eliminated and we end up with a *false* equation. Thus, if we obtain a false equation, such as $0 = 7$, when solving algebraically, we know that the system has **no solution**. The system is inconsistent, and the equations are independent.

Do Exercise 6.

Some systems have infinitely many solutions. How can we recognize such a situation when we are solving systems using an algebraic method?

EXAMPLE 5 Solve this system:

$$\begin{aligned} 3y - 2x &= 6, \quad &&\textbf{(1)} \\ -12y + 8x &= -24. \quad &&\textbf{(2)} \end{aligned}$$

The graphs are the same line. The system has an infinite number of solutions.

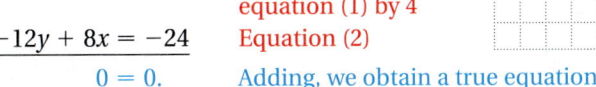

Suppose we try to solve this system by the elimination method:

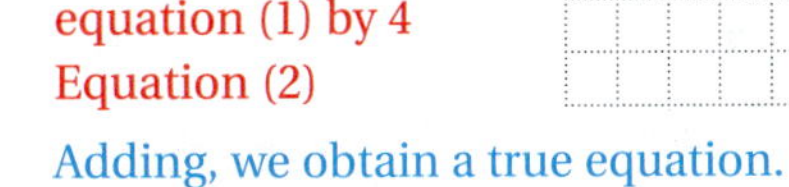

$$\begin{aligned} 12y - 8x &= 24 \quad &&\text{Multiplying equation (1) by 4} \\ -12y + 8x &= -24 \quad &&\text{Equation (2)} \\ \hline 0 &= 0. \quad &&\text{Adding, we obtain a true equation.} \end{aligned}$$

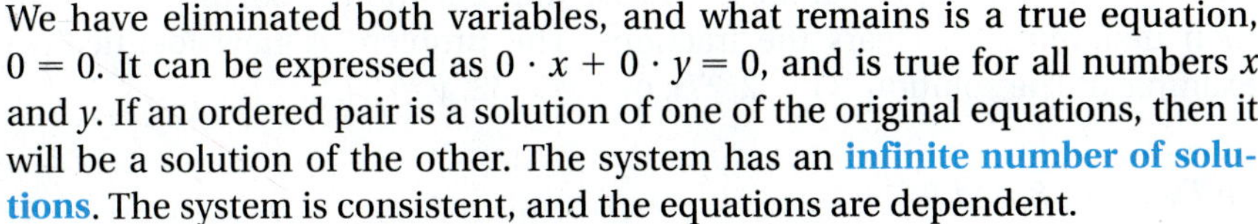

We have eliminated both variables, and what remains is a true equation, $0 = 0$. It can be expressed as $0 \cdot x + 0 \cdot y = 0$, and is true for all numbers x and y. If an ordered pair is a solution of one of the original equations, then it will be a solution of the other. The system has an **infinite number of solutions**. The system is consistent, and the equations are dependent.

SPECIAL CASES

When solving a system of two linear equations in two variables:

1. If a false equation is obtained, such as $0 = 7$, then the system has no solution. The system is inconsistent, and the equations are independent.
2. If a true equation is obtained, such as $0 = 0$, then the system has an infinite number of solutions. The system is consistent, and the equations are dependent.

Do Exercise 7.

6. Solve by the elimination method:

$$y + 2x = 3,$$
$$y + 2x = -1.$$

7. Solve by the elimination method:

$$2x - 5y = 10,$$
$$-6x + 15y = -30.$$

Answers on page A-48

8. Clear the decimals. Then solve.

$$0.02x + 0.03y = 0.01,$$
$$0.3x - 0.1y = 0.7$$

(*Hint*: Multiply the first equation by 100 and the second one by 10.)

9. Clear the fractions. Then solve.

$$\frac{3}{5}x + \frac{2}{3}y = \frac{1}{3},$$
$$\frac{3}{4}x - \frac{1}{3}y = \frac{1}{4}$$

Answers on page A-48

CALCULATOR CORNER

Solving Systems of Equations

1. Consider the system of equations in Example 4. Before we can enter these equations on a graphing calculator, each must be solved for y. What happens when we do this? What does this indicate about the nature of the solutions of the system of equations?
2. Consider the system of equations in Example 5. Before we can enter these equations on a graphing calculator, each must be solved for y. What happens when we do this? What does this indicate about the nature of the solutions of the system of equations?

When solving a system by the elimination method, it helps to first write the equations in the form $Ax + By = C$. When decimals or fractions occur, it also helps to *clear* before solving.

EXAMPLE 6 Solve this system:

$$0.2x + 0.3y = 1.7,$$
$$\tfrac{1}{7}x + \tfrac{1}{5}y = \tfrac{29}{35}.$$

We have

$$0.2x + 0.3y = 1.7, \longrightarrow \text{Multiplying by 10} \longrightarrow 2x + 3y = 17,$$
$$\tfrac{1}{7}x + \tfrac{1}{5}y = \tfrac{29}{35} \longrightarrow \text{Multiplying by 35} \longrightarrow 5x + 7y = 29.$$

We multiplied by 10 to clear the decimals. Multiplication by 35, the least common denominator, clears the fractions. The problem is now identical to Example 3. The solution is $(-32, 27)$, or $x = -32$, $y = 27$.

Do Exercises 8 and 9.

To use the elimination method to solve systems of two equations:

1. Write both equations in the form $Ax + By = C$.
2. Clear any decimals or fractions.
3. Choose a variable to eliminate.
4. Make a chosen variable's terms opposites by multiplying one or both equations by appropriate numbers if necessary.
5. Eliminate a variable by adding the sides of the equations and then solve for the remaining variable.
6. Substitute in either of the original equations to find the value of the other variable.

COMPARING METHODS

The following table is a summary that compares the graphical, substitution, and elimination methods for solving systems of equations.

When deciding which method to use, consider this table and directions from your instructor. The situation is analogous to having a piece of wood to cut and three different types of saws available. Although all three saws can cut the wood, the "best" choice depends on the particular piece of wood, the type of cut being made, and your level of skill with each saw.

METHOD	STRENGTHS	WEAKNESSES
Graphical	Can "see" solutions.	Inexact when solutions involve numbers that are not integers. Solutions may not appear on the part of the graph drawn.
Substitution	Yields exact solutions. Convenient to use when a variable has a coefficient of 1.	Can introduce extensive computations with fractions. Cannot "see" solutions quickly.
Elimination	Yields exact solutions. Convenient to use when no variable has a coefficient of 1. The preferred method for systems of 3 or more equations in 3 or more variables (see Section 8.5).	Cannot "see" solutions quickly.

b Solving Applied Problems Using Elimination

Let's now work a few fairly easy applied problems using elimination. In Section 13.4, we will do more complicated problems. We revisit Example 4 of Section 13.2, but this time we solve it using elimination.

EXAMPLE 7 *Architecture.* The ground floor of the John Hancock Building has a perimeter of 860 ft. The length is 100 ft more than the width. Find the length and the width of the ground floor of the building.

1., 2. Familiarize and **Translate.** Earlier, we translated the problem to the system of equations

$$2l + 2w = 860, \quad (1)$$
$$l = w + 100. \quad (2)$$

10. Architecture. The top floor of the John Hancock Building is also a rectangle, but its perimeter is 520 ft. The width is 60 ft less than the length. Find the length and the width.

a) Use elimination to solve the resulting system.

b) Compare the use of the substitution method in Margin Exercise 6 of Section 8.2 with the use of the elimination method here.

Answers on page A-48

3. Solve. To use *elimination*, we first rewrite both equations in the form $Ax + By = C$. Since equation (1) is already in this form, we need only rewrite equation (2):

$$l = w + 100 \qquad \text{Equation (2)}$$
$$l - w = 100. \qquad \text{Subtracting } w$$

Now we solve the system

$$2l + 2w = 860, \qquad \textbf{(1)}$$
$$l - w = 100. \qquad \textbf{(2)}$$

We multiply both sides of equation (2) by 2 and add:

$$2l + 2w = 860 \qquad \text{Equation (1)}$$
$$2l - 2w = 200 \qquad \text{Multiplying both sides of equation (2) by 2}$$
$$4l = 1060 \qquad \text{Adding}$$
$$l = 265. \qquad \text{Solving for } l$$

Next, we substitute 265 for l in the equation $l = w + 100$ and solve for w:

$$265 = w + 100$$
$$165 = w.$$

Which method, substitution or elimination, do you prefer to use?

4. Check. Consider the dimensions 265 ft and 165 ft. The length is 100 ft more than the width. The perimeter is 2(265 ft) + 2(165 ft), or 860 ft. The dimensions 265 ft and 165 ft check in the original problem.

5. State. The length is 265 ft and the width is 165 ft.

Do Exercise 10.

13.3 EXERCISE SET

For Extra Help

Digital Video Tutor CD 11 Videotape 15 | InterAct Math | Math Tutor Center | MathXL | MyMathLab

a Solve the system by the elimination method.

1. $x + 3y = 7,$
$-x + 4y = 7$

2. $x + y = 9,$
$2x - y = -3$

3. $9x + 5y = 6,$
$2x - 5y = -17$

4. $8x - 3y = 16,$
$8x + 3y = -8$

5. $5x + 3y = 19,$
$2x - 5y = 11$

6. $3x + 2y = 3,$
$9x - 8y = -2$

7. $5r - 3s = 24,$
$3r + 5s = 28$

8. $5x - 7y = -16,$
$2x + 8y = 26$

9. $0.3x - 0.2y = 4,$
$0.2x + 0.3y = 1$

10. $0.7x - 0.3y = 0.5,$
$-0.4x + 0.7y = 1.3$

11. $\frac{1}{2}x + \frac{1}{3}y = 4,$
$\frac{1}{4}x + \frac{1}{3}y = 3$

12. $\frac{2}{3}x + \frac{1}{7}y = -11,$
$\frac{1}{7}x - \frac{1}{3}y = -10$

13. $\frac{2}{5}x + \frac{1}{2}y = 2,$
$\frac{1}{2}x - \frac{1}{6}y = 3$

14. $\frac{1}{3}x + \frac{1}{5}y = 7,$
$\frac{1}{6}x - \frac{2}{5}y = -4$

15. $2x + 3y = 1,$
$4x + 6y = 2$

16. $3x - 2y = 1,$
$-6x + 4y = -2$

17. $2x - 4y = 5,$
$2x - 4y = 6$

18. $3x - 5y = -2,$
$5y - 3x = 7$

19. $5x - 9y = 7,$
$7y - 3x = -5$

20. $a - 2b = 16,$
$b + 3 = 3a$

21. $3(a - b) = 15,$
$4a = b + 1$

22. $10x + y = 306,$
$10y + x = 90$

23. $x - \frac{1}{10}y = 100,$
$y - \frac{1}{10}x = -100$

24. $\frac{1}{8}x + \frac{3}{5}y = \frac{19}{2},$
$-\frac{3}{10}x - \frac{7}{20}y = -1$

25. $0.05x + 0.25y = 22,$
$0.15x + 0.05y = 24$

26. $1.3x - 0.2y = 12,$
$0.4x + 17y = 89$

b Solve. Use the elimination method when solving the translated system.

27. *Soccer Field.* The perimeter of a soccer field is 340 m. The length exceeds the width by 50 m. Find the length and the width.

28. *Racquetball Court.* A regulation racquetball court has a perimeter of 120 ft, with a length that is twice the width. Find the length and the width of such a court.

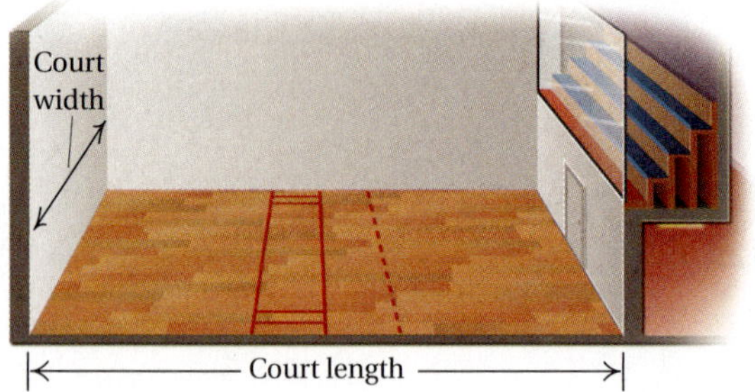

29. *Complementary Angles.* **Complementary angles** are angles whose sum is 90°. Two complementary angles are such that one angle is 6° more than five times the other. Find the measures of the angles.

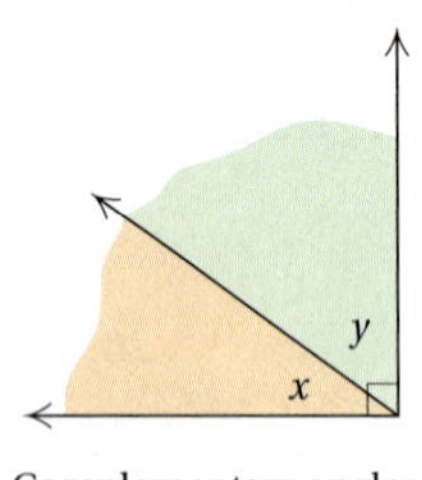

Complementary angles:
$x + y = 90°$

30. *Supplementary Angles.* **Supplementary angles** are angles whose sum is 180°. Two supplementary angles are such that one angle is 12° less than three times the other. Find the measures of the angles.

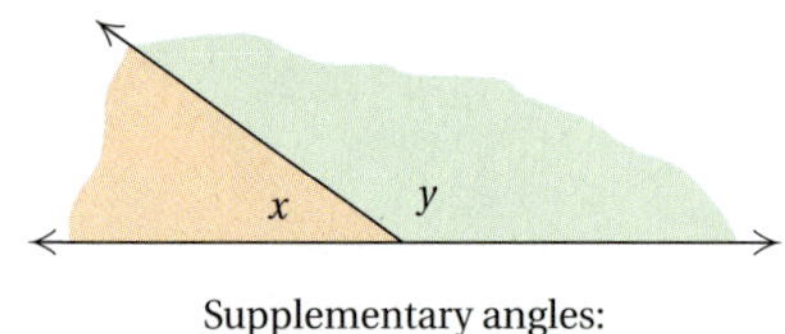

Supplementary angles:
$x + y = 180°$

31. *Airplane Seating.* An airplane has a total of 152 seats. The number of coach-class seats is 5 more than six times the number of first-class seats. How many of each type of seat are there on the plane?

32. *Hockey Points.* Hockey teams receive two points when they win a game and one point when they tie. One season, a team won a championship with 60 points. They won 9 more games than they tied. How many wins and how many ties did the team have?

33. DW Describe a method that could be used to create dependent systems of equations.

34. DW Write a problem for a classmate to solve that can be translated into a system of two equations. Devise the problem so that the solution is "Shelly gave 9 haircuts and 5 shampoos."

SKILL MAINTENANCE

Write an equation of the line containing the given point and perpendicular to the given line. [12.5d]

35. $(10, 1)$; $2x - 7y = 3$

36. $(-2, -3)$; $7y - 5x = 1$

Given the function $f(x) = 3x^2 - x + 1$, find each of the following function values. [12.1b]

37. $f(0)$

38. $f(-1)$

39. $f(1)$

40. $f(10)$

41. $f(-2)$

42. $f(2a)$

43. $f(-4)$

44. $f(1.8)$

SYNTHESIS

45. Use the INTERSECT feature to solve the following system of equations. You may need to first solve for y. Round answers to the nearest hundredth.

$$3.5x - 2.1y = 106.2,$$
$$4.1x + 16.7y = -106.28$$

46. Solve:

$$\frac{x+y}{2} - \frac{x-y}{5} = 1,$$
$$\frac{x-y}{2} + \frac{x+y}{6} = -2.$$

47. The solution of this system is $(-5, -1)$. Find A and B.

$$Ax - 7y = -3,$$
$$x - By = -1$$

48. Find an equation to pair with $6x + 7y = -4$ such that $(-3, 2)$ is a solution of the system.

49. The points $(0, -3)$ and $\left(-\frac{3}{2}, 6\right)$ are two of the solutions of the equation $px - qy = -1$. Find p and q.

50. Determine a and b for which $(-4, -3)$ will be a solution of the system

$$ax + by = -26,$$
$$bx - ay = 7.$$

13.4 SOLVING APPLIED PROBLEMS: TWO EQUATIONS

Objectives

a Solve applied problems involving total value and mixture using systems of two equations.

b Solve applied problems involving motion using systems of two equations.

a Total-Value and Mixture Problems

Systems of equations can be a useful tool in solving applied problems. Using systems often makes the *Translate* step easier than using a single equation. The first kind of problem we consider involves quantities of items sold and the total value of the items. We refer to this type of problem as a **total-value problem.**

EXAMPLE 1 *Retail Sales of Gloves.* In one day, Glovers, Inc., sold 20 pairs of gloves. Fleece gloves sold for \$24.95 a pair and Gore-Tex gloves for \$37.50 a pair. Receipts totaled \$687.25. How many of each kind of glove were sold?

1. **Familiarize.** To familiarize ourselves with the problem situation, let's guess that Glovers sold 12 pairs of fleece gloves and 8 pairs of Gore-Tex gloves. The total is 20. How much money was taken in? Since fleece gloves sold for \$24.95 a pair and Gore-Tex for \$37.50 a pair, the total received would then be

Money from fleece gloves plus Money from Gore-Tex gloves

$$12(\$24.95) + 8(\$37.50) = \$299.40 + \$300.00 = \$599.40.$$

Although the total number of pairs is correct, our guess is incorrect because the problem states that the total amount received was \$687.25. Since \$599.40 is less than \$687.25, more of the expensive gloves were sold than we had guessed. We could now adjust our guess accordingly. Instead, let's use an algebraic approach that avoids guessing.

We let $p =$ the number of pairs of fleece gloves sold and $q =$ the number of pairs of Gore-Tex gloves sold. It helps to organize the information in a table as follows:

	FLEECE GLOVE	GORE-TEX GLOVE	TOTAL	
Number Sold	p	q	20	$\rightarrow p + q = 20$
Price	\$24.95	\$37.50		
Amount Taken In	$24.95p$	$37.50q$	687.25	$\rightarrow 24.95p + 37.50q = 687.25$

Study Tips

TAPING YOUR LECTURES

Consider recording your notes and playing them back when convenient, for example, while commuting to campus. It can even be advantageous to record math lectures. (Be sure to get permission from your instructor before doing so, however.) Important points can be emphasized verbally.

2. **Translate.** The first row of the table and the first sentence of the problem tell us that a total of 20 pairs of gloves was sold. Thus we have one equation:

$$p + q = 20.$$

Since each pair of fleece gloves costs \$24.95 and p pairs were sold, $24.95p$ is the amount taken in from the sale of fleece gloves. Similarly, $37.50q$ is the amount taken in from the sale of q pairs of Gore-Tex gloves. From the third row of the table and the third sentence of the problem, we get the second equation:

$$24.95p + 37.50q = 687.25.$$

We can multiply both sides of this equation by 100 in order to clear the decimals. This gives us the following system of equations as a translation:

$$p + q = 20, \qquad \textbf{(1)}$$
$$2495p + 3750q = 68{,}725. \qquad \textbf{(2)}$$

3. **Solve.** We choose to use the elimination method to solve the system. We eliminate p by multiplying equation (1) by -2495 and adding it to equation (2):

$$\begin{aligned} -2495p - 2495q &= -49{,}900 && \text{Multiplying equation (1) by } -2495 \\ 2495p + 3750q &= 68{,}725 && \\ \hline 1255q &= 18{,}825 && \text{Adding} \\ q &= 15. && \text{Solving for } q \end{aligned}$$

To find p, we substitute 15 for q in equation (1) and solve for p:

$$\begin{aligned} p + q &= 20 && \text{Equation (1)} \\ p + 15 &= 20 && \text{Substituting 15 for } q \\ p &= 5. && \text{Solving for } p \end{aligned}$$

We obtain $(5, 15)$, or $p = 5$, $q = 15$.

4. **Check.** We check in the original problem. Remember that p is the number of pairs of fleece gloves and q the number of pairs of Gore-Tex gloves:

Number of pairs of gloves: $p + q = 5 + 15 = 20$
Money from fleece gloves: $\$24.95p = 24.95 \times 5 = \$124.75$
Money from Gore-Tex gloves: $\$37.50q = 37.50 \times 15 = \$562.50$
Total $= \$687.25$

The numbers check.

5. **State.** The store sold 5 pairs of fleece gloves and 15 pairs of Gore-Tex gloves.

Do Exercise 1.

1. **Retail Sales of Sweatshirts.** Sandy's Sweatshirt Shop sells college sweatshirts. White sweatshirts sell for \$18.95 each and red ones sell for \$19.50 each. If receipts for the sale of 30 sweatshirts total \$572.90, how many of each color did the shop sell? Complete the following table, letting $w =$ the number of white sweatshirts and $r =$ the number of red sweatshirts.

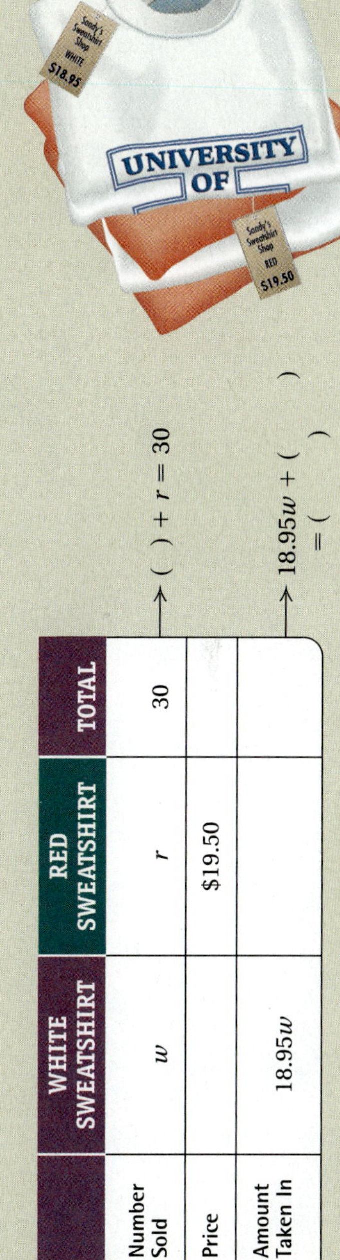

	WHITE SWEATSHIRT	RED SWEATSHIRT	TOTAL	
Number Sold	w	r	30	→ $(\quad) + r = 30$
Price		\$19.50		
Amount Taken In	$18.95w$			→ $18.95w + (\quad) = (\quad)$

Answers on page A-49

2. **Blending Coffees.** The Coffee Counter charges \$9.00 per pound for Kenyan French Roast coffee and \$8.00 per pound for Sumatran coffee. How much of each type should be used to make a 20-lb blend that sells for \$8.40 per pound?

Answer on page A-49

The following problem, similar to Example 1, is called a **mixture problem.**

EXAMPLE 2 *Blending Flower Seeds.* Tara's website, Garden Edibles, specializes in the sale of herbs and flowers for colorful meals and garnishes. Tara sells packets of nasturtium seeds for \$0.95 each and packets of Johnny-jump-up seeds for \$1.43 each. She decides to offer a 16-packet spring-garden combination, combining packets of both types of seeds at \$1.10 per packet. How many packets of each type of seed should be put in her garden mix?

1. **Familiarize.** To familiarize ourselves with the problem situation, we make a guess and do some calculations. The total number of packets of seed is 16. Let's try 12 packets of nasturtiums and 4 packets of Johnny-jump-ups.

   The sum of the number of packets is $12 + 4$, or 16.

   The value of these seed packets is found by multiplying the cost per packet by the number of packets and adding:

   $\$0.95(12) + \$1.43(4)$, or \$17.12.

   The desired cost is \$1.10 per packet. If we multiply \$1.10 by 16, we get 16(\$1.10), or \$17.60. This does not agree with \$17.12, but these calculations give us a basis for understanding how to translate.

   We let $a =$ the number of packets of nasturtium seeds and $b =$ the number of packets of Johnny-jump-up seeds. Next, we organize the information in a table, as follows.

	NASTURTIUM	JOHNNY-JUMP-UP	SPRING	
Number of Packets	a	b	16	$\longrightarrow a + b = 16$
Price per Packet	\$0.95	\$1.43	\$1.10	
Value of Packets	$0.95a$	$1.43b$	$16 \cdot 1.10$, or 17.60	$\longrightarrow 0.95a + 1.43b = 17.60$

2. **Translate.** The total number of packets is 16, so we have one equation:

$a + b = 16.$

The value of the nasturtium seeds is $0.95a$ and the value of the Johnny-jump-up seeds is $1.43b$. These amounts are in dollars. Since the total value is to be 16($1.10), or $17.60, we have

$0.95a + 1.43b = 17.60.$

We can multiply both sides of this equation by 100 to clear the decimals. Thus we have the translation, a system of equations:

$a + b = 16,$ **(1)**

$95a + 143b = 1760.$ **(2)**

3. **Solve.** We decide to use substitution, although elimination could be used as we did in Example 1. When equation (1) is solved for b, we get $b = 16 - a$. Substituting $16 - a$ for b in equation (2) and solving gives us

$95a + 143(16 - a) = 1760$ Substituting

$95a + 2288 - 143a = 1760$ Using the distributive law

$-48a = -528$ Subtracting 2288 and collecting like terms

$a = 11.$

We have $a = 11$. Substituting this value in the equation $b = 16 - a$, we obtain $b = 16 - 11$, or 5.

4. **Check.** We check in a manner similar to our guess in the *Familiarize* step. The total number of packets is $11 + 5$, or 16. The value of the packet mixture is

$0.95(11) + $1.43(5), or $17.60.

Thus the numbers of packets check.

5. **State.** The spring garden mixture can be made by combining 11 packets of nasturtium seeds with 5 packets of Johnny-jump-up seeds.

Do Exercise 2 on the preceding page.

EXAMPLE 3 *Student Loans.* Enid's student loans totaled $9600. Part was a Perkins loan made at 5% interest and the rest was a Federal Education loan made at 8% interest. After one year, Enid's loans accumulated $633 in interest. What was the amount of each loan?

1. **Familiarize.** Listing the given information in a table will help. The columns in the table come from the formula for simple interest: $I = Prt$. We let $x =$ the number of dollars in the Perkins loan and $y =$ the number of dollars in the Federal Education loan.

	PERKINS LOAN	FEDERAL LOAN	TOTAL	
Principal	x	y	$9600	→ $x + y = 9600$
Rate of Interest	5%	8%		
Time	1 yr	1 yr		
Interest	$0.05x$	$0.08y$	$633	→ $0.05x + 0.08y = 633$

3. Client Investments. Kaufman Financial Corporation makes investments for corporate clients. It makes an investment of $3700 for one year at simple interest, yielding $297. Part of the money is invested at 7% and the rest at 9%. How much was invested at each rate?

Do the *Familiarize* and *Translate* steps by completing the following table. Let $x =$ the number of dollars invested at 7% and $y =$ the number of dollars invested at 9%.

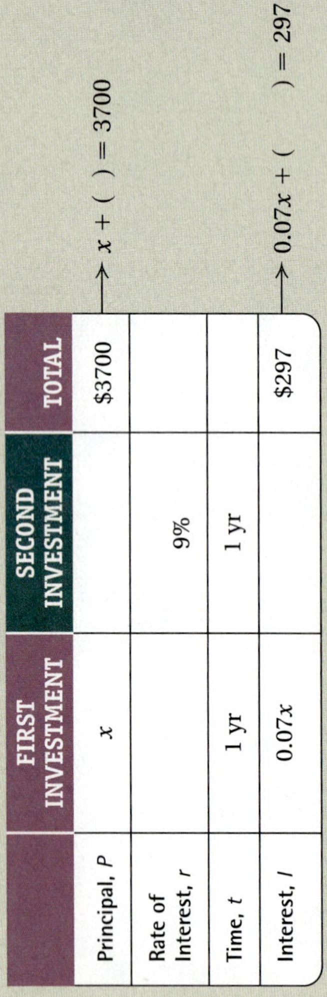

	FIRST INVESTMENT	SECOND INVESTMENT	TOTAL	
Principal, P	x		$3700	$\to x + (\ \ ) = 3700$
Rate of Interest, r		9%		
Time, t	1 yr	1 yr		
Interest, I	$0.07x$		$297	$\to 0.07x + (\ \ ) = 297$

Answer on page A-49

2. Translate. The total of the amounts of the loans is found in the first row of the table. This gives us one equation:

$$x + y = 9600.$$

Look at the last row of the table. The interest, or **yield,** totals $633. This gives us a second equation:

$$5\%x + 8\%y = 633, \quad \text{or} \quad 0.05x + 0.08y = 633.$$

After we multiply on both sides to clear the decimals, we have

$$5x + 8y = 63{,}300.$$

3. Solve. Using either elimination or substitution, we solve the resulting system:

$$x + y = 9600,$$
$$5x + 8y = 63{,}300.$$

We find that $x = 4500$ and $y = 5100$.

4. Check. The sum is $4500 + $5100, or $9600. The interest from $4500 at 5% for one year is 5%($4500), or $225. The interest from $5100 at 8% for one year is 8%($5100), or $408. The total interest is $225 + $408, or $633. The numbers check in the problem.

5. State. The Perkins loan was for $4500 and the Federal Education loan was for $5100.

Do Exercise 3.

EXAMPLE 4 *Mixing Fertilizers.* Yardbird Gardening carries two kinds of fertilizer containing nitrogen and water. "Gently Green" is 5% nitrogen and "Sun Saver" is 15% nitrogen. Yardbird Gardening needs to combine the two types of solution to make 90 L of a solution that is 12% nitrogen. How much of each brand should be used?

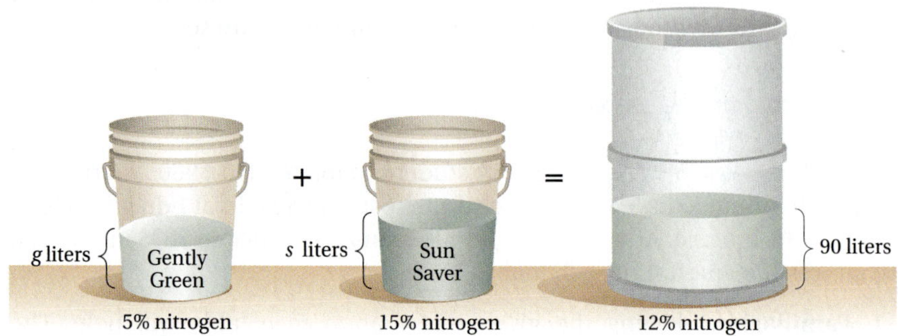

1. Familiarize. We first make a drawing and a guess to become familiar with the problem.

We choose two numbers that total 90 L—say, 40 L of Gently Green and 50 L of Sun Saver—for the amounts of each fertilizer. Will the resulting mixture have the correct percentage of nitrogen? To find out, we multiply as follows:

$$5\%(40 \text{ L}) = 2 \text{ L of nitrogen} \quad \text{and} \quad 15\%(50 \text{ L}) = 7.5 \text{ L of nitrogen}.$$

Thus the total amount of nitrogen in the mixture is 2 L + 7.5 L, or 9.5 L.

The final mixture of 90 L is supposed to be 12% nitrogen. Now

$$12\%(90\text{ L}) = 10.8\text{ L}.$$

Since 9.5 L and 10.8 L are not the same, our guess is incorrect. But these calculations help us to become familiar with the problem and to make the translation.

We let g = the number of liters of Gently Green and s = the number of liters of Sun Saver.

The information can be organized in a table, as follows.

	GENTLY GREEN	SUN SAVER	MIXTURE	
Number of Liters	g	s	90	$\longrightarrow g + s = 90$
Percent of Nitrogen	5%	15%	12%	
Amount of Nitrogen	$0.05g$	$0.15s$	0.12×90, or 10.8 liters	$\longrightarrow 0.05g + 0.15s = 10.8$

2. Translate. If we add g and s in the first row, we get 90, and this gives us one equation:

$$g + s = 90.$$

If we add the amounts of nitrogen listed in the third row, we get 10.8, and this gives us another equation:

$$5\%g + 15\%s = 10.8, \quad \text{or} \quad 0.05g + 0.15s = 10.8.$$

After clearing the decimals, we have the following system:

$$g + s = 90, \qquad (1)$$
$$5g + 15s = 1080. \qquad (2)$$

3. Solve. We solve the system using elimination. We multiply equation (1) by -5 and add the result to equation (2):

$$\begin{aligned} -5g - 5s &= -450 && \text{Multiplying equation (1) by } -5 \\ 5g + 15s &= 1080 && \\ \hline 10s &= 630 && \text{Adding} \\ s &= 63; && \text{Dividing by 10} \end{aligned}$$

$$\begin{aligned} g + 63 &= 90 && \text{Substituting in equation (1) of the system} \\ g &= 27. && \text{Solving for } g \end{aligned}$$

4. Mixing Cleaning Solutions. King's Service Station uses two kinds of cleaning solution containing acid and water. "Attack" is 2% acid and "Blast" is 6% acid. They want to mix the two to get 60 qt of a solution that is 5% acid. How many quarts of each should they use?

Do the *Familiarize* and *Translate* steps by completing the following table. Let a = the number of quarts of Attack and b = the number of quarts of Blast.

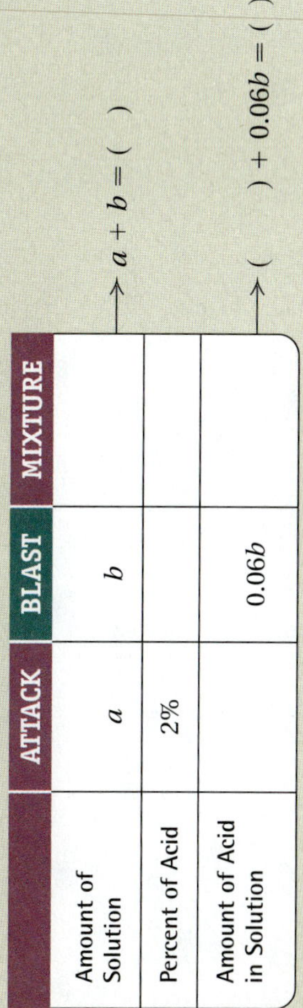

	ATTACK	BLAST	MIXTURE	
Amount of Solution	a	b		$\rightarrow a + b = (\quad)$
Percent of Acid	2%			
Amount of Acid in Solution		$0.06b$		$\rightarrow (\quad) + 0.06b = (\quad)$

Answer on page A-49

4. **Check.** Remember that g is the number of liters of Gently Green, with 5% nitrogen, and s is the number of liters of Sun Saver, with 15% nitrogen.

   Total number of liters of mixture: $g + s = 27 + 63 = 90$

   Amount of nitrogen: $5\%(27) + 15\%(63) = 1.35 + 9.45 = 10.8$ L

   Percentage of nitrogen in mixture: $\dfrac{10.8}{90} = 0.12 = 12\%$

   The numbers check in the original problem.

5. **State.** Yardbird Gardening should mix 27 L of Gently Green and 63 L of Sun Saver.

Do Exercise 4 on the preceding page.

b Motion Problems

When a problem deals with speed, distance, and time, we can expect to use the following *motion formula.*

THE MOTION FORMULA

Distance = Rate (or speed) · Time

$d = rt$

TIPS FOR SOLVING MOTION PROBLEMS

1. Draw a diagram using an arrow or arrows to represent distance and the direction of each object in motion.
2. Organize the information in a table or chart.
3. Look for as many things as you can that are the same, so you can write equations.

EXAMPLE 5 *Auto Travel.* Your brother leaves on a trip, forgetting his suitcase. You know that he normally drives at a speed of 55 mph. You do not discover the suitcase until 1 hr after he has left. If you follow him at a speed of 65 mph, how long will it take you to catch up with him?

1. **Familiarize.** We first make a drawing.

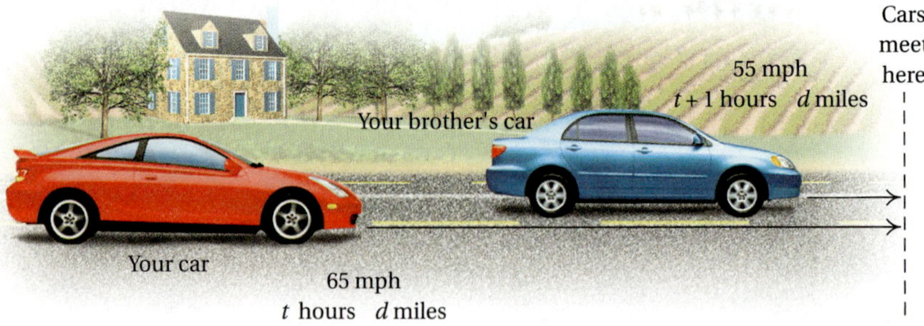

From the drawing, we see that when you catch up with your brother, the distances from home are the same. Let's call the distance d. If we let t = the time for you to catch your brother, then $t + 1$ = the time traveled by your brother at a slower speed.

We organize the information in a table.

	DISTANCE (d)	RATE (r)	TIME (t)	
Brother	d	55	$t + 1$	$\longrightarrow d = 55(t + 1)$
You	d	65	t	$\longrightarrow d = 65t$

2. Translate. Using $d = rt$ in each row of the table, we get an equation. Thus we have a system of equations:

$$d = 55(t + 1), \quad (1)$$
$$d = 65t. \quad (2)$$

3. Solve. We solve the system using the substitution method:

$65t = 55(t + 1)$ Substituting $65t$ for d in equation (1)

$65t = 55t + 55$ Multiplying to remove parentheses on the right

$10t = 55$, $t = 5.5.$ Solving for t

Your time is 5.5 hr, which means that your brother's time is 5.5 + 1, or 6.5 hr.

4. Check. At 65 mph, you will travel $65 \cdot 5.5$, or 357.5 mi, in 5.5 hr. At 55 mph, your brother will travel $55 \cdot 6.5$, or the same 357.5 mi, in 6.5 hr. The numbers check.

5. State. You will overtake your brother in 5.5 hr.

Do Exercise 5.

5. Train Travel. A train leaves Barstow traveling east at 35 km/h. One hour later, a faster train leaves Barstow, also traveling east on a parallel track at 40 km/h. How far from Barstow will the faster train catch up with the slower one?

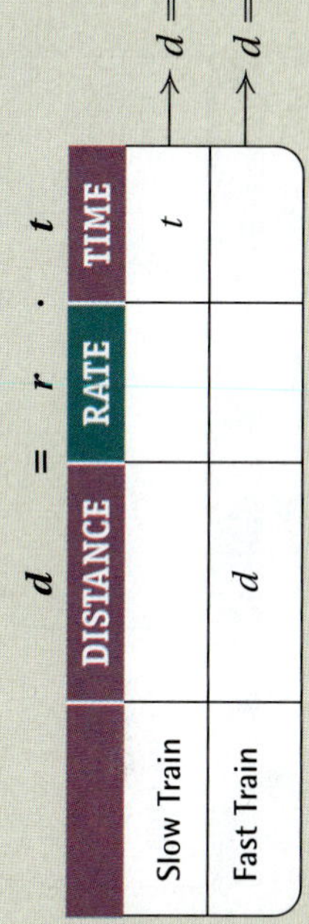

EXAMPLE 6 *Marine Travel.* A Coast-Guard patrol boat travels 4 hr on a trip downstream with a 6-mph current. The return trip against the same current takes 5 hr. Find the speed of the boat in still water.

Answer on page A-49

6. Air Travel. An airplane flew for 4 hr with a 20-mph tailwind. The return flight against the same wind took 5 hr. Find the speed of the plane in still air.

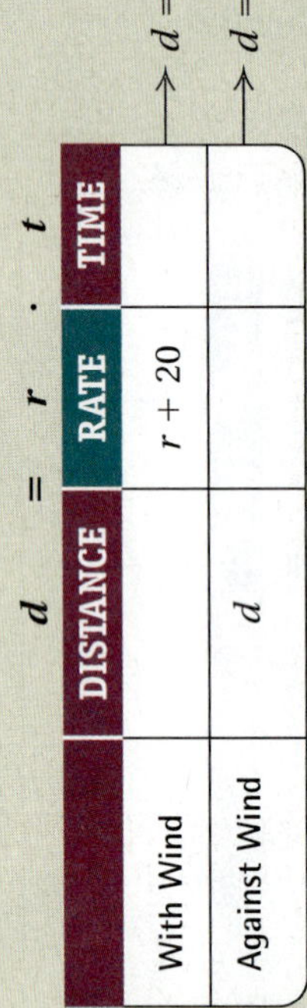

Answer on page A-49

1. **Familiarize.** We first make a drawing. From the drawing, we see that the distances are the same. We let $d =$ the distance, in miles, and $r =$ the speed of the boat in still water, in miles per hour. Then, when the boat is traveling downstream, its speed is $r + 6$ (the current helps the boat along). When it is traveling upstream, its speed is $r - 6$ (the current holds the boat back). We can organize the information in a table. We use the formula $d = rt$.

	DISTANCE (d)	RATE (r)	TIME (t)	
Downstream	d	$r + 6$	4	→ $d = (r + 6)4$
Upstream	d	$r - 6$	5	→ $d = (r - 6)5$

2. **Translate.** From each row of the table, we get an equation, $d = rt$:

$$d = 4r + 24, \quad (1)$$
$$d = 5r - 30. \quad (2)$$

3. **Solve.** We solve the system by the substitution method:

$$4r + 24 = 5r - 30 \quad \text{Substituting } 4r + 24 \text{ for } d \text{ in equation (2)}$$
$$\left.\begin{aligned} 24 &= r - 30 \\ 54 &= r. \end{aligned}\right\} \quad \text{Solving for } r$$

4. **Check.** If $r = 54$, then $r + 6 = 60$; and $60 \cdot 4 = 240$, the distance traveled downstream. If $r = 54$, then $r - 6 = 48$; and $48 \cdot 5 = 240$, the distance traveled upstream. The distances are the same. In this type of problem, a problem-solving tip to keep in mind is "Have I found what the problem asked for?" We could solve for a certain variable but still have not answered the question of the original problem. For example, we might have found speed when the problem wanted distance. In this problem, we want the speed of the boat in still water, and that is r.

5. **State.** The speed in still water is 54 mph.

Do Exercise 6.

For Extra Help

Digital Video Tutor CD 11 Videotape 15

InterAct Math

Math Tutor Center

MathXL

MyMathLab

a Solve.

1. *Retail Sales.* Paint Town sold 45 paintbrushes, one kind at \$8.50 each and another at \$9.75 each. In all, \$398.75 was taken in for the brushes. How many of each kind were sold?

2. *Retail Sales.* Mountainside Fleece sold 40 neckwarmers. Solid-color neckwarmers sold for \$9.90 each and print ones sold for \$12.75 each. In all, \$421.65 was taken in for the neckwarmers. How many of each type were sold?

3. *Sales of Pharmaceuticals.* The Diabetic Express recently charged \$15.75 for a vial of Humulin insulin and \$12.95 for a vial of Novolin insulin. If a total of \$959.35 was collected for 65 vials of insulin, how many vials of each type were sold?

4. *Fundraising.* The St. Mark's Community Barbecue served 250 dinners. A child's plate cost \$3.50 and an adult's plate cost \$7.00. A total of \$1347.50 was collected. How many of each type of plate was served?

5. *Radio Airplay.* Rudy must play 12 commercials during his 1-hr radio show. Each commercial is either 30 sec or 60 sec long. If the total commercial time during that hour is 10 min, how many commercials of each type does Rudy play?

6. *Nontoxic Floor Wax.* A nontoxic floor wax can be made by combining lemon juice and food-grade linseed oil. The amount of oil should be twice the amount of lemon juice. How much of each ingredient is needed in order to make 32 oz of floor wax? (The mix should be spread with a rag and buffed when dry.)

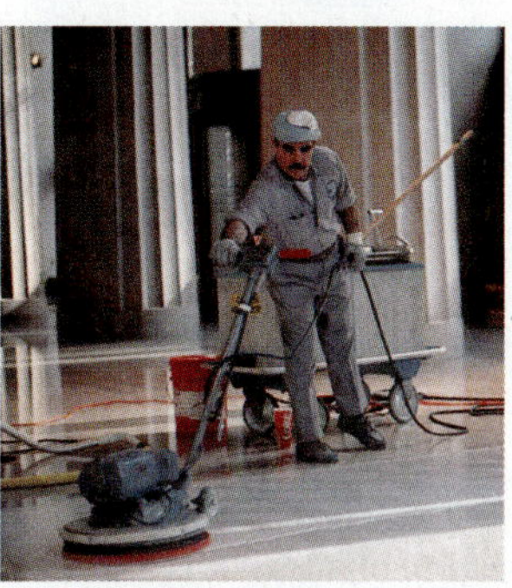

7. *Catering.* Stella's Catering is planning a wedding reception. The bride and groom would like to serve a nut mixture containing 25% peanuts. Stella has available mixtures that are either 40% or 10% peanuts. How much of each type should be mixed to get a 10-lb mixture that is 25% peanuts?

8. *Blending Granola.* Deep Thought Granola is 25% nuts and dried fruit. Oat Dream Granola is 10% nuts and dried fruit. How much of Deep Thought and how much of Oat Dream should be mixed to form a 20-lb batch of granola that is 19% nuts and dried fruit?

9. *Ink Remover.* Etch Clean Graphics uses one cleanser that is 25% acid and a second that is 50% acid. How many liters of each should be mixed to get 10 L of a solution that is 40% acid?

10. *Livestock Feed.* Soybean meal is 16% protein and corn meal is 9% protein. How many pounds of each should be mixed to get a 350-lb mixture that is 12% protein?

11. *Student Loans.* Sarah's two student loans totaled \$12,000. One of her loans was at 6% simple interest and the other at 9%. After one year, Sarah owed \$855 in interest. What was the amount of each loan?

12. *Investments.* An executive nearing retirement made two investments totaling \$15,000. In one year, these investments yielded \$1432 in simple interest. Part of the money was invested at 9% and the rest at 10%. How much was invested at each rate?

13. *Food Science.* The following bar graph shows the milk fat percentages in three dairy products. How many pounds each of whole milk and cream should be mixed to form 200 lb of milk for cream cheese?

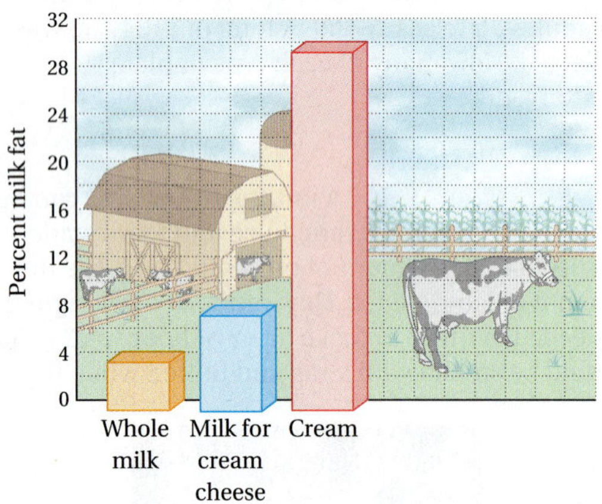

14. *Automotive Maintenance.* "Arctic Antifreeze" is 18% alcohol and "Frost No-More" is 10% alcohol. How many liters of each should be mixed to get 20 L of a mixture that is 15% alcohol?

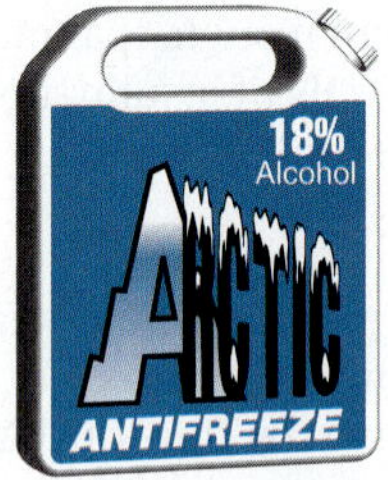

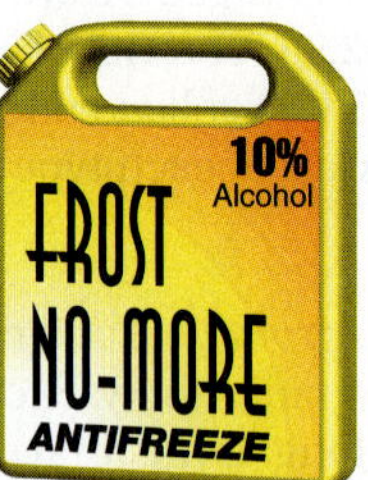

15. *Teller Work.* Juan goes to a bank and gets change for a \$50 bill consisting of all \$5 bills and \$1 bills. There are 22 bills in all. How many of each kind are there?

16. *Making Change.* Christina makes a \$9.25 purchase at a bookstore in Reno with a \$20 bill. The store has no bills and gives her the change in quarters and fifty-cent pieces. There are 30 coins in all. How many of each kind are there?

b Solve.

17. *Train Travel.* A train leaves Danville Junction and travels north at a speed of 75 mph. Two hours later, a second train leaves on a parallel track and travels north at 125 mph. How far from the station will they meet?

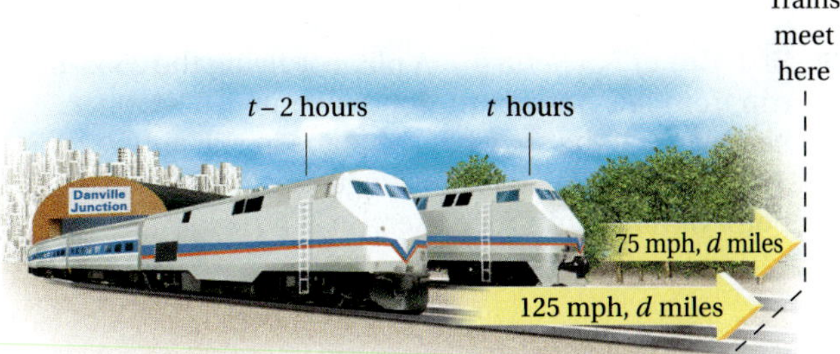

18. *Car Travel.* Two cars leave Denver traveling in opposite directions. One car travels at a speed of 80 km/h and the other at 96 km/h. In how many hours will they be 528 km apart?

19. *Canoeing.* Darren paddled for 4 hr with a 6-km/h current to reach a campsite. The return trip against the same current took 10 hr. Find the speed of Darren's canoe in still water.

20. *Boating.* Mia's motorboat took 3 hr to make a trip downstream with a 6-mph current. The return trip against the same current took 5 hr. Find the speed of the boat in still water.

21. *Car Travel.* Donna is late for a sales meeting after traveling from one town to another at a speed of 32 mph. If she had traveled 4 mph faster, she could have made the trip in $\frac{1}{2}$ hr less time. How far apart are the towns?

22. *Air Travel.* Rod is a pilot for Crossland Airways. He computes his flight time against a headwind for a trip of 2900 mi at 5 hr. The flight would take 4 hr and 50 min if the headwind were half as great. Find the headwind and the plane's air speed.

23. *Air Travel.* Two planes travel toward each other from cities that are 780 km apart at rates of 190 km/h and 200 km/h. They started at the same time. In how many hours will they meet?

24. *Motorcycle Travel.* Sally and Rocky travel on motorcycles toward each other from Chicago and Indianapolis, which are about 350 km apart, and they are biking at rates of 110 km/h and 90 km/h. They started at the same time. In how many hours will they meet?

25. *Air Travel.* Two airplanes start at the same time and fly toward each other from points 1000 km apart at rates of 420 km/h and 330 km/h. After how many hours will they meet?

26. *Truck and Car Travel.* A truck and a car leave a service station at the same time and travel in the same direction. The truck travels at 55 mph and the car at 40 mph. They can maintain CB radio contact within a range of 10 mi. When will they lose contact?

27. *Point of No Return.* A plane flying the 3458-mi trip from New York City to London has a 50-mph tailwind. The flight's *point of no return* is the point at which the flight time required to return to New York is the same as the time required to continue to London. If the speed of the plane in still air is 360 mph, how far is New York from the point of no return?

28. *Point of No Return.* A plane is flying the 2553-mi trip from Los Angeles to Honolulu into a 60-mph headwind. If the speed of the plane in still air is 310 mph, how far from Los Angeles is the plane's point of no return? (See Exercise 27.)

29. DW List three or four study tips for someone beginning this exercise set.

30. DW Write a problem similar to Example 1 for a classmate to solve. Design the problem so the answer is "The florist sold 14 hanging plants and 9 flats of petunias."

SKILL MAINTENANCE

Given the function $f(x) = 4x - 7$, find each of the following function values. [12.1b]

31. $f(0)$
32. $f(-1)$
33. $f(1)$
34. $f(10)$
35. $f(-2)$
36. $f(2a)$
37. $f(-4)$
38. $f(1.8)$
39. $f\left(\frac{3}{4}\right)$
40. $f(-2.5)$
41. $f(-3h)$
42. $f(1000)$

SYNTHESIS

43. *Automotive Maintenance.* The radiator in Michelle's car contains 16 L of antifreeze and water. This mixture is 30% antifreeze. How much of this mixture should she drain and replace with pure antifreeze so that there will be a mixture of 50% antifreeze?

44. *Physical Exercise.* Natalie jogs and walks to school each day. She averages 4 km/h walking and 8 km/h jogging. The distance from home to school is 6 km and Natalie makes the trip in 1 hr. How far does she jog in a trip?

45. *Fuel Economy.* Sally Cline's station wagon gets 18 miles per gallon (mpg) in city driving and 24 mpg in highway driving. The car is driven 465 mi on 23 gal of gasoline. How many miles were driven in the city and how many were driven on the highway?

46. *Gender.* Phil and Phyllis are siblings. Phyllis has twice as many brothers as she has sisters. Phil has the same number of brothers as sisters. How many girls and how many boys are in the family?

47. *Wood Stains.* Bennet Custom Flooring has 0.5 gal of stain that is 20% brown and 80% neutral. A customer orders 1.5 gal of a stain that is 60% brown and 40% neutral. How much pure brown stain and how much neutral stain should be added to the original 0.5 gal in order to make up the order?

48. See Exercise 47. Let $x =$ the amount of pure brown stain added to the original 0.5 gal. Find a function $P(x)$ that can be used to determine the percentage of brown stain in the 1.5-gal mixture. On a graphing calculator, draw the graph of P and use ZOOM and TRACE or the TABLE feature to confirm the answer to Exercise 47.

13.5 SYSTEMS OF EQUATIONS IN THREE VARIABLES

Objective

a Solve systems of three equations in three variables.

a Solving Systems in Three Variables

A **linear equation in three variables** is an equation equivalent to one of the type $Ax + By + Cz = D$. A **solution** of a system of three equations in three variables is an ordered triple (x, y, z) that makes *all three* equations true.

The substitution method can be used to solve systems of three equations, but it is not efficient unless a variable has already been eliminated from one or more of the equations. Therefore, we will use only the elimination method—essentially the same procedure for systems of three equations as for systems of two equations. The first step is to eliminate a variable and obtain a system of two equations in two variables.

EXAMPLE 1 Solve the following system of equations:

$$\begin{aligned} x + y + z &= 4, & (1) \\ x - 2y - z &= 1, & (2) \\ 2x - y - 2z &= -1. & (3) \end{aligned}$$

a) We first use *any* two of the three equations to get an equation in two variables. In this case, let's use equations (1) and (2) and add to eliminate z:

$$\begin{aligned} x + y + z &= 4 & (1) \\ x - 2y - z &= 1 & (2) \\ \hline 2x - y \quad &= 5. & (4) \quad \text{Adding} \end{aligned}$$

b) We use a *different* pair of equations and eliminate the **same variable** that we did in part (a). Let's use equations (1) and (3) and again eliminate z.

CAUTION!

A common error is to eliminate a different variable the second time.

$$\begin{aligned} x + y + z &= 4 & (1) \\ 2x - y - 2z &= -1 & (3) \end{aligned}$$

$$\begin{aligned} 2x + 2y + 2z &= 8 & \text{Multiplying equation (1) by 2} \\ 2x - y - 2z &= -1 & (3) \\ \hline 4x + y \quad &= 7 & (5) \quad \text{Adding} \end{aligned}$$

c) Now we solve the resulting system of equations, (4) and (5). That solution will give us two of the numbers. Note that we now have two equations in two variables. Had we eliminated two *different* variables in parts (a) and (b), this would not be the case.

$$\begin{aligned} 2x - y &= 5 & (4) \\ 4x + y &= 7 & (5) \\ \hline 6x \quad &= 12 & \text{Adding} \\ x &= 2 \end{aligned}$$

1. Solve. Don't forget to check.

$$4x - y + z = 6,$$
$$-3x + 2y - z = -3,$$
$$2x + y + 2z = 3$$

Answer on page A-49

We can use either equation (4) or (5) to find y. We choose equation (5):

$$\begin{aligned} 4x + y &= 7 && \textbf{(5)} \\ 4(2) + y &= 7 && \text{Substituting 2 for } x \\ 8 + y &= 7 \\ y &= -1. \end{aligned}$$

d) We now have $x = 2$ and $y = -1$. To find the value for z, we use any of the original three equations and substitute to find the third number, z. Let's use equation (1) and substitute our two numbers in it:

$$\begin{aligned} x + y + z &= 4 && \textbf{(1)} \\ 2 + (-1) + z &= 4 && \text{Substituting 2 for } x \text{ and } -1 \text{ for } y \\ 1 + z &= 4 && \text{Solving for } z \\ z &= 3. \end{aligned}$$

We have obtained the ordered triple $(2, -1, 3)$. We check as follows, substituting $(2, -1, 3)$ into each of the three equations using alphabetical order.

CHECK:

$$x + y + z = 4$$
$$2 + (-1) + 3 \; ? \; 4$$
$$4 \quad \text{TRUE}$$

$$x - 2y - z = 1$$
$$2 - 2(-1) - 3 \; ? \; 1$$
$$2 + 2 - 3$$
$$1 \quad \text{TRUE}$$

$$2x - y - 2z = -1$$
$$2(2) - (-1) - 2 \cdot 3 \; ? \; -1$$
$$4 + 1 - 6$$
$$-1 \quad \text{TRUE}$$

The triple $(2, -1, 3)$ checks and is the solution.

To use the elimination method to solve systems of three equations:

1. Write all equations in the standard form $Ax + By + Cz = D$.
2. Clear any decimals or fractions.
3. Choose a variable to eliminate. Then use *any* two of the three equations to eliminate that variable, getting an equation in two variables.
4. Next, use a different pair of equations and get another equation in *the same two variables*. That is, eliminate the same variable that you did in step (3).
5. Solve the resulting system (pair) of equations. That will give two of the numbers.
6. Then use any of the original three equations to find the third number.

Do Exercise 1.

EXAMPLE 2 Solve this system:

$$\begin{aligned} 4x - 2y - 3z &= 5, \quad &(1)\\ -8x - y + z &= -5, \quad &(2)\\ 2x + y + 2z &= 5. \quad &(3) \end{aligned}$$

a) The equations are in standard form and do not contain decimals or fractions.

b) We decide to eliminate the variable y since the y-terms are opposites in equations (2) and (3). We add:

$$\begin{aligned} -8x - y + z &= -5 \quad &(2)\\ 2x + y + 2z &= 5 \quad &(3)\\ \hline -6x + 3z &= 0. \quad &(4) \quad \text{Adding} \end{aligned}$$

c) We use another pair of equations to get an equation in the same two variables, x and z. That is, we eliminate the same variable y that we did in step (b). We use equations (1) and (3) and eliminate y:

$$\begin{aligned} 4x - 2y - 3z &= 5 \quad &(1)\\ 2x + y + 2z &= 5; \quad &(3) \end{aligned}$$

$$\begin{aligned} 4x - 2y - 3z &= 5 \quad &(1)\\ 4x + 2y + 4z &= 10 \quad &\text{Multiplying equation (3) by 2}\\ \hline 8x + z &= 15. \quad &(5) \quad \text{Adding} \end{aligned}$$

d) Now we solve the resulting system of equations (4) and (5). That will give us two of the numbers:

$$\begin{aligned} -6x + 3z &= 0, \quad &(4)\\ 8x + z &= 15. \quad &(5) \end{aligned}$$

We multiply equation (5) by -3. (We could also have multiplied equation (4) by $-\frac{1}{3}$.)

$$\begin{aligned} -6x + 3z &= 0 \quad &(4)\\ -24x - 3z &= -45 \quad &\text{Multiplying equation (5) by } -3\\ \hline -30x &= -45 \quad &\text{Adding}\\ x &= \tfrac{-45}{-30} = \tfrac{3}{2} \end{aligned}$$

We now use equation (5) to find z:

$$\begin{aligned} 8x + z &= 15 \quad &(5)\\ 8\left(\tfrac{3}{2}\right) + z &= 15 \quad &\text{Substituting } \tfrac{3}{2} \text{ for } x\\ 12 + z &= 15 \quad &\text{Solving for } z\\ z &= 3. \end{aligned}$$

2. Solve. Don't forget to check.

$$\begin{aligned} 2x + y - 4z &= 0, \\ x - y + 2z &= 5, \\ 3x + 2y + 2z &= 3 \end{aligned}$$

Answer on page A-49

e) Next, we use any of the original equations and substitute to find the third number, y. We choose equation (3) since the coefficient of y there is 1:

$$\begin{aligned} 2x + y + 2z &= 5 && \text{(3)} \\ 2\left(\tfrac{3}{2}\right) + y + 2(3) &= 5 && \text{Substituting } \tfrac{3}{2} \text{ for } x \text{ and 3 for } z \\ 3 + y + 6 &= 5 \\ y + 9 &= 5 && \text{Solving for } y \\ y &= -4. \end{aligned}$$

The solution is $\left(\frac{3}{2}, -4, 3\right)$. The check is as follows.

CHECK:

$$\begin{array}{c|l} 4x - 2y - 3z & 5 \\ \hline 4 \cdot \frac{3}{2} - 2(-4) - 3(3) \; ? & 5 \\ 6 + 8 - 9 & \\ 5 & \quad \text{TRUE} \end{array}$$

$$\begin{array}{c|l} -8x - y + z & -5 \\ \hline -8 \cdot \frac{3}{2} - (-4) + 3 \; ? & -5 \\ -12 + 4 + 3 & \\ -5 & \quad \text{TRUE} \end{array}$$

$$\begin{array}{c|l} 2x + y + 2z & 5 \\ \hline 2 \cdot \frac{3}{2} + (-4) + 2(3) \; ? & 5 \\ 3 - 4 + 6 & \\ 5 & \quad \text{TRUE} \end{array}$$

Do Exercise 2.

In Example 3, two of the equations have a missing variable.

EXAMPLE 3 Solve this system:

$$\begin{aligned} x + y + z &= 180, && \text{(1)} \\ x \qquad - z &= -70, && \text{(2)} \\ 2y - z &= 0. && \text{(3)} \end{aligned}$$

We note that there is no y in equation (2). In order to have a system of two equations in the variables x and z, we need to find another equation without a y. We use equations (1) and (3) to eliminate y:

$$\begin{aligned} x + y + z &= 180 && \text{(1)} \\ 2y - z &= 0 && \text{(3)} \end{aligned}$$

$$\begin{aligned} -2x - 2y - 2z &= -360 && \text{Multiplying equation (1) by } -2 \\ 2y - z &= 0 && \text{(3)} \\ \hline -2x \qquad - 3z &= -360. && \text{(4)} \quad \text{Adding} \end{aligned}$$

Now we solve the resulting system of equations (2) and (4):

$$\begin{aligned} x - z &= -70 && \text{(2)} \\ -2x - 3z &= -360 && \text{(4)} \end{aligned}$$

$$\begin{aligned} 2x - 2z &= -140 && \text{Multiplying equation (2) by 2} \\ -2x - 3z &= -360 && \text{(4)} \\ \hline -5z &= -500 && \text{Adding} \\ z &= 100. \end{aligned}$$

To find x, we substitute 100 for z in equation (2) and solve for x:

$$x - z = -70$$
$$x - 100 = -70$$
$$x = 30.$$

To find y, we substitute 100 for z in equation (3) and solve for y:

$$2y - z = 0$$
$$2y - 100 = 0$$
$$2y = 100$$
$$y = 50.$$

The triple (30, 50, 100) is the solution. The check is left to the student.

Do Exercise 3.

It is possible for a system of three equations to have no solution, that is, to be inconsistent. An example is the system

$$x + y + z = 14.$$
$$x + y + z = 11,$$
$$2x - 3y + 4z = -3.$$

Note the first two equations. It is not possible for a sum of three numbers to be both 14 and 11. Thus the system has no solution. We will not consider such systems here, nor will we consider systems with infinitely many solutions, which also exist.

3. Solve. Don't forget to check.

$$x + y + z = 100,$$
$$x - y = -10,$$
$$x - z = -30$$

Answer on page A-49

13.5 EXERCISE SET

For Extra Help

Digital Video Tutor CD 11 Videotape 15 | InterAct Math | Math Tutor Center | MathXL | MyMathLab

a Solve.

1. $x + y + z = 2,$
$2x - y + 5z = -5,$
$-x + 2y + 2z = 1$

2. $2x - y - 4z = -12,$
$2x + y + z = 1,$
$x + 2y + 4z = 10$

3. $2x - y + z = 5,$
$6x + 3y - 2z = 10,$
$x - 2y + 3z = 5$

4. $x - y + z = 4,$
$3x + 2y + 3z = 7,$
$2x + 9y + 6z = 5$

5. $2x - 3y + z = 5,$
$x + 3y + 8z = 22,$
$3x - y + 2z = 12$

6. $6x - 4y + 5z = 31,$
$5x + 2y + 2z = 13,$
$x + y + z = 2$

7. $3a - 2b + 7c = 13,$
$a + 8b - 6c = -47,$
$7a - 9b - 9c = -3$

8. $x + y + z = 0,$
$2x + 3y + 2z = -3,$
$-x + 2y - 3z = -1$

9. $2x + 3y + z = 17,$
$x - 3y + 2z = -8,$
$5x - 2y + 3z = 5$

10. $2x + y - 3z = -4,$
$4x - 2y + z = 9,$
$3x + 5y - 2z = 5$

11. $2x + y + z = -2,$
$2x - y + 3z = 6,$
$3x - 5y + 4z = 7$

12. $2x + y + 2z = 11,$
$3x + 2y + 2z = 8,$
$x + 4y + 3z = 0$

13. $x - y + z = 4,$
$5x + 2y - 3z = 2,$
$3x - 7y + 4z = 8$

14. $2x + y + 2z = 3,$
$x + 6y + 3z = 4,$
$3x - 2y + z = 0$

15. $4x - y - z = 4,$
$2x + y + z = -1,$
$6x - 3y - 2z = 3$

16. $a + 2b + c = 1,$
$7a + 3b - c = -2,$
$a + 5b + 3c = 2$

17. $2r + 3s + 12t = 4,$
$4r - 6s + 6t = 1,$
$r + s + t = 1$

18. $10x + 6y + z = 7,$
$5x - 9y - 2z = 3,$
$15x - 12y + 2z = -5$

19. $4a + 9b = 8,$
$8a + 6c = -1,$
$6b + 6c = -1$

20. $3p + 2r = 11,$
$q - 7r = 4,$
$p - 6q = 1$

21. $x + y + z = 57,$
$-2x + y = 3,$
$x - z = 6$

22. $x + y + z = 105,$
$10y - z = 11,$
$2x - 3y = 7$

23. $r + s = 5,$
$3s + 2t = -1,$
$4r + t = 14$

24. $a - 5c = 17,$
$b + 2c = -1,$
$4a - b - 3c = 12$

25. D_W Explain a procedure that could be used to solve a system of four equations in four variables.

26. D_W Is it possible for a system of three equations to have exactly two ordered triples in its solution set? Why or why not?

SKILL MAINTENANCE

Solve for the indicated letter. [7.4b]

27. $F = 3ab$, for a

28. $Q = 4(a + b)$, for a

29. $F = \frac{1}{2}t(c - d)$, for c

30. $F = \frac{1}{2}t(c - d)$, for d

31. $Ax - By = c$, for y

32. $Ax + By = c$, for y

Find the slope and the y-intercept. [12.3b]

33. $y = -\frac{2}{3}x - \frac{5}{4}$

34. $y = 5 - 4x$

35. $2x - 5y = 10$

36. $7x - 6.4y = 20$

SYNTHESIS

Solve.

37. $w + x + y + z = 2,$
$w + 2x + 2y + 4z = 1,$
$w - x + y + z = 6,$
$w - 3x - y + z = 2$

38. $w + x - y + z = 0,$
$w - 2x - 2y - z = -5,$
$w - 3x - y + z = 4,$
$2w - x - y + 3z = 7$

13.6 SOLVING APPLIED PROBLEMS: THREE EQUATIONS

Objective

a Solve applied problems using systems of three equations.

1. Triangle Measures. One angle of a triangle is twice as large as a second angle. The remaining angle is 20° greater than the first angle. Find the measure of each angle.

a Using Systems of Three Equations

Solving systems of three or more equations is important in many applications occurring in the natural and social sciences, business, and engineering.

EXAMPLE 1 *Architecture.* In a triangular cross-section of a roof, the largest angle is 70° greater than the smallest angle. The largest angle is twice as large as the remaining angle. Find the measure of each angle.

1. **Familiarize.** We first make a drawing. Since we do not know the size of any angle, we use x, y, and z for the measures of the angles. We let $x =$ the smallest angle, $z =$ the largest angle, and $y =$ the remaining angle.

2. **Translate.** In order to translate the problem, we need to make use of a geometric fact—that is, the sum of the measures of the angles of a triangle is 180°. This fact about triangles gives us one equation:

   $$x + y + z = 180.$$

   There are two statements in the problem that we can translate directly.

The largest angle	is	70°	greater than	the smallest angle.
z	$=$	70	$+$	x

The largest angle	is	twice as large as the remaining angle.
z	$=$	$2y$

   We now have a system of three equations:

   $$\begin{aligned} x + y + z &= 180, \\ x + 70 &= z, \\ 2y &= z; \end{aligned} \qquad \text{or} \qquad \begin{aligned} x + y + z &= 180, \\ x \qquad - z &= -70, \\ 2y - z &= 0. \end{aligned}$$

3. **Solve.** The system was solved in Example 3 of Section 8.5. The solution is $(30, 50, 100)$.
4. **Check.** The sum of the numbers is 180. The largest angle measures 100° and the smallest measures 30°. The largest angle is 70° greater than the smallest. The remaining angle measures 50°. The largest angle is twice as large as the remaining angle. We do have an answer to the problem.
5. **State.** The measures of the angles of the triangle are 30°, 50°, and 100°.

Do Exercise 1.

Answer on page A-49

EXAMPLE 2 *Cholesterol Levels.* Americans have become very conscious of their cholesterol levels. Recent studies indicate that a child's intake of cholesterol should be no more than 300 mg per day. By eating 1 egg, 1 cupcake, and 1 slice of pizza, a child consumes 302 mg of cholesterol. If the child eats 2 cupcakes and 3 slices of pizza, he or she takes in 65 mg of cholesterol. By eating 2 eggs and 1 cupcake, a child consumes 567 mg of cholesterol. How much cholesterol is in each item?

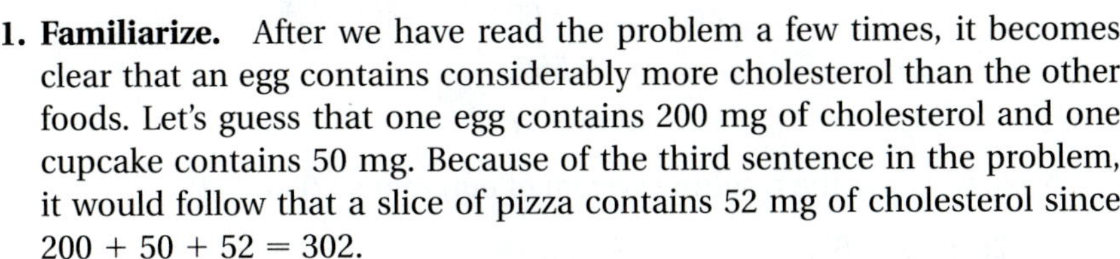

1. **Familiarize.** After we have read the problem a few times, it becomes clear that an egg contains considerably more cholesterol than the other foods. Let's guess that one egg contains 200 mg of cholesterol and one cupcake contains 50 mg. Because of the third sentence in the problem, it would follow that a slice of pizza contains 52 mg of cholesterol since $200 + 50 + 52 = 302$.

   To see if our guess satisfies the other statements in the problem, we find the amount of cholesterol that 2 cupcakes and 3 slices of pizza would contain: $2 \cdot 50 + 3 \cdot 52 = 256$. Since this does not match the 65 mg listed in the fourth sentence of the problem, our guess was incorrect. Rather than guess again, we examine how we checked our guess and let e, c, and s = the number of milligrams of cholesterol in an egg, a cupcake, and a slice of pizza, respectively.

2. **Translate.** By rewording some of the sentences in the problem, we can translate it into three equations.

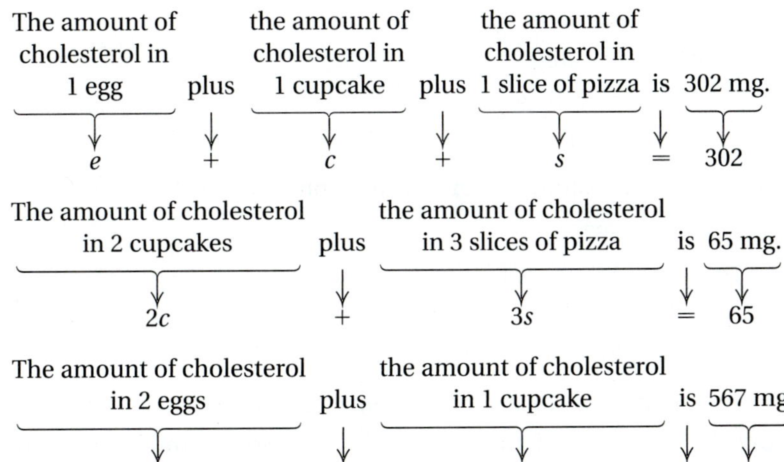

We now have a system of three equations:

$$
\begin{aligned}
e + c + s &= 302, && (1)\\
2c + 3s &= 65, && (2)\\
2e + c \phantom{{}+s} &= 567. && (3)
\end{aligned}
$$

2. Client Investments. Kaufman Financial Corporation makes investments for corporate clients. One year, a client receives \$1620 in simple interest from three investments that total \$25,000. Part is invested at 5%, part at 6%, and part at 7%. There is \$11,000 more invested at 7% than at 6%. How much was invested at each rate?

Answer on page A-49

3. Solve. To solve, we first note that the variable e does not appear in equation (2). In order to have a system of two equations in the variables c and s, we need to find another equation without the variable e. We use equations (1) and (3) to eliminate e:

$$\begin{aligned} e + c + s &= 302, && (1) \\ 2e + c \quad\; &= 567; && (3) \end{aligned}$$

$$\begin{aligned} -2e - 2c - 2s &= -604, && \text{Multiplying equation (1) by } -2 \\ 2e + c \qquad\; &= 567, && (3) \\ \hline -c - 2s &= -37. && (4) \quad \text{Adding} \end{aligned}$$

Next, we solve the resulting system of equations (2) and (4):

$$\begin{aligned} 2c + 3s &= 65, && (2) \\ -c - 2s &= -37; && (4) \end{aligned}$$

$$\begin{aligned} 2c + 3s &= 65 && (2) \\ -2c - 4s &= -74 && \text{Multiplying equation (4) by 2} \\ \hline -s &= -9 && \text{Adding} \\ s &= 9. \end{aligned}$$

To find c, we substitute 9 for s in equation (4) and solve for c:

$$\begin{aligned} -c - 2s &= -37 && (4) \\ -c - 2(9) &= -37 && \text{Substituting} \\ -c - 18 &= -37 \\ -c &= -19 \\ c &= 19. \end{aligned}$$

To find e, we substitute 19 for c in equation (3) and solve for e:

$$\begin{aligned} 2e + c &= 567 && (3) \\ 2e + 19 &= 567 && \text{Substituting} \\ 2e &= 548 \\ e &= 274. \end{aligned}$$

The solution is $c = 19$, $e = 274$, $s = 9$, or $(19, 274, 9)$.

4. Check. The sum of 19, 274, and 9 is 302 so the total cholesterol in 1 cupcake, 1 egg, and 1 slice of pizza checks. Two cupcakes and three slices of pizza would contain $2 \cdot 19 + 3 \cdot 9$, or 65 mg, while two eggs and one cupcake would contain $2 \cdot 274 + 19$, or 567 mg of cholesterol. The answer checks.

5. State. A cupcake contains 19 mg of cholesterol, an egg contains 274 mg of cholesterol, and a slice of pizza contains 9 mg of cholesterol.

Do Exercise 2.

13.6 EXERCISE SET

a Solve.

1. *Restaurant Management.* Reggie works at Dunkin® Donuts, where a 10-oz cup of coffee costs \$1.09, a 14-oz cup costs \$1.29, and a 20-oz cup costs \$1.49. During one busy period, Reggie served 34 cups of coffee, emptying five 96-oz pots while collecting a total of \$43.46. How many cups of each size did Reggie fill?
Source: Dunkin Donuts

2. *Restaurant Management.* McDonald's® recently sold small soft drinks for \$1.09, medium soft drinks for \$1.19, and large soft drinks for \$1.39. During a lunch-time rush, Maggie sold 45 soft drinks for a total of \$55.05. The number of small and large drinks, combined, was 15 fewer than the number of medium drinks. How many drinks of each size were sold?
Source: McDonald's Corporation

3. *Triangle Measures.* In triangle ABC, the measure of angle B is three times that of angle A. The measure of angle C is 20° more than that of angle A. Find the measure of each angle.

4. *Triangle Measures.* In triangle ABC, the measure of angle B is twice the measure of angle A. The measure of angle C is 80° more than that of angle A. Find the measure of each angle.

5. *Automobile Pricing.* A recent basic model of a particular automobile had a price of \$12,685. The basic model with the added features of automatic transmission and power door locks was \$14,070. The basic model with air conditioning (AC) and power door locks was \$13,580. The basic model with AC and automatic transmission was \$13,925. What was the individual cost of each of the three options?

6. *Telemarketing.* Sven, Tillie, and Isaiah can process 740 telephone orders per day. Sven and Tillie together can process 470 orders, while Tillie and Isaiah together can process 520 orders per day. How many orders can each person process alone?

7. *Lens Production.* When Sight-Rite's three polishing machines, A, B, and C, are all working, 5700 lenses can be polished in one week. When only A and B are working, 3400 lenses can be polished in one week. When only B and C are working, 4200 lenses can be polished in one week. How many lenses can be polished in a week by each machine alone?

8. *Welding Rates.* Elrod, Dot, and Wendy can weld 74 linear feet per hour when working together. Elrod and Dot together can weld 44 linear feet per hour, while Elrod and Wendy can weld 50 linear feet per hour. How many linear feet per hour can each weld alone?

9. *Investments.* A business class divided an imaginary investment of \$80,000 among three mutual funds. The first fund grew by 10%, the second by 6%, and the third by 15%. Total earnings were \$8850. The earnings from the first fund were \$750 more than the earnings from the third. How much was invested in each fund?

10. *Advertising.* In a recent year, companies spent a total of \$84.8 billion on newspaper, television, and radio ads. The total amount spent on television and radio ads was only \$2.6 billion more than the amount spent on newspaper ads alone. The amount spent on newspaper ads was \$5.1 billion more than what was spent on television ads. How much was spent on each form of advertising? (*Hint*: Let the variables represent numbers of billions of dollars.)

11. *Twin Births.* In the United States, the highest incidence of fraternal twin births occurs among Asian–Americans, then African–Americans, and then Caucasians. Of every 15,400 births, the total number of fraternal twin births for all three is 739, where there are 185 more for Asian–Americans than African–Americans and 231 more for Asian–Americans than Caucasians. How many births of fraternal twins are there for each group out of every 15,400 births?

12. *Crying Rate.* The sum of the average number of times a man, a woman, and a one-year-old child cry each month is 71.7. A one-year-old cries 46.4 more times than a man. The average number of times a one-year-old cries per month is 28.3 more than the average number of times combined that a man and a woman cry. What is the average number of times per month that each cries?

13. *Nutrition.* A dietician in a hospital prepares meals under the guidance of a physician. Suppose that for a particular patient a physician prescribes a meal to have 800 calories, 55 g of protein, and 220 mg of vitamin C. The dietician prepares a meal of roast beef, baked potatoes, and broccoli according to the data in the following table.

	CALORIES	PROTEIN (in grams)	VITAMIN C (in milligrams)
Roast Beef, 3 oz	300	20	0
Baked Potato	100	5	20
Broccoli, 156 g	50	5	100

How many servings of each food are needed in order to satisfy the doctor's orders?

14. *Nutrition.* Repeat Exercise 13 but replace the broccoli with asparagus, for which one 180-g serving contains 50 calories, 5 g of protein, and 44 mg of vitamin C. Which meal would you prefer eating?

15. *Golf.* On an 18-hole golf course, there are par-3 holes, par-4 holes, and par-5 holes. A golfer who shoots par on every hole has a total of 70. There are twice as many par-4 holes as there are par-5 holes. How many of each type of hole are there on the golf course?

16. *Golf.* On an 18-hole golf course, there are par-3 holes, par-4 holes, and par-5 holes. A golfer who shoots par on every hole has a total of 72. The sum of the number of par-3 holes and the number of par-5 holes is 8. How many of each type of hole are there on the golf course?

17. *Basketball Scoring.* The New York Knicks recently scored a total of 92 points on a combination of 2-point field goals, 3-point field goals, and 1-point foul shots. Altogether, the Knicks made 50 baskets and 19 more 2-pointers than foul shots. How many shots of each kind were made?

18. *History.* Find the year in which the first U.S. transcontinental railroad was completed. The following are some facts about the number. The sum of the digits in the year is 24. The ones digit is 1 more than the hundreds digit. Both the tens and the ones digits are multiples of 3.

19. DW Exercise 8 can be solved mentally after a careful reading of the problem. How is this possible?

20. DW *Ticket Revenue.* A pops-concert audience of 100 people consists of adults, senior citizens, and children. The ticket prices are \$10 each for adults, \$3 each for senior citizens, and \$0.50 each for children. The total amount of money taken in is \$100. How many adults, senior citizens, and children are in attendance? Does there seem to be some information missing? Do some careful reasoning and explain.

21. **DW** Consider Exercise 17. Suppose there were no foul shots made. Would there still be a solution? Why or why not?

22. **DW** Consider Exercise 1. Suppose Reggie collected \$46. Could the problem still be solved? Why or why not?

SKILL MAINTENANCE

Determine whether the correspondence is a function. [12.1a]

23. *Domain* (State) → *Range* (City)

California → Los Angeles
California → San Francisco
California → San Diego
Kansas → Kansas City
Kansas → Topeka
Kansas → Wichita

24. *Domain* (Boy's Age, in months) → *Range* (Average Daily Weight Gain, in grams)

2 → 24.3
9 → 11.7
16 → 8.2
23 → 7.0

Source: *American Family Physician*, December 1993, p. 1435

25. Find the domain of the function: [12.2a]

$$f(x) = \frac{x - 5}{x + 7}.$$

26. Find the domain and the range of the function: [12.2a]

$$g(x) = 5 - x^2.$$

27. Find an equation of the line with slope $-\frac{3}{5}$ and y-intercept $(0, -7)$. [12.5a]

28. Simplify: $\dfrac{(a^2b^3)^5}{a^7b^{16}}$. [9.1e], [9.2a, b]

SYNTHESIS

29. Find the sum of the angle measures at the tips of the star in this figure.

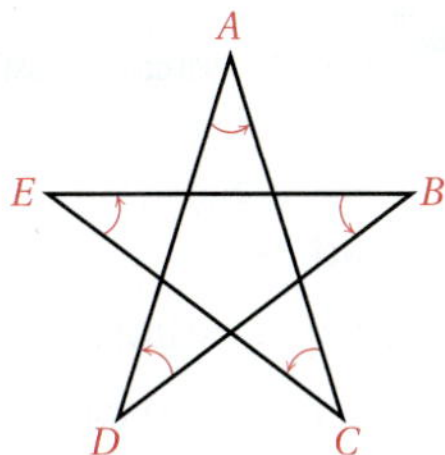

30. *Sharing Raffle Tickets.* Hal gives Tom as many raffle tickets as Tom has and Gary as many as Gary has. In like manner, Tom then gives Hal and Gary as many tickets as each then has. Similarly, Gary gives Hal and Tom as many tickets as each then has. If each finally has 40 tickets, with how many tickets does Tom begin?

31. *Digits.* Find a three-digit positive integer such that the sum of all three digits is 14, the tens digit is 2 more than the ones digit, and if the digits are reversed, the number is unchanged.

32. *Ages.* Tammy's age is the sum of the ages of Carmen and Dennis. Carmen's age is 2 more than the sum of the ages of Dennis and Mark. Dennis's age is four times Mark's age. The sum of all four ages is 42. How old is Tammy?

13.7 BUSINESS AND ECONOMICS APPLICATIONS

Objectives

a. Given total-cost and total-revenue functions, find the total-profit function and the break-even point.

b. Given supply and demand functions, find the equilibrium point.

a Break-Even Analysis

When a company manufactures x units of a product, it invests money. This is **total cost** and can be thought of as a function C, where $C(x)$ is the total cost of producing x units. When the company sells x units of the product, it takes in money. This is **total revenue** and can be thought of as a function R, where $R(x)$ is the total revenue from the sale of x units. **Total profit** is the money taken in less the money spent, or total revenue minus total cost. Total profit from the production and sale of x units is a function P given by

Profit = Revenue − Cost, or $P(x) = R(x) - C(x)$.

If $R(x)$ is greater than $C(x)$, the company has a profit. If $C(x)$ is greater than $R(x)$, the company has a loss. When $R(x) = C(x)$, the company breaks even.

There are two kinds of costs. First, there are costs like rent, insurance, machinery, and so on. These costs, which must be paid whether a product is produced or not, are called **fixed costs.** When a product is being produced, there are costs for labor, materials, marketing, and so on. These are called **variable costs,** because they vary according to the amount of the product being produced. The sum of the fixed costs and the variable costs gives the **total cost** of producing a product.

EXAMPLE 1 *Manufacturing Radios.* Ergs, Inc., is planning to make a new kind of radio. Fixed costs will be \$90,000, and it will cost \$15 to produce each radio (variable costs). Each radio sells for \$26.

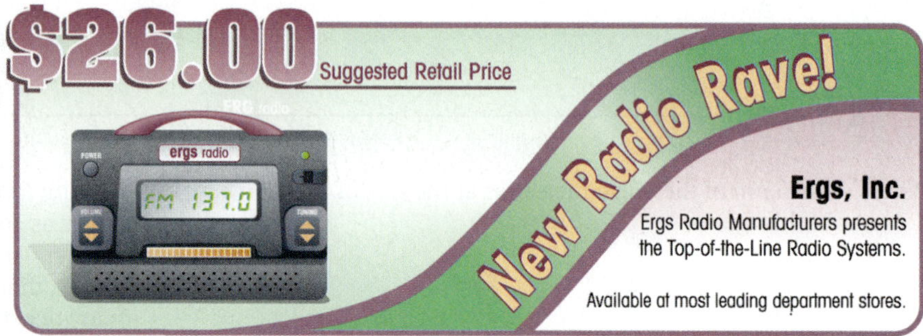

a) Find the total cost $C(x)$ of producing x radios.

b) Find the total revenue $R(x)$ from the sale of x radios.

c) Find the total profit $P(x)$ from the production and sale of x radios.

d) What profit or loss will the company realize from the production and sale of 3000 radios? of 14,000 radios?

e) Graph the total-cost, total-revenue, and total-profit functions using the same set of axes. Determine the break-even point.

a) Total cost is given by

$$C(x) = (\text{Fixed costs}) \text{ plus } (\text{Variable costs}),$$

or $$C(x) = 90{,}000 + 15x,$$

where x is the number of radios produced.

b) Total revenue is given by

$R(x) = 26x$. $26 times the number of radios sold. We assume that every radio produced is sold.

c) Total profit is given by

$$P(x) = R(x) - C(x)$$
$$= 26x - (90{,}000 + 15x)$$
$$= 11x - 90{,}000.$$

d) Profits will be

$$P(3000) = 11 \cdot 3000 - 90{,}000 = -\$57{,}000$$

when 3000 radios are produced and sold, and

$$P(14{,}000) = 11 \cdot 14{,}000 - 90{,}000 = \$64{,}000$$

when 14,000 radios are produced and sold. Thus the company loses $57,000 if only 3000 radios are sold, but makes $64,000 if 14,000 are sold.

e) The graphs of each of the three functions are shown below:

$C(x) = 90{,}000 + 15x$, **(1)**

$R(x) = 26x$, **(2)**

$P(x) = 11x - 90{,}000$. **(3)**

$R(x)$, $C(x)$, and $P(x)$ are all in dollars.

Equation (2) has a graph that goes through the origin and has a slope of 26. Equation (1) has an intercept on the y-axis of 90,000 and has a slope of 15. Equation (3) has an intercept on the y-axis of $-90{,}000$ and has a slope of 11. It is shown by the dashed line. The red dashed line shows a "negative" profit, which is a loss. (That is what is known as "being in the red.") The black dashed line shows a "positive" profit, or gain. (That is what is known as "being in the black.")

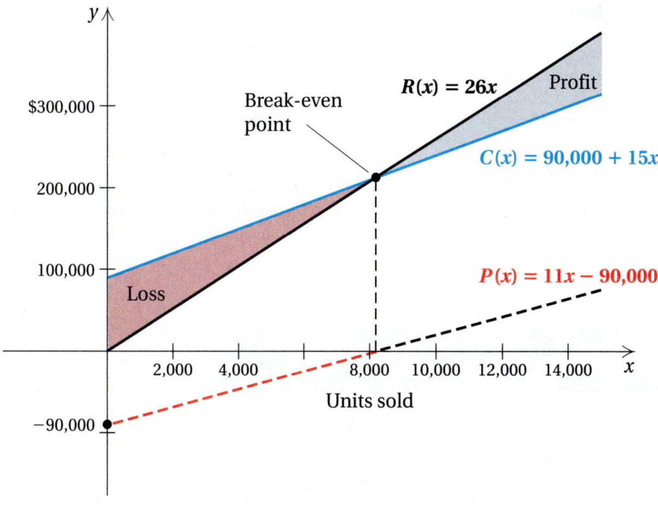

Profits occur when the revenue is greater than the cost. Losses occur when the revenue is less than the cost. The **break-even point** occurs where the graphs of R and C cross. Thus to find the break-even point, we solve a system:

$$C(x) = 90{,}000 + 15x,$$
$$R(x) = 26x.$$

1. Manufacturing Radios. Refer to Example 1. Suppose that fixed costs are $80,000, and it costs $20 to produce each radio. Each radio sells for $36.

a) Find the total cost $C(x)$ of producing x radios.

b) Find the total revenue $R(x)$ from the sale of x radios.

c) Find the total profit $P(x)$ from the production and sale of x radios.

d) What profit or loss will the company realize from the production and sale of 4000 radios? of 16,000 radios?

e) Graph the total-cost, total-revenue, and total-profit functions using the same set of axes. Determine the break-even point.

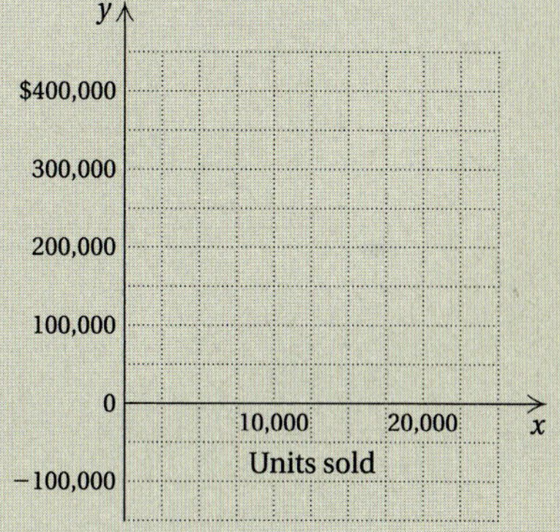

Answers on page A-49

Since both revenue and cost are in *dollars* and they are equal at the break-even point, the system can be rewritten as

$$d = 90{,}000 + 15x, \quad (1)$$
$$d = 26x \quad (2)$$

and solved using substitution:

$$26x = 90{,}000 + 15x \quad \text{Substituting } 26x \text{ for } d \text{ in equation (1)}$$
$$11x = 90{,}000$$
$$x \approx 8181.8.$$

The firm will break even if it produces and sells about 8182 radios (8181 will yield a tiny loss and 8182 a tiny gain), and takes in a total of $R(8182) = 26 \cdot 8182 = \$212{,}732$ in revenue. Note that the x-coordinate of the break-even point is also the x-intercept of the profit function. It can also be found by solving $P(x) = 0$.

Do Exercise 1 on the preceding page.

b Supply and Demand

As the price of coffee varies, the amount sold varies. The table and graph below both show that consumer demand goes down as the price goes up and the demand goes up as the price goes down.

DEMAND FUNCTION, D

PRICE, p, PER KILOGRAM	QUANTITY, $D(p)$ (in millions of kilograms)
\$ 8.00	25
9.00	20
10.00	15
11.00	10
12.00	5

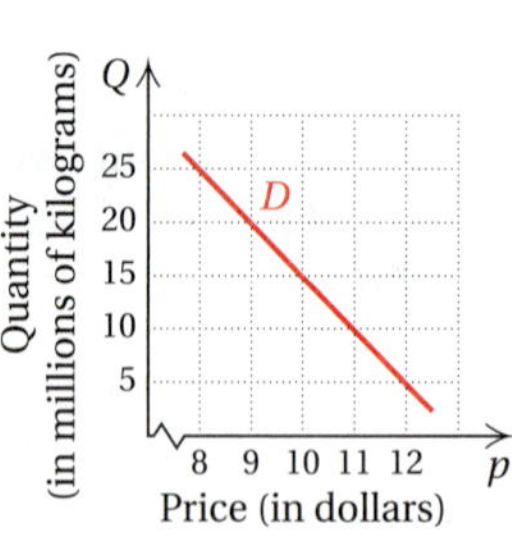

As the price of coffee varies, the amount available varies. The table and graph below both show that sellers will supply less as the price goes down, but will supply more as the price goes up.

SUPPLY FUNCTION, S

PRICE, p, PER KILOGRAM	QUANTITY, $S(p)$ (in millions of kilograms)
\$ 9.00	5
9.50	10
10.00	15
10.50	20
11.00	25

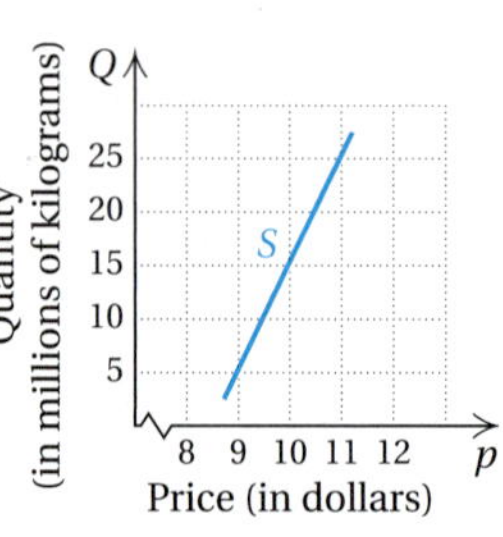

Let's look at the above graphs together. We see that as price increases, demand decreases. As price increases, supply increases. The point of intersection of the demand and supply functions is called the **equilibrium point.** At the equilibrium point, the amount that the seller will supply is the same amount that the consumer will buy. The situation is analogous to a buyer and a seller negotiating the price of an item. The equilibrium point is the price and quantity on which they finally agree.

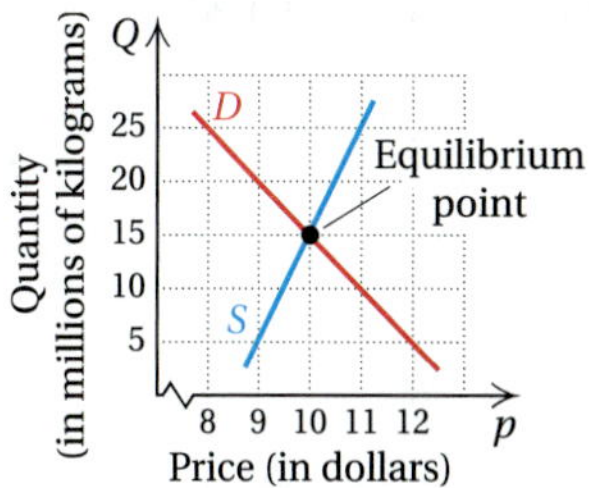

Any ordered pair of coordinates from the graph is (price, quantity), because the horizontal axis is the price axis and the vertical axis is the quantity axis. If D is a demand function and S is a supply function, then the equilibrium point is where demand equals supply:

$$D(p) = S(p).$$

EXAMPLE 2 Find the equilibrium point for the following demand and supply functions:

$$D(p) = 1000 - 60p, \quad (1)$$
$$S(p) = 200 + 4p. \quad (2)$$

Since both demand and supply are *quantities* and they are equal at the equilibrium point, we rewrite the system as

$$q = 1000 - 60p, \quad (1)$$
$$q = 200 + 4p. \quad (2)$$

We substitute $200 + 4p$ for q in equation (1) and solve:

$$200 + 4p = 1000 - 60p$$
$$200 + 64p = 1000 \quad \text{Adding } 60p \text{ on both sides}$$
$$64p = 800 \quad \text{Subtracting 200 on both sides}$$
$$p = \tfrac{800}{64} = 12.5.$$

Thus the equilibrium price is \$12.50 per unit.

To find the equilibrium quantity, we substitute \$12.50 into either $D(p)$ or $S(p)$. We use $S(p)$:

$$S(12.5) = 200 + 4(12.5)$$
$$= 200 + 50 = 250.$$

Thus the equilibrium quantity is 250 units, and the equilibrium point is (\$12.50, 250).

Do Exercise 2.

2. Find the equilibrium point for the following supply and demand functions:

$$D(p) = 1000 - 46p,$$
$$S(p) = 300 + 4p.$$

Answer on page A-50

13.7 EXERCISE SET

For Extra Help

Digital Video Tutor CD 11 Videotape 15 | InterAct Math | Math Tutor Center | MathXL | MyMathLab

a For each of the following pairs of total-cost and total-revenue functions, find **(a)** the total-profit function and **(b)** the break-even point.

1. $C(x) = 25x + 270{,}000$;
$R(x) = 70x$

2. $C(x) = 45x + 300{,}000$;
$R(x) = 65x$

3. $C(x) = 10x + 120{,}000$;
$R(x) = 60x$

4. $C(x) = 30x + 49{,}500$;
$R(x) = 85x$

5. $C(x) = 20x + 10{,}000$;
$R(x) = 100x$

6. $C(x) = 40x + 22{,}500$;
$R(x) = 85x$

7. $C(x) = 22x + 16{,}000$;
$R(x) = 40x$

8. $C(x) = 15x + 75{,}000$;
$R(x) = 55x$

9. $C(x) = 50x + 195{,}000$;
$R(x) = 125x$

10. $C(x) = 34x + 928{,}000$;
$R(x) = 128x$

Solve.

11. *Manufacturing Lamps.* City Lights is planning to manufacture a new type of lamp. For the first year, the fixed costs for setting up production are $22,500. The variable costs for producing each lamp are $40. The revenue from each lamp is $85. Find the following.

a) The total cost $C(x)$ of producing x lamps
b) The total revenue $R(x)$ from the sale of x lamps
c) The total profit $P(x)$ from the production and sale of x lamps
d) The profit or loss from the production and sale of 3000 lamps; of 400 lamps
e) The break-even point

12. *Computer Manufacturing.* Sky View Electronics is planning to introduce a new line of computers. For the first year, the fixed costs for setting up production are $125,100. The variable costs for producing each computer are $750. The revenue from each computer is $1050. Find the following.

a) The total cost $C(x)$ of producing x computers
b) The total revenue $R(x)$ from the sale of x computers
c) The total profit $P(x)$ from the production and sale of x computers
d) The profit or loss from the production and sale of 400 computers; of 700 computers
e) The break-even point

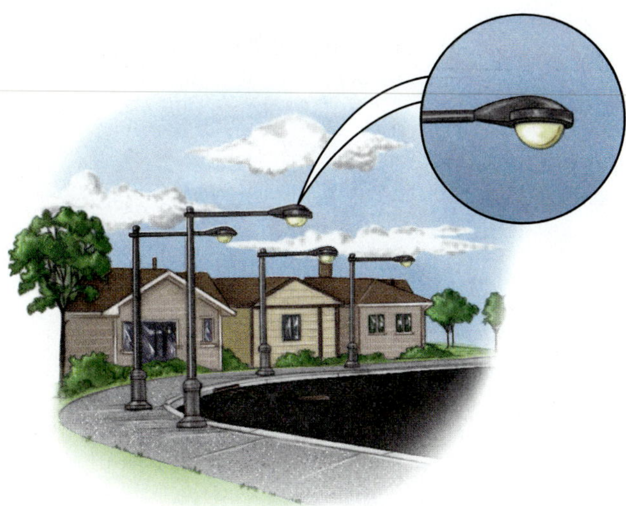

Solve.

13. *Manufacturing Caps.* Martina's Custom Printing is planning on adding painter's caps to its product line. For the first year, the fixed costs for setting up production are $16,404. The variable costs for producing a dozen caps are $6.00. The revenue on each dozen caps is $18.00. Find the following.

a) The total cost $C(x)$ of producing x dozen caps
b) The total revenue $R(x)$ from the sale of x dozen caps
c) The total profit $P(x)$ from the production and sale of x dozen caps
d) The profit or loss from the production and sale of 3000 dozen caps; of 1000 dozen caps
e) The break-even point

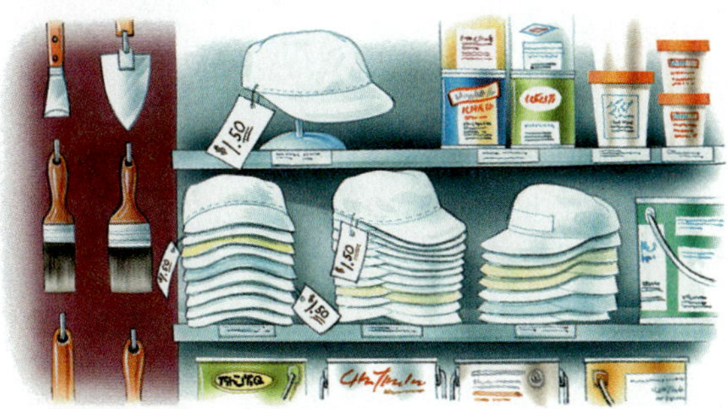

14. *Sport Coat Production.* Sarducci's is planning a new line of sport coats. For the first year, the fixed costs for setting up production are $10,000. The variable costs for producing each coat are $20. The revenue from each coat is $100. Find the following.

a) The total cost $C(x)$ of producing x coats
b) The total revenue $R(x)$ from the sale of x coats
c) The total profit $P(x)$ from the production and sale of x coats
d) The profit or loss from the production and sale of 2000 coats; of 50 coats
e) The break-even point

Find the equilibrium point for each of the following pairs of demand and supply functions.

15. $D(p) = 1000 - 10p;$
$S(p) = 230 + p$

16. $D(p) = 2000 - 60p;$
$S(p) = 460 + 94p$

17. $D(p) = 760 - 13p;$
$S(p) = 430 + 2p$

18. $D(p) = 800 - 43p;$
$S(p) = 210 + 16p$

19. $D(p) = 7500 - 25p;$
$S(p) = 6000 + 5p$

20. $D(p) = 8800 - 30p;$
$S(p) = 7000 + 15p$

21. $D(p) = 1600 - 53p;$
$S(p) = 320 + 75p$

22. $D(p) = 5500 - 40p;$
$S(p) = 1000 + 85p$

23. $^D{}_W$ Variable costs and fixed costs are often compared to the slope and the y-intercept, respectively, of an equation of a line. Explain why this analogy is valid.

24. $^D{}_W$ In this section, we examined supply and demand functions for coffee. Does it seem realistic to you for the graph of D to have a constant slope? Why or why not?

SKILL MAINTENANCE

Find the slope and the y-intercept. [12.3b]

25. $5y - 3x = 8$

26. $6x + 7y - 9 = 4$

27. $2y = 3.4x + 98$

28. $\frac{x}{3} + \frac{y}{4} = 1$

CHAPTER 13

Summary and Review

The review that follows is meant to prepare you for a chapter exam. It consists of two parts. The first part is a checklist of some of the Study Tips referred to in this and preceding chapters. The second part is the Review Exercises. These provide practice exercises for the exam, together with references to section objectives so you can go back and review. Before beginning, stop and look back over the skills you have obtained. What skills in mathematics do you have now that you did not have before studying this chapter?

STUDY TIPS CHECKLIST

The foundation of all your study skills is TIME!

- ☐ Have you tried taping your lectures, with your instructor's permission?
- ☐ Are you using the time-management suggestions we have given?
- ☐ Are you using the five steps for problem solving?
- ☐ Have you tried tutoring a fellow student?
- ☐ Are you finding quiet, nondistracting places to study?

REVIEW EXERCISES

Solve graphically. Then classify the system as consistent or inconsistent and the equations as dependent or independent. [13.1a]

1. $4x - y = -9,$
$x - y = -3$

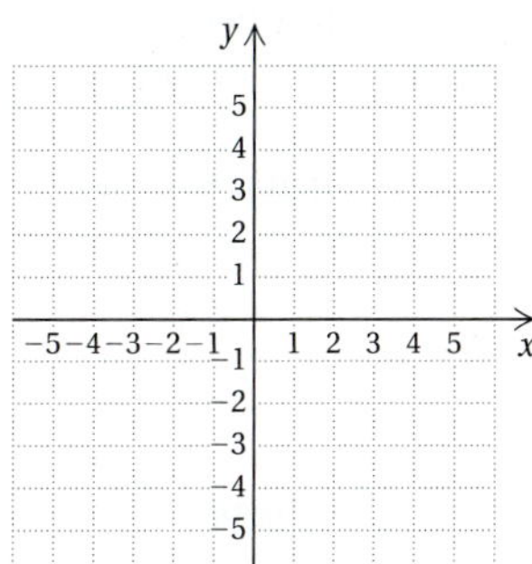

2. $15x + 10y = -20,$
$3x + 2y = -4$

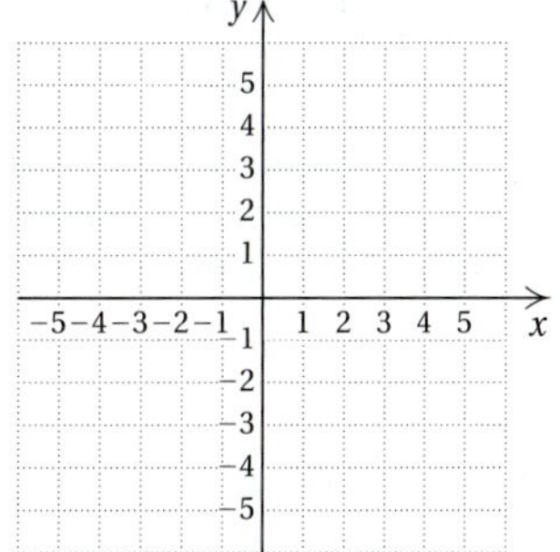

3. $y - 2x = 4,$
$y - 2x = 5$

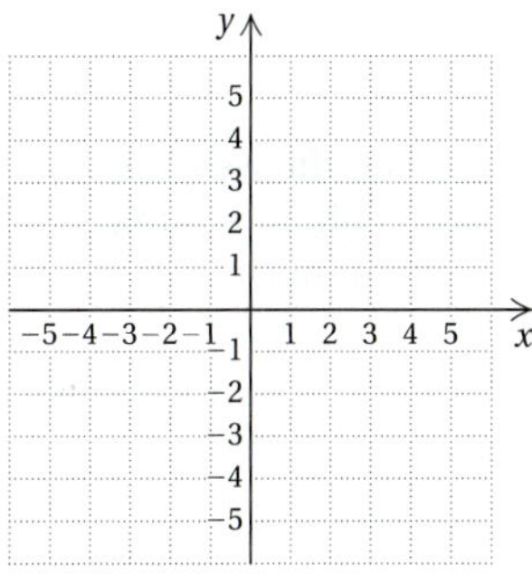

Solve by the substitution method. [13.2a]

4. $7x - 4y = 6,$
$y - 3x = -2$

5. $y = x + 2,$
$y - x = 8$

6. $9x - 6y = 2,$
$x = 4y + 5$

Solve by the elimination method. [13.3a]

7. $8x - 2y = 10,$
$-4y - 3x = -17$

8. $4x - 7y = 18,$
$9x + 14y = 40$

9. $3x - 5y = -4,$
$5x - 3y = 4$

10. $1.5x - 3 = -2y,$
$3x + 4y = 6$

11. *Music Spending.* Sean has \$37 to spend. He can spend all of it on two compact discs and a cassette, or he can buy one CD and two cassettes and have \$5.00 left over. What is the price of a CD? of a cassette? [13.4a]

12. *Orange Drink Mixtures.* "Orange Thirst" is 15% orange juice and "Quencho" is 5% orange juice. How many liters of each should be combined in order to get 10 L of a mixture that is 10% orange juice? [13.4a]

13. *Train Travel.* A train leaves Watsonville at noon traveling north at 44 mph. One hour later, another train, going 52 mph, travels north on a parallel track. How many hours will the second train travel before it overtakes the first train? [13.4b]

Solve. [13.5a]

14. $x + 2y + z = 10,$
$2x - y + z = 8,$
$3x + y + 4z = 2$

15. $3x + 2y + z = 3,$
$6x - 4y - 2z = -34,$
$-x + 3y - 3z = 14$

16. $2x - 5y - 2z = -4,$
$7x + 2y - 5z = -6,$
$-2x + 3y + 2z = 4$

17. $x + y + 2z = 1,$
$x - y + z = 1,$
$x + 2y + z = 2$

18. *Triangle Measure.* In triangle ABC, the measure of angle A is four times the measure of angle C, and the measure of angle B is 45° more than the measure of angle C. What are the measures of the angles of the triangle? [13.6a]

19. *Money Mixtures.* Elaine has \$194, consisting of \$20, \$5, and \$1 bills. The number of \$1 bills is 1 less than the total number of \$20 and \$5 bills. If she has 39 bills in her purse, how many of each denomination does she have? [13.6a]

20. *Bed Manufacturing.* Kregel Furniture is planning to produce a new type of bed. For the first year, the fixed costs for setting up production are \$35,000. The variable costs for producing each bed are \$175. The revenue from each bed is \$300. Find the following. [13.7a]

a) The total cost $C(x)$ of producing x beds
b) The total revenue $R(x)$ from the sale of x beds
c) The total profit from the production and sale of x beds
d) The profit or loss from the production and sale of 1200 beds; of 200 beds
e) The break-even point

21. Find the equilibrium point for the following demand and supply functions: [13.7b]

$$D(p) = 120 - 13p,$$
$$S(p) = 60 + 7p.$$

22. DW Briefly compare the strengths and the weaknesses of the graphical, substitution, and elimination methods as applied to the solution of two equations in two variables. [13.1a], [13.2a], [13.3a]

23. **DW** Explain the advantages of using a system of equations to solve an applied problem. [13.2b], [13.3b], [13.4a, b], [13.6a]

24. *Peanuts.* Refer to the figure at the bottom of the page. Help Peppermint Patty with her dilemma. Translate Peppermint Patty's problem to a system of equations. Then solve the problem. [13.3b]
Source: PEANUTS reprinted by permission of United Feature Syndicate, Inc.

SKILL MAINTENANCE

Certain objectives from four particular sections will be retested on the chapter test. The objectives are listed with the practice problems that follow.

Write an equation of the line containing the given point and parallel to the given line. [12.5d]

25. $(0, 5)$; $x - 2y = -3$

26. $(1, -1)$; $10x - 5y = 7$

Write an equation of the line containing the given point and perpendicular to the given line. [12.5d]

27. $(8, 0)$; $4x + y = 10$

28. $(2, -2)$; $y = 4 - x$

29. Solve $Q = at - 4t$ for t. [7.4b]

30. Given the function $f(x) = 8 - 3x$, find $f(0)$ and $f(-2)$. [12.1b]

31. For $5x - 8y = 40$, find the slope and the y-intercept. [12.3b]

SYNTHESIS

32. Solve graphically:

$$y = x + 2,$$
$$y = x^2 + 2.$$

[13.1a]

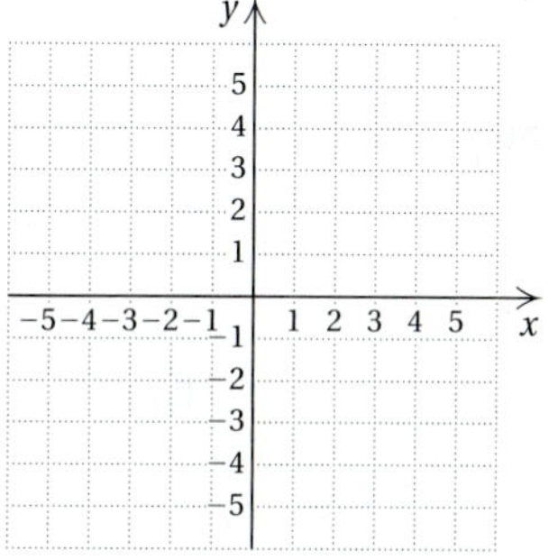

33. *Height Estimation in Anthropology.* An anthropologist can use linear functions to estimate the height of a male or female, given the length of certain bones. The *femur* is the large bone from the hip to the knee. Let x = the length of the femur, in centimeters. Then the height, in centimeters, of a male with a femur of length x is given by the function

$$M(x) = 1.88x + 81.31.$$

The height, in centimeters, of a female with a femur of length x is given by the function

$$F(x) = 1.95x + 72.85.$$

[13.1a]

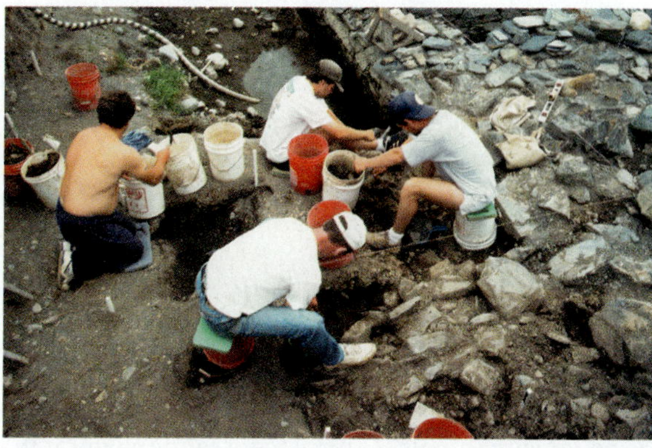

A 45-cm femur was uncovered at an anthropological dig.

a) If we assume that it was from a male, how tall was he?

b) If we assume that it was from a female, how tall was she?

c) Graph each equation and find the point of intersection of the graphs of the equations.

d) For what length of a male femur and a female femur, if any, would the height be the same?

CHAPTER 13

Chapter Test

Solve graphically. Then classify the system as consistent or inconsistent and the equations as dependent or independent.

1. $y = 3x + 7,$
$3x + 2y = -4$

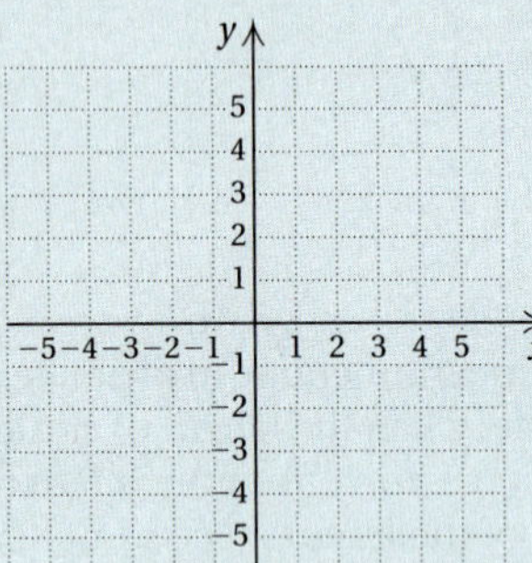

2. $y = 3x + 4,$
$y = 3x - 2$

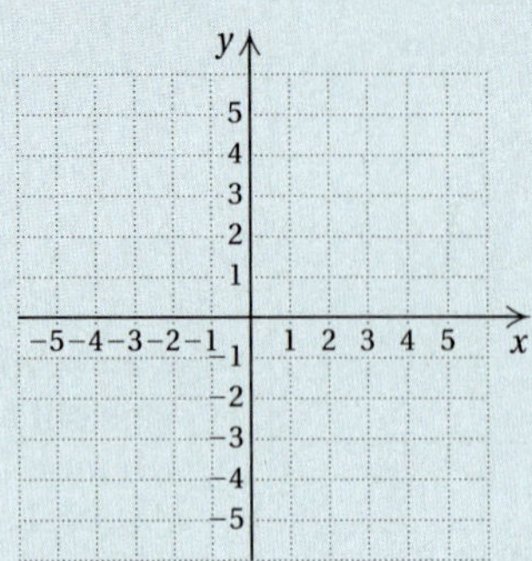

3. $y - 3x = 6,$
$6x - 2y = -12$

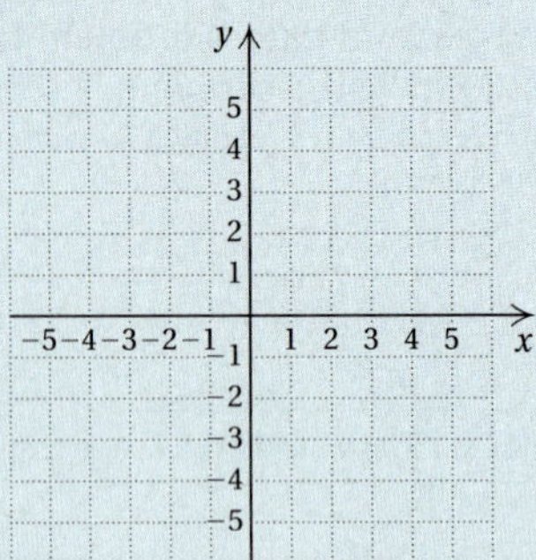

Solve by the substitution method.

4. $x + 3y = -8,$
$4x - 3y = 23$

5. $2x + 4y = -6,$
$y = 3x - 9$

Solve by the elimination method.

6. $4x - 6y = 3,$
$6x - 4y = -3$

7. $4y + 2x = 18,$
$3x + 6y = 26$

Solve.

8. *Saline Solutions.* Saline (saltwater) solutions are often used for sore throats. A nurse has a saline solution that is 34% salt and the rest water. She also has a 61% saline solution. She wants to use a 50% solution with a particular patient without wasting the existing solutions. How many milliliters (mL) of each solution would be needed in order to obtain 120 mL of a mixture that is 50% salt?

9. *Chicken Dinners.* High Flyin' Wings charges $12 for a bucket of chicken wings and $7 for a chicken dinner. After filling 28 orders for buckets and dinners during a football game, the waiters had collected $281. How many buckets and how many dinners did they sell?

10. *Air Travel.* An airplane flew for 5 hr with a 20-km/h tailwind and returned in 7 hr against the same wind. Find the speed of the plane in still air.

11. *Tennis Court.* The perimeter of a standard tennis court used for playing doubles is 288 ft. The width of the court is 42 ft less than the length. Find the length and the width.

12. Solve:

$$6x + 2y - 4z = 15,$$
$$-3x - 4y + 2z = -6,$$
$$4x - 6y + 3z = 8.$$

13. Find the equilibrium point for the following demand and supply functions:

$$D(p) = 79 - 8p,$$
$$S(p) = 37 + 6p.$$

Solve.

14. *Repair Rates.* An electrician, a carpenter, and a plumber are hired to work on a house. The electrician earns \$21 per hour, the carpenter \$19.50 per hour, and the plumber \$24 per hour. The first day on the job, they worked a total of 21.5 hr and earned a total of \$469.50. If the plumber worked 2 hr more than the carpenter did, how many hours did the electrician work?

15. *Manufacturing Tennis Rackets.* Sweet Spot Manufacturing is planning to produce a new type of tennis racket. For the first year, the fixed costs for setting up production are \$40,000. The variable costs for producing each racket are \$30. The sales department predicts that 1500 rackets can be sold during the first year. The revenue from each racket is \$80. Find the following.

a) The total cost $C(x)$ of producing x rackets
b) The total revenue $R(x)$ from the sale of x rackets
c) The total profit from the production and sale of x rackets
d) The profit or loss from the production and sale of 1200 rackets; of 200 rackets
e) The break-even point

SKILL MAINTENANCE

16. Write an equation of the line containing the given point and parallel to the given line.

$$(-1, -2);\ 3x + y = -4$$

17. Write an equation of the line containing the given point and perpendicular to the given line.

$$(0, 4);\ 5x + 4y = -2$$

18. Solve $P = 4a - 3b$ for a.

19. For $7x = 14 - 2y$, find the slope and the y-intercept.

20. Given the function $f(x) = x^2 - 8$, find $f(0)$ and $f(-3)$.

SYNTHESIS

21. The graph of the function $f(x) = mx + b$ contains the points $(-1, 3)$ and $(-2, -4)$. Find m and b.

More on Inequalities

Gateway to Chapter 14

In this chapter, we expand our coverage of inequalities by expressing solution sets in interval notation. We solve and graph conjunctions and disjunctions of inequalities and solve systems of inequalities. We also solve equations and inequalities with absolute-value expressions.

Real-World Application

The equation $y = 5.06x + 9.7$ can be used to predict the average daily volume y of e-mail messages, in billions, in North America x years after 2000. Predict the average daily volume of e-mail messages in 2005 and in 2010. Determine those years in which the average daily volume of e-mail messages will be greater than 110.9 billion.

Source: Pitney Bowes

This problem appears as Exercise 80 in Exercise Set 14.1.

CHAPTER 14

CHAPTER

14 Pretest

Solve.

1. $-7x < 21$ [14.1c]

2. $2x - 1 \le 5x + 8$ [14.1c]

3. $-1 \le 3x + 2 \le 4$ [14.2a]

4. $|2y - 7| = 9$ [14.3c]

5. $4a + 5 < -3$ *or* $4a + 5 > 3$ [14.2b]

6. $|3a + 5| > 1$ [14.3e]

7. $|3x + 4| = |x - 7|$ [14.3d]

8. $-8(x - 7) \ge 10(2x + 3) - 12$ [14.1c]

9. Find the distance between 8.2 and 10.6. [14.3b]

10. **Records in the Women's 100-m Dash.** Florence Griffith Joyner set a world record of 10.49 sec in the women's 100-m dash in 1988. The equation

$$R = -0.0433t + 10.49$$

can be used to predict the world record in the women's 100-m dash t years after 1988. Determine those years for which the world record will be less than 10.35 sec. [14.1d]

Source: International Amateur Athletics Federation

11. Find the intersection: [14.2a]

$$\{-4, -3, 1, 4, 9\} \cap \{-3, -2, 2, 4, 5\}.$$

12. Find the union: $\{-1, 1\} \cup \{0\}$. [14.2b]

13. Simplify: $\left|\frac{-5y}{2x}\right|$. [14.3a]

14. Graph on a plane: $6x - 2y < 12$. [14.4b]

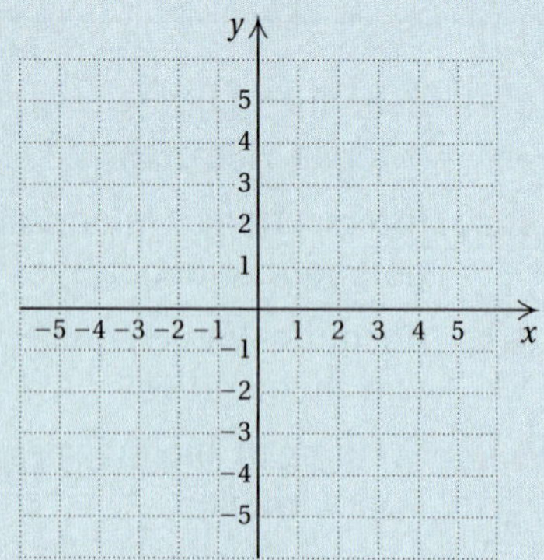

15. Graph the following system of inequalities. Find the coordinates of any vertices formed. [14.4c]

$$x + y \le 16,$$
$$3x + 6y \le 60,$$
$$x \ge 0,$$
$$y \ge 0$$

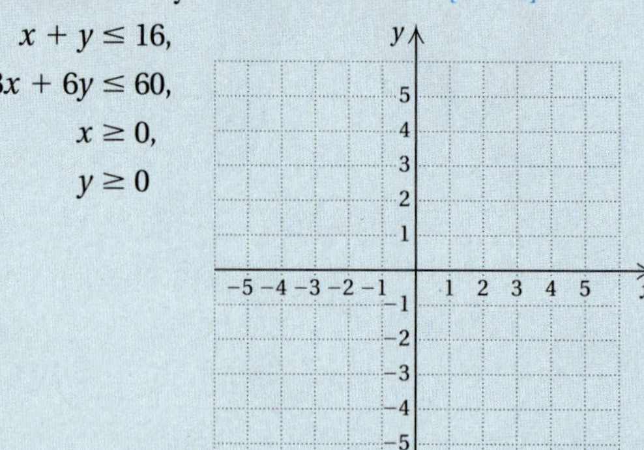

14.1 SETS, INEQUALITIES, AND INTERVAL NOTATION

Objectives

a Determine whether a given number is a solution of an inequality.

b Write interval notation for the solution set or graph of an inequality.

c Solve an inequality using the addition and multiplication principles and then graph the inequality.

d Solve applied problems by translating to inequalities.

We begin this chapter with a review of solving inequalities. In Chapter 7, we wrote solution sets of inequalities using *set-builder notation.* Here we will also write solution sets using *interval notation.*

a Inequalities

INEQUALITY

An **inequality** is any sentence containing $<$, $>$, $\leq$, $\geq$, or $\neq$.

Examples of inequalities are the following:

$-2 < a, \quad x > 4, \quad x + 3 \leq 6, \quad 6 - 7y \geq 10y - 4,$ and $5x \neq 10.$

SOLUTION OF AN INEQUALITY

Any replacement or value for the variable that makes an inequality true is called a **solution.** The set of all solutions is called the **solution set.** When all the solutions of an inequality have been found, we say that we have **solved** the inequality.

EXAMPLES Determine whether the given number is a solution of the inequality.

1. $x + 3 < 6$; 5

   We substitute and get $5 + 3 < 6$, or $8 < 6$, a *false* sentence. Therefore, 5 is not a solution.

2. $2x - 3 > -3$; 1

   We substitute and get $2(1) - 3 > -3$, or $-1 > -3$, a *true* sentence. Therefore, 1 is a solution.

3. $4x - 1 \leq 3x + 2$; 3

   We substitute and get $4(3) - 1 \leq 3(3) + 2$, or $11 \leq 11$, a *true* sentence. Therefore, 3 is a solution.

Do Exercises 1–3.

Determine whether the given number is a solution of the inequality.

1. $3 - x < 2$; 8

2. $3x + 2 > -1$; -2

3. $3x + 2 \leq 4x - 3$; 5

b Inequalities and Interval Notation

The **graph** of an inequality is a drawing that represents its solutions. An inequality in one variable can be graphed on the number line.

EXAMPLE 4 Graph $x < 4$ on the number line.

The solutions are all real numbers less than 4, so we shade all numbers less than 4 on the number line. To indicate that 4 is not a solution, we use a right parenthesis ")" at 4.

−7 −6 −5 −4 −3 −2 −1 0 1 2 3 4 5 6 7

Answers on page A-50

We can write the solution set for $x < 4$ using **set-builder notation:**

$$\{x \mid x < 4\}.$$

This is read "The set of all x such that x is less than 4."

Another way to write solutions of an inequality in one variable is to use **interval notation.** Interval notation uses parentheses ( ) and brackets [ ].

If a and b are real numbers such that $a < b$, we define the interval (a, b) as the set of all numbers between but not including a and b—that is, the set of all x for which $a < x < b$. Thus,

$$(a, b) = \{x \mid a < x < b\}.$$

The points a and b are the **endpoints.** The parentheses indicate that the endpoints are *not* included in the graph.

The interval $[a, b]$ is defined as the set of all numbers x for which $a \leq x \leq b$. Thus,

$$[a, b] = \{x \mid a \leq x \leq b\}.$$

The brackets indicate that the endpoints *are* included in the graph.*

CAUTION!

Do not confuse the *interval* (a, b) with the *ordered pair* (a, b) used in connection with an equation in two variables in the plane, as in Chapter 13. The context in which the notation appears usually makes the meaning clear.

The following intervals include one endpoint and exclude the other:

$(a, b] = \{x \mid a < x \leq b\}$. The graph excludes a and includes b.

$[a, b) = \{x \mid a \leq x < b\}$. The graph includes a and excludes b.

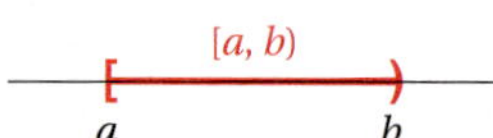

Some intervals extend without bound in one or both directions. We use the symbols ∞, read "infinity," and $-\infty$, read "negative infinity," to name these intervals. The notation (a, ∞) represents the set of all numbers greater than a—that is,

$$(a, \infty) = \{x \mid x > a\}.$$

Similarly, the notation $(-\infty, a)$ represents the set of all numbers less than a—that is,

$$(-\infty, a) = \{x \mid x < a\}.$$

*Some books use the representations and instead of, respectively, and .

The notations $[a, \infty)$ and $(-\infty, a]$ are used when we want to include the endpoints. The interval $(-\infty, \infty)$ names the set of all real numbers.

$(-\infty, \infty) = \{x | x \text{ is a real number}\}$

Interval notation is summarized in the following table.

INTERVALS: NOTATION AND GRAPHS

INTERVAL NOTATION	SET NOTATION	GRAPH
(a, b)	$\{x \| a < x < b\}$	a ( ) b
$[a, b]$	$\{x \| a \le x \le b\}$	a [ ] b
$[a, b)$	$\{x \| a \le x < b\}$	a [ ) b
$(a, b]$	$\{x \| a < x \le b\}$	a ( ] b
(a, ∞)	$\{x \| x > a\}$	a (→
$[a, \infty)$	$\{x \| x \ge a\}$	a [→
$(-\infty, b)$	$\{x \| x < b\}$	←) b
$[-\infty, b]$	$\{x \| x \le b\}$	←] b
$(-\infty, \infty)$	$\{x \| x \text{ is a real number}\}$	← →

CAUTION!

Whenever the symbol ∞ is included in interval notation, a right parenthesis ")" is used. Similarly, when $-\infty$ is included, a left parenthesis "(" is used.

EXAMPLES Write interval notation for the given set or graph.

5. $\{x | -4 < x < 5\} = (-4, 5)$
6. $\{x | x \ge -2\} = [-2, \infty)$
7. $(-2, 4]$

-6 -5 -4 -3 -2 -1 0 1 2 3 4 5 6

8. 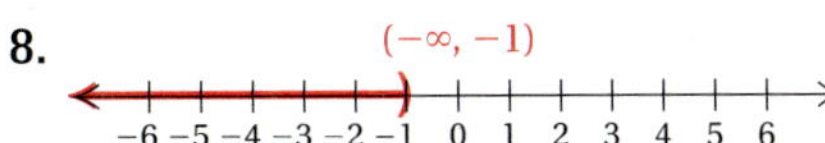

Do Exercises 4–7.

C Solving Inequalities

Two inequalities are **equivalent** if they have the same solution set. For example, the inequalities $x > 4$ and $4 < x$ are equivalent. Just as the addition principle for equations gives us equivalent equations, the addition principle for inequalities gives us equivalent inequalities.

Write interval notation for the given set or graph.

4. $\{x | -4 \le x < 5\}$

5. $\{x | x \le -2\}$

6.

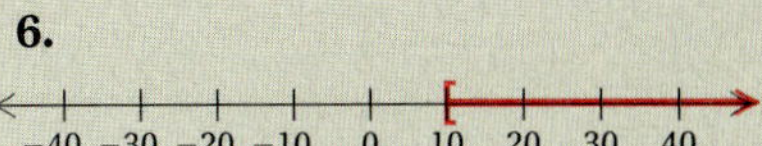

7.

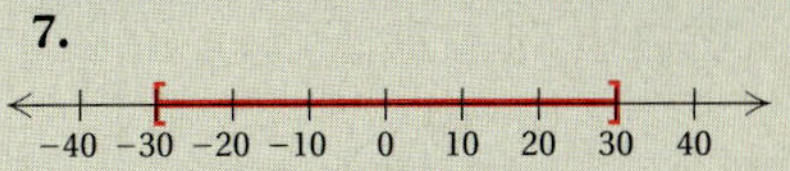

Answers on page A-50

Solve and graph.

8. $x + 6 > 9$

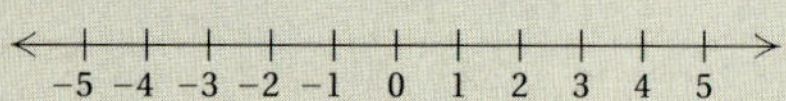

9. $x + 4 \leq 7$

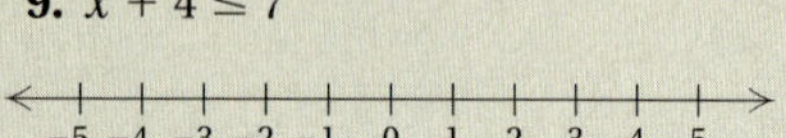

10. Solve and graph:

$$2x - 3 \geq 3x - 1.$$

−5 −4 −3 −2 −1 0 1 2 3 4 5

Answers on page A-50

THE ADDITION PRINCIPLE FOR INEQUALITIES

For any real numbers a, b, and c:

$a < b$ is equivalent to $a + c < b + c$;

$a > b$ is equivalent to $a + c > b + c$.

Similar statements hold for $\leq$ and $\geq$.

Since subtracting c is the same as adding $-c$, there is no need for a separate subtraction principle.

EXAMPLE 9 Solve and graph: $x + 5 > 1$.

We have

$$x + 5 > 1$$

$$x + 5 - 5 > 1 - 5 \qquad \text{Using the addition principle: adding } -5 \text{ or subtracting } 5$$

$$x > -4.$$

We used the addition principle to show that the inequalities $x + 5 > 1$ and $x > -4$ are equivalent. The solution set is $\{x \mid x > -4\}$ and consists of an infinite number of solutions. We cannot possibly check them all. Instead, we can perform a partial check by substituting one member of the solution set (here we use -1) into the original inequality:

$$\begin{array}{c|c} x + 5 & > 1 \\ \hline -1 + 5 & ?\ 1 \\ 4 & \end{array} \quad \text{TRUE}$$

Since $4 > 1$ is true, we have our check. The solution set is $\{x \mid x > -4\}$, or $(-4, \infty)$. The graph is as follows:

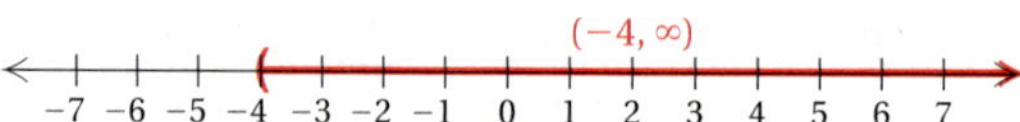

Do Exercises 8 and 9.

EXAMPLE 10 Solve and graph: $4x - 1 \geq 5x - 2$.

We have

$$4x - 1 \geq 5x - 2$$

$$4x - 1 + 2 \geq 5x - 2 + 2 \qquad \text{Adding 2}$$

$$4x + 1 \geq 5x \qquad \text{Simplifying}$$

$$4x + 1 - 4x \geq 5x - 4x \qquad \text{Subtracting } 4x$$

$$1 \geq x. \qquad \text{Simplifying}$$

The inequalities $1 \geq x$ and $x \leq 1$ have the same meaning and the same solutions. You can check that any number less than or equal to 1 is a solution. The solution set is $\{x \mid 1 \geq x\}$ or, more commonly, $\{x \mid x \leq 1\}$. Using interval notation, we write that the solution set is $(-\infty, 1]$. The graph is as follows:

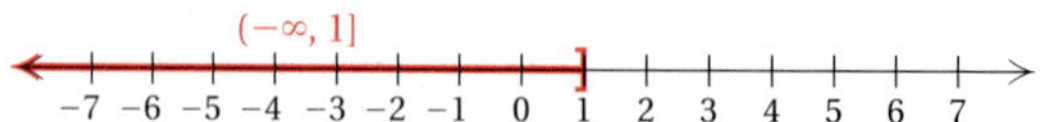

Do Exercise 10.

The multiplication principle for inequalities is somewhat different from the multiplication principle for equations. Consider this true inequality:

$-4 < 9.$ True

If we multiply both numbers by 2, we get another true inequality:

$-4(2) < 9(2),$ or $-8 < 18.$ True

If we multiply both numbers by -3, we get a false inequality:

$-4(-3) < 9(-3),$ or $12 < -27.$ False

However, if we now *reverse* the inequality symbol above, we get a true inequality:

$12 > -27.$ True

THE MULTIPLICATION PRINCIPLE FOR INEQUALITIES

For any real numbers a and b, and any *positive* number c:

$a < b$ is equivalent to $ac < bc$;

$a > b$ is equivalent to $ac > bc$.

For any real numbers a and b, and any *negative* number c:

$a < b$ is equivalent to $ac > bc$;

$a > b$ is equivalent to $ac < bc$.

Similar statements hold for $\leq$ and $\geq$.

Since division by c is the same as multiplication by $1/c$, there is no need for a separate division principle.

EXAMPLE 11 Solve and graph: $3y < \frac{3}{4}$.

We have

$$3y < \frac{3}{4}$$

$$\frac{3y}{3} < \frac{\frac{3}{4}}{3}$$ Dividing by 3. The symbol stays the same.

$$y < \frac{3}{4} \div 3$$ Simplifying; rewriting the division

$$y < \frac{3}{4} \cdot \frac{1}{3}$$ Dividing by multiplying by a reciprocal

$$y < \frac{1}{4}.$$ Simplifying

Any number less than $\frac{1}{4}$ is a solution. The solution set is $\{y \mid y < \frac{1}{4}\}$, or $(-\infty, \frac{1}{4})$. The graph is as follows:

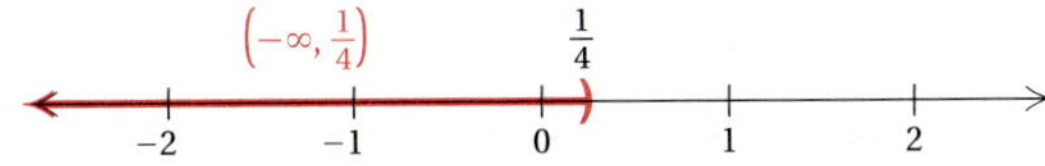

Solve and graph.

11. $5y \le \frac{3}{2}$

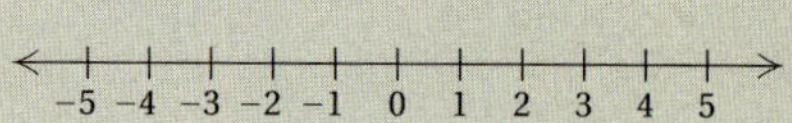

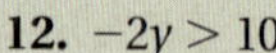

12. $-2y > 10$

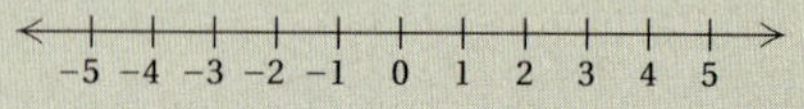

13. $-\frac{1}{3}x \le -4$

Answers on page A-50

EXAMPLE 12 Solve and graph: $-5x \ge -80$.

We have

$$-5x \ge -80$$

$$\frac{-5x}{-5} \le \frac{-80}{-5} \quad \text{Dividing by } -5. \text{ The symbol must be reversed.}$$

$$x \le 16.$$

The solution set is $\{x \mid x \le 16\}$, or $(-\infty, 16]$. The graph is as follows:

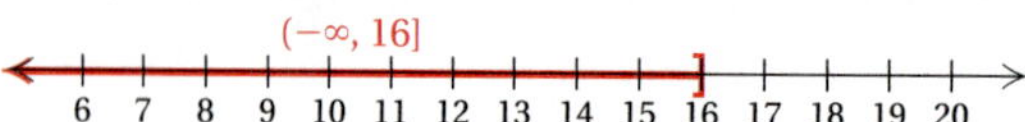

Do Exercises 11–13.

We use the addition and multiplication principles together in solving inequalities in much the same way as in solving equations.

EXAMPLE 13 Solve: $16 - 7y \ge 10y - 4$.

We have

$$16 - 7y \ge 10y - 4$$

$$-16 + 16 - 7y \ge -16 + 10y - 4 \quad \text{Adding } -16$$

$$-7y \ge 10y - 20 \quad \text{Collecting like terms}$$

$$-10y + (-7y) \ge -10y + 10y - 20 \quad \text{Adding } -10y$$

$$-17y \ge -20 \quad \text{Collecting like terms}$$

$$\frac{-17y}{-17} \le \frac{-20}{-17} \quad \text{Dividing by } -17. \text{ The symbol must be reversed.}$$

$$y \le \frac{20}{17}. \quad \text{Simplifying}$$

The solution set is $\left\{y \mid y \le \frac{20}{17}\right\}$, or $\left(-\infty, \frac{20}{17}\right]$.

In some cases, you can avoid the concern about multiplying or dividing by a negative number by using the addition principle in a different way. Let's rework Example 13 by adding $7y$ instead of $-10y$:

$$16 - 7y \ge 10y - 4$$

$$16 - 7y + 7y \ge 10y - 4 + 7y \quad \text{Adding } 7y$$

$$16 \ge 17y - 4 \quad \text{Collecting like terms}$$

$$16 + 4 \ge 17y - 4 + 4 \quad \text{Adding 4}$$

$$20 \ge 17y \quad \text{Collecting like terms}$$

$$\frac{20}{17} \ge \frac{17y}{17} \quad \text{Dividing by 17. The symbol stays the same.}$$

$$\frac{20}{17} \ge y.$$

Note that $\frac{20}{17} \ge y$ is equivalent to $y \le \frac{20}{17}$.

EXAMPLE 14 Solve: $-3(x+8)-5x>4x-9$.

We have

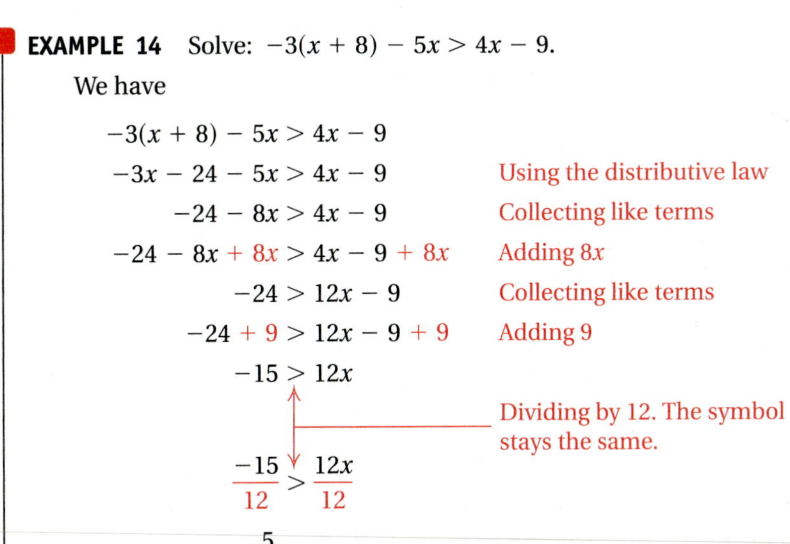

$$
\begin{aligned}
-3(x+8)-5x &> 4x-9 \\
-3x-24-5x &> 4x-9 && \text{Using the distributive law} \\
-24-8x &> 4x-9 && \text{Collecting like terms} \\
-24-8x+8x &> 4x-9+8x && \text{Adding } 8x \\
-24 &> 12x-9 && \text{Collecting like terms} \\
-24+9 &> 12x-9+9 && \text{Adding 9} \\
-15 &> 12x \\
\frac{-15}{12} &> \frac{12x}{12} && \text{Dividing by 12. The symbol stays the same.} \\
-\frac{5}{4} &> x.
\end{aligned}
$$

The solution set is $\{x \mid -\frac{5}{4} > x\}$, or $\{x \mid x < -\frac{5}{4}\}$, or $(-\infty, -\frac{5}{4})$.

Do Exercises 14–16.

d Applications and Problem Solving

Many problem-solving and applied situations translate to inequalities. In addition to “is less than” and “is more than,” other phrases are commonly used.

IMPORTANT WORDS	SAMPLE SENTENCE	TRANSLATION
is at least	Max is at least 5 years old.	$m \geq 5$
is at most	At most 6 people could fit in the elevator.	$n \leq 6$
cannot exceed	Total weight in the elevator cannot exceed 2000 pounds.	$w \leq 2000$
must exceed	The speed must exceed 15 mph.	$s > 15$
is between	Heather’s income is between \$23,000 and \$35,000.	$23{,}000 < h < 35{,}000$
no more than	Bing weighs no more than 90 pounds.	$w \leq 90$
no less than	Saul would accept no less than \$4000 for the piano.	$t \geq 4000$

The following phrases deserve special attention.

TRANSLATING “AT LEAST” AND “AT MOST”

A quantity x is **at least** some amount q: $x \geq q$.
(If x is at least q, it cannot be less than q.)

A quantity x is **at most** some amount q: $x \leq q$.
(If x is at most q, it cannot be more than q.)

Do Exercises 17–26.

Solve.

14. $6 - 5y \geq 7$

15. $3x + 5x < 4$

16. $17 - 5(y-2) \leq 45y + 8(2y-3) - 39y$

Translate.

17. Emma scored no less than 88 on her Chinese exam.

18. The average credit-card holder is at least \$4000 in debt.
Source: CBS television, "60 Minutes"

19. The price of that PT Cruiser is at most \$21,000.

20. The time of the test was between 50 and 60 min.

21. The University of Northern Kentucky is more than 25 miles away.

22. Sarah’s weight is less than 110 lb.

23. That number is greater than −8.

24. The costs of production of that DVD model cannot exceed \$135,000.

25. At most 11.6% of all deaths in Minnesota are from cancer.

26. Yesterday, at least 37 people got tickets for drunk driving.

Answers on page A-51

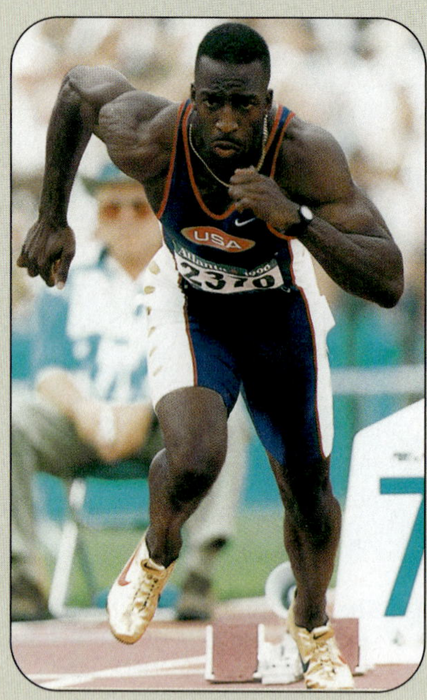

27. Records in the Men's 200-m Dash. Refer to Example 15. Determine (in terms of an inequality) those years for which the world record will be less than 18.6 sec.

EXAMPLE 15 *Records in the Men's 200-m Dash.* Michael Johnson set a world record of 19.32 sec in the men's 200-m dash in the 1996 Olympics. The equation

$$R = -0.0513t + 19.32$$

can be used to predict the world record in the men's 200-m dash t years after 1996. Determine (in terms of an inequality) those years for which the world record will be less than 19.0 sec.

Source: International Amateur Athletic Foundation

1. **Familiarize.** We already have a formula. To become more familiar with it, we might make a substitution for t. Suppose we want to know the record after 20 yr, in the year 2016. We substitute 20 for t:

   $$R = -0.0513(20) + 19.32 = 18.294 \text{ sec.}$$

   We see that by 2016, we expect that the record will be less than 19.0 sec. To predict the exact year in which the 19.0-sec mark will be broken, we could make other guesses that are less than 20. Instead, we proceed to the next step.

2. **Translate.** The record R is to be *less than* 19.0 sec. Thus we have

   $$R < 19.0.$$

   We replace R with $-0.0513t + 19.32$ to find the times t that solve the inequality:

   $$-0.0513t + 19.32 < 19.0.$$

3. **Solve.** We solve the inequality:

   $$\begin{aligned} -0.0513t + 19.32 &< 19.0 \\ -0.0513t &< -0.32 && \text{Subtracting 19.32} \\ t &> 6.24. && \text{Dividing by } -0.0513 \text{ and rounding} \end{aligned}$$

4. **Check.** A partial check is to substitute a value for t greater than 6.24. We did that in the *Familiarize* step.

5. **State.** The record will be less than 19.0 sec for races occurring more than 6.24 yr after 1996, or approximately after 2002.

Do Exercise 27.

Answer on page A-51

EXAMPLE 16 *Salary Plans.* On a new job, Rose can be paid in one of two ways: *Plan A* is a salary of \$600 per month, plus a commission of 4% of sales; and *Plan B* is a salary of \$800 per month, plus a commission of 6% of sales in excess of \$10,000. For what amount of monthly sales is plan A better than plan B, if we assume that sales are always more than \$10,000?

1. **Familiarize.** Listing the given information in a table will be helpful.

PLAN A: MONTHLY INCOME	PLAN B: MONTHLY INCOME
\$600 salary 4% of sales *Total*: \$600 + 4% of sales	\$800 salary 6% of sales over \$10,000 *Total*: \$800 + 6% of sales over \$10,000

Next, suppose that Rose sold \$12,000 in one month. Which plan would be better? Under plan A, she would earn \$600 plus 4% of \$12,000, or

$$600 + 0.04(12{,}000) = \$1080.$$

Since with plan B commissions are paid only on sales in excess of \$10,000, Rose would earn \$800 plus 6% of (\$12,000 − \$10,000), or

$$800 + 0.06(12{,}000 - 10{,}000) = \$920.$$

This shows that for monthly sales of \$12,000, plan A is better. Similar calculations will show that for sales of \$30,000 a month, plan B is better. To determine *all* values for which plan A pays more money, we must solve an inequality that is based on the calculations above.

2. **Translate.** We let S = the amount of monthly sales. If we examine the calculations in the *Familiarize* step, we see that the monthly income from plan A is $600 + 0.04S$ and from plan B is $800 + 0.06(S - 10{,}000)$. Thus we want to find all values of S for which

Income from plan A — is greater than — Income from plan B

$$600 + 0.04S \quad > \quad 800 + 0.06(S - 10{,}000).$$

3. **Solve.** We solve the inequality:

$600 + 0.04S > 800 + 0.06(S - 10{,}000)$

$600 + 0.04S > 800 + 0.06S - 600$ Using the distributive law

$600 + 0.04S > 200 + 0.06S$ Collecting like terms

$400 > 0.02S$ Subtracting 200 and $0.04S$

$20{,}000 > S$, or $S < 20{,}000$. Dividing by 0.02

4. **Check.** For $S = 20{,}000$, the income from plan A is

$600 + 4\% \cdot 20{,}000$, or \$1400.

The income from plan B is

$800 + 6\% \cdot (20{,}000 - 10{,}000)$, or \$1400.

This confirms that for sales of \$20,000, Rose's pay is the same under either plan.

In the *Familiarize* step, we saw that for sales of \$12,000, plan A pays more. Since $12{,}000 < 20{,}000$, this is a partial check. Since we cannot check all possible values of S, we will stop here.

5. **State.** For monthly sales of less than \$20,000, plan A is better.

Do Exercise 28.

28. **Salary Plans.** A painter can be paid in one of two ways:

Plan A: \$500 plus \$4 per hour;
Plan B: Straight \$9 per hour.

Suppose that the job takes n hours. For what values of n is plan A better for the painter?

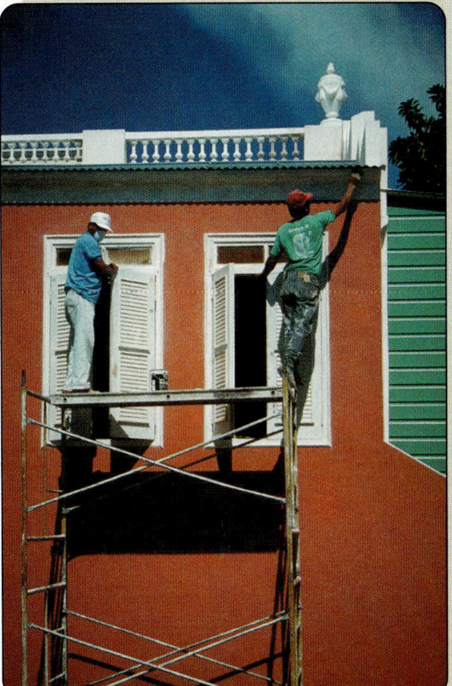

Answer on page A-51

14.1 EXERCISE SET

For Extra Help

Digital Video Tutor CD 12 Videotape 16 | InterAct Math | Math Tutor Center | MathXL | MyMathLab

a Determine whether the given numbers are solutions of the inequality.

1. $x - 2 \geq 6$; $-4, 0, 4, 8$

2. $3x + 5 \leq -10$; $-5, -10, 0, 27$

3. $t - 8 > 2t - 3$; $0, -8, -9, -3, -\frac{7}{8}$

4. $5y - 7 < 8 - y$; $2, -3, 0, 3, \frac{2}{3}$

b Write interval notation for the given set or graph.

5. $\{x \mid x < 5\}$

6. $\{t \mid t \geq -5\}$

7. $\{x \mid -3 \leq x \leq 3\}$

8. $\{t \mid -10 < t \leq 10\}$

9. $-6\ -5\ -4\ -3\ -2\ -1\ 0\ 1\ 2\ 3\ 4\ 5\ 6$

10.

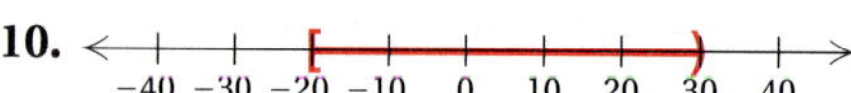

11.

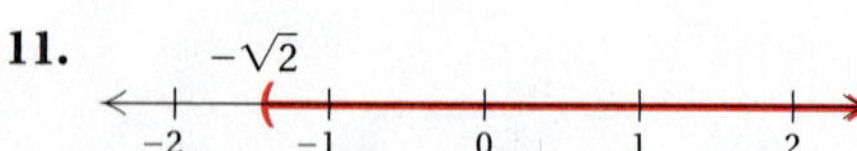

12.

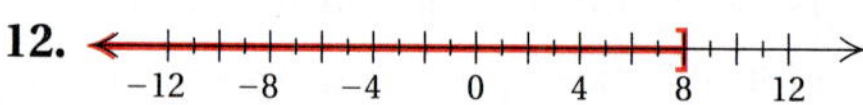

c Solve and graph.

13. $x + 2 > 1$

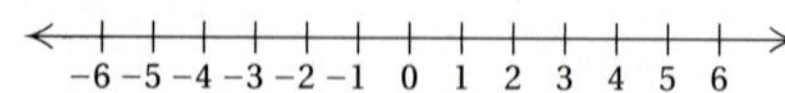

14. $x + 8 > 4$

$-6\ -5\ -4\ -3\ -2\ -1\ 0\ 1\ 2\ 3\ 4\ 5\ 6$

15. $y + 3 < 9$

$-6\ -5\ -4\ -3\ -2\ -1\ 0\ 1\ 2\ 3\ 4\ 5\ 6$

16. $y + 4 < 10$

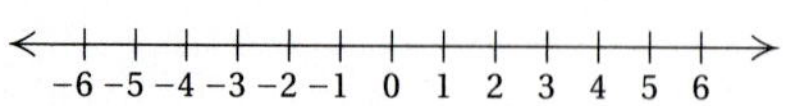

17. $a - 9 \leq -31$

$-40\ -30\ -20\ -10\ 0\ 10\ 20\ 30\ 40$

18. $a + 6 \leq -14$

$-40\ -30\ -20\ -10\ 0\ 10\ 20\ 30\ 40$

19. $t + 13 \geq 9$

$-6\ -5\ -4\ -3\ -2\ -1\ 0\ 1\ 2\ 3\ 4\ 5\ 6$

20. $x - 8 \leq 17$

$-40\ -30\ -20\ -10\ 0\ 10\ 20\ 30\ 40$

21. $y - 8 > -14$

$-6\ -5\ -4\ -3\ -2\ -1\ 0\ 1\ 2\ 3\ 4\ 5\ 6$

22. $y - 9 > -18$

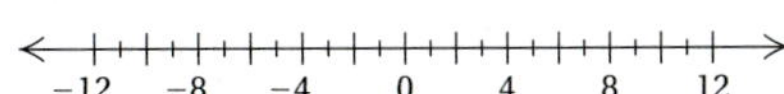

23. $x - 11 \le -2$

−12 −8 −4 0 4 8 12

24. $y - 18 \le -4$

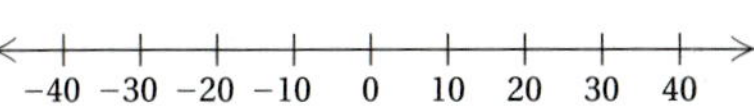

25. $8x \ge 24$

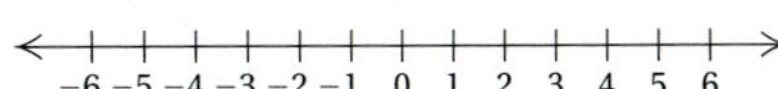

26. $8t < -56$

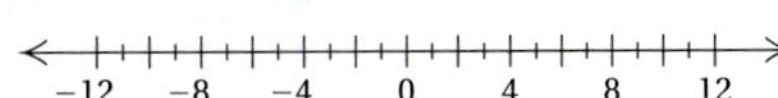

27. $0.3x < -18$

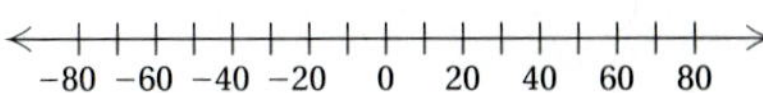

28. $0.6x < 30$

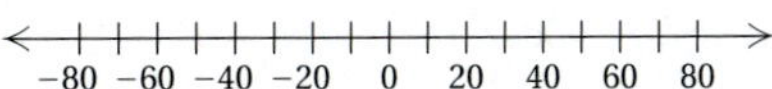

Solve.

29. $-9x \ge -8.1$

30. $-5y \le 3.5$

31. $-\frac{3}{4}x \ge -\frac{5}{8}$

32. $-\frac{1}{8}y \le -\frac{9}{8}$

33. $2x + 7 < 19$

34. $5y + 13 > 28$

35. $5y + 2y \le -21$

36. $-9x + 3x \ge -24$

37. $2y - 7 < 5y - 9$

38. $8x - 9 < 3x - 11$

39. $0.4x + 5 \le 1.2x - 4$

40. $0.2y + 1 > 2.4y - 10$

41. $5x - \frac{1}{12} \le \frac{5}{12} + 4x$

42. $2x - 3 < \frac{13}{4}x + 10 - 1.25x$

43. $4(4y - 3) \geq 9(2y + 7)$

44. $2m + 5 \geq 16(m - 4)$

45. $3(2 - 5x) + 2x < 2(4 + 2x)$

46. $2(0.5 - 3y) + y > (4y - 0.2)8$

47. $5[3m - (m + 4)] > -2(m - 4)$

48. $[8x - 3(3x + 2)] - 5 \geq 3(x + 4) - 2x$

49. $3(r - 6) + 2 > 4(r + 2) - 21$

50. $5(t + 3) + 9 < 3(t - 2) + 6$

51. $19 - (2x + 3) \leq 2(x + 3) + x$

52. $13 - (2c + 2) \geq 2(c + 2) + 3c$

53. $\frac{1}{4}(8y + 4) - 17 < -\frac{1}{2}(4y - 8)$

54. $\frac{1}{3}(6x + 24) - 20 > -\frac{1}{4}(12x - 72)$

55. $2[4 - 2(3 - x)] - 1 \geq 4[2(4x - 3) + 7] - 25$

56. $5[3(7 - t) - 4(8 + 2t)] - 20 \leq -6[2(6 + 3t) - 4]$

57. $\frac{4}{5}(7x - 6) < 40$

58. $\frac{2}{3}(4x - 3) > 30$

59. $\frac{3}{4}(3 + 2x) + 1 \geq 13$

60. $\frac{7}{8}(5 - 4x) - 17 \geq 38$

61. $\frac{3}{4}\left(3x - \frac{1}{2}\right) - \frac{2}{3} < \frac{1}{3}$

62. $\frac{2}{3}\left(\frac{7}{8} - 4x\right) - \frac{5}{8} < \frac{3}{8}$

63. $0.7(3x + 6) \geq 1.1 - (x + 2)$

64. $0.9(2x + 8) < 20 - (x + 5)$

65. $a + (a - 3) \leq (a + 2) - (a + 1)$

66. $0.8 - 4(b - 1) > 0.2 + 3(4 - b)$

d Solve.

Body Mass Index. The *body mass index I* can be used to determine an individual's risk of heart disease. An index less than 25 indicates a low risk. The body mass index is given by the formula, or model,

$$I = \frac{704.5W}{H^2},$$

where W is the weight, in pounds, and H is the height, in inches.

Source: National Center for Health Statistics

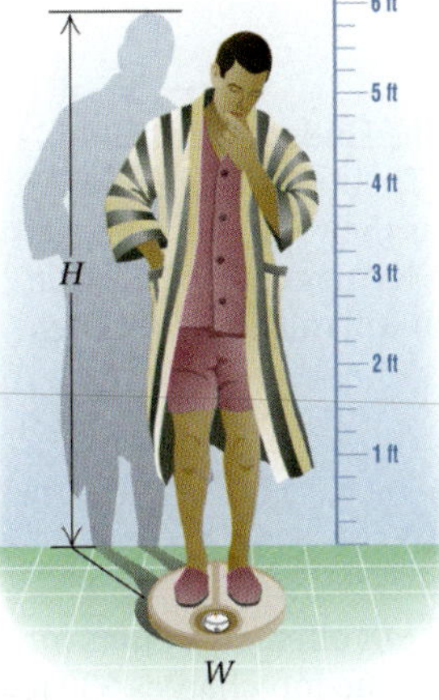

67. *Body Mass Index.*

a) Marv weighs 192 lb and his height is 73 in. What is his body mass index?

b) Determine (in terms of an inequality) those weights W for Marv that will keep him in the low-risk category.

68. *Body Mass Index.*

a) Elaine weighs 110 lb and her height is 67 in. What is her body mass index?

b) Determine (in terms of an inequality) those weights W for Elaine that will keep her in the low-risk category.

69. *Grades.* You are taking a European history course in which there will be 4 tests, each worth 100 points. You have scores of 89, 92, and 95 on the first three tests. You must make a total of at least 360 in order to get an A. What scores on the last test will give you an A?

70. *Grades.* You are taking a literature course in which there will be 5 tests, each worth 100 points. You have scores of 94, 90, and 89 on the first three tests. You must make a total of at least 450 in order to get an A. What scores on the fourth test will keep you eligible for an A?

71. *Insurance Claims.* After a serious automobile accident, most insurance companies will replace the damaged car with a new one if repair costs exceed 80% of the N.A.D.A., or "blue-book," value of the car. Miguel's car recently sustained $9200 worth of damage but was not replaced. What was the blue-book value of his car?

72. *Phone Rates.* A long-distance telephone call using Down East Calling costs 20 cents for the first minute and 16 cents for each additional minute. The same call, placed on Long Call Systems, costs 19 cents for the first minute and 18 cents for each additional minute. For what length phone calls is Down East Calling less expensive?

73. *Salary Plans.* Toni can be paid in one of two ways:

Plan A: A salary of \$400 per month plus a commission of 8% of gross sales;

Plan B: A salary of \$610 per month, plus a commission of 5% of gross sales.

For what amount of gross sales should Toni select plan A?

74. *Salary Plans.* Branford can be paid for his masonry work in one of two ways:

Plan A: \$300 plus \$9.00 per hour;

Plan B: Straight \$12.50 per hour.

Suppose that the job takes n hours. For what values of n is plan B better for Branford?

75. *Checking-Account Rates.* The Hudson Bank offers two checking-account plans. Their Anywhere plan charges 20¢ per check whereas their Acu-checking plan costs \$2 per month plus 12¢ per check. For what numbers of checks per month will the Acu-checking plan cost less?

76. *Insurance Benefits.* Bayside Insurance offers two plans. Under plan A, Giselle would pay the first \$50 of her medical bills and 20% of all bills after that. Under plan B, Giselle would pay the first \$250 of bills, but only 10% of the rest. For what amount of medical bills will plan B save Giselle money? (Assume that her bills will exceed \$250.)

77. *Wedding Costs.* The Arnold Inn offers two plans for wedding parties. Under plan A, the inn charges \$30 for each person in attendance. Under plan B, the inn charges \$1300 plus \$20 for each person in excess of the first 25 who attend. For what size parties will plan B cost less? (Assume that more than 25 guests will attend.)

78. *Investing.* Lillian is about to invest \$20,000, part at 6% and the rest at 8%. What is the most that she can invest at 6% and still be guaranteed at least \$1500 in interest per year?

79. *Converting Dress Sizes.* The formula

$$I = 2(s + 10)$$

can be used to convert dress sizes s in the United States to dress sizes I in Italy. For what dress sizes in the United States will dress sizes in Italy be larger than 36?

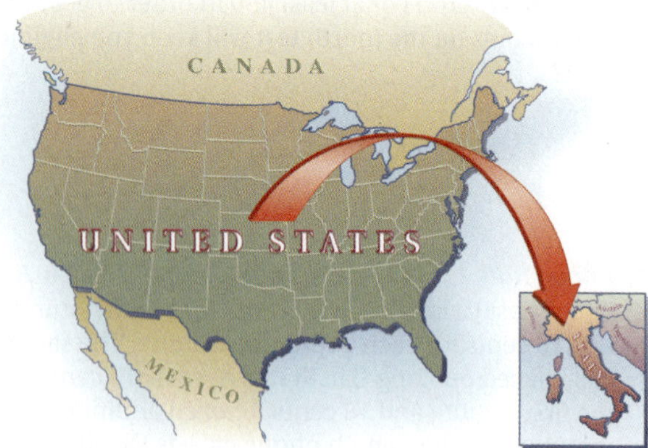

80. *E-mail Messages.* The equation $y = 5.06x + 9.7$ can be used to predict the average daily volume y of e-mail messages, in billions, in North America x years after 2000. Predict the average daily volume of e-mail messages in 2005 and in 2010. Determine those years in which the average daily volume of e-mail messages will be greater than 110.9 billion.

Source: Pitney Bowes

81. *Consumption of Bottled Water.* People are consuming more bottled water each year. The number N of gallons per year that each person drinks t years after 1990 is approximated by

$$N = 0.733t + 8.398.$$

a) How many gallons of bottled water did each person drink in 1990 ($t = 0$)? in 1995 ($t = 5$)? in 2000?
b) For what years will the amount of bottled water drunk be at least 15 gal?

82. *Dewpoint Spread.* Pilots use the **dewpoint spread,** or the difference between the current temperature and the dewpoint (the temperature at which dew occurs), to estimate the height of the cloud cover. Each 3° of dewpoint spread corresponds to an increased height of cloud cover of 1000 ft. A plane, flying with limited instruments, must have a cloud cover higher than 3500 ft. What dewpoint spreads will allow the plane to fly?

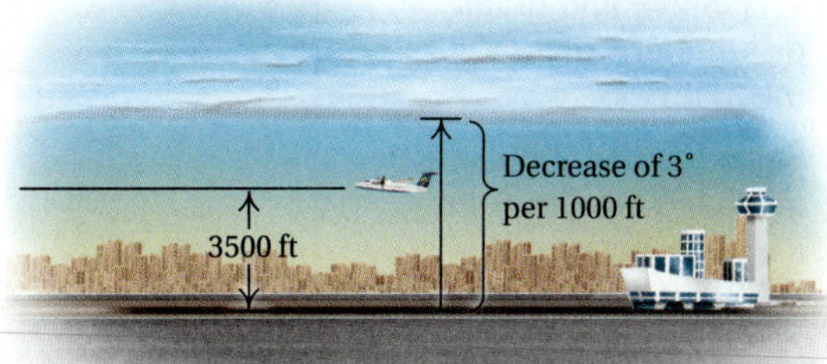

83. DW Explain in your own words why the inequality symbol must be reversed when both sides of an inequality are multiplied or divided by a negative number.

84. DW Graph the solution set of each of the following on a number line and compare:

$$x < 2, \quad x > 2, \quad x = 2, \quad x \leq 2, \quad \text{and} \quad x \geq 2.$$

SKILL MAINTENANCE

Multiply. [9.6a]

85. $(3x - 4)(x + 8)$

86. $(r - 4s)(6r + s)$

87. $(2a - 5)(3a + 11)$

88. $(t + 2s)(t - 9s)$

Find the domain. [12.2a]

89. $f(x) = \dfrac{-3}{x + 8}$

90. $f(x) = 3x - 5$

91. $f(x) = |x| - 4$

92. $f(x) = \dfrac{x + 7}{3x - 2}$

SYNTHESIS

93. *Supply and Demand.* The supply S and demand D for a certain product are given by

$$S = 460 + 94p \quad \text{and} \quad D = 2000 - 60p.$$

a) Find those values of p for which supply exceeds demand.
b) Find those values of p for which supply is less than demand.

Determine whether the statement is true or false. If false, give a counterexample.

94. For any real numbers x and y, if $x < y$, then $x^2 < y^2$.

95. For any real numbers a, b, c, and d, if $a < b$ and $c < d$, then $a + c < b + d$.

96. Determine whether the inequalities

$$x < 3 \quad \text{and} \quad 0 \cdot x < 0 \cdot 3$$

are equivalent. Give reasons to support your answer.

Solve.

97. $x + 5 \leq 5 + x$

98. $x + 8 < 3 + x$

99. $x^2 + 1 > 0$

Objectives

a Find the intersection of two sets. Solve and graph conjunctions of inequalities.

b Find the union of two sets. Solve and graph disjunctions of inequalities.

c Solve applied problems involving conjunctions and disjunctions of inequalities.

14.2 INTERSECTIONS, UNIONS, AND COMPOUND INEQUALITIES

The steroid cholesterol is a fat-related compound in tissues of cells. One commonly used measure of cholesterol is *total cholesterol.* The following table shows the health-risk levels of this measure of cholesterol.

Type of Cholesterol	Normal Risk	Borderline Risk	High Risk
Total	Total $\leq$ 200	200 $<$ Total $<$ 240	Total $\geq$ 240

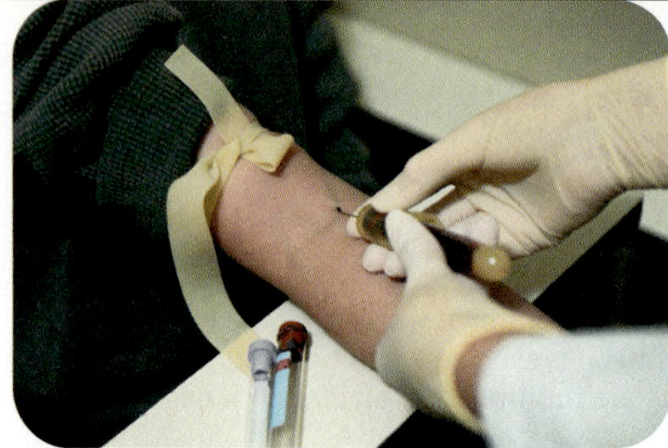

A total-cholesterol level T between 200 and 240 is considered borderline risk. We can express this by the sentence

$200 < T$ *and* $T < 240$

or more simply by

$200 < T < 240.$

This is an example of a *compound inequality.* **Compound inequalities** consist of two or more inequalities joined by the word *and* or the word *or.* We now "solve" such sentences—that is, we find the set of all solutions.

a Intersections of Sets and Conjunctions of Inequalities

INTERSECTION

The **intersection** of two sets A and B is the set of all members that are common to A and B. We denote the intersection of sets A and B as

$A \cap B.$

The intersection of two sets is often illustrated as shown at right.

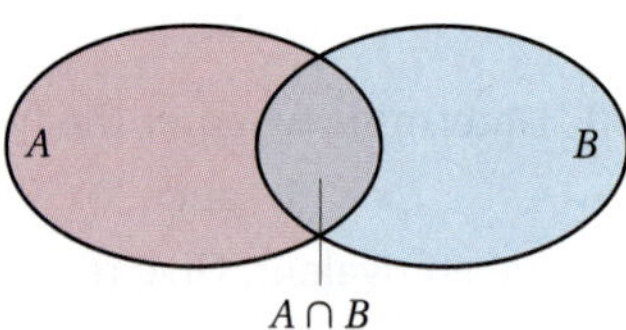

EXAMPLE 1 Find the intersection: $\{1,2,3,4,5\} \cap \{-2,-1,0,1,2,3\}$.

The numbers 1, 2, and 3 are common to the two sets, so the intersection is $\{1,2,3\}$.

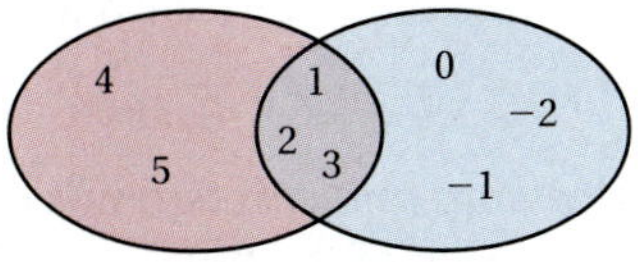

Do Exercises 1 and 2.

CONJUNCTION

When two or more sentences are joined by the word *and* to make a compound sentence, the new sentence is called a **conjunction** of the sentences.

The following is a conjunction of inequalities:

$$-2 < x \quad and \quad x < 1.$$

In order for a conjunction to be true, each individual sentence must be true. *The solution set of a conjunction is the intersection of the solution sets of the individual sentences.* Consider the conjunction

$$-2 < x \quad and \quad x < 1.$$

The graphs of each separate sentence are shown below, and the intersection is the last graph. We use both set-builder and interval notations.

$\{x \mid -2 < x\}$

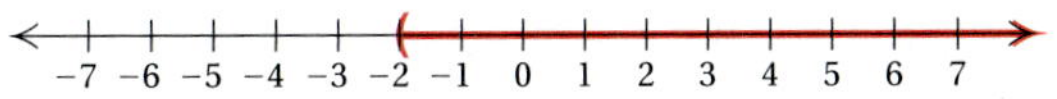

$(-2, \infty)$

$\{x \mid x < 1\}$

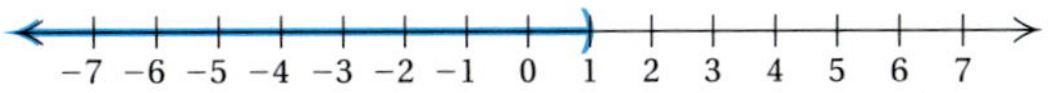

$(-\infty, 1)$

$\{x \mid -2 < x\} \cap \{x \mid x < 1\}$
$= \{x \mid -2 < x \text{ and } x < 1\}$

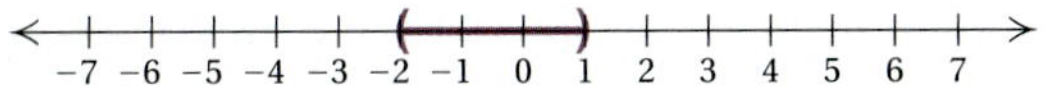

$(-2, 1)$

Because there are numbers that are both greater than -2 and less than 1, the conjunction $-2 < x$ *and* $x < 1$ can be abbreviated by $-2 < x < 1$. Thus the interval $(-2, 1)$ can be represented as $\{x \mid -2 < x < 1\}$, the set of all numbers that are *simultaneously* greater than -2 *and* less than 1. Note that, in general, for $a < b$,

$a < x$ and $x < b$ can be abbreviated $a < x < b$;
and $b > x$ and $x > a$ can be abbreviated $b > x > a$.

CAUTION!

"$a > x$ *and* $x < b$" cannot be abbreviated as "$a > x < b$".

"AND"; "INTERSECTION"

The word **"and"** corresponds to **"intersection"** and to the symbol "$\cap$". In order for a number to be a solution of a conjunction, it must be in *both* solution sets.

Do Exercise 3.

1. Find the intersection:

$\{0,3,5,7\} \cap \{0,1,3,11\}$.

2. Shade the intersection of sets A and B.

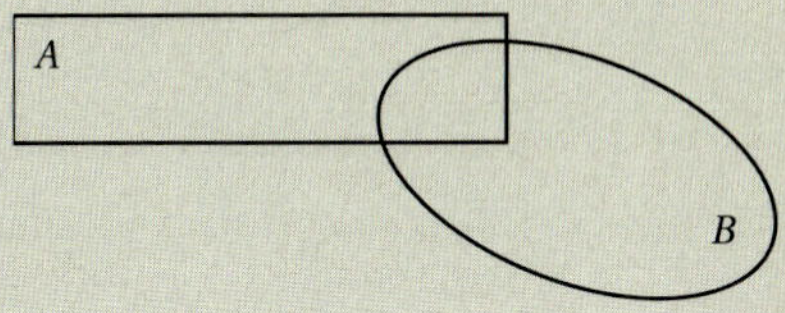

3. Graph and write interval notation:

$-1 < x$ *and* $x < 4$.

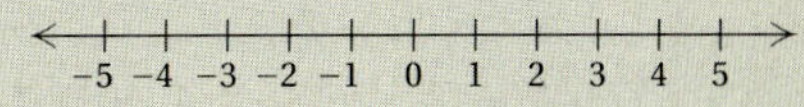

Answers on page A-51

4. Solve and graph:

$$-22 < 3x - 7 \le 23.$$

−12 −8 −4 0 4 8 12

Answer on page A-51

EXAMPLE 2 Solve and graph: $-1 \le 2x + 5 < 13$.

This inequality is an abbreviation for the conjunction

$$-1 \le 2x + 5 \quad \textit{and} \quad 2x + 5 < 13.$$

The word *and* corresponds to set *intersection*, $\cap$. To solve the conjunction, we solve each of the two inequalities separately and then find the intersection of the solution sets:

$-1 \le 2x + 5$	*and*	$2x + 5 < 13$	
$-6 \le 2x$	*and*	$2x < 8$	Subtracting 5
$-3 \le x$	*and*	$x < 4.$	Dividing by 2

We now abbreviate the answer:

$$-3 \le x < 4.$$

The solution set is $\{x \mid -3 \le x < 4\}$, or, in interval notation, $[-3, 4)$. The graph is the intersection of the two separate solution sets.

$\{x \mid -3 \le x\}$

$[-3, \infty)$

$\{x \mid x < 4\}$

$(-\infty, 4)$

$\{x \mid -3 \le x\} \cap \{x \mid x < 4\}$
$= \{x \mid -3 \le x < 4\}$

−7 −6 −5 −4 −3 −2 −1 0 1 2 3 4 5 6 7

$[-3, 4)$

The steps above are generally combined as follows:

$-1 \le 2x + 5 < 13$	$2x + 5$ appears in both inequalities.
$-6 \le 2x < 8$	Subtracting 5
$-3 \le x < 4.$	Dividing by 2

Such an approach saves some writing and will prove useful in Section 14.3. Check numbers such as −5, −2, and 4. Which are solutions?

Do Exercise 4.

EXAMPLE 3 Solve and graph: $2x - 5 \ge -3$ *and* $5x + 2 \ge 17$.

We first solve each inequality separately:

$2x - 5 \ge -3$	*and*	$5x + 2 \ge 17$
$2x \ge 2$	*and*	$5x \ge 15$
$x \ge 1$	*and*	$x \ge 3.$

Next, we find the intersection of the two separate solution sets:

$\{x \mid x \geq 1\}$  $[1, \infty)$

$\{x \mid x \geq 3\}$ −7 −6 −5 −4 −3 −2 −1 0 1 2 3 4 5 6 7 $[3, \infty)$

$\{x \mid x \geq 1\} \cap \{x \mid x \geq 3\}$
$= \{x \mid x \geq 3\}$ −7 −6 −5 −4 −3 −2 −1 0 1 2 3 4 5 6 7 $[3, \infty)$

The numbers common to both sets are those that are greater than or equal to 3. Thus the solution set is $\{x \mid x \geq 3\}$, or, in interval notation, $[3, \infty)$. You should check that any number in $[3, \infty)$ satisfies the conjunction whereas numbers outside $[3, \infty)$ do not. Check numbers such as -1, 2, and 4. Which are solutions?

Do Exercise 5.

5. Solve and graph:

$3x + 4 < 10$ *and* $2x - 7 < -13$.

−5 −4 −3 −2 −1 0 1 2 3 4 5

EMPTY SET; DISJOINT SETS

Sometimes two sets have no elements in common. In such a case, we say that the intersection of the two sets is the **empty set,** denoted { } or $\varnothing$. Two sets with an empty intersection are said to be **disjoint.**

$A \cap B = \varnothing$

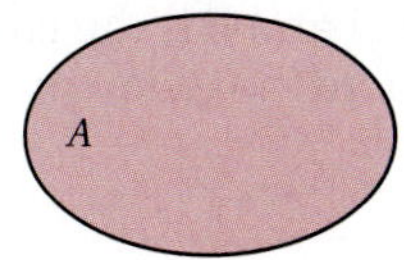

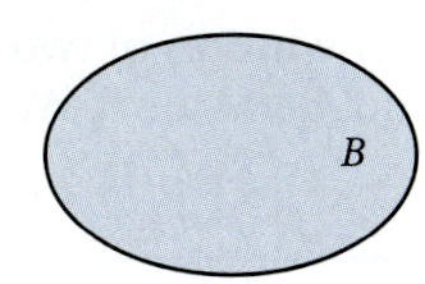

6. Solve and graph.

$3x - 7 \leq -13$ *and* $4x + 3 > 8$.

−6 −5 −4 −3 −2 −1 0 1 2 3 4 5 6

EXAMPLE 4 Solve and graph: $2x - 3 > 1$ *and* $3x - 1 < 2$.

We solve each inequality separately:

$$\begin{aligned} 2x - 3 > 1 \quad &\textit{and} \quad 3x - 1 < 2 \\ 2x > 4 \quad &\textit{and} \quad 3x < 3 \\ x > 2 \quad &\textit{and} \quad x < 1. \end{aligned}$$

The solution set is the intersection of the solution sets of the individual inequalities.

$\{x \mid x > 2\}$ 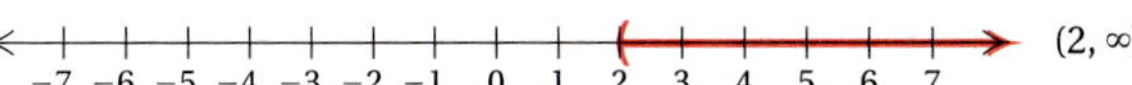 $(2, \infty)$

$\{x \mid x < 1\}$ −7 −6 −5 −4 −3 −2 −1 0 1 2 3 4 5 6 7 $(-\infty, 1)$

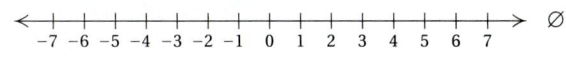

$\{x \mid x > 2\} \cap \{x \mid x < 1\}$
$= \{x \mid x > 2 \text{ and } x < 1\}$
$= \varnothing$ −7 −6 −5 −4 −3 −2 −1 0 1 2 3 4 5 6 7 $\varnothing$

Since no number is both greater than 2 and less than 1, the solution set is the empty set, $\varnothing$.

Do Exercise 6.

Answers on page A-51

7. Solve: $-4 \leq 8 - 2x \leq 4$.

8. Find the union:

$\{0, 1, 3, 4\} \cup \{0, 1, 7, 9\}$.

9. Shade the union of sets A and B.

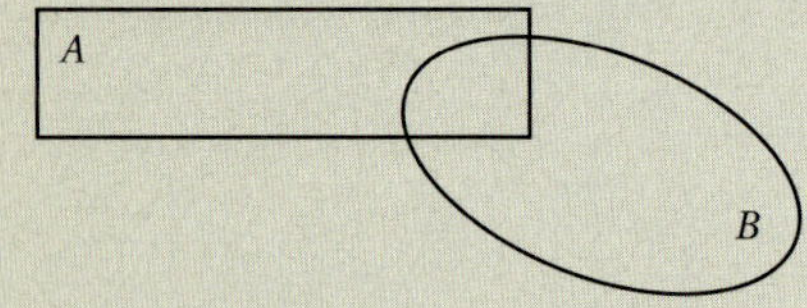

Answers on page A-51

EXAMPLE 5 Solve: $3 \leq 5 - 2x < 7$.

We have

$$3 \leq 5 - 2x < 7$$

$$3 - 5 \leq 5 - 2x - 5 < 7 - 5 \quad \text{Subtracting 5}$$

$$-2 \leq -2x < 2 \quad \text{Simplifying}$$

$$\frac{-2}{-2} \geq \frac{-2x}{-2} > \frac{2}{-2} \quad \text{Dividing by } -2\text{. The symbols must be reversed.}$$

$$1 \geq x > -1. \quad \text{Simplifying}$$

The solution set is $\{x \mid 1 \geq x > -1\}$, or $\{x \mid -1 < x \leq 1\}$, since the inequalities $1 \geq x > -1$ and $-1 < x \leq 1$ are equivalent. The solution, in interval notation, is $(-1, 1]$.

Do Exercise 7.

b Unions of Sets and Disjunctions of Inequalities

UNION

The **union** of two sets A and B is the collection of elements belonging to A and/or B. We denote the union of A and B by

$A \cup B$.

The union of two sets is often pictured as shown below.

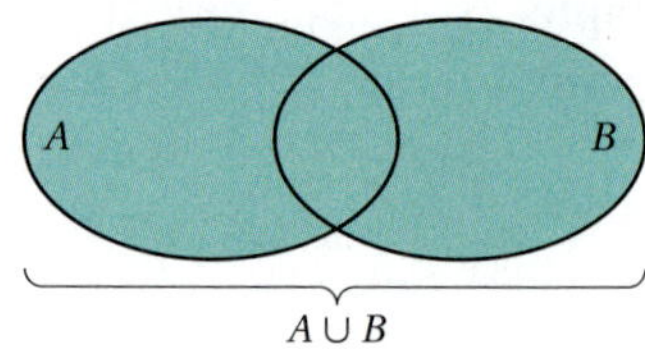

EXAMPLE 6 Find the union: $\{2, 3, 4\} \cup \{3, 5, 7\}$.

The numbers in either or both sets are 2, 3, 4, 5, and 7, so the union is $\{2, 3, 4, 5, 7\}$. We don't list the number 3 twice.

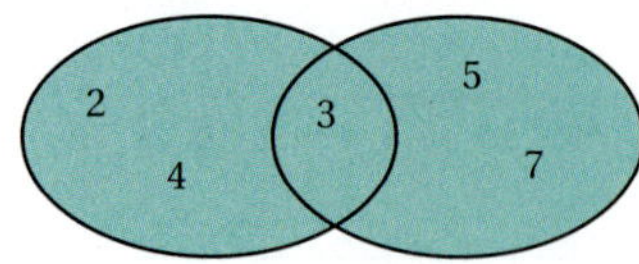

Do Exercises 8 and 9.

DISJUNCTION

When two or more sentences are joined by the word *or* to make a compound sentence, the new sentence is called a **disjunction** of the sentences.

Here are three examples of disjunctions:

$x < -3$ *or* $x > 3$;

y is an odd number *or* y is a prime number;

$x < 0$ *or* $x = 0$ *or* $x > 0$.

In order for a disjunction to be true, at least one of the individual sentences must be true. *The solution set of a disjunction is the union of the individual solution sets.* Consider the disjunction

$$x < -3 \quad or \quad x > 3.$$

The graphs of each separate sentence are shown below, and the union is the last graph. Again, we use both set-builder and interval notations.

$\{x \mid x < -3\}$

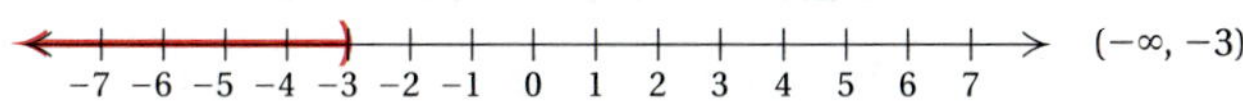

$(-\infty, -3)$

$\{x \mid x > 3\}$

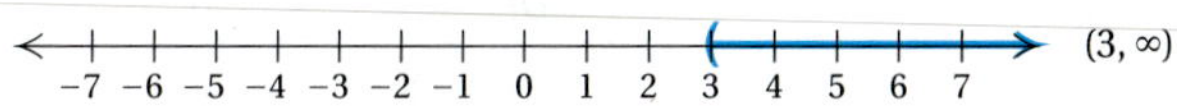

$(3, \infty)$

$\{x \mid x < -3\} \cup \{x \mid x > 3\}$
$= \{x \mid x < -3 \text{ or } x > 3\}$

−7 −6 −5 −4 −3 −2 −1 0 1 2 3 4 5 6 7

$(-\infty, -3) \cup (3, \infty)$

Answers to disjunctions can rarely be written in a shorter manner. The solution set of $x < -3$ *or* $x > 3$ is simply written $\{x \mid x < -3 \text{ or } x > 3\}$, or $(-\infty, -3) \cup (3, \infty)$.

"OR"; "UNION"

The word **"or"** corresponds to **"union"** and the symbol "$\cup$". In order for a number to be in the solution set of a disjunction, it must be in *at least one* of the solution sets.

Do Exercise 10.

10. Graph and write interval notation:

$x \le -2$ *or* $x > 4$.

−5 −4 −3 −2 −1 0 1 2 3 4 5

EXAMPLE 7 Solve and graph: $7 + 2x < -1$ *or* $13 - 5x \le 3$.

We solve each inequality separately, retaining the word *or*:

$$7 + 2x < -1 \quad or \quad 13 - 5x \le 3$$
$$2x < -8 \quad or \quad -5x \le -10$$

Dividing by -5. The symbol must be reversed.

$$x < -4 \quad or \quad x \ge 2.$$

To find the solution set of the disjunction, we consider the individual graphs. We graph $x < -4$ and then $x \ge 2$. Then we take the union of the graphs.

$\{x \mid x < -4\}$ −7 −6 −5 −4 −3 −2 −1 0 1 2 3 4 5 6 7 $(-\infty, -4)$

$\{x \mid x \geq 2\}$ −7 −6 −5 −4 −3 −2 −1 0 1 2 3 4 5 6 7 $[2, \infty)$

$\{x \mid x < -4 \text{ or } x \geq 2\}$ −7 −6 −5 −4 −3 −2 −1 0 1 2 3 4 5 6 7 $(-\infty, -4) \cup [2, \infty)$

The solution set is $\{x \mid x < -4 \text{ or } x \ge 2\}$, or, in interval notation, $(-\infty, -4) \cup [2, \infty)$.

Answer on page A-51

Solve and graph.

11. $x - 4 < -3$ *or* $x - 3 \geq 3$

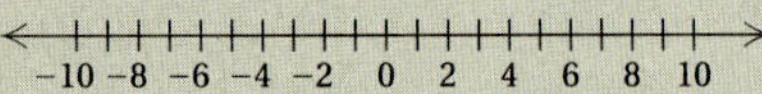

12. $-2x + 4 \leq -3$ *or* $x + 5 < 3$

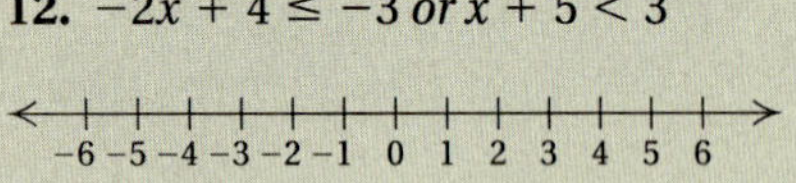

13. Solve:

$-3x - 7 < -1$ *or* $x + 4 < -1$.

14. Solve and graph:

$5x - 7 \leq 13$ *or* $2x - 1 \geq -7$.

←|-6 -5 -4 -3 -2 -1 0 1 2 3 4 5 6|→

Answers on page A-51

CAUTION!

A compound inequality like

$$x < -4 \quad or \quad x \geq 2,$$

as in Example 7, *cannot* be expressed as $2 \leq x < -4$ because to do so would be to say that x is *simultaneously* less than -4 and greater than or equal to 2. No number is both less than -4 *and* greater than or equal to 2, but many are less than -4 *or* greater than or equal to 2.

Do Exercises 11 and 12.

EXAMPLE 8 Solve: $-2x - 5 < -2$ *or* $x - 3 < -10$.

We solve the individual inequalities separately, retaining the word *or*:

$$-2x - 5 < -2 \quad or \quad x - 3 < -10$$
$$-2x < 3 \quad or \quad x < -7$$
$$x > -\tfrac{3}{2} \quad or \quad x < -7.$$

Reversing the symbol

Keep the word "or."

The solution set is $\{x \mid x < -7 \text{ or } x > -\frac{3}{2}\}$, or, in interval notation, $(-\infty, -7) \cup (-\frac{3}{2}, \infty)$.

Do Exercise 13.

EXAMPLE 9 Solve: $3x - 11 < 4$ *or* $4x + 9 \geq 1$.

We solve the individual inequalities separately, retaining the word *or*:

$$3x - 11 < 4 \quad or \quad 4x + 9 \geq 1$$
$$3x < 15 \quad or \quad 4x \geq -8$$
$$x < 5 \quad or \quad x \geq -2.$$

To find the solution set, we first look at the individual graphs.

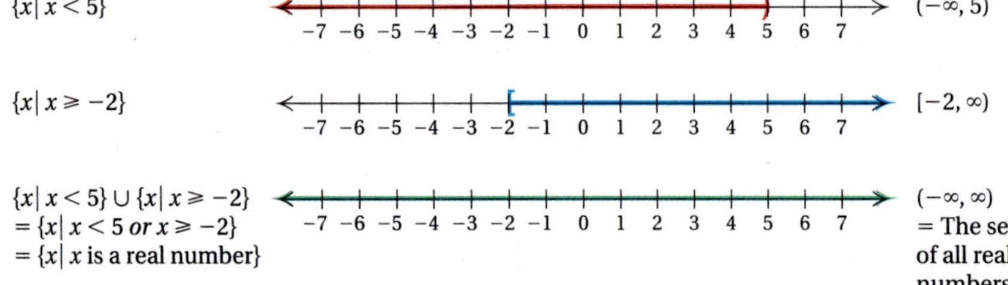

Since any number is either less than 5 or greater than or equal to -2, the two sets fill the entire number line. Thus the solution set is the set of all real numbers.

Do Exercise 14.

C Applications and Problem Solving

EXAMPLE 10 *Converting Dress Sizes.* The equation

$$I = 2(s + 10)$$

can be used to convert dress sizes s in the United States to dress sizes I in Italy. Which dress sizes in the United States correspond to dress sizes between 32 and 46 in Italy?

1. **Familiarize.** We have a formula for converting the dress sizes. Thus we can substitute a value into the formula. For a dress of size 6 in the United States, we get the corresponding dress size in Italy as follows:

   $$I = 2(6 + 10) = 2 \cdot 16 = 32.$$

   This familiarizes us with the formula and also tells us that the United States sizes that we are looking for must be larger than size 6.

2. **Translate.** We want the Italian sizes *between* 32 and 46, so we want to find those values of s for which

   $32 < I < 46$ I is between 32 and 46

   or

   $32 < 2(s + 10) < 46.$ Substituting $2(s + 10)$ for I

   Thus we have translated the problem to an inequality.

3. **Solve.** We solve the inequality:

   $32 < 2(s + 10) < 46$

   $\frac{32}{2} < \frac{2(s + 10)}{2} < \frac{46}{2}$ Dividing by 2

   $16 < s + 10 < 23$

   $6 < s < 13.$ Subtracting 10

4. **Check.** We substitute some values as we did in the *Familiarize* step.

5. **State.** Dress sizes between 6 and 13 in the United States correspond to dress sizes between 32 and 46 in Italy.

Do Exercise 15.

15. **Converting Dress Sizes.** Refer to Example 10. Which dress sizes in the United States correspond to dress sizes between 36 and 58 in Italy?

Answer on page A-51

14.2 EXERCISE SET

For Extra Help

Digital Video Tutor CD 12 Videotape 16 | InterAct Math | Math Tutor Center | MathXL | MyMathLab

a, b Find the intersection or union.

1. $\{9, 10, 11\} \cap \{9, 11, 13\}$

2. $\{1, 5, 10, 15\} \cap \{5, 15, 20\}$

3. $\{a, b, c, d\} \cap \{b, f, g\}$

4. $\{m, n, o, p\} \cap \{m, o, p\}$

5. $\{9, 10, 11\} \cup \{9, 11, 13\}$

6. $\{1, 5, 10, 15\} \cup \{5, 15, 20\}$

7. $\{a, b, c, d\} \cup \{b, f, g\}$

8. $\{m, n, o, p\} \cup \{m, o, p\}$

9. $\{2, 5, 7, 9\} \cap \{1, 3, 4\}$

10. $\{a, e, i, o, u\} \cap \{m, q, w, s, t\}$

11. $\{3, 5, 7\} \cup \varnothing$

12. $\{3, 5, 7\} \cap \varnothing$

a Graph and write interval notation.

13. $-4 < a$ *and* $a \le 1$

−6 −5 −4 −3 −2 −1 0 1 2 3 4 5 6

14. $-\frac{5}{2} \le m$ *and* $m < \frac{3}{2}$

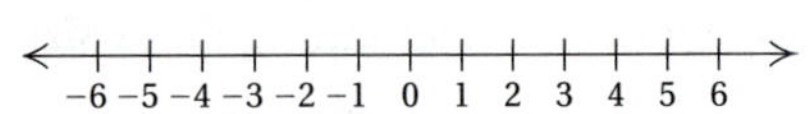

15. $1 < x < 6$

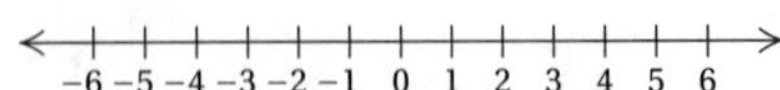

16. $-3 \le y \le 4$

−6 −5 −4 −3 −2 −1 0 1 2 3 4 5 6

Solve and graph.

17. $-10 \le 3x + 2$ *and* $3x + 2 < 17$

−6 −5 −4 −3 −2 −1 0 1 2 3 4 5 6

18. $-11 < 4x - 3$ *and* $4x - 3 \le 13$

−6 −5 −4 −3 −2 −1 0 1 2 3 4 5 6

19. $3x + 7 \ge 4$ *and* $2x - 5 \ge -1$

−6 −5 −4 −3 −2 −1 0 1 2 3 4 5 6

20. $4x - 7 < 1$ *and* $7 - 3x > -8$

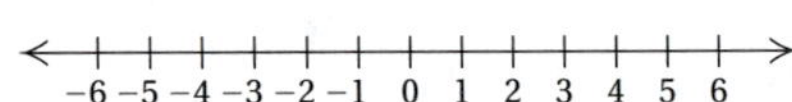

21. $4 - 3x \ge 10$ *and* $5x - 2 > 13$

−6 −5 −4 −3 −2 −1 0 1 2 3 4 5 6

22. $5 - 7x > 19$ *and* $2 - 3x < -4$

−6 −5 −4 −3 −2 −1 0 1 2 3 4 5 6

Solve.

23. $-4 < x + 4 < 10$

24. $-6 < x + 6 \le 8$

25. $6 > -x \ge -2$

26. $3 > -x \ge -5$

27. $1 < 3y + 4 \leq 19$

28. $5 \leq 8x + 5 \leq 21$

29. $-10 \leq 3x - 5 \leq -1$

30. $-18 \leq -2x - 7 < 0$

31. $2 < x + 3 \leq 9$

32. $-6 \leq x + 1 < 9$

33. $-6 \leq 2x - 3 < 6$

34. $4 > -3m - 7 \geq 2$

35. $-\frac{1}{2} < \frac{1}{4}x - 3 \leq \frac{1}{2}$

36. $-\frac{2}{3} \leq 4 - \frac{1}{4}x < \frac{2}{3}$

37. $-3 < \frac{2x - 5}{4} < 8$

38. $-4 \leq \frac{7 - 3x}{5} \leq 4$

b Graph and write interval notation.

39. $x < -2$ *or* $x > 1$

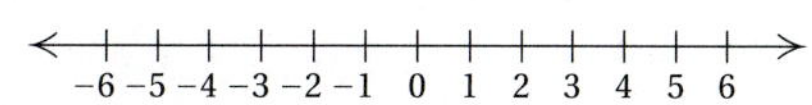

40. $x < -4$ *or* $x > 0$

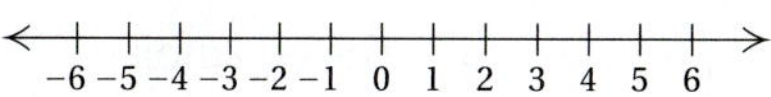

41. $x \leq -3$ *or* $x > 1$

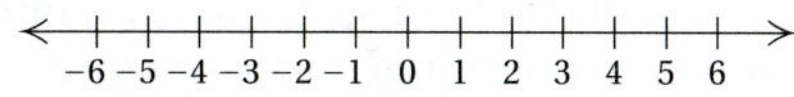

42. $x \leq -1$ *or* $x > 3$

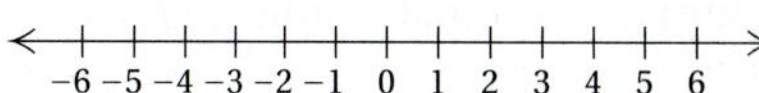

Solve and graph.

43. $x + 3 < -2$ *or* $x + 3 > 2$

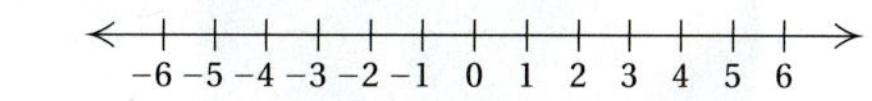

44. $x - 2 < -1$ *or* $x - 2 > 3$

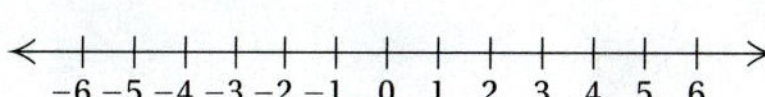

45. $2x - 8 \leq -3$ *or* $x - 1 \geq 3$

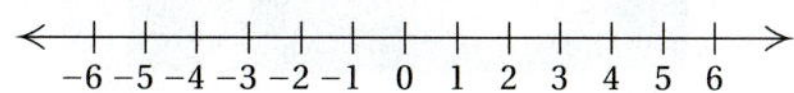

46. $x - 5 \leq -4$ *or* $2x - 7 \geq 3$

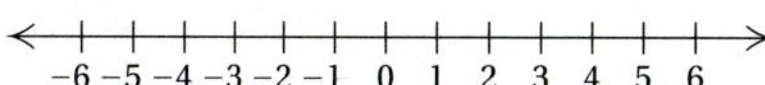

47. $7x + 4 \geq -17$ *or* $6x + 5 \geq -7$

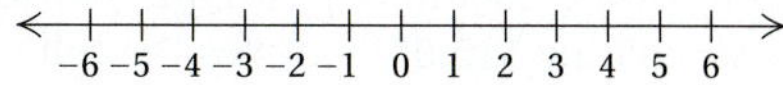

48. $4x - 4 < -8$ *or* $4x - 4 < 12$

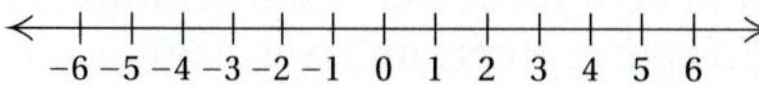

Solve.

49. $7 > -4x + 5$ *or* $10 \leq -4x + 5$

50. $6 > 2x - 1$ *or* $-4 \leq 2x - 1$

51. $3x - 7 > -10$ *or* $5x + 2 \leq 22$

52. $3x + 2 < 2$ *or* $4 - 2x < 14$

53. $-2x - 2 < -6$ *or* $-2x - 2 > 6$

54. $-3m - 7 < -5$ *or* $-3m - 7 > 5$

55. $\frac{2}{3}x - 14 < -\frac{5}{6}$ *or* $\frac{2}{3}x - 14 > \frac{5}{6}$

56. $\frac{1}{4} - 3x \leq -3.7$ *or* $\frac{1}{4} - 5x \geq 4.8$

57. $\frac{2x - 5}{6} \leq -3$ *or* $\frac{2x - 5}{6} \geq 4$

58. $\frac{7 - 3x}{5} < -4$ *or* $\frac{7 - 3x}{5} > 4$

C Solve.

59. *Pressure at Sea Depth.* The equation

$$P = 1 + \frac{d}{33}$$

gives the pressure P, in atmospheres (atm), at a depth of d feet in the sea. For what depths d is the pressure at least 1 atm and at most 7 atm?

60. *Temperatures of Liquids.* The formula

$$C = \tfrac{5}{9}(F - 32)$$

can be used to convert Fahrenheit temperatures F to Celsius temperatures C.

a) Gold is a liquid for Celsius temperatures C such that $1063° \leq C < 2660°$. Find such an inequality for the corresponding Fahrenheit temperatures.

b) Silver is a liquid for Celsius temperatures C such that $960.8° \leq C < 2180°$. Find such an inequality for the corresponding Fahrenheit temperatures.

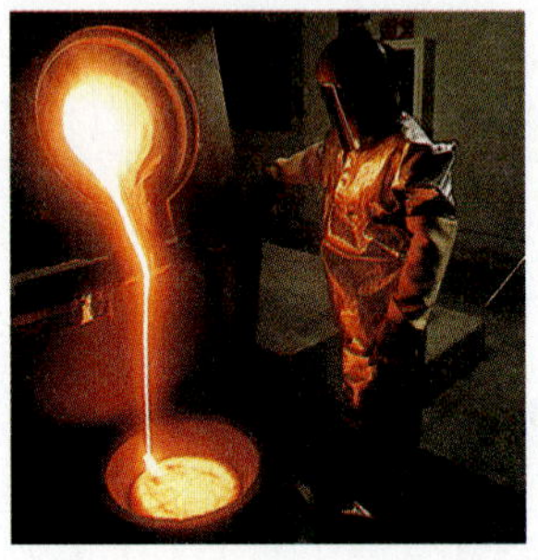

61. *Aerobic Exercise.* In order to achieve maximum results from aerobic exercise, one should maintain one's heart rate at a certain level. A 30-yr-old woman with a resting heart rate of 60 beats per minute should keep her heart rate between 138 and 162 beats per minute while exercising. She checks her pulse for 10 sec while exercising. What should the number of beats be?

62. *Minimizing Tolls.* A \$3.00 toll is charged to cross the bridge from Sanibel Island to mainland Florida. A six-month pass, costing \$15.00, reduces the toll to \$0.50. A one-year pass, costing \$150, allows for free crossings. How many crossings per year does it take, on average, for the two six-month passes to be the most economical choice? Assume a constant number of trips per month.

63. *Online Shoppers.* The equation

$$y = 14.57x + 62.91$$

can be used to predict the number y of online shoppers, in millions, x years after 2000. For what years will the number of online shoppers be between 106,620,000 and 194,040,000?
Source: eMarketer

64. *Young's Rule in Medicine.* Young's rule for determining the amount of a medicine dosage for a child is given by

$$c = \frac{ad}{a + 12},$$

where a is the child's age and d is the usual adult dosage, in milligrams. (*Warning*! Do not apply this formula without checking with a physician!) An 8-yr-old child needs medication. What adult dosage can be used if a child's dosage must stay between 100 mg and 200 mg?
Source: Olsen, June L., et al., *Medical Dosage Calculations*, 6th ed. Reading, MA: Addison Wesley Longman, p. A-31

65. D_W Explain why the conjunction $3 < x$ *and* $x < 5$ is equivalent to $3 < x < 5$, but the disjunction $3 < x$ *or* $x < 5$ is not.

66. D_W Describe the circumstances under which, for intervals, $[a, b] \cup [c, d] = [a, d]$.

SKILL MAINTENANCE

Solve. [13.2a], [13.3a]

67. $3x - 2y = -7,$
$2x + 5y = 8$

68. $4x - 7y = 23,$
$x + 6y = -33$

69. $x + y = 0,$
$x - y = 8$

Find an equation of the line containing the given pair of points. [12.5c]

70. $(2, 7), (3, -4)$

71. $(0, 7), (2, -1)$

72. $(4, -2), (-2, 4)$

Multiply. [9.6a]

73. $(2a - b)(3a + 5b)$

74. $(5y + 6)(5y + 1)$

75. $(7x - 8)(3x - 5)$

76. $(13x - 2y)(x + 3y)$

SYNTHESIS

Solve.

77. $-\frac{2}{15} \le \frac{2}{3}x - \frac{2}{5} \le \frac{2}{15}$

78. $4m - 8 > 6m + 5$ *or* $5m - 8 < -2$

79. $3x < 4 - 5x < 5 + 3x$

80. $2[5(3 - y) - 2(y - 2)] > y + 4$

81. $x + 4 < 2x - 6 \le x + 12$

82. $2x - \frac{3}{4} < -\frac{1}{10}$ *or* $2x - \frac{3}{4} > \frac{1}{10}$

Determine whether each sentence is true or false for all real numbers a, b, and c.

83. If $-b < -a$, then $a < b$.

84. If $a \le c$ and $c \le b$, then $b \ge a$.

85. If $a < c$ and $b < c$, then $a < b$.

86. If $-a < c$ and $-c > b$, then $a > b$.

87. What is the union of the set of all rational numbers with the set of all irrational numbers? the intersection?

14.3 ABSOLUTE-VALUE EQUATIONS AND INEQUALITIES

Objectives

- **a** Simplify expressions containing absolute-value symbols.
- **b** Find the distance between two points on a number line.
- **c** Solve equations with absolute-value expressions.
- **d** Solve equations with two absolute-value expressions.
- **e** Solve inequalities with absolute-value expressions.

Simplify, leaving as little as possible inside the absolute-value signs.

1. $|7x|$

2. $|x^8|$

3. $|5a^2b|$

4. $\left|\frac{7a}{b^2}\right|$

5. $|-9x|$

Answers on page A-52

a Properties of Absolute Value

We can think of the **absolute value** of a number as its distance from zero on the number line. Recall the formal definition from Section 6.2.

ABSOLUTE VALUE

The **absolute value** of x, denoted $|x|$, is defined as follows:

$x \geq 0 \longrightarrow |x| = x; \qquad x < 0 \longrightarrow |x| = -x.$

Some simple properties of absolute value allow us to manipulate or simplify algebraic expressions.

PROPERTIES OF ABSOLUTE VALUE

a) $|ab| = |a| \cdot |b|$, for any real numbers a and b.
(The absolute value of a product is the product of the absolute values.)

b) $\left|\frac{a}{b}\right| = \frac{|a|}{|b|}$, for any real numbers a and $b \neq 0$.
(The absolute value of a quotient is the quotient of the absolute values.)

c) $|-a| = |a|$, for any real number a.
(The absolute value of the opposite of a number is the same as the absolute value of the number.)

EXAMPLES Simplify, leaving as little as possible inside the absolute-value signs.

1. $|5x| = |5| \cdot |x| = 5|x|$
2. $|-3y| = |-3| \cdot |y| = 3|y|$
3. $|7x^2| = |7| \cdot |x^2| = 7|x^2| = 7x^2$ Since x^2 is never negative for any number x
4. $\left|\frac{6x}{-3x^2}\right| = \left|\frac{2}{-x}\right| = \frac{|2|}{|-x|} = \frac{2}{|x|}$

Do Exercises 1–5.

b Distance on a Number Line

The number line below shows that the distance between -3 and 2 is 5.

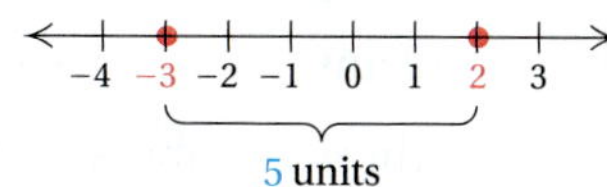

Another way to find the distance between two numbers on a number line is to take the absolute value of the difference, as follows:

$$|-3 - 2| = |-5| = 5, \quad \text{or} \quad |2 - (-3)| = |5| = 5.$$

Note that the order in which we subtract does not matter because we are taking the absolute value after we have subtracted.

DISTANCE AND ABSOLUTE VALUE

For any real numbers a and b, the **distance** between them is $|a - b|$.

We should note that the distance is also $|b - a|$, because $a - b$ and $b - a$ are opposites and hence have the same absolute value.

EXAMPLE 5 Find the distance between -8 and -92 on a number line.

$$|-8 - (-92)| = |84| = 84, \quad \text{or} \quad |-92 - (-8)| = |-84| = 84$$

EXAMPLE 6 Find the distance between x and 0 on the number line.

$$|x - 0| = |x|$$

Do Exercises 6–8.

c Equations with Absolute Value

EXAMPLE 7 Solve: $|x| = 4$. Then graph using the number line.

Note that $|x| = |x - 0|$, so that $|x - 0|$ is the distance from x to 0. Thus solutions of the equation $|x| = 4$, or $|x - 0| = 4$, are those numbers x whose distance from 0 is 4. Those numbers are -4 and 4. The solution set is $\{-4, 4\}$. The graph consists of just two points, as shown.

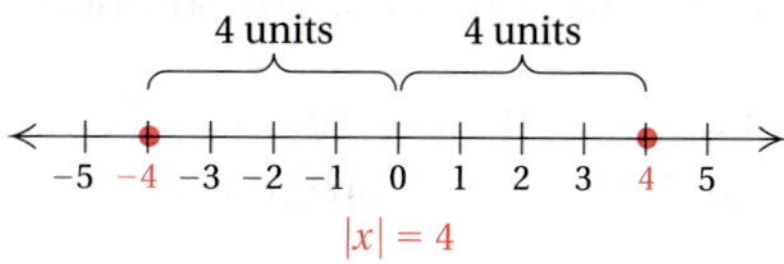

EXAMPLE 8 Solve: $|x| = 0$.

The only number whose absolute value is 0 is 0 itself. Thus the solution is 0. The solution set is $\{0\}$.

EXAMPLE 9 Solve: $|x| = -7$.

The absolute value of a number is always nonnegative. There is no number whose absolute value is -7. Thus there is no solution. The solution set is $\varnothing$.

Find the distance between the points.

6. $-6, -35$

7. 19, 14

8. $0, p$

9. Solve: $|x| = 6$. Then graph using a number line.

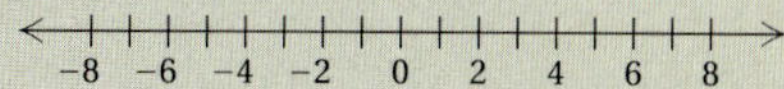

10. Solve: $|x| = -6$.

11. Solve: $|p| = 0$.

Answers on page A-52

Solve.

12. $|3x| = 6$

13. $4|x| + 10 = 27$

14. $3|x| - 2 = 10$

15. Solve: $|x - 4| = 1$. Use two methods as in Example 11.

Answers on page A-52

Examples 7–9 lead us to the following principle for solving linear equations with absolute value.

THE ABSOLUTE-VALUE PRINCIPLE

For any positive number p and any algebraic expression X:

a) The solution of $|X| = p$ are those numbers that satisfy $X = -p$ *or* $X = p$.

b) The equation $|X| = 0$ is equivalent to the equation $X = 0$.

c) The equation $|X| = -p$ has no solution.

Do Exercises 9–11 on the preceding page.

We can use the absolute-value principle with the addition and multiplication principles to solve equations with absolute value.

EXAMPLE 10 Solve: $2|x| + 5 = 9$.

We first use the addition and multiplication principles to get $|x|$ by itself. Then we use the absolute-value principle.

$$2|x| + 5 = 9$$

$$2|x| = 4 \qquad \text{Subtracting 5}$$

$$|x| = 2 \qquad \text{Dividing by 2}$$

$$x = -2 \quad or \quad x = 2 \qquad \text{Using the absolute-value principle}$$

The solutions are -2 and 2. The solution set is $\{-2, 2\}$.

Do Exercises 12–14.

EXAMPLE 11 Solve: $|x - 2| = 3$.

We can consider solving this equation in two different ways.

METHOD 1 This allows us to see the meaning of the solutions graphically. The solution set consists of those numbers that are 3 units from 2 on the number line.

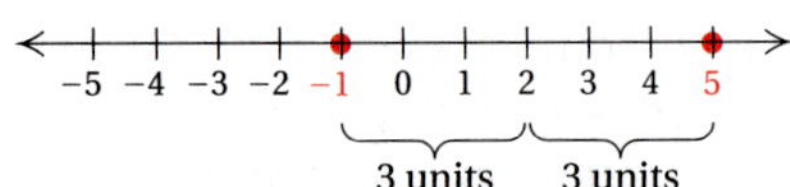

The solutions of $|x - 2| = 3$ are -1 and 5. The solution set is $\{-1, 5\}$.

METHOD 2 This method is more efficient. We use the absolute-value principle, replacing X with $x - 2$ and p with 3. Then we solve each equation separately.

$$|X| = p$$

$$|x - 2| = 3$$

$$x - 2 = -3 \quad or \quad x - 2 = 3 \qquad \text{Absolute-value principle}$$

$$x = -1 \quad or \quad x = 5$$

The solutions are -1 and 5. The solution set is $\{-1, 5\}$.

Do Exercise 15.

EXAMPLE 12 Solve: $|2x + 5| = 13$.

We use the absolute-value principle, replacing X with $2x + 5$ and p with 13:

$$\begin{aligned} |X| &= p \\ |2x + 5| &= 13 \\ 2x + 5 = -13 \quad &or \quad 2x + 5 = 13 \qquad \text{Absolute-value principle} \\ 2x = -18 \quad &or \quad 2x = 8 \\ x = -9 \quad &or \quad x = 4. \end{aligned}$$

The solutions are -9 and 4. The solution set is $\{-9, 4\}$.

Do Exercise 16.

EXAMPLE 13 Solve: $|4 - 7x| = -8$.

Since absolute value is always nonnegative, this equation has no solution. The solution set is $\varnothing$.

Do Exercise 17.

d Equations with Two Absolute-Value Expressions

Sometimes equations have two absolute-value expressions. Consider $|a| = |b|$. This means that a and b are the same distance from 0. If a and b are the same distance from 0, then either they are the same number or they are opposites of each other.

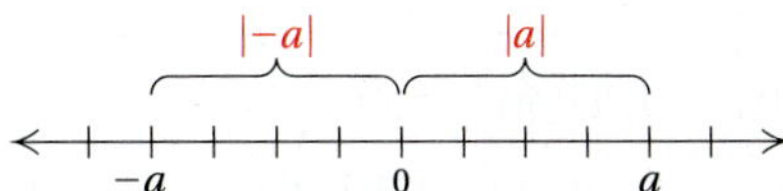

EXAMPLE 14 Solve: $|2x - 3| = |x + 5|$.

Either $2x - 3 = x + 5$ *or* $2x - 3 = -(x + 5)$. We solve each equation:

$$\begin{aligned} 2x - 3 = x + 5 \quad &or \quad 2x - 3 = -(x + 5) \\ x - 3 = 5 \quad &or \quad 2x - 3 = -x - 5 \\ x = 8 \quad &or \quad 3x - 3 = -5 \\ x = 8 \quad &or \quad 3x = -2 \\ x = 8 \quad &or \quad x = -\tfrac{2}{3}. \end{aligned}$$

The solutions are 8 and $-\frac{2}{3}$. The solution set is $\{8, -\frac{2}{3}\}$.

EXAMPLE 15 Solve: $|x + 8| = |x - 5|$.

$$\begin{aligned} x + 8 = x - 5 \quad &or \quad x + 8 = -(x - 5) \\ 8 = -5 \quad &or \quad x + 8 = -x + 5 \\ 8 = -5 \quad &or \quad 2x = -3 \\ 8 = -5 \quad &or \quad x = -\tfrac{3}{2} \end{aligned}$$

The first equation has no solution. The second equation has $-\frac{3}{2}$ as a solution. The solution set is $\{-\frac{3}{2}\}$.

Do Exercises 18 and 19.

16. Solve: $|3x - 4| = 17$.

17. Solve: $|6 + 2x| = -3$.

Solve.

18. $|5x - 3| = |x + 4|$

19. $|x - 3| = |x + 10|$

Answers on page A-52

20. Solve: $|x| = 5$. Then graph using a number line.

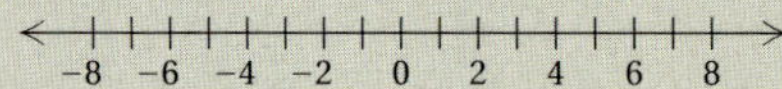

21. Solve: $|x| < 5$. Then graph.

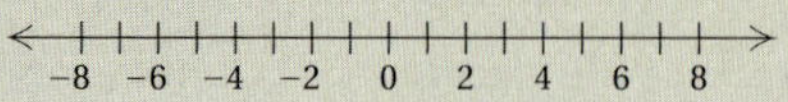

22. Solve: $|x| \geq 5$. Then graph.

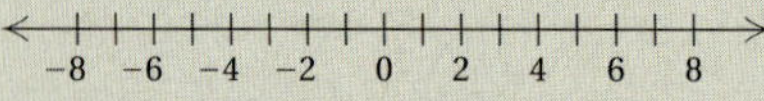

Answers on page A-52

e Inequalities with Absolute Value

We can extend our methods for solving equations with absolute value to those for solving inequalities with absolute value.

EXAMPLE 16 Solve: $|x| = 4$. Then graph using a number line.

From Example 7, we know that the solutions are -4 and 4. The solution set is $\{-4, 4\}$. The graph consists of just two points, as shown here.

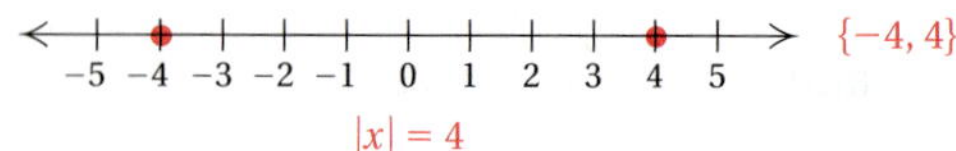

Do Exercise 20.

EXAMPLE 17 Solve: $|x| < 4$. Then graph.

The solutions of $|x| < 4$ are the solutions of $|x - 0| < 4$ and are those numbers x whose distance from 0 is less than 4. We can check by substituting or by looking at the number line that numbers like $-3, -2, -1, -\frac{1}{2}, -\frac{1}{4}, 0, \frac{1}{4}, \frac{1}{2}$, 1, 2, and 3 are all solutions. In fact, the solutions are all the real numbers x between -4 and 4. The solution set is $\{x \mid -4 < x < 4\}$ or, in interval notation, $(-4, 4)$. The graph is as follows.

Do Exercise 21.

EXAMPLE 18 Solve: $|x| \geq 4$. Then graph.

The solutions of $|x| \geq 4$ are solutions of $|x - 0| \geq 4$ and are those numbers whose distance from 0 is greater than or equal to 4—in other words, those numbers x such that $x \leq -4$ *or* $x \geq 4$. The solution set is $\{x \mid x \leq -4 \text{ or } x \geq 4\}$, or $(-\infty, -4] \cup [4, \infty)$. The graph is as follows.

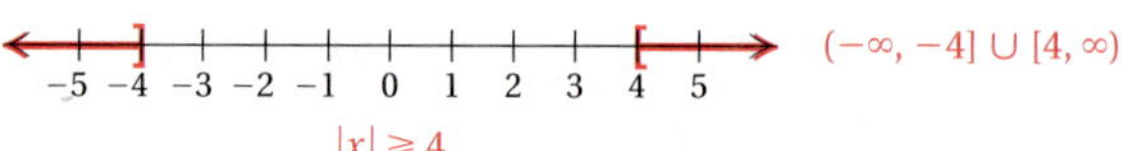

Do Exercise 22.

Examples 16–18 illustrate three cases of solving equations and inequalities with absolute value. The expression inside the absolute-value signs can be something besides a single variable. The following is a general principle for solving.

SOLUTIONS OF ABSOLUTE-VALUE EQUATIONS AND INEQUALITIES

For any positive number p and any algebraic expression X:

a) The solutions of $|X| = p$ are those numbers that satisfy $X = -p$ *or* $X = p$.

As an example, replacing X with $5x - 1$ and p with 8, we see that the solutions of $|5x - 1| = 8$ are those numbers x for which

$$5x - 1 = -8 \quad or \quad 5x - 1 = 8$$
$$5x = -7 \quad or \quad 5x = 9$$
$$x = -\tfrac{7}{5} \quad or \quad x = \tfrac{9}{5}.$$

The solution set is $\left\{-\frac{7}{5}, \frac{9}{5}\right\}$.

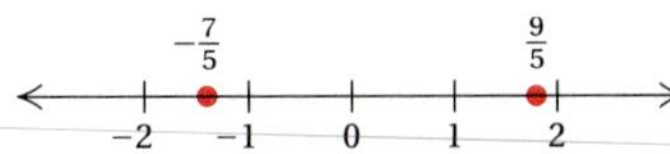

b) The solutions of $|X| < p$ are those numbers that satisfy $-p < X < p$.

As an example, replacing X with $6x + 7$ and p with 5, we see that the solutions of $|6x + 7| < 5$ are those numbers x for which

$$-5 < 6x + 7 < 5$$
$$-12 < 6x < -2$$
$$-2 < x < -\tfrac{1}{3}.$$

The solution set is $\left\{x \mid -2 < x < -\frac{1}{3}\right\}$, or $\left(-2, -\frac{1}{3}\right)$.

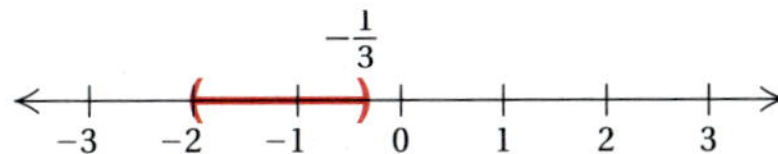

c) The solutions of $|X| > p$ are those numbers that satisfy $X < -p$ *or* $X > p$.

As an example, replacing X with $2x - 9$ and p with 4, we see that the solutions of $|2x - 9| > 4$ are those numbers x for which

$$2x - 9 < -4 \quad or \quad 2x - 9 > 4$$
$$2x < 5 \quad or \quad 2x > 13$$
$$x < \tfrac{5}{2} \quad or \quad x > \tfrac{13}{2}.$$

The solution set is $\left\{x \mid x < \frac{5}{2} \text{ or } x > \frac{13}{2}\right\}$, or $\left(-\infty, \frac{5}{2}\right) \cup \left(\frac{13}{2}, \infty\right)$.

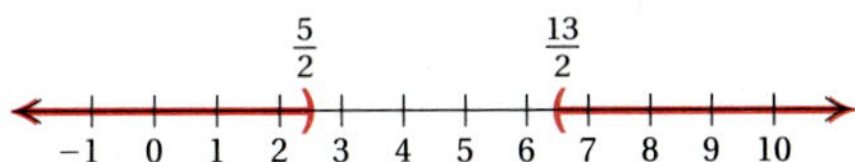

23. Solve: $|2x - 3| < 7$. Then graph.

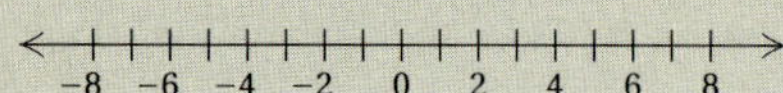

24. Solve: $|7 - 3x| \leq 4$.

25. Solve: $|3x + 2| \geq 5$. Then graph.

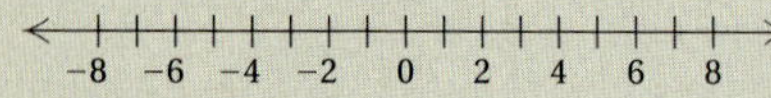

Answers on page A-52

EXAMPLE 19 Solve: $|3x - 2| < 4$. Then graph.

We use part (b). In this case, X is $3x - 2$ and p is 4:

$$|X| < p$$

$$|3x - 2| < 4 \qquad \text{Replacing } X \text{ with } 3x - 2 \text{ and } p \text{ with } 4$$

$$-4 < 3x - 2 < 4$$

$$-2 < 3x < 6$$

$$-\frac{2}{3} < x < 2.$$

The solution set is $\{x \mid -\frac{2}{3} < x < 2\}$, or $(-\frac{2}{3}, 2)$. The graph is as follows.

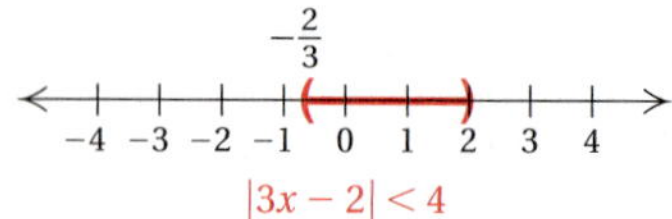

EXAMPLE 20 Solve: $|8 - 4x| \leq 5$.

We use part (b). In this case, X is $8 - 4x$ and p is 5:

$$|X| \leq p$$

$$|8 - 4x| \leq 5 \qquad \text{Replacing } X \text{ with } 8 - 4x \text{ and } p \text{ with } 5$$

$$-5 \leq 8 - 4x \leq 5$$

$$-13 \leq -4x \leq -3$$

$$\frac{13}{4} \geq x \geq \frac{3}{4}. \qquad \text{Dividing by } -4 \text{ and reversing the inequality symbols}$$

The solution set is $\{x \mid \frac{13}{4} \geq x \geq \frac{3}{4}\}$, or $\{x \mid \frac{3}{4} \leq x \leq \frac{13}{4}\}$, or $[\frac{3}{4}, \frac{13}{4}]$.

EXAMPLE 21 Solve: $|4x + 2| \geq 6$.

We use part (c). In this case, X is $4x + 2$ and p is 6:

$$|X| \geq p$$

$$|4x + 2| \geq 6 \qquad \text{Replacing } X \text{ with } 4x + 2 \text{ and } p \text{ with } 6$$

$$4x + 2 \leq -6 \quad \textit{or} \quad 4x + 2 \geq 6$$

$$4x \leq -8 \quad \textit{or} \quad 4x \geq 4$$

$$x \leq -2 \quad \textit{or} \quad x \geq 1.$$

The solution set is $\{x \mid x \leq -2 \textit{ or } x \geq 1\}$, or $(-\infty, -2] \cup [1, \infty)$.

Do Exercises 23–25.

14.3 EXERCISE SET

For Extra Help

Digital Video Tutor CD 12 Videotape 16 | InterAct Math | Math Tutor Center | MathXL | MyMathLab

a Simplify, leaving as little as possible inside absolute-value signs.

1. $|9x|$

2. $|26x|$

3. $|2x^2|$

4. $|8x^2|$

5. $|-2x^2|$

6. $|-20x^2|$

7. $|-6y|$

8. $|-17y|$

9. $\left|\frac{-2}{x}\right|$

10. $\left|\frac{y}{3}\right|$

11. $\left|\frac{x^2}{-y}\right|$

12. $\left|\frac{x^4}{-y}\right|$

13. $\left|\frac{-8x^2}{2x}\right|$

14. $\left|\frac{9y}{3y^2}\right|$

b Find the distance between the points on a number line.

15. $-8,\ -46$

16. $-7,\ -32$

17. $36,\ 17$

18. $52,\ 18$

19. $-3.9,\ 2.4$

20. $-1.8,\ -3.7$

21. $-5,\ 0$

22. $\frac{2}{3},\ -\frac{5}{6}$

c Solve.

23. $|x| = 3$

24. $|x| = 5$

25. $|x| = -3$

26. $|x| = -9$

27. $|q| = 0$

28. $|y| = 7.4$

29. $|x - 3| = 12$

30. $|3x - 2| = 6$

31. $|2x - 3| = 4$

32. $|5x + 2| = 3$

33. $|4x - 9| = 14$

34. $|9y - 2| = 17$

35. $|x| + 7 = 18$

36. $|x| - 2 = 6.3$

37. $574 = 283 + |t|$

38. $-562 = -2000 + |x|$

39. $|5x| = 40$

40. $|2y| = 18$

41. $|3x| - 4 = 17$

42. $|6x| + 8 = 32$

43. $7|w| - 3 = 11$

44. $5|x| + 10 = 26$

45. $\left|\frac{2x - 1}{3}\right| = 5$

46. $\left|\frac{4 - 5x}{6}\right| = 7$

47. $|m + 5| + 9 = 16$

48. $|t - 7| - 5 = 4$

49. $10 - |2x - 1| = 4$

50. $2|2x - 7| + 11 = 25$

51. $|3x - 4| = -2$

52. $|x - 6| = -8$

53. $|\frac{5}{9} + 3x| = \frac{1}{6}$

54. $|\frac{2}{3} - 4x| = \frac{4}{5}$

d Solve.

55. $|3x + 4| = |x - 7|$

56. $|2x - 8| = |x + 3|$

57. $|x + 3| = |x - 6|$

58. $|x - 15| = |x + 8|$

59. $|2a + 4| = |3a - 1|$

60. $|5p + 7| = |4p + 3|$

61. $|y - 3| = |3 - y|$

62. $|m - 7| = |7 - m|$

63. $|5 - p| = |p + 8|$

64. $|8 - q| = |q + 19|$

65. $\left|\frac{2x - 3}{6}\right| = \left|\frac{4 - 5x}{8}\right|$

66. $\left|\frac{6 - 8x}{5}\right| = \left|\frac{7 + 3x}{2}\right|$

67. $|\frac{1}{2}x - 5| = |\frac{1}{4}x + 3|$

68. $|2 - \frac{2}{3}x| = |4 + \frac{7}{8}x|$

e Solve.

69. $|x| < 3$

70. $|x| \leq 5$

71. $|x| \geq 2$

72. $|y| > 12$

73. $|x - 1| < 1$

74. $|x + 4| \leq 9$

75. $5|x + 4| \leq 10$

76. $2|x - 2| > 6$

77. $|2x - 3| \leq 4$

78. $|5x + 2| \leq 3$

79. $|2y - 7| > 10$

80. $|3y - 4| > 8$

81. $|4x - 9| \geq 14$

82. $|9y - 2| \geq 17$

83. $|y - 3| < 12$

84. $|p - 2| < 6$

85. $|2x + 3| \leq 4$

86. $|5x + 2| \leq 13$

87. $|4 - 3y| > 8$

88. $|7 - 2y| > 5$

89. $|9 - 4x| \geq 14$

90. $|2 - 9p| \geq 17$

91. $|3 - 4x| < 21$

92. $|-5 - 7x| \leq 30$

93. $\left|\frac{1}{2} + 3x\right| \geq 12$

94. $\left|\frac{1}{4}y - 6\right| > 24$

95. $\left|\frac{x - 7}{3}\right| < 4$

96. $\left|\frac{x + 5}{4}\right| \leq 2$

97. $\left|\frac{2 - 5x}{4}\right| \geq \frac{2}{3}$

98. $\left|\frac{1 + 3x}{5}\right| > \frac{7}{8}$

99. $|m + 5| + 9 \leq 16$

100. $|t - 7| + 3 \geq 4$

101. $7 - |3 - 2x| \geq 5$

102. $16 \leq |2x - 3| + 9$

103. $\left|\frac{2x - 1}{0.0059}\right| \leq 1$

104. $\left|\frac{3x - 2}{5}\right| \geq 1$

105. **DW** Explain in your own words why the solutions of the inequality $|x + 5| \le 2$ can be interpreted as "all those numbers x whose distance from -5 is at most 2 units."

106. **DW** Explain in your own words why the interval $[6, \infty)$ is only part of the solution set of $|x| \ge 6$.

SKILL MAINTENANCE

Find the domain. [12.2a]

107. $f(x) = |x - 2|$

108. $f(x) = \dfrac{2x - 4}{2x + 5}$

109. $f(x) = \dfrac{1}{x - 5}$

110. $f(x) = 10 - 2x$

Solve. [13.2a], [13.3a]

111. $2x + y = 7,$
$x - 3y = 21$

112. $10x + y = 16,$
$x - 10y = 42$

SYNTHESIS

113. *Motion of a Spring.* A weighted spring is bouncing up and down so that its distance d above the ground satisfies the inequality $|d - 6 \text{ ft}| \le \frac{1}{2}$ ft. Find all possible distances d.

114. *Container Sizes.* A container company is manufacturing rectangular boxes of various sizes. The length of any box must exceed the width by at least 3 in., but the perimeter cannot exceed 24 in. What widths are possible?

$$l \ge w + 3,$$
$$2l + 2w \le 24$$

Solve.

115. $|x + 5| = x + 5$

116. $1 - \left|\frac{1}{4}x + 8\right| = \frac{3}{4}$

117. $|7x - 2| = x + 4$

118. $|x - 1| = x - 1$

119. $|x - 6| \le -8$

120. $|3x - 4| > -2$

121. $|x + 5| > x$

122. $\left|\frac{5}{9} + 3x\right| < -\frac{1}{6}$

123. $|x| \ge 0$

124. $2 \le |x - 1| \le 5$

Find an equivalent inequality with absolute value.

125. $-3 < x < 3$

126. $-5 \le y \le 5$

127. $x \le -6$ *or* $x \ge 6$

128. $-5 < x < 1$

129. $x < -8$ *or* $x > 2$

14.4 SYSTEMS OF INEQUALITIES IN TWO VARIABLES

Objectives

a Determine whether an ordered pair of numbers is a solution of an inequality in two variables.

b Graph linear inequalities in two variables.

c Graph systems of linear inequalities and find coordinates of any vertices.

A **graph** of an inequality is a drawing that represents its solutions. An inequality in one variable can be graphed on the number line (see Section 14.1). An inequality in two variables can be graphed on a coordinate plane.

A **linear inequality** is one that we can get from a related linear equation by changing the equals symbol to an inequality symbol. The graph of a linear inequality is a region on one side of a line. This region is called a **half-plane.** The graph sometimes includes the graph of the related line at the boundary of the half-plane.

a Solutions of Inequalities in Two Variables

The solutions of an inequality in two variables are ordered pairs.

EXAMPLES Determine whether the ordered pair is a solution of the inequality $5x - 4y > 13$.

1. $(-3, 2)$

We have

$$\begin{array}{r|l} 5x - 4y & > 13 \\ \hline 5(-3) - 4 \cdot 2 & \overset{?}{} \ 13 \\ -15 - 8 & \\ -23 & \end{array}$$

We use alphabetical order to replace x with -3 and y with 2.

FALSE

Since $-23 > 13$ is false, $(-3, 2)$ is not a solution.

2. $(4, -3)$

We have

$$\begin{array}{r|l} 5x - 4y & > 13 \\ \hline 5(4) - 4(-3) & \overset{?}{} \ 13 \\ 20 + 12 & \\ 32 & \end{array}$$

Replacing x with 4 and y with -3

TRUE

Since $32 > 13$ is true, $(4, -3)$ is a solution.

Do Exercises 1 and 2.

1. Determine whether $(1, -4)$ is a solution of $4x - 5y < 12$.

$$\begin{array}{c|c} 4x - 5y & < 12 \\ \hline & ? \end{array}$$

2. Determine whether $(4, -3)$ is a solution of $3y - 2x \leq 6$.

$$\begin{array}{c|c} 3y - 2x & \leq 6 \\ \hline & ? \end{array}$$

Answers on page A-52

b Graphing Inequalities in Two Variables

Let's visualize the results of Examples 1 and 2. The equation $5x - 4y = 13$ is represented by the dashed line in the graphs below. The solutions of the inequality are shaded below that dashed line. The pair $(-3, 2)$ is not a solution of the inequality $5x - 4y > 13$ and is not in the shaded region. See the graph on the left at the top of the following page.

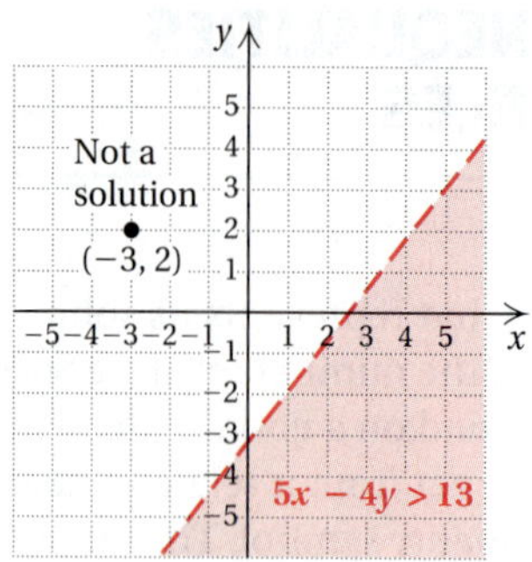

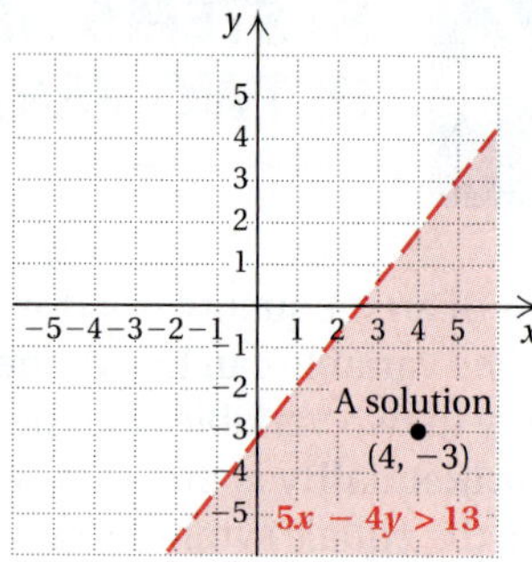

The pair $(4, -3)$ is a solution of the inequality $5x - 4y > 13$ and is in the shaded region. See the graph on the right above.

We now consider how to graph inequalities.

EXAMPLE 3 Graph: $y < x$.

We first graph the line $y = x$. Every solution of $y = x$ is an ordered pair like $(3, 3)$, where the first and second coordinates are the same. The graph of $y = x$ is shown on the left below. We draw it dashed because these points are *not* solutions of $y < x$.

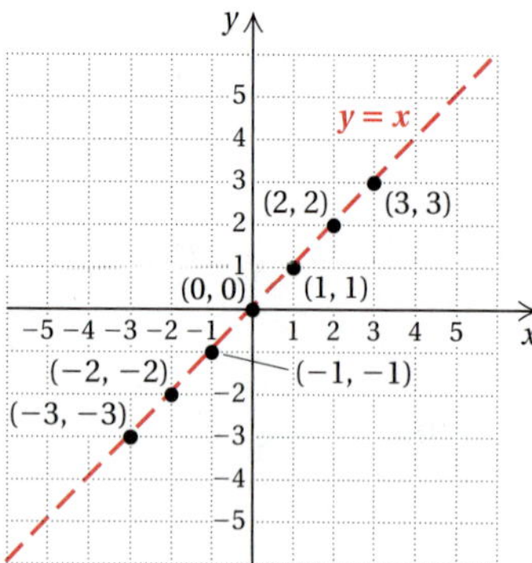

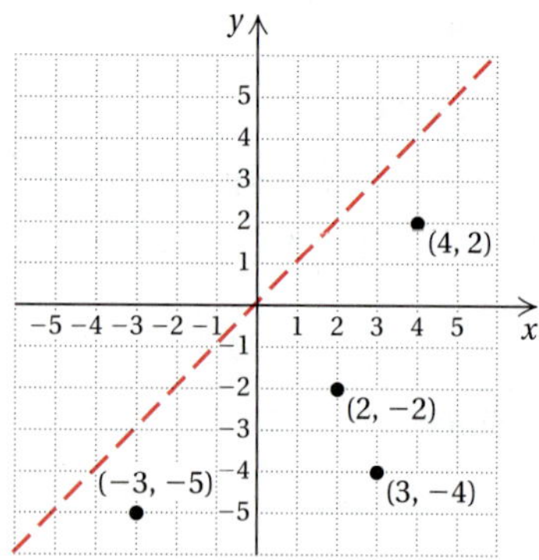

Now look at the graph on the right above. Several ordered pairs are plotted on the half-plane below $y = x$. Each is a solution of $y < x$. We can check the pair $(4, 2)$ as follows:

$$\begin{array}{c|c} y & < x \\ \hline 2 & ?\ 4 \end{array}\quad \text{TRUE}$$

It turns out that any point on the same side of $y = x$ as $(4, 2)$ is also a solution. Thus, *if you know that one point in a half-plane is a solution, then all points in that half-plane are solutions.* In this text, we will usually indicate this by color shading. We shade the half-plane below $y = x$.

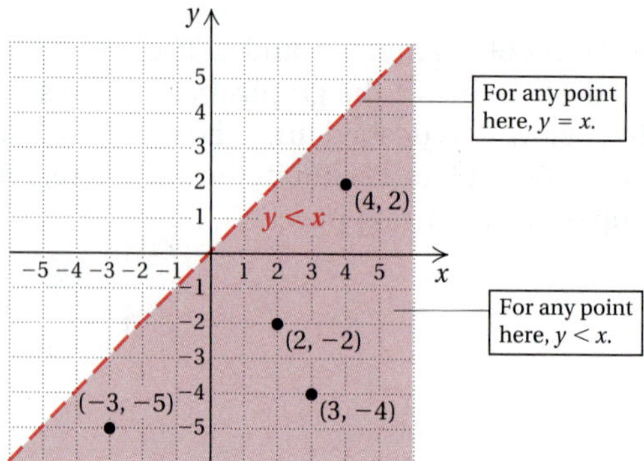

EXAMPLE 4 Graph: $8x + 3y \geq 24$.

First, we sketch the line $8x + 3y = 24$. Points on the line $8x + 3y = 24$ are also in the graph of $8x + 3y \geq 24$, so we draw the line solid. This indicates that all points on the line are solutions. The rest of the solutions are in the half-plane either to the left or to the right of the line. To determine which, we select a point that is not on the line and determine whether it is a solution of $8x + 3y \geq 24$. We try $(-3, 4)$ as a test point:

$$\begin{array}{r|l} 8x + 3y & \geq 24 \\ \hline 8(-3) + 3(4) & ?\ 24 \\ -24 + 12 & \\ -12 & \end{array} \quad \text{FALSE}$$

We see that $-12 \geq 24$ is *false*. Since $(-3, 4)$ is not a solution, none of the points in the half-plane containing $(-3, 4)$ is a solution. Thus the points in the opposite half-plane are solutions. We shade that half-plane and obtain the graph shown below.

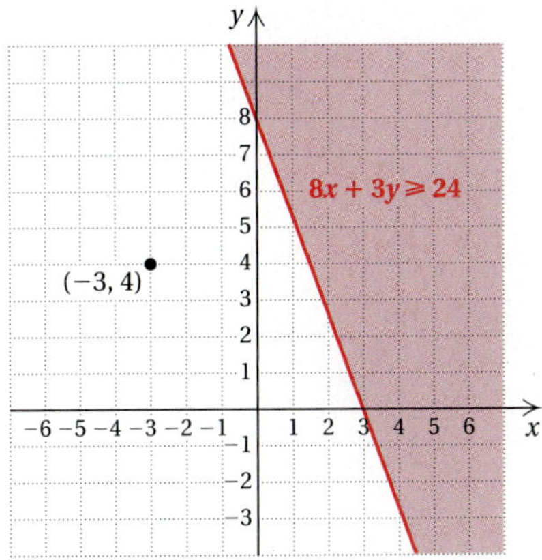

To graph an inequality in two variables:

1. Replace the inequality symbol with an equals sign and graph this related equation. This separates points that represent solutions from those that do not.
2. If the inequality symbol is $<$ or $>$, draw the line dashed. If the inequality symbol is $\leq$ or $\geq$, draw the line solid.
3. The graph consists of a half-plane that is either above or below or to the left or right of the line and, if the line is solid, the line as well. To determine which half-plane to shade, choose a point not on the line as a test point. Substitute to determine whether that point is a solution. If so, shade the half-plane containing that point. If not, shade the opposite half-plane.

Graph.

3. $6x - 3y < 18$

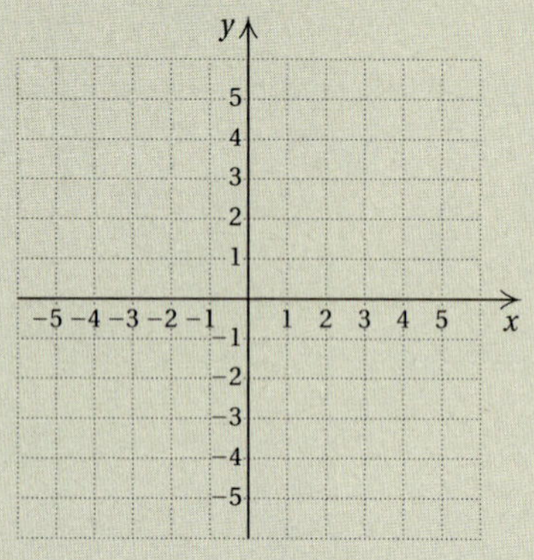

4. $4x + 3y \geq 12$

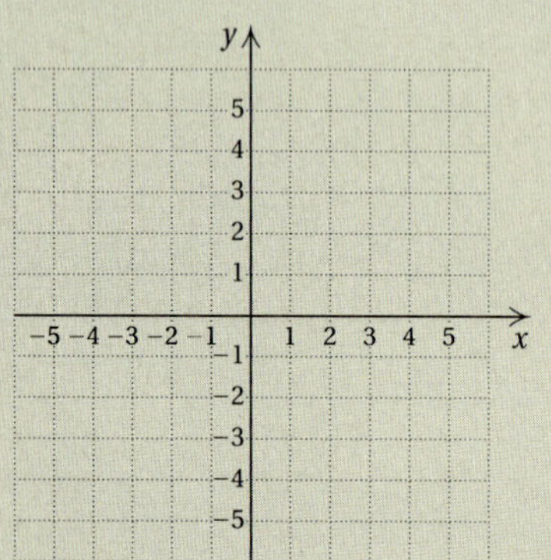

Answers on page A-52

EXAMPLE 5 Graph: $6x - 2y < 12$.

1. We first graph the related equation $6x - 2y = 12$.
2. Since the inequality uses the symbol $<$, points on the line are not solutions of the inequality, so we draw a dashed line.
3. To determine which half-plane to shade, we consider a test point *not* on the line. We try $(0, 0)$ and substitute:

$$\begin{array}{c|l} 6x - 2y < 12 & \\ \hline 6(0) - 2(0) \ ? \ 12 & \\ 0 - 0 & \\ 0 & \quad \text{TRUE} \end{array}$$

Since the inequality $0 < 12$ is *true*, the point $(0, 0)$ is a solution; each point in the half-plane containing $(0, 0)$ is a solution. Thus each point in the opposite half-plane is *not* a solution. The graph is shown below.

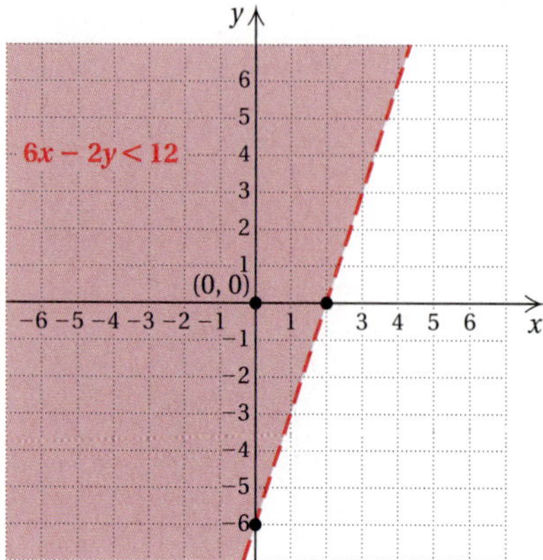

Do Exercises 3 and 4.

EXAMPLE 6 Graph $x > -3$ on a plane.

There is a missing variable in this inequality. If we graph the inequality on the number line, its graph is as follows:

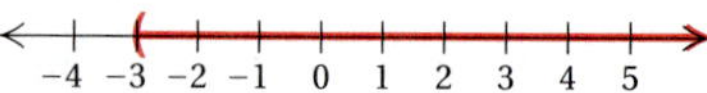

However, we can also write this inequality as $x + 0y > -3$ and consider graphing it in the plane. We are, in effect, determining which ordered pairs have x-values greater than -3. We use the same technique that we have used with the other examples. We first graph the related equation $x = -3$ in the plane. We draw the boundary with a dashed line. The rest of the graph is a half-plane to the right or left of the line $x = -3$. To determine which, we consider a test point, $(2, 5)$:

$$\begin{array}{c|l} x + 0y > -3 & \\ \hline 2 + 0(5) \ ? \ -3 & \\ 2 & \quad \text{TRUE} \end{array}$$

Since $(2, 5)$ is a solution, all the points in the half-plane containing $(2, 5)$ are solutions. We shade that half-plane.

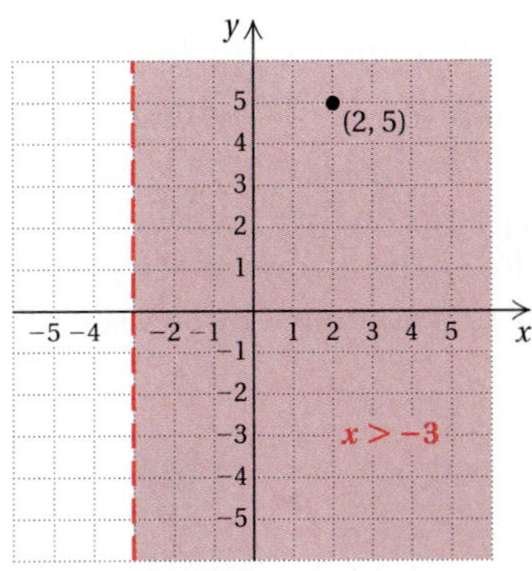

Graph on a plane.

5. $x < 3$

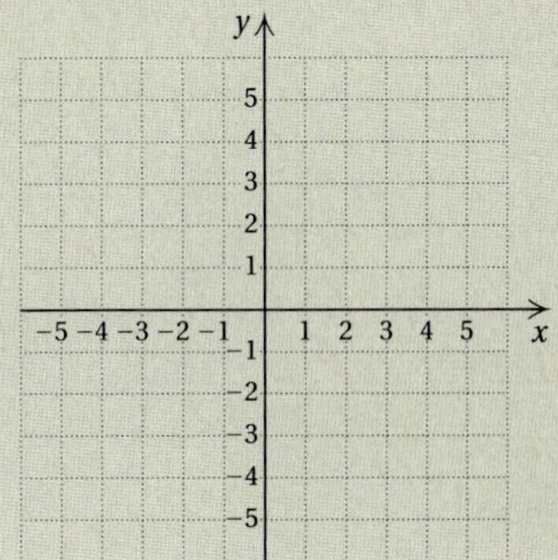

EXAMPLE 7 Graph $y \leq 4$ on a plane.

We first graph $y = 4$ using a solid line. We then use $(-2, 5)$ as a test point and substitute in $0x + y \leq 4$:

$$\begin{array}{r|l} 0x + y \leq 4 & \\ \hline 0(-2) + 5 \;?\; 4 & \\ 0 + 5 & \\ 5 & \end{array} \quad \text{FALSE}$$

We see that $(-2, 5)$ is *not* a solution, so all the points in the half-plane containing $(-2, 5)$ are not solutions. Thus each point in the opposite half-plane is a solution.

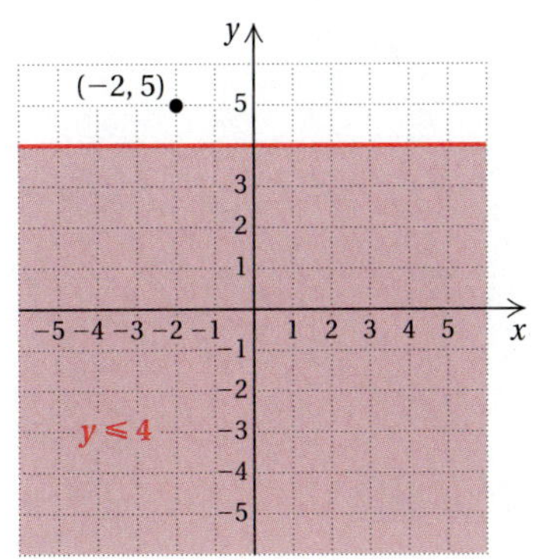

6. $y \geq -4$

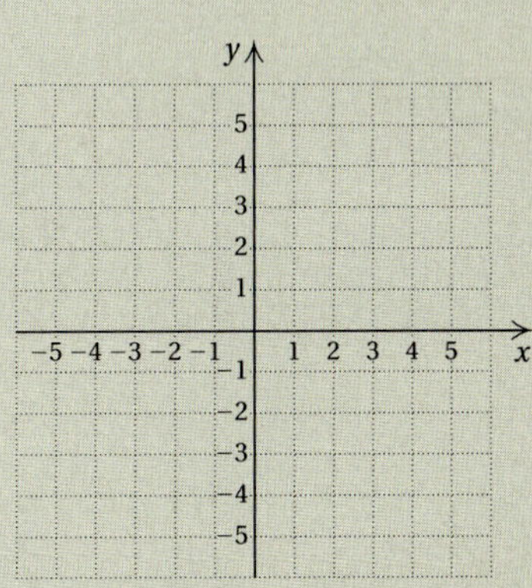

Do Exercises 5 and 6.

C Systems of Linear Inequalities

The following is an example of a system of two linear inequalities in two variables:

$$x + y \leq 4,$$
$$x - y < 4.$$

A **solution** of a system of linear inequalities is an ordered pair that is a solution of *both* inequalities. We now graph solutions of systems of linear inequalities. To do so, we graph each inequality and determine where the graphs overlap, or intersect. That will be a region in which the ordered pairs are solutions of both inequalities.

Answers on page A-52

7. Graph:

$$x + y \geq 1,$$
$$y - x \geq 2.$$

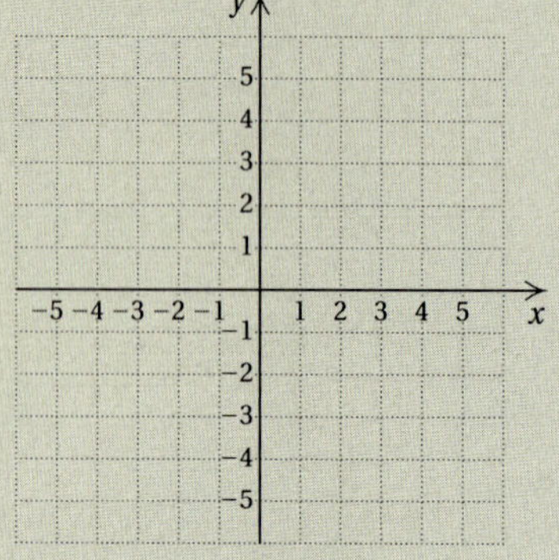

Answer on page A-53

EXAMPLE 8 Graph the solutions of the system

$$x + y \leq 4,$$
$$x - y < 4.$$

We graph the inequality $x + y \leq 4$ by first graphing the equation $x + y = 4$ using a solid red line. We consider $(0, 0)$ as a test point and find that it is a solution, so we shade all points on that side of the line using red shading. (See the graph on the left below.) The arrows at the ends of the line also indicate the half-plane, or region, that contains the solutions.

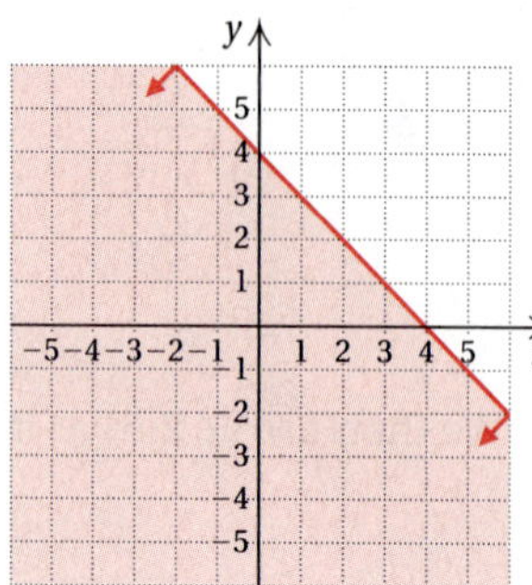

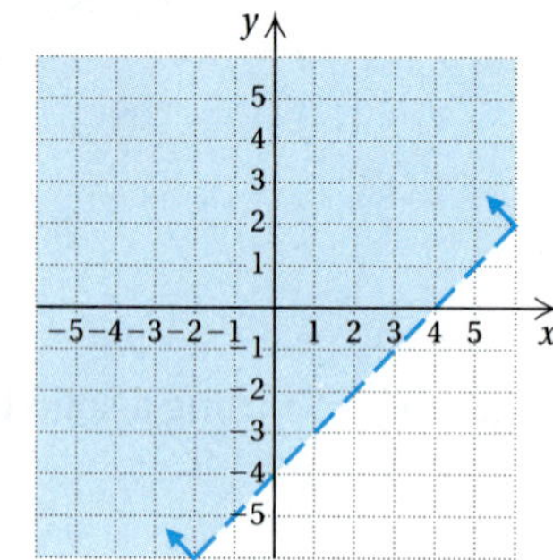

Next, we graph $x - y < 4$. We begin by graphing the equation $x - y = 4$ using a dashed blue line and consider $(0, 0)$ as a test point. Again, $(0, 0)$ is a solution so we shade that side of the line using blue shading. (See the graph on the right above.) The solution set of the system is the region that is shaded both red and blue and part of the line $x + y = 4$. (See the graph below.)

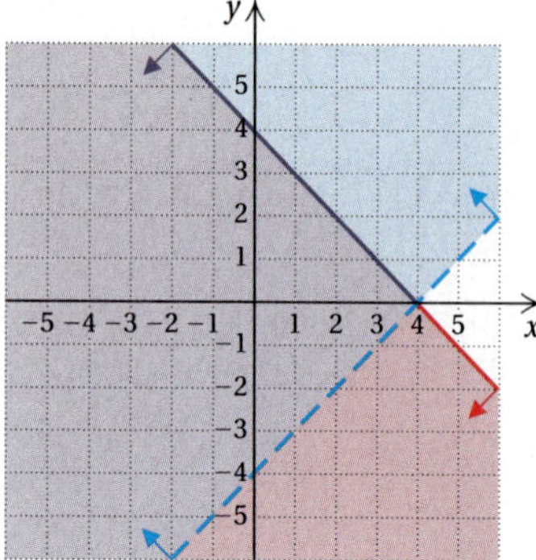

Do Exercise 7.

EXAMPLE 9 Graph: $-2 < x \leq 5$.

This is actually a system of inequalities:

$$-2 < x,$$
$$x \leq 5.$$

We graph the equation $-2 = x$ and see that the graph of the first inequality is the half-plane to the right of the line $-2 = x$ (see the graph on the left below).

Next, we graph the second inequality, starting with the line $x = 5$, and find that its graph is the line and also the half-plane to the left of it (see the graph on the right below).

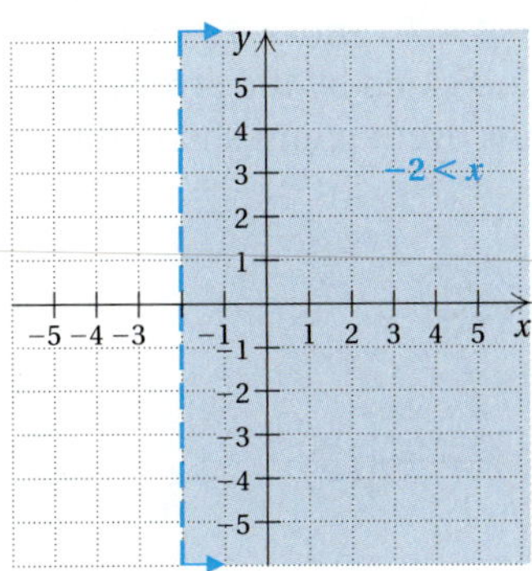

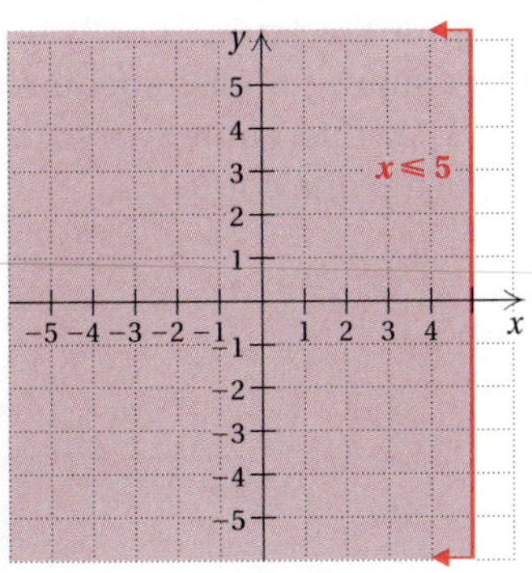

We shade the intersection of these graphs.

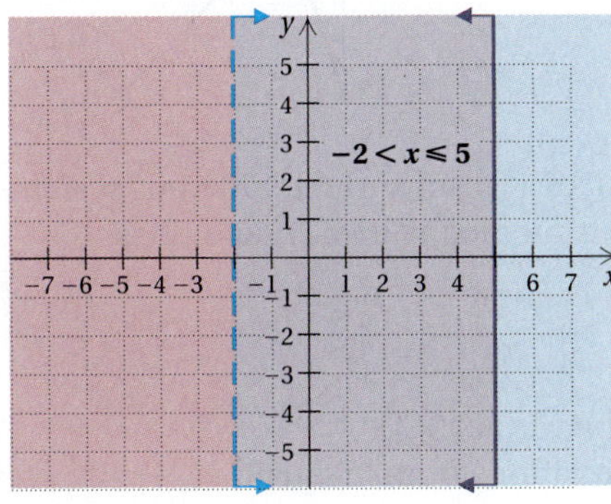

Do Exercise 8.

8. Graph: $-3 \leq y < 4$.

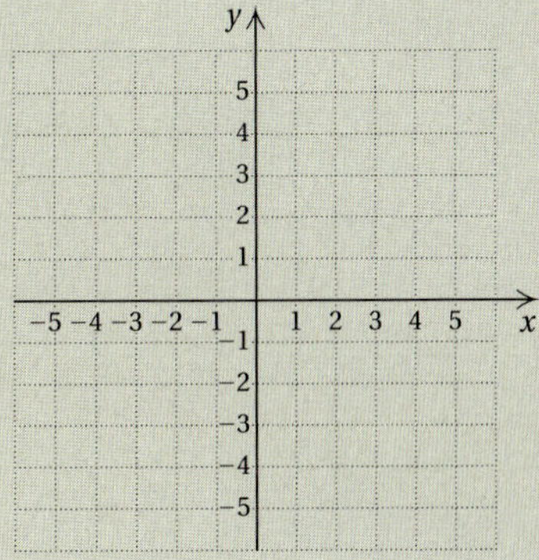

Answer on page A-53

9. Graph the system of inequalities. Find the coordinates of any vertices formed.

$$5x + 6y \le 30,$$
$$0 \le y \le 3,$$
$$0 \le x \le 4$$

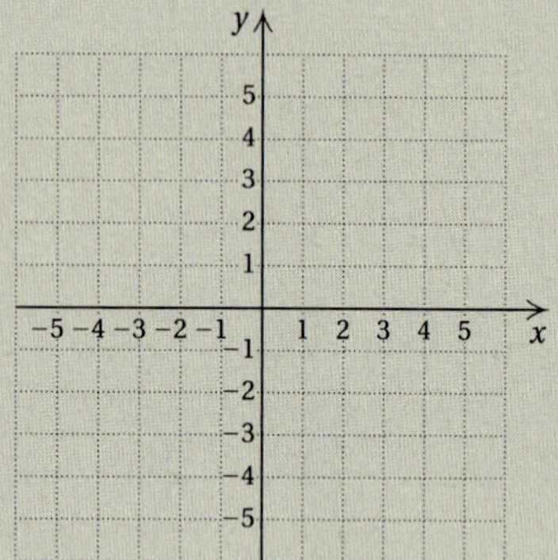

Answer on page A-53

A system of inequalities may have a graph that consists of a polygon and its interior. In *linear programming*, which is a topic rich in application that you may study in a later course, it is important to be able to find the vertices of such a polygon.

EXAMPLE 10 Graph the following system of inequalities. Find the coordinates of any vertices formed.

$$6x - 2y \le 12, \quad (1)$$
$$y - 3 \le 0, \quad (2)$$
$$x + y \ge 0 \quad (3)$$

We graph the lines $6x - 2y = 12$, $y - 3 = 0$, and $x + y = 0$ using solid lines. The regions for each inequality are indicated by the arrows at the ends of the lines. We then note where the regions overlap and shade the region of solutions using one color.

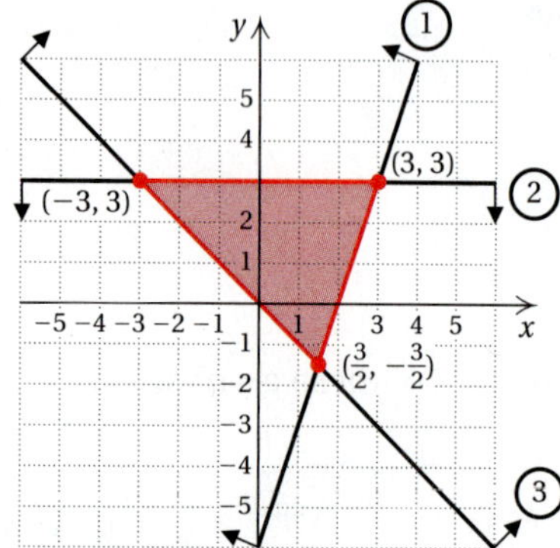

To find the vertices, we solve three different systems of equations. The system of equations from inequalities (1) and (2) is

$$6x - 2y = 12, \quad (1)$$
$$y - 3 = 0. \quad (2)$$

Solving, we obtain the vertex $(3, 3)$.

The system of equations from inequalities (1) and (3) is

$$6x - 2y = 12, \quad (1)$$
$$x + y = 0. \quad (3)$$

Solving, we obtain the vertex $\left(\frac{3}{2}, -\frac{3}{2}\right)$.

The system of equations from inequalities (2) and (3) is

$$y - 3 = 0, \quad (2)$$
$$x + y = 0. \quad (3)$$

Solving, we obtain the vertex $(-3, 3)$.

Do Exercise 9.

EXAMPLE 11 Graph the following system of inequalities. Find the coordinates of any vertices formed.

$$x + y \le 16, \quad (1)$$
$$3x + 6y \le 60, \quad (2)$$
$$x \ge 0, \quad (3)$$
$$y \ge 0 \quad (4)$$

We graph each inequality using solid lines. The regions for each inequality are indicated by the arrows at the ends of the lines. We then note where the regions overlap and shade the region of solutions using one color.

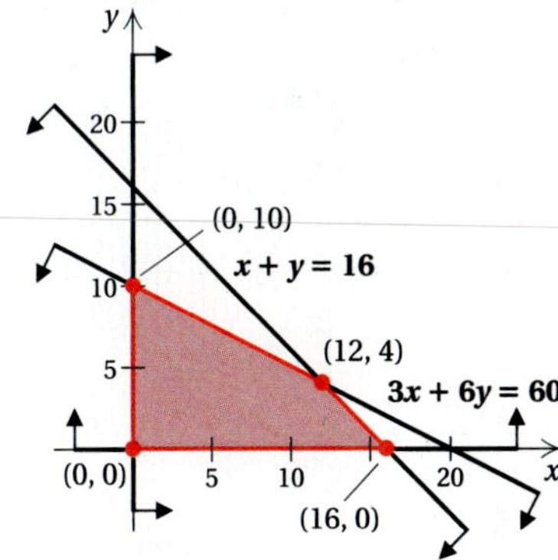

To find the vertices, we solve four different systems of equations. The system of equations from inequalities (1) and (2) is

$$x + y = 16, \quad (1)$$
$$3x + 6y = 60. \quad (2)$$

Solving, we obtain the vertex $(12, 4)$.

The system of equations from inequalities (1) and (4) is

$$x + y = 16, \quad (1)$$
$$y = 0. \quad (4)$$

Solving, we obtain the vertex $(16, 0)$.

The system of equations from inequalities (3) and (4) is

$$x = 0, \quad (3)$$
$$y = 0. \quad (4)$$

The vertex is $(0, 0)$.

The system of equations from inequalities (2) and (3) is

$$3x + 6y = 60, \quad (2)$$
$$x = 0. \quad (3)$$

Solving, we obtain the vertex $(0, 10)$.

Do Exercise 10.

10. Graph the system of inequalities. Find the coordinates of any vertices formed.

$$2x + 4y \le 8,$$
$$x + y \le 3,$$
$$x \ge 0,$$
$$y \ge 0$$

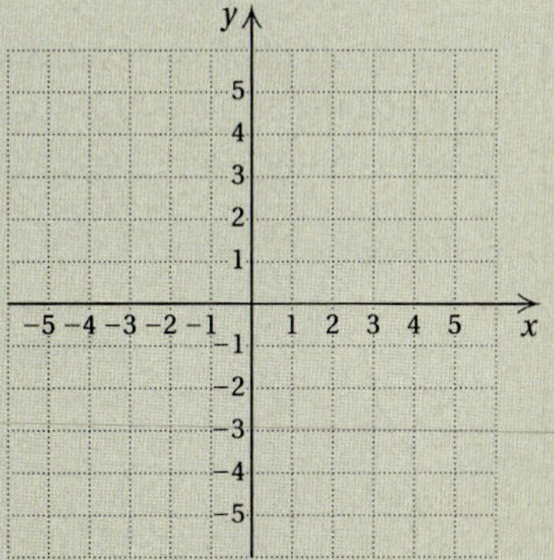

Answer on page A-53

CALCULATOR CORNER

Graphs of Inequalities We can graph inequalities on a graphing calculator, shading the region of the solution set. To graph the inequality in Example 4, $8x + 3y \geq 24$, we first graph the line $8x + 3y = 24$. Solving for y, we get $y = \frac{24 - 8x}{3}$, or $y = -\frac{8}{3}x + 8$. We enter this equation on the equation-editor screen. After determining algebraically that the solution set consists of all points above the line, we use the calculator's "shade above" GraphStyle to shade this region. To do this, we position the cursor over the graphstyle icon on the equation-editor screen to the left of the equation. Then we press ENTER repeatedly, until the "shade above" icon appears. We select a window and press GRAPH to display the graph of the inequality.

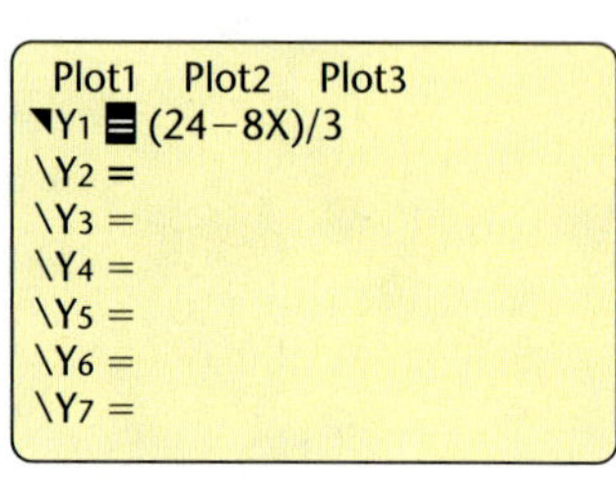

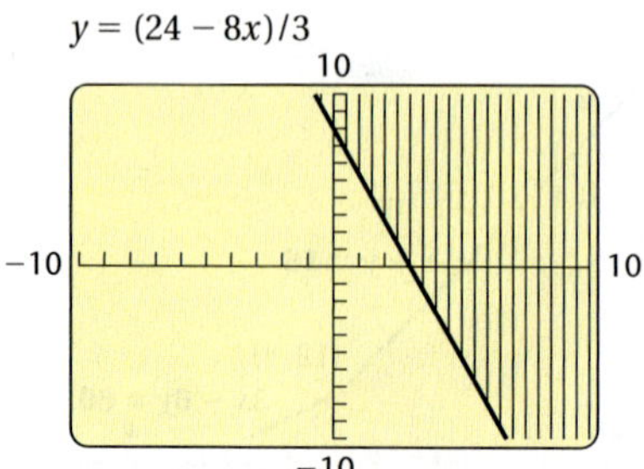

Note that we cannot use the GraphStyle option to graph an inequality like the one in Example 6, $x > -3$, because the related equation has no y-term and thus cannot be entered in "$y =$" form.

We can also graph a system of inequalities by shading the solution set of each inequality in the system with a different pattern. When the "shade above" or "shade below" GraphStyle options are selected, the TI-83 Plus rotates through four shading patterns. Vertical lines shade the graph of the first inequality, horizontal lines the second, negatively sloping diagonal lines the third, and positively sloping diagonal lines the fourth. These patterns repeat if more than four inequalities are graphed. For example, to graph the solution set of the system of inequalities

$$x + y \leq 4,$$
$$x - y \geq 2,$$

we first graph the equation $x + y = 4$, entering it in the form $y_1 = -x + 4$. We determine that the solution set of $x + y \leq 4$ consists of all points below the line $x + y = 4$, or $y_1 = -x + 4$, so we select the "shade below" GraphStyle for this function. Next, we graph $x - y = 2$, entering it in the form $y_2 = x - 2$. The solution set of $x - y \geq 2$ consists of all points below the line $x - y = 2$, or $y_2 = x - 2$, so we also choose the "shade below" GraphStyle for this function. Now we select a window and press GRAPH to display the solution sets of each inequality in the system and the region where they overlap. The region of overlap is the solution set of the system of inequalities. We could also use the INTERSECT feature to find the coordinates of the vertex formed by this system of inequalities.

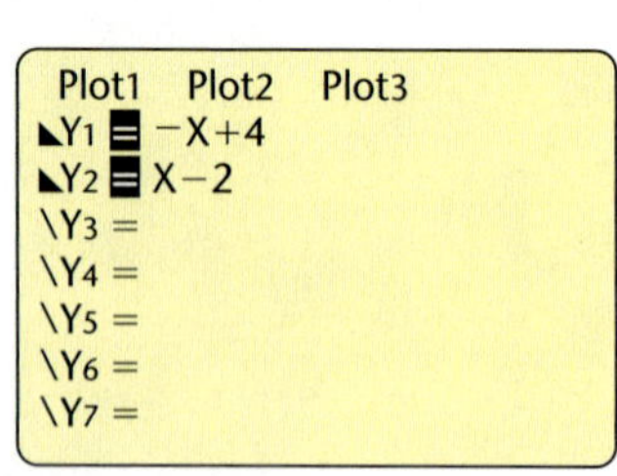

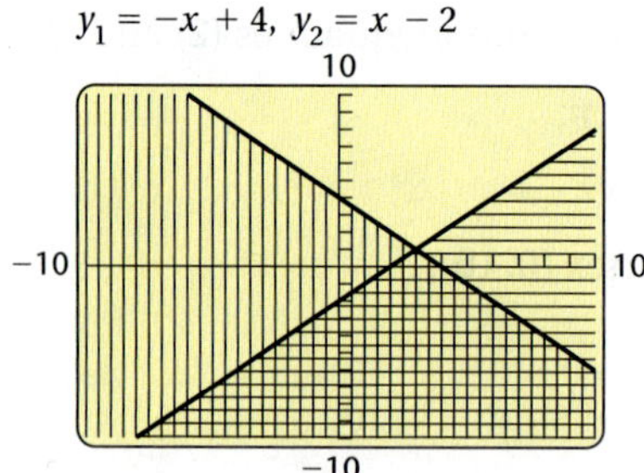

Exercises:

1. Use a graphing calculator to graph the inequalities in Margin Exercises 4 and 6.
2. Use a graphing calculator to graph the system of inequalities in Margin Exercise 7.

14.4 EXERCISE SET

For Extra Help

Digital Video Tutor CD 12 Videotape 16 | InterAct Math | Math Tutor Center | MathXL | MyMathLab

a Determine whether the given ordered pair is a solution of the given inequality.

1. $(-3, 3)$; $3x + y < -5$

2. $(6, -8)$; $4x + 3y \geq 0$

3. $(5, 9)$; $2x - y > -1$

4. $(5, -2)$; $6y - x > 2$

b Graph the inequality on a plane.

5. $y > 2x$

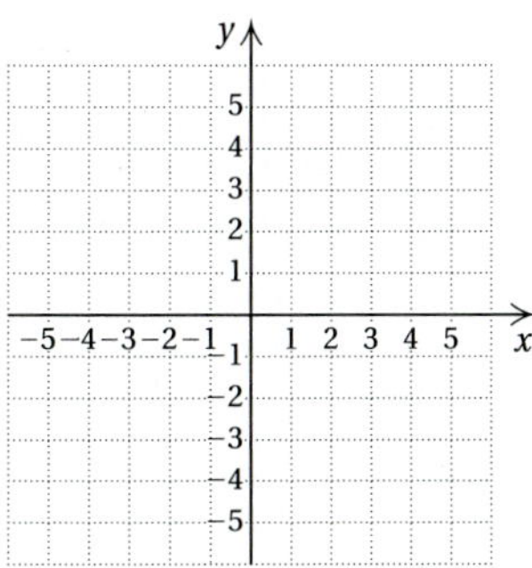

6. $y < 3x$

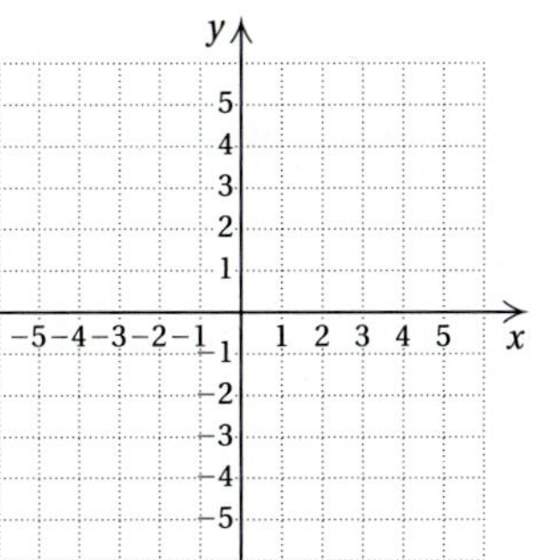

7. $y < x + 1$

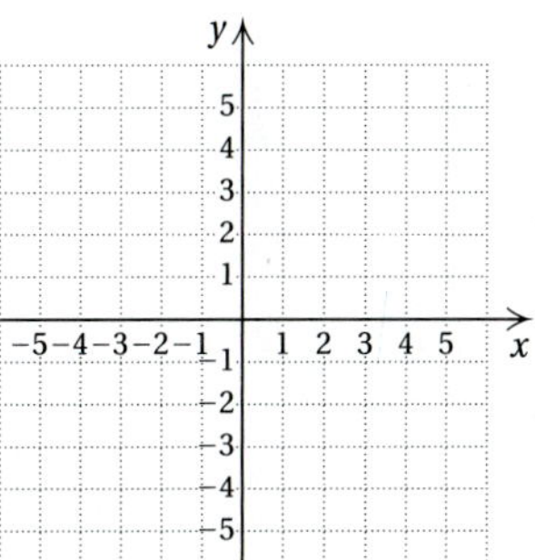

8. $y \leq x - 3$

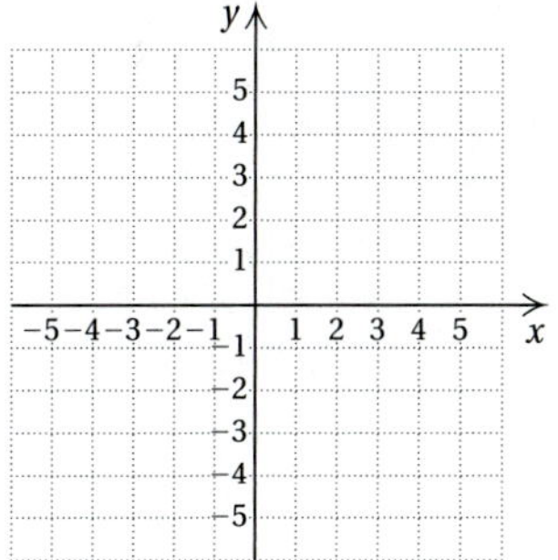

9. $y > x - 2$

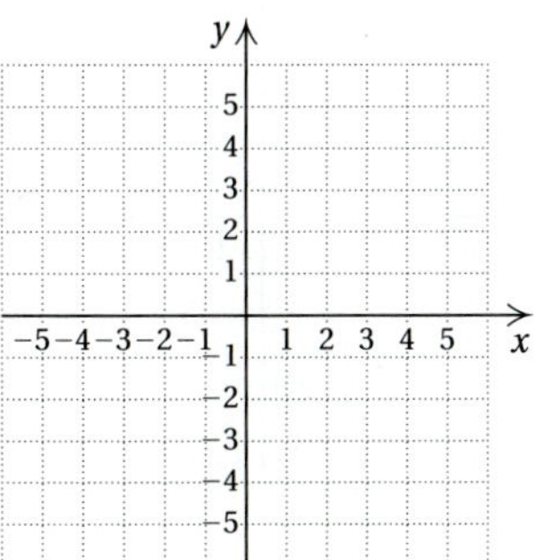

10. $y \geq x + 4$

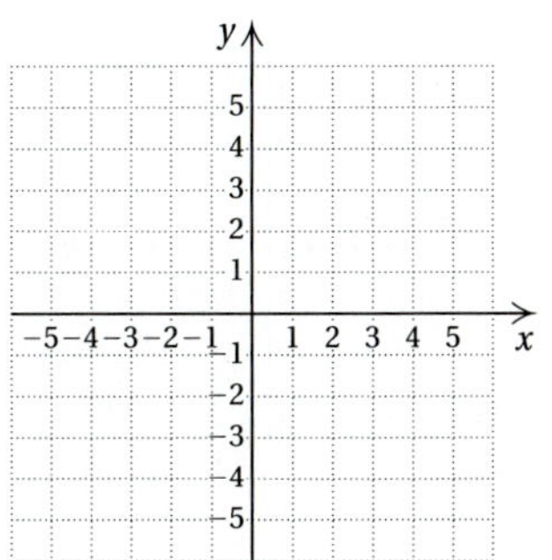

11. $x + y < 4$

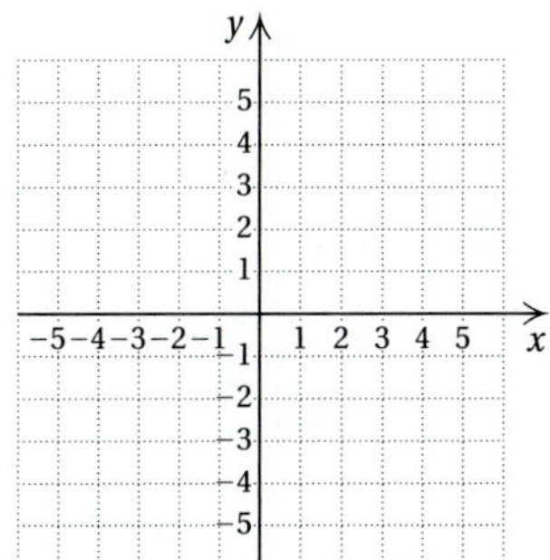

12. $x - y \geq 3$

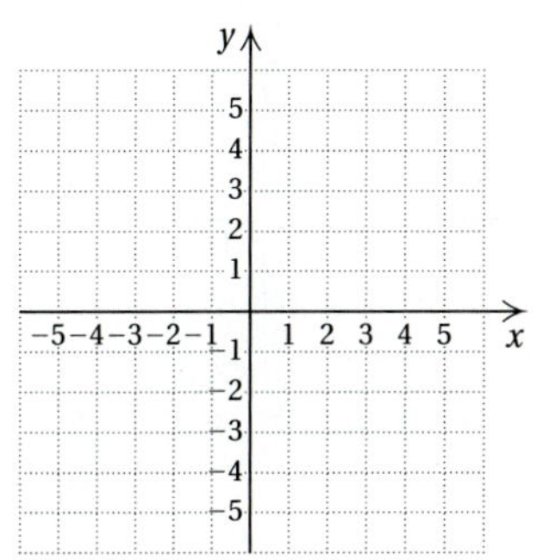

13. $3x + 4y \leq 12$

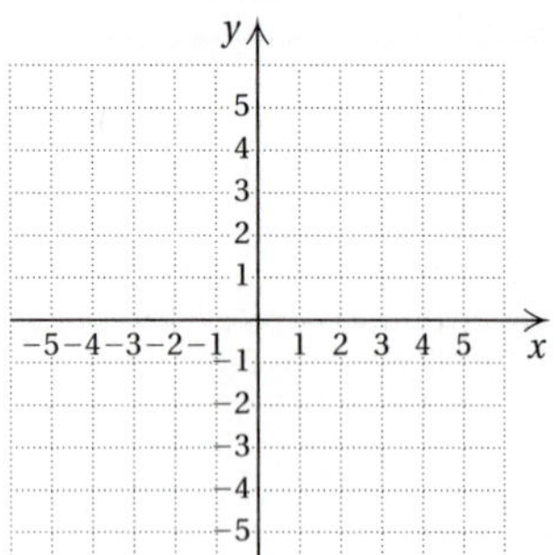

14. $2x + 3y < 6$

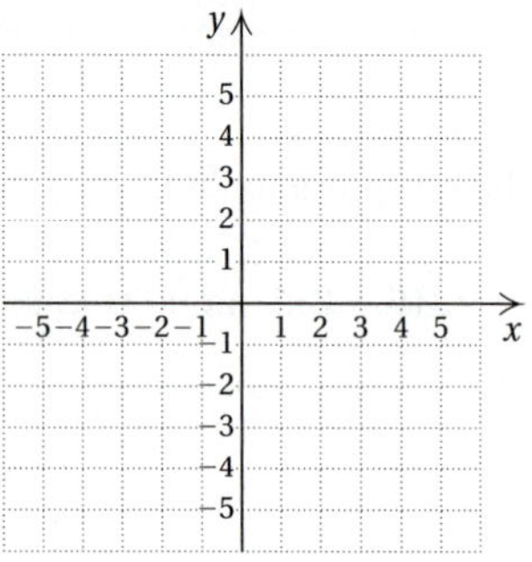

15. $2y - 3x > 6$

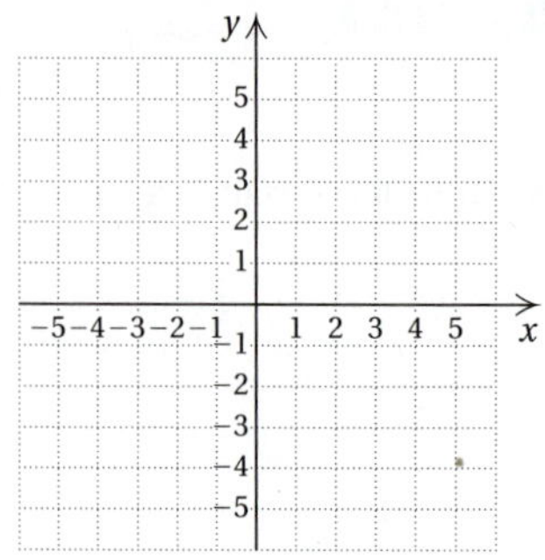

16. $2y - x \leq 4$

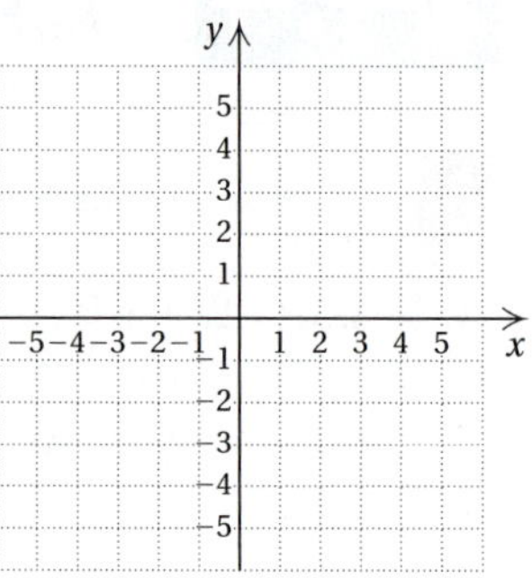

17. $3x - 2 \leq 5x + y$

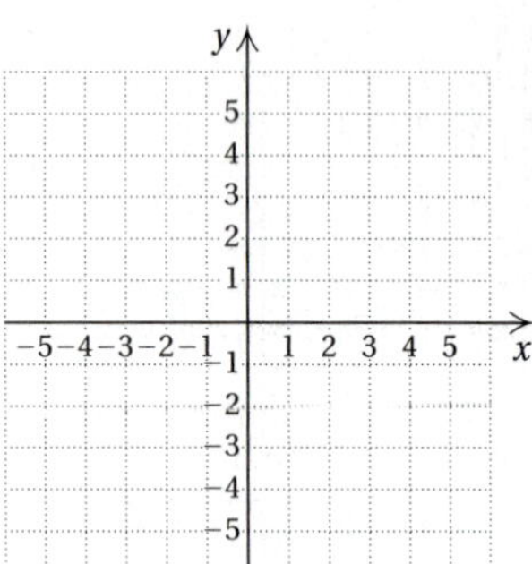

18. $2x - 2y \geq 8 + 2y$

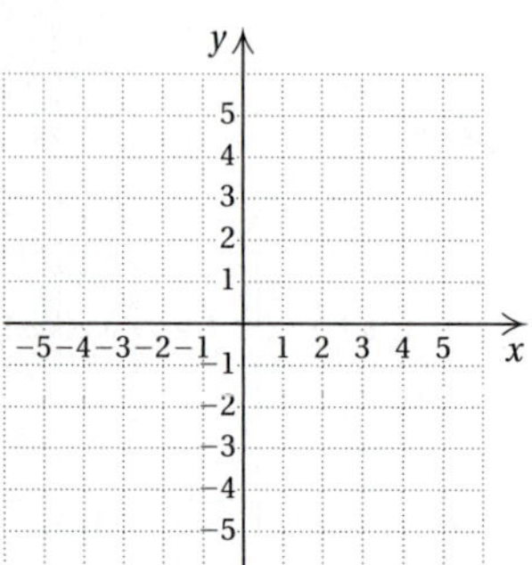

19. $x < 5$

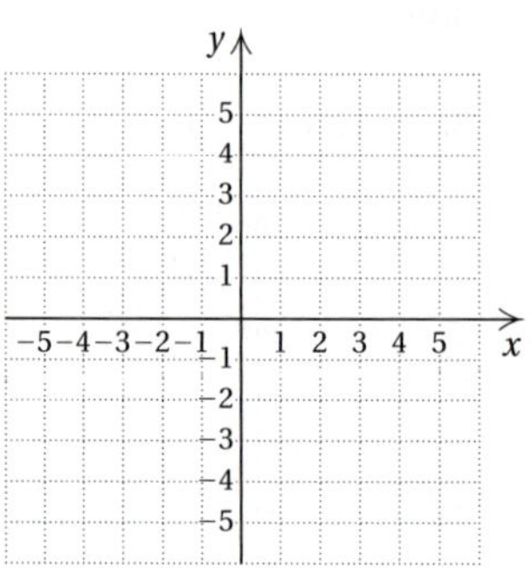

20. $y \geq -2$

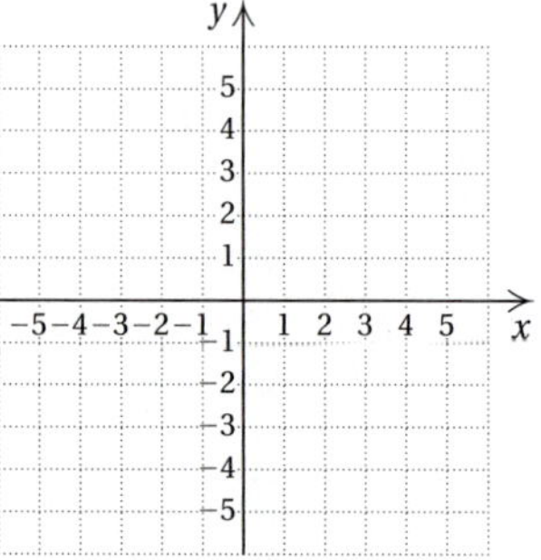

21. $y > 2$

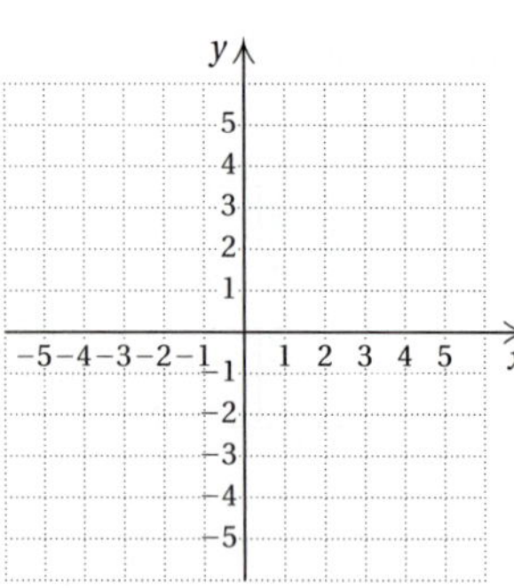

22. $x \leq -4$

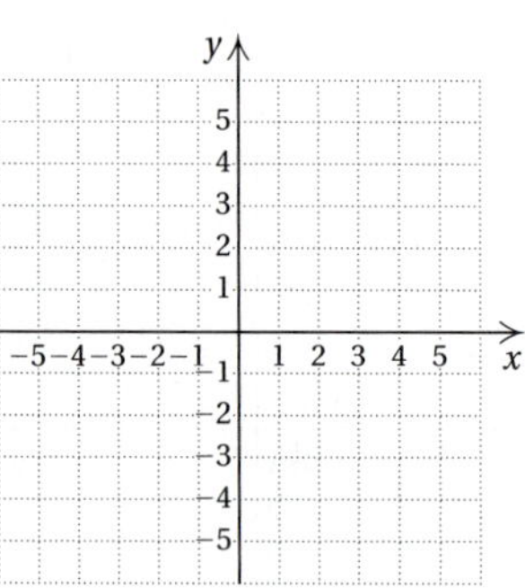

23. $2x + 3y \leq 6$

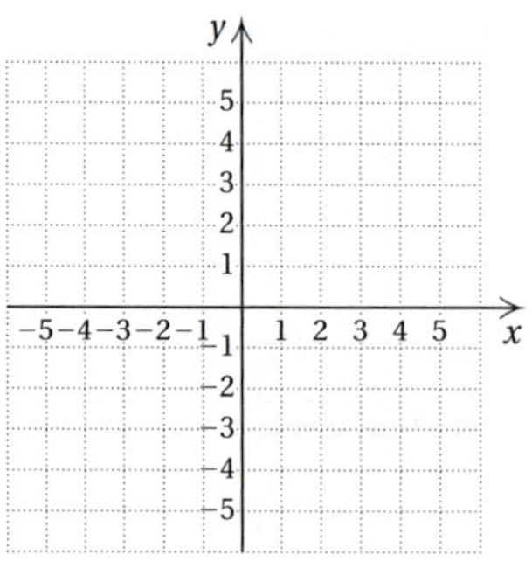

24. $7x + 2y \geq 21$

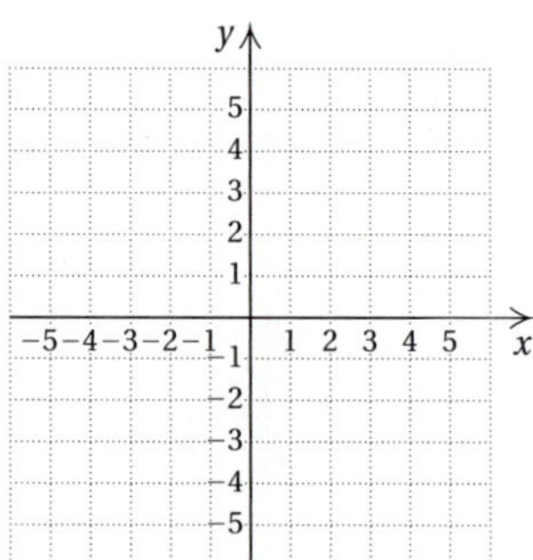

Matching. Each of Exercises 25–30 shows the graph of an inequality. Match the graph with one of the appropriate inequalities (A)–(F) that follow.

25.

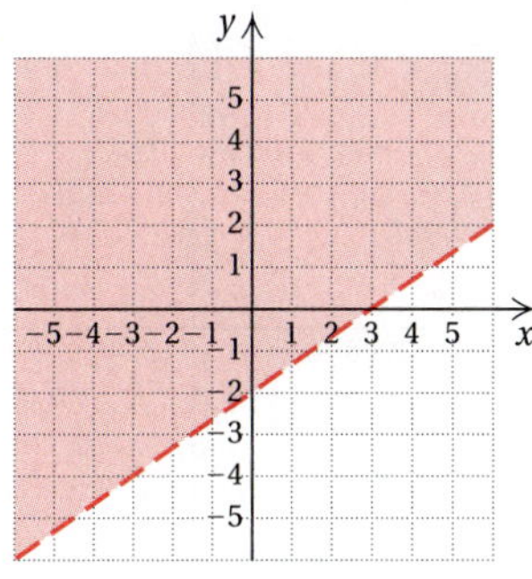

26.

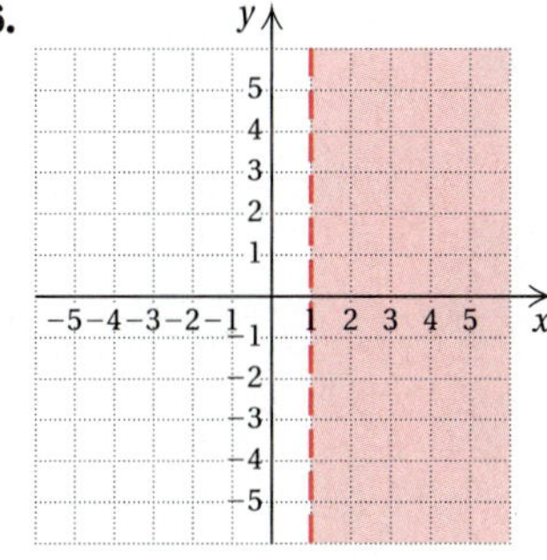

27.

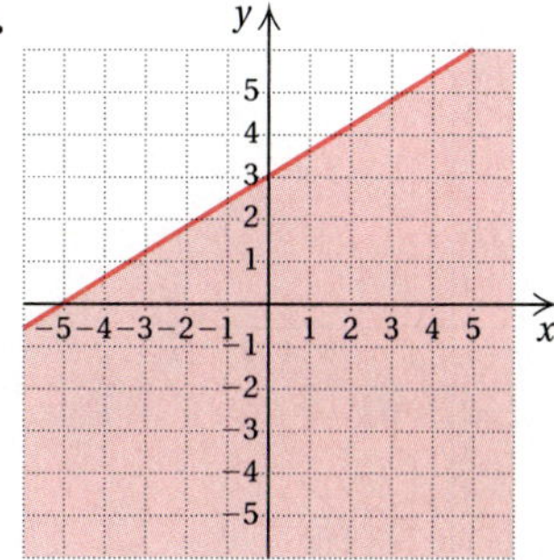

28.

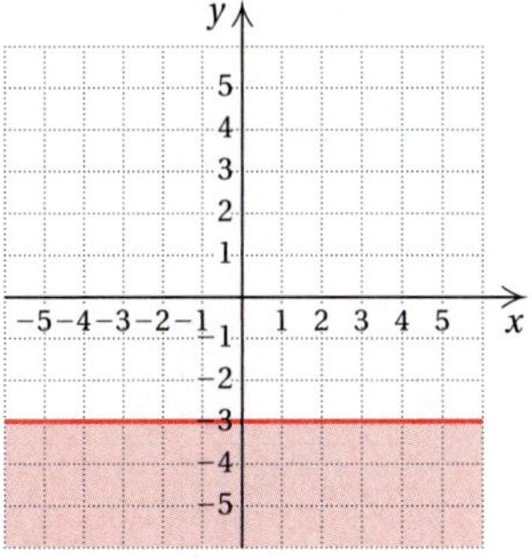

29.

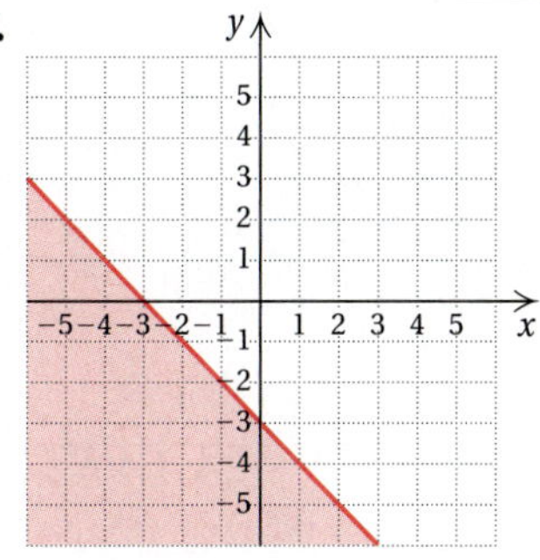

30.

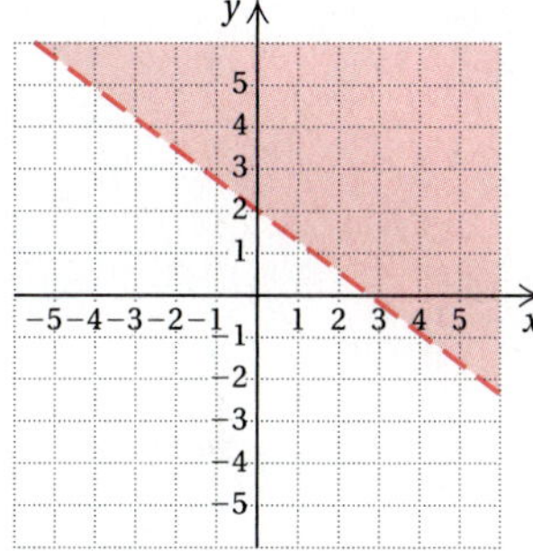

A. $4y > 8 - 3x$ **B.** $3x \geq 5y - 15$ **C.** $y + x \leq -3$ **D.** $x > 1$ **E.** $y \leq -3$ **F.** $2x - 3y < 6$

C Graph the system of inequalities. Find the coordinates of any vertices formed.

31. $y \geq x,$
$y \leq -x + 2$

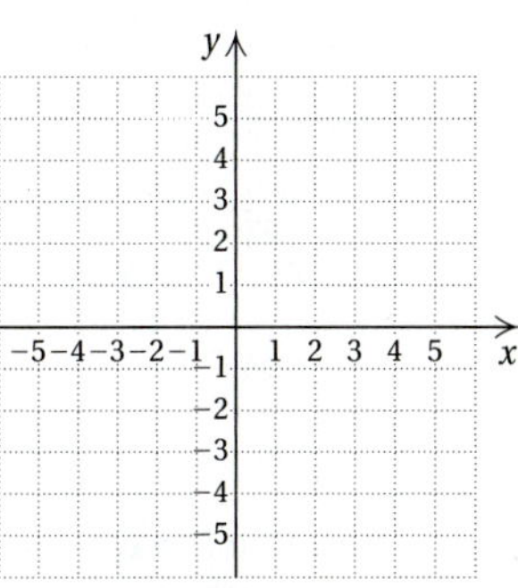

32. $y \geq x,$
$y \leq -x + 4$

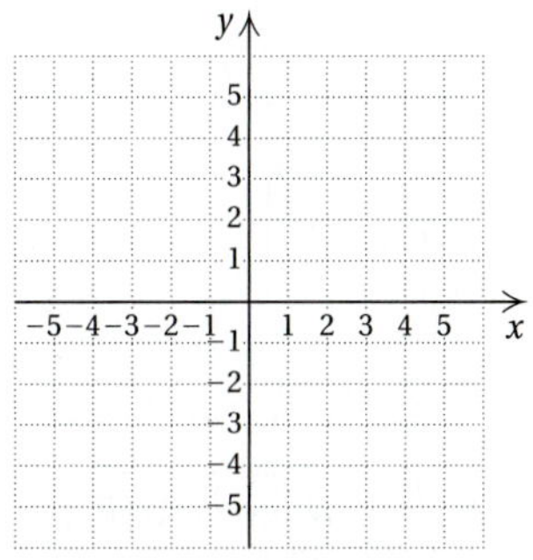

33. $y > x,$
$y < -x + 1$

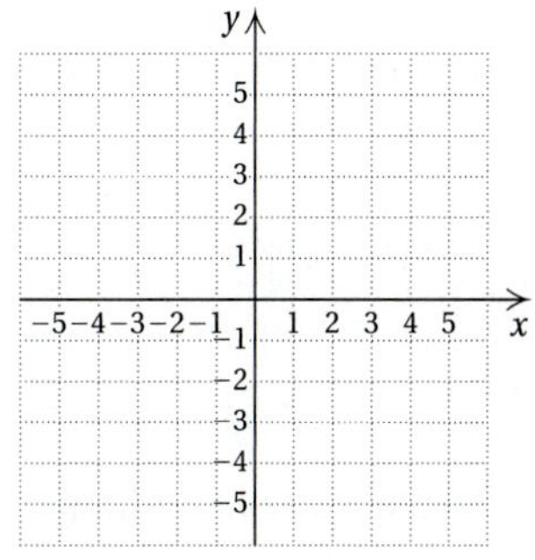

34. $y < x,$
$y > -x + 3$

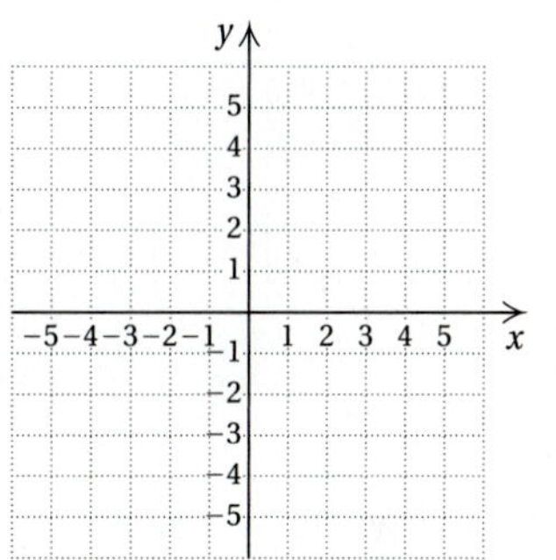

35. $y \geq -2,$
$x \geq 1$

36. $y \leq -2,$
$x \geq 2$

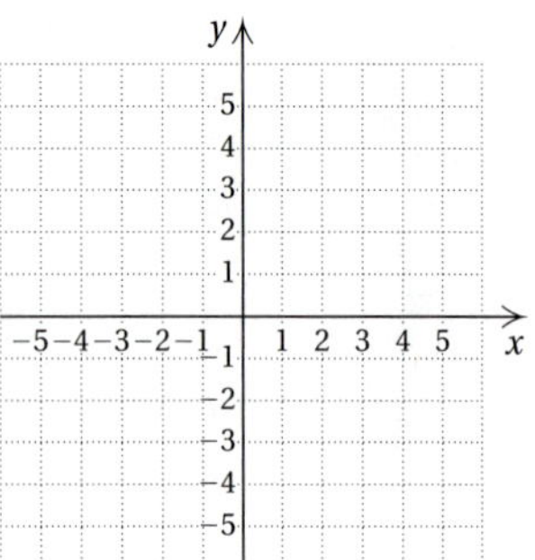

37. $x \leq 3,$
$y \geq -3x + 2$

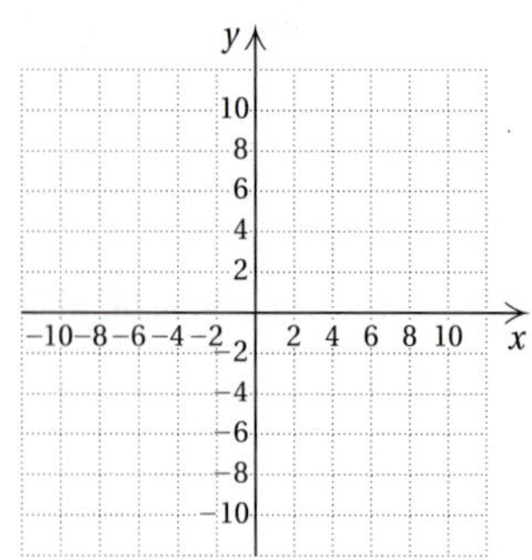

38. $x \geq -2,$
$y \leq -2x + 3$

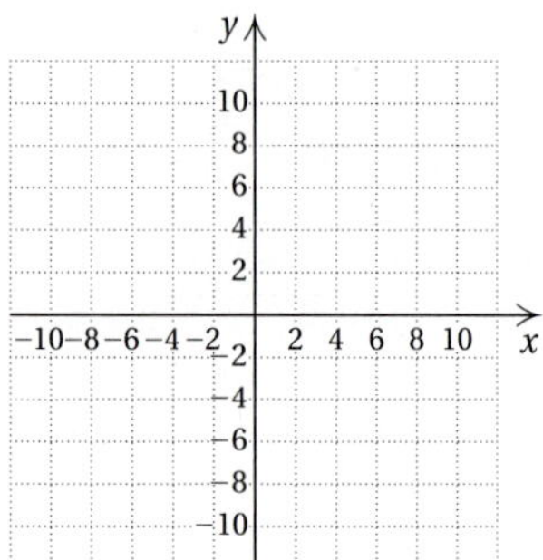

39. $y \leq 2x + 1,$
$y \geq -2x + 1,$
$x \leq 2$

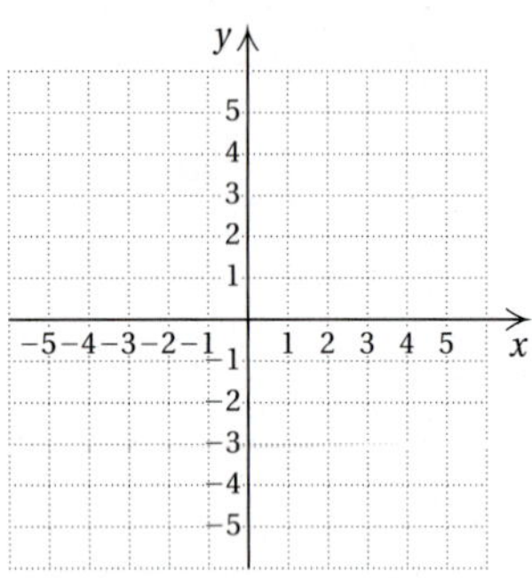

40. $x - y \leq 2,$
$x + 2y \geq 8,$
$y \leq 4$

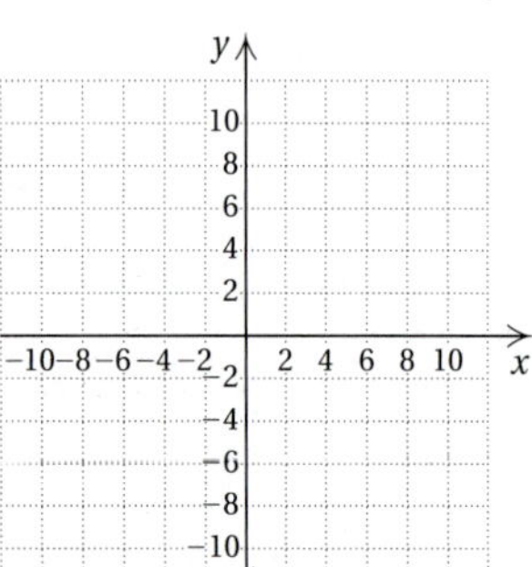

41. $x + y \leq 1,$
$x - y \leq 2$

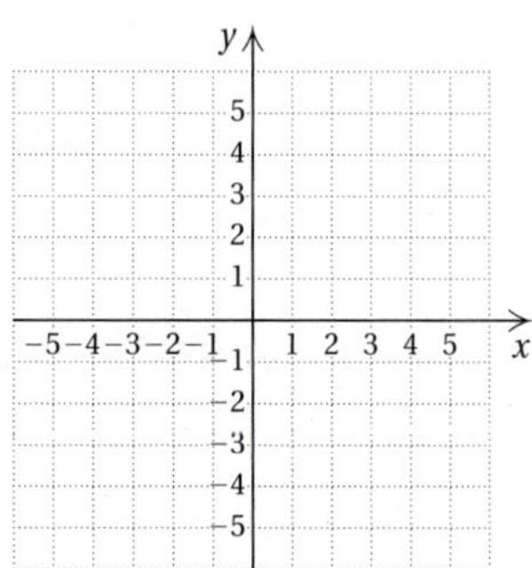

42. $x + y \leq 3,$
$x - y \leq 4$

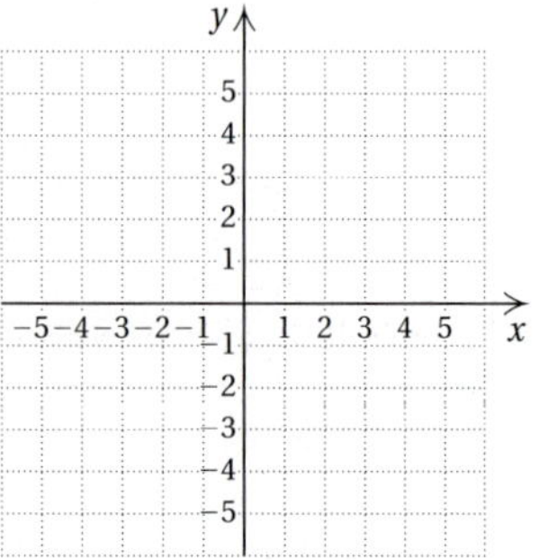

43. $x + 2y \leq 12,$
$2x + y \leq 12,$
$x \geq 0,$
$y \geq 0$

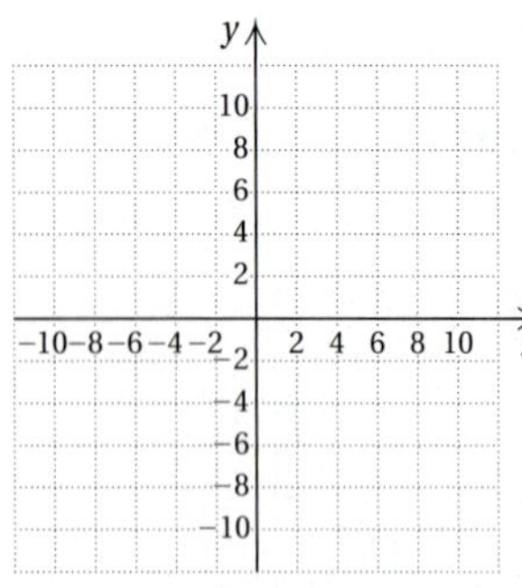

44. $4y - 3x \geq -12,$
$4y + 3x \geq -36,$
$y \leq 0,$
$x \leq 0$

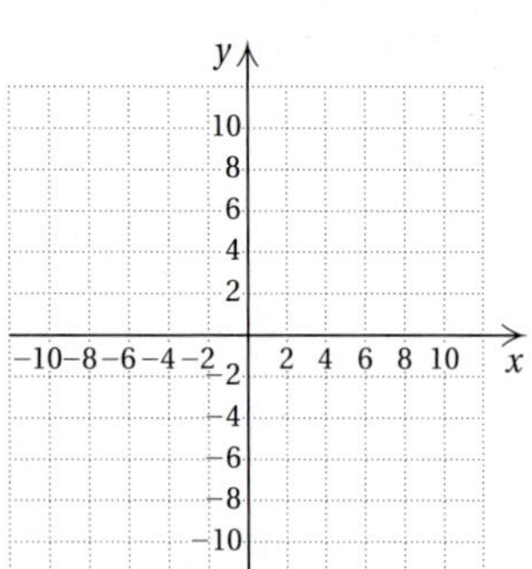

45. $8x + 5y \leq 40,$
$x + 2y \leq 8,$
$x \geq 0,$
$y \geq 0$

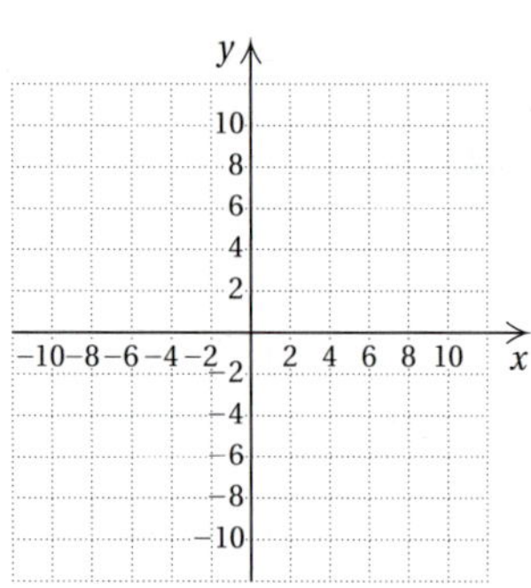

46. $y - x \geq 1,$
$y - x \leq 3,$
$2 \leq x \leq 5$

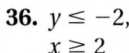

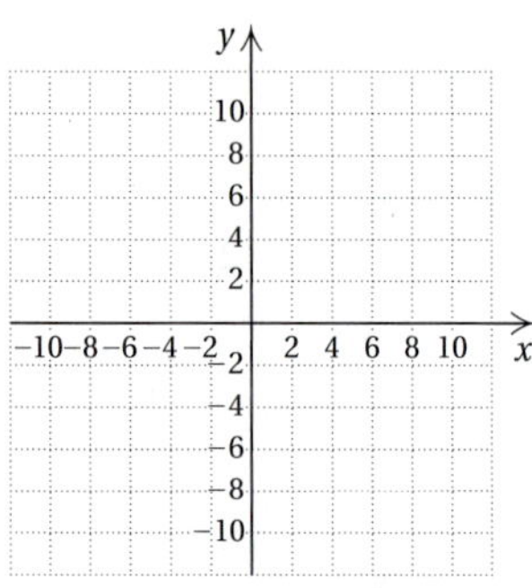

47. $^{\mathrm{D}}\mathrm{W}$ Do all systems of linear inequalities have solutions? Why or why not?

48. $^{\mathrm{D}}\mathrm{W}$ When graphing linear inequalities, Ron always shades above the line when he sees a $\geq$ symbol. Is this wise? Why or why not?

SKILL MAINTENANCE

Find an equation of the line containing the given pair of points. [12.5c]

49. $(-5, -6), (3, -2)$

50. $(0, 0), (8, 3)$

51. $(3, 6), (1, 12)$

52. $(-2, 0), (0, 9)$

53. $(-5, 5), (-1, -1)$

54. $(2, 6), (-3, 6)$

Given the function $f(x) = |2 - x|$, find each of the following function values. [12.1b]

55. $f(0)$

56. $f(-1)$

57. $f(1)$

58. $f(10)$

59. $f(-2)$

60. $f(2a)$

61. $f(-4)$

62. $f(1.8)$

SYNTHESIS

63. *Luggage Size.* Unless an additional fee is paid, most major airlines will not check any luggage that is more than 62 in. long. The U.S. Postal Service will ship a package only if the sum of the package's length and girth (distance around its midsection) does not exceed 108 in. Concert Productions is ordering several 62-in. long trunks that will be both mailed and checked as luggage. Using w and h for width and height (in inches), respectively, write and graph a system of inequalities that represents all acceptable combinations of width and height.

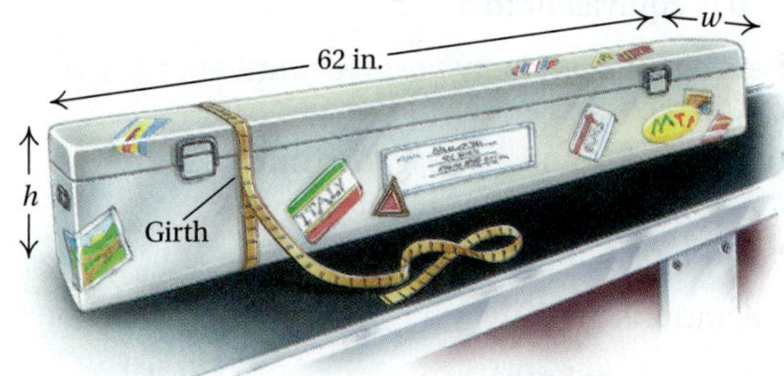

64. *Exercise Danger Zone.* It is dangerous to exercise when the weather is hot and humid. The solutions of the following system of inequalities give a "danger zone" for which it is dangerous to exercise intensely:

$$4H - 3F < 70,$$
$$F + H > 160,$$
$$2F + 3H > 390,$$

where F is the temperature, in degrees Fahrenheit, and H is the humidity.

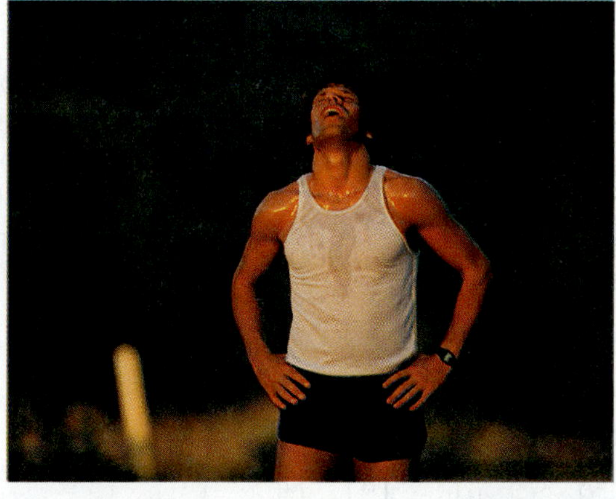

a) Draw the danger zone by graphing the system of inequalities.

b) Is it dangerous to exercise when $F = 80°$ and $H = 80\%$?

In Exercises 65–68, use a graphing calculator with a SHADE feature to graph the inequality.

65. $3x + 6y > 2$

66. $x - 5y \leq 10$

67. $13x - 25y + 10 \leq 0$

68. $2x + 5y > 0$

69. Use a graphing calculator with a SHADE feature to check your answers to Exercises 31 and 32. Then use the INTERSECT feature to determine any point(s) of intersection.

CHAPTER 14 Summary and Review

The review that follows is meant to prepare you for a chapter exam. It consists of two parts. The first part is a checklist of some of the Study Tips referred to in this and preceding chapters, as well as a list of important properties and formulas. The second part is the Review Exercises. These provide practice exercises for the exam, together with references to section objectives so you can go back and review. Before beginning, stop and look back over the skills you have obtained. What skills in mathematics do you have now that you did not have before studying this chapter?

STUDY TIPS CHECKLIST

The foundation of all your study skills is TIME!	
	☐ Are you attempting to keep a positive attitude as you study mathematics?
	☐ Are you using the five-step problem-solving strategy when doing applied problems?
	☐ Are you stopping to work the margin exercises when directed to do so?
	☐ Have you tried viewing the videotapes or the Digital Video Tutor?

IMPORTANT PROPERTIES AND FORMULAS

The Addition Principle for Inequalities: For any real numbers a, b, and c: $a < b$ is equivalent to $a + c < b + c$; $a > b$ is equivalent to $a + c > b + c$.

The Multiplication Principle for Inequalities: For any real numbers a and b, and any *positive* number c: $a < b$ is equivalent to $ac < bc$; $a > b$ is equivalent to $ac > bc$.
For any real numbers a and b, and any *negative* number c: $a < b$ is equivalent to $ac > bc$; $a > b$ is equivalent to $ac < bc$.
Similar statements hold for $\leq$ and $\geq$.

Set Intersection: $A \cap B = \{x \mid x \text{ is in } A \text{ and } x \text{ is in } B\}$

Set Union: $A \cup B = \{x \mid x \text{ is in } A \text{ or in } B\text{, or both}\}$

"$a < x$ *and* $x < b$" is equivalent to "$a < x < b$"

Properties of Absolute Value

$|ab| = |a| \cdot |b|$, $\left|\frac{a}{b}\right| = \frac{|a|}{|b|}$, $|-a| = |a|$, The distance between a and b is $|a - b|$.

Principles for Solving Equations and Inequalities Involving Absolute Value:
For any positive number p and any algebraic expression X:
a) The solutions of $|X| = p$ are those numbers that satisfy $X = -p$ *or* $X = p$.
b) The solutions of $|X| < p$ are those numbers that satisfy $-p < X < p$.
c) The solutions of $|X| > p$ are those numbers that satisfy $X < -p$ *or* $X > p$.

REVIEW EXERCISES

Write interval notation for the given set or graph. [14.1b]

1. $\{x \mid -8 \le x < 9\}$

2. 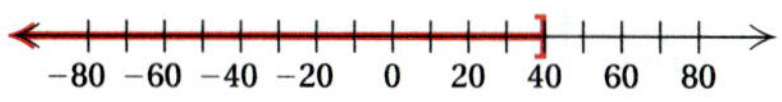

Solve and graph. Write interval notation for the solution set. [14.1c]

3. $x - 2 \le -4$

4. $x + 5 > 6$

Solve. [14.1c]

5. $a + 7 \le -14$

6. $y - 5 \ge -12$

7. $4y > -16$

8. $-0.3y < 9$

9. $-6x - 5 < 13$

10. $4y + 3 \le -6y - 9$

11. $-\frac{1}{2}x - \frac{1}{4} > \frac{1}{2} - \frac{1}{4}x$

12. $0.3y - 8 < 2.6y + 15$

13. $-2(x - 5) \ge 6(x + 7) - 12$

14. *Moving Costs.* Musclebound Movers charges \$85 plus \$40 an hour to move households across town. Champion Moving charges \$60 an hour for cross-town moves. For what lengths of time is Champion more expensive? [14.1d]

15. *Investments.* You are going to invest \$30,000, part at 13% and part at 15%, for one year. What is the most that can be invested at 13% in order to make at least \$4300 interest in one year? [14.1d]

Graph and write interval notation. [14.2a, b]

16. $-2 \le x < 5$

17. $x \le -2$ *or* $x > 5$

Solve. [14.2a, b]

18. $2x - 5 < -7$ *and* $3x + 8 \ge 14$

19. $-4 < x + 3 \le 5$

20. $-15 < -4x - 5 < 0$

21. $3x < -9$ *or* $-5x < -5$

22. $2x + 5 < -17$ *or* $-4x + 10 \le 34$

23. $2x + 7 \le -5$ *or* $x + 7 \ge 15$

Simplify. [14.3a]

24. $\left|-\frac{3}{x}\right|$

25. $\left|\frac{2x}{y^2}\right|$

26. $\left|\frac{12y}{-3y^2}\right|$

27. Find the distance between -23 and 39. [14.3b]

Solve. [14.3c, d]

28. $|x| = 6$

29. $|x - 2| = 7$

30. $|2x + 5| = |x - 9|$

31. $|5x + 6| = -8$

Solve. [14.3e]

32. $|2x + 5| < 12$

33. $|x| \geq 3.5$

34. $|3x - 4| \geq 15$

35. $|x| < 0$

36. *Records in the Women's 100-m Dash.* In 1988, Florence Griffith Joyner set a world record of 10.49 sec in the women's 100-m dash. The equation

$$R = -0.0433t + 10.49$$

can be used to predict the world record in the women's 100-m dash t years after 1988. For what years was the record between 10.15 and 10.35 sec? [14.2c]

Source: International Amateur Athletic Foundation

37. Find the intersection: [14.2a]

$\{1, 2, 5, 6, 9\} \cap \{1, 3, 5, 9\}$.

38. Find the union: [14.2b]

$\{1, 2, 5, 6, 9\} \cup \{1, 3, 5, 9\}$.

Graph. [14.4b]

39. $2x + 3y < 12$

40. 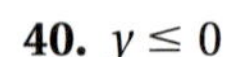 $y \leq 0$

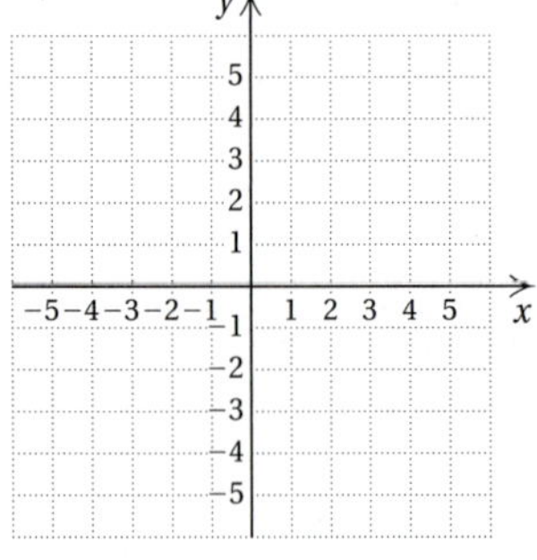

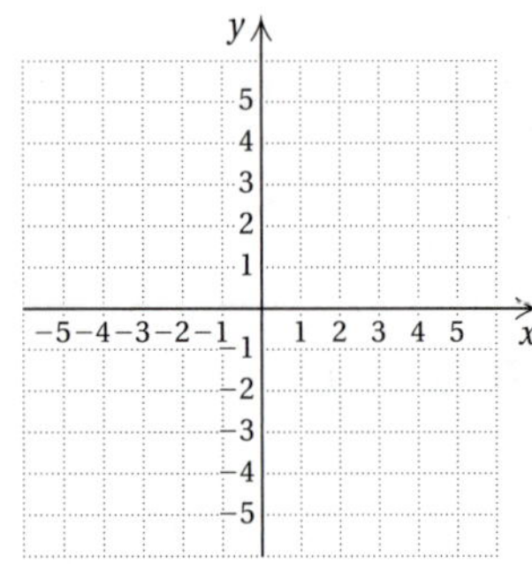

41. 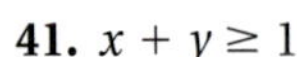 $x + y \geq 1$

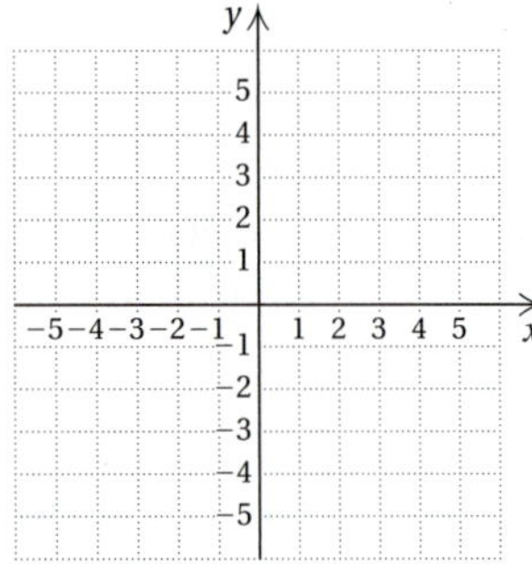

Graph. Find the coordinates of any vertices formed. [14.4c]

42. $y \geq -3,$
$x \geq 2$

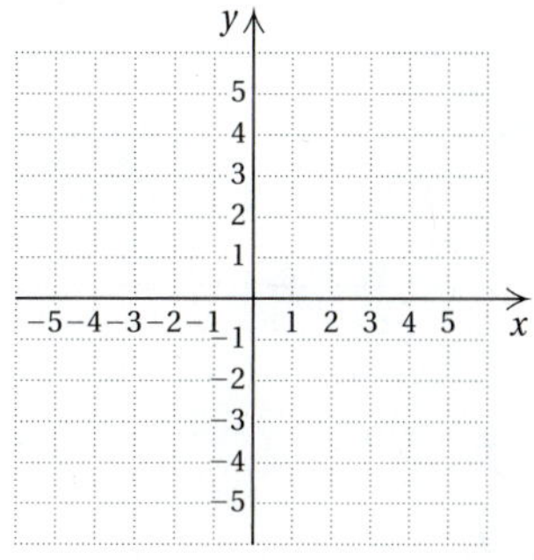

43. $x + 3y \geq -1,$
$x + 3y \leq 4$

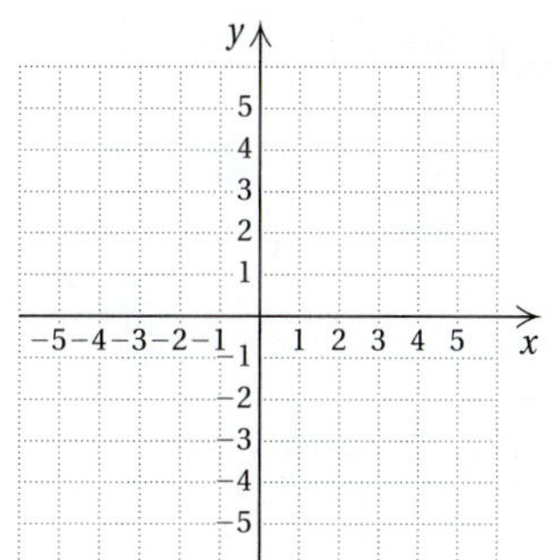

44. $x - 3y \leq 3,$
$x + 3y \geq 9,$
$y \leq 6$

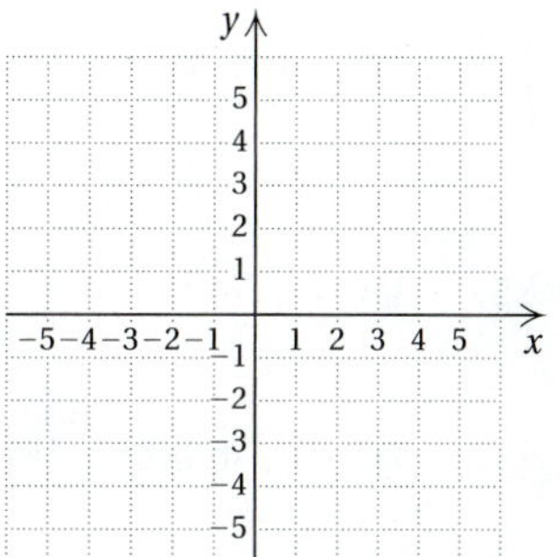

45. DW Find the error or errors in each of the following steps: [14.1c]

$$7 - 9x + 6x < -9(x + 2) + 10x$$
$$7 - 9x + 6x < -9x + 2 + 10x \quad (1)$$
$$7 + 6x > 2 + 10x \quad (2)$$
$$-4x > 8 \quad (3)$$
$$x > -2. \quad (4)$$

46. DW How does the word "solve" vary in meaning in this chapter? [14.1c, d], [14.2a, b, c], [14.3c, d, e]

SKILL MAINTENANCE

Certain objectives from three particular sections will be retested on the chapter test. The objectives are listed with the practice problems that follow.

47. Multiply: $(8y + 5)(3y - 7)$. [9.6a]

48. Find the domain: $f(x) = \dfrac{4}{x + 9}$. [12.2a]

49. Solve: $6x - y = 13,$ [13.2a], [13.3a]
$y - 5x = -11.$

50. Find an equation of the line containing the points $(3, -3)$ and $(-2, -1)$. [12.5c]

SYNTHESIS

51. Solve: $|2x + 5| \leq |x + 3|$. [14.3d, e]

CHAPTER 14

Chapter Test

Write interval notation for the given set or graph.

1. $\{x \mid -3 < x \leq 2\}$

2. 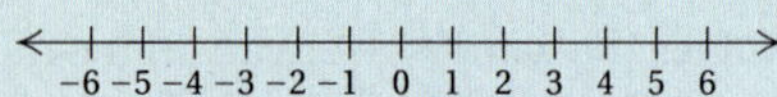

Solve and graph. Write interval notation for the solution set.

3. $x - 2 \leq 4$

4. $-4y - 3 \geq 5$

Solve.

5. $-0.6y < 30$

6. $3a - 5 \leq -2a + 6$

7. $-5y - 1 > -9y + 3$

8. $4(5 - x) < 2x + 5$

9. $-8(2x + 3) + 6(4 - 5x) \geq 2(1 - 7x) - 4(4 + 6x)$

Solve.

10. *Moving Costs.* Motivated Movers charges \$105 plus \$30 an hour to move households across town. Quick-Pak Moving charges \$80 an hour for cross-town moves. For what lengths of time is Quick-Pak more expensive?

11. *Pressure at Sea Depth.* The equation
$$P = 1 + \frac{d}{33}$$
gives the pressure P, in atmospheres (atm), at a depth of d feet in the sea. For what depths d is the pressure at least 2 atm and at most 8 atm?

Graph and write interval notation.

12. $-3 \leq x \leq 4$

13. $x < -3$ *or* $x > 4$

Solve.

14. $5 - 2x \leq 1$ *and* $3x + 2 \geq 14$

15. $-3 < x - 2 < 4$

16. $-11 \leq -5x - 2 < 0$

17. $-3x > 12$ *or* $4x > -10$

18. $x - 7 \leq -5$ *or* $x - 7 \geq -10$

19. $3x - 2 < 7$ *or* $x - 2 > 4$

Simplify.

20. $\left|\frac{7}{x}\right|$

21. $\left|\frac{-6x^2}{3x}\right|$

22. Find the distance between 4.8 and -3.6.

Solve.

23. $|x| = 9$

24. $|x| > 3$

25. $|4x - 1| < 4.5$

26. $|-5x - 3| \geq 10$

27. $|x + 10| = |x - 12|$

28. $|2 - 5x| = -10$

29. $\left|\dfrac{6 - x}{7}\right| \leq 15$

30. $|x - 3| = 9$

31. Find the intersection:
$\{1, 3, 5, 7, 9\} \cap \{3, 5, 11, 13\}$.

32. Find the union:
$\{1, 3, 5, 7, 9\} \cup \{3, 5, 11, 13\}$.

Graph. Find the coordinates of any vertices formed.

33. $x - 6y < -6$

34. $x + y \geq 3,$
$x - y \geq 5$

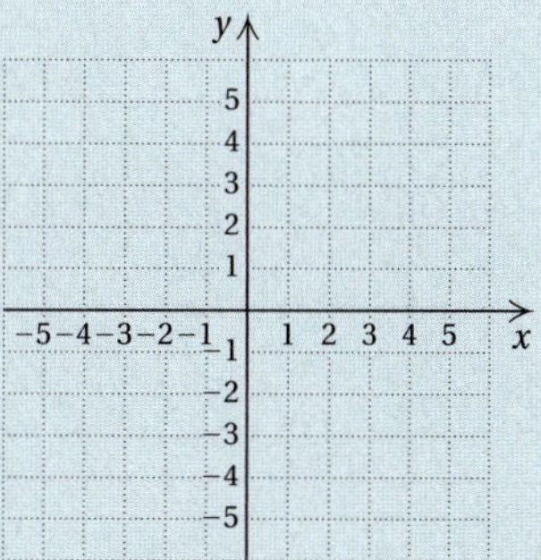

35. $2y - x \geq -7,$
$2y + 3x \leq 15,$
$y \leq 0,$
$x \leq 0$

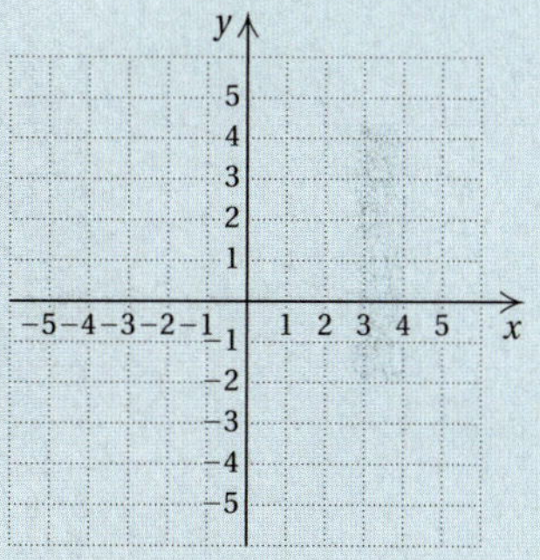

SKILL MAINTENANCE

36. Multiply: $(9x - 7)(8x + 5)$.

37. Find an equation of the line containing the points $(6, -8)$ and $(1, 2)$.

38. Solve: $4x - 5y = -12,$
$2x - 9y = 20.$

39. Find the domain: $f(x) = \dfrac{-20}{x - 20}$.

SYNTHESIS

Solve.

40. $|3x - 4| \leq -3$

41. $7x < 8 - 3x < 6 + 7x$

Radical Expressions, Equations, and Functions

Gateway to Chapter 15

In this chapter, we introduce square roots, cube roots, fourth roots, fifth roots, and so on. We study them in relation to radical expressions and functions and use them to solve applied problems like the one below. We also consider expressions with rational exponents and their related functions.

Real-World Application

The wind chill temperature is what the temperature would have to be with no wind in order to give the same chilling effect as with the wind. A formula for finding the wind chill temperature, T_W, is

$$T_W = 91.4 - (91.4 - T)(0.478 + 0.301\sqrt{v} - 0.02v),$$

where T is the actual temperature given by a thermometer, in degrees Fahrenheit, and v is the wind speed, in miles per hour. Find the wind chill temperature when $T = 10°\text{F}$ and $v = 20$ mph.

Source: The National Weather Service

This problem appears as Exercise 44 in Exercise Set 15.7.

CHAPTER 15

CHAPTER 15
Pretest

Simplify. Assume that letters can represent *any* real number.

1. $\sqrt{t^2}$ [15.1b]

2. $\sqrt[3]{27x^3}$ [15.1c]

3. $\sqrt[12]{y^{12}}$ [15.1d]

In Questions 4–20, assume that all expressions under the radicals represent positive numbers.

Simplify.

4. $(\sqrt[3]{9a^2b})^2$ [15.2d]

5. $\sqrt{45} - 3\sqrt{125} + 4\sqrt{80}$ [15.4a]

Multiply and simplify.

6. $\sqrt{18x^2}\,\sqrt{8x^3}$ [15.3a]

7. $(2\sqrt{6} - 1)^2$ [15.4b]

8. Divide and simplify: [15.3b]

$$\frac{\sqrt{52a^4}}{\sqrt{13a^3}}.$$

9. Rationalize the denominator: [15.5b]

$$\frac{3}{2 + \sqrt{5}}.$$

Solve.

10. $\sqrt{2x - 1} = 5$ [15.6a]

11. $\sqrt[3]{1 - 6x} = 2$ [15.6a]

12. $\sqrt{3x + 1} - \sqrt{2x} = 1$ [15.6b]

In Questions 13 and 14, give an exact answer and an approximation to three decimal places.

13. In a right triangle with leg $b = 5$ ft and hypotenuse $c = 8$ ft, find the length of leg a. [15.7a]

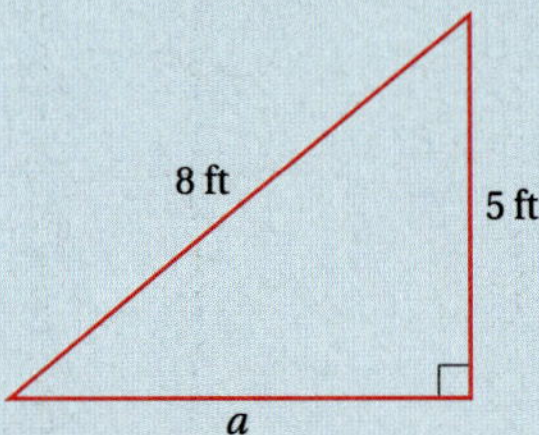

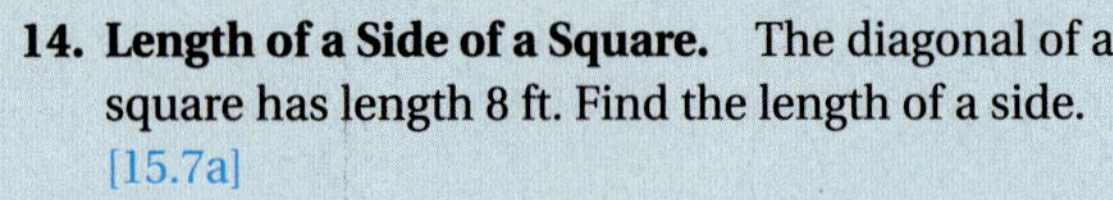

14. Length of a Side of a Square. The diagonal of a square has length 8 ft. Find the length of a side. [15.7a]

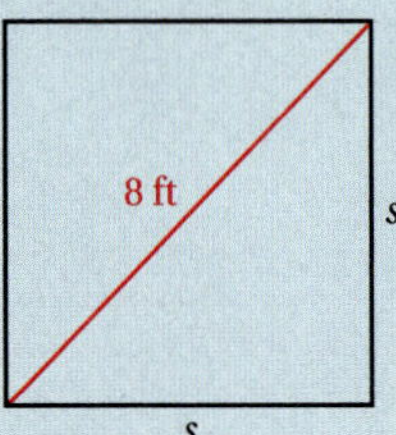

15. Use rational exponents to write a single radical expression: [15.2d]

$$\sqrt{x}\,\sqrt[3]{x}.$$

16. Determine whether $-1 + i$ is a solution of $x^2 + 2x + 4 = 0$. [15.8f]

17. Subtract: $(-2 + 7i) - (3 - 6i)$. [15.8b]

18. Multiply: $(4 + 3i)(3 - 4i)$. [15.8c]

19. Divide: $\dfrac{-4 + i}{3 - 2i}$. [15.8e]

20. Simplify: i^{87}. [15.8d]

15.1 RADICAL EXPRESSIONS AND FUNCTIONS

In this section, we consider roots, such as square roots and cube roots. We define the symbolism and consider methods of manipulating symbols to get equivalent expressions.

a Square Roots and Square-Root Functions

When we raise a number to the second power, we say that we have **squared** the number. Sometimes we may need to find the number that was squared. We call this process **finding a square root** of a number.

SQUARE ROOT

The number c is a **square root** of a if $c^2 = a$.

For example:

5 is a *square root* of 25 because $5^2 = 5 \cdot 5 = 25$;

-5 is a *square root* of 25 because $(-5)^2 = (-5)(-5) = 25$.

The number -4 does not have a real-number square root because there is no real number c such that $c^2 = -4$.

PROPERTIES OF SQUARE ROOTS

Every positive real number has two real-number square roots.

The number 0 has just one square root, 0 itself.

Negative numbers do not have real-number square roots.*

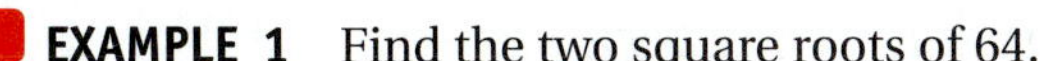

EXAMPLE 1 Find the two square roots of 64.

The square roots of 64 are 8 and -8 because $8^2 = 64$ and $(-8)^2 = 64$.

Do Exercises 1–3.

PRINCIPAL SQUARE ROOT

The **principal square root** of a nonnegative number is its nonnegative square root. The symbol $\sqrt{a}$ represents the principal square root of a. To name the negative square root of a, we can write $-\sqrt{a}$.

EXAMPLES Simplify.

2. $\sqrt{25} = 5$

Remember: $\sqrt{}$ indicates the principal (nonnegative) square root.

3. $-\sqrt{25} = -5$

*In Section 15.8, we will consider an expansion of the real-number system, in which negative numbers do have square roots.

Objectives

a Find principal square roots and their opposites, approximate square roots, find outputs of square-root functions, graph square-root functions, and find the domains of square-root functions.

b Simplify radical expressions with perfect-square radicands.

c Find cube roots, simplifying certain expressions, and find outputs of cube-root functions.

d Simplify expressions involving odd and even roots.

Find the square roots.

1. 9

2. 36

3. 121

Simplify.

4. $\sqrt{1}$

5. $\sqrt{36}$

6. $\sqrt{\dfrac{81}{100}}$

7. $\sqrt{0.0064}$

Answers on page A-55

Find the following.

8. a) $\sqrt{16}$

b) $-\sqrt{16}$

c) $\sqrt{-16}$

9. a) $\sqrt{49}$

b) $-\sqrt{49}$

c) $\sqrt{-49}$

10. a) $\sqrt{144}$

b) $-\sqrt{144}$

c) $\sqrt{-144}$

11. $\sqrt{\frac{25}{64}}$

12. $-\sqrt{0.81}$

13. $\sqrt{1.44}$

It would be helpful to memorize the following table of exact square roots.

TABLE OF COMMON SQUARE ROOTS	
$\sqrt{1} = 1$	$\sqrt{196} = 14$
$\sqrt{4} = 2$	$\sqrt{225} = 15$
$\sqrt{9} = 3$	$\sqrt{256} = 16$
$\sqrt{16} = 4$	$\sqrt{289} = 17$
$\sqrt{25} = 5$	$\sqrt{324} = 18$
$\sqrt{36} = 6$	$\sqrt{361} = 19$
$\sqrt{49} = 7$	$\sqrt{400} = 20$
$\sqrt{64} = 8$	$\sqrt{441} = 21$
$\sqrt{81} = 9$	$\sqrt{484} = 22$
$\sqrt{100} = 10$	$\sqrt{529} = 23$
$\sqrt{121} = 11$	$\sqrt{576} = 24$
$\sqrt{144} = 12$	$\sqrt{625} = 25$
$\sqrt{169} = 13$	

Use a calculator to approximate the square root to three decimal places.

14. $\sqrt{17}$

15. $\sqrt{40}$

16. $\sqrt{1138$}

17. $-\sqrt{867.6}$

18. $\sqrt{\frac{22}{35}}$

19. $-\sqrt{\frac{2103.4}{67.82}}$

Answers on page A-55

4. $\sqrt{\frac{81}{64}} = \frac{9}{8}$
5. $\sqrt{0.0049} = 0.07$
6. $-\sqrt{0.000001} = -0.001$
7. $\sqrt{0} = 0$
8. $\sqrt{-25}$ Does not exist as a real number. Negative numbers do not have real-number square roots.

Do Exercises 4–13. (Exercises 4–7 are on the preceding page.)

We found exact square roots in Examples 1–8. We often need to use rational numbers to approximate square roots that are irrational. Such expressions can be found using a calculator with a square-root key.

EXAMPLES Use a calculator to approximate each of the following.

	Number	*Using a calculator with a 10-digit readout*	*Rounded to three decimal places*
9.	$\sqrt{11}$	3.316624790	3.317
10.	$\sqrt{487}$	22.06807649	22.068
11.	$-\sqrt{7297.8}$	−85.42716196	−85.427
12.	$\sqrt{\frac{463}{557}}$	.9117229728	0.912

Do Exercises 14–19.

RADICAL; RADICAL EXPRESSION; RADICAND

The symbol $\sqrt{}$ is called a **radical.**

An expression written with a radical is called a **radical expression.**

The expression written under the radical is called the **radicand.**

These are radical expressions:

$$\sqrt{5}, \quad \sqrt{a}, \quad -\sqrt{5x}, \quad \sqrt{y^2 + 7}.$$

The radicands in these expressions are 5, a, $5x$, and $y^2 + 7$, respectively.

EXAMPLE 13 Identify the radicand in $\sqrt{x^2 - 9}$.

The radicand in $\sqrt{x^2 - 9}$ is $x^2 - 9$.

Do Exercises 20 and 21 on the following page.

Since each nonnegative real number x has exactly one principal square root, the symbol $\sqrt{x}$ represents exactly one real number and thus can be used to define a square-root function:

$$f(x) = \sqrt{x}.$$

The domain of this function is the set of nonnegative real numbers. In interval notation, the domain is $[0, \infty)$.

EXAMPLE 14 For the given function, find the indicated function values:

$$f(x) = \sqrt{3x - 2}; \quad f(1), f(5), \text{ and } f(0).$$

We have

$f(1) = \sqrt{3 \cdot 1 - 2}$ Substituting

$= \sqrt{3 - 2} = \sqrt{1} = 1;$ Simplifying and taking the square root

$f(5) = \sqrt{3 \cdot 5 - 2}$ Substituting

$= \sqrt{13} \approx 3.606;$ Simplifying and approximating

$f(0) = \sqrt{3 \cdot 0 - 2}$ Substituting

$= \sqrt{-2}.$ Negative radicand. No real-number function value exists; 0 is not in the domain of f.

Do Exercises 22 and 23.

EXAMPLE 15 Find the domain of $g(x) = \sqrt{x + 2}$.

The expression $\sqrt{x + 2}$ is a real number only when $x + 2$ is nonnegative. Thus the domain of $g(x) = \sqrt{x + 2}$ is the set of all x-values for which $x + 2 \geq 0$. We solve as follows:

$x + 2 \geq 0$

$x \geq -2.$ Adding -2

The domain of $g = \{x | x \geq -2\} = [-2, \infty)$.

EXAMPLE 16 Graph: **(a)** $f(x) = \sqrt{x}$; **(b)** $g(x) = \sqrt{x + 2}$.

We first find outputs as we did in Example 14. We can either select inputs that have exact outputs or use a calculator to make approximations. Once ordered pairs have been calculated, a smooth curve can be drawn.

a)

x	$f(x) = \sqrt{x}$	$(x, f(x))$
0	0	(0, 0)
1	1	(1, 1)
3	1.7	(3, 1.7)
4	2	(4, 2)
7	2.6	(7, 2.6)
9	3	(9, 3)

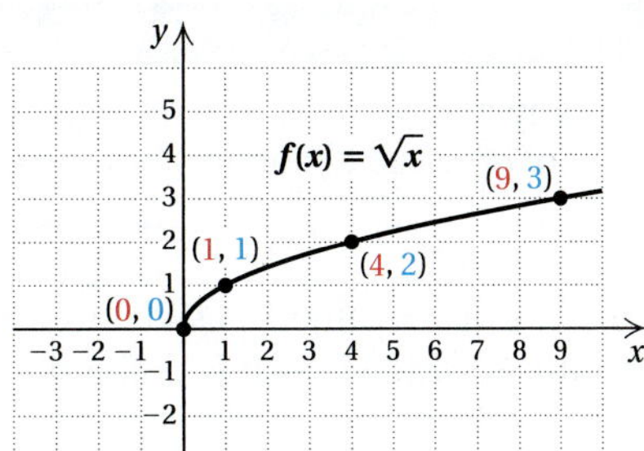

We can see from the table and the graph that the domain is $[0, \infty)$. The range is also the set of nonnegative real numbers $[0, \infty)$.

Identify the radicand.

20. $\sqrt{28 + x}$

21. $\sqrt{\dfrac{y}{y + 3}}$

For the given function, find the indicated function values.

22. $g(x) = \sqrt{6x + 4}$; $g(0)$, $g(3)$, and $g(-5)$

23. $f(x) = -\sqrt{x}$; $f(4)$, $f(7)$, and $f(-3)$

Answers on page A-55

b)

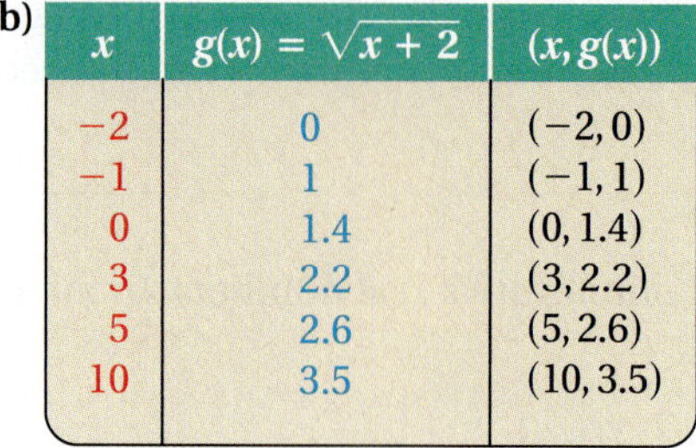

x	$g(x) = \sqrt{x+2}$	$(x, g(x))$
−2	0	(−2, 0)
−1	1	(−1, 1)
0	1.4	(0, 1.4)
3	2.2	(3, 2.2)
5	2.6	(5, 2.6)
10	3.5	(10, 3.5)

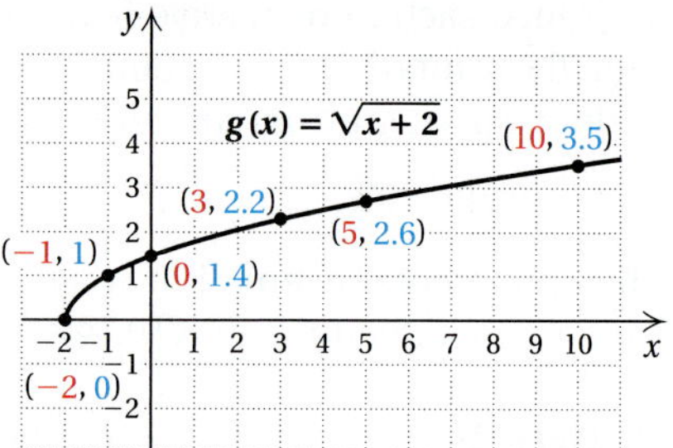

We can see from the table, the graph, and Example 15 that the domain is $[-2, \infty)$. The range is the set of nonnegative real numbers $[0, \infty)$.

Do Exercises 24–27.

b Finding $\sqrt{a^2}$

In the expression $\sqrt{a^2}$, the radicand is a perfect square. It is tempting to think that $\sqrt{a^2} = a$, but we see below that this is not the case.

Suppose $a = 5$. Then we have $\sqrt{5^2}$, which is $\sqrt{25}$, or 5.
Suppose $a = -5$. Then we have $\sqrt{(-5)^2}$, which is $\sqrt{25}$, or 5.
Suppose $a = 0$. Then we have $\sqrt{0^2}$, which is $\sqrt{0}$, or 0.

The symbol $\sqrt{a^2}$ never represents a negative number. It represents the principal square root of a^2. Note the following.

SIMPLIFYING $\sqrt{a^2}$

$a \geq 0 \longrightarrow \sqrt{a^2} = a$

If a is positive or 0, the principal square root of a^2 is a.

$a < 0 \longrightarrow \sqrt{a^2} = -a$

If a is negative, the principal square root of a^2 is the opposite of a.

In all cases, the radical expression represents the absolute value of a.

PRINCIPAL SQUARE ROOT OF a^2

For any real number a, $\sqrt{a^2} = |a|$. The principal (nonnegative) square root of a^2 is the absolute value of a.

The absolute value is used to ensure that the principal square root is nonnegative, which is as it is defined.

Find the domain of the function.

24. $f(x) = \sqrt{x-5}$

25. $g(x) = \sqrt{2x+3}$

Graph.

26. $g(x) = -\sqrt{x}$

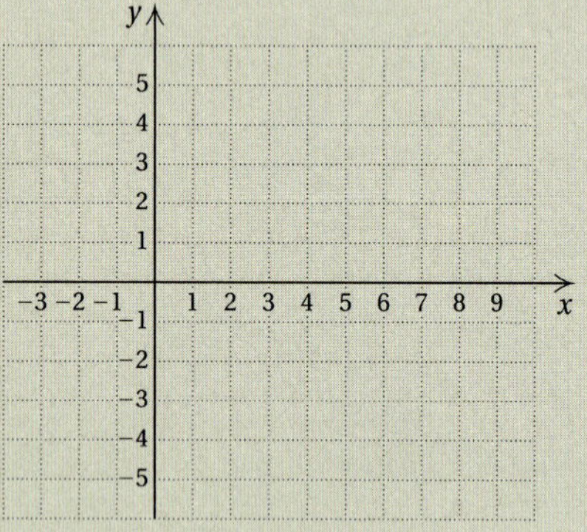

27. $f(x) = 2\sqrt{x+3}$

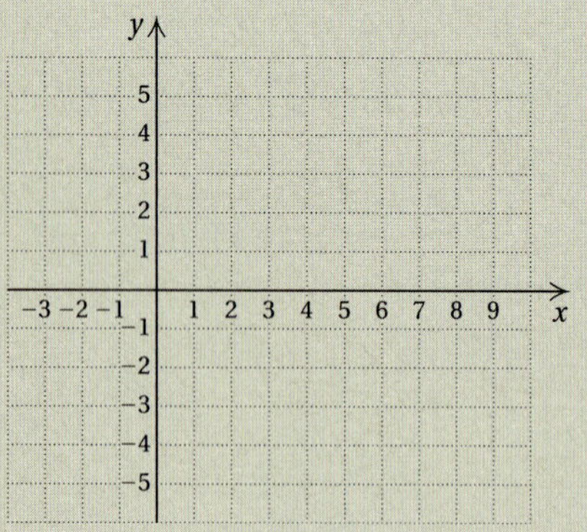

Answers on page A-55

EXAMPLES Find the following. Assume that letters can represent any real number.

17. $\sqrt{(-16)^2} = |-16|$, or 16

18. $\sqrt{(3b)^2} = |3b| = |3| \cdot |b| = 3|b|$

$|3b|$ can be simplified to $3|b|$ because the absolute value of any product is the product of the absolute values. That is, $|a \cdot b| = |a| \cdot |b|$.

19. $\sqrt{(x-1)^2} = |x-1|$

20. $\sqrt{x^2 + 8x + 16} = \sqrt{(x+4)^2}$
$= |x+4|$

CAUTION! $|x+4|$ is *not* the same as $|x| + 4$.

Do Exercises 28–35.

Find the following. Assume that letters can represent *any* real number.

28. $\sqrt{y^2}$

29. $\sqrt{(-24)^2}$

30. $\sqrt{(5y)^2}$

31. $\sqrt{16y^2}$

32. $\sqrt{(x+7)^2}$

33. $\sqrt{4(x-2)^2}$

34. $\sqrt{49(y+5)^2}$

35. $\sqrt{x^2 - 6x + 9}$

C Cube Roots

CUBE ROOT

The number c is the **cube root** of a if its third power is a—that is, if $c^3 = a$.

For example:

2 is the *cube root* of 8 because $2^3 = 2 \cdot 2 \cdot 2 = 8$;

-4 is the *cube root* of -64 because $(-4)^3 = (-4)(-4)(-4) = -64$.

We talk about *the* cube root of a number because of the following.

Every real number has exactly one cube root in the system of real numbers. The symbol $\sqrt[3]{a}$ represents *the* cube root of a.

EXAMPLES Find the following.

21. $\sqrt[3]{8} = 2$

22. $\sqrt[3]{-27} = -3$

23. $\sqrt[3]{-\frac{216}{125}} = -\frac{6}{5}$

24. $\sqrt[3]{0.001} = 0.1$

25. $\sqrt[3]{x^3} = x$

26. $\sqrt[3]{-8} = -2$

27. $\sqrt[3]{0} = 0$

28. $\sqrt[3]{-8y^3} = -2y$

When we are determining a cube root, no absolute-value signs are needed because a real number has just one cube root. The real-number cube root of a positive number is positive. The real-number cube root of a negative number is negative. The cube root of 0 is 0. That is, $\sqrt[3]{a^3} = a$ whether $a > 0$, $a < 0$, or $a = 0$.

Do Exercises 36–39.

Find the following.

36. $\sqrt[3]{-64}$

37. $\sqrt[3]{27y^3}$

38. $\sqrt[3]{8(x+2)^3}$

39. $\sqrt[3]{-\frac{343}{64}}$

Answers on page A-55

40. For the given function, find the indicated function values.

$g(x) = \sqrt[3]{x-4}$; $g(-23)$, $g(4)$, $g(-1)$, and $g(11)$

Find the following.

41. $\sqrt[5]{243}$

42. $\sqrt[5]{-243}$

43. $\sqrt[5]{x^5}$

44. $\sqrt[7]{y^7}$

45. $\sqrt[5]{0}$

46. $\sqrt[5]{-32x^5}$

47. $\sqrt[7]{(3x+2)^7}$

Answers on page A-55

Since the symbol $\sqrt[3]{x}$ represents exactly one real number, it can be used to define a cube-root function: $f(x) = \sqrt[3]{x}$.

EXAMPLE 29 For the given function, find the indicated function values.

$f(x) = \sqrt[3]{x}$; $f(125)$, $f(-8)$, $f(0)$, and $f(-10)$.

We have

$f(125) = \sqrt[3]{125} = 5$;

$f(0) = \sqrt[3]{0} = 0$;

$f(-8) = \sqrt[3]{-8} = -2$;

$f(-10) = \sqrt[3]{-10} \approx -2.1544$.

For calculator instructions for finding higher roots, see the Calculator Corner on p. 1126.

Do Exercise 40.

The graph of $f(x) = \sqrt[3]{x}$ is shown below for reference. Note that the domain and the range *each* consists of the entire set of real numbers, $(-\infty, \infty)$.

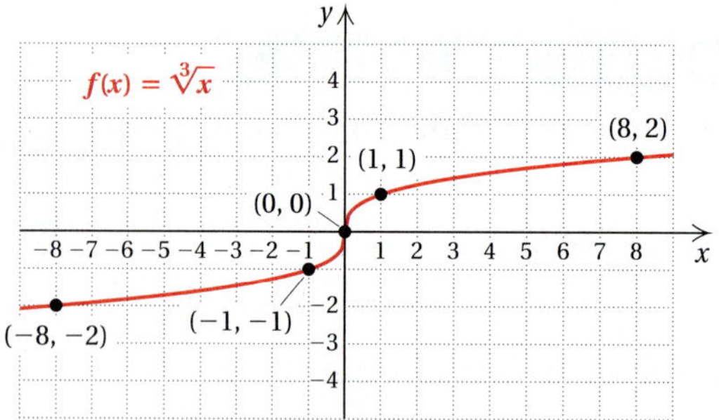

d Odd and Even kth Roots

In the expression $\sqrt[k]{a}$, we call k the **index** and assume $k \geq 2$.

ODD ROOTS

The 5th root of a number a is the number c for which $c^5 = a$. There are also 7th roots, 9th roots, and so on. Whenever the number k in $\sqrt[k]{}$ is an odd number, we say that we are taking an **odd root.**

Every number has just one real-number odd root. If the number is positive, then the root is positive. If the number is negative, then the root is negative. If the number is 0, then the root is 0. Absolute-value signs are *not* needed when we are finding odd roots.

> If k is an *odd* natural number, then for any real number a,
> $$\sqrt[k]{a^k} = a.$$

EXAMPLES Find the following.

30. $\sqrt[5]{32} = 2$

31. $\sqrt[5]{-32} = -2$

32. $-\sqrt[5]{32} = -2$

33. $-\sqrt[5]{-32} = -(-2) = 2$

34. $\sqrt[7]{x^7} = x$

35. $\sqrt[7]{128} = 2$

36. $\sqrt[7]{-128} = -2$

37. $\sqrt[7]{0} = 0$

38. $\sqrt[5]{a^5} = a$

39. $\sqrt[9]{(x-1)^9} = x - 1$

Do Exercises 41–47 on the preceding page.

EVEN ROOTS

When the index k in $\sqrt[k]{\ }$ is an even number, we say that we are taking an **even root.** When the index is 2, we do not write it. Every positive real number has two real-number kth roots when k is even. One of those roots is positive and one is negative. Negative real numbers do not have real-number kth roots when k is even. When we are finding even kth roots, absolute-value signs are sometimes necessary, as they are with square roots. For example,

$$\sqrt{64} = 8, \qquad \sqrt[6]{64} = 2, \qquad -\sqrt[6]{64} = -2, \qquad \sqrt[6]{64x^6} = |2x| = 2|x|.$$

Note that in $\sqrt[6]{64x^6}$, we need absolute-value signs because a variable is involved.

EXAMPLES Find the following. Assume that letters can represent any real number.

40. $\sqrt[4]{16} = 2$

41. $-\sqrt[4]{16} = -2$

42. $\sqrt[4]{-16}$ Does not exist as a real number.

43. $\sqrt[4]{81x^4} = 3|x|$

44. $\sqrt[6]{(y+7)^6} = |y + 7|$

45. $\sqrt{81y^2} = 9|y|$

The following is a summary of how absolute value is used when we are taking even or odd roots.

SIMPLIFYING $\sqrt[k]{a^k}$

For any real number a:

a) $\sqrt[k]{a^k} = |a|$ when k is an *even* natural number. We use absolute value when k is even unless a is nonnegative.

b) $\sqrt[k]{a^k} = a$ when k is an *odd* natural number greater than 1. We do not use absolute value when k is odd.

Do Exercises 48–56.

Find the following. Assume that letters can represent any real number.

48. $\sqrt[4]{81}$

49. $-\sqrt[4]{81}$

50. $\sqrt[4]{-81}$

51. $\sqrt[4]{0}$

52. $\sqrt[4]{16(x-2)^4}$

53. $\sqrt[6]{x^6}$

54. $\sqrt[8]{(x+3)^8}$

55. $\sqrt[7]{(x+3)^7}$

56. $\sqrt[5]{243x^5}$

Answers on page A-55

CALCULATOR CORNER

Approximating Roots We can use a graphing calculator to approximate square roots, cube roots, and higher roots of real numbers. To approximate $\sqrt{21}$, for example, we press [2nd] [$\sqrt{\ }$] [2] [1] [)] [ENTER]. ($\sqrt{\ }$ is the second operation associated with the [x^2] key.) To approximate $-\sqrt{6.95}$, we press [(–)] [2nd] [$\sqrt{\ }$] [6] [.] [9] [5] [)] [ENTER]. Although it is not necessary to include the right parenthesis in either of these entries, we do so here in order to close the set of parentheses that are opened when the calculator displays "$\sqrt{\ }$(". We see that $\sqrt{21} \approx 4.583$ and $-\sqrt{6.95} \approx -2.636$.

We can also find higher roots on a graphing calculator. To find $\sqrt[3]{-71}$, we will use the cube-root operation from the MATH menu. We press [MATH] [4] to select this operation. Then we press [(–)] [7] [1] [)] [ENTER] to enter the radicand and display the result. As with square roots, we choose to close the parentheses although it is not necessary. To find fourth, fifth, or higher roots, we use the *x*th-root operation from the MATH menu. To find $\sqrt[6]{178.4}$, we first press [6] to indicate that we are finding a sixth root. Then we press [MATH] [5] to select the *x*th-root operation. Finally, we press [1] [7] [8] [.] [4] [ENTER] to enter the radicand and display the result. Note that since this operation does not supply a left parenthesis, we do not enter a right parenthesis at the end. We see that $\sqrt[3]{-71} \approx -4.141$ and $\sqrt[6]{178.4} \approx 2.373$.

```
√(21)
              4.582575695
-√(6.95)
             -2.636285265
```

```
3√(-71)
             -4.140817749
6x√178.4
              2.372643426
```

Exercises: Use a graphing calculator to approximate each of the following to three decimal places.

1. $\sqrt{43}$
2. $\sqrt{10{,}467}$
3. $-\sqrt{9406}$
4. $-\sqrt{\dfrac{11}{17}}$
5. $\sqrt[3]{416.73}$
6. $-\sqrt[3]{-800}$
7. $\sqrt[4]{16.4}$
8. $\sqrt[7]{-1389.7}$

15.1 EXERCISE SET

For Extra Help

Digital Video Tutor CD 12 Videotape 17 | InterAct Math | Math Tutor Center | MathXL | MyMathLab

a Find the square roots.

1. 16 **2.** 225 **3.** 144 **4.** 9 **5.** 400 **6.** 81

Simplify.

7. $-\sqrt{\dfrac{49}{36}}$

8. $-\sqrt{\dfrac{361}{9}}$

9. $\sqrt{196}$

10. $\sqrt{441}$

11. $\sqrt{0.0036}$

12. $\sqrt{0.04}$

Use a calculator to approximate to three decimal places.

13. $\sqrt{347}$

14. $-\sqrt{1839.2}$

15. $\sqrt{\dfrac{285}{74}}$

16. $\sqrt{\dfrac{839.4}{19.7}}$

Identify the radicand.

17. $9\sqrt{y^2 + 16}$

18. $-3\sqrt{p^2 - 10}$

19. $x^4y^5\sqrt{\dfrac{x}{y-1}}$

20. $a^2b^2\sqrt{\dfrac{a^2-b}{b}}$

For the given function, find the indicated function values.

21. $f(x) = \sqrt{5x - 10}$; $f(6)$, $f(2)$, $f(1)$, and $f(-1)$

22. $t(x) = -\sqrt{2x + 1}$; $t(4)$, $t(0)$, $t(-1)$, and $t\left(-\frac{1}{2}\right)$

23. $g(x) = \sqrt{x^2 - 25}$; $g(-6)$, $g(3)$, $g(6)$, and $g(13)$

24. $F(x) = \sqrt{x^2 + 1}$; $F(0)$, $F(-1)$, and $F(-10)$

25. Find the domain of the function f in Exercise 21.

26. Find the domain of the function t in Exercise 22.

27. *Speed of a Skidding Car.* How do police determine how fast a car had been traveling after an accident has occurred? The function

$$S(x) = 2\sqrt{5x}$$

can be used to approximate the speed S, in miles per hour, of a car that has left a skid mark of length x, in feet. What was the speed of a car that left skid marks of length 30 ft? 150 ft?

28. *Parking-Lot Arrival Spaces.* The attendants at a parking lot park cars in temporary spaces before the cars are taken to permanent parking stalls. The number N of such spaces needed is approximated by the function

$$N(a) = 2.5\sqrt{a},$$

where a is the average number of arrivals in peak hours. What is the number of spaces needed when the average number of arrivals is 66? 100?

Graph.

29. $f(x) = 2\sqrt{x}$

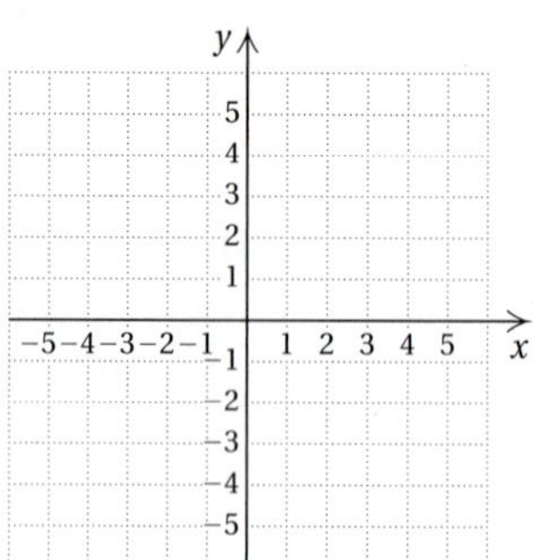

30. $g(x) = 3 - \sqrt{x}$

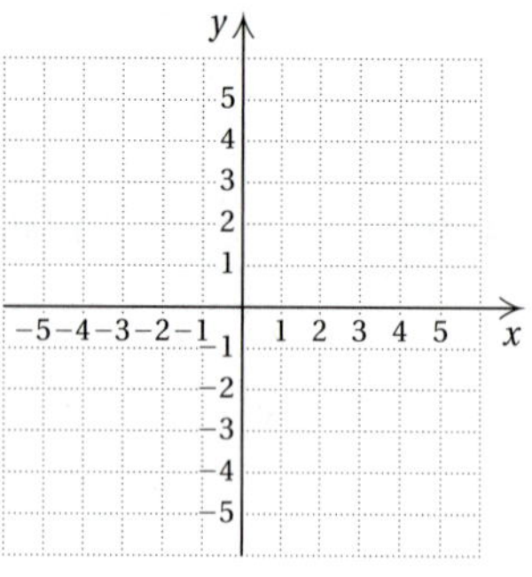

31. $F(x) = -3\sqrt{x}$

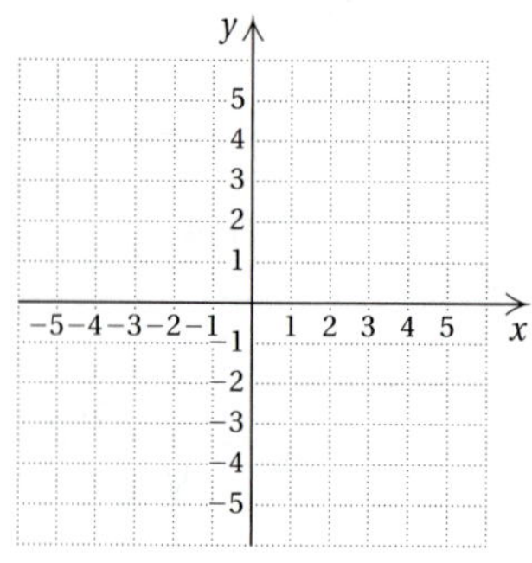

32. $f(x) = 2 + \sqrt{x - 1}$

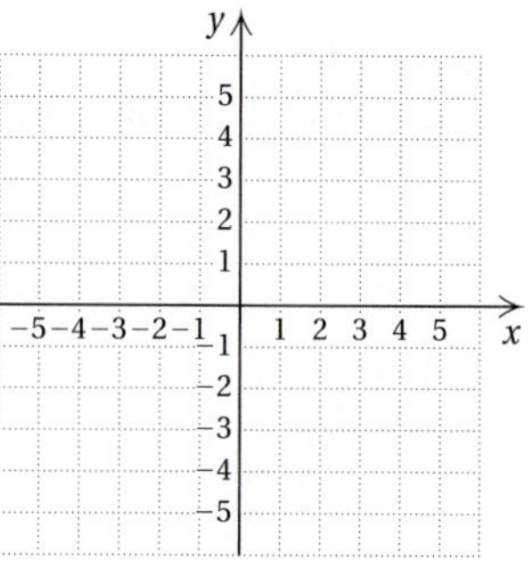

33. $f(x) = \sqrt{x}$

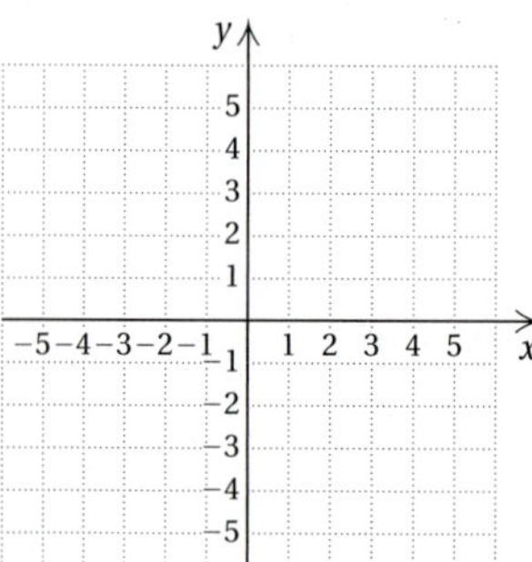

34. $g(x) = -\sqrt{x}$

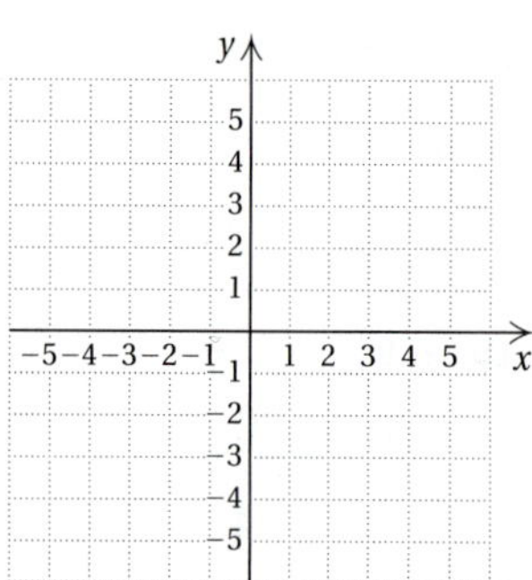

35. $f(x) = \sqrt{x - 2}$

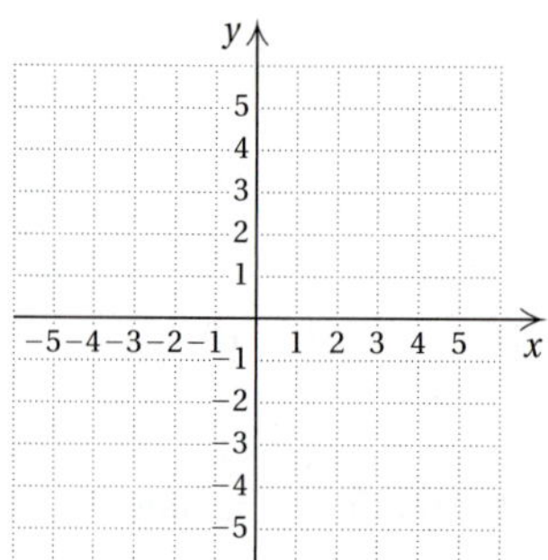

36. $g(x) = \sqrt{x + 3}$

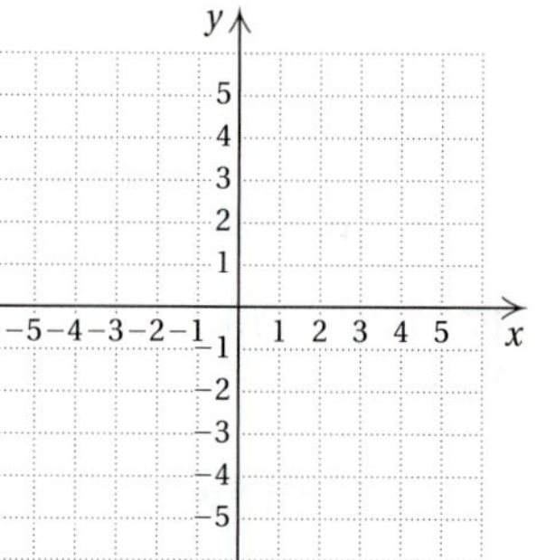

37. $f(x) = \sqrt{12 - 3x}$

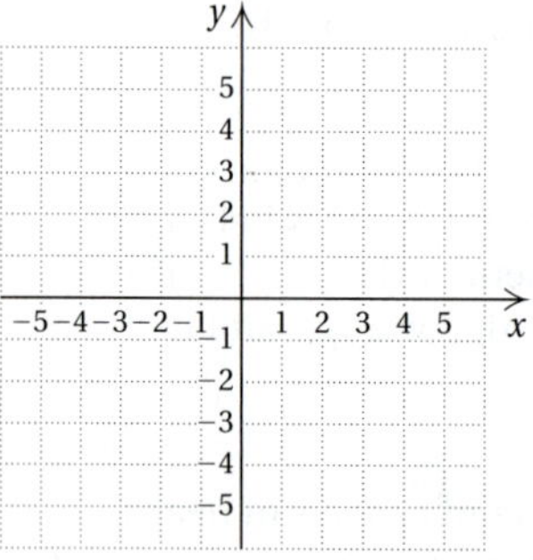

38. $g(x) = \sqrt{8 - 4x}$

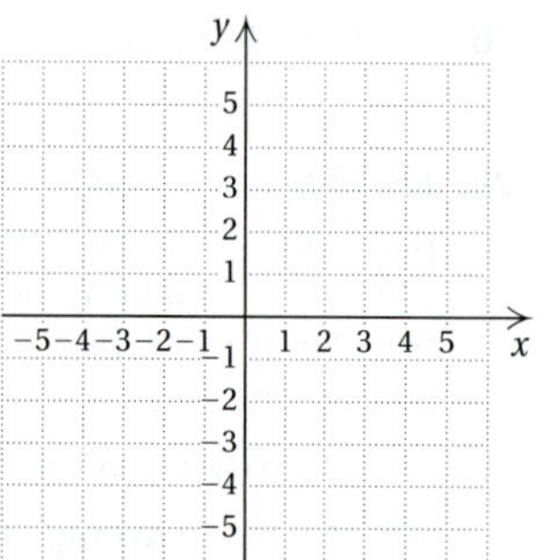

39. $g(x) = \sqrt{3x + 9}$

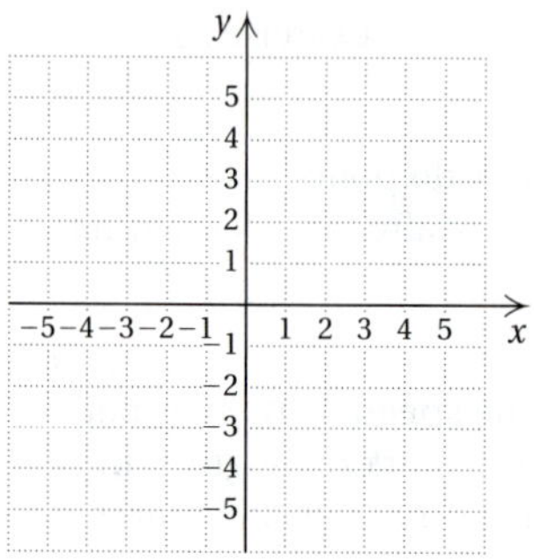

40. $f(x) = \sqrt{3x - 6}$

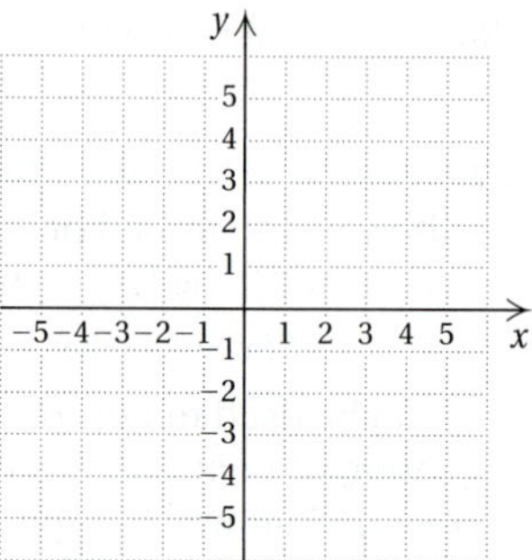

b Find the following. Assume that letters can represent *any* real number.

41. $\sqrt{16x^2}$

42. $\sqrt{25t^2}$

43. $\sqrt{(-12c)^2}$

44. $\sqrt{(-9d)^2}$

45. $\sqrt{(p + 3)^2}$

46. $\sqrt{(2 - x)^2}$

47. $\sqrt{x^2 - 4x + 4}$

48. $\sqrt{9t^2 - 30t + 25}$

c Simplify.

49. $\sqrt[3]{27}$

50. $-\sqrt[3]{64}$

51. $\sqrt[3]{-64x^3}$

52. $\sqrt[3]{-125y^3}$

53. $\sqrt[3]{-216}$

54. $-\sqrt[3]{-1000}$

55. $\sqrt[3]{0.343(x + 1)^3}$

56. $\sqrt[3]{0.000008(y - 2)^3}$

For the given function, find the indicated function values.

57. $f(x) = \sqrt[3]{x + 1}$; $f(7)$, $f(26)$, $f(-9)$, and $f(-65)$

58. $g(x) = -\sqrt[3]{2x - 1}$; $g(-62)$, $g(0)$, $g(-13)$, and $g(63)$

59. $f(x) = -\sqrt[3]{3x + 1}$; $f(0)$, $f(-7)$, $f(21)$, and $f(333)$

60. $g(t) = \sqrt[3]{t - 3}$; $g(30)$, $g(-5)$, $g(1)$, and $g(67)$

d Find the following. Assume that letters can represent *any* real number.

61. $-\sqrt[4]{625}$

62. $-\sqrt[4]{256}$

63. $\sqrt[5]{-1}$

64. $\sqrt[5]{-32}$

65. $\sqrt[5]{-\frac{32}{243}}$

66. $\sqrt[5]{-\frac{1}{32}}$

67. $\sqrt[6]{x^6}$

68. $\sqrt[8]{y^8}$

69. $\sqrt[4]{(5a)^4}$

70. $\sqrt[4]{(7b)^4}$

71. $\sqrt[10]{(-6)^{10}}$

72. $\sqrt[12]{(-10)^{12}}$

73. $\sqrt[414]{(a + b)^{414}}$

74. $\sqrt[1999]{(2a + b)^{1999}}$

75. $\sqrt[7]{y^7}$

76. $\sqrt[3]{(-6)^3}$

77. $\sqrt[5]{(x - 2)^5}$

78. $\sqrt[9]{(2xy)^9}$

79. $^{\text{D}}\text{W}$ Does the nth root of x^2 always exist? Why or why not?

80. $^{\text{D}}\text{W}$ Explain how to formulate a radical expression that can be used to define a function f with a domain of $\{x | x \leq 5\}$.

SKILL MAINTENANCE

Solve. [10.8b]

81. $x^2 + x - 2 = 0$

82. $x^2 + x = 0$

83. $4x^2 - 49 = 0$

84. $2x^2 - 26x + 72 = 0$

85. $3x^2 + x = 10$

86. $4x^2 - 20x + 25 = 0$

87. $4x^3 - 20x^2 + 25x = 0$

88. $x^3 - x^2 = 0$

Simplify.

89. $(a^3b^2c^5)^3$ [9.2b]

90. $(5a^7b^8)(2a^3b)$ [9.1d]

SYNTHESIS

91. Find the domain of

$$f(x) = \frac{\sqrt{x + 3}}{\sqrt{2 - x}}.$$

92. Use a graphing calculator to check your answers to Exercises 33, 37, and 39.

93. Use only the graph of $f(x) = \sqrt{x}$, shown below, to approximate $\sqrt{3}$, $\sqrt{5}$, and $\sqrt{10}$. Answers may vary.

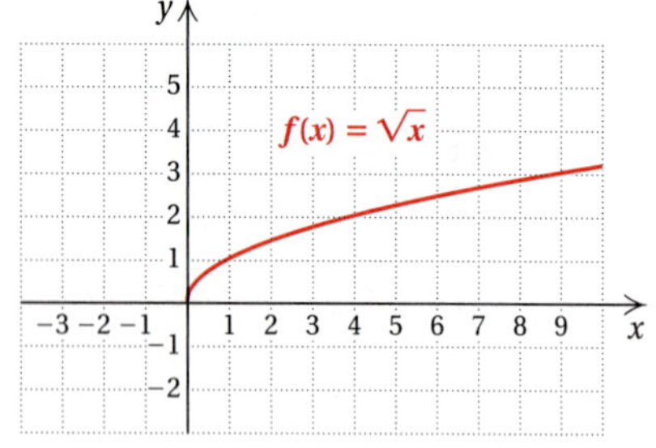

94. Use only the graph of $f(x) = \sqrt[3]{x}$, shown below, to approximate $\sqrt[3]{4}$, $\sqrt[3]{6}$, and $\sqrt[3]{-5}$. Answers may vary.

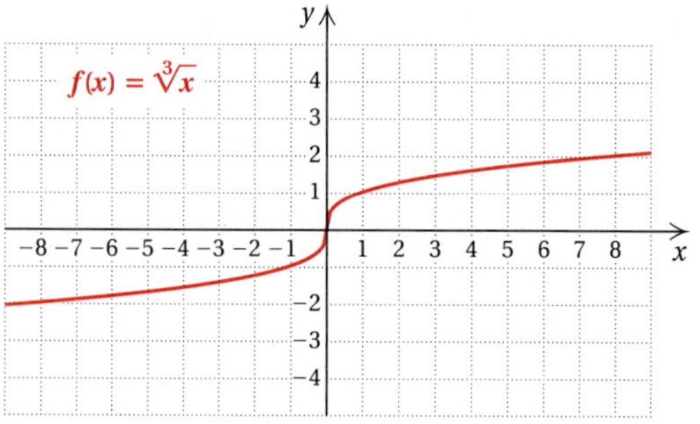

95. Use the TABLE, TRACE, and GRAPH features of a graphing calculator to find the domain and the range of each of the following functions.

a) $f(x) = \sqrt[3]{x}$

b) $g(x) = \sqrt[3]{4x - 5}$

c) $q(x) = 2 - \sqrt{x + 3}$

d) $h(x) = \sqrt[4]{x}$

e) $t(x) = \sqrt[4]{x - 3}$

15.2 RATIONAL NUMBERS AS EXPONENTS

Objectives

a Write expressions with or without rational exponents, and simplify, if possible.

b Write expressions without negative exponents, and simplify, if possible.

c Use the laws of exponents with rational exponents.

d Use rational exponents to simplify radical expressions.

In this section, we give meaning to expressions such as $a^{1/3}$, $7^{-1/2}$, and $(3x)^{0.84}$, which have rational numbers as exponents. We will see that using such notation can help simplify certain radical expressions.

a Rational Exponents

Expressions like $a^{1/2}$, $5^{-1/4}$, and $(2y)^{4/5}$ have not yet been defined. We will define such expressions so that the general properties of exponents hold.

Consider $a^{1/2} \cdot a^{1/2}$. If we want to multiply by adding exponents, it must follow that $a^{1/2} \cdot a^{1/2} = a^{1/2+1/2}$, or a^1. Thus we should define $a^{1/2}$ to be a square root of a. Similarly, $a^{1/3} \cdot a^{1/3} \cdot a^{1/3} = a^{1/3+1/3+1/3}$, or a^1, so $a^{1/3}$ should be defined to mean $\sqrt[3]{a}$.

$a^{1/n}$

For any *nonnegative* real number a and any natural number index n $(n \neq 1)$,

$a^{1/n}$ means $\sqrt[n]{a}$ (the nonnegative nth root of a).

Whenever we use rational exponents, we assume that the bases are nonnegative.

EXAMPLES Rewrite without rational exponents, and simplify, if possible.

1. $x^{1/2} = \sqrt{x}$ An index of 2 is not written.
2. $27^{1/3} = \sqrt[3]{27} = 3$
3. $(abc)^{1/5} = \sqrt[5]{abc}$

Do Exercises 1–5.

Rewrite without rational exponents, and simplify, if possible.

1. $y^{1/4}$
2. $(3a)^{1/2}$
3. $16^{1/4}$
4. $(125)^{1/3}$
5. $(a^3b^2c)^{1/5}$

EXAMPLES Rewrite with rational exponents.

4. $\sqrt[5]{7xy} = (7xy)^{1/5}$ We need parentheses around the radicand here.
5. $8\sqrt[3]{xy} = 8(xy)^{1/3}$
6. $\sqrt[7]{\frac{x^3y}{9}} = \left(\frac{x^3y}{9}\right)^{1/7}$

Do Exercises 6–9.

Rewrite with rational exponents.

6. $\sqrt[3]{19ab}$
7. $19\sqrt[3]{ab}$
8. $\sqrt[5]{\frac{x^2y}{16}}$
9. $7\sqrt[4]{2ab}$

How should we define $a^{2/3}$? If the general properties of exponents are to hold, we have $a^{2/3} = (a^{1/3})^2$, or $(a^2)^{1/3}$, or $(\sqrt[3]{a})^2$, or $\sqrt[3]{a^2}$. We define this accordingly.

$a^{m/n}$

For any natural numbers m and n $(n \neq 1)$ and any nonnegative real number a,

$a^{m/n}$ means $\sqrt[n]{a^m}$, or $(\sqrt[n]{a})^m$.

Answers on page A-56

Rewrite without rational exponents, and simplify, if possible.

10. $x^{3/5}$

11. $8^{2/3}$

12. $4^{5/2}$

Rewrite with rational exponents.

13. $(\sqrt[3]{7abc})^4$

14. $\sqrt[5]{6^7}$

Rewrite with positive exponents, and simplify, if possible.

15. $16^{-1/4}$

16. $(3xy)^{-7/8}$

17. $81^{-3/4}$

18. $7p^{3/4}q^{-6/5}$

19. $\left(\frac{11m}{7n}\right)^{-2/3}$

Answers on page A-56

EXAMPLES Rewrite without rational exponents, and simplify, if possible.

7. $(27)^{2/3} = \sqrt[3]{27^2} = (\sqrt[3]{27})^2 = 3^2 = 9$

8. $4^{3/2} = \sqrt[2]{4^3} = (\sqrt[2]{4})^3 = 2^3 = 8$

Do Exercises 10–12.

EXAMPLES Rewrite with rational exponents.

The index becomes the denominator of the rational exponent.

9. $\sqrt[3]{9^4} = 9^{4/3}$

10. $(\sqrt[4]{7xy})^5 = (7xy)^{5/4}$

Do Exercises 13 and 14.

b Negative Rational Exponents

Negative rational exponents have a meaning similar to that of negative integer exponents.

$a^{-m/n}$

For any rational number m/n and any positive real number a,

$$a^{-m/n} \quad \text{means} \quad \frac{1}{a^{m/n}},$$

that is, $a^{m/n}$ and $a^{-m/n}$ are reciprocals.

EXAMPLES Rewrite with positive exponents, and simplify, if possible.

11. $9^{-1/2} = \frac{1}{9^{1/2}} = \frac{1}{\sqrt{9}} = \frac{1}{3}$

12. $(5xy)^{-4/5} = \frac{1}{(5xy)^{4/5}}$

13. $64^{-2/3} = \frac{1}{64^{2/3}} = \frac{1}{(\sqrt[3]{64})^2} = \frac{1}{4^2} = \frac{1}{16}$

14. $4x^{-2/3}y^{1/5} = 4 \cdot \frac{1}{x^{2/3}} \cdot y^{1/5} = \frac{4y^{1/5}}{x^{2/3}}$

15. $\left(\frac{3r}{7s}\right)^{-5/2} = \left(\frac{7s}{3r}\right)^{5/2}$ Since $\left(\frac{a}{b}\right)^{-n} = \left(\frac{b}{a}\right)^n$

Do Exercises 15–19.

CALCULATOR CORNER

Rational Exponents We can use a graphing calculator to approximate rational roots of real numbers. To approximate $7^{2/3}$, we press [7] [^] [(] [2] [÷] [3] [)] [ENTER]. Note that the parentheses around the exponent are necessary. If they are not used, the calculator will read the expression as $7^2 \div 3$. To approximate $14^{-1.9}$, we press [1] [4] [^] [(−)] [1] [·] [9] [ENTER]. Parentheses are not required when a rational exponent is expressed in a single decimal number. The display indicates that $7^{2/3} \approx 3.659$ and $14^{-1.9} \approx 0.007$.

```
7^(2/3)
                 3.65930571
14^-1.9
                 .006642885
```

Exercises: Approximate each of the following.

1. $5^{3/4}$
2. $8^{4/7}$
3. $29^{-3/8}$
4. $73^{0.56}$
5. $34^{-2.78}$
6. $32^{0.2}$

C Laws of Exponents

The same laws hold for rational-number exponents as for integer exponents. We list them for review.

For any real number a and any rational exponents m and n:

1. $a^m \cdot a^n = a^{m+n}$ In multiplying, we can add exponents if the bases are the same.
2. $\dfrac{a^m}{a^n} = a^{m-n}$ In dividing, we can subtract exponents if the bases are the same.
3. $(a^m)^n = a^{m \cdot n}$ To raise a power to a power, we can multiply the exponents.
4. $(ab)^m = a^m b^m$ To raise a product to a power, we can raise each factor to the power.
5. $\left(\dfrac{a}{b}\right)^n = \dfrac{a^n}{b^n}$ To raise a quotient to a power, we can raise both the numerator and the denominator to the power.

EXAMPLES Use the laws of exponents to simplify.

16. $3^{1/5} \cdot 3^{3/5} = 3^{1/5+3/5} = 3^{4/5}$ Adding exponents

17. $\dfrac{7^{1/4}}{7^{1/2}} = 7^{1/4-1/2} = 7^{1/4-2/4} = 7^{-1/4} = \dfrac{1}{7^{1/4}}$ Subtracting exponents

18. $(7.2^{2/3})^{3/4} = 7.2^{2/3 \cdot 3/4} = 7.2^{6/12} = 7.2^{1/2}$ Multiplying exponents

19. $(a^{-1/3}b^{2/5})^{1/2} = a^{-1/3 \cdot 1/2} \cdot b^{2/5 \cdot 1/2}$ Raising a product to a power and multiplying exponents

$$= a^{-1/6}b^{1/5} = \frac{b^{1/5}}{a^{1/6}}$$

Do Exercises 20–23.

Use the laws of exponents to simplify.

20. $7^{1/3} \cdot 7^{3/5}$

21. $\dfrac{5^{7/6}}{5^{5/6}}$

22. $(9^{3/5})^{2/3}$

23. $(p^{-2/3}q^{1/4})^{1/2}$

Answers on page A-56

Use rational exponents to simplify.

24. $\sqrt[4]{a^2}$

25. $\sqrt[4]{x^4}$

26. $\sqrt[6]{8}$

Use rational exponents to simplify.

27. $\sqrt[12]{x^3y^6}$

28. $\sqrt[6]{a^{12}b^3}$

29. $\sqrt[5]{a^5b^{10}}$

30. Use rational exponents to write a single radical expression:

$$\sqrt[4]{7} \cdot \sqrt{3}.$$

Answers on page A-56

d Simplifying Radical Expressions

Rational exponents can be used to simplify some radical expressions. The procedure is as follows.

SIMPLIFYING RADICAL EXPRESSIONS

1. Convert radical expressions to exponential expressions.
2. Use arithmetic and the laws of exponents to simplify.
3. Convert back to radical notation when appropriate.

Important: This procedure works only when all expressions under radicals are *nonnegative* since rational exponents are not defined otherwise. With this assumption, no absolute-value signs will be needed.

EXAMPLES Use rational exponents to simplify.

20. $\sqrt[6]{x^3} = x^{3/6}$ Converting to an exponential expression
$= x^{1/2}$ Simplifying the exponent
$= \sqrt{x}$ Converting back to radical notation

21. $\sqrt[6]{4} = 4^{1/6}$ Converting to exponential notation
$= (2^2)^{1/6}$ Renaming 4 as 2^2
$= 2^{2/6}$ Using $(a^m)^n = a^{mn}$; multiplying exponents
$= 2^{1/3}$ Simplifying the exponent
$= \sqrt[3]{2}$ Converting back to radical notation

Do Exercises 24–26.

EXAMPLE 22 Use rational exponents to simplify: $\sqrt[8]{a^2b^4}$.

$\sqrt[8]{a^2b^4} = (a^2b^4)^{1/8}$ Converting to exponential notation
$= a^{2/8} \cdot b^{4/8}$ Using $(ab)^n = a^nb^n$
$= a^{1/4} \cdot b^{1/2}$ Simplifying the exponents
$= a^{1/4} \cdot b^{2/4}$ Rewriting $\frac{1}{2}$ with a denominator of 4
$= (ab^2)^{1/4}$ Using $a^nb^n = (ab)^n$
$= \sqrt[4]{ab^2}$ Converting back to radical notation

Do Exercises 27–29.

We can use properties of rational exponents to write a single radical expression for a product or a quotient.

EXAMPLE 23 Use rational exponents to write a single radical expression for $\sqrt[3]{5} \cdot \sqrt{2}$.

$\sqrt[3]{5} \cdot \sqrt{2} = 5^{1/3} \cdot 2^{1/2}$ Converting to exponential notation
$= 5^{2/6} \cdot 2^{3/6}$ Rewriting so that exponents have a common denominator
$= (5^2 \cdot 2^3)^{1/6}$ Using $a^nb^n = (ab)^n$
$= \sqrt[6]{5^2 \cdot 2^3}$ Converting back to radical notation
$= \sqrt[6]{200}$ Multiplying under the radical

Do Exercise 30 on the preceding page.

EXAMPLE 24 Write a single radical expression for $a^{1/2}b^{-1/2}c^{5/6}$.

$a^{1/2}b^{-1/2}c^{5/6} = a^{3/6}b^{-3/6}c^{5/6}$ — Rewriting so that exponents have a common denominator

$= (a^3b^{-3}c^5)^{1/6}$ — Using $a^nb^n = (ab)^n$

$= \sqrt[6]{a^3b^{-3}c^5}$ — Converting to radical notation

EXAMPLE 25 Write a single radical expression for $\dfrac{x^{5/6} \cdot y^{3/8}}{x^{4/9} \cdot y^{1/4}}$.

$\dfrac{x^{5/6} \cdot y^{3/8}}{x^{4/9} \cdot y^{1/4}} = x^{5/6-4/9} \cdot y^{3/8-1/4}$ — Subtracting exponents

$= x^{15/18-8/18} \cdot y^{3/8-2/8}$ — Finding common denominators so that exponents can be subtracted

$= x^{7/18} \cdot y^{1/8}$ — Carrying out the subtraction of exponents

$= x^{28/72} \cdot y^{9/72}$ — Rewriting so that all exponents have a common denominator

$= \sqrt[72]{x^{28}y^9}$ — Converting to radical notation

Do Exercises 31 and 32.

EXAMPLES Use rational exponents to simplify.

26. $\sqrt[6]{(5x)^3} = (5x)^{3/6}$ — Converting to exponential notation

$= (5x)^{1/2}$ — Simplifying the exponent

$= \sqrt{5x}$ — Converting back to radical notation

27. $\sqrt[5]{t^{20}} = t^{20/5}$ — Converting to exponential notation

$= t^4$ — Simplifying the exponent

28. $(\sqrt[3]{pq^2c})^{12} = (pq^2c)^{12/3}$ — Converting to exponential notation

$= (pq^2c)^4$ — Simplifying the exponent

$= p^4q^8c^4$ — Using $(ab)^n = a^nb^n$

29. $\sqrt{\sqrt[3]{x}} = \sqrt{x^{1/3}}$ — Converting the radicand to exponential notation

$= (x^{1/3})^{1/2}$ — Try to go directly to this step.

$= x^{1/6}$ — Multiplying exponents

$= \sqrt[6]{x}$ — Converting back to radical notation

Do Exercises 33–36.

Write a single radical expression.

31. $x^{2/3}y^{1/2}z^{5/6}$

32. $\dfrac{a^{1/2}b^{3/8}}{a^{1/4}b^{1/8}}$

Use rational exponents to simplify.

33. $\sqrt[14]{(5m)^2}$

34. $\sqrt[18]{m^3}$

35. $(\sqrt[6]{a^5b^3c})^{24}$

36. $\sqrt[5]{\sqrt{x}}$

Answers on page A-56

15.2 EXERCISE SET

For Extra Help

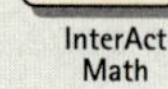

Digital Video Tutor CD 12 Videotape 17 | InterAct Math | Math Tutor Center | MathXL | MyMathLab

a Rewrite without rational exponents, and simplify, if possible.

1. $y^{1/7}$ **2.** $x^{1/6}$ **3.** $8^{1/3}$ **4.** $16^{1/2}$ **5.** $(a^3b^3)^{1/5}$

6. $(x^2y^2)^{1/3}$ **7.** $16^{3/4}$ **8.** $4^{7/2}$ **9.** $49^{3/2}$ **10.** $27^{4/3}$

Rewrite with rational exponents.

11. $\sqrt{17}$ **12.** $\sqrt{x^3}$ **13.** $\sqrt[3]{18}$ **14.** $\sqrt[3]{23}$ **15.** $\sqrt[5]{xy^2z}$

16. $\sqrt[7]{x^3y^2z^2}$ **17.** $(\sqrt{3mn})^3$ **18.** $(\sqrt[3]{7xy})^4$ **19.** $(\sqrt[7]{8x^2y})^5$ **20.** $(\sqrt[6]{2a^5b})^7$

b Rewrite with positive exponents, and simplify, if possible.

21. $27^{-1/3}$ **22.** $100^{-1/2}$ **23.** $100^{-3/2}$ **24.** $16^{-3/4}$ **25.** $x^{-1/4}$

26. $y^{-1/7}$ **27.** $(2rs)^{-3/4}$ **28.** $(5xy)^{-5/6}$ **29.** $2a^{3/4}b^{-1/2}c^{2/3}$ **30.** $5x^{-2/3}y^{4/5}z$

31. $\left(\frac{7x}{8yz}\right)^{-3/5}$ **32.** $\left(\frac{2ab}{3c}\right)^{-5/6}$ **33.** $\frac{1}{x^{-2/3}}$ **34.** $\frac{1}{a^{-7/8}}$ **35.** $2^{-1/3}x^4y^{-2/7}$

36. $3^{-5/2}a^3b^{-7/3}$ **37.** $\frac{7x}{\sqrt[3]{z}}$ **38.** $\frac{6a}{\sqrt[4]{b}}$ **39.** $\frac{5a}{3c^{-1/2}}$ **40.** $\frac{2z}{5x^{-1/3}}$

c Use the laws of exponents to simplify. Write the answers with positive exponents.

41. $5^{3/4} \cdot 5^{1/8}$ **42.** $11^{2/3} \cdot 11^{1/2}$ **43.** $\frac{7^{5/8}}{7^{3/8}}$ **44.** $\frac{3^{5/8}}{3^{-1/8}}$ **45.** $\frac{4.9^{-1/6}}{4.9^{-2/3}}$

46. $\frac{2.3^{-3/10}}{2.3^{-1/5}}$ **47.** $(6^{3/8})^{2/7}$ **48.** $(3^{2/9})^{3/5}$ **49.** $a^{2/3} \cdot a^{5/4}$ **50.** $x^{3/4} \cdot x^{2/3}$

51. $(a^{2/3} \cdot b^{5/8})^4$

52. $(x^{-1/3} \cdot y^{-2/5})^{-15}$

53. $(x^{2/3})^{-3/7}$

54. $(a^{-3/2})^{2/9}$

d Use rational exponents to simplify. Write the answer in radical notation if appropriate.

55. $\sqrt[6]{a^2}$

56. $\sqrt[6]{t^4}$

57. $\sqrt[3]{x^{15}}$

58. $\sqrt[4]{a^{12}}$

59. $\sqrt[6]{x^{-18}}$

60. $\sqrt[5]{a^{-10}}$

61. $(\sqrt[3]{ab})^{15}$

62. $(\sqrt[7]{cd})^{14}$

63. $\sqrt[14]{128}$

64. $\sqrt[6]{81}$

65. $\sqrt[6]{4x^2}$

66. $\sqrt[3]{8y^6}$

67. $\sqrt{x^4y^6}$

68. $\sqrt[4]{16x^4y^2}$

69. $\sqrt[5]{32c^{10}d^{15}}$

Use rational exponents to write a single radical expression.

70. $\sqrt[3]{3}\,\sqrt{3}$

71. $\sqrt[3]{7} \cdot \sqrt[4]{5}$

72. $\sqrt[7]{11} \cdot \sqrt[6]{13}$

73. $\sqrt[4]{5} \cdot \sqrt[5]{7}$

74. $\sqrt[3]{y}\,\sqrt[5]{3y}$

75. $\sqrt{x}\,\sqrt[3]{2x}$

76. $(\sqrt[3]{x^2y^5})^{12}$

77. $(\sqrt[5]{a^2b^4})^{15}$

78. $\sqrt[4]{\sqrt{x}}$

79. $\sqrt[3]{\sqrt[6]{m}}$

80. $a^{2/3} \cdot b^{3/4}$

81. $x^{1/3} \cdot y^{1/4} \cdot z^{1/6}$

82. $\dfrac{x^{8/15} \cdot y^{7/5}}{x^{1/3} \cdot y^{-1/5}}$

83. $\left(\dfrac{c^{-4/5}d^{5/9}}{c^{3/10}d^{1/6}}\right)^3$

84. $\sqrt[3]{\sqrt[4]{xy}}$

85. **DW** Find the domain of

$$f(x) = (x + 5)^{1/2}(x + 7)^{-1/2}$$

and explain how you found your answer.

86. **DW** Explain why $\sqrt[3]{x^6} = x^2$ for any value of x, but $\sqrt{x^6} = x^3$ only when $x \geq 0$.

SKILL MAINTENANCE

Solve. [14.3c]

87. $|7x - 5| = 9$

88. $|3x| = 120$

89. $8 - |2x + 5| = -2$

90. $\left|\dfrac{1}{2} + x\right| = \dfrac{7}{8}$

SYNTHESIS

91. Use the SIMULTANEOUS mode to graph

$$y_1 = x^{1/2},\quad y_2 = 3x^{2/5},\quad y_3 = x^{4/7},\quad y_4 = \tfrac{1}{5}x^{3/4}.$$

Then, looking only at coordinates, match each graph with its equation.

92. Simplify:

$$(\sqrt[10]{\sqrt[5]{x^{15}}})^5(\sqrt[5]{\sqrt[10]{x^{15}}})^5.$$

15.3 SIMPLIFYING RADICAL EXPRESSIONS

Objectives

a Multiply and simplify radical expressions.

b Divide and simplify radical expressions.

a Multiplying and Simplifying Radical Expressions

Note that $\sqrt{4}\,\sqrt{25} = 2 \cdot 5 = 10$. Also $\sqrt{4 \cdot 25} = \sqrt{100} = 10$. Likewise,

$$\sqrt[3]{27}\,\sqrt[3]{8} = 3 \cdot 2 = 6 \quad \text{and} \quad \sqrt[3]{27 \cdot 8} = \sqrt[3]{216} = 6.$$

These examples suggest the following.

THE PRODUCT RULE FOR RADICALS

For any nonnegative real numbers a and b and any index k,

$$\sqrt[k]{a} \cdot \sqrt[k]{b} = \sqrt[k]{a \cdot b}, \quad \text{or} \quad a^{1/k} \cdot b^{1/k} = (ab)^{1/k}.$$

(To multiply, multiply the radicands.)

The index must be the same throughout.

EXAMPLES Multiply.

1. $\sqrt{3} \cdot \sqrt{5} = \sqrt{3 \cdot 5} = \sqrt{15}$
2. $\sqrt{5a}\,\sqrt{2b} = \sqrt{5a \cdot 2b} = \sqrt{10ab}$
3. $\sqrt[3]{4}\,\sqrt[3]{5} = \sqrt[3]{4 \cdot 5} = \sqrt[3]{20}$
4. $\sqrt[4]{\dfrac{y}{5}}\,\sqrt[4]{\dfrac{7}{x}} = \sqrt[4]{\dfrac{y}{5} \cdot \dfrac{7}{x}} = \sqrt[4]{\dfrac{7y}{5x}}$

CAUTION! A common error is to omit the index in the answer.

Do Exercises 1–4.

Keep in mind that the product rule can be used only when the indexes are the same. When indexes differ, we can use rational exponents as we did in Section 15.2.

EXAMPLE 5 Multiply: $\sqrt{5x} \cdot \sqrt[4]{3y}$.

$\sqrt{5x} \cdot \sqrt[4]{3y} = (5x)^{1/2}(3y)^{1/4}$ — Converting to exponential notation

$= (5x)^{2/4}(3y)^{1/4}$ — Rewriting so that exponents have a common denominator

$= [(5x)^2(3y)]^{1/4}$ — Using $a^n b^n = (ab)^n$

$= [(25x^2)(3y)]^{1/4}$ — Squaring $5x$

$= \sqrt[4]{(25x^2)(3y)}$ — Converting back to radical notation

$= \sqrt[4]{75x^2y}$ — Multiplying under the radical

Do Exercises 5 and 6.

Multiply.

1. $\sqrt{19}\,\sqrt{7}$

2. $\sqrt{3p}\,\sqrt{7q}$

3. $\sqrt[4]{403}\,\sqrt[4]{7}$

4. $\sqrt[3]{\dfrac{5}{p}} \cdot \sqrt[3]{\dfrac{2}{q}}$

Multiply.

5. $\sqrt{5}\,\sqrt[3]{2}$

6. $\sqrt{x}\,\sqrt[3]{5y}$

Answers on page A-57

We can reverse the product rule to simplify a product. We simplify the root of a product by taking the root of each factor separately.

FACTORING RADICAL EXPRESSIONS

For any nonnegative real numbers a and b and any index k,

$$\sqrt[k]{ab} = \sqrt[k]{a} \cdot \sqrt[k]{b}, \quad \text{or} \quad (ab)^{1/k} = a^{1/k} \cdot b^{1/k}.$$

(Take the kth root of each factor separately.)

Compare the following:

$$\sqrt{50} = \sqrt{10 \cdot 5} = \sqrt{10}\,\sqrt{5};$$
$$\sqrt{50} = \sqrt{25 \cdot 2} = \sqrt{25}\,\sqrt{2} = 5\sqrt{2}.$$

In the second case, the radicand has the perfect-square factor 25. If you do not recognize perfect-square factors, try factoring the radicand into its prime factors. For example,

$$\sqrt{50} = \sqrt{2 \cdot \underbrace{5 \cdot 5}} = 5\sqrt{2}.$$

Perfect square (a pair of the same numbers)

Square-root radical expressions in which the radicand has no perfect-square factors, such as $5\sqrt{2}$, are considered to be in simplest form. A procedure for simplifying kth roots follows.

SIMPLIFYING kth ROOTS

To simplify a radical expression by factoring:

1. Look for the largest factors of the radicand that are perfect kth powers (where k is the index).
2. Then take the kth root of the resulting factors.
3. A radical expression, with index k, is *simplified* when its radicand has no factors that are perfect kth powers.

EXAMPLES Simplify by factoring.

6. $\sqrt{50} = \sqrt{25 \cdot 2} = \sqrt{25} \cdot \sqrt{2} = \sqrt{5 \cdot 5} \cdot \sqrt{2} = 5\sqrt{2}$

This factor is a perfect square.

7. $\sqrt[3]{32} = \sqrt[3]{8 \cdot 4} = \sqrt[3]{8} \cdot \sqrt[3]{4} = \sqrt[3]{2 \cdot 2 \cdot 2} \cdot \sqrt[3]{2 \cdot 2} = 2\sqrt[3]{4}$

This factor is a perfect cube (third power).

8. $\sqrt[4]{48} = \sqrt[4]{16 \cdot 3} = \sqrt[4]{16} \cdot \sqrt[4]{3} = \sqrt[4]{2 \cdot 2 \cdot 2 \cdot 2} \cdot \sqrt[4]{3} = 2\sqrt[4]{3}$

This factor is a perfect fourth power.

Do Exercises 7 and 8.

In many situations, expressions under radicals never represent negative numbers. In such cases, absolute-value notation is not necessary. For this reason, we will henceforth assume that *all expressions under radicals are nonnegative.*

Simplify by factoring.

7. $\sqrt{32}$

8. $\sqrt[3]{80}$

Answers on page A-57

Simplify by factoring. Assume that all expressions under radicals represent nonnegative numbers.

9. $\sqrt{300}$

10. $\sqrt{36y^2}$

11. $\sqrt{12a^2b}$

12. $\sqrt{12ab^3c^2}$

13. $\sqrt[3]{16}$

14. $\sqrt[3]{81x^4y^8}$

Answers on page A-57

EXAMPLES Simplify by factoring. Assume that all expressions under radicals represent nonnegative numbers.

9. $\sqrt{5x^2} = \sqrt{5 \cdot x^2}$ Factoring the radicand

$= \sqrt{5} \cdot \sqrt{x^2}$ Factoring into two radicals

$= \sqrt{5} \cdot x$ Taking the square root of x^2

Absolute-value notation is not needed because the expression under the radical is not negative.

10. $\sqrt{18x^2y} = \sqrt{9 \cdot 2 \cdot x^2 \cdot y}$ Factoring the radicand and looking for perfect-square factors

$= \sqrt{9 \cdot x^2 \cdot 2 \cdot y}$

$= \sqrt{9} \cdot \sqrt{x^2} \cdot \sqrt{2} \cdot \sqrt{y}$ Factoring into several radicals

$= 3x\sqrt{2y}$ Taking square roots

11. $\sqrt{216x^5y^3} = \sqrt{36 \cdot 6 \cdot x^4 \cdot x \cdot y^2 \cdot y}$ Factoring the radicand and looking for perfect-square factors

$= \sqrt{36 \cdot x^4 \cdot y^2 \cdot 6 \cdot x \cdot y}$

$= \sqrt{36}\,\sqrt{x^4}\,\sqrt{y^2}\,\sqrt{6xy}$ Factoring into several radicals

$= 6x^2y\sqrt{6xy}$ Taking square roots

Let's look at this example another way. We do a complete factorization and look for pairs of factors. Each pair of factors makes a square:

$\sqrt{216x^5y^3} = \sqrt{\underline{2 \cdot 2} \cdot 2 \cdot \underline{3 \cdot 3} \cdot 3 \cdot \underline{x \cdot x} \cdot \underline{x \cdot x} \cdot x \cdot \underline{y \cdot y} \cdot y}$ Each pair of factors makes a perfect square.

$= 2 \cdot 3 \cdot x \cdot x \cdot y \cdot \sqrt{2 \cdot 3 \cdot x \cdot y}$

$= 6x^2y\sqrt{6xy}.$

12. $\sqrt[3]{16a^7b^{11}} = \sqrt[3]{8 \cdot 2 \cdot a^6 \cdot a \cdot b^9 \cdot b^2}$ Factoring the radicand. The index is 3, so we look for the largest powers that are multiples of 3 because these are perfect cubes.

$= \sqrt[3]{8} \cdot \sqrt[3]{a^6} \cdot \sqrt[3]{b^9} \cdot \sqrt[3]{2ab^2}$ Factoring into radicals

$= 2a^2b^3\sqrt[3]{2ab^2}$ Taking cube roots

Let's look at this example another way. We do a complete factorization and look for triples of factors. Each triple of factors makes a cube:

$\sqrt[3]{16a^7b^{11}}$

$= \sqrt[3]{\underline{2 \cdot 2 \cdot 2} \cdot 2 \cdot \underline{a \cdot a \cdot a} \cdot \underline{a \cdot a \cdot a} \cdot a \cdot \underline{b \cdot b \cdot b} \cdot \underline{b \cdot b \cdot b} \cdot \underline{b \cdot b \cdot b} \cdot b \cdot b}$

Each triple of factors makes a cube.

$= 2 \cdot a \cdot a \cdot b \cdot b \cdot b \cdot \sqrt[3]{2 \cdot a \cdot b \cdot b}$

$= 2a^2b^3\sqrt[3]{2ab^2}.$

Do Exercises 9–14.

Sometimes after we have multiplied, we can then simplify by factoring.

EXAMPLES Multiply and simplify. Assume that all expressions under radicals represent nonnegative numbers.

13. $\sqrt{20}\sqrt{8} = \sqrt{20 \cdot 8} = \sqrt{\underline{4} \cdot 5 \cdot \underline{4} \cdot 2} = 4\sqrt{10}$

14. $3\sqrt[3]{25} \cdot 2\sqrt[3]{5} = 6 \cdot \sqrt[3]{25 \cdot 5}$
$= 6 \cdot \sqrt[3]{5 \cdot 5 \cdot 5}$
$= 6 \cdot 5 = 30$

15. $\sqrt[3]{18y^3}\,\sqrt[3]{4x^2} = \sqrt[3]{18y^3 \cdot 4x^2}$ Multiplying radicands
$= \sqrt[3]{2 \cdot 3 \cdot 3 \cdot \underline{y \cdot y \cdot y} \cdot 2 \cdot 2 \cdot x \cdot x}$
$= 2 \cdot y \cdot \sqrt[3]{3 \cdot 3 \cdot x \cdot x}$
$= 2y\sqrt[3]{9x^2}.$

Do Exercises 15–18.

Multiply and simplify. Assume that all expressions under radicals represent nonnegative numbers.

15. $\sqrt{3}\,\sqrt{6}$

16. $\sqrt{18y}\,\sqrt{14y}$

17. $\sqrt[3]{3x^2y}\,\sqrt[3]{36x}$

18. $\sqrt{7a}\,\sqrt{21b}$

b Dividing and Simplifying Radical Expressions

Note that $\dfrac{\sqrt[3]{27}}{\sqrt[3]{8}} = \dfrac{3}{2}$ and that $\sqrt[3]{\dfrac{27}{8}} = \dfrac{3}{2}$. This example suggests the following.

THE QUOTIENT RULE FOR RADICALS

For any nonnegative number a, any positive number b, and any index k,

$$\frac{\sqrt[k]{a}}{\sqrt[k]{b}} = \sqrt[k]{\frac{a}{b}}, \quad \text{or} \quad \frac{a^{1/k}}{b^{1/k}} = \left(\frac{a}{b}\right)^{1/k}.$$

(To divide, divide the radicands. After doing this, you can sometimes simplify by taking roots.)

EXAMPLES Divide and simplify. Assume that all expressions under radicals represent positive numbers.

16. $\dfrac{\sqrt{80}}{\sqrt{5}} = \sqrt{\dfrac{80}{5}} = \sqrt{16} = 4$ We divide the radicands.

17. $\dfrac{5\sqrt[3]{32}}{\sqrt[3]{2}} = 5\sqrt[3]{\dfrac{32}{2}} = 5\sqrt[3]{16} = 5\sqrt[3]{8 \cdot 2} = 5\sqrt[3]{8}\,\sqrt[3]{2} = 5 \cdot 2\sqrt[3]{2} = 10\sqrt[3]{2}$

18. $\dfrac{\sqrt{72xy}}{2\sqrt{2}} = \dfrac{1}{2}\,\dfrac{\sqrt{72xy}}{\sqrt{2}} = \dfrac{1}{2}\sqrt{\dfrac{72xy}{2}} = \dfrac{1}{2}\sqrt{36xy} = \dfrac{1}{2}\sqrt{36}\,\sqrt{xy}$
$= \dfrac{1}{2} \cdot 6\sqrt{xy} = 3\sqrt{xy}$

Do Exercises 19–22.

Divide and simplify. Assume that all expressions under radicals represent positive numbers.

19. $\dfrac{\sqrt{75}}{\sqrt{3}}$

20. $\dfrac{14\sqrt{128xy}}{2\sqrt{2}}$

21. $\dfrac{\sqrt{50a^3}}{\sqrt{2a}}$

22. $\dfrac{4\sqrt[3]{250}}{7\sqrt[3]{2}}$

Answers on page A-57

Simplify by taking the roots of the numerator and the denominator. Assume that all expressions under radicals represent positive numbers.

23. $\sqrt{\frac{25}{36}}$

24. $\sqrt{\frac{x^2}{100}}$

25. $\sqrt[3]{\frac{54x^5}{125}}$

26. Divide and simplify:

$$\frac{\sqrt[4]{x^3y^2}}{\sqrt[3]{x^2y}}.$$

Answers on page A-57

We can reverse the quotient rule to simplify a quotient. We simplify the root of a quotient by taking the roots of the numerator and of the denominator separately.

*k*th ROOTS OF QUOTIENTS

For any nonnegative number a, any positive number b, and any index k,

$$\sqrt[k]{\frac{a}{b}} = \frac{\sqrt[k]{a}}{\sqrt[k]{b}}, \quad \text{or} \quad \left(\frac{a}{b}\right)^{1/k} = \frac{a^{1/k}}{b^{1/k}}.$$

(Take the kth roots of the numerator and of the denominator separately.)

EXAMPLES Simplify by taking the roots of the numerator and the denominator. Assume that all expressions under radicals represent positive numbers.

19. $\sqrt[3]{\frac{27}{125}} = \frac{\sqrt[3]{27}}{\sqrt[3]{125}} = \frac{3}{5}$ — We take the cube root of the numerator and of the denominator.

20. $\sqrt{\frac{25}{y^2}} = \frac{\sqrt{25}}{\sqrt{y^2}} = \frac{5}{y}$ — We take the square root of the numerator and of the denominator.

21. $\sqrt{\frac{16x^3}{y^4}} = \frac{\sqrt{16x^3}}{\sqrt{y^4}} = \frac{\sqrt{16x^2 \cdot x}}{\sqrt{y^4}} = \frac{\sqrt{16x^2} \cdot \sqrt{x}}{\sqrt{y^4}} = \frac{4x\sqrt{x}}{y^2}$

22. $\sqrt[3]{\frac{27y^5}{343x^3}} = \frac{\sqrt[3]{27y^5}}{\sqrt[3]{343x^3}} = \frac{\sqrt[3]{27y^3 \cdot y^2}}{\sqrt[3]{343x^3}} = \frac{\sqrt[3]{27y^3} \cdot \sqrt[3]{y^2}}{\sqrt[3]{343x^3}} = \frac{3y\sqrt[3]{y^2}}{7x}$

We are assuming that no expression represents 0 or a negative number. Thus we need not be concerned about zero denominators.

Do Exercises 23–25.

When indexes differ, we can use rational exponents.

EXAMPLE 23 Divide and simplify: $\frac{\sqrt[3]{a^2b^4}}{\sqrt{ab}}$.

$$\frac{\sqrt[3]{a^2b^4}}{\sqrt{ab}} = \frac{(a^2b^4)^{1/3}}{(ab)^{1/2}}$$ Converting to exponential notation

$$= \frac{a^{2/3}b^{4/3}}{a^{1/2}b^{1/2}}$$ Using the product and power rules

$$= a^{2/3-1/2}b^{4/3-1/2}$$ Subtracting exponents

$$= a^{4/6-3/6}b^{8/6-3/6}$$ Finding common denominators so exponents can be subtracted

$$= a^{1/6}b^{5/6}$$

$$= (ab^5)^{1/6}$$ Using $a^nb^n = (ab)^n$

$$= \sqrt[6]{ab^5}$$ Converting back to radical notation

Do Exercise 26.

15.3 EXERCISE SET

For Extra Help

Digital Video Tutor CD 12 Videotape 17 | InterAct Math | Math Tutor Center |  MathXL |  MyMathLab 

a Simplify by factoring. Assume that all expressions under radicals represent nonnegative numbers.

1. $\sqrt{24}$

2. $\sqrt{20}$

3. $\sqrt{90}$

4. $\sqrt{18}$

5. $\sqrt[3]{250}$

6. $\sqrt[3]{108}$

7. $\sqrt{180x^4}$

8. $\sqrt{175y^6}$

9. $\sqrt[3]{54x^8}$

10. $\sqrt[3]{40y^3}$

11. $\sqrt[3]{80t^8}$

12. $\sqrt[3]{108x^5}$

13. $\sqrt[4]{80}$

14. $\sqrt[4]{32}$

15. $\sqrt{32a^2b}$

16. $\sqrt{75p^3q^4}$

17. $\sqrt[4]{243x^8y^{10}}$

18. $\sqrt[4]{162c^4d^6}$

19. $\sqrt[5]{96x^7y^{15}}$

20. $\sqrt[5]{p^{14}q^9r^{23}}$

Multiply and simplify. Assume that all expressions under radicals represent nonnegative numbers.

21. $\sqrt{10}\sqrt{5}$

22. $\sqrt{6}\sqrt{3}$

23. $\sqrt{15}\sqrt{6}$

24. $\sqrt{2}\sqrt{32}$

25. $\sqrt[3]{2}\sqrt[3]{4}$

26. $\sqrt[3]{9}\sqrt[3]{3}$

27. $\sqrt{45}\sqrt{60}$

28. $\sqrt{24}\sqrt{75}$

29. $\sqrt{3x^3}\sqrt{6x^5}$

30. $\sqrt{5a^7}\sqrt{15a^3}$

31. $\sqrt{5b^3}\sqrt{10c^4}$

32. $\sqrt{2x^3y}\sqrt{12xy}$

33. $\sqrt[3]{5a^2}\,\sqrt[3]{2a}$

34. $\sqrt[3]{7x}\,\sqrt[3]{3x^2}$

35. $\sqrt[3]{y^4}\,\sqrt[3]{16y^5}$

36. $\sqrt[3]{s^2t^4}\,\sqrt[3]{s^4t^6}$

37. $\sqrt[4]{16}\,\sqrt[4]{64}$

38. $\sqrt[5]{64}\,\sqrt[5]{16}$

39. $\sqrt{12a^3b}\,\sqrt{8a^4b^2}$

40. $\sqrt{30x^3y^4}\,\sqrt{18x^2y^5}$

41. $\sqrt{2}\,\sqrt[3]{5}$

42. $\sqrt{6}\,\sqrt[3]{5}$

43. $\sqrt[4]{3}\,\sqrt{2}$

44. $\sqrt[3]{5}\,\sqrt[4]{2}$

45. $\sqrt{a}\,\sqrt[4]{a^3}$

46. $\sqrt[3]{x^2}\,\sqrt[6]{x^5}$

47. $\sqrt[5]{b^2}\,\sqrt{b^3}$

48. $\sqrt[4]{a^3}\,\sqrt[3]{a^2}$

49. $\sqrt{xy^3}\,\sqrt[3]{x^2y}$

50. $\sqrt[4]{9ab^3}\,\sqrt{3a^4b}$

b Divide and simplify. Assume that all expressions under radicals represent positive numbers.

51. $\dfrac{\sqrt{90}}{\sqrt{5}}$

52. $\dfrac{\sqrt{98}}{\sqrt{2}}$

53. $\dfrac{\sqrt{35q}}{\sqrt{7q}}$

54. $\dfrac{\sqrt{30x}}{\sqrt{10x}}$

55. $\dfrac{\sqrt[3]{54}}{\sqrt[3]{2}}$

56. $\dfrac{\sqrt[3]{40}}{\sqrt[3]{5}}$

57. $\dfrac{\sqrt{56xy^3}}{\sqrt{8x}}$

58. $\dfrac{\sqrt{52ab^3}}{\sqrt{13a}}$

59. $\dfrac{\sqrt[3]{96a^4b^2}}{\sqrt[3]{12a^2b}}$

60. $\dfrac{\sqrt[3]{189x^5y^7}}{\sqrt[3]{7x^2y^2}}$

61. $\dfrac{\sqrt{128xy}}{2\sqrt{2}}$

62. $\dfrac{\sqrt{48ab}}{2\sqrt{3}}$

63. $\dfrac{\sqrt[4]{48x^9y^{13}}}{\sqrt[4]{3xy^5}}$

64. $\dfrac{\sqrt[5]{64a^{11}b^{28}}}{\sqrt[5]{2ab^2}}$

65. $\dfrac{\sqrt[3]{a}}{\sqrt{a}}$

66. $\dfrac{\sqrt{x}}{\sqrt[4]{x}}$

67. $\dfrac{\sqrt[3]{a^2}}{\sqrt[4]{a}}$

68. $\dfrac{\sqrt[3]{x^2}}{\sqrt[5]{x}}$

69. $\dfrac{\sqrt[4]{x^2y^3}}{\sqrt[3]{xy}}$

70. $\dfrac{\sqrt[5]{a^4b^2}}{\sqrt[3]{ab^2}}$

Simplify.

71. $\sqrt{\dfrac{25}{36}}$

72. $\sqrt{\dfrac{49}{64}}$

73. $\sqrt{\dfrac{16}{49}}$

74. $\sqrt{\dfrac{100}{81}}$

75. $\sqrt[3]{\dfrac{125}{27}}$

76. $\sqrt[3]{\dfrac{343}{1000}}$

77. $\sqrt{\dfrac{49}{y^2}}$

78. $\sqrt{\dfrac{121}{x^2}}$

79. $\sqrt{\dfrac{25y^3}{x^4}}$

80. $\sqrt{\dfrac{36a^5}{b^6}}$

81. $\sqrt[3]{\dfrac{27a^4}{8b^3}}$

82. $\sqrt[3]{\dfrac{64x^7}{216y^6}}$

83. $\sqrt[4]{\dfrac{81x^4}{16}}$

84. $\sqrt[4]{\dfrac{81x^4}{y^8z^4}}$

85. $\sqrt[5]{\dfrac{32x^8}{y^{10}}}$

86. $\sqrt[5]{\dfrac{32b^{10}}{243a^{20}}}$

87. $\sqrt[6]{\dfrac{x^{13}}{y^6z^{12}}}$

88. $\sqrt[6]{\dfrac{p^9q^{24}}{r^{18}}}$

89. D_W Is the quotient of two irrational numbers always an irrational number? Why or why not?

90. D_W Ron is puzzled. When he uses a graphing calculator to graph $y = \sqrt{x} \cdot \sqrt{x}$, he gets the following screen. Explain why Ron did not get the complete line $y_1 = x$.

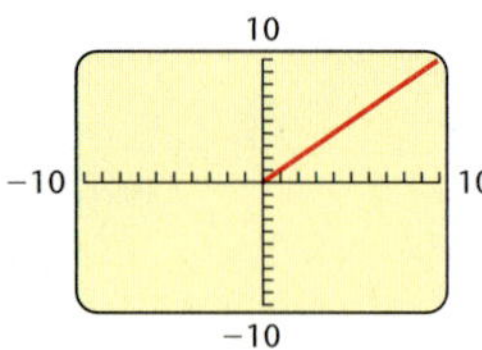

SKILL MAINTENANCE

Solve. [10.9a]

91. The sum of a number and its square is 90. Find the number.

92. *Triangle Dimensions.* The base of a triangle is 2 in. longer than the height. The area is 12 in^2. Find the height and the base.

Solve. [11.7a]

93. $\dfrac{12x}{x-4} - \dfrac{3x^2}{x+4} = \dfrac{384}{x^2-16}$

94. $\dfrac{2}{3} + \dfrac{1}{t} = \dfrac{4}{5}$

95. $\dfrac{18}{x^2-3x} = \dfrac{2x}{x-3} - \dfrac{6}{x}$

96. $\dfrac{4x}{x+5} + \dfrac{20}{x} = \dfrac{100}{x^2+5x}$

SYNTHESIS

97. *Pendulums.* The **period** of a pendulum is the time it takes to complete one cycle, swinging to and fro. For a pendulum that is L centimeters long, the period T is given by the function

$$T(L) = 2\pi\sqrt{\frac{L}{980}},$$

where T is in seconds. Find, to the nearest hundredth of a second, the period of a pendulum of length **(a)** 65 cm; **(b)** 98 cm; **(c)** 120 cm. Use a calculator's $\boxed{\pi}$ key if possible.

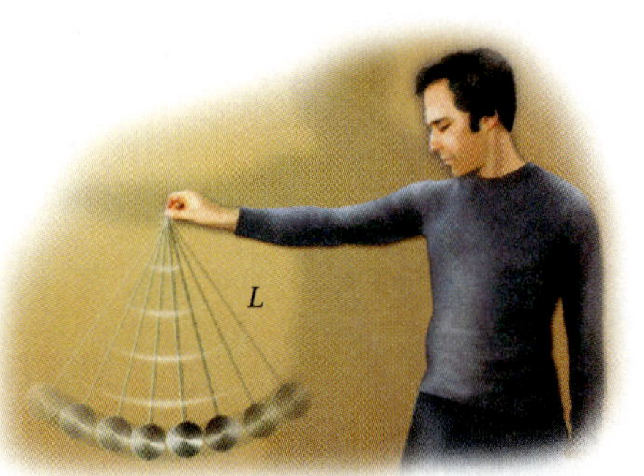

Simplify.

98. $\dfrac{\sqrt[3]{x^3 - y^3}}{\sqrt[3]{x-y}}$

99. $\dfrac{\sqrt{44x^2y^9z}\,\sqrt{22y^9z^6}}{(\sqrt{11xy^8z^2})^2}$

100.  Use a graphing calculator to check your answers to Exercises 7, 12, 30, and 54.

15.4 ADDITION, SUBTRACTION, AND MORE MULTIPLICATION

a Addition and Subtraction

Any two real numbers can be added. For example, the sum of 7 and $\sqrt{3}$ can be expressed as $7 + \sqrt{3}$. We cannot simplify this sum. However, when we have **like radicals** (radicals having the same index and radicand), we can use the distributive laws to simplify by collecting like radical terms. For example,

$$7\sqrt{3} + \sqrt{3} = 7\sqrt{3} + 1 \cdot \sqrt{3} = (7 + 1)\sqrt{3} = 8\sqrt{3}.$$

EXAMPLES Add or subtract. Simplify by collecting like radical terms, if possible.

1. $6\sqrt{7} + 4\sqrt{7} = (6 + 4)\sqrt{7}$ Using a distributive law (factoring out $\sqrt{7}$)
$= 10\sqrt{7}$

2. $8\sqrt[3]{2} - 7x\sqrt[3]{2} + 5\sqrt[3]{2} = (8 - 7x + 5)\sqrt[3]{2}$ Factoring out $\sqrt[3]{2}$
$= (13 - 7x)\sqrt[3]{2}$

These parentheses *are* necessary!

3. $6\sqrt[5]{4x} + 4\sqrt[5]{4x} - \sqrt[3]{4x} = (6 + 4)\sqrt[5]{4x} - \sqrt[3]{4x}$
$= 10\sqrt[5]{4x} - \sqrt[3]{4x}$

Note that these expressions have the *same* radicand, but they are *not* like radicals because they do not have the same index.

Do Exercises 1 and 2.

Sometimes we need to simplify radicals by factoring in order to obtain terms with like radicals.

EXAMPLES Add or subtract. Simplify by collecting like radical terms, if possible.

4. $3\sqrt{8} - 5\sqrt{2} = 3(\sqrt{4 \cdot 2}) - 5\sqrt{2}$ Factoring 8
$= 3\sqrt{4} \cdot \sqrt{2} - 5\sqrt{2}$ Factoring $\sqrt{4 \cdot 2}$ into two radicals
$= 3 \cdot 2\sqrt{2} - 5\sqrt{2}$ Taking the square root of 4
$= 6\sqrt{2} - 5\sqrt{2}$
$= (6 - 5)\sqrt{2}$ Collecting like radical terms
$= \sqrt{2}$

5. $5\sqrt{2} - 4\sqrt{3}$ No simplification possible

6. $5\sqrt[3]{16y^4} + 7\sqrt[3]{2y} = 5\sqrt[3]{8y^3 \cdot 2y} + 7\sqrt[3]{2y}$
$= 5\sqrt[3]{8y^3} \cdot \sqrt[3]{2y} + 7\sqrt[3]{2y}$ Factoring the first radical
$= 5 \cdot 2y \cdot \sqrt[3]{2y} + 7\sqrt[3]{2y}$ Taking the cube root of $8y^3$
$= 10y\sqrt[3]{2y} + 7\sqrt[3]{2y}$
$= (10y + 7)\sqrt[3]{2y}$ Collecting like radical terms

Do Exercises 3–5.

Objectives

a Add or subtract with radical notation and simplify.

b Multiply expressions involving radicals in which some factors contain more than one term.

Add or subtract. Simplify by collecting like radical terms, if possible.

1. $5\sqrt{2} + 8\sqrt{2}$

2. $7\sqrt[4]{5x} + 3\sqrt[4]{5x} - \sqrt{7}$

Add or subtract. Simplify by collecting like radical terms, if possible.

3. $7\sqrt{45} - 2\sqrt{5}$

4. $3\sqrt[3]{y^5} + 4\sqrt[3]{y^2} + \sqrt[3]{8y^6}$

5. $\sqrt{25x - 25} - \sqrt{9x - 9}$

Answers on page A-57

Multiply. Assume that all expressions under radicals represent nonnegative numbers.

6. $\sqrt{2}(5\sqrt{3} + 3\sqrt{7})$

7. $\sqrt[3]{a^2}(\sqrt[3]{3a} - \sqrt[3]{2})$

Multiply. Assume that all expressions under radicals represent nonnegative numbers.

8. $(\sqrt{3} - 5\sqrt{2})(2\sqrt{3} + \sqrt{2})$

9. $(\sqrt{a} + 2\sqrt{3})(3\sqrt{b} - 4\sqrt{3})$

10. $(\sqrt{2} + \sqrt{5})(\sqrt{2} - \sqrt{5})$

11. $(\sqrt{p} - \sqrt{q})(\sqrt{p} + \sqrt{q})$

Multiply.

12. $(2\sqrt{5} - y)^2$

13. $(3\sqrt{6} + 2)^2$

Answer on page A-57

b More Multiplication

To multiply expressions in which some factors contain more than one term, we use the procedures for multiplying polynomials.

EXAMPLES Multiply.

7. $\sqrt{3}(x - \sqrt{5}) = \sqrt{3} \cdot x - \sqrt{3} \cdot \sqrt{5}$ Using a distributive law
$= x\sqrt{3} - \sqrt{15}$ Multiplying radicals

8. $\sqrt[3]{y}(\sqrt[3]{y^2} + \sqrt[3]{2}) = \sqrt[3]{y} \cdot \sqrt[3]{y^2} + \sqrt[3]{y} \cdot \sqrt[3]{2}$ Using a distributive law
$= \sqrt[3]{y^3} + \sqrt[3]{2y}$ Multiplying radicals
$= y + \sqrt[3]{2y}$ Simplifying $\sqrt[3]{y^3}$.

Do Exercises 6 and 7.

EXAMPLE 9 Multiply: $(4\sqrt{3} + \sqrt{2})(\sqrt{3} - 5\sqrt{2})$.

F O I L

$(4\sqrt{3} + \sqrt{2})(\sqrt{3} - 5\sqrt{2}) = 4(\sqrt{3})^2 - 20\sqrt{3} \cdot \sqrt{2} + \sqrt{2} \cdot \sqrt{3} - 5(\sqrt{2})^2$
$= 4 \cdot 3 - 20\sqrt{6} + \sqrt{6} - 5 \cdot 2$
$= 12 - 20\sqrt{6} + \sqrt{6} - 10$
$= 2 - 19\sqrt{6}$ Collecting like terms

EXAMPLE 10 Multiply: $(\sqrt{a} + \sqrt{3})(\sqrt{b} + \sqrt{3})$. Assume that all expressions under radicals represent nonnegative numbers.

$$(\sqrt{a} + \sqrt{3})(\sqrt{b} + \sqrt{3}) = \sqrt{a}\sqrt{b} + \sqrt{a}\sqrt{3} + \sqrt{3}\sqrt{b} + \sqrt{3}\sqrt{3}$$
$$= \sqrt{ab} + \sqrt{3a} + \sqrt{3b} + 3$$

EXAMPLE 11 Multiply: $(\sqrt{5} + \sqrt{7})(\sqrt{5} - \sqrt{7})$.

$(\sqrt{5} + \sqrt{7})(\sqrt{5} - \sqrt{7}) = (\sqrt{5})^2 - (\sqrt{7})^2$ This is now a difference of two squares: $(A - B)(A + B) = A^2 - B^2$.
$= 5 - 7 = -2$

EXAMPLE 12 Multiply: $(\sqrt{a} + \sqrt{b})(\sqrt{a} - \sqrt{b})$. Assume that all expressions under radicals represent nonnegative numbers.

$$(\sqrt{a} + \sqrt{b})(\sqrt{a} - \sqrt{b}) = (\sqrt{a})^2 - (\sqrt{b})^2$$
$$= a - b$$ ← No radicals

Expressions of the form $\sqrt{a} + \sqrt{b}$ and $\sqrt{a} - \sqrt{b}$ are called **conjugates.** Their product is always an expression that has no radicals.

Do Exercises 8–11.

EXAMPLE 13 Multiply: $(\sqrt{3} + x)^2$.

$(\sqrt{3} + x)^2 = (\sqrt{3})^2 + 2x\sqrt{3} + x^2$ Squaring a binomial
$= 3 + 2x\sqrt{3} + x^2$

Do Exercises 12 and 13.

15.4 EXERCISE SET

For Extra Help

Digital Video Tutor CD 13 Videotape 17 | InterAct Math | Math Tutor Center | MathXL | MyMathLab

a Add or subtract. Then simplify by collecting like radical terms, if possible. Assume that all expressions under radicals represent nonnegative numbers.

1. $7\sqrt{5} + 4\sqrt{5}$

2. $2\sqrt{3} + 9\sqrt{3}$

3. $6\sqrt[3]{7} - 5\sqrt[3]{7}$

4. $13\sqrt[5]{3} - 8\sqrt[5]{3}$

5. $4\sqrt[3]{y} + 9\sqrt[3]{y}$

6. $6\sqrt[4]{t} - 3\sqrt[4]{t}$

7. $5\sqrt{6} - 9\sqrt{6} - 4\sqrt{6}$

8. $3\sqrt{10} - 8\sqrt{10} + 7\sqrt{10}$

9. $4\sqrt[3]{3} - \sqrt{5} + 2\sqrt[3]{3} + \sqrt{5}$

10. $5\sqrt{7} - 8\sqrt[4]{11} + \sqrt{7} + 9\sqrt[4]{11}$

11. $8\sqrt{27} - 3\sqrt{3}$

12. $9\sqrt{50} - 4\sqrt{2}$

13. $8\sqrt{45} + 7\sqrt{20}$

14. $9\sqrt{12} + 16\sqrt{27}$

15. $18\sqrt{72} + 2\sqrt{98}$

16. $12\sqrt{45} - 8\sqrt{80}$

17. $3\sqrt[3]{16} + \sqrt[3]{54}$

18. $\sqrt[3]{27} - 5\sqrt[3]{8}$

19. $2\sqrt{128} - \sqrt{18} + 4\sqrt{32}$

20. $5\sqrt{50} - 2\sqrt{18} + 9\sqrt{32}$

21. $\sqrt{5a} + 2\sqrt{45a^3}$

22. $4\sqrt{3x^3} - \sqrt{12x}$

23. $\sqrt[3]{24x} - \sqrt[3]{3x^4}$

24. $\sqrt[3]{54x} - \sqrt[3]{2x^4}$

25. $5\sqrt[3]{32} - \sqrt[3]{108} + 2\sqrt[3]{256}$

26. $3\sqrt[3]{8x} - 4\sqrt[3]{27x} + 2\sqrt[3]{64x}$

27. $\sqrt[3]{6x^4} + \sqrt[3]{48x} - \sqrt[3]{6x}$

28. $\sqrt[4]{80x^5} - \sqrt[4]{405x^9} + \sqrt[4]{5x}$

29. $\sqrt{4a - 4} + \sqrt{a - 1}$

30. $\sqrt{9y + 27} + \sqrt{y + 3}$

31. $\sqrt{x^3 - x^2} + \sqrt{9x - 9}$

32. $\sqrt{4x - 4} + \sqrt{x^3 - x^2}$

b Multiply.

33. $\sqrt{5}(4 - 2\sqrt{5})$

34. $\sqrt{6}(2 + \sqrt{6})$

35. $\sqrt{3}(\sqrt{2} - \sqrt{7})$

36. $\sqrt{2}(\sqrt{5} - \sqrt{2})$

37. $\sqrt{3}(2\sqrt{5} - 3\sqrt{4})$

38. $\sqrt{2}(3\sqrt{10} - 2\sqrt{2})$

39. $\sqrt[3]{2}(\sqrt[3]{4} - 2\sqrt[3]{32})$

40. $\sqrt[3]{3}(\sqrt[3]{9} - 4\sqrt[3]{21})$

41. $\sqrt[3]{a}(\sqrt[3]{2a^2} + \sqrt[3]{16a^2})$

42. $\sqrt[3]{x}(\sqrt[3]{3x^2} - \sqrt[3]{81x^2})$

43. $(\sqrt{3} - \sqrt{2})(\sqrt{3} + \sqrt{2})$

44. $(\sqrt{5} + \sqrt{6})(\sqrt{5} - \sqrt{6})$

45. $(\sqrt{8} + 2\sqrt{5})(\sqrt{8} - 2\sqrt{5})$

46. $(\sqrt{18} + 3\sqrt{7})(\sqrt{18} - 3\sqrt{7})$

47. $(7 + \sqrt{5})(7 - \sqrt{5})$

48. $(4 - \sqrt{3})(4 + \sqrt{3})$

49. $(2 - \sqrt{3})(2 + \sqrt{3})$

50. $(11 - \sqrt{2})(11 + \sqrt{2})$

51. $(\sqrt{8} + \sqrt{5})(\sqrt{8} - \sqrt{5})$

52. $(\sqrt{6} - \sqrt{7})(\sqrt{6} + \sqrt{7})$

53. $(3 + 2\sqrt{7})(3 - 2\sqrt{7})$

54. $(6 - 3\sqrt{2})(6 + 3\sqrt{2})$

For the following exercises, assume that all expressions under radicals represent nonnegative numbers.

55. $(\sqrt{a} + \sqrt{b})(\sqrt{a} - \sqrt{b})$

56. $(\sqrt{x} - \sqrt{y})(\sqrt{x} + \sqrt{y})$

57. $(3 - \sqrt{5})(2 + \sqrt{5})$

58. $(2 + \sqrt{6})(4 - \sqrt{6})$

59. $(\sqrt{3} + 1)(2\sqrt{3} + 1)$

60. $(4\sqrt{3} + 5)(\sqrt{3} - 2)$

61. $(2\sqrt{7} - 4\sqrt{2})(3\sqrt{7} + 6\sqrt{2})$

62. $(4\sqrt{5} + 3\sqrt{3})(3\sqrt{5} - 4\sqrt{3})$

63. $(\sqrt{a} + \sqrt{2})(\sqrt{a} + \sqrt{3})$

64. $(2 - \sqrt{x})(1 - \sqrt{x})$

65. $(2\sqrt[3]{3} + \sqrt[3]{2})(\sqrt[3]{3} - 2\sqrt[3]{2})$

66. $(3\sqrt[4]{7} + \sqrt[4]{6})(2\sqrt[4]{9} - 3\sqrt[4]{6})$

67. $(2 + \sqrt{3})^2$

68. $(\sqrt{5} + 1)^2$

69. $(\sqrt[5]{9} - \sqrt[5]{3})(\sqrt[5]{8} + \sqrt[5]{27})$

70. $(\sqrt[3]{8x} - \sqrt[3]{5y})^2$

71. **DW** Why do we need to know how to simplify radical expressions before we learn to add them?

72. **DW** In what way(s) is collecting like radical terms the same as collecting like monomial terms?

SKILL MAINTENANCE

Multiply or divide and simplify.

73. $\dfrac{x^3 + 4x}{x^2 - 16} \div \dfrac{x^2 + 8x + 15}{x^2 + x - 20}$ [11.2b]

74. $\dfrac{a^2 - 4}{a} \div \dfrac{a - 2}{a + 4}$ [11.2b]

75. $\dfrac{a^3 + 8}{a^2 - 4} \cdot \dfrac{a^2 - 4a + 4}{a^2 - 2a + 4}$ [11.1d]

76. $\dfrac{y^3 - 27}{y^2 - 9} \cdot \dfrac{y^2 - 6y + 9}{y^2 + 3y + 9}$ [11.1d]

Simplify. [11.6a]

77. $\dfrac{x - \frac{1}{3}}{x + \frac{1}{4}}$

78. $\dfrac{1 - \frac{1}{x}}{1 - \frac{1}{x^2}}$

79. $\dfrac{\frac{1}{p} - \frac{1}{q}}{\frac{1}{p^2} - \frac{1}{q^2}}$

80. $\dfrac{\frac{1}{a} + \frac{1}{b}}{\frac{1}{a^2} + \frac{1}{b^2}}$

SYNTHESIS

81. Graph the function $f(x) = \sqrt{(x - 2)^2}$. What is the domain?

82. Use a graphing calculator to check your answers to Exercises 5, 22, and 64.

Multiply and simplify.

83. $\sqrt{9 + 3\sqrt{5}}\sqrt{9 - 3\sqrt{5}}$

84. $(\sqrt{x + 2} - \sqrt{x - 2})^2$

85. $(\sqrt{3} + \sqrt{5} - \sqrt{6})^2$

86. $\sqrt[3]{y}(1 - \sqrt[3]{y})(1 + \sqrt[3]{y})$

87. $(\sqrt[3]{9} - 2)(\sqrt[3]{9} + 4)$

88. $\left[\sqrt{3 + \sqrt{2 + \sqrt{1}}}\right]^4$

15.5 MORE ON DIVISION OF RADICAL EXPRESSIONS

Objectives

a Rationalize the denominator of a radical expression having one term in the denominator.

b Rationalize the denominator of a radical expression having two terms in the denominator.

a Rationalizing Denominators

Sometimes in mathematics it is useful to find an equivalent expression without a radical in the denominator. This provides a standard notation for expressing results. The procedure for finding such an expression is called **rationalizing the denominator.** We carry this out by multiplying by 1.

EXAMPLE 1 Rationalize the denominator: $\sqrt{\frac{7}{3}}$.

We multiply by 1, using $\sqrt{3}/\sqrt{3}$. We do this so that the denominator of the radicand will be a perfect square.

$$\sqrt{\frac{7}{3}} = \frac{\sqrt{7}}{\sqrt{3}} \cdot \frac{\sqrt{3}}{\sqrt{3}}$$
$$= \frac{\sqrt{7} \cdot \sqrt{3}}{\sqrt{3} \cdot \sqrt{3}}$$
$$= \frac{\sqrt{21}}{\sqrt{3^2}} = \frac{\sqrt{21}}{3}$$

The radicand is a perfect square.

Historically, rationalizing a denominator made approximations of roots of rational expressions easier when a table of square roots was involved. For example, look at

$$\frac{\sqrt{21}}{3} \approx \frac{4.5826}{3} \approx 1.5275.$$

To find the approximation, we would look up $\sqrt{21}$ in a table of square roots of natural numbers and then divide by 3. However, what if we wanted to approximate

$$\sqrt{\frac{7}{3}}?$$

We could divide 7 by 3 and approximate it as

$$\sqrt{\frac{7}{3}} \approx \sqrt{2.3333}.$$

But most tables listed square roots of only natural numbers. Thus rationalizing the denominator allowed approximations using tables. Such a difficulty does not exist today with the advent of calculators, which we can use to go quickly to an approximate answer.

Do Exercise 1.

1. Rationalize the denominator:

$$\sqrt{\frac{2}{5}}.$$

Answer on page A-57

2. Rationalize the denominator:

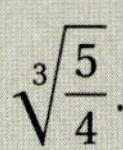

$$\sqrt[3]{\frac{5}{4}}.$$

EXAMPLE 2 Rationalize the denominator: $\sqrt[3]{\frac{7}{25}}$.

We first factor the denominator:

$$\sqrt[3]{\frac{7}{25}} = \sqrt[3]{\frac{7}{5 \cdot 5}}.$$

To get a perfect cube in the denominator, we consider the index 3 and the factors. We have 2 factors of 5, and we need 3 factors of 5. We achieve this by multiplying by 1, using $\sqrt[3]{5}/\sqrt[3]{5}$.

$$\sqrt[3]{\frac{7}{25}} = \sqrt[3]{\frac{7}{5 \cdot 5}} \cdot \frac{\sqrt[3]{5}}{\sqrt[3]{5}}$$ Multiplying by $\frac{\sqrt[3]{5}}{\sqrt[3]{5}}$ to make the denominator of the radicand a perfect cube

$$= \frac{\sqrt[3]{7} \cdot \sqrt[3]{5}}{\sqrt[3]{5 \cdot 5} \cdot \sqrt[3]{5}}$$

$$= \frac{\sqrt[3]{35}}{\sqrt[3]{5^3}}$$ The radicand is a perfect cube.

$$= \frac{\sqrt[3]{35}}{5}.$$

Do Exercise 2.

3. Rationalize the denominator:

$$\sqrt{\frac{4a}{3b}}.$$

EXAMPLE 3 Rationalize the denominator: $\sqrt{\frac{2a}{5b}}$. Assume that all expressions under radicals represent positive numbers.

$$\sqrt{\frac{2a}{5b}} = \frac{\sqrt{2a}}{\sqrt{5b}}$$ Converting to a quotient of radicals

$$= \frac{\sqrt{2a}}{\sqrt{5b}} \cdot \frac{\sqrt{5b}}{\sqrt{5b}}$$ Multiplying by 1

$$= \frac{\sqrt{10ab}}{\sqrt{5^2b^2}}$$ The radicand in the denominator is a perfect square.

$$= \frac{\sqrt{10ab}}{5b}$$

Do Exercise 3.

Rationalize the denominator.

4. $\frac{\sqrt[4]{7}}{\sqrt[4]{2}}$

5. $\sqrt[3]{\frac{3x^5}{2y}}$

EXAMPLE 4 Rationalize the denominator: $\frac{\sqrt[3]{a}}{\sqrt[3]{9x}}$.

We factor the denominator:

$$\frac{\sqrt[3]{a}}{\sqrt[3]{9x}} = \frac{\sqrt[3]{a}}{\sqrt[3]{3 \cdot 3 \cdot x}}.$$

To choose the symbol for 1, we look at $3 \cdot 3 \cdot x$. To make it a cube, we need another 3 and two more x's. Thus we multiply by 1, using $\sqrt[3]{3x^2}/\sqrt[3]{3x^2}$:

$$\frac{\sqrt[3]{a}}{\sqrt[3]{9x}} = \frac{\sqrt[3]{a}}{\sqrt[3]{3 \cdot 3 \cdot x}} \cdot \frac{\sqrt[3]{3x^2}}{\sqrt[3]{3x^2}}$$ Multiplying by 1

$$= \frac{\sqrt[3]{3ax^2}}{\sqrt[3]{3^3x^3}}$$ The radicand in the denominator is a perfect cube.

$$= \frac{\sqrt[3]{3ax^2}}{3x}.$$

Do Exercises 4 and 5.

Answers on page A-57

EXAMPLE 5 Rationalize the denominator: $\dfrac{3x}{\sqrt[5]{2x^2y^3}}$.

$$\begin{aligned}\frac{3x}{\sqrt[5]{2x^2y^3}} &= \frac{3x}{\sqrt[5]{2 \cdot x \cdot x \cdot y \cdot y \cdot y}}\\ &= \frac{3x}{\sqrt[5]{2x^2y^3}} \cdot \frac{\sqrt[5]{2^4x^3y^2}}{\sqrt[5]{2^4x^3y^2}}\\ &= \frac{3x\sqrt[5]{16x^3y^2}}{\sqrt[5]{2^5x^5y^5}} \quad \text{The radicand in the denominator is a perfect fifth power.}\\ &= \frac{3x\sqrt[5]{16x^3y^2}}{2xy}\\ &= \frac{x}{x} \cdot \frac{3\sqrt[5]{16x^3y^2}}{2y}\\ &= \frac{3\sqrt[5]{16x^3y^2}}{2y}\end{aligned}$$

Do Exercise 6.

b Rationalizing When There Are Two Terms

Do Exercises 7 and 8.

Certain pairs of expressions containing square roots, such as $c - \sqrt{b}$, $c + \sqrt{b}$ and $\sqrt{a} - \sqrt{b}$, $\sqrt{a} + \sqrt{b}$, are called **conjugates.** The product of such a pair of conjugates has no radicals in it. (See Example 12 of Section 15.4.) Thus when we wish to rationalize a denominator that has two terms and one or more of them involves a square-root radical, we multiply by 1 using the conjugate of the denominator to write a symbol for 1.

EXAMPLES What symbol for 1 would you use to rationalize the denominator?

	Expression	*Symbol for 1*
6.	$\dfrac{3}{x + \sqrt{7}}$	$\dfrac{x - \sqrt{7}}{x - \sqrt{7}}$
7.	$\dfrac{\sqrt{7} + 4}{3 - 2\sqrt{5}}$	$\dfrac{3 + 2\sqrt{5}}{3 + 2\sqrt{5}}$

Change the operation sign in the denominator to obtain the conjugate. Use the conjugate for the numerator and denominator of the symbol for 1.

Do Exercises 9 and 10.

6. Rationalize the denominator:

$$\frac{7x}{\sqrt[3]{4xy^5}}.$$

Multiply.

7. $(c - \sqrt{b})(c + \sqrt{b})$

8. $(\sqrt{a} + \sqrt{b})(\sqrt{a} - \sqrt{b})$

What symbol for 1 would you use to rationalize the denominator?

9. $\dfrac{\sqrt{5} + 1}{\sqrt{3} - y}$

10. $\dfrac{1}{\sqrt{2} + \sqrt{3}}$

Answers on page A-57

Rationalize the denominator.

11. $\dfrac{14}{3 + \sqrt{2}}$

12. $\dfrac{5 + \sqrt{2}}{1 - \sqrt{2}}$

Answers on page A-57

EXAMPLE 8 Rationalize the denominator: $\dfrac{4}{\sqrt{3} + x}$.

$$\frac{4}{\sqrt{3} + x} = \frac{4}{\sqrt{3} + x} \cdot \frac{\sqrt{3} - x}{\sqrt{3} - x}$$
$$= \frac{4(\sqrt{3} - x)}{(\sqrt{3} + x)(\sqrt{3} - x)}$$
$$= \frac{4\sqrt{3} - 4x}{3 - x^2}$$

EXAMPLE 9 Rationalize the denominator: $\dfrac{4 + \sqrt{2}}{\sqrt{5} - \sqrt{2}}$.

$$\frac{4 + \sqrt{2}}{\sqrt{5} - \sqrt{2}} = \frac{4 + \sqrt{2}}{\sqrt{5} - \sqrt{2}} \cdot \frac{\sqrt{5} + \sqrt{2}}{\sqrt{5} + \sqrt{2}}$$

Multiplying by 1, using the conjugate of $\sqrt{5} - \sqrt{2}$, which is $\sqrt{5} + \sqrt{2}$

$$= \frac{(4 + \sqrt{2})(\sqrt{5} + \sqrt{2})}{(\sqrt{5} - \sqrt{2})(\sqrt{5} + \sqrt{2})}$$

Multiplying numerators and denominators

$$= \frac{4\sqrt{5} + 4\sqrt{2} + \sqrt{2}\,\sqrt{5} + (\sqrt{2})^2}{(\sqrt{5})^2 - (\sqrt{2})^2}$$

Using $(A - B)(A + B) = A^2 - B^2$ in the denominator

$$= \frac{4\sqrt{5} + 4\sqrt{2} + \sqrt{10} + 2}{5 - 2}$$
$$= \frac{4\sqrt{5} + 4\sqrt{2} + \sqrt{10} + 2}{3}$$

Do Exercises 11 and 12.

15.5 EXERCISE SET

For Extra Help

Digital Video Tutor CD 13 Videotape 17 | InterAct Math | Math Tutor Center | MathXL | MyMathLab

a Rationalize the denominator. Assume that all expressions under radicals represent positive numbers.

1. $\sqrt{\frac{5}{3}}$

2. $\sqrt{\frac{8}{7}}$

3. $\sqrt{\frac{11}{2}}$

4. $\sqrt{\frac{17}{6}}$

5. $\frac{2\sqrt{3}}{7\sqrt{5}}$

6. $\frac{3\sqrt{5}}{8\sqrt{2}}$

7. $\sqrt[3]{\frac{16}{9}}$

8. $\sqrt[3]{\frac{3}{9}}$

9. $\frac{\sqrt[3]{3a}}{\sqrt[3]{5c}}$

10. $\frac{\sqrt[3]{7x}}{\sqrt[3]{3y}}$

11. $\frac{\sqrt[3]{2y^4}}{\sqrt[3]{6x^4}}$

12. $\frac{\sqrt[3]{3a^4}}{\sqrt[3]{7b^2}}$

13. $\frac{1}{\sqrt[4]{st}}$

14. $\frac{1}{\sqrt[3]{yz}}$

15. $\sqrt{\frac{3x}{20}}$

16. $\sqrt{\frac{7a}{32}}$

17. $\sqrt[3]{\frac{4}{5x^5y^2}}$

18. $\sqrt[3]{\frac{7c}{100ab^5}}$

19. $\sqrt[4]{\frac{1}{8x^7y^3}}$

20. $\frac{2x}{\sqrt[5]{18x^8y^6}}$

b Rationalize the denominator. Assume that all expressions under radicals represent positive numbers.

21. $\dfrac{9}{6-\sqrt{10}}$

22. $\dfrac{3}{8+\sqrt{5}}$

23. $\dfrac{-4\sqrt{7}}{\sqrt{5}-\sqrt{3}}$

24. $\dfrac{34\sqrt{5}}{2\sqrt{5}-\sqrt{3}}$

25. $\dfrac{\sqrt{5}-2\sqrt{6}}{\sqrt{3}-4\sqrt{5}}$

26. $\dfrac{\sqrt{6}-3\sqrt{5}}{\sqrt{3}-2\sqrt{7}}$

27. $\dfrac{2-\sqrt{a}}{3+\sqrt{a}}$

28. $\dfrac{5+\sqrt{x}}{8-\sqrt{x}}$

29. $\dfrac{5\sqrt{3}-3\sqrt{2}}{3\sqrt{2}-2\sqrt{3}}$

30. $\dfrac{7\sqrt{2}+4\sqrt{3}}{4\sqrt{3}-3\sqrt{2}}$

31. $\dfrac{\sqrt{x}-\sqrt{y}}{\sqrt{x}+\sqrt{y}}$

32. $\dfrac{\sqrt{a}+\sqrt{b}}{\sqrt{a}-\sqrt{b}}$

33. **D**W A student *incorrectly* claims that

$$\frac{5+\sqrt{2}}{\sqrt{18}}=\frac{5+\sqrt{1}}{\sqrt{9}}=\frac{5+1}{3}=2.$$

How could you convince the student that a mistake has been made? How would you explain the correct way of rationalizing the denominator?

34. **D**W A student considers the radical expression

$$\frac{11}{\sqrt[3]{4}-\sqrt[3]{5}}$$

and tries to rationalize the denominator by multiplying by

$$\frac{\sqrt[3]{4}+\sqrt[3]{5}}{\sqrt[3]{4}+\sqrt[3]{5}}.$$

Discuss the difficulties of such a plan.

SKILL MAINTENANCE

Solve. [11.7a]

35. $\dfrac{1}{2}-\dfrac{1}{3}=\dfrac{5}{t}$

36. $\dfrac{5}{x-1}+\dfrac{9}{x^2+x+1}=\dfrac{15}{x^3-1}$

Divide and simplify. [11.2b]

37. $\dfrac{1}{x^3-y^3}\div\dfrac{1}{(x-y)(x^2+xy+y^2)}$

38. $\dfrac{2x^2-x-6}{x^2+4x+3}\div\dfrac{2x^2+x-3}{x^2-1}$

SYNTHESIS

39. Use a graphing calculator to check your answers to Exercises 15, 16, and 28.

40. Express each of the following as the product of two radical expressions.

a) $x-5$

b) $x-a$

Simplify. (*Hint*: Rationalize the denominator.)

41. $\sqrt{a^2-3}-\dfrac{a^2}{\sqrt{a^2-3}}$

42. $\dfrac{1}{4+\sqrt{3}}+\dfrac{1}{\sqrt{3}}+\dfrac{1}{\sqrt{3}-4}$

15.6 SOLVING RADICAL EQUATIONS

a The Principle of Powers

A **radical equation** has variables in one or more radicands—for example,

$$\sqrt[3]{2x} + 1 = 5, \qquad \sqrt{x} + \sqrt{4x - 2} = 7.$$

To solve such an equation, we need a new equation-solving principle. Suppose that an equation $a = b$ is true. If we square both sides, we get another true equation: $a^2 = b^2$. This can be generalized.

THE PRINCIPLE OF POWERS

For any natural number n, if an equation $a = b$ is true, then $a^n = b^n$ is true.

However, if an equation $a^n = b^n$ is true, it *may not* be true that $a = b$, if n is even. For example, $3^2 = (-3)^2$ is true, but $3 = -3$ is not true. Thus we *must* make a check when we solve an equation using the principle of powers.

EXAMPLE 1 Solve: $\sqrt{x} - 3 = 4$.

$$\sqrt{x} - 3 = 4$$
$$\sqrt{x} = 7 \qquad \text{Adding to isolate the radical}$$
$$(\sqrt{x})^2 = 7^2 \qquad \text{Using the principle of powers (squaring)}$$
$$x = 49. \qquad \sqrt{x} \cdot \sqrt{x} = x$$

The number 49 is a possible solution. But we *must* make a check in order to be sure!

CHECK:

$$\sqrt{x} - 3 = 4$$
$$\sqrt{49} - 3 \ ? \ 4$$
$$7 - 3$$
$$4 \qquad \text{TRUE}$$

The solution is 49.

CAUTION!

The principle of powers does not always give equivalent equations. For this reason, a check is a must!

Objectives

a Solve radical equations with one radical term.

b Solve radical equations with two radical terms.

c Solve applied problems involving radical equations.

Study Tips

BEGINNING TO STUDY FOR THE FINAL EXAM (PART 1)

It is never too soon to begin to study for the final examination. Take a few minutes each week to review the highlighted information, such as formulas, properties, and procedures. Make special use of the Summary and Reviews and Chapter Tests, as well as the supplements such as Interact Math Tutorial software and MathXL.

"Practice does not make perfect; practice makes permanent."

Dr. Richard Chase, former president, Wheaton College

Solve.

1. $\sqrt{x} - 7 = 3$

2. $\sqrt{x} = -2$

Solve.

3. $x + 2 = \sqrt{2x + 7}$

4. $x + 1 = 3\sqrt{x - 1}$

Answers on page A-57

EXAMPLE 2 Solve: $\sqrt{x} = -3$.

We might observe at the outset that this equation has no solution because the principal square root of a number is never negative. Let's continue as above for comparison.

$$\sqrt{x} = -3$$
$$(\sqrt{x})^2 = (-3)^2$$
$$x = 9$$

CHECK: $\sqrt{x} = -3$

$\sqrt{9} \;?\; -3$

$3 \;|\;$ FALSE

The number 9 does *not* check. Thus the equation $\sqrt{x} = -3$ has no real-number solution. Note that the equation $x = 9$ has solution 9, but that $\sqrt{x} = -3$ has *no* solution. Thus the equations $x = 9$ and $\sqrt{x} = -3$ are *not* equivalent. That is, $\sqrt{9} \neq -3$.

Do Exercises 1 and 2.

To solve an equation with a radical term, we first isolate the radical term on one side of the equation. Then we use the principle of powers.

EXAMPLE 3 Solve: $x - 7 = 2\sqrt{x + 1}$.

The radical term is already isolated. We proceed with the principle of powers:

$$x - 7 = 2\sqrt{x + 1}$$
$$(x - 7)^2 = (2\sqrt{x + 1})^2 \quad \text{Using the principle of powers (squaring)}$$
$$(x - 7) \cdot (x - 7) = (2\sqrt{x + 1})(2\sqrt{x + 1})$$
$$x^2 - 14x + 49 = 2^2(\sqrt{x + 1})^2$$
$$x^2 - 14x + 49 = 4(x + 1)$$
$$x^2 - 14x + 49 = 4x + 4$$
$$x^2 - 18x + 45 = 0$$
$$(x - 3)(x - 15) = 0 \quad \text{Factoring}$$
$$x - 3 = 0 \quad or \quad x - 15 = 0 \quad \text{Using the principle of zero products}$$
$$x = 3 \quad or \quad x = 15.$$

The possible solutions are 3 and 5. We check.

For 3:

$x - 7 = 2\sqrt{x + 1}$

$3 - 7 \;?\; 2\sqrt{3 + 1}$

$-4 \;|\; 2\sqrt{4}$

$\;|\; 2(2)$

$\;|\; 4$ FALSE

For 15:

$x - 7 = 2\sqrt{x + 1}$

$15 - 7 \;?\; 2\sqrt{15 + 1}$

$8 \;|\; 2\sqrt{16}$

$\;|\; 2(4)$

$\;|\; 8$ TRUE

The number 3 does *not* check, but the number 15 does check. The solution is 15.

The number 3 in Example 3 is what is sometimes called an *extraneous solution*, but such terminology is risky at best because the number 3 is in *no way* a solution of the original equation.

Do Exercises 3 and 4.

ALGEBRAIC–GRAPHICAL CONNECTION

We can visualize or check the solutions of a radical equation graphically. Consider the equation of Example 3: $x - 7 = 2\sqrt{x + 1}$. We can examine the solutions by graphing the equations

$$y = x - 7 \quad \text{and} \quad y = 2\sqrt{x + 1}$$

using the same set of axes. A hand-drawn graph of $y = 2\sqrt{x + 1}$ would involve approximating square roots on a calculator.

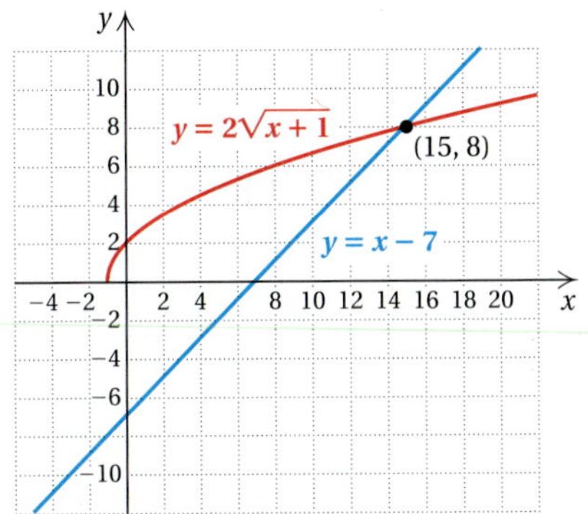

It appears from the graph that when $x = 15$, the values of $y = x - 7$ and $y = 2\sqrt{x + 1}$ are the same, 8. We can check this as we did in Example 3. Note too that the graphs *do not* intersect at $x = 3$, the "extraneous" solution.

CALCULATOR CORNER

Solving Radical Equations We can solve radical equations graphically. Consider the equation in Example 3,

$$x - 7 = 2\sqrt{x + 1}.$$

We first graph each side of the equation. We enter $y_1 = x - 7$ and $y_2 = 2\sqrt{x + 1}$ on the equation-editor screen and graph the equations using the window $[-5, 20, -10, 10]$. Note that there is one point of intersection. We use the INTERSECT feature to find its coordinates. (See the Calculator Corner on p. 991 for the procedure.) The first coordinate, 15, is the value of x for which $y_1 = y_2$, or $x - 7 = 2\sqrt{x + 1}$. It is the solution of the equation. Note that the graph shows a single solution whereas the algebraic solution in Example 3 yields two possible solutions, 3 and 15, that must be checked. The algebraic check shows that 15 is the only solution.

$y_1 = x - 7,\ y_2 = 2\sqrt{x + 1}$

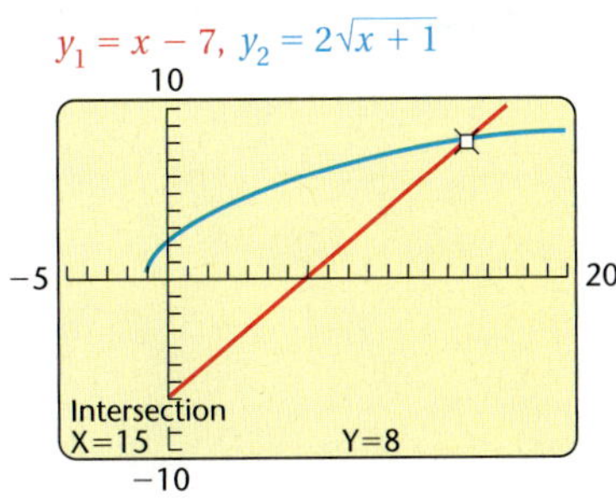

Exercises:

1. Solve the equations in Examples 1 and 4 graphically.
2. Solve the equations in Margin Exercises 1, 3, and 4 graphically.

Solve.

5. $x = \sqrt{x+5} + 1$

6. $\sqrt[4]{x-1} - 2 = 0$

Answers on page A-57

EXAMPLE 4 Solve: $x = \sqrt{x+7} + 5$.

$$x = \sqrt{x+7} + 5$$

$$x - 5 = \sqrt{x+7}$$ Subtracting 5 to isolate the radical term

$$(x-5)^2 = \left(\sqrt{x+7}\right)^2$$ Using the principle of powers (squaring both sides)

$$x^2 - 10x + 25 = x + 7$$

$$x^2 - 11x + 18 = 0$$

$$(x-9)(x-2) = 0$$ Factoring

$$x = 9 \quad \text{or} \quad x = 2$$ Using the principle of zero products

The possible solutions are 9 and 2. Let's check.

For 9:

$$\begin{array}{c|c} x = \sqrt{x+7} + 5 & \\ \hline 9 \ ? \ \sqrt{9+7} + 5 & \\ & 9 \end{array}$$ TRUE

For 2:

$$\begin{array}{c|c} x = \sqrt{x+7} + 5 & \\ \hline 2 \ ? \ \sqrt{2+7} + 5 & \\ & 8 \end{array}$$ FALSE

Since 9 checks but 2 does not, the solution is 9.

EXAMPLE 5 Solve: $\sqrt[3]{2x+1} + 5 = 0$.

$$\sqrt[3]{2x+1} + 5 = 0$$

$$\sqrt[3]{2x+1} = -5$$ Subtracting 5. This isolates the radical term.

$$\left(\sqrt[3]{2x+1}\right)^3 = (-5)^3$$ Using the principle of powers (raising to the third power)

$$2x + 1 = -125$$

$$2x = -126$$ Subtracting 1

$$x = -63$$

CHECK:

$$\begin{array}{r|l} \sqrt[3]{2x+1} + 5 = 0 & \\ \hline \sqrt[3]{2\cdot(-63)+1} + 5 \ ? & 0 \\ \sqrt[3]{-125} + 5 & \\ -5 + 5 & \\ 0 & \end{array}$$

The solution is -63.

Do Exercises 5 and 6.

b Equations with Two Radical Terms

A general strategy for solving radical equations, including those with two radical terms, is as follows.

SOLVING RADICAL EQUATIONS

To solve radical equations:

1. Isolate one of the radical terms.
2. Use the principle of powers.
3. If a radical remains, perform steps (1) and (2) again.
4. Check possible solutions.

CAUTION!

Equations with more than one radical term are especially likely to yield numbers that are not solutions.

EXAMPLE 6 Solve: $\sqrt{x-3} + \sqrt{x+5} = 4$.

$$\sqrt{x-3} + \sqrt{x+5} = 4$$

$$\sqrt{x-3} = 4 - \sqrt{x+5}$$ Subtracting $\sqrt{x+5}$. This isolates one of the radical terms.

$$\left(\sqrt{x-3}\right)^2 = \left(4 - \sqrt{x+5}\right)^2$$ Using the principle of powers (squaring both sides)

$$x - 3 = 16 - 8\sqrt{x+5} + (x+5)$$ Using $(A - B)^2 = A^2 - 2AB + B^2$. See this rule in Section 9.5.

$$-3 = 21 - 8\sqrt{x+5}$$ Subtracting x and collecting like terms

$$-24 = -8\sqrt{x+5}$$ Isolating the remaining radical term

$$3 = \sqrt{x+5}$$ Dividing by -8

$$3^2 = \left(\sqrt{x+5}\right)^2$$ Squaring

$$9 = x + 5$$

$$4 = x$$

The number 4 checks and is the solution.

Solve.

7. $\sqrt{x} - \sqrt{x-5} = 1$

8. $\sqrt{2x-5} - 2 = \sqrt{x-2}$

9. Solve:

$\sqrt{3x+1} - 1 - \sqrt{x+4} = 0.$

Answers on page A-57

EXAMPLE 7 Solve: $\sqrt{2x-5} = 1 + \sqrt{x-3}$.

$$\sqrt{2x-5} = 1 + \sqrt{x-3}$$

$$(\sqrt{2x-5})^2 = (1 + \sqrt{x-3})^2 \quad \text{One radical is already isolated. We square both sides.}$$

$$2x - 5 = 1 + 2\sqrt{x-3} + (\sqrt{x-3})^2$$

$$2x - 5 = 1 + 2\sqrt{x-3} + (x-3)$$

$$x - 3 = 2\sqrt{x-3} \quad \text{Isolating the remaining radical term}$$

$$(x-3)^2 = (2\sqrt{x-3})^2 \quad \text{Squaring both sides}$$

$$x^2 - 6x + 9 = 4(x-3)$$

$$x^2 - 6x + 9 = 4x - 12$$

$$x^2 - 10x + 21 = 0$$

$$(x-7)(x-3) = 0 \quad \text{Factoring}$$

$$x = 7 \quad or \quad x = 3 \quad \text{Using the principle of zero products}$$

The possible solutions are 7 and 3. We check.

For 7:

$\sqrt{2x-5}$	$= 1 + \sqrt{x-3}$	
$\sqrt{2(7)-5}$	? $1 + \sqrt{7-3}$	
$\sqrt{14-5}$	$1 + \sqrt{4}$	
$\sqrt{9}$	$1 + 2$	
3	3	TRUE

For 3:

$\sqrt{2x-5}$	$= 1 + \sqrt{x-3}$	
$\sqrt{2(3)-5}$	? $1 + \sqrt{3-3}$	
$\sqrt{6-5}$	$1 + \sqrt{0}$	
$\sqrt{1}$	$1 + 0$	
1	1	TRUE

The numbers 7 and 3 check and are the solutions.

Do Exercises 7 and 8.

EXAMPLE 8 Solve: $\sqrt{x+2} - \sqrt{2x+2} + 1 = 0$.

We first isolate one radical.

$$\sqrt{x+2} - \sqrt{2x+2} + 1 = 0$$

$$\sqrt{x+2} + 1 = \sqrt{2x+2} \quad \text{Adding } \sqrt{2x+2} \text{ to isolate a radical expression}$$

$$(\sqrt{x+2} + 1)^2 = (\sqrt{2x+2})^2 \quad \text{Squaring both sides}$$

$$x + 2 + 2\sqrt{x+2} + 1 = 2x + 2$$

$$2\sqrt{x+2} = x - 1$$

$$(2\sqrt{x+2})^2 = (x-1)^2$$

$$4(x+2) = x^2 - 2x + 1$$

$$4x + 8 = x^2 - 2x + 1$$

$$0 = x^2 - 6x - 7$$

$$0 = (x-7)(x+1) \quad \text{Factoring}$$

$$x - 7 = 0 \quad or \quad x + 1 = 0 \quad \text{Using the principle of zero products}$$

$$x = 7 \quad or \quad x = -1$$

The check is left to the student. The number 7 checks, but -1 does not. The solution is 7.

Do Exercise 9.

C Applications

Speed of Sound. Many applications translate to radical equations. For example, at a temperature of t degrees Fahrenheit, sound travels S feet per second, where

$$S = 21.9\sqrt{5t + 2457}. \quad (1)$$

EXAMPLE 9 *Orchestra Practice.* During orchestra practice, the temperature of a room was 72°F. How fast was the sound of the orchestra traveling through the room?

We substitute 72 for t in equation (1) and find an approximation using a calculator:

$$\begin{aligned} S &= 21.9\sqrt{5t + 2457} \\ &= 21.9\sqrt{5(72) + 2457} \\ &= 21.9\sqrt{360 + 2457} \\ &= 21.9\sqrt{2817} \\ &\approx 1162.4 \text{ ft/sec.} \end{aligned}$$

Do Exercise 10.

EXAMPLE 10 *Musical Performances.* The group *NSYNC regularly performs outdoors for large audiences. A scientific instrument at one of their concerts determined that the sound of the group was traveling at a rate of 1170 ft/sec. What was the air temperature at the concert?

We substitute 1170 for S in the formula $S = 21.9\sqrt{5t + 2457}$:

$$1170 = 21.9\sqrt{5t + 2457}.$$

Then we solve the equation for t:

$$\begin{aligned} 1170 &= 21.9\sqrt{5t + 2457} \\ \frac{1170}{21.9} &= \sqrt{5t + 2457} && \text{Dividing by 21.9} \\ \left(\frac{1170}{21.9}\right)^2 &= \left(\sqrt{5t + 2457}\right)^2 && \text{Squaring both sides} \\ 2854.2 &\approx 5t + 2457 && \text{Simplifying} \\ 397.2 &\approx 5t && \text{Subtracting 2457} \\ 79 &\approx t. && \text{Dividing by 5} \end{aligned}$$

The temperature at the concert was about 79°F.

Do Exercise 11.

10. Music During Surgery. It has been shown that playing music during surgery may hasten a patient's recovery after surgery. During a heart bypass surgery, doctors played music in the operating room. The temperature of the room was kept at 60°F. How fast did the music travel through the room?
Source: *Acta Anaesthesiologica Scandinavica*, 2001; 45: 812–817

11. Musical Performances. During an outdoor concert given by LeAnn Rimes, the speed of sound from the music was measured by a scientific instrument to be 1179 ft/sec. What was the air temperature at the concert?

Answers on page A-57

15.6 EXERCISE SET

For Extra Help

Digital Video Tutor CD 13 Videotape 17 | InterAct Math | Math Tutor Center | MathXL | MyMathLab

a Solve.

1. $\sqrt{2x-3}=4$

2. $\sqrt{5x+2}=7$

3. $\sqrt{6x}+1=8$

4. $\sqrt{3x}-4=6$

5. $\sqrt{y+7}-4=4$

6. $\sqrt{x-1}-3=9$

7. $\sqrt{5y+8}=10$

8. $\sqrt{2y+9}=5$

9. $\sqrt[3]{x}=-1$

10. $\sqrt[3]{y}=-2$

11. $\sqrt{x+2}=-4$

12. $\sqrt{y-3}=-2$

13. $\sqrt[3]{x+5}=2$

14. $\sqrt[3]{x-2}=3$

15. $\sqrt[4]{y-3}=2$

16. $\sqrt[4]{x+3}=3$

17. $\sqrt[3]{6x+9}+8=5$

18. $\sqrt[3]{3y+6}+2=3$

19. $8=\dfrac{1}{\sqrt{x}}$

20. $\dfrac{1}{\sqrt{y}}=3$

21. $x-7=\sqrt{x-5}$

22. $x-5=\sqrt{x+7}$

23. $2\sqrt{x+1}+7=x$

24. $\sqrt{2x+7}-2=x$

25. $3\sqrt{x-1}-1=x$

26. $x-1=\sqrt{x+5}$

27. $x-3=\sqrt{27-3x}$

28. $x-1=\sqrt{1-x}$

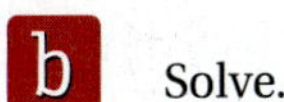 Solve.

29. $\sqrt{3y+1} = \sqrt{2y+6}$

30. $\sqrt{5x-3} = \sqrt{2x+3}$

31. $\sqrt{y-5} + \sqrt{y} = 5$

32. $\sqrt{x-9} + \sqrt{x} = 1$

33. $3 + \sqrt{z-6} = \sqrt{z+9}$

34. $\sqrt{4x-3} = 2 + \sqrt{2x-5}$

35. $\sqrt{20-x} + 8 = \sqrt{9-x} + 11$

36. $4 + \sqrt{10-x} = 6 + \sqrt{4-x}$

37. $\sqrt{4y+1} - \sqrt{y-2} = 3$

38. $\sqrt{y+15} - \sqrt{2y+7} = 1$

39. $\sqrt{x+2} + \sqrt{3x+4} = 2$

40. $\sqrt{6x+7} - \sqrt{3x+3} = 1$

41. $\sqrt{3x-5} + \sqrt{2x+3} + 1 = 0$

42. $\sqrt{2m-3} + 2 - \sqrt{m+7} = 0$

43. $2\sqrt{t-1} - \sqrt{3t-1} = 0$

44. $3\sqrt{2y+3} - \sqrt{y+10} = 0$

C Solve.

Sightings to the Horizon. How far can you see from a given height? There is a formula for this. At a height of h meters, you can see V kilometers to the horizon. These numbers are related as follows:

$$V = 3.5\sqrt{h}.$$

Use this formula for Exercises 45–52.

45. How far can you see to the horizon through an airplane window at a height, or altitude, of 9000 m?

46. How far can you see to the horizon through an airplane window at a height of 8000 m?

47. Elaine can see 50.4 km to the horizon from the top of a cliff. What is the height of Elaine's eyes?

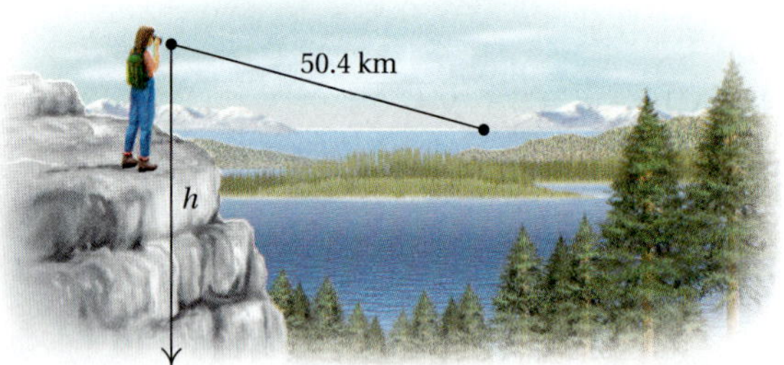

48. A technician can see 49 km to the horizon from the top of a radio tower. How high is the tower?

49. A steeplejack can see 21 km to the horizon from the top of a building. What is the height of the steeplejack's eyes?

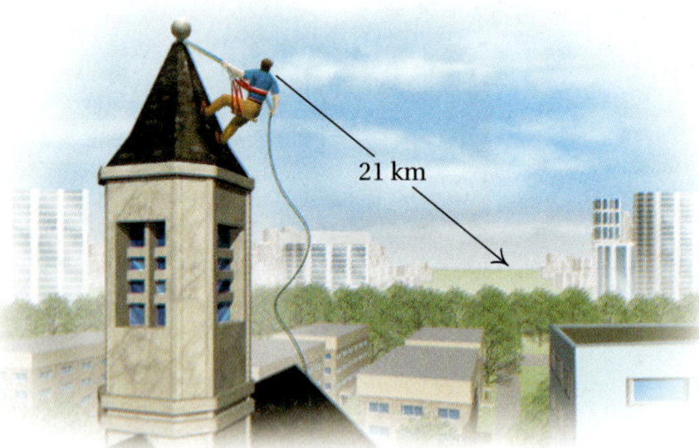

50. A person can see 371 km to the horizon from an airplane window. How high is the airplane?

51. How far can a sailor see to the horizon from the top of a mast that is 37 m high?

52. How far can you see to the horizon through an airplane window at a height of 9800 m?

Speed of a Skidding Car. After an accident, how do police determine the speed at which the car had been traveling? The formula

$$r = 2\sqrt{5L}$$

can be used to approximate the speed r, in miles per hour, of a car that has left a skid mark of length L, in feet. Use this formula for Exercises 53 and 54.

53. How far will a car skid at 55 mph? at 75 mph?

54. How far will a car skid at 65 mph? at 100 mph?

Temperature and the Speed of Sound. Solve Exercises 55 and 56 using the formula $S = 21.9\sqrt{5t + 2457}$ from Example 9.

55. During blasting for avalanche control in Utah's Wasatch Mountains, sound traveled at a rate of 1113 ft/sec. What was the temperature at the time?

56. At a recent concert by the Dave Matthews Band, sound traveled at a rate of 1176 ft/sec. What was the temperature at the time?

Period of a Swinging Pendulum. The formula $T = 2\pi\sqrt{L/32}$ can be used to find the period T, in seconds, of a pendulum of length L, in feet.

57. What is the length of a pendulum that has a period of 1.0 sec? Use 3.14 for π.

58. What is the length of a pendulum that has a period of 2.0 sec? Use 3.14 for π.

59. ^{D}W The principle of powers contains an "if–then" statement that becomes false when the parts are interchanged. Find another mathematical example of such an "if–then" statement.

60. ^{D}W Is checking necessary when the principle of powers is used with an odd power n? Why or why not?

SKILL MAINTENANCE

Solve. [11.8a]

61. *Painting a Room.* Julia can paint a room in 8 hr. George can paint the same room in 10 hr. How long will it take them, working together, to paint the same room?

62. *Delivering Leaflets.* Jeff can drop leaflets in mailboxes three times as fast as Grace can. If they work together, it takes them 1 hr to complete the job. How long would it take each to deliver the leaflets alone?

Solve. [11.8b]

63. *Bicycle Travel.* A cyclist traveled 702 mi in 14 days. At this same ratio, how far would the cyclist have traveled in 56 days?

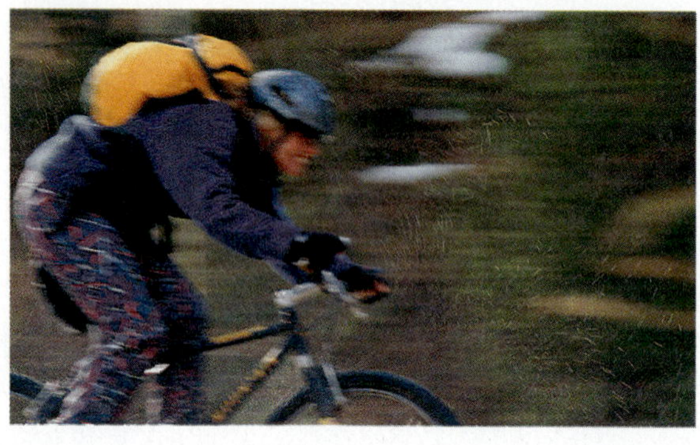

64. *Earnings.* Dharma earned \$696.64 working for 56 hr at a fruit stand. How many hours must she work in order to earn \$1044.96?

Solve. [10.8b]

65. $x^2 + 2.8x = 0$

66. $3x^2 - 5x = 0$

67. $x^2 - 64 = 0$

68. $2x^2 = x + 21$

SYNTHESIS

69. Use a graphing calculator to check your answers to Exercises 4, 9, 33, and 38.

70. Consider the equation
$$\sqrt{2x+1} + \sqrt{5x-4} = \sqrt{10x+9}.$$
a) Use a graphing calculator to solve the equation.
b) Solve the equation algebraically.
c) Explain the advantages and disadvantages of using each method. Which do you prefer?

Solve.

71. $\sqrt[3]{\frac{z}{4}} - 10 = 2$

72. $\sqrt[4]{z^2 + 17} = 3$

73. $\sqrt{\sqrt{y+49} - \sqrt{y}} = \sqrt{7}$

74. $\sqrt[3]{x^2 + x + 15} - 3 = 0$

75. $\sqrt{\sqrt{x^2 + 9x + 34}} = 2$

76. $\sqrt{8-b} = b\sqrt{8-b}$

77. $\sqrt{x-2} - \sqrt{x+2} + 2 = 0$

78. $6\sqrt{y} + 6y^{-1/2} = 37$

79. $\sqrt{a^2 + 30a} = a + \sqrt{5a}$

80. $\sqrt{\sqrt{x} + 4} = \sqrt{x} - 2$

81. $\frac{x-1}{\sqrt{x^2 + 3x + 6}} = \frac{1}{4}$

82. $\sqrt{x+1} - \frac{2}{\sqrt{x+1}} = 1$

83. $\sqrt{y^2 + 6} + y - 3 = 0$

84. $2\sqrt{x-1} - \sqrt{3x-5} = \sqrt{x-9}$

85. $\sqrt{y+1} - \sqrt{2y-5} = \sqrt{y-2}$

86. Evaluate: $\sqrt{7 + 4\sqrt{3}} - \sqrt{7 - 4\sqrt{3}}$.

15.7 APPLICATIONS INVOLVING POWERS AND ROOTS

a Applications

There are many kinds of applied problems that involve powers and roots. Many also make use of right triangles and the Pythagorean theorem: $a^2 + b^2 = c^2$.

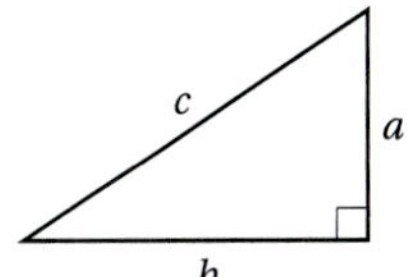

EXAMPLE 1 *Vegetable Garden.* Benito and Dominique are planting a vegetable garden in the backyard. They decide that it will be a 30-ft by 40-ft rectangle and begin to lay it out using string. They soon realize that it is difficult to form the right angles and that it would be helpful to know the length of a diagonal. Find the length of a diagonal.

Using the Pythagorean theorem, $a^2 + b^2 = c^2$, we substitute 30 for a and 40 for b and then solve for c:

$$a^2 + b^2 = c^2$$
$$30^2 + 40^2 = c^2 \quad \text{Substituting}$$
$$900 + 1600 = c^2$$
$$2500 = c^2$$
$$\sqrt{2500} = c$$
$$50 = c.$$

We consider only the positive root since length cannot be negative.

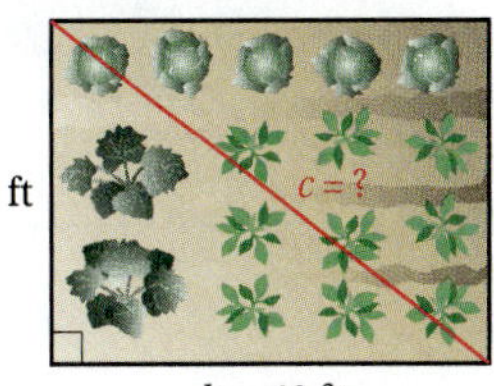

The length of the hypotenuse, or the diagonal, is 50 ft. Knowing this measurement would help in laying out the garden. Construction workers often use a procedure like this to lay out a right angle.

EXAMPLE 2 Find the length of the hypotenuse of this right triangle. Give an exact answer and an approximation to three decimal places.

$$7^2 + 4^2 = c^2 \quad \text{Substituting}$$
$$49 + 16 = c^2$$
$$65 = c^2$$

Exact answer: $c = \sqrt{65}$

Approximation: $c \approx 8.062$ Using a calculator

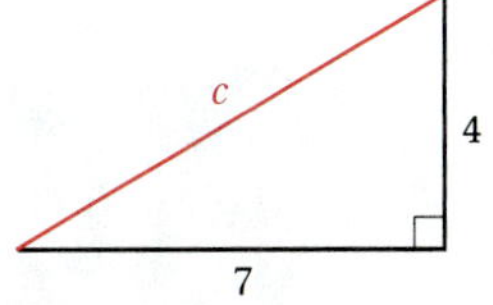

EXAMPLE 3 Find the missing length b in this right triangle. Give an exact answer and an approximation to three decimal places.

$$1^2 + b^2 = (\sqrt{11})^2 \quad \text{Substituting}$$
$$1 + b^2 = 11$$
$$b^2 = 10$$

Exact answer: $b = \sqrt{10}$

Approximation: $b \approx 3.162$ Using a calculator

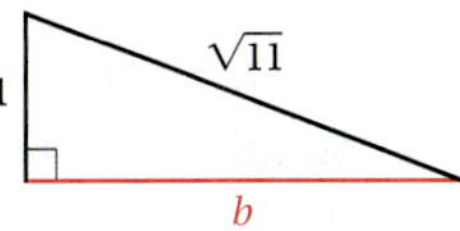

Do Exercises 1–3.

Objective

a Solve applied problems involving the Pythagorean theorem and powers and roots.

1. Find the length of the hypotenuse of this right triangle. Give an exact answer and an approximation to three decimal places.

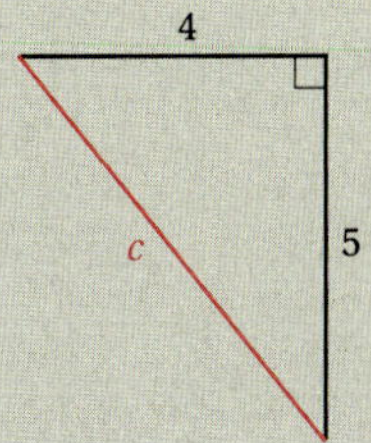

2. Find the length of the leg of this right triangle. Give an exact answer and an approximation to three decimal places.

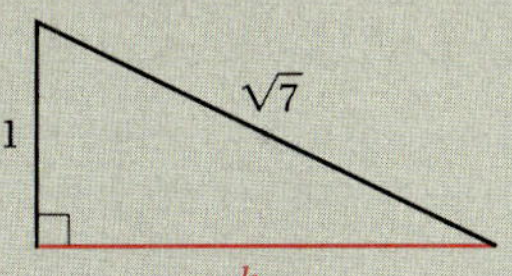

3. Find the length of the hypotenuse of this right triangle. Give an exact answer and an approximation to three decimal places.

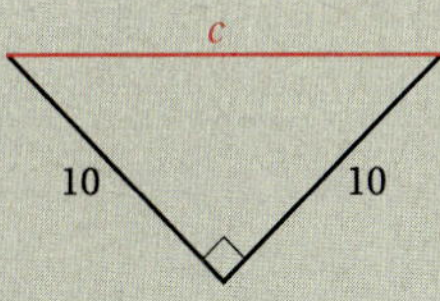

Answers on page A-58

4. Baseball Diamond. A baseball diamond is actually a square 90 ft on a side. Suppose a catcher fields a bunt along the third-base line 10 ft from home plate. How far would the catcher have to throw the ball to first base? Give an exact answer and an approximation to three decimal places.

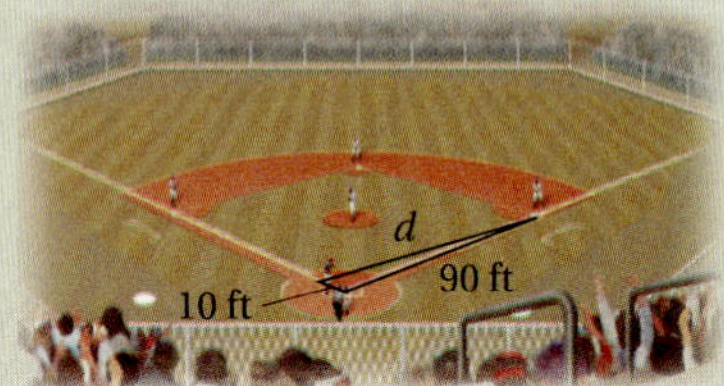

5. Referring to Example 5, find L given that $h = 3$ ft and $d = 180$ ft. You will need a calculator with an exponential key $\boxed{y^x}$, or $\boxed{\wedge}$.

Answers on page A-58

EXAMPLE 4 *Ramps for the Handicapped.* Laws regarding access ramps for the handicapped state that a ramp must be in the form of a right triangle, where every vertical length (leg) of 1 ft has a horizontal length (leg) of 12 ft. What is the length of a ramp with a 12-ft horizontal leg and a 1-ft vertical leg? Give an exact answer and an approximation to three decimal places.

We make a drawing and let $h =$ the length of the ramp. It is the length of the hypotenuse of a right triangle whose legs are 12 ft and 1 ft. We substitute these values into the Pythagorean theorem to find h.

$$h^2 = 12^2 + 1^2$$
$$h^2 = 144 + 1$$
$$h^2 = 145$$

Exact answer: $h = \sqrt{145}$ ft

Approximation: $h \approx 12.042$ ft Using a calculator

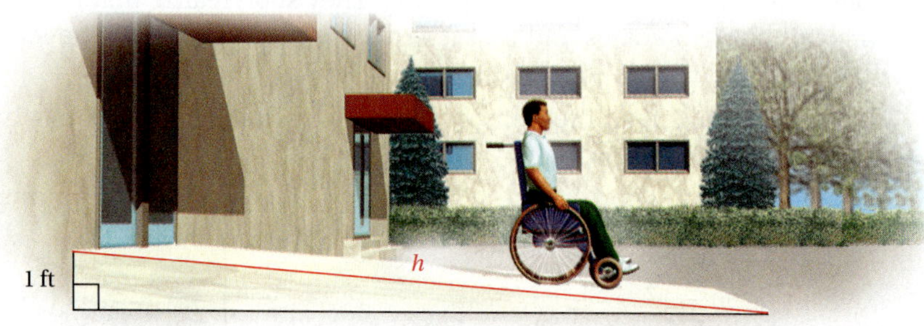

Do Exercise 4.

EXAMPLE 5 *Road-Pavement Messages.* In a psychological study, it was determined that the proper length L of the letters of a word painted on pavement is given by

$$L = \frac{0.000169d^{2.27}}{h},$$

where d is the distance of a car from the lettering and h is the height of the eye above the road. All units are in feet. For a person h feet above the road, a message d feet away will be the most readable if the length of the letters is L.

Find L, given that $h = 4$ ft and $d = 180$ ft.

We substitute 4 for h and 180 for d and calculate L using a calculator with an exponential key $\boxed{y^x}$, or $\boxed{\wedge}$:

$$L = \frac{0.000169(180)^{2.27}}{4} \approx 5.6 \text{ ft}.$$

Do Exercise 5.

15.7 EXERCISE SET

For Extra Help

Digital Video Tutor CD 13 Videotape 17 | InterAct Math | Math Tutor Center | MathXL | MyMathLab

a In a right triangle, find the length of the side not given. Give an exact answer and an approximation to three decimal places.

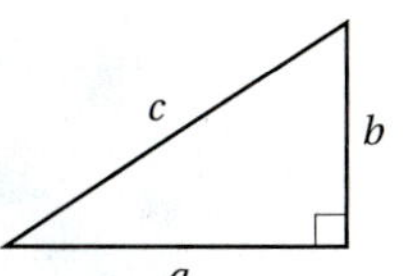

1. $a = 3, \quad b = 5$

2. $a = 8, \quad b = 10$

3. $a = 15, \quad b = 15$

4. $a = 8, \quad b = 8$

5. $b = 12, \quad c = 13$

6. $a = 5, \quad c = 12$

7. $c = 7, \quad a = \sqrt{6}$

8. $c = 10, \quad a = 4\sqrt{5}$

9. $b = 1, \quad c = \sqrt{13}$

10. $a = 1, \quad c = \sqrt{12}$

11. $a = 1, \quad c = \sqrt{n}$

12. $c = 2, \quad a = \sqrt{n}$

In the following problems, give an exact answer and, where appropriate, an approximation (using a calculator) to three decimal places.

13. *Bridge Expansion.* During the summer heat, a 2-mi bridge expands 2 ft in length. If we assume that the bulge occurs straight up the middle, how high is the bulge? (The answer may surprise you. In reality, bridges are build with expansion spaces to avoid such buckling.)

14. *Triangle Areas.* Triangle *ABC* has sides of lengths 25 ft, 25 ft, and 30 ft. Triangle *PQR* has sides of lengths 25 ft, 25 ft, and 40 ft. Which triangle has the greater area and by how much?

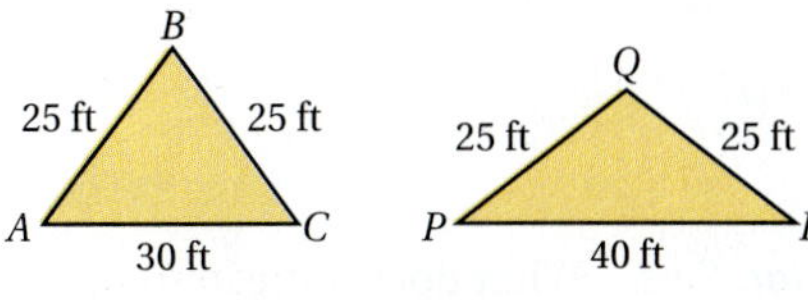

15. Each side of a regular octagon has length s. Find a formula for the distance d between the parallel sides of the octagon.

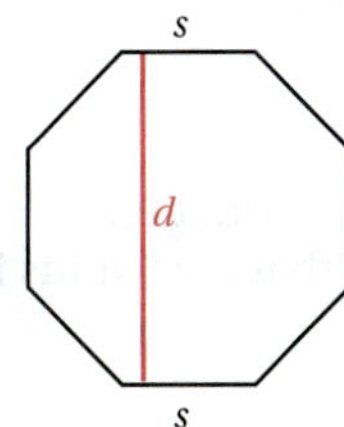

16. The two equal sides of an isosceles right triangle are of length s. Find a formula for the length of the hypotenuse.

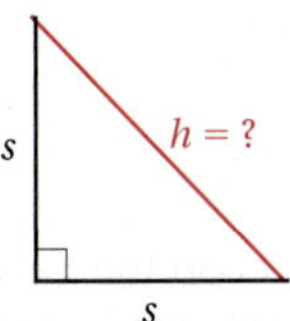

17. *Guy Wire.* How long is a guy wire reaching from the top of a 10-ft pole to a point on the ground 4 ft from the pole?

18. *Softball Diamond.* A slow-pitch softball diamond is actually a square 65 ft on a side. How far is it from home to second base?

19. *Road-Pavement Messages.* Using the formula of Example 5, find the length L of a road-pavement message when $h = 4$ ft and $d = 200$ ft.

20. *Road-Pavement Messages.* Using the formula of Example 5, find the length L of a road-pavement message when $h = 8$ ft and $d = 300$ ft.

21. The length and the width of a rectangle are given by consecutive integers. The area of the rectangle is 90 cm^2. Find the length of a diagonal of the rectangle.

22. The diagonal of a square has length $8\sqrt{2}$ ft. Find the length of a side of the square.

23. *Television Sets.* What does it mean to refer to a 20-in. TV set or a 25-in. TV set? Such units refer to the diagonal of the screen. A 20-in. TV set also has a width of 16 in. What is its height?

24. *Television Sets.* A 25-in. TV set has a screen with a height of 15 in. What is its width?

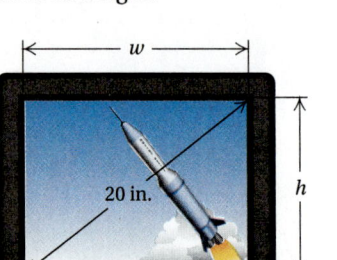

25. Find all ordered pairs on the x-axis of a Cartesian coordinate system that are 5 units from the point $(0, 4)$.

26. Find all ordered pairs on the y-axis of a Cartesian coordinate system that are 5 units from the point $(3, 0)$.

27. *Speaker Placement.* A stereo receiver is in a corner of a 12-ft by 14-ft room. Speaker wire will run under a rug, diagonally, to a speaker in the far corner. If 4 ft of slack is required on each end, how long a piece of wire should be purchased?

28. *Distance Over Water.* To determine the width of a pond, a surveyor locates two stakes at either end of the pond and uses instrumentation to place a third stake so that the distance across the pond is the length of a hypotenuse. If the third stake is 90 m from one stake and 70 m from the other, how wide is the pond?

29. *Plumbing.* Plumbers use the Pythagorean theorem to calculate pipe length. If a pipe is to be offset, as shown in the figure, the *travel*, or length, of the pipe, is calculated using the lengths of the *advance* and *offset*.

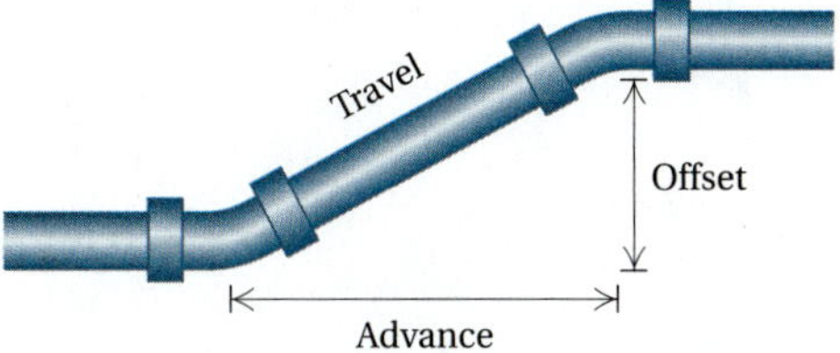

Find the travel if the offset is 17.75 in. and the advance is 10.25 in.

30. *Carpentry.* Darla is laying out the footer of a house. To see if the corner is square, she measures 16 ft from the corner along one wall and 12 ft from the corner along the other wall. How long should the diagonal be between those two points if the corner is a right angle?

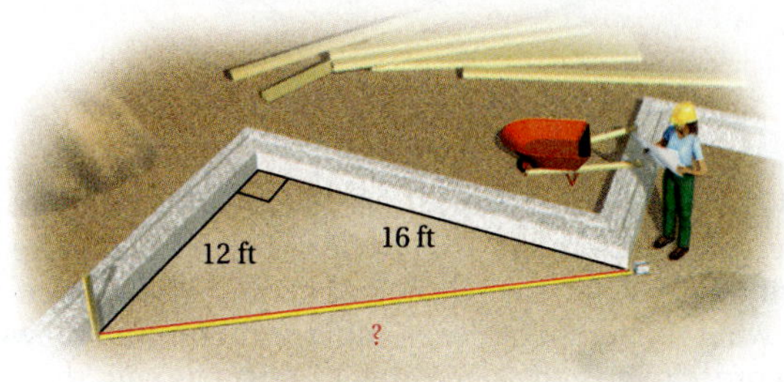

31. D$_W$ Write a problem for a classmate to solve in which the solution is "The height of the tepee is $5\sqrt{3}$ yd."

32. D$_W$ Write a problem for a classmate to solve in which the solution is "The height of the window is $15\sqrt{3}$ yd."

SKILL MAINTENANCE

Solve. [11.8a]

33. *Commuter Travel.* The speed of the Zionsville Flash commuter train is 14 mph faster than that of the Carmel Crawler. The Flash travels 290 mi in the same time that it takes the Crawler to travel 230 mi. Find the speed of each train.

34. *Marine Travel.* A motor boat travels three times as fast as the current in the Saskatee River. A trip up the river and back takes 10 hr, and the total distance of the trip is 100 mi. Find the speed of the current.

Solve.

35. $2x^2 + 11x - 21 = 0$ [10.8b]

36. $x^2 + 24 = 11x$ [10.8b]

37. $\dfrac{x+2}{x+3} = \dfrac{x-4}{x-5}$ [11.7a]

38. $3x^2 - 12 = 0$ [10.8b]

39. $\dfrac{x-5}{x-7} = \dfrac{4}{3}$ [11.7a]

40. $\dfrac{x-1}{x-3} = \dfrac{6}{x-3}$ [11.7a]

SYNTHESIS

41. *Roofing.* Kit's cottage, which is 24 ft wide and 32 ft long, needs a new roof. By counting clapboards that are 4 in. apart, Kit determines that the peak of the roof is 6 ft higher than the sides. If one packet of shingles covers $33\frac{1}{3}$ square feet, how many packets will the job require?

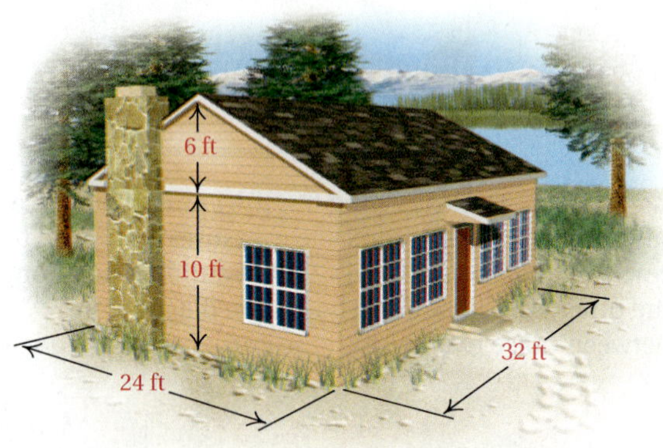

42. *Painting.* (Refer to Exercise 41.) A gallon of paint covers about 275 square feet. If Kit's first floor is 10 ft high, how many gallons of paint should be bought to paint the house? What assumption(s) is made in your answer?

43. *Cube Diagonal.* A cube measures 5 cm on each side. How long is the diagonal that connects two opposite corners of the cube? Give an exact answer.

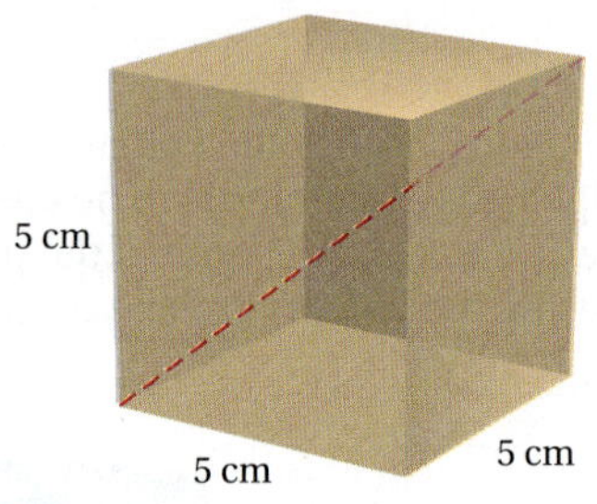

44. *Wind Chill Temperature.* Because wind enhances the loss of heat from the skin, we feel colder when there is wind than when there is not. The *wind chill temperature* is what the temperature would have to be with no wind in order to give the same chilling effect as with the wind. A formula for finding the wind chill temperature, T_W, is

$$T_W = 91.4 - (91.4 - T)(0.478 + 0.301\sqrt{v} - 0.02v),$$

where T is the actual temperature given by a thermometer, in degrees Fahrenheit, and v is the wind speed, in miles per hour.* Use a calculator to find the wind chill temperature in each case. Round to the nearest degree.

a) $T = 40°F$, $v = 25$ mph
b) $T = 20°F$, $v = 25$ mph
c) $T = 10°F$, $v = 20$ mph
d) $T = 10°F$, $v = 40$ mph
e) $T = -5°F$, $v = 35$ mph
f) $T = -16°F$, $v = 35$ mph

*This formula can be used only when the wind speed is *above* 4 mph.

15.8 THE COMPLEX NUMBERS

Objectives

a. Express imaginary numbers as bi, where b is a nonzero real number, and complex numbers as $a + bi$, where a and b are real numbers.

b. Add and subtract complex numbers.

c. Multiply complex numbers.

d. Write expressions involving powers of i in the form $a + bi$.

e. Find conjugates of complex numbers and divide complex numbers.

f. Determine whether a given complex number is a solution of an equation.

a Imaginary and Complex Numbers

Negative numbers do not have square roots in the real-number system. However, mathematicians have invented a larger number system that contains the real-number system, such that negative numbers have square roots. That system is called the **complex-number system.** We begin by defining a number that is a square root of -1. We call this new number i.

THE COMPLEX NUMBER i

We define the number i to be $\sqrt{-1}$. That is,

$$i = \sqrt{-1} \quad \text{and} \quad i^2 = -1.$$

To express roots of negative numbers in terms of i, we can use the fact that in the complex numbers, $\sqrt{-p} = \sqrt{-1 \cdot p} = \sqrt{-1}\,\sqrt{p}$ when p is a positive real number.

EXAMPLES Express in terms of i.

1. $\sqrt{-7} = \sqrt{-1 \cdot 7} = \sqrt{-1} \cdot \sqrt{7} = i\sqrt{7}$, or $\sqrt{7}\,i$

   i is *not* under the radical.

2. $\sqrt{-16} = \sqrt{-1 \cdot 16} = \sqrt{-1} \cdot \sqrt{16} = i \cdot 4 = 4i$
3. $-\sqrt{-13} = -\sqrt{-1 \cdot 13} = -\sqrt{-1} \cdot \sqrt{13} = -i\sqrt{13}$, or $-\sqrt{13}\,i$
4. $-\sqrt{-64} = -\sqrt{-1 \cdot 64} = -\sqrt{-1} \cdot \sqrt{64} = -i \cdot 8 = -8i$
5. $\sqrt{-48} = \sqrt{-1 \cdot 48} = \sqrt{-1} \cdot \sqrt{48} = i\sqrt{48} = i \cdot 4\sqrt{3} = 4\sqrt{3}\,i$, or $4i\sqrt{3}$

Do Exercises 1–5.

Express in terms of i.

1. $\sqrt{-5}$
2. $\sqrt{-25}$
3. $-\sqrt{-11}$
4. $-\sqrt{-36}$
5. $\sqrt{-54}$

Answers on page A-58

IMAGINARY NUMBER

An **imaginary* number** is a number that can be named

bi,

where b is some real number and $b \neq 0$.

To form the system of **complex numbers,** we take the imaginary numbers and the real numbers and all possible sums of real and imaginary numbers. These are complex numbers:

$$7 - 4i, \qquad -\pi + 19i, \qquad 37, \qquad i\sqrt{8}.$$

COMPLEX NUMBER

A **complex number** is any number that can be named

$a + bi$,

where a and b are any real numbers. (Note that either a or b or both can be 0.)

*Don't let the name "imaginary" fool you. The imaginary numbers are very important in such fields as engineering and the physical sciences.

Add or subtract.

6. $(7 + 4i) + (8 - 7i)$

7. $(-5 - 6i) + (-7 + 12i)$

8. $(8 + 3i) - (5 + 8i)$

9. $(5 - 4i) - (-7 + 3i)$

Answers on page A-58

Since $0 + bi = bi$, every imaginary number is a complex number. Similarly, $a + 0i = a$, so every real number is a complex number. The relationships among various real and complex numbers are shown in the following diagram.

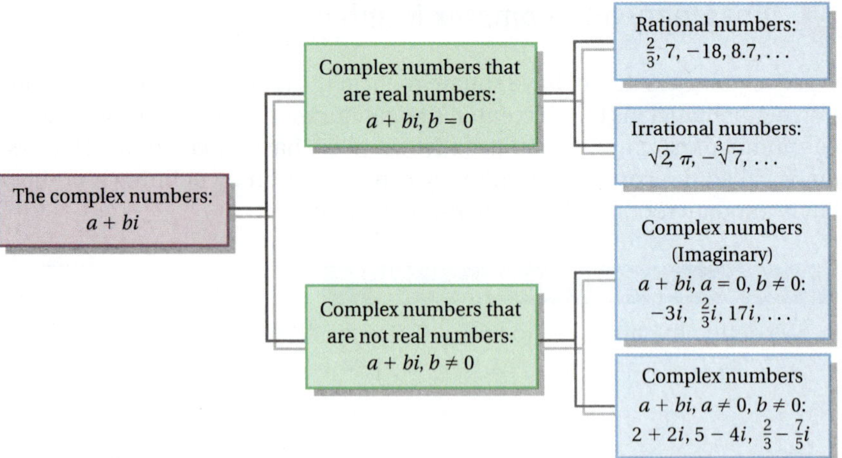

It is important to keep in mind some comparisons between numbers that have real-number roots and those that have complex-number roots that are not real. For example, $\sqrt{-48}$ is a complex number that is not a real number because we are taking the square root of a negative number. *But*, $\sqrt[3]{-125}$ is a real number because we are taking the cube root of a negative number and *any* real number has a cube root that is a real number.

b Addition and Subtraction

The complex numbers follow the commutative and associative laws of addition. Thus we can add and subtract them as we do binomials with real-number coefficients, that is, we collect like terms.

EXAMPLES Add or subtract.

6. $(8 + 6i) + (3 + 2i) = (8 + 3) + (6 + 2)i = 11 + 8i$

7. $(3 + 2i) - (5 - 2i) = (3 - 5) + [2 - (-2)]i = -2 + 4i$

Do Exercises 6–9.

c Multiplication

The complex numbers obey the commutative, associative, and distributive laws. But although the property $\sqrt{a}\sqrt{b} = \sqrt{ab}$ does *not* hold for complex numbers in general, it does hold when $a = -1$ and b is a positive real number. To multiply square roots of negative real numbers, we first express them in terms of i. For example,

$$\sqrt{-2} \cdot \sqrt{-5} = \sqrt{-1} \cdot \sqrt{2} \cdot \sqrt{-1} \cdot \sqrt{5} = i\sqrt{2} \cdot i\sqrt{5}$$
$$= i^2\sqrt{10} = -\sqrt{10} \quad \text{is correct!}$$

But $\sqrt{-2} \cdot \sqrt{-5} = \sqrt{(-2)(-5)} = \sqrt{10}$ is wrong!

Keeping this and the fact that $i^2 = -1$ in mind, we multiply in much the same way that we do with real numbers.

CAUTION!

The rule $\sqrt{a}\sqrt{b} = \sqrt{ab}$ holds only for nonnegative real numbers.

EXAMPLES Multiply.

8. $\sqrt{-49} \cdot \sqrt{-16} = \sqrt{-1} \cdot \sqrt{49} \cdot \sqrt{-1} \cdot \sqrt{16}$
$= i \cdot 7 \cdot i \cdot 4$
$= i^2(28)$
$= (-1)(28)$ $\quad$ $i^2 = -1$
$= -28$

9. $\sqrt{-3} \cdot \sqrt{-7} = \sqrt{-1} \cdot \sqrt{3} \cdot \sqrt{-1} \cdot \sqrt{7}$
$= i \cdot \sqrt{3} \cdot i \cdot \sqrt{7}$
$= i^2(\sqrt{21})$
$= (-1)\sqrt{21}$ $\quad$ $i^2 = -1$
$= -\sqrt{21}$

10. $-2i \cdot 5i = -10 \cdot i^2$
$= (-10)(-1)$ $\quad$ $i^2 = -1$
$= 10$

11. $(-4i)(3 - 5i) = (-4i) \cdot 3 - (-4i)(5i)$ $\quad$ Using a distributive law
$= -12i + 20i^2$
$= -12i + 20(-1)$ $\quad$ $i^2 = -1$
$= -12i - 20$
$= -20 - 12i$

12. $(1 + 2i)(1 + 3i) = 1 + 3i + 2i + 6i^2$ $\quad$ Multiplying each term of one number by every term of the other (FOIL)
$= 1 + 3i + 2i + 6(-1)$ $\quad$ $i^2 = -1$
$= 1 + 3i + 2i - 6$
$= -5 + 5i$ $\quad$ Collecting like terms

13. $(3 - 2i)^2 = 3^2 - 2(3)(2i) + (2i)^2$ $\quad$ Squaring the binomial
$= 9 - 12i + 4i^2$
$= 9 - 12i + 4(-1)$ $\quad$ $i^2 = -1$
$= 9 - 12i - 4$
$= 5 - 12i$

Do Exercises 10–17.

Multiply.

10. $\sqrt{-25} \cdot \sqrt{-4}$

11. $\sqrt{-2} \cdot \sqrt{-17}$

12. $-6i \cdot 7i$

13. $-3i(4 - 3i)$

14. $5i(-5 + 7i)$

15. $(1 + 3i)(1 + 5i)$

16. $(3 - 2i)(1 + 4i)$

17. $(3 + 2i)^2$

Answers on page A-58

Simplify.

18. i^{47}

19. i^{68}

20. i^{85}

21. i^{90}

Simplify.

22. $8 - i^5$

23. $7 + 4i^2$

24. $6i^{11} + 7i^{14}$

25. $i^{34} - i^{55}$

Answers on page A-58

d Powers of i

We now want to simplify certain expressions involving powers of i. To do so, we first see how to simplify powers of i. Simplifying powers of i can be done by using the fact that $i^2 = -1$ and expressing the given power of i in terms of even powers, and then in terms of powers of i^2. Consider the following:

$$i,$$
$$i^2 = -1,$$
$$i^3 = i^2 \cdot i = (-1)i = -i,$$
$$i^4 = (i^2)^2 = (-1)^2 = 1,$$
$$i^5 = i^4 \cdot i = (i^2)^2 \cdot i = (-1)^2 \cdot i = i,$$
$$i^6 = (i^2)^3 = (-1)^3 = -1.$$

Note that the powers of i cycle themselves through the values i, -1, $-i$, and 1.

EXAMPLES Simplify.

14. $i^{37} = i^{36} \cdot i = (i^2)^{18} \cdot i = (-1)^{18} \cdot i = 1 \cdot i = i$
15. $i^{58} = (i^2)^{29} = (-1)^{29} = -1$
16. $i^{75} = i^{74} \cdot i = (i^2)^{37} \cdot i = (-1)^{37} \cdot i = -1 \cdot i = -i$
17. $i^{80} = (i^2)^{40} = (-1)^{40} = 1$

Do Exercises 18–21.

Now let's simplify other expressions.

EXAMPLES Simplify to the form $a + bi$.

18. $8 - i^2 = 8 - (-1) = 8 + 1 = 9$
19. $17 + 6i^3 = 17 + 6 \cdot i^2 \cdot i = 17 + 6(-1)i = 17 - 6i$
20. $i^{22} - 67i^2 = (i^2)^{11} - 67(-1) = (-1)^{11} + 67 = -1 + 67 = 66$
21. $i^{23} + i^{48} = (i^{22}) \cdot i + (i^2)^{24} = (i^2)^{11} \cdot i + (-1)^{24} = (-1)^{11} \cdot i + (-1)^{24}$
$= -i + 1 = 1 - i$

Do Exercises 22–25.

e Conjugates and Division

Conjugates of complex numbers are defined as follows.

CONJUGATE

The **conjugate** of a complex number $a + bi$ is $a - bi$, and the **conjugate** of $a - bi$ is $a + bi$.

EXAMPLES Find the conjugate.

22. $5 + 7i$ — The conjugate is $5 - 7i$.
23. $14 - 3i$ — The conjugate is $14 + 3i$.
24. $-3 - 9i$ — The conjugate is $-3 + 9i$.
25. $4i$ — The conjugate is $-4i$.

Do Exercises 26–28.

When we multiply a complex number by its conjugate, we get a real number.

EXAMPLES Multiply.

26.
$$\begin{aligned}(5 + 7i)(5 - 7i) &= 5^2 - (7i)^2 && \text{Using } (A+B)(A-B) = A^2 - B^2\\ &= 25 - 49i^2\\ &= 25 - 49(-1) && i^2 = -1\\ &= 25 + 49\\ &= 74\end{aligned}$$

27.
$$\begin{aligned}(2 - 3i)(2 + 3i) &= 2^2 - (3i)^2\\ &= 4 - 9i^2\\ &= 4 - 9(-1) && i^2 = -1\\ &= 4 + 9\\ &= 13\end{aligned}$$

Do Exercises 29 and 30.

We use conjugates in dividing complex numbers.

EXAMPLE 28 Divide and simplify to the form $a + bi$: $\dfrac{-5 + 9i}{1 - 2i}$.

$$\begin{aligned}\frac{-5 + 9i}{1 - 2i} \cdot \frac{1 + 2i}{1 + 2i} &= \frac{(-5 + 9i)(1 + 2i)}{(1 - 2i)(1 + 2i)} && \text{Multiplying by 1 using the conjugate of the denominator in the symbol for 1}\\ &= \frac{-5 - 10i + 9i + 18i^2}{1^2 - 4i^2}\\ &= \frac{-5 - i + 18(-1)}{1 - 4(-1)} && i^2 = -1\\ &= \frac{-5 - i - 18}{1 + 4}\\ &= \frac{-23 - i}{5}\\ &= -\frac{23}{5} - \frac{1}{5}i\end{aligned}$$

Note the similarity between the preceding example and rationalizing denominators. In both cases, we used the conjugate of the denominator to write another name for 1. In Example 28, the symbol for the number 1 was chosen using the conjugate of the divisor, $1 - 2i$.

Find the conjugate.

26. $6 + 3i$

27. $-9 - 5i$

28. $\pi - \frac{1}{4}i$

Multiply.

29. $(7 - 2i)(7 + 2i)$

30. $(-3 - i)(-3 + i)$

Answers on page A-58

Divide and simplify to the form $a + bi$.

31. $\dfrac{6 + 2i}{1 - 3i}$

32. $\dfrac{2 + 3i}{-1 + 4i}$

Answers on page A-58

EXAMPLE 29 What symbol for 1 would you use to divide?

Division to be done	*Symbol for 1*
$\dfrac{3 + 5i}{4 + 3i}$	$\dfrac{4 - 3i}{4 - 3i}$

EXAMPLE 30 Divide and simplify to the form $a + bi$: $\dfrac{3 + 5i}{4 + 3i}$.

$$\frac{3 + 5i}{4 + 3i} \cdot \frac{4 - 3i}{4 - 3i} = \frac{(3 + 5i)(4 - 3i)}{(4 + 3i)(4 - 3i)} \qquad \text{Multiplying by 1}$$

$$= \frac{12 - 9i + 20i - 15i^2}{4^2 - 9i^2}$$

$$= \frac{12 + 11i - 15(-1)}{16 - 9(-1)} \qquad i^2 = -1$$

$$= \frac{27 + 11i}{25} = \frac{27}{25} + \frac{11}{25}i$$

Do Exercises 31 and 32.

CALCULATOR CORNER

Complex Numbers We can perform operations on complex numbers on a graphing calculator. To do so, we first set the calculator in complex, or $a + bi$, mode by pressing [MODE], using the [▽] and [▷] keys to position the blinking cursor over $a + bi$, and then pressing [ENTER]. We press [2nd] [QUIT] to go to the home screen. Now we can add, subtract, multiply, and divide complex numbers.

To find $(3 + 4i) - (7 - i)$, for example, we press [(] [3] [+] [4] [2nd] [i] [)] [−] [(] [7] [−] [2nd] [i] [)] [ENTER]. (i is the second operation associated with the [.] key.) Note that although the parentheses around $3 + 4i$ are optional, those around $7 - i$ are necessary to ensure that both parts of the second complex number are subtracted from the first number.

To find $\dfrac{5 - 2i}{-1 + 3i}$ and display the result using fraction notation, we press [(] [5] [−] [2] [2nd] [i] [)] [÷] [(] [(−)] [1] [+] [3] [2nd] [i] [)] [MATH] [1] [ENTER]. Since the fraction bar acts as a grouping symbol in the original expression, the parentheses must be used to group the numerator and the denominator when the expression is entered in the calculator. To find $\sqrt{-4} \cdot \sqrt{-9}$, we press [2nd] [√] [(−)] [4] [)] [×] [2nd] [√] [(−)] [9] [)] [ENTER]. Note that the calculator supplies the left parenthesis in each radicand and we supply the right parenthesis. The results of these operations are shown below.

```
(3+4i)-(7-i)
                 -4+5i
(5-2i)/(-1+3i)►Frac
        -11/10-13/10i
√(-4)*√(-9)
                    -6
```

Exercises: Carry out each operation.

1. $(9 + 4i) + (-11 - 13i)$
2. $(9 + 4i) - (-11 - 13i)$
3. $(9 + 4i) \cdot (-11 - 13i)$
4. $(9 + 4i) \div (-11 - 13i)$
5. $\sqrt{-16} \cdot \sqrt{-25}$
6. $\sqrt{-23} \cdot \sqrt{-35}$
7. $\dfrac{4 - 5i}{-6 + 8i}$
8. $(-3i)^4$
9. $(1 - i)^3 - (2 + 3i)^4$
10. $\dfrac{(1 - i)^3}{(2 + 3i)^2}$

f Solutions of Equations

The equation $x^2 + 1 = 0$ has no real-number solution, but it has *two* nonreal complex solutions.

EXAMPLE 31 Determine whether i is a solution of the equation $x^2 + 1 = 0$.

We substitute i for x in the equation.

$$\begin{array}{r|l} x^2 + 1 & 0 \\ \hline i^2 + 1 & ?\ 0 \\ -1 + 1 & \\ 0 & \end{array} \quad \text{TRUE}$$

The number i is a solution.

Do Exercise 33.

Any equation consisting of a polynomial in one variable on one side and 0 on the other has complex-number solutions (some may be real). It is not always easy to find the solutions, but they always exist.

EXAMPLE 32 Determine whether $1 + i$ is a solution of the equation $x^2 - 2x + 2 = 0$.

We substitute $1 + i$ for x in the equation.

$$\begin{array}{r|l} x^2 - 2x + 2 & 0 \\ \hline (1 + i)^2 - 2(1 + i) + 2 & ?\ 0 \\ 1 + 2i + i^2 - 2 - 2i + 2 & \\ 1 + 2i - 1 - 2 - 2i + 2 & \\ (1 - 1 - 2 + 2) + (2 - 2)i & \\ 0 + 0i & \\ 0 & \end{array} \quad \text{TRUE}$$

The number $1 + i$ is a solution.

EXAMPLE 33 Determine whether $2i$ is a solution of $x^2 + 3x - 4 = 0$.

$$\begin{array}{r|l} x^2 + 3x - 4 & 0 \\ \hline (2i)^2 + 3(2i) - 4 & ?\ 0 \\ 4i^2 + 6i - 4 & \\ -4 + 6i - 4 & \\ -8 + 6i & \end{array} \quad \text{FALSE}$$

The number $2i$ is not a solution.

Do Exercise 34.

33. Determine whether $-i$ is a solution of $x^2 + 1 = 0$.

$$\begin{array}{c|c} x^2 + 1 & 0 \\ \hline & ? \end{array}$$

34. Determine whether $1 - i$ is a solution of $x^2 - 2x + 2 = 0$.

$$\begin{array}{c|c} x^2 - 2x + 2 & 0 \\ \hline & ? \end{array}$$

Answers on page A-58

15.8 EXERCISE SET

For Extra Help

Digital Video Tutor CD 13 Videotape 17 | InterAct Math | Math Tutor Center | MathXL | MyMathLab

a Express in terms of i.

1. $\sqrt{-35}$
2. $\sqrt{-21}$
3. $\sqrt{-16}$
4. $\sqrt{-36}$
5. $-\sqrt{-12}$
6. $-\sqrt{-20}$
7. $\sqrt{-3}$
8. $\sqrt{-4}$
9. $\sqrt{-81}$
10. $\sqrt{-27}$
11. $\sqrt{-98}$
12. $-\sqrt{-18}$
13. $-\sqrt{-49}$
14. $-\sqrt{-125}$
15. $4 - \sqrt{-60}$
16. $6 - \sqrt{-84}$
17. $\sqrt{-4} + \sqrt{-12}$
18. $-\sqrt{-76} + \sqrt{-125}$

b Add or subtract and simplify.

19. $(7 + 2i) + (5 - 6i)$
20. $(-4 + 5i) + (7 + 3i)$
21. $(4 - 3i) + (5 - 2i)$
22. $(-2 - 5i) + (1 - 3i)$
23. $(9 - i) + (-2 + 5i)$
24. $(6 + 4i) + (2 - 3i)$
25. $(6 - i) - (10 + 3i)$
26. $(-4 + 3i) - (7 + 4i)$
27. $(4 - 2i) - (5 - 3i)$
28. $(-2 - 3i) - (1 - 5i)$
29. $(9 + 5i) - (-2 - i)$
30. $(6 - 3i) - (2 + 4i)$

c Multiply.

31. $\sqrt{-36} \cdot \sqrt{-9}$

32. $\sqrt{-16} \cdot \sqrt{-64}$

33. $\sqrt{-7} \cdot \sqrt{-2}$

34. $\sqrt{-11} \cdot \sqrt{-3}$

35. $-3i \cdot 7i$

36. $8i \cdot 5i$

37. $-3i(-8 - 2i)$

38. $4i(5 - 7i)$

39. $(3 + 2i)(1 + i)$

40. $(4 + 3i)(2 + 5i)$

41. $(2 + 3i)(6 - 2i)$

42. $(5 + 6i)(2 - i)$

43. $(6 - 5i)(3 + 4i)$

44. $(5 - 6i)(2 + 5i)$

45. $(7 - 2i)(2 - 6i)$

46. $(-4 + 5i)(3 - 4i)$

47. $(3 - 2i)^2$

48. $(5 - 2i)^2$

49. $(1 + 5i)^2$

50. $(6 + 2i)^2$

51. $(-2 + 3i)^2$

52. $(-5 - 2i)^2$

d Simplify.

53. i^7

54. i^{11}

55. i^{24}

56. i^{35}

57. i^{42}

58. i^{64}

59. i^9

60. $(-i)^{71}$

61. i^6

62. $(-i)^4$

63. $(5i)^3$

64. $(-3i)^5$

Simplify to the form $a + bi$.

65. $7 + i^4$

66. $-18 + i^3$

67. $i^{28} - 23i$

68. $i^{29} + 33i$

69. $i^2 + i^4$

70. $5i^5 + 4i^3$

71. $i^5 + i^7$

72. $i^{84} - i^{100}$

73. $1 + i + i^2 + i^3 + i^4$

74. $i - i^2 + i^3 - i^4 + i^5$

75. $5 - \sqrt{-64}$

76. $\sqrt{-12} + 36i$

77. $\frac{8 - \sqrt{-24}}{4}$

78. $\frac{9 + \sqrt{-9}}{3}$

e Divide and simplify to the form $a + bi$.

79. $\frac{4 + 3i}{3 - i}$

80. $\frac{5 + 2i}{2 + i}$

81. $\frac{3 - 2i}{2 + 3i}$

82. $\frac{6 - 2i}{7 + 3i}$

83. $\frac{8 - 3i}{7i}$

84. $\frac{3 + 8i}{5i}$

85. $\frac{4}{3 + i}$

86. $\frac{6}{2 - i}$

87. $\frac{2i}{5 - 4i}$

88. $\frac{8i}{6 + 3i}$

89. $\frac{4}{3i}$

90. $\frac{5}{6i}$

91. $\frac{2 - 4i}{8i}$

92. $\frac{5 + 3i}{i}$

93. $\frac{6 + 3i}{6 - 3i}$

94. $\frac{4 - 5i}{4 + 5i}$

f Determine whether the complex number is a solution of the equation.

95. $1 - 2i$;

$$x^2 - 2x + 5 = 0$$

?

96. $1 + 2i$;

$$x^2 - 2x + 5 = 0$$

?

97. $2 + i$;

$$x^2 - 4x - 5 = 0$$

?

98. $1 - i$;

$$x^2 + 2x + 2 = 0$$

?

99. **D_W** How are conjugates of complex numbers similar to the conjugates used in Section 11.5?

100. **D_W** Is every real number a complex number? Why or why not?

SKILL MAINTENANCE

Solve. [11.7a]

101. $\dfrac{196}{x^2 - 7x + 49} - \dfrac{2x}{x + 7} = \dfrac{2058}{x^3 + 343}$

102. $\dfrac{5}{t} - \dfrac{3}{2} = \dfrac{4}{7}$

Solve. [14.3c, d, e]

103. $|3x + 7| = 22$

104. $|3x + 7| < 22$

105. $|3x + 7| \geq 22$

106. $|3x + 7| = |2x - 5|$

SYNTHESIS

107. A complex function g is given by

$$g(z) = \frac{z^4 - z^2}{z - 1}.$$

Find $g(2i)$, $g(1 + i)$, and $g(-1 + 2i)$.

108. Evaluate $\dfrac{1}{w - w^2}$ when $w = \dfrac{1 - i}{10}$.

Express in terms of i.

109. $\dfrac{1}{8}\left(-24 - \sqrt{-1024}\right)$

110. $12\sqrt{-\dfrac{1}{32}}$

111. $7\sqrt{-64} - 9\sqrt{-256}$

Simplify.

112. $\dfrac{i^5 + i^6 + i^7 + i^8}{(1 - i)^4}$

113. $(1 - i)^3(1 + i)^3$

114. $\dfrac{5 - \sqrt{5}\,i}{\sqrt{5}\,i}$

115. $\dfrac{6}{1 + \dfrac{3}{i}}$

116. $\left(\dfrac{1}{2} - \dfrac{1}{3}i\right)^2 - \left(\dfrac{1}{2} + \dfrac{1}{3}i\right)^2$

117. $\dfrac{i - i^{38}}{1 + i}$

118. Find all numbers a for which the opposite of a is the same as the reciprocal of a.

CHAPTER

15 Summary and Review

The review that follows is meant to prepare you for a chapter exam. It consists of two parts. The first part is a checklist of some of the Study Tips referred to in this and preceding chapters, as well as a list of important properties and formulas. The second part is the Review Exercises. These provide practice exercises for the exam, together with references to section objectives so you can go back and review. Before beginning, stop and look back over the skills you have obtained. What skills in mathematics do you have now that you did not have before studying this chapter?

STUDY TIPS CHECKLIST

The foundation of all your study skills is TIME!

- ☐ Are you doing some even-numbered exercises, whether they have been assigned or not, to better prepare you to take tests?
- ☐ Have you begun preparing for the final examination?
- ☐ Are you keeping one section ahead in your syllabus?
- ☐ Are you approaching your study of mathematics with a positive attitude?
- ☐ Have you visited the learning resource centers on your campus, such as a math lab, tutor center, and your instructor's office?

IMPORTANT PROPERTIES AND FORMULAS

$\sqrt{a^2} = |a|$; $\quad \sqrt[k]{a^k} = |a|$, when k is even; $\quad \sqrt[k]{a^k} = a$, when k is odd;

$\sqrt[k]{ab} = \sqrt[k]{a} \cdot \sqrt[k]{b}$; $\quad \sqrt[k]{\dfrac{a}{b}} = \dfrac{\sqrt[k]{a}}{\sqrt[k]{b}}$;

$a^{1/n} = \sqrt[n]{a}$; $\quad a^{m/n} = \sqrt[n]{a^m} = \left(\sqrt[n]{a}\right)^m$; $\quad a^{-m/n} = \dfrac{1}{a^{m/n}}$

Principle of Powers: If $a = b$ is true, then $a^n = b^n$ is true.

Pythagorean Theorem: $a^2 + b^2 = c^2$, in a right triangle.

$i = \sqrt{-1}$, $\quad i^2 = -1$, $\quad i^3 = -1$, $\quad i^4 = 1$

Imaginary Numbers: bi, $i^2 = -1$, $b \neq 0$

Complex Numbers: $a + bi$, $i^2 = -1$

Conjugates: $a + bi$, $a - bi$

REVIEW EXERCISES

Use a calculator to approximate to three decimal places. [15.1a]

1. $\sqrt{778}$

2. $\sqrt{\dfrac{963.2}{23.68}}$

3. For the given function, find the indicated function values. [15.1a]

$f(x) = \sqrt{3x - 16}$; $f(0), f(-1), f(1)$, and $f\left(\frac{41}{3}\right)$

4. Find the domain of the function f in Exercise 3.

Simplify. Assume that letters represent *any* real number. [15.1b]

5. $\sqrt{81a^2}$

6. $\sqrt{(-7z)^2}$

7. $\sqrt{(c-3)^2}$

8. $\sqrt{x^2 - 6x + 9}$

Simplify. [15.1c]

9. $\sqrt[3]{-1000}$

10. $\sqrt[3]{-\dfrac{1}{27}}$

11. For the given function, find the indicated function values. [15.1c]

$f(x) = \sqrt[3]{x+2}$; $f(6), f(-10)$, and $f(25)$

Simplify. Assume that letters represent *any* real number. [15.1d]

12. $\sqrt[10]{x^{10}}$

13. $-\sqrt[13]{(-3)^{13}}$

Rewrite without rational exponents, and simplify, if possible. [15.2a]

14. $a^{1/5}$

15. $64^{3/2}$

Rewrite with rational exponents. [15.2a]

16. $\sqrt{31}$

17. $\sqrt[5]{a^2b^3}$

Rewrite with positive exponents, and simplify, if possible. [15.2b]

18. $49^{-1/2}$

19. $(8xy)^{-2/3}$

20. $5a^{-3/4}b^{1/2}c^{-2/3}$

21. $\dfrac{3a}{\sqrt[4]{t}}$

Use the laws of exponents to simplify. Write answers with positive exponents. [15.2c]

22. $(x^{-2/3})^{3/5}$

23. $\dfrac{7^{-1/3}}{7^{-1/2}}$

Use rational exponents to simplify. Write the answer in radical notation if appropriate. [15.2d]

24. $\sqrt[3]{x^{21}}$

25. $\sqrt[3]{27x^6}$

Use rational exponents to write a single radical expression. [15.2d]

26. $x^{1/3}y^{1/4}$

27. $\sqrt[4]{x}\sqrt[3]{x}$

Simplify by factoring. Assume that all expressions under radicals represent nonnegative numbers. [15.3a]

28. $\sqrt{245}$

29. $\sqrt[3]{-108}$

30. $\sqrt[3]{250a^2b^6}$

Simplify. Assume that all expressions under radicals represent positive numbers. [15.3b]

31. $\sqrt{\dfrac{49}{36}}$

32. $\sqrt[3]{\dfrac{64x^6}{27}}$

33. $\sqrt[4]{\dfrac{16x^8}{81y^{12}}}$

Perform the indicated operations and simplify. Assume that all expressions under radicals represent positive numbers. [15.3a, b]

34. $\sqrt{5x}\,\sqrt{3y}$

35. $\sqrt[3]{a^5b}\,\sqrt[3]{27b}$

36. $\sqrt[3]{a}\,\sqrt[5]{b^3}$

37. $\dfrac{\sqrt[3]{60xy^3}}{\sqrt[3]{10x}}$

38. $\dfrac{\sqrt{75x}}{2\sqrt{3}}$

39. $\dfrac{\sqrt[3]{x^2}}{\sqrt[4]{x}}$

Add or subtract. Assume that all expressions under radicals represent nonnegative numbers. [15.4a]

40. $5\sqrt[3]{x} + 2\sqrt[3]{x}$

41. $2\sqrt{75} - 7\sqrt{3}$

42. $\sqrt[3]{8x^4} + \sqrt[3]{xy^6}$

43. $\sqrt{50} + 2\sqrt{18} + \sqrt{32}$

Multiply. [15.4b]

44. $(\sqrt{5} - 3\sqrt{8})(\sqrt{5} + 2\sqrt{8})$

45. $(1 - \sqrt{7})^2$

46. $(\sqrt[3]{27} - \sqrt[3]{2})(\sqrt[3]{27} + \sqrt[3]{2})$

Rationalize the denominator. [15.5a, b]

47. $\sqrt{\dfrac{8}{3}}$

48. $\dfrac{2}{\sqrt{a} + \sqrt{b}}$

Solve. [15.6a, b]

49. $\sqrt[4]{x + 3} = 2$

50. $1 + \sqrt{x} = \sqrt{3x - 3}$

51. $x - 3 = \sqrt{5 - x}$

52. *Length of a Side of a Square.* The diagonal of a square has length $9\sqrt{2}$ cm. Find the length of a side of the square. [15.7a]

53. *Bookcase Width.* A bookcase is 5 ft tall and has a 7-ft diagonal brace, as shown. How wide is the bookcase? [15.7a]

Automotive Repair. For an engine with a displacement of 2.8 L, the function given by

$$d(n) = 0.75\sqrt{2.8n}$$

can be used to determine the diameter size of the carburetor's opening, $d(n)$, in millimeters, where n is the number of rpm's at which the engine achieves peak performance. [15.6c]

Source: macdizzy.com

54. If a carburetor's opening is 81 mm, for what number of rpm's will the engine produce peak power?

55. If a carburetor's opening is 84 mm, for what number of rpm's will the engine produce peak power?

In a right triangle, find the length of the side not given. Give an exact answer and an answer to three decimal places. [15.7a]

56. $a = 7, \quad b = 24$

57. $a = 2, \quad c = 5\sqrt{2}$

58. Express in terms of i: $\sqrt{-25} + \sqrt{-8}$. [15.8a]

Add or subtract. [15.8b]

59. $(-4 + 3i) + (2 - 12i)$

60. $(4 - 7i) - (3 - 8i)$

Multiply. [15.8c, d]

61. $(2 + 5i)(2 - 5i)$

62. i^{13}

63. $(6 - 3i)(2 - i)$

Divide. [15.8e]

64. $\dfrac{-3 + 2i}{5i}$

65. $\dfrac{6 - 3i}{2 - i}$

66. Determine whether $1 + i$ is a solution of $x^2 + x + 2 = 0$. [15.8f]

$x^2 + x + 2 = 0$	
	?

67. Graph: $f(x) = \sqrt{x}$. [15.1a]

68. DW We learned a new method of equation solving in this chapter. Explain how this procedure differs from others we have used. [15.6a, b]

SKILL MAINTENANCE

Certain objectives from four particular sections will be retested on the chapter test. The objectives are listed with the practice problems that follow.

Solve.

69. $\dfrac{7}{x + 2} + \dfrac{5}{x^2 - 2x + 4} = \dfrac{84}{x^3 + 8}$ [11.7a]

70. $3x^2 - 5x - 12 = 0$ [10.8b]

71. Multiply and simplify: [11.1d]

$$\frac{x^2 + 3x}{x^2 - y^2} \cdot \frac{x^2 - xy + 2x - 2y}{x^2 - 9}.$$

72. *Vehicle Speeds.* A motorcycle and a sport utility vehicle (SUV) are driven out of a car dealership at the same time. The SUV travels 8 mph faster than the motorcycle. The SUV travels 105 mi in the same time that the motorcycle travels 93 mi. Find the speed of each vehicle. [11.8a]

SYNTHESIS

73. Simplify: $i \cdot i^2 \cdot i^3 \cdots i^{99} \cdot i^{100}$. [15.8c, d]

74. Solve: $\sqrt{11x + \sqrt{6 + x}} = 6$. [15.6a]

CHAPTER 15 Chapter Test

1. Use a calculator to approximate $\sqrt{148}$ to three decimal places.

2. For the given function, find the indicated function values.
$f(x) = \sqrt{8 - 4x}$; $f(1)$ and $f(3)$

3. Find the domain of the function f in Question 2.

Simplify. Assume that letters represent *any* real number.

4. $\sqrt{(-3q)^2}$

5. $\sqrt{x^2 + 10x + 25}$

6. $\sqrt[3]{-\frac{1}{1000}}$

7. $\sqrt[5]{x^5}$

8. $\sqrt[10]{(-4)^{10}}$

Rewrite without rational exponents, and simplify, if possible.

9. $a^{2/3}$

10. $32^{3/5}$

Rewrite with rational exponents.

11. $\sqrt{37}$

12. $(\sqrt{5xy^2})^5$

Rewrite with positive exponents, and simplify, if possible.

13. $1000^{-1/3}$

14. $8a^{3/4}b^{-3/2}c^{-2/5}$

Use the laws of exponents to simplify. Write answers with positive exponents.

15. $(x^{2/3}y^{-3/4})^{12/5}$

16. $\frac{2.9^{-5/8}}{2.9^{2/3}}$

Use rational exponents to simplify. Write the answer in radical notation if appropriate. Assume that all expressions under radicals represent nonnegative numbers.

17. $\sqrt[8]{x^2}$

18. $\sqrt[4]{16x^6}$

Use rational exponents to write a single radical expression.

19. $a^{2/5}b^{1/3}$

20. $\sqrt[4]{2y}\sqrt[3]{y}$

Simplify by factoring. Assume that all expressions under radicals represent nonnegative numbers.

21. $\sqrt{148}$

22. $\sqrt[4]{80}$

23. $(\sqrt[3]{16a^2b})^2$

Simplify. Assume that all expressions under radicals represent positive numbers.

24. $\sqrt[3]{-\frac{8}{x^6}}$

25. $\sqrt{\frac{25x^2}{36y^4}}$

Perform the indicated operations and simplify. Assume that all expressions under radicals represent positive numbers.

26. $\sqrt[3]{2x}\,\sqrt[3]{5y^2}$

27. $\sqrt[4]{x^3y^2}\,\sqrt{xy}$

28. $\dfrac{\sqrt[5]{x^3y^4}}{\sqrt[5]{xy^2}}$

29. $\dfrac{\sqrt{300a}}{5\sqrt{3}}$

30. Add: $3\sqrt{128} + 2\sqrt{18} + 2\sqrt{32}$.

Multiply.

31. $(\sqrt{20} + 2\sqrt{5})(\sqrt{20} - 3\sqrt{5})$

32. $(3 + \sqrt{x})^2$

33. Rationalize the denominator: $\dfrac{1 + \sqrt{2}}{3 - 5\sqrt{2}}$.

Solve.

34. $\sqrt[5]{x - 3} = 2$

35. $\sqrt{x - 6} = \sqrt{x + 9} - 3$

36. $\sqrt{x - 1} + 3 = x$

37. *Length of a Side of a Square.* The diagonal of a square has length $7\sqrt{2}$ ft. Find the length of a side of the square.

38. *Sighting to the Horizon.* A person can see 247.49 km to the horizon from an airplane window. How high is the airplane? Use the formula $V = 3.5\sqrt{h}$.

In a right triangle, find the length of the side not given. Give an exact answer and an answer to three decimal places.

39. $a = 7, \quad b = 7$

40. $a = 1, \quad c = \sqrt{5}$

41. Express in terms of i: $\sqrt{-9} + \sqrt{-64}$.

42. Subtract: $(5 + 8i) - (-2 + 3i)$.

Multiply.

43. $(3 - 4i)(3 + 7i)$

44. i^{95}

45. Divide: $\dfrac{-7 + 14i}{6 - 8i}$.

46. Determine whether $1 + 2i$ is a solution of $x^2 + 2x + 5 = 0$.

SKILL MAINTENANCE

Solve.

47. $\dfrac{11x}{x + 3} + \dfrac{33}{x} + 12 = \dfrac{99}{x^2 + 3x}$

48. $6x^2 = 13x + 5$

49. Divide and simplify:
$$\frac{x^3 - 27}{x^2 - 16} \div \frac{x^2 + 3x + 9}{x + 4}.$$

50. *Mowing Time.* Fran and Juan do all the mowing for the local community college. It takes Juan 9 hr more than Fran to do the mowing. Working together, they can complete the job in 20 hr. How long would it take each, working alone, to do the mowing?

SYNTHESIS

51. Simplify: $\dfrac{1 - 4i}{4i(1 + 4i)^{-1}}$.

52. Solve: $\sqrt{2x - 2} + \sqrt{7x + 4} = \sqrt{13x + 10}$.

Quadratic Equations and Functions

Gateway to Chapter 16

We began our study of factoring quadratic equations in Chapter 10 in connection with polynomials. In this chapter, we extend our equation-solving skills to those quadratic equations that do not lend themselves to being solved easily by factoring. We consider both real-number and complex-number solutions found using the quadratic formula.

We also learn to graph quadratic functions in various forms and to solve related applied problems.

Real-World Application

Typically rivers are deepest in the middle, with the depth decreasing to 0 at the edges. A hydrologist measures the depths D, in feet, of a river at distances x, in feet, from one bank. Some of the data points found by the hydrologist are $(0, 0)$, $(50, 20)$, and $(100, 0)$. Use the data points to find a quadratic function that fits the data and graph the function.

This problem appears as Example 7 in Section 16.7.

CHAPTER 16

CHAPTER 16

Pretest

Solve.

1. $5x^2 + 15x = 0$ [16.1b], [16.2a]

2. $y^2 + 4y + 8 = 0$ [16.1b], [16.2a]

3. $x^2 - 10x + 25 = 0$ [16.1a, b]

4. $3x^4 - 7x^2 + 2 = 0$ [16.4c]

5. $\frac{2x}{2x+1} + \frac{x+1}{2x-1} = \frac{6}{4x^2-1}$ [16.2a]

6. $x + 3\sqrt{x} - 4 = 0$ [16.4c]

7. Solve. Give the exact solution and approximate the solutions to three decimal places. [16.2a]

$$2x^2 + 4x - 1 = 0$$

8. Solve for T: [16.3b]

$$W = \sqrt{\frac{1}{RT}}.$$

9. Write a quadratic equation having solutions $\frac{2}{3}$ and -2. [16.4b]

10. For $f(x) = 2x^2 - 12x + 16$: [16.6a]

a) Find the vertex.
b) Find the line of symmetry.
c) Graph the function.

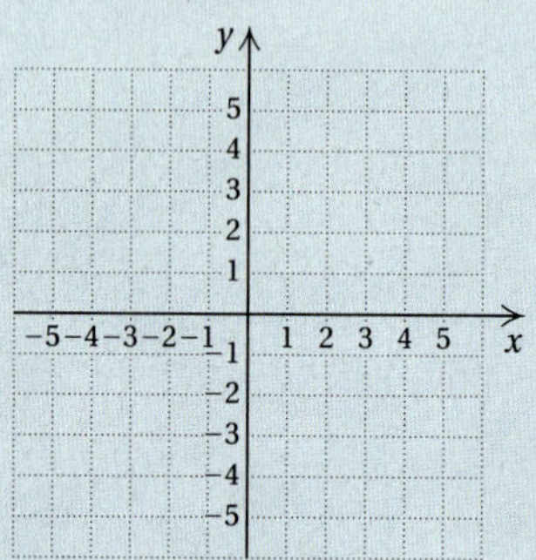

d) Find the minimum value of the function.

11. Find the x-intercepts: $f(x) = x^2 - 6x + 4$. [16.6b]

12. Find the quadratic function that fits the data points $(0, 1)$, $(1, 0)$, and $(2, 7)$. [16.7b]

Solve.

13. Box Construction. An open box is to be made from a 10-ft by 20-ft rectangular piece of cardboard by cutting a square from each corner. The area of the bottom of the box is to be 96 ft^2. What is the length of the sides of the squares that are cut from the corners? [16.3a]

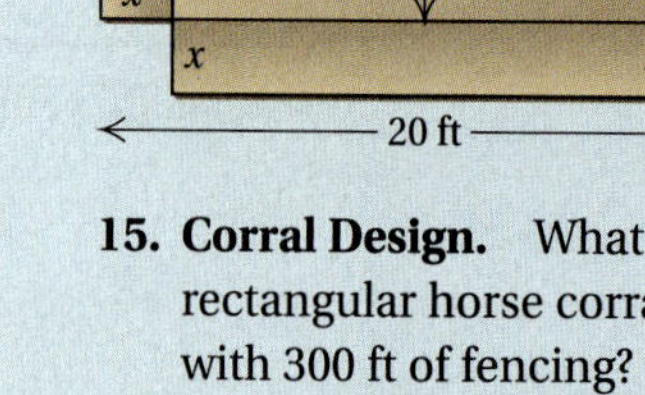

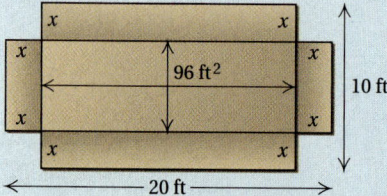

14. Car Travel. During the first part of a trip, a car travels 100 mi at a certain speed. It travels 120 mi on the second part of the trip at a speed that is 8 mph faster. The total time for the trip is 5 hr. Find the speed of the car during each part of the trip. [16.3a]

15. Corral Design. What is the area of the largest rectangular horse corral that a rancher can enclose with 300 ft of fencing? [16.7a]

Solve.

16. $x(x - 3)(x + 5) < 0$ [16.8a]

17. $\frac{x-2}{x+3} \geq 0$ [16.8b]

16.1 THE BASICS OF SOLVING QUADRATIC EQUATIONS

Objectives

a Solve quadratic equations using the principle of square roots and find the x-intercepts of the graph of a related function.

b Solve quadratic equations by completing the square.

c Solve applied problems using quadratic equations.

ALGEBRAIC–GRAPHICAL CONNECTION

Let's reexamine the graphical connections to the algebraic equation-solving concepts we have studied before.

In Chapter 12, we introduced the graph of a quadratic function:

$$f(x) = ax^2 + bx + c, \quad a \neq 0.$$

For example, the graph of the function $f(x) = x^2 + 6x + 8$ and its x-intercepts are shown below.

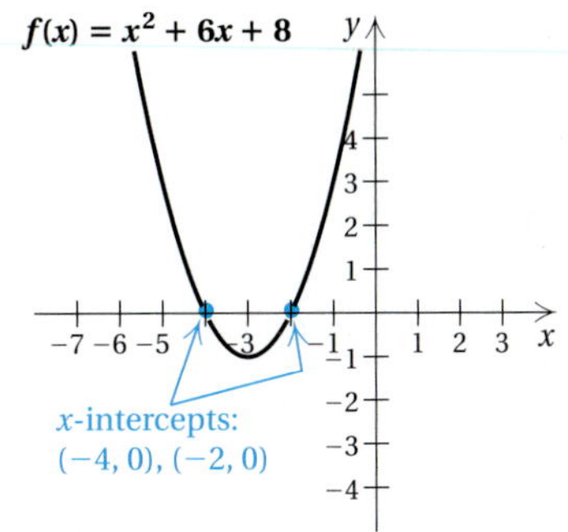

The x-intercepts are $(-4, 0)$ and $(-2, 0)$. These pairs are also the points of intersection of the graphs of $f(x) = x^2 + 6x + 8$ and $g(x) = 0$ (the x-axis). We will analyze the graphs of quadratic functions in greater detail in Sections 12.5–12.7.

In Chapter 10, we solved quadratic equations like $x^2 + 6x + 8 = 0$ using factoring, as here:

$$x^2 + 6x + 8 = 0$$

$$(x + 4)(x + 2) = 0 \qquad \text{Factoring}$$

$$x + 4 = 0 \quad or \quad x + 2 = 0 \qquad \text{Using the principle of zero products}$$

$$x = -4 \quad or \quad x = -2.$$

We see that the solutions of $0 = x^2 + 6x + 8$, -4 and -2, are the first coordinates of the x-intercepts, $(-4, 0)$ and $(-2, 0)$, of the graph of $f(x) = x^2 + 6x + 8$.

Do Exercise 1.

We now extend our ability to solve quadratic equations.

a The Principle of Square Roots

The quadratic equation

$$5x^2 + 8x - 2 = 0$$

is said to be in **standard form.** The quadratic equation

$$5x^2 = 2 - 8x$$

is equivalent to the preceding, but it is *not* in standard form.

1. Consider solving the equation

$$x^2 - 6x + 8 = 0.$$

Below is the graph of

$$f(x) = x^2 - 6x + 8.$$

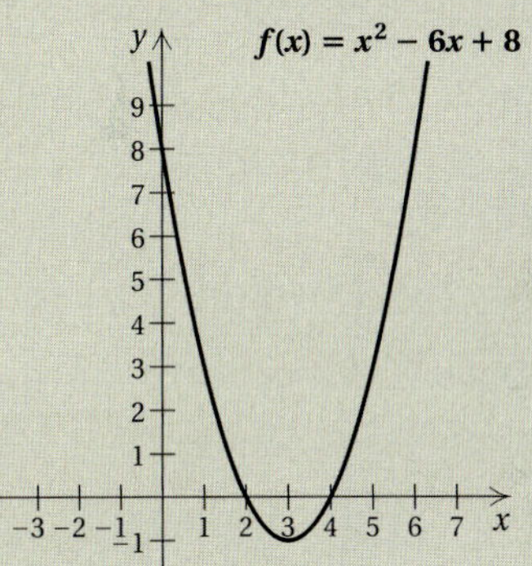

a) What are the x-intercepts of the graph?

b) What are the solutions of $x^2 - 6x + 8 = 0$?

c) What relationship exists between the answers to parts (a) and (b)?

Answers on page A-59

2. a) Solve: $x^2 = 16$.

b) Find the x-intercepts of $f(x) = x^2 - 16$.

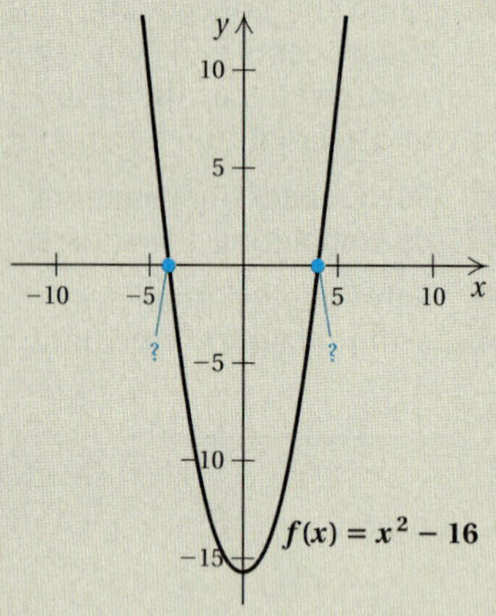

3. a) Solve: $4x^2 + 14x = 0$.

b) Find the x-intercepts of $f(x) = 4x^2 + 14x$.

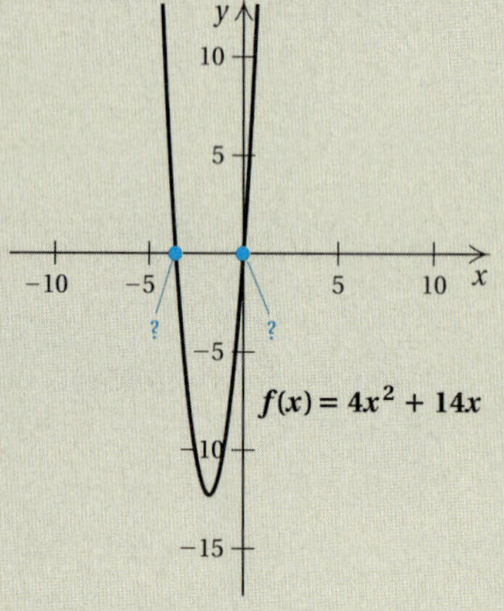

Answers on page A-59

CALCULATOR CORNER

Solving Quadratic Equations Use the ZERO feature to solve the equations in Example 2 and Margin Exercise 3. See the Calculator Corner on p. 809 to review the procedure.

QUADRATIC EQUATION

An equation of the type $ax^2 + bx + c = 0$, where a, b, and c are real-number constants and $a \neq 0$, is called the **standard form of a quadratic equation.**

For the equation $-5x^2 + 4x - 7 = 0$, we can find an equivalent equation by multiplying both sides by -1:

$$-1(-5x^2 + 4x - 7) = -1(0)$$
$$5x^2 - 4x + 7 = 0.$$

In Section 10.8, we studied the use of factoring and the principle of zero products to solve certain quadratic equations. Let's review that procedure and introduce a new one.

EXAMPLE 1

a) Solve: $x^2 = 25$.

b) Find the x-intercepts of $f(x) = x^2 - 25$.

a) We first find standard form and then factor:

$x^2 - 25 = 0$ Subtracting 25

$(x - 5)(x + 5) = 0$ Factoring

$x - 5 = 0$ *or* $x + 5 = 0$ Using the principle of zero products

$x = 5$ *or* $x = -5$.

The solutions are 5 and -5.

b) The x-intercepts of $f(x) = x^2 - 25$ are $(-5, 0)$ and $(5, 0)$. The solutions of the equation $x^2 = 25$ are the first coordinates of the x-intercepts of the graph of $f(x) = x^2 - 25$.

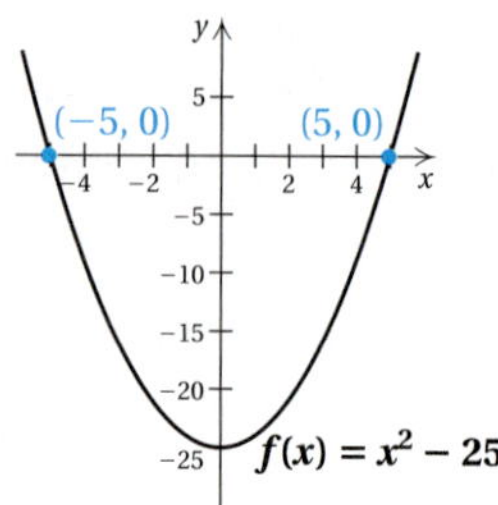

EXAMPLE 2 Solve: $6x^2 - 15x = 0$.

We factor and use the principle of zero products:

$$6x^2 - 15x = 0$$
$$3x(2x - 5) = 0$$
$$3x = 0 \quad or \quad 2x - 5 = 0$$
$$x = 0 \quad or \quad x = \tfrac{5}{2}.$$

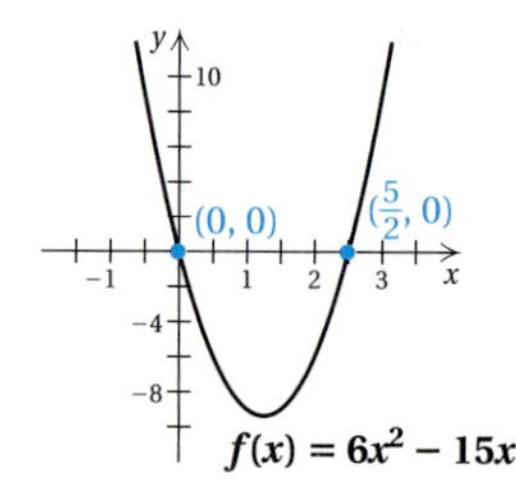

The solutions are 0 and $\frac{5}{2}$. The check is left to the student.

Do Exercises 2 and 3.

EXAMPLE 3

a) Solve: $3x^2 = 2 - x$.

b) Find the x-intercepts of $f(x) = 3x^2 + x - 2$.

a) We first find standard form. Then we factor and use the principle of zero products.

$$3x^2 = 2 - x$$

$$3x^2 + x - 2 = 0 \quad \text{Adding } x \text{ and subtracting 2 to get the standard form}$$

$$(3x - 2)(x + 1) = 0 \quad \text{Factoring}$$

$$3x - 2 = 0 \quad or \quad x + 1 = 0 \quad \text{Using the principle of zero products}$$

$$3x = 2 \quad or \quad x = -1$$

$$x = \tfrac{2}{3} \quad or \quad x = -1$$

CHECK: For $\frac{2}{3}$:

$$\begin{array}{c|c} 3x^2 & 2 - x \\ \hline 3\left(\frac{2}{3}\right)^2 & 2 - \left(\frac{2}{3}\right) \\ 3 \cdot \frac{4}{9} & \frac{6}{3} - \frac{2}{3} \\ \frac{4}{3} & \frac{4}{3} \end{array} \quad \text{TRUE}$$

For -1:

$$\begin{array}{c|c} 3x^2 & 2 - x \\ \hline 3(-1)^2 & 2 - (-1) \\ 3 \cdot 1 & 2 + 1 \\ 3 & 3 \end{array} \quad \text{TRUE}$$

The solutions are -1 and $\frac{2}{3}$.

b) The x-intercepts of $f(x) = 3x^2 + x - 2$ are $(-1, 0)$ and $\left(\frac{2}{3}, 0\right)$. The solutions of the equation $3x^2 = 2 - x$ are the first coordinates of the x-intercepts of the graph of $f(x) = 3x^2 + x - 2$.

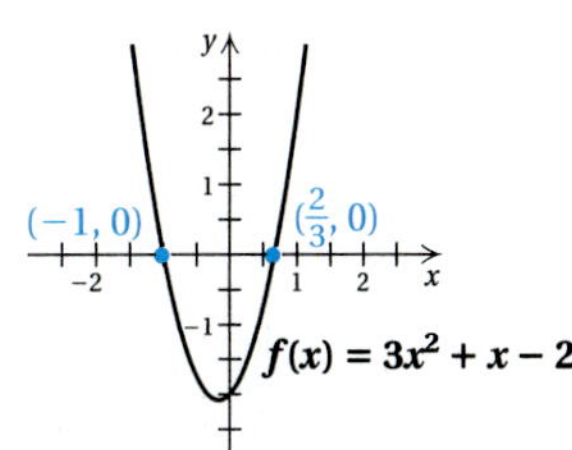

Do Exercise 4.

4. a) Solve: $5x^2 = 8x - 3$.

b) Find the x-intercepts of $f(x) = 5x^2 - 8x + 3$.

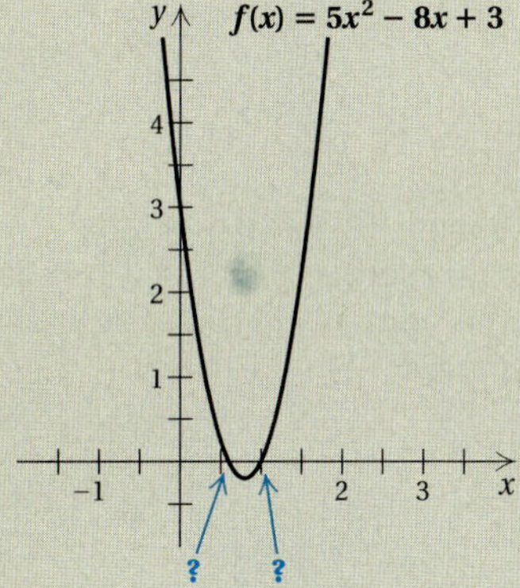

Answers on page A-59

SOLVING EQUATIONS OF THE TYPE $x^2 = d$

Consider the equation $x^2 = 25$ again. We know from Chapter 15 that the number 25 has two real-number square roots, namely, 5 and -5. Note that these are the solutions of the equation in Example 1. This exemplifies the principle of square roots, which provides a quick method for solving equations of the type $x^2 = d$.

THE PRINCIPLE OF SQUARE ROOTS

The solutions of the equation $x^2 = d$ are $\sqrt{d}$ and $-\sqrt{d}$.

When $d > 0$, the solutions are two real numbers.

When $d = 0$, the only solution is 0.

When $d < 0$, the solutions are two imaginary numbers.

5. Solve: $5x^2 = 15$. Give the exact solution and approximate the solutions to three decimal places.

EXAMPLE 4 Solve: $3x^2 = 6$. Give the exact solutions and approximate the solutions to three decimal places.

We have

$$3x^2 = 6$$
$$x^2 = 2$$
$$x = \sqrt{2} \quad or \quad x = -\sqrt{2}.$$

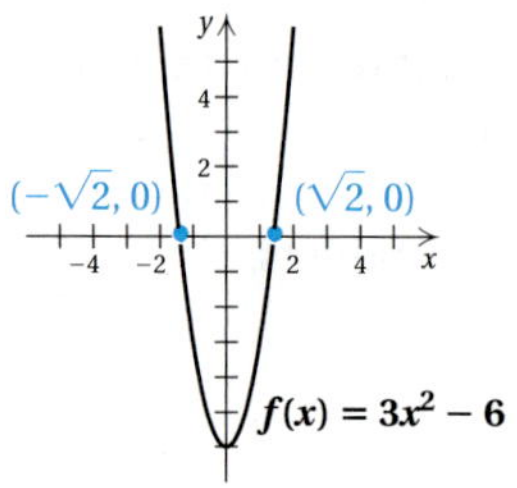

We often use the symbol $\pm\sqrt{2}$ to represent both of the solutions.

CHECK: For $\sqrt{2}$:

$$\begin{array}{r|l} 3x^2 & = 6 \\ \hline 3(\sqrt{2})^2 & ?\ 6 \\ 3 \cdot 2 & \\ 6 & \end{array} \quad \text{TRUE}$$

For $-\sqrt{2}$:

$$\begin{array}{r|l} 3x^2 & = 6 \\ \hline 3(-\sqrt{2})^2 & ?\ 6 \\ 3 \cdot 2 & \\ 6 & \end{array} \quad \text{TRUE}$$

The solutions are $\sqrt{2}$ and $-\sqrt{2}$, or $\pm\sqrt{2}$, which are about 1.414 and -1.414 when rounded to three decimal places.

Do Exercise 5.

Sometimes we rationalize denominators to simplify answers.

6. Solve: $-3x^2 + 8 = 0$. Give the exact solution and approximate the solutions to three decimal places.

EXAMPLE 5 Solve: $-5x^2 + 2 = 0$. Give the exact solutions and approximate the solutions to three decimal places.

$$-5x^2 + 2 = 0$$
$$x^2 = \frac{2}{5} \qquad \text{Subtracting 2 and dividing by } -5$$
$$x = \sqrt{\frac{2}{5}} \quad or \quad x = -\sqrt{\frac{2}{5}} \qquad \text{Using the principle of square roots}$$
$$x = \sqrt{\frac{2}{5} \cdot \frac{5}{5}} \quad or \quad x = -\sqrt{\frac{2}{5} \cdot \frac{5}{5}} \qquad \text{Rationalizing the denominators}$$
$$x = \frac{\sqrt{10}}{5} \quad or \quad x = -\frac{\sqrt{10}}{5}$$

CHECK: We check both numbers at once, since there is no x-term in the equation. We could have checked both numbers at once in Example 4 as well.

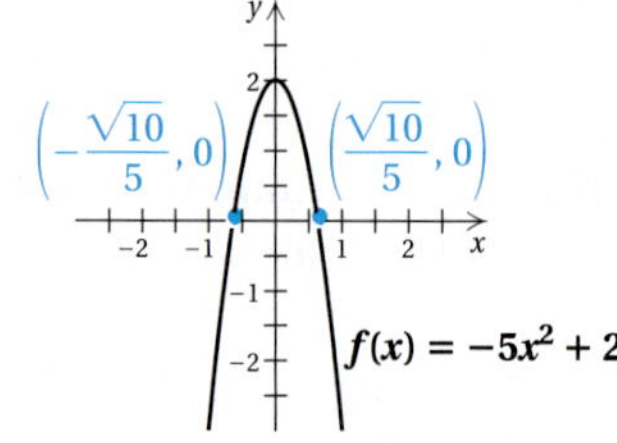

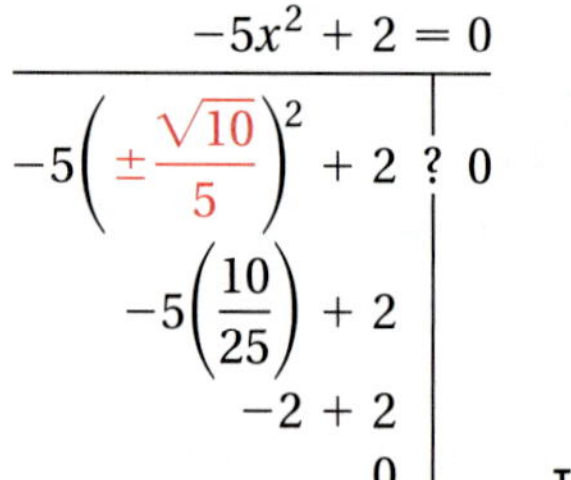

$$\begin{array}{r|l} -5x^2 + 2 & = 0 \\ \hline -5\left(\pm\frac{\sqrt{10}}{5}\right)^2 + 2 & ?\ 0 \\ -5\left(\frac{10}{25}\right) + 2 & \\ -2 + 2 & \\ 0 & \end{array} \quad \text{TRUE}$$

The solutions are $\frac{\sqrt{10}}{5}$ and $-\frac{\sqrt{10}}{5}$, or $\pm\frac{\sqrt{10}}{5}$. We can use a calculator for approximations:

$$\pm\frac{\sqrt{10}}{5} \approx \pm 0.632.$$

Answers on page A-59

Do Exercise 6 on the preceding page.

Sometimes we get solutions that are imaginary numbers.

EXAMPLE 6 Solve: $4x^2 + 9 = 0$.

$$4x^2 + 9 = 0$$

$$x^2 = -\frac{9}{4} \quad \text{Subtracting 9 and dividing by 4}$$

$$x = \sqrt{-\frac{9}{4}} \quad or \quad x = -\sqrt{-\frac{9}{4}} \quad \text{Using the principle of square roots}$$

$$x = \frac{3}{2}i \quad or \quad x = -\frac{3}{2}i \quad \text{Simplifying}$$

CHECK:

$$\begin{array}{c|c} 4x^2 + 9 & 0 \\ \hline 4\left(\pm\frac{3}{2}i\right)^2 + 9 \ ? & 0 \\ 4\left(-\frac{9}{4}\right) + 9 & \\ -9 + 9 & \\ 0 & \end{array} \quad \text{TRUE}$$

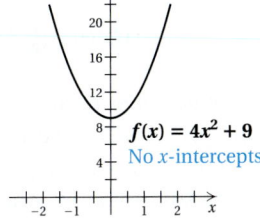

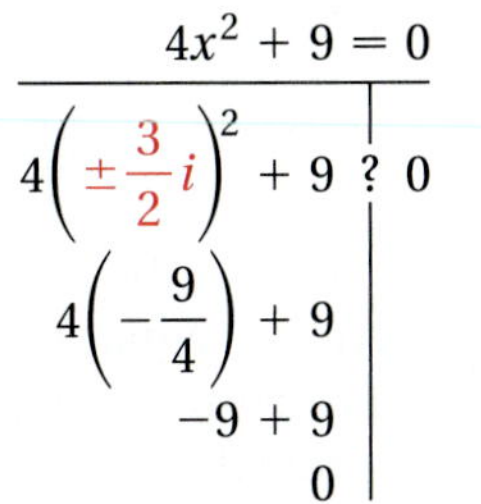

The solutions are $\frac{3}{2}i$ and $-\frac{3}{2}i$, or $\pm\frac{3}{2}i$.

We see that the graph of $f(x) = 4x^2 + 9$ does not cross the x-axis. This is true because the equation $4x^2 + 9 = 0$ has *imaginary* complex-number solutions.

Do Exercise 7.

SOLVING EQUATIONS OF THE TYPE $(x + c)^2 = d$

The equation $(x - 2)^2 = 7$ can also be solved using the principle of square roots.

EXAMPLE 7

a) Solve: $(x - 2)^2 = 7$.

b) Find the x-intercepts of $f(x) = (x - 2)^2 - 7$.

a) We have

$$(x - 2)^2 = 7$$

$$x - 2 = \sqrt{7} \quad or \quad x - 2 = -\sqrt{7} \quad \text{Using the principle of square roots}$$

$$x = 2 + \sqrt{7} \quad or \quad x = 2 - \sqrt{7}.$$

The solutions are $2 + \sqrt{7}$ and $2 - \sqrt{7}$, or $2 \pm \sqrt{7}$.

b) The x-intercepts of $f(x) = (x - 2)^2 - 7$ are $(2 - \sqrt{7}, 0)$ and $(2 + \sqrt{7}, 0)$.

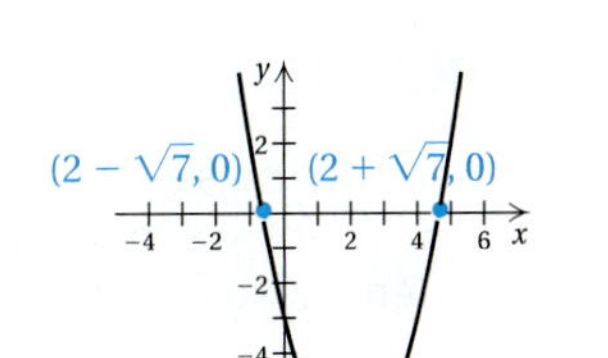

Do Exercise 8.

CALCULATOR CORNER

Imaginary Solutions of Quadratic Equations

What happens when you use the ZERO feature to solve the equation in Example 6? Explain why this happens.

7. Solve: $2x^2 + 1 = 0$.

8. a) Solve: $(x - 1)^2 = 5$.

b) Find the x-intercepts of $f(x) = (x - 1)^2 - 5$.

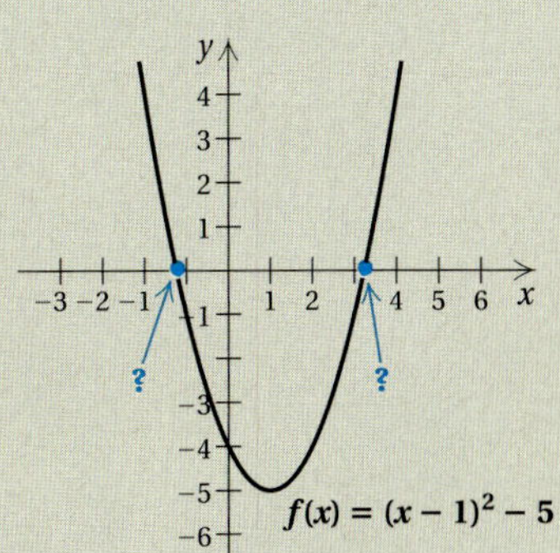

Answers on page A-59

9. Solve: $x^2 + 16x + 64 = 11$.

Answer on page A-59

If we can express the left side of an equation as the square of a binomial, we can proceed as we did in Example 7.

EXAMPLE 8 Solve: $x^2 + 6x + 9 = 2$.

We have

$$x^2 + 6x + 9 = 2 \quad \text{The left side is the square of a binomial.}$$
$$(x + 3)^2 = 2$$
$$x + 3 = \sqrt{2} \quad or \quad x + 3 = -\sqrt{2} \quad \text{Using the principle of square roots}$$
$$x = -3 + \sqrt{2} \quad or \quad x = -3 - \sqrt{2}.$$

The solutions are $-3 + \sqrt{2}$ and $-3 - \sqrt{2}$, or $-3 \pm \sqrt{2}$.

Do Exercise 9.

b Completing the Square

We can solve quadratic equations like $3x^2 = 6$ and $(x - 2)^2 = 7$ by using the principle of square roots. We can also solve an equation such as $x^2 + 6x + 9 = 2$ in like manner because the expression on the left side is the square of a binomial, $(x + 3)^2$. This second procedure is the basis for a method called **completing the square.** *It can be used to solve any quadratic equation.*

Suppose we have the following quadratic equation:

$$x^2 + 14x = 4.$$

If we could add on both sides of the equation a constant that would make the expression on the left the square of a binomial, we could then solve the equation using the principle of square roots.

How can we determine what to add to $x^2 + 14x$ to construct the square of a binomial? We want to find a number a such that the following equation is satisfied:

$$x^2 + 14x + a^2 = (x + a)(x + a) = x^2 + 2ax + a^2.$$

Thus a is such that $2a = 14$. Solving, we get $a = 7$. That is, a is half of the coefficient of x in $x^2 + 14x$. Since $a^2 = \left(\frac{14}{2}\right)^2 = 7^2 = 49$, we add 49 to our original expression:

$$x^2 + 14x + 49 \text{ is the square of } x + 7;$$

that is,

$$x^2 + 14x + 49 = (x + 7)^2.$$

COMPLETING THE SQUARE

When solving an equation, to **complete the square** of an expression like $x^2 + bx$, we take half the x-coefficient, which is $b/2$, and square it. Then we add that number, $(b/2)^2$, on both sides of the equation.

Returning to solving our original equation, we first add 49 on *both* sides to *complete the square* on the left. Then we solve:

$$x^2 + 14x = 4 \quad \text{Original equation}$$
$$x^2 + 14x + 49 = 4 + 49 \quad \text{Adding 49: } \left(\tfrac{14}{2}\right)^2 = 7^2 = 49$$
$$(x + 7)^2 = 53$$
$$x + 7 = \sqrt{53} \quad \text{or} \quad x + 7 = -\sqrt{53} \quad \text{Using the principle of square roots}$$
$$x = -7 + \sqrt{53} \quad \text{or} \quad x = -7 - \sqrt{53}.$$

The solutions are $-7 \pm \sqrt{53}$.

We have seen that a quadratic equation $(x + c)^2 = d$ can be solved using the principle of square roots. Any equation, such as $x^2 - 6x + 8 = 0$, can be put in this form by completing the square. Then we can solve as before.

EXAMPLE 9 Solve: $x^2 - 6x + 8 = 0$.

We have

$$x^2 - 6x + 8 = 0$$
$$x^2 - 6x = -8. \quad \text{Subtracting 8}$$

We take half of -6 and square it, to get 9. Then we add 9 on *both* sides of the equation. This makes the left side the square of a binomial, $x - 3$. We have now *completed the square.*

$$x^2 - 6x + 9 = -8 + 9 \quad \text{Adding 9: } \left(\tfrac{-6}{2}\right)^2 = (-3)^2 = 9$$
$$(x - 3)^2 = 1$$
$$x - 3 = 1 \quad \text{or} \quad x - 3 = -1 \quad \text{Using the principle of square roots}$$
$$x = 4 \quad \text{or} \quad x = 2$$

The solutions are 2 and 4.

Do Exercises 10 and 11.

EXAMPLE 10 Solve $x^2 + 4x - 7 = 0$ by completing the square.

We have

$$x^2 + 4x - 7 = 0$$
$$x^2 + 4x = 7 \quad \text{Adding 7}$$
$$x^2 + 4x + 4 = 7 + 4 \quad \text{Adding 4: } \left(\tfrac{4}{2}\right)^2 = (2)^2 = 4$$
$$(x + 2)^2 = 11$$
$$x + 2 = \sqrt{11} \quad \text{or} \quad x + 2 = -\sqrt{11} \quad \text{Using the principle of square roots}$$
$$x = -2 + \sqrt{11} \quad \text{or} \quad x = -2 - \sqrt{11}.$$

The solutions are $-2 \pm \sqrt{11}$.

Do Exercise 12.

Solve.

10. $x^2 + 6x + 8 = 0$

11. $x^2 - 8x - 20 = 0$

12. Solve by completing the square:

$x^2 + 6x - 1 = 0.$

Answers on page A-59

When the coefficient of x^2 is not 1, we can make it 1, as shown in the following example.

EXAMPLE 11 Solve $2x^2 = 3x - 7$ by completing the square.

$$2x^2 = 3x - 7$$

$$2x^2 - 3x = -7 \qquad \text{Subtracting } 3x$$

$$\frac{1}{2}(2x^2 - 3x) = \frac{1}{2} \cdot (-7) \qquad \text{Multiplying by } \tfrac{1}{2} \text{ to make the } x^2\text{-coefficient 1}$$

$$x^2 - \frac{3}{2}x = -\frac{7}{2} \qquad \text{Multiplying and simplifying}$$

$$x^2 - \frac{3}{2}x + \frac{9}{16} = -\frac{7}{2} + \frac{9}{16} \qquad \text{Adding } \frac{9}{16}: \left[\frac{1}{2}\left(-\frac{3}{2}\right)\right]^2 = \left[-\frac{3}{4}\right]^2 = \frac{9}{16}$$

$$\left(x - \frac{3}{4}\right)^2 = -\frac{56}{16} + \frac{9}{16} \qquad \text{Finding a common denominator}$$

$$\left(x - \frac{3}{4}\right)^2 = -\frac{47}{16}$$

$$x - \frac{3}{4} = \sqrt{-\frac{47}{16}} \quad or \quad x - \frac{3}{4} = -\sqrt{-\frac{47}{16}} \qquad \text{Using the principle of square roots}$$

$$x - \frac{3}{4} = i\sqrt{\frac{47}{16}} \quad or \quad x - \frac{3}{4} = -i\sqrt{\frac{47}{16}} \qquad \sqrt{-1} = i$$

$$x = \frac{3}{4} + \frac{i\sqrt{47}}{4} \quad or \quad x = \frac{3}{4} - \frac{i\sqrt{47}}{4}$$

The solutions are $\frac{3}{4} \pm i\frac{\sqrt{47}}{4}$.

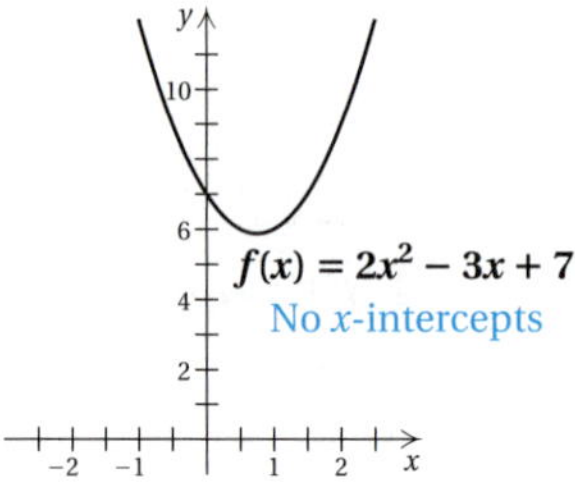

We see that the graph of $f(x) = 2x^2 - 3x + 7$ does not cross the x-axis. This is true because the equation $2x^2 = 3x - 7$ has nonreal complex-number solutions.

SOLVING BY COMPLETING THE SQUARE

To solve an equation $ax^2 + bx + c = 0$ by completing the square:

1. If $a \neq 1$, multiply by $1/a$ so that the x^2-coefficient is 1.
2. If the x^2-coefficient is 1, add or subtract so that the equation is in the form

$$x^2 + bx = -c, \quad \text{or} \quad x^2 + \frac{b}{a}x = -\frac{c}{a} \text{ if step (1) has been applied.}$$

3. Take half of the x-coefficient and square it. Add the result on both sides of the equation.
4. Express the side with the variables as the square of a binomial.
5. Use the principle of square roots and complete the solution.

Completing the square provides a base for proving the quadratic formula in Section 16.2.

Do Exercise 13.

13. Solve by completing the square:

$$3x^2 - 2x = 7.$$

C Applications and Problem Solving

EXAMPLE 12 *Hang Time.* One of the most exciting plays in basketball is the dunk shot. The amount of time T that passes from the moment a player leaves the ground, goes up, makes the shot, and arrives back on the ground is called the *hang time*. A function relating an athlete's vertical leap V, in inches, to hang time T, in seconds, is given by

$$V(T) = 48T^2.$$

a) Hall-of-Famer Michael Jordan has a hang time of about 0.889 sec. What is his vertical leap?

b) Although his height is only 5 ft 7 in., Spud Webb, formerly of the Sacramento Kings, had a vertical leap of about 44 in. What is his hang time?

Source: Peter Brancazio, "The Mechanics of a Slam Dunk," *Popular Mechanics*, November 1991. Courtesy of Professor Peter Brancazio, Brooklyn College.

a) To find Jordan's vertical leap, we substitute 0.889 for T in the function and compute V:

$$V(0.889) = 48(0.889)^2 \approx 37.9 \text{ in.}$$

Jordan's vertical leap is about 37.9 in. Surprisingly, Jordan does not have the vertical leap most fans would expect.

b) To find Webb's hang time, we substitute 44 for V and solve for T:

$$44 = 48T^2 \quad \text{Substituting 44 for } V$$

$$\frac{44}{48} = T^2 \quad \text{Solving for } T^2$$

$$0.91\overline{6} = T^2$$

$$\sqrt{0.91\overline{6}} = T \quad \text{Hang time is positive.}$$

$$0.957 \approx T. \quad \text{Using a calculator}$$

Webb's hang time is 0.957 sec. Note that his hang time is greater than Jordan's.

Do Exercises 14 and 15.

14. **Hang Time.** Anfernee Hardaway, of the Phoenix Suns, has a hang time of about 0.866 sec. What is his vertical leap?

15. **Hang Time.** The record for the greatest vertical leap in the NBA is held by Darryl Griffith of the Utah Jazz. It was 48 in. What was his hang time?

Answers on page A-59

16.1 EXERCISE SET

For Extra Help

Digital Video Tutor CD 13 Videotape 18 | InterAct Math | Math Tutor Center | MathXL | MyMathLab

a

1. a) Solve: $6x^2 = 30$.
   b) Find the x-intercepts of $f(x) = 6x^2 - 30$.

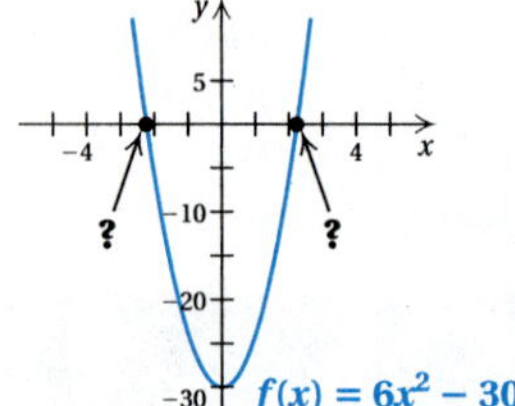

2. a) Solve: $5x^2 = 35$.
   b) Find the x-intercepts of $f(x) = 5x^2 - 35$.

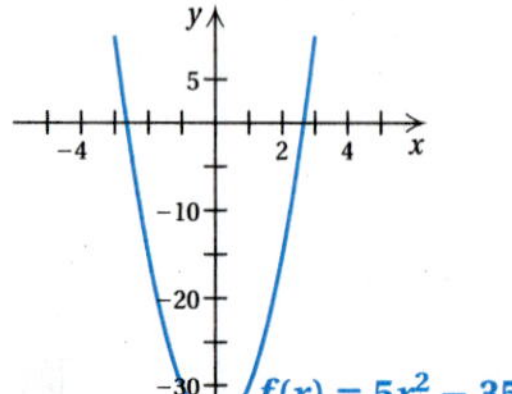

3. a) Solve: $9x^2 + 25 = 0$.
   b) Find the x-intercepts of $f(x) = 9x^2 + 25$.

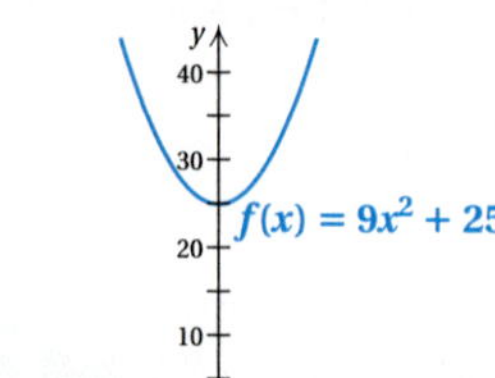

4. a) Solve: $36x^2 + 49 = 0$.
   b) Find the x-intercepts of $f(x) = 36x^2 + 49$.

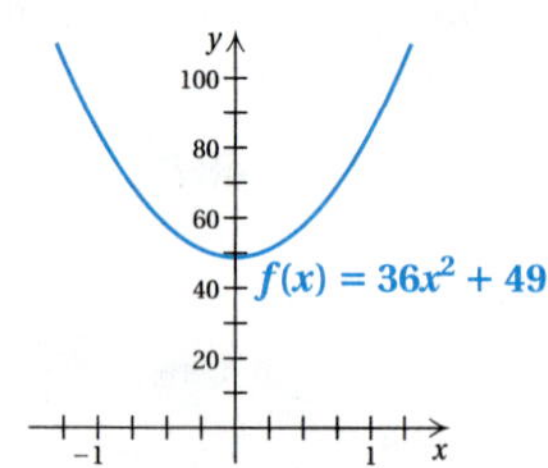

Solve. Give the exact solution and approximate solutions to three decimal places, when appropriate.

5. $2x^2 - 3 = 0$

6. $3x^2 - 7 = 0$

7. $(x + 2)^2 = 49$

8. $(x - 1)^2 = 6$

9. $(x - 4)^2 = 16$

10. $(x + 3)^2 = 9$

11. $(x - 11)^2 = 7$

12. $(x - 9)^2 = 34$

13. $(x - 7)^2 = -4$

14. $(x + 1)^2 = -9$

15. $(x - 9)^2 = 81$

16. $(t - 2)^2 = 25$

17. $\left(x - \frac{3}{2}\right)^2 = \frac{7}{2}$

18. $\left(y + \frac{3}{4}\right)^2 = \frac{17}{16}$

19. $x^2 + 6x + 9 = 64$

20. $x^2 + 10x + 25 = 100$

21. $y^2 - 14y + 49 = 4$

22. $p^2 - 8p + 16 = 1$

Solve by completing the square. Show your work.

23. $x^2 + 4x = 2$

24. $x^2 + 2x = 5$

25. $x^2 - 22x = 11$

26. $x^2 - 18x = 10$

27. $x^2 + x = 1$

28. $x^2 - x = 3$

29. $t^2 - 5t = 7$

30. $y^2 + 9y = 8$

31. $x^2 + \frac{3}{2}x = 3$

32. $x^2 - \frac{4}{3}x = \frac{2}{3}$

33. $m^2 - \frac{9}{2}m = \frac{3}{2}$

34. $r^2 + \frac{2}{5}r = \frac{4}{5}$

35. $x^2 + 6x - 16 = 0$

36. $x^2 - 8x + 15 = 0$

37. $x^2 + 22x + 102 = 0$

38. $x^2 + 18x + 74 = 0$

39. $x^2 - 10x - 4 = 0$

40. $x^2 + 10x - 4 = 0$

41. a) Solve:
$x^2 + 7x - 2 = 0$.
b) Find the x-intercepts of $f(x) = x^2 + 7x - 2$.

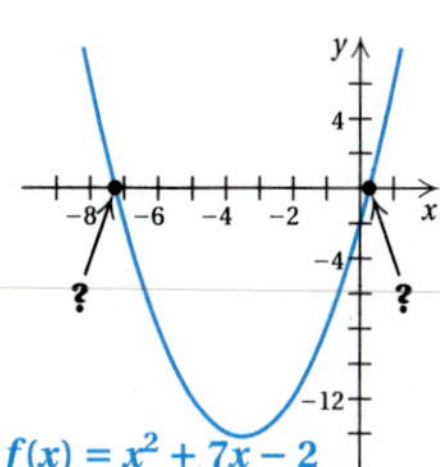

42. a) Solve:
$x^2 - 7x - 2 = 0$.
b) Find the x-intercepts of $f(x) = x^2 - 7x - 2$.

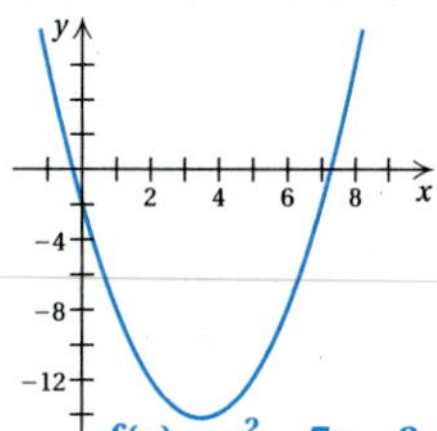

43. a) Solve:
$2x^2 - 5x + 8 = 0$.
b) Find the x-intercepts of $f(x) = 2x^2 - 5x + 8$.

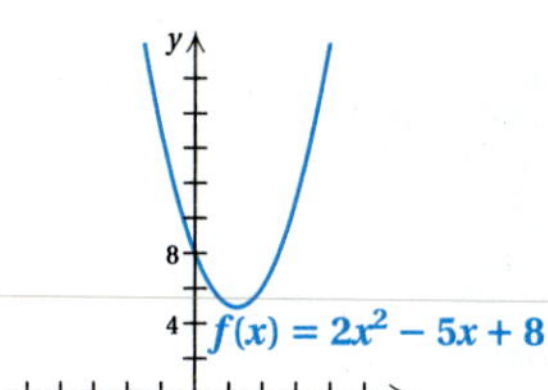

44. a) Solve:
$2x^2 - 3x + 9 = 0$.
b) Find the x-intercepts of $f(x) = 2x^2 - 3x + 9$.

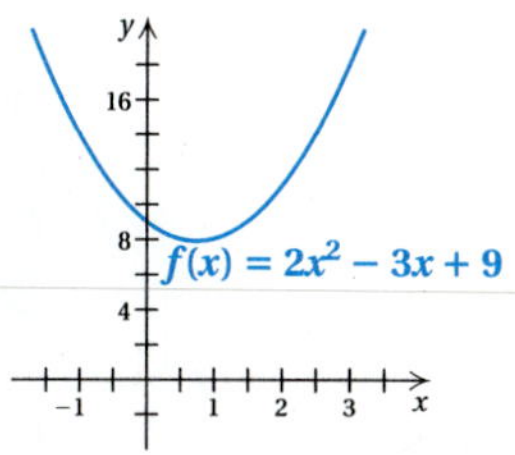

Solve by completing the square. Show your work.

45. $x^2 - \frac{3}{2}x - \frac{1}{2} = 0$

46. $x^2 + \frac{3}{2}x - 2 = 0$

47. $2x^2 - 3x - 17 = 0$

48. $2x^2 + 3x - 1 = 0$

49. $3x^2 - 4x - 1 = 0$

50. $3x^2 + 4x - 3 = 0$

51. $x^2 + x + 2 = 0$

52. $x^2 - x + 1 = 0$

53. $x^2 - 4x + 13 = 0$

54. $x^2 - 6x + 13 = 0$

C *Hang Time.* For Exercises 55 and 56, use the hang-time function $V(T) = 48T^2$, relating vertical leap to hang time.

55. The NBA's Vince Carter, of the Toronto Raptors, has a vertical leap of about 36 in. What is his hang time?

56. Tracy McGrady, of the Orlando Magic, has a vertical leap of 40 in. What is his hang time?

Free-Falling Objects. The function $s(t) = 16t^2$ is used to approximate the distance s, in feet, that an object falls freely from rest in t seconds. Use the formula for Exercises 57–60.

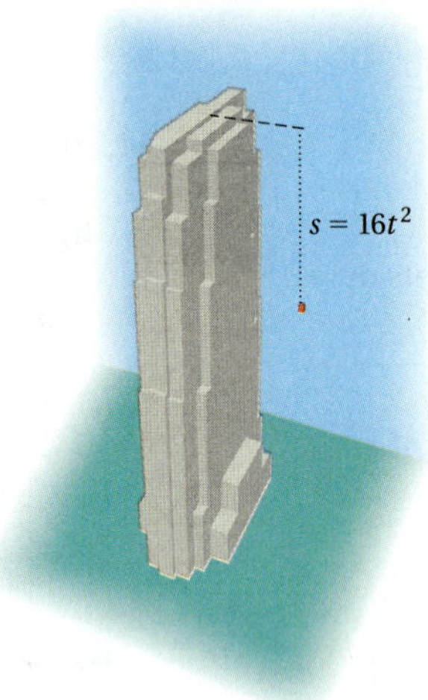

57. The RCA Building in New York City is 850 ft tall. How long will it take an object to fall from the top?

58. The CN Tower in Toronto, at 1815 ft, is the world's tallest self-supporting tower (no guy wires). How long would it take an object to fall freely from the top?
Source: *The Guinness Book of Records*

59. Reaching 745 ft above the water, the towers of California's Golden Gate Bridge are the world's tallest bridge towers. How long would it take an object to fall freely from the top?
Source: *The Guinness Book of Records*

60. The Gateway Arch in St. Louis is 640 ft high. How long would it take an object to fall freely from the top?

61. D_W Explain in your own words a sequence of steps that you might follow to solve any quadratic equation.

62. D_W Write a problem involving hang time for a classmate to solve. (See Example 12.) Devise the problem so that the solution is "The player's hang time is about 0.935 sec."

SKILL MAINTENANCE

63. *Tattoo Removal.* The following table lists data regarding the number of people who visited a doctor for tattoo removal in 1996 and 2000. [12.5e]

Number of years since 1995	Number of people who visited a doctor for tattoo removal (in thousands)
1	275
5	410

Source: *Star-Tribune*, Minneapolis–St. Paul

a) Use the two data points to find a linear function $T(t) = mt + b$ that fits the data.
b) Use the function to estimate the number of people who will visit a doctor in 2005 to have a tattoo removed.
c) In what year will 1,085,000 people visit a doctor for tattoo removal?

Graph. [12.1c], [12.4a]

64. $f(x) = 5 - 2x^2$

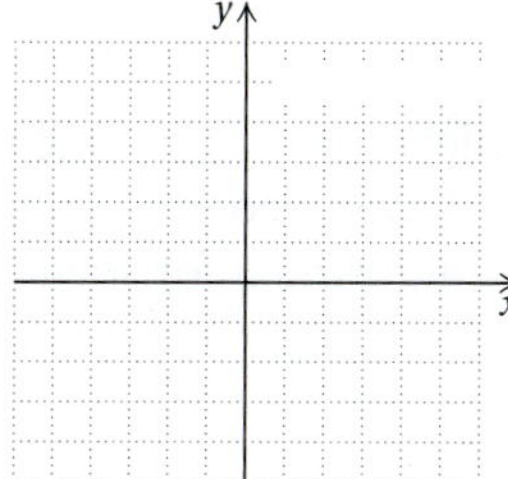

65. $f(x) = 5 - 2x$

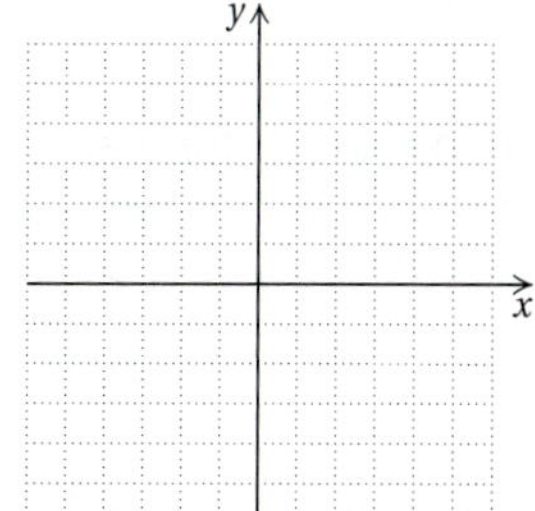

66. $2x - 5y = 10$

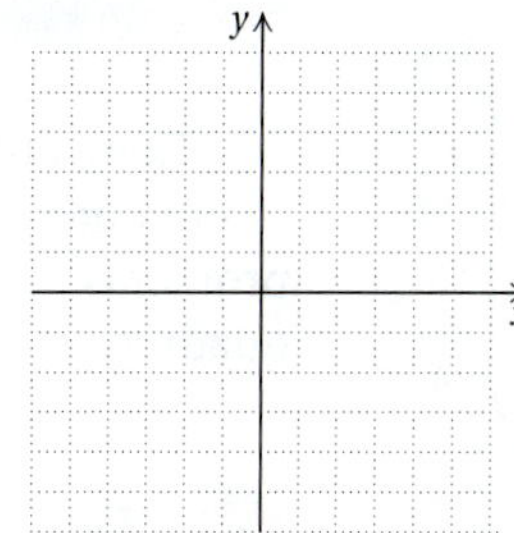

67. $f(x) = |5 - 2x|$

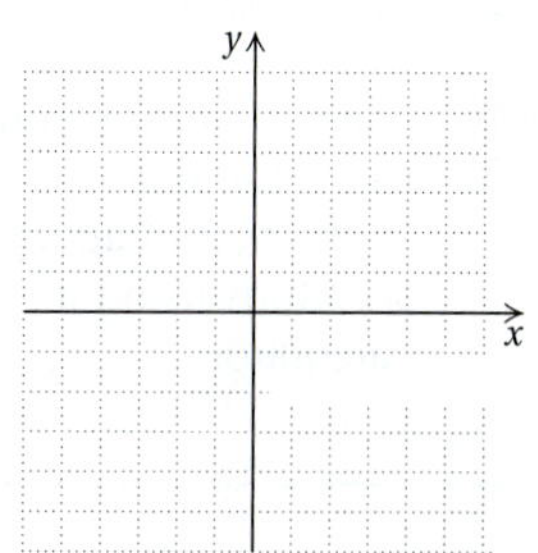

68. Simplify: $\sqrt{88}$. [15.3a]

69. Rationalize the denominator: $\sqrt{\frac{2}{5}}$. [15.5a]

SYNTHESIS

70. Use a graphing calculator to solve each of the following equations.

a) $25.55x^2 - 1635.2 = 0$
b) $-0.0644x^2 + 0.0936x + 4.56 = 0$
c) $2.101x + 3.121 = 0.97x^2$

71. Problems such as those in Exercises 17, 21, and 25 can be solved without first finding standard form by using the INTERSECT feature on a graphing calculator. We let $y_1 =$ the left side of the equation and $y_2 =$ the right side. Use a graphing calculator to solve Exercises 17, 21, and 25 in this manner.

Find b such that the trinomial is a square.

72. $x^2 + bx + 75$

73. $x^2 + bx + 64$

Solve.

74. $\left(x - \frac{1}{3}\right)\left(x - \frac{1}{3}\right) + \left(x - \frac{1}{3}\right)\left(x + \frac{2}{9}\right) = 0$

75. $x(2x^2 + 9x - 56)(3x + 10) = 0$

76. *Boating.* A barge and a fishing boat leave a dock at the same time, traveling at right angles to each other. The barge travels 7 km/h slower than the fishing boat. After 4 hr, the boats are 68 km apart. Find the speed of each vessel.

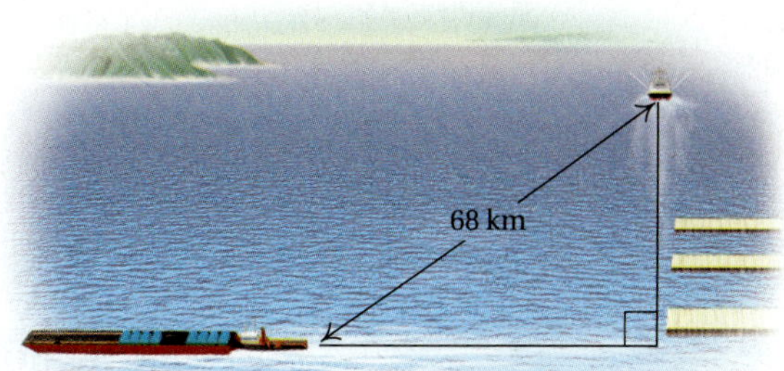

Objective

a Solve quadratic equations using the quadratic formula, and approximate solutions using a calculator.

16.2 THE QUADRATIC FORMULA

There are at least two reasons for learning to complete the square. One is to enhance your ability to graph certain equations that are needed to solve problems in Section 16.7. The other is to prove a general formula for solving quadratic equations.

a Solving Using the Quadratic Formula

Each time you solve by completing the square, the procedure is the same. When we do the same kind of procedure many times, we look for a formula to speed up our work. Consider

$$ax^2 + bx + c = 0, \quad a > 0.$$

Note that if $a < 0$, we can get an equivalent form with $a > 0$ by first multiplying by -1.

Let's solve by *completing the square*. As we carry out the steps, compare them with Example 11 in the preceding section.

$$x^2 + \frac{b}{a}x + \frac{c}{a} = 0 \qquad \text{Multiplying by } \frac{1}{a}$$

$$x^2 + \frac{b}{a}x = -\frac{c}{a} \qquad \text{Subtracting } \frac{c}{a}$$

Half of $\frac{b}{a}$ is $\frac{b}{2a}$. The square is $\frac{b^2}{4a^2}$. We add $\frac{b^2}{4a^2}$ on both sides:

$$x^2 + \frac{b}{a}x + \frac{b^2}{4a^2} = -\frac{c}{a} + \frac{b^2}{4a^2} \qquad \text{Adding } \frac{b^2}{4a^2}$$

$$\left(x + \frac{b}{2a}\right)^2 = -\frac{4ac}{4a^2} + \frac{b^2}{4a^2} \qquad \text{Factoring the left side and finding a common denominator on the right}$$

$$\left(x + \frac{b}{2a}\right)^2 = \frac{b^2 - 4ac}{4a^2}$$

$$x + \frac{b}{2a} = \sqrt{\frac{b^2 - 4ac}{4a^2}} \quad or \quad x + \frac{b}{2a} = -\sqrt{\frac{b^2 - 4ac}{4a^2}}. \qquad \text{Using the principle of square roots}$$

Since $a > 0$, $\sqrt{4a^2} = 2a$, so we can simplify as follows:

$$x + \frac{b}{2a} = \frac{\sqrt{b^2 - 4ac}}{2a} \quad \text{or} \quad x + \frac{b}{2a} = -\frac{\sqrt{b^2 - 4ac}}{2a}.$$

Thus,

$$x = -\frac{b}{2a} \pm \frac{\sqrt{b^2 - 4ac}}{2a}, \quad \text{or} \quad x = \frac{-b \pm \sqrt{b^2 - 4ac}}{2a}.$$

We now have the following.

THE QUADRATIC FORMULA

The solutions of $ax^2 + bx + c = 0$ are given by

$$x = \frac{-b \pm \sqrt{b^2 - 4ac}}{2a}.$$

The formula also holds when $a < 0$. A similar proof would show this, but we will not consider it here.

ALGEBRAIC–GRAPHICAL CONNECTION

The Quadratic Formula (Algebraic). The solutions of $ax^2 + bx + c = 0$, $a \neq 0$, are given by

$$x = \frac{-b \pm \sqrt{b^2 - 4ac}}{2a}.$$

The Quadratic Formula (Graphical). The x-intercepts of the graph of the function $f(x) = ax^2 + bx + c$, $a \neq 0$, if they exist, are given by

$$\left(\frac{-b \pm \sqrt{b^2 - 4ac}}{2a}, 0\right).$$

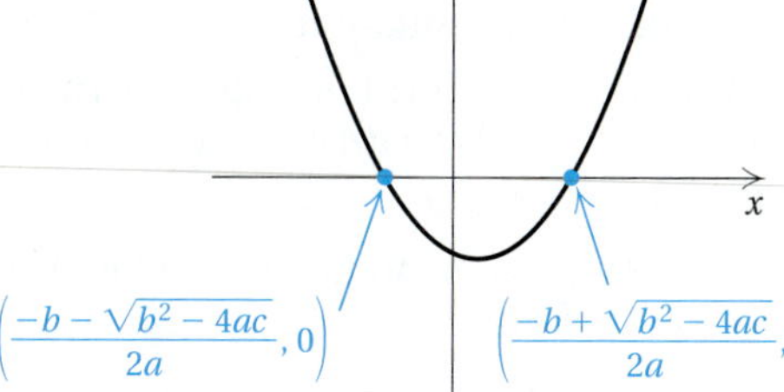

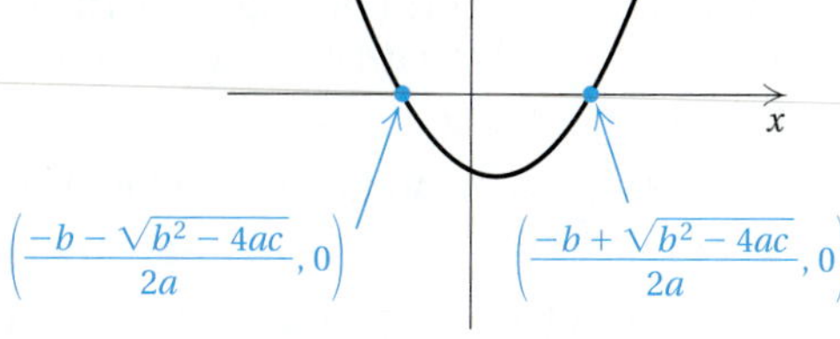

EXAMPLE 1 Solve $5x^2 + 8x = -3$ using the quadratic formula.

We first find standard form and determine a, b, and c:

$$5x^2 + 8x + 3 = 0;$$
$$a = 5, \quad b = 8, \quad c = 3.$$

We then use the quadratic formula:

$$x = \frac{-b \pm \sqrt{b^2 - 4ac}}{2a}$$

$$x = \frac{-8 \pm \sqrt{8^2 - 4 \cdot 5 \cdot 3}}{2 \cdot 5} \quad \text{Substituting}$$

Be sure to write the fraction bar all the way across.

$$x = \frac{-8 \pm \sqrt{64 - 60}}{10}$$

$$x = \frac{-8 \pm \sqrt{4}}{10}$$

$$x = \frac{-8 \pm 2}{10}$$

$$x = \frac{-8 + 2}{10} \quad or \quad x = \frac{-8 - 2}{10}$$

$$x = \frac{-6}{10} \quad or \quad x = \frac{-10}{10}$$

$$x = -\frac{3}{5} \quad or \quad x = -1.$$

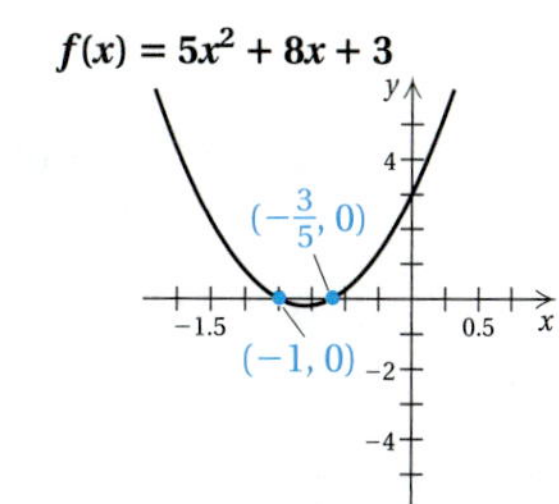

The solutions are $-\frac{3}{5}$ and -1.

It turns out that we could have solved the equation in Example 1 more easily by factoring as follows:

$$
\begin{aligned}
5x^2 + 8x + 3 &= 0 \\
(5x + 3)(x + 1) &= 0 \\
5x + 3 = 0 \quad &or \quad x + 1 = 0 \\
5x = -3 \quad &or \quad x = -1 \\
x = -\tfrac{3}{5} \quad &or \quad x = -1.
\end{aligned}
$$

To solve a quadratic equation:

1. Check for the form $x^2 = d$ or $(x + c)^2 = d$. If it is in this form, use the principle of square roots as in Section 16.1.
2. If it is not in the form of step (1), write it in standard form $ax^2 + bx + c = 0$ with a and b nonzero.
3. Then try factoring.
4. If it is not possible to factor or if factoring seems difficult, use the quadratic formula.

The solutions of a quadratic equation cannot always be found by factoring. They can always be found using the quadratic formula.

The solutions to all the exercises in this section could also be found by completing the square. However, the quadratic formula is the preferred method when solving applied problems because it is faster.

Do Exercise 1.

We will see in Example 2 that we cannot always rely on factoring.

EXAMPLE 2 Solve: $5x^2 - 8x = 3$. Give the exact solution and approximate the solutions to three decimal places.

We first find standard form and determine a, b, and c:

$$5x^2 - 8x - 3 = 0;$$

$$a = 5, \quad b = -8, \quad c = -3.$$

We then use the quadratic formula, $x = \dfrac{-b \pm \sqrt{b^2 - 4ac}}{2a}$:

$$x = \frac{-(-8) \pm \sqrt{(-8)^2 - 4 \cdot 5 \cdot (-3)}}{2 \cdot 5} \quad \text{Substituting}$$

$$= \frac{8 \pm \sqrt{64 + 60}}{10} = \frac{8 \pm \sqrt{124}}{10} = \frac{8 \pm \sqrt{4 \cdot 31}}{10}$$

$$= \frac{8 \pm 2\sqrt{31}}{10} = \frac{2(4 \pm \sqrt{31})}{2 \cdot 5} = \frac{2}{2} \cdot \frac{4 \pm \sqrt{31}}{5} = \frac{4 \pm \sqrt{31}}{5}.$$

CAUTION!

To avoid a common error in simplifying, remember to *factor the numerator and the denominator* and then remove a factor of 1.

We can use a calculator to approximate the solutions:

$$\frac{4 + \sqrt{31}}{5} \approx 1.914; \qquad \frac{4 - \sqrt{31}}{5} \approx -0.314.$$

1. Consider the equation

$$2x^2 = 4 + 7x.$$

a) Solve using the quadratic formula.

b) Solve by factoring.

Answers on page A-60

CALCULATOR CORNER

Approximating Solutions of Quadratic Equations In Example 2, we found that the solutions of the equation $5x^2 - 8x = 3$ are $\dfrac{4 + \sqrt{31}}{5}$ and $\dfrac{4 - \sqrt{31}}{5}$. We can use a calculator to approximate these solutions. To approximate $\dfrac{4 + \sqrt{31}}{5}$, we press

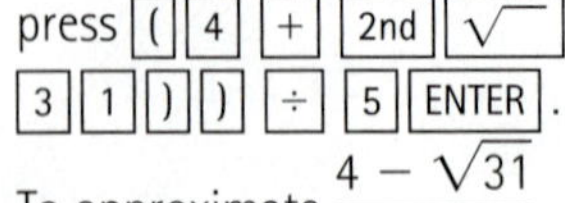

. To approximate $\dfrac{4 - \sqrt{31}}{5}$, we press

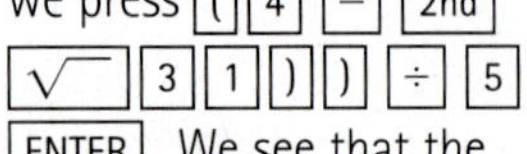

ENTER. We see that the solutions are approximately 1.914 and −0.314.

```
(4+√(31))/5
                1.913552873
(4−√(31))/5
               −.3135528726
```

Exercises: Use a calculator to approximate the solutions in each of the following. Round to three decimal places.

1. Example 4
2. Margin Exercise 2
3. Margin Exercise 4

CHECK: Checking the exact solutions $(4 \pm \sqrt{31})/5$ can be quite cumbersome. It could be done on a calculator or by using the approximations. Here we check 1.914; the check for -0.314 is left to the student.

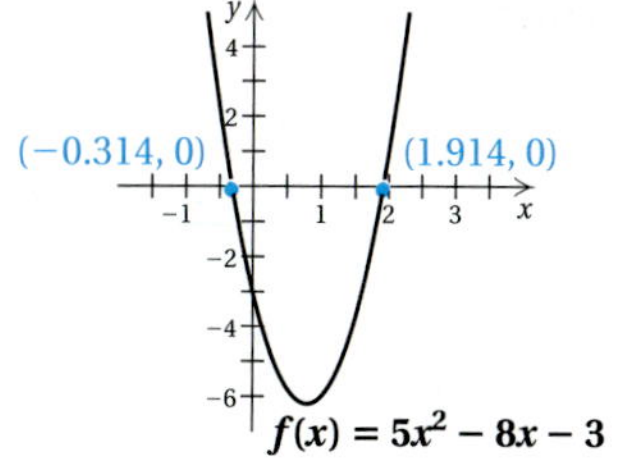

For 1.914:

$$\begin{array}{c|l} 5x^2 - 8x & 3 \\ \hline 5(1.914)^2 - 8(1.914) & ?\ 3 \\ 5(3.663396) - 15.312 & \\ 3.00498 & \end{array}$$

APPROXIMATELY TRUE

We do not have a perfect check due to the rounding error. But our check seems to confirm the solutions.

Do Exercise 2.

Some quadratic equations have solutions that are nonreal complex numbers.

EXAMPLE 3 Solve: $x^2 + x + 1 = 0$.

We have $a = 1$, $b = 1$, $c = 1$. We use the quadratic formula:

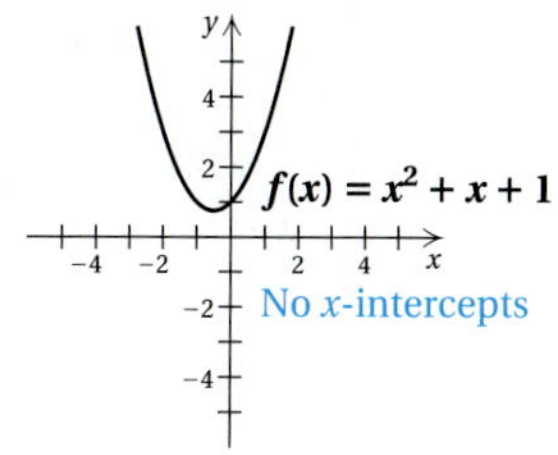

$$\begin{aligned} x &= \frac{-1 \pm \sqrt{1^2 - 4 \cdot 1 \cdot 1}}{2 \cdot 1} \\ &= \frac{-1 \pm \sqrt{1 - 4}}{2} \\ &= \frac{-1 \pm \sqrt{-3}}{2} \\ &= \frac{-1 \pm i\sqrt{3}}{2}. \end{aligned}$$

The solutions are

$$\frac{-1 + i\sqrt{3}}{2} \quad \text{and} \quad \frac{-1 - i\sqrt{3}}{2}.$$

The solutions can also be expressed in the form

$$-\frac{1}{2} + i\frac{\sqrt{3}}{2} \quad \text{and} \quad -\frac{1}{2} - i\frac{\sqrt{3}}{2}.$$

Do Exercise 3.

EXAMPLE 4 Solve: $2 + \frac{7}{x} = \frac{5}{x^2}$. Give the exact solution and approximate solutions to three decimal places.

We first find standard form:

$$\begin{aligned} x^2\left(2 + \frac{7}{x}\right) &= x^2 \cdot \frac{5}{x^2} && \text{Multiplying by } x^2 \text{ to clear fractions, noting that } x \neq 0 \\ 2x^2 + 7x &= 5 \\ 2x^2 + 7x - 5 &= 0. && \text{Subtracting 5} \end{aligned}$$

2. Solve using the quadratic formula:

$$3x^2 + 2x = 7.$$

Give the exact solution and approximate solutions to three decimal places.

3. Solve: $x^2 - x + 2 = 0$.

Answers on page A-60

4. Solve:

$$3 = \frac{5}{x} + \frac{4}{x^2}.$$

Give the exact solution and approximate solutions to three decimal places.

Answer on page A-60

Then

$$a = 2, \quad b = 7, \quad c = -5$$

$$x = \frac{-7 \pm \sqrt{7^2 - 4 \cdot 2 \cdot (-5)}}{2 \cdot 2} \qquad \text{Substituting}$$

$$x = \frac{-7 \pm \sqrt{49 + 40}}{4} = \frac{-7 \pm \sqrt{89}}{4}$$

$$x = \frac{-7 + \sqrt{89}}{4} \quad or \quad x = \frac{-7 - \sqrt{89}}{4}.$$

The quadratic formula always gives correct results when we begin with the standard form. In such cases, we need check only to detect errors in substitution and computation. In this case, since we began with a rational equation, which was *not* in standard form, we *do* need to check. We cleared the fractions before obtaining standard form, and this step could introduce numbers that do not check in the original equation. At the very least, we need to show that neither of the numbers makes a denominator 0. Since neither of them does, the solutions are

$$\frac{-7 + \sqrt{89}}{4} \quad \text{and} \quad \frac{-7 - \sqrt{89}}{4}.$$

We can use a calculator to approximate the solutions:

$$\frac{-7 + \sqrt{89}}{4} \approx 0.608;$$

$$\frac{-7 - \sqrt{89}}{4} \approx -4.108.$$

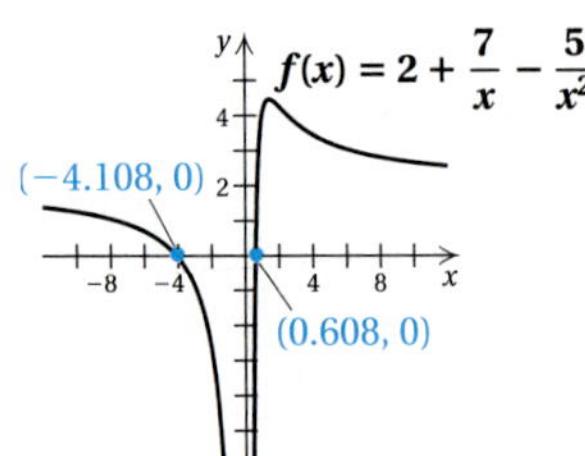

Do Exercise 4.

CALCULATOR CORNER

Visualizing Solutions of Quadratic Equations The graph of $f(x) = 5x^2 - 8x - 3$ in Example 2 has two x-intercepts. This indicates that the equation $5x^2 - 8x - 3 = 0$ (or, equivalently, $5x^2 - 8x = 3$) has two real-number solutions. The graph of $f(x) = x^2 + x + 1$ in Example 3 has no x-intercepts. This indicates that the equation $x^2 + x + 1 = 0$ has no real-number solutions.

Exercises:

1. Explain how the graph of $y = x^2 - x + 2$ shows that the equation $x^2 - x + 2 = 0$ has no real-number solutions.
2. Use a graph to determine whether the equation $x^2 + x = 1$ has real-number solutions. If so, how many are there?
3. Use a graph to determine whether the equation $x^2 - x - 2 = 0$ has real-number solutions. If so, how many are there?
4. Use a graph to determine whether the equation $4x^2 + 1 = 4x$ has real-number solutions. If so, how many are there?

CALCULATOR CORNER

Solving Quadratic Equations A quadratic equation written with 0 on one side of the equals sign can be solved using the ZERO feature of a graphing calculator. See the Calculator Corner on p. 809 for the procedure.

We can also use the INTERSECT feature to solve a quadratic equation. Consider the equation in Exercise 19 in Exercise Set 16.2: $4x(x - 2) - 5x(x - 1) = 2$. First, we enter $y_1 = 4x(x - 2) - 5x(x - 1)$ and $y_2 = 2$ on the equation-editor screen and graph the equations in a window that shows the point(s) of intersection of the graphs. We use the window $[-5, 3, -2, 4]$. We see that there are two points of intersection, so the equation has two solutions.

We use the INTERSECT feature to find the coordinates of the left-hand point of intersection. (See the Calculator Corner on p. 991 for the procedure.) The first coordinate of this point, -2, is one solution of the equation. We use the INTERSECT feature again to find the other solution, -1.

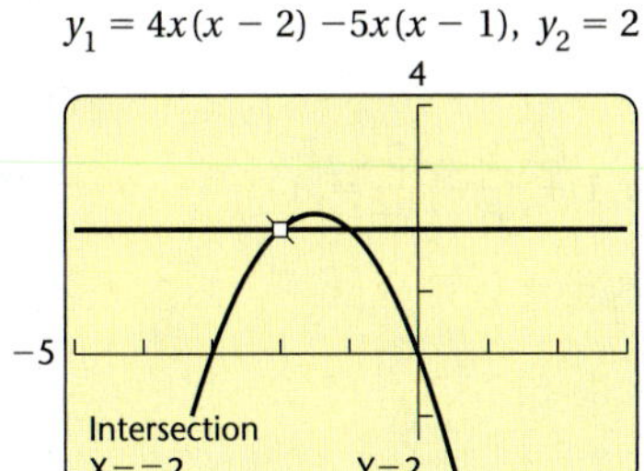

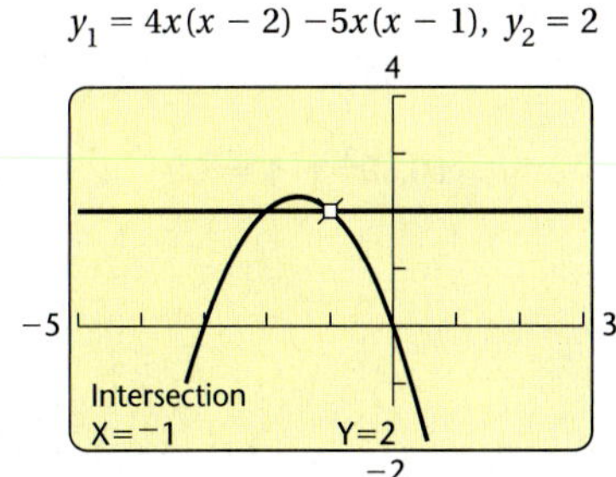

Note that we could use the ZERO feature to solve this equation if we first write it with 0 on one side: $4x(x - 2) - 5x(x - 1) - 2 = 0$.

Exercises: Solve.

1. $5x^2 = -11x + 12$
2. $2x^2 - 15 = 7x$
3. $6(x - 3) = (x - 3)(x - 2)$
4. $(x + 1)(x - 4) = 3(x - 4)$

Study Tips

BEGINNING TO STUDY FOR THE FINAL EXAM (PART 2)

The best scenario for preparing for a final exam is to do so over a period of at least two weeks. Work in a diligent, disciplined manner, doing some final-exam preparation *each* day. Here is a detailed plan that many find useful.

1. **Begin by browsing through each chapter, reviewing the highlighted or boxed information regarding important formulas in both the text and the Summary and Review.** There may be some formulas that you will need to memorize.
2. **Retake each chapter test that you took in class, assuming your instructor has returned it. Otherwise, use the chapter test in the book.** Restudy the objectives in the text that correspond to each question you missed.
3. **If you are still missing questions, use the supplements for extra review.** For example, you might check out the video- or audiotapes, the *Student's Solutions Manual*, the InterAct Math Tutorial Software, or MathXL.
4. **For remaining difficulties, see your instructor, go to a tutoring session, or participate in a study group.**

"The door of opportunity won't open unless you do some pushing."

Anonymous

16.2 EXERCISE SET

For Extra Help

Digital Video Tutor CD 13 Videotape 18 | InterAct Math | Math Tutor Center | MathXL | MyMathLab

a Solve.

1. $x^2 + 6x + 4 = 0$

2. $x^2 - 6x - 4 = 0$

3. $3p^2 = -8p - 1$

4. $3u^2 = 18u - 6$

5. $x^2 - x + 1 = 0$

6. $x^2 + x + 2 = 0$

7. $x^2 + 13 = 4x$

8. $x^2 + 13 = 6x$

9. $r^2 + 3r = 8$

10. $h^2 + 4 = 6h$

11. $1 + \frac{2}{x} + \frac{5}{x^2} = 0$

12. $1 + \frac{5}{x^2} = \frac{2}{x}$

13. a) Solve: $3x + x(x - 2) = 0$.
b) Find the x-intercepts of $f(x) = 3x + x(x - 2)$.

14. a) Solve: $4x + x(x - 3) = 0$.
b) Find the x-intercepts of $f(x) = 4x + x(x - 3)$.

15. a) Solve: $11x^2 - 3x - 5 = 0$.
b) Find the x-intercepts of $f(x) = 11x^2 - 3x - 5$.

16. a) Solve: $7x^2 + 8x = -2$.
b) Find the x-intercepts of $f(x) = 7x^2 + 8x + 2$.

17. a) Solve: $25x^2 = 20x - 4$.
b) Find the x-intercepts of $f(x) = 25x^2 - 20x + 4$.

18. a) Solve: $49x^2 - 14x + 1 = 0$.
b) Find the x-intercepts of $f(x) = 49x^2 - 14x + 1$.

Solve.

19. $4x(x - 2) - 5x(x - 1) = 2$

20. $3x(x + 1) - 7x(x + 2) = 6$

21. $14(x - 4) - (x + 2) = (x + 2)(x - 4)$

22. $11(x - 2) + (x - 5) = (x + 2)(x - 6)$

23. $5x^2 = 17x - 2$

24. $15x = 2x^2 + 16$

25. $x^2 + 5 = 4x$

26. $x^2 + 5 = 2x$

27. $x + \frac{1}{x} = \frac{13}{6}$

28. $\frac{3}{x} + \frac{x}{3} = \frac{5}{2}$

29. $\frac{1}{y} + \frac{1}{y + 2} = \frac{1}{3}$

30. $\frac{1}{x} + \frac{1}{x + 4} = \frac{1}{7}$

31. $(2t - 3)^2 + 17t = 15$

32. $2y^2 - (y + 2)(y - 3) = 12$

33. $(x - 2)^2 + (x + 1)^2 = 0$

34. $(x + 3)^2 + (x - 1)^2 = 0$

35. $x^3 - 1 = 0$
(*Hint*: Factor the difference of cubes. Then use the quadratic formula.)

36. $x^3 + 27 = 0$

Solve. Give the exact solution and approximate solutions to three decimal places.

37. $x^2 + 6x + 4 = 0$

38. $x^2 + 4x - 7 = 0$

39. $x^2 - 6x + 4 = 0$

40. $x^2 - 4x + 1 = 0$

41. $2x^2 - 3x - 7 = 0$

42. $3x^2 - 3x - 2 = 0$

43. $5x^2 = 3 + 8x$

44. $2y^2 + 2y - 3 = 0$

45. ^{D}W The list of steps on p. 1212 does not mention completing the square as a method of solving quadratic equations. Why not?

46. ^{D}W Given the solutions of a quadratic equation, is it possible to reconstruct the original equation? Why or why not?

SKILL MAINTENANCE

Solve. [15.6a, b]

47. $x = \sqrt{x + 2}$

48. $x = \sqrt{15 - 2x}$

49. $\sqrt{x + 2} = \sqrt{2x - 8}$

50. $\sqrt{x + 1} + 2 = \sqrt{3x + 1}$

51. $\sqrt{x + 5} = -7$

52. $\sqrt{2x - 6} + 11 = 2$

53. $\sqrt[3]{4x - 7} = 2$

54. $\sqrt[4]{3x - 1} = 2$

SYNTHESIS

55. Use a graphing calculator to solve the equations in Exercises 3, 16, 17, and 37 using the INTERSECT feature, letting $y_1 =$ the left side and $y_2 =$ the right side. Then solve $2.2x^2 + 0.5x - 1 = 0$.

56. Use a graphing calculator to solve the equations in Exercises 9, 27, and 30. Then solve $5.33x^2 = 8.23x + 3.24$.

Solve.

57. $2x^2 - x - \sqrt{5} = 0$

58. $\frac{5}{x} + \frac{x}{4} = \frac{11}{7}$

59. $ix^2 - x - 1 = 0$

60. $\sqrt{3}x^2 + 6x + \sqrt{3} = 0$

61. $\frac{x}{x + 1} = 4 + \frac{1}{3x^2 - 3}$

62. $(1 + \sqrt{3})x^2 - (3 + 2\sqrt{3})x + 3 = 0$

63. Let $f(x) = (x - 3)^2$. Find all inputs x such that $f(x) = 13$.

64. Let $f(x) = x^2 + 14x + 49$. Find all inputs x such that $f(x) = 36$.

16.3 APPLICATIONS INVOLVING QUADRATIC EQUATIONS

Objectives

a Solve applied problems involving quadratic equations.

b Solve a formula for a given letter.

1. Box Construction. An open box is to be made from a 10-ft by 20-ft rectangular piece of cardboard by cutting a square from each corner. The area of the bottom of the box is to be 96 ft^2. What is the length of the sides of the squares that are cut from the corners?

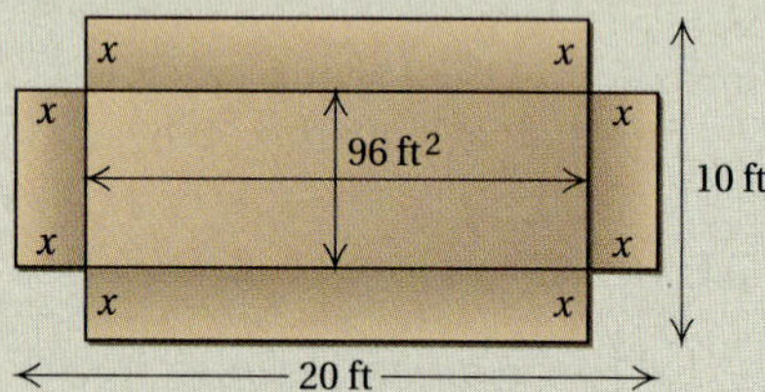

Answer on page A-60

a Applications and Problem Solving

Sometimes when we translate a problem to mathematical language, the result is a quadratic equation.

EXAMPLE 1 *Landscaping.* A rectangular garden is 60 ft by 80 ft. Part of the garden is torn up to install a sidewalk of uniform width around it. The area of the new garden is $\frac{1}{2}$ of the old area. How wide is the sidewalk?

1. **Familiarize.** We first make a drawing and label it with the known information. We don't know how wide the sidewalk is, so we have called its width x.

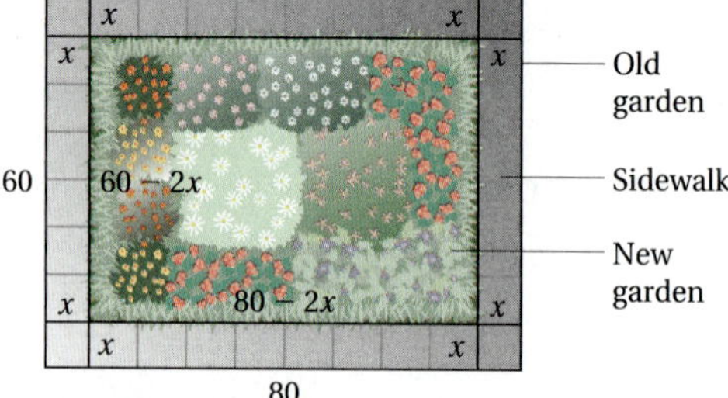

2. **Translate.** Remember, the area of a rectangle is lw (length times width). Then:

   Area of old garden $= 60 \cdot 80$;

   Area of new garden $= (60 - 2x)(80 - 2x)$.

   Since the area of the new garden is $\frac{1}{2}$ of the area of the old, we have

   $(60 - 2x)(80 - 2x) = \frac{1}{2} \cdot 60 \cdot 80.$

3. **Solve.** We solve the equation:

$$4800 - 120x - 160x + 4x^2 = 2400 \quad \text{Using FOIL on the left}$$
$$4x^2 - 280x + 2400 = 0 \quad \text{Collecting like terms}$$
$$x^2 - 70x + 600 = 0 \quad \text{Dividing by 4}$$
$$(x - 10)(x - 60) = 0 \quad \text{Factoring}$$
$$x = 10 \quad or \quad x = 60. \quad \text{Using the principle of zero products}$$

4. **Check.** We check in the original problem. We see that 60 is not a solution because when $x = 60$, $60 - 2x = -60$, and the width of the garden cannot be negative.

   If the sidewalk is 10 ft wide, then the garden itself will have length $80 - 2 \cdot 10$, or 60 ft. The width will be $60 - 2 \cdot 10$, or 40 ft. The new area is thus $60 \cdot 40$, or 2400 ft^2. The old area was $60 \cdot 80$, or 4800 ft^2. The new area of 2400 ft^2 is $\frac{1}{2}$ of 4800 ft^2, so the number 10 checks.

5. **State.** The sidewalk is 10 ft wide.

Do Exercise 1.

EXAMPLE 2 *Ladder Location.* A ladder leans against a building, as shown below. The ladder is 20 ft long. The distance to the top of the ladder is 4 ft greater than the distance d from the building. Find the distance d and the distance to the top of the ladder.

1. **Familiarize.** We first make a drawing and label it. We want to find d and $d + 4$.

2. **Translate.** As we look at the figure, we see that a right triangle is formed. We can use the Pythagorean equation, which we studied in Chapter 15: $c^2 = a^2 + b^2$. In this problem, we have

$$20^2 = d^2 + (d + 4)^2.$$

3. **Solve.** We solve the equation:

$$400 = d^2 + d^2 + 8d + 16 \quad \text{Squaring}$$
$$2d^2 + 8d - 384 = 0 \quad \text{Finding standard form}$$
$$d^2 + 4d - 192 = 0 \quad \text{Dividing by 2}$$
$$(d + 16)(d - 12) = 0 \quad \text{Factoring}$$
$$d + 16 = 0 \quad \text{or} \quad d - 12 = 0 \quad \text{Using the principle of zero products}$$
$$d = -16 \quad \text{or} \quad d = 12.$$

4. **Check.** We know that -16 is not an answer because distances are not negative. If $d = 12$, then $d + 4 = 16$, and

$$d^2 + (d + 4)^2 = 12^2 + 16^2 = 144 + 256 = 400.$$

Since $20^2 = 400$, the distance 12 ft checks.

5. **State.** The distance d is 12 ft, and the distance to the top of the ladder is 12 + 4, or 16 ft.

Do Exercise 2.

EXAMPLE 3 *Ladder Location.* Suppose that the ladder in Example 2 has length 10 ft. Find the distance d and the distance $d + 4$.

Using the same reasoning that we did in Example 2, we translate the problem to the equation

$$10^2 = d^2 + (d + 4)^2.$$

We solve as follows. Note that the quadratic equation we get is not easily factored, so we use the quadratic formula:

$$100 = d^2 + d^2 + 8d + 16 \quad \text{Squaring}$$
$$2d^2 + 8d - 84 = 0 \quad \text{Finding standard form}$$
$$d^2 + 4d - 42 = 0. \quad \text{Multiplying by } \tfrac{1}{2}\text{, or dividing by 2}$$

2. **Town Planning.** Three towns A, B, and C are situated as shown. The roads at A form a right angle. The distance from A to B is 2 mi less than the distance from A to C. The distance from B to C is 10 mi. Find the distance from A to B and the distance from A to C.

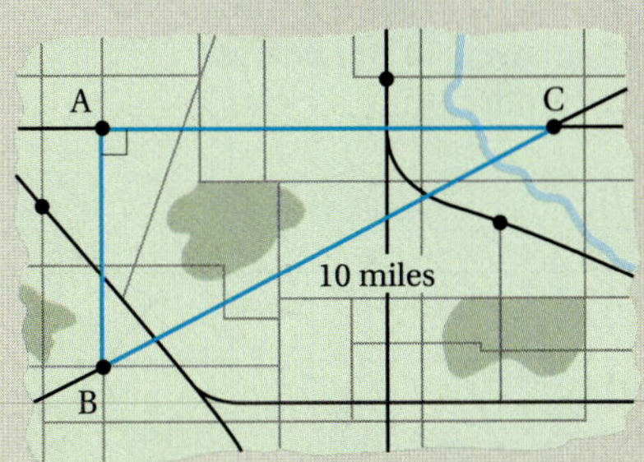

Answer on page A-60

3. Town Planning. Three towns A, B, and C are situated as shown. The roads at A form a right angle. The distance from A to B is 2 mi less than the distance from A to C. The distance from B to C is 8 mi. Find the distance from A to B and the distance from A to C. Find exact and approximate answers to the nearest hundredth of a mile.

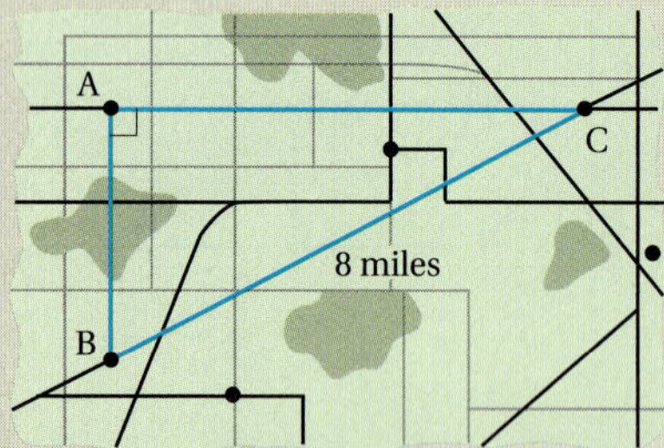

Answer on page A-60

Then

$$d = \frac{-b \pm \sqrt{b^2 - 4ac}}{2a}$$
$$= \frac{-4 \pm \sqrt{4^2 - 4(1)(-42)}}{2(1)}$$
$$= \frac{-4 \pm \sqrt{16 + 168}}{2} = \frac{-4 \pm \sqrt{184}}{2}$$
$$= \frac{-4 \pm \sqrt{4(46)}}{2} = \frac{-4 \pm 2\sqrt{46}}{2}$$
$$= -2 \pm \sqrt{46}.$$

Since $-2 - \sqrt{46} < 0$ and $\sqrt{46} > 2$, it follows that d is given by $d = -2 + \sqrt{46}$.Using a calculator, we find that $d \approx -2 + 6.782 \approx 4.782$ ft, and that $d + 4 \approx 4.782 + 4$, or 8.782 ft.

Do Exercise 3.

Some problems translate to rational equations. The solution of such rational equations can involve quadratic equations as well.

EXAMPLE 4 *Motorcycle Travel.* Karin's motorcycle traveled 300 mi at a certain speed. Had she gone 10 mph faster, she could have made the trip in 1 hr less time. Find her speed.

1. **Familiarize.** We first make a drawing, labeling it with known and unknown information. We can also organize the information, in a table, as we did in Section 11.8. We let $r =$ the speed, in miles per hour, and $t =$ the time, in hours.

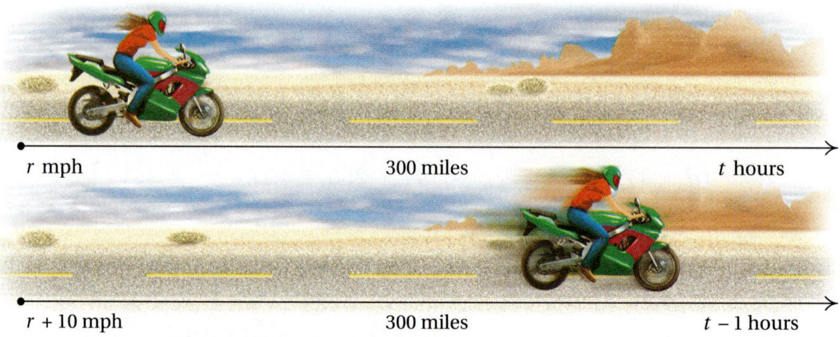

DISTANCE	SPEED	TIME	
300	r	t	$\rightarrow r = \frac{300}{t}$
300	$r + 10$	$t - 1$	$\rightarrow r + 10 = \frac{300}{t - 1}$

Recalling the motion formula $d = rt$ and solving for r, we get $r = d/t$. From the rows of the table, we obtain

$$r = \frac{300}{t} \quad \text{and} \quad r + 10 = \frac{300}{t - 1}.$$

2. **Translate.** We substitute for r from the first equation into the second and get a translation:

$$\frac{300}{t} + 10 = \frac{300}{t - 1}.$$

3. Solve. We solve as follows:

$$\frac{300}{t} + 10 = \frac{300}{t-1}$$

$$t(t-1)\left[\frac{300}{t} + 10\right] = t(t-1) \cdot \frac{300}{t-1} \quad \text{Multiplying by the LCM}$$

$$t(t-1) \cdot \frac{300}{t} + t(t-1) \cdot 10 = t(t-1) \cdot \frac{300}{t-1}$$

$$300(t-1) + 10(t^2 - t) = 300t$$

$$10t^2 - 10t - 300 = 0 \quad \text{Standard form}$$

$$t^2 - t - 30 = 0 \quad \text{Dividing by 10}$$

$$(t-6)(t+5) = 0 \quad \text{Factoring}$$

$$t = 6 \quad or \quad t = -5. \quad \text{Using the principle of zero products}$$

4. Check. Since negative time has no meaning in this problem, we try 6 hr. Remembering that $r = d/t$, we get $r = 300/6 = 50$ mph.

To check, we take the speed 10 mph faster, which is 60 mph, and see how long the trip would have taken at that speed:

$$t = \frac{d}{r} = \frac{300}{60} = 5 \text{ hr.}$$

This is 1 hr less than the trip actually took, so we have an answer.

5. State. Karin's speed was 50 mph.

Do Exercise 4.

b Solving Formulas

Recall that to solve a formula for a certain letter, we use the principles for solving equations to get that letter alone on one side.

EXAMPLE 5 *Period of a Pendulum.* The time T required for a pendulum of length L to swing back and forth (complete one period) is given by the formula $T = 2\pi\sqrt{L/g}$, where g is the gravitational constant. Solve for L.

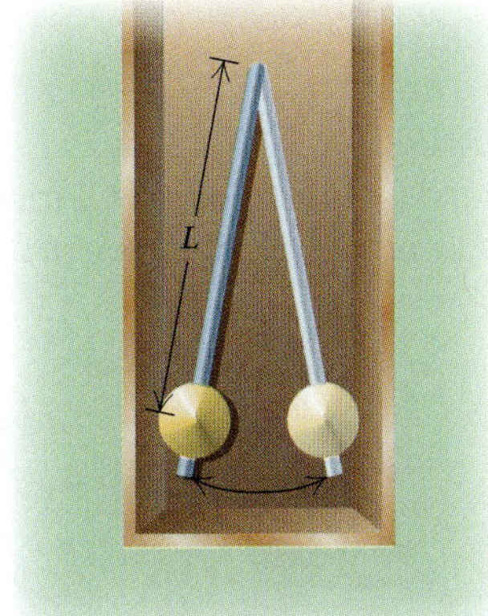

$$T = 2\pi\sqrt{\frac{L}{g}} \quad \text{This is a radical equation (see Section 15.6).}$$

$$T^2 = \left(2\pi\sqrt{\frac{L}{g}}\right)^2 \quad \text{Principle of powers (squaring)}$$

$$T^2 = 2^2\pi^2\frac{L}{g}$$

$$gT^2 = 4\pi^2L \quad \text{Clearing fractions}$$

$$\frac{gT^2}{4\pi^2} = L. \quad \text{Multiplying by } \frac{1}{4\pi^2}$$

We now have L alone on one side and L does not appear on the other side, so the formula is solved for L.

Do Exercise 5.

4. Marine Travel. Two ships make the same voyage of 3000 nautical miles. The faster ship travels 10 knots faster than the slower one (a *knot* is 1 nautical mile per hour). The faster ship makes the voyage in 50 hr less time than the slower one. Find the speeds of the two ships.

Complete this table to help with the familiarization.

	DISTANCE	SPEED	TIME
Faster Ship	3000		
Slower Ship	3000	r	t

5. Solve $A = \sqrt{\frac{w_1}{w_2}}$ for w_2.

Answers on page A-60

6. Solve $V = \pi r^2 h$ for r.

(Volume of a right circular cylinder)

7. Solve $s = gt + 16t^2$ for t.

Answers on page A-60

In most formulas, variables represent nonnegative numbers, so we do not need to use absolute-value signs when taking square roots.

EXAMPLE 6 *Hang Time.* An athlete's *hang time* is the amount of time that the athlete can remain airborne when jumping. A formula relating an athlete's vertical leap V, in inches, to hang time T, in seconds, is $V = 48T^2$. (See Example 12 in Section 16.1.) Solve for T.

We have

$$48T^2 = V$$

$$T^2 = \frac{V}{48} \qquad \text{Multiplying by } \tfrac{1}{48} \text{ to get } T^2 \text{ alone}$$

$$T = \sqrt{\frac{V}{48}} \qquad \text{Using the principle of square roots; note that } T \geq 0.$$

$$T = \sqrt{\frac{V}{2 \cdot 2 \cdot 2 \cdot 2 \cdot 3} \cdot \frac{3}{3}}$$

$$T = \frac{\sqrt{3V}}{2 \cdot 2 \cdot 3} = \frac{\sqrt{3V}}{12}.$$

Do Exercise 6.

EXAMPLE 7 *Falling Distance.* An object tossed downward with an initial speed (velocity) of v_0 will travel a distance of s meters, where $s = 4.9t^2 + v_0 t$ and t is measured in seconds. Solve for t.

Since t is squared in one term and raised to the first power in the other term, the equation is quadratic in t.

We have

$$4.9t^2 + v_0t = s$$

$$4.9t^2 + v_0t - s = 0 \quad \text{Writing standard form}$$

$$a = 4.9, \quad b = v_0, \quad c = -s$$

$$t = \frac{-v_0 \pm \sqrt{v_0^2 - 4(4.9)(-s)}}{2(4.9)} \quad \text{Using the quadratic formula: } t = \frac{-b \pm \sqrt{b^2 - 4ac}}{2a}$$

$$t = \frac{-v_0 \pm \sqrt{v_0^2 + 19.6s}}{9.8}.$$

Since the negative square root would yield a negative value for t, we use only the positive root:

$$t = \frac{-v_0 + \sqrt{v_0^2 + 19.6s}}{9.8}.$$

The steps listed in the margin should help you when solving formulas for a given letter. Try to remember that when solving a formula, you do the same things you would do to solve any equation.

Do Exercise 7 on the preceding page.

EXAMPLE 8 Solve $t = \frac{a}{\sqrt{a^2 + b^2}}$ for a.

In this case, we could either clear the fractions first or use the principle of powers first. Let's clear the fractions. Multiplying by $\sqrt{a^2 + b^2}$, we have

$$t\sqrt{a^2 + b^2} = a.$$

Now we square both sides and then continue:

$$\left(t\sqrt{a^2 + b^2}\right)^2 = a^2 \quad \text{Squaring}$$

Caution! Don't forget to square both t and $\sqrt{a^2 + b^2}$.

$$t^2(a^2 + b^2) = a^2$$

$$t^2a^2 + t^2b^2 = a^2$$

$$t^2b^2 = a^2 - t^2a^2 \quad \text{Getting all } a^2\text{-terms together}$$

$$t^2b^2 = a^2(1 - t^2) \quad \text{Factoring out } a^2$$

$$\frac{t^2b^2}{1 - t^2} = a^2 \quad \text{Dividing by } 1 - t^2$$

$$\sqrt{\frac{t^2b^2}{1 - t^2}} = a \quad \text{Taking the square root}$$

$$\frac{tb}{\sqrt{1 - t^2}} = a. \quad \text{Simplifying}$$

You need not rationalize denominators in situations such as this.

Do Exercise 8.

To solve a formula for a letter, say, b:

1. Clear the fractions and use the principle of powers, as needed, until b does not appear in any radicand or denominator. (In some cases, you may clear the fractions first, and in some cases you may use the principle of powers first.)
2. Collect all terms with b^2 in them. Also collect all terms with b in them.
3. If b^2 does not appear, you can finish by using just the addition and multiplication principles.
4. If b^2 appears but b does not, solve the equation for b^2. Then take square roots on both sides.
5. If there are terms containing both b and b^2, write the equation in standard form and use the quadratic formula.

8. Solve $\frac{b}{\sqrt{a^2 - b^2}} = t$ for b.

Answer on page A-60

16.3 EXERCISE SET

For Extra Help

Digital Video Tutor CD 13 Videotape 18

InterAct Math

Math Tutor Center

MathXL

MyMathLab

a Solve.

1. *Flower Bed.* The width of a rectangular flower bed is 7 ft less than the length. The area is 18 ft^2. Find the length and the width.

2. *Feed Lot.* The width of a rectangular feed lot is 8 m less than the length. The area is 20 m^2. Find the length and the width.

3. *Parking Lot.* The length of a rectangular parking lot is twice the width. The area is 162 yd^2. Find the length and the width.

4. *Computer Part.* The length of a rectangular computer part is twice the width. The area is 242 cm^2. Find the length and the width.

5. *Sailing.* The base of a triangular sail is 9 m less than its height. The area is 56 m^2. Find the base and the height of the sail.

6. *Parking Lot.* The width of a rectangular parking lot is 50 ft less than its length. Determine the dimensions of the parking lot if it measures 250 ft diagonally.

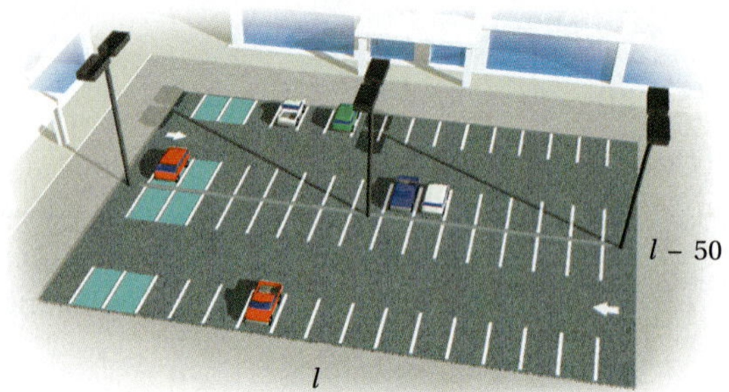

7. *Sailing.* The base of a triangular sail is 8 ft less than its height. The area is 56 ft^2. Find the base and the height of the sail.

8. *Parking Lot.* The width of a rectangular parking lot is 51 ft less than its length. Determine the dimensions of the parking lot if it measures 250 ft diagonally.

9. *Picture Framing.* The outside of a picture frame measures 12 cm by 20 cm; 84 cm^2 of picture shows. Find the width of the frame.

10. *Picture Framing.* The outside of a picture frame measures 14 in. by 20 in.; 160 in^2 of picture shows. Find the width of the frame.

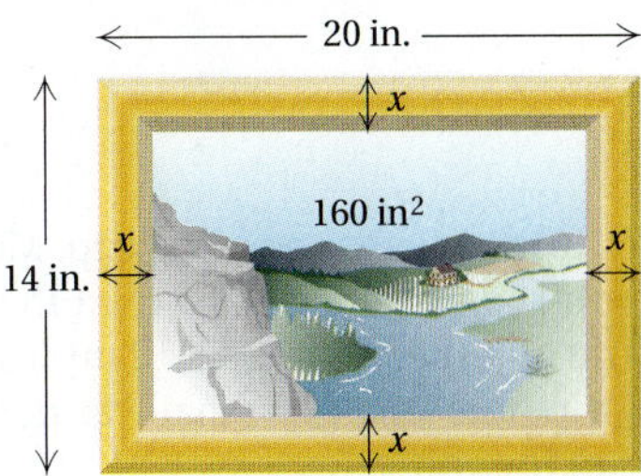

11. *Landscaping.* A landscaper is designing a flower garden in the shape of a right triangle. She wants 10 ft of a perennial border to form the hypotenuse of the triangle, and one leg is to be 2 ft longer than the other. Find the lengths of the legs.

12. The hypotenuse of a right triangle is 25 m long. The length of one leg is 17 m less than the other. Find the lengths of the legs.

13. *Page Numbers.* A student opens a literature book to two facing pages. The product of the page numbers is 812. Find the page numbers.

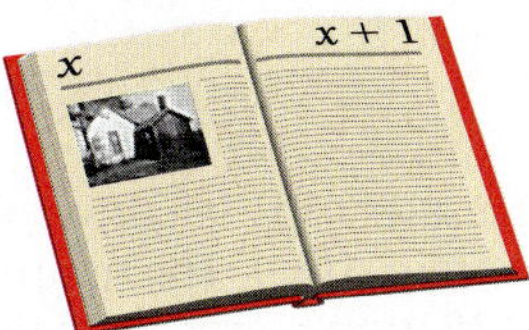

14. *Page Numbers.* A student opens a mathematics book to two facing pages. The product of the page numbers is 1980. Find the page numbers.

Solve. Find exact and approximate answers rounded to three decimal places.

15. The width of a rectangle is 4 ft less than the length. The area is 10 ft^2. Find the length and the width.

16. The length of a rectangle is twice the width. The area is 328 cm^2. Find the length and the width.

17. *Page Dimensions.* The outside of an oversized book page measures 14 in. by 20 in.; 100 in^2 of printed text shows. Find the width of the margin.

18. *Picture Framing.* The outside of a picture frame measures 13 cm by 20 cm; 80 cm^2 of picture shows. Find the width of the frame.

19. The hypotenuse of a right triangle is 24 ft long. The length of one leg is 14 ft more than the other. Find the lengths of the legs.

20. The hypotenuse of a right triangle is 22 m long. The length of one leg is 10 m less than the other. Find the lengths of the legs.

21. *Car Trips.* During the first part of a trip, Meira's Honda traveled 120 mi at a certain speed. Meira then drove another 100 mi at a speed that was 10 mph slower. If Meira's total trip time was 4 hr, what was her speed on each part of the trip?

DISTANCE	SPEED	TIME

22. *Canoeing.* During the first part of a canoe trip, Tim covered 60 km at a certain speed. He then traveled 24 km at a speed that was 4 km/h slower. If the total time for the trip was 8 hr, what was the speed on each part of the trip?

DISTANCE	SPEED	TIME

23. *Car Trips.* Petra's Plymouth travels 200 mi at a certain speed. If the car had gone 10 mph faster, the trip would have taken 1 hr less. Find Petra's speed.

24. *Car Trips.* Sandi's Subaru travels 280 mi at a certain speed. If the car had gone 5 mph faster, the trip would have taken 1 hr less. Find Sandi's speed.

25. *Air Travel.* A Cessna flies 600 mi at a certain speed. A Beechcraft flies 1000 mi at a speed that is 50 mph faster, but takes 1 hr longer. Find the speed of each plane.

26. *Air Travel.* A turbo-jet flies 50 mph faster than a super-prop plane. If a turbo-jet goes 2000 mi in 3 hr less time than it takes the super-prop to go 2800 mi, find the speed of each plane.

27. *Bicycling.* Naoki bikes the 40 mi to Hillsboro at a certain speed. The return trip is made at a speed that is 6 mph slower. Total time for the round trip is 14 hr. Find Naoki's speed on each part of the trip.

28. *Car Speed.* On a sales trip, Gail drives the 600 mi to Richmond at a certain speed. The return trip is made at a speed that is 10 mph slower. Total time for the round trip was 22 hr. How fast did Gail travel on each part of the trip?

29. *Navigation.* The current in a typical Mississippi River shipping route flows at a rate of 4 mph. In order for a barge to travel 24 mi upriver and then return in a total of 5 hr, approximately how fast must the barge be able to travel in still water?

30. *Navigation.* The Hudson River flows at a rate of 3 mph. A patrol boat travels 60 mi upriver and returns in a total time of 9 hr. What is the speed of the boat in still water?

b Solve the formula for the given letter. Assume that all variables represent nonnegative numbers.

31. $A = 6s^2$, for s
(Surface area of a cube)

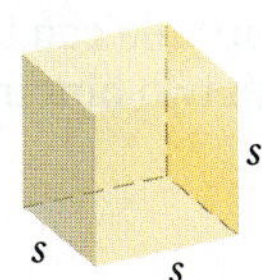

32. $A = 4\pi r^2$, for r
(Surface area of a sphere)

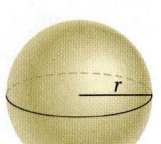

33. $F = \dfrac{Gm_1m_2}{r^2}$, for r

34. $N = \dfrac{kQ_1Q_2}{s^2}$, for s
(Number of phone calls between two cities)

35. $E = mc^2$, for c
(Einstein's energy–mass relationship)

36. $V = \frac{1}{3}s^2h$, for s
(Volume of a pyramid)

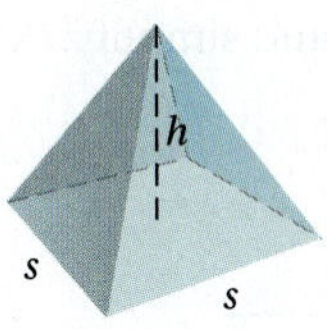

37. $a^2 + b^2 = c^2$, for b
(Pythagorean formula in two dimensions)

38. $a^2 + b^2 + c^2 = d^2$, for c
(Pythagorean formula in three dimensions)

39. $N = \dfrac{k^2 - 3k}{2}$, for k
(Number of diagonals of a polygon of k sides)

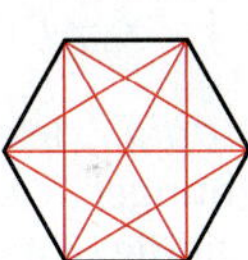

40. $s = v_0t + \dfrac{gt^2}{2}$, for t
(A motion formula)

41. $A = 2\pi r^2 + 2\pi rh$, for r
(Surface area of a cylinder)

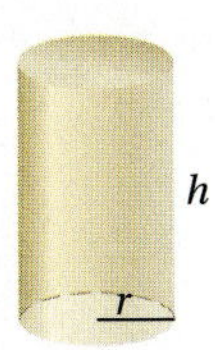

42. $A = \pi r^2 + \pi rs$, for r
(Surface area of a cone)

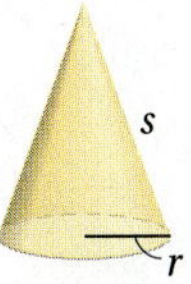

43. $T = 2\pi\sqrt{\dfrac{L}{g}}$, for g
(A pendulum formula)

44. $W = \sqrt{\dfrac{1}{LC}}$, for L
(An electricity formula)

45. $I = \dfrac{704.5W}{H^2}$, for H
(Body mass index; see Example 1 of Section 1.2)

46. $N + p = \frac{6.2A^2}{pR^2}$, for R

47. $m = \frac{m_0}{\sqrt{1 - \frac{v^2}{c^2}}}$, for v

(A relativity formula)

48. Solve the formula given in Exercise 47 for c.

49. $^{\text{D}}\text{W}$ Explain how Exercises 1–30 can be solved using a calculator but without using factoring, completing the square, or the quadratic formula.

50. $^{\text{D}}\text{W}$ Explain how the quadratic formula can be used to factor a quadratic polynomial into two binomials. Use it to factor $5x^2 + 8x - 3$.

SKILL MAINTENANCE

Add or subtract.

51. $\frac{1}{x - 1} + \frac{1}{x^2 - 3x + 2}$ [11.4a]

52. $\frac{x + 1}{x - 1} - \frac{x + 1}{x^2 + x + 1}$ [11.5a]

53. $\frac{2}{x + 3} - \frac{x}{x - 1} + \frac{x^2 + 2}{x^2 + 2x - 3}$ [11.5b]

54. Multiply and simplify: $\sqrt{3x^2}\,\sqrt{3x^3}$. [15.3a]

55. Express in terms of i: $\sqrt{-20}$. [15.8a]

Simplify. [11.6a]

56. $\dfrac{\frac{3}{x - 1}}{\frac{1}{x + 1} + \frac{2}{x - 1}}$

57. $\dfrac{\frac{4}{a^2b}}{\frac{3}{a} - \frac{4}{b^2}}$

SYNTHESIS

58. Solve: $\frac{4}{2x + i} - \frac{1}{x - i} = \frac{2}{x + i}$.

59. Find a when the reciprocal of $a - 1$ is $a + 1$.

60. *Bungee Jumping.* Jesse is tied to one end of a 40-m elasticized (bungee) cord. The other end of the cord is tied to the middle of a train trestle. If Jesse jumps off the bridge, for how long will he fall before the cord begins to stretch? (See Example 7 and let $v_0 = 0$.)

61. *Surface Area.* A sphere is inscribed in a cube as shown in the figure below. Express the surface area of the sphere as a function of the surface area S of the cube.

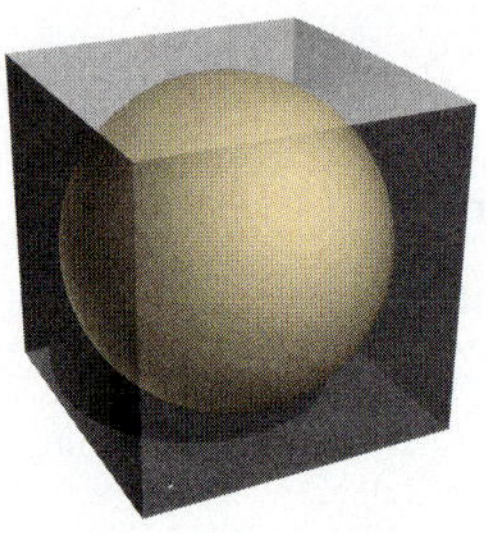

62. *Pizza Crusts.* At Pizza Perfect, Ron can make 100 large pizza crusts in 1.2 hr less than Chad. Together they can do the job in 1.8 hr. How long does it take each to do the job alone?

63. *The Golden Rectangle.* For over 2000 yr, the proportions of a "golden" rectangle have been considered visually appealing. A rectangle of width w and length l is considered "golden" if

$$\frac{w}{l} = \frac{l}{w + l}.$$

Solve for l.

16.4 MORE ON QUADRATIC EQUATIONS

Objectives

a Determine the nature of the solutions of a quadratic equation.

b Write a quadratic equation having two numbers specified as solutions.

c Solve equations that are quadratic in form.

a The Discriminant

From the quadratic formula, we know that the solutions x_1 and x_2 of a quadratic equation are given by

$$x_1 = \frac{-b + \sqrt{b^2 - 4ac}}{2a} \quad \text{and} \quad x_2 = \frac{-b - \sqrt{b^2 - 4ac}}{2a}.$$

The expression $b^2 - 4ac$ is called the **discriminant.** When using the quadratic formula, it is helpful to compute the discriminant first. If it is 0, there will be just one real solution. If it is positive, there will be two real solutions. If it is negative, we will be taking the square root of a negative number; hence there will be two nonreal complex-number solutions, and they will be complex conjugates.

DISCRIMINANT $b^2 - 4ac$	NATURE OF SOLUTIONS	x-INTERCEPTS
0	Only one solution; it is a real number	Only one
Positive	Two different real-number solutions	Two different
Negative	Two different nonreal complex-number solutions (complex conjugates)	None

If the discriminant is a perfect square, we can solve the equation by factoring, not needing the quadratic formula.

EXAMPLE 1 Determine the nature of the solutions of $9x^2 - 12x + 4 = 0$.

We have

$$a = 9, \quad b = -12, \quad c = 4.$$

We compute the discriminant:

$$\begin{aligned} b^2 - 4ac &= (-12)^2 - 4 \cdot 9 \cdot 4 \\ &= 144 - 144 \\ &= 0. \end{aligned}$$

There is just one solution, and it is a real number. Since 0 is a perfect square, the equation can be solved by factoring.

EXAMPLE 2 Determine the nature of the solutions of $x^2 + 5x + 8 = 0$.

We have

$$a = 1, \quad b = 5, \quad c = 8.$$

We compute the discriminant:

$$\begin{aligned} b^2 - 4ac &= 5^2 - 4 \cdot 1 \cdot 8 \\ &= 25 - 32 \\ &= -7. \end{aligned}$$

Since the discriminant is negative, there are two nonreal complex-number solutions. The equation cannot be solved by factoring because -7 is not a perfect square.

Study Tips

TEST PREPARATION: SYSTEMATIC VS. INTENSE STUDY

Let's consider two ways to prepare for an exam scheduled for November 15.

1. *Systematic Study*: You begin studying on November 1 and continue every day until the day of the test.
2. *Intense Study*: You wait until November 14 to begin studying and "cram" all day and through the night.

Which of these methods would produce a better result on the exam? Research has shown that best results come from a combination of these two methods: Study systematically well ahead of the test, but do intense study the day before. (This works so long as you don't stay up all night.) Try this on your next few exams.

Determine the nature of the solutions without solving.

1. $x^2 + 5x - 3 = 0$

2. $9x^2 - 6x + 1 = 0$

3. $3x^2 - 2x + 1 = 0$

Answers on page A-61

EXAMPLE 3 Determine the nature of the solutions of $x^2 + 5x + 6 = 0$.

We have

$$a = 1, \quad b = 5, \quad c = 6;$$
$$b^2 - 4ac = 5^2 - 4 \cdot 1 \cdot 6 = 1.$$

Since the discriminant is positive, there are two solutions, and they are real numbers. The equation can be solved by factoring since the discriminant is a perfect square.

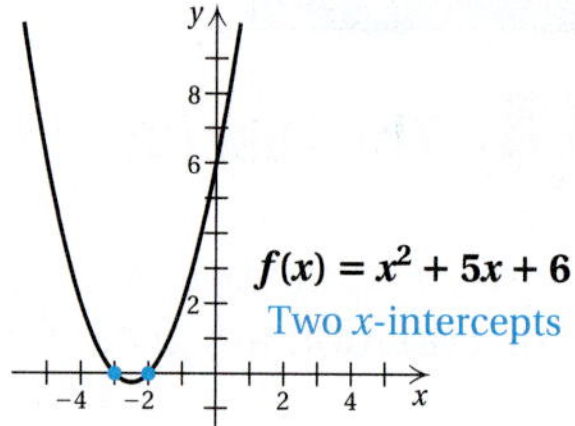

The discriminant, $b^2 - 4ac$, tells us how many real-number solutions the equation $ax^2 + bx + c = 0$ has, so it also indicates how many x-intercepts the graph of $f(x) = ax^2 + bx + c$ has. Compare the following.

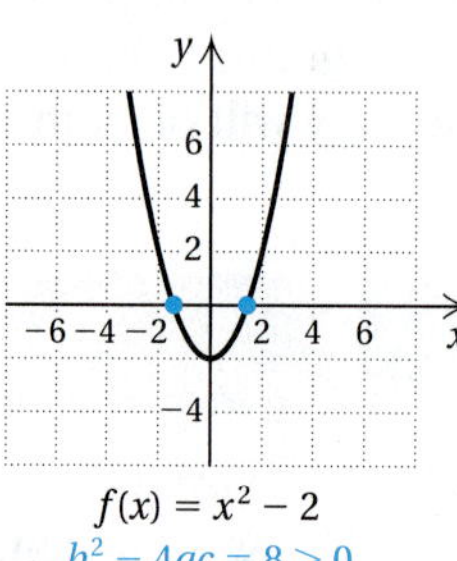

$f(x) = x^2 - 2$
$b^2 - 4ac = 8 > 0$
Two real solutions
Two x-intercepts

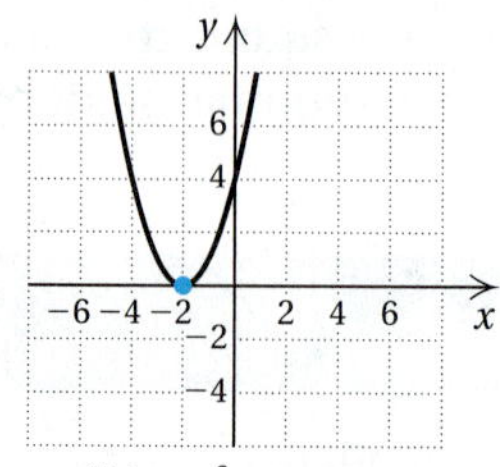

$f(x) = x^2 + 4x + 4$
$b^2 - 4ac = 0$
One real solution
One x-intercept

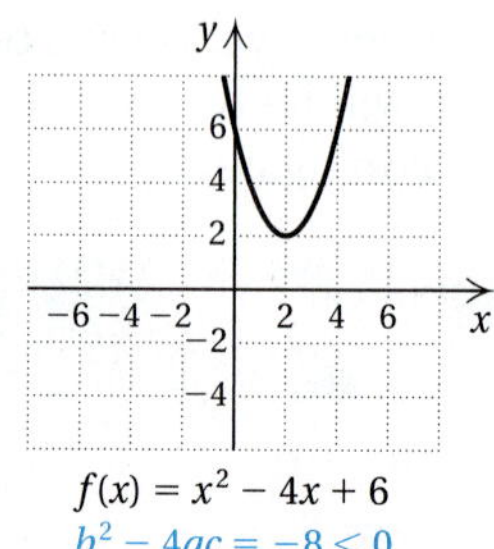

$f(x) = x^2 - 4x + 6$
$b^2 - 4ac = -8 < 0$
No real solutions
No x-intercepts

Do Exercises 1–3.

b Writing Equations from Solutions

We know by the principle of zero products that $(x - 2)(x + 3) = 0$ has solutions 2 and -3. If we know the solutions of an equation, we can write the equation, using this principle in reverse.

EXAMPLE 4 Find a quadratic equation whose solutions are 3 and $-\frac{2}{5}$.

We have

$$x = 3 \quad or \quad x = -\tfrac{2}{5}$$
$$x - 3 = 0 \quad or \quad x + \tfrac{2}{5} = 0 \qquad \text{Getting the 0's on one side}$$
$$x - 3 = 0 \quad or \quad 5x + 2 = 0 \qquad \text{Clearing the fraction}$$
$$(x - 3)(5x + 2) = 0 \qquad \text{Using the principle of zero products in reverse}$$
$$5x^2 - 13x - 6 = 0. \qquad \text{Using FOIL}$$

EXAMPLE 5 Write a quadratic equation whose solutions are $2i$ and $-2i$.

We have

$$x = 2i \quad or \quad x = -2i$$
$$x - 2i = 0 \quad or \quad x + 2i = 0 \qquad \text{Getting the 0's on one side}$$
$$(x - 2i)(x + 2i) = 0 \qquad \text{Using the principle of zero products in reverse}$$
$$x^2 - (2i)^2 = 0 \qquad \text{Using } (A - B)(A + B) = A^2 - B^2$$
$$x^2 - 4i^2 = 0$$
$$x^2 - 4(-1) = 0$$
$$x^2 + 4 = 0.$$

EXAMPLE 6 Write a quadratic equation whose solutions are $\sqrt{3}$ and $-2\sqrt{3}$.

We have

$$x = \sqrt{3} \quad or \quad x = -2\sqrt{3}$$
$$x - \sqrt{3} = 0 \quad or \quad x + 2\sqrt{3} = 0 \qquad \text{Getting the 0's on one side}$$
$$(x - \sqrt{3})(x + 2\sqrt{3}) = 0 \qquad \text{Using the principle of zero products}$$
$$x^2 + 2\sqrt{3}x - \sqrt{3}x - 2(\sqrt{3})^2 = 0 \qquad \text{Using FOIL}$$
$$x^2 + \sqrt{3}x - 6 = 0. \qquad \text{Collecting like terms}$$

Do Exercises 4–7.

Find a quadratic equation having the following solutions.

4. 7 and -2

5. -4 and $\frac{5}{3}$

6. $5i$ and $-5i$

7. $-2\sqrt{2}$ and $\sqrt{2}$

C Equations Quadratic in Form

Certain equations that are not really quadratic can still be solved as quadratic. Consider this fourth-degree equation.

$$x^4 - 9x^2 + 8 = 0$$
$$\downarrow$$
$$(x^2)^2 - 9(x^2) + 8 = 0 \qquad \text{Thinking of } x^4 \text{ as } (x^2)^2$$
$$\downarrow$$
$$u^2 - 9u + 8 = 0 \qquad \text{To make this clearer, write } u \text{ instead of } x^2.$$

The equation $u^2 - 9u + 8 = 0$ can be solved by factoring or by the quadratic formula. After that, we can find x by remembering that $x^2 = u$. Equations that can be solved like this are said to be **quadratic in form,** or **reducible to quadratic.**

EXAMPLE 7 Solve: $x^4 - 9x^2 + 8 = 0$.

Let $u = x^2$. Then we solve the equation found by substituting u for x^2:

$$u^2 - 9u + 8 = 0$$
$$(u - 8)(u - 1) = 0 \qquad \text{Factoring}$$
$$u - 8 = 0 \quad or \quad u - 1 = 0 \qquad \text{Using the principle of zero products}$$
$$u = 8 \quad or \quad u = 1.$$

Answers on page A-61

8. Solve: $x^4 - 10x^2 + 9 = 0$.

9. Solve $x + 3\sqrt{x} - 10 = 0$. Be sure to check.

Answers on page A-61

Next, we substitute x^2 for u and solve these equations:

$$x^2 = 8 \quad or \quad x^2 = 1$$
$$x = \pm\sqrt{8} \quad or \quad x = \pm 1$$
$$x = \pm 2\sqrt{2} \quad or \quad x = \pm 1.$$

To check, first note that when $x = 2\sqrt{2}$, $x^2 = 8$ and $x^4 = 64$. Also, when $x = -2\sqrt{2}$, $x^2 = 8$ and $x^4 = 64$. Similarly, when $x = 1$, $x^2 = 1$ and $x^4 = 1$, and when $x = -1$, $x^2 = 1$ and $x^4 = 1$. Thus, instead of making four checks, we need make only two.

CHECK:

For $\pm 2\sqrt{2}$:

$x^4 - 9x^2 + 8 = 0$		
$(\pm 2\sqrt{2})^4 - 9(\pm 2\sqrt{2})^2 + 8$	? 0	
$64 - 9 \cdot 8 + 8$		
0		TRUE

For ± 1:

$x^4 - 9x^2 + 8 = 0$		
$(\pm 1)^4 - 9(\pm 1)^2 + 8$	? 0	
$1 - 9 + 8$		
0		TRUE

The solutions are 1, -1, $2\sqrt{2}$, and $-2\sqrt{2}$.

CAUTION!

A common error is to solve for u and then forget to solve for x. Remember that you *must* find values for the *original* variable!

Do Exercise 8.

Solving equations quadratic in form can sometimes introduce numbers that are not solutions of the original equation. Thus a check by substitution in the original equation is necessary.

EXAMPLE 8 Solve: $x - 3\sqrt{x} - 4 = 0$.

Let $u = \sqrt{x}$. Then we solve the equation found by substituting u for $\sqrt{x}$ and u^2 for x.

$$u^2 - 3u - 4 = 0$$
$$(u - 4)(u + 1) = 0$$
$$u = 4 \quad or \quad u = -1$$

Next, we substitute $\sqrt{x}$ for u and solve these equations:

$$\sqrt{x} = 4 \quad or \quad \sqrt{x} = -1.$$

Squaring the first equation, we get $x = 16$. Squaring the second equation, we get $x = 1$. We check both solutions.

CHECK:

For 16:

$x - 3\sqrt{x} - 4 = 0$		
$16 - 3\sqrt{16} - 4$	? 0	
$16 - 3 \cdot 4 - 4$		
$16 - 12 - 4$		
0		TRUE

For 1:

$x - 3\sqrt{x} - 4 = 0$		
$1 - 3\sqrt{1} - 4$	? 0	
$1 - 3 \cdot 1 - 4$		
-6		FALSE

The solution is 16.

Do Exercise 9.

EXAMPLE 9 Solve: $y^{-2} - y^{-1} - 2 = 0$.

Let $u = y^{-1}$. Then we solve the equation found by substituting u for y^{-1} and u^2 for y^{-2}:

$$u^2 - u - 2 = 0$$
$$(u - 2)(u + 1) = 0$$
$$u = 2 \quad \text{or} \quad u = -1.$$

Next, we substitute y^{-1} or $1/y$ for u and solve these equations:

$$\frac{1}{y} = 2 \quad \text{or} \quad \frac{1}{y} = -1.$$

Solving, we get

$$y = \frac{1}{2} \quad \text{or} \quad y = \frac{1}{(-1)} = -1.$$

The numbers $\frac{1}{2}$ and -1 both check. They are the solutions.

Do Exercise 10.

EXAMPLE 10 Find the x-intercepts of the graph of $f(x) = (x^2 - 1)^2 - (x^2 - 1) - 2$.

The x-intercepts occur where $f(x) = 0$ so we must have

$$(x^2 - 1)^2 - (x^2 - 1) - 2 = 0.$$

Let $u = x^2 - 1$. Then we solve the equation found by substituting u for $x^2 - 1$:

$$u^2 - u - 2 = 0$$
$$(u - 2)(u + 1) = 0$$
$$u = 2 \quad \text{or} \quad u = -1.$$

Next, we substitute $x^2 - 1$ for u and solve these equations:

$$x^2 - 1 = 2 \quad \text{or} \quad x^2 - 1 = -1$$
$$x^2 = 3 \quad \text{or} \quad x^2 = 0$$
$$x = \pm\sqrt{3} \quad \text{or} \quad x = 0.$$

The numbers $\sqrt{3}$, $-\sqrt{3}$, and 0 check. They are the solutions of $(x^2 - 1)^2 - (x^2 - 1) - 2 = 0$. Thus the x-intercepts of the graph of $f(x)$ are $(-\sqrt{3}, 0)$, $(0, 0)$, and $(\sqrt{3}, 0)$.

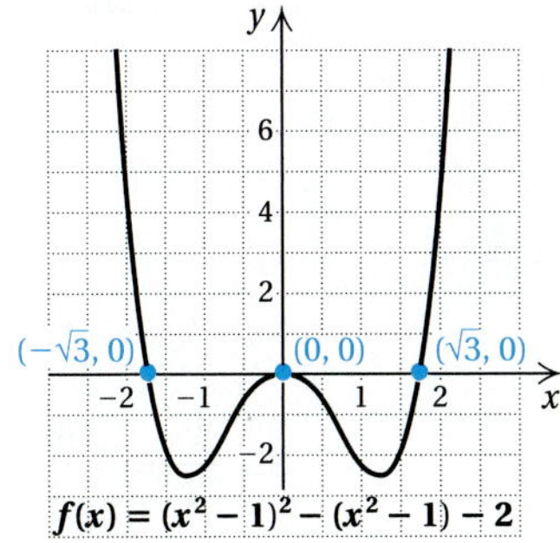

Do Exercise 11.

10. Solve: $x^{-2} + x^{-1} - 6 = 0$.

11. Find the x-intercepts of

$f(x) = (x^2 - x)^2 - 14(x^2 - x) + 24$.

Answers on page A-61

16.4 EXERCISE SET

For Extra Help

Digital Video Tutor CD 14 Videotape 18 | InterAct Math | Math Tutor Center | MathXL | MyMathLab

a Determine the nature of the solutions of the equation.

1. $x^2 - 8x + 16 = 0$

2. $x^2 + 12x + 36 = 0$

3. $x^2 + 1 = 0$

4. $x^2 + 6 = 0$

5. $x^2 - 6 = 0$

6. $x^2 - 3 = 0$

7. $4x^2 - 12x + 9 = 0$

8. $4x^2 + 8x - 5 = 0$

9. $x^2 - 2x + 4 = 0$

10. $x^2 + 3x + 4 = 0$

11. $9t^2 - 3t = 0$

12. $4m^2 + 7m = 0$

13. $y^2 = \frac{1}{2}y + \frac{3}{5}$

14. $y^2 + \frac{9}{4} = 4y$

15. $4x^2 - 4\sqrt{3}x + 3 = 0$

16. $6y^2 - 2\sqrt{3}y - 1 = 0$

b Write a quadratic equation having the given numbers as solutions.

17. -4 and 4

18. -11 and 9

19. -2 and -7

20. 3 and 10

21. 8, only solution [*Hint*: It must be a double solution, that is, $(x - 8)(x - 8) = 0$.]

22. -3, only solution

23. $-\frac{2}{5}$ and $\frac{6}{5}$

24. $-\frac{1}{4}$ and $-\frac{1}{2}$

25. $\frac{k}{3}$ and $\frac{m}{4}$

26. $\frac{c}{2}$ and $\frac{d}{2}$

27. $-\sqrt{3}$ and $2\sqrt{3}$

28. $\sqrt{2}$ and $3\sqrt{2}$

C Solve.

29. $x^4 - 6x^2 + 9 = 0$

30. $x^4 - 7x^2 + 12 = 0$

31. $x - 10\sqrt{x} + 9 = 0$

32. $2x - 9\sqrt{x} + 4 = 0$

33. $(x^2 - 6x)^2 - 2(x^2 - 6x) - 35 = 0$

34. $(x^2 + 5x)^2 + 2(x^2 + 5x) - 24 = 0$

35. $x^{-2} - 5x^{-1} - 36 = 0$

36. $3x^{-2} - x^{-1} - 14 = 0$

37. $(1 + \sqrt{x})^2 + (1 + \sqrt{x}) - 6 = 0$

38. $(2 + \sqrt{x})^2 - 3(2 + \sqrt{x}) - 10 = 0$

39. $(y^2 - 5y)^2 - 2(y^2 - 5y) - 24 = 0$

40. $(2t^2 + t)^2 - 4(2t^2 + t) + 3 = 0$

41. $t^4 - 6t^2 - 4 = 0$

42. $w^4 - 7w^2 + 7 = 0$

43. $2x^{-2} + x^{-1} - 1 = 0$

44. $m^{-2} + 9m^{-1} - 10 = 0$

45. $6x^4 - 19x^2 + 15 = 0$

46. $6x^4 - 17x^2 + 5 = 0$

47. $x^{2/3} - 4x^{1/3} - 5 = 0$

48. $x^{2/3} + 2x^{1/3} - 8 = 0$

49. $\left(\frac{x-4}{x+1}\right)^2 - 2\left(\frac{x-4}{x+1}\right) - 35 = 0$

50. $\left(\frac{x+3}{x-3}\right)^2 - \left(\frac{x+3}{x-3}\right) - 6 = 0$

51. $9\left(\frac{x+2}{x+3}\right)^2 - 6\left(\frac{x+2}{x+3}\right) + 1 = 0$

52. $16\left(\frac{x-1}{x-8}\right)^2 + 8\left(\frac{x-1}{x-8}\right) + 1 = 0$

53. $\left(\frac{x^2-2}{x}\right)^2 - 7\left(\frac{x^2-2}{x}\right) - 18 = 0$

54. $\left(\frac{y^2-1}{y}\right)^2 - 4\left(\frac{y^2-1}{y}\right) - 12 = 0$

Find the x-intercepts of the function.

55. $f(x) = 5x + 13\sqrt{x} - 6$

56. $f(x) = 3x + 10\sqrt{x} - 8$

57. $f(x) = (x^2 - 3x)^2 - 10(x^2 - 3x) + 24$

58. $f(x) = x^{2/5} + x^{1/5} - 6$

59. DW Describe a procedure that could be used to write an equation having the first seven natural numbers as solutions.

60. DW Describe a procedure that could be used to write an equation that is quadratic in $3x^2 + 1$ and has real-number solutions.

SKILL MAINTENANCE

Solve. [13.4a]

61. *Coffee Beans.* Twin Cities Roasters has Kenyan coffee worth \$6.75 per pound and Peruvian coffee worth \$11.25 per pound. How many pounds of each kind should be mixed in order to obtain a 50-lb mixture that is worth \$8.55 per pound?

62. *Solution Mixtures.* Solution A is 18% alcohol and solution B is 45% alcohol. How many liters of each should be mixed in order to get 12 L of a solution that is 36% alcohol?

Multiply and simplify. Assume that all expressions under radicals represent nonnegative numbers. [15.3a]

63. $\sqrt{8x}\,\sqrt{2x}$

64. $\sqrt[3]{x^2}\,\sqrt[3]{27x^4}$

65. $\sqrt[4]{9a^2}\,\sqrt[4]{18a^3}$

66. $\sqrt[5]{16}\,\sqrt[5]{64}$

Graph. [12.1c], [12.4a, c]

67. $f(x) = -\frac{3}{5}x + 4$

68. $5x - 2y = 8$

69. $y = 4$

70. $f(x) = -x - 3$

SYNTHESIS

71. Use a graphing calculator to check your answers to Exercises 30, 32, 34, and 37.

72. Use a graphing calculator to solve each of the following equations.

a) $6.75x - 35\sqrt{x} - 5.26 = 0$
b) $\pi x^4 - \pi^2 x^2 = \sqrt{99.3}$
c) $x^4 - x^3 - 13x^2 + x + 12 = 0$

For each equation under the given condition, **(a)** find k and **(b)** find the other solution.

73. $kx^2 - 2x + k = 0$; one solution is -3.

74. $kx^2 - 17x + 33 = 0$; one solution is 3.

75. Find a quadratic equation for which the sum of the solutions is $\sqrt{3}$ and the product is 8.

76. Find k given that $kx^2 - 4x + (2k - 1) = 0$ and the product of the solutions is 3.

77. The graph of a function of the form

$$f(x) = ax^2 + bx + c$$

is a curve similar to the one shown below. Determine a, b, and c from the information given.

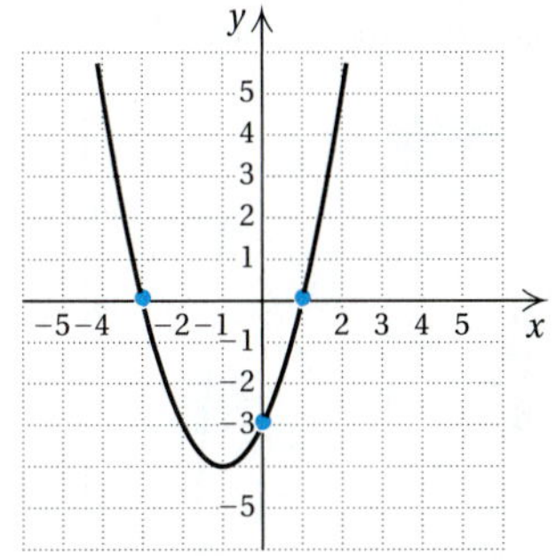

78. While solving a quadratic equation of the form $ax^2 + bx + c = 0$ with a graphing calculator, Shawn-Marie gets the following screen.

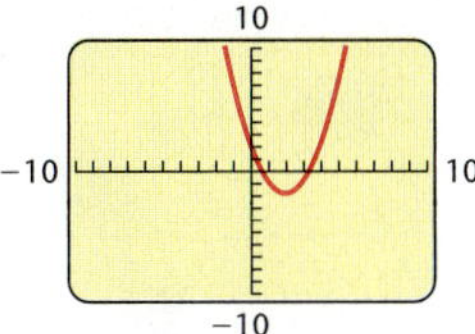

How could the discriminant help her check the graph?

Solve.

79. $\dfrac{x}{x-1} - 6\sqrt{\dfrac{x}{x-1}} - 40 = 0$

80. $\dfrac{x}{x-3} - 24 = 10\sqrt{\dfrac{x}{x-3}}$

81. $\sqrt{x-3} - \sqrt[4]{x-3} = 12$

82. $a^3 - 26a^{3/2} - 27 = 0$

83. $x^6 - 28x^3 + 27 = 0$

84. $x^6 + 7x^3 - 8 = 0$

Objectives

a Graph quadratic functions of the type $f(x) = ax^2$ and then label the vertex and the line of symmetry.

b Graph quadratic functions of the type $f(x) = a(x - h)^2$ and then label the vertex and the line of symmetry.

c Graph quadratic functions of the type $f(x) = a(x - h)^2 + k$, finding the vertex, the line of symmetry, and the maximum or minimum y-value.

16.5 GRAPHING $f(x) = a(x - h)^2 + k$

In this section and the next, we develop techniques for graphing quadratic functions.

a Graphs of $f(x) = ax^2$

The most basic quadratic function is $f(x) = x^2$.

EXAMPLE 1 Graph: $f(x) = x^2$.

We choose some values for x and compute $f(x)$ for each. Then we plot the ordered pairs and connect them with a smooth curve.

x	$f(x) = x^2$	$(x, f(x))$
-3	9	$(-3, 9)$
-2	4	$(-2, 4)$
-1	1	$(-1, 1)$
0	0	$(0, 0)$
1	1	$(1, 1)$
2	4	$(2, 4)$
3	9	$(3, 9)$

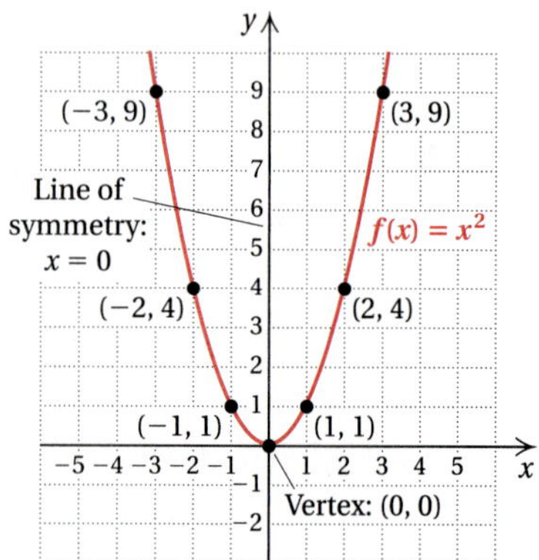

All quadratic functions have graphs similar to the one in Example 1. Such curves are called **parabolas.** They are cup-shaped curves that are symmetric with respect to a vertical line known as the parabola's **line of symmetry,** or **axis of symmetry.** In the graph of $f(x) = x^2$, shown above, the y-axis (or the line $x = 0$) is the line of symmetry. If the paper were to be folded on this line, the two halves of the curve would match. The point $(0, 0)$ is the **vertex** of this parabola.

Let's compare the graphs of $g(x) = \frac{1}{2}x^2$ and $h(x) = 2x^2$ with the graph of $f(x) = x^2$. We choose x-values and plot points for both functions.

x	$g(x) = \frac{1}{2}x^2$
-3	$\frac{9}{2}$
-2	2
-1	$\frac{1}{2}$
0	0
1	$\frac{1}{2}$
2	2
3	$\frac{9}{2}$

x	$h(x) = 2x^2$
-3	18
-2	8
-1	2
0	0
1	2
2	8
3	18

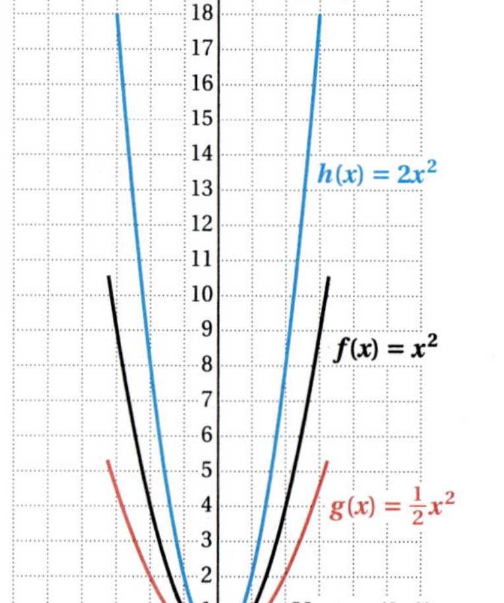

Note that the graph of $g(x) = \frac{1}{2}x^2$ is a wider parabola than the graph of $f(x) = x^2$, and the graph of $h(x) = 2x^2$ is narrower. The vertex and the line of symmetry, however, remain $(0, 0)$ and $x = 0$, respectively.

CALCULATOR CORNER

Graphing Quadratic Functions Use a graphing calculator to make a table of values for $f(x) = x^2$. (See the Calculator Corner on p. 610 for the procedure.)

CALCULATOR CORNER

Graphs of Quadratic Functions

1. Graph each of the following equations in the viewing window $[-5, 5, -10, 10]$:

$$y_1 = x^2, \qquad y_2 = 3x^2, \qquad y_3 = \frac{1}{3}x^2.$$

Determine a rule that describes the effect of the constant a on the graph of $y = ax^2$, where $a > 1$ and where $0 < a < 1$. Graph some other equations to test your rule.

2. Graph each of the following equations in the viewing window $[-5, 5 -10, 10]$:

$$y_1 = -x^2, \qquad y_2 = -4x^2, \qquad y_3 = -\frac{2}{3}x^2.$$

Determine a rule that describes the effect of the constant a on the graph of $y = ax^2$, where $a < -1$ and where $-1 < a < 0$. Graph some other equations to test your rule.

When we consider the graph of $k(x) = -\frac{1}{2}x^2$, we see that the parabola opens down and is the same shape as the graph of $g(x) = \frac{1}{2}x^2$.

x	$k(x) = -\frac{1}{2}x^2$
-3	$-\frac{9}{2}$
-2	-2
-1	$-\frac{1}{2}$
0	0
1	$-\frac{1}{2}$
2	-2
3	$-\frac{9}{2}$

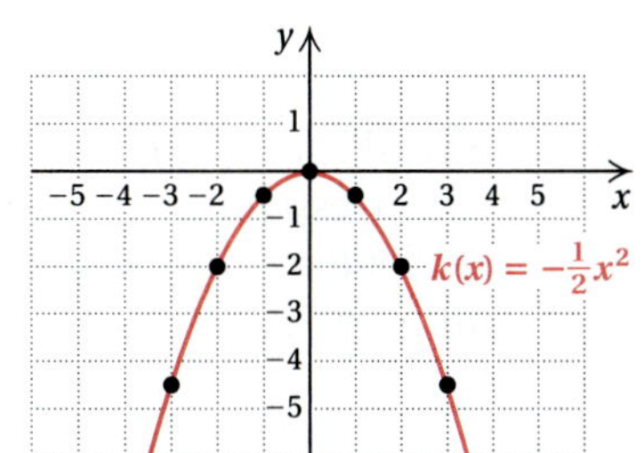

GRAPHS OF $f(x) = ax^2$

The graph of $f(x) = ax^2$, or $y = ax^2$, is a parabola with $x = 0$ as its line of symmetry; its vertex is the origin.

For $a > 0$, the parabola opens up; for $a < 0$, the parabola opens down.

If $|a|$ is greater than 1, the parabola is narrower than $y = x^2$.

If $|a|$ is between 0 and 1, the parabola is wider than $y = x^2$.

Do Exercises 1–3.

Graph.

1. $f(x) = -\frac{1}{3}x^2$

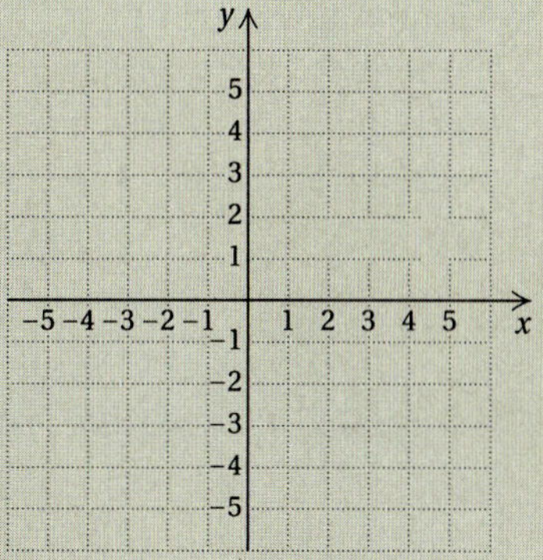

2. $f(x) = 3x^2$

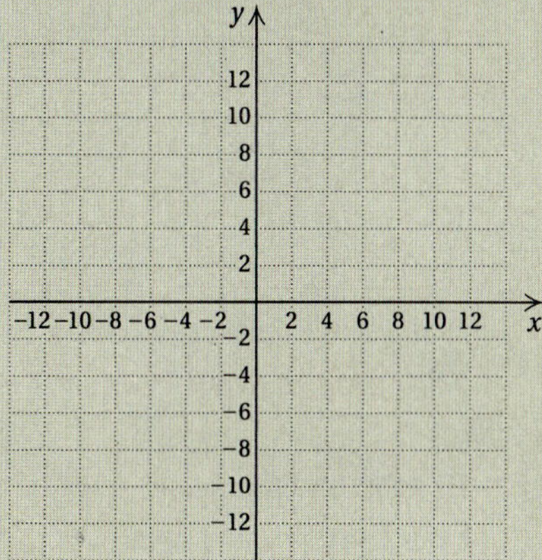

3. $f(x) = -2x^2$

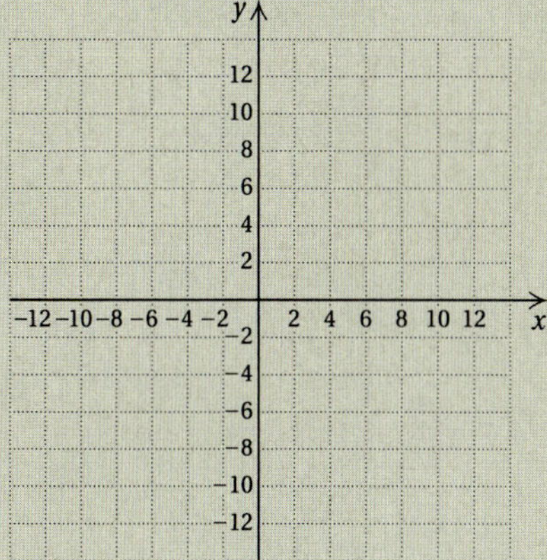

Answers on page A-61

CALCULATOR CORNER

Graphing Quadratic Functions Use a graphing calculator to graph the equations in Margin Exercises 1–3.

b Graphs of $f(x) = a(x - h)^2$

It would seem logical now to consider functions of the type

$$f(x) = ax^2 + bx + c.$$

We are heading in that direction, but it is convenient to first consider graphs of $f(x) = a(x - h)^2$ and then $f(x) = a(x - h)^2 + k$, where a, h, and k are constants.

EXAMPLE 2 Graph: $g(x) = (x - 3)^2$.

We choose some values for x and compute $g(x)$. Then we plot the points and draw the curve.

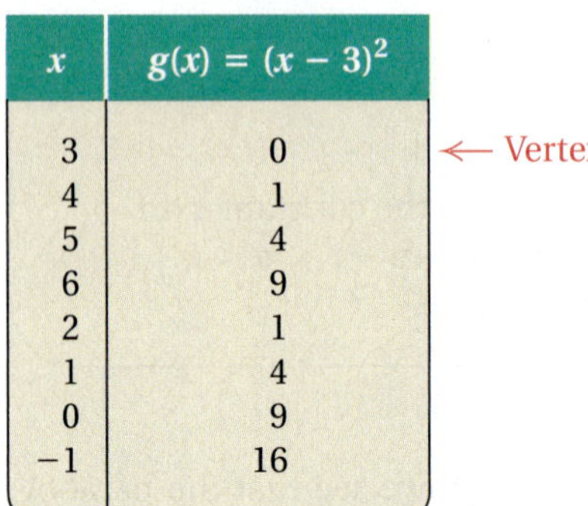

x	$g(x) = (x - 3)^2$
3	0
4	1
5	4
6	9
2	1
1	4
0	9
−1	16

← Vertex

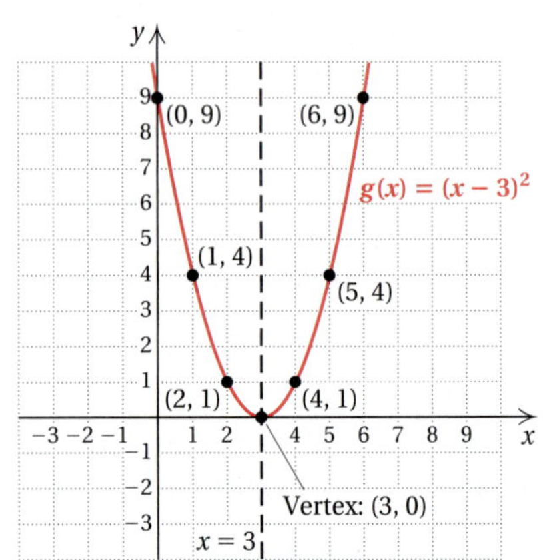

First note that for an x-value of 3, $g(3) = (3 - 3)^2 = 0$. As we increase x-values from 3, note that the corresponding y-values increase. Then as we decrease x-values from 3, note that the corresponding y-values increase again. The line $x = 3$ is a line of symmetry.

CALCULATOR CORNER

Graphs of Quadratic Functions Graph each of the following equations in the viewing window $[-5, 5, -5, 5]$:

$$y_1 = 7x^2,$$
$$y_2 = 7(x - 1)^2,$$
$$y_3 = 7(x + 2)^2.$$

Determine a rule that describes the effect of the constant h on the graph of $y = a(x - h)^2$. Graph some other equations to test your rule.

EXAMPLE 3 Graph: $t(x) = (x + 3)^2$.

We choose some values for x and compute $t(x)$. Then we plot the points and draw the curve.

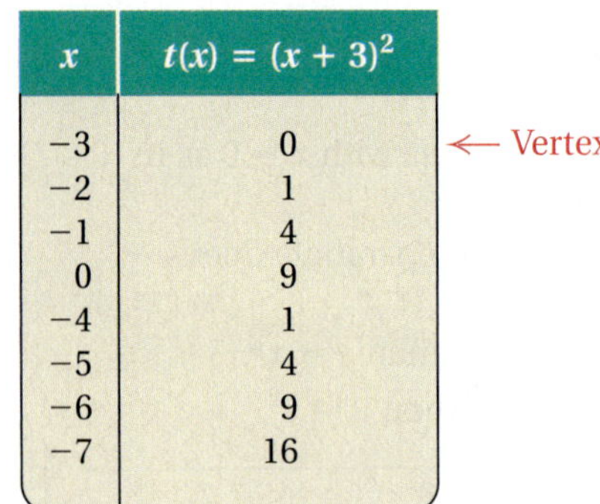

x	$t(x) = (x + 3)^2$
−3	0
−2	1
−1	4
0	9
−4	1
−5	4
−6	9
−7	16

← Vertex

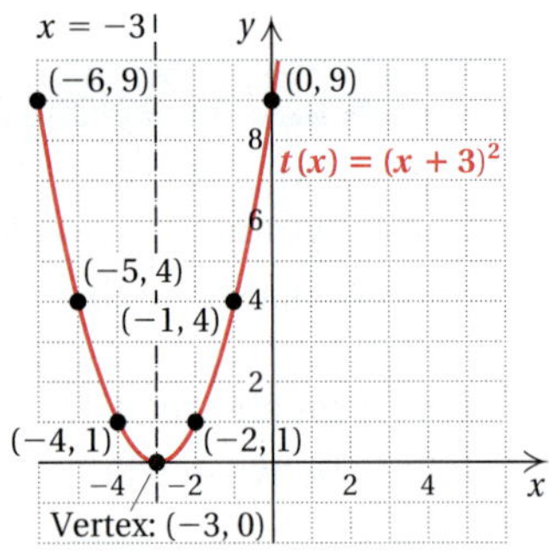

First note that for an x-value of -3, $t(-3) = (-3 + 3)^2 = 0$. As we increase x-values from -3, note that the corresponding y-values increase. Then as we decrease x-values from -3, note that the y-values increase again. The line $x = -3$ is a line of symmetry.

The graph of $g(x) = (x - 3)^2$ in Example 2 looks just like the graph of $f(x) = x^2$ in Example 1, except that it is moved, or translated, 3 units to the right. Comparing the pairs for $g(x)$ with those for $f(x)$, we see that when an input for $g(x)$ is 3 more than an input for $f(x)$, the outputs match.

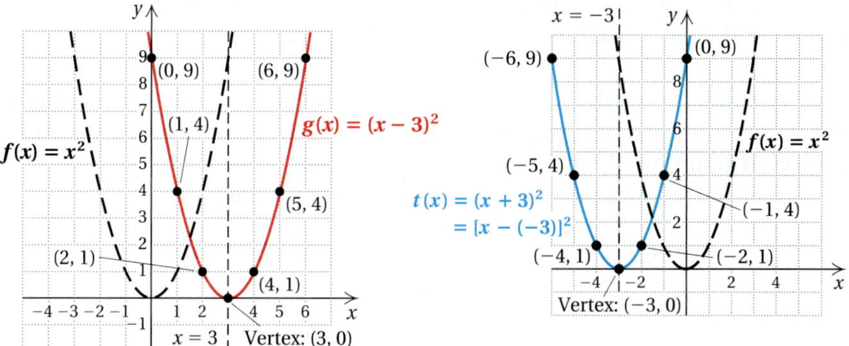

The graph of $t(x) = (x + 3)^2 = [x - (-3)]^2$ in Example 3 looks just like the graph of $f(x) = x^2$ in Example 1, except that it is moved, or translated, 3 units to the left. Comparing the pairs for $t(x)$ with those for $f(x)$, we see that when an input for $t(x)$ is 3 less than an input for $f(x)$, the outputs match.

GRAPHS OF $f(x) = a(x - h)^2$

The graph of $f(x) = a(x - h)^2$ has the same shape as the graph of $y = ax^2$.

If h is positive, the graph of $y = ax^2$ is shifted h units to the right.

If h is negative, the graph of $y = ax^2$ is shifted $|h|$ units to the left.

The vertex is $(h, 0)$ and the axis of symmetry is $x = h$.

EXAMPLE 4 Graph: $f(x) = -2(x + 3)^2$.

We first rewrite the equation as $f(x) = -2[x - (-3)]^2$. In this case, $a = -2$ and $h = -3$, so the graph looks like that of $g(x) = 2x^2$ translated 3 units to the left and, since $-2 < 0$, the graph opens down. The vertex is $(-3, 0)$, and the line of symmetry is $x = -3$. Plotting points as needed, we obtain the graph shown below.

x	$f(x) = -2(x + 3)^2$
−3	0 ← Vertex
−2	−2
−1	−8
0	−18
−4	−2
−5	−8
−6	−18
−7	−32

Do Exercises 4 and 5.

Graph. Find and label the vertex and the line of symmetry.

4. $f(x) = \frac{1}{2}(x - 4)^2$

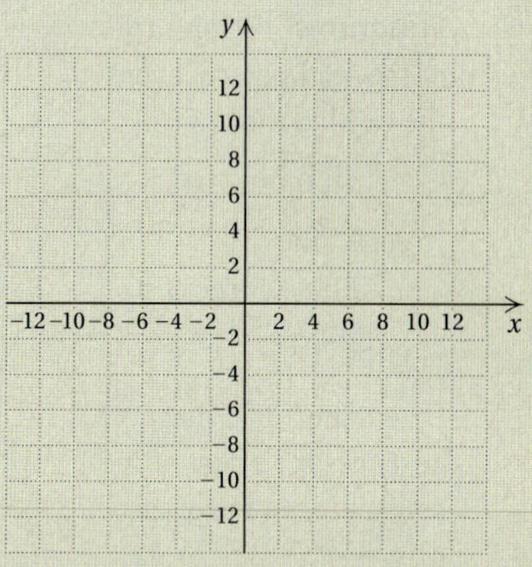

5. $f(x) = -\frac{1}{2}(x - 4)^2$

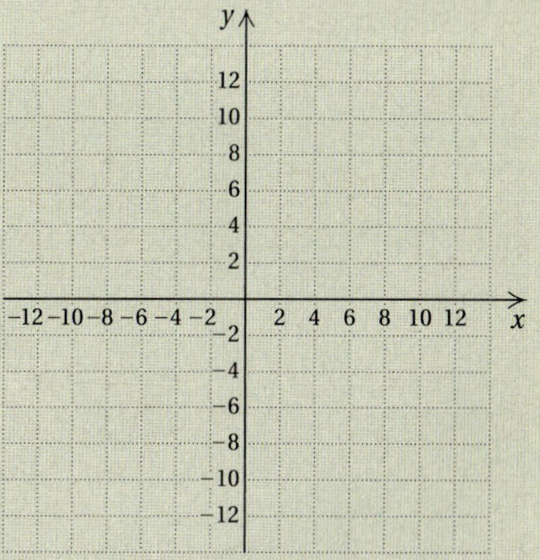

Answers on page A-61

CALCULATOR CORNER

Graphs of Quadratic Functions Graph each of the following equations in the viewing window $[-5, 5, -5, 5]$:

$y_1 = 7(x-1)^2,$

$y_2 = 7(x-1)^2 + 2,$

$y_3 = 7(x-1)^2 - 3.$

Determine a rule that describes the effect of the constant k on the graph of $y = a(x-h)^2 + k$. Graph some other equations to test your rule.

C Graphs of $f(x) = a(x-h)^2 + k$

Given a graph of $f(x) = a(x-h)^2$, what happens if we add a constant k? Suppose that we add 2. This increases each function value $f(x)$ by 2, so the curve is moved up. If k is negative, the curve is moved down. The line of symmetry for the parabola remains $x = h$, but the vertex will be at (h, k), or equivalently, $(h, f(h))$.

Note that if a parabola opens up $(a > 0)$, the function value, or y-value, at the vertex is a least, or **minimum,** value. That is, it is less than the y-value at any other point on the graph. If the parabola opens down $(a < 0)$, the function value at the vertex is a greatest, or **maximum,** value.

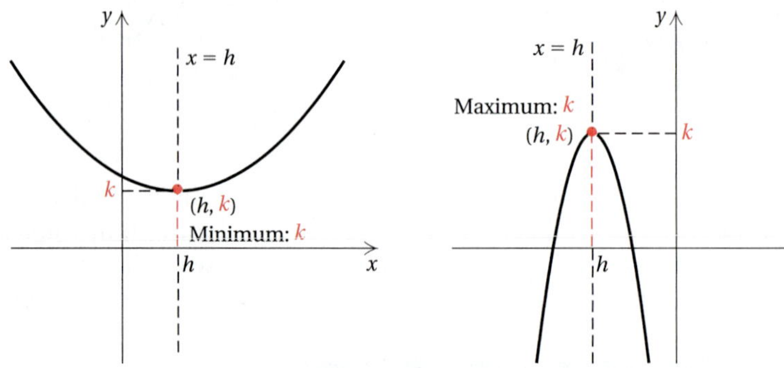

GRAPHS OF $f(x) = a(x-h)^2 + k$

The graph of $f(x) = a(x-h)^2 + k$ has the same shape as the graph of $y = a(x-h)^2$.

If k is positive, the graph of $y = a(x-h)^2$ is shifted k units up.

If k is negative, the graph of $y = a(x-h)^2$ is shifted $|k|$ units down.

The vertex is (h, k), and the line of symmetry is $x = h$.

For $a > 0$, k is the minimum function value. For $a < 0$, k is the maximum function value.

EXAMPLE 5 Graph $f(x) = (x-3)^2 - 5$, and find the minimum function value.

The graph will look like that of $g(x) = (x-3)^2$ (see Example 2) but translated 5 units down. You can confirm this by plotting some points. For instance,

$$f(4) = (4-3)^2 - 5 = -4,$$

whereas in Example 2,

$$g(4) = (4-3)^2 = 1.$$

Note that the vertex is $(h, k) = (3, -5)$, so we begin calculating points on both sides of $x = 3$. The line of symmetry is $x = 3$, and the minimum function value is -5.

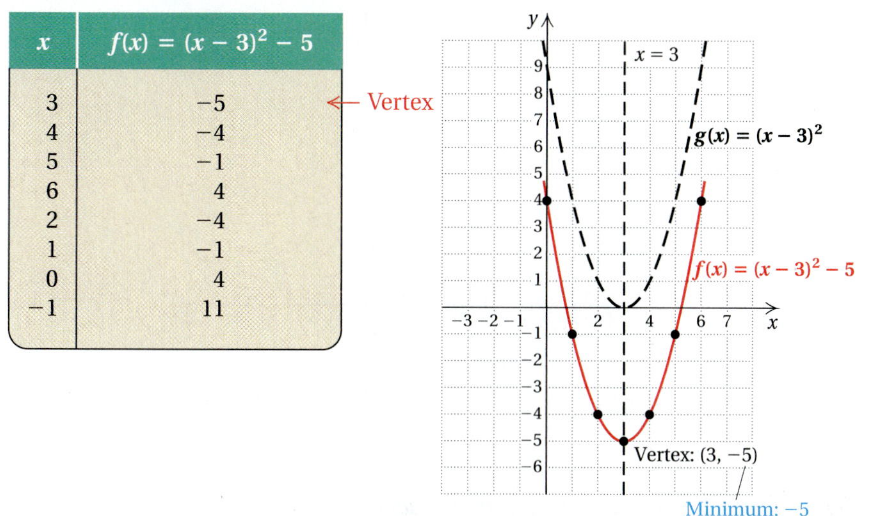

x	$f(x) = (x-3)^2 - 5$	
3	−5	← Vertex
4	−4	
5	−1	
6	4	
2	−4	
1	−1	
0	4	
−1	11	

EXAMPLE 6 Graph $t(x) = \frac{1}{2}(x-3)^2 + 5$, and find the minimum function value.

The graph looks just like that of $f(x) = \frac{1}{2}x^2$ but moved 3 units to the right and 5 units up. The vertex is $(3, 5)$, and the line of symmetry is $x = 3$. We draw $f(x) = \frac{1}{2}x^2$ and then shift the curve over and up. The minimum function value is 5. By plotting some points, we have a check.

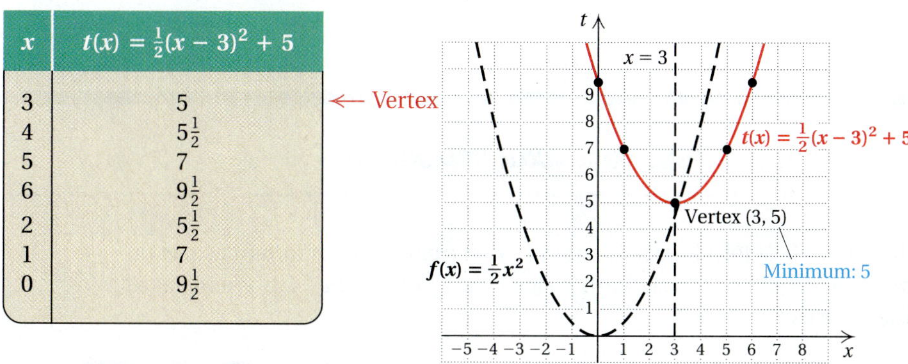

x	$t(x) = \frac{1}{2}(x-3)^2 + 5$	
3	5	← Vertex
4	$5\frac{1}{2}$	
5	7	
6	$9\frac{1}{2}$	
2	$5\frac{1}{2}$	
1	7	
0	$9\frac{1}{2}$	

EXAMPLE 7 Graph $f(x) = -2(x+3)^2 + 5$. Find the vertex, the line of symmetry, and the maximum or minimum value.

We first express the equation in the equivalent form

$$f(x) = -2[x - (-3)]^2 + 5.$$

The graph looks like that of $g(x) = -2x^2$ translated 3 units to the left and 5 units up. The vertex is $(-3, 5)$, and the line of symmetry is $x = -3$. Since $-2 < 0$, we know that the graph opens down so 5, the second coordinate of the vertex, is the maximum y-value.

We compute a few points as needed. The graph is shown on the following page.

x	$f(x) = -2(x+3)^2 + 5$	
−3	5	← Vertex
−2	3	
−1	−3	
−4	3	
−5	−3	

Graph. Find the vertex, the line of symmetry, and the maximum or minimum y-value.

6. $f(x) = \frac{1}{2}(x+2)^2 - 4$

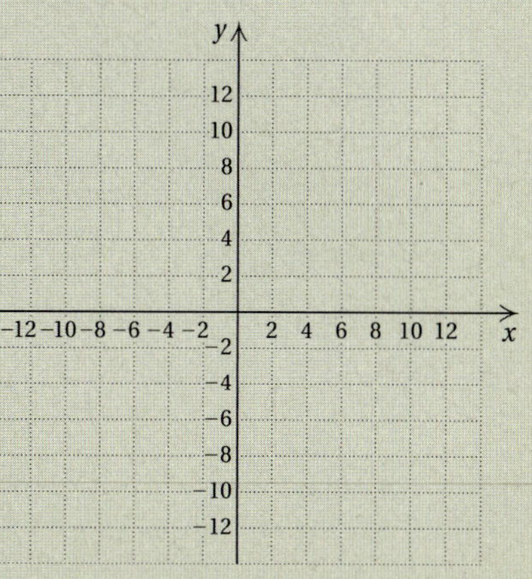

7. $f(x) = -2(x-5)^2 + 3$

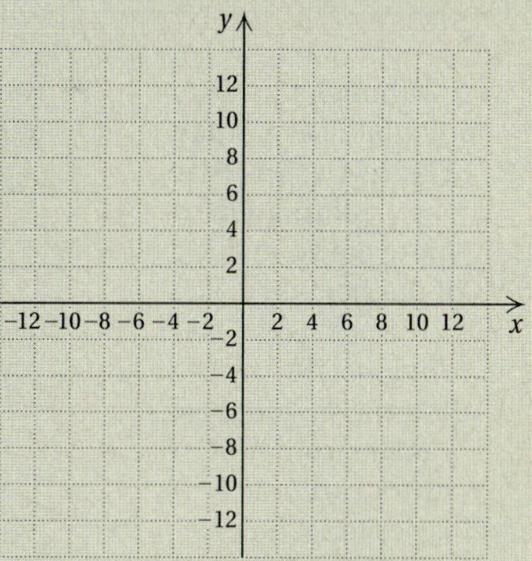

Answers on page A-61

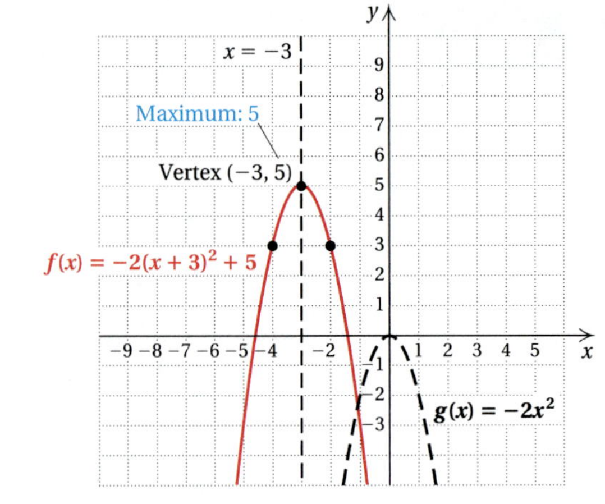

Do Exercises 6 and 7 on the preceding page.

Study Tips

THE FOURTEEN BEST JOBS: HOW MATH STACKS UP

Although this does not qualify as a Study Tip, you can use the information to motivate your study of mathematics. The book *Jobs Rated Almanac* 2001 by Les Krantz lists 250 jobs, ranked from best to worst, according to six criteria: income, stress, physical demands, potential growth, job security, and work environment.

We list the Top 14 best jobs here and note whether math is an important aspect or requirement of the job. Note that math is significant in 11 of the 14 jobs and has at least some use in all the jobs!

	JOB	MATH EMPHASIS	MID-LEVEL SALARY
1.	Financial planner	Yes	$107,000
2.	Web site manager	Yes	68,000
3.	Computer systems analyst	Yes	54,000
4.	Actuary	Yes	71,000
4.	Computer programmer (tie)	Yes	57,000
6.	Software engineer	Yes	53,000
7.	Meteorologist	Some	57,000
8.	Biologist	Some	53,000
9.	Astronomer	Yes	74,000
10.	Paralegal assistant	Some	37,000
11.	Statistician	Yes	55,000
12.	Hospital administrator	Yes	64,000
13.	Dietician	Yes	39,000
14.	Mathematician	Yes	45,000

Source: Les Krantz, *Jobs Rated Almanac* 2001. New York: St. Martin's Press, 2000

16.5 EXERCISE SET

For Extra Help

Digital Video Tutor CD 14 Videotape 19

InterAct Math

Math Tutor Center

MathXL

MyMathLab

a, **b** Graph. Find and label the vertex and the line of symmetry.

1. $f(x) = 4x^2$

x	$f(x)$
0	
1	
2	
−1	
−2	

Vertex: (____, ____)
Line of symmetry: $x =$ ____

2. $f(x) = 5x^2$

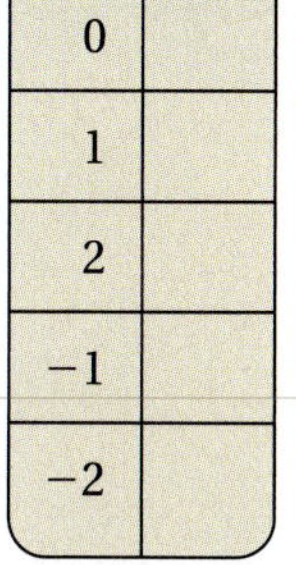

x	$f(x)$
0	
1	
2	
−1	
−2	

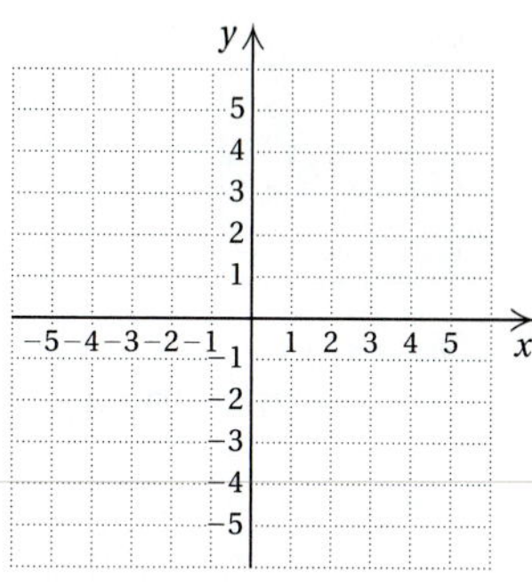

Vertex: (____, ____)
Line of symmetry: $x =$ ____

3. $f(x) = \frac{1}{3}x^2$

x	$f(x)$
0	
1	
2	
−1	
−2	

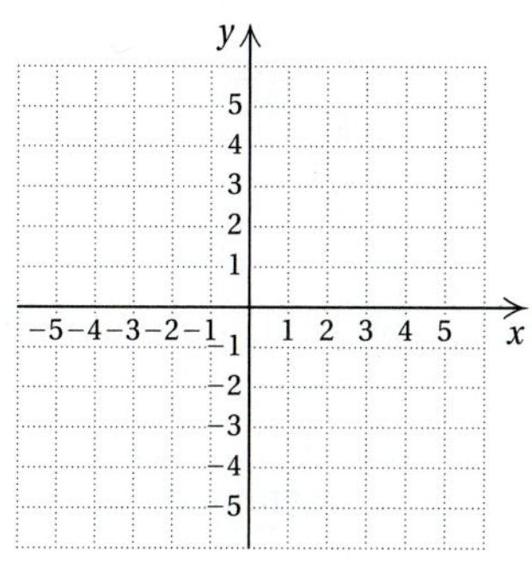

Vertex: (____, ____)
Line of symmetry: $x =$ ____

4. $f(x) = \frac{1}{4}x^2$

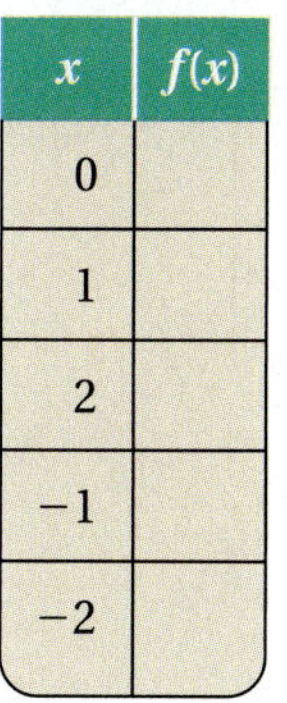

x	$f(x)$
0	
1	
2	
−1	
−2	

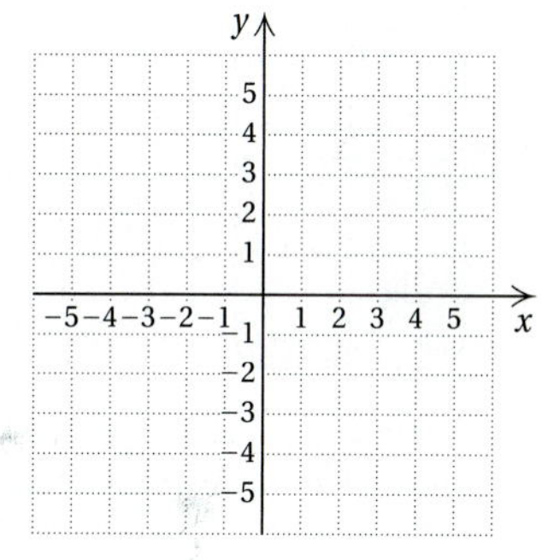

Vertex: (____, ____)
Line of symmetry: $x =$ ____

5. $f(x) = -\frac{1}{2}x^2$

x	$f(x)$

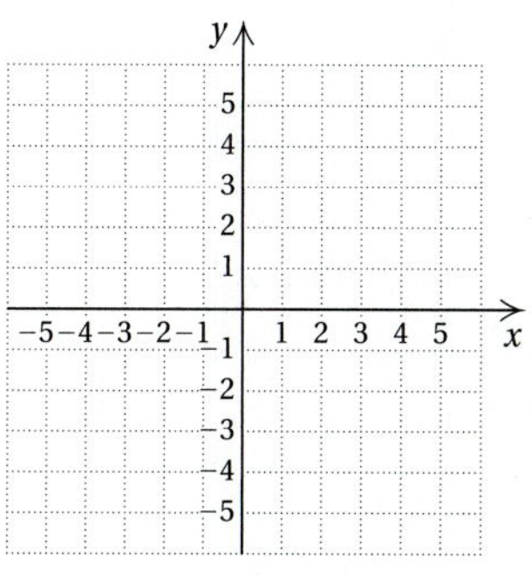

Vertex: (____, ____)
Line of symmetry: $x =$ ____

6. $f(x) = -\frac{1}{4}x^2$

x	$f(x)$

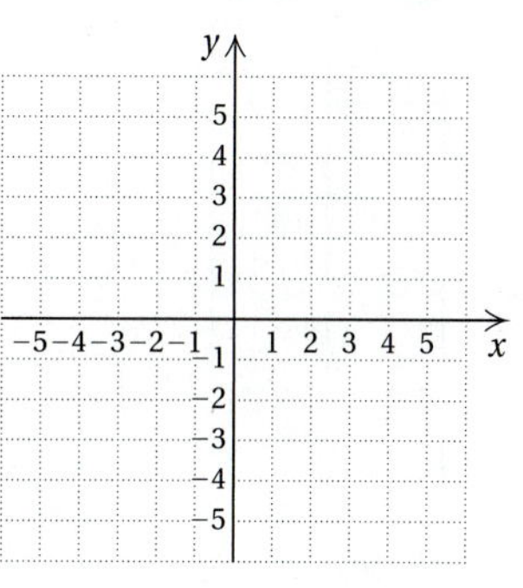

Vertex: (____, ____)
Line of symmetry: $x =$ ____

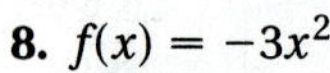

7. $f(x) = -4x^2$

x	$f(x)$

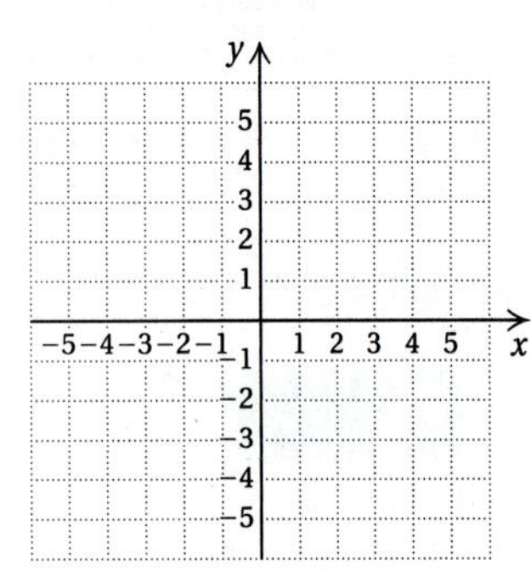

Vertex: (____, ____)
Line of symmetry: $x =$ ____

8. $f(x) = -3x^2$

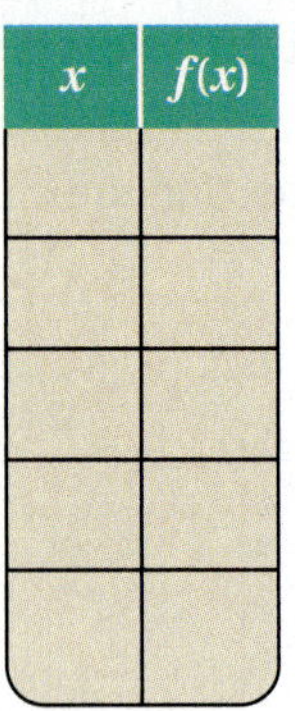

x	$f(x)$

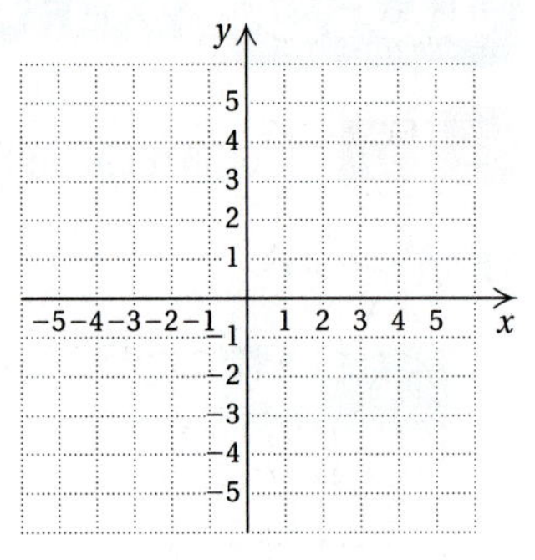

Vertex: (____, ____)
Line of symmetry: $x =$ ____

9. $f(x) = (x + 3)^2$

x	$f(x)$
−3	
−2	
−1	
−4	
−5	

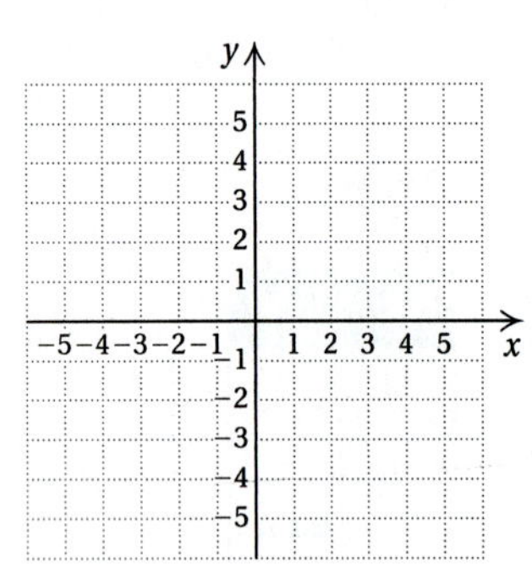

Vertex: (____, ____)
Line of symmetry: $x =$ ____

10. $f(x) = (x + 1)^2$

x	$f(x)$
−1	
0	
1	
−2	
−3	

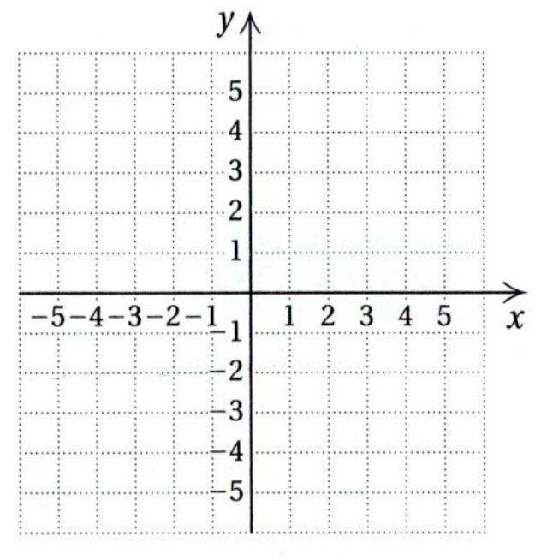

Vertex: (____, ____)
Line of symmetry: $x =$ ____

11. $f(x) = 2(x - 4)^2$

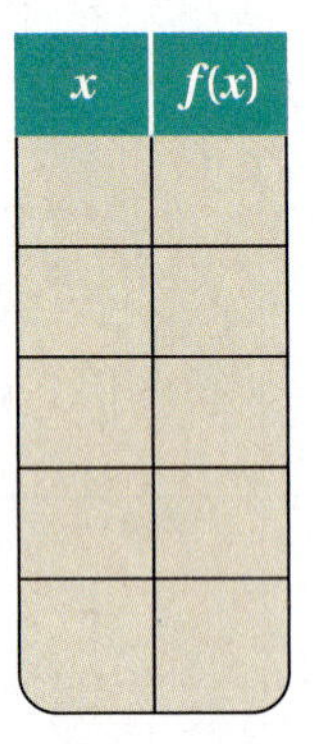

x	$f(x)$

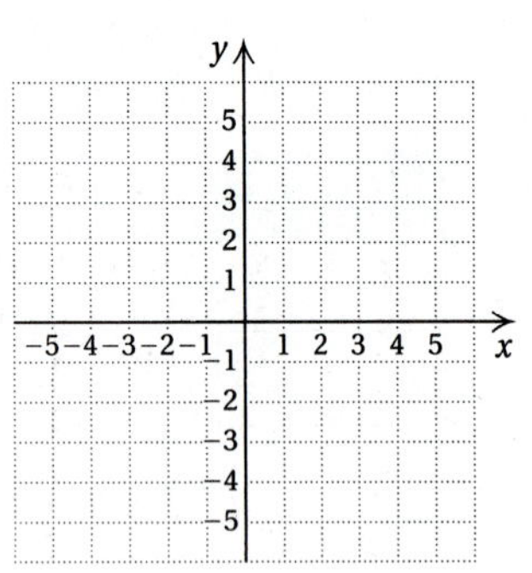

Vertex: (____, ____)
Line of symmetry: $x =$ ____

12. $f(x) = 4(x - 1)^2$

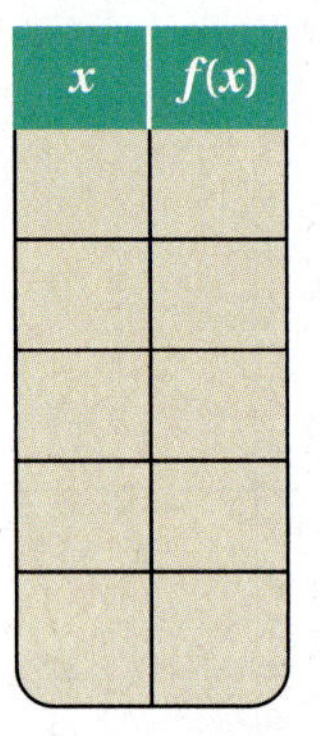

x	$f(x)$

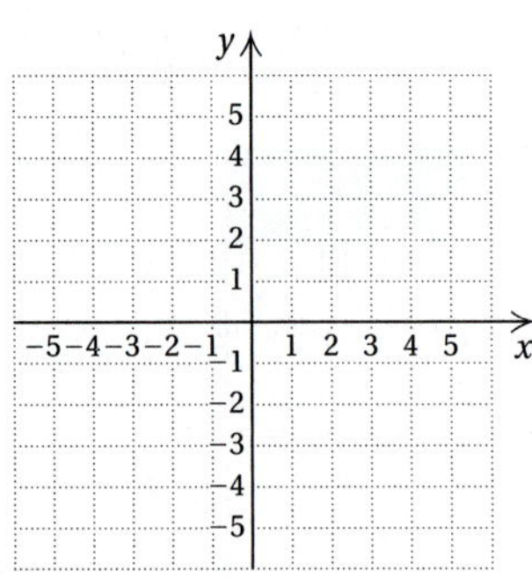

Vertex: (____, ____)
Line of symmetry: $x =$ ____

13. $f(x) = -2(x + 2)^2$

x	$f(x)$

Vertex: (____, ____)
Line of symmetry: $x =$ ____

14. $f(x) = -2(x + 4)^2$

x	$f(x)$

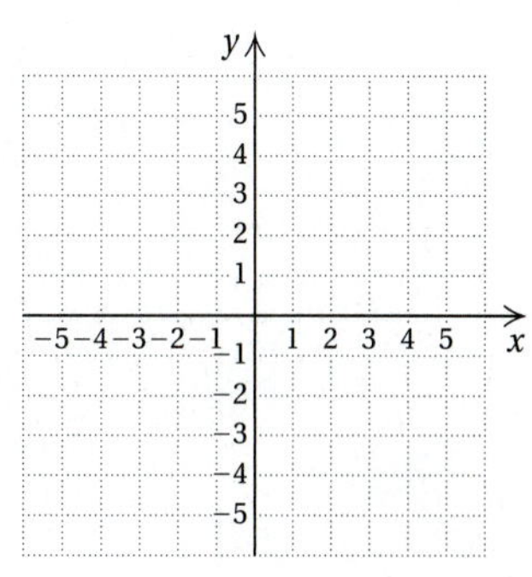

Vertex: (____, ____)
Line of symmetry: $x =$ ____

15. $f(x) = 3(x - 1)^2$

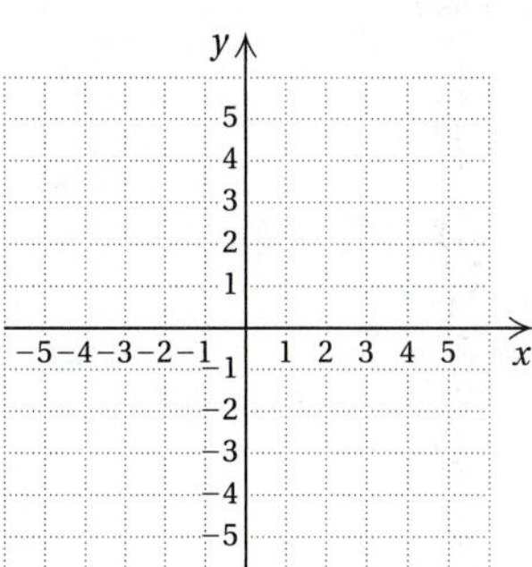

16. $f(x) = 4(x - 2)^2$

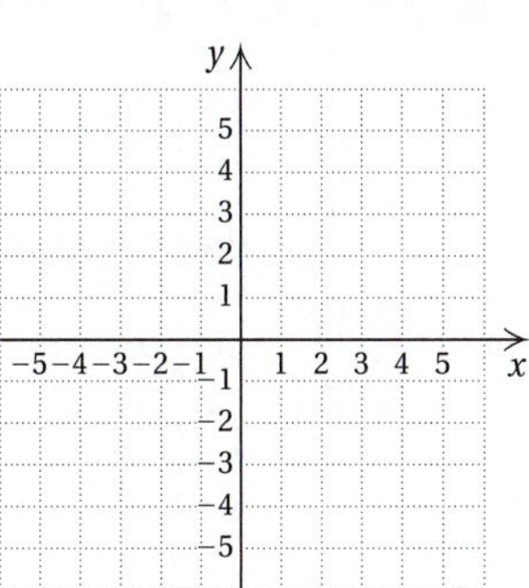

17. $f(x) = -\frac{3}{2}(x + 2)^2$

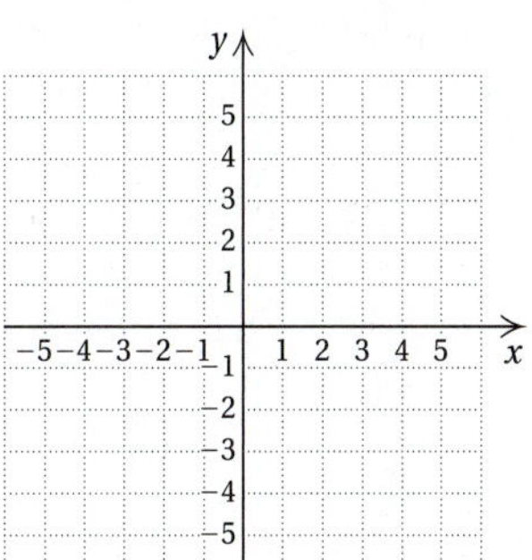

18. $f(x) = -\frac{5}{2}(x + 3)^2$

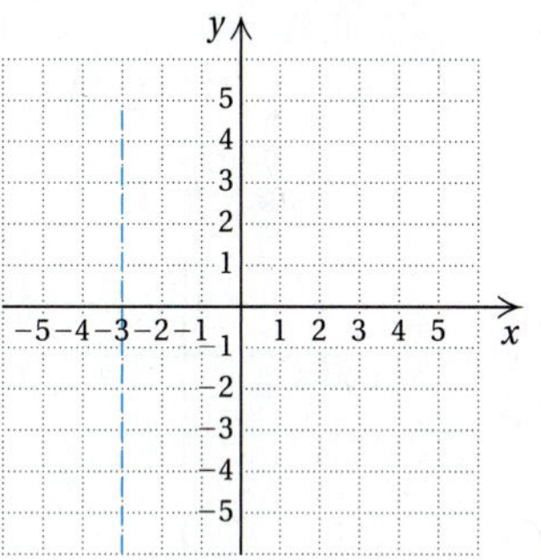

C Graph. Find and label the vertex and the line of symmetry. Find the maximum or minimum value.

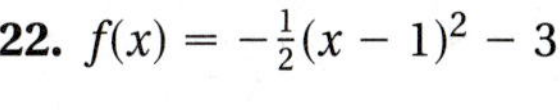

19. $f(x) = (x - 3)^2 + 1$

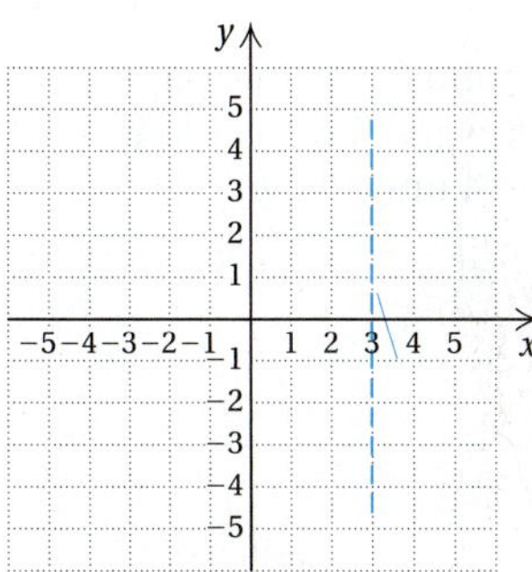

Vertex: (____, ____)
Line of symmetry: $x =$ ____
Minimum value: ____

20. $f(x) = (x + 2)^2 - 3$

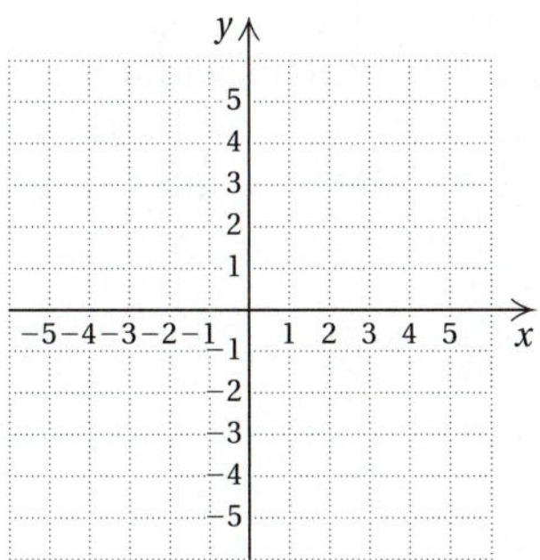

Vertex: (____, ____)
Line of symmetry: $x =$ ____
Minimum value: ____

21. $f(x) = -3(x + 4)^2 + 1$

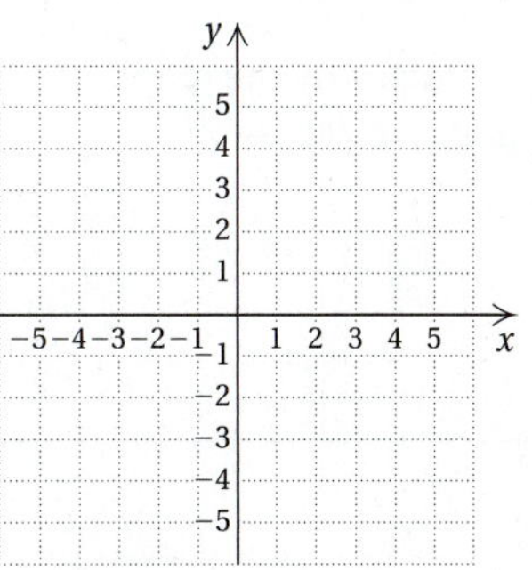

Vertex: (____, ____)
Line of symmetry: $x =$ ____
Maximum value: ____

22. $f(x) = -\frac{1}{2}(x - 1)^2 - 3$

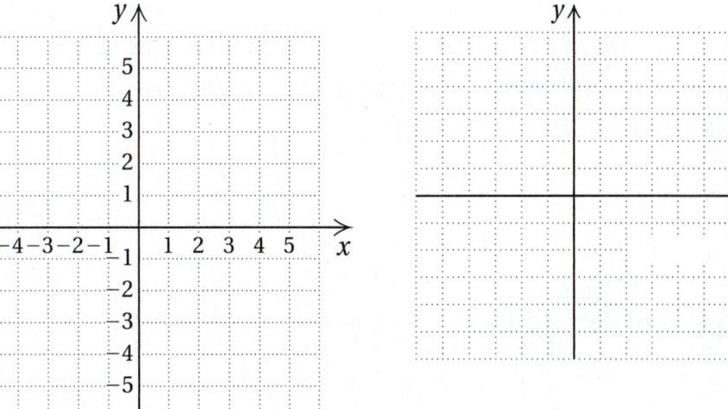

Vertex: (____, ____)
Line of symmetry: $x =$ ____
Maximum value: ____

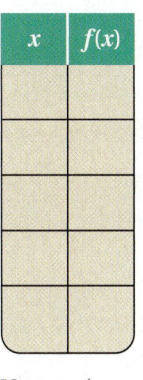

23. $f(x) = \frac{1}{2}(x + 1)^2 + 4$

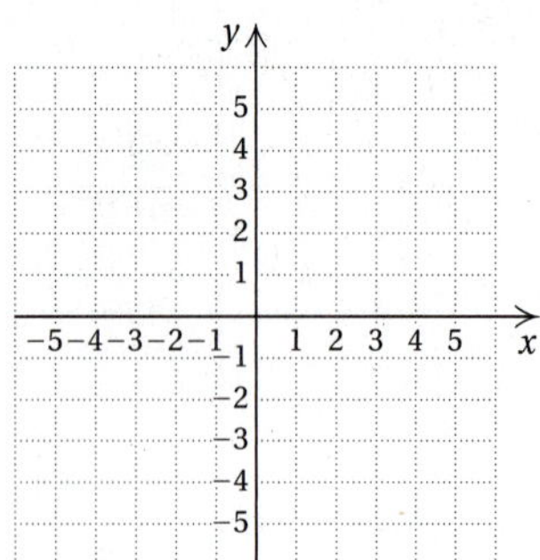

Vertex: (__,__)
Line of symmetry: $x =$ ____
__________ value: _____

24. $f(x) = -2(x - 5)^2 - 3$

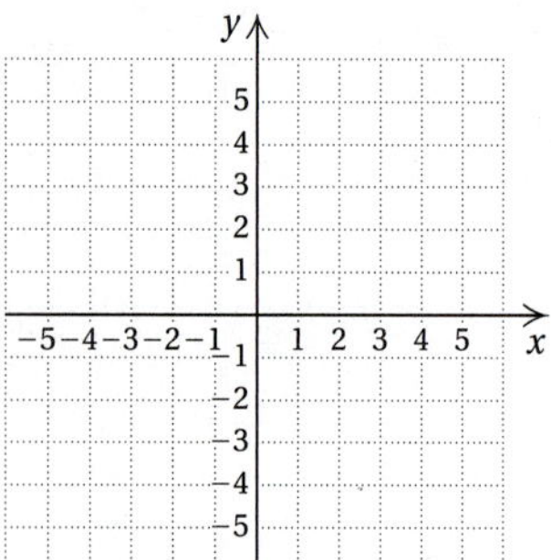

Vertex: (__,__)
Line of symmetry: $x =$ ____
__________ value: _____

25. $f(x) = -(x + 1)^2 - 2$

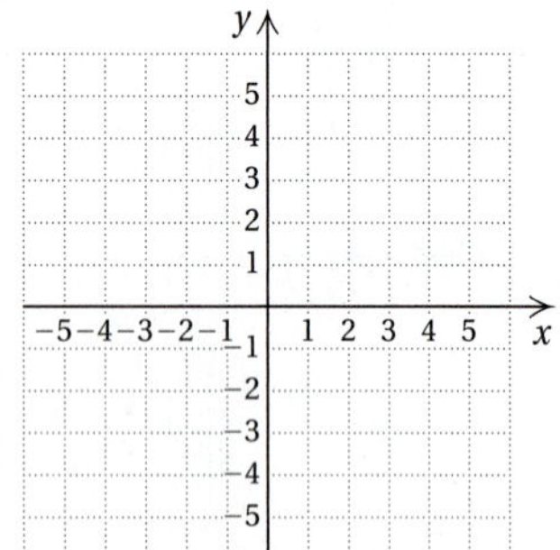

Vertex: (__,__)
Line of symmetry: $x =$ ____
__________ value: _____

26. $f(x) = 3(x - 4)^2 + 2$

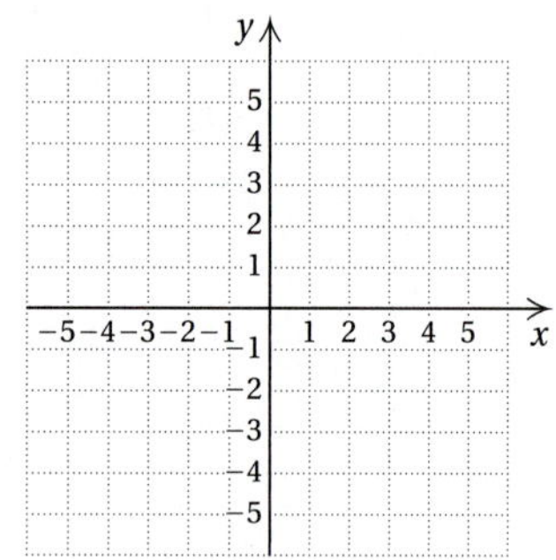

Vertex: (__,__)
Line of symmetry: $x =$ ____
__________ value: _____

27. ^{D}W Explain, without plotting points, why the graph of $f(x) = (x + 3)^2$ looks like the graph of $f(x) = x^2$ translated 3 units to the left.

28. ^{D}W Explain, without plotting points, why the graph of $f(x) = (x + 3)^2 - 4$ looks like the graph of $f(x) = x^2$ translated 3 units to the left and 4 units down.

SKILL MAINTENANCE

Solve. [15.6a, b]

29. $x - 5 = \sqrt{x + 7}$

30. $\sqrt{2x + 7} = \sqrt{5x - 4}$

31. $\sqrt{x + 4} = -11$

32. $x = 7 + 2\sqrt{x + 1}$

SYNTHESIS

33. Use the TRACE and/or TABLE features on a graphing calculator to confirm the maximum or minimum values given in Exercises 23, 24, and 26.

16.6 GRAPHING $f(x) = ax^2 + bx + c$

a Graphing and Analyzing $f(x) = ax^2 + bx + c$

By *completing the square,* we can begin with any quadratic polynomial $ax^2 + bx + c$ and find an equivalent expression $a(x - h)^2 + k$. This allows us to combine the skills of Sections 16.1 and 16.5 to graph and analyze any quadratic function $f(x) = ax^2 + bx + c$.

EXAMPLE 1 For $f(x) = x^2 - 6x + 4$, find the vertex, the line of symmetry, and the minimum value. Then graph.

We first find the vertex and the line of symmetry. To do so, we find the equivalent form $a(x - h)^2 + k$ by completing the square, beginning as follows:

$$f(x) = x^2 - 6x + 4 = (x^2 - 6x \qquad) + 4.$$

We complete the square inside the parentheses, but in a different manner than we did before. We take half the x-coefficient, $-6/2 = -3$, and square it: $(-3)^2 = 9$. Then we add 0, or $9 - 9$, inside the parentheses. (Because we are using function notation, instead of adding $(b/2)^2$ on both sides of an equation, we add and subtract it on the same side, effectively adding 0 and not changing the value of the expression.)

$$f(x) = (x^2 - 6x + 0) + 4 \quad \text{Adding 0}$$
$$= (x^2 - 6x + 9 - 9) + 4 \quad \text{Substituting } 9 - 9 \text{ for } 0$$
$$= (x^2 - 6x + 9) + (-9 + 4) \quad \text{Using the associative law of addition to regroup}$$
$$= (x - 3)^2 - 5. \quad \text{Factoring and simplifying}$$

(This equation was graphed in Example 5 of Section 16.5.) The vertex is $(3, -5)$, and the line of symmetry is $x = 3$. The coefficient of x^2 is 1, which is positive, so the graph opens up. This tells us that -5 is a minimum. We plot the vertex and draw the line of symmetry. We choose some x-values on both sides of the vertex and graph the parabola. Suppose we compute the pair $(5, -1)$:

$$f(5) = 5^2 - 6(5) + 4 = 25 - 30 + 4 = -1.$$

We note that it is 2 units to the right of the line of symmetry. There will also be a pair with the same y-coordinate on the graph 2 units to the *left* of the line of symmetry. Thus we get a second point, $(1, -1)$, without making another calculation.

x	$f(x)$	
3	-5	← Vertex
4	-4	
5	-1	
6	4	
2	-4	
1	-1	
0	4	

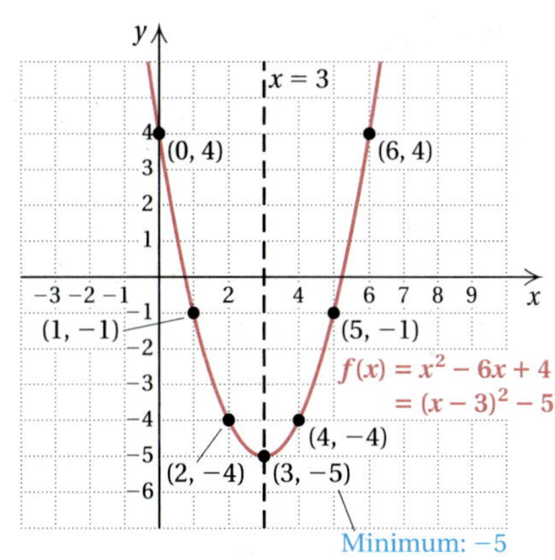

Objectives

a For a quadratic function, find the vertex, the line of symmetry, and the maximum or minimum value, and graph the function.

b Find the intercepts of a quadratic function.

1. For $f(x) = x^2 - 4x + 7$, find the vertex, the line of symmetry, and the minimum value. Then graph.

x	$f(x)$

Vertex: (____, ____)
Line of symmetry: $x =$ _____
Minimum value: _____

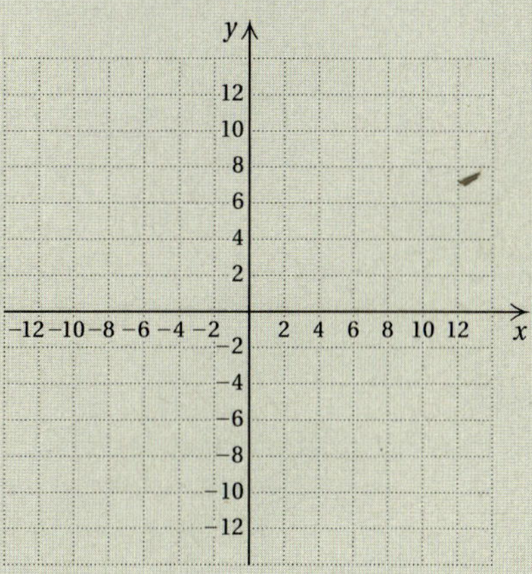

Answer on page A-63

2. For $f(x) = 3x^2 - 24x + 43$, find the vertex, the line of symmetry, and the minimum value. Then graph.

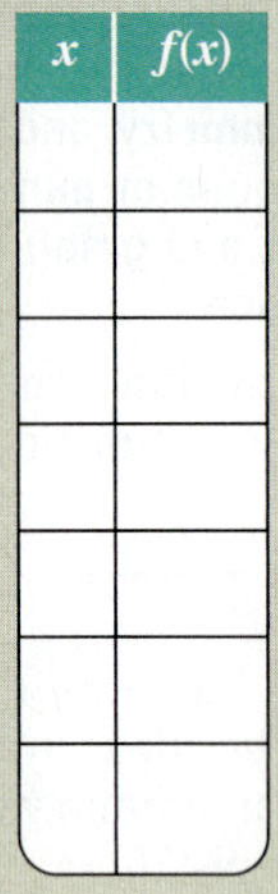

Vertex: (_____, _____)
Line of symmetry: $x =$ ______
Minimum value: ______

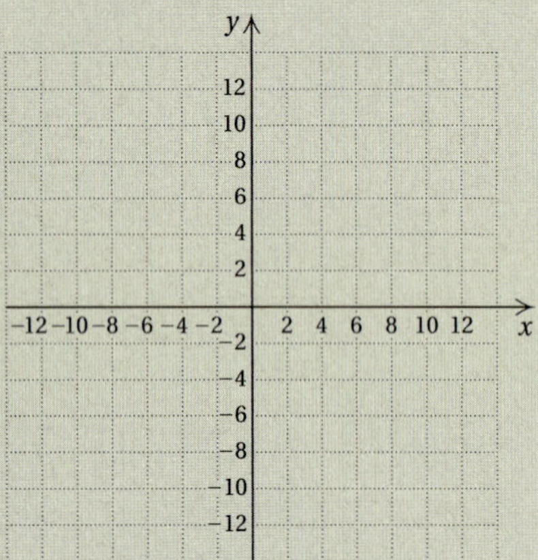

Answer on page A-63

Do Exercise 1 on the preceding page.

EXAMPLE 2 For $f(x) = 3x^2 + 12x + 13$, find the vertex, the line of symmetry, and the minimum value. Then graph.

Since the coefficient of x^2 is not 1, we factor out 3 from only the *first two* terms of the expression. Remember that we want to get to the form $f(x) = a(x - h)^2 + k$:

$$f(x) = 3x^2 + 12x + 13$$
$$= 3(x^2 + 4x) + 13. \qquad \text{Factoring 3 out of the first two terms}$$

Next, we complete the square inside the parentheses:

$$f(x) = 3(x^2 + 4x \qquad) + 13.$$

We take half the x-coefficient, $\frac{1}{2} \cdot 4 = 2$, and square it: $2^2 = 4$. Then we add 0, or $4 - 4$, inside the parentheses:

$$f(x) = 3(x^2 + 4x + 0) + 13 \qquad \text{Adding 0}$$
$$= 3(x^2 + 4x + 4 - 4) + 13 \qquad \text{Substituting } 4 - 4 \text{ for } 0$$
$$= 3(x^2 + 4x + 4 - 4) + 13$$
$$= 3(x^2 + 4x + 4) + 3(-4) + 13 \qquad \text{Using the distributive law to separate } -4 \text{ from the trinomial}$$
$$= 3(x + 2)^2 + 1. \qquad \text{Factoring and simplifying}$$

The vertex is $(-2, 1)$, and the line of symmetry is $x = -2$. The coefficient of x^2 is 3, so the graph is narrow and opens up. This tells us that 1 is a minimum. We choose a few x-values on one side of the line of symmetry, compute y-values, and use the resulting coordinates to find more points on the other side of the line of symmetry. We plot points and graph the parabola.

x	$f(x)$	
−2	1	← Vertex
−1	4	
−3	4	
0	13	
−4	13	

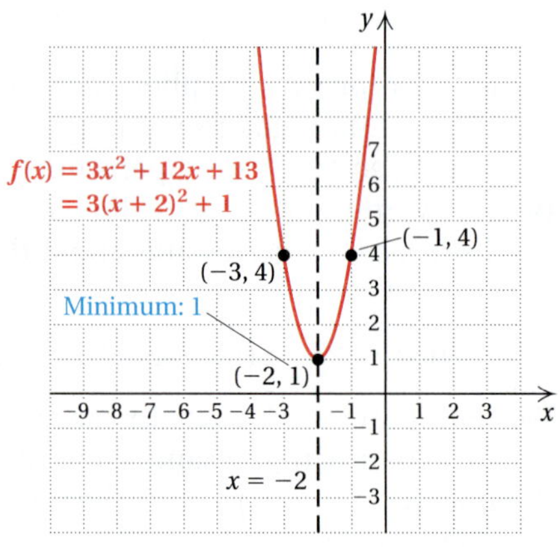

Do Exercise 2.

EXAMPLE 3 For $f(x) = -2x^2 + 10x - 7$, find the vertex, the line of symmetry, and the maximum value. Then graph.

Again, the coefficient of x^2 is not 1. We factor out -2 from only the *first two* terms of the expression. This makes the coefficient of x^2 inside the parentheses 1:

$$f(x) = -2x^2 + 10x - 7$$
$$= -2(x^2 - 5x) - 7.$$

Next, we complete the square as before:

$$f(x) = -2(x^2 - 5x \qquad\qquad) - 7.$$

We take half the x-coefficient, $\frac{1}{2}(-5) = -\frac{5}{2}$, and square it: $\left(-\frac{5}{2}\right)^2 = \frac{25}{4}$. Then we add 0, or $\frac{25}{4} - \frac{25}{4}$, inside the parentheses:

$$f(x) = -2\left(x^2 - 5x + \frac{25}{4} - \frac{25}{4}\right) - 7 \quad \text{Adding 0, or } \frac{25}{4} - \frac{25}{4}$$
$$= -2\left(x^2 - 5x + \frac{25}{4} - \frac{25}{4}\right) - 7$$
$$= -2\left(x^2 - 5x + \frac{25}{4}\right) + (-2)\left(-\frac{25}{4}\right) - 7 \quad \text{Using the distributive law to separate the } -\frac{25}{4} \text{ from the trinomial}$$
$$= -2\left(x^2 - 5x + \frac{25}{4}\right) + \frac{25}{2} - 7$$
$$= -2\left(x - \frac{5}{2}\right)^2 + \frac{11}{2}. \quad \text{Factoring and simplifying}$$

The vertex is $\left(\frac{5}{2}, \frac{11}{2}\right)$, and the line of symmetry is $x = \frac{5}{2}$. The coefficient of x^2 is -2, so the graph is narrow and opens down. This tells us that $\frac{11}{2}$ is a maximum. We choose a few x-values on one side of the line of symmetry, compute y-values, and use the resulting coordinates to find more points on the other side of the line of symmetry. We plot points and graph the parabola.

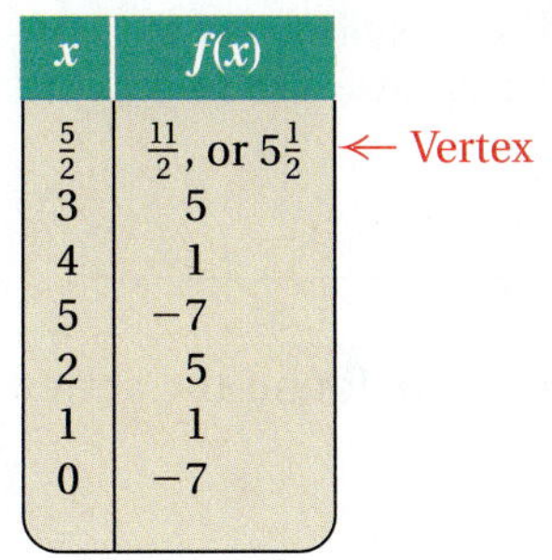

x	$f(x)$
$\frac{5}{2}$	$\frac{11}{2}$, or $5\frac{1}{2}$ ← Vertex
3	5
4	1
5	−7
2	5
1	1
0	−7

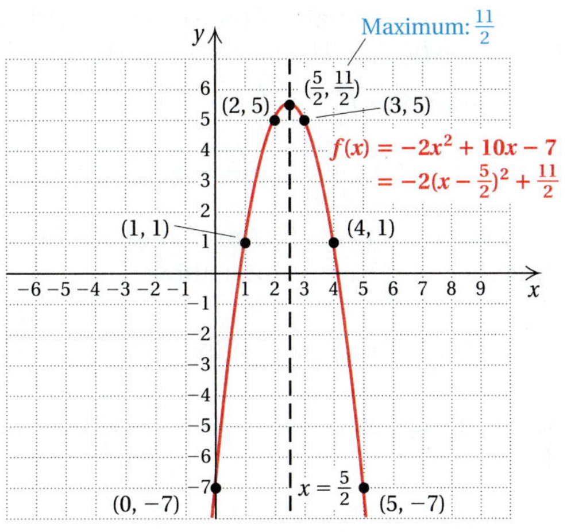

Do Exercise 3.

3. For $f(x) = -4x^2 + 12x - 5$, find the vertex, the line of symmetry, and the maximum value. Then graph.

x	$f(x)$

Vertex: (____, ____)
Line of symmetry: $x =$ _____
Minimum value: _____

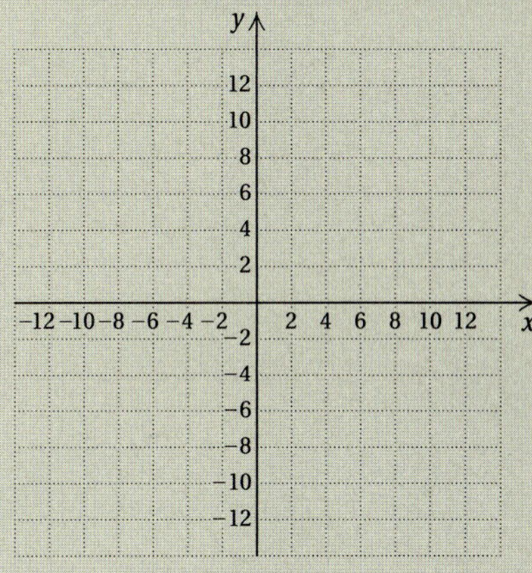

Answer on page A-63

Find the vertex of the parabola using the formula.

4. $f(x) = x^2 - 6x + 4$

5. $f(x) = 3x^2 - 24x + 43$

6. $f(x) = -4x^2 + 12x - 5$

Answers on page A-63

The method used in Examples 1–3 can be generalized to find a formula for locating the vertex. We complete the square as follows:

$$f(x) = ax^2 + bx + c$$
$$= a\left(x^2 + \frac{b}{a}x\right) + c. \quad \text{Factoring } a \text{ out of the first two terms. Check by multiplying.}$$

Half of the x-coefficient, $\frac{b}{a}$, is $\frac{b}{2a}$. We square it to get $\frac{b^2}{4a^2}$ and add $\frac{b^2}{4a^2} - \frac{b^2}{4a^2}$ inside the parentheses. Then we distribute the a:

$$f(x) = a\left(x^2 + \frac{b}{a}x + \frac{b^2}{4a^2} - \frac{b^2}{4a^2}\right) + c$$
$$= a\left(x^2 + \frac{b}{a}x + \frac{b^2}{4a^2}\right) + a\left(-\frac{b^2}{4a^2}\right) + c \quad \text{Using the distributive law}$$
$$= a\left(x + \frac{b}{2a}\right)^2 + \frac{-b^2}{4a} + \frac{4ac}{4a} \quad \text{Factoring and finding a common denominator}$$
$$= a\left[x - \left(-\frac{b}{2a}\right)\right]^2 + \frac{4ac - b^2}{4a}.$$

Thus we have the following.

VERTEX; LINE OF SYMMETRY

The **vertex** of a parabola given by $f(x) = ax^2 + bx + c$ is

$$\left(-\frac{b}{2a}, \frac{4ac - b^2}{4a}\right), \quad \text{or} \quad \left(-\frac{b}{2a}, f\left(-\frac{b}{2a}\right)\right).$$

The x-coordinate of the vertex is $-b/(2a)$. The **line of symmetry** is $x = -b/(2a)$. The second coordinate of the vertex is easiest to find by computing $f\left(-\frac{b}{2a}\right)$.

Let's reexamine Example 3 to see how we could have found the vertex directly. From the formula above,

$$\text{the } x\text{-coordinate of the vertex is } -\frac{b}{2a} = -\frac{10}{2(-2)} = \frac{5}{2}.$$

Substituting $\frac{5}{2}$ into $f(x) = -2x^2 + 10x - 7$, we find the second coordinate of the vertex:

$$f\left(\tfrac{5}{2}\right) = -2\left(\tfrac{5}{2}\right)^2 + 10\left(\tfrac{5}{2}\right) - 7$$
$$= -2\left(\tfrac{25}{4}\right) + 25 - 7$$
$$= -\tfrac{25}{2} + 18 = -\tfrac{25}{2} + \tfrac{36}{2} = \tfrac{11}{2}.$$

The vertex is $\left(\frac{5}{2}, \frac{11}{2}\right)$. The line of symmetry is $x = \frac{5}{2}$.

We have developed two methods for finding the vertex. One is by completing the square and the other is by using a formula. You should check with your instructor about which method to use.

Do Exercises 4–6.

b Finding the Intercepts of a Quadratic Function

The points at which a graph crosses an axis are called **intercepts.** We determine the y-intercept by finding $f(0)$. For $f(x) = ax^2 + bx + c$, the y-intercept is $(0, c)$.

To find the x-intercepts, we look for values of x for which $f(x) = 0$. For $f(x) = ax^2 + bx + c$, we solve

$$0 = ax^2 + bx + c.$$

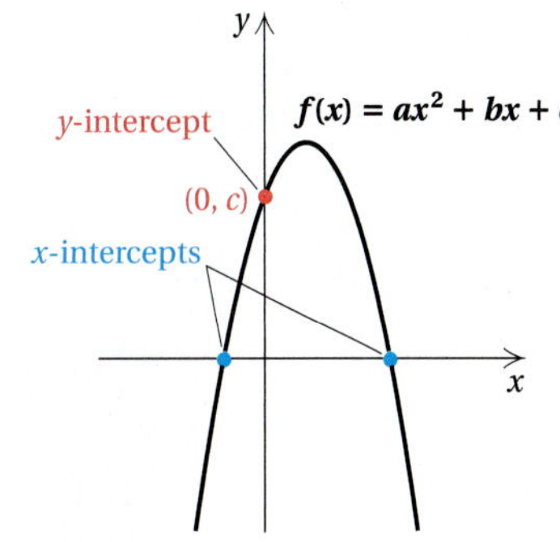

EXAMPLE 4 Find the intercepts of $f(x) = x^2 - 2x - 2$.

The y-intercept is $(0, f(0))$. Since $f(0) = 0^2 - 2 \cdot 0 - 2 = -2$, the y-intercept is $(0, -2)$. To find the x-intercepts, we solve

$$0 = x^2 - 2x - 2.$$

Using the quadratic formula gives us $x = 1 \pm \sqrt{3}$. Thus the x-intercepts are $(1 - \sqrt{3}, 0)$ and $(1 + \sqrt{3}, 0)$, or, approximately, $(-0.732, 0)$ and $(2.732, 0)$.

Do Exercises 7–9.

Find the intercepts.

7. $f(x) = x^2 + 2x - 3$

8. $f(x) = x^2 + 8x + 16$

9. $f(x) = x^2 - 4x + 1$

Answers on page A-63

16.6 EXERCISE SET

For Extra Help

Digital Video Tutor CD 14 Videotape 19

InterAct Math

Math Tutor Center

MathXL

MyMathLab

a For each quadratic function, find **(a)** the vertex, **(b)** the line of symmetry, and **(c)** the maximum or minimum value. Then **(d)** graph the function.

1. $f(x) = x^2 - 2x - 3$

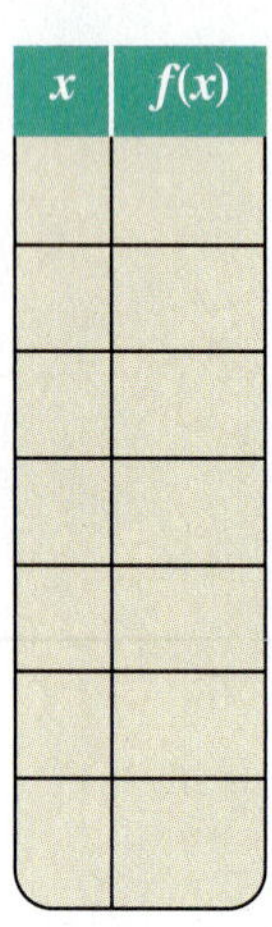

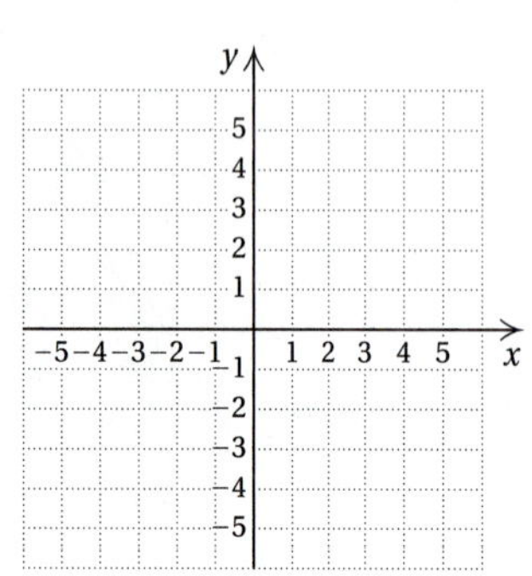

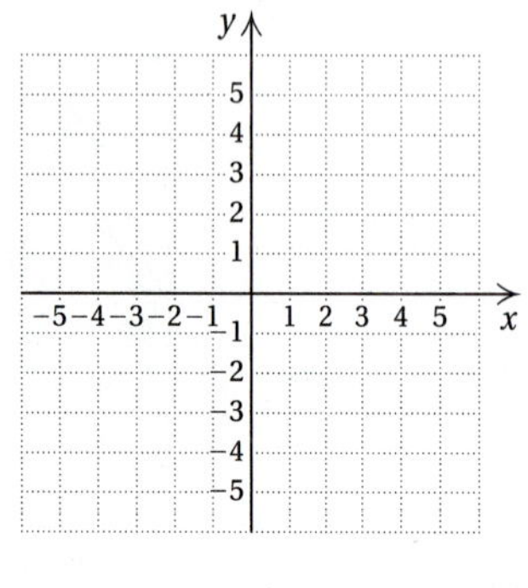

Vertex: (____, ____)
Line of symmetry: $x =$ ____
__________ value: ____

2. $f(x) = x^2 + 2x - 5$

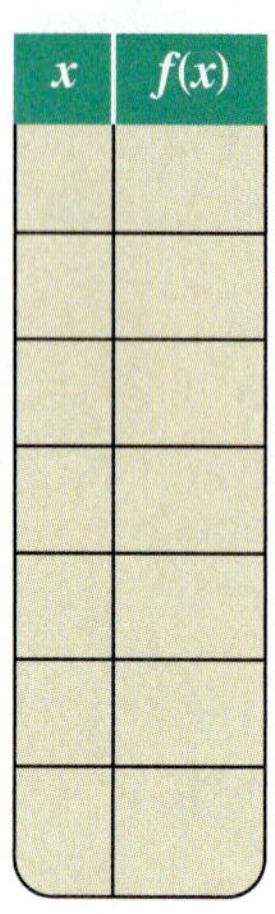

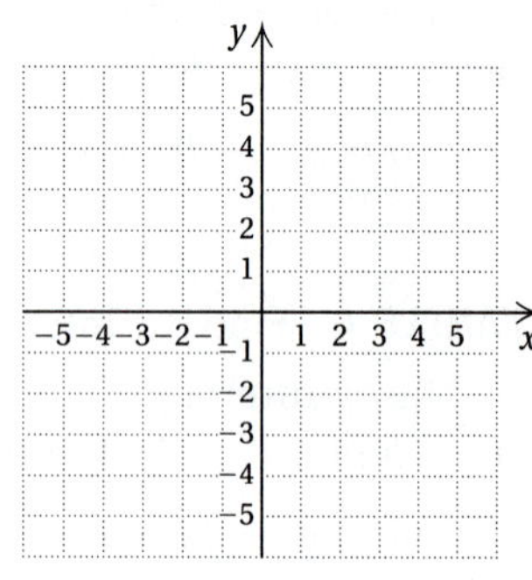

Vertex: (____, ____)
Line of symmetry: $x =$ ____
__________ value: ____

3. $f(x) = -x^2 - 4x - 2$

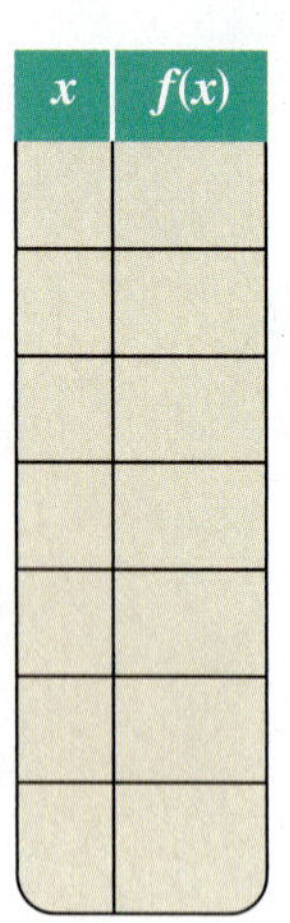

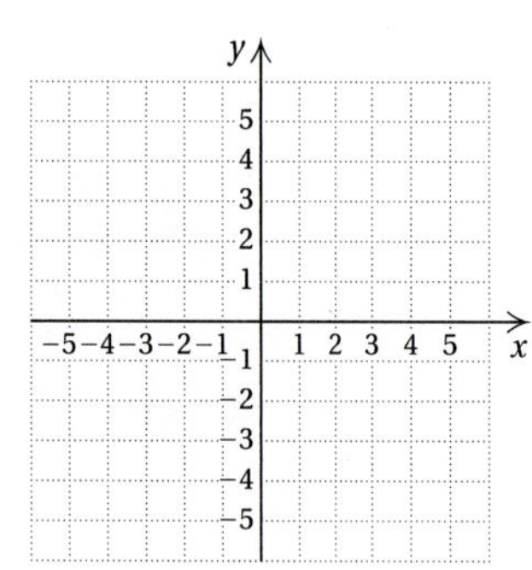

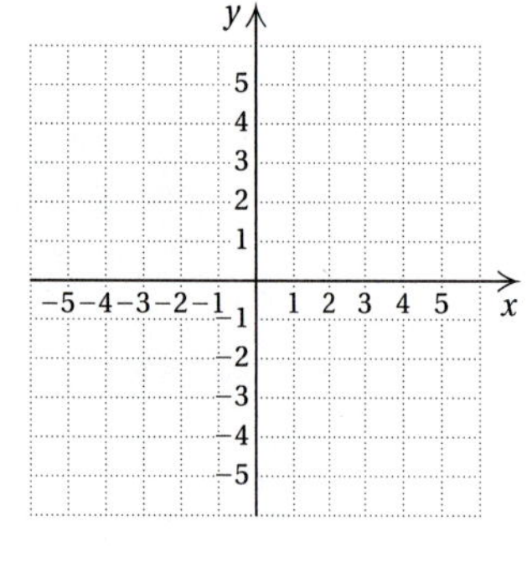

Vertex: (____, ____)
Line of symmetry: $x =$ ____
__________ value: ____

4. $f(x) = -x^2 + 4x + 1$

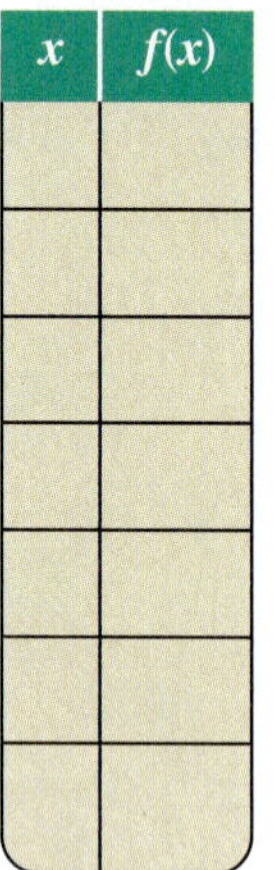

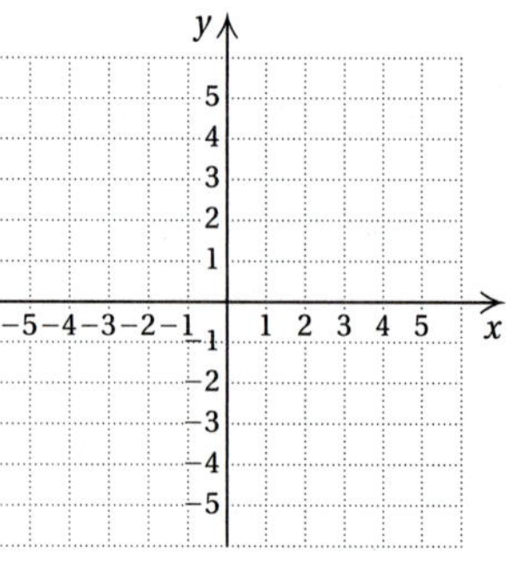

Vertex: (____, ____)
Line of symmetry: $x =$ ____
__________ value: ____

5. $f(x) = 3x^2 - 24x + 50$

x	$f(x)$

Vertex: (____,____)
Line of symmetry: $x =$ ____
__________ value: _____

6. $f(x) = 4x^2 + 8x + 1$

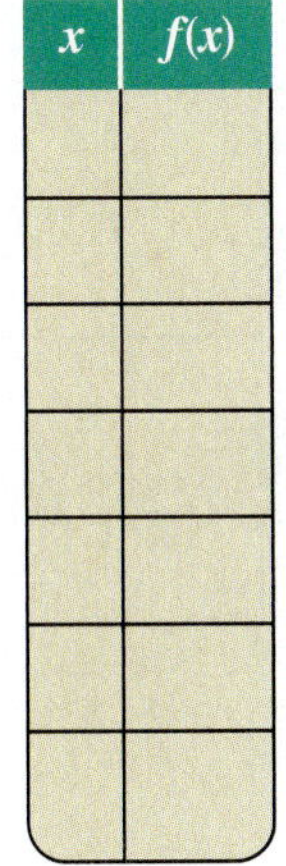

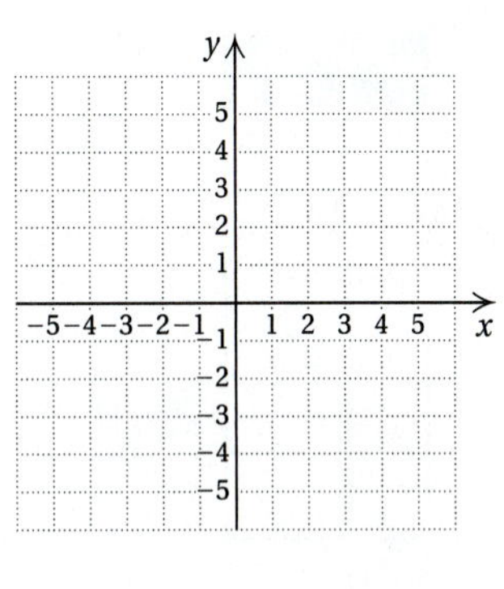

Vertex: (____,____)
Line of symmetry: $x =$ ____
__________ value: _____

7. $f(x) = -2x^2 - 2x + 3$

x	$f(x)$

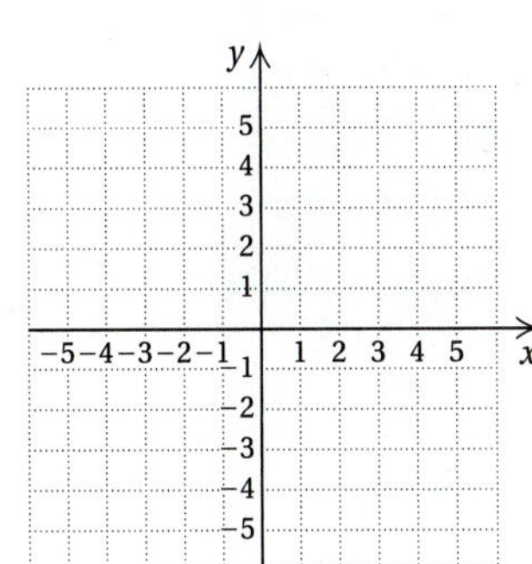

Vertex: (____,____)
Line of symmetry: $x =$ ____
__________ value: _____

8. $f(x) = -2x^2 + 2x + 1$

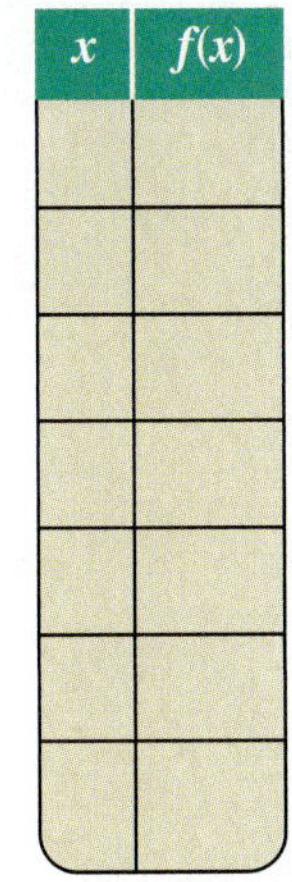

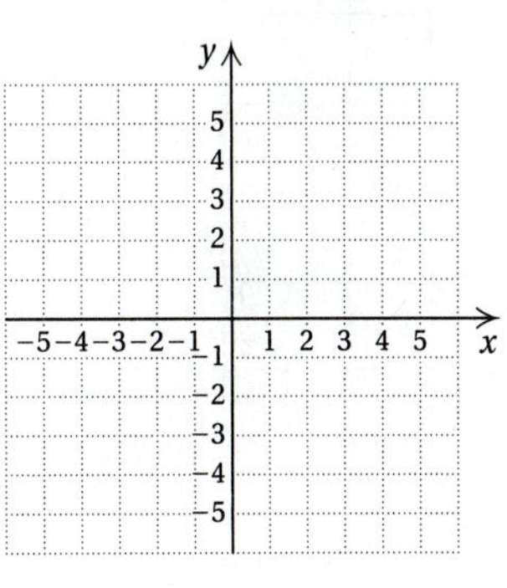

Vertex: (____,____)
Line of symmetry: $x =$ ____
__________ value: _____

9. $f(x) = 5 - x^2$

x	$f(x)$

Vertex: (____,____)
Line of symmetry: $x =$ ____
__________ value: _____

10. $f(x) = x^2 - 3x$

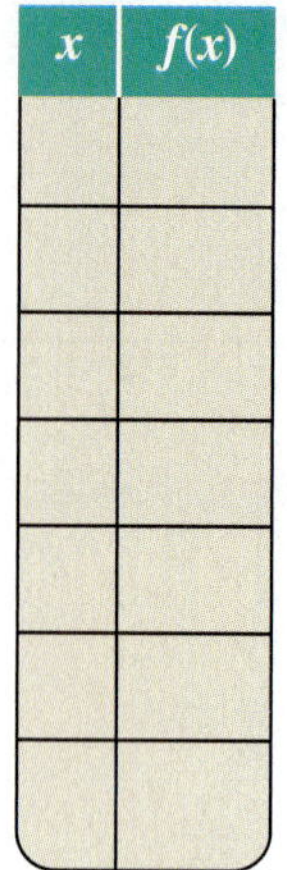

x	$f(x)$

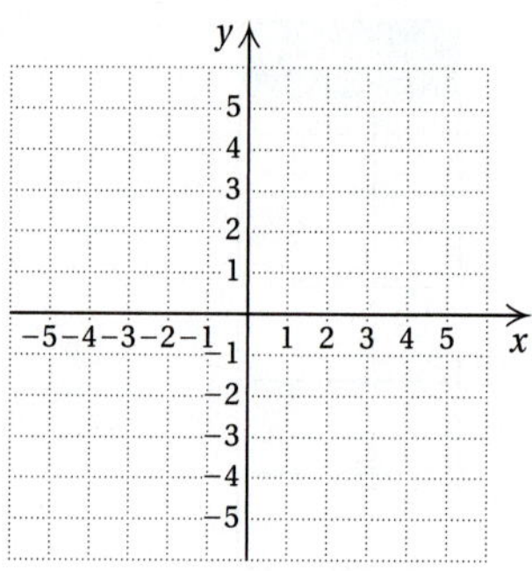

Vertex: (____,____)
Line of symmetry: $x =$ ____
__________ value: _____

11. $f(x) = 2x^2 + 5x - 2$

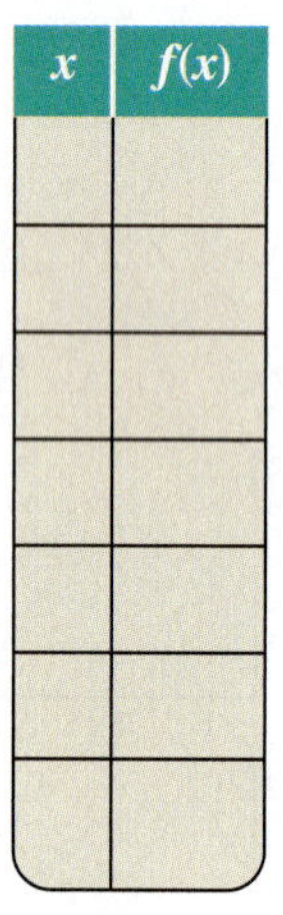

x	$f(x)$

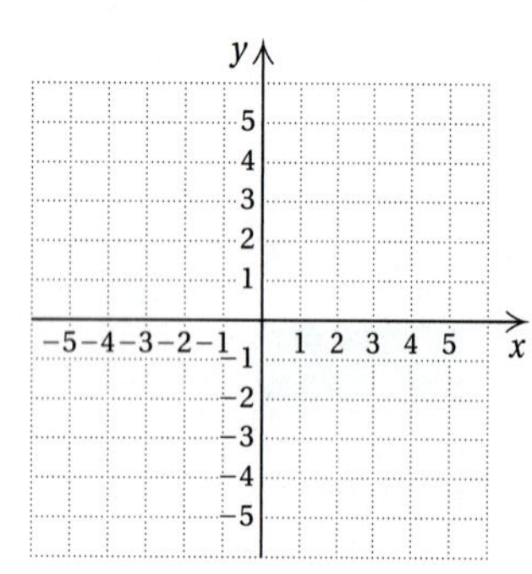

Vertex: (____,____)
Line of symmetry: $x =$ ____
__________ value: _____

12. $f(x) = -4x^2 - 7x + 2$

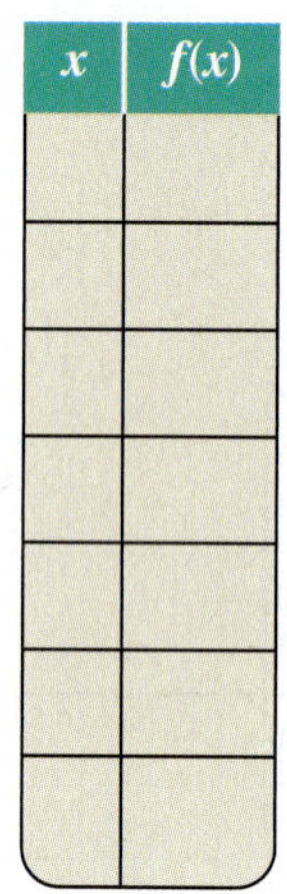

x	$f(x)$

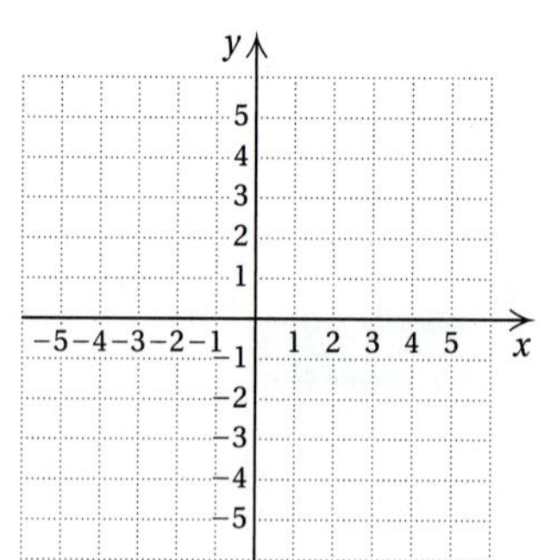

Vertex: (____,____)
Line of symmetry: $x =$ ____
__________ value: _____

b Find the x- and y-intercepts.

13. $f(x) = x^2 - 6x + 1$

14. $f(x) = x^2 + 2x + 12$

15. $f(x) = -x^2 + x + 20$

16. $f(x) = -x^2 + 5x + 24$

17. $f(x) = 4x^2 + 12x + 9$

18. $f(x) = 3x^2 - 6x + 1$

19. $f(x) = 4x^2 - x + 8$

20. $f(x) = 2x^2 + 4x - 1$

21. D_W Does the graph of every quadratic function have a y-intercept? Why or why not?

22. D_W Is it possible for the graph of a quadratic function to have only one x-intercept if the vertex is off the x-axis? Why or why not?

SKILL MAINTENANCE

Solve. [11.9b]

23. *Determining Medication Dosage.* A child's dosage D, in milligrams, of a medication varies directly as the child's weight w, in kilograms. To control a fever, a doctor suggests that a child who weighs 28 kg be given 420 mg of Tylenol.

a) Find an equation of variation.

b) How much Tylenol would be recommended for a child who weighs 42 kg?

24. *Calories Burned.* The number C of calories burned while exercising varies directly as the time t, in minutes, spent exercising. Harold exercises for 24 min on a stairmaster and burns 356 calories.

a) Find an equation of variation.

b) How many calories would he burn if he were to exercise for 48 min?

Find the variation constant and an equation of variation in which y varies inversely as x and the following are true. [11.9c]

25. $y = 125$ when $x = 2$

26. $y = 2$ when $x = 125$

Find the variation constant and an equation of variation in which y varies directly as x and the following are true. [11.9a]

27. $y = 125$ when $x = 2$

28. $y = 2$ when $x = 125$

SYNTHESIS

29. Use the TRACE and/or TABLE features of a graphing calculator to estimate the maximum or minimum values of the following functions.

a) $f(x) = 2.31x^2 - 3.135x - 5.89$

b) $f(x) = -18.8x^2 + 7.92x + 6.18$

30. Use the TRACE and/or TABLE features of a graphing calculator to confirm the maximum or minimum values given in Exercises 8, 11, and 12.

Graph.

31. $f(x) = |x^2 - 1|$

32. $f(x) = |x^2 + 6x + 4|$

33. $f(x) = |x^2 - 3x - 4|$

34. $f(x) = |2(x - 3)^2 - 5|$

35. A quadratic function has $(-1, 0)$ as one of its intercepts and $(3, -5)$ as its vertex. Find an equation for the function.

36. A quadratic function has $(4, 0)$ as one of its intercepts and $(-1, 7)$ as its vertex. Find an equation for the function.

37. Consider

$$f(x) = \frac{x^2}{8} + \frac{x}{4} - \frac{3}{8}.$$

Find the vertex, the line of symmetry, and the maximum or minimum value. Then draw the graph.

38. Use only the graph in Exercise 37 to approximate the solutions of each of the following equations.

a) $\frac{x^2}{8} + \frac{x}{4} - \frac{3}{8} = 0$

b) $\frac{x^2}{8} + \frac{x}{4} - \frac{3}{8} = 1$

c) $\frac{x^2}{8} + \frac{x}{4} - \frac{3}{8} = 2$

Use the INTERSECT feature of a graphing calculator to find the points of intersection of the graphs of each pair of functions.

39. $f(x) = x^2 - 4x + 2, \quad g(x) = 2 + x$

40. $f(x) = x^2 + 2x + 1, \quad g(x) = -2x^2 - 4x + 1$

16.7 MATHEMATICAL MODELING WITH QUADRATIC FUNCTIONS

Objectives

a Solve maximum–minimum problems involving quadratic functions.

b Fit a quadratic function to a set of data to form a mathematical model, and solve related applied problems.

We now consider some of the many situations in which quadratic functions can serve as mathematical models.

a Maximum–Minimum Problems

We have seen that for any quadratic function $f(x) = ax^2 + bx + c$, the value of $f(x)$ at the vertex is either a maximum or a minimum, meaning that either all outputs are smaller than that value for a maximum or larger than that value for a minimum.

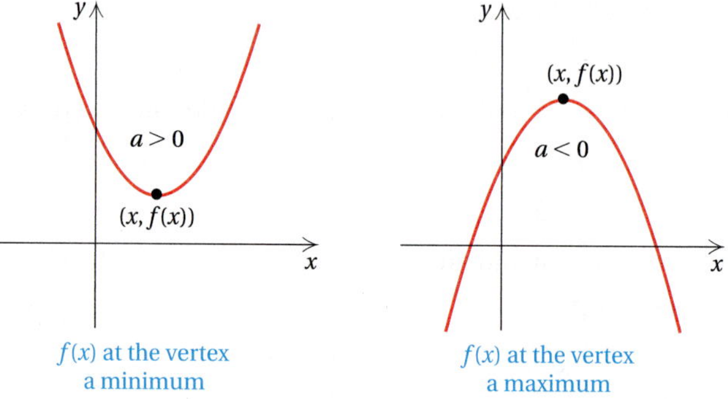

There are many types of applied problems in which we want to find a maximum or minimum value of a quantity. If a quadratic function can be used as a model, we can find such maximums or minimums by finding coordinates of the vertex.

EXAMPLE 1 *Fenced-In Land.* A farmer has 64 yd of fencing. What are the dimensions of the largest rectangular pen that the farmer can enclose?

1. **Familiarize.** We first make a drawing and label it. We let $l =$ the length of the pen and $w =$ the width. Recall the following formulas:

   Perimeter: $2l + 2w$;

   Area: $l \cdot w$.

l	w	A
22	10	220
20	12	240
18	14	252
18.5	13.5	249.75
12.4	19.6	243.04
15	17	255

To become familiar with the problem, let's choose some dimensions (shown at left) for which $2l + 2w = 64$ and then calculate the corresponding areas. What choice of l and w will maximize A?

2. **Translate.** We have two equations, one for perimeter and one for area:

$$2l + 2w = 64,$$
$$A = l \cdot w.$$

Let's use them to express A as a function of l or w, but not both. To express A in terms of w, for example, we solve for l in the first equation:

$$\begin{aligned} 2l + 2w &= 64 \\ 2l &= 64 - 2w \\ l &= \frac{64 - 2w}{2} \\ &= 32 - w. \end{aligned}$$

Substituting $32 - w$ for l, we get a quadratic function $A(w)$, or just A:

$$A = lw = (32 - w)w = 32w - w^2 = -w^2 + 32w.$$

3. **Carry out.** Note here that we are altering the third step of our five-step problem-solving strategy to "carry out" some kind of mathematical manipulation, because we are going to find the vertex rather than solve an equation. To do so, we complete the square as in Section 16.6:

$$\begin{aligned} A &= -w^2 + 32w && \text{This is a parabola opening down, so a maximum exists.} \\ &= -1(w^2 - 32w) && \text{Factoring out } -1 \\ &= -1(w^2 - 32w + 256 - 256) && \tfrac{1}{2}(-32) = -16;\ (-16)^2 = 256. \text{ We add 0, or } 256 - 256. \\ &= -1(w^2 - 32w + 256) + (-1)(-256) && \text{Using the distributive law} \\ &= -(w - 16)^2 + 256. \end{aligned}$$

The vertex is $(16, 256)$. Thus the maximum value is 256. It occurs when $w = 16$ and $l = 32 - w = 32 - 16 = 16$.

4. **Check.** We note that 256 is larger than any of the values found in the *Familiarize* step. To be more certain, we could make more calculations. We leave this to the student. We can also use the graph of the function to check the maximum value.

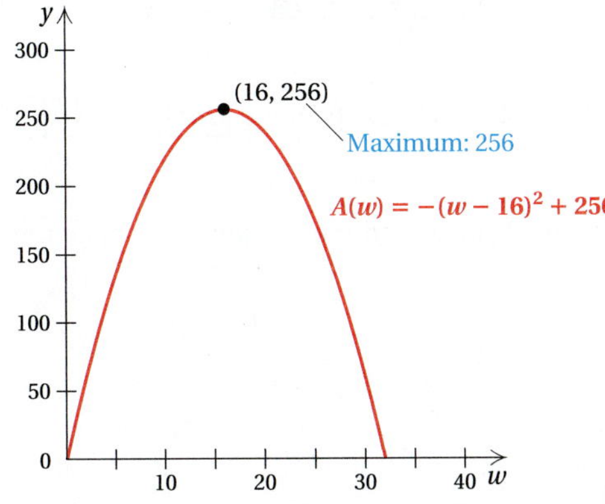

5. **State.** The largest rectangular pen that can be enclosed is 16 yd by 16 yd; that is, a square.

Do Exercise 1.

1. **Fenced-In Land.** A farmer has 100 yd of fencing. What are the dimensions of the largest rectangular pen that the farmer can enclose?

To familiarize yourself with the problem, complete the following table.

l	w	A
12	38	456
15	35	
24	26	
25	25	
26.2	23.8	

Answer on page A-64

CALCULATOR CORNER

Maximum and Minimum Values We can use a graphing calculator to find the maximum or minimum value of a quadratic function. Consider the quadratic function in Example 1, $A = -w^2 + 32w$. First, we replace w with x and A with y and graph the function in a window that displays the vertex of the graph. We choose $[0, 40, 0, 300]$, with Xscl $= 5$ and Yscl $= 20$. Now, we press [2nd] [CALC] 4 or [2nd] [CALC] [▽] [▽] [▽] [ENTER] to select the MAXIMUM feature from the CALC menu. We are prompted to select a left bound for the maximum point. This means that we must choose an x-value that is to the left of the x-value of the point where the maximum occurs. This can be done by using the left- and right-arrow keys to move the cursor to a point to the left of the maximum point or by keying in an appropriate value. Once this is done, we press [ENTER]. Now, we are prompted to select a right bound. We move the cursor to a point to the right of the maximum point or key in an appropriate value.

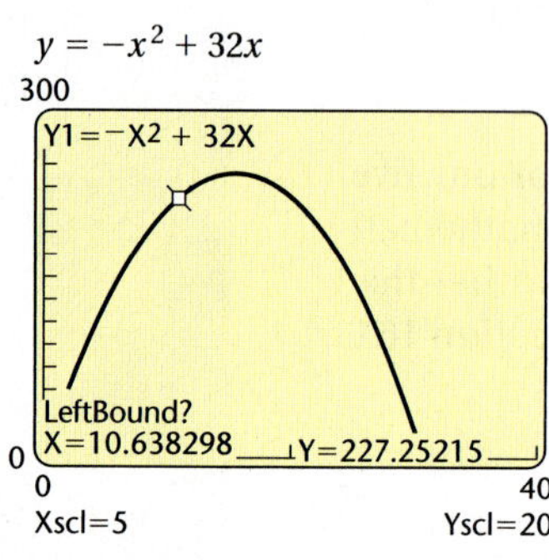

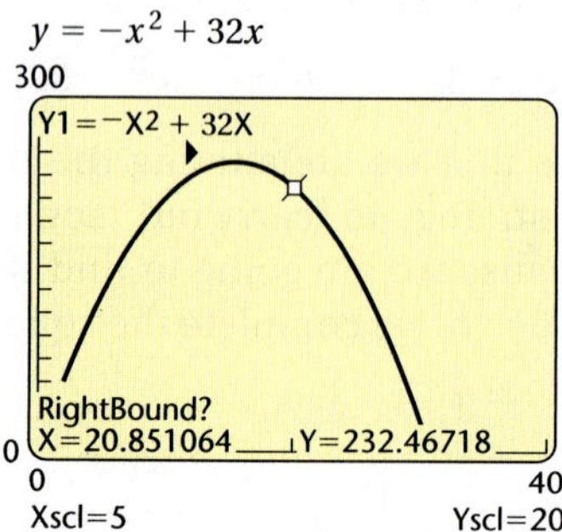

We press [ENTER] again. Finally, we are prompted to guess the x-value at which the maximum occurs. We move the cursor close to the maximum or key in an x-value. We press [ENTER] a third time and see that the maximum function value of 256 occurs when $x = 16$. (One or both coordinates of the maximum point might be approximations of the actual values, as shown with the x-value below, because of the method the calculator uses to find these values.)

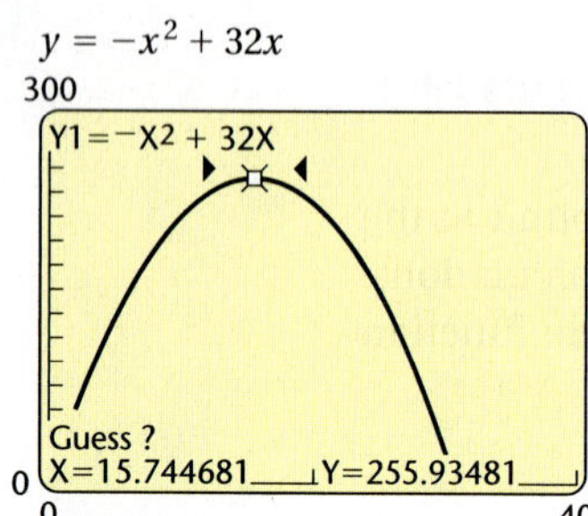

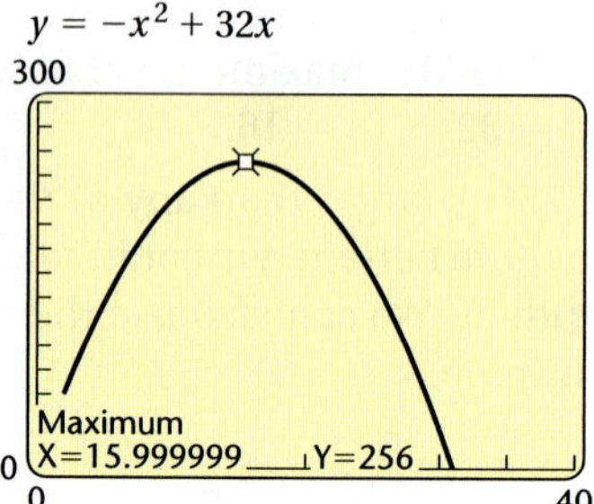

To find a minimum value, we select item 3, "minimum," from the CALC menu by pressing [2nd] [CALC] [3] [ENTER] or [2nd] [CALC] [▽] [▽] [ENTER].

Exercises: Use the maximum or minimum feature on a graphing calculator to find the maximum or minimum value of each function.

1. $y = 3x^2 - 6x + 4$
2. $y = 2x^2 + x + 5$
3. $y = -x^2 + 4x + 2$
4. $y = -4x^2 + 5x - 1$

b Fitting Quadratic Functions to Data

As we move through our study of mathematics, we develop a library of functions. These functions can serve as models for many applications. Some of them are graphed below. We have not considered the cubic or quartic functions in detail other than in the Calculator Corners (we leave that discussion to a later course), but we show them here for reference.

Linear function:

$f(x) = mx + b$

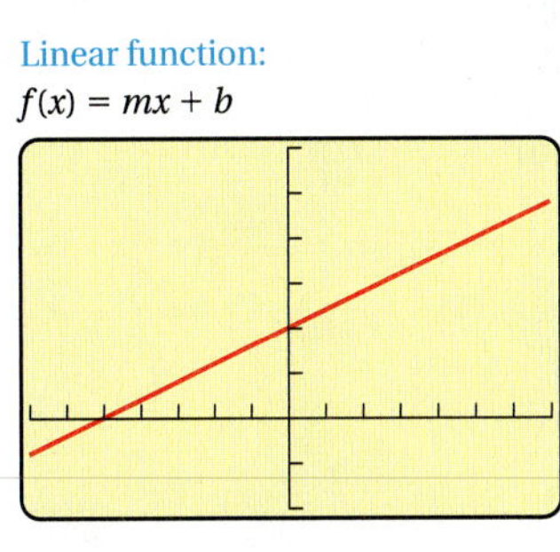

Quadratic function:

$f(x) = ax^2 + bx + c,\ a > 0$

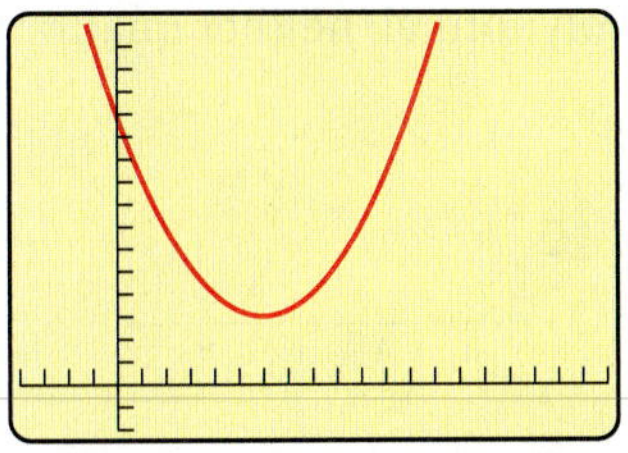

Quadratic function:

$f(x) = ax^2 + bx + c,\ a < 0$

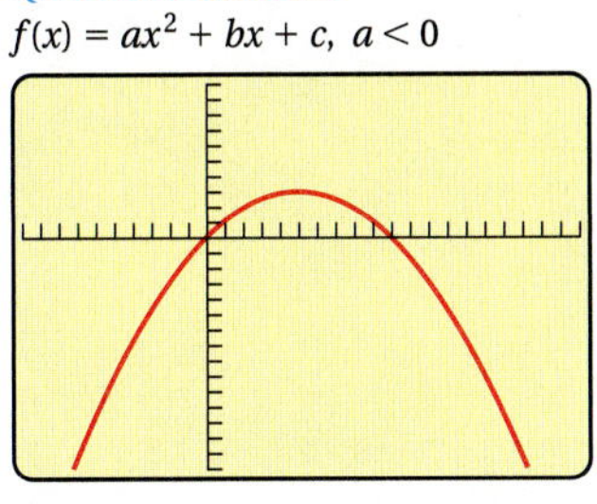

Absolute-value function:

$f(x) = |x|$

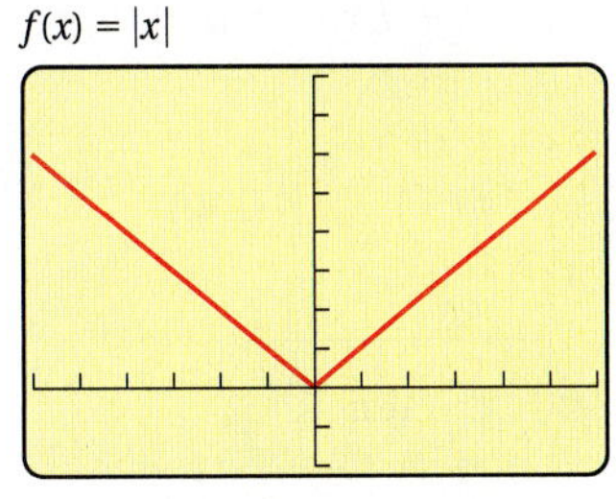

Cubic function:

$f(x) = ax^3 + bx^2 + cx + d,\ a > 0$

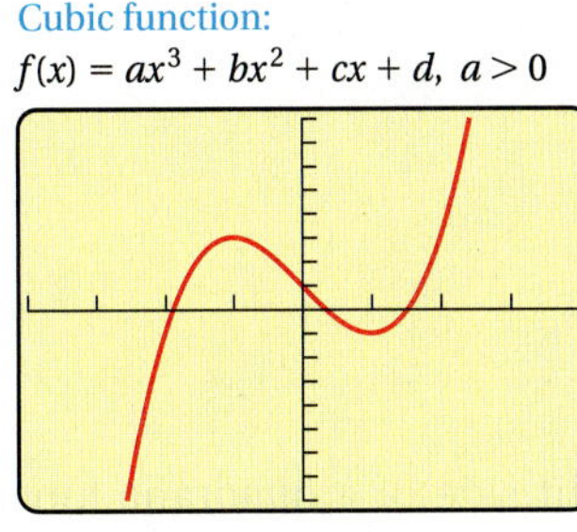

Quartic function:

$f(x) = ax^4 + bx^3 + cx^2 + dx + e,\ a > 0$

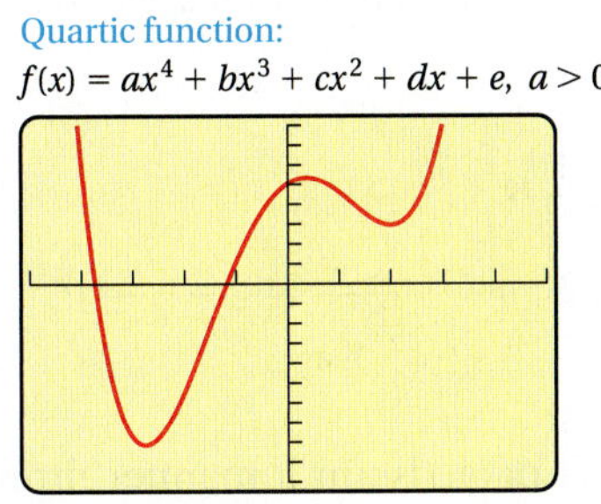

Now let's consider some real-world data. How can we decide which type of function might fit the data of a particular application? One simple way is to graph the data and look for a pattern resembling one of the graphs above. For example, data might be modeled by a linear function if the graph resembles a straight line. The data might be modeled by a quadratic function if the graph rises and then falls, or falls and then rises, in a curved manner resembling a parabola. For a quadratic, it might also just rise or fall in a curved manner as if following only one part of the parabola.

Choosing Models. For the scatterplots and graphs in Margin Exercises 2–5, determine which, if any, of the following functions might be used as a model for the data.

Linear, $f(x) = mx + b$;

Quadratic, $f(x) = ax^2 + bx + c$, $a > 0$;

Quadratic, $f(x) = ax^2 + bx + c$, $a < 0$;

Polynomial, neither quadratic nor linear

2. 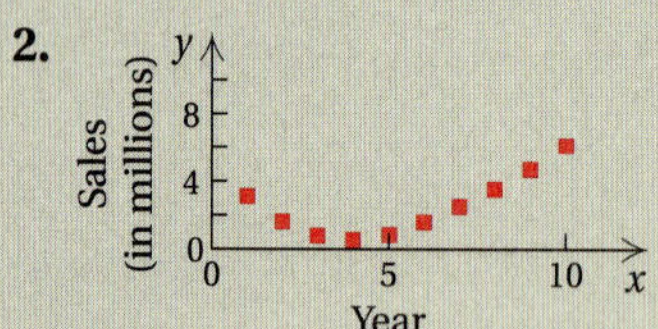

3. 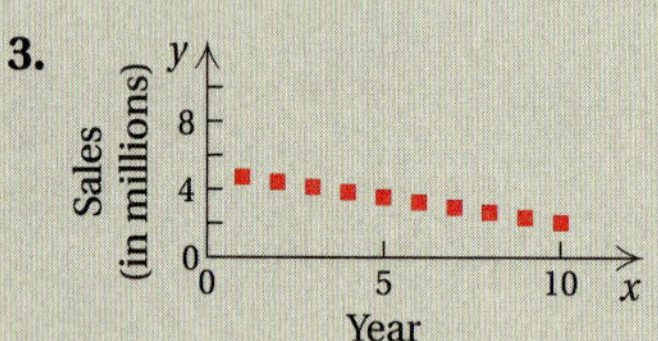

4. 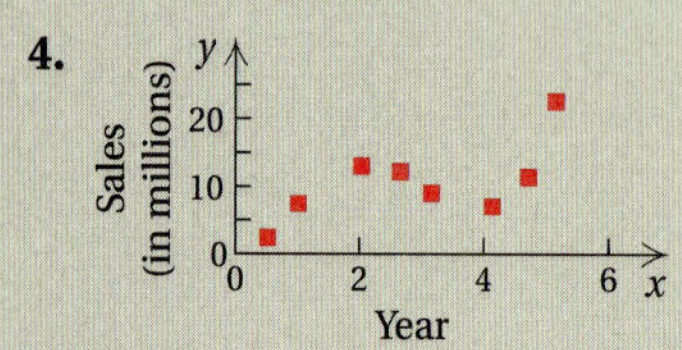

5. 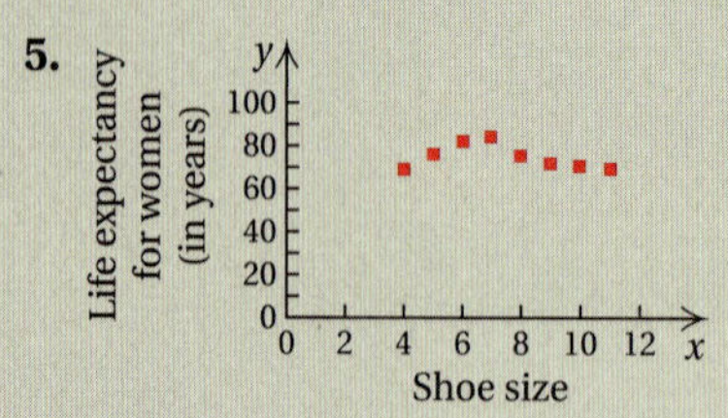

Source: Orthopedic Quarterly

Answers on page A-64

Let's now use our library of functions to see which, if any, might fit certain data situations.

EXAMPLES *Choosing Models.* For the scatterplots and graphs below, determine which, if any, of the following functions might be used as a model for the data.

Linear, $f(x) = mx + b$;

Quadratic, $f(x) = ax^2 + bx + c, a > 0$;

Quadratic, $f(x) = ax^2 + bx + c, a < 0$;

Polynomial, neither quadratic nor linear

2. 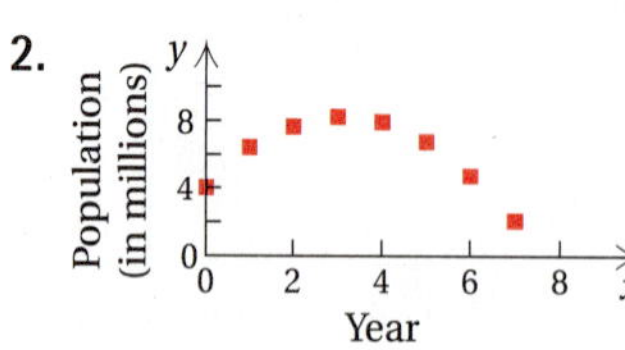

The data rise and then fall in a curved manner fitting a quadratic function $f(x) = ax^2 + bx + c, a < 0$.

3. 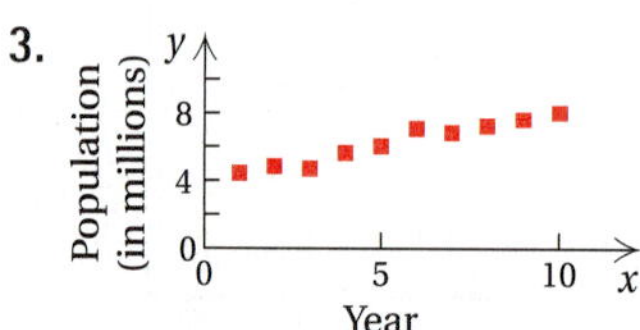

The data seem to fit a linear function $f(x) = mx + b$.

4. 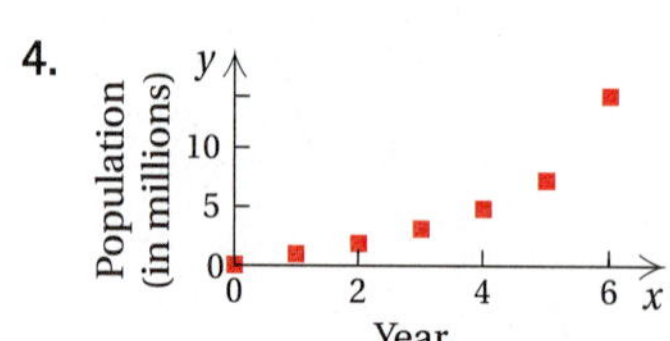

The data rise in a manner fitting the right side of a quadratic function $f(x) = ax^2 + bx + c, a > 0$.

5. **Driver Fatalities by Age**

Number of licensed drivers per 100,000 who died in motor vehicle accidents in 1990. The fatality rates for both the 70–79 group and the 80+ age group were lower than for the 15- to 24-year-olds.

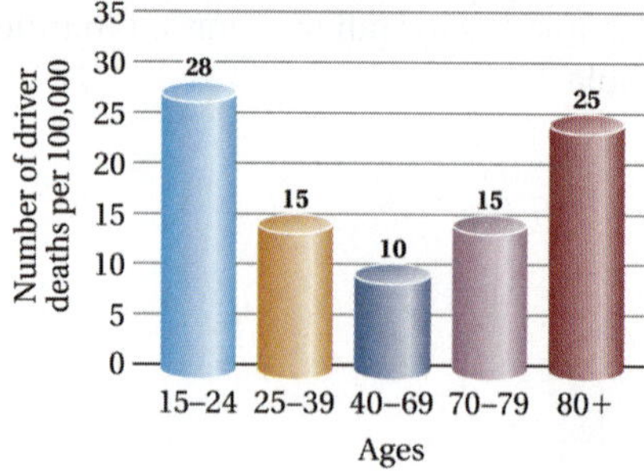

Source: National Highway Traffic Administration

The data fall and then rise in a curved manner fitting a quadratic function $f(x) = ax^2 + bx + c, a > 0$.

6.

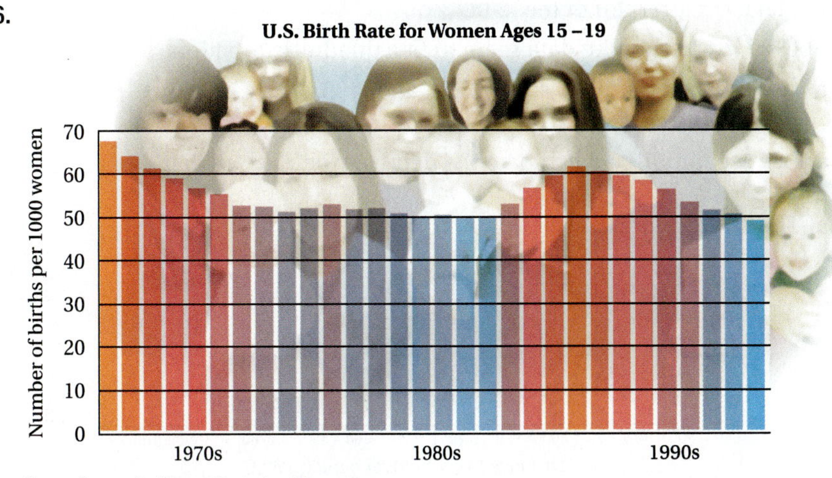

The data fall, then rise, then fall again. They do not appear to fit a linear or quadratic function but might fit a polynomial function that is neither quadratic nor linear.

Do Exercises 2–5 on the preceding page.

Whenever a quadratic function seems to fit a data situation, that function can be determined if at least three inputs and their outputs are known.

EXAMPLE 7 *River Depth.* The drawing below shows the cross section of a river. Typically rivers are deepest in the middle, with the depth decreasing to 0 at the edges. A hydrologist measures the depths D, in feet, of a river at distances x, in feet, from one bank. The results are listed in the table at right.

DISTANCE, x, FROM THE RIVERBANK (in feet)	DEPTH, D, OF THE RIVER (in feet)
0	0
15	10.2
25	17
50	20
90	7.2
100	0

6. Ticket Profits. Valley Community College is presenting a play. The profit P, in dollars, after x days is given in the following table. (Profit can be negative when costs exceed revenue. See Section 13.7.)

DAYS, x	PROFIT, P
0	$\$-100$
90	560
180	872
270	870
360	548
450	-100

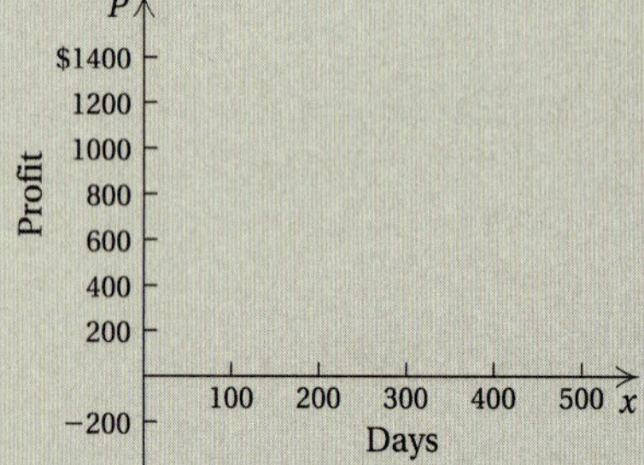

a) Make a scatterplot of the data.

b) Decide whether the data can be modeled by a quadratic function.

c) Use the data points $(0, -100)$, $(180, 872)$, and $(360, 548)$ to find a quadratic function that fits the data.

d) Use the function to estimate the profits after 225 days.

Answers on page A-64

a) Make a scatterplot of the data.

b) Decide whether the data seem to fit a quadratic function.

c) Use the data points $(0, 0)$, $(50, 20)$, and $(100, 0)$ to find a quadratic function that fits the data.

d) Use the function to estimate the depth of the river at 75 ft.

a) The scatterplot is as follows.

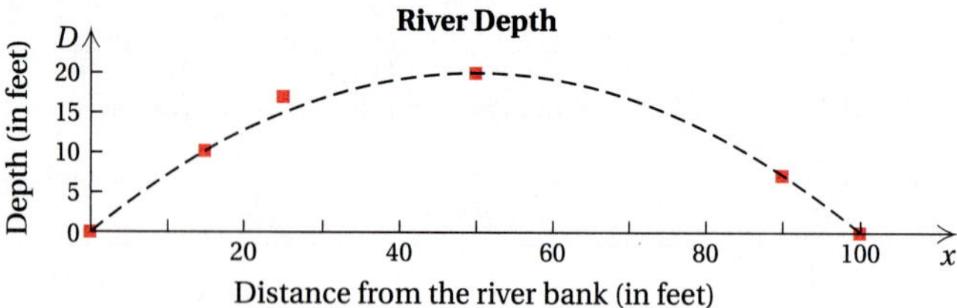

b) The data seem to rise and fall in a manner similar to a quadratic function. The dashed black line in the graph represents a sample quadratic function of fit. Note that it may not necessarily go through each point.

c) We are looking for a quadratic function

$$D(x) = ax^2 + bx + c.$$

We need to determine the constants a, b, and c. We use the three data points $(0, 0)$, $(50, 20)$, and $(100, 0)$ and substitute as follows:

$$0 = a \cdot 0^2 + b \cdot 0 + c,$$
$$20 = a \cdot 50^2 + b \cdot 50 + c,$$
$$0 = a \cdot 100^2 + b \cdot 100 + c.$$

After simplifying, we see that we need to solve the system

$$0 = c,$$
$$20 = 2{,}500a + 50b + c,$$
$$0 = 10{,}000a + 100b + c.$$

Since $c = 0$, the system reduces to a system of two equations in two variables:

$$20 = 2{,}500a + 50b, \quad (1)$$
$$0 = 10{,}000a + 100b. \quad (2)$$

We multiply equation (1) by -2, add, and solve for a (see Section 13.3):

$$-40 = -5{,}000a - 100b,$$
$$0 = 10{,}000a + 100b$$
$$-40 = 5000a \qquad \text{Adding}$$
$$\frac{-40}{5000} = a \qquad \text{Solving for } a$$
$$-0.008 = a.$$

Next, we substitute -0.008 for a in equation (2) and solve for b:

$$0 = 10{,}000(-0.008) + 100b$$
$$0 = -80 + 100b$$
$$80 = 100b$$
$$0.8 = b.$$

This gives us the quadratic function:

$$D(x) = -0.008x^2 + 0.8x.$$

d) To find the depth 75 ft from the riverbank, we substitute:

$$D(75) = -0.008(75)^2 + 0.8(75) = 15.$$

At a distance of 75 ft from the riverbank, the depth of the river is 15 ft.

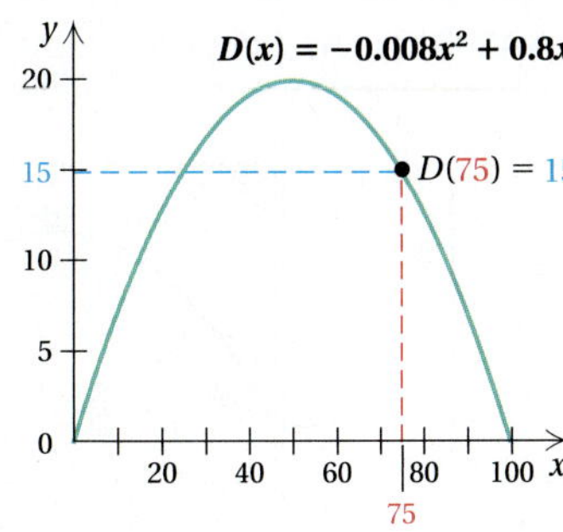

Do Exercise 6 on the preceding page.

CALCULATOR CORNER

Mathematical Modeling: Fitting a Quadratic Function to Data We can use the quadratic regression feature on a graphing calculator to fit a quadratic function to a set of data. The following table shows the average number of live births for women of various ages.

AGE	AVERAGE NUMBER OF LIVE BIRTHS PER 1000 WOMEN
16	34
18.5	86.5
22	111.1
27	113.9
32	84.5
37	35.4
42	6.8

a) Make a scatterplot of the data and verify that the data can be modeled with a quadratic function.

b) Fit a quadratic function to the data using the quadratic regression feature on a graphing calculator.

c) Use the function to estimate the average number of live births per 1000 women of age 20 and of age 30.

(continued)

a) We enter the data on the STAT list editor screen, turn on Plot 1, and plot the points as described in the Calculator Corner on p. 972. The data points rise and then fall in a manner consistent with a quadratic function.

L1	L2	L3 2
16	34	
18.5	86.5	
22	111.1	
27	113.9	
32	84.5	
37	35.4	
42	6.8	

L2(7)=6.8

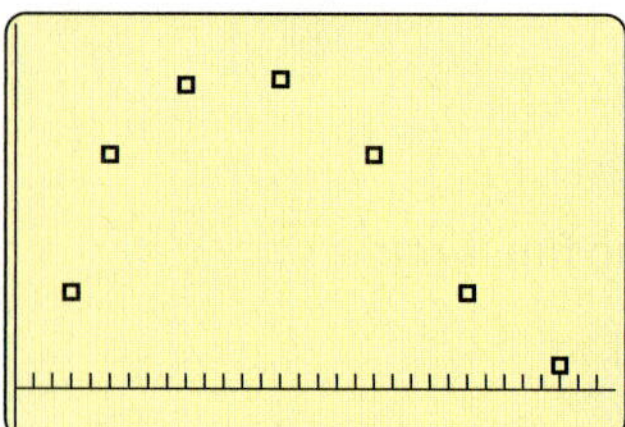

b) To fit a quadratic function to the data, we press [STAT] [▷] [5] [VARS] [▷] [1] [1] [ENTER]. The first three keystrokes select QuadReg from the STAT CALC menu and display the coefficients a, b, and c of the regression equation $y = ax^2 + bx + c$. The keystrokes [VARS] [▷] [1] [1] copy the regression equation to the equation-editor screen as y_1. We see that the regression equation is $y = -0.4868465035x^2 + 25.94985182x - 238.4892193$. We can press [ZOOM] [9] to see the regression equation graphed with the data points.

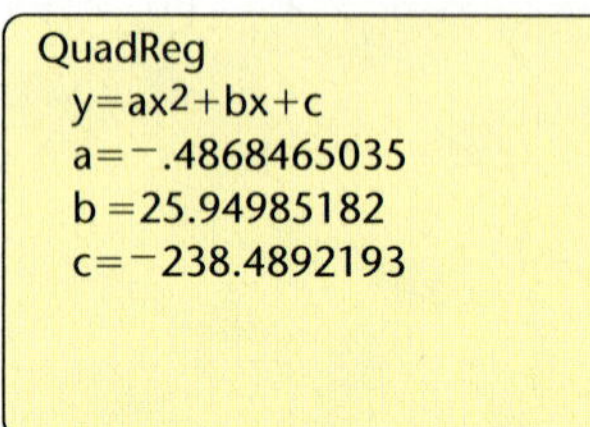

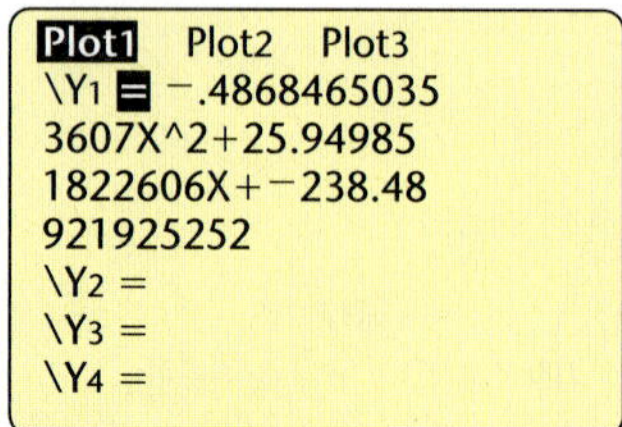

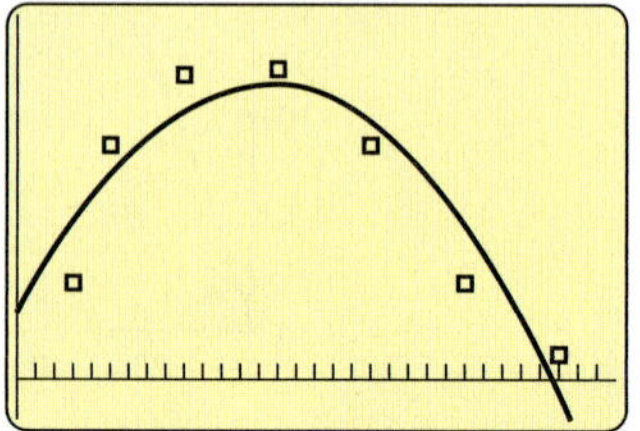

c) Since the equation is entered on the equation-editor screen, we can use a table set in ASK mode to estimate the average number of live births per 1000 women of age 20 and of age 30. We see that when $x = 20$, $y \approx 85.8$, and when $x = 30$, $y \approx 101.8$, so we estimate that there are about 85.8 live births per 1000 women of age 20 and about 101.8 live births per 1000 women of age 30.

X	Y1	
20	85.769	
30	101.84	

X=

Remember to turn off the STAT PLOT as described on p. 974 before you graph other equations.

Exercise:

1. Consider the data in the table in Example 7.
   a) Use a graphing calculator to make a scatterplot of the data.
   b) Use a graphing calculator to fit a quadratic function to the data. Compare this function with the one found in Example 7.
   c) Graph the quadratic function with the scatterplot.
   d) Use the function found in part (b) to estimate the depth of the river 75 ft from the riverbank. Compare this estimate with the one found in Example 7.

For Extra Help

Digital Video Tutor CD 14 Videotape 19 | InterAct Math | Math Tutor Center | MathXL | MyMathLab

a Solve.

1. *Architecture.* An architect is designing the floors for a hotel. Each floor is to be rectangular and is allotted 720 ft of security piping around walls outside the rooms. What dimensions of the atrium will allow an atrium at the bottom to have maximum area?

2. *Stained-Glass Window Design.* An artist is designing a rectangular stained-glass window with a perimeter of 84 in. What dimensions will yield the maximum area?

3. *Molding Plastics.* Economite Plastics plans to produce a one-compartment vertical file by bending the long side of an 8-in. by 14-in. sheet of plastic along two lines to form a U shape. How tall should the file be in order to maximize the volume that the file can hold?

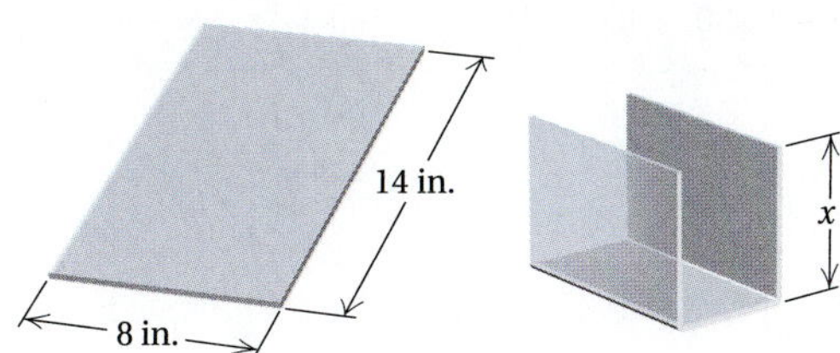

4. *Patio Design.* A stone mason has enough stones to enclose a rectangular patio with 60 ft of perimeter, assuming that the attached house forms one side of the rectangle. What is the maximum area that the mason can enclose? What should the dimensions of the patio be in order to yield this area?

5. *Minimizing Cost.* Aki's Bicycle Designs has determined that when x hundred bicycles are built, the average cost per bicycle is given by

$$C(x) = 0.1x^2 - 0.7x + 2.425,$$

where $C(x)$ is in hundreds of dollars. How many bicycles should the shop build in order to minimize the average cost per bicycle?

6. *Corral Design.* A rancher needs to enclose two adjacent rectangular corrals, one for sheep and one for cattle. If a river forms one side of the corrals and 180 yd of fencing is available, what is the largest total area that can be enclosed?

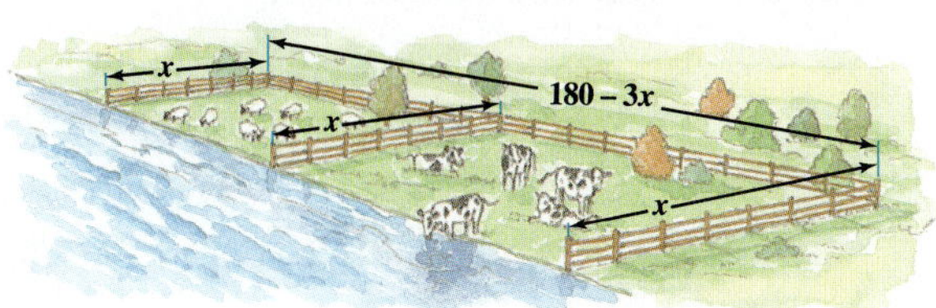

7. *Garden Design.* A farmer decides to enclose a rectangular garden, using the side of a barn as one side of the rectangle. What is the maximum area that the farmer can enclose with 40 ft of fence? What should the dimensions of the garden be in order to yield this area?

8. *Composting.* A rectangular compost container is to be formed in a corner of a fenced yard, with 8 ft of chicken wire completing the other two sides of the rectangle. If the chicken wire is 3 ft high, what dimensions of the base will maximize the volume of the container?

9. *Ticket Sales.* The number of tickets sold each day for an upcoming performance of Handel's Messiah is given by

$$N(x) = -0.4x^2 + 9x + 11,$$

where x is the number of days since the concert was first announced. When will daily ticket sales peak and how many tickets will be sold that day?

10. *Stock Prices.* The value of a share of R. P. Mugahti, in dollars, can be represented by $V(x) = x^2 - 6x + 13$, where x is the number of months after January 2003. What is the lowest value $V(x)$ will reach, and when did that occur?

Maximizing Profit. Recall (Section 13.7) that total profit P is the difference between total revenue R and total cost C. Given the following total-revenue and total-cost functions, find the total profit, the maximum value of the total profit, and the value of x at which it occurs.

11. $R(x) = 1000x - x^2$,
$C(x) = 3000 + 20x$

12. $R(x) = 200x - x^2$,
$C(x) = 5000 + 8x$

13. What is the maximum product of two numbers whose sum is 22? What numbers yield this product?

14. What is the maximum product of two numbers whose sum is 45? What numbers yield this product?

15. What is the minimum product of two numbers whose difference is 4? What are the numbers?

16. What is the minimum product of two numbers whose difference is 6? What are the numbers?

17. What is the maximum product of two numbers that add to -12? What numbers yield this product?

18. What is the minimum product of two numbers that differ by 9? What are the numbers?

Choosing Models. For the scatterplots and graphs in Exercises 19–26, determine which, if any, of the following functions might be used as a model for the data: Linear, $f(x) = mx + b$; quadratic, $f(x) = ax^2 + bx + c, a > 0$; quadratic, $f(x) = ax^2 + bx + c, a < 0$; polynomial, neither quadratic nor linear.

19.

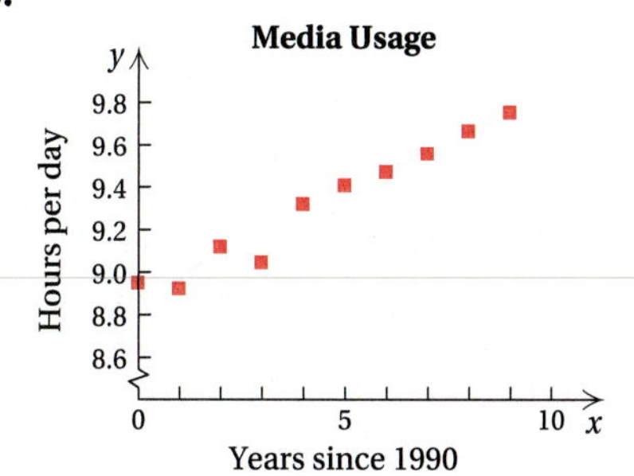

20.

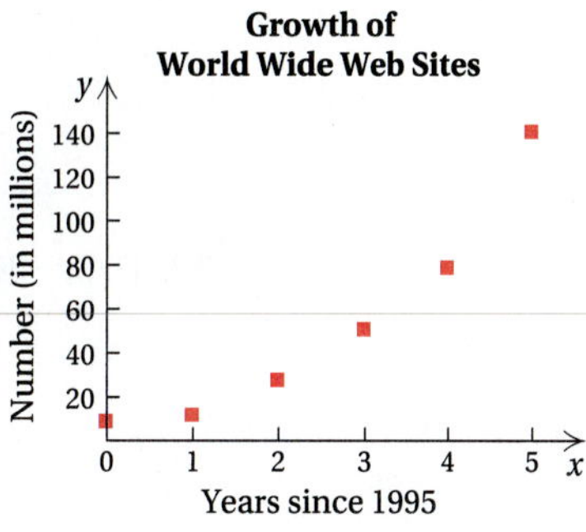

21.

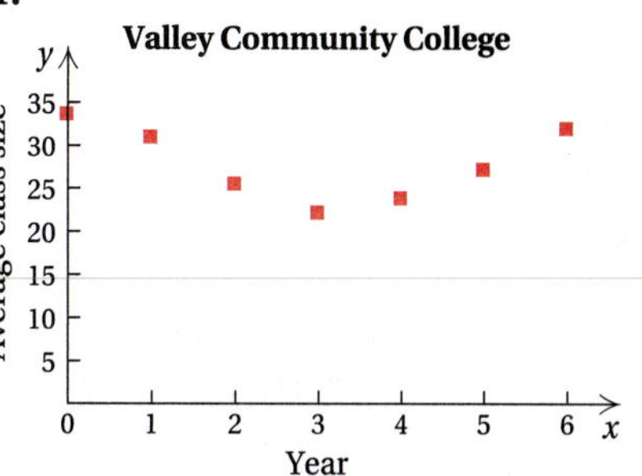

22.

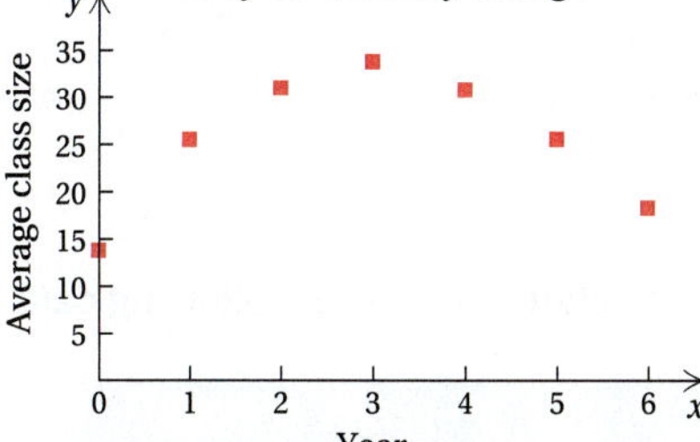

23.

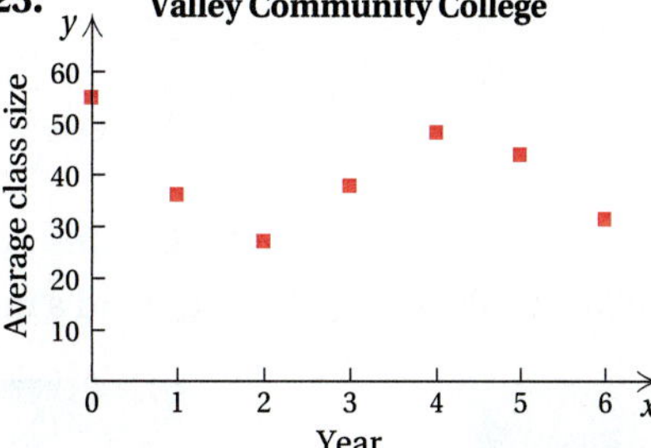

24.

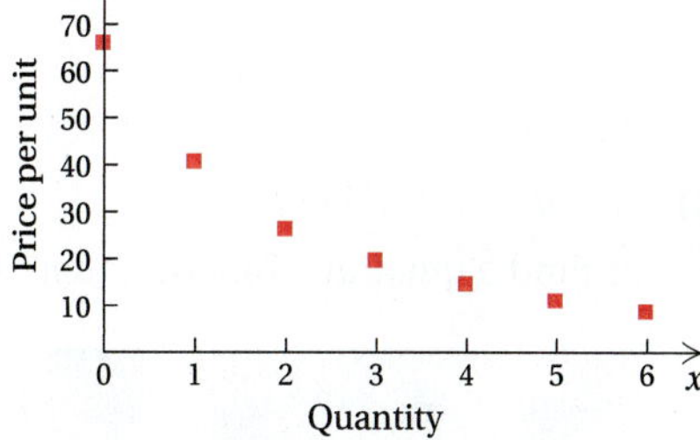

25.

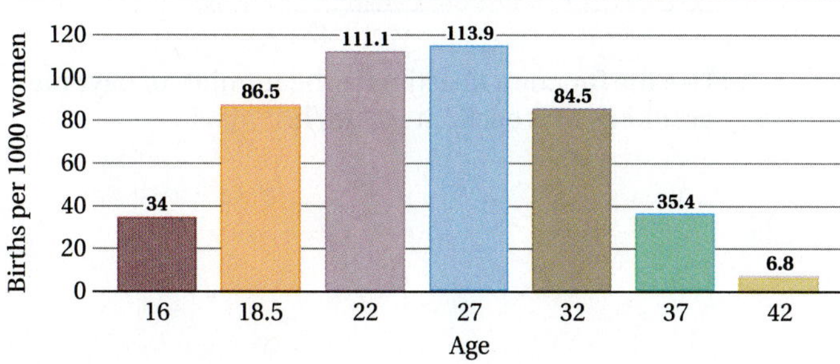

Source: U.S. Centers for Disease Control

26.

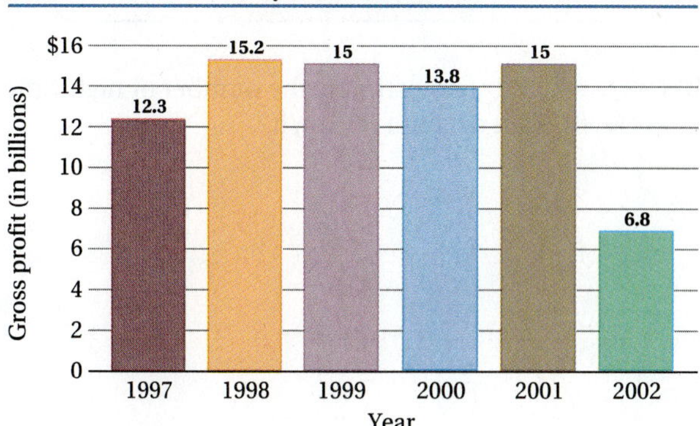

Source: The New York Stock Exchange

Find a quadratic function that fits the set of data points.

27. $(1, 4), (-1, -2), (2, 13)$

28. $(1, 4), (-1, 6), (-2, 16)$

29. $(2, 0), (4, 3), (12, -5)$

30. $(-3, -30), (3, 0), (6, 6)$

31. *Nighttime Accidents.*

a) Find a quadratic function that fits the following data.

TRAVEL SPEED (in kilometers per hour)	NUMBER OF NIGHTTIME ACCIDENTS (for every 200 million kilometers driven)
60	400
80	250
100	250

b) Use the function to estimate the number of nighttime accidents that occur at 50 km/h.

32. *Daytime Accidents.*

a) Find a quadratic function that fits the following data.

TRAVEL SPEED (in kilometers per hour)	NUMBER OF DAYTIME ACCIDENTS (for every 200 million kilometers driven)
60	100
80	130
100	200

b) Use the function to estimate the number of daytime accidents that occur at 50 km/h.

33. *Archery.* The Olympic flame tower at the 1992 Summer Olympics was lit at a height of about 27 m by a flaming arrow that was launched about 63 m from the base of the tower. If the arrow landed about 63 m beyond the tower, find a quadratic function that expresses the height h of the arrow as a function of the distance d that it traveled horizontally.

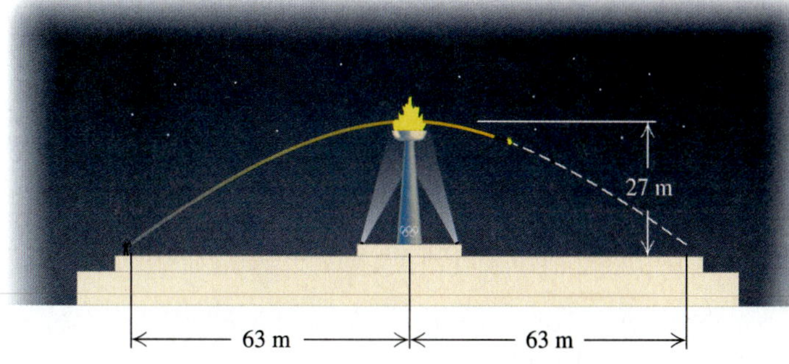

34. *Pizza Prices.* Pizza Unlimited has the following prices for pizzas.

DIAMETER	PRICE
8 in.	$ 6.00
12 in.	$ 8.50
16 in.	$11.50

Is price a quadratic function of diameter? It probably should be, because the price should be proportional to the area, and the area is a quadratic function of the diameter. (The area of a circular region is given by $A = \pi r^2$ or $(\pi/4) \cdot d^2$.)

a) Express price as a quadratic function of diameter using the data points $(8, 6)$, $(12, 8.50)$, and $(16, 11.50)$.

b) Use the function to find the price of a 14-in. pizza.

35. DW Explain the restrictions that should be placed on the domains of the quadratic functions found in Exercises 11 and 31 and why such restrictions are needed.

36. DW Explain how the leading coefficient of a quadratic function can be used to determine whether a maximum or minimum function value exists.

SKILL MAINTENANCE

Multiply and simplify. Assume that all expressions under radicals represent nonnegative numbers. [15.3a]

37. $\sqrt[4]{5x^3y^5}\,\sqrt[4]{125x^2y^3}$

38. $\sqrt{9a^3}\,\sqrt{16ab^4}$

Solve. [15.6a, b]

39. $\sqrt{4x - 4} = \sqrt{x + 4} + 1$

40. $\sqrt{5x - 4} + \sqrt{13 - x} = 7$

41. $-35 = \sqrt{2x + 5}$

42. $\sqrt{7x - 5} = \sqrt{4x + 7}$

SYNTHESIS

43. *Sony Electronics, Inc.* Use the REGRESSION feature on your graphing calculator to fit a quartic function to the data in Exercise 26.

44. The sum of the base and the height of a triangle is 38 cm. Find the dimensions for which the area is a maximum, and find the maximum area.

16.8 POLYNOMIAL AND RATIONAL INEQUALITIES

Objectives

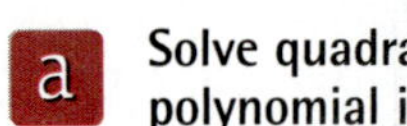

a Solve quadratic and other polynomial inequalities.

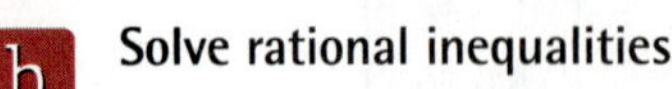

b Solve rational inequalities.

a Quadratic and Other Polynomial Inequalities

Inequalities like the following are called **quadratic inequalities:**

$$x^2 + 3x - 10 < 0, \qquad 5x^2 - 3x + 2 \geq 0.$$

In each case, we have a polynomial of degree 2 on the left. We will solve such inequalities in two ways. The first method provides understanding and the second yields the more efficient method.

The first method for solving a quadratic inequality, such as $ax^2 + bx + c > 0$, is by considering the graph of a related function, $f(x) = ax^2 + bx + c$.

EXAMPLE 1 Solve: $x^2 + 3x - 10 > 0$.

Consider the function $f(x) = x^2 + 3x - 10$ and its graph. The graph opens up since the leading coefficient ($a = 1$) is positive. We find the x-intercepts by setting the polynomial equal to 0 and solving:

$$x^2 + 3x - 10 = 0$$
$$(x + 5)(x - 2) = 0$$
$$x + 5 = 0 \quad or \quad x - 2 = 0$$
$$x = -5 \quad or \quad x = 2.$$

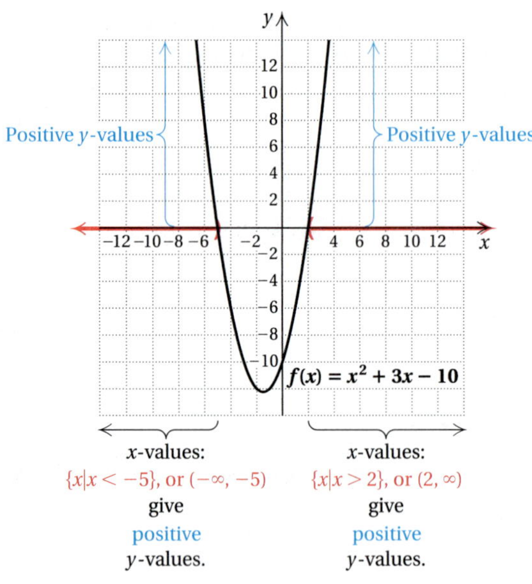

Values of y will be positive to the left and right of the intercepts, as shown. Thus the solution set of the inequality is

$$\{x \mid x < -5 \text{ or } x > 2\}, \quad \text{or} \quad (-\infty, -5) \cup (2, \infty).$$

Do Exercise 1.

We can solve any inequality by considering the graph of a related function and finding x-intercepts as in Example 1. In some cases, we may need to use the quadratic formula to find the intercepts.

1. Solve by graphing:

$$x^2 + 2x - 3 > 0.$$

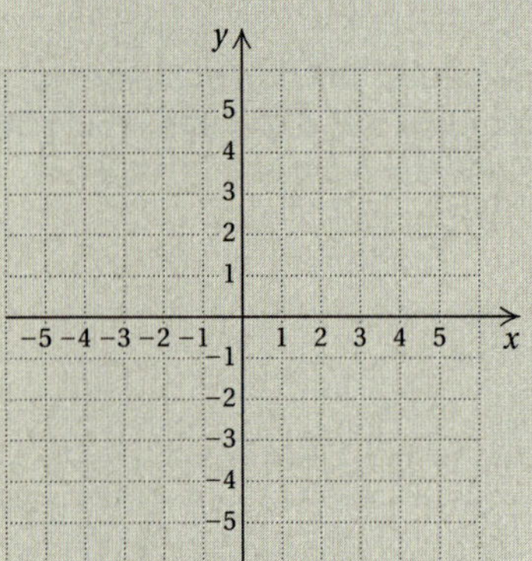

Answer on page A-64

EXAMPLE 2 Solve: $x^2 + 3x - 10 < 0$.

Looking again at the graph of $f(x) = x^2 + 3x - 10$ or at least visualizing it tells us that y-values are negative for those x-values between -5 and 2.

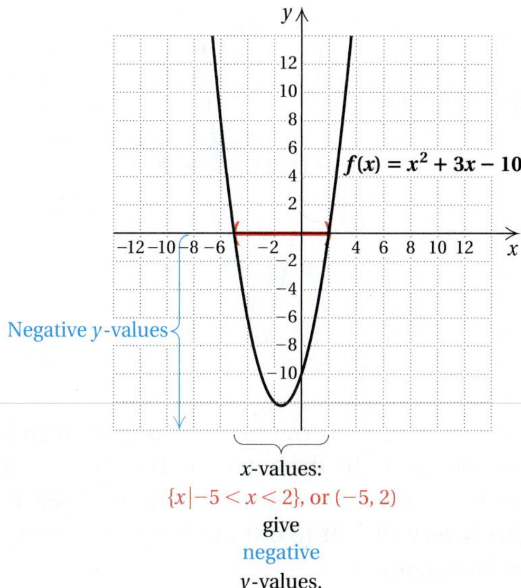

That is, the solution set is $\{x|-5 < x < 2\}$, or $(-5, 2)$.

When an inequality contains $\leq$ or $\geq$, the x-values of the x-intercepts must be included. Thus the solution set of the inequality $x^2 + 3x - 10 \leq 0$ is $\{x|-5 \leq x \leq 2\}$, or $[-5, 2]$.

Do Exercises 2 and 3. (Exercise 3 is on the following page.)

2. Solve by graphing:

$x^2 + 2x - 3 < 0$.

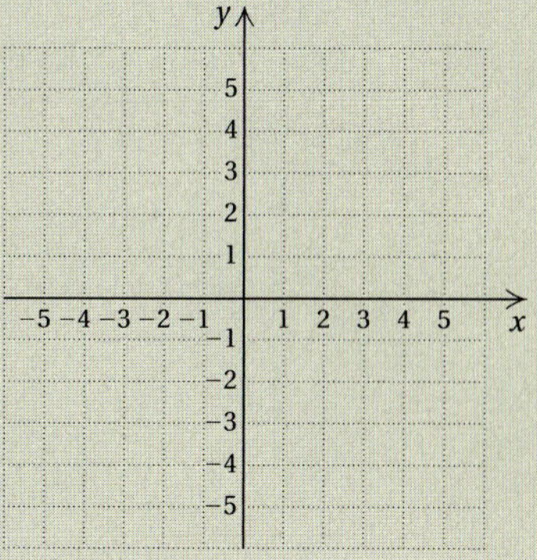

Answer on page A-64

CALCULATOR CORNER

Solving Polynomial Inequalities We can solve polynomial inequalities graphically. Consider the inequality in Example 2, $x^2 + 3x - 10 < 0$. We first graph the function $f(x) = x^2 + 3x - 10$. Then we use the ZERO feature to find the solutions of the equation $f(x) = 0$, or $x^2 + 3x - 10 = 0$. The solutions, -5 and 2, divide the number line into three intervals, $(-\infty, -5)$, $(-5, 2)$, and $(2, \infty)$.

Since we want to find the values of x for which $f(x) < 0$, we look for the interval(s) on which the function values are negative. That is, we note where the graph lies below the x-axis. This occurs in the interval $(-5, 2)$, so the solution set is $\{x|-5 < x < 2\}$, or $(-5, 2)$.

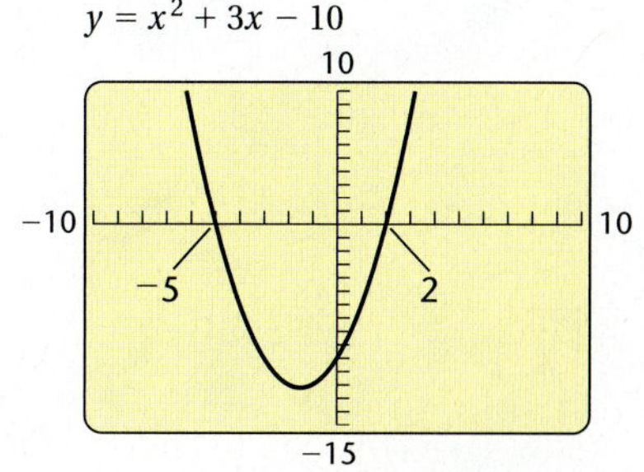

If we were solving the inequality $x^2 + 3x - 10 > 0$, we would look for the intervals on which the graph lies above the x-axis. We can see that $x^2 + 3x - 10 > 0$ for $\{x|x < -5 \text{ or } x > 2\}$, or $(-\infty, -5) \cup (2, \infty)$. If the inequality symbol were $\leq$ or $\geq$, we would include the endpoints of the intervals as well.

Exercises: Solve graphically.

1. $x^2 + 3x - 4 > 0$
2. $x^2 - x - 6 < 0$
3. $6x^3 + 9x^2 - 6x \leq 0$
4. $x^3 - 16x \geq 0$

3. Solve by graphing:

$x^2 + 2x - 3 \leq 0.$

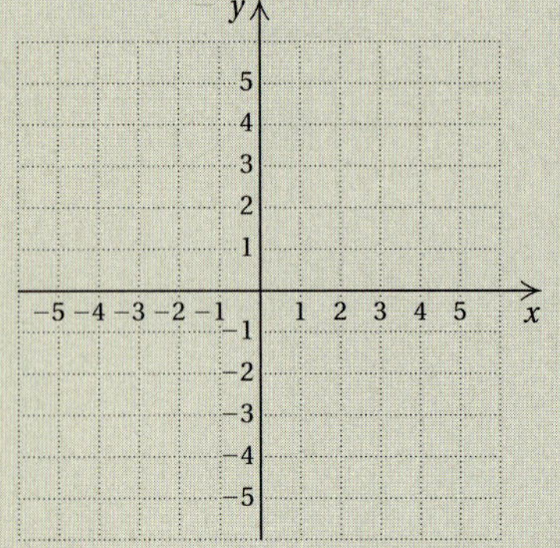

Answer on page A-64

We now consider a more efficient method for solving polynomial inequalities. The preceding discussion provides the understanding for this method. In Examples 1 and 2, we see that the x-intercepts divide the number line into intervals.

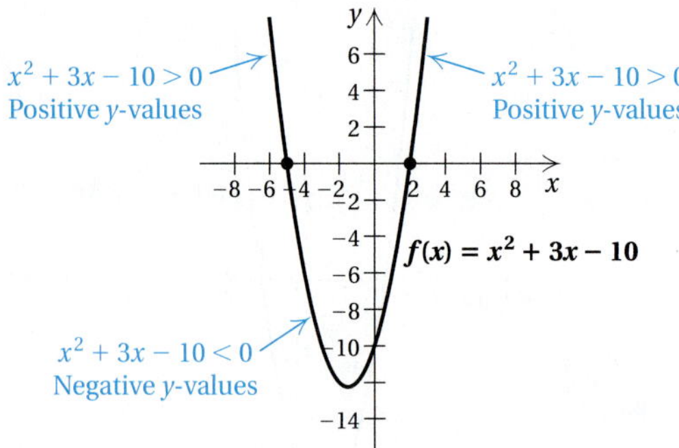

If a function has a positive output for one number in an interval, it will be positive for all the numbers in the interval. The same is true for negative outputs. Thus we can merely make a test substitution in each interval to solve the inequality. This is very similar to our method of using test points to graph a linear inequality in a plane.

EXAMPLE 3 Solve: $x^2 + 3x - 10 < 0$.

We set the polynomial equal to 0 and solve. The solutions of $x^2 + 3x - 10 = 0$, or $(x + 5)(x - 2) = 0$, are -5 and 2. We then locate them on a number line as follows. Note that the numbers divide the number line into three intervals, which we will call A, B, and C.

A B C

-5 2

We choose a test number in interval A, say -7, and substitute -7 for x in the function $f(x) = x^2 + 3x - 10$:

$$f(-7) = (-7)^2 + 3(-7) - 10$$
$$= 49 - 21 - 10 = 18. \qquad \text{Thus, } f(-7) > 0.$$

Note that $18 > 0$, so the function values will be positive for any number in interval A.

Next, we try a test number in interval B, say 1, and find the corresponding function value:

$$f(1) = 1^2 + 3(1) - 10$$
$$= 1 + 3 - 10 = -6. \qquad \text{Thus, } f(1) < 0.$$

Note that $-6 < 0$, so the function values will be negative for any number in interval B.

Next, we try a test number in interval C, say 4, and find the corresponding function value:

$$f(4) = 4^2 + 3(4) - 10$$
$$= 16 + 12 - 10 = 18. \qquad \text{Thus, } f(4) > 0.$$

Note that $18 > 0$, so the function values will be positive for any number in interval C.

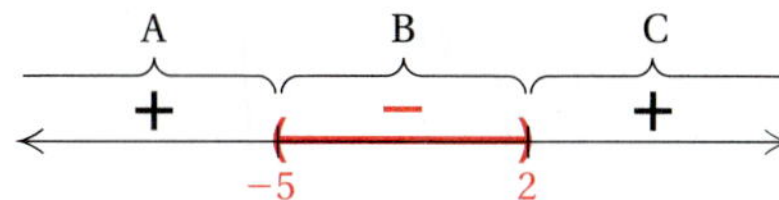

We are looking for numbers x for which $f(x) = x^2 + 3x - 10 < 0$. Thus any number x in interval B is a solution. If the inequality had been $\leq$, it would have been necessary to include the intercepts -5 and 2 in the solution set as well. The solution set is $\{x \mid -5 < x < 2\}$, or the interval $(-5, 2)$.

Solve using the method of Example 3.

4. $x^2 + 3x > 4$

5. $x^2 + 3x \leq 4$

To solve a polynomial inequality:

1. Get 0 on one side, set the expression on the other side equal to 0, and solve to find the x-intercepts.
2. Use the numbers found in step (1) to divide the number line into intervals.
3. Substitute a number from each interval into the related function. If the function value is positive, then the expression will be positive for all numbers in the interval. If the function value is negative, then the expression will be negative for all numbers in the interval.
4. Select the intervals for which the inequality is satisfied and write set-builder or interval notation for the solution set.

Do Exercises 4 and 5.

EXAMPLE 4 Solve: $5x(x + 3)(x - 2) \geq 0$.

The solutions of $f(x) = 0$, or $5x(x + 3)(x - 2) = 0$, are -3, 0, and 2. They divide the real-number line into four intervals, as shown below.

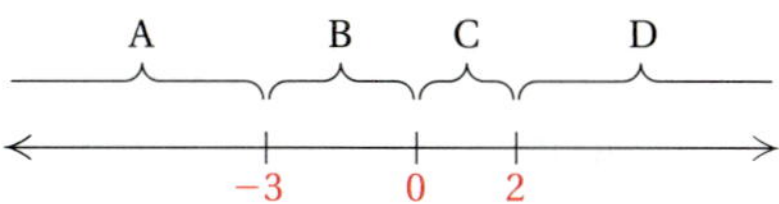

We try test numbers in each interval:

A: Test -5, $f(-5) = 5(-5)(-5 + 3)(-5 - 2) = -350 < 0.$

B: Test -2, $f(-2) = 5(-2)(-2 + 3)(-2 - 2) = 40 > 0.$

C: Test 1, $f(1) = 5(1)(1 + 3)(1 - 2) = -20 < 0.$

D: Test 3, $f(3) = 5(3)(3 + 3)(3 - 2) = 90 > 0.$

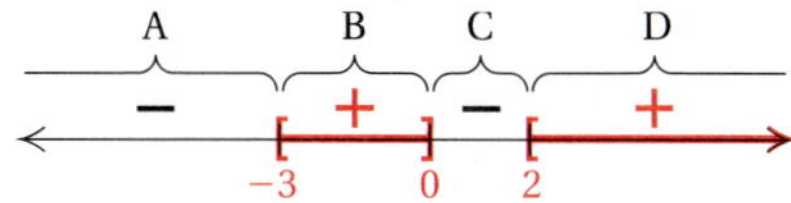

Answers on page A-64

6. Solve: $6x(x + 1)(x - 1) < 0$.

The expression is positive for values of x in intervals B and D. Since the inequality symbol is $\geq$, we need to include the x-intercepts. The solution set of the inequality is

$$\{x \mid -3 \leq x \leq 0 \text{ or } 2 \leq x\}, \quad \text{or} \quad [-3, 0] \cup [2, \infty).$$

We visualize this with the graph below.

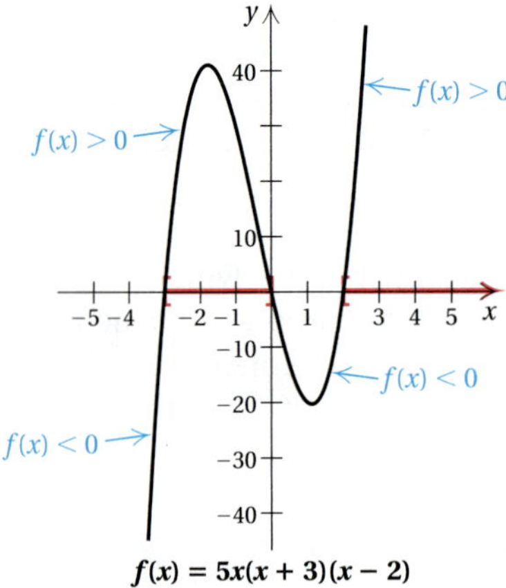

Do Exercise 6.

b Rational Inequalities

We adapt the preceding method when an inequality involves rational expressions. We call these **rational inequalities.**

EXAMPLE 5 Solve: $\dfrac{x - 3}{x + 4} \geq 2$.

We write a related equation by changing the $\geq$ symbol to $=$:

$$\frac{x - 3}{x + 4} = 2.$$

Then we solve this related equation. First, we multiply both sides of the equation by the LCM, which is $x + 4$:

$$(x + 4) \cdot \frac{x - 3}{x + 4} = (x + 4) \cdot 2$$
$$x - 3 = 2x + 8$$
$$-11 = x.$$

With rational inequalities, we also need to determine those numbers for which the rational expression is not defined—that is, those numbers that make the denominator 0. We set the denominator equal to 0 and solve: $x + 4 = 0$, or $x = -4$. Next, we use the numbers -11 and -4 to divide the number line into intervals, as shown below.

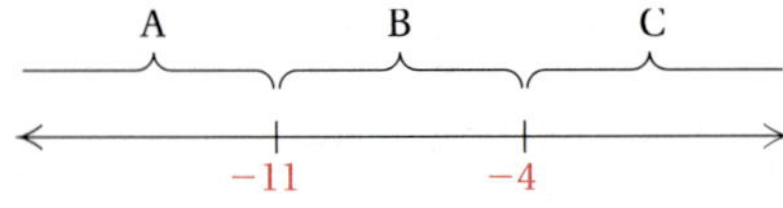

We try test numbers in each interval to see if each satisfies the original inequality.

Answer on page A-64

A: Test -15,

$$\frac{x-3}{x+4} \geq 2$$

$$\begin{array}{c|c} \dfrac{-15-3}{-15+4} \; ? \; 2 \\ \dfrac{18}{11} & \text{FALSE} \end{array}$$

Since the inequality is false for $x = -15$, the number -15 is not a solution of the inequality. Interval A is *not* part of the solution set.

B: Test -8,

$$\frac{x-3}{x+4} \geq 2$$

$$\begin{array}{c|c} \dfrac{-8-3}{-8+4} \; ? \; 2 \\ \dfrac{11}{4} & \text{TRUE} \end{array}$$

Since the inequality is true for $x = -8$, the number -8 is a solution of the inequality. Interval B *is* part of the solution set.

C: Test 1,

$$\frac{x-3}{x+4} \geq 2$$

$$\begin{array}{c|c} \dfrac{1-3}{1+4} \; ? \; 2 \\ -\dfrac{2}{5} & \end{array}$$

Since the inequality is false for $x = 1$, the number 1 is not a solution of the inequality. Interval C is *not* part of the solution set.

The solution set includes the interval B. The number -11 is also included since the inequality symbol is $\geq$ and -11 is a solution of the related equation. The number -4 is not included; it is not an allowable replacement because it results in division by 0. Thus the solution set of the original inequality is

$$\{x \mid -11 \leq x < -4\}, \quad \text{or} \quad [-11, -4).$$

To solve a rational inequality:

1. Change the inequality symbol to an equals sign and solve the related equation.
2. Find the numbers for which any rational expression in the inequality is not defined.
3. Use the numbers found in steps (1) and (2) to divide the number line into intervals.
4. Substitute a number from each interval into the inequality. If the number is a solution, then the interval to which it belongs is part of the solution set.
5. Select the intervals for which the inequality is satisfied and write set-builder or interval notation for the solution set.

Do Exercises 7 and 8.

Solve.

7. $\dfrac{x+1}{x-2} \geq 3$

8. $\dfrac{x}{x-5} < 2$

Answers on page A-64

ALGEBRAIC–GRAPHICAL CONNECTION

Let's compare the algebraic solution of Example 5 to a graphical solution. Look at the graph of the function

$$f(x) = \frac{x - 3}{x + 4}.$$

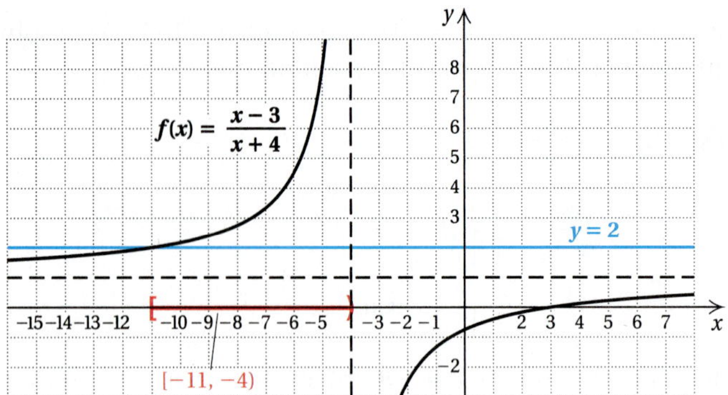

We see that the solutions of the inequality

$$\frac{x - 3}{x + 4} \geq 2$$

can be found by drawing the line $y = 2$ and determining all x-values for which $f(x) \geq 2$.

16.8 EXERCISE SET

For Extra Help

Digital Video Tutor CD 14 Videotape 19 | InterAct Math | Math Tutor Center | MathXL | MyMathLab

a Solve algebraically and verify results from the graph.

1. $(x - 6)(x + 2) > 0$

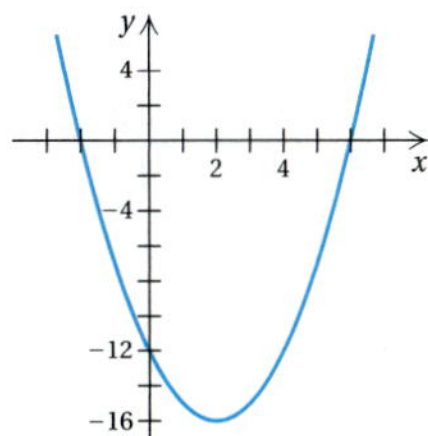

2. $(x - 5)(x + 1) > 0$

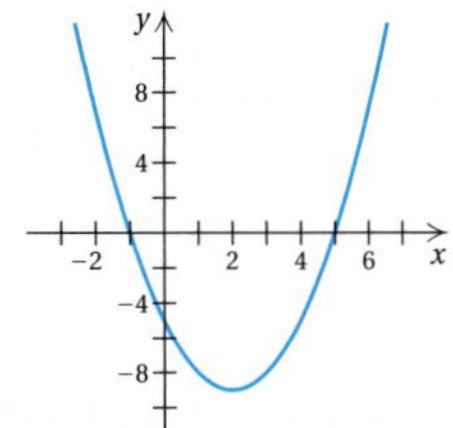

3. $4 - x^2 \geq 0$

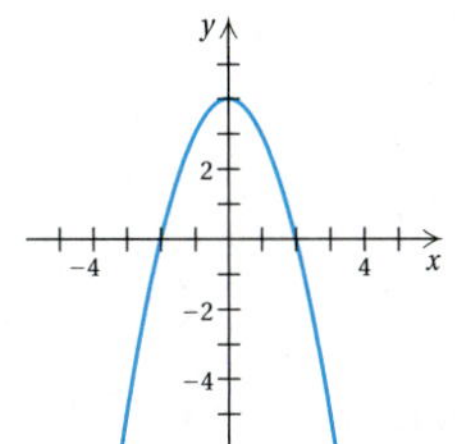

4. $9 - x^2 \leq 0$

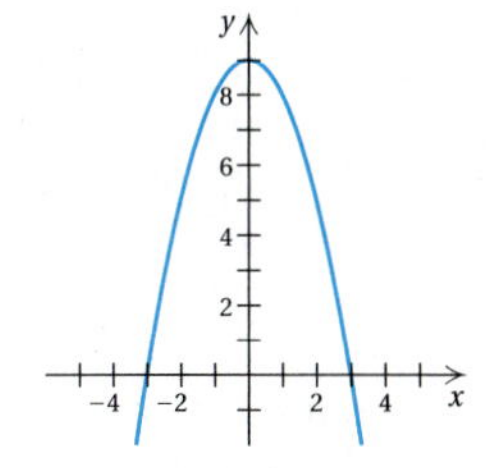

Solve.

5. $3(x + 1)(x - 4) \leq 0$

6. $(x - 7)(x + 3) \leq 0$

7. $x^2 - x - 2 < 0$

8. $x^2 + x - 2 < 0$

9. $x^2 - 2x + 1 \geq 0$

10. $x^2 + 6x + 9 < 0$

11. $x^2 + 8 < 6x$

12. $x^2 - 12 > 4x$

13. $3x(x + 2)(x - 2) < 0$

14. $5x(x + 1)(x - 1) > 0$

15. $(x + 9)(x - 4)(x + 1) > 0$

16. $(x - 1)(x + 8)(x - 2) < 0$

17. $(x + 3)(x + 2)(x - 1) < 0$

18. $(x - 2)(x - 3)(x + 1) < 0$

b Solve.

19. $\frac{1}{x - 6} < 0$

20. $\frac{1}{x + 4} > 0$

21. $\frac{x + 1}{x - 3} > 0$

22. $\frac{x - 2}{x + 5} < 0$

23. $\frac{3x + 2}{x - 3} \leq 0$

24. $\frac{5 - 2x}{4x + 3} \leq 0$

25. $\frac{x - 1}{x - 2} > 3$

26. $\frac{x + 1}{2x - 3} < 1$

27. $\dfrac{(x-2)(x+1)}{x-5} < 0$

28. $\dfrac{(x+4)(x-1)}{x+3} > 0$

29. $\dfrac{x+3}{x} \leq 0$

30. $\dfrac{x}{x-2} \geq 0$

31. $\dfrac{x}{x-1} > 2$

32. $\dfrac{x-5}{x} < 1$

33. $\dfrac{x-1}{(x-3)(x+4)} < 0$

34. $\dfrac{x+2}{(x-2)(x+7)} > 0$

35. $3 < \dfrac{1}{x}$

36. $\dfrac{1}{x} \leq 2$

37. $\dfrac{(x-1)(x+2)}{(x+3)(x-4)} > 0$

38. $\dfrac{x^2 - 11x + 30}{x^2 - 8x - 9} \geq 0$

39. **D**W Describe a method that could be used to create quadratic inequalities that have no solution.

40. **D**W Describe a method that could be used to create quadratic inequalities that have all real numbers as solutions.

SKILL MAINTENANCE

Simplify. [15.3b]

41. $\sqrt[3]{\dfrac{125}{27}}$

42. $\sqrt{\dfrac{25}{4a^2}}$

43. $\sqrt{\dfrac{16a^3}{b^4}}$

44. $\sqrt[3]{\dfrac{27c^5}{343d^3}}$

Add or subtract. [15.4a]

45. $3\sqrt{8} - 5\sqrt{2}$

46. $7\sqrt{45} - 2\sqrt{20}$

47. $5\sqrt[3]{16a^4} + 7\sqrt[3]{2a}$

48. $3\sqrt{10} + 8\sqrt{20} - 5\sqrt{80}$

SYNTHESIS

49. Use a graphing calculator to solve Exercises 11, 22, and 25 by graphing two curves, one for each side of an inequality.

50. Use a graphing calculator to solve each of the following.

a) $x + \dfrac{1}{x} < 0$

b) $x - \sqrt{x} \geq 0$

c) $\frac{1}{3}x^3 - x + \frac{2}{3} \leq 0$

Solve.

51. $x^2 - 2x \leq 2$

52. $x^2 + 2x > 4$

53. $x^4 + 2x^2 > 0$

54. $x^4 + 3x^2 \leq 0$

55. $\left|\dfrac{x+2}{x-1}\right| < 3$

56. *Total Profit.* A company determines that its total profit on the production and sale of x units of a product is given by

$$P(x) = -x^2 + 812x - 9600.$$

a) A company makes a profit for those nonnegative values of x for which $P(x) > 0$. Find the values of x for which the company makes a profit.

b) A company loses money for those nonnegative values of x for which $P(x) < 0$. Find the values of x for which the company loses money.

57. *Height of a Thrown Object.* The function

$$H(t) = -16t^2 + 32t + 1920$$

gives the height H of an object thrown from a cliff 1920 ft high, after time t seconds.

a) For what times is the height greater than 1920 ft?

b) For what times is the height less than 640 ft?

CHAPTER 16

Summary and Review

The review that follows is meant to prepare you for a chapter exam. It consists of two parts. The first part is a checklist of some of the Study Tips referred to in this and preceding chapters, as well as a list of important properties and formulas. The second part is the Review Exercises. These provide practice exercises for the exam, together with references to section objectives so you can go back and review. Before beginning, stop and look back over the skills you have obtained. What skills in mathematics do you have now that you did not have before studying this chapter?

STUDY TIPS CHECKLIST

The foundation of all your study skills is TIME!

- ☐ Are you beginning to prepare for the final exam?
- ☐ Are you studying in a systematic way for the test on this chapter?
- ☐ As you take a test, are you working a bit slower, but more deliberately, to avoid mistakes?
- ☐ As you take a test, are you using estimation or an alternate technique to check your answers?
- ☐ As you take a test and work the applied problems, are you checking possible answers in the original problems?

IMPORTANT PROPERTIES AND FORMULAS

Principle of Square Roots: $x^2 = d$ has solutions $\sqrt{d}$ and $-\sqrt{d}$.

Quadratic Formula: $x = \dfrac{-b \pm \sqrt{b^2 - 4ac}}{2a}$; *Discriminant:* $b^2 - 4ac$

The *vertex* of the graph of $f(x) = ax^2 + bx + c$ is $\left(-\dfrac{b}{2a}, \dfrac{4ac - b^2}{4a}\right)$, or $\left(-\dfrac{b}{2a}, f\left(-\dfrac{b}{2a}\right)\right)$.

The *line of symmetry* of the graph of $f(x) = ax^2 + bx + c$ is $x = -\dfrac{b}{2a}$.

REVIEW EXERCISES

1. a) Solve: $2x^2 - 7 = 0$. [16.1a]
b) Find the x-intercepts of $f(x) = 2x^2 - 7$.

Solve. [16.2a]

2. $14x^2 + 5x = 0$

3. $x^2 - 12x + 27 = 0$

4. $4x^2 + 3x + 1 = 0$

5. $x^2 - 7x + 13 = 0$

6. $4x(x - 1) + 15 = x(3x + 4)$

7. $x^2 + 4x + 1 = 0$. Give exact solutions and approximate solutions to three decimal places.

8. $\dfrac{x}{x - 2} + \dfrac{4}{x - 6} = 0$

9. $\dfrac{x}{4} - \dfrac{4}{x} = 2$

10. $15 = \dfrac{8}{x + 2} - \dfrac{6}{x - 2}$

11. Solve $x^2 + 4x + 1 = 0$ by completing the square. Show your work. [16.1b]

12. *Hang Time.* Use the function $V(T) = 48T^2$. A basketball player has a vertical leap of 39 in. What is his hang time? [16.1c]

13. *DVD Player Screen.* The width of a rectangular screen on a portable DVD player is 5 cm less than the length. The area is 126 cm^2. Find the length and the width. [16.3a]

14. *Picture Matting.* A picture mat measures 12 in. by 16 in.; 140 in^2 of picture shows. Find the width of the mat. [16.3a]

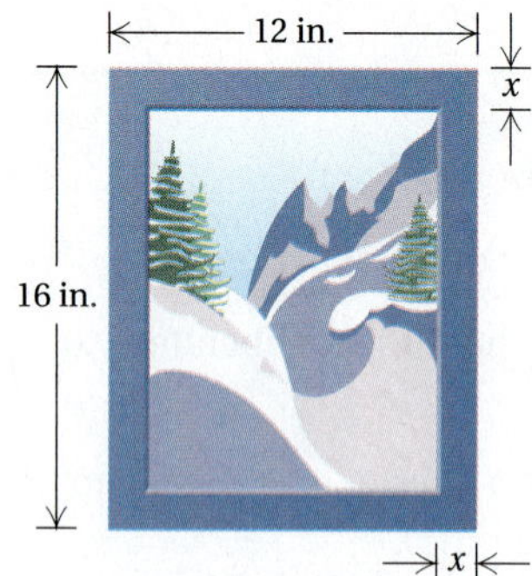

15. *Motorcycle Travel.* During the first part of a trip, a motorcyclist travels 50 mi at a certain speed. The rider travels 80 mi on the second part of the trip at a speed that is 10 mph slower. The total time for the trip is 3 hr. What is the speed on each part of the trip? [16.3a]

Determine the nature of the solutions of the equation. [16.4a]

16. $x^2 + 3x - 6 = 0$

17. $x^2 + 2x + 5 = 0$

Write a quadratic equation having the given solutions. [16.4b]

18. $\frac{1}{5}, -\frac{3}{5}$

19. -4, only solution

Solve for the indicated letter. [16.3b]

20. $N = 3\pi\sqrt{\frac{1}{p}}$, for p

21. $2A = \frac{3B}{T^2}$, for T

Solve. [16.4c]

22. $x^4 - 13x^2 + 36 = 0$

23. $15x^{-2} - 2x^{-1} - 1 = 0$

24. $(x^2 - 4)^2 - (x^2 - 4) - 6 = 0$

25. $x - 13\sqrt{x} + 36 = 0$

For each quadratic function in Exercises 26–28, find and label **(a)** the vertex, **(b)** the line of symmetry, and **(c)** the maximum or minimum value. Then **(d)** graph the function. [16.5c], [16.6a]

26. $f(x) = -\frac{1}{2}(x - 1)^2 + 3$

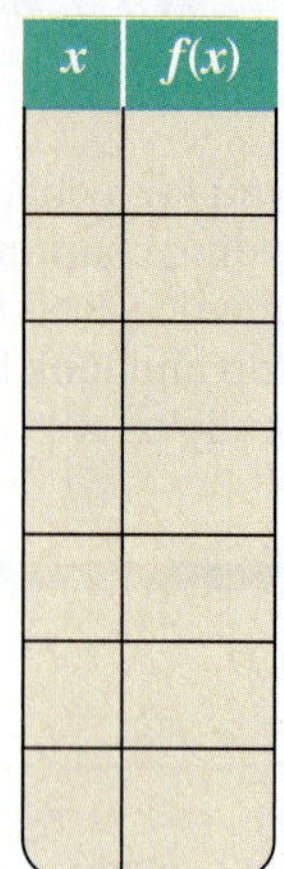

x	$f(x)$

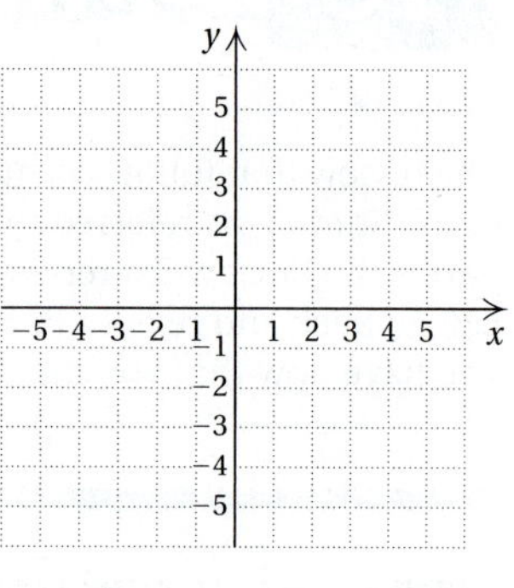

Vertex: (____, ____)
Line of symmetry: $x =$ ____
__________ value: _____

27. $f(x) = x^2 - x + 6$

x	$f(x)$

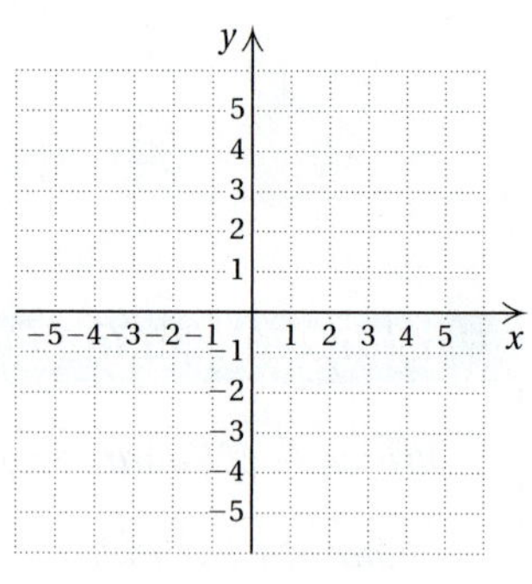

Vertex: (____, ____)
Line of symmetry: $x =$ ____
__________ value: _____

28. $f(x) = -3x^2 - 12x - 8$

Vertex: (____, ____)
Line of symmetry:
$x =$ ____
__________ value:
_____

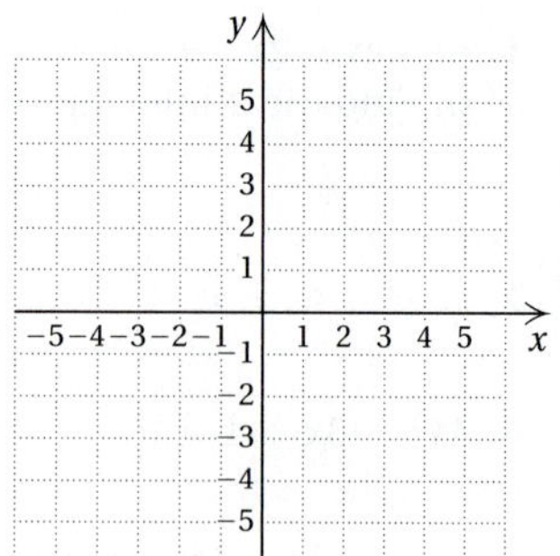

29. Find the x- and y-intercepts: [16.6b]

$$f(x) = x^2 - 9x + 14.$$

30. What is the minimum product of two numbers whose difference is 22? What numbers yield this product? [16.7a]

31. Find the quadratic function that fits the data points $(0, -2)$, $(1, 3)$, and $(3, 7)$. [16.7b]

32. *Live Births by Age.* The average number of live births per 1000 women rises and falls according to age, as seen in the following bar graph. [16.7b]

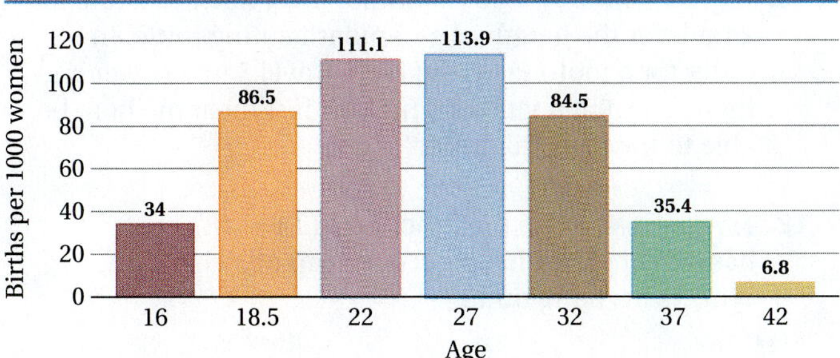

Source: U.S. Centers for Disease Control

a) Use the data points $(16, 34)$, $(27, 113.9)$, and $(37, 35.4)$ to fit a quadratic function to the data.
b) Use the quadratic function to estimate the number of live births per 1000 women of age 30.

Solve. [16.8a, b]

33. $(x + 2)(x - 1)(x - 2) > 0$

34. $\dfrac{(x + 4)(x - 1)}{(x + 2)} < 0$

35. DW Explain as many characteristics as you can of the graph of the quadratic function $f(x) = ax^2 + bx + c$. [16.5a, b, c], [16.6a, b]

36. DW Explain how the x-intercepts of a quadratic function can be used to help find the vertex of the function. What piece of information would still be missing? [16.6a, b]

SKILL MAINTENANCE

Certain objectives from four particular sections will be retested on the chapter test. The objectives are listed with the practice problems that follow.

37. *SAT Math Scores.* The following table lists SAT math scores for recent years. [12.5e]

YEARS SINCE 1994, x	SAT MATH SCORE
0	504
1	505
2	508
3	511
4	512
5	511
6	514

Source: The College Board

a) Use the data points $(1, 505)$ and $(6, 514)$ to fit a linear function to the data.
b) Use the linear function to predict SAT math scores in 2003 and in 2010.

38. Add and simplify: [11.4a]

$$\frac{x}{x^2 - 3x + 2} + \frac{2}{x^2 - 5x + 6}.$$

39. Multiply and simplify: [15.3a]

$$\sqrt[3]{9t^6}\,\sqrt[3]{3s^4t^9}.$$

40. Solve: $\sqrt{5x - 1} + \sqrt{2x} = 5.$ [15.6b]

SYNTHESIS

41. A quadratic function has x-intercepts $(-3, 0)$ and $(5, 0)$ and y-intercept $(0, -7)$. Find an equation for the function. What is its maximum or minimum value? [16.6a, b]

42. Find h and k such that $3x^2 - hx + 4k = 0$, the sum of the solutions is 20, and the product of the solutions is 80. [16.2a]

43. The average of two numbers is 171. One of the numbers is the square root of the other. Find the numbers. [16.3a]

CHAPTER 16

Chapter Test

1. a) Solve: $3x^2 - 4 = 0$.
   b) Find the x-intercepts of $f(x) = 3x^2 - 4$.

Solve.

2. $x^2 + x + 1 = 0$

3. $x - 8\sqrt{x} + 7 = 0$

4. $4x(x - 2) - 3x(x + 1) = -18$

5. $x^4 - 5x^2 + 5 = 0$

6. $x^2 + 4x = 2$. Give exact solutions and approximate solutions to three decimal places.

7. $\frac{1}{4 - x} + \frac{1}{2 + x} = \frac{3}{4}$

8. Solve $x^2 - 4x + 1 = 0$ by completing the square. Show your work.

9. *Free-Falling Objects.* The Peachtree Plaza in Atlanta, Georgia, is 723 ft tall. Use the function $s(t) = 16t^2$ to approximate how long it would take an object to fall from the top.

10. *Marine Travel.* The Columbia River flows at a rate of 2 mph for the length of a popular boating route. In order for a motorized dinghy to travel 3 mi upriver and then return in a total of 4 hr, how fast must the boat be able to travel in still water?

11. *Memory Board.* A computer-parts company wants to make a rectangular memory board that has a perimeter of 28 cm. What dimensions will allow the board to have a maximum area?

12. *Hang Time.* Use the function $V(T) = 48T^2$. A basketball player has a vertical leap of 35 in. What is his hang time?

13. Determine the nature of the solutions of the equation $x^2 + 5x + 17 = 0$.

14. Write a quadratic equation having the solutions $\sqrt{3}$ and $3\sqrt{3}$.

15. Solve $V = 48T^2$ for T.

For each quadratic function, find and label **(a)** the vertex, **(b)** the line of symmetry, and **(c)** the maximum or minimum value. Then **(d)** graph the function.

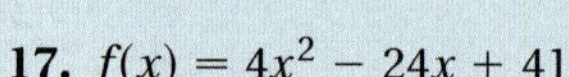

16. $f(x) = -x^2 - 2x$

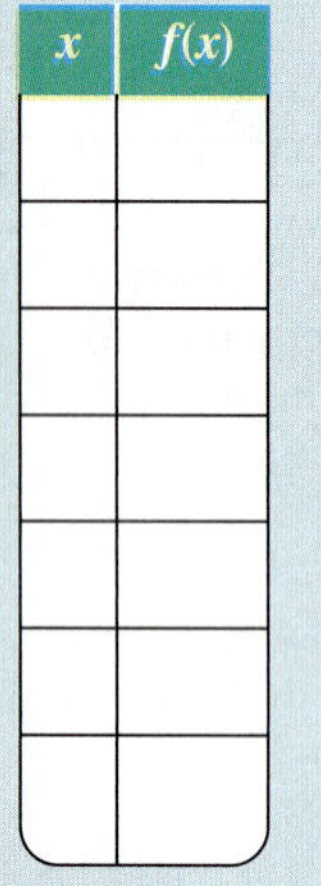

x	$f(x)$

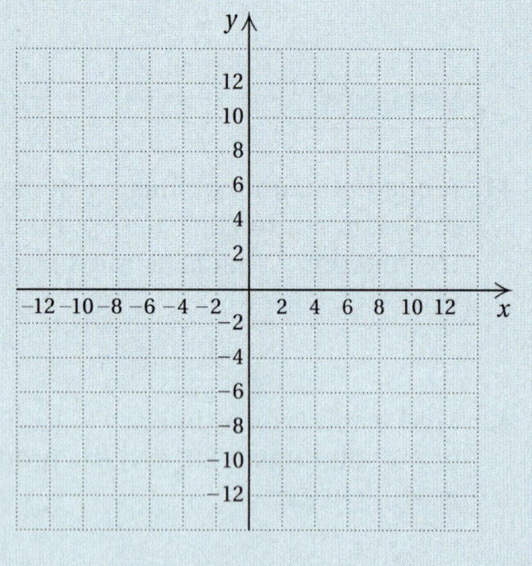

Vertex: (____, ____)
Line of symmetry: $x =$ ____
________ value: ____

17. $f(x) = 4x^2 - 24x + 41$

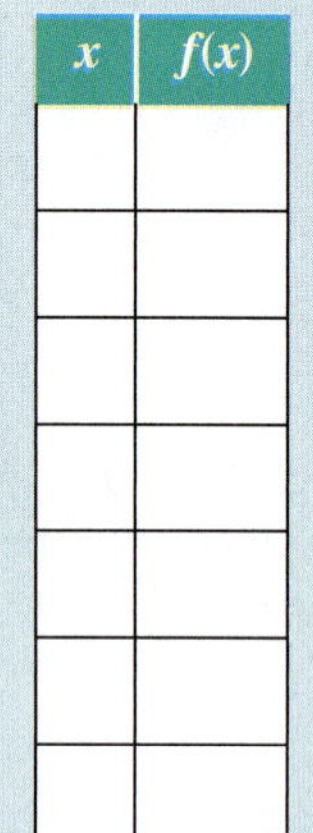

x	$f(x)$

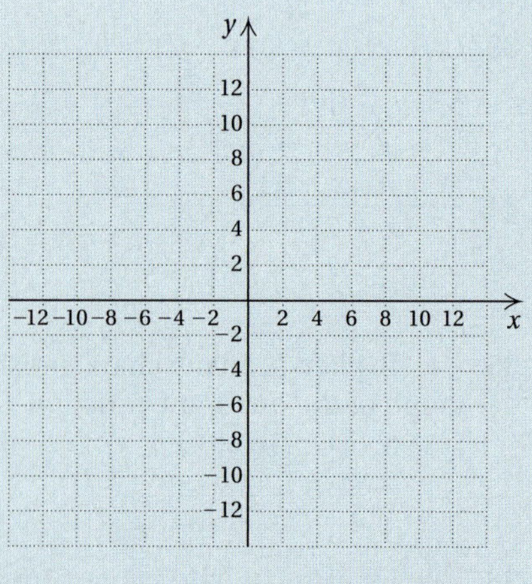

Vertex: (____, ____)
Line of symmetry: $x =$ ____
________ value: ____

18. Find the x- and y-intercepts:

$$f(x) = -x^2 + 4x - 1.$$

19. What is the minimum product of two numbers whose difference is 8? What numbers yield this product?

20. Find the quadratic function that fits the data points $(0, 0)$, $(3, 0)$, and $(5, 2)$.

21. *Internet Radio Stations.* The graph at right shows the number of radio stations broadcasting over the Internet for various years. It appears that the graph might be fit by one part or side of a quadratic function.
Source: BRS Media Inc.

a) Use the data points $(1, 56)$, $(4, 2261)$, and $(6, 5058)$ to fit a quadratic function $N(t) = at^2 + bt + c$ to the data, where N is the number of Internet radio stations t years after 1995 and $t = 0$ corresponds to 1995.

b) Use the quadratic function to predict the number of Internet radio stations in 2008 and in 2010.

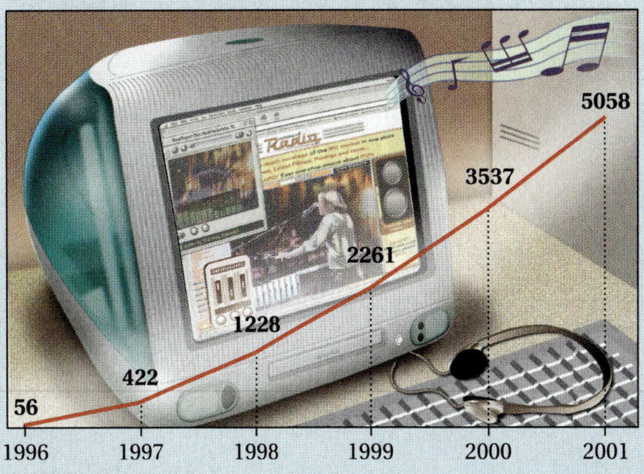

Source: BRS Media, Inc.

Solve.

22. $x^2 < 6x + 7$

23. $\dfrac{x - 5}{x + 3} < 0$

24. $\dfrac{(x - 2)}{(x + 3)(x - 1)} \geq 0$

SKILL MAINTENANCE

25. *SAT Verbal Scores.* The following table lists SAT verbal scores for recent years.

YEARS SINCE 1994, x	SAT VERBAL SCORE
0	499
1	504
2	505
3	505
4	505
5	505
6	505

Source: The College Board

a) Use the data points $(1, 504)$ and $(6, 505)$ to fit a linear function to the data.

b) Use the linear function to predict SAT verbal scores in 2004 and in 2010.

26. Subtract and simplify:

$$\frac{x}{x^2 + 15x + 56} - \frac{7}{x^2 + 13x + 42}.$$

27. Multiply and simplify:

$$\sqrt[4]{2a^2b^3}\,\sqrt[4]{a^4b}.$$

28. Solve: $\sqrt{x + 3} = x - 3$.

SYNTHESIS

29. A quadratic function has x-intercepts $(-2, 0)$ and $(7, 0)$ and y-intercept $(0, 8)$. FInd an equation for the function. What is its maximum or minimum value?

30. One solution of $kx^2 + 3x - k = 0$ is -2. Find the other solution.

31. Solve: $x^8 - 20x^4 + 64 = 0$.

Exponential and Logarithmic Functions

Gateway to Chapter 17

The functions that we consider in this chapter are important for their rich applications to many fields. We will look at such applications as compound interest, population growth, and radioactive decay, but there are many others.

Exponents are the basis of the theory in this chapter. We define some functions having variable exponents, called exponential functions. The logarithmic functions follow from the exponential functions.

Real-World Application

The restoration of antique automobiles is a popular hobby. The value of a 1929 Pierce-Arrow Roadster was $4800 when it was sold brand-new. Fully restored, the value of this automobile in 2001 was $150,000. **(a)** Fit an exponential growth function to the data. **(b)** Use the function found in part (a) to predict the value of the automobile in 2019. **(c)** Assuming exponential growth continues, predict the year in which the value will be $200,000.

Source: Dr. Rex B. Gosnell

This problem appears as Example 6 in Section 17.7.

CHAPTER 17

CHAPTER 17

Pretest

Graph.

1. $f(x) = 2^x$ [17.1a]

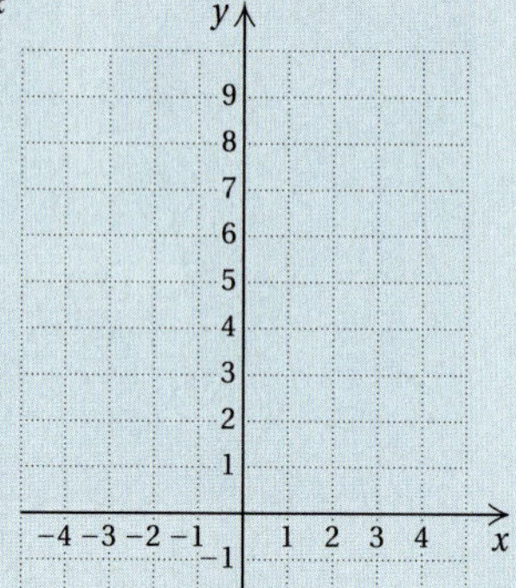

2. $f(x) = \log_2 x$ [17.3a]

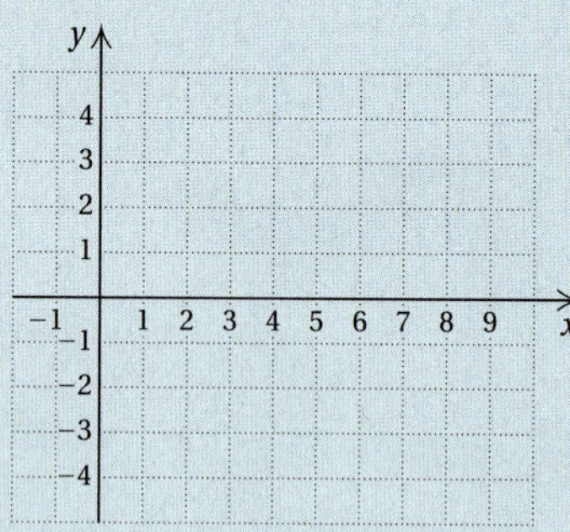

3. Convert to a logarithmic equation: [17.3b]

   $a^3 = 1000.$

4. Convert to an exponential equation: [17.3b]

   $\log_2 32 = t.$

5. Express as a single logarithm: [17.4d]

   $$2 \log_a M + \log_a N - \frac{1}{3} \log_a Q.$$

6. Express in terms of logarithms of x, y, and z: [17.4d]

   $$\log_a \sqrt[4]{\frac{x^3 y}{z^2}}.$$

Solve.

7. $\log_x \frac{1}{4} = -2$ [17.6b]

8. $\log_5 x = 2$ [17.6b]

9. $3^x = 8.6$ [17.6a]

10. $16^{x-3} = 4$ [17.6a]

11. $\log(x^2 - 4) - \log(x - 2) = 1$ [17.6b]

12. $\ln x = 0$ [17.6b]

Find each of the following using a calculator.

13. $\log 714$ [17.3d]

14. $\ln 0.0008464$ [17.5a]

15. $\ln 1$ [17.5a]

16. $e^{4.2}$ [17.5a]

17. $e^{-2.13}$ [17.5a]

18. $\log(0.0234)$ [17.3d]

19. **Credit-Card Spending.** The amount S, in billions of dollars, charged on credit cards between Thanksgiving and Christmas each year has been increasing exponentially according to the function

    $$S(t) = 52.4(1.16)^t,$$

    where t is the number of years since 1990. [17.7b]

    Source: RAM Research Group, National Credit Counseling Services

    a) Estimate the amount charged between Thanksgiving and Christmas in 2008.
    b) In what year will the amount charged be \$1 trillion?
    c) What is the doubling time?

20. **Carbon Dating.** How old is an animal bone that

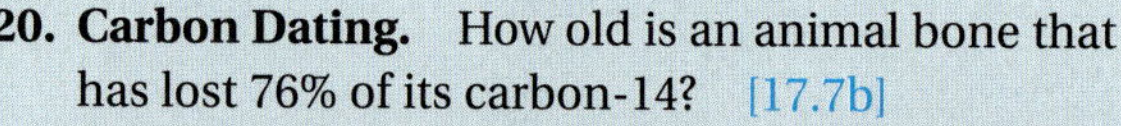

has lost 76% of its carbon-14? [17.7b]

17.1 EXPONENTIAL FUNCTIONS

Objectives

a Graph exponential equations and functions.

b Graph exponential equations in which x and y have been interchanged.

c Solve applied problems involving applications of exponential functions and their graphs.

The rapidly rising graph shown below approximates the graph of an *exponential function.* We will consider such functions and some of their applications.

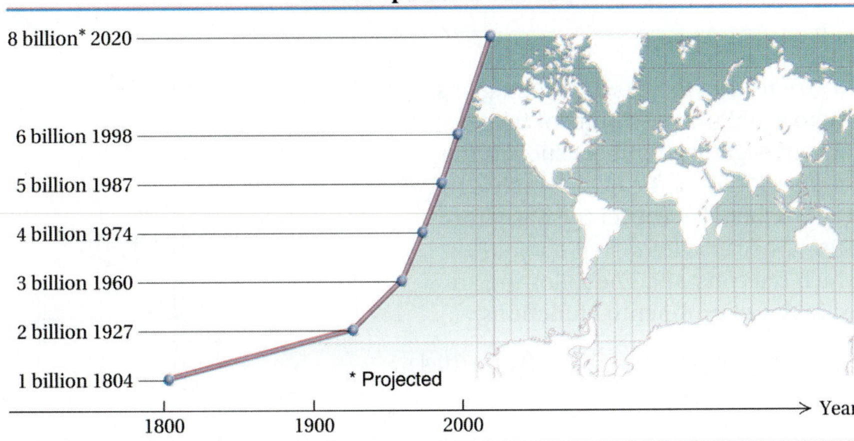

Source: U.S. Bureau of the Census

a Graphing Exponential Functions

In Chapter 15, we gave meaning to exponential expressions with rational-number exponents such as

$$8^{1/4},\quad 3^{-3/4},\quad 7^{2.34},\quad 5^{1.73}.$$

For example, $5^{1.73}$, or $5^{173/100}$, or $\sqrt[100]{5^{173}}$, means to raise 5 to the 173rd power and then take the 100th root. We now develop the meaning of exponential expressions with irrational exponents. Examples of expressions with irrational exponents are

$$5^{\sqrt{3}},\quad 7^{\pi},\quad 9^{-\sqrt{2}}.$$

Since we can approximate irrational numbers with decimal approximations, we can also approximate expressions with irrational exponents. For example, consider $5^{\sqrt{3}}$. We know that $5^{\sqrt{3}} \approx 5^{1.73} = \sqrt[100]{5^{173}}$. As rational values of r get close to $\sqrt{3}$, 5^r gets close to some real number. Note the following:

r closes in on $\sqrt{3}$. r	5^r closes in on some real number p. 5^r
$1 < \sqrt{3} < 2$	$5 = 5^1 < p < 5^2 = 25$
$1.7 < \sqrt{3} < 1.8$	$15.426 = 5^{1.7} < p < 5^{1.8} = 18.119$
$1.73 < \sqrt{3} < 1.74$	$16.189 = 5^{1.73} < p < 5^{1.74} = 16.452$
$1.732 < \sqrt{3} < 1.733$	$16.241 = 5^{1.732} < p < 5^{1.733} = 16.267$

As r closes in on $\sqrt{3}$, 5^r closes in on some real number p. We define $5^{\sqrt{3}}$ to be that number p. To seven decimal places, we have

$$5^{\sqrt{3}} \approx 16.2424508.$$

1. Graph: $f(x) = 3^x$.

a) Complete this table of solutions.

x	$f(x)$
0	
1	
2	
3	
−1	
−2	
−3	

b) Plot the points from the table and connect them with a smooth curve.

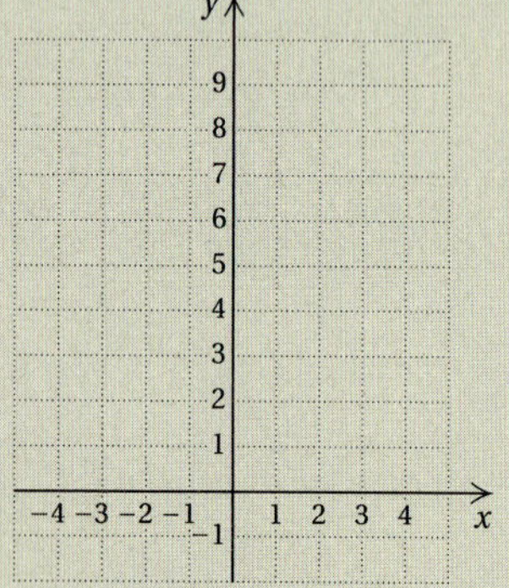

Answers on page A-66

Any positive irrational exponent can be defined in a similar way. Negative irrational exponents are then defined in the same way as negative integer exponents. Then the expression a^x has meaning for any real number x. The general laws of exponents still hold, but we will not prove that here.

We now define exponential functions.

EXPONENTIAL FUNCTION

The function $f(x) = a^x$, where a is a positive constant different from 1, is called an **exponential function,** base a.

We restrict the base a to being positive to avoid the possibility of taking even roots of negative numbers such as the square root of -1, $(-1)^{1/2}$, which is not a real number. We restrict the base from being 1 because for $a = 1$, $t(x) = 1^x = 1$, which is a constant. The following are examples of exponential functions:

$$f(x) = 2^x, \qquad f(x) = \left(\tfrac{1}{2}\right)^x, \qquad f(x) = (0.4)^x.$$

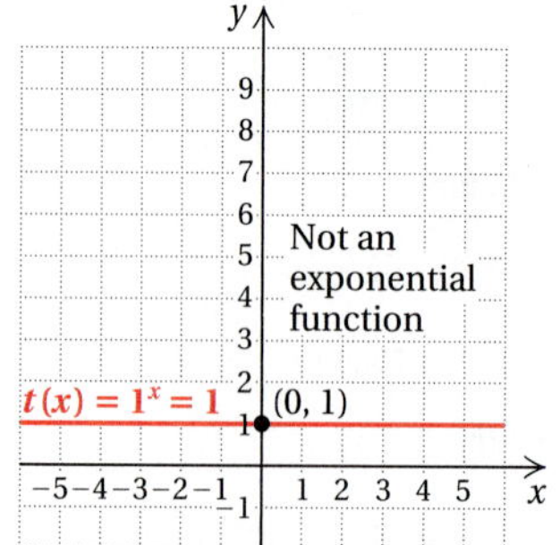

Note that in contrast to polynomial functions like $f(x) = x^2$ and $f(x) = x^3$, the variable is *in the exponent.* Let's consider graphs of exponential functions.

EXAMPLE 1 Graph the exponential function $f(x) = 2^x$.

We compute some function values and list the results in a table. It is a good idea to begin by letting $x = 0$.

$$f(0) = 2^0 = 1;$$
$$f(1) = 2^1 = 2;$$
$$f(2) = 2^2 = 4;$$
$$f(3) = 2^3 = 8;$$
$$f(-1) = 2^{-1} = \frac{1}{2^1} = \frac{1}{2};$$
$$f(-2) = 2^{-2} = \frac{1}{2^2} = \frac{1}{4};$$
$$f(-3) = 2^{-3} = \frac{1}{2^3} = \frac{1}{8}.$$

x	$f(x)$
0	1
1	2
2	4
3	8
−1	$\frac{1}{2}$
−2	$\frac{1}{4}$
−3	$\frac{1}{8}$

Next, we plot these points and connect them with a smooth curve.

In graphing, be sure to plot enough points to determine how steeply the curve rises.

The curve comes very close to the x-axis, but does not touch or cross it.

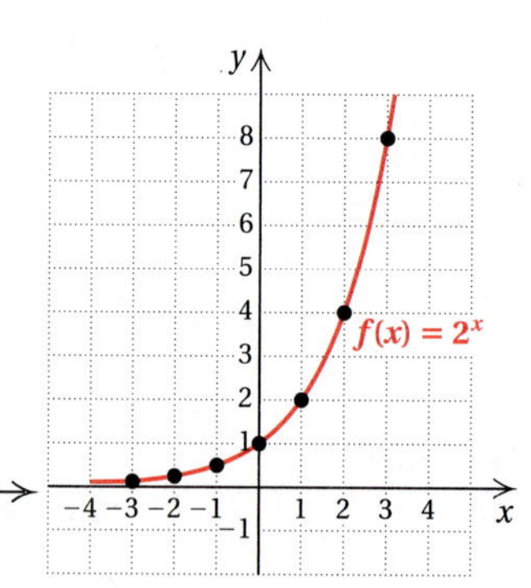

Note that as x increases, the function values increase indefinitely. As x decreases, the function values decrease, getting very close to 0. The x-axis, or the line $y = 0$, is an *asymptote,* meaning here that as x gets very small, the curve comes very close to but never touches the axis.

Do Exercise 1 on the preceding page.

EXAMPLE 2 Graph the exponential function $f(x) = \left(\frac{1}{2}\right)^x$.

We compute some function values and list the results in a table. Before we do so, note that

$$f(x) = \left(\tfrac{1}{2}\right)^x = (2^{-1})^x = 2^{-x}.$$

Then we have

$$f(0) = 2^{-0} = 1;$$
$$f(1) = 2^{-1} = \frac{1}{2^1} = \frac{1}{2};$$
$$f(2) = 2^{-2} = \frac{1}{2^2} = \frac{1}{4};$$
$$f(3) = 2^{-3} = \frac{1}{2^3} = \frac{1}{8};$$
$$f(-1) = 2^{-(-1)} = 2^1 = 2;$$
$$f(-2) = 2^{-(-2)} = 2^2 = 4;$$
$$f(-3) = 2^{-(-3)} = 2^3 = 8.$$

x	$f(x)$
0	1
1	$\frac{1}{2}$
2	$\frac{1}{4}$
3	$\frac{1}{8}$
−1	2
−2	4
−3	8

Next, we plot these points and draw the curve. Note that this graph is a reflection across the y-axis of the graph in Example 1. The line $y = 0$ is again an asymptote.

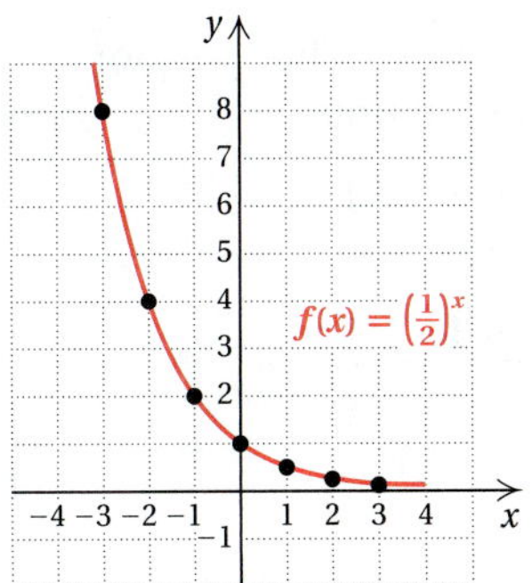

Do Exercise 2.

The preceding examples illustrate exponential functions with various bases. Let's list some of their characteristics. Keep in mind that the definition of an exponential function, $f(x) = a^x$, requires that the base be positive and different from 1.

2. Graph: $f(x) = \left(\dfrac{1}{3}\right)^x$.

a) Complete this table of solutions.

x	$f(x)$
0	
1	
2	
3	
−1	
−2	
−3	

b) Plot the points from the table and connect them with a smooth curve.

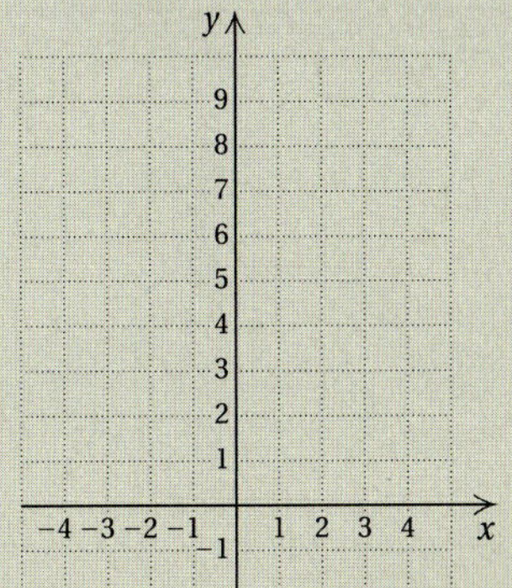

Answers on page A-66

Graph.

3. $f(x) = 4^x$

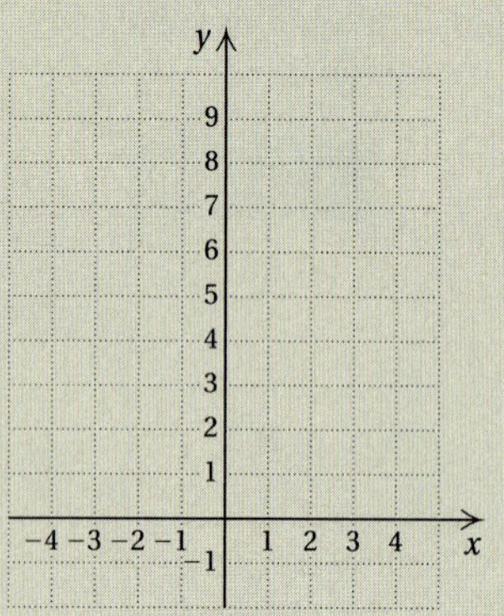

4. $f(x) = \left(\frac{1}{4}\right)^x$

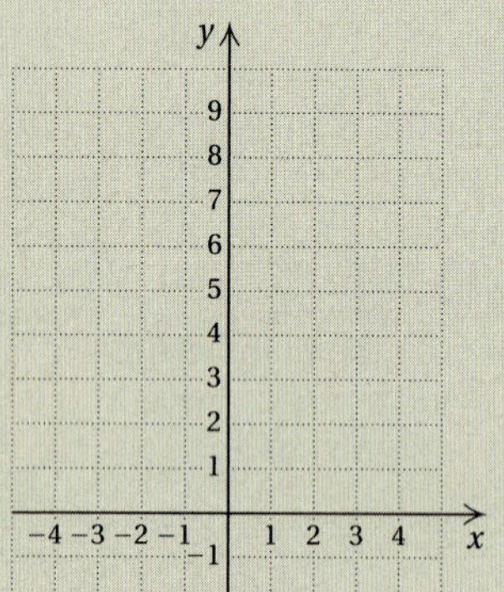

5. Graph: $f(x) = 2^{x+2}$.

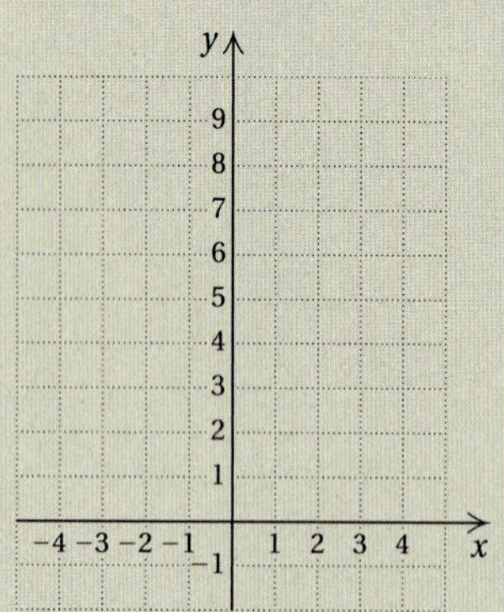

Answers on page A-66

When $a > 1$, the function $f(x) = a^x$ increases from left to right. The greater the value of a, the steeper the curve. As x gets smaller and smaller, the curve gets closer to the line $y = 0$: It is an asymptote.

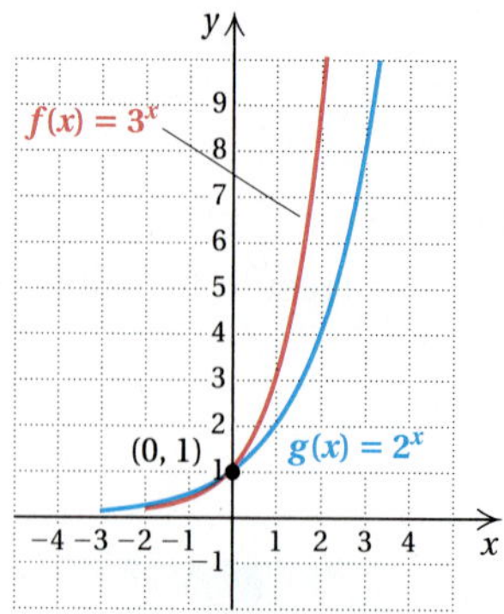

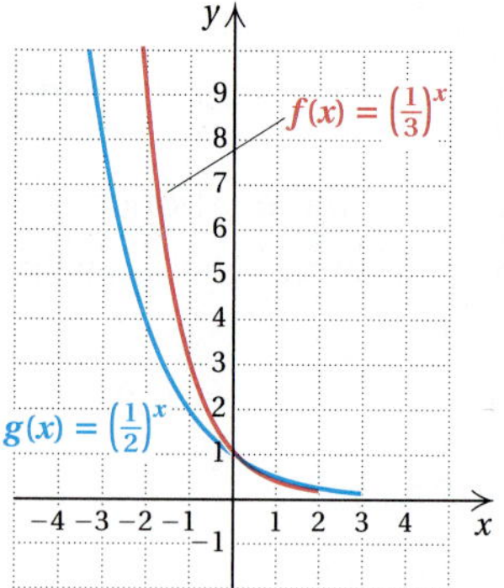

When $0 < a < 1$, the function $f(x) = a^x$ decreases from left to right. As a approaches 1, the curve becomes less steep. As x gets larger and larger, the curve gets closer to the line $y = 0$: It is an asymptote.

y-INTERCEPT OF AN EXPONENTIAL FUNCTION

All functions $f(x) = a^x$ go through the point $(0, 1)$. That is, the y-intercept is $(0, 1)$.

Do Exercises 3 and 4.

EXAMPLE 3 Graph: $f(x) = 2^{x-2}$.

We construct a table of values. Then we plot the points and connect them with a smooth curve. Be sure to note that $x - 2$ is the *exponent*.

$$f(0) = 2^{0-2} = 2^{-2} = \frac{1}{2^2} = \frac{1}{4};$$

$$f(1) = 2^{1-2} = 2^{-1} = \frac{1}{2^1} = \frac{1}{2};$$

$$f(2) = 2^{2-2} = 2^0 = 1;$$

$$f(3) = 2^{3-2} = 2^1 = 2;$$

$$f(4) = 2^{4-2} = 2^2 = 4;$$

$$f(-1) = 2^{-1-2} = 2^{-3} = \frac{1}{2^3} = \frac{1}{8};$$

$$f(-2) = 2^{-2-2} = 2^{-4} = \frac{1}{2^4} = \frac{1}{16}$$

x	$f(x)$
0	$\frac{1}{4}$
1	$\frac{1}{2}$
2	1
3	2
4	4
−1	$\frac{1}{8}$
−2	$\frac{1}{16}$

The graph looks just like the graph of $g(x) = 2^x$, but it is translated 2 units to the right.

The y-intercept of $g(x) = 2^x$ is $(0, 1)$. The y-intercept of $f(x) = 2^{x-2}$ is $\left(0, \frac{1}{4}\right)$. The line $y = 0$ is still an asymptote.

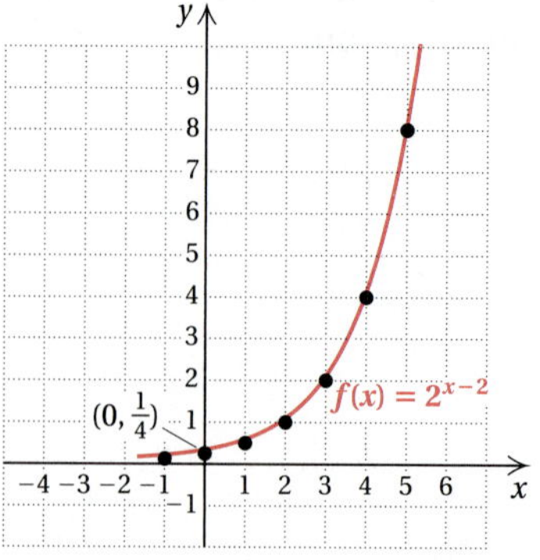

Do Exercise 5 on the preceding page.

EXAMPLE 4 Graph: $f(x) = 2^x - 3$.

We construct a table of values. Then we plot the points and connect them with a smooth curve. Note that the only expression in the exponent is x.

$$f(0) = 2^0 - 3 = 1 - 3 = -2;$$
$$f(1) = 2^1 - 3 = 2 - 3 = -1;$$
$$f(2) = 2^2 - 3 = 4 - 3 = 1;$$
$$f(3) = 2^3 - 3 = 8 - 3 = 5;$$
$$f(4) = 2^4 - 3 = 16 - 3 = 13;$$
$$f(-1) = 2^{-1} - 3 = \frac{1}{2} - 3 = -\frac{5}{2};$$
$$f(-2) = 2^{-2} - 3 = \frac{1}{4} - 3 = -\frac{11}{4}$$

x	$f(x)$
0	−2
1	−1
2	1
3	5
4	13
−1	$-\frac{5}{2}$
−2	$-\frac{11}{4}$

The graph looks just like the graph of $g(x) = 2^x$, but it is translated down 3 units. The y-intercept is $(0, -2)$. The line $y = -3$ is an asymptote. The curve gets closer to this line as x gets smaller and smaller.

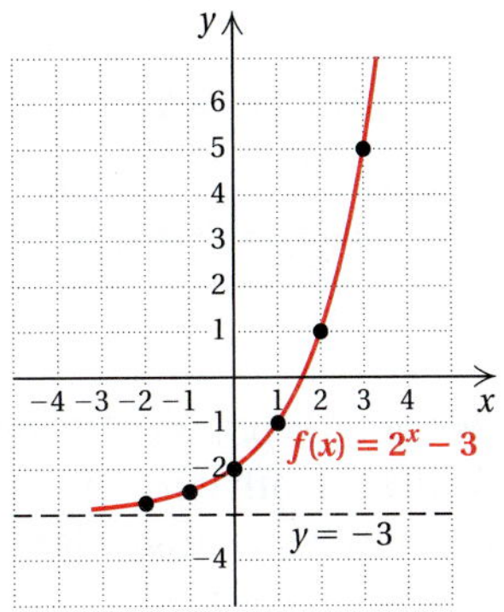

Do Exercise 6.

6. Graph: $f(x) = 2^x - 4$.

x	$f(x)$
0	
1	
2	
3	
4	
−1	
−2	

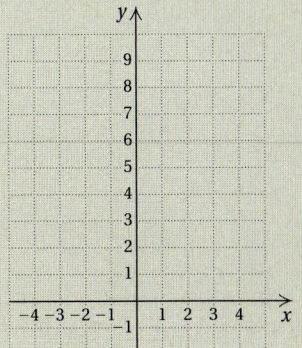

Answer on page A-66

CALCULATOR CORNER

Graphing Exponential Functions We can use a graphing calculator to graph exponential functions. It might be necessary to try several sets of window dimensions in order to find the ones that give a good view of the curve.

To graph $f(x) = 3^x - 1$, we enter the equation as y_1 by pressing [3] [∧] [X, T, θ, n] [−] [1]. We can begin graphing with the standard window $[-10, 10, -10, 10]$ by pressing [ZOOM] [6]. Although this window gives a good view of the curve, we might want to adjust it to show more of the curve in the first quadrant. Changing the dimensions to $[-10, 10, -5, 15]$ accomplishes this.

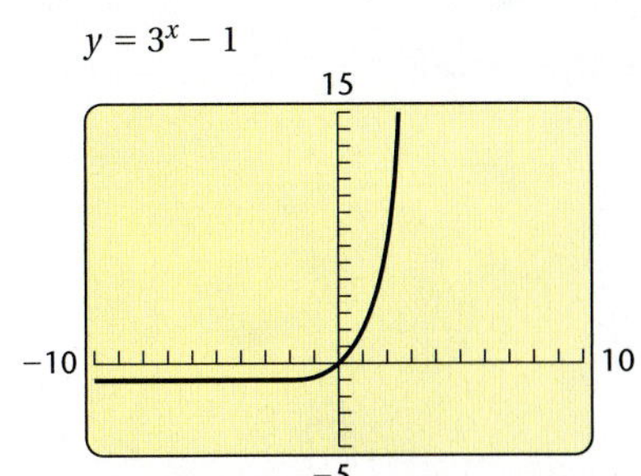

Exercises:

1. Use a graphing calculator to graph the functions in Examples 1–4.
2. Use a graphing calculator to graph the functions in Margin Exercises 1–6.

7. Graph: $x = 3^y$.

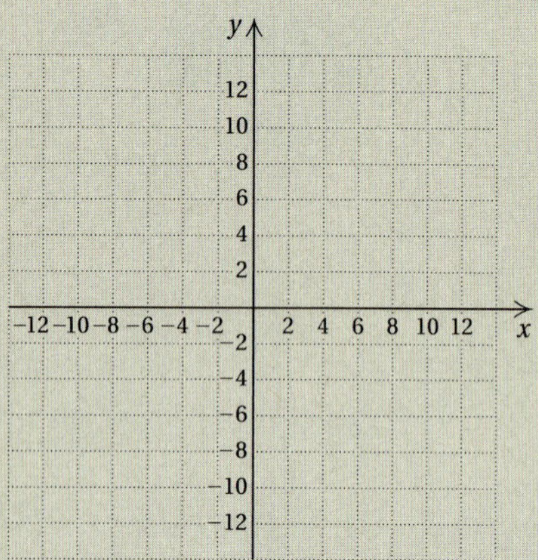

Answer on page A-66

b Equations with x and y Interchanged

It will be helpful in later work to be able to graph an equation in which the x and the y in $y = a^x$ are interchanged.

EXAMPLE 5 Graph: $x = 2^y$.

Note that x is alone on one side of the equation. We can find ordered pairs that are solutions more easily by choosing values for y and then computing the x-values.

For $y = 0$, $x = 2^0 = 1$.
For $y = 1$, $x = 2^1 = 2$.
For $y = 2$, $x = 2^2 = 4$.
For $y = 3$, $x = 2^3 = 8$.
For $y = -1$, $x = 2^{-1} = \frac{1}{2^1} = \frac{1}{2}$.
For $y = -2$, $x = 2^{-2} = \frac{1}{2^2} = \frac{1}{4}$.
For $y = -3$, $x = 2^{-3} = \frac{1}{2^3} = \frac{1}{8}$.

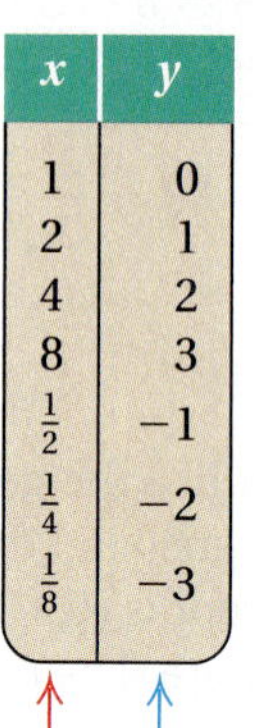

x	y
1	0
2	1
4	2
8	3
$\frac{1}{2}$	−1
$\frac{1}{4}$	−2
$\frac{1}{8}$	−3

(1) Choose values for y.
(2) Compute values for x.

We plot the points and connect them with a smooth curve. What happens as y-values become smaller?

This curve does not touch or cross the y-axis.

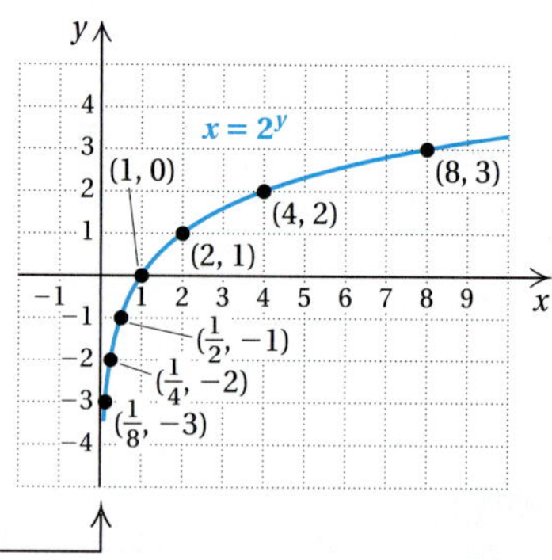

Note that this curve $x = 2^y$ looks just like the graph of $y = 2^x$, except that it is reflected, or flipped, across the line $y = x$, as shown below.

$y = 2^x$	
x	y
0	1
1	2
2	4
3	8
−1	$\frac{1}{2}$
−2	$\frac{1}{4}$
−3	$\frac{1}{8}$

$x = 2^y$	
x	y
1	0
2	1
4	2
8	3
$\frac{1}{2}$	−1
$\frac{1}{4}$	−2
$\frac{1}{8}$	−3

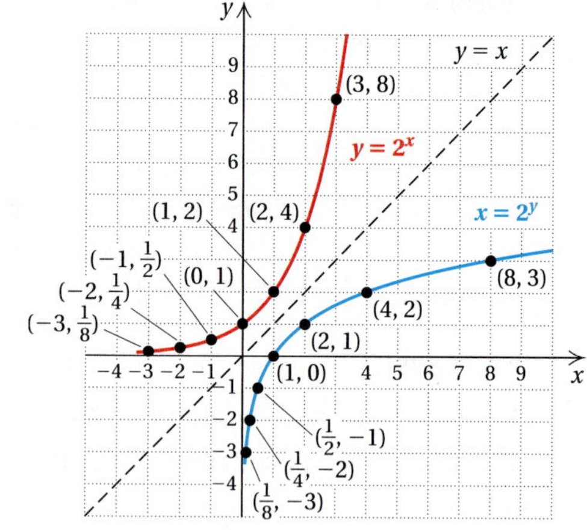

Do Exercise 7.

C Applications of Exponential Functions

When interest is paid on interest, we call it **compound interest.** This is the type of interest paid on investments and loans. Suppose you have \$100,000 in a savings account at an interest rate of 8%. This means that in 1 yr, the account will contain the original \$100,000 plus 8% of \$100,000. Thus the total in the account after 1 yr will be

$$\$100{,}000 \quad \text{plus} \quad \$100{,}000 \times 0.08.$$

This can also be expressed as

$$\begin{aligned}\$100{,}000 + \$100{,}000 \times 0.08 &= \$100{,}000 \times 1 + \$100{,}000 \times 0.08 \\ &= \$100{,}000(1 + 0.08) \qquad \text{Factoring out \$100,000 using the distributive law} \\ &= \$100{,}000(1.08) \\ &= \$108{,}000.\end{aligned}$$

Now suppose that the total of \$108,000 remains in the account for another year. At the end of the second year, the account will contain the \$108,000 plus 8% of \$108,000. The total in the account will be

$$\$108{,}000 \quad \text{plus} \quad \$108{,}000 \times 0.08,$$

or

$$\$108{,}000(1.08) = \$100{,}000(1.08)^2 = \$116{,}640.$$

Note that in the second year, interest is earned on the first year's interest as well as the original amount. When this happens, we say the interest is **compounded annually.** If the original amount of \$100,000 earned only simple interest for 2 yr, the interest would be

$$\$100{,}000 \times 0.08 \times 2, \quad \text{or} \quad \$16{,}000,$$

and the amount in the account would be

$$\$100{,}000 + \$16{,}000 = \$116{,}000,$$

less than the \$116,640 when interest is compounded annually.

Do Exercise 8.

The following table shows how the computation continues over 4 yr.

\$100,000 IN AN ACCOUNT

YEAR	INTEREST COMPOUNDED ANNUALLY	SIMPLE INTEREST
Beginning of 1st yr	\$100,000	
End of 1st yr	$\$108{,}000 = \$100{,}000(1.08)^1$	\$108,000
Beginning of 2nd yr	\$108,000	
End of 2nd yr	$\$116{,}640 = \$100{,}000(1.08)^2$	\$116,000
Beginning of 3rd yr	\$116,640	
End of 3rd yr	$\$125{,}971.20 = \$100{,}000(1.08)^3$	\$124,000
Beginning of 4th yr	\$125,971.20	
End of 4th yr	$\$136{,}048.90 = \$100{,}000(1.08)^4$	\$132,000

8. **Interest Compounded Annually.** Find the amount in an account after 1 yr and after 2 yr if \$40,000 is invested at 5%, compounded annually.

Answer on page A-66

9. **Interest Compounded Annually.** Suppose that $40,000 is invested at 5% interest, compounded annually.

a) Find a function for the amount in the account after t years.

b) Find the amount of money in the account at $t = 0$, $t = 4$, $t = 8$, and $t = 10$.

c) Graph the function.

Answers on page A-66

CALCULATOR CORNER

Evaluating Exponential Functions Use one of the methods for finding function values described in the Calculator Corner on p. 924 to find the function values in Example 6(b).

We can express interest compounded annually using an exponential function.

EXAMPLE 6 *Interest Compounded Annually.* The amount of money A that a principal P will grow to after t years at interest rate r, compounded annually, is given by the formula

$$A = P(1 + r)^t.$$

Suppose that $100,000 is invested at 8% interest, compounded annually.

a) Find a function for the amount in the account after t years.

b) Find the amount of money in the account at $t = 0$, $t = 4$, $t = 8$, and $t = 10$.

c) Graph the function.

We solve as follows:

a) If $P = \$100{,}000$ and $r = 8\% = 0.08$, we can substitute these values and form the following function:

$$A(t) = \$100{,}000(1 + 0.08)^t = \$100{,}000(1.08)^t.$$

b) To find the function values, you might find a calculator with a power key helpful.

$$A(0) = \$100{,}000(1.08)^0 = \$100{,}000;$$
$$A(4) = \$100{,}000(1.08)^4 \approx \$136{,}048.90;$$
$$A(8) = \$100{,}000(1.08)^8 \approx \$185{,}093.02;$$
$$A(10) = \$100{,}000(1.08)^{10} \approx \$215{,}892.50$$

c) We use the function values computed in (b) with others, if we wish, to draw the graph as follows. Note that the axes are scaled differently because of the large numbers.

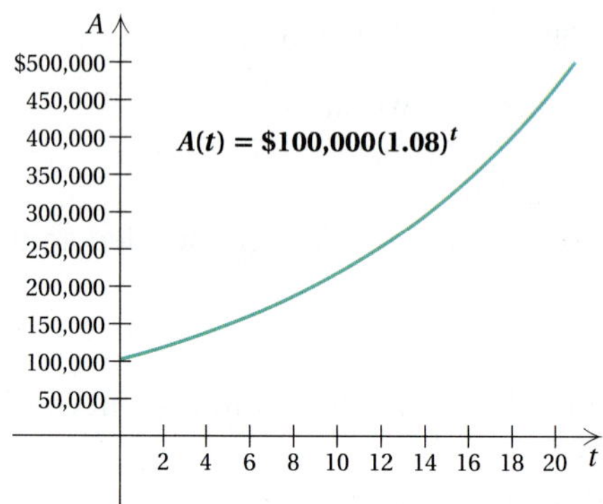

Do Exercise 9.

Suppose the principal of $100,000 we just considered were **compounded semiannually**—that is, every half year. Interest would then be calculated twice a year at a rate of 8% ÷ 2, or 4%, each time. The computations are as follows:

After the first $\frac{1}{2}$ year, the account will contain 104% of $100,000:

$$\$100{,}000 \times 1.04 = \$104{,}000.$$

After a second $\frac{1}{2}$ year (1 full year),the account will contain 104% of $104,000:

$$\$104{,}000 \times 1.04 = \$100{,}000 \times (1.04)^2 = \$108{,}160.$$

After a third $\frac{1}{2}$ year $\left(1\frac{1}{2}\text{ full years}\right)$, the account will contain 104% of $108,160:

$$\$108{,}160 \times 1.04 = \$100{,}000 \times (1.04)^3 = \$112{,}486.40.$$

After a fourth $\frac{1}{2}$ year (2 full years), the account will contain 104% of \$112,486.40:

$$\$112{,}486.40 \times 1.04 = \$100{,}000 \times (1.04)^4$$
$$\approx \$116{,}985.86. \quad \text{Rounded to the nearest cent}$$

Comparing these results with those in the table on p. 1295, we can see that by having more compounding periods, we increase the amount in the account.

We have illustrated the following result.

COMPOUND-INTEREST FORMULA

If a principal P has been invested at interest rate r, compounded n times a year, in t years it will grow to an amount A given by

$$A = P \cdot \left(1 + \frac{r}{n}\right)^{n \cdot t}.$$

EXAMPLE 7 The Ibsens invest \$4000 in an account paying $8\frac{5}{8}\%$, compounded quarterly. Find the amount in the account after $2\frac{1}{2}$ yr.

The compounding is quarterly—that is, four times a year—so in $2\frac{1}{2}$ yr, there are ten $\frac{1}{4}$-yr periods. We substitute \$4000 for P, $8\frac{5}{8}\%$, or 0.08625, for r, 4 for n, and $2\frac{1}{2}$, or $\frac{5}{2}$, for t and compute A:

$$\begin{aligned} A &= P \cdot \left(1 + \frac{r}{n}\right)^{n \cdot t} \\ &= 4000 \cdot \left(1 + \frac{8\frac{5}{8}\%}{4}\right)^{4 \cdot \frac{5}{2}} \\ &= 4000 \cdot \left(1 + \frac{0.08625}{4}\right)^{10} \\ &= 4000(1.0215625)^{10} \quad \text{Using a calculator} \\ &\approx \$4951.19. \end{aligned}$$

The amount in the account after $2\frac{1}{2}$ yr is \$4951.19.

Do Exercise 10.

10. A couple invests \$7000 in an account paying 6.4%, compounded quarterly. Find the amount in the account after $5\frac{1}{2}$ yr.

Answer on page A-66

CALCULATOR CORNER

The Compound-Interest Formula When interest is compounded quarterly, we can find a formula like the one considered in this section as follows:

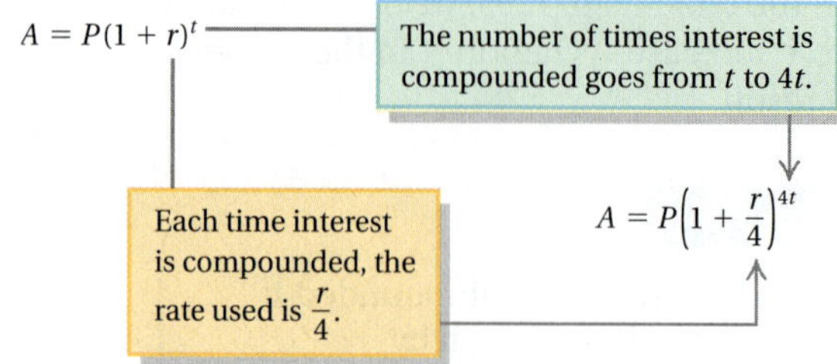

$A = P(1 + r)^t$
The number of times interest is compounded goes from t to $4t$.
Each time interest is compounded, the rate used is $\frac{r}{4}$.
$A = P\left(1 + \frac{r}{4}\right)^{4t}$

In general, the following formula for compound interest can be used. The amount of money A in an account is given by

$$A = P\left(1 + \frac{r}{n}\right)^{nt},$$

where P is the principal, r is the annual interest rate, n is the number of times per year that interest is compounded, and t is the time, in years.

Example Suppose that \$1000 is invested at 8%, compounded quarterly. How much is in the account at the end of 2 yr?

We use the equation above, substituting 1000 for P, 0.08 for r, 4 for n (compounding quarterly), and 2 for t. Then we get

$$A = P\left(1 + \frac{r}{n}\right)^{nt} = 1000\left(1 + \frac{0.08}{4}\right)^{4 \cdot 2}.$$

To do this computation on a calculator, we press [1] [0] [0] [0] [(] [1] [+] [.] [0] [8] [÷] [4] [)] [^] [(] [4] [×] [2] [)] [ENTER]. The result is approximately \$1171.66.

Exercises:

1. Suppose that \$1000 is invested at 6%, compounded semiannually. How much is in the account at the end of 2 yr?
2. Suppose that \$1000 is invested at 6%, compounded monthly. How much is in the account at the end of 2 yr?
3. Suppose that \$20,000 is invested at 4.2%, compounded quarterly. How much is in the account at the end of 10 yr?
4. Suppose that \$420,000 is invested at 4.2%, compounded daily. How much is in the account at the end of 30 yr?
5. Suppose that \$10,000 is invested at 5.4%. How much is in the account at the end of 1 yr, if interest is compounded **(a)** annually? **(b)** semiannually? **(c)** quarterly? **(d)** daily? **(e)** hourly?
6. Suppose that \$10,000 is invested at 3%. How much is in the account at the end of 1 yr, if interest is compounded **(a)** annually? **(b)** semiannually? **(c)** quarterly? **(d)** daily? **(e)** hourly?

17.1 EXERCISE SET

For Extra Help

Digital Video Tutor CD 14 Videotape 20

InterAct Math

Math Tutor Center

MathXL

MyMathLab

a Graph.

1. $f(x) = 2^x$

x	$f(x)$
0	
1	
2	
3	
−1	
−2	
−3	

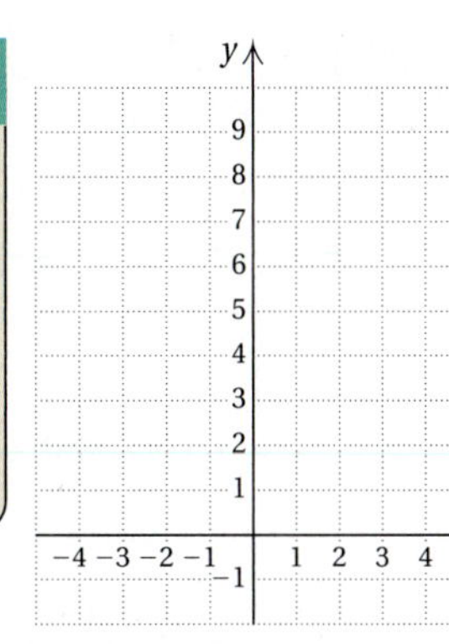

2. $f(x) = 3^x$

x	$f(x)$
0	
1	
2	
3	
−1	
−2	
−3	

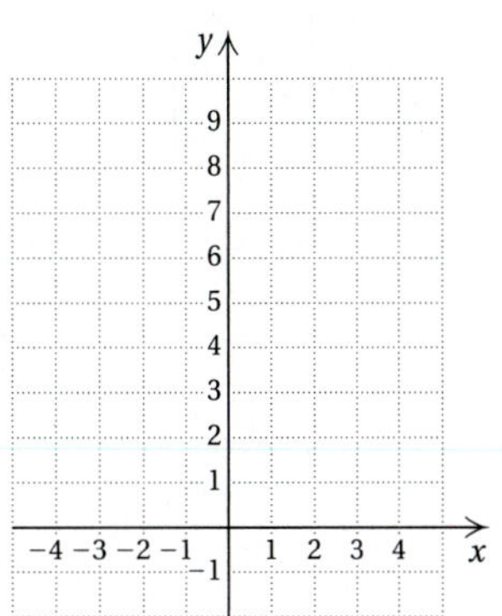

3. $f(x) = 5^x$

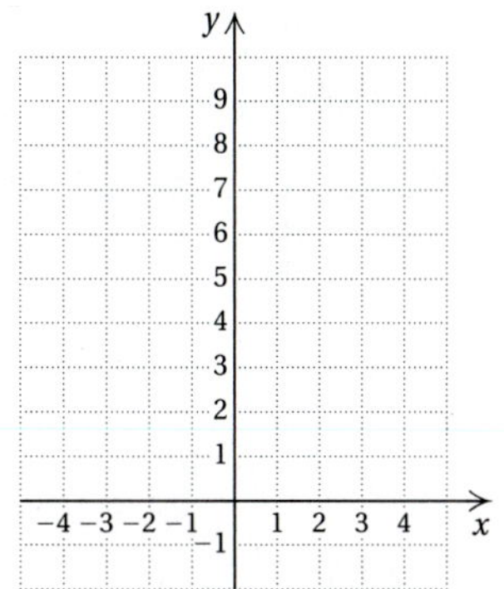

4. $f(x) = 6^x$

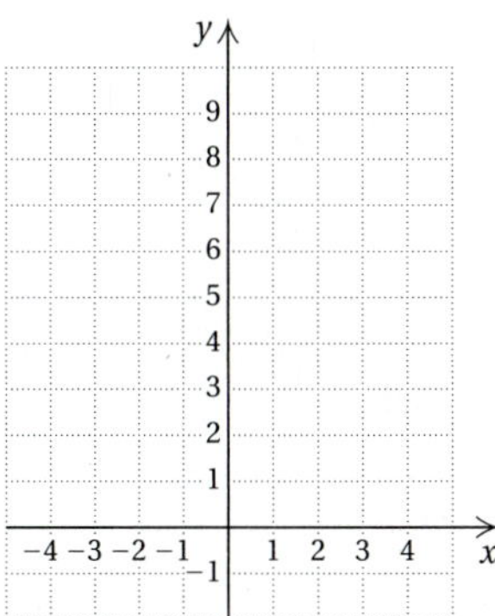

5. $f(x) = 2^{x+1}$

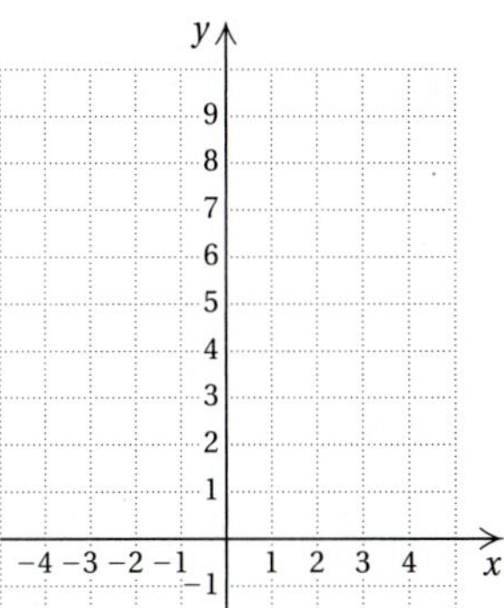

6. $f(x) = 2^{x-1}$

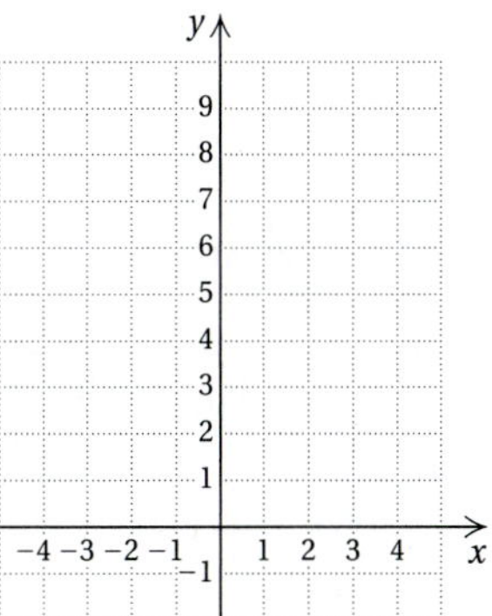

7. $f(x) = 3^{x-2}$

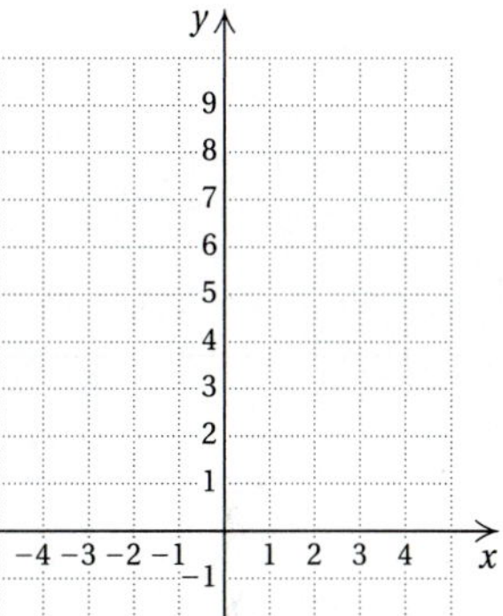

8. $f(x) = 3^{x+2}$

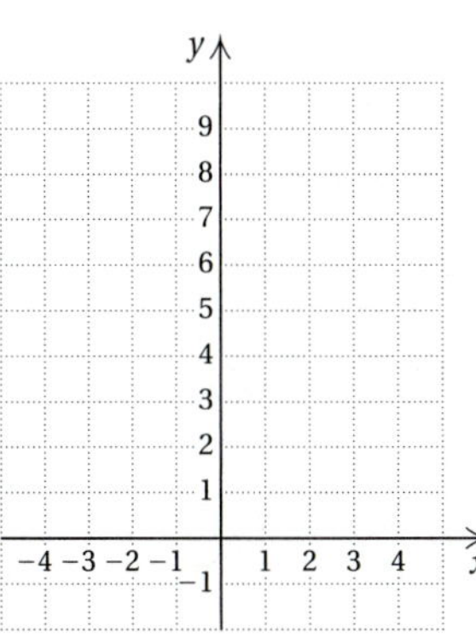

9. $f(x) = 2^x - 3$

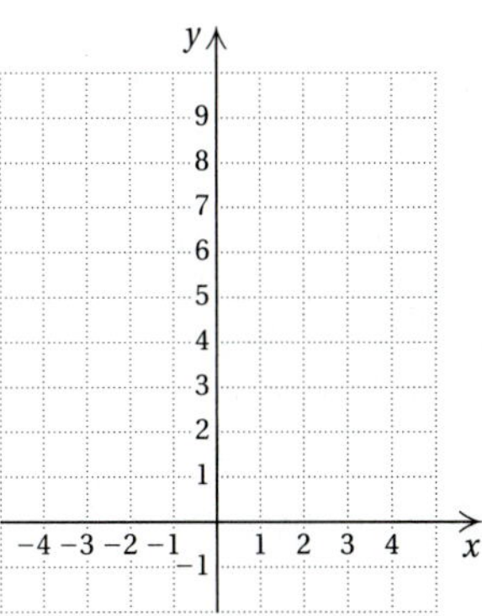

10. $f(x) = 2^x + 1$

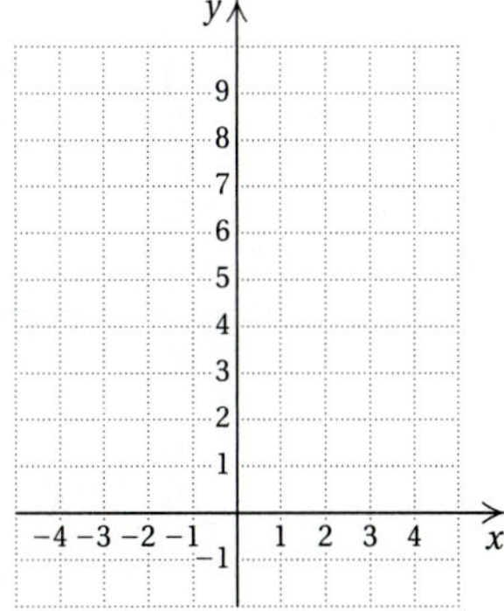

11. $f(x) = 5^{x+3}$

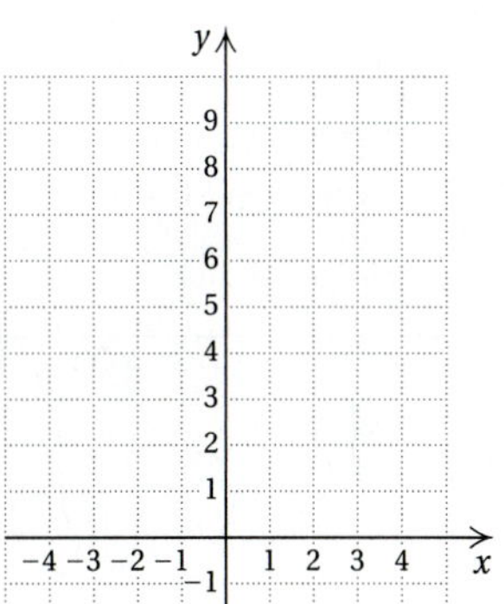

12. $f(x) = 6^{x-4}$

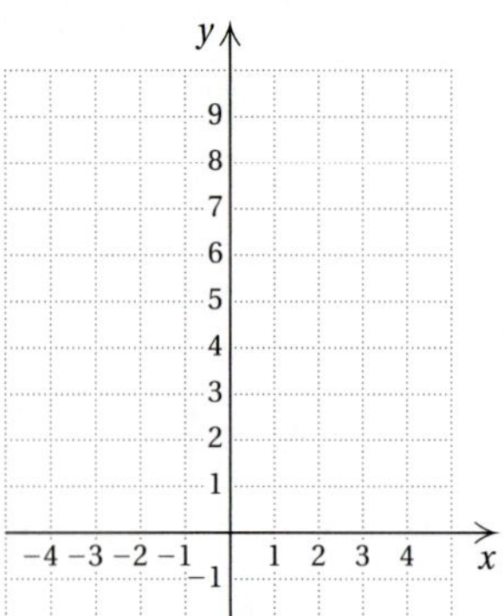

13. $f(x) = \left(\frac{1}{2}\right)^x$

x	$f(x)$
0	
1	
2	
3	
−1	
−2	
−3	

14. $f(x) = \left(\frac{1}{3}\right)^x$

x	$f(x)$
0	
1	
2	
3	
−1	
−2	
−3	

15. $f(x) = \left(\frac{1}{5}\right)^x$

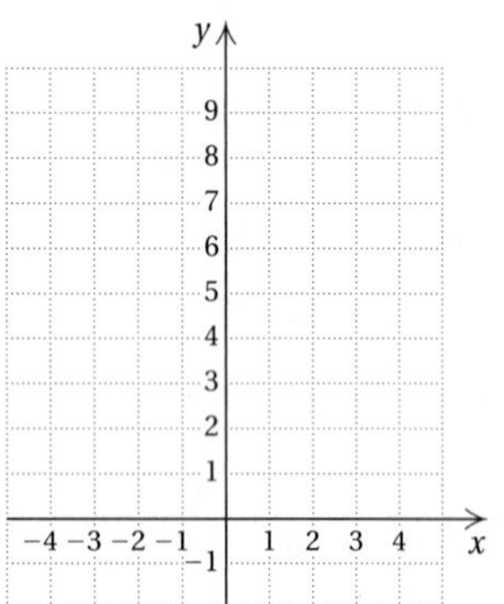

16. $f(x) = \left(\frac{1}{4}\right)^x$

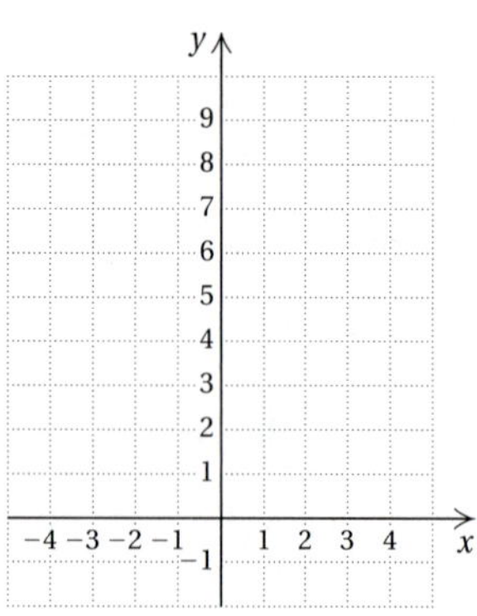

17. $f(x) = 2^{2x-1}$

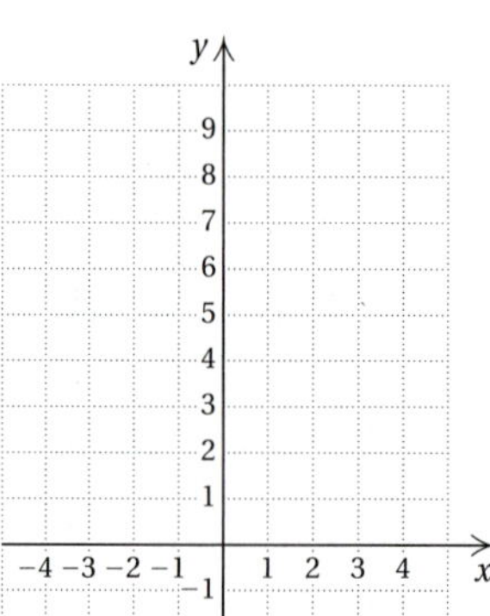

18. $f(x) = 3^{3-x}$

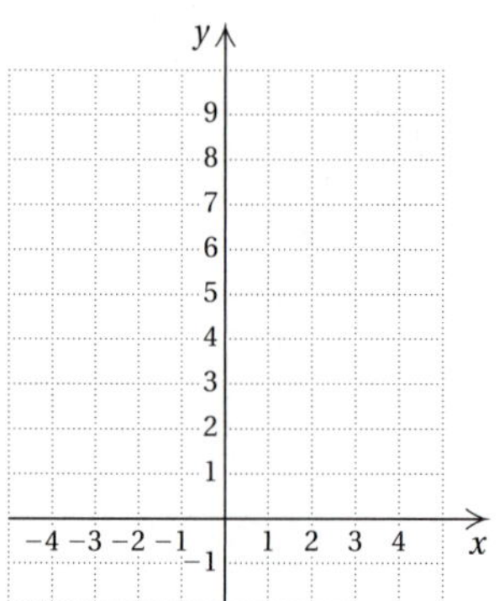

b Graph.

19. $x = 2^y$

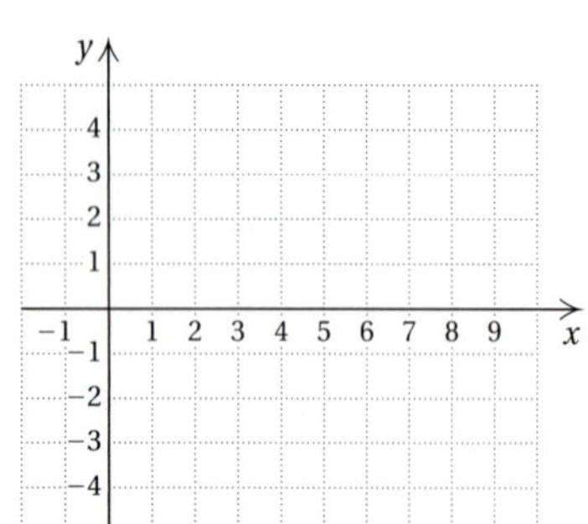

20. $x = 6^y$

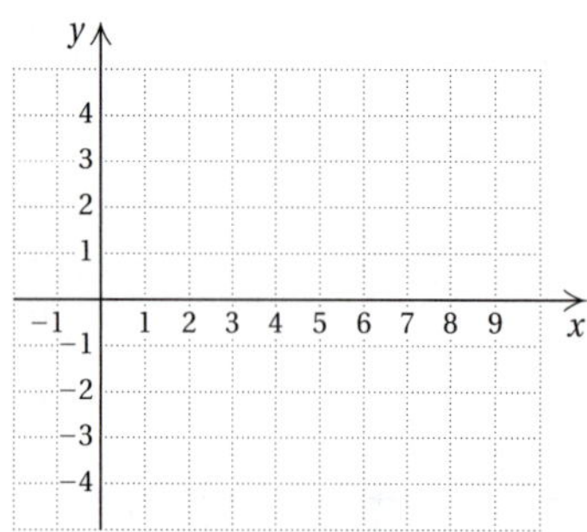

21. $x = \left(\frac{1}{2}\right)^y$

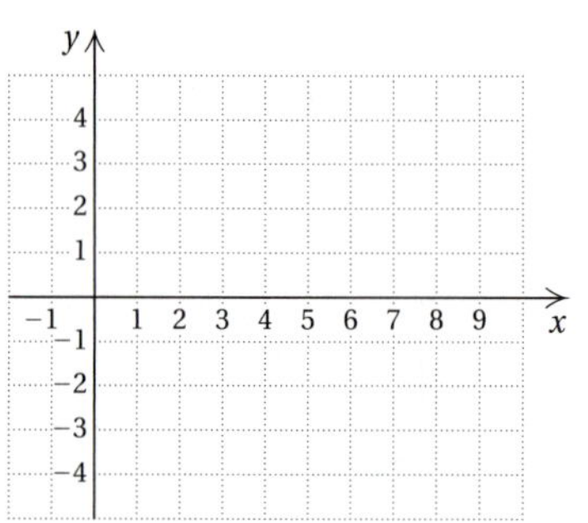

22. $x = \left(\frac{1}{3}\right)^y$

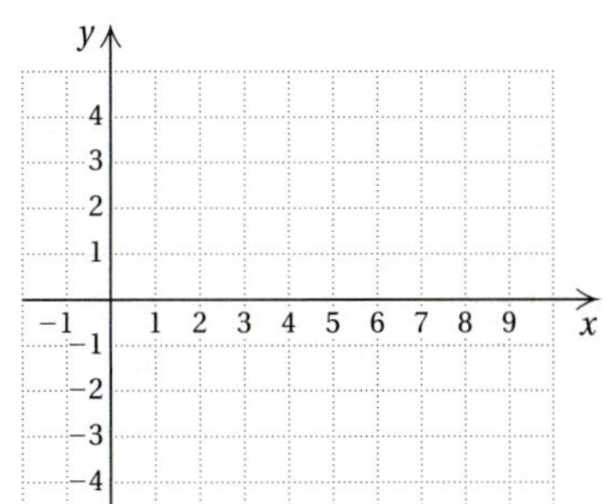

23. $x = 5^y$

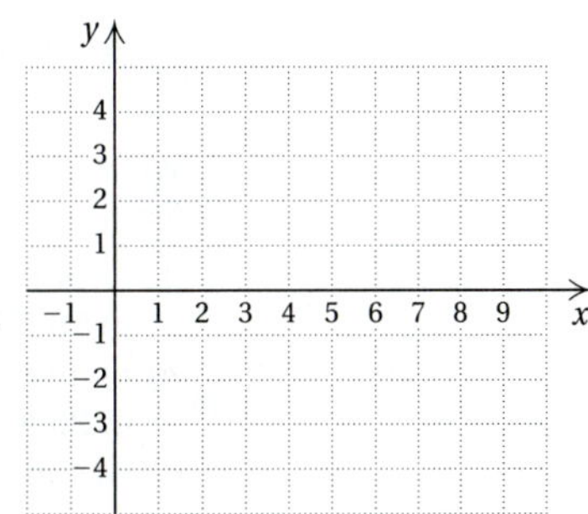

24. $x = \left(\frac{2}{3}\right)^y$

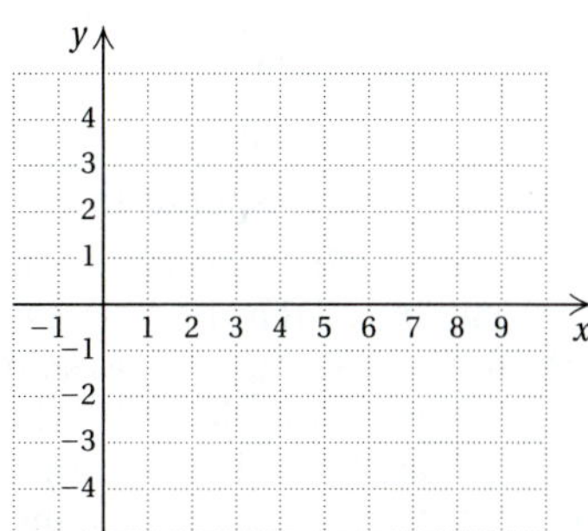

Graph both equations using the same set of axes.

25. $y = 2^x, \quad x = 2^y$

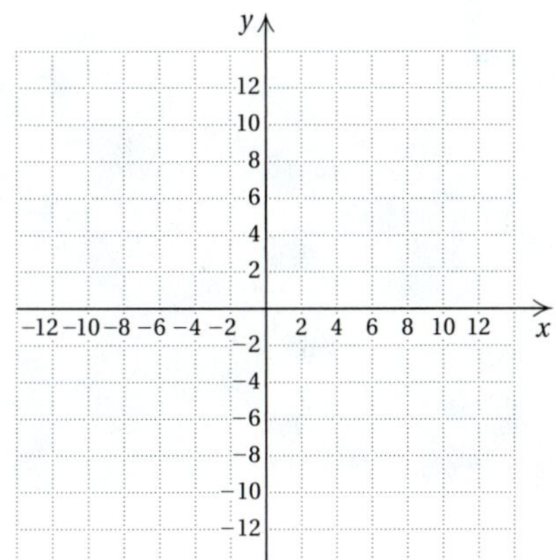

26. $y = \left(\frac{1}{2}\right)^x, \quad x = \left(\frac{1}{2}\right)^y$

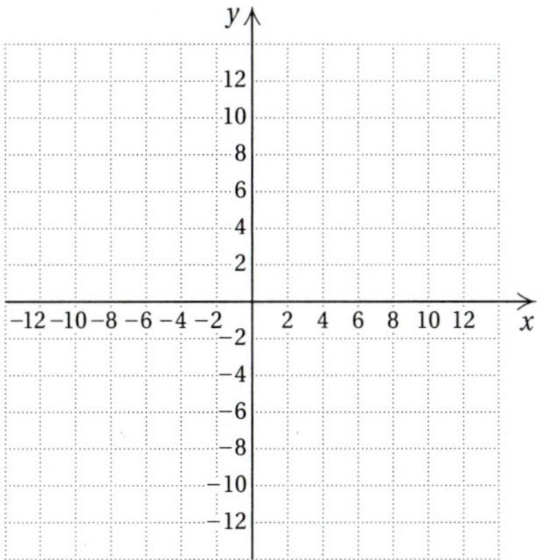

C Solve.

27. *Interest Compounded Annually.* Suppose that \$50,000 is invested at 6% interest, compounded annually.

a) Find a function A for the amount in the account after t years.

b) Complete the following table of function values.

t	$A(t)$
0	
1	
2	
4	
8	
10	
20	

c) Graph the function.

28. *Interest Compounded Annually.* Suppose that \$50,000 is invested at 8% interest, compounded annually.

a) Find a function A for the amount in the account after t years.

b) Complete the following table of function values.

t	$A(t)$
0	
1	
2	
4	
8	
10	
20	

c) Graph the function.

29. *Cliff Notes.* Most students are familiar with Cliff Notes, the booklets that offer abridged versions of classic literary works. The value V of the CliffNotes Company, t years after 1958 when it was founded, can be approximated by

$$V(t) = 4000(1.22)^t,$$

where $t = 0$ corresponds to 1998.

a) What was the value of the CliffNotes Company in 1958? in 1970? in 1980? in 1990? in 1998?
b) CliffNotes Company was sold to IDG Books Worldwide, Inc., in 1999. What was the value of the company when it was sold?
c) Cliff Hillegass, founder of the company, died on May 7, 2001. What might its value have been at the time of his death?
d) Graph the function.

30. *E-mail Boxes.* The growth in the number of e-mail boxes in North America is exponential. The number N of boxes, in millions, t years after 1995 can be approximated by

$$N(t) = 57(1.3)^t,$$

where $t = 0$ corresponds to 1995.

Source: IDC Research

a) How many e-mail boxes were there in North America in 1995? in 2000? in 2003?
b) Estimate the number of e-mail boxes in 2008 and in 2020.
c) Graph the function.

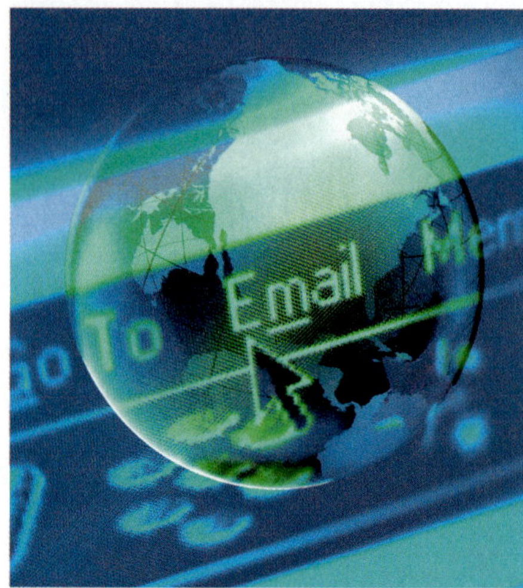

31. *Marine Biology.* Due to excessive whaling prior to the mid-1970s, the humpback whale is considered an endangered species. The worldwide population of humpbacks, $P(t)$, in thousands, t years after 1900 $(t < 70)$ can be approximated by

$$P(t) = 150(0.960)^t.$$

Source: American Cetecean Society, 2001; ASK Archive, 1998

a) How many humpback whales were alive in 1930? in 1960?
b) Graph the function.

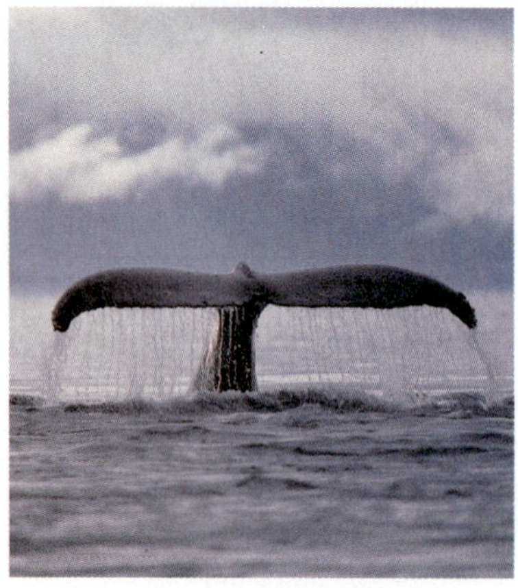

32. *Marine Biology.* As a result of preservation efforts in most countries in which whaling was common, the humpback whale population has grown since the 1970s. The worldwide population of humpbacks, $P(t)$, in thousands, t years after 1982 can be approximated by

$$P(t) = 5.5(1.047)^t.$$

Source: American Cetecean Society, 2001; ASK Archive, 1998

a) How many humpback whales were alive in 1992? in 2001?
b) Graph the function.

33. *Growth of Bacteria.* Bladder infections are often caused when the bacteria *Escherichia coli* reach the human bladder. Suppose that 3000 of the bacteria are present at time $t = 0$. Then t minutes later, the number of bacteria present will be

$$N(t) = 3000(2)^{t/20}.$$

Source: Chris Hayes, "Detecting a Human Health Risk: *E. coli*," *Laboratory Medicine* 29, no. 6, June 1998: 347–355

a) How many bacteria will be present after 10 min? 20 min? 30 min? 40 min? 60 min?

b) Graph the function.

34. *Salvage Value.* An office machine is purchased for \$5200. Its value each year is about 80% of the value the preceding year. Its value after t years is given by the exponential function

$$V(t) = \$5200(0.8)^t.$$

a) Find the value of the machine after 0 yr, 1 yr, 2 yr, 5 yr, and 10 yr.

b) Graph the function.

35. DW Why was it necessary to discuss irrational exponents before graphing exponential functions?

36. DW Suppose that \$1000 is invested for 5 yr at 6% interest, compounded annually. In what year will the greatest amount of interest be earned? Why?

SKILL MAINTENANCE

37. Multiply and simplify: $x^{-5} \cdot x^3$. [9.1d, f]

38. Simplify: $(x^{-3})^4$. [9.2a, f]

Simplify. [9.1b]

39. 9^0

40. $\left(\frac{2}{3}\right)^0$

41. $\left(\frac{2}{3}\right)^1$

42. 2.7^1

Divide and simplify. [9.1e, f]

43. $\dfrac{x^{-3}}{x^4}$

44. $\dfrac{x}{x^{11}}$

45. $\dfrac{x}{x^0}$

46. $\dfrac{x^{-3}}{x^{-4}}$

SYNTHESIS

47. Simplify: $(5^{\sqrt{2}})^{2\sqrt{2}}$.

48. Which is larger: $\pi^{\sqrt{2}}$ or $(\sqrt{2})^{\pi}$?

Graph.

49. $y = 2^x + 2^{-x}$

50. $y = |2^x - 2|$

51. $y = \left|\left(\frac{1}{2}\right)^x - 1\right|$

52. $y = 2^{-x^2}$

Graph both equations using the same set of axes.

53. $y = 3^{-(x-1)}$, $x = 3^{-(y-1)}$

54. $y = 1^x$, $x = 1^y$

55. Use a graphing calculator to graph each of the equations in Exercises 49–52.

17.2 INVERSE AND COMPOSITE FUNCTIONS

Objectives

a Find the inverse of a relation if it is described as a set of ordered pairs or as an equation.

b Given a function, determine whether it is one-to-one and has an inverse that is a function.

c Find a formula for the inverse of a function, if it exists, and graph inverse relations and functions.

d Find the composition of functions and express certain functions as a composition of functions.

e Determine whether a function is an inverse by checking its composition with the original function.

1. Consider the relation g given by

$$g = \{(2, 5), (-1, 4), (-2, 1)\}.$$

The graph of the relation is shown below in red. Find the inverse and draw its graph in blue.

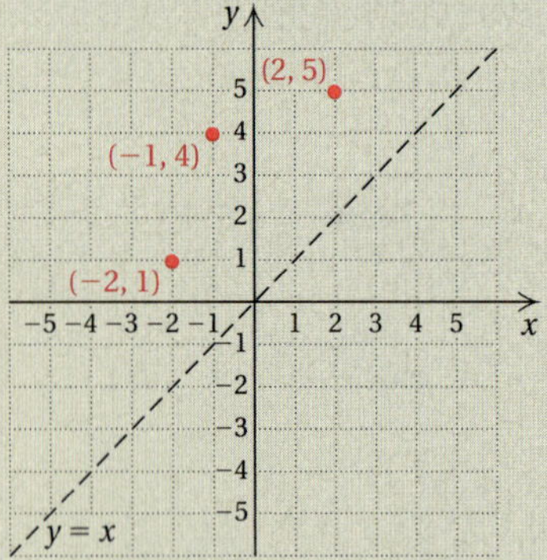

Answer on page A-68

When we go from an output of a function back to its input or inputs, we get an *inverse relation*. When that relation is a function, we have an *inverse function*. We now study such inverse functions and how to find formulas when the original function has a formula. We do so to understand the relationships among the special functions that we study in this chapter.

a Inverses

A set of ordered pairs is called a **relation.** When we consider the graph of a function, we are thinking of a set of ordered pairs. Thus a function can be thought of as a special kind of relation, in which to each first coordinate there corresponds one and only one second coordinate.

Consider the relation h given as follows:

$$h = \{(-7, 4), (3, -1), (-6, 5), (0, 2)\}.$$

Suppose we *interchange* the first and second coordinates. The relation we obtain is called the **inverse** of the relation h and is given as follows:

$$\text{Inverse of } h = \{(4, -7), (-1, 3), (5, -6), (2, 0)\}.$$

INVERSE RELATION

Interchanging the coordinates of the ordered pairs in a relation produces the **inverse relation.**

EXAMPLE 1 Consider the relation g given by

$$g = \{(2, 4), (-1, 3), (-2, 0)\}.$$

In the figure below, the relation g is shown in red. The inverse of the relation is

$$\{(4, 2), (3, -1), (0, -2)\}$$

and is shown in blue.

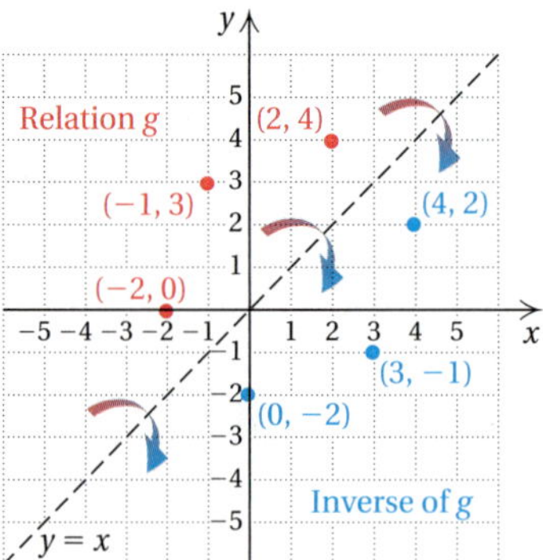

Do Exercise 1.

INVERSE RELATION

If a relation is defined by an equation, interchanging the variables produces an equation of the **inverse relation.**

EXAMPLE 2 Find an equation of the inverse of the relation

$$y = 3x - 4.$$

Then graph both the relation and its inverse.

We interchange x and y and obtain an equation of the inverse:

$$x = 3y - 4.$$

Relation: $y = 3x - 4$ → *Inverse*: $x = 3y - 4$

x	y
0	−4
1	−1
2	2
3	5

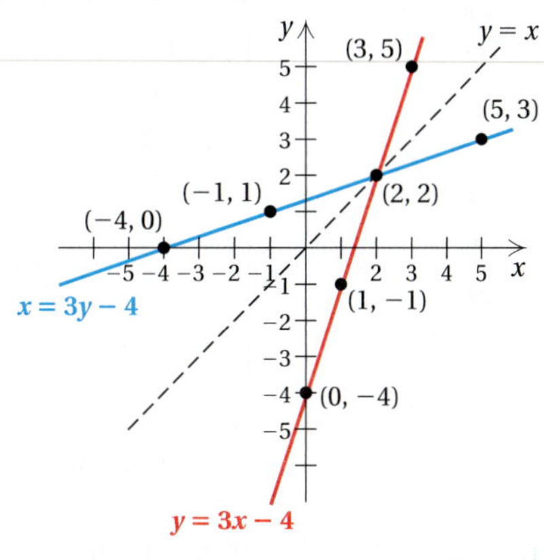

x	y
−4	0
−1	1
2	2
5	3

Note in Example 2 that the relation $y = 3x - 4$ is a function and its inverse relation $x = 3y - 4$ is also a function. Each graph passes the vertical-line test (see Section 12.1).

EXAMPLE 3 Find an equation of the inverse of the relation

$$y = 6x - x^2.$$

Then graph both the original relation and its inverse.

We interchange x and y and obtain an equation of the inverse:

$$x = 6y - y^2.$$

Relation: $y = 6x - x^2$ → *Inverse*: $x = 6y - y^2$

x	y
−1	−7
0	0
1	5
3	9
5	5

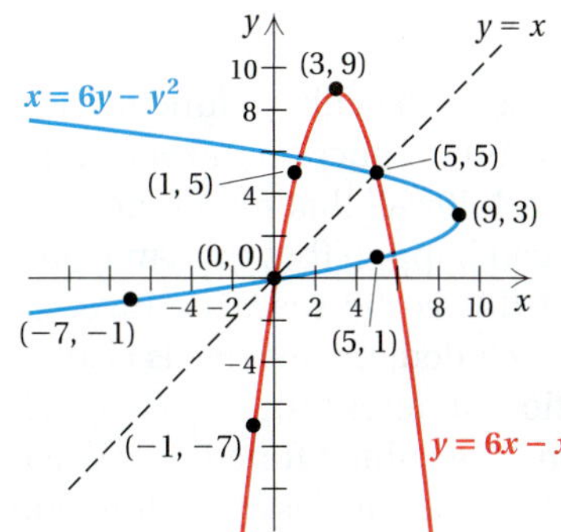

x	y
−7	−1
0	0
5	1
9	3
5	5

2. Find an equation of the inverse relation. Then complete the table and graph both the original relation and its inverse.

Relation: $y = 6 - 2x$

x	y
0	6
2	2
3	0
5	−4

Inverse: ____________

x	y
6	
2	
0	
−4	

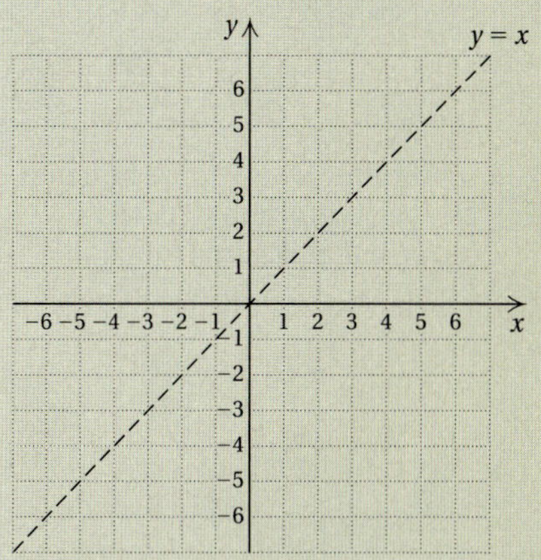

Answers on page A-68

3. Find an equation of the inverse relation. Then complete the table and graph both the original relation and its inverse.

Relation:
$y = x^2 - 4x + 7$

x	y
0	7
1	4
2	3
3	4
4	7

Inverse: ______

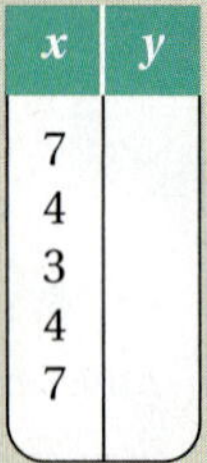

x	y
7	
4	
3	
4	
7	

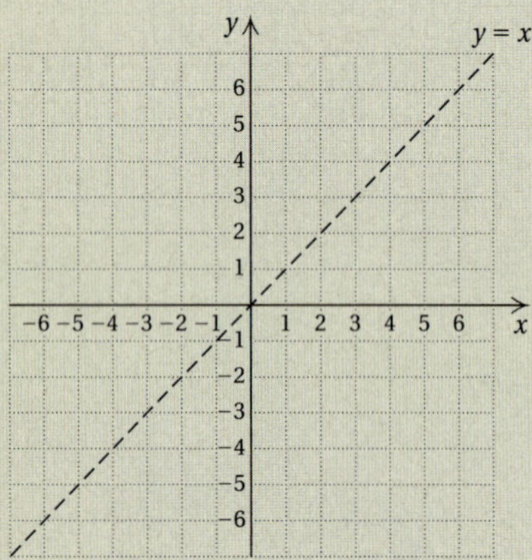

4. Determine whether the function is one-to-one and thus has an inverse that is also a function.

$f(x) = 4 - x$

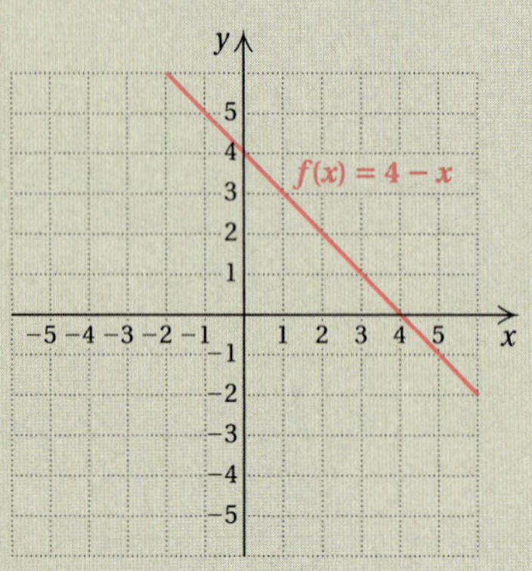

Answers on page A-68

Note in Example 3 that the relation $y = 6x - x^2$ is a function because it passes the vertical-line test. However, its inverse relation $x = 6y - y^2$ is not a function because its graph fails the vertical-line test. Therefore, the inverse of a function is *not* always a function.

Do Exercises 2 and 3. (Exercise 2 is on the preceding page.)

b Inverses and One-To-One Functions

Let's consider the following two functions.

NUMBER (Domain)	CUBE (Range)
−3 →	−27
−2 →	−8
−1 →	−1
0 →	0
1 →	1
2 →	8
3 →	27

YEAR (Domain)	FIRST-CLASS POSTAGE COST, IN CENTS (Range)
1983 →	20
1984 →	20
1989 →	25
1991 →	29
1995 →	32
1999 →	33
2001 →	34
2002 →	37

Source: U.S. Postal Service

Suppose we reverse the arrows. Are these inverse relations functions?

CUBE ROOT (Range)	NUMBER (Domain)
−3	← −27
−2	← −8
−1	← −1
0	← 0
1	← 1
2	← 8
3	← 27

YEAR (Range)	FIRST-CLASS POSTAGE COST, IN CENTS (Domain)
1983	← 20
1984	← 20
1989	← 25
1991	← 29
1995	← 32
1999	← 33
2001	← 34
2002	← 37

We see that the inverse of the cubing function is a function. The inverse of the postage function is not a function, however, because the input 20 has *two* outputs, 1983 and 1984. Recall that for a function, each input has exactly one output. However, it can happen that the same output comes from two or more different inputs. If this is the case, the inverse cannot be a function. When this possibility is excluded, the inverse is also a function.

In the cubing function, different inputs have different outputs. Thus its inverse is also a function. The cubing function is what is called a **one-to-one function.** If the inverse of a function f is also a function, it is named f^{-1} (read "f-inverse").

CAUTION!

The -1 in f^{-1} is *not* an exponent and f^{-1} does not represent a reciprocal!

ONE-TO-ONE FUNCTION AND INVERSES

A function f is **one-to-one** if different inputs have different outputs—that is,

$$\text{if} \quad a \neq b, \quad \text{then} \quad f(a) \neq f(b). \quad \text{Or,}$$

A function f is **one-to-one** if when the outputs are the same, the inputs are the same—that is,

$$\text{if} \quad f(a) = f(b), \quad \text{then} \quad a = b.$$

If a function is one-to-one, then its inverse is a function.

The domain of a one-to-one function f is the range of the inverse f^{-1}.

The range of a one-to-one function f is the domain of the inverse f^{-1}.

How can we tell graphically whether a function is one-to-one and thus has an inverse that is a function?

EXAMPLE 4 The graph of the exponential function $f(x) = 2^x$, or $y = 2^x$, is shown on the left below. The graph of the inverse $x = 2^y$ is shown on the right. How can we tell by examining only the graph on the left whether it has an inverse that is a function?

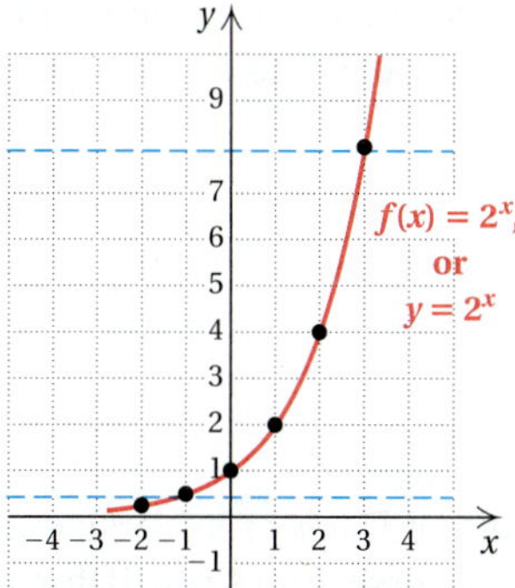

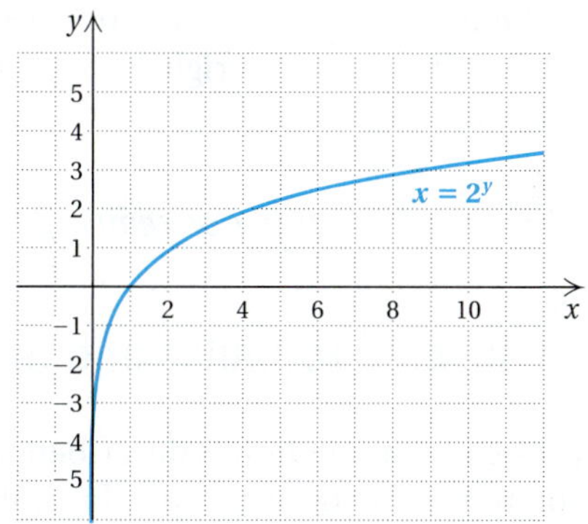

We see that the graph on the right passes the vertical-line test, so we know it is that of a function. However, if we look only at the graph on the left, we think as follows:

A function is one-to-one if different inputs have different outputs. In other words, no two x-values will have the same y-value. For this function, we cannot find two x-values that have the same y-value. Note also that no horizontal line can be drawn that will cross the graph more than once. The function is thus one-to-one and its inverse is a function.

THE HORIZONTAL-LINE TEST

If it is possible for a horizontal line to intersect the graph of a function more than once, then the function is not one-to-one and therefore its inverse is not a function.

Determine whether the function is one-to-one and thus has an inverse that is also a function.

5. $f(x) = x^2 - 1$

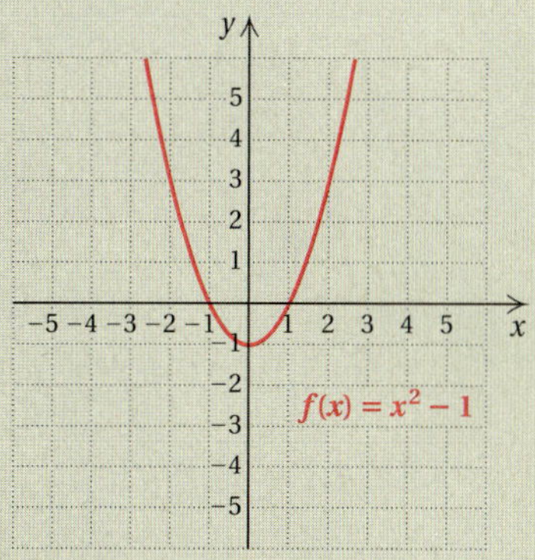

6. $f(x) = 4^x$
(Sketch this graph yourself.)

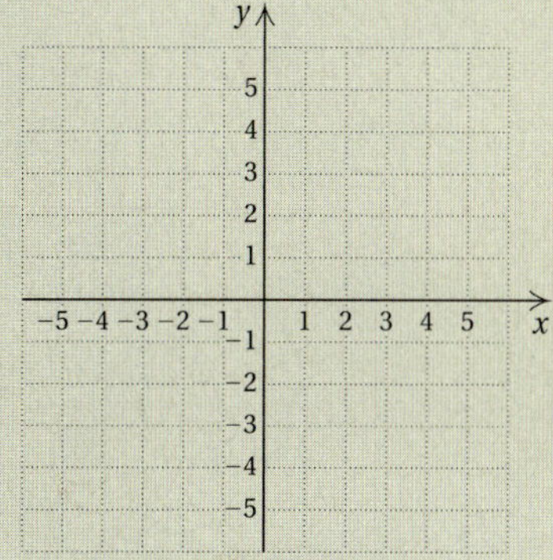

7. $f(x) = |x| - 3$
(Sketch this graph yourself.)

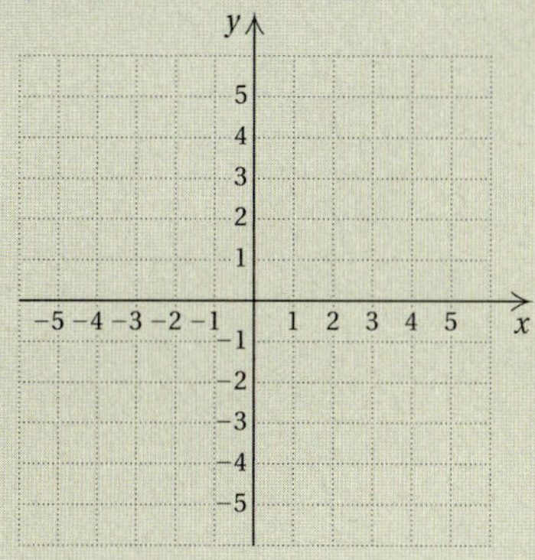

Answers on page A-68

A graph is that of a function if no vertical line crosses the graph more than once. A function has an inverse that is also a function if no horizontal line crosses the graph more than once.

EXAMPLE 5 Determine whether the function $f(x) = x^2$ is one-to-one and has an inverse that is also a function.

The graph of $f(x) = x^2$, or $y = x^2$, is shown on the left below. There are many horizontal lines that cross the graph more than once, so this function is not one-to-one and does not have an inverse that is a function.

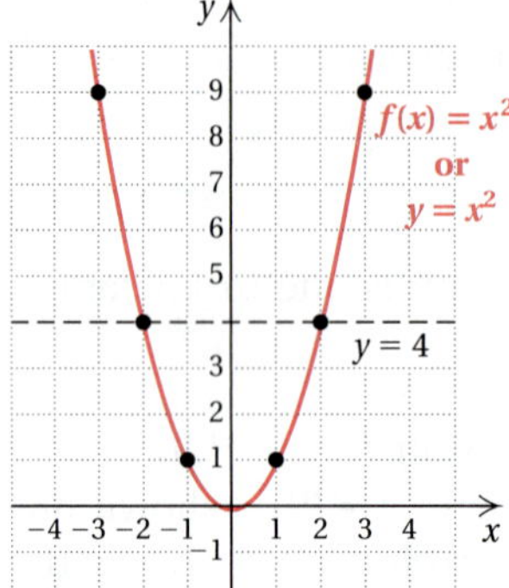

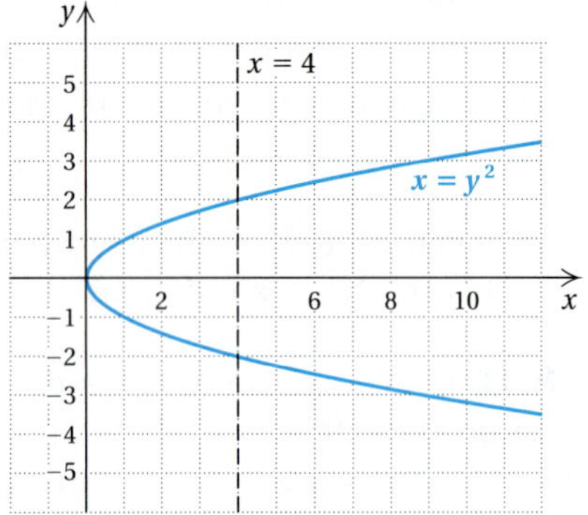

The inverse of the function $y = x^2$ is the relation $x = y^2$. The graph of $x = y^2$ is shown on the right above. It fails the vertical-line test and is not a function.

Do Exercises 4–7 on the preceding pages.

C Inverse Formulas and Graphs

Suppose that a function is described by a formula. If it has an inverse that is a function, how do we find a formula for the inverse function? If for any equation with two variables such as x and y we interchange the variables, we obtain an equation of the inverse relation. We proceed as follows to find a formula for f^{-1}.

If a function f is one-to-one, a formula for its inverse f^{-1} can be found as follows:

1. Replace $f(x)$ with y.
2. Interchange x and y. (This gives the inverse relation.)
3. Solve for y.
4. Replace y with $f^{-1}(x)$.

EXAMPLE 6 Given $f(x) = x + 1$:

a) Determine whether the function is one-to-one.

b) If it is one-to-one, find a formula for $f^{-1}(x)$.

c) Graph the inverse function, if it exists.

a) The graph of $f(x) = x + 1$ is shown below. It passes the horizontal-line test, so it is one-to-one. Thus its inverse is a function.

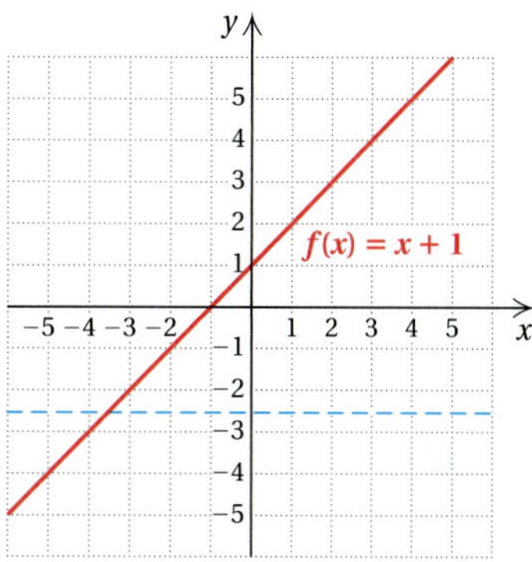

b) 1. Replace $f(x)$ with y: $y = x + 1$.

2. Interchange x and y: $x = y + 1$. This gives the inverse relation.

3. Solve for y: $x - 1 = y$.

4. Replace y with $f^{-1}(x)$: $f^{-1}(x) = x - 1$.

c) We graph $f^{-1}(x) = x - 1$, or $y = x - 1$. The graph is shown below.

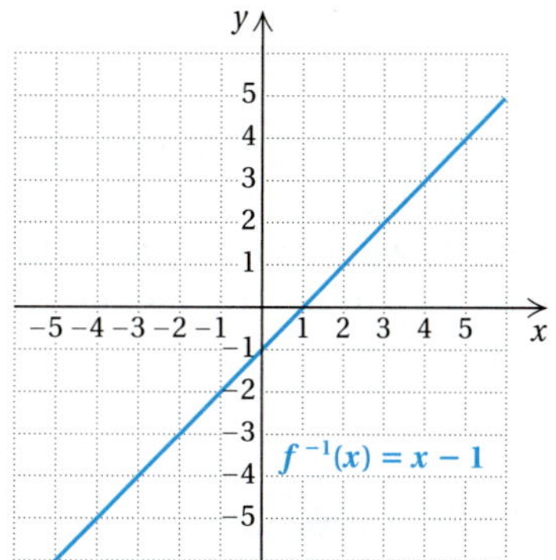

EXAMPLE 7 Given $f(x) = 2x - 3$:

a) Determine whether the function is one-to-one.

b) If it is one-to-one, find a formula for $f^{-1}(x)$.

c) Graph the inverse function, if it exists.

a) The graph of $f(x) = 2x - 3$ is shown below. It passes the horizontal-line test and is one-to-one.

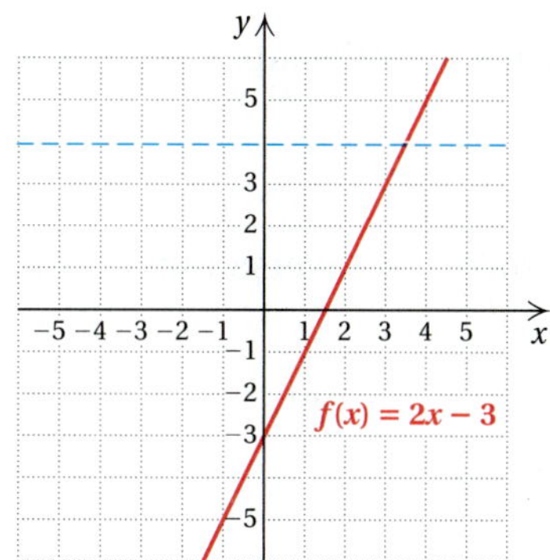

Given each function:

a) Determine whether it is one-to-one.

b) If it is one-to-one, find a formula for the inverse.

c) Graph the inverse function, if it exists.

8. $f(x) = 3 - x$

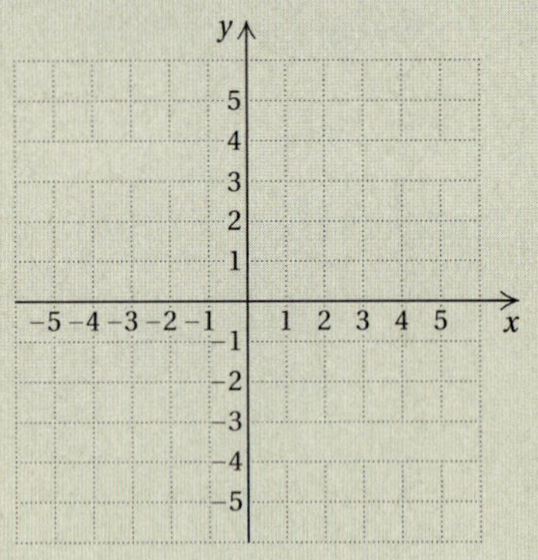

9. $g(x) = 3x - 2$

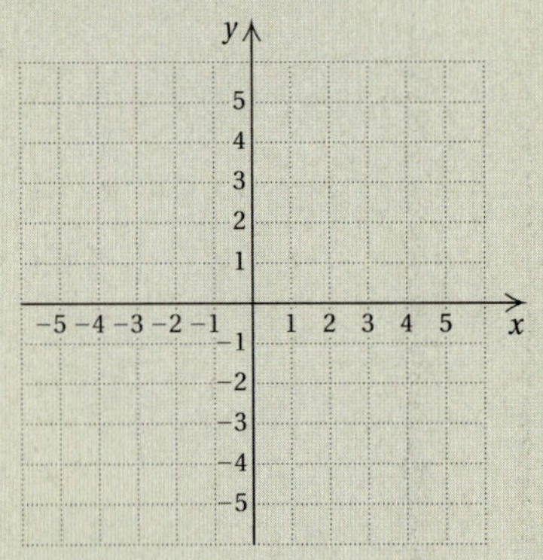

Answers on page A-68

b) 1. Replace $f(x)$ with y: $\quad y = 2x - 3.$

2. Interchange x and y: $\quad x = 2y - 3.$

3. Solve for y: $\quad x + 3 = 2y$

$$\frac{x+3}{2} = y.$$

4. Replace y with $f^{-1}(x)$: $\quad f^{-1}(x) = \dfrac{x+3}{2}.$

c) We graph

$$f^{-1}(x) = \frac{x+3}{2}, \quad \text{or}$$

$$y = \frac{1}{2}x + \frac{3}{2}.$$

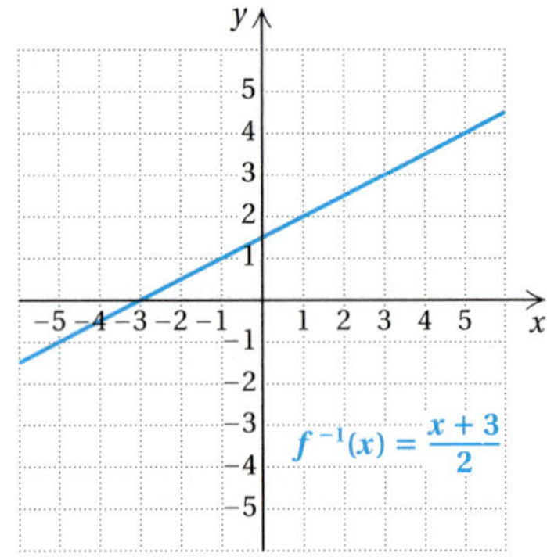

Do Exercises 8 and 9.

Let's now consider inverses of functions in terms of a function machine. Suppose that a one-to-one function f is programmed into a machine. If the machine has a reverse switch, when the switch is thrown, the machine performs the inverse function f^{-1}. Inputs then enter at the opposite end, and the entire process is reversed.

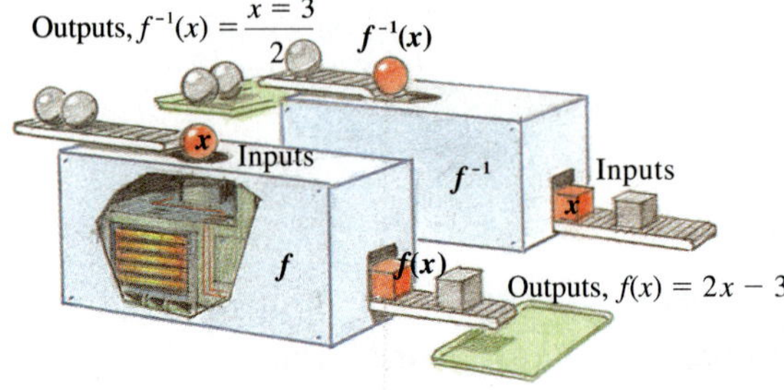

Consider $f(x) = 2x - 3$ and $f^{-1}(x) = (x + 3)/2$ from Example 7. For the input 5,

$$f(5) = 2 \cdot 5 - 3 = 10 - 3 = 7.$$

The output is 7. Now we use 7 for the input in the inverse:

$$f^{-1}(7) = \frac{7+3}{2} = \frac{10}{2} = 5.$$

The function f takes 5 to 7. The inverse function f^{-1} takes the number 7 back to 5.

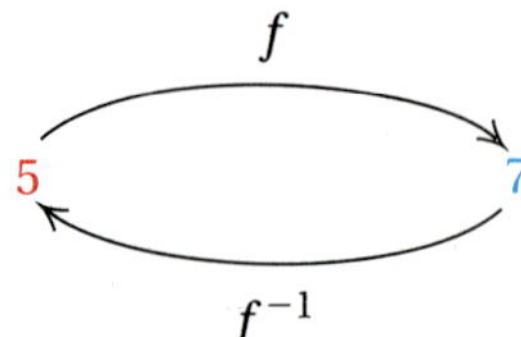

How do the graphs of a function and its inverse compare?

EXAMPLE 8 Graph $f(x) = 2x - 3$ and $f^{-1}(x) = (x + 3)/2$ using the same set of axes. Then compare.

The graph of each function follows. Note that the graph of f^{-1} can be drawn by reflecting the graph of f across the line $y = x$. That is, if we graph $f(x) = 2x - 3$ in wet ink and fold the paper along the line $y = x$, the graph of $f^{-1}(x) = (x + 3)/2$ will appear as the impression made by f.

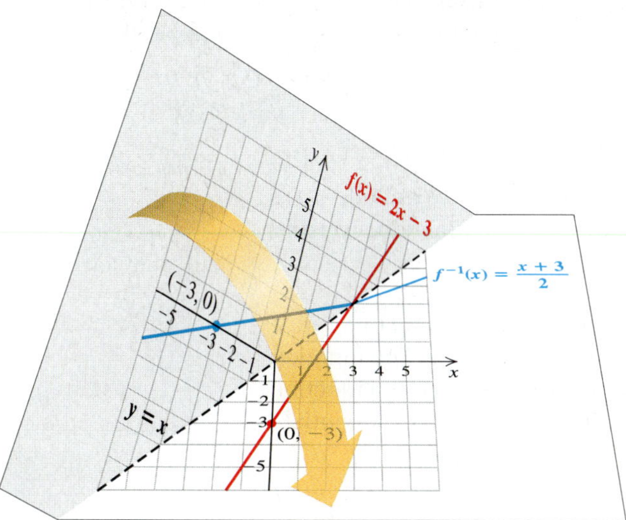

When x and y are interchanged to find a formula for the inverse, we are, in effect, flipping the graph of $f(x) = 2x - 3$ over the line $y = x$. For example, when the coordinates of the y-intercept of the graph of f, $(0, -3)$, are reversed, we get the x-intercept of the graph of f^{-1}, $(-3, 0)$.

The graph of f^{-1} is a reflection of the graph of f across the line $y = x$.

Do Exercise 10.

EXAMPLE 9 Consider $g(x) = x^3 + 2$.

a) Determine whether the function is one-to-one.

b) If it is one-to-one, find a formula for its inverse.

c) Graph the inverse, if it exists.

a) The graph of $g(x) = x^3 + 2$ is shown at right in red. It passes the horizontal-line test and thus is one-to-one.

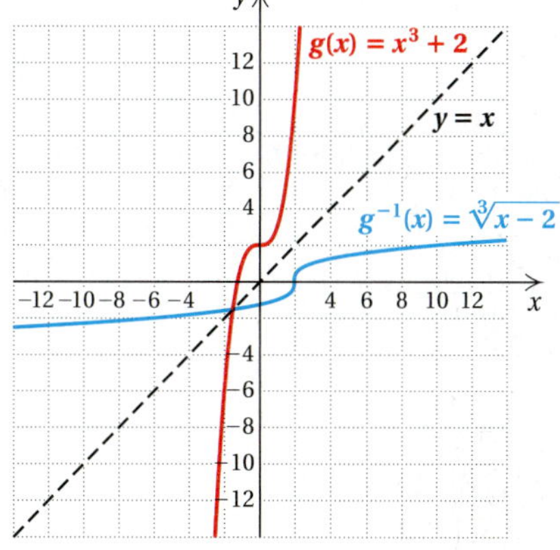

10. Graph $g(x) = 3x - 2$ and $g^{-1}(x) = (x + 2)/3$ using the same set of axes.

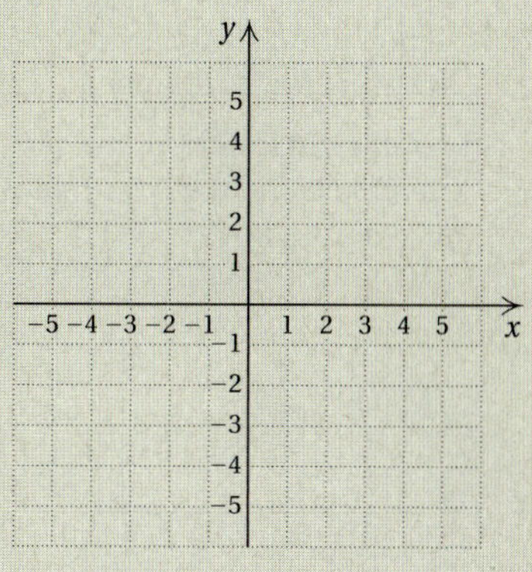

Answer on page A-69

11. Given $f(x) = x^3 + 1$:

a) Determine whether the function is one-to-one.

b) If it is one-to-one, find a formula for its inverse.

c) Graph the function and its inverse using the same set of axes.

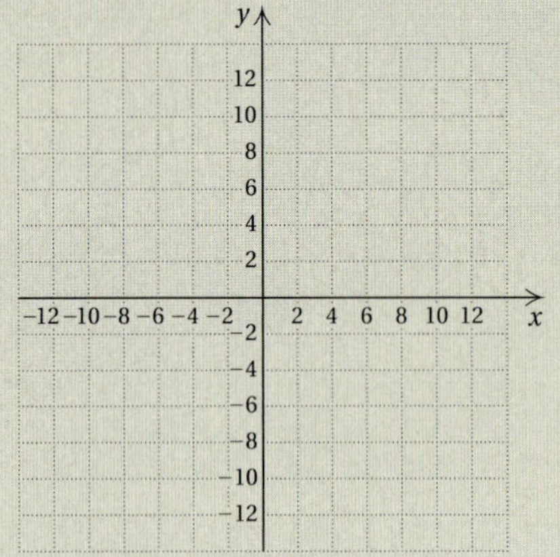

Answers on page A-69

b) 1. Replace $g(x)$ with y: $y = x^3 + 2$.

2. Interchange x and y: $x = y^3 + 2$.

3. Solve for y: $x - 2 = y^3$

$\sqrt[3]{x - 2} = y$. Since a number has only one cube root, we can solve for y.

4. Replace y with $g^{-1}(x)$: $g^{-1}(x) = \sqrt[3]{x - 2}$.

c) To find the graph, we reflect the graph of $g(x) = x^3 + 2$ across the line $y = x$, as we did in Example 8. It can also be found by substituting into $g^{-1}(x) = \sqrt[3]{x - 2}$ and plotting points. The graphs of g and g^{-1} are shown together above.

Do Exercise 11.

We can now see why we exclude 1 as a base for an exponential function. Consider

$$f(x) = a^x = 1^x = 1.$$

The graph of f is the horizontal line $y = 1$. The graph is not one-to-one. The function does not have an inverse that is a function. All other positive bases yield exponential functions that are one-to-one.

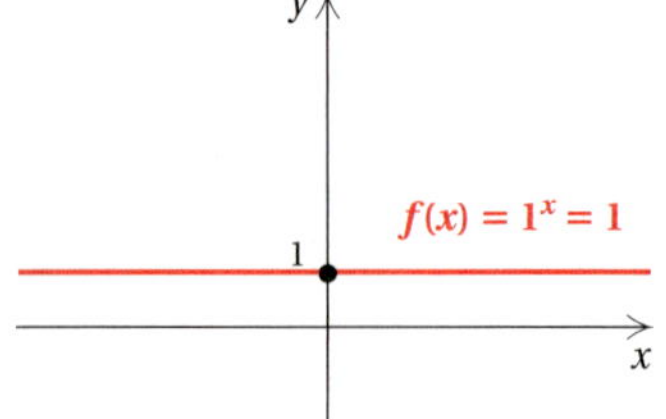

CALCULATOR CORNER

Graphing an Inverse Function The DRAWINV operation can be used to graph a function and its inverse on the same screen. A formula for the inverse function need not be found in order to do this. The graphing calculator must be set in FUNC mode when this operation is used.

To graph $f(x) = 2x - 3$ and $f^{-1}(x)$ using the same set of axes, we first clear any existing equations on the equation-editor screen and then enter $y_1 = 2x - 3$. Now, we press [2nd] [DRAW] 8 to select the DRAWINV operation. (DRAW is the second operation associated with the [PRGM] key.) We press [VARS] [▷] [1] [1] to indicate that we want to graph the inverse of y_1. Then we press [ENTER] to see the graph of the function and its inverse. The graphs are shown here in the standard window.

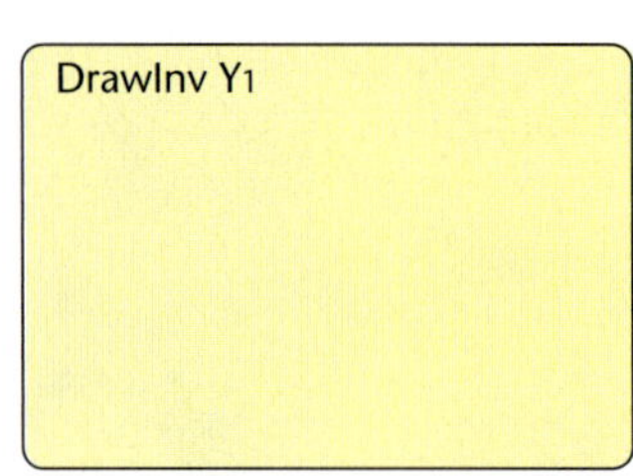

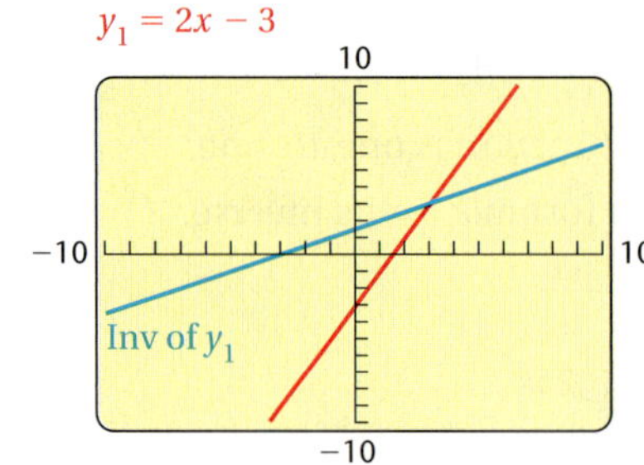

Exercises: Use the DRAWINV operation on a graphing calculator to graph each function with its inverse on the same screen.

1. $f(x) = x - 5$
2. $f(x) = \frac{2}{3}x$
3. $f(x) = x^2 + 2$
4. $f(x) = x^3 - 3$

d Composite Functions

In the real world, functions frequently occur in which some quantity depends on a variable that, in turn, depends on another variable. For instance, the number of employees hired by a firm may depend on the firm's profits, which may in turn depend on the number of items the firm produces. Functions like this are called **composite functions.**

For example, the function g that gives a correspondence between women's shoe sizes in the United States and those in Italy is given by $g(x) = 2x + 24$, where x is the U.S. size and $g(x)$ is the Italian size. Thus a U.S. size 4 corresponds to a shoe size of $g(4) = 2 \cdot 4 + 24$, or 32, in Italy.

There is also a function that gives a correspondence between women's shoe sizes in Italy and those in Britain. The function is given by $f(x) = \frac{1}{2}x - 14$, where x is the Italian size and $f(x)$ is the corresponding British size. Thus an Italian size 32 corresponds to a British size $f(32) = \frac{1}{2}(32) - 14$, or 2.

It seems reasonable to conclude that a shoe size of 4 in the United States corresponds to a size of 2 in Britain and that some function h describes this correspondence. Can we find a formula for h? If we look at the following tables, we might guess that such a formula is $h(x) = x - 2$, and that is indeed correct. But, for more complicated formulas, we would need to use algebra.

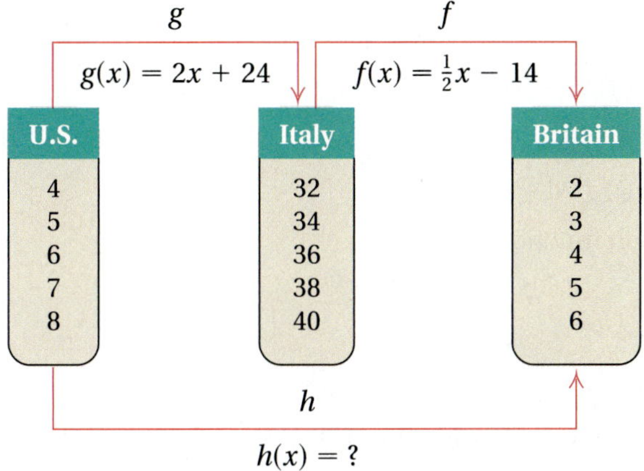

A shoe size x in the United States corresponds to a shoe size $g(x)$ in Italy, where

$$g(x) = 2x + 24.$$

Now $2x + 24$ is a shoe size in Italy. If we replace x in $f(x) = \frac{1}{2}x - 14$ with $g(x)$, or $2x + 24$, we can find the corresponding shoe size in Britain:

$$\begin{aligned} f(x) &= \tfrac{1}{2}x - 14 \\ f(g(x)) &= f(2x + 24) = \tfrac{1}{2}[2x + 24] - 14 \\ &= x + 12 - 14 = x - 2. \end{aligned}$$

This gives a formula for h:

$$h(x) = x - 2.$$

Thus a shoe size of 4 in the United States corresponds to a shoe size of $h(4) = 4 - 2$, or 2, in Britain. The function h is the **composition** of f and g, symbolized by $f \circ g$. To find $f \circ g(x)$, we substitute $g(x)$ for x in $f(x)$.

12. Given $f(x) = x + 5$ and $g(x) = x^2 - 1$, find $f \circ g(x)$ and $g \circ f(x)$.

Answer on page A-69

COMPOSITE FUNCTION

The **composite function** $f \circ g$, the **composition** of f and g, is defined as

$$f \circ g(x) = f(g(x)), \quad \text{or} \quad (f \circ g)(x) = f[g(x)].$$

We can visualize the composition of functions as follows.

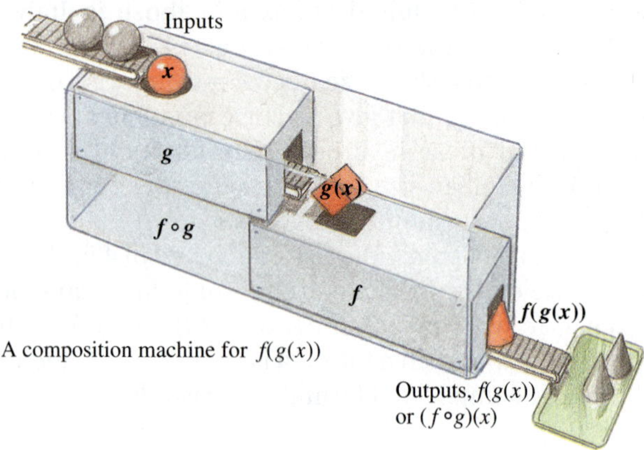

A composition machine for $f(g(x))$

EXAMPLE 10 Given $f(x) = 3x$ and $g(x) = 1 + x^2$:

a) Find $f \circ g(5)$ and $g \circ f(5)$.

b) Find $f \circ g(x)$ and $g \circ f(x)$.

We consider each function separately:

$f(x) = 3x$ — This function multiplies each input by 3.

and $g(x) = 1 + x^2$. — This function adds 1 to the square of each input.

a) $f \circ g(5) = f(g(5)) = f(1 + 5^2) = f(26) = 3(26) = 78;$

$g \circ f(5) = g(f(5)) = g(3 \cdot 5) = g(15) = 1 + 15^2 = 1 + 225 = 226$

b) $f \circ g(x) = f(g(x))$

$= f(1 + x^2)$ — Substituting $1 + x^2$ for x

$= 3(1 + x^2)$

$= 3 + 3x^2;$

$g \circ f(x) = g(f(x))$

$= g(3x)$ — Substituting $3x$ for x

$= 1 + (3x)^2$

$= 1 + 9x^2$

We can check the values in part (a) with the formulas found in part (b):

$f \circ g(x) = 3 + 3x^2$	$g \circ f(x) = 1 + 9x^2$
$f \circ g(5) = 3 + 3 \cdot 5^2$	$g \circ f(5) = 1 + 9 \cdot 5^2$
$= 3 + 3 \cdot 25$	$= 1 + 9 \cdot 25$
$= 3 + 75$	$= 1 + 225$
$= 78;$	$= 226.$

Example 10 shows that $f \circ g(5) \neq g \circ f(5)$ and, in general, $f \circ g(x) \neq g \circ f(x)$.

Do Exercise 12.

EXAMPLE 11 Given $f(x) = \sqrt{x}$ and $g(x) = x - 1$, find $f \circ g(x)$ and $g \circ f(x)$.

$$f \circ g(x) = f(g(x)) = f(x - 1) = \sqrt{x - 1};$$

$$g \circ f(x) = g(f(x)) = g(\sqrt{x}) = \sqrt{x} - 1$$

Do Exercise 13.

13. Given $f(x) = 4x + 5$ and $g(x) = \sqrt[3]{x}$, find $f \circ g(x)$ and $g \circ f(x)$.

It is important to be able to recognize how a function can be expressed, or "broken down," as a composition. Such a situation can occur in a study of calculus.

EXAMPLE 12 Find $f(x)$ and $g(x)$ such that $h(x) = f \circ g(x)$:

$$h(x) = (7x + 3)^2.$$

This is $7x + 3$ to the 2nd power. Two functions that can be used for the composition are $f(x) = x^2$ and $g(x) = 7x + 3$. We can check by forming the composition:

$$h(x) = f \circ g(x) = f(g(x)) = f(7x + 3) = (7x + 3)^2.$$

This is the most "obvious" answer to the question. There can be other less obvious answers. For example, if

$$f(x) = (x - 1)^2 \quad \text{and} \quad g(x) = 7x + 4,$$

then $h(x) = f \circ g(x) = f(g(x)) = f(7x + 4) = (7x + 4 - 1)^2 = (7x + 3)^2.$

Do Exercise 14.

14. Find $f(x)$ and $g(x)$ such that $h(x) = f \circ g(x)$. Answers may vary.

a) $h(x) = \sqrt[3]{x^2 + 1}$

b) $h(x) = \dfrac{1}{(x + 5)^4}$

e Inverse Functions and Composition

Suppose that we used some input x for the function f and found its output, $f(x)$. The function f^{-1} would then take that output back to x. Similarly, if we began with an input x for the function f^{-1} and found its output, $f^{-1}(x)$, the original function f would then take that output back to x.

> If a function f is one-to-one, then f^{-1} is the unique function for which
>
> $$f^{-1} \circ f(x) = x \quad \text{and} \quad f \circ f^{-1}(x) = x.$$

EXAMPLE 13 Let $f(x) = 2x - 3$. Use composition to show that

$$f^{-1}(x) = \frac{x + 3}{2}. \quad \text{(See Example 7.)}$$

We find $f^{-1} \circ f(x)$ and $f \circ f^{-1}(x)$ and check to see that each is x.

$$\begin{aligned} f^{-1} \circ f(x) &= f^{-1}(f(x)) \\ &= f^{-1}(2x - 3) \\ &= \frac{(2x - 3) + 3}{2} \\ &= \frac{2x}{2} \\ &= x; \end{aligned}$$

$$\begin{aligned} f \circ f^{-1}(x) &= f(f^{-1}(x)) \\ &= f\left(\frac{x + 3}{2}\right) \\ &= 2 \cdot \frac{x + 3}{2} - 3 \\ &= x + 3 - 3 \\ &= x \end{aligned}$$

Do Exercise 15.

15. Let $f(x) = \frac{2}{3}x - 4$. Use composition to show that

$$f^{-1}(x) = \frac{3x + 12}{2}.$$

Answers on page A-69

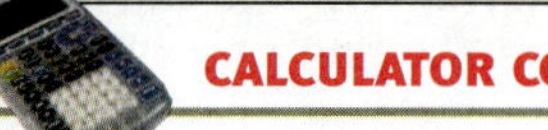

CALCULATOR CORNER

Composition of Functions We can evaluate composite functions on a graphing calculator. For example, given that $f(x) = 2x - 5$ and $g(x) = x^2 - 3x + 8$, we can find $f \circ g(7)$ and $g \circ f(7)$.

On the equation-editor screen, we enter $y_1 = 2x - 5$ and $y_2 = x^2 - 3x + 8$. Then $f \circ g(7) = y_1 \circ y_2(7)$, or $y_1(y_2(7))$ and $g \circ f(7) = y_2 \circ y_1(7)$, or $y_2(y_1(7))$. To find these function values, we press [2nd] [QUIT] to go to the home screen. Then we enter $y_1(y_2(7))$ by pressing [VARS] [▷] [1] [1] [(] [VARS] [▷] [1] [2] [(] [7] [)] [)] [ENTER]. Now we enter $y_2(y_1(7))$ by pressing [VARS] [▷] [1] [2] [(] [VARS] [▷] [1] [1] [(] [7] [)] [)] [ENTER]. We see that $y_1(y_2(7))$, or $f \circ g(7)$, is 67 and $y_2(y_1(7))$, or $g \circ f(7)$, is 62.

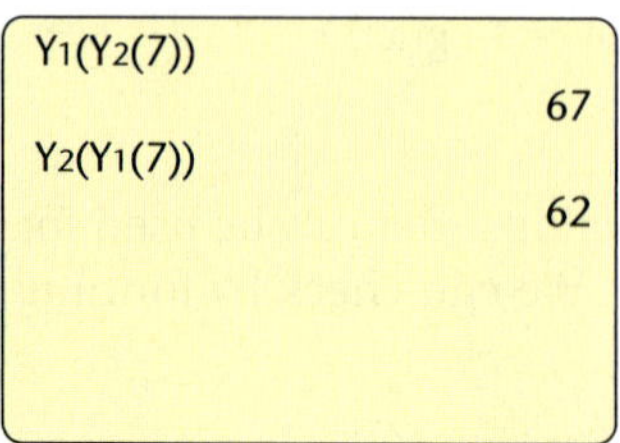

Exercises:

1. Use a graphing calculator to check Example 10(a).
2. Given $f(x) = x + 3$ and $g(x) = 2x^2 - 1$, use a graphing calculator to find $f \circ g(-1)$ and $g \circ f(-1)$.
3. Given $f(x) = \sqrt{x}$ and $g(x) = x^2$, use a graphing calculator to find $f \circ g(9)$ and $g \circ f(9)$.
4. Given $f(x) = x^3$ and $g(x) = x - 2$, use a graphing calculator to find $f \circ g(3)$ and $g \circ f(3)$.

Study Tips

BEGINNING TO STUDY FOR THE FINAL EXAM (PART 3): THREE DAYS TO TWO WEEKS OF STUDY TIME

1. **Begin by browsing through each chapter, reviewing the highlighted or boxed information regarding important formulas in both the text and the Summary and Review.** There may be some formulas that you will need to memorize. Summarize them on an index card and quiz yourself frequently.
2. **Retake each chapter test that you took in class, assuming your instructor has returned it. Otherwise, use the chapter test in the book.** Restudy the objectives in the text that correspond to each question you missed.
3. **For remaining difficulties, see your instructor, go to a tutoring session, or participate in a study group.**

"It is a great piece of skill to know how to guide your luck, even while waiting for it."

Baltasar Gracian,
seventeenth-century Spanish philosopher and writer

17.2 EXERCISE SET

For Extra Help

Digital Video Tutor CD 15 Videotape 20 | InterAct Math | Math Tutor Center | MathXL | MyMathLab

a Find the inverse of the relation. Graph the original relation in red and then graph the inverse relation in blue.

1. $\{(1, 2), (6, -3), (-3, -5)\}$

2. $\{(3, -1), (5, 2), (5, -3), (2, 0)\}$

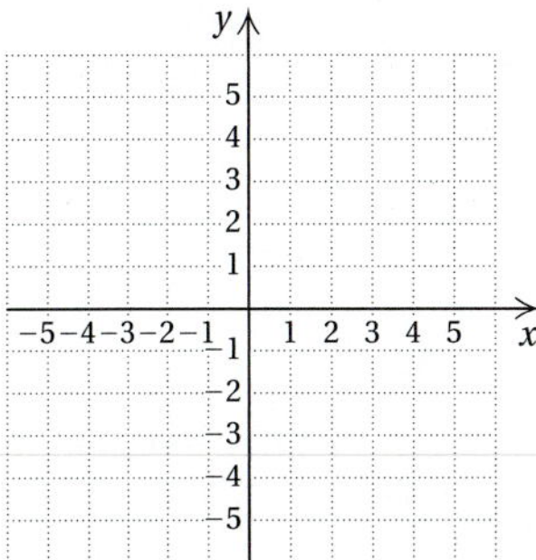

Find an equation of the inverse of the relation. Then complete the second table and graph both the original relation and its inverse.

3. $y = 2x + 6$

x	y
-1	4
0	6
1	8
2	10
3	12

x	y
4	
6	
8	
10	
12	

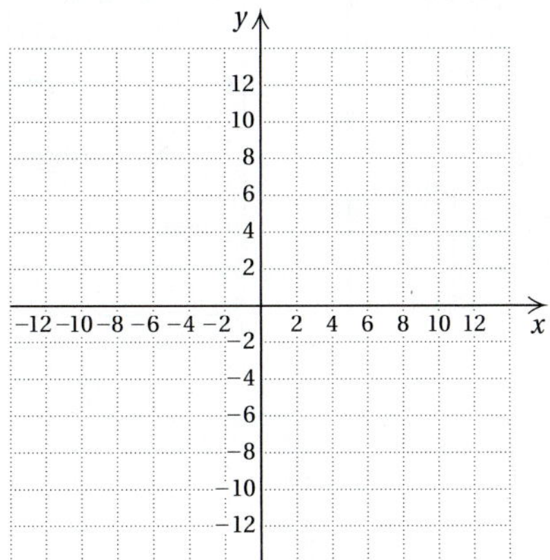

4. $y = \frac{1}{2}x^2 - 8$

x	y
-4	0
-2	-6
0	-8
2	-6
4	0

x	y
0	
-6	
-8	
-6	
0	

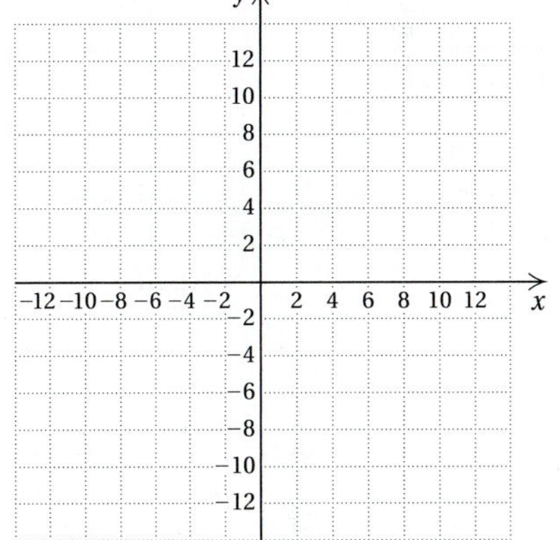

b Determine whether the function is one-to-one.

5. $f(x) = x - 5$

6. $f(x) = 3 - 6x$

7. $f(x) = x^2 - 2$

8. $f(x) = 4 - x^2$

9. $f(x) = |x| - 3$

10. $f(x) = |x - 2|$

11. $f(x) = 3^x$

12. $f(x) = \left(\frac{1}{2}\right)^x$

C Determine whether the function is one-to-one. If it is, find a formula for its inverse.

13. $f(x) = 5x - 2$

14. $f(x) = 4 + 7x$

15. $f(x) = \dfrac{-2}{x}$

16. $f(x) = \dfrac{1}{x}$

17. $f(x) = \frac{4}{3}x + 7$

18. $f(x) = -\frac{7}{8}x + 2$

19. $f(x) = \dfrac{2}{x + 5}$

20. $f(x) = \dfrac{1}{x - 8}$

21. $f(x) = 5$

22. $f(x) = -2$

23. $f(x) = \dfrac{2x + 1}{5x + 3}$

24. $f(x) = \dfrac{2x - 1}{5x + 3}$

25. $f(x) = x^3 - 1$

26. $f(x) = x^3 + 5$

27. $f(x) = \sqrt[3]{x}$

28. $f(x) = \sqrt[3]{x - 4}$

Graph the function and its inverse using the same set of axes.

29. $f(x) = \frac{1}{2}x - 3$,

$f^{-1}(x) =$ ________

x	$f(x)$

x	$f^{-1}(x)$

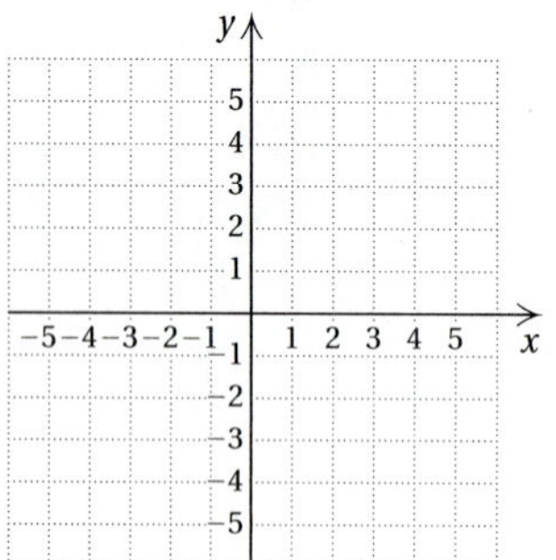

30. $g(x) = x + 4$,

$g^{-1}(x) =$ ________

x	$g(x)$

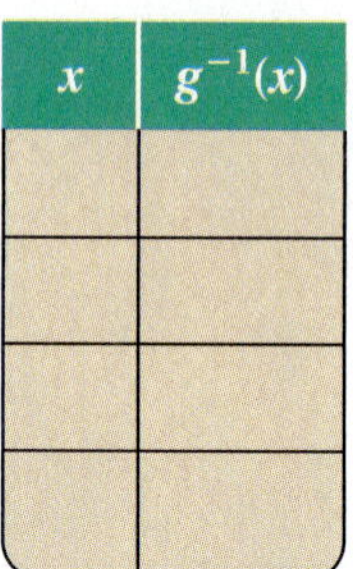

x	$g^{-1}(x)$

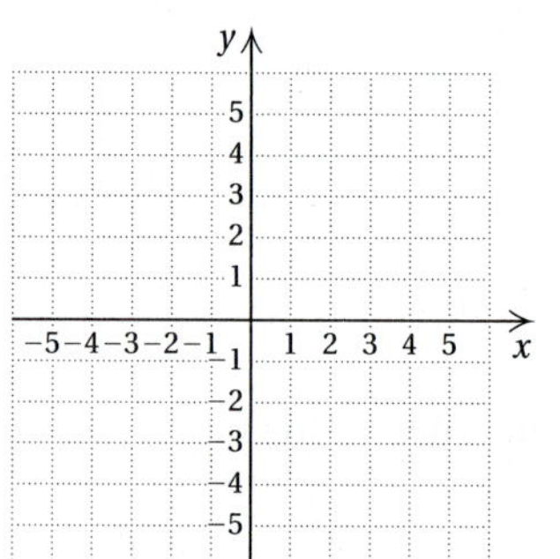

31. $f(x) = x^3$,

$f^{-1}(x) =$ ________

x	$f(x)$
0	
1	
2	
3	
−1	
−2	
−3	

x	$f^{-1}(x)$

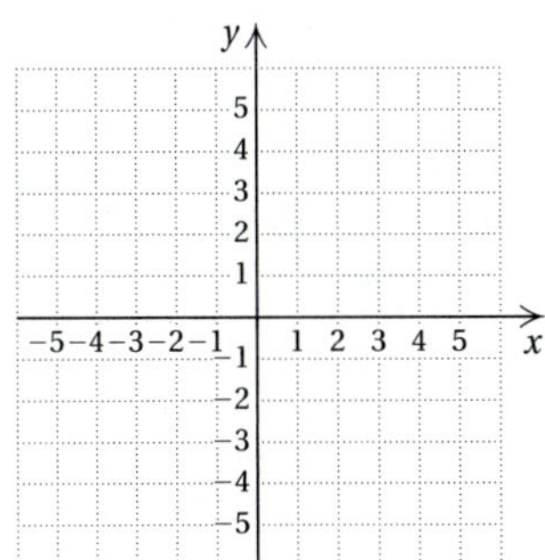

32. $f(x) = x^3 - 1$,

$f^{-1}(x) =$ ________

x	$f(x)$
0	
1	
2	
3	
−1	
−2	
−3	

x	$f^{-1}(x)$

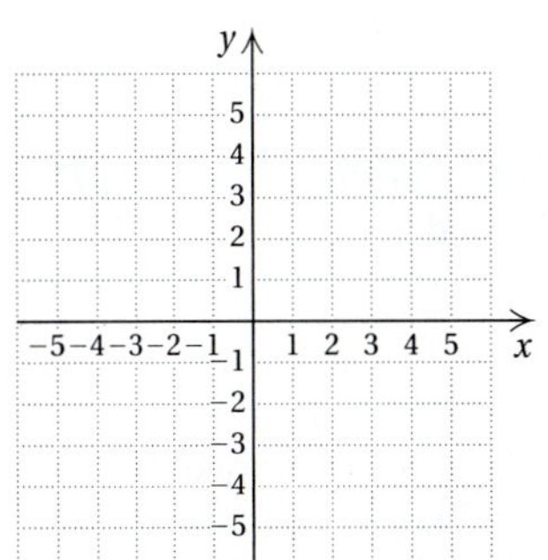

d Find $f \circ g(x)$ and $g \circ f(x)$.

33. $f(x) = 2x - 3$, $g(x) = 6 - 4x$

34. $f(x) = 9 - 6x$, $g(x) = 0.37x + 4$

35. $f(x) = 3x^2 + 2$, $g(x) = 2x - 1$

36. $f(x) = 4x + 3$, $g(x) = 2x^2 - 5$

37. $f(x) = 4x^2 - 1$, $g(x) = \dfrac{2}{x}$

38. $f(x) = \dfrac{3}{x}$, $g(x) = 2x^2 + 3$

39. $f(x) = x^2 + 5$, $g(x) = x^2 - 5$

40. $f(x) = \dfrac{1}{x^2}$, $g(x) = x - 1$

Find $f(x)$ and $g(x)$ such that $h(x) = f \circ g(x)$. Answers may vary.

41. $h(x) = (5 - 3x)^2$

42. $h(x) = 4(3x - 1)^2 + 9$

43. $h(x) = \sqrt{5x + 2}$

44. $h(x) = (3x^2 - 7)^5$

45. $h(x) = \dfrac{1}{x - 1}$

46. $h(x) = \dfrac{3}{x} + 4$

47. $h(x) = \dfrac{1}{\sqrt{7x + 2}}$

48. $h(x) = \sqrt{x - 7} - 3$

49. $h(x) = (\sqrt{x} + 5)^4$

50. $h(x) = \dfrac{x^3 + 1}{x^3 - 1}$

e For each function, use composition to show that the inverse is correct.

51. $f(x) = \frac{4}{5}x$,
$f^{-1}(x) = \frac{5}{4}x$

52. $f(x) = \dfrac{x+7}{2}$,
$f^{-1}(x) = 2x - 7$

53. $f(x) = \dfrac{1-x}{x}$,
$f^{-1}(x) = \dfrac{1}{x+1}$

54. $f(x) = x^3 - 5$,
$f^{-1}(x) = \sqrt[3]{x+5}$

Find the inverse of the given function by thinking about the operations of the function and then reversing, or undoing, them. Then use composition to show whether the inverse is correct.

	Function	*Inverse*
55.	$f(x) = 3x$	$f^{-1}(x) =$ ________
56.	$f(x) = \frac{1}{4}x + 7$	$f^{-1}(x) =$ ________
57.	$f(x) = -x$	$f^{-1}(x) =$ ________
58.	$f(x) = \sqrt[3]{x} - 5$	$f^{-1}(x) =$ ________
59.	$f(x) = \sqrt[3]{x-5}$	$f^{-1}(x) =$ ________
60.	$f(x) = x^{-1}$	$f^{-1}(x) =$ ________

61. *Dress Sizes in the United States and France.* A size-6 dress in the United States is size 38 in France. A function that converts dress sizes in the United States to those in France is

$$f(x) = x + 32.$$

a) Find the dress sizes in France that correspond to sizes of 8, 10, 14, and 18 in the United States.
b) Determine whether this function has an inverse that is a function. If so, find a formula for the inverse.
c) Use the inverse function to find dress sizes in the United States that correspond to sizes of 40, 42, 46, and 50 in France.

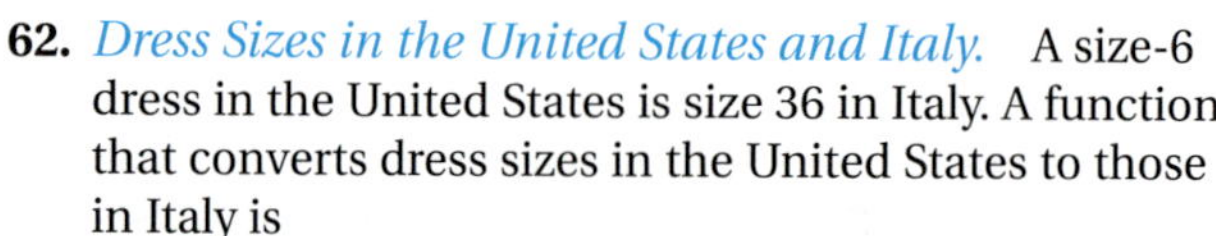

62. *Dress Sizes in the United States and Italy.* A size-6 dress in the United States is size 36 in Italy. A function that converts dress sizes in the United States to those in Italy is

$$f(x) = 2(x + 12).$$

a) Find the dress sizes in Italy that correspond to sizes of 8, 10, 14, and 18 in the United States.
b) Determine whether this function has an inverse that is a function. If so, find a formula for the inverse.
c) Use the inverse function to find dress sizes in the United States that correspond to sizes of 40, 44, 52, and 60 in Italy.

63. $\mathbf{D_W}$ The function $V(t) = 750(1.2)^t$ is used to predict the value V of a certain rare stamp t years from 1999. Do not calculate $V^{-1}(t)$ but explain how V^{-1} could be used.

64. $\mathbf{D_W}$ An organization determines that the cost per person of chartering a bus is given by the function

$$C(x) = \frac{100 + 5x}{x},$$

where x is the number of people in the group and $C(x)$ is in dollars. Determine $C^{-1}(x)$ and explain how this inverse function could be used.

SKILL MAINTENANCE

Use rational exponents to simplify. [15.2d]

65. $\sqrt[6]{a^2}$

66. $\sqrt[6]{x^4}$

67. $\sqrt{a^4b^6}$

68. $\sqrt[3]{8t^6}$

69. $\sqrt[8]{81}$

70. $\sqrt[4]{32}$

71. $\sqrt[12]{64x^6y^6}$

72. $\sqrt[8]{p^4t^2}$

73. $\sqrt[5]{32a^{15}b^{40}}$

74. $\sqrt[3]{1000x^9y^{18}}$

75. $\sqrt[4]{81a^8b^8}$

76. $\sqrt[3]{27p^3q^9}$

SYNTHESIS

In Exercises 77–80, use a graphing calculator to help determine whether or not the given functions are inverses of each other.

77. $f(x) = 0.75x^2 + 2$; $g(x) = \sqrt{\dfrac{4(x-2)}{3}}$

78. $f(x) = 1.4x^3 + 3.2$; $g(x) = \sqrt[3]{\dfrac{x-3.2}{1.4}}$

79. $f(x) = \sqrt{2.5x + 9.25}$; $g(x) = 0.4x^2 - 3.7, x \geq 0$

80. $f(x) = 0.8x^{1/2} + 5.23$; $g(x) = 1.25(x^2 - 5.23), x \geq 0$

81. Use a graphing calculator to help match each function in Column A with its inverse from Column B.

Column A	*Column B*
(1) $y = 5x^3 + 10$	**A.** $y = \dfrac{\sqrt[3]{x} - 10}{5}$
(2) $y = (5x + 10)^3$	**B.** $y = \sqrt[3]{\dfrac{x}{5}} - 10$
(3) $y = 5(x + 10)^3$	**C.** $y = \sqrt[3]{\dfrac{x-10}{5}}$
(4) $y = (5x)^3 + 10$	**D.** $y = \dfrac{\sqrt[3]{x-10}}{5}$

In Exercises 82 and 83, graph the inverse of f.

82.

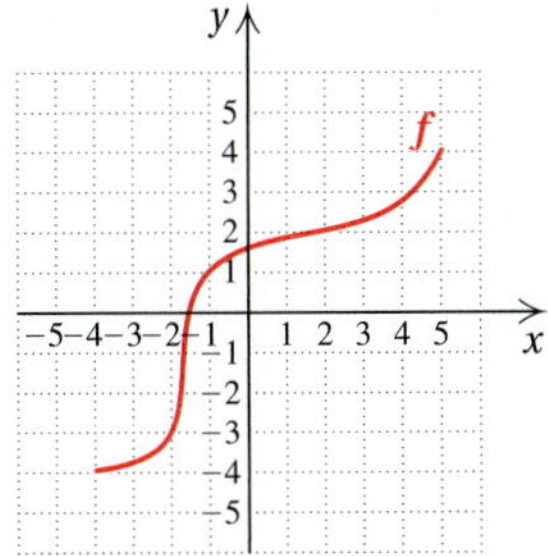

83.

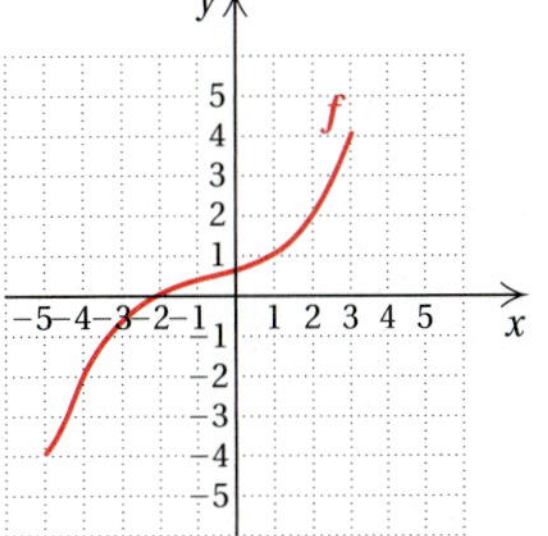

84. Examine the following table. Does it appear that f and g could be inverses of each other? Why or why not?

x	$f(x)$	$g(x)$
6	6	6
7	6.5	8
8	7	10
9	7.5	12
10	8	14
11	8.5	16
12	9	18

85. Assume in Exercise 84 that f and g are both linear functions. Find equations for $f(x)$ and $g(x)$. Are f and g inverses of each other?

17.3 LOGARITHMIC FUNCTIONS

Objectives

a Graph logarithmic functions.

b Convert from exponential equations to logarithmic equations and from logarithmic equations to exponential equations.

c Solve logarithmic equations.

d Find common logarithms on a calculator.

1. Write the meaning of $\log_2 64$. Then find $\log_2 64$.

We are now ready to study inverses of exponential functions. These functions have many applications and are referred to as *logarithm*, or *logarithmic functions.*

a Graphing Logarithmic Functions

Consider the exponential function $f(x) = 2^x$. Like all exponential functions, f is one-to-one. Can a formula for f^{-1} be found? To answer this, we use the method of Section 17.2:

1. Replace $f(x)$ with y: $\quad y = 2^x$.
2. Interchange x and y: $\quad x = 2^y$.
3. Solve for y: $\quad y =$ the power to which we raise 2 to get x.
4. Replace y with $f^{-1}(x)$: $\quad f^{-1}(x) =$ the power to which we raise 2 to get x.

We now define a new symbol to replace the words "the power to which we raise 2 to get x".

MEANING OF LOGARITHMS

$\log_2 x$, read "the logarithm, base 2, of x," or "log, base 2, of x," means "the power to which we raise 2 to get x."

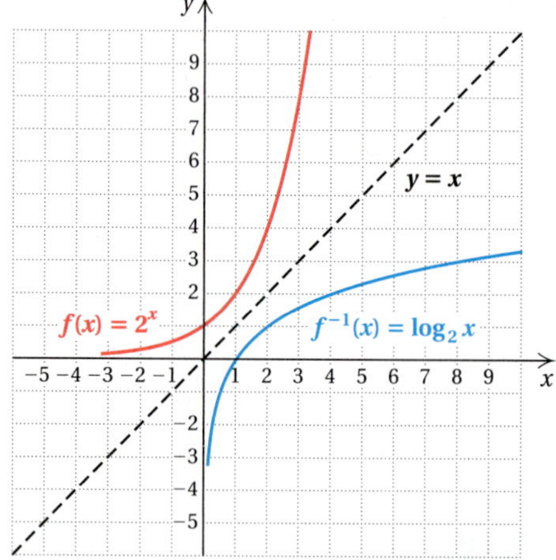

Thus if $f(x) = 2^x$, then $f^{-1}(x) = \log_2 x$. Note that $f^{-1}(8) = \log_2 8 = 3$, because 3 is the *power to which we raise* 2 *to get* 8; that is, $2^3 = 8$.

Although expressions like $\log_2 13$ can only be approximated, remember that $\log_2 13$ represents the ***power*** *to which we raise* 2 *to get* 13. That is, $2^{\log_2 13} = 13$.

Do Exercise 1.

Answer on page A-70

For any exponential function $f(x) = a^x$, the inverse is called a **logarithmic function, base *a*.** The graph of the inverse can, of course, be drawn by reflecting the graph of $f(x) = a^x$ across the line $y = x$. It will be helpful to remember that the inverse of $f(x) = a^x$ is given by $f^{-1}(x) = \log_a x$. Normally, we use a number a that is greater than 1 for the logarithm base.

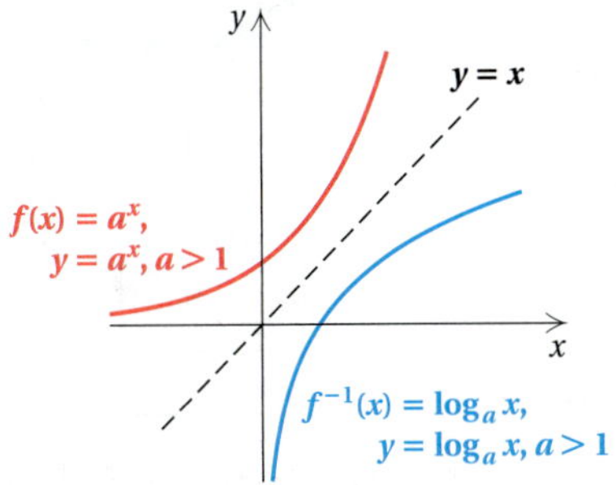

LOGARITHMS

The inverse of $f(x) = a^x$ is given by

$$f^{-1}(x) = \log_a x.$$

We read "$\log_a x$" as "the logarithm, base a, of x." We define $y = \log_a x$ as that number y such that $a^y = x$, where $x > 0$ and a is a positive constant other than 1.

It is helpful in dealing with logarithmic functions to remember that the logarithm of a number is an **exponent.** It is the exponent y in $x = a^y$. Keep thinking, "The logarithm, base a, of a number x is the power to which a must be raised in order to get x."

EXPONENTIAL FUNCTION	LOGARITHMIC FUNCTION
$y = a^x$	$x = a^y$
$f(x) = a^x$	$f^{-1}(x) = \log_a x$
$a > 0, a \neq 1$	$a > 0, a \neq 1$
Domain = The set of real numbers	Range = The set of real numbers
Range = The set of positive numbers	Domain = The set of positive numbers

Why do we exclude 1 from being a logarithm base? See the graph below. If we allow 1 as a logarithm base, the graph of the relation $y = \log_1 x$, or $x = 1^y = 1$, is a vertical line, which is not a function and therefore not a logarithmic function.

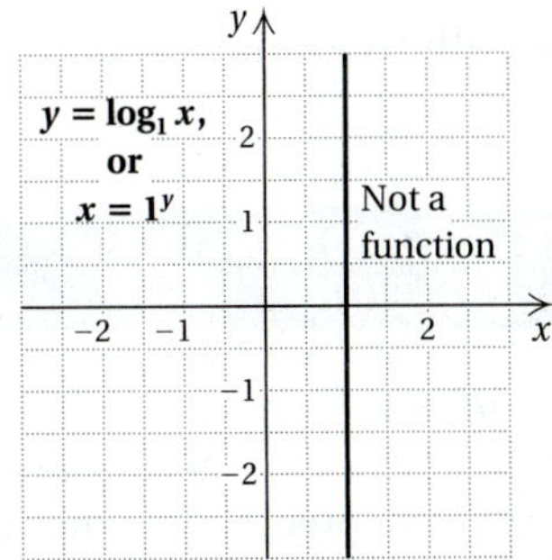

2. Graph: $y = f(x) = \log_3 x$.

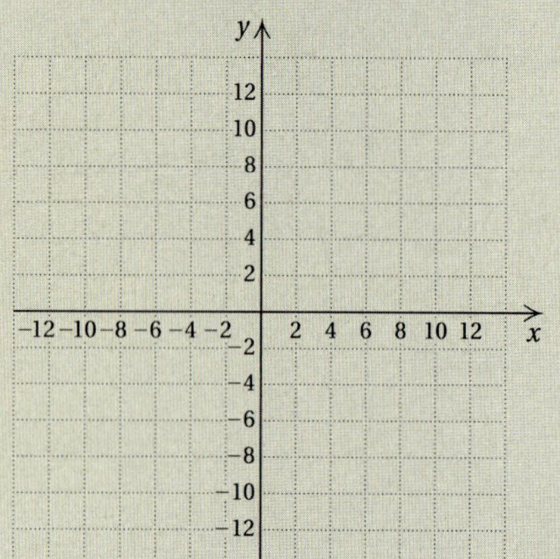

Answer on page A-70

EXAMPLE 1 Graph: $y = f(x) = \log_5 x$.

The equation $y = \log_5 x$ is equivalent to $5^y = x$. We can find ordered pairs that are solutions by choosing values for y and computing the corresponding x-values.

For $y = 0$, $x = 5^0 = 1$.
For $y = 1$, $x = 5^1 = 5$.
For $y = 2$, $x = 5^2 = 25$.
For $y = 3$, $x = 5^3 = 125$.
For $y = -1$, $x = 5^{-1} = \frac{1}{5}$.
For $y = -2$, $x = 5^{-2} = \frac{1}{25}$.

x, or 5^y	y
1	0
5	1
25	2
125	3
$\frac{1}{5}$	-1
$\frac{1}{25}$	-2

(1) Select y.
(2) Compute x.

The table shows the following:

$\log_5 1 = 0;$
$\log_5 5 = 1;$
$\log_5 25 = 2;$
$\log_5 \frac{1}{5} = -1;$
$\log_5 \frac{1}{25} = -2.$

These can all be checked using the equations above.

We plot the ordered pairs and connect them with a smooth curve. The graph of $y = 5^x$ has been shown only for reference.

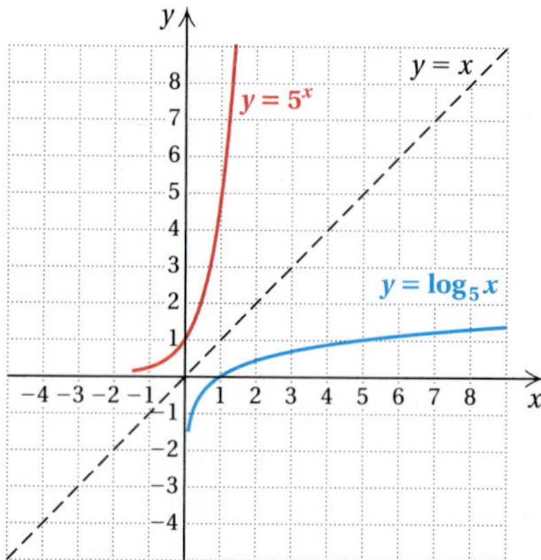

Do Exercise 2.

b Converting Between Exponential Equations and Logarithmic Equations

We use the definition of logarithms to convert from exponential equations to logarithmic equations.

CONVERTING BETWEEN EXPONENTIAL AND LOGARITHMIC EQUATIONS

$y = \log_a x \rightarrow a^y = x$; $\quad a^y = x \rightarrow y = \log_a x$

Be sure to memorize this relationship! It is probably the most important definition in the chapter. Many times this definition will be a justification for a proof or a procedure that we are considering.

EXAMPLES Convert to a logarithmic equation.

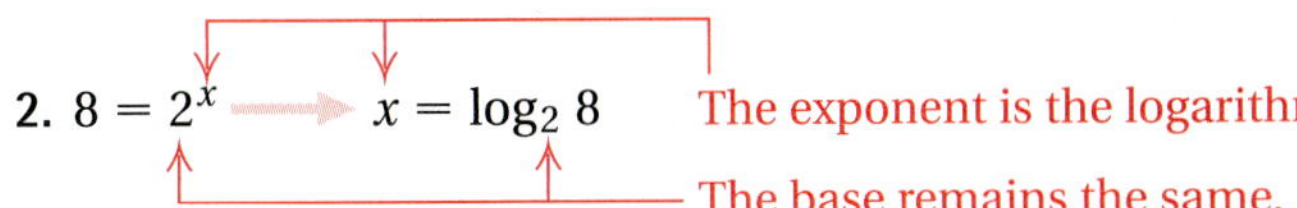

2. $8 = 2^x \rightarrow x = \log_2 8$ — The exponent is the logarithm. The base remains the same.

3. $y^{-1} = 4 \rightarrow -1 = \log_y 4$

4. $a^b = c \rightarrow b = \log_a c$

Do Exercises 3–6.

We also use the definition of logarithms to convert from logarithmic equations to exponential equations.

EXAMPLES Convert to an exponential equation.

5. $y = \log_3 5 \rightarrow 3^y = 5$ — The logarithm is the exponent. The base does not change.

6. $-2 = \log_a 7 \rightarrow a^{-2} = 7$

7. $a = \log_b d \rightarrow b^a = d$

Do Exercises 7–10.

C Solving Certain Logarithmic Equations

Certain equations involving logarithms can be solved by first converting to exponential equations. We will solve more complicated equations later.

EXAMPLE 8 Solve: $\log_2 x = -3$.

$$\log_2 x = -3$$
$$2^{-3} = x \quad \text{Converting to an exponential equation}$$
$$\frac{1}{2^3} = x$$
$$\frac{1}{8} = x$$

CHECK: $\log_2 \frac{1}{8}$ is the exponent to which we raise 2 to get $\frac{1}{8}$. Since $2^{-3} = \frac{1}{8}$, we know that $\frac{1}{8}$ checks and is the solution.

EXAMPLE 9 Solve: $\log_x 16 = 2$.

$$\log_x 16 = 2$$
$$x^2 = 16 \quad \text{Converting to an exponential equation}$$
$$x = 4 \quad or \quad x = -4 \quad \text{Using the principle of square roots}$$

CHECK: $\log_4 16 = 2$ because $4^2 = 16$. Thus, 4 is a solution. Since all logarithm bases must be positive, $\log_{-4} 16$ is not defined. Therefore, -4 is not a solution.

Do Exercises 11–13 on the following page.

Convert to a logarithmic equation.

3. $6^0 = 1$

4. $10^{-3} = 0.001$

5. $16^{0.25} = 2$

6. $m^T = P$

Convert to an exponential equation.

7. $\log_2 32 = 5$

8. $\log_{10} 1000 = 3$

9. $\log_a Q = 7$

10. $\log_t M = x$

Answers on page A-70

Solve.

11. $\log_{10} x = 4$

12. $\log_x 81 = 4$

13. $\log_2 x = -2$

Find each of the following.

14. $\log_{10} 10{,}000$

15. $\log_{10} 0.0001$

16. $\log_7 1$

Answers on page A-70

To think of finding logarithms as solving equations may help in some cases.

EXAMPLE 10 Find $\log_{10} 1000$.

METHOD 1. Let $\log_{10} 1000 = x$. Then

$$10^x = 1000 \quad \text{Converting to an exponential equation}$$
$$10^x = 10^3$$
$$x = 3. \quad \text{The exponents are the same.}$$

Therefore, $\log_{10} 1000 = 3$.

METHOD 2. Think of the meaning of $\log_{10} 1000$. It is the exponent to which we raise 10 to get 1000. That exponent is 3. Therefore, $\log_{10} 1000 = 3$.

EXAMPLE 11 Find $\log_{10} 0.01$.

METHOD 1. Let $\log_{10} 0.01 = x$. Then

$$10^x = 0.01 \quad \text{Converting to an exponential equation}$$
$$10^x = \frac{1}{100}$$
$$10^x = 10^{-2}$$
$$x = -2. \quad \text{The exponents are the same.}$$

Therefore, $\log_{10} 0.01 = -2$.

METHOD 2. $\log_{10} 0.01$ is the exponent to which we raise 10 to get 0.01. Noting that

$$0.01 = \frac{1}{100} = \frac{1}{10^2} = 10^{-2},$$

we see that the exponent is -2. Therefore, $\log_{10} 0.01 = -2$.

EXAMPLE 12 Find $\log_5 1$.

METHOD 1. Let $\log_5 1 = x$. Then

$$5^x = 1 \quad \text{Converting to an exponential equation}$$
$$5^x = 5^0$$
$$x = 0. \quad \text{The exponents are the same.}$$

Therefore, $\log_5 1 = 0$.

METHOD 2. $\log_5 1$ is the exponent to which we raise 5 to get 1. That exponent is 0. Therefore, $\log_5 1 = 0$.

Do Exercises 14–16.

THE LOGARITHM OF 1

For any base a,

$$\log_a 1 = 0.$$

The logarithm, base a, of 1 is always 0.

The proof follows from the fact that $a^0 = 1$. This is equivalent to the logarithmic equation $\log_a 1 = 0$.

Another property follows similarly. We know that $a^1 = a$ for any real number a. In particular, it holds for any positive number a. This is equivalent to the logarithmic equation $\log_a a = 1$.

THE LOGARITHM, BASE a, OF a

For any base a,

$$\log_a a = 1.$$

EXAMPLE 13 Simplify: $\log_m 1$ and $\log_t t$.

$\log_m 1 = 0$; $\quad \log_t t = 1$

Do Exercises 17–20.

d Finding Common Logarithms on a Calculator

Base-10 logarithms are called **common logarithms.** Before calculators became so widely available, common logarithms were used extensively to do complicated calculations. In fact, that is why logarithms were invented. The abbreviation **log,** with no base written, is used for the common logarithm, base-10. Thus,

$\log 29$ means $\log_{10} 29$.

Be sure to memorize $\log a = \log_{10} a$.

We can approximate log 29. Note the following:

$$\left.\begin{aligned} \log 100 &= \log_{10} 100 = 2; \\ \log 29 &= ?; \\ \log 10 &= \log_{10} 10 = 1. \end{aligned}\right\}$$

It seems reasonable that log 29 is between 1 and 2.

On a scientific or graphing calculator, the key for common logarithms is generally marked LOG. We find that

$$\log 29 \approx 1.462397998 \approx 1.4624,$$

rounded to four decimal places. This also tells us that $10^{1.4624} \approx 29$.

On some scientific calculators, the keystrokes for doing such a calculation might be

2 9 LOG =. The display would then read 1.462398.

Using a graphing calculator, the keystrokes might be

LOG 2 9 ENTER. The display would then read 1.462397998.

EXAMPLES Find each common logarithm, to four decimal places, on a scientific or graphing calculator.

Function Value	*Readout*	*Rounded*
14. log 287,523	5.458672591	5.4587
15. log 0.000486	−3.313363731	−3.3134
16. log (−5)	NONREAL ANS	NONREAL ANS

Simplify.

17. $\log_3 1$

18. $\log_3 3$

19. $\log_c c$

20. $\log_c 1$

Find the common logarithm, to four decimal places, on a scientific or graphing calculator.

21. log 78,235.4

22. log 0.0000309

23. log (−3)

24. Find

log 1000 and log 10,000

without using a calculator. Between what two whole numbers is log 9874? Then on a calculator approximate log 9874, rounded to four decimal places.

Answers on page A-70

25. Complete the following table to express each number in the first column as a power of 10. Round each exponent to the nearest ten-thousandth.

NUMBER	EXPRESSED AS A POWER OF 10
8	
947	
634,567	
0.00708	
0.000778899	
18,600,000	
1860	

26. Find $10^{4.8934}$ using a calculator. (Compare your computation to that of Margin Exercise 21.)

Answers on page A-70

In Example 16, $\log(-5)$ does not exist as a real number because there is no real-number power to which we can raise 10 to get -5. The number 10 raised to any power is nonnegative. The logarithm of a negative number does not exist as a real number (though it can be defined as a complex number).

Do Exercises 21–24 on the preceding page.

We can use common logarithms to express any positive number as a power of 10. We simply find the common logarithm of the number on a calculator. Considering very large or very small numbers as powers of 10 might be a helpful way to compare those numbers.

EXAMPLE 17 Complete the following table to express each number in the first column as a power of 10. Round each exponent to the nearest ten-thousandth.

We simply find the common logarithm of the number using a calculator.

NUMBER	EXPRESSED AS A POWER OF 10
4	$4 = 10^{0.6021}$
625	$625 = 10^{2.7959}$
134,567	$134{,}567 = 10^{5.1289}$
0.00567	$0.00567 = 10^{-2.2464}$
0.000374859	$0.000374859 = 10^{-3.4261}$
186,000	$186{,}000 = 10^{5.2695}$
186,000,000	$186{,}000{,}000 = 10^{8.2695}$

Do Exercise 25.

The inverse of a logarithmic function is an exponential function. Thus, if $f(x) = \log x$, then $f^{-1}(x) = 10^x$. Because of this, on many calculators, the [LOG] key doubles as the [10^x] key after a [2nd] or [SHIFT] key has been pressed. To find $10^{5.4587}$ on a scientific calculator, we might enter 5.4587 and press [10^x]. On a graphing calculator, we press [2nd] [10^x], followed by 5.4587. In either case, we get the approximation

$$10^{5.4587} \approx 287{,}541.1465.$$

Compare this computation to Example 14. Note that, apart from the rounding error, $10^{5.4587}$ takes us back to about 287,523.

Do Exercise 26.

Using the scientific keys on a calculator would allow us to construct a graph of $f(x) = \log_{10} x = \log x$ by finding function values directly, rather than converting to exponential form as we did in Example 1.

x	$f(x)$
0.5	-0.3010
1	0
2	0.3010
3	0.4771
5	0.6990
9	0.9542
10	1

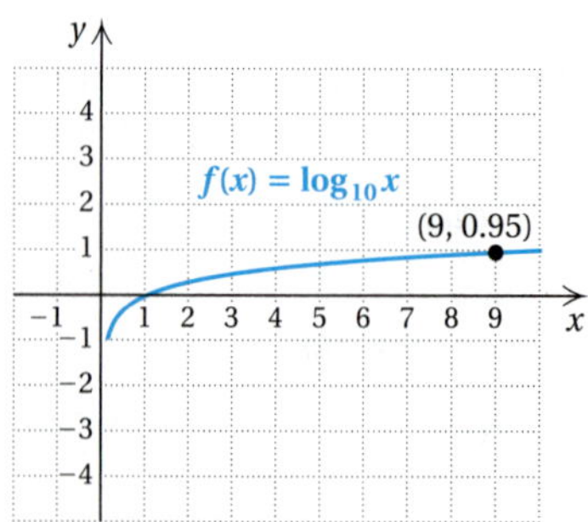

17.3 EXERCISE SET

For Extra Help

Digital Video Tutor CD 15 Videotape 20 | InterAct Math | Math Tutor Center | MathXL | MyMathLab

a Graph.

1. $f(x) = \log_2 x$, or $y = \log_2 x$
$y = \log_2 x \rightarrow x =$ ________

x, or 2^y	y
	0
	1
	2
	3
	−1
	−2
	−3

2. $f(x) = \log_{10} x$, or $y = \log_{10} x$
$y = \log_{10} x \rightarrow x =$ ________

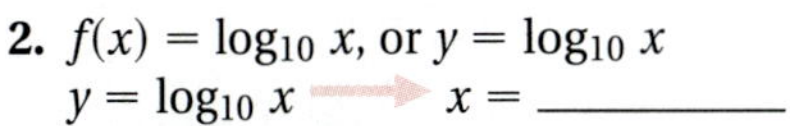

x, or 10^y	y
	0
	1
	2
	3
	−1
	−2
	−3

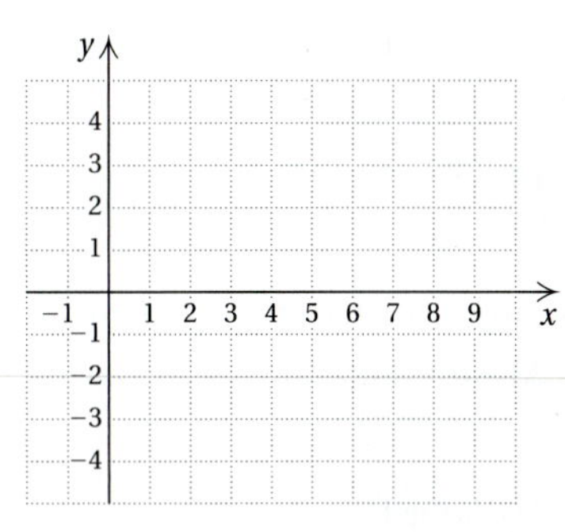

3. $f(x) = \log_{1/3} x$

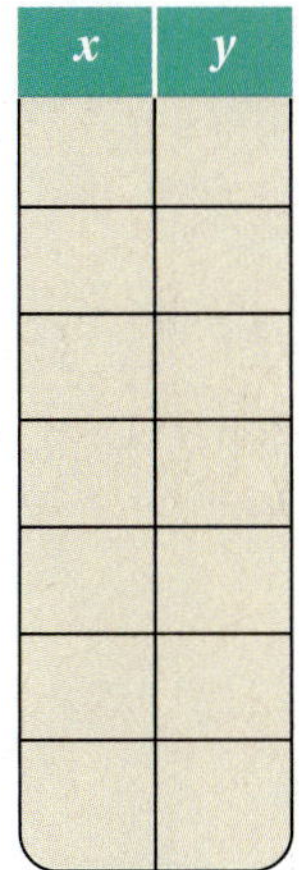

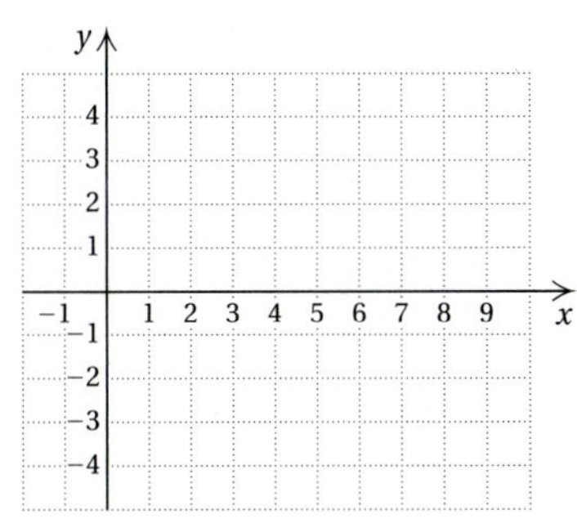

4. $f(x) = \log_{1/2} x$

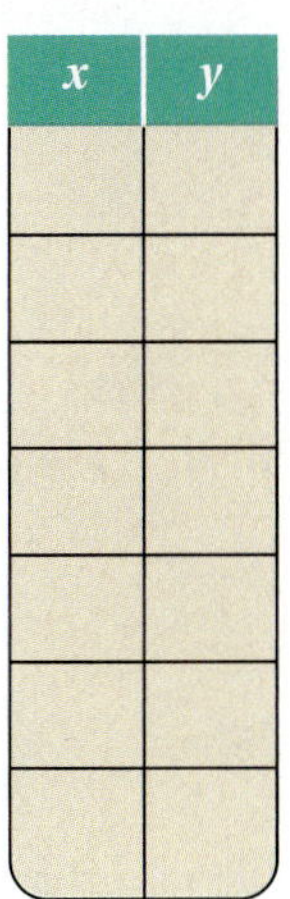

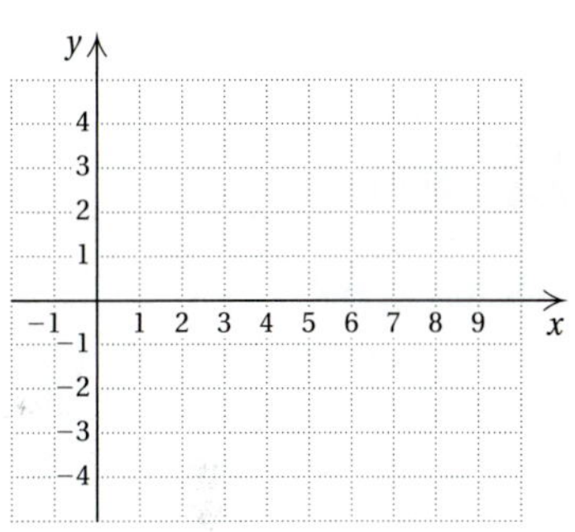

Graph both functions using the same set of axes.

5. $f(x) = 3^x, \quad f^{-1}(x) = \log_3 x$

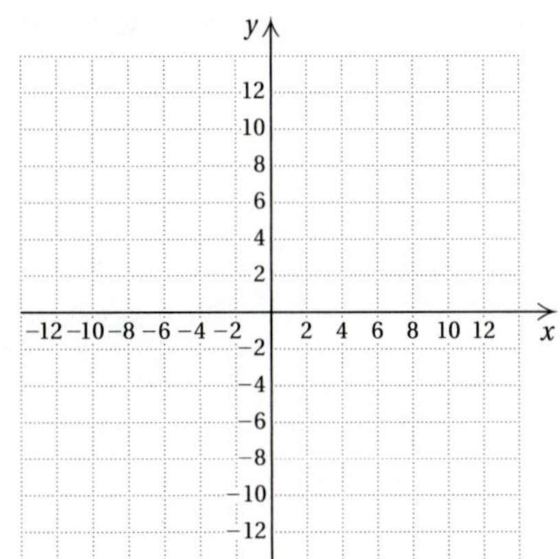

6. $f(x) = 4^x, \quad f^{-1}(x) = \log_4 x$

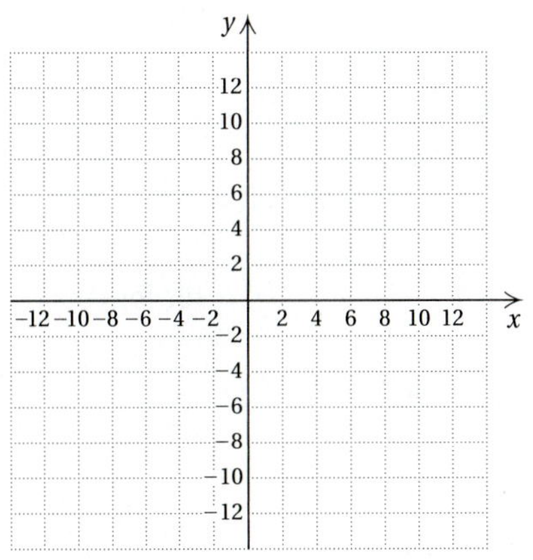

b Convert to a logarithmic equation.

7. $10^3 = 1000$

8. $10^2 = 100$

9. $5^{-3} = \frac{1}{125}$

10. $4^{-5} = \frac{1}{1024}$

11. $8^{1/3} = 2$

12. $16^{1/4} = 2$

13. $10^{0.3010} = 2$

14. $10^{0.4771} = 3$

15. $e^2 = t$

16. $p^k = 3$

17. $Q^t = x$

18. $P^m = V$

19. $e^2 = 7.3891$

20. $e^3 = 20.0855$

21. $e^{-2} = 0.1353$

22. $e^{-4} = 0.0183$

Convert to an exponential equation.

23. $w = \log_4 10$

24. $t = \log_5 9$

25. $\log_6 36 = 2$

26. $\log_7 7 = 1$

27. $\log_{10} 0.01 = -2$

28. $\log_{10} 0.001 = -3$

29. $\log_{10} 8 = 0.9031$

30. $\log_{10} 2 = 0.3010$

31. $\log_e 100 = 4.6052$

32. $\log_e 10 = 2.3026$

33. $\log_t Q = k$

34. $\log_m P = a$

c Solve.

35. $\log_3 x = 2$

36. $\log_4 x = 3$

37. $\log_x 16 = 2$

38. $\log_x 64 = 3$

39. $\log_2 16 = x$

40. $\log_5 25 = x$

41. $\log_3 27 = x$

42. $\log_4 16 = x$

43. $\log_x 25 = 1$

44. $\log_x 9 = 1$

45. $\log_3 x = 0$

46. $\log_2 x = 0$

47. $\log_2 x = -1$

48. $\log_3 x = -2$

49. $\log_8 x = \frac{1}{3}$

50. $\log_{32} x = \frac{1}{5}$

Find each of the following.

51. $\log_{10} 100$

52. $\log_{10} 100{,}000$

53. $\log_{10} 0.1$

54. $\log_{10} 0.001$

55. $\log_{10} 1$

56. $\log_{10} 10$

57. $\log_5 625$

58. $\log_2 64$

59. $\log_7 49$

60. $\log_5 125$

61. $\log_2 8$

62. $\log_8 64$

63. $\log_9 \frac{1}{81}$

64. $\log_5 \frac{1}{125}$

65. $\log_8 1$

66. $\log_6 6$

67. $\log_e e$

68. $\log_e 1$

69. $\log_{27} 9$

70. $\log_8 2$

d Find the common logarithm, to four decimal places, on a calculator.

71. $\log 78{,}889.2$

72. $\log 9{,}043{,}788$

73. $\log 0.67$

74. $\log 0.0067$

75. $\log(-97)$

76. $\log 0$

77. $\log\left(\frac{289}{32.7}\right)$

78. $\log\left(\frac{23}{86.2}\right)$

79. Complete the following table to express each number in the first column as a power of 10. Round each exponent to the nearest ten-thousandth.

NUMBER	EXPRESSED AS A POWER OF 10
6	
84	
987,606	
0.00987606	
98,760.6	
70,000,000	
7000	

80. Complete the following table to express each number in the first column as a power of 10. Round each exponent to the nearest ten-thousandth.

NUMBER	EXPRESSED AS A POWER OF 10
7	
314	
31.4	
31,400,000	
0.000314	
3.14	
0.0314	

81. DW Explain in your own words what is meant by $\log_a b = c$.

82. DW John Napier (1550–1617) of Scotland is credited by mathematicians as the inventor of logarithms. Make a report on Napier and his work.

SKILL MAINTENANCE

For each quadratic function, find and label **(a)** the vertex, **(b)** the line of symmetry, and **(c)** the maximum or minimum value. Then **(d)** graph the function. [16.5c], [16.6a]

83. $f(x) = 4 - x^2$

84. $f(x) = (x + 3)^2$

85. $f(x) = -2(x - 1)^2 - 3$

86. $f(x) = x^2 - 2x - 5$

Solve for the given letter. [16.3b]

87. $E = mc^2$, for c

88. $B = 3a^2 - 4a$, for a

89. $A = \sqrt{3ab}$, for b

90. $T = 2\pi\sqrt{\frac{L}{g}}$, for L

91. *Automobile Accidents by Age.* The following table lists the percentages of drivers of age a involved in automobile accidents in a recent year. [16.7b]

Age, a	Percentage Involved in an Accident
20	31
24	34
34	22
44	18

Source: National Safety Council

a) Use the data points (20, 31), (24, 34), and (34, 22) to fit a quadratic function to the data.
b) Use the quadratic function to predict the percentage of drivers involved in accidents at age 30 and at age 37.

SYNTHESIS

Graph.

92. $f(x) = \log_2 (x - 1)$

93. $f(x) = \log_3 |x + 1|$

Solve.

94. $|\log_3 x| = 3$

95. $\log_{125} x = \frac{2}{3}$

96. $\log_4 (3x - 2) = 2$

97. $\log_8 (2x + 1) = -1$

98. $\log_{10} (x^2 + 21x) = 2$

Simplify.

99. $\log_{1/4} \frac{1}{64}$

100. $\log_{81} 3 \cdot \log_3 81$

101. $\log_{10} (\log_4 (\log_3 81))$

102. $\log_2 (\log_2 (\log_4 256))$

103. $\log_{1/5} 25$

17.4 PROPERTIES OF LOGARITHMIC FUNCTIONS

The ability to manipulate logarithmic expressions is important in many applications and in more advanced mathematics. We now establish some basic properties that are useful in manipulating logarithmic expressions.

a Logarithms of Products

PROPERTY 1: THE PRODUCT RULE

For any positive numbers M and N,

$$\log_a (M \cdot N) = \log_a M + \log_a N.$$

(The logarithm of a product is the sum of the logarithms of the factors. The number a can be any logarithm base.)

EXAMPLE 1 Express as a sum of logarithms: $\log_2 (4 \cdot 16)$.

$$\log_2 (4 \cdot 16) = \log_2 4 + \log_2 16 \quad \text{By Property 1}$$

EXAMPLE 2 Express as a single logarithm: $\log_{10} 0.01 + \log_{10} 1000$.

$$\log_{10} 0.01 + \log_{10} 1000 = \log_{10} (0.01 \times 1000) \quad \text{By Property 1}$$
$$= \log_{10} 10$$

Do Exercises 1–4.

A PROOF OF PROPERTY 1 (*OPTIONAL*). Let $\log_a M = x$ and $\log_a N = y$. Converting to exponential equations, we have $a^x = M$ and $a^y = N$. Then we multiply to obtain

$$M \cdot N = a^x \cdot a^y = a^{x+y}.$$

Converting $M \cdot N = a^{x+y}$ back to a logarithmic equation, we get

$$\log_a M \cdot N = x + y.$$

Remembering what x and y represent, we get

$$\log_a M \cdot N = \log_a M + \log_a N.$$

b Logarithms of Powers

PROPERTY 2: THE POWER RULE

For any positive number M and any real number k,

$$\log_a M^k = k \cdot \log_a M.$$

(The logarithm of a power of M is the exponent times the logarithm of M. The number a can be any logarithm base.)

Objectives

a Express the logarithm of a product as a sum of logarithms, and conversely.

b Express the logarithm of a power as a product.

c Express the logarithm of a quotient as a difference of logarithms, and conversely.

d Convert from logarithms of products, quotients, and powers to expressions in terms of individual logarithms, and conversely.

e Simplify expressions of the type $\log_a a^k$.

Express as a sum of logarithms.

1. $\log_5 (25 \cdot 5)$

2. $\log_b PQ$

Express as a single logarithm.

3. $\log_3 7 + \log_3 5$

4. $\log_a J + \log_a A + \log_a M$

Answers on page A-71

Express as a product.

5. $\log_7 4^5$

6. $\log_a \sqrt{5}$

7. Express as a difference of logarithms:

$$\log_b \frac{P}{Q}.$$

8. Express as a single logarithm:

$$\log_2 125 - \log_2 25.$$

Answers on page A-71

CALCULATOR CORNER

Properties of Logarithms Use a table or a graph to determine whether each of the following is correct.

1. $\log(5x) = \log 5 \cdot \log x$
2. $\log(5x) = \log 5 + \log x$
3. $\log x^2 = \log x \cdot \log x$
4. $\log x^2 = 2 \log x$
5. $\log\left(\frac{x}{3}\right) = \frac{\log x}{\log 3}$
6. $\log\left(\frac{x}{3}\right) = \log x - \log 3$
7. $\log(x + 2) = \log x + \log 2$
8. $\log(x + 2) = \log x \cdot \log 2$

EXAMPLES Express as a product.

3. $\log_a 9^{-5} = -5 \log_a 9$ — By Property 2

4. $\log_a \sqrt[4]{5} = \log_a 5^{1/4}$ — Writing exponential notation
$= \frac{1}{4} \log_a 5$ — By Property 2

Do Exercises 5 and 6.

A PROOF OF PROPERTY 2 (*OPTIONAL*). Let $x = \log_a M$. Then we convert to an exponential equation to get $a^x = M$. Raising both sides to the kth power, we obtain

$$(a^x)^k = M^k, \quad \text{or} \quad a^{xk} = M^k.$$

Converting back to a logarithmic equation with base a, we get $\log_a M^k = xk$. But $x = \log_a M$, so

$$\log_a M^k = (\log_a M)k = k \cdot \log_a M.$$

C Logarithms of Quotients

PROPERTY 3: THE QUOTIENT RULE

For any positive numbers M and N,

$$\log_a \frac{M}{N} = \log_a M - \log_a N.$$

(The logarithm of a quotient is the logarithm of the numerator minus the logarithm of the denominator. The number a can be any logarithm base.)

EXAMPLE 5 Express as a difference of logarithms: $\log_t \frac{6}{U}$.

$$\log_t \frac{6}{U} = \log_t 6 - \log_t U \qquad \text{By Property 3}$$

EXAMPLE 6 Express as a single logarithm: $\log_b 17 - \log_b 27$.

$$\log_b 17 - \log_b 27 = \log_b \frac{17}{27} \qquad \text{By Property 3}$$

EXAMPLE 7 Express as a single logarithm: $\log_{10} 10{,}000 - \log_{10} 100$.

$$\log_{10} 10{,}000 - \log_{10} 100 = \log_{10} \frac{10{,}000}{100} = \log_{10} 100$$

Do Exercises 7 and 8.

A PROOF OF PROPERTY 3 (*OPTIONAL*). The proof makes use of Property 1 and Property 2.

$$\begin{aligned} \log_a \frac{M}{N} &= \log_a MN^{-1} \\ &= \log_a M + \log_a N^{-1} && \text{By Property 1} \\ &= \log_a M + (-1)\log_a N && \text{By Property 2} \\ &= \log_a M - \log_a N \end{aligned}$$

d Using the Properties Together

EXAMPLES Express in terms of logarithms.

8. $\log_a \dfrac{x^2y^3}{z^4} = \log_a (x^2y^3) - \log_a z^4$ — Using Property 3

$= \log_a x^2 + \log_a y^3 - \log_a z^4$ — Using Property 1

$= 2\log_a x + 3\log_a y - 4\log_a z$ — Using Property 2

9. $\log_a \sqrt[4]{\dfrac{xy}{z^3}} = \log_a \left(\dfrac{xy}{z^3}\right)^{1/4}$ — Writing exponential notation

$= \frac{1}{4}\log_a \dfrac{xy}{z^3}$ — Using Property 2

$= \frac{1}{4}(\log_a xy - \log_a z^3)$ — Using Property 3 (note the parentheses)

$= \frac{1}{4}(\log_a x + \log_a y - 3\log_a z)$ — Using Properties 1 and 2

$= \frac{1}{4}\log_a x + \frac{1}{4}\log_a y - \frac{3}{4}\log_a z$ — Distributive law

10. $\log_b \dfrac{xy}{m^3n^4} = \log_b xy - \log_b m^3n^4$ — Using Property 3

$= (\log_b x + \log_b y) - (\log_b m^3 + \log_b n^4)$ — Using Property 1

$= \log_b x + \log_b y - \log_b m^3 - \log_b n^4$ — Removing parentheses

$= \log_b x + \log_b y - 3\log_b m - 4\log_b n$ — Using Property 2

Do Exercises 9–11.

EXAMPLES Express as a single logarithm.

11. $\dfrac{1}{2}\log_a x - 7\log_a y + \log_a z$

$= \log_a x^{1/2} - \log_a y^7 + \log_a z$ — Using Property 2

$= \log_a \dfrac{\sqrt{x}}{y^7} + \log_a z$ — Using Property 3

$= \log_a \dfrac{z\sqrt{x}}{y^7}$ — Using Property 1

12. $\log_a \dfrac{b}{\sqrt{x}} + \log_a \sqrt{bx}$

$= \log_a b - \log_a \sqrt{x} + \log_a \sqrt{bx}$ — Using Property 3

$= \log_a b - \frac{1}{2}\log_a x + \frac{1}{2}\log_a (bx)$ — Using Property 2

$= \log_a b - \frac{1}{2}\log_a x + \frac{1}{2}(\log_a b + \log_a x)$ — Using Property 1

$= \log_a b - \frac{1}{2}\log_a x + \frac{1}{2}\log_a b + \frac{1}{2}\log_a x$

$= \frac{3}{2}\log_a b$ — Collecting like terms

$= \log_a b^{3/2}$ — Using Property 2

Example 12 could also be done as follows:

$\log_a \dfrac{b}{\sqrt{x}} + \log_a \sqrt{bx} = \log_a \dfrac{b}{\sqrt{x}}\sqrt{bx}$ — Using Property 1

$= \log_a \dfrac{b}{\sqrt{x}} \cdot \sqrt{b} \cdot \sqrt{x}$

$= \log_a b\sqrt{b}$, or $\log_a b^{3/2}$.

Express in terms of logarithms of x, y, z, and w.

9. $\log_a \sqrt{\dfrac{z^3}{xy}}$

10. $\log_a \dfrac{x^2}{y^3z}$

11. $\log_a \dfrac{x^3y^4}{z^5w^9}$

Express as a single logarithm.

12. $5\log_a x - \log_a y + \dfrac{1}{4}\log_a z$

13. $\log_a \dfrac{\sqrt{x}}{b} - \log_a \sqrt{bx}$

Answers on page A-71

Given

$\log_a 2 = 0.301,$
$\log_a 5 = 0.699,$

find each of the following.

14. $\log_a 4$

15. $\log_a 10$

16. $\log_a \frac{2}{5}$

17. $\log_a \frac{5}{2}$

18. $\log_a \frac{1}{5}$

19. $\log_a \sqrt{a^3}$

20. $\log_a 5a$

21. $\log_a 16$

Simplify.

22. $\log_2 2^6$

23. $\log_{10} 10^{3.2}$

24. $\log_e e^{12}$

Answers on page A-71

CAUTION!

Keep in mind that, in general,

$\log_a (M + N) \neq \log_a M + \log_a N,$

$\log_a (M - N) \neq \log_a M - \log_a N,$

$\log_a (MN) \neq (\log_a M)(\log_a N),$

and

$\log_a (M/N) \neq (\log_a M) \div (\log_a N).$

Do Exercises 12 and 13 on the preceding page.

EXAMPLES Given $\log_a 2 = 0.301$ and $\log_a 3 = 0.477$, find each of the following.

13. $\log_a 6 = \log_a (2 \cdot 3) = \log_a 2 + \log_a 3$ Property 1
$= 0.301 + 0.477 = 0.778$

14. $\log_a \frac{2}{3} = \log_a 2 - \log_a 3$ Property 3
$= 0.301 - 0.477 = -0.176$

15. $\log_a 81 = \log_a 3^4 = 4 \log_a 3$ Property 2
$= 4(0.477) = 1.908$

16. $\log_a \frac{1}{3} = \log_a 1 - \log_a 3$ Property 3
$= 0 - 0.477 = -0.477$

17. $\log_a \sqrt{a} = \log_a a^{1/2} = \frac{1}{2} \log_a a = \frac{1}{2} \cdot 1 = \frac{1}{2}$ Property 2

18. $\log_a 2a = \log_a 2 + \log_a a$ Property 1
$= 0.301 + 1 = 1.301$

19. $\log_a 5$ No way to find using these properties ($\log_a 5 \neq \log_a 2 + \log_a 3$)

20. $\frac{\log_a 3}{\log_a 2} = \frac{0.477}{0.301} \approx 1.58$ We simply divide the logarithms, not using any property.

Do Exercises 14–21.

e The Logarithm of the Base to a Power

PROPERTY 4

For any base a,

$$\log_a a^k = k.$$

(The logarithm, base a, of a to a power is the power.)

A PROOF OF PROPERTY 4 (*OPTIONAL*). The proof involves Property 2 and the fact that $\log_a a = 1$:

$$\log_a a^k = k(\log_a a) \quad \text{Using Property 2}$$
$$= k \cdot 1 \quad \text{Using } \log_a a = 1$$
$$= k.$$

EXAMPLES Simplify.

21. $\log_3 3^7 = 7$

22. $\log_{10} 10^{5.6} = 5.6$

23. $\log_e e^{-t} = -t$

Do Exercises 22–24.

17.4 EXERCISE SET

For Extra Help

Digital Video Tutor CD 15 Videotape 20 | InterAct Math | Math Tutor Center | MathXL | MyMathLab

a Express as a sum of logarithms.

1. $\log_2 (32 \cdot 8)$

2. $\log_3 (27 \cdot 81)$

3. $\log_4 (64 \cdot 16)$

4. $\log_5 (25 \cdot 125)$

5. $\log_a Qx$

6. $\log_r 8Z$

Express as a single logarithm.

7. $\log_b 3 + \log_b 84$

8. $\log_a 75 + \log_a 5$

9. $\log_c K + \log_c y$

10. $\log_t H + \log_t M$

b Express as a product.

11. $\log_c y^4$

12. $\log_a x^3$

13. $\log_b t^6$

14. $\log_{10} y^7$

15. $\log_b C^{-3}$

16. $\log_c M^{-5}$

c Express as a difference of logarithms.

17. $\log_a \frac{67}{5}$

18. $\log_t \frac{T}{7}$

19. $\log_b \frac{2}{5}$

20. $\log_a \frac{z}{y}$

Express as a single logarithm.

21. $\log_c 22 - \log_c 3$

22. $\log_d 54 - \log_d 9$

d Express in terms of logarithms.

23. $\log_a x^2y^3z$

24. $\log_a 5xy^4z^3$

25. $\log_b \frac{xy^2}{z^3}$

26. $\log_b \frac{p^2q^5}{m^4n^7}$

27. $\log_c \sqrt[3]{\frac{x^4}{y^3z^2}}$

28. $\log_a \sqrt{\frac{x^6}{p^5q^8}}$

29. $\log_a \sqrt[4]{\frac{m^8n^{12}}{a^3b^5}}$

30. $\log_a \sqrt{\frac{a^6b^8}{a^2b^5}}$

Express as a single logarithm and, if possible, simplify.

31. $\frac{2}{3}\log_a x - \frac{1}{2}\log_a y$

32. $\frac{1}{2}\log_a x + 3\log_a y - 2\log_a x$

33. $\log_a 2x + 3(\log_a x - \log_a y)$

34. $\log_a x^2 - 2\log_a \sqrt{x}$

35. $\log_a \frac{a}{\sqrt{x}} - \log_a \sqrt{ax}$

36. $\log_a (x^2 - 4) - \log_a (x - 2)$

Given $\log_b 3 = 1.099$ and $\log_b 5 = 1.609$, find each of the following.

37. $\log_b 15$

38. $\log_b \frac{3}{5}$

39. $\log_b \frac{5}{3}$

40. $\log_b \frac{1}{3}$

41. $\log_b \frac{1}{5}$

42. $\log_b \sqrt{b}$

43. $\log_b \sqrt{b^3}$

44. $\log_b 3b$

45. $\log_b 5b$

46. $\log_b 9$

e Simplify.

47. $\log_e e^t$

48. $\log_w w^8$

49. $\log_p p^5$

50. $\log_Y Y^{-4}$

Solve for x.

51. $\log_2 2^7 = x$

52. $\log_9 9^4 = x$

53. $\log_e e^x = -7$

54. $\log_a a^x = 2.7$

55. DW Find a way to express $\log_a (x/5)$ as a difference of logarithms without using the quotient rule. Explain your work.

56. DW A student incorrectly reasons that

$$\begin{aligned}\log_a \frac{1}{x} &= \log_a \frac{x}{x \cdot x} \\ &= \log_a x - \log_a x + \log_a x \\ &= \log_a x.\end{aligned}$$

What mistake has the student made? Explain what the answer should be.

SKILL MAINTENANCE

Compute and simplify. Express answers in the form $a + bi$, where $i^2 = -1$. [15.8b, c, d, e]

57. i^{29}

58. i^{34}

59. $(2 + i)(2 - i)$

60. $\frac{2 + i}{2 - i}$

61. $(7 - 8i) - (-16 + 10i)$

62. $2i^2 \cdot 5i^3$

63. $(8 + 3i)(-5 - 2i)$

64. $(2 - i)^2$

SYNTHESIS

65. Use the TABLE and GRAPH features to show that

$$\log x^2 \neq (\log x)(\log x).$$

66. Use the TABLE and GRAPH features to show that

$$\frac{\log x}{\log 4} \neq \log x - \log 4.$$

Express as a single logarithm and, if possible, simplify.

67. $\log_a (x^8 - y^8) - \log_a (x^2 + y^2)$

68. $\log_a (x + y) + \log_a (x^2 - xy + y^2)$

Express as a sum or a difference of logarithms.

69. $\log_a \sqrt{1 - s^2}$

70. $\log_a \frac{c - d}{\sqrt{c^2 - d^2}}$

Determine whether each is true or false.

71. $\frac{\log_a P}{\log_a Q} = \log_a \frac{P}{Q}$

72. $\frac{\log_a P}{\log_a Q} = \log_a P - \log_a Q$

73. $\log_a 3x = \log_a 3 + \log_a x$

74. $\log_a 3x = 3 \log_a x$

75. $\log_a (P + Q) = \log_a P + \log_a Q$

76. $\log_a x^2 = 2 \log_a x$

17.5 NATURAL LOGARITHMIC FUNCTIONS

Objectives

- **a** Find logarithms or powers, base e, using a calculator.
- **b** Use the change-of-base formula to find logarithms to bases other than e or 10.
- **c** Graph exponential and logarithmic functions, base e.

Any positive number other than 1 can serve as the base of a logarithmic function. Common, or base-10, logarithms, which were introduced in Section 17.3, are useful because they have the same base as our "commonly" used decimal system of naming numbers.

Today, another base is widely used. It is an irrational number named e. We now consider e and **natural logarithms,** or logarithms base e.

a The Base *e* and Natural Logarithms

When interest is computed n times per year, the compound-interest formula is

$$A = P\left(1 + \frac{r}{n}\right)^{nt},$$

where A is the amount that an initial investment P will grow to after t years at interest rate r. Suppose that \$1 could be invested at 100% interest for 1 year. (In reality, no financial institution would pay such an interest rate.) The preceding formula becomes a function A defined in terms of the number of compounding periods n:

$$A(n) = \left(1 + \frac{1}{n}\right)^{n}. \qquad \text{(See Section 17.1.)}$$

Let's find some function values, using a calculator and rounding to six decimal places. The numbers in the table at right approach a very important number called e. It is an irrational number, so its decimal representation neither terminates nor repeats.

n	$A(n) = \left(1 + \frac{1}{n}\right)^{n}$
1 (compounded annually)	\$2.00
2 (compounded semiannually)	\$2.25
3	\$2.370370
4 (compounded quarterly)	\$2.441406
5	\$2.488320
100	\$2.704814
365 (compounded daily)	\$2.714567
8760 (compounded hourly)	\$2.718127

THE NUMBER *e*

$e \approx 2.7182818284\ldots$

Logarithms, base e, are called **natural logarithms,** or **Naperian logarithms,** in honor of John Napier (1550–1617), a Scotsman who invented logarithms.

The abbreviation **ln** is commonly used with natural logarithms. Thus,

$\ln 29$ means $\log_e 29$. Be sure to memorize $\ln a = \log_e a$.

We usually read "ln 29" as "the natural log of 29," or simply "el en of 29."

On a calculator, the key for natural logarithms is generally marked [LN]. Using that key, we find that

$$\ln 29 \approx 3.36729583 \approx 3.3673,$$

rounded to four decimal places. This also tells us that $e^{3.3673} \approx 29$.

On some scientific calculators, the keystrokes for doing such a calculation might be

[2] [9] [LN] [=].

The display would then read 3.3672958.

Find the natural logarithm, to four decimal places, on a calculator.

1. ln 78,235.4

2. ln 0.0000309

3. ln (−3)

4. ln 0

5. ln 10

6. Find $e^{11.2675}$ using a calculator. (Compare this computation to that of Margin Exercise 1.)

7. Find e^{-2} using a calculator.

Answers on page A-71

If we were to use a graphing calculator, the keystrokes might be

[LN] [2] [9] [ENTER].

The display would then read 3.36729583.

EXAMPLES Find each natural logarithm, to four decimal places, on a calculator.

Function Value	*Readout*	*Rounded*
1. ln 287,523	12.56905814	12.5691
2. ln 0.000486	−7.629301934	−7.6293
3. ln (−5)	NONREAL ANS	NONREAL ANS
4. ln (e)	1	1
5. ln 1	0	0

Do Exercises 1–5.

The inverse of a logarithmic function is an exponential function. Thus, if $f(x) = \ln x$, then $f^{-1}(x) = e^x$. Because of this, on many calculators, the [LN] key doubles as the [e^x] key after a [2nd] or [SHIFT] key has been pressed.

EXAMPLE 6 Find $e^{12.5691}$ using a calculator.

On a scientific calculator, we might enter 12.5691 and press [e^x]. On a graphing calculator, we might press [2nd] [e^x], followed by 12.5691. In either case, we get the approximation

$$e^{12.5691} \approx 287{,}535.0371.$$

Compare this computation to Example 1. Note that, apart from the rounding error, $e^{12.5691}$ takes us back to about 287,523.

EXAMPLE 7 Find $e^{-1.524}$ using a calculator.

On a scientific calculator, we might enter −1.524 and press [e^x]. On a graphing calculator, we might press [2nd] [e^x], followed by −1.524. In either case, we get the approximation

$$e^{-1.524} \approx 0.2178.$$

Do Exercises 6 and 7.

b Changing Logarithm Bases

Most calculators give the values of both common logarithms and natural logarithms. To find a logarithm with some other base, we can use the following conversion formula.

THE CHANGE-OF-BASE FORMULA

For any logarithm bases a and b and any positive number M,

$$\log_b M = \frac{\log_a M}{\log_a b}.$$

A PROOF OF THE CHANGE-OF-BASE FORMULA (*OPTIONAL*). Let $x = \log_b M$. Then, writing an equivalent exponential equation, we have $b^x = M$. Next, we take the logarithm base a on both sides. This gives us

$$\log_a b^x = \log_a M.$$

By Property 2,

$$x \log_a b = \log_a M,$$

and solving for x, we obtain

$$x = \frac{\log_a M}{\log_a b}.$$

But $x = \log_b M$, so we have

$$\log_b M = \frac{\log_a M}{\log_a b},$$

which is the change-of-base formula.

EXAMPLE 8 Find $\log_4 7$ using common logarithms.

Let $a = 10$, $b = 4$, and $M = 7$. Then we substitute into the change-of-base formula:

$$\log_b M = \frac{\log_a M}{\log_a b}$$

$$\log_4 7 = \frac{\log_{10} 7}{\log_{10} 4} \qquad \text{Substituting 10 for } a\text{, 4 for } b\text{, and 7 for } M$$

$$= \frac{\log 7}{\log 4}$$

$$\approx 1.4037.$$

To check, we use a calculator with a power key $\boxed{y^x}$ or $\boxed{\wedge}$ to verify that

$$4^{1.4037} \approx 7.$$

We can also use base e for a conversion.

EXAMPLE 9 Find $\log_4 7$ using natural logarithms.

$$\log_b M = \frac{\log_a M}{\log_a b}$$

$$\log_4 7 = \frac{\log_e 7}{\log_e 4} \qquad \text{Substituting } e \text{ for } a\text{, 4 for } b\text{, and 7 for } M$$

$$= \frac{\ln 7}{\ln 4}$$

$$\approx 1.4037 \qquad \text{Note that this is the same answer as Example 8.}$$

EXAMPLE 10 Find $\log_5 29$ using natural logarithms.

Substituting e for a, 5 for b, and 29 for M, we have

$$\log_5 29 = \frac{\log_e 29}{\log_e 5} \qquad \text{Using the change-of-base formula}$$

$$= \frac{\ln 29}{\ln 5} \approx 2.0922.$$

Do Exercises 8 and 9.

8. a) Find $\log_6 7$ using common logarithms.

b) Find $\log_6 7$ using natural logarithms.

9. Find $\log_2 46$ using natural logarithms.

Answers on page A-71

CALCULATOR CORNER

The Change-of-Base Formula To find a logarithm with a base other than 10 or e, we use the change-of-base formula, $\log_b M = \dfrac{\log_a M}{\log_a b}$, where a and b are any logarithm bases and M is any positive number. For example, we can find $\log_5 8$ using common logarithms.

We let $a = 10$, $b = 5$, and $M = 8$ and substitute in the change-of-base formula. We press [LOG] 8 [)] [÷] [LOG] 5 [)] [ENTER]. Note that the parentheses must be closed in the numerator to enter the expression correctly. We also close the parentheses in the denominator for completeness. The result is about 1.2920. We could have let $a = e$ and used natural logarithms to find $\log_5 8$ as well.

```
log(8)/log(5)
                1.292029674
```

C Graphs of Exponential and Logarithmic Functions, Base e

EXAMPLE 11 Graph $f(x) = e^x$ and $g(x) = e^{-x}$.

We use a calculator with an [e^x] key to find approximate values of e^x and e^{-x}. Using these values, we can graph the functions.

x	e^x	e^{-x}
0	1	1
1	2.7	0.4
2	7.4	0.1
−1	0.4	2.7
−2	0.1	7.4

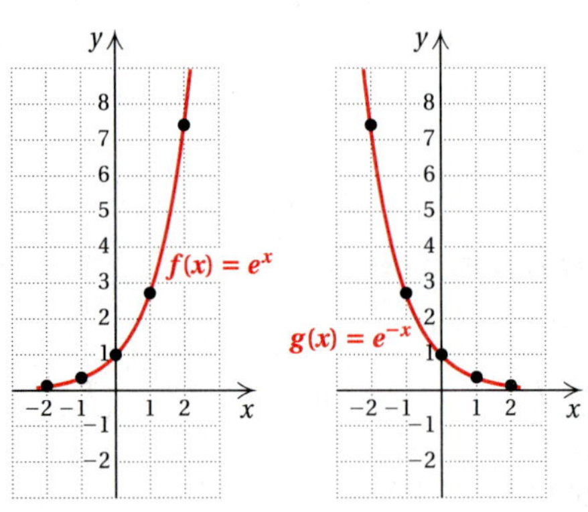

Note that each graph is the image of the other reflected across the y-axis.

EXAMPLE 12 Graph: $f(x) = e^{-0.5x}$.

We find some solutions with a calculator, plot them, and then draw the graph. For example,

$$f(2) = e^{-0.5(2)} = e^{-1} \approx 0.4.$$

x	$e^{-0.5x}$
0	1
1	0.6
2	0.4
3	0.2
−1	1.6
−2	2.7
−3	4.5

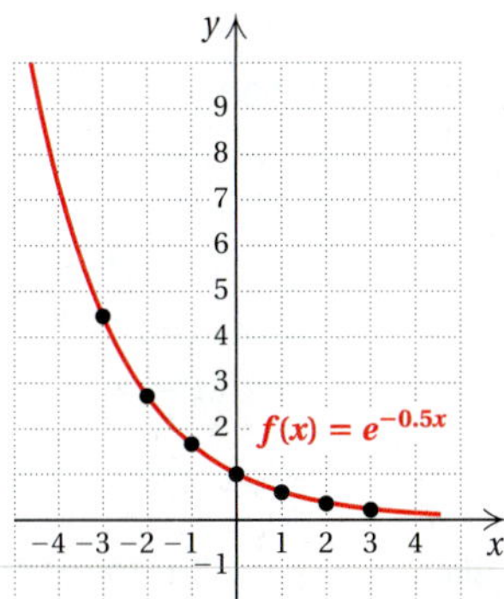

Do Exercises 10 and 11.

EXAMPLE 13 Graph: $g(x) = \ln x$.

We find some solutions with a calculator and then draw the graph. As expected, the graph is a reflection across the line $y = x$ of the graph of $y = e^x$.

x	$\ln x$
1	0
4	1.4
7	1.9
0.5	−0.7

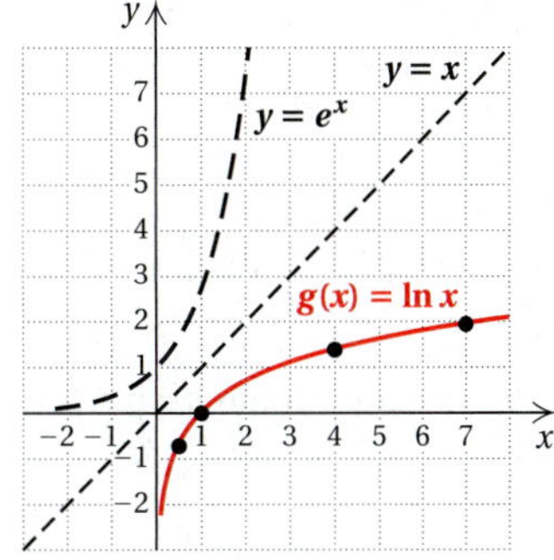

EXAMPLE 14 Graph: $f(x) = \ln (x + 3)$.

We find some solutions with a calculator, plot them, and then draw the graph.

x	$\ln (x + 3)$
0	1.1
1	1.4
2	1.6
3	1.8
4	1.9
−1	0.7
−2	0
−2.5	−0.7

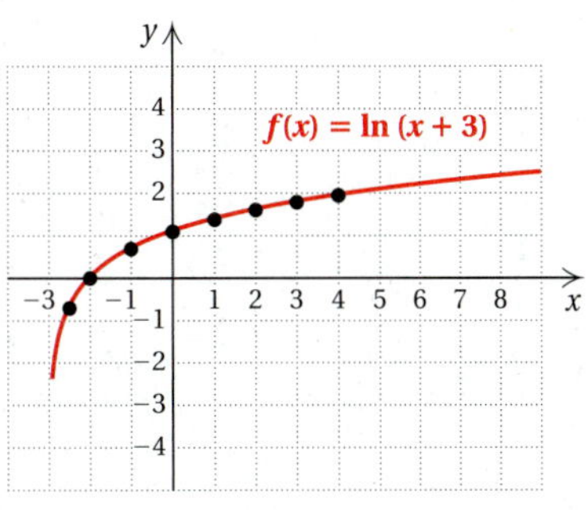

The graph of $y = \ln (x + 3)$ is the graph of $y = \ln x$ translated 3 units to the left.

Do Exercises 12 and 13.

Graph.

10. $f(x) = e^{2x}$

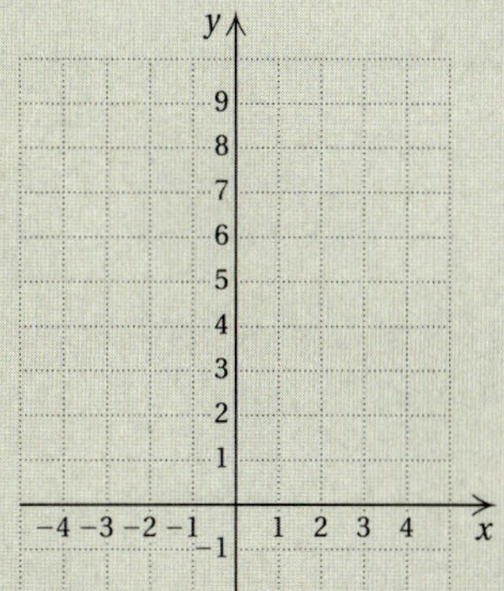

11. $g(x) = \frac{1}{2}e^{-x}$

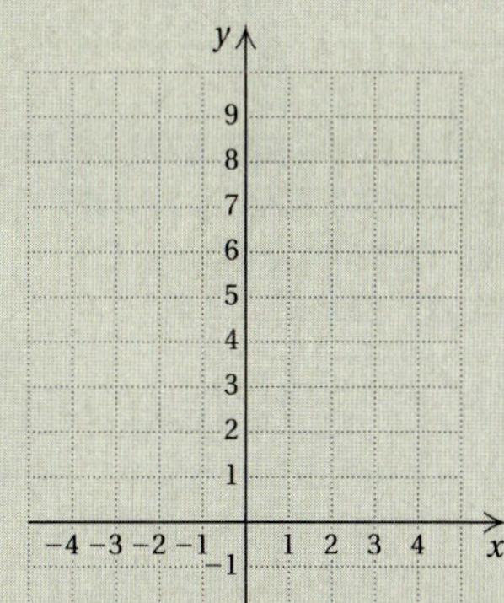

Graph.

12. $f(x) = 2 \ln x$

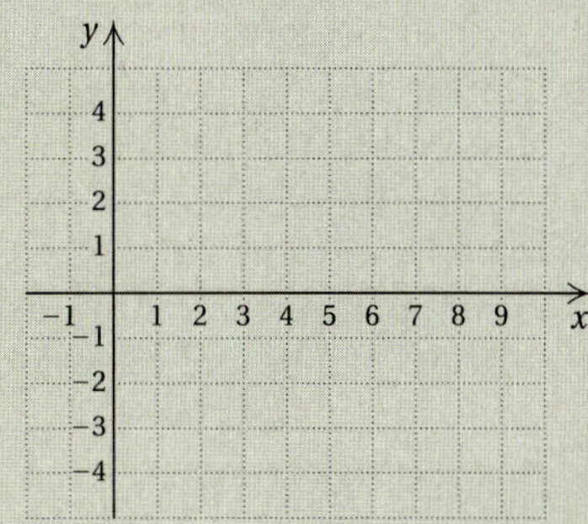

13. $g(x) = \ln (x - 2)$

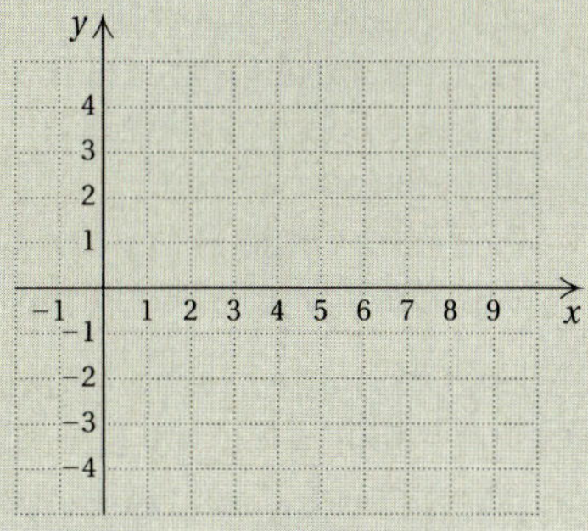

Answers on page A-71

CALCULATOR CORNER

Graphing Logarithmic Functions We can graph logarithmic functions with base 10 or base e by entering the function on the equation-editor screen using the LOG or LN key. To graph a logarithmic function with a base other than 10 or e, we must first use the change-of-base formula to change the base to 10 or e.

In Example 1 of Section 17.3, we graphed the function $y = \log_5 x$ by finding a table of x- and y-values and plotting points. We will now graph this function on a graphing calculator by first changing the base to e. We let $a = e$, $b = 5$, and $M = x$ and substitute in the change-of-base formula. We enter $y_1 = \dfrac{\ln x}{\ln 5}$ on the equation-editor screen, select a window, and press GRAPH. The TI-83+ forces the use of parentheses with the ln function, so the parentheses in the numerator must be closed: ln (x)/ln (5). The right parenthesis following the 5 is optional but we include it for completeness.

We could have let $a = 10$ and used base-10 logarithms to graph this function as well.

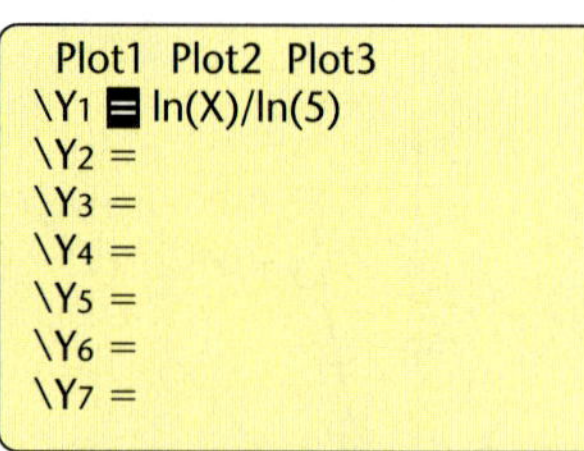

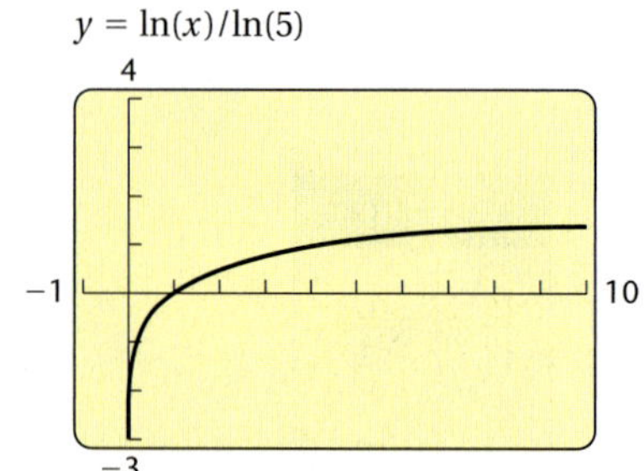

Exercises: Graph each of the following on a graphing calculator.

1. $y = \log_2 x$
2. $y = \log_3 x$
3. $y = \log_{1/2} x$
4. $y = \log_{2/3} x$

Study Tips

BEGINNING TO STUDY FOR THE FINAL EXAM (PART 4): ONE OR TWO DAYS OF STUDY TIME

1. **Begin by browsing through each chapter, reviewing the highlighted or boxed information regarding important formulas in both the text and the Summary and Review.** There may be some formulas that you will need to memorize.
2. **Do the Review Exercises at the end of the chapter during the last couple of days before the final.** Set aside approximately the same amount of time that you will have for the final. See how many exercises you can complete under test-like conditions. Be careful to avoid any questions corresponding to objectives not covered. Restudy the objectives in the text that correspond to each question you missed.
3. **Attend a final-exam review session if one is available.**

"Great is the art of beginning, but greater is the art of ending."

Henry Wadsworth Longfellow, nineteenth-century American poet

17.5 EXERCISE SET

For Extra Help

Digital Video Tutor CD 15 Videotape 20 | InterAct Math | Math Tutor Center | MathXL | MyMathLab

a Find each of the following logarithms or powers, base e, using a calculator. Round answers to four decimal places.

1. $\ln 2$ **2.** $\ln 5$ **3.** $\ln 62$ **4.** $\ln 30$ **5.** $\ln 4365$ **6.** $\ln 901.2$

7. $\ln 0.0062$ **8.** $\ln 0.00073$ **9.** $\ln 0.2$ **10.** $\ln 0.04$ **11.** $\ln 0$ **12.** $\ln(-4)$

13. $\ln\left(\frac{97.4}{558}\right)$ **14.** $\ln\left(\frac{786.2}{77.2}\right)$ **15.** $\ln e$ **16.** $\ln e^2$ **17.** $e^{2.71}$ **18.** $e^{3.06}$

19. $e^{-3.49}$ **20.** $e^{-2.64}$ **21.** $e^{4.7}$ **22.** $e^{1.23}$ **23.** $\ln e^5$ **24.** $e^{\ln 7}$

b Find each of the following logarithms using the change-of-base formula.

25. $\log_6 100$ **26.** $\log_3 100$ **27.** $\log_2 100$ **28.** $\log_7 100$ **29.** $\log_7 65$ **30.** $\log_5 42$

31. $\log_{0.5} 5$ **32.** $\log_{0.1} 3$ **33.** $\log_2 0.2$ **34.** $\log_2 0.08$ **35.** $\log_\pi 200$ **36.** $\log_\pi \pi$

c Graph.

37. $f(x) = e^x$

x	$f(x)$
0	
1	
2	
3	
−1	
−2	
−3	

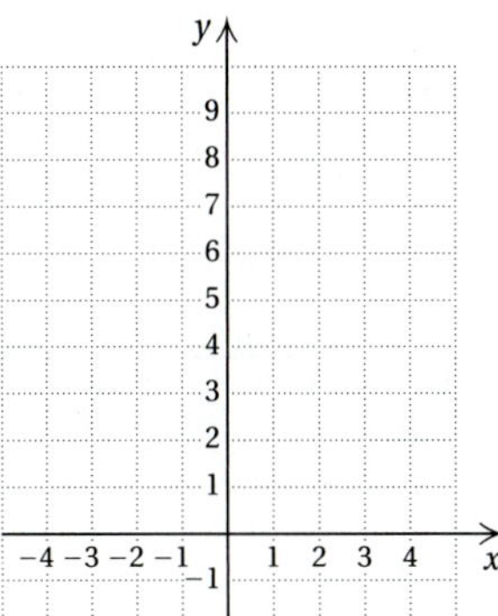

38. $f(x) = e^{0.5x}$

x	$f(x)$
0	
1	
2	
3	
−1	
−2	
−3	

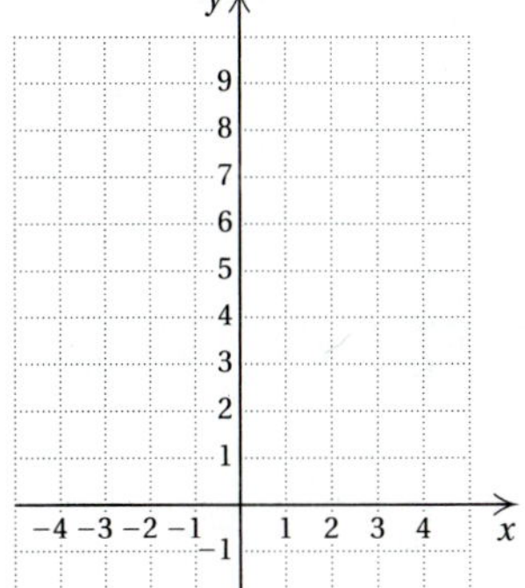

39. $f(x) = e^{-0.5x}$

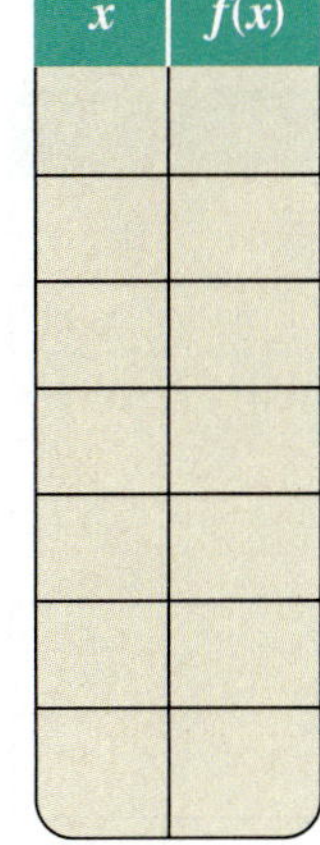

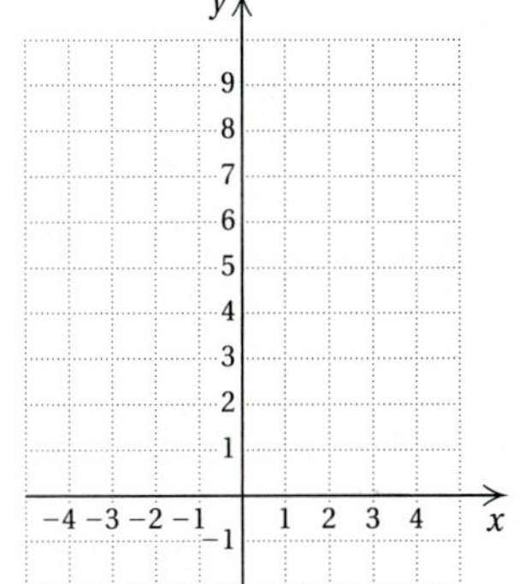

40. $f(x) = e^{-x}$

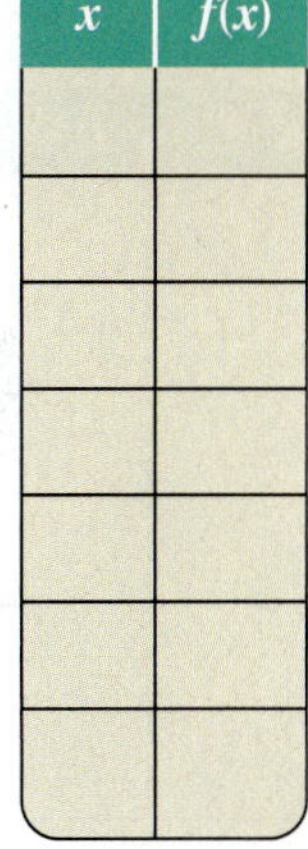

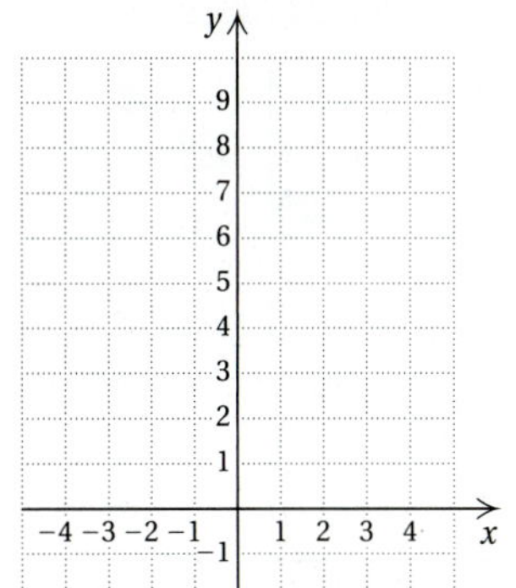

41. $f(x) = e^{x-1}$

x	$f(x)$

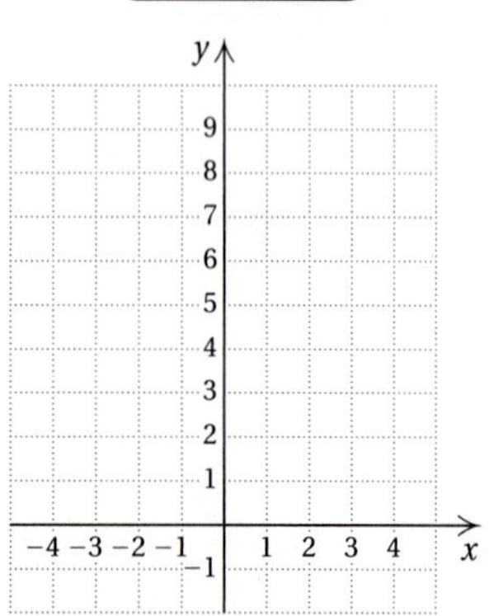

42. $f(x) = e^{-x} + 3$

x	$f(x)$

43. $f(x) = e^{x+2}$

x	$f(x)$

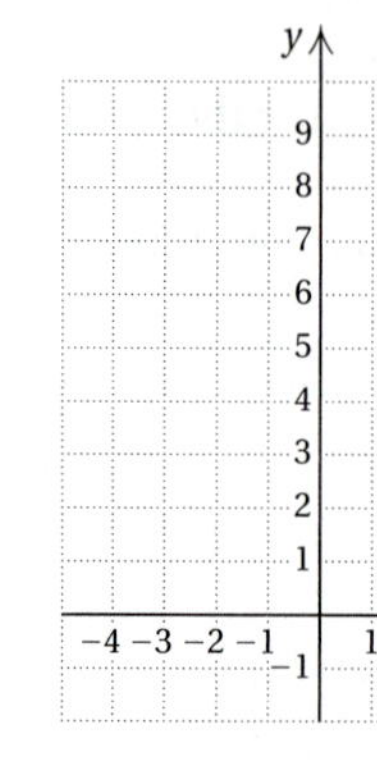

44. $f(x) = e^{x-2}$

x	$f(x)$

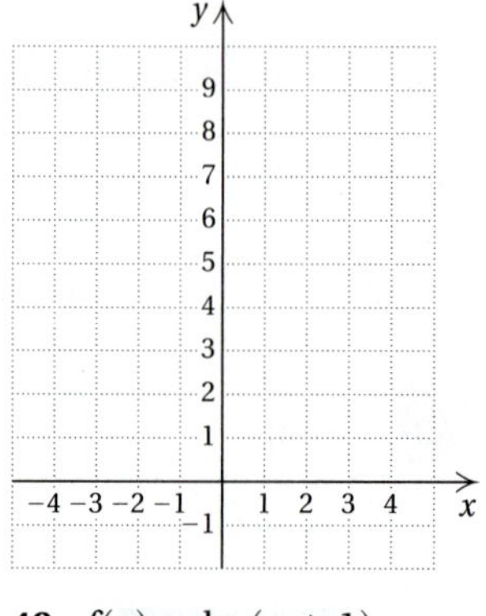

45. $f(x) = e^{x} - 1$

x	$f(x)$

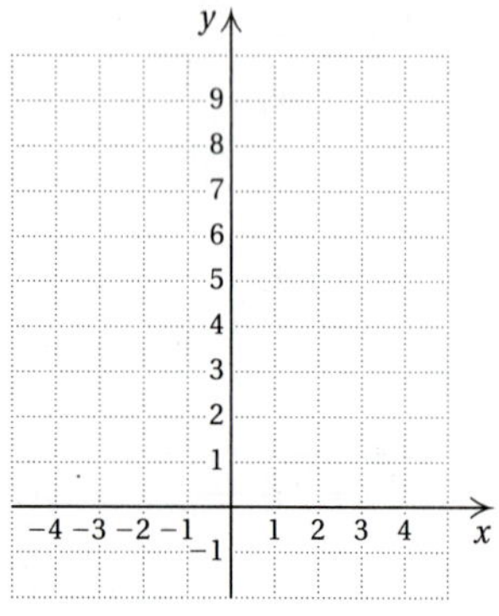

46. $f(x) = 2e^{0.5x}$

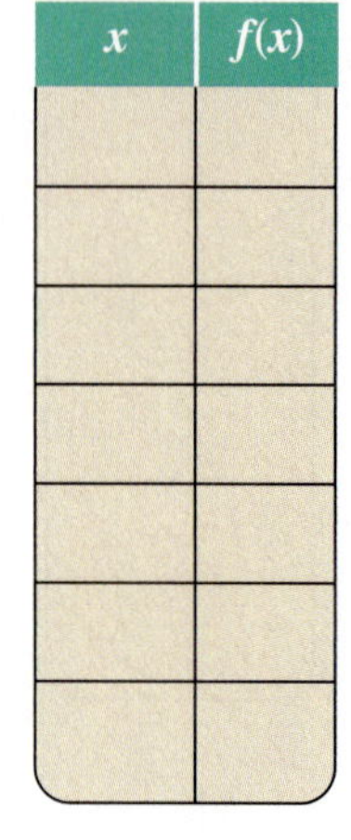

x	$f(x)$

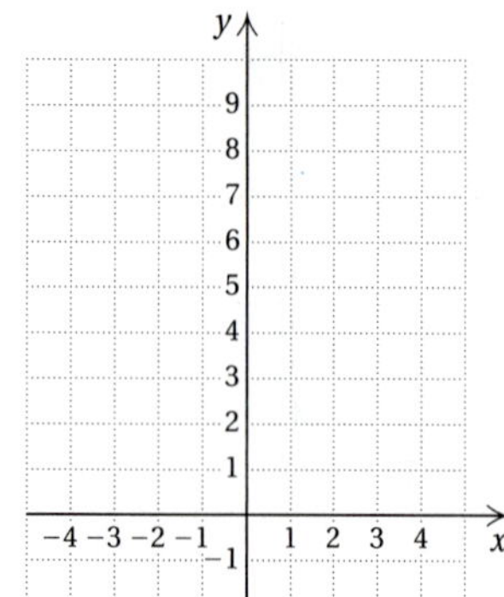

47. $f(x) = \ln(x + 2)$

x	$f(x)$
0	
1	
2	
3	
−0.5	
−1	
−1.5	

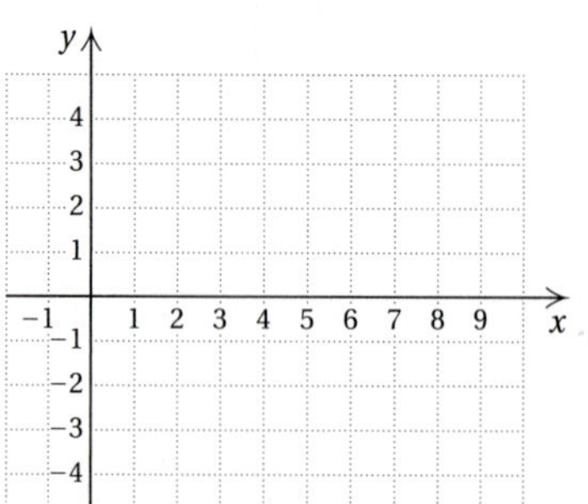

48. $f(x) = \ln(x + 1)$

x	$f(x)$
0	
1	
2	
3	
4	
−0.5	
−0.75	

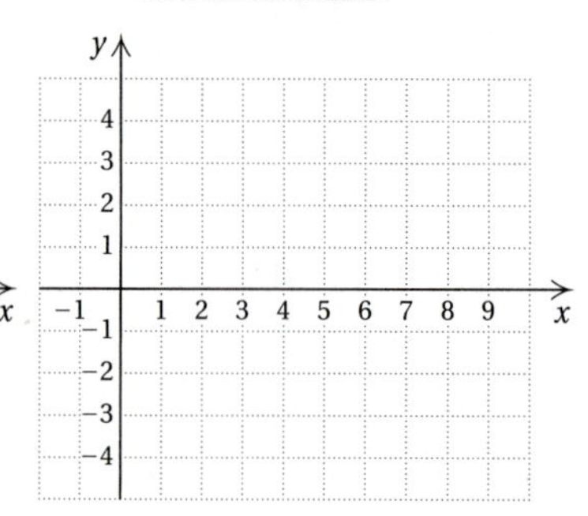

49. $f(x) = \ln (x - 3)$

x	$f(x)$

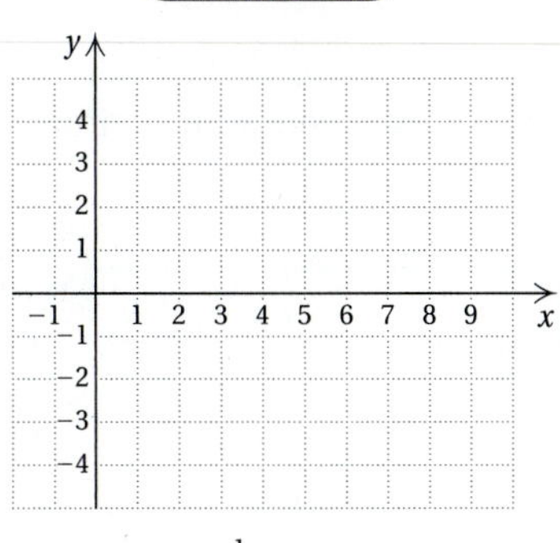

50. $f(x) = 2 \ln (x - 2)$

x	$f(x)$

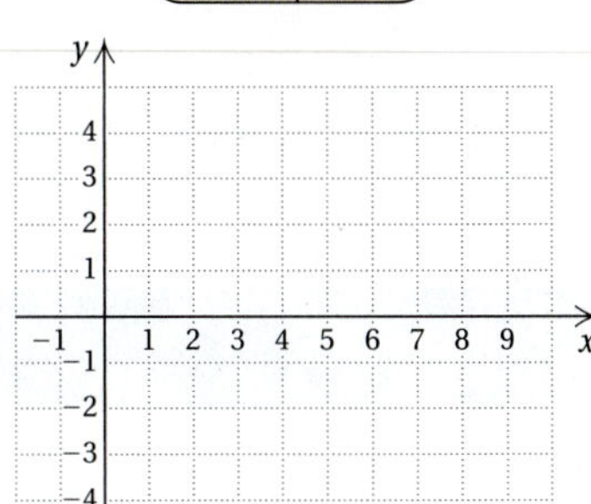

51. $f(x) = 2 \ln x$

x	$f(x)$

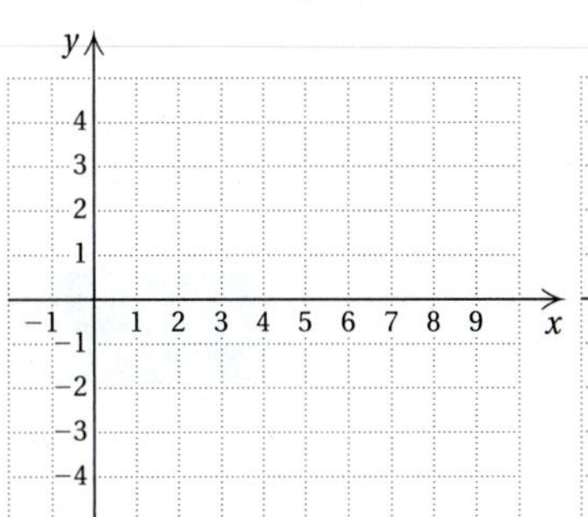

52. $f(x) = \ln x - 3$

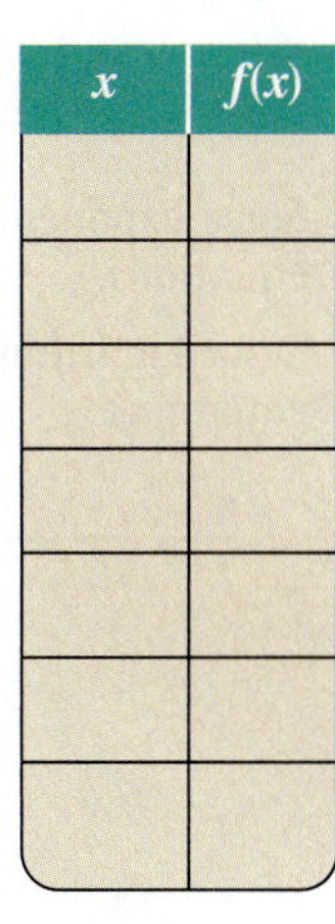

x	$f(x)$

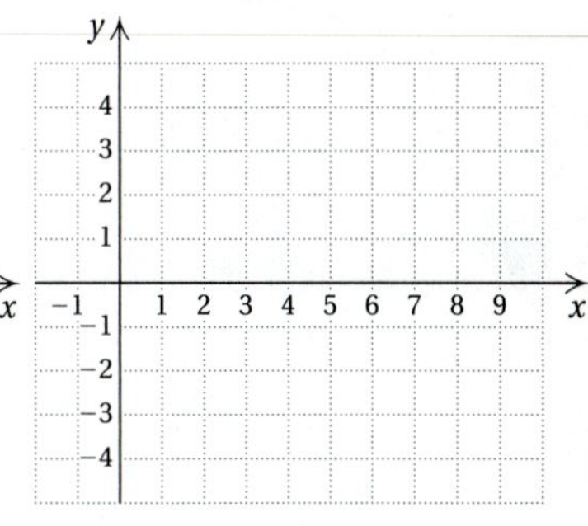

53. $f(x) = \frac{1}{2} \ln x + 1$

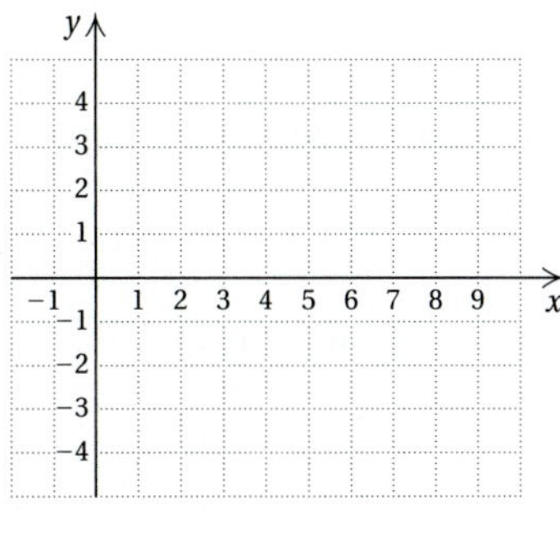

54. $f(x) = \ln x^2$

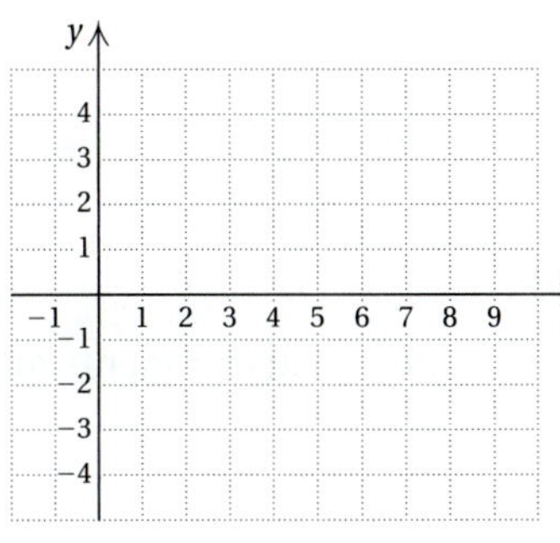

55. $f(x) = |\ln x|$

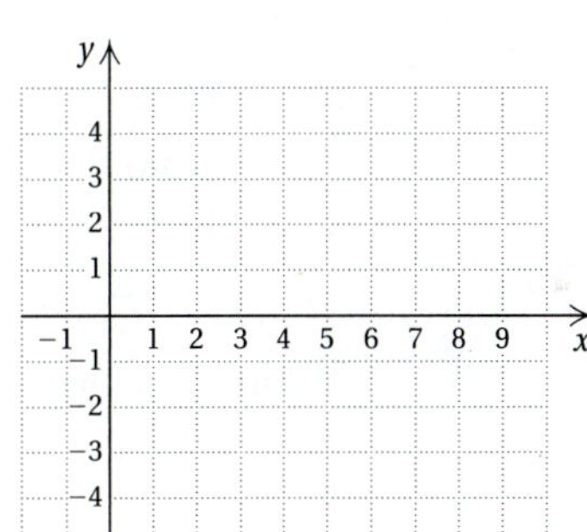

56. $f(x) = \ln |x|$

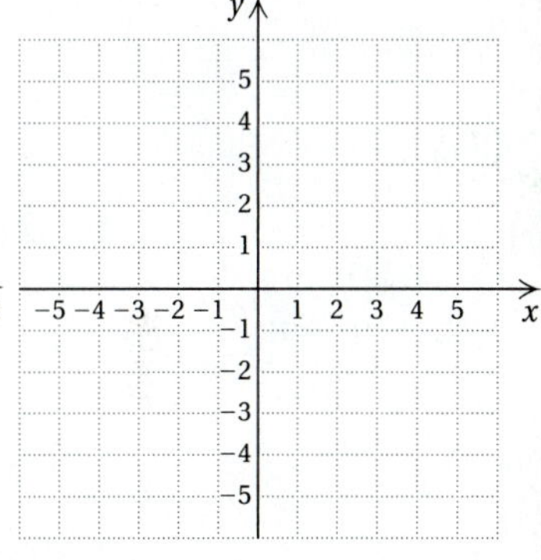

57. $\mathrm{D_W}$ Explain how the graph of $f(x) = e^x$ could be used to obtain the graph of $g(x) = 1 + \ln x$.

58. $\mathrm{D_W}$ Find a formula for converting common logarithms to natural logarithms. Then find a formula for converting natural logarithms to common logarithms.

SKILL MAINTENANCE

Solve. [16.4c]

59. $x^{1/2} - 6x^{1/4} + 8 = 0$

60. $2y - 7\sqrt{y} + 3 = 0$

61. $x - 18\sqrt{x} + 77 = 0$

62. $x^4 - 25x^2 + 144 = 0$

SYNTHESIS

Graph and find the domain and the range.

63. $f(x) = 10x^2e^{-x}$

64. $f(x) = 7.4e^x \ln x$

65. $f(x) = 100(1 - e^{-0.3x})$

Find the domain.

66. $f(x) = \log_3 x^2$

67. $f(x) = \log (2x - 5)$

17.6 SOLVING EXPONENTIAL AND LOGARITHMIC EQUATIONS

Objectives

a Solve exponential equations.

b Solve logarithmic equations.

a Solving Exponential Equations

Equations with variables in exponents, such as $5^x = 12$ and $2^{7x} = 64$, are called **exponential equations.** Sometimes, as is the case with $2^{7x} = 64$, we can write each side as a power of the *same* number:

$$2^{7x} = 2^6.$$

Since the base is the same, 2, the exponents are the same. We can set them equal and solve:

$$7x = 6$$
$$x = \tfrac{6}{7}.$$

We use the following property, which is true because exponential functions are one-to-one.

PRINCIPLE OF EXPONENTIAL EQUALITY

For any $a > 0$, $a \neq 1$,

$$a^x = a^y \longrightarrow x = y.$$

(When powers are equal, the exponents are equal.)

EXAMPLE 1 Solve: $2^{3x-5} = 16$.

Note that $16 = 2^4$. Thus we can write each side as a power of the same number:

$$2^{3x-5} = 2^4.$$

Since the base is the same, 2, the exponents must be the same. Thus,

$$3x - 5 = 4$$
$$3x = 9$$
$$x = 3.$$

CHECK:

$$\begin{array}{c|c} 2^{3x-5} & 16 \\ \hline 2^{3\cdot 3-5} \ ? & 16 \\ 2^{9-5} & \\ 2^4 & \\ 16 & \end{array} \quad \text{TRUE}$$

The solution is 3.

Do Exercises 1 and 2.

Solve.

1. $3^{2x} = 9$

2. $4^{2x-3} = 64$

Answers on page A-72

ALGEBRAIC–GRAPHICAL CONNECTION

The solution, 3, of the equation $2^{3x-5} = 16$ in Example 1 is the x-coordinate of the point of intersection of the graphs of $y = 2^{3x-5}$ and $y = 16$, as we see in the graph on the left below.

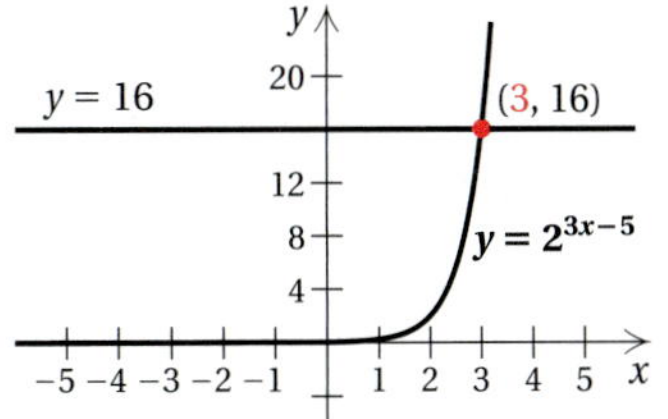

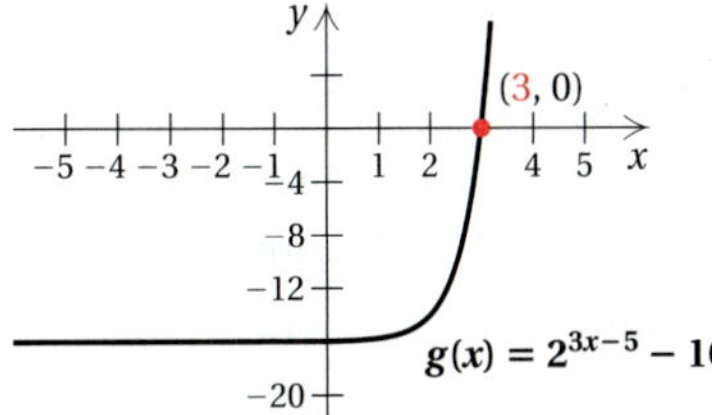

If we subtract 16 on both sides of $2^{3x-5} = 16$, we get $2^{3x-5} - 16 = 0$. The solution, 3, is then the x-coordinate of the x-intercept of the function $g(x) = 2^{3x-5} - 16$, as we see in the graph on the right above.

When it does not seem possible to write both sides of an equation as powers of the same base, we can use the following principle along with the properties developed in Section 17.4.

THE PRINCIPLE OF LOGARITHMIC EQUALITY

For any logarithm base a, and for $x, y > 0$,

$$\log_a x = \log_a y \rightarrow x = y.$$

(If the logarithms, base a, of two expressions are the same, then the expressions are the same.)

Because calculators can generally find only common or natural logarithms (without resorting to the change-of-base formula), we usually take the common or natural logarithm on both sides of the equation.

The principle of logarithmic equality is useful anytime a variable appears as an exponent.

EXAMPLE 2 Solve: $5^x = 12$.

$5^x = 12$

$\log 5^x = \log 12$ Taking the common logarithm on both sides

$x \log 5 = \log 12$ Property 2

$x = \dfrac{\log 12}{\log 5}$ ← **CAUTION!** This is not $\log \frac{12}{5}$!

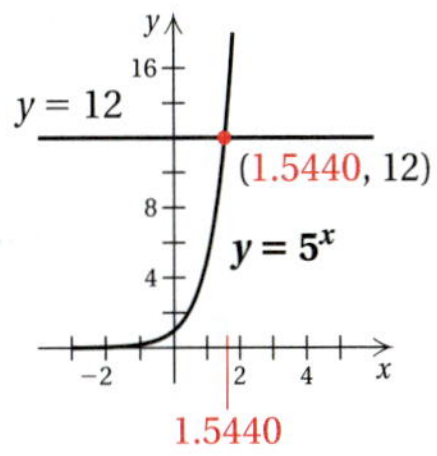

This is an exact answer. We cannot simplify further, but we can approximate using a calculator:

$$x = \frac{\log 12}{\log 5} \approx 1.5440.$$

We can also partially check this answer by finding $5^{1.5440}$ using a calculator:

$$5^{1.5440} \approx 12.00078587.$$

We get an answer close to 12, due to the rounding. This checks.

3. Solve: $7^x = 20$.

Answer on page A-72

4. Solve: $e^{0.3t} = 80$.

5. Solve: $\log_5 x = 2$.

Answers on page A-72

CALCULATOR CORNER

Solving Exponential Equations Use the INTERSECT feature to solve the equations in Examples 1–3 and Margin Exercises 1–4. (See the Calculator Corner on p. 991 for the procedure.)

Do Exercise 3 on the preceding page.

If the base is e, we can make our work easier by taking the logarithm, base e, on both sides.

EXAMPLE 3 Solve: $e^{0.06t} = 1500$.

We take the natural logarithm on both sides:

$$e^{0.06t} = 1500$$
$$\ln e^{0.06t} = \ln 1500 \quad \text{Taking ln on both sides}$$
$$\log_e e^{0.06t} = \ln 1500 \quad \text{Definition of natural logarithms}$$
$$0.06t = \ln 1500 \quad \text{Here we use Property 4: } \log_a a^k = k.$$
$$t = \frac{\ln 1500}{0.06}.$$

We can approximate using a calculator:

$$t = \frac{\ln 1500}{0.06} \approx \frac{7.3132}{0.06} \approx 121.89.$$

We can also partially check this answer using a calculator.

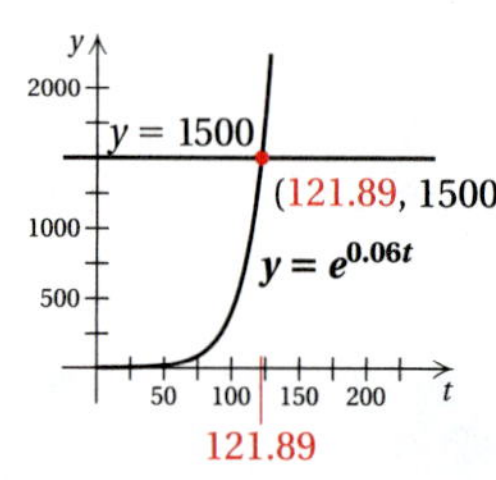

CHECK:

$e^{0.06t} = 1500$	
$e^{0.06(121.89)}$?	1500
$e^{7.3134}$	
1500.269444	TRUE

The solution is about 121.89.

Do Exercise 4.

b Solving Logarithmic Equations

Equations containing logarithmic expressions are called **logarithmic equations.** We solved some logarithmic equations in Section 17.3 by converting to equivalent exponential equations.

EXAMPLE 4 Solve: $\log_2 x = 3$.

We obtain an equivalent exponential equation:

$$x = 2^3$$
$$x = 8.$$

The solution is 8.

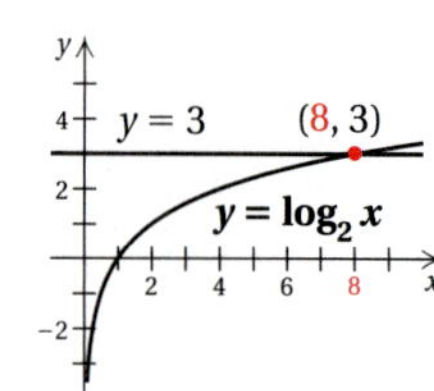

Do Exercise 5.

To solve a logarithmic equation, first try to obtain a single logarithmic expression on one side and then write an equivalent exponential equation.

EXAMPLE 5 Solve: $\log_4 (8x - 6) = 3$.

We already have a single logarithmic expression, so we write an equivalent exponential equation:

$$8x - 6 = 4^3 \quad \text{Writing an equivalent exponential equation}$$
$$8x - 6 = 64$$
$$8x = 70$$
$$x = \frac{70}{8}, \text{ or } \frac{35}{4}$$

CHECK:

$$\begin{array}{r|l} \log_4 (8x - 6) = 3 & \\ \hline \log_4 \left(8 \cdot \frac{35}{4} - 6\right) \ ? & 3 \\ \log_4 (70 - 6) & \\ \log_4 64 & \\ 3 & \end{array} \quad \text{TRUE}$$

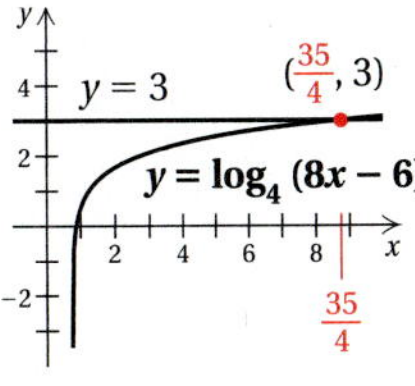

The solution is $\frac{35}{4}$.

Do Exercise 6.

6. Solve: $\log_3 (5x + 7) = 2$.

EXAMPLE 6 Solve: $\log x + \log (x - 3) = 1$.

Here we have common logarithms. It will help us follow the solution to first write in the 10's before we obtain a single logarithmic expression on the left.

$$\log_{10} x + \log_{10} (x - 3) = 1$$
$$\log_{10} [x(x - 3)] = 1 \quad \text{Using Property 1 to obtain a single logarithm}$$
$$x(x - 3) = 10^1 \quad \text{Writing an equivalent exponential expression}$$
$$x^2 - 3x = 10$$
$$x^2 - 3x - 10 = 0$$
$$(x + 2)(x - 5) = 0 \quad \text{Factoring}$$
$$x + 2 = 0 \quad or \quad x - 5 = 0 \quad \text{Using the principle of zero products}$$
$$x = -2 \quad or \quad x = 5$$

CHECK: For −2:

$$\begin{array}{r|l} \log x + \log (x - 3) = 1 & \\ \hline \log (-2) + \log (-2 - 3) \ ? & 1 \end{array}$$

The number −2 does *not* check because negative numbers do not have logarithms.

For 5:

$$\begin{array}{r|l} \log x + \log (x - 3) = 1 & \\ \hline \log 5 + \log (5 - 3) \ ? & 1 \\ \log 5 + \log 2 & \\ \log (5 \cdot 2) & \\ \log 10 & \\ 1 & \end{array} \quad \text{TRUE}$$

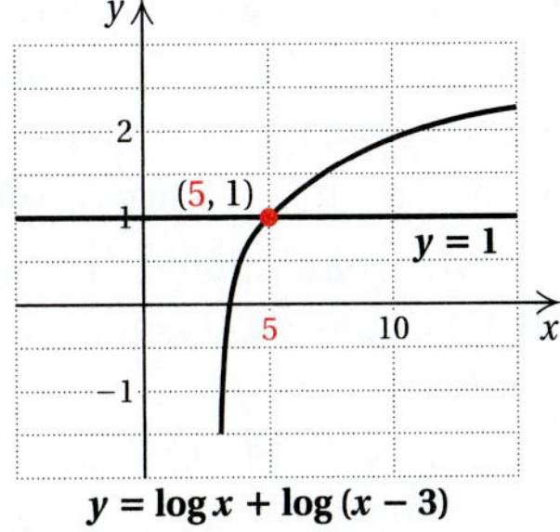

$y = \log x + \log (x - 3)$

The solution is 5.

Do Exercise 7.

7. Solve: $\log x + \log (x + 3) = 1$.

Answers on page A-72

8. Solve:

$\log_3 (2x - 1) - \log_3 (x - 4) = 2.$

Answer on page A-72

EXAMPLE 7 Solve: $\log_2 (x + 7) - \log_2 (x - 7) = 3.$

$$\log_2 (x + 7) - \log_2 (x - 7) = 3$$

$$\log_2 \frac{x+7}{x-7} = 3 \qquad \text{Using Property 3 to obtain a single logarithm}$$

$$\frac{x+7}{x-7} = 2^3 \qquad \text{Writing an equivalent exponential expression}$$

$$\frac{x+7}{x-7} = 8$$

$$x + 7 = 8(x - 7) \qquad \text{Multiplying by the LCM, } x - 7$$

$$x + 7 = 8x - 56 \qquad \text{Using a distributive law}$$

$$63 = 7x$$

$$\frac{63}{7} = x$$

$$9 = x$$

CHECK:

$$\log_2 (x + 7) - \log_2 (x - 7) = 3$$

$$\log_2 (9 + 7) - \log_2 (9 - 7) \;?\; 3$$

$$\log_2 16 - \log_2 2$$

$$\log_2 \tfrac{16}{2}$$

$$\log_2 8$$

$$3 \qquad \text{TRUE}$$

The solution is 9.

Do Exercise 8.

Study Tips

READ FOR SUCCESS

In his article "The Daily Dozen Disciplines for Massive Success in 2001 & Beyond," Jerry Clark comments, "Research has shown that 58% of high school graduates never read another book from cover to cover the rest of their adult lives, that 78% of the population have not been in a bookstore in the last 5 years, and that 97% of the population of the U.S. do not have library cards." Clark then suggests spending at least 15 minutes each day reading an empowering and uplifting book or article. The following books are some suggestions from your author. Their motivating words may empower you in your study of mathematics.

1. *Fish*, by Stephen C. Lundin, Harry Paul, and John Christensen (Hyperion). This was a Wall Street Journal Business Bestseller. Though it has a strange title, it discusses a remarkable way to boost morale and improve results.
2. *True Success: A New Philosophy of Excellence*, by Tom Morris (Grosset/Putnam). Morris was a well-loved philosophy professor at Notre Dame. Students, especially athletes, flocked to his classes.
3. *The Road Less Traveled, Abounding Grace*, by M. Scott Peck. Noted psychiatrist and author of many excellent books, Peck has amazing insights and wisdom about life.
4. *The Weight of Glory, The Great Divorce*, by C. S. Lewis. British author and scholar, noted for his philosophical exposition, Lewis also wrote science-fiction fantasies with moral overtones.

"Far worse than not reading books is not realizing that it matters!"

Jim Rohn, motivational speaker

17.6 EXERCISE SET

For Extra Help

Digital Video Tutor CD 15 Videotape 20 | InterAct Math | Math Tutor Center | MathXL | MyMathLab

a Solve.

1. $2^x = 8$

2. $3^x = 81$

3. $4^x = 256$

4. $5^x = 125$

5. $2^{2x} = 32$

6. $4^{3x} = 64$

7. $3^{5x} = 27$

8. $5^{7x} = 625$

9. $2^x = 11$

10. $2^x = 20$

11. $2^x = 43$

12. $2^x = 55$

13. $5^{4x-7} = 125$

14. $4^{3x+5} = 16$

15. $3^{x^2} \cdot 3^{4x} = \frac{1}{27}$

16. $3^{5x} \cdot 9^{x^2} = 27$

17. $4^x = 8$

18. $6^x = 10$

19. $e^t = 100$

20. $e^t = 1000$

21. $e^{-t} = 0.1$

22. $e^{-t} = 0.01$

23. $e^{-0.02t} = 0.06$

24. $e^{0.07t} = 2$

25. $2^x = 3^{x-1}$

26. $3^{x+2} = 5^{x-1}$

27. $(3.6)^x = 62$

28. $(5.2)^x = 70$

b Solve.

29. $\log_4 x = 4$

30. $\log_7 x = 3$

31. $\log_2 x = -5$

32. $\log_9 x = \frac{1}{2}$

33. $\log x = 1$

34. $\log x = 3$

35. $\log x = -2$

36. $\log x = -3$

37. $\ln x = 2$

38. $\ln x = 1$

39. $\ln x = -1$

40. $\ln x = -3$

41. $\log_3 (2x + 1) = 5$

42. $\log_2 (8 - 2x) = 6$

43. $\log x + \log (x - 9) = 1$

44. $\log x + \log (x + 9) = 1$

45. $\log x - \log (x + 3) = -1$

46. $\log (x + 9) - \log x = 1$

47. $\log_2 (x + 1) + \log_2 (x - 1) = 3$

48. $\log_2 x + \log_2 (x - 2) = 3$

49. $\log_4 (x + 6) - \log_4 x = 2$

50. $\log_4 (x + 3) - \log_4 (x - 5) = 2$

51. $\log_4 (x + 3) + \log_4 (x - 3) = 2$

52. $\log_5 (x + 4) + \log_5 (x - 4) = 2$

53. $\log_3 (2x - 6) - \log_3 (x + 4) = 2$

54. $\log_4 (2 + x) - \log_4 (3 - 5x) = 3$

55. **DW** Explain how Exercises 37–40 could be solved using only the graph of $f(x) = \ln x$.

56. **DW** Christina first determines that the solution of $\log_3 (x + 4) = 1$ is -1, but then rejects it. What mistake do you think she might be making?

SKILL MAINTENANCE

Solve. [16.4c]

57. $x^4 + 400 = 104x^2$

58. $x^{2/3} + 2x^{1/3} = 8$

59. $(x^2 + 5x)^2 + 2(x^2 + 5x) = 24$

60. $10 = x^{-2} + 9x^{-1}$

61. Simplify: $(125x^3y^{-2}z^6)^{-2/3}$. [15.2c]

62. Simplify: i^{79}. [15.8d]

SYNTHESIS

63. Find the value of x for which the natural logarithm is the same as the common logarithm.

64. Use a graphing calculator to check your answers to Exercises 4, 20, 36, and 54.

65. Use a graphing calculator to solve each of the following equations.

a) $e^{7x} = 14$ b) $8e^{0.5x} = 3$
c) $xe^{3x-1} = 5$ d) $4 \ln (x + 3.4) = 2.5$

66. Use the INTERSECT feature of a graphing calculator to find the points of intersection of the graphs of each pair of functions.

a) $f(x) = e^{0.5x-7}$, $g(x) = 2x + 6$
b) $f(x) = \ln 3x$, $g(x) = 3x - 8$
c) $f(x) = \ln x^2$, $g(x) = -x^2$

Solve.

67. $2^{2x} + 128 = 24 \cdot 2^x$

68. $27^x = 81^{2x-3}$

69. $8^x = 16^{3x+9}$

70. $\log_x (\log_3 27) = 3$

71. $\log_6 (\log_2 x) = 0$

72. $x \log \frac{1}{8} = \log 8$

73. $\log_5 \sqrt{x^2 - 9} = 1$

74. $2^{x^2+4x} = \frac{1}{8}$

75. $\log (\log x) = 5$

76. $\log_5 |x| = 4$

77. $\log x^2 = (\log x)^2$

78. $\log_3 |5x - 7| = 2$

79. $\log_a a^{x^2+4x} = 21$

80. $\sqrt{x} \cdot \sqrt[3]{x} \cdot \sqrt[4]{x} \cdot \sqrt[5]{x} = 146$

81. $3^{2x} - 8 \cdot 3^x + 15 = 0$

82. If $x = (\log_{125} 5)^{\log_5 125}$, what is the value of $\log_3 x$?

17.7 MATHEMATICAL MODELING WITH EXPONENTIAL AND LOGARITHMIC FUNCTIONS

Objectives

a Solve applied problems involving logarithmic functions.

b Solve applied problems involving exponential functions.

Exponential and logarithmic functions can now be added to our library of functions that can serve as models for many kinds of applications. Let's review some of their graphs.

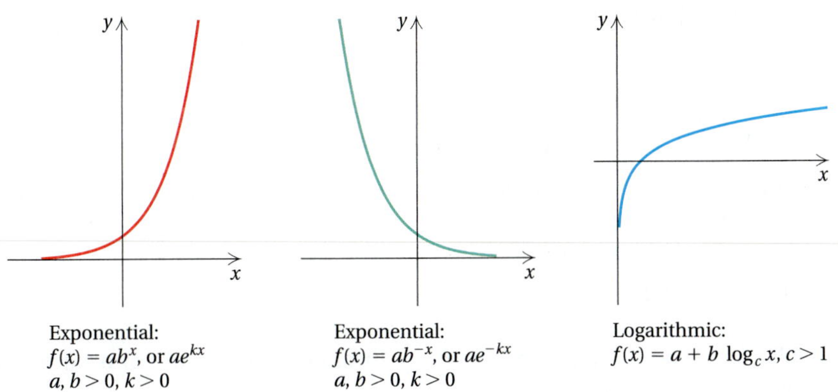

Exponential: $f(x) = ab^x$, or ae^{kx} $a, b > 0, k > 0$

Exponential: $f(x) = ab^{-x}$, or ae^{-kx} $a, b > 0, k > 0$

Logarithmic: $f(x) = a + b \log_c x, c > 1$

a Applications of Logarithmic Functions

EXAMPLE 1 *Sound Levels.* To measure the "loudness" of any particular sound, the decibel scale is used. The loudness L, in decibels (dB), of a sound is given by

$$L = 10 \cdot \log \frac{I}{I_0},$$

where I is the intensity of the sound, in watts per square meter ($\mathrm{W/m^2}$), and $I_0 = 10^{-12}\ \mathrm{W/m^2}$. ($I_0$ is approximately the intensity of the softest sound that can be heard.)

a) It is common for the intensity of sound at live performances of rock music to reach $10^{-1}\ \mathrm{W/m^2}$ (even higher close to the stage). How loud, in decibels, is this sound level?

b) Audiologists and physicians recommend that earplugs be worn when one is exposed to sounds in excess of 90 dB. What is the intensity of such sounds?

Sarah Fisher puts in her earplugs in the pits at the Indianapolis Motor Speedway on May 19, 2002. She is the youngest woman to qualify for the Indy 500.

a) To find the loudness, in decibels, we use the above formula:

$$\begin{aligned} L &= 10 \cdot \log \frac{I}{I_0} \\ &= 10 \cdot \log \frac{10^{-1}}{10^{-12}} && \text{Substituting} \\ &= 10 \cdot \log 10^{11} && \text{Subtracting exponents} \\ &= 10 \cdot 11 && \log 10^a = a \\ &= 110. \end{aligned}$$

The volume of the music is 110 decibels.

1. **Acoustics.** The intensity of sound in normal conversation is about $3.2 \times 10^{-6}\ \text{W/m}^2$. How high is this sound level in decibels?

2. **Audiology.** Overexposure to excessive sound levels can diminish one's hearing to the point where the softest sound that is audible is 28 dB. What is the intensity of such a sound?

Answers on page A-73

b) We substitute and solve for I:

$$L = 10 \cdot \log \frac{I}{I_0}$$

$$90 = 10 \cdot \log \frac{I}{10^{-12}} \qquad \text{Substituting}$$

$$9 = \log \frac{I}{10^{-12}} \qquad \text{Dividing by 10}$$

$$9 = \log I - \log 10^{-12} \qquad \text{Using Property 3}$$

$$9 = \log I - (-12) \qquad \log 10^a = a$$

$$-3 = \log I \qquad \text{Adding } -12$$

$$10^{-3} = I. \qquad \text{Converting to an exponential equation}$$

Earplugs are recommended for sounds with intensities that exceed $10^{-3}\ \text{W/m}^2$.

Do Exercises 1 and 2.

EXAMPLE 2 *Chemistry: pH of Liquids.* In chemistry the pH of a liquid is defined as:

$$\text{pH} = -\log\,[\text{H}^+],$$

where $[\text{H}^+]$ is the hydrogen ion concentration in moles per liter.

a) The hydrogen ion concentration of human blood is normally about 3.98×10^{-8} moles per liter. Find the pH.

b) The pH of seawater is about 8.3. Find the hydrogen ion concentration.

a) To find the pH of human blood, we use the above formula:

$$\text{pH} = -\log\,[\text{H}^+] = -\log\,[3.98 \times 10^{-8}]$$

$$\approx -(-7.400117) \qquad \text{Using a calculator}$$

$$\approx 7.4.$$

The pH of human blood is normally about 7.4.

b) We substitute and solve for $[\text{H}^+]$:

$$8.3 = -\log\,[\text{H}^+] \qquad \text{Using pH} = -\log\,[\text{H}^+]$$

$$-8.3 = \log\,[\text{H}^+] \qquad \text{Dividing by } -1$$

$$10^{-8.3} = [\text{H}^+] \qquad \text{Converting to an exponential equation}$$

$$5.01 \times 10^{-9} \approx [\text{H}^+]. \qquad \text{Using a calculator; writing scientific notation}$$

The hydrogen ion concentration of seawater is about 5.01×10^{-9} moles per liter.

Do Exercises 3 and 4.

b Applications of Exponential Functions

EXAMPLE 3 *Interest Compounded Annually.* Suppose that \$30,000 is invested at 8% interest, compounded annually. In t years, it will grow to the amount A given by the function

$$A(t) = 30{,}000(1.08)^t.$$

(See Example 6 in Section 17.1.)

a) How long will it take to accumulate \$150,000 in the account?

b) Let $T =$ the amount of time it takes for the \$30,000 to double itself; T is called the **doubling time.** Find the doubling time.

a) We set $A(t) = 150{,}000$ and solve for t:

$$150{,}000 = 30{,}000(1.08)^t$$

$$\frac{150{,}000}{30{,}000} = (1.08)^t \quad \text{Dividing by 30,000}$$

$$5 = (1.08)^t$$

$$\log 5 = \log (1.08)^t \quad \text{Taking the common logarithm on both sides}$$

$$\log 5 = t \log 1.08 \quad \text{Using Property 2}$$

$$\frac{\log 5}{\log 1.08} = t \quad \text{Dividing by log 1.08}$$

$$20.9 \approx t. \quad \text{Using a calculator}$$

It will take about 20.9 yr for the \$30,000 to grow to \$150,000.

b) To find the *doubling time T*, we replace $A(t)$ with 60,000 and t with T and solve for T:

$$60{,}000 = 30{,}000(1.08)^T$$

$$2 = (1.08)^T \quad \text{Dividing by 30,000}$$

$$\log 2 = \log (1.08)^T \quad \text{Taking the common logarithm on both sides}$$

$$\log 2 = T \log 1.08 \quad \text{Using Property 2}$$

$$T = \frac{\log 2}{\log 1.08} \approx 9.0. \quad \text{Using a calculator}$$

The doubling time is about 9 yr.

Do Exercise 5.

The function in Example 3 illustrates exponential growth. Populations often grow exponentially according to the following model.

3. Coffee. The hydrogen ion concentration of freshly brewed coffee is about 1.3×10^{-5} moles per liter. Find the pH.

4. Acidosis. When the pH of a patient's blood drops below 7.4, a condition called *acidosis* sets in. Acidosis can be fatal at a pH level of 7.0. What would the hydrogen ion concentration of the patient's blood be at that point?

5. Interest Compounded Annually. Suppose that \$40,000 is invested at 7% interest, compounded annually.

a) After what amount of time will there be \$250,000 in the account?

b) Find the doubling time.

Answers on page A-73

6. Population Growth of the United States. What will the population of the United States be in 2008? in 2025?

EXPONENTIAL GROWTH MODEL

An **exponential growth model** is a function of the form

$$P(t) = P_0e^{kt}, \quad k > 0,$$

where P_0 is the population at time 0, $P(t)$ is the population at time t, and k is the **exponential growth rate** for the situation. The **doubling time** is the amount of time necessary for the population to double in size.

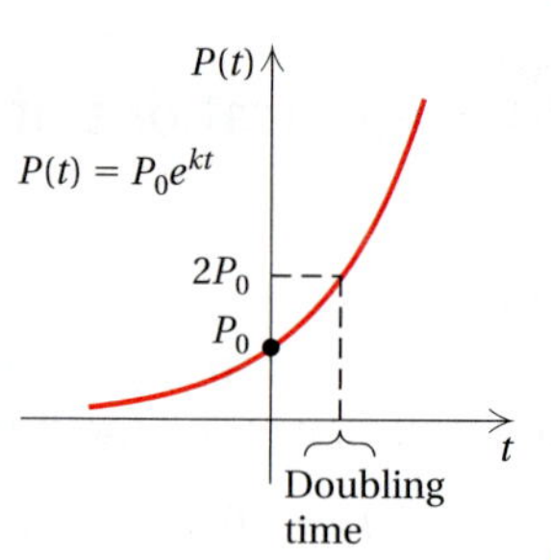

The exponential growth rate is the rate of growth of a population at any *instant* in time. Since the population is continually growing, the percent of total growth after one year will exceed the exponential growth rate.

EXAMPLE 4 *Population Growth of the United States.* In 2000, the population of the United States was 281 million, and the exponential growth rate was 0.8% per year.
Source: U.S. Bureau of the Census

a) Find the exponential growth function.

b) What will the population be in 2012?

a) We are trying to find a model. The given information allows us to create one. At $t = 0$ (2000), the population was 281 million. We substitute 281 for P_0 and 0.8%, or 0.008, for k to obtain the exponential growth function:

$$P(t) = P_0e^{kt}$$
$$P(t) = 281e^{0.008t}.$$

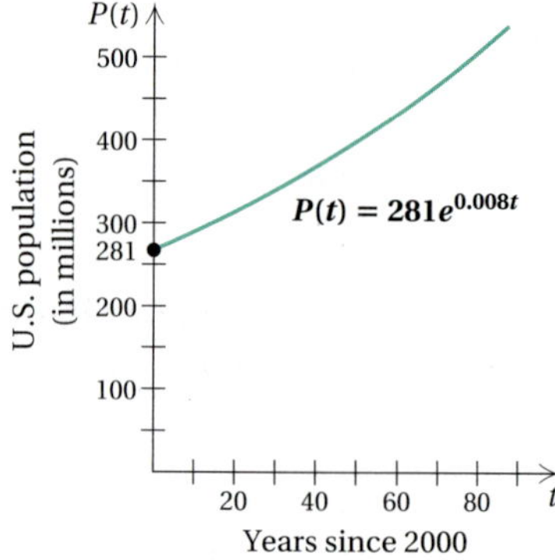

b) In 2012, we have $t = 12$. That is, 12 yr have passed since 2000. To find the population in 2012, we substitute 12 for t:

$$P(12) = 281e^{0.008(12)} \quad \text{Substituting 12 for } t$$
$$\approx 309 \text{ million.} \quad \text{Using a calculator}$$

The population of the United States will be about 309 million in 2012.

Do Exercise 6.

EXAMPLE 5 *Interest Compounded Continuously.* Suppose that an amount of money P_0 is invested in a savings account at interest rate k, compounded continuously. That is, suppose that interest is computed every "instant" and added to the amount in the account. The balance $P(t)$, after t years, is given by the exponential growth model

$$P(t) = P_0e^{kt}.$$

a) Suppose that \$30,000 is invested and grows to \$44,754.75 in 5 yr. Find the interest rate and then the exponential growth function.

b) What is the balance after 10 yr?

c) What is the doubling time?

Answer on page A-73

a) We have $P(0) = 30{,}000$. Thus the exponential growth function is

$$P(t) = 30{,}000e^{kt}, \quad \text{where } k \text{ must still be determined.}$$

We know that $P(5) = 44{,}754.75$. We substitute and solve for k:

$44{,}754.75 = 30{,}000e^{k(5)} = 30{,}000e^{5k}$

$\frac{44{,}754.75}{30{,}000} = e^{5k}$ Dividing by 30,000

$1.491825 = e^{5k}$

$\ln 1.491825 = \ln e^{5k}$ Taking the natural logarithm on both sides

$0.4 \approx 5k$ Finding ln 1.491825 on a calculator and simplifying $\ln e^{5k}$

$\frac{0.4}{5} = 0.08 \approx k.$

The interest rate is about 0.08, or 8%, compounded continuously. Note that since interest is being compounded continuously, the interest earned each year is more than 8%. The exponential growth function is

$$P(t) = 30{,}000e^{0.08t}.$$

b) We substitute 10 for t:

$$P(10) = 30{,}000e^{0.08(10)} = 66{,}766.23.$$

The balance in the account after 10 yr will be $66,766.23.

c) To find the doubling time T, we replace $P(t)$ with 60,000 and solve for T:

$60{,}000 = 30{,}000e^{0.08T}$

$2 = e^{0.08T}$ Dividing by 30,000

$\ln 2 = \ln e^{0.08T}$ Taking the natural logarithm on both sides

$\ln 2 = 0.08T$

$\frac{\ln 2}{0.08} = T$ Dividing

$8.7 \approx T.$

Thus the original investment of $30,000 will double in about 8.7 yr, as shown in the following graph of the growth function.

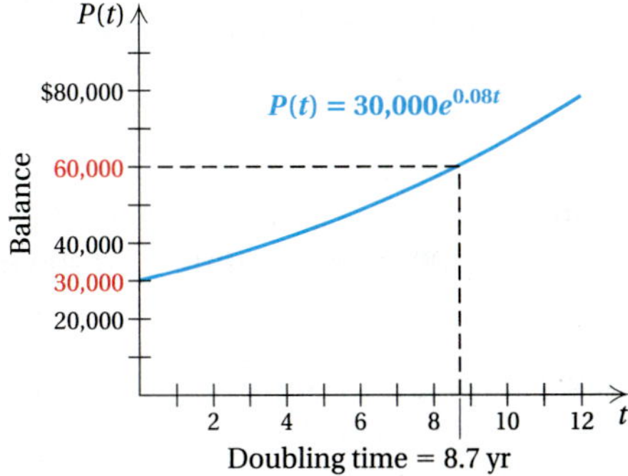

Do Exercise 7.

Compare the results of Examples 3(b) and 5(c). Note that continuous compounding gives a higher yield and shorter doubling time. Greater understanding about continuous compounding belongs to a course in business calculus.

7. Interest Compounded Continuously.

a) Suppose that $5000 is invested and grows to $6356.25 in 4 yr. Find the interest rate and then the exponential growth function.

b) What is the balance after 1 yr? 2 yr? 10 yr?

c) What is the doubling time?

Answers on page A-73

Take a look at the graphs in Example 6 and in Margin Exercise 8. Note that the data tend to be rising in a manner that might fit an exponential function. To fit an exponential function to the data, we choose a pair of data points and use them to determine P_0 and k.

EXAMPLE 6 *1929 Pierce-Arrow.* The restoration of antique automobiles is a popular hobby among many people. The value of a 1929 Pierce-Arrow Roadster was $4800 when it was sold brand-new. Fully restored, the value of this automobile in 2001 was $150,000.

Source: Dr. Rex B. Gosnell, www.pierce-arrow.org/specs.html

Dr. and Mrs. Rex B. Gosnell (the uncle and aunt of Marvin Bittinger, one of the authors of this textbook) donated this 1929 Pierce-Arrow Roadster to the Gilmore-Classic Car Club of America Museum in Hickory Corners, Michigan.

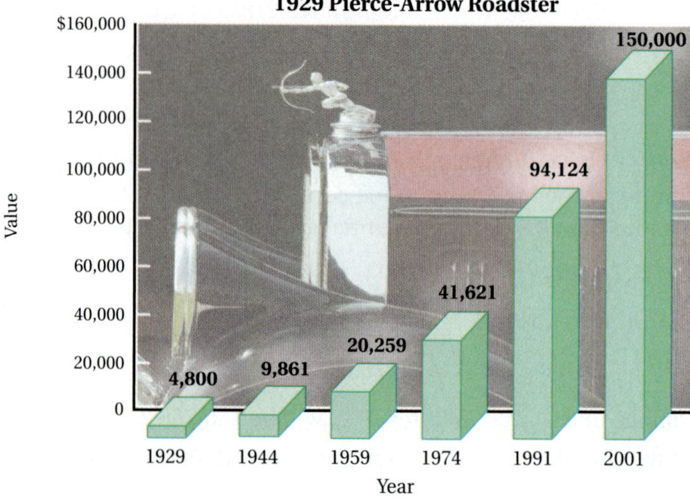

a) If we let $t = 0$ correspond to 1929 and $t = 72$ to 2001, then t is the number of years since 1929. Use the data points $(0, 4800)$ and $(72, 150{,}000)$ to find the exponential growth rate and fit an exponential growth function to the data.

b) Use the function found in part(a) to predict the value of the automobile in 2019.

c) Assuming exponential growth continues, predict the year in which the value of such an automobile will be $200,000.

a) We use the equation $P(t) = P_0e^{kt}$, where $P(t)$ is the value of the automobile, in dollars, t years after 1929. In 1929, at $t = 0$, the value was $4800. Thus we substitute 4800 for P_0:

$$P(t) = 4800e^{kt}.$$

To find the exponential growth rate k, we note that 72 years later, at the end of 2001, the value was $150,000. We substitute and solve for k:

$$\left.\begin{aligned} P(72) &= 4800e^{k(72)} \\ 150{,}000 &= 4800e^{k(72)} \end{aligned}\right\} \quad \text{Substituting}$$

$$\begin{aligned} 31.25 &= e^{72k} && \text{Dividing by 4800} \\ \ln 31.25 &= \ln e^{72k} && \text{Taking the natural logarithm on both sides} \\ 3.4420 &\approx 72k && \ln e^a = a \\ 0.048 &\approx k. && \text{Dividing by 72} \end{aligned}$$

The exponential growth rate was 0.048, or 4.8%, and the exponential growth function is $P(t) = 4800e^{0.048t}$.

b) Since 2019 is 90 yr from 1929, we substitute 90 for t:

$$P(90) = 4800e^{0.048(90)} \approx \$360{,}905.$$

The value of the automobile will be about \$360,905 in 2019.

c) To predict when the value of the automobile will be \$200,000, we substitute 200,000 for $P(t)$ and solve for t:

$$200{,}000 = 4800e^{0.048t}$$
$$41.6667 \approx e^{0.048t} \qquad \text{Dividing by 4800}$$
$$\ln 41.6667 \approx \ln e^{0.048t} \qquad \text{Taking the natural logarithm on both sides}$$
$$3.7297 \approx 0.048t \qquad \ln e^a = a$$
$$77.7 \approx t. \qquad \text{Dividing by 0.048}$$

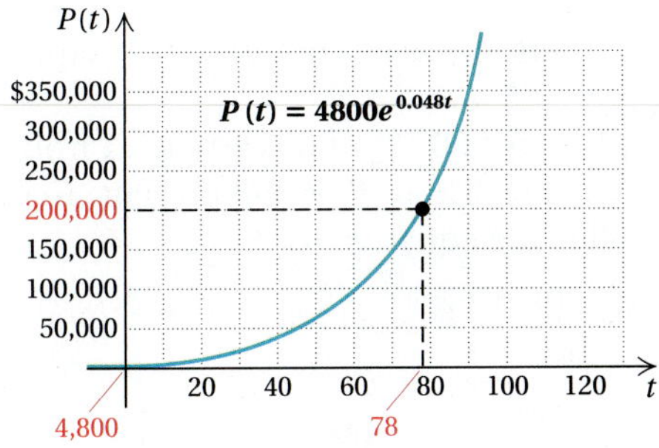

Rounding to 78 yr, we see that, according to this model, by the end of 1929 + 78, or 2007, the value of a 1929 Pierce-Arrow Roadster will be \$200,000.

Do Exercise 8.

In some real-life situations, a quantity or population is *decreasing* or *decaying* exponentially.

EXPONENTIAL DECAY MODEL

An **exponential decay model** is a function of the form

$$P(t) = P_0e^{-kt}, \quad k > 0,$$

where P_0 is the quantity present at time 0, $P(t)$ is the amount present at time t, and k is the **decay rate.** The **half-life** is the amount of time necessary for half of the quantity to decay.

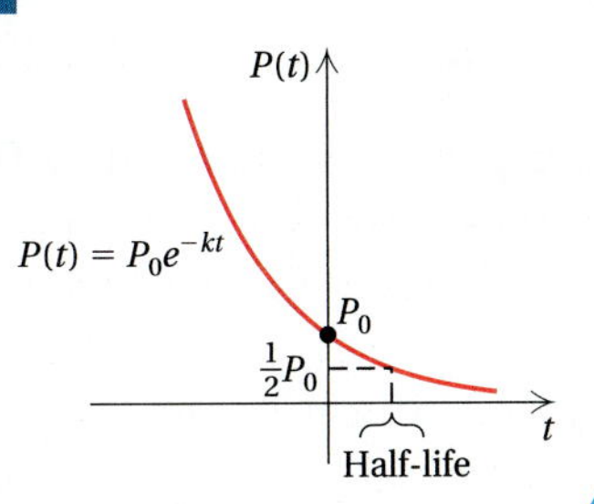

EXAMPLE 7 *Carbon Dating.* The radioactive element carbon-14 has a half-life of 5750 yr. The percentage of carbon-14 present in the remains of organic matter can be used to determine the age of that organic matter. Recently, while digging in Chaco Canyon, New Mexico, archaeologists found corn pollen that had lost 38.1% of its carbon-14. The age of this corn pollen was evidence that Native Americans had been cultivating crops in the Southwest centuries earlier than scientists had thought. What was the age of the pollen?
Source: *American Anthropologist*

8. Online Service Fees. The fees paid by small businesses for online services are projected to increase exponentially over the next few years, as shown in the graph below.

Online Fees for Small Business

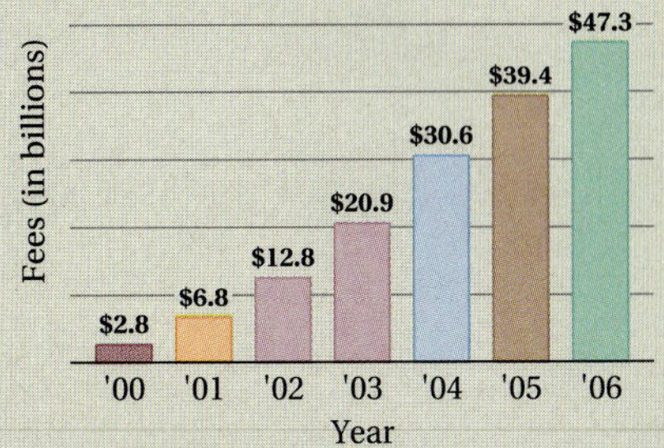

Source: Keenan Vision

a) Let $P(t) = P_0e^{kt}$, where $P(t)$ is the online fees, in billions of dollars, t years after 2000. Then $t = 0$ corresponds to 2000 and $t = 6$ corresponds to 2006. Use the data points $(0, 2.8)$ and $(6, 47.3)$ to find the exponential growth rate and fit an exponential growth function to the data.

b) Use the function found in part (a) to predict the online fees that will be paid in 2020.

c) Assuming exponential growth continues, predict the year in which the online fees will be \$100 billion.

Answers on page A-73

9. Carbon Dating. How old is an animal bone that has lost 30% of its carbon-14?

Answer on page A-73

We first find k. To do so, we use the concept of half-life. When $t = 5750$ (the half-life), $P(t)$ will be half of P_0. Then

$$0.5P_0 = P_0 e^{-k(5750)}$$

$$0.5 = e^{-5750k} \quad \text{Dividing by } P_0$$

$$\ln 0.5 = \ln e^{-5750k} \quad \text{Taking the natural logarithm on both sides}$$

$$\ln 0.5 = -5750k$$

$$\frac{\ln 0.5}{-5750} = k$$

$$0.00012 \approx k.$$

Now we have a function for the decay of carbon-14:

$$P(t) = P_0 e^{-0.00012t}. \quad \text{This completes the first part of our solution.}$$

(*Note*: This equation can be used for any subsequent carbon-dating problem.)

If the corn pollen has lost 38.1% of its carbon-14 from an initial amount P_0, then 100% − 38.1%, or 61.9%, of P_0 is still present. To find the age t of the pollen, we solve the following equation for t:

$$61.9\%P_0 = P_0 e^{-0.00012t} \quad \text{We want to find } t \text{ for which } P(t) = 0.619P_0.$$

$$0.619 = e^{-0.00012t} \quad \text{Dividing by } P_0$$

$$\ln 0.619 = \ln e^{-0.00012t} \quad \text{Taking the natural logarithm on both sides}$$

$$-0.4797 = -0.00012t \quad \ln e^a = a$$

$$\frac{-0.4797}{-0.00012} \approx t \quad \text{Dividing by } -0.00012$$

$$4000 \approx t. \quad \text{Rounding}$$

The pollen is about 4000 yr old.

Do Exercise 9.

CALCULATOR CORNER

Mathematical Modeling: Fitting an Exponential Function to Data We can use the exponential regression feature on a graphing calculator to fit an exponential function of the form $y = a \cdot b^x$ to a set of data. When we use the regression feature to find a mathematical model, we can use all the data points rather than only two points, as we did in Example 6.

The table at right lists the data from the graph in Margin Exercise 8 regarding fees paid by small businesses for online services.

YEARS SINCE 2000	FEES (in billions)
0	$ 2.8
1	6.8
2	12.8
3	20.9
4	30.6
5	39.4
6	47.3

Source: Keenan Vision

(*continued*)

a) Use the exponential regression feature on a graphing calculator to fit an exponential function to the data.

b) Make a scatterplot of the data. Then graph the function found in part (a) with the scatterplot.

c) Use the function found in part (a) to predict the fees that small businesses will pay for online services in 2020. Compare your answer with the one found in part (b) of Margin Exercise 8.

a) We enter the data on the STAT list editor screen as described in the Calculator Corner on p. 972. Then, to fit an exponential function to the data, we press [STAT] [▷] [0] [VARS] [▷] [1] [1] [ENTER]. The first three keystrokes display the STAT CALC menu and select EXPREG from that menu. The keystrokes [VARS] [▷] [1] [1] copy the regression equation to the equation-editor screen as y_1. We see that the regression equation is $y = 4.077027434 \cdot 1.583277276^x$.

L1	L2	L3 3
0	2.8	
1	6.8	
2	12.8	
3	20.9	
4	30.6	
5	39.4	
6	47.3	
L3(1)=		

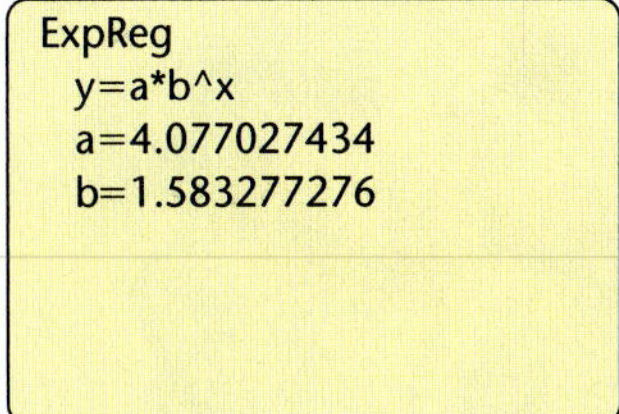

b) To make a scatterplot of the data and graph the regression equation along with the scatterplot, we first turn on Plot 1 as described in the Calculator Corner on pp. 972–974. Then we press [ZOOM] [9] to see the graph.

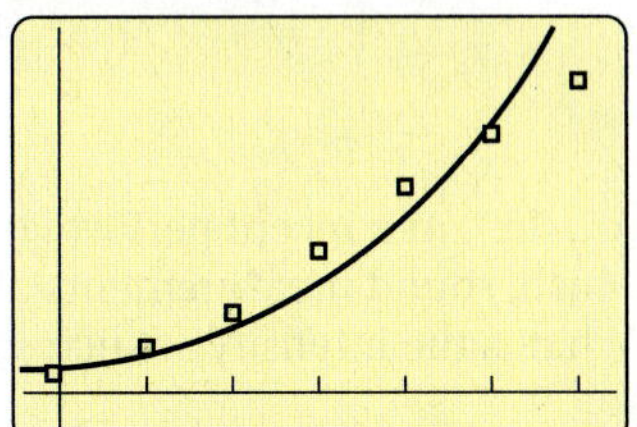

c) We can use any of the methods described in the Calculator Corner on p. 924 to predict the fees that will be paid by small businesses for online services in 2020. We will use a table set in ASK mode. The year 2020 is 20 years after the year 2000, so we enter 20 for x and find that $y \approx 39{,}947$. Thus we predict that in 2020 small businesses will pay about \$39,947 billion, or \$39.947 trillion, for online services. The prediction found with exponential regression is approximately \$5400 higher than the prediction found algebraically in Margin Exercise 8(b).

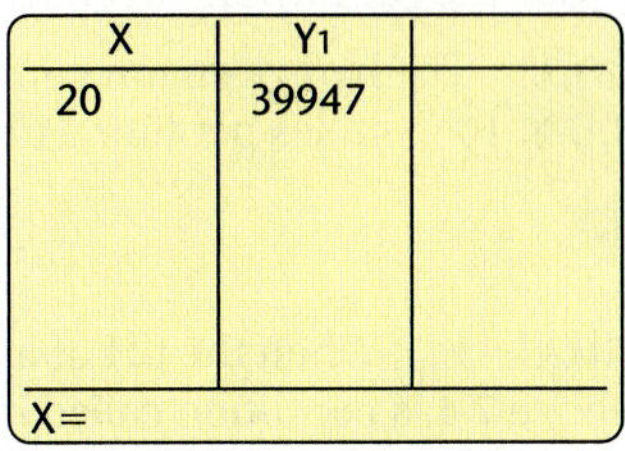

Exercises: As the following table shows, the number of Hispanic-owned businesses in the United States has soared in recent years.

YEARS SINCE 1982	NUMBER OF BUSINESSES
0	284,011
5	489,973
10	862,605
15	1,121,387
19	1,500,000

Source: U.S. Small Business Administration

1. Use the exponential regression feature on a graphing calculator to fit an exponential function to the data.
2. Make a scatterplot of the data. Then graph the function found in part (a) with the scatterplot.
3. Use the function found in part (a) to predict the number of Hispanic-owned businesses in 2010.

17.7 EXERCISE SET

For Extra Help

Digital Video Tutor CD 15 Videotape 20 | InterAct Math | Math Tutor Center | MathXL | MyMathLab

a Solve.

Sound Levels. Use the decibel formula from Example 1 for Exercises 1–4.

1. *Acoustics.* The intensity of sound of a riveter at work is about $3.2 \times 10^{-3}\ \text{W/m}^2$. How high is this sound level in decibels?

2. *Acoustics.* The intensity of sound of a dishwasher is about $2.5 \times 10^{-6}\ \text{W/m}^2$. How high is this sound level in decibels?

3. *Music.* At a recent performance of the rock group Phish, sound measurements of 105 dB were recorded. What is the intensity of such sounds?

4. *Jet Plane at Takeoff.* A jet plane at takeoff can generate sound measurements of 130 dB. What is the intensity of such sounds?

pH. Use the pH formula from Example 2 for Exercises 5–8.

5. *Milk.* The hydrogen ion concentration of milk is about 1.6×10^{-7} moles per liter. Find the pH.

6. *Mouthwash.* The hydrogen ion concentration of mouthwash is about 6.3×10^{-7} moles per liter. Find the pH.

7. *Alkalosis.* When the pH of a patient's blood rises above 7.4, a condition called *alkalosis* sets in. Alkalosis can be fatal at a pH level above 7.8. What would the hydrogen ion concentration of the patient's blood be at that point?

8. *Orange Juice.* The pH of orange juice is 3.2. What is its hydrogen ion concentration?

Walking Speed. In a study by psychologists Bornstein and Bornstein, it was found that the average walking speed w, in feet per second, of a person living in a city of population P, in thousands, is given by the function

$$w(P) = 0.37 \ln P + 0.05.$$

In Exercises 9–12, various cities and their populations are given. Find the walking speed of people in each city in 2000.
Source: *International Journal of Psychology*

9. Phoenix, Arizona: 3,251,876 (Metropolitan area)

10. Las Vegas, Nevada: 1,563,282 (Metropolitan area)

11. Fayetteville–Springdale–Rogers, Arkansas: 311,121
Source: U.S. Bureau of the Census, Census 2000

12. Naples, Florida: 251,377

b Solve.

13. *Pay It Forward.* The movie *Pay It Forward*, starring Kevin Spacey, Helen Hunt, and Haley Joel Osment, suggests an application of exponential functions. In the film, Osment's character does an act of kindness to each of three people. In return, each of these people must agree to "pay forward" three acts of kindness to other people, and so on. Let's assume that each person performs these acts of kindness within 1 month. Then the number N of people who receive the acts of kindness in month t is given by the function

$$N(t) = 3^t.$$

a) How many people would have received an act of kindness in month 5?
b) The population of the world is about 6.2 billion. After what amount of time will the acts of kindness reach the entire world?
c) What is the doubling time for the number of people who receive an act of kindness?

14. *Spread of Rumor.* The number of people who hear a rumor increases exponentially. If 20 people start a rumor and if each person who hears the rumor repeats it to two people a day, the number of people N who have heard the rumor after t days is given by the function

$$N(t) = 20(3)^t.$$

a) How many people have heard the rumor after 5 days?
b) After what amount of time will 1000 people have heard the rumor?
c) What is the doubling time for the number of people who have heard the rumor?

15. *Projected College Costs.* In the new millennium, the cost of tuition, books, room, and board at a state university is projected to follow the exponential function

$$C(t) = 11{,}054(1.06)^t,$$

where C is the cost, in dollars, and t is the number of years after 2000.

Source: College Board, Senate Labor Committee

a) Find the college costs in 2005.
b) In what year will the cost be $21,000?
c) What is the doubling time of the costs?

16. *Salvage Value.* A color photocopier is purchased for $5200. Its value each year is about 70% of its value in the preceding year. Its value in dollars after t years is given by the exponential function

$$V(t) = 5200(0.7)^t.$$

a) Find the salvage value of the copier after 3 yr.
b) After what amount of time will the salvage value be $1200?
c) After what amount of time will the salvage value be half the original value?

Growth. Use the exponential growth model $P(t) = P_0e^{kt}$ for Exercises 17–22.

17. *World Population Growth.* In 1998, the population of the world reached 6 billion, and the exponential growth rate was 1.5% per year.

a) Find the exponential growth function.
b) What will the population be in 2010?
c) In what year will the population be 10 billion?
d) What is the doubling time?

18. *Population Growth of Laredo, Texas.* In 2000, with a population of 193,117, Laredo, Texas, was the ninth fastest-growing metropolitan area in the United States, with an exponential growth rate of 3.7% per year.

Source: U.S. Bureau of the Census

a) Find the exponential growth function.
b) What will the population be in 2010?
c) In what year will the population be 600,000?
d) What is the doubling time?

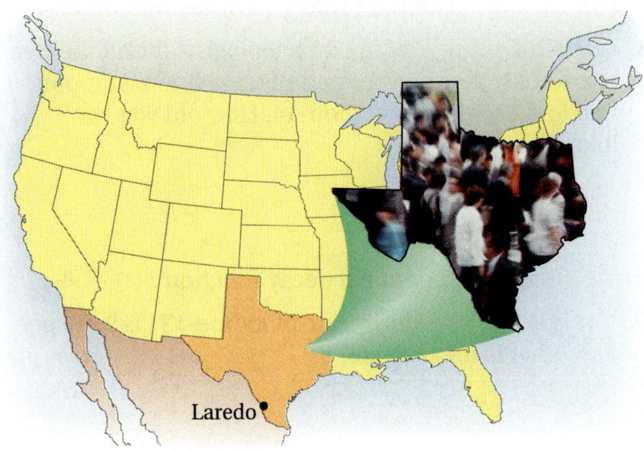

19. *Interest Compounded Continuously.* Suppose that P_0 is invested in a savings account in which interest is compounded continuously at 6% per year.

a) Express $P(t)$ in terms of P_0 and 0.06.
b) Suppose that \$5000 is invested. What is the balance after 1 yr? 2 yr? 10 yr?
c) When will the investment of \$5000 double itself?

20. *Interest Compounded Continuously.* Suppose that P_0 is invested in a savings account in which interest is compounded continuously at 5.4% per year.

a) Express $P(t)$ in terms of P_0 and 0.054.
b) Suppose that \$10,000 is invested. What is the balance after 1 yr? 2 yr? 10 yr?
c) When will the investment of \$10,000 double itself?

21. *Population Growth of Las Vegas, Nevada.* In 2000, the population of Las Vegas, Nevada, was 1,563,282. It had grown from a population of 852,737 in 1990, and was the fastest growing metropolitan area in the United States. Assume that the population increases according to an exponential growth function.
Source: U.S. Bureau of the Census

a) Let $t = 0$ correspond to 1990 and $t = 10$ correspond to 2000. Then t is the number of years since 1990. Use the data points (0, 852,737) and (10, 1,563,282) to find the exponential growth rate and fit an exponential growth function $P(t) = P_0e^{kt}$ to the data, where $P(t)$ is the population of Las Vegas t years after 1990.
b) Use the function found in part (a) to predict the population of Las Vegas in 2010.
c) When will the population reach 8 million?

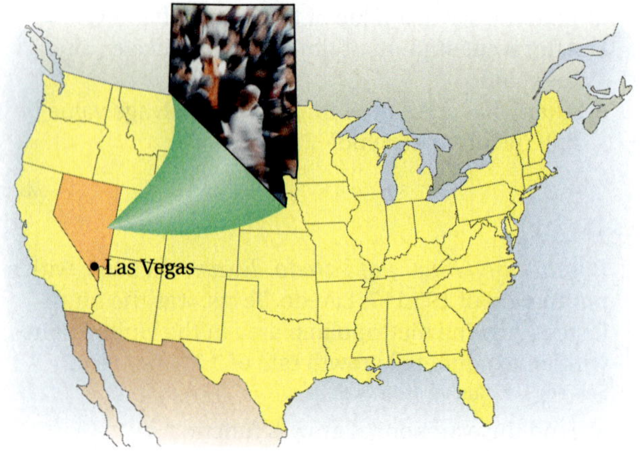

22. *First-Class Postage.* First-class postage (for the first ounce) was 20¢ in 1981 and 37¢ in 2002. Assume the cost increases according to an exponential growth function.
Source: U.S. Postal Service

a) Let $t = 0$ correspond to 1981 and $t = 21$ correspond to 2002. Then t is the number of years since 1981. Use the data points $(0, 20)$ and $(21, 37)$ to find the exponential growth rate and fit an exponential growth function $P(t) = P_0e^{kt}$ to the data, where $P(t)$ is the cost of first-class postage, in cents, t years after 1981.
b) Use the function found in part (a) to predict the cost of first-class postage in 2008.
c) When will the cost of first-class postage be \$1.00 or 100¢?

Carbon Dating. Use the carbon-14 decay function $P(t) = P_0e^{-0.00012t}$ for Exercises 23 and 24.

23. *Carbon Dating.* When archaeologists found the Dead Sea scrolls, they determined that the linen wrapping had lost 22.3% of its carbon-14. How old was the linen wrapping?

24. *Carbon Dating.* In 1998, researchers found an ivory tusk that had lost 18% of its carbon-14. How old was the tusk?

Decay. Use the exponential decay function $P(t) = P_0e^{-kt}$ for Exercises 25 and 26.

25. *Chemistry.* The decay rate of iodine-131 is 9.6% per day. What is the half-life?

26. *Chemistry.* The decay rate of krypton-85 is 6.3% per day. What is the half-life?

27. *Home Construction.* The chemical urea formaldehyde was found in some insulation used in houses built during the mid to late 1960s. Unknown at the time was the fact that urea formaldehyde emitted toxic fumes as it decayed. The half-life of urea formaldehyde is 1 yr. What is its decay rate?

28. *Plumbing.* Lead pipes and solder are often found in older buildings. Unfortunately, as lead decays, toxic chemicals can get in the water resting in the pipes. The half-life of lead is 22 yr. What is its decay rate?

29. *Decline of Discarded Yard Waste.* The amount of discarded yard waste has declined considerably in recent years because of increased recycling and composting. In 1996, 17.5 million tons were discarded, but by 1998 the figure dropped to 14.5 million tons. Assume the amount of discarded yard waste is decreasing according to the exponential decay model.
Source: *Statistical Abstract of the United States*, 2000
 a) Find the value k, and write an exponential function that describes the amount of yard waste discarded t years after 1996.
 b) Estimate the amount of discarded yard waste in 2010.
 c) In what year (theoretically) will only 1 ton of yard waste be discarded?

30. *Decline in Cases of Mumps.* The number of cases of mumps has dropped exponentially from 5300 in 1990 to 800 in 1996.
Source: *Statistical Abstract of the United States*, 2000
 a) Find the value k, and write an exponential function that can be used to estimate the number of cases t years after 1990.
 b) Estimate the number of cases of mumps in 2008.
 c) In what year (theoretically) will there be only 1 case of mumps?

31. *Value of a Sports Card.* Because he objected to smoking, and because his first baseball card was issued in cigarette packs, the great shortstop Honus Wagner halted production of his card before many were produced. One of these cards was sold in 1991 for \$451,000 and again in 1996 for \$640,500. For the following questions, assume that the card's value increases exponentially.

 a) Find an exponential function $V(t)$, where t is the number of years after 1991, if $V_0 = 451{,}000$.
 b) Predict the card's value in 2005.
 c) What is the doubling time of the value of the card?
 d) In what year will the value of the card first exceed \$1,000,000?
 e) In 2001, this card drew a bid of \$1.1 million on eBay. How does this correlate with what can be predicted from this function? Would you buy the card?

32. *Portrait of Dr. Gachet.* As of May 2001, the most ever paid for a painting is \$82.5 million, paid in 1990 for Vincent Van Gogh's *Portrait of Dr. Gachet*. The same painting sold for \$58 million in 1987. Assume that the growth in the value V of the painting is exponential.

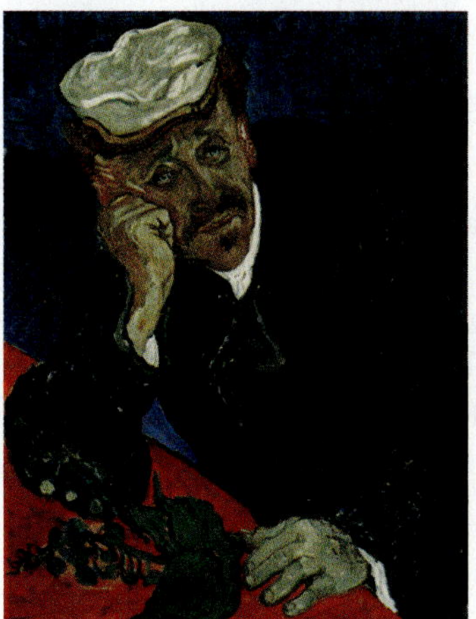

Van Gogh's *Portrait of Dr. Gachet*, oil on canvas.

 a) Find the exponential growth rate k, and determine the exponential growth function V, for which $V(t)$ is the painting's value, in millions of dollars, t years after 1987.
 b) Estimate the value of the painting in 2007.
 c) What is the doubling time of the value of the painting?
 d) How long after 1987 will the value of the painting be \$1 billion?

33. *Population Decay of Pittsburgh.* The population of the metropolitan area of Pittsburgh declined from 2,394,811 in 1990 to 2,242,798 in 2000. Assume the population decreases according to an exponential decay function.
Source: U.S. Bureau of the Census

a) Let $t = 0$ correspond to 1990 and $t = 10$ correspond to 2000. Then t is the number of years since 1990. Use the data points $(0, 2{,}394{,}811)$ and $(10, 2{,}242{,}798)$ to find the exponential decay rate and fit an exponential decay function $P(t) = P_0e^{-kt}$ to the data, where $P(t)$ is the population of Pittsburgh t years after 1990.
b) Use the function found in part (a) to predict the population of Pittsburgh in 2010.
c) When will the population decline to 1 million?

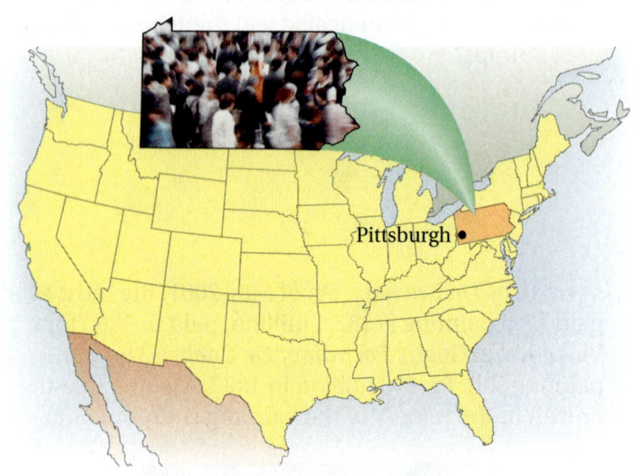

34. *Skateboarding.* The number of skateboarders of age x (with $x \geq 16$), in thousands, can be approximated by

$$N(x) = 600(0.873)^{x-16}.$$

Sources: Based on figures from the U.S. Bureau of the Census and *Statistical Abstract of the United States*, 2000

a) Estimate the number of 41-yr-old skateboarders.
b) At what age are there only 2000 skateboarders? 200 skateboarders?

35. D$_W$ Write a problem for a classmate to solve in which data that seem to fit an exponential growth function are provided. Try to find data in a newspaper to make the problem as realistic as possible.

36. D$_W$ Do some research or consult with a chemist to determine how carbon dating is carried out. Write a report.

SKILL MAINTENANCE

Compute and simplify. Express answers in the form $a + bi$, where $i^2 = -1$. [15.8c, d, e]

37. i^{46} **38.** i^{48} **39.** i^{53} **40.** i^{97} **41.** $i^{14} + i^{15}$

42. $i^{18} - i^{16}$ **43.** $\dfrac{8 - i}{8 + i}$ **44.** $\dfrac{2 + 3i}{5 - 4i}$ **45.** $(5 - 4i)(5 + 4i)$ **46.** $(-10 - 3i)^2$

SYNTHESIS

Use a graphing calculator to solve each of the following equations.

47. $2^x = x^{10}$ **48.** $(\ln 2)x = 10 \ln x$ **49.** $x^2 = 2^x$ **50.** $x^3 = e^x$

51. *Sports Salaries.* In 2001, Derek Jeter of the New York Yankees signed a $189 million 10-yr contract that will pay him $21 million in 2010. How much would Yankee owner George Steinbrenner need to invest in 2001 at 5% interest compounded continuously, in order to have the $21 million for Jeter in 2010?

52. *Nuclear Energy.* Plutonium-239 (Pu-239) is used in nuclear energy plants. The half-life of Pu-239 is 24,360 yr. How long will it take for a fuel rod of Pu-239 to lose 90% of its radioactivity?
Source: *Microsoft Encarta 97 Encyclopedia*

CHAPTER 17

Summary and Review

The review that follows is meant to prepare you for a chapter exam. It consists of two parts. The first part is a checklist of some of the Study Tips referred to in this and preceding chapters, as well as a list of important properties and formulas. The second part is the Review Exercises. These provide practice exercises for the exam, together with references to section objectives so you can go back and review. Before beginning, stop and look back over the skills you have obtained. What skills in mathematics do you have now that you did not have before studying this chapter?

STUDY TIPS CHECKLIST

The foundation of all your study skills is TIME!

- ☐ As you work the applied problems and take a test, are you careful to state the answers with appropriate units?
- ☐ Are you beginning a systematic preparation for the final exam?
- ☐ Are you using MyMathLab as preparation for the chapter test and the final exam?
- ☐ Have you used the videotapes to supplement your learning?
- ☐ Are you stopping to work the margin exercises when directed to do so?

IMPORTANT PROPERTIES AND FORMULAS

Exponential Functions: $f(x) = a^x$, $f(x) = e^x$

Composition of Functions: $f \circ g(x) = f(g(x))$

Definition of Logarithms: $y = \log_a x$ is that number y such that $x = a^y$, where $x > 0$ and a is a positive constant other than 1.

Properties of Logarithms:

$\log M = \log_{10} M$, $\log_a 1 = 0$, $\ln M = \log_e M$, $\log_a a = 1$,

$\log_a MN = \log_a M + \log_a N$, $\log_a a^k = k$, $\log_a M^k = k \cdot \log_a M$, $\log_b M = \dfrac{\log_a M}{\log_a b}$,

$\log_a \dfrac{M}{N} = \log_a M - \log_a N$, $e \approx 2.7182818284\ldots$

Growth: $P(t) = P_0 e^{kt}$

Decay: $P(t) = P_0 e^{-kt}$

Carbon Dating: $P(t) = P_0 e^{-0.00012t}$

Interest Compounded Annually: $A = P(1 + r)^t$

Interest Compounded n Times per Year: $A = P\left(1 + \dfrac{r}{n}\right)^{nt}$

Interest Compounded Continuously: $P(t) = P_0 e^{kt}$, where P_0 dollars are invested for t years at interest rate k

REVIEW EXERCISES

1. Find the inverse of the relation [17.2a]

$$\{(-4, 2), (5, -7), (-1, -2), (10, 11)\}.$$

Determine whether the function is one-to-one. If it is, find a formula for its inverse. [17.2b, c]

2. $f(x) = 4 - x^2$

3. $g(x) = \dfrac{2x - 3}{7}$

4. $f(x) = 8x^3$

5. $f(x) = \dfrac{4}{3 - 2x}$

6. Graph the function $f(x) = x^3 + 1$ and its inverse using the same set of axes. [17.2c]

Graph.

7. $f(x) = 3^{x-1}$ [17.1a]

x	$f(x)$
0	
1	
2	
3	
−1	
−2	
−3	

8. $f(x) = \log_3 x$, or $y = \log_3 x$ [17.3a]

$y = \log_3 x \rightarrow x =$ __________

x, or 3^y	y
	0
	1
	2
	3
	−1
	−2
	−3

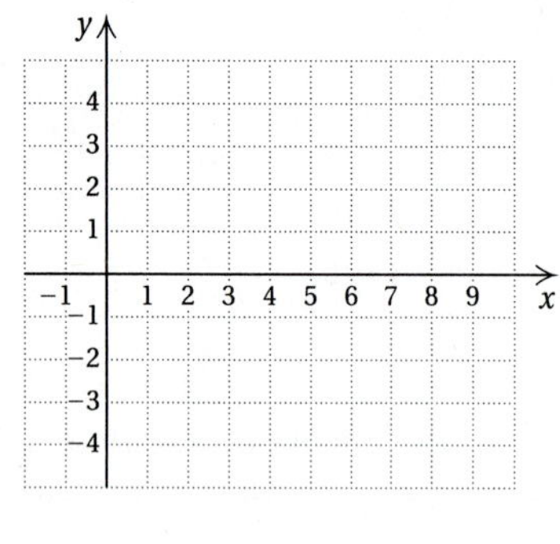

9. $f(x) = e^{x+1}$ [17.5c]

x	$f(x)$
0	
1	
2	
3	
−1	
−2	
−3	

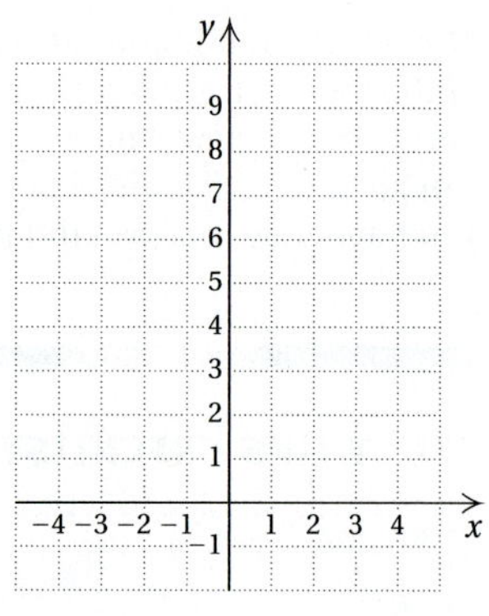

10. $f(x) = \ln(x - 1)$ [17.5c]

x	$f(x)$

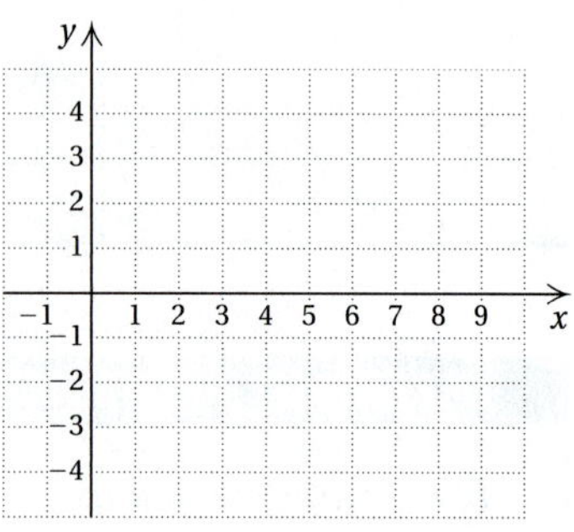

11. Find $f \circ g(x)$ and $g \circ f(x)$ if $f(x) = x^2$ and $g(x) = 3x - 5$. [17.2d]

12. If $h(x) = \sqrt{4 - 7x}$, find $f(x)$ and $g(x)$ such that $h(x) = f \circ g(x)$. [17.2d]

Convert to a logarithmic equation. [17.3b]

13. $10^4 = 10{,}000$

14. $25^{1/2} = 5$

Convert to an exponential equation. [17.3b]

15. $\log_4 16 = x$

16. $\log_{1/2} 8 = -3$

Find each of the following. [17.3c]

17. $\log_3 9$

18. $\log_{10} \frac{1}{10}$

19. $\log_m m$

20. $\log_m 1$

Find the common logarithm, to four decimal places, using a calculator. [17.3d]

21. $\log\left(\frac{78}{43{,}112}\right)$

22. $\log(-4)$

Express in terms of logarithms of x, y, and z. [17.4d]

23. $\log_a x^4y^2z^3$

24. $\log\sqrt[4]{\frac{z^2}{x^3y}}$

Express as a single logarithm. [17.4d]

25. $\log_a 8 + \log_a 15$

26. $\frac{1}{2}\log a - \log b - 2\log c$

Simplify. [17.4e]

27. $\log_m m^{17}$

28. $\log_m m^{-7}$

Given $\log_a 2 = 1.8301$ and $\log_a 7 = 5.0999$, find each of the following. [17.4d]

29. $\log_a 28$

30. $\log_a 3.5$

31. $\log_a \sqrt{7}$

32. $\log_a \frac{1}{4}$

Find each of the following, to four decimal places, using a calculator. [17.5a]

33. $\ln 0.06774$

34. $e^{-0.98}$

35. $e^{2.91}$

36. $\ln 1$

37. $\ln 0$

38. $\ln e$

Find the logarithm using the change-of-base formula. [17.5b]

39. $\log_5 2$

40. $\log_{12} 70$

Solve. Where appropriate, give approximations to four decimal places. [17.6a, b]

41. $\log_3 x = -2$

42. $\log_x 32 = 5$

43. $\log x = -4$

44. $3\ln x = -6$

45. $4^{2x-5} = 16$

46. $2^{x^2} \cdot 2^{4x} = 32$

47. $4^x = 8.3$

48. $e^{-0.1t} = 0.03$

49. $\log_4 16 = x$

50. $\log_4 x + \log_4 (x - 6) = 2$

51. $\log x + \log(x - 15) = 2$

52. $\log_3 (x - 4) = 3 - \log_3 (x + 4)$

Solve. [17.7a, b]

53. *Acoustics.* The intensity of the sound at the foot of Niagara Falls is about 10^{-3} W/m^2. How high is this sound level in decibels? [Use $L = 10 \cdot \log(I/I_0)$.]

54. *Major-League Baseball Salaries.* The salaries of major-league baseball players have been rising exponentially over recent years. The total amount S, in millions of dollars, paid out t years after 1997 can be approximated by the exponential function

$$S(t) = 37.18(1.15)^t.$$

Sources: Associated Press; Major League Baseball Players Association

a) Predict the amount of salaries paid in 2005, in 2008, and in 2010.
b) In what year will the amount of salaries paid reach $1 billion?
c) What is the doubling time of the amount of salaries?

55. *Aeronautics.* There were 6 commercial space launches in 1994 and 12 in 1996. Assume that the number N is growing exponentially according to the function

$$N(t) = N_0e^{kt}.$$

Sources: Federal Aviation Administration; U.S. Department of Transportation

a) Find k and write the exponential growth function.
b) Predict the number of launches in 2003.
c) When will there be 192 annual scheduled launches?

56. The population of Riverton doubled in 16 yr. Find the exponential growth rate.

57. How long will it take $7600 to double itself if it is invested at 8.4%, compounded continuously?

58. How old is a skeleton that has lost 34% of its carbon-14? (Use $P(t) = P_0e^{-0.00012t}$.)

59. D$_\text{W}$ Explain why you cannot take the logarithm of a negative number. [17.3a]

60. D$_\text{W}$ Explain why $\log_a 1 = 0$. [17.3a]

SKILL MAINTENANCE

Certain objectives from four particular sections will be retested on the chapter test. The objectives are listed with the practice problems that follow.

61. *Pizza Prices.* Pizza Unlimited has the following prices for pizza.

DIAMETER	PRICE
8 in.	$ 6.00
12 in.	8.50
16 in.	11.50

a) Use the data points $(8, 6)$, $(12, 8.5)$, and $(16, 11.5)$ to fit a quadratic function to the data.
b) Use the function to estimate the price of a 24-in. pizza. [16.7b]

62. Solve: $x^4 - 11x^2 - 80 = 0$. [16.4c]

63. For $f(x) = x^2 + 2x + 3$, find and label **(a)** the vertex, **(b)** the line of symmetry, and **(c)** the maximum or minimum value. Then **(d)** graph the function. [16.6a]

Compute and simplify. Express answers in the form $a + bi$, where $i^2 = -1$. [15.8b, c, d, e]

64. $(2 - 3i)(4 + i)$

65. i^{20}

66. $\dfrac{4 - 5i}{1 + 3i}$

67. $(4 - 5i) + (1 + 3i)$

SYNTHESIS

Solve. [17.6a, b]

68. $\ln(\ln x) = 3$

69. $5^{x+y} = 25,\ 2^{2x-y} = 64$

CHAPTER 17

Chapter Test

Graph.

1. $f(x) = 2^{x+1}$

x	$f(x)$

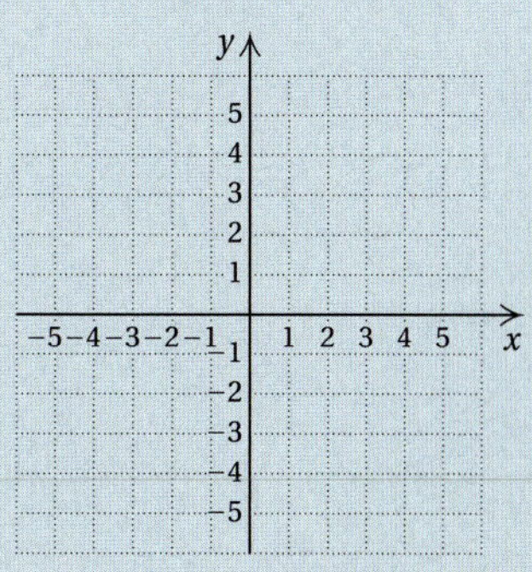

2. $y = \log_2 x$

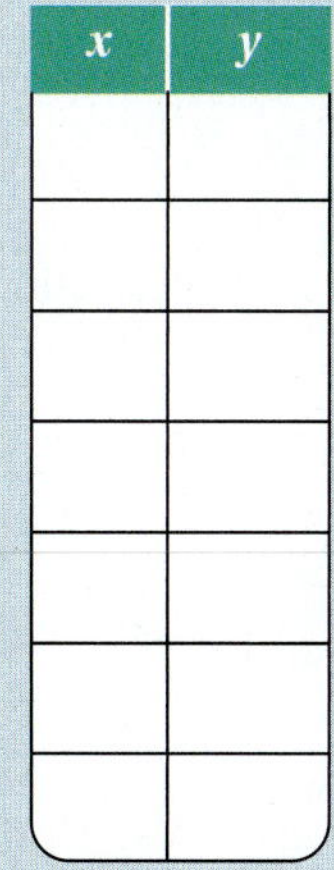

x	y

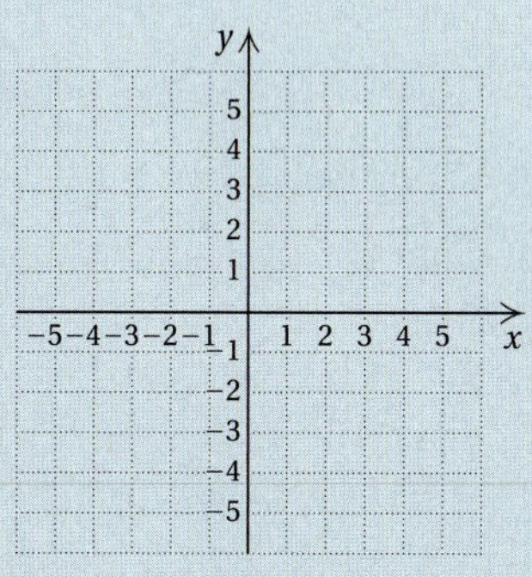

3. $f(x) = e^{x-2}$

x	$f(x)$

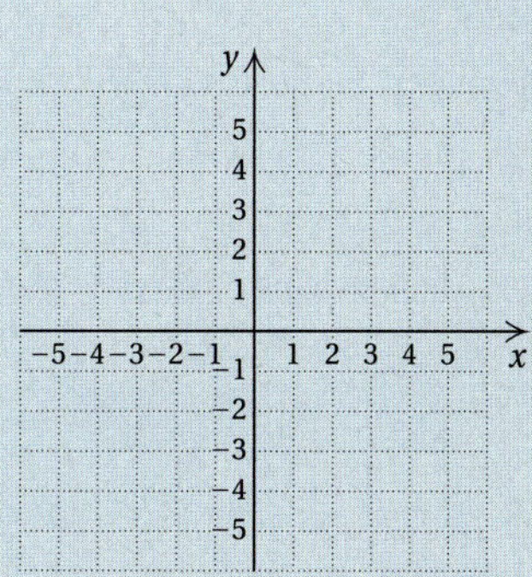

4. $f(x) = \ln(x - 4)$

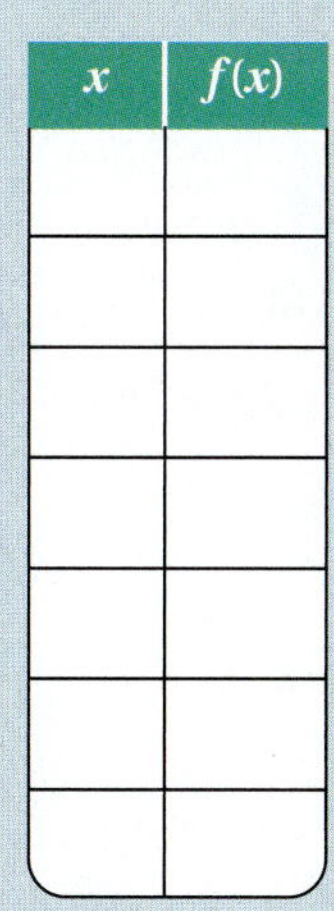

x	$f(x)$

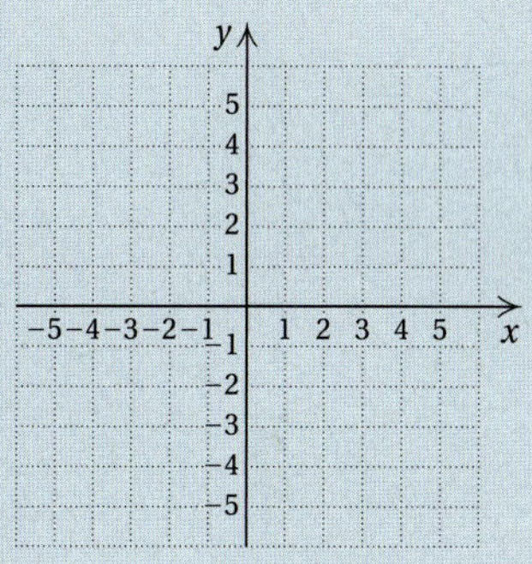

5. Find the inverse of the relation $\{(-4, 3), (5, -8), (-1, -3), (10, 12)\}$.

Determine whether the function is one-to-one. If it is, find a formula for its inverse.

6. $f(x) = 4x - 3$

7. $f(x) = (x + 1)^3$

8. $f(x) = 2 - |x|$

9. Find $f \circ g(x)$ and $g \circ f(x)$ if $f(x) = x + x^2$ and $g(x) = 5x - 2$.

10. Convert to a logarithmic equation:
$256^{1/2} = 16.$

11. Convert to an exponential equation:
$m = \log_7 49.$

Find each of the following.

12. $\log_5 125$

13. $\log_t t^{23}$

14. $\log_p 1$

Find the common logarithm, to four decimal places, using a calculator.

15. $\log 0.0123$

16. $\log(-5)$

17. Express in terms of logarithms of a, b, and c:
$\log \frac{a^3 b^{1/2}}{c^2}.$

18. Express as a single logarithm:
$\frac{1}{3}\log_a x - 3\log_a y + 2\log_a z.$

Given $\log_a 2 = 0.301$, $\log_a 6 = 0.778$, and $\log_a 7 = 0.845$, find each of the following.

19. $\log_a \frac{2}{7}$

20. $\log_a 12$

Find each of the following, to four decimal places, using a calculator.

21. $\ln 807.39$

22. $e^{4.68}$

23. $\ln 1$

24. Find $\log_{18} 31$ using the change-of-base formula.

Solve. Where appropriate, give approximations to four decimal places.

25. $\log_x 25 = 2$

26. $\log_4 x = \frac{1}{2}$

27. $\log x = 4$

28. $\ln x = \frac{1}{4}$

29. $7^x = 1.2$

30. $\log(x^2 - 1) - \log(x - 1) = 1$

31. $\log_5 x + \log_5 (x + 4) = 1$

32. *Tomatoes.* What is the pH of tomatoes if the hydrogen ion concentration is 6.3×10^{-5} moles per liter? (Use $\text{pH} = -\log[\text{H}^+]$.)

33. *Projected College Costs.* The cost of tuition, books, room, and board at a private university is projected to follow the exponential function

$$C(t) = 24{,}534(1.06)^t,$$

where C is the cost, in dollars, and t is the number of years after 2000.

Sources: College Board; Senate Labor Committee

a) Find the college costs in 2008.
b) In what year will the cost be \$43,937?
c) What is the doubling time of the costs?

34. *Population Growth of Canada.* The population of Canada was 30 million in 1996 and 30.753 million in 2000. Assume the number N is growing exponentially according to the equation

$$N(t) = N_0 e^{kt}.$$

Source: Census Bureau of Canada

a) Find k and write the exponential growth function.
b) Estimate the population in 2003 and in 2010.
c) When will the population be 50 million?
d) What is the doubling time?

35. An investment with interest compounded continuously doubled in 15 yr. What is the interest rate?

36. How old is an animal bone that has lost 43% of its carbon-14? (Use $P(t) = P_0 e^{-0.00012t}$.)

SKILL MAINTENANCE

37. *Audio Equipment.* The owner's manual for a top-rated cassette deck includes a table relating the time that a tape has to run to the counter reading.

a) Use the data points $(0, 0)$, $(400, 25)$, and $(700, 50)$ to fit a quadratic function to the data.
b) Use the function to estimate how long a tape has run when the counter reading is 200.

COUNTER READING	TIME TAPE HAS RUN (in minutes)
000	0
400	25
700	50

38. Solve: $y - 9\sqrt{y} + 8 = 0$.

39. For $f(x) = x^2 - 4x - 2$, find and label **(a)** the vertex, **(b)** the line of symmetry, and **(c)** the maximum or minimum value. Then **(d)** graph the function.

Compute and simplify. Express answers in the form $a + bi$, where $i^2 = -1$.

40. i^{29}

41. $\dfrac{4 - i}{1 + i}$

SYNTHESIS

42. Solve: $\log_3 |2x - 7| = 4$.

43. If $\log_a x = 2$, $\log_a y = 3$, and $\log_a z = 4$, find

$$\log_a \frac{\sqrt[3]{x^2 z}}{\sqrt[3]{y^2 z^{-1}}}.$$

Appendixes

A. **LINEAR MEASURES: AMERICAN AND METRIC UNITS**

B. **MEAN, MEDIAN, AND MODE**

C. **TABLES AND PICTOGRAPHS**

D. **THE DISTANCE FORMULA AND MIDPOINTS**

E. **NONLINEAR INEQUALITIES**

F. **SEQUENCES AND SERIES**

G. **APPLYING REASONING SKILLS**

H. **TASP PRACTICE TEST**

I. **TASP TEST MATHEMATICS SKILLS**

A LINEAR MEASURES: AMERICAN AND METRIC UNITS

Objectives

a Convert from one American unit of length to another.

b Convert from one metric unit of length to another.

c Convert between American units of length and metric units of length.

Use the unit below to measure the length of each segment or object.

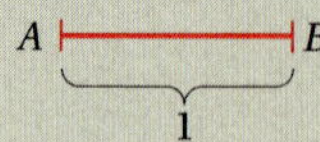

1.

2.

3.

4.

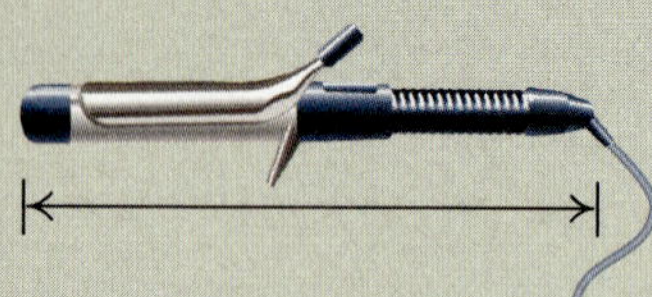

Answers on page A-74

Length, or distance, is one kind of measure. To find lengths, we start with some **unit segment** and assign to it a measure of 1. Suppose $\overline{AB}$ below is a unit segment.

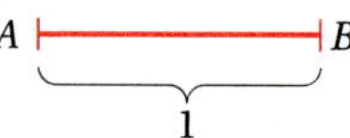

Let's measure segment $\overline{CD}$ below, using $\overline{AB}$ as our unit segment.

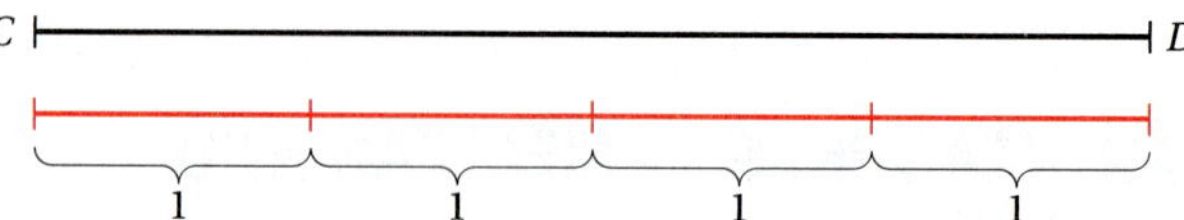

Since we can place 4 unit segments end to end along $\overline{CD}$, the measure of $\overline{CD}$ is 4.

Sometimes we have to use parts of units, called **subunits.** For example, the measure of the segment $\overline{MN}$ below is $1\frac{1}{2}$. We place one unit segment and one half-unit segment end to end.

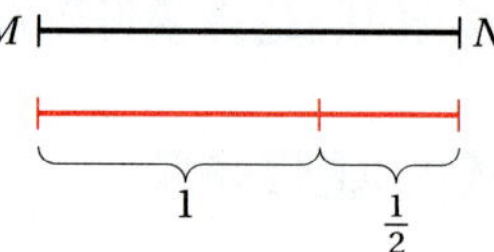

Do Exercises 1–4.

a American Measures

American units of length are related as follows.

(Actual size, in inches)

AMERICAN UNITS OF LENGTH

12 inches (in.) = 1 foot (ft)	3 feet = 1 yard (yd)
36 inches = 1 yard	5280 feet = 1 mile (mi)

We can visualize comparisons of the units as follows:

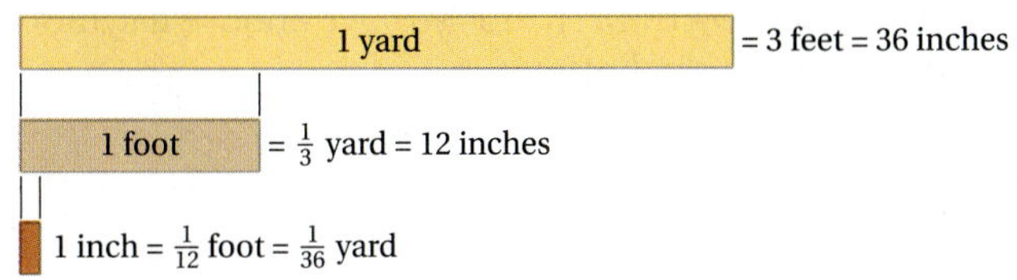

The symbolism 13 in. = 13″ and 27 ft = 27′ is also used for inches and feet. American units have also been called "English," or "British–American," because at one time they were used by both countries. Today, both Canada and England have officially converted to the metric system. However, if you travel in England, you will still see units such as "miles" on road signs.

To change from certain American units to others, we make substitutions. Such a substitution is usually helpful when we are converting from a *larger* unit to a *smaller* one.

EXAMPLE 1 Complete: $7\frac{1}{3}$ yd = _______ in.

$$
\begin{aligned}
7\frac{1}{3}\text{ yd} &= 7\frac{1}{3} \times 1\text{ yd} \\
&= 7\frac{1}{3} \times 3\text{ ft} && \text{Substituting 3 ft for 1 yd} \\
&= 7\frac{1}{3} \times 3 \times 1\text{ ft} \\
&= \frac{22}{3} \times 3 \times 12\text{ in.} && \text{Substituting 12 in. for 1 ft} \\
&= 264\text{ in.}
\end{aligned}
$$

Do Exercises 5–7.

Sometimes it helps to use multiplying by 1 in making conversions. For example, 12 in. = 1 ft, so

$$\frac{12\text{ in.}}{1\text{ ft}} = 1 \quad \text{and} \quad \frac{1\text{ ft}}{12\text{ in.}} = 1.$$

If we divide 12 in. by 1 ft or 1 ft by 12 in., we get 1 because the lengths are the same. Let's first convert from *smaller* to *larger* units.

EXAMPLE 2 Complete: 48 in. = _______ ft.

We want to convert from "in." to "ft." We multiply by 1 using a symbol for 1 with "in." on the bottom and "ft" on the top to eliminate inches and to convert to feet:

$$
\begin{aligned}
48\text{ in.} &= \frac{48\text{ in.}}{1} \times \frac{1\text{ ft}}{12\text{ in.}} && \text{Multiplying by 1 using } \frac{1\text{ ft}}{12\text{ in.}} \text{ to eliminate in.} \\
&= \frac{48\text{ in.}}{12\text{ in.}} \times 1\text{ ft} \\
&= \frac{48}{12} \times \frac{\text{in.}}{\text{in.}} \times 1\text{ ft} \\
&= 4 \times 1\text{ ft} && \text{The } \frac{\text{in.}}{\text{in.}} \text{ acts like 1, so we can omit it.} \\
&= 4\text{ ft.}
\end{aligned}
$$

Do Exercises 8 and 9.

Complete.

5. 8 yd = _______ in.

6. $2\frac{5}{6}$ yd = _______ ft

7. 3.8 mi = _______ in.

Complete.

8. 72 in. = _______ ft

9. 17 in. = _______ ft

Answers on page A-74

Complete.

10. 24 ft = ______ yd

11. 35 ft = ______ yd

Complete.

12. 26,400 ft = ______ mi

13. 2640 ft = ______ mi

Answers on page A-74

EXAMPLE 3 Complete: 25 ft = ______ yd.

Since we are converting from "ft" to "yd," we choose a symbol for 1 with "yd" on the top and "ft" on the bottom:

$$25 \text{ ft} = 25 \text{ ft} \times \frac{1 \text{ yd}}{3 \text{ ft}}$$

$3 \text{ ft} = 1 \text{ yd}$, so $\frac{3 \text{ ft}}{1 \text{ yd}} = 1$, and $\frac{1 \text{ yd}}{3 \text{ ft}} = 1$. We use $\frac{1 \text{ yd}}{3 \text{ ft}}$ to eliminate ft.

$$= \frac{25}{3} \times \frac{\text{ft}}{\text{ft}} \times 1 \text{ yd}$$

$$= 8\frac{1}{3} \times 1 \text{ yd}$$

The $\frac{\text{ft}}{\text{ft}}$ acts like 1, so we can omit it.

$$= 8\frac{1}{3} \text{ yd, or } 8.\overline{3} \text{ yd.}$$

We can also look at this conversion as "canceling" units:

$$25 \text{ ft} = 25 \cancel{\text{ft}} \times \frac{1 \text{ yd}}{3 \cancel{\text{ft}}}$$

$$= \frac{25}{3} \times 1 \text{ yd} = 8\frac{1}{3} \text{ yd, or } 8.\overline{3} \text{ yd.}$$

Do Exercises 10 and 11.

EXAMPLE 4 Complete: 23,760 ft = ______ mi.

We choose a symbol for 1 with "mi" on the top and "ft" on the bottom:

$$23{,}760 \text{ ft} = 23{,}760 \text{ ft} \times \frac{1 \text{ mi}}{5280 \text{ ft}}$$

$5280 \text{ ft} = 1 \text{ mi}$, so $\frac{1 \text{ mi}}{5280 \text{ ft}} = 1$.

$$= \frac{23{,}760}{5280} \times \frac{\text{ft}}{\text{ft}} \times 1 \text{ mi}$$

$$= 4.5 \times 1 \text{ mi} \qquad \text{Dividing}$$

$$= 4.5 \text{ mi.}$$

Let's also consider this example using canceling:

$$23{,}760 \text{ ft} = 23{,}760 \cancel{\text{ft}} \times \frac{1 \text{ mi}}{5280 \cancel{\text{ft}}}$$

$$= \frac{23{,}760}{5280} \times 1 \text{ mi}$$

$$= 4.5 \times 1 \text{ mi} = 4.5 \text{ mi.}$$

Do Exercises 12 and 13.

b Metric Measures

The **metric system** is used in most countries of the world, but very little in the United States. The metric system does not use inches, feet, pounds, and so on, although units for time and electricity are the same as those used now in the United States.

An advantage of the metric system is that it is easier to convert from one unit to another. That is because the metric system is based on the number 10.

The basic unit of length is the **meter.** It is just over a yard. In fact, 1 meter $\approx$ 1.1 yd.

(Comparative sizes are shown.)

1 Meter

1 Yard

The other units of length are multiples of the length of a meter:

10 times a meter, 100 times a meter, 1000 times a meter, and so on,

or fractions of a meter:

$\frac{1}{10}$ of a meter, $\frac{1}{100}$ of a meter, $\frac{1}{1000}$ of a meter, and so on.

METRIC UNITS OF LENGTH

1 *kilo*meter (km) = 1000 meters (m)
1 *hecto*meter (hm) = 100 meters (m)
1 *deka*meter (dam) = 10 meters (m)
1 meter (m)
1 *deci*meter (dm) = $\frac{1}{10}$ meter (m)
1 *centi*meter (cm) = $\frac{1}{100}$ meter (m)
1 *milli*meter (mm) = $\frac{1}{1000}$ meter (m)

dam and *dm* are not used often.

You should memorize these names and abbreviations. Think of *kilo*- for 1000, *hecto*- for 100, *deka*- for 10, *deci*- for $\frac{1}{10}$, *centi*- for $\frac{1}{100}$, and *milli*- for $\frac{1}{1000}$. We will also use these prefixes when considering units of area, capacity, and mass.

THINKING METRIC

To familiarize yourself with metric units, consider the following.

1 kilometer (1000 meters)	is slightly more than $\frac{1}{2}$ mile (0.6 mi).
1 meter	is just over a yard (1.1 yd).
1 centimeter (0.01 meter)	is a little more than the width of a paperclip (about 0.3937 inch).

1 cm

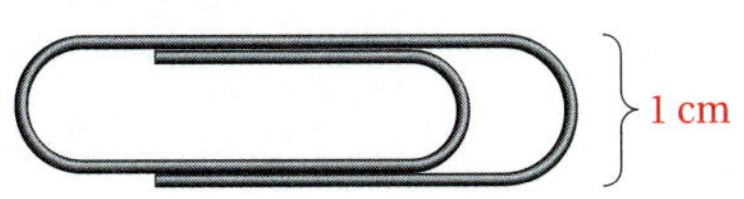

Use a centimeter ruler. Measure each object.

14.

15.

0 1 2 3

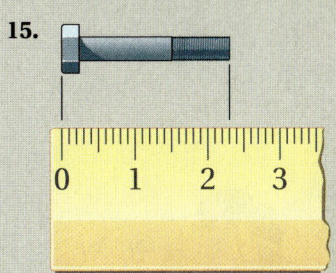

16.

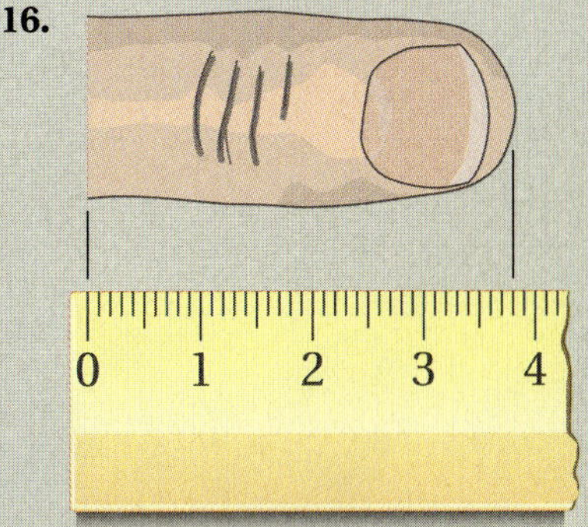

Answers on page A-74

1 inch is about 2.54 centimeters.

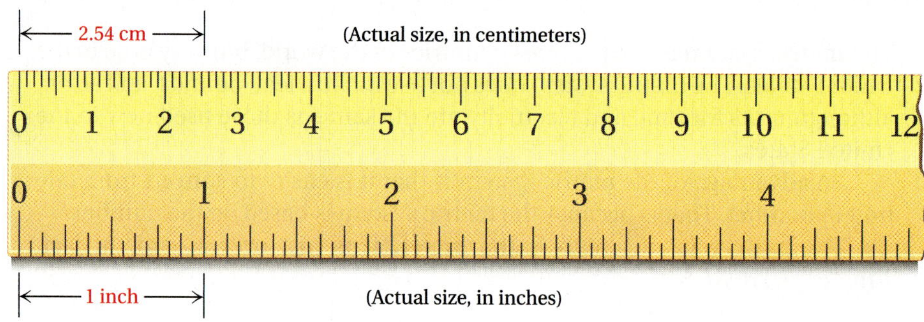

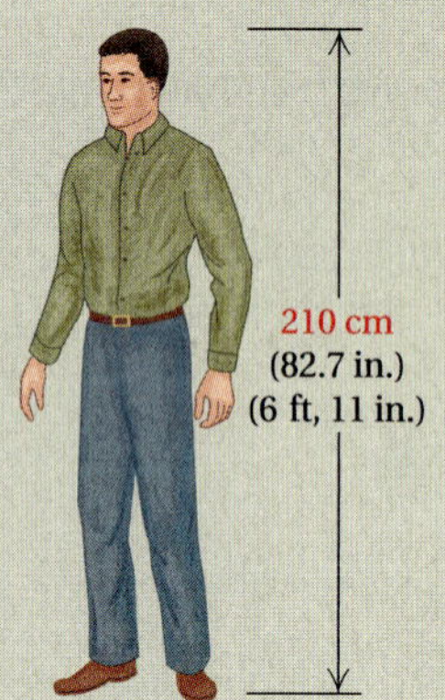

1 millimeter is about the diameter of a paperclip wire.

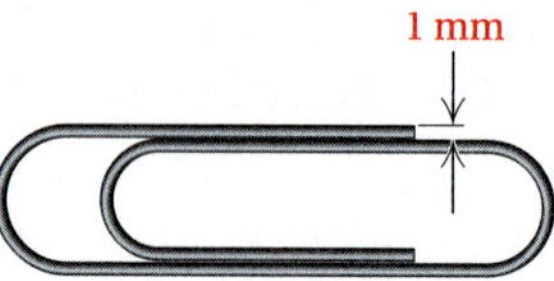

The millimeter (mm) is used to measure small distances, especially in industry. In many countries, the centimeter (cm) is used for body dimensions and clothing sizes.

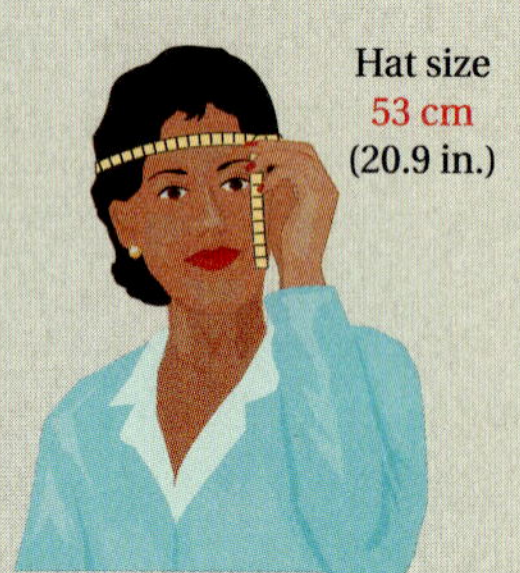

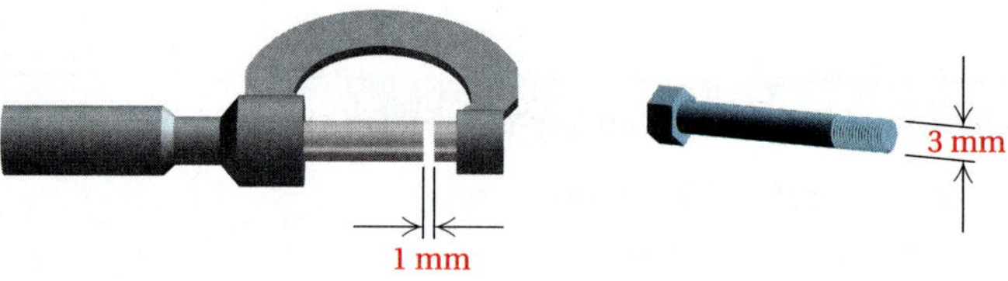

Do Exercises 14–16 on the preceding page.

The meter (m) is used for expressing dimensions of larger objects—say, the length of a building—and for shorter distances, such as the length of a rug.

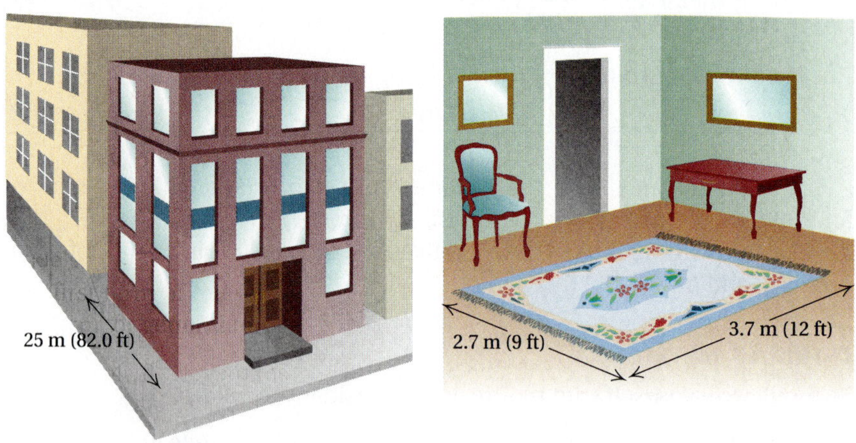

The kilometer (km) is used for longer distances, mostly in cases where miles are now being used.

1 mile is about 1.6 km.

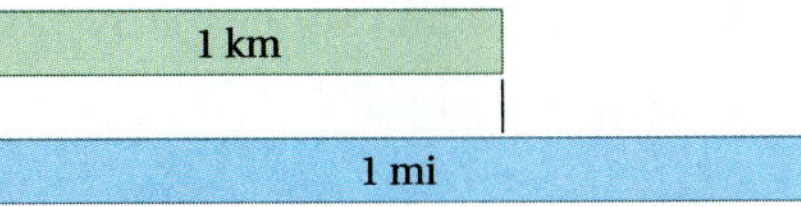

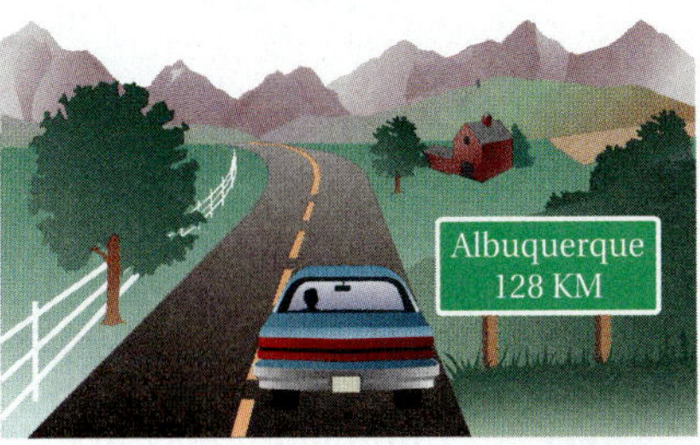

Do Exercises 17–22.

As with American units, when changing from a *larger* unit to a *smaller* unit, we usually make substitutions.

EXAMPLE 5 Complete: 4 km = ______ m.

We want to convert from "km" to "m." Since we are converting from a *larger* to a *smaller* unit, we use substitution.

$$\begin{aligned} 4 \text{ km} &= 4 \times 1 \text{ km} \\ &= 4 \times 1000 \text{ m} \quad \text{Substituting 1000 m for 1 km} \\ &= 4000 \text{ m} \end{aligned}$$

Do Exercises 23 and 24 on the following page.

Since

$$\frac{1}{10}\text{ m} = 1 \text{ dm}, \quad \frac{1}{100}\text{ m} = 1 \text{ cm}, \quad \text{and} \quad \frac{1}{1000}\text{ m} = 1 \text{ mm},$$

it follows that

$$1 \text{ m} = 10 \text{ dm}, \quad 1 \text{ m} = 100 \text{ cm}, \quad \text{and} \quad 1 \text{ m} = 1000 \text{ mm}.$$

EXAMPLE 6 Complete: 93.4 m = ______ cm.

We want to convert from "m" to "cm." Since we are converting from a *larger* to a *smaller* unit, we use substitution.

$$\begin{aligned} 93.4 \text{ m} &= 93.4 \times 1 \text{ m} \\ &= 93.4 \times 100 \text{ cm} \quad \text{Substituting 100 cm for 1 m} \\ &= 9340 \text{ cm} \end{aligned}$$

Complete with mm, cm, m, or km.

17. A stick of gum is 7 ______ long.

18. Minneapolis is 3213 ______ from San Francisco.

19. A penny is 1 ______ thick.

20. The halfback ran 7 ______ .

21. The book is 3 ______ thick.

22. The desk is 2 ______ long.

Answers on page A-74

Complete.

23. 23 km = ______ m

24. 4 hm = ______ m

Complete.

25. 1.78 m = ______ cm

26. 9.04 m = ______ mm

Complete.

27. 7814 m = ______ km

28. 7814 m = ______ dam

Complete.

29. 9.67 mm = ______ cm

30. 89 km = ______ cm

Answers on page A-74

EXAMPLE 7 Complete: 0.248 m = ______ mm.

We want to convert from "m" to "mm." Since we are again converting from a *larger* to a *smaller* unit, we use substitution.

$$0.248 \text{ m} = 0.248 \times 1 \text{ m}$$
$$= 0.248 \times 1000 \text{ mm} \qquad \text{Substituting 1000 mm for 1 m}$$
$$= 248 \text{ mm}$$

Do Exercises 25 and 26.

We now convert from "m" to "km." Since we are converting from a *smaller* unit to a *larger* unit, we use multiplying by 1. We choose a symbol for 1 with "km" in the numerator and "m" in the denominator.

EXAMPLE 8 Complete: 2347 m = ______ km.

$$2347 \text{ m} = 2347 \text{ m} \times \frac{1 \text{ km}}{1000 \text{ m}} \qquad \text{Multiplying by 1 using } \frac{1 \text{ km}}{1000 \text{ m}}$$
$$= \frac{2347}{1000} \times \frac{\text{m}}{\text{m}} \times 1 \text{ km} \qquad \text{The } \frac{\text{m}}{\text{m}} \text{ acts like 1, so we omit it.}$$
$$= 2.347 \text{ km}$$

Dividing by 1000 moves the decimal point three places to the left.

Using canceling, we can work this example as follows:

$$2347 \text{ m} = 2347 \cancel{\text{m}} \times \frac{1 \text{ km}}{1000 \cancel{\text{m}}}$$
$$= \frac{2347}{1000} \times 1 \text{ km} = 2.347 \text{ km}.$$

Do Exercises 27 and 28.

Sometimes we multiply by 1 more than once.

EXAMPLE 9 Complete: 8.42 mm = ______ cm.

$$8.42 \text{ mm} = 8.42 \text{ mm} \times \frac{1 \text{ m}}{1000 \text{ mm}} \times \frac{100 \text{ cm}}{1 \text{ m}} \qquad \text{Multiplying by 1 using } \frac{1 \text{ m}}{1000 \text{ mm}} \text{ and } \frac{100 \text{ cm}}{1 \text{ m}}$$
$$= \frac{8.42 \times 100}{1000} \times \frac{\text{mm}}{\text{mm}} \times \frac{\text{m}}{\text{m}} \times 1 \text{ cm}$$
$$= \frac{842}{1000} \text{ cm} = 0.842 \text{ cm}$$

Do Exercises 29 and 30.

MENTAL CONVERSION

Look back over the examples and exercises done so far and you will see that changing from one unit to another in the metric system amounts to only the movement of a decimal point. That is because the metric system is based on 10. Let's find a faster way to convert. Look at the following table.

1000 m	100 m	10 m	1 m	0.1 m	0.01 m	0.001 m
1 km	1 hm	1 dam	1 m	1 dm	1 cm	1 mm

Each place in the table has a value $\frac{1}{10}$ that to the left or 10 times that to the right. Thus moving one place in the table corresponds to moving one decimal place.

Let's convert mentally.

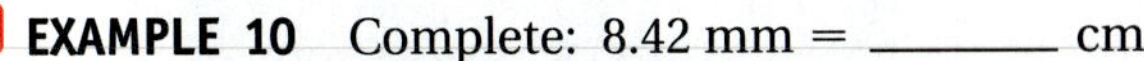

EXAMPLE 10 Complete: 8.42 mm = _______ cm.

Think: To go from mm to cm in the table is a move of one place to the left. Thus we move the decimal point one place to the left.

1000 m	100 m	10 m	1 m	0.1 m	0.01 m	0.001 m
1 km	1 hm	1 dam	1 m	1 dm	1 cm	1 mm

1 place to the left

8.42 0.8.42 8.42 mm = 0.842 cm

EXAMPLE 11 Complete: 1.886 km = _______ cm.

Think: To go from km to cm is a move of five places to the right. Thus we move the decimal point five places to the right.

1000 m	100 m	10 m	1 m	0.1 m	0.01 m	0.001 m
1 km	1 hm	1 dam	1 m	1 dm	1 cm	1 mm

5 places to the right

1.886 1.88600. 1.886 km = 188,600 cm

EXAMPLE 12 Complete: 3 m = _______ cm.

Think: To go from m to cm in the table is a move of two places to the right. Thus we move the decimal point two places to the right.

1000 m	100 m	10 m	1 m	0.1 m	0.01 m	0.001 m
1 km	1 hm	1 dam	1 m	1 dm	1 cm	1 mm

2 places to the right

3 3.00. 3 m = 300 cm

You should try to make metric conversions mentally as much as possible.

Complete. Try to do this mentally using the table.

31. 6780 m = _______ km

32. 9.74 cm = _______ mm

33. 1 mm = _______ cm

34. 845.1 mm = _______ dm

Answers on page A-74

Complete.

35. 100 yd = ________ m
(The length of a football field)

36. 500 mi = ________ km
(The Indianapolis 500-mile race)

37. 2383 km = ________ mi
(The distance from St. Louis to Phoenix)

Answers on page A-74

The fact that conversions can be done so easily is an important advantage of the metric system. The most commonly used metric units of length are km, m, cm, and mm. We have purposely used these more often than the others in the exercises.

Do Exercises 31–34 on the preceding page.

C Converting Between American and Metric Units

We can make conversions between American and metric units by using the following table. These listings are rounded approximations. Again, we either make a substitution or multiply by 1 appropriately.

AMERICAN	METRIC
1 in.	2.540 cm
1 ft	0.305 m
1 yd	0.914 m
1 mi	1.609 km
0.621 mi	1 km
1.094 yd	1 m
3.281 ft	1 m
39.370 in.	1 m

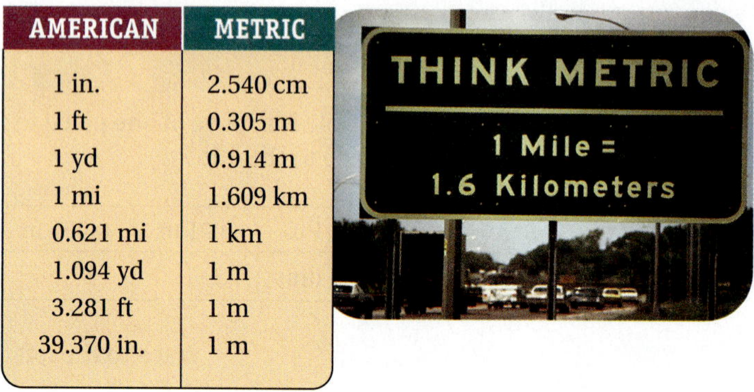

This table is set up to enable us to make all conversions by substitution.

EXAMPLE 13 Complete: 26.2 mi = ________ km.
(The length of the Olympic marathon)

$$\begin{aligned} 26.2 \text{ mi} &= 26.2 \times 1 \text{ mi} \\ &\approx 26.2 \times 1.609 \text{ km} \quad \text{Substituting 1.609 km for 1 mi} \\ &= 42.1558 \text{ km} \end{aligned}$$

EXAMPLE 14 Complete: 2.16 m = ________ in.
(The height of Shaquille O'Neal of the Los Angeles Lakers)

$$\begin{aligned} 2.16 \text{ m} &= 2.16 \times 1 \text{ m} \\ &\approx 2.16 \times 39.37 \text{ in.} \quad \text{Substituting 39.37 in. for 1 m} \\ &= 85.0392 \text{ in.} \end{aligned}$$

In an application like this one, the answer would probably be rounded to the nearest one, as 85 in.

EXAMPLE 15 Complete: 100 m = ________ ft.
(The length of the 100-meter dash)

$$\begin{aligned} 100 \text{ m} &= 100 \times 1 \text{ m} \\ &\approx 100 \times 3.281 \text{ ft} \quad \text{Substituting 3.281 ft for 1 m} \\ &= 328.1 \text{ ft} \end{aligned}$$

EXAMPLE 16 Complete: 4544 km = _______ mi.
(The distance from New York to Los Angeles)

$$4544 = 4544 \times 1\text{ km}$$
$$\approx 4544 \times 0.621\text{ mi} \qquad \text{Substituting 0.621 mi for 1 km}$$
$$= 2821.824\text{ mi}$$

In practical situations, we would probably round this answer to 2822 mi.

Do Exercises 35–37 on the preceding page.

EXAMPLE 17 *Petronas Towers.* The height of the Petronas Towers in Kuala Lumpur, Malaysia, is 1483 ft. Find the height in meters.
Source: *The New York Times Almanac*

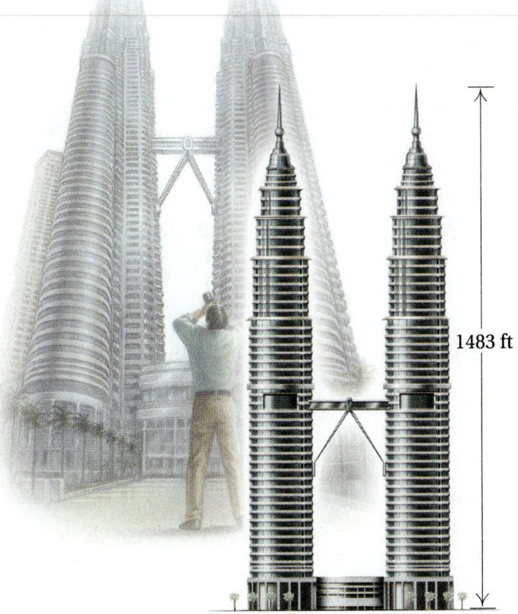

The height H, in meters, is given by

$$H = 1483\text{ ft} = 1483 \times 1\text{ ft}$$
$$\approx 1483 \times 0.305\text{ m} \qquad \text{Substituting 0.305 m for 1 ft}$$
$$= 452.315\text{ m.}$$

Do Exercises 38 and 39.

EXAMPLE 18 Complete: 0.0041 in. = _______ mm.
(The thickness of a $1 bill)

In this case, we must make two substitutions since the chart on the preceding page does not provide an easy way to convert from inches to millimeters.

$$0.0041\text{ in.} = 0.0041 \times 1\text{ in.}$$
$$\approx 0.0041 \times 2.54\text{ cm} \qquad \text{Substituting 2.54 cm for 1 in.}$$
$$= 0.0041 \times 2.54 \times 1\text{ cm}$$
$$= 0.0041 \times 2.54 \times 10\text{ mm} \qquad \text{Substituting 10 mm for 1 cm}$$
$$= 0.10414\text{ mm}$$

Do Exercise 40.

Complete.

38. 568 mi = _______ km
(The distance from San Francisco to Las Vegas)

39. The height of the John Hancock Building in Chicago is 1127 ft. Find the height in meters.

40. Complete:
0.125 in. = _______ mm.
(The thickness of a quarter)

Answers on page A-74

A EXERCISE SET

a Complete.

1. 1 ft = ______ in.

2. 1 yd = ______ ft

3. 1 in. = ______ ft

4. 1 mi = ______ yd

5. 1 mi = ______ ft

6. 1 ft = ______ yd

7. 3 yd = ______ in.

8. 10 yd = ______ ft

9. 84 in. = ______ ft

10. 48 ft = ______ yd

11. 18 in. = ______ ft

12. 29 ft = ______ yd

13. 5 mi = ______ ft

14. 5 mi = ______ yd

15. 63 in. = ______ ft

16. 11,616 ft = ______ mi

17. 10 ft = ______ yd

18. 9.6 yd = ______ ft

19. 7.1 mi = ______ ft

20. 31,680 ft = ______ mi

21. $4\frac{1}{2}$ ft = ______ yd

22. 48 in. = ______ ft

23. 45 in. = ______ yd

24. $6\frac{1}{3}$ yd = ______ in.

25. 330 ft = ______ yd

26. 5280 yd = ______ mi

27. 3520 yd = ______ mi

28. 25 mi = ______ ft

29. 100 yd = ______ ft

30. 480 in. = ______ ft

31. 360 in. = ______ ft

32. 720 in. = ______ yd

33. 1 in. = ______ yd

34. 25 in. = ______ ft

35. 2 mi = ______ in.

36. 63,360 in. = ______ mi

b Complete. Do as much as possible mentally.

37. **a)** 1 km = ________ m
b) 1 m = ________ km

38. **a)** 1 hm = ________ m
b) 1 m = ________ hm

39. **a)** 1 dam = ________ m
b) 1 m = ________ dam

40. **a)** 1 dm = ________ m
b) 1 m = ________ dm

41. **a)** 1 cm = ________ m
b) 1 m = ________ cm

42. **a)** 1 mm = ________ m
b) 1 m = ________ mm

43. **a)** 6.7 km = ________ m
b) This conversion is from a larger unit to a smaller. Did you substitute or multiply by 1?

44. 27 km = ________ m

45. **a)** 98 cm = ________ m
b) This conversion is from a smaller unit to a larger. Did you substitute or multiply by 1?

46. 0.789 cm = ________ m

47. 8921 m = ________ km

48. 8664 m = ________ km

49. 56.66 m = ________ km

50. 4.733 m = ________ km

51. 5666 m = ________ cm

52. 869 m = ________ cm

53. 477 cm = ________ m

54. 6.27 mm = ________ m

55. 6.88 m = ________ cm

56. 6.88 m = ________ dm

57. 1 mm = ________ cm

58. 1 cm = ________ km

59. 1 km = ________ cm

60. 2 km = ________ cm

61. 14.2 cm = ________ mm

62. 25.3 cm = ________ mm

63. 8.2 mm = ________ cm

64. 9.7 mm = ________ cm

65. 4500 mm = ________ cm

66. 8,000,000 m = ________ km

67. 0.024 mm = ________ m

68. 60,000 mm = ________ dam

69. 6.88 m = ________ dam

70. 7.44 m = ________ hm

71. 2.3 dam = ________ dm

72. 9 km = ________ hm

73. 392 dam = ________ km

74. 0.056 mm = ________ dm

Complete the following table.

	OBJECT	MILLIMETERS (mm)	CENTIMETERS (cm)	METERS (m)
75.	Length of a calculator		18	
76.	Width of a calculator	85		
77.	Length of a piece of typing paper			0.278
78.	Length of a football field			109.09
79.	Width of a football field		4844	
80.	Film size	33		
81.	Length of 4 meter sticks			4
82.	Length of 3 meter sticks		300	
83.	Thickness of an index card	0.27		
84.	Thickness of a piece of cardboard		0.23	
85.	Height of the Sears Tower			442
86.	Height of the CN Tower (Toronto)	553,000		

C Complete.

87. 330 ft = ________ m
(The length of most baseball foul lines)

88. 12 in. = ________ cm
(The length of a common ruler)

89. 1171.352 km = ________ mi
(The distance from Cleveland to Atlanta)

90. 2 m = ________ ft
(The length of a desk)

91. 65 mph = ________ km/h
(The common speed limit in the United States)

92. 100 km/h = ________ mph
(A common speed limit in Canada)

93. 180 mi = ________ km
(The distance from Indianapolis to Chicago)

94. 141,600,000 mi = ________ km
(The farthest distance of Mars from the sun)

95. 70 mph = ________ km/h
(An interstate speed limit in Arizona)

96. 60 km/h = ________ mph
(A city speed limit in Canada)

97. 10 yd = ________ m
(The length needed for a first down in football)

98. 450 ft = ________ m
(The length of a long home run in baseball)

99. 2.13 m = ________ in.
(The height of Tim Duncan of the San Antonio Spurs)

100. 87 in. = ________ m
(The height of Arvydas Sabonis of the Portland Trail Blazers)

101. 381 m = ________ ft
(The height of the Empire State Building)

102. 1127 ft = ________ m
(The height of the John Hancock Center)

103. 7.5 in. = ________ cm
(The length of a pencil)

104. 15.7 cm = ________ in.
(The length of a $1 bill)

105. 2216 km = ________ mi
(The distance from Chicago to Miami)

106. 1862 mi = ________ km
(The distance from Seattle to Kansas City)

107. 13 mm = ________ in.
(The thickness of a plastic case for a CD-ROM)

108. 0.25 in. = ________ mm
(The thickness of an eraser on a pencil)

Complete the following table.

	OBJECT	YARDS (yd)	CENTIMETERS (cm)	INCHES (in.)	METERS (m)	MILLIMETERS (mm)
109.	Length of a mousepad		23.8			
110.	Width of a mousepad		20.3			
111.	Width of a piece of typing paper			$8\frac{1}{2}$		
112.	Length of a football field	120 yd				
113.	Width of a football field		4844			
114.	Film size					33
115.	Length of 4 yard sticks	4				
116.	Length of 3 meter sticks		300			
117.	Thickness of an index card				0.00027	
118.	Thickness of a piece of cardboard		0.23			
119.	Height of the Sears Tower				442	
120.	Height of Central Plaza, Hong Kong	409				

SYNTHESIS

121. Develop a formula to convert from inches to millimeters.

122. Develop a formula to convert from millimeters to inches. How does it relate to the answer for Exercise 121?

Objective

a Find the mean (average), the median, and the mode of a set of data and solve related applied problems.

Find the mean. Round to the nearest tenth.

1. 28, 103, 39

2. 85, 46, 105.7, 22.1

3. A student scored the following on five tests:

   78, 95, 84, 100, 82.

   What was the average score?

Answers on page A-75

B MEAN, MEDIAN, AND MODE

a Mean, Median, and Mode

One way to analyze data is to look for a single representative number, called a **center point** or **measure of central tendency.** Those most often used are the **mean** (or **average**), the **median,** and the **mode.**

MEAN

Let's first consider the *mean*, or *average.*

MEAN, OR AVERAGE

The **mean,** or **average,** of a set of numbers is the sum of the numbers divided by the number of addends.

EXAMPLE 1 Consider the following data on revenue, in billions of dollars, at McDonald's restaurants in five recent years:

$12.5, $13.2, $14.2, $14.9, $15.9.

What is the mean of the numbers?
Source: McDonalds Corporation

First, we add the numbers:

$$12.5 + 13.2 + 14.2 + 14.9 + 15.9 = 70.7.$$

Then we divide by the number of addends, 5:

$$\frac{(12.5 + 13.2 + 14.2 + 14.9 + 15.9)}{5} = \frac{70.7}{5} = 14.14.$$

The mean, or average, revenue of McDonald's for those five years is $14.14 billion.

Note that $14.14 + 14.14 + 14.14 + 14.14 + 14.14 = 70.7$. If we use this center point, 14.14, repeatedly as the addend, we get the same sum that we do when adding individual data numbers.

Do Exercises 1–3.

MEDIAN

The *median* is useful when we wish to de-emphasize extreme scores. For example, suppose five workers in a technology company manufactured the following number of computers during one day's work:

Sarah:	88	Jen:	94
Matt:	92	Mark:	91
Pat:	66		

Let's first list the scores in order from smallest to largest:

66 88 91 92 94.

↑ Middle number (91)

The middle number—in this case, 91—is the **median.**

MEDIAN

Once a set of data has been arranged from smallest to largest, the **median** of the set of data is the middle number if there is an odd number of data numbers. If there is an even number of data numbers, then there are two middle numbers and the median is the *average* of the two middle numbers.

EXAMPLE 2 What is the median of the following set of yearly salaries?

\$76,000, \$58,000, \$87,000, \$32,500, \$64,800, \$62,500

We first rearrange the numbers in order from smallest to largest.

\$32,500, \$58,000, \$62,500, ↑ \$64,800, \$76,000, \$87,000

Median

There is an even number of numbers. We look for the middle two, which are \$62,500 and \$64,800. In this case, the median is the average of \$62,500 and \$64,800:

$$\frac{\$62{,}500 + \$64{,}800}{2} = \$63{,}650.$$

Do Exercises 4–6.

MODE

The last center point we consider is called the *mode.* A number that occurs most often in a set of data can be considered a representative number or center point.

MODE

The **mode** of a set of data is the number or numbers that occur most often. If each number occurs the same number of times, there is *no* mode.

EXAMPLE 3 Find the mode of the following data:

23, 24, 27, 18, 19, 27

The number that occurs most often is 27. Thus the mode is 27.

EXAMPLE 4 Find the mode of the following data:

83, 84, 84, 84, 85, 86, 87, 87, 87, 88, 89, 90.

There are two numbers that occur most often, 84 and 87. Thus the modes are 84 and 87.

EXAMPLE 5 Find the mode of the following data:

115, 117, 211, 213, 219.

Each number occurs the same number of times. The set of data has *no* mode.

Do Exercises 7–10.

Find the median.

4. 17, 13, 18, 14, 19

5. 17, 18, 16, 19, 13, 14

6. 122, 102, 103, 91, 83, 81, 78, 119, 88

Find any modes that exist.

7. 33, 55, 55, 88, 55

8. 90, 54, 88, 87, 87, 54

9. 23.7, 27.5, 54.9, 17.2, 20.1

10. In conducting laboratory tests, Carole discovers bacteria in different lab dishes grew to the following areas, in square millimeters:

25, 19, 29, 24, 28.

a) What is the mean?

b) What is the median?

c) What is the mode?

Answers on page A-75

B EXERCISE SET

a For each set of numbers, find the mean (average), the median, and any modes that exist.

1. 17, 19, 29, 18, 14, 29

2. 72, 83, 85, 88, 92

3. 5, 37, 20, 20, 35, 5, 25

4. 13, 32, 25, 27, 13

5. 4.3, 7.4, 1.2, 5.7, 7.4

6. 13.4, 13.4, 12.6, 42.9

7. 234, 228, 234, 229, 234, 278

8. $29.95, $28.79, $30.95, $29.95

9. *Atlantic Storms and Hurricanes.* The following bar graph shows the number of Atlantic storms or hurricanes that formed in various months from 1980 to 2000. What is the average number for the 9 months given? the median? the mode?

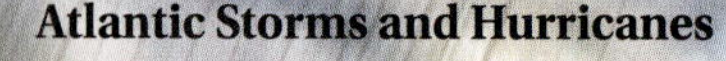

Source: Colorado State University

10. *Cheddar Cheese Prices.* The following prices per pound of sharp cheddar cheese were found at five supermarkets:

$5.99, $6.79, $5.99, $6.99, $6.79.

What was the average price per pound? the median price? the mode?

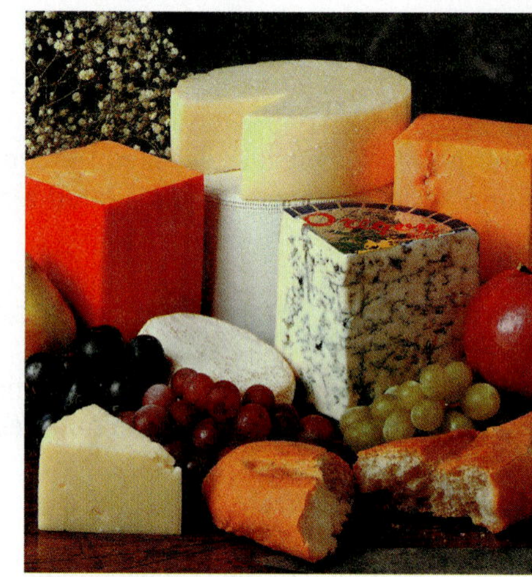

11. *Coffee Consumption.* The following lists the annual coffee consumption, in cups per person, for various countries. Find the mean, the median, and the mode.

Germany	1113
United States	610
Switzerland	1215
France	798
Italy	750

Source: Beverage Marketing Corporation

12. *NBA Tall Men.* The following is a list of the heights, in inches, of the tallest men in the NBA in a recent year. Find the mean, the median, and the mode.

Shaquille O'Neal	85
Gheorghe Muresan	91
Shawn Bradley	90
Priest Lauderdale	88
Rik Smits	88
David Robinson	85
Arvydas Sabonis	87

Source: National Basketball Association

13. *Salmon Prices.* The following prices per pound of Atlantic salmon were found at five fish markets:

$6.99, $8.49, $8.99, $6.99, $9.49.

What was the average price per pound? the median price? the mode?

14. *PBA Scores.* Chris Barnes rolled scores of 224, 224, 254, and 187 in a recent tournament of the Professional Bowlers Association. What was his average? his median? his mode?
Source: Professional Bowlers Association

SYNTHESIS

Grade Point Average. The tables in Exercises 15 and 16 show the grades of a student for one semester. In each case, find the grade point average. Assume that the grade point values are 4.0 for an A, 3.0 for a B, and so on. Round to the nearest tenth.

15.

GRADE	NUMBER OF CREDIT HOURS IN COURSE
B	4
A	5
D	3
C	4

16.

GRADE	NUMBER OF CREDIT HOURS IN COURSE
A	5
C	4
F	3
B	5

17. *Hank Aaron.* Hank Aaron averaged $34\frac{7}{22}$ home runs per year over a 22-yr career. After 21 yr, Aaron had averaged $35\frac{10}{21}$ home runs per year. How many home runs did Aaron hit in his final year?

18. The ordered set of data 18, 21, 24, a, 36, 37, b has a median of 30 and an average of 32. Find a and b.

19. *Length of Pregnancy.* Marta was pregnant 270 days, 259 days, and 272 days for her first three pregnancies. In order for Marta's average length of pregnancy to equal the worldwide average of 266 days, how long must her fourth pregnancy last?
Source: David Crystal (ed.), *The Cambridge Factfinder.* Cambridge CB2 1RP: Cambridge University Press, 1993, p. 84.

20. *Male Height.* Jason's brothers are 174 cm, 180 cm, 179 cm, and 172 cm tall. The average male is 176.5 cm tall. How tall is Jason if he and his brothers have an average height of 176.5 cm?

C TABLES AND PICTOGRAPHS

Objectives

a Extract and interpret data from tables.

b Extract and interpret data from pictographs.

a Reading and Interpreting Tables

A **table** is often used to present data in rows and columns.

EXAMPLE 1 *Nutrition Information.* The following table lists nutrition information for a 1-cup serving of five name-brand cereals (it does not consider the use of milk, sugar, or sweetener).

CEREAL	CALORIES	FAT (in grams)	TOTAL CARBOHYDRATES (in grams)	SODIUM (in milligrams)
Cinnamon Life	160	1.3	34.7	200
Life (Regular)	160	2.0	33.3	213.3
Lucky Charms	120	1.0	25.0	210
Kellogg's Complete	120	0.7	30.7	280
Wheaties	110	1.0	24.0	220

Sources: Quaker Oats; General Mills; Kellogg's

a) Which cereal has the least amount of sodium per serving?

b) Which cereal has the greatest amount of fat?

c) Which cereal has the least amount of fat?

d) Find the average total carbohydrates in the cereals listed.

Careful examination of the table will give the answers.

a) To determine which cereal has the least amount of sodium, look down the column headed "Sodium" and find the smallest number. That number is 200 mg. Then look across that row to find the brand of cereal, Cinnamon Life.

b) To determine which cereal has the greatest amount of fat, look down the column headed "Fat" and find the largest number. That number is 2.0 g. Then look across that row to find the cereal, Life (Regular).

c) To determine which cereal has the least amount of fat, look down the column headed "Fat" and find the smallest number. That number is 0.7 g. Then look across that row to find the cereal, Kellogg's Complete.

d) Find the average of all the numbers in the column headed "Total Carbohydrates":

$$\frac{34.7 + 33.3 + 25.0 + 30.7 + 24.0}{5} = \frac{147.7}{5} = 29.54 \text{ g.}$$

The average total carbohydrates is 29.54 g.

Do Exercises 1–7. (Exercises 5–7 are on the following page.)

Use the table in Example 1 to answer Margin Exercises 1–7.

1. Which cereal has the most total carbohydrates?

2. Which cereal has the least total carbohydrates?

3. Which cereal has the least number of calories?

4. Find the average number of grams of fat in the cereals listed.

Answers on page A-75

EXAMPLE 2 *Wheaties Nutrition Facts.* Most foods are required by law to provide factual information regarding nutrition, as shown in the following table of Nutrition Facts from a box of Wheaties cereal. Although this can be very helpful to the consumer, one must be careful in interpreting the data. The % Daily Value figures shown here are based on a 2000-calorie diet. Your daily values may be higher or lower, depending on your calorie needs or intake.

Nutrition Facts

Serving Size 1 cup (30g)
Servings Per Container 17

Amount Per Serving	Wheaties	with ½ cup skim milk
Calories	110	150
Calories from Fat	10	10
	% Daily value**	
Total Fat 1g*	**1%**	**2%**
Saturated Fat 0g	**0%**	**0%**
Cholesterol 0mg	**0%**	**1%**
Sodium 220mg	**9%**	**12%**
Potassium 110mg	**3%**	**9%**
Total Carbohydrate 24g	**8%**	**10%**
Dietary Fiber 3g	**12%**	**12%**
Sugars 4g		
Other Carbohydrate 17g		
Protein 3g		
Vitamin A	10%	15%
Vitamin C	10%	10%
Calcium	0%	15%
Iron	45%	45%
Vitamin D	10%	25%
Thiamin	50%	50%
Riboflavin	50%	60%
Niacin	50%	50%
Vitamin B_6	50%	50%
Folic Acid	50%	50%
Phosphorus	10%	20%
Magnesium	8%	10%
Zinc	50%	50%
Copper	4%	4%

*Amount in Cereal. A serving of cereal plus skim milk provides 1g fat, less than 5mg cholesterol, 280mg sodium, 310mg potassium, 30g carbohydrate (10g sugars), and 7g protein.

**Percent Daily Values are based on a 2,000 calorie diet. Your daily values may be higher or lower depending on your calorie needs.

Source: General Mills

Suppose your morning bowl of cereal consists of 2 cups of Wheaties with 1 cup of skim milk, with artificial sweetener containing 0 calories.

a) How many calories have you consumed?

b) What percent of the daily value of total fat have you consumed?

c) A nutritionist recommends that you look for foods that provide 10% or more of the daily value for vitamin C. Do you get that with your bowl of Wheaties?

d) Suppose you are trying to limit your daily caloric intake to 2500 calories. How many bowls of cereal would it take to exceed the 2500 calories, even though you probably would not eat just cereal?

Careful examination of the table of nutrition facts will give the answers.

a) Look at the column marked "with 1/2 cup skim milk" and note that 1 cup of cereal with 1/2 cup skim milk contains 150 calories. Since you are having twice that amount, you are consuming

$$2 \times 150, \quad \text{or} \quad 300 \text{ calories.}$$

b) Read across from "Total Fat" and note that in 1 cup of cereal with 1/2 cup skim milk, you get 2% of the daily value of fat. Since you are doubling that, you get 4% of the daily value of fat.

c) Find the row labeled "Vitamin C" on the left and look under the column labeled "with 1/2 cup skim milk." Note that you get 10% of the daily value for "1 cup with 1/2 cup of skim milk," and since you are doubling that, you are more than satisfying the 10% requirement.

d) From part (a), we know that you are consuming 300 calories per bowl. Dividing 2500 by 300 gives $\frac{2500}{300} \approx 8.33$. Thus if you eat 9 bowls of cereal in this manner, you will exceed the 2500 calories.

Do Exercises 8–12.

5. Find the average amount of sodium in the cereals.

6. Find the median of the amount of sodium in the cereals.

7. Find the average, the median, and the mode of the number of calories in the cereals.

Use the Nutrition Facts data from the Wheaties box and the bowl of cereal described in Example 2 to answer Margin Exercises 8–12.

8. How many calories from fat are in your bowl of cereal?

9. A nutritionist recommends that you look for foods that provide 10% or more of the daily value for iron. Do you get that with your bowl of Wheaties?

10. How much sodium have you consumed?

11. What daily value of sodium have you consumed?

12. How much protein have you consumed?

Answers on page A-75

Use the pictograph in Example 3 to answer Margin Exercises 13–15.

13. How many elephants are there in Tanzania?

14. How does the elephant population of Zimbabwe compare to that of Cameroon?

15. What is the average number of elephants in these six countries?

Answers on page A-75

b Reading and Interpreting Pictographs

Pictographs (or *picture graphs*) are another way to show information. Instead of actually listing the amounts to be considered, a **pictograph** uses symbols to represent the amounts. In addition, a *key* is given telling what each symbol represents.

EXAMPLE 3 *Elephant Population.* The following pictograph shows the elephant population of various countries in Africa. Located on the graph is a key that tells you that each symbol represents 10,000 elephants.

Elephant Population

= 10,000 elephants

Cameroon

Zimbabwe

Sudan

Zaire

Tanzania

Botswana

Source: National Geographic

a) Which country has the greatest number of elephants?

b) Which country has the least number of elephants?

c) How many more elephants are there in Zaire than in Botswana?

We can compute the answers by first reading the pictograph.

a) The country with the most symbols has the greatest number of elephants: Zaire, with $11 \times 10{,}000$, or 110,000 elephants.

b) The countries with the fewest symbols have the least number of elephants: Cameroon and Sudan, each with $2 \times 10{,}000$, or 20,000 elephants.

c) From part (a), we know that there are 110,000 elephants in Zaire. In Botswana there are $7 \times 10{,}000$, or 70,000 elephants. Thus there are $110{,}000 - 70{,}000$, or 40,000 more elephants in Zaire than in Botswana.

Do Exercises 13–15.

You have probably noticed that, although they seem to be very easy to read, pictographs are difficult to draw accurately because whole symbols reflect loose approximations due to significant rounding. In pictographs, you also need to use some mathematics to find the actual amounts.

EXAMPLE 4 *Coffee Consumption.* For selected countries, the following pictograph shows approximately how many cups of coffee each person (per capita) drinks annually.

Coffee Consumption

Source: Beverage Marketing Corporation

a) Determine the approximate annual coffee consumption per capital of Germany.

b) Which two countries have the greatest difference in coffee consumption? Estimate that difference.

We use the data from the pictograph as follows.

a) Germany's consumption is represented by 11 whole symbols (1100 cups) and, though it is visually debatable, about $\frac{1}{8}$ of another symbol (about 13 cups), for a total of 1113 cups.

b) Visually, we see that Switzerland has the most consumption and that the United States has the least consumption. Switzerland's annual coffee consumption per capita is represented by 12 whole symbols (1200 cups) and about $\frac{1}{5}$ of another symbol (20 cups), for a total of 1220 cups. U.S. consumption is represented by 6 whole symbols (600 cups) and about $\frac{1}{10}$ of another symbol (10 cups), for a total of 610 cups. The difference between these amounts is $1220 - 610$, or 610 cups.

One advantage of pictographs is that the appropriate choice of a symbol will tell you, at a glance, the kind of measurement being made. Another advantage is that the comparison of amounts represented in the graph can be expressed more easily by just counting symbols. For instance, in Example 3, the ratio of elephants in Zaire to those in Cameroon is 11:2.

There are at least three disadvantages of pictographs:

1. To make a pictograph easy to read, the amounts must be rounded significantly to the unit that a symbol represents. This makes it difficult to accurately represent an amount.
2. It is difficult to determine very accurately how much a partial symbol represents.
3. Some mathematics is required to finally compute the amount represented, since there is usually no explicit statement of the amount.

Do Exercises 16–18.

Use the pictograph in Example 4 to answer Margin Exercises 16–18.

16. Determine the approximate coffee consumption per capita of France.

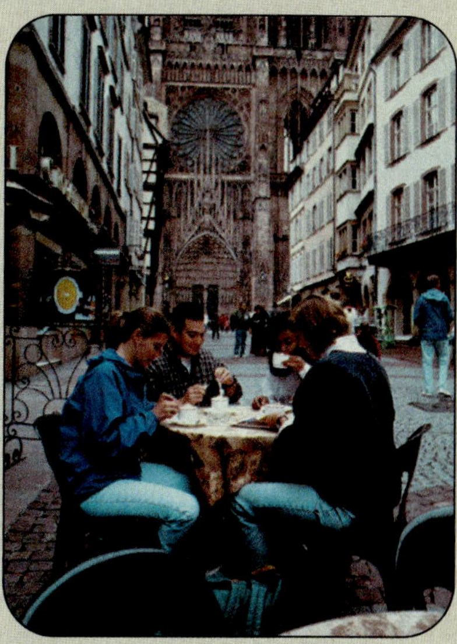

17. Determine the approximate coffee consumption per capita of Italy.

18. The approximate coffee consumption of Finland is about the same as the combined coffee consumptions of Switzerland and the United States. What is the approximate coffee consumption of Finland?

Answers on page A-75

C EXERCISE SET

a *Planets.* Use the following table, which lists information about the planets, for Exercises 1–10.

PLANET	AVERAGE DISTANCE FROM SUN (in miles)	DIAMETER (in miles)	LENGTH OF PLANET'S DAY IN EARTH TIME (in days)	TIME OF REVOLUTION IN EARTH TIME (in years)
Mercury	35,983,000	3,031	58.82	0.24
Venus	67,237,700	7,520	224.59	0.62
Earth	92,955,900	7,926	1.00	1.00
Mars	141,634,800	4,221	1.03	1.88
Jupiter	483,612,200	88,846	0.41	11.86
Saturn	888,184,000	74,898	0.43	29.46
Uranus	1,782,000,000	31,763	0.45	84.01
Neptune	2,794,000,000	31,329	0.66	164.78
Pluto	3,666,000,000	1,423	6.41	248.53

Source: *The Handy Science Answer Book*, Gale Research, Inc.

1. Find the average distance from the sun to Jupiter.

2. How long is a day on Venus?

3. Which planet has a time of revolution of 164.78 yr?

4. Which planet has a diameter of 4221 mi?

5. Which planets have an average distance from the sun that is greater than 1,000,000 mi?

6. Which planets have a diameter that is less than 100,000 mi?

7. About how many earth diameters would it take to equal one Jupiter diameter?

8. How much longer is the longest time of revolution than the shortest?

9. What are the average, the median, and the mode of the diameters of the planets?

10. What are the average, the median, and the mode of the average distances from the sun of the planets?

Heat Index. In warm weather, a person can feel hotter due to reduced heat loss from the skin caused by higher humidity. The **temperature–humidity index,** or **apparent temperature,** is what the temperature would have to be with no humidity in order to give the same heat effect. The following table lists the apparent temperatures for various actual temperatures and relative humidities. Use this table for Exercises 11–22.

ACTUAL TEMPERATURE (°F)	RELATIVE HUMIDITY									
	10%	20%	30%	40%	50%	60%	70%	80%	90%	100%
	APPARENT TEMPERATURE (°F)									
75°	75	77	79	80	82	84	86	88	90	92
80°	80	82	85	87	90	92	94	97	99	102
85°	85	88	91	94	97	100	103	106	108	111
90°	90	93	97	100	104	107	111	114	118	121
95°	95	99	103	107	111	115	119	123	127	131
100°	100	105	109	114	118	123	127	132	137	141
105°	105	110	115	120	125	131	136	141	146	151

In Exercises 11–14, find the apparent temperature for the given actual temperature and humidity combinations.

11. 80°, 60%

12. 90°, 70%

13. 85°, 90%

14. 95°, 80%

15. How many listed temperature–humidity combinations give an apparent temperature of 100°?

16. How many listed temperature–humidity combinations given an apparent temperature of 111°?

17. At a relative humidity of 50%, what actual temperatures give an apparent temperature above 100°?

18. At a relative humidity of 90%, what actual temperatures give an apparent temperature above 100°?

19. At an actual temperature of 95°, what relative humidities give an apparent temperature above 100°?

20. At an actual temperature of 85°, what relative humidities give an apparent temperature above 100°?

21. At an actual temperature of 85°, by how much would the humidity have to increase in order to raise the apparent temperature from 94° to 108°?

22. At an actual temperature of 80°, by how much would the humidity have to increase in order to raise the apparent temperature from 87° to 102°?

Global Warming. Ecologists are increasingly concerned about global warming, that is, the trend of average global temperatures to rise over recent years. One possible effect is the melting of the polar icecaps. Use the following table for Exercises 23–26.

YEAR	1990	1991	1992	1993	1994	1995	1996	1997	1998	1999
Global temperature (in degrees Fahrenheit)	59.85°	59.74°	59.23°	59.36°	59.56°	59.72°	59.58°	59.74°	60.26°	59.81°

Sources: Lester R. Brown et al., *Vital Signs*; the Council of Environmental Quality

23. Find the average global temperatures in 1997 and 1998. What was the percent of increase in the temperature from 1997 to 1998?

24. Find the average global temperatures in 1998 and 1999. What was the percent of decrease in the temperature from 1998 to 1999?

25. Find the average of the average global temperatures for the years 1990 to 1993. Find the average of the average global temperatures for the years 1997 to 1999. By how many degrees does the latter average exceed the former?

26. Find the average of the average global temperatures for the years 1994 to 1996. Find the ten-year average of the average global temperatures for the years 1990 to 1999. By how many degrees does the ten-year average exceed the average for the years 1994 to 1996?

b *World Population Growth.* The following pictograph shows world population in various years. Use the pictograph for Exercises 27–34.

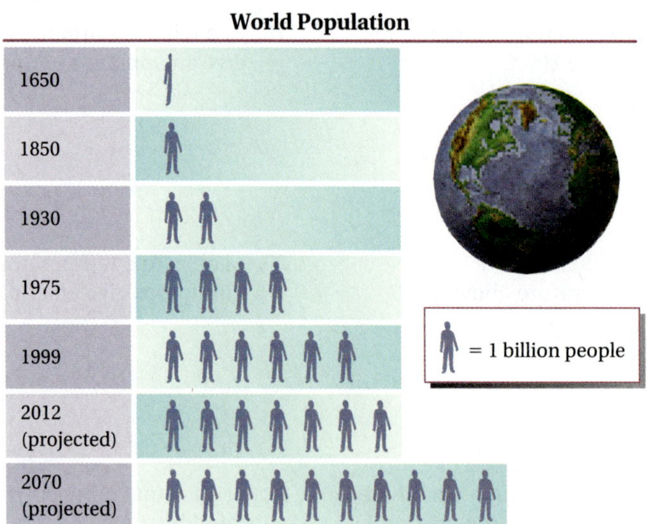

Source: U.S. Census Bureau, International Data Base

27. What was the world population in 1850?

28. What was the world population in 1975?

29. In which year will the population be the greatest?

30. In which year was the population the least?

31. Between which two years was the amount of growth the least?

32. Between which two years was the amount of growth the greatest?

33. How much greater will the world population in 2012 be than in 1975? What is the percent of increase?

34. How much greater was the world population in 1999 than in 1930? What is the percent of increase?

Water Consumption. The following pictograph shows water consumption, per person, in different regions of the world in a recent year. Use the pictograph for Exercises 35–40.

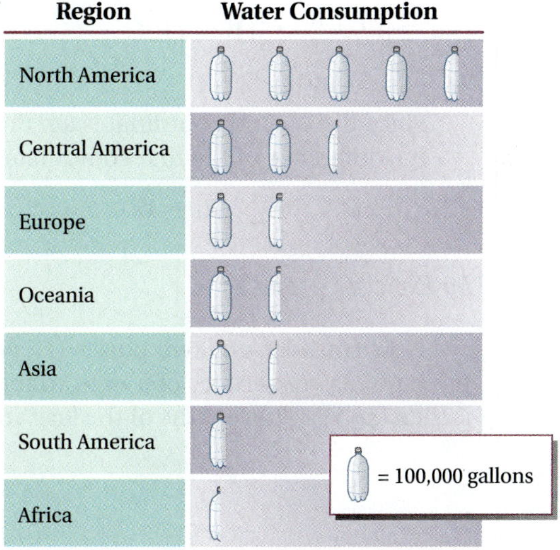

Sources: World Resources Institute; U.S. Energy Information Administration

35. What region consumes the least water?

36. Which region consumes the most water?

37. About how many gallons are consumed per person in North America?

38. About how many gallons are consumed per person in Europe?

39. Approximately how many more gallons are consumed per person in North America than in Asia?

40. Approximately how many more gallons are consumed per person in Central America than in Africa?

D THE DISTANCE FORMULA AND MIDPOINTS

Objectives

a Use the distance formula to find the distance between two points whose coordinates are known.

b Use the midpoint formula to find the midpoint of a segment when the coordinates of its endpoints are known.

Find the distance between the pair of points.

1. $(7, 12)$ and $(7, -2)$

2. $(6, 2)$ and $(-5, 2)$

Answers on page A-75

a The Distance Formula

Suppose that two points are on a horizontal line, and thus have the same second coordinate. We can find the distance between them by subtracting their first coordinates. This difference may be negative, depending on the order in which we subtract. So, to make sure we get a positive number, we take the absolute value of this difference. The distance between two points on a horizontal line (x_1, y_1) and (x_2, y_1) is thus $|x_2 - x_1|$. Similarly, the distance between two points on a vertical line (x_2, y_1) and (x_2, y_2) is $|y_2 - y_1|$.

EXAMPLES Find the distance between these points.

1. $(-5, 13)$ and $(-5, 2)$

   We take the absolute value of the difference of the second coordinates, since the first coordinates are the same. The distance is

   $$|13 - 2| = |11| = 11.$$

2. $(7, -3)$ and $(-5, -3)$

   Since the second coordinates are the same, we take the absolute value of the difference of the first coordinates. The distance is

   $$|-5 - 7| = |-12| = 12.$$

Do Exercises 1 and 2.

Now consider *any* two points (x_1, y_1) and (x_2, y_2). If $x_1 \neq x_2$ and $y_1 \neq y_2$, these points are vertices of a right triangle, as shown below. The other vertex is then (x_2, y_1). The lengths of the legs are $|x_2 - x_1|$ and $|y_2 - y_1|$.

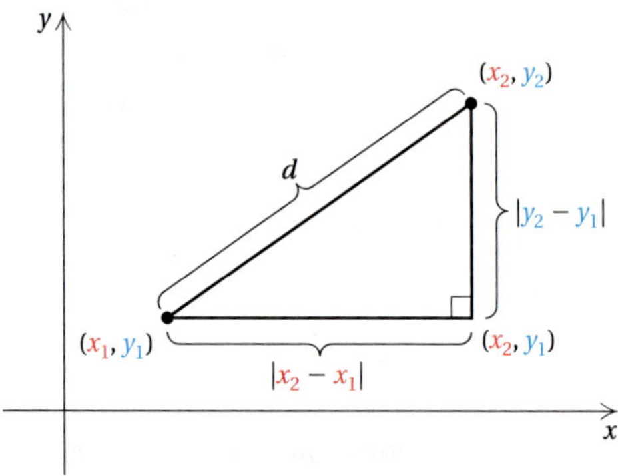

We find d, the length of the hypotenuse, by using the Pythagorean theorem (Sections 10.9 and 15.7):

$$d^2 = |x_2 - x_1|^2 + |y_2 - y_1|^2.$$

Since the square of a number is the same as the square of its opposite, we don't need these absolute-value signs. Thus,

$$d^2 = (x_2 - x_1)^2 + (y_2 - y_1)^2.$$

Taking the principal square root, we obtain the distance between two points.

THE DISTANCE FORMULA

The distance between any two points (x_1, y_1) and (x_2, y_2) is given by
$$d = \sqrt{(x_2 - x_1)^2 + (y_2 - y_1)^2}.$$

This formula holds even when the two points *are* on a vertical or a horizontal line.

EXAMPLE 3 Find the distance between $(4, -3)$ and $(-5, 4)$. Give an exact answer and an approximation to three decimal places.

We substitute into the distance formula:

$$d = \sqrt{(-5 - 4)^2 + [4 - (-3)]^2}$$
$$= \sqrt{(-9)^2 + 7^2}$$
$$= \sqrt{130} \approx 11.402.$$

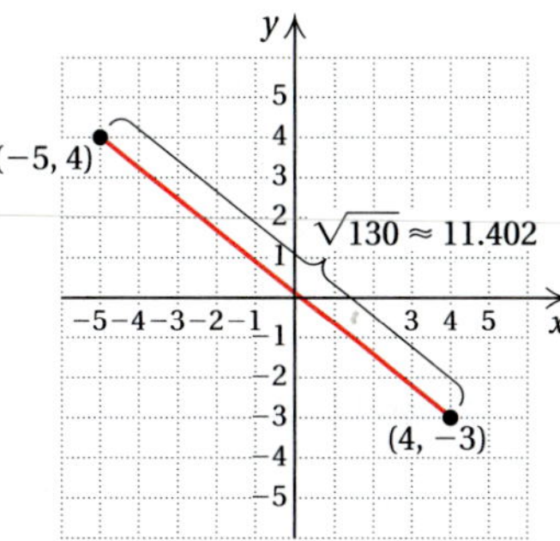

Do Exercises 3 and 4.

Find the distance between the pair of points. Where appropriate, give an approximation to three decimal places.

3. $(2, 6)$ and $(-4, -2)$

4. $(-2, 1)$ and $(4, 2)$

b Midpoints of Segments

The distance formula can be used to derive a formula for finding the midpoint of a segment when the coordinates of the endpoints are known.

THE MIDPOINT FORMULA

If the endpoints of a segment are (x_1, y_1) and (x_2, y_2), then the coordinates of the midpoints are
$$\left(\frac{x_1 + x_2}{2}, \frac{y_1 + y_2}{2}\right).$$
(To locate the midpoint, determine the average of the x-coordinates and the average of the y-coordinates.)

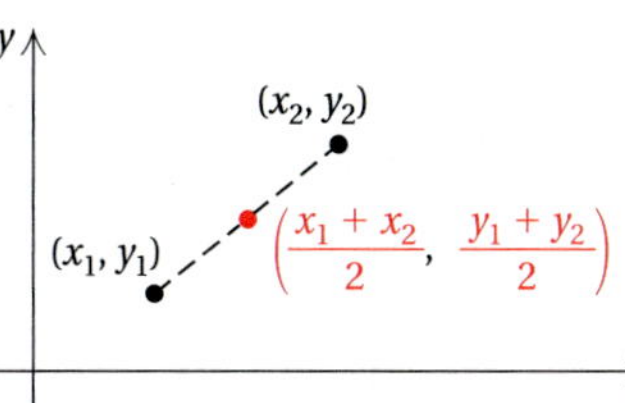

EXAMPLE 4 Find the midpoint of the segment with endpoints $(-2, 3)$ and $(4, -6)$.

Using the midpoint formula, we obtain
$$\left(\frac{-2 + 4}{2}, \frac{3 + (-6)}{2}\right), \quad \text{or} \quad \left(\frac{2}{2}, \frac{-3}{2}\right), \quad \text{or} \quad \left(1, -\frac{3}{2}\right).$$

Do Exercises 5 and 6.

Find the midpoint of the segment with the given endpoints.

5. $(-3, 1)$ and $(6, -7)$

6. $(10, -7)$ and $(8, -3)$

Answers on page A-75

D EXERCISE SET

a Find the distance between the pair of points. Where appropriate, give an approximation to three decimal places.

1. $(6, -4)$ and $(2, -7)$

2. $(1, 2)$ and $(-4, 14)$

3. $\left(-5, \frac{3}{4}\right)$ and $\left(-5, -\frac{3}{2}\right)$

4. $\left(-6, \frac{1}{2}\right)$ and $\left(-3, \frac{1}{2}\right)$

5. $(0, -4)$ and $(5, -6)$

6. $(8, 3)$ and $(8, -3)$

7. $(9, 9)$ and $(-9, -9)$

8. $(2, 22)$ and $(-8, 1)$

9. $(2.8, -3.5)$ and $(-4.3, -3.5)$

10. $(6.1, 2)$ and $(5.6, -4.4)$

11. $\left(\frac{5}{7}, \frac{1}{14}\right)$ and $\left(\frac{1}{7}, \frac{11}{14}\right)$

12. $(0, \sqrt{7})$ and $(\sqrt{6}, 0)$

13. $(-23, 10)$ and $(56, -17)$

14. $(34, -18)$ and $(-46, -38)$

15. (a, b) and $(0, 0)$

16. $(0, 0)$ and (p, q)

17. $(\sqrt{2}, -\sqrt{3})$ and $(-\sqrt{7}, \sqrt{5})$

18. $(\sqrt{8}, \sqrt{3})$ and $(-\sqrt{5}, -\sqrt{6})$

19. $(1000, -240)$ and $(-2000, 580)$

20. $(-3000, 560)$ and $(-430, -640)$

21. $\left(-\frac{2}{5}, 14\right)$ and $(0, 14)$

22. $(0, 7)$ and $(0, -7)$

b Find the midpoint of the segment with the given endpoints.

23. $(-1, 9)$ and $(4, -2)$

24. $(5, 10)$ and $(2, -4)$

25. $(3, 5)$ and $(-3, 6)$

26. $(7, -3)$ and $(4, 11)$

27. $(-10, -13)$ and $(8, -4)$

28. $(6, -2)$ and $(-5, 12)$

29. $(-3.4, 8.1)$ and $(2.9, -8.7)$

30. $(4.1, 6.9)$ and $(5.2, -6.9)$

31. $\left(\frac{1}{6}, -\frac{3}{4}\right)$ and $\left(-\frac{1}{3}, \frac{5}{6}\right)$

32. $\left(-\frac{4}{5}, -\frac{2}{3}\right)$ and $\left(\frac{1}{8}, \frac{3}{4}\right)$

33. $(\sqrt{2}, -1)$ and $(\sqrt{3}, 4)$

34. $(9, 2\sqrt{3})$ and $(-4, 5\sqrt{3})$

SYNTHESIS

Find the distance between the given points.

35. $(-1, 3k)$ and $(6, 2k)$

36. (a, b) and $(-a, -b)$

37. $(6m, -7n)$ and $(-2m, n)$

38. $(-3\sqrt{3}, 1 - \sqrt{6})$ and $(\sqrt{3}, 1 + \sqrt{6})$

If the sides of a triangle have lengths a, b, and c and $a^2 + b^2 = c^2$, then the triangle is a right triangle. Determine whether the given points are vertices of a right triangle.

39. $(-8, -5)$, $(6, 1)$, and $(-4, 5)$

40. $(9, 6)$, $(-1, 2)$, and $(1, -3)$

41. Find the midpoint of the segment with the endpoints $(2 - \sqrt{3}, 5\sqrt{2})$ and $(2 + \sqrt{3}, 3\sqrt{2})$.

42. Find the point on the y-axis that is equidistant from $(2, 10)$ and $(6, 2)$.

Objectives

a Solve quadratic and other polynomial inequalities.

b Graph quadratic inequalities.

1. Solve by graphing:

$$x^2 + 2x - 3 > 0.$$

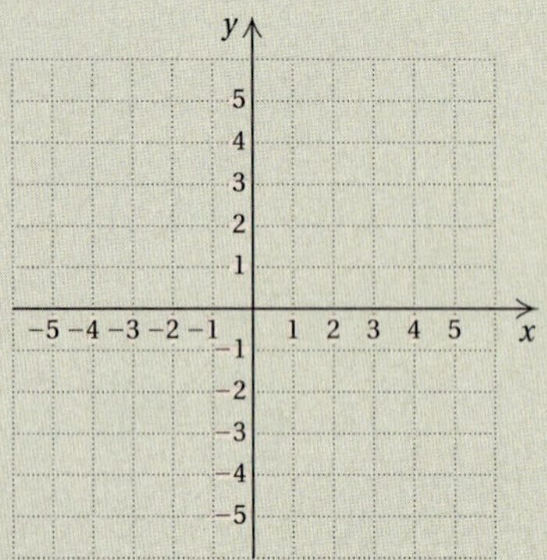

Answer on page A-75

E NONLINEAR INEQUALITIES

a Solving Polynomial Inequalities

Inequalities like the following are called **quadratic inequalities:**

$$x^2 + 3x - 10 < 0, \qquad 5x^2 - 3x + 2 \geq 0.$$

In each case, we have a polynomial of degree 2 on the left.

We will first consider solving a quadratic inequality, such as $ax^2 + bx + c > 0$, using the graph of a related equation, $y = ax^2 + bx + c$.

EXAMPLE 1 Solve: $x^2 + 3x - 10 > 0$.

Consider the equation $y = x^2 + 3x - 10$ and its graph. Its graph opens up since the leading coefficient $(a = 1)$ is positive. Values of y will be positive to the left and right of the intercepts, as shown below. We find the intercepts by setting the polynomial equal to 0 and solving:

$$x^2 + 3x - 10 = 0$$
$$(x + 5)(x - 2) = 0$$
$$x + 5 = 0 \quad or \quad x - 2 = 0$$
$$x = -5 \quad or \quad x = 2.$$

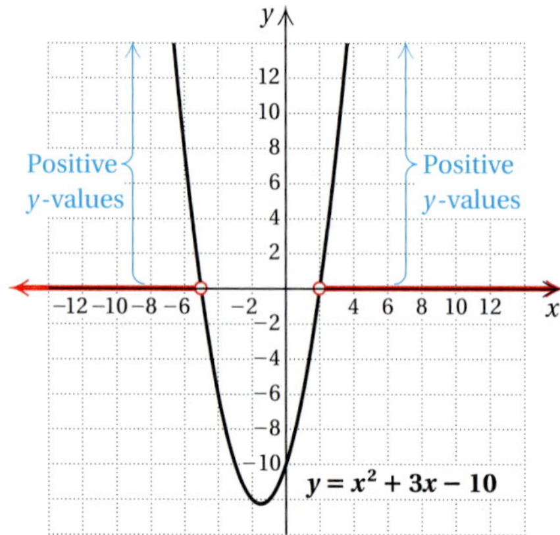

Thus the solution set of the inequality is

$$\{x \mid x < -5 \text{ or } x > 2\}, \text{ or } (-\infty, -5) \cup (2, \infty).$$

Do Exercise 1.

EXAMPLE 2 Solve: $x^2 + 3x - 10 < 0$.

Looking again at the graph of $y = x^2 + 3x - 10$ or at least visualizing it tells us that y-values are negative for those x-values between -5 and 2.

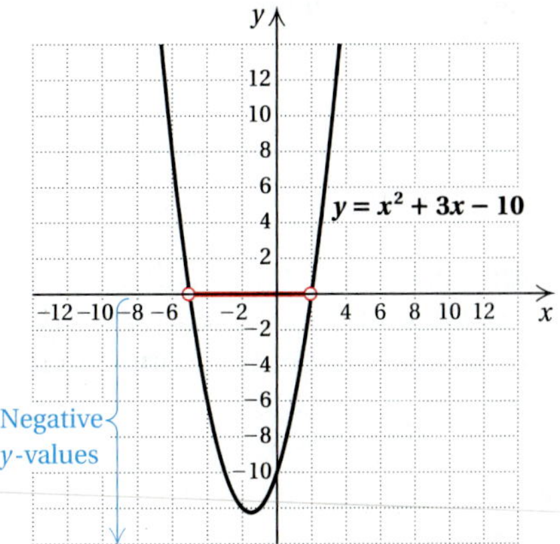

That is, the solution set is $\{x \mid -5 < x < 2\}$, or $(-5, 2)$.

Do Exercise 2.

When an inequality contains $\leq$ or $\geq$, the x-values of the intercepts must be included. Thus the solution set of the inequality $x^2 + 3x - 10 \leq 0$ is $\{x \mid -5 \leq x \leq 2\}$, or $[-5, 2]$.

Do Exercise 3.

In Examples 1 and 2, we see that the intercepts divide the number line into intervals. If a particular equation has a positive output for one number in an interval, it will be positive for all the numbers in the interval. Thus we can merely make a test substitution in each interval to solve the inequality. This is very similar to our method of using test points to graph a linear inequality in a plane.

EXAMPLE 3 Solve: $x^2 + 3x - 10 < 0$.

We set the polynomial equal to 0 and solve. The solutions of $x^2 + 3x - 10 = 0$, or $(x + 5)(x - 2) = 0$, are -5 and 2. We then locate them on a number line as follows. Note that the numbers divide the number line into three intervals A, B, and C.

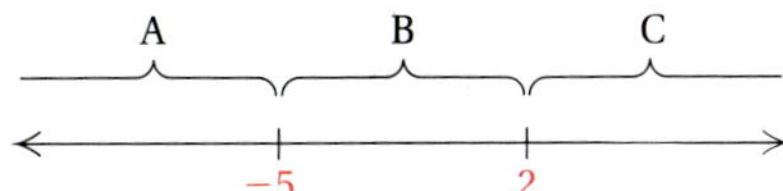

We choose a test number in interval A, say -7, and substitute -7 for x in the equation $y = x^2 + 3x - 10$:

$$y = (-7)^2 + 3(-7) - 10 = 49 - 21 - 10 = 18.$$

Note that $18 > 0$, so the y-values will be positive for any number in interval A.

Next, we try a test number in interval B, say 1, and find the corresponding y-value:

$$y = 1^2 + 3(1) - 10 = 1 + 3 - 10 = -6.$$

Note that $-6 < 0$, so the y-values will be negative for any number in interval B.

2. Solve by graphing:

$$x^2 + 2x - 3 < 0.$$

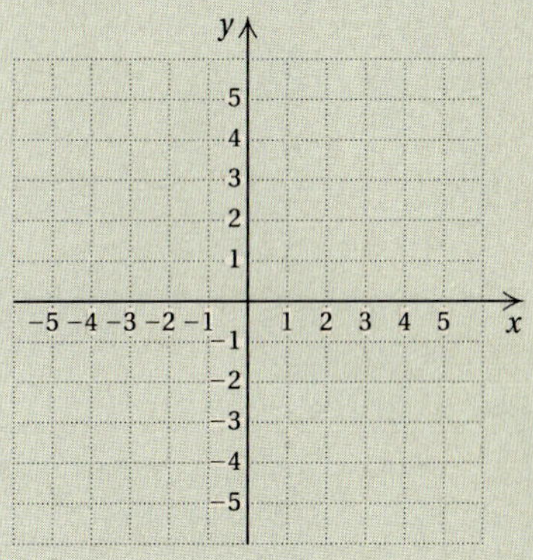

3. Solve by graphing:

$$x^2 + 2x - 3 \leq 0.$$

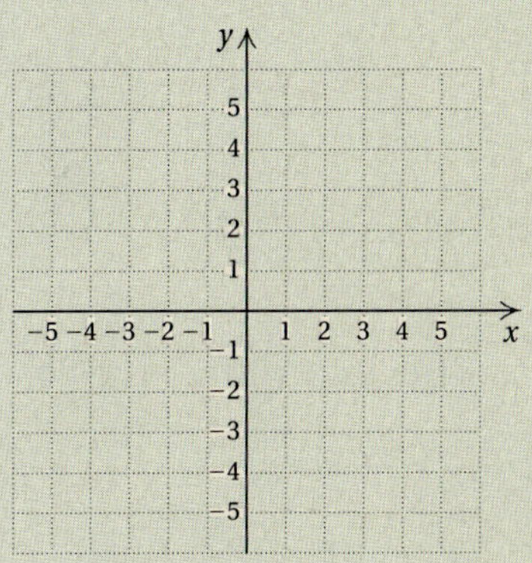

Answers on page A-75

Solve using the method of Example 3.

4. $x^2 + 3x > 4$

5. $x^2 + 3x \le 4$

6. Solve: $6x(x + 1)(x - 1) < 0$.

Answers on page A-75

Next, we try a test number in interval C, say 4, and find the corresponding y-value:

$$y = 4^2 + 3(4) - 10 = 16 + 12 - 10 = 18.$$

Note that $18 > 0$, so the y-values will be positive for any number in interval C.

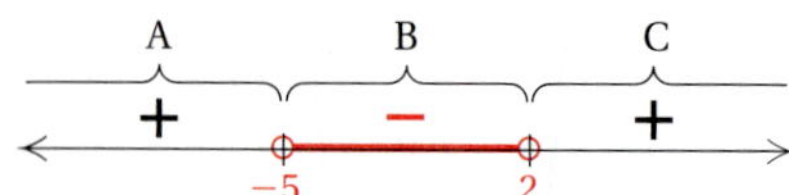

We are looking for numbers x for which $x^2 + 3x - 10 < 0$. Thus any number x in interval B is a solution. If the inequality had been $\le$ or $\ge$, we would also need to include the intercepts -5 and 2 in the solution set. The solution set is $\{x \mid -5 < x < 2\}$, or $(-5, 2)$.

Do Exercises 4 and 5.

EXAMPLE 4 Solve: $5x(x + 3)(x - 2) \ge 0$.

The solutions of $5x(x + 3)(x - 2) = 0$ are -3, 0, and 2. They divide the real-number line into four intervals, as follows.

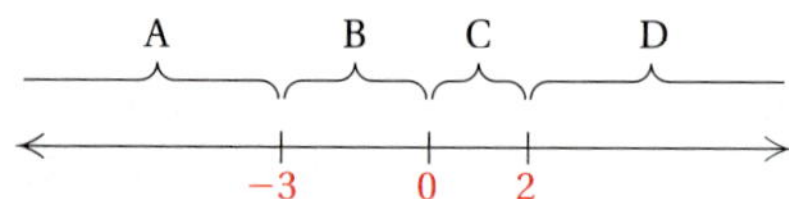

We try test numbers in each interval:

A: Test -5, $y = 5(-5)(-5 + 3)(-5 - 2) = -350$.

B: Test -2, $y = 5(-2)(-2 + 3)(-2 - 2) = 40$.

C: Test 1, $y = 5(1)(1 + 3)(1 - 2) = -20$.

D: Test 3, $y = 5(3)(3 + 3)(3 - 2) = 90$.

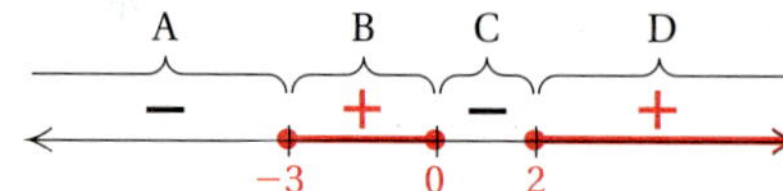

The expression is positive for values of x in intervals B and D. Since the inequality symbol is $\ge$, we will need to include the intercepts. The solution set of the inequality is

$$\{x \mid -3 \le x \le 0 \text{ or } x \ge 2\}, \text{ or } [-3, 0] \cup [2, \infty).$$

Do Exercise 6.

b Graphing Quadratic Inequalities

Graphing quadratic inequalities involves the same procedure as graphing quadratic equations, but the graph will also include the interior region enclosed by the parabola or the exterior region outside the parabola.

EXAMPLE 5 Graph: $y \le x^2 + 2x - 3$.

We first replace the inequality symbol with an equals sign and graph the equation:

$$y = x^2 + 2x - 3.$$

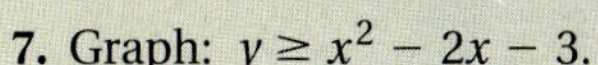

The x-coordinate of the vertex is

$$-\frac{b}{2a} = -\frac{2}{2 \cdot 1} = -1.$$

We substitute -1 for x in the equation to find the second coordinate of the vertex:

$$\begin{aligned} y &= x^2 + 2x - 3 \\ &= (-1)^2 + 2(-1) - 3 \\ &= 1 - 2 - 3 \\ &= -4. \end{aligned}$$

The vertex is $(-1, -4)$. The line of symmetry is $x = -1$. We choose some x-values on both sides of the vertex and graph the parabola.

For $x = 0$, $\quad y = x^2 + 2x - 3 = 0^2 + 2 \cdot 0 - 3 = -3$.

For $x = -2$, $\quad y = x^2 + 2x - 3 = (-2)^2 + 2(-2) - 3 = -3$.

For $x = 1$, $\quad y = x^2 + 2x - 3 = 1^2 + 2 \cdot 1 - 3 = 0$.

For $x = -3$, $\quad y = x^2 + 2x - 3 = (-3)^2 + 2(-3) - 3 = 0$.

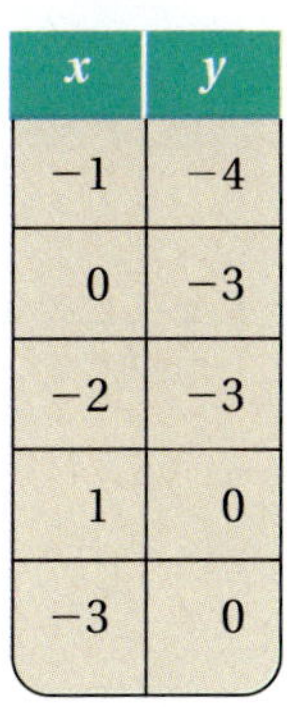

x	y
-1	-4
0	-3
-2	-3
1	0
-3	0

← This is the vertex.

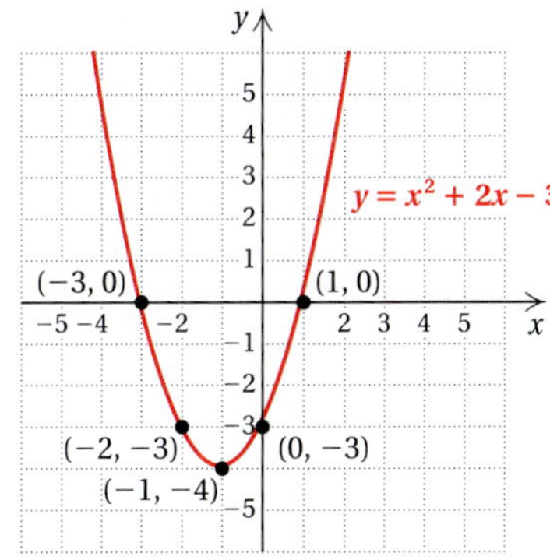

The inequality symbol is $\leq$, so we draw the curve solid.

To determine which region to shade, we choose a point not on the curve as a test point. The origin $(0, 0)$ is usually an easy one to use.

$$\begin{array}{c|l} \multicolumn{2}{c}{y \leq x^2 + 2x - 3} \\ \hline 0 & 0^2 + 2 \cdot 0 - 3 \\ & 0 + 0 - 3 \\ & -3 \end{array} \qquad \textbf{FALSE}$$

We see that $(0, 0)$ is *not* a solution, so we shade the exterior region. Had the substitution given us a true inequality, we would have shaded the interior.

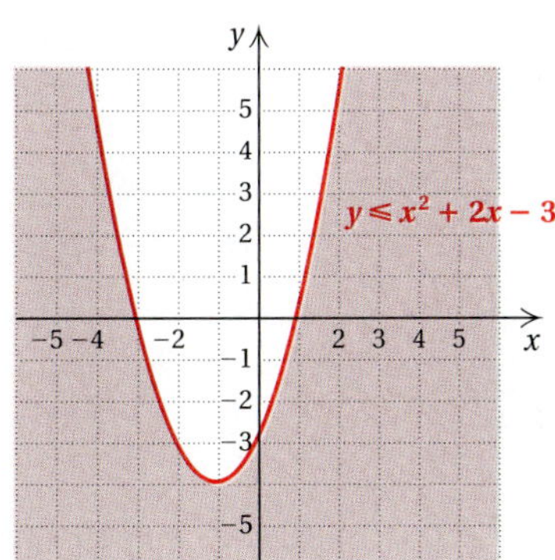

Do Exercises 7 and 8.

7. Graph: $y \geq x^2 - 2x - 3$.

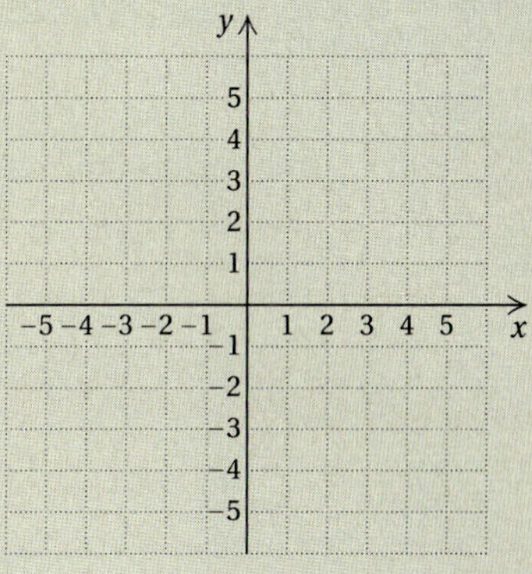

8. Graph: $y > -x^2 + 2x$.

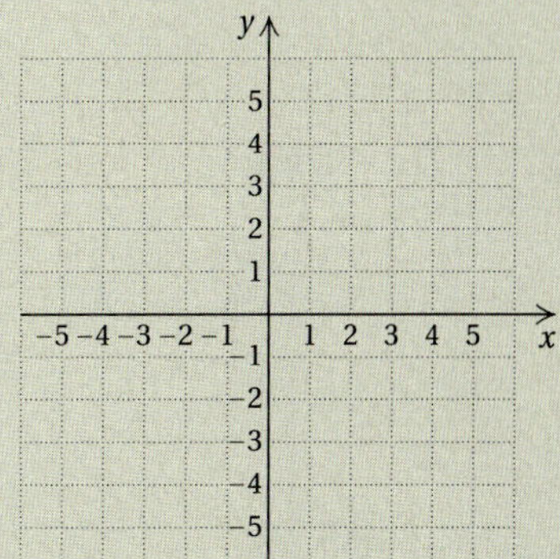

Answers on page A-75

E EXERCISE SET

a Solve algebraically and verify results from the graph.

1. $(x-6)(x+2)>0$

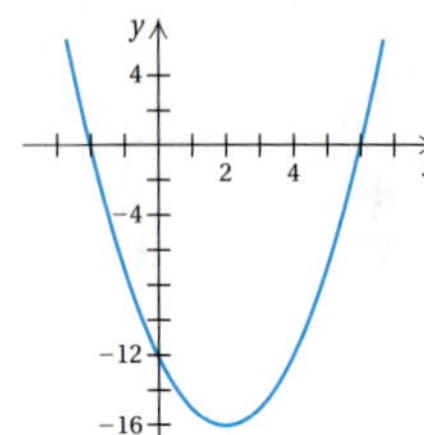

2. $(x-5)(x+1)>0$

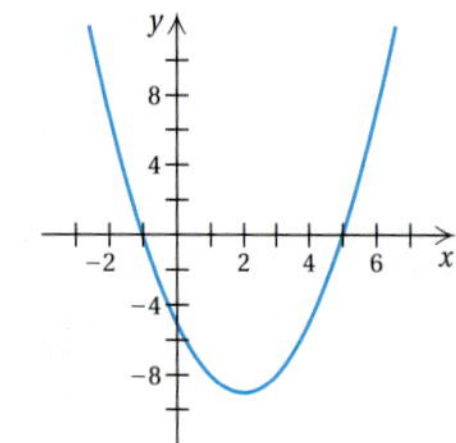

3. $4-x^2\geq 0$

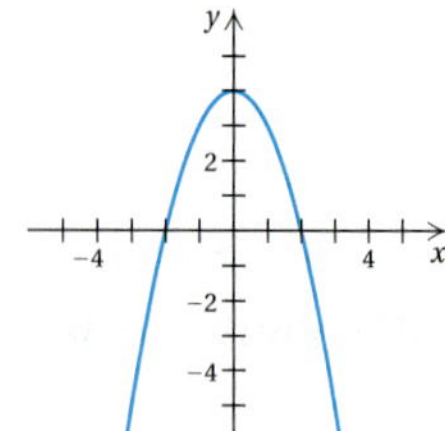

4. $9-x^2\leq 0$

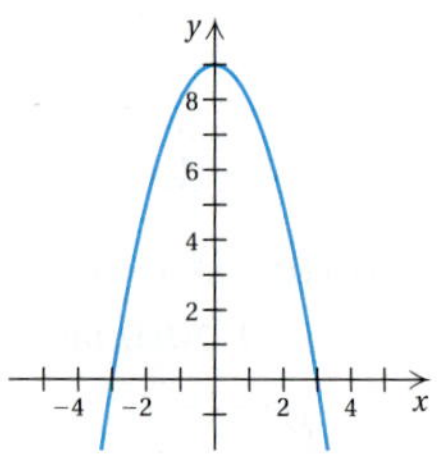

Solve.

5. $3(x+1)(x-4)\leq 0$

6. $(x-7)(x+3)\leq 0$

7. $x^2-x-2<0$

8. $x^2+x-2<0$

9. $x^2-2x+1\geq 0$

10. $x^2+6x+9<0$

11. $x^2+8<6x$

12. $x^2-12>4x$

13. $3x(x+2)(x-2)<0$

14. $5x(x+1)(x-1)>0$

15. $(x+9)(x-4)(x+1)>0$

16. $(x-1)(x+8)(x-2)<0$

17. $(x+3)(x+2)(x-1)<0$

18. $(x-2)(x-3)(x+1)<0$

b Graph.

19. $y \leq 3x^2$

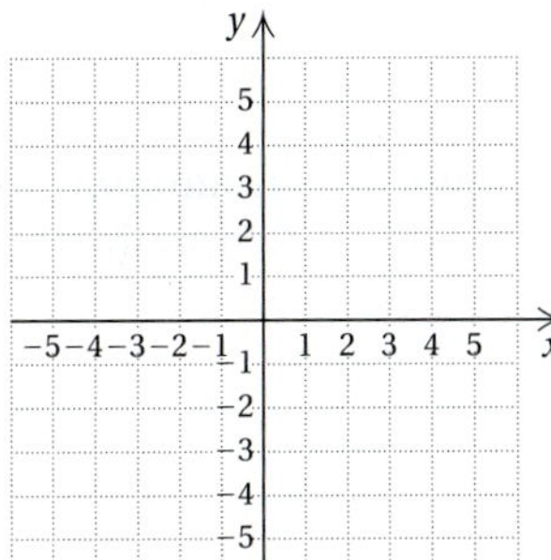

20. $y > -x^2$

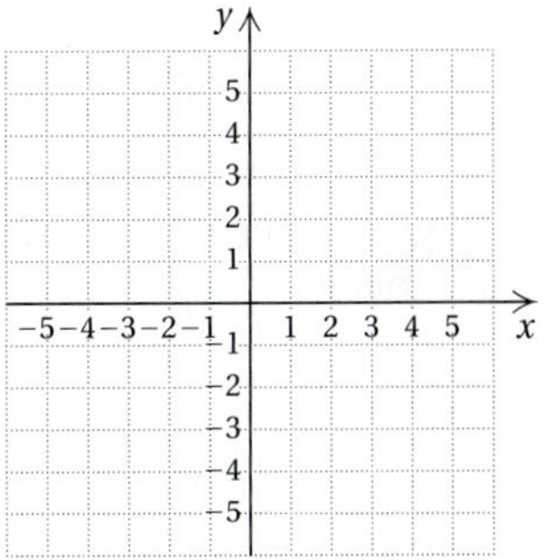

21. $y < 4 - x^2$

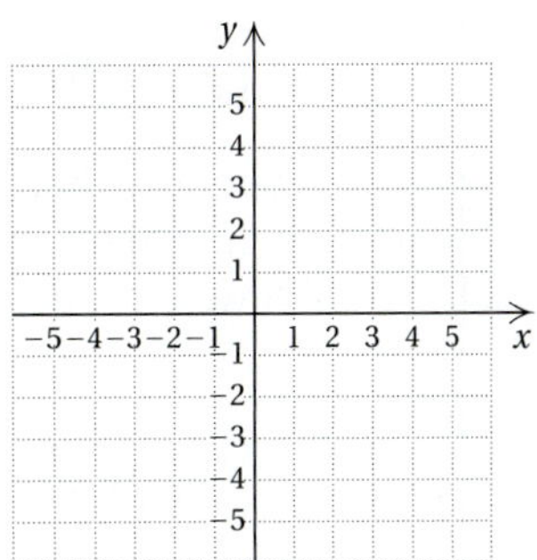

22. $y \leq 2 - x^2$

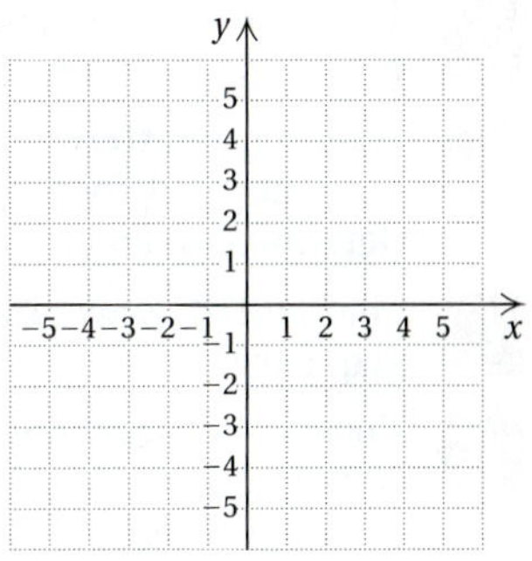

23. $y \leq \frac{1}{2}x^2 - 1$

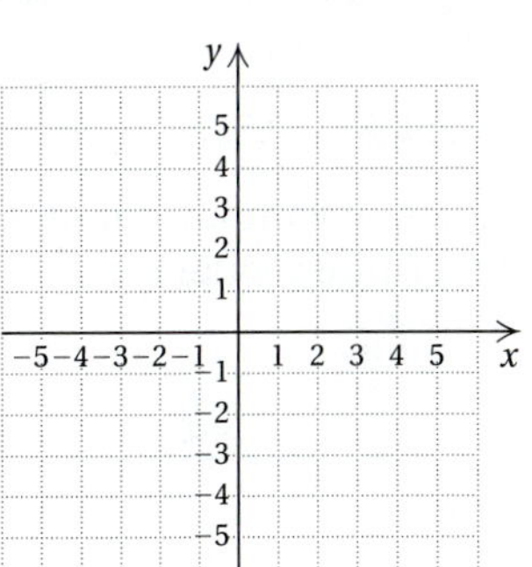

24. $y < -\frac{1}{2}x^2 + 2$

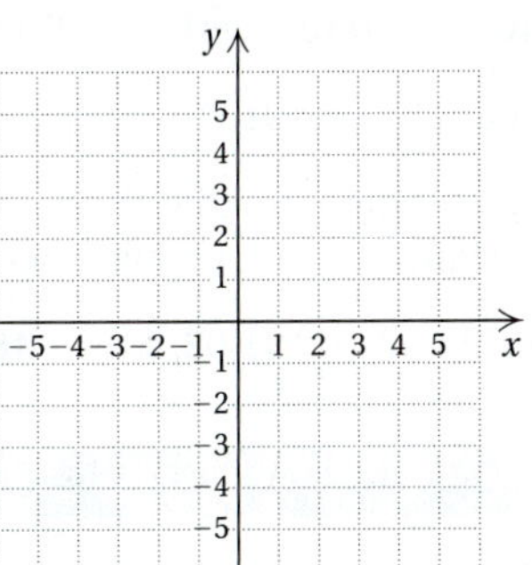

25. $y > x^2 + 4x - 1$

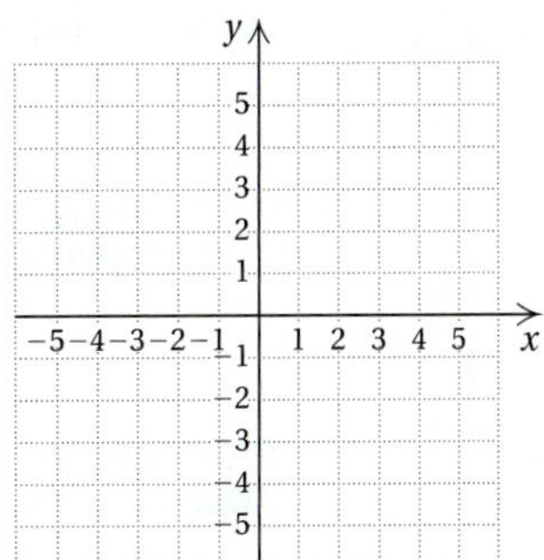

26. $y \leq x^2 - 2x - 3$

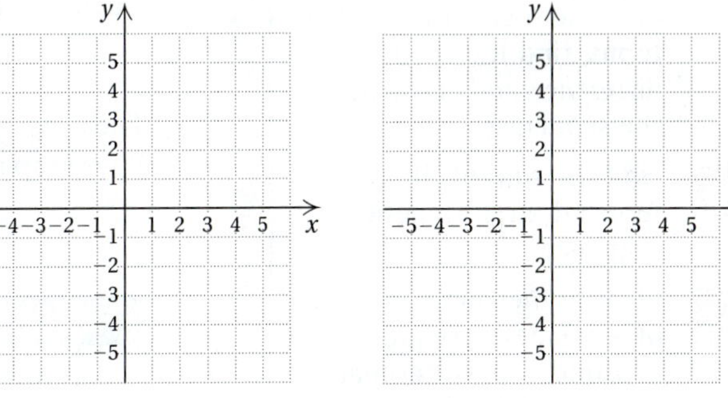

27. $y \geq x^2 + 2x + 1$

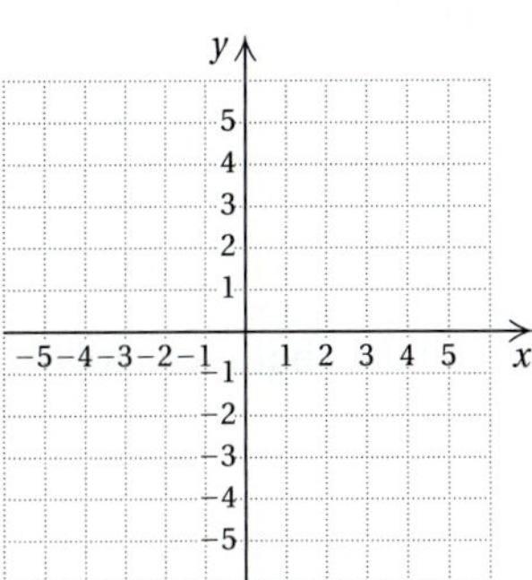

28. $y > x^2 + x - 6$

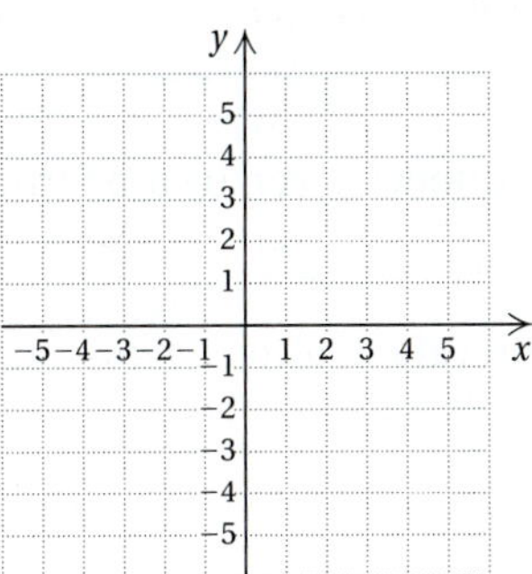

29. $y < 5 - x - x^2$

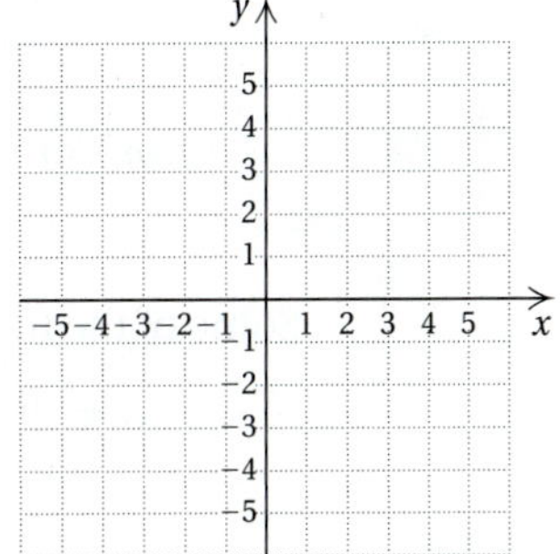

30. $y \geq -x^2 + 2x + 3$

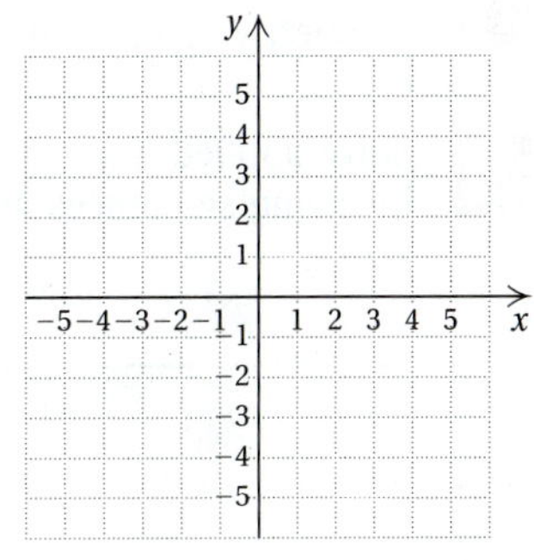

31. $y \leq -x^2 - 2x + 3$

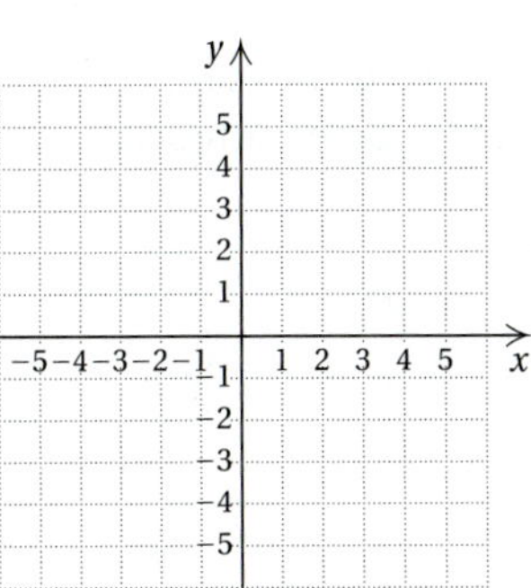

32. $y < -2x^2 - 4x + 1$

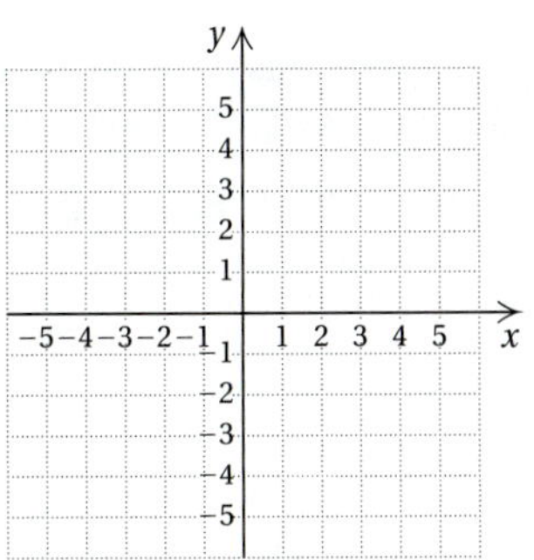

33. $y > 2x^2 + 4x - 1$

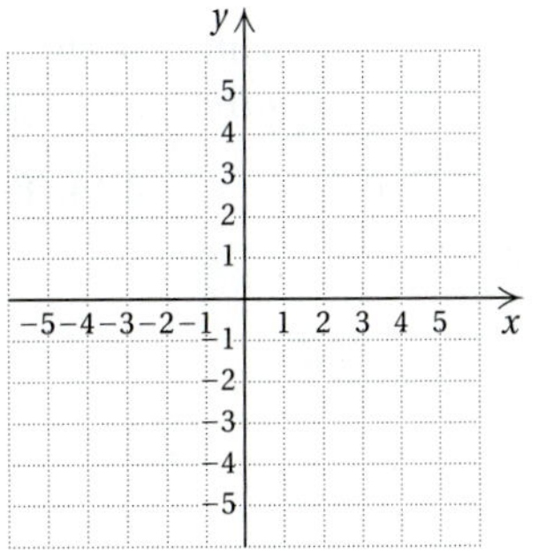

34. $y \leq x^2 + 2x - 5$

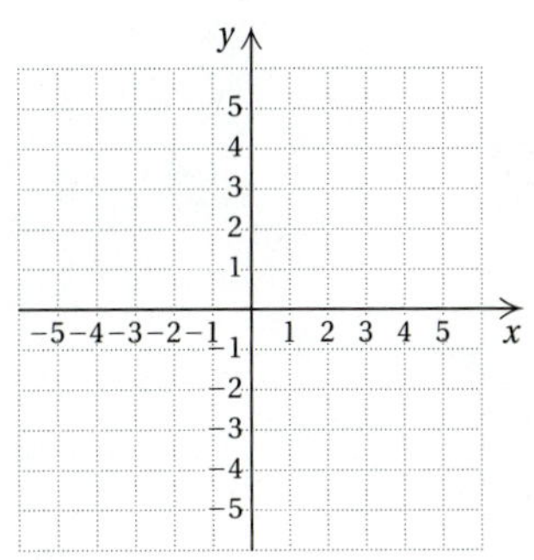

Objectives

a Find terms of sequences given the nth term.

b Look for a pattern in a sequence and try to determine a general term.

c Find partial sums of a sequence.

d Convert between sigma notation and other notation for a series.

e For any arithmetic sequence, find the nth term when n is given and n when the nth term is given, and given two terms, find the common difference and construct the sequence.

f Find the sum of the first n terms of an arithmetic sequence.

g Identify the common ratio of a geometric sequence, and find a given term and the sum of the first n terms.

h Find the sum of an infinite geometric series, if it exists.

i Solve applied problems involving sequences and series.

F SEQUENCES AND SERIES

a Sequences

Suppose that \$1000 is invested at 6%, compounded annually. The amounts to which the account will grow after 1 yr, 2 yr, 3 yr, 4 yr, and so on, form the following sequence of numbers:

$$\underset{\downarrow}{(1)}\quad \underset{\downarrow}{(2)}\quad \underset{\downarrow}{(3)}\quad \underset{\downarrow}{(4)}$$

\$1060.00, \$1123.60, \$1191.02, \$1262.48, ...

We can think of this as a function that pairs 1 with \$1060.00, 2 with \$1123.60, 3 with \$1191.02, and so on. A **sequence** is thus a *function*, where the domain is a set of consecutive positive integers beginning with 1.

If we continue to compute the amounts of money in the account forever, we obtain an **infinite sequence** with function values

\$1060.00, \$1123.60, \$1191.02, \$1262.48, \$1338.23, \$1418.52, ...

The dots ". . ." at the end indicate that the sequence goes on without stopping. If we stop after a certain number of years, we obtain a **finite sequence:**

\$1060.00, \$1123.60, \$1191.02, \$1262.48.

SEQUENCES

An **infinite sequence** is a function having for its domain the set of positive integers, $\{1, 2, 3, 4, 5, \ldots\}$.

A **finite sequence** is a function having for its domain a set of positive integers, $\{1, 2, 3, 4, 5, \ldots, n\}$, for some positive integer n.

Consider the sequence given by the formula

$$a(n) = 2^n, \quad \text{or} \quad a_n = 2^n.$$

Some of the function values, also known as the **terms** of the sequence, follow:

$$a_1 = 2^1 = 2,$$
$$a_2 = 2^2 = 4,$$
$$a_3 = 2^3 = 8,$$
$$a_4 = 2^4 = 16,$$
$$a_5 = 2^5 = 32.$$

The first term of the sequence is denoted as a_1, the fifth term as a_5, and the nth term, or **general term,** as a_n. This sequence can also be denoted as

$$2,\ 4,\ 8,\ \ldots, \quad \text{or as} \quad 2,\ 4,\ 8,\ \ldots,\ 2^n,\ \ldots.$$

EXAMPLE 1 Find the first 4 terms and the 23rd term of the sequence whose general term is given by $a_n = (-1)^n n^2$.

We have $a_n = (-1)^n n^2$, so

$$\begin{aligned} a_1 &= (-1)^1 \cdot 1^2 = -1, \\ a_2 &= (-1)^2 \cdot 2^2 = 4, \\ a_3 &= (-1)^3 \cdot 3^2 = -9, \\ a_4 &= (-1)^4 \cdot 4^2 = 16, \\ a_{23} &= (-1)^{23} \cdot 23^2 = -529. \end{aligned}$$

Do Exercises 1 and 2.

Note in Example 1 that the power $(-1)^n$ causes the signs of the terms to alternate between positive and negative, depending on whether n is even or odd. This kind of sequence is called an **alternating sequence.**

b Finding the General Term

When only the first few terms of a sequence are known, we do not know for sure what the general term is, but we might be able to make a prediction by looking for a pattern.

EXAMPLE 2 For each of the following sequences, predict the general term.

a) 1, 4, 9, 16, 25, ...

b) −1, 3, −9, 27, −81, ...

c) 2, 4, 8, ...

a) These are squares of numbers, so the general term might be n^2.

b) These are powers of 3 with alternating signs, so the general term might be $(-1)^n 3^{n-1}$.

c) If we see the pattern of powers of 2, we will see 16 as the next term and guess 2^n for the general term. Then the sequence could be written with more terms as

2, 4, 8, 16, 32, 64, 128, ...

If we see that we can get the second term by adding 2, the third term by adding 4, and the next term by adding 6, and so on, we will see 14 as the next term. A general term for the sequence is $n^2 - n + 2$, and the sequence can be written with more terms as

2, 4, 8, 14, 22, 32, 44, 58, ...

Do Exercises 3–6.

Example 2(c) illustrates that, in fact, you can never be certain about the general term. The fewer the given terms, the greater the uncertainty.

In each of the following, the nth term of a sequence is given. Find the first 4 terms and the 31st term.

1. $a_n = 2n + 1$

2. $a_n = \dfrac{(-1)^n}{n+5}$

Predict the general, or nth term, a_n, of the sequence. Answers may vary.

3. 3, 6, 9, 12, 15, ...

4. −1, 2, −3, 4, −5, ...

5. 1, 8, 27, 64, 125, ...

6. 1, 2, 4, 8, 16, ...

Answers on page A-76

7. For the sequence 1, −1, 2, −2, 3, −3, 4, −4, . . . , find each of the following.

a) S_2

b) S_3

c) S_7

Answers on page A-76

c Sums and Series

SERIES

Given the infinite sequence

$$a_1, a_2, a_3, a_4, \ldots, a_n, \ldots,$$

the sum of the terms

$$a_1 + a_2 + a_3 + \cdots + a_n + \cdots$$

is called an **infinite series.** A **partial sum** is the sum of the first n terms:

$$a_1 + a_2 + a_3 + \cdots + a_n.$$

A partial sum is also called a **finite series,** or ***n*th partial sum,** and is denoted S_n.

EXAMPLE 3 For the sequence −2, 4, −6, 8, −10, 12, −14, . . . , find each of the following.

a) S_1

b) S_4

c) S_5

a) $S_1 = -2$

b) $S_4 = -2 + 4 + (-6) + 8 = 4$

c) $S_5 = -2 + 4 + (-6) + 8 + (-10) = -6$

Do Exercise 7.

d Sigma Notation

The Greek letter Σ (sigma) can be used to simplify notation when the general term of a sequence is a formula. For example, the sum of the first four terms of the sequence 3, 5, 7, 9, . . . , $2k + 1$, . . . can be named as follows, using what is called **sigma notation,** or **summation notation:**

$$\sum_{k=1}^{4} (2k + 1).$$

This is read "the sum as k goes from 1 to 4 of $2k + 1$." The letter k is called the **index of summation.** Sometimes the index of summation starts at a number other than 1.

EXAMPLE 4 Find and evaluate each of the following sums.

a) $\sum_{k=1}^{5} k^3$ b) $\sum_{k=0}^{4} (-1)^k 5^k$ c) $\sum_{k=2}^{5} (2^k - 3)$

a) We replace k with 1, 2, 3, 4, and 5. Then we add the results.

$$\sum_{k=1}^{5} k^3 = 1^3 + 2^3 + 3^3 + 4^3 + 5^3$$
$$= 1 + 8 + 27 + 64 + 125$$
$$= 225$$

b) $\sum_{k=0}^{4} (-1)^k 5^k = (-1)^0 5^0 + (-1)^1 5^1 + (-1)^2 5^2 + (-1)^3 5^3 + (-1)^4 5^4$

$$= 1 - 5 + 25 - 125 + 625 = 521$$

c) $\sum_{k=2}^{5} (2^k - 3) = (2^2 - 3) + (2^3 - 3) + (2^4 - 3) + (2^5 - 3)$

$$= 1 + 5 + 13 + 29 = 48$$

Do Exercises 8–10.

EXAMPLE 5 Write sigma notation for each sum.

a) $1 + 2 + 4 + 8 + 16 + 32 + 64$

b) $-2 + 4 - 6 + 8 - 10$

c) $x + \frac{x^2}{2} + \frac{x^3}{3} + \frac{x^4}{4} + \cdots$

a) $1 + 2 + 4 + 8 + 16 + 32 + 64$

This is a sum of powers of 2, beginning with 2^0, or 1, and ending with 2^6, or 64. Sigma notation is $\Sigma_{k=0}^{6} 2^k$.

b) $-2 + 4 - 6 + 8 - 10$

Disregarding the alternating signs, we see that this is the sum of the first 5 even integers. Note that $2k$ is a formula for the kth positive even integer, and $(-1)^k = -1$ when k is odd and $(-1)^k = 1$ when k is even. Thus the general term is $(-1)^k(2k)$. The sum begins with $k = 1$ and ends with $k = 5$, so sigma notation is $\Sigma_{k=1}^{5} (-1)^k(2k)$.

c) $3 + 9 + 27 + 81 + \cdots$

This is a sum of powers of 3, and it is also an infinite series. We use the symbol ∞ to represent infinity and write the series using sigma notation: $\Sigma_{k=1}^{\infty} 3^k$.

Do Exercises 11–13.

Find and evaluate each sum.

8. $\sum_{k=1}^{5} k^2$

9. $\sum_{k=0}^{4} (-1)^k 2^k$

10. $\sum_{k=4}^{7} (k^2 - 12)$

Write sigma notation.

11. $2 + 4 + 6 + 8 + 10$

12. $1 - 8 + 27 - 64$

13. $4 + 9 + 16 + 25 + \cdots$

Answers on page A-76

For each of the following arithmetic sequences, identify the first term, a_1, and the common difference, d.

14. 3.1, 3.9, 4.7, 5.5, 6.3, ...

15. 14, 9, 4, −1, −6, ...

Answers on page A-76

e Arithmetic Sequences

In the sequence 2, 5, 8, 11, 14, 17, ..., note that adding 3 to any term produces the next term. In other words, the difference between any term and the preceding one is 3. This is an example of an **arithmetic sequence,** or **arithmetic progression.**

ARITHMETIC SEQUENCE

A sequence is **arithmetic** if there exists a number d, called the **common difference,** such that $a_{n+1} = a_n + d$ for any integer $n \geq 1$.

EXAMPLE 6 For each of the following arithmetic sequences, identify the first term, a_1, and the common difference, d.

a) 4, 9, 14, 19, 24, ...

b) 34, 27, 20, 13, 6, −1, −8, ...

The first term, a_1, is the first term listed. To find the common difference, d, we choose any term beyond the first and subtract the preceding term from it.

Sequence	*First Term, a_1*	*Common Difference, d*
a) 4, 9, 14, 19, 24, ...	4	5 $(9 - 4 = 5)$
b) 34, 27, 20, 13, 6, −1, −8, ...	34	−7 $(27 - 34 = -7)$

We obtained the common difference by subtracting a_1 from a_2. Had we subtracted a_2 from a_3 or a_3 from a_4, we would have obtained the same values for d. Thus we can check by adding d to each term in a sequence to see if we progress correctly to the next term.

CHECK:

a) $4 + 5 = 9$, $9 + 5 = 14$, $14 + 5 = 19$, $19 + 5 = 24$

b) $34 + (-7) = 27$, $27 + (-7) = 20$, $20 + (-7) = 13$, $13 + (-7) = 6$, $6 + (-7) = -1$, $-1 + (-7) = -8$

Do Exercises 14 and 15.

To find a formula for the general, or nth, term of any arithmetic sequence, we denote the common difference by d, write out the first few terms, and look for a pattern:

a_1,

$a_2 = a_1 + d$,

$a_3 = a_2 + d = (a_1 + d) + d = a_1 + 2d$ Substituting for a_2

$a_4 = a_3 + d = (a_1 + 2d) + d = a_1 + 3d$ Substituting for a_3

Note that the coefficient of d in each case is 1 less than the subscript.

Generalizing, we obtain the following formula.

nTH TERM OF AN ARITHMETIC SEQUENCE

The ***n*th term** of an arithmetic sequence is given by $a_n = a_1 + (n - 1)d$, for any integer $n \geq 1$.

EXAMPLE 7 Find the 14th term of the arithmetic sequence 4, 7, 10, 13, ...

We first note that $a_1 = 4$, $d = 7 - 4$, or 3, and $n = 14$. Then using the formula for the nth term, we obtain

$$a_n = a_1 + (n - 1)d$$

$$a_{14} = 4 + (14 - 1) \cdot 3 = 4 + 13 \cdot 3 = 4 + 39 = 43.$$

The 14th term is 43.

Do Exercise 16.

EXAMPLE 8 In the sequence of Example 7, which term is 301? That is, find n if $a_n = 301$.

We substitute 301 for a_n, 4 for a_1, and 3 for d in the formula for the nth term and solve for n:

$$a_n = a_1 + (n - 1)d$$

$$301 = 4 + (n - 1) \cdot 3 \quad \text{Substituting}$$

$$\left.\begin{aligned} 301 &= 4 + 3n - 3 \\ 301 &= 3n + 1 \\ 300 &= 3n \\ 100 &= n. \end{aligned}\right\} \quad \text{Solving for } n$$

The term 301 is the 100th term of the sequence.

Do Exercise 17.

16. Find the 13th term of the sequence 2, 6, 10, 14, ...

17. For the sequence given in Margin Exercise 16, which term is 298? That is, find n if $a_n = 298$.

Answers on page A-76

18. The 7th term of an arithmetic sequence is 79, and the 13th term is 151. Find a_1 and d and construct the sequence.

Answers on page A-76

Given two terms and their places in an arithmetic sequence, we can construct the sequence.

EXAMPLE 9 The 3rd term of an arithmetic sequence is 8, and the 16th term is 47. Find a_1 and d and construct the sequence.

We know that $a_3 = 8$ and $a_{16} = 47$. Thus we would have to add d 13 times to get 47 from 8. That is,

$$8 + 13d = 47. \qquad a_3 \text{ and } a_{16} \text{ are 13 terms apart.}$$

Solving $8 + 13d = 47$, we obtain

$$13d = 39$$
$$d = 3.$$

Since $a_3 = 8$, we subtract d twice to get a_1. Thus,

$$a_1 = 8 - 2 \cdot 3 = 2. \qquad a_1 \text{ and } a_3 \text{ are 2 terms apart.}$$

The sequence is 2, 5, 8, 11, ...

Do Exercise 18.

f Sum of the First *n* Terms of an Arithmetic Sequence

Consider the arithmetic sequence

$$3, 5, 7, 9, \ldots.$$

When we add the first 4 terms of the sequence, we get S_4, which is

$$3 + 5 + 7 + 9, \quad \text{or} \quad 24.$$

This sum is called an **arithmetic series.** To find a formula for the sum of the first n terms, S_n, of an arithmetic sequence, we first denote an arithmetic sequence, as follows:

This term is two terms back from the last. If you add d to this term, the result is the next-to-last term, $a_n - d$.

$$a_1, \quad (a_1 + d), \quad (a_1 + 2d), \quad \ldots, \quad (a_n - 2d), \quad (a_n - d), \quad a_n.$$

This is the next-to-last term. If you add d to this term, the result is a_n.

Then S_n is given by

$$S_n = a_1 + (a_1 + d) + (a_1 + 2d) + \cdots + (a_n - 2d) + (a_n - d) + a_n. \tag{1}$$

Reversing the order of the addition gives us

$$S_n = a_n + (a_n - d) + (a_n - 2d) + \cdots + (a_1 + 2d) + (a_1 + d) + a_1. \tag{2}$$

If we add corresponding terms of each side of equations (1) and (2), we get

$$2S_n = [a_1 + a_n] + [(a_1 + d) + (a_n - d)] + [(a_1 + 2d) + (a_n - 2d)]$$
$$+ \cdots + [(a_n - 2d) + (a_1 + 2d)]$$
$$+ [(a_n - d) + (a_1 + d)] + [a_n + a_1].$$

There are n expressions in square brackets.

This simplifies to

$$2S_n = [a_1 + a_n] + [a_1 + a_n] + [a_1 + a_n] + \cdots + [a_n + a_1]$$
$$+ [a_n + a_1] + [a_n + a_1].$$

Since $a_1 + a_n$ is being added n times, it follows that

$$2S_n = n(a_1 + a_n).$$

Dividing both sides by 2 leads to the following formula.

SUM OF THE FIRST n TERMS

The sum of the first n terms of an arithmetic sequence is given by

$$S_n = \frac{n}{2}(a_1 + a_n).$$

EXAMPLE 10 Find the sum of the first 100 natural numbers.

The sum is

$$1 + 2 + 3 + \cdots + 99 + 100.$$

This is the sum of the first 100 terms of the arithmetic sequence for which

$$a_1 = 1, \quad a_n = 100, \quad \text{and} \quad n = 100.$$

Thus substituting into the formula

$$S_n = \frac{n}{2}(a_1 + a_n),$$

we get

$$S_{100} = \frac{100}{2}(1 + 100) = 50(101) = 5050.$$

The sum of the first 100 natural numbers is 5050.

19. Find the sum of the even numbers 2 through 200.

20. Find the sum of the first 15 terms of the arithmetic sequence

$1, 3, 5, 7, 9, \ldots.$

Answers on page A-76

EXAMPLE 11 Find the sum of the first 15 terms of the arithmetic sequence 4, 7, 10, 13, ...

Note that $a_1 = 4$, $d = 3$, and $n = 15$. Before using the formula

$$S_n = \frac{n}{2}(a_1 + a_n),$$

we find the last term, a_{15}:

$$a_{15} = 4 + (15 - 1)3 \quad \text{Substituting into the formula } a_n = a_1 + (n - 1)d$$

$$= 4 + 14 \cdot 3 = 46.$$

Thus,

$$S_{15} = \frac{15}{2}(4 + 46) = \frac{15}{2}(50) = 375.$$

The sum of the first 15 terms is 375.

Do Exercises 19 and 20.

g Geometric Sequences

A sequence in which each term after the first is found by multiplying the preceding term by the same number is a **geometric sequence.** Consider this sequence:

$$2, 6, 18, 54, 162, \ldots.$$

Note that multiplying each term by 3 produces the next term. We call the number 3 the **common ratio** because it can be found by dividing any term by the preceding term. A geometric sequence is also called a *geometric progression.*

GEOMETRIC SEQUENCE

A sequence is **geometric** if there is a number r, called the **common ratio,** such that

$$\frac{a_{n+1}}{a_n} = r, \quad \text{or} \quad a_{n+1} = a_n r, \quad \text{for any integer } n \geq 1.$$

EXAMPLE 12 For each of the following geometric sequences, identify the common ratio.

a) 3, 6, 12, 24, 48, . . .

b) $1, -\frac{1}{2}, \frac{1}{4}, -\frac{1}{8}, \ldots$

c) \$5200, \$3900, \$2925, \$2193.75, . . .

Sequence	*Common Ratio*	
a) 3, 6, 12, 24, 48, . . .	2	$\left(\frac{6}{3} = 2, \frac{12}{6} = 2\right.$, and so on$\left.\right)$
b) $1, -\frac{1}{2}, \frac{1}{4}, -\frac{1}{8}, \ldots$	$-\frac{1}{2}$	$\left(\frac{-\frac{1}{2}}{1} = -\frac{1}{2}, \frac{\frac{1}{4}}{-\frac{1}{2}} = -\frac{1}{2}\right.$, and so on$\left.\right)$
c) \$5200, \$3900, \$2925, \$2193.75, . . .	0.75	$\left(\frac{\$3900}{\$5200} = 0.75, \frac{\$2925}{\$3900} = 0.75\right.$, and so on$\left.\right)$

For each of the following geometric sequences, identify the common ratio.

21. 1, 5, 25, 125, . . .

22. 3, −9, 27, −81, . . .

23. \$6000, \$5100, \$4335, \$3684.75, . . .

24. \$100, \$109, \$118.81, . . .

25. $1, \frac{1}{5}, \frac{1}{25}, \frac{1}{125}, \ldots$

Do Exercises 21–25.

We now find a formula for the general, or nth, term of a geometric sequence. Let a_1 be the first term and r the common ratio. The first few terms are as follows:

$$a_1,$$
$$a_2 = a_1 r,$$
$$a_3 = a_2 r = (a_1 r)r = a_1 r^2, \quad \text{Substituting } a_1 r \text{ for } a_2$$
$$a_4 = a_3 r = (a_1 r^2)r = a_1 r^3. \quad \text{Substituting } a_1 r^2 \text{ for } a_3$$

Note that the exponent is 1 less than the subscript.

Generalizing, we obtain the following.

nTH TERM OF A GEOMETRIC SEQUENCE

The ***n*th term** of a geometric sequence is given by

$$a_n = a_1 r^{n-1}, \quad \text{for any integer } n \geq 1.$$

EXAMPLE 13 Find the 7th term of the geometric sequence 4, 20, 100, ...

We first note that

$$a_1 = 4 \quad \text{and} \quad n = 7.$$

To find the common ratio, we can divide any term (other than the first) by the preceding term. Since the second term is 20 and the first is 4, we get

$$r = \frac{20}{4}, \quad \text{or} \quad 5.$$

Then using the formula $a_n = a_1 r^{n-1}$, we have

$$a_7 = 4 \cdot 5^{7-1} = 4 \cdot 5^6 = 4 \cdot 15{,}625 = 62{,}500.$$

Thus the 7th term is 62,500.

Answers on page A-76

26. Find the 9th term of the geometric sequence

$$2, 4, 8, 16, \ldots.$$

27. Find the 6th term of the geometric sequence

$$-3, 1, -\frac{1}{3}, \frac{1}{9}, \ldots.$$

28. Find the sum of the first 8 terms of the geometric sequence

$$4, 12, 36, 108, \ldots.$$

29. Find the sum of the first 10 terms of the geometric sequence

$$2, -1, \frac{1}{2}, -\frac{1}{4}, \ldots.$$

Answers on page A-76

EXAMPLE 14 Find the 10th term of the geometric sequence 64, -32, 16, $-8, \ldots$.

We first note that

$$a_1 = 64, \qquad n = 10, \quad \text{and} \quad r = \frac{-32}{64}, \text{ or } -\frac{1}{2}.$$

Then using the formula $a_n = a_1 r^{n-1}$, we have

$$a_{10} = 64 \cdot \left(-\frac{1}{2}\right)^{10-1} = 64 \cdot \left(-\frac{1}{2}\right)^9 = 2^6 \cdot \left(-\frac{1}{2^9}\right) = -\frac{1}{2^3} = -\frac{1}{8}.$$

Thus the 10th term is $-\frac{1}{8}$.

Do Exercises 26 and 27.

Next, we develop a formula for the sum S_n of the first n terms of a geometric sequence:

$$a_1, a_1 r, a_1 r^2, a_1 r^3, \ldots, a_1 r^{n-1}, \ldots.$$

The associated **geometric series** is given by

$$S_n = a_1 + a_1 r + a_1 r^2 + a_1 r^3 + \cdots + a_1 r^{n-1}. \qquad \textbf{(1)}$$

We want to find a formula for this sum. If we multiply on both sides of equation (1) by r, we have

$$rS_n = a_1 r + a_1 r^2 + a_1 r^3 + a_1 r^4 + \cdots + a_1 r^n. \qquad \textbf{(2)}$$

Subtracting equation (2) from equation (1), we see that the differences of the red terms are 0, leaving

$$S_n - rS_n = a_1 - a_1 r^n,$$

or $\quad S_n(1 - r) = a_1(1 - r^n). \quad$ Factoring

Dividing on both sides by $1 - r$ gives us the following formula.

SUM OF THE FIRST n TERMS

The sum of the first n terms of the geometric sequence is given by

$$S_n = \frac{a_1(1 - r^n)}{1 - r}, \quad \text{for any } r \neq 1.$$

EXAMPLE 15 Find the sum of the first 7 terms of the geometric sequence 3, 15, 75, 375, ...

We first note that

$$a_1 = 3, \qquad n = 7, \quad \text{and} \quad r = \tfrac{15}{3}, \text{ or } 5.$$

Then using the formula

$$S_n = \frac{a_1(1 - r^n)}{1 - r},$$

we have

$$S_7 = \frac{3(1 - 5^7)}{1 - 5} = \frac{3(1 - 78{,}125)}{-4} = 58{,}593.$$

Thus the sum of the first 7 terms is 58,593.

Do Exercises 28 and 29.

h Infinite Geometric Series

The sum of the terms of an infinite geometric sequence is an **infinite geometric series.** For some geometric sequences, S_n gets close to a specific number as n gets large. For example, consider the infinite series

$$\frac{1}{2} + \frac{1}{4} + \frac{1}{8} + \frac{1}{16} + \cdots + \frac{1}{2^n} + \cdots.$$

We examine some partial sums. Note that each of the partial sums is less than 1, but S_n gets very close to 1 as n gets large. We say that 1 is the **limit** of S_n and also that 1 is the **sum of the infinite geometric sequence.** The sum of an infinite geometric sequence is denoted S_∞. In this case, $S_\infty = 1$.

n	S_n
1	0.5
5	0.96875
10	0.9990234375
20	0.9999990463
30	0.9999999991

It can be shown (but we will not do so here) that the sum of the terms of an infinite geometric series exists if and only if $|r| < 1$ (that is, the absolute value of the common ratio is less than 1).

To find a formula for the sum of an infinite geometric series, we first consider the sum of the first n terms:

$$S_n = \frac{a_1(1 - r^n)}{1 - r} = \frac{a_1 - a_1r^n}{1 - r}. \qquad \text{Using the distributive law}$$

For $|r| < 1$, values of r^n get close to 0 as n gets large. As r^n gets close to 0, so does a_1r^n. Thus, S_n gets close to $a_1/(1 - r)$.

LIMIT OR SUM OF AN INFINITE GEOMETRIC SERIES

When $|r| < 1$, the limit or sum of an infinite geometric series is given by

$$S_\infty = \frac{a_1}{1 - r}.$$

Determine whether the infinite geometric series has a sum. If so, find it.

30. $1 + 7 + 49 + 343 + \cdots$

31. $1 + (-1) + 1 + (-1) + \cdots$

32. $\frac{1}{2} + \frac{1}{4} + \frac{1}{8} + \frac{1}{16} + \frac{1}{32} + \cdots$

33. $625 + 250 + 100 + 40 + \cdots$

Find fraction notation.

34. $0.222\overline{2}$

35. $0.13\overline{13}$

Answers on page A-76

EXAMPLE 16 Determine whether each of the following infinite geometric series has a limit. If a limit exists, find it.

a) $1 + 3 + 9 + 27 + \cdots$

b) $-2 + 1 - \frac{1}{2} + \frac{1}{4} - \frac{1}{8} + \cdots$

a) Here $r = 3$, so $|r| = |3| = 3$. Since $|r| > 1$, the series *does not* have a limit.

b) Here $r = -\frac{1}{2}$, so $|r| = \left|-\frac{1}{2}\right| = \frac{1}{2}$. Since $|r| < 1$, the series *does* have a limit. We find the limit:

$$S_\infty = \frac{a_1}{1 - r} = \frac{-2}{1 - \left(-\frac{1}{2}\right)} = \frac{-2}{\frac{3}{2}} = -\frac{4}{3}.$$

Do Exercises 30–33.

EXAMPLE 17 Find fraction notation for 0.78787878..., or $0.\overline{78}$.

We can express this as

$$0.78 + 0.0078 + 0.000078 + \cdots.$$

Then we see that this is an infinite geometric series, where $a_1 = 0.78$ and $r = 0.01$. Since $|r| < 1$, this series has a limit:

$$S_\infty = \frac{a_1}{1 - r} = \frac{0.78}{1 - 0.01} = \frac{0.78}{0.99} = \frac{78}{99}, \quad \text{or} \quad \frac{26}{33}.$$

Thus fraction notation for 0.78787878... is $\frac{26}{33}$. You can check this on your calculator.

Do Exercises 34 and 35.

i Applications

The translation of some applications and problem-solving situations may involve sequences or series. We consider some examples.

EXAMPLE 18 *Hourly Wages.* Gloria accepts a job, starting with an hourly wage of \$14.25, and is promised a raise of 15¢ per hour every 2 months for 5 yr. At the end of 5 yr, what will Gloria's hourly wage be?

It helps to first write down the hourly wage for several 2-month time periods:

Beginning:	\$14.25,
After two months:	\$14.40,
After four months:	\$14.55,
and so on.	

What appears is a sequence of numbers: 14.25, 14.40, 14.55, ... This sequence is arithmetic, because adding 0.15 each time gives us the next term.

We want to find the last term of an arithmetic sequence, so we use the formula $a_n = a_1 + (n - 1)d$. We know that $a_1 = 14.25$ and $d = 0.15$, but what is n? That is, how many terms are in the sequence? Each year there are 12/2, or 6 raises, since Gloria gets a raise every 2 months. There are 5 yr, so the total number of raises will be $5 \cdot 6$, or 30. Thus there will be 31 terms: the original wage and 30 increased rates.

Substituting in the formula $a_n = a_1 + (n - 1)d$ gives us

$$\begin{aligned} a_{31} &= 14.25 + (31 - 1) \cdot 0.15 \\ &= 18.75. \end{aligned}$$

Thus, at the end of 5 yr, Gloria's hourly wage will be \$18.75.

Do Exercise 36.

EXAMPLE 19 *A Daily Doubling Salary.* Suppose someone offered you a job for the month of September (30 days) under the following conditions. You will be paid \$0.01 for the first day, \$0.02 for the second, \$0.04 for the third, and so on, doubling your previous day's salary each day. How much would you earn?

You earn \$0.01 the first day, \$0.01(2) the second day, \$0.01(2)(2) the third day, and so on. The amount earned is the geometric series

$$\$0.01 + \$0.01(2) + \$0.01(2^2) + \$0.01(2^3) + \cdots + \$0.01(2^{29}),$$

where $a_1 = \$0.01$, $r = 2$, and $n = 30$. Using the formula

$$S_n = \frac{a_1(1 - r^n)}{1 - r},$$

we have

$$S_{30} = \frac{\$0.01(1 - 2^{30})}{1 - 2} = \$10{,}737{,}418.23.$$

The pay for the month would be \$10,737,418.23.

Do Exercise 37.

36. You take a job starting with an hourly rate of \$16.75. You are promised a raise of 25¢ per hour every 2 months for 7 years. At the end of 7 years, what will your hourly wage be?

37. Under the conditions of Example 19, how much would you make in October, which has 31 days?

Answers on page A-76

F EXERCISE SET

a In each of the following, the nth term of a sequence is given. Find the first 4 terms, a_{10}, and a_{15}.

1. $a_n = 4n - 1$

2. $a_n = (n - 1)(n - 2)(n - 3)$

3. $a_n = \frac{n}{n - 1}, n \geq 2$

4. $a_n = n^2 - 1, n \geq 3$

5. $a_n = \frac{n^2 - 1}{n^2 + 1}$

6. $a_n = \left(-\frac{1}{2}\right)^{n-1}$

7. $a_n = (-1)^n n^2$

8. $a_n = (-1)^{n-1}(3n - 5)$

9. $a_n = 5 + \frac{(-2)^{n+1}}{2^n}$

10. $a_n = \frac{2n - 1}{n^2 + 2n}$

b Predict the general, or nth term, a_n, of the sequence. Answers may vary.

11. 2, 4, 6, 8, 10, ...

12. 3, 9, 27, 81, 243, ...

13. −2, 6, −18, 54, ...

14. −2, 3, 8, 13, 18, ...

15. $\frac{2}{3}, \frac{3}{4}, \frac{4}{5}, \frac{5}{6}, \frac{6}{7}, \ldots$

16. $\sqrt{2}, 2, \sqrt{6}, 2\sqrt{2}, \sqrt{10}, \ldots$

17. $1 \cdot 2, 2 \cdot 3, 3 \cdot 4, 4 \cdot 5, \ldots$

18. −1, −4, −7, −10, −13, ...

19. 0, log 10, log 100, log 1000, ...

20. $\ln e^2, \ln e^3, \ln e^4, \ln e^5, \ldots$

c Find the indicated partial sums for the sequence.

21. $1, 2, 3, 4, 5, 6, 7, \ldots; S_3$ and S_7

22. $1, -3, 5, -7, 9, -11, \ldots; S_2$ and S_5

23. $2, 4, 6, 8, \ldots; S_4$ and S_5

24. $1, \frac{1}{4}, \frac{1}{9}, \frac{1}{16}, \frac{1}{25}, \ldots; S_1$ and S_5

d Find and evaluate the sum.

25. $\sum_{k=1}^{5} \frac{1}{2k}$

26. $\sum_{k=1}^{6} \frac{1}{2k+1}$

27. $\sum_{k=0}^{6} 2^k$

28. $\sum_{k=4}^{7} \sqrt{2k-1}$

29. $\sum_{k=7}^{10} \ln k$

30. $\sum_{k=1}^{4} \pi k$

31. $\sum_{k=1}^{8} \frac{k}{k+1}$

32. $\sum_{k=1}^{5} \frac{k-1}{k+3}$

33. $\sum_{k=1}^{5} (-1)^k$

34. $\sum_{k=0}^{5} (-1)^{k+1}$

Write sigma notation.

35. $5 + 10 + 15 + 20 + 25 + \cdots$

36. $7 + 14 + 21 + 28 + 35 + \cdots$

37. $2 - 4 + 8 - 16 + 32 - 64$

38. $3 + 6 + 9 + 12 + 15$

39. $-\frac{1}{2} + \frac{2}{3} - \frac{3}{4} + \frac{4}{5} - \frac{5}{6} + \frac{6}{7}$

40. $\frac{1}{1^2} + \frac{1}{2^2} + \frac{1}{3^2} + \frac{1}{4^2} + \frac{1}{5^2}$

41. $4 - 9 + 16 - 25 + \cdots + (-1)^n n^2$

42. $9 - 16 + 25 + \cdots + (-1)^{n+1} n^2$

43. $\frac{1}{1 \cdot 2} + \frac{1}{2 \cdot 3} + \frac{1}{3 \cdot 4} + \frac{1}{4 \cdot 5} + \cdots$

44. $\frac{1}{1 \cdot 2^2} + \frac{1}{2 \cdot 3^2} + \frac{1}{3 \cdot 4^2} + \frac{1}{4 \cdot 5^2} + \cdots$

e Find the first term and the common difference.

45. 3, 8, 13, 18, ...

46. \$1.08, \$1.16, \$1.24, \$1.32, ...

47. 9, 5, 1, -3, ...

48. -8, -5, -2, 1, 4, ...

49. $\frac{3}{2}, \frac{9}{4}, 3, \frac{15}{4}, \ldots$

50. $\frac{3}{5}, \frac{1}{10}, -\frac{2}{5}, \ldots$

51. \$316, \$313, \$310, \$307, ...

52. Find the 11th term of the arithmetic sequence 0.07, 0.12, 0.17, ...

53. Find the 12th term of the arithmetic sequence 2, 6, 10, ...

54. Find the 17th term of the arithmetic sequence 7, 4, 1, ...

55. Find the 14th term of the arithmetic sequence $3, \frac{7}{3}, \frac{5}{3}, \ldots$.

56. Find the 13th term of the arithmetic sequence \$1200, \$964.32, \$728.64, ...

57. In the sequence of Exercise 53, what term is the number 106?

58. In the sequence of Exercise 52, what term is the number 1.67?

59. In the sequence of Exercise 55, what term is -27?

60. In the sequence of Exercise 54, what term is -296?

f

61. Find a_1 when $d = 4$ and $a_8 = 33$.

62. Find a_{20} when $a_1 = 14$ and $d = -3$.

63. Find n when $a_1 = 25$, $d = -14$, and $a_n = -507$.

64. Find d when $a_1 = 8$ and $a_{11} = 26$.

65. In an arithmetic sequence, $a_{17} = \frac{25}{3}$ and $a_{32} = \frac{95}{6}$. Find a_1 and d. Write the first 5 terms of the sequence.

66. In an arithmetic sequence, $a_{17} = -40$ and $a_{28} = -73$. Find a_1 and d. Write the first 5 terms of the sequence.

67. Find the sum of the first 20 terms of the series $5 + 8 + 11 + 14 + \cdots$.

68. Find the sum of the first 14 terms of the series $11 + 7 + 3 + \cdots$.

69. Find the sum of the first 400 even natural numbers.

70. Find the sum of the first 300 natural numbers.

71. Find the sum of the multiples of 7 from 7 to 98, inclusive.

72. Find the sum of the odd numbers 1 to 199, inclusive.

g Find the common ratio.

73. $2, 4, 8, 16, \ldots$

74. $18, -6, 2, -\frac{2}{3}, \ldots$

75. $-1, 1, -1, 1, \ldots$

76. $-8, -0.8, -0.08, -0.008, \ldots$

77. $\frac{2}{3}, -\frac{4}{3}, \frac{8}{3}, -\frac{16}{3}, \ldots$

78. $75, 15, 3, \frac{3}{5}, \ldots$

79. $6.275, 0.6275, 0.06275, \ldots$

80. $\frac{1}{x}, \frac{1}{x^2}, \frac{1}{x^3}, \ldots$

81. $5, \frac{5a}{2}, \frac{5a^2}{4}, \frac{5a^3}{8}, \ldots$

82. \$780, \$858, \$943.80, \$1038.18, ...

Find the indicated term.

83. $2, 4, 8, 16, \ldots$; the 7th term

84. $2, -10, 50, -250, \ldots$; the 9th term

85. $2, 2\sqrt{3}, 6, \ldots$; the 9th term

86. $1, -1, 1, -1, \ldots$; the 57th term

87. $\frac{7}{625}, -\frac{7}{25}, 7, \ldots$; the 23rd term

88. \$1000, \$1060, \$1123.60, ...; the 5th term

89. Find the sum of the first 7 terms of the geometric series

$$6 + 12 + 24 + \cdots.$$

90. Find the sum of the first 10 terms of the geometric series

$$16 - 8 + 4 - \cdots.$$

91. Find the sum of the first 9 terms of the geometric series

$$\frac{1}{18} - \frac{1}{6} + \frac{1}{2} - \cdots.$$

92. Find the sum of the geometric series

$$-8 + 4 + (-2) + \cdots + \left(-\frac{1}{32}\right).$$

Find the sum, if it exists.

93. $4 + 2 + 1 + \cdots$

94. $7 + 3 + \frac{9}{7} + \cdots$

95. $25 + 20 + 16 + \cdots$

96. $100 - 10 + 1 - \frac{1}{10} + \cdots$

97. $8 + 40 + 200 + \cdots$

98. $-6 + 3 - \frac{3}{2} + \frac{3}{4} - \cdots$

99. $0.6 + 0.06 + 0.006 + \cdots$

100. $3 - 9 + 27 - 81 + \cdots$

Find fraction notation.

101. $0.131313\ldots$, or $0.\overline{13}$

102. $0.2222\ldots$, or $0.\overline{2}$

103. $8.999\overline{9}$

104. $6.1616\overline{16}$

105. $3.4125\overline{125}$

106. $12.7809\overline{809}$

i

107. *Pole Stacking.* How many poles will be in a stack of telephone poles if there are 50 in the first layer, 49 in the second, and so on, with 6 in the top layer?

108. *Investment Return.* Max is an investment counselor. He sets up an investment situation for a client that will return \$5000 the first year, \$6125 the second year, \$7250 the third year, and so on, for 25 yr. How much is received from the investment altogether?

109. *Garden Plantings.* A gardener is making a planting in the shape of a trapezoid. It will have 35 plants in the front row, 31 in the second row, 27 in the third row, and so on. If the pattern is consistent, how many plants will there be in the last row? How many plants are there altogether?

110. *Band Formation.* A formation of a marching band has 10 marchers in the front row, 12 in the second row, 14 in the third row, and so on, for 10 rows. How many marchers are in the last row? How many marchers are there altogether?

111. *Total Savings.* If 10¢ is saved on October 1, 20¢ is saved on October 2, 30¢ on October 3, and so on, how much is saved during the 31 days of October?

112. *Parachutist Free Fall.* It has been found that when a parachutist jumps from an airplane, the distances, in feet, which the parachutist falls in each successive second before pulling the ripcord to release the parachute are as follows:

16, 48, 80, 112, 144, ...

What is the total distance fallen after 10 sec?

113. *Bouncing Ping-Pong Ball.* A ping-pong ball is dropped from a height of 16 ft and always rebounds $\frac{1}{4}$ of the distance fallen.

a) How high does it rebound the 6th time?

b) Find the total sum of the rebound heights of the ball.

114. *Daily Doubling Salary.* Suppose someone offered you a job for the month of February (28 days) under the following conditions. You will be paid \$0.01 the 1st day, \$0.02 the 2nd, \$0.04 the 3rd, and so on, doubling your previous day's salary each day. How much would you earn altogether?

115. *Bungee Jumping.* A bungee jumper rebounds 60% of the height jumped. A bungee jump is made using a cord that stretches to 200 ft.

a) After jumping and then rebounding 9 times, how far has a bungee jumper traveled upward (the total rebound distance)?

b) About how far will a jumper have traveled upward (bounced) before coming to rest?

116. *Population Growth.* Gaintown has a present population of 100,000, and the population is increasing by 3% each year.

a) What will the population be in 15 yr?

b) How long will it take for the population to double?

117. *Loan Repayment.* A family borrows \$120,000. The loan is to be repaid in a single payment in 13 yr at 12% interest, compounded annually. How much will be repaid at the end of 13 yr?

118. *The Economic Multiplier.* The government is making a \$13,000,000,000 expenditure for educational improvement. If 85% of this is spent again, and so on, what is the total effect on the economy?

G APPLYING REASONING SKILLS

Objectives

a Draw conclusions using inductive reasoning.

b Draw conclusions using deductive reasoning.

The two basic types of reasoning commonly used to solve problems are inductive reasoning and deductive reasoning.

a Inductive Reasoning

Inductive reasoning is being used when conclusions are made on the basis of observations. This usually involves identifying a pattern exhibited by several items in a group and then drawing a general conclusion about the entire group. Four commonly occurring patterns are increasing, decreasing, alternating, and circular.

EXAMPLE 1 Find the next number in the sequence

2, 5, 8, 11, ...

The numbers increase in value in this sequence, so we have an increasing pattern. We observe that 3 is added to each number to get the next number. Thus the next number in the sequence is $11 + 3$, or 14.

EXAMPLE 2 Find the next number in the sequence

50, 45, 40, 35, ...

The numbers decrease in value in this sequence, so we have a decreasing pattern. We observe that 5 is subtracted from each number to get the next number. Thus the next number in the sequence is $35 - 5$, or 30.

EXAMPLE 3 Find the next number in the sequence

3, 4, 6, 9, 13, ...

The numbers increase in value in this sequence, so we have an increasing pattern. We observe that, first, 1 is added to 3 to get 4, then 2 is added to 4 to get 6, then 3 is added to 6 to get 9, and then 4 is added to 9 to get 13. To find the next number in the sequence, we must add 5 to 13. Thus the next number in the sequence is $13 + 5$, or 18.

Do Exercises 1–3.

EXAMPLE 4 Find the next number in the sequence

4, 9, 6, 11, 8, ...

Note that the numbers go from smaller to larger to smaller to larger and so on. This is an alternating pattern. Observe that first 5 is added to 4 to get 9. Then 3 is subtracted from 9 to get 6. Next 5 is added to 6 to get 11, and then 3 is subtracted from 11 to get 8. The pattern is to add 5 and then subtract 3. To find the next number in the sequence, we add 5 to 8. Thus the next number in the sequence is $8 + 5$, or 13.

Do Exercise 4.

Find the next number in each sequence.

1. 1, 6, 11, 16, ...

2. 35, 31, 27, 23, ...

3. 2, 4, 7, 11, 16, ...

4. Find the next number in the sequence

7, 13, 9, 15, 11, ...

Answers on page A-76

EXAMPLE 5 Find the next letter in the sequence

A, Z, B, Y, C, ...

First, we convert this sequence of letters to a sequence of numbers by assigning a numerical value to each letter. We let A = 1, B = 2, C = 3, and so on, ending with Z = 26. Rewriting the sequence in numerical form, we have

1, 26, 2, 25, 3, ...

The numbers alternate between small and large values, so we have an alternating sequence. We can use a diagram to find a pattern:

$$1 \xrightarrow{+25} 26 \xrightarrow{-24} 2 \xrightarrow{+23} 25 \xrightarrow{-22} 3$$

The diagram indicates that the next number in the sequence is 3 + 21, or 24. This corresponds to the letter X.

Do Exercises 5 and 6.

EXAMPLE 6 Find the next figure in the sequence.

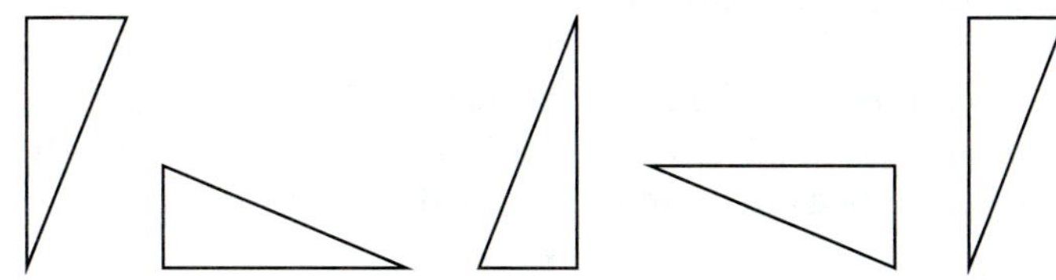

This pattern consists of a triangle that rotates one-fourth turn counterclockwise as the sequence moves from figure to figure. This is a circular pattern. To continue this pattern, we rotate the last triangle one-fourth turn counterclockwise to get the figure shown.

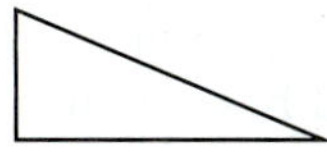

EXAMPLE 7 Find the next figure in the sequence.

This sequence consists of a square divided into four regions, one of which is color. We observe that the color rotates clockwise as the sequence moves from figure to figure, so we have a circular pattern. To continue this pattern, the next figure in the sequence should be color in the upper righthand region, as shown here:

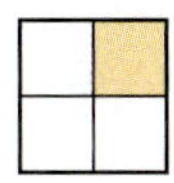

Do Exercises 7 and 8.

5. Find the next letter in the sequence

A, F, C, H, E, ...

6. Find the next number in the sequence

3, 13, 4, 12, 5, ...

Find the next figure in each sequence.

7.

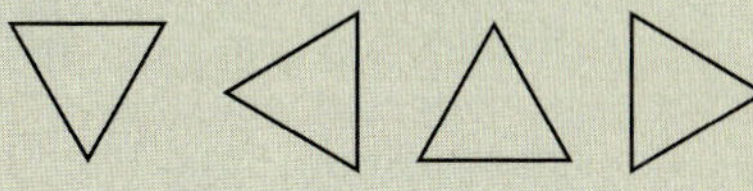

8.

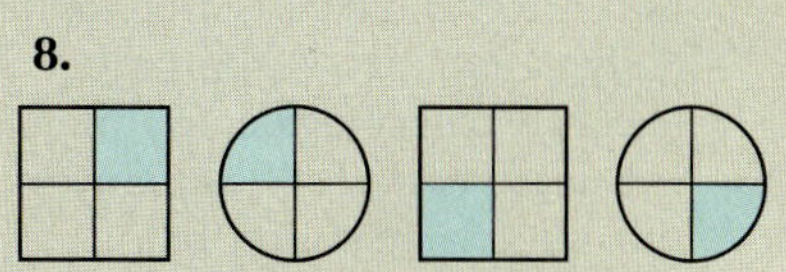

Answers on page A-76

EXAMPLE 8 Find the next figure in the sequence.

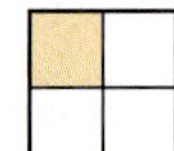

There are two patterns in this sequence. Circles and squares alternate to create one pattern. The second pattern is created by color that rotates counterclockwise as the sequence moves from figure to figure. The next figure in the shape pattern will be a circle. The next figure in the color sequence will be color in the lower lefthand region. Thus the next figure in the sequence is as shown below:

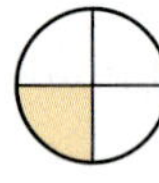

Do Exercises 9 and 10.

b Deductive Reasoning

Deductive reasoning is being used when conclusions are made on the basis of a set of facts. This usually involves deriving a series of new facts from those originally stated until a conclusion is reached.

EXAMPLE 9 A survey of 100 people shows that 65 own a dog, 25 own a cat, and 18 own both a dog and a cat. How many of those surveyed own neither a dog nor a cat?

First, we find that $65 + 25$, or 90 people, own a dog or a cat or both. From this number, we subtract the number of people who own both a dog and a cat, because they are counted twice—once as dog owners and once as cat owners. We have $90 - 18$, or 72. Finally, we subtract this number from 100, the number of people surveyed, to find how many own neither a dog nor a cat. We have $100 - 72$, or 28.

Do Exercises 11 and 12.

Find the next figure in each sequence.

9.

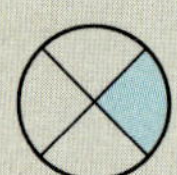

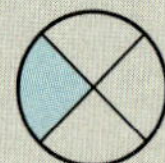
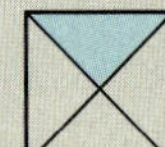

10.

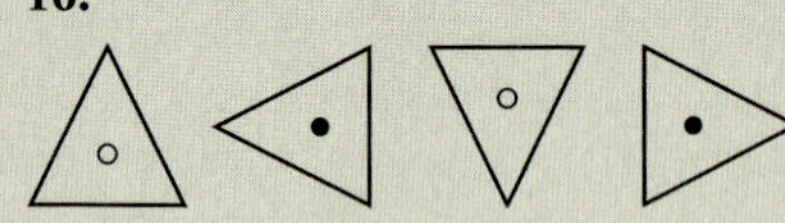

11. A poll of 125 people shows that 68 are on a low-fat diet, 39 are on a low-sodium diet, and 17 are on both a low-fat and a low-sodium diet. How many of those polled are on neither a low-fat nor a low-sodium diet?

12. A survey of 200 people shows that 125 of them subscribe to a daily newspaper, 70 subscribe to a weekly newsmagazine, and 65 subscribe to both. How many of those surveyed subscribe to neither a daily newspaper nor a weekly newsmagazine?

Answers on page A-76

EXAMPLE 10 Amy, Chloe, Marc, and Pete all work in the same building. One is a doctor, one is a lawyer, one is a computer analyst, and one is an architect. Use the statements below to determine who is the architect.

1. Chloe and Marc carpool with the computer analyst.
2. Amy and Pete work on the same floor of the building as the architect.
3. Chloe eats lunch with the lawyer and the architect.

We set up a table as shown below. We use the given statements to determine which jobs each person *cannot* hold. In statement 1, we learn that neither Chloe nor Marc is the computer analyst, so we put an X under "Analyst" in the rows labeled "Chloe" and "Marc." In statement 2, we learn that neither Amy not Pete is the architect, so we put an X under "Architect" in the rows labeled "Amy" and "Pete." Statement 3 tells us that Chloe is neither the lawyer nor the architect, so we put an X under both "Lawyer" and "Architect" in the row labeled "Chloe." Now we see that the only empty cell of the table under "Architect" corresponds to the row labeled "Marc," so Marc is the architect.

	DOCTOR	LAWYER	ANALYST	ARCHITECT
Amy				X
Chloe		X	X	X
Marc			X	
Pete				X

Do Exercise 13.

13. Maria, Tonya, Carlos, and Frank all live in the same building. One is a student, one is a teacher, one is a salesperson, and one is a paramedic. Use the statements and the table below to determine who is the teacher.

1. Tonya lives on the same floor as the teacher and the paramedic.
2. Maria and Frank jog with the student.
3. Carlos and Frank eat dinner with the teacher.

	STUDENT	TEACHER	SALESPERSON	PARAMEDIC
Maria				
Tonya				
Carlos				
Frank				

Answer on page A-76

G EXERCISE SET

a Find the next letter or number in each sequence.

1. 7, 10, 13, 16, ...

2. 25, 27, 21, 23, 17, ...

3. B, G, L, Q, ...

4. W, U, S, Q, ...

5. 40, 41, 39, 42, 38, ...

6. 1, 3, 4, 7, 11, ...

Find the next figure in each sequence.

7.

8.

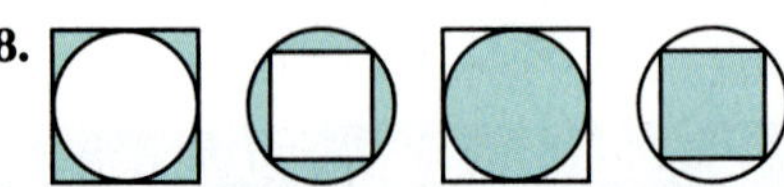

9.

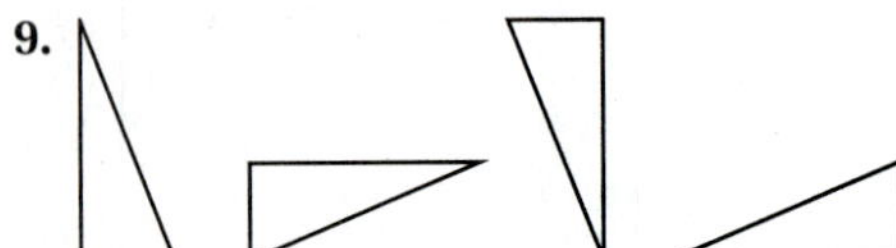

10.

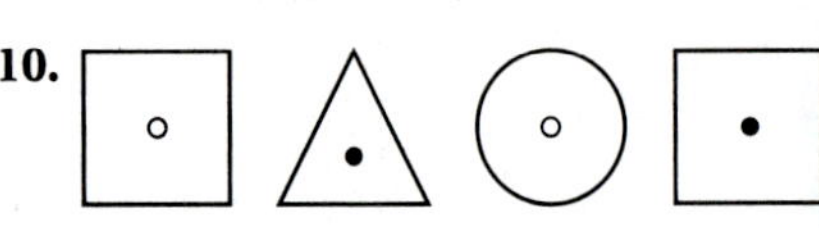

b

11. A survey of 100 households shows that 78 of them have a VCR, 42 have a CD player, and 37 have both a VCR and a CD player. How many of those surveyed have neither a VCR nor a CD player?

12. A poll of 80 library patrons shows that, in the past month, 60 of them checked out at least one novel, 25 checked out at least one audiotape, and 13 checked out both a novel and an audiotape. How many of those polled checked out neither a novel nor an audiotape in the past month?

13. Max, Jay, Bev, and Lea live on the same street. For exercise, one jogs, one walks, one swims, and one rides a bicycle. Use the statements and the table below to determine who is the swimmer.

1. Max and Lea carpool with the swimmer.
2. Jay and Bev live across the street from the walker.
3. Jay's next-door neighbors are the bike rider and the swimmer.

	JOGGER	WALKER	SWIMMER	BIKER
MAX				
JAY				
BEV				
LEA				

14. Antonio, Dale, Janet, and Lynn attend the same college. One majors in English, one in business, one in math, and one in Spanish. Use the statements and the table below to determine Lynn's major.

1. Antonio and Lynn eat lunch with the Spanish major.
2. Dale and Janet study with the business major.
3. Lynn is in the same speech class as the English and business majors.

	ENGLISH	BUSINESS	MATH	SPANISH
ANTONIO				
DALE				
JANET				
LYNN				

15. Pat, Chris, Casey, and Lee are in the same math class. One has brown hair, one has black hair, one has red hair, and one is a blond. Use the statements below to determine who has red hair.

1. Pat walks to class with the redhead and the blond.
2. Chris got a higher score on the last test than the students with black hair and blond hair.
3. Casey and Lee study with the redhead.

16. Joe, Sam, Fran, and Beth work in the same office. One is a supervisor, one is a typist, one is a receptionist, and one is a bookkeeper. Use the statements below to determine Joe's position.

1. Sam and Beth earn more than the typist.
2. Joe eats lunch with the receptionist and the bookkeeper.
3. The supervisor works later than Fran and Joe.

H THEA PRACTICE TEST

1. Green and White Landscaping employs 48 people during the summer. In August, $\frac{1}{3}$ of them return to school. Only $\frac{5}{8}$ of those remaining are employed for snow removal in the winter. How many people does Green and White employ during the winter?

   A. 10 **B.** 20 **C.** 46 **D.** 12

2. Debbie started a new job with an annual salary of $24,000. In January, she received a cost-of-living adjustment that raised her salary by 3%. In April, her salary was raised by 5%. What was her annual salary after both increases?

   A. $25,920 **B.** $25,956 **C.** $25,200 **D.** $24,800

3. Lindsay paid $4.60 for the use of 400 ft^3 of water. At that rate, how much would she have to pay for 700 ft^3 of water?

   A. $32.20 **B.** $2.63 **C.** $8.80 **D.** $8.05

4. Light travels at a speed of 3.0×10^5 km per second. Earth is, on average, 1.5×10^8 km from the sun. About how many minutes does it take for light to travel to Earth from the sun?

   A. 8.3 min **B.** 500 min **C.** 0.5 min **D.** 83 min

5. Jason began a hike at an elevation of 150 ft above sea level. He descended 265 ft, then climbed back up 75 ft. At what elevation was he at that point?

   A. 490 ft above sea level **B.** 340 ft above sea level **C.** 40 ft below sea level **D.** 190 ft below sea level

6. The following circle graph shows how college students, in general, use their spending money.

   College Spending Money

   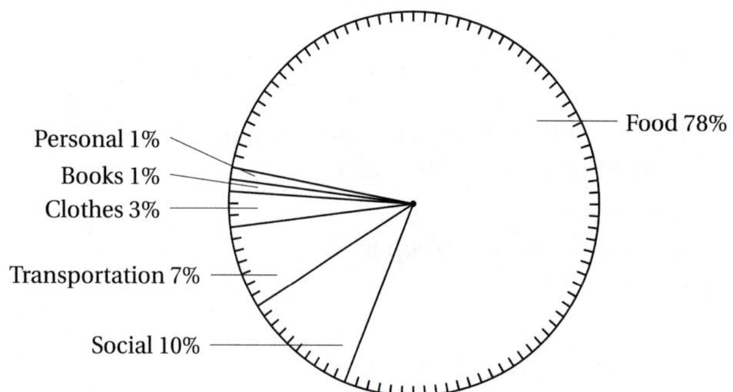

   After paying her fixed expenses, Carla has 25% of her income left for spending money. According to the circle graph, what percent of her income does she spend each month for social activities?

   A. 35%
   B. 2.5%
   C. 15%
   D. Not enough information given to determine the answer

7. Carol assembles 10 electronic parts per hour during the first half of her shift. She takes a half-hour lunch break. After lunch her rate drops to 8 parts per hour for the last half of her shift. Which of the following graphs represents the total number of parts assembled throughout the day?

A.

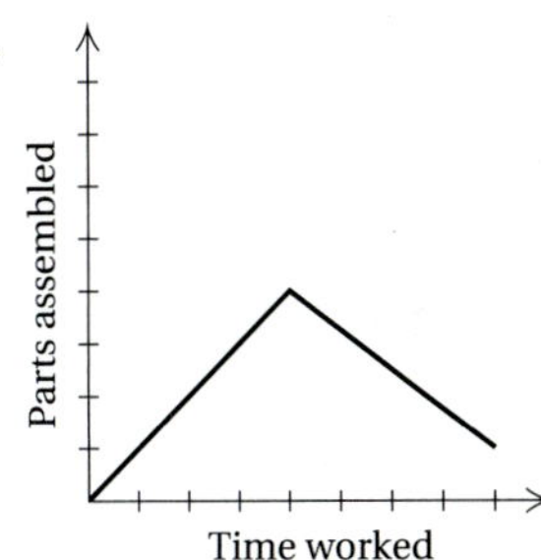

B.

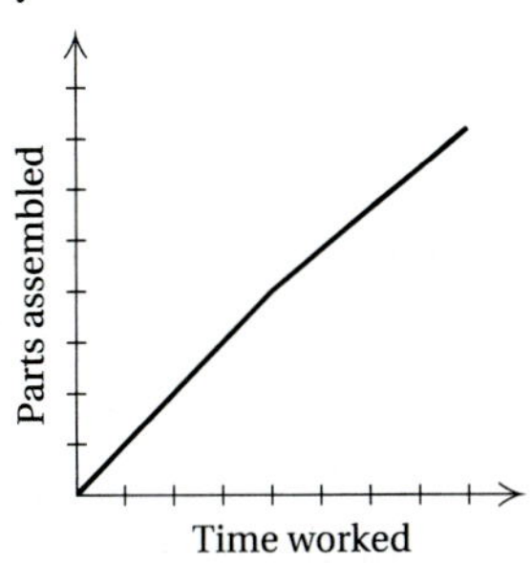

C.

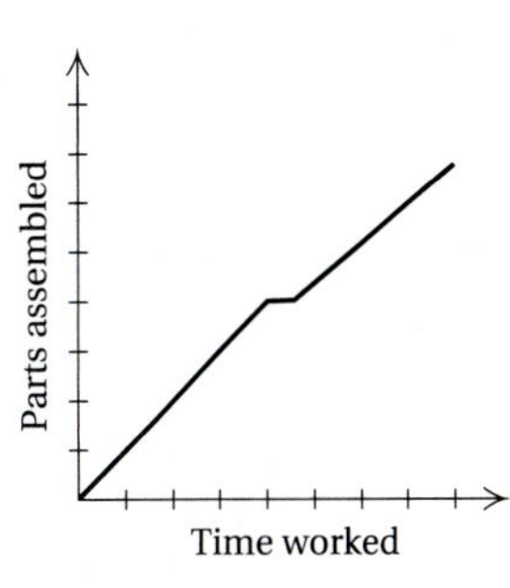

D.

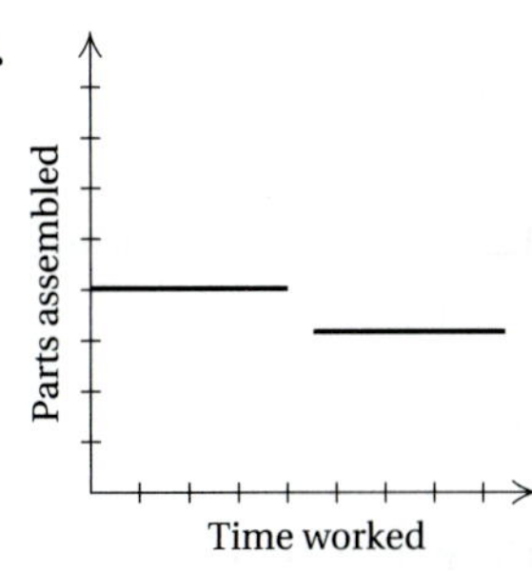

8. In order to estimate the market value of his home, Jordan needs to know the average selling price of other homes in his neighborhood. In the first half of the preceding year, three homes sold for an average selling price of $124,000. In the second half of the year, a home sold for $140,000 and another for $120,000. What was the average selling price of all the homes sold in the preceding year?

A. $124,000 **B.** $128,000 **C.** $130,200 **D.** $126,400

9. The curves below show the distribution of grades in two separate sections of a mathematics class.

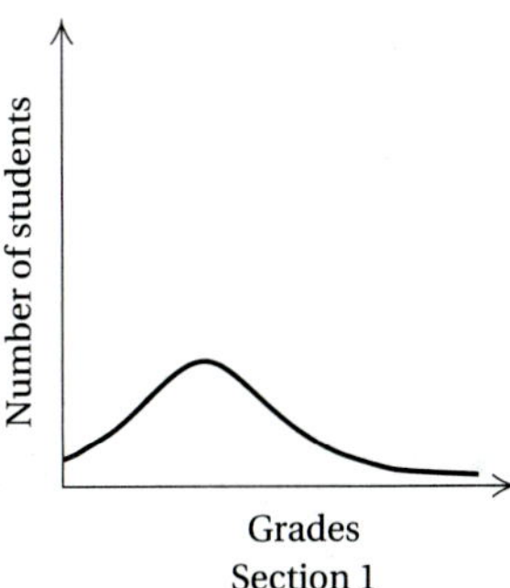

Which of the following statements correctly analyzes the information presented in these distributions?

A. The mean grade for both sections was the same.
B. There were more students in section 2 than in section 1.
C. There was more variability in the grades for section 1 than in those for section 2.
D. Students in section 2 received poorer grades than students in section 1.

10. Use the graph of the line *AB* at right to answer the question that follows.

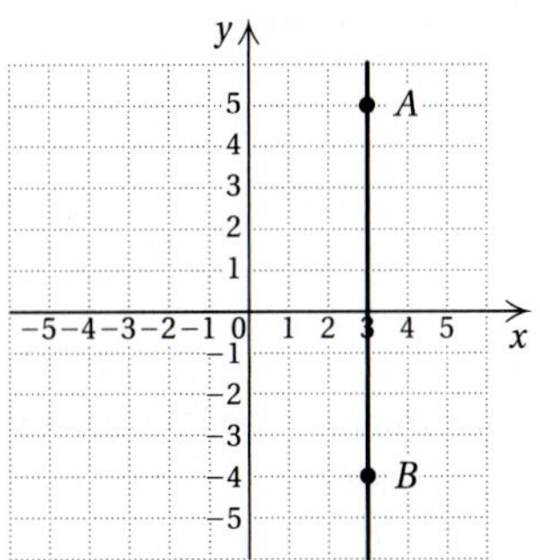

Which of the following equations represents the line *AB*?

A. $y = 3$ **B.** $x + y = 3$
C. $x = 3$ **D.** $y = x + 3$

11. Find the slope, if it exists, of the line containing the points $(0, -4)$ and $(-2, -1)$.

A. $-\frac{3}{2}$ B. $-\frac{2}{3}$ C. $\frac{3}{2}$ D. Undefined

12. What are the coordinates of the y-intercept of the line whose equation is $2x - 5y = 10$?

A. $(0, -2)$ B. $(0, 5)$ C. $(5, 0)$ D. $(-2, 0)$

13. Which of the following is an equation of the line containing the points $(-5, 6)$ and $(-2, 4)$?

A. $y = -\frac{3}{2}x - \frac{3}{2}$ B. $y = -\frac{2}{3}x + 6$ C. $y = \frac{3}{2}x + \frac{27}{2}$ D. $y = -\frac{2}{3}x + \frac{8}{3}$

14. The graph at right shows how the amount of time t that it takes to polish the rotor blade of a helicopter depends on the number of machinists n working on it. Which of the following statements about the relationship is true?

A. The more machinists working, the longer it takes to polish the blade.
B. It takes 4 machinists 10 hr to polish the blade.
C. If the blade must be polished in 2 hr, 5 machinists can get it done.
D. In 10 hr, 2 machinists can polish the blade.

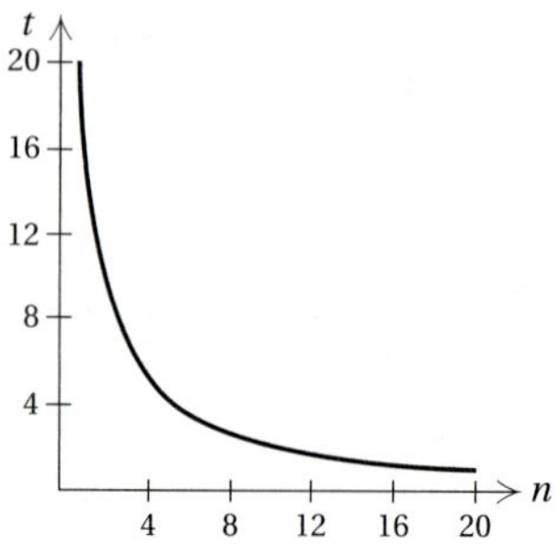

15. If $\frac{2}{5}x + 7 = 1$, what is the value of $10 - 3x$?

A. -15 B. 55 C. 17 D. -35

16. Solve $t = \frac{1}{5}(r + 5)$ for r.

A. $r = 5t - 5$ B. $r = 5t - 25$ C. $r = \frac{t - 1}{5}$ D. $r = \frac{1}{5}t - 1$

17. What is the solution of the system of equations $y = 2x^2 + 7x + 3$ and $4x + 3y = 9$?

A. $(-3, 7), \left(-\frac{1}{2}, \frac{11}{3}\right)$ B. $(0, 3), \left(-\frac{25}{6}, \frac{77}{9}\right)$

C. $(3\sqrt{3}, 3 - 4\sqrt{3}), (-3\sqrt{3}, 3 + 4\sqrt{3})$ D. No solution

18. Which of the following equations correctly translates this statement? The product of the length l of a fish and the square of the girth g of the fish is 280 times the weight w of the fish.

A. $(l \times g)^2 = 280w$ B. $280(l \times g^2) = w$ C. $l \times g^2 = 280w$ D. $280(l + g) = w$

19. Which of the following graphs shows the solution of the system of equations $y - 5 + x = 2x$ and $y = x^2 - 2$?

A.

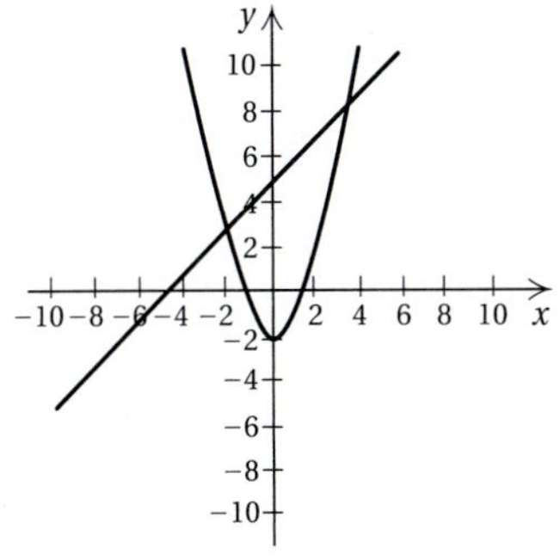

B.

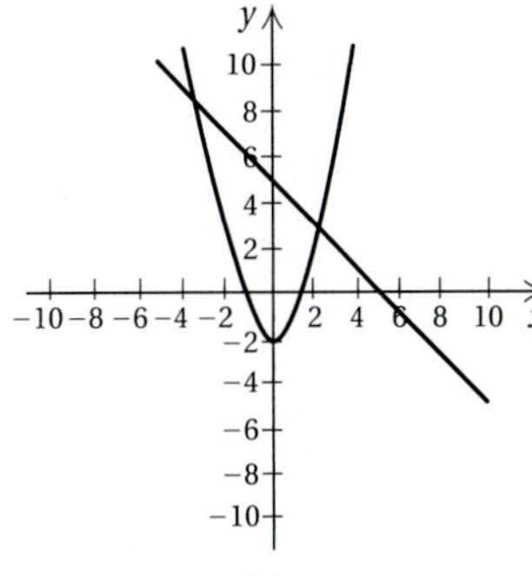

C.

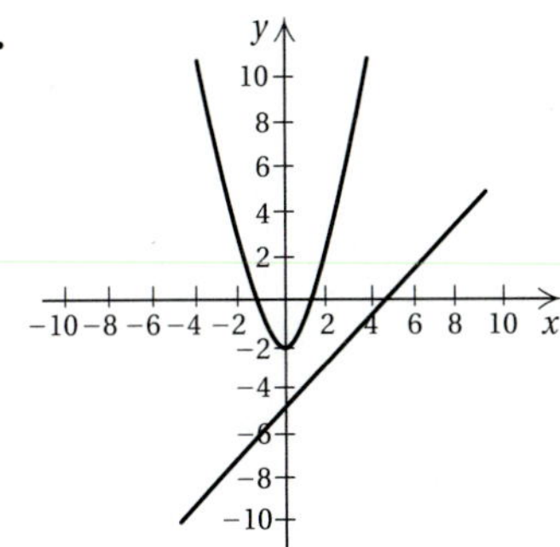

D.

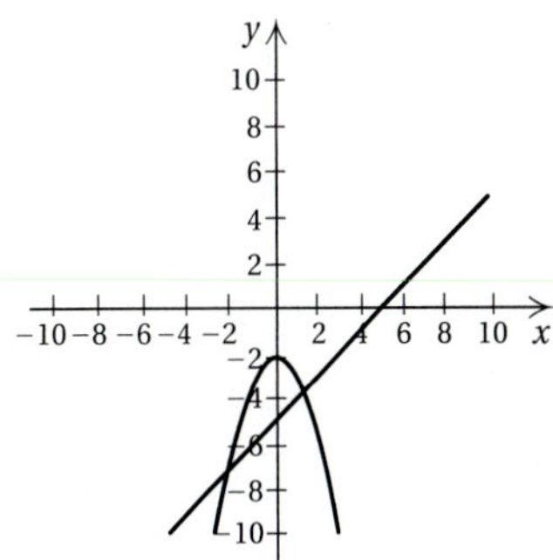

20. Karen earns 2% of the price of each home she sells as commission. Last week, she sold two houses and earned \$3500. The selling price of one house was $1\frac{1}{2}$ times the selling price of the other. What was the selling price of the more expensive house?

A. \$175,000 **B.** \$105,000 **C.** \$116,666 **D.** \$87,500

21. Party Time stores sell complete party packages. The Flyaway package contains paper products for 12 people plus 30 helium-filled balloons and sells for \$34.80. The Down-to-Earth package contains paper products for 15 people plus 1 balloon and sells for \$14.30. Party Time wants to begin selling a Party package that will contain paper products for 25 people plus 10 balloons. What should the selling price of the Party package be?

A. \$49.10 **B.** \$29.00 **C.** \$30.50 **D.** \$11.60

22. Which of the following is one factor of $3x^2 - 4x - 15$?

A. $(x - 5)$ **B.** $(3x - 5)$ **C.** $(x + 3)$ **D.** $(x - 3)$

23. Perform the multiplication: $(2x + 5)^2$.

A. $4x^2 + 25$ **B.** $4x^2 + 20x + 25$ **C.** $4x^2 + 10x + 25$ **D.** $2x^2 + 25$

24. Add and simplify, if possible: $\dfrac{x - 12}{x^2 + x - 6} + \dfrac{x}{x - 2}$.

A. $\dfrac{x + 6}{x + 3}$ **B.** $\dfrac{4x - 12}{x - 6}$ **C.** $\dfrac{2x - 12}{x^2 + 2x - 8}$ **D.** $\dfrac{2x - 12}{(x^2 + x - 6)(x - 2)}$

25. Add and simplify, if possible: $\sqrt{18x^3} - \sqrt{2x} + \sqrt{3x}$.

A. $\sqrt{18x^3 + x}$ B. $\sqrt{18x^3} + \sqrt{x}$ C. $(3x - 1)\sqrt{2x} + \sqrt{3x}$ D. $\sqrt{x}(2\sqrt{2} + \sqrt{3})$

26. If $f(x) = |2x - 7|$, find $f(2) \cdot f\left(\frac{1}{2}\right)$.

A. 18 B. 5 C. 25 D. 3

27. Shown at right is the graph of a quadratic function. Which of the following equations is represented by this graph?

A. $y = x^2 - 5x - 6$ B. $y = x^2 + 5x - 6$
C. $y = x^2 + 7x + 6$ D. $y = x^2 - 7x + 6$

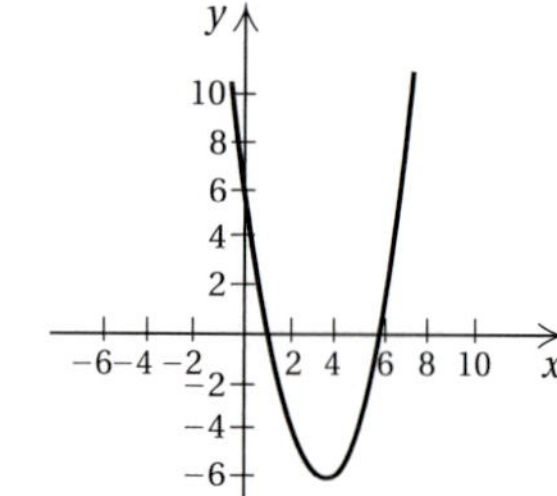

28. Shown at right is the graph of a quadratic inequality. Which of the following inequalities describes the shaded region?

A. $y \leq x^2 + 2x - 8$ B. $y \geq x^2 + 2x - 8$
C. $y \leq (x - 1)^2 - 9$ D. $y \geq (x - 1)^2 - 9$

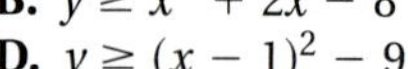

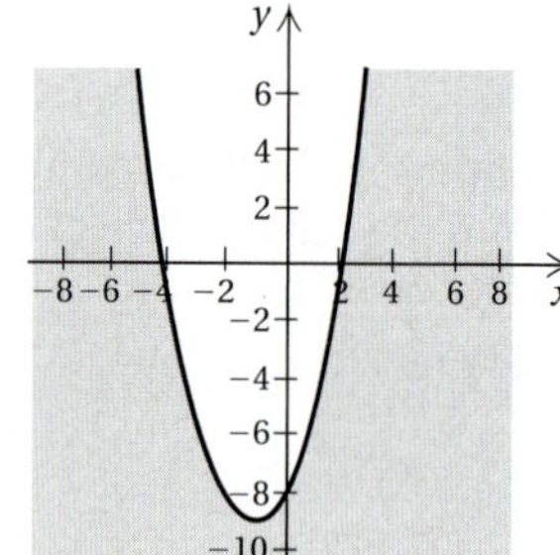

29. Which of the following numbers should be added to both sides of the equation $x^2 + 9x = 15$ in order to solve by completing the square?

A. 81 B. $\frac{81}{4}$ C. $\frac{9}{2}$ D. 125

30. Solve the equation $2x^2 - 5x = 8$ using the quadratic formula. In the solution, what is the number under the radical sign?

A. −39 B. 5 C. 41 D. 89

31. A rectangular piece of bound carpet lies in the center of a rectangular room. The distance between the carpet and the wall is the same all the way around the room. The room is 15 ft long and 12 ft wide. The carpet piece is 12 yd^2. What is the distance between the carpet and the wall?

A. 4 yd B. 4 ft C. $1\frac{1}{2}$ ft D. 3 ft

32. A window is square with a semicircular top section, as shown in the figure at right.

About how many square feet of glass will it take to make the window?

A. 16.1 ft^2　　**B.** 12.5 ft^2
C. 23.1 ft^2　　**D.** 28.3 ft^2

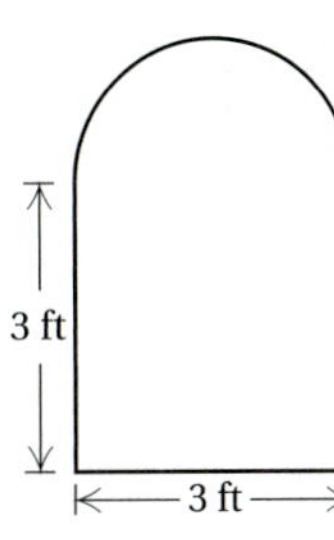

33. A decorative tin in the shape of a cylinder is full of candy. The tin is 8 in. high, and the diameter of the lid is 5 in. What is the volume of candy that the tin contains?

A. About 157 in^3　　**B.** About 165 in^3　　**C.** About 40 in^3　　**D.** About 126 in^3

34. Metal Manufacturing is making a hand rest for a medical supply company in the shape shown at right.

How much metal is used to make the hand rest?

A. 1700 cm^2　　**B.** 6500 cm^2
C. 2700 cm^2　　**D.** 2550 cm^2

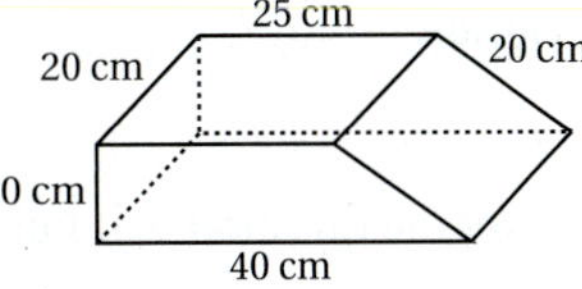

35. Buck Creek Township Fire Department has a fire truck with a ladder that extends straight to 60 ft. At a fire in an apartment building, the closest the truck could get to the burning building was 30 ft. About how high on the building did the ladder reach?

A. 52.0 ft　　**B.** 60.0 ft　　**C.** 30.0 ft　　**D.** 67.1 ft

36. Jessica is building a scale model of a house. The dining room is in the shape of a hexagon, as shown in the figure at right.

Polygon *ABCDEF*, the dining room of the house, is similar to polygon *PQRSTU*, the dining room of the model. *AB* measures 12 ft and *PQ* 4 in. How long should she make *UP* if *FA* is 5 ft?

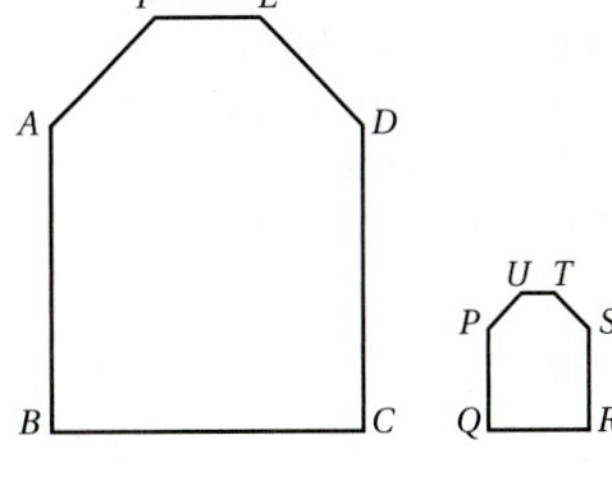

A. 15 in.　　**B.** $\frac{3}{5}$ in.

C. $1\frac{2}{3}$ in.　　**D.** 3 in.

37. Shown at right are two triangles, $\triangle ABC$ and $\triangle DEF$.

If you know that $\overline{AC} \cong \overline{DF}$ and $\angle ABC \cong \angle DEF$, which of the following is a valid conclusion about the triangles?

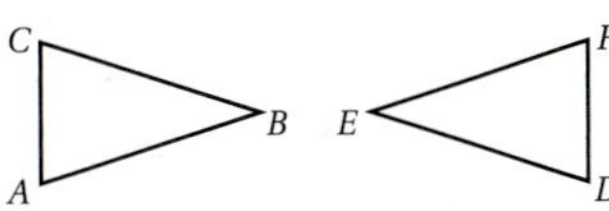

A. $\triangle ABC \cong \triangle DEF$ by SAS.
B. $\triangle ABC \cong \triangle DEF$ by ASA.
C. $\triangle ABC \cong \triangle DEF$ by SSS.
D. None of these is a valid conclusion.

38. In the hexagon shown at right, $\overline{AB} \parallel \overline{DE}$. $\overline{BE}$ divides the figure into two quadrilaterals, *ABEF* and *DEBC*.

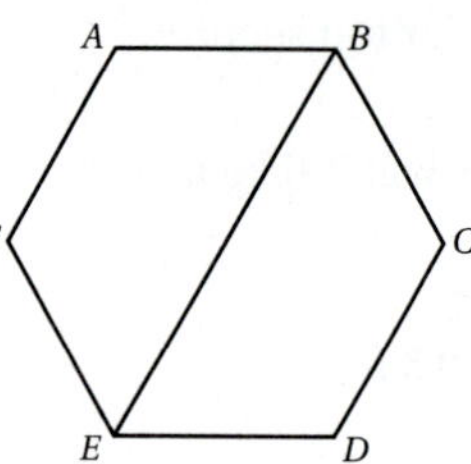

Which of the following is a valid conclusion about this figure?

A. $ABEF \cong DEBC$
B. $\angle CBE \cong \angle FEB$
C. $\angle ABE \cong \angle DEB$
D. None of these is a valid conclusion.

39. In the polygon shown at right, $\overline{BF}$ is perpendicular to both $\overline{BC}$ and $\overline{EG}$.

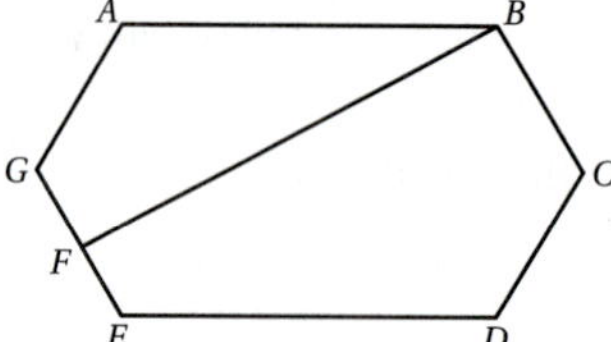

Which of the following is a valid conclusion?

A. $\overline{BC}$ is congruent to $\overline{EG}$.
B. $\overline{BC}$ is perpendicular to $\overline{EG}$.
C. $\overline{BC}$ is parallel to $\overline{EG}$.
D. None of these is a valid conclusion.

40. Use the statements below to answer the question that follows.

1. All employees of Poster Prints will get a pay raise this year.
2. Some employees of Poster Prints will be transferred to another location.
3. No management personnel will be transferred.
4. Jackie is an employee of Poster Prints.

Which of the following statements must be true?

A. Jackie will be transferred.
B. Jackie is in management.
C. Jackie will get a pay raise.
D. None of these statements must be true.

41. Mike, Linda, Jeff, and Paula are brothers and sisters. One is an engineer, one is an author, one is a consultant, and one is an accountant. Use the statements below to determine which one is an author.

1. Mike lives near the author and the consultant.
2. Linda is older than the engineer and the accountant.
3. Paula is younger than the author and the engineer.
4. Jeff does not live near any of his brothers or sisters.

Who is the author?

A. Mike **B.** Linda **C.** Jeff **D.** Paula

42. Use the pattern sequence below to answer the question that follows.

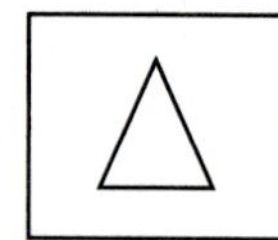
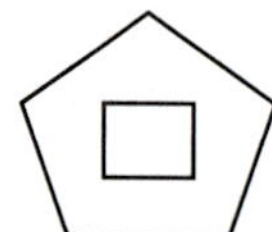
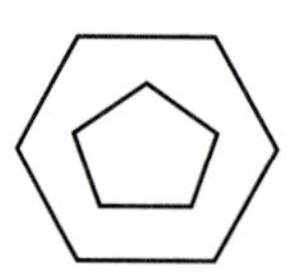

What figure comes next in the sequence?

A.
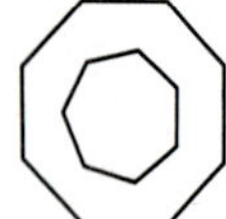

B.
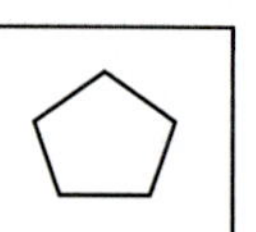

C.

D.
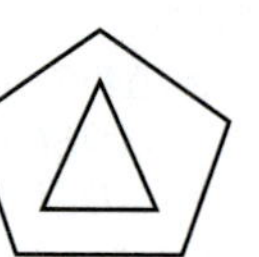

43. Use the pattern sequence below to answer the question that follows.

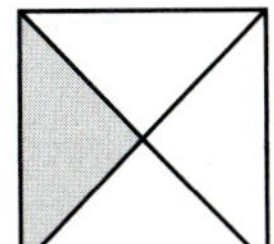 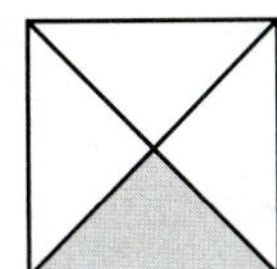 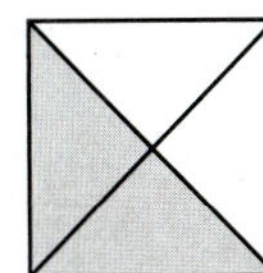 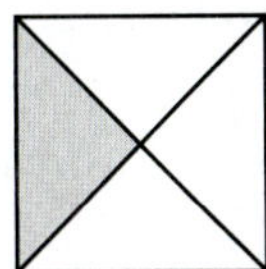 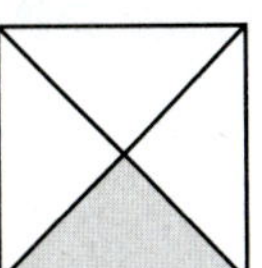 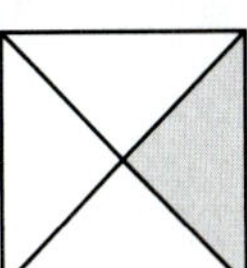  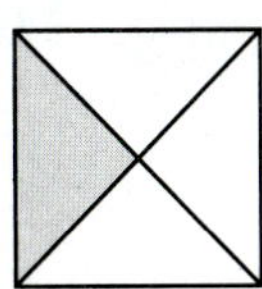 ?

What figure comes next in the sequence?

A. 

B. 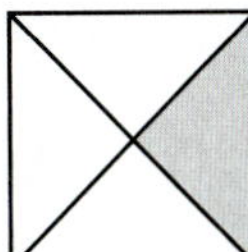

C. 

D. 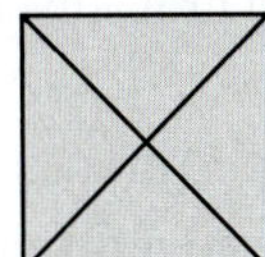

44. A flower garden is a square with a semicircle along each side, as shown in the figure at right.

A landscape designer wants to plant flowers along the perimeter of the garden. If there should be 4 plants for every meter, and each plant costs $0.70, how much will the flowers cost?

A. $13.30 **B.** $52.50
C. $26.60 **D.** $39.90

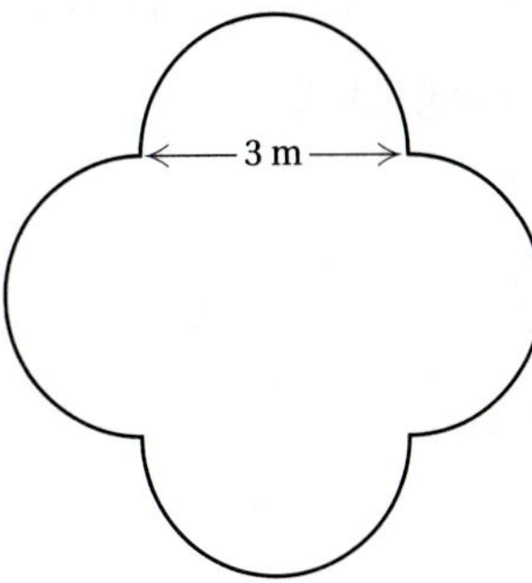

Use the information below for Questions 45 and 46.

Brenda is a graphics designer who has just accepted an entry-level job paying $24,000 a year. She can qualify for a mortgage of up to $2\frac{1}{2}$ times her annual salary. With the current interest rate, the monthly payment on a mortgage is about 1% of the amount borrowed.

45. If Brenda borrows as much as she can, what will be her monthly mortgage payment?

A. $600 **B.** $240 **C.** $480 **D.** $120

46. Brenda has other fixed expenses. If she buys the house, she will have to pay $100 a month for taxes and insurance. Income taxes take 20% of her earnings. Her car payment is $175 a month, and auto insurance is $100 a month. She spends 15% of her income on food. How much will she have left each month after paying her mortgage and other fixed expenses?

A. $1675 **B.** $325 **C.** $925 **D.** $1275

I THEA TEST MATHEMATICS SKILLS

*Skill**	*Text Reference*
FUNDAMENTAL MATHEMATICS	
1. Solve word problems involving integers, fractions, decimals, and units of measurement.	
• Solve word problems involving integers.	Sections 6.3–6.6
• Solve word problems involving fractions.	Section 2.5
• Solve word problems involving decimals (including percents).	Sections 3.7 and 4.6–4.8
• Solve word problems involving ratio and proportions.	Sections 4.1 and 11.8
• Solve word problems involving units of measurement and conversions (including scientific notation).	Section 9.2 and Appendixes A and B
2. Solve problems involving data interpretation and analysis.	
• Interpret information from line graphs, bar graphs, pictographs, and pie charts.	Section 8.1 and Appendix D
• Interpret data from tables.	Appendix D
• Recognize appropriate graphic representations of various data.	Section 8.1
• Analyze and interpret data using measures of central tendency (mean, median, and mode). Analyze and interpret data using the concept of variability.	Appendix C
ALGEBRA	
3. Graph numbers or number relationships.	
• Identify the graph of a given equation.	Sections 8.2 and 8.3
• Identify the graph of a given inequality.	Section 7.7 and Chapter 14
• Find the slope and/or intercepts of a given line.	Sections 8.3, 8.4, and 12.3–12.5
• Find the equation of a line.	Section 12.5
• Recognize and interpret information from the graph of a function (including direct and inverse variation).	Sections 11.9 and 12.1
4. Solve one- and two-variable equations.	
• Find the value of the unknown in a given one-variable equation.	Sections 7.1–7.3
• Express one variable in terms of a second variable in two-variable equations.	Sections 7.4 and 16.3
• Solve systems of two equations in two variables (including graphical solutions).	Sections 13.1–13.3
5. Solve word problems involving one and two variables.	
• Identify the algebraic equivalent of a stated relationship.	Sections 6.1 and 7.6
• Solve word problems involving one and two unknowns.	Sections 7.5, 7.6, 7.8, 10.9, 11.8, 13.2–13.4, 13.6, 13.7, 15.7, and 17.7

*Many of these skills, particularly those in problem solving and interpreting information, are not confined to the sections or chapters specified here.

*Skill**	*Text Reference*
6. Understand operations with algebraic expressions and functional notation.	
• Factor quadratics and polynomials.	Sections 10.1–10.7
• Perform operations on and simplify polynomial expressions.	Sections 9.3–9.8
• Perform operations on and simplify rational expressions.	Sections 11.1–11.6
• Perform operations on and simplify radical expressions.	Sections 15.1–15.5
• Apply principles of functions and functional notation.	Sections 12.1 and 12.2
7. Solve problems involving quadratic equations.	
• Graph quadratic functions.	Sections 16.5 and 16.6
• Graph quadratic inequalities.	Appendix F
• Solve quadratic equations using factoring, completing the square, or the quadratic formula.	Sections 10.8, 16.1, and 16.2
• Solve problems involving quadratic models.	Sections 10.9, 16.3, and 16.7
GEOMETRY	
8. Solve problems involving geometric figures.	
• Solve problems involving two-dimensional geometric figures (e.g., perimeter and area problems).	Sections 5.2–5.4
• Solve problems involving three-dimensional geometric figures (e.g., volume and surface area problems).	Section 5.5
• Solve problems using the Pythagorean theorem.	Sections 10.9 and 15.7
9. Solve problems involving geometric concepts.	
• Solve problems using principles of similarity and congruence.	Sections 5.7 and 5.8
• Solve problems using principles of parallelism and perpendicularity.	Sections 5.6 and 12.5
PROBLEM SOLVING	
10. Apply reasoning skills.	
• Draw conclusions using inductive reasoning.	Appendix H
• Draw conclusions using deductive reasoning.	Appendix H
11. Solve applied problems involving a combination of mathematical skills.	
• Apply combinations of mathematical skills to solve problems.	Throughout
• Apply combinations of mathematical skills to solve a series of related problems.	Throughout

*Many of these skills, particularly those in problem solving and interpreting information, are not confined to the sections or chapters specified here.

Answers

Chapter 1

Pretest: Chapter 1, p. 2

1. Three million, seventy-eight thousand, fifty-nine
2. 6 thousands + 9 hundreds + 8 tens + 7 ones
3. 2,047,398,589 **4.** 6 ten thousands **5.** 956,000
6. 60,000 **7.** 10,216 **8.** 4108 **9.** 22,976
10. 503 R 11 **11.** $<$ **12.** $>$ **13.** 5542 **14.** 22
15. 34 **16.** 25 **17.** 4500 sheets **18.** \$299 **19.** \$369
20. \$30,793 **21.** 64 **22.** 120 **23.** 0 **24.** 3
25. Prime **26.** $2 \cdot 2 \cdot 3 \cdot 5 \cdot 7$, or $2^2 \cdot 3 \cdot 5 \cdot 7$ **27.** Yes
28. Yes

Margin Exercises, Section 1.1, pp. 3–7

1. 2 ten thousands **2.** 2 hundred thousands
3. 2 millions **4.** 2 ten millions
5. 4 hundred thousands; 8 ten thousands; 6 thousands; 5 hundreds; 7 tens; 5 ones
6. 1 thousand + 8 hundreds + 9 tens + 5 ones
7. 2 ten thousands + 2 thousands + 1 hundred + 3 tens + 2 ones
8. 3 thousands + 0 hundreds + 3 tens + 1 one, or 3 thousands + 3 tens + 1 one
9. 4 thousands + 1 hundred + 0 tens + 0 ones, or 4 thousands + 1 hundred
10. 3 thousands + 8 hundreds + 6 tens + 0 ones, or 3 thousands + 8 hundreds + 6 tens **11.** 5689
12. 87,128 **13.** 9003 **14.** Eighty-eight **15.** Sixteen
16. Thirty-two **17.** Two hundred four
18. Forty-three thousand, seven hundred eighty-two
19. One million, eight hundred seventy-nine thousand, two hundred four **20.** Six billion, two hundred fifty-nine million, six hundred thousand **21.** 213,105,329 **22.** $<$
23. $>$ **24.** $>$ **25.** $<$ **26.** $<$ **27.** $>$

Exercise Set 1.1, p. 9

1. 5 thousands **3.** 5 hundreds **5.** The last 0 **7.** 1
9. 5 thousands + 7 hundreds + 0 tens + 2 ones, or 5 thousands + 7 hundreds + 2 ones
11. 9 ten thousands + 3 thousands + 9 hundreds + 8 tens + 6 ones
13. 2 thousands + 0 hundreds + 5 tens + 8 ones, or 2 thousands + 5 tens + 8 ones
15. 1 thousand + 2 hundreds + 6 tens + 8 ones
17. 2475 **19.** 68,939 **21.** 7304 **23.** 1009
25. Eighty-five **27.** Eighty-eight thousand
29. One hundred twenty-three thousand, seven hundred sixty-five **31.** Seven billion, seven hundred fifty-four million, two hundred eleven thousand, five hundred seventy-seven **33.** 2,233,812 **35.** 8,000,000,000
37. Five hundred sixty-six thousand, two hundred eighty
39. Four hundred sixty-seven million, three hundred twenty-two thousand, three hundred eighty-eight
41. 9,460,000,000,000 **43.** 64,186,000 **45.** $<$ **47.** $>$
49. $<$ **51.** $>$ **53.** $>$ **55.** $>$ **57.** Thrust SCC; 54 ft $>$ 47 ft **59.** 87 yr $>$ 81 yr, or 81 yr $<$ 87 yr **61.** DW
63. 138

Margin Exercises, Section 1.2, pp. 13–20

1. 97 **2.** 9745 **3.** 13,465 **4.** 16,182 **5.** 27,474
6. 16 in.; 26 in. **7.** 29 in. **8.** 22 ft **9.** $7 = 2 + 5$, or $7 = 5 + 2$ **10.** $17 = 9 + 8$, or $17 = 8 + 9$
11. $5 = 13 - 8$; $8 = 13 - 5$ **12.** $11 = 14 - 3$; $3 = 14 - 11$
13. 3801 **14.** 6328 **15.** 4747 **16.** 56 **17.** 205
18. 658 **19.** 2851 **20.** 1546

Calculator Corner, p. 21

1. 1602 **2.** 734 **3.** 1932 **4.** 864 **5.** 67 **6.** 119
7. 2128 **8.** 2593

Exercise Set 1.2, p. 22

1. 387 **3.** 5198 **5.** 8503 **7.** 5266 **9.** 100
11. 8310 **13.** 6608 **15.** 34,432 **17.** 18,424
19. 2320 **21.** 114 mi **23.** 570 ft **25.** $7 = 3 + 4$, or $7 = 4 + 3$ **27.** $43 = 27 + 16$, or $43 = 16 + 27$
29. $6 = 15 - 9$; $9 = 15 - 6$ **31.** $23 = 32 - 9$; $9 = 32 - 23$

33. 44 **35.** 1126 **37.** 5382 **39.** 3831 **41.** 2168 **43.** 43,028 **45.** 84 **47.** 454 **49.** 771 **51.** 7019 **53.** 5745 **55.** 12,134 **57.** 4206 **59.** 10,305 **61.** DW **63.** 8 ten thousands **64.** One hundred fourteen million, three hundred thirty-six thousand, six hundred ten **65.** $1 + 99 = 100, 2 + 98 = 100, \ldots, 49 + 51 = 100$. Then $49 \cdot 100 = 4900$ and $4900 + 50 + 100 = 5050$.

Margin Exercises, Section 1.3, pp. 26–40

1. 116 **2.** 148 **3.** 4938 **4.** 6740 **5.** 1035 **6.** 3024 **7.** 46,252 **8.** 205,065 **9.** 144,432 **10.** 287,232 **11.** 14,075,720 **12.** 391,760 **13.** 17,345,600 **14.** 56,200 **15.** 562,000 **16.** 45 sq ft **17.** $\frac{54}{6} = 9$, $6\overline{)54}$ with quotient 9 **18.** $15 = 5 \cdot 3$, or $15 = 3 \cdot 5$ **19.** $72 = 9 \cdot 8$, or $72 = 8 \cdot 9$ **20.** $6 = 12 \div 2; 2 = 12 \div 6$ **21.** $6 = 42 \div 7; 7 = 42 \div 6$ **22.** 6; $6 \cdot 9 = 54$ **23.** 6 R 7; $6 \cdot 9 = 54$, $54 + 7 = 61$ **24.** 4 R 5; $4 \cdot 12 = 48$, $48 + 5 = 53$ **25.** 6 R 13; $6 \cdot 24 = 144$, $144 + 13 = 157$ **26.** 59 R 3 **27.** 1475 R 5 **28.** 1015 **29.** 807 R 4 **30.** 1088 **31.** 360 R 4 **32.** 800 R 47 **33.** 40 **34.** 50 **35.** 70 **36.** 100 **37.** 40 **38.** 80 **39.** 90 **40.** 140 **41.** 470 **42.** 240 **43.** 290 **44.** 600 **45.** 800 **46.** 800 **47.** 9300 **48.** 8000 **49.** 8000 **50.** 19,000 **51.** 69,000 **52.** **(a)** \$19,500; **(b)** yes **53.** Eliminate driver's seat with 6-way power adjustment, sunroof, feature package, and sport package. Answers may vary. **54.** 200 **55.** 2600 **56.** 11,000 **57.** 210,000; 160,000

Calculator Corner, p. 28

1. 448 **2.** 21,970 **3.** 6380 **4.** 39,564 **5.** 180,480 **6.** 2,363,754

Calculator Corner, p. 37

1. 3 R 11 **2.** 28 **3.** 124 R 2 **4.** 131 R 18 **5.** 283 R 57 **6.** 843 R 187

Exercise Set 1.3, p. 41

1. 564 **3.** 2,340,000 **5.** 1527 **7.** 64,603 **9.** 4770 **11.** 3995 **13.** 46,080 **15.** 14,652 **17.** 207,672 **19.** 798,408 **21.** 20,723,872 **23.** 362,128 **25.** 20,064,048 **27.** 25,236,000 **29.** 529,984 sq mi **31.** 8100 sq ft **33.** $18 = 3 \cdot 6$, or $18 = 6 \cdot 3$ **35.** $22 = 22 \cdot 1$, or $22 = 1 \cdot 22$ **37.** $9 = 45 \div 5$; $5 = 45 \div 9$ **39.** $37 = 37 \div 1; 1 = 37 \div 37$ **41.** 12 **43.** 1 **45.** 22 **47.** Not defined **49.** 55 R 2 **51.** 108 **53.** 307 **55.** 92 R 2 **57.** 1703 **59.** 127 **61.** 29 R 5 **63.** 40 R 12 **65.** 90 R 22 **67.** 29 **69.** 1007 R 1 **71.** 23 **73.** 370 **75.** 609 R 15 **77.** 3508 R 219 **79.** 50 **81.** 470 **83.** 730 **85.** 900 **87.** 100 **89.** 1000 **91.** 9100 **93.** 32,900 **95.** 6000 **97.** 8000 **99.** 45,000 **101.** 373,000 **103.** 180 **105.** 5720 **107.** 16,500 **109.** 5200 **111.** 31,000 **113.** 69,000 **115.** $50 \cdot 70 = 3500$ **117.** $30 \cdot 30 = 900$ **119.** $900 \cdot 300 = 270,000$ **121.** $400 \cdot 200 = 80,000$ **123.** DW **125.** 7 thousands + 8 hundreds + 8 tens + 2 ones **126.** > **127.** $21 = 16 + 5$, or $21 = 5 + 16$ **128.** $56 = 14 + 42$, or $56 = 42 + 14$ **129.** $47 = 56 - 9$; $9 = 56 - 47$ **130.** $350 = 414 - 64; 64 = 414 - 350$ **131.** 54, 122; 33, 2772; 4, 8; 16, 19 **133.** 30 buses **135.** 247,464 sq ft

Margin Exercises, Section 1.4, pp. 46–50

1. 7 **2.** 5 **3.** No **4.** Yes **5.** 5 **6.** 10 **7.** 5 **8.** 22 **9.** 22,490 **10.** 9022 **11.** 570 **12.** 3661 **13.** 8 **14.** 45 **15.** 77 **16.** 3311 **17.** 6114 **18.** 8 **19.** 16 **20.** 644 **21.** 96 **22.** 94

Exercise Set 1.4, p. 51

1. 14 **3.** 0 **5.** 29 **7.** 0 **9.** 8 **11.** 14 **13.** 1035 **15.** 25 **17.** 450 **19.** 90,900 **21.** 32 **23.** 143 **25.** 79 **27.** 45 **29.** 324 **31.** 743 **33.** 37 **35.** 66 **37.** 15 **39.** 48 **41.** 175 **43.** 335 **45.** 104 **47.** 45 **49.** 4056 **51.** 17,603 **53.** 18,252 **55.** 205 **57.** DW **59.** $7 = 15 - 8; 8 = 15 - 7$ **60.** $6 = 48 \div 8; 8 = 48 \div 6$ **61.** < **62.** > **63.** > **64.** < **65.** 142 R 5 **66.** 142 **67.** 334 **68.** 334 R 11 **69.** 347

Margin Exercises, Section 1.5, pp. 54–61

1. 277 home runs **2.** 249 home runs **3.** Mark McGwire **4.** \$1874 **5.** \$223 **6.** \$34,776 **7.** 9180 sq in. **8.** 378 cartons with 1 can left over **9.** 34 gal **10.** 181 seats

Exercise Set 1.5, p. 62

1. 178,668 million, or 178,668,000,000 **3.** 12,113 million, or 12,113,000,000 **5.** 2000 **7.** 66,672 sq mi **9.** \$459,860 **11.** 211,000 people **13.** 240 mi **15.** 256 gal **17.** 480,000 pixels **19.** \$11,976 **21.** 216,000,000 CDs **23.** 2,545,000,000 CDs **25.** 35 weeks; 2 episodes left over **27.** 168 hr **29.** \$247 **31.** 2,959,500 **33.** 21 columns **35.** 563 packages; 7 bars left over **37.** **(a)** 4200 sq ft; **(b)** 268 ft **39.** 35 cartons **41.** 56 full cartons; 11 books left over. If 1355 books are shipped, it will take 57 cartons. **43.** 384 mi; 27 in. **45.** 18 rows **47.** 32 \$10 bills **49.** \$400 **51.** 280 min, or 4 hr 40 min **53.** 525 min, or 8 hr 45 min **55.** 106 bones **57.** \$704 **59.** DW **61.** 234,600 **62.** 234,560 **63.** 235,000 **64.** 22,000 **65.** 16,000 **66.** 8000 **67.** 4000 **68.** 320,000 **69.** 720,000 **70.** 46,800,000 **71.** 792,000 mi; 1,386,000 mi

Margin Exercises, Section 1.6, pp. 69–74

1. 5^4 **2.** 5^5 **3.** 10^2 **4.** 10^4 **5.** 10,000 **6.** 100 **7.** 512 **8.** 32 **9.** 51 **10.** 30 **11.** 584 **12.** 84

13. 4; 1 **14.** 52; 52 **15.** 29 **16.** 1880 **17.** 253 **18.** 93 **19.** 1880 **20.** 305 **21.** 75 **22.** 4 **23.** 86 in. **24.** 46 **25.** 4

Calculator Corner, p. 70

1. 243 **2.** 15,625 **3.** 20,736 **4.** 2048

Calculator Corner, p. 72

1. 49 **2.** 85 **3.** 36 **4.** 0 **5.** 73 **6.** 49

Exercise Set 1.6, p. 75

1. 3^4 **3.** 5^2 **5.** 7^5 **7.** 10^3 **9.** 49 **11.** 729 **13.** 20,736 **15.** 121 **17.** 22 **19.** 20 **21.** 100 **23.** 1 **25.** 49 **27.** 5 **29.** 434 **31.** 41 **33.** 88 **35.** 4 **37.** 303 **39.** 20 **41.** 70 **43.** 295 **45.** 32 **47.** 906 **49.** 62 **51.** 102 **53.** 256 **55.** $94 **57.** 110 **59.** 7 **61.** 544 **63.** 708 **65.** 27 **67.** D$_W$ **69.** 452 **70.** 835 **71.** 13 **72.** 37 **73.** 2342 **74.** 4898 **75.** 25 **76.** 100 **77.** 102,600 mi^2 **78.** 98 gal **79.** 24; $1 + 5 \cdot (4 + 3) = 36$ **81.** 7; $12 \div (4 + 2) \cdot 3 - 2 = 4$

Margin Exercises, Section 1.7, pp. 78–82

1. Yes **2.** No **3.** 1, 2, 5 **4.** 1, 3, 5, 9, 15, 45 **5.** 1, 2, 31, 62 **6.** 1, 2, 3, 4, 6, 8, 12, 24 **7.** $5 = 1 \cdot 5$; $45 = 9 \cdot 5$; $100 = 20 \cdot 5$ **8.** $10 = 1 \cdot 10$; $60 = 6 \cdot 10$; $110 = 11 \cdot 10$ **9.** 5, 10, 15, 20, 25, 30, 35, 40, 45, 50 **10.** Yes **11.** Yes **12.** No **13.** 2, 13, 19, 41, 73 are prime; 6, 12, 65, 99 are composite; 1 is neither **14.** $2 \cdot 3$ **15.** $2 \cdot 2 \cdot 3$ **16.** $3 \cdot 3 \cdot 5$ **17.** $2 \cdot 7 \cdot 7$ **18.** $2 \cdot 3 \cdot 3 \cdot 7$ **19.** $2 \cdot 2 \cdot 2 \cdot 2 \cdot 3 \cdot 3$

Calculator Corner, p. 80

1. Yes **2.** No **3.** No **4.** Yes **5.** No **6.** Yes **7.** Yes **8.** No **9.** Yes **10.** No

Exercise Set 1.7, p. 83

1. No **3.** Yes **5.** 1, 2, 3, 6, 9, 18 **7.** 1, 2, 3, 6, 9, 18, 27, 54 **9.** 1, 2, 4 **11.** 1, 7 **13.** 1 **15.** 1, 2, 7, 14, 49, 98 **17.** 4, 8, 12, 16, 20, 24, 28, 32, 36, 40 **19.** 20, 40, 60, 80, 100, 120, 140, 160, 180, 200 **21.** 3, 6, 9, 12, 15, 18, 21, 24, 27, 30 **23.** 12, 24, 36, 48, 60, 72, 84, 96, 108, 120 **25.** 10, 20, 30, 40, 50, 60, 70, 80, 90, 100 **27.** 9, 18, 27, 36, 45, 54, 63, 72, 81, 90 **29.** No **31.** Yes **33.** Yes **35.** No **37.** No **39.** Neither **41.** Composite **43.** Prime **45.** Prime **47.** $2 \cdot 2 \cdot 2$ **49.** $2 \cdot 7$ **51.** $2 \cdot 3 \cdot 7$ **53.** $5 \cdot 5$ **55.** $2 \cdot 5 \cdot 5$ **57.** $13 \cdot 13$ **59.** $2 \cdot 2 \cdot 5 \cdot 5$ **61.** $5 \cdot 7$ **63.** $2 \cdot 2 \cdot 2 \cdot 3 \cdot 3$ **65.** $7 \cdot 11$ **67.** $2 \cdot 2 \cdot 7 \cdot 103$ **69.** $3 \cdot 17$ **71.** D$_W$ **73.** 26 **74.** 256 **75.** 425 **76.** 4200 **77.** 0 **78.** 22 **79.** 1 **80.** 3 **81.** $612

82. 201 min, or 3 hr 21 min **83.** Row 1: 48, 90, 432, 63; row 2: 7, 2, 2, 10, 8, 6, 21, 10; row 3: 9, 18, 36, 14, 12, 11, 21; row 4: 29, 19, 42

Margin Exercises, Section 1.8, pp. 85–88

1. Yes **2.** No **3.** Yes **4.** No **5.** Yes **6.** No **7.** Yes **8.** No **9.** Yes **10.** No **11.** No **12.** Yes **13.** No **14.** Yes **15.** No **16.** Yes **17.** No **18.** Yes **19.** No **20.** Yes **21.** Yes **22.** No **23.** No **24.** Yes **25.** Yes **26.** No **27.** No **28.** Yes **29.** No **30.** Yes **31.** Yes **32.** No

Exercise Set 1.8, p. 89

1. 46, 224, 300, 36, 45,270, 4444, 256, 8064, 21,568 **3.** 224, 300, 36, 4444, 256, 8064, 21,568 **5.** 300, 36, 45,270, 8064 **7.** 36, 45,270, 711, 8064 **9.** 324, 42, 501, 3009, 75, 2001, 402, 111,111, 1005 **11.** 55,555, 200, 75, 2345, 35, 1005 **13.** 324 **15.** 200 **17.** D$_W$ **19.** 138 **20.** 139 **21.** 874 **22.** 56 **23.** 26 **24.** 13 **25.** 234 **26.** 4003 **27.** 45 gal **28.** 4320 min **29.** $2 \cdot 2 \cdot 2 \cdot 3 \cdot 5 \cdot 5 \cdot 13$ **31.** $2 \cdot 2 \cdot 3 \cdot 3 \cdot 7 \cdot 11$ **33.** 95,238

Margin Exercises, Section 1.9, pp. 91–94

1. 45 **2.** 40 **3.** 30 **4.** 24 **5.** 10 **6.** 80 **7.** 40 **8.** 360 **9.** 864 **10.** 2520 **11.** 18 **12.** 24 **13.** 36 **14.** 210

Exercise Set 1.9, p. 95

1. 4 **3.** 50 **5.** 40 **7.** 54 **9.** 150 **11.** 120 **13.** 72 **15.** 420 **17.** 144 **19.** 288 **21.** 30 **23.** 105 **25.** 72 **27.** 60 **29.** 36 **31.** 900 **33.** 48 **35.** 50 **37.** 143 **39.** 420 **41.** 378 **43.** 810 **45.** Every 60 yr **47.** Every 420 yr **49.** D$_W$ **51.** $<$ **52.** $539 **53.** 33,135 **54.** 6939 **55.** 2 ten thousands + 4 thousands + 6 hundreds + 5 ones **56.** One hundred two thousand, nine hundred sixty **57.** 18,900 **59.** 5 in. by 24 in.

Summary and Review: Chapter 1, p. 97

1. 2 thousands + 7 hundreds + 9 tens + 3 ones **2.** 5 ten thousands + 6 thousands + 0 hundreds + 7 tens + 8 ones, or 5 ten thousands + 6 thousands + 7 tens + 8 ones **3.** 8669 **4.** 90,844 **5.** Sixty-seven thousand, eight hundred nineteen **6.** Two million, seven hundred eighty-one thousand, four hundred twenty-seven **7.** 476,588 **8.** 2,000,400,000 **9.** 8 thousands **10.** 3 **11.** 14,272 **12.** 66,024 **13.** 22,098 **14.** 98,921 **15.** $10 = 6 + 4$, or $10 = 4 + 6$ **16.** $8 = 11 - 3$; $3 = 11 - 8$ **17.** 5148 **18.** 1153 **19.** 2274 **20.** 17,757 **21.** 345,800 **22.** 345,760 **23.** 346,000 **24.** $41,300 + 19,700 = 61,000$ **25.** $38,700 - 24,500 = 14,200$ **26.** $400 \cdot 700 = 280,000$

27. $>$ **28.** $<$ **29.** 420,000 **30.** 6,276,800 **31.** 506,748 **32.** 27,589 **33.** 5,331,810 **34.** $56 = 8 \cdot 7$, or $56 = 7 \cdot 8$ **35.** $4 = 52 \div 13$; $13 = 52 \div 4$ **36.** 12 R 3 **37.** 5 **38.** 913 R 3 **39.** 384 R 1 **40.** 4 R 46 **41.** 54 **42.** 452 **43.** 4389 **44.** 8 **45.** 45 **46.** 546 **47.** 4^3 **48.** 10,000 **49.** 36 **50.** 65 **51.** 233 **52.** 56 **53.** 32 **54.** 260 **55.** 165 **56.** \$502 **57.** \$484 **58.** 1982 **59.** 37 cartons **60.** 14 beehives **61.** \$2413 **62.** \$19,748 **63.** 137 beakers filled; 13 mL left over **64.** 98 ft^2; 42 ft **65.** $2 \cdot 5 \cdot 7$ **66.** $2 \cdot 3 \cdot 5$ **67.** $3 \cdot 3 \cdot 5$ **68.** $2 \cdot 3 \cdot 5 \cdot 5$ **69.** No **70.** No **71.** No **72.** Yes **73.** Prime **74.** 36 **75.** 90 **76.** 30 **77.** 1404 **78.** ^{D}W A vat contains 1152 oz of hot sauce. If 144 bottles are to be filled equally, how much will each bottle contain? Answers may vary. **79.** $d = 8$ **80.** $a = 8, b = 4$ **81.** 6 days **82.** 13, 11, 101, 73

Test: Chapter 1, p. 100

1. [1.1b] 8 thousands + 8 hundreds + 4 tens + 3 ones **2.** [1.1c] Thirty-eight million, four hundred three thousand, two hundred seventy-seven **3.** [1.1a] 5 **4.** [1.2a] 9989 **5.** [1.2a] 63,791 **6.** [1.2a] 34 **7.** [1.2a] 10,515 **8.** [1.2d] 3630 **9.** [1.2d] 1039 **10.** [1.2d] 6848 **11.** [1.2d] 5175 **12.** [1.3a] 41,112 **13.** [1.3a] 5,325,600 **14.** [1.3a] 2405 **15.** [1.3a] 534,264 **16.** [1.3d] 3 R 3 **17.** [1.3d] 70 **18.** [1.3d] 97 **19.** [1.3d] 805 R 8 **20.** [1.5a] 1852 12-packs; 7 cakes left over **21.** [1.5a] 1,285,156 sq mi **22.** **(a)** [1.2b, d], [1.3b] 300 in., 5000 in^2; 264 in., 3872 in^2; 228 in., 2888 in^2; **(b)** [1.5a] 2112 in^2 **23.** [1.5a] 99,076 patents **24.** [1.5a] 1808 lb **25.** [1.5a] 20 staplers **26.** [1.4b] 46 **27.** [1.4b] 13 **28.** [1.4b] 14 **29.** [1.3e] 35,000 **30.** [1.3e] 34,580 **31.** [1.3e] 34,600 **32.** [1.3f] $23{,}600 + 54{,}700 = 78{,}300$ **33.** [1.3f] $54{,}800 - 23{,}600 = 31{,}200$ **34.** [1.3f] $800 \cdot 500 = 400{,}000$ **35.** [1.1d] $>$ **36.** [1.1d] $<$ **37.** [1.6b] 343 **38.** [1.6b] 8 **39.** [1.6a] 12^4 **40.** [1.9a] 48 **41.** [1.6c] 64 **42.** [1.6c] 96 **43.** [1.6c] 2 **44.** [1.6d] 216 **45.** [1.6c] 18 **46.** [1.6c] 92 **47.** [1.7d] $2 \cdot 3 \cdot 3$ **48.** [1.7d] $2 \cdot 2 \cdot 3 \cdot 5$ **49.** [1.8a] Yes **50.** [1.8a] No **51.** [1.3b], [1.5a] 336 in^2 **52.** [1.5a] 80 payments **53.** [1.6c] 83 **54.** [1.6c] 9

Chapter 2

Pretest: Chapter 2, p. 104

1. 1 **2.** 68 **3.** 0 **4.** $\frac{1}{4}$ **5.** $<$ **6.** $\frac{8}{7}$ **7.** $\frac{61}{8}$ **8.** $5\frac{1}{2}$ **9.** $11\frac{31}{60}$ **10.** $1\frac{2}{3}$ **11.** $\frac{6}{5}$ **12.** $\frac{1}{40}$ **13.** 30 **14.** $\frac{2}{9}$ **15.** $21\frac{1}{4}$ lb **16.** $\frac{1}{24}$ m **17.** $351\frac{1}{5}$ mi **18.** $22\frac{1}{2}$ cups **19.** $\frac{49}{12}$, or $4\frac{1}{12}$ **20.** 10 **21.** 2

Margin Exercises, Section 2.1, pp. 105–112

1. 1, numerator; 6, denominator **2.** 5, numerator; 7, denominator **3.** 22, numerator; 3, denominator **4.** $\frac{1}{2}$ **5.** $\frac{1}{3}$ **6.** $\frac{1}{3}$ **7.** $\frac{2}{3}$ **8.** $\frac{3}{4}$ **9.** $\frac{15}{16}$ **10.** $\frac{5}{4}$ **11.** $\frac{7}{4}$ **12.** 1 **13.** 1 **14.** 1 **15.** 1 **16.** 1 **17.** 1 **18.** 0 **19.** 0 **20.** 0 **21.** 0 **22.** Not defined **23.** Not defined **24.** 8 **25.** 10 **26.** 346 **27.** 23 **28.** **29.** $\frac{15}{56}$ **30.** $\frac{32}{15}$ **31.** $\frac{3}{100}$ **32.** $\frac{14}{3}$ **33.** $\frac{8}{16}$ **34.** $\frac{30}{50}$ **35.** $\frac{52}{100}$ **36.** $\frac{200}{75}$ **37.** $\frac{12}{9}$ **38.** $\frac{18}{24}$ **39.** $\frac{90}{100}$ **40.** $\frac{9}{45}$ **41.** $\frac{56}{49}$ **42.** $\frac{1}{4}$ **43.** $\frac{5}{6}$ **44.** 5 **45.** $\frac{4}{3}$ **46.** $\frac{7}{8}$ **47.** $\frac{89}{78}$ **48.** $\frac{8}{7}$ **49.** $\frac{1}{4}$ **50.** $\frac{2}{100} = \frac{1}{50}$; $\frac{4}{100} = \frac{1}{25}$; $\frac{32}{100} = \frac{8}{25}$; $\frac{44}{100} = \frac{11}{25}$; $\frac{18}{100} = \frac{9}{50}$

Calculator Corner, p. 113

1. $\frac{14}{15}$ **2.** $\frac{7}{8}$ **3.** $\frac{138}{167}$ **4.** $\frac{7}{25}$

Exercise Set 2.1, p. 114

1. 3, numerator; 4, denominator **3.** 11, numerator; 20, denominator **5.** $\frac{2}{4}$ **7.** $\frac{1}{8}$ **9.** $\frac{4}{3}$ **11.** $\frac{12}{16}$ **13.** $\frac{3}{4}$ **15.** $\frac{4}{8}$ **17.** $\frac{6}{12}$ **19.** **(a)** $\frac{2}{8}$; **(b)** $\frac{6}{8}$ **21.** **(a)** $\frac{3}{8}$; **(b)** $\frac{5}{8}$ **23.** 18 **25.** 0 **27.** 1 **29.** Not defined **31.** Not defined **33.** 1 **35.** $\frac{1}{6}$ **37.** $\frac{5}{8}$ **39.** $\frac{2}{15}$ **41.** $\frac{4}{15}$ **43.** $\frac{9}{16}$ **45.** $\frac{14}{39}$ **47.** $\frac{21}{4}$ **49.** $\frac{49}{64}$ **51.** $\frac{5}{10}$ **53.** $\frac{20}{32}$ **55.** $\frac{75}{45}$ **57.** $\frac{42}{132}$ **59.** $\frac{3}{4}$ **61.** $\frac{1}{5}$ **63.** 3 **65.** $\frac{3}{4}$ **67.** $\frac{7}{8}$ **69.** 6 **71.** $\frac{1}{3}$ **73.** ^{D}W **75.** $\frac{1}{6}$ **77.** $\frac{2}{16}$, or $\frac{1}{8}$

Margin Exercises, Section 2.2, pp. 118–121

1. $\frac{7}{12}$ **2.** $\frac{1}{3}$ **3.** 6 **4.** $\frac{5}{2}$ **5.** $\frac{5}{2}$ **6.** $\frac{7}{10}$ **7.** $\frac{1}{9}$ **8.** 5 **9.** $\frac{8}{7}$ **10.** $\frac{8}{3}$ **11.** $\frac{1}{10}$ **12.** 100 **13.** 1 **14.** $\frac{14}{15}$ **15.** $\frac{4}{5}$ **16.** 32

Exercise Set 2.2, p. 122

1. $\frac{1}{3}$ **3.** $\frac{1}{6}$ **5.** $\frac{27}{10}$ **7.** $\frac{14}{9}$ **9.** 1 **11.** 1 **13.** 4 **15.** 9 **17.** $\frac{98}{5}$ **19.** 30 **21.** $\frac{1}{5}$ **23.** $\frac{9}{25}$ **25.** $\frac{11}{40}$ **27.** $\frac{5}{14}$ **29.** $\frac{6}{5}$ **31.** $\frac{1}{6}$ **33.** 6 **35.** $\frac{3}{10}$ **37.** $\frac{4}{5}$ **39.** $\frac{4}{15}$ **41.** 4 **43.** 2 **45.** $\frac{1}{8}$ **47.** $\frac{3}{7}$ **49.** 8 **51.** 35 **53.** 1 **55.** $\frac{2}{3}$ **57.** $\frac{9}{4}$ **59.** 144 **61.** 75 **63.** 2 **65.** $\frac{3}{5}$ **67.** 315 **69.** ^{D}W **71.** 67 **72.** 33 R 4 **73.** 285 R 2 **74.** 103 R 10 **75.** 67 **76.** 264 **77.** 8499 **78.** 4368 **79.** $\frac{3}{8}$

Margin Exercises, Section 2.3, pp. 124–129

1. $\frac{4}{5}$ **2.** 1 **3.** $\frac{1}{2}$ **4.** $\frac{3}{4}$ **5.** $\frac{5}{6}$ **6.** $\frac{29}{24}$ **7.** $\frac{5}{9}$ **8.** $\frac{413}{1000}$ **9.** $\frac{197}{210}$ **10.** $\frac{65}{72}$ **11.** $\frac{1}{2}$ **12.** $\frac{3}{8}$ **13.** $\frac{1}{2}$ **14.** $\frac{13}{18}$ **15.** $\frac{1}{2}$ **16.** $\frac{9}{112}$ **17.** $<$ **18.** $>$ **19.** $>$ **20.** $>$ **21.** $<$ **22.** $\frac{1}{6}$ **23.** $\frac{11}{40}$

Exercise Set 2.3, p. 130

1. 1 **3.** $\frac{3}{4}$ **5.** $\frac{3}{2}$ **7.** $\frac{7}{24}$ **9.** $\frac{3}{2}$ **11.** $\frac{19}{24}$ **13.** $\frac{9}{10}$ **15.** $\frac{29}{18}$ **17.** $\frac{31}{100}$ **19.** $\frac{41}{60}$ **21.** $\frac{189}{100}$ **23.** $\frac{7}{8}$ **25.** $\frac{13}{24}$ **27.** $\frac{17}{24}$ **29.** $\frac{3}{4}$ **31.** $\frac{437}{500}$ **33.** $\frac{53}{40}$ **35.** $\frac{391}{144}$ **37.** $\frac{2}{3}$ **39.** $\frac{3}{4}$ **41.** $\frac{5}{8}$ **43.** $\frac{1}{24}$ **45.** $\frac{1}{2}$ **47.** $\frac{9}{14}$ **49.** $\frac{3}{5}$ **51.** $\frac{7}{10}$ **53.** $\frac{17}{60}$ **55.** $\frac{53}{100}$ **57.** $\frac{26}{75}$ **59.** $\frac{9}{100}$ **61.** $\frac{13}{24}$ **63.** $\frac{1}{10}$ **65.** $\frac{1}{24}$ **67.** $\frac{13}{16}$ **69.** $\frac{31}{75}$ **71.** $\frac{13}{75}$ **73.** $<$ **75.** $>$ **77.** $<$ **79.** $<$ **81.** $>$ **83.** $>$ **85.** $<$ **87.** $\frac{1}{15}$ **89.** $\frac{2}{15}$ **91.** $\frac{1}{2}$ **93.** D_W **95.** 1 **96.** Not defined **97.** Not defined **98.** 4 **99.** \$73 **100.** \$11 **101.** \$3 **102.** \$9 **103.** \$1558 **104.** \$1684 **105.** $\frac{19}{24}$ **107.** $\frac{145}{144}$ **109.** $\frac{227}{420}$ km

Margin Exercises, Section 2.4, pp. 133–139

1. $1\frac{2}{3}$ **2.** $2\frac{3}{4}$ **3.** $8\frac{3}{4}$ **4.** $12\frac{2}{3}$ **5.** $\frac{22}{5}$ **6.** $\frac{61}{10}$ **7.** $\frac{29}{6}$ **8.** $\frac{37}{4}$ **9.** $\frac{62}{3}$ **10.** $2\frac{1}{3}$ **11.** $1\frac{1}{10}$ **12.** $18\frac{1}{3}$ **13.** $7\frac{2}{5}$ **14.** $12\frac{1}{10}$ **15.** $13\frac{7}{12}$ **16.** $1\frac{1}{2}$ **17.** $3\frac{1}{6}$ **18.** $3\frac{5}{18}$ **19.** $3\frac{2}{3}$ **20.** 20 **21.** $1\frac{7}{8}$ **22.** $12\frac{4}{5}$ **23.** $8\frac{1}{3}$ **24.** 16 **25.** $7\frac{3}{7}$ **26.** $1\frac{7}{8}$ **27.** $\frac{7}{10}$

Calculator Corner, p. 139

1. $\frac{7}{12}$ **2.** $\frac{11}{10}$ **3.** $\frac{35}{16}$ **4.** $\frac{3}{10}$ **5.** $10\frac{2}{15}$ **6.** $1\frac{1}{28}$ **7.** $10\frac{11}{15}$ **8.** $2\frac{91}{115}$

Exercise Set 2.4, p. 140

1. $\frac{29}{8}; \frac{11}{4}$ **3.** $\frac{17}{3}$ **5.** $\frac{59}{6}$ **7.** $\frac{51}{4}$ **9.** $3\frac{3}{5}$ **11.** $5\frac{7}{10}$ **13.** $43\frac{1}{8}$ **15.** $6\frac{1}{2}$ **17.** $2\frac{11}{12}$ **19.** $14\frac{7}{12}$ **21.** $21\frac{1}{2}$ **23.** $27\frac{7}{8}$ **25.** $27\frac{13}{24}$ **27.** $1\frac{3}{5}$ **29.** $4\frac{1}{10}$ **31.** $15\frac{3}{8}$ **33.** $7\frac{5}{12}$ **35.** $13\frac{3}{8}$ **37.** $11\frac{5}{18}$ **39.** $22\frac{2}{3}$ **41.** $2\frac{5}{12}$ **43.** $8\frac{1}{6}$ **45.** $9\frac{31}{40}$ **47.** $24\frac{91}{100}$ **49.** $975\frac{4}{5}$ **51.** $6\frac{1}{4}$ **53.** $1\frac{1}{5}$ **55.** $3\frac{9}{16}$ **57.** $1\frac{1}{8}$ **59.** $1\frac{8}{43}$ **61.** $\frac{9}{40}$ **63.** D_W **65.** 45,800 **66.** 45,770 **67.** Yes **68.** No **69.** No **70.** Yes **71.** No **72.** Yes **73.** Yes **74.** Yes **75.** 47 **76.** 889 **77.** 25 **78.** 477,978 **79.** $360\frac{60}{473}$ **81.** $35\frac{57}{64}$

Margin Exercises, Section 2.5, pp. 144–149

1. $\frac{3}{8}$ **2.** $\frac{63}{100}$ cm^2 **3.** 320 loops **4.** 200 gal **5.** $\frac{5}{24}$ mi **6.** $\frac{3}{8}$ in. **7.** $227\frac{1}{2}$ mi **8.** 20 mpg **9.** $240\frac{3}{4}$ ft^2

Exercise Set 2.5, p. 151

1. $\frac{9}{20}$ **3.** $\frac{12}{25}$ m^2 **5.** $\frac{5}{8}$ in. **7.** $\frac{1}{16}$ in. **9.** $\frac{1}{1521}$ **11.** 625 addresses **13.** 160 mi **15.** $\frac{1}{3}$ cup **17.** 75 times **19.** 690 kg; $\frac{14}{23}$ cement; $\frac{5}{23}$ stone; $\frac{4}{23}$ sand; 1 **21.** $\frac{51}{32}$ in. **23.** 32 pairs **25.** 24 bowls **27.** 16 L **29.** $\frac{5}{6}$ lb **31.** $\frac{23}{12}$ mi **33.** $\frac{1}{32}$ in. **35.** $\frac{1}{4}$ of the business **37.** $\frac{2}{3}$ cup **39.** $5\frac{3}{8}$ yd **41.** $7\frac{5}{12}$ lb **43.** $95\frac{1}{5}$ mi **45.** $3\frac{4}{5}$ hr **47.** 15 mpg **49.** $28\frac{3}{4}$ yd **51.** 68°F **53.** $7\frac{3}{8}$ ft **55.** $13\frac{1}{3}$ tsp **57.** $16\frac{1}{2}$ servings **59.** $343\frac{3}{4}$ lb **61.** $35\frac{115}{256}$ sq in. **63.** $59{,}538\frac{1}{8}$ sq ft **65.** D_W **67.** \$27,729 **68.** 130 gal **69.** 4992 ft^2; 284 ft **70.** \$928 **71.** 2203 **72.** 11,851 **73.** 204 **74.** 700 **75.** 5 **76.** 89 **77.** 3520 **78.** 9001 **79.** 50 **80.** 6399 **81.** $\frac{4}{15}$; \$320 **83.** $\frac{1}{12}$

Margin Exercises, Section 2.6, pp. 158–161

1. $\frac{1}{2}$ **2.** $\frac{3}{10}$ **3.** $20\frac{2}{3}$, or $\frac{62}{3}$ **4.** $9\frac{7}{8}$ in. **5.** $\frac{5}{9}$ **6.** $\frac{31}{40}$ **7.** $\frac{27}{56}$ **8.** 0 **9.** 1 **10.** $\frac{1}{2}$ **11.** 1 **12.** 12; answers may vary **13.** 32; answers may vary **14.** 27; answers may vary **15.** 15; answers may vary **16.** 1; answers may vary **17.** 1,000,000; answers may vary **18.** $22\frac{1}{2}$ **19.** 132 **20.** 37

Exercise Set 2.6, p. 162

1. $\frac{1}{24}$ **3.** $\frac{2}{5}$ **5.** $\frac{4}{7}$ **7.** $\frac{59}{30}$, or $1\frac{29}{30}$ **9.** $\frac{3}{20}$ **11.** $\frac{211}{8}$, or $26\frac{3}{8}$ **13.** $\frac{7}{16}$ **15.** $\frac{1}{36}$ **17.** $\frac{3}{8}$ **19.** $\frac{37}{48}$ **21.** $\frac{25}{72}$ **23.** $\frac{103}{16}$, or $6\frac{7}{16}$ **25.** $2\frac{41}{128}$ lb **27.** $18\frac{3}{5}$ days **29.** $\frac{17}{6}$, or $2\frac{5}{6}$ **31.** $\frac{8395}{84}$, or $99\frac{79}{84}$ **33.** 0 **35.** 0 **37.** $\frac{1}{2}$ **39.** $\frac{1}{2}$ **41.** 0 **43.** 1 **45.** 6 **47.** 12 **49.** 19 **51.** 6 **53.** 12 **55.** 16 **57.** 3 **59.** 13 **61.** 2 **63.** 2 **65.** $\frac{1}{2}$ **67.** $271\frac{1}{2}$ **69.** 3 **71.** 100 **73.** $29\frac{1}{2}$ **75.** D_W **77.** 3402 **78.** 1,038,180 **79.** 59 R 77 **80.** 348 **81.** 783 **82.** $\frac{8}{3}$ **83.** $\frac{3}{8}$ **84.** Prime: 5, 7, 23, 43; composite: 9, 14; neither: 1 **85.** 16 people **86.** 43 mg **87.** **(a)** $13 \cdot 9\frac{1}{4} + 8\frac{1}{4} \cdot 7\frac{1}{4}$; **(b)** $\frac{2881}{16}$, or $180\frac{1}{16}$ in^2; **(c)** Multiply before adding. **89.** $a = 2, b = 8$ **91.** The largest is $\frac{4}{3} + \frac{5}{2} = \frac{23}{6}$.

Summary and Review: Chapter 2, p. 166

1. 2, numerator; 7, denominator **2.** $\frac{3}{5}$ **3.** $\frac{3}{100}; \frac{8}{100} = \frac{2}{25}; \frac{10}{100} = \frac{1}{10}; \frac{15}{100} = \frac{3}{20}; \frac{21}{100}; \frac{43}{100}$ **4.** 0 **5.** 1 **6.** 48 **7.** 6 **8.** $\frac{2}{5}$ **9.** Not defined **10.** $\frac{1}{3}$ **11.** $\frac{1}{4}$ **12.** $\frac{2}{3}$ **13.** $\frac{3}{2}$ **14.** 24 **15.** $\frac{1}{14}$ **16.** $\frac{2}{3}$ **17.** $\frac{5}{4}$ **18.** $\frac{1}{3}$ **19.** $\frac{9}{2}$ **20.** 2 **21.** $\frac{11}{6}$ **22.** $\frac{1}{4}$ **23.** $\frac{9}{4}$ **24.** 300 **25.** 1 **26.** $\frac{4}{9}$ **27.** $\frac{63}{40}$ **28.** $\frac{19}{48}$ **29.** $\frac{29}{15}$ **30.** $\frac{7}{16}$ **31.** $\frac{1}{3}$ **32.** $\frac{1}{8}$ **33.** $\frac{5}{27}$ **34.** $\frac{11}{18}$ **35.** $>$ **36.** $>$ **37.** $\frac{15}{2}$ **38.** $\frac{67}{8}$ **39.** $2\frac{1}{3}$ **40.** $6\frac{3}{4}$ **41.** $10\frac{2}{5}$ **42.** $11\frac{11}{15}$ **43.** $10\frac{2}{3}$ **44.** $8\frac{1}{4}$ **45.** $7\frac{7}{9}$ **46.** $4\frac{11}{15}$ **47.** $4\frac{3}{20}$ **48.** $13\frac{3}{8}$ **49.** 16 **50.** $3\frac{1}{2}$ **51.** $2\frac{21}{50}$ **52.** 6 **53.** 12 **54.** $1\frac{7}{17}$ **55.** $\frac{1}{8}$ **56.** $\frac{9}{10}$ **57.** $\frac{3}{10}$ **58.** 240 **59.** $\frac{19}{40}$ **60.** $\frac{2}{5}$ **61.** 9 days **62.** 100 km **63.** $\frac{1}{3}$ cup; 2 cups **64.** \$15 **65.** $4\frac{1}{4}$ yd **66.** 24 lb **67.** About $69\frac{3}{8}$ kg **68.** $1\frac{73}{100}$ in. **69.** $177\frac{3}{4}$ in^2 **70.** $50\frac{1}{4}$ in^2 **71.** $8\frac{3}{8}$ cups **72.** $63\frac{2}{3}$ pies; $19\frac{1}{3}$ pies **73.** 1 **74.** $\frac{77}{240}$ **75.** $\frac{1}{2}$ **76.** 0 **77.** 1 **78.** 7 **79.** 10 **80.** 2 **81.** $28\frac{1}{2}$ **82.** D_W Taking $\frac{1}{2}$ of a number is equivalent to multiplying the number by $\frac{1}{2}$. Dividing by $\frac{1}{2}$ is equivalent to multiplying by the reciprocal of $\frac{1}{2}$, or 2. Thus taking $\frac{1}{2}$ of a number is not the same as dividing by $\frac{1}{2}$. **83.** D_W It might be necessary to find the least common denominator before adding or subtracting. The least common denominator is the least common multiple of the denominators. **84.** 24 **85.** 469 **86.** \$912 **87.** 774 mi **88.** 17 days **89.** 408 R 9 **90.** 3607 **91.** $\frac{6}{3} + \frac{5}{4} = 3\frac{1}{4}$ **92.** $a = 11{,}176$; $b = 9887$

Test: Chapter 2, p. 170

1. [2.1a] $\frac{3}{4}$ **2.** [2.3c] > **3.** [2.1b] 1 **4.** [2.1b] 0
5. [2.1e] $\frac{1}{14}$ **6.** [2.1b] Not defined **7.** [2.2a] 32
8. [2.2a] $\frac{5}{2}$ **9.** [2.2a] $\frac{1}{10}$ **10.** [2.2b] $\frac{8}{5}$ **11.** [2.2b] $\frac{1}{18}$
12. [2.2c] $\frac{3}{10}$ **13.** [2.2c] $\frac{8}{5}$ **14.** [2.2c] 18 **15.** [2.3a] 3
16. [2.3a] $\frac{37}{24}$ **17.** [2.3a] $\frac{79}{100}$ **18.** [2.3b] $\frac{1}{3}$ **19.** [2.3b] $\frac{1}{12}$
20. [2.3b] $\frac{1}{12}$ **21.** [2.2d] 64 **22.** [2.3d] $\frac{1}{4}$ **23.** [2.4a] $\frac{7}{2}$
24. [2.4a] $8\frac{2}{9}$ **25.** [2.4b] $14\frac{1}{5}$ **26.** [2.4c] $4\frac{7}{24}$
27. [2.4d] $4\frac{1}{2}$ **28.** [2.4e] 2 **29.** [2.5a] 4375 students
30. [2.5a] $\frac{3}{40}$ m **31.** [2.5a] 80 books **32.** [2.5a] $\frac{1}{16}$ in.
33. [2.5a] 105 kg **34.** [2.5a] $6\frac{11}{36}$ ft **35.** [2.6a] $3\frac{1}{2}$
36. [2.6b] 0 **37.** [2.6b] 1 **38.** [2.6b] $18\frac{1}{2}$ **39.** [2.6b] 16
40. [1.4b] 1805 **41.** [1.4b] 101 **42.** [1.5a] 3635 mi
43. [1.3d] 380 R 7 **44.** [1.2d] 4434 **45.** [2.5a] $\frac{15}{8}$ tsp
46. [2.5a] Dolores runs $\frac{17}{56}$ mi farther.

Chapter 3

Pretest: Chapter 3, p. 174

1. Two and three hundred forty-seven thousandths
2. Three thousand, two hundred sixty-four and $\frac{78}{100}$ dollars
3. $\frac{21}{100}$ **4.** $\frac{5408}{1000}$ **5.** 0.379 **6.** 28.439 **7.** 3.2 **8.** 0.099
9. 21.0 **10.** 21.045 **11.** 607.219 **12.** 39.0901
13. 0.6179 **14.** 0.32456 **15.** 30.4 **16.** 0.57698
17. 84.26 **18.** 6345.157 **19.** 1081.6 mi **20.** $285.95
21. $159.84 **22.** $3397.71 **23.** 224 **24.** 1.4
25. 0.925 **26.** 2.75 **27.** $4.\overline{142857}$ **28.** 4.6 **29.** 4.62
30. 4.616 **31.** $9.49 **32.** 490,000,000,000,000
33. 1548.8836 **34.** 58.17

Margin Exercises, Section 3.1, pp. 176–181

1. Fifteen and three tenths **2.** Two and five thousand three hundred thirty-three hundred-thousandths
3. Two hundred forty-five and eighty-nine hundredths
4. Thirty-four and sixty-four ten-thousandths
5. Thirty-one thousand, seventy-nine and seven hundred sixty-four thousandths **6.** Four thousand, two hundred seventeen and $\frac{56}{100}$ dollars **7.** Thirteen and $\frac{98}{100}$ dollars
8. $\frac{896}{1000}$ **9.** $\frac{2378}{100}$ **10.** $\frac{56,789}{10,000}$ **11.** $\frac{19}{10}$ **12.** 7.43
13. 0.406 **14.** 6.7089 **15.** 0.9 **16.** 0.057 **17.** 0.083
18. 4.3 **19.** 283.71 **20.** 456.013 **21.** 2.04 **22.** 0.06
23. 0.58 **24.** 1 **25.** 0.8989 **26.** 21.05 **27.** 2.8
28. 13.9 **29.** 234.4 **30.** 7.0 **31.** 0.64 **32.** 7.83
33. 34.68 **34.** 0.03 **35.** 0.943 **36.** 8.004
37. 43.112 **38.** 37.401 **39.** 7459.355 **40.** 7459.35
41. 7459.4 **42.** 7459 **43.** 7460 **44.** 7500 **45.** 7000

Exercise Set 3.1, p. 182

1. Two hundred forty-nine and ninety-four hundredths
3. Ninety-six and four thousand three hundred seventy-five ten-thousandths **5.** Thirty-four and eight hundred ninety-one thousandths **7.** Three hundred twenty-six and $\frac{48}{100}$ dollars **9.** Thirty-six and $\frac{72}{100}$ dollars
11. $\frac{83}{10}$ **13.** $\frac{356}{100}$ **15.** $\frac{4603}{100}$ **17.** $\frac{13}{100,000}$ **19.** $\frac{10,008}{10,000}$
21. $\frac{20,003}{1000}$ **23.** 0.8 **25.** 8.89 **27.** 3.798 **29.** 0.0078
31. 0.00019 **33.** 0.376193 **35.** 99.44 **37.** 3.798
39. 2.1739 **41.** 8.953073 **43.** 0.58 **45.** 0.91
47. 0.001 **49.** 235.07 **51.** $\frac{4}{100}$ **53.** 0.4325 **55.** 0.1
57. 0.5 **59.** 2.7 **61.** 123.7 **63.** 0.89 **65.** 0.67
67. 1.00 **69.** 0.09 **71.** 0.325 **73.** 17.002
75. 10.101 **77.** 9.999 **79.** 800 **81.** 809.473
83. 809 **85.** 34.5439 **87.** 34.54 **89.** 35 **91.** DW
93. 6170 **94.** 6200 **95.** 6000 **96.** 54 **97.** $6\frac{3}{5}$
98. $2 \cdot 2 \cdot 2 \cdot 2 \cdot 5 \cdot 5 \cdot 5$, or $2^4 \cdot 5^3$
99. $2 \cdot 3 \cdot 3 \cdot 5 \cdot 17$, or $2 \cdot 3^2 \cdot 5 \cdot 17$ **100.** $2 \cdot 7 \cdot 11 \cdot 13$
101. $2 \cdot 2 \cdot 2 \cdot 7 \cdot 7 \cdot 11$, or $2^3 \cdot 7^2 \cdot 11$
103. 2.000001, 2.0119, 2.018, 2.0302, 2.1, 2.108, 2.109
105. 6.78346 **107.** 0.03030

Margin Exercises, Section 3.2, pp. 185–189

1. 10.917 **2.** 34.2079 **3.** 4.969 **4.** 3.5617
5. 9.40544 **6.** 912.67 **7.** 2514.773 **8.** 10.754
9. 0.339 **10.** 0.5345 **11.** 0.5172 **12.** 7.36992
13. 1194.22 **14.** 4.9911 **15.** 38.534 **16.** 14.164
17. 2133.5
18. The "balance forward" column should read:
$3078.92
2738.23
2659.67
2890.47
2877.33
2829.33
2868.91
2766.04
2697.45
2597.45

Calculator Corner, p. 187

1. 317.645 **2.** 506.553 **3.** 17.15 **4.** 49.08 **5.** 4.4
6. 33.83 **7.** 454.74 **8.** 0.99

Exercise Set 3.2, p. 190

1. 334.37 **3.** 1576.215 **5.** 132.560 **7.** 84.417
9. 50.0248 **11.** 40.007 **13.** 771.967 **15.** 20.8649
17. 227.4680 **19.** 8754.8221 **21.** 1.3 **23.** 49.02
25. 45.61 **27.** 85.921 **29.** 2.4975 **31.** 3.397
33. 8.85 **35.** 3.37 **37.** 1.045 **39.** 3.703 **41.** 0.9902
43. 99.66 **45.** 4.88 **47.** 0.994 **49.** 17.802
51. 51.13 **53.** 2.491 **55.** 32.7386 **57.** 1.6666
59. 2344.90886 **61.** 11.65 **63.** 19.251 **65.** 384.68
67. 582.97 **69.** 15,335.3

71. The balance forward should read:

$ 9704.56	10416.72
9677.12	12916.72
10677.12	12778.94
10553.17	12797.82
10429.15	9997.82

73. ^{D}W **75.** 2720 **76.** $2 \cdot 2 \cdot 3 \cdot 19$ **77.** $\frac{1}{6}$ **78.** $\frac{34}{45}$
79. 6166 **80.** 5366 **81.** $16\frac{1}{2}$ servings **82.** $60\frac{1}{5}$ mi
83. 345.8

Margin Exercises, Section 3.3, pp. 195–198

1. 529.48 **2.** 5.0594 **3.** 34.2906 **4.** 0.348 **5.** 0.0348
6. 0.00348 **7.** 0.000348 **8.** 34.8 **9.** 348 **10.** 3480
11. 34,800 **12.** 4,400,000,000 **13.** 3,700,000
14. 1569¢ **15.** 17¢ **16.** $0.35 **17.** $5.77

Calculator Corner, p. 197

1. 48.6 **2.** 6930.5 **3.** 142.803 **4.** 0.5076 **5.** 7916.4
6. 20.4153

Exercise Set 3.3, p. 199

1. 60.2 **3.** 6.72 **5.** 0.252 **7.** 0.522 **9.** 237.6
11. 583,686.852 **13.** 780 **15.** 8.923 **17.** 0.09768
19. 0.782 **21.** 521.6 **23.** 3.2472 **25.** 897.6
27. 322.07 **29.** 55.68 **31.** 3487.5 **33.** 50.0004
35. 114.42902 **37.** 13.284 **39.** 90.72 **41.** 0.0028728
43. 0.72523 **45.** 1.872115 **47.** 45,678 **49.** 2888¢
51. 66¢ **53.** $0.34 **55.** $34.45 **57.** 93,000,000
59. 7,200,000,000 **61.** ^{D}W **63.** $11\frac{1}{5}$ **64.** $\frac{35}{72}$ **65.** $2\frac{7}{15}$
66. $7\frac{2}{15}$ **67.** 342 **68.** 87 **69.** 4566 **70.** 1257
71. 87 **72.** 1176 R 14 **73.** $10^{21} = 1$ sextillion
75. $10^{24} = 1$ septillion

Margin Exercises, Section 3.4, pp. 202–208

1. 0.6 **2.** 1.5 **3.** 0.47 **4.** 0.32 **5.** 3.75 **6.** 0.25
7. (a) 375; **(b)** 15 **8.** 4.9 **9.** 12.8 **10.** 15.625
11. 12.78 **12.** 0.001278 **13.** 0.09847 **14.** 67.832
15. 0.78314 **16.** 1105.6 **17.** 0.04 **18.** 0.2426
19. 593.44 **20.** 5967.5 m

Calculator Corner, p. 205

1. 14.3 **2.** 2.56 **3.** 200 **4.** 0.75 **5.** 20 **6.** 0.064
7. 15.7 **8.** 75.8

Exercise Set 3.4, p. 209

1. 2.99 **3.** 23.78 **5.** 7.48 **7.** 7.2 **9.** 1.143
11. 4.041 **13.** 0.07 **15.** 70 **17.** 20 **19.** 0.4
21. 0.41 **23.** 8.5 **25.** 9.3 **27.** 0.625 **29.** 0.26
31. 15.625 **33.** 2.34 **35.** 0.47 **37.** 0.2134567
39. 21.34567 **41.** 1023.7 **43.** 9.3 **45.** 0.0090678
47. 45.6 **49.** 2107 **51.** 303.003 **53.** 446.208
55. 24.14 **57.** 13.0072 **59.** 19.3204 **61.** 473.188278
63. 10.49 **65.** 911.13 **67.** 205 **69.** $1288.36
71. 16,249.6 **73.** ^{D}W **75.** $\frac{6}{7}$ **76.** $\frac{7}{8}$ **77.** $\frac{19}{73}$ **78.** $\frac{19}{73}$
79. $2 \cdot 2 \cdot 3 \cdot 3 \cdot 19$, or $2^2 \cdot 3^2 \cdot 19$
80. $2 \cdot 3 \cdot 3 \cdot 3 \cdot 3$, or $2 \cdot 3^4$ **81.** $3 \cdot 3 \cdot 223$, or $3^2 \cdot 223$
82. $5 \cdot 401$ **83.** $15\frac{1}{8}$ **84.** $5\frac{7}{8}$ **85.** 6.254194585
87. 1000 **89.** 100

Margin Exercises, Section 3.5, pp. 213–217

1. 0.8 **2.** 0.45 **3.** 0.275 **4.** 1.32 **5.** 0.4 **6.** 0.375
7. $0.1\overline{6}$ **8.** $0.\overline{6}$ **9.** $0.\overline{45}$ **10.** $1.\overline{09}$ **11.** $0.\overline{428571}$
12. 0.7; 0.67; 0.667 **13.** 0.8; 0.81; 0.808 **14.** 6.2; 6.25; 6.245 **15.** 0.510 **16.** 24.2 mpg **17.** 42.1 models
18. 0.72 **19.** 0.552 **20.** 9.6575

Exercise Set 3.5, p. 218

1. 0.23 **3.** 0.6 **5.** 0.325 **7.** 0.2 **9.** 0.85 **11.** 0.375
13. 0.975 **15.** 0.52 **17.** 20.016 **19.** 0.25 **21.** 1.16
23. 1.1875 **25.** $0.2\overline{6}$ **27.** $0.\overline{3}$ **29.** $1.\overline{3}$ **31.** $1.1\overline{6}$
33. $0.\overline{571428}$ **35.** $0.91\overline{6}$ **37.** 0.3; 0.27; 0.267
39. 0.3; 0.33; 0.333 **41.** 1.3; 1.33; 1.333 **43.** 1.2; 1.17; 1.167 **45.** 0.6; 0.57; 0.571 **47.** 0.9; 0.92; 0.917
49. 0.2; 0.18; 0.182 **51.** 0.3; 0.28; 0.278 **53. (a)** 0.429; **(b)** 0.75; **(c)** 0.571; **(d)** 1.333 **55.** 15.8 mpg
57. 17.8 mpg **59.** 15.2 mph **61.** $41.6875; $41.69
63. $25.875; $25.88 **65.** $19.046875; $19.05 **67.** 11.06
69. 8.4 **71.** $417.51\overline{6}$ **73.** 0 **75.** 2.8125 **77.** 0.20425
79. 317.14 **81.** 0.1825 **83.** 18 **85.** 2.736
87. ^{D}W **89.** 21 **90.** $238\frac{7}{8}$ **91.** 10 **92.** $\frac{43}{52}$
93. $50\frac{5}{24}$ **94.** $30\frac{7}{10}$ **95.** $1\frac{1}{2}$ **96.** $14\frac{13}{24}$ **97.** $1\frac{1}{24}$ cups
98. $1\frac{33}{100}$ in. **99.** 270 **100.** 792 **101.** $0.\overline{142857}$
103. $0.\overline{428571}$ **105.** $0.\overline{714285}$ **107.** $0.\overline{1}$ **109.** $0.\overline{001}$

Margin Exercises, Section 3.6, pp. 222–224

1. (b) **2.** (a) **3.** (d) **4.** (b) **5.** (a) **6.** (d) **7.** (b)
8. (c) **9.** (b) **10.** (b) **11.** (c) **12.** (a) **13.** (c)
14. (c)

Exercise Set 3.6, p. 225

1. (d) **3.** (c) **5.** (a) **7.** (c) **9.** 1.6 **11.** 6 **13.** 60
15. 2.3 **17.** 180 **19.** (a) **21.** (c) **23.** (b) **25.** (b)
27. $1500 \div 0.5 = 3000$; answers may vary **29.** ^{D}W
31. $2 \cdot 2 \cdot 3 \cdot 3 \cdot 3$, or $2^2 \cdot 3^3$
32. $2 \cdot 2 \cdot 2 \cdot 2 \cdot 5 \cdot 5$, or $2^4 \cdot 5^2$ **33.** $5 \cdot 5 \cdot 13$, or $5^2 \cdot 13$
34. $2 \cdot 3 \cdot 3 \cdot 37$, or $2 \cdot 3^2 \cdot 37$
35. $2 \cdot 2 \cdot 2 \cdot 2 \cdot 2 \cdot 2 \cdot 3 \cdot 3 \cdot 3$, or $2^6 \cdot 3^3$ **36.** $\frac{5}{16}$ **37.** $\frac{129}{251}$
38. $\frac{8}{9}$ **39.** $\frac{13}{25}$ **40.** $\frac{25}{19}$ **41.** Yes **43.** No
45. (a) $+, \times$; **(b)** $+, \times, -$

Margin Exercises, Section 3.7, pp. 228–235

1. 8.4° **2.** 148.1 gal **3.** $51.26 **4.** $368.75
5. 96.52 cm^2 **6.** $1.33 **7.** 28.6 mpg **8.** $289,683
9. $582,278

Exercise Set 3.7, p. 236

1. \$10.50 **3.** \$3.87 **5.** 102.8°F **7.** \$21,219.17 **9.** Area: 8.125 cm^2; perimeter: 11.5 cm **11.** 22,691.5 mi **13.** 20.2 mpg **15.** \$10 **17.** 11.9752 ft^3 **19.** 78.1 cm **21.** 28.5 cm **23.** 2.31 cm **25.** 876 calories **27.** \$1171.74 **29.** 227.75 ft^2 **31.** 0.372 **33.** \$69.24 **35.** \$906.50 **37.** 5.8¢, or \$0.058 **39.** 2152.56 yd^2 **41.** 4.229 billion **43.** 1.4°F **45.** \$142,989 **47.** \$112,762 **49.** \$75,097 **51.** DW **53.** 6335 **54.** $\frac{31}{24}$ **55.** $6\frac{5}{6}$ **56.** 2803 **57.** $\frac{1}{24}$ **58.** $1\frac{5}{6}$ **59.** 8 **60.** 11 **61.** $3\frac{2}{5}$ **62.** $19\frac{1}{21}$ **63.** $2\frac{8}{9}$ **64.** 4 **65.** 28 min **66.** $7\frac{1}{5}$ min **67.** \$17.28

Summary and Review: Chapter 3, p. 244

1. 6,590,000 **2.** 6,900,000 **3.** Three and forty-seven hundredths **4.** Thirty-one thousandths **5.** Five hundred ninety-seven and $\frac{25}{100}$ dollars **6.** Zero and $\frac{96}{100}$ dollars **7.** $\frac{9}{100}$ **8.** $\frac{4561}{1000}$ **9.** $\frac{89}{1000}$ **10.** $\frac{30,227}{10,000}$ **11.** 0.034 **12.** 4.2603 **13.** 27.91 **14.** 867.006 **15.** 0.034 **16.** 0.91 **17.** 0.741 **18.** 1.041 **19.** 17.4 **20.** 17.43 **21.** 17.429 **22.** 17 **23.** 574.519 **24.** 0.6838 **25.** 229.1 **26.** 45.551 **27.** 29.2092 **28.** 790.29 **29.** 29.148 **30.** 70.7891 **31.** 12.96 **32.** 0.14442 **33.** 4.3 **34.** 0.02468 **35.** 7.5 **36.** 0.45 **37.** 45.2 **38.** 1.022 **39.** 0.2763 **40.** 1389.2 **41.** 496.2795 **42.** 6.95 **43.** 42.54 **44.** 4.9911 **45.** 24.36 cups; 104.4 cups **46.** \$15.52 **47.** \$5788.56 **48.** \$224.99 **49.** 14.5 mpg **50.** **(a)** 54.6 lb; **(b)** 13.65 lb **51.** 272 **52.** 216 **53.** 4 **54.** \$125 **55.** 2.6 **56.** 1.28 **57.** 2.75 **58.** 3.25 **59.** $1.1\overline{6}$ **60.** $1.\overline{54}$ **61.** 1.5 **62.** 1.55 **63.** 1.545 **64.** \$82.73 **65.** \$4.87 **66.** 2493¢ **67.** 986¢ **68.** 1.8045 **69.** 57.1449 **70.** 15.6375 **71.** $41.537\overline{3}$

72. DW Multiply by 1 to get a denominator that is a power of 10:

$$\frac{44}{125} = \frac{44}{125} \cdot \frac{8}{8} = \frac{352}{1000} = 0.352.$$

We can also divide to find that $\frac{44}{125} = 0.352$.

73. DW Each decimal place in the decimal notation corresponds to one zero in the power of ten in the fraction notation. When the fractions are multiplied, the number of zeros in the denominator of the product is the sum of the number of zeros in the denominators of the factors. So the number of decimal places in the product is the sum of the number of decimal places in the factors.

74. $3\frac{3}{4}$ **75.** $6\frac{3}{5}$ **76.** $\frac{1}{2}$ **77.** $2 \cdot 2 \cdot 2 \cdot 2 \cdot 2 \cdot 2 \cdot 3$, or $2^6 \cdot 3$ **78.** 3300 **79.** **(a)** $2.56 \times 6.4 \div 51.2 - 17.4 + 89.7 = 72.62$; **(b)** $(11.12 - 0.29) \times 3^4 = 877.23$

80. $\frac{1}{3} + \frac{2}{3} = 0.33333333\ldots + 0.66666666\ldots$
$= 0.99999999\ldots.$

Therefore, $1 = 0.99999999\ldots$ because $\frac{1}{3} + \frac{2}{3} = 1$.

81. $2 = 1.\overline{9}$

Test: Chapter 3, p. 247

1. [3.3b] 8,900,000,000 **2.** [3.3b] 3,756,000 **3.** [3.1a] Two and thirty-four hundredths **4.** [3.1a] One thousand, two hundred thirty-four and $\frac{78}{100}$ dollars **5.** [3.1b] $\frac{91}{100}$ **6.** [3.1b] $\frac{2769}{1000}$ **7.** [3.1b] 0.074 **8.** [3.1b] 3.7047 **9.** [3.1b] 756.09 **10.** [3.1b] 91.703 **11.** [3.1c] 0.162 **12.** [3.1c] 0.078 **13.** [3.1c] 0.9 **14.** [3.1d] 6 **15.** [3.1d] 5.68 **16.** [3.1d] 5.678 **17.** [3.1d] 5.7 **18.** [3.2a] 0.7902 **19.** [3.2a] 186.5 **20.** [3.2a] 1033.23 **21.** [3.2b] 48.357 **22.** [3.2b] 19.0901 **23.** [3.2b] 152.8934 **24.** [3.3a] 0.03 **25.** [3.3a] 0.21345 **26.** [3.3a] 73,962 **27.** [3.4a] 4.75 **28.** [3.4a] 30.4 **29.** [3.4a] 0.19 **30.** [3.4a] 0.34689 **31.** [3.4a] 34,689 **32.** [3.4b] 84.26 **33.** [3.2c] 8.982 **34.** [3.7a] \$120.49 **35.** [3.7a] 28.3 mpg **36.** [3.7a] \$6572.45 **37.** [3.7a] \$1199.94 **38.** [3.7a] 53.9 million passengers **39.** [3.6a] 198 **40.** [3.6a] 4 **41.** [3.5a] 1.6 **42.** [3.5a] 0.88 **43.** [3.5a] 5.25 **44.** [3.5a] 0.75 **45.** [3.5a] $1.\overline{2}$ **46.** [3.5a] $2.\overline{142857}$ **47.** [3.5b] 2.1 **48.** [3.5b] 2.14 **49.** [3.5b] 2.143 **50.** [3.3b] \$9.49 **51.** [3.4c] 40.0065 **52.** [3.4c] 384.8464 **53.** [3.5c] 302.4 **54.** [3.5c] $52.339\overline{4}$ **55.** [2.4c] $26\frac{1}{2}$ **56.** [2.4e] $1\frac{1}{8}$ **57.** [2.1e] $\frac{11}{18}$ **58.** [1.7d] $2 \cdot 2 \cdot 2 \cdot 3 \cdot 3 \cdot 5$, or $2^3 \cdot 3^2 \cdot 5$ **59.** [1.9a] 360 **60.** [3.7a] \$35 **61.** [3.1b, c] $\frac{2}{3}, \frac{5}{7}, \frac{15}{19}, \frac{11}{13}, \frac{17}{20}, \frac{13}{15}$

Chapter 4

Pretest: Chapter 4, p. 252

1. $\frac{35}{43}$ **2.** $\frac{0.079}{1.043}$ **3.** 22.5 **4.** 25.5 mpg **5.** 12 min **6.** 22 packs **7.** 0.133 **8.** 50.4% **9.** 4% **10.** $\frac{19}{100}$ **11.** $x = 60\% \times 75$; 45 **12.** $\frac{n}{100} = \frac{35}{50}$; 70% **13.** 90 lb **14.** 19.2% **15.** \$14.30; \$300.30 **16.** \$5152 **17.** \$1112.50; \$3337.50 **18.** \$99.60 **19.** \$20 **20.** \$7128.60

Margin Exercises, Section 4.1, pp. 253–262

1. $\frac{5}{11}$, or 5:11 **2.** $\frac{57.3}{86.1}$, or 57.3:86.1 **3.** $\frac{6\frac{3}{4}}{7\frac{2}{5}}$, or $6\frac{3}{4}:7\frac{2}{5}$ **4.** $\frac{739}{12}$ **5.** $\frac{12}{14}$ **6.** $\frac{73}{248\frac{2}{3}}$; $\frac{248\frac{2}{3}}{73}$ **7.** $\frac{38.2}{56.1}$ **8.** 18 is to 27 as 2 is to 3 **9.** 3.6 is to 12 as 3 is to 10 **10.** 1.2 is to 1.5 as 4 is to 5 **11.** $\frac{3}{4}$ **12.** 5 mi/hr, or 5 mph **13.** 12 mi/hr, or 12 mph **14.** $\frac{89}{13}$ km/h, or 6.85 km/h **15.** 1100 ft/sec **16.** 4 ft/sec **17.** $\frac{121}{8}$ ft/sec, or 15.125 ft/sec **18.** 250 ft/sec **19.** 2 gal/day **20.** Yes **21.** No **22.** No **23.** 14 **24.** $11\frac{1}{4}$ **25.** 10.5 **26.** 2.64 **27.** 10.8 **28.** 15 gal **29.** 9.5 in. **30.** 2074 deer

Calculator Corner, p. 259

1. Left to the student **2.** Left to the student **3.** 27.5625 **4.** 25.6 **5.** 15.140625 **6.** 40.03952941 **7.** 39.74857143 **8.** 119

Exercise Set 4.1, p. 263

1. $\frac{4}{5}$ **3.** $\frac{56.78}{98.35}$ **5.** $\frac{4}{1}$ **7.** $\frac{3}{4}$ **9.** $\frac{7}{9}$ **11.** $\frac{478}{213}$; $\frac{213}{478}$ **13.** 40 km/h **15.** 7.48 mi/sec **17.** 28 mpg **19.** 0.623 gal/ft^2 **21.** 124 km/h **23.** 25 beats/min **25.** No **27.** Yes **29.** Yes **31.** No **33.** 45 **35.** 10 **37.** 20 **39.** 18 **41.** 0.06 **43.** 5 **45.** 1 **47.** 12.5725 **49.** 168.6 million, or 168,600,000 **51.** 9.75 gal **53.** 175 bulbs **55.** 2975 ft^2 **57.** **(a)** 450 Australian dollars; **(b)** \$27.78 **59.** **(a)** About 96 gal; **(b)** 3920 mi **61.** 64 gal **63.** 954 deer **65.** 58.1 mi **67.** 120 lb **69.** D_W **71.** 65 **72.** 39.5 **73.** 290.5 **74.** $1523.\overline{1}$ **75.** 17 positions

Margin Exercises, Section 4.2, pp. 267–270

1. $\frac{70}{100}$; $70 \times \frac{1}{100}$; 70×0.01 **2.** $\frac{23.4}{100}$; $23.4 \times \frac{1}{100}$; 23.4×0.01 **3.** $\frac{100}{100}$; $100 \times \frac{1}{100}$; 100×0.01 **4.** 0.34 **5.** 0.789 **6.** 0.06625 **7.** 0.63 **8.** 0.0008 **9.** 24% **10.** 347% **11.** 100% **12.** 60% **13.** 25.3%

Calculator Corner, p. 268

1. 0.14 **2.** 0.00069 **3.** 0.438 **4.** 1.25

Exercise Set 4.2, p. 271

1. $\frac{90}{100}$; $90 \times \frac{1}{100}$; 90×0.01 **3.** $\frac{12.5}{100}$; $12.5 \times \frac{1}{100}$; 12.5×0.01 **5.** 0.67 **7.** 0.456 **9.** 0.5901 **11.** 0.1 **13.** 0.01 **15.** 2 **17.** 0.001 **19.** 0.0009 **21.** 0.0018 **23.** 0.2319 **25.** 0.14875 **27.** 0.565 **29.** 0.4 **31.** 0.186 **33.** 0.29 **35.** 47% **37.** 3% **39.** 870% **41.** 33.4% **43.** 75% **45.** 40% **47.** 0.6% **49.** 1.7% **51.** 27.18% **53.** 2.39% **55.** 52.6% **57.** 17% **59.** 41.1% **61.** D_W **63.** $33\frac{1}{3}$ **64.** $37\frac{1}{2}$ **65.** $9\frac{3}{8}$ **66.** $18\frac{9}{16}$ **67.** $5\frac{11}{14}$ **68.** $111\frac{2}{3}$ **69.** $0.\overline{6}$ **70.** $0.\overline{3}$ **71.** $0.8\overline{3}$ **72.** $1.41\overline{6}$ **73.** $2.\overline{6}$ **74.** 0.9375

Margin Exercises, Section 4.3, pp. 275–277

1. 25% **2.** 62.5%, or $62\frac{1}{2}$% **3.** $66.\overline{6}$%, or $66\frac{2}{3}$% **4.** $83.\overline{3}$%, or $83\frac{1}{3}$% **5.** 57% **6.** 76% **7.** $\frac{3}{5}$ **8.** $\frac{13}{400}$ **9.** $\frac{2}{3}$

10.

Fraction Notation	$\frac{1}{5}$	$\frac{5}{6}$	$\frac{3}{8}$
Decimal Notation	0.2	$0.83\overline{3}$	0.375
Percent Notation	20%	$83.\overline{3}$%, or $83\frac{1}{3}$%	$37\frac{1}{2}$%

Calculator Corner, p. 275

1. 52% **2.** 38.46% **3.** 110.26% **4.** 171.43% **5.** 59.62% **6.** 28.31%

Calculator Corner, p. 278

1. 30.54; 1.31% **2.** 32.05; 1.20% **3.** 34.47; 1.19% **4.** 26.47; 1.00% **5.** 11.98; 4.32% **6.** 17.52; 0.89%

Exercise Set 4.3, p. 279

1. 41% **3.** 5% **5.** 20% **7.** 30% **9.** 50% **11.** 87.5%, or $87\frac{1}{2}$% **13.** 80% **15.** $66.\overline{6}$%, or $66\frac{2}{3}$% **17.** $16.\overline{6}$%, or $16\frac{2}{3}$% **19.** 18.75%, or $18\frac{3}{4}$% **21.** 81.25%, or $81\frac{1}{4}$% **23.** 16% **25.** 5% **27.** 34% **29.** 8% **31.** 21% **33.** 24% **35.** $\frac{17}{20}$ **37.** $\frac{5}{8}$ **39.** $\frac{1}{3}$ **41.** $\frac{1}{6}$ **43.** $\frac{29}{400}$ **45.** $\frac{1}{125}$ **47.** $\frac{203}{800}$ **49.** $\frac{176}{225}$ **51.** $\frac{711}{1100}$ **53.** $\frac{3}{2}$ **55.** $\frac{13}{40{,}000}$ **57.** $\frac{1}{3}$ **59.** $\frac{13}{50}$ **61.** $\frac{1}{20}$ **63.** $\frac{3}{50}$ **65.** $\frac{9}{20}$ **67.** $\frac{47}{100}$

69.

Fraction Notation	Decimal Notation	Percent Notation
$\frac{1}{8}$	0.125	12.5%, or $12\frac{1}{2}$%
$\frac{1}{6}$	$0.1\overline{6}$	$16.\overline{6}$%, or $16\frac{2}{3}$%
$\frac{1}{5}$	0.2	20%
$\frac{1}{4}$	0.25	25%
$\frac{1}{3}$	$0.\overline{3}$	$33.\overline{3}$%, or $33\frac{1}{3}$%
$\frac{3}{8}$	0.375	37.5%, or $37\frac{1}{2}$%
$\frac{2}{5}$	0.4	40%
$\frac{1}{2}$	0.5	50%

71.

Fraction Notation	Decimal Notation	Percent Notation
$\frac{1}{2}$	0.5	50%
$\frac{1}{3}$	$0.\overline{3}$	$33.\overline{3}\%$, or $33\frac{1}{3}\%$
$\frac{1}{4}$	0.25	25%
$\frac{1}{6}$	$0.1\overline{6}$	$16.\overline{6}\%$, or $16\frac{2}{3}\%$
$\frac{1}{8}$	0.125	12.5%, or $12\frac{1}{2}\%$
$\frac{3}{4}$	0.75	75%
$\frac{5}{6}$	$0.8\overline{3}$	$83.\overline{3}\%$, or $83\frac{1}{3}\%$
$\frac{3}{8}$	0.375	37.5%, or $37\frac{1}{2}\%$

73. DW **75.** 70 **76.** 5 **77.** 400 **78.** 18.75 **79.** 4 **80.** $\frac{3}{44}$ **81.** $33\frac{1}{3}$ **82.** $37\frac{1}{2}$ **83.** $83\frac{1}{3}$ **84.** $20\frac{1}{2}$ **85.** $43\frac{1}{8}$ **86.** $62\frac{1}{6}$ **87.** $18\frac{3}{4}$ **88.** $7\frac{4}{9}$ **89.** $11.\overline{1}\%$ **91.** $257.\overline{46317}\%$ **93.** $0.01\overline{5}$ **95.** $1.04\overline{142857}$

Margin Exercises, Section 4.4, pp. 283–286

1. $12\% \times 50 = a$ **2.** $a = 40\% \times 60$ **3.** $45 = 20\% \times t$ **4.** $120\% \times y = 60$ **5.** $16 = n \times 40$ **6.** $b \times 84 = 10.5$ **7.** 6 **8.** \$35.20 **9.** 225 **10.** \$50 **11.** 40% **12.** 12.5%

Calculator Corner, p. 287

1. 1.2 **2.** \$5.04 **3.** 48.64 **4.** \$22.40 **5.** 0.0112 **6.** \$29.70 **7.** Left to the student **8.** Left to the student

Exercise Set 4.4, p. 288

1. $y = 32\% \times 78$ **3.** $89 = a \times 99$ **5.** $13 = 25\% \times y$ **7.** 234.6 **9.** 45 **11.** \$18 **13.** 1.9 **15.** 78% **17.** 200% **19.** 50% **21.** 125% **23.** 40 **25.** \$40 **27.** 88 **29.** 20 **31.** 6.25 **33.** \$846.60 **35.** DW **37.** $\frac{9}{100}$ **38.** $\frac{179}{100}$ **39.** $\frac{875}{1000}$, or $\frac{7}{8}$ **40.** $\frac{125}{1000}$, or $\frac{1}{8}$ **41.** $\frac{9375}{10{,}000}$, or $\frac{15}{16}$ **42.** $\frac{6875}{10{,}000}$, or $\frac{11}{16}$ **43.** 0.89 **44.** 0.07 **45.** 0.3 **46.** 0.017 **47.** \$800 (can vary); \$843.20 **49.** \$10,000 (can vary); \$10,400 **51.** \$1875

Margin Exercises, Section 4.5, pp. 291–293

1. $\frac{12}{100} = \frac{a}{50}$ **2.** $\frac{40}{100} = \frac{a}{60}$ **3.** $\frac{130}{100} = \frac{a}{72}$ **4.** $\frac{20}{100} = \frac{45}{b}$ **5.** $\frac{120}{100} = \frac{60}{b}$ **6.** $\frac{P}{100} = \frac{16}{40}$ **7.** $\frac{P}{100} = \frac{10.5}{84}$ **8.** \$225 **9.** 35.2 **10.** 6 **11.** 50 **12.** 30% **13.** 12.5%

Exercise Set 4.5, p. 294

1. $\frac{37}{100} = \frac{a}{74}$ **3.** $\frac{P}{100} = \frac{4.3}{5.9}$ **5.** $\frac{25}{100} = \frac{14}{b}$ **7.** 68.4 **9.** 462 **11.** 40 **13.** 2.88 **15.** 25% **17.** 102% **19.** 25% **21.** 93.75% **23.** \$72 **25.** 90 **27.** 88 **29.** 20 **31.** 25 **33.** \$780.20 **35.** DW **37.** 8 **38.** 4000 **39.** 8 **40.** 2074 **41.** 100 **42.** 15 **43.** $8.0\overline{4}$ **44.** $\frac{3}{16}$, or 0.1875 **45.** $\frac{43}{48}$ qt **46.** $\frac{1}{8}$ T **47.** \$1134 (can vary); \$1118.64

Margin Exercises, Section 4.6, pp. 297–302

1. About 9.5% **2.** 750 mL **3.** **(a)** \$1475; **(b)** \$38,350 **4.** **(a)** \$9218.75; **(b)** \$27,656.25 **5.** About 30% **6.** About 16.2%

Exercise Set 4.6, p. 303

1. **(a)** About 67.7%; **(b)** 90% **3.** 134 passes **5.** Overweight: 168.6 million; obese: 70.25 million **7.** 20.4 mL; 659.6 mL **9.** 637 field goals **11.** 95%; 5% **13.** 36.4 correct; 3.6 incorrect **15.** 95 items **17.** 25% **19.** 166; 156; 146; 140; 122 **21.** 8% **23.** 20% **25.** About 30.6% **27.** \$30,030 **29.** \$16,174.50; \$12,130.88 **31.** About 25% **33.** 34.375%, or $34\frac{3}{8}\%$ **35.** 5.7%; 539,452 **37.** 71% **39.** \$1560 **41.** 80% **43.** 46,069; 8.2% **45.** 1,027,429; 21.1% **47.** 4,891,769; 9.6% **49.** \$36,400 **51.** 40% **53.** DW **55.** $2.\overline{27}$ **56.** 0.44 **57.** 3.375 **58.** $4.\overline{7}$ **59.** 0.92 **60.** $0.8\overline{3}$ **61.** 0.4375 **62.** 2.317 **63.** 3.4809 **64.** 0.675 **65.** About 5 ft 6 in. **67.** $83\frac{1}{3}\%$

Margin Exercises, Section 4.7, pp. 309–317

1. \$53.52; \$722.47 **2.** \$9.43; \$144.18 **3.** 6% **4.** \$420 **5.** \$5628 **6.** 12.5%, or $12\frac{1}{2}\%$ **7.** \$1675 **8.** \$180; \$360 **9.** 20% **10.** \$602 **11.** \$451.50 **12.** \$37.48; \$4837.48 **13.** \$2464.20 **14.** \$8146.86

Calculator Corner, p. 317

1. \$16,357.18 **2.** \$12,764.72

Exercise Set 4.7, p. 318

1. \$2.14 **3.** 5% **5.** \$18.55; \$283.55 **7.** 4% **9.** \$2000 **11.** \$800 **13.** \$719.86 **15.** 5.6% **17.** \$2700 **19.** 5% **21.** \$980 **23.** \$5880 **25.** 12% **27.** \$420 **29.** \$387; 30.4% **31.** \$30; \$270 **33.** \$2.55; \$14.45 **35.** \$125; \$112.50 **37.** 40%; \$360

39. \$16 **41.** \$84 **43.** \$113.52 **45.** \$1525 **47.** \$671.88 **49.** **(a)** \$128.22; **(b)** \$6628.22 **51.** \$484 **53.** \$2802.50 **55.** \$7853.38 **57.** \$125,562.26 **59.** \$4284.90 **61.** \$28,225.00 **63.** \$9270.87 **65.** \$129,871.09 **67.** \$4101.01 **69.** **DW** **71.** $\frac{93}{100}$ **72.** 37 **73.** $1.\overline{18}$ **74.** $2\frac{7}{11}$

Margin Exercises, Section 4.8, pp. 323–327

1. **(a)** \$97; **(b)** interest: \$86.40; amount applied to principal: \$10.60; **(c)** interest: \$55.17; amount applied to principal: \$41.83; **(d)** At 13.6%, the principal was decreased by \$31.23 more than at the 21.3% rate. The interest at 13.6% is \$31.23 less than at 21.3%. **2.** **(a)** Interest: \$197.44; amount applied to principal: \$186.56; **(b)** \$34.04; **(c)** \$2520 **3.** Interest: \$909.18; amount applied to principal: \$69.46 **4.** **(a)** Interest: \$909.69; amount applied to principal: \$328.73 **(b)** \$99,915.60; **(c)** The Sawyers will pay \$129,394.80 less in interest with the 15-yr loan than with the 30-yr loan. **5.** **(a)** Interest: \$666.25; amount applied to principal: \$405.21 **(b)** \$69,862.80; **(c)** The Sawyers will pay \$30,052.80 less in interest with the 15-yr loan at $6\frac{1}{2}$% than with the 15-yr loan at $8\frac{7}{8}$%.

Exercise Set 4.8, p. 328

1. **(a)** \$98; **(b)** interest: \$86.56; amount applied to principal: \$11.44 **(c)** interest: \$51.20; amount applied to principal: \$46.80 **(d)** At 12.6%, the principal is decreased by \$35.36 more than at the 21.3% rate. The interest at 12.6% is \$35.36 less than at 21.3%. **3.** **(a)** Interest: \$241.37; amount applied to principal: \$264.60; **(b)** \$74.26; **(c)** \$16,156.40, \$21,737.60, \$5581.20 **5.** **(a)** Interest: \$872.50; amount applied to principal: \$123.44; **(b)** \$208,538.40; **(c)** new principal: \$149,876.56; interest: \$871.78; amount applied to principal: \$124.16 **7.** **(a)** Interest: \$872.50; amount applied to principal: \$474.07; **(b)** \$92,382.60; **(c)** The Martinez family will pay \$116,155.80 less in interest with the 15-yr loan than with the 30-yr loan. **9.** \$99,917.71; \$99,834.94 **11.** \$99,712.04; \$99,422.15 **13.** \$149,882.75; \$149,764.79 **15.** \$199,382.07; \$198,760.41 **17.** **(a)** Interest: \$112.38; amount applied to principal: \$260.82; **(b)** new principal: \$14,739.18; \$1.96 less interest in second payment; **(c)** \$2913.60 **19.** **(a)** \$790; \$7110; **(b)** interest: \$74; amount applied to principal: \$163.82; **(c)** \$1451.52 **21.** **DW** **23.** **DW** **25.** 18 **26.** $\frac{22}{7}$ **27.** 265.625 **28.** 1.113 **29.** $0.\overline{5}$ **30.** $2.\overline{09}$ **31.** $0.91\overline{6}$ **32.** $1.\overline{857142}$ **33.** $2.\overline{142857}$ **34.** $1.58\overline{3}$ **35.** 4,030,000,000,000 **36.** 5,800,000 **37.** 42,700,000 **38.** 6,090,000,000,000

Summary and Review: Chapter 4, p. 331

1. $\frac{47}{84}$ **2.** $\frac{46}{1.27}$ **3.** $\frac{83}{100}$ **4.** $\frac{0.72}{197}$ **5.** $\frac{3}{4}$ **6.** $\frac{9}{16}$ **7.** 26 mpg **8.** 6300 rpm **9.** 0.638 gal/ft^2 **10.** 0.72 serving/lb **11.** No **12.** No **13.** 32 **14.** 7 **15.** $\frac{1}{40}$ **16.** 24 **17.** \$4.45 **18.** 351 circuits **19.** **(a)** 270 Euros; **(b)** \$46.30 **20.** 832 mi **21.** 27 acres **22.** Approximately 3,173,732 kg **23.** 6 in. **24.** Approximately 9906 lawyers **25.** 1.7% **26.** 56% **27.** 37.5% **28.** $33.\overline{3}$%, or $33\frac{1}{3}$% **29.** 0.735 **30.** 0.065 **31.** $\frac{6}{25}$ **32.** $\frac{63}{1000}$ **33.** $30.6 = p \times 90$; 34% **34.** $63 = 84\% \times n$; 75 **35.** $y = 38\frac{1}{2}\% \times 168$; 64.68 **36.** $\frac{24}{100} = \frac{16.8}{b}$; 70 **37.** $\frac{42}{30} = \frac{P}{100}$; 140% **38.** $\frac{10.5}{100} = \frac{a}{84}$; 8.82 **39.** 223 students; 105 students **40.** 44% **41.** 2500 mL **42.** 12% **43.** 93.15 **44.** \$14.40 **45.** 5% **46.** 11% **47.** \$42; \$308 **48.** \$42.70; \$262.30 **49.** \$2940 **50.** Approximately 25% **51.** \$36 **52.** **(a)** \$394.52; **b)** \$24,394.52 **53.** \$121 **54.** \$7727.26 **55.** \$9504.80 **56.** **(a)** \$129; **(b)** interest: \$100.18; amount applied to principal: \$28.82; **(c)** interest: \$70.72; amount applied to principal: \$58.28; **(d)** At 13.2%, the principal is decreased by \$29.46 more than at the 18.7% rate. The interest at 13.2% is \$29.46 less than at 18.7%. **57.** **DW** No; the 10% discount was based on the original price rather than on the sale price. **58.** **DW** A 40% discount is better. When successive discounts are taken, each is based on the previous discounted price rather than on the original price. A 20% discount followed by a 22% discount is the same as a 37.6% discount off the original price. **59.** $\frac{3107}{1000}$ **60.** $\frac{29}{100}$ **61.** 64 **62.** 7.6123 **63.** $3.\overline{6}$ **64.** $1.\overline{571428}$ **65.** $3\frac{2}{3}$ **66.** $17\frac{2}{7}$ **67.** 105 min, or 1 hr 45 min **68.** $66\frac{2}{3}$% **69.** \$168

Test: Chapter 4, p. 335

1. [4.1a] $\frac{85}{97}$ **2.** [4.1a] $\frac{0.34}{124}$ **3.** [4.1b] About 23.5 mpg **4.** [4.1b] $1\frac{1}{3}$ servings/lb **5.** [4.1c] Yes **6.** [4.1c] No **7.** [4.1d] 12 **8.** [4.1d] 360 **9.** [4.1e] 525 mi **10.** [4.1e] 4.8 min **11.** [4.1e] **(a)** 684 Canadian dollars; **(b)** \$368.42 **12.** [4.1e] About 86,151 arrests **13.** [4.2b] 90.5% **14.** [4.2b] 0.2 **15.** [4.3a] 137.5% **16.** [4.3b] $\frac{13}{20}$ **17.** [4.4a, b] $a = 40\% \cdot 55$; 22 **18.** [4.5a, b] $\frac{P}{100} = \frac{65}{80}$; 81.25% **19.** [4.6a] 400; 575 **20.** [4.6a] About 539 at-bats **21.** [4.6b] $50.\overline{90}$% **22.** [4.6a] 5.5% **23.** [4.7a] \$16.20; \$340.20 **24.** [4.7b] \$630 **25.** [4.7c] \$40; \$160 **26.** [4.7d] \$8.52 **27.** [4.7d] \$5356 **28.** [4.7e] \$1110.39 **29.** [4.7e] \$13,086.45 **30.** [4.6b] Dental assistant: 326, 42.4%; nurse/psychiatric aide: 333, 22.8%; child-care worker: 905, 26.1%; hairdresser/hair stylist/cosmetologist: 608, 62 **31.** [4.7c] \$131.95; 52.8% **32.** [4.8a] \$119,909.14; \$119,817.72 **33.** [3.4b] 222 **34.** [3.1b] $\frac{447}{10}$ **35.** [3.5a] $1.41\overline{6}$ **36.** [2.4a] $3\frac{21}{44}$ **37.** [4.7b] \$194,600 **38.** [4.1e] 5888

Chapter 5

Pretest: Chapter 5, p. 340

1. 60° **2.** 131 mm **3.** 92 in^2 **4.** 22 cm^2 **5.** $32\frac{1}{2}$ ft^2 **6.** 4 m^2 **7.** 9.6 m **8.** 30.144 m; 72.3456 m^2 **9.** 160 cm^3 **10.** 1256 ft^3 **11.** 33,493.$\overline{3}$ yd^3 **12.** 150.72 cm^3 **13.** $m\angle 1 = m\angle 7 = m\angle 3 = m\angle 5 = 151°$, $m\angle 2 = m\angle 6 = m\angle 4 = 29°$ **14.** $\overline{PQ} \cong \overline{ST}$, $\overline{QR} \cong \overline{TV}$, $\overline{RP} \cong \overline{VS}$; $\angle P \cong \angle S$, $\angle Q \cong \angle T$, $\angle R \cong \angle V$ **15.** $MA = 7$, $GT = 8$

Margin Exercises, Section 5.1, pp. 341–347

1. (a) and (b) E•——•F (c) $\overline{EF}, \overline{FE}$ **2.** •P •Q
3. •P——•Q **4.** •P——•Q——→, P
5. ←——•P——•Q, Q **6.** •R •S
7. •R——•S, R and S **8.** •R——•S——→, R
9. ←——•R——•S, S **10.** ←——•R——•S——→,
no endpoints **11.** $\overleftrightarrow{RS}, \overleftrightarrow{SR}, \overleftrightarrow{RT}, \overleftrightarrow{TR}, \overleftrightarrow{ST}, \overleftrightarrow{TS}, n$
12. Angle DEF, angle FED, $\angle DEF$, $\angle FED$, or $\angle E$
13. Angle PQR, angle RQP, $\angle PQR$, $\angle RQP$, or $\angle Q$ **14.** 127° **15.** 33° **16.** Right **17.** Acute **18.** Obtuse **19.** Straight **20.** Not perpendicular **21.** Perpendicular **22.** (a) $\triangle ABC$; (b) $\triangle ABC$, $\triangle MPN$; (c) $\triangle DEF$, $\triangle GHI$, $\triangle JKL$, $\triangle QRS$ **23.** Yes **24.** No **25.** (a) $\triangle DEF$; (b) $\triangle GHI$, $\triangle QRS$; (c) $\triangle ABC$, $\triangle MPN$, $\triangle JKL$ **26.** Quadrilateral **27.** Hexagon **28.** Triangle **29.** Quadrilateral **30.** Dodecagon **31.** Octagon **32.** 180° **33.** 64° **34.** (a) 3; (b) 180°; (c) 3, 540° **35.** 1080° **36.** 4140°

Exercise Set 5.1, p. 348

1. •G——•H, $\overline{GH}, \overline{HG}$ **3.** •Q——•D——→, $\overrightarrow{QD}$ **5.** $\overleftrightarrow{DE}, \overleftrightarrow{ED}, \overleftrightarrow{DF}, \overleftrightarrow{FD}, \overleftrightarrow{EF}, \overleftrightarrow{FE}, l$ **7.** Angle GHI, angle IHG, $\angle GHI$, $\angle IHG$, or $\angle H$ **9.** 10° **11.** 180° **13.** 130° **15.** Obtuse **17.** Acute **19.** Straight **21.** Obtuse **23.** Acute **25.** Obtuse **27.** Not perpendicular **29.** Perpendicular **31.** Scalene; obtuse **33.** Scalene; right **35.** Equilateral; acute **37.** Scalene; obtuse **39.** Quadrilateral **41.** Pentagon **43.** Triangle **45.** Pentagon **47.** Hexagon **49.** 1440° **51.** 900° **53.** 2160° **55.** 3240° **57.** 46° **59.** 120° **61.** 43° **63.** DW **65.** \$160 **66.** \$22.50 **67.** \$148 **68.** \$1116.67 **69.** \$33,597.91 **70.** \$413,458.31 **71.** \$641,566.26 **72.** \$684,337.34 **73.** $m\angle 2 = 67.13°$; $m\angle 3 = 33.07°$; $m\angle 4 = 79.8°$; $m\angle 5 = 67.13°$ **75.** $m\angle ACB = 50°$; $m\angle CAB = 40°$; $m\angle EBC = 50°$; $m\angle EBA = 40°$; $m\angle AEB = 100°$; $m\angle ADB = 50°$

Margin Exercises, Section 5.2, pp. 353–355

1. 26 cm **2.** 46 in. **3.** 12 cm **4.** 17.5 yd **5.** $27\frac{5}{6}$ in. **6.** 40 km **7.** 21 yd **8.** 31.2 km **9.** 70 ft; \$346.50

Exercise Set 5.2, p. 356

1. 17 mm **3.** 15.25 in. **5.** 18 km **7.** 30 ft **9.** 79.14 cm **11.** 88 ft **13.** 182 mm **15.** 826 m; \$1197.70 **17.** 122 cm **19.** (a) 228 ft; (b) \$1046.52 **21.** DW **23.** 0.561 **24.** 67.34% **25.** 112.5% **26.** 25 **27.** 100 **28.** 961 **29.** 4,700,000 **30.** 4,300,000,000 **31.** 9 ft

Margin Exercises, Section 5.3, pp. 358–363

1. 8 cm^2 **2.** 56 km^2 **3.** $18\frac{3}{8}$ yd^2 **4.** 144 km^2 **5.** 118.81 m^2 **6.** $12\frac{1}{4}$ yd^2 **7.** 43.8 cm^2 **8.** 12.375 km^2 **9.** 96 m^2 **10.** 18.7 cm^2 **11.** 100 m^2 **12.** 88 cm^2 **13.** 228 in^2

Exercise Set 5.3, p. 364

1. 15 km^2 **3.** 1.4 in^2 **5.** $6\frac{1}{4}$ yd^2 **7.** 8100 ft^2 **9.** 50 ft^2 **11.** 169.883 cm^2 **13.** $41\frac{2}{9}$ in^2 **15.** 484 ft^2 **17.** 3237.61 km^2 **19.** $28\frac{57}{64}$ yd^2 **21.** 32 cm^2 **23.** 60 in^2 **25.** 104 ft^2 **27.** 45.5 in^2 **29.** 8.05 cm^2 **31.** 297 cm^2 **33.** 7 m^2 **35.** 1197 m^2 **37.** (a) 5599.75 ft^2; (b) about \$45 **39.** 630.36 m^2 **41.** (a) 819.75 ft^2; (b) 10 gal; (c) \$179.50 **43.** 80 cm^2 **45.** 675 cm^2 **47.** 21 cm^2 **49.** 852.04 ft^2 **51.** DW **53.** $\frac{7}{20}$ **54.** $\frac{171}{200}$ **55.** $\frac{3}{8}$ **56.** $\frac{2}{3}$ **57.** $\frac{5}{6}$ **58.** $\frac{1}{6}$ **59.** 7500 **60.** 46; 2 cc left over **61.** 16,914 in^2

Margin Exercises, Section 5.4, pp. 369–373

1. 9 in. **2.** 5 ft **3.** 62.8 m **4.** 88 m **5.** 34.296 yd **6.** $78\frac{4}{7}$ km^2 **7.** 339.62 cm^2 **8.** 12-ft diameter flower bed, by about 13.04 ft^2

Exercise Set 5.4, p. 374

1. 14 cm; 44 cm; 154 cm^2 **3.** $1\frac{1}{2}$ in.; $4\frac{5}{7}$ in.; $1\frac{43}{56}$ in^2 **5.** 16 ft; 100.48 ft; 803.84 ft^2 **7.** 0.7 cm; 4.396 cm; 1.5386 cm^2 **9.** 3 cm; 18.84 cm; 28.26 cm^2 **11.** 153.86 ft^2 **13.** 2.5 cm; 1.25 cm; 4.90625 cm^2 **15.** 3.454 ft **17.** 65.94 yd^2 **19.** 45.68 ft **21.** 26.84 yd **23.** 45.7 yd **25.** 100.48 m^2 **27.** 6.9972 cm^2 **29.** 64.4214 in^2 **31.** DW **33.** 87.5% **34.** 58% **35.** 66.$\overline{6}$% **36.** 43.61% **37.** 37.5% **38.** 62.5% **39.** 66.$\overline{6}$% **40.** 20% **41.** 4 **42.** $8\frac{1}{2}$ **43.** 13 **44.** $39\frac{1}{2}$ **45.** 5 **46.** 0 **47.** 2 **48.** 3 **49.** $\frac{1}{2}$ **50.** 3 **51.** $275\frac{1}{2}$ **52.** $7\frac{1}{2}$ **53.** 3.1416 **55.** $3d$; πd; circumference of one ball, since $\pi > 3$

Margin Exercises, Section 5.5, pp. 378–383

1. 12 cm^3 **2.** 20 ft^3 **3.** 128 ft^3 **4.** 38.4 m^3; 75.2 m^2 **5.** $1\frac{7}{8}$ ft^3; $10\frac{1}{4}$ ft^2 **6.** 785 ft^3 **7.** 67,914 m^3 **8.** $91{,}989\frac{1}{3}$ ft^3 **9.** 38.77272 cm^3 **10.** 1695.6 m^3 **11.** 528 in^3 **12.** 83.7$\overline{3}$ mm^3

Calculator Corner, p. 382

1. Left to the student **2.** Left to the student

Exercise Set 5.5, p. 384

1. 768 cm^3; 512 cm^2 **3.** 45 in^3; 87 in^2 **5.** 75 m^3; 145 m^2 **7.** $357\frac{1}{2}$ yd^3; $311\frac{1}{2}$ yd^2 **9.** 803.84 in^3 **11.** 353.25 cm^3 **13.** 41,580,000 yd^3 **15.** $4{,}186{,}666\frac{2}{3}$ in^3 **17.** 124.72 m^3 **19.** $1950\frac{101}{168}$ ft^3 **21.** 113,982 ft^3 **23.** 24.64 cm^3 **25.** 0.423115 yd^3 **27.** 367.38 m^3 **29.** 143.72 cm^3 **31.** 32,993,440,000 mi^3 **33.** 646.74 cm^3 **35.** 61,600 cm^3 **37.** 5832 yd^3 **39.** DW **41.** \$19.20 **42.** \$96 **43.** 1000 **44.** 225 **45.** 49 **46.** 64 **47.** 5% **48.** 11% **49.** About 57,480 in^3 **51.** 9.425 L **53.** The diameter of the earth at the equator is about 7930 mi. The diameter of the earth between the north and south poles is about 7917 mi. If we use the average of these two diameters (7923.5 mi), the volume of the earth is $\frac{4}{3} \cdot \pi \cdot \left(\frac{7923.5}{2}\right)^3$, or about 260,000,000,000 mi^3. **55.** 0.331 m^3

Margin Exercises, Section 5.6, pp. 389–396

1. $\angle 1$ and $\angle 2$; $\angle 1$ and $\angle 4$; $\angle 2$ and $\angle 3$; $\angle 3$ and $\angle 4$ **2.** 45° **3.** 72° **4.** 5° **5.** $\angle 1$ and $\angle 2$; $\angle 1$ and $\angle 4$; $\angle 2$ and $\angle 3$; $\angle 3$ and $\angle 4$ **6.** 142° **7.** 23° **8.** 90° **9.** Not congruent **10.** Congruent **11.** Not congruent **12.** Congruent **13.** $m\angle 1 = 10°, m\angle 3 = 129°, m\angle 5 = 41°, m\angle 6 = 129°$ **14.** $\angle 1$ and $\angle 3$, $\angle 2$ and $\angle 4$, $\angle 5$ and $\angle 7$, $\angle 6$ and $\angle 8$ **15.** $\angle 2$, $\angle 3$, $\angle 6$, and $\angle 7$ **16.** $\angle 2$ and $\angle 7$, $\angle 6$ and $\angle 3$ **17.** $m\angle 7 = m\angle 1 = m\angle 5 = 51°, m\angle 8 = m\angle 2 = m\angle 6 = m\angle 4 = 129°$ **18.** $\angle CED \cong \angle BEA, \angle ECD \cong \angle EBA, \angle EDC \cong \angle EAB, \angle CEA \cong \angle BED$ **19.** $\angle TPQ \cong \angle TRS, \angle TQP \cong \angle TSR$

Exercise Set 5.6, p. 397

1. 79° **3.** 23° **5.** 32° **7.** 61° **9.** 177° **11.** 41° **13.** 95° **15.** 78° **17.** Not congruent **19.** Congruent **21.** $m\angle 2 = 67°, m\angle 3 = 33°, m\angle 4 = 80°, m\angle 6 = 33°$ **23.** **(a)** $\angle 1$ and $\angle 3$, $\angle 2$ and $\angle 4$, $\angle 8$ and $\angle 6$, $\angle 7$ and $\angle 5$; **(b)** $\angle 2$, $\angle 3$, $\angle 6$, and $\angle 7$; **(c)** $\angle 2$ and $\angle 6$, $\angle 3$ and $\angle 7$ **25.** $m\angle 6 = m\angle 2 = m\angle 8 = 125°, m\angle 5 = m\angle 3 = m\angle 7 = m\angle 1 = 55°$ **27.** $\angle ABE \cong \angle DCE$, 95°; $\angle BAE \cong \angle CDE$; $\angle AEB \cong \angle DEC$; $\angle BED \cong \angle AEC$ **29.** $\angle AEC \cong \angle DCE$, 50°; $\angle BED \cong \angle EDC$, 41° **31.** $11\frac{1}{4}$ **32.** $16\frac{1}{8}$ **33.** 118 **34.** $14\frac{2}{3}$

Margin Exercises, Section 5.7, pp. 400–405

1. $\angle A \cong \angle D, \angle B \cong \angle E, \angle C \cong \angle F; \overline{AB} \cong \overline{DE}, \overline{AC} \cong \overline{DF}, \overline{BC} \cong \overline{EF}$ **2.** $\angle N \cong \angle P, \angle M \cong \angle R, \angle O \cong \angle Q; \overline{NM} \cong \overline{PR}, \overline{NO} \cong \overline{PQ}, \overline{MO} \cong \overline{RQ}$ **3.** (a), (c) **4.** (a) **5.** (b) **6.** None **7.** SAS **8.** ASA **9.** SSS **10.** SAS **11.** $\triangle SRW \cong \triangle STV$ by ASA; $\angle RSW \cong \angle TSV, \overline{RS} \cong \overline{TS}, \overline{SW} \cong \overline{SV}$ **12.** $\triangle GKP \cong \triangle PTR$ by ASA. Thus corresponding parts $\overline{GP}$ and $\overline{PR}$ are congruent, and P is the midpoint of $\overline{GR}$. **13.** $m\angle C = 27°, m\angle B = m\angle D = 153°$ **14.** $m\angle S = 114°, m\angle P = m\angle R = 66°$ **15.** $QR = 10, SR = 8$ **16.** $EF = 13.6, DE = GF = 20.4$

Exercise Set 5.7, p. 406

1. $\angle A \cong \angle R, \angle B \cong \angle S, \angle C \cong \angle T; \overline{AB} \cong \overline{RS}, \overline{AC} \cong \overline{RT}, \overline{BC} \cong \overline{ST}$ **3.** $\angle D \cong \angle G, \angle E \cong \angle H, \angle F \cong \angle K; \overline{DE} \cong \overline{GH}, \overline{DF} \cong \overline{GK}, \overline{EF} \cong \overline{HK}$ **5.** $\angle X \cong \angle U, \angle Y \cong \angle V, \angle Z \cong \angle W; \overline{XY} \cong \overline{UV}, \overline{XZ} \cong \overline{UW}, \overline{YZ} \cong \overline{VW}$ **7.** $\angle A \cong \angle F, \angle C \cong \angle D, \angle B \cong \angle E; \overline{AC} \cong \overline{FD}, \overline{AB} \cong \overline{FE}, \overline{CB} \cong \overline{DE}$ **9.** $\angle M \cong \angle Q, \angle N \cong \angle P, \angle O \cong \angle S; \overline{MN} \cong \overline{QP}, \overline{MO} \cong \overline{QS}, \overline{NO} \cong \overline{PS}$ **11.** No **13.** Yes **15.** Yes **17.** No **19.** Yes **21.** Yes **23.** Yes **25.** Yes **27.** Yes **29.** ASA **31.** SAS **33.** SSS or SAS **35.** $\overline{PR} \cong \overline{TR}, \overline{SR} \cong \overline{QR}, \angle PRQ \cong \angle TRS$ (vertical angles); $\triangle PRQ \cong \triangle TRS$ by SAS **37.** $m\angle GLK = m\angle GLM = 90°, \angle GLK \cong \angle GLM, \overline{GL} \cong \overline{GL}, \overline{KL} \cong \overline{ML}$; $\triangle KLG \cong \triangle MLG$ by SAS **39.** $\overline{AE} \cong \overline{CD}, \overline{AB} \cong \overline{CB}, \overline{EB} \cong \overline{DB}$; $\triangle AEB \cong \triangle CDB$ by SSS **41.** $\triangle LKH \cong \triangle GKJ$ by SAS; $\angle HLK \cong \angle JGK, \angle LHK \cong \angle GJK, \overline{LH} \cong \overline{GJ}$ **43.** $\triangle PED \cong \triangle PFG$ by ASA. As corresponding parts, $\overline{EP} \cong \overline{FP}$; thus P is the midpoint of $\overline{EF}$. **45.** $m\angle A = 70°, m\angle D = m\angle B = 110°$ **47.** $m\angle M = 71°, m\angle J = m\angle L = 109°$ **49.** $TU = 9, NU = 15$ **51.** $KL = 3\frac{1}{2}, ML = JK = 7\frac{1}{2}$ **53.** $AC = 28, ED = 38$ **55.** 45.2% **56.** $33\frac{1}{3}\%$ **57.** 55% **58.** 88% **59.** $\frac{2.7}{13.1}; \frac{13.1}{2.7}$ **60.** $\frac{1}{4}; \frac{3}{4}$ **61.** 1.75 **62.** 2.34 **63.** 0.234 **64.** 0.0234 **65.** 13.85

Margin Exercises, Section 5.8, pp. 411–414

1. (a), (b), (d) **2.** $PQ \leftrightarrow GH, QR \leftrightarrow HK, PR \leftrightarrow GK, \angle P \leftrightarrow \angle G, \angle Q \leftrightarrow \angle H, \angle R \leftrightarrow \angle K$ **3.** $\angle J \cong \angle A, \angle K \cong \angle B, \angle L \cong \angle C; \frac{JK}{AB} = \frac{JL}{AC} = \frac{KL}{BC}$ **4.** $\frac{PN}{TS} = \frac{PM}{TR} = \frac{MN}{RS}$ **5.** $BT = 6\frac{3}{4}, CT = 9$ **6.** $QR = 10$ **7.** 24.75 ft **8.** 34.9 ft

Exercise Set 5.8, p. 415

1. $\angle R \leftrightarrow \angle A, \angle S \leftrightarrow \angle B, \angle T \leftrightarrow \angle C, \overline{RS} \leftrightarrow \overline{AB}, \overline{RT} \leftrightarrow \overline{AC}, \overline{ST} \leftrightarrow \overline{BC}$ **3.** $\angle C \leftrightarrow \angle W, \angle B \leftrightarrow \angle J, \angle S \leftrightarrow \angle Z, \overline{CB} \leftrightarrow \overline{WJ}, \overline{CS} \leftrightarrow \overline{WZ}, \overline{BS} \leftrightarrow \overline{JZ}$ **5.** $\angle A \cong \angle R, \angle B \cong \angle S, \angle C \cong \angle T; \frac{AB}{RS} = \frac{AC}{RT} = \frac{BC}{ST}$ **7.** $\angle M \cong \angle C, \angle E \cong \angle L, \angle S \cong \angle F; \frac{ME}{CL} = \frac{MS}{CF} = \frac{ES}{LF}$ **9.** $\frac{PS}{ND} = \frac{SQ}{DM} = \frac{PQ}{NM}$ **11.** $\frac{TA}{GF} = \frac{TW}{GC} = \frac{AW}{FC}$ **13.** $QR = 10, PR = 8$ **15.** $EC = 18$ **17.** 36 ft **19.** 100 ft **21.** DW **23.** $\frac{147}{5}$, or $29\frac{2}{5}$ **24.** 0.244 **25.** 78 **26.** 61.1611

Summary and Review: Chapter 5, p. 418

1. 54° **2.** 180° **3.** 140° **4.** 90° **5.** Acute **6.** Straight **7.** Obtuse **8.** Right **9.** 60° **10.** Scalene **11.** Right **12.** 720° **13.** 23 m **14.** 4.4 m **15.** 228 ft; 2808 ft^2 **16.** 36 ft; 81 ft^2 **17.** 17.6 cm; 12.6 cm^2 **18.** 60 cm^2 **19.** 35 mm^2

20. $22.5\ m^2$ **21.** $29.64\ cm^2$ **22.** $88\ m^2$ **23.** $145\frac{5}{9}\ in^2$ **24.** $840\ ft^2$ **25.** 8 m **26.** $\frac{14}{11}$ in., or $1\frac{3}{11}$ in. **27.** 14 ft **28.** 20 cm **29.** 50.24 m **30.** 8 in. **31.** $200.96\ m^2$ **32.** $5\frac{1}{11}\ in^2$ **33.** $1038.555\ ft^2$ **34.** $93.6\ m^3$ **35.** $193.2\ cm^3$ **36.** $31{,}400\ ft^3$ **37.** $4.71\ in^3$ **38.** $33.49\overline{3}\ cm^3$ **39.** $942\ cm^3$ **40.** 8° **41.** 85° **42.** 147° **43.** 47° **44.** $m\angle 2 = 105°$, $m\angle 3 = 37°$, $m\angle 4 = 38°$, $m\angle 6 = 37°$ **45.** **(a)** $\angle 1$ and $\angle 5$, $\angle 4$ and $\angle 8$, $\angle 3$ and $\angle 7$, $\angle 2$ and $\angle 6$; **(b)** $\angle 4$, $\angle 5$, $\angle 2$, and $\angle 7$; **(c)** $\angle 4$ and $\angle 7$, $\angle 2$ and $\angle 5$ **46.** $m\angle 1 = m\angle 3 = m\angle 7 = m\angle 5 = 45°$, $m\angle 6 = m\angle 2 = m\angle 8 = 135°$ **47.** $\angle D \cong \angle R$, $\angle H \cong \angle Z$, $\angle J \cong \angle K$; $\overline{DH} \cong \overline{RZ}$, $\overline{DJ} \cong \overline{RK}$, $\overline{HJ} \cong \overline{ZK}$ **48.** $\angle A \cong \angle G$, $\angle B \cong \angle D$, $\angle C \cong \angle F$, $\overline{AB} \cong \overline{GD}$, $\overline{AC} \cong \overline{GF}$, $\overline{BC} \cong \overline{DF}$ **49.** ASA **50.** SSS **51.** None **52.** $\overline{IJ} \cong \overline{KJ}$, $\angle HJI \cong \angle LJK$, $\angle HIJ \cong \angle LKJ$; $\triangle JIH \cong \triangle JKL$ by ASA **53.** $m\angle C = 63°$, $m\angle B = m\angle D = 117°$; $BC = 23$, $CD = 13$ **54.** $\angle C \cong \angle F$, $\angle Q \cong \angle A$, $\angle W \cong \angle S$; $\dfrac{CQ}{FA} = \dfrac{CW}{FS} = \dfrac{QW}{AS}$ **55.** $MO = 14$ **56.** D_W Volume of two spheres, each with radius r: $2\left(\frac{4}{3}\pi r^3\right) = \frac{8}{3}\pi r^3$; volume of one sphere with radius $2r$: $\frac{4}{3}\pi(2r)^3 = \frac{32}{3}\pi r^3$. The volume of the sphere with radius $2r$ is four times the volume of the two spheres, each with radius r: $\frac{32}{3}\pi r^3 = 4 \cdot \frac{8}{3}\pi r^3$. **57.** D_W Linear measure is one-dimensional, area is two-dimensional, and volume is three-dimensional. **58.** $54\frac{5}{8}$ **59.** 103.823 **60.** $\frac{1}{16}$ **61.** $\frac{73}{100}$ **62.** 47% **63.** 92% **64.** $100\ ft^2$ **65.** $7.83998704\ m^2$ **66.** $42.05915\ cm^2$

Test: Chapter 5, p. 422

1. [5.1b] 90° **2.** [5.1b] 35° **3.** [5.1b] 180° **4.** [5.1b] 113° **5.** [5.1c] Right **6.** [5.1c] Acute **7.** [5.1c] Straight **8.** [5.1c] Obtuse **9.** [5.1f] 35° **10.** [5.1e] Isosceles **11.** [5.1e] Obtuse **12.** [5.1f] 540° **13.** [5.2a], [5.3a] 32.82 cm; $65.894\ cm^2$ **14.** [5.2a], [5.3a] $19\frac{1}{2}$ in.; $23\frac{49}{64}\ in^2$ **15.** [5.3b] $25\ cm^2$ **16.** [5.3b] $12\ m^2$ **17.** [5.3b] $18\ ft^2$ **18.** [5.4a] $\frac{1}{4}$ in. **19.** [5.4a] 9 cm **20.** [5.4b] $\frac{11}{14}$ in. **21.** [5.4c] $254.34\ cm^2$ **22.** [5.3c], [5.4d] 65.46 km; $103.815\ km^2$ **23.** [5.5a] $84\ cm^3$; $142\ cm^2$ **24.** [5.5e] $420\ in^3$ **25.** [5.5b] $1177.5\ ft^3$ **26.** [5.5c] $4186.\overline{6}\ yd^3$ **27.** [5.5d] $113.04\ cm^3$ **28.** [5.6a] 149° **29.** [5.6a] 11° **30.** [5.1f], [5.6c] $m\angle 2 = 110°$, $m\angle 3 = 8°$, $m\angle 4 = 62°$, $m\angle 6 = 8°$ **31.** [5.6d] $m\angle 6 = m\angle 2 = m\angle 8 = 120°$, $m\angle 5 = m\angle 3 = m\angle 7 = m\angle 1 = 60°$ **32.** [5.7a] $\angle C \cong \angle A$, $\angle W \cong \angle T$, $\angle S \cong \angle Z$, $\overline{CW} \cong \overline{AT}$, $\overline{WS} \cong \overline{TZ}$, $\overline{SC} \cong \overline{ZA}$ **33.** [5.7a] SAS **34.** [5.7a] None **35.** [5.7a] ASA **36.** [5.7a] None **37.** [5.7b] $m\angle G = 105°$, $m\angle D = m\angle F = 75°$; $EF = 11$, $DE = GF = 20$ **38.** [5.7b] $LJ = 6.4$, $KM = 6$ **39.** [5.8a] $\angle E \cong \angle T$, $\angle R \cong \angle G$, $\angle S \cong \angle F$; $\dfrac{ER}{TG} = \dfrac{RS}{GF} = \dfrac{SE}{FT}$ **40.** [5.8b] $EK = 18$, $ZK = 27$ **41.** [1.6b] 1000 **42.** [1.6b] $\frac{1}{16}$ **43.** [4.3a] 81.25% **44.** [4.2b] 0.932 **45.** [4.3b] $\frac{1}{3}$ **46.** [2.4d] 22 **47.** [5.3b] $1.875\ ft^2$ **48.** [5.5a] $0.65\ ft^3$

Chapter 6

Pretest: Chapter 6, p. 428

1. $\frac{5}{16}$ **2.** $78\%x$, or $0.78x$ **3.** $360\ ft^2$ **4.** 12 **5.** $>$ **6.** $>$ **7.** $>$ **8.** $<$ **9.** 12 **10.** 2.3 **11.** 0 **12.** -5.4 **13.** $\frac{2}{3}$ **14.** $\frac{1}{10}$ **15.** $-\frac{3}{2}$ **16.** -17 **17.** 38.6 **18.** $-\frac{17}{15}$ **19.** -5 **20.** 63 **21.** $-\frac{5}{12}$ **22.** -98 **23.** 8 **24.** 24 **25.** 26 **26.** $9z - 18$ **27.** $-4a - 2b + 10c$ **28.** $4(x - 3)$ **29.** $3(2y - 3z - 6)$ **30.** $-y - 13$ **31.** $y + 18$ **32.** $12 < x$ **33.** 50°C higher

Margin Exercises, Section 6.1, pp. 429–432

1. $8 + x = 21$; 13 **2.** 64 **3.** 28 **4.** 60 **5.** $192\ ft^2$ **6.** 25 **7.** 16 **8.** 12 hr **9.** $x - 8$ **10.** $y + 8$, or $8 + y$ **11.** $m - 4$ **12.** $\frac{1}{2}p$ **13.** $6 + 8x$, or $8x + 6$ **14.** $a - b$ **15.** $59\%x$, or $0.59x$ **16.** $xy - 200$ **17.** $p + q$

Calculator Corner, p. 431

1. 56 **2.** 11.9 **3.** 1.8 **4.** 34,427.16 **5.** 20.1 **6.** 29.9

Exercise Set 6.1, p. 435

1. \$20,400; \$46,800; \$150,000 **3.** $1935\ m^2$ **5.** 260 mi **7.** $24\ ft^2$ **9.** 56 **11.** 8 **13.** 1 **15.** 6 **17.** 2 **19.** $b + 7$, or $7 + b$ **21.** $c - 12$ **23.** $4 + q$, or $q + 4$ **25.** $a + b$, or $b + a$ **27.** $x \div y$, or $\dfrac{x}{y}$, or x/y, or $x \cdot \dfrac{1}{y}$ **29.** $x + w$, or $w + x$ **31.** $n - m$ **33.** $x + y$, or $y + x$ **35.** $2z$ **37.** $3m$ **39.** $89\%s$, or $0.89s$, where s is the salary **41.** $65t$ miles **43.** \$50 − x **45.** D_W **47.** $357.5\ ft^2$ **48.** $45.51\ ft^2$ **49.** $54{,}756\ mi^2$ **50.** $1.4196\ mm^2$ **51.** $66.97\ ft^2$ **52.** $7976.1\ m^2$ **53.** $x + 3y$ **55.** $2x - 3$

Margin Exercises, Section 6.2, pp. 440–445

1. 8; -5 **2.** 950,000,000 **3.** -6 **4.** -10; 156 **5.** -120; 50; -80 **6.**

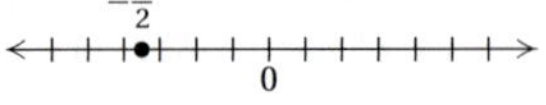

7.

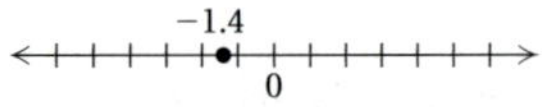

8.

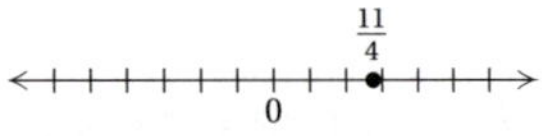

9. -0.375 **10.** $-0.\overline{54}$ **11.** $1.\overline{3}$ **12.** $<$ **13.** $<$ **14.** $>$ **15.** $>$ **16.** $>$ **17.** $<$ **18.** $<$ **19.** $>$ **20.** $7 > -5$ **21.** $4 < x$ **22.** False **23.** True **24.** True **25.** 8 **26.** 9 **27.** $\frac{2}{3}$ **28.** 5.6

Calculator Corner, p. 441

1. -0.75 **2.** -0.45 **3.** -0.125 **4.** -1.8 **5.** -0.675 **6.** -0.6875 **7.** -3.5 **8.** -0.76

Calculator Corner, p. 442

1. 8.717797887 **2.** 17.80449381 **3.** 67.08203932 **4.** 35.4807407 **5.** 3.141592654 **6.** 91.10618695 **7.** 530.9291585 **8.** 138.8663978

Calculator Corner, p. 445

1. 5 **2.** 17 **3.** 0 **4.** 6.48 **5.** 12.7 **6.** 0.9 **7.** $\frac{5}{7}$ **8.** $\frac{4}{3}$

Exercise Set 6.2, p. 446

1. −1286; 14,410 **3.** 24; −2 **5.** −5,600,000,000,000 **7.** Alley Cats: −34; Strikers: 34

9.

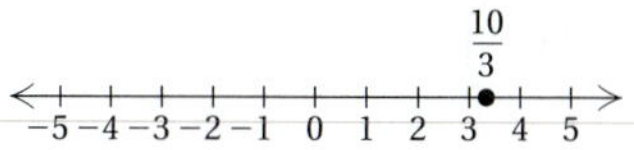

11.

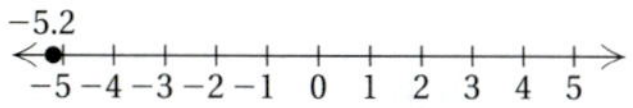

13. −0.875 **15.** $0.8\overline{3}$ **17.** $-1.1\overline{6}$ **19.** $0.\overline{6}$ **21.** −0.5 **23.** 0.1 **25.** > **27.** < **29.** < **31.** < **33.** > **35.** < **37.** > **39.** < **41.** < **43.** < **45.** True **47.** False **49.** $x < -6$ **51.** $y \geq -10$ **53.** 3 **55.** 10 **57.** 0 **59.** 24 **61.** $\frac{2}{3}$ **63.** 0 **65.** $3\frac{5}{8}$ **67.** D_W **69.** 70 **70.** 168 **71.** 180 **72.** $\frac{7}{6}$ **73.** $\frac{29}{24}$ **74.** $\frac{41}{20}$ **75.** $-\frac{5}{6}, -\frac{3}{4}, -\frac{2}{3}, \frac{1}{6}, \frac{3}{8}, \frac{1}{2}$ **77.** $\frac{1}{9}$ **79.** $5\frac{5}{9}$, or $\frac{50}{9}$

Margin Exercises, Section 6.3, pp. 449–453

1. −3 **2.** −3 **3.** −5 **4.** 4 **5.** 0 **6.** −2 **7.** −11 **8.** −12 **9.** 2 **10.** −4 **11.** −2 **12.** 0 **13.** −22 **14.** 3 **15.** 0.53 **16.** 2.3 **17.** −7.7 **18.** −6.2 **19.** $-\frac{2}{9}$ **20.** $-\frac{19}{20}$ **21.** −58 **22.** −56 **23.** −14 **24.** −12 **25.** 4 **26.** −8.7 **27.** 7.74 **28.** $\frac{8}{9}$ **29.** 0 **30.** −12 **31.** −14; 14 **32.** −1; 1 **33.** 19; −19 **34.** 1.6; −1.6 **35.** $-\frac{2}{3}; \frac{2}{3}$ **36.** $\frac{9}{8}; -\frac{9}{8}$ **37.** 4 **38.** 13.4 **39.** 0 **40.** $-\frac{1}{4}$ **41.** 24 students

Exercise Set 6.3, p. 454

1. −7 **3.** −6 **5.** 0 **7.** −8 **9.** −7 **11.** −27 **13.** 0 **15.** −42 **17.** 0 **19.** 0 **21.** 3 **23.** −9 **25.** 7 **27.** 0 **29.** 35 **31.** −3.8 **33.** −8.1 **35.** $-\frac{1}{5}$ **37.** $-\frac{7}{9}$ **39.** $-\frac{3}{8}$ **41.** $-\frac{19}{24}$ **43.** $\frac{1}{24}$ **45.** 37 **47.** 50 **49.** −1409 **51.** −24 **53.** 26.9 **55.** −8 **57.** $\frac{13}{8}$ **59.** −43 **61.** $\frac{4}{3}$ **63.** 24 **65.** $\frac{3}{8}$ **67.** 13,796 ft **69.** −3°F **71.** −$20,300 **73.** −$85 **75.** D_W **77.** 87,450 **78.** 87,500 **79.** 87,000 **80.** $\frac{17}{10}$ **81.** $\frac{17}{12}$ **82.** $\frac{13}{24}$ **83.** All positive **85.** (b)

Margin Exercises, Section 6.4, pp. 457–459

1. −10 **2.** 3 **3.** −5 **4.** −1 **5.** 2 **6.** −4 **7.** −2 **8.** −11 **9.** 4 **10.** −2 **11.** −6 **12.** −16 **13.** 7.1 **14.** 3 **15.** 0 **16.** $\frac{3}{2}$ **17.** −8 **18.** 7 **19.** −3 **20.** −23.3 **21.** 0 **22.** −9 **23.** 17 **24.** 12.7 **25.** 214°F higher

Exercise Set 6.4, p. 460

1. −7 **3.** −4 **5.** −6 **7.** 0 **9.** −4 **11.** −7 **13.** −6 **15.** 0 **17.** 0 **19.** 14 **21.** 11 **23.** −14 **25.** 5 **27.** −7 **29.** −1 **31.** 18 **33.** −10 **35.** −3 **37.** −21 **39.** 5 **41.** −8 **43.** 12 **45.** −23 **47.** −68 **49.** −73 **51.** 116 **53.** 0 **55.** −1 **57.** $\frac{1}{12}$ **59.** $-\frac{17}{12}$ **61.** $\frac{1}{8}$ **63.** 19.9 **65.** −8.6 **67.** −0.01 **69.** −193 **71.** 500 **73.** −2.8 **75.** −3.53 **77.** $-\frac{1}{2}$ **79.** $\frac{6}{7}$ **81.** $-\frac{41}{30}$ **83.** $-\frac{2}{15}$ **85.** 37 **87.** −62 **89.** −139 **91.** 6 **93.** 107 **95.** 219 **97.** 2385 m **99.** $347.94 **101.** **(a)** 77; **(b)** −41 **103.** 383 ft **105.** D_W **107.** $\frac{2}{3}$ **108.** $\frac{1}{6}$ **109.** $\frac{9}{40}$ **110.** 15% **111.** 72% **112.** $66.\overline{6}\%$, or $66\frac{2}{3}\%$ **113.** −309,882 **115.** False; $3 - 0 \neq 0 - 3$ **117.** True **119.** True **121.** **(a)** −2; **(b)** yes

Margin Exercises, Section 6.5, pp. 465–468

1. 20; 10; 0; −10; −20; −30 **2.** −18 **3.** −100 **4.** −80 **5.** $-\frac{5}{9}$ **6.** −30.033 **7.** $-\frac{7}{10}$ **8.** −10; 0; 10; 20; 30 **9.** 27 **10.** 32 **11.** 35 **12.** $\frac{20}{63}$ **13.** $\frac{2}{3}$ **14.** 13.455 **15.** −30 **16.** 30 **17.** 0 **18.** $-\frac{8}{3}$ **19.** 0 **20.** 0 **21.** −30 **22.** −30.75 **23.** $-\frac{5}{3}$ **24.** 120 **25.** −120 **26.** 6 **27.** 4; −4 **28.** 9; −9 **29.** 48; 48 **30.** 55°C

Exercise Set 6.5, p. 469

1. −8 **3.** −48 **5.** −24 **7.** −72 **9.** 16 **11.** 42 **13.** −120 **15.** −238 **17.** 1200 **19.** 98 **21.** −72 **23.** −12.4 **25.** 30 **27.** 21.7 **29.** $-\frac{2}{5}$ **31.** $\frac{1}{12}$ **33.** −17.01 **35.** $-\frac{5}{12}$ **37.** 420 **39.** $\frac{2}{7}$ **41.** −60 **43.** 150 **45.** $-\frac{2}{45}$ **47.** 1911 **49.** 50.4 **51.** $\frac{10}{189}$ **53.** −960 **55.** 17.64 **57.** $-\frac{5}{784}$ **59.** 0 **61.** −720 **63.** −30,240 **65.** 441; −147 **67.** 20; 20 **69.** −20 lb **71.** −54°C **73.** $12.71 **75.** −32 m **77.** D_W **79.** 16 **80.** $\frac{9}{10}$ **81.** $\frac{1}{36}$ **82.** 23% **83.** 140% **84.** 4.6% **85.** (a) **87.** Answer found at bottom of page

87.

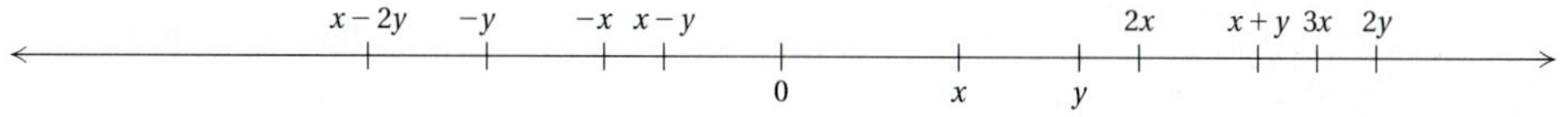

Margin Exercises, Section 6.6, pp. 472–477

1. -2 **2.** 5 **3.** -3 **4.** 8 **5.** -6 **6.** $-\frac{30}{7}$
7. Not defined **8.** 0 **9.** $\frac{3}{2}$ **10.** $-\frac{4}{5}$ **11.** $-\frac{1}{3}$
12. -5 **13.** $\frac{1}{1.6}$ **14.** $\frac{2}{3}$
15.

Number	Opposite	Reciprocal
$\frac{2}{3}$	$-\frac{2}{3}$	$\frac{3}{2}$
$-\frac{5}{4}$	$\frac{5}{4}$	$-\frac{4}{5}$
0	0	Not defined
1	-1	1
-8	8	$-\frac{1}{8}$
-4.5	4.5	$-\frac{1}{4.5}$

16. $\frac{4}{7} \cdot \left(-\frac{5}{3}\right)$ **17.** $5 \cdot \left(-\frac{1}{8}\right)$ **18.** $(a - b) \cdot \left(\frac{1}{7}\right)$
19. $-23 \cdot a$ **20.** $-5 \cdot \left(\frac{1}{7}\right)$ **21.** $-\frac{20}{21}$ **22.** $-\frac{12}{5}$ **23.** $\frac{16}{7}$
24. -7 **25.** $\frac{5}{-6}, -\frac{5}{6}$ **26.** $\frac{-8}{7}, \frac{8}{-7}$ **27.** $\frac{-10}{3}, -\frac{10}{3}$
28. -3.4°F per minute

Calculator Corner, p. 477

1. -4 **2.** -0.3 **3.** -12 **4.** -9.5 **5.** -12 **6.** 2.7
7. -2 **8.** -5.7 **9.** -32 **10.** -1.8 **11.** 35
12. 14.44 **13.** -2 **14.** -0.8 **15.** 1.4 **16.** 4

Exercise Set 6.6, p. 478

1. -8 **3.** -14 **5.** -3 **7.** 3 **9.** -8 **11.** 2
13. -12 **15.** -8 **17.** Not defined **19.** $\frac{23}{2}$ **21.** $\frac{7}{15}$
23. $-\frac{13}{47}$ **25.** $\frac{1}{13}$ **27.** $\frac{1}{4.3}$ **29.** -7.1 **31.** $\frac{q}{p}$ **33.** $4y$
35. $\frac{3b}{2a}$ **37.** $4 \cdot \left(\frac{1}{17}\right)$ **39.** $8 \cdot \left(-\frac{1}{13}\right)$ **41.** $13.9 \cdot \left(-\frac{1}{1.5}\right)$
43. $x \cdot y$ **45.** $(3x + 4)\left(\frac{1}{5}\right)$ **47.** $(5a - b)\left(\frac{1}{5a + b}\right)$
49. $-\frac{9}{8}$ **51.** $\frac{5}{3}$ **53.** $\frac{9}{14}$ **55.** $\frac{9}{64}$ **57.** -2 **59.** $\frac{11}{13}$
61. -16.2 **63.** Not defined **65.** -7.4% **67.** 42.4%
69. DW **71.** $0.\overline{09}$ **72.** $0.91\overline{6}$ **73.** \$3278.18
74. 84.6 cm **75.** $\frac{1}{-10.5}$; -10.5, the reciprocal of the reciprocal is the original number **77.** Negative
79. Positive **81.** Negative

Margin Exercises, Section 6.7, pp. 481–489

1.

Value	$x + x$	$2x$
$x = 3$	6	6
$x = -6$	-12	-12
$x = 4.8$	9.6	9.6

2.

Value	$x + 3x$	$5x$
$x = 2$	8	10
$x = -6$	-24	-30
$x = 4.8$	19.2	24

3. $\frac{6}{8}$ **4.** $\frac{3t}{4t}$ **5.** $\frac{3}{4}$ **6.** $-\frac{4}{3}$ **7.** 1; 1 **8.** -10; -10
9. $9 + x$ **10.** qp **11.** $t + xy$, or $yx + t$, or $t + yx$
12. 19; 19 **13.** 150; 150 **14.** $(r + s) + 7$ **15.** $(9a)b$
16. $(4t)u$, $(tu)4$, $t(4u)$; answers may vary
17. $(2 + r) + s$, $(r + s) + 2$, $s + (r + 2)$; answers may vary
18. **(a)** 63; **(b)** 63 **19.** **(a)** 80; **(b)** 80 **20.** **(a)** 28; **(b)** 28
21. **(a)** 8; **(b)** 8 **22.** **(a)** -4; **(b)** -4 **23.** **(a)** -25; **(b)** -25
24. $5x, -8y, 3$ **25.** $-4y, -2x, 3z$ **26.** $3x - 15$
27. $5x + 5$ **28.** $\frac{3}{5}p + \frac{3}{5}q - \frac{3}{5}t$ **29.** $-2x + 6$
30. $5x - 10y + 20z$ **31.** $-5x + 10y - 20z$ **32.** $6(x - 2)$
33. $3(x - 2y + 3)$ **34.** $b(x + y - z)$
35. $2(8a - 18b + 21)$ **36.** $\frac{1}{8}(3x - 5y + 7)$
37. $-4(3x - 8y + 4z)$ **38.** $3x$ **39.** $6x$ **40.** $-8x$
41. $0.59x$ **42.** $3x + 3y$ **43.** $-4x - 5y - 7$
44. $-\frac{2}{3} + \frac{1}{10}x + \frac{7}{9}y$

Exercise Set 6.7, p. 490

1. $\frac{3y}{5y}$ **3.** $\frac{10x}{15x}$ **5.** $-\frac{3}{2}$ **7.** $-\frac{7}{6}$ **9.** $8 + y$ **11.** nm
13. $xy + 9$, or $9 + yx$ **15.** $c + ab$, or $ba + c$
17. $(a + b) + 2$ **19.** $8(xy)$ **21.** $a + (b + 3)$
23. $(3a)b$
25. $2 + (b + a)$, $(2 + a) + b$, $(b + 2) + a$; answers may vary **27.** $(5 + w) + v$; $(v + 5) + w$; $(w + v) + 5$; answers may vary
29. $(3x)y$, $y(x \cdot 3)$, $3(yx)$; answers may vary
31. $a(7b)$, $b(7a)$, $(7b)a$; answers may vary **33.** $2b + 10$
35. $7 + 7t$ **37.** $30x + 12$ **39.** $7x + 28 + 42y$
41. $7x - 21$ **43.** $-3x + 21$ **45.** $\frac{2}{3}b - 4$
47. $7.3x - 14.6$ **49.** $-\frac{3}{5}x + \frac{3}{5}y - 6$
51. $45x + 54y - 72$ **53.** $-4x + 12y + 8z$
55. $-3.72x + 9.92y - 3.41$ **57.** $4x, 3z$ **59.** $7x, 8y, -9z$
61. $2(x + 2)$ **63.** $5(6 + y)$ **65.** $7(2x + 3y)$
67. $5(x + 2 + 3y)$ **69.** $8(x - 3)$ **71.** $4(8 - y)$
73. $2(4x + 5y - 11)$ **75.** $a(x - 1)$ **77.** $a(x - y - z)$
79. $6(3x - 2y + 1)$ **81.** $\frac{1}{3}(2x - 5y + 1)$ **83.** $19a$
85. $9a$ **87.** $8x + 9z$ **89.** $7x + 15y^2$ **91.** $-19a + 88$
93. $4t + 6y - 4$ **95.** b **97.** $\frac{13}{4}y$ **99.** $8x$ **101.** $5n$

103. $-16y$ **105.** $17a - 12b - 1$ **107.** $4x + 2y$ **109.** $7x + y$ **111.** $0.8x + 0.5y$ **113.** $\frac{35}{6}a + \frac{3}{2}b - 42$ **115.** **DW** **117.** 9 **118.** 12 **119.** 64 **120.** 32 **121.** 225 **122.** 81 **123.** Not equivalent; $3 \cdot 2 + 5 \neq 3 \cdot 5 + 2$ **125.** Equivalent; commutative law of addition **127.** $q(1 + r + rs + rst)$

Margin Exercises, Section 6.8, pp. 494–499

1. $-x - 2$ **2.** $-5x - 2y - 8$ **3.** $-6 + t$ **4.** $-x + y$ **5.** $4a - 3t + 10$ **6.** $-18 + m + 2n - 4z$ **7.** $2x - 9$ **8.** $3y + 2$ **9.** $2x - 7$ **10.** $3y + 3$ **11.** $-2a + 8b - 3c$ **12.** $-9x - 8y$ **13.** $-16a + 18$ **14.** $-26a + 41b - 48c$ **15.** $3x - 7$ **16.** 2 **17.** 18 **18.** 6 **19.** 17 **20.** $5x - y - 8$ **21.** -1237 **22.** 8 **23.** 4 **24.** 381 **25.** -12

Calculator Corner, p. 498

1. -11 **2.** 9 **3.** 114 **4.** 117,649 **5.** $-1{,}419{,}857$ **6.** $-1{,}124{,}864$ **7.** $-117{,}649$ **8.** $-1{,}419{,}857$ **9.** $-1{,}124{,}864$ **10.** -4 **11.** -2 **12.** 787

Exercise Set 6.8, p. 500

1. $-2x - 7$ **3.** $-8 + x$ **5.** $-4a + 3b - 7c$ **7.** $-6x + 8y - 5$ **9.** $-3x + 5y + 6$ **11.** $8x + 6y + 43$ **13.** $5x - 3$ **15.** $-3a + 9$ **17.** $5x - 6$ **19.** $-19x + 2y$ **21.** $9y - 25z$ **23.** $-7x + 10y$ **25.** $37a - 23b + 35c$ **27.** 7 **29.** -40 **31.** 19 **33.** $12x + 30$ **35.** $3x + 30$ **37.** $9x - 18$ **39.** $-4x - 64$ **41.** -7 **43.** -7 **45.** -16 **47.** -334 **49.** 14 **51.** 1880 **53.** 12 **55.** 8 **57.** -86 **59.** 37 **61.** -1 **63.** -10 **65.** -67 **67.** -7988 **69.** -3000 **71.** 60 **73.** 1 **75.** 10 **77.** $-\frac{13}{45}$ **79.** $-\frac{23}{18}$ **81.** -118 **83.** **DW** **85.** $\frac{13}{20}$ **86.** $\frac{3}{2}$ **87.** $\frac{3}{8}$ **88.** $\frac{1}{3}$ **89.** 30 yd; 94.2 yd; 706.5 yd^2 **90.** 16.4 m; 51.496 m; 211.1336 m^2 **91.** 19 mi; 59.66 mi; 283.385 mi^2 **92.** 4800 cm; 15,072 cm; 18,086,400 cm^2 **93.** $6y - (-2x + 3a - c)$ **95.** $6m - (-3n + 5m - 4b)$ **97.** $-2x - f$ **99. (a)** 52; 52; 28.130169; **(b)** -24; -24; -108.307025 **101.** -6

Summary and Review: Chapter 6, p. 504

1. 4 **2.** 19%x, or $0.19x$ **3.** -45, 72 **4.** 38

5.

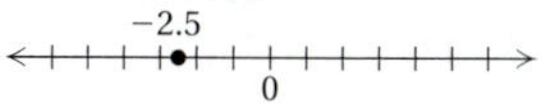

6.

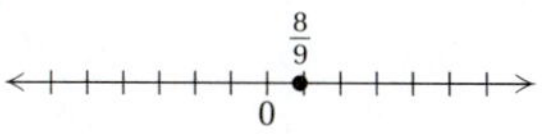

7. $<$ **8.** $>$ **9.** $>$ **10.** $<$ **11.** -3.8 **12.** $\frac{3}{4}$ **13.** $\frac{8}{3}$ **14.** $-\frac{1}{7}$ **15.** 34 **16.** 5 **17.** -3 **18.** -4 **19.** -5 **20.** 4 **21.** $-\frac{7}{5}$ **22.** -7.9 **23.** 54 **24.** -9.18 **25.** $-\frac{2}{7}$ **26.** -210 **27.** -7 **28.** -3 **29.** $\frac{3}{4}$ **30.** 40.4 **31.** -2 **32.** 2 **33.** -180 **34.** 8-yd gain **35.** $-\$130$ **36.** \$4.64 **37.** \$18.95 **38.** $15x - 35$ **39.** $-8x + 10$ **40.** $4x + 15$ **41.** $-24 + 48x$ **42.** $2(x - 7)$ **43.** $6(x - 1)$ **44.** $5(x + 2)$ **45.** $3(4 - x)$ **46.** $7a - 3b$ **47.** $-2x + 5y$ **48.** $5x - y$ **49.** $-a + 8b$ **50.** $-3a + 9$ **51.** $-2b + 21$ **52.** 6 **53.** $12y - 34$ **54.** $5x + 24$ **55.** $-15x + 25$ **56.** True **57.** False **58.** $x > -3$ **59.** **DW** If the sum of two numbers is 0, they are opposites, or additive inverses of each other. For every real number a, the opposite of a can be named $-a$, and $a + (-a) = (-a) + a = 0$. **60.** **DW** No; $|0| = 0$, and 0 is not positive. **61.** $-\frac{5}{8}$ **62.** -2.1 **63.** 1000 **64.** $4a + 2b$

Test: Chapter 6, p. 507

1. [6.1a] 6 **2.** [6.1b] $x - 9$ **3.** [6.1a] 240 ft^2 **4.** [6.2d] $<$ **5.** [6.2d] $>$ **6.** [6.2d] $>$ **7.** [6.2d] $<$ **8.** [6.2e] 7 **9.** [6.2e] $\frac{9}{4}$ **10.** [6.2e] 2.7 **11.** [6.3b] $-\frac{2}{3}$ **12.** [6.3b] 1.4 **13.** [6.3b] 8 **14.** [6.6b] $-\frac{1}{2}$ **15.** [6.6b] $\frac{7}{4}$ **16.** [6.4a] 7.8 **17.** [6.3a] -8 **18.** [6.3a] $\frac{7}{40}$ **19.** [6.4a] 10 **20.** [6.4a] -2.5 **21.** [6.4a] $\frac{7}{8}$ **22.** [6.5a] -48 **23.** [6.5a] $\frac{3}{16}$ **24.** [6.6a] -9 **25.** [6.6c] $\frac{3}{4}$ **26.** [6.6c] -9.728 **27.** [6.8d] -173 **28.** [6.8d] -5 **29.** [6.4b] 14°F **30.** [6.3c], [6.4b] Up 15 points **31.** [6.5b] 16,080 **32.** [6.6d] $\frac{33}{35}$°C per minute **33.** [6.7c] $18 - 3x$ **34.** [6.7c] $-5y + 5$ **35.** [6.7d] $2(6 - 11x)$ **36.** [6.7d] $7(x + 3 + 2y)$ **37.** [6.4a] 12 **38.** [6.8b] $2x + 7$ **39.** [6.8b] $9a - 12b - 7$ **40.** [6.8c] $68y - 8$ **41.** [6.8d] -4 **42.** [6.8d] 448 **43.** [6.2d] $-2 \geq x$ **44.** [2.3a] $\frac{17}{30}$ **45.** [4.6b] \$234 **46.** [5.2a] $21\frac{1}{2}$ in. **47.** [5.3a] 64 mm^2 **48.** [6.2e], [6.8d] 15 **49.** [6.8c] $4a$ **50.** [6.7e] $4x + 4y$

Chapter 7

Pretest: Chapter 7, p. 512

1. 8 **2.** -7 **3.** 2 **4.** -1 **5.** -5 **6.** $\frac{135}{32}$ **7.** 1 **8.** $\{y | y > -4\}$ **9.** $\{x | x \geq -6\}$ **10.** $\{a | a > -1\}$ **11.** $\{x | x \geq 3\}$ **12.** $\{y | y < -\frac{9}{4}\}$ **13.** No solution **14.** $x = \dfrac{y}{A}$ **15.** $a = \dfrac{A + b}{3}$ **16.** Width: 34 in.; length: 39 in. **17.** \$460 **18.** 81, 82, 83 **19.** $\{l | l \geq 174 \text{ yd}\}$

20.

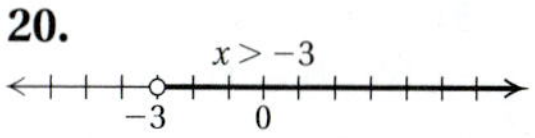

21.

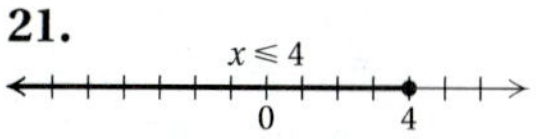

22. 20.4 **23.** 54 **24.** 20% **25.** About 26.5%

Margin Exercises, Section 7.1, pp. 513–516

1. False **2.** True **3.** Neither **4.** Yes **5.** No **6.** No **7.** 9 **8.** -13 **9.** 22 **10.** 13.2 **11.** -6.5 **12.** -2 **13.** $\frac{31}{8}$

Exercise Set 7.1, p. 517

1. Yes **3.** No **5.** No **7.** Yes **9.** No **11.** No **13.** 4 **15.** -20 **17.** -14 **19.** -18 **21.** 15

23. -14 25. 2 27. 20 29. -6 31. $6\frac{1}{2}$ 33. 19.9
35. $\frac{7}{3}$ 37. $-\frac{7}{4}$ 39. $\frac{41}{24}$ 41. $-\frac{1}{20}$ 43. 5.1 45. 12.4
47. -5 49. $1\frac{5}{6}$ 51. $-\frac{10}{21}$ 53. D_W 55. -11 56. 5
57. $-\frac{5}{12}$ 58. $\frac{1}{3}$ 59. $-\frac{3}{2}$ 60. -5.2 61. $-\frac{1}{24}$
62. 172.72 63. $\$83 - x$ 64. $65t$ miles 65. 342.246
67. $-\frac{26}{15}$ 69. -10 71. All real numbers 73. $-\frac{5}{17}$
75. 13, -13

Margin Exercises, Section 7.2, pp. 519–522

1. 15 2. $-\frac{7}{4}$ 3. -18 4. 10 5. 10 6. $-\frac{4}{5}$
7. 7800 8. -3 9. 28

Exercise Set 7.2, p. 523

1. 6 3. 9 5. 12 7. -40 9. 1 11. -7 13. -6
15. 6 17. -63 19. 36 21. -21 23. $-\frac{3}{5}$ 25. $-\frac{3}{2}$
27. $\frac{9}{2}$ 29. 7 31. -7 33. 8 35. 15.9 37. -50
39. -14 41. D_W 43. $7x$ 44. $-x + 5$ 45. $8x + 11$
46. $-32y$ 47. $x - 4$ 48. $-5x - 23$ 49. $-10y - 42$
50. $-22a + 4$ 51. $8r$ miles 52. $\frac{1}{2}b \cdot 10\text{ m}^2$, or $5b\text{ m}^2$
53. -8655 55. No solution 57. No solution
59. $\dfrac{b}{3a}$ 61. $\dfrac{4b}{a}$

Margin Exercises, Section 7.3, pp. 525–531

1. 5 2. 4 3. 4 4. 39 5. $-\frac{3}{2}$ 6. -4.3 7. -3
8. 800 9. 1 10. 2 11. 2 12. $\frac{17}{2}$ 13. $\frac{8}{3}$
14. $-\frac{43}{10}$, -4.3 15. 2 16. 3 17. -2 18. $-\frac{1}{2}$
19. Yes 20. Yes 21. Yes 22. Yes 23. No
24. No 25. No 26. No 27. All real numbers
28. No solution

Calculator Corner, p. 532

1. Left to the student 2. Left to the student

Exercise Set 7.3, p. 533

1. 5 3. 8 5. 10 7. 14 9. -8 11. -8 13. -7
15. 15 17. 6 19. 4 21. 6 23. -3 25. 1
27. 6 29. -20 31. 7 33. 2 35. 5 37. 2
39. 10 41. 4 43. 0 45. -1 47. $-\frac{4}{3}$ 49. $\frac{2}{5}$
51. -2 53. -4 55. $\frac{4}{5}$ 57. $-\frac{28}{27}$ 59. 6 61. 2
63. No solution 65. All real numbers 67. 6 69. 8
71. 1 73. All real numbers 75. No solution
77. 17 79. $-\frac{5}{3}$ 81. -3 83. 2 85. $\frac{4}{7}$
87. No solution 89. All real numbers 91. $-\frac{51}{31}$
93. D_W 95. -6.5 96. -75.14 97. $7(x - 3 - 2y)$
98. $8(y - 11x + 1)$ 99. 4.4233464 101. $-\frac{5}{32}$

Margin Exercises, Section 7.4, pp. 537–540

1. 2.8 mi 2. 341 mi 3. $q = 3B$ 4. $r = \dfrac{d}{t}$ 5. $I = \dfrac{E}{R}$
6. $x = y - 5$ 7. $x = y + 7$ 8. $x = y + b$
9. $y = \dfrac{5x}{9}$, or $\dfrac{5}{9}x$ 10. $p = \dfrac{bq}{a}$ 11. $x = \dfrac{y - b}{m}$
12. $Q = \dfrac{a + p}{t}$ 13. $D = \dfrac{C}{\pi}$ 14. $c = 4A - a - b - d$

Exercise Set 7.4, p. 541

1. (a) 57,000 Btu's; (b) $a = \dfrac{B}{30}$ 3. (a) $1\frac{3}{5}$ mi; (b) $t = 5M$
5. (a) 1423; (b) $n = 15f$ 7. 10.5 calories per ounce
9. 42 games 11. $x = \dfrac{y}{5}$ 13. $c = \dfrac{a}{b}$ 15. $x = y - 13$
17. $x = y - b$ 19. $x = 5 - y$ 21. $x = a - y$
23. $y = \dfrac{5x}{8}$, or $\dfrac{5}{8}x$ 25. $x = \dfrac{By}{A}$ 27. $t = \dfrac{W - b}{m}$
29. $x = \dfrac{y - c}{b}$ 31. $b = 3A - a - c$ 33. $t = \dfrac{A - b}{a}$
35. $h = \dfrac{A}{b}$ 37. $w = \dfrac{P - 2l}{2}$, or $\dfrac{1}{2}P - l$ 39. $a = 2A - b$
41. $a = \dfrac{F}{m}$ 43. $c^2 = \dfrac{E}{m}$ 45. $x = \dfrac{c - By}{A}$ 47. $t = \dfrac{3k}{v}$
49. D_W 51. 1 52. -90 53. -9.325 54. 44
55. -13.2 56. $-21a + 12b$ 57. $\frac{1}{6}$ 58. $-\frac{3}{2}$
59. (a) 1901 calories;
(b) $a = \dfrac{917 + 6w + 6h - K}{6}$;
$h = \dfrac{K - 917 - 6w + 6a}{6}$;
$w = \dfrac{K - 917 - 6h + 6a}{6}$
61. $b = \dfrac{2A - ah}{h}$; $h = \dfrac{2A}{a + b}$
63. A quadruples. 65. A increases by $2h$ units.

Margin Exercises, Section 7.5, pp. 545–548

1. $13\% \cdot 80 = a$ 2. $a = 60\% \cdot 70$ 3. $43 = 20\% \cdot b$
4. $110\% \cdot b = 30$ 5. $16 = n \cdot 80$ 6. $n \cdot 94 = 10.5$
7. 1.92 8. 115 9. 36% 10. 111,416 mi^2
11. About 3.9 million 12. About 58%

Exercise Set 7.5, p. 549

1. 20% 3. 150 5. 546 7. 24% 9. 2.5 11. 5%
13. 25% 15. 84 17. 24% 19. 16% 21. $46\frac{2}{3}$
23. 0.8 25. 5 27. 40 29. \$198 31. \$1584
33. \$528 35. U.S.: 68.4%; Asia: 25.9%; Europe: 5.7%
37. About 603 at-bats 39. \$280 41. (a) 16%; (b) \$29
43. (a) \$3.75; (b) \$28.75 45. (a) \$28.80; (b) \$33.12
47. 200 women 49. About 31.5 lb 51. \$305; 71%
53. \$1560; \$780 55. \$1310; 80% 57. D_W
59. -11 60. -100 61. $-\frac{2}{5}$ 62. 2 63. $a + c$
64. $7x - 9y$ 65. -3.9 66. $-6\frac{1}{8}$ 67. 6 ft 7 in.

Margin Exercises, Section 7.6, pp. 554–562

1. $62\frac{2}{3}$ mi **2.** Top: 24 ft; middle: 72 ft; bottom: 144 ft **3.** 313 and 314 **4.** 60,417 copies **5.** Length: 84 ft; width: 50 ft **6.** First: 30°; second: 90°; third: 60° **7.** \$8400 **8.** \$658

Exercise Set 7.6, p. 563

1. 180 in.; 60 in. **3.** \$3.67 **5.** \$6.3 billion **7.** $699\frac{1}{3}$ mi **9.** 273, 274 **11.** 41, 42, 43 **13.** 61, 63, 65 **15.** Length: 48 ft; width: 14 ft **17.** \$75 **19.** \$85 **21.** 11 visits **23.** 28°, 84°, 68° **25.** 33°, 38°, 109° **27.** \$350 **29.** \$852.94 **31.** 12 mi **33.** \$36 **35.** D_W **37.** $-\frac{47}{40}$ **38.** $-\frac{17}{40}$ **39.** $-\frac{3}{10}$ **40.** $-\frac{32}{15}$ **41.** -10 **42.** 1.6 **43.** 409.6 **44.** -9.6 **45.** -41.6 **46.** 0.1 **47.** 120 apples **49.** About 0.65 in. **51.** \$9.17, not \$9.10

Margin Exercises, Section 7.7, pp. 568–575

1. (a) No; **(b)** no; **(c)** no; **(d)** yes; **(e)** no; **(f)** no **2. (a)** Yes; **(b)** yes; **(c)** yes; **(d)** no; **(e)** yes; **(f)** yes

3. $x \le 4$ (0, 4)

4. $x > -2$ (−2, 0)

5. $-2 < x \le 4$ (−2, 0, 4)

6. $\{x|x > 2\}$; (0, 2)

7. $\{x|x \le 3\}$; (0, 3)

8. $\{x|x < -3\}$; (−3, 0)

9. $\left\{x \middle| x \ge \frac{2}{15}\right\}$ **10.** $\{y|y \le -3\}$

11. $\{x|x < 8\}$; (0, 8)

12. $\{y|y \ge 32\}$; (0, 10, 32)

13. $\{x|x \ge -6\}$ **14.** $\left\{y \middle| y < -\frac{13}{5}\right\}$ **15.** $\left\{x \middle| x > -\frac{1}{4}\right\}$ **16.** $\left\{y \middle| y \ge \frac{19}{9}\right\}$ **17.** $\left\{y \middle| y \ge \frac{19}{9}\right\}$ **18.** $\{x|x \ge -2\}$ **19.** $\{x|x \ge -4\}$ **20.** $\left\{x \middle| x > \frac{8}{3}\right\}$

Exercise Set 7.7, p. 576

1. (a) Yes; **(b)** yes; **(c)** no; **(d)** yes; **(e)** yes **3. (a)** No; **(b)** no; **(c)** no; **(d)** yes; **(e)** no

5. $x > 4$ (0, 4)

7. $t < -3$ (−3, 0)

9. $m \ge -1$ (−1, 0)

11. $-3 < x \le 4$ (−3, 0, 4)

13. $0 < x < 3$ (0, 3)

15. $\{x|x > -5\}$; (−5, 0)

17. $\{x|x \le -18\}$; (−18, −20, 0)

19. $\{y|y > -5\}$

21. $\{x|x > 2\}$ **23.** $\{x|x \le -3\}$ **25.** $\{x|x < 4\}$ **27.** $\{t|t > 14\}$ **29.** $\left\{y \middle| y \le \frac{1}{4}\right\}$ **31.** $\left\{x \middle| x > \frac{7}{12}\right\}$

33. $\{x|x < 7\}$; (0, 7)

35. $\{x|x < 3\}$; (0, 3)

37. $\left\{y \middle| y \ge -\frac{2}{5}\right\}$ **39.** $\{x|x \ge -6\}$ **41.** $\{y|y \le 4\}$ **43.** $\left\{x \middle| x > \frac{17}{3}\right\}$ **45.** $\left\{y \middle| y < -\frac{1}{14}\right\}$ **47.** $\left\{x \middle| x \le \frac{3}{10}\right\}$ **49.** $\{x|x < 8\}$ **51.** $\{x|x \le 6\}$ **53.** $\{x|x < -3\}$ **55.** $\{x|x > -3\}$ **57.** $\{x|x \le 7\}$ **59.** $\{x|x > -10\}$ **61.** $\{y|y < 2\}$ **63.** $\{y|y \ge 3\}$ **65.** $\{y|y > -2\}$ **67.** $\{x|x > -4\}$ **69.** $\{x|x \le 9\}$ **71.** $\{y|y \le -3\}$ **73.** $\{y|y < 6\}$ **75.** $\{m|m \ge 6\}$ **77.** $\left\{t \middle| t < -\frac{5}{3}\right\}$ **79.** $\{r|r > -3\}$ **81.** $\left\{x \middle| x \ge -\frac{57}{34}\right\}$ **83.** $\{x|x > -2\}$ **85.** D_W **87.** -74 **88.** 4.8 **89.** $-\frac{5}{8}$ **90.** -1.11 **91.** -38 **92.** $-\frac{7}{8}$ **93.** -9.4 **94.** 1.11 **95.** 140 **96.** 41 **97.** $-2x - 23$ **98.** $37x - 1$ **99. (a)** Yes; **(b)** yes; **(c)** no; **(d)** no; **(e)** no; **(f)** yes; **(g)** yes **101.** All real numbers

Margin Exercises, Section 7.8, pp. 580–583

1. $m \ge 92$ **2.** $c \ge 4000$ **3.** $p \le 21{,}900$ **4.** $45 < t < 55$ **5.** $d > 15$ **6.** $w < 110$ **7.** $n > -2$ **8.** $c \le 12{,}500$ **9.** $d \le 11.4\%$ **10.** $s \ge 23$ **11.** $\frac{9}{5}C + 32 < 88; \left\{C \middle| C < 31\frac{1}{9}^\circ\right\}$

12. $\dfrac{91 + 86 + 89 + s}{4} \ge 90; \{s|s \ge 94\}$

Exercise Set 7.8, p. 584

1. $n \ge 7$ **3.** $w > 2$ kg **5.** 90 mph $< s <$ 110 mph **7.** $a \le 1{,}200{,}000$ **9.** $c \ge \$1.50$ **11.** $x > 8$ **13.** $y \le -4$ **15.** $n \ge 1300$ **17.** $A \le 500$ L **19.** $2 + 3x < 13$ **21.** $\{x|x \ge 84\}$ **23.** $\{C|C < 1063^\circ\}$ **25.** $\{Y|Y \ge 1935\}$ **27.** $\{L|L \ge 5 \text{ in.}\}$ **29.** 15 or fewer copies **31.** 5 min or more **33.** 2 courses **35.** 4 servings or more **37.** Lengths greater than or equal to 92 ft; lengths less than or equal to 92 ft **39.** Lengths less than 21.5 cm **41.** The blue-book value is greater than or equal to \$10,625. **43.** It has at least 16 g of fat. **45.** Dates at least 6 weeks after July 1 **47.** Heights greater than or equal to 4 ft **49.** 21 calls or more **51.** D_W **53.** -160 **54.** $-17x + 18$ **55.** $91x - 242$ **56.** 0.25 **57.** Temperatures between -15°C and $-9\frac{4}{9}$°C **59.** They contain at least 7.5 g of fat per serving.

Summary and Review: Chapter 7, p. 589

1. -22 **2.** 1 **3.** 25 **4.** 9.99 **5.** $\frac{1}{4}$ **6.** 7 **7.** -192 **8.** $-\frac{7}{3}$ **9.** $-\frac{15}{64}$ **10.** -8 **11.** 4 **12.** -5 **13.** $-\frac{1}{3}$ **14.** 3 **15.** 4 **16.** 16 **17.** All real numbers **18.** 6 **19.** -3 **20.** 28 **21.** 4 **22.** No solution **23.** Yes **24.** No **25.** Yes **26.** $\left\{y \middle| y \ge -\frac{1}{2}\right\}$ **27.** $\{x|x \ge 7\}$ **28.** $\{y|y > 2\}$ **29.** $\{y|y \le -4\}$ **30.** $\{x|x < -11\}$ **31.** $\{y|y > -7\}$ **32.** $\left\{x \middle| x > -\frac{9}{11}\right\}$ **33.** $\left\{x \middle| x \ge -\frac{1}{12}\right\}$

34. $x < 3$ (0, 3)

35. $-2 < x \le 5$ (−2, 0, 5)

36. $y > 0$ (0) **37.** $d = \dfrac{C}{\pi}$ **38.** $B = \dfrac{3V}{h}$

39. $a = 2A - b$ **40.** $x = \dfrac{y - b}{m}$ **41.** Length: 365 mi; width: 275 mi **42.** 345, 346 **43.** \$2117 **44.** 27 **45.** 35°, 85°, 60° **46.** 15 **47.** 18.75% **48.** 600 **49.** About 26% **50.** \$220 **51.** \$53,400 **52.** \$138.95 **53.** 86 **54.** $\{w \mid w > 17 \text{ cm}\}$ **55.** $\mathbf{D_W}$ The end result is the same either way. If s is the original salary, the new salary after a 5% raise followed by an 8% raise is $1.08(1.05s)$. If the raises occur the other way around, the new salary is $1.05(1.08s)$. By the commutative and associative laws of multiplication, we see that these are equal. However, it would be better to receive the 8% raise first, because this increase yields a higher salary initially than a 5% raise. **56.** $\mathbf{D_W}$ The inequalities are equivalent by the multiplication principle for inequalities. If we multiply both sides of one inequality by -1, the other inequality results. **57.** $\frac{41}{4}$ **58.** $58t$ **59.** -45 **60.** $-43x + 8y$ **61.** 23, -23 **62.** 20, -20 **63.** $a = \dfrac{y - 3}{2 - b}$

Test: Chapter 7, p. 592

1. [7.1b] 8 **2.** [7.1b] 26 **3.** [7.2a] -6 **4.** [7.2a] 49 **5.** [7.3b] -12 **6.** [7.3a] 2 **7.** [7.3a] -8 **8.** [7.1b] $-\frac{7}{20}$ **9.** [7.3c] 7 **10.** [7.3c] $\frac{5}{3}$ **11.** [7.3b] $\frac{5}{2}$ **12.** [7.3c] No solution **13.** [7.3c] All real numbers **14.** [7.7c] $\{x \mid x \le -4\}$ **15.** [7.7c] $\{x \mid x > -13\}$ **16.** [7.7d] $\{x \mid x \le 5\}$ **17.** [7.7d] $\{y \mid y \le -13\}$ **18.** [7.7d] $\{y \mid y \ge 8\}$ **19.** [7.7d] $\{x \mid x \le -\frac{1}{20}\}$ **20.** [7.7e] $\{x \mid x < -6\}$ **21.** [7.7e] $\{x \mid x \le -1\}$ **22.** [7.7b] $y \le 9$ (0, 4, 9) **23.** [7.7b, e] $x < 1$ (0, 1)

24. [7.7b] $-2 \le x \le 2$ (-2, 0, 2) **25.** [7.5a] 18

26. [7.5a] 16.5% **27.** [7.5a] 40,000 **28.** [7.5a] About 30.3% **29.** [7.6a] Width: 7 cm; length: 11 cm **30.** [7.5a] About \$141.5 billion **31.** [7.6a] 2509, 2510, 2511 **32.** [7.6a] \$880 **33.** [7.6a] 3 m, 5 m **34.** [7.8b] $\{l \mid l \ge 174 \text{ yd}\}$ **35.** [7.8b] $\{b \mid b \le \$105\}$ **36.** [7.8b] $\{c \mid c \le 143{,}750\}$ **37.** [7.4b] $r = \dfrac{A}{2\pi h}$ **38.** [7.4b] $x = \dfrac{y - b}{8}$ **39.** [6.3a] $-\frac{2}{9}$ **40.** [6.1a] $\frac{8}{3}$ **41.** [6.1b] 73% p, or $0.73p$ **42.** [6.8b] $-18x + 37y$ **43.** [7.4b] $d = \dfrac{1 - ca}{-c}$, or $\dfrac{ca - 1}{c}$ **44.** [6.2e], [7.3a] 15, -15 **45.** [7.6a] 60 tickets

Chapter 8

Pretest: Chapter 8, p. 596

1.

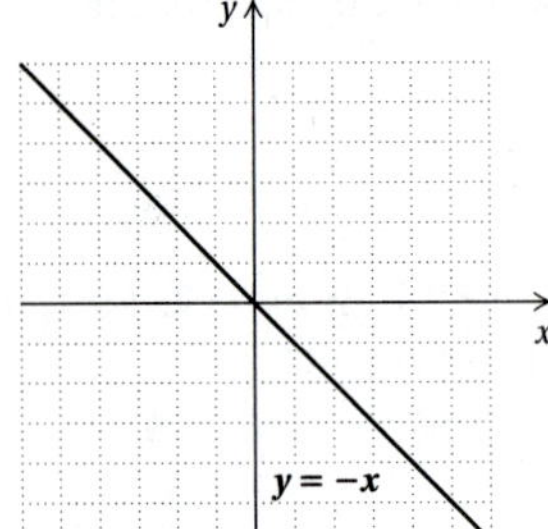

2.

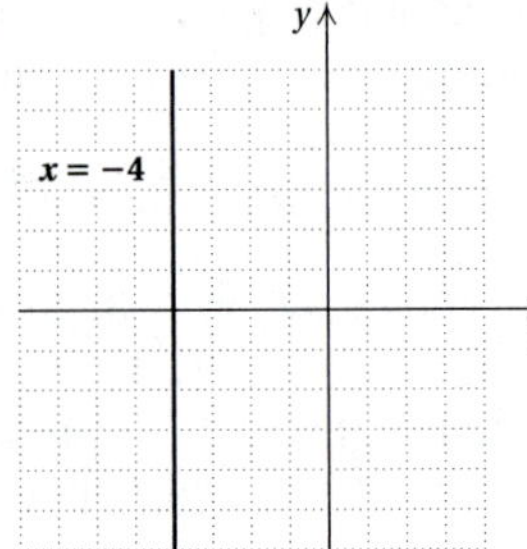

3.

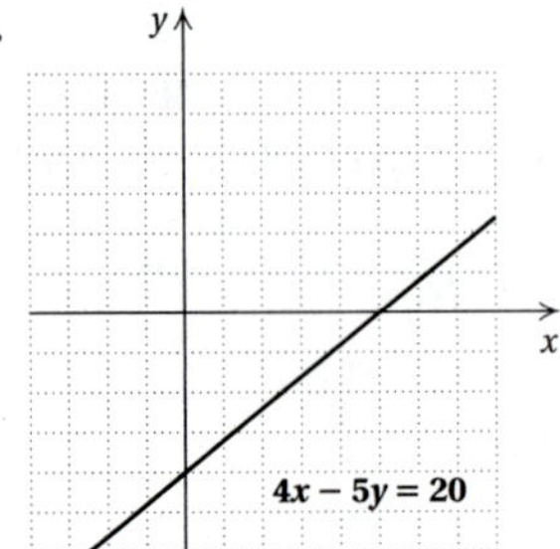

4.

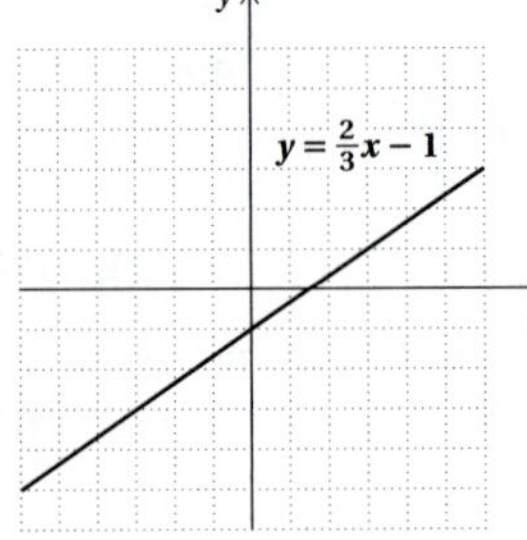

5. III **6.** No **7.** y-intercept: $(0, -4)$; x-intercept: $(5, 0)$ **8.** $(0, -8)$ **9.** 320¢, or \$3.20

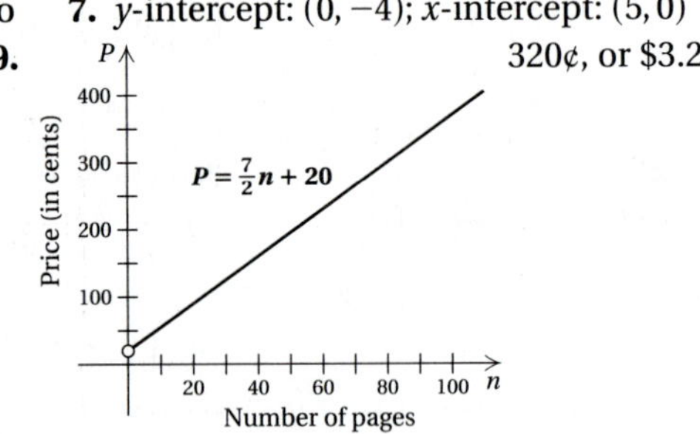

10. $-\frac{2}{3}$ **11.** $2\frac{1}{2}$ cars per hour **12.** 15 calories per minute **13.** $\frac{3}{5}$

Margin Exercises, Section 8.1, pp. 597–601

1. 270 travelers **2.** **(a)** 3 A.M.–6 A.M.; **(b)** midnight–3 A.M.; 3 A.M.–6 A.M.; 6 A.M.–9 A.M.; 9 A.M.–noon **3.** **(a)** 2 months; **(b)** 60 beats per minute **4.–11.**

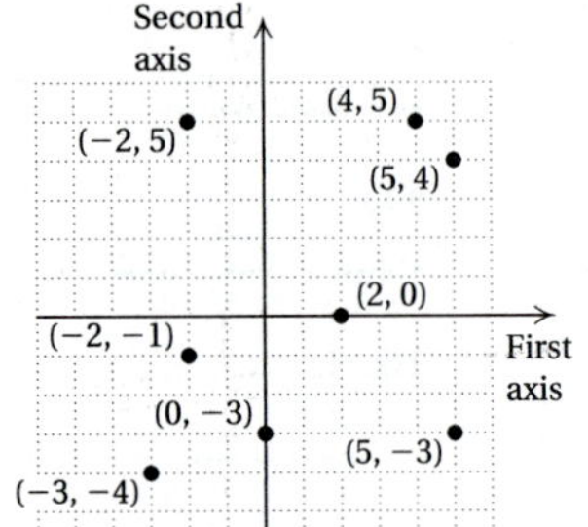

12. Both are negative numbers. **13.** First, positive; second, negative **14.** I **15.** III **16.** IV **17.** II **18.** Not in any quadrant **19.** A: $(-5, 1)$; B: $(-3, 2)$; C: $(0, 4)$; D: $(3, 3)$; E: $(1, 0)$; F: $(0, -3)$; G: $(-5, -4)$

Exercise Set 8.1, p. 602

1. 47.6% **3.** 44.6% **5.** 49,500 **7.** 18,900
9. 6 drinks **11.** The weight is greater than 200 lb.
13. The weight is greater than 120 lb.
15. 17,000 **17.** 1998 **19.** About 1000
21.

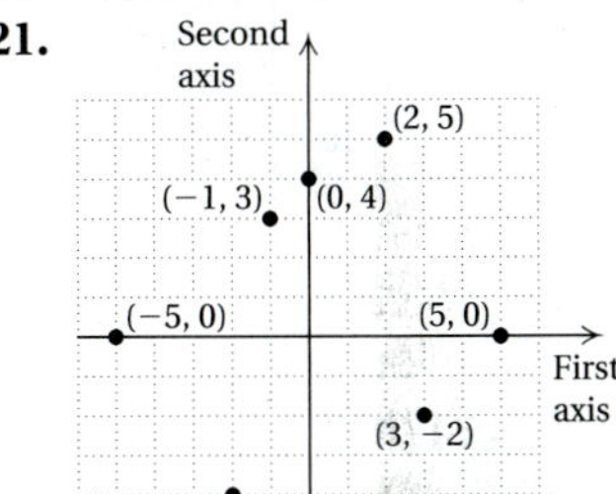

23. II **25.** IV
27. III **29.** I **31.** II
33. IV **35.** I
37. Positive **39.** I, IV
41. I, III

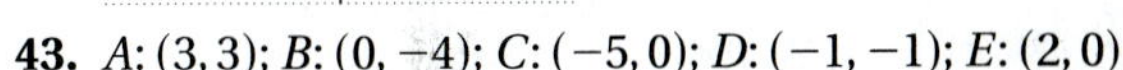

43. A: (3, 3); B: (0, −4); C: (−5, 0); D: (−1, −1); E: (2, 0)
45. $\mathbf{D_W}$ **47.** 12 **48.** 4.89 **49.** 0 **50.** $\frac{4}{5}$ **51.** 3.4
52. $\sqrt{2}$ **53.** $\frac{2}{3}$ **54.** $\frac{7}{8}$ **55.** $28.32 **56.** $18.40
57. (−1, −5) **59.**

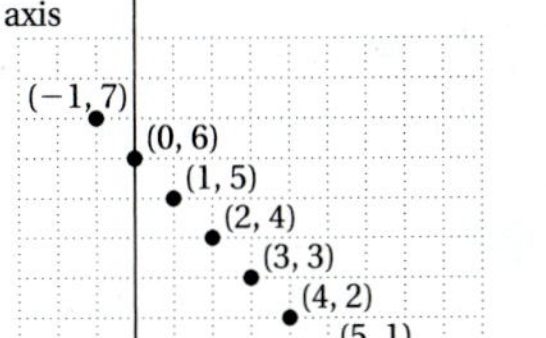

61. 26

Margin Exercises, Section 8.2, pp. 606–614

1. No **2.** Yes **3.** (−2, −3), (1, 3); answers may vary

4.

x	y	(x, y)
−3	6	(−3, 6)
−1	2	(−1, 2)
0	0	(0, 0)
1	−2	(1, −2)
3	−6	(3, −6)

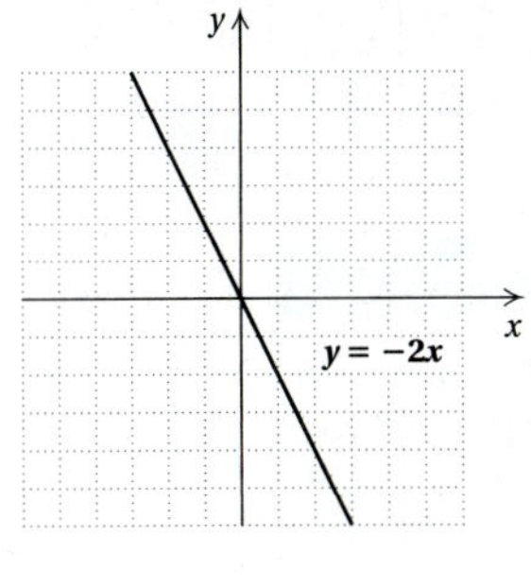

5.

x	y	(x, y)
4	2	(4, 2)
2	1	(2, 1)
0	0	(0, 0)
−2	−1	(−2, −1)
−4	−2	(−4, −2)
−1	$-\frac{1}{2}$	$(-1, -\frac{1}{2})$

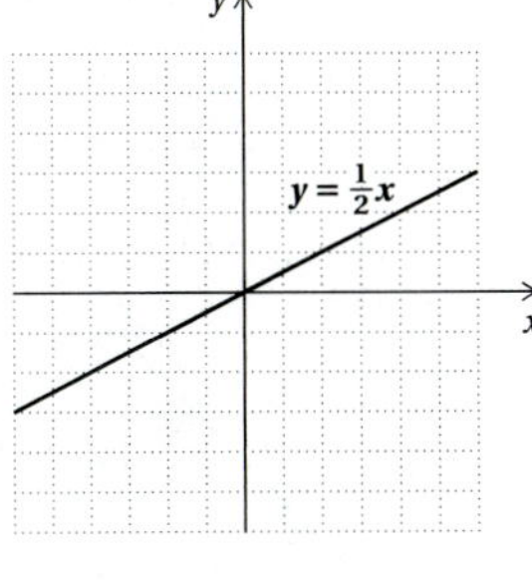

6.

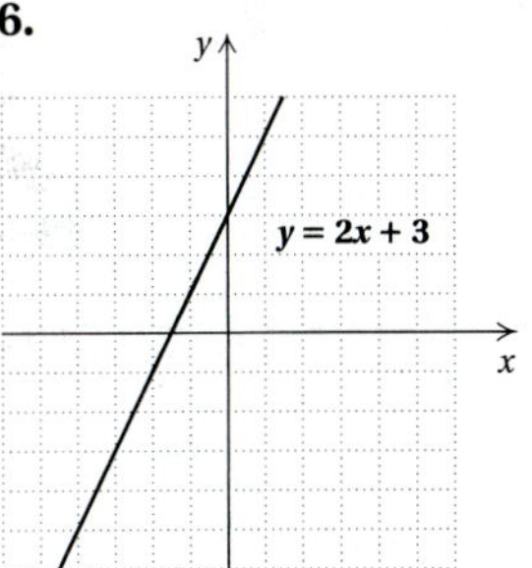

7.

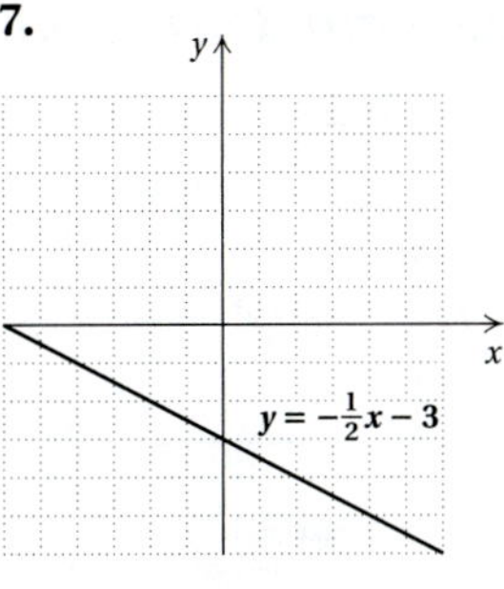

8.

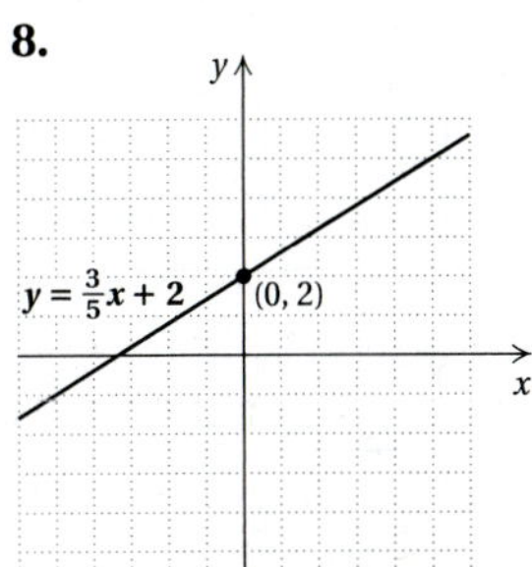

9.

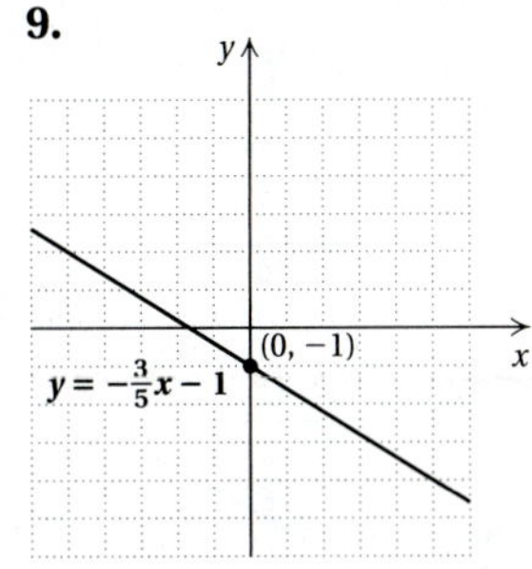

10.

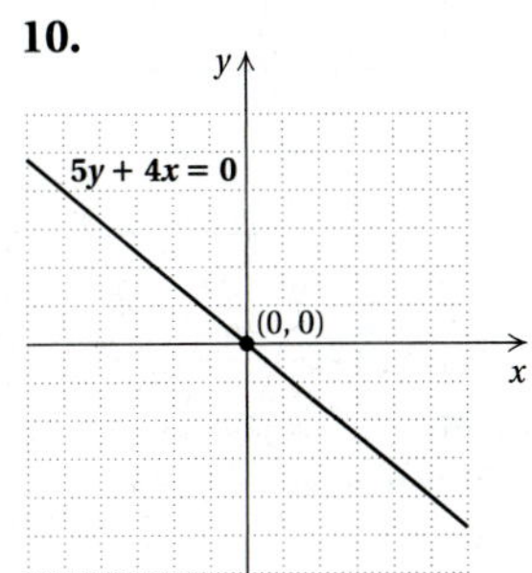

11.

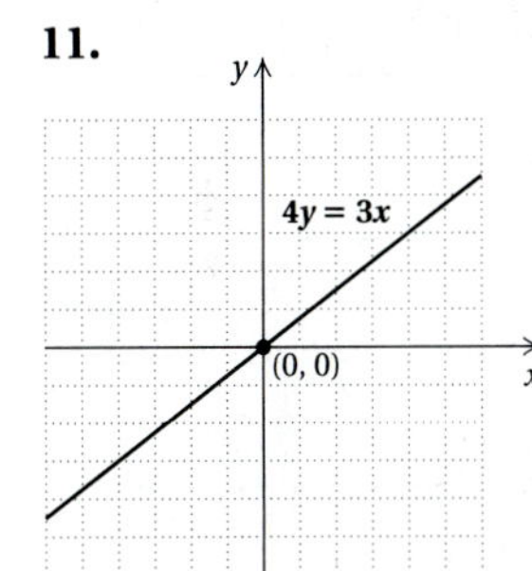

12.

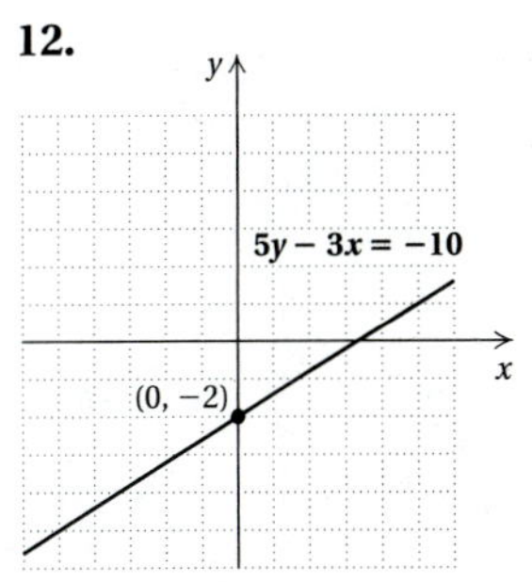

13.

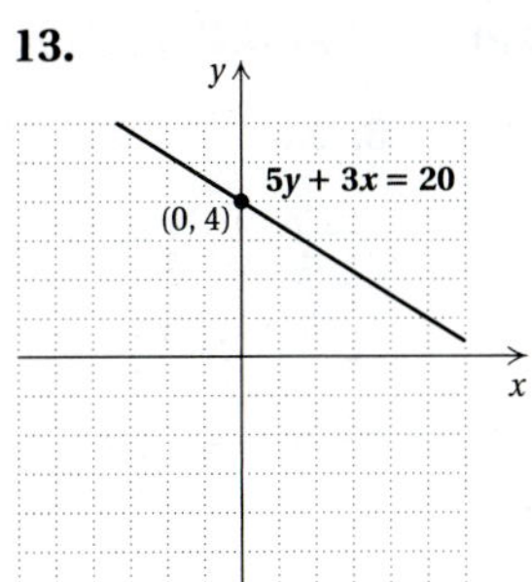

14. **(a)** $2720; $2040; $680; $0;
(b) about $1700;
(c) about 2.8 yr

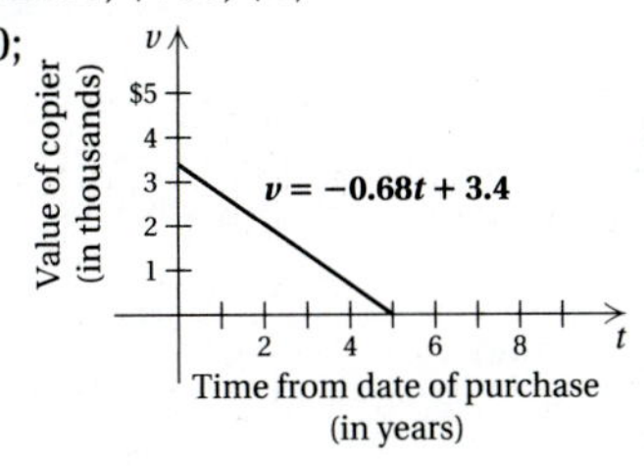

Calculator Corner, p. 610

1. Left to the student

Calculator Corner, p. 616

1. $y = 2x + 1$

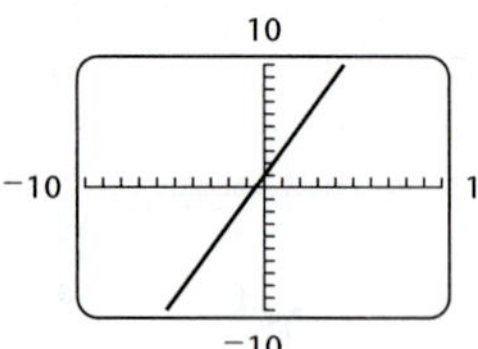
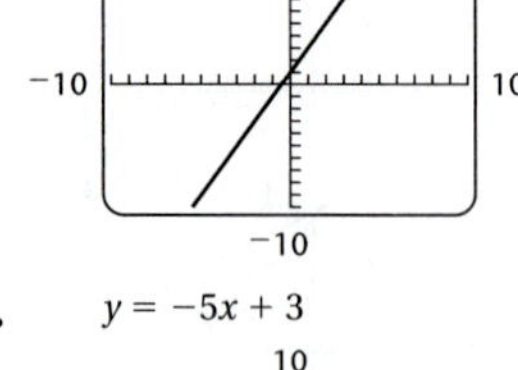

2. $y = -3x + 1$

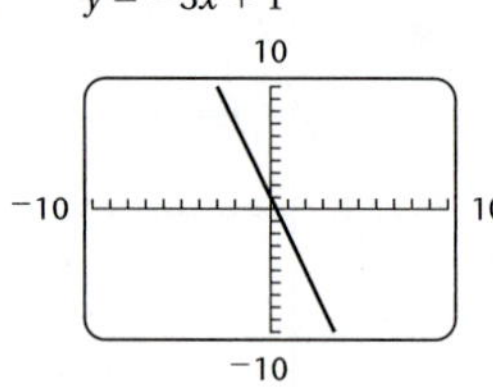

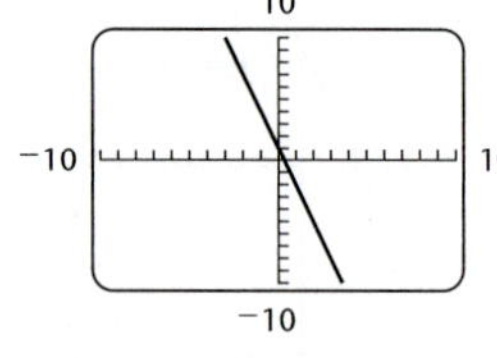

3. $y = -5x + 3$

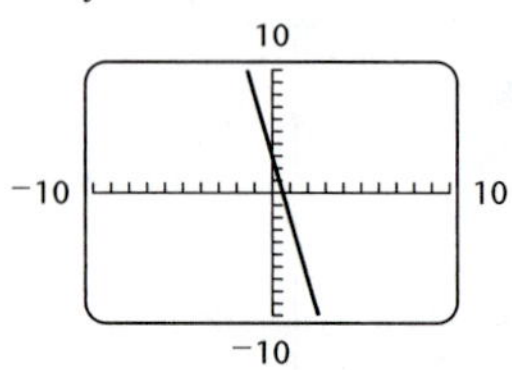

4. $y = 4x - 5$

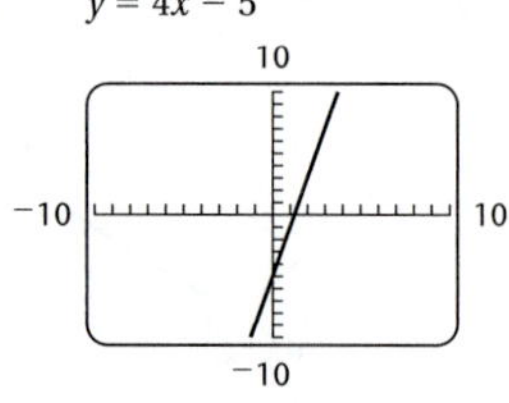
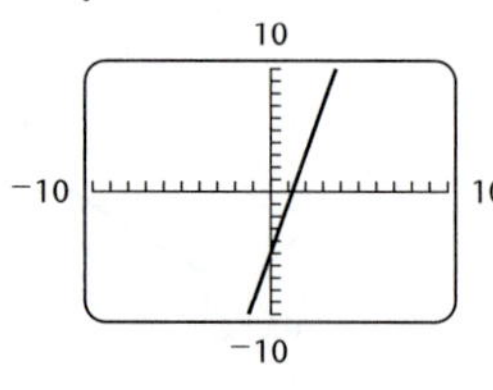

5. $y = \frac{4}{5}x + 2$

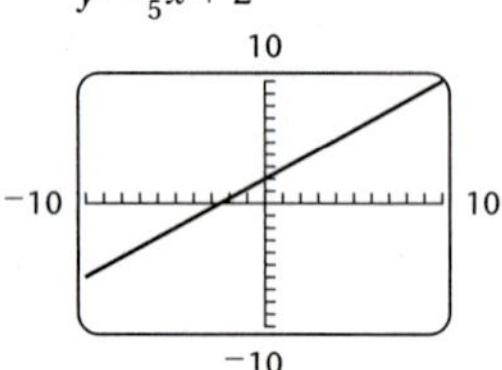

6. $y = -\frac{3}{5}x - 1$

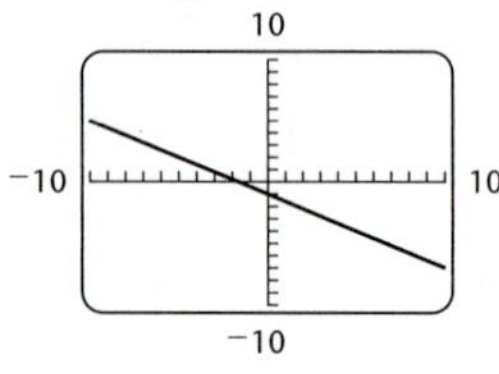

7. $y = 2.085x + 5.08$

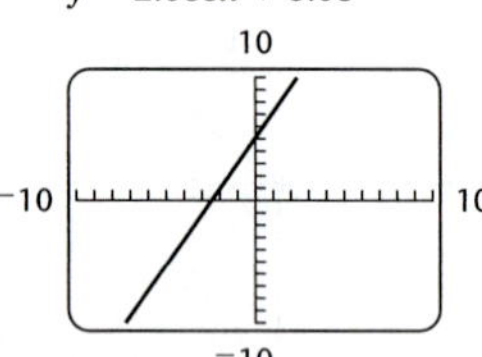

8. $y = -3.45x - 1.68$

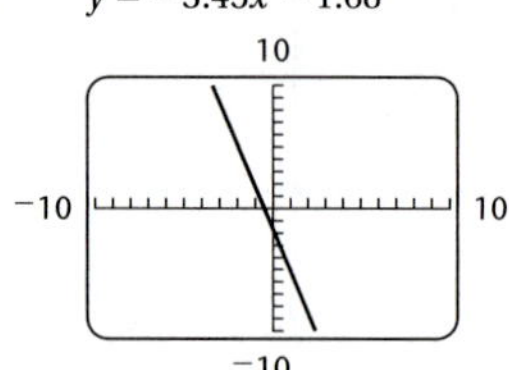
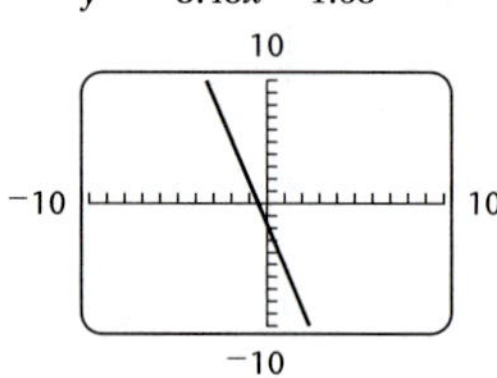

Exercise Set 8.2, p. 617

1. No **3.** No **5.** Yes

7.
$$\begin{array}{c|l} \multicolumn{2}{c}{y = x - 5} \\ \hline -1 \;?\; & 4 - 5 \\ & -1 \end{array} \quad \text{TRUE}$$

$$\begin{array}{c|l} \multicolumn{2}{c}{y = x - 5} \\ \hline -4 \;?\; & 1 - 5 \\ & -4 \end{array} \quad \text{TRUE}$$

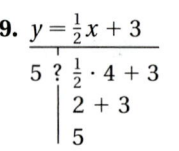

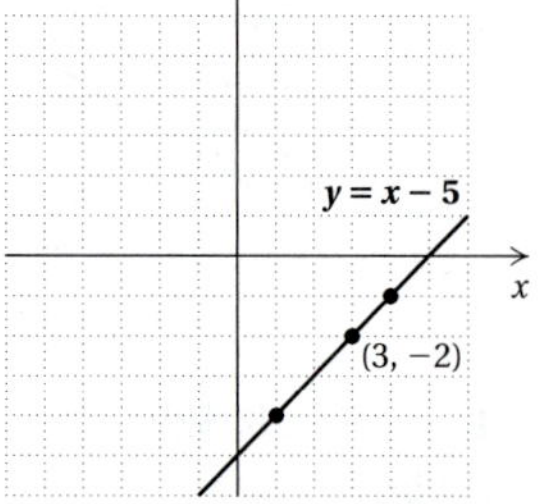

9.
$$\begin{array}{c|l} \multicolumn{2}{c}{y = \frac{1}{2}x + 3} \\ \hline 5 \;?\; & \frac{1}{2} \cdot 4 + 3 \\ & 2 + 3 \\ & 5 \end{array} \quad \text{TRUE}$$

$$\begin{array}{c|l} \multicolumn{2}{c}{y = \frac{1}{2}x + 3} \\ \hline 2 \;?\; & \frac{1}{2}(-2) + 3 \\ & -1 + 3 \\ & 2 \end{array} \quad \text{TRUE}$$

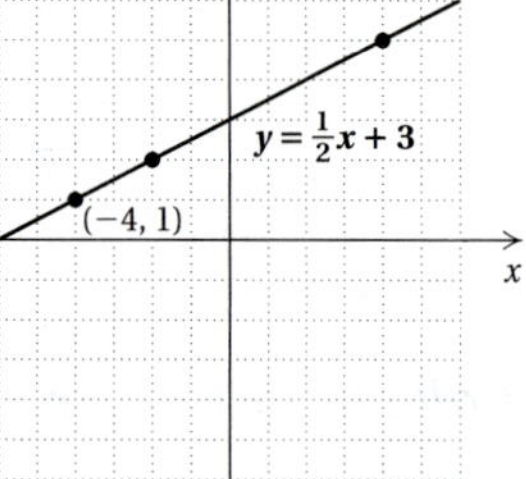

11.
$$\begin{array}{c|l} \multicolumn{2}{c}{4x - 2y = 10} \\ \hline 4 \cdot 0 - 2(-5) \;?\; & 10 \\ 0 + 10 & \\ 10 & \end{array} \quad \text{TRUE}$$

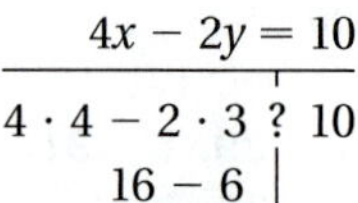

$$\begin{array}{c|l} \multicolumn{2}{c}{4x - 2y = 10} \\ \hline 4 \cdot 4 - 2 \cdot 3 \;?\; & 10 \\ 16 - 6 & \\ 10 & \end{array} \quad \text{TRUE}$$

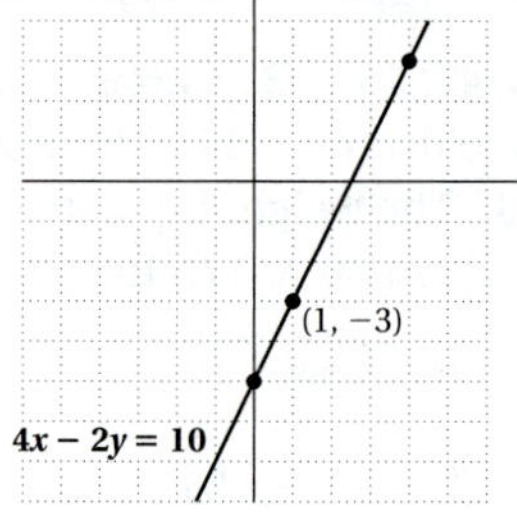

13.

x	y
−2	−1
−1	0
0	1
1	2
2	3
3	4

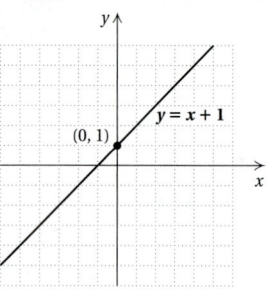

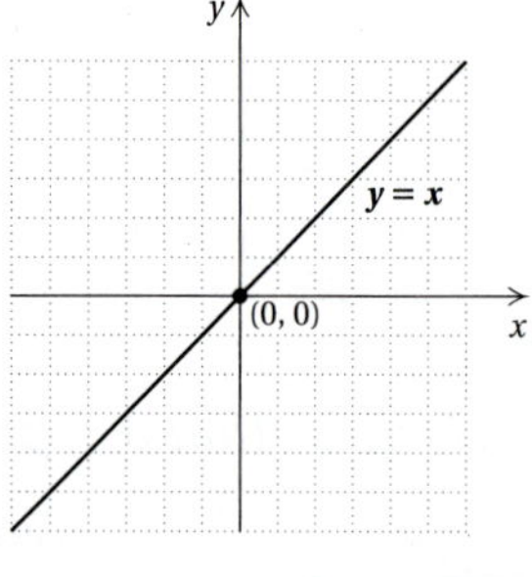

15.

x	y
−2	−2
−1	−1
0	0
1	1
2	2
3	3

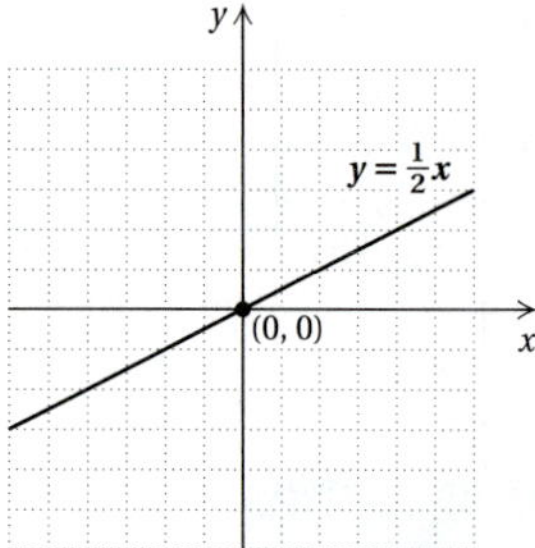

17.

x	y
−2	−1
0	0
4	2

19.

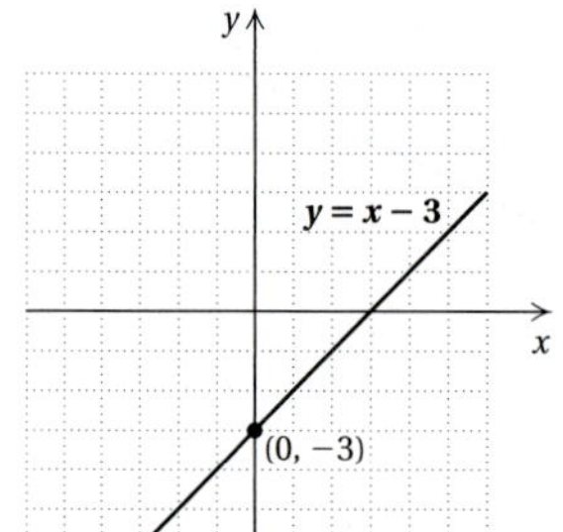

21.

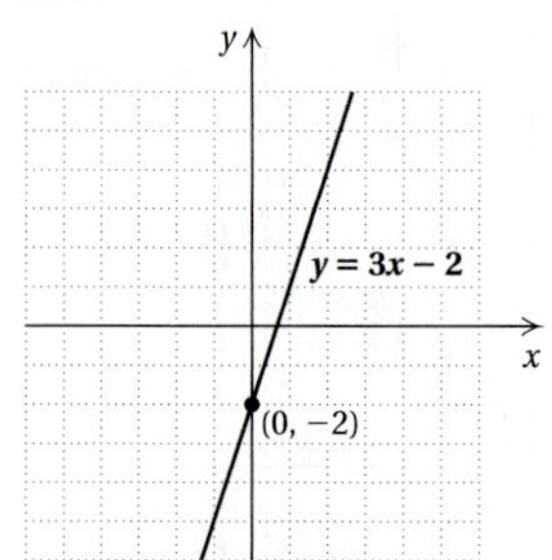

23.

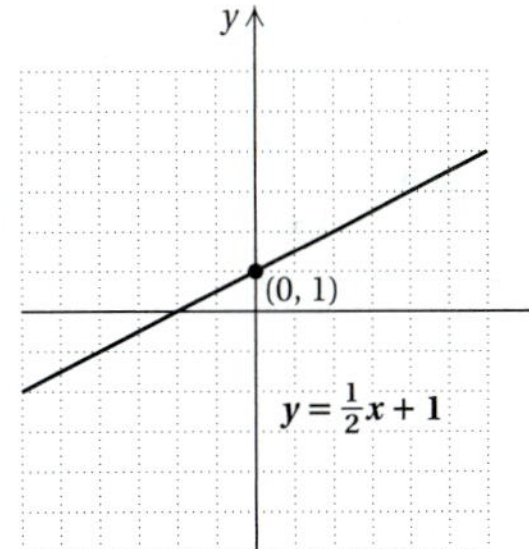

25.

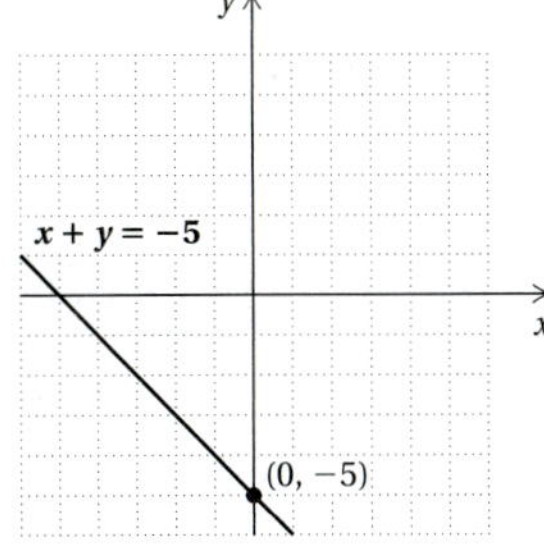

27.

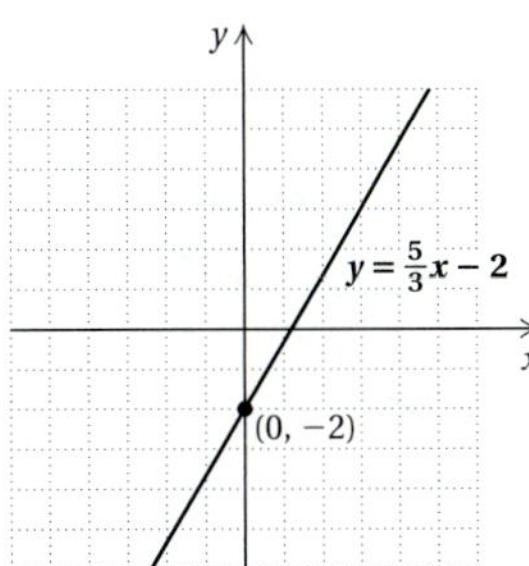

29.

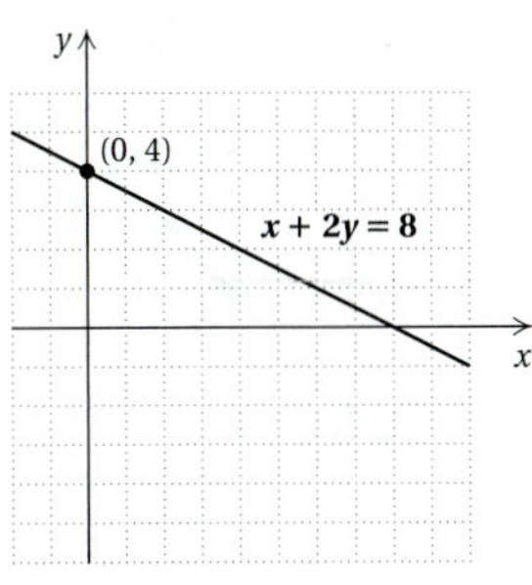

31.

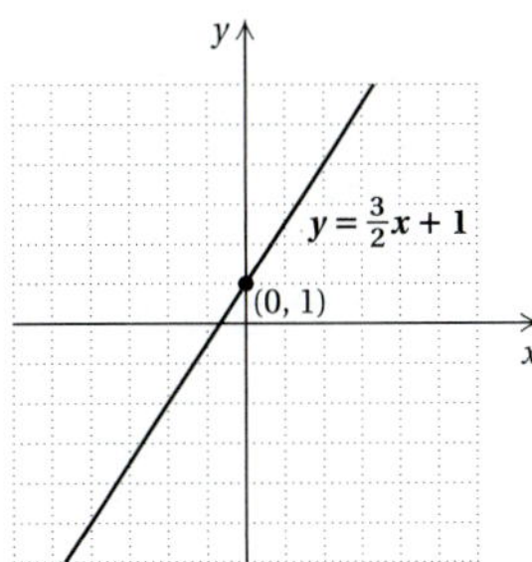

33.

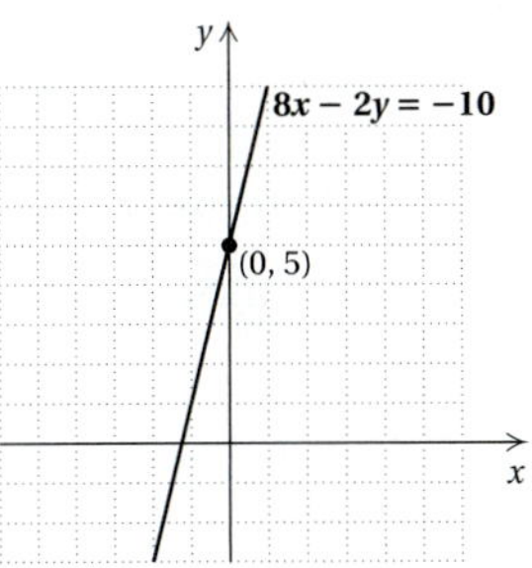

35.

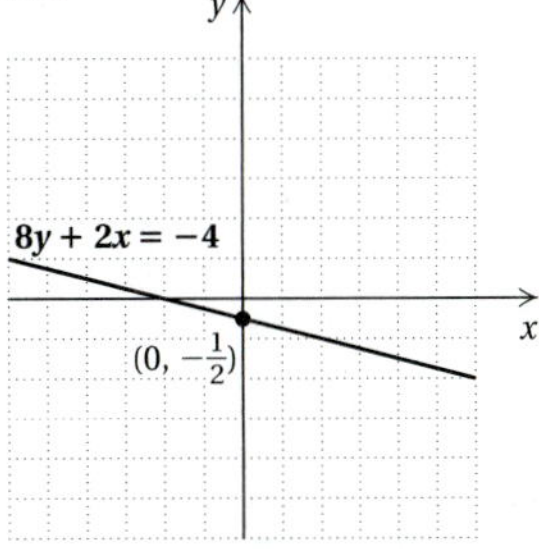

37. **(a)** \$300, \$100, \$0; **(b)**

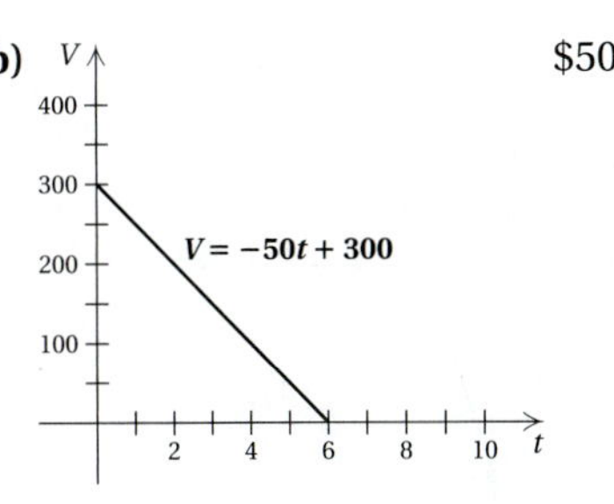

\$50; **(c)** 3 yr

39. **(a)** 7.1 gal, 7.5 gal, 8 gal, 9 gal; **(b)**

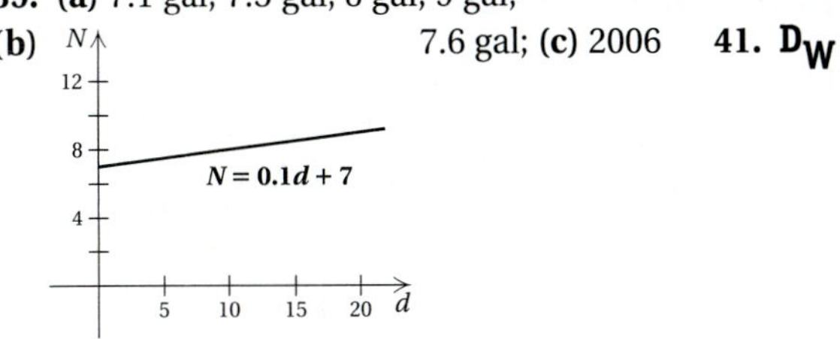

7.6 gal; **(c)** 2006 **41.** D_W

43. 7 **44.** $-\frac{1}{2}$ **45.** -4 **46.** -16 **47.** 4
48. 0.71875 **49.** -0.875 **50.** -2.25 **51.** 1.828125
52. -1.0625 **53.** $y = -x + 5$ **55.** $y = x + 2$

Margin Exercises, Section 8.3, pp. 623–627

1. **(a)** (0, 3); **(b)** (4, 0)

2.

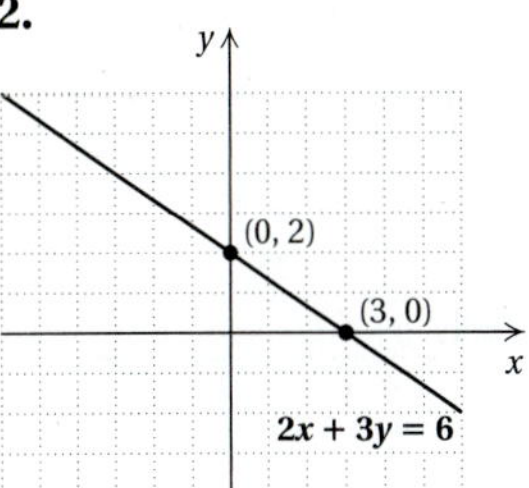

3.

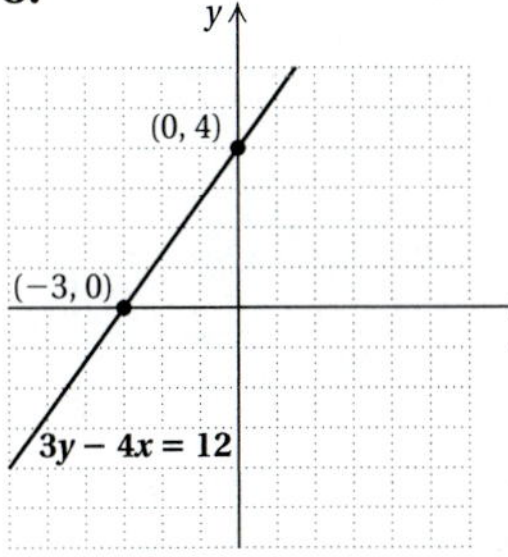

4.

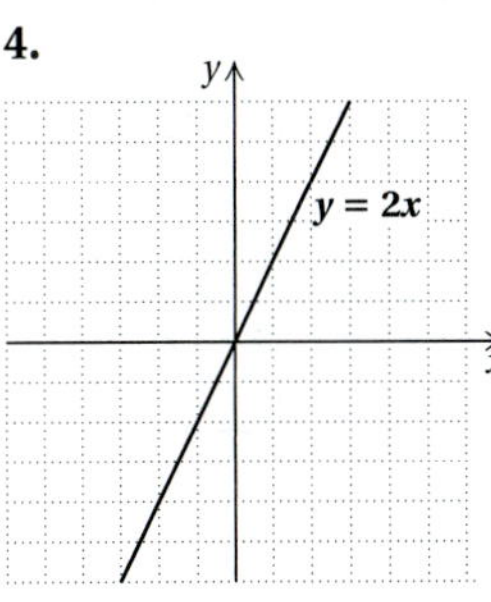

5.

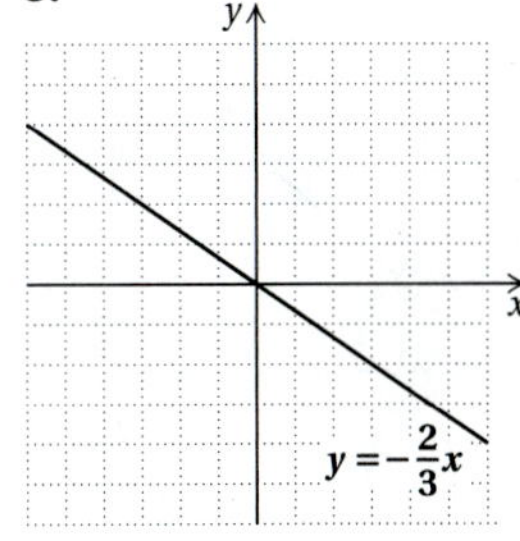

6.

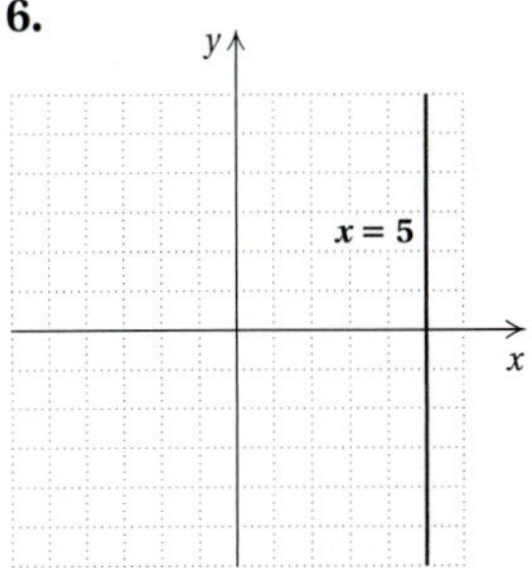

7.

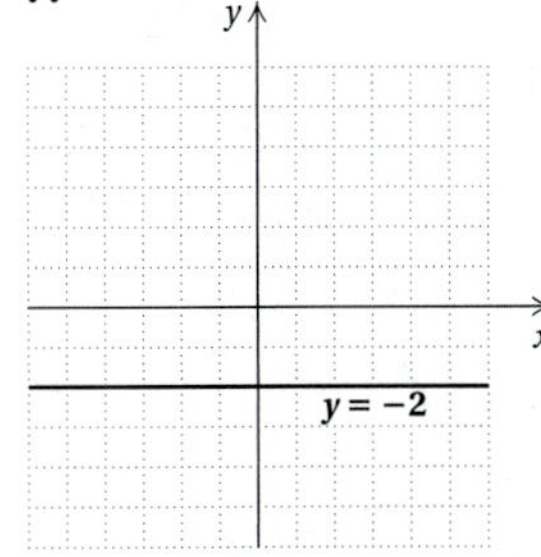

8.

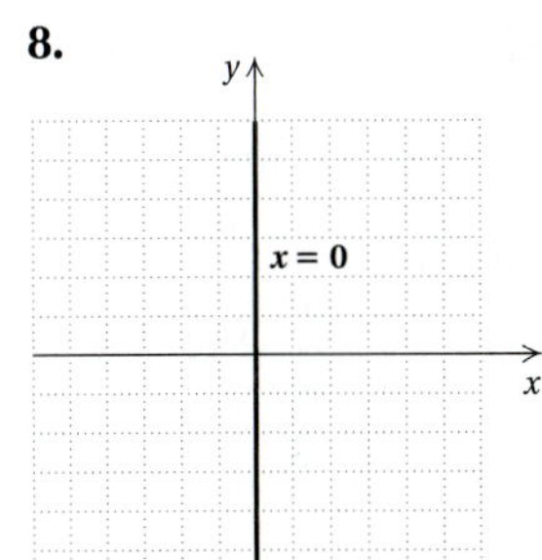

9.

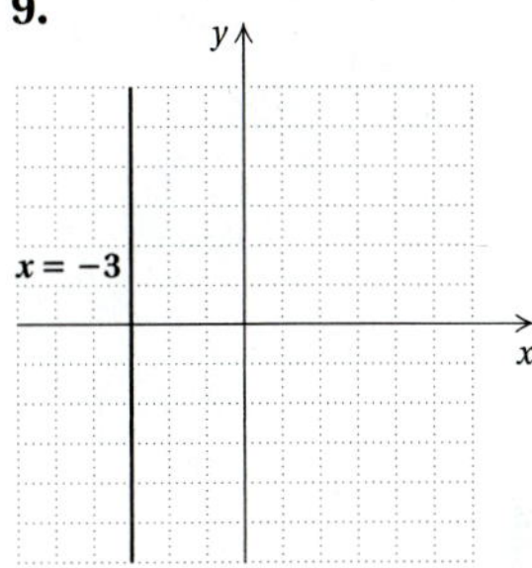

Calculator Corner, p. 625

1. y-intercept: $(0, -15)$; x-intercept: $(-2, 0)$;

$y = -7.5x - 15$

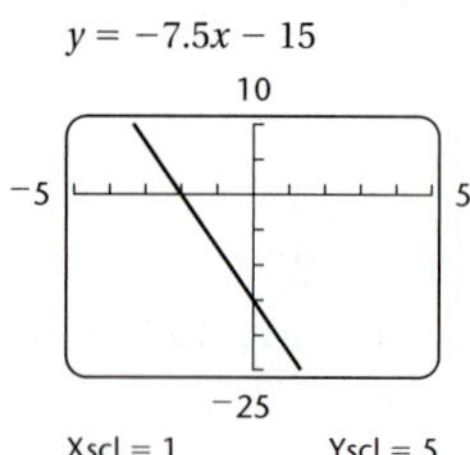

Xscl = 1 Yscl = 5

2. y-intercept: $(0, 43)$; x-intercept: $(-20, 0)$;

$y = 2.15x + 43$

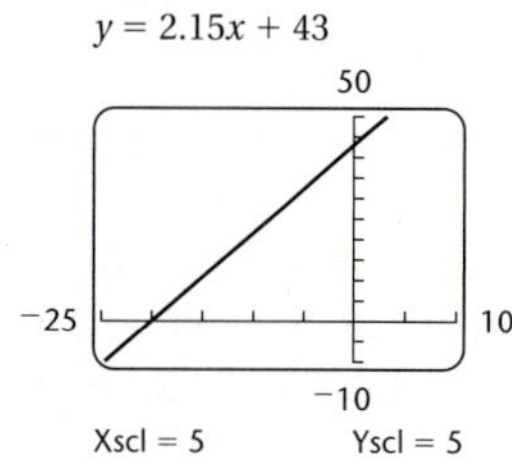

Xscl = 5 Yscl = 5

3. y-intercept: $(0, -30)$; x-intercept: $(25, 0)$;

$y = (6x - 150)/5$

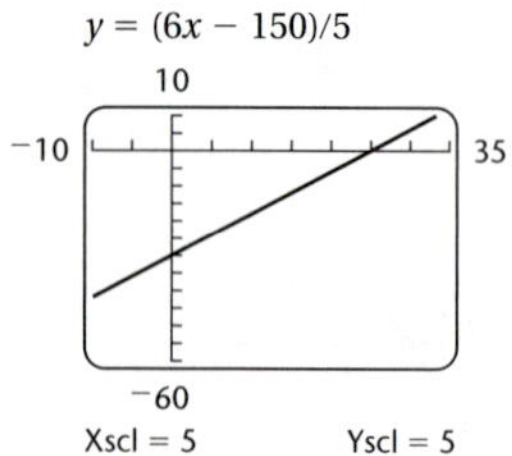

Xscl = 5 Yscl = 5

4. y-intercept: $(0, -4)$; x-intercept: $(20, 0)$;

$y = 0.2x - 4$

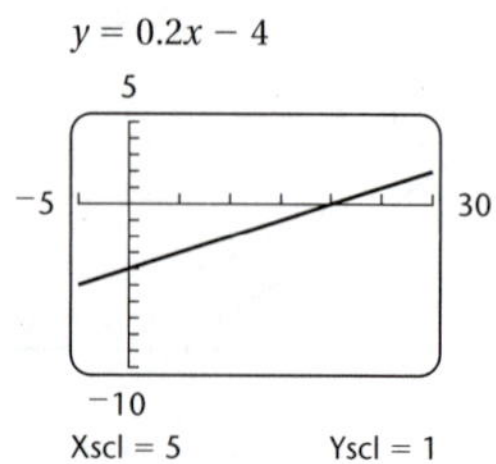

Xscl = 5 Yscl = 1

5. y-intercept: $(0, -15)$; x-intercept: $(10, 0)$;

$y = 1.5x - 15$

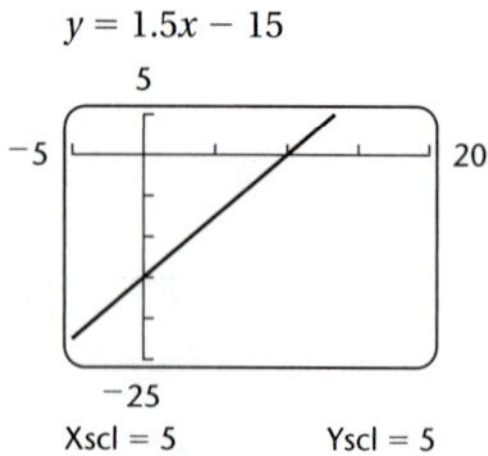

Xscl = 5 Yscl = 5

6. y-intercept: $\left(0, -\frac{1}{2}\right)$; x-intercept: $\left(\frac{2}{5}, 0\right)$;

$y = (5x - 2)/4$

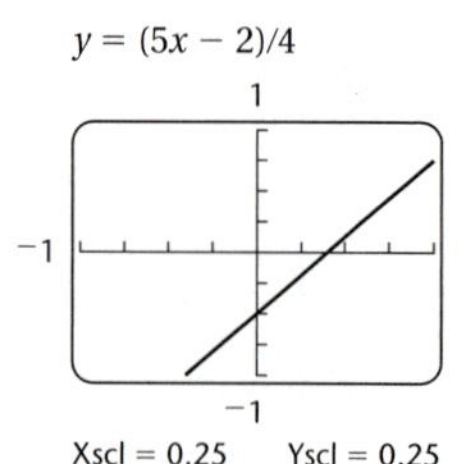

Xscl = 0.25 Yscl = 0.25

Exercise Set 8.3, p. 628

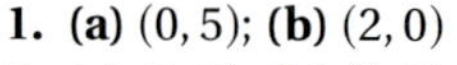

1. **(a)** $(0, 5)$; **(b)** $(2, 0)$ **3.** **(a)** $(0, -4)$; **(b)** $(3, 0)$
5. **(a)** $(0, 3)$; **(b)** $(5, 0)$ **7.** **(a)** $(0, -14)$; **(b)** $(4, 0)$
9. **(a)** $\left(0, \frac{10}{3}\right)$; **(b)** $\left(-\frac{5}{2}, 0\right)$ **11.** **(a)** $\left(0, -\frac{1}{3}\right)$; **(b)** $\left(\frac{1}{2}, 0\right)$

13.

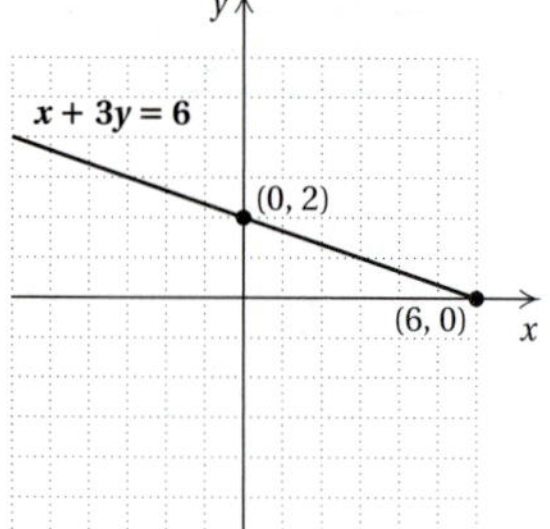

15.

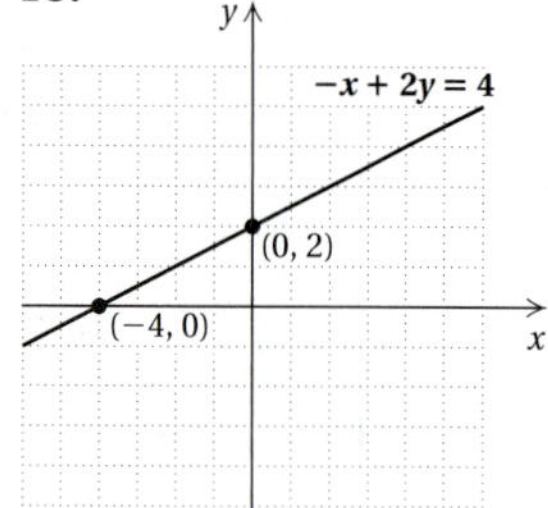

17.

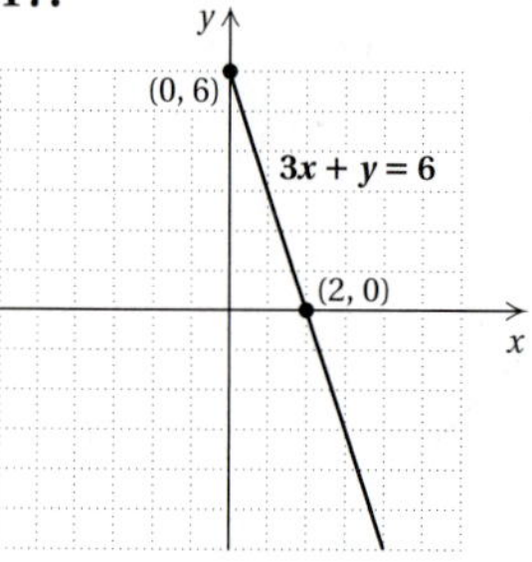

19.

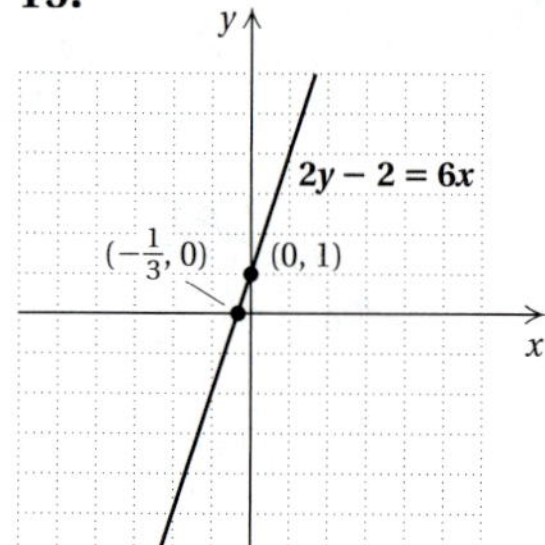

21.

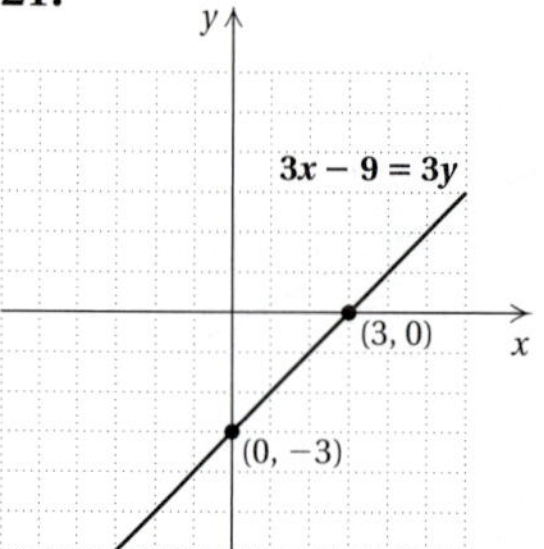

23.

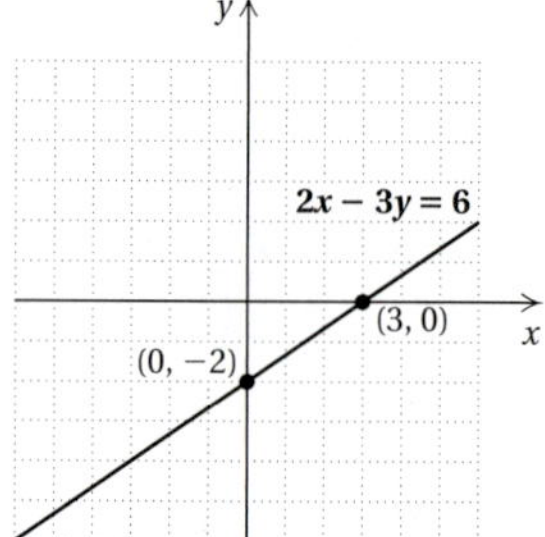

25.

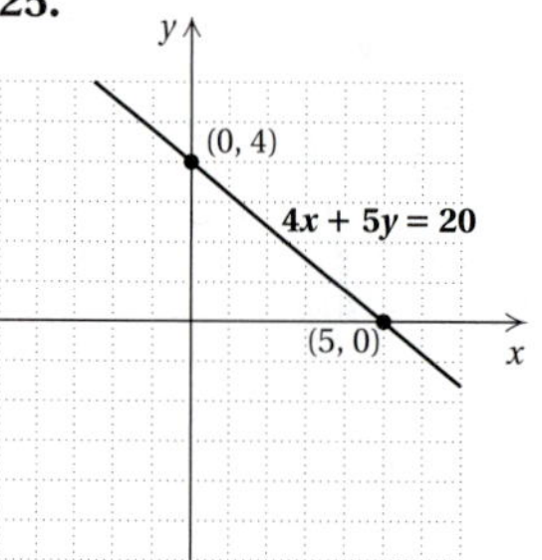

27.

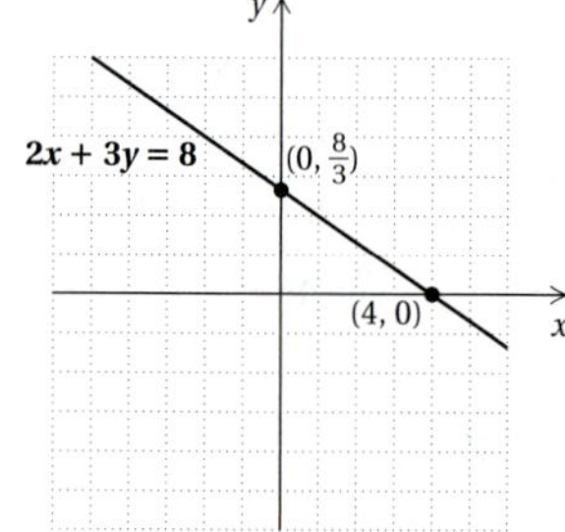

29.

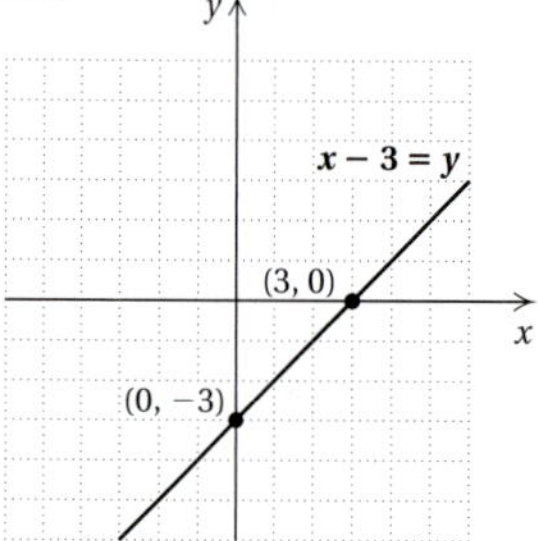

31.

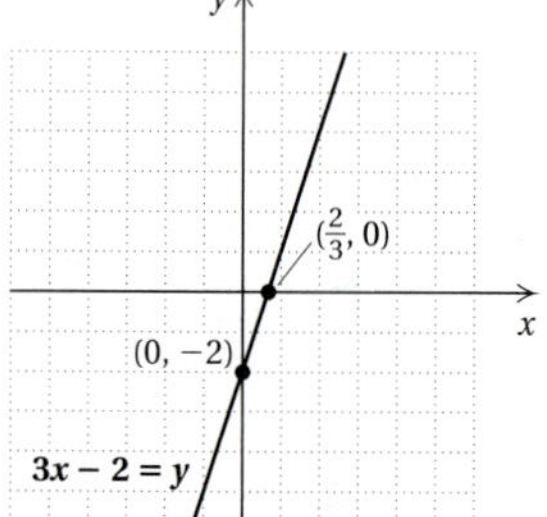

33.

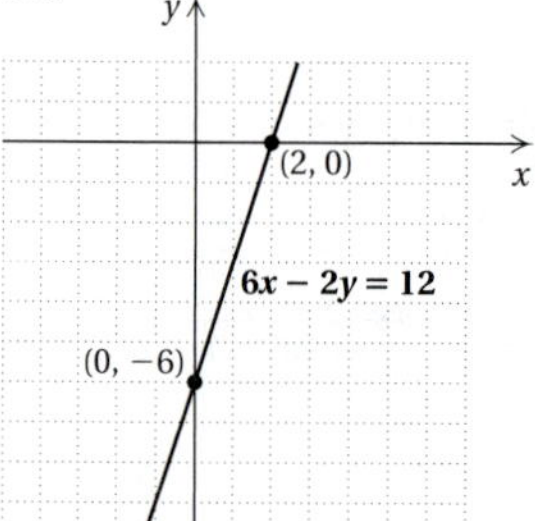

35.

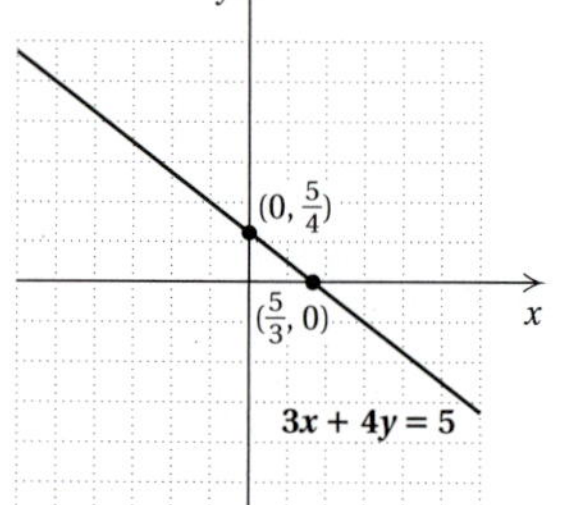

37.

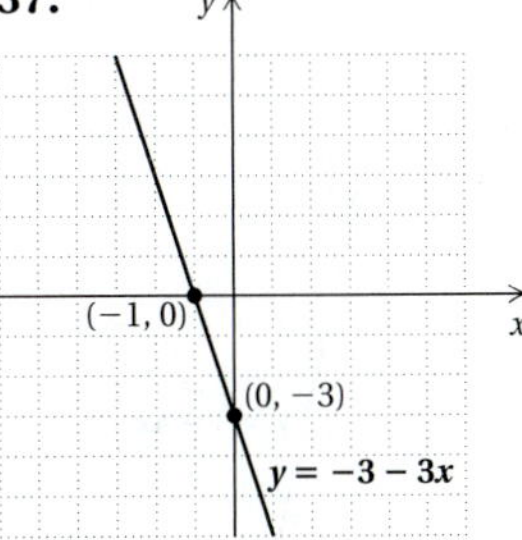

39.

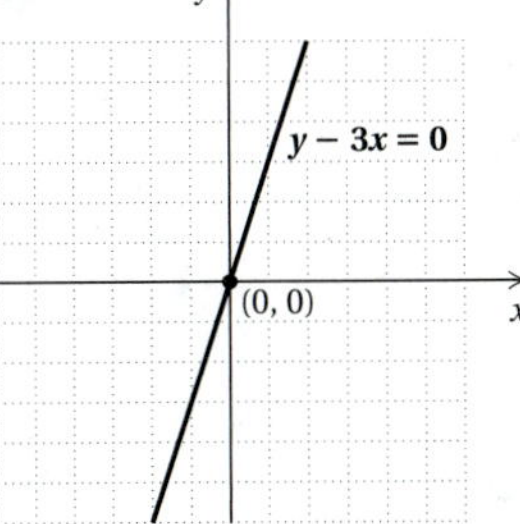

41.

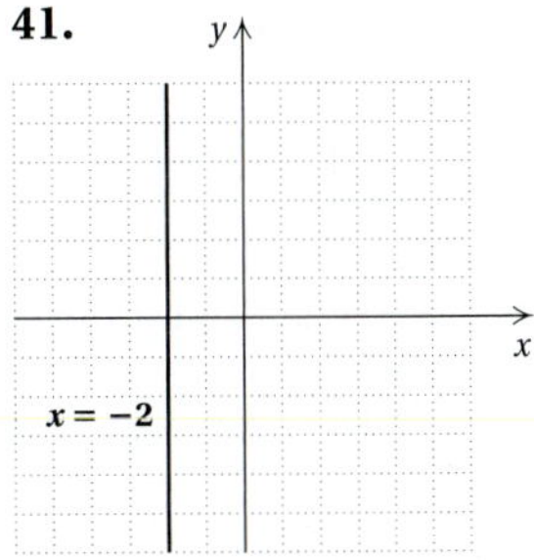

43.

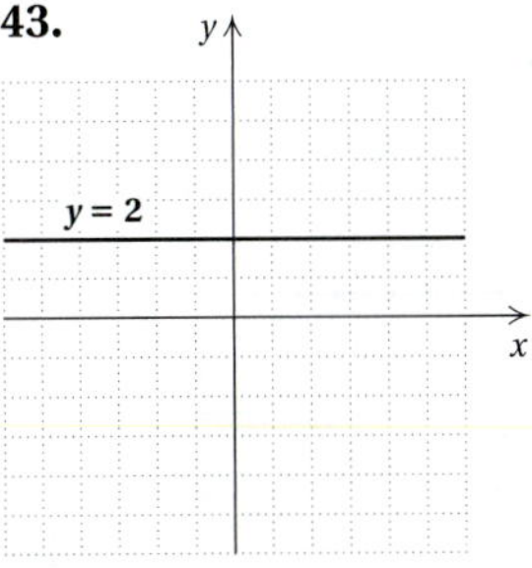

45.

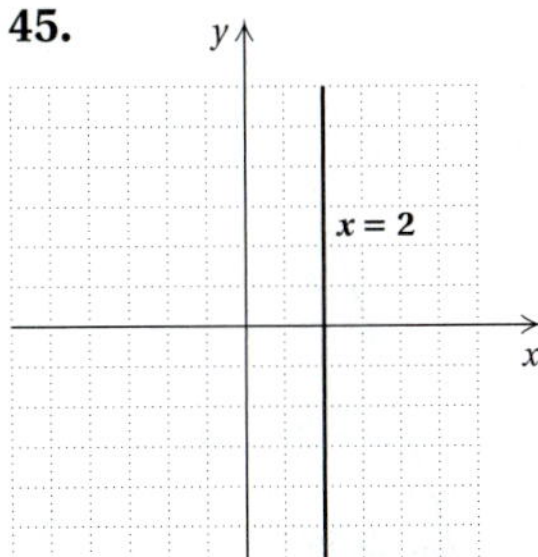

47.

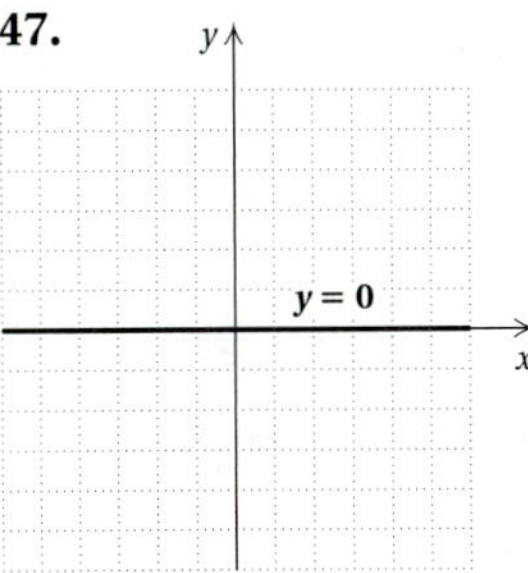

49.

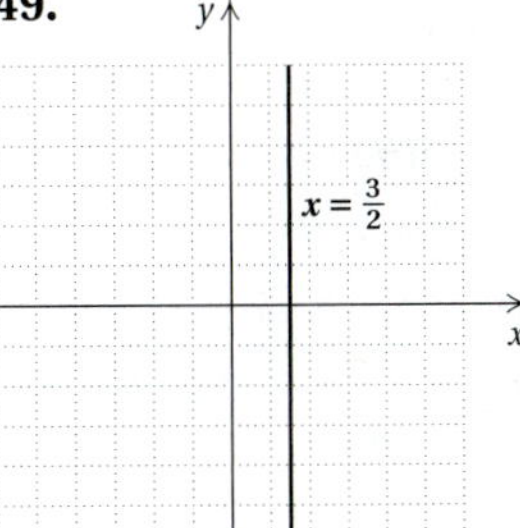

51.

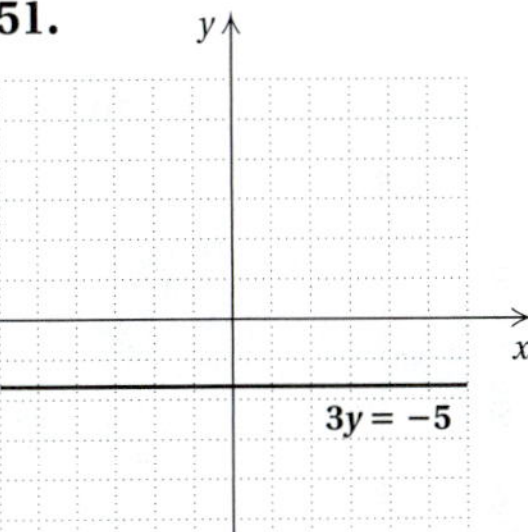

53.

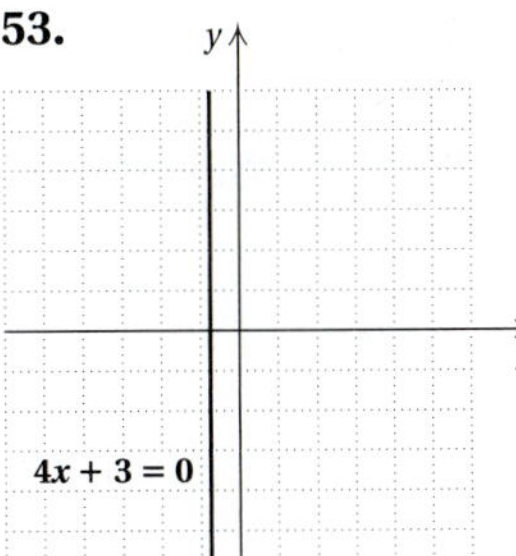

55.

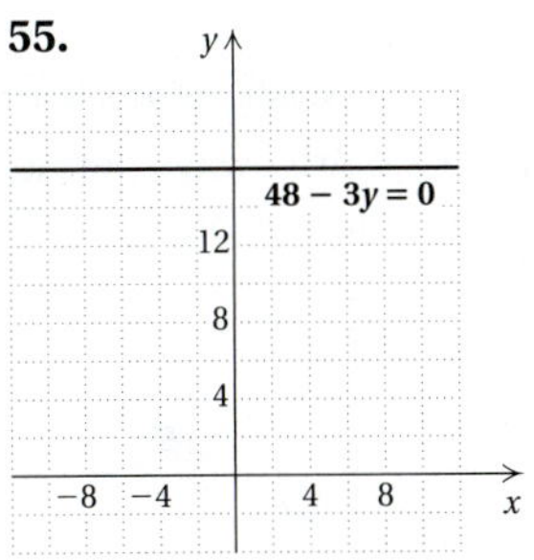

57. $y = -1$ **59.** $x = 4$ **61.** $\mathbf{D_W}$ **63.** 16% **64.** \$32.50 **65.** $\{x|x > -40\}$ **66.** $\{x|x \le -7\}$ **67.** $\{x|x < 1\}$ **68.** $\{x|x \ge 2\}$ **69.** $y = -4$ **71.** $k = 12$

Margin Exercises, Section 8.4, pp. 635–640

1. $\frac{2}{5}$

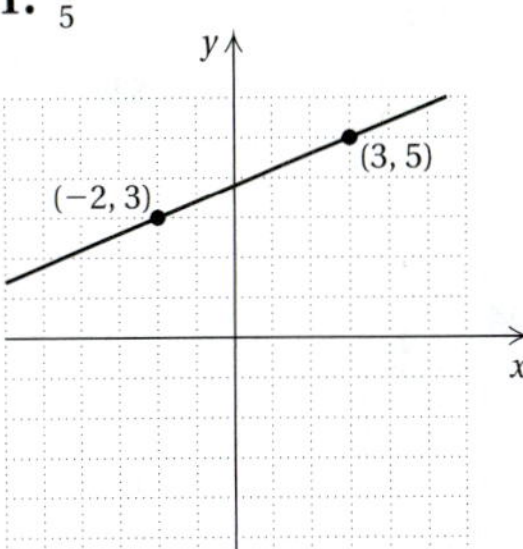

2. $-\frac{5}{3}$

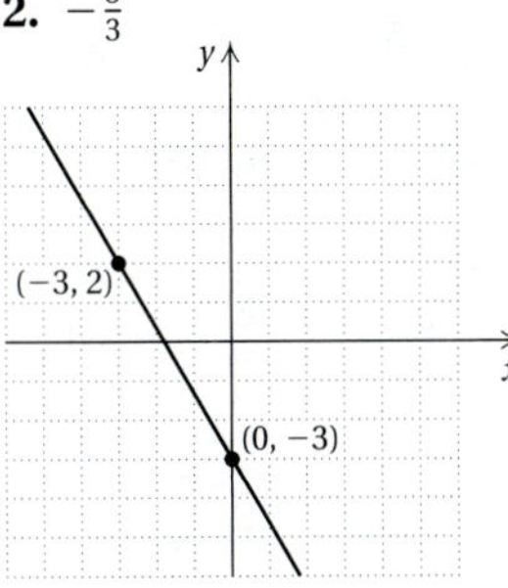

3. $63\frac{7}{11}\%$, or $63.\overline{63}\%$ **4.** 7 cents per minute **5.** −0.4% per year **6.** 4 **7.** −17 **8.** −1 **9.** $\frac{2}{3}$ **10.** −1 **11.** $\frac{5}{4}$ **12.** Not defined **13.** 0

Calculator Corner, p. 639

1. This line will pass through the origin and slant up from left to right. This line will be steeper than $y = 10x$.
2. This line will pass through the origin and slant up from left to right. This line will be less steep than $y = \frac{5}{32}x$.

Calculator Corner, p. 640

1. This line will pass through the origin and slant down from left to right. This line will be steeper than $y = -10x$.
2. This line will pass through the origin and slant down from left to right. This line will be less steep than $y = -\frac{5}{32}x$.

Exercise Set 8.4, p. 641

1. $-\frac{3}{7}$ **3.** $\frac{2}{3}$ **5.** $\frac{3}{4}$ **7.** 0

9. $-\frac{4}{5}$;

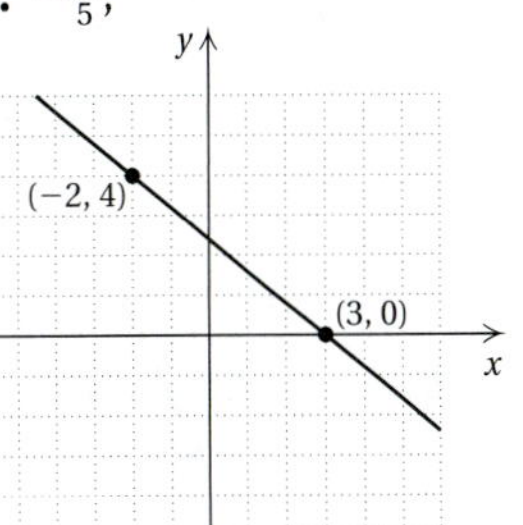

11. 3;

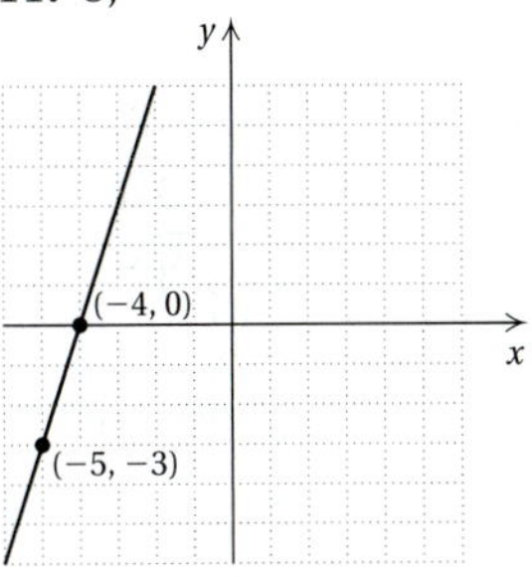

13. $-\frac{5}{6}$;

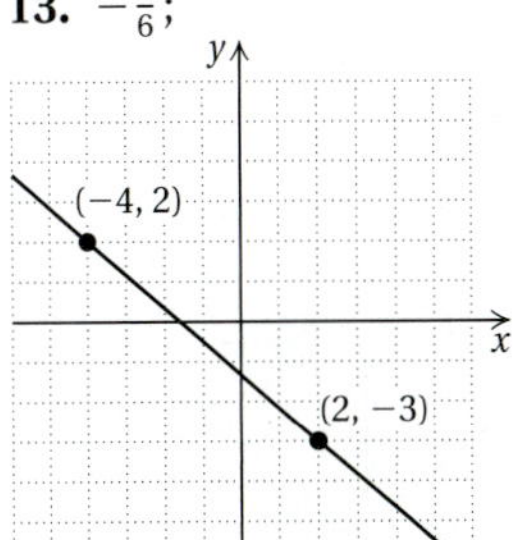

15. $\frac{7}{8}$;

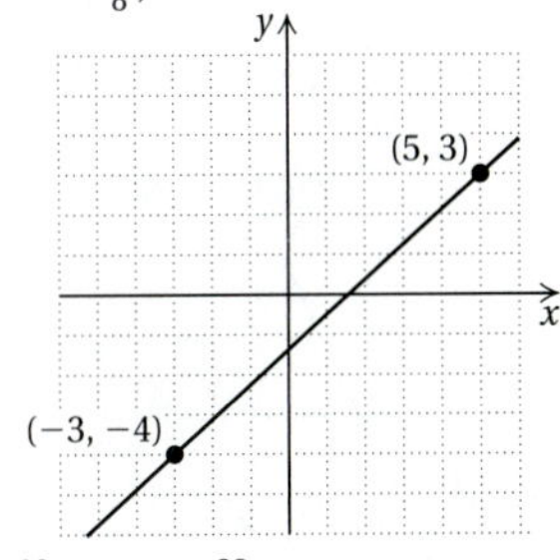

17. $\frac{2}{3}$ **19.** Not defined **21.** $\frac{12}{41}$ **23.** $\frac{28}{129}$ **25.** About 29.4% **27.** 25 miles per gallon **29.** −\$500 per year

31. About 7689 people per year **33.** -10 **35.** 3.78
37. 3 **39.** $-\frac{1}{5}$ **41.** $-\frac{3}{2}$ **43.** $\frac{5}{7}$ **45.** -2.74 **47.** 3
49. $\frac{5}{4}$ **51.** 0 **53.** $\mathbf{D_W}$ **55.** -4 **56.** -20 **57.** -37
58. $-\frac{3}{4}$ **59.** $3.57 **60.** $48.60 **61.** 20% **62.** $18
63. $45.15 **64.** $55

65. $y = 0.35x - 7$

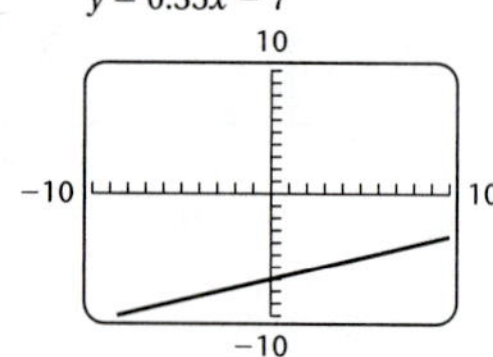

X	Y1
−10	−10.5
−9.9	−10.47
−9.8	−10.43
−9.7	−10.4
−9.6	−10.36
−9.5	−10.33
−9.4	−10.29

X = −10

67. $y = x^3 - 5$

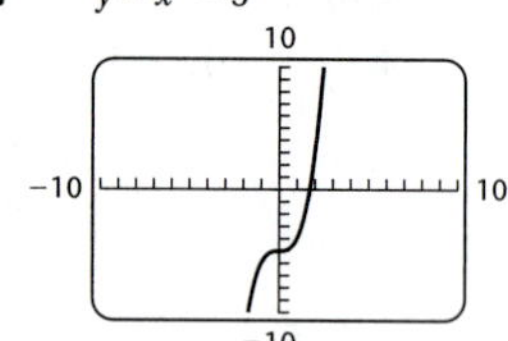

X	Y1
−10	−1005
−9.9	−975.3
−9.8	−946.2
−9.7	−917.7
−9.6	−889.7
−9.5	−862.4
−9.4	−835.6

X = −10

Summary and Review: Chapter 8, p. 647

1. $775.50; $634.50 **2.** 47 lb **3.** 80 lb **4.** 33 lb **5.** 1993 **6.** 1990–1995 **7.** One shower **8.** One toilet flush **9.** One shave, wash dishes, one shower **10.** One toilet flush **11.** $(-5, -1)$ **12.** $(-2, 5)$ **13.** $(3, 0)$

14.–16.

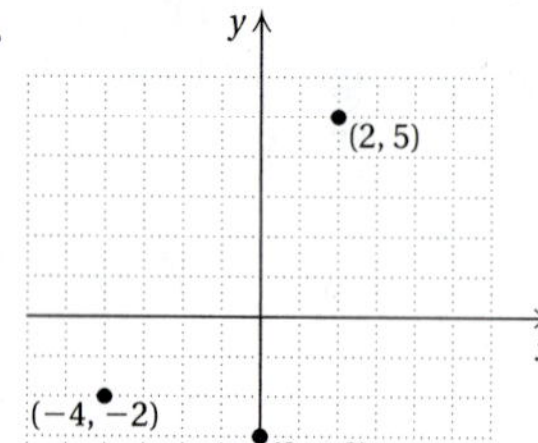

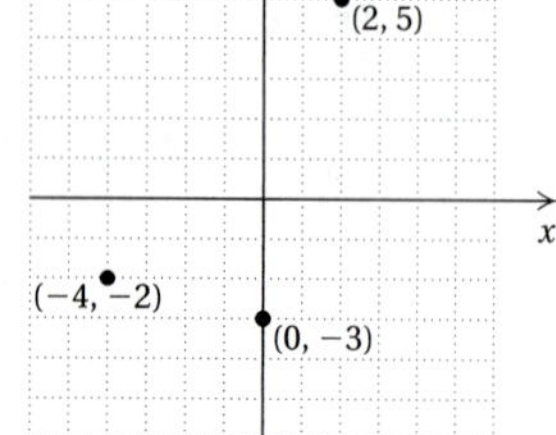

17. IV **18.** III **19.** I **20.** No **21.** Yes

22.

$$\begin{array}{c|c} 2x - y & 3 \\ \hline 2 \cdot 0 - (-3) & ?\ 3 \\ 0 + 3 & \\ 3 & \end{array} \quad \text{TRUE}$$

$$\begin{array}{c|c} 2x - y & 3 \\ \hline 2 \cdot 2 - 1 & ?\ 3 \\ 4 - 1 & \\ 3 & \end{array} \quad \text{TRUE}$$

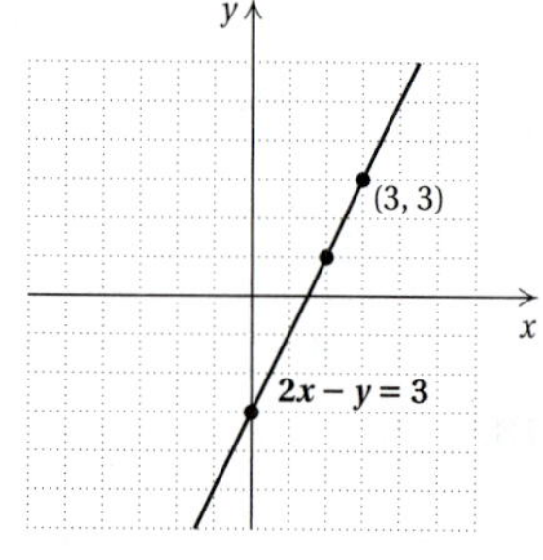

23.

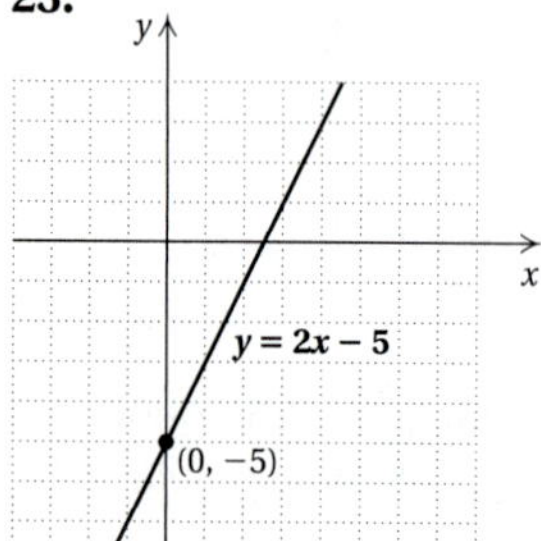

24.

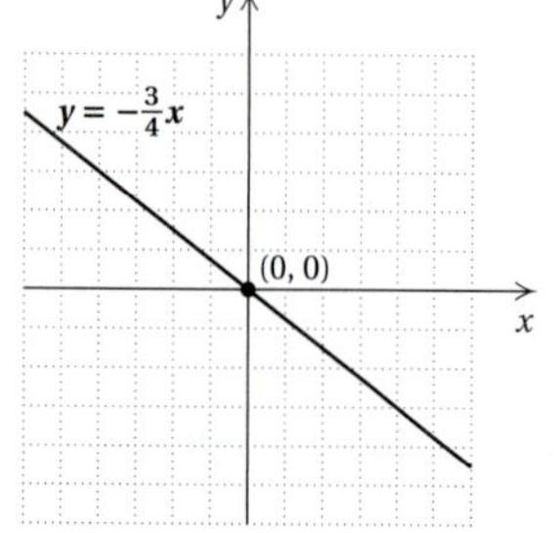

25.

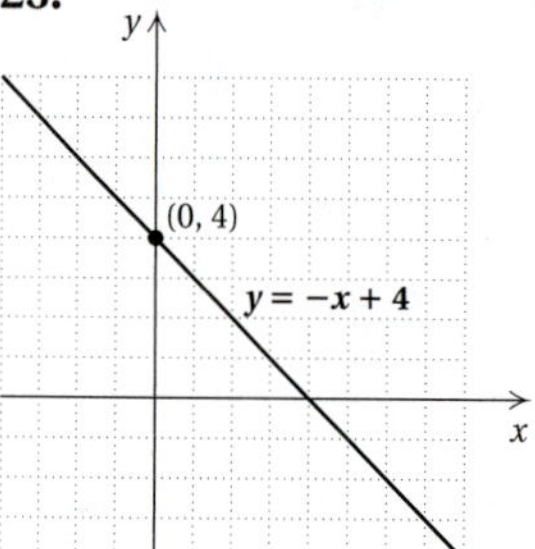

26.

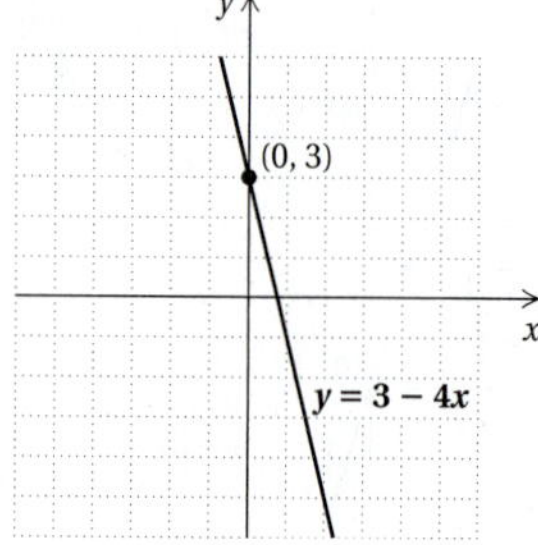

27.

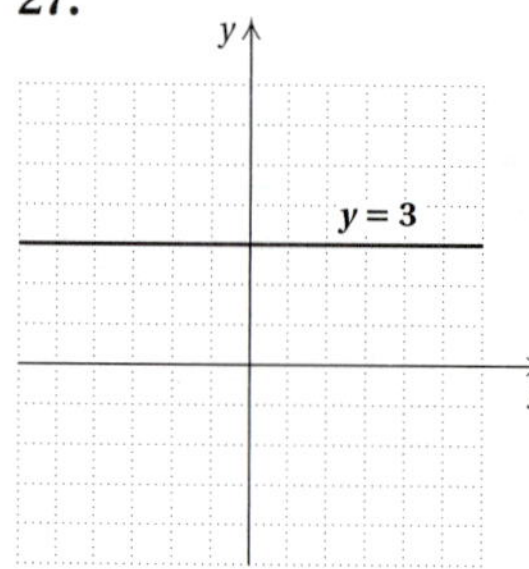

28.

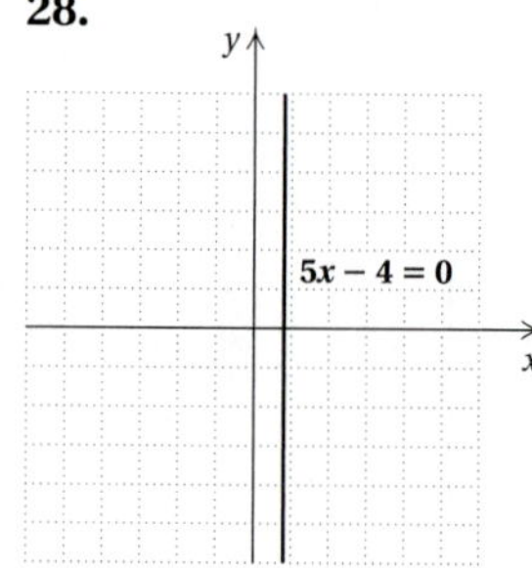

29.

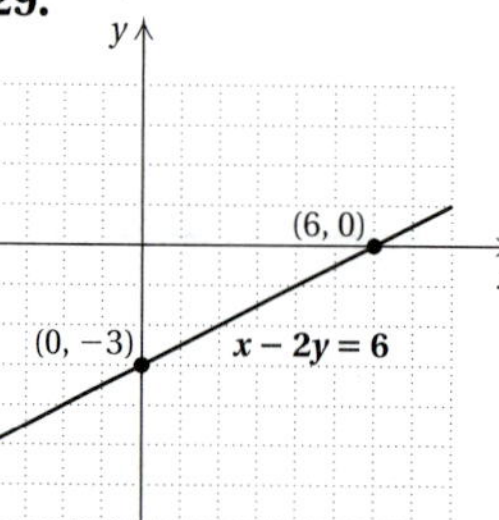

30.

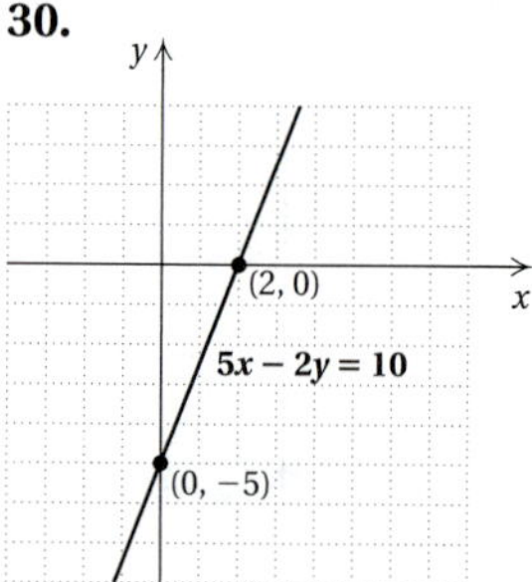

31. **(a)** $14\frac{1}{2}$ ft^3, 16 ft^3, $20\frac{1}{2}$ ft^3, 28 ft^3;
(b)

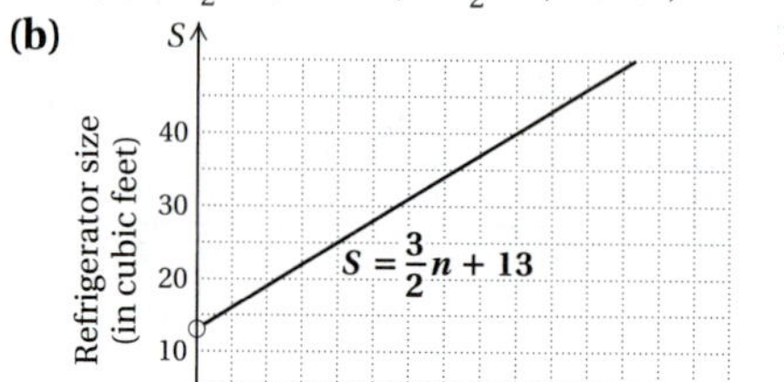

$17\frac{1}{2}$ ft^3; **(c)** 6

32. **(a)** 2.4 driveways per hour; **(b)** 25 minutes per driveway **33.** 4 manicures per hour **34.** $\frac{1}{3}$ **35.** $-\frac{1}{3}$
36. $\frac{3}{5}$;

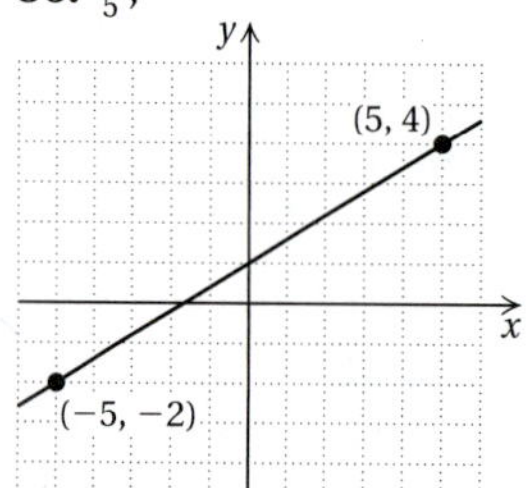

37. -1;

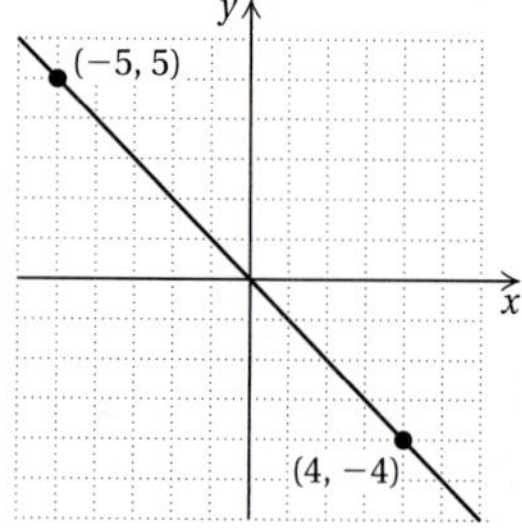

38. 7% **39.** $-\frac{5}{8}$ **40.** $\frac{1}{2}$ **41.** Not defined **42.** 0
43. $\mathbf{D_W}$ A small business might use a graph to look up prices quickly (as in the FedEx mailing costs example) or to plot change in sales over a period of time. Many other applications exist. **44.** $\mathbf{D_W}$ The *y*-intercept is the point at

which the graph crosses the y-axis. Since a point on the y-axis is neither left nor right of the origin, the first or x-coordinate of the point is 0. **45.** -0.34375 **46.** $0.\overline{8}$ **47.** 3.2 **48.** $\frac{17}{19}$ **49.** 11 **50.** $\frac{3}{4}$ **51.** \$9755.09 **52.** \$79.95 **53.** $m = -1$ **54.** 45 square units; 28 linear units **55.** **(a)** 3.709 feet per minute; **(b)** about 0.2696 minute per foot

Test: Chapter 8, p. 652

1. [8.1a] \$495,000,000 **2.** [8.1a] Crest and Colgate **3.** [8.1a] Crest **4.** [8.1a] Arm & Hammer **5.** [8.1a] June **6.** [8.1a] January **7.** [8.1a] March, April, May, June, July **8.** [8.1a] August **9.** [8.1a] 2001 **10.** [8.1a] 1996 **11.** [8.1a] 5002 **12.** [8.1a] 2000 and 2001 **13.** [8.1a] 3481 **14.** [8.1a] 2309 **15.** [8.1c] II **16.** [8.1c] III **17.** [8.1d] (3, 4) **18.** [8.1d] (0, −4)
19. [8.2a]

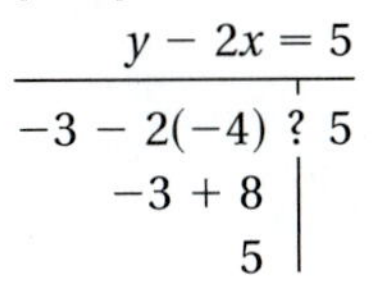

$y - 2x = 5$	
$-3 - 2(-4)$	? 5
$-3 + 8$	
5	TRUE

$y - 2x = 5$	
$3 - 2(-1)$	? 5
$3 + 2$	
5	TRUE

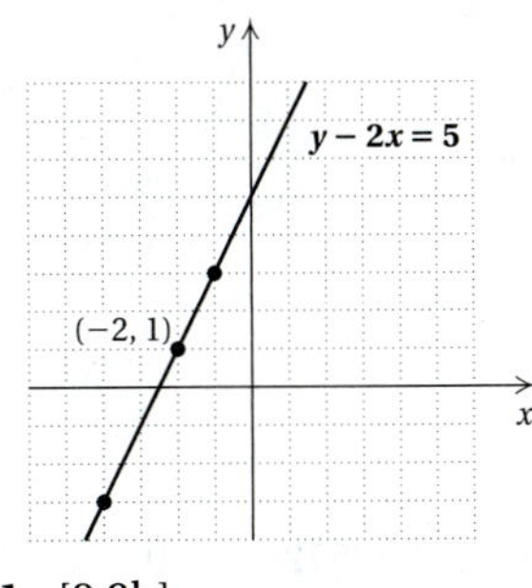

20. [8.2b]

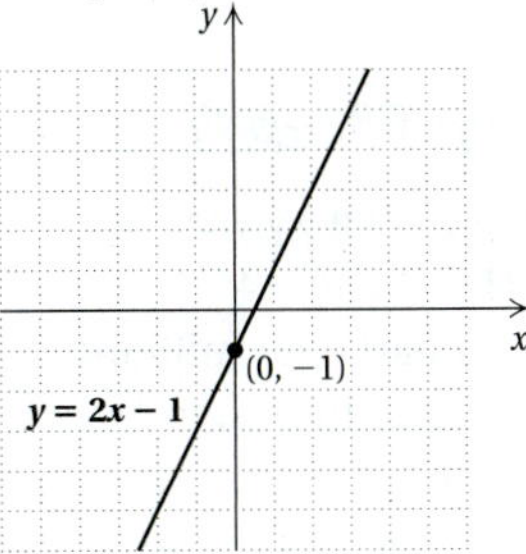

21. [8.2b]

22. [8.3b]

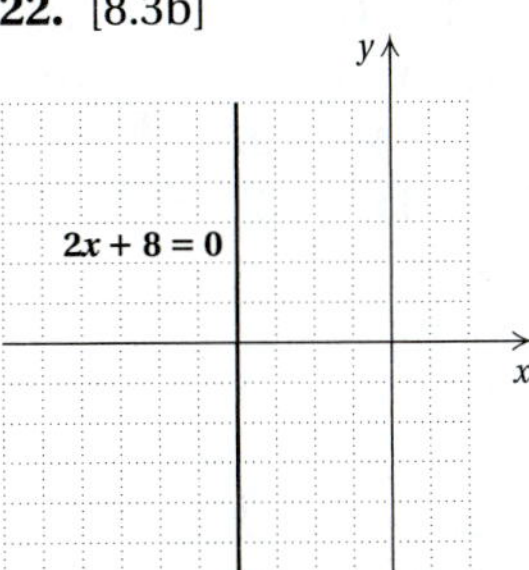

23. [8.3b]

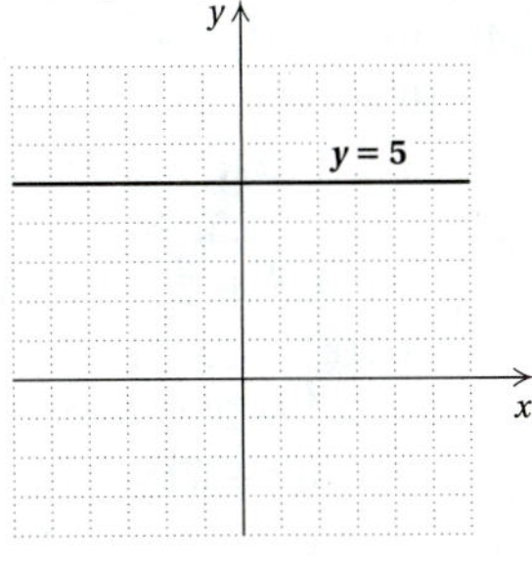

24. [8.3a]

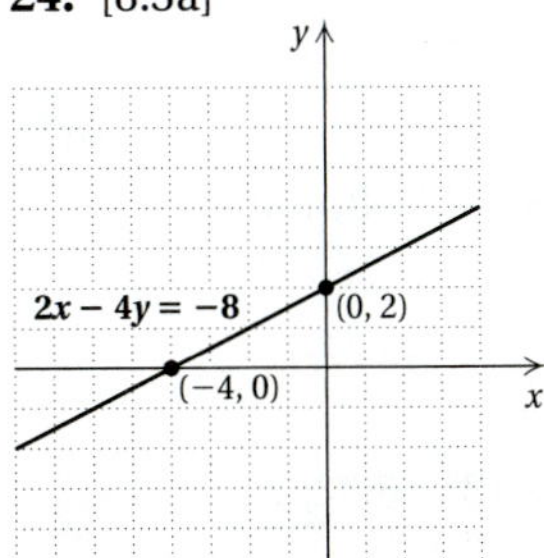

25. [8.3a]

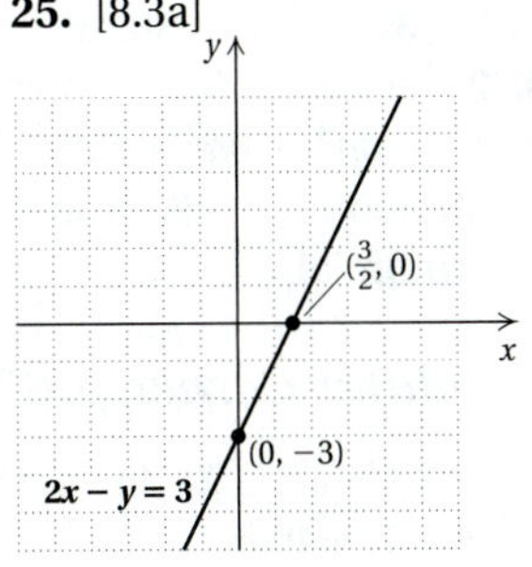

26. [8.2c] **(a)** \$17,000; \$19,400; \$22,600; \$24,200; \$27,500; **(b)** **(c)** 2002

27. [8.4b] **(a)** 14.5 floors per minute; **(b)** $4\frac{4}{29}$ seconds per floor **28.** [8.4b] 87.5 miles per hour **29.** [8.4a] −2 **30.** [8.4a] $\frac{3}{8}$;

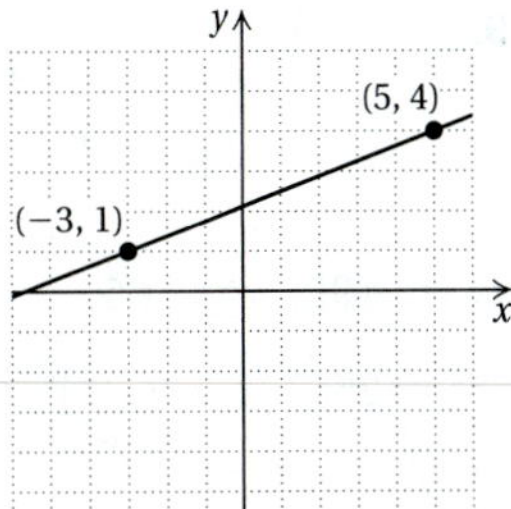

31. [8.4b] $-\frac{1}{20}$ **32.** [8.4c] **(a)** $\frac{2}{5}$; **(b)** not defined **33.** [6.2c] 0.975 **34.** [6.2c] $-1.08\overline{3}$ **35.** [6.2e] 71.2 **36.** [6.2e] $\frac{13}{47}$ **37.** [7.2a] $-\frac{3}{5}$ **38.** [7.2a] 13 **39.** [7.5a] \$84.50 **40.** [7.5a] \$36,400 **41.** [8.1b] 25 square units, 20 linear units **42.** [8.3b] $y = 3$

Chapter 9

Pretest: Chapter 9, p. 658

1. x^2 **2.** $\frac{1}{x^7}$ **3.** $\frac{16x^4}{y^6}$ **4.** $\frac{1}{p^3}$ **5.** 3.47×10^{-4} **6.** 3,400,000 **7.** 1.395×10^3 **8.** 8×10^{-2} **9.** 3, 2, 1, 0; 3 **10.** $-3a^3b - 2a^2b^2 + ab^3 + 12b^3 + 9$ **11.** $11x^2 + 4x - 11$ **12.** $-x^2 - 18x + 27$ **13.** $15x^4 - 20x^3 + 5x^2$ **14.** $x^2 + 10x + 25$ **15.** $x^2 - 25$ **16.** $4x^6 + 19x^3 - 30$ **17.** $4x^2 - 12xy + 9y^2$ **18.** $x^2 + x + 3$, R 8; or $x^2 + x + 3 + \frac{8}{x - 2}$ **19.** **(a)** $4w + 16$; **(b)** $w^2 + 8w$

Margin Exercises, Section 9.1, pp. 659–664

1. $5 \cdot 5 \cdot 5 \cdot 5$ **2.** $x \cdot x \cdot x \cdot x \cdot x$ **3.** $3t \cdot 3t$ **4.** $3 \cdot t \cdot t$ **5.** $(-x) \cdot (-x) \cdot (-x) \cdot (-x)$ **6.** 6 **7.** 1 **8.** 8.4 **9.** 1 **10.** 125 **11.** 3215.36 cm^2 **12.** 119 **13.** 3; −3 **14.** **(a)** 144; **(b)** 36; **(c)** no **15.** 3^{10} **16.** x^{10} **17.** p^{24} **18.** x^5 **19.** a^9b^8 **20.** 4^3 **21.** y^4 **22.** p^9 **23.** a^4b^2 **24.** $\frac{1}{4^3} = \frac{1}{64}$ **25.** $\frac{1}{5^2} = \frac{1}{25}$ **26.** $\frac{1}{2^4} = \frac{1}{16}$ **27.** $\frac{1}{(-2)^3} = -\frac{1}{8}$ **28.** $\frac{4}{p^3}$ **29.** x^2 **30.** 5^2 **31.** $\frac{1}{x^7}$ **32.** $\frac{1}{7^5}$ **33.** b **34.** t^6

Exercise Set 9.1, p. 665

1. $3 \cdot 3 \cdot 3 \cdot 3$ **3.** $(-1.1)(-1.1)(-1.1)(-1.1)(-1.1)$
5. $(\frac{2}{3})(\frac{2}{3})(\frac{2}{3})(\frac{2}{3})$ **7.** $(7p)(7p)$ **9.** $8 \cdot k \cdot k \cdot k$ **11.** 1
13. b **15.** 1 **17.** 1 **19.** ab **21.** ab **23.** 27
25. 19 **27.** 256 **29.** 93 **31.** 10; 4 **33.** 3629.84 ft^2
35. $\frac{1}{3^2} = \frac{1}{9}$ **37.** $\frac{1}{10^3} = \frac{1}{1000}$ **39.** $\frac{1}{7^3} = \frac{1}{343}$ **41.** $\frac{1}{a^3}$
43. $8^2 = 64$ **45.** y^4 **47.** z^n **49.** 4^{-3} **51.** x^{-3}
53. a^{-5} **55.** 2^7 **57.** 8^{14} **59.** x^7 **61.** 9^{38} **63.** $(3y)^{12}$
65. $(7y)^{17}$ **67.** 3^3 **69.** $\frac{1}{x}$ **71.** x^{17} **73.** $\frac{1}{x^{13}}$
75. $\frac{1}{a^{10}}$ **77.** 1 **79.** 7^3 **81.** 8^6 **83.** y^4 **85.** $\frac{1}{16^6}$
87. $\frac{1}{m^6}$ **89.** $\frac{1}{(8x)^4}$ **91.** 1 **93.** x^2 **95.** x^9 **97.** $\frac{1}{z^4}$
99. x^3 **101.** 1 **103.** $5^2 = 25; 5^{-2} = \frac{1}{25}; (\frac{1}{5})^2 = \frac{1}{25};$
$(\frac{1}{5})^{-2} = 25; -5^2 = -25; (-5)^2 = 25; -(-\frac{1}{5})^2 = -\frac{1}{25};$
$(-\frac{1}{5})^{-2} = 25$ **105.** $\mathbf{D_W}$ **107.** $64\%t$, or $0.64t$ **108.** 1
109. 64 **110.** 1579.5 **111.** $\frac{4}{3}$ **112.** $8(x-7)$
113. 8 in., 4 in. **114.** 228, 229 **115.** No **117.** No
119. y^{5x} **121.** a^{4t} **123.** 1 **125.** > **127.** <
129. Let $x = 2$; then $3x^2 = 12$, but $(3x)^2 = 36$.

Margin Exercises, Section 9.2, pp. 669–676

1. 3^{20} **2.** $\frac{1}{x^{12}}$ **3.** y^{15} **4.** $\frac{1}{x^{32}}$ **5.** $\frac{16x^{20}}{y^{12}}$ **6.** $\frac{25x^{10}}{y^{12}z^6}$
7. x^{74} **8.** $\frac{27z^{24}}{y^6x^{15}}$ **9.** $\frac{x^{12}}{25}$ **10.** $\frac{8t^{15}}{w^{12}}$ **11.** $\frac{9}{x^8}$
12. 5.17×10^{-4} **13.** 5.23×10^8 **14.** 689,300,000,000
15. 0.0000567 **16.** 5.6×10^{-15} **17.** 7.462×10^{-13}
18. 2.0×10^3 **19.** 5.5×10^2 **20.** 1.884672×10^{11} L
21. The mass of Saturn is 9.5×10 times the mass of Earth.

Calculator Corner, p. 674

1. 1.3545×10^{-4} **2.** 9.044×10^5 **3.** 3.2×10^5
4. 3.6×10^{12} **5.** 3×10^{-6} **6.** 4×10^5 **7.** 8×10^{-26}
8. 3×10^{13}

Exercise Set 9.2, p. 677

1. 2^6 **3.** $\frac{1}{5^6}$ **5.** x^{12} **7.** $\frac{1}{a^{18}}$ **9.** t^{18} **11.** $\frac{1}{t^{12}}$
13. x^8 **15.** a^3b^3 **17.** $\frac{1}{a^3b^3}$ **19.** $\frac{1}{m^3n^6}$ **21.** $16x^6$
23. $\frac{9}{x^8}$ **25.** $\frac{1}{x^{12}y^{15}}$ **27.** $x^{24}y^8$ **29.** $\frac{a^{10}}{b^{35}}$ **31.** $\frac{25t^6}{r^8}$
33. $\frac{b^{21}}{a^{15}c^6}$ **35.** $\frac{9x^6}{y^{16}z^6}$ **37.** $\frac{y^6}{4}$ **39.** $\frac{a^8}{b^{12}}$ **41.** $\frac{8}{y^6}$
43. $49x^6$ **45.** $\frac{x^6y^3}{z^3}$ **47.** $\frac{c^2d^6}{a^4b^2}$ **49.** 2.8×10^{10}
51. 9.07×10^{17} **53.** 3.04×10^{-6} **55.** 1.8×10^{-8}
57. 10^{11} **59.** 2.81×10^8 **61.** 10^{-7} **63.** 87,400,000
65. 0.00000005704 **67.** 10,000,000 **69.** 0.00001
71. 6×10^9 **73.** 3.38×10^4 **75.** 8.1477×10^{-13}
77. 2.5×10^{13} **79.** 5.0×10^{-4} **81.** 3.0×10^{-21}
83. Approximately 1.325×10^{14} ft^3 **85.** The mass of Jupiter is 3.18×10^2 times the mass of Earth. **87.** 1×10^{22}
89. The mass of the sun is 3.33×10^5 times the mass of Earth.
91. 4.375×10^2 days **93.** $\mathbf{D_W}$ **95.** $9(x-4)$
96. $2(2x - y + 8)$ **97.** $3(s + t + 8)$ **98.** $-7(x+2)$
99. $\frac{7}{4}$ **100.** 2 **101.** $-\frac{12}{7}$ **102.** $-\frac{11}{2}$
103.

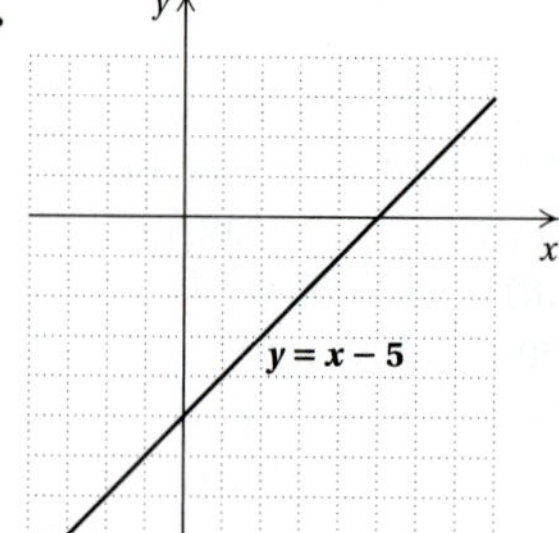

104.

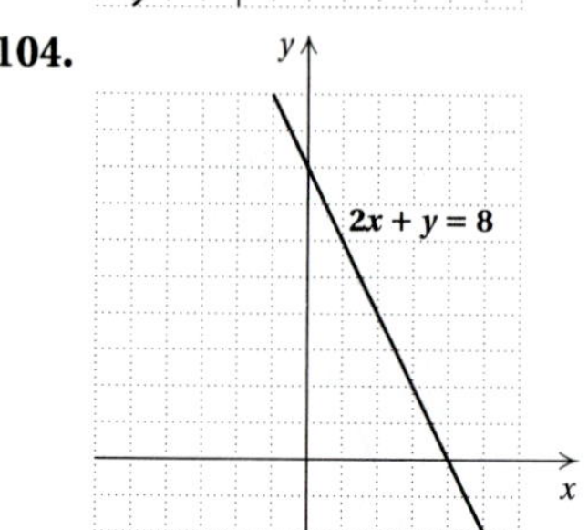

105. 2.478125×10^{-1} **107.** $\frac{1}{5}$ **109.** 3^{11} **111.** 7
113. $\frac{1}{0.4}$, or 2.5 **115.** False **117.** False

Margin Exercises, Section 9.3, pp. 682–689

1. $4x^2 - 3x + \frac{5}{4}; 15y^3; -7x^3 + 1.1$; answers may vary
2. -19 **3.** -104 **4.** -18 **5.** 21 **6.** 6; -4
7. 132 games **8.** 360 ft **9.** **(a)** 7.55 parts per million; **(b)** When $t = 3$, $C \approx 7.5$; so the value found in part (a) appears to be correct. **10.** 20 parts per million
11. $-9x^3 + (-4x^5)$ **12.** $-2y^3 + 3y^7 + (-7y)$
13. $3x^2, 6x, \frac{1}{2}$ **14.** $-4y^5, 7y^2, -3y, -2$ **15.** $4x^3$ and $-x^3$
16. $4t^4$ and $-7t^4$; $-9t^3$ and $10t^3$ **17.** $5x^2$ and $7x^2$; $3x$ and $-8x$; -10 and 11 **18.** 2, -7, -8.5, 10, -4 **19.** $8x^2$
20. $2x^3 + 7$ **21.** $-\frac{1}{4}x^5 + 2x^2$ **22.** $-4x^3$ **23.** $5x^3$
24. $25 - 3x^5$ **25.** $6x$ **26.** $4x^3 + 4$
27. $-\frac{1}{4}x^3 + 4x^2 + 7$ **28.** $3x^2 + x^3 + 9$
29. $6x^7 + 3x^5 - 2x^4 + 4x^3 + 5x^2 + x$
30. $7x^5 - 5x^4 + 2x^3 + 4x^2 - 3$
31. $14t^7 - 10t^5 + 7t^2 - 14$ **32.** $-2x^2 - 3x + 2$
33. $10x^4 - 8x - \frac{1}{2}$ **34.** 4, 2, 1, 0; 4 **35.** x **36.** x^3, x^2, x, x^0 **37.** x^2, x **38.** x^3
39. $2x^3 + 4x^2 + 0x - 2$; $2x^3 + 4x^2 \qquad - 2$
40. $a^4 + 0a^3 + 0a^2 + 0a + 10$; $a^4 \qquad + 10$
41. Monomial **42.** None of these **43.** Binomial
44. Trinomial

Calculator Corner, p. 685

1. 3; 2.25; -27 **2.** 44; 0; 9.28 **3.** 13; -3.32; 7
4. -1; -7; -40.6

Exercise Set 9.3, p. 690

1. $-18; 7$ **3.** $19; 14$ **5.** $-12; -7$ **7.** $-1; 5$ **9.** $9; 1$ **11.** $56; -2$ **13.** 1112 ft **15.** **(a)** 3.93 million gigawatt hours, 4.12 million gigawatt hours, 4.5 million gigawatt hours, 4.88 million gigawatt hours, 5.45 million gigawatt hours, 5.83 million gigawatt hours; **(b)** left to the student
17. \$18,750; \$24,000 **19.** $-4, 4, 5, 2.75, 1$
21. 1,820,000; 3,660,000 **23.** 9 words **25.** 6 **27.** 15
29. $2, -3x, x^2$ **31.** $6x^2$ and $-3x^2$
33. $2x^4$ and $-3x^4$; $5x$ and $-7x$ **35.** $3x^5$ and $14x^5$; $-7x$ and $-2x$; 8 and -9 **37.** $-3, 6$ **39.** $5, 3, 3$
41. $-5, 6, -3, 8, -2$ **43.** $-3x$ **45.** $-8x$ **47.** $11x^3 + 4$
49. $x^3 - x$ **51.** $4b^5$ **53.** $\frac{3}{4}x^5 - 2x - 42$ **55.** x^4
57. $\frac{15}{16}x^3 - \frac{7}{6}x^2$ **59.** $x^5 + 6x^3 + 2x^2 + x + 1$
61. $15y^9 + 7y^8 + 5y^3 - y^2 + y$ **63.** $x^6 + x^4$
65. $13x^3 - 9x + 8$ **67.** $-5x^2 + 9x$ **69.** $12x^4 - 2x + \frac{1}{4}$
71. $1, 0; 1$ **73.** $2, 1, 0; 2$ **75.** $3, 2, 1, 0; 3$ **77.** $2, 1, 6, 4; 6$
79.

Term	Coefficient	Degree of the Term	Degree of the Polynomial
$-7x^4$	-7	4	
$6x^3$	6	3	
$-3x^2$	-3	2	4
$8x$	8	1	
-2	-2	0	

81. x^2, x **83.** x^3, x^2, x^0 **85.** None missing
87. $x^3 + 0x^2 + 0x - 27$; $x^3 \qquad\qquad - 27$
89. $x^4 + 0x^3 + 0x^2 - x + 0x^0$; $x^4 \qquad\qquad - x$
91. None missing **93.** Trinomial **95.** None of these
97. Binomial **99.** Monomial **101.** D_W
103. 27 apples **104.** -19 **105.** $-\frac{17}{24}$ **106.** $\frac{5}{8}$
107. -2.6 **108.** $\frac{15}{2}$ **109.** $b = \dfrac{C + r}{a}$ **110.** 45%; 37.5%; 17.5% **111.** $3(x - 5y + 21)$ **113.** $3x^6$ **115.** 10
117. $-4, 4, 5, 2.75, 1$ **119.** 1,820,000; 3,660,000

Margin Exercises, Section 9.4, pp. 696–699

1. $x^2 + 7x + 3$ **2.** $-4x^5 + 7x^4 + x^3 + 2x^2 + 4$
3. $24x^4 + 5x^3 + x^2 + 1$ **4.** $2x^3 + \frac{10}{3}$ **5.** $2x^2 - 3x - 1$
6. $8x^3 - 2x^2 - 8x + \frac{5}{2}$ **7.** $-8x^4 + 4x^3 + 12x^2 + 5x - 8$
8. $-x^3 + x^2 + 3x + 3$ **9.** $-4x^3 + 6x - 3$
10. $-5x^4 - 3x^2 - 7x + 5$
11. $-14x^{10} + \frac{1}{2}x^5 - 5x^3 + x^2 - 3x$ **12.** $2x^3 + 2x + 8$
13. $x^2 - 6x - 2$ **14.** $-8x^4 - 5x^3 + 8x^2 - 1$
15. $x^3 - x^2 - \frac{4}{3}x - 0.9$ **16.** $2x^3 + 5x^2 - 2x - 5$
17. $-x^5 - 2x^3 + 3x^2 - 2x + 2$ **18.** Sum of perimeters: $13x$; sum of areas: $\frac{7}{2}x^2$ **19.** $x^2 - 64$ ft^2

Calculator Corner, p. 699

1. Yes **2.** Yes **3.** No **4.** Yes **5.** No **6.** Yes

Exercise Set 9.4, p. 700

1. $-x + 5$ **3.** $x^2 - 5x - 1$ **5.** $2x^2$ **7.** $5x^2 + 3x - 30$
9. $-2.2x^3 - 0.2x^2 - 3.8x + 23$ **11.** $6 + 12x^2$
13. $-\frac{1}{2}x^4 + \frac{2}{3}x^3 + x^2$
15. $0.01x^5 + x^4 - 0.2x^3 + 0.2x + 0.06$
17. $9x^8 + 8x^7 - 6x^4 + 8x^2 + 4$
19. $1.05x^4 + 0.36x^3 + 14.22x^2 + x + 0.97$ **21.** $5x$
23. $x^2 - 10x + 2$ **25.** $-12x^4 + 3x^3 - 3$ **27.** $-3x + 7$
29. $-4x^2 + 3x - 2$ **31.** $4x^4 - 6x^2 - \frac{3}{4}x + 8$
33. $7x - 1$ **35.** $-x^2 - 7x + 5$ **37.** -18
39. $6x^4 + 3x^3 - 4x^2 + 3x - 4$
41. $4.6x^3 + 9.2x^2 - 3.8x - 23$ **43.** $\frac{3}{4}x^3 - \frac{1}{2}x$
45. $0.06x^3 - 0.05x^2 + 0.01x + 1$ **47.** $3x + 6$
49. $11x^4 + 12x^3 - 9x^2 - 8x - 9$ **51.** $x^4 - x^3 + x^2 - x$
53. $5x^2 + 4x$ **55.** $\frac{23}{2}a + 12$ **57.** $(r + 11)(r + 9)$; $9r + 99 + 11r + r^2$, or $r^2 + 20r + 99$
59. $(x + 3)(x + 3)$, or $(x + 3)^2$; $x^2 + 3x + 9 + 3x$, or $x^2 + 6x + 9$ **61.** $\pi r^2 - 25\pi$ **63.** $18z - 64$ **65.** D_W
67. 6 **68.** -19 **69.** $-\frac{7}{22}$ **70.** 5 **71.** 5 **72.** 1
73. $\frac{39}{2}$ **74.** $\frac{37}{2}$ **75.** $\{x \mid x \geq -10\}$ **76.** $\{x \mid x < 0\}$
77. $20w + 42$ **79.** $2x^2 + 20x$ **81.** $y^2 - 4y + 4$
83. $12y^2 - 23y + 21$ **85.** $-3y^4 - y^3 + 5y - 2$

Margin Exercises, Section 9.5, pp. 704–707

1. $-15x$ **2.** $-x^2$ **3.** x^2 **4.** $-x^5$ **5.** $12x^7$ **6.** $-8y^{11}$
7. $7y^5$ **8.** 0 **9.** $8x^2 + 16x$ **10.** $-15t^3 + 6t^2$
11. $-5x^6 - 25x^5 + 30x^4 - 40x^3$
12. **(a)**
$$\begin{aligned}(y + 2)(y + 7) &= y \cdot (y + 7) + 2(y + 7)\\ &= y \cdot y + y \cdot 7 + 2 \cdot y + 2 \cdot 7\\ &= y^2 + 7y + 2y + 14\\ &= y^2 + 9y + 14;\end{aligned}$$
(b) $y^2 + 2y + 7y + 14$
13. $x^2 + 13x + 40$ **14.** $x^2 + x - 20$
15. $5x^2 - 17x - 12$ **16.** $6x^2 - 19x + 15$
17. $x^4 + 3x^3 + x^2 + 15x - 20$
18. $6y^5 - 20y^3 + 15y^2 + 14y - 35$
19. $3x^3 + 13x^2 - 6x + 20$
20. $20x^4 - 16x^3 + 32x^2 - 32x - 16$
21. $6x^4 - x^3 - 18x^2 - x + 10$

Calculator Corner, p. 707

1. Correct **2.** Correct **3.** Not correct **4.** Not correct

Exercise Set 9.5, p. 708

1. $40x^2$ **3.** x^3 **5.** $32x^8$ **7.** $0.03x^{11}$ **9.** $\frac{1}{15}x^4$ **11.** 0
13. $-24x^{11}$ **15.** $-2x^2 + 10x$ **17.** $-5x^2 + 5x$
19. $x^5 + x^2$ **21.** $6x^3 - 18x^2 + 3x$ **23.** $-6x^4 - 6x^3$
25. $18y^6 + 24y^5$ **27.** $x^2 + 9x + 18$ **29.** $x^2 + 3x - 10$
31. $x^2 - 7x + 12$ **33.** $x^2 - 9$ **35.** $25 - 15x + 2x^2$
37. $4x^2 + 20x + 25$ **39.** $x^2 - \frac{21}{10}x - 1$

41. $x^2 + 2.4x - 10.81$ **43.**

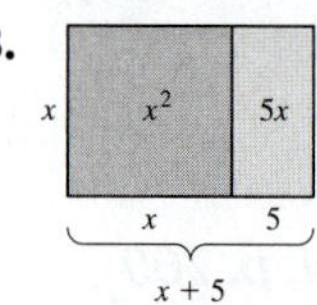

45.

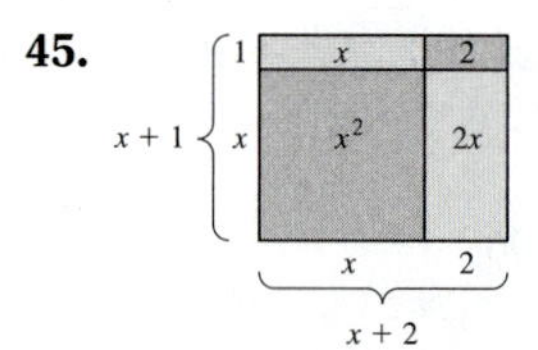

47.

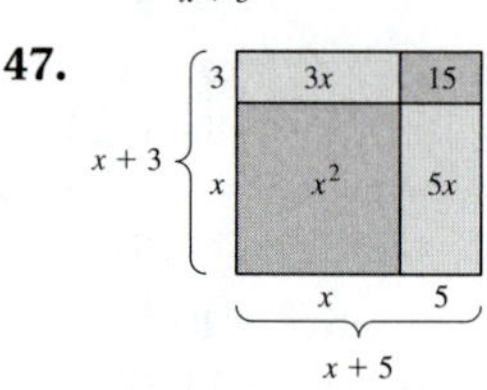

49.

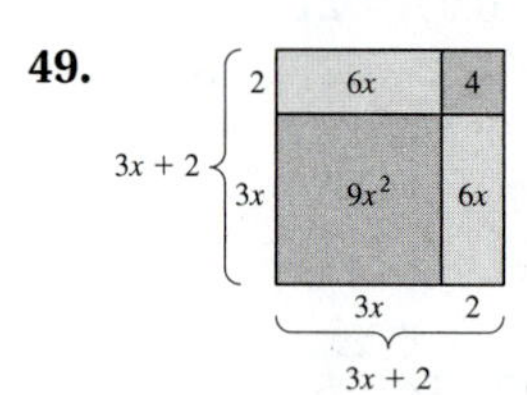

51. $x^3 - 1$

53. $4x^3 + 14x^2 + 8x + 1$ **55.** $3y^4 - 6y^3 - 7y^2 + 18y - 6$ **57.** $x^6 + 2x^5 - x^3$ **59.** $-10x^5 - 9x^4 + 7x^3 + 2x^2 - x$ **61.** $-1 - 2x - x^2 + x^4$ **63.** $6t^4 + t^3 - 16t^2 - 7t + 4$ **65.** $x^9 - x^5 + 2x^3 - x$ **67.** $x^4 - 1$ **69.** $x^4 + 8x^3 + 12x^2 + 9x + 4$ **71.** $2x^4 - 5x^3 + 5x^2 - \frac{19}{10}x + \frac{1}{5}$ **73.** D_W **75.** $-\frac{3}{4}$ **76.** 6.4 **77.** 96 **78.** 32 **79.** $3(5x - 6y + 4)$ **80.** $4(4x - 6y + 9)$ **81.** $-3(3x + 15y - 5)$ **82.** $100(x - y + 10a)$ **83.**

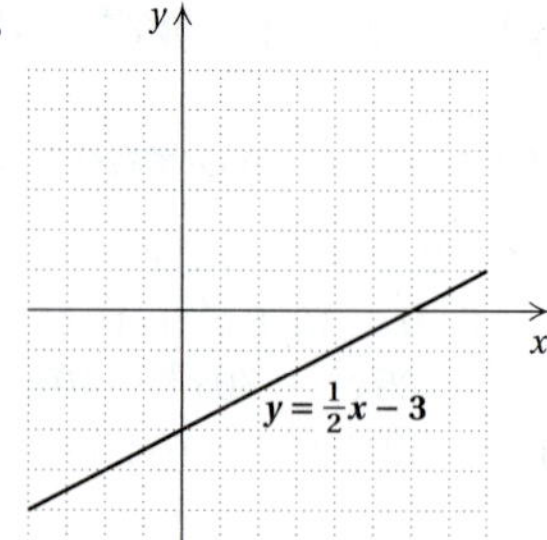

84. $\frac{23}{19}$ **85.** $75y^2 - 45y$ **87.** $V = 4x^3 - 48x^2 + 144x$ in^3; $S = -4x^2 + 144$ in^2 **89.** 5 **91.** $x^3 + 2x^2 - 210$ m^3 **93.** 0 **95.** 0

Margin Exercises, Section 9.6, pp. 712–716

1. $x^2 + 7x + 12$ **2.** $x^2 - 2x - 15$ **3.** $2x^2 - 9x + 4$ **4.** $2x^3 - 4x^2 - 3x + 6$ **5.** $12x^5 + 10x^3 + 6x^2 + 5$ **6.** $y^6 - 49$ **7.** $t^2 + 8t + 15$ **8.** $-2x^7 + x^5 + x^3$ **9.** $x^2 - \frac{16}{25}$ **10.** $x^5 + 0.5x^3 - 0.5x^2 - 0.25$ **11.** $8 + 2x^2 - 15x^4$ **12.** $30x^5 - 27x^4 + 6x^3$ **13.** $x^2 - 25$ **14.** $4x^2 - 9$ **15.** $x^2 - 4$ **16.** $x^2 - 49$ **17.** $36 - 16y^2$ **18.** $4x^6 - 1$ **19.** $x^2 - \frac{4}{25}$ **20.** $x^2 + 16x + 64$ **21.** $x^2 - 10x + 25$ **22.** $x^2 + 4x + 4$ **23.** $a^2 - 8a + 16$ **24.** $4x^2 + 20x + 25$ **25.** $16x^4 - 24x^3 + 9x^2$ **26.** $60.84 + 18.72y + 1.44y^2$ **27.** $9x^4 - 30x^2 + 25$ **28.** $x^2 + 11x + 30$ **29.** $t^2 - 16$ **30.** $-8x^5 + 20x^4 + 40x^2$ **31.** $81x^4 + 18x^2 + 1$ **32.** $4a^2 + 6a - 40$ **33.** $25x^2 + 5x + \frac{1}{4}$ **34.** $4x^2 - 2x + \frac{1}{4}$ **35.** $x^3 - 3x^2 + 6x - 8$

Exercise Set 9.6, p. 717

1. $x^3 + x^2 + 3x + 3$ **3.** $x^4 + x^3 + 2x + 2$ **5.** $y^2 - y - 6$ **7.** $9x^2 + 12x + 4$ **9.** $5x^2 + 4x - 12$ **11.** $9t^2 - 1$ **13.** $4x^2 - 6x + 2$ **15.** $p^2 - \frac{1}{16}$ **17.** $x^2 - 0.01$ **19.** $2x^3 + 2x^2 + 6x + 6$ **21.** $-2x^2 - 11x + 6$ **23.** $a^2 + 14a + 49$ **25.** $1 - x - 6x^2$ **27.** $x^5 + 3x^3 - x^2 - 3$ **29.** $3x^6 - 2x^4 - 6x^2 + 4$ **31.** $13.16x^2 + 18.99x - 13.95$ **33.** $6x^7 + 18x^5 + 4x^2 + 12$ **35.** $8x^6 + 65x^3 + 8$ **37.** $4x^3 - 12x^2 + 3x - 9$ **39.** $4y^6 + 4y^5 + y^4 + y^3$ **41.** $x^2 - 16$ **43.** $4x^2 - 1$ **45.** $25m^2 - 4$ **47.** $4x^4 - 9$ **49.** $9x^8 - 16$ **51.** $x^{12} - x^4$ **53.** $x^8 - 9x^2$ **55.** $x^{24} - 9$ **57.** $4y^{16} - 9$ **59.** $\frac{25}{64}x^2 - 18.49$ **61.** $x^2 + 4x + 4$ **63.** $9x^4 + 6x^2 + 1$ **65.** $a^2 - a + \frac{1}{4}$ **67.** $9 + 6x + x^2$ **69.** $x^4 + 2x^2 + 1$ **71.** $4 - 12x^4 + 9x^8$ **73.** $25 + 60t^2 + 36t^4$ **75.** $x^2 - \frac{5}{4}x + \frac{25}{64}$ **77.** $9 - 12x^3 + 4x^6$ **79.** $4x^3 + 24x^2 - 12x$ **81.** $4x^4 - 2x^2 + \frac{1}{4}$ **83.** $9p^2 - 1$ **85.** $15t^5 - 3t^4 + 3t^3$ **87.** $36x^8 + 48x^4 + 16$ **89.** $12x^3 + 8x^2 + 15x + 10$ **91.** $64 - 96x^4 + 36x^8$ **93.** $t^3 - 1$ **95.** 25; 49 **97.** 56; 16 **99.** $a^2 + 2a + 1$ **101.** $t^2 + 10t + 24$ **103.** D_W **105.** Lamps: 500 watts; air conditioner: 2000 watts; television: 50 watts **106.** $\frac{28}{27}$ **107.** $-\frac{41}{7}$ **108.** $\frac{27}{4}$ **109.** $y = \dfrac{3x - 12}{2}$, or $y = \dfrac{3}{2}x - 6$ **110.** $a = \dfrac{5d + 4}{3}$, or $a = \dfrac{5}{3}d + \dfrac{4}{3}$ **111.** $30x^3 + 35x^2 - 15x$ **113.** $a^4 - 50a^2 + 625$ **115.** $81t^{16} - 72t^8 + 16$ **117.** -7 **119.** First row: 90, -432, -63; second row: 7, -18, -36, -14, 12, -6, -21, -11; third row: 9, -2, -2, 10, -8, -8, -8, -10, 21; fourth row: -19, -6 **121.** Yes **123.** No

Margin Exercises, Section 9.7, pp. 721–724

1. -7940 **2.** -176 **3.** 1889 calories **4.** $-3, 3, -2, 1, 2$ **5.** 3, 7, 1, 1, 0; 7 **6.** $2x^2y + 3xy$ **7.** $5pq - 8$ **8.** $-4x^3 + 2x^2 - 4y + 2$ **9.** $14x^3y + 7x^2y - 3xy - 2y$ **10.** $-5p^2q^4 + 2p^2q^2 + 3p^2q + 6pq^2 + 3q + 5$ **11.** $-8s^4t + 6s^3t^2 + 2s^2t^3 - s^2t^2$ **12.** $-9p^4q + 9p^3q^2 - 4p^2q^3 - 9q^4 + 5$ **13.** $x^5y^5 + 2x^4y^2 + 3x^3y^3 + 6x^2$ **14.** $p^5q - 4p^3q^3 + 3pq^3 + 6q^4$ **15.** $3x^3y + 6x^2y^3 + 2x^3 + 4x^2y^2$ **16.** $2x^2 - 11xy + 15y^2$ **17.** $16x^2 + 40xy + 25y^2$ **18.** $9x^4 - 12x^3y^2 + 4x^2y^4$ **19.** $4x^2y^4 - 9x^2$ **20.** $16y^2 - 9x^2y^4$ **21.** $9y^2 + 24y + 16 - 9x^2$ **22.** $4a^2 - 25b^2 - 10bc - c^2$

Exercise Set 9.7, p. 725

1. -1 **3.** -15 **5.** 240 **7.** -145 **9.** 3.715 liters **11.** 110.4 m **13.** 44.46 in^2 **15.** 63.78125 in^2 **17.** Coefficients: 1, -2, 3, -5; degrees: 4, 2, 2, 0; 4 **19.** Coefficients: 17, -3, -7; degrees: 5, 5, 0; 5

21. $-a - 2b$ **23.** $3x^2y - 2xy^2 + x^2$ **25.** $20au + 10av$
27. $8u^2v - 5uv^2$ **29.** $x^2 - 4xy + 3y^2$ **31.** $3r + 7$
33. $-a^3b^2 - 3a^2b^3 + 5ab + 3$ **35.** $ab^2 - a^2b$
37. $2ab - 2$ **39.** $-2a + 10b - 5c + 8d$
41. $6z^2 + 7zu - 3u^2$ **43.** $a^4b^2 - 7a^2b + 10$
45. $a^6 - b^2c^2$ **47.** $y^6x + y^4x + y^4 + 2y^2 + 1$
49. $12x^2y^2 + 2xy - 2$ **51.** $12 - c^2d^2 - c^4d^4$
53. $m^3 + m^2n - mn^2 - n^3$
55. $x^9y^9 - x^6y^6 + x^5y^5 - x^2y^2$ **57.** $x^2 + 2xh + h^2$
59. $r^6t^4 - 8r^3t^2 + 16$ **61.** $p^8 + 2m^2n^2p^4 + m^4n^4$
63. $4a^6 - 2a^3b^3 + \frac{1}{4}b^6$ **65.** $3a^3 - 12a^2b + 12ab^2$
67. $4a^2 - b^2$ **69.** $c^4 - d^2$ **71.** $a^2b^2 - c^2d^4$
73. $x^2 + 2xy + y^2 - 9$ **75.** $x^2 - y^2 - 2yz - z^2$
77. $a^2 - b^2 - 2bc - c^2$ **79.** $\mathbf{D_W}$ **81.** IV **82.** III
83. I **84.** II **85.**

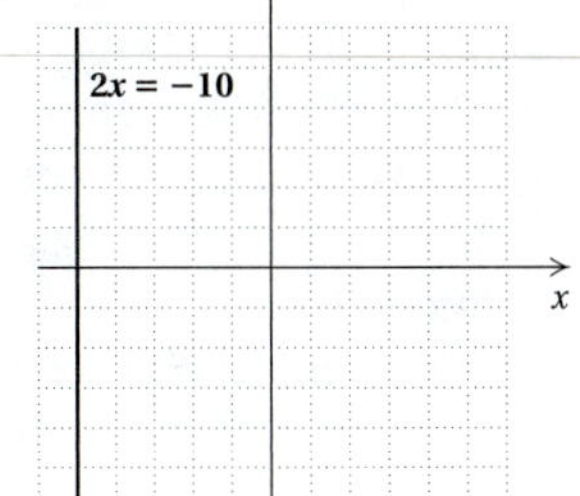

86. 

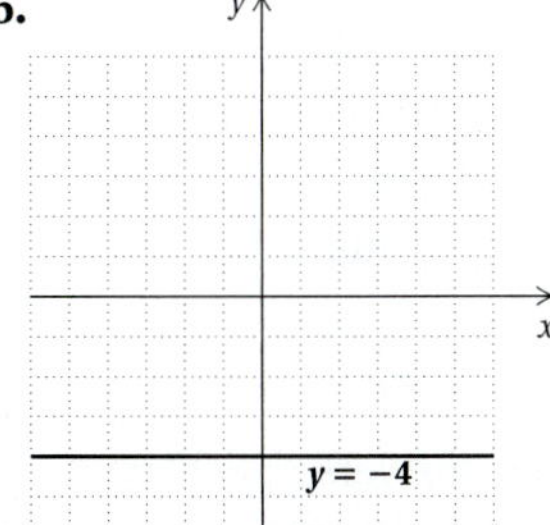

87.

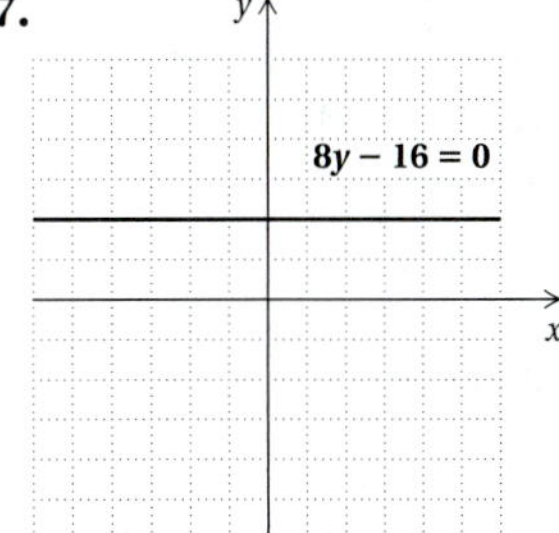

88.

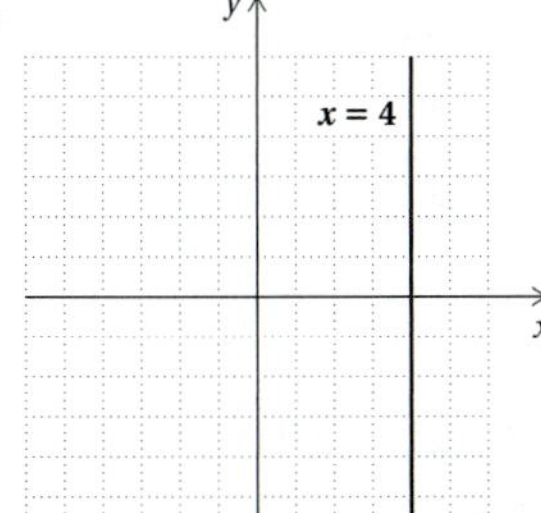

89. $4xy - 4y^2$

91. $2xy + \pi x^2$ **93.** $2\pi nh + 2\pi mh + 2\pi n^2 - 2\pi m^2$
95. 16 gal **97.** \$15,638.03

Margin Exercises, Section 9.8, pp. 730–735

1. $4x^2$ **2.** $-7x^{11}$ **3.** $-28p^3q$ **4.** $\frac{1}{4}x^4$ **5.** $7x^4 + 8x^2$
6. $x^2 + 3x + 2$ **7.** $2x^2 + x - \frac{2}{3}$ **8.** $4x^2 - \frac{3}{2}x + \frac{1}{2}$
9. $2x^2y^4 - 3xy^2 + 5y$ **10.** $x - 2$ **11.** $x + 4$
12. $x + 4$, R -2, or $x + 4 + \dfrac{-2}{x + 3}$ **13.** $x^2 + x + 1$
14. $2x^2 + 2x + 14$, R 34, or $2x^2 + 2x + 14 + \dfrac{34}{x - 3}$
15. $x^2 - 4x + 13$, R -30, or $x^2 - 4x + 13 + \dfrac{-30}{x + 2}$
16. $y^2 - y + 1$

Exercise Set 9.8, p. 736

1. $3x^4$ **3.** $5x$ **5.** $18x^3$ **7.** $4a^3b$
9. $3x^4 - \frac{1}{2}x^3 + \frac{1}{8}x^2 - 2$ **11.** $1 - 2u - u^4$
13. $5t^2 + 8t - 2$ **15.** $-4x^4 + 4x^2 + 1$
17. $6x^2 - 10x + \frac{3}{2}$ **19.** $9x^2 - \frac{5}{2}x + 1$
21. $6x^2 + 13x + 4$ **23.** $3rs + r - 2s$ **25.** $x + 2$
27. $x - 5 + \dfrac{-50}{x - 5}$ **29.** $x - 2 + \dfrac{-2}{x + 6}$ **31.** $x - 3$
33. $x^4 - x^3 + x^2 - x + 1$ **35.** $2x^2 - 7x + 4$
37. $x^3 - 6$ **39.** $x^3 + 2x^2 + 4x + 8$ **41.** $t^2 + 1$
43. $x^2 - x + 1 + \dfrac{-4}{x - 1}$ **45.** $a + 7 + \dfrac{-47}{a + 4}$
47. $x^2 - 5x - 23 + \dfrac{-43}{x - 2}$ **49.** $3x^2 - 2x + 2 + \dfrac{-3}{x + 3}$
51. $y^2 + 2y + 1 + \dfrac{12}{y - 2}$ **53.** $3x^3 + 9x^2 + 2x + 6$
55. $x^2 + 2x + 4$ **57.** $y^3 + 2y^2 + 4y + 8$ **59.** $\mathbf{D_W}$
61. -28 **62.** -59 **63.** 6.8 **64.** $-\frac{11}{8}$
65. 25, 543.75 ft^2 **66.** 51°, 27°, 102° **67.** $\frac{23}{14}$ **68.** $\frac{11}{10}$
69. $4(x - 3 + 6y)$ **70.** $2(128 - a - 2b)$ **71.** $x^2 + 5$
73. $a + 3 + \dfrac{5}{5a^2 - 7a - 2}$ **75.** $2x^2 + x - 3$
77. $a^5 + a^4b + a^3b^2 + a^2b^3 + ab^4 + b^5$ **79.** -5 **81.** 1

Summary and Review: Chapter 9, p. 740

1. $\dfrac{1}{7^2}$ **2.** y^{11} **3.** $(3x)^{14}$ **4.** t^8 **5.** 4^3 **6.** $\dfrac{1}{a^3}$ **7.** 1
8. $9t^8$ **9.** $36x^8$ **10.** $\dfrac{y^3}{8x^3}$ **11.** t^{-5} **12.** $\dfrac{1}{y^4}$
13. 3.28×10^{-5} **14.** 8,300,000 **15.** 2.09×10^4
16. 5.12×10^{-5} **17.** 4.4676×10^9 gal **18.** 10
19. $-4y^5, 7y^2, -3y, -2$ **20.** x^2, x^0 **21.** 3, 2, 1, 0; 3
22. Binomial **23.** None of these **24.** Monomial
25. $-2x^2 - 3x + 2$ **26.** $10x^4 - 7x^2 - x - \frac{1}{2}$
27. $x^5 - 2x^4 + 6x^3 + 3x^2 - 9$
28. $-2x^5 - 6x^4 - 2x^3 - 2x^2 + 2$ **29.** $2x^2 - 4x$
30. $x^5 - 3x^3 - x^2 + 8$ **31.** Perimeter: $4w + 6$; area: $w^2 + 3w$ **32.** $(t + 3)(t + 4)$, $t^2 + 7t + 12$
33. $x^2 + \frac{7}{6}x + \frac{1}{3}$ **34.** $49x^2 + 14x + 1$
35. $12x^3 - 23x^2 + 13x - 2$ **36.** $9x^4 - 16$
37. $15x^7 - 40x^6 + 50x^5 + 10x^4$ **38.** $x^2 - 3x - 28$
39. $9y^4 - 12y^3 + 4y^2$ **40.** $2t^4 - 11t^2 - 21$ **41.** 49
42. Coefficients: 1, −7, 9, −8; degrees: 6, 2, 2, 0; 6
43. $-y + 9w - 5$
44. $m^6 - 2m^2n + 2m^2n^2 + 8n^2m - 6m^3$
45. $-9xy - 2y^2$ **46.** $11x^3y^2 - 8x^2y - 6x^2 - 6x + 6$
47. $p^3 - q^3$ **48.** $9a^8 - 2a^4b^3 + \frac{1}{9}b^6$ **49.** $5x^2 - \frac{1}{2}x + 3$
50. $3x^2 - 7x + 4 + \dfrac{1}{2x + 3}$ **51.** $x^2 + 9x + 40 + \dfrac{153}{x - 4}$

52. $3x^3 - 8x^2 + 8x - 6 + \frac{-1}{x+1}$ **53.** 0, 3.75, −3.75, 0, 2.25
54. $\mathbf{D_W}$ 578.6×10^{-7} is not in scientific notation because 578.6 is larger than 10. **55.** $\mathbf{D_W}$ A monomial is an expression of the type ax^n, where n is a whole number and a is a real number. A binomial is a sum of two monomials and has two terms. A trinomial is a sum of three monomials and has three terms. A general polynomial is a monomial or a sum of monomials and has one or more terms.
56. $25(t - 2 + 4m)$ **57.** $\frac{9}{4}$ **58.** −12 **59.** −11.2
60. Width: 125.5 m; length: 144.5 m **61.** $\frac{1}{2}x^2 - \frac{1}{2}y^2$
62. $400 - 4a^2$ **63.** $-28x^8$ **64.** $\frac{94}{13}$
65. $x^4 + x^3 + x^2 + x + 1$ **66.** 16 ft by 8 ft

Test: Chapter 9, p. 743

1. [9.1d, f] $\frac{1}{6^5}$ **2.** [9.1d] x^9 **3.** [9.1d] $(4a)^{11}$
4. [9.1e] 3^3 **5.** [9.1e, f] $\frac{1}{x^5}$ **6.** [9.1b, e] 1 **7.** [9.2a] x^6
8. [9.2a, b] $-27y^6$ **9.** [9.2a, b] $16a^{12}b^4$ **10.** [9.2b] $\frac{a^3b^3}{c^3}$
11. [9.1d], [9.2a, b] $-216x^{21}$ **12.** [9.1d], [9.2a, b] $-24x^{21}$
13. [9.1d], [9.2a, b] $162x^{10}$ **14.** [9.1d], [9.2a, b] $324x^{10}$
15. [9.1f] $\frac{1}{5^3}$ **16.** [9.1f] y^{-8} **17.** [9.2c] 3.9×10^9
18. [9.2c] 0.00000005 **19.** [9.2d] 1.75×10^{17}
20. [9.2d] 1.296×10^{22} **21.** [9.2e] 1.5×10^4
22. [9.3a] −43 **23.** [9.3d] $\frac{1}{3}$, −1, 7 **24.** [9.3g] 3, 0, 1, 6; 6
25. [9.3i] Binomial **26.** [9.3e] $5a^2 - 6$
27. [9.3e] $\frac{7}{4}y^2 - 4y$ **28.** [9.3f] $x^5 + 2x^3 + 4x^2 - 8x + 3$
29. [9.4a] $4x^5 + x^4 + 2x^3 - 8x^2 + 2x - 7$
30. [9.4a] $5x^4 + 5x^2 + x + 5$
31. [9.4c] $-4x^4 + x^3 - 8x - 3$
32. [9.4c] $-x^5 + 0.7x^3 - 0.8x^2 - 21$
33. [9.5b] $-12x^4 + 9x^3 + 15x^2$ **34.** [9.6c] $x^2 - \frac{2}{3}x + \frac{1}{9}$
35. [9.6b] $9x^2 - 100$ **36.** [9.6a] $3b^2 - 4b - 15$
37. [9.6a] $x^{14} - 4x^8 + 4x^6 - 16$
38. [9.6a] $48 + 34y - 5y^2$ **39.** [9.5d] $6x^3 - 7x^2 - 11x - 3$
40. [9.6c] $25t^2 + 20t + 4$
41. [9.7c] $-5x^3y - y^3 + xy^3 - x^2y^2 + 19$
42. [9.7e] $8a^2b^2 + 6ab - 4b^3 + 6ab^2 + ab^3$
43. [9.7f] $9x^{10} - 16y^{10}$ **44.** [9.8a] $4x^2 + 3x - 5$
45. [9.8b] $2x^2 - 4x - 2 + \frac{17}{3x+2}$
46. [9.8c] $x^2 + 6x + 20 + \frac{54}{x-3}$
47. [9.8c] $4x^2 - 26x + 130 + \frac{-659}{x-5}$
48. [9.3a] 3, 1.5, −3.5, −5, −5.25 **49.** [9.4d] $28a + 90$
50. [9.4d] $(5t + 2)(5t + 2)$, $25t^2 + 20t + 4$ **51.** [7.3b] 13
52. [7.3c] −3 **53.** [6.7d] $16(4t - 2m + 1)$ **54.** [6.4a] $\frac{23}{20}$
55. [7.6a] 100°, 25°, 55° **56.** [9.5b], [9.6a] $V = l^3 - 3l^2 + 2l$
57. [7.3b], [9.6b, c] $-\frac{61}{12}$

Chapter 10

Pretest: Chapter 10, p. 748

1. $4(-5x^6)$, $(-2x^3)(10x^3)$, $x^2(-20x^4)$; answers may vary
2. $2(x + 1)^2$ **3.** $(x + 4)(x + 2)$ **4.** $4a(2a^4 + a^2 - 5)$
5. $(5x + 2)(x - 3)$ **6.** $(9 + z^2)(3 + z)(3 - z)$
7. $(y^3 - 2)^2$ **8.** $(x^2 + 4)(3x + 2)$ **9.** $(p - 6)(p + 5)$
10. $(x^2y + 8)(x^2y - 8)$ **11.** $(2p - q)(p + 4q)$
12. $(y - 4)(y^2 + 4y + 16)$ **13.** 0, 5 **14.** 4, $\frac{3}{5}$
15. $\frac{2}{3}$, −4 **16.** −3, 3 **17.** Base: 8 cm; height: 11 cm
18. Length: 18 in.; width: 9 in. **19.** 5 ft

Margin Exercises, Section 10.1, pp. 749–752

1. **(a)** $12x^2$; **(b)** $(3x)(4x)$, $(2x)(6x)$; answers may vary
2. **(a)** $-16x^3$; **(b)** $(2x)(-8x^2)$, $(-4x)(4x^2)$; answers may vary
3. $(8x)(x^3)$, $(4x^2)(2x^2)$, $(2x^3)(4x)$; answers may vary
4. $(-7x)(3x)$, $(7x)(-3x)$, $(-21x)(x)$; answers may vary
5. $(6x^4)(x)$, $(-2x^3)(-3x^2)$, $(3x^3)(2x^2)$; answers may vary
6. **(a)** $3x + 6$; **(b)** $3(x + 2)$
7. **(a)** $2x^3 + 10x^2 + 8x$; **(b)** $2x(x^2 + 5x + 4)$ **8.** $x(x + 3)$
9. $y^2(3y^4 - 5y + 2)$ **10.** $3x^2(3x^2 - 5x + 1)$
11. $\frac{1}{4}(3t^3 + 5t^2 + 7t + 1)$ **12.** $7x^3(5x^4 - 7x^3 + 2x^2 - 9)$
13. $2.8(3x^2 - 2x + 1)$ **14.** $(x^2 + 3)(x + 7)$
15. $(x^2 + 2)(a + b)$ **16.** $(x^2 + 3)(x + 7)$
17. $(2t^2 + 3)(4t + 1)$ **18.** $(3m^3 + 2)(m^2 - 5)$
19. $(3x^2 - 1)(x - 2)$ **20.** $(2x^2 - 3)(2x - 3)$
21. Not factorable using factoring by grouping

Exercise Set 10.1, p. 753

1. $(4x^2)(2x)$, $(-8)(-x^3)$, $(2x^2)(4x)$; answers may vary
3. $(-5a^5)(2a)$, $(10a^3)(-a^3)$, $(-2a^2)(5a^4)$; answers may vary
5. $(8x^2)(3x^2)$, $(-8x^2)(-3x^2)$, $(4x^3)(6x)$; answers may vary
7. $x(x - 6)$ **9.** $2x(x + 3)$ **11.** $x^2(x + 6)$
13. $8x^2(x^2 - 3)$ **15.** $2(x^2 + x - 4)$
17. $17xy(x^4y^2 + 2x^2y + 3)$ **19.** $x^2(6x^2 - 10x + 3)$
21. $x^2y^2(x^3y^3 + x^2y + xy - 1)$
23. $2x^3(x^4 - x^3 - 32x^2 + 2)$
25. $0.8x(2x^3 - 3x^2 + 4x + 8)$
27. $\frac{1}{3}x^3(5x^3 + 4x^2 + x + 1)$ **29.** $(x^2 + 2)(x + 3)$
31. $(5a^3 - 1)(2a - 7)$ **33.** $(x^2 + 2)(x + 3)$
35. $(2x^2 + 1)(x + 3)$ **37.** $(4x^2 + 3)(2x - 3)$
39. $(4x^2 + 1)(3x - 4)$ **41.** $(5x^2 - 1)(x - 1)$
43. $(x^2 - 3)(x + 8)$ **45.** $(2x^2 - 9)(x - 4)$ **47.** $\mathbf{D_W}$
49. $\{x | x > -24\}$ **50.** $\{x | x \le \frac{14}{5}\}$ **51.** 27
52. $p = 2A - q$ **53.** $y^2 + 12y + 35$ **54.** $y^2 + 14y + 49$
55. $y^2 - 49$ **56.** $y^2 - 14y + 49$

57. 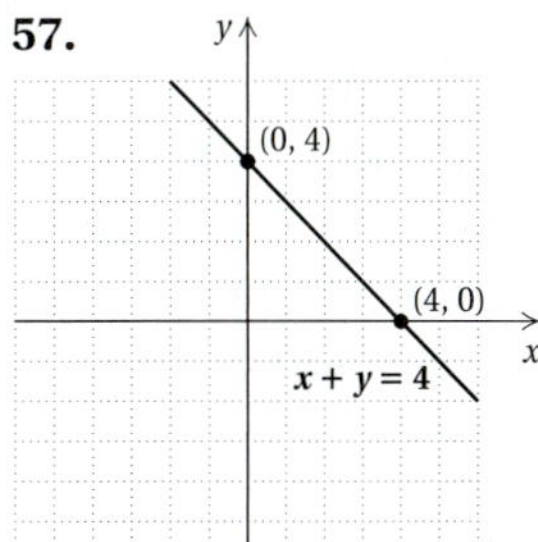

58. 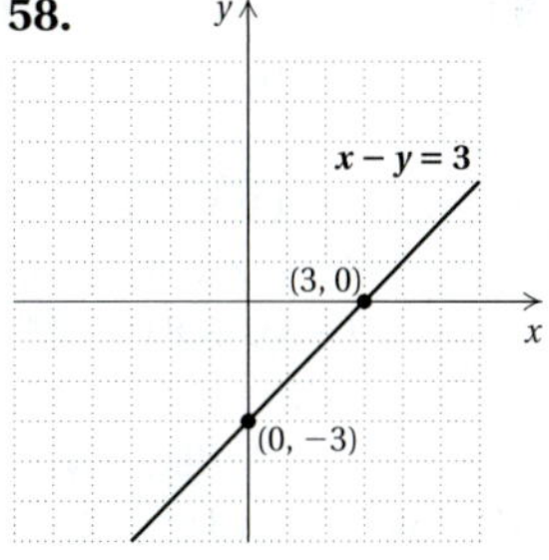

59. 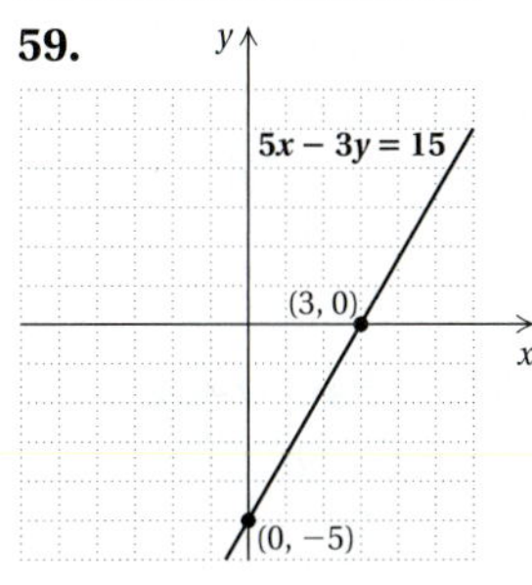

60. 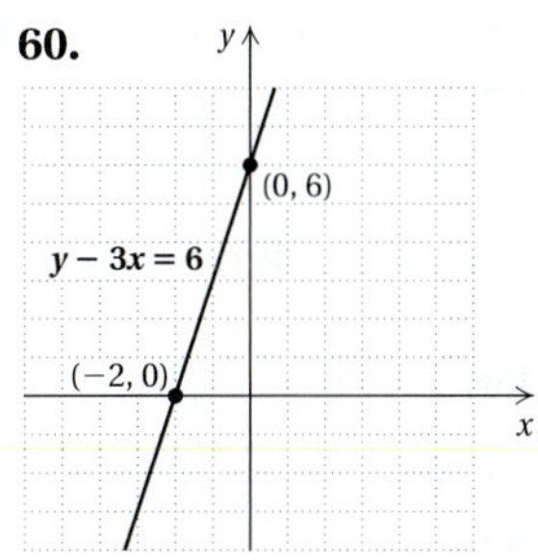

61. $(2x^3 + 3)(2x^2 + 3)$ **63.** $(x^7 + 1)(x^5 + 1)$
65. Not factorable by grouping

Margin Exercises, Section 10.2, pp. 755–760

1. (a) $-13, 8, -8, 7, -7$; **(b)** 13, 8, 7; both 7 and 12 are positive; **(c)** $(x + 3)(x + 4)$ **2.** $(x + 9)(x + 4)$
3. The coefficient of the middle term of the trinomial, -8, is negative and the sum of each pair of factors is positive.
4. $(x - 5)(x - 3)$ **5.** $(t - 5)(t - 4)$ **6. (a)** 23, 10, 5, 2; the positive factor has the larger absolute value and the coefficient of the middle term of the trinomial is negative; **(b)** $-23, -10, -5, -2$; the negative factor has the larger absolute value and the coefficient of the middle term of the trinomial is negative; **(c)** $(x + 3)(x - 8)$ **7. (a)** $-23, -10, -5, -2$; the negative factor has the larger absolute value and the coefficient of the middle term of the trinomial is positive; **(b)** 23, 10, 5, 2; the positive factor has the larger absolute value and the coefficient of the middle term of the trinomial is positive;
(c) $(x - 2)(x + 12)$ **8.** $(a - 2)(a + 12)$
9. $(t + 2)(t - 12)$ **10.** $(y - 6)(y + 2)$
11. $(t^2 + 7)(t^2 - 2)$ **12.** Prime **13.** $x(x + 6)(x - 2)$
14. $p(p - q - 3q^2)$ **15.** $3x(x + 4)^2$
16. $-1(x + 2)(x - 7)$, or $(-x - 2)(x - 7)$, or $(x + 2)(-x + 7)$ **17.** $-1(x + 3)(x - 6)$, or $(-x - 3)(x - 6)$, or $(x + 3)(-x + 6)$

Exercise Set 10.2, p. 761

1.

Pairs of Factors	Sums of Factors
1, 15	16
−1, −15	−16
3, 5	8
−3, −5	−8

$(x + 3)(x + 5)$

3.

Pairs of Factors	Sums of Factors
1, 12	13
−1, −12	−13
2, 6	8
−2, −6	−8
3, 4	7
−3, −4	−7

$(x + 3)(x + 4)$

5.

Pairs of Factors	Sums of Factors
1, 9	10
−1, −9	−10
3, 3	6
−3, −3	−6

$(x - 3)^2$

7.

Pairs of Factors	Sums of Factors
−1, 14	13
1, −14	−13
−2, 7	5
2, −7	−5

$(x + 2)(x - 7)$

9.

Pairs of Factors	Sums of Factors
1, 4	5
−1, −4	−5
2, 2	4
−2, −2	−4

$(b + 1)(b + 4)$

11.

Pairs of Factors	Sums of Factors
$\frac{1}{3}, \frac{1}{3}$	$\frac{2}{3}$
$-\frac{1}{3}, -\frac{1}{3}$	$-\frac{2}{3}$

$\left(x + \frac{1}{3}\right)^2$

13. $(d - 2)(d - 5)$ **15.** $(y - 1)(y - 10)$ **17.** Prime
19. $(x - 9)(x + 2)$ **21.** $x(x - 8)(x + 2)$
23. $y(y - 9)(y + 5)$ **25.** $(x - 11)(x + 9)$
27. $(c^2 + 8)(c^2 - 7)$ **29.** $(a^2 + 7)(a^2 - 5)$
31. $(x - 6)(x + 7)$ **33.** Prime **35.** $(x + 10)^2$
37. $-1(x - 10)(x + 3)$, or $(-x + 10)(x + 3)$, or $(x - 10)(-x - 3)$ **39.** $-1(a - 2)(a + 12)$, or $(-a + 2)(a + 12)$, or $(a - 2)(-a - 12)$
41. $x^2(x - 25)(x + 4)$ **43.** $(x - 24)(x + 3)$
45. $(x - 9)(x - 16)$ **47.** $(a + 12)(a - 11)$
49. $(x - 15)(x - 8)$ **51.** $-1(x + 12)(x - 9)$, or $(-x - 12)(x - 9)$, or $(x + 12)(-x + 9)$
53. $(y - 0.4)(y + 0.2)$ **55.** $(p + 5q)(p - 2q)$
57. $-1(t + 14)(t - 6)$, or $(-t - 14)(t - 6)$, or $(t + 14)(-t + 6)$ **59.** $(m + 4n)(m + n)$

61. $(s + 3t)(s - 5t)$ **63.** $6a^8(a + 2)(a - 7)$ **65.** DW
67. DW **69.** $16x^3 - 48x^2 + 8x$ **70.** $28w^2 - 53w - 66$
71. $49w^2 + 84w + 36$ **72.** $16w^2 - 88w + 121$
73. $16w^2 - 121$ **74.** $27x^{12}$ **75.** $\frac{8}{3}$ **76.** $-\frac{7}{2}$
77. 29,555 **78.** 100°, 25°, 55° **79.** 15, −15, 27, −27, 51, −51 **81.** $\left(x + \frac{1}{4}\right)\left(x - \frac{3}{4}\right)$ **83.** $(x + 5)\left(x - \frac{5}{7}\right)$
85. $(b^n + 5)(b^n + 2)$ **87.** $2x^2(4 - \pi)$
89. First consider all the factorizations of 36 that contain three factors. We also find the sum of the factors in each factorization.

Factorization	Sum of Factors
1 · 1 · 36	38
1 · 2 · 18	21
1 · 3 · 12	16
1 · 4 · 9	14
1 · 6 · 6	13
2 · 2 · 9	13
2 · 3 · 6	11
3 · 3 · 4	10

We can conclude that the number on the house next door is 13, because two sums are 13. This is what causes the census taker to be puzzled. She cannot determine which trio of factors gives the children's ages. When the mother supplies the additional information that there is an oldest child, the census taker knows that the ages of the children cannot be 1, 6, and 6 because there is not an oldest child in this group. Therefore, the children's ages must be 2, 2, and 9.

Margin Exercises, Section 10.3, pp. 766–769

1. $(2x + 5)(x - 3)$ **2.** $(4x + 1)(3x - 5)$
3. $(3x - 4)(x - 5)$ **4.** $2(5x - 4)(2x - 3)$
5. $(2x + 1)(3x + 2)$ **6.** $-1(2x - 1)(3x + 2)$, or $(2x - 1)(-3x - 2)$, or $(-2x + 1)(3x + 2)$
7. $-2(3x - 4)(x + 1)$, or $2(-3x + 4)(x + 1)$, or $2(3x - 4)(-x - 1)$ **8.** $(2a - b)(3a - b)$
9. $3(2x + 3y)(x + y)$

Calculator Corner, p. 770

1. Correct **2.** Correct **3.** Not correct **4.** Not correct
5. Not correct **6.** Correct **7.** Not correct **8.** Correct

Exercise Set 10.3, p. 771

1. $(2x + 1)(x - 4)$ **3.** $(5x + 9)(x - 2)$
5. $(3x + 1)(2x + 7)$ **7.** $(3x + 1)(x + 1)$
9. $(2x - 3)(2x + 5)$ **11.** $(2x + 1)(x - 1)$
13. $(3x - 2)(3x + 8)$ **15.** $(3x + 1)(x - 2)$
17. $(3x + 4)(4x + 5)$ **19.** $(7x - 1)(2x + 3)$
21. $(3x + 2)(3x + 4)$ **23.** $(3x - 7)^2$
25. $(24x - 1)(x + 2)$ **27.** $(5x - 11)(7x + 4)$
29. $-2(x - 5)(x + 2)$, or $2(-x + 5)(x + 2)$, or $2(x - 5)(-x - 2)$ **31.** $4(3x - 2)(x + 3)$
33. $6(5x - 9)(x + 1)$ **35.** $2(3y + 5)(y - 1)$
37. $(3x - 1)(x - 1)$ **39.** $4(3x + 2)(x - 3)$
41. $(2x + 1)(x - 1)$ **43.** $(3x + 2)(3x - 8)$
45. $5(3x + 1)(x - 2)$ **47.** $p(3p + 4)(4p + 5)$
49. $-1(3x + 2)(3x - 8)$, or $(-3x - 2)(3x - 8)$, or $(3x + 2)(-3x + 8)$ **51.** $-1(5x - 3)(3x - 2)$, or $(-5x + 3)(3x - 2)$, or $(5x - 3)(-3x + 2)$
53. $x^2(7x - 1)(2x + 3)$ **55.** $3x(8x - 1)(7x - 1)$
57. $(5x^2 - 3)(3x^2 - 2)$ **59.** $(5t + 8)^2$
61. $2x(3x + 5)(x - 1)$ **63.** Prime **65.** Prime
67. $(4m + 5n)(3m - 4n)$ **69.** $(2a + 3b)(3a - 5b)$
71. $(3a + 2b)(3a + 4b)$ **73.** $(5p + 2q)(7p + 4q)$
75. $6(3x - 4y)(x + y)$ **77.** DW **79.** $q = \frac{A + 7}{p}$
80. $x = \frac{y - b}{m}$ **81.** $y = \frac{6 - 3x}{2}$ **82.** $q = p + r - 2$
83. $\{x|x > 4\}$ **84.** $\left\{x|x \le \frac{8}{11}\right\}$
85.

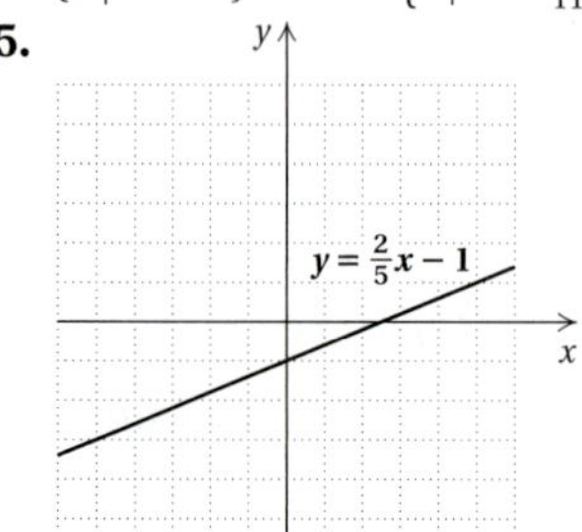

86. y^8 **87.** $9x^2 - 25$ **88.** $16a^2 - 24a + 9$
89. $(2x^n + 1)(10x^n + 3)$ **91.** $(x^{3a} - 1)(3x^{3a} + 1)$
93.–101. Left to the student

Margin Exercises, Section 10.4, p. 775

1. $(2x + 1)(3x + 2)$ **2.** $(4x + 1)(3x - 5)$
3. $3(2x + 3)(x + 1)$ **4.** $2(5x - 4)(2x - 3)$

Exercise Set 10.4, p. 776

1. $(x + 7)(x + 2)$ **3.** $(x - 1)(x - 4)$
5. $(2x + 3)(3x + 2)$ **7.** $(x - 4)(3x - 4)$
9. $(5x + 3)(7x - 8)$ **11.** $(2x - 3)(2x + 3)$
13. $(2x^2 + 5)(x^2 + 3)$ **15.** $(2x - 1)(x + 4)$
17. $(3x + 5)(x - 3)$ **19.** $(2x + 7)(3x + 1)$
21. $(3x - 1)(x - 1)$ **23.** $(2x + 3)(2x - 5)$
25. $(2x - 1)(x + 1)$ **27.** $(3x + 2)(3x - 8)$
29. $(3x - 1)(x + 2)$ **31.** $(3x - 4)(4x - 5)$
33. $(7x + 1)(2x - 3)$ **35.** $(3x + 2)(3x + 4)$
37. $(3x - 7)^2$ **39.** $(24x + 1)(x - 2)$
41. $-1(3a - 1)(3a + 5)$, or $(-3a + 1)(3a + 5)$, or $(3a - 1)(-3a - 5)$ **43.** $-2(x - 5)(x + 2)$, or $2(-x + 5)(x + 2)$, or $2(x - 5)(-x - 2)$
45. $4(3x - 2)(x + 3)$ **47.** $6(5x - 9)(x + 1)$
49. $2(3y + 5)(y - 1)$ **51.** $(3x - 1)(x - 1)$
53. $4(3x + 2)(x - 3)$ **55.** $(2x + 1)(x - 1)$
57. $(3x - 2)(3x + 8)$ **59.** $5(3x + 1)(x - 2)$
61. $p(3p + 4)(4p + 5)$ **63.** $-1(5x - 4)(x + 1)$, or $(-5x + 4)(x + 1)$, or $(5x - 4)(-x - 1)$

65. $-3(2t-1)(t-5)$, or $3(-2t+1)(t-5)$, or $3(2t-1)(-t+5)$ **67.** $x^2(7x-1)(2x+3)$ **69.** $3x(8x-1)(7x-1)$ **71.** $(5x^2-3)(3x^2-2)$ **73.** $(5t+8)^2$ **75.** $2x(3x+5)(x-1)$ **77.** Prime **79.** Prime **81.** $(4m+5n)(3m-4n)$ **83.** $(2a+3b)(3a-5b)$ **85.** $(3a-2b)(3a-4b)$ **87.** $(5p+2q)(7p+4q)$ **89.** $6(3x-4y)(x+y)$ **91.** $-6x(x-5)(x+2)$, or $6x(-x+5)(x+2)$, or $6x(x-5)(-x-2)$ **93.** $x^3(5x-11)(7x+4)$ **95.** D_W **97.** $\{x|x<-100\}$ **98.** $\{x|x\geq 217\}$ **99.** $\{x|x\leq 8\}$ **100.** $\{x|x<2\}$ **101.** $\{x|x\geq \frac{20}{3}\}$ **102.** $\{x|x>17\}$ **103.** $\{x|x>\frac{26}{7}\}$ **104.** $\{x|x\geq \frac{77}{17}\}$ **105.** About 6369 km, or 3949 mi **106.** 40° **107.** $(3x^5-2)^2$ **109.** $(4x^5+1)^2$ **111.–119.** Left to the student

Margin Exercises, Section 10.5, pp. 781–785

1. Yes **2.** No **3.** No **4.** Yes **5.** No **6.** Yes **7.** No **8.** Yes **9.** $(x+1)^2$ **10.** $(x-1)^2$ **11.** $(t+2)^2$ **12.** $(5x-7)^2$ **13.** $(7-4y)^2$ **14.** $3(4m+5)^2$ **15.** $(p^2+9)^2$ **16.** $z^3(2z-5)^2$ **17.** $(3a+5b)^2$ **18.** Yes **19.** No **20.** No **21.** No **22.** Yes **23.** Yes **24.** Yes **25.** $(x+3)(x-3)$ **26.** $4(4+t)(4-t)$ **27.** $(a+5b)(a-5b)$ **28.** $x^4(8+5x)(8-5x)$ **29.** $5(1+2t^3)(1-2t^3)$ **30.** $(9x^2+1)(3x+1)(3x-1)$ **31.** $(7p^2+5q^3)(7p^2-5q^3)$

Exercise Set 10.5, p. 786

1. Yes **3.** No **5.** No **7.** No **9.** $(x-7)^2$ **11.** $(x+8)^2$ **13.** $(x-1)^2$ **15.** $(x+2)^2$ **17.** $(q^2-3)^2$ **19.** $(4y+7)^2$ **21.** $2(x-1)^2$ **23.** $x(x-9)^2$ **25.** $3(2q-3)^2$ **27.** $(7-3x)^2$ **29.** $5(y^2+1)^2$ **31.** $(1+2x^2)^2$ **33.** $(2p+3q)^2$ **35.** $(a-3b)^2$ **37.** $(9a-b)^2$ **39.** $4(3a+4b)^2$ **41.** Yes **43.** No **45.** No **47.** Yes **49.** $(y+2)(y-2)$ **51.** $(p+3)(p-3)$ **53.** $(t+7)(t-7)$ **55.** $(a+b)(a-b)$ **57.** $(5t+m)(5t-m)$ **59.** $(10+k)(10-k)$ **61.** $(4a+3)(4a-3)$ **63.** $(2x+5y)(2x-5y)$ **65.** $2(2x+7)(2x-7)$ **67.** $x(6+7x)(6-7x)$ **69.** $(7a^2+9)(7a^2-9)$ **71.** $(a^2+4)(a+2)(a-2)$ **73.** $5(x^2+9)(x+3)(x-3)$ **75.** $(1+y^4)(1+y^2)(1+y)(1-y)$ **77.** $(x^6+4)(x^3+2)(x^3-2)$ **79.** $\left(y+\frac{1}{4}\right)\left(y-\frac{1}{4}\right)$ **81.** $\left(5+\frac{1}{7}x\right)\left(5-\frac{1}{7}x\right)$ **83.** $(4m^2+t^2)(2m+t)(2m-t)$ **85.** D_W **87.** -11 **88.** 400 **89.** $-\frac{5}{6}$ **90.** -0.9 **91.** 2 **92.** -160 **93.** $x^2-4xy+4y^2$ **94.** $\frac{1}{2}\pi x^2+2xy$ **95.** y^{12} **96.** $25a^4b^6$ **97.**

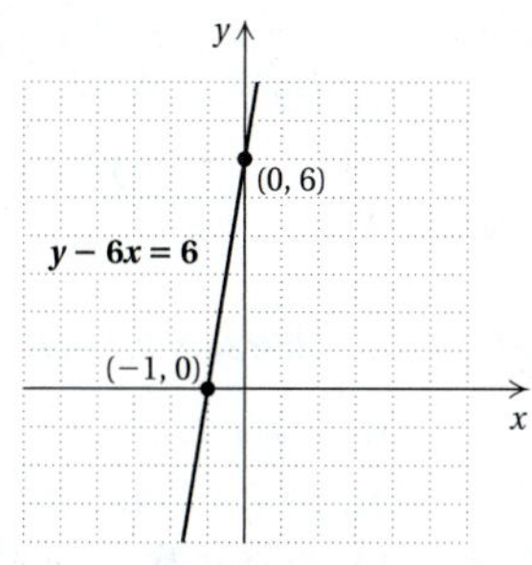

98.

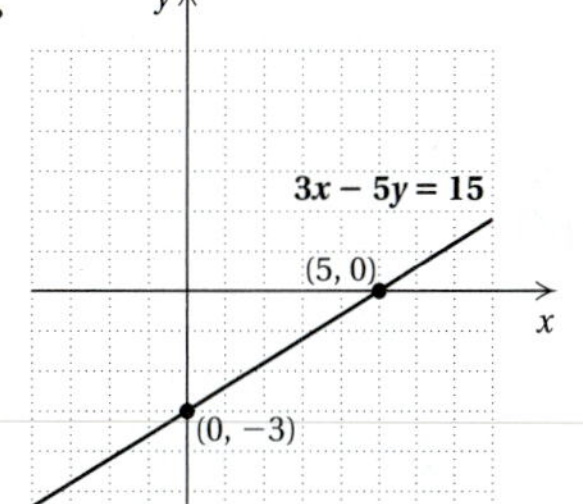

99. Prime **101.** $(x+11)^2$ **103.** $2x(3x+1)^2$ **105.** $(x^4+2^4)(x^2+2^2)(x+2)(x-2)$ **107.** $3x^3(x+2)(x-2)$ **109.** $2x\left(3x+\frac{2}{5}\right)\left(3x-\frac{2}{5}\right)$ **111.** $p(0.7+p)(0.7-p)$ **113.** $(0.8x+1.1)(0.8x-1.1)$ **115.** $x(x+6)$ **117.** $\left(x+\frac{1}{x}\right)\left(x-\frac{1}{x}\right)$ **119.** $(9+b^{2k})(3-b^k)(3+b^k)$ **121.** $(3b^n+2)^2$ **123.** $(y+4)^2$ **125.** 9 **127.** Not correct **129.** Not correct

Margin Exercises, Section 10.6, pp. 791–792

1. $(x-2)(x^2+2x+4)$ **2.** $(4-y)(16+4y+y^2)$ **3.** $(3x+y)(9x^2-3xy+y^2)$ **4.** $(2y+z)(4y^2-2yz+z^2)$ **5.** $(m+n)(m^2-mn+n^2)(m-n)(m^2+mn+n^2)$ **6.** $2xy(2x^2+3y^2)(4x^4-6x^2y^2+9y^4)$ **7.** $(3x+2y)(9x^2-6xy+4y^2)(3x-2y)(9x^2+6xy+4y^2)$ **8.** $(x-0.3)(x^2+0.3x+0.09)$

Exercise Set 10.6, p. 793

1. $(z+3)(z^2-3z+9)$ **3.** $(x-1)(x^2+x+1)$ **5.** $(y+5)(y^2-5y+25)$ **7.** $(2a+1)(4a^2-2a+1)$ **9.** $(y-2)(y^2+2y+4)$ **11.** $(2-3b)(4+6b+9b^2)$ **13.** $(4y+1)(16y^2-4y+1)$ **15.** $(2x+3)(4x^2-6x+9)$ **17.** $(a-b)(a^2+ab+b^2)$ **19.** $\left(a+\frac{1}{2}\right)\left(a^2-\frac{1}{2}a+\frac{1}{4}\right)$ **21.** $2(y-4)(y^2+4y+16)$ **23.** $3(2a+1)(4a^2-2a+1)$ **25.** $r(s+4)(s^2-4s+16)$ **27.** $5(x-2z)(x^2+2xz+4z^2)$ **29.** $(x+0.1)(x^2-0.1x+0.01)$ **31.** $8(2x^2-t^2)(4x^4+2x^2t^2+t^4)$ **33.** $2y(y-4)(y^2+4y+16)$ **35.** $(z-1)(z^2+z+1)(z+1)(z^2-z+1)$ **37.** $(t^2+4y^2)(t^4-4t^2y^2+16y^4)$ **39.** $\frac{343}{y^{15}}$ **40.** a^8b^{18} **41.** $\frac{16}{x^6}$ **42.** $4y^{10}-9$ **43.** $w^2-\frac{2}{3}w+\frac{1}{9}$ **44.** $x^2+0.04x-0.05$ **45.** 1; 19; 19; 7; 1

47. $(x^{2a} + y^b)(x^{4a} - x^{2a}y^b + y^{2b})$
49. $3(x^a + 2y^b)(x^{2a} - 2x^ay^b + 4y^{2b})$
51. $\frac{1}{3}(\frac{1}{2}xy + z)(\frac{1}{4}x^2y^2 - \frac{1}{2}xyz + z^2)$ 53. $y(3x^2 + 3xy + y^2)$
55. $4(3a^2 + 4)$

Margin Exercises, Section 10.7, pp. 796–799

1. $3(m^2 + 1)(m + 1)(m - 1)$ 2. $(x^3 + 4)^2$
3. $2x^2(x + 1)(x + 3)$ 4. $(3x^2 - 2)(x + 4)$
5. $8x(x - 5)(x + 5)$ 6. $x^2y(x^2y + 2x + 3)$
7. $2p^4q^2(5p^2 + 2pq + q^2)$ 8. $(a - b)(2x + 5 + y^2)$
9. $(a + b)(x^2 + y)$ 10. $(x^2 + y^2)^2$ 11. $(xy + 1)(xy + 4)$
12. $(p^2 + 9q^2)(p + 3q)(p - 3q)$
13. $15(a - 2b)(a^2 + 2ab + 4b^2)$

Exercise Set 10.7, p. 800

1. $3(x + 8)(x - 8)$ 3. $(a - 5)^2$ 5. $(2x - 3)(x - 4)$
7. $x(x + 12)^2$ 9. $(x + 2)(x - 2)(x + 3)$
11. $3(4x + 1)(4x - 1)$ 13. $3x(3x - 5)(x + 3)$
15. Prime 17. $x(x - 3)(x^2 + 7)$ 19. $x^3(x - 7)^2$
21. $-2(x - 2)(x + 5)$, or $2(-x + 2)(x + 5)$, or $2(x - 2)(-x - 5)$ 23. Prime
25. $4(x^2 + 4)(x + 2)(x - 2)$
27. $(1 + y^4)(1 + y^2)(1 + y)(1 - y)$ 29. $x^3(x - 3)(x - 1)$
31. $\frac{1}{9}(\frac{1}{3}x^3 - 4)^2$ 33. $m(x^2 + y^2)$ 35. $9xy(xy - 4)$
37. $2\pi r(h + r)$ 39. $(a + b)(2x + 1)$
41. $(x + 1)(x - 1 - y)$ 43. $(n + p)(n + 2)$
45. $(3q + p)(2q - 1)$ 47. $(2b - a)^2$, or $(a - 2b)^2$
49. $(4x + 3y)^2$ 51. $(7m^2 - 8n)^2$ 53. $(y^2 + 5z^2)^2$
55. $(\frac{1}{2}a + \frac{1}{3}b)^2$ 57. $(a + b)(a - 2b)$
59. $(m + 20n)(m - 18n)$ 61. $(mn - 8)(mn + 4)$
63. $r^3(rs - 2)(rs - 8)$ 65. $a^3(a - b)(a + 5b)$
67. $(a + \frac{1}{5}b)(a - \frac{1}{5}b)$
69. $7(x + y)(x^2 - xy + y^2)(x - y)(x^2 + xy + y^2)$
71. $(4 + p^2q^2)(2 + pq)(2 - pq)$
73. $(1 + 4x^6y^6)(1 + 2x^3y^3)(1 - 2x^3y^3)$
75. $(q + 1)(q - 1)(q + 8)$ 77. $(7x + 8y)^2$ 79. DW
81. 1999 82. 1990 83. 1996 84. 847 million
85. 10.9% 86. 2.6% 87. $-\frac{14}{11}$ 88. $25x^2 - 10xt + t^2$
89. $X = \dfrac{A + 7}{a + b}$ 90. $\{x|x < 32\}$ 91. $(a + 1)^2(a - 1)^2$
93. $(3.5x - 1)^2$ 95. $(5x + 4)(x + 1.8)$
97. $(y + 3)(y - 3)(y - 2)$ 99. $(a^2 + 1)(a + 4)$
101. $(x + 2)(x - 2)(x - 1)$ 103. $(y - 1)^3$
105. $(y + 4 + x)^2$

Margin Exercises, Section 10.8, pp. 805–808

1. 3, −4 2. 7, 3 3. $-\frac{1}{4}, \frac{2}{3}$ 4. $0, \frac{17}{3}$ 5. −2, 3
6. −4, 7 7. 3 8. 0, 4 9. $-\frac{4}{3}, \frac{4}{3}$ 10. $3, \frac{7}{2}$ 11. −5, 2
12. −3, 3 13. (−5, 0), (1, 0) 14. 0, 3

Calculator Corner, p. 809

1. Left to the student

Exercise Set 10.8, p. 810

1. −4, −9 3. −3, 8 5. −12, 11 7. 0, −3 9. 0, −18
11. $-\frac{5}{2}, -4$ 13. $-\frac{1}{5}, 3$ 15. $4, \frac{1}{4}$ 17. $0, \frac{2}{3}$ 19. $-\frac{1}{10}, \frac{1}{27}$
21. $\frac{1}{3}, -20$ 23. $0, \frac{2}{3}, \frac{1}{2}$ 25. −5, −1 27. −9, 2
29. 3, 5 31. 0, 8 33. 0, −18 35. −4, 4 37. $-\frac{2}{3}, \frac{2}{3}$
39. −3 41. 4 43. $0, \frac{6}{5}$ 45. $-1, \frac{5}{3}$ 47. $-\frac{1}{4}, \frac{2}{3}$
49. $-1, \frac{2}{3}$ 51. $-\frac{7}{10}, \frac{7}{10}$ 53. −2, 9 55. $\frac{4}{5}, \frac{3}{2}$
57. (−4, 0), (1, 0) 59. $(-\frac{5}{2}, 0)$, (2, 0) 61. (−3, 0), (5, 0)
63. −1, 4 65. −1, 3 67. DW 69. $(a + b)^2$
70. $a^2 + b^2$ 71. −16 72. −4.5 73. $-\frac{10}{3}$ 74. $\frac{3}{10}$
75. −5, 4 77. −3, 9 79. $-\frac{1}{8}, \frac{1}{8}$ 81. −4, 4
83. Answers may vary. (a) $x^2 - x - 12 = 0$; (b) $x^2 + 7x + 12 = 0$; (c) $4x^2 - 4x + 1 = 0$; (d) $x^2 - 25 = 0$; (e) $40x^3 - 14x^2 + x = 0$ 85. 2.33, 6.77
87. −9.15, −4.59 89. 0, 2.74

Margin Exercises, Section 10.9, pp. 813–818

1. Length: 24 in.; width: 12 in. 2. Height: 25 ft; width: 10 ft
3. (a) 342 games; (b) 9 teams 4. 22 and 23 5. 24 ft
6. 3 m, 4 m

Exercise Set 10.9, p. 819

1. Length: 12 ft; width: 2 ft 3. Length: 18 cm; width: 8 cm
5. Height: 4 cm; base: 14 cm 7. Base: 8 m; height: 16 m
9. 182 games 11. 12 teams 13. 4950 handshakes
15. 25 people 17. 20 people 19. 14 and 15
21. 12 and 14; −12 and −14 23. 15 and 17; −15 and −17
25. Hypotenuse: 17 ft; leg: 15 ft 27. 32 ft 29. 25 ft
31. Dining room: 12 ft by 12 ft; kitchen: 12 ft by 10 ft
33. 4 sec 35. 5 and 7 37. DW 39. $9x^2 - 25y^2$
40. $9x^2 - 30xy + 25y^2$ 41. $9x^2 + 30xy + 25y^2$
42. $6x^2 + 11xy - 35y^2$ 43. y-intercept: (0, −4); x-intercept: (16, 0) 44. y-intercept: (0, 4); x-intercept: (16, 0) 45. y-intercept: (0, −5); x-intercept: (6.5, 0)
46. y-intercept: $(0, \frac{2}{3})$; x-intercept: $(\frac{5}{8}, 0)$ 47. y-intercept: (0, 4); x-intercept: $(\frac{4}{5}, 0)$ 48. y-intercept: (0, −5); x-intercept: $(\frac{5}{2}, 0)$ 49. 35 ft 51. 5 ft 53. 30 cm by 15 cm 55. 39 cm

Summary and Review, Chapter 10, p. 825

1. $(-10x)(x)$; $(-5x)(2x)$; $(5x)(-2x)$; answers may vary
2. $(6x)(6x^4)$; $(4x^2)(9x^3)$; $(-2x^4)(-18x)$; answers may vary
3. $5(1 + 2x^3)(1 - 2x^3)$ 4. $x(x - 3)$
5. $(3x + 2)(3x - 2)$ 6. $(x + 6)(x - 2)$ 7. $(x + 7)^2$
8. $3x(2x^2 + 4x + 1)$ 9. $(x^2 + 3)(x + 1)$
10. $(3x - 1)(2x - 1)$ 11. $(x^2 + 9)(x + 3)(x - 3)$
12. $3x(3x - 5)(x + 3)$ 13. $2(x + 5)(x - 5)$
14. $(x^3 - 2)(x + 4)$ 15. $(4x^2 + 1)(2x + 1)(2x - 1)$
16. $4x^4(2x^2 - 8x + 1)$ 17. $3(2x + 5)^2$ 18. Prime
19. $x(x - 6)(x + 5)$ 20. $(2x - 5)(4x^2 + 10x + 25)$
21. $(3x - 5)^2$ 22. $2(3x + 4)(x - 6)$ 23. $(x - 3)^2$
24. $(2x + 1)(x - 4)$ 25. $2(3x - 1)^2$

26. $3(x+3)(x-3)$ **27.** $(x-5)(x-3)$ **28.** $(5x-2)^2$ **29.** $(7b^5-2a^4)^2$ **30.** $(xy+4)(xy-3)$ **31.** $3(2a+7b)^2$ **32.** $(m+t)(m+5)$ **33.** $32(x^2-2y^2z^2)(x^2+2y^2z^2)$ **34.** $1, -3$ **35.** $-7, 5$ **36.** $-4, 3$ **37.** $\frac{2}{3}, 1$ **38.** $-4, \frac{3}{2}$ **39.** $-2, 8$ **40.** Height: 6 cm; base: 5 cm **41.** -18 and -16; 16 and 18 **42.** -19 and -17; 17 and 19 **43.** 3 ft **44.** 6 km **45.** $(-5, 0), (-4, 0)$ **46.** $\left(-\frac{3}{2}, 0\right), (5, 0)$
47. DW Answers may vary. Because Sheri did not first factor out the largest common factor, 4, her factorization will not be "complete" until she removes a common factor of 2 from each binomial. Awarding 5 to 7 points seems reasonable.
48. DW The equations solved in this chapter have an x^2-term (are quadratic), whereas those solved previously have no x^2-term (are linear). The principle of zero products is used to solve quadratic equations; it is not used to solve linear equations.
49. $\frac{8}{35}$ **50.** $\{x|x \le \frac{4}{3}\}$ **51.** $4a^2-9$
52.

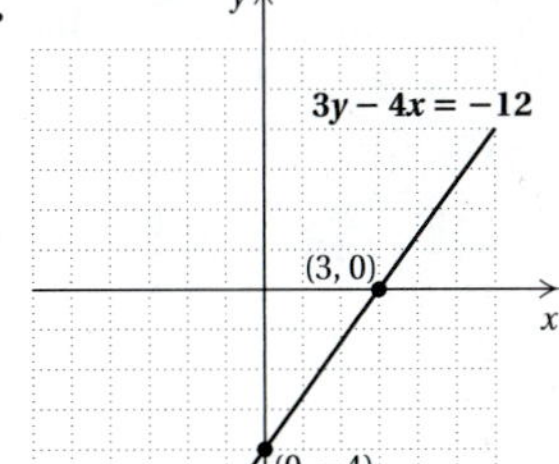

53. 2.5 cm **54.** 0, 2
55. Length: 12; width: 6 **56.** No solution **57.** $2, -3, \frac{5}{2}$ **58.** $-2, \frac{5}{4}, 3$ **59.** $(\pi-2)x^2$

Test: Chapter 10, p. 828

1. [10.1a] $(4x)(x^2)$; $(2x^2)(2x)$; $(-2x)(-2x^2)$; answers may vary **2.** [10.2a] $(x-5)(x-2)$ **3.** [10.5b] $(x-5)^2$
4. [10.1b] $2y^2(2y^2-4y+3)$ **5.** [10.1c] $(x^2+2)(x+1)$
6. [10.1b] $x(x-5)$ **7.** [10.2a] $x(x+3)(x-1)$
8. [10.3a], [10.4a] $2(5x-6)(x+4)$
9. [10.5d] $(2x+3)(2x-3)$ **10.** [10.2a] $(x-4)(x+3)$
11. [10.3a], [10.4a] $3m(2m+1)(m+1)$
12. [10.5d] $3(w+5)(w-5)$ **13.** [10.5b] $5(3x+2)^2$
14. [10.5d] $3(x^2+4)(x+2)(x-2)$ **15.** [10.5b] $(7x-6)^2$
16. [10.3a], [10.4a] $(5x-1)(x-5)$
17. [10.1c] $(x^3-3)(x+2)$
18. [10.5d] $5(4+x^2)(2+x)(2-x)$
19. [10.3a], [10.4a] $(2x+3)(2x-5)$
20. [10.3a], [10.4a] $3t(2t+5)(t-1)$
21. [10.2a] $3(m+2n)(m-5n)$
22. [10.6a] $(10a-3b)(100a^2+30ab+9b^2)$
23. [10.8b] $-4, 5$ **24.** [10.8b] $-5, \frac{3}{2}$ **25.** [10.8b] $-4, 7$
26. [10.9a] Length: 8 m; width: 6 m **27.** [10.9a] Height: 4 cm; base: 14 cm **28.** [10.9b] 5 ft
29. [10.8b] $(-5, 0), (7, 0)$ **30.** [10.8b] $\left(\frac{2}{3}, 0\right), (1, 0)$
31. [6.6c] $-\frac{10}{11}$ **32.** [7.7e] $\{x|x < \frac{19}{3}\}$

33. [8.3a]

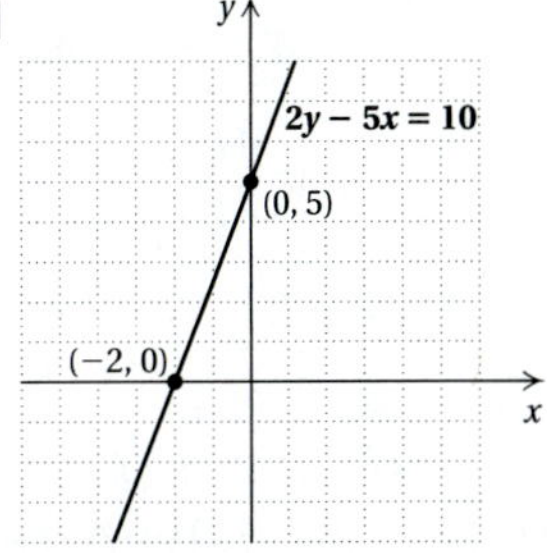

34. [9.6c] $25x^4-70x^2+49$ **35.** [10.9a] Length: 15; width: 3 **36.** [10.2a] $(a-4)(a+8)$ **37.** [10.8b] $-\frac{8}{3}, 0, \frac{2}{5}$
38. [9.6b], [10.5d] (d)

Chapter 11

Pretest: Chapter 11, p. 832

1. $(x+2)(x+3)^2$ **2.** $\frac{-b-1}{b^2-4}$, or $\frac{b+1}{4-b^2}$ **3.** $\frac{1}{y-2}$
4. $\frac{7a+6}{a(a+2)}$ **5.** $\frac{2x}{x+1}$ **6.** $\frac{2(x-3)}{x-2}$ **7.** $\frac{x-3}{x+3}$
8. $\frac{y+x}{y-x}$ **9.** -5 **10.** 0 **11.** About 55.6 gal
12. 10.5 hr **13.** $2\frac{8}{11}$ hr **14.** 60 mph, 80 mph

Margin Exercises, Section 11.1, pp. 833–838

1. 3 **2.** $-8, 3$ **3.** None **4.** $\frac{(2x+1)x}{(3x-2)x}$
5. $\frac{(x+1)(x+2)}{(x-2)(x+2)}$ **6.** $\frac{(x-8)(-1)}{(x-y)(-1)}$ **7.** 5 **8.** $\frac{x}{4}$
9. $\frac{2x+1}{3x+2}$ **10.** $\frac{x+1}{2x+1}$ **11.** $x+2$ **12.** $\frac{y+2}{4}$
13. -1 **14.** -1 **15.** -1 **16.** $\frac{a-2}{a-3}$ **17.** $\frac{x-5}{2}$

Calculator Corner, p. 839

1. Correct **2.** Correct **3.** Not correct **4.** Not correct
5. Not correct **6.** Not correct **7.** Correct **8.** Correct

Exercise Set 11.1, p. 840

1. 0 **3.** 8 **5.** $-\frac{5}{2}$ **7.** $-4, 7$ **9.** $-5, 5$ **11.** None
13. $\frac{(4x)(3x^2)}{(4x)(5y)}$ **15.** $\frac{2x(x-1)}{2x(x+4)}$ **17.** $\frac{-1(3-x)}{-1(4-x)}$
19. $\frac{(y+6)(y-7)}{(y+6)(y+2)}$ **21.** $\frac{x^2}{4}$ **23.** $\frac{8p^2q}{3}$ **25.** $\frac{x-3}{x}$
27. $\frac{m+1}{2m+3}$ **29.** $\frac{a-3}{a+2}$ **31.** $\frac{a-3}{a-4}$ **33.** $\frac{x+5}{x-5}$
35. $a+1$ **37.** $\frac{x^2+1}{x+1}$ **39.** $\frac{3}{2}$ **41.** $\frac{6}{t-3}$
43. $\frac{t+2}{2(t-4)}$ **45.** $\frac{w^2+wz+z^2}{w+z}$ **47.** -1 **49.** -1

51. -6 53. $-x-1$ 55. $\frac{56x}{3}$ 57. $\frac{2}{dc^2}$ 59. $\frac{x+2}{x-2}$
61. $\frac{(a+3)(a-3)}{a(a+4)}$ 63. $\frac{2a}{a-2}$ 65. $\frac{(t+2)(t-2)}{(t+1)(t-1)}$
67. $\frac{x+4}{x+2}$ 69. $\frac{5(a+6)}{a-1}$ 71. $c-2$ 73. DW
75. 18 and 20; -18 and -20 76. 3.125 L
77. $(x-8)(x+7)$ 78. $(a-8)^2$ 79. $x^3(x-7)(x+5)$
80. $(2y^2+1)(y-5)$ 81. $(2-t)(2+t)(4+t^2)$
82. $10(x+7)(x+1)$ 83. $(x-7)(x-2)$ 84. Prime
85. $(4x-5y)^2$ 86. $(a-7b)(a-2b)$ 87. $x+2y$
89. $\frac{(t-9)^2(t-1)}{(t^2+9)(t+1)}$ 91. $\frac{x-y}{x-5y}$
93. $$\frac{5(2x+5)-25}{10}=\frac{10x+25-25}{10}=\frac{10x}{10}=x$$
You get the same number you selected. To do a number trick, ask someone to select a number and then perform these operations. The person will probably be surprised that the result is the original number.

Margin Exercises, Section 11.2, pp. 844–846

1. $\frac{2}{7}$ 2. $\frac{2x^3-1}{x^2+5}$ 3. $\frac{1}{x-5}$ 4. x^2-3 5. $\frac{6}{7}$
6. $\frac{5}{8}$ 7. $\frac{(x-3)(x-2)}{(x+5)(x+5)}$ 8. $\frac{x-3}{x+2}$ 9. $\frac{(x-3)(x-2)}{x+2}$
10. $\frac{y+1}{y-1}$

Exercise Set 11.2, p. 847

1. $\frac{x}{4}$ 3. $\frac{1}{x^2-y^2}$ 5. $a+b$ 7. $\frac{x^2-4x+7}{x^2+2x-5}$ 9. $\frac{3}{10}$
11. $\frac{1}{4}$ 13. $\frac{b}{a}$ 15. $\frac{(a+2)(a+3)}{(a-3)(a-1)}$ 17. $\frac{(x-1)^2}{x}$
19. $\frac{1}{2}$ 21. $\frac{15}{8}$ 23. $\frac{15}{4}$ 25. $\frac{a-5}{3(a-1)}$ 27. $\frac{(x+2)^2}{x}$
29. $\frac{3}{2}$ 31. $\frac{c+1}{c-1}$ 33. $\frac{y-3}{2y-1}$ 35. $\frac{x^2+4x+16}{(x+4)^2}$
37. DW 39. $\{x|x \geq 77\}$ 40. Height: 4 cm; base: 14 cm
41. $8x^3-11x^2-3x+12$ 42. $-2p^2+4pq-4q^2$
43. $\frac{4y^8}{x^6}$ 44. $\frac{125x^{18}}{y^{12}}$ 45. $\frac{4x^6}{y^{10}}$ 46. $\frac{1}{a^{15}b^{20}}$ 47. $-\frac{1}{b^2}$
49. $\frac{(x-7)^2}{x+y}$

Margin Exercises, Section 11.3, pp. 849–850

1. 144 2. 12 3. 10 4. 120 5. $\frac{35}{144}$ 6. $\frac{1}{4}$ 7. $\frac{11}{10}$
8. $\frac{9}{40}$ 9. $60x^3y^2$ 10. $(y+1)^2(y+4)$
11. $7(t^2+16)(t-2)$ 12. $3x(x+1)^2(x-1)$

Exercise Set 11.3, p. 851

1. 108 3. 72 5. 126 7. 360 9. 500 11. $\frac{65}{72}$
13. $\frac{29}{120}$ 15. $\frac{23}{180}$ 17. $12x^3$ 19. $18x^2y^2$
21. $6(y-3)$ 23. $t(t+2)(t-2)$
25. $(x+2)(x-2)(x+3)$ 27. $t(t+2)^2(t-4)$
29. $(a+1)(a-1)^2$ 31. $(m-3)(m-2)^2$
33. $(2+3x)(2-3x)$ 35. $10v(v+4)(v+3)$
37. $18x^3(x-2)^2(x+1)$ 39. $6x^3(x+2)^2(x-2)$
41. DW 43. $(x-3)^2$ 44. $2x(3x+2)$
45. $(x+3)(x-3)$ 46. $(x+7)(x-3)$ 47. $(x+3)^2$
48. $(x-7)(x+3)$ 49. 54% 50. 64% 51. 74%
52. 98%; this seems unreasonably high. 53. 1965
54. 1999 55. 24 min

Margin Exercises, Section 11.4, pp. 853–856

1. $\frac{7}{9}$ 2. $\frac{3+x}{x-2}$ 3. $\frac{6x+4}{x-1}$ 4. $\frac{10x^2+9x}{48}$ 5. $\frac{9x+10}{48x^2}$
6. $\frac{4x^2-x+3}{x(x-1)(x+1)^2}$ 7. $\frac{2x^2+16x+5}{(x+3)(x+8)}$
8. $\frac{8x+88}{(x+16)(x+1)(x+8)}$ 9. $\frac{x-5}{4}$ 10. $\frac{x-1}{x-3}$
11. $\frac{-2x-11}{3(x+4)(x-4)}$

Exercise Set 11.4, p. 857

1. 1 3. $\frac{6}{3+x}$ 5. $\frac{2x+3}{x-5}$ 7. $\frac{2x+5}{x^2}$ 9. $\frac{41}{24r}$
11. $\frac{4x+6y}{x^2y^2}$ 13. $\frac{4+3t}{18t^3}$ 15. $\frac{x^2+4xy+y^2}{x^2y^2}$
17. $\frac{6x}{(x-2)(x+2)}$ 19. $\frac{11x+2}{3x(x+1)}$ 21. $\frac{x^2+6x}{(x+4)(x-4)}$
23. $\frac{6}{z+4}$ 25. $\frac{3x-1}{(x-1)^2}$ 27. $\frac{11a}{10(a-2)}$
29. $\frac{2x^2+8x+16}{x(x+4)}$ 31. $\frac{7a+6}{(a-2)(a+1)(a+3)}$
33. $\frac{2x^2-4x+34}{(x-5)(x+3)}$ 35. $\frac{3a+2}{(a+1)(a-1)}$ 37. $\frac{1}{4}$
39. $-\frac{1}{t}$ 41. $\frac{-x+7}{x-6}$ 43. $y+3$ 45. $\frac{2b-14}{b^2-16}$
47. $a+b$ 49. $\frac{5x+2}{x-5}$ 51. -1 53. $\frac{-x^2+9x-14}{(x-3)(x+3)}$
55. $\frac{2x+6y}{(x+y)(x-y)}$ 57. $\frac{a^2+7a+1}{(a+5)(a-5)}$
59. $\frac{5t-12}{(t+3)(t-3)(t-2)}$ 61. DW 63. x^2-1
64. $13y^3-14y^2+12y-73$ 65. $\frac{1}{8x^{12}y^9}$ 66. $\frac{x^6}{25y^2}$

67. $\frac{1}{x^{12}y^{21}}$ **68.** $\frac{25}{x^4y^6}$ **69.**

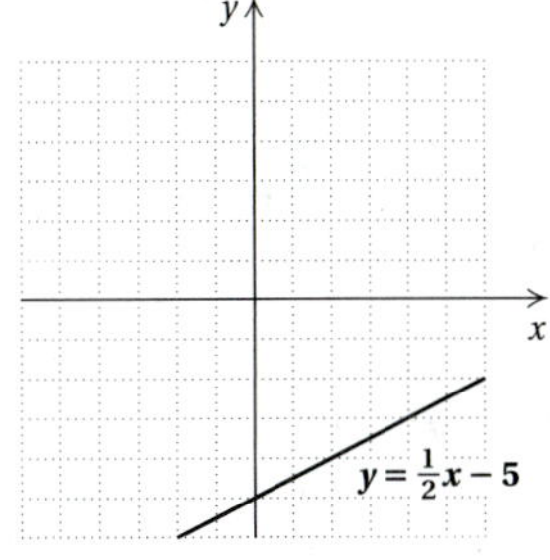

70.

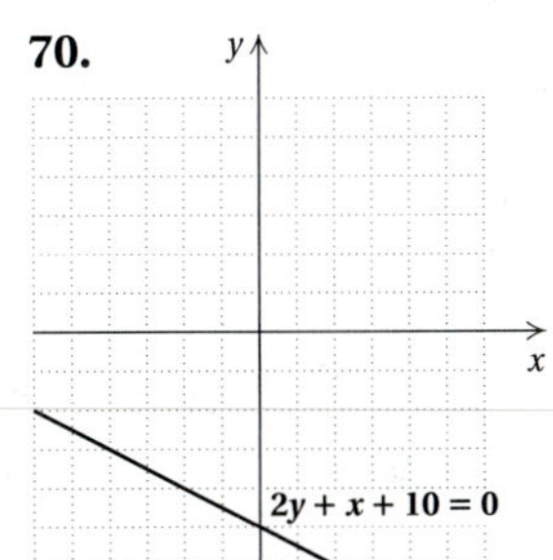

71.

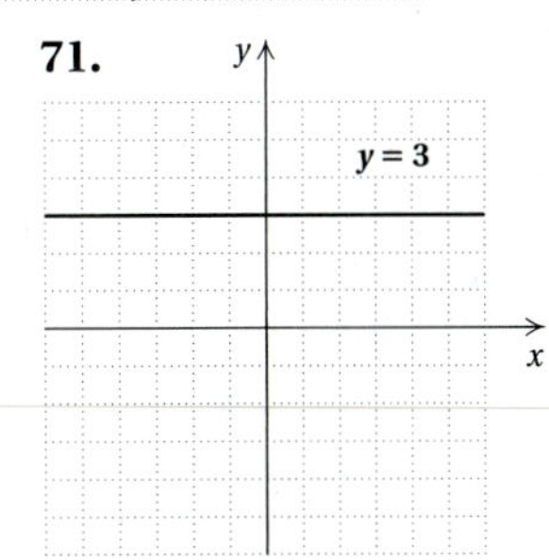

72.

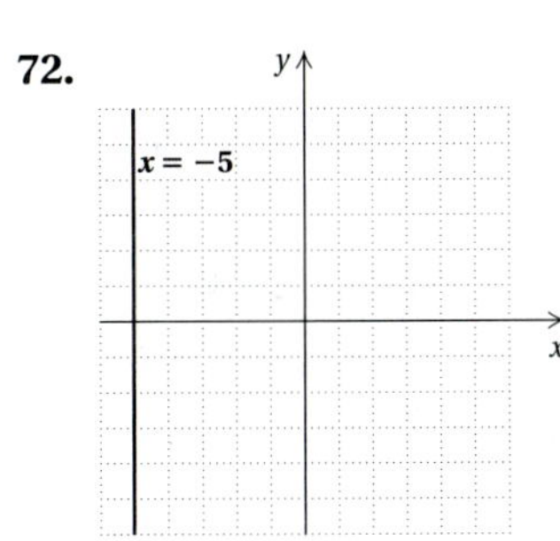

73. -8 **74.** $\frac{5}{6}$
75. 3, 5 **76.** $-2, 9$
77. Perimeter: $\frac{16y + 28}{15}$; area: $\frac{y^2 + 2y - 8}{15}$
79. $\frac{(z + 6)(2z - 3)}{(z + 2)(z - 2)}$

81. $\frac{11z^4 - 22z^2 + 6}{(z^2 + 2)(z^2 - 2)(2z^2 - 3)}$
83.–85. Left to the student

Margin Exercises, Section 11.5, pp. 861–864

1. $\frac{4}{11}$ **2.** $\frac{5}{y}$ **3.** $\frac{x^2 + 2x + 1}{2x + 1}$ **4.** $\frac{-x - 7}{15x}$
5. $\frac{x^2 - 48}{(x + 7)(x + 8)(x + 6)}$ **6.** $\frac{3x - 1}{3}$ **7.** $\frac{4x - 3}{x - 2}$
8. $\frac{-8y - 28}{(y + 4)(y - 4)}$ **9.** $\frac{x - 13}{(x + 3)(x - 3)}$ **10.** $\frac{6x^2 - 2x - 2}{3x(x + 1)}$

Exercise Set 11.5, p. 865

1. $\frac{4}{x}$ **3.** 1 **5.** $\frac{1}{x - 1}$ **7.** $\frac{-a - 4}{10}$ **9.** $\frac{7z - 12}{12z}$
11. $\frac{4x^2 - 13xt + 9t^2}{3x^2t^2}$ **13.** $\frac{2x - 40}{(x + 5)(x - 5)}$ **15.** $\frac{3 - 5t}{2t(t - 1)}$
17. $\frac{2s - st - s^2}{(t + s)(t - s)}$ **19.** $\frac{y - 19}{4y}$ **21.** $\frac{-2a^2}{(x + a)(x - a)}$
23. $\frac{8}{3}$ **25.** $\frac{13}{a}$ **27.** $\frac{8}{y - 1}$ **29.** $\frac{x - 2}{x - 7}$ **31.** $\frac{4}{a^2 - 25}$
33. $\frac{2x - 4}{x - 9}$ **35.** $\frac{9x + 12}{(x + 3)(x - 3)}$ **37.** $\frac{1}{2}$
39. $\frac{x - 3}{(x + 3)(x + 1)}$ **41.** $\frac{18x + 5}{x - 1}$ **43.** 0 **45.** $\frac{-9}{2x - 3}$

47. $\frac{20}{2y - 1}$ **49.** $\frac{2a - 3}{2 - a}$ **51.** $\frac{z - 3}{2z - 1}$ **53.** $\frac{2}{x + y}$
55. $\mathbf{D_W}$ **57.** x^5 **58.** $30x^{12}$ **59.** $\frac{b^{20}}{a^8}$ **60.** $18x^3$
61. $\frac{6}{x^3}$ **62.** $\frac{10}{x^3}$ **63.** $x^2 - 9x + 18$ **64.** $(4 - \pi)r^2$
65. $\frac{30}{(x - 3)(x + 4)}$ **67.** $\frac{x^2 + xy - x^3 + x^2y - xy^2 + y^3}{(x^2 + y^2)(x + y)^2(x - y)}$
69. Missing side: $\frac{-2a - 15}{a - 6}$; area: $\frac{-2a^3 - 15a^2 + 12a + 90}{2(a - 6)^2}$
71.–73. Left to the student

Margin Exercises, Section 11.6, pp. 870–872

1. $\frac{136}{5}$ **2.** $\frac{7x^2}{3(2 - x^2)}$ **3.** $\frac{x}{x - 1}$ **4.** $\frac{136}{5}$ **5.** $\frac{7x^2}{3(2 - x^2)}$
6. $\frac{x}{x - 1}$

Exercise Set 11.6, p. 873

1. $\frac{25}{4}$ **3.** $\frac{1}{3}$ **5.** -6 **7.** $\frac{1 + 3x}{1 - 5x}$ **9.** $\frac{2x + 1}{x}$ **11.** 8
13. $x - 8$ **15.** $\frac{y}{y - 1}$ **17.** $-\frac{1}{a}$ **19.** $\frac{ab}{b - a}$
21. $\frac{p^2 + q^2}{q + p}$ **23.** $\frac{2a^2 + 4a}{5 - 3a^2}$ **25.** $\frac{60 - 15a^3}{126a^2 + 28a^3}$
27. $\frac{ac}{bd}$ **29.** 1 **31.** $\frac{zw(w - z)}{w^2 - wz + z^2}$ **33.** $\mathbf{D_W}$
35. $4x^4 + 3x^3 + 2x - 7$ **36.** 0 **37.** $(p - 5)^2$
38. $(p + 5)^2$ **39.** $50(p^2 - 2)$ **40.** $5(p + 2)(p - 10)$
41. 14 yd **42.** 12 ft, 5 ft **43.** $\frac{(x - 1)(3x - 2)}{5x - 3}$
45. $\frac{5x + 3}{3x + 2}$

Margin Exercises, Section 11.7, pp. 875–879

1. $\frac{33}{2}$ **2.** $\frac{3}{2}$ **3.** 3 **4.** $-\frac{1}{8}$ **5.** 1 **6.** 2 **7.** 4

Calculator Corner, p. 878

1. Left to the student **2.** Left to the student

Study Tip, p. 880

1. Rational expression **2.** Solutions **3.** Rational expression **4.** Rational expression **5.** Rational expression **6.** Solutions **7.** Rational expression **8.** Solutions **9.** Solutions **10.** Solutions **11.** Rational expression **12.** Solutions **13.** Rational expression

Exercise Set 11.7, p. 881

1. $\frac{6}{5}$ **3.** $\frac{40}{29}$ **5.** $\frac{47}{2}$ **7.** -6 **9.** $\frac{24}{7}$ **11.** $-4, -1$
13. $-4, 4$ **15.** 3 **17.** $\frac{14}{3}$ **19.** 5 **21.** 5 **23.** $\frac{5}{2}$

25. -2 **27.** $-\frac{13}{2}$ **29.** $\frac{17}{2}$ **31.** No solution **33.** -5
35. $\frac{5}{3}$ **37.** $\frac{1}{2}$ **39.** No solution **41.** No solution
43. 4 **45.** $\mathbf{D_W}$ **47.** $\frac{1}{a^6b^{15}}$ **48.** x^8y^{12} **49.** $\frac{16x^4}{t^8}$
50. $\frac{w^4}{y^6}$ **51.** $32x^6$ **52.** $\frac{64x^{10}}{y^8}$

53.

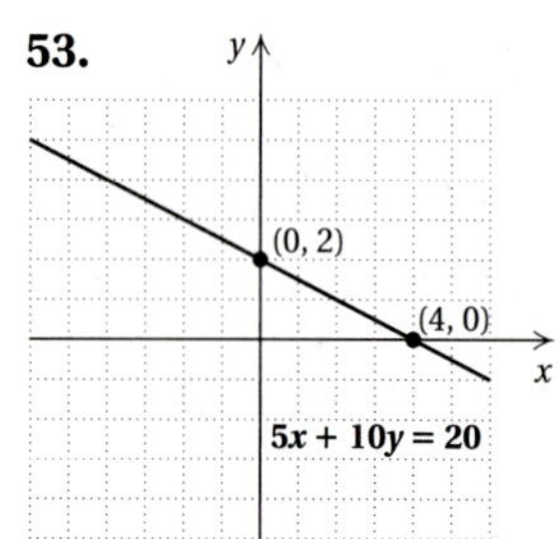

54.

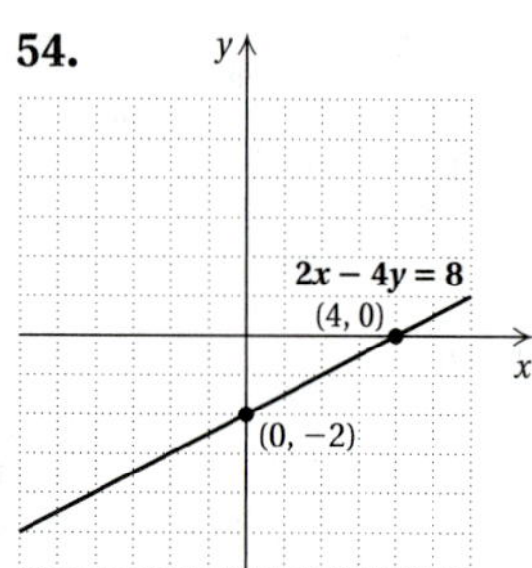

55.

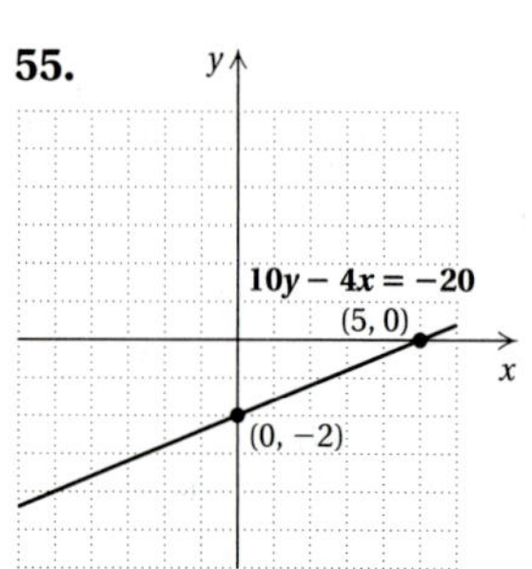

56.

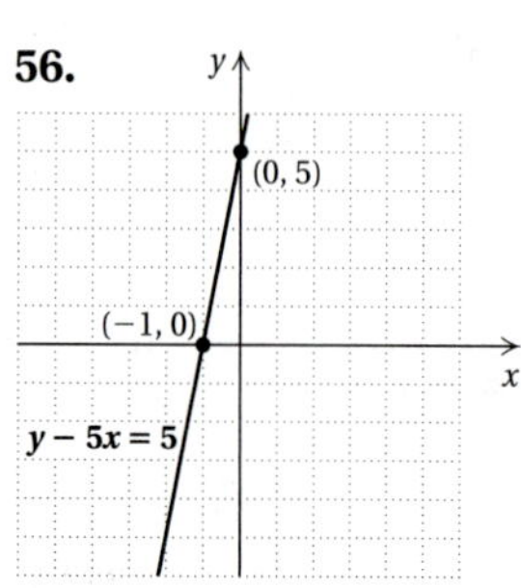

57. 7 **59.** No solution **61.** $-2, 2$ **63.** 4
65. Left to the student

Margin Exercises, Section 11.8, pp. 887–892

1. $3\frac{3}{7}$ hr **2.** Greg: 40 mph; Nancy: 60 mph **3.** 58 km/L
4. 0.280 **5.** 124 km/h **6.** 2.4 fish/yd^2 **7.** About 34.6 gal **8.** 90 whales **9.** No **10.** 24.75 ft
11. About 34.9 ft

Exercise Set 11.8, p. 893

1. $2\frac{2}{9}$ hr **3.** $25\frac{5}{7}$ min **5.** $3\frac{15}{16}$ hr **7.** $22\frac{2}{9}$ min
9. $7\frac{1}{2}$ min **11.** Sarah: 30 km/h; Rick: 70 km/h
13. Passenger: 80 mph; freight: 66 mph **15.** 20 mph
17. Hank: 14 km/h; Kelly: 19 km/h **19.** Ralph: 5 km/h; Bonnie: 8 km/h **21.** 3 hr **23.** $\frac{5}{9}$ divorce/marriage
25. 2.3 km/h **27.** 66 g **29.** 1.92 g **31.** 1.75 lb
33. $1\frac{11}{39}$ kg **35.** (a) 0.361; (b) 268 hits; (c) 202 hits
37. 22 in.; 55.8 cm **39.** $7\frac{1}{4}$; 57.9 cm **41.** $7\frac{1}{2}$; $23\frac{3}{5}$ in.
43. 10,000 blue whales **45.** (a) 4.8 tons; (b) 48 lb
47. 200 duds **49.** $\frac{21}{2}$ **51.** $\frac{8}{3}$ **53.** $\frac{35}{3}$ **55.** 15 ft
57. $\mathbf{D_W}$ **59.** x^{11} **60.** x **61.** $\frac{1}{x^{11}}$ **62.** $\frac{1}{x}$

63.

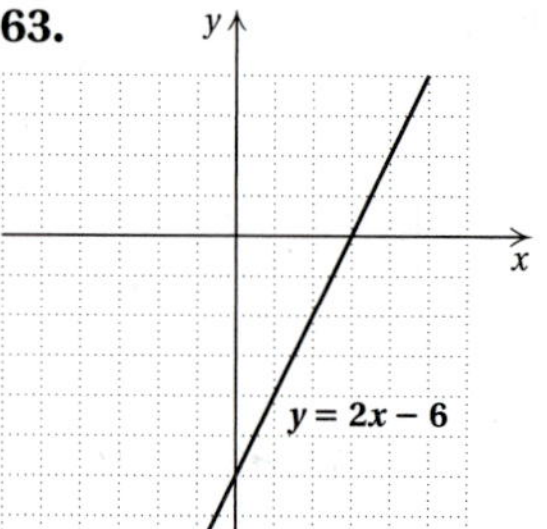

64.

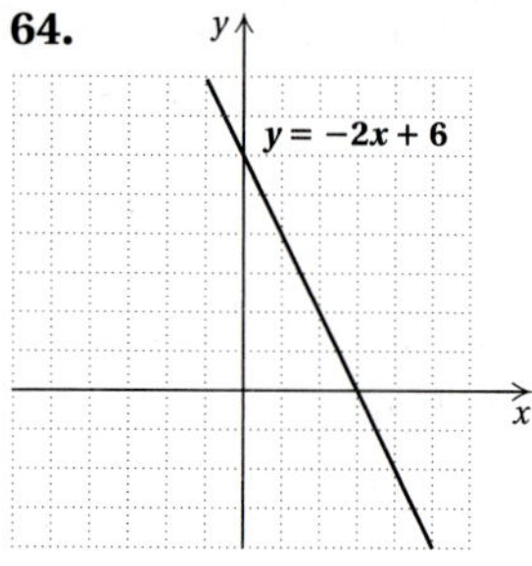

65.

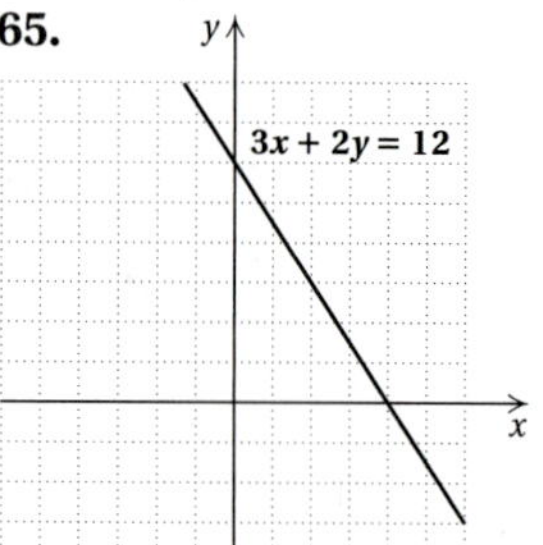

66.

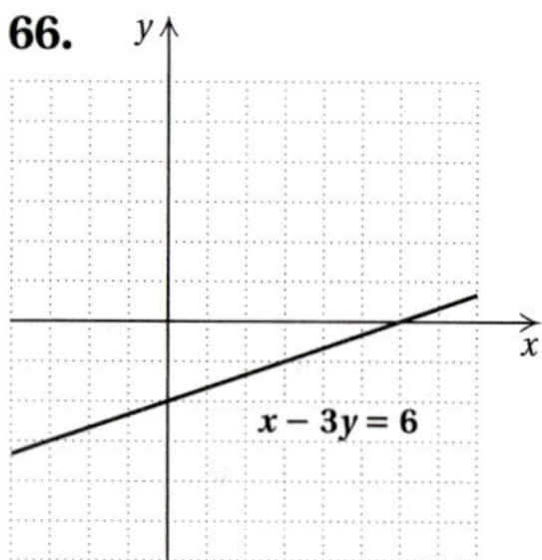

67.

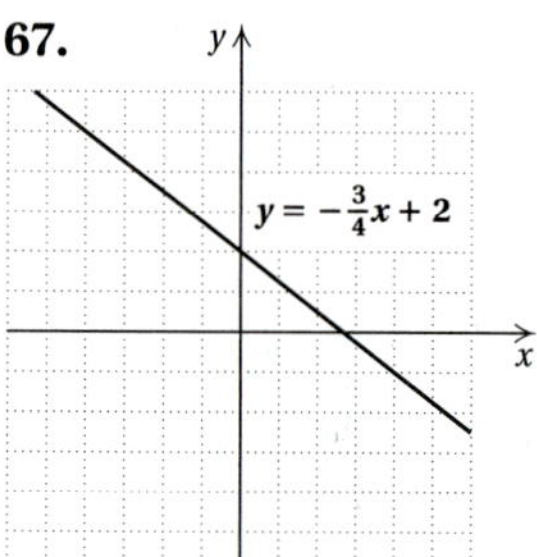

68.

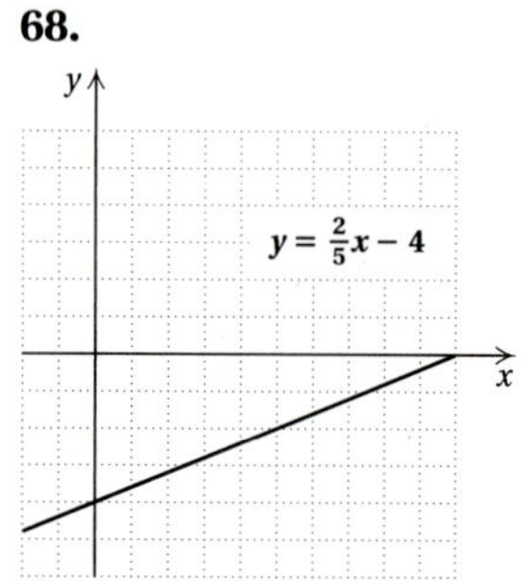

69. Ann: 6 hr; Betty: 12 hr **71.** $27\frac{3}{11}$ min **73.** $t = \frac{ab}{b + a}$

Margin Exercises, Section 11.9, pp. 901–906

1. $\frac{2}{5}$; $y = \frac{2}{5}x$ **2.** 0.7; $y = 0.7x$ **3.** 50 volts
4. 1,575,000 tons **5.** 0.6; $y = \frac{0.6}{x}$ **6.** 16 hr
7. $y = 7x^2$ **8.** $y = \frac{9}{x^2}$ **9.** $y = \frac{1}{2}xz$ **10.** $y = \frac{5xz^2}{w}$
11. 490 m **12.** 1 ohm

Exercise Set 11.9, p. 907

1. 5; $y = 5x$ **3.** $\frac{2}{15}$; $y = \frac{2}{15}x$ **5.** $\frac{9}{4}$; $y = \frac{9}{4}x$
7. 285,360,000 cans **9.** $66\frac{2}{3}$ cm **11.** 90 g **13.** 40 kg
15. 98; $y = \frac{98}{x}$ **17.** 36; $y = \frac{36}{x}$ **19.** 0.05; $y = \frac{0.05}{x}$
21. 3.5 hr **23.** $\frac{2}{9}$ ampere **25.** 1.92 ft **27.** 160 cm^3
29. $y = 15x^2$ **31.** $y = \frac{0.0015}{x^2}$ **33.** $y = xz$
35. $y = \frac{3}{10}xz^2$ **37.** $y = \frac{xz}{5wp}$ **39.** 36 mph **41.** 2.5 m
43. 106 earned runs **45.** 729 gal **47.** $\mathbf{D_W}$
49. $(x - 8)(x + 7)$ **50.** $(a - 8)^2$ **51.** $x^3(x - 7)(x + 5)$
52. $(2y^2 + 1)(y - 5)$ **53.** $(2 - t)(2 + t)(4 + t^2)$
54. $10(x + 7)(x + 1)$ **55.** $(x - 7)(x - 2)$

56. Not factorable **57.** $(4x - 5y)^2$
58. $(a - 7b)(a - 2b)$ **59.** $3(x - y)(x^2 + xy + y^2)$
60. $(w + t)(w^2 - wt + t^2)(w - t)(w^2 + wt + t^2)$

61. $\frac{\pi}{4}$ **63.** Q varies directly as the square of p and inversely as the cube of q. **65.** \$7.20

Summary and Review: Chapter 11, p. 913

1. 0 **2.** 6 **3.** $-6, 6$ **4.** $-6, 5$ **5.** -2 **6.** 0, 3, 5
7. $\frac{x-2}{x+1}$ **8.** $\frac{7x+3}{x-3}$ **9.** $\frac{1}{a-2}$ **10.** $\frac{a-6}{5}$
11. $\frac{6}{2t-1}$ **12.** $\frac{x^2+4x+16}{x+2}$ **13.** $-20t$ **14.** $\frac{2x^2-2x}{x+1}$
15. $30x^2y^2$ **16.** $4(a-2)$ **17.** $(y-2)(y+2)(y+1)$
18. $\frac{-3x+18}{x+7}$ **19.** -1 **20.** $\frac{2a}{a-1}$ **21.** $d + c$
22. $\frac{4}{x-4}$ **23.** $\frac{x+5}{2x}$ **24.** $\frac{2x+3}{x-2}$
25. $\frac{-x^2+x+26}{(x-5)(x+5)(x+1)}$ **26.** $\frac{2(x-2)}{x+2}$ **27.** $\frac{z}{1-z}$
28. $c - d$ **29.** 8 **30.** $-5, 3$ **31.** $y = 4x$
32. $y = \frac{2500}{x}$ **33.** 20 min **34.** $5\frac{1}{7}$ hr
35. 240 km/h, 280 km/h **36.** 95 mph, 175 mph
37. 160 defective calculators **38.** **(a)** $\frac{12}{13}$ c; **(b)** $4\frac{1}{5}$ c; **(c)** $9\frac{1}{3}$ c
39. 7200 frogs **40.** 6 **41.** **DW** $\frac{5x+6}{(x+2)(x-2)}$; used to find an equivalent expression for each rational expression with the LCM as the least common denominator
42. **DW** $\frac{3x+10}{(x-2)(x+2)}$; used to find an equivalent expression for each rational expression with the LCM as the least common denominator **43.** **DW** 4; used to clear fractions **44.** **DW** $\frac{4(x-2)}{x(x+4)}$; method 1: used to multiply by 1 using LCM/LCM; method 2: LCM of the denominators in the numerator used to subtract in the numerator and LCM of the denominators in the denominator used to add in the denominator **45.** $(5x^2 - 3)(x + 4)$ **46.** $\frac{1}{125x^9y^6}$
47. $-2x^3 + 3x^2 + 12x - 18$ **48.** Length: 5 cm; width: 3 cm; perimeter: 16 cm **49.** $\frac{5(a+3)^2}{a}$
50. $\frac{10a}{(a-b)(b-c)}$ **51.** They are equivalent proportions.

Test: Chapter 11, p. 916

1. [11.1a] 0 **2.** [11.1a] -8 **3.** [11.1a] $-7, 7$
4. [11.1a] 1, 2 **5.** [11.1a] 1 **6.** [11.1a] $-5, -3, 0$
7. [11.1c] $\frac{3x+7}{x+3}$ **8.** [11.1d] $\frac{a+5}{2}$
9. [11.2b] $\frac{(5x+1)(x+1)}{3x(x+2)}$
10. [11.3c] $(y-3)(y+3)(y+7)$ **11.** [11.4a] $\frac{23-3x}{x^3}$
12. [11.5a] $\frac{8-2t}{t^2+1}$ **13.** [11.4a] $\frac{-3}{x-3}$ **14.** [11.5a] $\frac{2x-5}{x-3}$
15. [11.4a] $\frac{8t-3}{t(t-1)}$ **16.** [11.5a] $\frac{-x^2-7x-15}{(x+4)(x-4)(x+1)}$
17. [11.5b] $\frac{x^2+2x-7}{(x-1)^2(x+1)}$ **18.** [11.6a] $\frac{3y+1}{y}$
19. [11.7a] 12 **20.** [11.7a] $-3, 5$
21. [11.8b] 16 defective spark plugs **22.** [11.8b] 50 zebras
23. [11.8a] 12 min **24.** [11.8a] Craig: 65 km/h; Marilyn: 45 km/h **25.** [11.9e] $Q = \frac{5}{2}xy$ **26.** [11.9c] $y = \frac{250}{x}$
27. [11.9b] \$495 **28.** [11.8b] 15
29. [10.7a] $(4a+7)(4a-7)$ **30.** [9.2a, b] $\frac{y^{12}}{81x^8}$
31. [9.4c] $13x^2 - 29x + 76$ **32.** [10.9a] 21 and 22; -22 and -21 **33.** [11.8a] Rema: 4 hr; Reggie: 10 hr
34. [11.6a] $\frac{3a+2}{2a+1}$

Chapter 12

Pretest: Chapter 12, p. 920

1.

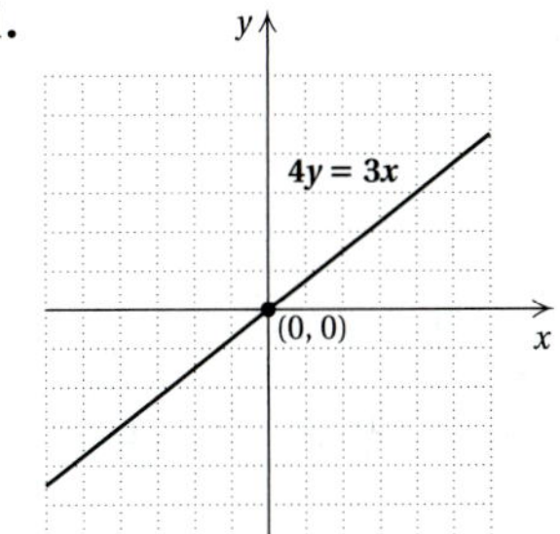

2.

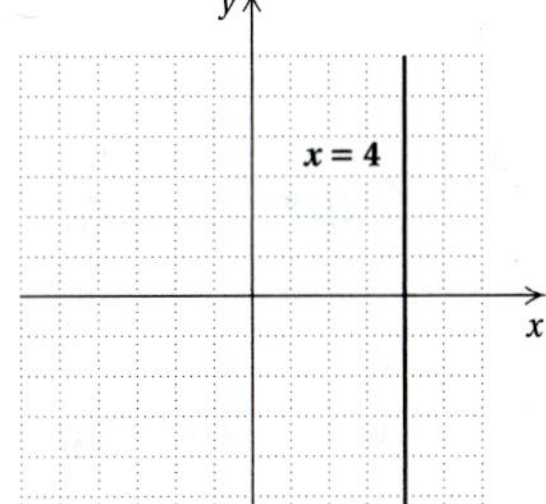

3.

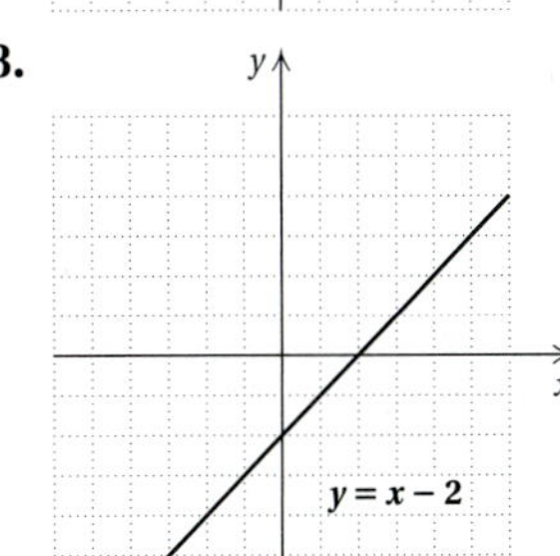

4.

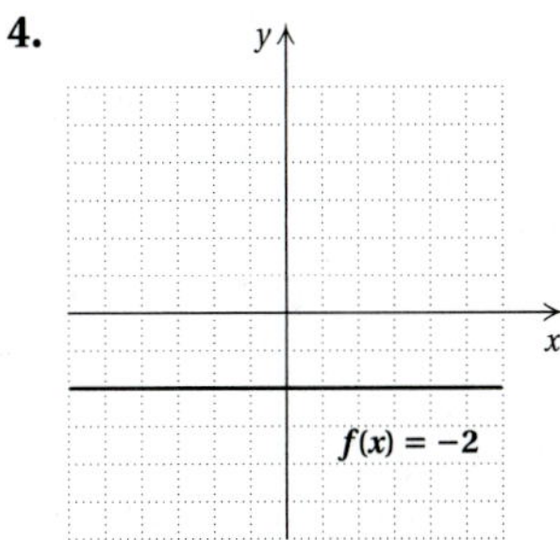

5.

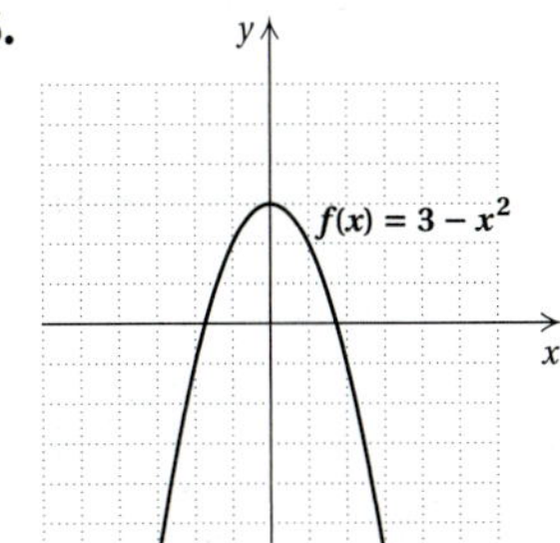

6. **(a)** Yes; **(b)** no

7. 5; $(0, -3)$ **8.** 0 **9.** $y = x + 10$ **10.** $y = 2x + 7$
11. $y = \frac{2}{7}x + \frac{31}{7}$ **12.** $y = -\frac{7}{2}x + 12$ **13.** Perpendicular

14. Parallel **15.** y-intercept: $(0, -4)$; x-intercept: $(6, 0)$
16. -3; -1; 1 **17.** $\{x \mid x \text{ is a real number } and\ x \neq \frac{5}{2}\}$
18. **(a)** $m(t) = 3.383t + 37.73$, where t is the number of years since 1991; **(b)** \$105.39

Margin Exercises, Section 12.1, pp. 922–928

1. Yes **2.** No **3.** Yes **4.** No **5.** Yes **6.** Yes
7. **(a)** -4; **(b)** 148; **(c)** 31; **(d)** $98a^2 + 21a - 4$
8. **(a)** -33; **(b)** 2; **(c)** 7; **(d)** $5a + 2$

9. 

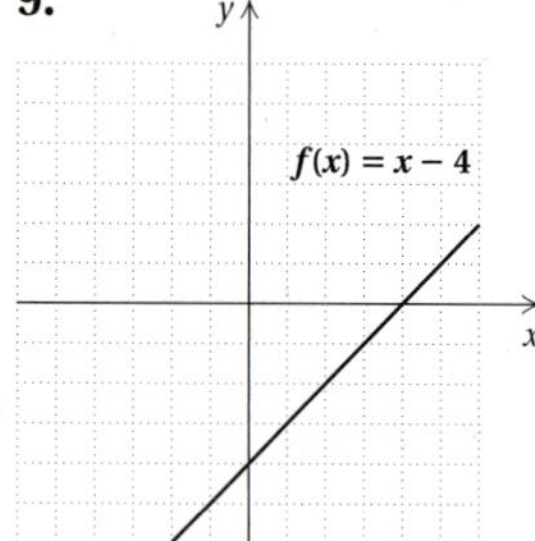

10. 

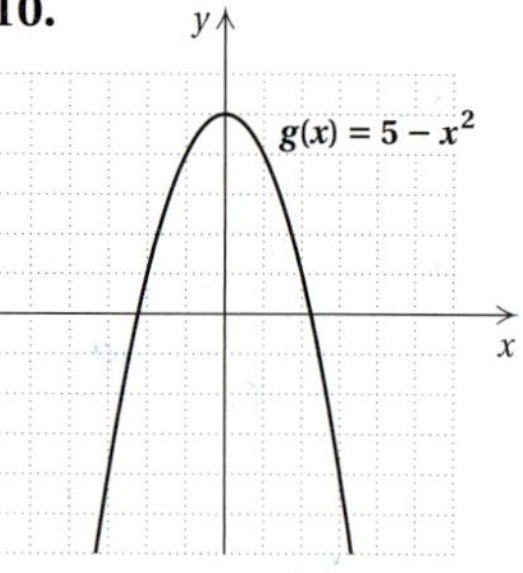

11.

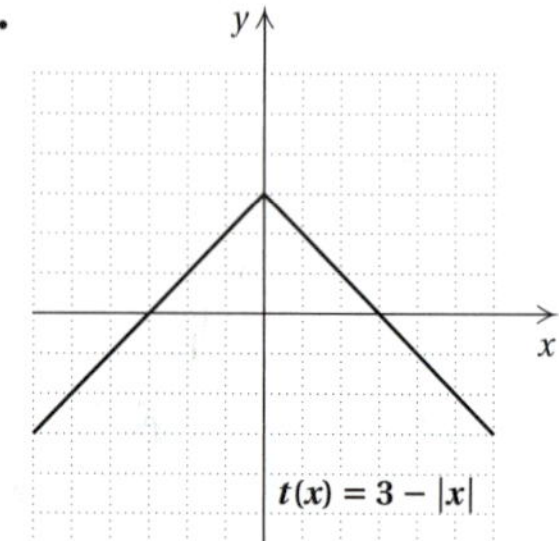

12. Yes **13.** No
14. No **15.** Yes
16. \$28 million
17. About \$8 million

Calculator Corner, p. 924

1. -13.3 **2.** -14.4 **3.** 14 **4.** 34

Calculator Corner, p. 927

1.

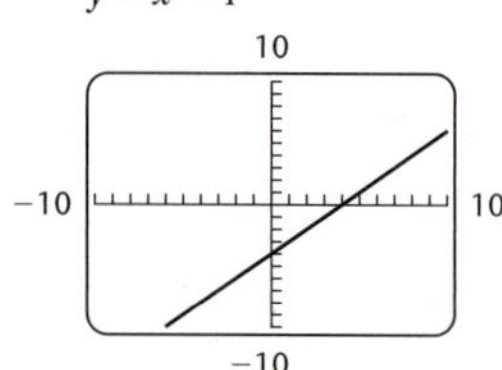

2.

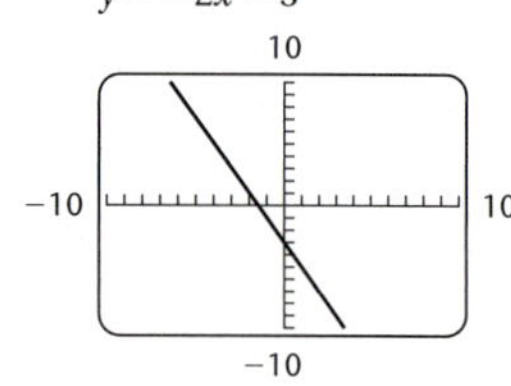

3.

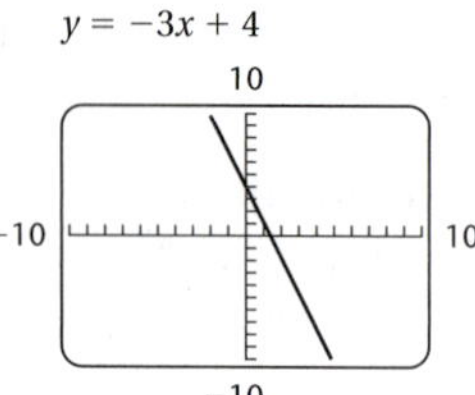

4.

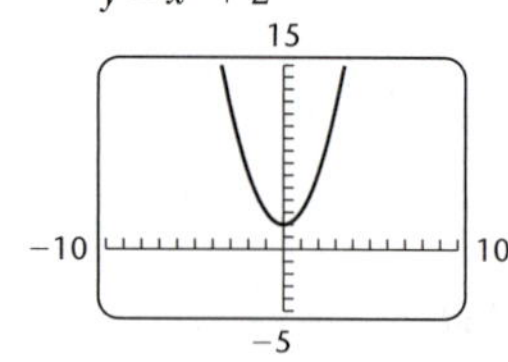

5.

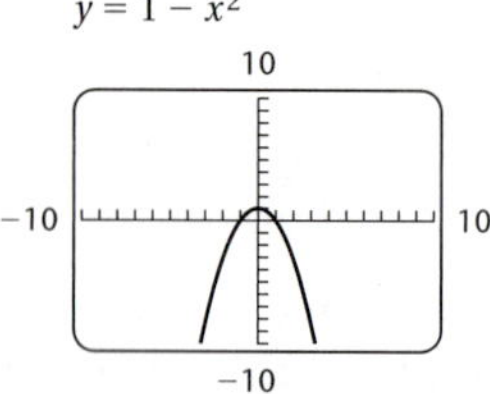

6.

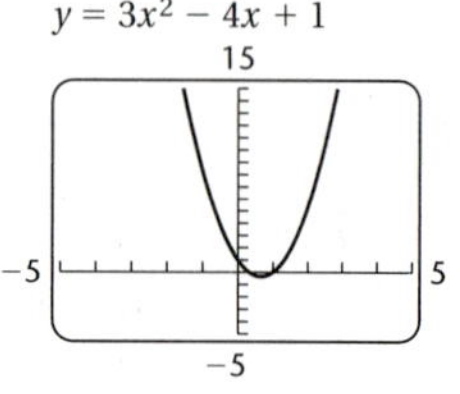

7.

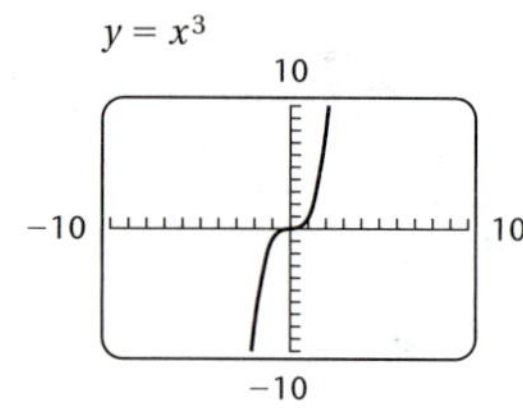

8.

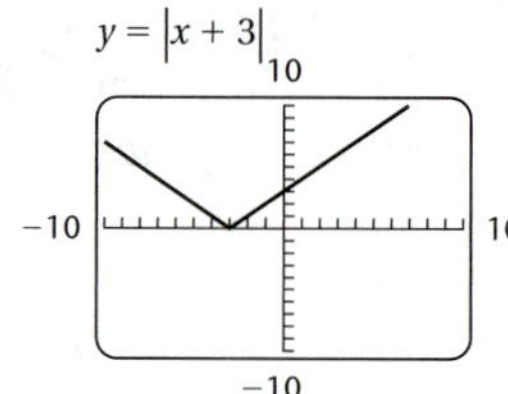

9.

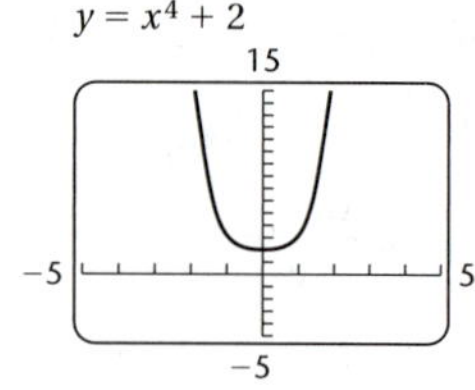

Exercise Set 12.1, p. 929

1. Yes **3.** Yes **5.** Yes **7.** No **9.** Yes **11.** No
13. Yes **15.** **(a)** 9; **(b)** 12; **(c)** 2; **(d)** 5; **(e)** 7.4; **(f)** $5\frac{2}{3}$
17. **(a)** -21; **(b)** 15; **(c)** 42; **(d)** 0; **(e)** 2; **(f)** $3a + 3$
19. **(a)** 7; **(b)** -17; **(c)** 24.1; **(d)** 4; **(e)** -26; **(f)** 6
21. **(a)** 0; **(b)** 5; **(c)** 2; **(d)** 170; **(e)** 65; **(f)** $32a^2 - 12a$
23. **(a)** 1; **(b)** 3; **(c)** 3; **(d)** 4; **(e)** 11; **(f)** $|a - 1| + 1$
25. **(a)** 0; **(b)** -1; **(c)** 8; **(d)** 1000; **(e)** -125; **(f)** -1000
27. **(a)** 159.48 cm; **(b)** 167.73 cm **29.** $1\frac{20}{33}$ atm; $1\frac{10}{11}$ atm; $4\frac{1}{33}$ atm **31.** 1.792 cm; 2.8 cm; 11.2 cm

33.

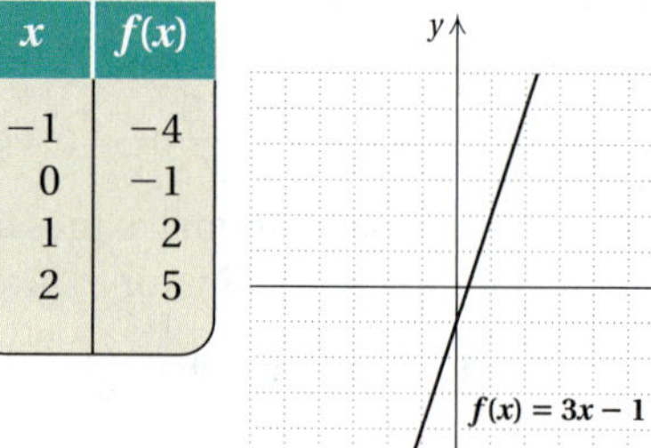

x	$f(x)$
-1	-4
0	-1
1	2
2	5

35.

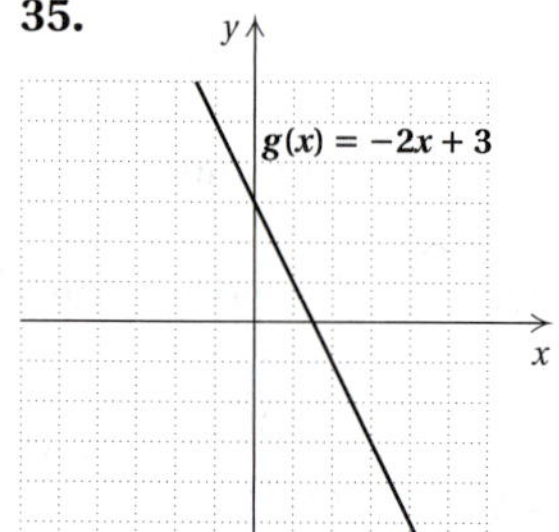

37.

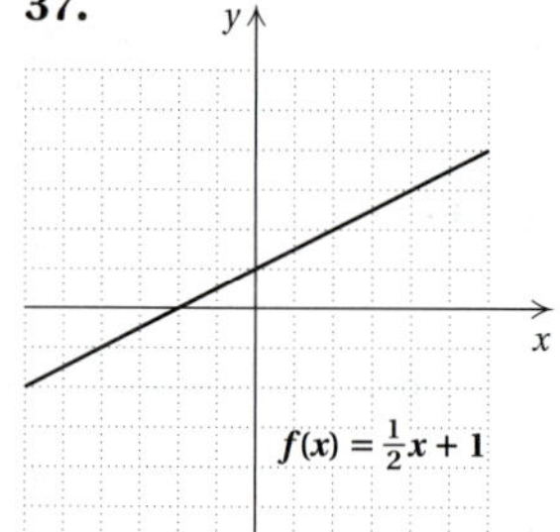

39.

x	$f(x)$
-3	-1
-2	0
-1	1
0	2
1	1
2	0
3	-1

y

$f(x) = 2 - |x|$

x

41.

x	$f(x)$
−3	9
−2	4
−1	1
0	0
1	1
2	4
3	9

$f(x) = x^2$

43.

x	$f(x)$
−3	10
−2	4
−1	0
0	−2
1	−2
2	0
3	4

$f(x) = x^2 - x - 2$

45. Yes **47.** Yes **49.** No **51.** No **53.** 75
55. About \$25 million **57.** $\mathbf{D_W}$ **59.** $\mathbf{D_W}$
61. 18, 20, 22 **62.** 12 ft; 20.8 ft **63.** $l = \dfrac{S - 2wh}{2h + 2w}$
64. $w = \dfrac{S - 2lh}{2l + 2h}$ **65.** 0.0000000093 **66.** 3,040,000
67. 1.075×10^9 **68.** 7.03×10^{-3}
69. $\left(w + \frac{1}{3}\right)\left(w^2 - \frac{1}{3}w + \frac{1}{9}\right)$ **70.** $(a - b)(a^2 + ab + b^2)$
71. $x(y + 4)(y^2 - 4y + 16)$
72. $(c - 0.2)(c^2 + 0.2c + 0.04)$ **73.** 26; 99
75. $g(x) = \frac{15}{4}x - \frac{13}{4}$

Margin Exercises, Section 12.2, pp. 936–938

1. Domain $= \{-3, -2, 0, 2, 5\}$; range $= \{-3, -2, 2, 3\}$
2. (a) -4; **(b)** $\{x \mid -3 \le x \le 3\}$; **(c)** $-3, 3$;
(d) $\{y \mid -5 \le y \le 4\}$ **3.** All real numbers
4. $\left\{x \mid x \text{ is a real number } \textit{and } x \ne -\frac{2}{3}\right\}$

Exercise Set 12.2, p. 939

1. (a) 3; **(b)** $\{-4, -3, -2, -1, 0, 1, 2\}$; **(c)** $-2, 0$; **(d)** $\{1, 2, 3, 4\}$
3. (a) 2; **(b)** $\{-6, -4, -2, 0, 1, 3, 4\}$; **(c)** 1, 3;
(d) $\{-5, -2, 0, 2, 5\}$ **5. (a)** $2\frac{1}{2}$; **(b)** $\{x \mid -3 \le x \le 5\}$; **(c)** $2\frac{1}{4}$;
(d) $\{y \mid 1 \le y \le 4\}$ **7. (a)** $2\frac{1}{4}$; **(b)** $\{x \mid -4 \le x \le 3\}$; **(c)** 0;
(d) $\{y \mid -5 \le y \le 4\}$ **9. (a)** 2; **(b)** $\{x \mid -5 \le x \le 4\}$;
(c) $\{x \mid 1 \le x \le 4\}$; **(d)** $\{y \mid -3 \le y \le 2\}$ **11. (a)** -1;
(b) $\{x \mid -6 \le x \le 5\}$; **(c)** $-4, 0, 3$; **(d)** $\{y \mid -2 \le y \le 2\}$
13. $\{x \mid x \text{ is a real number } \textit{and } x \ne -3\}$ **15.** All real numbers **17.** All real numbers **19.** $\left\{x \mid x \text{ is a real number } \textit{and } x \ne \frac{14}{5}\right\}$ **21.** All real numbers **23.** $\left\{x \mid x \text{ is a real number } \textit{and } x \ne \frac{3}{2}\right\}$ **25.** $\{x \mid x \text{ is a real number } \textit{and } x \ne 1\}$ **27.** All real numbers **29.** All real numbers
31. $\left\{x \mid x \text{ is a real number } \textit{and } x \ne \frac{5}{2}\right\}$ **33.** All real numbers
35. $\left\{x \mid x \text{ is a real number } \textit{and } x \ne -\frac{5}{4}\right\}$ **37.** $-8; 0; -2$
39. $\mathbf{D_W}$ **41.** $a - 1$ **42.** $\dfrac{2(y + 2)}{7(y + 7)}$ **43.** $\dfrac{5}{x + 2}$

44. $t - 4$ **45.** $w + 1$, R 2; or $w + 1 + \dfrac{2}{w + 3}$
46. $x^4 + x^3 - 2x^2 + x - 2$, R $2x + 3$; or
$x^4 + x^3 - 2x^2 + x - 2 + \dfrac{2x + 3}{x^2 - 1}$ **47.** $14x^2 + 57x - 27$
48. $a^2 - 1$ **49.** $81y^2 + 180y + 100$ **50.** $8w^2 - w - \frac{1}{4}$
51. $\{y \mid y \text{ is a real number } \textit{and } y \ne 0\}$; $\{y \mid y \ge 2\}$; $\{y \mid y \ge -4\}$; $\{y \mid y \ge 0\}$

Margin Exercises, Section 12.3, pp. 942–949

1.

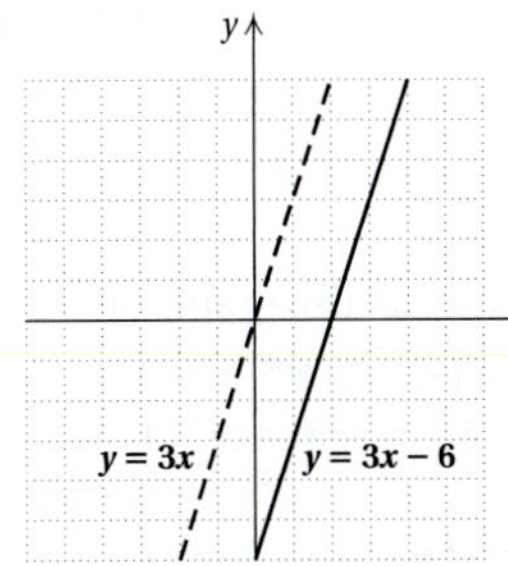

The graph of $y = 3x - 6$ looks just like the graph of $y = 3x$, but it is moved down 6 units.

2.

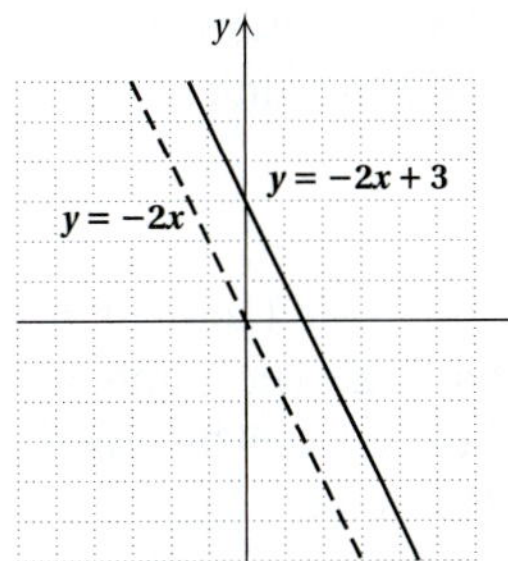

The graph of $y = -2x + 3$ looks just like the graph of $y = -2x$, but it is moved up 3 units.

3.

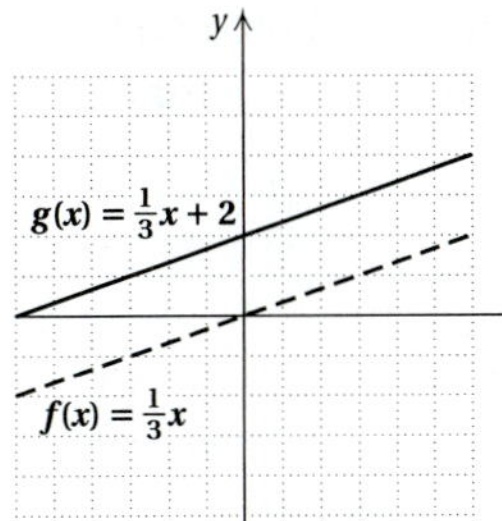

The graph of $g(x)$ looks just like the graph of $f(x)$, but it is moved up 2 units.

4. $(0, 8)$ **5.** $\left(0, -\frac{2}{3}\right)$

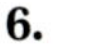

6.

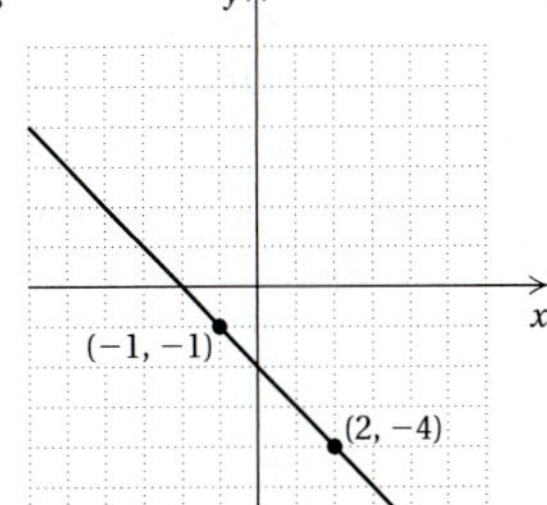

; $m = -1$

7. 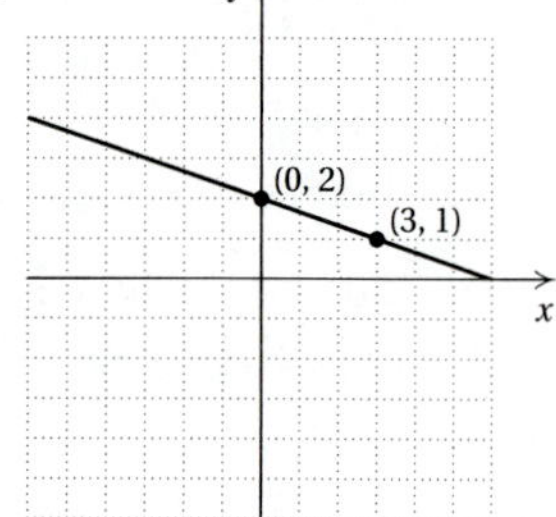

; $m = -\frac{1}{3}$ **8.** $m = -\frac{2}{3}$

9. Slope: -8; y-intercept: $(0, 23)$ **10.** Slope: $\frac{1}{2}$; y-intercept: $\left(0, -\frac{5}{2}\right)$ **11.** The rate of change is 3 haircuts per hour. **12.** The rate of change is $-\frac{3}{4}\%$ per year.

Calculator Corner, p. 941

1. The graph of $y_2 = x + 4$ is the same as the graph of $y_1 = x$, but it is moved up 4 units. **2.** The graph of $y_3 = x - 3$ is the same as the graph of $y_1 = x$, but it is moved down 3 units. **3.** The graph of $y = x + 8$ will be the same as the graph of $y_1 = x$, but it will be moved up 8 units. **4.** The graph of $y = x - 5$ will be the same as the graph of $y_1 = x$, but it will be moved down 5 units.

Calculator Corner, p. 945

1. The graph of $y = 10x$ will slant up from left to right. It will be steeper than the other graphs. **2.** The graph of $y = 0.005x$ will slant up from left to right. It will be less steep than the other graphs. **3.** The graph of $y = -10x$ will slant down from left to right. It will be steeper than the other graphs. **4.** The graph of $y = -0.005x$ will slant down from left to right. It will be less steep than the other graphs.

Exercise Set 12.3, p. 950

1. $m = 4$; y-intercept: $(0, 5)$ **3.** $m = -2$; y-intercept: $(0, -6)$ **5.** $m = -\frac{3}{8}$; y-intercept: $\left(0, -\frac{1}{5}\right)$ **7.** $m = 0.5$; y-intercept: $(0, -9)$ **9.** $m = \frac{2}{3}$; y-intercept: $\left(0, -\frac{8}{3}\right)$ **11.** $m = 3$; y-intercept: $(0, -2)$ **13.** $m = -8$; y-intercept: $(0, 12)$ **15.** $m = 0$; y-intercept: $\left(0, \frac{4}{17}\right)$ **17.** $m = -\frac{1}{2}$ **19.** $m = \frac{1}{3}$ **21.** $m = 2$ **23.** $m = \frac{2}{3}$ **25.** $m = -\frac{1}{3}$ **27.** $\frac{2}{25}$, or 8% **29.** 3.5% **31.** The rate of change is 5.06 billion messages daily per year. **33.** The rate of change is $-\$900$ per year. **35.** The rate of change is one point per \$1000 of family income. **37.** D_W **39.** -1323 **40.** $45x + 54$ **41.** $350x - 60y + 120$ **42.** 25 **43.** Square: 15 yd; triangle: 20 yd **44.** $(2 - 5x)(4 + 10x + 25x^2)$ **45.** $(c - d)(c^2 + cd + d^2)(c + d)(c^2 - cd + d^2)$ **46.** $7(2x - 1)(4x^2 + 2x + 1)$ **47.** $a - 10$, R -4; or $a - 10 + \dfrac{-4}{a - 1}$

Margin Exercises, Section 12.4, pp. 953–958

1. 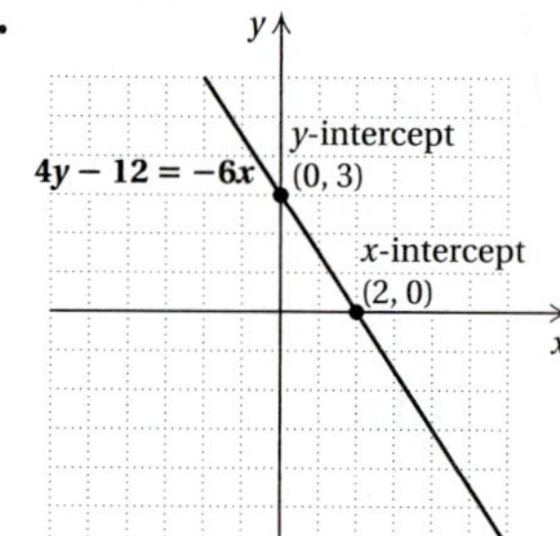

2. 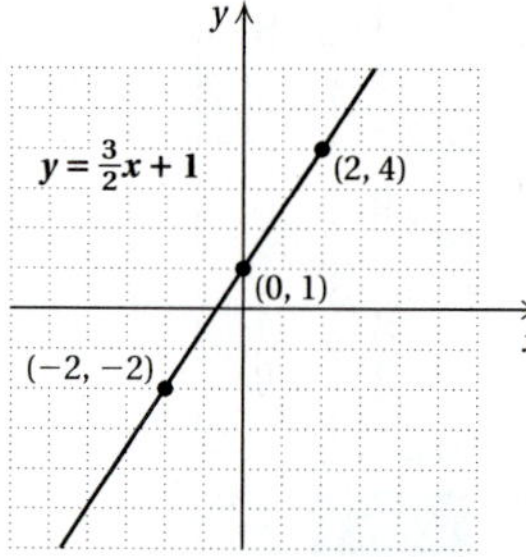

3. 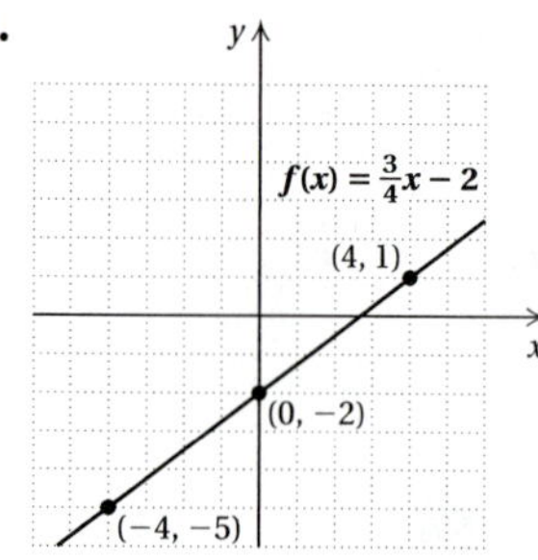

4. 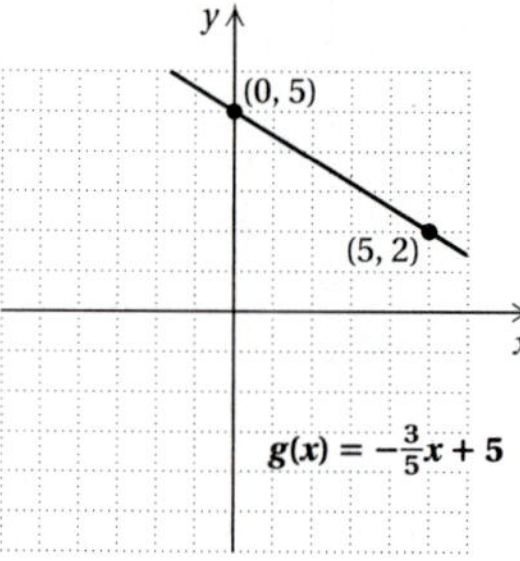

5. 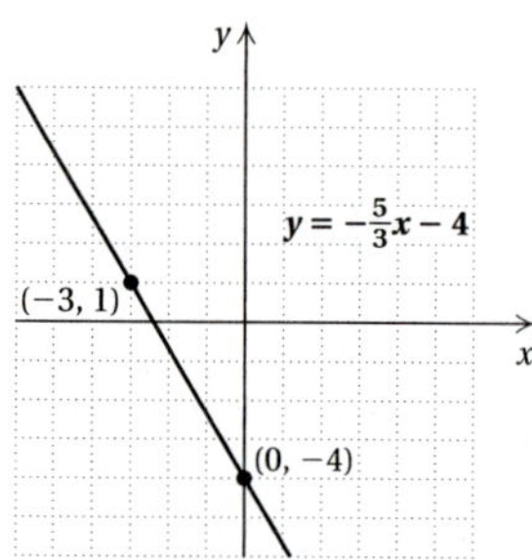

6. $m = 0$ 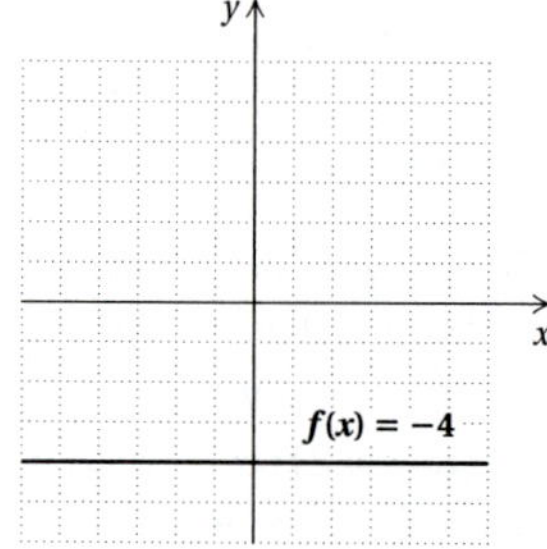

7. $m = 0$ 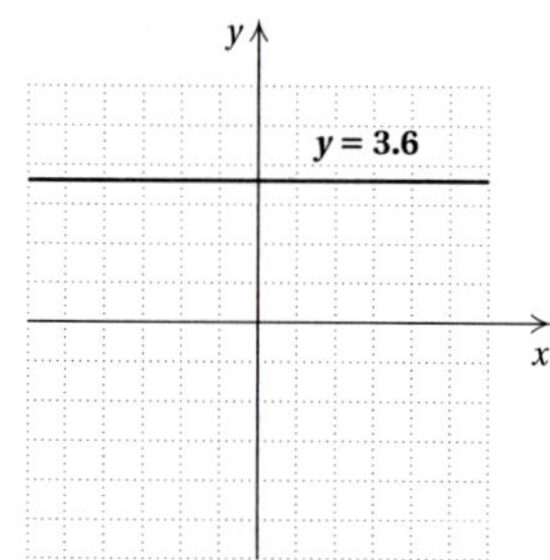

8. 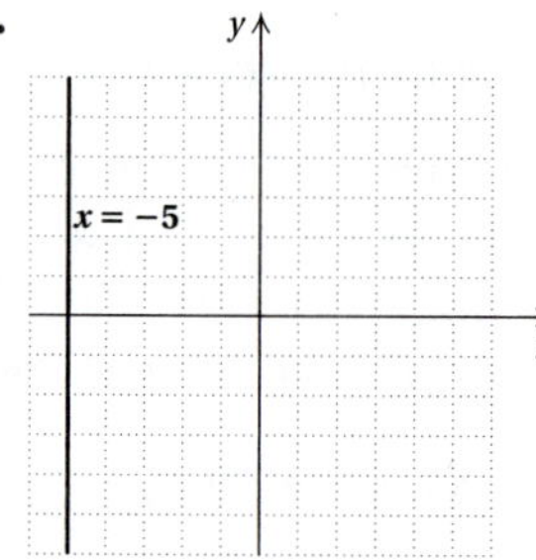

9. 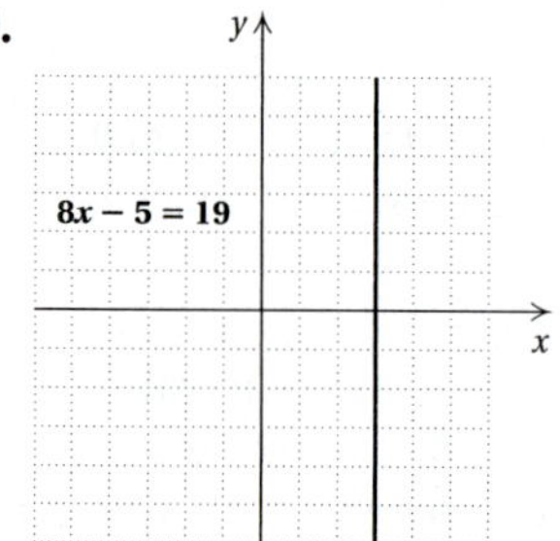

10. **(a)** Not defined; **(b)** $m = 0$; **(c)** $m = 0$; **(d)** not defined; **(e)** $m = 0$; **(f)** not defined **11.** Yes **12.** No **13.** No **14.** Yes **15.** No

Calculator Corner, p. 953

1. $y = -3.2x - 16$

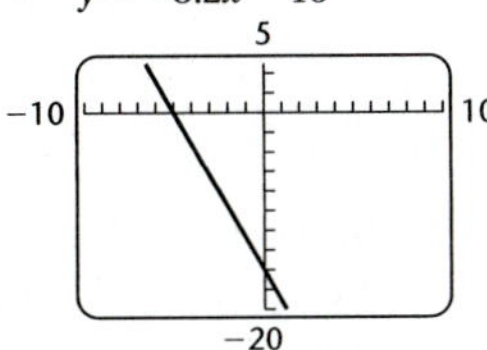

Xscl = 1   Yscl = 2

2. $y = 4.25x + 85$

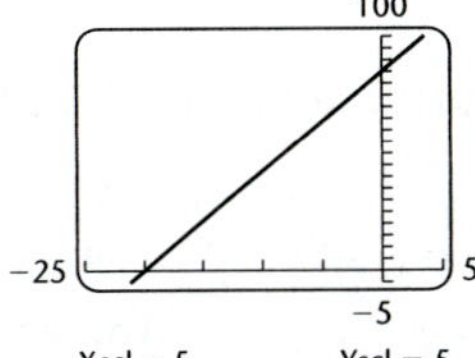

Xscl = 5   Yscl = 5

3. $y = (-6x + 90)/5$

25, −5, 25, −5

Xscl = 5   Yscl = 5

4. $y = (5x - 30)/6$

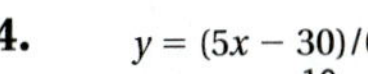

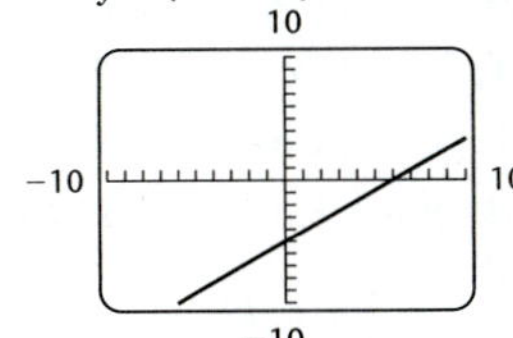

5. $y = (-8x + 9)/3$

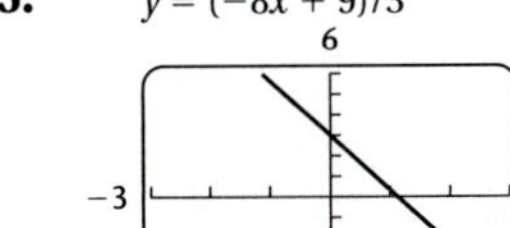

6. $y = 0.4x - 5$

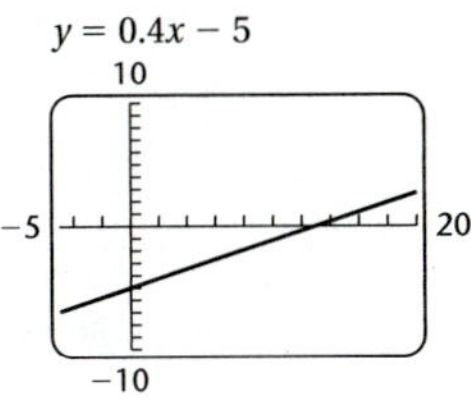

Xscl = 2   Yscl = 1

7. $y = 1.2x - 12$

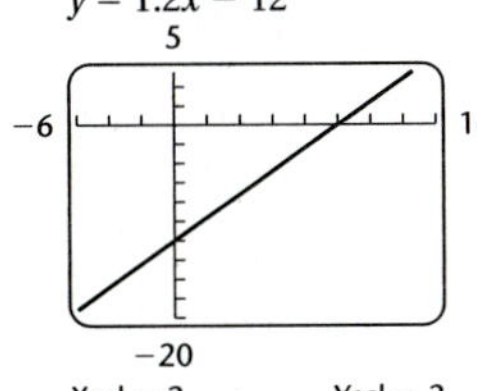

Xscl = 2   Yscl = 2

8. $y = (4x - 2)/5$

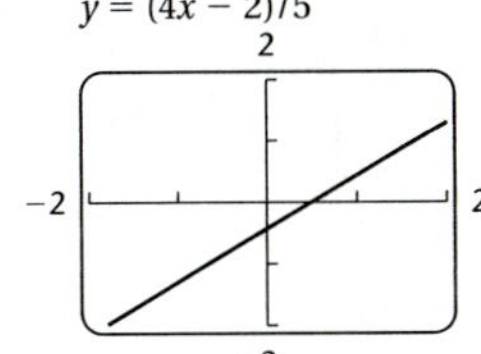

Calculator Corner, p. 959

1. Left to the student

Exercise Set 12.4, p. 960

1.

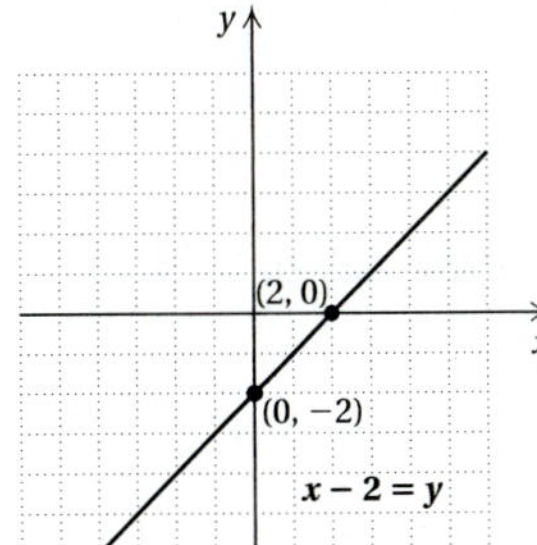

3.

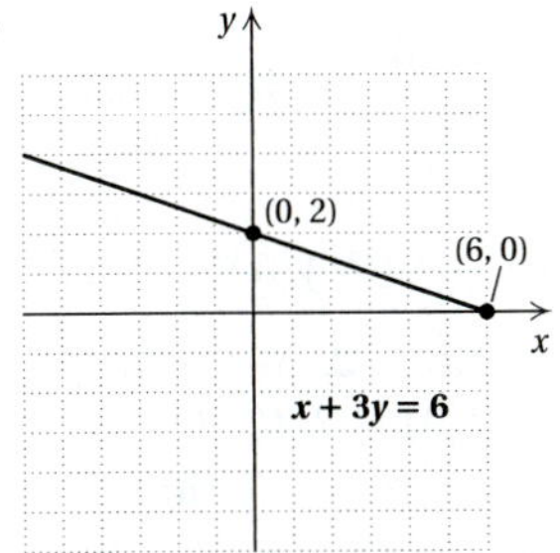

5.

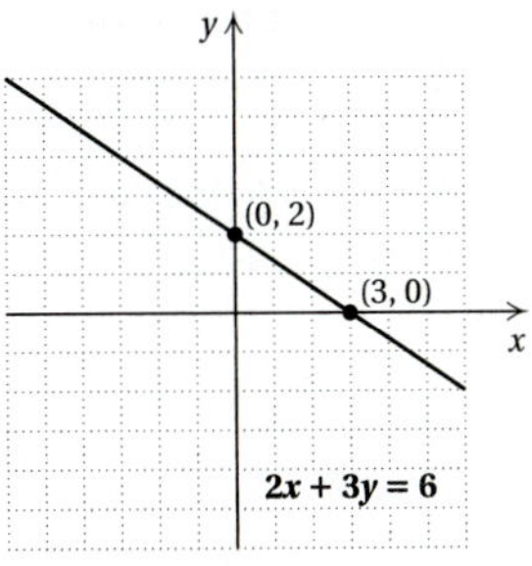

7.

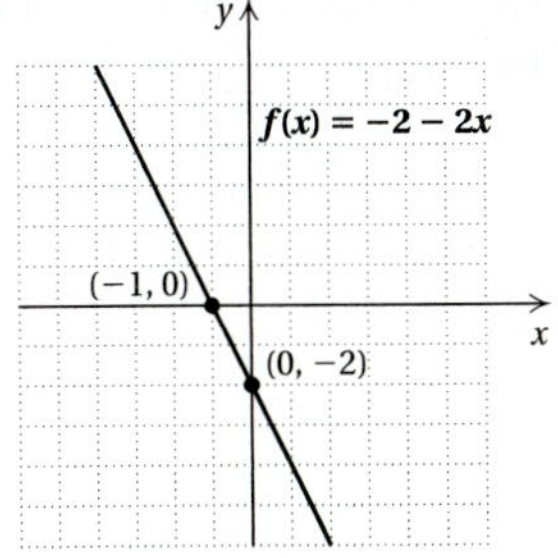

9.

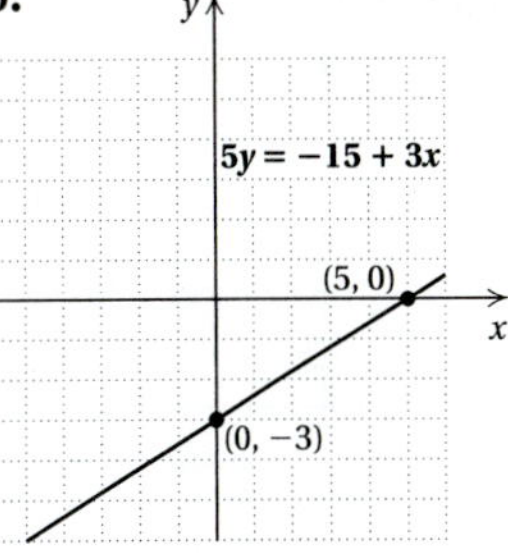

11.

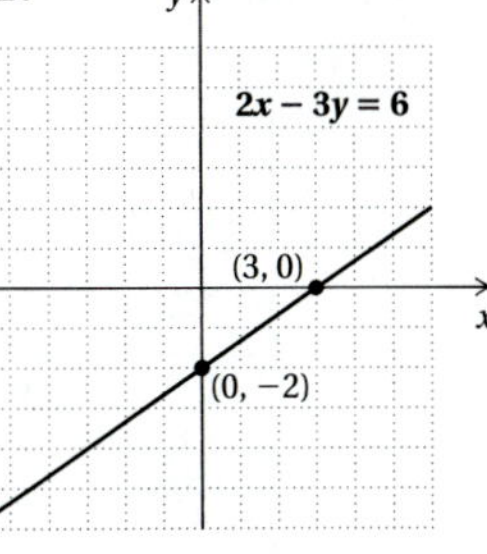

13.

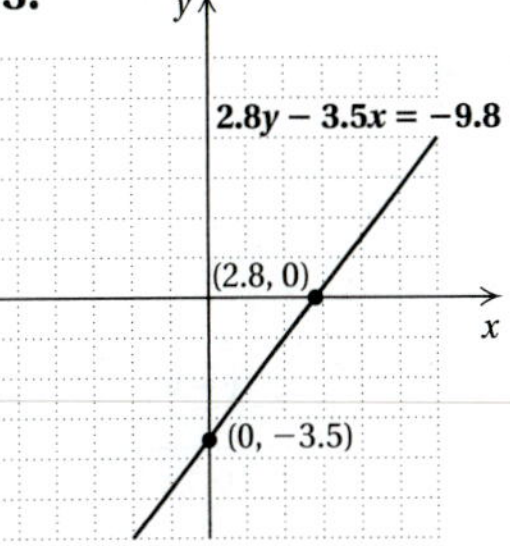

15.

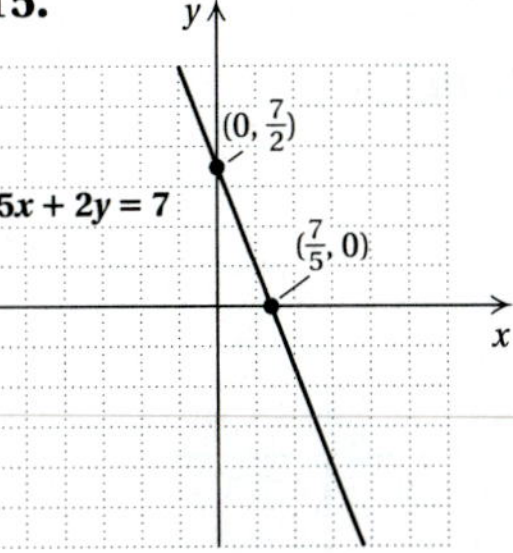

17.

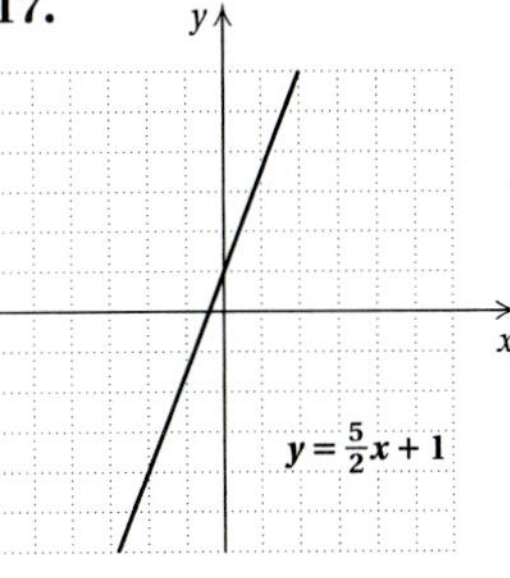

19.

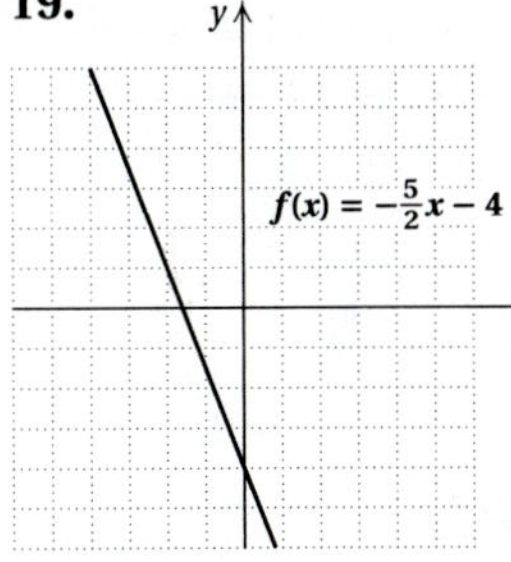

21.

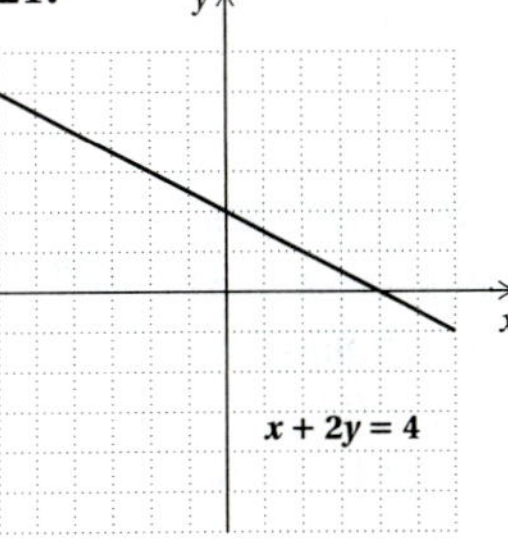

23.

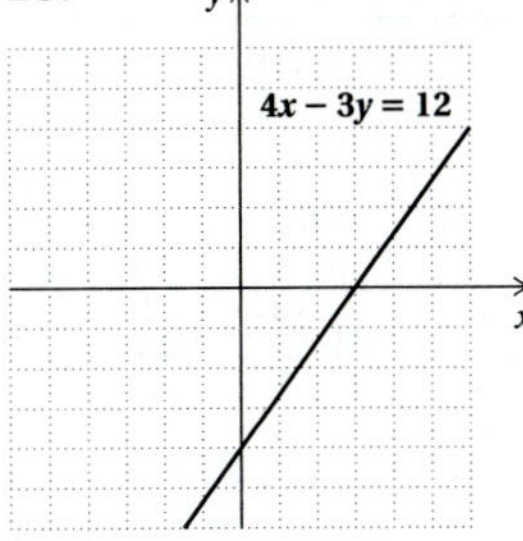

25.

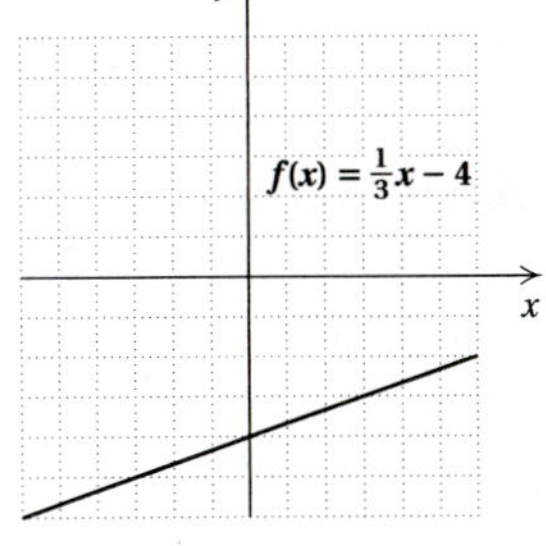

27.

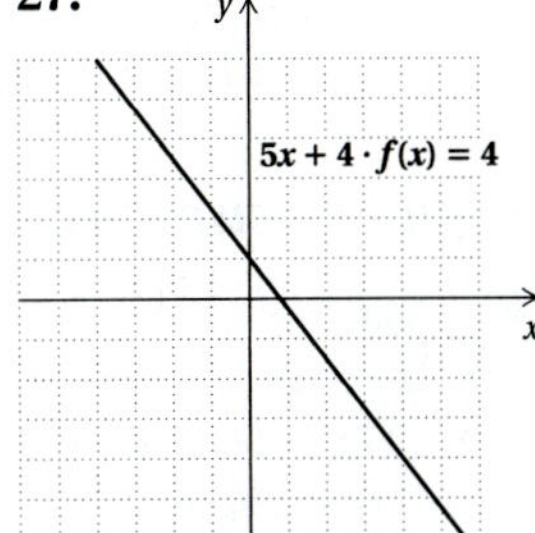

29. Not defined

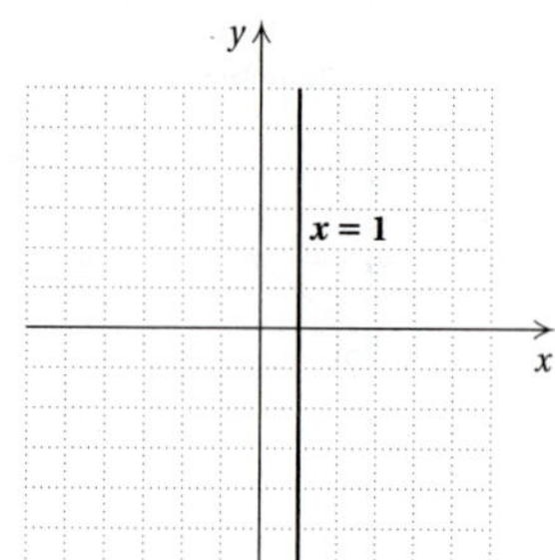

31. $m = 0$

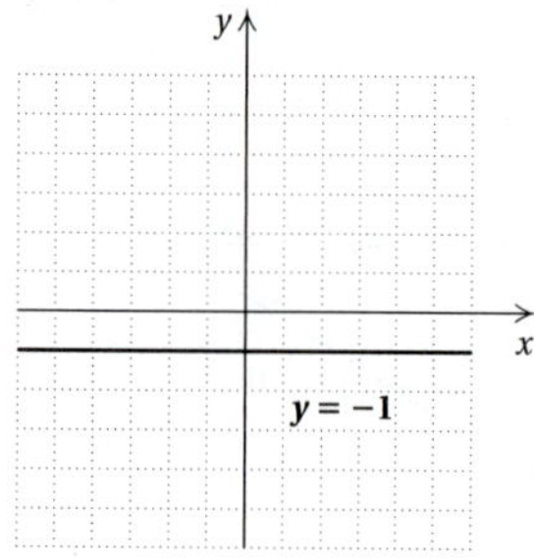

33. $m = 0$

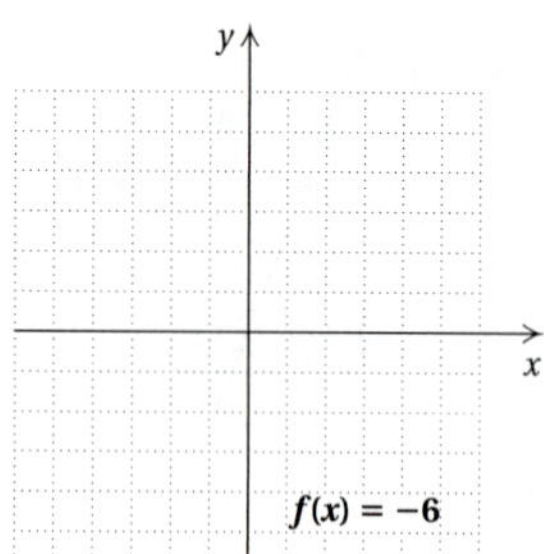

35. $m = 0$

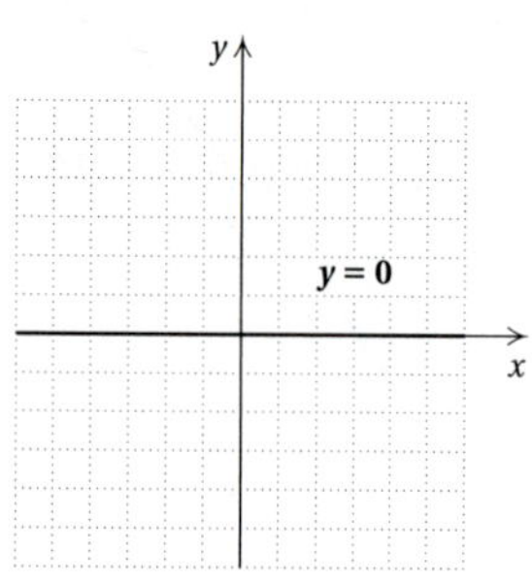

37. $m = 0$

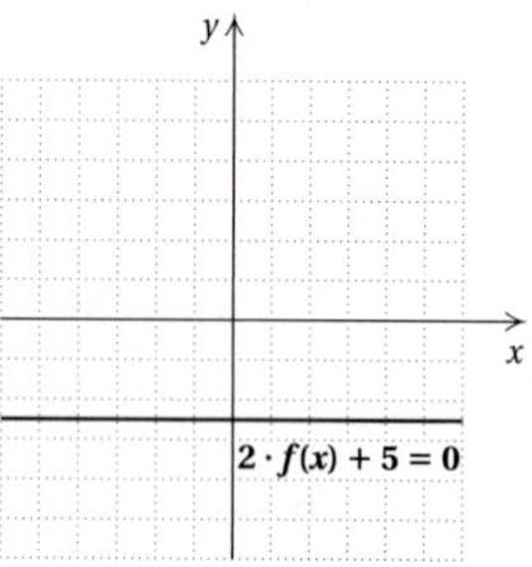

39. Not defined

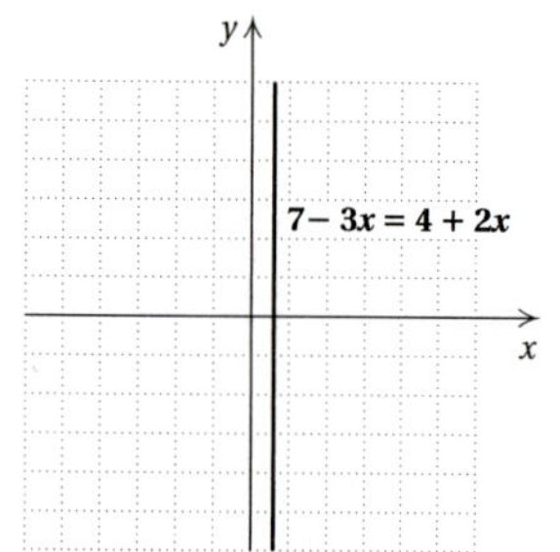

41. Yes **43.** No **45.** Yes **47.** Yes **49.** Yes **51.** No **53.** No **55.** Yes **57.** D_W **59.** 5.3×10^{10} **60.** 4.7×10^{-5} **61.** 1.8×10^{-2} **62.** 9.9902×10^{7} **63.** 0.0000213 **64.** 901,000,000 **65.** 20,000 **66.** 0.085677 **67.** $3(3x - 5y)$ **68.** $3a(4 + 7b)$ **69.** $7p(3 - q + 2)$ **70.** $64(x - 2y + 4)$ **71.** $y = 3$ **73.** $a = 2$ **75.** $y = \frac{2}{15}x + \frac{2}{5}$ **77.** $y = 0$; yes **79.** $m = -\frac{3}{4}$ **81.** **(a)** II; **(b)** IV; **(c)** I; **(d)** III

Margin Exercises, Section 12.5, pp. 965–971

1. $y = 3.4x - 8$ **2.** $y = -5x - 18$ **3.** $y = 3x - 5$ **4.** $y = 8x - 19$ **5.** $y = -\frac{2}{3}x + \frac{14}{3}$ **6.** $y = -\frac{5}{3}x + \frac{11}{3}$ **7.** $y = -17x - 56$ **8.** $y = \frac{8}{7}x - \frac{23}{7}$ **9.** $y = -2x + 14$ **10.** **(a)** $C(t) = 20t + 25$; **(b)**

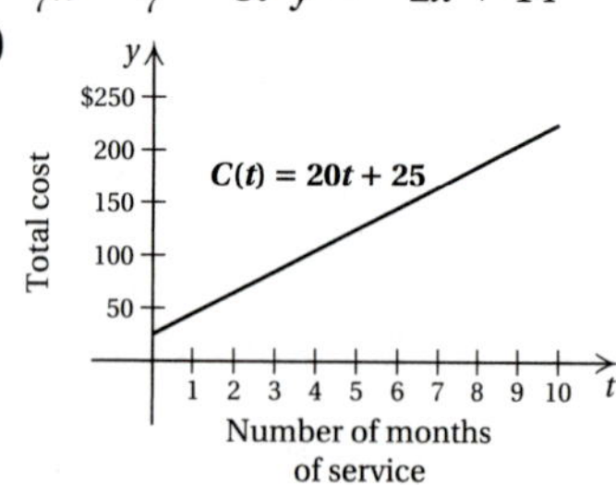

(c) $195 **11.** **(a)** $R(x) = 30.457x - 13.848$; **(b)** 5; **(c)** 0.832 in.

Calculator Corner, pp. 972–974

1. 5 **2.** **(a)** $y = 2.747311828x + 62.32258065$; **(b)** 90%

Exercise Set 12.5, p. 975

1. $y = -8x + 4$ **3.** $y = 2.3x - 1$ **5.** $f(x) = -\frac{7}{3}x - 5$ **7.** $f(x) = \frac{2}{3}x + \frac{5}{8}$ **9.** $y = 5x - 17$ **11.** $y = -3x + 33$ **13.** $y = x - 6$ **15.** $y = -2x + 16$ **17.** $y = -7$ **19.** $y = \frac{2}{3}x - \frac{8}{3}$ **21.** $y = \frac{1}{2}x + \frac{7}{2}$ **23.** $y = x$ **25.** $y = \frac{7}{4}x + 7$ **27.** $y = \frac{3}{2}x$ **29.** $y = \frac{1}{6}x$ **31.** $y = 13x - \frac{15}{4}$ **33.** $y = -\frac{1}{2}x + \frac{17}{2}$ **35.** $y = \frac{5}{7}x - \frac{17}{7}$ **37.** $y = \frac{1}{3}x + 4$ **39.** $y = \frac{1}{2}x + 4$ **41.** $y = \frac{4}{3}x - 6$ **43.** $y = \frac{5}{2}x + 9$ **45.** **(a)** $C(t) = 40t + 85$;
(b)

y
$500
400
300
200
100
Total cost
C(t) = 40t + 85
1 2 3 4 5 6 7 8 9 10 t
Number of hours of moving

; **(c)** $345

47. **(a)** $V(t) = 750 - 25t$;
(b)

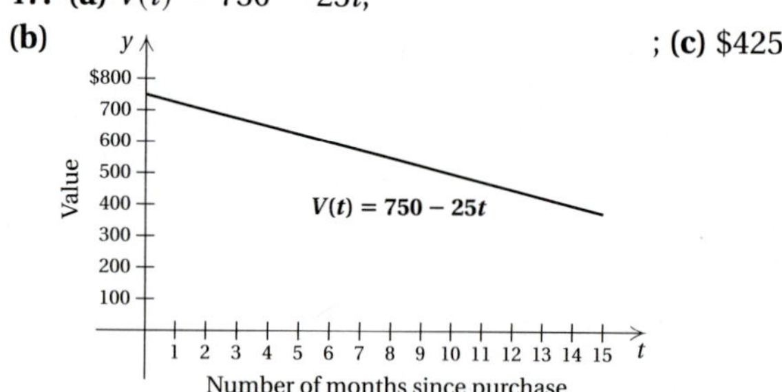

; **(c)** $425

49. **(a)** $S(x) = 115{,}900x + 155{,}100$; **(b)** $1,893,600; $2,473,100 **51.** **(a)** $R(t) = -0.075t + 46.8$; **(b)** 41.325 sec; 41.1 sec; **(c)** 2021 **53.** **(a)** $E(t) = \frac{9}{35}t + 71.8$; **(b)** 76.2 yr **55.** D_W **57.** -1 **58.** $b + 1$ **59.** $\dfrac{x - 3}{2(x - 5)}$ **60.** $\dfrac{4}{y - 9}$ **61.** $1350

Summary and Review: Chapter 12, p. 979

1. No **2.** Yes **3.** $g(0) = 5$; $g(-1) = 7$ **4.** $f(0) = 7$; $f(-1) = 12$ **5.** $14,315 **6.** Yes **7.** No **8.** **(a)** $f(2) = 3$; **(b)** $\{x \mid -2 \le x \le 4\}$; **(c)** -1; **(d)** $\{y \mid 1 \le y \le 5\}$ **9.** $\{x \mid x$ is a real number *and* $x \neq 4\}$ **10.** All real numbers **11.** Slope: -3; *y*-intercept: $(0, 2)$ **12.** Slope: $-\frac{1}{2}$; *y*-intercept: $(0, 2)$ **13.** $m = \frac{11}{3}$

14.

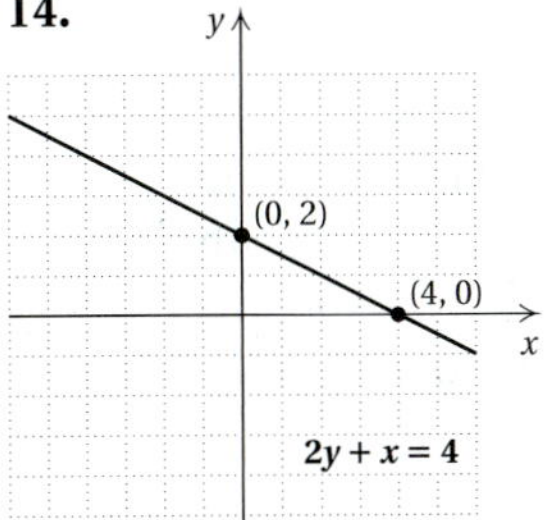

15.

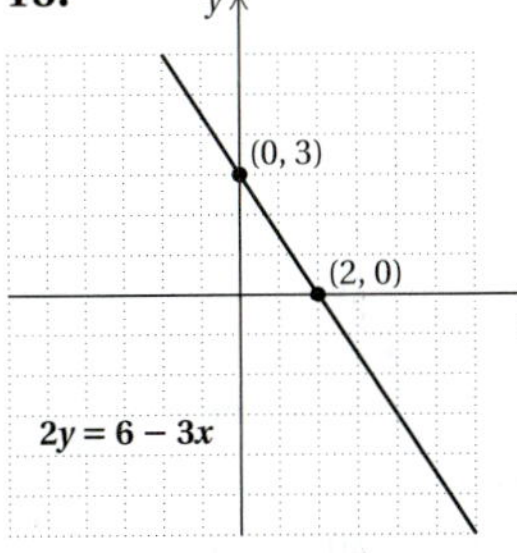

16.

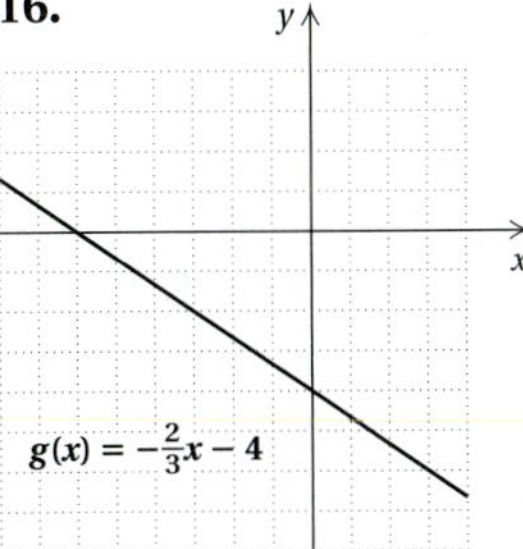

17.

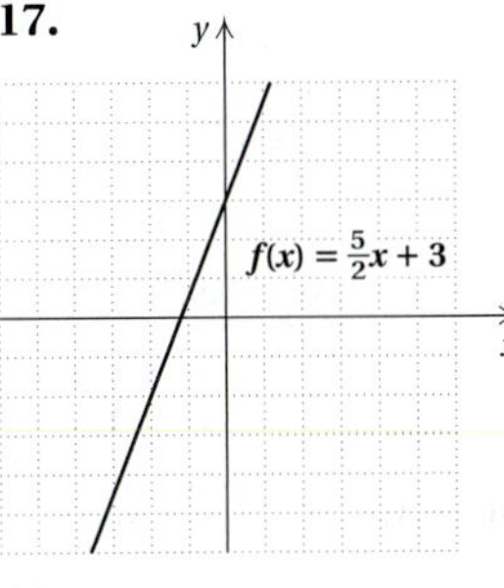

18.

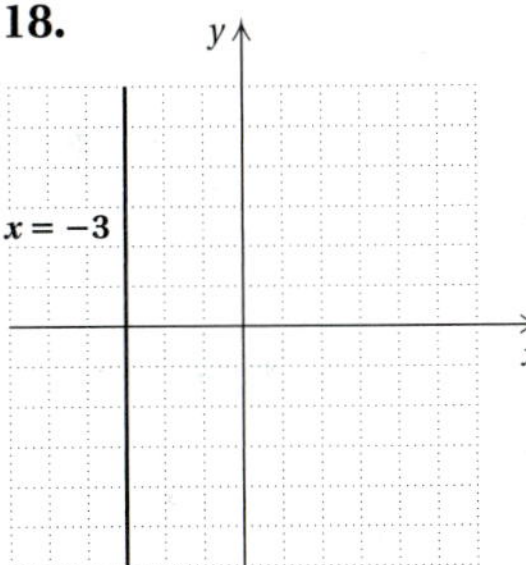

19.

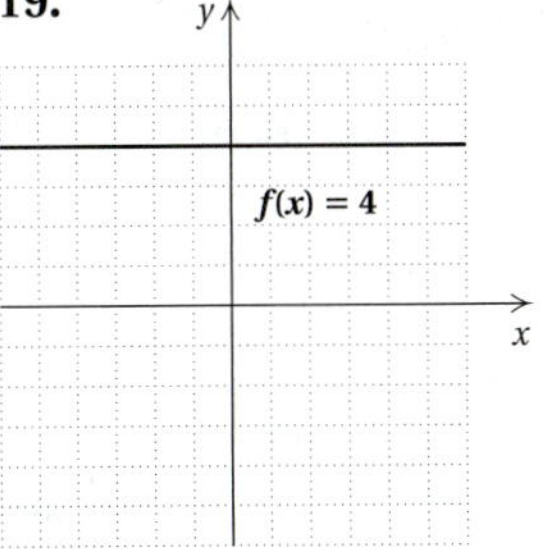

20. Perpendicular **21.** Parallel **22.** Parallel **23.** Perpendicular **24.** $f(x) = 4.7x - 23$ **25.** $y = -3x + 4$ **26.** $y = -\frac{3}{2}x$ **27.** $y = -\frac{5}{7}x + 9$ **28.** $y = \frac{1}{3}x + \frac{1}{3}$ **29.** **(a)** $R(x) = -0.07x + 46.8$; **(b)** 44.35 sec; 44 sec **30.** **DW** The concept of slope is useful in describing how a line slants. A line with positive slope slants up from left to right. A line with negative slope slants down from left to right. The larger the absolute value of the slope, the steeper the slant. **31.** **DW** The notation $f(x)$ can be read "f of x" or "f at x" or "the value of f at x." It represents the output of the function f for the input x. The notation $f(a) = b$ provides a concise way to indicate that for the input a, the output of the function f is b.
32. 0.000000033 **33.** 1.37×10^{-10} **34.** $3x - 1$
35. $r + 1$, R 10; or $r + 1 + \dfrac{10}{r - 4}$
36. $3(3y - 10)(9y^2 + 30y + 100)$
37. $9(1 + 2z)(1 - 2z + 4z^2)$ **38.** $\dfrac{a - 4}{3a - 1}$
39. $f(x) = 3.09x + 3.75$

Test: Chapter 12, p. 982

1. [12.1c]

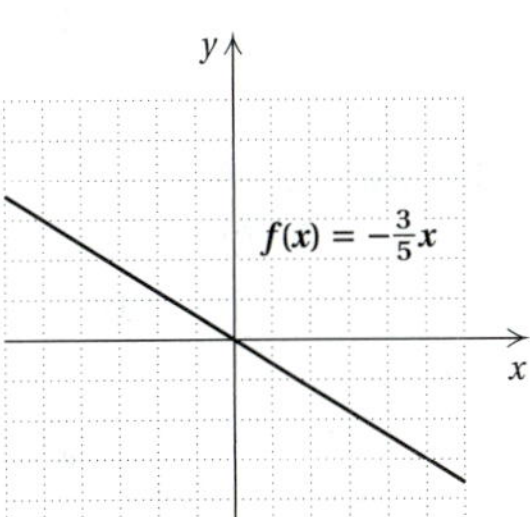

2. [12.1c]

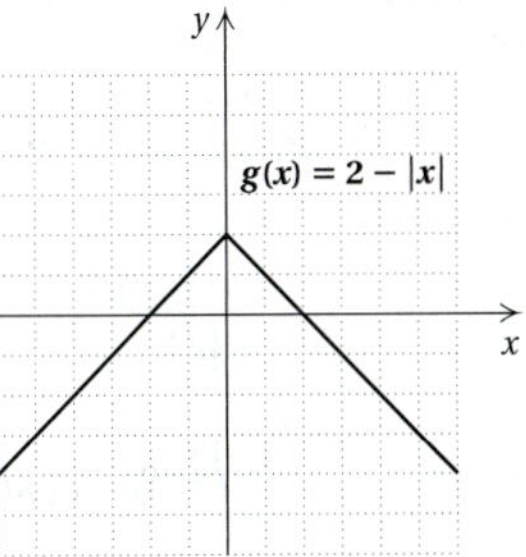

3. [12.1e] **(a)** 8.666 yr; **(b)** 1998 **4.** [12.1a] Yes **5.** [12.1a] No **6.** [12.1b] −4; 2 **7.** [12.1b] 7; 8 **8.** [12.1d] Yes **9.** [12.1d] No **10.** [12.1e] **(a)** \$105 million; **(b)** \$20 million **11.** [12.2a] **(a)** 1.2; **(b)** $\{x \mid -3 \le x \le 4\}$; **(c)** −3; **(d)** $\{y \mid -1 \le y \le 2\}$ **12.** [12.2a] All real numbers **13.** [12.2a] $\left\{x \mid x \text{ is a real number } and\ x \ne -\frac{3}{2}\right\}$ **14.** [12.3b] Slope: $-\frac{3}{5}$; y-intercept: (0, 12) **15.** [12.3b] Slope: $-\frac{2}{5}$; y-intercept: $\left(0, -\frac{7}{5}\right)$ **16.** [12.3b] $m = \frac{5}{8}$ **17.** [12.3b] $m = 0$ **18.** [12.3c] m (or rate of change) $= \frac{4}{5}$ km/min
19. [12.4a]

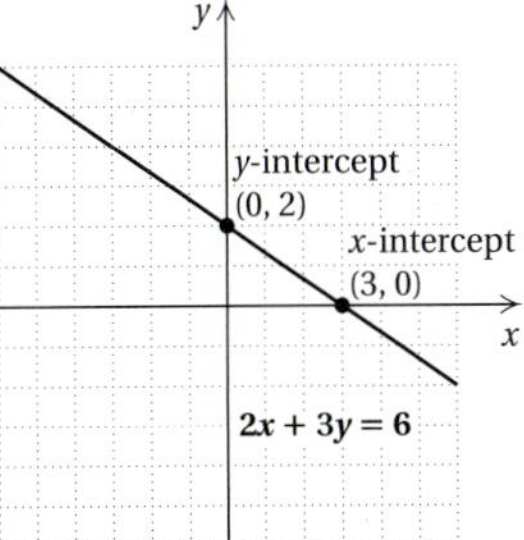

20. [12.4b]

21. [12.4c]

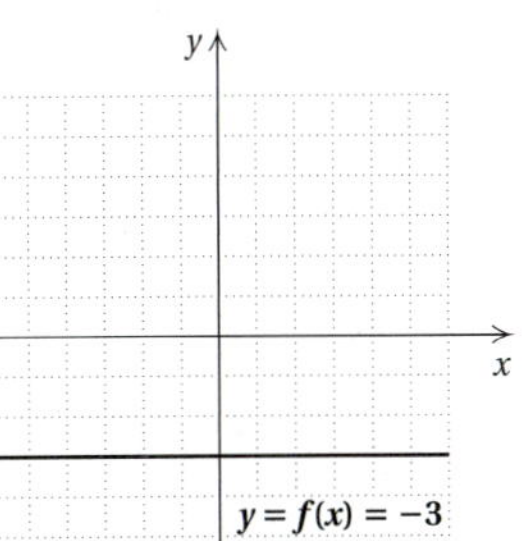

22. [12.4c]

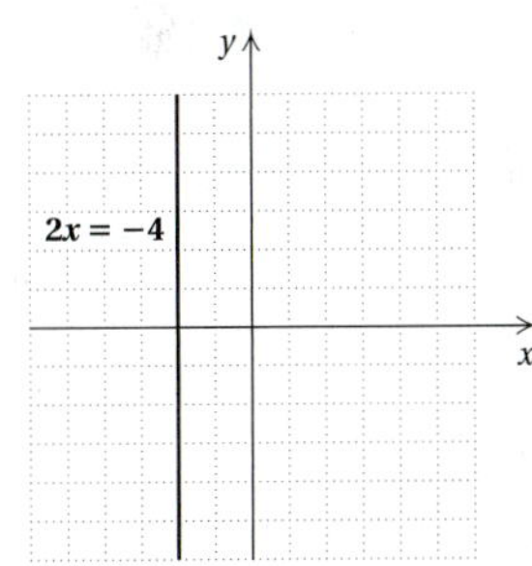

23. [12.4d] Parallel **24.** [12.4d] Perpendicular **25.** [12.5a] $y = -3x + 4.8$ **26.** [12.5a] $f(x) = 5.2x - \frac{5}{8}$ **27.** [12.5b] $y = -4x + 2$ **28.** [12.5c] $y = -\frac{3}{2}x$ **29.** [12.5d] $y = \frac{1}{2}x - 3$ **30.** [12.5d] $y = 3x - 1$ **31.** [12.5e] **(a)** $A(x) = 0.125x + 23.2$; **(b)** 27.95; 28.825 **32.** [9.2c] 8.204×10^{-7} **33.** [9.8b, c] $y + 2$, R −11; or $y + 2 + \dfrac{-11}{y + 4}$ **34.** [10.6a] $(5a + 3)(25a^2 - 15a + 9)$ **35.** [11.1c] $\dfrac{3m - 5}{7m + 4}$ **36.** [12.5d] $\frac{24}{5}$ **37.** [12.1b] $f(x) = 3$; answers may vary

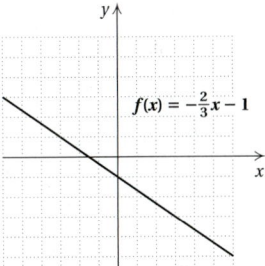

Chapter 13

Pretest: Chapter 13, p. 988

1. (1, 2) **2.** (1, 2) **3.** (−3, 2) **4.** Consistent **5.** Independent **6.** (1, 2, 3) **7.** Cashews: 20 lb; Brazil nuts: 30 lb **8.** \$3300 at 8%; \$4200 at 6% **9.** A: $64\frac{8}{13}$ lb; B: $55\frac{5}{13}$ lb **10.** 54 mph **11.** **(a)** $P(x) = 11x - 90{,}000$; **(b)** (8182, \$212,732)

Margin Exercises, Section 13.1, pp. 990–994

1. (0, 1) **2.** (2, 1) **3.** No solution **4.** Consistent: 1, 2; inconsistent: 3 **5.** Infinitely many solutions **6.** Independent: 1, 2, 3; dependent: 5 **7.** **(a)** −3; **(b)**

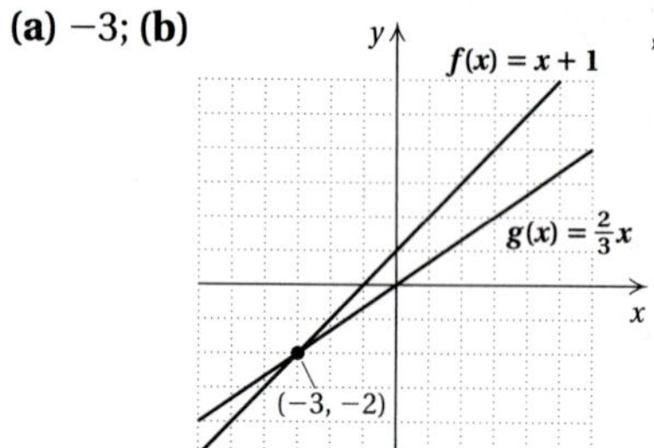

, −3; **(c)** the same, −3

8.

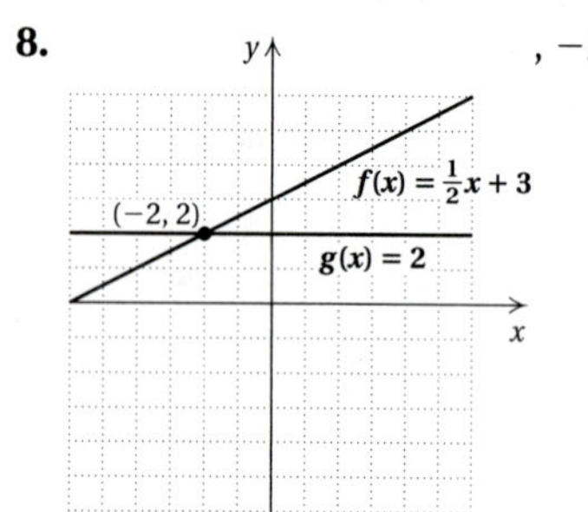

, −2

9. **(a)**

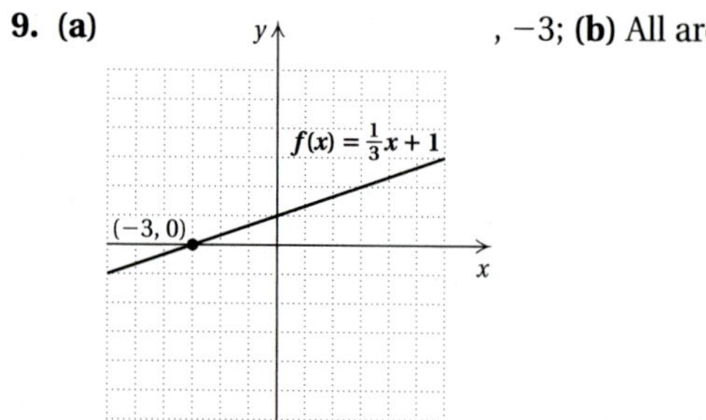

, −3; **(b)** All are −3.

10.

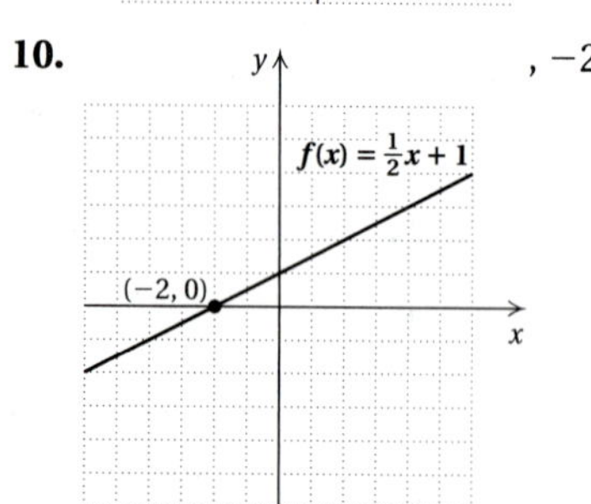

, −2

Calculator Corner, p. 991

1. (2, 3) **2.** (−4, −1) **3.** (−1, 5) **4.** (3, −1)

Exercise Set 13.1, p. 995

1. (3, 1); consistent; independent **3.** (1, −2); consistent; independent **5.** (4, −2); consistent; independent **7.** (2, 1); consistent; independent **9.** $\left(\frac{5}{2}, -2\right)$; consistent; independent **11.** (3, −2); consistent; independent **13.** No solution; inconsistent; independent **15.** Infinitely many solutions; consistent; dependent **17.** (4, −5); consistent; independent **19.** (2, −3); consistent; independent **21.** Consistent; independent; F **23.** Consistent; dependent; B **25.** Inconsistent; independent; D **27.** D_W **29.** $y = \frac{3}{5}x + \frac{22}{5}$ **30.** $y = \frac{3}{8}x + \frac{9}{4}$ **31.** (2.23, 1.14) **33.** (3, 3), (−5, 5)

Margin Exercises, Section 13.2, pp. 999–1001

1. (2, 4) **2.** (5, 2) **3.** (−3, 2) **4.** (13, 16) **5.** **(a)** No solution; **(b)** the same, no solution **6.** Length: 160 ft; width: 100 ft

Calculator Corner, p. 999

Left to the student

Exercise Set 13.2, p. 1002

1. (2, −3) **3.** $\left(\frac{21}{5}, \frac{12}{5}\right)$ **5.** (2, −2) **7.** (−2, −6) **9.** (−2, 1) **11.** $\left(\frac{1}{2}, \frac{1}{2}\right)$ **13.** $\left(\frac{19}{8}, \frac{1}{8}\right)$ **15.** No solution **17.** Length: 40 ft; width: 20 ft **19.** 48° and 132° **21.** Wins: 23; ties: 14 **23.** D_W **25.** 1.3 **26.** $-15y - 39$ **27.** $p = \dfrac{7A}{q}$ **28.** $\frac{7}{3}$ **29.** $m = -\frac{1}{2}$; $b = \frac{5}{2}$ **31.** Length: 57.6 in.; width: 20.4 in.

Margin Exercises, Section 13.3, pp. 1005–1010

1. (1, 4) **2.** $\left(\frac{1}{3}, \frac{1}{2}\right)$ **3.** (2, 3) **4.** (−127, 100) **5.** (−138, −118) **6.** No solution **7.** Infinitely many solutions **8.** $2x + 3y = 1$, $3x - y = 7$; (2, −1) **9.** $9x + 10y = 5$, $9x - 4y = 3$; $\left(\frac{25}{63}, \frac{1}{7}\right)$ **10.** **(a)** Length: 160 ft; width: 100 ft; **(b)** the solutions are the same.

Calculator Corner, p. 1008

1. We get $y = -3x + 5$ and $y = -3x - 2$. Since the graphs are parallel lines, the system of equations has no solution.

2. Each equation is equivalent to $y = \dfrac{2x + 6}{3}$, or $y = \frac{2}{3}x + 2$. Since the equations are equivalent, the graphs are the same and every solution of one is also a solution of the other. Thus we have an infinite number of solutions.

Exercise Set 13.3, p. 1011

1. (1, 2) **3.** (−1, 3) **5.** $\left(\frac{128}{31}, -\frac{17}{31}\right)$ **7.** (6, 2) **9.** $\left(\frac{140}{13}, -\frac{50}{13}\right)$ **11.** (4, 6) **13.** $\left(\frac{110}{19}, -\frac{12}{19}\right)$ **15.** Infinitely many solutions **17.** No solution **19.** $\left(\frac{1}{2}, -\frac{1}{2}\right)$ **21.** $\left(-\frac{4}{3}, -\frac{19}{3}\right)$ **23.** $\left(\frac{1000}{11}, -\frac{1000}{11}\right)$ **25.** (140, 60)

27. Length: 110 m; width: 60 m **29.** 14° and 76° **31.** Coach-class: 131; first-class: 21 **33.** ^{D}W **35.** $y = -\frac{7}{2}x + 36$ **36.** $y = -\frac{7}{5}x - \frac{29}{5}$ **37.** 1 **38.** 5 **39.** 3 **40.** 291 **41.** 15 **42.** $12a^2 - 2a + 1$ **43.** 53 **44.** 8.92 **45.** $(23.12, -12.04)$ **47.** $A = 2$, $B = 4$ **49.** $p = 2, q = -\frac{1}{3}$

Margin Exercises, Section 13.4, pp. 1015–1022

1. White: 22; red: 8

White	Red	Total	
w	r	30	→ $w + r = 30$
\$18.95	\$19.50		
$18.95w$	$19.50r$	572.90	→ $18.95w + 19.50r = 572.90$

2. Kenyan: 8 lb; Sumatran: 12 lb
3. \$1800 at 7%; \$1900 at 9%

First Investment	Second Investment	Total	
x	y	\$3700	→ $x + y = 3700$
7%	9%		
1 yr	1 yr		
$0.07x$	$0.09y$	\$297	→ $0.07x + 0.09y = 297$

4. Attack: 15 qt; Blast: 45 qt

Attack	Blast	Mixture	
a	b	60	→ $a + b = 60$
2%	6%	5%	
$0.02a$	$0.06b$	0.05×60, or 3	→ $0.02a + 0.06b = 3$

5. 280 km

Distance	Rate	Time	
d	35 km/h	t	→ $d = 35t$
d	40 km/h	$t - 1$	→ $d = 40(t - 1)$

6. 180 mph

Distance	Rate	Time	
d	$r + 20$	4 hr	→ $d = 4(r + 20)$
d	$r - 20$	5 hr	→ $d = 5(r - 20)$

Exercise Set 13.4, p. 1023

1. 32 at \$8.50; 13 at \$9.75 **3.** Humulin: 42; Novolin: 23 **5.** 30-sec: 4; 60-sec: 8 **7.** 5 lb of each **9.** 25% acid: 4 L; 50% acid: 6 L **11.** \$7500 at 6%; \$4500 at 9% **13.** Whole milk: $169\frac{3}{13}$ lb; cream: $30\frac{10}{13}$ lb **15.** \$5 bills: 7; \$1 bills: 15 **17.** 375 mi **19.** 14 km/h **21.** 144 mi **23.** 2 hr **25.** $1\frac{1}{3}$ hr **27.** About 1489 mi **29.** ^{D}W **31.** -7 **32.** -11 **33.** -3 **34.** 33 **35.** -15 **36.** $8a - 7$ **37.** -23 **38.** 0.2 **39.** -4 **40.** -17 **41.** $-12h - 7$ **42.** 3993 **43.** $4\frac{4}{7}$ L **45.** City: 261 mi; highway: 204 mi **47.** Brown: 0.8 gal; neutral: 0.2 gal

Margin Exercises, Section 13.5, pp. 1028–1031

1. $(2, 1, -1)$ **2.** $\left(2, -2, \frac{1}{2}\right)$ **3.** $(20, 30, 50)$

Exercise Set 13.5, p. 1032

1. $(1, 2, -1)$ **3.** $(2, 0, 1)$ **5.** $(3, 1, 2)$ **7.** $(-3, -4, 2)$ **9.** $(2, 4, 1)$ **11.** $(-3, 0, 4)$ **13.** $(2, 2, 4)$ **15.** $\left(\frac{1}{2}, 4, -6\right)$ **17.** $\left(\frac{1}{2}, \frac{1}{3}, \frac{1}{6}\right)$ **19.** $\left(\frac{1}{2}, \frac{2}{3}, -\frac{5}{6}\right)$ **21.** $(15, 33, 9)$ **23.** $(4, 1, -2)$ **25.** ^{D}W **27.** $a = \frac{F}{3b}$ **28.** $a = \frac{Q - 4b}{4}$, or $\frac{Q}{4} - b$ **29.** $c = \frac{2F + td}{t}$, or $\frac{2F}{t} + d$ **30.** $d = \frac{tc - 2F}{t}$, or $c - \frac{2F}{t}$ **31.** $y = \frac{Ax - c}{B}$ **32.** $y = \frac{c - Ax}{B}$ **33.** Slope: $-\frac{2}{3}$; y-intercept: $\left(0, -\frac{5}{4}\right)$ **34.** Slope: -4; y-intercept: $(0, 5)$ **35.** Slope: $\frac{2}{5}$; y-intercept: $(0, -2)$ **36.** Slope: 1.09375; y-intercept: $(0, -3.125)$ **37.** $(1, -2, 4, -1)$

Margin Exercises, Section 13.6, pp. 1034–1036

1. 64°, 32°, 84° **2.** \$4000 at 5%; \$5000 at 6%; \$16,000 at 7%

Exercise Set 13.6, p. 1037

1. 10-oz: 8; 14-oz: 20; 20-oz: 6 **3.** 32°, 96°, 52° **5.** Automatic transmission: \$865; power door locks: \$520; air conditioning: \$375 **7.** A: 1500; B: 1900; C: 2300 **9.** First fund: \$45,000; second fund: \$10,000; third fund: \$25,000 **11.** Asian–American: 385; African–American: 200; Caucasian: 154 **13.** Roast beef: 2; baked potato: 1; broccoli: 2 **15.** Par-3: 6; par-4: 8; par-5: 4 **17.** Two-pointers: 32; three-pointers: 5; foul shots: 13 **19.** ^{D}W **21.** ^{D}W **23.** No **24.** Yes **25.** $\{x \mid x \text{ is a real number and } x \neq -7\}$ **26.** Domain: all real numbers; range: $\{y \mid y \leq 5\}$ **27.** $y = -\frac{3}{5}x - 7$ **28.** $\frac{a^3}{b}$ **29.** 180° **31.** 464

Margin Exercises, Section 13.7, pp. 1043–1045

1. (a) $C(x) = 80{,}000 + 20x$; **(b)** $R(x) = 36x$; **(c)** $P(x) = 16x - 80{,}000$; **(d)** a loss of \$16,000; a profit of \$176,000;

(e)

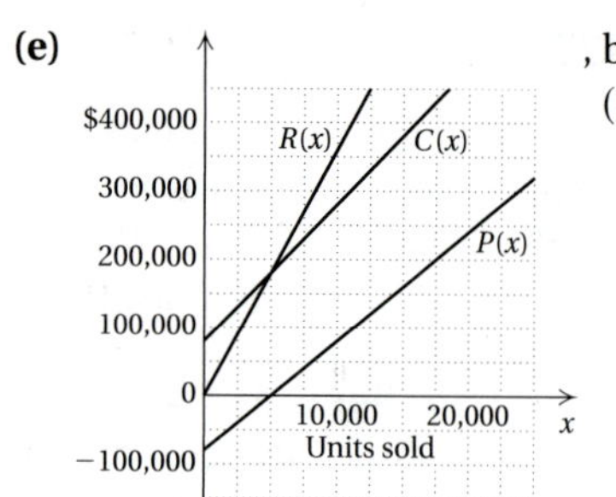

, break-even point: (5000, \$180,000)

2. (\$14, 356)

Exercise Set 13.7, p. 1046

1. (a) $P(x) = 45x - 270{,}000$; **(b)** (6000, \$420,000)
3. (a) $P(x) = 50x - 120{,}000$; **(b)** (2400, \$144,000)
5. (a) $P(x) = 80x - 10{,}000$; **(b)** (125, \$12,500)
7. (a) $P(x) = 18x - 16{,}000$; **(b)** (889, \$35,560)
9. (a) $P(x) = 75x - 195{,}000$; **(b)** (2600, \$325,000)
11. (a) $C(x) = 22{,}500 + 40x$; **(b)** $R(x) = 85x$; **(c)** $P(x) = 45x - 22{,}500$; **(d)** \$112,500 profit; \$4500 loss; **(e)** (500, \$42,500) **13. (a)** $C(x) = 16{,}404 + 6x$; **(b)** $R(x) = 18x$; **(c)** $P(x) = 12x - 16{,}404$; **(d)** \$19,596 profit; \$4404 loss; **(e)** (1367, \$24,606) **15.** (\$70, 300) **17.** (\$22, 474) **19.** (\$50, 6250) **21.** (\$10, 1070) **23.** $\mathbf{D_W}$ **25.** Slope: $\frac{3}{5}$; y-intercept: $\left(0, \frac{8}{5}\right)$ **26.** Slope: $-\frac{6}{7}$; y-intercept: $\left(0, \frac{13}{7}\right)$ **27.** Slope: 1.7; y-intercept: (0, 49) **28.** Slope: $-\frac{4}{3}$; y-intercept: (0, 4)

Summary and Review: Chapter 13, p. 1049

1. (−2, 1); consistent; independent **2.** Infinitely many solutions; consistent; dependent **3.** No solution; inconsistent; independent **4.** $\left(\frac{2}{5}, -\frac{4}{5}\right)$ **5.** No solution **6.** $\left(-\frac{11}{15}, -\frac{43}{30}\right)$ **7.** $\left(\frac{37}{19}, \frac{53}{19}\right)$ **8.** $\left(\frac{76}{17}, -\frac{2}{119}\right)$ **9.** (2, 2) **10.** Infinitely many solutions **11.** CD: \$14; cassette: \$9 **12.** 5 L of each **13.** $5\frac{1}{2}$ hr **14.** (10, 4, −8) **15.** $\left(-\frac{7}{3}, \frac{125}{27}, \frac{20}{27}\right)$ **16.** (2, 0, 4) **17.** $\left(2, \frac{1}{3}, -\frac{2}{3}\right)$ **18.** 90°, $67\frac{1}{2}$°, $22\frac{1}{2}$° **19.** \$20 bills: 5; \$5 bills: 15; \$1 bills: 19 **20. (a)** $C(x) = 35{,}000 + 175x$; **(b)** $R(x) = 300x$; **(c)** $P(x) = 125x - 35{,}000$; **(d)** \$115,000 profit; \$10,000 loss; **(e)** (280, \$84,000) **21.** (\$3, 81) **22.** $\mathbf{D_W}$ The comparison is summarized in the table in Section 8.3. **23.** $\mathbf{D_W}$ Many problems that deal with more than one unknown quantity are easier to translate to a system of equations than to a single equation. Problems involving complementary or supplementary angles, the dimensions of a geometric figure, mixtures, and the measures of the angles of a triangle are examples.

24. $d + q = 20,$
$25d + 10q = 90 + 10d + 25q;$
or $d + q = 20$
$d - q = 6;$
dimes: 13; quarters: 7

25. $y = \frac{1}{2}x + 5$ **26.** $y = 2x - 3$ **27.** $y = \frac{1}{4}x - 2$ **28.** $y = x - 4$ **29.** $t = \dfrac{Q}{a - 4}$ **30.** $f(0) = 8$; $f(-2) = 14$ **31.** Slope: $\frac{5}{8}$; y-intercept: (0, −5) **32.** (0, 2) and (1, 3) **33. (a)** 165.91 cm; **(b)** 160.60 cm; **(c)** (120.857, 308.521); **(d)** 120.857 cm

Test: Chapter 13, p. 1052

1. [13.1a] (−2, 1); consistent; independent **2.** [13.1a] No solution; inconsistent; independent **3.** [13.1a] Infinitely many solutions; consistent; dependent **4.** [13.2a] $\left(3, -\frac{11}{3}\right)$ **5.** [13.2a] $\left(\frac{15}{7}, -\frac{18}{7}\right)$ **6.** [13.3a] $\left(-\frac{3}{2}, -\frac{3}{2}\right)$ **7.** [13.3a] No solution **8.** [13.4a] 34% solution: $48\frac{8}{9}$ mL; 61% solution: $71\frac{1}{9}$ mL **9.** [13.4a] Buckets: 17; dinners: 11 **10.** [13.4b] 120 km/h **11.** [13.2b], [13.3b] Length: 93 ft; width: 51 ft **12.** [13.5a] $\left(2, -\frac{1}{2}, -1\right)$ **13.** [13.7b] (\$3, 55) **14.** [13.6a] 3.5 hr **15.** [13.7a] **(a)** $C(x) = 40{,}000 + 30x$; **(b)** $R(x) = 80x$; **(c)** $P(x) = 50x - 40{,}000$; **(d)** \$20,000 profit; \$30,000 loss; **(e)** (800, \$64,000) **16.** [12.5d] $y = -3x - 5$ **17.** [12.5d] $y = \frac{4}{5}x + 4$ **18.** [7.4b] $a = \dfrac{P + 3b}{4}$ **19.** [12.3b] Slope: $-\frac{7}{2}$; y-intercept: (0, 7) **20.** [12.1b] $f(0) = -8$; $f(-3) = 1$ **21.** [13.2a], [13.3a] $m = 7$; $b = 10$

Chapter 14

Pretest: Chapter 14, p. 1056

1. $\{x \mid x > -3\}$ **2.** $\{x \mid x \geq -3\}$ **3.** $\left\{x \mid -1 \leq x \leq \frac{2}{3}\right\}$ **4.** −1, 8 **5.** $\left\{a \mid a < -2 \text{ or } a > -\frac{1}{2}\right\}$ **6.** $\left\{a \mid a < -2 \text{ or } a > -\frac{4}{3}\right\}$ **7.** $-\frac{11}{2}, \frac{3}{4}$ **8.** $\left\{x \mid x \leq \frac{19}{14}\right\}$ **9.** 2.4 **10.** $\{t \mid t > 3.23\}$, 3.23 yr after 1988 **11.** $\{-3, 4\}$ **12.** $\{-1, 0, 1\}$ **13.** $\dfrac{5}{2}\left|\dfrac{y}{x}\right|$

14.

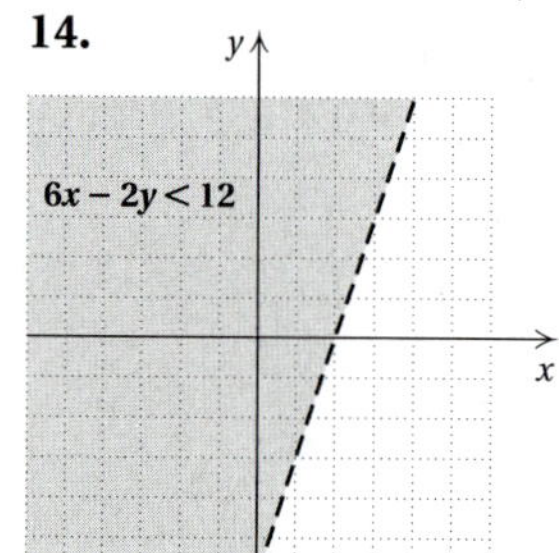

15.

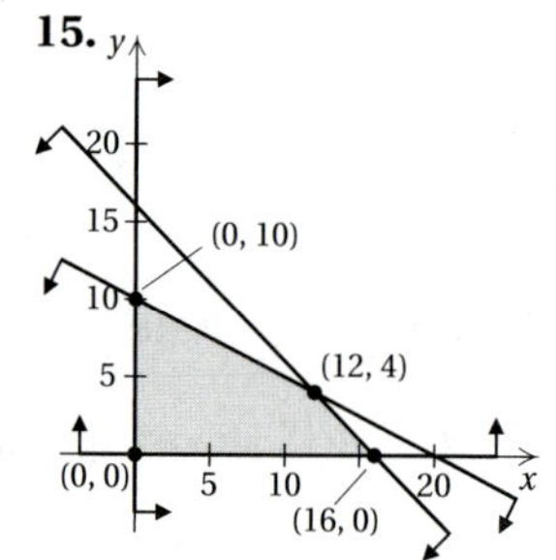

Margin Exercises, Section 14.1, pp. 1057–1065

1. Yes **2.** No **3.** Yes **4.** [−4, 5) **5.** (−∞, −2] **6.** [10, ∞) **7.** [−30, 30]

8. $\{x \mid x > 3\}$, or $(3, \infty)$;

0 3

9. $\{x \mid x \leq 3\}$, or $(-\infty, 3]$;

0 3

10. $\{x \mid x \leq -2\}$, or $(-\infty, -2]$;

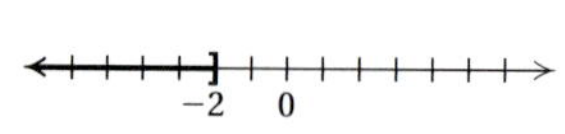

11. $\left\{y \mid y \leq \frac{3}{10}\right\}$, or $\left(-\infty, \frac{3}{10}\right]$;

$\frac{3}{10}$
0

12. $\{y \mid y < -5\}$, or $(-\infty, -5)$ **13.** $\{x \mid x \geq 12\}$, or $[12, \infty)$

14. $\left\{y \mid y \leq -\frac{1}{5}\right\}$, or $\left(-\infty, -\frac{1}{5}\right]$ **15.** $\left\{x \mid x < \frac{1}{2}\right\}$, or $\left(-\infty, \frac{1}{2}\right)$

16. $\left\{y \mid y \geq \frac{17}{9}\right\}$, or $\left[\frac{17}{9}, \infty\right)$ **17.** $s \geq 88$ **18.** $d \geq \$4000$

19. $p \leq \$21{,}000$ **20.** 50 min $< t <$ 60 min

21. $d > 25$ mi **22.** $w < 110$ lb **23.** $n > -8$

24. $c \leq \$135{,}000$ **25.** $d \leq 11.6\%$ **26.** $p \geq 37$

27. $\{t \mid t > 14.04\}$, or approximately after 2010

28. For $\{n \mid n < 100 \text{ hr}\}$, plan A is better.

Exercise Set 14.1, p. 1066

1. No, no, no, yes **3.** No, yes, yes, no, no **5.** $(-\infty, 5)$
7. $[-3, 3]$ **9.** $(-2, 5)$ **11.** $(-\sqrt{2}, \infty)$
13. $\{x \mid x > -1\}$, or $(-1, \infty)$ **15.** $\{y \mid y < 6\}$, or $(-\infty, 6)$
17. $\{a \mid a \leq -22\}$, or $(-\infty, -22]$
19. $\{t \mid t \geq -4\}$, or $[-4, \infty)$
21. $\{y \mid y > -6\}$, or $(-6, \infty)$ **23.** $\{x \mid x \leq 9\}$, or $(-\infty, 9]$
25. $\{x \mid x \geq 3\}$, or $[3, \infty)$
27. $\{x \mid x < -60\}$, or $(-\infty, -60)$
29. $\{x \mid x \leq 0.9\}$, or $(-\infty, 0.9]$
31. $\left\{x \mid x \leq \frac{5}{6}\right\}$, or $\left(-\infty, \frac{5}{6}\right]$ **33.** $\{x \mid x < 6\}$, or $(-\infty, 6)$
35. $\{y \mid y \leq -3\}$, or $(-\infty, -3]$ **37.** $\left\{y \mid y > \frac{2}{3}\right\}$, or $\left(\frac{2}{3}, \infty\right)$
39. $\{x \mid x \geq 11.25\}$, or $[11.25, \infty)$
41. $\left\{x \mid x \leq \frac{1}{2}\right\}$, or $\left(-\infty, \frac{1}{2}\right]$ **43.** $\left\{y \mid y \leq -\frac{75}{2}\right\}$, or $\left(-\infty, -\frac{75}{2}\right]$
45. $\left\{x \mid x > -\frac{2}{17}\right\}$, or $\left(-\frac{2}{17}, \infty\right)$ **47.** $\left\{m \mid m > \frac{7}{3}\right\}$, or $\left(\frac{7}{3}, \infty\right)$
49. $\{r \mid r < -3\}$, or $(-\infty, -3)$ **51.** $\{x \mid x \geq 2\}$, or $[2, \infty)$
53. $\{y \mid y < 5\}$, or $(-\infty, 5)$ **55.** $\left\{x \mid x \leq \frac{4}{7}\right\}$, or $\left(-\infty, \frac{4}{7}\right]$
57. $\{x \mid x < 8\}$, or $(-\infty, 8)$ **59.** $\left\{x \mid x \geq \frac{13}{2}\right\}$, or $\left[\frac{13}{2}, \infty\right)$
61. $\left\{x \mid x < \frac{11}{18}\right\}$, or $\left(-\infty, \frac{11}{18}\right)$ **63.** $\left\{x \mid x \geq -\frac{51}{31}\right\}$, or $\left[-\frac{51}{31}, \infty\right)$
65. $\{a \mid a \leq 2\}$, or $(-\infty, 2]$ **67. (a)** 25.38;
(b) $\{W \mid W <$ (approximately) 189.1 lb$\}$ **69.** $\{S \mid S \geq 84\}$
71. $\{B \mid B \geq \$11{,}500\}$ **73.** $\{S \mid S > \$7000\}$
75. $\{n \mid n > 25\}$ **77.** $\{p \mid p > 80\}$ **79.** $\{s \mid s > 8\}$
81. (a) 8.398 gal, 12.063 gal, 15.728 gal; **(b)** years after 1999
83. D_W **85.** $3x^2 + 20x - 32$ **86.** $6r^2 - 23rs - 4s^2$
87. $6a^2 + 7a - 55$ **88.** $t^2 - 7st - 18s^2$
89. $\{x \mid x$ is a real number *and* $x \neq -8\}$, or
$(-\infty, -8) \cup (-8, \infty)$ **90.** All real numbers **91.** All real numbers **92.** $\left\{x \mid x \text{ is a real number } \textit{and } x \neq \frac{2}{3}\right\}$, or
$\left(-\infty, \frac{2}{3}\right) \cup \left(\frac{2}{3}, \infty\right)$ **93. (a)** $\{p \mid p > 10\}$; **(b)** $\{p \mid p < 10\}$
95. True **97.** All real numbers **99.** All real numbers

Margin Exercises, Section 14.2, pp. 1073–1079

1. $\{0, 3\}$ **2.**

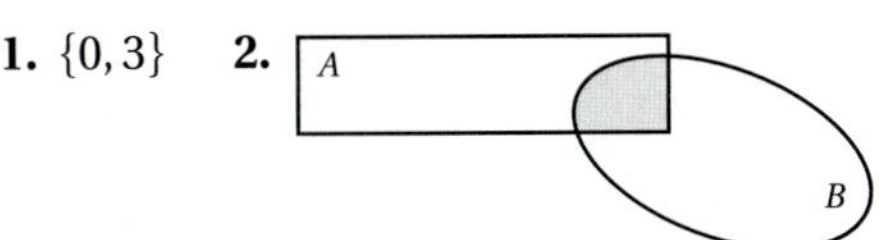

3. ; $(-1, 4)$

4. $\{x \mid -5 < x \leq 10\}$, or $(-5, 10]$;

5. $\{x \mid x < -3\}$; **6.** $\varnothing$

7. $\{x \mid 2 \leq x \leq 6\}$, or $[2, 6]$ **8.** $\{0, 1, 3, 4, 7, 9\}$

9.

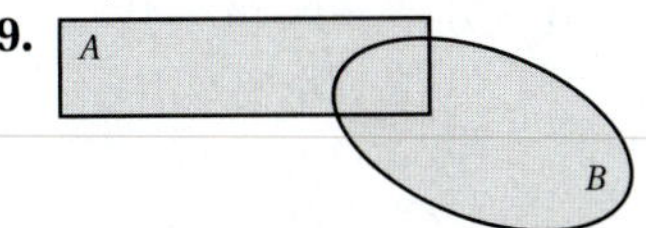

10. ; $(-\infty, -2] \cup (4, \infty)$

11. $\{x \mid x < 1 \textit{ or } x \geq 6\}$, or $(-\infty, 1) \cup [6, \infty)$;

12. $\left\{x \mid x \geq \frac{7}{2} \textit{ or } x < -2\right\}$, or $(-\infty, -2) \cup \left[\frac{7}{2}, \infty\right)$;

13. $\{x \mid x > -2 \textit{ or } x < -5\}$, or $(-\infty, -5) \cup (-2, \infty)$

14. All real numbers; **15.** $\{s \mid 8 < s < 19\}$

Exercise Set 14.2, p. 1080

1. $\{9, 11\}$ **3.** $\{b\}$ **5.** $\{9, 10, 11, 13\}$ **7.** $\{a, b, c, d, f, g\}$
9. $\varnothing$ **11.** $\{3, 5, 7\}$ **13.** ; $(-4, 1]$
15. ; $(1, 6)$
17. $\{x \mid -4 \leq x < 5\}$, or $[-4, 5)$;
19. $\{x \mid x \geq 2\}$, or $[2, \infty)$; **21.** $\varnothing$
23. $\{x \mid -8 < x < 6\}$, or $(-8, 6)$
25. $\{x \mid -6 < x \leq 2\}$, or $(-6, 2]$
27. $\{y \mid -1 < y \leq 5\}$, or $(-1, 5]$
29. $\left\{x \mid -\frac{5}{3} \leq x \leq \frac{4}{3}\right\}$, or $\left[-\frac{5}{3}, \frac{4}{3}\right]$
31. $\{x \mid -1 < x \leq 6\}$, or $(-1, 6]$
33. $\left\{x \mid -\frac{3}{2} \leq x < \frac{9}{2}\right\}$, or $\left[-\frac{3}{2}, \frac{9}{2}\right)$
35. $\{x \mid 10 < x \leq 14\}$, or $(10, 14]$
37. $\left\{x \mid -\frac{7}{2} < x < \frac{37}{2}\right\}$, or $\left(-\frac{7}{2}, \frac{37}{2}\right)$
39. ; $(-\infty, -2) \cup (1, \infty)$
41. ; $(-\infty, -3] \cup (1, \infty)$
43. $\{x \mid x < -5 \textit{ or } x > -1\}$, or $(-\infty, -5) \cup (-1, \infty)$;

45. $\{x|x \le \frac{5}{2}$ *or* $x \ge 4\}$, or $(-\infty, \frac{5}{2}] \cup [4, \infty)$;

47. $\{x|x \ge -3\}$, or $[-3, \infty)$;

49. $\{x|x \le -\frac{5}{4}$ *or* $x > -\frac{1}{2}\}$, or $(-\infty, -\frac{5}{4}] \cup (-\frac{1}{2}, \infty)$
51. All real numbers, or $(-\infty, \infty)$
53. $\{x|x < -4$ *or* $x > 2\}$, or $(-\infty, -4) \cup (2, \infty)$
55. $\{x|x < \frac{79}{4}$ *or* $x > \frac{89}{4}\}$, or $(-\infty, \frac{79}{4}) \cup (\frac{89}{4}, \infty)$
57. $\{x|x \le -\frac{13}{2}$ *or* $x \ge \frac{29}{2}\}$, or $(-\infty, -\frac{13}{2}] \cup [\frac{29}{2}, \infty)$
59. $\{d|0 \text{ ft} \le d \le 198 \text{ ft}\}$ 61. Between 23 and 27 beats
63. $\{t|3 \text{ yr} < t < 9 \text{ yr}\}$, or between 2003 and 2009
65. DW 67. $(-1, 2)$ 68. $(-3, -5)$ 69. $(4, -4)$
70. $y = -11x + 29$ 71. $y = -4x + 7$ 72. $y = -x + 2$
73. $6a^2 + 7ab - 5b^2$ 74. $25y^2 + 35y + 6$
75. $21x^2 - 59x + 40$ 76. $13x^2 + 37xy - 6y^2$
77. $\{x|\frac{2}{5} \le x \le \frac{4}{5}\}$, or $[\frac{2}{5}, \frac{4}{5}]$ 79. $\{x|-\frac{1}{8} < x < \frac{1}{2}\}$, or $(-\frac{1}{8}, \frac{1}{2})$
81. $\{x|10 < x \le 18\}$; $(10, 18]$ 83. True 85. False
87. All real numbers; $\varnothing$

Margin Exercises, Section 14.3, pp. 1084–1090

1. $7|x|$ 2. x^8 3. $5a^2|b|$ 4. $\frac{7|a|}{b^2}$ 5. $9|x|$ 6. 29
7. 5 8. $|p|$ 9. $\{6, -6\}$; 10. $\varnothing$

11. $\{0\}$ 12. $\{2, -2\}$ 13. $\{\frac{17}{4}, -\frac{17}{4}\}$ 14. $\{4, -4\}$
15. $\{3, 5\}$ 16. $\{-\frac{13}{3}, 7\}$ 17. $\varnothing$ 18. $\{\frac{7}{4}, -\frac{1}{6}\}$
19. $\{-\frac{7}{2}\}$ 20. $\{5, -5\}$;

21. $\{x|-5 < x < 5\}$, or $(-5, 5)$;

22. $\{x|x \le -5$ *or* $x \ge 5\}$, or $(-\infty, -5] \cup [5, \infty)$;

23. $\{x|-2 < x < 5\}$, or $(-2, 5)$;

24. $\{x|1 \le x \le \frac{11}{3}\}$, or $[1, \frac{11}{3}]$
25. $\{x|x \le -\frac{7}{3}$ *or* $x \ge 1\}$, or $(-\infty, -\frac{7}{3}] \cup [1, \infty)$;

Exercise Set 14.3, p. 1091

1. $9|x|$ 3. $2x^2$ 5. $2x^2$ 7. $6|y|$ 9. $\frac{2}{|x|}$ 11. $\frac{x^2}{|y|}$
13. $4|x|$ 15. 38 17. 19 19. 6.3 21. 5
23. $\{3, -3\}$ 25. $\varnothing$ 27. $\{0\}$ 29. $\{15, -9\}$
31. $\{\frac{7}{2}, -\frac{1}{2}\}$ 33. $\{\frac{23}{4}, -\frac{5}{4}\}$ 35. $\{11, -11\}$
37. $\{291, -291\}$ 39. $\{8, -8\}$ 41. $\{7, -7\}$ 43. $\{2, -2\}$
45. $\{8, -7\}$ 47. $\{2, -12\}$ 49. $\{\frac{7}{2}, -\frac{5}{2}\}$ 51. $\varnothing$
53. $\{-\frac{13}{54}, -\frac{7}{54}\}$ 55. $\{\frac{3}{4}, -\frac{11}{2}\}$ 57. $\{\frac{3}{2}\}$ 59. $\{5, -\frac{3}{5}\}$
61. All real numbers 63. $\{-\frac{3}{2}\}$ 65. $\{\frac{24}{23}, 0\}$ 67. $\{32, \frac{8}{3}\}$
69. $\{x|-3 < x < 3\}$, or $(-3, 3)$
71. $\{x|x \le -2$ *or* $x \ge 2\}$, or $(-\infty, -2] \cup [2, \infty)$
73. $\{x|0 < x < 2\}$, or $(0, 2)$
75. $\{x|-6 \le x \le -2\}$, or $[-6, -2]$
77. $\{x|-\frac{1}{2} \le x \le \frac{7}{2}\}$, or $[-\frac{1}{2}, \frac{7}{2}]$
79. $\{y|y < -\frac{3}{2}$ *or* $y > \frac{17}{2}\}$, or $(-\infty, -\frac{3}{2}) \cup (\frac{17}{2}, \infty)$
81. $\{x|x \le -\frac{5}{4}$ *or* $x \ge \frac{23}{4}\}$, or $(-\infty, -\frac{5}{4}] \cup [\frac{23}{4}, \infty)$
83. $\{y|-9 < y < 15\}$, or $(-9, 15)$
85. $\{x|-\frac{7}{2} \le x \le \frac{1}{2}\}$, or $[-\frac{7}{2}, \frac{1}{2}]$
87. $\{y|y < -\frac{4}{3}$ *or* $y > 4\}$, or $(-\infty, -\frac{4}{3}) \cup (4, \infty)$
89. $\{x|x \le -\frac{5}{4}$ *or* $x \ge \frac{23}{4}\}$, or $(-\infty, -\frac{5}{4}] \cup [\frac{23}{4}, \infty)$
91. $\{x|-\frac{9}{2} < x < 6\}$, or $(-\frac{9}{2}, 6)$
93. $\{x|x \le -\frac{25}{6}$ *or* $x \ge \frac{23}{6}\}$, or $(-\infty, -\frac{25}{6}] \cup [\frac{23}{6}, \infty)$
95. $\{x|-5 < x < 19\}$, or $(-5, 19)$
97. $\{x|x \le -\frac{2}{15}$ *or* $x \ge \frac{14}{15}\}$, or $(-\infty, -\frac{2}{15}] \cup [\frac{14}{15}, \infty)$
99. $\{m|-12 \le m \le 2\}$, or $[-12, 2]$
101. $\{x|\frac{1}{2} \le x \le \frac{5}{2}\}$, or $[\frac{1}{2}, \frac{5}{2}]$
103. $\{x|0.49705 \le x \le 0.50295\}$, or $[0.49705, 0.50295]$
105. DW 107. All real numbers 108. $\{x|x$ is a real number *and* $x \ne -\frac{5}{2}\}$, or $(-\infty, -\frac{5}{2}) \cup (-\frac{5}{2}, \infty)$
109. $\{x|x$ is a real number *and* $x \ne 5\}$, or $(-\infty, 5) \cup (5, \infty)$
110. All real numbers 111. $(6, -5)$ 112. $(2, -4)$
113. $\{d|5\frac{1}{2} \text{ ft} \le d \le 6\frac{1}{2} \text{ ft}\}$ 115. $\{x|x \ge -5\}$, or $[-5, \infty)$
117. $\{1, -\frac{1}{4}\}$ 119. $\varnothing$ 121. All real numbers
123. All real numbers 125. $|x| < 3$ 127. $|x| \ge 6$
129. $|x + 3| > 5$

Margin Exercises, Section 14.4, pp. 1095–1103

1. No 2. Yes 3.

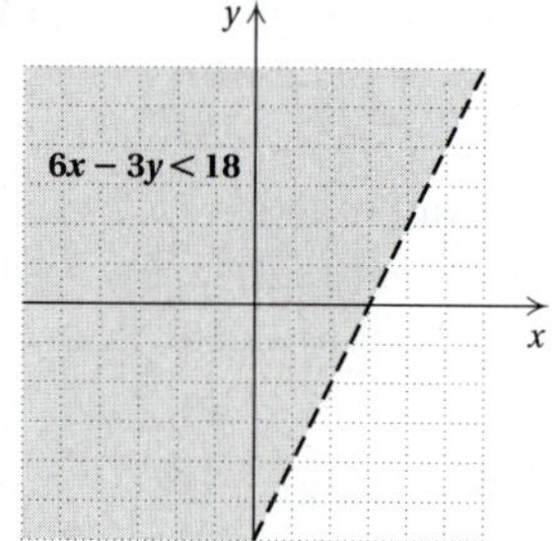

4.

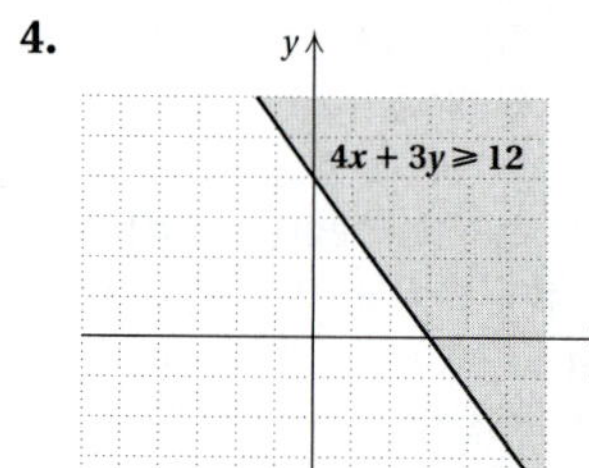

5.

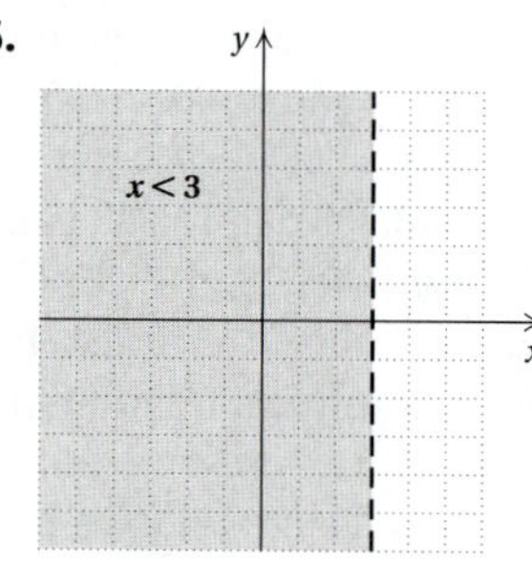

6.

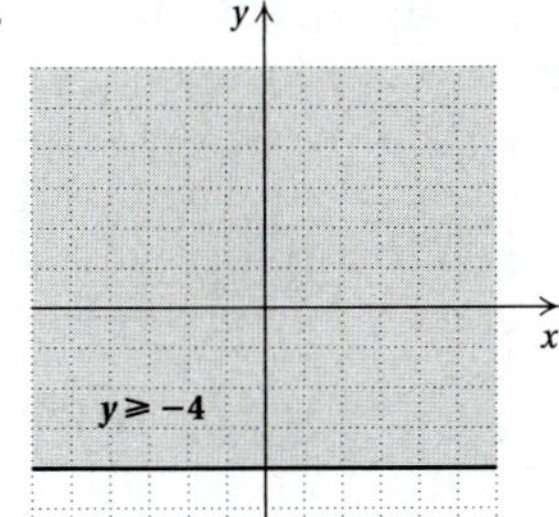

7.

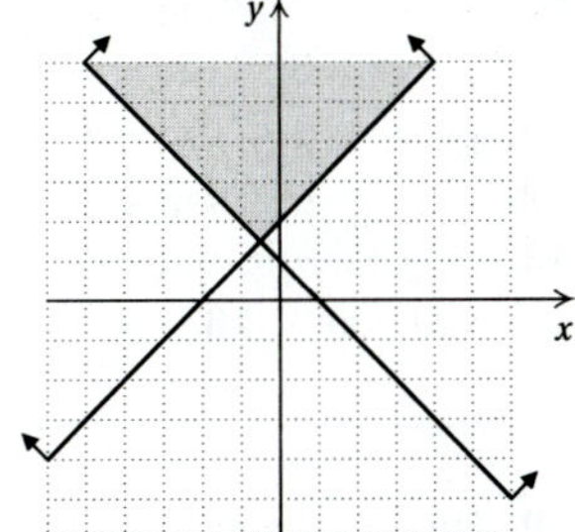

8.

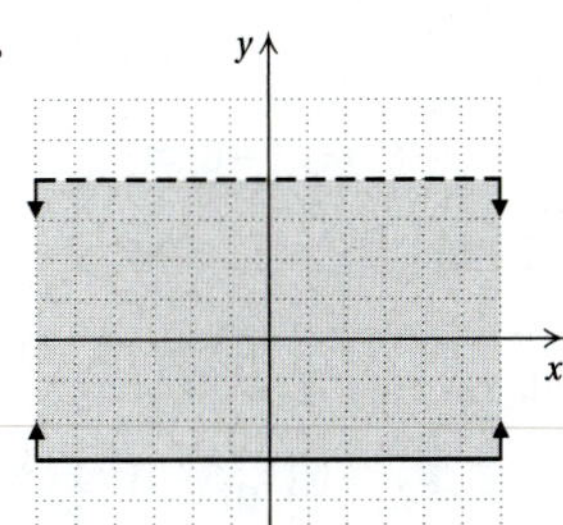

9.

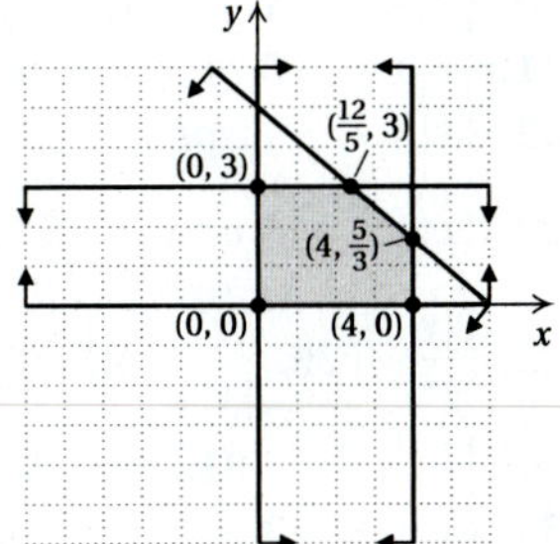

10.

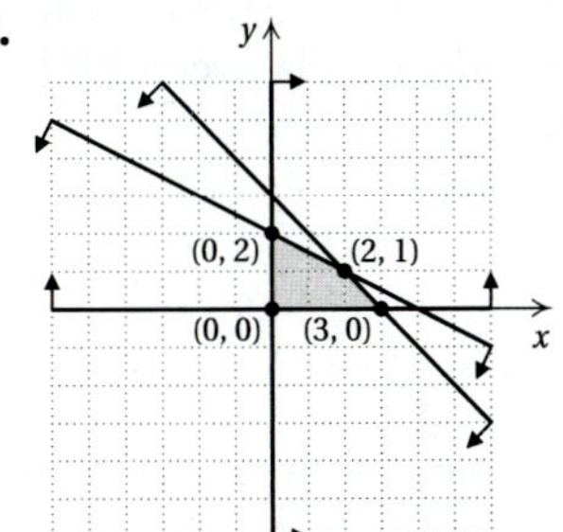

Calculator Corner, p. 1104

1. Left to the student **2.** Left to the student

Exercise Set 14.4, p. 1105

1. Yes **3.** Yes **5.**

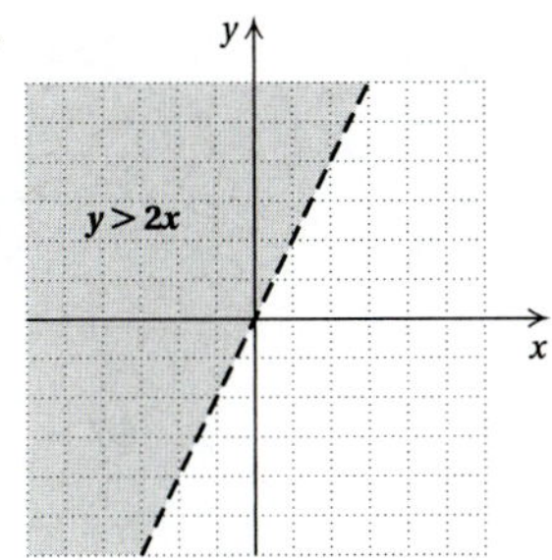

7.

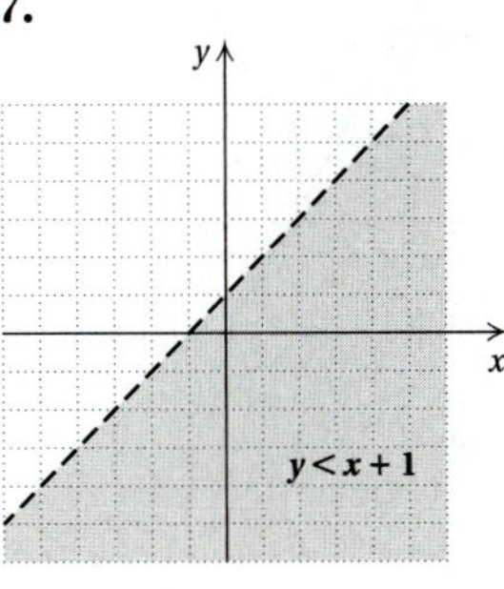

9.

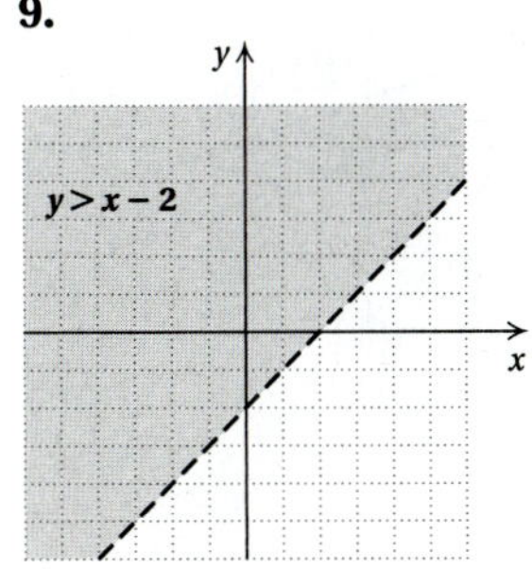

11.

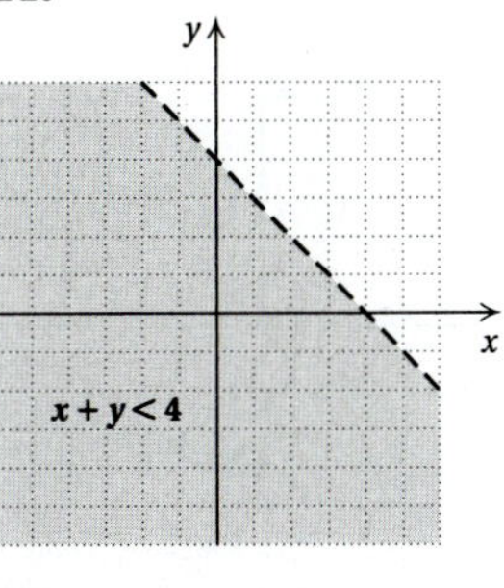

13.

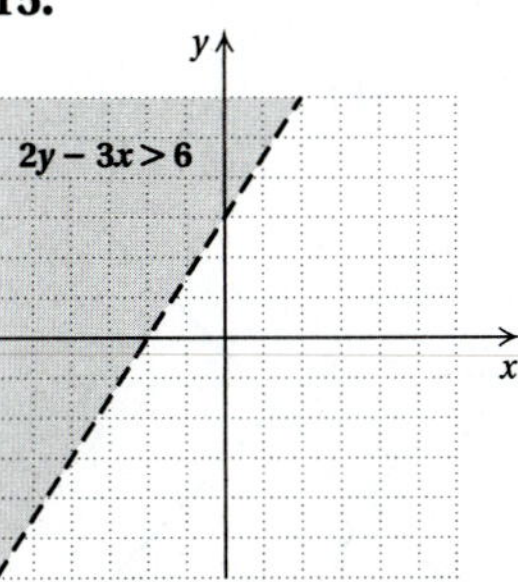

15.

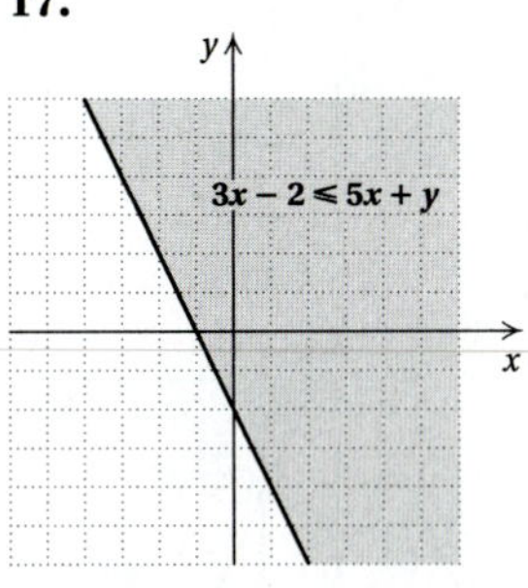

17.

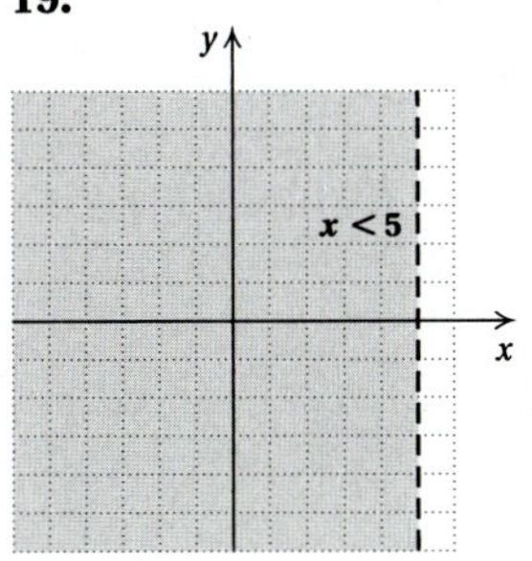

19.

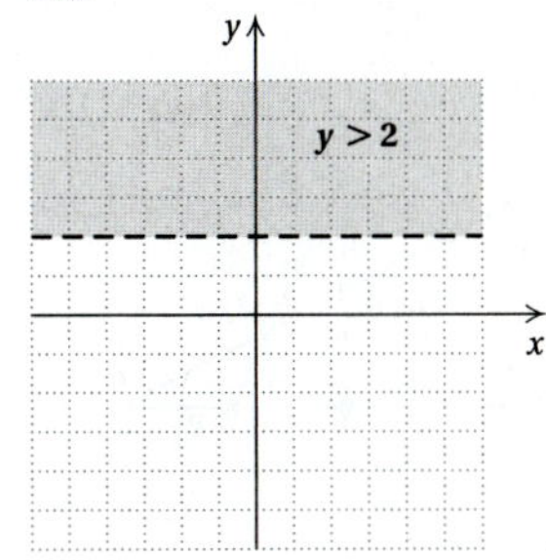

21.

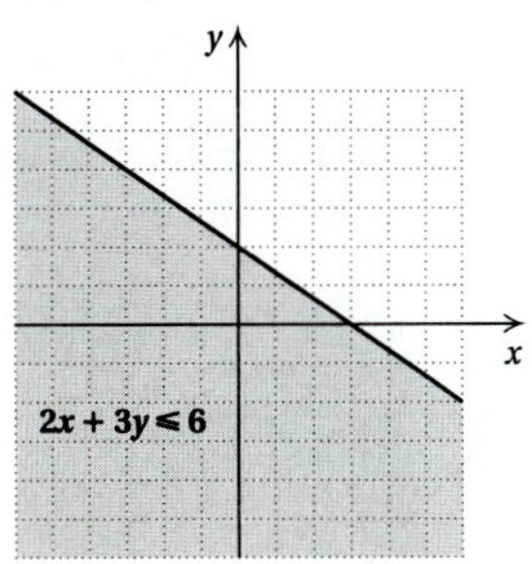

23.

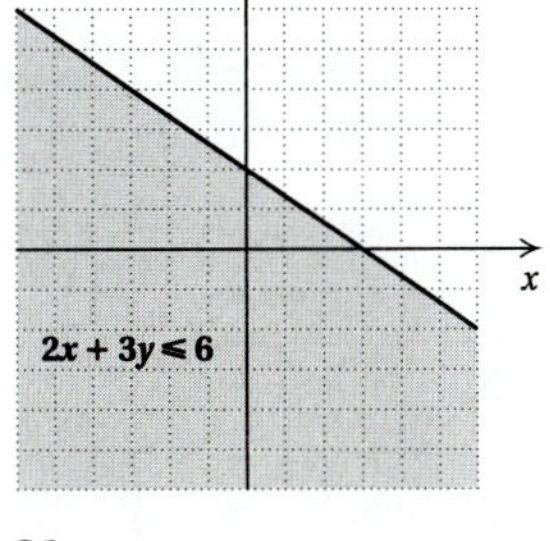

25. F
27. B
29. C

31.

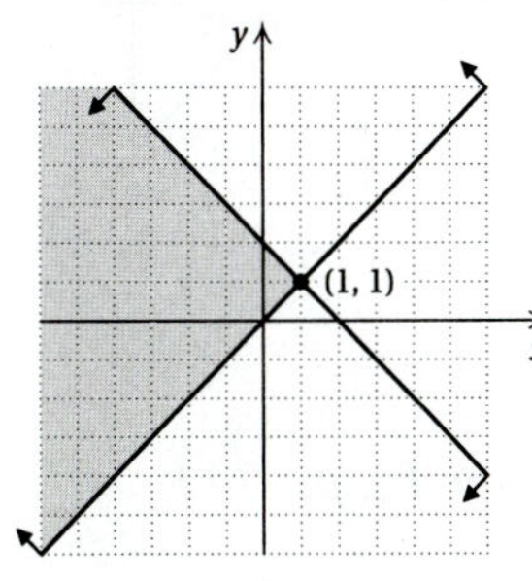

33.

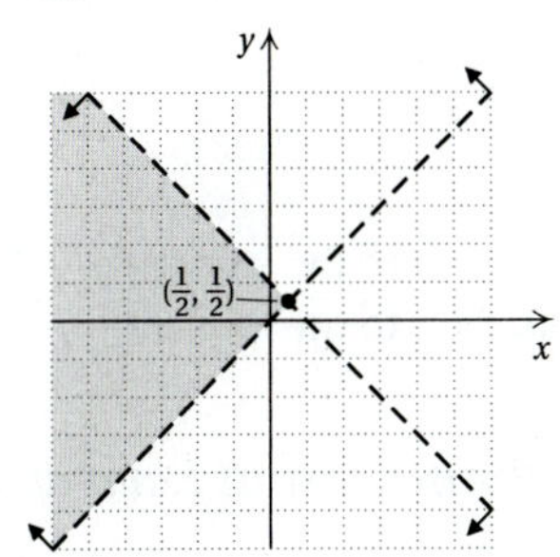

35.

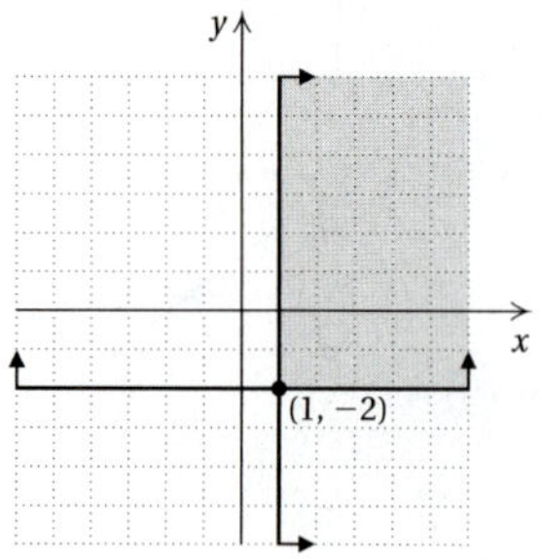

37.

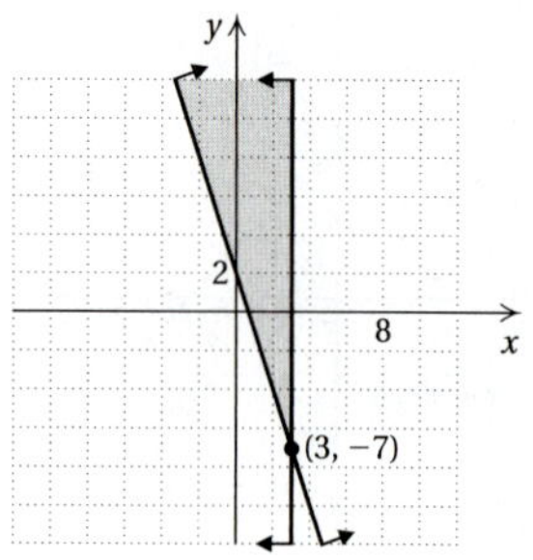

39.

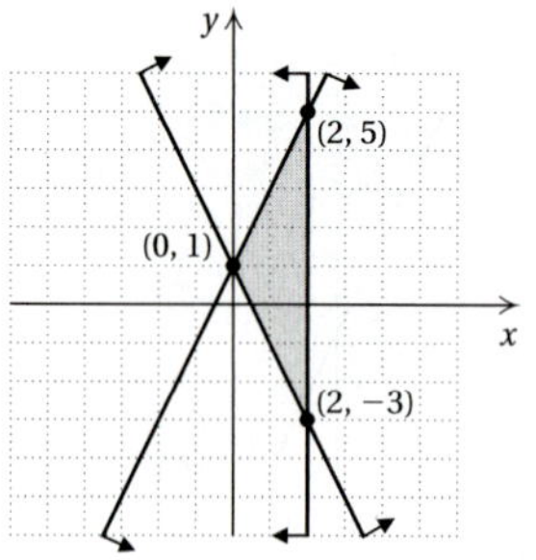

41.

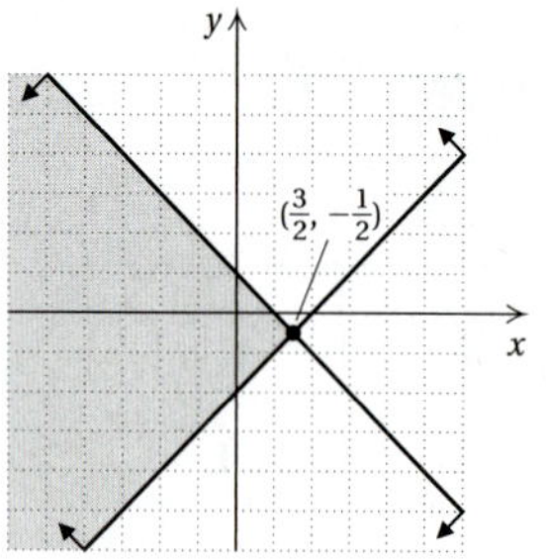

43.

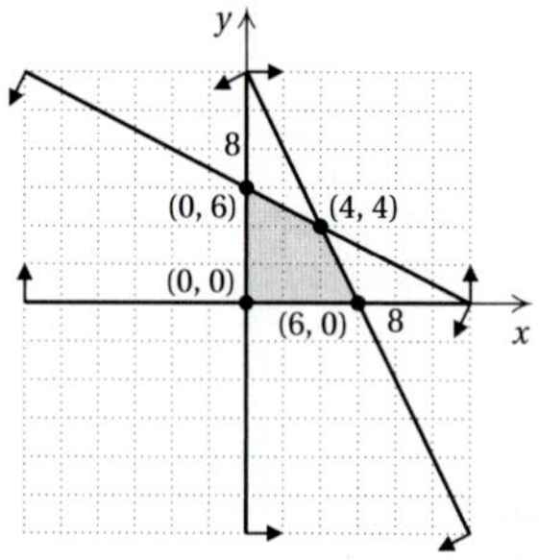

45.

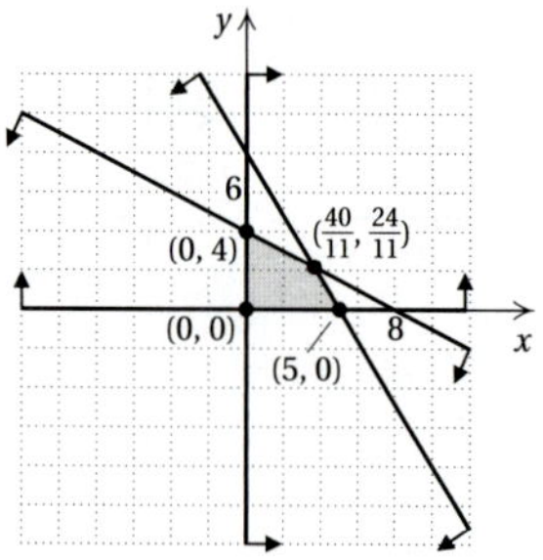

47. DW **49.** $y = \frac{1}{2}x - \frac{7}{2}$ **50.** $y = \frac{3}{8}x$
51. $y = -3x + 15$ **52.** $y = \frac{9}{2}x + 9$ **53.** $y = -\frac{3}{2}x - \frac{5}{2}$
54. $y = 6$ **55.** 2 **56.** 3 **57.** 1 **58.** 8 **59.** 4
60. $|2 - 2a|$, or $2|1 - a|$ **61.** 6 **62.** 0.2
63. $0 < w \leq 62$,
$0 < h \leq 62$,
$62 + 2w + 2h \leq 108$, or
$w + h \leq 23$;

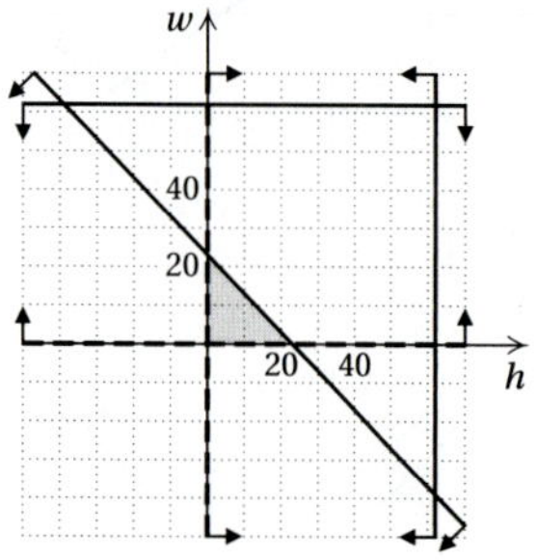

65. Left to the student **67.** Left to the student
69. Left to the student

Summary and Review: Chapter 14, p. 1111

1. $[-8, 9)$ **2.** $(-\infty, 40]$
3. 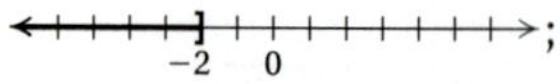; $(-\infty, -2]$
4. 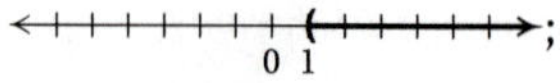; $(1, \infty)$
5. $\{a \mid a \leq -21\}$, or $(-\infty, -21]$

6. $\{y \mid y \geq -7\}$, or $[-7, \infty)$ **7.** $\{y \mid y > -4\}$, or $(-4, \infty)$
8. $\{y \mid y > -30\}$, or $(-30, \infty)$ **9.** $\{x \mid x > -3\}$, or $(-3, \infty)$
10. $\{y \mid y \leq -\frac{6}{5}\}$, or $(-\infty, -\frac{6}{5}]$ **11.** $\{x \mid x < -3\}$, or $(-\infty, -3)$
12. $\{y \mid y > -10\}$, or $(-10, \infty)$
13. $\{x \mid x \leq -\frac{5}{2}\}$, or $(-\infty, -\frac{5}{2}]$ **14.** $\{t \mid t > 4\frac{1}{4}\text{ hr}\}$
15. \$10,000 **16.** 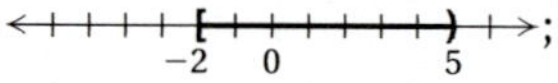; $[-2, 5)$
17. 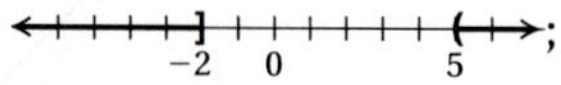; $(-\infty, -2] \cup (5, \infty)$ **18.** $\varnothing$
19. $\{x \mid -7 < x \leq 2\}$, or $(-7, 2]$
20. $\{x \mid -\frac{5}{4} < x < \frac{5}{2}\}$, or $(-\frac{5}{4}, \frac{5}{2})$
21. $\{x \mid x < -3 \text{ or } x > 1\}$, or $(-\infty, -3) \cup (1, \infty)$
22. $\{x \mid x < -11 \text{ or } x \geq -6\}$, or $(-\infty, -11) \cup [-6, \infty)$
23. $\{x \mid x \leq -6 \text{ or } x \geq 8\}$, or $(-\infty, -6] \cup [8, \infty)$ **24.** $\frac{3}{|x|}$
25. $\frac{2|x|}{y^2}$ **26.** $\frac{4}{|y|}$ **27.** 62 **28.** 6, −6 **29.** $\{9, -5\}$
30. $\{-14, \frac{4}{3}\}$ **31.** $\varnothing$ **32.** $\{x \mid -\frac{17}{2} < x < \frac{7}{2}\}$, or $(-\frac{17}{2}, \frac{7}{2})$
33. $\{x \mid x \leq -3.5 \text{ or } x \geq 3.5\}$, or $(-\infty, -3.5] \cup [3.5, \infty)$
34. $\{x \mid x \leq -\frac{11}{3} \text{ or } x \geq \frac{19}{3}\}$, or $(-\infty, -\frac{11}{3}] \cup [\frac{19}{3}, \infty)$ **35.** $\varnothing$
36. $\{t \mid 3.23 < t < 7.85\}$, or approximately between 1991 and 1996 **37.** $\{1, 5, 9\}$ **38.** $\{1, 2, 3, 5, 6, 9\}$

39.

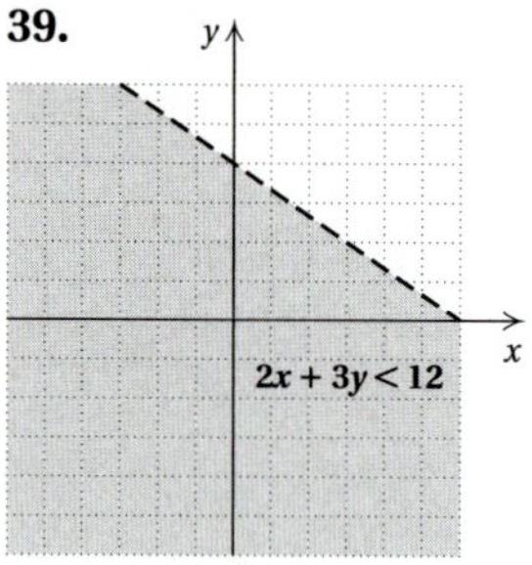

40.

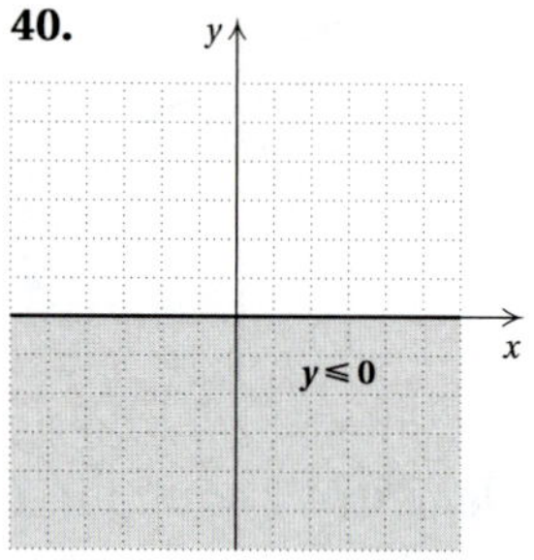

41.

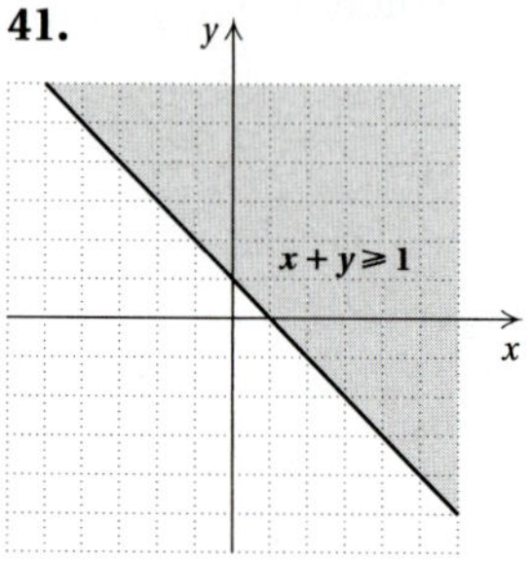

42.

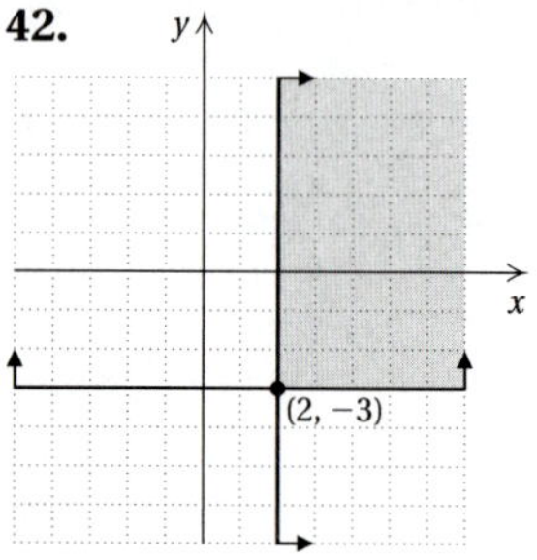

43.

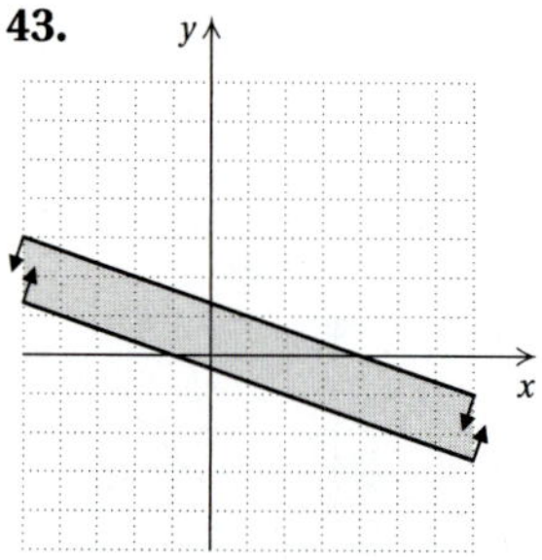

44.

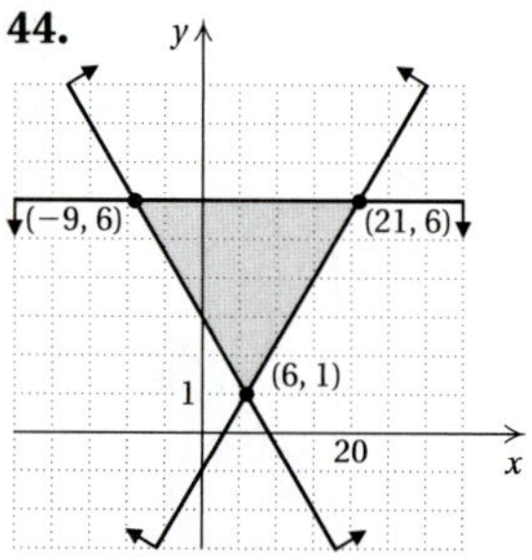

45. $\mathbf{D_W}$ **(1)** $-9(x + 2) = -9x - 18$, not $-9x + 2$. **(2)** This would be correct if (1) were correct. **(3)** If (2) were correct, the right-hand side would be -5, not 8. **(4)** The inequality symbol should be reversed. The correct solution is

$$\begin{aligned} 7 - 9x + 6x &< -9(x + 2) + 10x \\ 7 - 9x + 6x &< -9x - 18 + 10x \\ 7 - 3x &< x - 18 \\ -4x &< -25 \\ x &> \tfrac{25}{4}. \end{aligned}$$

46. $\mathbf{D_W}$ "Solve" can mean to find all the replacements that make an equation or inequality true. It can also mean to express a formula as an equivalent equation with a given variable alone on one side. **47.** $24y^2 - 41y - 35$ **48.** $\{x \mid x$ is a real number *and* $x \neq -9\}$, or $(-\infty, -9) \cup (-9, \infty)$ **49.** $(2, -1)$ **50.** $y = -\frac{2}{5}x - \frac{9}{5}$ **51.** $\left\{x \middle| -\frac{8}{3} \le x \le -2\right\}$, or $\left[-\frac{8}{3}, -2\right]$

Test: Chapter 14, p. 1114

1. [14.1b] $(-3, 2]$ **2.** [14.1b] $(-4, \infty)$
3. [14.1c] 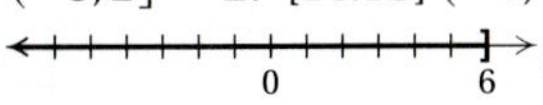

; $(-\infty, 6]$
4. [14.1c] 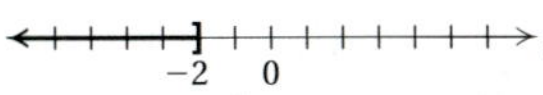

; $(-\infty, -2]$
5. [14.1c] $\{y \mid y > -50\}$, or $(-50, \infty)$
6. [14.1c] $\left\{a \middle| a \le \frac{11}{5}\right\}$, or $\left(-\infty, \frac{11}{5}\right]$
7. [14.1c] $\{y \mid y > 1\}$, or $(1, \infty)$
8. [14.1c] $\left\{x \middle| x > \frac{5}{2}\right\}$, or $\left(\frac{5}{2}, \infty\right)$
9. [14.1c] $\left\{x \middle| x \le \frac{7}{4}\right\}$, or $\left(-\infty, \frac{7}{4}\right]$ **10.** [14.1d] $\left\{h \middle| h > 2\frac{1}{10}\text{ hr}\right\}$
11. [14.2c] $\{d \mid 33 \text{ ft} \le d \le 231 \text{ ft}\}$
12. [14.2a] 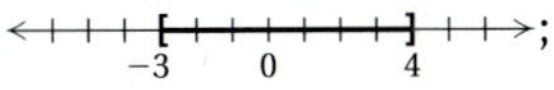

; $[-3, 4]$
13. [14.2b] 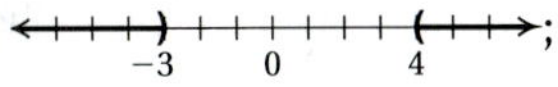

; $(-\infty, -3) \cup (4, \infty)$
14. [14.2a] $\{x \mid x \ge 4\}$, or $[4, \infty)$
15. [14.2a] $\{x \mid -1 < x < 6\}$, or $(-1, 6)$
16. [14.2a] $\left\{x \middle| -\frac{2}{5} < x \le \frac{9}{5}\right\}$, or $\left(-\frac{2}{5}, \frac{9}{5}\right]$
17. [14.2b] $\left\{x \middle| x < -4 \text{ or } x > -\frac{5}{2}\right\}$, or $(-\infty, -4) \cup \left(-\frac{5}{2}, \infty\right)$
18. [14.2b] All real numbers, or $(-\infty, \infty)$
19. [14.2b] $\{x \mid x < 3 \text{ or } x > 6\}$, or $(-\infty, 3) \cup (6, \infty)$
20. [14.3a] $\dfrac{7}{|x|}$ **21.** [14.3a] $2|x|$ **22.** [14.3b] 8.4
23. [14.3c] $\{9, -9\}$
24. [14.3e] $\{x \mid x < -3 \text{ or } x > 3\}$, or $(-\infty, -3) \cup (3, \infty)$
25. [14.3e] $\left\{x \middle| -\frac{7}{8} < x < \frac{11}{8}\right\}$, or $\left(-\frac{7}{8}, \frac{11}{8}\right)$
26. [14.3e] $\left\{x \middle| x \le -\frac{13}{5} \text{ or } x \ge \frac{7}{5}\right\}$, or $\left(-\infty, -\frac{13}{5}\right] \cup \left[\frac{7}{5}, \infty\right)$
27. [14.3d] $\{1\}$ **28.** [14.3c] $\varnothing$
29. [14.3e] $\{x \mid -99 \le x \le 111\}$, or $[-99, 111]$
30. [14.3c] $\{-6, 12\}$ **31.** [14.2a] $\{3, 5\}$
32. [14.3b] $\{1, 3, 5, 7, 9, 11, 13\}$

33. [14.4b]

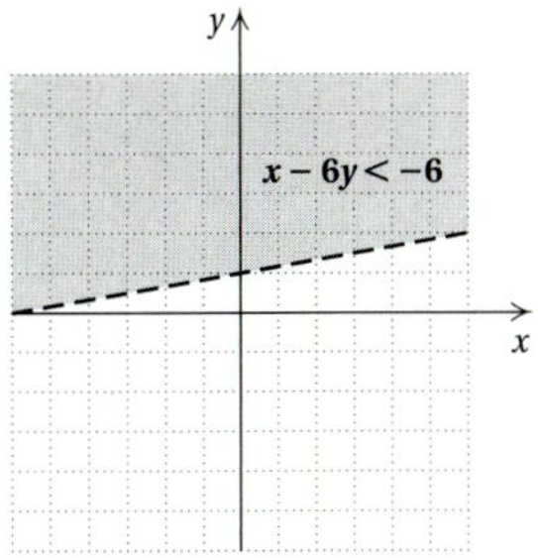

34. [14.4c]

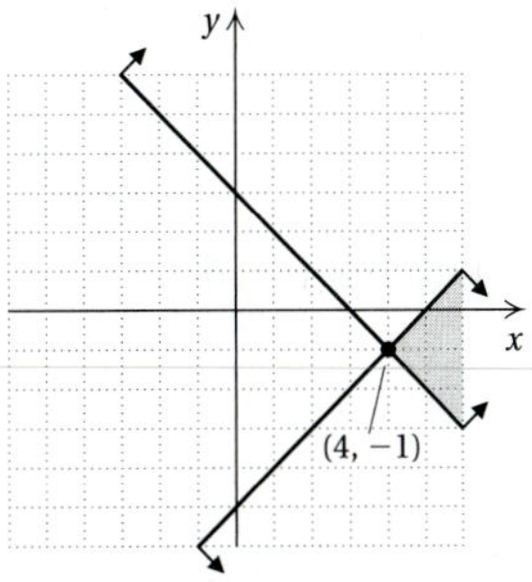

35. [14.4c]

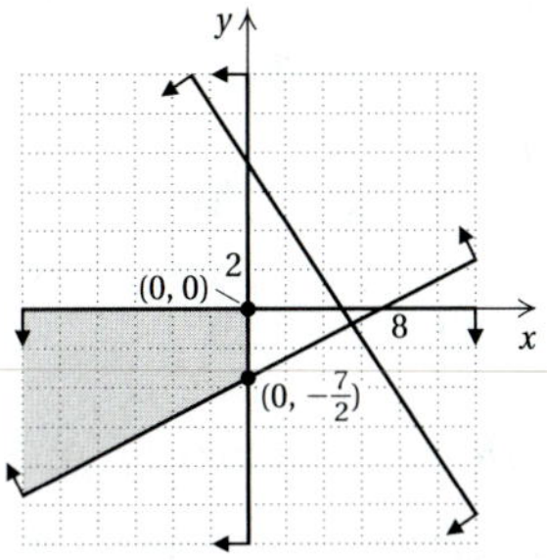

36. [9.6a] $72x^2 - 11x - 35$ **37.** [12.5c] $y = -2x + 4$
38. [13.2a], [13.3a] $(-8, -4)$ **39.** [12.2a] $\{x \mid x$ is a real number *and* $x \neq 20\}$, or $(-\infty, 20) \cup (20, \infty)$ **40.** [14.3e] $\varnothing$
41. [14.2a] $\left\{x \middle| \frac{1}{5} < x < \frac{4}{5}\right\}$, or $\left(\frac{1}{5}, \frac{4}{5}\right)$

Chapter 15

Pretest: Chapter 15, p. 1118

1. $|t|$ **2.** $3x$ **3.** $|y|$ **4.** $3a\sqrt[3]{3ab^2}$ **5.** $4\sqrt{5}$
6. $12x^2\sqrt{x}$ **7.** $25 - 4\sqrt{6}$ **8.** $2\sqrt{a}$ **9.** $-6 + 3\sqrt{5}$
10. 13 **11.** $-\frac{7}{6}$ **12.** 0, 8 **13.** $\sqrt{39}$ ft; 6.245 ft
14. $\sqrt{32}$ ft; 5.657 ft **15.** $\sqrt[6]{x^5}$ **16.** No **17.** $-5 + 13i$
18. $24 - 7i$ **19.** $-\frac{14}{13} - \frac{5}{13}i$ **20.** $-i$

Margin Exercises, Section 15.1, pp. 1119–1125

1. 3, −3 **2.** 6, −6 **3.** 11, −11 **4.** 1 **5.** 6 **6.** $\frac{9}{10}$
7. 0.08 **8.** **(a)** 4; **(b)** −4; **(c)** does not exist as a real number **9.** **(a)** 7; **(b)** −7; **(c)** does not exist as a real number **10.** **(a)** 12; **(b)** −12; **(c)** does not exist as a real number **11.** $\frac{5}{8}$ **12.** −0.9 **13.** 1.2 **14.** 4.123
15. 6.325 **16.** 33.734 **17.** −29.455 **18.** 0.793
19. −5.569 **20.** $28 + x$ **21.** $\dfrac{y}{y + 3}$ **22.** 2; $\sqrt{22} \approx 4.690$; does not exist as a real number
23. −2; $-\sqrt{7} \approx -2.646$; does not exist as a real number
24. Domain $= \{x \mid x \ge 5\} = [5, \infty)$
25. Domain $= \left\{x \middle| x \ge -\frac{3}{2}\right\} = \left[-\frac{3}{2}, \infty\right)$

26.

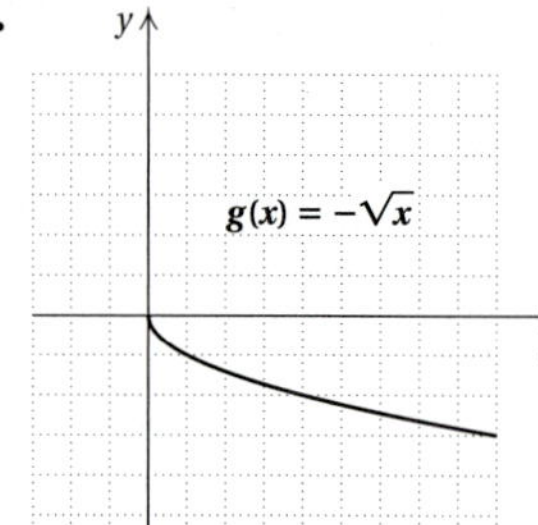

27.

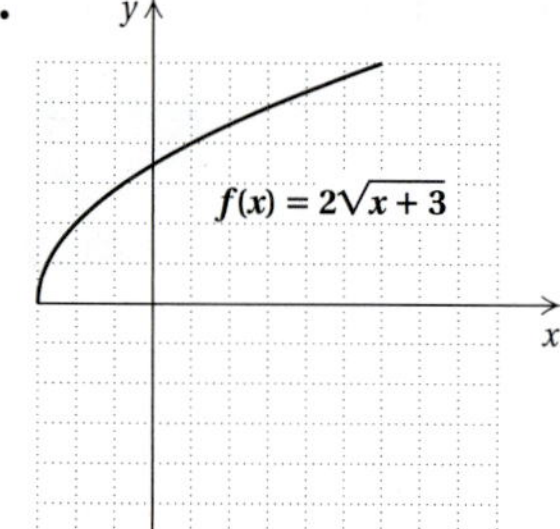

28. $|y|$ **29.** 24 **30.** $5|y|$ **31.** $4|y|$ **32.** $|x+7|$
33. $2|x-2|$ **34.** $7|y+5|$ **35.** $|x-3|$ **36.** -4
37. $3y$ **38.** $2(x+2)$ **39.** $-\frac{7}{4}$
40. -3; 0; $\sqrt[3]{-5} \approx -1.710$; $\sqrt[3]{7} \approx 1.913$ **41.** 3 **42.** -3
43. x **44.** y **45.** 0 **46.** $-2x$ **47.** $3x+2$ **48.** 3
49. -3 **50.** Does not exist as a real number **51.** 0
52. $2|x-2|$ **53.** $|x|$ **54.** $|x+3|$ **55.** $x+3$ **56.** $3x$

Calculator Corner, p. 1126

1. 6.557 **2.** 102.308 **3.** -96.985 **4.** -0.804
5. 7.469 **6.** 9.283 **7.** 2.012 **8.** -2.812

Exercise Set 15.1, p. 1127

1. 4, -4 **3.** 12, -12 **5.** 20, -20 **7.** $-\frac{7}{6}$ **9.** 14
11. 0.06 **13.** 18.628 **15.** 1.962 **17.** $y^2 + 16$
19. $\dfrac{x}{y-1}$ **21.** $\sqrt{20} \approx 4.472$; 0; does not exist as a real number; does not exist as a real number
23. $\sqrt{11} \approx 3.317$; does not exist as a real number; $\sqrt{11} \approx 3.317$; 12 **25.** Domain $= \{x \mid x \geq 2\} = [2, \infty)$
27. About 24.5 mph; about 54.8 mph

29.

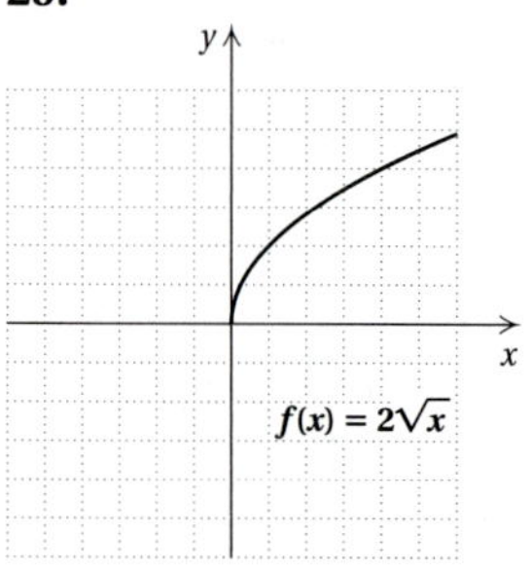

31.

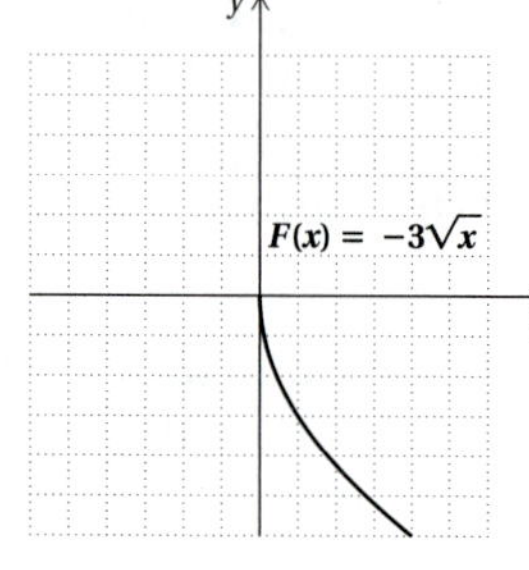

33.

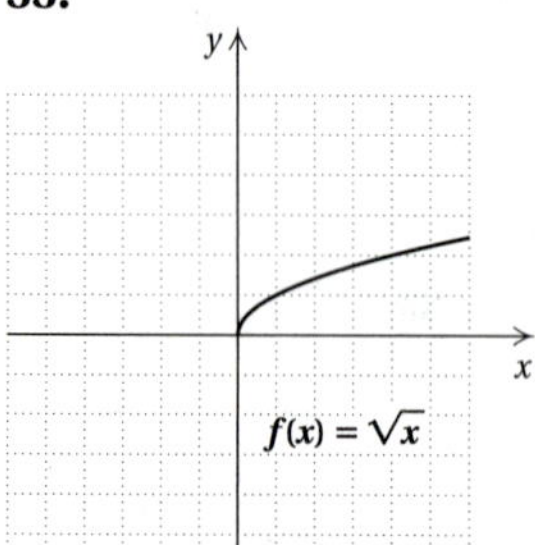

35.

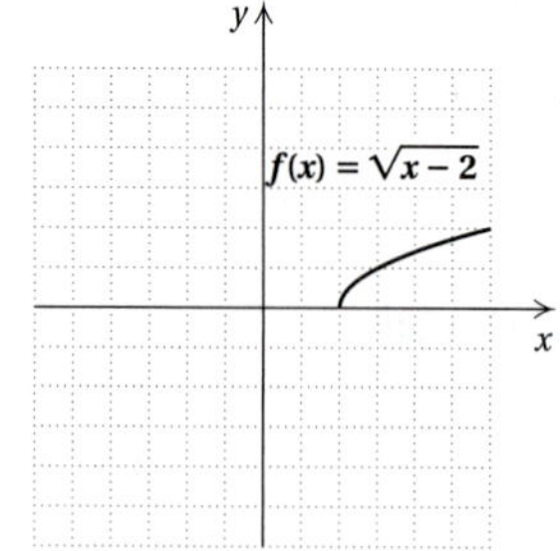

37.

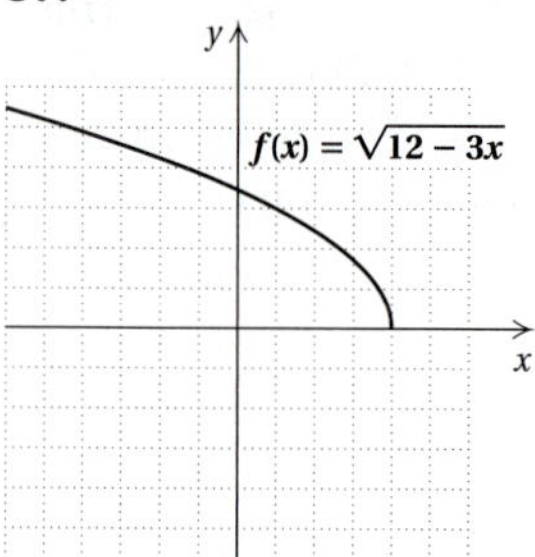

39.

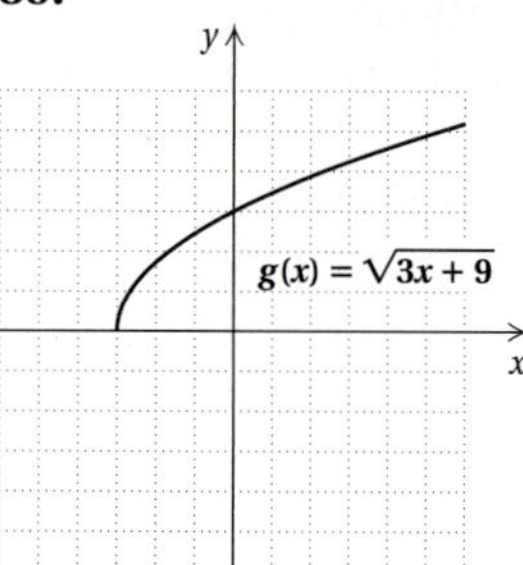

41. $4|x|$ **43.** $12|c|$ **45.** $|p+3|$ **47.** $|x-2|$ **49.** 3
51. $-4x$ **53.** -6 **55.** $0.7(x+1)$ **57.** 2; 3; -2; -4
59. -1; $-\sqrt[3]{-20}$, or $\sqrt[3]{20} \approx 2.7144$; -4; -10 **61.** -5
63. -1 **65.** $-\frac{2}{3}$ **67.** $|x|$ **69.** $5|a|$ **71.** 6
73. $|a+b|$ **75.** y **77.** $x-2$ **79.** $\mathbf{D_W}$ **81.** -2, 1
82. -1, 0 **83.** $-\frac{7}{2}, \frac{7}{2}$ **84.** 4, 9 **85.** $-2, \frac{5}{3}$ **86.** $\frac{5}{2}$
87. $0, \frac{5}{2}$ **88.** 0, 1 **89.** $a^9b^6c^{15}$ **90.** $10a^{10}b^9$
91. Domain $= \{x \mid -3 \leq x < 2\} = [-3, 2)$ **93.** 1.7; 2.2; 3.2
95. (a) Domain: $(-\infty, \infty)$; range: $(-\infty, \infty)$; **(b)** domain: $(-\infty, \infty)$; range: $(-\infty, \infty)$; **(c)** domain: $[-3, \infty)$; range: $(-\infty, 2]$; **(d)** domain: $[0, \infty)$; range: $[0, \infty)$; **(e)** domain: $[3, \infty)$; range: $[0, \infty)$

Margin Exercises, Section 15.2, pp. 1131–1135

1. $\sqrt[4]{y}$ **2.** $\sqrt{3a}$ **3.** 2 **4.** 5 **5.** $\sqrt[5]{a^3b^2c}$
6. $(19ab)^{1/3}$ **7.** $19(ab)^{1/3}$ **8.** $\left(\dfrac{x^2y}{16}\right)^{1/5}$ **9.** $7(2ab)^{1/4}$
10. $\sqrt[5]{x^3}$ **11.** 4 **12.** 32 **13.** $(7abc)^{4/3}$ **14.** $6^{7/5}$
15. $\frac{1}{2}$ **16.** $\dfrac{1}{(3xy)^{7/8}}$ **17.** $\frac{1}{27}$ **18.** $\dfrac{7p^{3/4}}{q^{6/5}}$ **19.** $\left(\dfrac{7n}{11m}\right)^{2/3}$
20. $7^{14/15}$ **21.** $5^{1/3}$ **22.** $9^{2/5}$ **23.** $\dfrac{q^{1/8}}{p^{1/3}}$ **24.** $\sqrt{a}$
25. x **26.** $\sqrt{2}$ **27.** $\sqrt[4]{xy^2}$ **28.** $a^2\sqrt{b}$ **29.** ab^2
30. $\sqrt[4]{63}$ **31.** $\sqrt[6]{x^4y^3z^5}$ **32.** $\sqrt[4]{ab}$ **33.** $\sqrt[7]{5m}$
34. $\sqrt[6]{m}$ **35.** $a^{20}b^{12}c^4$ **36.** $\sqrt[10]{x}$

Calculator Corner, p. 1133

1. 3.344 **2.** 3.281 **3.** 0.283 **4.** 11.053
5. 5.527×10^{-5} **6.** 2

Exercise Set 15.2, p. 1136

1. $\sqrt[7]{y}$ **3.** 2 **5.** $\sqrt[5]{a^3b^3}$ **7.** 8 **9.** 343 **11.** $17^{1/2}$
13. $18^{1/3}$ **15.** $(xy^2z)^{1/5}$ **17.** $(3mn)^{3/2}$ **19.** $(8x^2y)^{5/7}$
21. $\frac{1}{3}$ **23.** $\frac{1}{1000}$ **25.** $\dfrac{1}{x^{1/4}}$ **27.** $\dfrac{1}{(2rs)^{3/4}}$ **29.** $\dfrac{2a^{3/4}c^{2/3}}{b^{1/2}}$
31. $\left(\dfrac{8yz}{7x}\right)^{3/5}$ **33.** $x^{2/3}$ **35.** $\dfrac{x^4}{2^{1/3}y^{2/7}}$ **37.** $\dfrac{7x}{z^{1/3}}$
39. $\dfrac{5ac^{1/2}}{3}$ **41.** $5^{7/8}$ **43.** $7^{1/4}$ **45.** $4.9^{1/2}$
47. $6^{3/28}$ **49.** $a^{23/12}$ **51.** $a^{8/3}b^{5/2}$ **53.** $\dfrac{1}{x^{2/7}}$
55. $\sqrt[3]{a}$ **57.** x^5 **59.** $\dfrac{1}{x^3}$ **61.** a^5b^5 **63.** $\sqrt{2}$

65. $\sqrt[3]{2x}$ 67. x^2y^3 69. $2c^2d^3$ 71. $\sqrt[12]{7^4 \cdot 5^3}$
73. $\sqrt[20]{5^5 \cdot 7^4}$ 75. $\sqrt[6]{4x^5}$ 77. a^6b^{12} 79. $\sqrt[18]{m}$
81. $\sqrt[12]{x^4y^3z^2}$ 83. $\sqrt[30]{\dfrac{d^{35}}{c^{99}}}$ 85. **DW** 87. $\{-\frac{4}{7}, 2\}$
88. $\{-40, 40\}$ 89. $\{-\frac{15}{2}, \frac{5}{2}\}$ 90. $\{-\frac{11}{8}, \frac{3}{8}\}$
91. Left to the student

Margin Exercises, Section 15.3, pp. 1138–1142

1. $\sqrt{133}$ 2. $\sqrt{21pq}$ 3. $\sqrt[4]{2821}$ 4. $\sqrt[3]{\dfrac{10}{pq}}$
5. $\sqrt[6]{500}$ 6. $\sqrt[6]{25x^3y^2}$ 7. $4\sqrt{2}$ 8. $2\sqrt[3]{10}$
9. $10\sqrt{3}$ 10. $6y$ 11. $2a\sqrt{3b}$ 12. $2bc\sqrt{3ab}$
13. $2\sqrt[3]{2}$ 14. $3xy^2\sqrt[3]{3xy^2}$ 15. $3\sqrt{2}$ 16. $6y\sqrt{7}$
17. $3x\sqrt[4]{4y}$ 18. $7\sqrt{3ab}$ 19. 5 20. $56\sqrt{xy}$ 21. $5a$
22. $\frac{20}{7}$ 23. $\frac{5}{6}$ 24. $\dfrac{x}{10}$ 25. $\dfrac{3x\sqrt[3]{2x^2}}{5}$ 26. $\sqrt[12]{xy^2}$

Exercise Set 15.3, p. 1143

1. $2\sqrt{6}$ 3. $3\sqrt{10}$ 5. $5\sqrt[3]{2}$ 7. $6x^2\sqrt{5}$
9. $3x^2\sqrt[3]{2x^2}$ 11. $2t^2\sqrt[3]{10t^2}$ 13. $2\sqrt[4]{5}$ 15. $4a\sqrt{2b}$
17. $3x^2y^2\sqrt[4]{3y^2}$ 19. $2xy^3\sqrt[5]{3x^2}$ 21. $5\sqrt{2}$ 23. $3\sqrt{10}$
25. 2 27. $30\sqrt{3}$ 29. $3x^4\sqrt{2}$ 31. $5bc^2\sqrt{2b}$
33. $a\sqrt[3]{10}$ 35. $2y^3\sqrt[3]{2}$ 37. $4\sqrt[4]{4}$ 39. $4a^3b\sqrt{6ab}$
41. $\sqrt[6]{200}$ 43. $\sqrt[4]{12}$ 45. $a\sqrt[4]{a}$ 47. $b\sqrt[10]{b^9}$
49. $xy\sqrt[6]{xy^5}$ 51. $3\sqrt{2}$ 53. $\sqrt{5}$ 55. 3 57. $y\sqrt{7y}$
59. $2\sqrt[3]{a^2b}$ 61. $4\sqrt{xy}$ 63. $2x^2y^2$ 65. $\dfrac{1}{\sqrt[6]{a}}$
67. $\sqrt[12]{a^5}$ 69. $\sqrt[12]{x^2y^5}$ 71. $\frac{5}{6}$ 73. $\frac{4}{7}$ 75. $\frac{5}{3}$ 77. $\dfrac{7}{y}$
79. $\dfrac{5y\sqrt{y}}{x^2}$ 81. $\dfrac{3a\sqrt[3]{a}}{2b}$ 83. $\dfrac{3x}{2}$ 85. $\dfrac{2x\sqrt[5]{x^3}}{y^2}$
87. $\dfrac{x^2\sqrt[6]{x}}{yz^2}$ 89. **DW** 91. $-10, 9$ 92. Height: 4 in.; base: 6 in. 93. 8 94. $\frac{15}{2}$ 95. No solution 96. No solution 97. **(a)** 1.62 sec; **(b)** 1.99 sec; **(c)** 2.20 sec
99. $2yz\sqrt{2z}$

Margin Exercises, Section 15.4, pp. 1147–1148

1. $13\sqrt{2}$ 2. $10\sqrt[4]{5x} - \sqrt{7}$ 3. $19\sqrt{5}$
4. $(3y + 4)\sqrt[3]{y^2} + 2y^2$ 5. $2\sqrt{x - 1}$ 6. $5\sqrt{6} + 3\sqrt{14}$
7. $a\sqrt[3]{3} - \sqrt[3]{2a^2}$ 8. $-4 - 9\sqrt{6}$
9. $3\sqrt{ab} - 4\sqrt{3a} + 6\sqrt{3b} - 24$ 10. -3 11. $p - q$
12. $20 - 4y\sqrt{5} + y^2$ 13. $58 + 12\sqrt{6}$

Exercise Set 15.4, p. 1149

1. $11\sqrt{5}$ 3. $\sqrt[3]{7}$ 5. $13\sqrt[3]{y}$ 7. $-8\sqrt{6}$ 9. $6\sqrt[3]{3}$
11. $21\sqrt{3}$ 13. $38\sqrt{5}$ 15. $122\sqrt{2}$ 17. $9\sqrt[3]{2}$
19. $29\sqrt{2}$ 21. $(1 + 6a)\sqrt{5a}$ 23. $(2 - x)\sqrt[3]{3x}$
25. $15\sqrt[3]{4}$ 27. $(x + 1)\sqrt[3]{6x}$ 29. $3\sqrt{a - 1}$
31. $(x + 3)\sqrt{x - 1}$ 33. $4\sqrt{5} - 10$ 35. $\sqrt{6} - \sqrt{21}$
37. $2\sqrt{15} - 6\sqrt{3}$ 39. -6 41. $3a\sqrt[3]{2}$ 43. 1
45. -12 47. 44 49. 1 51. 3 53. -19
55. $a - b$ 57. $1 + \sqrt{5}$ 59. $7 + 3\sqrt{3}$ 61. -6
63. $a + \sqrt{3a} + \sqrt{2a} + \sqrt{6}$ 65. $2\sqrt[3]{9} - 3\sqrt[3]{6} - 2\sqrt[3]{4}$
67. $7 + 4\sqrt{3}$ 69. $3 - \sqrt[5]{24} - \sqrt[5]{81} + \sqrt[5]{72}$ 71. **DW**
73. $\dfrac{x(x^2 + 4)}{(x + 4)(x + 3)}$ 74. $\dfrac{(a + 2)(a + 4)}{a}$ 75. $a - 2$
76. $\dfrac{(y - 3)(y - 3)}{y + 3}$ 77. $\dfrac{4(3x - 1)}{3(4x + 1)}$ 78. $\dfrac{x}{x + 1}$
79. $\dfrac{pq}{q + p}$ 80. $\dfrac{a^2b^2}{b^2 - ab + a^2}$ 81. Domain $= (-\infty, \infty)$
83. 6 85. $14 + 2\sqrt{15} - 6\sqrt{2} - 2\sqrt{30}$
87. $3\sqrt[3]{3} + 2\sqrt[3]{9} - 8$

Margin Exercises, Section 15.5, pp. 1153–1156

1. $\dfrac{\sqrt{10}}{5}$ 2. $\dfrac{\sqrt[3]{10}}{2}$ 3. $\dfrac{2\sqrt{3ab}}{3b}$ 4. $\dfrac{\sqrt[4]{56}}{2}$ 5. $\dfrac{x\sqrt[3]{12x^2y^2}}{2y}$
6. $\dfrac{7\sqrt[3]{2x^2y}}{2y^2}$ 7. $c^2 - b$ 8. $a - b$ 9. $\dfrac{\sqrt{3} + y}{\sqrt{3} + y}$
10. $\dfrac{\sqrt{2} - \sqrt{3}}{\sqrt{2} - \sqrt{3}}$ 11. $6 - 2\sqrt{2}$ 12. $-7 - 6\sqrt{2}$

Exercise Set 15.5, p. 1157

1. $\dfrac{\sqrt{15}}{3}$ 3. $\dfrac{\sqrt{22}}{2}$ 5. $\dfrac{2\sqrt{15}}{35}$ 7. $\dfrac{2\sqrt[3]{6}}{3}$ 9. $\dfrac{\sqrt[3]{75ac^2}}{5c}$
11. $\dfrac{y\sqrt[3]{9yx^2}}{3x^2}$ 13. $\dfrac{\sqrt[4]{s^3t^3}}{st}$ 15. $\dfrac{\sqrt{15x}}{10}$ 17. $\dfrac{\sqrt[3]{100xy}}{5x^2y}$
19. $\dfrac{\sqrt[4]{2xy}}{2x^2y}$ 21. $\dfrac{54 + 9\sqrt{10}}{26}$ 23. $-2\sqrt{35} - 2\sqrt{21}$
25. $\dfrac{\sqrt{15} + 20 - 6\sqrt{2} - 8\sqrt{30}}{-77}$ 27. $\dfrac{6 - 5\sqrt{a} + a}{9 - a}$
29. $\dfrac{3\sqrt{6} + 4}{2}$ 31. $\dfrac{x - 2\sqrt{xy} + y}{x - y}$ 33. **DW** 35. 30
36. $-\frac{19}{5}$ 37. 1 38. $\dfrac{x - 2}{x + 3}$ 39. Left to the student
41. $-\dfrac{3\sqrt{a^2 - 3}}{a^2 - 3}$

Margin Exercises, Section 15.6, pp. 1160–1165

1. 100 2. No solution 3. 1 4. 2, 5 5. 4 6. 17
7. 9 8. 27 9. 5 10. About 1149.9 ft/sec
11. About 88°F

Calculator Corner, p. 1161

1. Left to the student 2. Left to the student

Exercise Set 15.6, p. 1166

1. $\frac{19}{2}$ 3. $\frac{49}{6}$ 5. 57 7. $\frac{92}{5}$ 9. -1 11. No solution
13. 3 15. 19 17. -6 19. $\frac{1}{64}$ 21. 9 23. 15
25. 2, 5 27. 6 29. 5 31. 9 33. 7 35. $\frac{80}{9}$
37. 2, 6 39. -1 41. No solution 43. 3 45. About 332 km 47. 207.36 m 49. 36 m 51. About 21 km
53. 151.25 ft; 281.25 ft 55. About 25°F 57. About 0.81 ft 59. **DW** 61. $4\frac{4}{9}$ hr 62. Jeff: $1\frac{1}{3}$ hr; Grace: 4 hr

63. 2808 mi **64.** 84 hr **65.** 0, -2.8 **66.** 0, $\frac{5}{3}$ **67.** $-8, 8$ **68.** $-3, \frac{7}{2}$ **69.** Left to the student **71.** 6912 **73.** 0 **75.** $-6, -3$ **77.** 2 **79.** $0, \frac{125}{4}$ **81.** 2 **83.** $\frac{1}{2}$ **85.** 3

Margin Exercises, Section 15.7, pp. 1171–1172

1. $\sqrt{41}$; 6.403 **2.** $\sqrt{6}$; 2.449 **3.** $\sqrt{200}$; 14.142 **4.** $\sqrt{8200}$ ft; 90.554 ft **5.** 7.4 ft

Exercise Set 15.7, p. 1173

1. $\sqrt{34}$; 5.831 **3.** $\sqrt{450}$; 21.213 **5.** 5 **7.** $\sqrt{43}$; 6.557 **9.** $\sqrt{12}$; 3.464 **11.** $\sqrt{n-1}$ **13.** $\sqrt{10{,}561}$ ft; 102.767 ft **15.** $s + s\sqrt{2}$ **17.** $\sqrt{116}$ ft; 10.770 ft **19.** 7.1 ft **21.** $\sqrt{181}$ cm; 13.454 cm **23.** 12 in. **25.** $(3, 0), (-3, 0)$ **27.** $\sqrt{340} + 8$ ft; 26.439 ft **29.** $\sqrt{420.125}$ in. $\approx$ 20.497 in. **31.** DW **33.** Flash: $67\frac{2}{3}$ mph; Crawler: $53\frac{2}{3}$ mph **34.** $3\frac{3}{4}$ mph **35.** $-7, \frac{3}{2}$ **36.** 3, 8 **37.** 1 **38.** $-2, 2$ **39.** 13 **40.** 7 **41.** 26 packets **43.** $\sqrt{75}$ cm

Margin Exercises, Section 15.8, pp. 1177–1183

1. $i\sqrt{5}$, or $\sqrt{5}i$ **2.** $5i$ **3.** $-i\sqrt{11}$, or $-\sqrt{11}i$ **4.** $-6i$ **5.** $3i\sqrt{6}$, or $3\sqrt{6}i$ **6.** $15 - 3i$ **7.** $-12 + 6i$ **8.** $3 - 5i$ **9.** $12 - 7i$ **10.** -10 **11.** $-\sqrt{34}$ **12.** 42 **13.** $-9 - 12i$ **14.** $-35 - 25i$ **15.** $-14 + 8i$ **16.** $11 + 10i$ **17.** $5 + 12i$ **18.** $-i$ **19.** 1 **20.** i **21.** -1 **22.** $8 - i$ **23.** 3 **24.** $-7 - 6i$ **25.** $-1 + i$ **26.** $6 - 3i$ **27.** $-9 + 5i$ **28.** $\pi + \frac{1}{4}i$ **29.** 53 **30.** 10 **31.** $2i$ **32.** $\frac{10}{17} - \frac{11}{17}i$

33.

$x^2 + 1 = 0$	
$(-i)^2 + 1$	? 0
$i^2 + 1$	
$-1 + 1$	
0	TRUE

Yes

34.

$x^2 - 2x + 2 = 0$	
$(1 - i)^2 - 2(1 - i) + 2$	? 0
$1 - 2i + i^2 - 2 + 2i + 2$	
$1 - 2i - 1 - 2 + 2i + 2$	
0	TRUE

Yes

Calculator Corner, p. 1182

1. $-2 - 9i$ **2.** $20 + 17i$ **3.** $-47 - 161i$ **4.** $-\frac{151}{290} + \frac{73}{290}i$ **5.** -20 **6.** -28.373 **7.** $-\frac{16}{25} - \frac{1}{50}i$ **8.** 81 **9.** $117 + 118i$ **10.** $-\frac{14}{169} + \frac{34}{169}i$

Exercise Set 15.8, p. 1184

1. $i\sqrt{35}$, or $\sqrt{35}i$ **3.** $4i$ **5.** $-2i\sqrt{3}$, or $-2\sqrt{3}i$ **7.** $i\sqrt{3}$, or $\sqrt{3}i$ **9.** $9i$ **11.** $7i\sqrt{2}$, or $7\sqrt{2}i$ **13.** $-7i$ **15.** $4 - 2\sqrt{15}i$, or $4 - 2i\sqrt{15}$ **17.** $(2 + 2\sqrt{3})i$ **19.** $12 - 4i$ **21.** $9 - 5i$ **23.** $7 + 4i$ **25.** $-4 - 4i$ **27.** $-1 + i$ **29.** $11 + 6i$ **31.** -18 **33.** $-\sqrt{14}$ **35.** 21 **37.** $-6 + 24i$ **39.** $1 + 5i$ **41.** $18 + 14i$ **43.** $38 + 9i$ **45.** $2 - 46i$ **47.** $5 - 12i$ **49.** $-24 + 10i$ **51.** $-5 - 12i$ **53.** $-i$ **55.** 1 **57.** -1 **59.** i **61.** -1 **63.** $-125i$ **65.** 8 **67.** $1 - 23i$ **69.** 0 **71.** 0 **73.** 1 **75.** $5 - 8i$ **77.** $2 - \frac{\sqrt{6}}{2}i$ **79.** $\frac{9}{10} + \frac{13}{10}i$ **81.** $-i$ **83.** $-\frac{3}{7} - \frac{8}{7}i$ **85.** $\frac{6}{5} - \frac{2}{5}i$ **87.** $-\frac{8}{41} + \frac{10}{41}i$ **89.** $-\frac{4}{3}i$ **91.** $-\frac{1}{2} - \frac{1}{4}i$ **93.** $\frac{3}{5} + \frac{4}{5}i$

95.

$x^2 - 2x + 5 = 0$	
$(1 - 2i)^2 - 2(1 - 2i) + 5$	? 0
$1 - 4i + 4i^2 - 2 + 4i + 5$	
$1 - 4i - 4 - 2 + 4i + 5$	
0	TRUE

Yes

97.

$x^2 - 4x - 5 = 0$	
$(2 + i)^2 - 4(2 + i) - 5$	? 0
$4 + 4i + i^2 - 8 - 4i - 5$	
$4 + 4i - 1 - 8 - 4i - 5$	
-10	FALSE

No

99. DW **101.** 7 **102.** $\frac{70}{29}$ **103.** $-\frac{29}{3}, 5$ **104.** $\{x \mid -\frac{29}{3} < x < 5\}$, or $(-\frac{29}{3}, 5)$ **105.** $\{x \mid x \le -\frac{29}{3}$ *or* $x \ge 5\}$, or $(-\infty, -\frac{29}{3}] \cup [5, \infty)$ **106.** $-12, -\frac{2}{5}$ **107.** $-4 - 8i$; $-2 + 4i$; $8 - 6i$ **109.** $-3 - 4i$ **111.** $-88i$ **113.** 8 **115.** $\frac{3}{5} + \frac{9}{5}i$ **117.** 1

Summary and Review: Chapter 15, p. 1189

1. 27.893 **2.** 6.378 **3.** $f(0)$, $f(-1)$, and $f(1)$ do not exist as real numbers; $f(\frac{41}{3}) = 5$ **4.** Domain $= \{x \mid x \ge \frac{16}{3}\}$, or $[\frac{16}{3}, \infty)$ **5.** $9|a|$ **6.** $7|z|$ **7.** $|c - 3|$ **8.** $|x - 3|$ **9.** -10 **10.** $-\frac{1}{3}$ **11.** 2; -2; 3 **12.** $|x|$ **13.** 3 **14.** $\sqrt[5]{a}$ **15.** 512 **16.** $31^{1/2}$ **17.** $(a^2b^3)^{1/5}$ **18.** $\frac{1}{7}$ **19.** $\frac{1}{4x^{2/3}y^{2/3}}$ **20.** $\frac{5b^{1/2}}{a^{3/4}c^{2/3}}$ **21.** $\frac{3a}{t^{1/4}}$ **22.** $\frac{1}{x^{2/5}}$ **23.** $7^{1/6}$ **24.** x^7 **25.** $3x^2$ **26.** $\sqrt[12]{x^4y^3}$ **27.** $\sqrt[12]{x^7}$ **28.** $7\sqrt{5}$ **29.** $-3\sqrt[3]{4}$ **30.** $5b^2\sqrt[3]{2a^2}$ **31.** $\frac{7}{6}$ **32.** $\frac{4}{3}x^2$ **33.** $\frac{2x^2}{3y^3}$ **34.** $\sqrt{15xy}$ **35.** $3a\sqrt[3]{a^2b^2}$ **36.** $\sqrt[15]{a^5b^9}$ **37.** $y\sqrt[3]{6}$ **38.** $\frac{5}{2}\sqrt{x}$ **39.** $\sqrt[12]{x^5}$ **40.** $7\sqrt[3]{x}$ **41.** $3\sqrt{3}$ **42.** $(2x + y^2)\sqrt[3]{x}$ **43.** $15\sqrt{2}$ **44.** $-43 - 2\sqrt{10}$ **45.** $8 - 2\sqrt{7}$ **46.** $9 - \sqrt[3]{4}$ **47.** $\frac{2\sqrt{6}}{3}$ **48.** $\frac{2\sqrt{a} - 2\sqrt{b}}{a - b}$ **49.** 13 **50.** 4 **51.** 4 **52.** 9 cm **53.** $\sqrt{24}$ ft; 4.899 ft **54.** About 4166 rpm **55.** 4480 rpm **56.** 25 **57.** $\sqrt{46}$; 6.782 **58.** $(5 + 2\sqrt{2})i$ **59.** $-2 - 9i$ **60.** $1 + i$ **61.** 29 **62.** i **63.** $9 - 12i$ **64.** $\frac{2}{5} + \frac{3}{5}i$ **65.** 3 **66.** No

67. 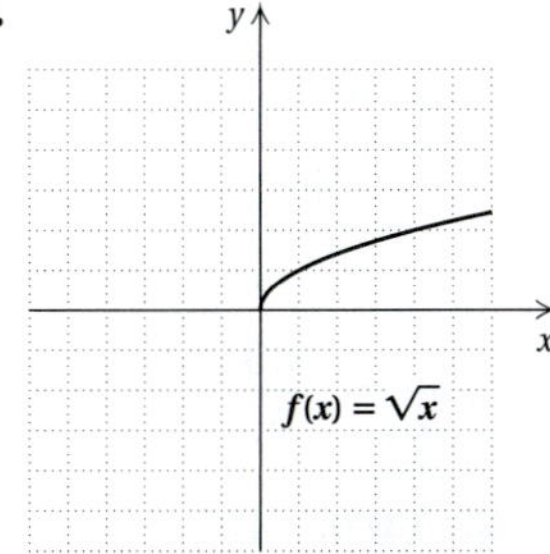

68. DW The procedure for solving radical equations is to isolate one of the radical terms, use the principle of powers, repeat these steps if necessary until all radicals are eliminated, and then check the possible solutions. A check is necessary since the principle of powers does not always yield equivalent equations. **69.** $\frac{23}{7}$ **70.** $-\frac{4}{3}, 3$

71. $\frac{x(x+2)}{(x+y)(x-3)}$ **72.** Motorcycle: 62 mph; SUV: 70 mph
73. -1 **74.** 3

Test: Chapter 15, p. 1192

1. [15.1a] 12.166 **2.** [15.1a] 2; does not exist as a real number **3.** [15.1a] Domain $= \{x \mid x \le 2\}$, or $(-\infty, 2]$
4. [15.1b] $3|q|$ **5.** [15.1b] $|x+5|$ **6.** [15.1c] $-\frac{1}{10}$
7. [15.1d] x **8.** [15.1d] 4 **9.** [15.2a] $\sqrt[3]{a^2}$
10. [15.2a] 8 **11.** [15.2a] $37^{1/2}$ **12.** [15.2a] $(5xy^2)^{5/2}$
13. [15.2b] $\frac{1}{10}$ **14.** [15.2b] $\frac{8a^{3/4}}{b^{3/2}c^{2/5}}$ **15.** [15.2c] $\frac{x^{8/5}}{y^{9/5}}$
16. [15.2c] $\frac{1}{2.9^{31/24}}$ **17.** [15.2d] $\sqrt[4]{x}$ **18.** [15.2d] $2x\sqrt{x}$
19. [15.2d] $\sqrt[15]{a^6b^5}$ **20.** [15.2d] $\sqrt[12]{8y^7}$ **21.** [15.3a] $2\sqrt{37}$
22. [15.3a] $2\sqrt[4]{5}$ **23.** [15.3a] $4a\sqrt[3]{4ab^2}$ **24.** [15.3b] $-\frac{2}{x^2}$
25. [15.3b] $\frac{5x}{6y^2}$ **26.** [15.3a] $\sqrt[3]{10xy^2}$ **27.** [15.3a] $xy\sqrt[4]{x}$
28. [15.3b] $\sqrt[5]{x^2y^2}$ **29.** [15.3b] $2\sqrt{a}$ **30.** [15.4a] $38\sqrt{2}$
31. [15.4b] -20 **32.** [15.4b] $9 + 6\sqrt{x} + x$
33. [15.5b] $\frac{13 + 8\sqrt{2}}{-41}$ **34.** [15.6a] 35 **35.** [15.6b] 7
36. [15.6a] 5 **37.** [15.7a] 7 ft **38.** [15.6c] About 5000 m
39. [15.7a] $\sqrt{98}$; 9.899 **40.** [15.7a] 2 **41.** [15.8a] $11i$
42. [15.8b] $7 + 5i$ **43.** [15.8c] $37 + 9i$ **44.** [15.8d] $-i$
45. [15.8e] $-\frac{77}{50} + \frac{7}{25}i$ **46.** [15.8f] No **47.** [11.7a] No solution **48.** [10.8b] $-\frac{1}{3}, \frac{5}{2}$ **49.** [11.2b] $\frac{x-3}{x-4}$
50. [11.8a] Fran: 36 hr; Juan: 45 hr **51.** [15.8c, e] $-\frac{17}{4}i$
52. [15.6b] 3

Chapter 16

Pretest: Chapter 16, p. 1196

1. $0, -3$ **2.** $-2 + 2i, -2 - 2i$ **3.** 5
4. $\frac{\sqrt{3}}{3}, -\frac{\sqrt{3}}{3}, \sqrt{2}, -\sqrt{2}$ **5.** $-1, \frac{5}{6}$ **6.** 1 **7.** $\frac{-2 \pm \sqrt{6}}{2}$; 0.225, -2.225 **8.** $T = \frac{1}{RW^2}$ **9.** $3x^2 + 4x - 4 = 0$
10. **(a)** $(3, -2)$; **(b)** $x = 3$; **(c)** 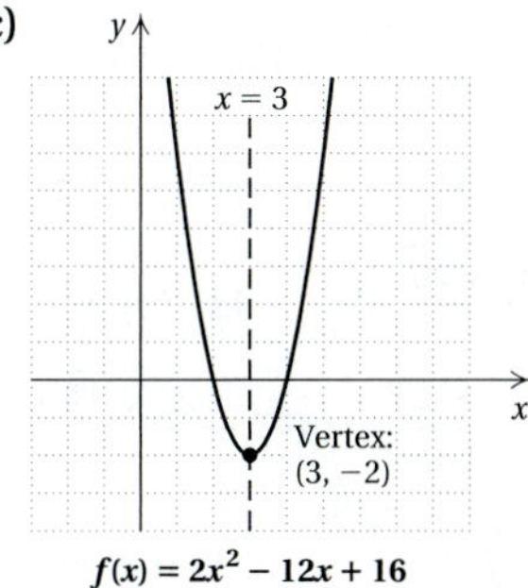

;

(d) -2 **11.** $(3 + \sqrt{5}, 0), (3 - \sqrt{5}, 0)$
12. $f(x) = 4x^2 - 5x + 1$ **13.** 2 ft **14.** 40 mph; 48 mph
15. 5625 ft^2 **16.** $\{x \mid x < -5 \text{ or } 0 < x < 3\}$, or $(-\infty, -5) \cup (0, 3)$ **17.** $\{x \mid x < -3 \text{ or } x \ge 2\}$, or $(-\infty, -3) \cup [2, 0)$

Margin Exercises, Section 16.1, pp. 1197–1205

1. **(a)** $(2, 0), (4, 0)$; **(b)** 2, 4; **(c)** The solutions of $x^2 - 6x + 8 = 0$, 2 and 4, are the first coordinates of the x-intercepts, $(2, 0)$ and $(4, 0)$, of the graph of $f(x) = x^2 - 6x + 8$. **2.** **(a)** 4 and -4, or ± 4; **(b)** $(-4, 0), (4, 0)$ **3.** **(a)** $0, -\frac{7}{2}$; **(b)** $(-\frac{7}{2}, 0), (0, 0)$
4. **(a)** $\frac{3}{5}, 1$; **(b)** $(\frac{3}{5}, 0), (1, 0)$ **5.** $\sqrt{3}$ and $-\sqrt{3}$, or $\pm\sqrt{3}$; 1.732 and -1.732, or ± 1.732 **6.** $\frac{2\sqrt{6}}{3}$ and $-\frac{2\sqrt{6}}{3}$, or $\pm\frac{2\sqrt{6}}{3}$; 1.633 and -1.633, or ± 1.633 **7.** $\frac{\sqrt{2}}{2}i$ and $-\frac{\sqrt{2}}{2}i$, or $\pm\frac{\sqrt{2}}{2}i$ **8.** **(a)** $1 \pm \sqrt{5}$; **(b)** $(1 - \sqrt{5}, 0), (1 + \sqrt{5}, 0)$
9. $-8 \pm \sqrt{11}$ **10.** $-2, -4$ **11.** $10, -2$
12. $-3 \pm \sqrt{10}$ **13.** $\frac{1 \pm \sqrt{22}}{3}$ **14.** About 36.0 in.
15. 1 sec

Calculator Corner, p. 1198

Left to the student

Calculator Corner, p. 1201

The calculator returns an ERROR message because the graph of $y = 4x^2 + 9$ has no x-intercepts. This indicates that the equation $4x^2 + 9 = 0$ has no real-number solutions.

Exercise Set 16.1, p. 1206

1. **(a)** $\sqrt{5}, -\sqrt{5}$, or $\pm\sqrt{5}$; **(b)** $(-\sqrt{5}, 0), (\sqrt{5}, 0)$ **3.** **(a)** $\frac{5}{3}i$, $-\frac{5}{3}i$, or $\pm\frac{5}{3}i$; **(b)** no x-intercepts **5.** $\pm\frac{\sqrt{6}}{2}$; ± 1.225
7. $5, -9$ **9.** $8, 0$ **11.** $11 \pm \sqrt{7}$; 13.646, 8.354
13. $7 \pm 2i$ **15.** $18, 0$ **17.** $\frac{3 \pm \sqrt{14}}{2}$; 3.371, -0.371
19. $5, -11$ **21.** $9, 5$ **23.** $-2 \pm \sqrt{6}$ **25.** $11 \pm 2\sqrt{33}$

27. $\frac{-1 \pm \sqrt{5}}{2}$ 29. $\frac{5 \pm \sqrt{53}}{2}$ 31. $\frac{-3 \pm \sqrt{57}}{4}$
33. $\frac{9 \pm \sqrt{105}}{4}$ 35. 2, −8 37. $-11 \pm \sqrt{19}$
39. $5 \pm \sqrt{29}$ 41. (a) $\frac{-7 \pm \sqrt{57}}{2}$; (b) $\left(\frac{-7 - \sqrt{57}}{2}, 0\right)$, $\left(\frac{-7 + \sqrt{57}}{2}, 0\right)$ 43. (a) $\frac{5}{4} \pm i\frac{\sqrt{39}}{4}$; (b) no x-intercepts
45. $\frac{3 \pm \sqrt{17}}{4}$ 47. $\frac{3 \pm \sqrt{145}}{4}$ 49. $\frac{2 \pm \sqrt{7}}{3}$
51. $-\frac{1}{2} \pm i\frac{\sqrt{7}}{2}$ 53. $2 \pm 3i$ 55. 0.866 sec
57. About 7.3 sec 59. About 6.8 sec 61. $\mathbf{D_W}$
63. (a) $T(t) = 33.75t + 241.25$, where t is the number of years since 1995; (b) 578,750; (c) 2020
64.

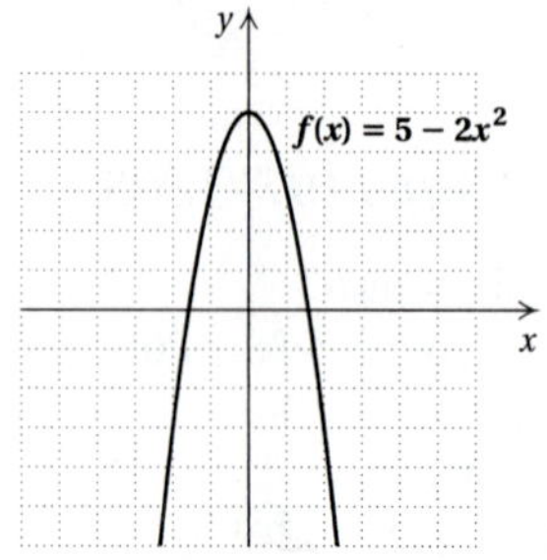

65.

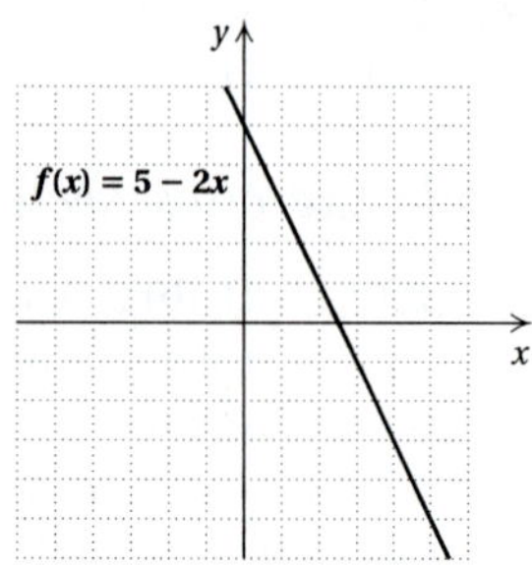

66.

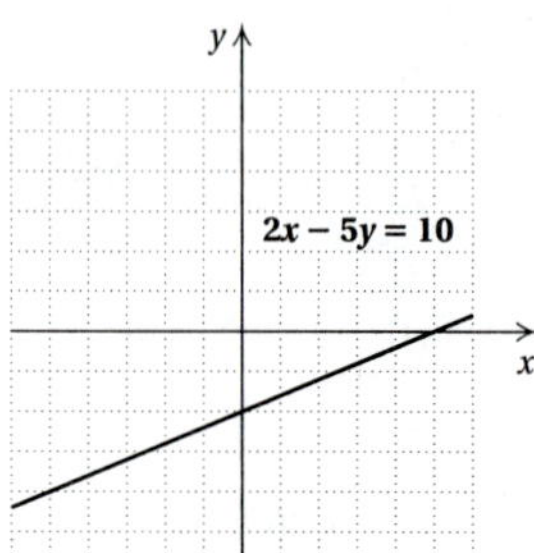

67.

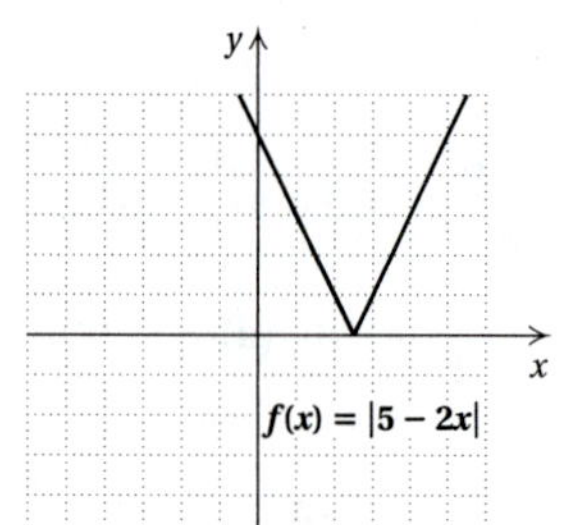

68. $2\sqrt{22}$ 69. $\frac{\sqrt{10}}{5}$ 71. Left to the student
73. 16, −16 75. $0, \frac{7}{2}, -8, -\frac{10}{3}$

Margin Exercises, Section 16.2, pp. 1212–1214

1. (a) $-\frac{1}{2}$, 4; (b) $-\frac{1}{2}$, 4 2. $\frac{-1 \pm \sqrt{22}}{3}$; 1.230, −1.897
3. $\frac{1 \pm i\sqrt{7}}{2}$, or $\frac{1}{2} \pm i\frac{\sqrt{7}}{2}$ 4. $\frac{5 \pm \sqrt{73}}{6}$; 2.257, −0.591

Calculator Corner, p. 1212

1.–3. Left to the student

Calculator Corner, p. 1214

1. The graph has no x-intercepts. 2. Yes; 2 3. Yes; 2
4. Yes; 1

Calculator Corner, p. 1215

1. −3, 0.8 2. −1.5, 5 3. 3, 8 4. 2, 4

Exercise Set 16.2, p. 1216

1. $-3 \pm \sqrt{5}$ 3. $\frac{-4 \pm \sqrt{13}}{3}$ 5. $\frac{1}{2} \pm i\frac{\sqrt{3}}{2}$ 7. $2 \pm 3i$
9. $\frac{-3 \pm \sqrt{41}}{2}$ 11. $-1 \pm 2i$ 13. (a) 0, −1; (b) (0, 0), (−1, 0) 15. (a) $\frac{3 \pm \sqrt{229}}{22}$; (b) $\left(\frac{3 + \sqrt{229}}{22}, 0\right)$, $\left(\frac{3 - \sqrt{229}}{22}, 0\right)$ 17. (a) $\frac{2}{5}$; (b) $\left(\frac{2}{5}, 0\right)$ 19. −1, −2
21. 5, 10 23. $\frac{17 \pm \sqrt{249}}{10}$ 25. $2 \pm i$ 27. $\frac{2}{3}, \frac{3}{2}$
29. $2 \pm \sqrt{10}$ 31. $\frac{3}{4}, -2$ 33. $\frac{1}{2} \pm \frac{3}{2}i$
35. $1, -\frac{1}{2} \pm i\frac{\sqrt{3}}{2}$ 37. $-3 \pm \sqrt{5}$; −0.764, −5.236
39. $3 \pm \sqrt{5}$; 5.236, 0.764 41. $\frac{3 \pm \sqrt{65}}{4}$; 2.766, −1.266
43. $\frac{4 \pm \sqrt{31}}{5}$; 1.914, −0.314 45. $\mathbf{D_W}$ 47. 2 48. 3
49. 10 50. 8 51. No solution 52. No solution
53. $\frac{15}{4}$ 54. $\frac{17}{3}$ 55. Left to the student; −0.797, 0.570
57. $\frac{1 \pm \sqrt{1 + 8\sqrt{5}}}{4}$ 59. $\frac{-i \pm i\sqrt{1 + 4i}}{2}$
61. $\frac{-1 \pm 3\sqrt{5}}{6}$ 63. $3 \pm \sqrt{13}$

Margin Exercises, Section 16.3, pp. 1218–1223

1. 2 ft 2. A to B is 6 mi; A to C is 8 mi 3. A to B is $\sqrt{31} - 1$ mi, or 4.57 mi; A to C is $1 + \sqrt{31}$ mi, or 6.57 mi
4.

	Distance	Speed	Time
Faster Ship	3000	$r + 10$	$t - 50$
Slower Ship	3000	r	t

20 knots, 30 knots
5. $w_2 = \frac{w_1}{A^2}$ 6. $r = \sqrt{\frac{V}{\pi h}}$ 7. $t = \frac{-g + \sqrt{g^2 + 64s}}{32}$
8. $b = \frac{ta}{\sqrt{1 + t^2}}$

Exercise Set 16.3, p. 1224

1. Length: 9 ft; width: 2 ft 3. Length: 18 yd; width: 9 yd
5. Height: 16 m; base: 7 m 7. Height: $4 + 8\sqrt{2}$ ft; base: $8\sqrt{2} - 4$ ft 9. 3 cm 11. 6 ft, 8 ft 13. 28 and 29

15. Length: $2 + \sqrt{14}$ ft ≈ 5.742 ft; width: $\sqrt{14} - 2$ ft ≈ 1.742 ft
17. $\dfrac{17 - \sqrt{109}}{2}$ in. ≈ 3.280 in.
19. $7 + \sqrt{239}$ ft ≈ 22.460 ft; $\sqrt{239} - 7$ ft ≈ 8.460 ft
21. First part: 60 mph; second part: 50 mph
23. 40 mph **25.** Cessna: 150 mph; Beechcraft: 200 mph; or Cessna: 200 mph; Beechcraft: 250 mph **27.** To Hillsboro: 10 mph; return trip: 4 mph **29.** About 11 mph
31. $s = \sqrt{\dfrac{A}{6}}$ **33.** $r = \sqrt{\dfrac{Gm_1m_2}{F}}$ **35.** $c = \sqrt{\dfrac{E}{m}}$
37. $b = \sqrt{c^2 - a^2}$ **39.** $k = \dfrac{3 + \sqrt{9 + 8N}}{2}$
41. $r = \dfrac{-\pi h + \sqrt{\pi^2h^2 + 2\pi A}}{2\pi}$ **43.** $g = \dfrac{4\pi^2 L}{T^2}$
45. $H = \sqrt{\dfrac{704.5W}{I}}$ **47.** $v = \dfrac{c\sqrt{m^2 - m_0^2}}{m}$ **49.** DW
51. $\dfrac{1}{x-2}$ **52.** $\dfrac{x^3 + x^2 + 2x + 2}{(x-1)(x^2+x+1)}$ **53.** $\dfrac{-x}{(x+3)(x-1)}$
54. $3x^2\sqrt{x}$ **55.** $2i\sqrt{5}$ **56.** $\dfrac{3x+3}{3x+1}$ **57.** $\dfrac{4b}{a(3b^2 - 4a)}$
59. $\pm\sqrt{2}$ **61.** $A(S) = \dfrac{\pi S}{6}$ **63.** $l = \dfrac{w + w\sqrt{5}}{2}$

Margin Exercises, Section 16.4, pp. 1230–1233

1. Two real **2.** One real **3.** Two nonreal
4. $x^2 - 5x - 14 = 0$ **5.** $3x^2 + 7x - 20 = 0$
6. $x^2 + 25 = 0$ **7.** $x^2 + \sqrt{2}x - 4 = 0$ **8.** $\pm 3, \pm 1$
9. 4 **10.** $-\frac{1}{3}, \frac{1}{2}$ **11.** $(-3, 0), (-1, 0), (2, 0), (4, 0)$

Exercise Set 16.4, p. 1234

1. One real **3.** Two nonreal **5.** Two real **7.** One real **9.** Two nonreal **11.** Two real **13.** Two real **15.** One real **17.** $x^2 - 16 = 0$ **19.** $x^2 + 9x + 14 = 0$
21. $x^2 - 16x + 64 = 0$ **23.** $25x^2 - 20x - 12 = 0$
25. $12x^2 - (4k + 3m)x + km = 0$
27. $x^2 - \sqrt{3}x - 6 = 0$ **29.** $\pm\sqrt{3}$ **31.** 1, 81
33. $-1, 1, 5, 7$ **35.** $-\frac{1}{4}, \frac{1}{9}$ **37.** 1 **39.** $-1, 1, 4, 6$
41. $\pm\sqrt{3 + \sqrt{13}}, \pm\sqrt{3 - \sqrt{13}}$ **43.** $-1, 2$
45. $\pm\dfrac{\sqrt{15}}{3}, \pm\dfrac{\sqrt{6}}{2}$ **47.** $-1, 125$ **49.** $-\frac{11}{6}, -\frac{1}{6}$ **51.** $-\frac{3}{2}$
53. $\dfrac{9 \pm \sqrt{89}}{2}, -1 \pm \sqrt{3}$ **55.** $\left(\frac{4}{25}, 0\right)$ **57.** $(4, 0), (-1, 0),$ $\left(\dfrac{3 + \sqrt{33}}{2}, 0\right), \left(\dfrac{3 - \sqrt{33}}{2}, 0\right)$ **59.** DW **61.** Kenyan: 30 lb; Peruvian: 20 lb **62.** Solution A: 4 L; solution B: 8 L
63. $4x$ **64.** $3x^2$ **65.** $3a\sqrt[4]{2a}$ **66.** 4

67.
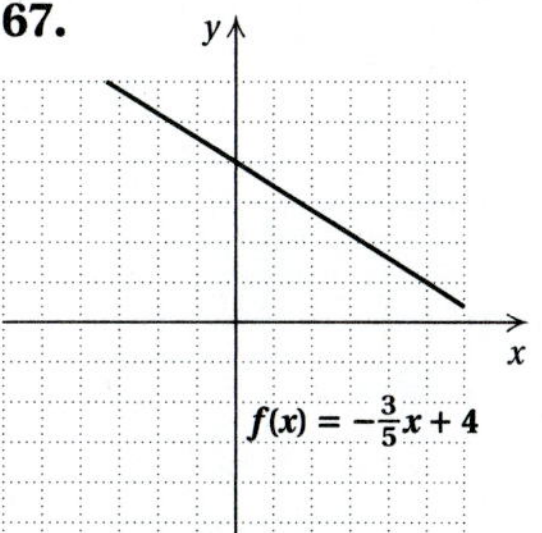

68.
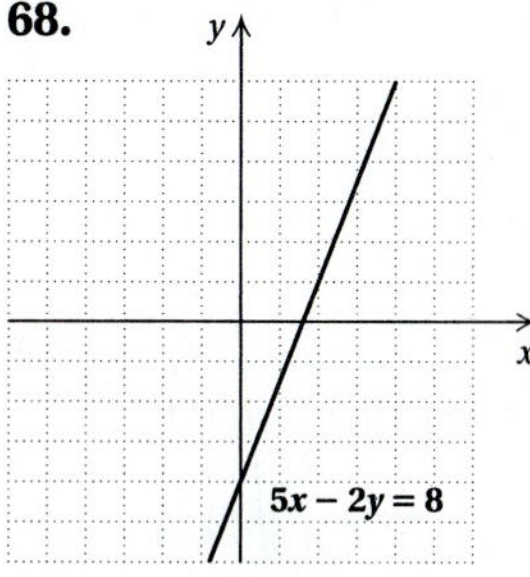

69.
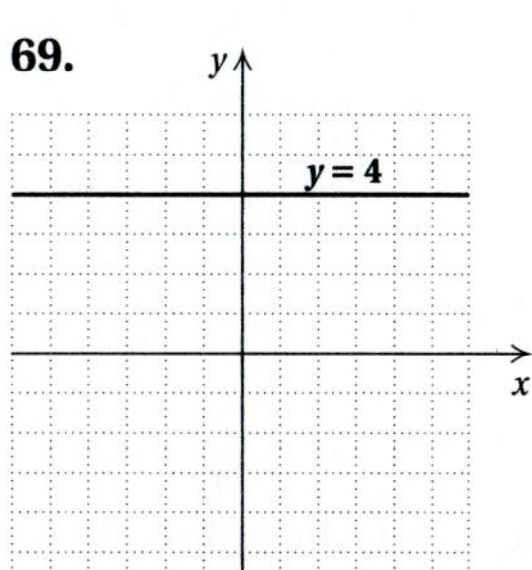

70.
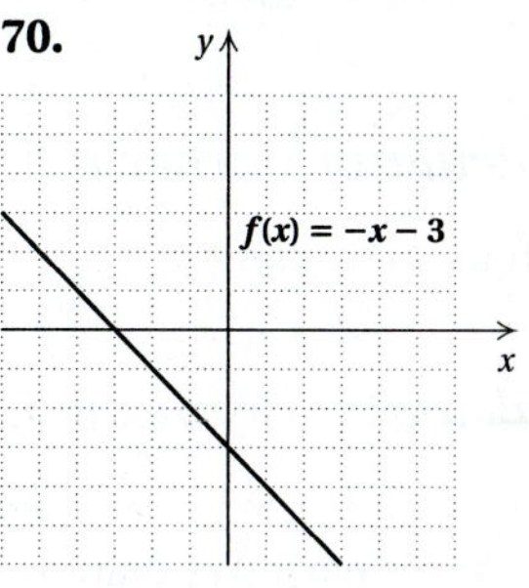

71. Left to the student **73.** (a) $-\frac{3}{5}$; (b) $-\frac{1}{3}$
75. $x^2 - \sqrt{3}x + 8 = 0$ **77.** $a = 1, b = 2, c = -3$
79. $\frac{100}{99}$ **81.** 259 **83.** 1, 3

Margin Exercises, Section 16.5, pp. 1239–1243

1.
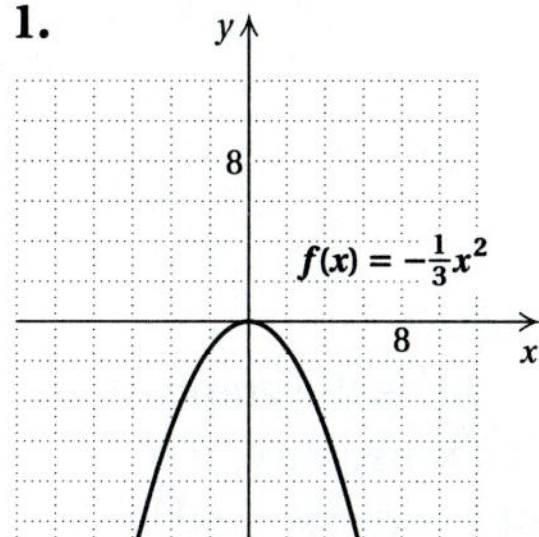

2.
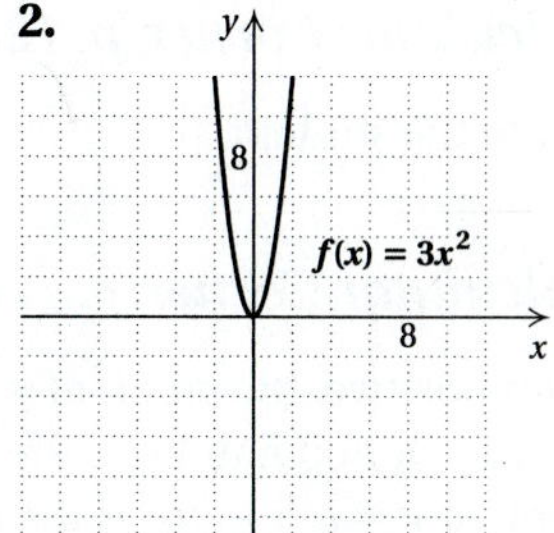

3.
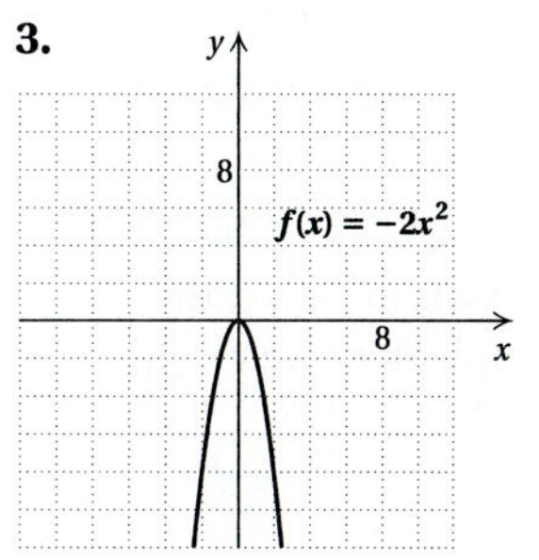

4.
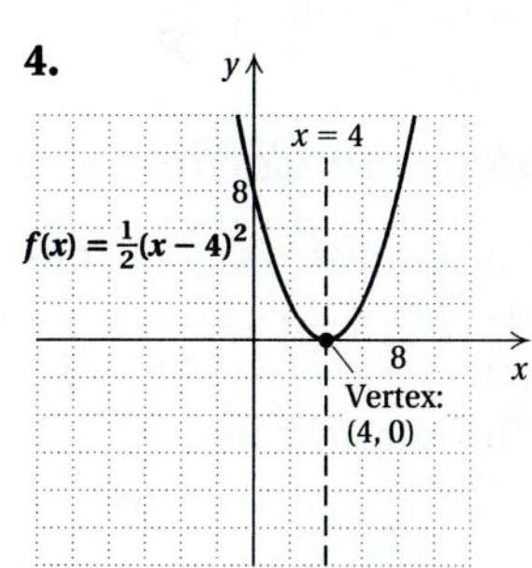

5.
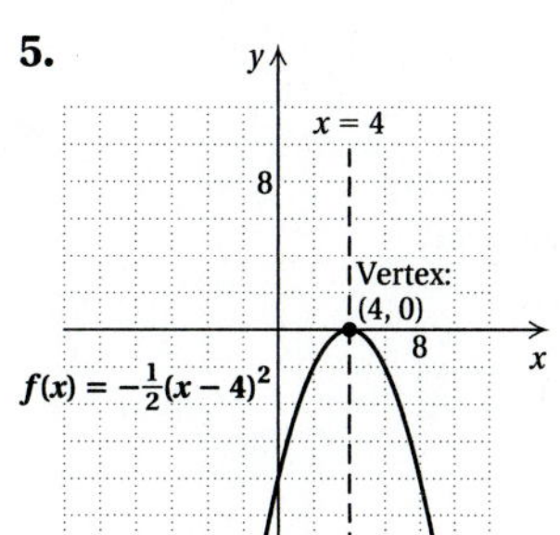

6.
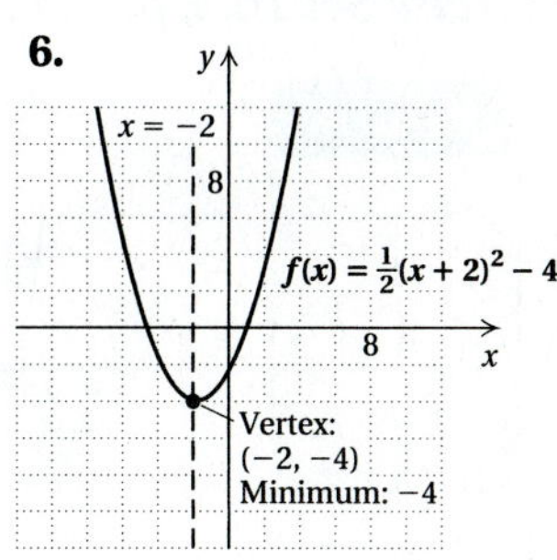

7.

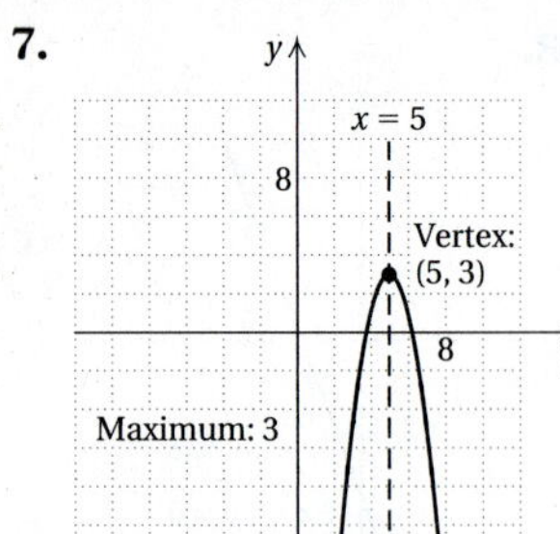

Calculator Corner, p. 1238

Left to the student

Calculator Corner, p. 1239

1. For $a > 0$, the graph opens up. For $a > 1$, the graph of $y = ax^2$ is narrower than the graph of $y = x^2$. For $0 < a < 1$, the graph of $y = ax^2$ is wider than the graph of $y = x^2$.
2. For $a < 0$, the graph opens down. For $a < -1$, the graph of $y = ax^2$ is narrower than the graph of $y = x^2$. For $-1 < a < 0$, the graph of $y = ax^2$ is wider than the graph of $y = x^2$.

Calculator Corner, p. 1240

Left to the student

Calculator Corner, p. 1240

If h is positive, the graph of $y = a(x - h)^2$ is the graph of $y = ax^2$ shifted h units to the right. If h is negative, the graph of $y = a(x - h)^2$ is the graph of $y = ax^2$ shifted $|h|$ units to the left.

Calculator Corner, p. 1242

If k is positive, the graph of $y = a(x - h)^2 + k$ is the graph of $y = a(x - h)^2$ shifted up k units. If k is negative, the graph of $y = a(x - h)^2 + k$ is the graph of $y = a(x - h)^2$ shifted down $|k|$ units.

Exercise Set 16.5, p. 1245

1.

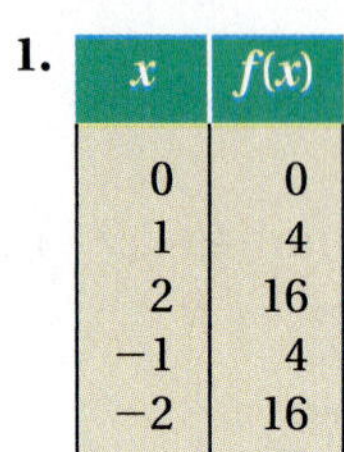

x	$f(x)$
0	0
1	4
2	16
−1	4
−2	16

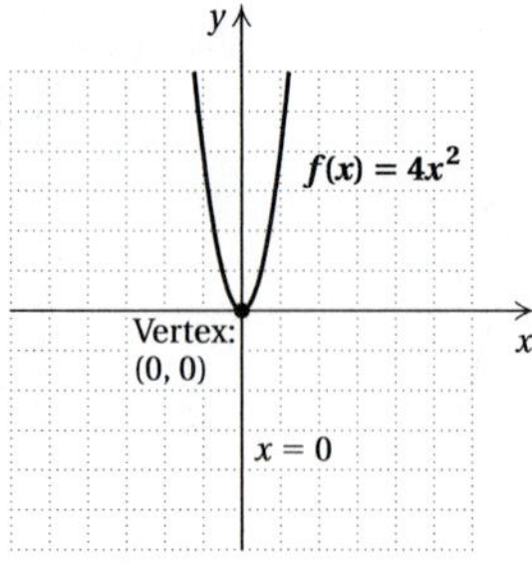

3.

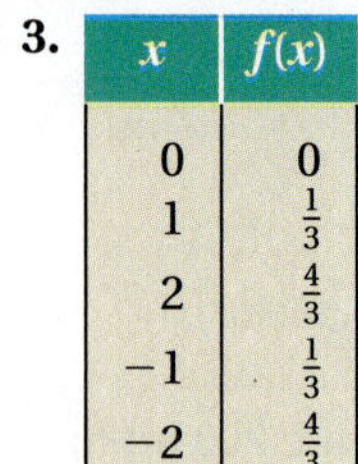

x	$f(x)$
0	0
1	$\frac{1}{3}$
2	$\frac{4}{3}$
−1	$\frac{1}{3}$
−2	$\frac{4}{3}$

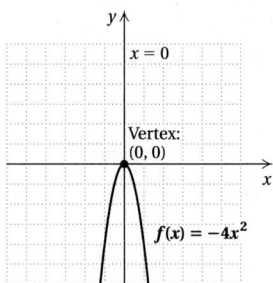

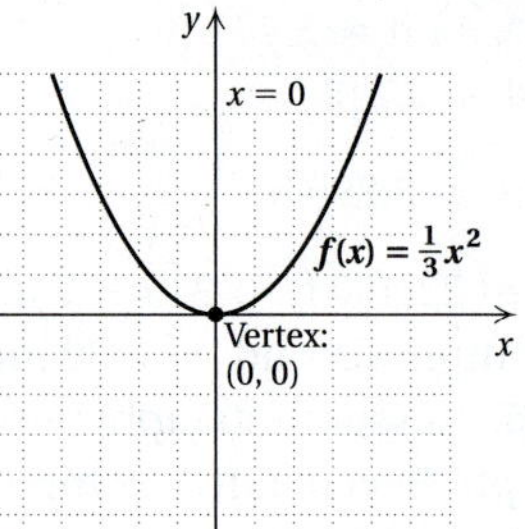

5.

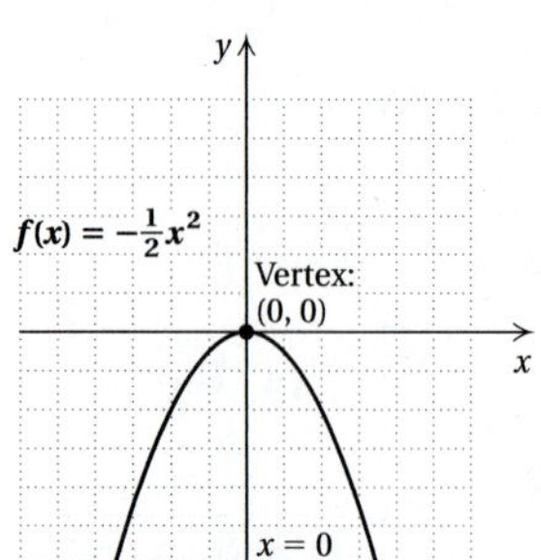

7.

y
x = 0
Vertex:
(0, 0)
x
$f(x) = -4x^2$

9.

x	$f(x)$
−3	0
−2	1
−1	4
−4	1
−5	4

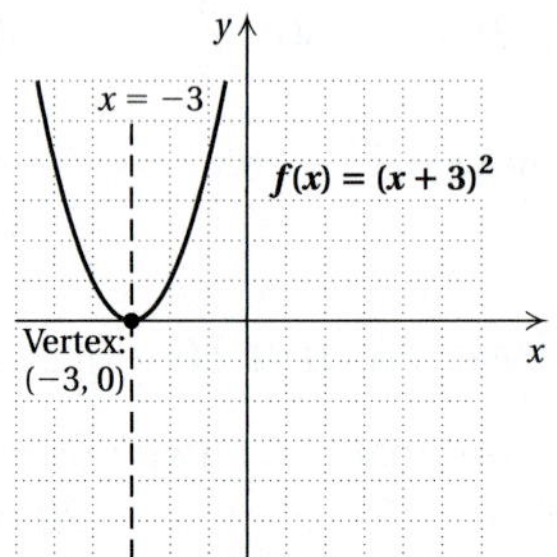

11.

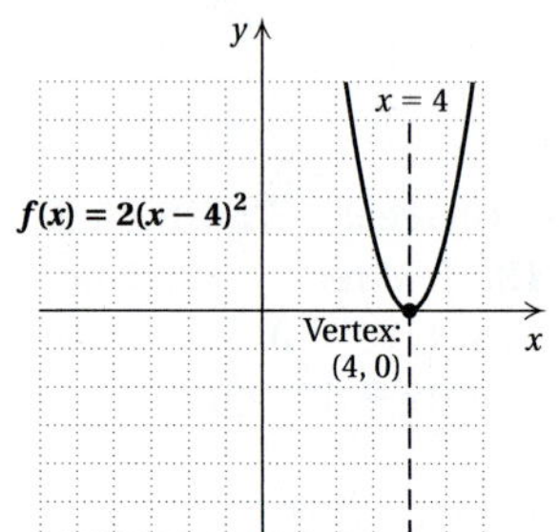

13.

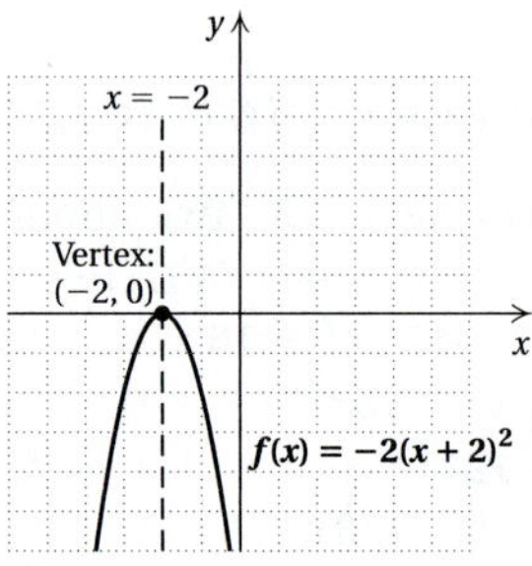

15.

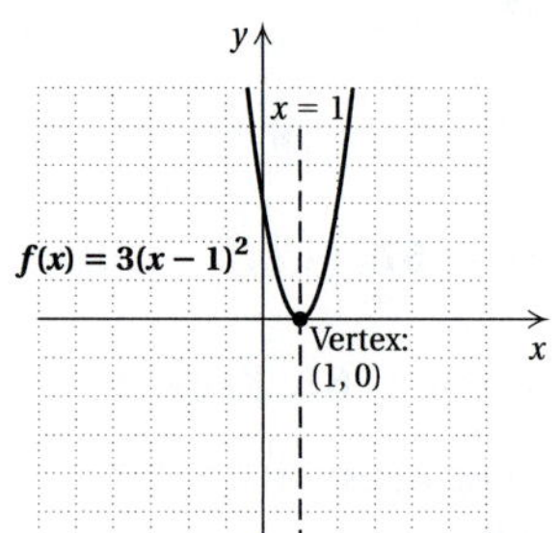

17.

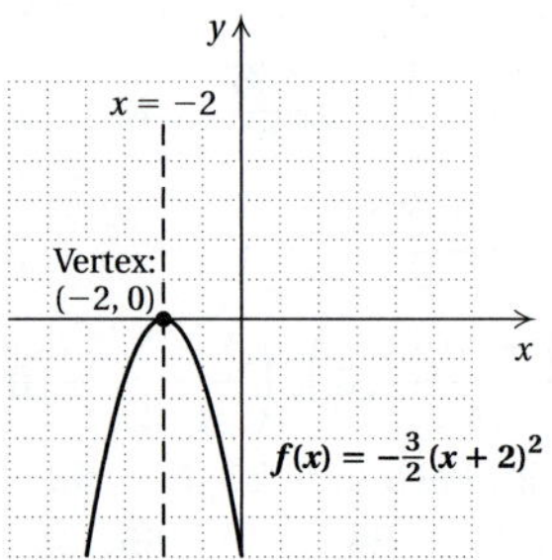

19.

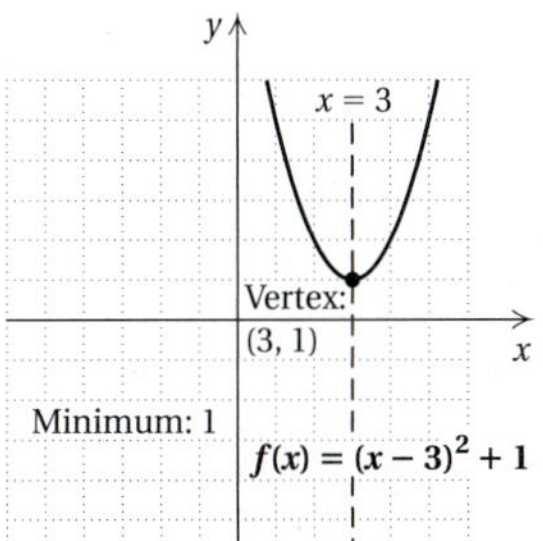

21.

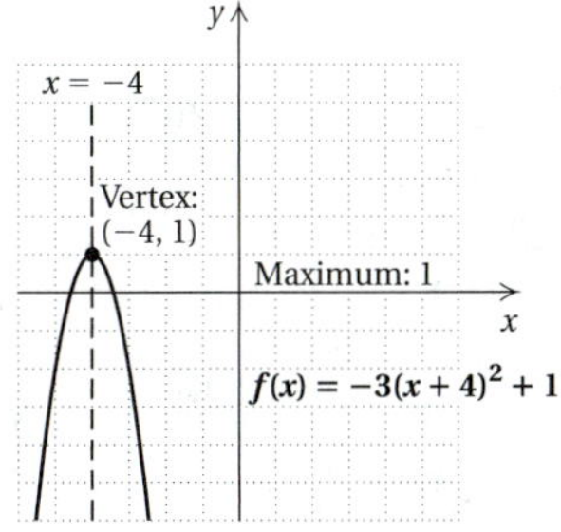

23.

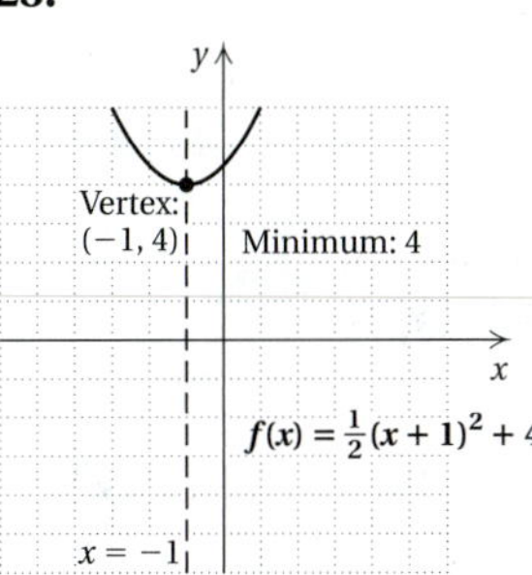

25.

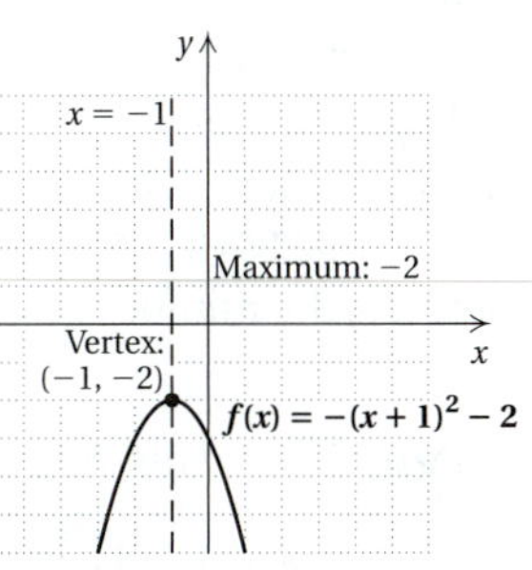

27. **DW** **29.** 9 **30.** $\frac{11}{3}$ **31.** No solution **32.** 15
33. Left to the student

Margin Exercises, Section 16.6, pp. 1249–1253

1.

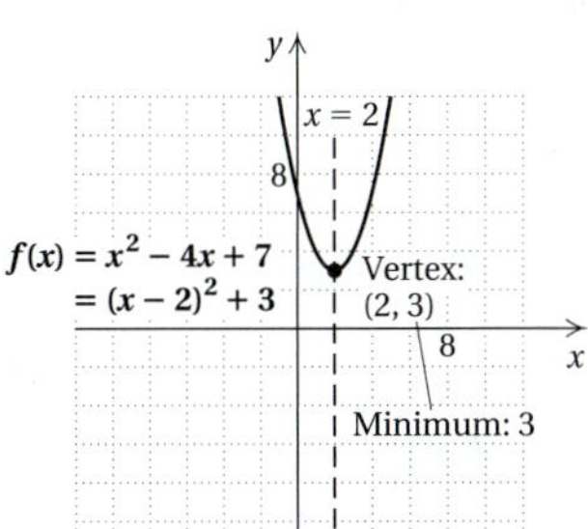

2.

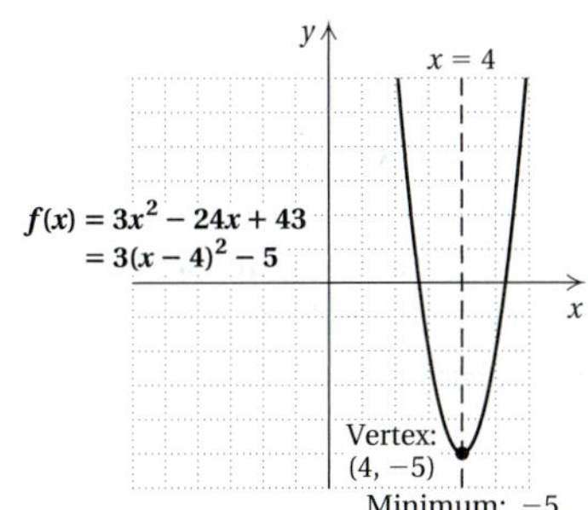

3.

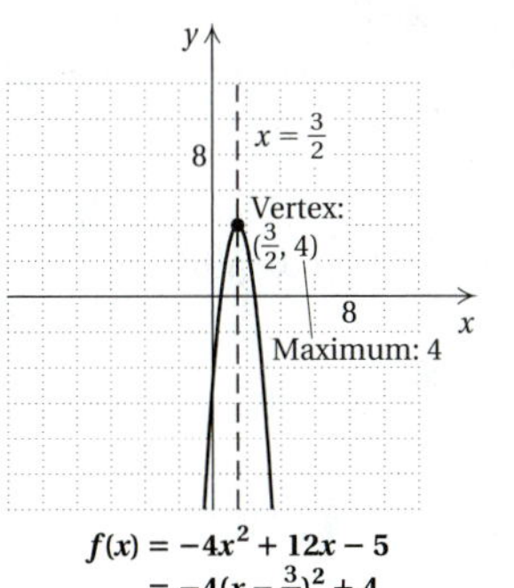

4. $(3, -5)$ **5.** $(4, -5)$

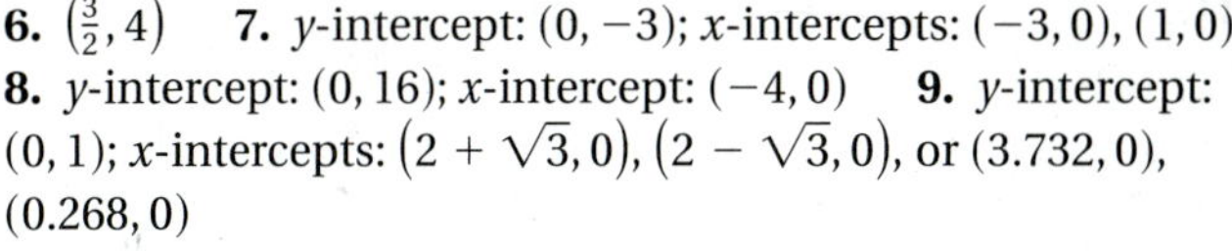

6. $\left(\frac{3}{2}, 4\right)$ **7.** y-intercept: $(0, -3)$; x-intercepts: $(-3, 0)$, $(1, 0)$
8. y-intercept: $(0, 16)$; x-intercept: $(-4, 0)$ **9.** y-intercept: $(0, 1)$; x-intercepts: $(2 + \sqrt{3}, 0)$, $(2 - \sqrt{3}, 0)$, or $(3.732, 0)$, $(0.268, 0)$

Exercise Set 16.6, p. 1254

1.

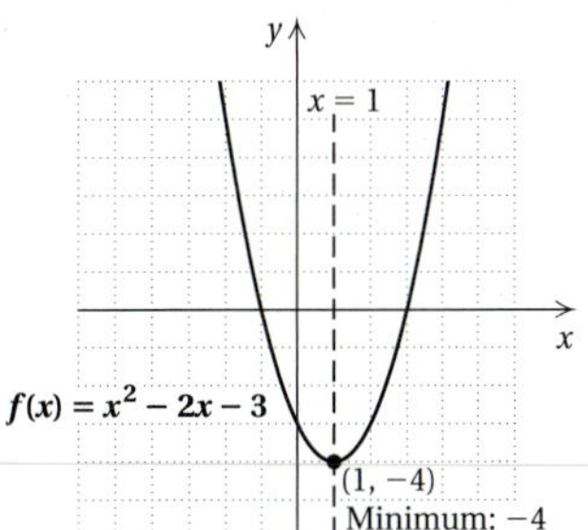

3.

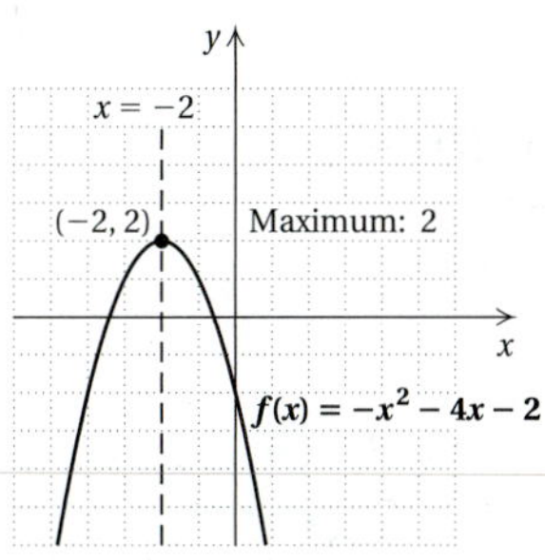

5.

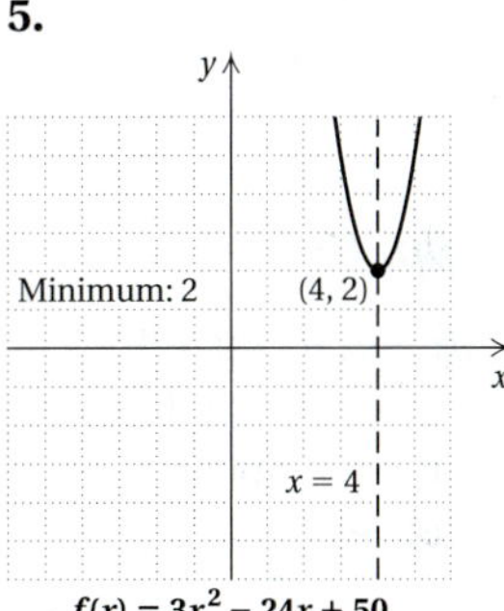

7.

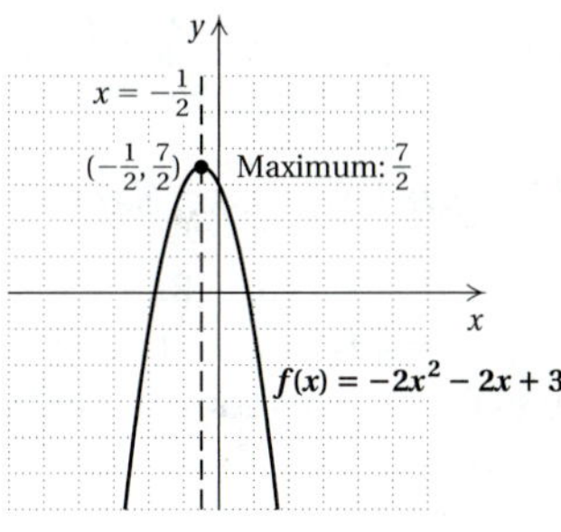

9.

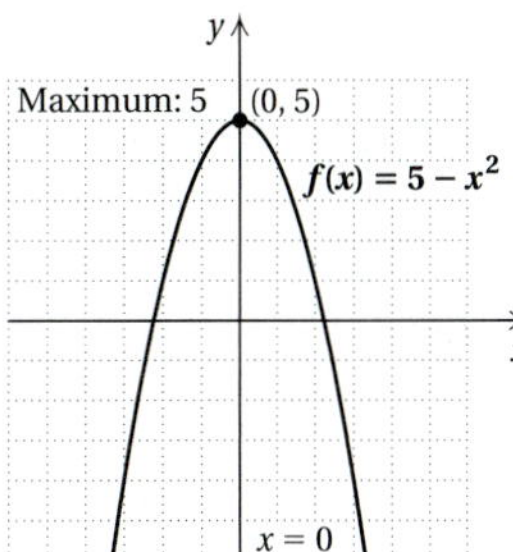

11.

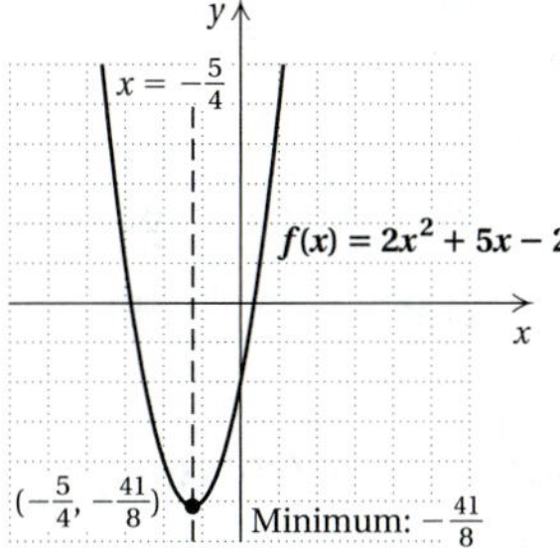

13. y-intercept: $(0, 1)$; x-intercepts: $(3 + 2\sqrt{2}, 0)$, $(3 - 2\sqrt{2}, 0)$ **15.** y-intercept: $(0, 20)$; x-intercepts: $(5, 0)$, $(-4, 0)$ **17.** y-intercept: $(0, 9)$; x-intercept: $\left(-\frac{3}{2}, 0\right)$
19. y-intercept: $(0, 8)$; x-intercepts: none **21.** **DW**
23. **(a)** $D = 15w$; **(b)** 630 mg **24.** **(a)** $C = \frac{89}{6}t$; **(b)** 712 calories **25.** 250; $y = \dfrac{250}{x}$ **26.** 250; $y = \dfrac{250}{x}$
27. $\frac{125}{2}$; $y = \frac{125}{2}x$ **28.** $\frac{2}{125}$; $y = \frac{2}{125}x$ **29.** **(a)** Minimum: -6.954; **(b)** maximum: 7.014

31.

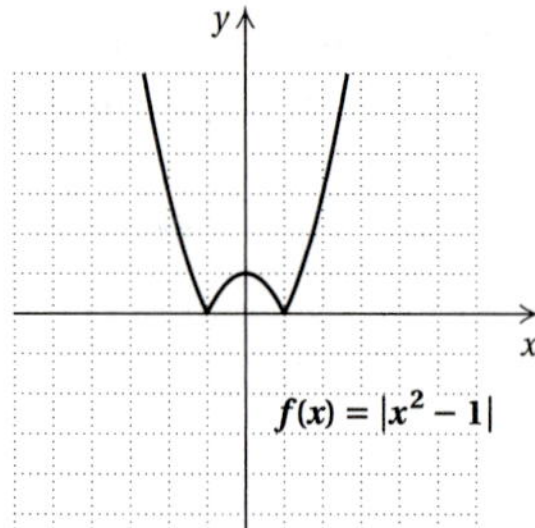

33.

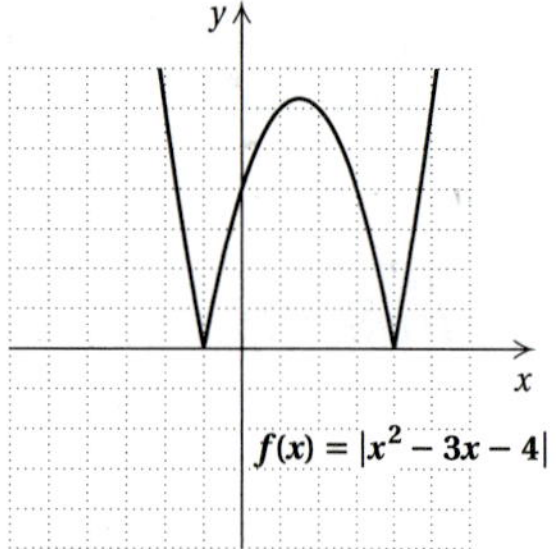

35. $f(x) = \frac{5}{16}x^2 - \frac{15}{8}x - \frac{35}{16}$

37.

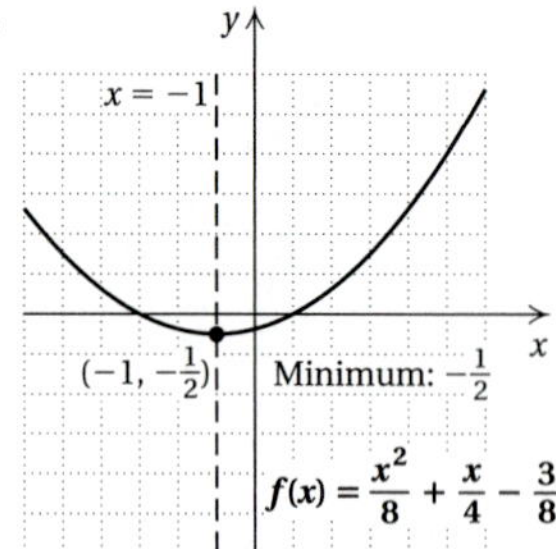

39. $(0, 2), (5, 7)$

Margin Exercises, Section 16.7, pp. 1259–1264

1. 25 yd by 25 yd **2.** $f(x) = ax^2 + bx + c, a > 0$ **3.** $f(x) = mx + b$ **4.** Polynomial, neither quadratic nor linear **5.** Polynomial, neither quadratic nor linear **6.** **(a)**

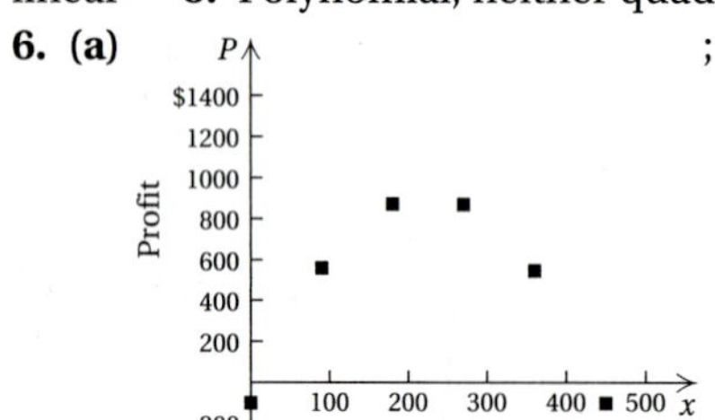

; **(b)** yes; **(c)** $f(x) = -0.02x^2 + 9x - 100$; **(d)** $912.50

Calculator Corner, p. 1260

1. Minimum: 1 **2.** Minimum: 4.875 **3.** Maximum: 6 **4.** Maximum: 0.5625

Calculator Corner, pp. 1265–1266

1. **(a)**

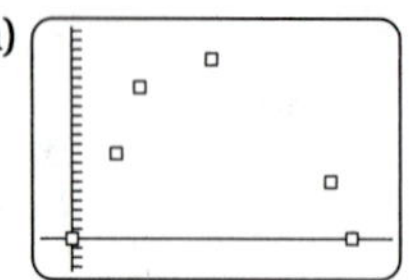

(b) $y = -0.0082833093x^2 + 0.8242996891x + 0.2121786608$;
(c)

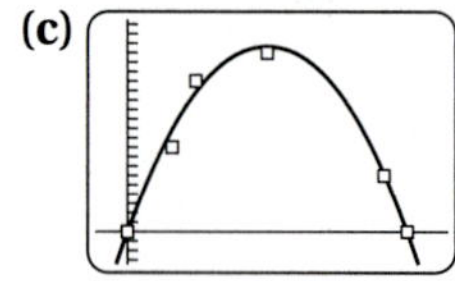

(d) 15.4 ft; the estimates of 15.4 ft and 15 ft are relatively close.

Exercise Set 16.7, p. 1267

1. 180 ft by 180 ft **3.** 3.5 in. **5.** 3.5 hundred, or 350 **7.** 200 ft^2; 10 ft by 20 ft **9.** 11 days after the concert was announced; about 62 **11.** $P(x) = -x^2 + 980x - 3000$; $237,100 at $x = 490$ **13.** 121; 11 and 11 **15.** −4; 2 and −2 **17.** 36; −6 and −6 **19.** $f(x) = mx + b$ **21.** $f(x) = ax^2 + bx + c, a > 0$ **23.** Polynomial, neither quadratic nor linear **25.** $f(x) = ax^2 + bx + c, a < 0$ **27.** $f(x) = 2x^2 + 3x - 1$ **29.** $f(x) = -\frac{1}{4}x^2 + 3x - 5$ **31.** **(a)** $A(s) = \frac{3}{16}s^2 - \frac{135}{4}s + 1750$; **(b)** about 531 per 200,000,000 kilometers driven **33.** $h(d) = -\frac{1}{147}d^2 + \frac{6}{7}d \approx -0.0068d^2 + 0.8571d$ **35.** D$_W$ **37.** $5xy^2\sqrt[4]{x}$ **38.** $12a^2b^2$ **39.** 5 **40.** 4 **41.** No solution **42.** 4 **43.** $y = -0.290x^4 + 2.699x^3 - 8.306x^2 + 9.190x + 12.235$, where x is the number of years after 1997

Margin Exercises, Section 16.8, pp. 1272–1277

1. $\{x \mid x < -3 \text{ or } x > 1\}$, or $(-\infty, -3) \cup (1, \infty)$
2. $\{x \mid -3 < x < 1\}$, or $(-3, 1)$ **3.** $\{x \mid -3 \le x \le 1\}$, or $[-3, 1]$ **4.** $\{x \mid x < -4 \text{ or } x > 1\}$, or $(-\infty, -4) \cup (1, \infty)$
5. $\{x \mid -4 \le x \le 1\}$, or $[-4, 1]$ **6.** $\{x \mid x < -1 \text{ or } 0 < x < 1\}$, or $(-\infty, -1) \cup (0, 1)$ **7.** $\{x \mid 2 < x \le \frac{7}{2}\}$, or $\left(2, \frac{7}{2}\right]$
8. $\{x \mid x < 5 \text{ or } x > 10\}$, or $(-\infty, 5) \cup (10, \infty)$

Calculator Corner, p. 1273

1. $\{x \mid x < -4 \text{ or } x > 1\}$, or $(-\infty, -4) \cup (1, \infty)$
2. $\{x \mid -2 < x < 3\}$, or $(-2, 3)$
3. $\{x \mid x \le -2 \text{ or } 0 \le x \le 0.5\}$, or $(-\infty, -2] \cup [0, 0.5]$
4. $\{x \mid -4 \le x \le 0 \text{ or } x \ge 4\}$, or $[-4, 0] \cup [4, \infty)$

Exercise Set 16.8, p. 1279

1. $\{x \mid x < -2 \text{ or } x > 6\}$, or $(-\infty, -2) \cup (6, \infty)$
3. $\{x \mid -2 \le x \le 2\}$, or $[-2, 2]$ **5.** $\{x \mid -1 \le x \le 4\}$, or $[-1, 4]$ **7.** $\{x \mid -1 < x < 2\}$, or $(-1, 2)$ **9.** All real numbers, or $(-\infty, \infty)$ **11.** $\{x \mid 2 < x < 4\}$, or $(2, 4)$
13. $\{x \mid x < -2 \text{ or } 0 < x < 2\}$, or $(-\infty, -2) \cup (0, 2)$
15. $\{x \mid -9 < x < -1 \text{ or } x > 4\}$, or $(-9, -1) \cup (4, \infty)$
17. $\{x \mid x < -3 \text{ or } -2 < x < 1\}$, or $(-\infty, -3) \cup (-2, 1)$
19. $\{x \mid x < 6\}$, or $(-\infty, 6)$ **21.** $\{x \mid x < -1 \text{ or } x > 3\}$, or $(-\infty, -1) \cup (3, \infty)$ **23.** $\{x \mid -\frac{2}{3} \le x < 3\}$, or $\left[-\frac{2}{3}, 3\right)$
25. $\{x \mid 2 < x < \frac{5}{2}\}$, or $\left(2, \frac{5}{2}\right)$ **27.** $\{x \mid x < -1 \text{ or } 2 < x < 5\}$, or $(-\infty, -1) \cup (2, 5)$ **29.** $\{x \mid -3 \le x < 0\}$, or $[-3, 0)$
31. $\{x \mid 1 < x < 2\}$, or $(1, 2)$ **33.** $\{x \mid x < -4 \text{ or } 1 < x < 3\}$, or $(-\infty, -4) \cup (1, 3)$ **35.** $\{x \mid 0 < x < \frac{1}{3}\}$, or $\left(0, \frac{1}{3}\right)$
37. $\{x \mid x < -3 \text{ or } -2 < x < 1 \text{ or } x > 4\}$, or $(-\infty, -3) \cup (-2, 1) \cup (4, \infty)$ **39.** D$_W$ **41.** $\frac{5}{3}$ **42.** $\frac{5}{2a}$
43. $\frac{4a}{b^2}\sqrt{a}$ **44.** $\frac{3c}{7d}\sqrt[3]{c^2}$ **45.** $\sqrt{2}$ **46.** $17\sqrt{5}$
47. $(10a + 7)\sqrt[3]{2a}$ **48.** $3\sqrt{10} - 4\sqrt{5}$ **49.** Left to the

student **51.** $\{x \mid 1 - \sqrt{3} \le x \le 1 + \sqrt{3}\}$, or $[1 - \sqrt{3}, 1 + \sqrt{3}]$ **53.** All real numbers except 0, or $(-\infty, 0) \cup (0, \infty)$ **55.** $\{x \mid x < \frac{1}{4} \text{ or } x > \frac{5}{2}\}$, or $(-\infty, \frac{1}{4}) \cup (\frac{5}{2}, \infty)$ **57.** **(a)** $\{t \mid 0 < t < 2\}$, or $(0, 2)$; **(b)** $\{t \mid t > 10\}$, or $(10, \infty)$

Summary and Review: Chapter 16, p. 1281

1. **(a)** $\pm\frac{\sqrt{14}}{2}$; **(b)** $\left(-\frac{\sqrt{14}}{2}, 0\right), \left(\frac{\sqrt{14}}{2}, 0\right)$ **2.** $0, -\frac{5}{14}$ **3.** 3, 9 **4.** $-\frac{3}{8} \pm i\frac{\sqrt{7}}{8}$ **5.** $\frac{7}{2} \pm i\frac{\sqrt{3}}{2}$ **6.** 3, 5 **7.** $-2 \pm \sqrt{3}$; $-0.268, -3.732$ **8.** 4, -2 **9.** $4 \pm 4\sqrt{2}$ **10.** $\frac{1 \pm \sqrt{481}}{15}$ **11.** $-2 \pm \sqrt{3}$ **12.** 0.90 sec **13.** Length: 14 cm; width: 9 cm **14.** 1 in. **15.** First part: 50 mph; second part: 40 mph **16.** Two real **17.** Two nonreal **18.** $25x^2 + 10x - 3 = 0$ **19.** $x^2 + 8x + 16 = 0$ **20.** $p = \frac{9\pi^2}{N^2}$ **21.** $T = \sqrt{\frac{3B}{2A}}$ **22.** 2, -2, 3, -3 **23.** 3, -5 **24.** $\pm\sqrt{7}, \pm\sqrt{2}$ **25.** 81, 16 **26.** **(a)** $(1, 3)$; **(b)** $x = 1$; **(c)** maximum: 3;

(d)

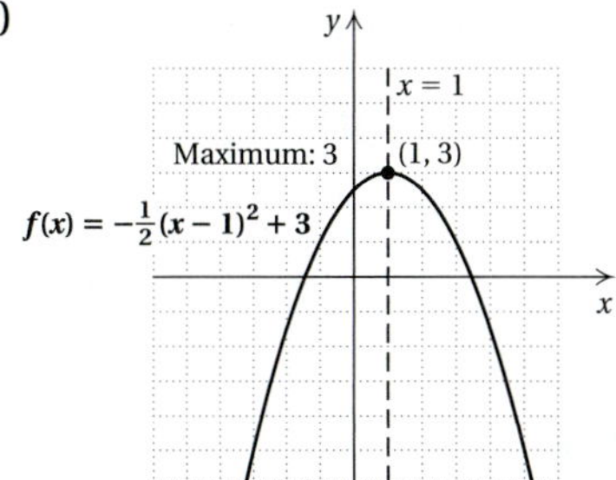

27. **(a)** $\left(\frac{1}{2}, \frac{23}{4}\right)$; **(b)** $x = \frac{1}{2}$; **(c)** minimum: $\frac{23}{4}$;

(d)

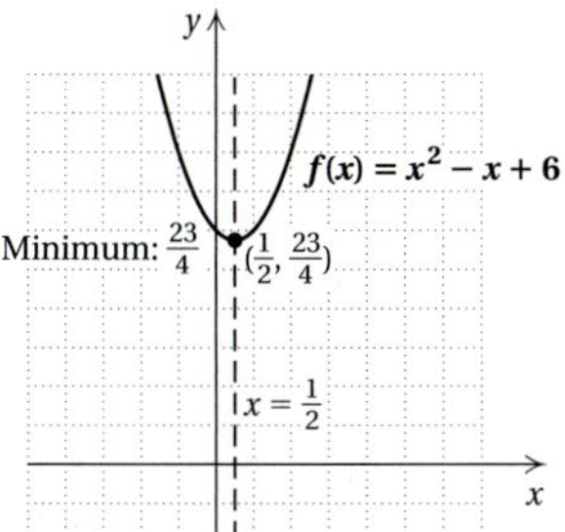

28. **(a)** $(-2, 4)$; **(b)** $x = -2$; **(c)** maximum: 4;

(d)

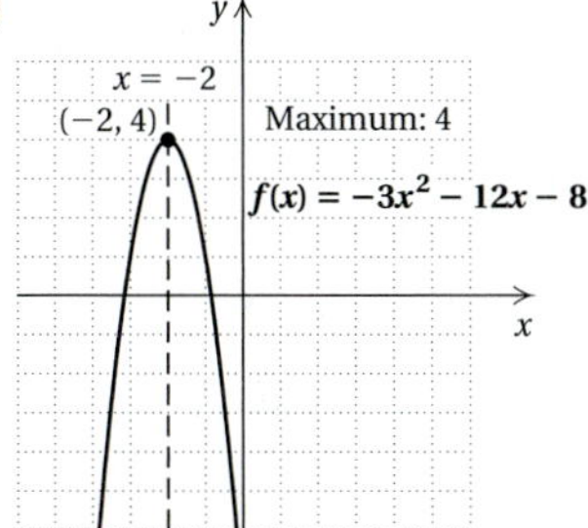

29. y-intercept: $(0, 14)$; x-intercepts: $(2, 0)$, $(7, 0)$ **30.** -121; 11 and -11 **31.** $f(x) = -x^2 + 6x - 2$ **32.** **(a)** $N(x) = -0.720x^2 + 38.211x - 393.127$; **(b)** about 105 **33.** $\{x \mid -2 < x < 1 \text{ or } x > 2\}$, or $(-2, 1) \cup (2, \infty)$ **34.** $\{x \mid x < -4 \text{ or } -2 < x < 1\}$, or $(-\infty, -4) \cup (-2, 1)$ **35.** $\mathbf{D_W}$ The graph of $f(x) = ax^2 + bx + c$ is a parabola with vertex $\left(-\frac{b}{2a}, \frac{4ac - b^2}{4a}\right)$. The line of symmetry is $x = -\frac{b}{2a}$. If $a < 0$, then the parabola opens down and $\frac{4ac - b^2}{4a}$ is the maximum function value. If $a > 0$, the parabola opens up and $\frac{4ac - b^2}{4a}$ is the minimum function value. The x-intercepts are $\left(\frac{-b - \sqrt{b^2 - 4ac}}{2a}, 0\right)$ and $\left(\frac{-b + \sqrt{b^2 - 4ac}}{2a}, 0\right)$, for $b^2 - 4ac > 0$. When $b^2 - 4ac = 0$, there is just one x-intercept, $\left(-\frac{b}{2a}, 0\right)$. When $b^2 - 4ac < 0$, there are no x-intercepts. **36.** $\mathbf{D_W}$ The x-coordinate of the maximum or minimum point lies halfway between the x-coordinates of the x-intercepts. The function must be evaluated for this value of x in order to determine the maximum or minimum value. **37.** **(a)** $f(x) = 1.8x + 503.2$; **(b)** $f(9) = 519.4$, $f(16) = 532$ **38.** $\frac{x + 1}{(x - 3)(x - 1)}$ **39.** $3t^5s\sqrt[3]{s}$ **40.** 2 **41.** $f(x) = \frac{7}{15}x^2 - \frac{14}{15}x - 7$; minimum: $-\frac{112}{15}$ **42.** $h = 60$, $k = 60$ **43.** 18 and 324

Test: Chapter 16, p. 1284

1. [16.1a] **(a)** $\pm\frac{2\sqrt{3}}{3}$; **(b)** $\left(\frac{2\sqrt{3}}{3}, 0\right), \left(-\frac{2\sqrt{3}}{3}, 0\right)$ **2.** [16.2a] $-\frac{1}{2} \pm i\frac{\sqrt{3}}{2}$ **3.** [16.4c] 49, 1 **4.** [16.2a] 9, 2 **5.** [16.4c] $\pm\sqrt{\frac{5 + \sqrt{5}}{2}}, \pm\sqrt{\frac{5 - \sqrt{5}}{2}}$ **6.** [16.2a] $-2 \pm \sqrt{6}$; 0.449, -4.449 **7.** [16.2a] 0, 2 **8.** [16.1b] $2 \pm \sqrt{3}$ **9.** [16.1c] About 6.7 sec **10.** [16.3a] About 2.89 mph **11.** [16.7a] 7 cm by 7 cm **12.** [16.1c] About 0.85 sec **13.** [16.4a] Two nonreal **14.** [16.4b] $x^2 - 4\sqrt{3}x + 9 = 0$ **15.** [16.3b] $T = \sqrt{\frac{V}{48}}$, or $\frac{\sqrt{3V}}{12}$ **16.** [16.6a] **(a)** $(-1, 1)$; **(b)** $x = -1$; **(c)** maximum: 1;

(d)

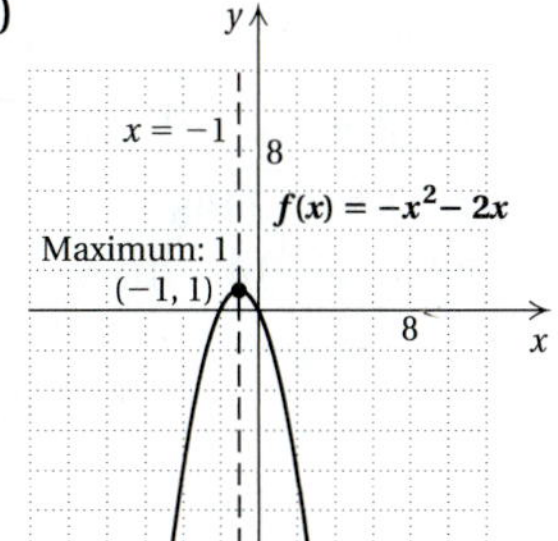

17. [16.6a] **(a)** $(3, 5)$; **(b)** $x = 3$; **(c)** minimum: 5;
(d)

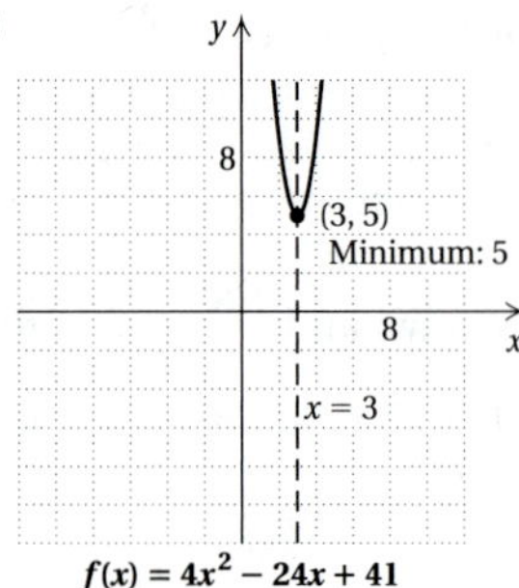

18. [16.6b] y-intercept: $(0, -1)$; x-intercepts: $(2 - \sqrt{3}, 0), (2 + \sqrt{3}, 0)$ **19.** [16.7a] -16; 4 and -4
20. [16.7b] $f(x) = \frac{1}{5}x^2 - \frac{3}{5}x$
21. [16.7b] **(a)** $N(x) = 132.7x^2 + 71.5x - 148.2$; **(b)** about 23,208; about 30,782
22. [16.8a] $\{x \mid -1 < x < 7\}$, or $(-1, 7)$
23. [16.8b] $\{x \mid -3 < x < 5\}$, or $(-3, 5)$
24. [16.8b] $\{x \mid -3 < x < 1 \text{ or } x \geq 2\}$, or $(-3, 1) \cup [2, \infty)$
25. [12.5e] **(a)** $f(x) = 0.2x + 503.8$; **(b)** about 506; about 507
26. [11.5a] $\dfrac{x - 8}{(x + 6)(x + 8)}$ **27.** [15.3a] $ab\sqrt[4]{2a^2}$
28. [15.6a] 6 **29.** [16.6a, b], [16.7b] $f(x) = -\frac{4}{7}x^2 + \frac{20}{7}x + 8$; maximum: $\frac{81}{7}$ **30.** [16.2a] $\frac{1}{2}$
31. [16.4c] $\pm 2, \pm\sqrt{2}, \pm 2i, \pm\sqrt{2}i$

Chapter 17

Pretest: Chapter 17, p. 1288

1.

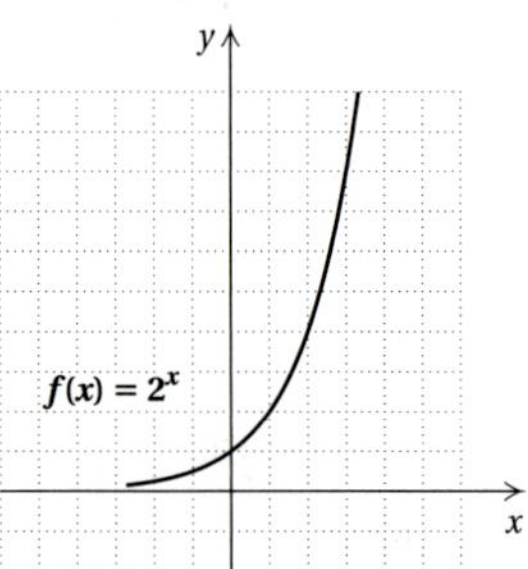

2.

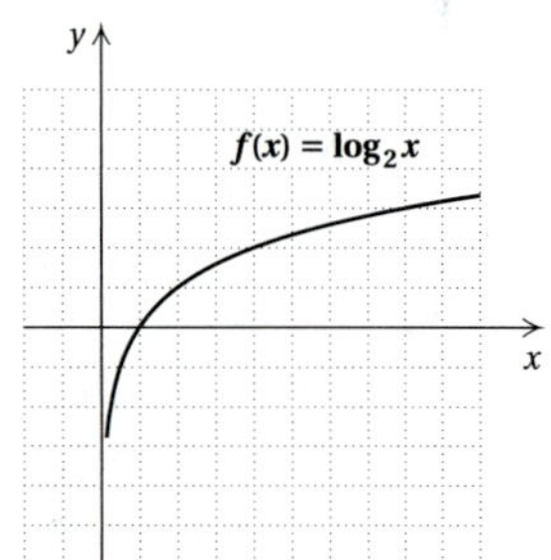

3. $\log_a 1000 = 3$ **4.** $2^t = 32$ **5.** $\log_a \dfrac{M^2N}{\sqrt[3]{Q}}$
6. $\frac{3}{4}\log_a x + \frac{1}{4}\log_a y - \frac{1}{2}\log_a z$ **7.** 2 **8.** 25
9. $\dfrac{\log 8.6}{\log 3} \approx 1.9586$ **10.** $\frac{7}{2}$ **11.** 8 **12.** 1 **13.** 2.8537
14. -7.0745 **15.** 0 **16.** 66.6863 **17.** 0.1188
18. -1.6308 **19.** **(a)** About \$757.8 billion; **(b)** 2010; **(c)** 4.7 yr **20.** About 11,893 yr

Margin Exercises, Section 17.1, pp. 1290–1297

1. (a)

x	$f(x)$
0	1
1	3
2	9
3	27
−1	$\frac{1}{3}$
−2	$\frac{1}{9}$
−3	$\frac{1}{27}$

(b)

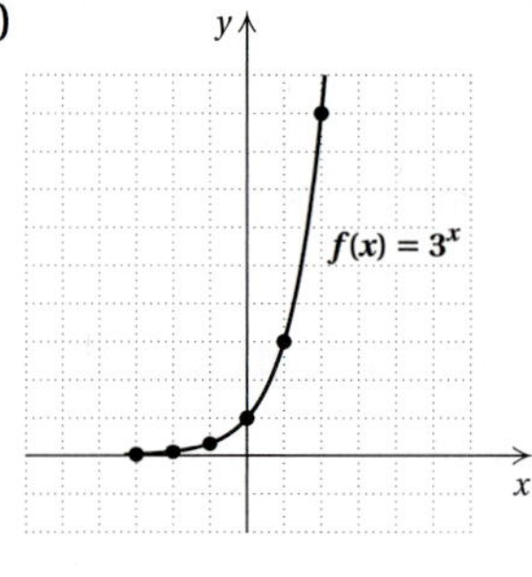

2. (a)

x	$f(x)$
0	1
1	$\frac{1}{3}$
2	$\frac{1}{9}$
3	$\frac{1}{27}$
−1	3
−2	9
−3	27

(b)

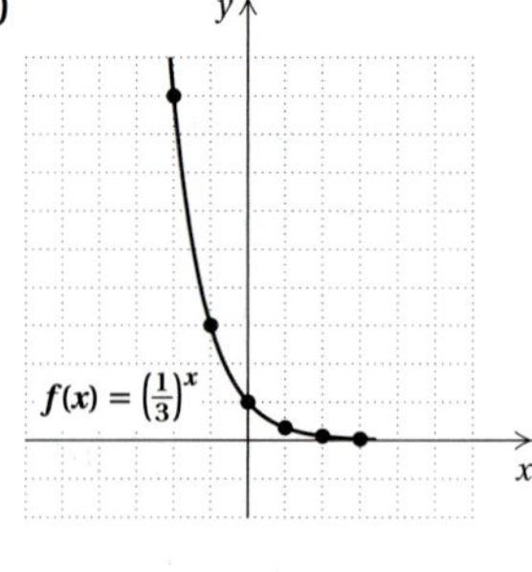

3.

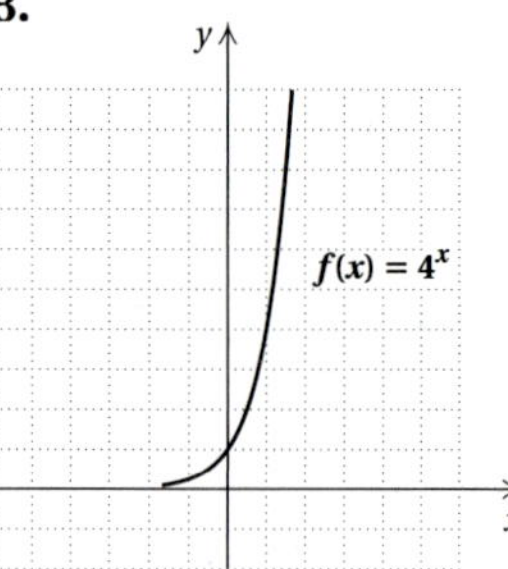

4.

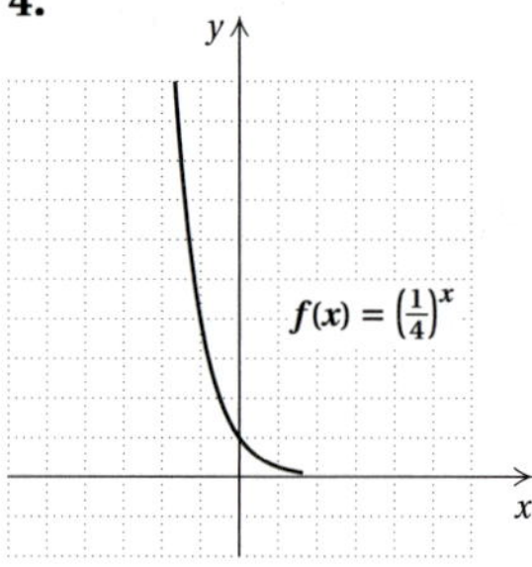

5.

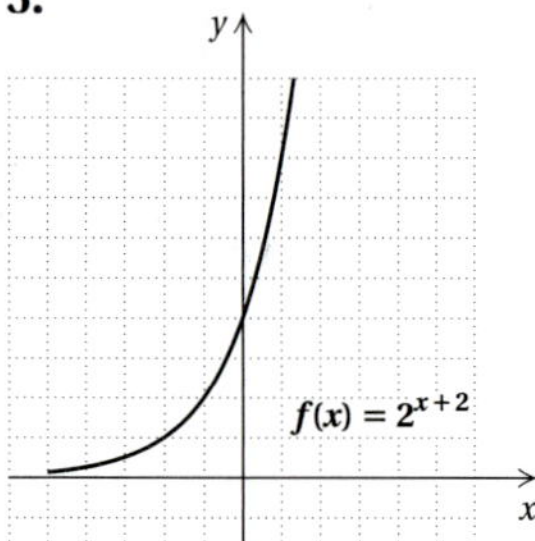

6.

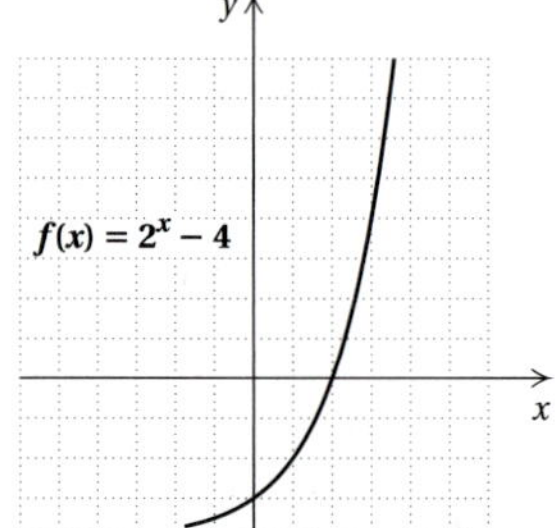

7.

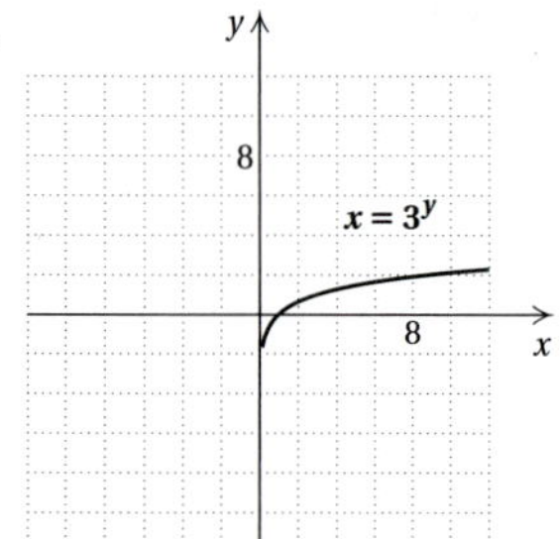

8. \$42,000; \$44,100 **9.** **(a)** $A(t) = \$40{,}000(1.05)^t$; **(b)** \$40,000, \$48,620.25, \$59,098.22, \$65,155.79;

(c)

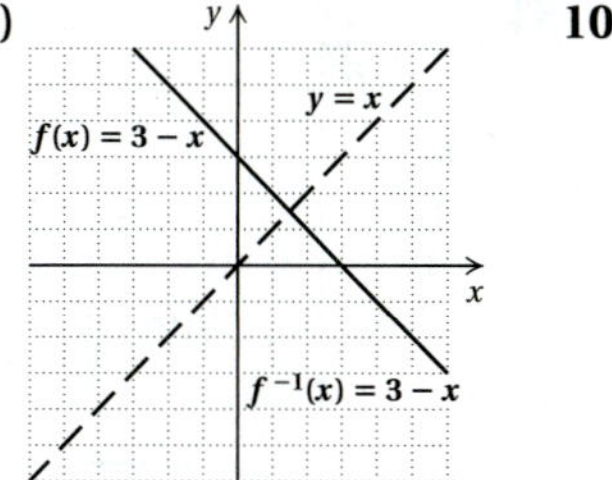

10. \$9925.67

Calculator Corner, p. 1293

1. Left to the student **2.** Left to the student

Calculator Corner, p. 1296

Left to the student

Calculator Corner, p. 1298

1. \$1125.51 **2.** \$1127.16 **3.** \$30,372.65
4. \$1,480,569.70 **5.** **(a)** \$10,540; **(b)** \$10,547.29; **(c)** \$10,551.03; **(d)** \$10,554.80; **(e)** \$10,554.84
6. **(a)** \$10,300; **(b)** \$10,302.25; **(c)** \$10,303.39; **(d)** \$10,304.53; **(e)** \$10,304.54

Exercise Set 17.1, p. 1299

1.

x	$f(x)$
0	1
1	2
2	4
3	8
−1	$\frac{1}{2}$
−2	$\frac{1}{4}$
−3	$\frac{1}{8}$

$f(x) = 2^x$

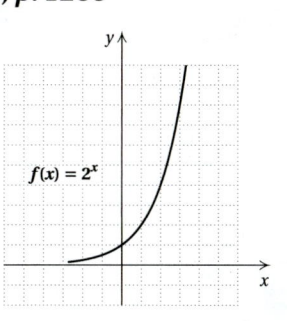

3.

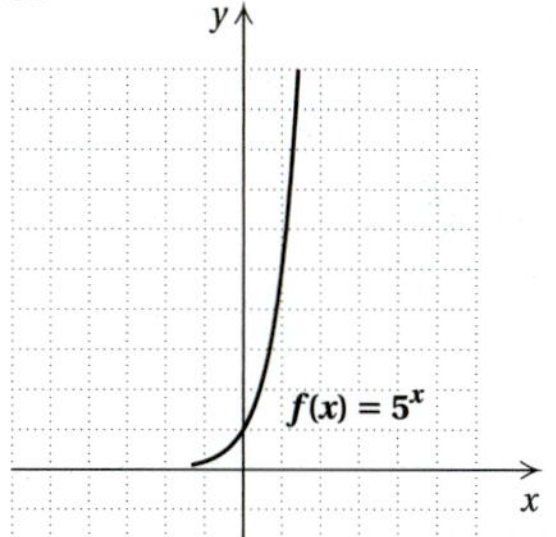

5.

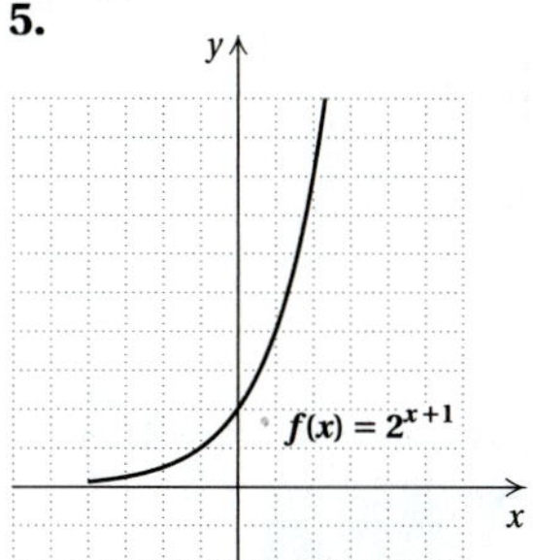

7.

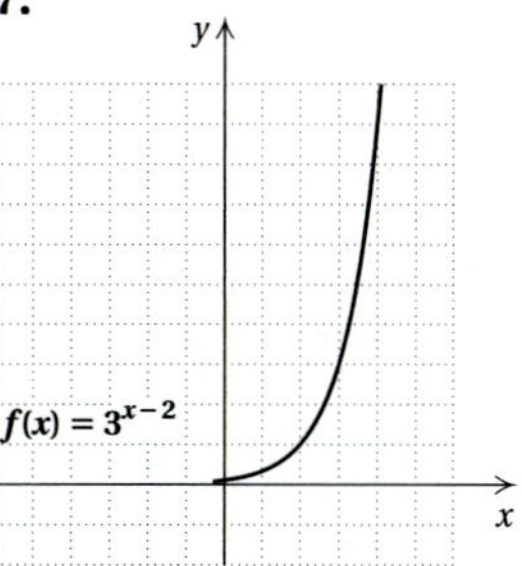

9.

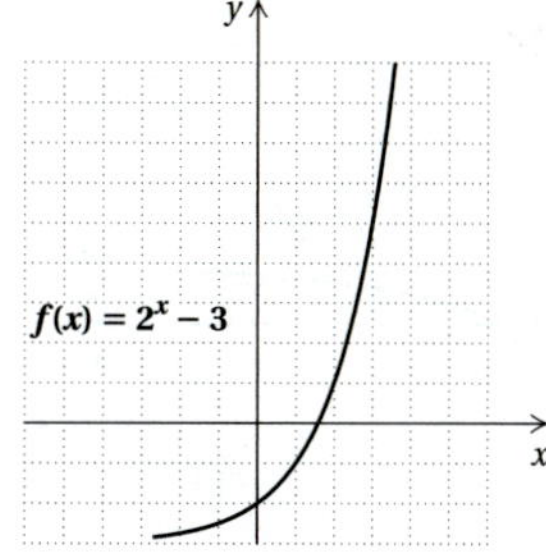

11.

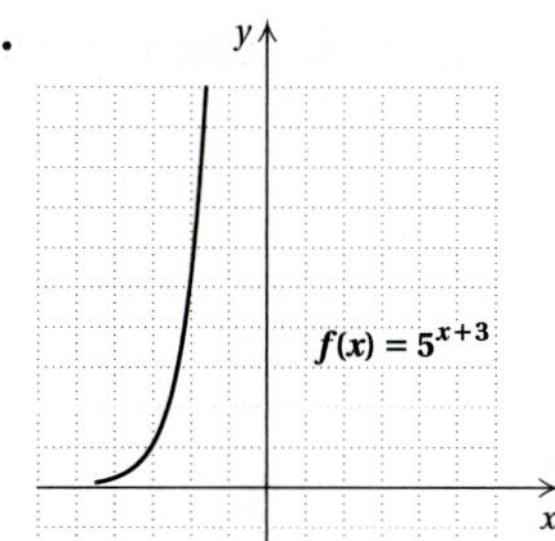

13.

x	$f(x)$
0	1
1	$\frac{1}{2}$
2	$\frac{1}{4}$
3	$\frac{1}{8}$
−1	2
−2	4
−3	8

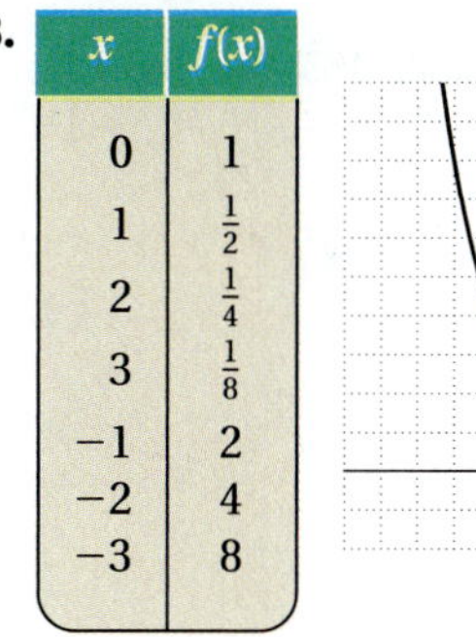

15.

$f(x) = \left(\frac{1}{5}\right)^x$

17.

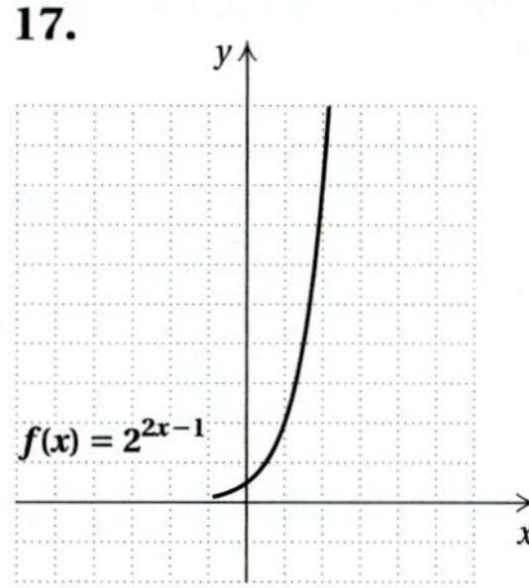

19.

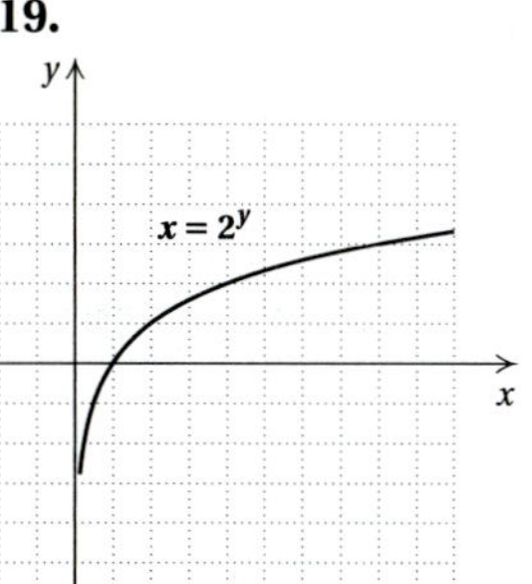

21.

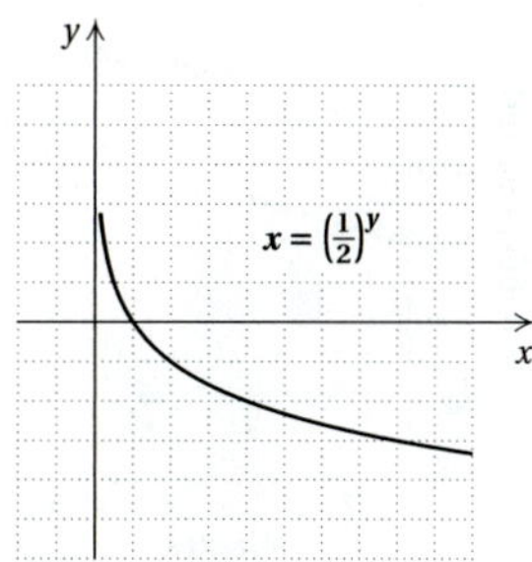

23.

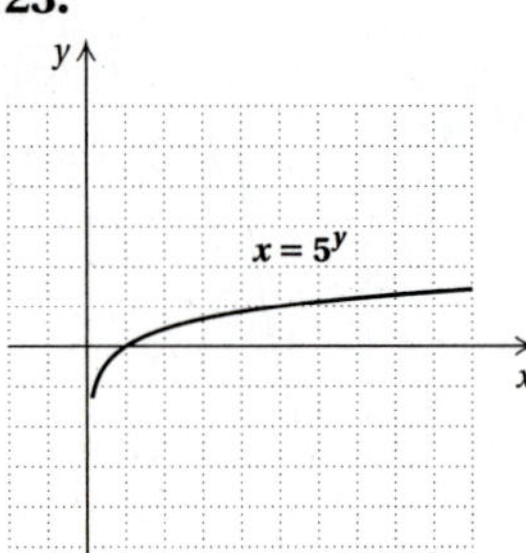

25.

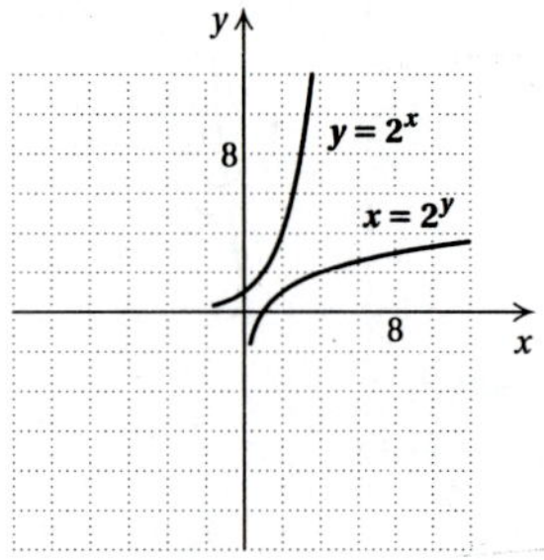

27. **(a)** $A(t) = \$50{,}000(1.06)^t$; **(b)** \$50,000; \$53,000; \$56,180; \$63,123.85; \$79,692.40; \$89,542.38; \$160,356.77

(c)

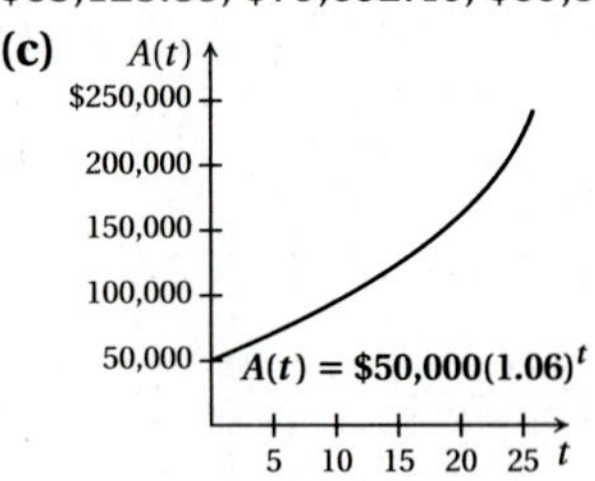

29. **(a)** \$4000; \$43,489; \$317,670; \$2,320,463; \$11,388,151; **(b)** \$13,893,544; **(c)** \$20,679,151;

(d)

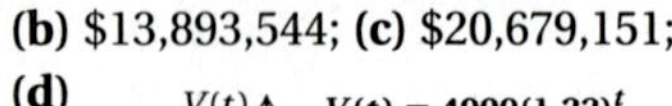

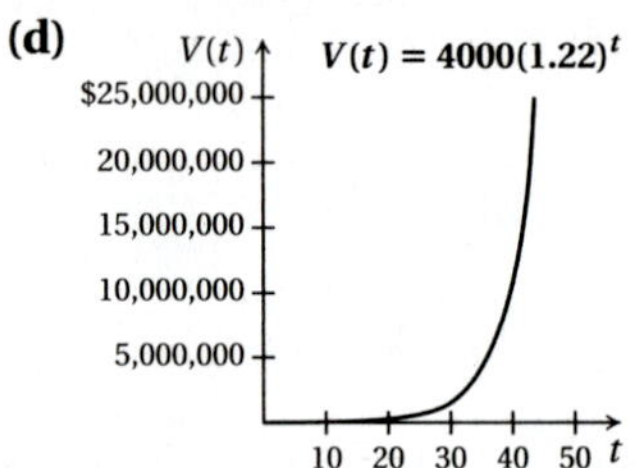

31. **(a)** About 44,079 whales; about 12,953 whales;

(b)

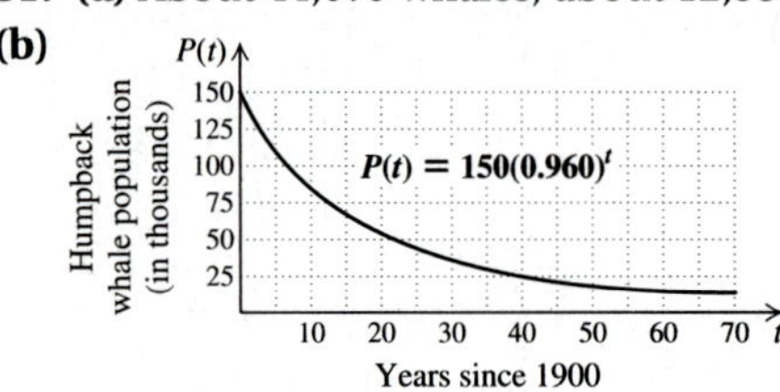

33. **(a)** 4243; 6000; 8485; 12,000; 24,000;

(b)

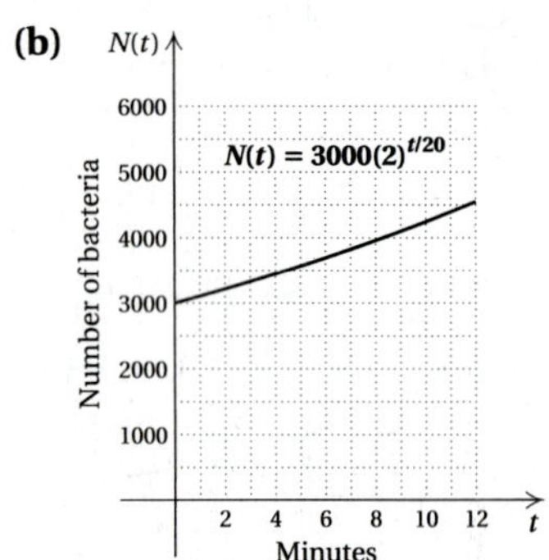

35. DW **37.** $\frac{1}{x^2}$

38. $\frac{1}{x^{12}}$ **39.** 1 **40.** 1 **41.** $\frac{2}{3}$ **42.** 2.7 **43.** $\frac{1}{x^7}$

44. $\frac{1}{x^{10}}$ **45.** x **46.** x **47.** 5^4, or 625

49.

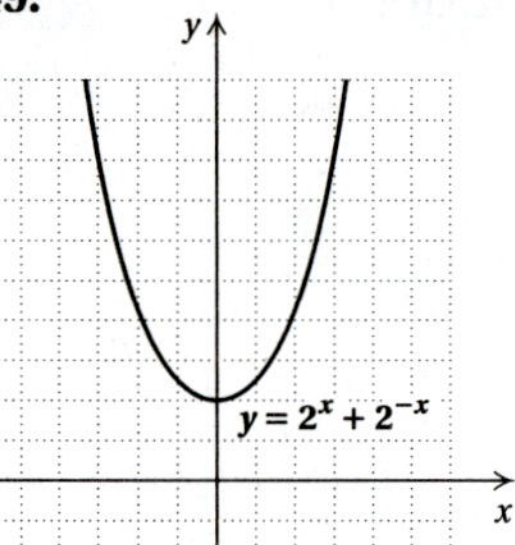

51.

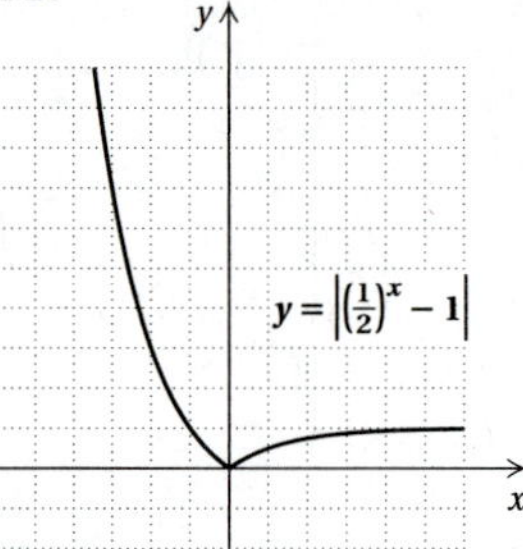

53.

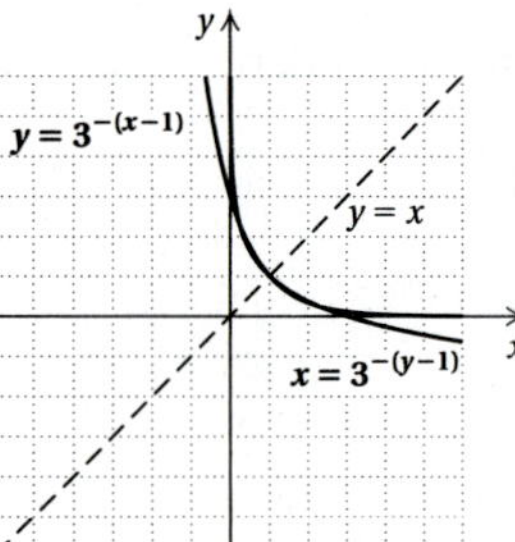

55. Left to the student

Margin Exercises, Section 17.2, pp. 1304–1315

1. Inverse of $g = \{(5, 2), (4, -1), (1, -2)\}$

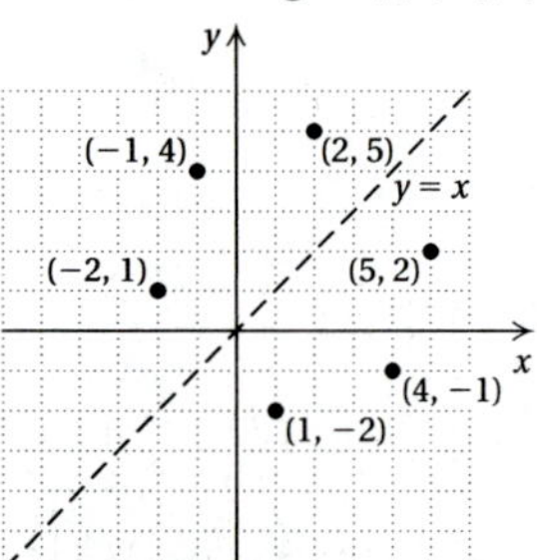

2. Inverse: $x = 6 - 2y$;

x	y
6	0
2	2
0	3
−4	5

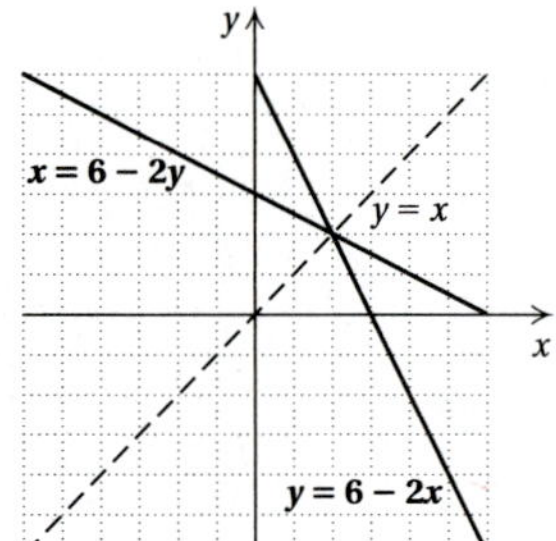

3. Inverse: $x = y^2 - 4y + 7$;

x	y
7	0
4	1
3	2
4	3
7	4

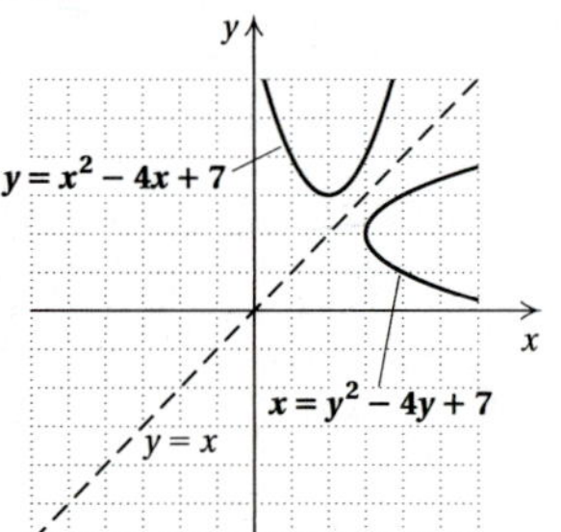

4. Yes **5.** No **6.** Yes **7.** No **8.** **(a)** Yes;

(b) $f^{-1}(x) = 3 - x$; **(c)**

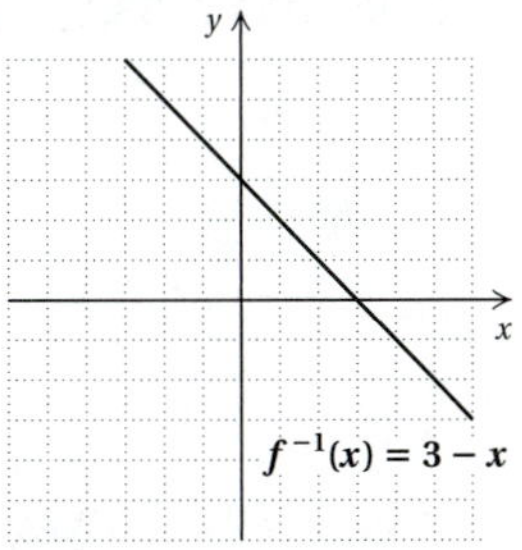

9. **(a)** Yes; **(b)** $g^{-1}(x) = \dfrac{x + 2}{3}$;

(c)

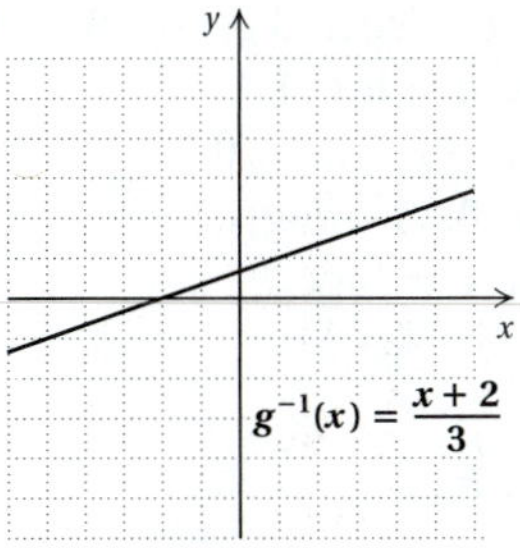

10.

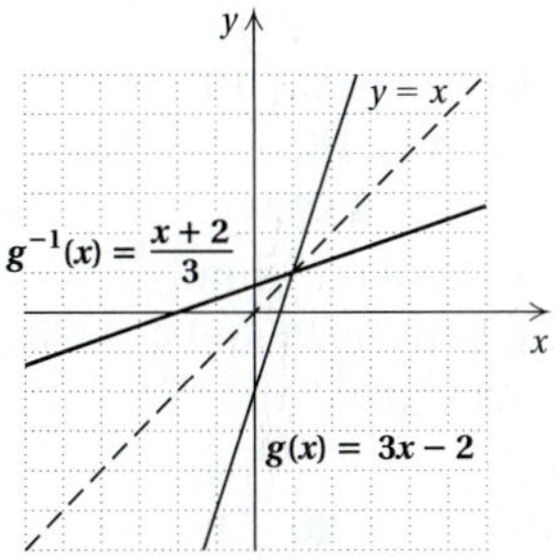

11. **(a)** Yes; **(b)** $f^{-1}(x) = \sqrt[3]{x - 1}$;
(c)

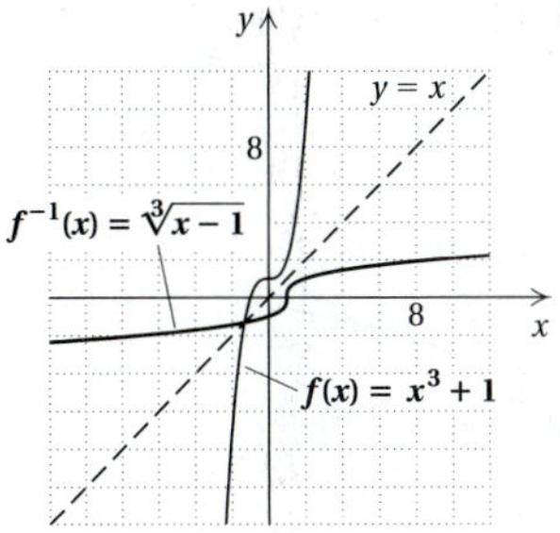

12. $x^2 + 4$; $x^2 + 10x + 24$

13. $4\sqrt[3]{x} + 5$; $\sqrt[3]{4x + 5}$ **14.** **(a)** $f(x) = \sqrt[3]{x}$, $g(x) = x^2 + 1$;
(b) $f(x) = \dfrac{1}{x^4}$, $g(x) = x + 5$

15. $f^{-1} \circ f(x) = f^{-1}(f(x)) = f^{-1}\left(\frac{2}{3}x - 4\right)$

$$= \frac{3\left(\frac{2}{3}x - 4\right) + 12}{2} = \frac{2x - 12 + 12}{2}$$

$$= \frac{2x}{2} = x;$$

$$f \circ f^{-1}(x) = f(f^{-1}(x)) = f\left(\frac{3x + 12}{2}\right)$$

$$= \frac{2}{3}\left(\frac{3x + 12}{2}\right) - 4 = \frac{6x + 24}{6} - 4$$

$$= x + 4 - 4 = x$$

Calculator Corner, p. 1312

1.

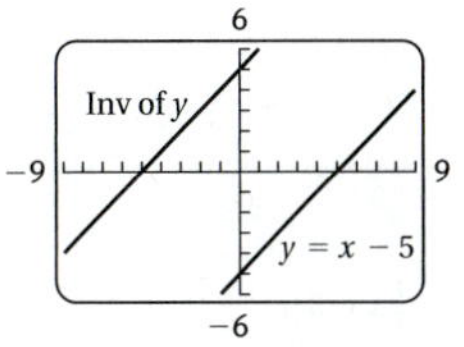

2.

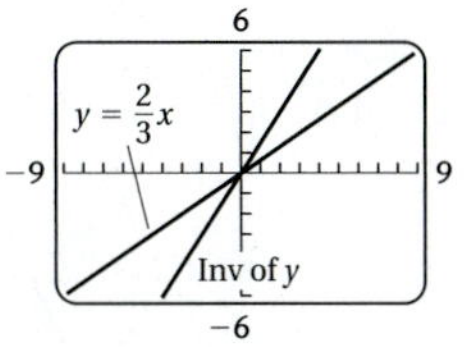

3.

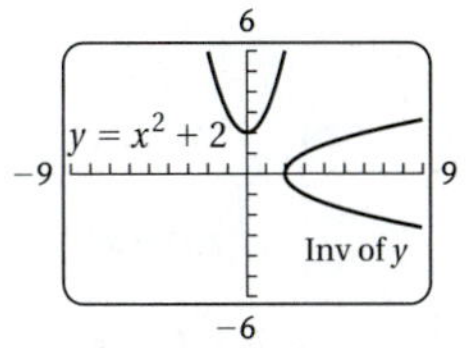

4.

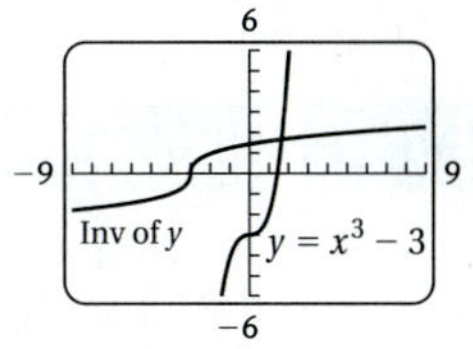

Calculator Corner, p. 1316

1. Left to the student **2.** 4; 7 **3.** 9; 9 **4.** 1; 25

Exercise Set 17.2, p. 1317

1. Inverse: $\{(2, 1), (-3, 6), (-5, -3)\}$

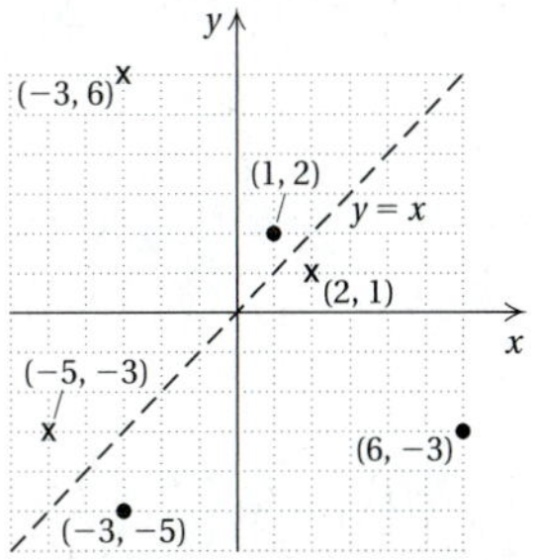

3. Inverse: $x = 2y + 6$

x	y
4	−1
6	0
8	1
10	2
12	3

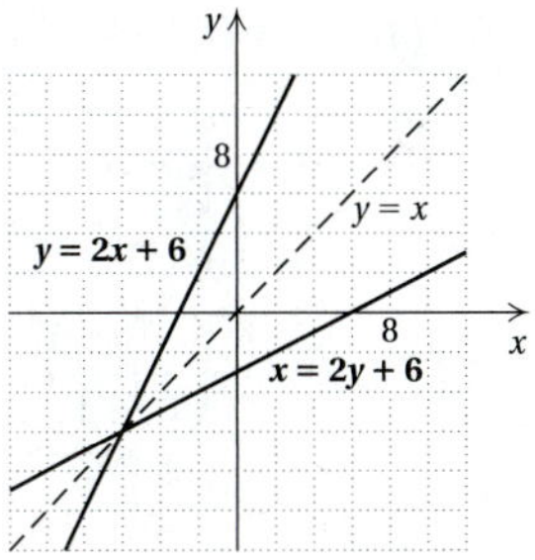

5. Yes **7.** No **9.** No **11.** Yes **13.** $f^{-1}(x) = \dfrac{x + 2}{5}$

15. $f^{-1}(x) = \dfrac{-2}{x}$ **17.** $f^{-1}(x) = \frac{3}{4}(x - 7)$

19. $f^{-1}(x) = \dfrac{2}{x} - 5$ **21.** Not one-to-one

23. $f^{-1}(x) = \dfrac{1 - 3x}{5x - 2}$ **25.** $f^{-1}(x) = \sqrt[3]{x + 1}$

27. $f^{-1}(x) = x^3$

29. $f^{-1}(x) = 2x + 6$

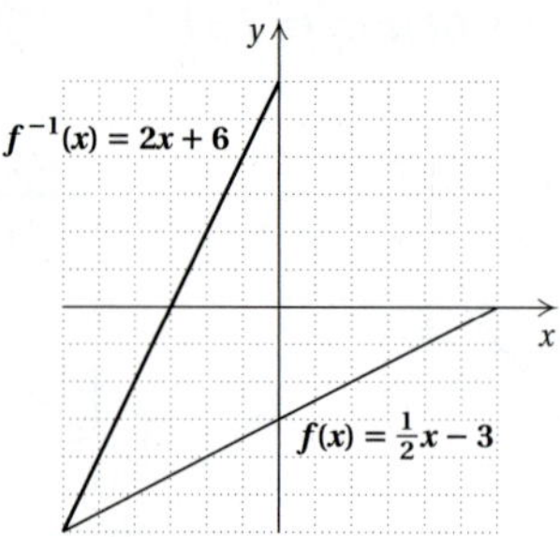

31. $f^{-1}(x) = \sqrt[3]{x}$

x	$f(x)$
0	0
1	1
2	8
3	27
−1	−1
−2	−8
−3	−27

x	$f^{-1}(x)$
0	0
1	1
8	2
27	3
−1	−1
−8	−2
−27	−3

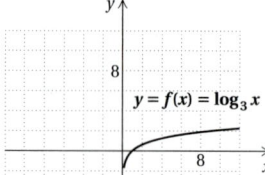

33. $-8x + 9; -8x + 18$ **35.** $12x^2 - 12x + 5; 6x^2 + 3$

37. $\frac{16}{x^2} - 1; \frac{2}{4x^2 - 1}$ **39.** $x^4 - 10x^2 + 30; x^4 + 10x^2 + 20$

41. $f(x) = x^2, g(x) = 5 - 3x$

43. $f(x) = \sqrt{x}, g(x) = 5x + 2$ **45.** $f(x) = \frac{1}{x}, g(x) = x - 1$

47. $f(x) = \frac{1}{\sqrt{x}}, g(x) = 7x + 2$

49. $f(x) = x^4, g(x) = \sqrt{x} + 5$

51. $f^{-1} \circ f(x) = f^{-1}(f(x)) = f^{-1}\left(\frac{4}{5}x\right) = \frac{5}{4}\left(\frac{4}{5}x\right) = x;$
$f \circ f^{-1}(x) = f(f^{-1}(x)) = f\left(\frac{5}{4}x\right) = \frac{4}{5}\left(\frac{5}{4}x\right) = x$

53. $f^{-1} \circ f(x) = f^{-1}(f(x)) = f^{-1}\left(\frac{1 - x}{x}\right)$

$$= \frac{1}{\frac{1 - x}{x} + 1} = \frac{1}{\frac{1}{x}} = x;$$

$$f \circ f^{-1}(x) = f(f^{-1}(x)) = f\left(\frac{1}{x + 1}\right)$$

$$= \frac{1 - \frac{1}{x + 1}}{\frac{1}{x + 1}} = \frac{\frac{x}{x + 1}}{\frac{1}{x + 1}} = x$$

55. $f^{-1}(x) = \frac{1}{3}x$ **57.** $f^{-1}(x) = -x$ **59.** $f^{-1}(x) = x^3 + 5$

61. **(a)** 40, 42, 46, 50; **(b)** $f^{-1}(x) = x - 32$; **(c)** 8, 10, 14, 18

63. D$_W$ **65.** $\sqrt[3]{a}$ **66.** $\sqrt[3]{x^2}$ **67.** a^2b^3 **68.** $2t^2$

69. $\sqrt{3}$ **70.** $2\sqrt[4]{2}$ **71.** $\sqrt{2xy}$ **72.** $\sqrt[4]{p^2t}$ **73.** $2a^3b^8$

74. $10x^3y^6$ **75.** $3a^2b^2$ **76.** $3pq^3$ **77.** No **79.** Yes

81. (1) C; (2) A; (3) B; (4) D **83.**

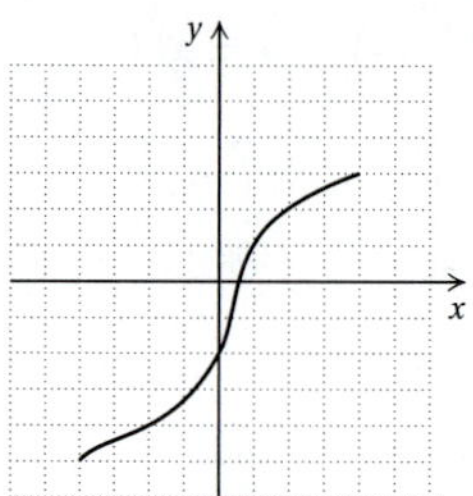

85. $f(x) = \frac{1}{2}x + 3$; $g(x) = 2x - 6$; yes

Margin Exercises, Section 17.3, pp. 1322–1328

1. $\log_2 64$ is the power to which we raise 2 to get 64; 6

2.

3. $0 = \log_6 1$

4. $-3 = \log_{10} 0.001$

5. $0.25 = \log_{16} 2$

6. $T = \log_m P$

7. $2^5 = 32$ **8.** $10^3 = 1000$ **9.** $a^7 = Q$ **10.** $t^x = M$

11. 10,000 **12.** 3 **13.** $\frac{1}{4}$ **14.** 4 **15.** −4 **16.** 0

17. 0 **18.** 1 **19.** 1 **20.** 0 **21.** 4.8934

22. −4.5100 **23.** Does not exist as a real number

24. log 1000 = 3, log 10,000 = 4; between 3 and 4; 3.9945

25. $8 = 10^{0.9031}$; $947 = 10^{2.9763}$; $634{,}567 = 10^{5.8025}$; $0.00708 = 10^{-2.1500}$; $0.000778899 = 10^{-3.1085}$; $18{,}600{,}000 = 10^{7.2695}$; $1860 = 10^{3.2695}$ **26.** 78,234.8042

Exercise Set 17.3, p. 1329

1. $x = 2^y$

x, or 2^y	y
1	0
2	1
4	2
8	3
$\frac{1}{2}$	−1
$\frac{1}{4}$	−2
$\frac{1}{8}$	−3

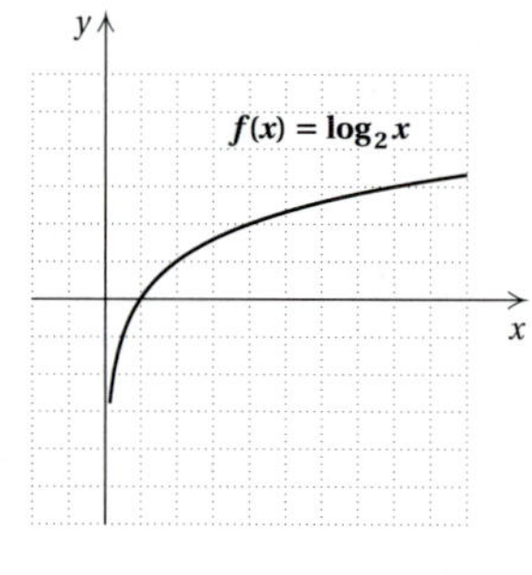

3.

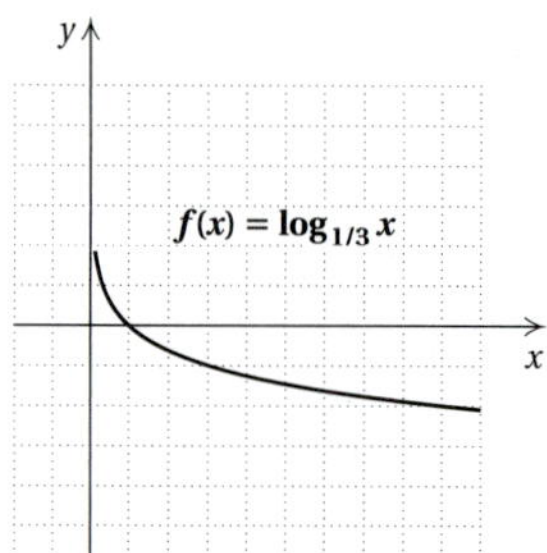

5.

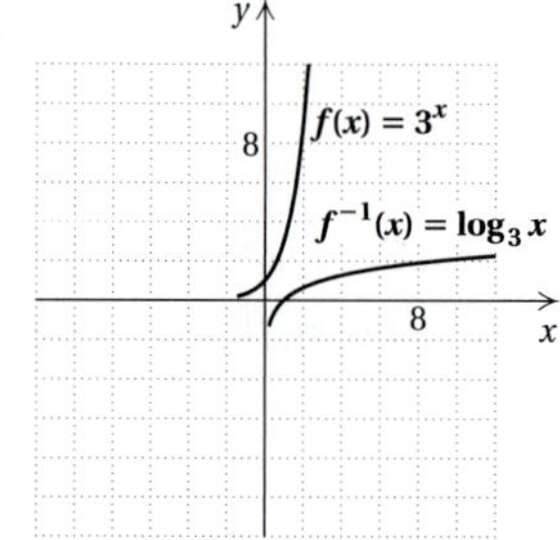

7. $3 = \log_{10} 1000$ **9.** $-3 = \log_5 \frac{1}{125}$ **11.** $\frac{1}{3} = \log_8 2$

13. $0.3010 = \log_{10} 2$ **15.** $2 = \log_e t$ **17.** $t = \log_Q x$ **19.** $2 = \log_e 7.3891$ **21.** $-2 = \log_e 0.1353$ **23.** $4^w = 10$ **25.** $6^2 = 36$ **27.** $10^{-2} = 0.01$ **29.** $10^{0.9031} = 8$ **31.** $e^{4.6052} = 100$ **33.** $t^k = Q$ **35.** 9 **37.** 4 **39.** 4 **41.** 3 **43.** 25 **45.** 1 **47.** $\frac{1}{2}$ **49.** 2 **51.** 2 **53.** -1 **55.** 0 **57.** 4 **59.** 2 **61.** 3 **63.** -2 **65.** 0 **67.** 1 **69.** $\frac{2}{3}$ **71.** 4.8970 **73.** -0.1739 **75.** Does not exist as a real number **77.** 0.9464 **79.** $6 = 10^{0.7782}$; $84 = 10^{1.9243}$; $987{,}606 = 10^{5.9946}$; $0.00987606 = 10^{-2.0054}$; $98{,}760.6 = 10^{4.9946}$; $70{,}000{,}000 = 10^{7.8451}$; $7000 = 10^{3.8451}$ **81.** DW

83.

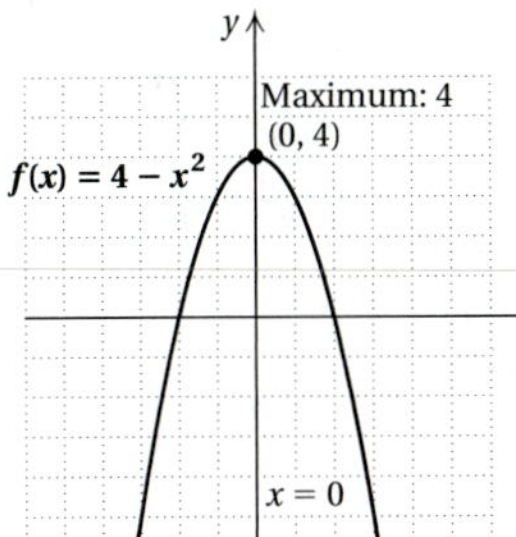

84.

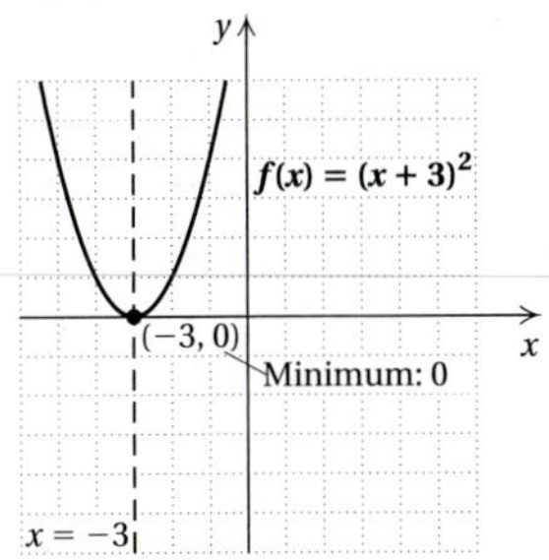

85.

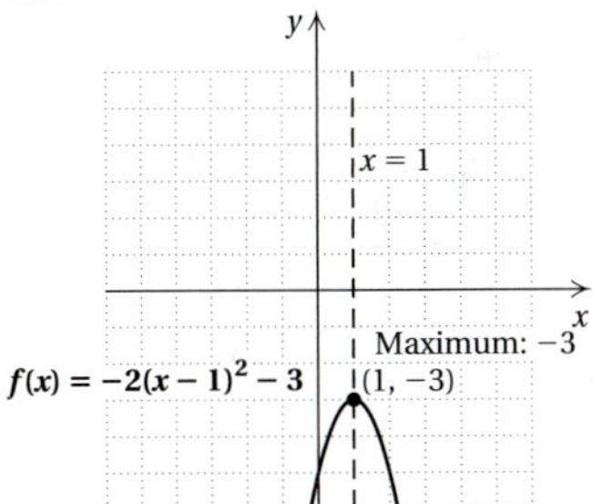

86.

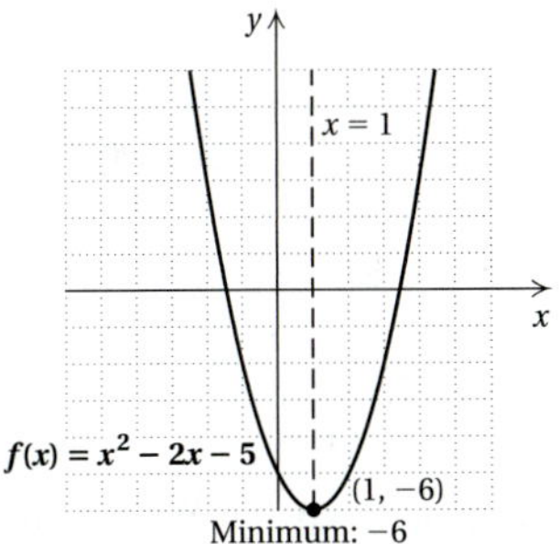

87. $c = \sqrt{\dfrac{E}{m}}$ **88.** $a = \dfrac{2 + \sqrt{4 + 3B}}{3}$ **89.** $b = \dfrac{A^2}{3a}$ **90.** $L = \dfrac{T^2 g}{4\pi^2}$ **91.** (a) $f(x) = -\frac{39}{280}x^2 + \frac{963}{140}x - \frac{356}{7}$; (b) 30%, 13% **93.**

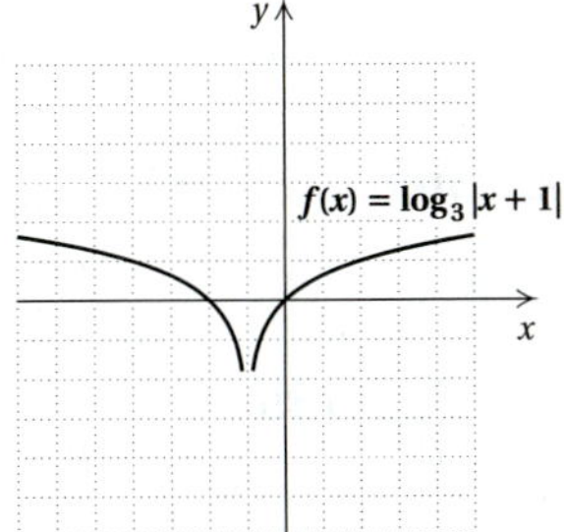

95. 25 **97.** $-\frac{7}{16}$ **99.** 3 **101.** 0 **103.** -2

Margin Exercises, Section 17.4, pp. 1333–1336

1. $\log_5 25 + \log_5 5$ **2.** $\log_b P + \log_b Q$ **3.** $\log_3 35$ **4.** $\log_a (JAM)$ **5.** $5 \log_7 4$ **6.** $\frac{1}{2} \log_a 5$ **7.** $\log_b P - \log_b Q$ **8.** $\log_2 5$ **9.** $\frac{3}{2} \log_a z - \frac{1}{2} \log_a x - \frac{1}{2} \log_a y$ **10.** $2 \log_a x - 3 \log_a y - \log_a z$ **11.** $3 \log_a x + 4 \log_a y - 5 \log_a z - 9 \log_a w$ **12.** $\log_a \dfrac{x^5 z^{1/4}}{y}$ **13.** $\log_a \dfrac{1}{b\sqrt{b}}$, or $\log_a b^{-3/2}$ **14.** 0.602 **15.** 1 **16.** -0.398 **17.** 0.398 **18.** -0.699 **19.** $\frac{3}{2}$ **20.** 1.699 **21.** 1.204 **22.** 6 **23.** 3.2 **24.** 12

Calculator Corner, p. 1334

1. Not correct **2.** Correct **3.** Not correct **4.** Correct **5.** Not correct **6.** Correct **7.** Not correct **8.** Not correct

Exercise Set 17.4, p. 1337

1. $\log_2 32 + \log_2 8$ **3.** $\log_4 64 + \log_4 16$ **5.** $\log_a Q + \log_a x$ **7.** $\log_b 252$ **9.** $\log_c Ky$ **11.** $4 \log_c y$ **13.** $6 \log_b t$ **15.** $-3 \log_b C$ **17.** $\log_a 67 - \log_a 5$ **19.** $\log_b 2 - \log_b 5$ **21.** $\log_c \frac{22}{3}$ **23.** $2 \log_a x + 3 \log_a y + \log_a z$ **25.** $\log_b x + 2 \log_b y - 3 \log_b z$ **27.** $\frac{4}{3} \log_c x - \log_c y - \frac{2}{3} \log_c z$ **29.** $2 \log_a m + 3 \log_a n - \frac{3}{4} - \frac{5}{4} \log_a b$ **31.** $\log_a \dfrac{x^{2/3}}{y^{1/2}}$, or $\log_a \dfrac{\sqrt[3]{x^2}}{\sqrt{y}}$ **33.** $\log_a \dfrac{2x^4}{y^3}$ **35.** $\log_a \dfrac{\sqrt{a}}{x}$ **37.** 2.708 **39.** 0.51 **41.** -1.609 **43.** $\frac{3}{2}$ **45.** 2.609 **47.** t **49.** 5 **51.** 7 **53.** -7 **55.** DW **57.** i **58.** -1 **59.** 5 **60.** $\frac{3}{5} + \frac{4}{5}i$ **61.** $23 - 18i$ **62.** $10i$ **63.** $-34 - 31i$ **64.** $3 - 4i$ **65.** Left to the student **67.** $\log_a (x^6 - x^4y^2 + x^2y^4 - y^6)$ **69.** $\frac{1}{2} \log_a (1 - s) + \frac{1}{2} \log_a (1 + s)$ **71.** False **73.** True **75.** False

Margin Exercises, Section 17.5, pp. 1340–1343

1. 11.2675 **2.** -10.3848 **3.** Does not exist as a real number **4.** Does not exist **5.** 2.3026 **6.** 78,237.1596 **7.** 0.1353 **8.** (a) 1.086; (b) 1.086 **9.** 5.5236

10.

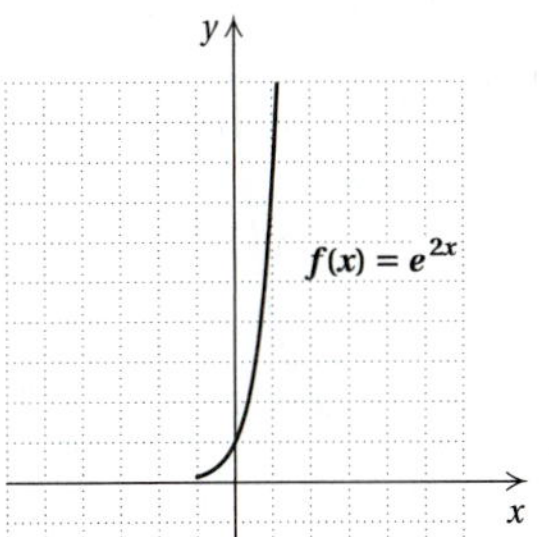

11.

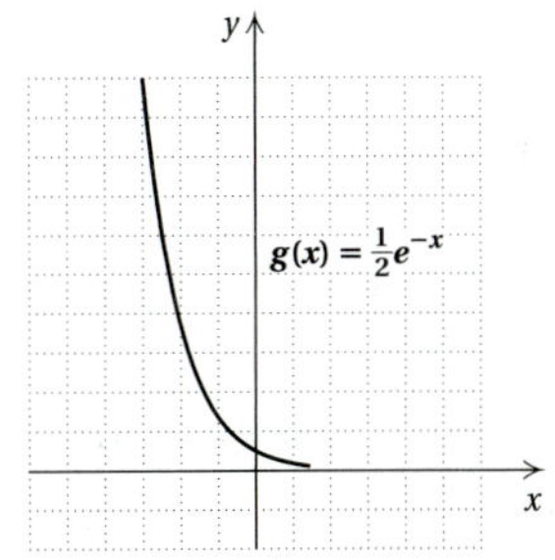

12.

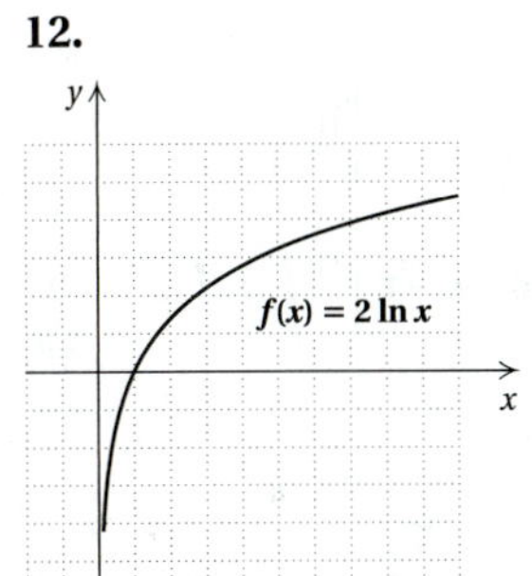

13.

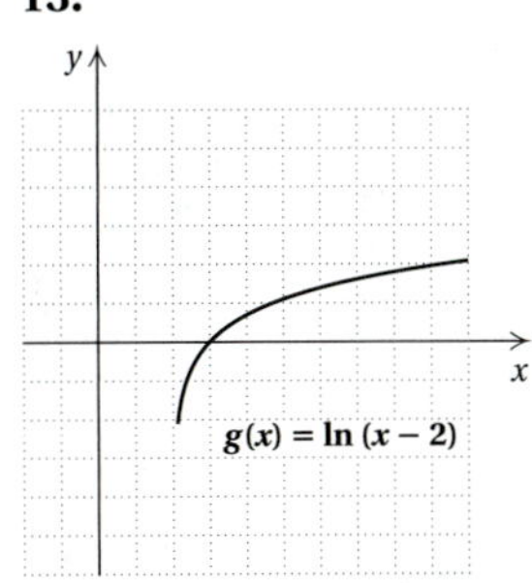

Calculator Corner, p. 1344

1.

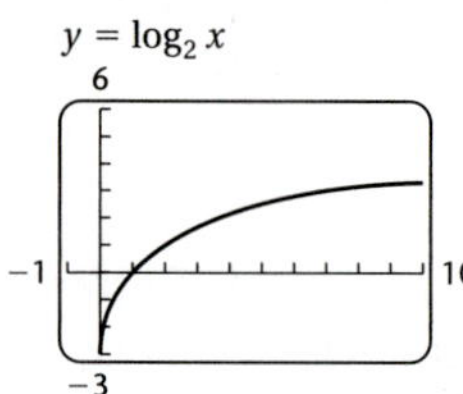

2.

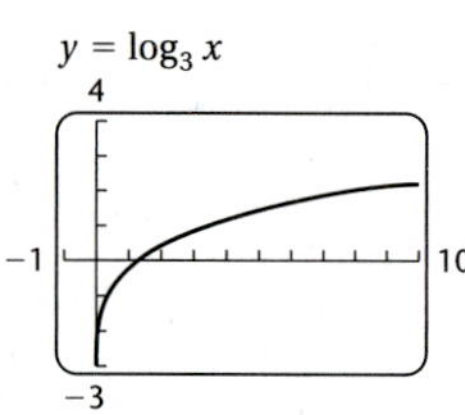

3.

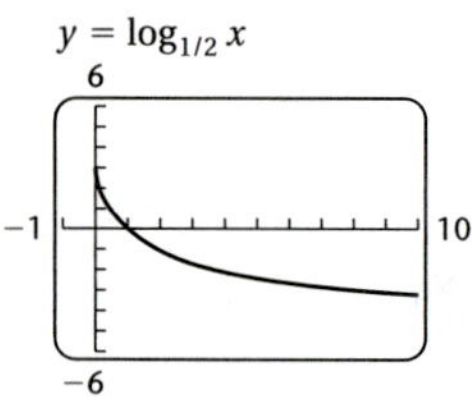

4.

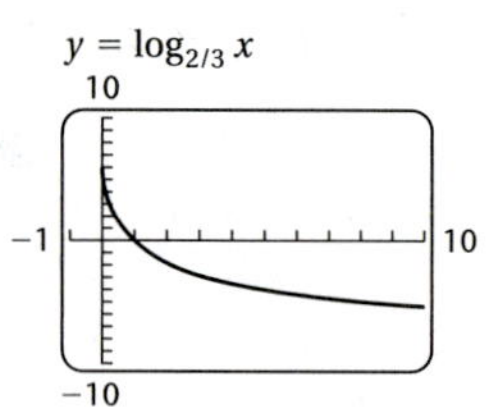

Exercise Set 17.5, p. 1345

1. 0.6931 **3.** 4.1271 **5.** 8.3814 **7.** −5.0832 **9.** −1.6094 **11.** Does not exist **13.** −1.7455 **15.** 1 **17.** 15.0293 **19.** 0.0305 **21.** 109.9472 **23.** 5 **25.** 2.5702 **27.** 6.6439 **29.** 2.1452 **31.** −2.3219 **33.** −2.3219 **35.** 4.6284

37.

x	$f(x)$
0	1
1	2.7
2	7.4
3	20.1
−1	0.4
−2	0.1
−3	0.05

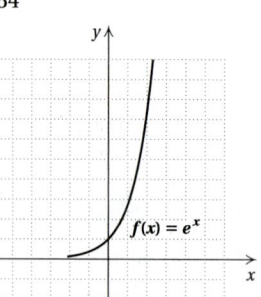

$f(x) = e^x$

39.

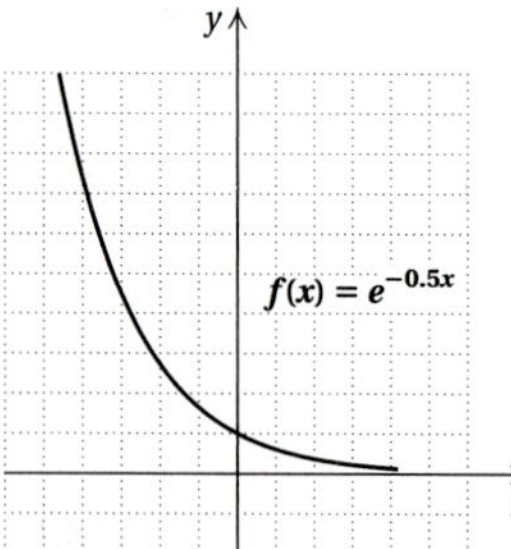

41.

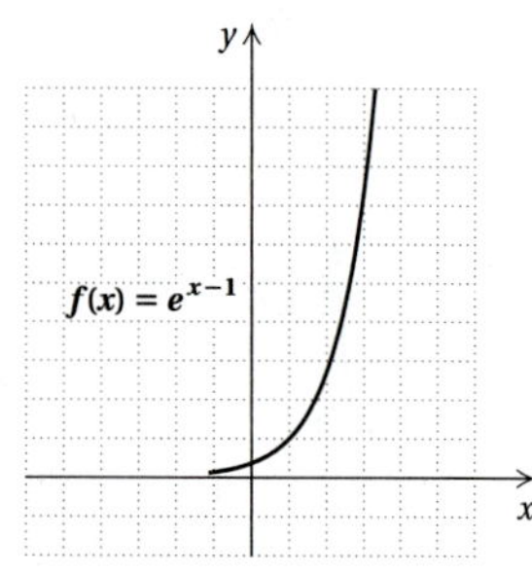

43.

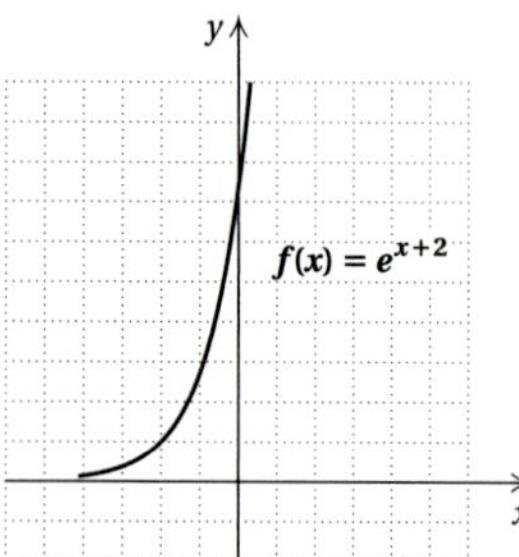

45.

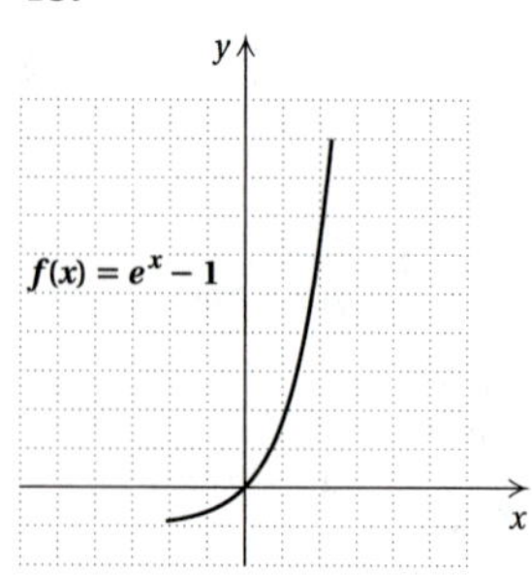

47.

x	$f(x)$
0	0.7
1	1.1
2	1.4
3	1.6
−0.5	0.4
−1	0
−1.5	−0.7

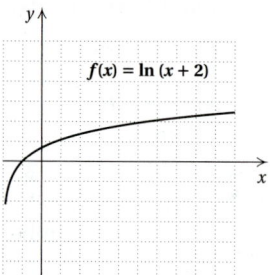

$f(x) = \ln(x + 2)$

49.

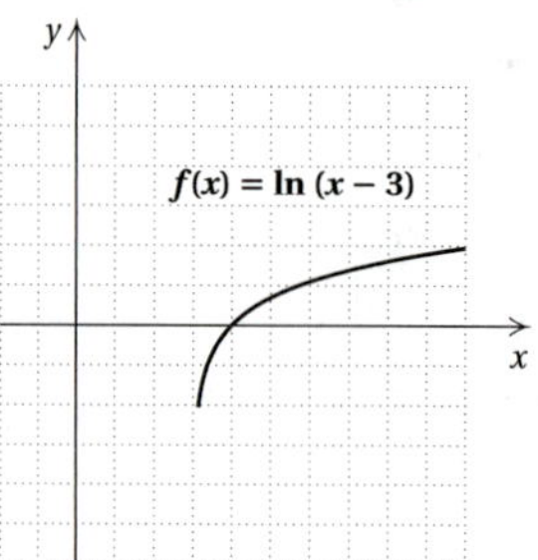

51.

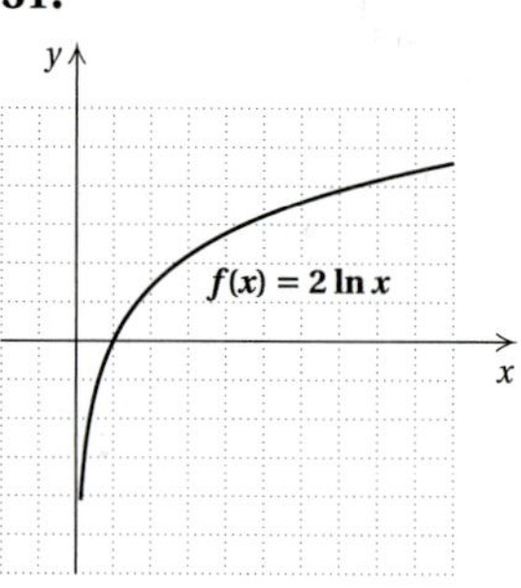

53.

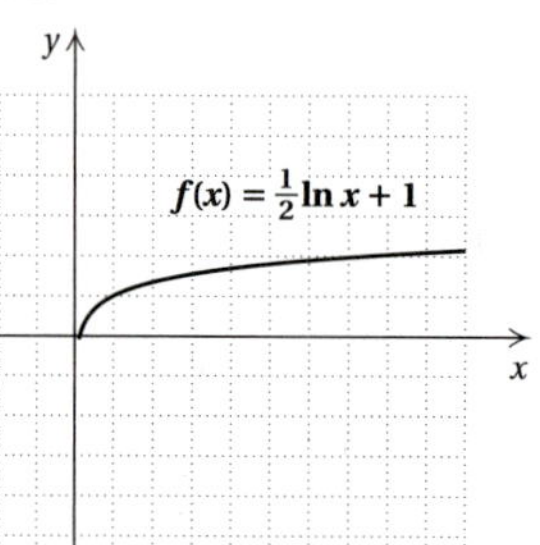

55.

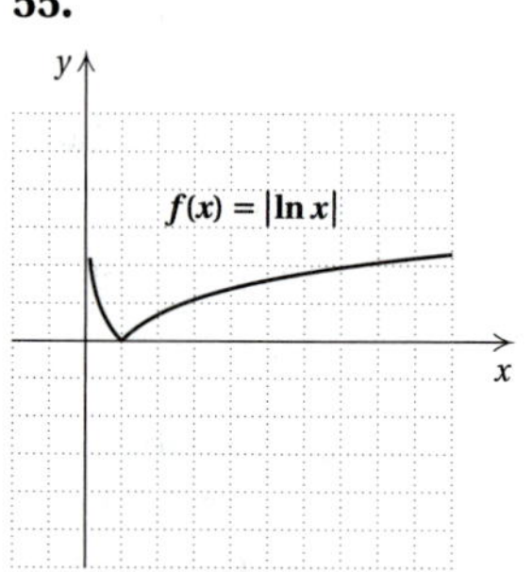

57. $\mathbf{D_W}$ **59.** 16, 256 **60.** $\frac{1}{4}$, 9 **61.** 49, 121 **62.** $\pm 3, \pm 4$ **63.** Domain: $(-\infty, \infty)$; range: $[0, \infty)$ **65.** Domain: $(-\infty, \infty)$; range: $(-\infty, 100)$ **67.** $\left(\frac{5}{2}, \infty\right)$

Margin Exercises, Section 17.6, pp. 1348–1352

1. 1 **2.** 3 **3.** 1.5395 **4.** 14.6068 **5.** 25 **6.** $\frac{2}{5}$ **7.** 2 **8.** 5

Calculator Corner, p. 1350

Left to the student

Exercise Set 17.6, p. 1353

1. 3 **3.** 4 **5.** $\frac{5}{2}$ **7.** $\frac{3}{5}$ **9.** 3.4594 **11.** 5.4263 **13.** $\frac{5}{2}$ **15.** −3, −1 **17.** $\frac{3}{2}$ **19.** 4.6052 **21.** 2.3026 **23.** 140.6705 **25.** 2.7095 **27.** 3.2220 **29.** 256 **31.** $\frac{1}{32}$ **33.** 10 **35.** $\frac{1}{100}$ **37.** $e^2 \approx 7.3891$ **39.** $\frac{1}{e} \approx 0.3679$ **41.** 121 **43.** 10 **45.** $\frac{1}{3}$ **47.** 3 **49.** $\frac{2}{5}$ **51.** 5 **53.** No solution **55.** $\mathbf{D_W}$ **57.** $\pm 10, \pm 2$ **58.** −64, 8 **59.** $-2, -3, \frac{-5 \pm \sqrt{41}}{2}$ **60.** $-\frac{1}{10}$, 1

61. $\dfrac{y^{4/3}}{25x^2z^4}$ 62. $-i$ 63. 1 65. (a) 0.3770; (b) -1.9617; (c) 0.9036; (d) -1.5318 67. 3, 4 69. -4 71. 2 73. $\pm\sqrt{34}$ 75. $10^{100,000}$ 77. 1, 100 79. 3, -7 81. 1, 1.465

Margin Exercises, Section 17.7, pp. 1356–1362

1. About 65 decibels **2.** $10^{-9.2}$ W/m^2 **3.** About 4.9 **4.** 10^{-7} moles per liter **5.** (a) 27.1 yr; (b) 10.2 yr **6.** About 300 million; about 343 million **7.** (a) $k \approx 6\%$, $P(t) = 5000e^{0.06t}$; (b) \$5309.18; \$5637.48; \$9110.59; (c) about 11.6 yr **8.** (a) $k \approx 0.471$, $P(t) = 2.8e^{0.471t}$, where $P(t)$ is in billions of dollars and t is the number of years after 2000; (b) about \$34,531 billion, or \$34.531 trillion; (c) 2008 **9.** 2972 yr

Calculator Corner, pp. 1362–1363

1. $y = 309{,}870.3567 \cdot 1.090787419^x$
2. 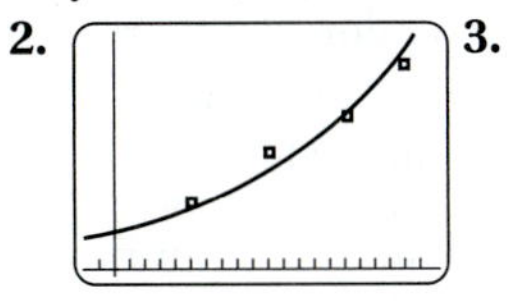 **3.** 3,531,046

Exercise Set 17.7, p. 1364

1. About 95 dB **3.** $10^{-1.5}$ W/m^2, or about 3.2×10^{-2} W/m^2 **5.** About 6.8 **7.** 1.58×10^{-8} moles per liter **9.** 3.04 ft/sec **11.** 2.17 ft/sec **13.** (a) 243 people; (b) about 20.5 months; (c) about 0.6 month **15.** (a) \$14,793; (b) 2011; (c) 11.9 yr **17.** (a) $P(t) = 6e^{0.015t}$; (b) 7.2 billion; (c) 2032; (d) 46.2 yr **19.** (a) $P(t) = P_0e^{0.06t}$; (b) \$5309.18; \$5637.48; \$9110.59; (c) in 11.6 yr **21.** (a) $k \approx 0.061$, $P(t) = 852{,}737e^{0.061t}$; (b) 2,888,380; (c) 2027 **23.** About 2103 yr **25.** About 7.2 days **27.** 69.3% per year **29.** (a) $k \approx 0.094$, $W(t) = 17.5e^{-0.094t}$; (b) 4.7 million tons; (c) 2173 **31.** (a) $V(t) = 451{,}000e^{0.07t}$; (b) \$1,201,670; (c) 9.9 yr; (d) 2002; (e) The function predicts that the card's value will be about \$908,202 in 2001. According to this, the card would not be a good buy at \$1.1 million. **33.** (a) $k \approx 0.007$, $P(t) = 2{,}394{,}811e^{-0.007t}$; (b) about 2,081,949; (c) 2115 **35.** D_W **37.** -1 **38.** 1 **39.** i **40.** i **41.** $-1 - i$ **42.** -2 **43.** $\frac{63}{65} - \frac{16}{65}i$ **44.** $-\frac{2}{41} + \frac{23}{41}i$ **45.** 41 **46.** $91 + 60i$ **47.** -0.937, 1.078, 58.770 **49.** -0.767, 2, 4 **51.** \$13.4 million

Summary and Review: Chapter 17, p. 1370

1. $\{(2, -4), (-7, 5), (-2, -1), (11, 10)\}$ **2.** Not one-to-one **3.** $g^{-1}(x) = \dfrac{7x + 3}{2}$ **4.** $f^{-1}(x) = \frac{1}{2}\sqrt[3]{x}$ **5.** $f^{-1}(x) = \dfrac{3x - 4}{2x}$

6. 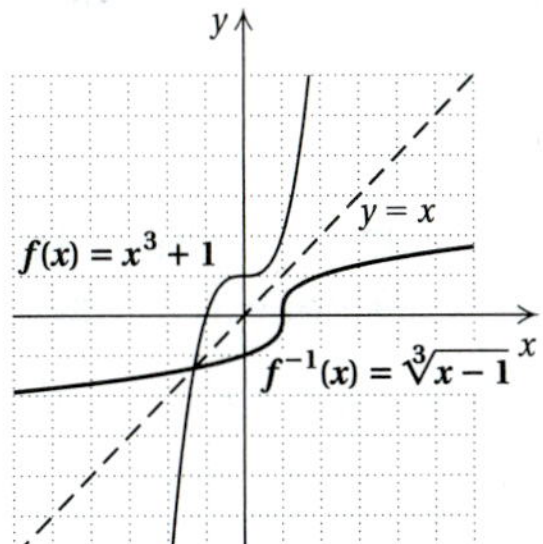

7.

x	$f(x)$
0	$\frac{1}{3}$
1	1
2	3
3	9
-1	$\frac{1}{9}$
-2	$\frac{1}{27}$
-3	$\frac{1}{81}$

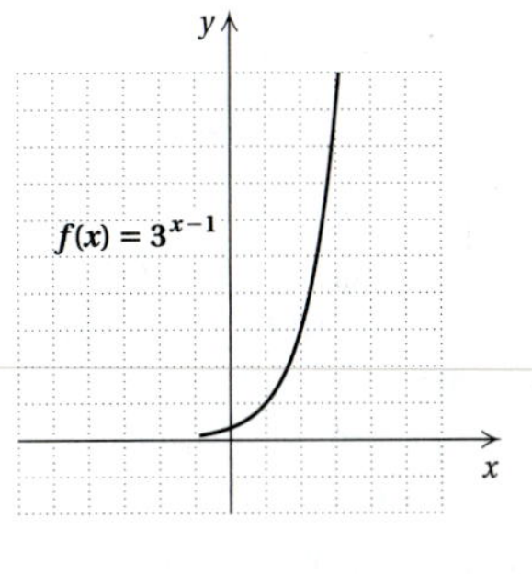

8. $x = 3^y$

x, or 3^y	y
1	0
3	1
9	2
27	3
$\frac{1}{3}$	-1
$\frac{1}{9}$	-2
$\frac{1}{27}$	-3

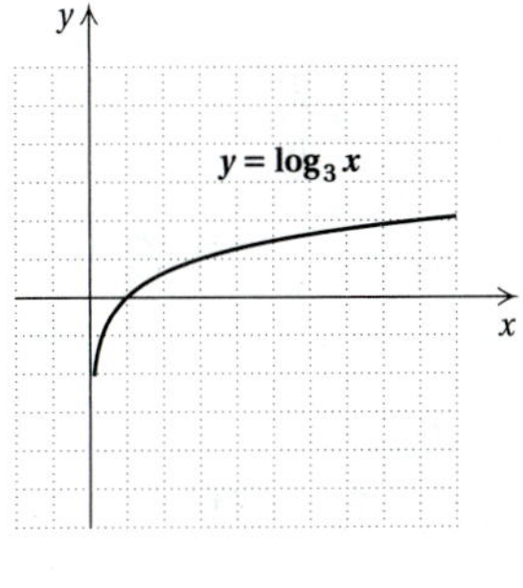

9.

x	$f(x)$
0	2.7
1	7.4
2	20.1
3	54.6
-1	1
-2	0.4
-3	0.1

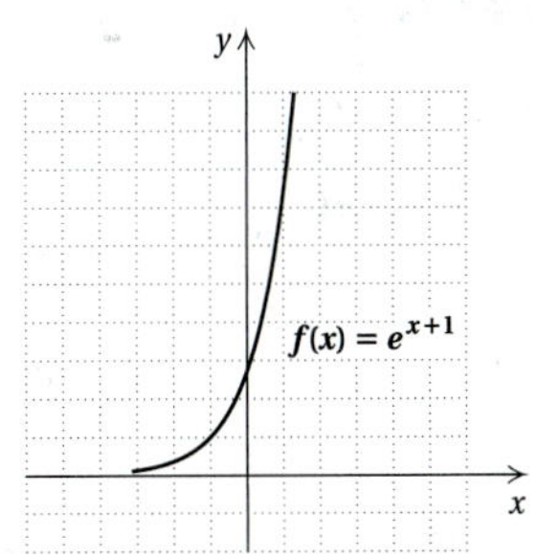

10. 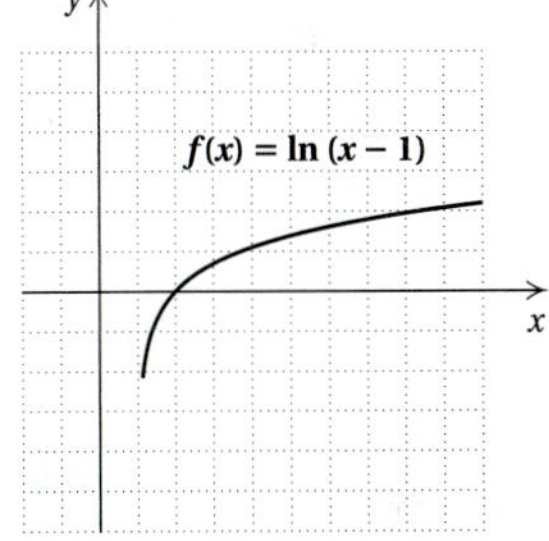

11. $f \circ g(x) = 9x^2 - 30x + 25$; $g \circ f(x) = 3x^2 - 25$ **12.** $f(x) = \sqrt{x}$, $g(x) = 4 - 7x$; answers may vary **13.** $4 = \log 10{,}000$ **14.** $\frac{1}{2} = \log_{25} 5$ **15.** $4^x = 16$ **16.** $\left(\frac{1}{2}\right)^{-3} = 8$ **17.** 2 **18.** -1 **19.** 1 **20.** 0 **21.** -2.7425 **22.** Does not exist as a real number

23. $4 \log_a x + 2 \log_a y + 3 \log_a z$

24. $\frac{1}{2}\log z - \frac{3}{4}\log x - \frac{1}{4}\log y$ 25. $\log_a 120$ 26. $\log \frac{a^{1/2}}{bc^2}$

27. 17 28. −7 29. 8.7601 30. 3.2698 31. 2.54995
32. −3.6602 33. −2.6921 34. 0.3753 35. 18.3568
36. 0 37. Does not exist 38. 1 39. 0.4307

40. 1.7097 41. $\frac{1}{9}$ 42. 2 43. $\frac{1}{10{,}000}$

44. $e^{-2} \approx 0.1353$ 45. $\frac{7}{2}$ 46. 1, −5 47. 1.5266
48. 35.0656 49. 2 50. 8 51. 20 52. $\sqrt{43}$
53. 90 dB 54. **(a)** \$113.73 million, \$172.98 million, \$228.76 million; **(b)** 2021; **(c)** about 5.0 yr
55. **(a)** $k \approx 0.347$; $N(t) = 6e^{0.347t}$, where t is the number of years since 1994; **(b)** 136; **(c)** 2004 56. $k \approx 0.043$, or 4.3%
57. About 8.25 yr 58. About 3463 yr 59. $\mathbf{D_W}$ You cannot take the logarithm of a negative number because logarithm bases are positive and there is no real-number power to which a positive number can be raised to yield a negative number. 60. $\mathbf{D_W}$ $\log_a 1 = 0$ because $a^0 = 1$.
61. **(a)** $f(x) = \frac{1}{64}x^2 + \frac{5}{16}x + \frac{5}{2}$; **(b)** \$19 62. $\pm 4, \pm i\sqrt{5}$
63.

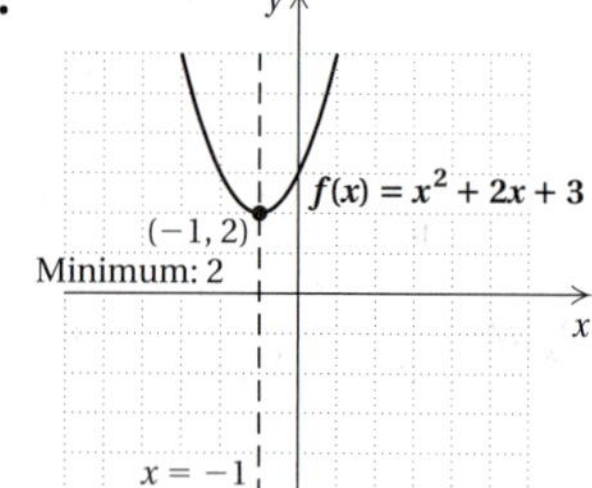

64. $11 - 10i$ 65. 1

66. $-\frac{11}{10} - \frac{17}{10}i$ 67. $5 - 2i$ 68. e^{e^3} 69. $\left(\frac{8}{3}, -\frac{2}{3}\right)$

Test: Chapter 17, p. 1373

1. [17.1a]

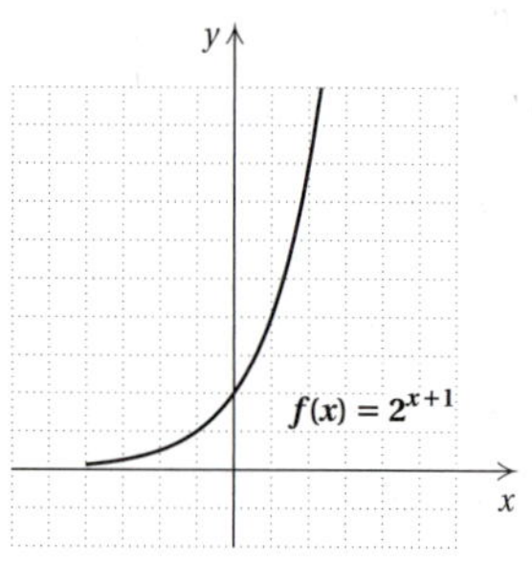

2. [17.3a]

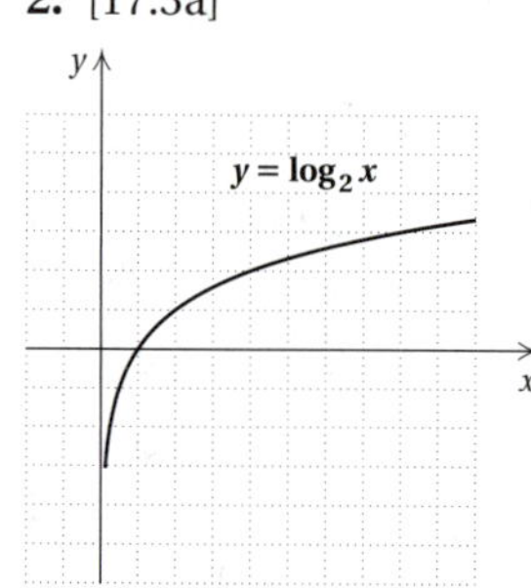

3. [17.5c]

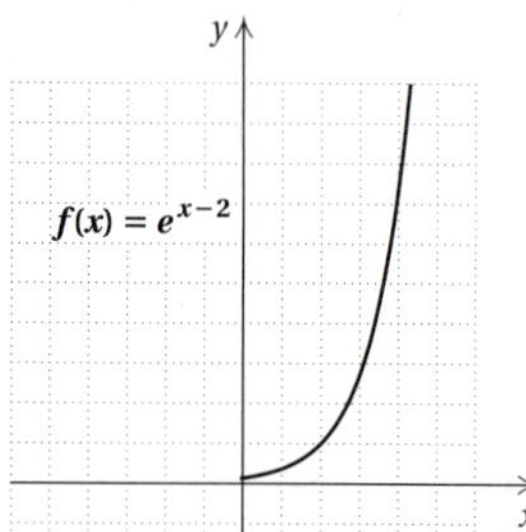

4. [17.5c]

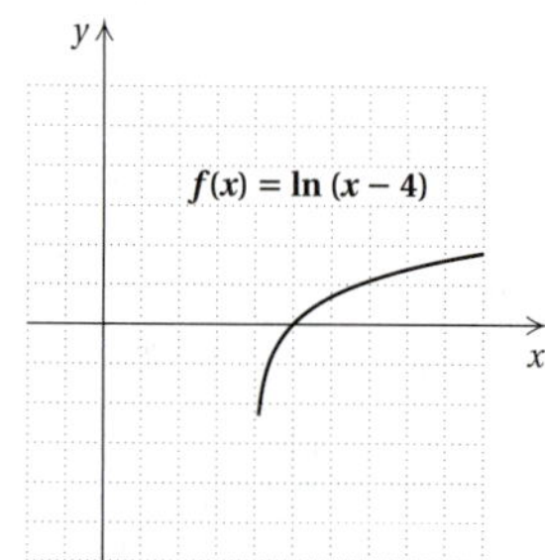

5. [17.2a] $\{(3, -4), (-8, 5), (-3, -1), (12, 10)\}$

6. [17.2b, c] $f^{-1}(x) = \frac{x + 3}{4}$

7. [17.2b, c] $f^{-1}(x) = \sqrt[3]{x} - 1$
8. [17.2b, c] Not one-to-one
9. [17.2d] $f \circ g(x) = 25x^2 - 15x + 2$, $g \circ f(x) = 5x^2 + 5x - 2$ 10. [17.3b] $\log_{256} 16 = \frac{1}{2}$
11. [17.3b] $7^m = 49$ 12. [17.3c] 3 13. [17.3c] 23
14. [17.3c] 0 15. [17.3d] −1.9101
16. [17.3d] Does not exist as a real number
17. [17.4d] $3 \log a + \frac{1}{2}\log b - 2 \log c$

18. [17.4d] $\log_a \frac{x^{1/3}z^2}{y^3}$ 19. [17.4d] −0.544

20. [17.4d] 1.079 21. [17.5a] 6.6938
22. [17.5a] 107.7701 23. [17.5a] 0 24. [17.5b] 1.1881
25. [17.6b] 5 26. [17.6b] 2 27. [17.6b] 10,000
28. [17.6b] $e^{1/4} \approx 1.2840$ 29. [17.6a] 0.0937
30. [17.6b] 9 31. [17.6b] 1 32. [17.7a] 4.2
33. [17.7b] **(a)** \$39,103; **(b)** 2010; **(c)** 11.9 yr
34. [17.7b] **(a)** $k \approx 0.006$, $N(t) = 30e^{0.006t}$, where t is the number of years since 1996 and N is in millions; **(b)** 31.287 million, 32.629 million; **(c)** 2081; **(d)** 115.5 yr
35. [17.7b] About 4.6% 36. [17.7b] About 4684 yr

37. [16.7b] **(a)** $y = \frac{1}{33{,}600}x^2 + \frac{17}{336}x$; **(b)** about 11.3 min

38. [16.4c] 64, 1 39. [16.6a]

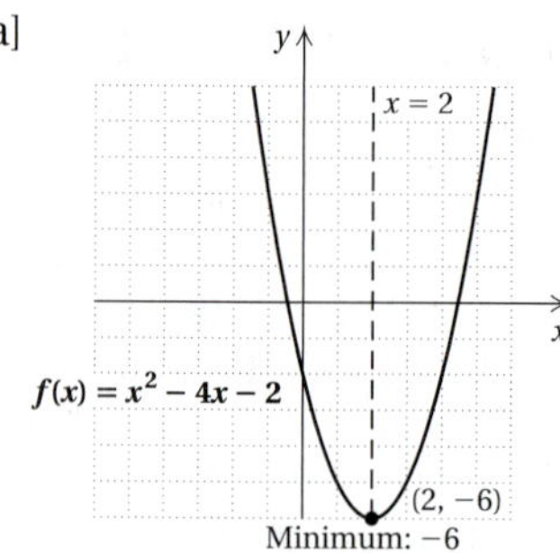

40. [15.8d] i 41. [15.8e] $\frac{3}{2} - \frac{5}{2}i$ 42. [17.6b] 44, −37
43. [17.4d] 2

Appendixes

Margin Exercises, Appendix A, pp. 1378–1387

1. 2 2. 3 3. $1\frac{1}{2}$ 4. $2\frac{1}{2}$ 5. 288 6. $8\frac{1}{2}$
7. 240,768 8. 6 9. $1\frac{5}{12}$ 10. 8 11. $11\frac{2}{3}$, or $11.\overline{6}$
12. 5 13. 0.5 14. 2 cm, or 20 mm 15. 2.3 cm, or 23 mm 16. 3.8 cm, or 38 mm 17. cm 18. km
19. mm 20. m 21. cm 22. m 23. 23,000
24. 400 25. 178 26. 9040 27. 7.814 28. 781.4
29. 0.967 30. 8,900,000 31. 6.78 32. 97.4 33. 0.1
34. 8.451 35. 91.4 36. 804.5 37. 1479.843
38. 913.912 39. 343.735 m 40. 3.175

Exercise Set A, p. 1388

1. 12 3. $\frac{1}{12}$ 5. 5280 7. 108 9. 7 11. $1\frac{1}{2}$
13. 26,400 15. $5\frac{1}{4}$, or 5.25 17. $3\frac{1}{3}$ 19. 37,488

21. $1\frac{1}{2}$ **23.** $1\frac{1}{4}$, or 1.25 **25.** 110 **27.** 2 **29.** 300 **31.** 30 **33.** $\frac{1}{36}$ **35.** 126,720 **37.** **(a)** 1000; **(b)** 0.001 **39.** **(a)** 10; **(b)** 0.1 **41.** **(a)** 0.01; **(b)** 100 **43.** **(a)** 6700; **(b)** Substitute **45.** **(a)** 0.98; **(b)** Multiply by 1 **47.** 8.921 **49.** 0.05666 **51.** 566,600 **53.** 4.77 **55.** 688 **57.** 0.1 **59.** 100,000 **61.** 142 **63.** 0.82 **65.** 450 **67.** 0.000024 **69.** 0.688 **71.** 230 **73.** 3.92 **75.** 180; 0.18 **77.** 278; 27.8 **79.** 48,440; 48.44 **81.** 4000; 400 **83.** 0.027; 0.00027 **85.** 442,000; 44,200 **87.** 100.65 **89.** 727.409592 **91.** 104.585 **93.** 289.62 **95.** 112.63 **97.** 9.14 **99.** 83.8581 **101.** 1250.061 **103.** 19.05 **105.** 1376.136 **107.** 0.51181

109.–119. Answers may vary, depending on the conversion factor used.

	yd	cm	in.	m	mm
109.	0.2604	23.8	9.37006	0.238	238
111.	0.2361	21.59	$8\frac{1}{2}$	0.2159	215.9
113.	52.9934	4844	1907.0828	48.44	48,440
115.	4	365.6	144	3.656	3656
117.	0.000295	0.027	0.0106299	0.00027	0.27
119.	483.548	44,200	17,401.54	442	442,000

121. 1 in. = 25.4 mm

Margin Exercises, Appendix B, pp. 1392–1393

1. 56.7 **2.** 64.7 **3.** 87.8 **4.** 17 **5.** 16.5 **6.** 91 **7.** 55 **8.** 54, 87 **9.** No mode exists. **10.** **(a)** 25 mm^2; **(b)** 25 mm^2; **(c)** No mode exists.

Exercise Set B, p. 1394

1. Mean: 21; median: 18.5; mode: 29 **3.** Mean: 21; median: 20; modes: 5, 20 **5.** Mean: 5.2; median: 5.7; mode: 7.4 **7.** Mean: 239.5; median: 234; mode: 234 **9.** Mean: $23.\overline{8}$; median: 15; mode: 1 **11.** Mean: 897.2; median: 798; no mode exists **13.** Mean: \$8.19; median: \$8.49; mode: \$6.99 **15.** 2.7 **17.** 10 home runs **19.** 263 days

Margin Exercises, Appendix C, pp. 1396–1399

1. Cinnamon Life **2.** Wheaties **3.** Wheaties **4.** 1.2 g **5.** 224.66 mg **6.** 213.3 mg **7.** Average: 134; median: 120; mode: 120, 160 **8.** 20 calories **9.** Yes **10.** 560 mg **11.** 24% **12.** 14 g **13.** 60,000 elephants **14.** Two and one half as many in Zimbabwe as in Cameroon **15.** 55,000 elephants **16.** 795 cups; answers may vary **17.** 750 cups; answers may vary **18.** 1830 cups; answers may vary

Exercise Set C, p. 1400

1. 483,612,200 mi **3.** Neptune **5.** All **7.** 11 **9.** Average: $27{,}884.\overline{1}$ mi; median: 7926 mi; no mode exists **11.** 92° **13.** 108° **15.** 3 **17.** 90° and higher **19.** 30% and higher **21.** 50% **23.** 1997: 59.74°; 1998: 60.26°; 0.87% **25.** 59.545°; 59.937°; 0.392° **27.** 1.0 billion **29.** 2070 **31.** 1650 and 1850 **33.** 3 billion; 75% **35.** Africa **37.** 475,000 gal **39.** 325,000 gal

Margin Exercises, Appendix D, pp. 1404–1405

1. 14 **2.** 11 **3.** 10 **4.** $\sqrt{37} \approx 6.083$ **5.** $\left(\frac{3}{2}, -3\right)$ **6.** $(9, -5)$

Exercise Set D, p. 1406

1. 5 **3.** $\frac{9}{4}$ **5.** $\sqrt{29} \approx 5.385$ **7.** $\sqrt{648} \approx 25.456$ **9.** 7.1 **11.** $\frac{\sqrt{41}}{7} \approx 0.915$ **13.** $\sqrt{6970} \approx 83.487$ **15.** $\sqrt{a^2 + b^2}$ **17.** $\sqrt{17 + 2\sqrt{14} + 2\sqrt{15}} \approx 5.677$ **19.** $\sqrt{9{,}672{,}400} \approx 3110.048$ **21.** $\frac{2}{5}$ **23.** $\left(\frac{3}{2}, \frac{7}{2}\right)$ **25.** $\left(0, \frac{11}{2}\right)$ **27.** $\left(-1, -\frac{17}{2}\right)$ **29.** $(-0.25, -0.3)$ **31.** $\left(-\frac{1}{12}, \frac{1}{24}\right)$ **33.** $\left(\frac{\sqrt{2} + \sqrt{3}}{2}, \frac{3}{2}\right)$ **35.** $\sqrt{49 + k^2}$ **37.** $8\sqrt{m^2 + n^2}$ **39.** Yes **41.** $(2, 4\sqrt{2})$

Margin Exercises, Appendix E, pp. 1408–1411

1. $\{x \mid x < -3 \text{ or } x > 1\}$, or $(-\infty, -3) \cup (1, \infty)$ **2.** $\{x \mid -3 < x < 1\}$, or $(-3, 1)$ **3.** $\{x \mid -3 \le x \le 1\}$, or $[-3, 1]$ **4.** $\{x \mid x < -4 \text{ or } x > 1\}$, or $(-\infty, -4) \cup (1, \infty)$ **5.** $\{x \mid -4 \le x \le 1\}$, or $[-4, 1]$ **6.** $\{x \mid x < -1 \text{ or } 0 < x < 1\}$, or $(-\infty, -1) \cup (0, 1)$

7.

$y \ge x^2 - 2x - 3$

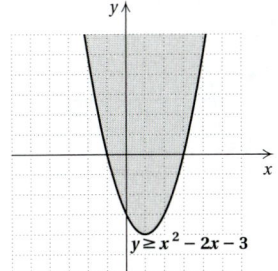

8.

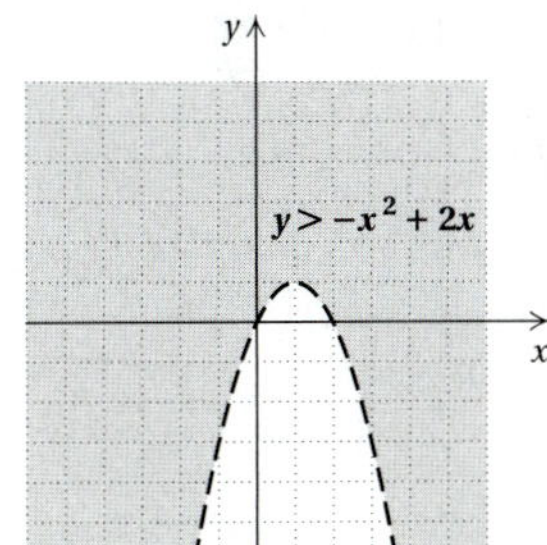

Exercise Set E, p. 1412

1. $\{x \mid x < -2 \text{ or } x > 6\}$, or $(-\infty, -2) \cup (6, \infty)$ **3.** $\{x \mid -2 \le x \le 2\}$, or $[-2, 2]$ **5.** $\{x \mid -1 \le x \le 4\}$, or $[-1, 4]$ **7.** $\{x \mid -1 < x < 2\}$, or $(-1, 2)$ **9.** All real numbers, or $(-\infty, \infty)$ **11.** $\{x \mid 2 < x < 4\}$, or $(2, 4)$ **13.** $\{x \mid x < -2 \text{ or } 0 < x < 2\}$, or $(-\infty, -2) \cup (0, 2)$ **15.** $\{x \mid -9 < x < -1 \text{ or } x > 4\}$, or $(-9, -1) \cup (4, \infty)$ **17.** $\{x \mid x < -3 \text{ or } -2 < x < 1\}$, or $(-\infty, -3) \cup (-2, 1)$

19.

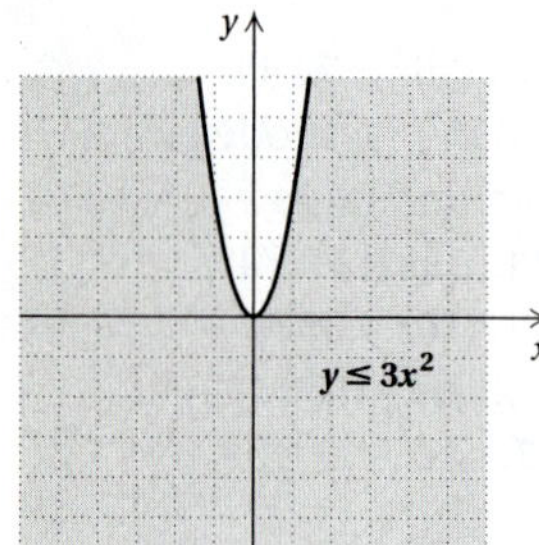

21.

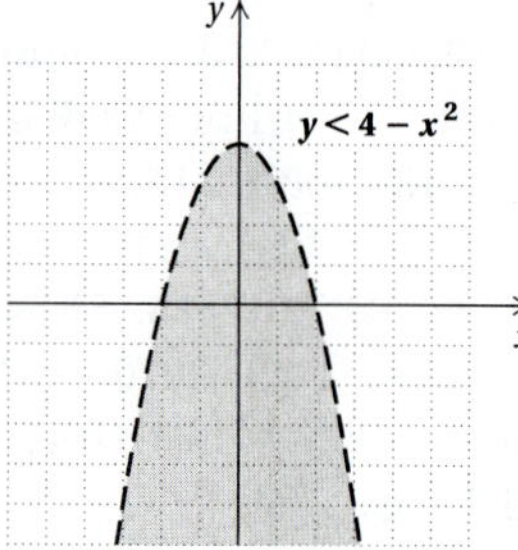

23.

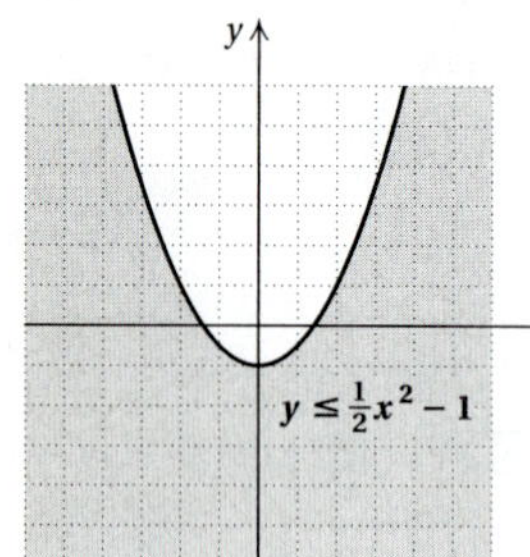

25.

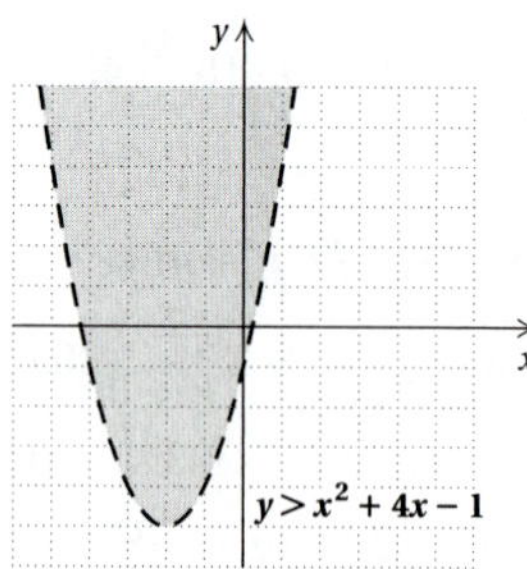

27.

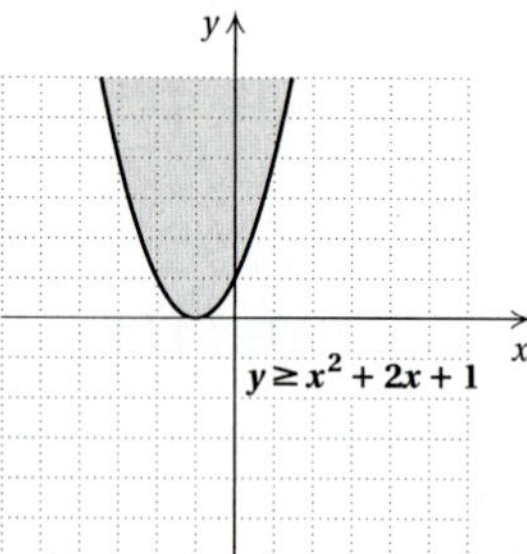

29.

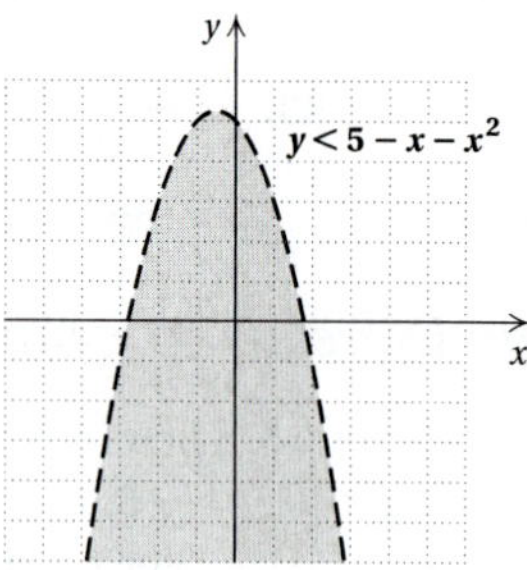

31.

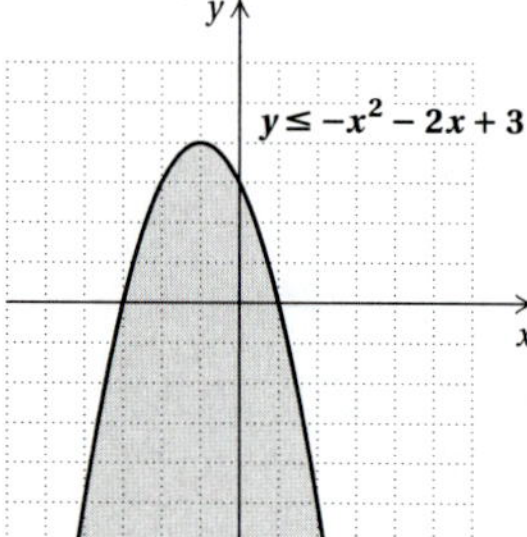

33.

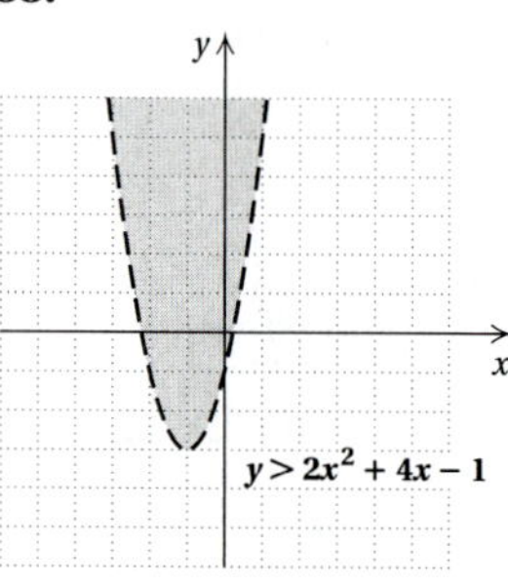

Margin Exercises, Appendix F, pp. 1415–1427

1. 3, 5, 7, 9; 63 **2.** $-\frac{1}{6}, \frac{1}{7}, -\frac{1}{8}, \frac{1}{9}; -\frac{1}{36}$ **3.** $3n$ **4.** $(-1)^n n$ **5.** n^3 **6.** 2^{n-1} **7.** **(a)** 0; **(b)** 2; **(c)** 4 **8.** $1 + 4 + 9 + 16 + 25 = 55$ **9.** $1 - 2 + 4 - 8 + 16 = 11$ **10.** $4 + 13 + 24 + 37 = 78$ **11.** $\sum_{k=1}^{5} 2k$ **12.** $\sum_{k=1}^{4} (-1)^{k+1}k^3$ **13.** $\sum_{k=2}^{5} k^2$ **14.** $a_1 = 3.1; d = 0.8$ **15.** $a_1 = 14; d = -5$ **16.** 50 **17.** 75th **18.** $a_1 = 7$; $d = 12$; 7, 19, 31, 43, 55, ... **19.** 10,100 **20.** 225 **21.** 5 **22.** −3 **23.** 0.85 **24.** 1.09 **25.** $\frac{1}{5}$ **26.** 512 **27.** $\frac{1}{81}$ **28.** 13,120 **29.** $\frac{1025}{768}$ **30.** Does not exist **31.** Does not exist **32.** 1 **33.** $\frac{3125}{3}$ **34.** $\frac{2}{9}$ **35.** $\frac{13}{99}$ **36.** \$27.25 **37.** \$21,474,836.47

Exercise Set F, p. 1428

1. 3, 7, 11, 15; 39; 59 **3.** $2, \frac{3}{2}, \frac{4}{3}, \frac{5}{4}; \frac{10}{9}; \frac{15}{14}$ **5.** $0, \frac{3}{5}, \frac{4}{5}, \frac{15}{17}, \frac{99}{101}; \frac{112}{113}$ **7.** $-1, 4, -9, 16; 100; -225$ **9.** 7, 3, 7, 3; 3; 7 **11.** $2n$ **13.** $(-1)^n \cdot 2 \cdot 3^{n-1}$ **15.** $\frac{n+1}{n+2}$ **17.** $n(n+1)$ **19.** $\log 10^{n-1}$, or $n - 1$ **21.** 6; 28 **23.** 20; 30 **25.** $\frac{1}{2} + \frac{1}{4} + \frac{1}{6} + \frac{1}{8} + \frac{1}{10} = \frac{137}{120}$ **27.** $1 + 2 + 4 + 8 + 16 + 32 + 64 = 127$ **29.** $\ln 7 + \ln 8 + \ln 9 + \ln 10 = \ln(7 \cdot 8 \cdot 9 \cdot 10) = \ln 5040 \approx 8.5252$ **31.** $\frac{1}{2} + \frac{2}{3} + \frac{3}{4} + \frac{4}{5} + \frac{5}{6} + \frac{6}{7} + \frac{7}{8} + \frac{8}{9} = \frac{15{,}551}{2520}$ **33.** $-1 + 1 - 1 + 1 - 1 = -1$ **35.** $\sum_{k=1}^{\infty} 5k$ **37.** $\sum_{k=1}^{6} (-1)^{k+1}2^k$ **39.** $\sum_{k=1}^{6} (-1)^k \frac{k}{k+1}$ **41.** $\sum_{k=2}^{n} (-1)^k k^2$ **43.** $\sum_{k=1}^{\infty} \frac{1}{k(k+1)}$ **45.** $a_1 = 3, d = 5$ **47.** $a_1 = 9, d = -4$ **49.** $a_1 = \frac{3}{2}, d = \frac{3}{4}$ **51.** $a_1 = \$316$, $d = -\$3$ **53.** $a_{12} = 46$ **55.** $a_{14} = -\frac{17}{3}$ **57.** 27th **59.** 46th **61.** $a_1 = 5$ **63.** $n = 39$ **65.** $a_1 = \frac{1}{3}$; $d = \frac{1}{2}; \frac{1}{3}, \frac{5}{6}, \frac{4}{3}, \frac{11}{6}, \frac{7}{3}$ **67.** 670 **69.** 160,400 **71.** 735 **73.** 2 **75.** −1 **77.** −2 **79.** 0.1 **81.** $\frac{a}{2}$ **83.** 128 **85.** 162 **87.** $7(5)^{40}$ **89.** 762 **91.** $\frac{4921}{18}$ **93.** 8 **95.** 125 **97.** Does not exist **99.** $\frac{2}{3}$ **101.** $\frac{13}{99}$ **103.** 9 **105.** $\frac{34{,}091}{9990}$ **107.** 1260 **109.** 3; 171 **111.** 4960¢, or \$49.60 **113.** **(a)** $\frac{1}{256}$ ft; **(b)** $5\frac{1}{3}$ ft **115.** **(a)** About 297 ft; **(b)** 300 ft **117.** \$523,619.17

Margin Exercises, Appendix G, pp. 1434–1437

1. 21 **2.** 19 **3.** 22 **4.** 17 **5.** J **6.** 11 **7.** **8.** **9.** **10.**

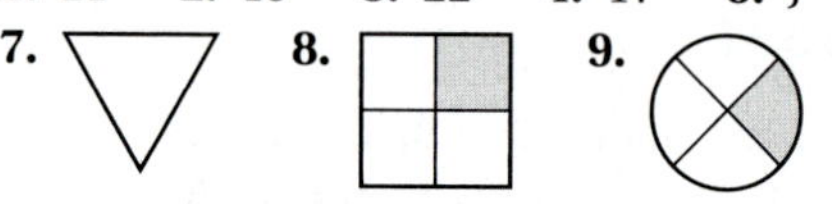

11. 35 **12.** 70 **13.** Maria

Exercise Set G, p. 1438

1. 19 **3.** V **5.** 43 **7.** **9.** **11.** 17 **13.** Bev **15.** Chris

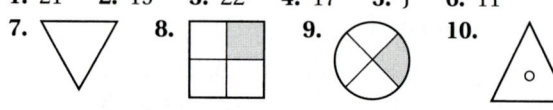

Exercise Set H, p. 1440

1. B **2.** B **3.** D **4.** A **5.** C **6.** B **7.** C **8.** D **9.** C **10.** C **11.** A **12.** A **13.** D **14.** D **15.** B **16.** A **17.** B **18.** C **19.** A **20.** B **21.** C **22.** D **23.** B **24.** A **25.** C **26.** A **27.** D **28.** A **29.** B **30.** D **31.** C **32.** B **33.** A **34.** D **35.** A **36.** C **37.** D **38.** C **39.** C **40.** C **41.** B **42.** C **43.** A **44.** B **45.** A **46.** B

Index

Photo Credits

9, The Image Bank **10, 11,** Tony Stone Images **29,** Camel **53,** UPI Newspictures **63,** EyeWire **100,** Dick Morton **105 (left),** © 2001 Corbis **105 (right top),** PhotoDisc **105 (right bottom),** Karen Bittinger **140,** PhotoDisc **145,** AP/Wide World Photos **148,** Karen Bittinger **153,** PhotoDisc **154,** FoodPix **156,** PhotoDisc **159,** David Young-Wolff/ PhotoEdit **168,** Corbis **176,** © AFP/Robert Sullivan/CORBIS **182,** Creative Labs **197,** © Bettmann/Corbis **215,** Daniel J. Cox, Tony Stone Images **219,** © Chuck Savage/Corbis Stockmarket **220,** Chris Thomaidis, Tony Stone Images **223,** Gary Gay, The Image Bank **237 (left),** Russ Einhorn, UPI Photo **237 (right),** © Lee Celano, AFB/Corbis **240,** © Janette Beckman/CORBIS **244,** © Tom Nebbia/Corbis Stockmarket **245,** Bob Gage, FPG International **253,** © AFP Photo, Bruce Weaver/Corbis **255,** EyeWire Collection **260,** Buccina Studios, PhotoDisc **262,** Nicholas Devore III/Photographers Aspen **264,** Jeremy Woodhouse, PhotoDisc **265,** Barry Rosenthal, FPG International **267,** FPG International **268,** Kenneth Chen/Envision **272 (top),** PhotoDisc **272 (bottom),** Hayden Roger Celestin, UPI Photo Service **273 (top),** PhotoDisc **273 (bottom),** Hayden Roger Celestin, UPI Photo Service **275, 284,** PhotoDisc **285,** Corbis **286, 290,** PhotoDisc **301,** Charles Gupton, Stone/Getty Images **303,** Jessie Cohen, National Zoo, UPI Photo Service **304,** Christine Chew, UPI Photo Service **318,** Ezio Petersen, UPI Photo Service **319,** Christopher Bissell, Stone/Getty Images **322,** PhotoLink **324,** Walter Hodges/Tony Stone Images **325,** PhotoDisc **332,** Patrick Clark, PhotoDisc **333,** Chad Baker/Ryan McVay, PhotoDisc **375,** © Nancy Dudley/Stock Boston **388,** © 1988 Al Satterwhite **431,** © Dale O'Dell/Corbis Stock Market **446 (left),** © Michel Tcherevkoff Ltd/The Image Bank **446 (right),** David Cannon/Allsport **455,** © Ruth Dixon/Stock Boston **463,** AP/Wide World Photos **464,** John W. Banagan/The Image Bank **505,** © Comstock Images **512,** Jack Hollingsworth, PhotoDisc **547,** Sharp Electronics Corporation **548 (left),** FPG International **548 (right),** Compaq **551 (top),** Doug Meneuz, PhotoDisc **551 (bottom),** Brian Spurlock **559,** AP/Wide World Photos **561,** Ryan McVay, PhotoDisc **563,** EyeWire Collection **564,** © Kevin Horan, Tony Stone **566,** EyeWire Collection **581,** Christine D. Tuff **585,** Tom Tracy/FPG International **591,** © Lawrence Migdale, Tony Stone **597,** EyeWire Collection **637,** The Image Bank **643 (left),** Mary Clay/Tom Stack and Associates **643 (right),** Terje Rakke/The Image Bank **650,** Federal Highway Administration and Washington Infrastructure Services, Inc. **651,** Copyright © by Klev Schoening **671 (left),** © Dave G. Houser/CORBIS **671 (middle),** FPG International **671 (right),** EyeWire Collection **678,** Jamie Squire/Allsport **679 (right),** Hoby Finn, PhotoDisc **680,** Bob Daemmrich/The Image Works **691,** Ron Chapple, FPG International **721,** Brian Spurlock **748,** Karen Bittinger **779,** FPG International **813,** Karen Bittinger **815,** Chris Salvo, FPG International **818,** Judy Gelles/Stock Boston **820 (top),** Duncan Smith/PhotoDisc **820 (bottom), 885,** EyeWire Collection **889,** American Honda **890 (left),** © John Warden/Index Stock Image **890 (right),** © 1990 Glenn Randall **895,** PhotoLink/PhotoDisc **896 (left),** Geostock/PhotoDisc **896 (right),** Tom Turpin, Purdue University **897 (left),** Reuters/Anthony P. Bolante/Reuters NewMedia Inc./CORBIS **897 (right),** REUTERS/Sue Ogrocki/Reuters NewMedia Inc./CORBIS **898,** Joseph Sohm, ChromoSohn Inc./CORBIS

902, Richard Shock/Corbis **903,** PhotoDisc **906,** EyeWire Collection **907,** George D. Lepp/Corbis **909,** PhotoDisc **914,** Hisham E. Ibrahim, PhotoDisc **915 (top),** Steve Mason, PhotoDisc **915 (bottom),** PhotoLink/PhotoDisc **963,** Mark Hamel/Tony Stone **971,** EyeWire Collection **976,** Yellow Dog Productions/The Image Bank **989,** EyeWire Collection **1000,** Angelo Hornak/Corbis **1023,** Robert Brenner/PhotoEdit **1038,** Stephanie Maze/Corbis **1039,** Michael Newman/PhotoEdit **1040,** Karen Bittinger **1051 (left),** PEANUTS reprinted by permission of United Feature Syndicate, Inc. **1051 (right),** PhotoDisc **1064,** Reuters/Wolfgang Rattay/Archive Photos **1065,** Macduff Everton/Corbis **1072,** Keith Brofsky/PhotoDisc **1082,** Keith Wood/Tony Stone Images **1109,** Warren Morgan/Corbis **1111,** Archive Photos **1127,** © 1999 D & I MacDonald/Envision **1165,** © 1999 Bill Ellzey **1170,** EyeWire Collection **1205,** Reuters/Ira Strikstein/Archive **1267,** Thomas A. Heinz/Corbis **1288,** PhotoDisc **1302 (left),** © John Warden/Index Stock Image **1302 (right),** Crowther and Carter/Tony Stone **1313,** Akos Szilvasi/Stock Boston **1355,** Reuters NewMedia Inc./Corbis **1357,** Reed Kaestner/Corbis **1358,** Alan Schein Photography/Corbis **1360,** John Lacko **1361,** Mark E. Gibson/Mira **1364,** AP/Wide World Photos **1365,** David Zimmerman/Corbis **1367 (left),** from *Classic Baseball Cards,* by Bert Randolph Sugar, copyright © 1977 by Dover Publishing, Inc. **1367 (right),** AP/Wide World Photos **1368,** Corbis **1371,** © Dave G. Houser/Corbis **1372,** PhotoDisc **1386,** Owen Franken/Stock, Boston **1392,** AP/Wide World Photos **1394,** © Corbis Images **1395,** PBA **1399,** Bob Handelman, Tony Stone Images

Index of Applications

Biology, Health, and Life Science

Business and Economics

Construction and Engineering

Consumer Applications

Environment and Geography

General Interest

Geometry

Physical Science

Sports, Leisure, and Entertainment

Statistics and Demographics

Transportation

Index of Study Tips